INSTRUMENT AND AUTOMATION ENGINEERS' HANDBOOK
FIFTH EDITION

VOLUME II
Analysis and Analyzers

INSTRUMENT AND AUTOMATION ENGINEERS' HANDBOOK
FIFTH EDITION

VOLUME II
Analysis and Analyzers

BÉLA G. LIPTÁK, Editor-in-Chief

KRISZTA VENCZEL, Volume Editor

International Society of Automation
67 Alexander Drive
P.O. Box 12277
Research Triangle Park, NC 27709
PHONE (919) 549-8411
FAX (919) 549-8288
E-MAIL info@isa.org
www.isa.org

CRC Press
Taylor & Francis Group
Boca Raton London New York

CRC Press is an imprint of the
Taylor & Francis Group, an **informa** business

CRC Press
Taylor & Francis Group
6000 Broken Sound Parkway NW, Suite 300
Boca Raton, FL 33487-2742

© 2017 by Bela G. Liptak
CRC Press is an imprint of Taylor & Francis Group, an Informa business

No claim to original U.S. Government works

Printed on acid-free paper
Version Date: 20160725

International Standard Book Number-13: 978-1-4987-2764-8 (Hardback)

This book contains information obtained from authentic and highly regarded sources. Reasonable efforts have been made to publish reliable data and information, but the author and publisher cannot assume responsibility for the validity of all materials or the consequences of their use. The authors and publishers have attempted to trace the copyright holders of all material reproduced in this publication and apologize to copyright holders if permission to publish in this form has not been obtained. If any copyright material has not been acknowledged please write and let us know so we may rectify in any future reprint.

Except as permitted under U.S. Copyright Law, no part of this book may be reprinted, reproduced, transmitted, or utilized in any form by any electronic, mechanical, or other means, now known or hereafter invented, including photocopying, microfilming, and recording, or in any information storage or retrieval system, without written permission from the publishers.

For permission to photocopy or use material electronically from this work, please access www.copyright.com (http://www.copyright.com/) or contact the Copyright Clearance Center, Inc. (CCC), 222 Rosewood Drive, Danvers, MA 01923, 978-750-8400. CCC is a not-for-profit organization that provides licenses and registration for a variety of users. For organizations that have been granted a photocopy license by the CCC, a separate system of payment has been arranged.

Trademark Notice: Product or corporate names may be trademarks or registered trademarks, and are used only for identification and explanation without intent to infringe.

Library of Congress Cataloging-in-Publication Data

Names: Liptáak, Bâela G., editor.
Title: Instrument and automation engineers' handbook : measurement and safety / editor, Bela Liptak.
Other titles: Instrument engineers' handbook.
Description: Fifth edition. | Boca Raton : Taylor & Francis, CRC Press, 2017. | Revision of: Instrument engineers' handbook. | Includes bibliographical references and index.
Identifiers: LCCN 2016025938 | ISBN 9781498727648 (hard back : alk. paper)
Subjects: LCSH: Process control--Handbooks, manuals, etc. | Measuring instruments--Handbooks, manuals, etc. | Automatic control--Handbooks, manuals, etc. | Plant engineering--Safety measures--Handbooks, manuals, etc.
Classification: LCC TS156.8 .I56 2017 | DDC 658.2/8--dc23
LC record available at https://lccn.loc.gov/2016025938

Visit the Taylor & Francis Web site at
http://www.taylorandfrancis.com

and the CRC Press Web site at
http://www.crcpress.com

This handbook is dedicated to the next generation of automation engineers working in the fields of analysis, measurement, control, and safety. I hope that learning from these pages will increase their professional standing around the world. It is also my hope that our knowledge accumulated during the last half century will speed the coming of the age of full automation. I hope that what we have learned in optimizing industrial processes will be used to improve the understanding of all processes.
I hope that this knowledge will also help overcome our environmental ills and will smooth the conversion of our lifestyle into a sustainable, safe, and clean one.

Béla G. Lipták

CONTENTS

Introduction   ix
Contributors   xvii

1   Analytical Measurement   1

1.1   Analyzer Selection and Application   19
1.2   Analyzer Sampling   53
1.3   Analyzer Sampling: Stack Monitoring   79
1.4   Analyzer Sampling: Air Quality Monitoring   91
1.5   Ammonia Analyzers   109
1.6   Biometers to Quantify Microorganisms   117
1.7   Carbon Dioxide   123
1.8   Carbon Monoxide   132
1.9   Chlorine Analyzers   142
1.10   Chromatographs: Gas   154
1.11   Chromatographs: Liquid   190
1.12   Coal Analyzers   200
1.13   Colorimeters   210
1.14   Combustible Gas or Vapor Sensors   219
1.15   Conductivity Measurement   235
1.16   Consistency Measurement   248
1.17   Corrosion Monitoring   258
1.18   Cyanide Analyzers: Weak-Acid Dissociable (WAD)   269
1.19   Differential Vapor Pressure   277
1.20   Dioxin and Persistent Organic Pollutants Analyzers   284
1.21   Electrochemical Analyzers   290
1.22   Elemental Analyzers   303
1.23   Fiber-Optic (FO) Probes and Cables   312
1.24   Flame, Fire, and Smoke Detectors   330
1.25   Fluoride Analyzers   343
1.26   Hazardous and Toxic Gas Monitoring   352
1.27   Heating Value Calorimeters   378
1.28   Hydrocarbon Analyzers   390
1.29   Hydrogen Cyanide [HCN] Detectors   403
1.30   Hydrogen in Steam or Air Analyzers   411
1.31   Hydrogen Sulfide Detectors   421
1.32   Infrared and Near-Infrared Analyzers   429
1.33   Ion-Selective Electrodes (ISE)   458
1.34   Leak Detectors   474

1.35 Mass Spectrometers 486
1.36 Mercury in Ambient Air 497
1.37 Mercury in Water 507
1.38 Moisture in Air: Humidity and Dew Point 519
1.39 Moisture in Gases and Liquids 540
1.40 Moisture in Solids 561
1.41 Molecular Weight of Liquids 580
1.42 Natural Gas Measurements 597
1.43 Nitrogen, Ammonia, Nitrite and Nitrate 606
1.44 Nitrogen Oxide (NO_x) Analyzers 615
1.45 Odor Detection 625
1.46 Oil in or on Water 634
1.47 Oxidation–Reduction Potential (ORP) 652
1.48 Oxygen Demands (BOD, COD, TOD) 665
1.49 Oxygen in Gases 682
1.50 Oxygen in Liquids (Dissolved Oxygen) 700
1.51 Ozone in Gas 714
1.52 Ozone in Water 724
1.53 Particle Size Distribution (PSD) Monitors 731
1.54 Particulate, Opacity, Air and Emission Monitoring 742
1.55 pH Measurement 764
1.56 Phosphate Analyzer 792
1.57 Physical Properties Analyzers for Petroleum Products 800
1.58 Raman Analyzers 825
1.59 Refractometers 849
1.60 Rheometers 864
1.61 Sand Concentration and Subsea Pipeline Erosion Detectors 878
1.62 Spectrometers, Open Path (OP) 886
1.63 Streaming Current Particle Charge Analyzer 909
1.64 Sulfur Dioxide and Trioxide 920
1.65 Sulfur in Oil and Gas 931
1.66 Thermal Conductivity Detectors 941
1.67 Total Carbon and Total Organic Carbon (TOC) Analyzers 951
1.68 Turbidity, Sludge and Suspended Solids 966
1.69 Ultraviolet and Visible Analyzers 981
1.70 Viscometers: Application and Selection 1002
1.71 Viscometers: Industrial 1016
1.72 Viscometers: Laboratory 1045
1.73 Water Quality Monitoring 1069
1.74 Wet Chemistry and Autotitrator Analyzers 1087

Appendix 1101

A.1 Definitions 1103
A.2 Abbreviations, Acronyms, and Symbols 1150
A.3 Organizations 1170
A.4 Flowsheet and Functional Diagrams Symbols 1173
A.5 Conversion among Engineering Units 1220
A.6 Chemical Resistance of Materials 1257
A.7 Composition and Properties of Metallic and Other Materials 1280
A.8 Steam and Water Tables 1287

Index 1295

INTRODUCTION

I started to work on the first edition of this handbook when I was 25 years old. Today, when you start turning the pages of this fifth edition, I am 80. This book started out as an American handbook on analytical instrumentation, while today it is a reference source used on all five continents. When I started writing the first edition, most composition analysis was done using manual samples that were analyzed by chromatographs in the laboratory and only a few density and pH controllers operated online. Even in the few cases where the analyzers were located close to the process, their sampling systems were complicated and required much maintenance (Figure I.1).

Most of today's analyzers have been moved out of analyzer houses and are mounted online, miniaturized, or are modular, and if they use sampling systems, they are *smart* and automated. They are also provided with wired or wireless communication between the sample system components, the analyzer–sensor, and the control system that controls the unit operation involved.

The role of analyzers in our everyday life is becoming increasingly important and their capabilities and sophistication are exploding. We have found that while using grab samples might be acceptable for product quality control purposes, because of the time it takes to get a sample, transport it to the laboratory, and wait for the results, it is unacceptable for safety or for process control, optimization, and energy conservation purposes. In these applications, the analysis must not only be continuous and online but also smart, rugged, often explosion proof with local display, and low maintenance (Figure I.2).

FIG. I.1
Analyzer installation 50–60 years ago.

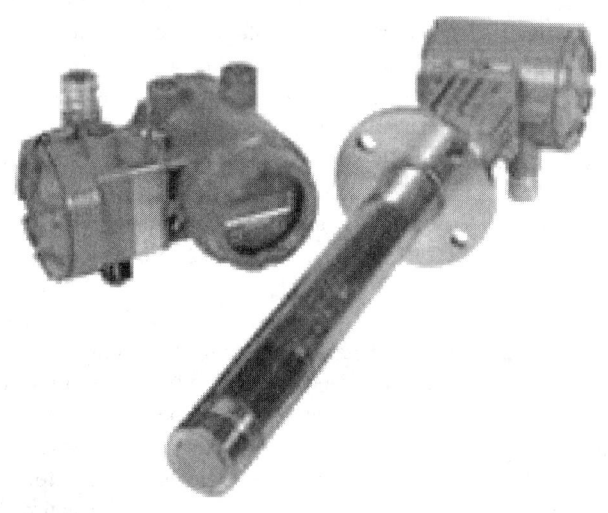

FIG. I.2
A typical state-of-the-art, online analyzer transmitter used for the measurement of oxygen. (Courtesy of Emerson, Rosemount Analytical.)

It is not only the control portion of the analyzer loop that has to be automatic, but where sampling systems are required, they too must automatically bring the process sample to the analyzer (Figure 1.5a). Also, the new self-diagnosing and self-calibrating analyzers are major advancements and so are the designs based on lower cost and less maintenance. Orientation Table 1.1a will help the reader to select the best analyzer for the application at hand. It is complemented by Table 1.1v that compares key attributes (cost, complexity, type of sample, etc.) of the many analyzers discussed in this chapter. These tables are good starting points in the selection process.

Our professions have also changed. At the time of the first edition, I was teaching process control in the *chemical* engineering department of Yale University and my handbook was published by the *electrical* engineering division of its publisher. Why? It was not because Yale or Chilton had something against our profession! No, it was because *they did not even know* that an automation profession existed!

Things have changed! Today, it is the basic know-how of chemical or electrical engineers that is taken for granted, and the focus is on the almost daily advances in automation, robotics, artificial intelligence and optimization. The increased importance of automation is complemented by the increased availability of self-checking digital components, wireless transmission, redundant safety backups, ease of configuring complex algorithms, and generating dynamic displays. It is being realized that automation can simultaneously maximize safety and product quality while minimizing operating and energy costs.

THE BIRTH OF THIS HANDBOOK

In 1956, we Hungarians rebelled against both Communism and Soviet occupation. Our revolution was crushed and 250,000 young and educated Hungarians (2.5% of the population) escaped. I was one of them. I received a scholarship at the Stevens Institute of Technology and graduated there as a naval architect in 1958, but I could not get a job, because all ship design firms had some connection with the Navy and I was considered to be a *security risk*, because my family lived behind the Iron Curtain. So I had to pick a new profession and, luckily, I picked automation.

At this point I got lucky, because an engineer named Sam Russell was just starting an engineering firm and he hired me. Sam worked for President Roosevelt during World War II in the effort to replace the natural rubber supplies that were blocked by the Japanese with synthetic rubber. Sam knew how to get things done! His engineering design firm—focusing on plastics—was a success, and I, as his chief instrument engineer, had to hire more people.

With my thick Hungarian accent and the age of 25, I did not feel comfortable hiring experienced engineers twice my age, so I asked Sam to let me hire smart, fresh graduates from schools like MIT and Caltech and to use one day a week to teach them our profession. He agreed, and in a couple of years, we had one of the best automation engineering department.

I kept the notes I used in my weekly classes—a foot-high pile of paper accumulated at the corner on my desk. At this point I got lucky again, because an old-fashioned publisher named Nick Gronevelt visited me. He reminded me of my grandfather, with his hair parted in the middle and the gold chain of his watch hanging out of his vest pocket. Nick asked about the pile of notes on my desk, and when I explained, he decided to publish it.

It took me nearly five years to complete the three volumes of the first edition. My goal was to produce an experience-based practical and reliable book, written by users for users. The coauthors included representatives of suppliers (Hans Baumann and Greg Shinskey) and of academia (Paul Murrill, Cecil Smith), but my focus was on authors who could share their personal experience and these people were busy and not used to writing. Edward Teller wrote the preface and the 100+ authors of the first edition were all respected professionals in their fields.

Recent Changes

This fifth edition is the first one that is written for a global audience. In order to convert the *Instrument and Automation Engineers' Handbook* (*IEAH*) from an American to a universal handbook, overseas products are also included and authors were invited from all five continents. The *IEAH* is now available not only in printed form but also on DVD.

To take advantage of the digital age, web addresses of the suppliers of all analyzers discussed in this book are provided, so that the reader can gain access to the specifications of any product of a particular manufacturer with a single click. Similarly, web addresses of reading material are included so that the interested reader can get more details on a particular analyzer with just another click.

Finally, we have paid special attention to provide a very thorough index that should help the reader with quick access to specific information.

SELECTING THE RIGHT ANALYZER

Reasons for installing process analyzers include the improvement of safety and product quality, reduction of by-products, decrease in analysis time, tightening of specifications, and monitoring of contaminants, toxicants, or pollutants. The analyzer selection process begins with deciding on how the measurement will be used. A number of questions must be answered such as the following: Will the analyzer be used for closed-loop control or for monitoring purposes only? Must the measurement be performed *online*, or could 80% of the benefits be obtained for 20% of the cost and effort by making the measurement using grab samples? What is the benefit of the measurement? How will the cost be justified? Will staffing be reduced by moving the measurement from the lab to the pipeline? Will it increase production, reduce energy consumption, or yield product quality improvement? In addition, it is important to understand the process to be monitored. What is the response time of the process? Obviously, if a process takes 5 minutes to start responding and 60 minutes to fully change a material or heat balance, the type of analyzer needed is quite different from the one that serves a fast process. These issues are important and should be resolved early.

The critical question to be answered during this information gathering phase is "What attribute of the analyte would be easiest to measure?" For example, if oxygen concentration measurement is required (gaseous or dissolved oxygen), the engineer has to evaluate several oxygen measurement approaches, each of which uses a different property of the oxygen molecule. Depending on the process conditions, the engineer could select a paramagnetic oxygen analyzer, which makes use of the fact that elemental oxygen is one of the few gases attracted to a magnet, or an electrochemical cell device, which makes use of oxygen's electroactivity. After the appropriate analyzer type is selected, it is also important to resolve how the analyzer will be calibrated. Will it be evaluated against a laboratory measurement? If yes will that measurement be reliable and more accurate than that of the online analyzer?

This book describes the analyzers that are used online and/or in the laboratory for the determination of the compositions of process fluids and gases. The chapters are in alphabetical order, starting with application- and selection-related general topics. When one is in the process of selecting the best analyzer for a particular application, it is recommended to start by reading these first chapters, which give an overview of the capabilities of the various designs and also guidance concerning the selection of the right ones for various applications.

In alphabetical listing, two types of chapters can be found. The subject matter of one category is *application specific*, such as the measurement of mercury in air or oil in water, while the other category of chapters is *method specific*, such as the description of chromatographs or infrared analyzers. In the *application-specific* chapters, the capabilities, advantages, and disadvantages of several types of analyzers are discussed that can be considered for the measurement of the particular component. Inversely, in the *method-specific* chapters, the capabilities of such analyzer categories are discussed, which are suitable for measuring the concentration of several components. The chapter of interest can be quickly found, because the chapters are in alphabetic order.

Orientation Tables

In the selection process, the first step is to review the orientation tables provided in Chapter 1.1. The orientation table is a bird's-eye view of the capabilities of all the analyzers, and thereby it can quickly guide the reader to the best detector for the application at hand. In these tables, I have placed check marks in the rows and columns, indicating the sensor categories that can be considered for a particular measurement application. After studying the table, the reader should select two or three best candidates after reviewing their ranges, accuracies, costs, etc., and select the one that is closest to the requirements of the application.

In making the selection, one must first balance the maintenance costs against the value of the measurement data obtained. Orientation Table 1.1a sums up the capabilities of the different analyzers. The information in Orientation Table 1.1a is complemented by that in Table 1.1v, which compares the key attributes (cost, complexity, type of sample, etc.) of the many analyzers discussed in this chapter.

In addition to these overall orientation tables, other useful information and listings can be found elsewhere in Chapter 1.1:

Tables 1.1d and 1.1e: UV and visible radiation absorbing compounds
Table 1.1f: Fluorescent compounds
Table 1.1g and Figure 1.1a: IR and NIR absorbing compounds
Table 1.1i: Refractive index of various compounds
Table 1.1j: Ion-selective electrode (ISE) detectable compounds

All of these tables should be used with the proverbial grain of salt, because analyzers are constantly changing; electronics, intelligence, and sensing elements are developing every day. Therefore, the use of these tables is only a starting point, which should be followed by careful reading of the chapter(s) selected.

Feature Summaries

If the information provided in the orientation tables is insufficient and more details are needed, the reader should proceed to the front pages of the corresponding chapters, which provide Feature Summaries of the analyzers of interest. At the beginning of each chapter, a summary is provided listing the basic features of the instruments described in that chapter. This summary allows the reader to quickly determine if that sensor category is worth further consideration for the particular application in question. These summaries include data on range, accuracy, design pressure, design temperature, materials of construction, cost, suppliers, and in most cases advantages and disadvantages.

A partial alphabetical list of suppliers is also provided, and in some cases, where such information is available, the *most popular* manufacturers are also identified. Next to the names of each manufacturer are their web addresses, where the reader can find the specifications of the analyzer of that supplier.

The most experienced automation engineers can make a selection after just checking the orientation table. Those who want to refresh their memories concerning the many analyzer options and receive selection guidance or review their relative merits for many different applications should read Chapter 1.1. After that, the reader can turn to the Feature Summary at the beginning of the chapter discussing the selected analyzer to check their range, accuracy, rangeability, cost, supplier, and other relevant information. If time is available, it is also advisable to read not only the feature summary, but the whole chapter. Finally, if one wants to study the most recent developments in their features and capabilities, which have occurred after the publication of this handbook, one can visit the web addresses given at the end of the feature summary, where detailed specifications are provided.

Preparing the Specifications and Bidding

Once the analyzer type is selected, it is time to prepare the specifications to obtain bids. Typical specification forms are included at the end of each chapter. In addition, specification format are included at the end of the chapters. The forms in most cases were prepared by the International Society of Automation (ISA), but any other form can also be used as long as it fully describes the process condition and the requirements of the application. Once the specification form is completed, it is time to obtain quotations from a number of suppliers.

Over the years, I found that obtaining early bids is valuable, because the preparation of the quotations usually bring up additional questions and looking into those can help make the right selection, because they will ask questions, which could have been overlooked earlier, and by answering those questions, the designers and their clients will better understand their requirements.

Another advantage of early bidding is obtaining accurate cost estimates. Once the bids are in, and the bid analysis is prepared, the costs, accuracies, rangeabilities, calibration and maintenance requirements, etc., are known and guaranteed. This does not mean that the supplier will be selected as soon as the quotations are received. No, one's options should be kept open at this stage, but it means that from this point on, the design will be based on guaranteed facts instead of assumptions and estimates.

Appendix

The appendix contains information that engineers have to look up daily. The appendix entries are provided in the back part of this handbook and cover the following topics:

Chapter A.1: Definitions
Chapter A.2: Abbreviations, Acronyms, and Symbols
Chapter A.3: Organizations
Chapter A.4: Flowsheet and Functional Diagrams Symbols
Chapter A.5: Conversions among Engineering Units
Chapter A.6: Chemical Resistance of Materials
Chapter A.7: Composition and Properties of Metallic and Other Materials
Chapter A.8: Steam Tables

ROLE OF ANALYZERS IN INDUSTRIAL SAFETY

Analyzers play a critical role in the prevention of industrial accidents as they detect the presence of combustible, explosive, or toxic materials. In the oil industry, the detection of flammable concentrations of methane at the drilling rig or, in the case of the nuclear industry, the detection of hydrogen should trigger automatic plant shutdown. Yet in most cases, they only actuate alarms and it is up to the operators to decide on the action that will be taken. In other words, manual control still exists in many of our industries. We live at a time when cultural attitudes concerning automation are changing as we debate the proper role of machines in our lives, but it is still the operator who usually has the *last word* and automation is seldom used to prevent human errors.

Human Error Prevention

Safety statistics tell us that the number one cause of all industrial accidents is human error and automation can provide increased safety, if it is designed to overrule the actions of panicked or badly trained operators who often make the wrong decisions at the time of an emergency. In other words, the culture of trusting man better than machines has to change. It has to be recognized that safety provided by automation is also *man-made*, but it is made by different men rather than the panicked, rooky operators running around in the dark at 2 AM. Properly designed safety automation is made by professional control engineers who have spent months identifying all potential *what-if* sources of possible accidents, evaluating their potential consequences before deciding on the actions that are to be triggered automatically when they arise.

Safety Standards

There is ongoing effort to develop national and international standards that the safety of control loops (Safety Instrumented Systems [SIS]*) and also define how one might determine the level of safety (often referred to as the safety integrity level [SIL]), which these safety control loops provide. The safety control loop, which the SIL applies to (often referred to as safety instrumented function [SIF]), consists only of three components: (1) the analyzer or process variable sensor/transmitter, (2) the controller/logic system, and (3) the control valve or other final control element.

The reliability of the three loop components is quantified only by their probability of failure when called upon to operate (often called probability of failure on demand [PFD]).

---

* International standard IEC 61511 was published in 2003 to provide guidance to end users on the application of SIS in process industries. This standard is based on IEC 61508, a generic standard for design, construction, and operation of electrical/electronic/programmable electronic systems. Other industry sectors may also have standards that are based on IEC 61508, such as IEC 62061 (for machinery systems), IEC 62425 (for railway signaling systems), IEC 61513 (for nuclear systems), and ISO 26262 (for road vehicles, currently a draft international standard).

TABLE I.1
Control Loop (SIF) Safety (SIL) Levels

SIL Number	Minimum Hardware Fault Tolerance and Backup (Backup Architectures)	Required Safety Availability (RSA)	Probability of Failure on Demand (PFD)
SIL1	0 (1001 or 2002)	90%–99%	0.1–0.01
SIL2	1 (1002)	99%–99.9%	0.01–0.001
SIL3	2 (2003)	99.9%–99.99%	0.001–0.0001
SIL4	See IEC 611508	99.99%–99.999%	0.0001–0.00001

The PFD value of an instrument is determined by the vendor and is then reviewed by external certification agencies (e.g., Technischer Überwachungsverein [TÜV], exida*). This certification applies only to the particular loop component, but does not determine the SIL of the full control loop (SIF). It only certifies that the component is *suitable for use* at a particular level of safety (SIL) if all other components in that control loop (SIF) and the fault tolerance (hardware fault tolerance [HFT]) of the loop itself are both suitable for use at that SIL. Table I.1 provides the quantitative definition of what the four levels of control loop safety (SIL) could be.

Fault tolerance is the property that enables a control loop (SIF) to continue operating properly in the event of failure (or one or more faults within) some of its components (analyzer–sensor, logic, or valve). If the minimum hardware fault tolerance (HFT) is satisfied, a small failure will not cause total breakdown but can decrease the operating quality of the loop (SIF). Fault tolerance is required in high-availability or life-critical control loops.

The required safety availability (RSA) value refers to the suitability of a component for use in a particular safety control loop (SIF) at the listed level of safety (SIL). Conversely, the PFD is the mathematical complement of RSA (PFD = 1 − RSA), expressing the probability that the particular component in the safety loop (SIF) will fail to do its job when called upon.

Unfortunately, it is much easier to insert three zeros in a table than to increase the safety of a real process a thousand-fold. This is because most control loops (SIF) consist of more than the three components and they are not considered by SIS (various accessories, power supplies, etc.) and we know that "a chain is just as strong as its weakest link!"

Other limitations of the value of the present SIS standards include the following:

- Their complexity, their excessive use of acronyms, excessive focus on terminology, statistics, and calculations based on complex equations instead of clear and simple rules of verification and application
- Their vagueness, leaving key aspects of design safety requirements to interpretation
- Their lack of clear definitions of what components can be shared between SIS and BPCS (Business Planning and Control System) systems
- Leaving up to the manufacturer and its certification agency to decide if its product is suitable ("SIL certified") for use at a particular SIL
- Not considering the need for positive feedback to signal if final control elements have properly responded to SIS signal
- Treating loop component reliabilities as if they were the same, when in fact their probabilities of failure greatly differ (my estimate is valve ~ 80%, analyzer–sensor ~ 20%, and logic ~ 1%)
- Not considering that cyber interference through the Internet is possible
- Not defining the limits of what the operator is allowed to do and what the operator must be prevented from doing

Overrule Safety Control (OSC)

From the previous paragraphs, one can see that I consider the present state of safety standards insufficient. This is not only because they only provide vague statements, pseudo-quantitative statistics, and complicated equations, instead of clearly verifying and justifying safety systems, but I also consider them insufficient because they do not provide protection against cyber attacks and operator errors.

Overrule safety control (OSC) is a layer of safety instrumentation, which *cannot be turned off or overruled by anything or anybody.* When the plant conditions enter a highly accident-prone, life-safety region, such as the presence of hydrogen in a nuclear power plant or the presence of methane on an oil drilling rig, uninterruptible safe shutdown must be automatically triggered. The functioning of this layer is not subject to possible cyber attacks, because it is not connected to the Internet at all. The OSC final control elements and analyzers or other sensors are not shared with any of the other layers and are redundant (two out of three). The OSC logic overrules any and all actions of all other control layers or of the operators and cannot be modified or turned off by the operator. In short, once the OSC layer is activated, the plant shuts down under preplanned, totally automatic control and nothing and nobody can prevent that.

---

* Exida is considered by many to be the leading authorities in the field of functional safety. They serve to meet the growing need for companies to become more knowledgeable about the requirements for safety related applications. exida has evolved from two offices in the US & Germany to a global support network. They provide consulting, product testing, certification, assessment, cybersecurity, and alarm management services.

I hope that in the coming years SIS standards will be expanded to include the OSC layer and will deemphasize complicated equations and acronyms, will reduce vagueness and focus on the simple quantification of the safety levels of complete loops; after all, a chain is just as strong as its weakest link! I also hope that SIS will grow out of the *manual control culture* by incorporating protection against both operator errors and cyber terrorism.

Standardization of Digital Protocols

Just as it occurred in the *analog age*, a global standard is now evolving for digital communication, which could link all digital instruments, including analyzers, and could also act as a *translator* for those that were not designed to *speak the same language*. Naturally, this standardization should apply to both wired and wireless systems, thereby eliminating *captive markets* and should allow easy mixing of the products of different manufacturers in the same control loop.

In this regard, a major step was taken in 2011, when the five major automation foundations, FDT (field device tool) Group, Fieldbus Foundation, HART (highway addressable remote transducer) Communication Foundation, PROFIBUS and PROFINET International, and OPC* (Object Linking and Embedding for Process Control) Foundation, have developed a single common solution for Field Device Integration (FDI). These foundations decided to combine their efforts and form a joint company. The new company is named FDI Cooperation, LLC (a limited liability company under U.S. law). This was made formal by all representatives signing the contract documents in Karlsruhe, Germany, on September 26, 2011. The FDI Cooperation is headed by a board of managers. It is composed of the representatives of the involved organizations, as well as managers of global automation suppliers, including ABB, Emerson, Endress+Hauser, Honeywell, Invensys, Siemens, and Yokogawa.

FDI Cooperation originated from efforts at the Electronic Device Description Language (EDDL) Cooperation Team (ECT) to accelerate deployment of the FDI solution, which kicked off at the 2007 Hanover. FDI is a unified solution for simple as well as the most advanced field devices and for tasks associated with all phases of their life cycle such as configuration, commissioning, diagnostics, and calibration.

This trend is most welcome, because once completed, it will allow the automation and process control engineers to once again concentrate on designing safe and optimized control systems and not worry about the possibility that the analyzers or other instruments supplied by different suppliers might not be able to talk to each other. Thereby, it is hoped that the "Babel of communication protocols" will shortly be over.

Standardizing Accuracy and Rangeability Statements

The ISA should issue standards that would define the same basis for all accuracy statements. It is important that when manufacturers state the accuracy of their analyzers, the users know *the percentage of what* is being stated, such as % full scale (%FS), % upper range value (%URV), or % actual reading (%AR). In the case of analyzers, this means uniformity in the definition not only of the guaranteed accuracy but also of the range over which that accuracy guarantee applies. This would mean the end of giving accuracy statements, without stating their basis.

It is also time for professional societies and independent testing laboratories to make their findings widely available and also time for technical magazines and lecturers at technical conferences to clearly state their findings, so far as indicating if a particular manufacturer's products lived up to the performance specifications and which did not.

In addition, suppliers should always state the range over which their inaccuracy statements are guaranteed. This *rangeability* or turndown should always be defined as the ratio between the maximum and minimum readings for which the inaccuracy statement is guaranteed. In their detailed specifications, they should also state if the analyzers were individually calibrated or not and if the inaccuracy statement is based only on the linearity, hysteresis, and repeatability errors or if it also includes the effects of drift, ambient temperature, overrange, supply voltage, humidity, RFI, and vibration.

Smarter Analyzers

In the case of transmitters, the overall performance is largely defined by the internal reference used, and this is particularly the case in multiple-range units that should be provided with multiple references. Analyzer improvements should also include the ability to switch between wired and wireless data transmission, provide local displays of both the present reading and its past history, be able to average over various time periods, and give minimum and maximum values experienced during those periods. They should also be provided with multivariable sensing capability, not only for compensation and redundant safety backup purposes, but also for signaling the need for maintenance, recalibration, or self-diagnosing of failure.

Analyzer Probes and Sampling Systems

In the area of continuous online analysis, further development is needed to extend the capabilities of probe-type

---

* OPC is the interoperability standard for the secure and reliable exchange of data in the industrial automation, space and in other industries. It is platform that ensures the seamless flow of information among devices from multiple vendors. The OPC Foundation is responsible for the development and maintenance of standards. The OPC standards are a series of specifications developed by industry vendors, end-users and software developers. These specifications define the interface between Clients and Servers, as well as Servers and Servers, including access to real-time data, monitoring of alarms and events, access to historical data and other applications.

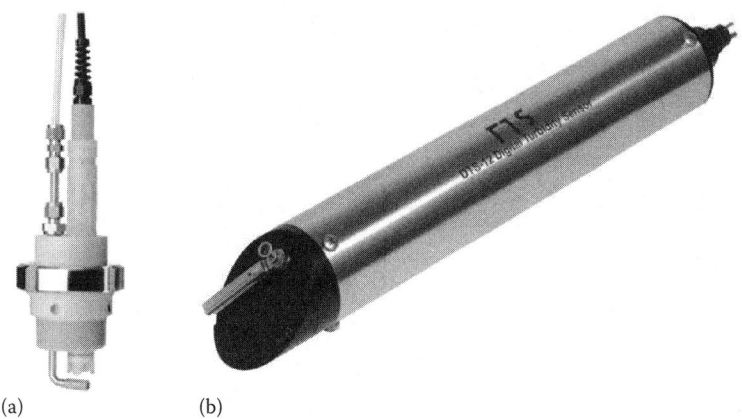

FIG. I.3
Automatic probe cleaners. (a) Jet-type probe cleaner. (Courtesy of Yokogawa.) (b) Mechanical probe cleaner. (Courtesy of GSA.)

analyzers. The needs include the changing of probe shapes to achieve better self-cleaning or using *flat tips* for ease of cleaning (Figure I.3). The automatically cleaned probes should be installed in sight flow indicators (see Figure 1.1r) so that the operator can visually check if the cleaner is operating properly. Wide varieties of probe-cleaning devices are available on the market. Their features and capabilities for the removal of various types of coatings and a list of their suppliers are provided in Table 1.1s.

In-line, probe-type analyzers are also used in stack gas composition measurement. Some of the microprocessor-based designs also provide self-calibration for both zero and span. In addition, they can provide automatic cleaning and drying to accommodate wet scrubber applications, in which it is necessary to periodically force hot and pressurized instrument air into the probe cavity to thoroughly dry the diffuser. The measurement is made by analyzing the measurement beam as it is returned by the retroreflector within the probe's gas measurement cavity (Figure 1.1p). These probe-type stack gas analyzers can operate in locations with high water vapor and particulate loadings and at temperatures up to 800°F (427°C).

Other advances including the wider use of fiber-optic probes, multiplexing and sharing the costs of analyzer electronics among several probes, and providing self-calibrating and self-diagnostics features are all important.

Sampling

Perhaps the most critical aspect of any analyzer installation is the sampling mechanism. The reading that an analyzer generates can be no better than the sample it has been presented for analysis. Sampling is such an important area that entire books have been published on that subject alone.* In this book, there are several chapters on sampling, including Chapters 3 through 5.

Conclusion

Analyzers play a key role in the automation profession. This profession can simultaneously increase GDP and industrial profitability without building a single new plant, just by optimizing the existing ones. We can achieve that goal, while also reducing both pollution and energy consumption, solely through applying the state of the art of automation. We can increase productivity without using a single pound of additional raw material and without spending a single additional BTU of energy. We can also protect our industries by replacing manual control with automatic safety controls.

I hope that you will find this fifth edition of the IAEH useful in your daily work and that what started out 50 years ago as a pile of notes at the corner on my desk will continue to contribute to the progress and recognition of our fine profession.

Béla G. Lipták
http://belaliptakpe.com/

---

* Houser, E. A., *Principles of Sample Handling and Sampling Systems Design for Process Analysis*, Research Triangle Park, NC: ISA, 1977; Cornish, D. C., Jepson, G., and Smurthwaite, M. J., *Sampling Systems for Process Analysers*, London, U.K.: Butterworths, 1981; Sherman, R. E., *Process Analyzer Sample-Conditioning System Technology*, New York: John Wiley & Sons, 2002.

CONTRIBUTORS

Here, only the authors who contributed or updated chapters for the 5th edition of this volume are listed in alphabetic order, together with their academic degrees and their titles or positions that they held at the time of making their contributions to this 5th edition.

In front of each chapter, the names of all contributors to all editions of that chapter are listed. The dates after their names indicate the edition to which they contributed. The author whose name is followed by the date (2017) is the author who has contributed or updated the chapter for the 5th edition.

L. JOSEPH BOLLYKY	PhD, PE, President, Bollyky Associates, Inc., Stamford, Connecticut
AERENTON FERREIRA BUENO	PhD, Chemistry, Petrobras, Sao Jose dos Campos, Brazil
AJAY V. DESHMUKH	PhD, ME, Principal, AGTI's Dr. Daulatro Aher College of Engineering Karad, Pune, India
DAVID CALVETE GALLEGO	I&CS Engineer, Fluor Corporation, Madrid, Spain
HARVINDAR SINGH GAMBHIR	AMIETE, FSCP (TUV), Vice President I&C, Reliance Industries, Ltd, Mumbai, India
RICHARD A. GILBERT	PhD, Professor Chemical and Biomedical Engineering, University of South Florida, Tampa, Florida
JAMES R. GRAY	BS Chemistry, MBA, Application Manager, Rosemount Analytical, Retired, Torrance, California
RICHARD S. HARNER	PhD, Senior Research Specialist, The Dow Chemical Company, Midland, Michigan
DUŠAN KOPECKÝ	PhD, Assistant Professor, University of Chemistry and Technology, Prague, Czech Republic
BÉLA G. LIPTÁK	MME, PE, ISA Fellow, Consultant, inducted into Control Process Automation Hall of Fame, in 2011, President, Béla Lipták Associates, Stamford, Connecticut
DAVID S. NYCE	BSEE, MBA, Owner, Revolution Sensor Company, Apex, North Carolina
STEPHEN HAROLD ORTON	ASEE, Cap, CCST, Master Electrician, Augusta Utilities, Grovetown, Georgia

MORTON W. REED	PhD, PE, Environmental Coordinator, Columbus Water Works, Columbus, Georgia
RAJAGOPALAN SREENEVASAN	BE, ME, MBA, CPEng, RPEQ, Principal Instrument Engineer, TetraTech Proteus, India
SURY VEDULA	MSc, Process Analyzer Specialist, Fluor Canada Ltd, Calgary, Alberta, Canada
IAN VERHAPPEN	BscEnv, BScCh, PE, CAP, Director, Orbis Engineering Field Services, Industrial Automatio Networks, Inc., Calgary, Alberta, Canada
VICTOR WEGELIN	BS, MBA, PE, Consulting Engineer, PMA Concepts, Redondo Beach, California
STEVEN J. WEST	BS, Senior Manager, Analytical and Temperature Development, Schneider Electric, Foxboro Field Devices Division, Foxborough, Massachusetts
SATISH YADAV	BE-Instrumentation & Controls, Manager Subsea Controls & Instrumentation, Reliance Industries Ltd, Mumbai, India
MIHA ZAVRŠNIK	PhD, Electrical Engineering, Product Manager, Anton Paar GmbH, Graz, Austria

Analytical Measurement

1

1.1
ANALYZER SELECTION AND APPLICATION 19

Introduction 19
Preparation Phase 19
   The Orientation Table 20
   Obtaining Preliminary Quotations 26
   Defining the Purpose of the Analysis 26
   Information Gathering 27
Selection Considerations 27
   Specificity 27
   Accuracy and Precision 28
   Calibration 28
   Analysis Frequency 28
Analyzer Types 29
   Chromatographs 29
   Spectrometers 29
   Electrochemical Analyzers 34
   Wet Chemistry Analyzers 36
   Miscellaneous Techniques 36
   Analyte-Specific Techniques 37
   Sampling Techniques 37
   Fiber-Optic Probes (FOP) 38
   Probe Cleaners 41
Analyzer Location 42
Data Handling 43
Maintenance 44
Cost 44
Conclusions 45
Specification Forms 45
Definitions 50
Abbreviations 50
Organizations 50
Bibliography 50
Basics 51
Measurement Applications 51

1.2
ANALYZER SAMPLING 53

Introduction 54
General Considerations 54
   Feasibility Evaluation 54
   Sample System Specification 55
   Sampling Point Location 56
   Sample Transport 56
   Transport Lag 57
   Sample Disposal 57
   Ambient Effects 58
   Test and Calibration 58
Sampling System Component Selection 58
   Filtering and Separation 59
   Separating Liquids from Gases 59
   Spargers, Packed Towers, and Strippers 60
   Separating Two Liquid Phases 62
   Gas Removal from Liquids 62
   Bypass Filters 63
   Self-Cleaning Filters 63
   Automatic Probe Cleaning 65
   Homogenizers 65
Sample Conditioning 66
   Vaporizing Samples 67
   Entrainment Removal 68
   Selection of Component Materials 68

Applications 68
　Gas Sampling 68
　Stack Gas Sampling 69
　Automatic Stack Sampling 69
　Sampling at High Pressure 71
　Chemical Reactor Samplers 71
　Duckbill Samplers 72
Solids Sampling 72
Difficult Process Sampling 73
　Trace Analysis Sampling 73
　Multistream Sampling 74
　Sampling for Mass Spectrometers 75
　New Sample System Initiative (NeSSI™) 75
Conclusions 75
Specification Forms 75
Definition 78
Abbreviations 78
Organizations 78
Bibliography 78

1.3
ANALYZER SAMPLING 79

Introduction 79
The EPA Particulate Sampling System 80
　Microprocessor-Controlled Stack Sampling 80
　Pitot Tube Assembly 81
　Heated Compartment (Hot Box) 84
　Ice-Bath Compartment (Cold Box) 84
　Control Unit 84
Automatic Sampling Trains 85
　Sampling for Gases and Vapors 86
Specification Forms 86
Abbreviations 90
Organizations 90
Bibliography 90

1.4
ANALYZER SAMPLING 91

Introduction 93
Air Quality Monitoring 93
　Air Quality Standards 93
　Single Source (Stacks) Monitoring 94
　Urban Area Monitoring 94
　Sampling Site Selection 95
Monitoring Methods 95
　Static Monitoring 95
　Laboratory Analyses 96
　Automatic Monitoring 97
　Sensors 97
　Data Transmission 97
　Data Processing 97

Audits 98
　Automatic Analyzers 98
Sampling 100
　Sampling Problems 100
Gas and Vapor Sampling 101
　Bag Sampling 101
Particulate Sampling 103
Specification Forms 105
Definitions 108
Abbreviations 108
Organizations 108
Bibliography 108

1.5
AMMONIA ANALYZERS 109

Introduction 110
Ammonia Measurement in Liquids 110
Colorimetric Method 110
Ion-Selective Method (ISE) 111
UV Spectrophotometry Method 112
Ammonia in Gases 112
Tunable Diode Laser (TDL) 113
Conclusion 113
Specification Forms 114
Abbreviations 116
Bibliography 116

1.6
BIOMETERS TO QUANTIFY MICROORGANISMS 117

Introduction 117
ATP and Living Organisms 118
ATP Analysis 118
Luminescence Biometer 119
Specification Forms 120
Abbreviations 122
Organization 122
Bibliography 122

1.7
CARBON DIOXIDE 123

Introduction 124
Role in Global Warming 124
Emissions 125
Ambient Air Measurement 126
　Nondispersive Infrared Sensors 126
　Gas Filter Correlation Type 127
Source Measurement 127
Specification Forms 128
Abbreviations 131
Bibliography 131

1.8
CARBON MONOXIDE 132

Introduction 133
Calibration Techniques 133
Analyzer Types 134
    Nondispersive Infrared (NDIR) 134
    Interferences 135
    Gas Filter Correlation (GFC) 135
    Mercury Vapor Analyzer 135
    Gas Chromatograph 136
    Electrochemical 137
Portable Monitors 137
    Catalytic Analysis 138
Spot Sampling of Ambient Air 138
Conclusions 138
Specification Forms 138
Abbreviations 141
Organizations 141
Bibliography 141

1.9
CHLORINE ANALYZERS 142

Introduction 143
Chlorine in Gas and Water 144
Analyzer Types 145
    Iodometric 145
    Colorimetric 145
    Amperometric 146
Regulations 150
Conclusions 150
Specification Forms 151
Definitions 153
Organizations 153
Bibliography 153

1.10
CHROMATOGRAPHS 154

Introduction 155
The Separation Process 155
The System Structure 156
The Process Gas Chromatograph (PGC) 157
Components of a PGC 158
    Analyzer 158
    Oven 159
    Valves 160
    Rotary Valve 161
    Sliding Plate Valve 161
    Diaphragm Valve 161
    Columns 162
    Column and Valve Configurations 163
    Hardware 164
    Sample Injection 164

    Back-Flush 165
    Heart-Cutting 166
Detectors 166
    Thermal Conductivity 167
    Flame Ionization Detector (FID) 168
    Flame Photometric Detector (FPD) 169
    Pulsed FPD 170
    Orifice-Capillary Detector (OCD) 172
    Miscellaneous Detectors 172
    Photoionization Detector (PID) 172
    Electron Capture Detector (ECD) 173
    Discharge Ionization Detectors 174
    Pulsed Discharge Detector (PDD) 174
Carrier Gas Flow Control 175
Programmer Controller 175
    Programmer 176
    Peak Processor 176
Microprocessor Operated PGC 176
    Input–Output 176
    Communication 177
Operator Interface 178
    Alarms and Diagnostics 180
    Quantitation 180
Sample Handling 180
    Sample Probe 180
    Sample Transport 181
    Sample Conditioning 181
    Multistream Analysis 183
    Sample Disposal 183
Installation 183
Advances 184
Summary and Acknowledgments 185
Specification Forms 185
Definitions 188
Abbreviations 188
Bibliography 189

1.11
CHROMATOGRAPHS 190

Introduction 191
Comparison with Gas Chromatographs 192
Main Components 192
    Carrier Supply 192
    Supply Pumps 193
    Pressure and Flow Controls 193
    Valves 193
    Columns 193
    Selectivity and Resolution 194
    Liquid–Partition Columns 194
    Liquid–Adsorption Columns 194
    Gel Permeation Columns 195
    Ion Exchange Columns 195
    Electrophoresis 195

Detectors 195
    Optical Absorbance 195
    Refractive Index 195
    Dielectric Constant 196
Applications 196
Specification Forms 196
Definitions 199
Abbreviations 199
Bibliography 199

1.12
COAL ANALYZERS 200

Introduction 200
Coal Property Measurement 201
Laboratory Techniques 202
    Thermogravimetry 202
    Bituminous Coal Analysis 202
    Gross Calorific Value 202
    Total Sulfur 203
    Ash Analysis 204
On-Line Monitors 204
    Prompt Gamma Neutron Activation
        Analyzers (PGNAA) 204
    Laser-Induced Breakdown Spectroscopy
        (LIBS) 206
Specification Forms 206
Definitions 209
Abbreviations 209
Bibliography 209

1.13
COLORIMETERS 210

Introduction 211
Color Measurement 211
    Absorbance and Transmittance
        Colorimetry 211
Spectrophotometric Analyzers 212
    Spectrophotometer Design 212
Tristimulus Method (Reflectance) 213
    Lab Algorithm of the Textile
        Industry 213
Continuous Color Monitors 214
    In-Line Liquid Color Measurement 214
    Online Shade Monitors 214
    Optical Fluorescence and Luminescence
        Sensors 214
    New Developments: Multispectral
        Analysis 215
Specification Forms 215
Abbreviations 218
Organizations 218
Bibliography 218

1.14
COMBUSTIBLE GAS OR VAPOR SENSORS 219

Introduction 220
Properties of Flammable Materials 221
Detector Types 221
Catalytic Combustion Type 222
    Limitations 222
    Measuring Circuits 222
    Diffusion Head Design 224
    Sampling System 225
    Accessories 225
    Total System Design 225
    Advantages and Disadvantages 227
Flame Ionization and Photo Ionization Types 227
    Flame Ionization Detectors 227
    Photo-Ionization Detectors 227
Infrared Types 228
    Point Monitoring System 228
    Open Path System 229
    Hydrocarbon Gases in the Atmosphere 230
    Point Measurement 231
Specification Forms 231
Definitions 234
Abbreviations 234
Bibliography 234

1.15
CONDUCTIVITY MEASUREMENT 235

Introduction 236
Theory of Operation 236
Cell Constant 236
Cell Dimensions 237
Two-Electrode Cells 238
Four-Electrode Measurement 238
Electrodeless Cells 239
Measurement Applications 239
Concentration Measurements 239
High-Purity Water Measurements 239
Corrosive and Fouling Applications 240
Calibration and Maintenance 241
    Calibration of Conductivity Sensors 241
    Maintenance of Conductivity Cells 241
New Developments 241
    Digital Conductivity Sensors 241
    New Conductivity Transmitter
        Developments 241
    Smart Conductivity Transmitters 243
    Asset Management 243
    Wireless HART™ Transmitters 244
Conclusion 244
Specification Forms 245
Definitions 247
Abbreviations 247
Organizations 247

Bibliography 247
Digital Conductivity Sensors 247
Inferred pH 247
Conductance Data for Commonly Used
    Chemicals 247

1.16
CONSISTENCY MEASUREMENT 248

Introduction 249
Sensor Designs 249
    Mechanical Devices 250
    Optical Designs 252
    Microwave Sensors 254
Conclusions 254
Specification Forms 254
Definitions 257
Abbreviations 257
Organization 257
Bibliography 257

1.17
CORROSION MONITORING 258

Introduction 258
Corrosion-Monitoring Techniques 259
    Coupon Monitoring 260
    Electrical Resistance (ER) Monitors 261
    Linear Polarization Resistance (LPR)
        Monitors 263
    LPR and ER Combination Designs 264
    Down-Hole Corrosion Monitoring
        (DCMS) 264
    Wall Thickness Monitoring 265
Specification Forms 265
Definition 268
Abbreviations 268
Bibliography 268

1.18
CYANIDE ANALYZERS: WEAK-ACID DISSOCIABLE (WAD) 269

Introduction 269
Cyanide Species 270
Cyanide Analysis: Laboratory Methods 270
Free Cyanide (CN_F) 270
    Available Cyanide (CN_{WAD}) 271
    Total Cyanide (CN_T): Manual
        Distillation 271
    Total Cyanide (CN_T): Automated 272
Primary and Alternate Analytical
    Methods 272
Online Cyanide Analysis Methods 273
Comparison of Online Cyanide
    Analyzers 273

Conclusions 273
Specification Forms 273
Abbreviations 276
Organizations 276
Bibliography 276
ASTM International Standards 276
U.S. EPA Standards 276
EN ISO Standards 276

1.19
DIFFERENTIAL VAPOR PRESSURE 277

Introduction 277
Applications 277
d/p Vapor Pressure Transmitter 280
Conclusions 280
Specification Forms 280
Definitions 283
Abbreviation 283
Bibliography 283

1.20
DIOXIN AND PERSISTENT ORGANIC POLLUTANTS ANALYZERS 284

Introduction 284
Sampling Systems 285
    Sample Recovery 286
    Sample Extraction 286
    Analysis 286
Conclusions 286
Specification Forms 286
Definition 289
Abbreviations 289
Bibliography 289

1.21
ELECTROCHEMICAL ANALYZERS 290

Introduction 291
Voltammetric Analysis 292
    Current, Voltage, and Time 293
    Potentiometry 293
    Galvanic and Electrolytic
        Probes 294
    Amperometry 295
    Polarography 298
    Coulometry 298
Conclusions 299
    Limitations 299
    Advances 299
Specification Forms 299
Abbreviations 302
Bibliography 302

1.22
ELEMENTAL ANALYZERS 303

Introduction   304
Atomic Absorption Spectrometer   305
Inductively Coupled Plasma Detector   306
   Operating Principle   306
Relative Merits of AA and ICP   307
X-Ray Fluorescence Spectrometer   307
   Instrumentation   307
Specification Forms   309
Abbreviations   311
Organization   311
Bibliography   311

1.23
FIBER-OPTIC (FO) PROBES AND CABLES 312

Introduction   313
Fiber-Optic Cables   313
   Internet Supporting Cables   313
Fiber-Optic Probes   315
   System Operation   315
   System Components   315
   Probe Designs   317
   Sealing Techniques   318
   Process Interfaces   320
Safety Considerations   322
   Fiber-Optic Ignition   322
Environmental Effects   323
Applications   324
   Absorption Measurement   324
   Scattering   324
   Critical Angle and Other Sensors   325
Conclusions   326
Specification Form   326
Definitions   328
Abbreviations   328
Organizations   328
Bibliography   328

1.24
FLAME, FIRE, AND SMOKE DETECTORS 330

Introduction   331
Smoke Detectors   331
   Ionization Chamber Sensors   331
   Photoelectric Sensors   332
Flame and Fire Detectors   332
   Thermal Sensors   332
   Optical Flame Detectors   332
   Fire Detectors   334
Burner Flame Safeguards   335
   Flame Rods   336
   Visible Radiation   337
   Flame Safeguard Installation   339

Conclusions   339
Specification Forms   340
Definition   342
Abbreviations   342
Organizations   342
Bibliography   342

1.25
FLUORIDE ANALYZERS 343

Introduction   344
Types of Fluoride Compounds   345
Detector Types   345
   Detector Tubes   345
   Electrochemical Cells   346
   Paper Tape   346
   Ion Mobility Spectrometry (IMS)   346
   Infrared Spectroscopy   347
   Ion-Specific Electrodes (ISE)   347
   Silicon Dioxide Sensors   348
   Other Methods   348
Laboratory Methods   348
Specification Forms   348
Abbreviations   351
Bibliography   351

1.26
HAZARDOUS AND TOXIC GAS MONITORING 352

Introduction   353
Hazardous Area Classification   356
   Safety Instrument Performance Standards   357
   Levels of Toxicity   357
   Exposure Limits   357
Detector Types   360
   Electrochemical Sensors   361
   Oxygen Detectors   361
   Amperometric Sensors   362
   Electrodes and Electrolytes   362
   Fixed Detectors   363
   Portable Detectors   364
   Advantages and Limitations   365
   Spectrometers   365
   Applications   365
   Semiconductor Detectors   366
   Catalytic Bead Detectors   366
   Photo and Flame Ionization Detectors   366
   Open-Path Detectors   366
Dosage Sensors   366
   Color-Change Badges   366
   Color Detector (Dosimeter) Tubes   368
   Sorption-Type Dosimeters   370
Calibration   371
   Dynamic Calibrators   372

Selection 372
Application Considerations 372
Communication and Maintenance 373
Specification Form 373
Definitions 375
Abbreviations 375
Organizations 375
Bibliography 376
Web Links to Government
    Organizations 376

1.27
HEATING VALUE CALORIMETERS 378

Introduction 379
Units and Wobbe Index 381
Design Variations 381
    Bomb Calorimeters 381
    Water Temperature Rise
        Calorimeter 382
        Air Temperature Rise Calorimeter 383
        Airflow Calorimeter 383
        Residual Oxygen Calorimeter 384
        Chromatographic Calorimeter 384
        Expansion Tube Calorimeter 384
Temperature-Based Designs 384
    Adiabatic Flame Temperature 384
    Thermopile 385
    Other Temperature-Based Designs 385
    Optical Calorimeters 385
Applications 386
Sample Conditioning 386
Conclusion 386
Specification Form 386
Definitions 388
Abbreviations 388
Organization 388
Bibliography 388

1.28
HYDROCARBON ANALYZERS 390

Introduction 392
Analyzer Types 393
    Flame Ionization Detectors 393
    Gas Chromatography 395
    Spectrometric Methods 396
    Ion Mobility Spectroscopy 398
    Hydrocarbon Dew-Point Meter 398
Calibration 399
Conclusions 399
Specification Forms 400
Definition 402
Abbreviations 402
Bibliography 402

1.29
HYDROGEN CYANIDE [HCN] DETECTORS 403

Introduction 404
Cyanide Chemistry 404
Exposure Limits 404
Detector Designs 405
    Fourier-Transform Infrared
        (FTIR) 405
    Performance Standards 406
Conclusion 407
Specification Forms 407
Definition 410
Abbreviations 410
Organizations 410
Bibliography 410

1.30
HYDROGEN IN STEAM OR AIR ANALYZERS 411

Introduction 411
    Hydrogen and Its Future Role 411
Properties and Safety 412
Detector Types 413
    Thermal Conductivity 413
    Solid State 414
    Electrochemical 414
    Palladium-Based 414
    Catalytic Bead (CB) 414
    Chromatography 415
    Mass Spectrometry 415
Sampling and Installations 415
    Specialized Applications 416
Conclusions 417
Specification Forms 417
Abbreviations 420
Organizations 420
Bibliography 420

1.31
HYDROGEN SULFIDE DETECTORS 421

Introduction 422
Analyzer Types 422
    Electrochemical 422
    Gold Film 423
    Solid State 423
    Lead Acetate Tape 423
    Ultraviolet 424
    Chromatograph 425
Applications 425
Specification Forms 425
Abbreviations 428
Bibliography 428

1.32 INFRARED AND NEAR-INFRARED ANALYZERS 429

Introduction 431
Principles of Analysis 432
    Beer–Lambert Law 434
    Laboratory Spectrophotometers 435
Analyzer Design Categories 436
    Single-Beam Configuration 436
    Dual-Beam Configuration 437
Laboratory Analyzers 438
    Grating Spectrophotometers 438
    Filter Spectrometers 440
    Fourier Transform Spectrometers 440
    Tunable Lasers 440
    Process Control Applications 441
Online Applications 441
    Single-Component Analyzers 441
    Dual-Chamber Luft Detector 442
    Design Variations 442
    Multiple-Component Fixed Filter Analyzer 444
    Programmed Circular Variable Filter Analyzer 444
Portable Analyzers 444
Infrared Sources 445
Infrared Detectors 445
    NDIR Detectors 445
    Thermal Detectors 445
    Photoconductive Detectors 446
Selecting the Cell 446
    Path Length Selection 446
    Gas Cells 446
    Liquid Cells: Transmission Type 447
    Liquid Cells: Reflection Type 447
    Solid Samples 448
Calibration 448
    Linearity 448
Packaging 449
Applications and Advances 449
Near-Infrared Analyzers 451
    Interpreting the Absorption Bands 451
    Sample Temperature Control 451
    Fiber-Optics 451
    Types of NIRs 452
    Optical Sources 452
    Gases 452
    Liquids 453
    Solids 453
    Calibration Transfer 453
Specification Forms 453
Definitions 456
Abbreviations 456
Organizations 456
Bibliography 456

1.33 ION-SELECTIVE ELECTRODES (ISE) 458

Introduction 459
Theory of Operation 460
    Reference Electrode 460
    Concentration and Activity 461
    Ionic Strength Adjustment, Buffers 462
    Temperature Effects 463
    System Accuracy 464
Electrode Types 464
    Glass 464
    Solid State 465
    Liquid Ion Exchange 466
Measurement Range 467
Interferences 467
    Solution Interferences 467
Calibration 468
    Multiple Electrode Heads 469
Advantages 469
Potential Disadvantages 469
Specification Forms 470
Definitions 473
Abbreviations 473
Bibliography 473

1.34 LEAK DETECTORS 474

Introduction 475
Above Ground Detection 476
    Pressure- and Vacuum-Based Testing 476
    Combustible or Toxic Leaks 476
    Personal Alarms 477
Detector Types 477
    Ultrasonic Detectors 477
    Thermal Conductivity Detectors 478
    Halogen Detectors 478
    Thermography 479
    Mass Spectrometry 479
Underground Detection 479
    Level Monitoring 479
    Soil Concentration Detection 480
    Aspirated Sensors 481
    Standpipe Detector 481
New Developments 481
Specification Forms 482
Abbreviations 485
Organizations 485
Bibliography 485

1.35 MASS SPECTROMETERS 486

Introduction 487
Operation Principles 487
    Sample Input 487
    Sample Ionization 487
    Ion Separation 489
    Ion Detection 491
    Vacuum Environment 491
    Data Reduction and Presentation 492
Residual Gas Analyzers 492
Conclusions 493
Specification Forms 493
Bibliography 496

1.36 MERCURY IN AMBIENT AIR 497

Introduction 498
Sampling and Concentration 498
    Impinger Collection Methods 498
    Amalgamation on Wettable Metals 500
    Activated Absorption 500
    Conversion of Mercury Vapors 501
Detection Methods 501
    Ultraviolet Absorption 501
    Colorimetric Methods 502
    Gas Chromatography for Mercury Organics 503
    Atomic Fluorescence Spectroscopy 503
    Other Analytical Procedures 503
Regulations 503
Specification Forms 504
Definitions 506
Abbreviations 506
Organizations 506
Bibliography 506

1.37 MERCURY IN WATER 507

Introduction 508
Total Mercury Detection 508
    Sample Treatment 508
    Colorimetric Detection 509
    Atomic Absorption Spectrophotometry 509
    Online Measurement 511
Organic Mercury Detection 513
    Sample Treatment 513
    Gas Chromatography 513
    Thin-Layer Chromatography 514
Direct (Thermal) Decomposition 515
Specification Forms 516
Abbreviations 518
Organization 518
Bibliography 518

1.38 MOISTURE IN AIR 519

Introduction 521
Operating Principles 521
    Enthalpy in the Air 522
Applications 523
    Using the Psychrometric Chart 523
    HVAC: Heating, Ventilating, and Air Conditioning 523
    Clean Rooms, Incubators, Microenvironments, Respiration Therapy 524
    Breathing Air 524
    Airline Environment 524
    Humidity Effect of Combustion 524
    Calculation of Relative Humidity from First Thermodynamic Principles 524
    Internet Options for the Calculation of Relative Humidity 525
Sensor Technology 525
    Spectroscopic Absorption 526
    Electrolytic Cells 526
Thin-Film Capacitance Probes 527
    Chilled Mirror 528
    Cooling Methods 528
    Legacy Designs: Historical Perspective 529
Additional Considerations 531
    Installation 531
    Condensing Atmospheres 531
Calibration 532
    Autocalibration 532
    Handheld Calibration 532
    Reference to Standards 533
    Two-Pressure Humidity Generator 533
Calibration Method 534
Conclusions 535
Selection Criteria 535
Specification Forms 535
Definitions 538
Abbreviations 538
Standards Agencies 538
Bibliography 538

1.39 MOISTURE IN GASES AND LIQUIDS 540

Introduction 542
Laboratory Analyzers 542
Process Analyzers 542
    Sampling Systems 543
    Electrolytic Hygrometer 543
    Capacitance Hygrometer 545
    Impedance Hygrometer 548
    Piezoelectric Hygrometer 549

Heat-of-Adsorption Hygrometer   550
Spectroscopy   551
Microwave Absorption Hygrometer   552
Dipole Polarization Effect Moisture Sensor   553
Cavity Ring-Down Spectroscopy (CRDS): Moisture Analysis   554
Neutron Backscatter Moisture Online Analyzer   554
Refraction-Based Moisture Analyzer   555
Calibration of Moisture Analyzers   555
Specification Forms   558
Abbreviations   560
Bibliography   560

1.40 MOISTURE IN SOLIDS   561

Introduction   563
Chemical Methods   563
Gravimetric Methods   563
   Oven Drying (Loss on Drying)   563
   Desiccants   565
   Thermogravimetry   565
   Lyophilization   565
   Absorption and Condensation   566
Fast Neutron Moderation   566
   Neutron Probe   566
Combined Neutron/Gamma Probes   567
Infrared Absorption and Reflection   569
Time Domain Reflectometry   570
Microwave Attenuation   571
   Basis Weight Compensation   571
   Moisture in Coal   571
   Sugar Industry Applications   572
Capacitance Moisture Gauge   572
   Sensor Designs   572
Capacitance Measurement   573
   Limitations   573
Resistance Moisture Gauge   573
   Limitations   573
Impedance Moisture Gauge   574
Nuclear Magnetic Resonance   574
Radio Frequency Absorption   575
Conclusions   575
Specification Forms   575
Definitions   578
Abbreviations and Symbols   578
Bibliography   578

1.41 MOLECULAR WEIGHT OF LIQUIDS   580

Introduction   581
Polymers   581
Analyzer Designs   582
   Osmometers   582
   Light-Scattering Photometers   585
   Viscometers   587
   Gel-Permeation Chromatography   588
   End Group Determination   591
   Electron Microscope   591
   Ultracentrifuge   592
Conclusions   592
Specification Forms   593
Definitions   595
Abbreviations   595
Bibliography   595

1.42 NATURAL GAS MEASUREMENTS   597

Introduction   597
Natural Gas Instrumentation   597
   Flow Measurement   597
   Analyzers   600
Specification Forms   603
Abbreviations   605
Organizations   605
Bibliography   605

1.43 NITROGEN, AMMONIA, NITRITE AND NITRATE   606

Introduction   607
Nitrogen and the Environment   607
   Nitrification and Denitrification   608
Analysis Goals   608
   Ammonium Detectors   608
   Nitrite Measurement   609
   Nitrate Measurement   609
   Total Nitrogen   610
Summary   611
Specification Forms   611
Definitions   614
Abbreviations   614
Organizations   614
Bibliography   614

1.44 NITROGEN OXIDE (NO_x) ANALYZERS   615

Introduction   616
Designs For Industrial Emission Applications   616
   Paramagnetic Analyzers   616
   Thermal Conductivity Analyzers   616
   Nondispersive Infrared Analyzers   617
   Ultraviolet Analyzers   617
   Chemiluminescent Analyzers   618
   Electrochemical Sensors   618
   Gas Chromatographs   618

Ambient Air Monitoring   618
　　Calibration Methods   618
　　NO–NO$_2$ Combination Analysis   619
　　Colorimetric Analyzers   620
Portable Monitors   620
Conclusions   620
Specification Forms   621
Definitions   624
Abbreviations   624
Bibliography   624

1.45
ODOR DETECTION   625

Introduction   625
Odor Measurement   626
　　Human Olfactory System   626
Odor Testing   626
　　Odor Panels   627
Odor Detectors   627
　　Gas Chromatography   627
　　Electronic Nose   628
　　Polymeric Film Sensors   629
　　Metal Oxide Sensors   629
　　Other Sensors   629
Applications   630
　　E-Nose Applications   630
Specification Forms   630
Definitions   633
Abbreviations   633
Organization   633
Bibliography   633

1.46
OIL IN OR ON WATER   634

Introduction   635
Environmental Pollution Sensors   635
Water-in-Oil Measurements   635
　　Capacitance   635
　　Radio Frequency (Microwave)   636
　　Conductivity-Based Interface   638
　　Ultrasonic Pipeline Interface   638
　　Nuclear   638
　　Near Infrared   639
Oil-in-Water Measurements   640
　　Ultraviolet   640
Oil-on-Water Measurements   641
　　Nephelometers Using Lasers   641
　　Oil Slick Thickness Detection   642
Conclusions   642
Specification Forms   643
Definitions   651
Abbreviations   651
Bibliography   651

1.47
OXIDATION–REDUCTION POTENTIAL (ORP)   652

Introduction   653
ORP Measurement Principles   654
　　ORP   654
　　ORP Scales   654
　　Oxidation and Reduction   655
　　Redox Couples and Reversibility   655
　　ORP Reaction   655
　　Practical Measurements   656
ORP Sensors   657
　　Nernst Equation   657
　　Electrode Mounting   657
ORP Applications   658
　　Chromium Reduction   658
　　Corrosion Protection   659
ORP Maintenance and Calibration   660
　　Maintenance   660
　　Calibration   660
ORP Control   660
　　Residence Time   661
Recent Developments   661
Specification Forms   661
Abbreviations   664
Organizations   664
Bibliography   664

1.48
OXYGEN DEMANDS (BOD, COD, TOD)   665

Introduction   666
Oxygen Demand Testing   666
　　Seed   666
　　pH   666
　　Temperature   667
　　Toxicity   667
　　Incubation Time   667
　　Nitrification   667
BOD Tests   667
　　Five-Day BOD Tests   667
　　Extended BOD Test   669
　　Manometric BOD Test   669
　　BOD Determination in Minutes   670
COD Measurement   671
　　Dichromate COD Test   671
　　Automatic On-Line Measurement   672
TOD Measurement   673
　　Sampling Valves   674
　　TOD Oxygen Detector   675
　　Calibration   675
　　Interferences   676
　　On-Line Monitoring   676
　　Applications   676
Correlation among BOD, COD, AND TOD   677
　　BOD and COD Correlation   677

Specification Forms 677
Definitions 681
Abbreviations 681
Organizations 681
Bibliography 681

1.49
OXYGEN IN GASES 682

Introduction 683
Paramagnetic Oxygen Sensors 684
   Deflection Analyzer 684
   Thermal Analyzer 685
   Dual-Gas Analyzer 686
Catalytic Combustion Oxygen Sensors 686
Electrochemical Oxygen Sensors 686
   High-Temperature Zirconium Dioxide Fuel Cells 687
   High-Temperature Current-Mode Oxygen Sensors 689
   Galvanic Sensors (Fuel Cells) 689
   Coulometric Sensor 692
   Polarographic Sensor 692
Spectroscopic Oxygen Detection 692
   Mass Spectroscopy 693
   Near-Infrared Spectroscopy 693
Other Oxygen Detection Methods 695
Developments Over the Last Decade 695
Specification Forms 695
Abbreviations 699
Bibliography 699

1.50
OXYGEN IN LIQUIDS (DISSOLVED OXYGEN) 700

Introduction 701
Electrochemical Detectors 702
   Polarographic (Voltametric) Cells 703
   Galvanic Cells 704
   Thallium Cells 707
Fluorescence Sensors 708
Wet Chemistry Analyzers 709
Considerations Applicable to All Sensors 709
   Calibration Methods 709
   Temperature Compensation 709
   Pressure Effects 710
   Salinity Effects 711
Applications 711
Installations 711
Specification Form 711
Definitions 713
Abbreviations 713
Bibliography 713

1.51
OZONE IN GAS 714

Introduction 715
   Ozone Monitoring 715
   Monitor Designs 716
Ultraviolet Analyzer 716
   Single-Beam Design 717
   Double-Beam Design 718
Amperometric Design 719
   Calibration 719
Thin-Film Semiconductor Design 719
Chemiluminescence Design 720
Specification Forms 721
Abbreviations 723
Organizations 723
Bibliography 723

1.52
OZONE IN WATER 724

Introduction 725
   Drinking Water Disinfection 725
Amperometric Sensors 725
   Bare Metal Electrodes 725
   Membrane-Type Designs 726
Stripping and Gas Phase Monitors 727
Ultraviolet Absorption 727
Colorimetric Method 727
The Indigo Method 727
Specification Forms 727
Abbreviations 730
Organizations 730
Bibliography 730

1.53
PARTICLE SIZE DISTRIBUTION (PSD) MONITORS 731

Introduction 732
Application Objectives 733
   Particle Characteristics and Sampling 734
Laboratory Measurements 734
   Small Samples 734
   Medium Size Samples 735
   Large Samples 735
Online Measurement 736
   Optical Analyzer 736
   Airborne-Particulate Counter 737
Particle Distribution Monitors 737
   Laser Diffraction 737
Specification Forms 738
Definitions 741
Abbreviations 741
Organizations 741
Bibliography 741

1.54
PARTICULATE, OPACITY, AIR AND EMISSION MONITORING 742

Introduction 744
Theory 744
    Opacity Measurement 744
    Units and Definitions 745
Stack Emission Monitoring 746
    Particulate Sampling 746
    Stack Gas Analysis and Sampling 747
Light Attenuation and
        Transmissometers 748
    Single-Pass Configuration 749
    Double-Pass Configuration 749
    Relative Performance Factors 750
Light Scattering Designs 751
    Laser-based Designs 751
Stack Samplers 752
    Manual Stack Samplers 752
    Automatic Stack Samplers 753
    Broken Bag Monitors 753
Ambient Air Opacity Monitoring 754
    Particulate Concentration 754
    Ambient Air Sampling 754
    Soiling Index 757
    Nephelometers 757
    Piezoelectric Crystal Mass
        Balance 757
    Impaction Devices 757
    Radiometric Devices 758
    Charge Transfer (Triboelectricity) 758
    Surface Ionization 759
    Visual Observation 759
    Remote Sensing 759
Conclusions 759
Specification Forms 760
Definitions 762
Abbreviations 762
Organizations 762
Standards 762
Bibliography 763

1.55
pH MEASUREMENT 764

Introduction 765
    Measurement Error 765
    Measurement Range 766
    Applications 766
Theoretical Review 766
    Ion Concentrations 767
pH Measurement 767
    Temperature Effects 769
    pH Electrodes and Sensors 770
    Reference Electrodes 773

Electrode Cleaners 775
    Shrouds and Filters 775
    Automatic Cleaners 775
Application Problems 778
    High Salt Errors 778
    High Acid Errors 779
    Solution Temperature Effects 779
    Water Concentration Errors 779
    Nonaqueous Solutions 780
    Probe Coating and Low Conductivity 780
Installation Methods 781
    Submersion Assemblies 781
    Retractable Units 782
    Median Selector 783
    Self-Diagnostics 783
Calibration 783
    Buffer Calibration Errors 784
    Cleaning and Calibration Systems 784
Conclusion 788
Specification Forms 788
Definitions 791
Abbreviations 791
Organizations 791
Bibliography 791

1.56
PHOSPHATE ANALYZER 792

Introduction 793
Phosphorus in Wastewater 793
Laboratory Analysis 793
Continuous Colorimeter 793
    Pumpless Design 794
Total Phosphate Analysis 795
Flame Photometric Analysis 795
    Operation 795
Chromatography 796
Sample Handling Systems 796
Specification Forms 796
Definitions 799
Abbreviation 799
Organizations 799
Bibliography 799

1.57
PHYSICAL PROPERTIES ANALYZERS FOR PETROLEUM PRODUCTS 800

Introduction 801
Advantages of Continuous Analyzers 803
Distillation Analyzers 803
    Laboratory Measurement Standards 803
    Online Distillation Analyzers 804
Vapor Pressure Analyzers 808
    Reid Method (ASTM Method D 323-90) 808

Dry Reid Method (ASTM Method D 4953-90) 809
Liquefied Petroleum Gases (ASTM Method D 1267-89) 809
Air-Saturated Vapor Pressure Analyzer—Continuous 809
Air-Saturated Vapor Pressure Analyzer—Cyclic 810
Dynamic Vapor Pressure Analyzer 810
Calibration 811
Applications 811
Vapor–Liquid Ratio Analyzers 811
Volatility Test (ASTM Method D 2533-90) 812
Continuous Vapor–Liquid Ratio Analyzer 812
Calibration and Applications 813
Pour-Point Temperature Analyzers 813
Pour-Point Test (ASTM Method D 97-87) 813
Pour Point by Differential Pressure 813
Viscous-Drag Pour Pointer 814
Calibration 814
Applications 814
Cloud-Point Analyzers 814
Cloud-Point Tests (ASTM Method D 2500-88) 814
Optical Cloud-Point Analyzer 815
Calibration and Applications 816
Freezing-Point Analyzer 816
Aviation Fuel Tests (ASTM Method D 2386-88) 816
Freezing-Point Analyzer for Aviation Fuel 816
Calibration and Applications 816
Flash-Point Analyzer 816
Flash-Point Tests (ASTM Methods D 56-90 and D 93-90) 817
Low-Temperature Flash-Point Analyzer 817
High-Temperature Flash-Point Analyzer 818
Calibration and Applications 818
Octane Analyzers 818
Octane Tests (ASTM Methods D 2699 and D 2700) 818
Octane Comparator Analyzer (ASTM Method D-2885) 818
Reactor-Tube Continuous Octane Analyzer 819
Calibration and Applications 819
Near-Infrared Analyzers 819
Chemometrics 820
Calibration and Applications 820
New Developments 820
Conclusions 820
Specification Forms 821

Definitions 823
Abbreviations 823
Organizations 823
Standards 823
Bibliography 824

1.58
RAMAN ANALYZERS 825

Introduction 826
Principles of Raman Spectroscopy 827
Optical Spectroscopy 827
Raman Scattering 827
Spectral Information 829
Instrumentation 830
Fourier Transform Raman Instrumentation 831
Dispersive Raman Instrumentation 832
Raman Process Analyzers 833
Components 833
Data Analysis 838
Diagnostics and Maintenance 840
Laser Safety 840
Outputs and Communication 840
Packaging 840
Installation and Maintenance 841
Probe Designs 842
Advantages and Disadvantages 842
Advantages 843
Limitations 843
Applications 843
Conclusion 844
Specification Forms 844
Definitions 847
Abbreviations 847
Organization 847
Bibliography 847

1.59
REFRACTOMETERS 849

Introduction 850
UNITS (RI and BRIX) 851
Theory of Operation 852
Critical Angle of Refraction 852
Refractometer Designs 853
Differential 853
Critical-Angle 854
Automatic 855
Design Variations 856
Reflected Light Measurement 859
Conclusions 860
Specification Forms 860
Definitions 863
Abbreviation 863
Bibliography 863

Contents of Section 1   15

1.60
RHEOMETERS   864

Introduction   865
Polymer Characterization   865
    Solid Polymer Measurements   865
Tests to Distinguish Liquids from Solids   866
    Shear-Strain Tests   866
    Shear Creep Tests   867
    Nonlinear Shear Flow   867
Selection   867
Detector Designs   868
    Cone and Plate   868
    Shear Flow Properties   868
    Parallel Disk   871
    Rectangular Torsion   872
    Coaxial Cylinder   873
    Tension/Compression and Bending   873
    Extensional Flow   873
    Capillary   874
Conclusions   874
Specification Forms   875
Definitions   877
Abbreviations   877
Organization   877
Bibliography   877

1.61
SAND CONCENTRATION AND SUBSEA PIPELINE EROSION DETECTORS   878

Introduction   878
Background   878
Passive Acoustic Sand Detector   879
    Principle of Operation   879
    Advantages   879
    Limitations   879
Intrusive and Erosion Based Sand Detector   880
    Principle of Operation   880
Typical Sand Monitoring System/Data
    Flow   882
Conclusion   882
Specification Forms   883
Definitions   885
Abbreviations   885
Organizations   885

1.62
SPECTROMETERS, OPEN PATH (OP)   886

Introduction   887
Types of Designs   887
    IR, UV, and TDLAS   888
    Blackbody Radiation Interference   889
    Interferometers   889
Open-Path Absorption   889

    Beer's Law   890
    Path-Integrated Concentration   890
OP-FTIR Spectrometry   891
    Interferometer   891
    Sample Cell   891
    Transfer Optics and the Detector
        Types   892
    Sample Cell   893
    Data System/Controller   893
    Instrument Configurations   893
OP-UV Spectrometry   897
    The Spectrometer   897
OP-TDLAS Spectrometry   898
    Diode Lasers   899
    Applications   900
    Principle of Operation   900
    OP-CRDC Spectroscopy   902
Applications   902
    Toxic Gas Measurement   903
    Combustible Vapor Measurement   903
Specification Forms   904
Definitions   907
Abbreviations   907
Organizations   907
Government Agencies   907
Bibliography   908

1.63
STREAMING CURRENT PARTICLE CHARGE ANALYZER   909

Introduction   910
Operating Principle   910
    The Streaming Current   910
    Treatment Chemical Selection   912
Calibration   912
Applications   913
    Sampling   913
    Control System   914
Specification Forms   915
Definitions   918
Abbreviations   918
Organization   918
Bibliography   918

1.64
SULFUR DIOXIDE AND TRIOXIDE   920

Introduction   921
    Applications   922
Industrial Analyzer Types   923
    Nondispersive Infrared (NDIR)   923
    Ultraviolet   924
    Correlation Spectrometry   925
    Thermal Conductivity   925

16 Analytical Measurement

Ambient Air Analyzers   925
   Calibration   925
   Colorimetric Analyzers   925
   Conductimetric Analyzers   926
   Coulometric Analyzers   927
   Flame Photometric Analyzers   927
   Electrochemical Analyzers   927
Specification Forms   928
Abbreviations   930
Bibliography   930

1.65 SULFUR IN OIL AND GAS   931

Introduction   932
Gas Chromatography with Flame Photometric
   Detection   932
Hydrogenolysis and Rateometric
   Colorimetry   933
Sulfur Chemiluminescence   934
Pyrolysis and Ultraviolet Fluorescence   935
X-Ray Fluorescence   936
   Sampling and Calibration   937
   New Developments   937
Conclusions   937
Specification Forms   937
Abbreviations   940
Organizations   940
Bibliography   940

1.66 THERMAL CONDUCTIVITY DETECTORS   941

Introduction   942
Thermal Conductivity   942
   Measurement Ranges   942
TCD Analyzer   943
   Detectors   943
   TCD Cells   944
   Bridge Circuits   944
   Operation   946
   Packaging and Calibration   946
Conclusions   947
Specification Forms   947
Definitions   950
Abbreviations   950
Bibliography   950

1.67 TOTAL CARBON AND TOTAL ORGANIC CARBON (TOC) ANALYZERS   951

Introduction   952
   Advantages and Limitations   952
   Carbon Measurement Techniques   952
   Analyzer Development   953
   Official Methods of TOC Determination   954
Detector Types   954
   Nondispersive Infrared Analyzers   954
   Oxidation Methods   955
   Aqueous Conductivity   958
   Coulometric Analysis   961
   Colorimetric Analysis   961
   Flame Ionization Detector   961
Conclusion   962
Specification Forms   962
Abbreviations   965
Organizations   965
Bibliography   965

1.68 TURBIDITY, SLUDGE AND SUSPENDED SOLIDS   966

Introduction   967
   Light Absorption and Scattering   967
   Units of Turbidity   967
Turbidity Analyzer Designs   969
   Forward Scattering or Transmission Type   969
   Scattered Light Detectors
      (Nephelometers)   974
   Backscatter Turbidity Sensors   976
   Density-Based Sensors   976
Conclusions   976
Specification Forms   977
Definitions   980
Abbreviations   980
Organizations   980
Bibliography   980

1.69 ULTRAVIOLET AND VISIBLE ANALYZERS   981

Introduction   982
UV Theory   983
   UV Absorption   983
   Radiation Spectrum   984
   Beer–Lambert Law   985
   UV-Absorbing Compounds   986
   UV Absorption Spectrum   987
   Applications   987
UV Analyzer Components   987
   Radiation Sources   988
   Emission Types   988
   Selecting the Wavelength   989
   Monochromator   989
   Optical Dispersion   990
   Measuring Cell   990
   Detectors   990
   Readouts   991

UV Analyzer Designs 991
   Single-Beam 991
   Split-Beam 992
   Dual-Beam, Single-Detector 992
   Dual-Beam, Dual-Detector 993
   Flicker Photometer 993
   Photodiode Array (PDA)
      Spectrophotometers 994
   Scanning Spectrophotometers 995
   Retroreflector Probes 996
Visible and Nir Analyzers 997
   Visible Light Photometers 997
   Near-Infrared Photometers 997
Conclusions 997
Specification Forms 998
Definitions 1001
Abbreviations 1001
Organizations 1001
Bibliography 1001

1.70
VISCOMETERS 1002

Introduction 1002
Theory of Viscous Behavior 1002
   Stokes' Law 1003
   Hagen–Poiseuille Law 1003
   Intrinsic Viscosity 1004
   Newtonian and Non-Newtonian
      Fluids 1004
   Typical Viscosities and Conversion among
      Units 1006
Viscometer Selection and
   Application 1008
   Selection 1010
   Applications 1010
Specification Forms 1011
Definitions 1014
Abbreviations 1014
Organizations 1014
Bibliography 1014

1.71
VISCOMETERS 1016

Introduction 1018
Capillary Viscometers 1020
   Limitations 1020
   Calibration 1020
   Differential Pressure Type 1021
   Backpressure Type 1023
Falling Element Viscometers 1023
   Falling-Piston Viscometer 1023
   Falling-Slug or Falling-Ball
      Viscometers 1024

Float Viscometers 1025
   Precautions 1025
   Single-Float Design 1026
   Two-Float Design 1026
   Concentric Design 1027
Plastometers 1028
   Plastic Behavior 1028
   Cone-and-Plate Plastometer 1028
   Kneader Plastometer 1029
   Capillary Plastometer 1029
Rotational Element Viscometers 1030
   Rotary Spindle Design 1030
   Gyrating-Element Viscometer 1032
   Agitator Power 1032
   Dynamic Liquid Pressure 1033
Oscillating Viscometers 1034
   Oscillating Blade 1034
   Oscillating Piston 1035
Vibrational Viscometers 1036
   Torsional Oscillation Design 1036
   Vibrating-Reed/Rod Viscometer 1037
Coriolis Mass Flowmeter 1038
   Measuring Tube Torsional
      Movement 1038
New Developments 1039
   Trends 1039
Conclusions 1040
Specification Forms 1040
Definitions 1043
Abbreviations 1043
Organizations 1043
Bibliography 1043

1.72
VISCOMETERS 1045

Introduction 1046
Laboratory Viscometer Designs 1048
   Bubble-Time Viscometers 1048
   Capillary Viscometers 1048
   Falling/Rolling Element Viscometers 1053
   Orifice/Efflux Cup Viscometers 1055
   Oscillating Piston Viscometer 1057
   Rotational Viscometers 1057
   Vibrational Viscometers 1063
Trends 1063
   Automation 1063
   Lab Management Systems 1063
   Adaptability 1064
   Smart User Interfaces 1064
   Smart Accessories 1064
Specification Forms 1064
Definitions 1067
Abbreviation 1067
Organizations 1067
Bibliography 1067

1.73
WATER QUALITY MONITORING 1069

Introduction 1070
    Related Information Sources 1071
    Regulations 1071
Monitoring System Components 1071
    Sampling Systems 1071
    Analyzer Types 1073
    Ion-Selective Electrodes 1073
    Chromatographs 1079
    Portable Spectrophotometer 1081
Conclusions 1082
Specification Forms 1083
Definitions 1086
Abbreviations 1086
Organizations 1086
Bibliography 1086

1.74
WET CHEMISTRY AND AUTOTITRATOR ANALYZERS 1087

Introduction 1087
Titration End Points 1088
    End Point Detectors 1089
    Continuous and Batch Designs 1089

Volumetric Titrators 1089
    High Precision Designs 1089
    Simpler Designs 1091
    Applications 1092
Colorimetric Analyzers 1092
    High Precision Designs 1092
    Simpler Design 1094
    On-Off Batch Design 1094
    Applications 1095
Flow Injection Analyzers (FIA) 1095
    Laboratory Design 1096
    Industrial Designs 1096
    Application Example 1 1096
    Application Example 2 1097
    Calibration 1097
Conclusions 1097
Specification Form 1098
Definitions 1100
Abbreviations 1100
Organizations 1100
Bibliography 1100

1.1 Analyzer Selection and Application

A. H. ULLMAN (2003) **B. G. LIPTÁK** (1972, 1982, 1995, 2017)

INTRODUCTION

The role of analyzers in our everyday life is becoming more and more important and their capabilities and sophistication are exploding. Here, I will give just one practical example, involving the monitoring of methane emitted into the atmosphere by fracking for natural gas. I could have picked dozens of other examples, but this one combines some very important new ideas and solutions, which illustrate how the analyzer technology is evolving.

This analyzer, built by Picarro, is portable (mounted on a truck), it measures the methane concentration of the atmosphere (by laser technology) while the truck is moving. It combines that information with data on wind velocity from an anemometer and GPS data for location in an onboard computer, which continuously turns the readings into a 3D model of a gas plume—a funnel-shaped flow of contamination—and calculates the location and size of the origin. Another new capability of this analyzer is that it can measure the fraction of the total methane, which leaked out from natural gas wells. This is important, because landfills and sewers also emit methane.

I thought that by mentioning this example, I would encourage some of the readers, to use the contents of the coming chapters to combine the devices described into more complex systems that solve unique applications.

In the past, the compositions of process streams were often analyzed manually, by taking grab samples to a laboratory. This approach might be acceptable for product quality control purposes, but because of the time it takes to get a sample, transport it to the laboratory, and wait for the results, it is certainly unacceptable for safety or for process optimization and energy conservation purposes. In these applications, the analysis must be continuous and online.

For an analyzer system to fulfill its expectations, careful planning and evaluation must precede its purchase, and the users must realize that if an expensive analyzer is worth purchasing, it must also be worth calibrating and maintaining. The operator's acceptance, which depends largely on training and familiarization, is also crucial. New self-diagnosing and self-calibrating analyzers are a major contribution to improved operator acceptance, but people issues are still more important and should not be discounted.

From a process control point of view, it is equally important to analyze how the information gained from the analyzer is going to be used. Is it just a safety constraint or is it a controlled variable? If it is the latter, is the analyzer's dead time and time constant suitable and is its inaccuracy good enough for the task at hand? I have often seen that because these questions were not carefully evaluated during the design phase of the plant, it was only after start-up that the answer to some of the preceding questions was found to be a resounding no. It is for this reason that when on a project a process or project engineer asks for an analyzer, the instrument engineer (somewhere called process control or automation engineer) asks for a detailed explanation of how the measurement is intended to be used.

The purpose of this chapter is to give the reader guidance for selecting the best analyzer for the application at hand. Therefore, in this chapter, the main selection considerations will be discussed, so that once the choice is made, the reader can proceed to reading the chapter that discusses the nature of the selected analyzer in detail.

PREPARATION PHASE

Selecting the right analyzer for a particular application involves a sequence of logical steps:

1. Clearly defining the purpose of the analyzer.
2. Determining if a continuous online analyzer is needed or an intermittent and manual laboratory analysis is sufficient.
3. For online units, determining the re-calibration requirements. The question usually is if the instrument can be removed for calibration or if automatic, online self-calibration is needed. It is also desirable to resolve if the re-calibration should be triggered by self-diagnostics or its frequency be set by time.
4. The next step is to locate the analyzer, determine if off-line or in-line units are needed. If a probe type design is required, should it be self-cleaning? If a sampling system is to be used, what types of accessories are required for filtering, etc.?

5. Only after an in-depth review of these four questions should the process control engineer decide on the type of analyzer to be used and on the types of accessories to be utilized with them. At this point, the use of an orientation table (such as Table 1.1a) is useful, because it can quickly identify the types of analyzers to consider.
6. After this, the specifications can be prepared and sent out to the potential suppliers for competitive bidding. One can use the general analyzer specification forms of the International Society of Automation (ISA) that are reproduced at the end of this chapter.
7. While waiting for the bids and also after they are received, it can also be useful to combine the data that define the analyzer application in a single tabulation. For this purpose too, the ISA specification forms or the user's own specification forms can be used. Both are provided at the end of this chapter.
8. Once the vendor is selected, the preparation of the installation drawings can start and the communication between the analyzer and its remote displays and/or controllers can be resolved.
9. The last step comes after installation and check-out, when the maintenance specifications are prepared.

It is not uncommon to revise the specifications or the selection during the design phase of the plant as new information is obtained. This is normal and a sign of good engineering, while the need to make changes after start-up is not.

The Orientation Table

For the convenience of the readers of my handbook, I have prepared Table 1.1a, which should help to get quickly the oriented available options for the various applications. The orientation table lists over 100 components and over 25 analyzer categories and indicates (by check marks) the analyzer types, which can detect the concentration of the component of interest. Once these types have been identified, the reader can proceed to the corresponding chapters, where the feature summary on the front pages gives an idea about costs, inaccuracies, advantages, and disadvantages, while the body of the chapter provides a detailed description of the device.

Naturally, because of space constraints, not all compounds of interest are listed in the table and not all categories and subcategories of analyzer types are separated into individual columns. For example, practically all substances can be analyzed by chromatography, but the cost of that approach is so high that in most cases one would not even consider its use for an application that can be handled otherwise. Similarly, in some columns, I have combined several analyzer subcategories. For example, under "Electrochemical Analyzers," I have included amperometric, coulometric, galvanometric, polarographic, and potentiometric analyzers. Similarly, under "Fluorescence" I included not only atomic fluorescence analyzers but also some ultraviolet (UV) ones, and under infrared (IR), I also included nondispersive infrared (NDIR) and radio frequency (RF) wavelengths. Under "Ultraviolet" I also included photometric or light reflection designs. Under "Wet Chemistry" I also included titration and chemiluminescence systems. In short, the orientation table is not as detailed or as complete as it could be, but it is more complete than any table I have seen to date.

TABLE 1.1a
Orientation Table for Analyzer Selection

Suited for the Analysis of the Compounds[a] and Properties Listed in the Following Text	1-Atomic Absorption (AA)	2-Capacitance	3-Catalytic Combustion	4-Chilled Surface (Mirror)	5-Chromatography, Gas[b]	6-Chromatography, Liquid	7-Colorimetric, Titrimetric	8-Conductance, Resistance	9-Density	10-Electrochemical[c]	11-Electrolytic Hygrometer	12-Flame Ionization (FID), (FPD)	13-Fluorescence[d]	14-Infrared (IR), NIR, NDIR[e]	15-Ion Selective, Specific	16-Laser	17-Mass Spectrometer	18-Nuclear–Neutron Moderation	19-Paramagnetic	20-Phototape	21-pH, ORP, ISE[f]	22-Refractometer[g]	23-Thermal Conduct. (TCD)	24-Ultraviolet (UV) and Visible[h]	25-Wet Chemistry[i], FIA	26-Zirconium Oxide	27-XRF (X-Ray Fluorescence)
Acetaldehyde					✓																			✓			
Acetic anhydride																								✓			
Acetone																								✓			
Acid in water								✓													✓	✓			✓		
Acrylonitrile																								✓			
Alcohol in water									✓													✓					
Aldehydes														✓										✓			
Alkalinity of water																									✓		
Alkyl chloride										✓																	
Aluminum																								✓			
Amines, ppm												✓															
Ammonia					✓		✓			✓				✓	✓		✓				✓			✓			
Ammonium in water															✓						✓				✓		
Aniline												✓												✓			
Argon					✓												✓						✓				
Aromatics												✓															

[a] For the proper operation of probe type analyzers, it is essential to keep them clean. For the available types of probe cleaner designs and their suppliers, refer to Table 1.1i.
[b] A very wide range of materials can be analyzed by chromatgraphy. In that regard, the listing in this tabulatoin is rather incomplete, because while practically all materials listed, could be so analyzed, economics usually limits the number of it's application.
[c] This category includes battery operated gas diffusion sensors, amperometric, (also referred to as galvanometric or polarographic), potentiometric and coulometric detectors..
[d] This category includes UV and atomic fluorescent sensors. A list of fluorescent compounds can be found in Table 1.1e.
[e] This category of analyzers includes Raman, near infrared (NIR), nondispersive infrared (NDIR), microwave, and radio frequency-type sensors.
[f] These are basically potentiometric sensors. ORP is the abbreviation of oxidation–reduction potential and ISE stand for ion-selective electrode. pH is also an ISE, which is sensitive to the activity of the hydrogen ion. Table 1.1h lists the compounds that can be detected by ISEs.
[g] Table 1.1g provides the refractiv indexes (RI) of a number or compounds.
[h] Includes photometric and light reflection sensors. For a complete list of compounds that are absorbing within the range of UV and visible wavelength, see Table 1.1c.
[i] Includes chemiluminescence and titration type sensors. In case of an auto-titrator, the end point may be determined potentiometrically, conductimetrically, or photometrically.

(Continued)

TABLE 1.1a (Continued) Orientation Table for Analyzer Selection

Suited for the Analysis of the Compounds[a] and Properties Listed in the Following Text	1-Atomic Absorption (AA)	2-Capacitance	3-Catalytic Combustion	4-Chilled Surface (Mirror)	5-Chromatography, Gas[b]	6-Chromatography, Liquid	7-Colorimetric, Titrimetric	8-Conductance, Resistance	9-Density	10-Electrochemical[c]	11-Electrolytic Hygrometer	12-Flame Ionization (FID), (FPD)	13-Fluorescence[d]	14-Infrared (IR), NIR, NDIR[e]	15-Ion Selective, Specific	16-Laser	17-Mass Spectrometer	18-Nuclear–Neutron Moderation	19-Paramagnetic	20-Phototape	21-pH, ORP, ISE[f]	22-Refractometer[g]	23-Thermal Conduct. (TCD)	24-Ultraviolet (UV) and Visible[h]	25-Wet Chemistry[i], FIA	26-Zirconium Oxide	27-XRF (X-Ray Fluorescence)
Benzene												✓		✓										✓			
Brine								✓	✓													✓					
Bromide																					✓			✓			
Bromine																						✓		✓			
Butane		✓									✓																
Butadiene					✓									✓								✓		✓			
Cadmium																					✓						
Caffeine																								✓			
Calcium																					✓				✓		
Carbon disulfide										✓				✓											✓		
Carbon dioxide										✓				✓									✓		✓		
Carbon monoxide					✓					✓				✓											✓		
Carbon tetrachloride																											
Caustic									✓													✓					
Chloride ion										✓											✓				✓		
Chlorine							✓			✓										✓	✓				✓		
Chlorine residual										✓											✓				✓		
Chloroform																					✓				✓		
Chromium in water																									✓		
COD (Chem. O₂ Dem.)																									✓		
Color												✓												✓			
Combustibles										✓																	
Copper in water										✓												✓			✓		
Cyanide in water										✓												✓			✓		
Cyclohexane																✓						✓					
Divalent ions in water																									✓		
Diolefin vapors																								✓			

(Continued)

TABLE 1.1a (Continued)
Orientation Table for Analyzer Selection

Suited for the Analysis of the Compounds[c] and Properties Listed in the Following Text	1-Atomic Absorption (AA)	2-Capacitance	3-Catalytic Combustion	4-Chilled Surface (Mirror)	5-Chromatography, Gas[b]	6-Chromatography, Liquid	7-Colorimetric, Titrimetric	8-Conductance, Resistance	9-Density	10-Electrochemical[c]	11-Electrolytic Hygrometer	12-Flame Ionization (FID), (FPD)	13-Fluorescence[d]	14-Infrared (IR), NIR, NDIR[e]	15-Ion Selective, Specific	16-Laser	17-Mass Spectrometer	18-Nuclear–Neutron Moderation	19-Paramagnetic	20-Phototape	21-pH, ORP, ISE[f]	22-Refractometer[g]	23-Thermal Conduct. (TCD)	24-Ultraviolet (UV) and Visible[h]	25-Wet Chemistry[i], FIA	26-Zirconium Oxide	27-XRF (X-Ray Fluorescence)
Elements (all incl. metals)	✓																										✓
Ethane									✓					✓			✓										
Ethanol										✓												✓					
Ethyl chloride and bromide					✓																						
Ethyl benzene					✓									✓													
Ethylene									✓					✓			✓					✓					
Ethylene glycol					✓					✓				✓			✓										
Ethylene oxide					✓									✓							✓				✓		
Fluoride in water															✓						✓	✓			✓		
Freon									✓	✓																	
Furfural																								✓			
Glycerin and salt in water																						✓	✓		✓		
Hardness (water)				✓																					✓		
Hazardous gases					✓					✓				✓													
Helium																	✓						✓				
Hexane					✓									✓													
Hexavalent chromium															✓								✓		✓		
Humidity		✓									✓												✓				
Hydrocarbons					✓					✓		✓		✓													
Hydrazine in water					✓					✓		✓												✓			
Hydrocarbon in air					✓					✓							✓								✓		
Hydrogen is air or steam										✓							✓						✓				
Hydrogen cyanide					✓					✓															✓		
Hydrogen impurities					✓												✓			✓			✓				
Hydrogen chloride, bromide, cyanide										✓															✓		
Hydrogen sulfide										✓											✓				✓		
Iodide ions in water															✓										✓		
Iron in water																								✓	✓		
Isobutane					✓																	✓					

(Continued)

TABLE 1.1a (Continued)
Orientation Table for Analyzer Selection

Suited for the Analysis of the Compounds[a] and Properties Listed in the Following Text	1-Atomic Absorption (AA)	2-Capacitance	3-Catalytic Combustion	4-Chilled Surface (Mirror)	5-Chromatography, Gas[b]	6-Chromatography, Liquid	7-Colorimetric, Titrimetric	8-Conductance, Resistance	9-Density	10-Electrochemical[c]	11-Electrolytic Hygrometer	12-Flame Ionization (FID), (FPD)	13-Fluorescence[d]	14-Infrared (IR), NIR, NDIR[e]	15-Ion Selective, Specific	16-Laser	17-Mass Spectrometer	18-Nuclear–Neutron Moderation	19-Paramagnetic	20-Photoiape	21-pH[d], ORP, ISE	22-Refractometer[g]	23-Thermal Conduct. (TCD)	24-Ultraviolet (UV) and Visible[h]	25-Wet Chemistry[i], FIA	26-Zirconium Oxide	27-XRF (X-Ray Fluorescence)
Isoprene																								✓			
Leads	✓																										
Lead ions in water	✓																										
Mercury in air					✓	✓																		✓	✓		
Mercury in water					✓		✓																	✓	✓		
Methane												✓					✓										
Methanol in water						✓							✓														
Methyl bromide, chloride										✓	✓			✓													
Methylene chloride										✓	✓			✓													
Moisture in air or gas		✓		✓							✓			✓													
Moisture in liquids		✓					✓							✓													
Moisture in solids[j]		✓						✓										✓									
Nitrates in water								✓							✓									✓	✓		
Nitric acid in water																					✓				✓		
Nitric oxide in air														✓										✓	✓		
Nitrogen compounds												✓															
Nitrogen dioxide							✓																	✓	✓		
Nitrogen oxide, peroxide																	✓			✓							
Nitrous fumes																								✓			
Octane of gasoline														✓							✓	✓					
Oil in or on water		✓						✓					✓											✓			
Olefins																											
Oxygen in water[k] (DO)								✓		✓															✓		
Oxygen in gases			✓							✓				✓					✓				✓	✓		✓	
Ozone in gas										✓														✓			
Ozone in water							✓			✓														✓			

[j] Analyzers using IR and NIR, microwave, Rf, neutron, capacitance, impedance, and electrical resistance techniques are applicable for the measurement of moisture in both solids and liquids.

[k] Also includes the measurement of biological, chemical, and total oxygen demands (BOD, COD, and TOD). Total organic carbon (TOC) measurements give closely correlateable results to COD readings.

(Continued)

TABLE 1.1a (Continued)
Orientation Table for Analyzer Selection

Suited for the Analysis of the Compounds[a] and Properties Listed in the Following Text	1-Atomic Absorption (AA)	2-Capacitance	3-Catalytic Combustion	4-Chilled Surface (Mirror)	5-Chromatography, Gas[b]	6-Chromatography, Liquid	7-Colorimetric, Titrimetric	8-Conductance, Resistance	9-Density	10-Electrochemical[c]	11-Electrolytic Hygrometer	12-Flame Ionization (FID), (FPD)	13-Fluorescence[d]	14-Infrared (IR), NIR, NDIR[e]	15-Ion Selective, Specific	16-Laser	17-Mass Spectrometer	18-Nuclear–Neutron Moderation	19-Paramagnetic	20-Phototape	21-pH, ORP, ISE	22-Refractometer[g]	23-Thermal Conduct. (TCD)	24-Ultraviolet (UV) and Visible[h]	25-Wet Chemistry[i], FIA	26-Zirconium Oxide	27-XRF (X-Ray Fluorescence)
Particulates						✓																					
Phenol in water																								✓	✓		
Phosgene in air																				✓					✓		
Phosgene in water										✓														✓	✓		
Phosphoric acid in water					✓					✓														✓	✓		✓
Phosphate in water																								✓	✓		
Polymer-solvent mix.									✓																		
Potassium in water						✓					✓													✓			
Propane					✓				✓					✓			✓										
Proteins													✓	✓													
Silicon in paper													✓	✓													
Sodium hydroxide in water								✓	✓												✓				✓		
Sodium ions in water								✓													✓						
Solids (in Capsup, etc.)									✓					✓								✓					
Steam in air														✓				✓									
Sugar in juice, jam, syrup									✓					✓								✓		✓			
Sulfates							✓			✓											✓				✓		
Sulfur in oil					✓		✓			✓												✓		✓			✓
Sulfur dioxide trioxide					✓			✓		✓									✓				✓		✓		
Sulfuric acid								✓		✓													✓		✓		
Toluene in hydrocarbons														✓										✓	✓		
Total carbon					✓									✓													
Total nitrogen																									✓		
Toxic gases					✓									✓													
Vinyl chloride					✓																						
Xylenes in hydrocarbons					✓																						

Obtaining Preliminary Quotations

At this point, it is appropriate to obtain bids from the analyzer suppliers. Some would argue that this is too early, that further evaluation is needed, but my experience is contrary to that view. Over the years I found that obtaining early bids from the vendors provides valuable inputs for the selection process in two ways.

One valuable contribution is that while several suppliers prepare their quotations, they will ask a number of questions, which could have been overlooked earlier, and by answering those questions, the designers and their clients will better understand the requirements for a successful installation.

The other valuable contribution that early bidding provides is guaranteed data, because until the quotations are received, one is dealing only with estimates. Once the bids are in, one can prepare a bid analysis, which is based on guaranteed costs, accuracies, rangeabilities, clearly specified calibration, and maintenance requirements.

For the preceding reasons, it is my recommendation to prepare the bid specifications at this early point in the project. This does not mean that the supplier will be selected soon after the quotations are received. No, one's options should be kept open at this stage, but it does mean that from this point on, guaranteed facts and not estimates will be used in the selection process. In preparing the bid specification, it is important to provide all relevant information to the vendors, which can be guaranteed, if time-tested specification forms are used. For examples of such specification forms, refer to the tables provided at the end of this chapter.

After sending out the bid invitations and while waiting for the quotations, an in-depth evaluation of the particular analyzer application can continue, during which the purpose of the measurement and the process information is further refined.

Defining the Purpose of the Analysis

Defining the purpose of the application involves the answering of a number of questions and understanding how the measurement will be used. Is it for closed-loop control or for monitoring purposes only? Must the measurement be performed "online," or could 80% of the benefits be obtained for 20% of the cost and effort by making the measurement using grab samples? What is the benefit of the measurement? How will the cost be justified? Will staffing be reduced by moving the measurement from the lab to the pipe? Will it increase production, reduce energy consumption, or yield product quality improvement? In addition, it is important to understand the process to be monitored. What is the response time of the process? Obviously, if a process takes 5 min to start to respond and 60 min to fully change in material or heat balance, the type of analyzer needed to control is quite different from the one that serves a fast process. These issues are important and must be resolved early.

Before continuing our discussion of problem definition, it is helpful to define some commonly used terms. In analytical chemistry, we usually define the location of measurements as laboratory, at-line, online, and in-line. In addition, measurements can be invasive or noninvasive.

An *at-line* measurement is generally a manual measurement, instrumental or chemical, performed near the process, generally in or adjoining the control room. The method may be identical to the laboratory method, but normally is simpler and faster. The at-line approach is ideal when the resulting information is used to control slow response processes, or for quality control, or when true real-time measurements are prohibitively expensive, compared to the benefit. An example of an at-line measurement might be a caliper measurement of a paper product or a multiphase titration of a surfactant. In both cases, production samples are manually taken to the measurement station and tested. A typical frequency for an at-line measurement might be hourly.

Online measurements are those installed on the process by obtaining a sample or slipstream, which is brought to the analyzer, frequently after some sample conditioning. In general, this method is used when data are needed very frequently and the measurement approach is not conducive to in-line use. The slipstream approach is needed when the measurement instrument requires sample conditioning, such as filtration. This will be discussed in more detail later. Process gas chromatographs are examples of instruments that normally use the online method and require sample conditioning to allow proper sampling of the small volumes needed for injection. (Of course, chromatographs are expensive and provide only intermittent measurements at substantial time delays.)

In-line measurements are those in which a probe is inserted directly into the process. No sample conditioning is required. Common examples include pH and other electrode measurements. In a gray zone between the online and in-line categories, we find approaches in which the instrument is able to look into the process through a window (e.g., a spectroscopic measurement) or directly on a moving web or conveyor belt. These are real-time approaches, but not on a slipstream and not really in-line either.

An *invasive* measurement is one that in some way penetrates or changes the process. Any in-line approach would be considered invasive, as would an online approach that modified the process flow with a slipstream. A near-infrared (NIR) photometer positioned above a web would be *noninvasive* since it does not enter or change the process or material. Similarly, a clamp-on ultrasonic flow or density meter is noninvasive.

When an adequate discussion of the measurement need has occurred, we can assume that the problem has been defined sufficiently, at least in terms of the kind of process measurement needed and how it will be used. It may be necessary to return to this subject and refine the problem definition further, based on additional information that will be gathered in the next steps.

Information Gathering

Reasons for installing process analyzers include the improvement of safety as well as of product quality, reduction of by-products, decrease in analysis time, tightening of specifications, and monitoring of contaminants, toxicants, or pollutants.

Additional requirements and characteristics of the measurement must be understood, including the needed frequency of analysis, sample availability, cost, and safety considerations.

In gathering the data for defining the nature of the analyzer application, it is useful to combine that information in a tabulation. This tabulation can be started early in the project and can continue and be refined after the bids are received. The Specification Forms at the end of this chapter can be used in this information gathering phase. Table 1.11 is an example of such a tabulation form.

In selecting an analysis system, one must ask whether the need for the measurement justifies the cost. If the answer is yes, the study progresses until several possible types of analyzers are identified (Table 1.1a), at which time the question of cost is again raised, and some of the more expensive analyzers may be eliminated from further consideration. After the complete system has been defined, the estimated costs can be compared with the expected benefits and a decision reached.

Frequently, this information gathering step involves discussions with research and development analytical chemists, plant quality control laboratory personnel, and library research. While gathering information, the instrument engineer should be open-minded to the methods, while carefully evaluating each one in light of the specific problem at hand. If the results of the measurement will be compared to the laboratory's results, it becomes crucial to understand any differences between the methods to be used in each and whenever possible, similar methods should be used. However, lab methods are not typically suitable to online measurement, so other approaches will probably be needed.

The critical question to be answered during this information gathering phase is: "What attribute of the analyte would be easiest to measure?" For example, if oxygen concentration measurement is required (gaseous or dissolved oxygen), the engineer has to evaluate several oxygen measurement approaches, each of which uses a different property of the oxygen molecule. Depending on the process conditions, the engineer could select a paramagnetic oxygen analyzer, which makes use of the fact that elemental oxygen is one of the few gases that is attracted by magnets, or an electrochemical cell device, which makes use of oxygen's electroactivity. After the appropriate analyzer type is selected, it is also important to resolve how the analyzer will be calibrated. Will it be evaluated against a laboratory measurement and if it will, will that measurement be reliable and more accurate than the online one?

Frequently, while gathering information, one must return to seek further clarification. This is an iterative process and once the requirements are fully defined, the analyzer selection can be finalized.

SELECTION CONSIDERATIONS

The most important factors that must be weighed by the instrument engineer are discussed in the remainder of this chapter and in the chapters that follow. Unfortunately, no single analyzer combines all the desirable features of providing on-stream, specific, continuous, unattended high-sensitivity readings without drift, noise, or need for a sampling system. Therefore, the selection is always a compromise, which is likely to give satisfactory results only if it is preceded by careful evaluation.

If the problem has been defined as one requiring the measurement of one material in the presence of another, it is necessary first to look for the handle or unique property of the material to be measured. Usually, a first step in determining the suitable methods of analysis is to investigate laboratory methods used to determine the desired property. As previously noted, a meeting with laboratory personnel is a good first step. Industry methods are another good starting point. For example, the American Society for Testing Materials (ASTM) has established certain test methods for the determination of various properties and materials. Their methods are widely available in book form and on the web. Other organizations such as the American Oil Chemists Society (AOCS) and the Technical Association of the Pulp and Paper Industry (TAPPI) have also published compendia of analytical methods. Suppliers of the material to be analyzed are possible sources of information for methods of analysis, as are suppliers of online analyzers.

Once prospective methods of analysis have been selected, the search for appropriate hardware begins. Ordinarily, details of the desired measurement are given to the potential analyzer suppliers, and estimates of performance are obtained. The analysis system can often be purchased on a guaranteed performance basis. For proprietary installations, the user himself must determine how the general analyzer specifications apply for the particular application, and since the user is buying hardware only, the supplier will guarantee only the quality of materials and workmanship. Alternatively, once a single supplier has been selected, a confidentiality agreement can be signed allowing information sharing by both parties—a true partnership to assure a successful installation.

When comparing different analysis approaches, the following criteria are critical.

Specificity

Specificity is the characteristic of responding only to the property or component of interest. The specificity of the analyzer will not exceed that of the analysis. Specificity is often a function of the range to be measured, the sample background, and the process conditions (solid, liquid, or gas; pressure and temperature). The more selective a measurement is, the less we need be concerned with interferences from other constituents of the sample matrix. Obtaining a high degree of selectivity frequently means compromising other analytical attributes. For example, a gas chromatographic separation with baseline resolution of the analyte of interest may require a longer analysis time than is desired or allowable.

Accuracy and Precision

Frequently, absolute accuracy of a process measurement cannot be established owing to the lack of a suitable calibration standard or the inability to draw an unchanged sample from the process and analyze it by an "official" laboratory method. For this reason, other terms, such as repeatability, take on added importance. Often, the terms do not have industry-wide significance, so the definition should be agreed upon with each supplier. (Note: With proper calibration, an analyzer will be "accurate.")

In general, in process measurement, precision or repeatability is more important than accuracy. Since the measurement is being made continuously or repeatedly, we are much more interested in changes in a value than whether that value is "absolutely correct" or accurate. Changes in the process are what we are trying to measure and understand (and control).

For the purpose of this discussion, *precision* is defined as the ability of an analyzer to produce the same output each time the sample contains the same quantity of component or property being measured. The terms *stability*, *reliability*, and *reproducibility* are sometimes used synonymously with *repeatability*. However, *reliability* is also used to describe the instrument's "up time" and should not be used as a synonym for precision. Lack of repeatability may be caused by the analyzer, the sampling system, or the effects of temperature, power supply, composition, pressure, and flow rate. Poor precision or low repeatability can come in two forms: short term and long term. Short term may mean the measurement system has lots of randomness (noise). Such random noise degrades the utility of any given reading, but may be tolerable since many data points are collected. Long-term drift can be a more significant problem requiring frequent instrument recalibration or adjustment. Repeatability and inaccuracy are normally expressed either as a percentage of the full measurement range or as a percentage of the actual reading.

When discussing specifications for an analyzer, especially with prospective vendors, be very careful to define all technical terms and understand exactly what the literature or the sales engineer means by terms such as accuracy, drift, precision, etc. If there is any doubt, ask for a clarifying example with actual numbers, such as "The refractometer is stable to $\pm 1\%$ of the actual reading over 14 days, without any adjustments. That is, it will read 1.102 ± 0.1 for 2 weeks, if the temperature changes no more than $\pm 5°C$." Sales literature is notorious for specifications that are unclear and frequently misleading.

Calibration

The ability to calibrate an analyzer properly usually depends on the availability of a reliable reference sample, or the ability to perform reliable laboratory analysis on the actual sample that is entering the analyzer. Never lose sight of these inherent difficulties (challenges) in calibration: (1) Samples can change after removal from the process and hence are different when they reach the lab. (2) Highly variable process streams, in which the lag time between the process measurement point and the sampling valve cause the sample drawn from the stream and the sample measured in the process to be different. (In some cases, the sample itself may be inhomogeneous, which further complicates comparison measurements.) (3) The two measurement methods may not be measuring the identical analyte. For example, a near-infrared moisture analyzer may be calibrated against a loss-on-drying lab method, but these are not identical, and there may be a bias between them caused by other volatile constituents of the sample.

Interferences with the analysis can be caused by physical or chemical effects that cause a deviation in the analyzer output. If the effect of the interfering substance remains constant, a compensation factor can usually be applied in the calibration procedure. During checkout of the installed system, one should verify the effect of interference by calibrating the analyzer for each suspected substance.

When interferences cannot be predetermined, it may be both necessary and practical to send known samples to prospective suppliers for evaluation. In any case, known and suspected interferences and their concentrations should be made known to prospective suppliers when requesting quotations.

Both the method of calibration and the accuracy of the calibration procedure must be established before one purchases a process analyzer.

Analysis Frequency

As discussed in the Problem Definition section, one should first determine whether continuous analysis, automatic-repetitive batch sampling, or an occasional spot check is required. The information from the analyzer and the rate of the dynamic changes in the sample are the main factors to be considered.

The need for process control suggests either a continuous analyzer or an analyzer with a continuous output signal, because a continuous analyzer provides a better chance for the system to reach and maintain equilibrium with the sample. Also, the mechanical design of a continuous analyzer ordinarily is less complicated than that of a discontinuous system. However, the discontinuous analyzer may be more attractive if automatic zero checks are frequently needed, if reagents are blended with the sample, or if the sample is corrosive. Some analyzers, for example, chromatographs, are inherently discontinuous.

The rate at which the measured variable changes in the process is an important factor in determining the frequency of discontinuous analyses. If the sample has to be withdrawn and transported to the analyzer, the time lag factor must also be considered. If several different samples are to share the analyzer (e.g., stream switching), additional time allowances are required.

ANALYZER TYPES

Analytical techniques suitable for process measurement may be organized into several arbitrary categories. Not one approach is inherently superior to another, although for a specific project, the constraints covered previously will narrow the choice.

Chromatographs

Separation techniques literally involve the separation of different components in the sample with appropriate detection. Gas chromatography, which dates to the 1950s, is probably the most widely used technique in process analysis, with the possible exception of pH.

As early as in 1855, the German chemist Friedlieb Runge passed a mixture through an inert material to create separation of the solution components based on differential adsorption. It took nearly a century, when Archer Martin and Richard Synge received the Nobel Prize for developing the first gas chromatograph. By the 1960s, the work of Martin and Synge also set the stage for the development of liquid chromatography, where small sorbent particles and pressure are used to achieve separation.

In all chromatographic techniques, a small portion of the sample stream is injected into a flowing carrier fluid that passes through a column in which it contacts a separation medium prior to reaching a detector. In gas chromatography, the sample is vaporized, if it is a liquid, and carried with helium or nitrogen (most commonly) through the column. The column may be a tube packed with adsorbent media or a thin capillary column coated with an appropriate phase to cause a separation. If the sample was a gas to begin with, it is carried through the column without the need for vaporization. Depending on the analytes to be separated, one chooses the appropriate length of column and coating or packing material. Many different separation materials are available.

Similarly, many different gas chromatography detectors exist to provide specificity or broad detectability. For fixed gases, the thermal conductivity detector (TCD) is mandatory; for organic compounds, either the TCD or the flame ionization detector (FID) is acceptable, but the FID is more sensitive. Added sensitivity (lower detection limits) comes at a price: the FID needs additional gases and extra safety precautions. Later chapters will address these subjects in more detail. In liquid chromatography, the sample is injected into a flowing stream of solvent (mobile phase), separated, and detected. Common detectors include the refractive index (RI) detector, which monitors the eluting mobile phase for RI changes; absorbance detectors, which monitor the spectral absorbance of the mobile phase at one or more wavelength, commonly in the ultraviolet (UV) or visible (Vis) areas of the spectrum; and the electrochemical detector, which monitors an electrochemical property such as conductance of the mobile phase.

Other chromatographic techniques that are available for process measurement include ion chromatography, which can measure either the anions or cations of a stream; gel permeation chromatography, which measures the molecular weight distribution, typically of polymer samples; and supercritical fluid chromatography, which operates in the pressure and temperature regime where a carrier fluid, such as carbon dioxide, is neither liquid nor vapor.

The disadvantages of the chromatographic techniques are the length of time many of the separations take, the fact that these are not continuous techniques (samples are injected only after the previous run is complete and the system has stabilized), aging of the columns, consumption of gases or solvents, and maintenance of the frequently complex and costly sampling systems. In fact, the cost of the chromatograph itself can be less than half of the total installation cost and, if the expense of maintenance is also included, the fraction of the instrument's cost is reduced even further. Some reports indicate that for every successful chromatograph installation in the chemical industry, one has failed and been abandoned. For this reason, some engineers favor simple inexpensive chromatographs, monitoring one component in a single sample to increase reliability. Proponents maintain that an overburdened chromatograph is more a liability than an asset, and that even when it is operating, the volume of information generated is more likely to swamp than to assist the operator. Therefore, their target is to achieve a degree of standardization and simplicity similar to that found in flow or temperature detectors. It is reported that these simple chromatographs can accommodate the majority of present applications.

New developments in the simplification of sampling systems may reduce their cost and complexity while increasing reliability. Recent developments in the field of sampling systems using a small, industry standard platform of modular components show much potential. Recent research in the use of chemometrics to extract information from poorly resolved peaks may provide advantages in analysis time by allowing faster run times. Parallel columns of different separation power combined with chemometrics will provide additional benefits. Chemometrics can help address the issue of swamping the operator with information.

Developments in nonchromatographic separation techniques, such as capillary electrophoresis, in which an applied electric potential is used to drive the separation of charged species, are occurring rapidly. The combination or merger of chromatography and flow injection analysis (FIA) is also on the horizon.

Spectrometers

This family of sensors uses the interaction of electromagnetic radiation with matter. Most commonly, the technique operates on either absorption or reflection principles, but fluorescence and scattering are also used.

If radiation at different wavelengths (Figure 1.1a) is passed through a process material, the amount of absorption may be an indicator of sample identity or composition. The output of such an analyzer is related to the absorption of light in a linear way, as shown in Beer's law (more correctly, the Lambert–Beer law) of radiation absorption. (Transmission is logarithmic with respect to light transmission, but linear with absorption and concentration.)

$$A = abc = \log I_o/I \qquad 1.1(1)$$

where
- A is the absorbance
- a is the molar absorption or extinction coefficient (a constant for any given compound and wavelength)
- b is the sample path length
- c is the absorber (sample) concentration
- I_o is the incident radiation
- I is the radiation leaving sample

The region of the spectrum used for the measurement varies with the kind of compound and information desired

TABLE 1.1b
Absorption/Emission of Electromagnetic Energy Useful for Measurements

Type of Radiation	Wavelength Range	Characteristic Process Probed
Gamma-rays	<10^{-12} m	Nuclear transitions
X-rays	1 nm–1 pm	Inner-shell electron transitions in atoms
UV	400 nm–1 nm	n, π (valence) electron transitions in molecules
Vis	750 nm–400 nm	n, π (valence) electron transitions in molecules
NIR	2.5 µ/m–750 nm	Molecular vibrations
IR	25 µ/m–2.5 µ/m	Molecular vibrations
Microwaves	1 mm–25 µ/m	Rotations in molecules
Radio waves	>1 mm	Rotations in molecules, electron spin flips[a]

[a] NMR uses a magnetic field to split the energy levels and the Rf energy to probe the spin state.

(Table 1.1b). In addition, the instrument itself will be quite different in specific components since the properties of the light vary. Instruments to make these measurements can vary from single-wavelength photometers to grating instruments to multiplex designs, depending on the spectral region and the needs of the measurement.

Ultraviolet and Visible Spectrums The ultraviolet and visible spectrums are similar and are generally grouped together. In fact, most laboratory instruments for this region include both. The energy of this region corresponds to electronic transitions in compounds. UV analyzers are most often used to measure the concentrations of compounds with double bonds (Table 1.1c).

TABLE 1.1c
Partial List of UV-Absorbing Compounds

Acetic acid	Formic acid
Acetone	Hydrogen peroxide
Ammonia	Hydrogen sulfide
Benzene	Iodine
Bromine	Isoprene
Butadiene (1,3)	Mercury
Carbon disulfide	Naphthalene
Carbon tetrachloride	Nitric acid
Chlorine	Ozone
Chlorobenzene	Perchloroethane
Crotonaldehyde	Phenol
Cumene	Phosgene
Cyclohexane	Styrene
Cyclohexanol	Sulfur
Cyclohexanone	Sulfur dioxide
1,3-Cyclopentadiene	Sulfuric acid
Ethylbenzene	Toluene
Fluorine	Trichlorobenzene
Formaldehyde	Xylene (*ortho*, *meta*, and *para*)

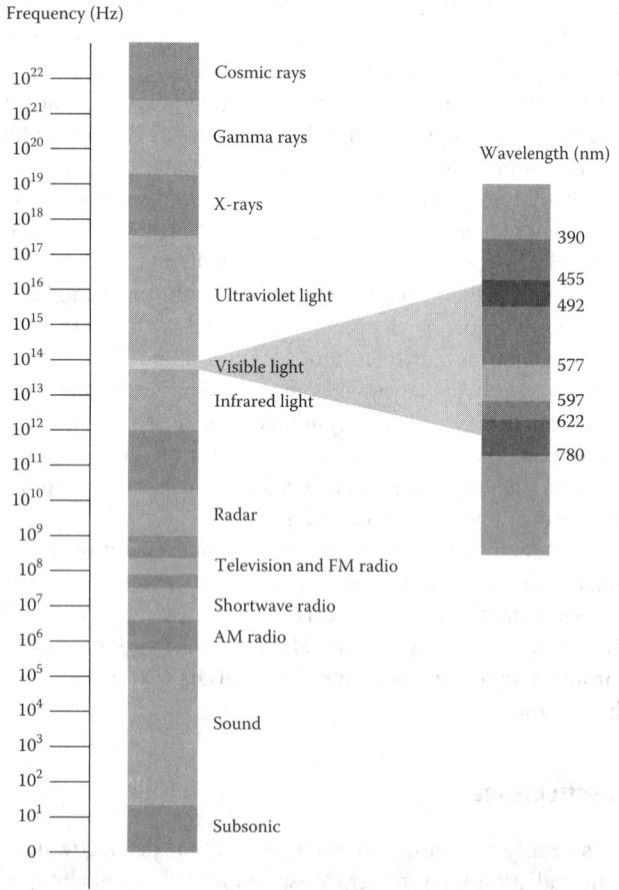

FIG. 1.1a
The frequency and wavelength ranges of various forms of radiation.

TABLE 1.1d
Absorption Bands of Some UV–Vis Absorbing Chromophores (Functional Groups)

Chromophore	Structure	Wavelength Maximum (nm)	Molar Absorptivity [l/(mol·cm)]
Aldehyde	CHO	280–300	11–18
Amine	NH_2	195	2800
Bromide	Br	208	300
Carbonyl	C=O	195, 270–285	1,000, 18–30
Disulfide	S—S	194, 255	5,500, 400
Ester	COOR	205	50
Ether	O	185	1,000
Nitrite	ONO	220–230	1,000–2,000
Oxime	NOH	190	5,000
Thioether	S	194, 215	4,600, 1,600
Carbon–carbon double bonds	C=C	190	8,000
	$(C=C)_2$	210–230	21,000
	$(C=C)_3$	260	35,000
	$(C=C)_4$	300	52,000
Carbonyl with C=C	C=C—C=O	210–250	10,000–20,000
Benzene	Aromatic ring	184, 204, 255	46,700, 6900, 170
Diphenyl	Linked aromatic rings	246	20,000
Anthracene	Fused aromatic rings	252	199,000

Other functional groups with π* electrons are also excellent absorbers (Table 1.1d). Vis absorbers are compounds that are colored to the human eye. Obviously, dyes, inks, etc., are readily measured in the visible portion of the spectrum. One major thing to watch out for is that compounds that absorb in this region may be such strong absorbers that a short path length is needed to get a good response, and only very low concentrations can be measured. Reducing the path length may help, but too small a path length in a typical process stream will act as a filter and cause the measurement to fail.

Fluorescent When radiation of one wavelength is absorbed by a molecule with the appropriate molecular structure, it can be remitted at a higher wavelength (lower energy). This process, called fluorescence, is an extremely useful measurement technique, provided that the analyte has the appropriate structure. Molecules with many double bonds, especially conjugated double bonds, will fluoresce. Many biomolecules, including some vitamins, proteins, cofactors, and the like, fluoresce, allowing the technique to be used in fermentation and other bioprocessing. Crude oil also contains fluorescent molecules; low levels of oil on water are frequently detected using fluorescence. Table 1.1e

TABLE 1.1e
Partial List of Some Fluorescent Compounds and Their Excitation and Fluorescence Wavelength Maxima

Compound	Excitation Maximum (nm)	Fluorescence (Emission) Maximum (nm)	Compound	Excitation Maximum (nm)	Fluorescence (Emission) Maximum (nm)
Adenine	280	375	Proteins	280	313, 350
Adenosine	285	395	p-Terphenyl	284	338
Adrenaline	285	325	Pyrene	330	382
Anilines	280, 290	344, 360	Pyrene, 4-methyl	338	386
Anthracene	420	430	Quinine	250, 350	450
Adenosine triphosphate (ATP)	285	395	Salicylic acid	310	435
Dibenzo[a,e]pyrene	370	400	Terramycin	390	520
Estradiol	285	330	Tetracycline	390	515
Folic acid	365	450	Thymol	265	300
Indole	280	355	Tryptophan	285	365
LSD	325	465	Vitamin A (carotene)	325	470
Neosynephrin	270	305	Vitamin B_{12} (cobalamine)	275	305
Pentabarbital	265	440	Vitamin B_2 (riboflavin)	270, 370, 445	520
Pentothal	315	530	Vitamin B_6 (pyridoxine)	340	400
Phenanthrene	252	362	Vitamin E (tocopherol)	285	330
Procaine	275	345			

Note: Excitation and fluorescence bands are broad and maxima may shift with pH or solvents.

---

* **Pi electron (π electron)** is an electron that resides in the Pi bond of a double or triple bond.

provides a list of some fluorescent compounds along with their excitation and fluorescence (emission) wavelengths maxima.

Infrared and Near-Infrared Infrared (IR) instruments were one of the first analyzers to be moved from the laboratory to the pipeline, and the technology is available for use with gas, liquid, or solid samples. In the absorption mode of operation on liquid samples, the path length has to be very short, though not as short as in the UV. Noncontacting backscatter designs readily measure moisture in solids (e.g., conveyor belts of flour) or composition with fiber-optic probes (FOPs) in liquids. The probe version of the IR analyzer eliminates the need for a sampling system and can be used in both liquid and gas phase processes. NIR is widely used for moisture measurement. Probably every paper machine in the world has an IR or NIR analyzer to measure and control drying. NIR instruments can be filter photometers or spectral instruments using gratings, interferometers, or acousto-optic tunable filters (AOTFs) to select the wavelengths needed for the measurement. Table 1.1f and Figure 1.1b give some absorption wavelengths for various organic functional groups in the IR and NIR regions.

TABLE 1.1f

Absorption Bands of Some Raman and Infrared-Absorbing Chromophores

Chromophore (Functional Group)	Region (cm^{-1})	Intensity Raman	Intensity Infrared
CC aliphatic chains	250–400	Strong	Weak
S—S	430–550	Strong	Weak
Si—O—Si	450–550	Strong	Weak
C—Cl	550–800	Strong	Strong
C—O—C	800–970	Medium	Weak
O—O	845–900	Strong	Weak
C—O—C asymmetric	1060–1150	Weak	Strong
CH$_3$	1380	Medium	Strong
CH$_2$, CH$_3$ asymmetric	1400–1470	Medium	Medium
N=N aromatic	1410–1440	Medium	—
C=C	1500–1900	Strong	Weak
C=N	1610–1680	Strong	Medium
H$_2$O	1640	Weak	Strong
C=O	1680–1820	Medium	Strong
C≡C	2100–2250	Strong	Weak
C≡N	2220–2255	Medium	Strong
C—H	2800–3000	Strong	Strong
O—H	3100–3650	Weak	Strong

Note: Intensities and wave number ranges are approximate and may vary with exact compounds.

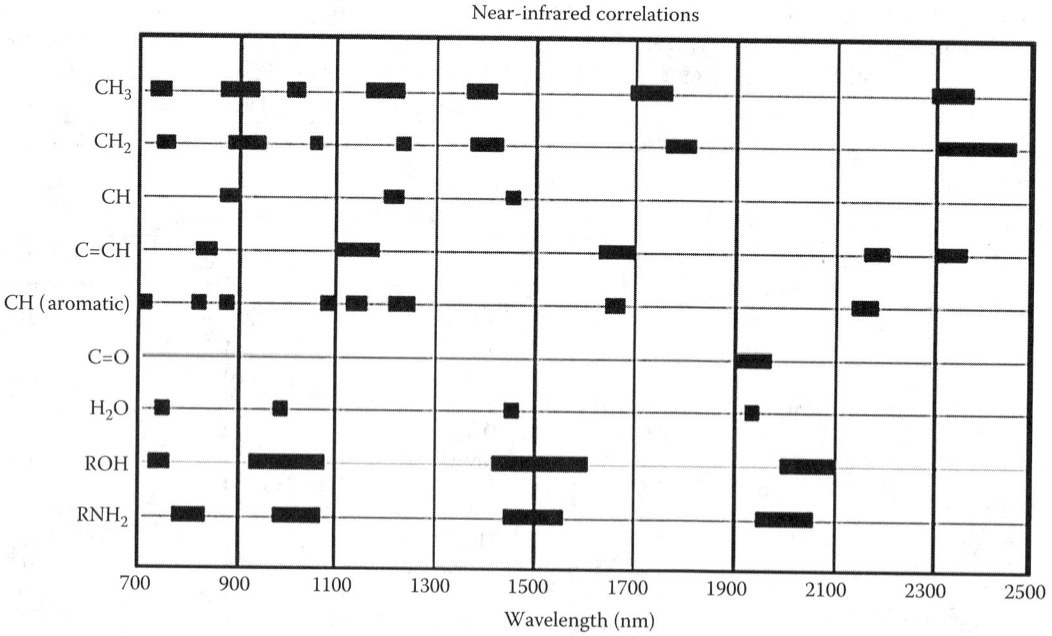

FIG. 1.1b

Absorption bands of some NIR-absorbing chromophores.

Raman Spectrometers In recent years, the Raman spectrometer has moved from the laboratory to the process floor. This has been driven by the combination of improved lasers, fiber optics, and compact spectrometer designs. Raman spectroscopy is complementary to IR and provides similar functional group information. Raman spectroscopy obtains its information from the scattering of light from molecules. It is amenable to direct, noninvasive sampling of solids and liquids. Water and glass are weak Raman scatters, providing the opportunity to investigate samples in aqueous media and through glass site windows, or a glass container, or via FOPs. Reports of Raman being used to monitor polymerization reactions and distillation processes are among the growing number of published applications. The use of NIR laser diodes has reduced the problem of background fluorescence. Table 1.1f compares some molecular absorption bands in the IR and Raman.

Microwave and Radio Frequency Microwave and radio frequency (Rf) analyzers operate in both the absorption and reflection modes. Their most common application is in the measurement of moisture in solids—without requiring physical contact with the sample. In these wavelength regions, dielectric properties are probed by the radiation; hence water, with its high dielectric constant (~78), compared to many common organic compounds (<5), is readily determined (e.g., moisture in wood). The technique is also used for binary solutions. Compared to the previously described spectral regions, this is a relative newcomer to process measurement.*

Refractometry (RI) The RI is a unique property of a chemical compound, based on the fact that light travels at different speeds in different media, and as such can be used for composition determination in binary systems and for various nonspecific concentration measurements, for example, dissolved solids, black liquor, and Brix (sugar). Snell's law expresses the relationship between RI, the angle of incidence (α), and the angle of refraction (β) as light of a given wavelength passes through the interface between two materials.

$$RI = \frac{\sin \alpha}{\sin \beta} \qquad 1.1(2)$$

Most refractometers use visible light, but at least one uses a diode source that emits just beyond the red. Differential refractometers utilize this relationship by keeping α constant; therefore, a measurement of β expresses RI. At a critical value of α, the light is totally reflected; the measurement of this angle can also be related to RI as follows:

$$\alpha \text{ critical} = \arcsin \frac{\text{Variable sample RI}}{\text{Fixed prism RI}} \qquad 1.1(3)$$

Although less sensitive, this latter technique (critical angle refractometry) requires no sampling system and, therefore, is preferred for on-stream applications. In fact, it has practically replaced the differential refractometer in all but the most demanding applications. It should be noted, though, that the critical angle refractometer only measures the RI of that portion of the sample that is in direct contact with the crystal. If even a thin coating develops on the surface, the readings will be useless. A variety of cleaning systems may be used if this is a problem. Turbulent flow has been proven useful in minimizing fouling. (For a listing of refractive indexes, see Table 1.1g.)

TABLE 1.1g
Refractive Index Table

Compound	RI	Compound	RI
Acetic acid	1.3718	Formic acid	1.3714
Acetone	1.3588	Glycerol	1.4729
Acrylic acid	1.4224	Glycol	1.4318
Amyl acetate	1.4012	Heptane	1.3876
Benzene	1.5011	Hexane	1.3749
Butyl acetate	1.3951	Hexanol	1.4135
Butyl alcohol	1.3993	Hydrazine	1.470
Butylene	1.3962	Hydrogen chloride	1.256
Carbon disulfide	1.6295	Lead tetraethyl	1.5198
Carbon tetrachloride	1.4631	Menthol	1.458
Chlorobenzene	1.5248	Methyl alcohol	1.3288
Chloroform	1.4464	Methyl-ethyl ketone	1.3807
Cycloheptane	1.4440	Nitric acid	1.397
Cyclohexane	1.4262	Nonane	1.4055
Cyclohexanone	1.4503	Octane	1.3975
Cyclopentane	1.4065	Pentane	1.3575
Decane	1.41203	Perchloroethylene	1.5053
Di-ethyl benzene	1.4955	Phenol	1.5425
Di-ethyl ether	1.3497	Propanol(iso)	1.3776
Di-methyl benzene	1.4972	Propanol(n)	1.3851
Ethyl acetate	1.3722	Styrene	1.5434
Ethyl alcohol	1.3624	Toluene	1.4969
Ethylbenzene	1.4952	Water	1.3330

Note: All data based on 20°C (68°F).

Turbidity and Particle Size As with refractometers, turbidity sensors also work with light, but make use of light scattering to measure the suspended solids in the liquid. Depending on the concentration of the particles, turbidity can be measured at 90°, 180°, or complete backscatter. (Technically, turbidity is the term applied to the straight-through approach, and nephelometry is used for the scattering measurement.) In addition to scattered light, some light may be absorbed by the liquid. Since there is a particle size dependence to light scatter, instruments are also available that use scattering to measure particle size distribution. Turbidity instruments are quite commonly used to monitor filtering systems, centrifuges, etc., for breakthrough, whereas the more expensive

---

* Lide, D. R. Ed., *CRC Handbook of Chemistry and Physics*, 76th edn., New York: CRC Press, 1995, and in Dean, J. A., Ed., *Lange's Handbook of Chemistry*, 14th edn., New York: McGraw-Hill, 1992.

particle size instruments find use monitoring crystallization and polymerization processes.

Nuclear Techniques A variety of techniques utilizing x-rays and other radiation sources are used in process measurements. X-ray fluorescence (XRF) is widely used to determine metal concentrations in slurries and to measure silicon coatings on paper. XRF works by exciting an inner-shell electron of an element, causing an x-ray to be emitted. The energy [wavelength] of the emitted x-ray is indicative of the element, and the number of x-rays is proportional to concentration. The excitation radiation can come from an x-ray tube or a radioactive source.

The absorption or backscatter of neutron or gamma radiation can also be correlated to composition in some systems. Neutrons have been used to measure the moisture content of solids in processes in which hydrogen is present only in the free water and is not bonded to the other molecules (or is relatively constant in the nonwater part of the matrix). For example, the moisture of wood chips in pulping has been determined by this technique. Gamma rays, which are extremely energetic, are absorbed nonspecifically, and their attenuation can be used as a measure of density. Since gamma rays can penetrate metallic walls, they can give composition data for binary systems without contacting the process stream. These techniques are also used extensively as a noninvasive way to measure tank or pipe level, density within a pipe, and the like.

Electrochemical Analyzers

Several kinds of measurements make direct use of electricity in the measurement. These are potentiometric, wherein an electric potential is measured and the solution remains unchanged; conductive, in which a minute current is measured, but the system is essentially unchanged; and amperometric, in which a chemical reaction occurs during the course of the measurement.

The basic principle of their operation for a system in which a reactant, R, goes to a product, P, is the Nernst equation:

$$E = E^\circ - \frac{RT}{nF} \ln \frac{C_P^0}{C_R^0} \qquad 1.1(4)$$

where
  E is the applied potential
  E° is the standard potential for the system (the chemistry of the reaction being measured)
  R is the molar gas constant
  T is the absolute temperature (°K)
  n is the number of electrons involved in the reaction
  F is Faraday constant
  C_P^0 is the concentration of product at the electrode surface
  C_R^0 is the concentration of reactant at the electrode surface

Potentiometric Sensors The family of analytical sensors that detect the electrical potential generated in response to the presence of dissolved, ionized solids in a solution include pH, oxidation–reduction potential (ORP), and ion-selective electrodes (ISEs) or probes. In potentiometric analysis, the potential difference between two electrodes (reference and measurement) is determined. As these two electrodes must be electrically connected, another potential, the junction potential, is also inherent to the measurement. (In fact, most of the problems of process pH measurement occur because of fouling of the junction.)

For these measurements, the Nernst equation can be rewritten as:

$$E_{cell} = K + \frac{RT}{z_i F} \ln \left(a_i + k_{ij} a_j^{\left(\frac{z_i}{z_j}\right)} \right) + E_{LJ} \qquad 1.1(5)$$

where
  E_{cell} is the measured potential
  K is a constant for the measurement system (electrode, etc.)
  R is the molar gas constant
  T is the absolute temperature (°K)
  F is the Faraday constant
  a_i is the activity of the analyte ion, i
  a_j is the activity of an interfering ion, j
  k_{ij} is the selectivity coefficient
  z_i and z_j are the charges of ions i and j, respectively
  E_{LJ} is the liquid junction potential

It can be further simplified in *dilute* solutions (where activity and concentration are approximately equal) to

$$E_{cell} = K + \frac{0.0592 \log c}{n} \qquad 1.1(6)$$

where
  c is the concentration of analyte
  n is the charge of the analyte ion
  0.0592 is the combination of the RT/F at 25°C

ORP sensors are available in probe designs to detect the ratio of reducing agent to oxidizing agent—an important parameter in effluent treatment controls, for example.

ISEs are also based on the Nernst law. If total ionized solids (conductivity) are constant, a correlation can be drawn between the activity of a specific ion and its concentration in the process stream. The ideal reference electrode produces a constant potential that is independent of the composition of the solution. A perfect measuring electrode gives a 59-mV change in potential for each 10-fold change in the *activity* of a monovalent ion. It is important to emphasize that it is the *activity* of free ions that the electrodes respond to and *not* the concentration. It is important to understand that according to the Nernst equation, concentration, c, can be determined by the measurement of activity (activity = γ c, where γ is the activity coefficient of the ion) *only if* the other variables of the equation are *constant*. To achieve this involves scrupulous design; occasionally, it also requires sample preparation. Frequently, in real-world applications, calibration of the electrode in the matrix, over a limited analyte range, can be used to minimize concerns over activity–concentration discrepancies.

Unless the interferences to which ion-selective measurements are subject can be recognized and eliminated in the potential installations, misapplications are likely. The available ISEs are listed in Table 1.1h, where they are also grouped by the type of membrane utilized. Coating or material buildup on the membranes calls for the same degree of maintenance as required for pH electrodes.

pH sensors are ISEs, sensitive to the activity of hydrogen ions in the process stream, and as such, reflect the acidity or alkalinity of the sample. pH is one of the most commonly made measurements in industry. On the surface, it appears to be easy to measure pH, since all that is needed is the insertion of the probe into the process, but in practice, it is not so simple. Probes may become fouled on the sensing surface; most are made of glass and quite fragile; concentrated solutions, or ones that are not aqueous, stretch the definition of pH (which is for dilute aqueous solutions), and high-purity water (e.g., boilers) requires specially designed probes, if it is to work at all. Recent improvements in electronics have reduced drift, instability, and transmission distance limitations. Combination electrodes (measurement and reference), gel fill *solutions*, and the development of various electrode-cleaning devices, such as air and water jets, and mechanical and ultrasonic cleaners have reduced many of the problems. New nonglass electrodes based on ion-sensitive field effect transistors (ISFETs) have opened up new applications where glass has been problematic (e.g., food and beverage processes). Automation of the removal, cleaning, and reinsertion of pH probes is now available "off-the-shelf" for those applications that require nearly continuous pH measurement and cannot tolerate downtime for maintenance. Nevertheless, despite all this activity, pH remains a difficult measurement for which installation of standby spare sensors and scheduled periodic maintenance are likely to be necessary.

Conductivity Sensors Conductivity sensors measure a solution's ability to conduct electricity, which is a function of all dissolved ionized solids in the solution. Two basic kinds of conductance measurement instrument are available: direct contact and inductive. The direct device makes direct, electrical contact with the process solution with two (or sometimes four) electrodes and measures the flow of electricity from one electrode to the other. Any fouling of the probes, which affects electric contact, requires cleaning of the probes. The inductive type is less sensitive to fouling because it does not make electrical contact with the process fluid. Toroidal (donut shapes are common) probes use several electric coils to probe the electric properties of the fluid. The probe is coated with an inert material such as a fluorocarbon. Another design builds the inductive probe directly into a pipe spool piece. Inductive conductance probes are widely used to measure the concentration of

TABLE 1.1h
Ion-Selective Electrodes: Types and Applications

Ion	Type of Electrode	Lower Detectable Limit (ppm)	Principal Interferences
Bromide	Solid state	0.4	CN^-, I^-, S^-
Cadmium	Solid state	0.01	Ag^+, Hg^{++}, Cu^{++}, Fe^{++}, Pb^{++}
Calcium	Liquid–ion exchange	0.4	Zn^{++}, Fe^{++}, Pb^{++}, Cu^{++}, Ni^{++}, Sr^{++}, Mg^{++}, Ba^{++}
Chloride	Liquid–ion exchange	0.4	ClO_4^-, I^-, OH^-, NO_3^-, Br^-, OAc^-, HCO_3^-, F^-, SO_4^-
Chloride	Solid state	1.8	Br^-, CN^-, SCN^-, I^-, NH_3, S^-
Copper (II)	Solid state	0.006	Ag^+, Hg^{++}, Fe^{+++}
Cyanide	Solid state	0.3	S^-, I^-
Fluoride	Solid state	0.02	OH^-
Fluoroborate (boron)	Liquid–ion exchange	0.11	I^-, HCO_3^-, NO_2^-, NO_3^-, F^-, Br^-, OAc^-, OH^-, Cl^-, SO_4^-
Iodide	Solid state	0.007	S^-, CN^-, $S_2O_3^-$
Lead	Solid state	0.02	Ag^+, Hg^{++}, Cu^{++}, Cd^{++}, Fe^{++}
Nitrate	Liquid–ion exchange	0.6	ClO_4^-, I^-, ClO_3^-, S^-, Br^-, NO_2^-, CN^-, HCO_3^-
Perchlorate	Liquid–ion exchange	1.0	OH^-, I^-, NO_3^-, MnO_4^-, IO_4^-, Cr_2O^-
Potassium	Liquid–ion exchange	2.0	H^+, NH_4^+, Ag^+, Na^+, Li^+, Cs^+
Redox	Solid state	Varies	All redox systems
Silver	Solid state	0.01	Hg^{++}
Sodium	Solid state	0.02	Ag^+, H^+, Li^+, K^+
Sulfide	Solid state	0.003	
Thiocyanate	Solid state	0.6	I^-, $S_2O_3^-$, Br^-, Cl^-, NH_3
Water hardness	Liquid–ion exchange	0.001	Zn^{++}, Fe^{++}, Cu^{++}, Ni^{++}, Ba^{++}, Sr^{++}

strong acid and base solutions. Plots of conductivity vs. concentration of various electrolytes are used to determine if an application is suitable for these devices. Conductance data for various compounds are available in various chemical handbooks and an excellent compilation from a vendor.

Amperometric Sensors Amperometric, also known as galvanic and polarographic, cells induce a reaction in the sample by applying a high enough voltage. The resulting current is measured and is proportional to the concentration of the species reacting. The electrolyte and electrodes are separated from the process stream by a membrane, a feature that eliminates the need for a sample system and can perform well *if* the membrane is kept clean. (Of course, the analyte molecules must diffuse through the membrane into the electrolyte to the electrodes if there is to be a reaction and resulting signal.) In the world of process measurements, such instruments are commonly used to measure oxygen, including both dissolved oxygen in liquids and vapor phase oxygen in gas samples. Membrane-separated amperometric sensors are also used for other gases, such as CO, H_2S, and Cl_2. Sensors using this concept are also being developed for bioanalytes such as glucose, where a specific enzyme is part of the measurement system.

Wet Chemistry Analyzers

In-process measurement using wet chemical methods can be performed in two different ways: batch, or discontinuous and continuous.

Process autotitrators draw a sample from the process at a regular interval and deliver the sample to a titration vessel. Once in the vessel, the sample may be diluted or prepared in other ways for a titration, very similar to a titration performed in the lab. An autoburet delivers the titrant until an end point is sensed. The titration cell is then drained and rinsed, and a new sample is drawn. Cycle times depend on the chemistry, but are typically 5–15 min. The end point may be determined potentiometrically, conductometrically, or photometrically. Methods to measure acid value (number) are common examples.

In continuously flowing wet chemistry methods, a very small sample is injected into a flowing stream that can be a solvent or a reagent. (In some applications, a small portion of the process stream is the carrier and reagent is injected into it.) Additional reagent(s) can be added, heating, mixing, and the time to allow a reaction, etc., can be increased until the reacted sample reaches a detector. Detectors are most commonly photometers, but flow-through pH and other electrochemical detectors are also available. The flowing system can be continuous (flow injection analysis), bubble segmented (as in auto-analyzers), or a variant of FIA, sequential injection, in which a rotary valve takes the place of much of the tubing.

Miscellaneous Techniques

Quartz Crystal Microbalances Certain crystals are known to generate an electric signal when they are deformed; conversely, these crystals can be caused to move when a voltage is applied. These phenomena are known as the piezoelectric effect, and it has been used to create sensors. The former effect has been used to detect pressure, acceleration, temperature, force, and thickness; the latter effect has been applied to measure various analytes, including moisture, in gases. The frequency of crystal oscillation is affected by the mass of material on the crystal, so that a crystal coated with an absorbent material will vibrate differently when the coating absorbs more or less analyte. If the specificity of the coating is high, a sensitive and specific analytical sensor can be developed. Surface acoustic wave (SAW) sensors, in which the speed of sound transmission through the coating is affected by the absorbed analyte, can be even more sensitive.

Mass Spectrometry Previously only available for gas streams, mass spectrometry (MS) is becoming available for liquids as well. Various approaches to the method are used, but all require a gaseous sample (vaporized, if necessary) that is injected into a vacuum and ionized (usually by an electron beam). The ionized sample molecules and the charged fragments of their decomposition are then sorted and measured by their mass-to-charge ratio. This last process can be accomplished by the time of flight of the ions down an evacuated tube, their trajectory through a magnetic field, or selection by an electrostatic field. Detection is by electron multipliers or other devices. The most common application of process MS is in gas plants, where many gases can be detected simultaneously, or multiple streams can be examined sequentially.

Nuclear Magnetic Resonance (NMR) Nuclear magnetic resonance (NMR) is an excellent technique for compositional information. In the laboratory, it is widely used to identify unknown materials. Its use in process analysis is a relatively recent occurrence. NMR makes use of a fundamental property of all matter—its nuclear structure. (Despite the fears of the public, NMR is not a technique involving radiation. If an application is contemplated using NMR, this will need to be explained to plant personnel.) NMR probes the molecules by applying both a magnetic and Rf field. Depending on the instrument, this can yield a simple number or a relatively high-resolution spectrum. NMR can only measure nuclei of atoms with a magnetic moment (caused by an odd number of neutrons or protons). Therefore, the technique can only be used to measure species containing ^1H, ^{13}C, ^{14}N, ^{19}F, and ^{31}P. From a practical standpoint, process instruments have not yet been built that are sensitive enough to be used with anything other than proton NMR. Low-resolution (wideband or pulsed) NMR units are useful for the measurement of free (unbound) proton-rich compounds such as water or oil.

Analyte-Specific Techniques

Many analytes have unique characteristics that have led to instruments designed specifically for their measurement. Water and oxygen stand out as examples.

Humidity and Moisture The measurement of water in the gas phase (humidity) is one of the most common. There may be more patents for humidity sensors than any other measurement. With so many different approaches for the measurement, selection of the appropriate one for a given application is both complicated and critical.

In the liquid or solid phase, moisture measurement can also be made with a wide variety of techniques. Already mentioned in this chapter were IR and NIR, microwave and Rf, and neutron techniques. In addition, there are techniques based on capacitance, impedance, and electrical resistance.

Several chapters of this book are devoted to water measurement in more detail.

Oxygen Oxygen is an important and unique analyte in industry. It is measured as the element in both gas and liquid samples. In the gas phase, it is important for breathing (environmental sensing), combustion control, and chemical stability. As mentioned previously, oxygen is paramagnetic and can be measured using this property. Such instruments operate best in relatively clean, noncondensing environments with oxygen levels from ~0.5% to ambient levels. Oxygen can also be measured electrochemically, and several varieties of instruments are available using different electrochemical techniques. High-temperature catalytic instruments are most commonly used in combustion control applications where the gas stream is already hot and the oxygen level is low. Fuel cell (amperometric) devices are also available. They require periodic replacement of the fuel cell, as it is a consumable part of the measurement.

Oxygen dissolved in water (and other liquids) can also be measured with an electrochemical probe. The oxygen diffuses through a membrane to the active electrode.

One of the parameters of interest in pollution control is the oxygen demand of the effluent. This is a measure of the organic load in the system, which will potentially use oxygen as it is consumed by microorganisms (biological oxygen demand [BOD]) or reacts chemically (chemical oxygen demand [COD]). Laboratory measurement of BOD takes several days of incubation and requires the sample to be inoculated with the appropriate organisms. Instruments are available that can reduce this time. Total organic carbon (TOC) analyzers give a close correlation to COD. Several different approaches are taken by the commercially available instruments. However, in all cases, the sample stream is first digested, and then the carbon dioxide formed from this combustion step is measured. The digestion may be strictly chemical or UV light induced (with catalyst). The resulting CO_2 is measured either directly by infrared photometry or by the change in conductivity of water into which the CO_2 is dissolved. The entire cycle is completed in a few minutes.

SAMPLING TECHNIQUES

Perhaps the most critical aspect of any analyzer installation is the sampling mechanism. The data an instrument generates can be no better than the sample it has been presented for analysis. Sampling is such an important area that entire books have been published on that subject alone.[*,†] In this volume, there are several chapters on sampling, including Chapters 1.2 and 1.3. Figure 1.1c illustrates a sampling system, which serves the removal of particles and water from the sample, prior to it being analyzed.

In the discussion earlier, mention has been made of ana-

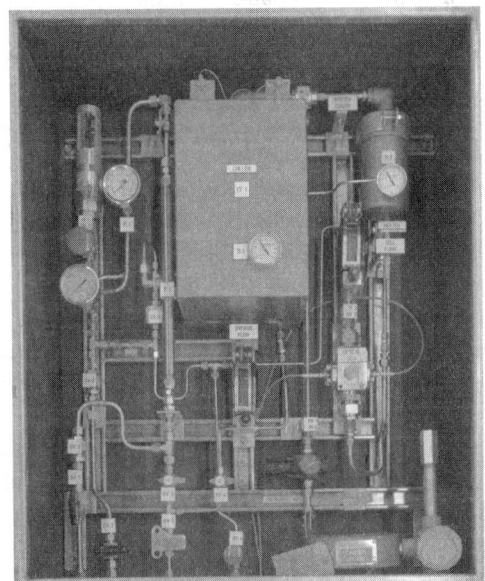

FIG. 1.1c
A sampling system serving the removal of particles and water from the sample, prior to being analyzed. (Courtesy of Guided Wave, Inc., http://www.guided-wave.com/products/sampling-systems.html.)

lyzers, such as chromatographs, that require the sample to be withdrawn from the process and delivered to the instrument. For such instruments, there is no option, but for many of the spectroscopic and electrochemical analyzers, the option of using directly inserted probes is available. Probe-type analyzers have been used for a long time to measure pH, conductivity, selective ion, dissolved oxygen, relative

---

[*] Ruzicka, J. and Marshall, G. D., Sequential injection: A new concept for chemical sensors, Process Analysis and Laboratory Assays, (1990) http://www.sciencedirect.com/science/article/pii/S0003267000839379.

[†] Sherman, R. E., Process analyzer sample-conditioning system technology, (2002) http://www.wiley.com/WileyCDA/WileyTitle/productCd-0471293644.html.

humidity, residual chlorine, oxidation–reduction potential, differential vapor pressure, corrosion, and many other variables. Improvements in quality and cost of fiber-optic probes have made them widely available for in-line measurement.

Fiber-Optic Probes (FOP)

Fiber-optic probes use waveguides made of glass or quartz or other more esoteric material to deliver and return the process-modified light from the probe to the instrument located some distance away, frequently in a control room (Figure 1.1d). FOPs can acquire data on spectral absorbance, diffuse reflectance, fluorescence, or scattering, and several FOPs can be multiplexed to the same computer-controlled analyzer. Performing spectroscopic measurements over FOPs first became popular with near-infrared instruments, but their use in the UV–Vis and IR regions has grown as well.

Insertion probe critical angle refractometers are available. RI units are also available with fiber-optic sensing and transmission (Figure 1.1e). They claim the same advantages of all FOP instruments—minimal consumption of space on the pipe and ease of access to hard-to-reach locations—and they are unaffected by electromagnetic interference, dust,

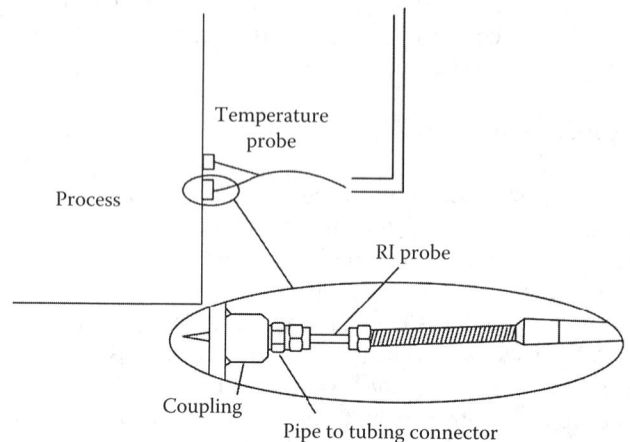

FIG. 1.1e
Temperature-compensated RI-detecting fiber-optic probe.

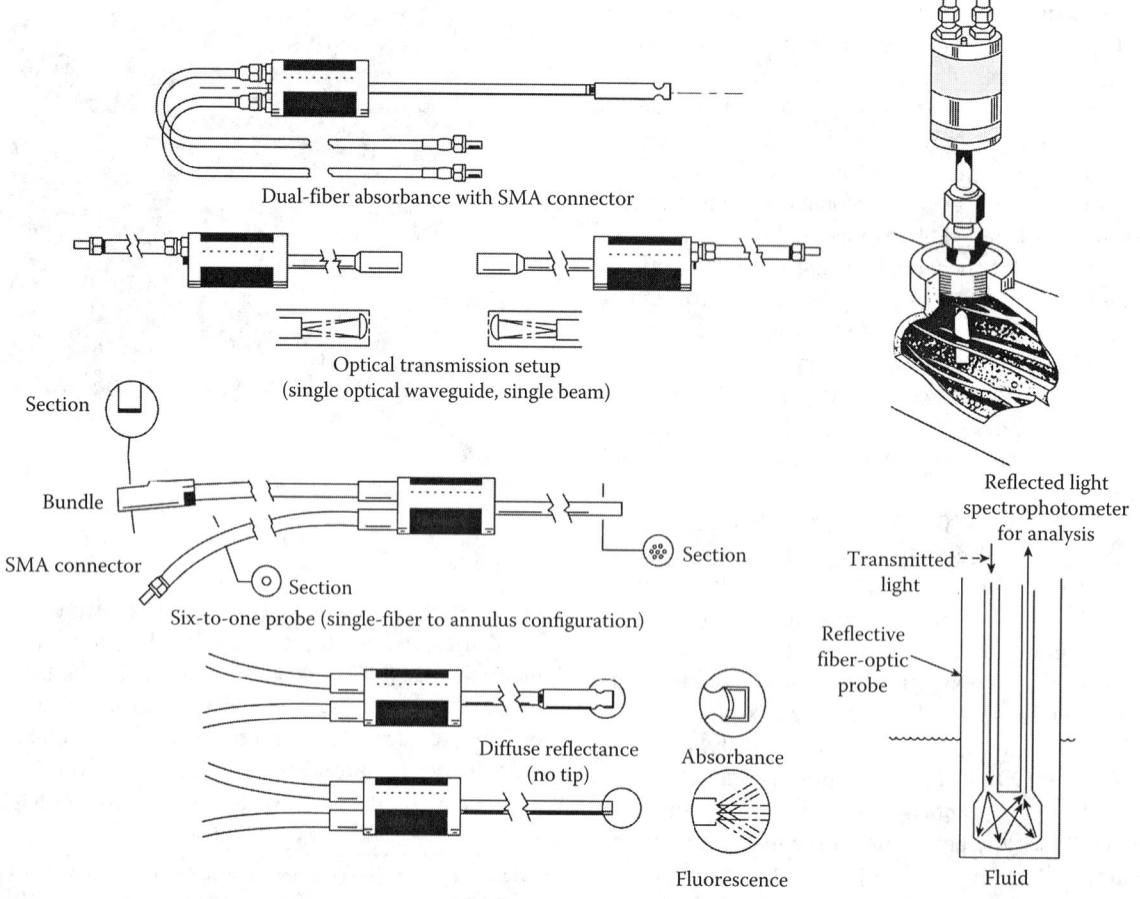

FIG. 1.1d
Fiber-optic probes can provide data on absorbance, diffuse reflectance, fluorescence, and scattering. (Courtesy of Guided Wave, Inc., http://www.guided-wave.com/products/probes.html.)

humidity, or hazardous environments. Measurements can be transmitted over large distances, up to 1 km (0.625 mile).

Probe-type sensors have also been developed for sludge and slurry measurement applications. These include the detection of sludge levels, interfaces, and densities in clarifiers and digesters (Figure 1.1f).

Figure 1.1g illustrates a mechanically self-cleaning probe design used to measure the suspended solids concentration in biological sludge. In this design, a motor-driven reciprocating piston located within a glass tube is stroked once every 15 sec. This action serves to refresh the sample while at the same time eliminating deposits and wiping the optical surfaces clean of dirt or slime. The result is a low-maintenance installation that provides reliable measurements of suspended solids concentration.

In-line, probe-type analyzers are also used in stack gas composition measurement. Some of the microprocessor-based designs also provide self-calibration for both zero and span. In addition, they can provide automatic cleaning and drying

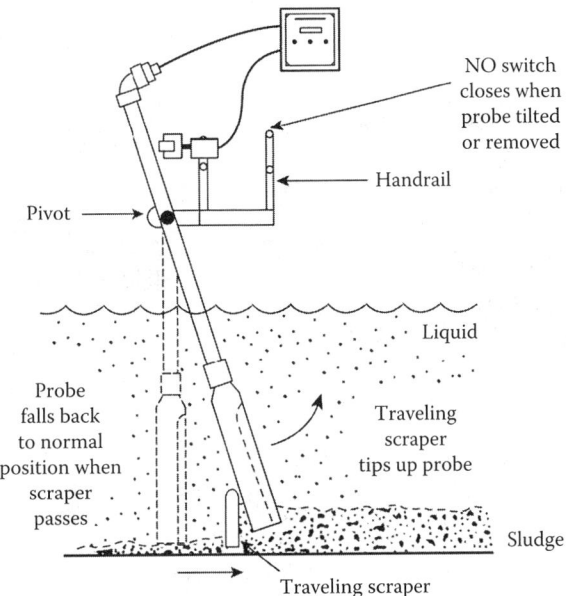

FIG. 1.1f
Probe-type sensors are used to detect the location and density of sludge and slurry layers in clarifiers. (Courtesy of Markland Specialty Engineering, Ltd., http://www.sludgecontrols.com/our-products/automatic-sludge-level-detector/.)

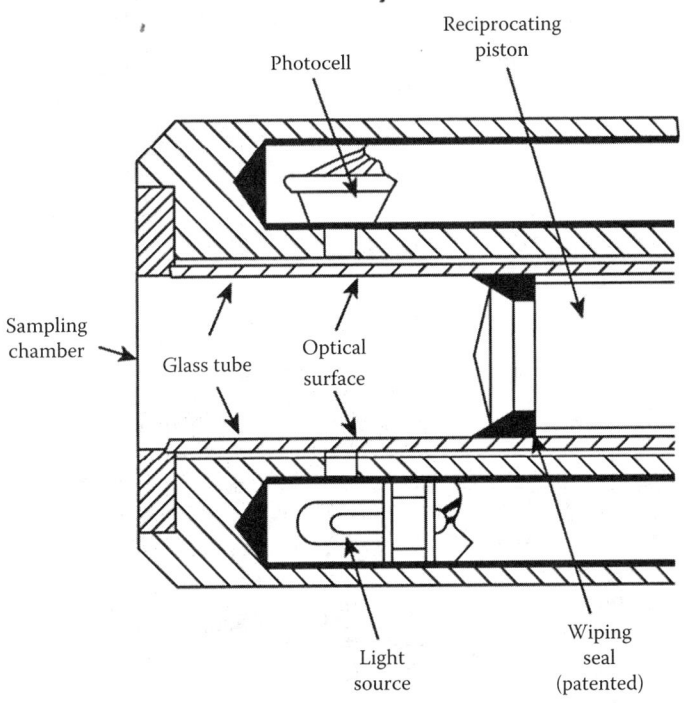

FIG. 1.1g
Probe-type self-cleaning suspended solids detector.

to accommodate wet-scrubber applications, in which it is necessary to periodically force hot and pressurized instrument air into the probe cavity to thoroughly dry the diffuser. The measurement is made by analyzing the measurement beam as it is returned by the retroreflector within the probe's gas measurement cavity (Figure 1.1h). These probe-type stack gas analyzers can operate in locations with high water vapor and particulate loadings and at temperatures up to 427°C (800°F).

A relatively new development in the area of in-stack analyzers is the introduction of acousto-optic tunable filters. Their main advantage is speed and the elimination of maintenance-prone mechanical elements such as filter wheels, moving gas cells and mirrors, diffraction gratings, and mechanical light choppers. The AOTF acts as an electronically controllable narrow-band filter that can be tuned to any desired frequency in milliseconds. As shown in Figure 1.1i, the resulting beam

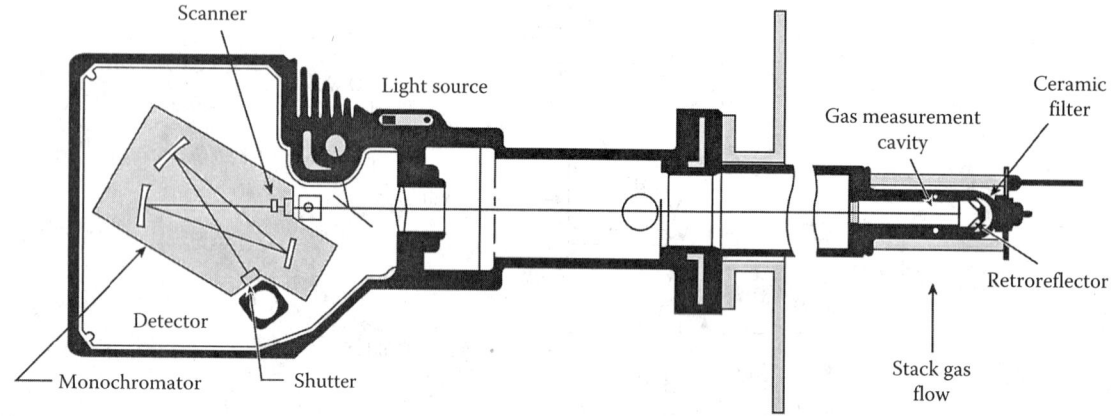

FIG. 1.1h
Probe-type stack gas analyzer. (Courtesy of Teledyne Monitor Labs, Inc., http://www.monitorlabs.com/sm8200.asp.)

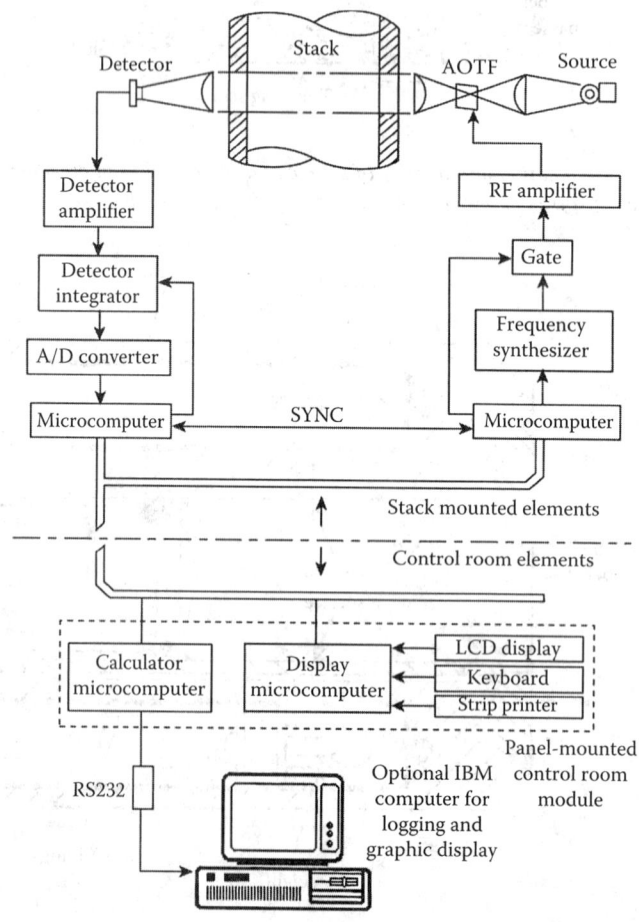

FIG. 1.1i
AOTFs can enhance the speed and increase the reliability of multicomponent analysis in stack gases.

at the selected wavelength is directed across the stack and can simultaneously measure the concentrations of CO, CO_2, C_2H_6, CH_4, NO, NO_2, H_2O, etc. This is achieved by the computer selecting the frequencies required for the measurement of the concentration of each component and the AOTF being tuned to each of the selected frequencies one at a time, but at a high speed. This results in a "scanning" spectrometer without moving parts.

Probe Cleaners

While probe-type in-line analyzers can eliminate the transportation lag and sample deterioration problems associated with online analysis, they are not without problems of their own, the largest of which is fouling. Various mechanisms are available to clean the probe automatically. It is recommended that when a probe cleaner is used, it be placed inside a sight glass so that the operator can continuously observe the performance of the cleaner (Figure 1.1j).

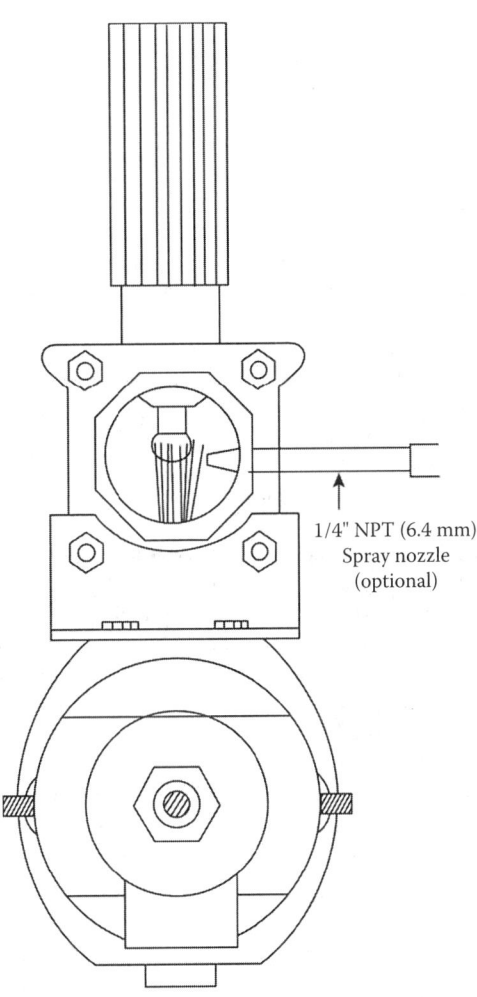

FIG. 1.1j
Probe cleaners should be mounted in sight-flow glasses for good visibility. (Courtesy of Aimco Instruments, Inc.)

TABLE 1.1i

Selection of Automatic Probe Cleaners[a]

	Applicable Choice of Probe Cleaner						
	Mechanical		Chemical			Hydrodynamic	Acoustical
	Brush	Rotary Scraper	Acid	Base	Emulsifier	(Self-Cleaning)	Ultrasonic
Service							
Oils, fats		√	√			√	
Resins (wood, pulp)			√		√		
Emulsion of latex		√					
Fibers (paper, textile)				√			
Solid suspensions				√			
Crystalline precipitation (carbonates)	√	√	√				
Amorphous precipitations (hydroxide)	√	√	√				√
Material of construction	Stainless steel (brush pH 7–14)	Stainless steel	PVC	PVC	PVC	Stainless steel	Polypropylene, stainless steel
Temperature °F	40–140	40–140	40–140	40–140	40–140	40–250	40–195
°C	4–60	4–60	4–60	4–60	4–60	4–120	4–90

[a] Probe cleaner suppliers:

Cole-Rarmer (Self-cleaning, http://www.coleparmer.com/buy/category/self-cleaning-ph-probe)

Emerson-Branson (ultrasonic cleaners, http://www.emersonindustrial.com/en-US/branson/Products/precision-cleaning/industrial-cleaning/Pages/default.aspx)

Fetterolf Corp. (spray rinse valve, http://www.fetterolfvalves.com/spray_rinse.htm)

Foxboro-Invensys (Clean in place, http://iom.invensys.com/EN/pdfLibrary/AppSolution_Foxboro_ConductivitySensorsCleanInPlaceProcedures_09–10.pdf9

Gamajet Cleaning Systems (tank spray washers, http://www.gamajet.com)

Helios Research Corp. (tank spray washers, http://www.heliojet.com)

Polymetron, Division of Danaher Corp. (probe cleaners, http://www.polymetron.com)

Spraying Systems Co. (tank spray washers, http://www.spray.com)

Toftejorg Inc. (tank spray washers, http://www.alfalaval.com/campaigns/tankequipment/tank-cleaning/pages/tank-cleaning.aspx)

Yokogawa (Chemical or mechanical, http://c15181565.r65.cf2.rackcdn.com/product_FC20_GS_11.pdf)

Many or most vendors of electrodes can provide self-cleaning or auto-cleaning capability for their probes. Many operate on a timer, cleaning the electrode periodically. Some can automatically extract a probe from the process, clean, and even recalibrate before reinserting it into the line. For example, pH, ORP, and ISE probe vendors have a variety of systems to clean electrodes, including ultrasonic, brushes, and Teflon balls.

Wide varieties of probe-cleaning devices are available on the market. Their features and capabilities for various types of coatings and a list of their suppliers are given in Table 1.1i. Probe cleaners for pH and other electrodes have been available for many years.

If there is no sampling system, the integrity of the sample is automatically guaranteed, and preference should be given to sensors that penetrate the process pipe with a retractable, cleanable probe. Probe sensors, either solid or membrane, require periodic cleaning, which can be done manually, by withdrawing the probe through an isolating valve so that the process is not opened when the electrode is cleaned, or automatically. Automatic probe-cleaning devices may be pressurized liquid or gas jets, or thermal, mechanical, or ultrasonic cleaning and scraping devices.

ANALYZER LOCATION

The location of the analyzer can be critical to the success of the installation. If the instrument is to be installed at some distance from the process, the lines bringing the sample to the instrument must be considered. Many questions will need to be answered. For example, what is the lag time to reach the instrument? Do the lines need heat tracing? Will the sample be returned to the process?

Ambient conditions for the analyzer must be considered. Should the instrument or several instruments be housed in an analyzer house or shelter? Instruments require power and other utilities that must be included in the planning. Access to the instrument for maintenance, local readout or display of the measured value, or of diagnostic information should

TABLE 1.1j
Electrical Area Classifications

Flammable Gas Class I	Combustible Dust Class II	Ignitable Fibers Class III
Group A: acetylene	Group E: metal dust, aluminum, magnesium, etc.	No groups listed
Group B: hydrogen or manufactured gas	Group F: carbon black, coal or coke dust	
Group C: ethyl ether, ethylene, cyclopropane	Group G: flour, starch, grain dust	
Group D: gasoline, hexane, naphtha, benzene, propane, butane, alcohol, acetone, benzol, lacquer solvent, natural gas		
Class I flammable gases		
Division 1 locations	Division 2 locations	
Locations in which:	Locations in which:	
(1) Hazardous concentrations exist continuously, intermittently, or periodically under normal operating conditions.	(1) Gases are handled, processed, or used, but will normally be confined to closed containers or systems from which they can escape only in accidental rupture or abnormal operation of equipment.	
(2) Gases exist frequently because of repair, maintenance, operations, or leakage.	(2) Gases are normally prevented by positive ventilation, but might become hazardous through failure of ventilating equipment.	
(3) Breakdown or faulty operation might release gases that might cause simultaneous failure of electrical equipment.	(3) Adjacent to Class I, Division I locations in which gases might occasionally be communicated.	

be considered. The quality of the electrical power can be critical, as many microprocessors in analyzers can be reset or damaged by voltage spikes.

If a sampling system is used, variation in pressure in vents and drains, and oil or moisture, or both, in air lines can wreak havoc on the system. The temperature and humidity of the atmosphere surrounding the analyzer may also contribute to unreliable performance.

Fire and explosion are the major safety hazards of analyzer installations (see Table 1.1j). The inability of most laboratory instruments to meet electrical requirements limits their use in online applications. Often, however, purging an analyzer that is not explosion proof allows it to be used in hazardous areas. Samples or stored reagents containing toxic, flammable, or noxious substances are also safety hazards. This can be a major issue when considering liquid chromatography analyzers, for example.

DATA HANDLING

Another issue that must be considered during the analyzer selection process regards the data. How will the valuable data your instrument and sampling system have generated be transferred from the analyzer to the people and control system computer that can make use of it? What kind of output is there? Will there be a local display?

For years, most analyzers provided only an analog output of current (most common, and typically 4–20 mA) or voltage. In recent years, this has changed so that more and more analyzers provide a digital output, as well as, or instead of, the analog output. In the case of the analog output, analyzers generally need to be calibrated in two ways. First, as already mentioned, calibration standards (samples that are independently measured by the lab) are determined and used to set the instrument, frequently just a zero and span setting. The second calibration is more of an adjustment. The output of the analog signal must be set. In other words, the low and high output must be made to read correctly. Frequently, this must also be matched to the input on the control system. For example, an NIR calibrated to read percent moisture of wheat flour from 0% to 25% might have its 4–20 mA output set so that 4 mA equals 10% moisture and 20 mA equals 25%, thus providing higher resolution for the output.

Newer-style analyzers provide digital outputs. These may be a relatively simple ASCII output using an RS232 connection or Ethernet, TCP/IP (Transmission Control Protocol/Internet Protocol), etc. If an analyzer provides a single piece of information, such as moisture, the issue is quite different from another analyzer that measures a process stream and transmits the concentrations of five different constituents, for example, a gas chromatograph measuring three alkanes, two alkenes, and several aromatic compounds.

Even more complex is the question of chemometric models that are used by some spectral techniques (e.g., IR, NIR) and are beginning to be used in chromatography as well.

When selecting the analyzer for an installation, the data handling must be considered. What inputs are available on the control system? Will chemometric calculations be done on the analyzer system or a separate computer system or by the control system? What kind of cable is needed for the data transmission, and over what distance can it be used without degrading the signal? How does that affect installation cost?

MAINTENANCE

Analyzer hardware is likely to receive better care if it is accessible and housed in pleasant surroundings, such as in air-conditioned buildings. Spare parts and special testing components should be ordered at the same time as the system to avoid future delays; the availability and most effective use of qualified maintenance personnel are also factors of major importance.

Analyzers already used in the plant are helpful in determining maintenance and spare parts requirements. Major subassemblies not stocked as spare parts can also be interchanged and can help to obviate waiting for parts for the duplicated system. Standardizing on one model of a particular analyzer within a site will reduce training and spare parts costs.

COST

For the approximate cost ranges of the various analyzer categories, refer to Table 1.1k, and for more accurate costs of the various analyzers, refer to the feature summaries at the

TABLE 1.1k
Summary of Some Attributes of Major Analyzer Approaches

Use this table as a basic guide. It provides you with a starting point, to help you think about the analyte's measurable attribute. Usually, several methods will be applicable to any given analyte, depending on concentration, matrix, etc.

Technique	Cost	Complexity	Calibration	Sample Phase	Sampling System	Selectivity	Mode	Range
Gas chromatography	2, 3	2, 3	1	1, 2	3	1	1a, 3	1, 2, 3
Liquid chromatography	3	3	1	2	3	1	1a, 3	1, 2, 3
IC	3	3	1	2	3	1	1a, 3	1, 2, 3
UV–Vis								
Filters (1–2 wavelengths)	1	1	1	1, 2	1	1, 2, 3	1a, 2	1, 2, 3
Full spectrum	2, 3	1	2, 3	1, 2, 3	1	1, 2, 3	1a, 2	1, 2, 3
Fluorescence (filter)	2	1	2	2, 3	1	1, 2	1a	1, 2
NIR								
Filters (2–6 wavelengths)	2	1	2	2, 3	1, 2	2	1a, 1b, 2	2, 3
Full spectrum	3	2, 3	3	2, 3	1, 2	2	1a, 1b, 2	2, 3
Raman, with fiber optics	3	2, 3	2, 3	2	2	2	1b, 2	2, 3
Microwave/Rf	2, 3	2	2	2, 3	2	3	1a, 1b, 2	2, 3
RI	1, 2	1	1	2	1	3	1b, 3	2, 3
Turbidity	1, 2	2	1	2	1	3	1b, 2	2, 3
Particle size	3	2	1	1, 2, 3	1, 2	3	1b, 2, 3	N/A
XRF	3	3	1	2, 3	1, 2	1	1a, 1b, 2	2, 3
Potentiometric (pH/ISE/ORP)	1	1	1	2	1, 2	1, 2	1b, 2	1, 2, 3
Conductance	1	1	1	2	1	3	1a, 1b, 2	2, 3
Amperometric (fuel cell, galvanic/polarographic)	1	1	1	1, 2	1	1, 2	1b, 2	1a, 2, 3
Autotitration	2, 3	2	2	2	2	2	3	1, 2, 3
FIA	3	3	2	2	2	2	2, 3	1, 2, 3
MS	3	2	2	1	2	1	2	1, 2, 3
NMR	3	3	2	2	2	2	3	3

Explanation of Codes:

Attribute	Code 1	Code 2	Code 3
Cost	<$5000	$5,000–20,000	>$20,000
Complexity	Low	Moderate	High
Calibration	Two points (or does not need)	Several points	Chemometrics
Sample phase	Gas	Liquid	Solid
Sampling system	None	Simple	Complex
Selectivity	Specific	Varies with system	Nonspecific
Mode	Noninvasive (a); invasive (b)	Continuous	Intermittent
Range	Trace (<0.1%)	0.1%–5%	>5%

Note: N/A, not applicable. As with all generalizations, these comparisons should be taken as a guide. Different installation schemes, different analytes, etc., can dramatically change costs, complexity, and the other attributes in the preceding table. The addition of fiber optics to any of the optical techniques will provide increased accessibility to the process from remote locations and eliminate the need for a sampling system, but the cost of the fiber, especially in the mid-infrared region, is considerable.

beginning of each chapter that follows. The total cost of an analyzer installation is highly variable, as it is the sum of the cost of the following:

- Engineering study
- Analyzer
- Recorder or other display hardware
- Sample system
- Startup and checkout in the laboratory
- Calibration standard
- Installation costs
- Spare parts
- Startup and checkout in the plant
- Utilities and reagents
- Training repairmen
- Maintenance

CONCLUSIONS

In an analyzer project, there are many different options and choices to be made. The skilled analyzer engineer or chemist must make many decisions and, in general, none of them are right or wrong; as in most of life, compromise is the order of the day. We must balance the cost, complexity, maintainability, etc., against the value of the measurement data to the process and make our choices. Table 1.1a provides an orientation table that sums up the capabilities of the different analytical techniques and helps the reader to select the best system for the application at hand.

The information in orientation Table 1.1a at the beginning of this chapter is complemented by the contents of Table 1.1k, which compares key attributes (cost, complexity, type of sample, etc.) of the many analyzers discussed in this chapter. Both of these tables should be used with the proverbial grain of salt. Analyzers are constantly changing; developments in electronics, sensing elements, and all other aspects of measurement improve almost daily. In addition, very few applications are exactly the same. Therefore, use these tables as a starting point, and carefully evaluate all the criteria discussed in this chapter and the chapters that follow.

SPECIFICATION FORMS

Analyzers should be carefully specified. Some users prefer to use their own specification forms (Table 1.1l); others use the forms prepared by the International Society of Automation describing the operating parameters of their installations (Forms 20A1001 and 20A1002). Both types of forms are reproduced on the following pages. The ISA documents are included with the permission of the International Society of Automation.

TABLE 1.1l
Typical Analyzer Specification Form Prepared by one of the Users

Project:... Date: ....

Specification no.: .... Code: ....

Information compiled by: ....

A. GENERAL INFORMATION:

1. Plant:.. 2. Unit: ....

3. Process: ....

B. CONDITIONS AT PLANNED ANALYZER LOCATION:

1. Ambient temperature range: .... to.. Normal ....

☐°C   ☐°F

2. Protected from weather: Yes ☐   ☐ No

3. Unusual ambient conditions ....

(Corrosive or explosive atmosphere, excessive moisture, dust, etc.)

4. Power available: .... V .... Hz

(a) Voltage variation:.. to V

(b) Frequency variation:... to ....Hz

(c) Grounding facilities available: ☐ Yes   ☐ No

5. Lighting level: *Good*   *Average*   *Poor*

(a) Front of instrument   ....

(b) Back of instrument   ....

(c) Direct sunlight will strike instrument: ☐ Yes   ☐ No

6. Steam lines near location: ....

7. Instrument air available:

(a) Pressure range: from....................................... to............................... Normal .... PSIA

(b) Temperature range: from........................... to............................... Normal ....☐C ☐F

(c) Contaminants: ....

(d) Size of header: .... Volume: .... ft^3/min

(Use separate page for each stream to be analyzed.)

C. SAMPLING INFORMATION:

1. Form of sample: Gas ☐   Liquid ☐   Other..

2. Temperature range: from....................................... to................. Normal................................ °C   °F

3. Pressure range: from....................................... to .... Normal............................ PSIA

4. Dew point:..°C At.. PSIG

5. Quantity available:... per hour

6. Low-pressure return line available:

Yes ☐   No ☐   Back-pressure... PSIG

7. Specific gravity: ....

8. Contaminants in sample:

Oil ☐   Wax ☐   Solids ☐   Particle Size ....

(Include identity and concentration in list of components below.)

9. Corrosive nature: Acidic ☐   Basic ☐   Other..

10. Other data (viscosity, unusual surges, etc.): ....

11. Materials of construction that may be used in contact with sample: ....

12. Distance from sample tap to analyzer location:... ft

13. Size of tap at process line, if any: ....

(Continued)

TABLE 1.1I (Continued)
Typical Analyzer Specification Form Prepared by one of the Users

14. Concentration ranges of all components (even if only traces) in stream: (Specify unit of measurement:% by volume,% by weight, or ppm for each component.)

Components to Be Analyzed				Other Components				
Maximum	Minimum	Normal	Unit	Water (Liquid, Vapor)	Maximum	Minimum	Normal	Unit

15. The above stream composition information is considered proprietary:
    Yes ☐   No ☐
16. Method of lab analysis used to measure sample: ....
17. Desired response time of analyzer:... min... sec..........................
18. Accuracy required:% of full-scale reading

D. INSTALLATION REQUIREMENTS:
1. Type of installation: Permanent ☐   Temporary ☐   Portable ☐
2. Type of mounting: None ☐   Rack ☐   Panel ☐
Other: ....
3. Electrical code: Class.............................. Group............................... Division ....
4. Recorder or indicator required: ....
    (a) To be supplied with analyzer: Yes ☐   No ☐
5. Location of recorder: At analyzer ☐   Distance from analyzer... ft
6. Recorder mounting: None ☐   Rack ☐   Panel ☐
7. Accessories:
    (a) Alarms: High ☐   Low ☐   (b) Controls: ....
    (c) Others: ....
8. Date required: ....
9. Sketch of system indicating sample points and distances to analyzer:

48 Analytical Measurement

ANALYSIS DEVICE — Operating Parameters

#	RESPONSIBLE ORGANIZATION			ANALYSIS DEVICE	#	SPECIFICATION IDENTIFICATIONS			
1					6				
2	ISA			Operating Parameters	7	Document no			
3					8	Latest revision		Date	
4					9	Issue status			
5					10				

#	ADMINISTRATIVE IDENTIFICATIONS				#	SERVICE IDENTIFICATIONS Continued			
11					40				
12	Project number		Sub project no		41	Return conn matl type			
13	Project				42	Inline hazardous area cl		Div/Zon	Group
14	Enterprise				43	Inline area min ign temp		Temp ident number	
15	Site				44	Remote hazardous area cl		Div/Zon	Group
16	Area		Cell	Unit	45	Remote area min ign temp		Temp ident number	
17					46				
18	SERVICE IDENTIFICATIONS				47				
19	Tag no/Functional ident				48	COMPONENT DESIGN CRITERIA			
20	Related equipment				49	Component type			
21	Service				50	Component style			
22					51	Output signal type			
23	P&ID/Reference dwg				52	Characteristic curve			
24	Process line/nozzle no				53	Compensation style			
25	Process conn pipe spec				54	Type of protection			
26	Process conn nominal size		Rating		55	Criticality code			
27	Process conn termn type		Style		56	Max EMI susceptibility		Ref	
28	Process conn schedule no		Wall thickness		57	Max temperature effect		Ref	
29	Process connection length				58	Max sample time lag			
30	Process line matl type				59	Max response time			
31	Fast loop line number				60	Min required accuracy		Ref	
32	Fast loop pipe spec				61	Avail nom power supply			Number wires
33	Fast loop conn nom size		Rating		62	Calibration method			
34	Fast loop conn termn type		Style		63	Testing/Listing agency			
35	Fast loop schedule no		Wall thickness		64	Test requirements			
36	Fast loop estimated lg				65	Supply loss failure mode			
37	Fast loop material type				66	Signal loss failure mode			
38	Return conn nominal size		Rating		67				
39	Return conn termn type		Style		68				

#	PROCESS VARIABLES	MATERIAL FLOW CONDITIONS			#	PROCESS DESIGN CONDITIONS		
69					101			
70	Flow Case Identification			Units	102	Minimum	Maximum	Units
71	Process pressure				103			
72	Process temperature				104			
73	Process phase type				105			
74	Process liquid actl flow				106			
75	Process vapor actl flow				107			
76	Process vapor std flow				108			
77	Process liquid density				109			
78	Process vapor density				110			
79	Process liquid viscosity				111			
80	Sample return pressure				112			
81	Sample vent/drain press				113			
82	Sample temperature				114			
83	Sample phase type				115			
84	Fast loop liq actl flow				116			
85	Fast loop vapor actl flow				117			
86	Fast loop vapor std flow				118			
87	Fast loop vapor density				119			
88	Conductivity/Resistivity				120			
89	pH/ORP				121			
90	RH/Dewpoint				122			
91	Turbidity/Opacity				123			
92	Dissolved oxygen				124			
93	Corrosivity				125			
94	Particle size				126			
95					127			
96	CALCULATED VARIABLES				128			
97	Sample lag time				129			
98	Process fluid velocity				130			
99	Wake/natural freq ratio				131			
100					132			

#	MATERIAL PROPERTIES			#	MATERIAL PROPERTIES Continued		
133				137			
134	Name			138	NFPA health hazard	Flammability	Reactivity
135	Density at ref temp	At		139			
136				140			

Rev	Date	Revision Description	By	Appv1	Appv2	Appv3	REMARKS

Form: 20A1001 Rev 0 © 2004 ISA

	RESPONSIBLE ORGANIZATION	ANALYSIS DEVICE COMPOSITION OR PROPERTY Operating Parameters (Continued)		SPECIFICATION IDENTIFICATIONS		
1			6	Document no		
2	ISA		7	Latest revision		Date
3			8	Issue status		
4			9			
5			10			

	PROCESS COMPOSITION OR PROPERTY			MEASUREMENT DESIGN CONDITIONS				
	Component/Property Name	Normal	Units	Minimum	Units	Maximum	Units	Repeatability
11								
12								
13								
...								
75								

Rev	Date	Revision Description	By	Appv1	Appv2	Appv3	REMARKS

Form: 20A1002 Rev 0

Definitions

Amperometry: The measurement of current for analytical purposes as an indication of concentration at constant potential difference (voltage). This measurement technology uses electrodes and electrode potential to measure current (electron generation). It can measure titration end point by measuring the electric current produced by the titration reaction or can measure the concentration of dissolved oxygen as it is reduced at the measuring electrode. In case of dissolved oxygen measurement, the oxygen is reduced at the cathode, which is usually made of a platinum wire, while the anode supplies the electrons for the current, which is directly proportional to the oxygen concentration.

Coulometry: A measurement technique in which the amount of a substance transformed (set free or deposited) is determined quantitatively by measuring the total amount of electricity (coulombs) that is consumed or produced. There are two basic categories of coulometric techniques. Potentiostatic coulometry involves holding the electric potential constant during the reaction and coulometric titration or amperostatic coulometry that is carried out by keeping the current (measured in amperes) constant during the reaction.

Galvanometry: A method for detecting or measuring small electric currents. This measurement can be done by means of evaluating the current's mechanical effect on a current-carrying coil in a magnetic field.

Nephelometer: An instrument for the measurement of the density or concentration of suspended particles in a liquid by measuring the degree to which the suspension scatters light.

Polarography: An amperometric method of analysis. It is used in electro-analytical chemistry to find the concentration of a solute in solution. In potentiometric measurements, the potential between two electrodes is measured using a high-impedance voltmeter.

Potentiometry: A method of analysis where the electromotive force (the difference in voltage) between a known voltage and the voltage output of a measurement cell is used to determine the concentration of a substance. One of the electrodes is the reference electrode, whose potential is known, the other is the measuring electrode, which usually is either a metal immersed into the solution whose concentration is to be measured or a carbon rod that contains the ions of interest.

Raman spectroscopy: It is a spectroscopic technique used to observe vibrational, rotational, and other low-frequency modes. It relies on Raman scattering, of monochromatic light, usually from a laser in the visible, near infrared, or near ultraviolet range. The laser light interacts with molecular vibrations, phonons, or other excitations, resulting in shift of the energy of the laser photons.

Refractive Index (RI): The refractive index or index of refraction of a substance is a dimensionless number that describes how light, or any other form of radiation, propagates through that substance. It is calculated as the ration of the speed of light through a vacuum divided by the speed of light through the particular substance. For example, the refractive index of water is 1.33, meaning that light travels 1.33 times as fast in vacuum as it does in water.

Refractometer: An instrument that measures the refractive index of a substance.

Abbreviations

AOTF	Acousto-optic tunable filters
BOD	Biological oxygen demand
CIP	Clean-in-place
COD	Chemical oxygen demand
FIA	Flow injection analysis
FID	Flame ionization detector
FOP	Fiber-optic probe
ISE	Ion-selective electrodes
ISFET	Ion-sensitive field effect transistors
MS	Mass spectrometry
NIR	Near infrared
NMR	Nuclear magnetic resonance
ORP	Oxidation–reduction potential
RF	Radio frequency
RI	Rrefractive index
SAW	Surface acoustic wave
TDC	Thermal conductivity detector
TOC	Total organic carbon
TOD	Total oxygen demand
UV	Ultra violet
XRF	X-ray fluorescence

Organizations

AOCS	American Oil Chemists Society
ASTM	American Society for Testing Materials
IEC	International Electrotechnical Commission
TAPPI	Technical Association of the Pulp and Paper Industry

Bibliography

(For the convenience of the reader, this listing is broken into three segments, separating the literature into the categories of dealing with dealing with basics and measurement applications.)

Basics

ACOS, Firestone, D., Official Methods and Recommended Practices of the AOCS, (2013) https://aocs.personifycloud.com/PersonifyEBusiness/Default.aspx?TabID=251&productId=114014.

ASTM, Bibliografia, (2005) http://www.tec.ac.cr/sitios/Docencia/forestal/kuru/Documents/2010/Kuru%20marzo%202010/Bibliografia.pdf.

Ball, D. W., Theory of Raman Spectroscopy, (2001) http://works.bepress.com/david_ball/115/.

Brown, H. W., Organic chemistry, (2011) http://www.amazon.com/Organic-Chemistry-William-H-Brown/dp/084005498X/ref=pd_sim_b_11.

Davis, R., Practical numerical methods for chemical engineers, (2012) http://www.amazon.com/Practical-Numerical-Methods-Chemical-Engineers/dp/1479146439/ref=sr_1_14?s=books&ie=UTF8&qid=1373046648&sr=1-14&keywords=chemical+analysis+methods.

Haldatec, Analyzer control systems, (2015) http://www.iceweb.com.au/Analyzer/sample_systems.htm.

Harris, D. C., Quantitative chemical analysis, (2010) http://www.amazon.com/s/ref=nb_sb_noss_2?url=search-alias%3Dstripbooks&field-keywords=Chemical+analysis.

Harrold, D., New sampling/sensor initiative, (2010) http://www2.emersonprocess.com/siteadmincenter/PM%20DeltaV%20Documents/Articles/ControlMagazine/Slowly-but-Surely.pdf.

Haynes, G. H., CRC handbook of chemistry, (2012) http://www.amazon.com/Handbook-Chemistry-Physics-93rd-Edition/dp/1439880492/ref=pd_sim_b_2.

Houser, E. A., Principles of sample handling and sampling systems design for process analysis, (1977) https://books.google.com/books/about/Principles_of_sample_handling_and_sampli.html?id=Mc__AAAAIAAJ.

Khandpur, R. S., Handbook of analytical instruments, (2006) http://books.google.com/books/about/Handbook_of_Analytical_Instruments.html?id=NcL_w2LQ8ZcC.

Lide, D. R., Ed., CRC handbook of chemistry and physics, 76th edn., (1995) http://www.amazon.ca/Handbook-Chemistry-Physics-76th-Edition/dp/0849304768.

Miller, B., Advanced organic chemistry, (2003) http://www.amazon.com/Chemical-Analysis-Instrumentation-Methods-Techniques/dp/0470859032/ref=sr_1_2?s=books&ie=UTF8&qid=1373045536&sr=1-2&keywords=Chemical+analysis.

Pande, A., Handbook of moisture determination and control, (1975) http://www.cabdirect.org/abstracts/19770432504.html;jsessionid=225030E1E01C5927B32405BBFE17490F.

Pedersen, S. F., Understanding the principles of organic chemistry, (2010) http://www.amazon.com/Understanding-Principles-Organic-Chemistry-Laboratory/dp/1111428166/ref=pd_sim_b_22.

Perry, R. and Green, D., Perry's chemical engineers' handbook, (2007) http://www.amazon.com/Perrys-Chemical-Engineers-Handbook-Edition/dp/0071422943/ref=pd_sim_b_23.

PIP, Process analyzer system acceptance testing, (2003) ftp://ip184-184-174-233.br.br.cox.net/My_Passport/SMHaik/Documents/MBA%20Computer/JOB/Specifications/EGGS/General%20Specifications/16%20Process%20Control/16GSTPA001.pdf.

Roberts, G. W., Chemical reactions and chemical reactors, (2008) http://www.amazon.com/Chemical-Reactions-Reactors-George-Roberts/dp/0471742201/ref=pd_sim_b_2.

Rosemount, Conductance data of commonly used materials, (2010) http://www2.emersonprocess.com/siteadmincenter/PM%20Rosemount%20Analytical%20Documents/LIQ_MAN_6039_Conductance_Data_Commonly_Used_Chemicals.pdf.

Rouessac, F., Chemical analysis: Modern instrumentation methods, (2007) http://www.amazon.com/Chemical-Analysis-Instrumentation-Methods-Techniques/dp/0470859032/ref=sr_1_2?s=books&ie=UTF8&qid=1373045536&sr=1-2&keywords=Chemical+analysis.

Ruzicka, J. and Marshall, G. D., Sequential injection: A new concept for chemical sensors, process analysis and laboratory assays, (1990) http://www.sciencedirect.com/science/article/pii/S0003267000839379.

Scott, W. W., Standard methods of chemical analysis, (2010) http://www.amazon.com/Standard-Methods-Chemical-Analysis-Analytical/dp/1143264983/ref=sr_1_5?s=books&ie=UTF8&qid=1373046300&sr=1-5&keywords=chemical+analysis+methods.

Sherman, R. E., Process analyzer sample-conditioning system technology, (2002) http://www.wiley.com/WileyCDA/WileyTitle/productCd-0471293644.html.

Speight, J., Lange's handbook of chemistry, (2005) http://www.amazon.com/Langes-Handbook-Chemistry-Anniversary-Edition/dp/0071432205.

TAPPI, Test methods, (2015) http://www.tappi.org/Standards-TIPs.aspx.

Towler, G., Chemical engineering design, (2012) http://www.amazon.com/Chemical-Engineering-Design-Second-Edition/dp/0080966594/ref=pd_sim_b_2.

Wolfbeis, O. S., Ed., Fiber optic chemical sensors, (2004) http://pubs.acs.org/doi/abs/10.1021/ac040049d?journalCode=ancham.

Measurement Applications

Annino, R. and Villalobos, R., *Fundamentals of Process Gas Chromatography*, Research Triangle Park, NC: ISA, 1991.

Biannual process analytical chemistry reviews in analytical chemistry, e.g.: Workman, J., Jr., et al., Process analytical chemistry, *Analytical Chemistry*, 71(12), 121R–180R, 1999. Workman, J., Jr., et al., Process analytical chemistry, *Analytical Chemistry*, 73(12), 2705R–2718R, 2001.

Carr-Brion, K., *Moisture Sensors in Process Control*, New York: Elsevier, 1986.

Cessna, G. D., Process mass spectrometry: History, fundamentals, and applications, *ISA Conference*, Houston, TX, October 1992.

Clevett, K. J., *Process Analyzer Technology*, New York: John John Wiley & Sons, 1986.

Cuonauito, R. P., "Oxygen Analyzers," Measurements and Control, February 1990.

Danigel, H., Fiber optical process measurement technique, *Optical Engineering*, 34(9), 2665, 1995.

Gill, A., Analyzer installation and maintenance, InTech, February 1994.

Huskins, D. J., *Gas Chromatographs as Industrial Process Analysers*, Bristol, U.K.: Hilger, 1977.

Huskins, D. J., *General Handbook of on Line Process Analyzers*, Chichester, U.K.: Ellis Horwood Ltd., 1981.

Huskins, D. J., *Quality Measuring Instruments in On-Line Process Analysis*, Chichester, U.K.: Ellis Horwood Ltd., 1982.

Kalis, G., How accurate is your on-line pH analyzers, *InTech*, 37(6), June 1990.

Khandpur, R. S., *Handbook of Analytical Instruments*, 2006, http://books.google.com/books/about/Handbook_of_Analytical_Instruments.html?id=NcL_w2LQ8ZcC.

Light, T. S., Industrial analysis and control, Chapter 10 in *Ion Selective Electrodes*, Durst, R. A., Ed., National Bureau of Standards Special Publication 314, Washington, DC, 1991.

Lodge, J. P., *Methods of Air Sampling and Analysis*, Chelsea, MI: Lewis, 1988.

Long, R. L., Sampling valves help analyze difficult process streams, InTech, May 1993.

Manka, D. P., Ed., *Automated Stream Analysis for Process Control*, Vols. 1 and 2, New York: Academic Press, 1982 and 1984.

McMillan, G. K., Understand some basic truths of pH measurement, *Chemical Engineering Progress*, 87(10), October 1991.

McMillan, G. K., *pH Measurement and Control*, 2nd edn., Research Triangle Park, NC: ISA, 1995.

Merritt, R., Hardware vs. virtual analyzers, *Control*, XV(2), 35–41, February 2002.

Miller, B., Advanced organic chemistry, 2003, http://www.amazon.com/Chemical-Analysis-Instrumentation-Methods-Techniques/dp/0470859032/ref=sr_1_2?s=books&ie=UTF8&qid=1373045536&sr=1–2&keywords=Chemical+analysis.

Mix, P. E., *The Design and Application of Process Analyzer Systems*, New York: John Wiley & Sons, 1984.

Ness, S. A., *Air Monitoring for Toxic Exposures: An Integrated Approach*, New York: Van Nostrand Reinhold, 1991.

Nichols, G. D., *On-Line Process Analyzers*, New York: John Wiley & Sons, 1988.

Norton, H. N., *Sensor and Analyzer Handbook*, Englewood Cliffs, NJ: Prentice Hall, 1982.

Ramanujam, R. S. and Fitzgibbon, P., X-ray fluorescence for on-line elemental analysis, *Control*, March 1990.

Riebe, M. T. and Eustace, D. J., Process analytical chemistry, *Analytical Chemistry*, 62, 65A–71A, 1990.

Rouessac, F., Chemical analysis: Modern instrumentation methods, 2007, http://www.amazon.com/Chemical-Analysis-Instrumentation-Methods-Techniques/dp/0470859032/ref=sr_1_2?s=books&ie=UTF8&qid=1373045536&sr=1–2&keywords=Chemical+analysis.

Schirmer, R. E., On-line fiber-optic-based near infrared absorption spectrophotometry for process control, *Proceedings of the ISA*, 1986, pp. 1229–1235.

Sherman, R. E. and Rhodes, L. J., Eds., *Analytical Instrumentation*, Research Triangle Park, NC: ISA, 1996.

Vána, J., *Gas and Liquid Analyzers*, New York: Elsevier, 1982.

Waller, M. H., *Measurement and Control of Paper Stock Consistency*, Research Triangle Park, NC: ISA, 1983.

Weaver, R., *Continuous Emissions Monitoring*, Measurements and Control, June 1992.

1.2 Analyzer Sampling

W. M. BARROWS (1972) **A. P. FOUNDOS** (1982)
B. G. LIPTÁK (1995, 2003) **B. G. LIPTÁK and I. VERHAPPEN** (2017)

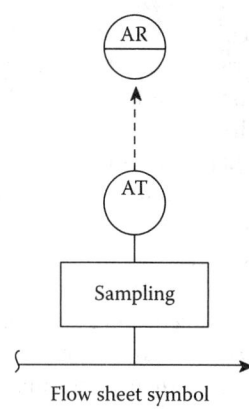

Flow sheet symbol

Costs Compressed air filters: $60; Stainless steel bypass filters: $300–$600; continuous-flow ultrasonic homogenizers, $3000; sludge centrifuge, $600; interval sampling pump only, $750; automatic liquid sampler system, $1500 and up; complete sampling system for single-process gas stream, $3000–$6000; complete sampling system for single-process liquid stream, $3500–$7000. The per-stream cost in a multi-stream sampling system drops as the number of streams increases.

Partial list of suppliers ABB Automation Analytical Div. http://new.abb.com/products/measurement-products/analytical/continuous-gas-analyzers/continuous-gas-analyzers-sample-handling, http%3A%2F%2Fwww.abb.com%2Fproduct%2Fus%2F9AAC100057.aspx%3Fcountry%3DUS
Air Dimensions Inc. (corrosion-resistant sampling pumps) (www.airdimensions.com)
Ametek Process Instruments http://www.ametekpi.com/searchresults.aspx?Keywords = sampling%20systems
APEX Instruments (Heated sample probe assembly) http://www.apexinst.com/products/iem-probes/pc4–4-s
Cole-Parmer Instrument Co. (homogenizers, composite samplers) (www.coleparmer.com)
Conax Technologies (sampling probe assembly) http://www.conaxtechnologies.com/details.aspx?cid = SGSA&pid = SPA
Cuno Filters a 3M company (self-cleaning filters) http://www.process-controls.com/John_Brooks_Company/filters_cleanable_cartridge.htm
Draeger Safety (www.draeger.com)
Fluid Metering Inc. (www.fmipump.com)
Hofi (stack gas sampler) http://www.environnement.it/public/articoli/171/Files/HOFI_uk_S_p(1).pdf
Horiba Instruments (www.horiba.com)
Kahn Instruments (www.kahn.com)
King Engineering Corp. (filters) (www.king-gage.com)
KNF Neuberber (www.knf.com)
Kurz Instruments Inc. (stack sampling) (www.kurz-instruments.com)
Markland Specialty Engineering (www.sludgecontrols.com)
Neutronics Inc. (www.neutronicsinc.com)
Orbital Sciences http://www.urc.cc/pubs/URC-1999d.pdf
Parker Filtration, formerly Balston http://www.balstonfilters.com/products/sample-analyzer-filters
Perma Pure (www.permapure.com)
PGI International (www.pgiint.com)
RotorFlush (self-cleaning backwash filter) http://www.rotorflush.com/
Scientific Instrument Services (Mass spectrometer probe) http://www.sisweb.com/referenc/applnote/app-44.htm
Sensidyne (www.sensidyne.com)
Sentry Equipment Corporation (liquid samplers) (www.sentry-equip.com) (formerly Bristol Equipment Co.)
Sentry Equipment Corp. http://sentry-equip.com/Products/ISOLOK-bulk-solids-samplers.htmSiemens Energy & Automation, Process & (www.sea.siemens.com) Analytical Div./Applied Automation (process samplers)
Servomex Applied Automation (www.servomex.com)
Teledyne Analytical Instruments (process sampling systems) (www.teledyne-ai.com)
Thermo Ramsey/Thermo Onix/Thermo Scientific (www.thermoscientific.com)
Tornado (self cleaning filters) http://soclema.com/wp-content/uploads/SOCLEMA_GENIEFILTERS_TORNADO602-PS.pdf
TracErase/Trace Technology (http://www.tracetechnology.com/?q = node/11)
Tyco Valves & Controls (www.tycovalves.com)
Zellweger Analytics (www.honeywellanalytics.com)

INTRODUCTION

This chapter starts with a discussion of the general considerations that the designer needs to evaluate before starting the actual hardware selection for a sampling system. Next, the various sample system components are described. The chapter is concluded by a detailed evaluation of the various sampling applications, considering gas, liquid, solids, stack, reactor, slurry, trace, and other types of process samples.

In the past, the analyzer sampling systems that were marketed, were designed to serve the analyzers they manufactured. In other cases, it was necessary for the users to design their own sampling system, combining the various components on the market. Complete, automatic, and self-diagnosing sampling systems (Figure 1.2a) that can provide samples to a variety of analyzers were not common. Today, there are automatic units that can be placed near the sample tap on the process pipeline and handle hot or corrosive samples and can clean particulate laden ones. Some of these units are also self-cleaning and are provided with temperature control and remote access in both analog or digital form.

FIG. 1.2a
Automatic, online analyzer sampling system. (Courtesy of ABB Analyzer Division.)

Sampling in general is also discussed in Chapter 1.1 and under the individual analyzer categories, such as in Chapter 1.10 for chromatographs or in Chapter 1.36 for mass spectrometers. The subject of sampling systems used in stack particulate concentration measurement is separately discussed in Chapter 1.4.

GENERAL CONSIDERATIONS

The sampling system is an integral and key part of an analyzer system and should be designed to obtain a representative sample; transport the sample to the analyzer; condition the sample without altering the composition or concentration of the component of interest; accomplish sample stream switching, if necessary; provide facilities for return and disposal of the sample; and provide not only calibration facilities, but also preventive maintenance features, and alarm functions for online reliability and operator alerts.

If a sample has to be brought to an analyzer, a sample transportation delay and a potential for interference with the integrity of the sample will be introduced. If the measured composition is to be controlled, the transportation lag can seriously deteriorate the closed-loop control stability of the system.

Even more serious is the potential for interference with the integrity of the sample, due to the effects of filtration, condensation, leakage, evaporation, and so on, because these operations can not only delay, but also change information that is to be measured: the composition of the process fluid.

Consequently, the best solution is to eliminate all sampling systems and place the analyzer directly into the process. The "in-pipe" analyzer designs are becoming more and more available. Particularly well suited for this design are the various radiant energy and probe-type sensors.

Feasibility Evaluation

Sample systems are rarely duplicated, and thus each system must be debugged as a new entity. In deciding the design of a sampling system, the following questions should be raised:

1. How will the result of the analysis be used (i.e., monitoring or control)?
2. Will the sample be adversely affected by sample transport and conditioning?
3. From what stage in the process will the sample be taken?
4. How can the sample be transported to the analyzer?
5. Is the sample a solid, liquid, gas, or mixture?
6. In what phase must the sample be for analysis?
7. Must the sample be altered (filtered)?
8. Is sufficient sample available?
9. What will be the time lag introduced by transporting the sample?
10. Where is the excess sample returned?
11. Will the analyzer be shared by one or more samples?

These considerations are applicable both to the continuously flowing sample systems and to the less frequently used grab samples. Grab sampling is limited by variations in cleanliness of the sample containers, changes in the method of collecting the sample, delays in transporting the sample, and deviations in withdrawing the sample from the sample container.

The piped-in sample system usually includes the hardware for calibrating the analysis system. It also provides a tap for obtaining samples for laboratory testing. The laboratory sample point must be located so that when used it does not adversely affect the operation through change in pressure or flow of the sample to the process analyzer. Sample conditioning systems are usually costly to maintain.

The installation price for the sampling system frequently exceeds the cost of the analyzer, but its importance overrides this economic consideration because a second-class analyzer can still furnish usable data if it operates with an efficient sample system, whereas a poor sample invalidates the entire measurement.

Therefore, the most important criterion is to keep the samples representative, both in time (short sample lines guarantee minimum transportation delays) and in composition. Whenever possible, the sample should not be tampered with because the steps of sample preparation (drying, vaporizing, condensing, filtering, and diluting) always degrade the sample.

Sample System Specification

A comprehensive listing of the characteristics of each sample stream, including any abnormal conditions, should be prepared before an analyzer sample system is designed. Data are usually available from process conditions and laboratory analysis in existing plants and from design data for new plants. If using plant design data for new plants be cautious, especially with characteristics of the "heavy ends," which are only estimates of those characteristics using "black oil" models rather than individual components. Table 1.2a is a form to be used in summarizing the characteristics of a sample. When purchasing an analyzer together with its sampling system, the filled out Table 1.2a should be attached to the applicable ISA analyzer specification form that can be found at the end of each chapter in this volume (Figure 1.2b).

TABLE 1.2a
Sample System Specification Form

Stream Name or Identification			
Stream Composition Data	Concentration in mol%, wt%, ppm		Range of Component to Be Measured
Component 1			
Component 2			
Component n			
Operating Process Data			
Temperature _____		Pressure _____	
Phase: Liquid _____		Vapor _____	
Sample bubble point		Dew point	
Corrosive components/solids			
Stability (polymerize, decompose, etc.)			
SAMPLE CONDITIONS			
Maximum distance: Tap to analyzer _____		Analyzer to return _____	
Speed loop required: Yes		No	
Sample return pressure point			
Sample probe requirements: Connection size		Orientation	
Materials of construction:			
Stainless steel Teflon Viton		Glass Other	
Electrical areas classification (shelter)		Sample point(s)	
Power supply			
Output signal			
Utilities available: Steam Air		Cooling water	

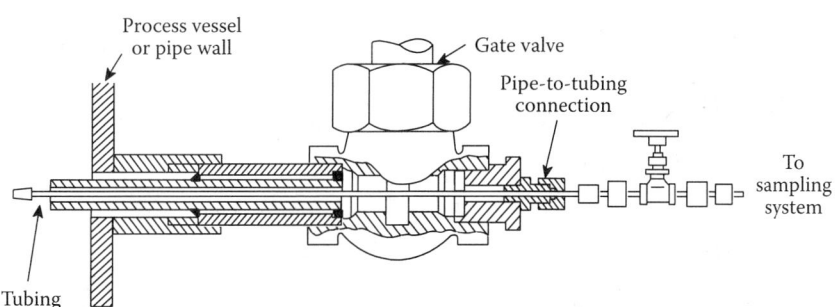

FIG. 1.2b
Sample probe assembly with process shutoff value.

Sampling Point Location

Sample conditioning begins with the location of a suitable sample takeoff point. To obtain a representative sample, the takeoff is usually located at the side of a process line, especially in the case of liquid samples where there is the possibility of vapor on the top of a horizontal line and dirt or solids on the bottom of the line. For sampling vapors, the connection may be located in the side or top of the process line, but in both cases with due consideration to accessibility for maintenance.

Ideally, the sample at the appropriately selected takeoff point will require little or no conditioning; however, it is good practice to install a sampling probe (Figure 1.2b) for most applications as a precautionary measure to prevent particulates from entering the sample transport system. Sampling processes that are still reacting chemically or pyrolysis gases may require reaction quenching, or fractionation, at the sample takeoff. This is done by cooling or back-flushing with an inert gas or liquid to keep the sample takeoff clean and reliably active (Figure 1.2c), while drawing off a reproducible sample for analysis.

With the advent of in situ analyzer detectors, the sample takeoff becomes the point of analysis and, for locating in situ analyzers, the preceding considerations must still be carefully evaluated.

Sample Transport

A representative sample extract from a process line must be continuously conditioned while in transport to avoid compromising the sample integrity. Thus, provision must be made to heat or cool the line as necessary for the specific condition (Figure 1.2d).

The sample is normally transported in one of three ways (Figure 1.2e).

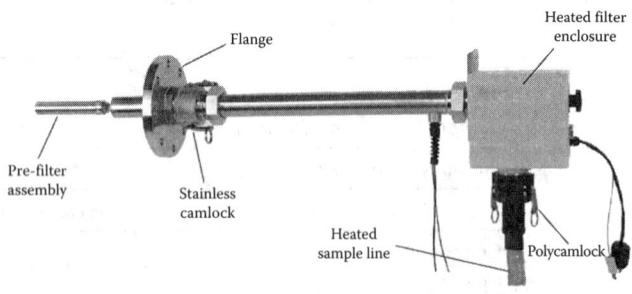

FIG. 1.2d
Heated sample probe assembly. (Courtesy of Apex Instruments.)

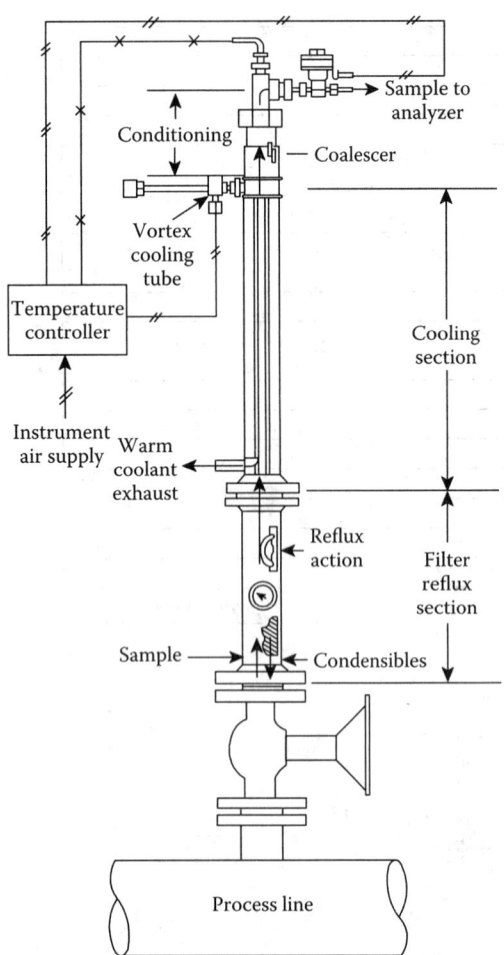

FIG. 1.2c
Pyrolysis gas sample fractionation and conditioning unit.

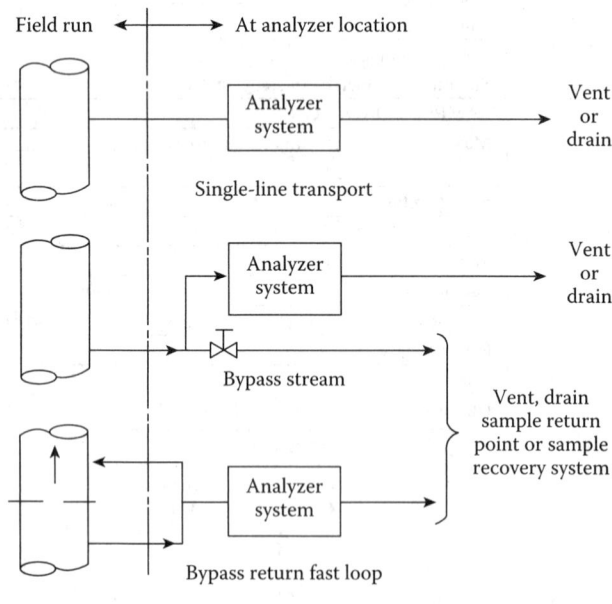

FIG. 1.2e
Sample transport methods.

Single-line transport is the most direct approach and is used when the sample line volume is small in relation to the analyzer sample consumption so that the transport time lag is reasonably short. It is usually used when the analyzer is field-mounted close to the sample point and sample exhaust facilities are available.

Bypass-stream transport is a commonly used method for maintaining a high sample transport velocity that provides minimum transportation lag. This method is used when samples are vaporized at the tap and no facilities exist for returning the vapor to the process, or in similar situations. Consideration must be given to the fact that the sample bypass must be piped to a drain or vent. This procedure may have a negative impact on the environment, despite the fact that the cost of the sample being disposed of may economically justify a sample recovery system.

Bypass-return fast loop is the most commonly used transport loop for an analyzer mounted away from the sample take-off point. By circulating a continuous high velocity across a device, to create a differential pressure, and drawing off a bypass stream to the analyzer, a fast loop is obtained, which is adjustable with no waste in the product. In most cases, such devices as pumps, control valves, and process equipment exist in the process for this purpose. Because a control valve is not a constant pressure drop device, it should only be used as a final resort to obtaining a differential pressure source. Otherwise, it may be necessary to install an orifice in the process line, a circulating sample pump, or an eductor. With a circulation sample pump, care must be taken to prevent cavitation by locating the pump close to the sample takeoff.

Transport Lag

After selecting the appropriate sample transport method for each analyzer system, a calculation of the sample time lag should be made, using conventional flow equations based on the following:

1. Available differential pressures.
2. Total length of the fast loop from the sample takeoff point to the analyzer location and back to the sample return point. Any restrictions on this loop should be excluded from available differential pressures.
3. Line sizes used.
4. Viscosity of the sample.

Table 1.2b should be used as an aid in establishing volumes and pressure drops for some typical tubing and pipes used in sample systems.

Another caution with tubing is that all metals are derated as temperature increases and in the case of stainless steel tubing the derating begins at approximately 150°C (300°F).

TABLE 1.2b
Dimensions and Volumes of Tubing and Pipe Used in Sample Systems

	Nominal Diameter, in.	Inner Diameter, in.	Internal Area		Internal Volume	
			in.^{2}	cc^2	ft^3/ft	cc^3/m
316 stainless steel tubing Wall thickness 0.035"	$1/8$	0.0787	0.055	0.0095	0.0613	6.6E-05
	$1/4$	0.1850	0.180	0.1018	0.6570	0.00071
	$3/8$	0.0253	0.305	0.2924	1.8862	0.00203
	$1/2$	0.4055	0.430	0.5811	3.7491	0.00404
Schedule 40 pipe	$1/4$	0.088	0.088	0.0243	0.1570	0.00017
	$3/8$	0.091	0.091	0.0260	0.1679	0.00018
	$1/2$	0.109	0.109	0.0373	0.2409	0.00026
	$3/4$	0.113	0.113	0.0401	0.2589	0.00028

Sample Disposal

Sample disposal is a critical area, both from the economic point of view that would preserve any quantity of sample not used and from the environmental side that would prevent the emission of most hydrocarbons into the air.

When there is an economic justification for saving the sample, for example, liquids in boiling point and viscometer analyzers, a sample collection and return systems must be furnished to collect the sample at atmospheric pressures and pump it back at high pressure into the process. For gases with no sample return point, the sample can be pressurized back into the process, or as is most frequently done, the sample can be vented into the flare system. However, except in rare cases, venting is done directly into the atmosphere.

When this is not possible, extreme caution should be taken to control the backpressure when venting the sample into a flare or other collection system with varying backpressures as this can affect the operation of the analyzer—especially for gas chromatographs that could see the carrier flow rates in the columns affected. As an alternative, that also "eliminates" hazardous environmental emissions, and a catalytic combustion process can be provided to oxidize the vented sample while maintaining the atmospheric pressure reference. This design is available from TRACErase technology. The vent mounted unit uses a continuous heat source to allow the oxidation process to be effective on intermittent fugitive emission streams at a constant backpressure.

Ambient Effects

Sample systems condition a sample suitable for introduction into an analyzer while maintaining the integrity of the sample. As has been emphasized throughout this chapter, once a sample is conditioned, it must be preserved in the conditioned state; therefore, provisions for heating, or in some cases cooling, the sample lines and system must be furnished for the integrity of the sample. Thus, the entire system must be protected from varying ambient temperatures, which could condense or flash the sample.

Furthermore, as a general rule, the sample should be located in enclosures that provide limited access to unauthorized personnel and also protect the equipment from any corrosive environment. It is accepted practice that systems be completely preassembled and tested in conjunction with analyzers prior to installation in the field.

Test and Calibration

The reliability of each analyzer is measured by its ability to check the analyzer calibration as recommended by the manufacturer. Therefore, each system must include provisions for testing and calibration. As is shown in Figure 1.2f, a means of isolating the inlet is needed in order to allow a calibration sample to be introduced manually or automatically.

Also necessary is a suitable calibration manifold with gas or liquid to furnish a reliable calibration sample to the analyzer. Storage of the test sample may be a consideration, especially for unstable liquids or gases with low dew points. Treatment of the containers is very important for trace analysis samples. One technique that has been successfully used to help reduce the separation of calibration gases while they sit in the cylinder is to apply a heat strip to the bottom of the cylinder as this induces thermal gradients in the fluid, which help reduce the stratification of calibration fluids over time due to differences in specific gravity between the components.

In all cases, the calibration provisions must be incorporated into the systems and a sample provided that is compatible with the desired stream composition and suitable for analysis by the analyzer used for the system. It is desirable, but not essential, that the calibration sample be introduced automatically from a remote location so that the instrument can be periodically checked; however, in most systems, the introduction of the sample is done by manual switching at the analyzer.

SAMPLING SYSTEM COMPONENT SELECTION

A number of devices are discussed here that are components of analyzer sampling systems. These devices include gas and liquid filters, bypass filters, liquid homogenizers, liquid grab sample collectors, chemical reactor sampling systems, and solids samplers. Later, under Applications, the various sampling probe packages and single-stream and multistream process sampling systems are described.

The more components there are in a sampling system, the less reliable it is likely to be. The mean time between failure and maintenance needs of the overall system will improve as the number of pumps, ejectors, regulating valves, coolers, heaters, filters, coalescers, dryers, knockout traps, manifolds, timers, and other components that comprise the system are reduced.

Sampling systems require certain components, which are commercially available. One source is the analyzer's manufacturer, who through the years has developed systems compatible with its analyzer, such as filter coalescers, condensers, and washing and treating systems.

A second source is the analyzer systems vendor, who had designed special components, such as kinetic separators, filter probes, and the like, for use with analyzers for applications in rather hard service.

A third source is the specialty vendor, who has developed unique sampling components, such as pyrolysis gas sample conditioners; permeation devices for water removal systems; high-efficiency, self-cleaning filters; and so on. It is desirable to check whether specialty items are available before trying to design new components. Most of the specialty components have taken years of field testing to develop and modify for successful application.

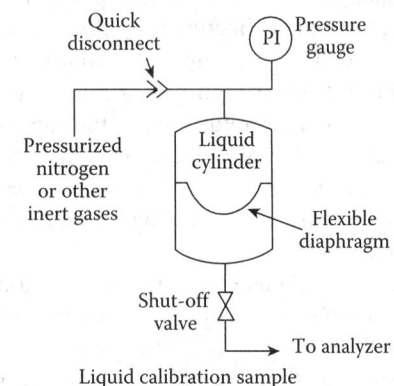

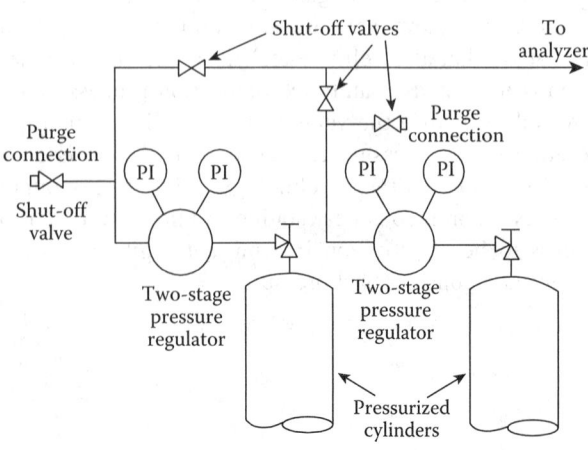

FIG. 1.2f
Calibration provisions for liquid.

When a large number of analyzers are used, the components selected must be of the same type and manufacture for interchangeability and stocking of spare parts. Documenting a sample system with complete flow schematics, part identification, and manufacturing of various components is an essential part of being able to properly start up a system and maintain it successfully over a long period.

The successful design of sample systems requires careful analysis of the physical and chemical conditions of the stream, as well as serious consideration of the ambient and transport conditions, to ensure integrity of the sample. Therefore, care should be taken in evaluating the preceding considerations with respect to a given stream and in applying the correct analyzer to provide the desired measurement.

Sample system design is based on experience, and whenever possible, previous experience should be given prime consideration in the selection and application of components. A successful sample system normally results in a successful analyzer system. Therefore, no effort must be spared in proper sample systems design, which requires a careful selection of the system components. These are discussed in the following text, starting with filters.

Filtering and Separation

Most analyzer sampling systems will include a filter with at least one wire mesh strainer (100 mesh or finer) to remove larger particles that might cause plugging. Available filter materials include cellulose, which should only be considered in applications where it does not absorb components of interest. Sintered metallic filters can remove particles as fine as 2 μm; cellulose filters can remove down to 3 μm; and ceramic or porous metallic elements can trap particles of 13 μm or larger. When the solids content is high, two filters can be installed in parallel, with isolation valves on each. Motorized self-cleaning filters (Figure 1.2g) are also available for such services.

Separating Liquids from Gases

Glass microfiber filter tubes efficiently separate suspended liquids from gases. The filter tubes capture the fine droplets suspended in the gas and cause the droplets to run together to form large drops within the depth of the filter tube. The large droplets are then forced by the gas flow to the downstream surface of the filter tube, from which the liquid drains by gravity. This process is called coalescing.

The coalesced liquid drains from the tubes at the same rate as liquid droplets enter the tubes. Therefore, the tubes have an unlimited life when coalescing liquids from relatively clean gases. The filters operate at their initial retention efficiency even when wet with liquid. The flow direction is inside to outside to permit the liquid to drip from the outside of the filter to the housing drain (Figure 1.2h).

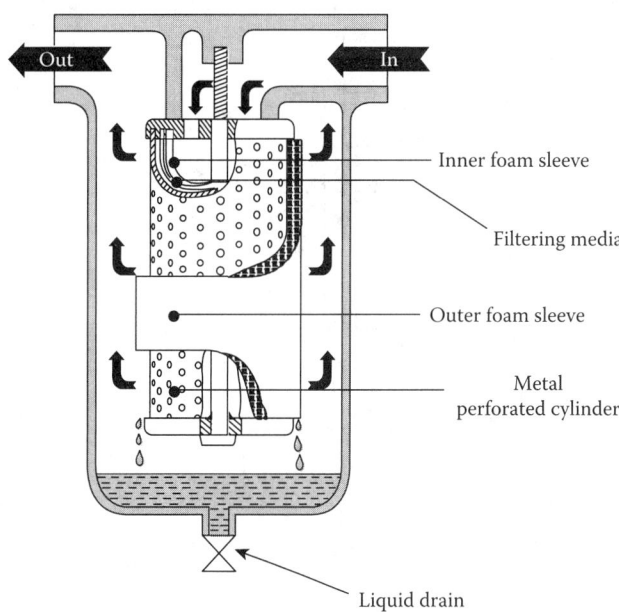

FIG. 1.2h
Coalescing gas filter serves to remove liquid entrainment. (Courtesy of Parker Filtration, formerly Balston Inc.)

The filter tube grade should be selected for maximum liquid drainage rate, rather than maximum filtration efficiency rating, because a liquid drainage rate decreases with increasing filtration efficiency rating. If liquid is carried into the filter in slugs rather than dispersed as droplets in the gas, a filter that is properly sized for steady-state conditions can be flooded and permit liquid carried over.

FIG. 1.2g
Self-cleaning analyzer filter with automatic backwash. (Courtesy of RotorFlush.)

If slugging of liquid is expected, a filter with a relatively large bowl should be selected to provide adequate liquid-holding capacity, and provisions should be made to drain the liquid automatically from the bowl of the housing as it accumulates.

An automatic float drain can be used if the pressure is in the 10–400-psig (28 bars) range. Above 400 psig, which is the upper limit for commercially available float drains (Figure 1.2i), the possibilities are (1) a constant bleed drain, (2) a valve with an automatic timed actuator, or (3) an external reservoir with manual valves. The external reservoir can be constructed of pipe or tubing with sufficient volume to hold all the liquid that is expected to be collected during any period of unattended operation. To drain liquid while the filter is operating at pressure or vacuum conditions, the reservoir inlet valve is closed and the outlet valve is opened.

Spargers, Packed Towers, and Strippers

In some analyzer applications, it is necessary to remove corrosive gases or condensable vapors from the sample. This removal can be done by bubbling the sample stream through a liquid solution.

In the case of a sparger (Figure 1.2j), the gas enters below the scrubbing medium through a sintered disc, which breaks it into small bubbles. This increases the liquid–gas contact surface area and thereby provides improved scrubbing efficiency. At the same time, the small bubbles also increase the tendency for foaming and liquid carryover, and therefore the sample flow velocities should be kept low.

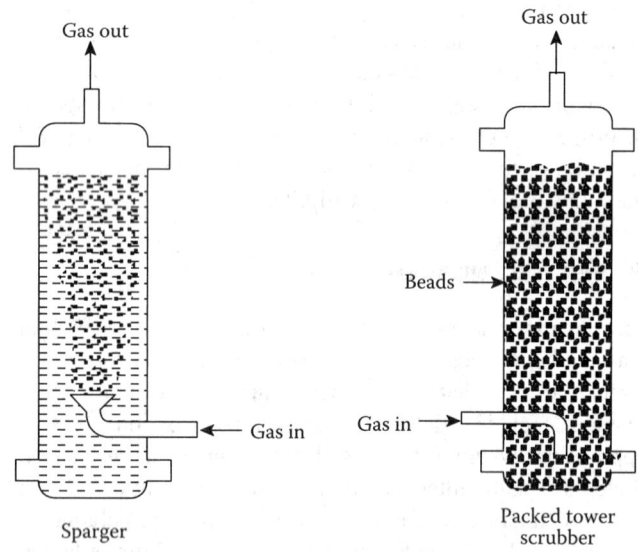

FIG. 1.2j
Removal of corrosive gases or of condensable vapors.

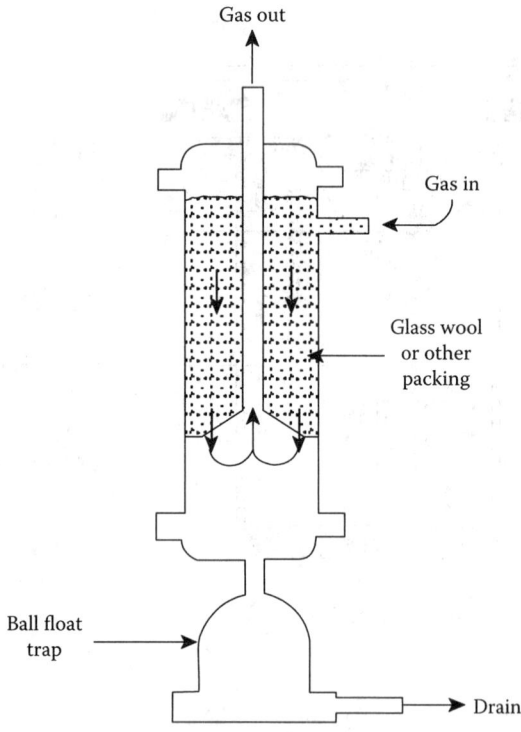

FIG. 1.2i
Entrained liquid separator with ball float trap. (From API RP 550, Manual on Installation of Refinery Instruments and Control Systems; Part II-Process Stream Analyzers (1984) http://www.techstreet.com/products/1821117.)

Packed towers (Figure 1.2j) can also be used as scrubbers. In these designs, the gas sample bubbles up through a nonreactive packing of beads that are wetted by a liquid solution. Foaming and liquid carryover are problems in packed towers as in a sparger.

Similar technology is used to analyze water streams for volatile components; however, in this case, the water flows through the vessel with appropriate packing and a nitrogen gas or similar stream is used to strip the components of interest from the fluid and the resulting vapor stream is then submitted to the process analyzer, typically a gas chromatograph for analysis.

When the impurities in the process gas stream are both solids and liquids, such as in particulate matter and mist—carryover problems in chlorine plants—the fiber mist

If the filter is under vacuum, the external reservoir is a practical method of collecting coalesced liquid for periodic manual draining. Alternatively, if an external vacuum source is available, such as an aspirator, the liquid may be drained continuously from the housing drain port.

1.2 Analyzer Sampling

If polymers represent a substantial portion of the process stream, the need for filter replacements becomes excessive and therefore impractical. A better technique is to vaporize the unreacted monomers through pressure reduction while keeping the polymers in a molten state through heating. This technique (Figure 1.2m) not only discharges polymers continuously but also provides a usable vapor sample.

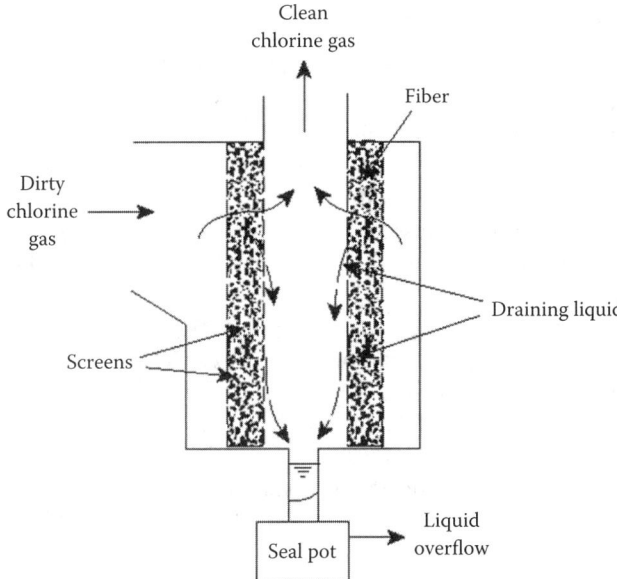

FIG. 1.2k
Fiber mist eliminator.

eliminator (Figure 1.2k) should be considered. The liquid particles form a film on the fiber surface, and the drag of the gas moves this film and the dissolved solids radially, while gravity causes them to move downward, resulting in a self-cleaning action.

Dissolved solids or polymer-forming compounds in a process stream would leave a residue and eventually plug the liquid sample valve. The logical solution is to force the residue formation to take place in a controlled area, such as in the fiberglass filter of the spray-stripping chamber shown in Figure 1.2l.

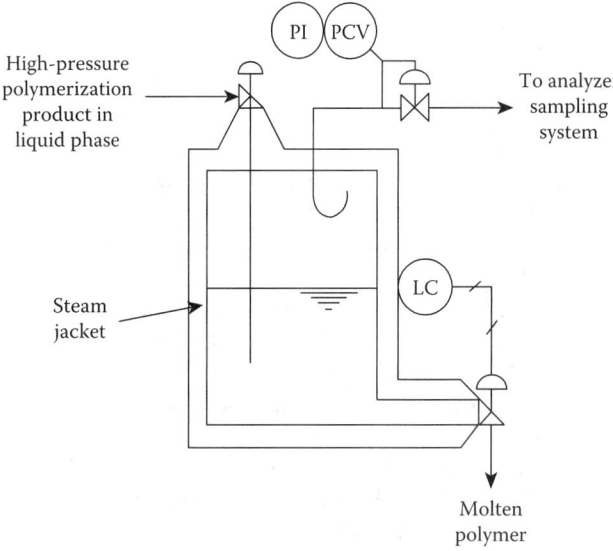

FIG. 1.2m
Flash chamber makes the analysis of unreacted monomers possible.

When it is necessary to clean the windows on the various photometers operating on gas samples, a warm air purge can be used, keeping the window compartment isolated from the sample.

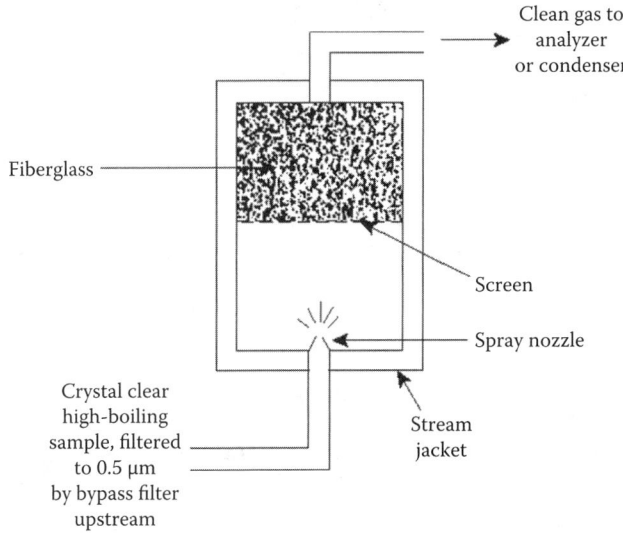

FIG. 1.2l
When dissolved inorganic solids or polymer-forming compounds are present, stripping the liquid sample may be the answer.

Separating Two Liquid Phases

Theoretically, glass micro-fiber filter tubes can separate suspended droplets of a liquid that are immiscible in another liquid using the same process by which they separate droplets of liquid from a gas. The liquid droplets suspended in the continuous liquid phase are trapped on the fibers and run together to form large drops, which are then forced through the filter to the downstream surface. The large drops separate from the continuous liquid phase by gravity difference, settling if heavier than the continuous phase and rising if lighter. The coalescing action of glass microfiber filters is effective with aqueous droplets suspended in oil or other hydrocarbons, and also with oil-in-water suspensions.

In practice, however, liquid–liquid separations are much more difficult to achieve than are liquid–gas separations. The specific gravity difference between two liquids is always less than that between a liquid and a gas and, therefore, a longer phase separation time is needed. Either the filter housing must be oversized or the flow rate greatly reduced to avoid carryover of the coalesced phase.

As a rule of thumb, flow rate for liquid–liquid separation should be no more than one-fifth the flow rate for solid–liquid separation. Even at low flow rates, if the specific gravity difference between the two liquids is less than 0.1 (e.g., if an oil suspended in water has a specific gravity between 0.9 and 1.1), the separation time for the coalesced phase may be too long to be practical. In that case, if there is only a small quantity of suspended liquid, the filter tube can be used until saturated with suspended liquid and then changed.

Another practical problem with liquid–liquid separation is that small quantities of impurities can act as surface-active agents and interfere with the coalescing action. For this reason, it is not possible to predict accurately the performance of a liquid–liquid coalescing filter, and each system must be tested on-site. Testing can be started with 25-μm filter tubes and with inside-out flows at very low flow rates. If the suspended liquid is lighter than the continuous phase, the housing should be oriented so that the drain port is up.

Gas Removal from Liquids

Glass microfiber filter tubes readily remove suspended gas bubbles from liquid, eliminating the need for deaeration tanks, baffles, or other separation devices. Flow direction through the filter is outside to inside, and the separated gas bubbles rise to the top of the housing and are vented through the drain port. If slipstream sampling is used, the separated bubbles are swept out of the housing with the bypassed liquid. Filter tubes rated at 25 μm are a good choice for gas bubble separation.

Columns with glass wool packing can also be used to remove entrained bubbles (Figure 1.2n). Here, the gas bubbles collect and agglomerate on the packing and form large bubbles. These bubbles break away and rise to the top, where they are separated from the liquid and vented.

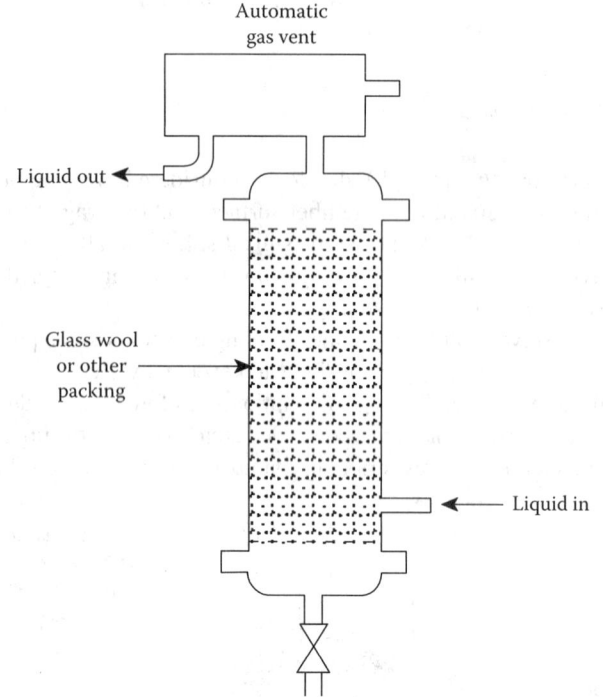

FIG. 1.2n
Entrained gas separator.

Bypass Filters

In order to minimize the transportation lag, a relatively large slipstream is usually taken from the process and brought near the analyzer. As the actual sample flow to the analyzer is small, only a small portion of this slipstream is sent to the analyzer; the bulk is returned to the process (Figure 1.2o). This arrangement permits the main volume of the filter to be swept continuously by the high-flow-rate system, thus minimizing lag time; at the same time, only the low-flow stream to the analyzer is filtered, thus maximizing filter life.

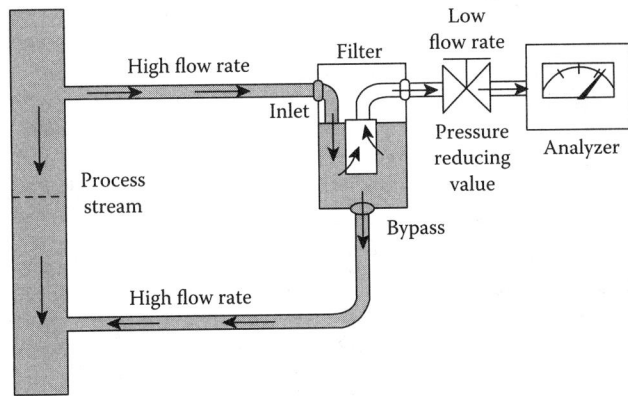

FIG. 1.2o
Slipstream or bypass filtration.

A slipstream filter requires that its inlet-to-outlet ports be located at opposite ends of the filter element to allow the high flow rate of the bypassed material to sweep the surface of the filter element. A third port connecting the low-flow-rate lines to the analyzer allows filtered samples to be withdrawn from the filter reservoir.

If bubble removal from a liquid is a requirement, this function may be combined with slipstream filtration, since the recommended flow direction for bubble removal is outside to inside and the separated bubbles will be swept out of the housing by the bypass stream. In this case, the liquid feed should enter at the bottom of the housing and the bypass liquid should exit at the top of the housing.

A special problem arises in slipstream sampling if the filter is to coalesce and continuously drain suspended liquid from a gas stream or to coalesce liquid droplets from a liquid stream. The coalescing filter requires two outlet ports, one for the dry gas and one for the liquid drain. To combine coalescing and slipstream filtration, a filter housing would need four ports—two for inlet and bypass and two for filtered gas and coalesced liquid—which is not a practical design.

Therefore, slipstreaming plus coalescing requires two stages of filtration (Figure 1.2p). The second (coalescing) stage must be located in the sample line into the analyzer and should be as small as possible to minimize lag time. If the quantity of suspended liquid is not large, a miniature inline disposable filter unit may be used as a trap for the suspended liquid, to be replaced when saturated.

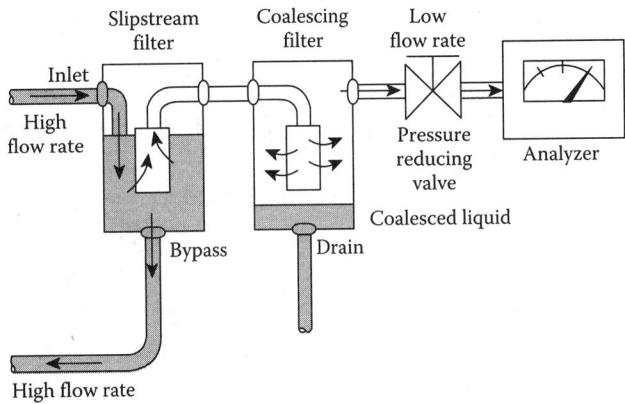

FIG. 1.2p
Slipstream plus coalescing filtration requires two stages.

Self-Cleaning Filters

If the material to be removed is dust, the self-cleaning bypass filter (Figure 1.2q) with automatic blowback constitutes a potential solution. In some instances, cyclone separators

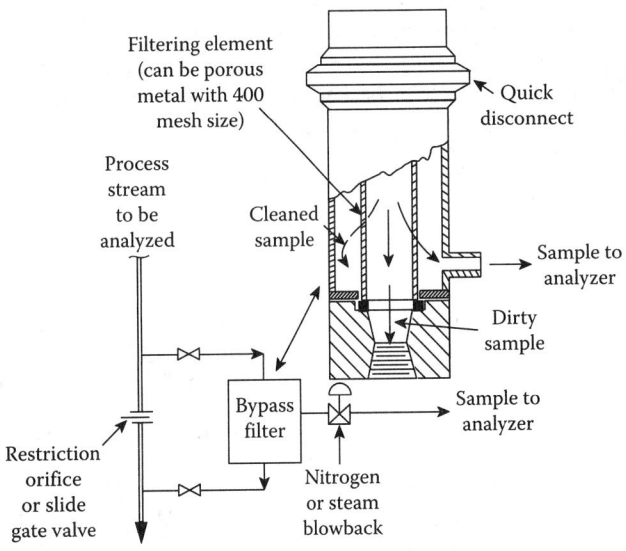

FIG. 1.2q
Self-cleaning bypass filter and its installation.

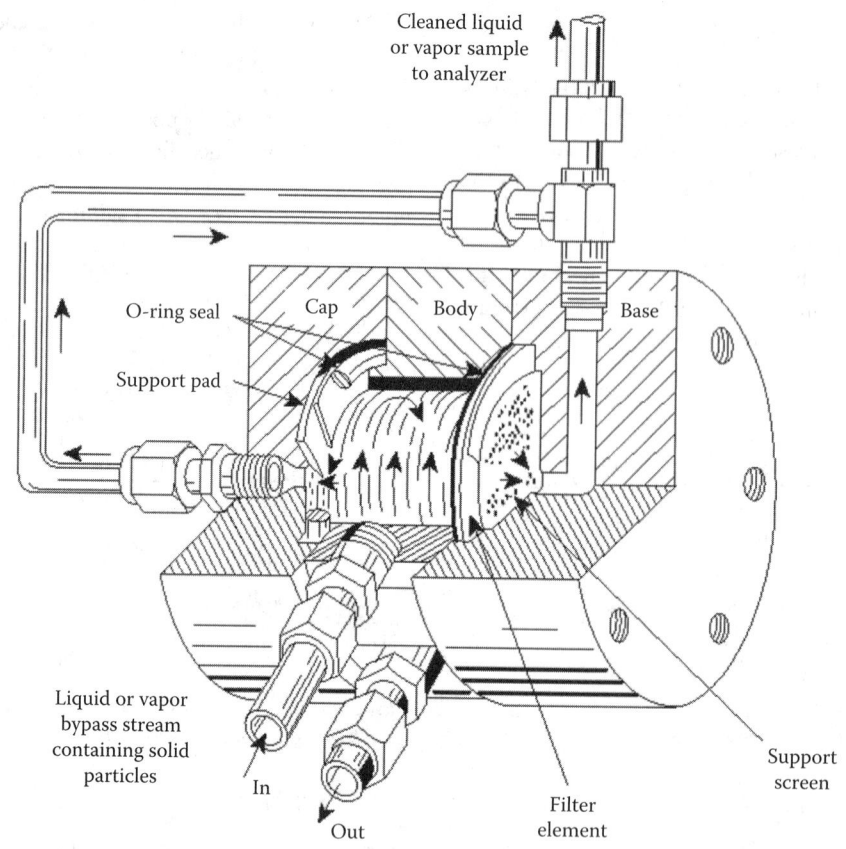

FIG. 1.2r
Bypass filter with its cleaning action amplified by the swirling of the tangentially entering sample.

should be considered. In the latter device (Figure 1.2r), the process stream enters tangentially to provide a swirling action, and the cleaned sample is taken near the center. Transportation lag can be kept to less than 1 min, and the unit is applicable to both gas and liquid samples. This type of centrifuge can also separate streams by gravity into their aqueous and organic constituents.

Another good filter design is the rotary disc filter (Figure 1.2s). Here, the filtered liquid enters through the small pores in the self-cleaning disc surfaces. The sample

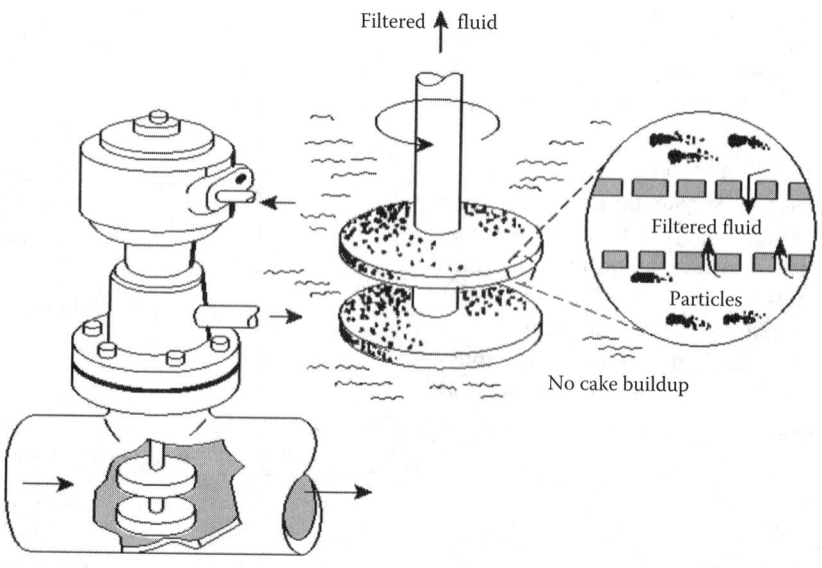

FIG. 1.2s
Rotary disc filter.

liquid is drawn by the sample pump through the hollow shaft and is transported to the analyzer.

For the removal of small amounts of polymer dust in vapor samples, there are melt filters with removable, heated metallic surfaces that melt and collect the polymer dust from the sample.

Liquid analyzer sample streams usually have high solids content. In addition, the analyzers are often located in remote areas of the plant and are infrequently serviced. Therefore, the sample filter system must have a long life between filter tube changes even when solid loading is high. The recommendation for this type of application is a two-stage filter system, a 75 μm prefilter followed by a 25 μm final filter. The filters should be oversized as much as possible without causing excessive lag time. Plastic filter housings are usually a good choice.

For installations with high solids loading, self-cleaning metal edge filters (like those of Cuno Inc.) consist of an all metal stack of alternating disks and spacers with spacings from 0.003" to 0.015" on a rotatable metal shaft. Cleaning blades are mounted on a rod adjacent to the cartridge stack with extensions into the disk spacings such that when the cartridge rotates either manually or continuously with an attached motor the particles are combed out from stack.

Automatic Probe Cleaning

One approach to keeping probes clean grew out of the mining industry, where gypsum buildup on pH and oxidation–reduction potential (ORP) probes was blinding these sensors.

The filtrate master (Figure 1.2t) consists of a filter cup at the tip of a 2 in. (50 mm) diameter standpipe, which is inserted into the process. The pH or other electrode is inside this cup, which is made out of porous sintered stainless steel. A pump housed outside the process draws a vacuum inside the cup, thereby drawing in the filtrate of the process fluid. The excess filtrate can be continuously returned by the pump, while the filtered-out solids accumulate on the outside surface of the filter cup.

Operating experience shows that solids can build up to over an inch (25 mm) thickness without blocking the filtrate flow. The timer is set to periodically reverse the filtrate pump. When the flow direction is reversed, the filtrate flows back into the filter cup and out through the porous filter. During back-flush, the buildup is quickly removed and another filtering cycle is initiated automatically.

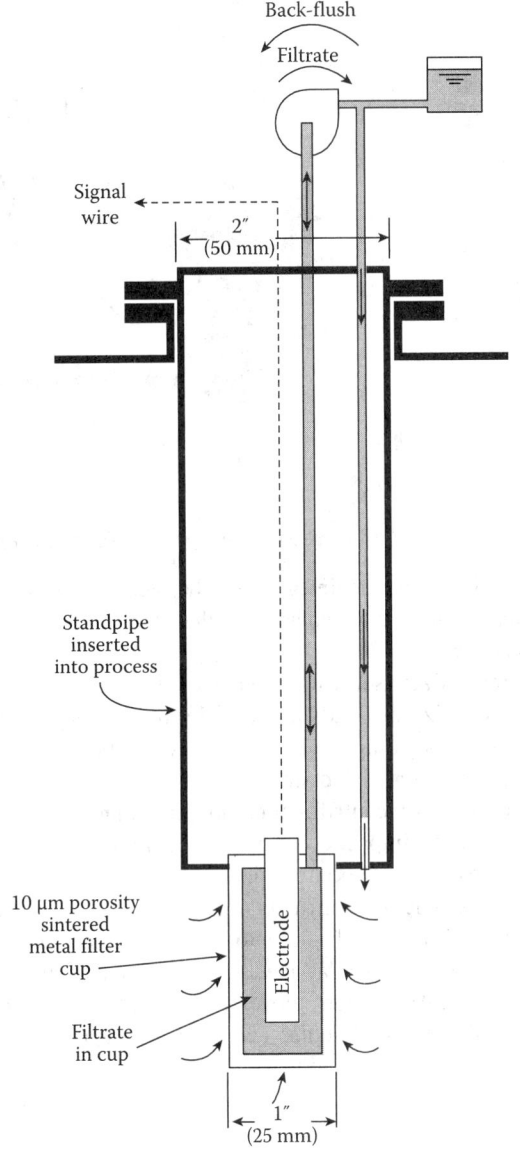

FIG. 1.2t
Electrode is protected from material buildup by back-flushed porous filter cup.

Homogenizers

A frequent problem of sampling systems is plugging. There are two ways to eliminate it. The older, more traditional approach is filtering. Unfortunately, as the filters remove materials that might otherwise plug the system, they also remove process constituents and make the sample less representative.

The newer approach is to eliminate the potential for plugging by reducing the size of solid particles (homogenization)

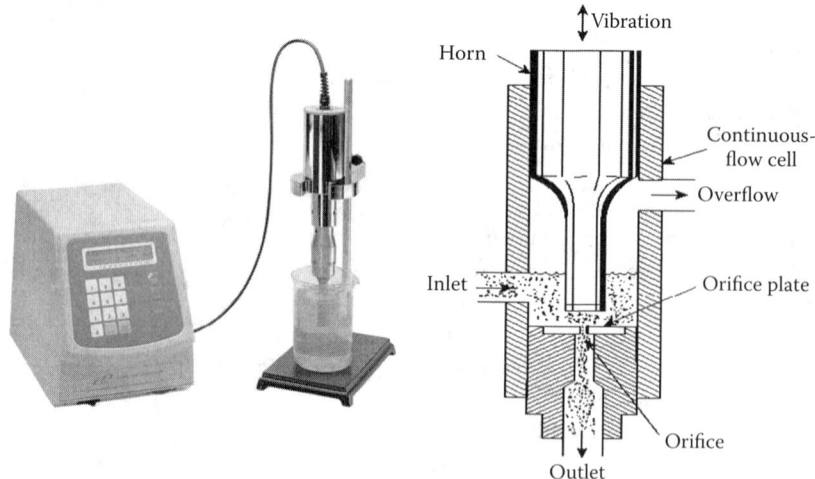

FIG. 1.2u
Laboratory and online ultrasonic homogenizers. (Courtesy of Cole-Parmer Instrument Co.)

while maintaining the integrity of the sample. Thus, when a pulverizer is used to replace the filter, the sample becomes representative.

Homogenizers serve to disperse, disintegrate, and reduce the particle size of solids and thereby reduce agglomerates and liquefy the sample. Homogenizers can be mechanical, using rotor-stator-type disintegrator heads. In this design, the rotor acts as a centrifugal pump, recirculating the slurry while the shear, impact collision, and cavitation at the disintegrator head provide homogenization.

In ultrasonic homogenizers, high-frequency mechanical vibration is introduced into the probe (horn), which creates pressure waves as it vibrates in front of an orifice (Figure 1.2u). As the horn moves away, it creates large numbers of microscopic bubbles (cavities), and when it moves forward, these bubbles implode, producing powerful shearing action and agitation due to cavitation.

Such homogenizers are available with continuous-flow cells for flow rates up to 4 g/hr (16 l/hr) and can homogenize liquids to less than 0.1 μm particle sizes. The flow cell is made of stainless steel and can operate at sample pressures of up to 100 psig (7 bars).

SAMPLE CONDITIONING

The extracted sample begins conditioning at the takeoff point, continues through the transport, and finishes conditioning at the analyzer location prior to entering the analyzer (Figure 1.2v). All samples require some form of conditioning

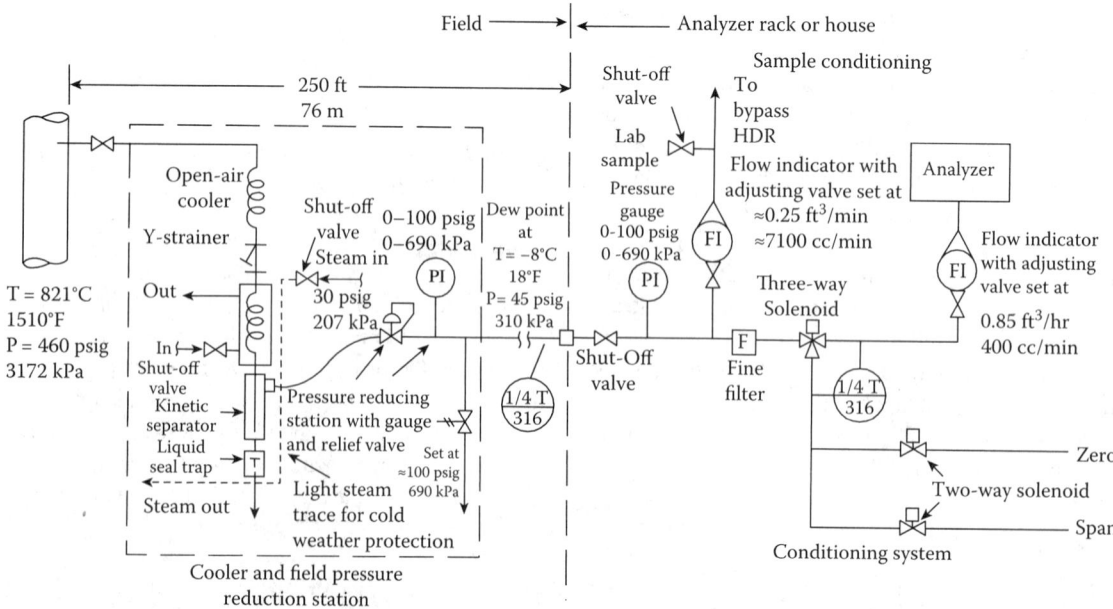

FIG. 1.2v
Typical sample conditioning system with remote preconditioning unit.

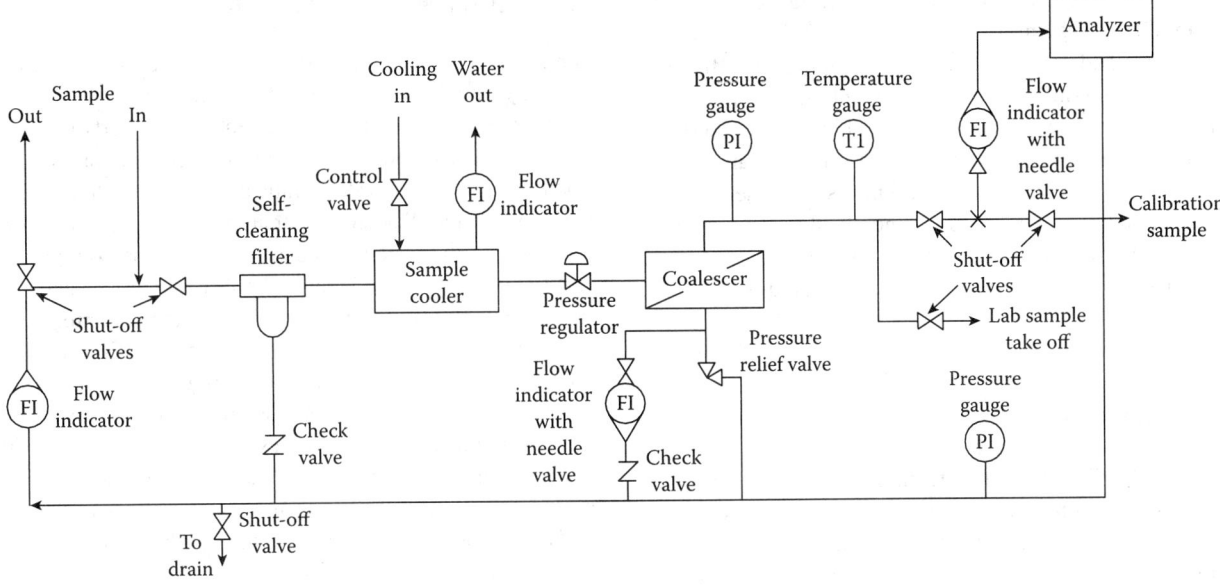

FIG. 1.2w
Typical liquid product sample system for refinery applications.

to make them suitable for the analyzer and to assure reliable on-stream operation. The conditioning is done at the appropriate location in the sample system loop in order to maintain the integrity of the sample (Figure 1.2w).

Sample washing is usually limited to dirty, particle-laden streams whose composition will not be affected by the solubility of the components in the liquid used to wash the sample. The conditions of flow, temperatures, and pressures must be controlled to maintain a relatively constant predetermined relationship of the composition. When washing, care must be taken to keep the sample in the vapor phase by providing heated transport lines or by making provisions for final moisture removal as the analyzer may require.

Vaporizing Samples

Vaporizing the sample can be necessary when the liquids are in equilibrium or when the analyzer requires a vapor sample. The vaporization is usually done by a vaporizer regulator (Figure 1.2x). In such a regulator, the sample is vaporized simply by pressure reduction across a capillary or, more often, by using a heater as well. Care must be exercised to avoid partial vaporization and fractionation

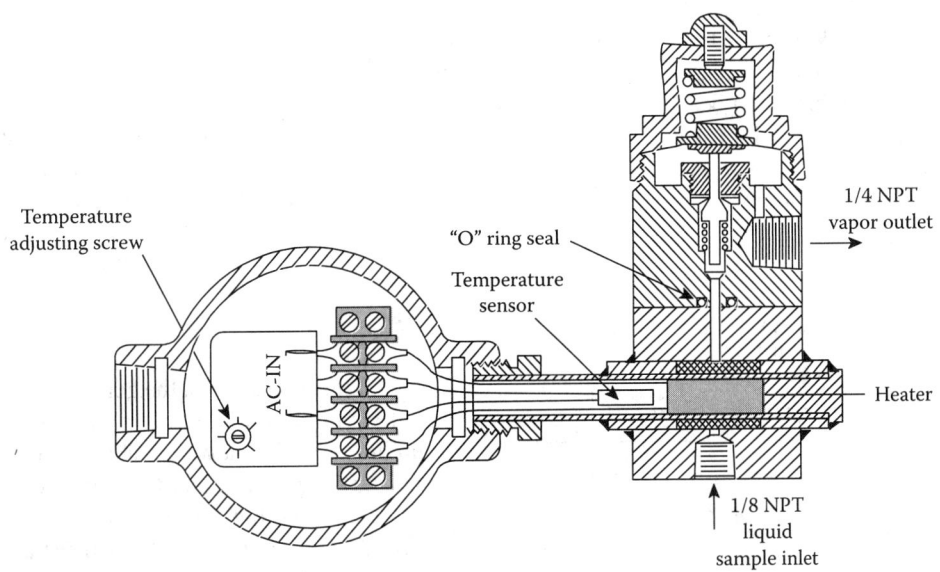

FIG. 1.2x
Vaporizing regulator assembly (electrically heated).

by selecting a suitably heated vaporizing regulator to accommodate the sample.

If a vaporized sample is desired or required for the analyzer and it is possible to do so, the sample should be vaporized as close to the sample point as possible as this will result in the minimum transport time to the analyzer and hence reduce the risk of contamination. This is because a gas is always several times the volume (typically almost an order of magnitude) of the same volume of liquid and because it has a lower density as well as viscosity than the liquid phase. This results in the vapor having a lower pressure drop than the associated liquid stream.

Entrainment Removal

The removal of entrainment from liquid or gas samples normally starts with locating a sample tap and providing a sample probe at the takeoff point. Filtration is normally used for both gases and liquids, because it removes both liquid and particulate entrainment from gases and particulate matter from liquids.

For gases requiring further conditioning of heavy loading, cyclone filters can be used if the sample has adequate velocity. For liquids, coalescers are frequently used to remove both undesirable gases and liquids by gravity. The removal of free water from a hydrocarbon stream is usually accomplished by passing the liquid through a hydrophilic element, causing the water droplets to accumulate on the element. The hydrocarbon stream is then passed through a hydrophobic element that rejects the water, removing it from the bottom with a hydrocarbon bypass stream.

Another method of removing moisture from a stream uses a selective permeation device with a drying medium that creates differential pressures to drive the water through the permeable materials, thus removing it from the flowing stream (Figure 1.2y). When using permeation devices, it is always important to check what materials in the process stream could affect the membrane material as some hydrocarbons tend to cause the membranes to become brittle, which could lead to their premature failure and introduction of either drying medium to the process or process fluid to drying medium at higher concentrations or pressures for which they are designed.

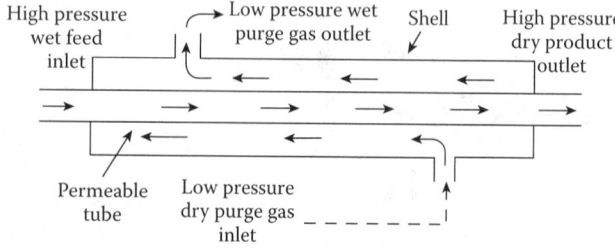

FIG. 1.2y
Permeable tube sample dryer.

Selection of Component Materials

Adsorption of the sample components of interest on the walls or surface with which the sample comes in contact will affect the analysis, especially for measurements in the parts per million (ppm) range. Therefore, proper selection of materials and conditioning is essential for establishing an equilibrium that will make analysis reliable. Selection of materials must also consider the finish of polish of the metal surfaces since, for example, stainless steel tubing with a "regular" polish can absorb and desorb water vapor as a function of pressure and temperature.

Specifically, water vapor samples reach equilibrium more rapidly in stainless steel lines than in copper or plastic tubing. Diffusion is another consideration. Samples system design should assure that gases do not permeate the walls of the sample system. This is especially important in high-pressure systems and ppm analysis. Another problem with diffusion is leakage, which can change the sample composition because gas molecules will flow in both directions of the leak and can significantly affect ppm analysis. This is best illustrated by the fact that in an oxygen ppm analyzer system, the slightest leak will create a full-scale reading on the instrument even though the leak is from a high-pressure sample to atmospheric pressure.

Some sample streams may contain corrosive gases or water vapor that influence the accuracy of the measurements or potentially damage the analyzer. The removal of such undesirable matter or components can be accomplished by passing the sample through a packed bed of soil chemicals or desiccant. Further, a liquid treating agent may be applied with provisions that gas streams be broken up into small bubbles to assure proper contact between the liquid and gas phases. Care should be taken in such systems to avoid alteration of the sample and to condition the sample to a desired and reproducible form for analysis.

APPLICATIONS

Here, a brief discussion will follow the various sample probe and sampling system package designs that are used for gas, liquid, and solids sampling. The discussion will also cover the more special cases of stack, reactor, condensate, trace component, and multistream sampling systems.

Gas Sampling

When taking gas samples, the goal is to obtain representative samples with minimum time lag, using short, small-volume sampling lines. Whenever possible, it is preferred to draw dry and clean samples in order to minimize the need for filters, dryers, knockout traps, or steam tracing.

The sample tap should be located on the side of the pipeline to minimize liquid or dirt entrainment, and the sample should be taken from the center of the pipe. If it is necessary to periodically remove the probe for cleaning or to perform a sampling

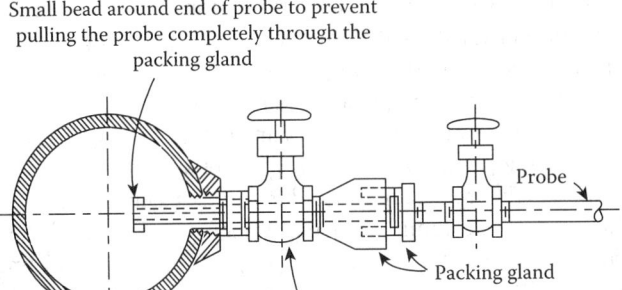

FIG. 1.2z
Gas sampling probe should be inserted from the side of the pipe to near the center of the pipe.

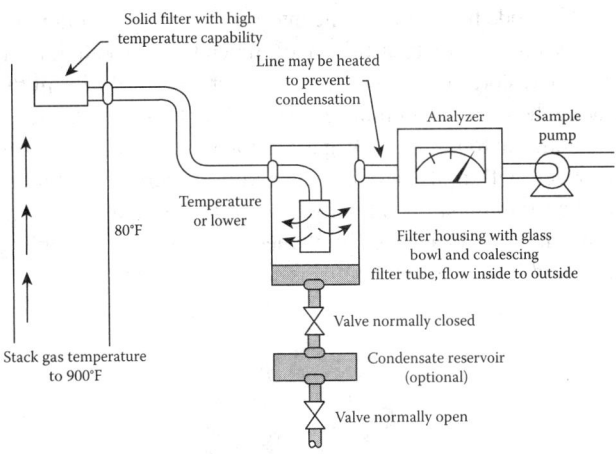

FIG. 1.2ab
Stack gas sampling.

traverse, it is desirable that the probe be inserted through a packing gland and block valve, as shown in Figure 1.2z.

The volume of the sampling system should be kept to a minimum, while the velocities through the sample lines should be high to protect against the settling of entrained liquids or particulates that can cause plugging. Sample line tubing can be as small as 1/8 in. (3 mm) in diameter, and sample flow velocities should be between 5 and 10 ft/sec (1.5–3 m/sec). When the sample is taken from a combustion zone or other dirty processes, a filter is usually provided at the tip of the sample probe, and a high-pressure filter blowback line is provided for periodic filter cleaning (Figure 1.2aa).

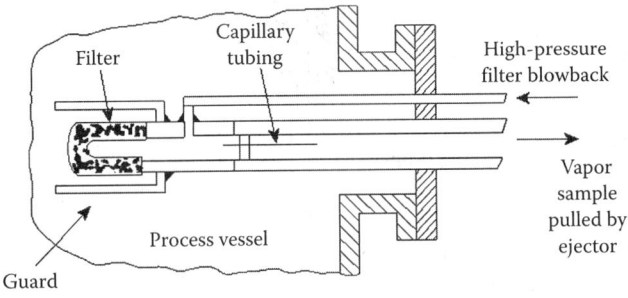

FIG. 1.2aa
Gas sampling probes used on combustion processes are usually provided with probe-tip filters and blowback lines for filter cleaning.

Stack Gas Sampling

This method of sampling is discussed in detail in Chapter 1.3. When sampling hot, wet stack gas, a filter capable of withstanding the gas temperature should be installed in the stack at the tip of the sample line to prevent solids from entering the gas sample line. After the sample is cooled, a coalescing filter is used to remove suspended liquids before the sample goes to the analyzer (Figure 1.2ab).

The sample flow direction is from inside to outside. Filter housings with Pyrex glass bowls are often used in this application to permit visual check of the liquid level in the filter housing. Since there is often a considerable amount of liquid present in the sample, steps must be taken to drain the housing to ensure that liquid does not build up and carry downstream to the analyzer.

The liquid coalescing filter should be located as close to the analyzer as possible to minimize the chance of condensation between the filter and the analyzer. Additional precautions that can be taken to avoid downstream condensation are to cool the sample below ambient temperature upstream of the coalescing filter and to heat the line gently between the filter and the analyzer.

Automatic Stack Sampling

In these sampling packages, a microprocessor directs the automatic sampling method, which can be selected to follow U.S. Environmental Protection Agency (EPA) Methods 1–6, 8, 17, and 23 or international methods specified by VDE, BSI, or ISO. The microprocessor stores all measurements, reviews and diagnoses all inputs, controls the required parameters, calculates isokinetic conditions, and either reports the results in a printed form or transfers them to permanent storage media.

Besides the controller, such a package usually consists of a probe, a filter (hot) box, a cold box, a flexible sample line, glassware, a node box, and a monorail system. The probe is usually 3, 5, 7, or 10 ft (0.9, 1.5, 2.1, or 3 m) and made of stainless steel with a glass liner. Most probes are jacket heated and are provided with both a liner thermocouple and a stack temperature thermocouple.

The measured variables include the temperatures of the stack, probe liner, filter box, condenser outlet, and the dry gas meter. The pressures are detected by an absolute and a differential pressure transducer and are used to measure the pressure of the stack gas, the barometric pressure, and the velocity pressure of the stack gas. The normal capacity of the vacuum pump that draws the sample is 0.75 CFM (21 l/min), and the dry gas meter has an operating range of 0.1–1.5 CFM (2.8–42 l/min).

The node box provides the interface between the filter box and the cold box by measuring the temperatures in both. It measures and stores the temperature, pressure, and velocity in the stack. The monorail eliminates the need for bulky supports.

Some automatic stack gas samplers (Figure 1.2ac) are provided with multiple channels and use microcomputers to do the sampling automatically. The units are capable of sampling multiple channels simultaneously while the sampling time and flow of each channel is set independently.

of sample every time the actuator piston is stroked. In the time-proportional mode, this sampling frequency is constant, while in the flow-proportional mode, this unit would vary the sampling frequency as a function of flow.

In some automatic liquid samplers, the sampling frequency is adjusted by pneumatic pulse relays or by electronic controls (Figure 1.2ae). Pulse duration is usually adjustable from 0.25 sec to 1 min, while pulse frequency can be adjusted from a few seconds to up to an hour. The intermittent in-line sampler illustrated in Figure 1.2ad can take samples at pressure up to 1500 psig (105 bars).

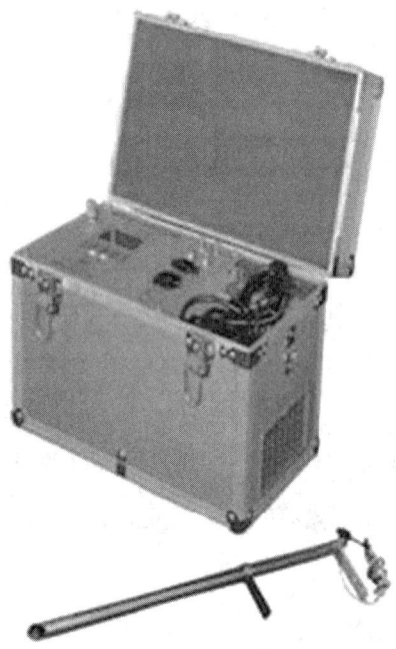

FIG. 1.2ac
Automatic stack gas sampler. (Courtesy of Wuhan Tianhong Instrument Co.)

Liquid Samplers Automatic liquid samplers collect intermittent samples from pressurized pipelines and deposit them in sample containers. The sample can be collected on a time-proportional or on a flow-proportional basis. Figure 1.2ad illustrates a sampler that withdraws a predetermined volume

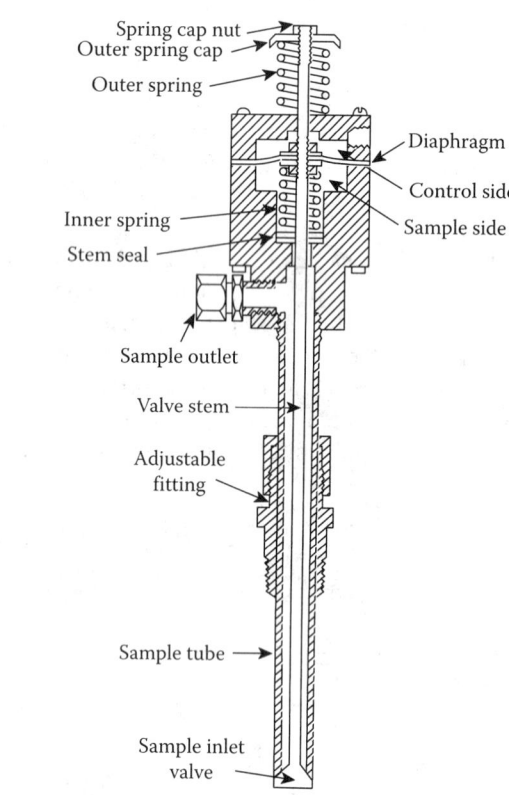

FIG. 1.2ae
Adjustable automatic load sampler.

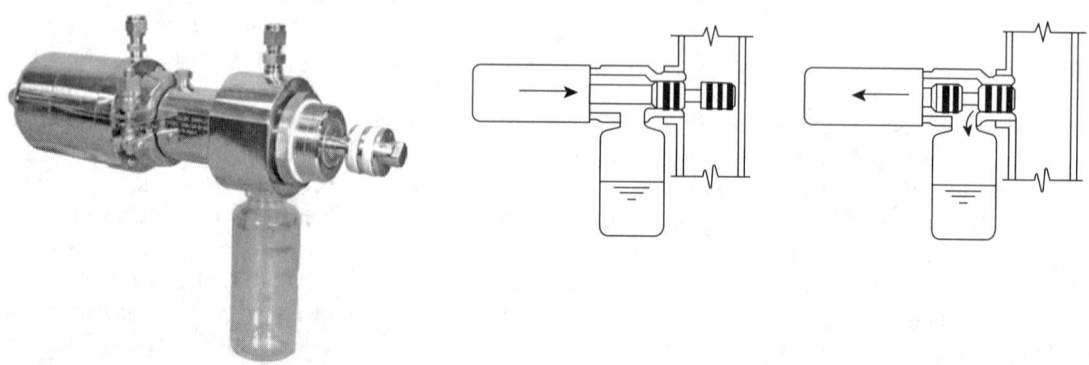

FIG. 1.2ad
Intermittent collection of samples. (Courtesy of Sentry Equipment Co.)

Sampling at High Pressure

Condensate in high-pressure boiler systems, measurements of condensate conductivity, specific ion concentration, and feed water additive concentrations are often required. In a continuous sampling system, the high-pressure steam or condensate is cooled below 38°C (100°F) and then reduced to near atmosphere pressure for metering to the analyzers. Filtration is required upstream of the pressure-reducing valves to prevent pitting to the valve seats by suspended particles and to eliminate variations in flow rate to the analyzers.

A stainless steel filter housing with the appropriate pressure rating and 25 μm filter tube is recommended. Since the analyzer system is often located some distance from the sampling point, slipstream filtration is usually required. Figure 1.2af shows a sampling system in operation at a steam generation facility.

Chemical Reactor Samplers

When samples are taken from chemical reactors by opening hatchways, the operators can be exposed to dangerous fumes while the product can be degraded or cause explosions. To eliminate these problems, reactor sampling systems have been designed that allow the safe taking of samples.

The design illustrated in Figure 1.2ag requires only one nozzle (3 in. or 75 mm) and utilizes a Teflon-coated sampler assembly that can be used up to 150 psig (10.5 bars) and 177°C (350°F). For continuous sampling applications, such as for closed-loop pH control, a Teflon diaphragm pump is used to continuously return the analyzed sample.

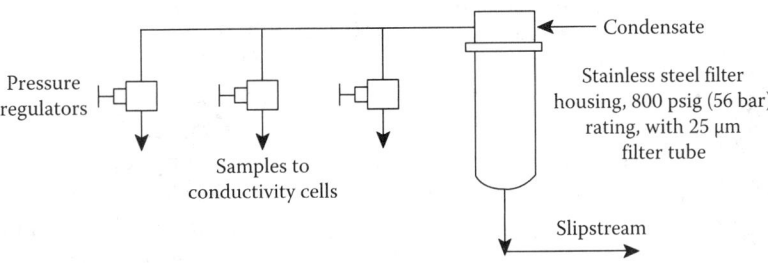

FIG. 1.2af
High-pressure stream sampling.

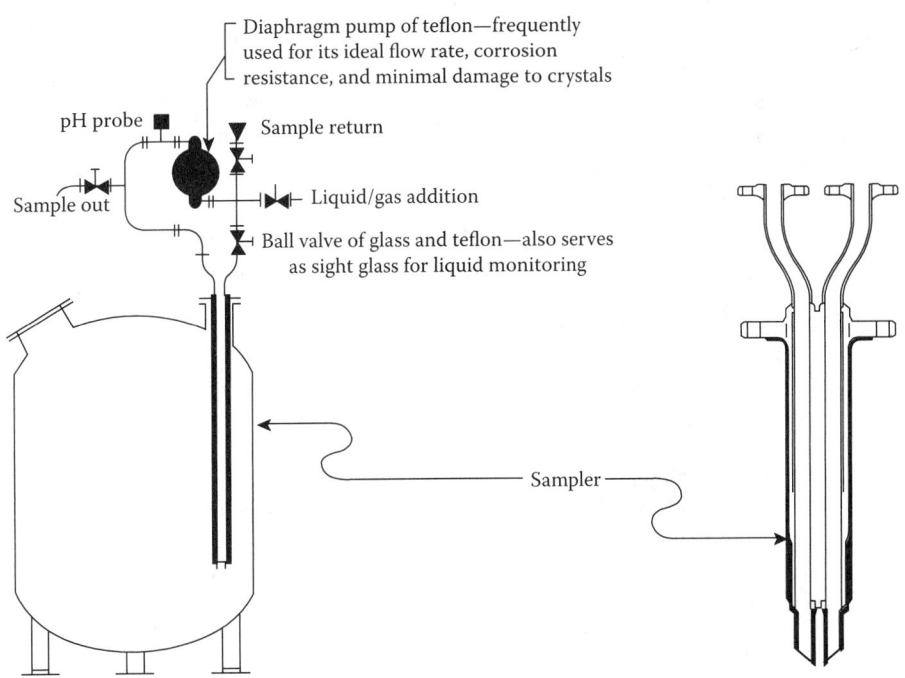

FIG. 1.2ag
Continuous or intermittent sampling of chemical reactor.

Duckbill Samplers

This sampler should be considered when liquid or sludge samples are to be collected at remote locations, from below the level in tanks, sewers, channels, sumps, lakes, or rivers. As shown in Figure 1.2ah, this device has no moving mechanical components, only a rubber (EPT, Buna-N, or Viton) bucket-shaped Duckbill®, which is inside a housing made of polyvinyl chloride (PVC), aluminum, or stainless steel. This rubber insert closes around fibers or particulate matter without jamming. It is operated by compressed air.

The sample enters by gravity through the bottom connected Duckbill inside the sampling chamber, which is installed below the process liquid (or sludge) level and traps some air at the top of the chamber as it fills with the process fluid. When a sample is required, compressed air is introduced, which closes the Duckbill inlet and discharges the sample from the bottom of the chamber. When a new sample is to be drawn into the sample chamber, the compressed air is vented and a new fill cycle is initiated.

An automatic controller is provided on which the user can adjust the frequency at which samples are to be taken into a composite sample collection bottle.

SOLIDS SAMPLING

When solids are to be sampled while flowing by gravity or while pneumatically conveyed under the pressure, the choice of sampling devices becomes more limited. The screw-type solids sampler can be used on these applications (Figure 1.2ai), but it is limited in the amount of pressure or vacuum it can seal against, and it does not provide an intermixed, representative sample. Screw samplers also introduce large transportation lags and cannot easily return the sample into the process.

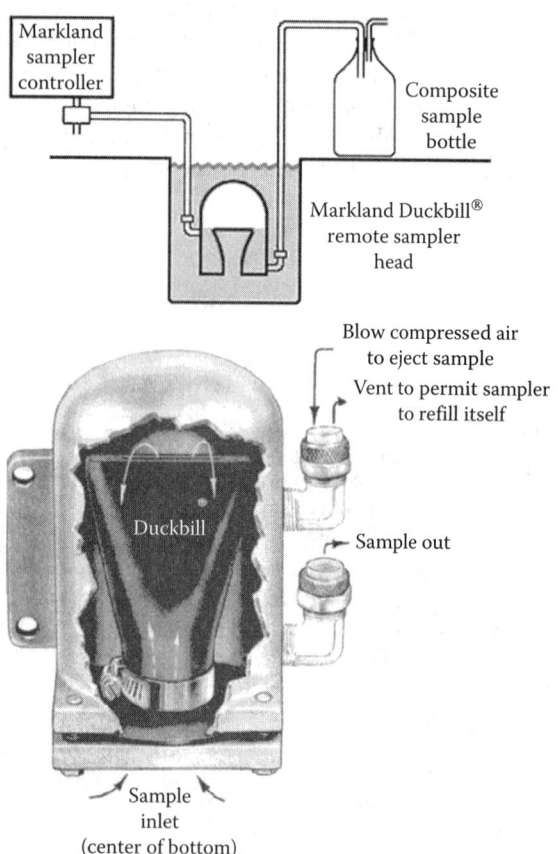

FIG. 1.2ah
Duckbill sampler. (Courtesy of Markland Specialty Engineering Ltd., www.sludgecontrols.com.)

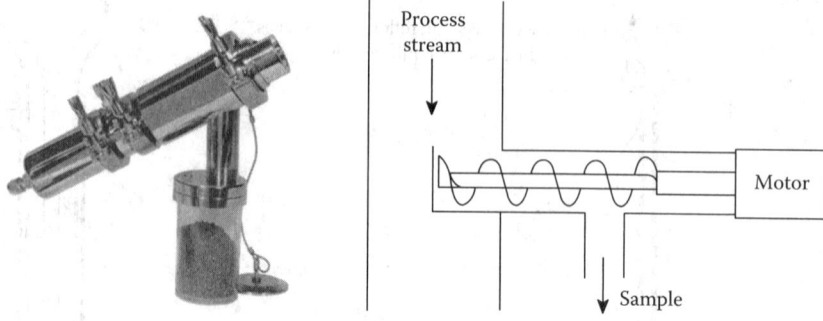

FIG. 1.2ai
Archimedean screw solid sampler. (Courtesy of Sentry Equipment Corp.)

An improved sampler design is described in Figure 1.2aj. This sampler can be sealed against small pressures or vacuums. It is operated by two actuators. During the sample collection phase, both actuators are extended into the process line. When a representative sample has been collected, it is withdrawn and simultaneously intermixed by pulling back actuator 1. As the analyzer is right above the withdrawn sample, the composition can be measured as soon as the sample has been taken. After analysis, actuator 1 is once again extended to return the old sample into the process. When it is time to take a new sample, actuator 2 is also extended and another sample is collected.

DIFFICULT PROCESS SAMPLING

An analyzer system enhances process control by providing a specific measurement of physical or compositional data of a process or ambient condition. With the advent of computer technology, the need for such analysis has increased substantially to provide online data that the computers use to optimize process control. Some of these advances have also improved the handling of such difficult tasks as sampling for trace component analysis or sampling multiple streams.

Trace Analysis Sampling

Trace analysis sampling systems necessitate more stringent requirements than normal analyzer sample systems because of contamination, adsorption, and desorption (Figure 1.2ak). Care should be taken in the selection of construction materials and proper applications of design criteria to avoid alteration of the sample. The following is a list of recommended practices for such systems:

1. Stainless steel seamless tubing is the preferred material because it provides inertness, smooth surfaces, and low porosity.
2. All components should be thoroughly cleaned of oil, gas, or other contaminating materials.
3. Dense fluorocarbons or other soft inert materials may be used as diaphragms as required.
4. Tubing sizes are critical, especially where low flows are used, because they limit the amount of increase in the adsorption–desorption phenomenon. A rule of

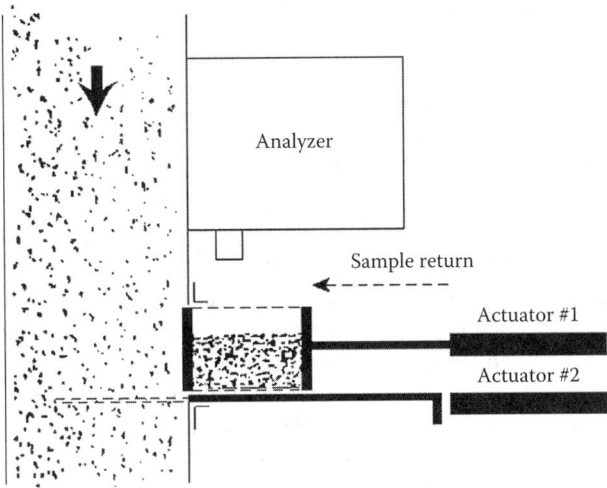

FIG. 1.2aj
The tray-type solids sampler.

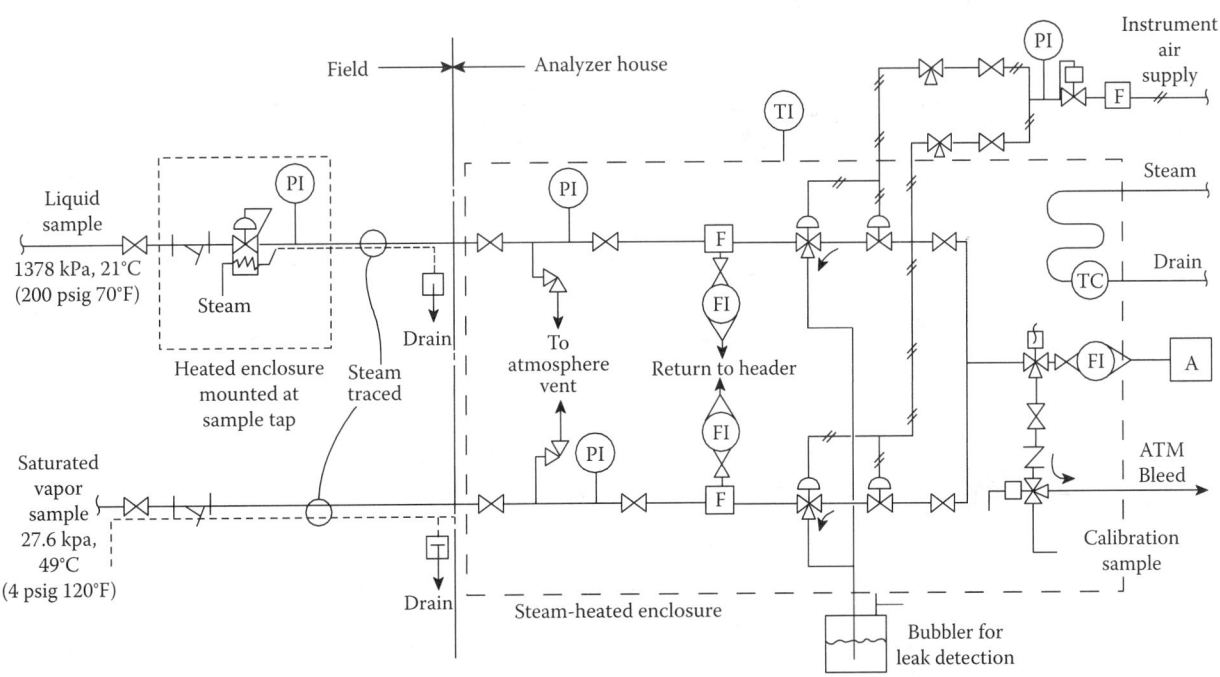

FIG. 1.2ak
Two-stream sampling system for trace analysis with a double-block, double-bleed configuration.

thumb is to use the smallest possible tubing to achieve maximum flow to accommodate the sample loop design.

5. Packless shutoff valves with a diaphragm or bellows seals should be used; however, due to their high cost, serious consideration must be given to this area, and conventional valves are suitable for most applications.
6. Filtration of the sample in ppm analysis can create significant problems unless the filter is totally inert. Therefore, stainless steel filters using a high flow rate or dense fluorocarbon inert materials are recommended for such applications.
7. Conditioning of the lines and system to accommodate the ambient temperature requirements for preventing condensation of components of interest in the sample must be considered; if necessary, heat tracing of the lines must be furnished.
8. When it is necessary to provide an aid for transporting the sample from its sample point to the sample system through the analyzer, as is frequently done in ambient monitoring systems, a sample pump, ejector, or aspirator is necessary. In such cases, the pump must be a diaphragm. If practical, an ejector to pressurize the sample or an aspirator to aspirate it through the measuring device can be used. Both are more desirable than a sample pump.

Multistream Sampling

Switching is usually used when it is practical to analyze several streams using one analyzer. However, because this system is more complex than single-stream sample conditioning, the following considerations should be reviewed to determine if multistreaming is feasible:

1. The potential problem of cross-contamination among multiple streams
2. The importance of each analysis and frequency of analysis with the associated potential impact on the control algorithm and its response time.
3. The loss of information from more than one analysis in case of analyzer failure
4. The cost of an additional analyzer vs. the cost of multistreaming
5. Maintenance requirements

After reviewing the preceding, one can decide if multistreaming is feasible and whether it should be manual or automatic. Whether manual or automatic, multistreaming requires good-quality valves for stream switching.

A typical multistream sampling system is shown in Figure 1.2al. A common and important requirement in all such systems is that a continuous bypass be provided for each sampling point to avoid dead-end sample lines. The sample system should be laid out in such a manner that contamination between streams is avoided. This is best accomplished by arranging the solenoid valve in a double-block double-bleed configuration, which is rather expensive.

More often, a three-way solenoid valve is used for each stream, with the venting port always at low pressures to create a relief in case of a leak. To prevent contamination, dead volume of the sample system should be considered, as well as equalization of the pressures upstream of the

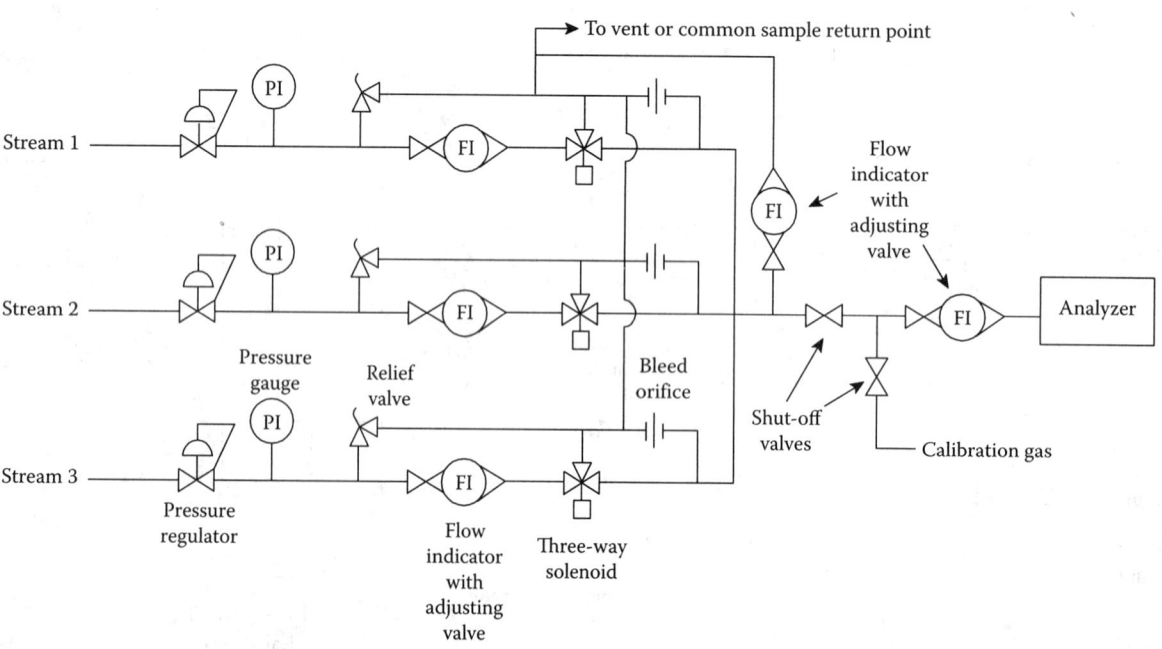

FIG. 1.2al
Typical multistream automatic sampling system.

three-way valves. The problem is more severe in ppm sampling systems because of the adsorption–desorption effects, and careful consideration should be given to the design criteria described earlier.

Sampling for Mass Spectrometers

The sample system and sample conditioning for a mass spectrometer are similar to those used for gas chromatographs (Chapter 1.10). In fact, it is very common to connect a gas chromatography system to the sample input port of a mass spectrometer. In any event, vapor samples are drawn by suction into the mass spectrometer through several possible types of inlet leaks. A sintered metal leak device is a commonly used input port because it allows a uniform molecular flow into the instrument based on the differential pressure across the sintered metal plate. Thin diaphragms made from Teflon may be used for corrosive service, and ultrafine metering valves can also be used.

Direct introduction of samples into mass spectrometers are suited for fast analysis, and one of the difficulties of using probes for introducing the samples is the interface that allows the probe to be inserted into the mass spectrometer. These interfaces must provide a sealed intermediate pumping area where a partial vacuum can be pulled before a valve is opened to allow the insertion of the probe into the process. This requires that the seals that are used in these inlets are designed to withstand the wear due to repetitive insertion and removal of the probe and that there be some method of insuring that the valve is closed and opened at the proper times so that venting of the mass spectrometer does not occur.

In the sampling probe assembly in Figure 1.2am, the valve isolates the intermediate pumping region from the high vacuum of the sample source for the mass spectrometer. This valve is a precision designed ball valve specifically designed for vacuum applications.

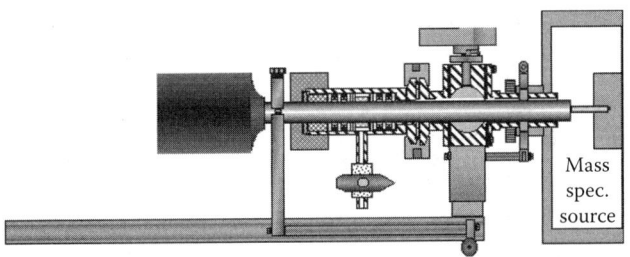

FIG. 1.2am
Mass spectrometer sampling probe. (Courtesy of Scientific Instrument Services.)

New Sample System Initiative (NeSSI™)

NeSSI™ (New Sampling/Sensor Initiative) is an industry-driven effort to define and promote a new standardized alternative to sample conditioning systems for analyzers and sensors in which rather than building a sample system from components connected together by tubing it is instead mounted on a substrate with predefined spacing as defined in international standards. The majority of the work being conducted in this area is led by a team at the Center for Process Analytical Chemistry (CPAC) at the University of Washington.

The group has defined three generations of products in the NeSSI family as follows:

GEN I. Standard fluidic interface for modular surface-mount components ISA S76.00.01
GEN II. Standard wiring and communications interfaces
GEN III. Standard platform for micro analytics

ANSI/ISA-76.00.02 and the corresponding international standard is IEC 62339–1:2006 for the modular sample systems that specifies the footprint, fluid ports, and mounting ports for components that mount to the surface of a mechanical "backbone," or substrate. Any manufacturer can supply components meeting these specifications and they will fit on any substrate that conforms to this standard.

A number of manufacturers provide these substrates as well as the associated surface mount components such as filters, pressure reducing elements, isolation valves, as well as sensors.

CONCLUSIONS

Due to the complexity and misunderstanding associated with process sample systems, many manufacturers are trying to develop analyzers that can be directly connected to the process with no or minimal associated sample systems. The NeSSI initiative is one such activity to reduce sample system complexity while providing better real-time understanding of the sample conditioning activity.

Manufacturers continue to refine their products with periodic new entries to the market.

Process analyzer sample systems continue to be a critical component of any process analytical measurement since if the sample does not represent the process the resulting measurement cannot provide a signal on which reliable control can be based. Similarly, the timeliness of the complete analysis cycle from sample collection, transport, conditioning, and actual analysis will have an impact on the ability of the resulting signal to control a process.

SPECIFICATION FORMS

Usually the analyzer sampling systems are specified and purchased together with the analyzers themselves. Figure 1.2b provided a form that allows the reader to specify the sampling system independently, while the specification forms ISA20A1001 and ISA20A1002 can be used to specify the analyzer and its sampling system both. These forms are reproduced on the following pages with the permission of the International Society of Automation.

Analysis Device — Operating Parameters

#	RESPONSIBLE ORGANIZATION		#	SPECIFICATION IDENTIFICATIONS	
1			6		
2			7	Document no	
3	(ISA)		8	Latest revision	Date
4			9	Issue status	
5			10		

ADMINISTRATIVE IDENTIFICATIONS

11		
12	Project number	Sub project no
13	Project	
14	Enterprise	
15	Site	
16	Area	Cell / Unit
17		

SERVICE IDENTIFICATIONS

18		
19	Tag no/Functional ident	
20	Related equipment	
21	Service	
22		
23	P&ID/Reference dwg	
24	Process line/nozzle no	
25	Process conn pipe spec	
26	Process conn nominal size	Rating
27	Process conn termn type	Style
28	Process conn schedule no	Wall thickness
29	Process connection length	
30	Process line matl type	
31	Fast loop line number	
32	Fast loop pipe spec	
33	Fast loop conn nom size	Rating
34	Fast loop conn termn type	Style
35	Fast loop schedule no	Wall thickness
36	Fast loop estimated lg	
37	Fast loop material type	
38	Return conn nominal size	Rating
39	Return conn termn type	Style

SERVICE IDENTIFICATIONS Continued

40		
41	Return conn matl type	
42	Inline hazardous area cl	Div/Zon / Group
43	Inline area min ign temp	Temp ident number
44	Remote hazardous area cl	Div/Zon / Group
45	Remote area min ign temp	Temp ident number
46		
47		

COMPONENT DESIGN CRITERIA

48		
49	Component type	
50	Component style	
51	Output signal type	
52	Characteristic curve	
53	Compensation style	
54	Type of protection	
55	Criticality code	
56	Max EMI susceptibility	Ref
57	Max temperature effect	Ref
58	Max sample time lag	
59	Max response time	
60	Min required accuracy	Ref
61	Avail nom power supply	Number wires
62	Calibration method	
63	Testing/Listing agency	
64	Test requirements	
65	Supply loss failure mode	
66	Signal loss failure mode	
67		
68		

PROCESS VARIABLES — MATERIAL FLOW CONDITIONS / PROCESS DESIGN CONDITIONS

#	Field	Value	Units	#	Minimum	Maximum	Units
69				101			
70	Flow Case Identification			102			
71	Process pressure			103			
72	Process temperature			104			
73	Process phase type			105			
74	Process liquid actl flow			106			
75	Process vapor actl flow			107			
76	Process vapor std flow			108			
77	Process liquid density			109			
78	Process vapor density			110			
79	Process liquid viscosity			111			
80	Sample return pressure			112			
81	Sample vent/drain press			113			
82	Sample temperature			114			
83	Sample phase type			115			
84	Fast loop liq actl flow			116			
85	Fast loop vapor actl flow			117			
86	Fast loop vapor std flow			118			
87	Fast loop vapor density			119			
88	Conductivity/Resistivity			120			
89	pH/ORP			121			
90	RH/Dewpoint			122			
91	Turbidity/Opacity			123			
92	Dissolved oxygen			124			
93	Corrosivity			125			
94	Particle size			126			
95				127			
96	CALCULATED VARIABLES			128			
97	Sample lag time			129			
98	Process fluid velocity			130			
99	Wake/natural freq ratio			131			
100				132			

MATERIAL PROPERTIES

#	Field	Value	#	Field	Flammability	Reactivity
133			137			
134	Name		138	NFPA health hazard		
135	Density at ref temp	At	139			
136			140			

Rev	Date	Revision Description	By	Appv1	Appv2	Appv3	REMARKS

Form: 20A1001 Rev 0

© 2004 ISA

1.2 Analyzer Sampling

	RESPONSIBLE ORGANIZATION	ANALYSIS DEVICE COMPOSITION OR PROPERTY Operating Parameters (Continued)		SPECIFICATION IDENTIFICATIONS		
1			6			
2			7	Document no		
3	(ISA)		8	Latest revision		Date
4			9	Issue status		
5			10			

	PROCESS COMPOSITION OR PROPERTY			MEASUREMENT DESIGN CONDITIONS				
	Component/Property Name	Normal	Units	Minimum	Units	Maximum	Units	Repeatability
11								
12								
...								
37								

(rows 38–75 blank)

Rev	Date	Revision Description	By	Appv1	Appv2	Appv3	REMARKS

Form: 20A1002 Rev 0 © 2004 ISA

Definition

Sparger: In chemistry, sparging, also known as gas flushing in metallurgy, is a technique that involves bubbling a chemically inert gas, such as nitrogen, argon, or helium, through a liquid. This can be used to remove dissolved gases from the liquid.

Abbreviations

CFM	Cubic Feet per Minute
EPT	Ethylene-propylene terepolymer
GEN	Generation
Mol%	Percent by moles of a process stream
NeSSI	New Sample System Initiative
NTU	Nephelometric Turbidity Units
ORP	Oxidation–Reduction Potential
ppm	Parts per million
SOGAT	Sour Oil & Gas Advanced Technology
Wt%	Percent by weight of a process stream

Organizations

ANSI	American National Standards Institute
ASTM	American Society for Testing and Materials
BSI	British Standards Institute
CPAC	Center for Process Analytical Chemistry
EPA	Environmental Protection Agency
IEC	International Electrotechnical Committee
ISA	International Society of Automation
ISO	International Organization for Standardization
VDE	Verband der Elektrotechnik

Bibliography

Anderson, R., *Sample Pretreatment and Separation*, 1987. http://www.amazon.com/Pretreatment-Separation-Analytical-Chemistry-Learning/dp/047191360X.

ANSI/ISA 76 and IEC 62339-1:2006, Modular Sample Systems (2006) http://www.circortech.com/CircorAssets/Products/Documents/77212_CT76_v13.pdf.

ANSI/ISA-76.00.02-2002, *Modular Component Interfaces for Surface-Mount Fluid Distribution Components—Part 1: Elastomeric Seals* (2002) http://www.bookdepository.com/ANSI-ISA-76–00–02–2002-Modular-Component-Interfaces-for-Surface-mount-Fluid-Distribution-Components-Pt-1/9781556178108.

API RP 550, Manual on Installation of Refinery Instruments and Control Systems; Part II-Process Stream Analyzers (1984) http://www.techstreet.com/products/1821117.

ASTM D3685/D3685M-13, *Standard Test Methods for Sampling and Determination of Particulate Matter in Stack Gases* (2013) http://www.astm.org/Standards/D3685.htm.

ASTM D7278-11, Standard Guide for Prediction of Analyzer Sample System Lag Times (2011) http://www.astm.org/Standards/D7278.htm.

ASTM D3764-13, Standard Practice for Validation of the Performance of Process Stream Analyzer Systems (2013) http://www.astm.org/Standards/D3764.htm.

Davidson, T., VanLandingham, T., Ruhl, G., Rodriguez, T., and Khidkikar, N., Development of modular refinery analyzer systems, *ISA Analysis Division Symposia* (2012) http://www.adsymposium.org/symposium/ad_2012.htm.

Dworetsky, S. H., Feasibility of automatic isokinetic stack sampler, *Environmental Science & Technology*, 8(5): 464–464, 1974, http://pubs.acs.org/doi/abs/10.1021/es60090a003.

Harris, P. and Pelligrini, M., Mass transport in sample system transport lines adsorption desorption effects and their influence on process analytical measurements, *ISA Analysis Division Symposium*, 2011 http://www.adsymposium.org/symposium/ad_2012.htm.

Haupt, O. et al. Automated monitoring of stack gas emissions (2002), http://www.icdd.com/resources/axa/vol42/v42_06.pdf.

Hofi, Stack gas samplers (2014) http://www.environnement.it/public/articoli/171/Files/HOFI_uk_S_p(1).pdf.

Mayeaux, D., Modern software for the analyzer engineer, *ISA 2014 Analysis Division Symposia, Analyzer Sample System References* (2014) www.iceweb.com.au/Analyzer/sample_systems.htm, 2014–09.

Sherman, R.E., *Process Analyzer Sample-Conditioning System Technology*, Wiley, New York, ISBN: 978–0–471–29364–4, (2002) http://search.aol.com/aol/search?enabled_terms=&s_it=client97_searchbox&q = ISBN%3A+978–0–471–29364–4.

Verhappen, I., The basics of analyzer sample systems. (2012) http://www.iceweb.com.au/Analyzer/basics_of_analyzer_sample_system.htm.

1.3 Analyzer Sampling
Stack Monitoring

D. H. F. LIU (1982, 1995)   **B. G. LIPTÁK** (2003, 2017)

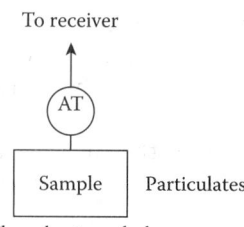

Flow sheet symbol

Types of sample	Gas-containing particulates
Standard design pressure	Generally atmospheric or near atmospheric
Standard design temperature	−32°C to 815°C (−25°F to 1500°F)
Sample velocity	120–3000 m (400–10,000 ft) per min
Materials of construction	316 or 304 stainless steel for pitot tubes; 316 or 304 stainless, quartz, or Incoloy for sample probes
Costs	Probes only in 1–3 m (3–10 ft) lengths with glass, quartz, or stainless steel lining— from $1,300 to $2,500; $10,000 to $15,000 for a complete EPA particulate sampling system (Reference Method 5)
Partial list of suppliers	AMP, EPA Method Sampling System, (2015) http://www.ampcherokee.com/v/vspfiles/photos/C002.0002-2T.jpg
	Apex Instruments, Clean Air Europe: http://www.alpha.cleanaireurope.com/page.php?u=produit_ventes&idprod=235
	Environmental Monitoring, http://www.em-monitor.com/IsokineticSamplers.html
	Inventys Inc. http://www.inventys.in/particle8.html
	Keika Ventures, http://www.keikaventures.com/productinfo.php?product_id=1174
	Mirion Technologies (https://www.mirion.com/search-results/?q=stack+samplers)
	New Star Environmental, http://www.newstarenvironmental.com/product/nsm9096-mini-stack-sampler-a
	Perma Pure, Baldwin, http://www.permapure.com/products/gas-drying-systems/baldwin-probes/baldwin-series-direct-extractive-filter-probes/
	Rupprecht & Patashnick Co. http://www.envirosource.com/domino/thielen/envrsrc.nsf/SearchAll/8CA1D1FE12B4E6FC8625662100762CAE?OpenDocument
	Sensidyne, Inc. (http://sensidyne.com/sensidyne-search-results.php?advsearch=oneword&search=particulate+sampler+stacks&sub=+%C2%A0+%C2%A0++++Search)
	Sierra Monitor Corp. (http://www.sierramonitor.com/protect/all-fire-gas-products/gas-sensor-accessories)
	Teledyne Analytical Instruments, http://www.teledyne-api.com/manuals/07318b_602.pdf
	Thermo Andersen, http://pine-environmental.com/product/instruments/thermo_andersen_tsp_sampler/
	Thermo Scientific, Particulate Monitoring (2015) http://www.thermoscientific.com/en/products/particulate-monitoring.html

INTRODUCTION

Stack gas sampling has already been discussed in Chapter 1.2 in connection with Figures 1.2ae and 1.2af. In this chapter, the emphasis will be on particulate sampling by EPA Method 5 and in that connection, the topics of traverse point locations and pitot tube designs will be emphasized.

Some stack gas samplers (Figure 1.3a) are provided with microcomputer controls and perform the sampling automatically.

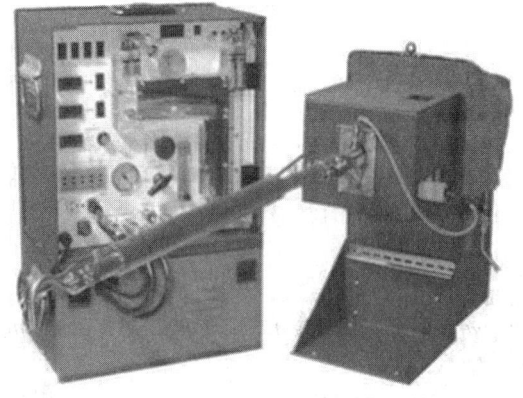

FIG. 1.3a
Isokinetic sampler. (Courtesy of Environmental Monitoring.)

THE EPA PARTICULATE SAMPLING SYSTEM

A complete Environmental Protection Agency (EPA) particulate sampling system (Reference Method 5)* is comprised of four major subsystems:

1. A pitot tube probe or *pitobe* assembly used for temperature and velocity measurements and for sampling
2. A two-module sampling unit that consists of a separate heated compartment with provision for a filter assembly, and a separate ice-bath compartment for the impinger train and bubblers
3. An operating–control unit with a vacuum pump and a standard dry gas meter
4. An integrated, modular umbilical cord that connects the sample unit and pitobe to the control unit

Figure 1.3b is a schematic of an EPA particulate sampling train (Method 5). As shown in the figure, the system can be readily adapted for sampling sulfur dioxide (SO_2), sulfur trioxide (SO_3), and sulfuric acid (H_2SO_4) mist (Method 8).

Microprocessor-Controlled Stack Sampling

In these sampling packages, a microprocessor directs the automatic sampling method, which can be selected to follow U.S. EPA Method 5† or other international methods specified by Verein Deutscher Ingenieure (VDI), British Standards Institution (BSI), or International Standards Organization (ISO). The microprocessor stores all measurements, reviews and diagnoses all inputs, controls the required parameters, calculates isokinetic conditions, and either reports the results in a printed form or transfers them to a floppy disk.

Besides the controller, such a package usually consists of a probe, a filter (hot) box, a cold box, a flexible sample line, glassware, a node box, and a monorail system. The probe is usually 0.9, 1.5, 2.1, or 3 m (3, 5, 7, or 10 ft) and made of stainless steel with a glass liner. Most probes are jacket heated and are provided with both a liner thermocouple and a stack temperature thermocouple.

This chapter will give a detailed description of each of the four subsystems: the pitot assembly, the heated and ice-bath compartments, and the control unit.

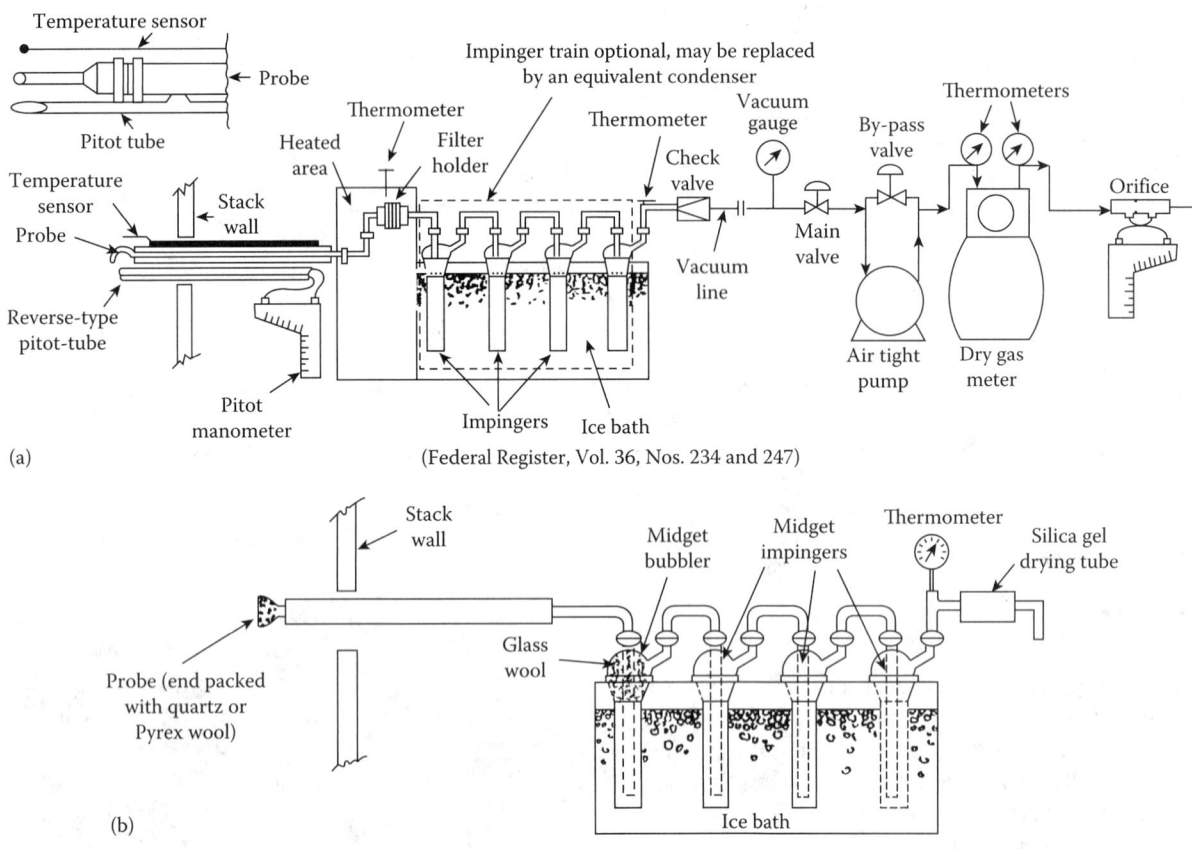

FIG. 1.3b
(a) EPA particulate sampling train (Method 5). (b) Sampling train adopted for SO_2, SO_3, and H_2SO_4 mist (Method 8).

---

* EPA, Method 5—Determination of particulate matter emissions from stationary sources, (2011) http://www3.epa.gov/ttnemc01/promgate/m-05.pdf.

† EPA, Method 5—Determination of particulate matter emissions from stationary sources, (2011) http://www3.epa.gov/ttnemc01/promgate/m-05.pdf.

Pitot Tube Assembly

The procurement of representative samples of particulates suspended in gas streams demands that the velocity at the entrance to the sampling probe be precisely equal to the stream velocity at that point. This is accomplished *by regulating the rate of sample withdrawal* so that the static pressure within the probe is equal to the static pressure in the fluid stream at the point of sampling.

A pitot tube of special design is used for such purposes with means for measuring the pertinent pressures. The pressure difference can be maintained at zero by automatically controlling the sample draw-off rate. Figure 1.3c shows a pitot tube manometer assembly for measuring stack gas velocity.

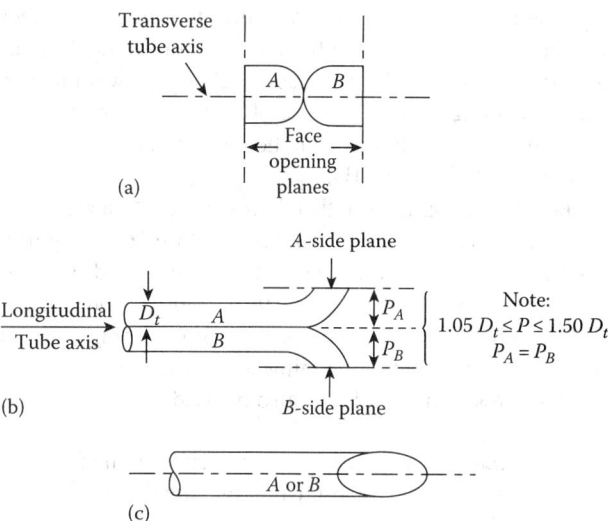

FIG. 1.3d
Properly constructed Type S pitot tube. (a) End view: face-opening planes perpendicular to transverse axis. (b) Top view: face-opening planes parallel to longitudinal axis. (c) Side view: both legs of equal length and center lines coincident, when viewed from both sides; baseline coefficient values of 0.84 may be assigned to pitot tubes constructed this way.

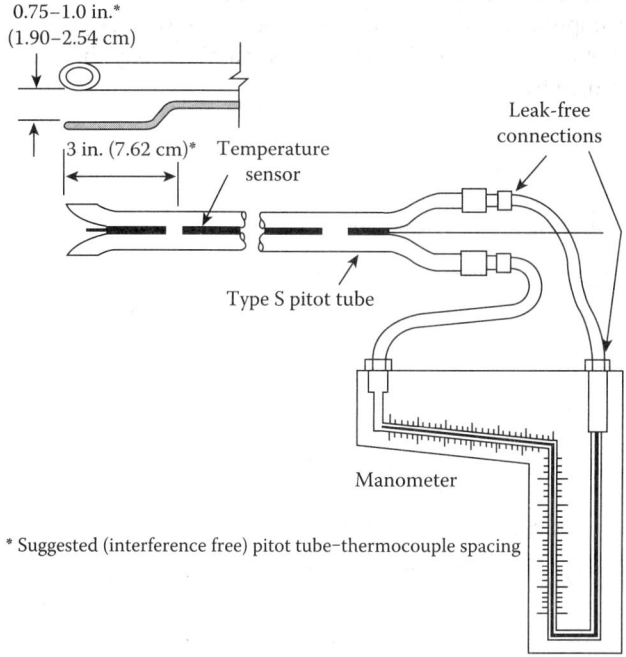

FIG. 1.3c
Type S pitot tube manometer assembly.

The Type S (Stauscheibe, or reverse) pitot tube consists of two opposing openings: one made to face upstream and the other downstream during the measurement. The pressure difference detected between the impact pressure (measured against the gas flow) and the static pressure is related to the stack velocity.

Type S Pitot and the Sampling Probe: Figure 1.3d illustrates the construction of the Type S pitot tube. The external tubing diameter is normally between 3/16 and 3/8 in. (4.8 and 9.5 mm). As can be seen, there is an equal distance from the base of each leg of the tube to its respective face-opening planes. This distance (P_A and P_B) is between 1.05 and 1.50 times the external tube diameter. The face openings of the pitot tube should be aligned as shown.

Figure 1.3e shows the pitot tube in combination with the sampling probe. The relative placement of these components

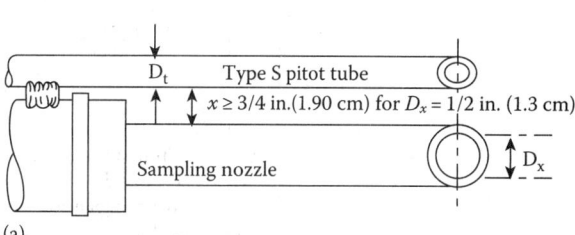

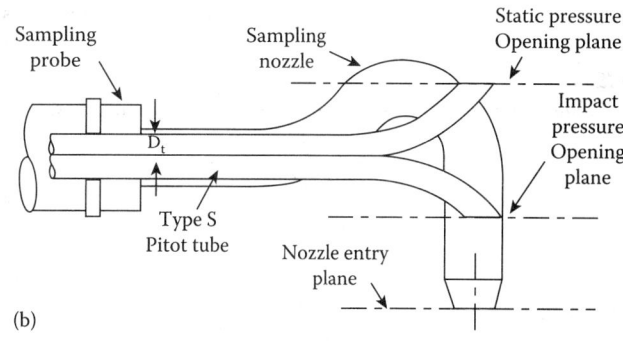

FIG. 1.3e
Proper pitot tube with sampling probe nozzle configuration to prevent aerodynamic interference. (a) Bottom view: minimum pitot nozzle separation. (b) Side view: to prevent pitot tube from interfering with gas flow streamlines approaching the nozzle, the impact pressure-opening plane of the pitot tube shall be even with or above the nozzle entry plane.

eliminates the major aerodynamic interference effects. The probe nozzle is of the bottom hook or elbow design. It is made of seamless 316 stainless steel or glass with a sharp, tapered leading edge. The angle of taper should be less than 30°, and the taper should be on the outside to preserve a constant internal diameter (ID).

For probe lining of either borosilicate or quartz glass, probe liners are used for stack temperatures up to approximately 482°C (900°F); quartz liners are used for temperatures between 482°C and 899°C (900°F and 1650°F). Although borosilicate or quartz glass probe linings are generally recommended, 316 stainless steel, Incoloy, or other corrosion-resistant metal may also be used.

Selecting the Sampling Point: The specific points of stack sampling are selected to ensure that the samples collected are representative of the material being discharged or controlled. These points are determined after examination of the process of the sources of emissions and their variation with time.

In general, the sampling point should be located at a distance equal to at least eight stack or duct diameters downstream and two diameters upstream from any source of flow disturbance, such as expansion, bend, contraction, valve, fitting, or visible flame. (*Note*: This eight and two criterion is adopted to ensure the presence of stable, fully developed flow patterns at the test section.) For rectangular stacks, the equivalent diameter is calculated from the following equation:

Equivalent diameter = 2(length × width)/(length + width)

1.3(1)

Traversing Point Locations: Next, provisions must be made to traverse the stack. The number of traverse points is 12. If the eight- and two-diameter criterion is not met, the required number of traverse points depends on the sampling point distance from the nearest upstream and downstream disturbances. This number may be determined by using Figure 1.3f.

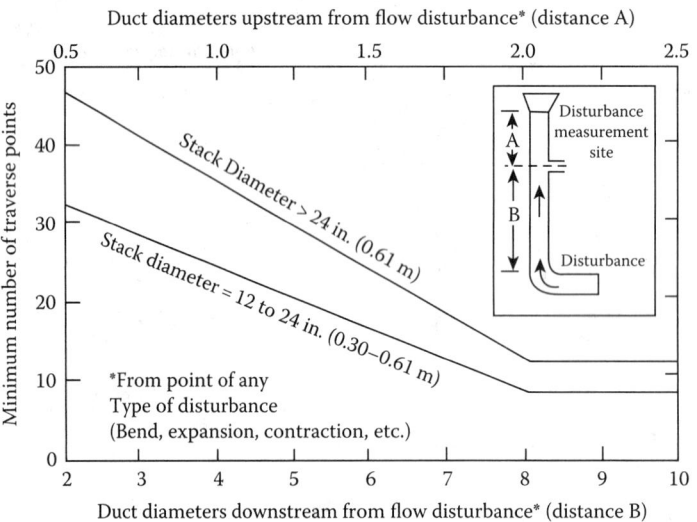

FIG. 1.3f
Minimum number of traverse points for particulate traverses.

The cross-sectional layout and location of traverse points are as follows:

1. For circular stacks, the traverse points should be located on two perpendicular diameters, as shown in Figure 1.3g and Table 1.3a.
2. For rectangular stacks, the cross section is divided into as many equal rectangular areas as traverse points, such that the ratio of the length to width of the elemental area is between 1 and 2. The traverse points are to be located at the center of at least nine and preferably more equal areas, as shown in Figure 1.3g.

Pitot Tube Calculation Form: The velocity head at various traverse points is measured using the pitot tube assembly shown in Figure 1.3c. The gas samples are collected at a rate proportional to the stack gas velocity and analyzed for carbon monoxide (CO), carbon dioxide (CO_2), and oxygen (O_2).

The pitot tube is calibrated by measuring the velocity head at some point in the flowing gas stream with both the Type S pitot tube and a standard pitot tube with a known coefficient. Other data also needed for calculation of the volumetric flow are stack temperature, stack and barometric pressures, and wet-bulb and dry-bulb temperatures of the gas sample at each traverse.

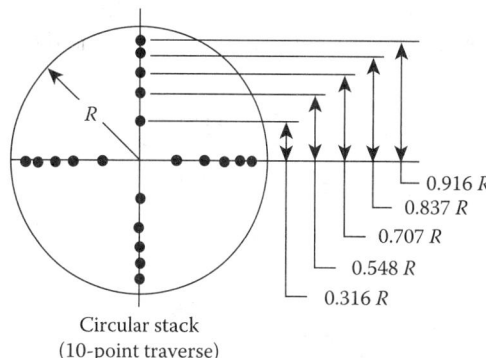

FIG. 1.3g
Traverse point locations for velocity measurement or for multipoint sampling.

TABLE 1.3a
Location of Traverse Points in Circular Stacks[a]

Traverse Point Number on a Diameter	Number of Traverse Points on a Diameter											
	2	4	6	8	10	12	14	16	18	20	22	24
1	14.6	6.7	4.4	3.2	2.6	2.1	1.8	1.6	1.4	1.3	1.1	1.1
2	85.4	25.0	14.6	10.5	8.2	6.7	5.7	4.9	4.4	3.9	3.5	3.2
3		75.6	29.6	19.4	14.6	11.8	9.9	8.5	7.5	6.7	6.0	5.5
4		93.3	70.4	32.3	22.6	17.7	14.6	12.5	10.9	9.7	8.7	7.9
5			85.4	67.7	34.2	25.0	20.1	16.9	14.6	12.9	11.6	10.5
6			95.6	80.6	65.8	35.6	26.9	22.0	18.8	16.5	14.6	13.2
7				89.5	77.4	64.4	36.6	28.3	23.6	20.4	18.0	16.1
8				96.8	85.4	75.0	63.4	37.5	29.6	25.0	21.8	19.4
9					91.8	82.3	73.1	62.5	38.2	30.6	26.2	23.0
10					97.4	88.2	79.9	71.7	61.8	38.8	31.5	27.2
11						93.3	85.4	78.0	70.4	61.2	39.3	32.3
12						97.9	90.1	83.1	76.4	69.4	60.7	39.8
13							94.3	87.5	81.2	75.0	68.5	60.2
14							98.4	91.5	85.4	79.6	73.8	67.7
15								95.1	89.1	83.5	78.2	72.8
16								98.4	92.5	87.1	82.0	77.0
17									95.6	90.3	85.4	80.6
18									98.6	93.3	88.4	83.9
19										96.1	91.3	86.8
20										98.7	94.0	89.5
21											96.5	92.1
22											98.9	94.5
23												96.8
24												99.9

Source: Courtesy of Clean Air.
[a] Percent of stack diameter from inside wall to traverse point.

Table 1.3b gives the equations for converting pitot tube readings into velocity and mass flow, and a typical data sheet for stack flow measurements.

Sampling Velocity for Particle Collection: Based on the range of velocity heads, a probe with a properly sized nozzle is selected to maintain isokinetic sampling of particulate matter. As shown in Figure 1.3h, a converging stream will be developed at the nozzle face if the sampling velocity is too high. Under this subisokinetic sampling condition, an excessive amount of lighter particles enters the probe.

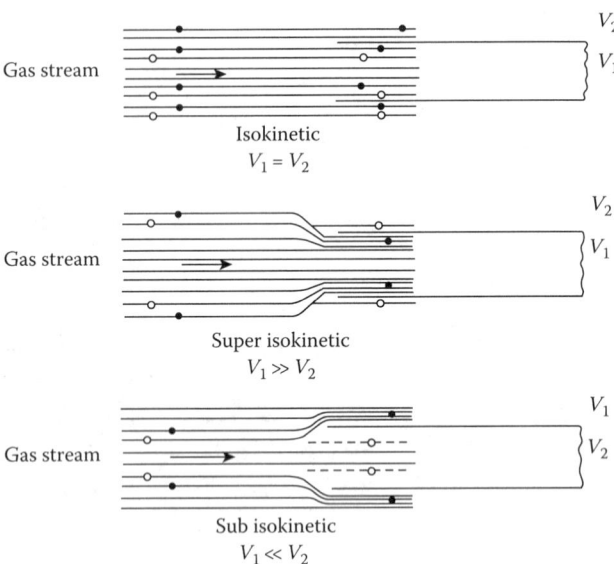

FIG. 1.3h
Particle collection and sampling velocity.

Because of the inertia effect, the heavier particles, especially those in the range of 3 μm or greater, travel around the edge of the nozzle and are not collected. The result is a sample indicating an excessively high concentration of lighter particles, and the weight of the solid sample is in error on the low side. Conversely, portions of the gas stream approaching at a higher velocity are deflected if the sampling velocity is below that of the flowing gas stream.

Under this superisokinetic sampling condition, the lighter particles follow the deflected stream and are not collected, while the heavier particles, because of their inertia, continue into the probe. The result is a sampling indicating high concentration of heavier particles, and the weight of solid sample is in error on the high side.

Isokinetic Sampling: Isokinetic sampling requires the precise adjustment of the sampling rate with the aid of the pitot tube manometer readings and nomographs such as APTD-0576 and nomographs. If the pressure drop across the filter in the sampling unit becomes too high, making isokinetic sampling difficult to maintain, the filter may be replaced in the middle of a sample run.

To measure the concentration of particulate matter, the sampling time for each run should be at least 60 min, and the minimum volumetric rate of sampling should be 30 dry scfm (51 m^3/hr).

Heated Compartment (Hot Box)

As shown in Figure 1.3b, the probe is connected to the heated compartment that contains the filter holder and other particulate-collecting devices, such as cyclone and flask. The filter holder is made of borosilicate glass, with a frit filter support and a silicone rubber gasket.

The compartment is insulated and equipped with a heating system capable of maintaining a temperature around the filter holder during sampling at 120°C ± 14°C (248°F ± 25°F), or such other temperature as specified by the EPA. The thermometer should measure temperature to within 3°C (5.4°F). The compartment should be provided with a circulating fan to minimize thermal gradients.

Ice-Bath Compartment (Cold Box)

The ice-bath compartment contains a number of impingers and bubblers. The system for determining stack gas moisture content consists of four impingers connected in series, as shown in Figure 1.3b. The first, third, and fourth impingers are of the Greenburg–Smith design.* To reduce the pressure drop, the tips are removed and replaced with a 0.5 in. (12.5 mm) ID glass tube extending to 0.5 in. (12.5 mm) from the bottom of the flask.

The second impinger is of the Greenburg–Smith design with a standard tip. During sampling for particulates, the first and second impingers are filled with 100 mL (3.4 oz) of distilled and deionized water. The third impinger is left dry to separate entrained water. The last impinger is filled with 200–300 g (7–10.5 oz) of precisely weighed silica gel (6–16 mesh) that has been dried at 177°C (350°F) for 2 hr to completely remove any remaining water.

A thermometer capable of measuring temperature to within 1.1°C (2°F) is placed at the outlet of the last impinger for monitoring purposes. Crushed ice should be added during the run to maintain the temperature of the gas, leaving the last impinger at 16°C (60°F) or less.

Control Unit

As shown in Figure 1.3b, the control unit consists of the system's vacuum pump, valves, switches, thermometers, and totalizing dry gas meter. This system is connected by a vacuum line to the last Greenburg–Smith impinger. The pump intake vacuum is monitored with a vacuum gauge just after the quick disconnect.

A bypass valve parallel with the vacuum pump provides fine control and permits recirculation of gases at a low sampling rate so that the pump motor is not overloaded. Downstream from the pump and bypass valve are

---

* Impinger, Greenburg-Smith, https://us.vwr.com/store/catalog/product.jsp?product_id=9256289.

thermometers, a dry gas meter, and calibrated orifice and inclined/vertical manometers.

The calibrated orifice and inclined manometer indicate the instantaneous sampling rate. The totalizing dry gas meter gives an integrated gas volume. The average of the two temperatures on each side of the dry gas meter gives the temperature at which the sample is collected. The addition of atmospheric pressure to orifice pressure gives meter pressure.

AUTOMATIC SAMPLING TRAINS

In the automatic sampling train (AST) packages (Figure 1.3i), a microprocessor stores all measurements, reviews and diagnoses all inputs, controls the required parameters, calculates isokinetic conditions, and either reports the results in a printed form or transfers them to a floppy disk.

The measured variables include the temperatures of the stack, probe liner, filter box, condenser outlet, and dry gas meter (Figure 1.3j). The pressures are detected by an absolute and a differential pressure transducer and are used to measure the pressure of the stack gas, the barometric pressure, and the velocity pressure of the stack gas. The normal capacity of the vacuum pump that draws the sample is 0.75 cfm (21 l/min), and the dry gas meter has an operating range of 0.1–1.5 cfm (2.8–42 l/min).

The node box provides the interface between the filter box and the cold box by measuring the temperatures in both. It measures and stores the temperature, pressure, and velocity in the stack. The monorail eliminates the need for bulky supports.

Precise measurements require that the thermometers be capable of measuring the temperature to within 3°C (5.4°F); the dry gas meter is inaccurate to within 2% of the volume; the barometer is inaccurate within 0.25 mmHg (torr) (0.035 kPa); and the manometer is inaccurate within 0.25 mmHg (torr) (0.035 kPa).

The umbilical cord is an integrated multiconductor assembly containing both pneumatic and electrical conductors. It connects the two-module sampling unit to the control unit, as well as the pitot tube stack velocity signals to the manometers or differential pressure gauges.

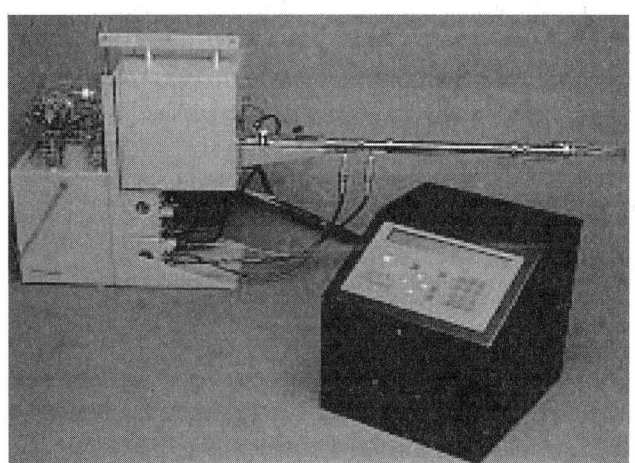

FIG. 1.3i
Automatic stack train. (Courtesy of Thermo Anderson.)

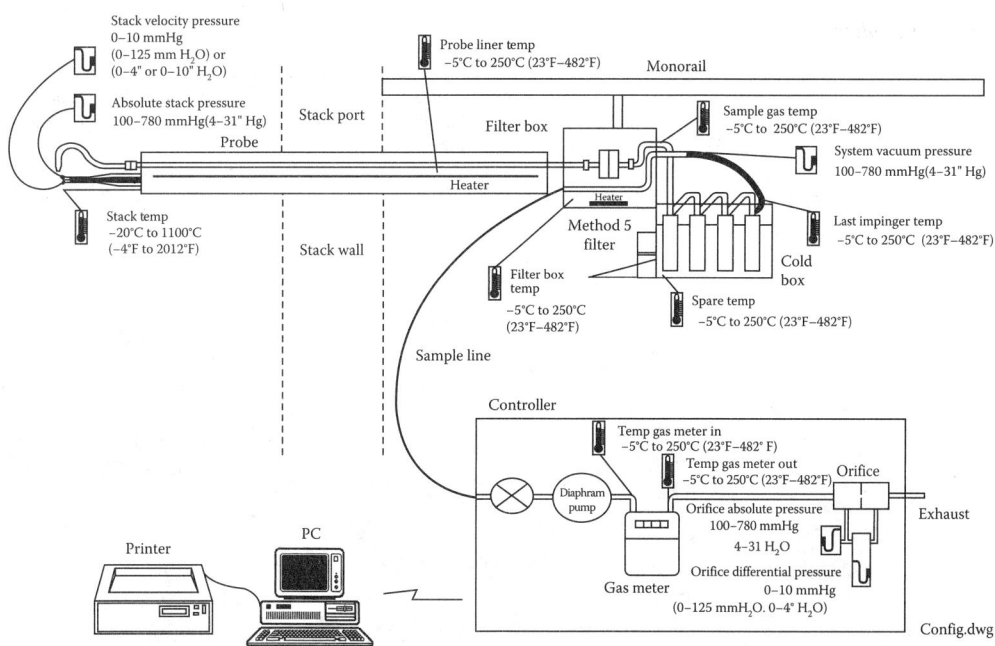

FIG. 1.3j
The components of an automatic stack train. (Courtesy of Thermo Andersen.)

Sampling for Gases and Vapors

Some commonly used components in stack sampling systems are illustrated in Figure 1.3k. If ball-and-socket joints and compression fittings are used, any arrangement of components is readily set up for field use. The stack sampling components are selected on the basis of the source to be sampled, the substances involved, and the data needed.

A summary of sampling procedure outlines was developed by industrial hygienists* for specific substances. The procedural outlines serve as a starting point in assembling a stack sampling system, after consideration has been given to the complications that might arise because of the presence of interfering substances in the gas samples.

SPECIFICATION FORMS

When specifying stack sampling systems, it is advisable to attach the Pitot Tube calculations that have been made for the system. To show these calculations, the form on the next page can be used.

The total sampling system can be specified by filling out the form Figure 1.2b, while the specification forms, found on the next 2 pages, ISA 20A1001 and ISA 20A1002 can be used to specify the combination of the analyzer and its sampling system. These forms are reproduced with the permission of the International Society of Automation.

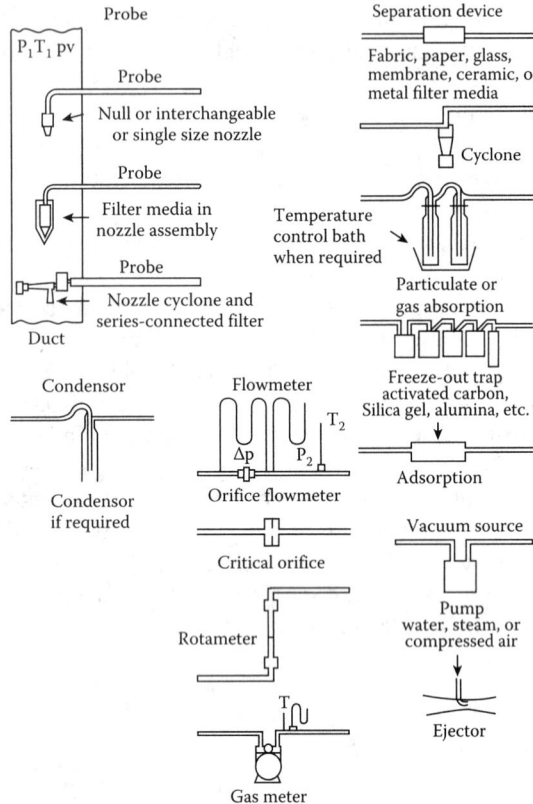

FIG. 1.3k
Components of common sampling systems.

---

* Ron, J.J., Environmental Calibration and Operation of Isokinetic Source-Sampling Equipment, (1972) http://nepis.epa.gov/Adobe/PDF/20013PMH.PDF.

TABLE 1.3b
Pitot Tube Calculation Sheet

Stack Volume Data

Stack no. _____ Station _____
Date _____ Page _____
Name of the firm _____

Point	Position, in.	Reading, in. of H_2O	$\sqrt{H}$	Temperature t_3, °F	Velocity V_3, ft/sec
1					
2					
3					
4					
5					
6					
7					
8					
9					
10					
11					
12					
13					
14					
15					
16					
	Totals				
	Average				
	Absolute temperature: $T_s = t_s + 460 =$ °R				

Dry-bulb temperature: $t_d =$ _____ °F Barometer: $P_b =$ _____ in Hg
Wet-bulb temperature: $t_w =$ _____ °F Stack gauge pressure: _____ in., H_2O
Absolute humidity: $W =$ ___ lb H_2O/lb dry gas Stack absolute pressure: $P_s =$ ___ in., H_2O/13.6 $\pm P_h$ ___ in., Hg
Stack area: $A_s =$ _____ ft² Pitot correction factor: $F_s =$ _____

Component	Volume Fraction, Dry Basis × Molecular Weight = Weight Fraction, Dry Basis	
Carbon dioxide		44 =
Carbon monoxide		28 =
Oxygen		32 =
Nitrogen		28 =

Average dry gas molecular weight: $M =$ _____
Specific gravity of stack gas: $G_S =$ <1083_Un8.3(1).eps> _____
(Reference dry air at same conditions)
Velocity: $V_s = 2.9\ F_s <$1083_un8.3(2).eps$> =$ _____ ft./sec.
Volume = _____ ft./sec × _____ ft² × 60 _____ = _____ cfm.
Standard volume = cfm. × <1083_υν8.3(3).εποσ>

Note: Sampling velocity for particle collection based on the range of velocity heads: a probe with a properly sized nozzle is selected to maintain isokinetic sampling of particulate matter. As shown in Figure 1.3h, a converging stream will be developed at the nozzle face if the sampling velocity is too high. Under this subisokinetic sampling condition, an excessive amount of lighter particles enters the probe.

Analytical Measurement

ANALYSIS DEVICE — Operating Parameters

RESPONSIBLE ORGANIZATION (ISA)

1–5	RESPONSIBLE ORGANIZATION
6	SPECIFICATION IDENTIFICATIONS
7	Document no
8	Latest revision / Date
9	Issue status
10	

ADMINISTRATIVE IDENTIFICATIONS
11	ADMINISTRATIVE IDENTIFICATIONS
12	Project number / Sub project no
13	Project
14	Enterprise
15	Site
16	Area / Cell / Unit
17	

SERVICE IDENTIFICATIONS
18	SERVICE IDENTIFICATIONS
19	Tag no/Functional ident
20	Related equipment
21	Service
22	
23	P&ID/Reference dwg
24	Process line/nozzle no
25	Process conn pipe spec
26	Process conn nominal size / Rating
27	Process conn termn type / Style
28	Process conn schedule no / Wall thickness
29	Process connection length
30	Process line matl type
31	Fast loop line number
32	Fast loop pipe spec
33	Fast loop conn nom size / Rating
34	Fast loop conn termn type / Style
35	Fast loop schedule no / Wall thickness
36	Fast loop estimated lg
37	Fast loop material type
38	Return conn nominal size / Rating
39	Return conn termn type / Style

SERVICE IDENTIFICATIONS Continued
40	SERVICE IDENTIFICATIONS Continued
41	Return conn matl type
42	Inline hazardous area cl / Div/Zon / Group
43	Inline area min ign temp / Temp ident number
44	Remote hazardous area cl / Div/Zon / Group
45	Remote area min ign temp / Temp ident number
46	
47	

COMPONENT DESIGN CRITERIA
48	COMPONENT DESIGN CRITERIA
49	Component type
50	Component style
51	Output signal type
52	Characteristic curve
53	Compensation style
54	Type of protection
55	Criticality code
56	Max EMI susceptibility / Ref
57	Max temperature effect / Ref
58	Max sample time lag
59	Max response time
60	Min required accuracy / Ref
61	Avail nom power supply / Number wires
62	Calibration method
63	Testing/Listing agency
64	Test requirements
65	Supply loss failure mode
66	Signal loss failure mode
67	
68	

PROCESS VARIABLES / MATERIAL FLOW CONDITIONS / PROCESS DESIGN CONDITIONS
#	PROCESS VARIABLES	MATERIAL FLOW CONDITIONS (Units)	#	PROCESS DESIGN CONDITIONS (Minimum / Maximum / Units)
69	PROCESS VARIABLES	MATERIAL FLOW CONDITIONS	101	PROCESS DESIGN CONDITIONS
70	Flow Case Identification	Units	102	Minimum / Maximum / Units
71	Process pressure		103	
72	Process temperature		104	
73	Process phase type		105	
74	Process liquid actl flow		106	
75	Process vapor actl flow		107	
76	Process vapor std flow		108	
77	Process liquid density		109	
78	Process vapor density		110	
79	Process liquid viscosity		111	
80	Sample return pressure		112	
81	Sample vent/drain press		113	
82	Sample temperature		114	
83	Sample phase type		115	
84	Fast loop liq actl flow		116	
85	Fast loop vapor actl flow		117	
86	Fast loop vapor std flow		118	
87	Fast loop vapor density		119	
88	Conductivity/Resistivity		120	
89	pH/ORP		121	
90	RH/Dewpoint		122	
91	Turbidity/Opacity		123	
92	Dissolved oxygen		124	
93	Corrosivity		125	
94	Particle size		126	
95			127	
96	CALCULATED VARIABLES		128	
97	Sample lag time		129	
98	Process fluid velocity		130	
99	Wake/natural freq ratio		131	
100			132	

MATERIAL PROPERTIES
133	MATERIAL PROPERTIES
134	Name
135	Density at ref temp / At
136	
137	MATERIAL PROPERTIES Continued
138	NFPA health hazard / Flammability / Reactivity
139	
140	

Rev	Date	Revision Description	By	Appv1	Appv2	Appv3	REMARKS

Form: 20A1001 Rev 0 © 2004 ISA

1.3 Analyzer Sampling: Stack Monitoring

1	RESPONSIBLE ORGANIZATION		ANALYSIS DEVICE COMPOSITION OR PROPERTY Operating Parameters (Continued)		6	SPECIFICATION IDENTIFICATIONS				
2					7	Document no				
3	(ISA logo)				8	Latest revision		Date		
4					9	Issue status				
5					10					

	PROCESS COMPOSITION OR PROPERTY			MEASUREMENT DESIGN CONDITIONS				
	Component/Property Name	Normal	Units	Minimum	Units	Maximum	Units	Repeatability
13								
14								
15								
...								
37								

(Rows 38–75: blank notes/remarks area)

Rev	Date	Revision Description	By	Appv1	Appv2	Appv3	REMARKS

Form: 20A1002 Rev 0 © 2004 ISA

Abbreviations

APTD   Air pollution technical data
AST   Automatic sampling train

Organizations

BSI   British Standards Institution
ISO   International Standards Organization
VDI   Verein Deutscher Ingenieure

Bibliography

AMP, EPA method sampling system, (2015) http://www.ampcherokee.com/v/vspfiles/photos/C002.0002-2T.jpg.

ASTM D6831-11, Standard test method for sampling and determining particulate matter in stack gases using an in-stack, inertial microbalance, (2011) http://www.astm.org/Standards/D6831.htm.

ASTM D3685/D3685M-13, Standard test methods for sampling and determination of particulate matter in stack gases, ASTM International, (2013) http://www.astm.org/Standards/D3685.htm.

ASTM D6331-14, Standard test method for determination of mass concentration of particulate matter from stationary sources at low concentrations (manual gravimetric method), (2014) http://www.astm.org/Standards/D6331.htm.

BACHARACH: Combustion analyzers, (2015) http://www.bacharach-inc.com/PDF/Instructions/24-9438.pdf.

Clean Air Engineering, Stack and source emission testing, (2014) http://www.cleanair.com/services/StackSourceTesting/index.htm.

Cooper, C. D. and Alley, F. C., Air pollution, (2002) http://www.amazon.com/Coopers-Pollution-Control-Edition-Hardcover/dp/B003WUI3I6.

EPA, current knowledge of particulate matter(PM) continuous emission monitoring, (2002) http://www3.epa.gov/ttnemc01/cem/pmcemsknowfinalrep.pdf.

EPA, method 5—Determination of particulate matter emissions from stationary sources, (2011) http://www3.epa.gov/ttnemc01/promgate/m-05.pdf.

Federal Register, Standards of performance for new stationary sources, (2011) https://www.federalregister.gov/articles/2011/03/21/2011-4495/standards-of-performance-for-new-stationary-sources-and-emission-guidelines-for-existing-sources.

ICS: Stack sampler, (2015) http://www.indiamart.com/industrial-commercial-services/environment-equipments.html.

Ron, J. J., Environmental calibration and operation of isokinetic source-sampling equipment, (1972) http://nepis.epa.gov/Adobe/PDF/20013PMH.PDF.

Sherman, R. E., Process analyzer sample-conditioning system technology, (2002) http://www.wiley.com/WileyCDA/WileyTitle/productCd-0471293644.html.

Teledyne Analytical Instruments, Particle measurement system, (2012) http://www.teledyne-api.com/manuals/07318b_602.pdf.

Thermo Scientific, Particulate monitoring, (2015) http://www.thermoscientific.com/en/products/particulate-monitoring.html.

1.4 Analyzer Sampling
Air Quality Monitoring

R. A. HERRICK and R. G. SMITH (1974)   **B. G. LIPTÁK** (1995, 2003, 2017)

Costs	Handheld portable units are available for a few hundred dollars (http://www.amazon.com/Dylos-DC1100-Laser-Quality-Monitor/dp/B000XG8XCI)
	A battery-operated adjustable-flow air sampling pump costs $500; equipment costs for an audit system might include $6,000 for an ozone analyzer, $10,000 for a CO analyzer, $6,000 for a pure air generator, $1,500 for a methane reactor, and $15,000 for a gas calibrator; a microprocessor-based portable infrared spectrometer with preprogrammed multicomponent identification capability and with space for 10 user-defined standards for calibration, an AC/DC converter, a sample probe, and a carrying case costs $17,000.
Partial list of suppliers	ABB (http://www.abb.com/analytical)
	AMC (Armstrong Monitoring Corp.) (www.armstrongmonitoring.com)
	Ametek/Thermox (http://www.environmental-expert.com/companies/ametek-process-instruments-thermox-9026/products)
	California Analytical Instruments Inc. (http://www.ierents.com/Spec%20Pages/1312%2520Lit.pdf)
	Ecotech (http://www.ecotech.com/systems/aaqms)
	EMS (Environmental Monitoring Systems) (www.emssales.net)
	Environmental Expert (http://www.environmental-expert.com/products/?keyword=air+quality+monitoring)
	Esensors Inc. (http://www.eesensors.com/air-quality-monitors/personal-air-quality-monitor.html?gclid=CNzMpuzD2bcCFUmY4AodgAoAEQ)
	Foxboro-Invensys (www.foxboro.com)
	Horiba Instrument Inc. (http://www.horiba.com/us/en/process-environmental/products/ambient/)
	Honeywell (http://www.instrumart.com/products/21650/honeywell-iaqprobe-air-quality-monitor)
	International Sensor Technology (http://www.intlsensor.com/pdf/4–20IQmanual.pdf)
	MSA Instrument Div. (http://ca.msasafety.com/Portable-Gas-Detection/c/114?N=10138&Ne=10198&isLanding=true)
	Pine Environmental Service: (http://www.pine-environmental.com/air-monitoring.htm)
	Purafil Inc. (http://www.purafil.com/products/monitoring/air_quality.aspx)
	Sensidyne Inc. (http://www.sensidyne.com/air-sampling-equipment/calibration-equipment.php)
	Siemens Energy & Automation (http://www.industry.usa.siemens.com/automation/us/en/process-instrumentation-and-analytics/process-analytics/pa-brochures/Documents/TotalSolutionsPIABR-00006–0509.pdf)
	Sierra Monitor Corp. (http://www.sierramonitor.com/markets/transportation.php)
	Teledyne Analytical Instruments (http://www.teledyne-api.com/)
	Thermo Fisher Scientific (http://www.thermofisher.com/global/en/markets/environmental.asp)
	ThermoScientific (http://www.thermoscientific.com/ecomm/servlet/productscatalog?storeId=11152&categoryId=80424&ca=air
	TSI (http://www.tsi.com/Air-Quality-Monitoring/)
	Thermo Gas Tech (http://www.pine-environmental.com/other-combustible-gas-indicators/thermo-gastech-402.htm)
	Yokogawa Corp. of America (www.yokogawa.com)
Orientation table	See Table 1.4a
Air quality standards	See Table 1.4b

TABLE 1.4a
Air Quality Parameters, Their Measurement Ranges, Accuracies, and Methods of Operation

Compound	Range	Accuracy	Technique Employed
Oxides of nitrogen	0–20 ppm	0.5 ppb	Chemiluminescence or DOAS open path
Sulfur dioxide	0–20 ppm	0.5 ppb	Fluorescence dual-channel ratiometric phase detection or DOAS open path
Ozone	0–20 ppm	0.5 ppb	Ultraviolet photometrics or DOAS open path
Carbon monoxide	0–200 ppm	50 ppb	Gas filter correlation or DOAS open path
Carbon dioxide	0–100%	50 ppb	Gas filter correlation or DOAS open path
Benzene, toluene, xylene	0–5 ppm	0.5 ppb	Gas chromatography or DOAS open path
Nonmethane hydrocarbons	0–1,000 ppm	0.01 ppm	FID or DOAS open path
Methane	0–1,000 ppm	0.01 ppm	FID or DOAS open path
Particulates (PM10, TSP, PM2.5)	0–5 g/m^3	0.1 µg/m^3	Tapered element oscillating microbalance
Carbon particulates	0–5 g/m^3	0.25 µg/m^3	Thermal CO_2 method
Local visual distance	0–16 km	±10%	Nephelometer
Wind speed	0–70 m/sec	0.22 m/sec	Anemometer
Wind direction	0°–540°	±3°	Airfoil vane
Ambient temperature	–50°C to 100°C	±0.1°C	Solid-state thermistor
Relative humidity	0%–100%	±2%	Thin film capacitor
Barometric pressure	800–1,200 mbar	±1.3 mbar	Solid-state transducer
Precipitation	NA	0.1 mm	Net radiometer
Solar radiation	250–2,800 nm	9 mV/kWm^2	Pyranometer
Net radiation	250–60,000 nm	8 mV/kWm^2	Net radiometer

Source: Courtesy of Ecotech, www.ecotech.com.au.
Abbreviations: DOAS, dedicated outdoor air system; FID, flame ionization detector; NA, nonapplicable; PM, particulate matter; PM10, particles smaller than 10 µm; PM2.5, particles smaller than 2.5 µm; TSP, total suspended particles.

TABLE 1.4b
National Ambient Air Quality Standards (NAAQS) of the United States

Pollutant [Final Rule Cite]		Primary/Secondary	Averaging Time	Level	Form
Carbon monoxide [76 FR 54294, Aug 31, 2011]		Primary	8 hr	9 ppm	Not to be exceeded more than once per year
			1 hr	35 ppm	
Lead [73 FR 66964, Nov 12, 2008]		Primary and secondary	Rolling 3 month average	0.15 µg/m^3	Not to be exceeded
Nitrogen dioxide [75 FR 6474, Feb 9, 2010] [61 FR 52852, Oct 8, 1996]		Primary	1 hr	100 ppb	Ninety-eighth percentile of 1 hr daily maximum concentrations, averaged over 3 years
		Primary and secondary	Annual	53 ppb	Annual mean
Ozone [73 FR 16436, Mar 27, 2008]		Primary and secondary	8 hr	0.075 ppm	Annual fourth highest daily maximum 8 hr concentration, averaged over 3 years
Particle pollution Dec 14, 2012	PM$_{2.5}$	Primary	Annual	12 µg/m^3	Annual mean, averaged over 3 years
		Secondary	Annual	15 µg/m^3	Annual mean, averaged over 3 years
		Primary and secondary	24 hr	35 µg/m^3	Ninety-eighth percentile, averaged over 3 years
	PM$_{10}$	Primary and secondary	24 hr	150 µg/m^3	Not to be exceeded more than once per year on average over 3 years
Sulfur dioxide [75 FR 35520, Jun 22, 2010] [38 FR 25678, Sept 14, 1973]		Primary	1 hr	75 ppb	Ninety-nineth percentile of 1 hr daily maximum concentrations, averaged over 3 years
		Secondary	3-hr	0.5 ppm	Not to be exceeded more than once per year

INTRODUCTION

Several Chapters of this handbook deal with some aspects of air quality monitoring and air sampling. The general topic of sampling was discussed in Chapter 1.2, and the specific subject of particulate sampling was covered in Chapter 1.3. Similarly, the various detectors and analyzers used to identify the concentrations of particular substances in a sample are discussed in the many chapters that follow this one.

This chapter starts with a general discussion of air quality–monitoring systems and continues with a description of air quality–detecting sensors, automatic monitoring packages, and microprocessor-based portable ambient air analyzers. The chapter is concluded with a discussion of ambient air sampling for both gases and particulate matter.

AIR QUALITY MONITORING

There are many options for the types of air quality information that can be collected, and the cost of air quality–monitoring systems can vary by orders of magnitude. Only with a thorough understanding of the decisions that must be made based on the information received from the air quality–monitoring system can an appropriate selection be made. Regardless of the type of instruments used to measure air quality, the data are only as good as they are representative of the sampling site selected. See Table 1.4a for a list of air quality and meteorological parameters and their measurement ranges, accuracy, and method of analysis.

The simplest air quality monitors are static sensors, which are exposed for a given length of time and are later analyzed in the laboratory. In some cases, these devices provided all the information required. More commonly, a system of automatic instruments measuring several different air quality parameters will be used. With more than a few instruments, the signals from these instruments can be retained on magnetic tape rather than on recorder charts.

The most common errors in the design of air quality–monitoring systems are poor site location and the acquisition of more data than necessary to accomplish the purpose of the installation.

Air Quality Standards

The principal purpose of air quality monitoring is frequently the acquisition of data for comparison to regulated standards. In the United States, standards have been promulgated by the federal government and by many of the states. Where possible, these standards have been based on the physiological effect of the air pollutant in question, so that human health is protected. The averaging time over which various concentration standards must be maintained is different for each pollutant.

The Clean Air Act* of the United States congress, which was last amended in 1990, requires the Environmental Protection Agency (EPA) to set National Ambient Air Quality Standards† (40 CFR part 50) for pollutants considered harmful to public health and the environment. The Clean Air Act identifies two types of national ambient air quality standards. Primary standards provide public health protection, including protecting the health of "sensitive" populations such as asthmatics, children, and the elderly. Secondary standards provide public welfare protection, including protection against decreased visibility and damage to animals, crops, vegetation, and buildings.

I will base this discussion on the National Ambient Air Quality Standards (NAAQS) of the United States, while emphasizing that similar standards exist in all nations and even in the individual states of the United States. In this Environmental Protection Agency (EPA) standard, as of December 2014, the maximum allowed average of fine particulate (PM2.5)‡ concentration, over a 3-year period, was 12.0 μg/m³. The regulation also limits the maximum allowable concentration of coarse particulates (PM10)§ not to be exceeded more than once per year on average over 3 years to 150 μg/m³.

EPA has set National Ambient Air Quality Standards for six principal pollutants, which are called "criteria" pollutants. They are listed in Table 1.4b. Units of measure for the standards are parts per million (ppm) by volume, parts per billion (ppb) by volume, and micrograms per cubic meter of air (μg/m³).

---

* http://www2.epa.gov/clean-air-act-overview
† http://www3.epa.gov/ttn/naaqs
‡ PM2.5 refers to particles smaller than 2.5 μm. They include combustion particles, organic compounds and metals. The World Health Organization (WHO) recommends that the PM2.5 exposure should not exceed 25 μg/m³ mean in any 24-hour period and 10 μg/m³ mean annually.
§ PM10 refers to particles smaller than 10μm. They include dust, pollen, and mold spores. The World Health Organization (WHO) recommends that the PM10 exposure should not exceed 50 μg/m³ mean in any 24-hour period and 20 μg/m³ mean annually.

Single Source (Stacks) Monitoring

Some air quality–monitoring systems are operated to determine the impact of a single source or a concentrated group of sources of emission on the surrounding area. In this case, it is important to determine the background level, the maximum ground-level concentration in the area, and the geographical extent of the air pollutant impact of the source. When the source is isolated, such as a single industrial plant in a rural area, the design is straightforward. Utilizing meteorological records, which are normally available from nearby airports or from government meteorological reporting stations, a wind rose can be prepared to estimate the direction of principal drift of the air pollutant from the source.

Dispersion calculations can be performed to estimate the location of the expected point of maximum ground-level concentration. As a rule of thumb, with stacks between about 50 and 350 ft tall, this point of maximum concentration will be approximately 10 stack heights downwind. The air quality–monitoring systems (which have been described in Chapter 1.3) should include at least one sensor at the point of expected maximum ground-level concentration. Additional sensors should be placed not less than 100 stack heights upwind (prevailing) to provide a background reading and at least two or three sensors should be placed between 100 and 200 stack heights downwind to determine the extent of the travel of the pollutants from the source in question. If adequate resources are available, sampling at the intersection points of a rectilinear grid with its center at the source in question is advisable.

With such a system for an isolated source, adequate data can be obtained in 1 year to determine the impact of the sources on the air quality of the area. There are very few instances where less than 1 year of data collection will provide adequate information because of the variability in climatic conditions on an annual basis. If a study of this type is performed to determine the effect on air quality of process changes, it may be necessary to continue the study for 2–5 years to develop information that is statistically reliable.

Some air quality monitoring is designed for the specific purpose of investigating complaints concerning an unidentified source. This usually happens in urban situations for odor complaints. In these cases, a triangulation technique is used. By the use of this technique, human observers over a period of days can correlate the location of the observed odor and the direction of the wind. Plotting on a map can pinpoint the offending source in most cases. While this is not an air pollution–monitoring system in an instrumental sense, it is a useful tool in certain situations.

Research Needs: The research needs of air pollution call for a completely different approach to air quality monitoring. Here, the purpose is to define some known variable or combination of variables. This can be either a new atmospheric phenomenon or the evaluation of a new air pollution sensor. In the former case, the most important consideration is the proper operation of the instrument used. In the latter case, the most important factor is the availability of a reference determination against which the results of the new instruments can be compared.

Urban Area Monitoring

Urban situations are the areas of major interest in air pollution in the United States, since most of the population lives in urban areas. The most sophisticated and expensive air quality–monitoring systems in the United States are those of large cities (and one or two of the largest states) where data collection and analysis are centralized at a single location through the use of telemetry. Online computer facilities provide data reduction.

Three philosophies can be utilized in the design of an urban air quality–monitoring system: (1) using location of sensors on a uniform area basis (rectilinear grid), (2) installing location of sensors in areas where pollutant concentrations are expected to be high, and (3) installing location sensors in proportion to population distribution. The operation of these systems is nearly identical, but the interpretation of the results can be radically different.

The most easily designed systems are those where the sensors are located uniformly on a geographical basis according to a rectilinear grid. Adequate coverage of an urban area frequently requires at least 100 sensors, so this concept is usually applied only with static or manual methods of air quality monitoring.

The location of air quality sensors at points of maximum concentration indicates the highest levels of air pollutants that are encountered throughout the area. Typically, this will include the central business district and the industrial areas on the periphery of the community. Data of this type are extremely useful when interpreted in the context of total system design. One or two sensors are usually placed in clean or background locations, so the average concentration of air pollutants over the entire area can be estimated. The basis of this philosophy is that if the concentrations in the dirtiest areas are below the air quality standards, there will certainly be no problem in the cleaner areas.

The design of air quality–monitoring systems based on population distribution calls for the placement of air quality sensors in the most populous areas. While this may not include all the high-pollutant-concentration areas in the urban region, it generally encompasses the central business district and is a measure of the levels of air pollutants to which the bulk of the population is exposed. The average concentrations from a sampling network of this type are an adequate description from a public health standpoint. However, this system design may miss some localized high-concentration areas.

Agreement on the purpose of air quality monitoring must be reached between the system designer and those responsible for interpreting the data before the system is designed.

Sampling Site Selection

Once the initial layout has been developed for an air quality–monitoring system, specific sampling sites must be located as close as practical to the ideal locations. The major considerations are the lack of obstruction from local interference and the adequacy of the site to represent the air mass of interest, accessibility, and security.

Local interference can cause major disruptions to air quality sensor sites. A sampler inlet placed at a sheltered interior corner of a building is most undesirable because of poor air motion. Tall buildings or trees immediately adjacent to the sampling site can also invalidate most readings.

The selection of sampling sites in urban areas is complicated by the canyon effect of streets and by the high density of pollutants, both gaseous and particulate, immediately at street level. In order to be representative of the air mass to be sampled, the purpose of the study must again be reviewed. If the data are to be collected for the determination of area-wide pollutant averages, it may be best to locate the sampler inlet in a city park, vacant lot, or other completely open area. If this is not possible, the sampler inlet could be at the roof level of a one- or two-storey building so that street-level effects could be minimized. On the other hand, if the physiological impact of the air pollutants is a prime consideration, the samplers should be at or near the breathing level of the people exposed. As a rule of thumb, an elevation of 3–6 m aboveground is suggested as optimum.

Sampling site location can be different for the same pollutant depending upon the purpose of the sampling. Carbon monoxide sampling is an example. The federally promulgated air quality standards for carbon monoxide include both an 8 and 1 hr concentration limit. Maximum 1 hr concentrations are likely to be found in a high-traffic-density center-city location. It is unlikely that people would ordinarily be exposed to these concentrations for 8 hr periods. When sampling for comparison with the 1 hr standard, the sensor should be located within about 20 ft of a major traffic intersection. When sampling for comparison with the 8 hr standard, the sensor should be located near a major thoroughfare in either the center-city area or the suburban area, with the sample less than about 50 ft from the intersection.

The purpose for two different locations of sampling sites is to be consistent with the physiological effects noted from carbon monoxide exposure and with the living pattern of the bulk of the population. If only one site can be selected, the location described for the 8 hr averaging time is preferred.

When the sampling instruments are located inside a building and an air sample is to be drawn in from the outside, it is frequently advantageous to utilize a sampling pipe with a small blower to bring outside air to the instrument inlets. This improves sampler line response time. An air velocity of approximately 700 ft/min in the pipe is a good choice to balance the problems of gravitational and inertial deposition of particular matter, where particulates are to be sampled.

The sampling site should be accessible to the operation and maintenance personnel. Since most air pollution–monitoring sites are unattended much of the time, sample site security is a very important consideration; the risk of vandalism is high in many areas.

MONITORING METHODS

Static Monitoring

Minimum capital cost is attained when static sensors are used to monitor air quality. While averaging times are in terms of weeks and sensitivity is frequently low, there are many cases where static monitors provide the most information for the amount of investment. They should not be rejected out of hand, but should be considered as useful adjuncts to more sophisticated systems.

Dust-Fall Jars: The simplest of all air quality–monitoring devices is the dust fall jar (Figure 1.4a). This device measures the fallout rate of coarse particulate matter, generally above about 10 μm in size. Dust fall and odor are two of the major reasons for citizen complaints concerning air pollution. Dust fall is offensive, because it builds up on porches, automobiles, and such, and it is highly visible and gritty to walk upon.

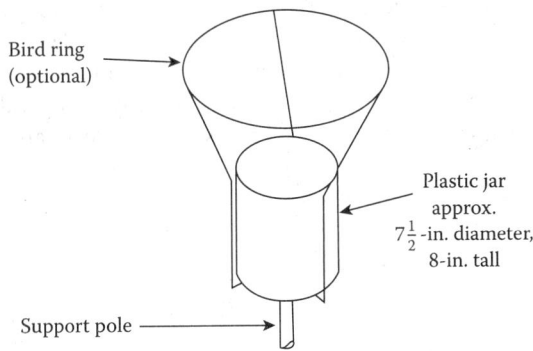

FIG. 1.4a
Dust fall jar.

Dust fall seldom carries for distances in excess of 1/2 mile, because these large particles are subject to strong gravitational effects. For this reason, dust fall stations must be more closely spaced than other air pollution sensors, if detailed study of an area is desired.

Dust measurements conducted in large cities in the United States in the 1920s and 1930s commonly indicated dust fall rates in hundreds of tons per square mile per month. These levels would be considered excessive today, as evidenced by the dust fall standards of 25–30 tons/mile^2/month promulgated by many of the states. While the measurement of low or moderate values of dust fall does not indicate freedom from air pollution problems, measured dust fall values in excess of 50–100 tons/mile²/month in an area are a sure indication of the existence of excessive air pollution.

The large size of the particulate matter found in dust fall jars makes it amenable to chemical analysis or to physical analysis by such techniques as microscopy. These analyses can be useful to identify specific sources.

Lead Peroxide Candles: For many years, lead peroxide candles had been used in measuring the concentration of sulfur dioxide. These devices are known as candles, because they are a mixture of lead peroxide paste spread on a porcelain cylinder that is about the size and shape of a candle. They are normally exposed to the ambient air in the location of interest for 1 month. Sulfur gases in the air react with lead peroxide to form lead sulfate. Sulfate is analyzed according to standard laboratory procedures to give an indication of the atmospheric levels of sulfur gases during the period of exposure.

The laboratory procedure can be simplified by a modification of this technique, which uses a fiber filter cemented to the inside of a plastic Petri dish (a flat-bottom dish with shallow walls, used for biological cultures). The filter in the dish is saturated with an aqueous mixture of lead peroxide and a gel (commonly gum tragacanth) and is allowed to dry. These dishes, or plates, are exposed to the ambient air in an inverted position for 1–4 weeks.

Lead peroxide estimation of sulfur dioxide suffers from inherent weakness. All sulfur gases, including reduced sulfur, react with lead peroxide to form lead sulfate. More importantly, the reactivity of lead peroxide is dependent upon its particle size distribution. For this reason, the results from different investigators are not directly comparable. Nevertheless, a network of lead peroxide plates over an area provides a good indication of the relative exposure to sulfur gases during the particular exposure period. This is useful for determining the geographical extent of sulfur pollution.

Other Static Methods: The use of a fiber filter cemented to a Petri dish has been modified by the use of sodium carbonate rather than lead peroxide for the measurement of sulfur gases. This method also shows some success in indicating the relative concentration of other gases, including nitrogen oxides and chlorides. Relative levels of gaseous fluoride air pollution have been measured using larger filters, for example, 3 in. in diameter, dipped in sodium carbonate, and placed in shelters to protect them from the rain. With all these static methods, the accuracy is extremely low and the data cannot be converted directly into ambient air concentrations. They do, however, provide a low-cost indicator of relative levels of pollution in a given area.

The corrosive nature of the atmosphere has been evaluated using standardized steel exposure plates for extended periods to measure the corrosion rate. This provides a gross indication of the corrosive nature of the atmosphere. As with other static samplers, the results are not directly related to the concentration of air pollutants.

Laboratory Analyses

Manual analyses for air quality measurements are those that require the sample to first be collected and then analyzed in the laboratory. Manual instruments provide no automatic indication of pollution levels.

The manual air sampling instrument, which is in widest use, is the high-volume sampler shown in Figure 1.4b.

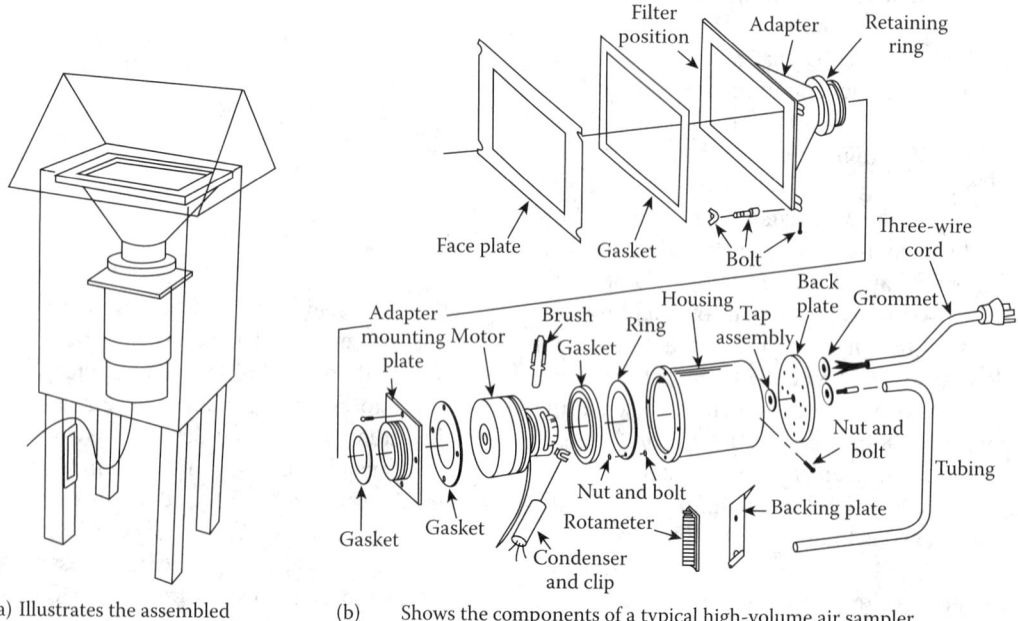

(a) Illustrates the assembled sampler and its shelter.

(b) Shows the components of a typical high-volume air sampler.

FIG. 1.4b
The high-volume air sampler.

With this method, ambient air is drawn through a preweighed filter at a rate of approximately 50 ACFM for 24 hr. The filter is then removed from the sampler, returned to the laboratory, and weighed. The gain in weight, in combination with the measured air volume through the sampler, allows for the calculation of particulate mass concentration, expressed in micrograms per cubic meter.

Reference methods for nearly all gaseous air pollutants involve the use of a wet sampling train in which air is drawn through a collecting medium for some period of time. The exposed collecting medium is then returned to the laboratory for chemical analysis. Sampling trains have been developed that allow the sampling of five or more gases simultaneously in separate bubblers. There are also sequential samplers, which automatically divert the airflow from one bubbler to another at preset time intervals.

These methods of sampling can be accomplished with a modest initial investment, but the manpower required to distribute and pick up the samples and to analyze them in the laboratory raises the total cost to a point where automated systems may be more economical for long-term studies.

Automatic Monitoring

As the need for accurate data that can be statistically reduced in a convenient manner becomes more the rule than the exception, automated sampling systems become more and more desirable. The elements of an automated system include the airflow handling system, the sensors, the data transmission storage and display apparatus, and the data processing facility. The overall system is no more valuable than the weakest link of this chain.

Sensors

Some of the sensor categories used have been listed in Table 1.4a. The detectors that are available to measure the various individual air pollutants are discussed in the chapters that follow.

The reliability of the output from an air quality sensor is dependent upon its inherent accuracy, sensitivity, zero drift, and calibration. The inherent accuracy and sensitivity are a function of the design of the instrument and of the principle upon which it operates. Zero drift can be either an electronic phenomenon or an indication of difficulties with the instrument. In those instruments using an optical path, lenses can become dirty. In wet chemical analyzers, the flow rates of reagents can vary, causing a change in both the zero and the span (range) of the instrument. Because of these potential problems, every instrument should have routine field calibrations at intervals determined by in-field practices to be reasonable for the sensor.

The calibration of an air quality sensor is frequently accomplished using either a standard gas mixture or a prepared, diluted gas mixture using permeation tubes. In some cases, the airstream entering the sensor is concurrently sampled by a reference wet chemical technique.

The operator of air quality sensors should always have an adequate supply of spare parts and tools so that downtime can be held to a minimum. Operator training should, at a minimum, include instruction to recognize the symptoms of equipment malfunction and vocabulary to describe the symptoms to the individual responsible for instrument repair. Ideally, each operator would receive a short training session from the instrument manufacturer or someone trained in the use and maintenance of the instrument, so that he or she can make repairs on-site. Since this is seldom possible in actual practice, the recognition of symptoms of malfunction becomes increasingly important.

Data Transmission

The output signal from a continuous monitor used in an air quality–monitoring system is typically the input to a strip-chart recorder, magnetic tape data storage, or an online data transmission bus or highway system. If the output of the air quality transmitter is in analog form, it is suitable for direct input to a strip-chart recorder, but it has to be converted into a digital signal before it can be sent over a data highway. In the case of sensors with a logarithmic output, it may be desirable to first convert this signal into a linear form.

Many of the first automated air quality–monitoring systems in the United States experienced major difficulties with their data transmission systems. In some cases, this was caused by attempts to overextend the lower range of the sensors, which resulted in an unfavorable signal-to-noise ratio. In other cases, the matching of sensor signal output to the data transmission system was poor. With developments of the new data highway protocols, these early difficulties have been overcome.

Continuous detectors tend to increase the complexity of the air quality–monitoring systems, but data from the continuous monitors can be stored on magnetic tape for later processing and statistical reduction. In case of an online system, this can be done instantaneously. When decisions with substantial community impact must be made within a very short time, this real-time capability is likely to be necessary.

Data Processing

The concentration of many air pollutants has been found to follow a lognormal rather than a normal distribution. In a lognormal distribution, a plot of the logarithm of the measured value more closely approximates the bell-shaped Gaussian distribution curve than does a plot of the numerical data. Suspended particulate concentrations are a prime example of this type of distribution.

When this is the case, the geometric mean is the statistical parameter that best describes the population of data. The arithmetic mean is of limited value, because it is dominated by a few occurrences of high values. The geometric mean, in combination with the geometric standard deviation, is a complete description of a frequency distribution for a lognormally distributed pollutant.

The averaging time over which the sample results are reported is a consideration in the processing and interpretation of air quality data. For sulfur dioxide, air quality standards have been promulgated by various agencies based on annual, monthly, weekly, 24 hr, 3 hr, and 1 hr arithmetic average concentrations.

The output of a continuous analyzer can be averaged over nearly any discrete time interval. In order to reduce the computation time, the time interval over which the continuous analyzer output is averaged to obtain a discrete input for the calculations must be considered. If a 1 hr average concentration is the shortest time interval of value in interpreting the study results, it is not economical to use a 1 or 2 min averaging time for inputs to the computation program.

Caution must be exercised in using strip-chart recorders for the acquisition of air quality data. The experience of many organizations, both governmental and industrial, is that the reduction of data from strip charts is a tedious chore at best. Many organizations have decided that they did not really need all that data in the first place, once they find a large backlog of unreduced strip charts. Two cautions are suggested by this experience. First, if you do not plan to use the data, do not collect it. Second, be aware of the advantages of magnetic tape data storage followed by computer processing.

The visual display of air quality data has considerable appeal to many nontechnical personnel. Long columns of numbers can be deceptive if only one or two important trends are to be shown. The use of bar charts or graphs is frequently advantageous, even though they do not show the complete history of the air quality over the time span of interest.

Audits

Periodic performance audits are required to validate the accuracy of the air monitoring system. The Code of Federal Regulations (CFR) requires that performance audits be conducted at least once a year for criteria pollutant analyzers operated at State and Local Air Monitoring Stations (SLAMS). The U.S. Environmental Protection Agency (EPA) recommends that each analyzer be disconnected from the monitoring station manifold and be individually connected to the audit, from which it will receive the audit gas of known concentration.

The audit gas concentrations are usually generated in a van, using a gas calibrator to dilute multiblend gases with zero air. Each state approaches air pollution monitoring differently. The California Air Resources Board (CARB), for example, has been conducting through-the-probe performance audits of continuous ambient air analyzers since 1981. CARB has the responsibility of overseeing the implementation of the California Clean Air Act and the Federal Clean Air Act in California.

Automatic Analyzers

The automatic sampling train (AST) packages have been described in Chapter 1.2. They can automate any of the U.S. EPA methods or international methods specified by VDI, BS, or ISO. In these packages, the microprocessor stores all measurements, diagnoses all inputs, controls the manipulated variables, calculates isokinetic conditions, and reports the results in either a printed form or over the data bus.

Infrared Spectrometers: Infrared and near infrared analyzers are discussed in detail in Chapter 1.33. Microprocessor-controlled spectrometers are also available to measure the concentrations of a variety of gases and vapors in ambient air. These units can be portable or permanently installed and can serve compliance with environmental and occupational safety regulations. In the infrared spectrometer design, the ambient air is drawn into the test cell by an integral air pump, operating at a flow rate of 0.88 ACFM (25 l/min).

The microprocessor selects the appropriate wavelengths for the components of interest, and the filter wheel in the analyzer allows the selected wavelengths to pass through the ambient air sample in the cell. The microprocessor automatically adjusts the path length through the cell to give the required sensitivity. Because of the folded path length design, the path length can be increased to 20 m (60 ft), and the resulting measurement sensitivity can be better than 1 ppm in many cases.

As shown in Table 1.4c, practically all organic and also some of the inorganic vapors and gases can be monitored by these infrared spectrometers. The advantage of the microprocessor-based operation is that the monitor is precalibrated for the analysis of over 100 Occupational Safety and Health Administration (OSHA)-cited compounds. The memory capacity of the microprocessor is sufficient to accommodate another 10 user-selected and user-calibrated gases.

Analysis time is minimized, because the microprocessor automatically sets the measurement wavelengths and parameters for any of the compounds in its memory. A general scan for contaminant in the atmosphere takes about 5 min, while the analysis of a specific compound can be completed in just a few minutes. The portable units are battery operated for 4 hr of continuous operation and are approved for use in hazardous areas.

TABLE 1.4c
Compounds That Can Be Analyzed by the Microprocessor-Controlled Portable Infrared Spectrometer

Compound	Range of Calibration (ppm)	Compound	Range of Calibration (ppm)
Acetaldehyde	0–400	Ethane	0–1000
Acetic acid	0–50	Ethanolamine	0–100
Acetone	0–2000	Fluorotrichloromethane (Freon 11)	0–2000
Acetonitrile	0–200	Formaldehyde	0–20
Acetophenone	0–100	Formic acid	0–20
Acetylene	0–200	Halothane	0–20
Acetylene tetrabromide	0–200	Heptane	0–10 and 0–100
Acrylonitrile	0–20 and 0–100	Hexane	0–1000
Ammonia	0–100 and 0–500	Hydrazine	0–1000
Aniline	0–20	Hydrogen cyanide	0–100
Benzaldehyde	0–500	Isoflurane	0–10 and 0–100
Benzene	0–50 and 0–200	Isopropyl alcohol	0–1000 and 0–2000
Benzyl chloride	0–100	Isopropyl ether	0–1000
Bromoform	0–10	Methane	0–100 and 0–1000
Butadiene	0–2000	Methoxyflurane	0–10 and 0–100
Butane	0–200 and 0–2000	Methyl acetate	0–500
2-Butanone (MEK)	0–250 and 0–1000	Methyl acetylene	0–1000 and 0–5000
Butyl acetate	0–300 and 0–600	Methyl acrylate	0–50
n-Butyl alcohol	0–200 and 0–1000	Methyl alcohol	0–500 and 0–1000
Carbon dioxide	0–2000	Methylamine	0–50
Carbon disulfide	0–50	Methyl bromide	0–50
Carbon monoxide	0–100 and 0–250	Methyl cellosolve	0–50
Carbon tetrachloride	0–20 and 0–200	Methyl chloride	0–200 and 0–1000
Chlorobenzene	0–150	Methyl chloroform	0–500
Chlorobromomethane	0–500	Methylene chloride	0–1000
Chlorodifluoromethane	0–1000	Methyl iodide	0–40
Chloroform	0–100 and 0–500	Methyl mercaptan	0–100
m-Cresol	0–20	Methyl methacrylate	0–250
Cumene	0–100	Morpholine	2–50
Cyclohexane	0–500	Nitrobenzene	0–20
Cyclopentane	0–500	Nitromethane	0–200
Diborane	0–10	Nitrous oxide	0–100 and 0–2000
m-Dichlorobenzene	0–150	Octane	0–100 and 0–1000
o-Dichlorobenzene	0–100	Pentane	0–1500
p-Dichlorobenzene	0–150	Perchloroethylene	0–200 and 0–500
Dichlorodifluoromethane (Freon 12)	0–5 and 0–800	Phosgene	0–5
1,1-Dichloroethane	0–200	Propane	0–2000
1,2-Dichloroethylene	0–500	n-Propyl alcohol	0–500
Dichloroethyl ether	0–50	Propylene oxide	0–200
Dichloromonofluoroethane (Freon 21)	0–1000	Pyridine	0–100
Dichlorotetrafluoromethane (Freon 114)	0–1000	Styrene	0–200 and 0–500
Diethylamine	0–50	Sulfur dioxide	0–100 and 0–250
Dimethylacetamide	0–50	Sulfur hexafluoride	0–5 and 0–500
Dimethylamine	0–50	1,1,2,2-Tetrachloro 1,2-difluoroethane (Freon 112)	0–2000
Dimethylformamide	0–50	1,1,2,2-Tetrachloroethane	0–50
Dioxane	0–100 and 0–500	Tetrahydrofuran	0–500
2-Ethoxyethyl acetate	0–200	Toluene	0–1000
Ethyl acetate	0–400 and 0–1000	Total hydrocarbons	0–1000
Ethyl alcohol	0–1000 and 0–2000	1,1,2-Trichloroethane	0–50
Ethylbenzene	0–200	Trichloroethylene	0–200 and 0–2000

(Continued)

TABLE 1.4c (Continued)
Compounds That Can Be Analyzed by the Microprocessor-Controlled Portable Infrared Spectrometer

Compound	Range of Calibration (ppm)	Compound	Range of Calibration (ppm)
Ethyl chloride	0–1500	1,1,2-Trichloro 1,2,2-trifluoroethane (Freon 113)	0–2000
Ethylene	0–100	Trifluoromonobromomethane (Freon 13B1)	0–1000
Ethylene dibromide	0–10 and 0–50	Vinyl acetate	0–1000
Ethylene dichloride	0–100	Vinyl chloride	0–10
Ethylene oxide	0–10 and 0–100	Vinylidene chloride	0–20
Ethyl ether	0–1000 and 0–2000	Xylene (xylol)	0–200 and 0–2000
Enflurane	0–10 and 0–100		

Handheld Indoor Air Quality Monitors: Battery-operated, handheld indoor air quality (IAQ) monitors are available for monitoring the air quality in schools, offices, meeting rooms, and greenhouses, and to service heating, ventilation, and air conditioning (HVAC) systems. These units monitor and record the temperature (range: 0°C–50°C = 32°F–122°F), relative humidity (range: 0%–100%), and the concentration of different selected gases. Plug-in sensors are available for gases, such as CO_2, with a range of 0–10,000 ppm.

Dual-beam infrared radiation is used for making the measurement. The sample flow is drawn by diffusion or is maintained at 100–300 ml/m. Sample rates are adjustable from 10 sec to one per day, and data loggers are also available with capacities for up to 50,000 samples.

SAMPLING

The sampling of process gases and vapors has been discussed in Chapter 1.2, and some of that discussion is also applicable to ambient air monitoring. Yet ambient air sampling is sufficiently different to justify a separate discussion in addition to what has already been said.

All substances in the ambient air exist as either particulate matter or gases and vapors. In general, the distinction is easily made; gases and vapors consist of substances dispersed as molecules in the atmosphere, while particulate matter consists of aggregates of molecules sufficiently large that they are said to behave like particles. Particulate matter (or particulates) are filterable, may be precipitated, and, in still air, may be expected to settle out. By contrast, gases and vapors do not behave in this fashion and are homogeneously mixed with the air molecules.

A substance such as carbon monoxide may exist only as a gas; an inorganic compound like iron oxide may exist only as a particle. Many substances may exist as either particles or vapors, however; additionally, substances that are gases can also be attached by some means to the particulate matter in the air. If sampling is to be intelligently conducted, prior knowledge of the physical state in which a substance exists must be available or else a judgment must be made. Particulate matter, opacity, dust, and smoke analyzers are discussed in a separate section in this chapter. They do not usually collect gases or vapors; hence, an incorrect selection of sampling method may lead to erroneous results. Fortunately, a considerable body of experience has evolved concerning the more common pollutants, and it is not difficult in most cases to select a sampling method reasonably suitable for the substances of interest.

Sampling Problems

Certain general observations relative to sampling ambient air must be recognized. For example, the quantity of a given substance contained in a volume of air is likely to be extremely small, and it will necessitate a sample of sufficient size for the analytical method employed to be adequate. Even air, which is heavily polluted, is not likely to contain more than a few milligrams per cubic meter of most contaminants and, more frequently, the amount present will best be measured in micrograms, or even nanograms, per cubic meter.

Consider, for example, the air quality standard for particulates, which is 75 g/m³. A cubic meter of air, or 35.3 ft³, is a relatively large volume for many sampling devices, and a considerable sampling period may be required to draw such a quantity of air through the sampler. If atmospheric mercury analyses are to be attempted, then it must be realized that background levels are likely to be as low as several nanograms per cubic meter and, in general, most substances tend to be of concern at quite low levels in the ambient air.

In addition to problems arising from the low concentrations of substances being sampled, other problems include those caused by the reactivity of the substances, changes after collection, and necessitating special measures to minimize such changes.

Whenever something is removed from a volume of air by sampling procedures, some alteration of the substance of interest may take place and analysis may be less informative than desired, or may even be misleading. Therefore, it would be ideal to perform analyses of the unchanged atmosphere if possible, using direct-reading devices, which could give accurate information concerning the chemical and physical states of contaminants as well as concentration information. Such instruments do exist for some substances, and many more are being developed, but conventional air sampling methods are still used in many instances and doubtlessly will continue to be required for some time to come.

GAS AND VAPOR SAMPLING

The simplest method of collecting a sample of air for subsequent analysis is to fill a bottle or other rigid container with it or, more conveniently, to use a bag made of suitable material. Although it may be relatively easy to sample by this method, the sample size is distinctly limited, and it may not be possible to collect a sufficiently large sample to permit subsequent analysis. Bottles larger than several liters in capacity are awkward to transport, and while bags of any size are conveniently transported when empty, they may be difficult to deal with when inflated. Nevertheless, it may prove more convenient to collect a number of samples in small bags than to take more complex sampling apparatus to a number of sampling sites. If it is possible to analyze for the contaminant of interest by means of gas chromatographic procedures, or by gas phase infrared spectroscopy, for example, then samples as small as a liter or less may be adequate and can be quickly and easily collected by bags.

There are several methods of filling rigid containers such as a bottle. One is to evacuate the bottle beforehand and then fill it at the sampling site by drawing air into the bottle and sealing it again (Figure 1.4c). Alternatively, a bottle may be filled with water, which is then allowed to drain and fill with the air. A third method consists of passing a sufficient amount of air through the bottle by using a pumping device until the original air is completely displaced by the air being sampled.

Bag Sampling

Plastic bags are frequently filled by means of a simple, hand-operated squeeze bulb with values on each end (Figure 1.4c) and then connected to a piece of tubing attached to the sampling inlet of the bag. In most cases, this procedure is satisfactory, but care should be taken to avoid contamination of the sampled air by the sampling bulb or possible losses of the constituent on the walls of the sampling bulb. To avoid problems of this kind, it is possible to place the bag in a rigid container such as a box and then withdraw air from the box so that a negative pressure is created, which results in air being drawn into the bag.

Bag materials must be selected with care, for some will permit losses of contaminants by diffusion through the walls, and others to the air being sampled. A number of polymers have been studied, and several have been found to be suitable for air sampling purposes. Materials suitable for use as sampling bags include Mylar, Saran, Scotchpak (a laminate of polyethylene, aluminum foil, and teflon), and teflon.

Even though the bag may be made of relatively inert materials, it is always possible that gas phase chemical reactions will occur, so that after a period of time, the contents of the bag may not be identical in composition with the air originally sampled. Thus, a reactive gas like sulfur dioxide or nitric oxide can be expected to gradually oxidize, depending upon the storage temperature. It is generally advisable to perform analyses as soon as possible after collecting the samples. Losses by adsorption or diffusion will also tend to be greater with the passage of time and will occur to some extent even though the best available bag materials have been used.

The use of small bags may permit the collection of samples to be analyzed for a relatively stable gas such as carbon monoxide at a number of locations throughout a community of interest, thus permitting routine air quality measurements that might otherwise be inordinately expensive.

Absorption: The gases or vapors of interest can also be absorbed in a suitable sampling medium. Ordinarily, this medium is a liquid of some kind, but absorption may also take place in solid absorbents or upon supporting materials such as filter papers, which are impregnated with suitable absorbents. Carbon dioxide, for example, may be absorbed in a granular bed of alkaline material, and sulfur dioxide is frequently measured by absorption of reactive chemicals placed on a cloth or ceramic support.

A number of gases are also detected by passing them through filter papers or glass tubes containing reactive chemicals, with the immediate production of a color change, which can be evaluated by eye to give a measure of the concentration of the substance of interest.

Most commonly, however, gases and vapors are absorbed by passing them through a liquid in which they are soluble,

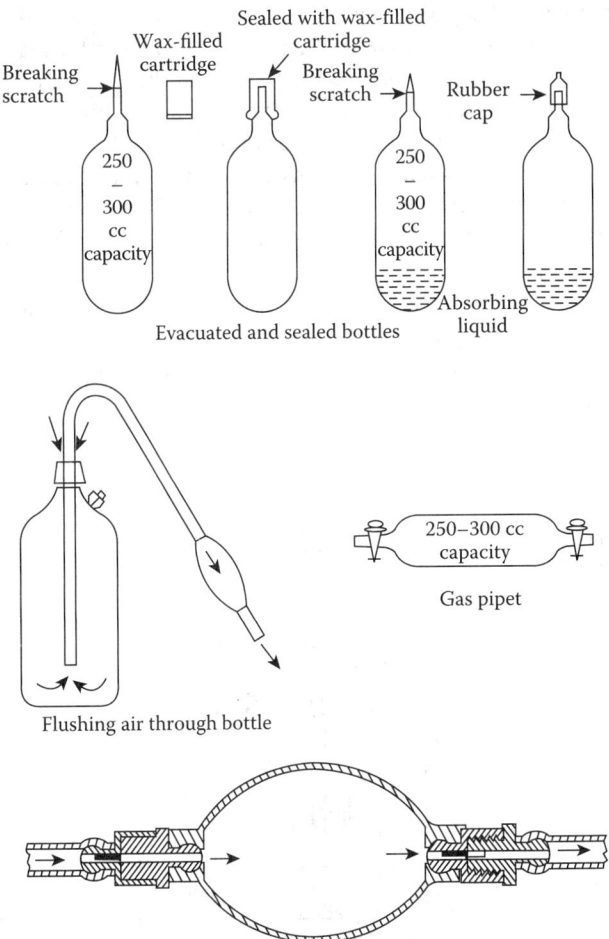

FIG. 1.4c
Devices for obtaining grab samples.

or which contains reactive chemicals that will combine with the substance being sampled. Many different absorption vessels have been designed, ranging from simple bubblers made by inserting a piece of tubing beneath the surface of a liquid to rather complex gas-washing devices that are designed to increase the length of time that the air and liquid are in contact with each other (Figure 1.4d).

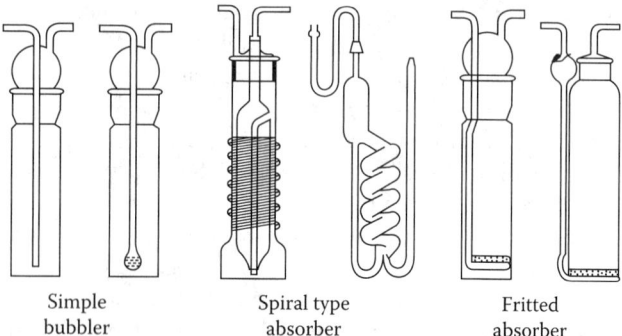

Simple bubbler   Spiral type absorber   Fritted absorber

FIG. 1.4d
Gas-absorbing vessels.

Impingers: Probably the most widely used contacting device is the impinger, which is available in several sizes and configurations. An impinger consists of an entrance tube terminating in a small orifice, causing the velocity of the air passing through the orifice to greatly increase. When this jet of air strikes a plate or the bottom of the sampling vessel, an intense impingement or bubbling action occurs, which results in much more efficient absorption of gases from the airstream than would take place if the air was simply bubbled through at low velocity.

The two impingement devices most frequently used are the standard impinger and the midget impinger (Figure 1.4e). They are designed to operate best at airflow rates of 1 and 0.1 ft^3/min, or 28.3 and 2.8 l/min, respectively. Using such devices for sampling periods of 10 or 30 min will result in the passage of a substantial amount of air through the devices, thus permitting low concentrations of trace substances to be determined with improved sensitivity and accuracy. Many relatively insoluble gases, such as nitrogen dioxide, are not quantitatively removed by passage through an impinger containing the usual sampling solutions.

Fritted Absorber: The most useful sampling devices for absorbing trace gases from air are those in which a gas dispersion tube made of fritted or sintered glass, ceramic, or other materials is immersed in a suitable vessel containing the absorption liquid (Figure 1.4d). This device causes the gas stream to be broken into thousands of small bubbles, thus promoting contact between the gas and the liquid with resulting high collection efficiencies in most cases.

In general, fritted absorbers are more widely applicable to sampling gases and vapors than are impingers and, in addition, are not as dependent upon flow rates as are impingers. They are standard items available from scientific supply houses and come in various sizes suitable for many sampling tasks. It is often advisable to prefilter the air prior to sampling with a fritted absorber in order to prevent the gradual accumulation of dirt within the pores of the frit.

The use of solid absorbents is not widely practiced in ambient air sampling, because the quantity of absorbed gases is usually determined by gravimetric means. With the

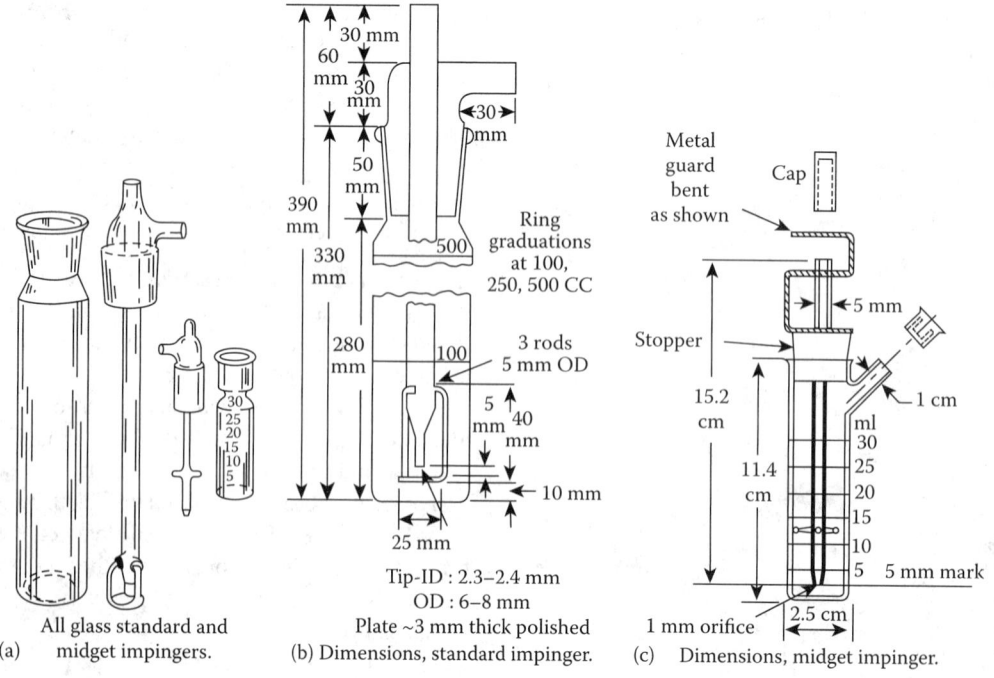

(a) All glass standard and midget impingers.   (b) Dimensions, standard impinger.   (c) Dimensions, midget impinger.

FIG. 1.4e
Standard and midget impingers.

exception of carbon dioxide, relatively few gases of interest in the atmosphere lend themselves to this type of analysis.

Adsorption: Adsorption, by contrast with absorption, consists of the retention of gaseous substances by solid adsorbents that, in most cases, do not chemically combine with the gases. Instead, the gases or vapors are held by adsorptive forces and may subsequently be removed unchanged. Any solid substance will adsorb a small amount of most gases, but to be useful as an adsorbent, a substance must have a large surface area and be able to concentrate a substantial amount of gas in a small volume of adsorbent.

Most widely used for this purpose are activated carbon or charcoal and activated silica gel. A small quantity of either adsorbent placed in a U-tube or other container through which air is passed will quantitatively remove many vapors and gases from a large volume of air and may subsequently be taken to the laboratory where desorption will remove the collected substance for analysis. Desorption is commonly achieved by heating the adsorbent and collecting the effluent gases or by eluting the collected substance with a suitable organic liquid.

In the case of most organic vapors, subsequent analysis by gas chromatographic, infrared, or ultraviolet spectroscopic means is usually most convenient. For some purposes, either silica gel or activated carbon may be used. In other instances, the use of silica gel is undesirable, because it also adsorbs water vapor, and a relatively short sampling period in humid air can result in saturation of the silica gel before sufficient contaminant is adsorbed. Charcoal does not adsorb water and hence may be used in humid environments and may be sampled for days or even weeks if the concentration of contaminant is low.

The ease of sampling by using adsorbents is offset somewhat by the difficulty of quantitatively desorbing the samples for analysis. When literature data are not available to assist in predicting the behavior of a new substance, it is advisable to perform tests in the laboratory to determine both the collection efficiency and the success of desorption procedures after sample collection.

Freeze-Out Sampling: Vapors or gases, which are condensable at low temperature, may be removed from the sampled airstream by passage through a vessel that is immersed in a refrigerating liquid. Table 1.4d lists liquids that are commonly used for this purpose, and usually, it is desirable to form a sampling train in which two or three coolant liquids

TABLE 1.4d
Coolant Solutions for Freeze-Out Sampling

Coolant	Temperature (°C)
Ice water	0
Ice and salt (NaCl)	−21
Dry ice and acetone	−78.5
Liquid air	−147
Liquid oxygen	−183
Liquid nitrogen	−196

progressively lower the air temperature in its passage through the system. All freeze-out systems are hampered somewhat by the accumulation of ice resulting from water vapor and may eventually become plugged with ice.

Flow rates through a freeze-out train are necessarily limited also; in order for the air to be cooled to the required degree, a sufficient residence time in the system must be provided. For these reasons, and because of the general inconvenience of assembling freeze-out sampling trains, they are not generally used for routine sampling purposes unless no other approach is feasible. However, freeze-out sampling is an excellent means of collecting substances for research studies, inasmuch as the low temperatures tend to arrest further chemical changes and ensure that the material being analyzed will remain in the sampling container ready for analysis after warming. Analysis is most frequently conducted by means of gas chromatographic, infrared, or ultraviolet spectrophotometry, or by mass spectrometric means.

PARTICULATE SAMPLING

Particulate matter is most conveniently removed from air by passage through a filter (Figure 1.4b). Before filtration is used to obtain a sample, however, consideration should be given to the purpose for which the sample is being taken. Many filters collect particulates efficiently, but thereafter, it may be impossible to remove the collected matter except by chemical treatment. If the samples were initially collected for the purpose of examining the particles and measuring their size, or noting morphological characteristics, then many filters are not suitable, because the particles are imbedded in the fibrous web of the filter and cannot readily be viewed or removed.

If the sample is collected for the purpose of performing a chemical analysis, then it is important that the filter itself does not contain significant quantities of the substance for which analysis is required. If the purpose of sampling is to collect a sufficient amount of particulate matter to permit weighing, then it is necessary to select a filter that can be weighed. This can be a problem, because many filtration materials are hygroscopic and change weight appreciably in response to changes in the relative humidity.

Filters may be made of many substances and, in fact, almost any solid substance could probably be made into a filter. In practice, however, fibrous substances, such as cellulose or paper, fabrics, asbestos, and a number of plastics or polymerized materials, are generally used. The most readily available filters are those made of cellulose or paper used in chemical laboratories for filtering liquids. Such filter papers come in a variety of sizes and range in efficiency from rather loose filters, which remove only the larger particles, to paper, which will remove very fine particles with high efficiency. All filters display similar behavior, and ordinarily, a high collection efficiency is accompanied by increased resistance to airflow.

Air Filters: Certain kinds of filtration media are more suited to air sampling than most paper or fibrous filters. Of these, membrane filters are of greatest utility. The commercially available membrane filters combine extremely high collection efficiencies with relatively low resistance to flow. Such filters are not made up of a fibrous mat, but instead are usually composed of gels of cellulose esters or other polymeric substances in such fashion that a smooth surface of predictable characteristics is formed.

Such filters contain many small holes, or pores, and may be made to exacting specifications in this regard so that their performance characteristics may be predicted. In addition, the filters are usually of high chemical purity and are well suited to trace metal analyses. Some of the membrane filters can also be rendered transparent and thus permit direct observation of collected particles with a microscope. Alternatively, the filters can be dissolved in an organic solvent, and the particles may be isolated and studied. Most membrane filters are not greatly affected by relative humidity changes and may be weighed before and after use to obtain reliable gravimetric data.

Fiberglass Filters: Another kind of filter that is widely used in sampling ambient air is the fiberglass filter. These filters are originally made of glass fibers in an organic binder; subsequently, the organic binder may be removed by firing, leaving a web of glass, which is very efficient in collecting fine particles from the air. The principal advantage of using this type of filter is its relatively low resistance to airflow and its virtually unchanging weight regardless of relative humidity.

The filters are not well suited to particle size studies, however; additionally, they are not chemically pure, and care must be taken to be certain that a substance for which analysis is to be made will not be contributed in unknown quantities by the filter itself. In the United States, most of the data relating to suspended particulate matter in our cities have been obtained on filters made of fiberglass and used in conjunction with a sampling device referred to as a high-volume sampler (Figure 1.4b).

Many other kinds of filters are available, but most sampling needs are well met by membrane or fiberglass filters. Extensive data concerning the types of filters available and their performance characteristics will be found in the bibliography cited at the end of this chapter.

Impingers: The impingers previously described in relation to sampling gases and vapors (Figure 1.4e) may also be used for the collection of particles and, in fact, they were originally developed for that purpose. However, in ambient air sampling, they are not used, inasmuch as the efficiency of collection tends to be low and unpredictable for the fine particles that may be present in ambient air. The relatively low sampling rates also make them less attractive than filters for general air sampling, but instances do arise when impingers can be satisfactorily used. When they are used, it is important that the correct sampling rates be maintained, for the collection efficiency of impingers for particles may vary drastically when flow rates are other than optimal.

Impactors: Of more widespread use in ambient air sampling are devices known as impactors, in which air is passed through small holes or orifices and made to impinge or impact against a solid surface. When such devices are constructed so that the air passing through one stage is subsequently directed onto another stage containing smaller holes, the resulting device is known as a cascade impactor and has the capability of separating particles according to their sizes. Various commercial devices are available.

Figure 1.4f portrays one that is widely used and consists of several layers of perforated plates through which the air must pass. Each plate contains a constant number of holes, but the hole size is progressively decreased so that the same volume of air passing through each stage will impinge at an increased velocity. The result is that coarse particles are deposited on the first stage and successively finer particles are removed at each subsequent stage. While the particle size fractions obtained by such instruments are not very accurate, they do perform predictably when the characteristics of the aerosol being sampled are known.

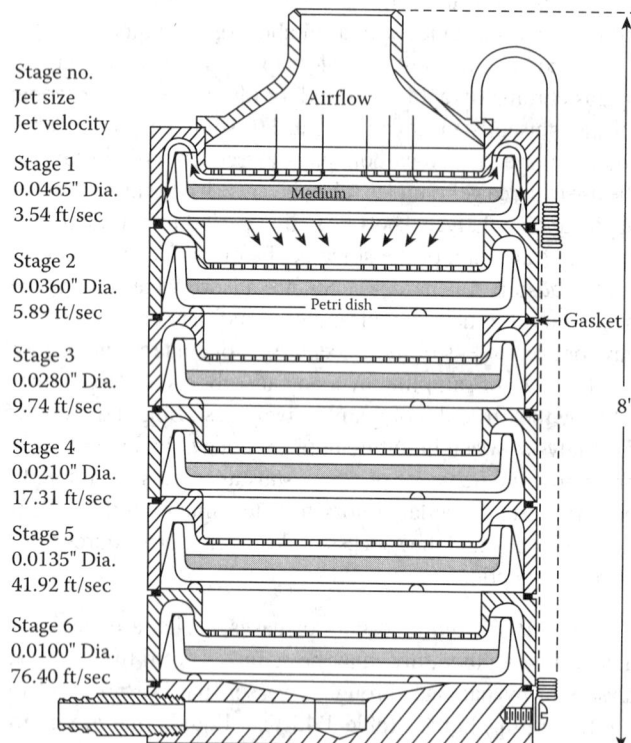

FIG. 1.4f
Cascade impactor. (Andersen sampler.)

In use, a cascade impactor is assembled after scrupulously cleaning each stage and applying a sticky substance or a removable surface on which the particles are to be deposited. After a suitably long period of sampling, during which time the volume of air is metered, the stages may be removed and the total weight of each fraction determined, as well as the fractions' chemical composition. Such information may be more useful than a single weight or chemical analysis of the total suspended particulate matter without regard to its particle size.

Electrostatic Precipitators: Particulate matter can be quantitatively removed from air by means of instruments known as electrostatic precipitators. They operate on the same principle as the devices used to remove particulate matter from stack gases prior to discharging into the atmosphere. There are several commercially available electrostatic precipitators that may be used for air sampling. They operate on the same general principle that passing the air between charged surfaces imparts a charge to the solid particles in the air. Therefore, these particles can be collected on an oppositely charged surface or plate.

In one of the more widely used commercial devices (Figure 1.4g), a high-voltage discharge is made to occur along a central wire; the collecting electrode is a metallic cylinder that is placed around the central wire, while the air to be sampled is passing through the tube. An intense corona discharge takes place on the central wire; the particles entering the tube are charged and are promptly swept to the walls of the tube where they remain firmly attached.

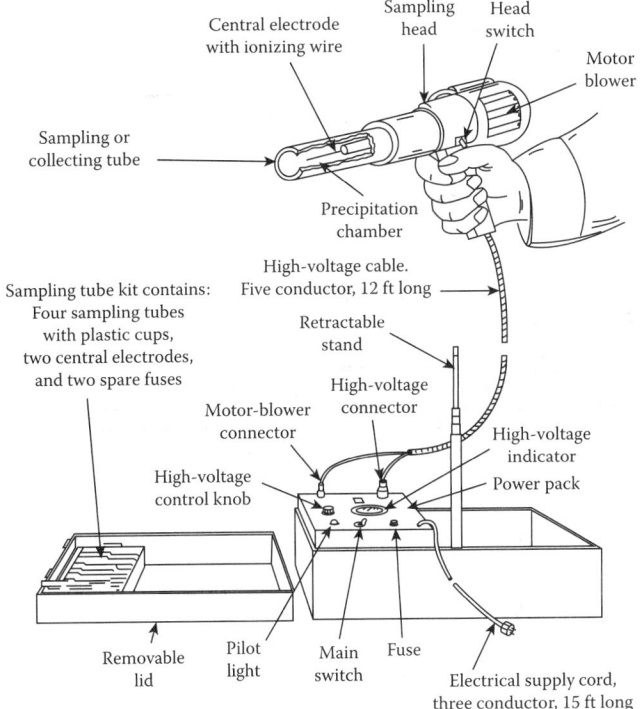

FIG. 1.4g
Electrostatic precipitator. (Courtesy of MSA Instrument Div.)

By this means, it is possible to collect a sample for subsequent weighing or chemical analysis. It is also possible to examine the particles and study their particle size and shape, and the intense electrical forces may produce aggregates of particles that are different than those that existed in the sampled air.

Electrostatic precipitators are not as widely used as filters for ambient air sampling, because they are generally less convenient and tend to be heavy due to the power pack necessary to generate the required high voltage. Nevertheless, when available, they are excellent instruments for obtaining samples for subsequent analysis, and samples at relatively high flow rates and very low resistance.

Thermal Precipitators: Whenever a strong temperature gradient exists between two adjacent surfaces, there will be a tendency for particles to be deposited on the colder of these surfaces. Collection of aerosols by this means is termed thermal precipitation and, in practice, several such commercial devices are available.

Because thermal forces are so weak, it is necessary to have a rather large temperature difference maintained in a small area. Additionally, the rate of airflow between the two surfaces must be low in order not to destroy the temperature gradient and to permit particles to be deposited before moving out of the collection area. As a result of these requirements, most devices use a heated wire as the source of the temperature differential and deposit a narrow ribbon of particles on the cold surface. Airflows are very small, being on the order of 10–25 ml/min. At such low rates, the amount of material collected will normally be insufficient for chemical analysis or weight determinations, but will be ample for examination by optical or electron microscopy.

Collection for particle size studies is, in fact, the principal use for thermal precipitation units, and they are well suited to collecting samples for such investigations. Because the collecting forces are gentle, it is generally believed that the particles are deposited unchanged. The microscopic examination gives information that can be translated into data concerning the number of particles and their morphological characteristics. It is convenient also to use a small grid suitable for insertion into an electron microscope as the collecting surface, thus making it unnecessary to perform additional manipulations of the sample prior to examination by electron microscopy.

SPECIFICATION FORMS

The air quality monitoring system can be specified by filling out the the specification forms, found on the next two pages, ISA 20A1001 and ISA 20A1002. These forms are reproduced with the permission of the International Society of Automation on the next pages.

Analytical Measurement

ANALYSIS DEVICE — Operating Parameters

#	RESPONSIBLE ORGANIZATION		#	SPECIFICATION IDENTIFICATIONS	
1	(ISA)		6	Document no	
2			7	Latest revision	Date
3			8	Issue status	
4			9		
5			10		

#	ADMINISTRATIVE IDENTIFICATIONS			#	SERVICE IDENTIFICATIONS Continued			
11				40				
12	Project number		Sub project no	41	Return conn matl type			
13	Project			42	Inline hazardous area cl		Div/Zon	Group
14	Enterprise			43	Inline area min ign temp		Temp ident number	
15	Site			44	Remote hazardous area cl		Div/Zon	Group
16	Area	Cell	Unit	45	Remote area min ign temp		Temp ident number	
17				46				

#	SERVICE IDENTIFICATIONS			#	COMPONENT DESIGN CRITERIA		
18				47			
19	Tag no/Functional ident			48			
20	Related equipment			49	Component type		
21	Service			50	Component style		
22				51	Output signal type		
23	P&ID/Reference dwg			52	Characteristic curve		
24	Process line/nozzle no			53	Compensation style		
25	Process conn pipe spec			54	Type of protection		
26	Process conn nominal size		Rating	55	Criticality code		
27	Process conn termn type		Style	56	Max EMI susceptibility		Ref
28	Process conn schedule no		Wall thickness	57	Max temperature effect		Ref
29	Process connection length			58	Max sample time lag		
30	Process line matl type			59	Max response time		
31	Fast loop line number			60	Min required accuracy		Ref
32	Fast loop pipe spec			61	Avail nom power supply		Number wires
33	Fast loop conn nom size		Rating	62	Calibration method		
34	Fast loop conn termn type		Style	63	Testing/Listing agency		
35	Fast loop schedule no		Wall thickness	64	Test requirements		
36	Fast loop estimated lg			65	Supply loss failure mode		
37	Fast loop material type			66	Signal loss failure mode		
38	Return conn nominal size		Rating	67			
39	Return conn termn type		Style	68			

#	PROCESS VARIABLES	MATERIAL FLOW CONDITIONS			#	PROCESS DESIGN CONDITIONS		
69					101			
70	Flow Case Identification			Units	102	Minimum	Maximum	Units
71	Process pressure				103			
72	Process temperature				104			
73	Process phase type				105			
74	Process liquid actl flow				106			
75	Process vapor actl flow				107			
76	Process vapor std flow				108			
77	Process liquid density				109			
78	Process vapor density				110			
79	Process liquid viscosity				111			
80	Sample return pressure				112			
81	Sample vent/drain press				113			
82	Sample temperature				114			
83	Sample phase type				115			
84	Fast loop liq actl flow				116			
85	Fast loop vapor actl flow				117			
86	Fast loop vapor std flow				118			
87	Fast loop vapor density				119			
88	Conductivity/Resistivity				120			
89	pH/ORP				121			
90	RH/Dewpoint				122			
91	Turbidity/Opacity				123			
92	Dissolved oxygen				124			
93	Corrosivity				125			
94	Particle size				126			
95					127			
96	CALCULATED VARIABLES				128			
97	Sample lag time				129			
98	Process fluid velocity				130			
99	Wake/natural freq ratio				131			
100					132			

#	MATERIAL PROPERTIES			#	MATERIAL PROPERTIES Continued		
133				137			
134	Name			138	NFPA health hazard	Flammability	Reactivity
135	Density at ref temp		At	139			
136				140			

Rev	Date	Revision Description	By	Appv1	Appv2	Appv3	REMARKS

Form: 20A1001 Rev 0 © 2004 ISA

1.4 Analyzer Sampling: Air Quality Monitoring

	RESPONSIBLE ORGANIZATION	ANALYSIS DEVICE COMPOSITION OR PROPERTY Operating Parameters (Continued)		SPECIFICATION IDENTIFICATIONS	
1			6		
2	ISA		7	Document no	
3			8	Latest revision	Date
4			9	Issue status	
5			10		

	PROCESS COMPOSITION OR PROPERTY			MEASUREMENT DESIGN CONDITIONS				
	Component/Property Name	Normal	Units	Minimum	Units	Maximum	Units	Repeatability
11								
12								
13								
...								
37								

Rev	Date	Revision Description	By	Appv1	Appv2	Appv3	REMARKS

Form: 20A1002 Rev 0 © 2004 ISA

Definitions

Particulate air pollution is defined by the US EPA as an air-suspended mixture of both solid and liquid particles. They are often separated into three classifications: coarse, fine, and ultrafine particles. Coarse particles have a diameter of between 10 µm and 2.5 µm and settle relatively quickly whereas fine (0.1–2.5 µm in diameter) and ultrafine (<0.1 µm in diameter) particles remain in suspension for longer. To put these sizes into perspective, human hair has a diameter of 50–70 µm and a grain of sand has a diameter of 90 µm.

PM2.5 refers to particles smaller than 2.5 µm. They include combustion particles, organic compounds, and metals. The World Health Organization (WHO) recommends that the PM2.5 exposure should not exceed 25 µg/m^3 mean in any 24 hr period and 10 µg/m^3 mean annually.

PM10 refers to particles smaller than 10 µm. They include dust, pollen, and mold spores. The World Health Organization (WHO) recommends that the PM10 exposure should not exceed 50 µg/m^3 mean in any 24 hr period and 20 µg/m^3 mean annually.

Total Suspended Particles (TSP) is an archaic regulatory measure of the mass concentration of particulate matter (PM) in air. It was a affected by the size selectivity of the inlet to the filter that collected the particles. The size cut varied with wind speed and direction and was from 20 to 50 µm in aerodynamic diameter. Under windy conditions the mass tended to be dominated by large wind-blown soil particles of relatively low toxicity.

Abbreviations

AAQS	Ambient Air Quality Standards
ACFM	Actual cubic feet per minute
AST	Automatic sampling train
CFR	Code of Federal Regulations
DOAS	Dedicated outdoor air system
FID	Flame ionization detector
HVAC	Heating, ventilation, and air conditioning
NAAQS	National Ambient Air Quality Standards
PM	Particulate matter
PM2.5	Particles smaller than 2.5 µm
PM10	Particles smaller than 10 µm
SLAMS	State and Local Air Monitoring Stations
TSP	Total suspended particles

Organizations

BS	British Standard
CARB	California Air Resources Board
ISO	International Standards Organization
OSHA	Occupational Safety and Health Administration
VDI	German Commission of Air Pollution Prevention, Verein Deutscher Ingenieure
WHO	World Health Organization

Bibliography

AMP, EPA method sampling system, (2015) http://www.ampcherokee.com/v/vspfiles/photos/C002.0002-2T.jpg.

California Environmental Protection Agency, Ambient air quality standards, (2013) http://www.arb.ca.gov/research/aaqs/aaqs.htm.

Cooper, C. D. and Alley, F. C., Air pollution, (2002) http://www.amazon.com/Coopers-Pollution-Control-Edition-Hardcover/dp/B003WUI3I6.

EPA, National ambient air quality standards, (2012) http://www3.epa.gov/ttn/naaqs/criteria.html.

EPA, Quality assurance handbook for air pollution measurement systems, (2013) http://www3.epa.gov/ttnamti1/files/ambient/pm25/qa/QA-Handbook-Vol-II.pdf.

Harrison, R. M. et al., Air quality in urban environments, (2015) http://www.slideshare.net/SothyrupanThiruchitt/057air-quality-in-urban-environments-issues-in-environmental-science-and-technologyronald-e.

New York State Department of Environmental Conservation, Air quality monitoring, (2015) http://www.dec.ny.gov/chemical/8406.html.

Ron, J. J., Environmental calibration and operation of isokinetic source-sampling equipment, (1972) http://nepis.epa.gov/Adobe/PDF/20013PMH.PDF.

Sherman, R. E., Process analyzer sample-conditioning system technology, (2002) http://www.wiley.com/WileyCDA/WileyTitle/productCd-0471293644.html.

Singal, S. P., Air quality monitoring and control strategy, (2012) http://www.amazon.com/Air-Quality-Monitoring-Control-Strategy/dp/1842657305.

Teledyne Analytical Instruments, Particle measurement system, (2012) http://www.teledyne-api.com/manuals/07318b_602.pdf.

Thermo Scientific, Particulate monitoring, (2015) http://www.thermoscientific.com/en/products/particulate-monitoring.html.

TSI, Air quality monitoring, (2015) http://www.tsi.com/Air-Quality-Monitoring/.

1.5 Ammonia Analyzers

D. C. GALLEGO (2017)

Methods of detection	Industrial Automatic Methods
	Liquids:
	A. Colorimetric
	B. ISE (Ion Selective)
	C. UV Spectrophotometry
	Gases:
	D. NIR (Infrared Spectrophotometry)
	E. Mass Spectrophotometry
	F. TDL (Tunable Diode Laser)
Measurement cycle	Liquids:
	I. Colorimetric and UV methods: between 3 and 15 min, when calibration cycle required between 15 and 40 min
	II. ISE method: continuous measurement
	Gases:
	III. IR and Mass spectrophotometer: long cycles due to sample conditioning before analysis cycle (10–15 min)
	IV. TDL open path: continuous measurement
	V. TDL extractive: less than 15 min
Accuracy	Liquids:
	a. Colorimetric: ±0.005 ppm
	b. ISE: ±0.02 to 0.5 ppm to 500ppm if the range is higher 7.5% of the full scale
	c. UV: 10% of full scale
Ranges	Liquids:
	a. Colorimetric: 0–3 ppm
	b. ISE: 0.5–1000 ppm of NH_3
	c. UV: FFT 0–1000 mg/l and UV chloramine method 0–500 mg/l
Price	Liquid:
	a. Colorimetric: Approximate price between $10,000 and $15,000
	b. ISE: Approximate price between $10,000 and $15,000
	c. UV: Approximate price between $35,000 and $40,000
	Gases:
	d. TDL: Approximate price between $50,000 (in situ) and $120,000 (extractive)
Partial list of suppliers	ISE:
	ABB (http://www.abb.com/product/us/9AAC100044.aspx)
	Emerson (http://www2.emersonprocess.com/en-US/brands/rosemountanalytical/Liquid/Systems/WQP/Pages/index.aspx)
	Endress + Hauser (http://www.endress.com/eh/home.nsf/#page/~product-news-nitrification-denitrification-control-isemax)
	Yokogawa (www.yokogawa.com)
	Colorimetric:
	ABB (http://www.abb.com/product/us/9AAC100044.aspx)
	ASA Analytics (http://www.asaanalytics.com/chemscan2150S.php)
	UV:
	Chemitec (http://www.chemitec.it/images/catalogo/Chemitec_Catalogo_Inglese_Analysers.pdf)
	Dtli (http://www.datalink-instruments.com/index.php?nav=3)

INTRODUCTION

The measurement of ammonia and ammonium ion concentration or traces (in gases or liquids) is very important in many industrial and environmental processes.

This chapter includes industrial measurement techniques for concentration determination. These techniques may be separated into measurements in liquids and measurements in gases.

Measurement in liquids is usually performed with colorimetric, ion selective, and/or UV spectrophotometric methods. However, measurement in gases is typically carried out using NIR spectrophotometry, mass spectrophotometers, or tunable diode lasers (TDL).

AMMONIA MEASUREMENT IN LIQUIDS

Ammonia measurement in liquids other than water can be performed by using chromatographs although these measurements are rarely required. The most common measurement is ammonia in water.

Ammonia is very soluble in water and other liquids. In water, ammonia exists simultaneously in two different conditions: as free unionized gaseous ammonia (NH_3) and as the ionized ammonium ion (NH_4^+). The ratio of NH_3 to NH_4^+ that exists in water depends mostly on the pH and in a lesser degree on the temperature:

- At 25°C a sample with pH 7.25 is mostly NH_4^+.
- At 25°C a sample with pH 9.25 contains similar amounts of NH_3 and NH_4^+.
- At 25°C a sample with pH 11.25 is mostly NH_3.

The most common application is the measurement in water treatment plants in order to determine water quality. The ammonia concentration, even in low concentrations, determines the presence of fecal bacteria or pathogens and, therefore, a poor water treatment or water pollution.

To determine the ammonia concentration in water the following methods can be used:

- *Colorimetric*: The ammonia concentration is determined by using a phenate that produces the reaction of a phenol and hypochlorite with ammonia to create a blue-colored compound. The color intensity is proportional to the ammonia concentration.
- *ISE Ammonia Selective Electrode*: This method is probably the easiest one to perform. After the pH has been adjusted to 11, the ammonia solution diffuses through a special membrane at the tip of the electrode. The change in electrical potential at the electrode is proportional to the ammonia concentration.

UV Spectrophotometry analysis. See Chapter 1.69.

These analysis methods can be either continuous or discontinuous with analysis cycle time of up to 50 min.

COLORIMETRIC METHOD

This method is based on the Berthelot reaction or indophenol method (see Chapter 1.15). The ammonia and all the ammonia compounds react with phenol giving place to indophenol. This reaction is described with Equations 1.5(1) through 1.5(3). Ammonia (NH_3) reacts with hypochlorite (NaOCl) to form a monochloramine, see Equation 1.5(1), which in turn reacts with two phenates to form an indophenol, see Equations 1.5(2) and 1.5(3). The stoichiometry of the overall reaction tells us that 1 mol of NH_3 reacts with 2 mol of phenate and 3 mol of NaOCl to form 1 mol of indophenol. The concentration of NH_3 is determined by measuring the absorbance spectrum of indophenol at $\lambda = 630$ nm. The obtained color intensity determines the ammonia concentration. In these reactions, NH_2Cl, NaOH, and phenol are important reagents for the indophenol formation.

$$NH_3 + NaOCl \longrightarrow NH_2Cl + NaOH \qquad 1.5(1)$$

$$NH_2Cl + \text{(phenate-ONa)} + 2NaOCl \longrightarrow Cl-N=\text{(quinone)}=O + 2NaCl + H_2O + NaOH \qquad 1.5(2)$$

$$Cl-N=\text{(quinone)}=O + \text{(phenate-ONa)} \longrightarrow HO-\text{(ring)}-N=\text{(ring)}=O + NaCl \qquad 1.5(3)$$

A previous step of membrane filtration is required in order to avoid possible interferences caused by turbidity in the sample.

For the analysis, three reagents are necessary, requiring that online analyzers must have a periodic maintenance in order to refill the tanks with the reagents.

These online analyzers have an analysis cycle time between 3 and 15 min.

The calibration and cleaning cycle varies between 15 and 45 min.

The typical measurement range for these analyzers is 0–3 ppm.

In Figure 1.5a, a typical configuration for an ammonia colorimeter in water is shown. Figure 1.5b shows the measurement cell.

ION-SELECTIVE METHOD (ISE)

This analyzer uses the ISE methods (see Chapter 1.33) for laboratory use ASTM D1426. In order to measure dissolved ammonia, this method requires an alkali (sodium hydroxide) reagent to treat the sample and measure all ammonia compounds (NH_3 and NH_4^+). The sodium hydroxide releases the available ammonia in a free form by adjusting the pH of the sample solution to a value greater than 11.

A permeable membrane is in contact with the water and this membrane allows the free ammonia to pass into contact with the NH_4Cl. This last compound is the liquid fill inside the membrane that reacts with the free ammonia and changes the pH of the ammonium chloride NH_4Cl solution.

The ISE probe measures the change of pH produced in the ammonium chloride solution. These pH changes are proportional to the concentration of free ammonia in the solution.

Figure 1.5c shows a complete system with a peristaltic pump and a reagent dosing system.

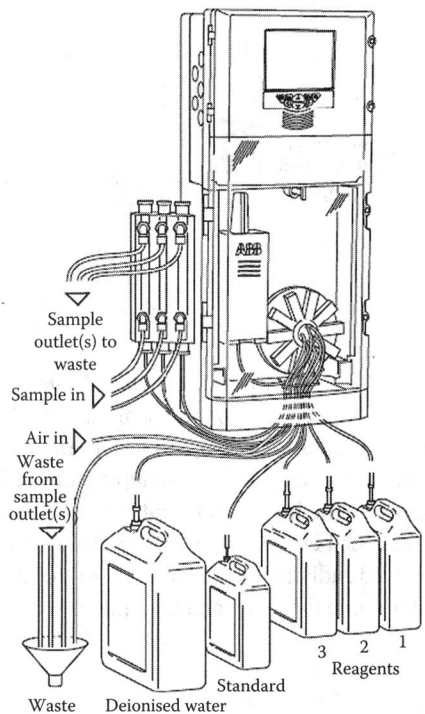

FIG. 1.5a
Multistream ammonia in water colorimeter analyzer. (Courtesy of ABB Analytics.)

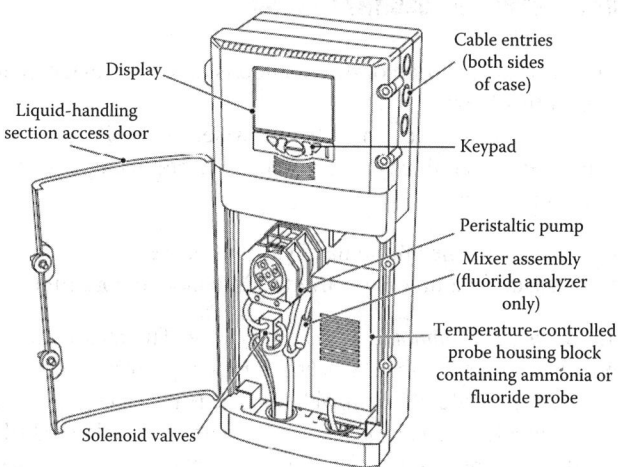

FIG. 1.5c
ISE ammonia water analyzer. (Courtesy of ABB Analytics.)

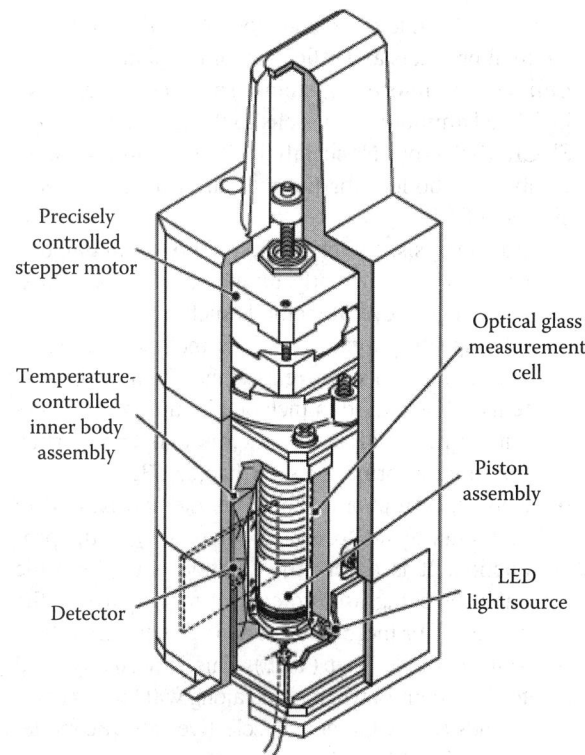

FIG. 1.5b
Measurement cell of colorimeter. (Courtesy of ABB Analytics.)

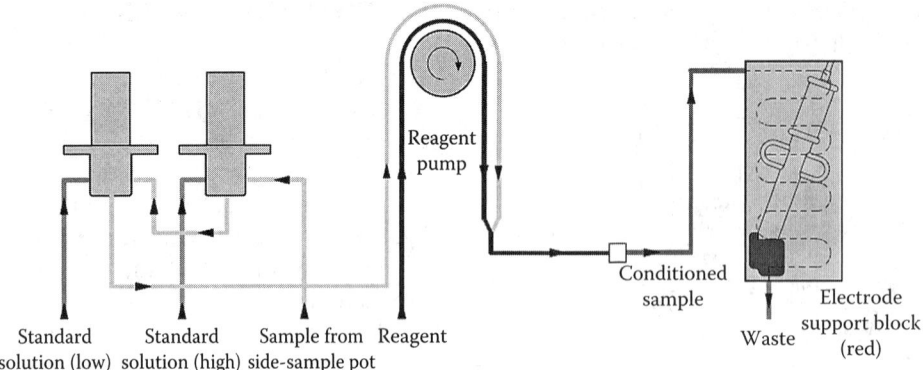

FIG. 1.5d
Flow schematic of ISE ammonia analyzer. (Courtesy of ABB Analytics.)

The flow schematic of this system is shown in Figure 1.5d.

This type of analyzer requires less maintenance than the analyzers using a colorimetric method as the only reagent used is the NaOH.

This type of analyzer is intended for continuous measurement but it is very important that the sample does not contain elements that can saturate the permeable membrane.

UV SPECTROPHOTOMETRY METHOD

The general operation of this method is described in Chapter 1.69 of this book.

The ammonia determination in water using UV spectrophotometry techniques can be accomplished using two different methods:

- Measuring the ammonia in gaseous phase
- Measuring ammonia converted into monochloramine

Measuring the ammonia in gaseous phase: The measurement principle is based on the UV light absorption spectrum of ammonia gas NH_3 in equilibrium with dissolved ammonia gas in the water sample. A small quantity of sodium hydroxide (NaOH) is added to the sample to increase the pH for transforming NH_4^+ into NH_3. A fast Fourier transform (FFT) is applied to the spectrum in order to extract the typical absorption signal of the ammonia gas. This method is very selective and it is suitable for river or waste water applications where other compounds might be present. Furthermore, turbidity or water coloring has no influence since the measurement is performed in gaseous phase.

Measuring ammonia converted into monochloramine: This type of analyzer is based on the UV light absorbance of monochloramine for ammonia concentration determination. Its absorbance is located in wavelengths between 200 and 450 nm. The analyzer comprises, like all UV spectrophotometers, a UV radiation source, a sample cell, a dispersion device, and a photodiode array (detector). The difference with a general UV spectrophotometer can be found in the treatment of the sample since for the measurement of the ammonia it is required that it has previously been converted into chloramines. This can be achieved first by adding NaOH to adjust the pH to a value greater than 12 and then by adding hypochlorite to convert all the remaining ammonia into monochloramine. The monochloramine is the substance that will absorb the UV radiation. Some UV ammonia analyzers use multiple paths of UV light in order to eliminate the impact of turbidity in the measurement.

The difference between the emitted radiation by the source and the received radiation is proportional to the concentration of monochloramine that is related to ammonia concentration.

AMMONIA IN GASES

The measurement of ammonia in gases is very important in many industries. Some clear examples are as follows:

- Ammonia production plants: to control the quality of the final products and efficiency of the plant.
- Emission monitoring systems: some countries have legislated ammonia emissions to the atmosphere.
- SNCR (Selective Noncatalytic Reductions): used in combustion boilers for the measurement of the efficiency of SCR units.
  - Gas analysis: to verify the quality and composition of a gas. Currently, ammonia in gases is measured using several methods such as:
  - NIR spectrophotometry. This method is used in ammonia plants and emissions monitoring systems. They are often included in analyzer shelters with other analyzers for emissions monitoring. For more information see Chapter 1.32.
  - Mass spectrophotometer. This method is used in ammonia plants to control the quality of the production. It is a versatile solution with the same analyzer being used to control many parts of the process. For more information see Chapter 1.35.
  - Tunable diode laser (TDL). This is normally used for NH_3 control when found along with others compounds (CO, CO_2, etc.) in selective catalytic reduction units (SCR). These analyzers can be either open path systems or in extractive arrangements, the extractive type being the more expensive.

TUNABLE DIODE LASER (TDL)

The TDL is an optical method that uses a laser to produce a specific wavelength of light tuned to a specific absorption line, the one corresponding to the light frequency of the target gas. This laser radiation stimulates the vibration and rotation in the gas molecules resulting in energy absorption. The principles used are the same as in the rest of the photometers (UV, IR). The concentration is directly proportional to the difference between the wavelengths of the emitted and received radiations.

The main difference between the TDL and other photometers (UV and IR) is that the TDL can be used mounted directly on equipment while the UV and IR photometers (see Chapters 1.32 and 1.69) are extractive types. Another difference is that the TDL has a greater precision than the UV or IR photometers.

In industry, two types of designs can be used for tunable diode lasers. These are as follows:

- *Open path TDL* (Figure 1.5e). The open path system is installed in situ (in the stream) and it is generally used to measure gases in either cross-stack or off-gas (atmospheric vent) applications. Cross-stack use is commonly applied to flue gas stacks to measure contaminants contained in gases emitted from a furnace or a boiler. Some examples of off-gas applications include the measurement of gases at the exhaust of furnaces, ducts, or stacks. These types of emission measurements are commonly referred to as continuous emissions monitoring systems (CEMS). CEMS analyzers monitor hazardous and prohibited contaminants such as NO_x. Another type of open path TDL is used to measure gas emissions across long distances (approximately 500 m). This type can be used to measure ammonia and CH_4 emissions of livestock.

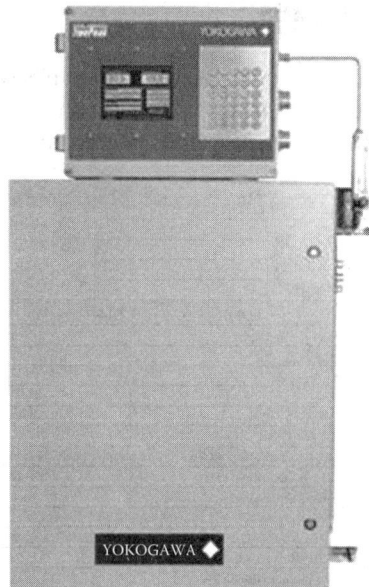

FIG. 1.5f
Extractive TDL for stack applications. (Courtesy of Yokogawa.)

- *Extractive TDL* (Figure 1.5f). The other major branch of TDL absorption spectroscopy is the extractive technology, or online/at-line approach. This technology samples gases extracted from a stream into a compact tubular chamber with mirrors. The mirrors reflect an incoming laser beam, between 1 and 100 times, before exiting the cell located inside the chamber. Typical cell diameters vary from 25 to 75 mm, and their length can be from 100 to 500 mm. Due to its compact design, the extractive TDL analyzers offer several advantages. They are small enough to permit convenient installation anywhere in a plant. The long optical path of the cell enables high sensitive measurements of trace gases in small volumes. Measurements of contaminants such as H_2S, and ammonia detecting concentrations as ppm and ppb levels. Because they operate outside the stream being measured, the extractive TDL is not subject to particulate matter in the flowing stream.

The main application of TDL in NH_3 measurements is emissions monitoring. When the required measurement must be very accurate and the concentration to be measured is very low, the extractive method is required (normally between 0 and 5 ppm). However, if the concentration is in the range of 0–5000 ppm, open path systems are normally used.

CONCLUSION

The method of measurement of ammonium depends on the state of the fluid, the range needed, and the accuracy required. The method must be revised according the measurement requirement to perform.

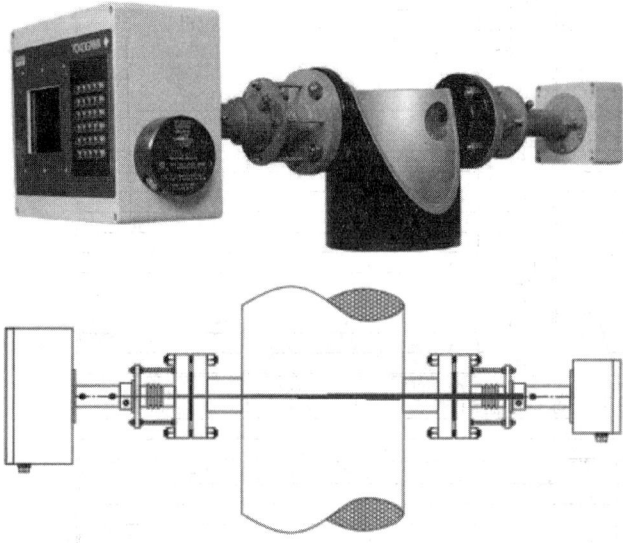

FIG. 1.5e
Open patch TDL for stack applications. (Courtesy of Yokogawa.)

SPECIFICATION FORMS

When specifying ammonia analyzers only, please use the ISA form 20A1001, and when specifying both the analyzer and the composition or the properties of the process material of interest, use ISA form 20A1002. The following are the forms that reproduced with the permission of the International Society of Automation.

#	RESPONSIBLE ORGANIZATION	ANALYSIS DEVICE	#	SPECIFICATION IDENTIFICATIONS	
1			6		
2	ISA		7	Document no	
3		Operating Parameters	8	Latest revision	Date
4			9	Issue status	
5			10		
11	ADMINISTRATIVE IDENTIFICATIONS		40	SERVICE IDENTIFICATIONS Continued	
12	Project number	Sub project no	41	Return conn matl type	
13	Project		42	Inline hazardous area cl	Div/Zon Group
14	Enterprise		43	Inline area min ign temp	Temp ident number
15	Site		44	Remote hazardous area cl	Div/Zon Group
16	Area	Cell Unit	45	Remote area min ign temp	Temp ident number
17			46		
18	SERVICE IDENTIFICATIONS		47		
19	Tag no/Functional ident		48	COMPONENT DESIGN CRITERIA	
20	Related equipment		49	Component type	
21	Service		50	Component style	
22			51	Output signal type	
23	P&ID/Reference dwg		52	Characteristic curve	
24	Process line/nozzle no		53	Compensation style	
25	Process conn pipe spec		54	Type of protection	
26	Process conn nominal size	Rating	55	Criticality code	
27	Process conn termn type	Style	56	Max EMI susceptibility	Ref
28	Process conn schedule no	Wall thickness	57	Max temperature effect	Ref
29	Process connection length		58	Max sample time lag	
30	Process line matl type		59	Max response time	
31	Fast loop line number		60	Min required accuracy	Ref
32	Fast loop pipe spec		61	Avail nom power supply	Number wires
33	Fast loop conn nom size	Rating	62	Calibration method	
34	Fast loop conn termn type	Style	63	Testing/Listing agency	
35	Fast loop schedule no	Wall thickness	64	Test requirements	
36	Fast loop estimated lg		65	Supply loss failure mode	
37	Fast loop material type		66	Signal loss failure mode	
38	Return conn nominal size	Rating	67		
39	Return conn termn type	Style	68		

#	PROCESS VARIABLES	MATERIAL FLOW CONDITIONS			Units	#	PROCESS DESIGN CONDITIONS		
69						101	Minimum	Maximum	Units
70	Flow Case Identification					102			
71	Process pressure					103			
72	Process temperature					104			
73	Process phase type					105			
74	Process liquid actl flow					106			
75	Process vapor actl flow					107			
76	Process vapor std flow					108			
77	Process liquid density					109			
78	Process vapor density					110			
79	Process liquid viscosity					111			
80	Sample return pressure					112			
81	Sample vent/drain press					113			
82	Sample temperature					114			
83	Sample phase type					115			
84	Fast loop liq actl flow					116			
85	Fast loop vapor actl flow					117			
86	Fast loop vapor std flow					118			
87	Fast loop vapor density					119			
88	Conductivity/Resistivity					120			
89	pH/ORP					121			
90	RH/Dewpoint					122			
91	Turbidity/Opacity					123			
92	Dissolved oxygen					124			
93	Corrosivity					125			
94	Particle size					126			
95						127			
96	CALCULATED VARIABLES					128			
97	Sample lag time					129			
98	Process fluid velocity					130			
99	Wake/natural freq ratio					131			
100						132			

#	MATERIAL PROPERTIES			#	MATERIAL PROPERTIES Continued		
133				137			
134	Name			138	NFPA health hazard	Flammability	Reactivity
135	Density at ref temp		At	139			
136				140			

Rev	Date	Revision Description	By	Appv1	Appv2	Appv3	REMARKS

Form: 20A1001 Rev 0 © 2004 ISA

1.5 Ammonia Analyzers

	RESPONSIBLE ORGANIZATION	ANALYSIS DEVICE COMPOSITION OR PROPERTY Operating Parameters (Continued)		SPECIFICATION IDENTIFICATIONS		
1			6			
2	ISA		7	Document no		
3			8	Latest revision		Date
4			9	Issue status		
5			10			

	PROCESS COMPOSITION OR PROPERTY			MEASUREMENT DESIGN CONDITIONS				
	Component/Property Name	Normal	Units	Minimum	Units	Maximum	Units	Repeatability
13								
14								
15								
16								
17								
18								
19								
20								
21								
22								
23								
24								
25								
26								
27								
28								
29								
30								
31								
32								
33								
34								
35								
36								
37								

(Rows 38–75: blank notes area)

Rev	Date	Revision Description	By	Appv1	Appv2	Appv3	REMARKS

Form: 20A1002 Rev 0 © 2004 ISA

Abbreviations

CEMS	Continuous emissions monitoring system
FFT	Fast Fournier transform
NaOCl	Sodium hypochlorite
NaOH	Sodium hydroxide
NH_3	Ammonia
NH_4Cl	Ammonium chloride
NIR	Near infrared
SNCR	Selective non-catalytic reductions
TDL	Tunable diode laser
UV	Ultraviolet

Bibliography

ABB. Ammonium analyzer catalog (http://www.abb.com/product/us/9AAC100044.aspx).

American Society for Testing and Materials. 1979. Method 1426–79. American Society fo Testing & Materials, Philadelphia, PA.

ITA Publications. 2001. Online ammonia analyzers for water and wastewater treatment applications. (http://www.instrument.org/ePubs/site_map.htm).

U.S. Environmental Protection Agency. 1979. Methods for chemical analysis of water and wastes. EPA-600/4-79-020, National Environmental Research Center, Cincinnati, OH.

1.6 Biometers to Quantify Microorganisms

I. G. YOUNG (1974, 1982)   **B. G. LIPTÁK** (1995, 2017)

Applications	Sludge dewatering, composting, product sanitation
Method of detection	Photometric measurement of light emitted by chemical reaction
Sample pressure	Atmospheric
Sample temperature	Ambient
Sample type	Grab sample
Materials of construction	Glass
Range	10^{-7} to 10^{-2} μg of ATP per 10 ml sample of bacterial extract. Sensitively to 10^{-7} μg per 10 μl sample. It can be calibrated for number of bacteria per microgram of ATP.
Response	Laboratory method: minutes after starting reaction
Cost	About $10,000
Partial list of suppliers	Bio-Tek Instruments: http://www.biotek.com/resources/articles/luminescent-atp-concentrations.html CE Water Management: http://www.cewater.com/A27_Monitoring.html Optofine Instruments, http://search.aol.com/aol/image?page=1&v_t=na&q=biometers&oreq=572bd7df9d954609b9af1e14e25f28a3 Sigma-Aldrich, http://www.sigmaaldrich.com/catalog/product/roche/11699695001?lang=en&region=US ThermoFisher Scientific, https://www.thermofisher.com/order/catalog/product/A22066 TOC Biometrics, https://www.tocbiometrics.com/ Tomey, http://www.tomey.com/Products/OCT/OA-1000.html

INTRODUCTION

Biometers measure the quantity and activity of microorganisms though the measurement of some of their byproducts. One such method is the measurement of carbon dioxide which is given off in proportion to the quantity of living matter in the sample (Figure 1.6a).

Other biosensors utilize a physicochemical, optical, piezoelectricl or electrochemical methods to detect microorganisms, bacteria, organelles, cell receptors, etc. The focus of this chapter is on adenosine triphosphate (ATP) biometers. ATP is an energy molecule found in living cells, which gives a direct measure of their health and biological concentration. ATP can be quantified by using a luminometer for measuring the light produced through its reaction with the enzyme luciferase. The amount of light produced

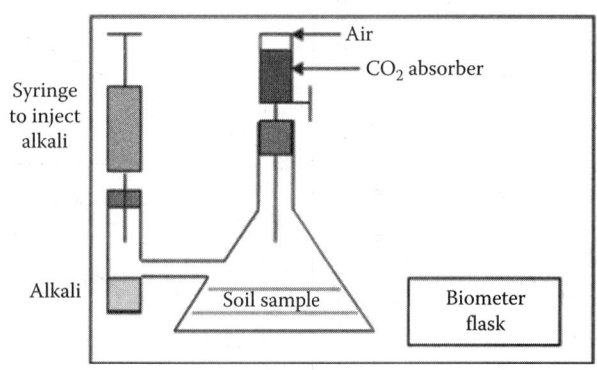

FIG. 1.6a
The Biometer flask is used to measure the activity of soil samples. Soil is incubated at the required temperature, moisture content, etc. in the main flask and air is added through the CO_2 absorber. (Courtesy of Colin Nayfield.)

is directly proportional to the amount of ATP present in the sample.

The sensors and measurement techniques discussed in this chapter are laboratory techniques used to measure the biological population and the biological oxidative activity by detecting the amount of adenosine triphosphate (ATP) and the changes in its concentration. These techniques can be used for the measurement and control of biological treatment, drinking water cleanliness, to manage fermentation processes.

For the key wastewater-related measurements, which detect the concentration of the discharged waste stream in terms of its oxygen demand (BOD, COD, and TOD), refer to Chapter 1.74.

ATP AND LIVING ORGANISMS

Adenosine 5′-triphosphate (ATP) is the primary energy currency of all living organisms and participates in a variety of cellular processes. All living things utilize ATP as a means for storing metabolic energy. Because of this, the detection and quantitation of ATP can be used as a means to detect and/or quantitate microorganisms such as bacteria and somatic cells. The assay used in these studies relies on the ATP dependence of the firefly luciferase reaction to detect live organisms.

ATP is a molecule found in and around living cells, and as such it gives a direct measure of biological concentration and health. ATP is quantified by measuring the light produced through its reaction with the naturally occurring firefly enzyme luciferase using a luminometer. The amount of light produced is directly proportional to the amount of living organisms present in the sample.

ATP ANALYSIS

In a detailed study of the control parameters for the activated sludge process, measurements of great interest are biochemical oxygen demand (BOD) and chemical oxygen demand (COD). BOD and COD reduction, biological population density, and biological oxidative activity are important indicators of this process.

It has been found that the amount of ATP is proportional to the viable biomass in a sample. It has also been found that changes in ATP concentration measure the oxidative capability of the biomass. Thus, it is of great interest to measure the ATP content of samples in the activated sludge process as well as in rivers, lakes, and other receiving waters.

Sensitive methods for ATP analysis have been developed based on the observation that the luminescent reaction in fireflies is absolutely dependent on the presence of ATP. The in vitro light-yielding reactions are given in Equations 1.6(1) and 1.6(2):

$$LH_2 + E + ATP(Mg_2^+)E - LH_2 - AMP + PP \quad 1.6(1)$$

$$E - LH_2 - AMP + O_2 \rightarrow E + Product + CO_2 + AMP + H\nu \quad 1.6(2)$$

where
 LH_2 is the luciferin
 E is the luciferase enzyme
 $E-LH_2-AMP$ is the enzyme–luciferin–adenosine monophosphate complex
 PP is the pyrophosphate

Figure 1.6b shows that the yield of light quanta (Hv) is in proportion to the amount of ATP present in the sample.

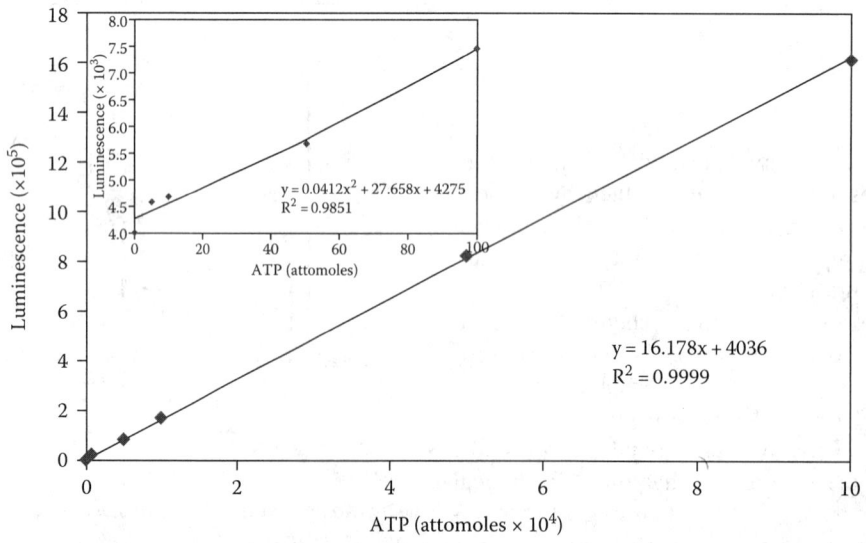

FIG 1.6b
The relationship between luminescence and ATP concentration is linear. (Courtesy of BioTek.)

LUMINESCENCE BIOMETER

ATP assay procedures have been developed based on the reactions just described. Briefly, the procedure involves rapid killing of the live bacterial cells and immediate extraction of ATP into aqueous solution. The latter is then treated with firefly lantern extract, and the light emission of the resultant solution is measured with a photometer. The firefly lantern extract and the ATP required for calibration are commercially available.

Manually operated instruments are available for the ATP measurement. It is supplied with all the required reagents. A tablet containing buffer and magnesium sulfate is dissolved in water, after which a homogeneous powder of luciferin and luciferase is added. The sample is filtered through a coarse filter to remove solid matter, and the latter is discarded. The filtrate is passed through a bacterial filter to catch all the living bacteria. The bacteria on the filter are treated with butanol, which ruptures the cell walls and releases the ATP.

The filtrate is made up to volume with water, and a microliter aliquot is added to the prepared reagent already in a cuvette. The cuvette is then placed in the instrument for reading of its light emission. The light flash is automatically converted to ATP or microorganism concentration per milliliter, depending on how the instrument is calibrated.

Bioluminescence ATP determination kits are suited for sensitive quantification of ATP in various applications such as ATP detection in biological samples or monitoring of ATP-dependent enzyme assays. The ATP kits are usually available in two different formats either for highly sensitive determination of low ATP concentrations or as a time-stable assay where ATP can be measured over a time scale of about 4 hr. Ten milliliters of the assay mix is sufficient to perform at least 200 analyses depending on assay volume.

For a luminescence reader, refer to Figure 1.6c.

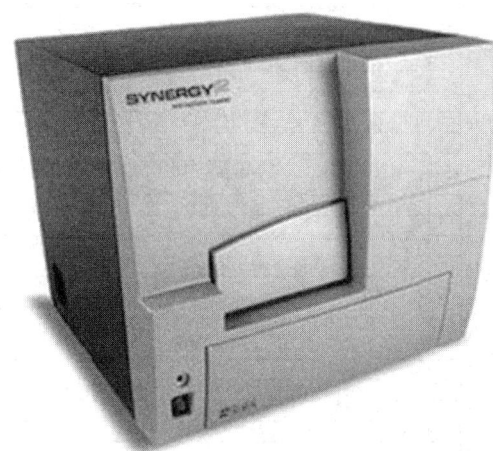

FIG. 1.6c
Synergy 2 SL luminescence microplate reader. (Courtesy of BioTek.)

120 Analytical Measurement

SPECIFICATION FORMS

When specifying biometers to quantify microorganisms only, please use the ISA form 20A1001 and when specifying both the analyzer and the composition or the properties of the process material of interest, use ISA form 20A1002. Both forms are reproduced with the permission of the International Society of Automation on the next pages.

#					#				
1	RESPONSIBLE ORGANIZATION		ANALYSIS DEVICE		6	SPECIFICATION IDENTIFICATIONS			
2					7	Document no			
3	(ISA)		Operating Parameters		8	Latest revision		Date	
4					9	Issue status			
5					10				
11	ADMINISTRATIVE IDENTIFICATIONS				40	SERVICE IDENTIFICATIONS Continued			
12	Project number		Sub project no		41	Return conn matl type			
13	Project				42	Inline hazardous area cl		Div/Zon	Group
14	Enterprise				43	Inline area min ign temp		Temp ident number	
15	Site				44	Remote hazardous area cl		Div/Zon	Group
16	Area		Cell	Unit	45	Remote area min ign temp		Temp ident number	
17					46				
18	SERVICE IDENTIFICATIONS				47				
19	Tag no/Functional ident				48	COMPONENT DESIGN CRITERIA			
20	Related equipment				49	Component type			
21	Service				50	Component style			
22					51	Output signal type			
23	P&ID/Reference dwg				52	Characteristic curve			
24	Process line/nozzle no				53	Compensation style			
25	Process conn pipe spec				54	Type of protection			
26	Process conn nominal size		Rating		55	Criticality code			
27	Process conn termn type		Style		56	Max EMI susceptibility		Ref	
28	Process conn schedule no		Wall thickness		57	Max temperature effect		Ref	
29	Process connection length				58	Max sample time lag			
30	Process line matl type				59	Max response time			
31	Fast loop line number				60	Min required accuracy		Ref	
32	Fast loop pipe spec				61	Avail nom power supply			Number wires
33	Fast loop conn nom size		Rating		62	Calibration method			
34	Fast loop conn termn type		Style		63	Testing/Listing agency			
35	Fast loop schedule no		Wall thickness		64	Test requirements			
36	Fast loop estimated lg				65	Supply loss failure mode			
37	Fast loop material type				66	Signal loss failure mode			
38	Return conn nominal size		Rating		67				
39	Return conn termn type		Style		68				
69	PROCESS VARIABLES	MATERIAL FLOW CONDITIONS			101	PROCESS DESIGN CONDITIONS			
70	Flow Case Identification			Units	102	Minimum	Maximum	Units	
71	Process pressure				103				
72	Process temperature				104				
73	Process phase type				105				
74	Process liquid actl flow				106				
75	Process vapor actl flow				107				
76	Process vapor std flow				108				
77	Process liquid density				109				
78	Process vapor density				110				
79	Process liquid viscosity				111				
80	Sample return pressure				112				
81	Sample vent/drain press				113				
82	Sample temperature				114				
83	Sample phase type				115				
84	Fast loop liq actl flow				116				
85	Fast loop vapor actl flow				117				
86	Fast loop vapor std flow				118				
87	Fast loop vapor density				119				
88	Conductivity/Resistivity				120				
89	pH/ORP				121				
90	RH/Dewpoint				122				
91	Turbidity/Opacity				123				
92	Dissolved oxygen				124				
93	Corrosivity				125				
94	Particle size				126				
95					127				
96	CALCULATED VARIABLES				128				
97	Sample lag time				129				
98	Process fluid velocity				130				
99	Wake/natural freq ratio				131				
100					132				
133	MATERIAL PROPERTIES				137	MATERIAL PROPERTIES Continued			
134	Name				138	NFPA health hazard		Flammability	Reactivity
135	Density at ref temp		At		139				
136					140				

Rev	Date	Revision Description	By	Appv1	Appv2	Appv3	REMARKS

Form: 20A1001 Rev 0 © 2004 ISA

		RESPONSIBLE ORGANIZATION	ANALYSIS DEVICE COMPOSITION OR PROPERTY Operating Parameters (Continued)			SPECIFICATION IDENTIFICATIONS		
1		ISA		6				
2				7	Document no			
3				8	Latest revision		Date	
4				9	Issue status			
5				10				

		PROCESS COMPOSITION OR PROPERTY			MEASUREMENT DESIGN CONDITIONS				
11		Component/Property Name	Normal	Units	Minimum	Units	Maximum	Units	Repeatability
12									
13									
...									
75									

Rev	Date	Revision Description	By	Appv1	Appv2	Appv3	REMARKS

Form: 20A1002 Rev 0 © 2004 ISA

Abbreviations

ATP   Adenosine triphosphate
BOD   Biochemical oxygen demand
COD   Chemical oxygen demand
PP   Pyrophosphate
TOD   Total oxygen demand

Organization

ASM   American Society for Microbiology

Bibliography

American Chemistry Council, Biomonitoring, (2015) http://www.americanchemistry.com/Policy/Chemical-Safety/Biomonitoring?gclid=CK2h3ZH5xsgCFcUUHwodZTcO_w.

ATP Testing, What is ATP testing?, (2013) http://atptesting.org/.

BioTek, Luminescent determination of ATP concentrations, (2015) http://www.biotek.com/resources/articles/luminescent-atp-concentrations.html.

Hach Products, Units of measurement for microbes, (2015) https://www.boundless.com/microbiology/textbooks/boundless-microbiology-textbook/microscopy-3/looking-at-microbes-28/units-of-measurement-for-microbes-239-6417/.

Held, P., BioTek, Luminescent ATP concentration, (2004) http://www.biotek.com/resources/articles/luminescent-atp-concentrations.html.

ThermoFisher Technology, ATP determination kit, (2015) https://tools.thermofisher.com/content/sfs/manuals/mp22066.pdf

1.7 Carbon Dioxide

R. A. HERRICK (1974, 1982)   **B. G. LIPTÁK** (1995, 2003, 2017)

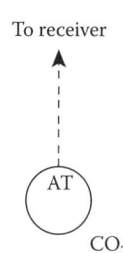

Flow sheet symbol

Types of sensors	a. Nondispersive infrared (NDIR) b. Gas filter correlation (GFC) c. Orsat (grab sample for flue gas measurement)
Applications	Ambient measurement and source (usually combustion) measurement
Inaccuracy	a. Can be as high as ±0.2 ppm; typically ±1% to ±2% of full-scale range b. 1%–2% of full scale, including drift c. Laboratory procedure
Ranges	a. 0–2000 ppm, 0–3000 ppm, 0–5000 ppm, 0%–1%, 0%–2%, 0%–5%, 0%–10%, 0%–20%, and 0%–100% b. 0–5 ppm, 0–10 ppm, 0–20 ppm, 0–50 ppm, 0–100 ppm, 0–500 ppm, 0–1000 ppm, 0–2000 ppm (low detectable limit is 0.1 ppm)
Sample pressure	Up to 15 psig (104 kPa), but atmospheric or near atmospheric is normal
Sample temperature	Up to approximately 49°C (120°F) not a consideration when freeze-out trap is used
Sample flow rate	Generally less than 0.5 acfm (2.35×10^{-4} m^3/sec); typically 1–2 l/min
Response	a. Determined by cell volume and sampling rate; typically less than 30 sec b. 90 sec with 30 sec signal averaging time
Costs	Detector tubes with 0.1%–5% range cost about $100. Portable, battery-operated, diffusion type monitor with two alarm settings, digital display, and 4–20 mA output is $1500; permanently installed, explosion-proof NDIR analyzer with recorder is about $10,000
Partial list of suppliers	Ametek/Thermox (http://www.ametekpi.com/products/Thermox-WDG-V-Combustion-Analyzer.aspx) Brasch, http://www.braschmfg.com/products/detectors/bgs-cd.htm CEA Instruments (http://www.ceainstr.com/pdf_datasheets/optimiser_Info.pdf) CO2Meter.com, http://www.co2meter.com/ Ecotech (http://www.ecotech.com.au/gas-analyzers-categories/stack) E&E Process (http://www.process-controls.com/EEProcess/Alnor/alnor-indoor-air-quality-meters-CF910.htm) EMS (Environmental Monitoring Systems) (http://www.emssales.net/store/cart.php?m=product_detail&slug=tsi-iaq-calc-7515-carbon-dioxide-co2-meter) Enviro Technology (http://www.et.co.uk/products/scientific-research/los-gatos-research-lgr-off-axis-icos-analysers/isotopic/carbon-dioxide-isotope-analyser/) Foxboro-Invensys (http://iom.invensys.com/EN/Pages/SearchResults.aspx?k=carbon%20dioxide) General Monitors, http://www.generalmonitors.com/Gas-Detectors/IR700-Point-IR-Carbon-Dioxide-Gas-Detector/p/000140006800001002 Horiba Instrument Inc. (http://www.horiba.com/process-environmental/products/ambient/details/apma-370-ambient-carbon-monoxide-monitor-270) International Sensor Technology (http://www.intlsensor.com/pdf/infrared.pdf) MSA Instrument Div. (http://ca.msasafety.com/search?text=carbon+dioxide&_requestConfirmationToken=6bcf37b5242d5d56f641b0a30ee4d448e8827ed5) Nova, http://www.nova-gas.com/analyzers/carbon-dioxide PCE Instruments, https://www.pce-instruments.com/us/measuring-instruments/test-meters/co2-analyser-kat_41831_1.htm

(Continued)

Sable Systems, http://www.sablesys.com/products/classic-line/ca-10-carbon-dioxide-analyzer/

Sensidyne Inc. (http://www.sensidyne.com/colorimetric-gas-detector-tubes/detector-tubes/sensidyne_gas_detector_tube_number-126SA_CARBON_DIOXIDE0.1–5.2.php)

Servomex, http://www.servomex.com/carbon_dioxide_gas_analyzer.html

Sieger Gasalarm; Siemens Energy & Automation (http://w3.usa.siemens.com/buildingtechnologies/us/en/building-automation-and-energy-management/sensors/air-quality-sensors/Pages/air-quality-sensors.aspx)

Sierra Monitor Corp. (http://www.sierramonitor.com/gas/IT/Toxic_Gas_Detector.php)

Systek Illinois, http://www.systechillinois.com/en/oxygen-and-carbon-dioxide-headspace-gas-analyzer-GS6000

Teledyne API http://www.teledyne-api.com/products/T360M.asp

Teledyne API (http://www.teledyne-api.com/manuals/07272B_T360-M.pdf)

Thermo Scientific (http://www.thermoscientific.com/ecomm/servlet/productsdetail?productId=11961827&groupType=PRODUCT&ssearchType=0&storeId=11152&from=search)

Topac (http://www.topac.com/guardianCO2.html)

University of Utah, co2.utah.edu/co2tutorial.php?site=7&id=4

INTRODUCTION

Carbon dioxide concentration can be measured in liquids or gases, they can be performed by simple Orsat equipment (patented in 1873 by Mr. H Orsat) in the laboratory, and they can serve ambient monitoring or industrial composition analysis purposes. In this chapter, we will focus on the industrial sensors, but before describing them, a brief discussion will be given of the role of carbon dioxide and its effects on the global climate.

The precise measurement of the carbon dioxide content of the atmosphere is of significant concern in determining long-term changes in the composition of the atmosphere. The measurement techniques used by geophysicists are highly precise compared to the techniques used for air quality– or air pollution–related measurements.

Air quality–related measurements can be used to monitor the return air quality from occupied spaces and, based on those measurements, to modulate the rate at which fresh air is being introduced. In air pollution–related applications, carbon dioxide is hardly ever measured in the ambient air. It is usually measured at emission points since some combustion equipment regulations are stated in terms of allowable pollutant discharges corrected to 50% excess air.

ROLE IN GLOBAL WARMING

Figure 1.7a—prepared by the U.S. Energy Information Administration in 2001—describes the global carbon cycle. In November 2007, the American National Academy of Science

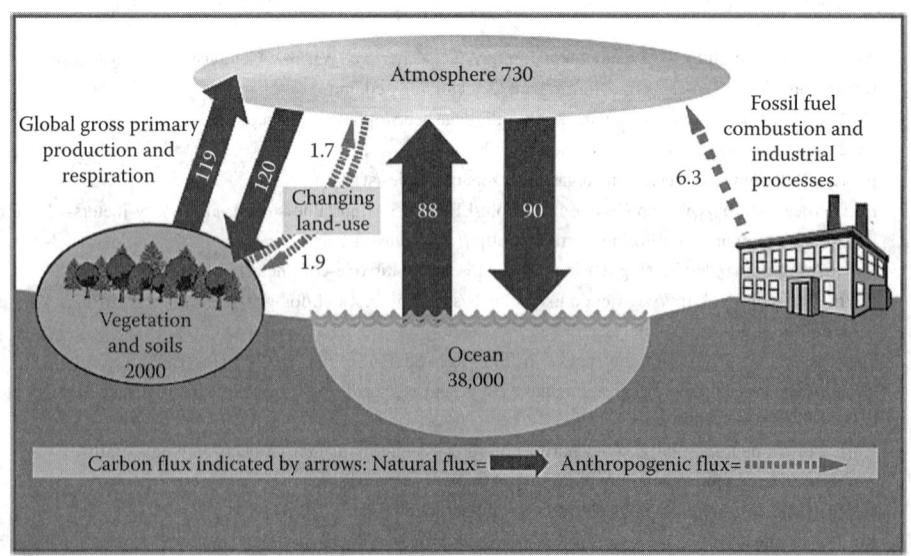

FIG. 1.7a
Global carbon cycle in billion metric tons carbon (U.S. Energy Information Administration).

reported the actual emission for 2006 as 8.4 billion tons and in 2015 it is approaching 10 billion tons.

The total quantity of carbon on earth is about 41,000 billion metric tons (92% in the oceans, 6% on land, and 2% in the atmosphere). Prior to the industrial age, the concentration of carbon dioxide in the atmosphere was stable. There was a balance as 210 billion tons (bt) of carbon dioxide entered the atmosphere and approximately the same amount was taken from it by the photosynthesis of the plants. That balance has been upset by fuel combustion, deforestation, and changing land use as the population increased.

Since the beginning of the industrial age, the atmospheric concentration of carbon dioxide increased from 280 to over 400 parts per million (ppm). The atmospheric concentration of carbon dioxide is the highest in 650,000 years.

World carbon dioxide emissions are expected to increase by 1.9% annually. Much of the increase in these emissions is expected to occur in the developing world where emerging economies, such as China and India, fuel their economic development with fossil energy. The carbon dioxide emissions of China are already approaching that of the United States and emissions of the developing countries are expected to surpass that of the industrialized countries around 2018.

Today the average American generates 21 tons of carbon dioxide a year, while the global per capita average is only 4 tons a year. The generation of each kWh of electricity from fossil fuels releases from 270 to 1050 g (0.27–1.05 kg) of carbon dioxide into the atmosphere, in addition to other pollutants such as NO_x, SO_2, and particulates. On average, an automobile generates about 160 g of carbon dioxide per kilometer (250 g/mile). Even a hybrid automobile generates 3 tons of carbon dioxide each year.

In comparison, the average tree consumes only 1 ton of carbon dioxide in a lifetime and an acre of rainforest consumes about 500 tons of carbon dioxide yearly. When the agribusiness and ethylene industry or pulp and paper corporations turn forests or rain forests into farmland, they also destroy an effective consumer of carbon dioxide.

In the United States, greenhouse gas emissions come from the following sources: power plants (33%), transportation (28%), industry (20%), agriculture (7%), commerce (7%), and households (5%). A 500 MW coal-fired power plant releases 100 tons of CO_2 every hour. The electric power generation capacity of the United States is around 1000 GW (the equivalent of 2000 such power plants).

EMISSIONS

The 840 million people in Africa generate only 3% of the global emission of carbon dioxide, yet they face the greatest risk of drought and disruption of their water supplies. Deforestation, soil erosion, storms, drought, and the devastation of agriculture are likely to result as temperatures exceed the heat tolerance of crops. These trends can combine to cause migration, ethnic strife, social destabilization, and wars. In the industrialized parts of the world, the effect of global warming is projected to be increased precipitation in Europe and the East coast of the United States as the increased ocean evaporation reaches the cooler land area. Some suggest this cooling is caused by the slowing of the Gulf Current, which in turn is caused by the melting of the ice on Greenland.

Figure 1.7b shows the global trend in carbon emissions. As a result of these emissions, the global temperature has already risen by 0.74°C (1.3°F). If this continues, it will result in the rise of ocean levels by about 60 cm (23″) just because of the thermal expansion of the ocean's

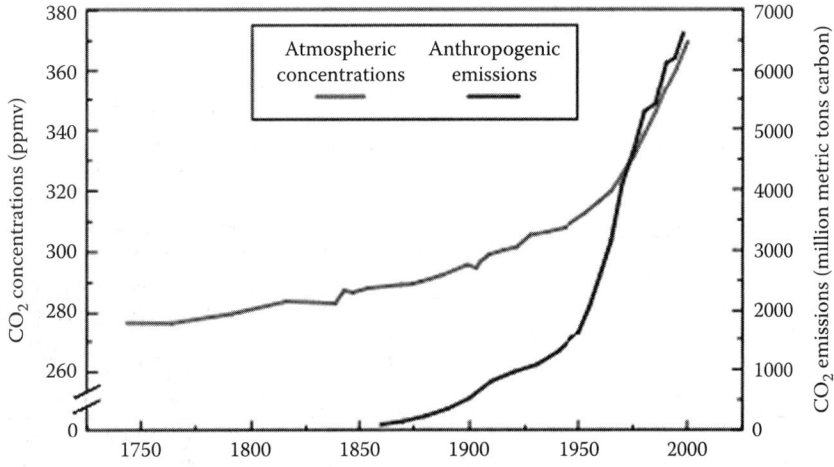

FIG. 1.7b
Trends in atmospheric concentrations and anthropogenic (manmade) emissions of carbon dioxide.

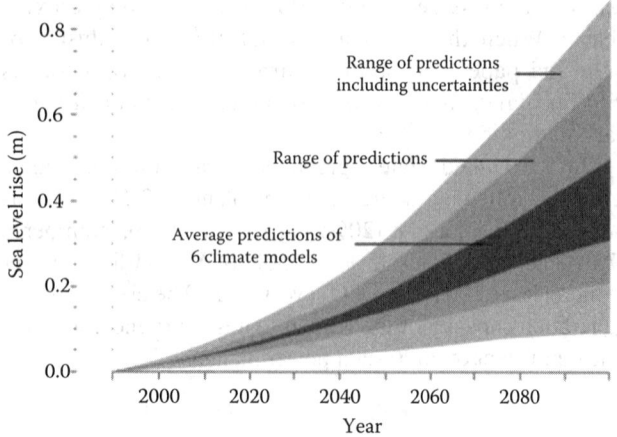

FIG. 1.7c
Sea level rise predictions. (Courtesy of NASA.)

waters (Figure 1.7c). In addition to these direct effects, there are also indirect consequences. One consequence is that as the oceans warm, they absorb less carbon dioxide, and the other is that as the Canadian, Siberian, and Alaskan permafrost melts, the rotting organic matter will release vast amounts of carbon dioxide and methane. The rising carbon dioxide concentration of the atmosphere also reduces the pH of the oceans. In addition, about 10 teratons of carbon (tera = 10^{12}) are stored in the frozen methane hydrates of the arctic regions, which could also be released, if the ice melts.

There is no question that the goal is to stabilize the climate, but it is also true that we can live with a few degrees of temperature rise and a few inches of sea level increase, while we convert our economy to a renewable one. Therefore, during the coming decade, the accurate monitoring of both the release of carbon dioxide from fossil fuel combustion sources and of the atmospheric concentration of carbon dioxide will gain importance.

The environmental cost of climate change will be paid by all of mankind. Sir Nicholas Stern, former chief economist at the World Bank, estimated that if no changes occur, this cost by 2020 will amount to 20% of the gross world product (GWP). The Cato Institute estimates that the environmental cost of climate change is between 5% and 10% of the gross world product (GWP) http://www.cato-at-liberty.org/2006/11/03/global-warming-costs-benefits/. The GWP of the United States is nearing 16 trillion dollars and the global GWP today is about 46 trillion dollars.

AMBIENT AIR MEASUREMENT

For the reasons discussed earlier, the precise knowledge of the carbon dioxide concentration in the atmosphere is necessary. Even a 1% increase in global carbon dioxide concentration (about 3–4 ppm) has significant influence on the weather. The instruments used to measure atmospheric carbon dioxide concentrations must, of necessity, be highly precise, such as the gas filter correlation (GFC) units discussed later.

Nondispersive Infrared Sensors

These instruments are discussed in detail in Chapter 1.33. Since the absorption bands of water and carbon dioxide somewhat overlap, a freeze-out trap (–80°C or –112°F) is often used in the sample preparation system to remove the water prior to the measurement. For some of the applications of infrared analyzers (IR), refer to Table 1.7a.

TABLE 1.7a
IR Analyzer Applications Summary

| Analyzer | Carbon Monoxide | Carbon Dioxide | Organic Vapors | | Organic Liquids | Solids (Reflection) | Comments |
			Simple Molecules	Complex Molecules			
NDIR	✓	✓	✓				Single-component analysis: ethylene, CO, acetylene, methane, etc.
	✓	✓	✓	✓	✓		Single-component analysis: same as earlier, including ammonia, vinyl chloride, carbon tetrachloride, methyl ethyl ketone, ethylene dichloride, etc.
Near-IR filter					✓	✓	Single-component analysis: ethylene dichloride, water, phenol, methyl alcohol, etc.; moisture in solids
Correlation spectrometer	✓						Stack analysis, single-component gas analysis
Multiple-filter near-IR						✓	Multiple components for cereal, meat, and paper analysis
Multiple-filter mid-IR	✓	✓	✓	✓	✓		Automotive exhaust analysis (CO, CO_2, –CH); multiple components for mike analysis, multiple components of gases using a programmable circular variable filter

When using prepared calibration standard mixtures of carbon dioxide in nitrogen, an inaccuracy of ±0.2 to 0.3 ppm is attainable by NDIR analyzers (Table 1.7b). At normal atmospheric CO_2 concentration levels of about 380 ppm, the error is less than ±0.1%.

When using prepared calibration standard mixtures of carbon dioxide in nitrogen, an inaccuracy of ±0.2 to 0.3 ppm is attainable by NDIR analyzers (Table 1.7b). At normal atmospheric CO_2 concentration levels of about 380 ppm, the error is less than ±0.1%.

TABLE 1.7b
Typical Applications for NDIR Analyzers

Gas	Minimum Range (ppm)	Maximum Range (%)
Ammonia (NH_3)	0–300	0–10
Butane (C_4H_{10})	0–300	0–100
Carbon dioxide (CO_2)	0–10	0–100
Carbon monoxide (CO)	0–50	0–100
Ethane (C_2H_6)	0–20,000	0–10
Ethylene (C_2H_4)	0–500	0–100
Hexane (C_6H_{14})	0–200	0–5
Methane (CH_4)	0–2,000	0–100
Nitrogen oxide (NO)	0–500	0–10
Propane (C_3H_8)	0–300	0–100
Sulfur dioxide (SO_2)	0–500	0–30
Water vapor (H_2O)	0–3,000	0–5

Nondispersive infrared (NDIR)-type CO_2 monitors for heating, ventilation, and air conditioning (HVAC) and industrial applications are available in both portable and permanently installed designs. The portable units are usually battery operated, and their ambient sample is received by a combination of diffusion and convection effects in the sensor head, without any pumps or filtering. These units are usually provided with digital displays, one or two alarm settings, and analog or digital output signals.

Figure 1.7d illustrates a microprocessor-based, infrared carbon dioxide gas detector that continuously monitors the CO_2 levels in the ppm range. All electronics are contained within an explosion-proof housing. It generates a 4–20 mA output signal and Modbus or HART configurations and predictive maintenance are also available. Figure 1.7d also shows another microprocessor based CO_2 analyzer which is available for measurement in the 0–100% range and can operate at pressures up to 1,500 psig, being suitable for natural gas applications.

The more expensive, permanently installed, or wall-mounted NDIR units often include data loggers, which can store about 1000 readings along with their times and dates. Some of these units can also detect other gases, such as CO, H_2S, or O_2.

Gas Filter Correlation Type

When very accurate low-level measurements are needed, or when background gases that have the potential to interfere with the measurement are present, GFC is used. In these designs, the measuring and reference filters are replaced by gas-filled cuvettes. The reference cuvette is filled with CO_2 and the measuring cuvette usually with nitrogen.

In addition to being unaffected by the presence of background gases, both the accuracy and the response time of these instruments are better than those using filters. If GFC is used in combination with single-beam, dual-wavelength technology, it is virtually immune to obstruction of the optics. This, in turn, prevents drift and thereby reduces the frequency at which recalibration is needed.

SOURCE MEASUREMENT

Measurement of the carbon dioxide, carbon monoxide, and oxygen concentration of flue gases from boilers has been done for many years. These measurements allow the precise setting of boiler operating variables for maximum fuel economy.

Before the use of the infrared analyzer became accepted practice, mechanical instruments were used to continuously determine the carbon dioxide content of flue gases.

CO_2 analyzer for ppm measurement range.
(Courtesy of General Monitors.)

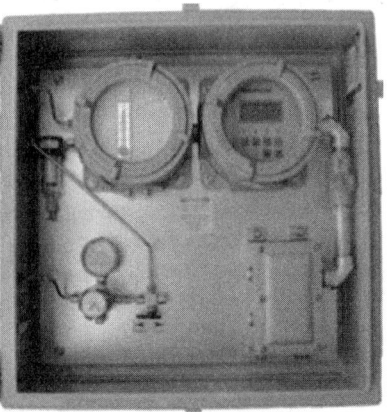

CO_2 analyzer for % range applications.
(Courtesy of Permtech.)

FIG. 1.7d
Microprocessor-based, infrared carbon dioxide gas detectors.

Their operation was based on the reduction in gas volume resulting from the absorption of carbon dioxide in a strong alkaline solution. This is the principle of the Orsat analyzer, still used as the standard method for manual determination of combustion gas composition.

The Orsat apparatus was patented before 1873 by Mr. H Orsat. It consists essentially of a calibrated water-jacketed gas burette connected by glass tubing to two or three absorption pipettes containing chemical solutions that absorb the gases it is required to measure. The gas to be analyzed is drawn into the burette and flushed through several times. Typically, 100 ml is withdrawn for ease of calculation. It is then passed into the caustic potash burette, left to stand for about 2 min, and then withdrawn, isolating the remaining gas via the stopcock arrangements After leveling the liquid in the bottle and burette, the remaining volume of gas in the burette indicates the percentage of carbon dioxide absorbed.

For air pollution testing purposes, the carbon dioxide content of the flue gas is determined only during the few hours of the test. The usual procedure is to slowly withdraw a low-volume integrated sample into a plastic bag over the duration of the test. The bag sample is then analyzed manually using an Orsat analyzer or instrumentally using a NDIR analyzer.

SPECIFICATION FORMS

For the specification of carbon dioxide analyzers, the forms prepared by the International Society of Automation, (Forms 20A1001 and 20A1002) can be used to describe the details and operating conditions of the application. These forms can be found on the next pages.

1	RESPONSIBLE ORGANIZATION		ANALYSIS DEVICE	6	SPECIFICATION IDENTIFICATIONS	
2				7	Document no	
3	ISA		Operating Parameters	8	Latest revision	Date
4				9	Issue status	
5				10		

11	ADMINISTRATIVE IDENTIFICATIONS			40	SERVICE IDENTIFICATIONS Continued		
12	Project number		Sub project no	41	Return conn matl type		
13	Project			42	Inline hazardous area cl	Div/Zon	Group
14	Enterprise			43	Inline area min ign temp	Temp ident number	
15	Site			44	Remote hazardous area cl	Div/Zon	Group
16	Area	Cell	Unit	45	Remote area min ign temp	Temp ident number	
17				46			
18	SERVICE IDENTIFICATIONS			47			
19	Tag no/Functional ident			48	COMPONENT DESIGN CRITERIA		
20	Related equipment			49	Component type		
21	Service			50	Component style		
22				51	Output signal type		
23	P&ID/Reference dwg			52	Characteristic curve		
24	Process line/nozzle no			53	Compensation style		
25	Process conn pipe spec			54	Type of protection		
26	Process conn nominal size		Rating	55	Criticality code		
27	Process conn termn type		Style	56	Max EMI susceptibility	Ref	
28	Process conn schedule no		Wall thickness	57	Max temperature effect	Ref	
29	Process connection length			58	Max sample time lag		
30	Process line matl type			59	Max response time		
31	Fast loop line number			60	Min required accuracy	Ref	
32	Fast loop pipe spec			61	Avail nom power supply		Number wires
33	Fast loop conn nom size		Rating	62	Calibration method		
34	Fast loop conn termn type		Style	63	Testing/Listing agency		
35	Fast loop schedule no		Wall thickness	64	Test requirements		
36	Fast loop estimated lg			65	Supply loss failure mode		
37	Fast loop material type			66	Signal loss failure mode		
38	Return conn nominal size		Rating	67			
39	Return conn termn type		Style	68			

69	PROCESS VARIABLES	MATERIAL FLOW CONDITIONS			101	PROCESS DESIGN CONDITIONS		
70	Flow Case Identification			Units	102	Minimum	Maximum	Units
71	Process pressure				103			
72	Process temperature				104			
73	Process phase type				105			
74	Process liquid actl flow				106			
75	Process vapor actl flow				107			
76	Process vapor std flow				108			
77	Process liquid density				109			
78	Process vapor density				110			
79	Process liquid viscosity				111			
80	Sample return pressure				112			
81	Sample vent/drain press				113			
82	Sample temperature				114			
83	Sample phase type				115			
84	Fast loop liq actl flow				116			
85	Fast loop vapor actl flow				117			
86	Fast loop vapor std flow				118			
87	Fast loop vapor density				119			
88	Conductivity/Resistivity				120			
89	pH/ORP				121			
90	RH/Dewpoint				122			
91	Turbidity/Opacity				123			
92	Dissolved oxygen				124			
93	Corrosivity				125			
94	Particle size				126			
95					127			
96	CALCULATED VARIABLES				128			
97	Sample lag time				129			
98	Process fluid velocity				130			
99	Wake/natural freq ratio				131			
100					132			

133	MATERIAL PROPERTIES			137	MATERIAL PROPERTIES Continued		
134	Name			138	NFPA health hazard	Flammability	Reactivity
135	Density at ref temp		At	139			
136				140			

Rev	Date	Revision Description	By	Appv1	Appv2	Appv3	REMARKS

Form: 20A1001 Rev 0

		RESPONSIBLE ORGANIZATION	ANALYSIS DEVICE COMPOSITION OR PROPERTY Operating Parameters (Continued)		SPECIFICATION IDENTIFICATIONS		
1				6			
2				7	Document no		
3		(ISA)		8	Latest revision	Date	
4				9	Issue status		
5				10			

		PROCESS COMPOSITION OR PROPERTY			MEASUREMENT DESIGN CONDITIONS				
		Component/Property Name	Normal	Units	Minimum	Units	Maximum	Units	Repeatability
11									
12									
13									
14									
15									
16									
17									
18									
19									
20									
21									
22									
23									
24									
25									
26									
27									
28									
29									
30									
31									
32									
33									
34									
35									
36									
37									
38–75									

Rev	Date	Revision Description	By	Appv1	Appv2	Appv3	REMARKS

Form: 20A1002 Rev 0

© 2004 ISA

Abbreviations

Bt	Billion tons
GFC	Gas filter correlation
Gt/year	Giga tons per year
GW	Giga Watt
GWP	Gross World Product
HVAC	Heating, ventilation, and air conditioning
MW	Mega Watt
ppm	Parts per billion

Bibliography

Ewing, G., Analytical Instrumentation Handbook, 3rd edn., (2013) http://www.scribd.com/doc/127912015/Ewing-s-Analytical-Instrumentation-Handbook.

International Sensor Technology, http://www.intlsensor.com/pdf/infrared.pdf.

Lipták, B. G. (Ed.), Environmental Engineers' Handbook, 2nd edn., (1997) http://www.amazon.com/Environmental-Engineers-Handbook-Second-Edition/dp/0849399718.

Lodge, J. P., Methods of air sampling and analysis, (1988) http://www.amazon.com/Methods-Sampling-Analysis-James-Lodge/dp/0873711416.

Marzouk, S. A. M., Gas analyzer for continuous monitoring of carbon dioxide in gas streams, (2010) http://www.sciencedirect.com/science/article/pii/S0925400509009988.

Sherman, R. E., Process analyzer sample-conditioning system technology, (2002) http://www.wiley.com/WileyCDA/WileyTitle/productCd-0471293644.html.

University of Utah, Atmospheric carbon dioxide measurement, (2015) co2.utah.edu/co2tutorial.php?site=7&id=4.

Yokogawa, Renewable energy, (2015) http://www.yokogawa.com/success/newenergy/index.htm.

1.8 Carbon Monoxide

R. J. GORDON (1974, 1982)　　**B. G. LIPTÁK** (1995, 2003, 2017)

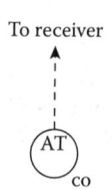

Flow sheet symbol

Detector types	A. Nondispersive infrared (NDIR); see Table 1.7b for IR analyzer designs and applications
	A1. Gas filter correlation (GFC)
	B. Mercury vapor
	C. Gas chromatography
	D. Electrochemical fuel cell
	E. Catalytic oxidation
	F. Others include, from color change badges, dosimeter tubes, and radon canisters to mass spectrometers
Reference method	Infrared
Inaccuracy	For NDIR sensors, ±1% of full scale for up to 1000 ppm and ±2% to 3% of full scale for higher ranges. For infrared stack gas analyzers, 2%–4% of reading can be expected, whereas for electrochemical ambient monitors, 1%–3% of full scale is usual.
Sensitivity	Generally 1 ppm; chromatographs can provide 0.1 ppm and mercury vapor analyzers 0.05 ppm.
Application	For ambient air monitoring, electrochemical sensors are used most often; for detecting stack gas concentration, infrared sensors are the most popular.
Ranges (see Table 1.8a)	A. From 0 to 1, 0 to 5, 0 to 10, 0 to 20, 0 to 50, and 0 to 100 ppm for ambient and 0 to 200, 0 to 500, 0 to 1000, 0 to 2000, 0 to 5000, and 0 to 10,000 ppm for other applications, including stack gas. (See Table 8.9b for overall range capability.)
	A1. 0–100 ppb to 0–1,000 ppm.
	B. 0–50 ppm
	C. 0–200 ppm
	D. 0–50 to 0–500 ppm or more
	E. 0–500 ppm
Response times	Infrared units are usually adjustable down to a few seconds, whereas electrochemical sensors require 30–60 sec.
Costs	Manual spot sensor tubes: Under $100. Pocket-size, battery-operated personal toxic gas monitor—$200–$700; continuous industrial electrochemical or infrared monitor/alarm/transmitter— $400–$2000; portable, battery-operated flue gas analyzer—$2000–$5000; mercury vapor analyzer—$7500; NDIR with recorder included—about $10,000; and gas chromatograph (see Chapter 227)—$25,000 and up.
Partial list of suppliers	Alnor, http://www.tsi.com/alnor-cga-801-carbon-monoxide-analyzer/
	AMC (Armstrong Monitoring Corp.) (http://www.armstrongmonitoring.com/cm2.cfm?fid=2&&extranet=0&sid=1&lang=1&html=carbon-monoxide-gas-detection-equipment.html)
	Ametek/Thermox (http://www.ametekpi.com/searchresults.aspx?Keywords=carbon%20monoxide)
	Bacharach, http://www.mybacharach.com/wp-content/uploads/pdf/Snifit-50/Snifit50%20datasheet.pdf
	Bran & Luebbe (http://www.spx.com)
	CEA Instruments (http://www.ceainstr.com/CO.html)
	Ecotech (http://www.ecotech.com.au/gas-analyzers-categories/stack)
	EMS (Environmental Monitoring Systems) (http://www.emssales.net/store/cart.php?m=product_detail&slug=extech-co10-carbon-monoxide-meter)

(Continued)

Enviro Technology (http://www.et.co.uk/products/scientific-research/los-gatos-research-lgr-off-axis-icos-analysers/trace-gas/carbon-monoxide-co-analyser/)
Fluke (http://www.fluke.com/fluke/usen/electrical-test-tools/gas-measurement-testers/fluke-co-220.htm?PID=56159)
Horiba Instrument Inc. (http://www.horiba.com/us/en/process-environmental/products/ambient/details/apma-370-ambient-carbon-monoxide-monitor-270/)
Inspector Tools, http://www.inspectortools.com/CO-Detectors-s/48.htm
Ketech, http://ketech.hu/non-destructive-equipments-and-materials/hands-free-single-gas-monitor/3120
MSA Instrument Div. (http://ca.msasafety.com/Fixed-Gas-%26-Flame-Detection/Gas-Detectors/Chemgard%26reg%3B-Photoacoustic-Infrared-Gas-Monitor-Series/p/00007000050000610)
Nova, http://www.nova-gas.com/analyzers/carbon-monoxide
OxyCheq, http://www.oxycheq.com/analyzers-sensors/analyzers/carbon-monoxide-analyzers.html
Sensidyne Inc. (http://www.sensidyne.com/colorimetric-gas-detector-tubes/detector-tubes/sensidyne_gas_detector_tube_number-106SC_CARBON_MONOXIDE1-50ppm.php)
SensIt, http://www.gasleaksensors.com/products/sensit_co.html
Servomex Co. (http://www.servomex.com/servomex/web/web.nsf/en/ir1520)
Siemens Energy & Automation (http://www.industry.usa.siemens.com/automation/us/en/process-instrumentation-and-analytics/process-analytics/continuous-gas-analytics/flk-probe/Pages/flk-probe.aspx)
Sierra Monitor Corp. (http://www.sierramonitor.com/gas/about/Gas-Detection/Carbon-Monoxide-CO-gas-detection.php)
Teledyne/API (http://www.teledyne-api.com/manuals/04033A_300M.pdf)
Thermo Scientific (http://www.thermoscientific.com/ecomm/servlet/productscatalog?categoryId=89577&storeId=11152&from=search&taxonomytype=4)
Topac (http://topac.com/safetypalmt.html)
TPI, http://www.testproductsintl.com/707.html#.ViQKVcsrpF0
Yokogawa Corp. of America (http://www.yokogawa.com/an/nr800/an-nr-01en.htm)

INTRODUCTION

Carbon monoxide (CO) is a colorless, odorless, and tasteless gas that is slightly less dense than air. It is toxic to humans and animals because it interferes with the normal hemoglobin function of carrying oxygen to body tissues.

Chapter 1.26 discusses the measurement of all toxic and hazardous gases while the coverage of this chapter is limited to the monitoring of carbon dioxide.

Carbon monoxide (CO) is a toxic gas and, as such, it is monitored in the ambient air for personal safety reasons. The threshold limit value (TLV) adopted by the American Conference of Governmental Industrial Hygienists for CO is 50 ppm. In industrial applications, lower limits are often used for alarm set points. Electrochemical and catalytic sensors are used in personal toxic gas monitors.

Carbon monoxide is also an indicator of incomplete combustion and, therefore, it is measured to optimize boilers and other combustion processes. For these applications, the most frequently used analyzer is the nondispersive infrared (NDIR) sensor (Chapter 1.33).

In the chemical processing industries, when higher sensitivity (0.1 ppm) measurements of carbon monoxide are required, mercury vapor (Chapter 1.37) and gas chromatographic analyzers (Chapter 1.10) are also used. Each of these techniques will be briefly described in the discussion that follows.

CALIBRATION TECHNIQUES

Ambient carbon monoxide analyzers (Table 1.8a) are calibrated with gas mixtures of known concentrations. Such mixtures may be prepared by volumetric dilution of pure

TABLE 1.8a
Atmospheric Carbon Monoxide Analyzers

Detection Method	Range, ppm	Sensitivity, ppm	Advantages	Disadvantages
NDIR	0–25, 50, 100	0.5–1	U.S. reference method, accurate, stable, dry gases	Sensitive, water, and CO_2 interferences (correctable), zero gas problems
Mercury vapor (hot HgO + CO releasing Hg vapor)	0–50	0.05	Sturdy, accurate, dry gases, high sensitivity	Interferences by water and other gases
Gas chromatography (reduction of CO to CH_4, flame ionization detection)	0–200	0.1	Accurate, high sensitivity, also read CH_4, dry gases	Complex and expensive

carbon monoxide with nitrogen or helium, which are free of carbon monoxide. If the volumes used in dilution are accurately known, the concentration may be calculated.

A known small volume of pure carbon monoxide is placed in an evacuated tank of known volume, and the tank is then filled with the dilutor gas. The pressures of the carbon monoxide volume and the final mixture must be known (or else known to be equal). Smaller samples may be prepared in plastic bags by injecting pure carbon monoxide from a gas syringe into a stream of dilutor gas metered accurately into the bag with a flow device such as a wet test meter.

Because cylinder nitrogen commonly has small amounts of carbon monoxide present in it, it must not be assumed to be pure without verification. Pure helium is reliably free of carbon monoxide. A useful procedure is to zero the detector using helium. After that, one can measure the CO content of the available cylinder nitrogen and can make the required correction. This permits using the less expensive nitrogen for most calibrations.

A reference gas mixture whose carbon monoxide content is not accurately known can also be analyzed gravimetrically or volumetrically by various methods, but these techniques are usually less convenient and often require large amounts of gas.

ANALYZER TYPES

Carbon monoxide analyzers can be distinguished by their applications and by their operating principles. Their main applications in the industry serve personal protection and combustion process optimization, and these analyzers most often utilize infrared detectors.

In nonindustrial applications, their purpose is personal protection, and the operating principles used include biomedical (a gel changes color when it absorbs carbon monoxide, and this color change triggers the alarm) metal oxide semiconductor (a silica chip's lowers its electrical resistance, and this change triggers the alarm) and electrochemical (an electrode in a chemical solution senses the change in electrical current and triggers the alarm.

In the discussion in this chapter, all of these designs will be discussed, but the reader is reminded that separate chapters are also provided in this volume giving more detailed descriptions of each of these instruments.

Nondispersive Infrared (NDIR)

NDIR analysis* is the reference method for the U.S. National Air Quality Standard for carbon monoxide and is discussed in detail in Chapter 1.33. It allows continuous analysis, because carbon monoxide absorbs the infrared radiation at a wavelength of 4.6 μm. Because infrared absorption is a nonlinear measurement, it is necessary for the analyzer to accurately linearize its output signal.

---

* EPA, Carbon Monoxide (CO) Standards, (2015) http://www3.epa.gov/ttn/naaqs/standards/co/s_co_index.html.

A schematic diagram of a typical NDIR analyzer is shown in Figure 1.8a. Infrared radiation from a hot filament is chopped to pass alternately through sample and reference cells, to be absorbed in the detector cell divided by a pressure-sensitive diaphragm. If the sample contains carbon monoxide, it will absorb part of the radiation, causing that half of the detector to exert less pressure on the diaphragm, whose distortion is converted to an electrical signal for rectification and amplification.

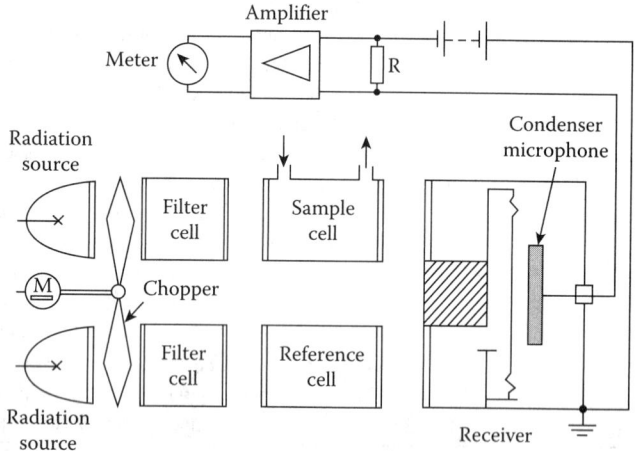

FIG. 1.8a
NDIR carbon monoxide analyzer.

Sample airflow is continuous at or sometimes above atmospheric pressure. The cell is commonly 0.5 m long. The measuring range usually extends from a minimum of 0.5–1 ppm up to a full-scale range of 25–100 ppm. Response times are in the range of less than 1–5 min. Although the NDIR response is nonlinear, it is assumed to be linear over the limited calibration range in use. Some instruments correct for nonlinearity in the output amplifier. These analyzers can be operated by nontechnical personnel.

Infrared carbon monoxide analyzers are also available for direct duct mounting in industrial combustion processes as long as the gas temperature is under 600°C (1112°F) and the concentration of CO is between 100 and 10,000 ppm (Figure 1.8b)

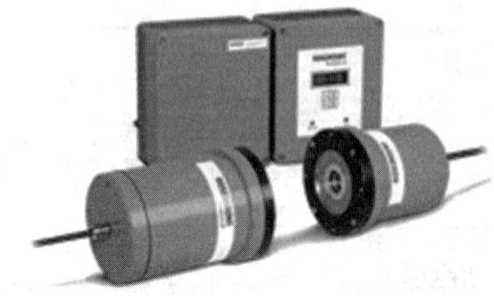

FIG. 1.8b
Direct duct mounted infrared analyzer to monitor combustion processes. (Courtesy of Emerson.)

Interferences

Carbon dioxide and water vapor in the sample interfere with the measurement. Filter cells filled with these gases, or optical filters when placed in front of the cells, can minimize effects of normal atmospheric levels of these interfering gases. The control of water vapor interference by its removal with desiccants (e.g., silica gel) or by a refrigerator condenser is preferable in many cases, even at the cost of some increase in response time.

Gas Filter Correlation (GFC)

When very accurate low-level measurements are needed, or when background gases that have the potential to interfere with the measurement are present, gas filter correlation (GFC) is used (Figure 1.8c). In these designs, the measuring and reference filters are replaced by gas-filled cuvettes. The reference cuvette is filled with CO and the measuring cuvette usually with nitrogen.

In addition to being unaffected by the presence of background gases, both the accuracy and the response time of these instruments are better than those using filters. If GFC is used in combination with single-beam, dual-wavelength technology, it is virtually immune to obstruction of the optics. This in turn prevents drift and thereby reduces the frequency at which recalibration is needed.

Mercury Vapor Analyzer

Carbon monoxide is oxidized by hot mercuric oxide as follows:

$$CO + HgO\,(s) \xrightarrow{(210°C)} CO_2 + Hg\,(g) \qquad 1.8(1)$$

The mercury vapor released may be measured photometrically. An analyzer based on this principle (Figure 1.8d) can be operated continuously. It has higher sensitivity than

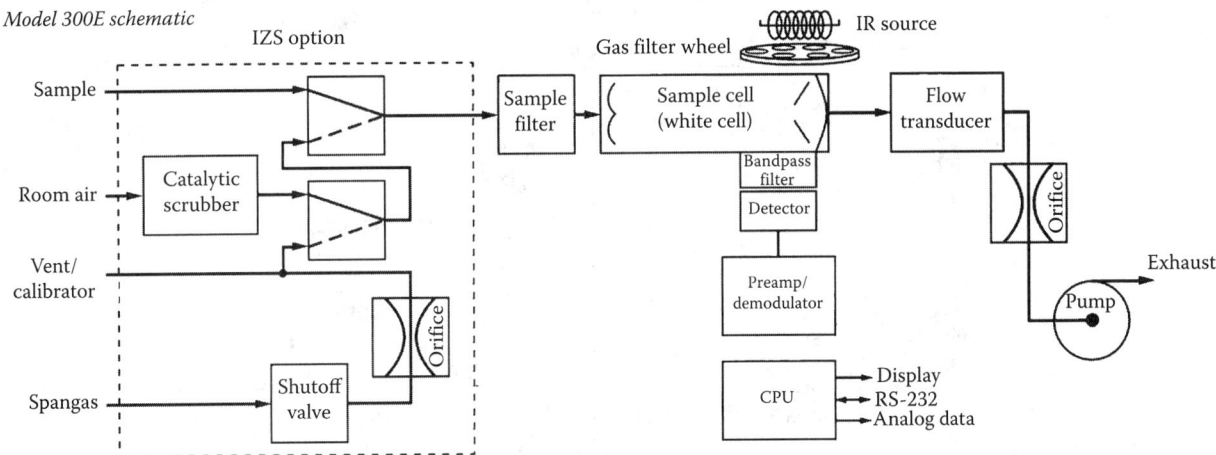

FIG. 1.8c
GFC-type carbon monoxide analyzer. (Courtesy of Teledyne Technologies Co.)

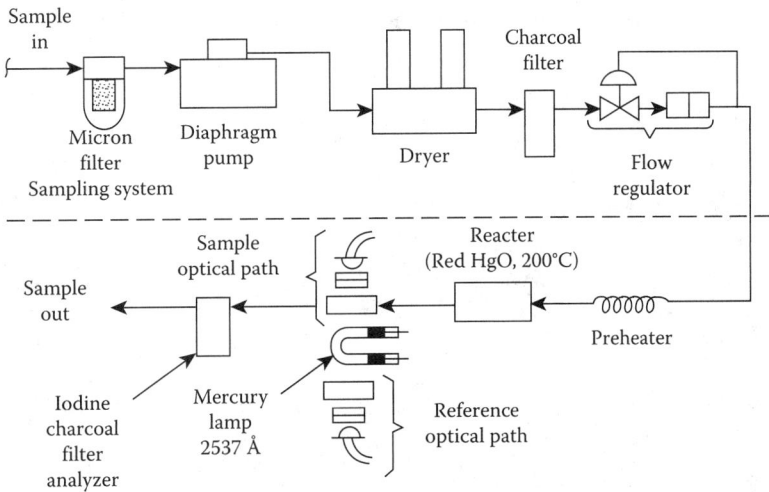

FIG. 1.8d
Mercury vapor carbon monoxide analyzer.

does the NDIR type, but it also suffers from some interference. Detection levels go down to 0.025 ppm, and changes of a tenth of that can be observed. Oxygenated hydrocarbons, olefins, and hydrogen interfere with the measurement, but all are normally at much lower concentrations in air than carbon monoxide. Water also interferes and should be removed by a dryer. This instrument has found particular use in nonurban measurements where carbon monoxide levels are low.

Gas Chromatograph

Gas chromatographs are discussed in detail in Chapter 1.10. Figure 1.8e illustrates an automated gas chromatograph that is the heart of a high precision and specificity system measuring methane and carbon monoxide.

A precolumn prevents carbon dioxide, water, and hydrocarbons other than methane from reaching the molecular sieve separation column. After separation, a catalytic nickel reactor converts carbon monoxide to methane, which is detected by flame ionization. The system permits determination of both methane and carbon monoxide about once every 5 min. The output is linear for both components and can be read from 0.1 to 200 ppm. The instrument is relatively complex and expensive, however, and requires technically trained operators.

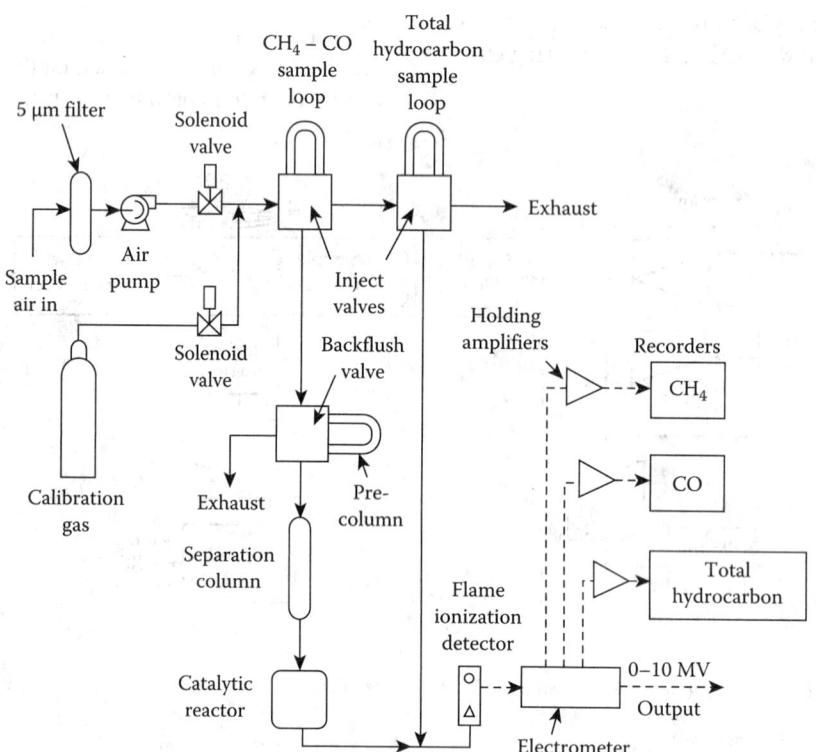

FIG. 1.8e
Methane and carbon monoxide analyzer.

Electrochemical

A galvanic cell for continuous carbon monoxide analysis* is based on the reaction of carbon monoxide with iodine pentoxide:

$$5CO + I_2O_5 \xrightarrow{(150°C)} 5CO_2 + I_2 \qquad 1.8(2)$$

The iodine liberated is absorbed by an electrolyte and reaches the cathode of a galvanic cell where it is reduced. The resulting current is measured by a galvanometer. Interference by mercaptan, hydrogen sulfide, hydrogen, olefins, and acetylene may be minimized by sampling through an absorption tube of mercuric sulfate on silica gel. Water vapor interference can be eliminated by the use of a drying column.

The same reaction is used in a coulometric method with a modified Hersch-type cell. The iodine is passed into the cell, and the current flow is measured by an electrometer. The interference possibilities are the same as those for the galvanic analyzer.

The minimum detectable concentration is 1 ppm with good precision if flow rates and temperature are controlled. Careful column preparation is required, and the response time is relatively slow.

PORTABLE MONITORS

For purposes of personal protection, battery-operated portable units are available. These units are usually provided with one or two alarm set points and with memory for some thousands of data points, together with their times and dates.

These pocket-size, battery-operated, portable electrochemical detectors are usually provided with digital displays and audible alarms. They can be configured for one or more monitoring channels. Figure 1.8f illustrates a unit with four individual channels of detection.

For industrial applications, carbon monoxide alarms are also available in microprocessor controlled "hands free" designs (Figure 1.8g), that can be worn on the wrist like a watch or attached to the belt or hard hat. They provide distinctive audible and visual alarms when the presence of carbon monoxide or other dangerous gases are detected. They are battery operated for about 3000 hr or about 1 year of normal use.

---

*EPA, Air Quality Criteria for Carbon Monoxide, http://nepis.epa.gov/Exe/ZyNET.exe/9100E2UA.TXT?ZyActionD=ZyDocument&Client=EPA&Index=Prior+to+1976&Doc.

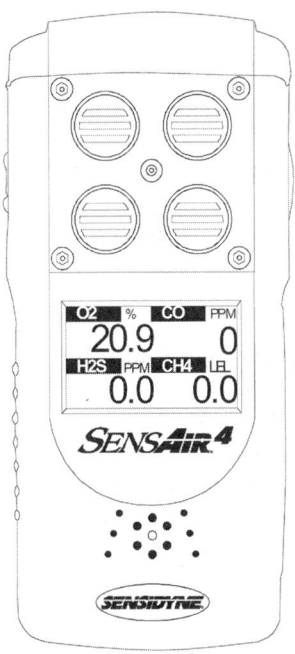

FIG. 1.8f
Typical pocket-size portable electrochemical detector configured for four individual channels of detection. (Courtesy of Sensidyne Inc.)

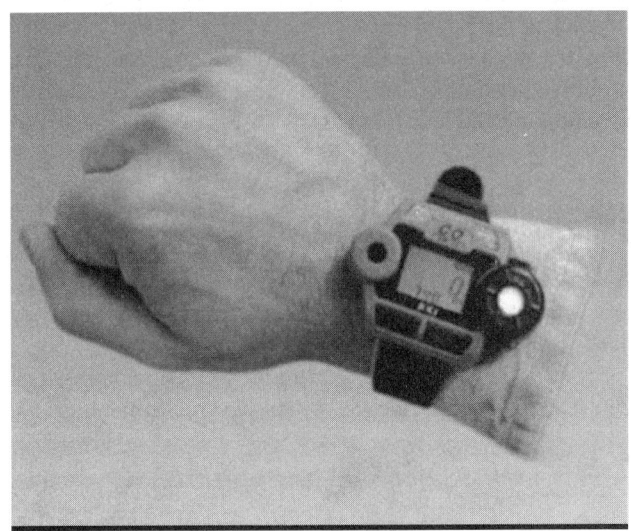

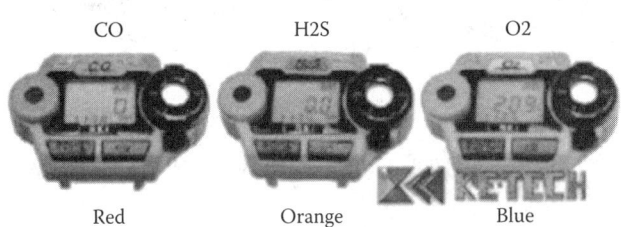

FIG 1.8g
"Hands free" alarm monitors for carbon monoxide or other dangerous gas applications. (Courtesy of Ketech.)

Catalytic Analysis

The catalyst Hopcalite will oxidize carbon monoxide to carbon dioxide. The resultant temperature rise may be recorded continuously as a measure of carbon monoxide concentration. The catalyst temperature and residence time must be controlled to avoid interference by hydrocarbons. The method is not suitable for most air monitoring applications because of low sensitivity.

SPOT SAMPLING OF AMBIENT AIR

When only intermittent analyses are required, it is convenient to collect samples in the field for later analysis in the laboratory. Rigid glass bulbs or stainless steel tanks may be evacuated and then simply opened briefly to collect the air sample. Plastic bags may be filled by means of a small air pump.

The samples may be analyzed later by various means, including use of a continuous analyzer at some other location. The samples may be analyzed in a central laboratory by an infrared spectrophotometer with a long-path gas cell or by suitable gas chromatographic apparatus.

Some colorimetric methods are also available for carbon monoxide analysis, although in general, their sensitivity and precision are low for atmospheric work. An NBS colorimetric indicating gel, if freshly prepared, will limit errors to 5%–10%, with detectability down to 0.1 ppm. The technique is simple but time consuming and tedious, with interference by oxidizing and reducing gases.

CONCLUSIONS

As discussed in more detail in Chapter 1.26, for simple and less accurate measurements of short-term carbon monoxide levels, gel tubes can be used (Figure 1.8h).

Similarly, simple devices are the breath CO monitors which detect carbon monoxide gas by an electrochemical sensor.

FIG. 1.8h
Carbon monoxide measuring tube with a range of 1–50 ppm. (Courtesy of Sensidyne Inc.)

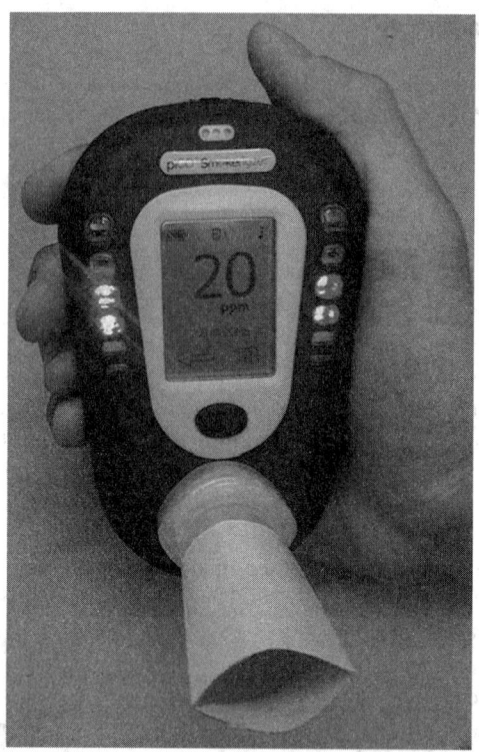

FIG. 1.8i
Commercial breath CO monitor, displaying a breath concentration together with the corresponding blood concentration.

These monitors typically incorporate an LCD display that provides a concentration level in a numeric format and/or colored indicators that correspond to various concentration ranges (Figure 1.8i).

For continuous and higher precision measurements, nondispersive infrared analysis is the most common.

If it is necessary to also measure methane, the combination gas chromatograph with carbon monoxide analyzer is worthy of consideration, but this is an expensive choice. The mercury vapor analyzer is suitable where the carbon monoxide levels are low.

SPECIFICATION FORMS

For the specification of carbon monoxide analyzers, the forms prepared by the International Society of Automation, (Forms 20A1001 and 20A1002) can be used to describe the details and operating conditions of the application. These forms can be found on the next pages.

1.8 Carbon Monoxide

1	RESPONSIBLE ORGANIZATION	ANALYSIS DEVICE	6	SPECIFICATION IDENTIFICATIONS	
2			7	Document no	
3	(ISA logo)	Operating Parameters	8	Latest revision	Date
4			9	Issue status	
5			10		

11	ADMINISTRATIVE IDENTIFICATIONS			40	SERVICE IDENTIFICATIONS Continued			
12	Project number		Sub project no	41	Return conn matl type			
13	Project			42	Inline hazardous area cl		Div/Zon	Group
14	Enterprise			43	Inline area min ign temp		Temp ident number	
15	Site			44	Remote hazardous area cl		Div/Zon	Group
16	Area	Cell	Unit	45	Remote area min ign temp		Temp ident number	
17				46				
18	SERVICE IDENTIFICATIONS			47				
19	Tag no/Functional ident			48	COMPONENT DESIGN CRITERIA			
20	Related equipment			49	Component type			
21	Service			50	Component style			
22				51	Output signal type			
23	P&ID/Reference dwg			52	Characteristic curve			
24	Process line/nozzle no			53	Compensation style			
25	Process conn pipe spec			54	Type of protection			
26	Process conn nominal size	Rating		55	Criticality code			
27	Process conn termn type	Style		56	Max EMI susceptibility		Ref	
28	Process conn schedule no	Wall thickness		57	Max temperature effect		Ref	
29	Process connection length			58	Max sample time lag			
30	Process line matl type			59	Max response time			
31	Fast loop line number			60	Min required accuracy		Ref	
32	Fast loop pipe spec			61	Avail nom power supply			Number wires
33	Fast loop conn nom size	Rating		62	Calibration method			
34	Fast loop conn termn type	Style		63	Testing/Listing agency			
35	Fast loop schedule no	Wall thickness		64	Test requirements			
36	Fast loop estimated lg			65	Supply loss failure mode			
37	Fast loop material type			66	Signal loss failure mode			
38	Return conn nominal size	Rating		67				
39	Return conn termn type	Style		68				

69	PROCESS VARIABLES	MATERIAL FLOW CONDITIONS			101	PROCESS DESIGN CONDITIONS			
70	Flow Case Identification			Units	102	Minimum	Maximum		Units
71	Process pressure				103				
72	Process temperature				104				
73	Process phase type				105				
74	Process liquid actl flow				106				
75	Process vapor actl flow				107				
76	Process vapor std flow				108				
77	Process liquid density				109				
78	Process vapor density				110				
79	Process liquid viscosity				111				
80	Sample return pressure				112				
81	Sample vent/drain press				113				
82	Sample temperature				114				
83	Sample phase type				115				
84	Fast loop liq actl flow				116				
85	Fast loop vapor actl flow				117				
86	Fast loop vapor std flow				118				
87	Fast loop vapor density				119				
88	Conductivity/Resistivity				120				
89	pH/ORP				121				
90	RH/Dewpoint				122				
91	Turbidity/Opacity				123				
92	Dissolved oxygen				124				
93	Corrosivity				125				
94	Particle size				126				
95					127				
96	CALCULATED VARIABLES				128				
97	Sample lag time				129				
98	Process fluid velocity				130				
99	Wake/natural freq ratio				131				
100					132				

133	MATERIAL PROPERTIES			137	MATERIAL PROPERTIES Continued		
134	Name			138	NFPA health hazard	Flammability	Reactivity
135	Density at ref temp	At		139			
136				140			

Rev	Date	Revision Description	By	Appv1	Appv2	Appv3	REMARKS

Form: 20A1001 Rev 0 © 2004 ISA

Analytical Measurement

	RESPONSIBLE ORGANIZATION	ANALYSIS DEVICE COMPOSITION OR PROPERTY Operating Parameters (Continued)		SPECIFICATION IDENTIFICATIONS	
1			6		
2	ISA		7	Document no	
3			8	Latest revision	Date
4			9	Issue status	
5			10		

	PROCESS COMPOSITION OR PROPERTY			MEASUREMENT DESIGN CONDITIONS				
	Component/Property Name	Normal	Units	Minimum	Units	Maximum	Units	Repeatability
11								
12								
13								
14								
15								
16								
17								
18								
19								
20								
21								
22								
23								
24								
25								
26								
27								
28								
29								
30								
31								
32								
33								
34								
35								
36								
37								
38–75								

Rev	Date	Revision Description	By	Appv1	Appv2	Appv3	REMARKS

Form: 20A1002 Rev 0

© 2004 ISA

Abbreviations

GFC   Gas filter correlation
NDIR   Nondispersive infrared
TLV   Threshold limit value

Organizations

ACGIH   American Conference of Governmental Industrial Hygienists
AIHA   American Industrial Hygiene Association
ASHRAE   American Society of Heating, Refrigerating, and Air-Conditioning Engineers
ASTM   American Society for Testing and Materials

Bibliography

(For a more complete listing of reading material on toxic and hazardous gas detection refer to Chapter 1.26)

Bittoun, R., Carbon monoxide meter, (2008) http://www.bedfont.com/file.php?f=ZmlsZSMjMTAzNQ==.

Burns, D. A., Handbook of near-infrared analysis, (2001) http://books.google.com/books/about/Handbook_of_Near_Infrared_Analysis_Secon.html?id=XkALgZVXxQQC.

Dwyer, Leatherman, Manclark, Kimball, and Rasmussen, Carbon monoxide: A clear and present danger, (2004) http://www.amazon.com/Carbon-Monoxide-Clear-Present-Danger/dp/1930044208.

Emerson-Rosemount: Continuous emission monitoring, (2013) http://www2.emersonprocess.com/siteadmincenter/PM%20Rosemount%20Analytical%20Documents/PGA_AN_42-PGA-AN-POWER-CEMS-Cogen.pdf.

EPA, Air quality criteria for carbon monoxide, http://nepis.epa.gov/Exe/ZyNET.exe/9100E2UA.TXT?ZyActionD=ZyDocument&Client=EPA&Index=Prior+to+1976&Doc.

EPA, Ambient air sampling, (2011) http://www.epa.gov/region4/sesd/fbqstp/Ambient-Air-Sampling.pdf.

EPA, An introduction to indoor air quality (IAQ), (2015) http://www.epa.gov/iaq/co.html.

EPA, Carbon monoxide (CO) standards, (2015) http://www3.epa.gov/ttn/naaqs/standards/co/s_co_index.html

Nova, Carbon monoxide, (2015) http://www.nova-gas.com/analyzers/carbon-monoxide

OSHA, Carbon monoxide in workplace atmospheres, (1991) https://www.osha.gov/dts/sltc/methods/inorganic/id210/id210.html.

Siesler, H. W., Near-infrared spectroscopy: Principles, instruments, applications, (2008) http://www.amazon.com/Near-Infrared-Spectroscopy-Principles-Instruments-Applications/dp/3527301496.

Smith, A. L., Applied infrared spectroscopy, http://books.google.com/books/about/Applied_infrared_spectroscopy.html?id=yiRRAAAAMAAJ.

Thermo Electron, Near infrared sensor for online measurement, (2005) http://www.thermo.com/eThermo/CMA/PDFs/Various/File_27448.pdf.

Van der Maas, J. H., Basic infrared spectroscopy, London: Heyden & Sons, Ltd., 1972. https://archive.org/details/BasicInfraredSpectroscopy.

Yokogawa, Carbon monoxide measurement in coal-fired power boilers, (2008) http://cdn2.us.yokogawa.com/TDLS_A_001.pdf.

1.9 Chlorine Analyzers

G. P. WHITTLE (1974, 1982) **B. G. LIPTÁK** (1995, 2003)
B. G. LIPTÁK and S. H. ORTON (2017)

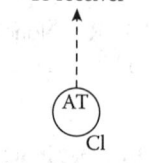

To receiver

Flow sheet symbol

Methods of detection in water	A: Requires reagent
	A1: Colorimetric: Visual
	A2: Spectrophotometric
	B: Reagent free
	B1: Amperometric
	B2: Polarographic (B2)
	C: Iodometric titration
Related chapters	Chapter 1.36: Leak detection in ambient air
	Chapter 1.10: Gas sample chromatography
	Chapter 1.70: Gas sample analysis by UV
	Chapter 1.75: Wet chemistry and auto-titration
Inaccuracy	A: Generally ±2% to 5% of full scale for ranges up to 20 ppm
	B: Unbuffered—2%FS or ±0.01 10 ppb, whichever is the greater at pH <7.5
	C: Buffered—2%FS or ±10 ppb, whichever is the greater at <pH 10
	D: Excellent accuracy
Ranges	A: Available ranges include 0–1 ppm free chlorine, and 0–3, 0–5, and 0–10 ppm total chlorine.
	B: 0–0.1, 0–1, 0–2, 0–5, 0–10, and 0–20 ppm or higher are available. The three-electrode units can measure chlorine residuals from the parts per billion (ppb) range to as high as 60 mg/l.
	C: The measurement range is from 0.001 to 10 mg/l.
Response time	A2: Responds to concentration change in 3 min
	B: Generally less than 10 (continuous analyzer)
Sampling	All three methods can be used both on grab samples and as automatic continuous analyzers.
Sample pressure	Generally atmospheric or near atmospheric. For continuous analyzer, water pressure is usually reduced to atmospheric.
Sample temperature	All methods are generally limited to the range of 0°C–49°C (32°F–120°F). Method B employs automatic temperature compensation within this range. Method A may or may not require precise temperature control, depending on the reagents employed.
Sample size of flow rate	A1: Grab samples as small as 5 ml are sufficient.
	A2: Flow rates of 10–75 ml/min are generally specified.
	B1: Flow rates of 100–750 ml/min are required.
	B2: Design probes are available for in situ installations without sampling.
Materials of construction	Enclosures for units are available in fiberglass, styrene, urethane-painted steel, vinyl-covered aluminum, and other corrosion-resistant construction, suitable for modular or control panel installation. Wetted parts are constructed of polyvinyl chloride (PVC), Teflon, Lucite, polyethylene, or glass. In method B, gold or platinum measuring and copper reference electrodes generally are employed.

(Continued)

Readout	All designs have indicating meters (usually digital) or transmitted output signals for remote display, recording, and control. High–low alarm actuation or chlorinator controls are also available.
Specificity	A, B: Determine either free or total residual chlorine. C: Measures total residual chlorine.
Interferences	A: Interfering substances may include other oxidants, for example, manganese, nitrite, and chlorine dioxide, as well as turbidity and color. B: Nitrogen trichloride and chlorine dioxide may interfere with free chlorine determinations.
Cost	A1: Visual test kits are available from $50 to $200. A2: Automatic colorimetric analyzer $3000 to $4000. B1: Portable amperometric titrator costs about $1800. B1: Corrosion-resistant amperometric transmitter system with probe cleaner costs about $6000, with the yearly cost of chemicals and maintenance amounting to $1200 or more. B2: Polarographic transmitter costs about $6000, but the cost of chemicals is low. C: Free chlorine electrode in acrylic flow cell and with residual chlorine controller is about $5000.
Partial list of suppliers	Applied Analytics Inc. (A2), http://www.a-a-inc.com/product.php?recordID=6 Emerson Process Management, Rosemount Analytical (Free and Total Chlorine), http://www2.emersonprocess.com/en-us/brands/rosemountanalytical/liquid/sensors/chlorine/fcl/pages/index.aspx Foxcroft Equipment and Science (B1), http://www.foxcroft.com/analyzers/chlorine-analyzer-controller-FX1000p-cs/ Hach (A1, B1), http://www.hach.com/chlorine/cl17-colorimetric-chlorine-analyzer/family?productCategoryId=22219801222 Hanna Instruments (A1), http://hannainst.com/usa/search_results.cfm?q=chlorine%20analyzer Honeywell, http://honeywell.com/Pages/Search.aspx?k=chlorine+analyzer Hydro Instruments (B1 and Leak detector), http://www.hydroinstruments.com/images/sce/RAH-210%20new%20[800x600].jpg; http://www.hydroinstruments.com/page.aspx?page_id=31 Omega, http://www.omega.com/pptst/CLDTX_FCLTX.html Siemens (A1), http://www.siemens.com/press/en/pressrelease/?press=/en/pressrelease/2010/industry_solutions/iis201006469.htm Teledyne Analytical Instruments (B), http://www.teledyne-ai.com/products/fca220.asp Thermo Scientific (A1, B, C), http://www.thermoscientific.com/en/product/orion-aq3070-aquafast-chlorine-meter.html; http://www.thermoscientific.com/content/tfs/en/products/chlorine-detection.html?wt.srch=1&wt.mc_id=led_chlorinedetection_adwords&adwordskeyword=%2Bchlorine+%2Banalyzer YSI, http://www.ysi.com/search.php?q=chlorine%2520 Yokogawa (B2), http://www.yokogawa.com/an/chlorine/an-fc400g-001en.htm

INTRODUCTION

Chlorine (Cl) is is a yellow-green gas under standard conditions, where it forms diatomic molecules, having an atomic mass of 35.5. It is a strong oxidizing agent. Its most common compound is sodium chloride, which is table salt (NaCl). Elemental chlorine is produced from brine by electrolysis. The high oxidizing potential of free chlorine makes it ideal for bleaching and disinfection to keep clean and sanitary conditions. Artificially produced chlorinated organics can be toxic. Elemental chlorine at high concentrations is extremely dangerous and poisonous to all living organisms and in the past was used as warfare agent.

The concentration of chlorine is of interest in both liquid and gas samples. The discussion in this chapter concentrates on the measurement of total or free residual chlorine in water, which can be measured by colorimetric, amperometric, polarographic, or iodometric analyzers. Because the most often used designs are the colorimetric and the amperometric

TABLE 1.9a
Relative Merits of Colorimetric and Amperometric Analyzers

Consideration	Colorimetric	Amperometric
Type of sample	Better suited for clarified natural or treated waters than for highly turbid or colored waters and wastewaters	Turbidity and color generally not a problem; applicable to both treated water and wastewater
Interference	Interfering ions should be absent; oxidized manganese compounds produce serious interference	Copper and silver ions may interfere by plating out on electrodes
Sample temperature	Temperature control may or may not be required, depending on reagent employed	Manual or automatic temperature compensation required
Speed of response	Generally 3 min or more required to detect a change in chlorine concentration	Chlorine concentration change detected in 10 sec or less
Calibration	Analyzer precalibrated; periodic standardization requires only simple manipulations	Periodic calibration required by separate analytical technique
Reagents required	External reagent solution required	External buffer may be required for varying sample pH
Maintenance stability	Cell staining may require periodic cleaning	Electrodes may require periodic cleaning
Stability	Drift compensated for by relatively simple standardization step	Drift not a problem when electrodes are kept clean
Initial cost	Generally less expensive	Generally more expensive

ones, the orientation table (Table 1.9a) compares the features of those two categories of chlorine analyzers.

The relative merits of the continuous colorimetric and amperometric analyzers are listed in Table 1.9a.

CHLORINE IN GAS AND WATER

The measurement of chlorine concentration in gas samples is not discussed here, except to note that ultraviolet (UV) analyzers (Chapter 1.70) and chromatographs (Chapter 1.10) with thermal conductivity or flame ionization detectors are most often used to make this measurement.

Chlorine gas leak detectors are discussed in detail in Chapter 1.35 and one microprocessor operated electrochemical unit, which is used as a chlorine leak detector is shown in Figure 1.9a. These units are capable of detecting concentrations as low as 0.5 ppm by volume.

Chlorine does not exist as Cl_2 in water. Therefore, the amount of free available chlorine is defined as the amount of chlorine that exists in the form of HOCl. Total available chlorine, on the other hand, is defined as the chlorine that exists in any of the following forms: HOCl, NH_2Cl, NCl_3, or $NHCl_3$. The combined available (residual) chlorine is obtained as the difference between total and free chlorine.

FIG. 1.9a
Leak detector for chlorine and other gases. (Courtesy of Hydro Instruments.)

ANALYZER TYPES

Iodometric

Standard laboratory methods for determining aqueous chlorine concentrations involve iodometry, in which free iodine liberated from potassium iodide is titrated with sodium thiosulfate, using starch as an indicator. All active forms of chlorine will liberate free iodine, and therefore the iodometric methods of analysis are limited to the determination of total active chlorine. For an automatic total residual chlorine analyzer with a 4–20 mADC logarithmic output, refer to Figure 1.9b.

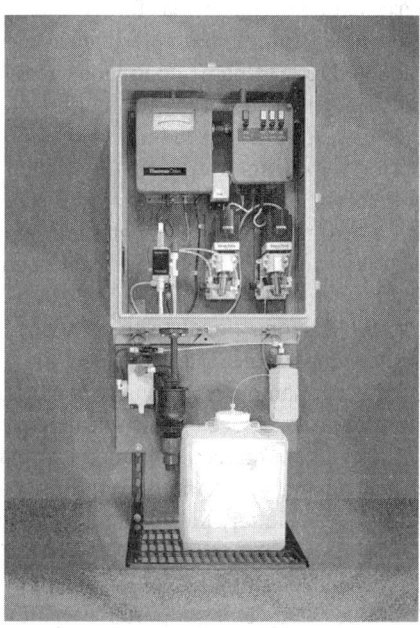

FIG. 1.9b
Iodometric total residual chlorine analyzer with a range of 0.001–10 mg/l and a 4- to 20-mADC logarithmic output. (Courtesy of Rosemount Inc.)

Colorimetric

Various colorimetric procedures are available in which a colorless indicator is oxidized to a colored product, the color intensity of which is proportional to the chlorine concentration. Many of the colorimetric procedures allow the determination of the various forms of chlorine. Colorimetric procedures are adaptable to field determination by the use of kits containing the necessary reagents, color standards for visual estimation, and portable, battery-operated colorimeters (Figure 1.9c).

In the Environmental Protection Agency (EPA) recommended DPD 330.5 method of chlorine analysis, a chemical reaction causes the intensity of the color of magenta to change in proportion to the concentration of chlorine. In the automatic analyzer, the color intensity of the sample is measured twice: once before reagents are added, for automatic zeroing and to compensate for any turbidity or natural color in the sample; and again after addition of the reagents and stirring of the sample. The second measured wavelength, around 510 nm, is compared with the reference. The chlorine concentration is calculated on the basis of the difference between the two readings.

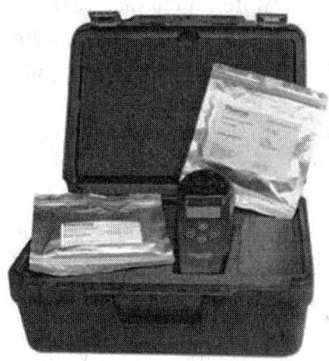

Hand-held colorimeter for calibrating in-line chlorine analyzers. (Courtesy of Siemens.)

Expanded range for free and total chlorine measurements. (Courtesy of Thermo Scientific.)

FIG. 1.9c
Portable, battery-operated visual colorimeters measure free and total chlorine per US EPA approved methods.

Figure 1.9d illustrates the basic components of a chlorine colorimetric analyzer. The water sample is introduced into a head regulator where the sample flow rate is regulated by an overflow arrangement and a capillary delivery tube. The sample is passed into a channel and is mixed with the colorimetric reagent metered through another capillary tube from a storage container. The treated sample flows over a sample heater, if required, and into the sample cell, which fills and overflows at a predetermined frequency.

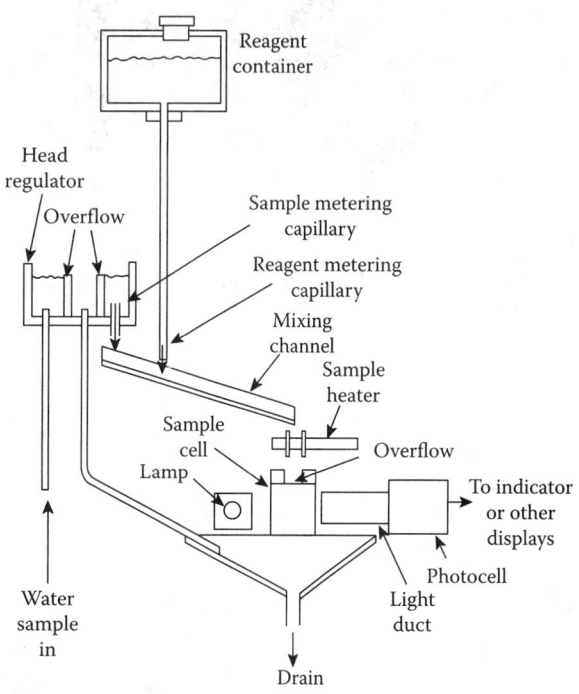

FIG. 1.9d
Chlorine colorimetric analyzer. (Courtesy of Hach Inc.)

If chlorine is present in the water, a characteristic color develops with an intensity in proportion to the amount of chlorine present. A filtered photocell develops an output signal, which is in proportion to the reduction in intensity of the transmitted light through the sample. Periodic standardization of the analyzer is required and is done by adjustment of the indicator to read 0 ppm chlorine on an untreated water sample and to read on the extreme upper end of the scale when the photocell is completely shielded.

Automatic Colorimetric Automatic free or total chlorine analyzers can control either in the on–off mode by stopping and starting the sample feed pump and sounding alarms or can provide throttling control by modulating the pumping rate (Figure 1.9e). The output signals can be analog or digital. The available analog signal ranges include 0–10 mV, 0–100 mV, 0–1 V, and 4–20 mA. Digital signals can be sent through RS232 interfaces, or thru Hart, Fieldbus or even by wireless transmission. These analyzers use about 500 ml of reagents and color indicators per month. Therefore, monthly refilling is sufficient. The cell should be cleaned at the same time of reagent refilling. Although requiring about 1 hr of maintenance every month, these are relatively inexpensive and stable analyzers that are insensitive to sample pH variations or other interferences.

Amperometric

Other titrimetric procedures employing indicators other than starch have also been developed for the determination and differentiation of the active chlorine forms. The amperometric titration method is based on polarographic principles and allows determination of the various forms of active chlorine. The amperometric titration and colorimetric methods have also been adapted to automated continuous analyzers.

In amperometric analyzers, the composition of the electrodes is so selective that the polarization of the measuring electrode prevents a current flow when no strong oxidizing agent is present in the sample. On the other hand, when a strong oxidizing agent such as chlorine is present, the electrodes are depolarized and a current flow is generated.

This current is proportional to the amount of chlorine residual in the sample. Because pH affects the dissociation constants, the water samples are usually conditioned with a buffer solution. Similarly, because of sensitivity to temperature variations, temperature compensation is provided.

Amperometric analyzers are used in three basic forms: as titrators, as free residual chlorine analyzers, and as total residual chlorine analyzers. The amperometric titrator determines the end point of a 200 ml sample in a manually filled sample jar by using a portable, battery-operated instrument.

During titration for free residual chlorine, a buffer solution maintains the sample pH at neutrality and a titrant (phenylarsine oxide) is added until no more current is produced by the electrode cell. This is considered the end point, and the amount of titrant used is an indication of free residual chlorine in the sample.

When total residual chlorine is measured, the titration process is similar, but the buffer is at a pH of 4 and potassium iodide is also added to the sample.

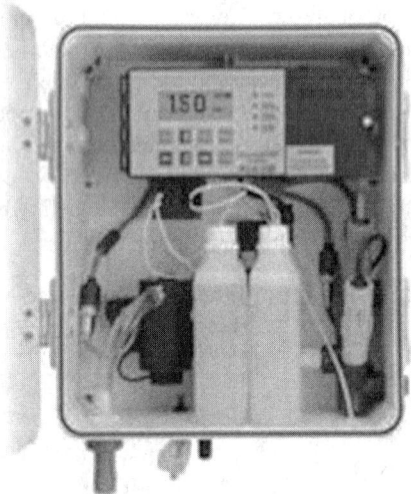

FIG. 1.9e

Automatic, EPA-compliant colorimetric chlorine analyzer (PCA 310). (Courtesy of Hach Co.)

Free Residual Chlorine Type For free residual chlorine analysis, the sample pH is held between 5 and 9. Simple flow-through units are available that do not require the use of buffers or reagents (Figure 1.9f). These analyzers are temperature compensated and operate in a range of 0–5 ppm of free chlorine, with an inaccuracy of about 0.1 ppm.

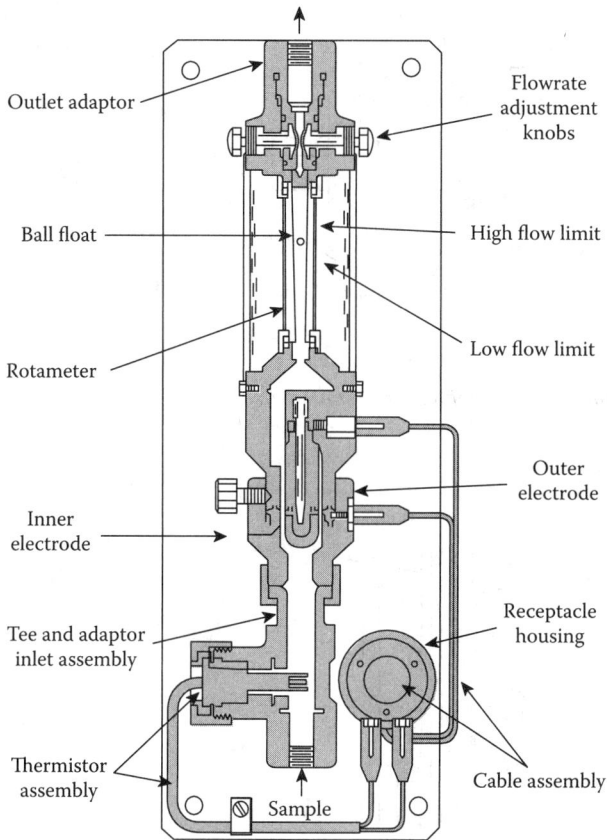

FIG. 1.9f
Measurement cell of a free residual chlorine analyzer.

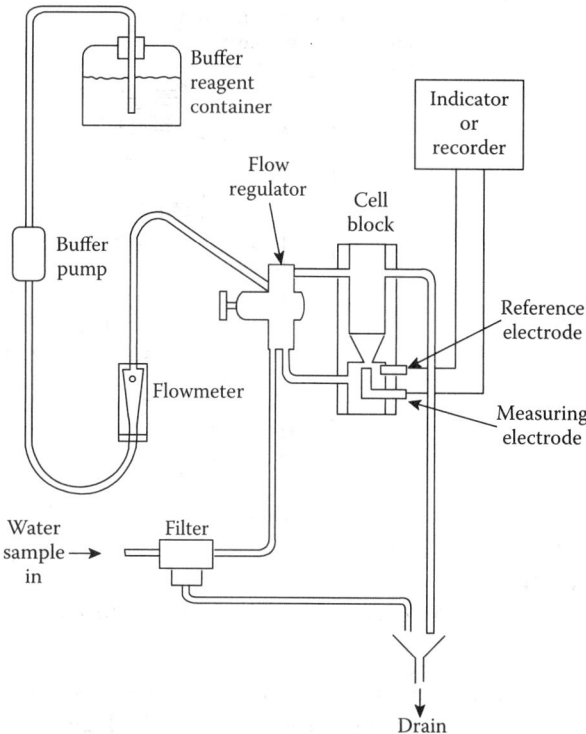

FIG. 1.9g
Chlorine amperometric analyzer.

Figure 1.9g shows the components of a typical amperometric analyzer, which does require buffers or reagents. The water sample is delivered to a diaphragm-type regulator for flow rate control. If required for pH control, a metered amount of buffer solution is also pumped to the regulator and is mixed with the sample prior to its reaching the cell block.

In the cell block, contact is made between the electrodes and the sample, and a direct current is generated in proportion to the chlorine in the sample. The cell block usually contains grit, which cleans the electrodes by means of sample velocity agitation. Periodic calibration is required through a separate determination of chlorine and is usually performed on a laboratory amperometric titrator using a standardized phenylarsine oxide solution.

Membrane Probes Microprocessor-based chlorine analyzers are available with built-in diagnostics and simulator features. They can automatically detect chlorine and temperature and can also be equipped with a pH detector probe.

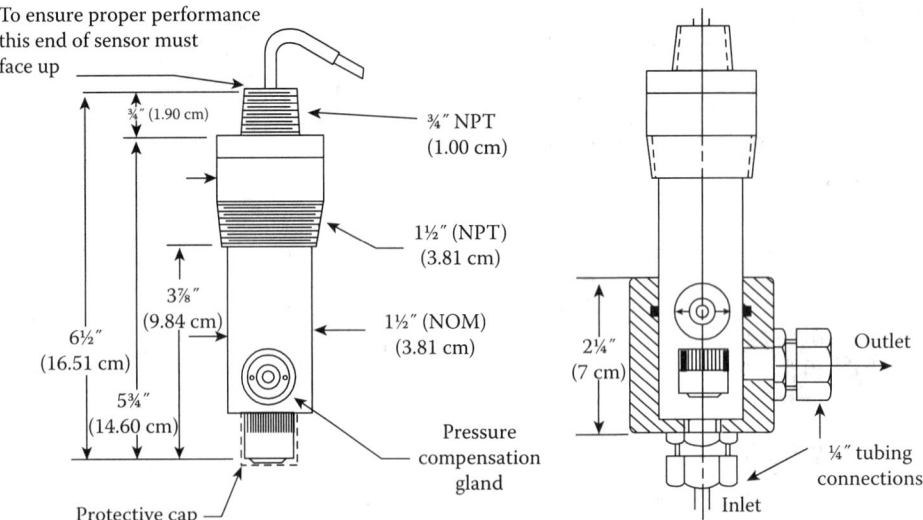

FIG. 1.9h
Amperometric chlorine probe for in-line or sample bypass-type installation. (Courtesy of Rosemount Analytical Inc.)

The amperometric probe consists of a gas-permeable membrane, a gold cathode, and a silver anode (Figure 1.9h). The probe is filled with a salt electrolyte. As chlorine penetrates the membrane, it is reduced at the gold cathode, which causes a current to flow that is proportional to the chlorine concentration in the water.

The main advantages of the immersion probe designs include the elimination of the sampling system and the elimination of the continuous need for chemical reagents and buffers. The elimination of sample pumps and other sampling system components not only lowers costs and maintenance, but also reduces transportation lag, which makes the installation better suited for closed-loop operation.

The elimination of the need for the continuous feed of chemicals reduces the yearly operating cost by thousands of dollars. The maintenance that is still needed is usually limited to replacing the membrane and refilling the probe with fresh electrolyte once every 2 months.

The membrane probe is also shown in Figure 1.9i. It can be provided with an in situ nutating sensor or with paddle-oscillation-type cleaner agitators to provide the required apparent sample velocity at the membrane. The probe can be inserted in pipes at pressures of up to 150 psig (10.6 bars) or can be handrail mounted (Figure 1.9i) in open tanks, with the local chlorine transmitter mounted on or above the handrail.

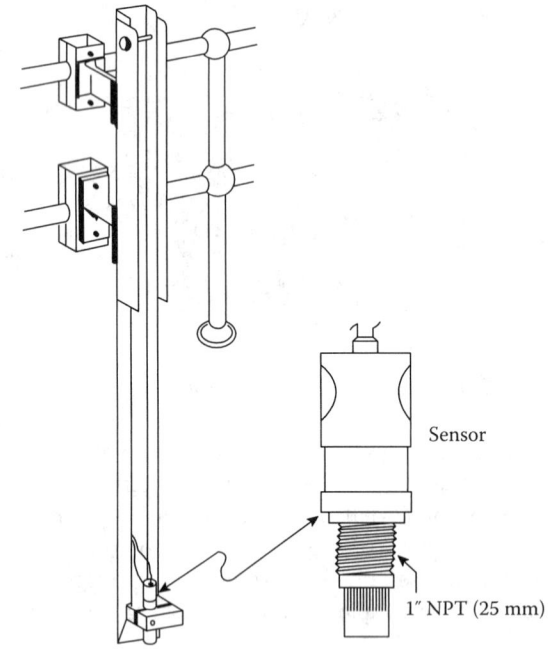

FIG. 1.9i
Immersion probe-type polarographic chlorine detector. (Courtesy of Rosemount Analytical Inc.)

Buffers and Reagents The main disadvantage of some amperometric chlorine analyzers is the need for the periodic replacement of substantial quantities of buffering and reagent chemicals.

When measuring free residual chlorine, the buffer maintains a pH of 7 and the reagent that is added to the buffer is usually potassium bromide. When measuring total residual chlorine, the buffer maintains the pH at around 4 and the reagent added to the buffer is usually potassium iodide. Other chemicals that are sometimes used to adjust the pH include acetic acid and sodium hydroxide.

The cost of these chemicals can reach several thousand dollars per year, and the scheduled replenishment of these materials also adds to the maintenance cost of operating the plant. It is one of the reasons why some users prefer the reagentless chlorine analyzers (Figure 1.9j).

Electrode Cleaners Keeping the chlorine electrodes clean also requires attention. In some designs, the measuring electrode rotates at 1550 rpm to maintain ideal electrolysis conditions (relative velocity) between the sample and the electrode surface. In other designs, the space between electrodes contains plastic pellets, which are continuously agitated by the swirling water or by a rotating scrubber (Figure 1.9k).

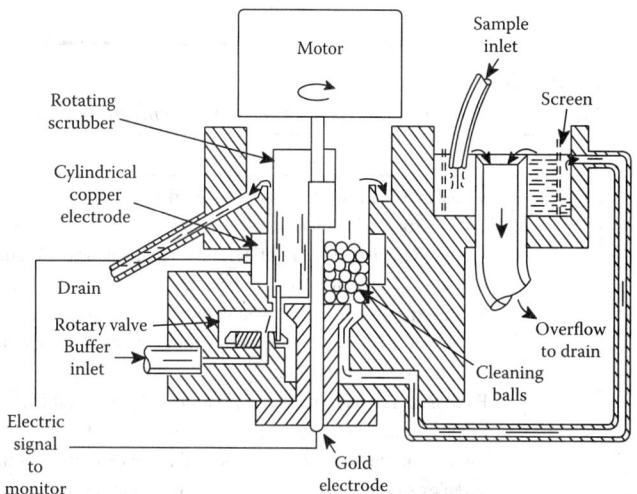

FIG. 1.9k
Electrode cleaner used in chlorine or ozone analyzers.

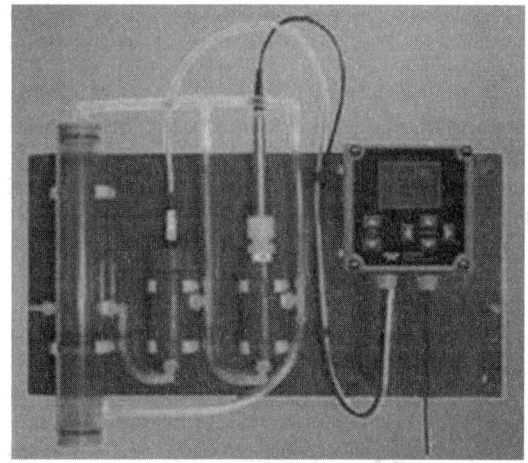

Reagent free chlorine analyzer.
(Courtesy of Teledyne Analytical Instruments.)

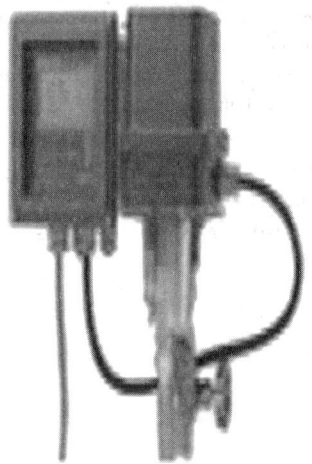

Reagent free, online, polarographic analyzer.
(Courtesy of Yokogawa.)

FIG. 1.9j
Reagent-free chlorine analyzer designs.

REGULATIONS

The U.S. Department of health has established some quality control procedures for the handling of both grab sample based and for continuous chlorine analysis. These are summarized in Table 1.9b.

CONCLUSIONS

Colorimetric and amperometric analyzers may be obtained with diverse accessories serving various functions, including alarm actuation measurement transmission, data logging, and PID control. Automatic temperature compensation is also available for the amperometric designs. An added advantage of the three-electrode amperometric designs is that they do not require zero calibration and are capable of detecting chlorine residuals as low as 1 μg/l (1 ppb).

TABLE 1.9b
Quality Control Procedures for Chlorine Concentration Measurement (Department of Health)

	Grab Sample	Online Chlorine Analyzer
Initial	1. Verify internal calibration curve according to the manufacturer's instructions, or prepare calibration curve using a method blank[a] and a set of at least three aqueous calibration standards. The standards must span the concentration range that is expected to be observed in the grab samples with the minimum at 0.2 mg/l. Each calibration point must be within ± 15% of its expected value. 2. Each field sampler must perform an IDC by analyzing a method blank[a] and five replicate independent reference samples at the same concentration. The concentration of the samples should be in the mid range of the calibration curve or near the expected concentration of the water samples. The average concentration for the five replicates must be within ± 15% of the expected value to demonstrate accuracy. The relative standard deviation (RSD)[b] must be ±15% to demonstrate precision.	1. Install the analyzer according to the manufacturer's specifications. 2. After the accuracy of the grab sample measurement is verified, collect and analyze a grab sample collected as close as feasible to the location where the sample enters the online chlorine analyzer. Compare the results from the grab sample analysis to the measurement made by the online chlorine analyzer. Adjust analyzer to have analyzer readings agree with grab sample measurements within 0.1 mg/l or ±15%. 3. Perform the IDC for the each online chlorine analyzer[c] by comparing online chlorine analyzer readings with grab sample analyses collected at least daily for consecutive 14 days. The analyzer reading must be within ±0.1 mg/l or ±15% (whichever is larger) of the grab sample measurement for each data pair. When multiple online chlorine analyzers are being installed, the IDC may be shortened if the same model analyzer is installed at each location and the water quality characteristics and treatment processes are equivalent at each location.
Routine	Prepare an aqueous calibration check standard at a concentration near the expected concentration of the water samples. The grab sample measured concentration of the calibration check standard must be within ±15% of the expected value. The results from analyses of calibration check standards must be recorded and maintained according to item 8 of this document. Min. calibration check frequency: quarterly.	1. Compare online chlorine analyzer readings with grab sample analyses. The analyzer reading must be within ±0.1 mg/l or ±15% (whichever is larger) of the grab sample measurement for each data pair. 2. Adjust the calibration of the analyzer so it gives the same value as the grab sample analysis. Follow the manufacturer's instructions. 3. After a major repair or after replacement of the online chlorine analyzer with an equivalent model, perform the initial calibration procedure. Return to the routine schedule for grab sample comparisons after verifying the accuracy of the analyzer on a daily basis for 7 consecutive days (or business days). Minimum calibration check frequency: every 5 days.

Source: https://www.health.ny.gov/environmental/water/drinking/chlorine_analyzer_guidance.htm.

[a] The method blank concentration must be ≤ ⅓ the concentration of the lowest standard used to prepare/verify the calibration curve.

[b] The standard deviation (S) and RSD are calculated using the following equations:

$$S = \sqrt{\frac{1}{N-1}\sum_{i=1}^{N}(X_i - \overline{X})^2} \quad \text{1.9(1)}$$

$$RSD = \frac{S}{\overline{X}} \times 100\% \quad \text{1.9(2)}$$

Where S is the standard deviation for the replicate values, $\overline{X}$ is the average value for the replicates, N is the number of replicates, X_i is the individual observation result and RSD is the Relative Standard Deviation.

[c] The IDC is not required if historical operating data for the on-line chlorine analyzer demonstrate the criterion are being met on an on-going basis. Historical data must show that the analyzer remains in agreement with the grab sample method over a period of two consecutive weeks without analyzer maintenance or calibration adjustment.

SPECIFICATION FORMS

For the specification of chlorine analyzers, the forms prepared by the International Society of Automation, (Forms 20A1001 and 20A1002) can be used to describe the details and operating conditions of the application. These forms are the following.

#	RESPONSIBLE ORGANIZATION	ANALYSIS DEVICE	#	SPECIFICATION IDENTIFICATIONS
1			6	
2	ISA	Operating Parameters	7	Document no
3			8	Latest revision — Date
4			9	Issue status
5			10	

#	ADMINISTRATIVE IDENTIFICATIONS	#	SERVICE IDENTIFICATIONS Continued
11		40	
12	Project number — Sub project no	41	Return conn matl type
13	Project	42	Inline hazardous area cl — Div/Zon — Group
14	Enterprise	43	Inline area min ign temp — Temp ident number
15	Site	44	Remote hazardous area cl — Div/Zon — Group
16	Area — Cell — Unit	45	Remote area min ign temp — Temp ident number
17		46	
18	SERVICE IDENTIFICATIONS	47	
19	Tag no/Functional ident	48	COMPONENT DESIGN CRITERIA
20	Related equipment	49	Component type
21	Service	50	Component style
22		51	Output signal type
23	P&ID/Reference dwg	52	Characteristic curve
24	Process line/nozzle no	53	Compensation style
25	Process conn pipe spec	54	Type of protection
26	Process conn nominal size — Rating	55	Criticality code
27	Process conn termn type — Style	56	Max EMI susceptibility — Ref
28	Process conn schedule no — Wall thickness	57	Max temperature effect — Ref
29	Process connection length	58	Max sample time lag
30	Process line matl type	59	Max response time
31	Fast loop line number	60	Min required accuracy — Ref
32	Fast loop pipe spec	61	Avail nom power supply — Number wires
33	Fast loop conn nom size — Rating	62	Calibration method
34	Fast loop conn termn type — Style	63	Testing/Listing agency
35	Fast loop schedule no — Wall thickness	64	Test requirements
36	Fast loop estimated lg	65	Supply loss failure mode
37	Fast loop material type	66	Signal loss failure mode
38	Return conn nominal size — Rating	67	
39	Return conn termn type — Style	68	

#	PROCESS VARIABLES — MATERIAL FLOW CONDITIONS — Units	#	PROCESS DESIGN CONDITIONS — Minimum — Maximum — Units
69		101	
70	Flow Case Identification	102	
71	Process pressure	103	
72	Process temperature	104	
73	Process phase type	105	
74	Process liquid actl flow	106	
75	Process vapor actl flow	107	
76	Process vapor std flow	108	
77	Process liquid density	109	
78	Process vapor density	110	
79	Process liquid viscosity	111	
80	Sample return pressure	112	
81	Sample vent/drain press	113	
82	Sample temperature	114	
83	Sample phase type	115	
84	Fast loop liq actl flow	116	
85	Fast loop vapor actl flow	117	
86	Fast loop vapor std flow	118	
87	Fast loop vapor density	119	
88	Conductivity/Resistivity	120	
89	pH/ORP	121	
90	RH/Dewpoint	122	
91	Turbidity/Opacity	123	
92	Dissolved oxygen	124	
93	Corrosivity	125	
94	Particle size	126	
95		127	
96	CALCULATED VARIABLES	128	
97	Sample lag time	129	
98	Process fluid velocity	130	
99	Wake/natural freq ratio	131	
100		132	

#	MATERIAL PROPERTIES	#	MATERIAL PROPERTIES Continued
133		137	
134	Name	138	NFPA health hazard — Flammability — Reactivity
135	Density at ref temp — At	139	
136		140	

Rev	Date	Revision Description	By	Appv1	Appv2	Appv3	REMARKS

Form: 20A1001 Rev 0 © 2004 ISA

ANALYSIS DEVICE COMPOSITION OR PROPERTY
Operating Parameters (Continued)

	RESPONSIBLE ORGANIZATION			SPECIFICATION IDENTIFICATIONS		
1	ISA		6			
2			7	Document no		
3			8	Latest revision		Date
4			9	Issue status		
5			10			

	PROCESS COMPOSITION OR PROPERTY			MEASUREMENT DESIGN CONDITIONS				
	Component/Property Name	Normal	Units	Minimum	Units	Maximum	Units	Repeatability
13								
14								
15								
... (rows 13–37)								

Rows 38–75: open area

Rev	Date	Revision Description	By	Appv1	Appv2	Appv3	REMARKS

Form: 20A1002 Rev 0 © 2004 ISA

Definitions

Free chlorine is the amount of chlorine that exists in the form of HOCl in the water.

Iodometric titration is a method of volumetric chemical analysis where the appearance or disappearance of elementary iodine indicates the end point. Iodometry involves indirect titration of iodine liberated by reaction with the analyte, whereas iodimetry involves direct titration using iodine as the titrant.

Residual chlorine is the difference between total and free chlorine.

Total available chlorine is the chlorine that exists in any of the following forms: $HOCl$, NH_2Cl, NCl_3, or $NHCl_3$ in the water.

Organizations

APHA	American Public Health Association
AWWA	American Water Works Association
IDC	Initial Demonstration of Capability
ITA	Instrumentation Testing Association
NIST	National Institute of Standards and Technology
RSD	Relative Standard Deviation
WEF	Water Environment Federation
WHO	World Health Organization

Bibliography

APHA, *Standard Methods for the Examination of Water and Wastewater*, 22nd edn., Washington, DC: APHA (2012), http://www.amazon.com/Standard-Methods-Examination-Water-Wastewater/dp/0875530133.

AWWA, Chlorine Safety DVD (2014), http://www.awwa.org/store/productdetail.aspx?ProductId=36027583.

Department of Health, Guidance for using an on-line chlorine analyzer for compliance monitoring (2014), https://www.health.ny.gov/environmental/water/drinking/chlorine_analyzer_guidance.htm.

Emerson-Rosemount, Chlorine measurement by amperometric sensor application (2009), http://www2.emersonprocess.com/siteadmincenter/pm%20rosemount%20analytical%20documents/liq_ads_43-6063.pdf.

EPA, Determination of residual chlorine in drinking water using residual chlorine analyzers (2009), http://water.epa.gov/scitech/drinkingwater/labcert/upload/met334_0.pdf.

EPA, Method 334.0: Determination of residual chlorine in drinking water using an on-line chlorine analyzer, (2009) http://water.epa.gov/scitech/drinkingwater/labcert/upload/met334_0.pdf.

EPA, Method 330.5: Total residual chlorine by spectrophotometer (2013), http://www.caslab.com/EPA-Method-330_5/.

HDR Engineering Inc., *Handbook of Public Water Systems*, New York: Wiley (2001), http://www.awwa.org/store/product-detail.aspx?ProductId=6464.

Instrumentation Testing Association, Chlorine residual analyzers, http://www.amazon.com/Residual-Analyzers-Wastewater-Treatment-Applications/dp/1583460063.

Metzger, M. R., Measuring chlorine (2007), http://www.wqpmag.com/measuring-chlorine.

Myron L. Co., Free chlorine equivalent (2012), http://www.myronl.com/PDF/fcetr.pdf.

Rosemount, Chlorine measurement by amperometric sensor, (2009) http://www2.emersonprocess.com/siteadmincenter/PM%20Rosemount%20Analytical%20Documents/Liq_ADS_43-6063.pdf.

Sherman, R. E., *Process Analyzer Sample-Conditioning System Technology*, New York: John Wiley & Sons (2002), http://www.amazon.com/Process-Analyzer-Sample-Conditioning-Technology/dp/0471293644.

WHO, Chlorite in drinking water http://www.who.int/water_sanitation_health/dwq/chemicals/chlorateandchlorite0505.pdf.

WHO, How to measure chlorine residual in water (2005), http://www.who.int/water_sanitation_health/hygiene/envsan/chlorineresid.pdf.

Yokogawa (2008) http://c15181565.r65.cf2.rackcdn.com/ORP_A_001.pdf.

1.10 Chromatographs
Gas

R. ANNINO (1995, 2003)   **B. G. LIPTÁK** (2017)

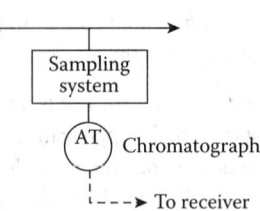

Flow sheet symbol

Types of sample	Gas, vapor, or vaporizable liquid
Detector options	Thermal conductivity (TCD), flame ionization (FID), flame photometric (FPD), and helium ionization (HID). Column options include capillary, micropacked, and packed columns
Sample pressure	0–100 psig (1–7 bar)
Ambient temperature	−20°C to 50°C (−4°F to 122°F); however, a sheltered environment is recommended
Analysis zone	60°C–180°C (140°F–356°F) provided that the desired oven temperature is at least 10°C (18°F) above ambient (stable oven temperature ±0.05°C (±0.1°F) at steady ambient ±0.5°C (±1°F) for −20°C to 50°C (−4°F to 122°F) ambient change)
Contacting material	Stainless steel or surface-deactivated steel for trace analysis of reactive compounds, teflon
Auxiliary utilities	Instrument air-dry, oil-free, available at 50 psig min (3.5 bar) and 2–3 scfm (1–1.5 sl/sec); carrier gas, zero air, and hydrogen for FID; possible steam tracing or electrical heating for sample lines
Cycle time	2–20 min, depending on application and packed or capillary column operation
Special features	Accepts other inputs (e.g., flow rate and density in the calculation of output, e.g., BTU/hr); built-in diagnostics with local and remote indication; multicomponent readout; stand-alone or networked; communication link (fieldbus or Ethernet directly to the plant or local LAN and local control system); local and remote operator interface; single-stream or multistream analysis
Location	Class 1, Groups B, C, and D, Division 1 hazardous areas
Analyzer cost	Basic analyzer, $25,000–$45,000
Installed cost	$45,000–$100,000 (depending on type, sampling system, and application)
Partial list of suppliers	ABB Process Analytics (http://www.abb.com/product/hu/9AAC128815.aspx?country=US) Agilent Technologies, https://www.agilent.com/cs/library/brochures/5989-9384EN.pdf Analytical Instruments, http://www.alpha-mos.com/analytical-instruments/pr2100-gas-chromatograph.php Applied Instrument Technologies (http://www.aitanalyzers.com/process-analyzer-product.php?id=13) ChemInfo, http://www.chem.info/product-releases/2012/05/high-speed-gas-chromatographs Emerson Process Management (http://www2.emersonprocess.com/en-US/brands/rosemountanalytical/GC/PGC/Pages/index.aspx) EquipNet, http://www.equipnet.com/gas-chromatographs-(gc)-equipment-47432/ Gow-Mac, http://www.uam.es/docencia/quimcursos/Scimedia/chem-ed/sep/gc/gcpics.htm Hewlett-Packard, GMI, http://www.gmi-inc.com/hewlett-packard-hp-5890-series-gc.html HNU Process Analyzers (http://www.hnu.com/products.php?v=322) Interlink Scientific Services Ltd., http://iss-store.co.uk/catalog/reconditioned-instruments/reconditioned-gas-chromatographs Questar Baseline Industries (http://products.baseline-mocon.com/category/on-line-gas-chromatographs) Rosemont, Emerson (http://www2.emersonprocess.com/en-US/brands/rosemountanalytical/GC/Pages/index.aspx) Service Technology Inc. http://www.st2-service.com/6890_GC.htm Siemens Applied Automation (http://www.industry.usa.siemens.com/automation/us/en/process-instrumentation-and-analytics/process-analytics/pa-brochures/Documents/PIABR-00017-0313-Gas-Chromatograph.pd) SRA Instruments (http://www.srainstruments.com/PN_93_AllinOne.html) Thermo Scientific (Thermo Onix) (http://www.thermoscientific.com/en/products/gas-chromatography.html) VALCO Instruments Co., http://www.chromtech.net.au/pdf2/VICI09_94-178A.pdf Varian, http://www.labx.com/product/varian-gas-chromatograph Wasson-ECE Instrumentation (http://www.wasson-ece.com/education/class_basic_gc.html) Yokagawa (http://www.yokogawa.com/an/gc1000m2/an-gc-01en.htm) Among these suppliers, about 65% of the market is shared by ABB and Siemens AA. About 30% of the market is shared among AIT (supplying the Foxboro 931D), Thermo Scientific, Emerson (supplying Daniel Industries and Rosemont units), and Yokagawa

INTRODUCTION

Chromatography is a method to separate the components of a sample stream consisting of a mixture of gas or vapor constituents, after being dissolved in a moving stream of fluid (called the mobile phase), which transports these components through the *stationary* phase, inside a "column." In this process, the components in the sample stream travel at different speeds, causing the stream to separate into its constituents. The detectors on the system serve to determine the relative proportions of *analytes* in a mixture.

In the 1950s, the work of A.J. Porter and Martin led to the development of both gas chromatography (GC) and high-performance liquid chromatography (Chapter 1.11). Since then, the technology has advanced rapidly and by today, it has become a mature field of analysis, which will be discussed in this chapter.

THE SEPARATION PROCESS

GC is a method for separating the components of a sample that contains a mixture of volatile compounds. The separations are made in order to determine the quantity of each of the sample components of interest. It has become one of the most often used procedures in analytical chemistry for separation and analysis. The reasons for its popularity can be traced to its ease of use for the separation of complex mixtures, its high sensitivity, and the small sample required for the analysis.

In industrial applications, its intermittent readings, high cost, and maintenance represent some of its limitations.

In the elution form of GC, the sample mixture to be separated is vaporized and injected into a flowing stream of carrier gas (the mobile phase) that carries it into a column containing a so-called stationary phase. The stationary phase is a high-boiling nonvolatile liquid that is suspended on an inert solid with a large surface area or coated in a thin film on the walls of a small-bore tube (capillary, or wall-coated open tubular [WCOT] column). The support-coated solid can be packed into a fairly large-bore column (packed column) or attached to the wall of a small-bore capillary (support-coated open tubular [SCOT] column). Similarly, a solid stationary phase (such as silica gel, alumina, or charcoal) can be packed into a column or suspended on the walls of a small-bore tube.

Separations occur, because the sample components have different solubilities in the liquid stationary phase or different adsorptivities on a solid stationary phase. Therefore, each sample component is retarded a different amount by the stationary phase and is carried down the column by the mobile phase at a different rate. Provided that a stationary phase has been selected that maximizes the solubility or absorptivity differences, complete separation will occur with each sample component emerging from the column at a different time. A detector that responds to some property difference between the carrier gas and the sample components is placed at the end of the column. It yields a signal that, when recorded as a function of time, produces the familiar chromatogram such as that shown in Figure 1.10a. The observed peak separation

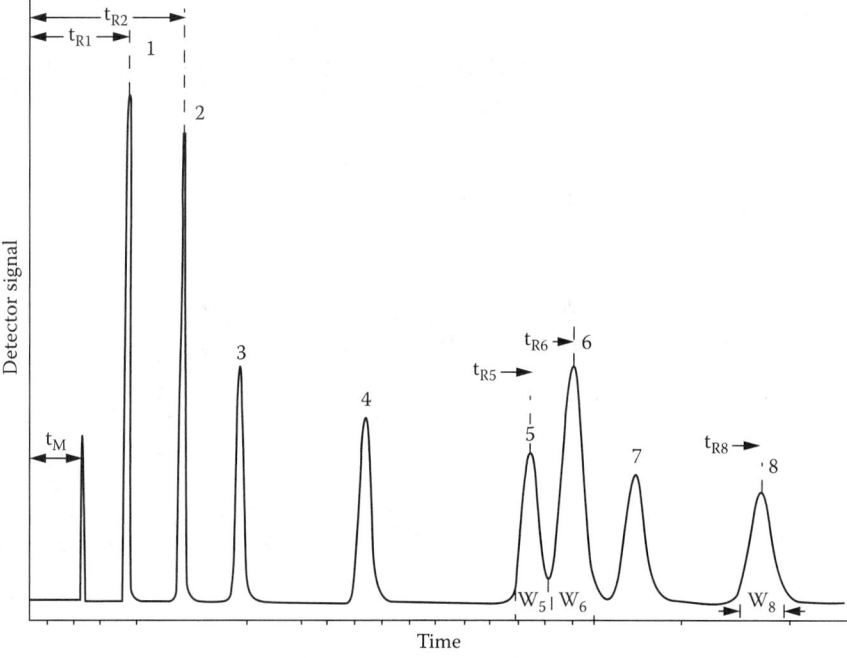

FIG. 1.10a
A typical packed column chromatogram showing fully and partially resolved peaks. Retention time of an unretained sample component (such as air) measured from the time of injection = t_M. Retention time of sample components 1–8 = t_{R1} to t_{R8}. The resolution of peaks 5 and 6 = 1.1 (see Equation 1.10(1)). The number of theoretical plates calculated using peak 8 = 4158 (see Equation 1.10(2)).

(resolution, R) is a function of the above-mentioned solubility or adsorptivity differences and the efficiency of the column, that is, its ability to produce narrow peaks. The efficiency (hetp) of a GC column is expressed as height equivalent to a theoretical plate (N). The smaller the hetp, the more efficient is the column (i.e., there are more theoretical plates per unit length of column).

It is important to remember that the plates do not really exist; they are a figment of the imagination that helps us understand the processes at work in the column. They also serve as a way of measuring column efficiency, either by stating the number of theoretical plates in a column, N (the more plates the better), or by stating the plate height; the HEIP—Height Equivalent to a Theoretical Plate (the smaller the better). If the length of the column is L, then the HETP is = L/N.

The resolution of peaks 5 and 6 in Figure 1.10a at retention times of t_{R5} and t_{R6} is calculated from the chromatogram as follows:

$$R = \frac{2(t_{R6} - t_{R5})}{w_5 + w_6} \qquad 1.10(1)$$

where w_5 and w_6 are the respective peak widths measured at the base of the peak. In this case, the resolution of the two peaks is 1.1. Two compounds of equal concentration are baseline resolved at a resolution of 1.5.

The theoretical plate as calculated from a peak (number 8 in Figure 1.10a) using Equation 1.10(2) is found to be 4158 and was obtained using a packed column.

$$N = 16 \frac{t_{R8}^2}{w_8^2} \qquad 1.10(2)$$

The efficiency of a column of length, L, is

$$\text{hetp} = \frac{N}{L} \qquad 1.10(3)$$

The time required for each component to emerge is called the retention time, t_R, and the magnitude of the detector signal (peak height or peak area) is related to the amount of the compound present in the injected sample. If these individual retention times and detector responses are made to remain constant (by adequate control of carrier gas flow rate, column temperature, amount of stationary phase, and detector variables), the process can be automated so that the instrument will provide this information quite precisely and cyclically on a repetitive basis.

THE SYSTEM STRUCTURE

The basic instrumentation necessary to accomplish the aforementioned chromatography is extremely simple (see Figure 1.10b).

It consists of a supply of carrier gas (commonly helium, hydrogen, or nitrogen) with appropriate pressure regulation, a sample introduction device (described later on) to inject a fixed volume of sample into the carrier gas stream, a column containing the stationary phase, and a detector. Since the degree of solubility of the sample components in the stationary phase is temperature dependent (and thus so will be the retention times), a precisely regulated temperature environment must be provided for the column. In some cases, to minimize the analysis time for samples that contain a wide boiling range of compounds, the temperature of the oven is

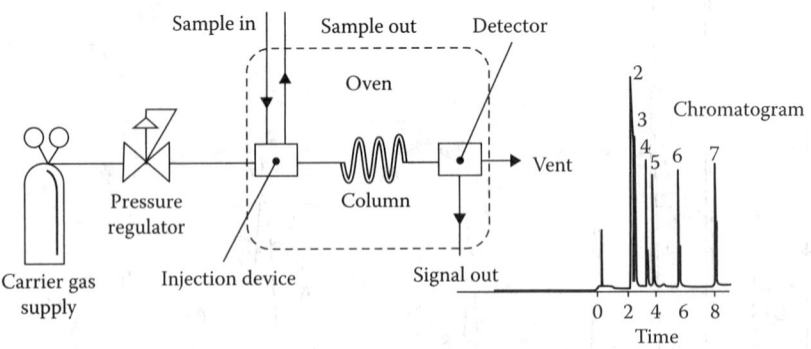

FIG. 1.10b
Basic elements of a chromatograph.

varied linearly with time. This procedure is called temperature-programmed chromatography.

THE PROCESS GAS CHROMATOGRAPH (PGC)

Although they employ the same principles, the process gas chromatograph (PGC) is quite a different instrument than its laboratory counterpart. These differences appear as the result of a number of factors, not the least of which include:

- The need for a continuous, reliable operation of the unit
- The need for the cycle time of the analysis to be shorter than the time required for the process control system to achieve proper control action
- The nature of the environment in which the PGC is placed
- The diverse user attitudes regarding maintenance, and their capability regarding the operation of the unit
- The necessity to interface with modern-day computer-controlled systems

Thus, a premium is placed on simple, reliable design to ensure a low mean failure rate, but straightforward, rapid repair when required, and the strategic placement of appropriate sensors to aid in diagnosing problems and to provide alarms with regard to the operation of the unit.

To summarize, the PGC has the following distinguishing attributes:

- Located in the plant as close as possible to the sample point to minimize sample transport time
- Dedicated to monitor one or more components in one or more process streams
- Designed for a continuous, unattended operation
- Designed for operation in hazardous environments
- Designed to withstand exposure to weather, humidity, dust, and corrosive atmospheres
- Contains integral hardware and software to allow the communication with the process control system as well as remote maintenance stations
- Contains alarms and diagnostic aids to continuously monitor the health of the instrument and aid in diagnosing problems

A process gas chromatographic system (as shown in Figure 1.10c) consists of both the PGC and a sample handling system (SHS). They will be discussed separately in the following paragraphs.

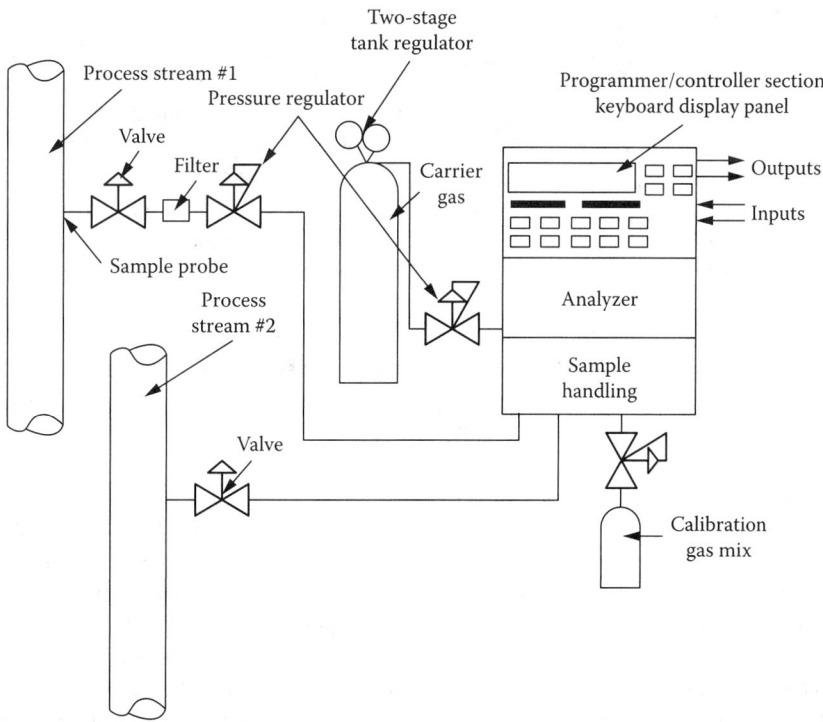

FIG. 1.10c
Basic elements of a multistream process gas chromatographic system.

COMPONENTS OF A PGC

The architecture of a PGC can be broken into two principal parts: the analyzer, containing the hardware that makes up the basic chromatograph, and the programmable controller, containing the electronics for the control and operation of the chromatograph plus a communication package. What follows is a detailed discussion of these various parts and specifications of the PGC.

The traditionally designed unit, such as illustrated in Figure 1.10d, is fairly large (typically 40″ × 26″ × 18″ in. HWD) and is mounted in a protected analyzer enclosure. An alternative so-called transmitter design, focused on a narrow market such

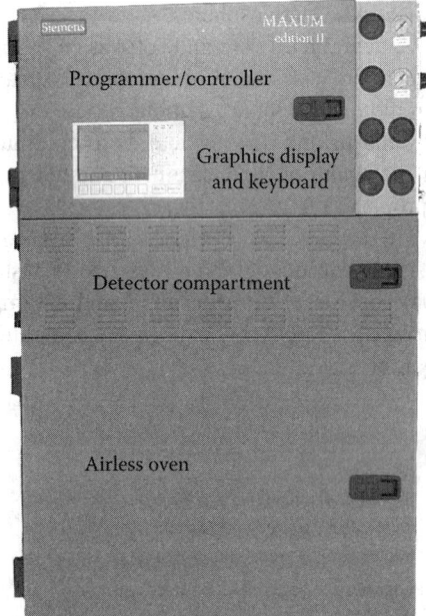

FIG. 1.10d
Typical PGC for mounting in an analyzer shelter. This unit can be obtained as shown with a single large oven, or with this space divided to provide two separate ovens each with its own thermostat. (Photo courtesy of Siemens Applied Automation.)

as natural gas analysis in custody transfer applications, has been available for some time. However, the development of a full-featured transmitter to compete with the more traditional design in a variety of applications has a checkered history. A totally pneumatic PGC transmitter was introduced in the mid-1970s and was on the market until 1989. More recently, a detailed transmitter design appeared in the literature but was never commercialized, and in the mid-1990s, a unit was introduced to the U.S. market (designated as the SGC 3000) that has since disappeared but may be offered in Japan and Europe. Finally, a transmitter design in the late 1990s, a transmitter design (illustrated in Figure 1.10e) was marketed that is still current and appears to be quite successful. All of the functions that will be discussed in the following text also pertain to this latter design.

The advantages of the transmitter are that it can be mounted close to the sample point (thereby decreasing sample transport time and thus cycle time) and it does not require

FIG. 1.10e
State of the art chromatograph transmitter mounted in a rugged fieldmountable enclosure. Among other features, it can provide hydrocarbon dew point (HCDP) calculations at four different pressures. (Courtesy of Rosemount, Emerson.)

an analyzer shelter (thus decreasing installed cost). The overall size of the transmitter and its sample handling hardware is much smaller than the shelter-mounted units for ease of field mounting. In previous designs, the oven space necessary to mount the complex valve and column arrangements was sacrificed along with its ability to handle many applications. This severely limited the market for such a design and may account for its rather limited success over the years. However, the newer design appears to be much more flexible, offering parallel chromatography, complex column valving options, many detector options, and a full communication package. In addition, efforts have been made to design the units for easy maintenance. In the absence of a complex sample-handling package, however, the transmitter, with its simple modular sample-handling unit close-coupled to it, is basically a single- or, at most, a dual-process stream analyzer. However, to accommodate users who wish to spread the cost of the analyzer by using it for the analysis of several streams, the transmitter equipped with an external sample-handling package can handle up to 16 streams. The only advantage left for this design then is the decreased cost due to the elimination of the analyzer shack required for the traditional PGC.

Analyzer

The analyzer section contains all of the basic elements of the chromatograph, namely, columns, sample and column-switching valves, and detectors, all enclosed in a precisely temperature-controlled oven. In some designs, the associated pneumatic components are also placed in the oven, and in others, they are not. Which option is elected depends on the quality of the components and robustness of the software to tolerate retention time changes that may occur when pressure and flow controllers experience a varying temperature environment. The oven may be designed for isothermal and temperature programming operation.

As mentioned previously, temperature programming is a procedure for decreasing the analysis time for a sample consisting of compounds of widely divergent boiling points and polarity. The temperature of the column is usually maintained at some initial value for a fixed period of time while the low boilers elute from the column, and is then raised in a linear fashion (at a suitable rate) during the remainder of the cycle so as to ensure the separation of the remaining sample components in a much shorter period of time than if the analysis had been run isothermally. The oven then must be cooled rapidly to exactly the same starting temperature and programmed at the same rate if one is to obtain reproducible retention times for the various sample constituents—clearly not a trivial design problem, especially if the unit is to be certified for use in Division 1 areas.

Oven

For economy and design simplicity, a single isothermal temperature zone is most frequently used. However, units with two separated temperature zones are now available. This provides the application engineer with tremendous flexibility, since parallel chromatography can be run on the same sample at the same or different temperatures with columns containing the same or different stationary phases. In this manner, the solution to difficult applications can be considerably simplified and the cycle time shortened.

Although isothermal oven operation is usually the choice, temperature-programmed units are offered to extend GC applications to other volatility measurements such as simulated distillation, Reid vapor pressure, etc.

The majority of PGCs utilize air-bath ovens for both isothermal and temperature-programmed applications. In this design, each of the analyzer elements, such as the oven heater and detectors, is built to be explosion proof for mounting in a basically non–explosion-proof air-bath enclosure. A continuous flow of instrument air (at 3–4 scfm) is passed through a heater (see Figure 1.10f) and the oven, and then vented inside or outside the analyzer shelter. The temperature of the oven is usually in the range of 50°C–225°C (±0.1°C), but may be limited by the area safety classification regarding the permitted surface temperature (the T rating). The T rating may also limit heat-up rates.

Also available for the analyzer shelter-mounted units is an airless heat-sink oven of about the same size as the air-bath units. The heating elements in this oven are not exposed, but cast into the oven walls, providing conductive and convective heat transfer with heat-up rates not much slower for the same oven volume than that of the air-bath variety. Also, higher temperatures can be achieved with this design within a given area T rating than with the air-bath oven. The advantages of such a design are that operational costs are decreased because

- It does not require instrument air for its operation.
- Power consumption is less than half that of the air-bath unit, since only conductive heat loss must be compensated, while with the air-bath design, new ambient air must be heated continuously.

In addition, with this design, the explosion-proof detectors can be mounted outside the oven (similar to laboratory GC design), providing much more working space within the oven (see Figure 1.10g).

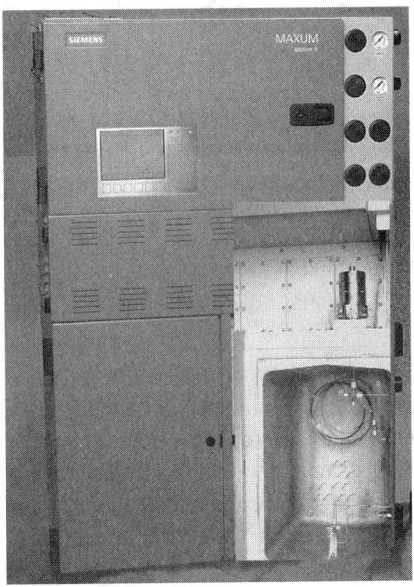

FIG. 1.10g
One-half of a dual airless oven assembly showing the placement of the explosion-proof detector outside of the oven. Space for the explosion-proof heater assembly is not required in an airless oven and an application using "valveless" column switching has further reduced the oven space needed for the application. (Photo courtesy of Siemens Applied Automation.)

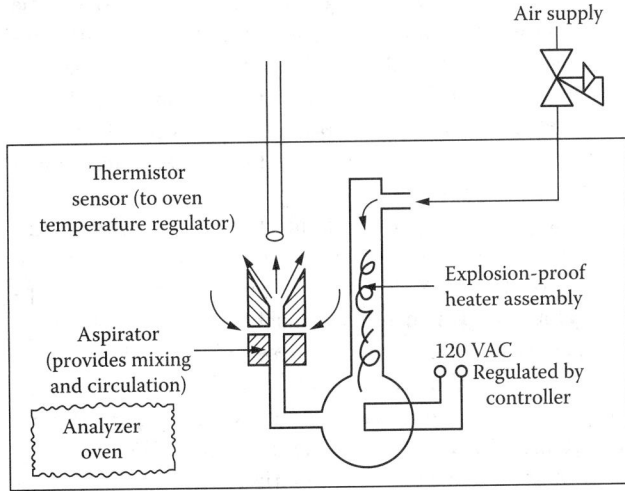

FIG. 1.10f
Basic elements of an explosion-proof isothermal air bath oven heater assembly.

The second and oldest design that has evolved through the years is achieved by placing all the elements of the analyzer in an explosion-proof enclosure. This design also does not require purge air for the oven. Historically, this has led to a very heavy and unwieldy package, but now, using the

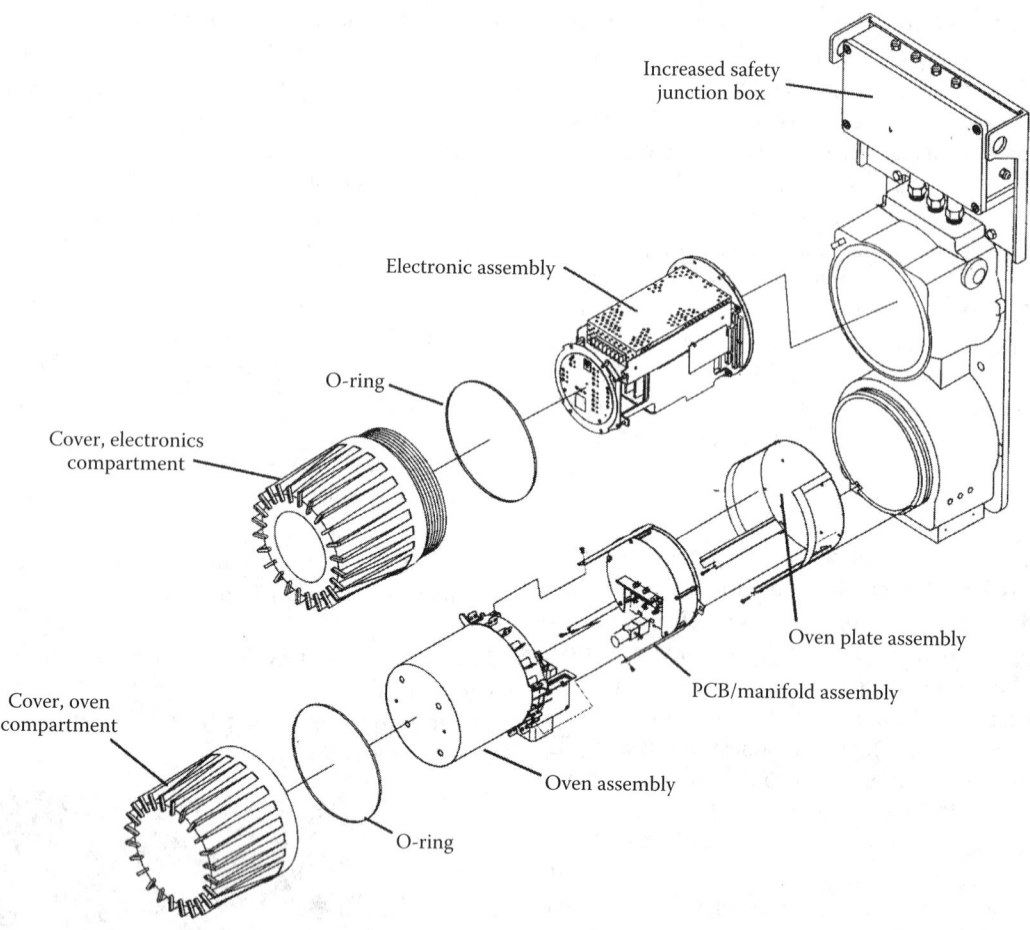

FIG. 1.10h
An exploded view of a "transmitter" illustrates the modular construction of the unit. Rapid field maintenance of the analyzer is met by replacement of the total oven assembly, which includes columns, detectors, valves, actuators, heaters, temperature sensors, and electronic pressure/flow controller components. In the electronics section, the printed circuit board assemblies are "euro-card" slide rail mounted and may be removed/replaced without the use of tools. (Courtesy of Rosemont Analytical Inc.)

smaller, modern engineered analyzer and electronic components, compact units such as the PGC transmitter are available. Field maintenance of the analyzer portion of these instruments consists of replacement of the total oven–GC assembly (see Figure 1.10h for details of the transmitter construction). Similarly, as with modern traditional design, the electronics package has been modularized to allow quick replacement of electronic components.

Valves

Valves are used for the injection of sample and as column switches in the various column configurations that may be required in the application. The most popular of these are the rotary, sliding plate, diaphragm mechanical, and no-moving-part valves, and the so-called valveless or Deans switch.

The sample valve is designed to deliver a fixed volume of sample to the head of the column. For very small samples (less than a microliter), the sample volume is defined internally by a slot machined into the movable member of the valve (sliding plate or rotor), and for larger volumes, by an external loop made from a length of deactivated tubing attached to the valve.

In the past, the smaller volumes were used to inject a liquid volume of sample (which was rapidly vaporized in the carrier gas). However, with the increasing popularity of capillary columns (which require much smaller samples than their packed column analogs), these valves are being used for the injection of gaseous samples. In any case, because liquid samples pose particular problems with regard to sample valve reliability, sample handling, and safety, liquid sample injection is best avoided by vaporizing the sample before presenting it to the sample valve.

Rotary Valve

The most popular rotary valve (especially for capillary chromatography) contains a conical rotor with machined grooves interconnecting the outer ports that are drilled into the outer body. Two positions of the rotor define different flow paths, as shown in Figure 1.10i.

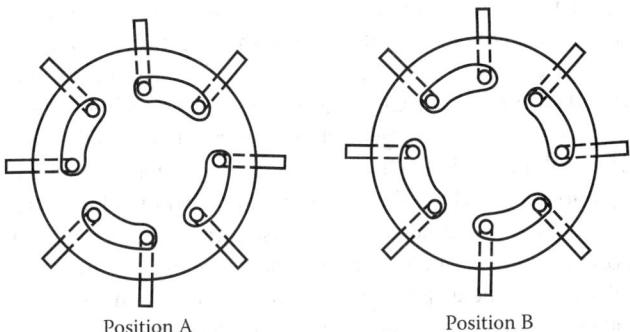

FIG. 1.10i
Rotary valve.

The rotor faces are of filled teflon or similar polymer that provides chemical resistance as well as sealing and lubricity. The valve body is made of stainless steel, or other corrosion-resistant alloy, with highly polished conical faces to form a mating seal with the rotor. The rotor is held against the stator by a factory-calibrated threaded cap that applies the correct force. An exploded view of a manually operated version of such a valve is shown in Figure 1.10j.

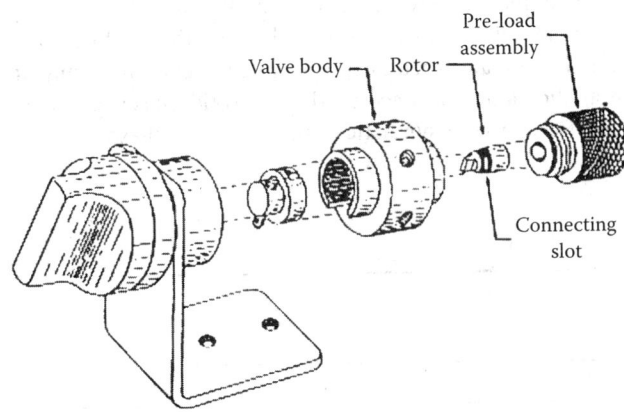

FIG. 1.10j
Exploded view of a manually actuated rotary valve. (Courtesy of Valco Instrument Co. Inc.)

The PGC version of this valve is actuated either by a compressed air operator or by an electrically driven solenoid. Although these valves meet the stringent demands for capillary chromatography, there are disadvantages in that they require regular maintenance, since the rotors wear and must be replaced after about 200,000 cycles. In addition, they require a much more complicated driver, since the simple linear motion commonly available must be converted to a rotary one.

Sliding Plate Valve

The sliding plate valve comes in both linear and rotary versions, the former being the most popular by virtue of the simple operator required for lateral rather than rotary movement of the slider.

In a manner similar to that of the rotary valve, the switching is done by a plate with channels machined in its face that is switched over holes that are drilled into the valve body. Again, the slide is made of filled teflon or similar polymer, for the reasons outlined earlier. The slider is moved between its two positions either by a diaphragm operator mechanically linked to it (with compressed air, released by an external solenoid, supplying the necessary force) or directly with a solenoid whose shaft is linked to it.

Diaphragm Valve

The diaphragm valve comes in 10-, 6-, and 4-port versions and operates on a different principle than the sliding plate or rotary valves. The flow of gases between adjacent ports arranged in a circular pattern in a machined and polished metal surface is controlled by a teflon diaphragm. Two versions of the diaphragm valve are available. In one, the force is applied to the diaphragm by a series of plungers that are in turn moved by two pneumatically actuated pistons. This is illustrated in Figure 1.10k for a six-port sample valve.

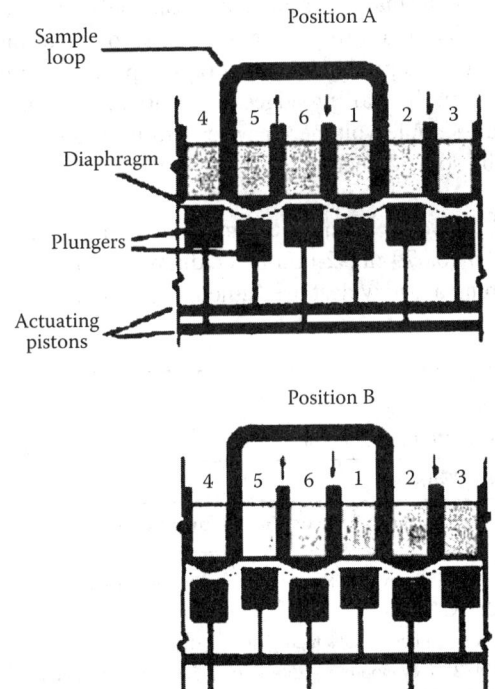

FIG. 1.10k
Schematic view of a six-port diaphragm sample inject valve. Carrier gas enters at port 3 and exits at port 4. Sample in/out ports are 1 and 6. In Position A, the sample loop is being filled with sample. In Position B, sample is swept from the loop by carrier gas. (Courtesy of Valco Instrument Co. Inc.)

The other version uses pressure-on-diaphragm activation with no moving parts. Both versions are highly reliable (1-million-cycle rating for the piston-operated one, and 16 million for the no-moving-part variety).

Although the sample valve example shown in Figure 1.10k utilizes an external sample loop, these valves are also available with small internal-volume loops.

Columns

Packed columns are still the most commonly used in most PGC applications. However, open tubular (OT) capillary column solutions are making inroads in many cases. WCOT columns are often preferred, because they are more efficient than packed columns, and thus, it is possible to simplify the column configuration necessary to achieve the desired separations. In addition, they often allow for faster analysis cycles. There is a downside, however, to using capillary columns in PGC applications in that many of the older designed PGCs will not adequately support them. Also, because of the ease of plugging the small-diameter columns and connecting plumbing, sample handling to produce extremely clean samples is of paramount importance. Very small sample volumes must be injected to maintain the efficiency of the WCOT columns. Thus, special sample valves and additional hardware to split the sample delivered from the sample valve are necessary. Also, detectors designed with small internal volumes are necessary, as well as critical attention to plumbing details, to avoid unswept volumes that degrade chromatographic peaks by creating long tails on the trailing edges. In addition, in some cases, the small sample volumes may lead to inadequate detectability by some detectors.

Packed Columns Packed columns in 1/8-in.-outer diameter (OD) (0.079-in., 2-mm-inner diameter (ID)) stainless steel tubing are in widest use, although the smaller-diameter 1/16-in.-OD. (0.039-in., 1-mm-ID) micropacked (smaller-diameter particles) are increasingly being used. Occasionally, the larger 3/16-in.-OD columns are used in special applications. The length of these columns will vary with the difficulty in making the separations. Longer columns produce more plates and thus have more separating power. However, the maximum column length is ultimately determined by the pneumatic equipment. Large pressure drops are avoided by keeping column lengths to less than 12 ft. The stationary phase is coated on support particles (sieved to 80–100 or 100–120 mesh) of Chromosorb™ (a diatomaceous earth product that has the large surface area needed to make efficient columns). Some popular stationary phases and the separations for which they are used are listed in Table 1.10a.

Additionally, it has been suggested in the literature that most separations can be accomplished by using mixtures of a polar and nonpolar phase in the proper proportions. Indeed, computer-aided series-connected capillary column design procedures have been published that produce column systems optimized for the analysis restraints (such as analysis time and detectability) that are operative for that particular application.

Active solid supports such as alumina, activated charcoals, and silica gels have been used in the past for the separation of fixed gases and light hydrocarbons. However, their use is now avoided if possible because their activity is difficult to stabilize, leading to changes in retention time and separating power. Similarly, molecular sieves (synthetic alkali metal aluminosilicates) once used primarily for the separation of hydrogen, oxygen, nitrogen, methane, and carbon monoxide must be protected from carbon dioxide (by using a trapping column ahead of it) and from other hydrocarbon gases that are adsorbed and deactivate the column. Many of the applications that once used these solid phases can now be performed with synthetic porous polymer phases such as Porapaks™ and HaySep™.

TABLE 1.10a
Popular Stationary Phases

Stationary Phase	Comments
OV-101™ (100% methyl silicone) useful from 0°C to 330°C	Most frequently used GC phase. Low polarity, separates homologous series of compounds according to their boiling points.
OV-17™ (50% phenyl–50% methyl silicone) useful from 50°C–250°C	Moderately polar silicone phase.
OV-25™ (75% phenyl–25% methyl silicone)	Highly polar silicone phase. Selective for aromatics over aliphatic hydrocarbons.
OV-225™ (25% cyanopropyl–25% phenyl–50% methyl silicone useful from 0°C to 250°C	Increased polarity over OV-25. Increased retention of aromatics over aliphatics, alcohols over ethers, ketones over primary alcohols.
Carbowax 20M (polyethylene glycol, mol. wt 15,000–20,000) useful from 60 to 225°C	Generally useful polar phase.

WCOT (Capillary) Columns In the past, capillary columns were avoided because of their fragility (they were made of glass, since metal capillaries had active surfaces that produced less efficient columns due to tailing peaks) and also the stationary phase tended to bleed off over a period of time. However, these problems are no longer present because of two technological breakthroughs. The fragility problem has been solved by coating the outside of the fused silica capillary with a polyamide (similar to the coating used for optical fibers), which turns it into a tough, bendable capillary tube. In addition, the stationary phase stability problem has been eliminated by performing the appropriate chemistry on the stationary phase after it is coated on the walls of the capillary tubing to produce partial cross-linking of the phase to itself and to the silica surface. Columns of this type are classified as having stabilized stationary phases and are readily available from all GC column vendors.

Although capillaries of 0.25-mm ID are commonly used in laboratory GCs, it is much more common to see the so-called megabores (0.32- to 0.50-mm ID) used in PGC applications as a direct replacement to the packed column. Minimum hardware modifications are necessary on older PGCs to accommodate these larger capillary columns. They provide comparable or better efficiency than their packed column analogs and, in many cases, better separations, since longer columns can be used (because of their high permeability).

The characteristics of the various columns used in process GC are summarized in Table 1.10b.

Column and Valve Configurations

One of the most severe restraints operative in PGC applications is that of timely analysis cycle time. Since the cycle time includes the sample transport time as well as analysis time, this restraint has implications from both a chromatographic and a sample handling point of view. The latter concern will be discussed later on in this text. The solution to analysis time has historically been to use multiple columns in various configurations. Multiple columns and column-switching systems serve several important functions, namely:

- Housekeeping: To ensure that all components are removed from the columns during each analysis cycle. The method commonly used to achieve this objective is called back-flushing.
- Discard unmeasured components: In many process control applications, a rapid analysis of only one or two components is required, not the total analysis. Both back-flushing and heart-cutting are used.
- Simplify the separation problem: A single column that will separate all of the measured sample components within the desired cycle time may be difficult to find. However, it may be possible to optimize separate columns for different portions of the analysis and then combine them in such a fashion to achieve the desired analysis. This simplifies the application problem. Trap-and-store methods, as well as parallel and series columns, are frequently used for this purpose.

TABLE 1.10b
Characteristics of Columns Used in PGCs

Type	Length, m	OD mm	ID mm	Flow, ml/min
Packed				
Conventional	0.1–5	3.18 (1/8")	2	10–50
Micropacked	0.02–1	1.59 (1/16")	1	1–10
Capillary	1–100	0.4–0.8 (1/64–1/32)	0.1–0.5	0.5–10

Type	Advantages	Disadvantages
Conventional	Ease of fabrication	Large carrier gas consumption
	Large sample capacity affords	Modest efficiency
	High detectability	
	Relatively inexpensive	
Micropacked	Lower gas consumption	Higher head pressure reqd.
	Higher efficiency	Lower sample capacity
	Faster analysis	Difficult to fabricate
Capillary	Lower gas consumption	Lowest sample capacity
	Higher efficiency	Reduced detectability
	Fast analysis	Expensive
	Best separating power	

While the solutions to some applications may have changed with the advent of more instrumentation that can support capillary, temperature-programmed, and parallel chromatography, many remain that can only be solved with the use of the various column configurations discussed in the following text. While there are literally hundreds of possible column configurations, they are all built from a few basic ones.

Hardware

Six- and ten-port valves are often used in multicolumn chromatography. However, an alternative switching procedure, the so-called valveless or Deans switching, is used extensively in process GC, particularly in Europe. Basically, this valve depends on applying a pressure differential to change the flow pattern of carrier gas. The columns are connected in series by a T, the center connection of which is connected to a second source of carrier gas through a solenoid located outside the heated oven. The solenoid controls the supply of carrier gas to the T to shift the flow pattern (see Figure 1.10l). The main hardware advantages of this arrangement are that solenoids that satisfy the lower temperature specification are much easier to find; it is a less expensive solution to the problem, since a solenoid and a T are less expensive than a high-temperature-rated chromatographic grade-switching valve; and, finally, the assembly takes up much less oven space than a conventional rotary or diaphragm valve.

Sample Injection

The primary function of the sample inject valve is to place a sharply defined fixed volume of sample into the carrier gas stream so that it can be carried into the column for separation into its individual components. The sample volume is defined by an external length of tubing connecting two of the valve ports, as shown in Figure 1.10k, or internally by a groove machined into one of the valve faces (see Figure 1.10m for an internal sample loop in a rotary valve). The latter configuration is used for small-volume gaseous or liquid injections.

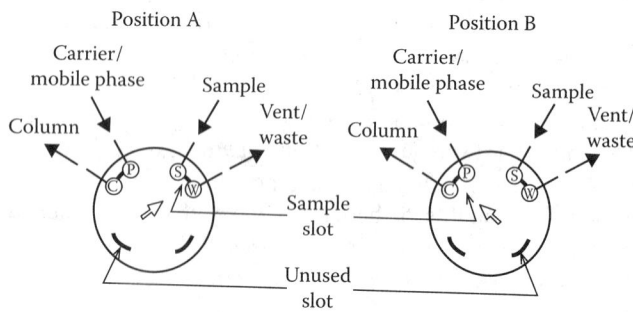

FIG. 1.10m
Rotary valve with internal sample loop for the injection of small volume liquid or vaporized sample.

In the case of some liquid samples, a special valve is used that is designed to have two distinct temperature zones: (1) a cool zone near ambient through which the sample circulates, and (2) an internally heated vaporizing zone where the liquid sample is injected into the carrier gas stream, vaporized, and carried into the column. This type of sample valve is called a vaporizing liquid sample valve. An important feature of this design is that the circulating liquid sample does not contact the heated zone. Hence, it can be used for samples that will not tolerate heat, for example:

- Samples that tend to polymerize (such as styrene)
- Samples with a high vapor pressure such that heating them to the temperature of the column would increase their vapor pressure above the line pressure and result in partial vaporization (flashing) of the sample

The internally heated vaporizing valve is also used if local or company safety codes prohibit the introduction of flammable liquids into the heated analyzer zone.

Capillary columns require much smaller injection volumes than do packed columns. For gas samples, the appropriate volumes (0.5–10 ml) can be obtained using the external or internal

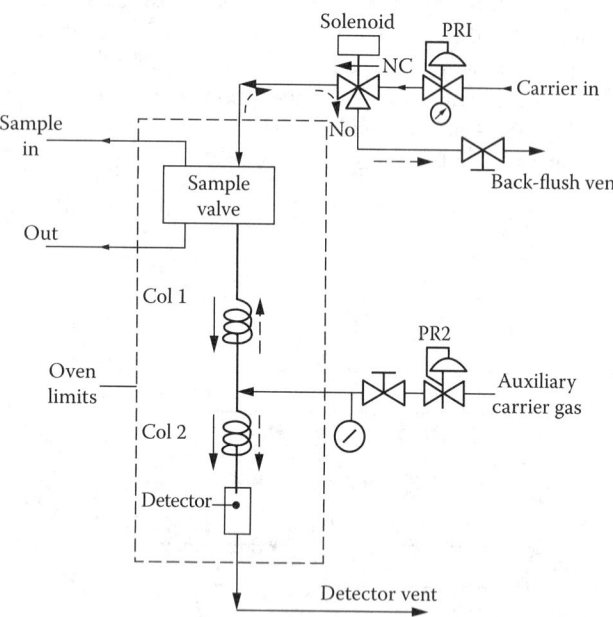

FIG. 1.10l
Back-flush to vent configuration using a Deans switch.* Column 1 is back-flushed to vent in the "fail-safe" un-energized position of the solenoid. The auxiliary regulator, PR2, supplies carrier gas to both columns in its back-flush mode. Except for the sample valve, there are no moving parts within the heated analyzer oven. (From Annino, R. and Villalobos, R., Process Gas Chromatography: Fundamentals and Applications, ISA. Copyright 1992 ISA. Used with permission. All rights reserved.)

---

* Agilent Technologies, Capillary Flow Technology: Deans Switch, (2013) https://www.agilent.com/cs/library/brochures/5989-9384EN.pdf.

sample loop. Acceptable liquid sample volumes for capillary chromatography, however, are a thousand times smaller than this and are obtained by adding a splitter between the sample inject valve and the column. The splitter is essentially a Y specially designed so as not to produce any discrimination of the vaporized sample according to some function of composition; that is, it provides only a volume split of the sample. Since one leg of the Y may be operating at 200–500 times the column flow rate, it potentially constitutes an increased operating cost for the instrument. For this reason, it is normally dead-ended, except when injecting sample.

Back-Flush

The column configuration most frequently used in process GC is the back-flush precut to vent. Both this and the sample inject operation are shown combined in the 10-port valve configuration shown in Figure 1.10n, using a rotary valve, and in Figure 1.10l, using a Deans switch. The back-flush procedure performs the all-important housekeeping function of removing all unmeasured heavy components or those that may degrade the analytical column (such as water) from the column system each cycle. The system consists of at least two columns: a precut (or stripper) column, C1, and an analysis column, C2. The precut column is back-flushed shortly after the sample is injected, thus allowing only the components that are to be measured to enter the analytical column.

Typically, the precut column is approximately one-third the length of the analytical column. This allows sufficient time for all components to back-flush during the time necessary to chromatograph the rest of the sample on the analytical column.

Another version of back-flush, called back-flush to detector, is illustrated in Figure 1.10o. It is used when it is desired to have a measure of the "heavies" that are back-flushed.

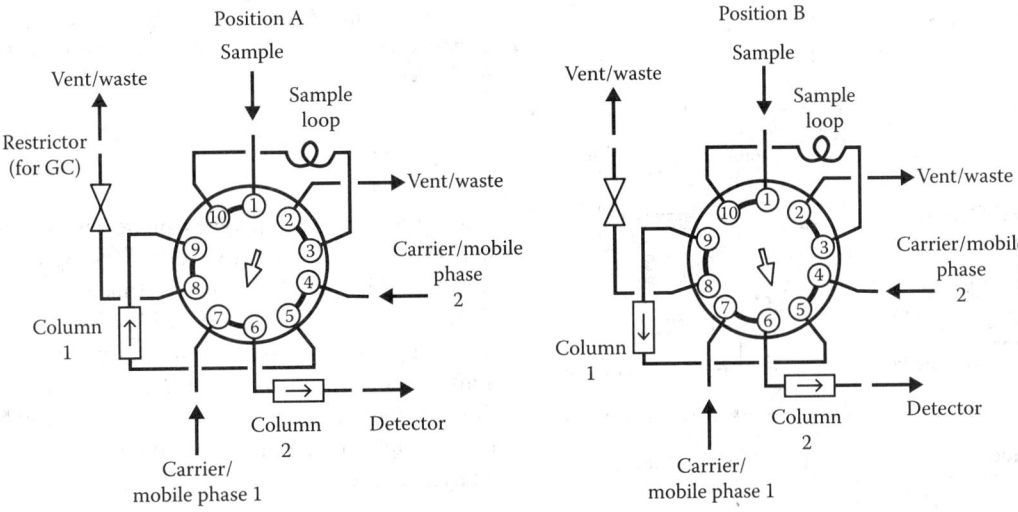

FIG. 1.10n
A 10-port rotary valve with an external sample loop configured in Position B for the injection of sample into column 1 and column 2 and, in Position A, for back-flush to vent of the sample components trapped on column 1, while the rest of the sample is separated on column 2 and the sample loop is refilled with fresh sample. (Courtesy of Valco Instrument Co. Inc.)

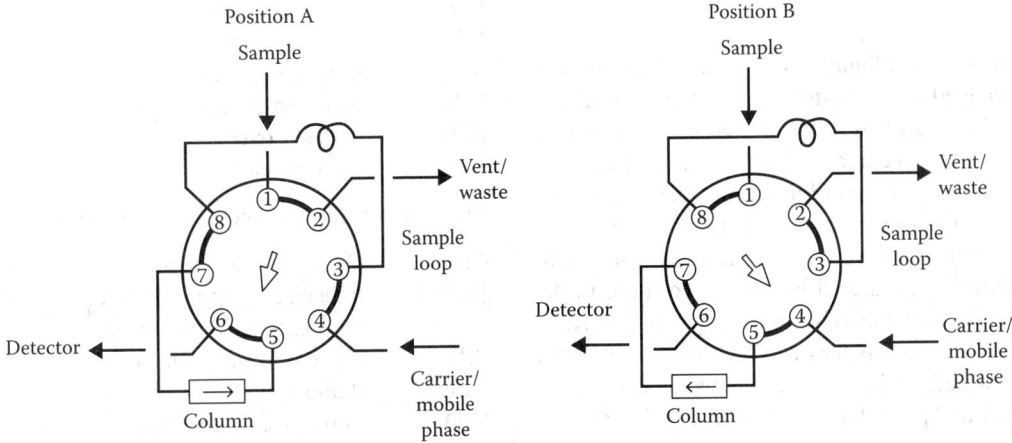

FIG. 1.10o
Back-flush to measure configuration: an 8-port rotary valve is configured so that the back-flushed components can be flushed through the detector and measured as a group. (Courtesy of Valco Instrument Co. Inc.)

Heart-Cutting

Heart-cutting is one of the most useful column configurations. It is frequently used for the analysis of trace components that elute immediately after a large concentration of another sample component that interferes with its determination (e.g., trace amounts of acetylene in ethylene). The design consists of two columns, a heart-cut column and an analysis column, as shown in Figure 1.10p, using a six-port sliding-plate valve. In the normal position of the heart-cut valve, the effluent of the heart-cut column is diverted to vent. When the component of interest is about to elute, the heart-cut valve diverts it into the analysis column and then returns to its venting position. A heart-cut of only the compound of interest and a narrow band or *tail* from the major component are thus introduced into the analysis column. This operation may be repeated for each component of interest, although it is usually not practical to measure more than three heart-cut components per analysis.

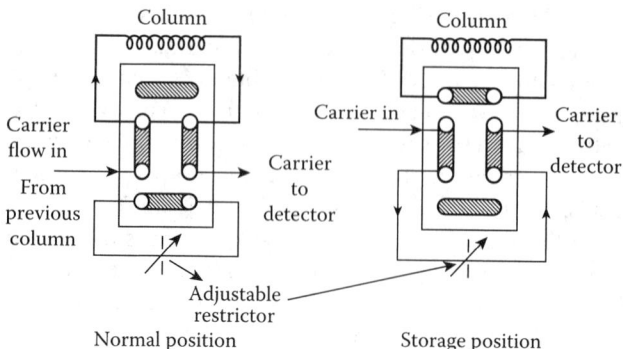

FIG. 1.10q
A six-port plate valve configured so that a group of early eluting sample components pass into a storage column, which is then isolated from carrier gas flow (the group of components remain stationary on the column except for a small amount of longitudinal diffusion) and stored until the rest of the sample is separated and measured by the detector. When the valve is again switched, the components on the storage column are developed and measured by the detector.

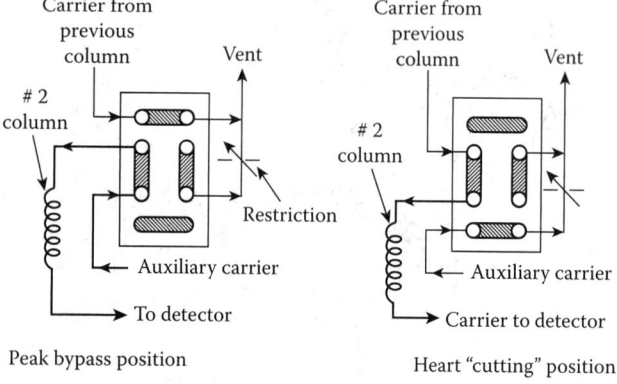

FIG. 1.10p
Heart-cut configuration: a six-port plate valve is configured so as to divert a large portion of a major component in order to simplify the separation of a small concentration of a sample component riding on its tail.

The trap-and-store configuration consists of two analytical columns arranged in series through a column-switching valve, as shown in Figure 1.10q. The switching valve is used to direct the effluent from column 1 into column 2 or through a bypass restrictor to the detector for measurement. In a manner similar to that used in heart-cutting, two or more components that are not separated on column 1 are diverted to column 2, as shown in Figure 1.10q, normal position. In the other storage position of the valve, carrier gas is not available for this leg. Column 2 is thus isolated, and the diverted compounds are stored, while the rest of the sample is developed in column 1 and passed on through the restrictor to the detector for measurement. The trapped compounds are then passed to the detector for measurement.

DETECTORS

There are two general types of detectors: destructive and nondestructive. Destructive detectors perform continuous transformation of the column effluent (burning, evaporation, or mixing with reagents) with subsequent measurement of some physical property of the resulting material (plasma, aerosol, or reaction mixture). Nondestructive detectors are directly measuring some property of the column effluent (for example UV absorption) and thus affords for the further analyte recovery.

The destructive detector designs include

AED	Atomic-emission
ELSD	Evaporative light scattering
FID	Flame ionization
FPD	Flame photometric
MD	Mira detector
MS	Mass spectrometer
NPD	Nitrogen phosphorous
NQA	Aerosol-based
SMSD	Sumon detector

The nondestructive detectors designs include

CD	Conductivity detector
ECD	RElectron capture detector
FD	Fluorescence detector
PID	Photoionizaton detector
RFD	Radio flow detector
RID	Refractive index detector
TCD	Thermal conductivity detector
UV/DAD	Ultraviolet diode array detector

This list of detectors exceeds the list of the ones that are discussed here, because they include both gas and liquid detectors (liquid chromatographs are discussed in Chapter 1.11) and also because some of these detector designs are discontinued or not widely used.

Of the many detectors that have been proposed for use in PGC over the years, two emerge as the most popular and have been applied to a far-ranging number of applications. They are the thermal conductivity detector (TCD) and the flame ionization detector (FID). Their popularity stems from their basic simplicity and ruggedness coupled with their acceptable sensitivity for most applications. There are other detectors used in PGC applications (such as the flame photometric and other ionization detector), but only under special circumstances.

Thermal Conductivity

The TCD is a concentration-responsive detector of moderate sensitivity (MDQ = 10^{-9} g) and dynamic range (10^5). It is a reliable, simple, easy-to-maintain, and relatively inexpensive detector with a universal response. The unit shown in Figure 1.10r is of classical design and consists of a cavity in a metal block with a filament (made of a metal with a large resistance/temperature coefficient such as tungsten or an alloy such as tungsten-rhenium) suspended in the carrier gas flow stream. In support of capillary chromatography, the newer detector designs strive to minimize detector cavity volume. The filament is heated electrically and reaches some equilibrium temperature (and corresponding resistance) based on the power supplied to it and the thermal conductance of the gas passing over it, and the temperature of the cavity wall. The cavity wall temperature is maintained constant by the oven environment, and the carrier gas is selected for its large thermal conductivity (helium or hydrogen), compared to the components of the sample. Thus, as sample components–carrier gas mixtures enter the detector, the consequent thermal conductance change experienced by the filament environment leads to a change in the equilibrium temperature and resistance of the filament.

Traditionally, the filament is made to be one arm of a Wheatstone bridge that is supplied with a constant voltage. The change in null voltage experienced by the bridge,

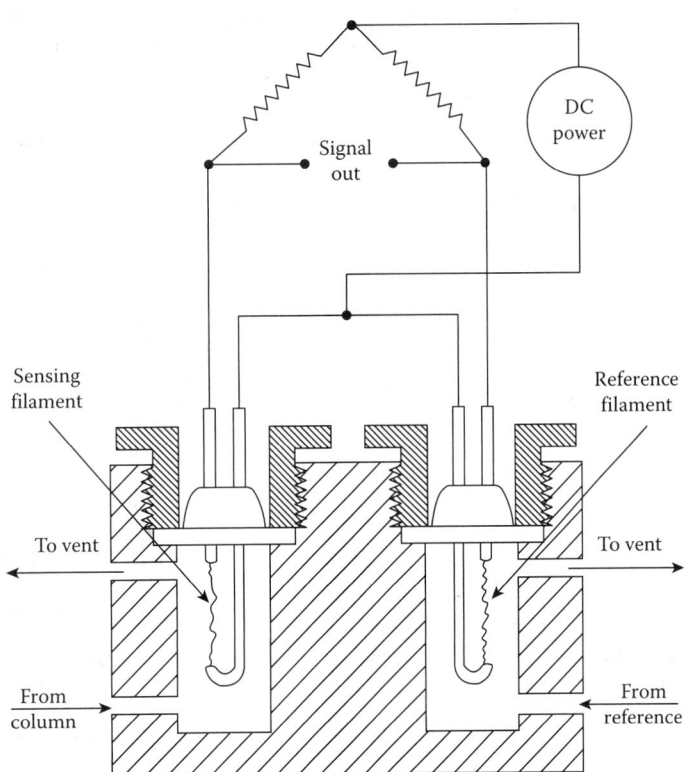

FIG. 1.10r
Single sensing filament, single reference filament, flow-through TCD with replaceable elements shown in a classical Wheatstone bridge circuit. In a properly designed TCD block assembly, the inclusion of reference filaments imparts greater stability to the signal by balancing out small temperature variations in the cell block.

when the filament changes resistance, is recorded as the chromatographic signal. Modern TCD electronics, however, favor constant-current (Figure 1.10s) or constant-resistance designs (Figure 1.10t) for faster response and also for filament protection in the presence of large sample concentrations and cessation of carrier gas flow.

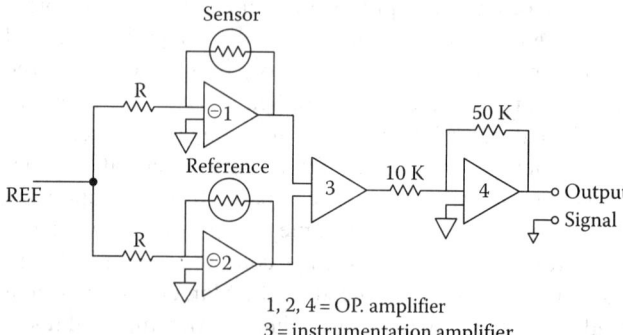

FIG. 1.10s
Simple constant current circuit for a thermistor bead TCD with a gain of 5. (From Annino, R. and Villalobos, R., Process Gas Chromatography: Fundamentals and Applications, ISA. Copyright 1992 ISA. Used with permission. All rights reserved.)

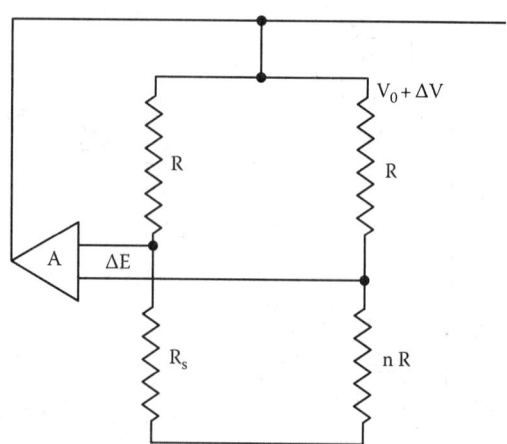

FIG. 1.10t
Basic constant resistance/temperature circuit for a TCD. (From Annino, R. and Villalobos, R., Process Gas Chromatography: Fundamentals and Applications, ISA. Copyright 1992 ISA. Used with permission. All rights reserved.)

Alternatively, the filaments can be replaced with thermistors (sintered mixtures of manganese, cobalt, and nickel oxides along with trace elements). They are mounted in the form of a bead on platinum wire leads and coated with glass to make them inert. Thermistors have a very large negative temperature coefficient, making them very sensitive sensors. However, they have a limited temperature range and their sensitivity decreases with increasing temperatures

Thus, they are most useful in low-temperature applications. Also, their use is commonly restricted to helium carrier gas as the glass covering becomes embrittled and cracks when hydrogen is used.

Flame Ionization Detector (FID)

Even though the use of a FID involves increased operating cost and increased analyzer complexity, its very large dynamic range (10^7), high sensitivity (0.015 coulombs/g C), low detection limits (10^{-12} gC/sec), and vanishing small dead volume (10^{-2} μl) have made it probably one of the most popular GC detectors in use today. With optimized operating parameters, it is possible to determine as low as 20 pg, or about 5 ppb, of a sample.

The FID consists of an oxygen-rich hydrogen flame that burns organic molecules, producing ionized fragments. These ions are subjected to an electrical field produced by impressing a potential (usually 150–300 V) across a jet and the collecting electrode (the jet is usually made negative, and the collector positive).

The essential components of a FID are shown schematically in Figure 1.10u. The hydrogen fuel is mixed with the column effluent and then fed to the jet. An outer sheath of air supports the flame. In some designs, the flame points downward. This ensures removal of water from the flame vicinity together with any solid particles that may be present. The presence of either of these substances will contribute to

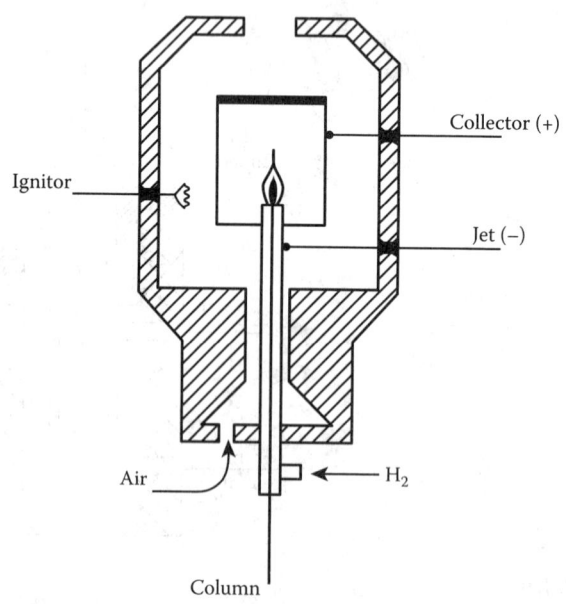

FIG. 1.10u
Schematic illustration of an FID assembly. The capillary column is run up inside the jet to the tip (thus minimizing detector dead volume). The collector is positively polarized usually with +300 volts. The flame is supported by an external sheath of air.

the background noise of the system and thus reduce minimum detection levels. The response of the detector depends on the flow rate of all three gases—with the hydrogen flow rate being the most critical (the ratio of air to hydrogen is usually around 10).

The response of the FID, which varies with the identity of the component in the flame, depends primarily on the number and type of carbon atoms being oxidized. However, the identity and manner in which other atoms are combined with the carbon also influences the response. The concept of effective carbon number (ECN) has been introduced as a means of relating the various responses. The ECN is the sum of the contributions made by the individual carbon atoms modified by the functional group contributions (a sample of these ECNs is given in Table 1.10c).

TABLE 1.10c
FID ECN

Atom	Type	ECN
Carbon	Aliphatic	1.0
Carbon	Aromatic	1.0
Carbon	Olefin	0.95
Carbon	Acetylenic	1.30
Oxygen	Ether	−1.00
Oxygen	Primary Alcohol	−0.60

A number of compounds cannot be detected by the FID. These include the fixed gases, oxides of nitrogen, H_2S, SO_2, COS, CS, CO, CO_2, H_2O, NH_3, and the noble gases. However, if the carrier gas is doped with an ionizable gas (such as methane), these gases can be detected (yielding negative peaks as they dilute the concentration of ionizing gas during the elution). Also, in the case of carbon dioxide, monoxide, or formaldehyde, chemically reacting the gases as they emerge from the column to form methane makes them FID responsive. Thus, most suppliers of PGCs offer a methanizer, which is placed between the end of the column and the FID. Hydrogen is also "teed in" at this point. The methanizer consists of a length of 1/8-in.-OD tubing filled with a catalyst of 10% (wt/wt) nickel on Chromosorb P,

over which the reducing reaction occurs. The reaction is nonselective in that all hydrocarbons are converted to methane.

The FID is a very inefficient ionizing source, and it is only its very low noise level (10^{-14} A), allowing large amplifications, that makes this detector so sensitive. As shown in Figure 1.10v, high-gain, low-noise, electrometers mounted as close to the detector as possible are required to exploit the full range of this detector. In addition, to maintain this low noise level, highly purified gases should be used.

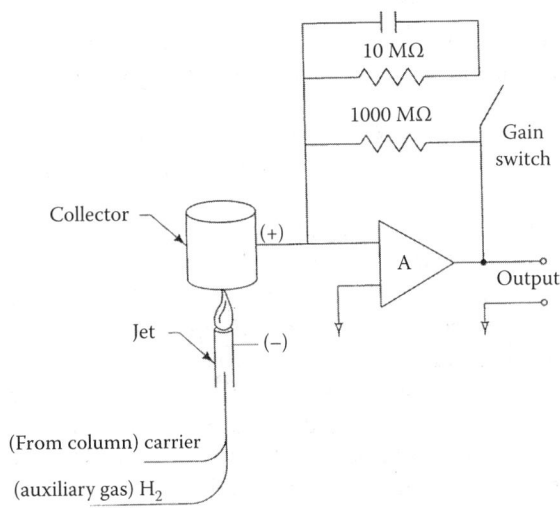

FIG. 1.10v
Schematic view of a simple high-impedance current amplifier circuit used for the measurement of the small currents generated by the FID. (From Annino, R. and Villalobos, R., Process Gas Chromatography: Fundamentals and Applications, ISA. Copyright 1992 ISA. Used with permission. All rights reserved.)

Flame Photometric Detector (FPD)

The flame photometric detector (FPD), illustrated in Figure 1.10w, is essentially a flame emission spectrometer design optimized for use as a GC detector. It is an element-specific detector and is used primarily for the determination

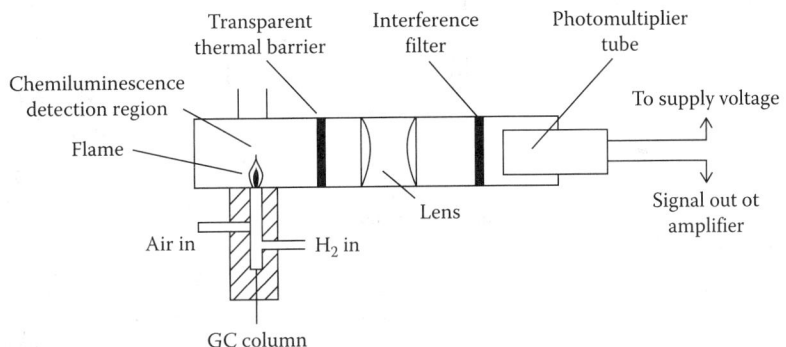

FIG. 1.10w
Basic FPD.

of sulfur- or phosphorus-bearing compounds (e.g., H_2S, CS_2, SO_2, COS, mercaptans, and alkali sulfides in various pulp milling processes and in petroleum fractions).

The column effluent is fed to a hydrogen-/oxygen-rich flame where individual atoms contained in the sample or the reactants of these atoms with hydrogen and oxygen are excited to higher electronic states by the energy of the flame. These excited atoms and molecular fragments subsequently return to the ground state with the emission of characteristic atomic or molecular band spectra. For sulfur, the S_2 species at 394 nm is used, while the 526-nm emission from HPO is used for phosphorous. A narrow-band-pass filter is used to isolate the appropriate wavelength, and its intensity is measured with a photomultiplier tube (PMT).

This detector has a number of severe limitations. These include response dependency on the O_2/H_2 ratio, the H_2 flow rate, the type of sulfur compound, quenching of the sulfur emission by large concentrations of other organic compounds, and even the length of time that the flame is lit.

The quenching of the sulfur emission by other organic compounds and the flame extinguishing problem caused by large concentrations of organics such as the solvent peak are largely overcome by the dual-flame assembly shown in Figure 1.10x, but at the expense of detection limits as the sample is diluted in passing to the second flame. In addition, the sulfur compound response dependency is eliminated with this detector, because all the sulfur compounds are converted to SO_2 in the lower flame.

The FPD response to sulfur and phosphorous is 10^4 greater than its response to hydrocarbons, and its minimum detectable quantity (MDQ) is at the nanogram level. However, since the detector signal is due to the emission of the S_2 species, it is not a linear function of concentration. Theoretically, it should be proportional to the square of the sulfur concentration (in practice, a factor of 1.8 provides a better fit to the calibration curve), and thus become much more sensitive to changes in sulfur concentration as the concentration increases. The reduced detectability and response characteristic at trace levels of the sulfur sample due to the detector's quadratic response can be improved by doping the carrier gas with a volatile sulfur compound (methyl mercaptan is a good choice) to bring the signal further up on the calibration curve. Finally, the PFD response is greatly dependent on carrier gas identity and flow rate, as well as oxygen (or air) flow rate to the flame. Clearly, it is not a very robust detector.

Pulsed FPD

In a conventional FPD, the detectivity is limited by light emissions of the continuous flame combustion products. Narrow-band-pass filters used to isolate the wavelength also limit the fraction of the element-specific light that reaches the PMT, and are not completely effective in eliminating the flame background and hydrocarbon interference. In a pulsed FPD (PFPD), the fuel gas (H_2) flow is set so low that a continuous flame cannot be sustained. By inserting a constant ignition source into the gas flow path, the fuel gas ignites and propagates back through a quartz combustion tube to a constriction in its path where it is extinguished; then it refills the detector, ignites, and repeats the cycle. Carbon light emissions and the emissions from the hydrogen–oxygen combustion flame are complete in 2–3 msec, after which a number of heteroatomic

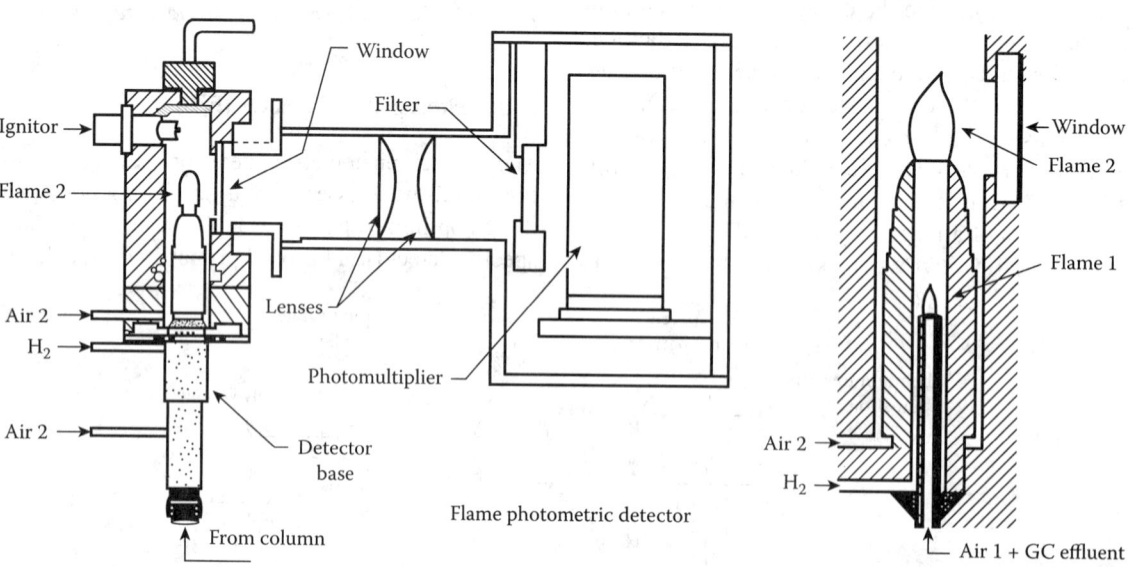

FIG. 1.10x

Schematic view of an FPD showing details of a dual flame assembly. The dual flame scheme is used to minimize the quenching of the sulfur emission by other organic compounds and also to eliminate the flame quenching problem when large concentrations of organics are present. It also provides the same sulfur response for all compounds, since they are all oxidized to SO_2 in the lower flame. (Courtesy of Variant/Agilent Technologies.)

species give delayed emissions that can last up to 20 msec. These delayed emissions are filtered with a wide-band-pass filter, detected by a PMT. By using the leading edge of the flame background emission to trigger a gated amplifier with an adjustable delay, heteroatomic emission can be amplified to the virtual exclusion of the hydrocarbon background emission. The PFPD is thus uniquely characterized by the addition of this time domain information (illustrated in Figure 1.10y), which allows the detector to selectively detect many other elements (such as As, Sn, Se, Ge, Te, Sb, Br, Ga, In, Cu, etc.) with no hydrocarbon interference.

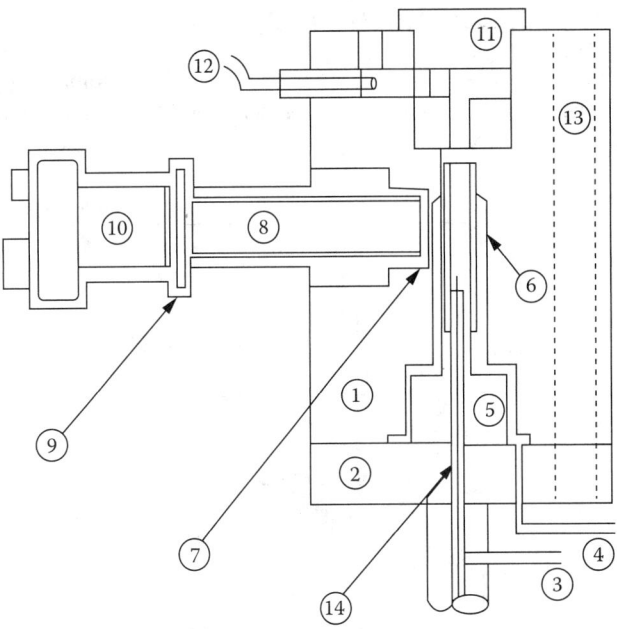

FIG. 1.10z
Schematic view of a PFPD. (Reprinted with permission from Amirav, A. and Jing, H., Anal. Chem., 67(18), 3305. Copyright 1995 American Chemical Society.)

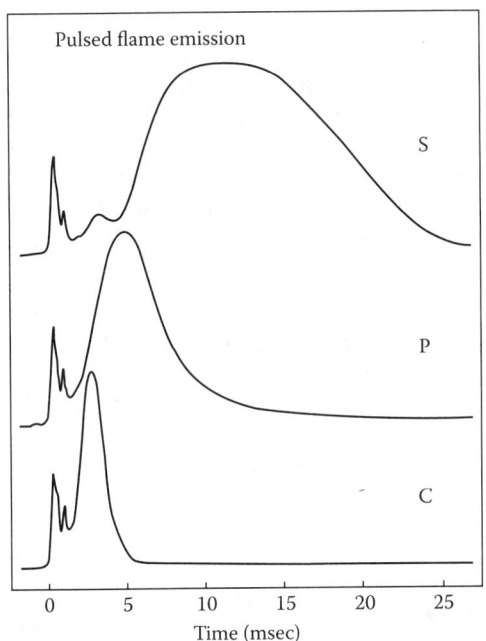

FIG. 1.10y
Emission spectra of sulfur, phosphorous, and carbon from a PFPD showing their separation in the time domain. The initial hydrocarbon emission spike serves as an excellent reference from which to time the appearance of the selected emission. This provides infinite selectivity against hydrocarbon emission as well as unique heteroatom identification capability. (From Amirav, A. et al., Pulsed Flame Photometric Detector for Gas Chromatography Report, Tel Aviv University, 2001.)

In the PFPD illustrated in Figure 1.10z, hydrogen and air (3) is continuously fed into the small pulsed flame chamber (6), together with sample molecules eluted from the GC column (14). The combustible gas mixture is also separately flowing into (4), a light-shielded, continuously heated wire igniter (12). The ignited flame is propagated back to the gas source through the pulsed flame chamber (6) and is self-terminated in a few milliseconds, since the pulsed flame cannot propagate through the small hole at the bottom of the combustion chamber. The continuous combustible gas flow creates another ignition after a few hundred milliseconds in a pulsed (~3 Hz) periodic fashion. The emitted light is transferred with a light pipe (8) through a broad (not narrow)-band-pass filter (9) and detected with a PMD (10).

The PFPD is much more sensitive than the continuous flame version (2×10^{-13} gS/sec, 1×10^{-14} gP/sec, and 2×10^{-12} gN/sec) and much more selective from unwanted hydrocarbon emission, with total discrimination against hydrocarbon compounds (selectivity greater than 10^7). The PFPD sulfur mode has similar detection limits as that of the sulfur chemiluminescence detector (SCD), but its detection of signal to noise is better at practical detection levels (due to its quadratic response). Quenching is still a problem, however. It can be reduced by injecting smaller samples, but at the expense of sensitivity. The real advantage of this detector is with mass limited samples, such as when using capillary columns rather than packed columns, and for trace analysis.

Orifice-Capillary Detector (OCD)

The orifice-capillary detector (OCD) schematically illustrated in Figure 1.10aa was used in a fully functional, commercially available PGC whose operation was based totally on pneumatic power. Designed as a transmitter primarily for control applications, this PGC measured one or two components of the sample. It traded sensitivity for increased reliability and lower installed cost. An intrinsically safe product was ensured by eliminating all electrical power for its operation (the oven was heated by steam, and the temperature was regulated by a pneumatic temperature regulator).

The chromatogram generated from the detector was produced by the variation in differential pressure generated across an orifice by a density change in the carrier gas resulting from the elution of sample components from the column. This differential pressure was amplified by a pneumatic amplifier and fed to a pneumatic computer programmed to detect and measure the peak height of the two desired sample components and transmit these measurements to a controller as updated analog trend signals.

There has been a proposal in the literature[15] for increasing the minimum detection levels of such a device. The differential pressure from the OCD is converted directly to a frequency-modulated optical signal, and the complete chromatogram is transmitted over optical fiber to a remote station for more sophisticated computer data processing. In addition, the capability of the pneumatic PGC is also expanded to that of a multicomponent analyzer. This design modification changes this PGC to one of a split architecture with the nonelectric analyzer unit in the field and the electronic programmer controller (PC) remotely positioned. Such an instrument has not yet been commercialized.

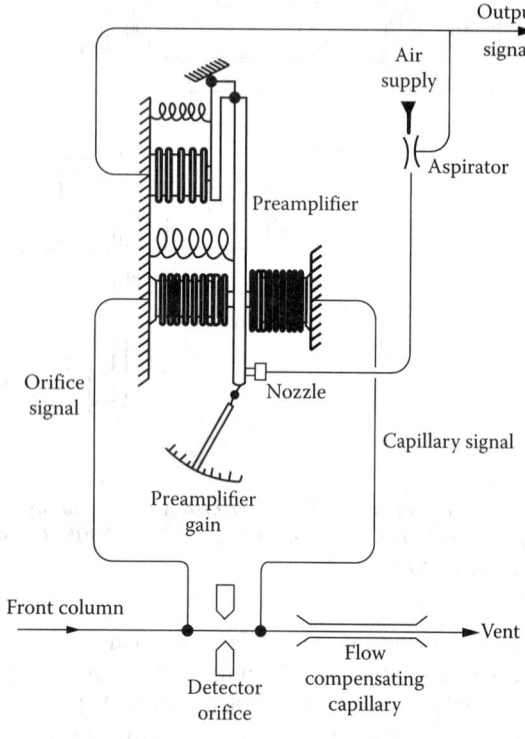
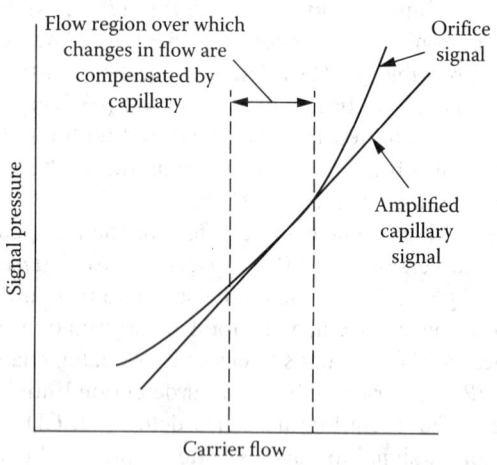

FIG. 1.10aa
OCD and pneumatic preamplifier. A properly designed capillary compensates for any differential pressure fluctuations across the orifice due to changes in flow rate so that the final differential pressure signal output of the amplifier is the result only of changes in density of the carrier gas due to the elution of sample components from the column.

Miscellaneous Detectors

There are a number of other detectors that have been introduced since the birth of GC, but because of issues such as high maintenance requirements, high customer burden, and unreliable operation over extended periods in the field, they have been limited primarily to laboratory applications. However, there are times when the minimum detection limits of the application cannot be satisfied by the basic detectors normally used in process GC. In those cases, some of these less-than-robust detectors have been used.

Photoionization Detector (PID)

The photoionization detector (PID) functions by irradiating the column effluent with high-energy ultraviolet (UV) light generated by a high-voltage discharge lamp containing

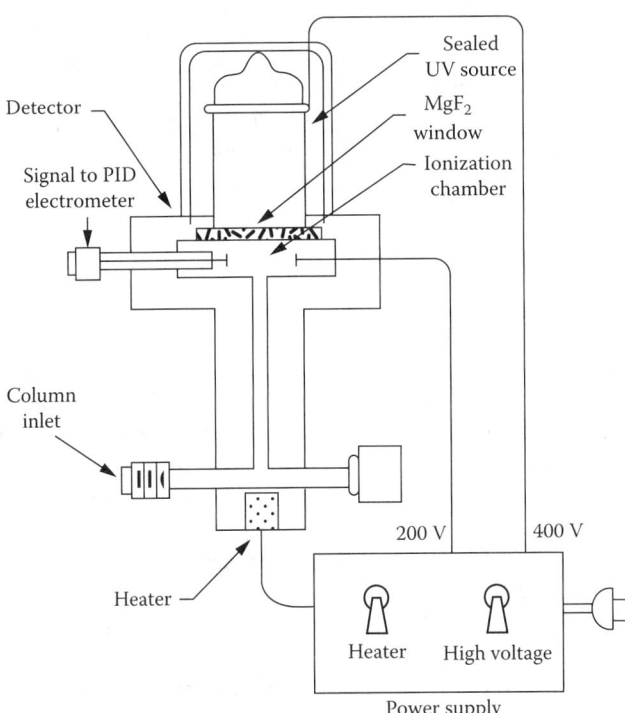

FIG. 1.10ab
Principal elements of a PID showing a 200 V supply for the collecting electrodes and a separate 400 V supply for the UV lamp. (From Driscol, J. N. and Spaziani, F. F., PID Development Gives New Performance Levels, Research/Development, May 1976.)

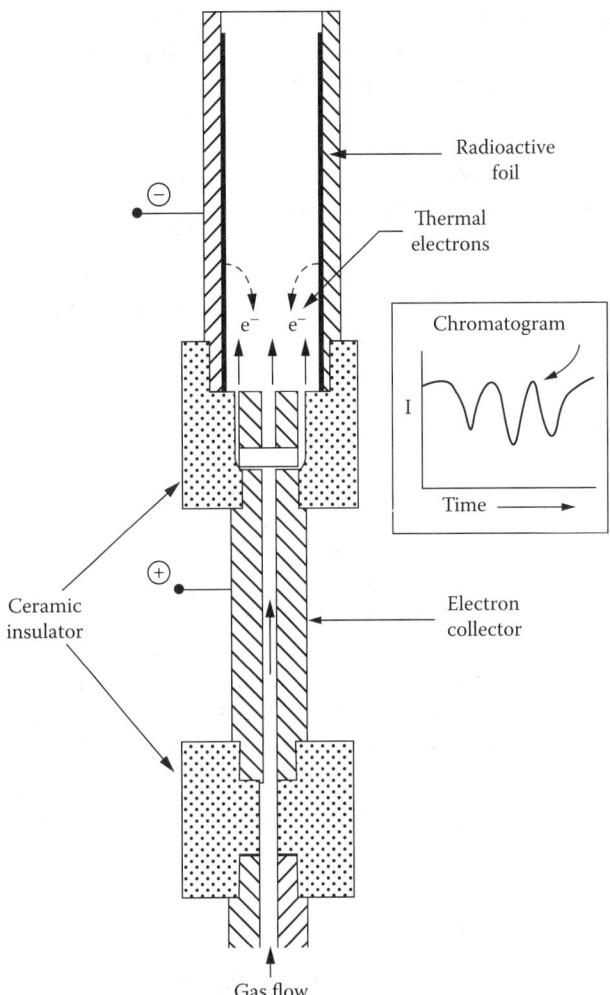

FIG. 1.10ac
Schematic view of an ECD arranged for constant voltage operation. Also shown is the chromatogram produced by the reduction of the standing current when compounds of higher electron affinity than the carrier gas are eluted from the column and pass through the detector. (Courtesy of Variant/Agilent Technologies.)

a noble gas (e.g., krypton) (Figure 1.10ab). The ions are collected at a polarized electrode, and the resulting current is measured with an FID-type electrometer. This detector is selective in that only compounds whose ionization potential is less than the UV radiation will be ionized. Compounds with aromatic ring structures give the highest sensitivities (as much as 10 times the FID), while compounds like methane and ethane with ionization potentials greater than 12.98 eV give no response with the commonly used PID tubes (9.5, 10.0, 10.2, 10.9, and 11.7 eV). The PID has a dynamic range of 10^7, extending from 2 pg to 30 mg. An advantage to this detector is that it does not require auxiliary gases, as does the FID. This advantage is offset by the fact that it does not respond sensitively to many of the compounds of interest, and it requires much maintenance to maintain quantitative accuracy.

Electron Capture Detector (ECD)

The electron capture detector (ECD) (see Figure 1.10ac) is yet another type of ionization detector. The column effluent passes between two electrodes, one of which has been treated with a radioactive source (tritium or nickel-63; the latter is preferred because of its extended detector stability) that emits high-energy electrons. These electrons produce large quantities of low-energy thermal electrons in the carrier gas, which are in turn collected by the other electrode to produce a steady-state current in the presence of pure carrier gas. Compounds eluting from the column that have an affinity for thermal electrons reduce this steady-state current, thereby producing the chromatogram. The detector is thus highly selective, with halogenated compounds being the most responsive (detection at the picogram level). Other groups exhibiting good selectivity include anhydrides, peroxides, conjugated carbonyls, nitriles and nitrates, and sulfur-containing compounds.

Maintenance is of critical importance with this detector, more so than with any other GC detector (except the helium ionization detector, discussed next). It responds exceptionally well to oxygen, necessitating leak-free systems and oxygen-free carrier gases. Also, response to water vapor can cause unstable baselines so that molecular sieve traps (which require periodic maintenance) are required in the carrier gas lines. Finally, care must be taken to use this detector only with columns of very low bleed, as the condensed stationary phase in the detector can easily be polymerized (by radiation and electron bombardment) to a hard insoluble deposit, which is almost impossible to eliminate and which interferes with the proper functioning of the detector.

Although a pulsed mode of operation has improved the linear range of the ECD, it has not improved its robustness, and all of the aforementioned concerns still apply.

Discharge Ionization Detectors

In contrast to the older-design ionization detectors that required a radioactive source of strontium, nickel, or tritium to produce metastable helium atoms by collisions with the high-energy electrons generated by the source, the newer designs require only helium. Two electrodes support a low-current arc through the helium makeup gas flow. The helium molecules between the electrodes are elevated from their ground state to form a helium plasma cloud. As the helium molecules collapse back down to the ground state, they give off a high-energy photon that will ionize all compounds having an ionization potential lower than 17.7 eV. Thus, the ionization detector will respond to volatile inorganics that the FID does not (such as NO_x, CO, CO_2, O_2, N_2, H_2S, and H_2). The ions are collected at polarized electrodes to produce a current, as in any ionization detector.

Pulsed Discharge Detector (PDD)

One version of this detector utilizes a steady-state helium plasma cloud, while a more recent design (see Figure 1.10ad) pulses the discharge (pulsed discharge detector [PDD]). Performance of the discharge ionization detectors is equal to or better than that of the older-design ionization detectors that require radioactive sources. In its helium photoionization mode of operation, it is a universal, nondestructive, high-sensitivity detector responding to fixed gases in the low parts per billion range. Its response to both inorganic and organic compounds is linear over a wide range (10^5).

The PDD is extremely versatile. If the carrier gas, at the column exit, is doped with methane (see Figure 1.10ad), the PDD can operate as an ECD with sensitivity and

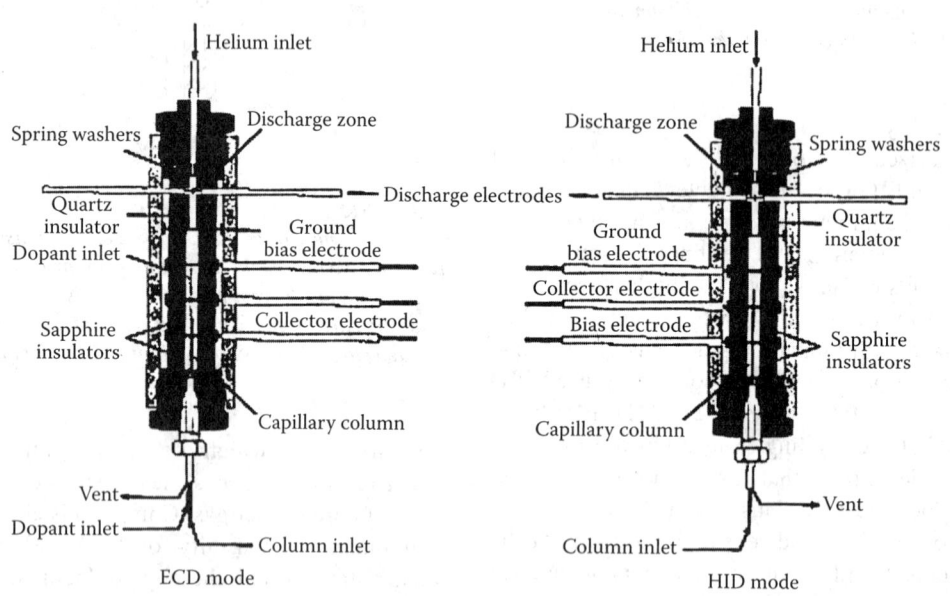

FIG. 1.10ad
Schematic view of a pulsed discharge detector configured as a PHID and as a PECD. A stable low-power, pulsed DC discharge, in helium is utilized as the ionization source. Eluents from the GC column flowing counter to the flow of helium from the discharge zone are ionized by photons from the helium discharge and the resulting electrons are focused toward the collector electrode by the two bias electrodes. The principal mode of ionization is photoionization in the range of 13.5–17.7 eV. (Courtesy of Valco Instrument Co. Inc.)

response similar to those of a conventional radioactive ECDs. Its MDQ for halogenated compounds ranges from 10^{-15} to 10^{-12} g.

In addition, when the helium discharge gas is doped with a suitable noble gas such as argon, krypton, or xenon (depending on the desired final ionizing power), the PDD can function as a specific PID for selective determination of aliphatics, aromatics, etc., as discussed in the section devoted to PIDs.

The HID is a high-maintenance item. It is the only detector that can measure permanent gases at the 1-ppb level. Therefore, leak-free plumbing is an absolute necessity. It responds to trace impurities, including water, in the carrier gas, thus requiring traps in the carrier gas stream to ensure the removal of these trace impurities. Clearly, this is a disadvantage for users not willing to devote maintenance time to monitor and frequently replace these traps. In addition, the HID is easily fouled by deposits from column bleed, which also contributes to a high noise level. Thus, its lowest detection limit is reached only with active solid supports such as molecular sieves.

CARRIER GAS FLOW CONTROL

Controlling carrier gas flow rate is of critical importance if one is to reproduce peak areas and retention times, and thus automate the analysis. Two methods, pressure control and flow control, may be used, but pressure control is by far the favored procedure used in process gas chromatographic instrumentation, for the following reasons:

1. A pressure regulator has an almost immediate response to downstream upsets (such as valve switching). A flow controller responds to flow upset only as fast as its flow set point allows.
2. A pressure regulator has the ability to provide some compensation for variation in retention time caused by changes in temperature.
3. A pressure regulator has the ability to supply the proper carrier gas flow rate to the column even in the presence of downstream leaks between it and the sample valve.

All pressure regulators are, to some degree, sensitive to temperature and input pressure. Tank regulators will not provide sufficiently stable output pressures under varying conditions of temperature and tank pressure. Thus, another regulator is mounted in the line close-coupled to the analyzer. Ideally, it should be mounted in the analyzer oven compartment to ensure a stable temperature environment, but few if any can withstand the 200°C (398°F) temperatures that are sometimes used. Therefore, it is important that this regulator has a minimum temperature coefficient (not as important a requirement when the analyzer is mounted in an analyzer shelter that has a fairly stable temperature environment) and that it can tolerate the input pressure changes it will experience from the tank regulator that is likely to be located outside, exposed to the varying temperatures of the environment.

For pressure drops across the column of 75 psi (5.3 bars), the relative error in peak area will be 1.7 times the relative error in the pressure drop. Therefore, to limit the error from this source to 0.5%, one must control pressure to within 0.4 psig (0.03 bars). In addition, assuming one is using a pressure regulator that is referenced to ambient pressure, the error caused by variations in ambient pressure will be 1.4 times the relative barometric pressure variation. Thus, errors of 0.2%–0.5% due to changes in weather are difficult to eliminate.

Recently, electronically controlled pressure regulators have become available and have been incorporated into PGCs. The heart of such a regulator is a solid-state pressure transducer that produces an electrical output proportional to pressure. This output is used to operate the integral control valve and to produce an output signal. In the case of the transmitter, miniature fluidic thermistors (fluistors) are used as pressure/flow controllers. Such an addition to the PGC is a tremendous advantage for maintenance staff, since it allows for control of carrier gas flow rate from the remote analyzer maintenance station (AMS) rather than at the unit.

PROGRAMMER CONTROLLER

The PC section is an extremely important part of a modern PGC, containing all the electronics to power the system, the controller, the data reduction hardware and software, and, of increasing importance, the communications package. Its design has changed radically over the past few years with the advances in electronic technology.

The PC unit can in theory be located in the field at the analyzer (stand-alone) or in a remote area such as the control room or instrument room close by (split architecture). The stand-alone design is by far the most popular. Although in this option the analyzer electronics are also usually located within the PC section, the primary purpose of the PC is to control all the functions in the analyzer. These include sample injection and column switching as well as various housekeeping tasks, such as modification of the detector signal, autozeroing, peak gating, integration, conversion of area units to concentration, and data transfer to the proper location. The transmitter utilizes what looks like a split-architecture design in that the electronics are packaged separately from the analyzer, but the two are close-coupled to each other in the field to yield a stand-alone PGC.

Most PC designs allow for operation in a Division 1 area when used with X or Y* air purge, and in Division 2 areas without purge.

Programmer

Programming tasks are managed by an on-board microprocessor-based computer. In older microprocessor-based PGCs, the PC was contained on several plug-in circuit boards located in a separate compartment from the analyzer. Current designs reflect the advances in electronics and focus on completely modular designs that allow for complete replacement of large portions of the electronic package (plug and play) for rapid repair and ease of maintenance. A number of provisions are made in the design to ensure the unit's rapid return-to-service in case of power failure.

Peak Processor

The peak processor section of the programmable controller has as its primary functions the detection of chromatographic peaks produced by the sample constituents of interest and the determination of the appropriate peak areas or peak heights. Peak detection is performed by comparing the instantaneous slope of the original (i.e., the rate of change) to some reference value. The objective is to differentiate between noise and the true onset of a peak. The peak detector can be disabled except for distinct time intervals (defined by the so-called gates) and integration performed only if a peak is found within these gates, or alternatively, integration can be forced between the gates.

Most systems have the capacity of measuring different types of peaks and allocating areas on the basis of some internal logic. This peak allocation is usually determined as follows:

1. Incompletely resolved peaks: Areas are allocated by dropping a perpendicular from the valley minimum to the baseline.
2. Proportional area allocation: Areas are allocated in proportion to the gross areas of the peaks.
3. Tangent skimming: Used for small rider peaks on the tail of major peaks. A tangent is drawn from the valley point to the back of the rider peak.
4. Forced integration: Bypasses slope detector logic and forced integration between start and stop points.
5. Peak height: Value of the peak maximum used as direct measure of concentration.

The peak areas are then converted to concentration using information obtained previously from the analysis of calibration standards. Both the peak selection and peak measurement procedures are selected by the user from a menu of choices at the time of setup.

MICROPROCESSOR OPERATED PGC

In modern microprocessor-based PGCs, the amplified detector signal is sampled by an analog-to-digital (A/D) converter and the digital values are stored in the random access memory (RAM) for processing. This is in contrast to the older procedure for acquiring chromatographic data: that is, monitoring the detector signal as a function of time with a potentiometer recorder. Recorder outputs are still present in some currently available PGCs to assist in setup and maintenance purposes. However, the record that is produced may not be as diagnostically useful as in the older designs, since, in many cases, it is not a record of the raw detector signal but an analog signal reconstructed from the RAM-stored filtered digital values.

Input–Output

The term process control generally implies the direct control of a process variable by an online instrument. Commonly, adjustments are made continuously in process conditions in order to reduce to zero the difference between the measured value of this variable and a desired (set point) value. For instruments whose output is a continuous trend representation of the measured variable (e.g., pH, conductivity, pressure, temperature, or flow), there is no problem in using this signal directly in a control scheme by comparing it to the set point and generating an error signal that is used to drive the process back to the set point. However, if the process variable is the concentration of one or two process stream constituents, the use of a chromatograph for the measurement presents special problems. The chromatographic signal is not continuous, but rather is transient. Furthermore, more than one component is represented in the information contained in the chromatogram, and this information is encoded as a function of time. Therefore, to be useful as a control instrument, the output must be deciphered and presented to the control loop in a form that is useful. For the traditional analog controller, this consists of a continuous analog signal (4–20 mA) whose magnitude is proportional to the concentration of the desired process variable. For an intelligent controller, this consists of a digital signal encoding the desired

---

* These are International Society of Automation (ISA) definitions of purge systems. X purge allows a general-purpose instrument to be placed in a Division 1 area, while Y purge allows for a Division 2-rated instrument to be used in a Division 1 area.

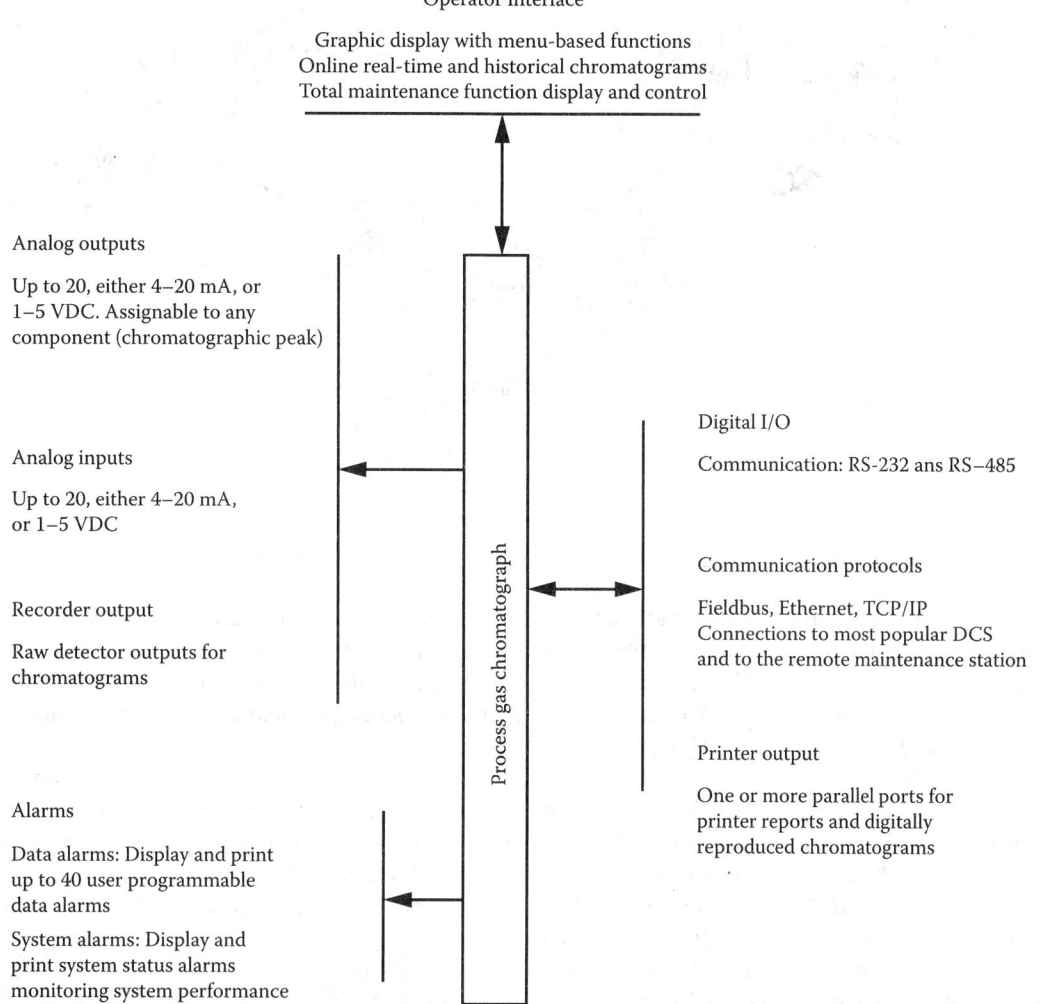

FIG. 1.10ae
I/O characteristics of a typical microprocessor-based PGC.

value of the process variable. Thus, a modern chromatograph must be able to deliver either of these outputs.

The input–output (I/O) characteristics of a typical PGC are summarized in Figure 1.10ae. Cyclically updated 4–20-mA analog trend signals proportional to the concentration of the analyzed peaks are available, as well as digital communication lines, RS-232 or RS-485. Various alarm signals are also generated. Additionally, provision is sometimes made to accept inputs from other devices where these data are required for a calculation of the desired process variable (e.g., flow rate and density measurements for BTU analysis).

Communication

Communication with the plant's distributed control system (DCS) is a must for modern PGCs. However, because of the large number of different DCSs in place—all using communication systems proprietary to the manufacturer of that system—direct interfacing with all networks is all but impossible for the user to ascertain that a gateway is available to interface the PGC to his or her control system.

Over the years, there has been much discussion and research into the development of a digital communication system to replace the traditional 4–20-mA analog standard. Worldwide adoption of an industry standard is still in the future. However, emerging from the various discussions within standards groups is a consensus that the local PGC will never be connected directly to the main data highway, but rather, communications will take place on a so-called fieldbus.

Fieldbus is a generic term that describes a new digital communication network that will be used in industry to replace the existing 4–20 mA analog signal standard. The network is a digital, bidirectional, multidrop, serial-bus

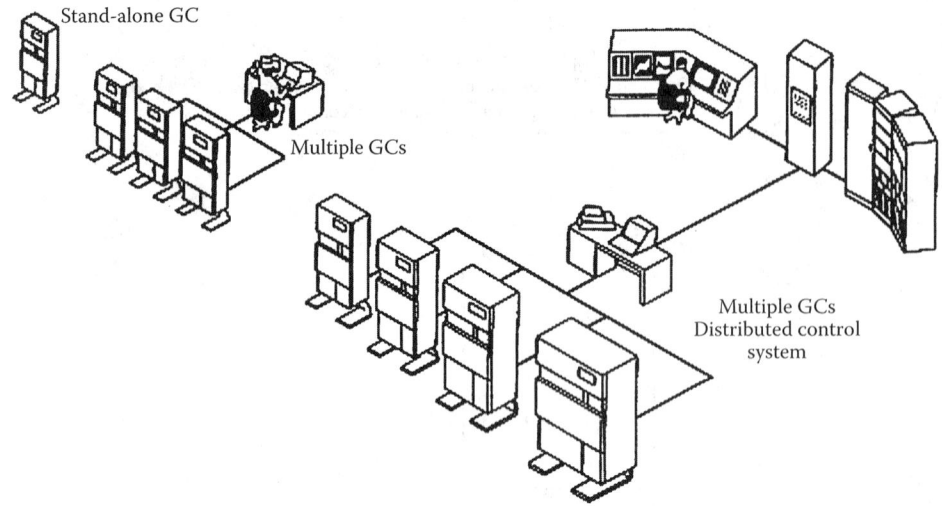

FIG. 1.10af
Illustration of the various hardware and communication configurations that must be supported by a PGC.

communication network used to link isolated field devices. The bidirectional specification allows data to be transmitted in two directions at the same time, multidrop can be interpreted as a single bus with many nodes connected to it, and serial means that the data are transmitted serially according to RS-232 or RS-485 protocol.

Currently, there are two advanced protocols for all digital communication with field devices for measurement and control of continuous processes. These are the Foundation Fieldbus (FF) and the Profibus-PA (also in the field is a hybrid analog/digital protocol called HART, which uses a frequency-based digital signal superimposed on the 4–20-mA analog signal).

In the middle of this fast-changing environment, current PGC design must allow for operation in a number of modes—stand-alone or multiple—reporting either to a single process measurement, control computer, and compound computer, or to the DCS (Figure 1.10af). Thus, many modern designs support the popular local area network (LAN) Ethernet protocol, as well as the new FF and Profibus-PA and older Modbus protocol.

OPERATOR INTERFACE

The operator interface has undergone many changes during the years following introduction of the gas chromatograph as a process control measurement device. In the original, rather crude chromatograph (from a modern technology viewpoint), the operator interface for control data also served as the maintenance technician's interface. However, in the modern system, the two functions have been largely separated. GC data, status, and validation inputs necessary to control and evaluate the process are sent to the control room, and the DCS console thus becomes the process or plant operator's interface. Maintenance and management data required to achieve maximum uptime are sent to a separate I/O interface called an AMS. This station can be positioned at the analyzer or at a remote location close by, with access to the system made with the ubiquitous industrial-hardened personal computer, or through a modem at a remote location some miles away. At the PC-based network workstation, any analyzer can be programmed and monitored. Graphical displays are available as an operator aid for simple operation, maintenance, and diagnostics. In addition, most PCs have a close-coupled I/O interface that includes a keyboard or display panel (illustrated in Figure 1.10ag)

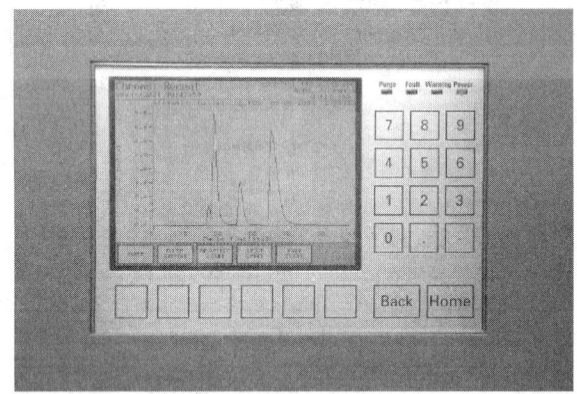

FIG. 1.10ag
Close-coupled full function AMS panel. (Photo courtesy of Siemens Applied Automation.)

providing access to the application and data reduction programs. Real-time chromatograms can be displayed here as well as hours of stored chromatograms complete with voltage and cycle times for future comparisons to simplify ongoing maintenance. Thus, from either of these locations, the maintenance technician can monitor the performance of the analyzer and reprogram the unit if necessary.

The full-featured close-coupled I/O interface described earlier is not available for the transmitter. Instead, a panel of LED indicator lights visible through a window in the electronics enclosure (see Figure 1.10ah) provides the technician with an overview of the instrument's operational status. Further information and programming options are available through a PC-based AMS.

A communication structure that combines all of the earlier discussed options is shown in Figure 1.10ai. The user requirements will vary depending on the size and complexity of the plant or process under consideration. Thus, the PGC is designed to satisfy simple as well as complex requirements.

FIG. 1.10ah
Close-up view of the see-through LED panel in the electronic enclosure. (Courtesy of Rosemont Analytical Inc of Emerson.)

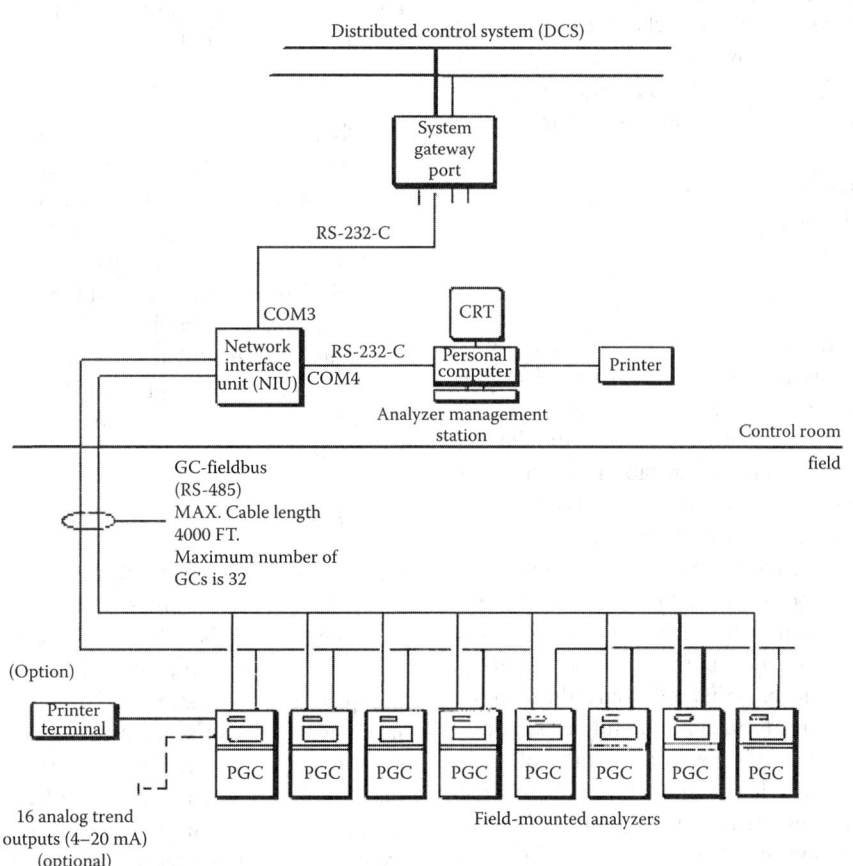

FIG. 1.10ai
A typical multi-PGC interface to the DCS. Some variation will exist among the various vendors. In this configuration, only PGC data, validation, system alarms, and stream sequence control are available to the DCS operator. All control and data communication is available to the AMS operator.

Alarms and Diagnostics

The amount of self-checking and diagnostic features that will continuously monitor the health of the instrument and alert personnel to problems and their source will continue to increase in importance as an integral part of the PGC design.

Present units have two types of alarms: system alarms and data alarms. System alarms are concerned with the pneumatic, mechanical, and electrical operation of the system. They include low carrier gas pressure, low calibration gas pressure, low sample flow, detector failure, oven temperature high/low, PROM error, and EEPROM error.

The data alarms (Hi/Lo) are used to monitor the results of the analysis. In some cases, these alarms alert only the operator; in other cases, action is taken automatically (such as not updating the trend signal with a new value when larger- or smaller-than-expected deviations occur). Retention time drifts and low area counts obtained for a calibration sample are examples of this type of alarm. Additionally, continuous read and write tests on the RAM memory are performed by the central processing unit (CPU) to detect the failure of a particular cell.

Quantitation

Since peak area is less sensitive to instrument variables than peak height, it has become the method of choice for translating the chromatographic record into process sample composition.

There are four principal methods used to relate the record of detector response to the sample composition. They are

1. Reference to calibration standards
2. Relative response factors
3. Internal normalization
4. Internal standard

All of these methods rely on some form of calibration of the detector by comparing its response to a calibration standard containing a known concentration of selected compounds. It is not always possible to prepare a standard of the process components (because of instability of the compounds, dew point problems, expense, etc.) However, the preparation of stable mixtures of reference compounds for which the relative detector responses to the compounds of interest are known may be possible. Process sample composition is then calculated using this information. The advantage over direct calibration is that all components of the sample do not need to be separated, only the components of interest. This leads to faster and simpler solutions to some application problems. The disadvantage of these direct calibration procedures is the direct effect that instrument variables (such as pressure of the sample loop) have on the accuracy and precision of the analysis. Normalization, on the other hand, eliminates sample volumes and pressure as variables, but requires that all sample components be measured. The normalization procedure involves taking the individual component measurements and adjusting each according to its response factor, and then dividing by the sum of these individual adjusted areas to yield the percent composition of the sample.

SAMPLE HANDLING

The importance of the sample handling system (SHS) and sample conditioning system (SCS) for ensuring reliability to the operation of the PGC and validity to its output cannot be overemphasized. In the author's experience, over 70% of the problems encountered in maintaining gas chromatographs can be traced to failures in the SHS/SCS. The reader who wishes to read detailed information on this subject should refer to Chapter 1.2 and the bibliography at the end of this chapter.

Unlike the laboratory chromatograph, when samples are usually held in an appropriate container and hand-carried to the laboratory where the chromatograph is situated, the PGC must receive a sample straight from the sampling point, untouched by human hands. Moreover, for the PGC to provide an accurate analysis of the process stream, a sample must be made available that is representative of the process composition. This is no trivial task, since the process sample may be quite hot; may be under considerable pressure; or may contain water vapor, solids, condensed liquids, and so on. Thus, the PGC requires a sample-handling front end to obtain, transport, and condition the sample.

The sample handling and conditioning system must

1. Obtain a sample of the process stream that is representative of its bulk composition.
2. Transport this sample to the analyzer within a time period such that the transport time plus the analysis time (i.e., the turnaround time) is short enough to satisfy the requirements of the control algorithm.
3. Condition the sample (i.e., clean, vaporize, condense, adjust pressure, dilute, etc.) and present it to the analyzer in a condition appropriate to the analyzer specifications.
4. Return the unused sample either to the process or to a waste disposal system.

Sample Probe

Because the low-velocity portion of the sample that is found along the walls of a process pipe is not representative, and also because debris is likely to accumulate along the walls, a simple tap into the process line to obtain a sample is not acceptable. A sample probe is required, which by virtue of its design can also serve quite effectively as a preliminary stage of conditioning. This is accomplished by (1) removing the sampling point from the walls, (2) orienting the probe so that the sample must turn 180° at the point of entry, thus excluding larger particles that cannot make the turn due to their momentum, (3) including a large-porosity filter in the probe, and (4) designing the probe to be large with respect to the sample line and allowing gravity to assist in separating particles from the gas phase.

Sample Transport

The sampling point and analyzer may be separated by hundreds of feet. It is necessary, therefore, to divert a portion of the process stream (through an appropriate probe), transport it to the analyzer, and return it either to the process stream at a lower pressure point or to a suitable waste disposal system. The time necessary to transport the sample to the analyzer can, in some cases, constitute the largest share of the system dead time or turnaround time. Mounting the analyzer closer to the sampling point is the most direct way of decreasing this time. Lacking this alternative, the next best solution is to increase the flow sample in the line. An added plus to using a so-called fast loop is that it provides additional mixing due to turbulent flow, and thus further ensures a representative sample.

Sample transport time is a function of sample line length, line diameter, the absolute pressure on the line, and the sample flow rate. Neglecting compressibility, the transport time for a gas sample is given as

$$t_{lag} \frac{VL(p+14.7)(530)}{F(14.7)(t+460)} \quad \quad 1.10(4)$$

where
- V is the volume of sample per unit length, ml/cm
- L is the length of sample line, cm
- F is the volume flow rate (ml/min) of sample under standard conditions (14.7 psia, 70°F)
- t is the temperature of sample, °F
- p is the sample line pressure, psig

Additional time must be allowed to purge sample conditioning equipment and any unswept volumes that are present due to poor design. Commonly, systems are designed to provide sample flow rates between 0.5 and 15 GPM (2 and 50 l/min).

Sample Conditioning

Conditioning of the sample either to allow easier handling or to put it into a form acceptable to the analyzer is always required to some degree. Conditioning may be as simple as pressure reduction and filtering or as complex as scrubbing and drying. Pressure reduction is the most effective way to prevent condensation of a condensable gas sample, and maintenance of a minimum pressure is the most effective way to prevent a liquid sample from vaporizing. Thus, pressure regulators for vaporization and pressure reduction are a common element of a conditioning system. In all cases, conditioning must be accomplished without affecting the sample composition. Some common sample conditioning elements are given in Table 1.10d.

The basic technology of sample conditioning has not changed much over the last 40 years. Currently, an SCS is constructed using components (listed in Table 1.10d) that have evolved from industries such as pneumatic and hydraulic service. These are mounted in a cabinet (which is usually secured to the outside wall of the analyzer shelter) and interconnected with tubing and fittings. Not being designed specifically to fit process analyzer needs, they have large dead volumes that are difficult to purge, thus leading to decreased sample system performance.

TABLE 1.10d
Common SHS/SCS Elements

Element	Function/Comments
Filter	Remove particulates from the sample. Filtering below 10 μm requires vigilant preventive maintenance unless some sort of self-cleaning filter can be devised (e.g., cyclone separation with automatic cleanout).
Vaporizing regulator	If possible, at least in the case of a PGC, it is best to present the sample as a vapor. The vaporizing regulator is best located at the sampling point (to provide short transport times for the sample), but it can be placed at the analyzer.
Condensers/separators	Used in applications where condensables are to be removed from the vapor sample
Coalescers	Used to force finely divided liquid droplets to combine into larger droplets so that they can be separated by gravity
Knockout pots	Used to collect liquid that has been separated from the sample. Should be provided with automatic drains.
Aspirators, ejectors	This is the preferred pump (by creating a vacuum) if the sample pressure is not sufficient to drive the sampling system. It is inexpensive and quite reliable because of the absence of moving parts
Rotameter	Consists of a ball or float in a tapered tube. Commonly used to measure flow. It is quite unreliable as the float tends to stick when the steam becomes dirty. It is a good indirect indicator of filter integrity.
Pressure regulators	Nonbleeding type, using corrosion resistant stainless steel and teflon. Used to regulate pressure between sampling point and the analyzer as a means of controlling sample flow rate. Since the quantity of sample is directly proportional to the sample in the loop, in most cases, a "block and bleed" configuration is used at the analyzer to equilibrate the sample to atmospheric pressure (this still does not eliminate inaccuracies due to changes in atmospheric pressure).
Pressure gauges	Installed downstream from regulators as an aid in setting regulator pressure and as an aid in maintenance checks. Installing in high-speed bypass lines and downstream of the analyzer eliminates extended turnaround problems that can occur because of the time required to flush sample from the pressure gauge bourdon tubes.
Steam or electrical heat tracing	To avoid condensation in the sample lines.

Recently, there has been some initiative directed toward changing the previously discussed conventional design to fit the specific requirements of the process analyzer. The potential of a smart modular miniature process analyzer sampling system integral to the analyzer is being explored with the idea of decreasing the installed cost of the analyzer, increasing reliability of the system, and, at the same time, decreasing cost of manufacture. One problem with sampling systems, however, is that they are application dependent. Thus, the design must allow for the convenient connection of simple modules to form the more complex one required for the particular application. A proposed design (borrowed from the gas management systems used in the semiconductor industry) shown in Figure 1.10aj uses a substrate assembly that provides the flow path for a process sample and consists of a variety of sample conditioning components. In turn, it is attached to a manifold system that will also accept various flow components and provide a flow path between two or more substrates.

Thus far, implementation of these ideas has been restricted to fairly simple and straightforward applications such as clean, dry, light hydrocarbon streams.

The PGC transmitter described previously is offered as a single- or dual-process stream analyzer (with two sample calibration streams) complete with a modular SHS for close-coupling to the sample tap.

Unfortunately, sampling systems usually fail during process upset conditions (just when the analysis is most needed to bring the process back into control). Thus, whenever possible, these systems should be overdesigned to be able to handle the upset condition. A process data sheet containing the information summarized in Table 1.10e is required to design a proper SCS. This definition of all the process conditions such as pressure, temperature, phase, particulates, chemical composition, physical properties (viscosity, density, etc.), and any possible process upsets that might occur, such as breakthrough of contaminants and runaway thermal reactions, ensures a design that will guarantee a correct sample presentation to the analyzer or, alternatively, a controlled shutdown or process hold with accompanying alarm signal.

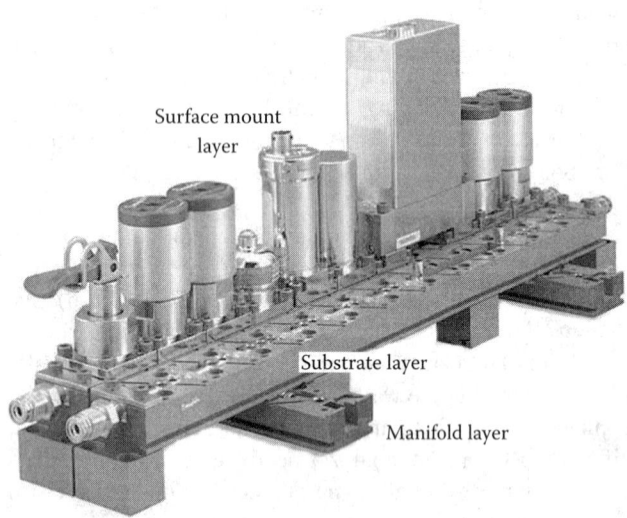

FIG. 1.10aj
Illustration of the architecture of a modular SHS with the various miniature SHS components surface mounted and interconnected on a so-called substrate. Individual substrates can then be manifolded together to form the complete SHS. (Photograph ©2002 Swagelok Company; text ©2002 R. Annino.)

TABLE 1.10e
Minimum Process Sample Information Required for Proper SHS Design

Liquid	Vapor
Process pressure and temperature	Process pressure and temperature
Bubble point at the process or highest ambient temperature (whichever is highest) and atmospheric pressure	Dew point and lowest ambient temperature
Viscosity	Average molecular weight
Specific gravity	Specific gravity
Sample composition under normal and upset conditions	Sample composition under normal and upset conditions
Sample return point pressure	Sample return point pressure
Distance between sampling point and the analyzer	Distance between sampling point and the analyzer
Required sample turnaround time (lag time of the SHS/SCS)	Required sample turnaround time (lag time of the SHS/SCS)
Maximum pressure for which the PGC sample valve is specified	Maximum pressure for which the PGC sample valve is specified
Whether the PGC valve can accept either liquid or vapor sample	Whether the PGC valve can accept either liquid or vapor sample
Minimum flow rate required through the PGC sample loop	Minimum flow rate required through the PGC sample loop
Corrosiveness of the sample (related to material compatibility)	Corrosiveness of the sample (related to material compatibility)
Sample toxicity	Sample toxicity

Multistream Analysis

The PGC must be able to accommodate at least two streams: the process stream and the calibration or standard line. The clean calibration sample can bypass the SHS and be plumbed directly into a stream-switching valve that will allow the process stream to be diverted, while the standard sample is flowing through the analyzer sample loop (Figure 1.10ak).

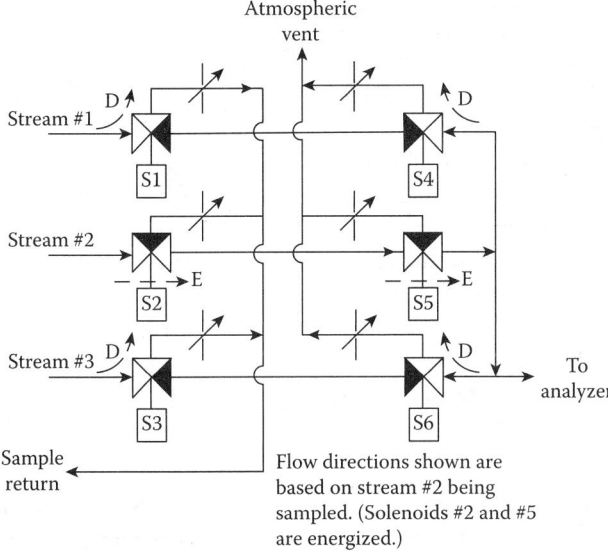

FIG. 1.10ak
Dual solenoid sampling system with block, bleed, and back purge.

The sharing of one PGC among a number of process streams is not recommended but, unfortunately, is common practice in the industry as a way of justifying the cost of the PGC. It is questionable if the savings are real when one considers the increased complexity of the SHS/SCS (one for each stream) and the maintenance problems that are incurred. Add to this the fact that if the analyzer goes down, the analysis of more than one stream is affected—a result that may have more impact on product cost than purchasing separate analyzers. The availability of a lower installed cost PGC with its simpler modular SHS may affect a shift in this paradigm in the future.

In any case, stream switching requires a design that prevents mixing of the process stream samples. The double block-and-bleed with bypass purge illustrated in Figure 1.10ak is probably the most efficient circuit for accomplishing this objective. In this circuit, all sample streams are flowing continuously, even in the unselected lines, thus ensuring up-to-date sample composition when selected. Sample stream 2 is selected in this case and continuously flushes the PGC sample loop and all other sample lines.

Sample Disposal

The vent of the PGC is almost totally carrier gas and is usually vented to the atmosphere, as is the sample loop vent line, unless the sample is toxic or otherwise environmentally harmful. In such cases, the sample must be vented to a specific waste disposal area. Alternatively, if a low-pressure point in the process stream is available, the sample can be returned to the process. Since the accuracy of the analysis is directly affected by the pressure of the sample loop (unless one is using normalization procedures), provision must be made to either measure this pressure or provide hardware to maintain a constant pressure.

In summary, the SHS/SCS is a very important part of PGC hardware. Proper design and maintenance of this part of the system will ensure a large uptime for the analyzer.

There has been some attempt to utilize the micromachined type of GC that has been used with some success, in the laboratory and as portable GC devices, as the analyzer portion of a PGC. The advantages of adopting such a design are, analytically, to achieve faster analysis and, engineering-wise, to design a smaller analyzer. Such a design puts an even greater burden on the SHS to provide an absolutely particle-free sample.

INSTALLATION

Typical utility provisions for PGCs are outlined in Table 1.10f. Additional support services may be required, such as refrigerated or heated sample runs, an analyzer shelter, etc. As a result, installation costs account for the principal portion of a traditionally designed PGC system. One of the prime advantages offered by the PGC transmitter is a significant reduction in installation and operating costs. However, severe environmental conditions can prevail at many industrial locations that can have an impact not only on the instrument but also on the maintenance personnel. Thus, mounting the PGC in a walk-in environmentally controlled analyzer shelter may be necessary for the protection of maintenance personnel while they are working on the instrument.

TABLE 1.10f
Typical Utility Requirement

Item	Specification
Instrument air	4 scfm per heated zone, 2 scfm purge
Instrument air quality	Clean, dry, −40°C dew point, oil free, particles ≤5 ppm microns, ISA grade hydrocarbon free
Carrier and other gases	Compressed gases, application dependent, may include helium, nitrogen, hydrogen, air, etc.
Compressed gas quality	Hydrogen, 99.995% ultrapure grade burner air, ≤1 ppm hydrocarbons, and ≤5 ppm water
	Helium, ≤99.995% ultrapure grade
	Nitrogen, 99.995% ultrapure grade
AC power	115 VAC, 50/60 Hz 10 amp service (per oven) or 220/230 VAC, 50/60 Hz

TABLE 1.10g
Breakdown of Installation Costs

Item	Traditional PGC	PGC Transmitter
Analyzer with basic SHS	45,000	33,000
Analyzer shelter (1/4 pro rata)	35,000	NA
Installation and startup	20,000	5,000
Total[a]	100,000	38,000
Sample/utility connection costs		
Electrical		
1 in. conduit	$28/ft	
2 in. conduit	$35/ft	
Sample lines		
Bare	$24/ft	
Heated	$85/ft	
Instrument air		
1 in. carbon steel		
Steam line (insulated 1.5 in.)	$30/ft	

[a] This is a minimum installation cost as each unit may require specific hardware/software depending on the specific PGC and the application.

A very approximate breakdown of these installation costs is given in Table 1.10g. The difference between the installed transmitter cost and that of the traditional PGC may be more striking when one considers the cost of running heat-traced sample lines. However, one must not be misled by claims that in all cases, the sample lines to a transmitter will be shorter, since it will be mounted close to the sample point: where the transmitter is mounted will always be a compromise between short sample lines and accessibility for maintenance purposes.

In addition, provision must be made for space to house the AMS. This unit may be located close by in a general-purpose area or located off-site and connected to the analyzer through a modem. Required gases (such as carrier gas, calibration gas, FID fuel, etc.) are stored as compressed gases in cylinders that should be placed in an easily accessible area outside, but fairly close to the analyzer shack.

ADVANCES

GC is used in all branches of the chemical industry, refining and hydrocarbon processing to provide the most flexible chemical composition analysis in all phases of production. GC is a mature technique of analysis, but advances in this technology are still being made and are applied in various fields. Research is still continuing to develop new stationary phases and columns, although this area has been much investigated and developed during the last six decades. Today, investigations are mostly focused on comprehensive 2D GC (GC × GC), fast GC, portable GC, and micro-GC.

In the area of stationary phase researches, development continues in the studying of ionic liquids and modified cyclodextrins. The GC columns are usually commercial capillary ones, and most of the stationary phases are 5% phenyl polydimethylsiloxane.

In order to measure trace impurities in pure products, multidimensional capillary GC has been developed to isolate very small amounts of analytes and to improve chromatographic resolution. This technology is complex and involves instability. In order to improve these limitations, septum-less sample handling, valve-less switching, cold trapping, microprocessor control, and mass flow controllers were introduced.

Much improvement has also been made in the area of developing intelligent displays and workstations, enabling easier monitoring or making modifications. Alarm management has been made simpler and can require only the clicking on the icon, automatically calling up the required screens on which the needed performance parameters are displayed. Such workstation portals also

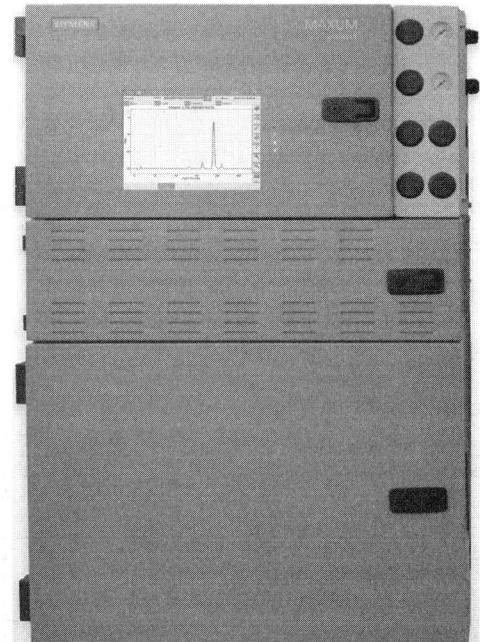

FIG. 1.10al
New workstation displays combine the previous menu-driven options with additional intuitive icons and graphical elements to increase the speed and simplicity of interactions with the GC during maintenance and operation. (Courtesy of Siemens.)

provide automatic data logging and reporting functions (Figure 1.10al).

The color touch screen user interface replaced the previously used black and white LCD panels. They are suitable for use in hazardous locations and meet the regulations of ATEX in Europe and CSA, FM in Canada and the US for Division 1/Zone 1 environments. These displays provide both the traditional menu-driven options and the new intuitive icons and graphical elements to increase the speed and simplicity of interactions for maintenance and operation.

SUMMARY AND ACKNOWLEDGMENTS

Although the PGC is perhaps the most expensive of the commonly used process analyzers, it provides the compositional information that is required for tight quality control. The introduction of the truly field-mountable PGC with the remote diagnostic and maintenance procedures that involve one technician doing a quick field replacement of the designated module may lead to an even greater increase in the use of PGCs in process control applications.

The author thanks Stephen Staphanos of Rosemont Analytical and Robert Bade of Siemens/Applied Automation for their fruitful discussions concerning PGC design, and Peter Vanvuren of Exxon/Mobil and the Center for Analytical Process Control for information regarding modular SCS design.

SPECIFICATION FORMS

When specifying gas chromatographs only, use the ISA form 20A1001 and when specifying both the analyzer and the composition or the properties of the process material of interest, use ISA form 20A1002. Both forms are reproduced with the permission of the International Society of Automation on the next pages.

As of this writing, the 2 forms mentioned below are still being prepared and when ready, they are recommended for use. For specifying the chromatograph components, the ISA form 20A2142 is recommended, while for the chromatograph itself use form 20A2141.

ANALYSIS DEVICE
Operating Parameters

#	RESPONSIBLE ORGANIZATION		#	SPECIFICATION IDENTIFICATIONS		
1			6			
2	ISA		7	Document no		
3			8	Latest revision	Date	
4			9	Issue status		
5			10			

#	ADMINISTRATIVE IDENTIFICATIONS			#	SERVICE IDENTIFICATIONS Continued			
11				40				
12	Project number	Sub project no		41	Return conn matl type			
13	Project			42	Inline hazardous area cl		Div/Zon	Group
14	Enterprise			43	Inline area min ign temp		Temp ident number	
15	Site			44	Remote hazardous area cl		Div/Zon	Group
16	Area	Cell	Unit	45	Remote area min ign temp		Temp ident number	
17				46				
18	SERVICE IDENTIFICATIONS			47				
19	Tag no/Functional ident			48	COMPONENT DESIGN CRITERIA			
20	Related equipment			49	Component type			
21	Service			50	Component style			
22				51	Output signal type			
23	P&ID/Reference dwg			52	Characteristic curve			
24	Process line/nozzle no			53	Compensation style			
25	Process conn pipe spec			54	Type of protection			
26	Process conn nominal size	Rating		55	Criticality code			
27	Process conn termn type	Style		56	Max EMI susceptibility		Ref	
28	Process conn schedule no	Wall thickness		57	Max temperature effect		Ref	
29	Process connection length			58	Max sample time lag			
30	Process line matl type			59	Max response time			
31	Fast loop line number			60	Min required accuracy		Ref	
32	Fast loop pipe spec			61	Avail nom power supply			Number wires
33	Fast loop conn nom size	Rating		62	Calibration method			
34	Fast loop conn termn type	Style		63	Testing/Listing agency			
35	Fast loop schedule no	Wall thickness		64	Test requirements			
36	Fast loop estimated lg			65	Supply loss failure mode			
37	Fast loop material type			66	Signal loss failure mode			
38	Return conn nominal size	Rating		67				
39	Return conn termn type	Style		68				

#	PROCESS VARIABLES	MATERIAL FLOW CONDITIONS			#	PROCESS DESIGN CONDITIONS		
69					101			
70	Flow Case Identification			Units	102	Minimum	Maximum	Units
71	Process pressure				103			
72	Process temperature				104			
73	Process phase type				105			
74	Process liquid actl flow				106			
75	Process vapor actl flow				107			
76	Process vapor std flow				108			
77	Process liquid density				109			
78	Process vapor density				110			
79	Process liquid viscosity				111			
80	Sample return pressure				112			
81	Sample vent/drain press				113			
82	Sample temperature				114			
83	Sample phase type				115			
84	Fast loop liq actl flow				116			
85	Fast loop vapor actl flow				117			
86	Fast loop vapor std flow				118			
87	Fast loop vapor density				119			
88	Conductivity/Resistivity				120			
89	pH/ORP				121			
90	RH/Dewpoint				122			
91	Turbidity/Opacity				123			
92	Dissolved oxygen				124			
93	Corrosivity				125			
94	Particle size				126			
95					127			
96	CALCULATED VARIABLES				128			
97	Sample lag time				129			
98	Process fluid velocity				130			
99	Wake/natural freq ratio				131			
100					132			

#	MATERIAL PROPERTIES			#	MATERIAL PROPERTIES Continued		
133				137			
134	Name			138	NFPA health hazard	Flammability	Reactivity
135	Density at ref temp	At		139			
136				140			

Rev	Date	Revision Description	By	Appv1	Appv2	Appv3	REMARKS

Form: 20A1001 Rev 0 © 2004 ISA

	RESPONSIBLE ORGANIZATION	ANALYSIS DEVICE COMPOSITION OR PROPERTY Operating Parameters (Continued)		SPECIFICATION IDENTIFICATIONS			
1			6				
2			7	Document no			
3	ISA		8	Latest revision		Date	
4			9	Issue status			
5			10				

	PROCESS COMPOSITION OR PROPERTY			MEASUREMENT DESIGN CONDITIONS				
11								
12	Component/Property Name	Normal	Units	Minimum	Units	Maximum	Units	Repeatability
13								
14								
15								
16								
17								
18								
19								
20								
21								
22								
23								
24								
25								
26								
27								
28								
29								
30								
31								
32								
33								
34								
35								
36								
37								

Rev	Date	Revision Description	By	Appv1	Appv2	Appv3	REMARKS

Form: 20A1002 Rev 0 © 2004 ISA

Definitions

EEPROM: Stands for Electrically Erasable Programmable Read-Only Memory and is a type of nonvolatile memory used in computers and other electronic devices to store small amounts of data that must be saved when power is removed, for example, calibration tables or device configuration.

HART: The HART Communications Protocol (Highway Addressable Remote Transducer Protocol) is an early implementation of Fieldbus, a digital industrial automation protocol. Its most notable advantage is that it can communicate over analog instrumentation wiring, sharing the pair of wires used by the older system.

MDQ: In chromatography, MDQ is the minimum amount of solute that will produce a detector signal twice the peak-to-peak noise of the detector. The dynamic range of the detector is the range of concentration of the test solute over which a change in concentration results in a measurable change in detector signal.

Mobile phase: The mobile phase is the part of the chromatographic system, which carries the solutes through the stationary phase. The temperature of the gas mobile phase is used to adjust the retention in GC.

RAM: Random access memory (RAM/) is a form of computer data storage. A random access device allows stored data to be accessed directly in any random order. In contrast, other data storage media such as hard disks, CDs, DVDs, and magnetic tape, as well as early primary memory types such as drum memory, read and write data only in a predetermined order, consecutively, because of mechanical design limitations.

Retention time: Retention time (tR) is the time it takes for a solute to travel through the column. The retention time is assigned to the corresponding solute peak. The retention time is a measure of the amount of time that a solute spends in a column. It is the sum of the time spent in the stationary phase and the mobile phase.

Solute: A substance dissolved in another substance, usually the component of a solution present in the lesser amount. In case of GC, the components of the sample.

Stationary phase: The stationary phase is the part of the chromatographic system though which the mobile phase flows where distribution of the solutes between the phases occurs. The stationary phase may be a solid or a liquid that is immobilized or adsorbed on a solid. In general, immobilization by absorbtion on a solid can be used in GC. The stationary phase may consist of particles (porous or solid), the walls of a tube (e.g., capillary) or a fibrous material (e.g., paper).

X and Y PURGE: X purge allows a general-purpose instrument to be placed in a Division 1 area, while Y purge allows for a Division 2-rated instrument to be used in a Division 1 area. These are Instrument Society of America (ISA) definitions of purge systems.

Abbreviations

A/D	Analog to digital
AED	Atomic emission detector
AMS	Analyzer maintenance station
ATEX	Atmosphères Explosibles
BTU	British thermal unit
CD	Conductivity detector
CPU	Central processing unit
ECD	Electron capture detector
ECN	Effective carbon number
ELSD	Evaporative light scattering detector
FD	Fluorescence detector
FF	Fieldbus
FID	Flame ionization detector
FPD	Flame photometric detector
GC	Gas chromatograph
HART	Highway Accessible Remote ransducer protocol
HCDP	Hydrocarbon dew point
HETP	Height equivalent to a theoretical plate
HID	Helium ionization detector
hwd	Height, width, depth
ID	Inside diameter
LAN	Local area network
LED	Light-emitting diodes
MD	Mira detector
MDQ	Minimum detectable quantity
MS	Mass spectrometer
NPD	Nitrogen phosphorous detector
NQA	Aerosol-based detector
OCD	Orifice-capillary detector
OD	Outside diameter
OT	Open tubular
PDD	Pulsed discharge detector
PECD	Pulsed electron capture detector
PFPD	Pulsed flame photometric detector
PGC	Process gas chromatograph
PHID	Pulsed helium ionization detector
PID	Photo ionizaton detector
PMT	Photomultiplier tube
PROM	Programmable read-only memory
RAM	Random access memory
RFD	Tadio flow detector
RID	Refractive index detector
SCD	Sulfur chemiluminescence detector
SCOT	Support-coated open tubular
SCS	Sample conditioning system
SGC	Smart gas chromatograph
SHS	Sample handling system
SMSD	Sumon detector
TCD	Thermal conductivity
UV/DAD	Ultraviolet diode array detector
WCOT	Wall-coated open tubular

Bibliography

Agilent Technologies, Capillary flow technology: Deans switch, (2013) https://www.agilent.com/cs/library/brochures/5989-9384EN.pdf.

Allgood, C., Correlations of relative sensitivities in gas chromatography, (1990) http://www.sciencedirect.com/science/article/pii/104403059085020M.

Annino, R., Industrial-grade field-mountable gas chromatograph for process monitoring and control, (1994) http://www.sciencedirect.com/science/article/pii/0021967394804753.

ANSI/ISA 76 and IEC 62339-1:2006, Modular sample systems, (2006) http://www.circortech.com/CircorAssets/Products/Documents/77212_CT76_v13.pdf.

ASTM D7278-11, Standard guide for prediction of analyzer sample system lag times, (2011) http://www.astm.org/Standards/D7278.htm.

Bruchner, M. Z., Ion chropatography, (2015) http://serc.carleton.edu/microbelife/research_methods/biogeochemical/ic.html.

Clevett, K. J., Process analyzer technology, (1986). http://www.amazon.com/Process-Analyzer-Technology-Kenneth-Clevett/dp/0471883166.

Grob, R. L., Modern practice of gas chromatography, 2004, http://www.amazon.com/Modern-Practice-Chromatography-Robert-Grob/dp/0471229830/ref=pd_sim_b_5.

Hoffmann, A., Bremer, R., and Nawrot, N., New developments in multi-dimensional capillary gas chromatography, http://www.gerstel.com/pdf/p-gc-an-1998–04.pdf.

LabTraining, Types of gas chromatography columns, (2011) http://lab-training.com/landing/gc-module-6/.

Mayeaux, D., Modern software for the analyzer engineer, ISA 2014 Analysis Division Symposia, Analyzer Sample System References, (2014) www.iceweb.com.au/Analyzer/sample_systems.htm, 2014–09.

McNair, H. M., Basic gas chromatography, (2009) http://www.amazon.com/Basic-Gas-Chromatography-Harold-McNair/dp/0470439548.

Meyers, R. A., Ed., Encyclopedia of Analytical Chemistry: Instrumentation and Applications, (2011) http://www.wiley.com/WileyCDA/WileyTitle/productCd-0470973331.html.

Poole, C., Gas chromatography, 2011, http://www.amazon.com/Gas-Chromatography-Colin-Poole/dp/0123855403/ref=pd_sim_b_12.

Rood, D., The troubleshooting and maintenance guide for gas chromatographers, (2007) http://www.amazon.com/The-Troubleshooting-Maintenance-Guide-Chromatographers/dp/3527313737/ref=pd_sim_b_1-#.

Rosemount Emerson, Gas chromatograph solutions, (2015) http://www2.emersonprocess.com/siteadmincenter/PM%20Danalyzer%20Documents/NGC_BRO_XA_Series.pdf.

Sherman, R. E., Process analyzer sample-conditioning system technology, (2002) http://www.wiley.com/WileyCDA/WileyTitle/productCd-0471293644.html.

Siemens, Gas chromatographs, (2015) http://news.usa.siemens.biz/press-release/gas-chromatograph/siemens-enhances-maxum-gas-chromatograph-color-touchscreen-display.

Sparkman, D., Gas chromatography and mass spectrometry: A practical guide, 2nd edn., (2011) http://www.amazon.com/Gas-Chromatography-Mass-Spectrometry-Practical/dp/0123736285/ref=pd_sim_b_3.

Thomson, M., Interfacing sample handling systems for on-line process analyzers, (2002) www.measurementation.com.au/tp-1.htm.

U.S. National Library of Medicine, Recent advances of gas chromatography, http://www.ncbi.nlm.nih.gov/pubmed/20073193.

Verhappen, I., The basics of analyzer sample systems, (2012) http://www.iceweb.com.au/Analyzer/basics_of_analyzer_sample_system.htm.

1.11 Chromatographs
Liquid

L. B. ROOF (1982) **B. G. LIPTÁK** (1995, 2003, 2017)

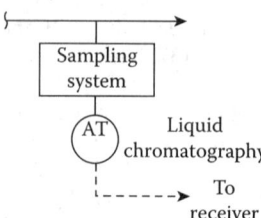

Flow sheet symbol

Types	Laboratory: thin-layer and paper chromatography
	Industrial: layer chromatography, column chromatography, or HPLC
	HPLC (High-Pressure Liquid Chromatography) can be normal or reversed phase. Reversed phase HPLC is the most commonly used form
Column	Liquid–solid absorption, liquid–liquid partition, gel permeation (exclusion), and ion exchange
Detectors	Fiber-optic probes, differential refractive index, fixed- or variable-wavelength ultraviolet, dielectric constant, electrical conductivity
Type of sample	Liquid
Sample pressure	5–1000 psig (0.35–70 barg)
Sample temperature	16°C–149°C (60°F–300°F)
Ambient temperature	−18°C to 50°C (0°F to 122°F)
Contacting materials	Stainless steel, Teflon standard; all conventional materials available
Utilities required	Electrical power, carrier solvent, air at 100 psi (7 barg)
Repeatability	±0.5% for most applications
Cycle time	3–20 min for most applications
Special features	Multicomponent readout, molecular weight readout
Costs	Laboratory system component costs: HPLC columns cost from $500 to $2000; solvent delivery pumps range from $2000 to $5000; fixed-wavelength ultraviolet–visible (UV-VIS) detectors cost $2500; variable-wavelength ones are about $5000; microprocessor-based data acquisition integrator costs about $3000
	Complete online process HPLC units cost about $50,000, and their installed cost is about $100,000
Partial list of suppliers	Beckman Coulter (http://www.gmi-inc.com/beckman-system-gold-high-performance-liquid-chromatograph.html)
	Bruker (http://www.bruker.com/products/mass-spectrometry-and-separations/lcms/liquid-chromatography.html)
	Buck Scientific (http://www.bucksci.com/atomic-absorption-spectrophotometers/category/6/high-performance-liquid-chromatographs.html)
	Cole-Parmer (http://www.coleparmer.com/Category/High_Performance_Liquid_Chromatography_HPLC/50089)
	Daniel (http://www.southeastern-automation.com/PDF/Emerson/Chromatograph%20Notes/NGL%20Fractionation%20Part%202%2027KB.pdf)
	EMD Millipore (http://www.emdmillipore.com/chemicals/analytical-hplc/c_BOab.s1LVnkAAAEWm.AfVhTl?cid=1USEN2330&ppc_kw=liquid%20chromatography)
	Hitachi (http://www.hitachi-hta.com/products/life-sciences-chemical-analysis/liquid-chromatography?gclid=CJHuxYDZ67cCFRSi4AodeVsAeA)
	Labcompare (http://www.labcompare.com/Laboratory-Analytical-Instruments/35-Liquid-Chromatography-HPLC/)
	Regis Technologies Inc. (http://www.registech.com/chromatography)

(Continued)

Siemens Energy & Automation (http://news.usa.siemens.biz/press-release/gas-chromatograph/siemens-enhances-maxum-gas-chromatograph-color-touch-screen-display)

Thermo Scientific (http://www.dionex.com/en-us/products/liquid-chromatography/lp-71340.html)

Thomson Instrument (http://www.htslabs.com/catalog/?d=flash)

Waters, http://www.waters.com/waters/en_US/HPLC-Columns/nav.htm?cid=511505&locale=en_US

Yamazen (http://yamazenusa.com/wp-content/themes/openair/images/green_flash_chromatography.pdf)

INTRODUCTION

High-performance liquid chromatography (HPLC) was formerly referred to as high-pressure liquid chromatography. It is a technique used to separate, identify, and quantify components in a liquid mixture. It relies on pumps to pass a pressurized liquid solvent containing the sample mixture through a column filled with a solid adsorbent material where each component in the sample interacts slightly differently with the adsorbent, causing each component to flow at a different velocity. This leads to the separation of the components as they flow out the column.

Liquid chromatography is often considered for the analysis of process samples that cannot be easily vaporized and therefore gas chromatography cannot be used. Liquid chromatography is a method to separate, identify, and quantify the components of a liquid sample after it being injected into a moving stream of fluid (called the mobile phase), which transports the sample through the "stationary" phase inside a "column." In this process, the components in the sample travel at different speeds, causing the constituents to separate. The detectors at the outlet of the column serve to determine the relative proportions of the constituents ("analytes") in the mixture (Figure 1.11a).

In a liquid chromatograph column, the stationary phase can consist of a finely powdered solid adsorbent packed into a thin metal column, and the mobile phase can be an eluting solvent that is forced through the column by a high-pressure pump.

The sample to be analyzed is injected into the column inlet and, after the separation of its constituents in the column, they are identified and quantified by a detector. A wide variety of liquid chromatograph packing, eluting solvents, and detectors are available. Their combinations are selected to obtain the desired resolution.

High-pressure liquid chromatography relies on pumps to pass a pressurized liquid and a sample mixture through such a column filled with a sorbent and thereby achieve separation of the sample components. The active component of the column, the sorbent, is typically a granular material made of solid particles (e.g., silica, polymers, etc.), 2–50 μm in size. The components of the sample mixture are separated from each other due to their different degrees of interaction with the sorbent particles. The pressurized liquid is typically a mixture of solvents and is referred to as the "mobile phase."

HPLC is distinguished from traditional ("low pressure") liquid chromatography because operational pressures are significantly higher (50–400 bars), whereas ordinary liquid chromatography typically relies on the force of gravity to pass the mobile phase through the column. Due to the small sample amount separated in analytical HPLC, typical column dimensions are 2.1–4.6 mm diameter, and 30–250 mm length. Also, HPLC columns are made with smaller sorbent particles (2–5 μm in average particle size). This gives HPLC superior resolving power when separating mixtures.

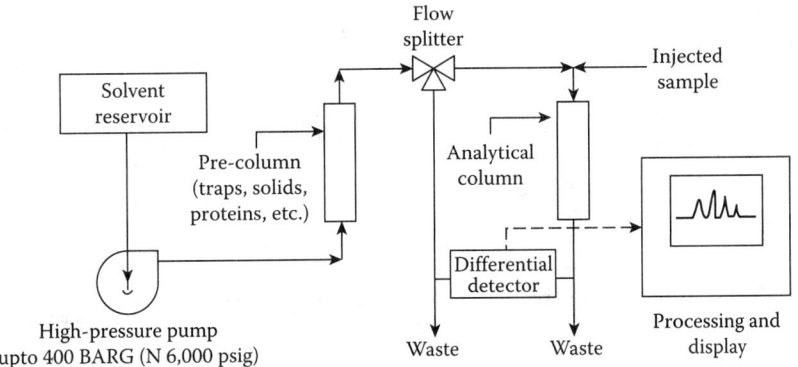

FIG. 1.11a
The main components of an HPLC loop.

COMPARISON WITH GAS CHROMATOGRAPHS

In many ways the liquid chromatograph is similar to the gas chromatograph. The basic difference is that the carrier is a liquid instead of a gas. All the other changes in the instrumentation used result from this difference. In fact, the only components that are significantly different are those that are in direct contact with the carrier solvent. Chapter 1.10 describes the gas chromatographs and should be referred to for a basic understanding of the operation of chromatographs in general.

MAIN COMPONENTS

The liquid chromatographic system is divided into two parts: the analysis section and the control section (Figure 1.11b). The analysis section is located near the process stream and is often contained in an instrument house. This section includes (1) the chromatographic oven containing the valves, columns, and detector; (2) the carrier supply; (3) an electronics compartment containing the circuitry for the detector, temperature control, valve operators, and local data handling; and (4) the sample preparation system, which may be in a separate temperature-controlled oven.

The control section is often located in the control room and includes the programmer to control the instrument and process the data, and the data display, which may include a strip chart recorder, a digital display, or a digital computer for further data processing. The programmer discussed here includes stream selection, which may be a physically separate unit.

However, the liquid chromatograph operating conditions are different from those of the gas chromatograph. In particular, the carrier pressure is typically 1000 psig (~70 MPa) at a flow rate of 1 ml/min, and the chromatographic cycle time is typically 3–20 min.

Carrier Supply

A typical carrier flow diagram of the liquid chromatograph is illustrated in Figure 1.11c. The sample valve injects a measured volume of sample into the controlled flow of the carrier liquid, which transports it through the columns and

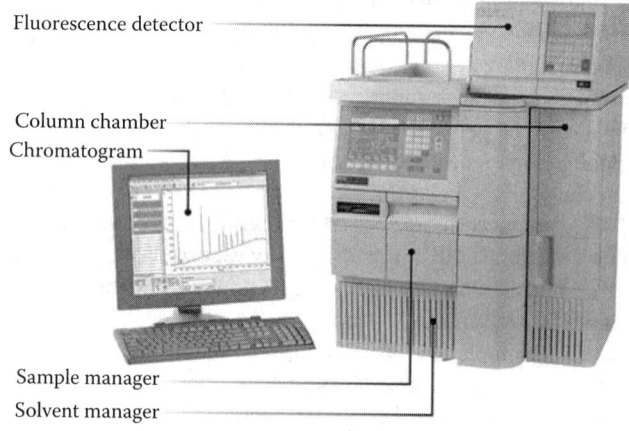

FIG. 1.11c
State-of-the-art high-pressure liquid chromatograph. (Courtesy of Waters.)

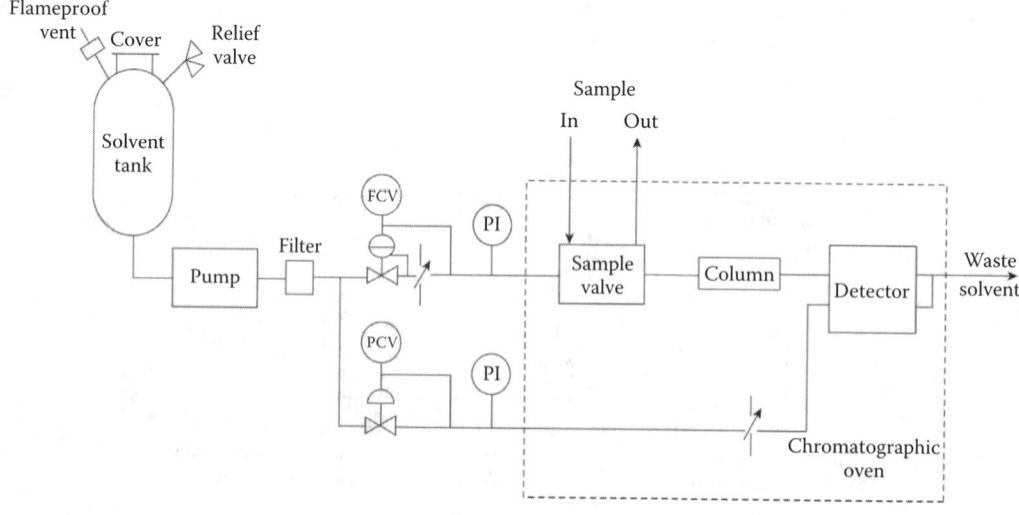

FIG. 1.11b
Carrier flow diagram of the liquid chromatograph.

into the detector. Interaction between the stationary column packing material and the flowing liquid carrier causes the sample components of interest to move through the column at different velocities, causing the separation. For each sample component, the detector provides an electrical signal proportional to its concentration in the carrier. The electrical signal is recorded as a chromatogram or otherwise displayed in any of the standard chromatographic forms.

The carrier supply is made up of (1) the liquid carrier, (2) a reservoir to hold it, (3) a carrier pump, (4) pressure and flow controls, and (5) a filter and gauges. The carrier is either a single-component solvent or a blend of solvents. The selection and uniformity of solvents and solvent blends are critical to the liquid chromatograph, affecting both component separation and detector response.

A typical carrier reservoir is a 15-gal (57-l) stainless steel tank equipped with a flameproof vent and a relief valve for flammable solvents. Some solvents are affected by oxygen or moisture in the air, and an inert gas, such as nitrogen, is either bubbled through or blanketed over the carrier in the tank.

Supply Pumps

There are two types of carrier pumps: air driven and motor driven. Both are reciprocating pumps containing check valves to divert the liquid through the pump. Air-driven pumps are powered by a large-diameter air piston, which is connected to a small-diameter liquid piston. The pump recycles at the end of each stroke. The air-driven pump provides a pressure, relatively independent of flow, that is nominally some multiple of the air pressure.

Motor-driven pumps are powered with an electric motor connected to an eccentric that drives one or more liquid pistons. The motor-driven pump provides a flow, relatively independent of pressure that is controlled by motor speed or piston stroke length.

Pressure and Flow Controls

Pressure and flow control depend on the type of pump used. Air-driven pumps provide a pressure that is reduced to the desired value with a diaphragm pressure regulator. Flow control is provided with a diaphragm-type differential pressure regulator that controls a preset pressure difference across a fixed restriction.

This system has the advantages of flexibility and of providing a continuous, not pulsating, flow, but requires numerous components. Carrier control with motor-driven pumps is generally limited to flow control. Two flow control methods are in use. The first option is to manually set the motor speed or pump stroke to the desired flow. This has the advantage of simplicity, but the quality of control is often insufficient. If automatic, closed-loop control is provided, the quality of control is improved, but the system becomes more complex and expensive.

Valves

Liquid chromatographic valves operate on the same principles as gas chromatographic valves. The air-operated diaphragm valves, plug valves, and rotary valves are also used (see Chapter 1.10). These valves are strengthened and use higher seal forces due to the high operating pressures. They also have smaller internal flow passage to reduce sample carrier mixing in the valve.

COLUMNS

Liquid columns are packed in a $\frac{1}{4}$ in. or $\frac{3}{8}$ in. (6- or 9.6-mm) outer-diameter straight stainless steel tube 2–12 in. (50–305 mm) long. The total column length in an instrument rarely exceeds 1.2 m (4 ft) and is more often 0.3 m (1 ft) or less. The columns are filled with particulate packing that is typically in the 5–10-µm range. There are four types of liquid columns classified by the principle of separation that is utilized in them: liquid–liquid columns, liquid–solid columns, size exclusion columns, and ion exchange columns. Table 1.11a lists a few typical column applications.

TABLE 1.11a
Typical Column Applications

Column Type	Applications
Liquid–liquid, normal phase	Phenols
	Esters
	Pigments
	Cooking oils
Liquid–liquid, reverse phase	Pesticides
	Herbicides
	Organic aromatics
	Water pollutants
Liquid–solid	Plasticizers
	Antioxidants
	Organic acids
	Water pollutants
Size exclusion	Polymers
	Resins
	Carbohydrates
	Hydrocarbons
Ion exchange	Inorganic ions
	Dies
	Detergents
	Sugars

Selectivity and Resolution

As the constituents of interest pass through high-pressure liquid chromatograph (HPLC) columns, they are retarded due to the selective absorption or adsorption effects of the column packing. The more efficient a column, the more able it is to produce narrow peak bands during the elution of the sample.

Figure 1.11d illustrates how a process sample containing constituents A and B enters three different columns, which produce three different separations between the peaks of the constituents. For example, column 1 in Figure 1.11e is inefficient (low resolution), whereas columns 2 and 3 are more efficient. Column 2 provides moderate and column 3 provides high resolution. In other words, although the efficiencies of columns 2 and 3 are similar, their selectivities—the relative affinity of the sample constituents to the packing—are different, as the selectivity of column 3 is better, or higher.

The coefficient, which expresses selectivity (Table 1.11b), is called by different names as a function of the type of column used. Selectivity can be improved by increasing column size or by raising the ratio of distribution factors. The combined effect of efficiency and selectivity is called resolution. In Figure 1.11d, the resolution of column 3 is the best. Unfortunately, the steps that will improve resolution (smaller tubing diameter, smaller injection volume, finer packing, and larger or multiple columns) also tend to extend analysis time and decrease downtime and maintenance due to plugging.

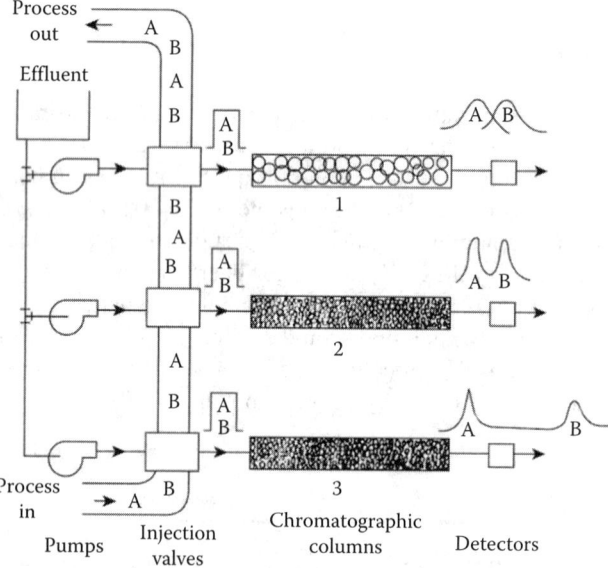

FIG. 1.11d
Relative efficiency, selectivity, and resolution of chromatographic columns.

TABLE 1.11b
Selectivity Coefficient Names Used in Case of Different Columns

Column Type	Name Used for Selectivity Coefficient
Liquid–solid adsorption	Adsorption
Liquid–liquid partition	Partition
Gel permeation (exclusion)	Permeation
Ion exchange	Distribution

Liquid–Partition Columns

Liquid–liquid columns (liquid–partition columns) are filled with solid support, usually silica, coated with a liquid stationary phase. To prevent loss of the liquid, it is usually chemically bonded to the support. The separation depends on the relative solubility of the sample components in the stationary and mobile, or carrier, phases.

Liquid–liquid separations are subdivided into normal and reverse phase separations. In normal phase separation, the stationary phase is more polar (in the chromatographic sense, not dipole moment) than the mobile phase. While in reverse phase separation, the stationary phase is less polar than the mobile phase. When sample components can be separated with either type, the components elute from the column in reverse order when changing from the normal to the reverse phase.

To improve separation, the binary mixture of solvents can be changed in the carrier. This is called solvent programming or gradient elution and is comparable to temperature programming in gas chromatography. Its use is limited almost exclusively to laboratory analysis.

Liquid–Adsorption Columns

In this case, the liquid–solid columns (liquid–adsorption columns) are filled with an adsorbent. Silica is the most common, but alumina and even charcoal are also used

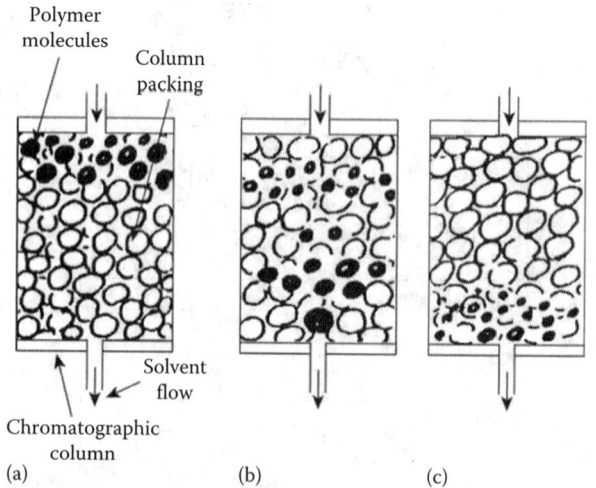

FIG. 1.11e
Three stages in the chromatographic separation of polymers: (a) at sample injection; (b) during separation; and (c) at sample elution.

occasionally. Separation occurs as the sample components are retained by the adsorption sites on the packing. The mobile phase competes with the sample for adsorption sites and displaces the sample so it can move through the column. As with liquid–liquid columns, the mobile phase is chosen or blended to provide the desired separation of sample components. Solvent programming can also be used.

Gel Permeation Columns

In gel permeation chromatography, sample components are separated on the basis of their molecular size. Large molecules move straight down the column, whereas small molecules stick in the pores of the porous beads of the gel and are retarded.

Size exclusion columns (steric exclusion columns or gel permeation columns) are filled with a porous solid, such as silica, or a porous polymer, such as cross-linked polystyrene (Figure 1.11e). The pore size in the column varies uniformly over a specific range, for example, 300–600 Å.

While moving through the column in the mobile phase, a sample component with a chain length (or more correctly, hydrodynamic volume) of 400 Å can diffuse into any of the pores that are 400 Å and larger. On the other hand, a sample component with a chain length of 500 Å can diffuse into only those pores 500 Å and larger. Since it enters fewer of the stationary pores, the 500 Å component moves through the column faster than the 400 Å component. Thus, long-chain or high-molecular-weight compounds elute from the column first. The mobile phase has no direct effect on the separation process in the columns.

Ion Exchange Columns

For process samples containing ions, ion exchange chromatography can be used. These columns can be packed with ion exchange resins, which contain exchangeable ions and therefore can separate the ions in the process sample from the neutral or oppositely charged components.

Ion exchange columns are filled with a cross-linked polystyrene resin containing charge-bearing functional groups on its surface. The stationary phase is called anion exchange if the functional groups are positively charged, and cation exchange if negatively charged. For the sample ions to be retained, they must be of opposite charge to the functional groups. The mobile phase is generally water containing a fixed quantity of ions of the same charge as the sample and buffered to a specific pH. The sample ions are separated, depending on how strongly they interact with the functional groups on the stationary phase in competition with the ions in the mobile phase. The quantity, the type of ion, and the pH in the mobile phase control the separation.

Electrophoresis

Electrophoresis is the motion of particles in a fluid under the influence of an electric field. This phenomenon was observed for the first time in 1807 by Ferdinand Frederic Reuss, who noticed that the application of a constant electric field caused clay particles in water to migrate. This is the basis for a number of analytical techniques for separating molecules by size, charge, or binding affinity.

Mixtures of ions can also be analyzed by using a column of polymeric gel, which is saturated with an electrolyte. In this case, the sample to be analyzed is spotted onto the gel and two electrodes are connected (5000 V), which cause the migration of positive ions toward one and the negative ones toward the other. This method is used in analyzing mixtures of proteins.

DETECTORS

All detectors used in the liquid chromatograph are simplified, small internal-volume versions of other analyzers. The most common ones are based on the measurement of optical absorbance, refractive index, or the dielectric constant.

Optical Absorbance

The optical absorbance detector measures the absorption of a fixed wavelength in the ultraviolet, visible, or infrared spectrum. (See Chapters 1.33 and 1.70.) It is highly sensitive to many absorbing sample compounds and relatively insensitive to external effects, such as temperature, flow, and carrier composition. It can only measure absorbing compounds in a nonabsorbing carrier.

Refractive Index

The refractive index detector measures the difference between the refractive index of the sample compounds and the carrier. With the proper choice of carrier, it is sensitive to all sample compounds, but the sensitivity is generally lower than that of the optical absorbance detector. It is also quite sensitive to temperature and carrier composition variations. Refractometers are described in Chapter 1.60.

Dielectric Constant

The dielectric constant detector measures the difference between the dielectric constants of the sample compounds and the carrier. Because, for compounds with no dipole moment, refractive index and dielectric constant are related, the advantages and disadvantages of this detector and the refractive index detector are similar. However, if, for example, the sample compounds have a dipole moment and the carrier does not, the dielectric constant detector has a higher sensitivity and more uniform response than the refractive index detector.

APPLICATIONS

The liquid chromatograph extends the advantages of the chromatograph to nonvolatile and thermally unstable samples, as well as polymers and inorganic salts. The prime advantages are the ability to both qualitatively and quantitatively analyze multicomponent streams with the versatility to analyze a wide range of samples.

When other instruments provide a satisfactory analysis, the liquid chromatograph is rarely used because it is more complex, more expensive, and sometimes less sensitive, and it requires a longer analysis time. However, there are many sample streams that can be analyzed satisfactorily only with the liquid chromatograph.

SPECIFICATION FORMS

When specifying liquid chromatographs only, please use the ISA form 20A1001, and when specifying both the analyzer and the composition or the properties of the process material of interest, use ISA form 20A1002. Both forms are reproduced with the permission of the International Society of Automation on the next pages.

1.11 Chromatographs: Liquid

1	RESPONSIBLE ORGANIZATION		ANALYSIS DEVICE	6	SPECIFICATION IDENTIFICATIONS		
2				7	Document no		
3	ISA		Operating Parameters	8	Latest revision	Date	
4				9	Issue status		
5				10			
11	ADMINISTRATIVE IDENTIFICATIONS			40	SERVICE IDENTIFICATIONS Continued		
12	Project number		Sub project no	41	Return conn matl type		
13	Project			42	Inline hazardous area cl	Div/Zon	Group
14	Enterprise			43	Inline area min ign temp	Temp ident number	
15	Site			44	Remote hazardous area cl	Div/Zon	Group
16	Area	Cell	Unit	45	Remote area min ign temp	Temp ident number	
17				46			
18	SERVICE IDENTIFICATIONS			47			
19	Tag no/Functional ident			48	COMPONENT DESIGN CRITERIA		
20	Related equipment			49	Component type		
21	Service			50	Component style		
22				51	Output signal type		
23	P&ID/Reference dwg			52	Characteristic curve		
24	Process line/nozzle no			53	Compensation style		
25	Process conn pipe spec			54	Type of protection		
26	Process conn nominal size	Rating		55	Criticality code		
27	Process conn termn type	Style		56	Max EMI susceptibility	Ref	
28	Process conn schedule no	Wall thickness		57	Max temperature effect	Ref	
29	Process connection length			58	Max sample time lag		
30	Process line matl type			59	Max response time		
31	Fast loop line number			60	Min required accuracy	Ref	
32	Fast loop pipe spec			61	Avail nom power supply		Number wires
33	Fast loop conn nom size	Rating		62	Calibration method		
34	Fast loop conn termn type	Style		63	Testing/Listing agency		
35	Fast loop schedule no	Wall thickness		64	Test requirements		
36	Fast loop estimated lg			65	Supply loss failure mode		
37	Fast loop material type			66	Signal loss failure mode		
38	Return conn nominal size	Rating		67			
39	Return conn termn type	Style		68			

69	PROCESS VARIABLES	MATERIAL FLOW CONDITIONS			101	PROCESS DESIGN CONDITIONS		
70	Flow Case Identification			Units	102	Minimum	Maximum	Units
71	Process pressure				103			
72	Process temperature				104			
73	Process phase type				105			
74	Process liquid actl flow				106			
75	Process vapor actl flow				107			
76	Process vapor std flow				108			
77	Process liquid density				109			
78	Process vapor density				110			
79	Process liquid viscosity				111			
80	Sample return pressure				112			
81	Sample vent/drain press				113			
82	Sample temperature				114			
83	Sample phase type				115			
84	Fast loop liq actl flow				116			
85	Fast loop vapor actl flow				117			
86	Fast loop vapor std flow				118			
87	Fast loop vapor density				119			
88	Conductivity/Resistivity				120			
89	pH/ORP				121			
90	RH/Dewpoint				122			
91	Turbidity/Opacity				123			
92	Dissolved oxygen				124			
93	Corrosivity				125			
94	Particle size				126			
95					127			
96	CALCULATED VARIABLES				128			
97	Sample lag time				129			
98	Process fluid velocity				130			
99	Wake/natural freq ratio				131			
100					132			

133	MATERIAL PROPERTIES			137	MATERIAL PROPERTIES Continued		
134	Name			138	NFPA health hazard	Flammability	Reactivity
135	Density at ref temp	At		139			
136				140			

Rev	Date	Revision Description	By	Appv1	Appv2	Appv3	REMARKS

Form: 20A1001 Rev 0 © 2004 ISA

	RESPONSIBLE ORGANIZATION	ANALYSIS DEVICE COMPOSITION OR PROPERTY Operating Parameters (Continued)		SPECIFICATION IDENTIFICATIONS		
1			6			
2			7	Document no		
3	(ISA)		8	Latest revision		Date
4			9	Issue status		
5			10			

	PROCESS COMPOSITION OR PROPERTY			MEASUREMENT DESIGN CONDITIONS				
	Component/Property Name	Normal	Units	Minimum	Units	Maximum	Units	Repeatability
11								
12								
13								
14								
15								
16								
17								
18								
19								
20								
21								
22								
23								
24								
25								
26								
27								
28								
29								
30								
31								
32								
33								
34								
35								
36								
37								

Rev	Date	Revision Description	By	Appv1	Appv2	Appv3	REMARKS

Form: 20A1002 Rev 0

© 2004 ISA

Definitions

Electrophoresis is the motion of particles in a fluid under the influence of an electric field. This phenomenon was observed for the first time in 1807 by Ferdinand Frederic Reuss who noticed that the application of a constant electric field caused clay particles in water to migrate. This is the basis for a number of analytical techniques for separating molecules by size, charge, or binding affinity.

High-pressure liquid chromatography: HPLC is distinguished from traditional ("low pressure") liquid chromatography because operational pressures are significantly higher (50–350 bar), whereas ordinary liquid chromatography typically relies on the force of gravity to pass the mobile phase through the column.

Micellar liquid chromatography: MLC is a form of reversed-phase liquid chromatography that uses an aqueous micellar solutions as the mobile phase.

Mobile phase: The mobile phase is the part of the chromatographic system which carries the solutes through the stationary phase. The liquid mobile phases are used to adjust the chromatographic separation and retention in liquid chromatography.

Retention time: Retention time (tR) is the time it takes a solute to travel through the column. The retention time is assigned to the corresponding solute peak. The retention time is a measure of the amount of time a solute spends in a column. It is the sum of the time spent in the stationary phase and the mobile phase.

Reversed-phase chromatography: RPC (also called hydrophobic chromatography) includes any liquid chromatographic method that uses a hydrophobic stationary phase. In the 1970s, most liquid chromatography was performed using a solid support stationary phase (also called a "column") containing unmodified silica or alumina resins.

Solute: A substance dissolved in another substance, usually the component of a solution present in the lesser amount. In case of liquid chromatography, it is the components in the sample, which is injected into the mobile phase (the solvent).

Stationary phase: The stationary Phase is the part of the chromatographic system though which the mobile phase flows where distribution of the solutes between the phases occurs. The stationary phase may be a solid or a liquid that is immobilized or adsorbed on a solid. In general, immobilization by reaction of a liquid with a solid is used in liquid chromatography. The stationary phase may consist of particles (porous or solid), the walls of a tube (e.g., capillary), or a fibrous material (e.g., paper).

Abbreviations

HPLC	High-Pressure Liquid Chromatograph or High-Performance Liquid Chromatograph
MLC	Micellar liquid chromatography
RPC	Reverse Phase Chromatography

Bibliography

Ahuja, S. and Dong, M. W. (eds.), Handbook of Pharmaceutical Analysis by HPLC, Elsevier/Academic Press, 2005. http://store.elsevier.com/Handbook-of-Pharmaceutical-Analysis-by-HPLC/isbn-9780120885473/.

Ahuja, S. and Rasmussen, H. T. (eds.), HPLC Method Development for Pharmaceuticals, Academic Press, 2007. http://store.elsevier.com/HPLC-Method-Development-for-Pharmaceuticals/isbn-9780123705402/.

Bender, T. A., Fast Analytical Separations with High-Pressure Liquid Chromatography, 2013. http://pubs.acs.org/doi/abs/10.1021/ed300757a.

Dong, M. W., Modern HPLC for Practicing Scientists, Hoboken, NJ: John Wiley & Sons, 2006. http://www.amazon.com/Modern-HPLC-Practicing-Scientists-Michael/dp/047172789X/ref=pd_sim_b_4.

Fanali, S., Haddad, P. R., Poole, C., and Shoenmakers, P., Liquid Chromatography: Fundamentals and Instrumentation, Philadelphia, PA: Elsevier, 2013. http://www.amazon.com/Liquid-Chromatography-Instrumentation-Salvatore-Fanali/dp/0124158072.

Grob, R. L. Modern Practice of Gas Chromatography, 2004. http://www.amazon.com/Modern-Practice-Chromatography-Robert-Grob/dp/0471229830/ref=pd_sim_b_5.

Kazakevich, Y. V. and LoBrutto, R. (eds.), HPLC for Pharmaceutical Scientists, Wiley, 2007. http://www.amazon.com/HPLC-Pharmaceutical-Scientists-Yuri-Kazakevich/dp/0471681628.

Majors, R., Fast and Ultrafast HPLC on sub-2 μm Porous Particles, 2010. http://www.chromatographyonline.com/lcgc/article/articleDetail.jsp?id=333246&pageID=3#.

McMaster, M. C., HPLC, a Practical User's Guide, Wiley, 2007. http://onlinelibrary.wiley.com/doi/10.1002/jssc.200790030/abstract.

Meyer, V. R., Practical High-Performance Liquid Chromatography, Hoboken, NJ: John Wiley & Sons, 2010. http://www.amazon.com/Introduction-Modern-Liquid-Chromatography-Snyder/dp/0470167548.

Snyder, L. R., Kirkland, J. J., and Dolan, J. W., Introduction to Modern Liquid Chromatography, Hoboken, NJ: John Wiley & Sons, 2009, http://www.amazon.com/Introduction-Modern-Liquid-Chromatography-Snyder/dp/0470 167548.

Snyder, L.R., Kirkland, J.J., and Glajch, J.L., Practical HPLC Method Development, New York: John Wiley & Sons, 1997.

Snyder, L.R. et al., Practical HPLC Method Development, New York: John Wiley & Sons, 1988.

Thomson, M., Interfacing sample handling systems for on-line process analyzers, January 14, 2002, www.measurementation.com.au/tp-1.htm.

Uwe, D.N. et al., HPLC Columns: Theory, Technology, and Practice, New York: Wiley, 1997.

Waksmudzka-Hajnos, M., High performance liquid chromatography in phytochemical analysis, 2010, http://www.amazon.com/Performance-Chromatography-Phytochemical-Analysis-Chromatographic/dp/142009260X/ref=pd_sim_b_43.

Waters, High Performance Liquid Chromatography, 2013, http://www.waters.com/waters/en_US/HPLC—High-Performance-Liquid-Chromatography/nav.htm?cid=10048919&locale=en_US

Xiang, Y., Liu, Y., and Lee, M.L., Ultrahigh pressure liquid chromatography using elevated temperature. *Journal of Chromatography A* 1104(1–2): 198–202, 2006. http://www.ncbi.nlm.nih.gov/pubmed/16376355

1.12 Coal Analyzers

D. H. F. LIU (1995) **B. G. LIPTÁK** (2017)

Types	AA: Atomic absorption spectrophotometry
	CPA: Coal particle analyzer
	LIBS: Laser-induced breakdown spectroscopy
	LPSA: Laser particle size analyzer
	MA: Moisture analyzer
	OCB: Oxygen combustion bomb
	PGNAA: Prompt/pulsed gamma neutron activation analyzers
	TG: Thermogravimetry
	TS: Total sulfur analysis
	XRF: X-ray fluorescence
Accuracy of measurement	Heating value, ±200 to 300 BTU/lb; ash content, ±0.05 wt%; and moisture content, ±0.05 wt% moisture
Coal properties	See Table 1.12a
Partial list of suppliers	Dandong Dongfang Measurement& Control Technology Co., Ltd. (PGNAA, LIBS) http://en.dfmc.cc/product/520.html
	IKA (OCB) http://www.oxygenbombcalorimeter.com/how-does-a-calorimeter-work.htm
	Kanawha Scales (MA) http://www.kanawhascales.com/moisture/index.php
	Leco Corp. (TG) http://www.leco.com/products/analytical-sciences/moisture-ash-volatile-content-and-loss-on-ignition/tga701-thermogravimetric-analyzer
	Liaoning Dongfang Measurement & Control Group Co., Ltd.(PGNAA) http://www.dfmc.cc/en/product/520.html
	Parr Instrument Co. (OCB) http://www.parrinst.com/wp-content/uploads/downloads/2013/05/6750MB_Parr_6750_Proximate-Interface_Literature.pdf
	Perkin Elmer (TG) http://www.perkinelmer.com/CMSResources/Images/44–142549APP_Proximate_Analysis_Coal_Coke.pdf
	Progression Inc. (LIBS) http://www.progression-systems.com/products/laser-induced-breakdown-spectroscopy/titancca.php
	Real Time Instruments (RTI) http://www.moistscan.com/RealTime-Instruments/Products-by-Name/Online-Coal-Ash-Analysers/AshScan-(1).aspx
	Standard Laboratories Inc.
	Thermo Fisher Scientific (PGNAA) http://www.thermo.com.cn/Resources/201007/2711240530.pdf
	Thermo Scientific (PGNAA) http://www.thermoscientific.com/ecomm/servlet/productscatalog_11152_L10583_93020_-1_4

INTRODUCTION

Coal analysis techniques are specific analytical methods designed to measure the particular physical and chemical properties of coals. These methods are used primarily to determine the suitability of coal for "Coke (fuel)" coking, power generation or for "Smelting" smelting in the manufacture of steel.

Coal comes in four main types: lignite or "Brown coal" brown coal, "Bituminous coal" bituminous coal or black coal, anthracite and graphite. Each type of coal has a certain set of physical parameters that are often dependent on its moisture and volatile content (in terms of aliphatic aromatic hydrocarbons).

Coal analyzers were first introduced in the early 1980s, with the United States and Australia leading the way. The demand for coal analyzers increased in the United States, owing to the need to control sulfur as mandated by the Clean Air Act Amendments of 1977. By 2013, more than 600 coal analyzers were in use throughout the world. Most of these analyzers are mounted around existing conveyor belts, although a significant minority of analyzers analyze sample streams taken from the main process stream.

TABLE 1.12a
Comparison of Coal Characteristics

Fuel Characteristics	Units	Eastern Bituminous	30% PRB Coal Blend
Proximate Analysis			
a. Moisture	%	5.15	10.29
b. Ash	%	9.80	7.46
c. Volatile Matter	%	35.79	35.29
d. Fixed Carbon	%	49.59	46.95
Ultimate Analysis			
a. Hydrogen	%	5.21	5.70
b. Carbon	%	71.54	67.70
c. Nitrogen	%	1.31	1.23
d. Sulfur	%	2.39	1.57
e. Oxygen	%	9.44	16.35
f. Ash	%	9.80	7.46
Heating Value	BTU/lb	12855	12053
Free Swelling Index		7.8	7.2
Hardgrove Grind Index		57	51
Ash Fusion Temperature Reducing Atmosphere			
a. Initial Deform	°F	2115	2124
b. Softening	°F	2190	2178
c. Hemi	°F	2340	2236
d. Fluid	°F	2400	2337
Slag Viscosity Factor (T_{250} value)	°F	>2500	2494
Chlorine	%	0.06	0.05
Fouling Index		0.17	0.22
Slagging Index		0.83	0.60
Base/Acid Ratio		0.34	0.39

Coal is widely used as a source of power and heat by the chemical, paper, cement, and metal industries. In the United States, there are some over 1000 coal preparation plants and coal-fired power plants, and some 600 coal analyzers have already been installed in them. One key consideration in operating coal-burning facilities is the control of SO_2 emissions to the atmosphere from coal-fired power plants.* Because the most economical method of reducing the sulfur content of coal is through the blending of various coals, on-line coal analyzers are often needed. The characteristics of coal are monitored for environmental protection, quality assurance, and process control.

For some typical values of coal properties, see Table 1.12a: This chapter will describe the instrumental methods for

1. *Proximate analysis*: Used to establish the rank of coals, to show the ratio of combustible to incombustible constituents, to provide the basis for buying and selling, to evaluate the benefits, or for other purposes
2. *Gross calorific value*: Provides the basis for buying and selling
3. *Sulfur analysis*: Used in coal preparation and in the determination of potential sulfur emissions from coal combustion or conversion processes, or in the determination of coal quality against contract specifications
4. *Composition analysis of major and minor elements in coal and coke ash*: Predicts slagging and fouling characteristics of combusted materials and potential use of ash by-products (see American Society for Testing Materials (ASTM) Standards D346 and D2013 on methods for collection and preparation of coal samples for laboratory analysis)

COAL PROPERTY MEASUREMENT

These analyzers are used to determine coal quality parameters such as ash, moisture, sulfur, and energy value (also known as heat content). Although most coal operations can obtain this information about coal quality by taking physical samples, then preparing the samples and analyzing them with laboratory equipment, these processes often involve a time lag of up to 24 hr from gathering the sample to final analysis results. In contrast, coal analyzers provide analysis information almost immediately as it is being transported by conveyor either at the mine or in the power plant. Such fast information of coal quality allows the optimization of sorting, blending, homogenization, and plant control.

---

* More than 50% of the mercury, 60% of the SO_2, 62% of the arsenic, and 77% of the acid gases come from power plants.

One of the most sophisticated analyzers uses a technique known as prompt gamma neutron activation analysis (PGNAA) to determine the elemental content of the coal. Another emerging technology for elemental analysis is laser-induced breakdown spectroscopy (LIBS). PGNAA and LIBS enable the analysis of sulfur and ash (the latter, by summing the ash constituents), and when combined with a second type of analyzer, such as a moisture monitor, it can also determine heating value (energy content). Moisture meters are often found in conjunction with the elemental analyzers, but sometimes are used alone, or in conjunction with ash gauges. Most moisture meters use microwave technology while the emerging technology of magnetic resonance (MR) offers a more direct measurement. For a detailed description of solids moisture analyzers, see Chapter 1.41. Most ash gauges employ gamma attenuation principles.

LABORATORY TECHNIQUES

Thermogravimetry

Due to the recent addition of microcomputer control and dedicated data reduction, the thermogravimetry (TG) technique has become popular for routine approximate analysis of coal and coal products. The TG unit performs a multistep analytical sequence automatically and unattended.

The sample is loaded into a two-arm furnace tube (Figure 1.12a) and is sequentially dried and burned; the residue is then weighed. The tube allows the active gas (air or oxygen) to enter near the top of the furnace rather than through the balance mechanism. The low-mass furnace provides both rapid heating rates and a short cool down time. The microcomputer controller provides automatic switching between the purge gas and the active gas to obtain the measurements (Figure 1.12b).

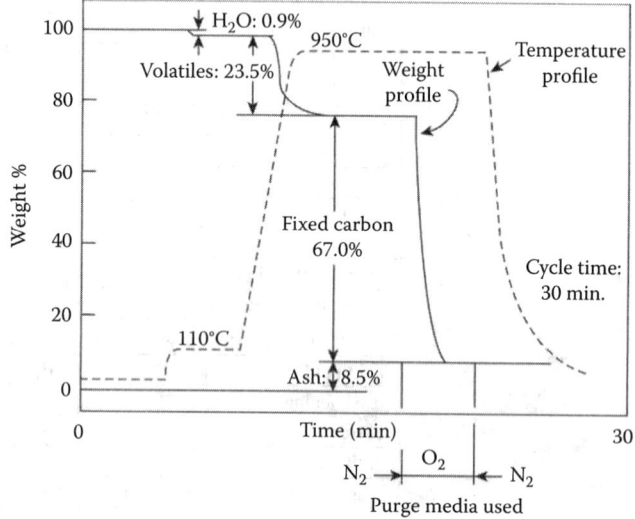

FIG. 1.12b
A typical proximate analysis of coal by microcomputer-controlled TG.

Bituminous Coal Analysis

Figure 1.12b shows a typical proximate analysis of a bituminous coal using the automated TG system. The furnace is heated to 110°C (230°F) at a rate of 60°C/min (140°F/min) and is held isothermally for 5 min, while water is vaporized off from the coal sample. The furnace is then heated at a rate of 80°C/min (176°F/min) to 950°C (1742°F) and held for 7 min while nitrogen is flowing through the TG, until all volatile matter is expelled from the sample. After the nitrogen purge, the purge gas is switched to either air or oxygen and the fixed carbon content of the char is oxidized, leaving the ash content as the residue.

The values from these determinations are read directly in weight percent from the chart recorder. The air or oxygen purge is switched back to nitrogen, and the system automatically cools back to load temperature. The elapsed time of the proximate analysis program and cooling of the tube back to load temperature totals 30 min.

Gross Calorific Value

Gross calorific value is determined by burning a weighed sample of coal. A calibrated isoperibol oxygen bomb calorimeter is used for that purpose under controlled conditions

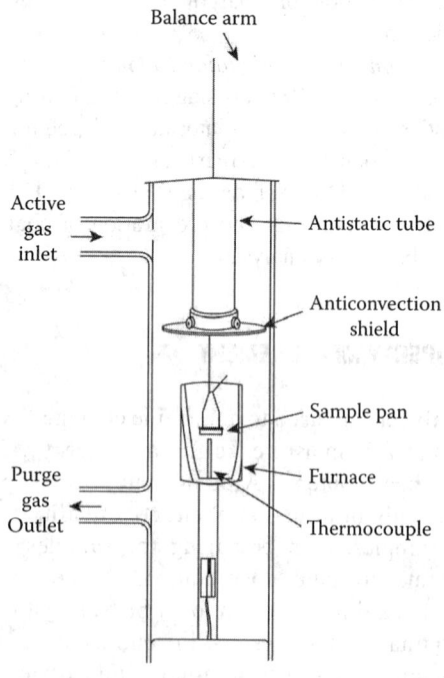

FIG. 1.12a
Two-arm furnace tube for active gas introduction in the microcomputer-controlled TG of coal and coal products. (Courtesy of the Perkin-Elmer Corp.)

A microprocessor control system monitors and controls the jacket temperature, fires the bomb, and monitors the temperature in the bucket. The test continues until the controller determines that equilibrium has been reached.

A microcomputer uses the sample weight and temperature data—applying correction for acid, sulfur, fuse, and any added combustion aids—to calculate the gross calorific value.

Total Sulfur

An ASTM method for determining sulfur in coal uses the washings from the oxygen bomb calorimeter. Sulfur is precipitated as barium sulfate from the washings. The precipitate is filtered, *ashed*, and weighed. An automatic titrimetric system is also available for rapid sulfur determination using the oxygen bomb washings. The washings are titrated with a lead perchlorate solution to obtain a lead precipitate. The titration takes place in a nonaqueous medium to ensure complete precipitation and a sharp end point with a lead ion-selective electrode.

Other procedures use high-temperature tube furnace combustion methods for rapid determination of sulfur in coal and coke using automated equipment. The instrumental analysis provides a reliable and rapid method for determining sulfur contents of coal or coke.

Figure 1.12d illustrates the high-temperature combustion method of sulfur detection using infrared absorption detection procedures. The sample is burned in a tube furnace at a minimum operating temperature of 1350°C (2462°F) in a stream of oxygen to oxidize the sulfur.

Moisture and particulate matter are removed from the gas by traps filled with magnesium perchlorate. Sulfur dioxide is measured, using an infrared absorption detector (For details on infrared analyzers, see Chapter 1.33 and for

FIG. 1.12c
Oxygen combustion bomb and bucket used in the isoperibol calorimetry. (Courtesy of Parr Instrument Co.)

(Figure 1.12c). For a complete discussion of heating value calorimeters, refer to Chapter 1.27.

The bucket, which holds the oxygen bomb, provides good circulation and rapid thermal equilibrium for the bomb. Thermal jacketing is provided by a circulating water system, which maintains cooling water flow around the bucket.

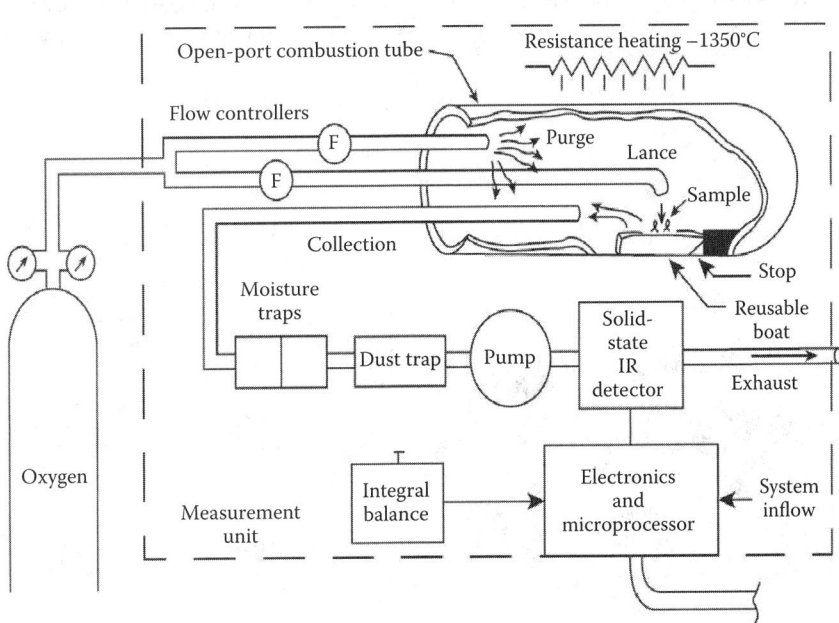

FIG. 1.12d
Apparatus for the determination of sulfur by the infrared detection method.

sulfur dioxide analyzers, see Chapter 1.65). Sulfur dioxide absorbs infrared energy at a precisely known wavelength within the infrared spectrum. Total sulfur as sulfur dioxide is determined on a continuous basis.

Ash Analysis

The major or minor elements in coal ash can be determined using x-ray fluorescence (XRF) techniques. The ash is fused with lithium tetraborate or other suitable flux and either ground or pressed into a glass disk. After that, the pellet or disk is irradiated, using an x-ray beam of short wavelength (high energy).

The characteristic x-rays of the atom that are emitted or fluoresced upon receiving the primary rays are dispersed, and their intensities are measured at selected wavelengths by sensitive detectors. Detector output is converted into concentration by computerized data-handling equipment. All elements are determined and are reported as oxides. They include Fe, Ca, K, Al, Si, P, Mg, Ti, and Na.

These major and minor elements can also be determined by atomic absorption (AA) spectrophotometry. See Chapter 1.22 for more details on elemental monitors.

ON-LINE MONITORS

Continuous coal analyzers can combine the monitoring of moisture, ash, and BTU in coal. In this system, microwave analyzers can be used to measures the moisture content of the coal, without requiring physical contact with the solids. (See Chapter 1.41 for the description of the various moisture analyzers used on solids, including microwave absorption hygrometer.)

The fingerprint of a given type of coal is its distinctive gamma spectrogram. This is produced by the detection and counting of photons released from atomic nuclei in the coal as it passes over a small source of neutron emissions.

The precision of the measurements of heating value, ash, and moisture content varies with the type of coal, typically 200–300 BTU/lb, 0.05 wt% ash, and 0.05 wt% moisture.

Prompt Gamma Neutron Activation Analyzers (PGNAA)

Prompt-gamma neutron activation analysis (PGAA) is a technique to simultaneously measure the amount of elements in small samples by irradiating them with a beam of neutrons. The elements in the sample absorb some of these neutrons and emit prompt gamma rays. The energies of these rays identify the elements that have captured the neutrons, while the intensities of the peaks reveal their concentrations.

The Harwell spectrometer is used in the PGNAA inline coal analyzer. This package also measures coal density and moisture content (Figure 1.12e). This multiple detector package also includes a feed hopper to direct the coal into the analyzer and a discharge conveyor, which takes away the coal under variable speed control. The system also includes an RS422–RS232 interface and remote displays for reporting the readings made by the analyzer.

An analyzer is also available for determining the ash and solids contents of coal–water mixtures or coal slurries. This analysis is made by the use of three probes, which are immersed into the coal slurry stream. The ash probe uses a source of low-energy x-rays to measure not only the ash content but also the concentration of iron. The density probe used gamma rays to detect the percent solids (slurry density) of the mixture. The low-intensity neutron source in the aeration probe detects the amount of air in the slurry, so that the density measurement can be corrected for aeration effects.

Western Kentucky University has developed a prototype coal analyzer operated with microsecond-wide 14-MeV neutron pulses and containing several gamma ray detectors. This analyzer measures the density and sulfur content of coal

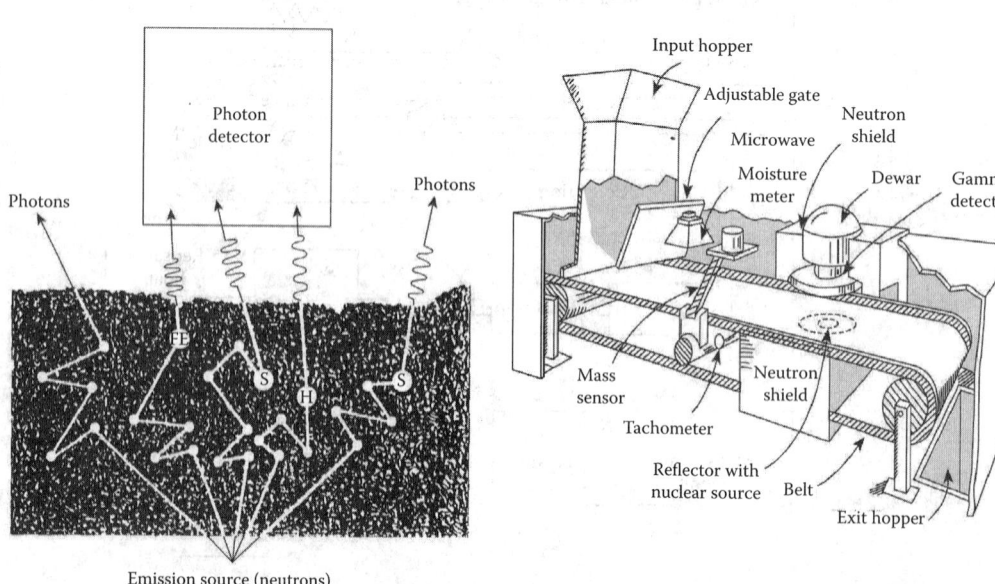

FIG. 1.12e
Neutron emission coal analyzer combined with moisture measurement.

along with its BTU, moisture, and volatile matter content. This pulsed fast/thermal neutron analyzer is self-calibrating and provides improved accuracy in the determination of elements such as carbon, oxygen, and sodium.

The most popular today are these inline neutron activation coal analyzers that can simultaneously provide a full elemental analysis of the coal, in addition to measuring, moisture, sulfur, ash contents, and calorific value of the coal passing through on the belt. Such units can be installed around the belt, without cutting it, and do the analysis without touching the coal (Figure 1.12f). These units can usually report the test results in practically real time (every minute) and can also provide blending controls where required. As such, they usually improve product quality while lowering production costs.

When the coal is the fuel to a power plant and several coal sources are available and it is desired to produce a more-or-less uniform blend from them, the PGNAA coal analyzer-controller can be used to manipulate the coal blend (Figure 1.12g).

When the coal is a mining product and its quality varies, the PGNAA coal analyzer-controller can be used to separate the coal products and send them to several destinations according to their quality (Figure 1.12h).

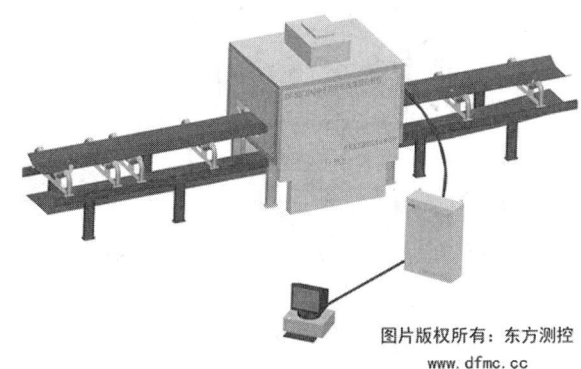

FIG. 1.12f
Inline neutron activation coal analyzer. (Courtesy of Dongfang Measurement & Control Technology Co.)

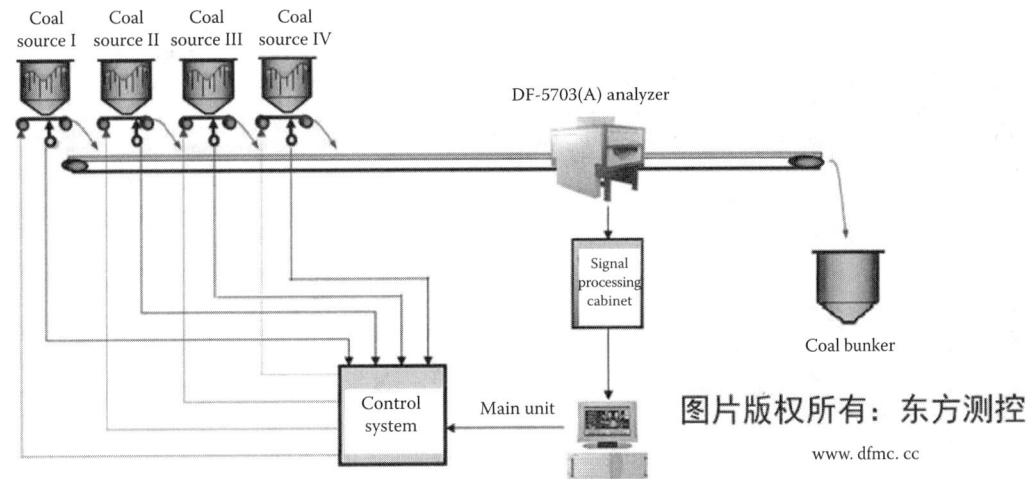

FIG. 1.12g
Inline neutron activation coal analyzer, serving the control of coal blending. (Dandong Dongfang Measurement & Control Technology Co., Ltd.)

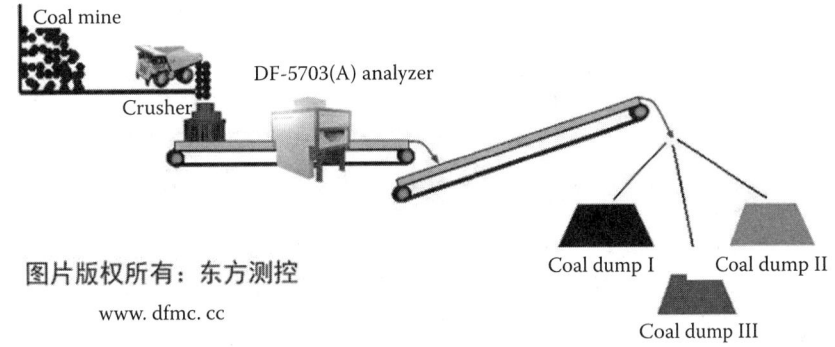

FIG. 1.12h
Inline neutron activation coal analyzer used to separate coal products according to their quality. (Dandong Dongfang Measurement & Control Technology Co., Ltd.)

Laser-Induced Breakdown Spectroscopy (LIBS)

LIBS is a new technology (Figure 1.12i). It uses a high-energy laser pulse that hits the coal sample and forms a plasma. Because all elements emit a light of a frequency characteristic of that element when excited, LIBS can detect all elements. In this design, the laser pulse that has vaporized a small part of the coal sample and caused the light emission identifies both the elements present and their concentrations, because the resulting spectrum intensities resulting from the pulse provide concentration data, while the wavelengths identify the elements.

LIBS is usually used with moisture analyzer (MA) sensors, discussed in Chapter 1.40, for the measurement of the moisture content of the coal, while LIBS provides the measurement of ash, sulfur, heating value, sodium, magnesium, phosphorus, mercury, and all other elements.

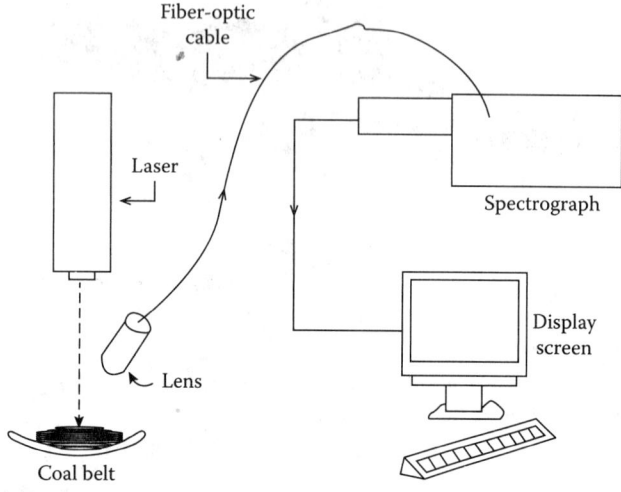

FIG. 1.12i
Main components of LIBS include a high-intensity laser source focused on the coal sample. The resulting light emission is sent to a spectrograph, provided with the required software to convert it into concentration data on all elements present in the sample.

SPECIFICATION FORMS

When specifying coal analyzers only, please use the ISA form 20A1001, and when specifying both the analyzer and the composition or the properties of the process material of interest, use ISA form 20A1002. Both forms are reproduced with the permission of the International Society of Automation on the next pages.

1.12 Coal Analyzers

1	RESPONSIBLE ORGANIZATION		ANALYSIS DEVICE		6	SPECIFICATION IDENTIFICATIONS			
2					7	Document no			
3	(ISA)		Operating Parameters		8	Latest revision		Date	
4					9	Issue status			
5					10				
11	ADMINISTRATIVE IDENTIFICATIONS				40	SERVICE IDENTIFICATIONS Continued			
12	Project number		Sub project no		41	Return conn matl type			
13	Project				42	Inline hazardous area cl		Div/Zon	Group
14	Enterprise				43	Inline area min ign temp		Temp ident number	
15	Site				44	Remote hazardous area cl		Div/Zon	Group
16	Area		Cell	Unit	45	Remote area min ign temp		Temp ident number	
17					46				
18	SERVICE IDENTIFICATIONS				47				
19	Tag no/Functional ident				48	COMPONENT DESIGN CRITERIA			
20	Related equipment				49	Component type			
21	Service				50	Component style			
22					51	Output signal type			
23	P&ID/Reference dwg				52	Characteristic curve			
24	Process line/nozzle no				53	Compensation style			
25	Process conn pipe spec				54	Type of protection			
26	Process conn nominal size		Rating		55	Criticality code			
27	Process conn termn type		Style		56	Max EMI susceptibility		Ref	
28	Process conn schedule no		Wall thickness		57	Max temperature effect		Ref	
29	Process connection length				58	Max sample time lag			
30	Process line matl type				59	Max response time			
31	Fast loop line number				60	Min required accuracy		Ref	
32	Fast loop pipe spec				61	Avail nom power supply			Number wires
33	Fast loop conn nom size		Rating		62	Calibration method			
34	Fast loop conn termn type		Style		63	Testing/Listing agency			
35	Fast loop schedule no		Wall thickness		64	Test requirements			
36	Fast loop estimated lg				65	Supply loss failure mode			
37	Fast loop material type				66	Signal loss failure mode			
38	Return conn nominal size		Rating		67				
39	Return conn termn type		Style		68				
69	PROCESS VARIABLES		MATERIAL FLOW CONDITIONS		101	PROCESS DESIGN CONDITIONS			
70	Flow Case Identification			Units	102	Minimum	Maximum	Units	
71	Process pressure				103				
72	Process temperature				104				
73	Process phase type				105				
74	Process liquid actl flow				106				
75	Process vapor actl flow				107				
76	Process vapor std flow				108				
77	Process liquid density				109				
78	Process vapor density				110				
79	Process liquid viscosity				111				
80	Sample return pressure				112				
81	Sample vent/drain press				113				
82	Sample temperature				114				
83	Sample phase type				115				
84	Fast loop liq actl flow				116				
85	Fast loop vapor actl flow				117				
86	Fast loop vapor std flow				118				
87	Fast loop vapor density				119				
88	Conductivity/Resistivity				120				
89	pH/ORP				121				
90	RH/Dewpoint				122				
91	Turbidity/Opacity				123				
92	Dissolved oxygen				124				
93	Corrosivity				125				
94	Particle size				126				
95					127				
96	CALCULATED VARIABLES				128				
97	Sample lag time				129				
98	Process fluid velocity				130				
99	Wake/natural freq ratio				131				
100					132				
133	MATERIAL PROPERTIES				137	MATERIAL PROPERTIES Continued			
134	Name				138	NFPA health hazard		Flammability	Reactivity
135	Density at ref temp		At		139				
136					140				

Rev	Date	Revision Description	By	Appv1	Appv2	Appv3	REMARKS

Form: 20A1001 Rev 0 © 2004 ISA

ANALYSIS DEVICE
COMPOSITION OR PROPERTY
Operating Parameters (Continued)

	RESPONSIBLE ORGANIZATION			SPECIFICATION IDENTIFICATIONS	
			Document no		
	ISA		Latest revision		Date
			Issue status		

PROCESS COMPOSITION OR PROPERTY			MEASUREMENT DESIGN CONDITIONS				
Component/Property Name	Normal	Units	Minimum	Units	Maximum	Units	Repeatability

Rev	Date	Revision Description	By	Appv1	Appv2	Appv3	REMARKS

Form: 20A1002 Rev 0 © 2004 ISA

Definitions

Gross calorific value is the amount of heat liberated by the complete combustion of unit weight of coal (or other fuel) under specified conditions; the water vapor produced during combustion is assumed to be completely condensed.

Prompt-gamma neutron activation analysis (PGAA) is a technique to simultaneously measure the amount of elements in small samples by irradiating them with a beam of neutrons. The elements in the sample absorb some of these neutrons and emit prompt gamma rays. The energies of these rays identify the elements that have captured the neutrons, while the intensities of the peaks reveal their concentrations.

Abbreviations

AA	Atomic absorption spectrophotometry
CPA	Coal particle analyzer
LIBS	Laser-induced breakdown spectroscopy
LPSA	Laser particle size analyzer
MA	Moisture analyzer
OCB	Oxygen combustion bomb
PGNAA	Prompt/pulsed gamma neutron activation analyzers
TG	Thermogravimetry
TS	Total sulfur analysis
XRF	X-ray fluorescence

Bibliography

ASTM, Standard Classification of Coals, (2015) http://www.astm.org/Standards/D388.htm.

Bhamidipati, V. N., Compliance blending of PRB Coal, (2004) http://www.thermo.com/eThermo/CMA/PDFs/Articles/articles-File_1029.pdf.

IEA, Clean coal center, on-line analysis and coal fired power plant, (2005) http://www.iea-coal.org.uk/documents/81375/5936/Online-analysis-and-coal-fired-power-plant.

Lehigh Energy, On-line lazer measurement of coal, (2011) http://www.lehigh.edu/~inenr/leu/leu_71.pdf.

Perkin Elmer, Thermographic coal analysis, (2013) http://www.perkinelmer.com/CMSResources/Images/44–142549APP_.

Progression Inc., Laser induced breakdown spectrograph coal analyzer, (2013) http://www.progression-systems.com/products/laserinduced-breakdown-spectroscopy/titancca.php.

SGS, Coal analysis, (2015) http://www.sgs.com/en/Energy/Energy-Sources/Coal/Coal-Analysis.aspx.

Snider, Kurt, Using an on-line elemental coal analyzer to reduce lost generation due to slagging, *International On-Line Coal Analyzer Technical Conference*, St. Louis, MO, November 8–10, 2004. http://www.thermo.com/com/CMA/Files/articles-File_20869.pdf.

Speight, J. G., *Handbook of Coal Analysis*, (2005) http://www.amazon.com/Handbook-Coal-Analysis-James-Speight/dp/0471522732.

Stallard, S. More from your online coal analyzer, (2007) http://www.energycentral.net/article/07/10/more-your-online-coal-analyzer.

Standard Laboratories Inc., Coal analysis, sampling, and testing, (2015) http://standardlabs.thomasnet-navigator.com/item/all-categories/coal-analysis-sampling-testing/item-1006?

Thermo Fisher Scientific, Online coal analyzer, (2007) http://www.thermo.com.cn/Resources/201007/2711240530.pdf.

Woodward, R., *A Promising New Development in the Coal Analyzer Industry*, (2013) http://www.coalage.com/index.php/features/2609-a-promising-new-development-in-the-coal-analyzer-industry.html.

1.13 Colorimeters

J. E. BROWN (1969, 1982)   **B. G. LIPTÁK** (1995)   **M. W. REED** (2003, 2017)

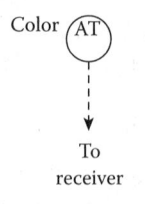

Flow sheet symbol

Types	Colorimeters (A) and spectrophotometers (B) (1D and multispectral)
Types of samples	Liquid, gas, or solid
Wavelength ranges	200–1000 nm is typical
Spectral resolution	2–20 nm
Source lamp	Halogen, xenon flash, LED, IR, fluorescent, fiber-optics
Detectors	CCD and photodiode linear arrays or photocells
Materials of construction	Standard materials, ordinary or quartz glass, ball lens
Cell length	From 0.5 to 1000 mm
Sample pressure	Vacuum to 150 psig (10.6 bars)
Inaccuracy	1 nm for spectrophotometers, 1%AR in transmittance for spectrometers and colorimeters, 5% precision for colorimeters
Costs	$500 for a single-channel colorimetric analog channel $1000 for fiber-optic spectrophotometers; on-line scanning systems cost $20,000–$60,000
Partial list of suppliers: ((A) denotes colorimeters, and (B) denotes spectrometers and spectrophotometers)	Agilent Technologies (B), http://www.chem.agilent.com/en-US/products-services/Parts-Supplies/Spectroscopy/8453-UV-Vis/Pages/default.aspx Applied Biosystems (B), http://www.lifetechnologies.com/us/en/home/life-science/cell-analysis/labeling-chemistry/fluorescence-spectraviewer.html Bayer Diagnostics (A), http://www.bayercontour.com/CMSPages/PortalTemplate.aspx?aliaspath=%2fBlood-Glucose-Monitoring%2fDiabetes-Management-Software%2fGlucofacts-Deluxe Bio-Rad,(B) (www.BioRad.com), http://www.bio-rad.com/en-us/spectroscopy Byk-Gardner, Inc. (B).https://www.byk.com/en/instruments/products/index.php?a=2&b=17&f=0&action=15&direct=1 Cole-Parmer, (B), http://www.coleparmer.com/Search/spectrometers Data Color International (B), http://industrial.datacolor.com/portfolio-view/datacolor-600/ Dolan Jenner Industries, light sources, http://www.dolan-Jenner.com/Pro/product.htm EMX Industries (A), http://www.emxinc.com/colormax.html Eppendorf North America (A,B), http://eshop.eppendorfna.com/categories/13 Excelitas, light sources http://www.excelitas.com/Pages/Product/Photonic-Detectors.aspx Fisher Scientific Co. (B), http://www.fishersci.com/ecomm/servlet/Search?keyWord=Spectrometers+%5BFT-IR%5D&store=&nav=10399987&offSet=0&storeId=10652&langId=-1&fromSearchPage=1&searchType=PROD&typeAheadCat=prodCat Guided Wave(B), http://www.guided-wave.com/products/spectrometers.html Hach Chemical Co, (A, B) (www.hach.com), http://www.hach.com/browse.by.category-list.jsa?productCategoryId=14371221977&parameterId=7639974407&productType=INSTRUMENT&secondPageNumber=1&pimContext=USen Hunter Labs (B), http://www.hunterlab.com/color-measurement-software.html

(Continued)

Konica Minolta Sensing, Inc. (A,B), http://sensing.konicaminolta.us/products/cr-10-color, http://sensing.konicaminolta.us/applications/color-measurement/
Mahlo America, Inc., (B), http://www.mahloamerica.com/pdf/CIS_en.pdf
Ocean Optics Inc. (B), http://www.oceanoptics.com/products/usb2000+.asp
Omega Engineering (A), http://www.omega.com/pptst/DC1200.html
Orbeco-Hellige Co. (A), http://www.grainger.com/product/ORBECO-Orbeco-SC450-Colorimeter-
Starna, optical cells, http://www.starnacells.com/
ThermoScientific Corp. (A,B), http://www.thermoscientific.com/en/products/colorimetry-turbidity-uv-vis-spectrophotometry.html
Turner Designs (A). http://www.turnerdesigns.com/products/submersible-fluorometer/c3-submersible-fluorometer
YSI, Inc., (A), http://www.ysi.com/productsdetail.php?6600V2-1
ZAPS Technologies, Inc. (A, B), http://www.zapstechnologies.com/about-liquid/

INTRODUCTION

Colorimetry involves the transmission, absorption, or reflectance of visible light and can be extended into the ultraviolet (UV) and near-infrared (NIR) spectrum. Gratings used in current spectrophotometers can be specified over a range of 200–1200 nm, with the visible range being 400–700 nm. Visible color can be measured as an indication of a concentration or as an indication of pH in titration using dyes. See Chapter 1.74.

Visual colorimetry involves the matching of the sample color against standards in cuvettes, Nessler tubes, or thin films. Simple color wheel kits are used in water quality analysis to match depth of color in the liquid sample with a reference color calibrated in the concentration of the specific chemical end group for measurements such as chlorine in drinking water. These are being replaced with low-cost handheld colorimeters and spectrometers, which can be programmed for field use.

If colored filters such as red, green, and blue (RGB) are used, the device is a colorimeter. If a grating or prism is used, the device is a spectrophotometer. Another term, spectrometer, is a general term applying to either design. Due to the current technology of matching of CCD and diode arrays to entire spectral images, an entire spectral curve may be measured continuously. This is usually the case when the color of paper, textiles, films, or solids is being measured. New techniques based on fluorescence, which can also use filters and spectral gratings, are being developed for rapid determination of pathogens in recreational waters such as by polymerase chain reaction (PCR*) systems.

COLOR MEASUREMENT

Color measurements involve that part of the electromagnetic spectrum that is sensed by the human eye and brain. This region is approximately 400–700 nm. The colors of the rainbow are associated with specific wavelengths of visible light as listed in Table 1.13a.

TABLE 1.13a
Color and Wavelength Association

Approximate Wavelength (nm)	Associated Color
400–450	Violet
450–500	Blue
500–570	Green
570–590	Yellow
590–610	Orange
610–700	Red

Industrial standards involve color matching in which color standards are compared either by eye or with the assistance of a spectrophotometer. In the world of art, separate color-matching methods have evolved that are based on color wheels that could be correlated to machine vision by taking spectral measurements of the subjective reference colors.[†,‡]

Absorbance and Transmittance Colorimetry

Absorbance spectra from spectrophotometers are used in the field of colorimetry of liquids where spectral absorbance curves can be compared one-dimensionally. The reason for converting to absorbance is that a logarithmic relation typically exists between transmittance at a certain wavelength and the concentration of colorants in liquid solutions. This is known as the Beer–Lambert law.[§] See Chapter 1.69.

Absorbance curves are generated from transmittance data with absorbance defined as the negative of the natural logarithm of T/T_o, where T_o is the transmittance of the reference standard. This standard might be a white reference

---

* Jochen, W. et al., Real-time PCR-based method for the estimation of Genome sizes, Nucleic Acids Research 2003, 31(10), Oxford University Press. http://www.rocheapplied-science.com/shop/CategoryDisplay?identifier=Real-Time+PCR+Overview#tab-0.

† Pearcy, B., *Decorative Painting Color Match Sourcebook*, Cincinnati, OH: North Light Books, 1997.
‡ TCS Digital Color Wheel, Tru-Color Systems 2 Dandridge, Danville, IN 46122, 2013 www.gotcs.com.
§ Haynes, W. M, Ed., *Handbook of Chemistry and Physics*, CRC Press, 2012, http://www.crcpress.com/product/isbn/9781439880494.

tile or a cuvette of distilled water, depending on the sample application. If the spectral curve is from light reflected from a solid, these reflectance curves can also be curve-fit and compared one-dimensionally. This is suitable for quality control of materials. When consumer products are accepted or rejected based on their color perception by humans, tristimulus model parameters such as XYZ, Lch, or Lab are generated from spectral data and tolerances are set.

SPECTROPHOTOMETRIC ANALYZERS

Interference filters are selected for desired wavelengths as determined from the spectral relationship curves. Detector photoconductors are chosen to give spectral response in the visible region. All known photodetectors are least sensitive in the blue end of the spectrum. This can be dealt with by using prefilters or by using narrow spectral ranges, which are calibrated for more sensitivity.

Analyzers have to be built to carry out colorimetric analysis based on existing standards. Hundreds of 1D color standards are used in the process industries. Other organizational databases include the American Public Health Association (APHA), the National Bureau of Standards (NBS), the National Institute of Standards and Technology (NIST), the American Association of Textile Chemists and Colorists (AATCC), and the American Society for Testing and Materials (ASTM). All are accessible on the Internet. The Saybolt Color scale in Table 1.13b is similar to ASTM D156, "Test for Color of Petroleum Products." The instrument that is used is called a chromometer. See http://www.thomassci.com/Instruments/Oil-Testing-Colorimeters/_/SAYBOLT-CHROMOMETER.

TABLE 1.13b
Saybolt Color Scale

Color Standard Filter Number	Depth of Oil, in. (mm)	Color Number
½	20 (508)	+30
½	16 (406)	+28
½	12 (305)	+26
1	20 (508)	+25
1	16 (406)	+23
1	10.75 (273.05)	+20
1	8.25 (209.55)	+18
1	6.25 (158.75)	+16
2	10.50 (266.7)	+15
2	9.00 (228.60)	+13
2	7.25 (184.15)	+10
2	5.75 (146.05)	+5
2	4.50 (114.30)	0
2	3.50 (88.90)	−5
2	3.00 (76.20)	−9
2	2.50 (63.50)	−13
2	2.125 (53.975)	−16

Spectrophotometer Design

The terms spectrometer and spectrophotometer are used interchangeably, but there is a fundamental design difference. See Chapter 1.32. A true spectrophotometer will use a diffraction grating to generate a spectrum and an array of sensors for each wavelength of the spectrum. The design of fiber-optic spectrophotometers is shown in Figure 1.13a. These units are small enough to fit in one's hand and cost about $5000. Improvements in spectrophotometers include a flashed xenon light source with dual-beam measurement. This is used for flashed light sources due to the variability of flashes. If filters are used instead of a grating, the design is referred to as a spectrometer. One has to get the instrument specifications to determine this.

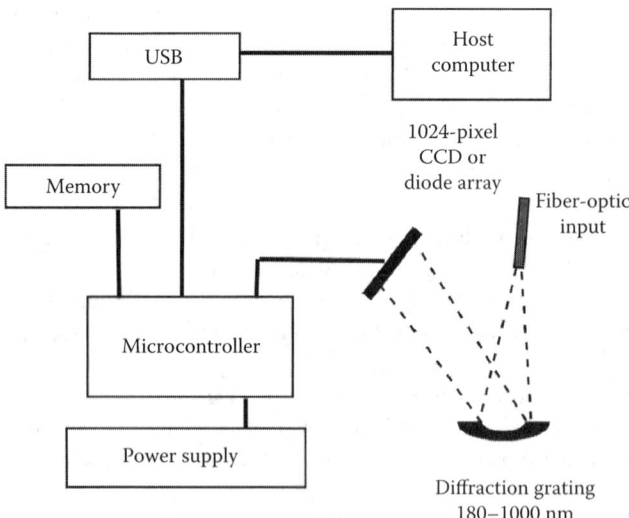

FIG. 1.13a
Fiber-optic spectrophotometer.

Dual-beam machines measure the spectrum of both the light source and the reflected light for each measurement. Spectrophotometers are characterized by the geometry of the light source and by the view of the photosensitive device. A common geometry is the so-called 45/0 design. This is achieved by illuminating the sample with a ring of light or fiber-optic cables at an angle of 45°. The reflected light is viewed at an angle of 0° normal to the surface. Other geometries are 0/45 and 0/0.

The processed reflected light may be captured with an integrating sphere. An integrating sphere may or may not be used. Such a device is needed to compare the diffuse and specular components of the reflected light. The number of spectral measurement points has increased from about 16 in the 1980s to about 1000 currently. Sometimes, colorimeters are marketed as 0/45 spectrophotometers by using up to 16 different filtered fiber-optic receptors in a ring for a 16-point spectral curve. This is inappropriate for materials having a directional shade. A true spectrophotometer collects the reflected light and then forms a spectrum. The buyer needs to make sure of the basic design inside the machine. The fiber-optic input can be from a liquid transmittance cell or from a fluorescence cell with the excitation wavelength filtered out.

TRISTIMULUS METHOD (REFLECTANCE)

Reflectance color measurements utilize the reflected light from a sample to monitor the surface color of the sample. This measurement can be done with a 1D color standard such as those discussed earlier, but in applications requiring better color definition, a tristimulus method must be used.

This method uses the Commission Internationale de l'Eclairage (CIE) model of human color perception. The method requires that the reflected spectrum from 400 to 700 nm be masked by the three functions shown in Figure 1.13b. The areas under the resulting curves measured as X, Y, and Z are correlated to human color perception. These values are normalized against standard values for different lighting conditions and types of observers. The algorithm has variations, depending on the vision model used.

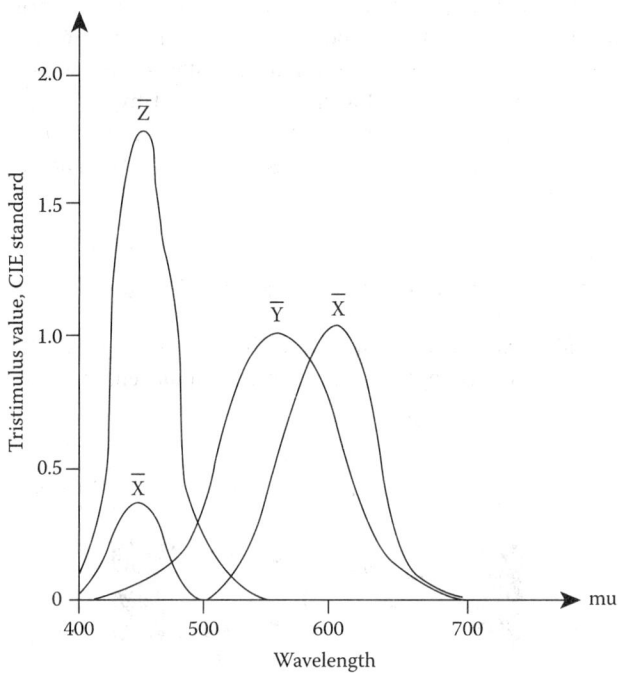

FIG. 1.13b
Spectral response of CIE standard values.

The parameters X0, Y0, and Z0 are defined to be the areas under the white reference reflectance mask curves. Values of X0, Y0, and Z0 will vary, depending on the design of the spectrophotometer. To model the human color space, the Lab shade model is used. L is the measurement of lightness and is calibrated to have the values of 0 to 100. Lch (cylindrical coordinates for lightness, chroma, and hue) and Lab (rectangular coordinates) are two common three-parameter models.

Lab Algorithm of the Textile Industry

Following is a section of the computer code, written in BASIC, by the author that takes a 16-point color spectrum and computes Lab. The PX(I), PY(I), and PZ(I) functions are those shown in Figure 1.13b. For tables of spectral data and more information on computing Lab under specific illuminants and angles of observation, see Judd and Wyszecki.*

Section of BASIC code for online L, a, and b determination:

```
X0 = 97.9392:Y0 = 100:Z0 = 117.9648  '2DEG ILLUM C
values of X, Y, Z for
'white reference tile determined at calibration
  PRINT #1, "m"  'serial port command to spectrophoto-
meter to send 16
  'ASCII spectral values from 400 to 700 nm every 20 nm
  FOR I = 1 TO 16
    INPUT #1, T(I)  'The 16 values make up the raw transmit-
tance or reflectance curve for the sample
  NEXT I
  FOR I = 1 TO 16
    R(I) = (T(I)-D(I))/((W(I)-D(I))
  'W(I) and D(I) are the white and dark tile values, respec-
tively, determined at
  'calibration of the system. Program must read white and
dark calibration tile
  'spectral curves and store in arrays W and D. Tristimulus
Data files are read for PX, PY, and PZ arrays.
    X = X + PX(I)*R(I)  'numerically integrates areas
    Y = Y + PY(I)*R(I)
    Z = Z + PZ(I)*R(I)
  NEXT I
  IF X/X0< = 0.008856 OR Y/Y0< =.008856 OR Z/Z0<
=.008856 THEN
    L = 903.3*(Y/Y0)  'low light condition
    A = 500*((7.787*(X/X0)+16/116) –(7.787*(Y/Y0)+16/116))
    B = 200*((7.787*(Y/Y0)+16/116) –(7.787*(Z/Z0)+16/116))
  ELSE
    L = 116*(Y/Y0)^.33333 – 16  'note that if Y = Y0 then
L = 100
    `A = 500*((X/X0)^.333333 –(Y/Y0)^.333333)
    B = 200*((Y/Y0)^.333333 –(Z/Z0)^.333333)
  `be sure to set all sums to zero for repeating this
```

The Lab algorithm is commonly used in the textile, automotive, and plastics industry. The parameter <u>a</u> ranges from 0 to ±128 (red to green), and the parameter <u>b</u> ranges from 0 to ±128 (yellow to blue). A white reference sample is taken by calibration to have the value 100, 0, 0, while the dark reference is taken by calibration to have the value 0, 0, 0.

Calibration is a very important issue, and vendors employ many methods. The Lab or Lch measurement is often called the shade of the material. Shade measurements can be made from different angles and areas of view. A recent patent* that

---
* Judd, D. B. and Wyszecki, G., *Color in Business, Science, and Industry*, 2nd edn., New York: John Wiley & Sons, 1963.

measures shade from different directions both online and off-line is given as a reference. Figure 1.13c illustrates how fiber-optics can be combined with robotics.

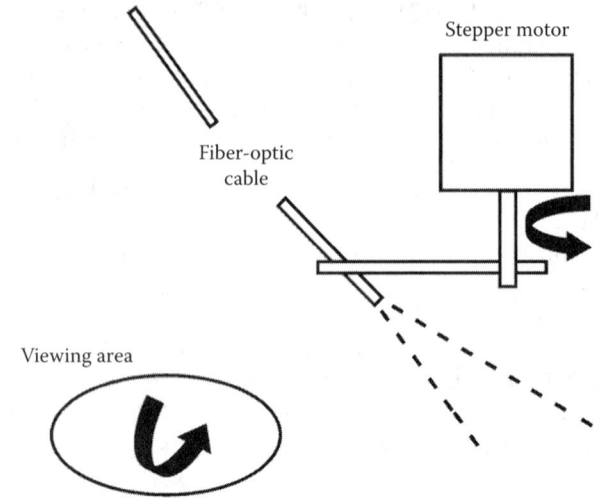

FIG. 1.13c
Fiber-optics robotically positioned.

CONTINUOUS COLOR MONITORS

In-Line Liquid Color Measurement

In-line liquid color measurement is illustrated by Figure 1.13d. Recent advances in fiber-optics have allowed the light source and the photosensitive device to be remotely located from the actual sensor windows. Various cells can be installed in flow lines from 0.5 to 6 in. in diameter, or in situ probes can be used for batch vessels. Temperature and pressure must be specified.

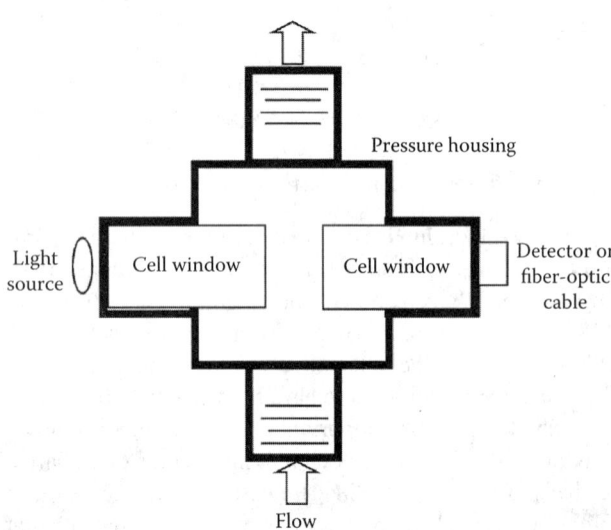

FIG. 1.13d
In-line transmittance cell.

The length of the light path in the cell is a key consideration due to the absorptivity of the liquid. Quartz must be used for UV analysis. Pressures may range up to 150 psig and temperatures to 250°C. The cost is lower if the sample stream can be reduced in size, pressure, and temperature. Cells can be in housings rated for NEMA 1, 4, 7 or Class 1, Group D, Division 1 explosion proof. Solvents such as chloroform that are used to extract and concentrate water-soluble analytes such as phenols can be eliminated by using longer cell path lengths if the spectrophotometer can accommodate them.*

Online Shade Monitors

Online shade monitors that feature robotics for scanning webs may also use fiber-optics to enable scanning while keeping the light source and spectrophotometer stationary. Fiber-optic-based spectrophotometers can be interfaced to host computers that also control stepper motors for positioning the fiber-optic for viewing at different angles.† Driver software is usually available for the spectrophotometer that will run in an application for data storage in large data files for statistical process control and trending. The output information from continuous systems can be used for alarming or for analog data transmitted by 4–20 mA current loop. Spectral information is transmitted serially by RS 232, RS 422, RS 485, USB (IEEE 1394), and I/P (Internet Protocol).

Optical Fluorescence and Luminescence Sensors

Optical measurement of turbidity is illustrated in Figure 1.13e. For field monitoring of surface waters, multisensor sondes

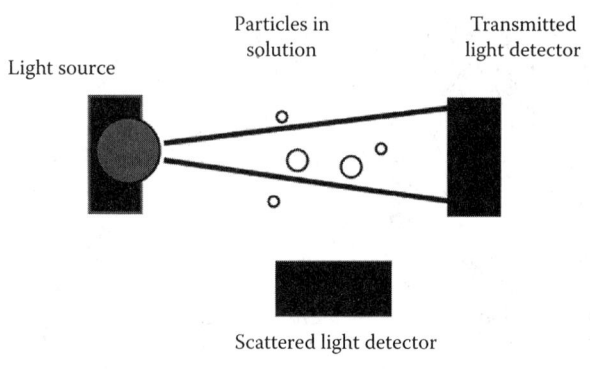

FIG. 1.13e
Laboratory turbidity measurement.

---

* Reed, M. W, et.al., Phenol Determination by the Direct Photometric Method, GAWP Spring Conference. April 13, 2010, Columbus, GA, www.mwreed@cwwga.org.
† United States Patent 5,559,605, Method and Apparatus for Determining the Directional Variation of Shade of Pile and Napped Materials, M. W. Reed, assigned to Milliken and Co., September 24, 1996.

such as that shown in Figures 1.13f and 1.13g combine a suite of sensors such as pH, temperature, and conductivity with optical sensors for phycocyanin and phycoerythrin (pigments in blue green algae), chlorophyll-a, turbidity, dissolved oxygen, etc. as summarized in Table 1.13c. The optical sensors have integrated motor-driven wipers. These sondes are installed remotely in streams and on buoys, and data are transmitted by cellular modems to cloud-based hosting services.

New Developments: Multispectral Analysis

For complex analysis of water, wastewater samples, and surface water,* bench-top spectrophotometers can have built-in algorithms for processing spectral curves. Advanced methods such as PCR methods using fluorescent dyes such as ethidium bromide attached to target DNA strands combine the fluorescence measurement with thermal cycling needed for DNA amplification. The American Water Works Association (AWWA) Source Book† has a list of analyzers by categories. Algorithms for combining fluorescence and spectral data for BOD determination are being developed.‡

Recently, multispectral analysis is also used for real-time process analysis in the LiquID™ system made by ZAPS, Inc. The concept of hybrid multispectral analysis (HMA) is possible in their hybrid fiber-optic spectrophotometer that is capable of UV/VIS absorption, fluorescence, and reflectance measurements. The ZAPS fiber-optic spectrophotometer was developed as an in situ probe for work in the oceans, which led to the development of the LiquID (LID) fiber-optic spectrophotometer. LID couples a xenon flash lamp (Excelitas Model FX 1165) and photon-counting detector (proprietary). The common end of the bifurcated fiber bundle terminates in a flow cell with a watertight seal. The fiber bundle is potted to withstand 300 psi. Source light enters the sample in the flow cell directly where it is collected and collimated by a submerged optical lens system.§ The lens returns light to the detector elements of the common end of the fiber bundle; this two-way distance is the effective path length. Wavelengths for each measurement are selected using two filter wheels, one placed before the lamp limb of the fiber bundle and one between the detector limb and the detector. Measurements can be changed by swapping filters in the wheels or by selecting different filter pairs through the software.

FIG. 1.13f
Multisensor sonde. (With permission from YSI Incorporated, a Xylem brand.)

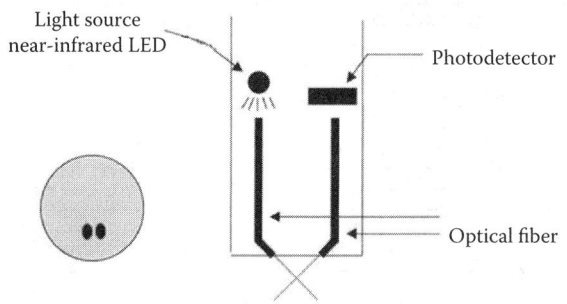

FIG. 1.13g
Optical field sensor for turbidity. (With permission from YSI Incorporated, a Xylem brand.)

SPECIFICATION FORMS

When specifying colorimeters only, please use the ISA form 20A1001, and when specifying both the analyzer and the composition or the properties of the process material of interest, use ISA form 20A1002. Both forms are reproduced with the permission of the International Society of Automation on the next pages.

TABLE 1.13c
Typical Optical Sensor Applications

Illuminant or Color	Fluorescence Color or Response	Analyte
UV, 200–400 nm	Blue or absorbed	Organic bond absorbance, E. coli metabolite fluorescence
Blue	Red	Chlorophyll-a
Blue	Red (chemiluminescence)	Dissolved oxygen
Green	Red	Phycoerythrin
Green	Yellow	Rhodamine WT
Orange	Red	Phycocyanin
NIR	NIR transmitted at 90° angle	Turbidity

---

* United States Patent 5,304,492, A spectrophotometer for Chemical Analysis of Fluids; G. Klinkhammer, 1994.
† American Water Works Association, AWWA, Source Book, 2013, http://sourcebook.awwa.org/CategoryResult.asp?scid=2
‡ Kwak, J. et al., Estimation of biochemical oxygen demand based on dissolved organic carbon, UV absorption, and fluorescence measurements, Journal of Chemistry, 2013, Article 243769, 9 pages.
§ Klinkhammer, G. *Light Returning Target For a Photometer*; 2008, U.S. Patent: 7,411,668.

Analytical Measurement

ANALYSIS DEVICE — Operating Parameters

1	RESPONSIBLE ORGANIZATION			6	SPECIFICATION IDENTIFICATIONS		
2				7	Document no		
3	(ISA)			8	Latest revision		Date
4				9	Issue status		
5				10			

11	ADMINISTRATIVE IDENTIFICATIONS			40	SERVICE IDENTIFICATIONS Continued		
12	Project number		Sub project no	41	Return conn matl type		
13	Project			42	Inline hazardous area cl	Div/Zon	Group
14	Enterprise			43	Inline area min ign temp	Temp ident number	
15	Site			44	Remote hazardous area cl	Div/Zon	Group
16	Area	Cell	Unit	45	Remote area min ign temp	Temp ident number	
17				46			
18	SERVICE IDENTIFICATIONS			47			
19	Tag no/Functional ident			48	COMPONENT DESIGN CRITERIA		
20	Related equipment			49	Component type		
21	Service			50	Component style		
22				51	Output signal type		
23	P&ID/Reference dwg			52	Characteristic curve		
24	Process line/nozzle no			53	Compensation style		
25	Process conn pipe spec			54	Type of protection		
26	Process conn nominal size	Rating		55	Criticality code		
27	Process conn termn type	Style		56	Max EMI susceptibility	Ref	
28	Process conn schedule no	Wall thickness		57	Max temperature effect	Ref	
29	Process connection length			58	Max sample time lag		
30	Process line matl type			59	Max response time		
31	Fast loop line number			60	Min required accuracy	Ref	
32	Fast loop pipe spec			61	Avail nom power supply		Number wires
33	Fast loop conn nom size	Rating		62	Calibration method		
34	Fast loop conn termn type	Style		63	Testing/Listing agency		
35	Fast loop schedule no	Wall thickness		64	Test requirements		
36	Fast loop estimated lg			65	Supply loss failure mode		
37	Fast loop material type			66	Signal loss failure mode		
38	Return conn nominal size	Rating		67			
39	Return conn termn type	Style		68			

69	PROCESS VARIABLES	MATERIAL FLOW CONDITIONS			101	PROCESS DESIGN CONDITIONS		
70	Flow Case Identification			Units	102	Minimum	Maximum	Units
71	Process pressure				103			
72	Process temperature				104			
73	Process phase type				105			
74	Process liquid actl flow				106			
75	Process vapor actl flow				107			
76	Process vapor std flow				108			
77	Process liquid density				109			
78	Process vapor density				110			
79	Process liquid viscosity				111			
80	Sample return pressure				112			
81	Sample vent/drain press				113			
82	Sample temperature				114			
83	Sample phase type				115			
84	Fast loop liq actl flow				116			
85	Fast loop vapor actl flow				117			
86	Fast loop vapor std flow				118			
87	Fast loop vapor density				119			
88	Conductivity/Resistivity				120			
89	pH/ORP				121			
90	RH/Dewpoint				122			
91	Turbidity/Opacity				123			
92	Dissolved oxygen				124			
93	Corrosivity				125			
94	Particle size				126			
95					127			
96	CALCULATED VARIABLES				128			
97	Sample lag time				129			
98	Process fluid velocity				130			
99	Wake/natural freq ratio				131			
100					132			

133	MATERIAL PROPERTIES			137	MATERIAL PROPERTIES Continued		
134	Name			138	NFPA health hazard	Flammability	Reactivity
135	Density at ref temp	At		139			
136				140			

Rev	Date	Revision Description	By	Appv1	Appv2	Appv3	REMARKS

Form: 20A1001 Rev 0 © 2004 ISA

	RESPONSIBLE ORGANIZATION	ANALYSIS DEVICE COMPOSITION OR PROPERTY Operating Parameters (Continued)		SPECIFICATION IDENTIFICATIONS		
1			6			
2	ISA		7	Document no		
3			8	Latest revision		Date
4			9	Issue status		
5			10			

	PROCESS COMPOSITION OR PROPERTY			MEASUREMENT DESIGN CONDITIONS				
	Component/Property Name	Normal	Units	Minimum	Units	Maximum	Units	Repeatability
11								
12								
13								
...								
37								

(rows 38–75: blank remarks/notes area)

Rev	Date	Revision Description	By	Appv1	Appv2	Appv3	REMARKS

Form: 20A1002 Rev 0 © 2004 ISA

Abbreviations

%AR	Percent actual reading
BOD	Biological oxygen demand
CCD	Charge-coupled device
DNA	Deoxyribonucleic acid
HMA	Hybrid multispectral analysis
IR	Infrared
Lab	CIE color space (rectangular coordinates) for human color recognition
Lch	CIE color space (cylindrical coordinates for lightless, chroma, and hue)
LED	Light-emitting diode
NIR	Near infrared
PCR	Polymerase chain reaction
RGB	Red, green, and blue color space
RS232	Recommended Standard 232 for voltage level serial data
RS422	Recommended Standard 422 for current loop serial data
RS485	Recommended Standard 485 for multidrop communications
UV	Ultraviolet

Organizations

AATCC	American Association of Textile Chemists and Colorists
APHA	American Public Health Association
ASTM	American Society for Testing and Materials
AWWA	American Water Works Association
CIE	Commission Internationale de l'Eclairage
GAWP	Georgia Association of Water Professionals
IEEE	Institute of Electrical and Electronics Engineers
NBS	National Bureau of Standards
NEMA	National Electrical Manufacturers Association
NIST	National Institute of Standards and Technology: http://www.nist.gov/pml/data/
WEF	Water Environment Federation

Bibliography

American Public Health Association, APHA, American Water Works Association, AWWA, Water Environment Federation, WEF, *Standard Methods for the Examination of Water and Wastewater*, 21st edn. 2005, 22nd edn., 2013, http://www.standardmethods.org

Ewing, G., *Analytical Instrumentation Handbook*, 1990.

Grayson, M., Ed., *Kirk-Othmer Concise Encyclopedia of Chemical Technology*, 4th edn., New York: John Wiley & Sons, 1999.

Nassau, K., *The Physics and Chemistry of Color*, New York: John Wiley & Sons, 1983.

Photonics Spectra Magazine, Laurin Publishing Co., Inc. ISSN 0731–1230, http://www.photonics.com.

YSI Incorporated, *6-SeriesMultiparameter Water Quality Sondes User Manual Revision F, 2009,* YSI Incorporated, a Xylem Brand, 1700/1725 Brannum. Lane, Yellow Springs, Ohio 45387 USA, environmental@ysi.com.

1.14 Combustible Gas or Vapor Sensors

R. NUSSBAUM (1969, 1982) **J. F. TATERA** (2003)
B. G. LIPTÁK (1995, 2017)

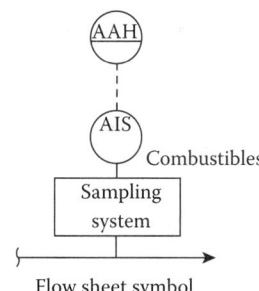

Flow sheet symbol

Types	A. Measurement of filament temperature or resistance in catalytic combustion sensors is most common. Thermal conductivity is used at higher concentrations. Electrochemical and semiconductor sensors can be used when hydrogen and other known gases are to be detected. B. Flame ionization and photoionization with or without a chromatograph can be used for accurate hydrocarbon detection. Response varies and gases of concern need to be known in the design and selection phase of the project. C. Infrared (IR) can be used for both point and area (open-path) applications. It cannot detect hydrogen. *Note:* All three types are available as portable and fixed devices, and Type A is also frequently found in a personal (pocket) device version.
Materials of construction	Many choices exist and offer the opportunity to select an appropriate one for a given application. Stainless steel and polymer sensor heads with ceramic and metal sensors are usually offered. Various polymer and metal constructions with the appropriate optical window selections for photoionization and IR applications are available.
Inaccuracy	A. 5% of lower explosive limit (LEL); linearity and repeatability from 2% to 3% of LEL B. Concentrations in the ppm range can be detected and monitored C. ppm and low % LEL levels achievable, but vary dramatically and usually more a function of the application than the instrument
Drift	A. 1% to 3% of LEL per month B. No generally accepted drift range per value C. No generally accepted drift range per value
Cost	For a large number of designs, see the websites of http://www.grainger.com/Grainger/combustible-gas-detectors/andindoor-air-quality/test-instruments/ecatalog/N-bbi or http://www.professionalequipment.com/combustible-gas-detectors/. Battery-operated portable gas leak detectors with sensing probes cost from $100 to $1000 (http://www.amazon.com/Reed-GD-3300-Combustible-Detector-Length/dp/B007WVCD4G); a combined oxygen and combustibles sensor, microprocessor based, portable with diffusion sampling, costs $2,500. For a permanently installed single-channel monitor with alarm or for a multichannel system, the cost per channel is $1,000 to $2,500. With sampled remote head installations, the installation cost of tubing can increase the per-channel cost to $3,000 to $5,000, and when a flame ionization or photoionization detector is used, the cost is in the $5,000 to $10,000 range. A portable chromatograph with electrochemical detector and 50 ppb sensitivity costs about $15,000 to $20,000. IR systems cost from $1,500 to $3,000 for a point system and $7,500 to $20,000 for an open-path system.
Partial list of suppliers	ABB (A, continuous, industrial) http://www.abb.com/product/seitp330/bf8f6ce43b17afbac1256d2600486d17.aspx?productLanguage=us&country=US American Gas and Chemical Co. Ltd. (A, wall mount alarm monitor) http://www.amgas.com/sbgpage.htm Bacharach Inc. (A, portable monitor) http://www.bacharach-inc.com/combustion-analyzers.htm Bascom-Turner Instruments (A, portable monitor and alarm) http://www.bascomturner.com/products.php B W Technologies (A, portable monitor) http://www.amazon.com/UEi-Instruments-CD100A-Combustible-Detector/dp/B000HHTY7Q/ref = pd_sim_sbs_indust_6 Cole-Parmer Instrument Co. (A, portable leak detector) http://www.coleparmer.com/Product/UEI_Combustible_Gas_Leak_Detector/EW-05500–35 Control Instruments Corp. (A, B, transmitter) http://www.controlinstruments.com/sites/default/files/product-brochure/SmartMaxIIBrochure.pdf

(Continued)

Detector Electronics Corp. (A, transmitter) http://www.detronics.com/utcfs/Templates/Pages/Template-53/0,8062,pageId%3D2720%26siteId%3D462,00.html

(C, open path) http://www.detronics.com/utcfs/Templates/Pages/Template-53/0,8062,pageId%3D2653%26siteId%3D462,00.html

Draeger Safety Inc. (B, flameproof) http://www.draeger.com/sites/enus_us/Pages/Fire/Infrared-sensors.aspx?navID=1947

GDS Corp., (C) http://www.gdscorp.com/wireless-gas-detectors

General Monitors (A, four channel) http://www.generalmonitors.com/Gas-Detectors/610A-Four-Channel-Combustible-Gas-Monitor/p/000140006500001008

(transmitter) http://www.generalmonitors.com/Gas-Detectors/S4000CH-Combustible-Gas-Detector/p/000140006500001028

Honeywell Analytics (acquired Zellweger) (A, C, portable monitor) http://www.honeywellanalytics.com/en-US/products/portablegasdetection/PHD6/Pages/PHD6.aspx

International Sensor Technology (A, B, C) http://www.intlsensor.com/

Macurco (A, wall-mounted alarm) http://www.aerionics.info/combustible_gas.html

MSA Instrument Div. (C, transmitter) http://ca.msasafety.com/Fixed-Gas-%26-Flame-Detection/Gas-Detectors/Ultima%26reg%3B-XIR-Gas-Monitor/p/000070001800003211

(32-point sampler) http://ca.msasafety.com/search?text = combustible+gas+analyzer&_requestConfirmationToken=1f15a88c656cc5e136bcdc53568a819f7f35c455

Sensidyne Inc. (C, transmitter, open path) http://www.sensidynegasdetection.com/products/open-path-gas-detection/safeye-700-series-open-path-combustible-gas-detector.html (wireless: http://www.sensidynegasdetection.com/)

Sick Maihak Inc. (B) http://www.sick.com/us1/en-us/home/products/product_portfolio/analyzers_systems/Pages/fid3006.aspx

Sierra Monitor Corp. (A, transmitter) http://www.sierramonitor.com/gas/IT/Combustible_gas_detector-Catalytic_Bead.php

Teledyne Analytical Instruments (A, B) http://www.teledyne-ai.com/industrialsensors.asp

Thermo Scientific (A) http://www.thermoscientific.com/ecomm/servlet/productscatalog?categoryId=89581&storeId=11152&from=search&taxonomytype=4

INTRODUCTION

Combustible analyzers will be discussed in this chapter. These instruments utilize a variety of operating principles that are covered in more detail in the following chapters:

Chapter 1.10—Gas Chromatography
Chapter 1.21—Electrochemical
Chapter 1.26—Toxic Monitoring
Chapter 1.28—Hydrocarbon Analyzers
Chapter 1.32—Infrared
Chapter 1.62—Open Path Spectrometry
Chapter 1.66—Thermal Conductivity

These instruments are designed to detect the presence and measure the concentration of combustible gases and vapors on a continuous basis. The methods of detecting the presence of combustible gases and vapors can utilize the phenomena of catalytic combustion, electrical resistance, luminosity, thermal conductivity, infrared (IR) absorption, or gas ionization.

Of the previously mentioned methods, the most widely used is catalytic combustion, where a change in the resistance or temperature of the sensing filament is caused by the catalytic combustion of the flammable gases, and this change is measured to detect the concentration of combustibles.

The second most widely used and a much newer technique is IR. Most suppliers offer a variety of designs, so that the user might select the best choice for his application. The selection process usually considers cost, robustness, selectivity, poison resistance, speed of response, etc.

PROPERTIES OF FLAMMABLE MATERIALS

In order to sustain combustion, each combustible gas or vapor requires a particular amount of oxygen. Some combustible gas mixtures ignite more easily than others (Table 1.14a). Additionally, the energy that is required to spark combustion also varies with the composition of mixtures.

The vaporization rates of the various liquids are a function of their vapor pressures, and vaporization rate increases with increased temperature. Flammable liquids are therefore more combustible at higher temperatures.

As can be seen from Table 1.14a, the ranges of air percentages within which some liquids and gases are flammable are extremely wide. In detecting the presence of such vapors or gases, their lower explosive limits (LELs) are usually of most interest, and in order to maintain safety, flammable gas and vapor concentrations must be kept below those limits.

Since air is usually the diluent and is almost always present, all concentrations above LEL are usually dangerous. Instruments are commonly calibrated with ranges in LEL units. LEL is selected as a limit on acceptable safety, because in order to reach a buildup of atmospheric concentration of flammables, which is above the upper explosive limit (UEL), the concentration must have necessarily passed through the full hazardous explosive range. Similarly, bringing the concentration back down to a safe level below the LEL, the concentration must pass again through the full hazardous explosive range.

DETECTOR TYPES

The most commonly used combustibles detectors are the catalytic filament units, which use a self-heated platinum wire as the catalytic surface to initiate combustion. A special portable variation of this unit is one that can be pinpointed at leaks by pointing a sample probe at the seals on manholes, tanks, or other containers that are likely to leak. In some instruments, two filaments are provided: a catalytic combustion filament for low ranges, and a thermal conductivity filament for higher ranges.

When the goal of the measurement is the detection of total hydrocarbons, or if the presence of lead, silicone, chlorinated compounds, or sulfur compounds could otherwise poison the catalytic filament, IR and flame or photo-ionization analyzers should be considered. Flame ionization instruments (as discussed in Chapters 1.24, 1.26 and 1.28) use the burning of the sample in a hydrogen flame. Since the flame of pure hydrogen contains practically no ions, even traces of organic material can be detected by the drastic rise in the number of ions in the flame. Measuring circuits for catalytic bead-type sensors usually include the Wheatstone bridge for resistance and null-balance potentiometers with thermocouples for temperature measurements.

In addition to the discussion of the measuring means, complete loops consisting of measuring, readout, and alarm devices and their applicability are covered in the following paragraphs.

TABLE 1.14a
Properties of Some Flammable Liquids and Gases

Material	Chemical Formula	Specific Gravity Air = 1	Ignition Temperature in Air (°F)	Ignition Temperature in Air (°C)	Flammability Limits in Air (% Vol.) Lower	Flammability Limits in Air (% Vol.) Upper
Methane	CH_4	0.55	1193	645	5.3	15.0
Natural gas	Blend	0.65	1163	628	4.5	14.5
Ethane	C_2H_6	1.04	993–1101	534–596	3.0	12.5
Propane	C_3H_8	1.56	957–1090	514–588	2.2	9.5
Butane	C_4H_{10}	2.01	912–1056	489–569	1.9	8.5
Toluene	C_7H_8	3.14	1026–1031	552–555	1.3	6.7
Gasoline	A blend	3–4.00	632	333	1.4	7.6
Acetone	C_3HO	2.00	1042	561	2.6	12.8
Benzene	C_6H_6	2.77	968	520	1.4	6.7
Carbon monoxide	CO	0.97	1191–1216	644–658	12.5	74.0
Hydrogen	H_2	0.07	1076–1094	580–590	4.0	75.0
Hydrogen sulfide	H_2S	1.18	655–714	346–379	4.3	45.0

CATALYTIC COMBUSTION TYPE

When mixtures of flammable gases or vapors in air come in contact with a heated and catalytically treated, fine, uniform, homogeneous platinum filament, combustion is induced at a temperature considerably below the normal ignition temperature of the particular gas or vapor. The heat generated by the combustion is measured by sensing the change of temperature of the filament by using thermocouples or by measuring the change of resistance of the filament.

Limitations

One of the common limitations of catalytic combustion-type analyzers is the poisoning of the filament by silicon, sulfur, chlorinated compounds, or lead compounds. When detecting the concentration of leaded gasoline vapors, which contain tetraethyllead, a solid lead combustion product can form (by condensation) on the filament surface, which reduces its catalytic activity. One way to protect the filament against lead condensation is to maintain the filament at a temperature that is high enough to prevent this condensation. Compounds containing silicone can also poison the filaments.

These effects impair the life of the sensor to different extents, depending on sensor packaging. Specially packaged diffusion head sensors (to be discussed shortly) are more likely to last longer on such services than do the flowing sample–type systems. Filament poisoning by chlorinated or sulfur compounds is also a serious problem.

In addition to special catalytic bead protective measures, ionization and IR detectors should be considered as an alternate means of measurement where sensor poisoning is an issue.

A variety of filament protection means have been added to increase the poison resistance of the sensors. Figure 1.14a illustrates one such design, in which the catalyst support consists of a low-density macroporous structure that surrounds the platinum wire deep within the bead assembly. This provides both protection and an increased catalyst surface area. The reported result is a 10-fold or better increase in sensor life expectancy on services such as hexamethyldisiloxane (HMDS), leaded gasoline, Freon-12, and ethyl mercaptan.

Life expectancies are usually defined in terms of exposure concentration hours. One high-concentration exposure of a poison has been known to knock out a sensor, and many do not respond in a fail-safe way. For this reason, nonpoisoning techniques should be considered, when poisoning is an issue.

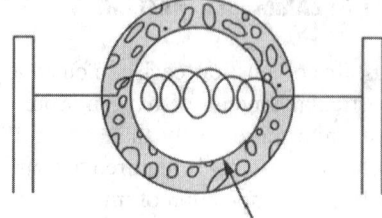

Porous structure to maximize support area for Poison-resistant catalyst

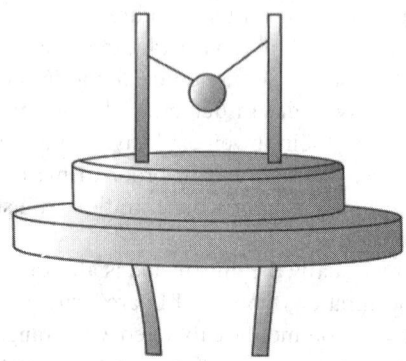

FIG. 1.14a
Porous bead construction provides poison resistance to catalytic combustion-type sensor.

Measuring Circuits

Whether the measurement is based on the change of temperature or resistance, it is convenient to use two filaments. One filament is constantly exposed to the sample (detector filament). The other is hermetically sealed in an inert atmosphere (reference filament). The reference filament is not activated with a catalyst, but its temperature resistance characteristics are similar to those of the detector filament. Its inert surface is usually exposed to the sample in a way that simplifies measurement compensation for changes in sample temperature, flow, and other potentially interfering characteristics.

The active detector filament and often the inert reference filament are mounted in a measuring chamber that is relatively large with respect to the size of the filaments. This permits a relatively large volume of sample to pass through the instrument, which ensures that the measurement filament is in contact with the sample and is measuring the current sample conditions. This design still only allows a relatively small portion of the sample to come in contact with the sensor, thereby increasing its useful life.

Thermocouple Detector In this design, two thermocouples are used. One thermocouple is bonded to the reference filament, and the other to the detector filament. The two thermocouples are connected in series opposition, so that a differential electromotive force (emf) is developed and applied at the terminals of the potentiometric circuit (see Figure 1.14b).

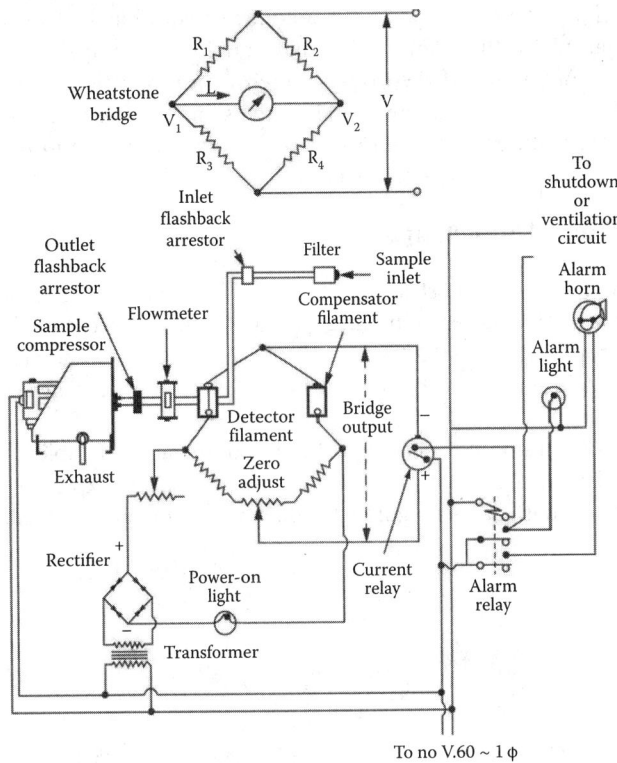

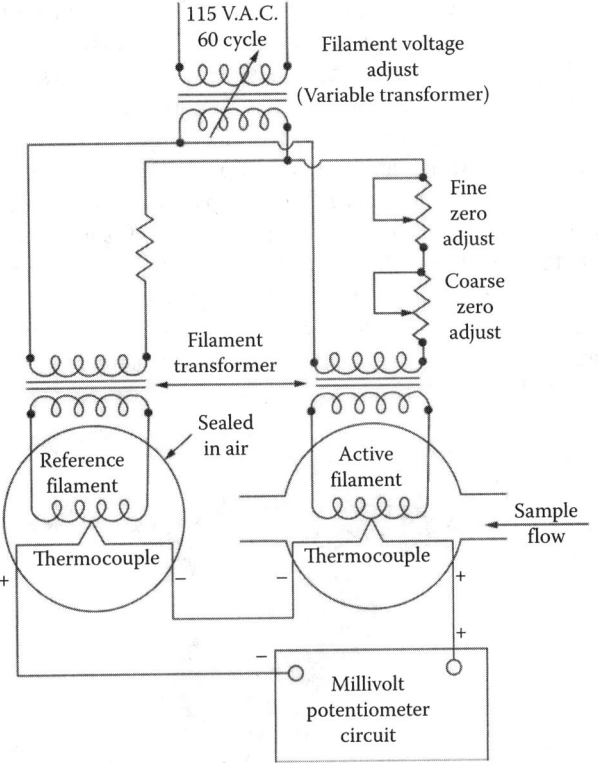

FIG. 1.14b
Thermocouple detector.

FIG. 1.14c
Wheatstone bridge detector with accessories.

When a combustible gas or vapor is admitted to the measuring chamber, combustion increases the temperature of the detector filament, resulting in an increased emf for the thermocouple bonded to it. The temperature of the reference filament remains the same as the sample temperature, since no combustion occurs on its bead. The potentiometric transmitter or the indicating, recording, and alarming instruments respond to the resultant differential emf.

Wheatstone Bridge Detector A Wheatstone bridge is typically used for resistance measurement. Its operation is based on the comparison of an unknown resistance to a resistor of known value, as shown in Figure 1.14c.

In this figure,
$R_1 = R_2 =$ constant
$R_3 =$ reference
$R_4 =$ sensor's measured resistance (compared to R_3)

For current I to be zero,

$$V_1 = V_2$$

$$V_1 = \frac{R_3}{R_1 + R_3} V$$

$$V_2 = \frac{R_4}{R_2 + R_4} V$$

$$\frac{R_3}{R_1 + R_3} = \frac{R_4}{R_2 + R_4} \qquad 1.14(1)$$

$$R_3 R_2 + R_3 R_4 = R_4 R_1 + R_4 R_3$$

$$R_3 R_2 = R_4 R_1$$

$$R_3 = R_4$$

In the catalytic bead-type combustibles detectors, R_3 is the reference filament and R_4 the detector filament. If the sample contains no combustibles, the bridge circuit remains in balance. If, however, there are combustibles in the sample, the combustion will cause heating of the detector filament. The change of resistance of the detector filament due to heating will result in unbalancing the bridge in proportion to the amount of additional heating caused by the combustible material in the sample. The output voltage of the bridge, which is in proportion to the concentration of combustibles in the sample, is detected by a transmitter or is used to operate indicating or recording instruments or to actuate alarms.

Diffusion Head Design

In contrast to most analyzers, the diffusion head analyzer does not require a sampling pump or a controlled sample flow. Rather, the diffusion head–type analyzer generates sample movement by diffusion, density difference, convection, or similar effects.

Diffusion-type catalytic bead sensors are available in poison-resistant designs and in intrinsically safe or explosion-proof construction. Figure 1.14d illustrates a conduit-mounted, diffusion-type transmitter with 4–20 mADC output. This unit is provided with a stainless steel sensor head and a polyvinyl chloride (PVC)-coated anodized aluminum conduit. Diffusion sensors can be used in still air or provided with plant air aspirators or pumps to draw a sample flow over the sensor.

Semiconductor sensors are also available in diffusion head designs. Semiconductor sensors respond to a combustible (target) gas that has been absorbed onto the doped surface of a metal oxide semiconductor, by displaying a change in the resistance of the semiconductor surface. By varying the doping layer, the manufacturer can vary the responsiveness of the detector to various materials. As with the catalyst bead surface effect sensors, poisoning is an issue.

Diffusion-type electrochemical and semiconductor sensors are also available to detect hydrogen, using a sensor that is not responsive to other hydrocarbons. This is desirable in semiconductor manufacturing plants, where it is a continual task to monitor for hydrogen leaks.

More recently, the combustible gas detectors have also become available in wireless designs (Figure 1.14e). They can be stationary or portable and can be battery or solar powered.

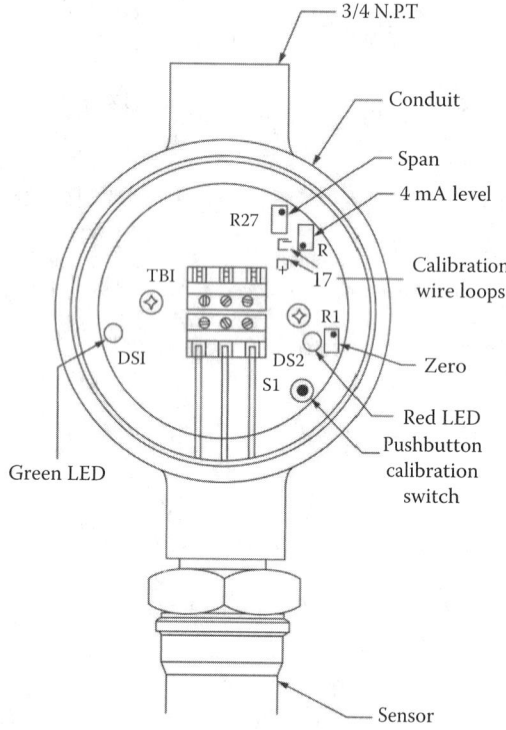

FIG. 1.14d
Diffusion-type sensor in combustible gas transmitter. (Courtesy of Sensidyne Inc.)

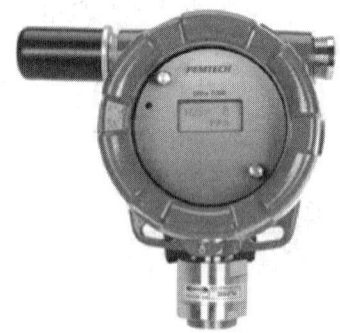

Stationary design. (Courtesy of Pemtech.)

Portable design. (Courtesy of REA.)

FIG. 1.14e
Wireless gas detectors are available in both permanently mounted and portable designs.

Sampling System

When diffusion-based systems are not adequate, active sampling systems may be required. The sampling systems should be carefully designed. Most importantly, the sample admitted into the analyzing cell should be wholly representative of the combustible components that are present in the monitored area. The sample should also be free of particulate matter and moisture.

In applications where the sample is at excessively high or low temperatures, it is advisable to use sample conditioners. This is particularly important if the sample is hot and humid and tends to cool while passing through the sampling line. The reason is that cooling would result in condensation, which, in turn, could block the sample line or introduce a time lag in the analyzer response. The sampling system should permit transport of the sample to the analyzer cell at the proper rate and minimum transportation time lag.

Since the vapors of all flammable liquids are heavier than air, detection of such vapors requires that the probes be located near the ground. Gases like hydrogen are lighter than air and require elevated probe locations. In dealing with gases, their molecular weight (heavier or lighter than air) will decide whether sampling probes should be near the ground or the ceiling of the monitored area. This may seem trivial, but is still worth mentioning.

Accessories

It is important to make sure to avoid the propagation of flame when the sampled air containing an explosive mixture of gas is ignited on the detector filament. This is not a problem when the concentrations are low or below the LEL, but a leak or spill can result in concentrations exceeding the LEL (within the explosive concentration envelope), and this leak or spill can become a source of ignition in that area of the plant.

Flashback arrestors of coiled copper screen or sintered metal are usually provided at the inlets and outlets of filament chambers. These prevent the energy that is liberated by combustion from propagating to the outside.

Samples containing hydrogen or acetylene, with concentrations of oxygen in excess of that found in normal air, have high rates of flame propagation. In such mixtures, standard flame arrestors cannot dissipate the energy liberated by combustion and, therefore, special flame arrestors have to be used.

To ensure safe operation of the detectors, a variety of alarms are provided. These alarms can signal filament failure, power failure, alarm relay failure, and low sample flow rates (not available for diffusion head–type designs). Many alarms do not detect sensor poisoning as a sensor failure. If in a particular application poisoning is likely, one should make sure to thoroughly understand the functioning of the alarms before issuing a purchase order.

To ensure that an adequate amount of sample passes through the measuring chamber, flowmeters (rotameters) and needle valves can be provided for all except the diffusion head type of units.

Total System Design

The selection is usually made from among three basic systems, and the choice is based on the plant layout, the required speed of response, and economic considerations. The three choices are (1) remote head (continuous measurement, continuous readout); (2) multiple head (continuous measurement, sequential readout); and (3) tube sampling system (sequential measurement, continuous readout).

Remote Head System   The remote head system offers the maximum application flexibility, but it does that at the highest initial cost. As shown in Figure 1.14f, this system typically consists of a number of locally mounted analyzer heads (suitable for hazardous areas) and an equal number of panel-mounted control and readout devices. The maximum number of areas monitored from one central panel is a function of the capacity of the sample pumps (or aspirators) and of the physical size of the panel. Because the analyzer heads are located in the monitored areas, the speed of response is fast.

Samples are continuously drawn, and the electrical signal corresponding to the measured combustible concentration is instantaneously transmitted to the control unit. The used

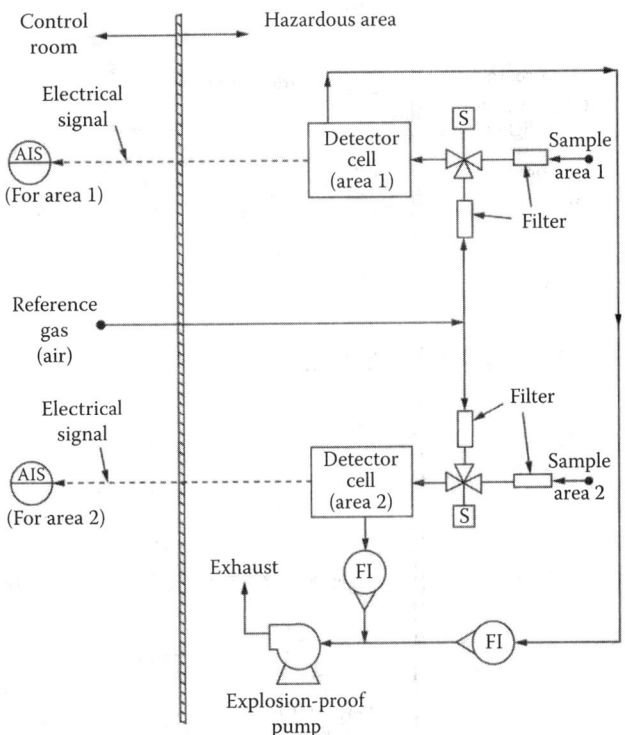

FIG. 1.14f
Remote head system.

sample is continuously withdrawn from the analyzer head through the tubing to the aspirator and is exhausted. Since the analyzer head is in the monitored area, it can be temperature-controlled to prevent condensation. The remote head system should be selected where fast response is essential and justifies the cost.

Multiple Head System Multiple head systems are used where at least four or more areas are monitored and a cyclic readout with the accompanying time delay can be tolerated. The multiple head system consists of a number of analyzer heads (one in each area to be monitored), one control unit with readout, and one or more sample pumps. The electrical circuit incorporates a single readout device common to all analyzing cells.

A separate alarm unit is associated with each detecting unit. The sample is drawn continuously to each sample chamber. The pump continuously withdraws the expended sample. The electrical output of each unit is transmitted to the panel, where sequential readout is provided. The dwell time for each area is typically 10 sec; that is, if four areas are being monitored, 40 sec elapses between subsequent readings for a given area.

This system is less costly than the remote head arrangement, but it can be used only where the combustible concentration buildup is likely to occur at a slow rate (see Figure 1.14g).

Tube Sampling System The tube sampling system consists of one analyzer head, one readout device, and a sample pump. This may be the least expensive arrangement, but sometimes the tubing cost (purchase and installation) can exceed the amount saved on the instrumentation. Samples from different areas are sequentially admitted to the common analyzer head. The electrical signal is then transmitted to the readout device.

A sample selector unit, consisting of time-sequenced solenoid valves, is arranged to admit one sample to the detector and connect all other sample lines to the sample pump. The sample is drawn continuously through each line. The sample selector is located at the analyzer head; thus, lag time between successive analysis and delay due to sample travel is minimized, since a fresh sample is always present at the sample selector.

One possible means of improving the system is to use separate pumps for the sample analyzed and for those that are bypassing the detector. A clean gas purge can be provided after each analysis to prevent an erroneous reading caused by residual carryover in this type of system (see Figure 1.14h).

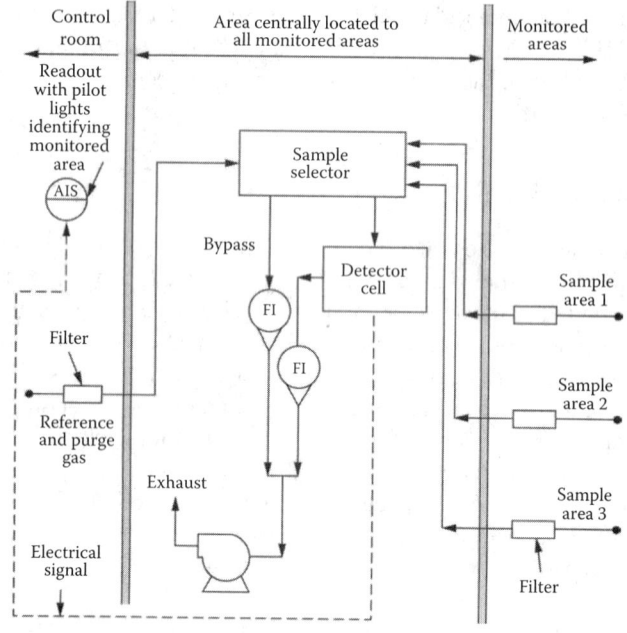

FIG. 1.14h
Tube sampling system.

In order to eliminate the problems associated with condensation in the sample tubes, these arrangements should be used only where true gases and vapors with boiling points well below ambient temperatures are to be detected.

Tube sampling systems usually have a 30 sec dwell time per hour. Therefore, they should be considered only if such slow response can be tolerated. For additional safety, readout devices can be calibrated with full-scale ranges as low as 0%–20% LEL. The alarm switches contained in the measuring circuit are used to actuate alarms, start ventilation, shut down sparking devices, and so on.

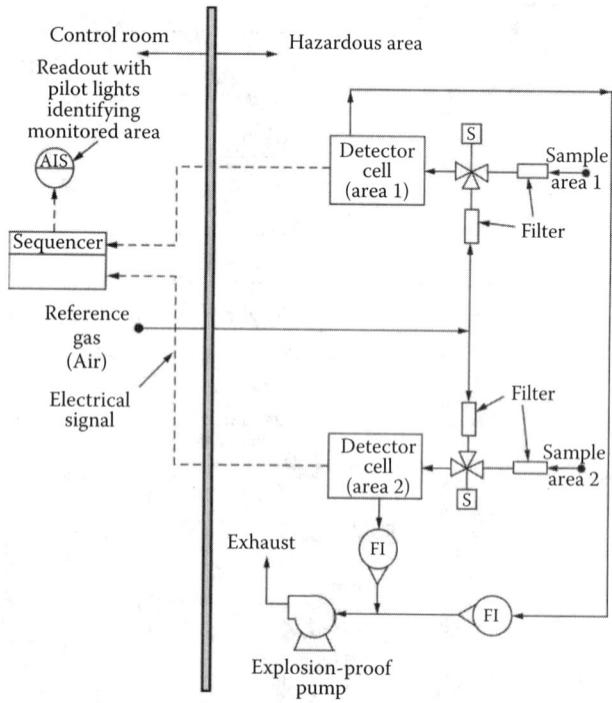

FIG. 1.14g
Multiple head system.

These systems are found in coating ovens, solvent recovery, and soybean extraction plants, just to mention a few typical applications.

Advantages and Disadvantages

In the diffusion head–type analyzer, the use of sample pump or aspirator is eliminated. Dispensing with any moving part increases reliability. Therefore, the use of a diffusion head analyzer is recommended wherever a flowing sample is not needed or where clean, dry samples are to be analyzed.

Large amounts of particular matter, moisture, and dust can and will cause plugging, which is difficult to detect in diffusion head analyzers since they cannot be furnished with low-sample-flow alarms.

In addition to the aforementioned facts, the selection parameters should include the considerations of plant layout, required speed of response, rate of gas buildup, and economy.

Comparing the Wheatstone and the thermocouple cells, the following should be considered. Whereas Wheatstone bridge cells use a fine, helical filament, the thermocouple cell uses a heavy, straight filament with a much longer useful life. Further, the evaporation on the exposed filament results in a constant change of base resistance of the filament.

In the Wheatstone bridge circuit, this change of base resistance produces a shifting of zero and requires frequent rebalancing of the bridge. The temperature change measured by the thermocouple is independent of filament deterioration. Thus, for the thermocouple detector, the zero drift is reduced to a negligible amount even over long periods of time. Therefore, the thermocouple detector is often superior to the Wheatstone bridge–type detector.

FLAME IONIZATION AND PHOTO IONIZATION TYPES

The theory and operation of flame ionization detectors (FIDs) and photoionization detectors (PIDs) are described in greater detail in the chapters describing chromatography (Chapter 1.10), hydrocarbon analyzers (Chapter 1.28). Both detectors are commonly used in chromatography and have been utilized for combustibles monitoring in both portable and fixed installation (Figure 1.14f).

Flame Ionization Detectors

The FID actually burns the sample in a hydrogen flame. In a simple combustibles application, no columns or carrier gases are used and the sample is used as an oxygen/air source. The sample is consumed during the combustion process. In a chromatography application, extremely clean air (as the source of oxygen) is introduced into the chromatographic column's effluent (which contains the sample) and is sent into the flame.

In these configurations, the only variable sources for ion formation in the flame are the components of interest in the column effluent or combustible contaminants in the combustibles sample. The appropriate combustible materials in the sample form ions in the flame. An oxygen-rich hydrogen and air flame basically exhausts water, nitrogen, and unconsumed oxygen. None of these are ionic in nature.

A charged electrical field is positioned across the flame, and it can conduct a current utilizing available ions in the flame as its conductor. When most combustible materials are introduced into the flame, they produce ions in their combustion products, and these are detected by the increased flow of current across the electric field (flame).

This detection method has been called a carbon counter, because of its response profile. It essentially responds to each carbon atom in the sample that has been consumed and used to form an ion in the flame. For example, one molecule of ethane has nearly twice the response of one molecule of methane.

This has both advantages and disadvantages in combustibles monitoring applications. The sensor is very sensitive to larger organic molecules. Its response to a mixture that may vary in composition can be difficult to calibrate, since different components have different LEL concentrations (see Table 1.14a) and different detector responses. Specific and unique calibrations may be needed for each sample or application.

The instrument cannot detect hydrogen (no ions are formed in the flame). These advantages and disadvantages are listed only as examples and are by no means exhaustive. Each application needs to be studied in full detail prior to selecting an appropriate measurement method.

Photo-Ionization Detectors

The PID utilizes a high-energy light source (normally ultraviolet (UV) radiation) as its source of ionization and measures the resulting flow of a current through the ionized sample, across a charged electric field. This detector also has several advantages and disadvantages.

It does not require auxiliary utilities (fuel gases). It can easily be made portable. It does not necessarily respond (depending on the ionization source chosen) to many potential components of interest. It requires frequent calibration and maintenance (as radiation sources deteriorate). Several different lamp strengths are available, and an appropriate one needs to be selected for a given sample. The ionization potential (IP) (eV)* of each molecule needs to be matched to the strength of the ionization source being used.

For example, acetone has an IP of 9.71 and can be ionized by most common lamps having IPs of 9.8, 10.6, or 11.7 eV. Of course, each different lamp has a different response factor for acetone, while methanol has an IP of 10.85 and therefore responds only to the 11.7 eV lamp. Methane has an IP of 12.51 and would not respond to any of these lamps.

These advantages and disadvantages are only given as an example and by no means are exhaustive. As mentioned

---

* Electronvolt (eV) is a unit of energy equal to approximately 160 zeptojoules (symbol zJ) or 1.6×10^{-19}. By definition, it is the amount of energy gained (or lost) by the charge of a single electron moved across an electric potential difference of one volt.

previously, each application needs to be fully studied prior to selecting an appropriate measurement method.

INFRARED TYPES

IR combustibles monitors are primarily a simplified and special-purpose version of an IR filter photometer. In cases of very simple applications, they have even been used as a substitute for an IR photometer and actually used to monitor a process sample that was introduced to them. For this reason, they have sometimes been called the poor man's IR analyzer.

IR photometers and spectrometers, and the technologies that they are based on, are discussed in great detail in Chapter 187, "Infrared and Near-Infrared Analyzers." Basically, ann IR beam of radiation that will excite the target gas molecules is used to measure the concentration of combustible gas molecules in the sample. For combustible gas monitoring, the radiation wavelength chosen is usually one that is absorbed by the C–H bond of most hydrocarbon molecules.

When the beam of radiation excites the molecules, a portion of its energy is absorbed and the amount of energy absorbed (lost to the beam) can be correlated to the amount of the target gas in the sample. Because many other factors could impact the intensity of the selected beam of IR radiation, these instruments usually also monitor a reference (another) wavelength of radiation that is not absorbed by the combustible gas, but is influenced by several of the other factors that could affect the measured beam's intensity.

IR combustibles monitoring instruments are available as both point and open-path (area) monitors. Even the point monitors are sometimes called open, because the IR measurement cell is actually open to the atmosphere. They typically rely on atmospheric diffusion to supply the sample and, consequently, the cell must be open to allow the diffusion of sample into the measurement area to take place.

Open-path instruments, on the other hand, actually use a large, open atmospheric path as their measurement cell (tens to hundreds of meters). IR combustibles monitors are a relatively new innovation in the field of combustibles monitoring, but they have already gained wide acceptance as a niche technology. They perform well on many samples that other technologies have problems with. This is especially true for many gases that can poison other combustibles sensors and for monitoring requirements where the likely points of leakage are difficult or impossible to predict.

Diatomic molecules like hydrogen, oxygen, and nitrogen have no usable IR absorbance and cannot be detected by these IR monitors. Consequently, IR combustibles monitoring systems should not be used for hydrogen or hydrogen-containing combustibles mixtures. The response of each potential gas or mixture to the detection method needs to be considered when selecting a monitor.

Point Monitoring System

Point IR systems monitor the sample at the measuring head, just like the other previously discussed point style combustibles monitors. If it is intended to monitor a sample that is not diffusing into the sensor head and is not located immediately adjacent to it, the sample must be transported to the sample head using a sample transport system. A couple of point IR designs are shown in Figures 1.14i and 1.14j.

Figure 1.14i depicts a reflector style point sensor design, where the IR source and detector are both located on the same side of the sample chamber. The measurement and reference

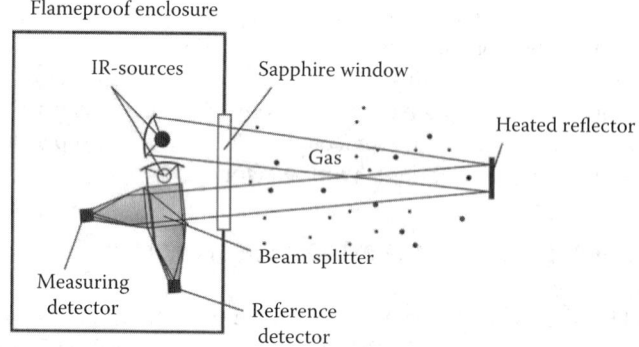

FIG. 1.14i
IR reflector style point sensor. (Courtesy of Draeger Safety Inc.)

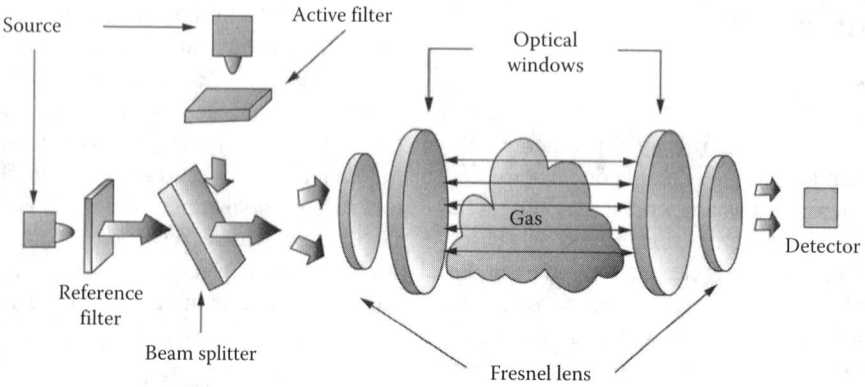

FIG. 1.14j
IR one-pass point sensor. (Courtesy of General Monitors Inc.)

IR beams are transmitted, reflected off of a mirrored reflector, and pass through the sample twice during the analysis. With this type of sensor, there is no chemical reaction of the gas, and as such, the materials that poisoned catalytic beads cannot poison these sensors.

But nothing is perfect or without its own Achilles' heel. For an IR point monitor to do its job, the IR radiation beam must pass through the sample and be partially absorbed by the sample of interest before reaching the detector. If the sample becomes opaque to the IR measurement beam or if the optical path is otherwise blocked (condensation or dirt on the windows, heavy fog, dust, etc.), the instrument can be rendered inoperable. Typically, the mirror and windowed instrument compartment are purged or maintained at a temperature that is intended to prevent condensation on either the window or mirror.

Figure 1.14j depicts a one-pass point IR sensor design, where the IR source and detector are located on opposite sides of the sample chamber. This analyzer is very much analogous to the reflector style sensor head depicted in Figure 1.14i, except that the beam of radiation only passes through the sample once.

In the reflector style, the beam has the opportunity to interact with the sample twice, and if all else was equal and good, the instrument should depict twice the sensitivity (or twice the interference to things like fog and dust). In real life, the sensitivities are usually not that different between the different designs, and the benefits are more often expressed in the form of a smaller sensor head or other geometric benefits.

Open Path System

All of the previously covered combustibles monitoring technologies can be classified as point monitoring systems. They only measure the atmosphere at the points where they have been located (the gas of interest diffuses to a sensor) or sampled from. To monitor a large area, one would have to locate many monitors (points) and hope that they represent the area's general atmosphere.

Open-path IR combustibles monitors project their IR beams in a path that is typically 10–200 m in length and monitor all of the combustibles in that path. This is not really an area (more of a line or path) detector, but the value it determines can be viewed as more representative of an area value, and few instruments could provide a value that more nearly represents an area than would be needed with point monitors.

Figure 1.14k depicts some of the conceptual differences between point and open-path applications. The figure shows examples of leak detection applications under both no-wind and mild-wind conditions. It can be seen that the leak cloud shape varies as a function of atmospheric conditions. Leak cloud shapes also vary as a function of the composition and the conditions of the sample.

Lighter and hotter gases rise faster, and samples under different pressures produce different rates of release and dispersion. Therefore, it is very difficult to locate a sensor (point or open path) in a way that will always accurately measure and detect a leak, unless it is located almost exactly at the source of the leak.

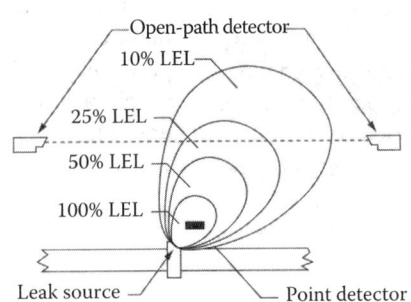

FIG. 1.14k

IR open-path vs. point monitoring concept. (Courtesy of General Monitors Inc.)

Both leak examples in Figure 1.14k show the use of both a point and an open-path monitor. It is easy to see the benefit one can gain by locating a good point monitor at a potential leak source. By doing so, one would get both an earlier and quicker detection, as the sample concentration is always higher, closer to the leak.

It is also easy to see the benefits of locating a single open-path monitor along a pipeline that may contain many potential leak sources, as opposed to installing many point sensors, which would be very costly and impractical. Open-path sensors can also better cover a general area, where the positions of potential leak sources may be difficult to predict. Similar benefits can be visualized regarding other applications, like perimeter monitoring, room monitoring, fence line monitoring, or other general-area monitoring tasks.

It is important to note that the point and open-path techniques utilize different reporting values. Point techniques utilize parts per million (ppm) or percent LEL (%LEL) values, as described in the definitions at the beginning of this chapter and as listed in Table 1.14a. Open-path techniques utilize the units of parts per million or LEL meters (ppm.m or LEL.m). These will be discussed in more detail later in this chapter. It is fair to say that in most applications, open-path monitors are used more to detect leaks than to determine the absolute degree of hazard associated with the leak. This is because of the various leak cloud shapes that could exist and the way the instruments add or average the concentrations along their path.

Hydrocarbon Gases in the Atmosphere

Properly applied open-path monitors can be effective in monitoring combustible hydrocarbon gases in the atmosphere. Figures 1.14l and 1.14m shows an example of a typical open-path monitor. This installation involves installing two field

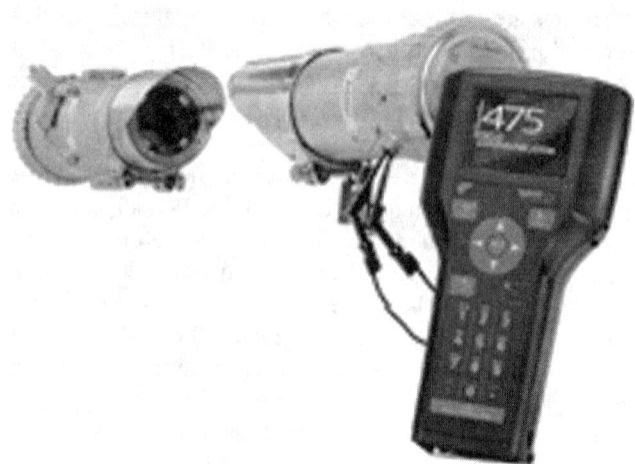

FIG. 1.14m
Open-path IR detector. (Courtesy of Detector Electronics Corporation.)

devices. In this example, they are the source and the receiver or detector sections of the monitoring instrument. In other examples, they may consist of the instrument (source and detector) and a reflector.

In all cases, proper positioning and alignments are crucial to the success of the application. The beam must be positioned in a way that will enable it to detect the leaks of interest. The alignment is typically done with the aid of a vendor-supplied or recommended rifle scope that is mounted on one of the sections of the unit; using it helps in precisely aligning the beam, so that it properly hits the other unit.

The instrument needs to be located not only where it can make the best measurement of the leak, but also where it can perform. It must not be located where it can be exposed to shock or vibration, because this could make the alignment unstable or impossible. The monitor also must not be located where people, cars, or other equipment can block the beam.

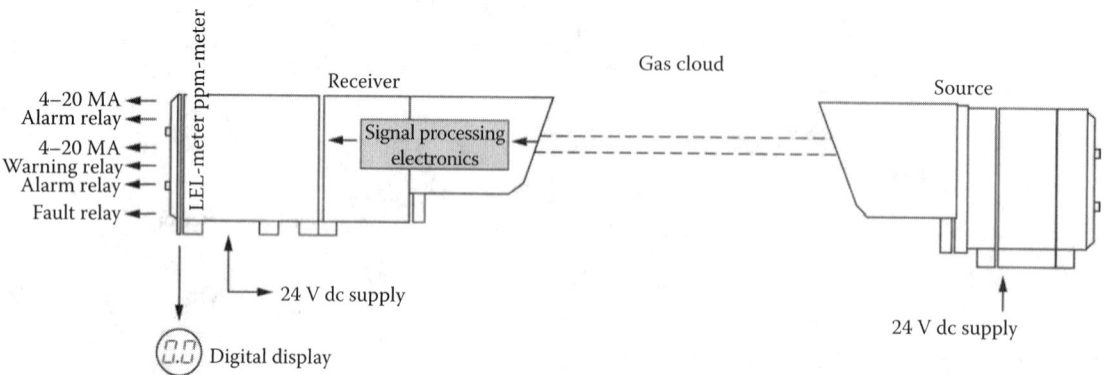

FIG. 1.14l
A single-pass, open-path IR system. (Courtesy of General Monitors Inc.)

Figure 1.14n depicts how an open-path instrument can compensate for partial blockages of its beam by light-obscuring interference, such as rain, fog, and dust. Essentially, it calculates the ratio of the measurement and the reference radiation signal. Most partially obscuring interferences will reduce both signals to the same extent. Therefore, the ratio of the signals is relatively unaffected by the interfering obstruction. On the other hand, the presence of a combustible gas will reduce only the measurement signal and therefore will result in a change in the ratio of the two radiation signals. This naturally is not the case if the signals are totally or nearly totally blocked. In that case, the instrument sensitivity and ability to detect a combustible gas are partially or completely lost.

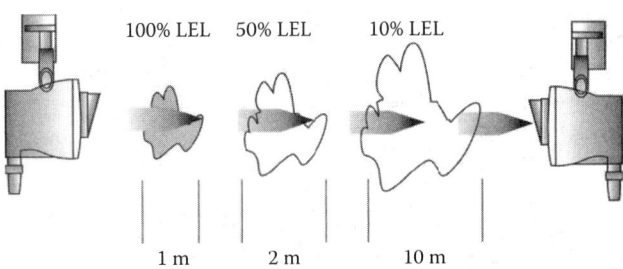

FIG. 1.14o
Open-path IR measurement units. (Courtesy of Honeywell Analytics–Zellweger Analytics Inc.)

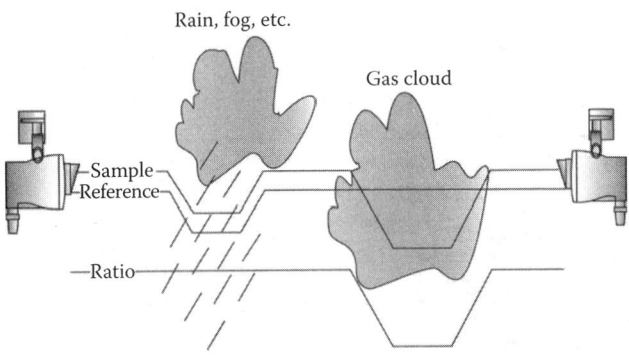

FIG. 1.14n
Open-path IR signal response. (Courtesy of Honeywell Analytics–Zellweger Analytics Inc.)

Sunlight, flames, and many other light sources also produce IR radiation. To reduce or eliminate their effects on the performance of IR instruments, choppers, filters, focusing optics, digital signal processing techniques, and other aids are utilized. In general, these miscellaneous IR sources do not interfere with the performance of today's IR type combustible monitors. Yet, under extreme situations, they can still reduce sensitivity, by swamping the detector with too much radiation, and consequently, they should be considered when designing the field installation.

Point Measurement

The open-path IR systems monitor the concentration over the length of their optical path, but how can they be used to measure the concentration at any one point along this path? Do they operate like radar and actually monitor the concentrations at various points along the optical path? No, they do not. They essentially measure the number (concentration) of combustible gas molecules along the path in an integrated or cumulative fashion and report a number that incorporates the dimensions of both the concentration and distance.

Figure 1.14o illustrates how these instruments measure and report their readings. It should be kept in mind that the measurement is along a beam or path and, therefore, it is not detecting an area or a point. Percent or parts per million (ppm) readings are obtained by integrating the product of the concentration of the gas (along the length of the IR beam) by the length of the cloud (along the optical path).

In Figure 1.14o, there are three clouds in the optical path of the monitor. Each cloud has a different gas concentration and size, but all three are equal in LEL-meter units. In terms of their open-path reporting dimensions (100% × 1 m = 50% × 2 m = 10% × 10 m = 1 LEL·m each), if all three clouds were in the optical path simultaneously, the instrument would report the total presence of combustible gases as 3 LEL·m.

Clearly, the 1 m 100% LEL cloud is potentially explosive and the most dangerous of the three, but this method of monitoring does allow one to distinguish between them. It cannot distinguish between small clouds of high concentration and large clouds of low concentration. It simply measures the total amount of target gas in its optical path. For this reason, the use and applicability of open-path IR combustibles monitoring is limited. Yet, this sensor still fills a niche market, and it can make some difficult monitoring applications feasible and practical.

All of the combustibles monitoring technologies reviewed in this chapter have strengths and weaknesses. Each has some advantages and disadvantages relative to the others. It is up to the user and the supplier to work together in evaluating these differences and picking the most appropriate technology for a given application.

SPECIFICATION FORMS

When specifying combustible gas and vapor detection systems, one can describe the devices required using the ISA Form 20A1001 and the probable composition at the target area being monitored by using form 20A1002. These forms can be found on the next pages where they are reproduced with the permission of the International Society of Automation.

Analytical Measurement

ANALYSIS DEVICE — Operating Parameters

#	RESPONSIBLE ORGANIZATION		#	SPECIFICATION IDENTIFICATIONS	
1	ISA		6		
2			7	Document no	
3			8	Latest revision	Date
4			9	Issue status	
5			10		

ADMINISTRATIVE IDENTIFICATIONS

11				
12	Project number		Sub project no	
13	Project			
14	Enterprise			
15	Site			
16	Area		Cell	Unit
17				

SERVICE IDENTIFICATIONS

18			
19	Tag no/Functional ident		
20	Related equipment		
21	Service		
22			
23	P&ID/Reference dwg		
24	Process line/nozzle no		
25	Process conn pipe spec		
26	Process conn nominal size	Rating	
27	Process conn termn type	Style	
28	Process conn schedule no	Wall thickness	
29	Process connection length		
30	Process line matl type		
31	Fast loop line number		
32	Fast loop pipe spec		
33	Fast loop conn nom size	Rating	
34	Fast loop conn termn type	Style	
35	Fast loop schedule no	Wall thickness	
36	Fast loop estimated lg		
37	Fast loop material type		
38	Return conn nominal size	Rating	
39	Return conn termn type	Style	

SERVICE IDENTIFICATIONS Continued

40			
41	Return conn matl type		
42	Inline hazardous area cl	Div/Zon	Group
43	Inline area min ign temp	Temp ident number	
44	Remote hazardous area cl	Div/Zon	Group
45	Remote area min ign temp	Temp ident number	
46			
47			

COMPONENT DESIGN CRITERIA

48			
49	Component type		
50	Component style		
51	Output signal type		
52	Characteristic curve		
53	Compensation style		
54	Type of protection		
55	Criticality code		
56	Max EMI susceptibility	Ref	
57	Max temperature effect	Ref	
58	Max sample time lag		
59	Max response time		
60	Min required accuracy	Ref	
61	Avail nom power supply		Number wires
62	Calibration method		
63	Testing/Listing agency		
64	Test requirements		
65	Supply loss failure mode		
66	Signal loss failure mode		
67			
68			

PROCESS VARIABLES — MATERIAL FLOW CONDITIONS

#	Field				Units
69					
70	Flow Case Identification				
71	Process pressure				
72	Process temperature				
73	Process phase type				
74	Process liquid actl flow				
75	Process vapor actl flow				
76	Process vapor std flow				
77	Process liquid density				
78	Process vapor density				
79	Process liquid viscosity				
80	Sample return pressure				
81	Sample vent/drain press				
82	Sample temperature				
83	Sample phase type				
84	Fast loop liq actl flow				
85	Fast loop vapor actl flow				
86	Fast loop vapor std flow				
87	Fast loop vapor density				
88	Conductivity/Resistivity				
89	pH/ORP				
90	RH/Dewpoint				
91	Turbidity/Opacity				
92	Dissolved oxygen				
93	Corrosivity				
94	Particle size				
95					

CALCULATED VARIABLES

96					
97	Sample lag time				
98	Process fluid velocity				
99	Wake/natural freq ratio				
100					

PROCESS DESIGN CONDITIONS

#	Minimum	Maximum	Units
101			
102			
103			
104			
105			
106			
107			
108			
109			
110			
111			
112			
113			
114			
115			
116			
117			
118			
119			
120			
121			
122			
123			
124			
125			
126			
127			
128			
129			
130			
131			
132			

MATERIAL PROPERTIES

133			
134	Name		
135	Density at ref temp		At
136			

MATERIAL PROPERTIES Continued

#	NFPA health hazard	Flammability	Reactivity
137			
138			
139			
140			

Rev	Date	Revision Description	By	Appv1	Appv2	Appv3	REMARKS

Form: 20A1001 Rev 0

© 2004 ISA

	RESPONSIBLE ORGANIZATION	ANALYSIS DEVICE COMPOSITION OR PROPERTY Operating Parameters (Continued)		SPECIFICATION IDENTIFICATIONS		
1			6			
2			7	Document no		
3	(ISA logo)		8	Latest revision	Date	
4			9	Issue status		
5			10			

	PROCESS COMPOSITION OR PROPERTY			MEASUREMENT DESIGN CONDITIONS				
11								
12	Component/Property Name	Normal	Units	Minimum	Units	Maximum	Units	Repeatability
13								
14								
15								
16								
17								
18								
19								
20								
21								
22								
23								
24								
25								
26								
27								
28								
29								
30								
31								
32								
33								
34								
35								
36								
37								

(Rows 38–75: blank open area)

Rev	Date	Revision Description	By	Appv1	Appv2	Appv3	REMARKS

Form: 20A1002 Rev 0 © 2004 ISA

Definitions

Electronvolt (eV) is a unit of energy equal to approximately 160 zeptojoules (symbol zJ) or 1.6×10^{-19}. By definition, it is the amount of energy gained (or lost) by the charge of a single electron moved across an electric potential difference of one volt.

Flash point: The lowest temperature at which a flammable liquid gives off enough vapors to form a flammable or ignitable mixture with air near the surface of the liquid or within the container used. Many hazardous liquids have flash points at or below room temperatures. They are normally covered by a layer of flammable vapors that will ignite in the presence of a source of ignition.

Lower explosive limit (LEL): The lowest concentration of gas or vapor in air where, once ignition occurs, the gas or vapor will continue to burn after the source of ignition has been removed.

Upper explosive limit (UEL): The highest concentration of gas or vapor in air in which a flame will continue to burn after the source of ignition has been removed.

Abbreviations

eV	Electronvolt
FID	Flame ionization detector
HMDS	Hexamethyldisiloxane
IP	Ionization potential
IR	Infrared
LEL	Lower explosive limit
mADC	Milliampere, direct current
PID	Photoionization detector
PVC	Polyvinyl chloride
UEL	Upper explosive limit

Bibliography

ACGIH, Air sampling instruments for evaluation of atmospheric contaminants, (2009), http://www.acgih.org/store/ProductDetail.cfm?id=2100.

Angle, J., Occupational safety and health in the emergency services, (2012) http://www.amazon.com/Occupational-Safety-Health-Emergency-Services/dp/1439057508/ref=sr_1_2?s=books&ie=UTF8&qid=1380310289&sr=1–2&keywords=combustible+gas+monitoring.

Bretherick, L., Bretherick's Handbook of Reactive Chemical Hazards, Butterworth-Heinemann, (2006), http://www.amazon.com/Brethericks-Handbook-Reactive-Chemical-Edition-Two/dp/0123725631.

General Monitors, Fundamentals of combustible gas detection, (2010) http://s7d9.scene7.com/is/content/minesafetyappliances/Combustible%20Gas%20Detection%20White%20Paper.

Honeywell, Gas Detection Book, (2013) http://www.honeywellanalytics.com/~/media/honeywell-analytics/documents/english/11296_gas-book_v5_0413_lr_en.pdf?la=en.

International Sensor Technology, (2014) Catalytic combustible gas sensors, http://www.intlsensor.com/pdf/catalyticbead.pdf.

International Sensor Technology, Sensor selection overview, (2015) http://www.intlsensor.com/pdf/sensorSelectionGuide.pdf.

Klinoff, R., Introduction to fire protection, (2011), http://www.amazon.com/Introduction-Fire-Protection-Robert-Klinoff/dp/1439058423/ref=pd_bxgy_b_img_y.

Lodge, J. P., Ed., Methods of Air Sampling and Analysis, Lewis Publishers, Boca Raton, FL, http://www.amazon.com/Methods-Sampling-Analysis-James-Lodge/dp/0873711416.

MSA, Gas Detection Handbook, (2007) http://www.gilsoneng.com/reference/gasdetectionhandbook.pdf.

New Jersey, Department of Health, (2014) Hazardous Substance Fact Sheets, http://web.doh.state.nj.us/rtkhsfs/indexfs.aspx.

OSHA, Use of combination oxygen and combustible gas detectors, (2013) https://www.osha.gov/dts/hib/hib_data/hib19900118.html.

Shackelford, R., Fire behavior and combustion processes, (2008) http://www.amazon.com/Fire-Behavior-Combustion-Processes-Shackelford/dp/1401880169/ref=pd_bxgy_b_img_y.

Yaws, C. L., Baker, W., Matheson Gas Data Book, (2001) https://books.google.com/books/about/Matheson_Gas_Data_Book.html?id=Sfvbgvu9OQMC.

Zabetakis, M. G., Flammability characteristics of combustible gases and vapors, https://books.google.com/books/about/Flammability_Characteristics_of_Combusti.html?id=nolWGQAACAAJ.

1.15 Conductivity Measurement

A. BRODGESELL (1969, 1982) **K. S. FLETCHER** (1995)

J. R. GRAY (2003, 2017)

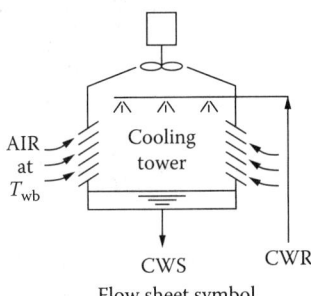

Flow sheet symbol

Standard design pressure	To 500 psig (3.5 MPa)
Standard design temperature	To 200°C (390°F)
Element materials	Cells: glass, epoxy, and stainless steel. Electrodes: platinum, nickel, titanium, and carbon. Electrodeless: epoxy, Noryl, PFA, polyether ether ketone (PEEK), and polypropylene
Cost	Conductivity sensors range in price from $400 for contacting sensors to $700 for toroidal sensors; transmitters, $1000 (two two-wire transmitters) to $2000 or more (line-powered transmitter)
Range	0–0.05 µS/cm minimum; 0–2 S/cm
Inaccuracy	Up to ±0.5% of full scale
Partial list of suppliers	ABB, http://new.abb.com/products/measurement-products Advanced Sensor Technologies Inc., http://www.astisensor.com/conductivity_sensors.htm Analytical Technology, http://www.analyticaltechnology.com/public/productlist.aspx?CategoryID=1006 Electro-Chemical Devices, http://www.ecdi.com/products/t80.html Endress+Hauser, http://www.us.endress.com/en/products/analytics#/en/products/analytics?filter1.product-measurementmethod=conductivity&filter1.product-family=&filter1.measuring-task=&filter1.internal-industry-relevance=&filter1.product-type=&filter1.design=&filter1.digital-process-communication=&filter1.product-approval=Mettler–Toledo, http://us.mt.com/us/en/home/products/Process-Analytics/transmitter.html Hach Co., http://www.hach.com/sc200-models/category-products?productCategoryId=14371221969&secondPageNumber=1&isNew=false&pimContext=USen Hamilton Co., http://www.hamiltoncompany.com/products/sensors/c/15/ Hanna Instruments, http://www.hannainst.com/usa/prods2.cfm?id=017005 Honeywell, https://www.honeywellprocess.com/en-US/explore/products/instrumentation/analytical-instruments-andsensors/default/Pages/conductivity.aspx Horiba Instruments Inc., http://www.horiba.com/us/en/process-environmental/products/water-treatment-environment/on-site-h-1-series/ Knick, http://www.knick-international.com/products/measuring_systems/ Rosemount Analytical, http://www2.emersonprocess.com/en-US/brands/rosemountanalytical/Liquid/Instruments/Pages/index.aspx Sensorex, http://www.sensorex.com/products/categories/category/conductivity_sensors Swan Analytical Instruments, http://www.swan.ch/Catalog/en/ProductOverview.aspx Teledyne Analytical Instruments, http://www.teledyne-ai.com/products/lxt.asp The Foxboro Company, http://www.fielddevices.foxboro.com/en-gb/products/analytical-echem/contacting-conductivity-and-resistivity/+GF+ Signet, http://www.gfps.com/content/gfps/signet/en_US/sensors/conduct.html Thermo Scientific, http://www.thermoscientific.com/en/products/ph-measurement.html Van London—Phoenix Co., http://www.vl-pc.com/default/index.cfm/conductivity/all-conductivity-products Yokogawa, http://www.yokogawa.com/us/products/analytical-products/ph-orp/index.htm

Conductivity sensors measure a solution's ability to conduct electricity, which is a function of all dissolved ionized solids in the solution. These detectors are packaged either as probes (with isolating valves for removal, without opening up the process) or in the flow-through designs.

INTRODUCTION

Conductivity analyzers measure ionic concentration of electrolyte samples. Cells and instrumentation are designed to measure the electrical resistance (or its reciprocal, the conductance) in a volume element of the electrolyte and to limit electrode–solution interfacial contributions to this measurement. A variety of sensors have been developed: some using electrodes in contact with the sample and others not. These sensors can be combined with modern microelectronics, often with integrated software programs, which improves the quality of measurement of concentration of ionic components in process samples.

THEORY OF OPERATION

The unit of conductance, the reciprocal of the ohm, is Siemens (S). This unit impresses a measurement of the mobility or velocity of ions in an electrolyte under an imposed electric field. The value of this unit depends on the number and hence on the concentration of ions present, which provides its value in analytical measurements. However, since the mobilities of dissimilar ions are different, the measurement response provided by the cell is useful only if the detected component is the sole, or at least the major, contributor to the measured conductivity. *Equivalent conductivity* is defined as the conductance that is reported for one gram-equivalent weight of the conducting ion.

The mobility of ions is affected by temperature and by the total concentration of all ions in the solution. Mobility of ions and hence conductance increases with temperature (about 2% per °C) and also with dilution. Table 1.15a shows

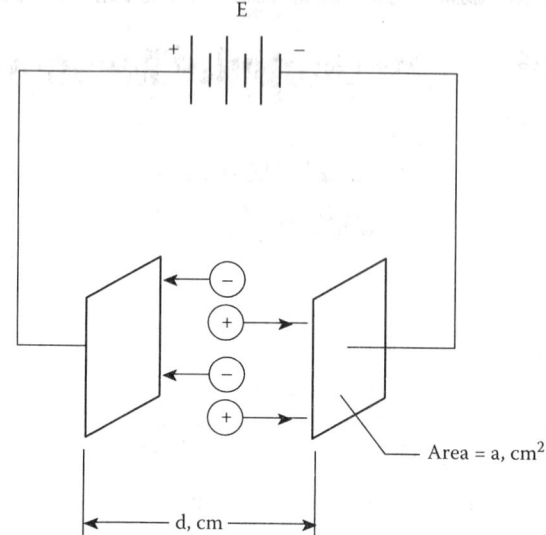

FIG. 1.15a
A simplified schematic view of two-electrode conductivity circuit.

values of conductance, expressed as equivalent conductivity, of several ions at 25°C corrected to infinite dilution, together with temperature coefficients.

Figure 1.15a shows the relationship of cell geometry to the measured conductance of the solution. The electric field applied to the cell is E/d. The current density i/A is the sum of the individual charge carriers in the field, and therefore the conductivity L (which from Ohm's law is the current divided by the voltage) is given by

$$L = (a/d)\sum_i z_i c_i \Gamma_i \qquad 1.15(1)$$

where
  L is the conductance in ohm^{-1} or Siemens
  a is the area of electrodes in cm^2
  d is the distance between electrodes in cm
  c_i is the concentrations of the participating ions in equiv./cm^3
  Γ_i is the equivalent conductivity of the participating ion in S/cm^2/equiv.
  z_i is the charge on the participating ion

CELL CONSTANT

It is not convenient to measure the ratio d/a geometrically for each cell, but since it is constant for any given cell, it may be assigned a value θ, in cm^{-1}, termed the *cell constant*. It is determined experimentally using solutions of accurately known concentrations of potassium chloride, for which values of specific conductivity (the conductance of a cube of the solution, 1 cm on each side), denoted by k (in S/cm), have been precisely determined as per American Society

TABLE 1.15a
Equivalent Conductivity of Several Ions at Infinite Dilution at 25°C[a] (S/cm^2/mol)

Cations[b]	Γ, degrees	Tempco, Degrees	Anions[b]	Γ, degrees	Tempco[c]
H^+	349.8	0.0139	OH^-	198.6	0.018
K^+	43.5	0.0193	Cl^-	76.4	0.0202
Na^+	50.11	0.0220	SO_3^-	71.42	0.020
Ca^{+2}	59.50	0.0230	SO_4^{-2}	80.0	0.022
Mg^{+2}	53.06	0.022	CO_3^{-2}	69.3	0.02
Cu^{+2}	53.6	0.02	HCO_3^-	44.5	—
$(n-Bu)_4N^+$	19.5	0.02	Picrate⁻	30.4	0.025

[a] Data from the measurement of electrolytic conductance, in *Ewing's Analytical Instrumentation Handbook*, 3rd edn., Cazes, J., Dekker, M., Ed., New York, NY, 2005.
[b] Data are on an equivalent basis.
[c] Tempco = $(1/\Gamma)(d\Gamma/dT)$.

TABLE 1.15b
Specific Conductivity of Potassium Chloride Solutions Used for Determination of Cell Constants

Approximate Normality	Weight, KCl in g/l of Solution	Temperature (°C)	k μS/cm
1.0	72.2460	0	65,176
		18	97,838
		25	113.342
0.1	7.4265	0	7138
		18	11,167
		25	12,856
0.01	0.7440	0	773.6
		18	1220.5
		25	1408.8

for Testing Materials (ASTM) Standard D1125–14*. A few specific conductivity values are listed in Table 1.15b.

The cell constant is readily determined using the expression $\theta = k/L$, where k (the specific conductivity) is known from the tabulated values, and L (the conductance) is measured using the cell being calibrated. Note that conductance and resistance are values read by measuring instruments and have the units of Siemens and ohms, respectively. Conductivity and resistivity are intrinsic properties of solutions; they are obtained after application of the cell constant and have the units of Siemens-centimeter and ohm-centimeter, respectively.

CELL DIMENSIONS

As the dimensions of the cell are changed, the cell constant varies as the ratio of d to a. For solutions of low conductivity (about 0.05–200 μS/cm), the electrodes can be placed closer together, giving cell constants in the range of 0.1–0.01 cm^{-1}. Similarly, for more conductive solutions (about 10–20,000 μS/cm), electrode separations can be increased to give cell constants of 1, 10, or sometimes 100 cm^{-1}.

This has the effect of adjusting the actual conductance read by the instrument to a conveniently measured range. Signal-to-noise considerations limit resistance measurements to less than about 2 Mohm. Conductance measurements, which had typically been limited by signal level magnitudes to less than about 0.2 mS, can be as high as 20 mS due to recent improvements in transmitter electronics. This makes accurate conductivity measurements possible over a wider range. For example, conductivity can now be measured from 1.0 to 20,000 μS/cm with an accuracy of ±0.6% of actual reading using a 1.0 cm^{-1} cell. The specifications of the conductivity transmitter should always be checked when choosing the cell constant for a particular conductivity range.

Composition measurement using conductivity is popular in industrial process measurement and control applications, because of the inherent simplicity and reliability of the technique. Cells are available that cover a resistivity range of $1–10^8$ in aqueous electrolytes (Figure 1.15b).

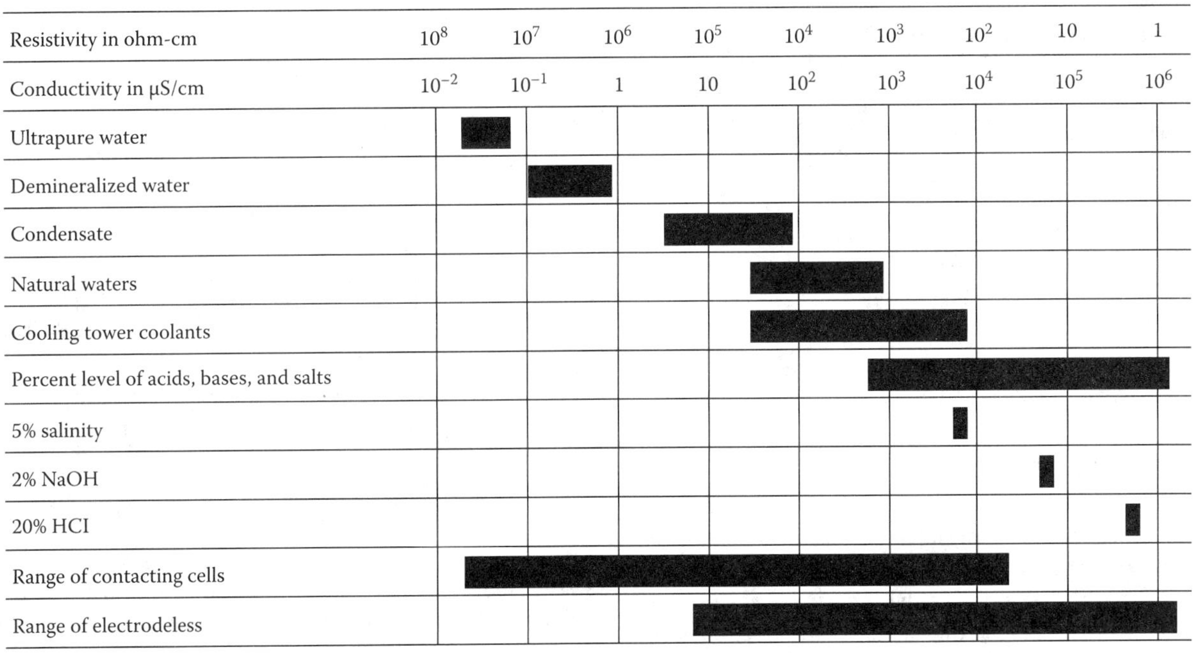

Resistivity in ohm-cm	10^8	10^7	10^6	10^5	10^4	10^3	10^2	10	1
Conductivity in μS/cm	10^{-2}	10^{-1}	1	10	10^2	10^3	10^4	10^5	10^6
Ultrapure water	■								
Demineralized water		■							
Condensate			■						
Natural waters				■					
Cooling tower coolants				■	■				
Percent level of acids, bases, and salts						■	■	■	■
5% salinity							■		
2% NaOH								■	
20% HCl								■	
Range of contacting cells		■	■	■	■	■	■	■	■
Range of electrodeless				■	■	■	■	■	■

FIG. 1.15b
Resistivity/conductivity spectrum of aqueous electrolytes. (From Light, T.S., Chemtech, 4960, 1990.)

---

* ASTM D1125-14, Standard Test Methods for Electrical Conductivity and Resistivity of Water, ASTM International, West Conshohocken, PA, 2014, www.astm.org.

Three types of cells are used: two electrode, four electrode, and electrodeless. The four-electrode and electrodeless cells and their associated instrumentation are shown in Figures 1.15c and 1.15d, respectively.

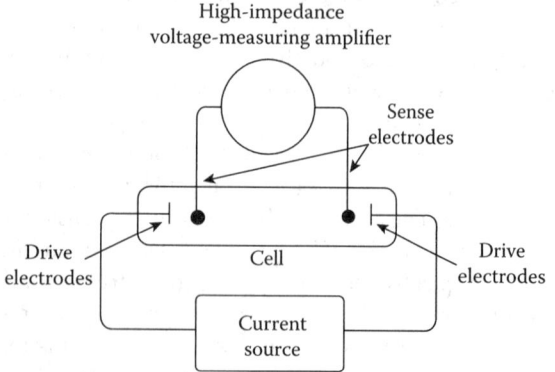

FIG. 1.15c
Four-electrode conductivity measuring circuit.

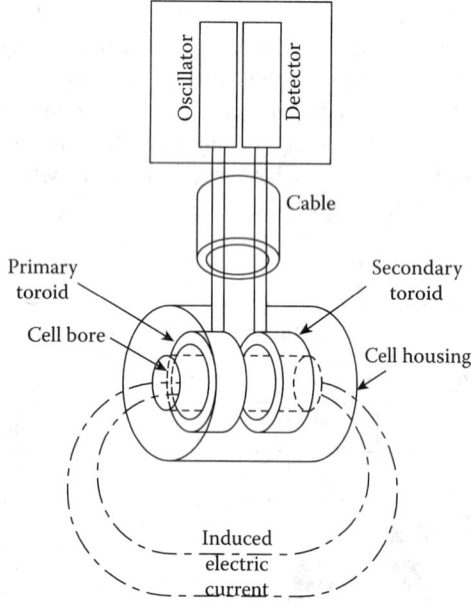

FIG. 1.15d
The electrodeless conductivity cell and instrument. (From Light, T.S. et al., Talanta, 36, 235, 1989.)

TWO-ELECTRODE CELLS

Two-electrode cells are best suited for measurement in clean solutions to avoid errors caused by the formation of coatings and films on the electrodes. In these designs, it is desirable to minimize the interfacial impedance of the electrodes with the solution, because the goal is to measure the bulk conductivity of the electrolyte.

Derivation of Equation 1.15(1) assumed no iR (potential) loss at the electrodes or in the leads to the cell. The cell and instrument are designed accordingly. In order to avoid significant electrolysis, a small-amplitude (usually sinusoidal) waveform having a frequency in the range of 100–1000 Hz is used for excitation.

In addition, the electrode materials are selected to reduce polarization or iR (potential) drops at electrode–solution interfaces.* Ideally, the electrodes are made of platinum and are coated with a layer of platinum black. As the conductance of the measured solution decreases, polarization and coating effects become less significant, and metals other than platinum, such as Monel and titanium, are considered as inert electrode materials.

Particularly noteworthy is the class of two-electrode conductivity applications, called resistivity measurements, which employ titanium two-electrode cells in monitoring the high-purity water used in semiconductor manufacturing, steam turbine applications, and nuclear reactors.

FOUR-ELECTRODE MEASUREMENT

Four-electrode conductivity is useful for high conductance when coating and fouling of electrodes are a concern. Current is imposed across two drive electrodes, and the potential drop through the electrolyte is detected between two points in the cell using two sense electrodes (Figure 1.5e). The sense electrodes are monitored with a high-input impedance, voltage-measuring amplifier to minimize the current drawn and electrode polarization.

Polarization at the drive electrodes has no effect on the measurement, provided the drive voltage is able to maintain the control current through the cell.† This voltage increases

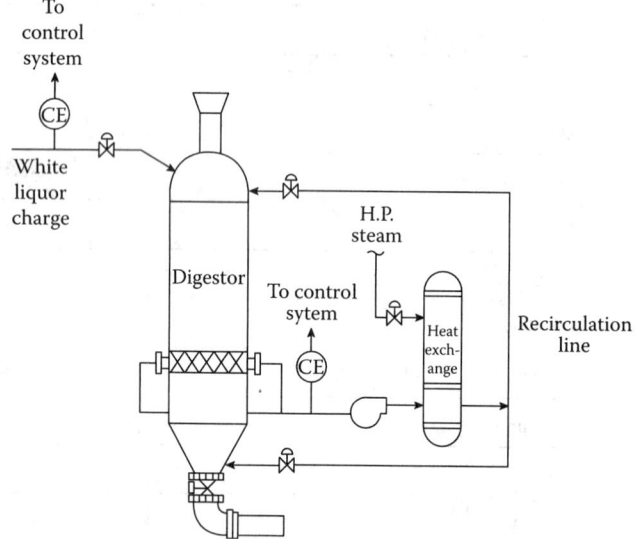

FIG. 1.15e
Example of conductivity measurement used in control of batch digestor for paper making.

---

* Braunstein, J. and Robbins, G. D., Electrolytic conductance measurements and capacitive balance, *Journal of Chemical Education*, 48, 52–59, 1971.
† Conductivity and Conductometry, in *Laboratory Techniques in Electroanalytical Chemistry*, Kissinger, P. T. and Heineman, W. R., Eds., New York: Marcel Dekker, 1984.

with fouling and can be used as a diagnostic tool to signal the user when cleaning is required.

Because of geometrical considerations, four-electrode designs are not suited to probe configurations. Precise measurements require flow-through cells that allow linear distribution of current across the sense electrodes. When used in probes, four-electrode measurements are best applied to set-point control.

Typical applications include measurement of salts, acids, and alkalis in chemical processes in the mining, metallurgy, pulp and paper, and aluminum industries, where samples often contain solids, oils, or other materials that form insulating coatings on the electrodes.*

ELECTRODELESS CELLS

One way to eliminate electrode polarization effects is to eliminate the electrodes. Techniques to do this are referred to as electrodeless conductivity measurement or, alternately, as inductive or toroidal sensing of conductivity.[†,‡] The probe shown in Figure 1.15d consists of two encapsulated toroids. When immersed in the electrolyte, the solution forms a conductive loop shared by both toroids. One toroid radiates an electric field in this loop, and the other detects a small, induced electric current. Practically speaking, the two toroids form a transformer whose coils are interconnected by the resistance of the electrolyte.

The radiated field is typically 20 kHz, and the induced current, which is proportional to the conductivity, is amplified, rectified, and displayed. These probes are encapsulated in nonconductive, temperature-stable, and chemically resistant materials such as the fluorocarbon polymers.

Electrodeless sensors can be completely covered by a nonconductive coating without a noticeable loss of signal. It is not until coating becomes thicker that signal is lost; when the center of the toroid is plugged, the entire conductivity signal is lost. In spite of these limitations, electrodeless conductivity is the most fouling resistant conductivity measurement technology available.

The effective cell constant of an electrodeless sensor can be changed by wall effects. Wall effects are due to the proximity of the sensor to the walls of a pipe or vessel. A metal wall near the sensor provides a conductive path and increases the measured conductivity, while proximity to plastic pipe reduces the volume of electrolyte measured by the sensor and reduces the measured conductivity. A toroidal sensor should be mounted at a distance of three-fourth of the toroid diameter to eliminate wall effects. When a sensor cannot be mounted to avoid wall effects, the measurement should be calibrated with the sensor in the process using a grab sample to take these effects into account.

---

* Anderson, F. P., Brookes, H. C., Hotz, M. C. B., and Spong, A. H., Measurement of electrolytic conductance with a fourelectrode alternating current, *Journal of Scientific Instruments* (*Journal of Physics E*), 2(2), 491–502, 1969.
† Light, T. S., *Chemtech*, August, 4960–4501, 1990.
‡ Light, T. S., McHale, E. J., and Fletcher, K. S., Electrodeless conductivity, *Talanta*, 36, 235–241, 1989.

The temperature element used in an electrodeless sensor is most often encapsulated by plastic within the sensor, and as a result, the response of the temperature measurement can have a time constant of 5 min or more. When there is a significant change in temperature, there will be an error in the conductivity measurement until the sensor and its temperature element reaches the temperature of the process. In applications where this creates a problem, a separate RTD (resistance temperature detector) and thermowell can be used to provide faster temperature response, or in some cases, when digital communications are used, the temperature measurement from a nearby temperature transmitter can be input and used by the conductivity transmitter for temperature compensation.

MEASUREMENT APPLICATIONS

Modern conductivity analyzers, with on-board computers, provide essential features such as temperature correction to reference values, digital display of concentration data from measured values, controller functionality, self-diagnostics, and calculations such as water subtraction and percent rejection.

CONCENTRATION MEASUREMENTS

Temperature compensation and concentration computation data for common acids, bases, or salts are often embedded in the instrument's software. Data for less common materials may often be loaded by the end user. Concentration can be derived from the conductivity of an electrolyte, when there is a significant increase or decrease in conductivity with increasing concentration.

While concentration measurements are most often applied to a single electrolyte in solution, they can also be applied to mixtures of electrolytes, when the ratio of the components of the mixture is constant. Concentration measurement is also used in batch reactions, where the progress of the reaction is accompanied by a significant increase or decrease in conductivity. Output signals, either digital or analog, are used in control systems for measurement and control of processes such as boiler feedwater or the monitoring of gas scrubber solutions and pickling and plating liquors.

HIGH-PURITY WATER MEASUREMENTS

A special class of temperature compensation has evolved for measuring high-purity water. Monitoring of high-purity water is used in semiconductor processing. In power and pharmaceutical applications, it is necessary to distinguish the temperature coefficient of pure water from that of ionic purity.

Today's *intelligent* instruments can measure both conductivity or resistivity and temperature. They can therefore compute the value of conductivity or resistivity of pure water at the measured temperature. They can also calculate the measurement contribution due to the impurities and output the conductivity or resistivity of the process water, referenced to a standard temperature, or the concentration of impurity after subtraction of the contribution due to the water (e.g., as total dissolved solids [TDSs] or particles per million [ppm] NaCl).

Multiple sensors are used in conjunction with ion-exchange columns or reverse osmosis systems to monitor and control inlet and outlet resistivity across the bed and the breakthrough of unwanted ions and, as a predictive tool, to signal the need for bed regeneration. If the sample temperature is not controlled, water temperature compensation is necessary for accurate measurements for conductivity values at or below 1 μS/cm, or resistivity readings at or above 1 Mohm/cm. A variant of this temperature compensation, called cation conductivity temperature compensation, is used for high-purity water samples, such as the effluent of a cation exchange bed, which has acid as the major impurity.

CORROSIVE AND FOULING APPLICATIONS

The other extreme, that is, highly conductive solutions—those that are highly concentrated, corrosive, and contaminated with fouling materials—is best measured with the electrodeless designs. Here, coatings only affect the measurement response to the extent that they alter the geometry, and hence the cell constant, of the probe. The relatively large size of these probes renders this effect small or negligible. Examples of applications include online analysis of oleum in H_2SO_4–SO_3–H_2O, measurement and control of alkalinity and solution strength in many industries using lime slaking, industrial dishwashing rinse control, and gas scrubber solution concentration control.*

Pulp processing uses extremely corrosive chemicals, high temperatures and pressures, and samples entrained with solids and particles; such processing provides an example of how conductance analyzers are applied. Figure 1.15e shows how the conductivity sensors are integrated into a continuous Kraft digester commonly used in paper pulp making.†

Temperature, flow, and alkali concentration data are used by the control systems to control the uniformity of the pulp by manipulating the residual alkali strength in the cooking liquor in response to changing feed properties such as wood chip composition, species, chip moisture, and uniformity of concentration of makeup chemicals. For a summary of conductivity analyzer applications, refer to Table 1.15c.

TABLE 1.15c
Conductivity Applications

Process	Application (Usage) and Comments
Chemical streams	To measure and control solution strength.
Steam boilers	*Blowdown* is a method of lowering the amount of dissolved solids in a boiler by dilution. To control buildup of dissolved solids to prevent scaling and corrosion.
	Condensate return is usually checked for quality before being returned to the boiler. If out of limits, it is dumped.
Waste streams	A means of determining the amount of dissolved salts being discharged.
Cooling towers	*Bleed control* is a method of reducing the total dissolved solids in a tower by dilution (similar to blowdown). To prevent scaling and corrosion. For bleed control, the electrodeless conductivity system works best to minimize maintenance and failure.
Fruit peeling	Strong caustic is used, and its strength can be determined by conductivity.
Rinse water	Plating shop running rinse water is monitored for dissolved salts—a method of reducing water consumption.
Semiconductor rinse water	Requires ultrapure water, usually measured in megohms/centimeter.
Interface determination	Usually used in food processing, e.g., dairy and brewing. Most commonly used in cleaning in place (CIP); interfaces in pipes are easily determined and can be diverted by valves controlled by conductivity.
Demineralizer output	Determination of ion-exchange exhaustion.
Reverse osmosis	Efficiency of reverse osmosis (RO) operations is usually monitored by comparing inlet and outlet conductivity or TDS ratio (cell 1/cell 2). The inlet conductivity is installed upstream of the RO feed pump to avoid high-pressure requirements. Also, abnormal readings can be used to diagnose membrane fouling, improper flow rate, membrane failure, etc.
Desalination	Similar to reverse osmosis and demineralization process.
Deionization process	Conductivity or resistivity measurement provides capability for monitoring and controlling the acid and caustic dilution. Regeneration of deionizers requires consistent application of acid and caustic to obtain repeatable results. Savings is provided by consistent regeneration, which assures deionized water availability, less frequent regeneration, long resin life, and conservation of costly reagents.
	More precise control can be obtained by using conductivity ratio measurement. A comparison of inlet and outlet (ratio of cell 1/cell 2) conductivity across the bed can determine the unwanted ions and the need for bed regeneration, which can compensate and control for variations in mineral concentration of feedwater.
Ion exchange	Occasionally loses resin. If a resin bead or fines are trapped between the electrodes of a cell, it is shorted and produces a very low resistivity (or high conductivity) reading. This feature is a great help in troubleshooting.

* Light, T. S., *Chemtech*, August, 4960–4501, 1990.
† Gingerella, M. and Jacanin, J. A., Is there an accurate low conductivity standard solution? *Cal Lab*, July–August, 29–36, 2000. http://www2.emersonprocess.com/siteadmincenter/PM%20Rosemount%20Analytical%20Documents/LIQ_TWP_Accurate_Low_Conductivity_Standard_Solution.pdf.

CALIBRATION AND MAINTENANCE

Calibration of Conductivity Sensors

Conductivity measurements can be calibrated using conductivity standards or online, by grab sample analysis. The conductivity sensor should be given sufficient time to reach the temperature of the standard solution, or in the case of online calibration, the sensors should be calibrated only after the process has been at a stable temperature for some time. By doing so, temperature compensation errors are eliminated, because the temperature element in the conductivity sensor will have had time to reach the same temperature as that of the standard or the process. This is especially important when using electrodeless sensors, which typically have a much larger mass than contacting sensors, and therefore require more time to reach thermal equilibrium, up to 30 min in some cases.

Conductivity or resistivity measurements in high-purity water applications cannot be calibrated by using standards or calibrated online by using grab samples. This is because of the extreme sensitivity of the samples to contamination by trace amounts of electrolytes and even to atmospheric CO_2. It has been argued that the accuracy of calibrations, when using conductivity standards of less than 100 µS/cm, can be questionable.* A subsequent study showed that two standards, both 5 µS/cm and below, can remain within their specified accuracy if packaged, handled, and stored correctly.†

In general, the conductivity sensors used for high-purity water applications are calibrated by determining the cell constant in a conductive standard solution (usually done by the manufacturer). The input to the analyzer is calibrated with precision resistors (also usually done by the manufacturer). In the field, the predetermined cell constant of the conductivity cell is entered into the software of the analyzer by the user. Further calibrations are done using a certified reference conductivity system.

Maintenance of Conductivity Cells

Conductivity measuring systems may be designed to be trouble free and produce reliable measurements; however, some maintenance is required, especially for the electrodes. In addition, the cell may require periodic cleaning depending on the type of application, the quality of the water passing through it, and the type of cell used. Some types of contaminants may not interfere directly with the measured conductivity—for example, organic materials, rust, and suspended solids—but may form deposits on the electrode surfaces. In most cases, these surfaces can be cleaned with a bristle brush and a weak detergent solution.

Problems may also occur in hard water applications, where gradual formation of scale will reduce the active area of the electrodes, which over a period of time will result in an apparent decrease in conductivity. For this type of fouling, simple brush cleaning is insufficient, as it will not remove scale from the cell. To remove the scaling, the electrode should be treated with a 10% solution of formic or hydrochloric acid. The presence of bubbles will indicate that the scale is being dissolved. It takes about 2 or 3 min and is complete when the bubble formation ceases. Then the cell should be thoroughly rinsed to remove all traces of acid before it is returned into the process.

Cells with stainless steel electrodes are generally used in applications where a low conductivity is combined with low concentrations of organic contamination. For high-purity water applications, titanium electrodes are used, because of their better performance characteristics in very low conductivity samples.

In applications where fouling or corrosion is anticipated, the need for cleaning can best be minimized by the use of electrodeless conductivity sensors.

NEW DEVELOPMENTS

Digital Conductivity Sensors

Digital conductivity sensors provide a digitized raw conductivity and temperature signal to the conductivity transmitter, which provides improved signal integrity, especially over long cable lengths. It also makes installation more convenient by using connectors, which avoids having to connect individual wires to a terminal strip.

Digital conductivity sensors are also smart in the sense that they have a memory, which allows them to be precalibrated by the manufacturer or in the lab or shop before installation. Other information stored in the memory can include calibration dates, changes in calibration constants, exposure to high temperature, as well as serial code and manufacturing information. This information aids in maintenance and troubleshooting, especially when combined with sensor maintenance software.

New Conductivity Transmitter Developments

Conductivity transmitters have been developed with larger, often multilingual displays and more descriptive menu systems and keypads, which has made configuration and troubleshooting easier. Calibration at the local display is a stepwise process using prompts. Diagnostics in transmitters include overrange and negative reading alarms, as well as alerts related to temperature measurement and problems with the transmitter's electronics and memory. The use of a larger display makes it possible to more fully describe problems and can even offer corrective action.

---

* Gingerella, M. and Jacanin, J. A., Is there an accurate low conductivity standard solution? *Cal Lab*, July–August, 29–36, 2000. http://www2.emersonprocess.com/siteadmincenter/PM%20Rosemount%20Analytical%20Documents/LIQ_TWP_Accurate_Low_Conductivity_Standard_Solution.pdf.
† Barron, J. and Ashton, C., Stable low level conductivity standards Cal Lab. *International Journal of Metrology*, 12(1), 1–8, 2005. http://reagecon.com/pdf/technicalpapers/stable-low-levelcond-stds-v11.pdf.

These advances have been applied to two-wire transmitters and line-powered transmitters. In some two-wire transmitters, a second externally powered 4–20 mA output has been added to provide a second conductivity output with a different range or to output a secondary measurement such as temperature.

Line-powered transmitters have seen even more developments. Data logging can also be available to track alarms and changes to the calibration constants, and log runtime data including conductivity and secondary data such as temperature, raw conductivity, temperature, RTD resistance, and analog outputs. In many cases, these data can be plotted on screen or downloaded to a portable memory device for a more detailed examination using a PC or laptop. Even though a plant or batch historian may be available, data logging can provide a better tool for troubleshooting because it captures secondary measurements, which can show the root cause of a measurement problem. When digital conductivity sensors are used, the data stored in the sensor can be viewed on the analyzer display.

Line-powered transmitters have become multiparameter and can include one or more analytical measurements in addition to conductivity. In some cases, an additional measurement can be provided by 4–20 mA input to the transmitter, which allows it to receive, scale, and even power pressure, temperature, flow, or any other transmitter with a 4–20 mA output. This allows the transmitter to provide a more complete solution beyond a simple conductivity measurement.

The input from a temperature transmitter can be used for temperature compensation of a toroidal conductivity measurement. This avoids errors resulting from the slow temperature response of the temperature element mounted inside the polymeric body of the toroidal sensor.

Using a transmitter with two conductivity inputs makes it possible to apply a single analyzer in monitoring reverse osmosis, measuring the feedwater and permeate, and calculating the % passage and % rejection. In some applications, it is useful to calculate the conductivity ratio. In high-pressure steam plants, inferred pH can be calculated from straight conductivity and cation conductivity, using a line-powered transmitter.

In addition to monitoring, analytical transmitters can provide control functionality. They have more than one analog output, and one or more of these can be assigned to PID (proportional, integral, and derivative) control for the measurements made. The user display and keypad are used for setup and tuning of the PID loop, and if a data logger is present, it can capture the PV of the control application and the control output. If the transmitters have relays, discrete control functions can include simple high or low relay control, or possibly relay control with a delay timer, typically used to allow time chemical additions to mix. More sophisticated control can include pulse width modulation (see Figure 1.15f).

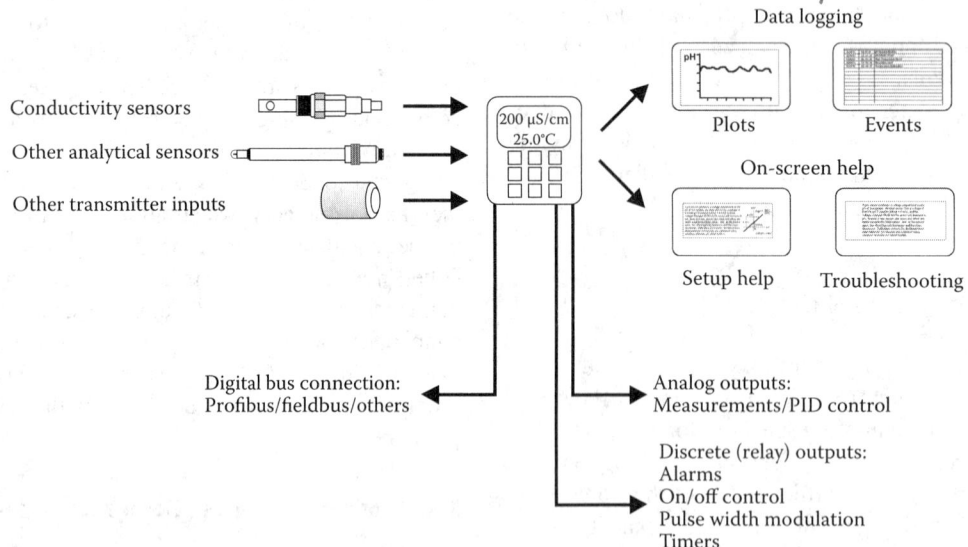

FIG. 1.15f
Features available in line-powered transmitters.

Smart Conductivity Transmitters

The term "smart" is used to describe a transmitter that has a digital output, in addition to or in place of an analog 4–20 mA output. This allows a user to have access to the transmitter's configuration parameters, diagnostic information, secondary measurements, such as raw conductivity and temperature, and control parameters if they are present. The most common digital protocols used are HART®, PROFIBUS®, and FOUNDATION® fieldbus. HART typically employs a digital signal superimposed on a 4–20mA output, while PROFIBUS and FOUNDATION fieldbus use digital physical layers.

One of the advantages of using smart transmitters is their ability to transmit more than just a conductivity measurement. In the case of reverse osmosis applications using dual conductivity measurements, smart transmitters can often transmit the conductivity of feedwater and permeate, as well as the % passage and the % rejection.

Asset Management

The advantage of smart transmitters to a conductivity user is the ability to access a transmitter remotely at a PC to check for diagnostic alarms, review configuration parameters, and even do calibrations. While PC software applications have been developed that simply list parameters, the preferred presentation of transmitter data is a Windows environment with the parameters organized according to function. The first of the two technologies used to create these windows is electronic device descriptive language (EDDL), which is a file created for a certain model of transmitter by its manufacturer that contains instructions for an asset management program to create windows. The second technology is FDT/DTM, where the field device tool (FDT) provides a frame application to display the information, and a device type manager (DTM), which is an executable program for each type of transmitter that creates the windows for it.

Beyond the ease of accessing and quickly reviewing transmitter, asset management software can also provide a historian for changes to the transmitter's parameters, including its calibration constants. It can also be used to create a configuration template and then download it to a transmitter, making the configuration of a number of transmitters faster and easier than using the local displays of each (see Figure 1.15g).

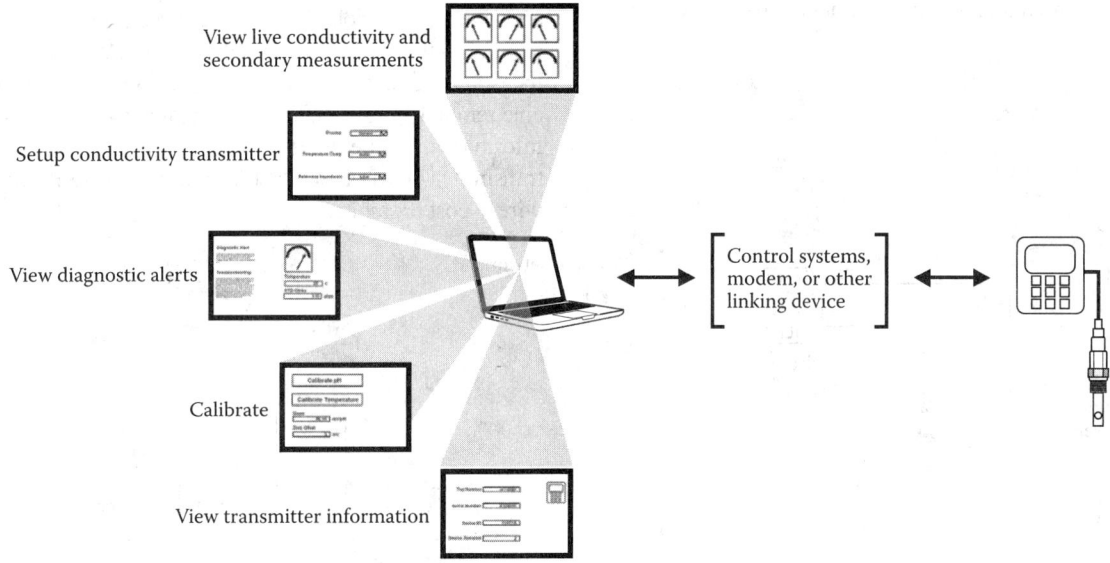

FIG. 1.15g
Asset management.

Wireless HART™ Transmitters

Wireless HART is part of the new version 7 of HART that defines wireless communications using wireless transmitters in a self-organizing, mesh network configuration. This provides multiple paths for the wireless signal among the transmitters in the network and results in a more reliable network, less susceptible to a loss of the wireless signal due to the failure or blockage of any single transmitter. The network communicates to a gateway, which communicates to the control system using Modbus, Ethernet, or even wirelessly. The advantage of wireless transmitters is that they can be installed without the time and expense of running power and signal wire to the sample point.

A wireless conductivity transmitter is battery powered, and depending on the transmission rate, battery life can be as long as 4 years at a one minute update rate. Wireless conductivity transmitters have essentially the same feature set as a two-wire transmitter.

When power is available at the sample point, a line-powered transmitter can be used in the network using a wireless HART adaptor, which can connect from one to three wired transmitters to the network. This makes it possible to use all the advanced features of a line-powered transmitter, such as multiple analytical and nonanalytical measurements and control, while maintaining communications with the control room and asset management software wirelessly (see Figure 1.15h).

CONCLUSION

Selecting the right conductivity cell includes having information on the cell constant used by the analyzer, the conductivity range, the materials of construction, selected to resist corrosion, and the appropriate mounting of the sensor.

When designing a conductivity measurement system, the first consideration is the conductivity range of the sample. For applications below 20,000 μS/cm, contacting conductivity sensors are typically used. For higher conductivity ranges, or in samples that can foul or corrode the metal electrodes, electrodeless detectors are the best choice. For conductivity measurements at 1 μS/cm and below, contacting conductivity sensors should be used in conjunction with a conductivity analyzer, which is provided with high-purity temperature compensation. It never hurts to consult sensor suppliers as to the suitability of their various designs to a particular process application.

Among the new developments for conductivity, digital conductivity sensors streamline sensor connections and improve signal integrity. On-board sensor memory provides calibration, process exposure, and factory history, which aids in maintenance and troubleshooting.

Conductivity transmitters can bring more capability to the sample point, especially in the form of multiparameter transmitters with added analytical or other measurements, and PID and discrete control. When the conductivity transmitters are smart, all the measurements, diagnostics, and parameters for measurement configuration and controller tuning are available remotely, organized by an asset management system. This information can be made available wirelessly, using a wireless transmitter or wireless adaptor, in cases where running signal wire is cost prohibitive.

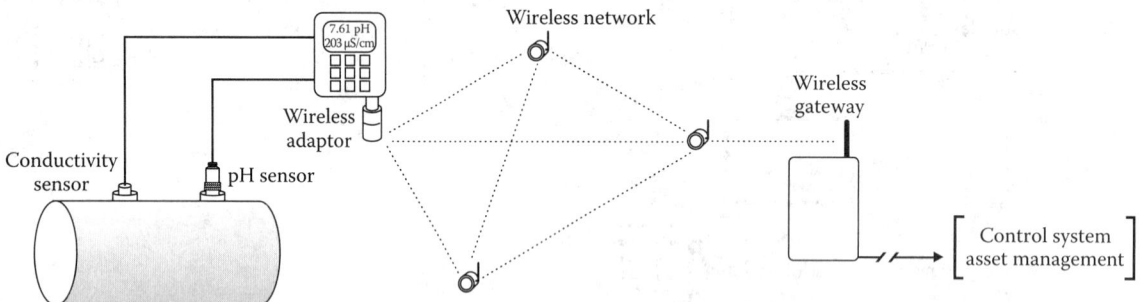

FIG. 1.15h
Remote conductivity and pH monitoring using a line-powered multiparameter transmitter and wireless adaptor.

1.15 Conductivity Measurement

SPECIFICATION FORMS

When specifying the required features and the operating conditions under which conductivity sensors are to be uses, utilize ISA Form 20A2381, and when specifying conductivity transmitters, use ISA Form 20A2342. The following are the forms that are used with the permission of the International Society of Automation.

#	Field		#	Field	
1	RESPONSIBLE ORGANIZATION	CONDUCTIVITY SENSOR w/wo INSERTION ASSEMBLY Device Specification	6	SPECIFICATION IDENTIFICATIONS	
2			7	Document no	
3	(ISA)		8	Latest revision	Date
4			9	Issue status	
5			10		
11	BODY HOLDER OR FITTING		60	INSERTION ASSEMBLY continued	
12	Body/Fitting type		61	Chamber wetted material	
13	Process conn nominal size	Rating	62	Compression ferrule matl	
14	Process conn termn type	Style	63		
15	Body/Fitting material		64		
16	Flange material		65	PERFORMANCE CHARACTERISTICS	
17	Seal/O ring material		66	Max press at design temp	At
18			67	Min working temperature	Max
19			68	Cell constant accuracy	
20			69	Measurement LRL	URL
21	SENSING ELEMENT		70	Temp compensation LRL	URL
22	Sensor type		71	Max temp response time	
23	Construction style		72	Max sensor to receiver lg	
24	Nominal cell constant		73		
25	Temperature sensor type		74		
26	Bore diameter		75		
27	Insertion/Immersion lg		76		
28	Temp sensor location		77		
29	Insulator material		78		
30	Electrode material		79		
31	Ext shaft/support matl		80		
32	Sheath material		81		
33			82		
34			83		
35			84		
36	LEAD WIRE AND EXTENSION		85	ACCESSORIES	
37	Extension type		86	Conductivity standard	
38	Cable length		87		
39	Max cable operating temp		88		
40	Signal termination type		89		
41	Cable jacket material		90		
42			91		
43	CONNECTION HEAD		92	SPECIAL REQUIREMENTS	
44	Housing type		93	Custom tag	
45	Enclosure type no/class		94	Reference specification	
46	Signal termination type		95	Compliance standard	
47	Cert/Approval type		96	Calibration report	
48	Mounting location/type		97		
49	Enclosure material		98		
50			99		
51	INSERTION ASSEMBLY		100	PHYSICAL DATA	
52	Assembly type		101	Estimated weight	
53	Isolation valve style		102	Overall height	
54	Process conn nominal size	Rating	103	Removal clearance	
55	Process conn termn type	Style	104	Signal conn nominal size	Style
56	Purge conn nominal size	Style	105	Mfr reference dwg	
57	Insertion/Immersion length		106		
58	Valve body material		107		
59	Valve seat material		108		

#	CALIBRATIONS AND TEST		INPUT OR TEST		OUTPUT	
110						
111	TAG NO/FUNCTIONAL IDENT	MEAS/SIGNAL/SCALE	LRV	URV	LRV	URV
112		Measurement-Output signal				
113		Temp-Output signal				
114						
115						
116						
117						

#	COMPONENT IDENTIFICATIONS		
118			
119	COMPONENT TYPE	MANUFACTURER	MODEL NUMBER
120			
121			
122			
123			
124			
125			

Rev	Date	Revision Description	By	Appv1	Appv2	Appv3	REMARKS

Form: 20A2381 Rev 0 © 2004 ISA

pH/ORP/CONDUCTIVITY/RESISTIVITY TRANSMITTER/ANALYZER/MONITOR Device Specification

#				#		
1	RESPONSIBLE ORGANIZATION			6	SPECIFICATION IDENTIFICATIONS	
2				7	Document no	
3				8	Latest revision	Date
4				9	Issue status	
5				10		

#	TRANSMITTER OR ANALYZER		#	PERFORMANCE CHARACTERISTICS	
11			56		
12	Housing type		57	Measurement accuracy	
13	Input sensor type		58	Temperature accuracy	
14	Input sensor style		59	Measurement LRL	URL
15	Output signal type		60	Temp compensation LRL	URL
16	Enclosure type no/class		61	Minimum span	Max
17	Control mode		62	Min ambient working temp	Max
18	Local operator interface		63	Contacts ac rating	At max
19	Characteristic curve		64	Contacts dc rating	At max
20	Digital communication std		65	Max sensor to receiver lg	
21	Signal power source		66		
22	Temperature sensor type		67		
23	Quantity of input sensors		68		
24	Preamplifier location		69		
25	Contacts arrangement	Quantity	70		
26	Integral indicator style		71		
27	Signal termination type		72		
28	Cert/Approval type		73		
29	Mounting location/type		74		
30	Failure/Diagnostic action		75		
31	Sensor diagnostics		76	ACCESSORIES	
32	Data history log		77	Remote indicator style	
33	Temperature compensation		78	Indicator enclosure	
34	Calibration function		79	Communicator style	
35	Enclosure material		80		
36			81		
37			82		
38			83	SPECIAL REQUIREMENTS	
39			84	Custom tag	
40			85	Reference specification	
41			86	Compliance standard	
42			87	Software configuration	
43			88	Software program	
44			89		
45			90		
46			91		
47			92	PHYSICAL DATA	
48			93	Estimated weight	
49			94	Overall height	
50			95	Removal clearance	
51			96	Signal conn nominal size	Style
52			97	Mfr reference dwg	
53			98		
54			99		
55			100		

#	CALIBRATIONS AND TEST		INPUT OR TEST			OUTPUT OR SCALE	
110							
111	TAG NO/FUNCTIONAL IDENT	MEAS/SIGNAL/SCALE	LRV	URV	ACTION	LRV	URV
112		Meas-Analog output 1					
113		Meas-Analog output 2					
114		Meas-Digital output					
115		Meas-Scale					
116		Temp-Scale					
117		Temp-Digital output					
118		Meas setpoint 1-Output					
119		Meas setpoint 2-Output					
120		Meas setpoint 3-Output					
121		Meas setpoint 4-Output					
122		Failure signal-Output					
123							

#	COMPONENT IDENTIFICATIONS		
124			
125	COMPONENT TYPE	MANUFACTURER	MODEL NUMBER
126			
127			
128			
129			

Rev	Date	Revision Description	By	Appv1	Appv2	Appv3	REMARKS

Form: 20A2342 Rev 0 © 2004 ISA

Definitions

Anion is a negatively charged ion such as chloride, carbonate, or sulfate.

Cation is a positively charged ion such as sodium, potassium, or hydrogen.

Cation conductivity is the conductivity of a water sample that has been passed through a column containing a cation exchange resin, which replaces cations with hydrogen ions. It is used in steam plant water analysis.

Inferred pH is a calculation that uses the standard and cation conductivity of a water sample in a steam plant that provides an estimate of the pH of the water based on models developed from the type of water treatment used.

% Passage % passage = 100 – % rejection; it is a measure of how much ionic contamination passes through the membrane with the permeate.

Raw conductivity is a measured conductivity that is not compensated for temperature changes.

% Rejection % rejection = 100 × [C(feedwater) – C(permeate)]/C(feedwater), where C() is conductivity; it is a measure of how much ionic contamination is being removed by reverse osmosis.

Reverse osmosis is a technique for water purification that forces the feedwater at elevated pressure through a semipermeable membrane that holds back ions and other molecules, resulting in permeate that is purified and less conductive than the feedwater.

Abbreviations

HART   stands for Highway Addressable Remote Transmitter, which is the protocol of the HART Communication Foundation.
PID   proportional, integral, and derivative
RTD   resistance temperature detector

Organizations

Electronic Device Descriptive Language (EDDL) www.eddl.org/Pages/default.aspx

FDT/DTM www.fdtgroup.org/

Fieldbus FOUNDATION™ www.fieldbus.org/

HART Communication Foundation www.hartcomm.org/

PROFIBUS® PROFIBUS International www.profibus.com/

Bibliography

Bevilacqua, A. C., Ultrapure water: The standard for resistivity measurements of ultrapure water, *Semiconductor Pure Water and Chemicals Conference*, Santa Clara, CA, March 2–5, 1998. http://www.snowpure.com/docs/thornton-upw-resistivity-measurement.pdf.

Gray, D. M. and Bevilacqua, A. C., Cation conductivity temperature compensation, *International Water Conference*, Pittsburgh, PA, November 1997. http://us.mt.com/dam/mt_ext_files/Editorial/Generic/4/Paper-THOR-Cation-Cond-Temp-Bev-Gray-11–97_Editorial-eneric_1161617581924_files/cation_cond_tempcompensation.pdf.

Morash, K. R., Thornton, R. D., Saunders, C. H., Bevilacqua, A. C., and Light, T. S., Measurement of the resistivity of ultrapure water at elevated temperatures, *Ultrapure Water Journal*, 11(9), 18–26, December 1994. http://www.gatewayequipment.com/whitepapers/resistivity_high_elevations.pdf.

Digital Conductivity Sensors

Gray, D. and Raabe, S., Benefits of integrated conductivity sensors in water treatment systems, *71st Annual International Water Conference*, San Antonio, TX, October, 2010. http://www.waterworld.com/articles/iww/print/volume-11/issue-1/feature-editorial/advanced-conductivity-measurement-for-water-treatment-systems.html.

Memosens, http://www.memosens.org/en/.

Inferred pH

Covey, J., Inferring pH from conductivity and cation conductivity, *Power Magazine*, November 1, 2007. http://www.power-eng.com/articles/print/volume-111/issue-11/features/inferring-ph-from-conductivity-and-cation-conductivity.html.

http://www2.emersonprocess.com/siteadmincenter/pm%20rosemount%20analytical%20documents/liq_ads_4900–87.pdf.

Conductance Data for Commonly Used Chemicals

http://www2.emersonprocess.com/siteadmincenter/PM%20Rosemount%20Analytical%20Documents/LIQ_MAN_6039_Conductance_Data_Commonly_Used_Chemicals.pdf.

1.16 Consistency Measurement

A. BRODGESELL (1969, 1982)   **M. H. WALLER** (2003)
B. G. LIPTÁK (1995, 2017)

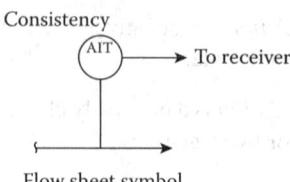

Flow sheet symbol

Types	A. Blade, rotary static (a fixed probe or blade) or moving (blades, rotating disks, or propeller)
	B. Optical fiber
	C. Microwave
	D. Ultra high-frequency electromagnetic wave
	E. Radiological, gamma
	F. *Laboratory*, grab sample testing
Element materials	Stainless steel and titanium
Normal design temperature	Up to 120°C (250°F)
Normal design pressure	Up to 125 psig (8.6 bars)
Range	0.01%–15% consistency
Sensitivity	0.01%–15% consistency
Repeatability	0.5% of reading
Inaccuracy	Function of empirical calibration, usually 1% of reading
Cost	Laboratory units, $500–$3,000; continuous industrial units, $5,000–$10,000 for most; higher for gamma radiation types
Partial list of suppliers	ABB (C): http://www.abb.us/industries/db0003db001873/0b91c983670d5790c12579d500375c58.aspx
	Aquar System (C): http://www.aquar-system.com/catalog/pulp-consistency-meter/
	Berthold Technologies: www.berthold.com
	BTG (A, B): http://www.btg.com/subapplication.asp?cat=prod&langage=1&appli=7&numProd=461 and http://www.btg.com/en/products/products-by-type/inline-instrumentation/consistency-detail/279/
	CCS Scientific (F): http://www.cscscientific.com/consistometer/bostwick-consistometer
	Cerlic (B): http://www.cerlic.com/ctx-consistency-measurements.aspx
	Coliy Technology GmbH: http://www.coliy.de/CA_c300.pdf
	DeZurik/Copes-Vulcan (A): http://www.dezurik.com/products/product-line/other-products-specialty-valves/accutrax-consistency-transmitters-sbc/11/48/
	Lorentzen and Wettre, ABB (A, B, C): http://www.lorentzen-wettre.com/index.php?page=shop.product_details&flypage=flypage.tpl&product_id=86&category_id=8&option=com_virtuemart&Itemid=53&lang=en and http://l-w.com/produkt/optical-consistency-transmitters/
	Metso Automation (B,C) http://www.metso.com/Automation/magazinebank.nsf/Resource/automation_0305_p26-27/$File/automation_0305_p26.pdf
	MX Servicios Para (A, B): http://www.mxchile.com/en/products/specialized-process-instrumentation/consistency-transmitters/
	Ronan Engineering (E): http://www.ronanmeasure.com/ppa5.html
	TEMCO Engineering (A): http://www.temcoeng.com/products/product-detail?ID=1
	Thermo Electron Corp. (E): http://www.cerlic.com/ctx-consistency-measurements.aspx
	Thermo Scientific (E): http://www.thermoscientific.com/en/product/consistencypro-density-gauge-consistency-pulp-paper.html and http://thermo.com.cn/Resources/200802/productPDF_14149.pdf
	Thompson Equipment Co. (A,C): http://www.teco-inc.com/site56.php
	Toshiba (C): http://www.toshiba.com/ind/product_display.jsp?id1=9&id2=90&id3=158

INTRODUCTION

Consistency is the term used to describe the solid content of a solid–liquid mixture. It is calculated as the ratio of the dry weight of the solids, divided by the total weight of the mixture. Mixtures having under 5% consistency are considered to be low, between 5% and 15%, medium, and over 15% as high consistency mixtures. Knowing the specific gravity of the solids and of the liquid phase, consistency can be determined (or approximated) on the basis of density measurement. The various density sensors are covered in full detail in Chapters 6.1 to 6.9 in the first volume of this handbook.

By definition, consistency is expressed as a percentage by dividing the mass of solid material by the total mass of a wet sample, resulting in units of mass per unit mass. Therefore, consistency is the weight percentage of the dry solid material in a slurry or pulp. It is calculated by the following equation:

$$C = \left(\frac{F}{W}\right)(100) \qquad 1.16(1)$$

where

C is the consistency of the pulp or slurry in %
F is the the weight of the dry solid or fibrous material in the sample
W is the total weight of the sample

When measured mechanically, consistency is the resistance to deformation or shear by fibrous or slurry materials, and thus is related to apparent viscosity. Such materials include wood pulp, dough, tomato paste, paint, gelatin, or drilling mud. This chapter discusses those methods that were used for the measurement of consistency in the paper and other industries. In laboratories, consistency is measured using a gravimetric method described in TAPPI Test Method T 240 om-93.* General industrial methods used for consistency measurement are described in this chapter and in the Bibliography at the end of the chapter.

As far as the paper industry applications are concerned, in order to have good control over the basis weight of the paper product, it is necessary to maintain the consistency of the feed constant. An increase in temperature or in the inorganic materials content will reduce the viscosity, and thus the apparent consistency, while increasing freeness (ability of the suspension to release water). In addition, increasing fiber length, or increasing pH, will also increase the apparent consistency. Pipeline velocity also can influence the consistency readings of mechanical sensors and, therefore, it is advisable to measure consistency in turbulent locations at constant flow rate.

In addition to density, which was mentioned earlier, the measurement of consistency can also be based on the detection of turbidity (Chapter 1.68) and infrared (Chapter 1.32) measurements.

The measurement of consistency is also related to the measurement of density (see Chapter 6.1 in Volume 1), suspended solids, and turbidity (Chapter 1.68). The operating principles of consistency analyzers are also discussed in other chapters, such as nuclear radiation in Chapter 7.8 in Volume 1 and optical and infrared in Chapters 1.23 and 1.32.

SENSOR DESIGNS

Some fifty years ago, consistency was considered to be a mature measurement, almost totally relying on mechanical devices for shear force measurements in the 2%–5% range. Today, we still rely on mechanical shear force measurements, but in addition, we have devices utilizing light scattering, light transmission, nuclear radiation, radio waves, microwave, and ultra-high-frequency sensors. Mechanical devices might be categorized as either static (a fixed probe or blade) or moving (blades, rotating disks, or propeller). The development of high-intensity light-emitting diodes 25 years ago allowed further development of optical consistency sensors. These gauges relied on scattered or transmitted light for measurement up to 4% consistency.

Several other approaches to consistency measurement have been attempted with varying degrees of success. Gamma attenuation devices measure consistency on the basis of density changes. Recent developments in plastic scintillation detector technology have improved sensitivity and stability. Because the density difference between fibers and water is very small, high sensitivity is a must. Unfortunately, fillers are quite dense and, if present in the pulp, will yield a false high reading. Similarly, the presence of air will yield false low readings.

Microwave measurement techniques are independent of pulp type, fiber length, brightness, color, and flow rate. The most prevalent commercial technique is the measurement of propagation velocity, or time of flight through the stock, which is a function of the relative permittivity of the material. Because of the factor of 10 difference in the dielectric constant of water and wood fiber, velocity is a strong function of consistency. These microwave devices measure both fiber and filler, and compensation must be made for the filler amount and type. In addition, this method is sensitive to air, conductivity, and temperature, for which compensation must be made.

While consistency measurement is a mature technology, new sensors are still being introduced. The newer optical

---

* TAPPI Test Methods, Consistency (Concentration) of Pulp Suspensions, T 240 om-93, (1993) http://grayhall.co.uk/BeloitResearch/tappi/t240.pdf.

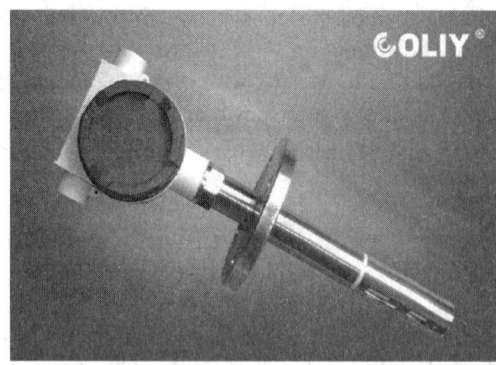 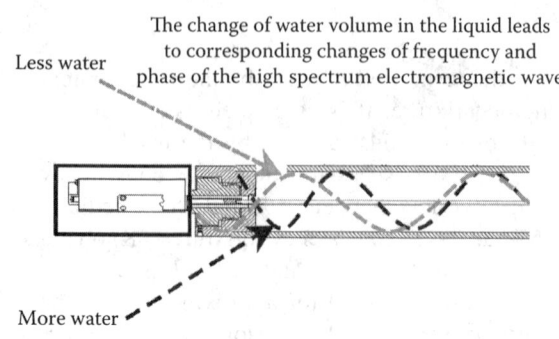

FIG. 1.16a
The resonant frequency of electromagnetic waves can measure concentration. (Courtesy of Coliy Technology GmbH.)

sensors have found great utility at low consistencies, while microwave measurements have the advantages of being able to measure total consistency without being affected by flow rate, brightness, color, fiber length, freeness, wood species, or blends that are present in the pulp stream.

One of the more recently introduced designs is the ultra-high-frequency electromagnetic wave type detector (Figure 1.16a). It is of the probe design using the resonance characteristics of an ultra-high-frequency electromagnetic wave to detect concentration. By detecting the resonant frequency and peak width, while simultaneously also providing temperature compensation, according to the supplier, the medium's concentration can be measured with a resolution of 0.1% within a range of up to 0.1%–100% concentration.

Mechanical Devices

Mechanical sensors detect consistency by measuring the shear forces acting on the sensing element, when the consistencies exceed about 1%. In rotary devices, the shear force is measured as the torque required to maintain the sensor at constant speed, as the imbalance of a strain gauge resistance bridge, or as a turning moment. The instruments are calibrated in-line; thus, the output is not in terms of bone-dry consistency, but rather some arbitrary, reproducible value that is related to consistency.

Stationary sensors depend on the process flow for measurement, and for such instruments, the output is affected by the velocity of the flow. For blade sensors, the sensor contour is designed to minimize flow effects on the output over the operating flow range. On the other hand, rotating sensors do not depend on process flow for a measurement. While these units are also sensitive to flow velocity variations, they generally can be used over wider flow ranges. In addition, the rotary motion of the sensor produces some self-cleaning action, whereas the fixed sensors depend on either a properly designed contour or an upstream deflector to prevent material hang-up (Figure 1.16b).

FIG. 1.16b
Rotoforce transmitter. (Courtesy of BTG.)

One supplier's specifications give a measuring range up to 8% consistency (C), with a sensitivity and accuracy of 0.0005% C at a flow velocity in the range of 0.3–5 m/s (1–16 ft/s).

A similar microwave propagation technique uses the phase difference between an original wave and one that passed through the stock to determine consistency. This device is claimed to be resistant to the effects of contamination and bubbles, with a range of 1%–10% consistency.

Ideally, the complete process stream should be exposed to the sensor, but in very large flows, this is not practical and, therefore, samples are taken. The sample should be taken from the center of the pipe, usually from the discharge of a centrifugal pump so that separation or settling of solids is minimized (Figure 1.16c).

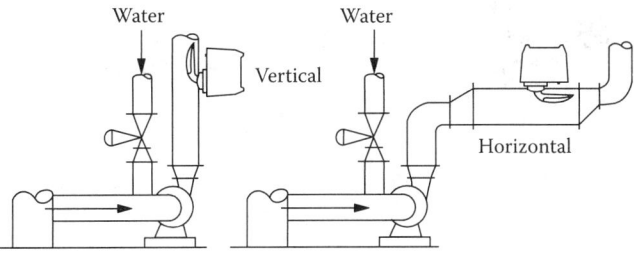

FIG. 1.16c
Installation of blade-type consistency transmitter in vertical and horizontal pipelines.

Probe Types This sensor transmitter functions as a resistance bridge strain gauge. The bridge elements are bonded to the inner wall of a hollow cylinder that is inserted into the process. The shear force acting in the cylinder, due to the consistency of the process fluid, causes an imbalance of the resistance bridge. The amount of imbalance is proportional to the shear force and the consistency of the process fluid. The resistance bridge is powered from a recorder that also contains the AC potentiometer electronics. Pipeline velocity must be measured as compensating information, and a deflector in the pipe upstream of the probe prevents accumulation of strings and like material.

The sensor is mounted through a threaded bushing furnished with the unit. Flowing velocity must be between 0.15 and 1.5 m/s (0.5 and 5.0 ft/s) in order to obtain repeatability, of around 0.1% of bone-dry consistency, up to 16%.

Blade Types The sensing element of this instrument is a fixed blade, specially shaped to minimize the effects of velocity. As shown in Figure 1.16d, material flowing past the blade, which is positioned along the line of flow, creates a shear force. Velocity of the process produces two drag forces, F_1 and F_2, whose resultant F_3 acts through the fulcrum. The moment arm of F_3 is therefore zero, and the effect of velocity on the measurement is negligible over a range of 0.23–1.5 m/s (0.75–5 ft/s).

Changes in consistency up to the 12% level are transmitted through the blade to the force bar, causing small changes in the relationship between flapper and nozzle. Therefore, the relay unit output pressure changes until the force due to the feedback unit balances the shear force. The instrument can be mounted on any line 4 in. (100 mm) or larger. Mounting is through a 2 in. (50 mm) flange supplied with the instrument. Most new instruments use electronic systems for force measurement, replacing pneumatic devices.

Moving-blade devices stroke the blade, cutting across the flow in the plane of the blade, measuring the time required to complete the stroke. Higher consistencies will require a longer time, and vice versa. Compensation for velocity is affected by a deflector mounted upstream.

Rotating Sensors This unit consists of a motor-driven, ribbed disk immersed in the process fluid. The disk is rotated at constant speed, and variations in torque output by the motor are sensed by a torque arm. The motor is suspended by flexure bearings and anchored to the torque arm, which senses motor reaction torque (Figure 1.16e). The detection of the torque force in the earlier designs was done pneumatically; today's versions are electrical. In some installations, problems have been introduced by the shaft seal required for this design. Also, because the torque variations should only

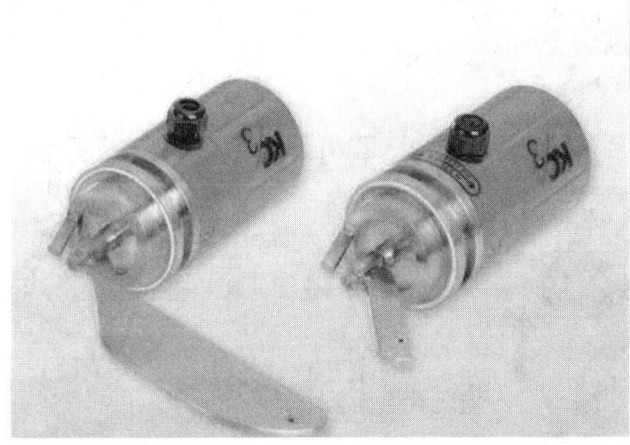

FIG. 1.16d
Stationary blade-type consistency detector. (Courtesy of MX, Servicios Para.)

FIG. 1.16e
Air schematic of rotating sensor. (Courtesy of TEMCO Engineering.)

be caused by consistency changes, shaft friction variations can be detrimental to the accuracy of the measurement.

Optical Designs

The range of these measurements is generally 1% and below for transmission devices, and up to 4% for reflection sensors. Accordingly, optical devices, either in transmission or scatter mode, are the sensors of choice for low consistencies, relying on the fiber's interaction with light, as shown in Figure 1.16f for three types of sensors.

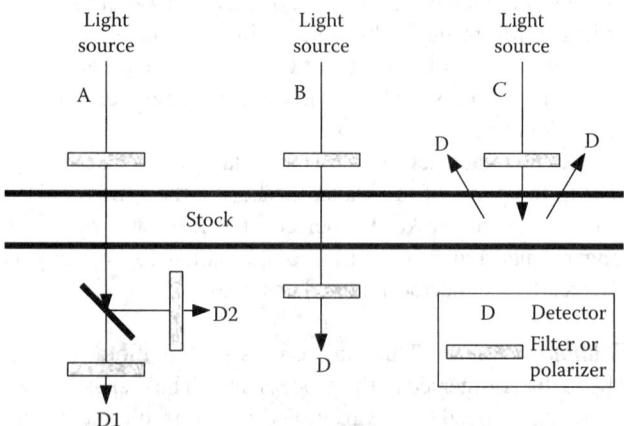

FIG. 1.16f
A variety of optical sensors.

Sensor A uses linearly polarized light from either a halogen bulb or a semiconductor laser, which is passed through the measurement cell. The transmitted light is split into two beams, one passing through a second transverse-plane polarizing filter, the other passing through a third in-plane polarizing filter. The beams are detected by photodiodes and combined to produce a relative depolarization signal, which is a function of the total fiber and filler. The signal is insensitive to brightness, color, freeness, or soluble additives.

Sensor B is based on the transmittance of light as a function of consistency. Unfortunately, this sensor is relatively sensitive to changes in freeness and color, exhibiting nonlinear behavior with changes in filler and dissolved solids.

Sensor C uses forward- and backscattered light to produce a signal combined from the several detectors that is proportional to consistency. This type of sensor can be used at much higher consistencies (ca. 4%), and in general, its sensitivity to variations in the content of nonfibrous substance lies between that of sensors A and B. The exception to this rule is filler, for which this sensor is the most sensitive.

Wood-Free Pulp Measurement Optical sensors are frequently used to manage retention on a paper machine by measuring the total consistency at the head box and in the machine white water early and late in the forming zone. One such device, the kajaaniRM-200 C, is illustrated in Figure 1.16g.

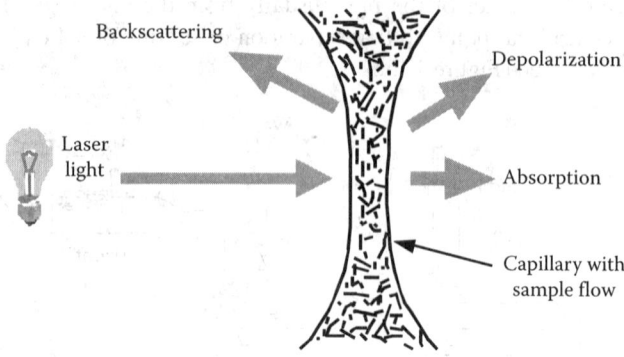

FIG. 1.16g
Consistency measurement used to control retention on paper machines. (Courtesy of Metso Automation.)

This device is similar to sensor A in Figure 1.16f, in that a polarized light beam is directed through a glass capillary cell, where the sample continuously flows. The transmitted light is directed through a special aperture disk for scattering measurements, and then through a second polarizer, which splits the light into cross-polarized and parallel-polarized components that are detected by photodiodes. The depolarization signal mainly indicates the total consistency of the sample, and the attenuation of light is affected by the total consistency and filler consistency. Attenuation is strongly affected by scattering and light absorption. Since backscattering and attenuation are influenced by small particles, filler consistency is calculated from these signals.

Pulp Containing Wood For pulp containing a considerable amount of fibers, and thus a large fraction of lignin, the depolarization scheme loses effectiveness. Another sensor has been developed that uses two light sources and a combination of optical measurement principles, including depolarization, absorption, and scattering at several wavelengths from the ultraviolet (UV) to the near infrared (NIR). An outline of this sensor is shown in Figure 1.16h.

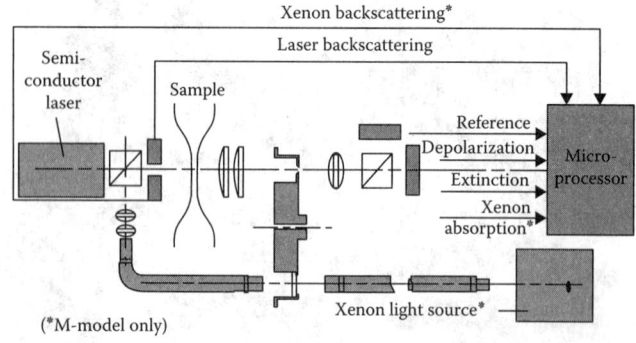

FIG. 1.16h
Consistency measurement with two light sources used on wood-based pulp. (Courtesy of Metso Automation.)

The NIR semiconductor laser light is polarized, passed through the cell, and then depolarized, as before, in Figure 1.16g. UV light from the xenon lamp is directed through the cell via a filter and polarizing prism. The forward-scattered light is directed through the lens and aperture disk to photodiodes. Backward scattering is also measured for both the UV and IR lights by detection with a photodiode before the cell. Light extinction, as well as backward and forward scattering, is measured at several different wavelengths. The signals are processed to monitor total solids and filler consistencies and flocculation in the sample.

A flow-through design of an optical transmitter is shown in Figure 1.16i. Here, the sensor measures the transmission of light through the suspended solids or wood fibers. The light source is an LED, which is generating pulses of NIR light at high intensity. The installation of this analyzer is relatively simple, its maintenance needs are likely to be low, and its consistency range is 0%–5%.

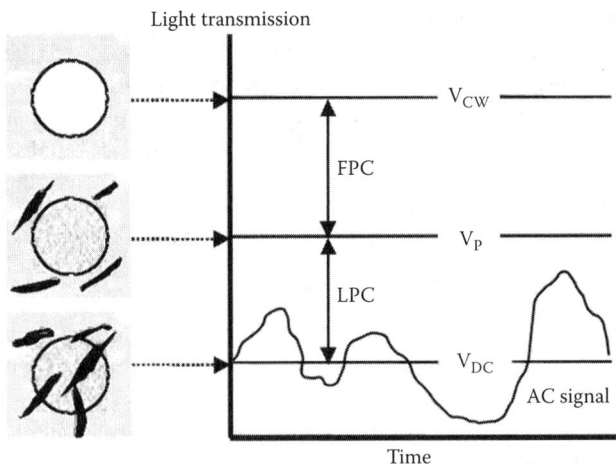

FIG. 1.16j
A sample signal trace from the BTG wet-end consistency analyzer. (From Wold, D., The peak method of optical analysis realizes the benefits of low-consistency measurement, in UpTimes, No. 5, BTG Pulp and Paper Technology, Säffle, 1999, pp. 24–25.)

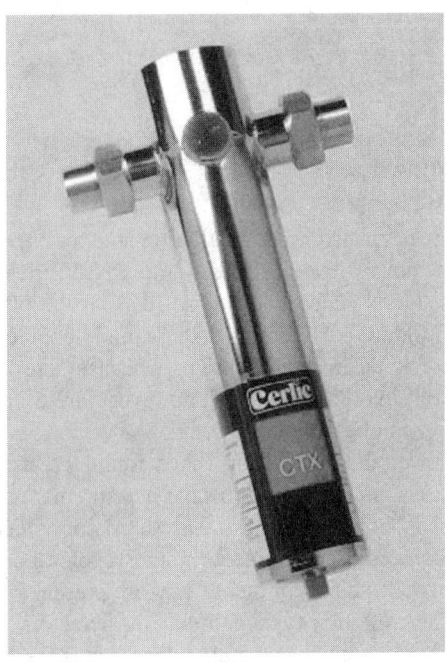

FIG. 1.16i
Flow-through optical consistency transmitter. (Courtesy of Lorentzen and Wettre, ABB Group.)

Figure 1.16j describes a typical signal trace from an analyzer using the measurement scheme that is similar to the one shown as A in Figure 1.16f. A light beam is directed at the suspension, and a photodetector senses the transmitted light. Three independent filters process the detector signal. The first filter determines the mean value V_{DC} of the transmitted light, the second determines the peak value V_P, and the third extracts the AC component V_{AC} of the signal.

The peak method used in the analysis assumes that the suspension is substantially characterized by large and small particles. The large particles (fibers) form a relatively transparent network within which the much greater number of smaller particles (fillers and fines) floats freely. Observation of a typical suspension over time reveals that the great number of small particles is relatively constant, whereas the number of large particles is few and variable. The average value of the transmitted light determines V_{DC}.

Deviations from this mean value are mainly due to the large particles passing through the light beam. The highest light intensity and V_P occur when no fibers are passing through the beam and the light is being dimmed only by the suspended fine particles. Thus, the respective amounts of large and small particles in the suspension can be determined by the mean and peak values.*

Referring to Figure 1.16j, V_{CW} is the detector signal for clear water and is used as a reference value. The AC signal, V_{AC}, is plotted along with V_{CW}, V_P, and V_{DC}. The large-particle content (LPC) is the difference between V_P and V_{DC}, while the fine-particle content (FPC) is the difference between V_{CW} and V_P. The total consistency is obtained by summing LPC and FPC.

The consistency measurement at 30% for TMP or CTMP pulp is based on dielectric measurement of the water content

---

* Shaw, P. and Fladda, G., A modern approach to retention measurement and control, (1993) http://cat.inist.fr/?aModele=afficheN&cpsidt=4668223.

and an optical measurement using an NIR technique with reflectance spectroscopy. This technique is based on the resonance vibration of water, which appears as absorption bands in the infrared region of the spectrum. A typical sensor uses four fixed wavelengths of the spectrum—two located in the absorption bands of water and two in a region where the effect of water is minimal.

Microwave Sensors

In connection with Figure 1.16a, the ultra-high-frequency electromagnetic wave-type consistency detectors have already been briefly discussed.

For the control of measurement of the total consistency in pulp and paper processes, these microwave-based sensors are widely used. Their main advantage is that they measure total consistency without being affected by fiber length, freeness, wood species, or blends that are present in the pulp stream.

Their operation is based on the phenomenon that solids such as fibers and fillers conduct microwaves faster than water, and therefore, the transmission times of microwaves shorten as consistencies increase. This relationship is linear, making it easy to calibrate the consistency sensor and therefore the consistency in recycled fiber, or other difficult pulp compositions can be measured.

Additionally, the measurement is not affected by flow rate, brightness, or color, which reduces the likelihood of off-grade product. The installation of microwave sensors is usually simple, and their maintenance needs are relatively low. These consistency detectors are also available in retractable probe designs (Figure 1.16k) and in on-line, flow through designs (Figure 1.16l).

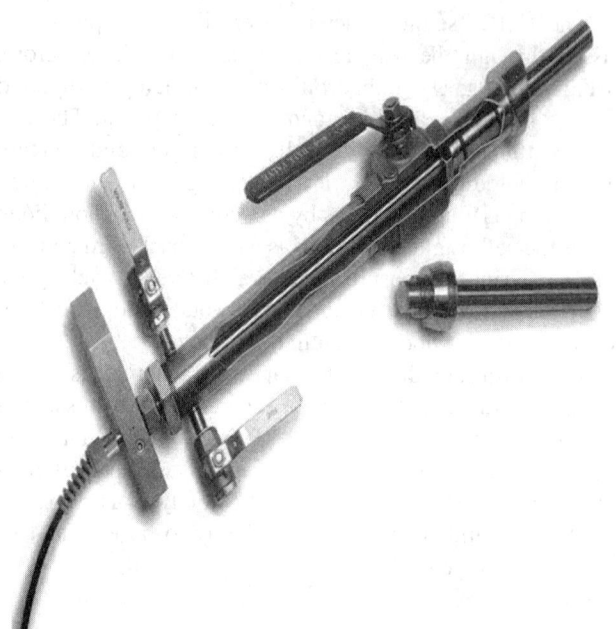

FIG. 1.16l
Consistency sensor in the retractable probe design. (Courtesy of the Thompson Equipment Company.)

CONCLUSIONS

While convenient from an installation standpoint, mechanical in-line consistency detectors are all sensitive to flow variations. Fixed sensors are more likely to be plagued by material buildup, particularly if the sample contains fibers. Rotating sensors are self-cleaning, because sensor motion will tend to spin off any material; however, variations in shaft seal friction can be troublesome.

The newer optical sensors have found great utility at low consistencies, while microwave measurements have the advantages of being able to measure total consistency without being affected by flow rate, brightness, color, fiber length, freeness, wood species, or blends that are present in the pulp stream. The radiation-type gamma ray detectors are basically density sensors. Their main advantage is their noncontact design, while their main limitation is cost.

SPECIFICATION FORMS

When specifying consistency detectors only, please use the ISA form 20A1001, and when specifying both the analyzer and the composition or the properties of the process material of interest, use ISA form 20A1002. Both forms are reproduced with the permission of the International Society of Automation on the next pages.

FIG. 1.16k
Online consistency meter using microwave phase shift technology to determine the solids concentration in the flowing fluid. (Courtesy of Toshiba.)

1.16 Consistency Measurement

	RESPONSIBLE ORGANIZATION	ANALYSIS DEVICE		SPECIFICATION IDENTIFICATIONS
1			6	
2			7	Document no
3	ISA	Operating Parameters	8	Latest revision — Date
4			9	Issue status
5			10	

	ADMINISTRATIVE IDENTIFICATIONS			SERVICE IDENTIFICATIONS Continued
11			40	
12	Project number	Sub project no	41	Return conn matl type
13	Project		42	Inline hazardous area cl — Div/Zon — Group
14	Enterprise		43	Inline area min ign temp — Temp ident number
15	Site		44	Remote hazardous area cl — Div/Zon — Group
16	Area	Cell — Unit	45	Remote area min ign temp — Temp ident number
17			46	
18	SERVICE IDENTIFICATIONS		47	
19	Tag no/Functional ident		48	COMPONENT DESIGN CRITERIA
20	Related equipment		49	Component type
21	Service		50	Component style
22			51	Output signal type
23	P&ID/Reference dwg		52	Characteristic curve
24	Process line/nozzle no		53	Compensation style
25	Process conn pipe spec		54	Type of protection
26	Process conn nominal size	Rating	55	Criticality code
27	Process conn termn type	Style	56	Max EMI susceptibility — Ref
28	Process conn schedule no	Wall thickness	57	Max temperature effect — Ref
29	Process connection length		58	Max sample time lag
30	Process line matl type		59	Max response time
31	Fast loop line number		60	Min required accuracy — Ref
32	Fast loop pipe spec		61	Avail nom power supply — Number wires
33	Fast loop conn nom size	Rating	62	Calibration method
34	Fast loop conn termn type	Style	63	Testing/Listing agency
35	Fast loop schedule no	Wall thickness	64	Test requirements
36	Fast loop estimated lg		65	Supply loss failure mode
37	Fast loop material type		66	Signal loss failure mode
38	Return conn nominal size	Rating	67	
39	Return conn termn type	Style	68	

	PROCESS VARIABLES	MATERIAL FLOW CONDITIONS			PROCESS DESIGN CONDITIONS
69				101	
70	Flow Case Identification	Units		102	Minimum — Maximum — Units
71	Process pressure			103	
72	Process temperature			104	
73	Process phase type			105	
74	Process liquid actl flow			106	
75	Process vapor actl flow			107	
76	Process vapor std flow			108	
77	Process liquid density			109	
78	Process vapor density			110	
79	Process liquid viscosity			111	
80	Sample return pressure			112	
81	Sample vent/drain press			113	
82	Sample temperature			114	
83	Sample phase type			115	
84	Fast loop liq actl flow			116	
85	Fast loop vapor actl flow			117	
86	Fast loop vapor std flow			118	
87	Fast loop vapor density			119	
88	Conductivity/Resistivity			120	
89	pH/ORP			121	
90	RH/Dewpoint			122	
91	Turbidity/Opacity			123	
92	Dissolved oxygen			124	
93	Corrosivity			125	
94	Particle size			126	
95				127	
96	CALCULATED VARIABLES			128	
97	Sample lag time			129	
98	Process fluid velocity			130	
99	Wake/natural freq ratio			131	
100				132	

	MATERIAL PROPERTIES			MATERIAL PROPERTIES Continued
133			137	
134	Name		138	NFPA health hazard — Flammability — Reactivity
135	Density at ref temp	At	139	
136			140	

Rev	Date	Revision Description	By	Appv1	Appv2	Appv3	REMARKS

Form: 20A1001 Rev 0 © 2004 ISA

	RESPONSIBLE ORGANIZATION	ANALYSIS DEVICE COMPOSITION OR PROPERTY Operating Parameters (Continued)		SPECIFICATION IDENTIFICATIONS		
1			6	Document no		
2			7	Latest revision	Date	
3	ISA		8	Issue status		
4			9			
5			10			

	PROCESS COMPOSITION OR PROPERTY			MEASUREMENT DESIGN CONDITIONS				
	Component/Property Name	Normal	Units	Minimum	Units	Maximum	Units	Repeatability
11								
12								
...								
37								

(rows 38–75 blank)

Rev	Date	Revision Description	By	Appv1	Appv2	Appv3	REMARKS

Form: 20A1002 Rev 0 © 2004 ISA

Definitions

Consistency is the term used to describe the solid content of a solid–liquid mixture. It is calculated as the ratio of the dry weight of the solids, divided by the total weight of the mixture. Mixtures having under 5% consistency are considered to be low, between 5% and 15%, medium, and over 15% as high consistency mixtures. Knowing the specific gravity of the solids and of the liquid phase, consistency can be determined (or approximated) on the basis of density measurement.

Freeness is the ability of pulp and water mixture to release or retain water on drainage.

Abbreviations

CTMP	Chemithermomechanical pulping
FPC	Fine-particle content
LED	Light-emitting diode
LPC	Large-particle content
NIR	Near infrared
TMP	Thermomechanical pulping

Organization

TAPPI	Technical Association of the Pulp and Paper Industry

Bibliography

Biermann, C. J., Handbook of pulping and paper making, (1996) http://www.sciencedirect.com/science/book/9780120973620.

Cumberbach, E., The vibrating element densitometer, (2012) https://ebooks.cambridge.org/chapter.jsf?bid=CBO9780511626326&cid=CBO9780511626326A037&tabName=Chapter.

Ewing, G., Analytical Instrumentation Handbook, (1997) http://books.google.com/books/about/Analytical_Instrumentation_Handbook_Seco.html?id=5VApgsCUL_IC.

Helmentstine, A. M., How to measure volume and density, (2007) http://chemistry.about.com/b/2007/08/05/how-to-measure-volume-density-a-tale-of-archimedes.htm.

Illinois Department of Transportation, Radiation safety and density by nuclear method, (2012) http://www.dot.state.il.us/materials/radiationsafetys34.pdf.

Innovative Technology DOE/EM-0490, Testing of pipeline slurry monitors, (1998) http://costperformance.org/pdf/itsr1547.pdf.

Shaw, P. and Fladda, G., A modern approach to retention measurement and control, (1993) http://cat.inist.fr/?aModele=afficheN&cpsidt=4668223.

TAPPI Test Methods, Consistency (concentration) of pulp suspensions, T 240 om-93, (1993) http://grayhall.co.uk/BeloitResearch/tappi/t240.pdf.

The Engineering Toolbox, Slurry/density, (2015) http://www.engineeringtoolbox.com/slurry-density-d_1188.html.

Thermo Fisher Scientific, Gamma based consistency measurement, (2015), http://thermo.com.cn/Resources/200802/productPDF_14149.pdf.

Tippet, J., Consistency measurement, (2005) http://www2.emersonprocess.com/siteadmincenter/PM%20PSS%20Services%20Documents/Consulting%20Services/ConsistencyControlSystematicandScientificDesignleadstoManyDifferentStrategies.pdf.

Waller, M. H., Papermaking retention aid control, (1998) http://citeseerx.ist.psu.edu/viewdoc/download?doi=10.1.1.524.7895&rep=rep1&type=pdf.

1.17 Corrosion Monitoring

D. H. F. LIU (1995)   **B. G. LIPTÁK** (2003, 2017)

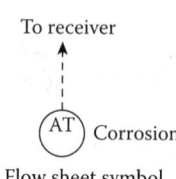
Flow sheet symbol

Type	A. Electric resistance and coupon monitoring
	B. Linear polarization resistance (LPR)
	C. Ultrasound
	D. Guided wave
Functions	Can be total corrosion, corrosion rate, and the state of corrosion
Accuracy	A. Usually 30- or 60-day coupons
	B. Cumulative metal loss, corrosion rate in μm, 0.1% resolution of span
	C. Repeatability to 0.2 mm (0.008″) at −40°C to 120°C (−40°F to 248°F) or 0.0025 mm (0.0001″) (high temp—350°C)
Design pressure	A. Up to 10,000 psig (690 bars)
Design temperature	A. Up to 560°C (1000°F)
Materials of construction	Wide range of corrosion-resistant metals or alloys
Partial list of suppliers	ACM Instruments (A): http://www.acminstruments.com/index.php?option=com_content&task=view&id=63&Itemid=85
	Cormon-Teledyne (A): http://www.cormon.com/products/ohlcd.aspx
	Emerson Process/Roxar (A): http://www2.emersonprocess.com/en-US/brands/roxar/corrosion/Pages/Corrosion.aspx
	GE (C): http://www.ge-mcs.com/en/ultrasound/corrosion-monitoring/rightrax-automate-lt.html
	ION Science: http://www.ionscience.com/products/hydrosteel-7000-fixed-corrosion-monitor
	Metal Samples (A–Monitoring Products): http://www.alspi.com/ms.htm
	Pepperl+Fuchs (A): http://www.pepperl-fuchs.com/global/en/classid_1685.htm?view=productgroupoverview
	Purafil Inc. (Corrosion in scrubbers for gas and air): http://www.corrosionmonitor.com/
	Rohrback Cosasco Systems (A, B): http://www.cosasco.com/corrosion-monitoring-instrument.html and http://www.cosasco.com/wireless-corrosion-monitoring-transmitters.html
	SGS (D) http://www.sgs.com/~/media/Global/Documents/Flyers%20and%20Leaflets/SGS-IND-NDT-Guided%20Wave-A4-EN-10.pdf
	Siemens (A): http://www.water.siemens.com/SiteCollectionDocuments/Product_Lines/Electrocatalytic_Products/Brochures/EPS-ELCM2-DS-0408.pdf
	Walton–Corrpro Ltd. (A): http://www.corrpro.com/content/142/corrpro-protects-neptune-spar-hull.aspx

INTRODUCTION

Corrosion monitoring provides critical information on the condition industrial equipment in the plants. It serves to identify the location, rate, and underlying causes of corrosion. Corrosion monitoring can provide significant advantages when integrated into both preventative maintenance and the processes inherent to safety management programs. Based on the results of the corrosion monitoring programs, informed decisions can be made, not only regarding the expected life of the equipment that is affected by corrosion but also regarding life extension strategies, prospective material selection, and cost-effective methods for selecting the proper remedies of the problems.

A detailed tabulation of the chemical resistance of materials is given in Chapter 2.6 of this volume. A more condensed

TABLE 1.17a
Corrosion and Cavitation Resistance of Various Materials

Trim or Valve Body Material	Relative Cavitation Resistance Index	Approximate Rockwell C Hardness Values[a]	Corrosion Resistance	Cost
Aluminum	1	0	Fair	Low
Synthetic sapphire	5	Very high	Excellent	High
Brass	12	2	Poor	Low
Carbon steel, AISI C1213	28	30	Fair	Low
Carbon steel, WCB	60	40	Fair	Low
Nodular iron	70	3	Fair	Low
Cast iron	120	25	Poor	Low
Tungsten carbide	140	72	Good	High
Stellite 1	150	54	Good	Medium
Stainless steel, type 316	160	35	Excellent	Medium
Stainless steel, type 410	200	40	Good	Medium
Aluminum oxide	200	72	Fair	High
K-Monel	300	32	Excellent	High
Stainless steel, type 17-4 PH	340	44	Excellent	Medium
Stellite 12	350	47	Excellent	Medium
Stainless steel, type 440C	400	55	Fair	High
Stainless steel, type 329, annealed	1000	45	Excellent	Medium
Stellite 6	3500	44	Excellent	Medium
Stellite 6B	3500	44	Excellent	High

[a] Rockwell scale is a hardness scale based on the indentation of materials. The Rockwell test determines the hardness by measuring the depth of penetration of an indenter under a large load compared to the penetration made by a preload. There are different scales, denoted by single letters (Rockwell A, B, C, etc.), corresponding to the different loads or indenters. The result is a dimensionless number noted as HRA, HRB, HRC, etc., where the last letter refers to the respective Rockwell scale.

summary of the corrosion resistance of some widely used materials is also given in Table 1.17a. In selecting the materials of construction for instruments, one should also keep in mind that most process fluids are not pure and that the rate of corrosion is also a function of flow velocity and of dissolved oxygen content.

Nondestructive testing, such as ultrasonic scanning, is also used in inspecting storage tanks, pressure vessels, and piping. If the plant's atmosphere contains corrosive gases, it is also desirable to enclose, purge, or otherwise protect the more sensitive instruments or their components. In such installations, it is advisable to protect the stems of control valves by boots.

Corrosion monitoring can be based on 30- or 60-day coupons, which provide data only on the average rate of corrosion. Mobile monitoring laboratories can provide spot checks, or in the more critical cases, permanent monitoring equipment can be installed to provide continuous corrosion rate and pitting tendency readings. In cooling systems, where chlorine is used to control biological deposition, corrosion monitoring is needed, because if the chlorine residual is too high, corrosion will occur.

CORROSION-MONITORING TECHNIQUES

A variety of corrosion-monitoring techniques are listed in Table 1.17b. These can monitor total corrosion, corrosion rate, and the state of corrosion. Some of them can also

TABLE 1.17b
Instrumentation for Corrosion Monitoring

Method	Measures or Detects
Corrosion coupon	Average corrosion rate over a known exposure period by weight loss.
Linear polarization	Corrosion is measured by electrochemical polarization resistance method.
Electric resistance	Metal loss is measured by the resistance change of corroding metal element.
Analytical	pH of process stream.
Radiography	Flaws and cracks by penetration of radiation and detection on film.
Ultrasonics	Thickness of metal and presence of cracks, pits, etc., by changes in response to ultrasonic waves.
Eddy current testing	Use of a magnetic probe to scan surface.
Hydrogen sensing	Hydrogen probe used to measure hydrogen gas liberated by corrosion.
Analytical	Concentration of corroded metal ions or concentration of inhibitor; oxygen concentration in process stream.
Potential monitoring	Potential change of monitoring metal with respect to a reference electrode.

determine the composition of products and detect the presence of defects or changes in physical parameters. However, any monitoring technique can provide only limited information, and the techniques should be regarded as complementary rather than competitive. Where more than one technique can provide the information required, such a cross-check can be very valuable, and their differences can add to their value.

This section focuses on coupon monitoring and on two frequently used online corrosion monitors: (1) the electric resistance monitors and (2) the linear polarization resistance (LPR) monitors.

Coupon monitoring provides long-term performance data on general corrosion and information on localized corrosion. Electric resistance monitors give medium- to long-term data on general corrosion or erosion. LPR monitors provide real-time operational data on general corrosion and enable engineers to study the dynamics of the corroding system and to observe the effects of the addition of various corrosion inhibitors.

Coupon Monitoring

Coupons are the simplest and most frequently used devices in the monitoring of corrosion. Coupons are small pieces of metals, usually of a rectangular shape, which are inserted in the process stream and are removed after a period for study. Specimens for standard corrosion, stress corrosion, crevice corrosion, and galvanic corrosion tests are shown in Figure 1.17a.

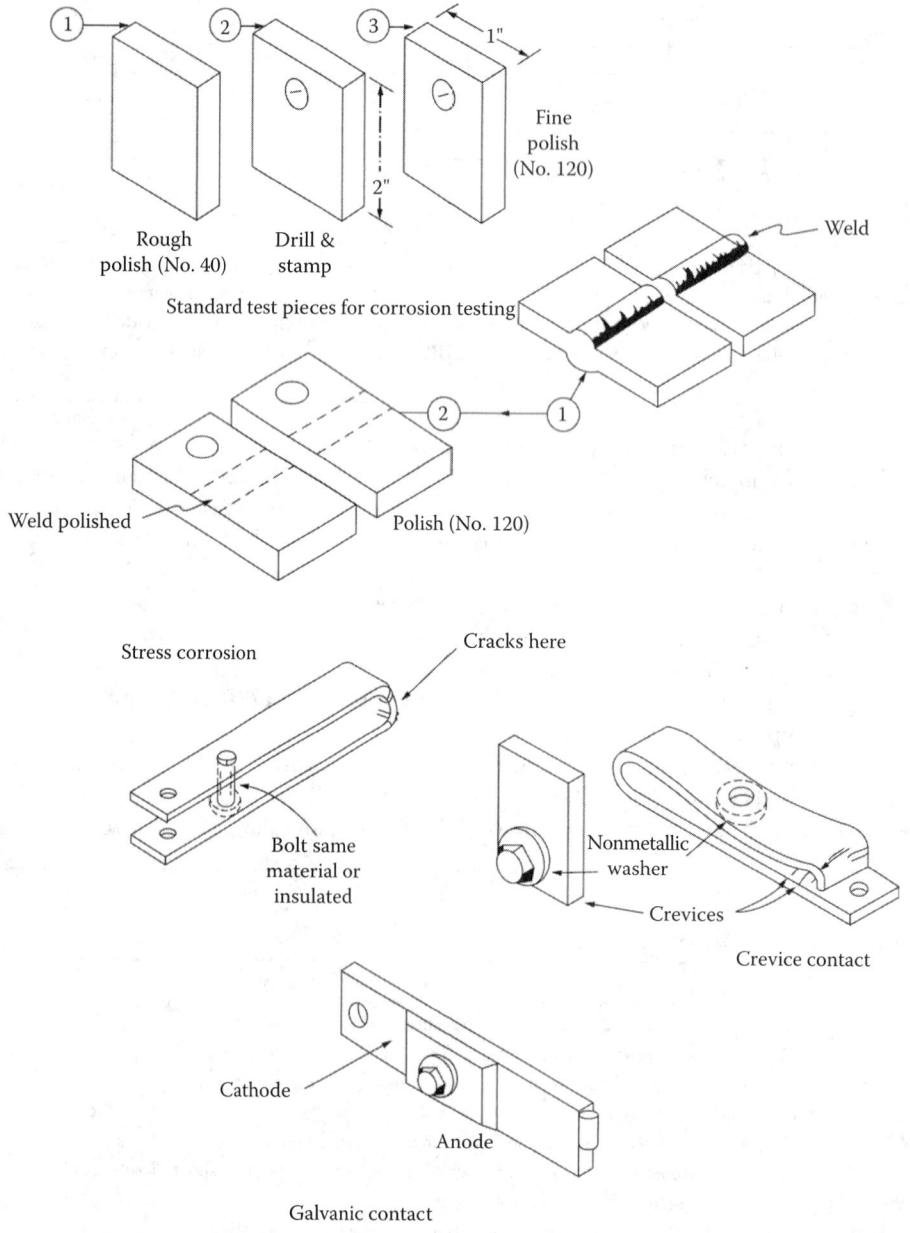

FIG. 1.17a
Monitoring coupon specimens for standard corrosion, stress corrosion, crevice corrosion, and galvanic corrosion tests.

The most common use of coupons is to determine average rate of uniform corrosion over the period of exposure. The coupon or specimen is weighed before and after exposure. The average corrosion rate is calculated from the weight loss, the initial surface area, and the time exposed. The rate is usually expressed in mils per year (miles/year) (1 mile = 0.001 in.) or millimeters per year (mm/year), since corrosion is generally a long-term effect.

Improper cleaning of coupons or specimens after exposure is the main source of error in determining the corrosion rate. American Society for Testing and Materials (ASTM) Standard 61–81* describes the recommended practice for preparing, cleaning, and evaluating test specimens. In addition, Fontana and Greene (see bibliography) have useful information, particularly on techniques for sample cleaning after exposure.

The measurement of the depths even of extremely shallow pits is important. Pit depth should be measured to an accuracy of 1.0 mil using a micrometer.

Exposure Time Proper selection of the time that the sample will be exposed is critical. In batch processes that involve cyclic exposure conditions, the coupon should undergo all of the batch conditions, including periods of shutdown. Exposure should be made for two or more complete cycles of operations to ensure that an equivalent of 2 weeks exposure time is obtained. Usually, a 2-week test is acceptable, but a 1-month test is preferable.

Normally, initial corrosion rates are high, because as the corrosion products form, they tend to protect the surface and, as a consequence, the corrosion rate will drop below the initial rate. Therefore, it will be necessary to conduct tests of sufficient duration to compensate for this effect.

Advantages and Limitations

Advantages of the coupon monitoring technique are as follows:

1. The technique is suitable for all environments.
2. Coupons provide information about the type of corrosion present. Coupons can be examined for evidence of pitting and other localized forms of attack.
3. There are a variety of coupons available for specialized analysis.

---
* ASTM, Corrosion Standards, (2015) http://www.astm.org/Standards/corrosion-and-wear-standards.html.

Limitations of the technique are as follows:

1. High corrosion rates for short periods of time may be undetectable and cannot be correlated to process upset conditions.
2. The technique can require plant shutdowns for installation or removal. Highly qualified personnel and reasonably sophisticated test procedures are required for the interpretation of the results.

Electrical Resistance (ER) Monitors

Figure 1.17b is a simplified diagram of an electric resistance corrosion monitor. It shows a probe that has both an exposed measuring element and a protected reference element. They are connected to a Wheatstone bridge measuring an output circuit, and a power supply.

The measuring element is a loop of wire that is exposed to the corrosive environment. The reference element is encapsulated inside the probe body by a thermally conductive plastic so that it will be at the process temperature. The

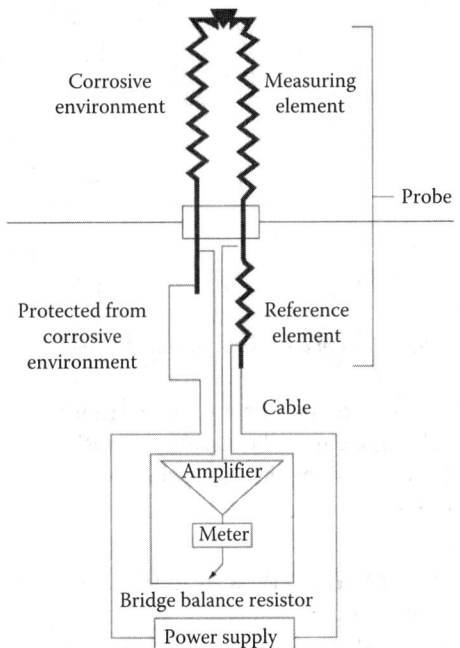

FIG. 1.17b
Basic diagram of electric resistance corrosion monitor with wire loop probe. (Courtesy of RCS—Rohrback Cosasco Systems.)

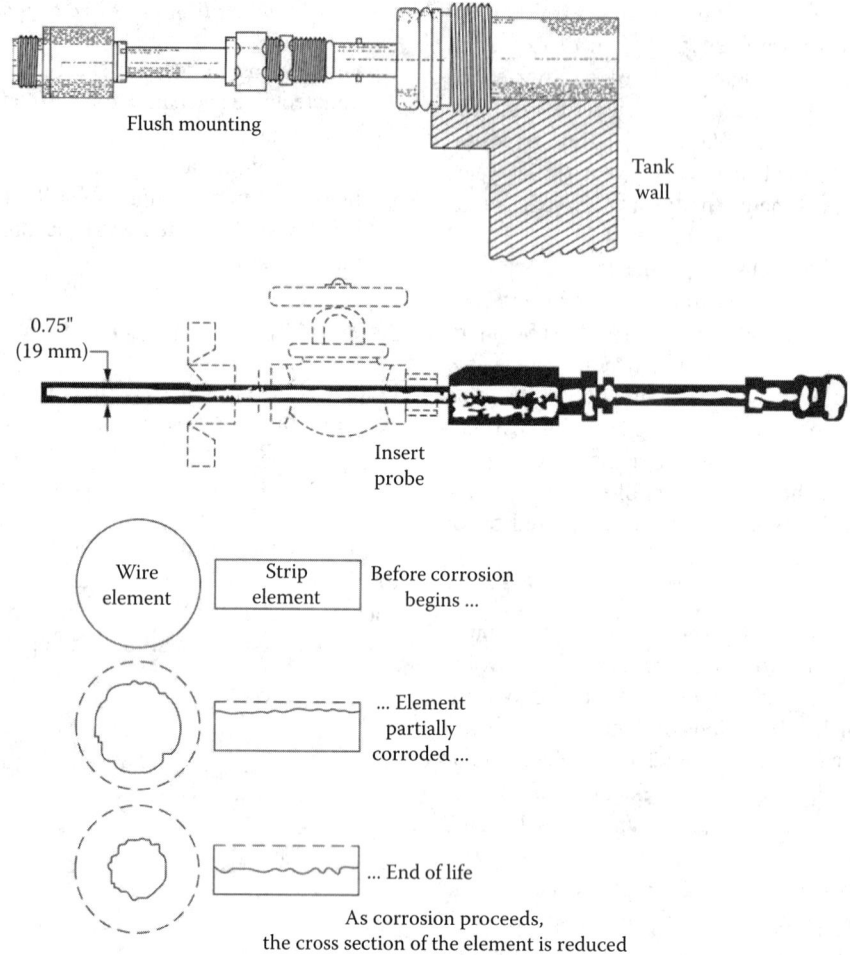

FIG. 1.17c
Probe configurations of corrosion sensors.

output is directly proportional to the resistance of the thin-wire measuring element, because as corrosion increases, it causes the thickness of the measuring element to decrease (Figure 1.17c).

Advantages and Limitations Electrical resistance (ER) monitors are basically automatic coupons, and they share many characteristics with coupons when it comes to advantages and limitations. The resistance probe must be allowed to corrode for a period of time before accurate corrosion measurements can be made.

Advantages of the electric resistance technique are as follows:

1. The technique is suitable for all environments except liquid metals or some conductive molten salts. The process material, which causes the corrosion, need not be an electrolyte (in fact, it need not be a liquid).
2. A corrosion measurement can be made without having to see or remove the test sample.
3. Corrosion measurements can be made quickly—in a few hours or days—or continuously. Sudden increases or decreases in corrosion rate can be detected, so that the user can modify the process to reduce the corrosion.
4. The method can detect low corrosion rates that would take a long time to detect with weight loss methods. Its accuracy is comparable to the coupon method.

Limitations of the technique are as follows:

1. The technique is usually limited to the measurement of uniform corrosion only and is not generally satisfactory for localized corrosion.
2. The probe design includes provisions for temperature variations. This feature is not totally successful. The most reliable results are obtained in constant-temperature systems.
3. The technique does not provide an instantaneous corrosion rate, so any corrective action must be delayed until an average corrosion rate can be determined.
4. The resistance method measures a combination of chemical and physical erosion without distinguishing between the two.

Linear Polarization Resistance (LPR) Monitors

Figure 1.17d shows a three-electrode and a two-electrode linear polarization system. In addition to the electrodes, the three-electrode system includes an ammeter, a voltmeter, and an adjustable current source. The meters have adjustable null features to compensate for normal variations in the surface conditions of the test and reference electrodes.

A small electrical current flows between the test and auxiliary electrodes. Because corrosion occurs at the anode, the test electrode is protected when it is the cathode. The direction of current flow is then reversed, and the test electrode becomes the anode and its corrosion rate is accelerated. The change in corrosion rate caused by the reversal in current causes a change in potential of the test electrode when compared to the freely corroding reference electrode. The relationship between the change in current flow (dI), the change in polarization voltage (dE), and the corrosion rate (CR) of the test electrode is as follows:

$$CR = K \frac{dI}{dE} \qquad 1.17(1)$$

Since all corrosion measurements are made at a constant polarization potential of 10 mV, the voltage term becomes constant, so the corrosion equation can be reduced to

$$CR = KI \qquad 1.17(2)$$

Under constant polarization voltage, the corrosion rate is directly proportional to the current required to produce that polarization voltage. A value representing the K factor is designed into the equipment so that the readings are linear and calibrated directly in mils per year.

Advantages and Limitations

Advantages of the LPR technique are as follows:

1. The polarization probes measure corrosion rate almost instantaneously. They measure the instantaneous corrosion rate instead of the average corrosion rate.
2. The probes are useful for comparing relative corrosion rates. Thus, it is possible to use the data to determine the process variables that give the lowest corrosion rates.
3. The commercially available polarization probes supply pitting tendency information.

Limitations of the technique are as follows:

1. The probes will not work in nonconductive fluids or fluids containing compounds that coat the electrodes (e.g., crude oil).
2. The absolute accuracy of the corrosion measurement is not as reliable as the one obtained from corrosion coupons.
3. The method measures the combined rate of any electrochemical reactions at the surface of the test sample. If reactions other than corrosion reactions are possible at comparable or greater rates, the measured rate will also include these other reactions.

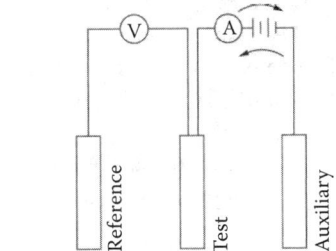

Three-electrode system polarized by 10 mv anodically and cathodically compared to reference electrode.

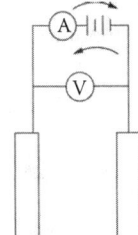

Two-electrode systems impress a 2 mv difference between the two electrodes. It is assumed that one electrode is shifted 10 mv anodically and the other 10 mv cathodically. Measurements can be made in either direction.

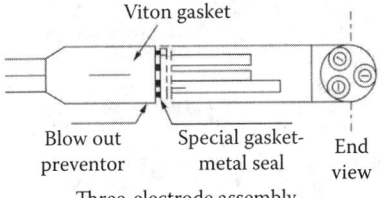

Three-electrode assembly

FIG. 1.17d
Linear polarization probes.

LPR and ER Combination Designs

Newer LPR designs have been developed to substantially increase the speed of response over conventional corrosion-monitoring techniques, such as coupons, ER probes, and traditional LPR designs by combining the rapid response of LPR and the ER designs. This monitoring system can be 50 times more sensitive and faster in response than ER systems. Such detector probes can be connected to transmitters and data loggers as shown in Figure 1.17e.

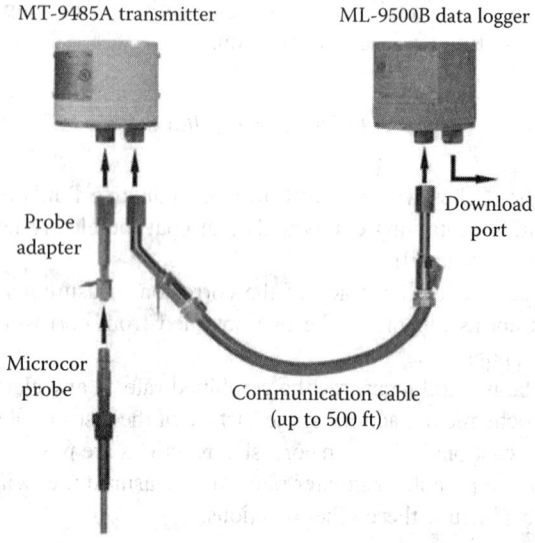

FIG. 1.17e
Configuration of a fast and sensitive corrosion detector. (Courtesy of RCS—Rohrback Cosasco Systems.)

As shown in Figure 1.17f, handheld corrosion monitors are also available.

Other advances include the availability of wireless transmission of galvanic corrosion and erosion rate information, which increases the flexibility and economy of both new and existing installations (Figure 1.17f). These units are rated intrinsically safe for operation in hazardous locations (Class I, Zone 1). Their intrinsically safe approvals include FM, ATEX, and CSA.

As shown in Figure 1.17g, handheld corrosion monitors are also available.

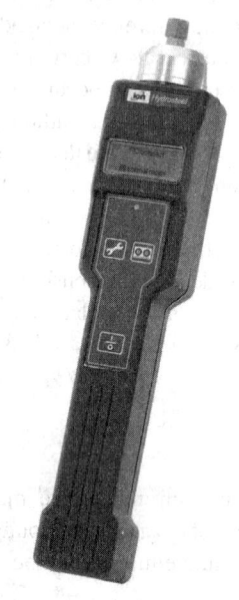

FIG 1.17g
Handheld corrosion monitor. (Courtesy of ION Science.)

Down-Hole Corrosion Monitoring (DCMS)

A Downhole Corrosion Monitoring System (DCMS™) system provides continuous corrosion history. It is a tool that provides recorded corrosion and temperature data even in the hostile environment of the downhole operation. The sensitivity of the system enables the film persistence of inhibitors to be evaluated in actual operating conditions (Figure 1.17h).

The LPR method used has the advantage of fast response to changes in corrosive conditions, so that

FIG 1.17f
Wireless corrosion monitor. (Courtesy of Rochback Cosasco Systems.)

FIG. 1.17h
DCMS. (Courtesy of Rohrback Cosasco.)

inhibition and process changes can limit corrosion damage quickly. It is generally advantageous to continuously monitor the corrosion rate, so that system upsets can be recorded for further evaluation. This can be used to determine cumulative metal loss. Normally, the DCMS monitor is used only to provide corrosion rate data and alarms, but not direct control.

Wall Thickness Monitoring

Corrosion and erosion can cause unscheduled plant or pipeline shutdowns, resulting in lost production, high repair costs, or possible fines. Therefore, improved corrosion and erosion monitoring can help in reducing operating costs. The design shown in Figure 1.17i describes an ultrasonic wall thickness measuring and monitoring system, which provides continuous measurement on the wall thickness of pipes, vessels, or other equipment.

This technology helps to increase safety and productivity while reducing overall inspection costs, because the permanently installed ultrasonic sensors permit remote monitoring of restricted or high-temperature areas (up to 350°C). The system provides online remote access, eliminating the need to erect scaffolding, remove insulation, or even shut the plant down. Online monitoring also reduces labor-intensive maintenance and traditional inspection routines.

SPECIFICATION FORMS

When specifying corrosion monitors only, please use the ISA form 20A1001, and when specifying both the analyzer and the composition or the properties of the process material of interest, use ISA form 20A1002. Both forms are reproduced with the permission of the International Society of Automation on the next pages.

FIG. 1.17i
Ultrasonic pipe thickness monitoring permits remote monitoring of restricted, hard-to-access, and/or high-temperature areas up to 350°C (662°F). (Courtesy of GE Rightrax HT.)

1	RESPONSIBLE ORGANIZATION		ANALYSIS DEVICE	6	SPECIFICATION IDENTIFICATIONS		
2				7	Document no		
3	ISA		Operating Parameters	8	Latest revision	Date	
4				9	Issue status		
5				10			

11	ADMINISTRATIVE IDENTIFICATIONS			40	SERVICE IDENTIFICATIONS Continued		
12	Project number		Sub project no	41	Return conn matl type		
13	Project			42	Inline hazardous area cl	Div/Zon	Group
14	Enterprise			43	Inline area min ign temp	Temp ident number	
15	Site			44	Remote hazardous area cl	Div/Zon	Group
16	Area	Cell	Unit	45	Remote area min ign temp	Temp ident number	
17				46			
18	SERVICE IDENTIFICATIONS			47			
19	Tag no/Functional ident			48	COMPONENT DESIGN CRITERIA		
20	Related equipment			49	Component type		
21	Service			50	Component style		
22				51	Output signal type		
23	P&ID/Reference dwg			52	Characteristic curve		
24	Process line/nozzle no			53	Compensation style		
25	Process conn pipe spec			54	Type of protection		
26	Process conn nominal size		Rating	55	Criticality code		
27	Process conn termn type		Style	56	Max EMI susceptibility	Ref	
28	Process conn schedule no		Wall thickness	57	Max temperature effect	Ref	
29	Process connection length			58	Max sample time lag		
30	Process line matl type			59	Max response time		
31	Fast loop line number			60	Min required accuracy	Ref	
32	Fast loop pipe spec			61	Avail nom power supply		Number wires
33	Fast loop conn nom size		Rating	62	Calibration method		
34	Fast loop conn termn type		Style	63	Testing/Listing agency		
35	Fast loop schedule no		Wall thickness	64	Test requirements		
36	Fast loop estimated lg			65	Supply loss failure mode		
37	Fast loop material type			66	Signal loss failure mode		
38	Return conn nominal size		Rating	67			
39	Return conn termn type		Style	68			

69	PROCESS VARIABLES	MATERIAL FLOW CONDITIONS				101	PROCESS DESIGN CONDITIONS		
70	Flow Case Identification				Units	102	Minimum	Maximum	Units
71	Process pressure					103			
72	Process temperature					104			
73	Process phase type					105			
74	Process liquid actl flow					106			
75	Process vapor actl flow					107			
76	Process vapor std flow					108			
77	Process liquid density					109			
78	Process vapor density					110			
79	Process liquid viscosity					111			
80	Sample return pressure					112			
81	Sample vent/drain press					113			
82	Sample temperature					114			
83	Sample phase type					115			
84	Fast loop liq actl flow					116			
85	Fast loop vapor actl flow					117			
86	Fast loop vapor std flow					118			
87	Fast loop vapor density					119			
88	Conductivity/Resistivity					120			
89	pH/ORP					121			
90	RH/Dewpoint					122			
91	Turbidity/Opacity					123			
92	Dissolved oxygen					124			
93	Corrosivity					125			
94	Particle size					126			
95						127			
96	CALCULATED VARIABLES					128			
97	Sample lag time					129			
98	Process fluid velocity					130			
99	Wake/natural freq ratio					131			
100						132			

133	MATERIAL PROPERTIES			137	MATERIAL PROPERTIES Continued		
134	Name			138	NFPA health hazard	Flammability	Reactivity
135	Density at ref temp		At	139			
136				140			

Rev	Date	Revision Description	By	Appv1	Appv2	Appv3	REMARKS

Form: 20A1001 Rev 0

1	RESPONSIBLE ORGANIZATION		ANALYSIS DEVICE COMPOSITION OR PROPERTY Operating Parameters (Continued)	6	SPECIFICATION IDENTIFICATIONS			
2				7	Document no			
3	(ISA)			8	Latest revision		Date	
4				9	Issue status			
5				10				
11	PROCESS COMPOSITION OR PROPERTY				MEASUREMENT DESIGN CONDITIONS			
12	Component/Property Name	Normal	Units	Minimum	Units	Maximum	Units	Repeatability
13–75								

Rev	Date	Revision Description	By	Appv1	Appv2	Appv3	REMARKS

Form: 20A1002 Rev 0

© 2004 ISA

Definition

Rockwell scale is a hardness scale based on the indentation of materials. The Rockwell test determines the hardness by measuring the depth of penetration of an indenter under a large load compared to the penetration made by a preload. There are different scales, denoted by single letters (Rockwell A, B, C, etc.), corresponding to the different loads or indenters. The result is a dimensionless number noted as HRA, HRB, HRC, etc., where the last letter refers to the respective Rockwell scale.

Abbreviations

CR	Corrosion rate
DCMS	Downhole corrosion-monitoring system
DE	Change in polarization voltage
DI	Change in current flow
LPR	Linear polarization resistance
mmpy	Millimeters per year

Bibliography

ASTM, Corrosion Standards, (2015) http://www.astm.org/Standards/corrosion-and-wear-standards.html.

Fontana, M. G., Corrosion engineering, (2011) http://www.amazon.com/Corrosion-Engineering-Materials-Science/dp/0071003606.

ION Science, Handheld corrosion monitor, (2015) http://www.ionscience.com/products/hydrosteel-6000-handheld-corrosion-monitor.

Leygraf, C. and Graedel, T., Atmospheric corrosion, (2001) http://www.electrochem.org/dl/interface/wtr/wtr01/IF12-01-Pages24-30.pdf.

McCafferty, E. Introduction to corrosion science, (2010) http://www.springer.com/us/book/9781441904546.

PDF: Schweityer Corriosion Tables, (2012) http://pdfbit.com/sc/schweitzer-corrosion-tables-pdf.html.

Perry, R. H., Ed., *Chemical Engineers Handbook*, (2008) http://www.amazon.com/Perrys-Chemical-Engineers-Handbook-Edition/dp/0071422943.

RCS, Corrosion monitoring sensors, (2013) http://www.cosasco.com/.

Roberge, P., Corrosion engineering, (2008) http://www.amazon.com/Corrosion-Engineering-Principles-Pierre-Roberge/dp/0071482431.

Rohrback Cosasco Systems, Inc., Corrosion monitoring primer, 2013 http://www.cosasco.com/.

Zakipour, S. and Leygraf, C., Quartz crystal microbalance applied to studies of atmospheric corrosion of metals (1992) http://www.maneyonline.com/doi/abs/10.1179/bcj.1992.27.4.295.

1.18 Cyanide Analyzers: Weak-Acid Dissociable (WAD)

R. SREENEVASAN (2017)

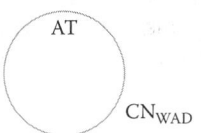

Applications	Free cyanide (CN_F)/weak acid dissociable (CN_{WAD}) limit compliance monitoring, effluent monitoring, control cyanide destruction/cyanidation process
Types	A. Laboratory analyzer (gas diffusion + amperometry)
	B. Online analyzer [(picric acid + colorimetry) or (gas diffusion + amperometry)], U.S. EPA standard)
	C. Online analyzer (distillation + colorimetry, EN I SO standard)
Range	0–50 ppm to 0–3000 ppm CN_{WAD} (online analyzers)
	2–5000 ppb available cyanide (laboratory analyzers)
	2–5000 ppb total cyanide (laboratory analyzers)
	5–500 ppb postdistillation cyanide (laboratory analyzers)
Inaccuracy	<5% for samples analyzed at the midpoint of full scale (laboratory analyzers)
	±3% of full scale (online analyzers)
Cost	Free or available cyanide laboratory analyzer ~$35,000
	Total cyanide laboratory analyzer ~$39,000
	Free cyanide process analyzer ~$44,000 to $70,000
Partial list of suppliers	CyPlus GmbH (B, C) http://www.cyplus.com/Navigation/Products/CyPlus_CCS.asp
	Metrohm Applikon (B) http://www.metrohm-applikon.com/Products/ADI2045TI.html
	Mintek (B) http://www.mintek.co.za/wp-content/uploads/2011/10/WAD-Cynoprobe.pdf
	NextChem Process Analyzers (A) http://nextchem-analyzers.com/wp-content/themes/nextchem/pdf/Cyanide.pdf
	O I Analytical (A, B) http://www.oico.com/default.aspx?id=product&productid=131, http://www.oico.com/default.aspx?id=product&productid=94
	Process Analytical Systems (B) http://www.process-analytical.co.za/prod05.html, http://www.process-analytical.co.za/prod01.html
	Skalar (A) http://www.skalar.com/analyzers/basic-colorimetric-analysis-for-the-environmental-field

INTRODUCTION

These days, the treatment of mine effluent and discharges for cyanide removal and/or destruction is one of the most important conditions to obtain a permit to operate a gold extraction plant.* Prior to 1980, gold extraction plants relied on natural degradation of cyanide compounds in the tailings ponds or dams. The explosion of gold mining and extraction since the early 1980s has caused regulators around the world to demand additional cyanide destruction in the tailings dam to avoid environmental problems.

A tailings impoundment facility at Baia Mare, Romania, breached in 2000, released a cyanide plume, which was detectable as far down as 2000 km downstream, killed a large number of fish in Tisza and Danube Rivers, and disrupted water supply. This and other calls to ban cyanide use in gold extraction plants resulted in an unprecedented effort by the global mining industry, United Nations Environmental Programme, the International Council of Mining and Metals,

---
* Process and Environmental Chemistry of Cyanidation, 2001. CyPlus GmbH.

the Gold Institute, and the World Wildlife Fund for Nature to develop the International Cyanide Management Code. This new code is administered by the International Cyanide Management Institute (ICMI).

CYANIDE SPECIES

Cyanides are present as three (3) different species, namely, free or reactive cyanide (HCN, CN$^-$), weak acid dissociable or weak to moderately strong metal complexes (Ag, Cd, Cu, Hg, Ni, Zn), and strong metal complexes (Fe, Co, Pt, Au).

Free cyanide (hydrogen cyanide [HCN] and cyanide ion [CN$^-$]) is usually released at pH 6. It is hundreds of times more toxic to aquatic organisms than terrestrial organisms.

Weak acid dissociable cyanide (available and amenable) consists of free, simple (Na, K), weak metal compounds (Cd, Zn) and moderate metal compounds (Ag, Cu, Hg, Ni) and is released under mild acid condition pH 3–6.

Total cyanide consists of free, weak acid dissociable, and strong metal cyanide compounds (Au, Fe, Co, Pt) and is released under strong acid conditions (pH < 2) plus high heat and catalyst or UV dissociation. Cyanate (CNO$^-$) and thiocyanate (SCN$^-$) are excluded from the total cyanide definition.

CYANIDE ANALYSIS: LABORATORY METHODS

All cyanide measurement methods follow these generalized steps:

- Generate HCN from sample matrix.
- Method parameters define CN "species."

- HCN separation is achieved by manual distillation or automated distillation or gas diffusion (gas permeable membrane).
- Measure CN.

FREE CYANIDE (CN$_F$)

The only viable method for the determination of CN$_F$ at a gold mine is the silver nitrate titration.* The differences occurring in practice are the methods used for endpoint determination, which is the crucial point of this analytical process. When silver nitrate is added to the titration sample containing free cyanide, all silver ions are immediately complexed. Only when all free cyanides in the solution are complexed, excess silver ions will be present. The presence of the surplus silver ions in the solution is used to determine the endpoint of titration.

Methods used to determine the endpoint of silver nitrate titration include the following:

- Potassium iodide indicator (least reliable)
- Rhodanine indicator (requires experienced operators, prone to interferences)
- Potentiometric detection (most reliable)

Automated method to determine free cyanide in aqueous solutions is defined by ASTM D7237. A sample is injected into a flowing stream of buffer solution (Lipps W). The generated HCN$^-$ is allowed to diffuse through a membrane into a flowing stream of base solution (assume KOH is the base solution, then HCN + KOH = KCN + H$_2$O). The resulting CN$^-$ strength is measured by a very sensitive amperometric detector (see Figure 1.18a).

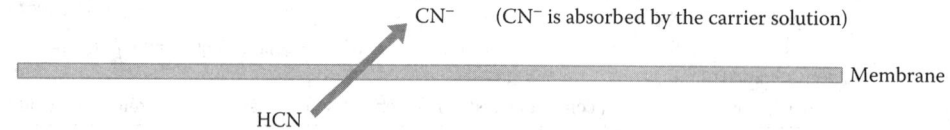

FIG. 1.18a
Gas diffusion through a membrane.

---

* Process and Environmental Chemistry of Cyanidation, 2001. CyPlus GmbH.

Available Cyanide (CN_{WAD})

The traditional ASTM D2036 method requires 3–4 hr to determine CN_{WAD} present in a sample, while the OIA 1677 or ASTM D6888 method requires only 2 min to determine CN_{WAD} present in the same sample.

In the pretreatment step, ligand exchange reagents are added at room temperature to 100 ml cyanide-containing sample (U.S. EPA OIA-1677: 2004). The ligand exchange reagents form very stable complexes with various transition metals resulting in the release cyanide ion from metal cyanide complexes.

Cyanide analysis is accomplished using a flow injection analysis (FIA) system. Pretreated sample is injected into the flow injection manifold. The addition of hydrochloric acid converts cyanide ion to HCN, which passes under a gas diffusion membrane. The HCN diffuses through the membrane into an alkaline receiving solution, where it is converted back to cyanide ion. The cyanide ion is monitored with a silver working electrode, silver/silver chloride reference electrode, and platinum/stainless steel counter electrode at an applied potential of zero volt. The current generated is proportional to the cyanide concentration present in the original sample (see Figure 1.18b).

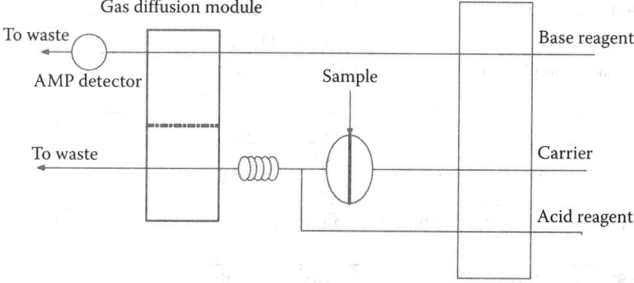

FIG. 1.18b
OIA 1677 configuration.

Picric acid method allows CN_{WAD} to be determined in a much shorter time (~25 min) than the time required for ISO/DIS or ASTM methods.* Since the picric acid procedure is very sensitive, only CN_{WAD} concentrations <~3 ppm can be measured directly. Higher concentrations, however, can be measured reliably by dilution.

The basic chemistry of the picric acid method is a straightforward example of a colorimetric reaction. The dissolved metal alkali picrate (a salt of picric acid obtained by dissolving in aqueous sodium hydroxide solution) is converted into the colored salt of the isopurpuric acid when reacting with cyanide. The intensity of the formed color is directly proportional to the concentration of CN_{WAD} in the sample. The intensity of the color is measured spectrophotometrically at 520 nm and the result is evaluated by using a calibration curve.

A further colorimetric procedure for CN_{WAD} determination is the pyridine/barbituric acid method. This method is designed to trace cyanide concentrations in the range of 2 ppb–5 ppm. Since a number of the typical constituents of the solutions from a typical cyanidation plant interfere with the pyridine/barbituric acid method, it is rarely used for direct CN_{WAD} analysis* (see Table 1.18a).

TABLE 1.18a
Comparison of ASTM Methods for Available Cyanide

	ASTM D2036	OIA 1677 or ASTM D6888
Step 1	Add Cl_2 to half sample. Spin for 1 hr.	Add ligands and run on analyzer.
Step 2	Distill Cl_2 half. Distill total half.	—
Step 3	Run on analyzer. Subtract results.	—
Sample preparation	Two distillations. 2–3 hr	Gas diffusion.
Analysis	1–2 min	1–2 min
Total time	3–4 hr	1–2 min

Total Cyanide (CN_T): Manual Distillation

For the analytical determination of total cyanide, strong acid reaction conditions and elevated temperatures are required to liberate the cyanide ions from the stable ferrocyanide and ferricyanide complexes. Since no colorimetric reaction for the indication of cyanides is known to operate under such harsh conditions, the determination of total cyanide is only possible by using distillation methods in a typically equipped laboratory at a cyanidation plant* (see Table 1.18b).

TABLE 1.18b
Available Manual Methods to Determine Total Cyanide

Method Number	Description	Measurement
EPA 335.4	Midi distillation $MgCl_2$	Automated colorimetry
ASTM D2036	Midi/micro or macro distillation $MgCl_2$	Colorimetry/ion-sensitive electrode (ISE)/ amperometry
ASTM D7284	Midi/micro distillation $MgCl_2$	Gas diffusion—amperometry

All manual distillation methods for total cyanide are essentially the same and will obtain the same result. All share the same interferences that occur during the distillation process.

---

*Process and Environmental Chemistry of Cyanidation, 2001. CyPlus GmbH.

Total Cyanide (CN_T): Automated (Table 1.18c)

TABLE 1.18c
Available Automated Methods to Determine Total Cyanide

Method Number	Description	Measurement
ASTM D4374 (Kelada 01)	High-power UV—auto distillation Alkaline pH	Automated colorimetry
EPA 335.3	Low-power UV—auto distillation pH < 2	Automated colorimetry
ASTM D7511	Low-power UV pH < 2	Gas diffusion—amperometry

PRIMARY AND ALTERNATE ANALYTICAL METHODS (TABLE 1.18d)

TABLE 1.18d
Primary and Alternate Methods to Determine Cyanides

Analyte	Method	Comments
Free cyanide (CN_F)	$AgNO_3$ titration	Preferred method
	$AgNO_3$ titration with potentiometric endpoint determination	Alternate method with precise endpoint determination
	Microdiffusion of HCN from static sample into NaOH (ASTM D4282)	Alternate method
	Ion-selective electrode	Alternate method
	Direct colorimetry	Alternate method
	Amperometric determination	Alternate method
Weak acid dissociable cyanide (CN_{WAD})	Manual distillation pH 4.5 + potentiometric or colorimetric finish (ISO 6703-2)	Preferred method
	Amenable to chlorination (ASTM D2036-B/U.S. EPA 9010)	Alternate method
	Segmented flow injection analysis (SFIA) in-line micro distillation pH 4.5 + colorimetric finish (ASTM D4374)	Alternate method
	Flow injection analysis (FIA), in-line ligand exchange + amperometric finish (U.S. EPA OIA-1677)	Alternate method
	Picric acid, colorimetric determination	Alternate method
Total cyanide (CN_T)	Manual batch distillation + titration/potentiometric or colorimetric finish (ISO 6703-1)	Preferred method
	SFIA, in-line UV irradiation, micro distillation + colorimetric finish (ASTM D4374)	Alternate method

Source: ICMI sampling and analysis.

ONLINE CYANIDE ANALYSIS METHODS

Cyanide is clearly the lixiviant of choice for gold leaching, with gold cyanidation being the most commonly used metallurgical technique throughout the world (van der Merwe 2010). Thiosulfate was first proposed as a way to recover precious metals in the early 1900s, but due to technical uncertainties, it has never been adopted by the gold industry (Rumball 2006). The cyanidation process has, however, been the topic of much controversial debate due to the highly toxic nature of cyanide and the occurrence of environmental disasters over the years.

Risks associated with cyanide handling can be minimized by taking proper precautions. Gold processing plants are required to enforce correct practices. Governments around the world have introduced stricter regulations to prevent potential environmental incidents and disasters.

The Cyanide Code specifies the following WAD cyanide concentration limits:

- 50 ppm CN_{WAD} in tailings storage facility (TSF)
- 0.5 ppm CN_{WAD} for plant effluents
- 22 ppb CN_F or 22 ppb CN_{WAD} as surrogate in the postmixing zone of river discharge

First, while laboratory-type cyanide analyzers have been in use for a long time, analyzers capable of surviving the rigors of an operating process plant environment have become available only in recent times (early 2000). Second, the reagent preparation time and shelf life of prepared reagents affected the widespread acceptance of online cyanide analyzers. Adoption of the ligand exchange method (OIA-1677) by the U.S. Environmental Protection Agency (EPA) as an international standard accelerated the acceptance of online cyanide analyzers by gold plant operators.

At the present time, two types of online cyanide analyzers are available: one based on picric acid/colorimetric measurement and the other based on gas diffusion/amperometric measurement. A very recent development is a method involving distillation with subsequent colorimetric measurement using barbituric acid. This method has the additional advantage of offering an online method to determine the weak acid dissociable cyanide according to the internationally recognized EN ISO standard.

COMPARISON OF ONLINE CYANIDE ANALYZERS

Amperometric analyzer capabilities have been increased to monitor and control an entire small-to-medium gold processing plant (van der Merwe 2010) (see Table 1.18e).

TABLE 1.18e
Comparison of Online Cyanide Analyzers

Features	Colorimetric Analyzers	Amperometric Analyzers
Cyanide species	CN_{WAD}	CN_F and CN_{WAD}
Range	0–50 ppm (up to 1000 ppm and more by dilution)	0–3000 ppm
Measurement cycle time	~20 min	~5 min
Application areas	Compliance monitoring for CN_{WAD} limits, cyanide destruction	Compliance monitoring including effluent limits, cyanide destruction, cyanidation control
Number of known suppliers/manufacturers	Three (3)	Two (2)

CONCLUSIONS

The adoption of simplified and enhanced international cyanide measurement standards has enabled manufacturers to offer online cyanide analyzers. Similarly, the widespread adoption of voluntary "Cyanide Code [ICMI]" by gold plant operators is facilitating the installation and acceptance of online cyanide analyzers to both minimize cyanide consumption and prevent environmental incidents.

SPECIFICATION FORMS

When specifying cyanide analyzers only, please use the ISA form 20A1001, and when specifying both the analyzer and the composition or the properties of the process material of interest, use ISA form 20A1002. Both forms are reproduced with the permission of the International Society of Automation on the next pages.

Analytical Measurement

1	RESPONSIBLE ORGANIZATION		ANALYSIS DEVICE		6	SPECIFICATION IDENTIFICATIONS		
2					7	Document no		
3	(ISA)		Operating Parameters		8	Latest revision	Date	
4					9	Issue status		
5					10			

11	ADMINISTRATIVE IDENTIFICATIONS			40	SERVICE IDENTIFICATIONS Continued			
12	Project number		Sub project no	41	Return conn matl type			
13	Project			42	Inline hazardous area cl		Div/Zon	Group
14	Enterprise			43	Inline area min ign temp		Temp ident number	
15	Site			44	Remote hazardous area cl		Div/Zon	Group
16	Area	Cell	Unit	45	Remote area min ign temp		Temp ident number	
17				46				
18	SERVICE IDENTIFICATIONS			47				
19	Tag no/Functional ident			48	COMPONENT DESIGN CRITERIA			
20	Related equipment			49	Component type			
21	Service			50	Component style			
22				51	Output signal type			
23	P&ID/Reference dwg			52	Characteristic curve			
24	Process line/nozzle no			53	Compensation style			
25	Process conn pipe spec			54	Type of protection			
26	Process conn nominal size	Rating		55	Criticality code			
27	Process conn termn type	Style		56	Max EMI susceptibility		Ref	
28	Process conn schedule no	Wall thickness		57	Max temperature effect		Ref	
29	Process connection length			58	Max sample time lag			
30	Process line matl type			59	Max response time			
31	Fast loop line number			60	Min required accuracy		Ref	
32	Fast loop pipe spec			61	Avail nom power supply			Number wires
33	Fast loop conn nom size	Rating		62	Calibration method			
34	Fast loop conn termn type	Style		63	Testing/Listing agency			
35	Fast loop schedule no	Wall thickness		64	Test requirements			
36	Fast loop estimated lg			65	Supply loss failure mode			
37	Fast loop material type			66	Signal loss failure mode			
38	Return conn nominal size	Rating		67				
39	Return conn termn type	Style		68				

69	PROCESS VARIABLES	MATERIAL FLOW CONDITIONS			101	PROCESS DESIGN CONDITIONS		
70	Flow Case Identification			Units	102	Minimum	Maximum	Units
71	Process pressure				103			
72	Process temperature				104			
73	Process phase type				105			
74	Process liquid actl flow				106			
75	Process vapor actl flow				107			
76	Process vapor std flow				108			
77	Process liquid density				109			
78	Process vapor density				110			
79	Process liquid viscosity				111			
80	Sample return pressure				112			
81	Sample vent/drain press				113			
82	Sample temperature				114			
83	Sample phase type				115			
84	Fast loop liq actl flow				116			
85	Fast loop vapor actl flow				117			
86	Fast loop vapor std flow				118			
87	Fast loop vapor density				119			
88	Conductivity/Resistivity				120			
89	pH/ORP				121			
90	RH/Dewpoint				122			
91	Turbidity/Opacity				123			
92	Dissolved oxygen				124			
93	Corrosivity				125			
94	Particle size				126			
95					127			
96	CALCULATED VARIABLES				128			
97	Sample lag time				129			
98	Process fluid velocity				130			
99	Wake/natural freq ratio				131			
100					132			

133	MATERIAL PROPERTIES			137	MATERIAL PROPERTIES Continued		
134	Name			138	NFPA health hazard	Flammability	Reactivity
135	Density at ref temp		At	139			
136				140			

Rev	Date	Revision Description	By	Appv1	Appv2	Appv3	REMARKS

Form: 20A1001 Rev 0 © 2004 ISA

1.18 Cyanide Analyzers: Weak-Acid Dissociable (WAD)

	RESPONSIBLE ORGANIZATION		ANALYSIS DEVICE COMPOSITION OR PROPERTY Operating Parameters (Continued)			SPECIFICATION IDENTIFICATIONS		
1					6			
2					7	Document no		
3	ISA				8	Latest revision		Date
4					9	Issue status		
5					10			

	PROCESS COMPOSITION OR PROPERTY			MEASUREMENT DESIGN CONDITIONS				
	Component/Property Name	Normal	Units	Minimum	Units	Maximum	Units	Repeatability
13								
14								
15								
16								
17								
18								
19								
20								
21								
22								
23								
24								
25								
26								
27								
28								
29								
30								
31								
32								
33								
34								
35								
36								
37								

Rev	Date	Revision Description	By	Appv1	Appv2	Appv3	REMARKS

Form: 20A1002 Rev 0 © 2004 ISA

Abbreviations

CNF   Free cyanide
CNT   Total cyanide
CNWAD   Weak acid dissociable cyanide
FIA   Flow injection analysis
HCN   Hydrogen cyanide
TSF   Tailings storage facility

Organizations

ASTM   American Society for Testing and Materials
EPA   Environmental Protection Agency
ICMI   International Cyanide Management Institute

Bibliography

Berthold, M. Available Cyanide Sampling and Analysis. Michigan Department of Natural Resources and Environment. http://www.mi-wea.org/docs/Available%20Cyanide%20Sampling%20and%20Analysis.pdf. Accessed June 18, 2012.

Cyanide Management. Environment Australia. Department of the Environment. Australian Government. http://www.ret.gov.au/resources/Documents/LPSDP/BPEMCyanide.pdf. Accessed June 11, 2012.

Cyanide Management Handbook. 2008. Australian Government. http://www.ret.gov.au/resources/documents/lpsdp/lpsdp-cyanidehandbook.pdf. Accessed June 11, 2012.

Lipps, W. An overview and comparison of methods for cyanide analysis. OI analytical. http://ezinearticles.com/?Free,-Available,-and-Total-Cyanide-Analysis-Using-Gas-Diffusion-Amperometry&id=6321765.

Lipps, W. Free, available and total cyanide analysis using gas diffusion amperometry.

Lipps, W. Fundamentals of cyanide chemistry.

Mulpeter, T. and Kotzen, D. 2007. Online cyanide analyzers optimize cyanidation and ensure cyanide code compliance for plant tailings. *World Gold Congress 2007*, Cairns, QLD, Australia, 2007.

Rumball, J. 2006. *Thiosulfate Shows Potential for Gold*. Parker CRC for Integrated Hydrometallurgy Solutions, CSIRO.

The International Cyanide Management Code. 2011. International Cyanide Management Institute. http://www.cyanidecode.org/cyanide-facts/sample-analysis.

van der Merwe, W. and Breuer, P. Cyanide analysis for complex cyanide solutions. http://ezinearticles.com/?Free,-Available,-and-Total-Cyanide-Analysis-Using-Gas-Diffusion-Amperometry&id=6321765.

van der Merwe, W. A. M., Lotz, P., and Smith, H.S. 2008. Recent refinements to an online weak acid dissociable cyanide measurement device. *MetPlant 2008*, Perth, WA, Australia, 2008.

van der Merwe, W., Lotz, P., and Smith, V.C. 2007. A case study for the development of an online weak acid dissociable cyanide measurement device. *World Gold Congress 2007*, Santiago, Chile, 2010.

van der Merwe, W. and Smith, S. 2010. Cyanide monitoring, control and destruction facilitated by an online weak acid dissociable cyanide measurement device. *AutoMining 2010*, Santiago, Chile, 2010.

ASTM International Standards

http://www.astm.org/bookstore/comps/cyanidecd10.htm.

D4374: Standard test method for cyanides in water—automated methods for total cyanide, weak acid dissociable cyanide and thiocyanate.

D6888: Standard test method for available cyanide with ligand displacement and flow injection analysis (FIA) utilizing gas diffusion separation and amperometric detection. http://www.astm.org/Standards/D6888.htm.

D6996: Standard guide for understanding cyanide species.

D7237: Standard test method for free cyanide with flow injection analysis (FIA) utilizing gas diffusion separation and amperometric detection.

D7284: Standard test method for total cyanide in water by micro distillation followed by flow injection analysis with gas diffusion separation and amperometric detection.

D7511: Standard test method for total cyanide by segmented flow injection analysis, in-line ultraviolet digestion and amperometric detection.

D7728: Standard guide for selection of ASTM analytical methods for implementation of international cyanide management code guidance.

U.S. EPA Standards

335.5:   Determination of total cyanide by semi-automated colorimetry.

9010C: Total and amenable cyanide—distillation. http://www.epa.gov/osw/hazard/testmethods/sw846/pdfs/9010c.pdf.

9012B: Total and amenable cyanide—automated colorimetric, with off-line distillation. http://www.epa.gov/osw/hazard/testmethods/sw846/pdfs/9012b.pdf.

9213:   Potentiometric determination of cyanide in aqueous samples and distillates with ion-selective electrode. http://www.epa.gov/osw/hazard/testmethods/sw846/pdfs/9213.pdf.

OIA-1677: Determination of available cyanide by flow injection, ligand exchange and amperometry. http://water.epa.gov/scitech/methods/cwa/metals/cyanide/upload/2007_08_14_methods_method_cyanide_1677-2004.pdf.

EN ISO Standards

6703-1: Determination of total cyanide. http://www.iso.org/iso/home/store/catalogue_tc/catalogue_detail.htm?csnumber=13141.

6703-2: Determination of easily liberatable cyanide. http://www.iso.org/iso/home/store/catalogue_tc/catalogue_detail.htm?csnumber=13142.

6703-3: Determination of cyanogen chloride. http://www.iso.org/iso/home/store/catalogue_tc/catalogue_detail.htm?csnumber=13143.

14403-1: Determination of cyanide in water by flow injection analysis (FIA). http://www.iso.org/iso/home/store/catalogue_tc/catalogue_detail.htm?csnumber=51083.

14403-2: Determination of cyanide in water by continuous flow analysis(CFA).http://www.iso.org/iso/home/store/catalogue_tc/catalogue_detail.htm?csnumber=52208.

1.19 Differential Vapor Pressure

B. G. LIPTÁK (1995–2017)

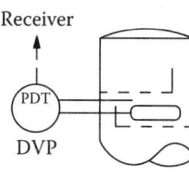

Flow sheet symbol

Application	Product composition control of such binary products as refrigerants or binary processes involving distillation, drying, or evaporation
Inaccuracy	The best d/p cells can be 0.1% FS accurate but are measuring at the operating and not at standard conditions
Design	Any d/p cell can be used as long as its accuracy and range match the requirements of the application and it is provided with a reference bulb, which is connected to one side of the d/p cell and is filled with the appropriate reference fluid
Typical ranges	From 1–0–1 in. H_2O (250–0–250 Pa) up to 500–0–500 in. H_2O (100–0–100 kPa)
Maximum working pressure	1500 psig (105 bars)
Costs	Cost is the same as that of the d/p cell used plus about $500 for the addition of the reference bulb
Partial list of suppliers	Practically any d/p cell can be used if it has a low enough range and sufficient accuracy and one side of it is connected to a bulb containing the reference fluid

INTRODUCTION

Physical property analyzers are covered in detail in Chapter 1.57. Vapor pressure is one of these physical properties of fluids, and they, including the Reid vapor pressure analyzers, are also covered in Chapter 1.57. This chapter concentrates only on the vapor pressure detectors which utilize a d/p detector to measure the vapor pressure difference between that of the process fluid and that of a reference fluid.

For more accurate measurements, the Reid Vapor Pressure (RVP) detectors are used, which operate on the basis of the ASTM D323-99a standard test method for vapor pressure of petroleum products. It is the vapor pressure of a chilled sample of gasoline or other fuel as measured in a test bomb at 100 F. The Reid vapor pressure is applicable only for gasoline, volatile crude oil, and other volatile petroleum products. It is not applicable for liquefied petroleum gases.

These analyzers can be used in processes where the concentration of a binary product is to be controlled. In such cases, the reference bulb is filled with a sample of the desired product. This way, if the vapor pressures of the process and of the reference fluid are closed, the product is good, while when the two move apart (the ΔP is rising), the control system acts to return the product concentration to set point. Such a control configuration is applicable only to processes where either a single product is made or it is acceptable to replace the reference fluid when the product changes.

APPLICATIONS

One common application of vapor pressure analyzers is the production of refrigerants. Vapor pressure is the pressure that a vapor exerts when it is in equilibrium with its own liquid. In some processes, such as refrigeration, it is important that working fluid (such as Freon-21) have a known and accurately predictable relationship between its temperature and its vapor pressure. Therefore, in the production of refrigerants, one of the carefully analyzed characteristics of the product is its vapor pressure. At a constant temperature, the vapor pressure is just as unique characteristic of a liquid as is its boiling point (Table 1.19a). Therefore, by measuring the vapor pressure of a fluid, one can determine the composition of that fluid.

TABLE 1.19a
Refrigerant Characteristics

Refrigerant		Feature						
	Applicable Compressor (R = Reciprocating, RO = Rotary, C = Centrifugal)	Boiling Point in °F[a] at Atmospheric Pressure	Evaporator Pressure in psia[b] if Operating Temperature Is −15°C (5°F)	Condenser Pressure in psia[b] if Operating Temperature Is 30°C (86°F)	Latent Heat in BTU/lbm[c] at −7.8°C (18°F)	Toxic (T), Flammable (F), Irritating (I)	Mixes and/or Compatible with the Lubricating Oil	Chemically Inert and Noncorrosive	Remarks
Ethane C_2H_6	R	−127	236	675	148	T, F	No	Yes	For low-temperature service
Carbon dioxide CO_2	R	−108	334	1039	116	No	Yes	Yes	Low-efficiency refrigerant
Propane C_3H_8	R	−48	42	155	132	T, F	No	Yes	For low-temperature service
Freon-22 $CHClF_2$	R	−41	43	175	92	No	(1)	Yes	High-efficiency refrigerant
Ammonia NH_3	R	−28	34	169	555	T, F	No	(2)	Most recommended
Freon-12 CCl_2F_2	R	−22	26	108	67	No	Yes	Yes	Expansion valve may freeze if water is present
Methyl chloride CH_3Cl	R	−11	21	95	178	(3)	Yes	(4)	Common to these refrigerants
Sulfur dioxide SO_2	R	+14	12	66	166	T, I	No	(4)	a. Evaporator under vacuum
Freon-21 $CHCl_2F$	RO	+48	5	31	108	No	Yes	Yes	b. Low compressor discharge pressure
Ethyl chloride C_2H_5Cl	RO	+54	5	27	175	F, I	No	(5)	c. High volume-to-mass ratio across compressor
Freon-11 CCl_3F	C	+75	3	18	83	No	Yes	Yes	
Dichloromethane CH_2Cl_2	C	+105	1	10	155	No	Yes	Yes	

Notes: (1) Oil floats on it at low temperature; (2) corrosive to copper-bearing alloys; (3) anesthetic; (4) corrosive in the presence of water; (5) attacks rubber compounds.

[a] $°C = \dfrac{°F - 32}{1.8}$
[b] psia = 6.9 kPa.
[c] BTU/lbm = 232.6 J/kg.

If the temperature of the process varies, the measurement of the vapor pressure requires temperature compensation. This compensation can be obtained either by filling a reference bulb with the desired product and comparing its vapor pressure with that of the product being produced. The other option available with microprocessor-based smart instruments (Figure 1.19a) is to provide the needed compensation in software that has stored the relationship between vapor pressure and temperature of the product. Figure 1.19b shows some of the temperature–vapor pressure relationships of a number of liquids.

FIG. 1.19a
Any d/p cell can be used if a reference bulb is connected to one side containing the right reference fluid. Intelligent transmitters are preferred, because they can provide temperature compensation needed. (Courtesy of Honeywell.)

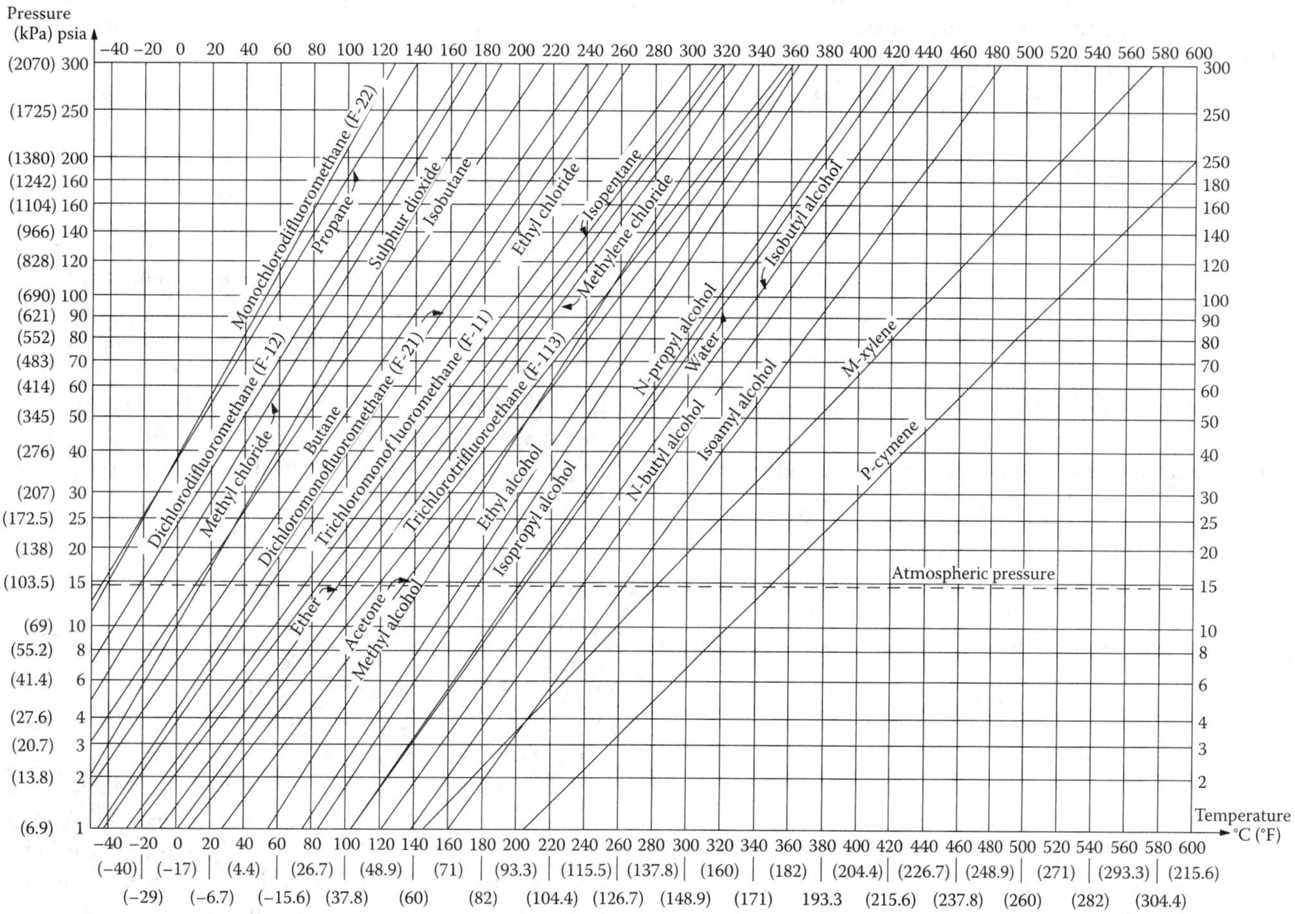

FIG. 1.19b
Relationship between vapor pressure and the temperature of some liquids.

d/p VAPOR PRESSURE TRANSMITTER

The differential vapor pressure transmitter is a regular d/p cell with one of its pressure ports connected to a temperature bulb filled with a reference fluid (Figure 1.19c), while the other side is connected to the process. This way, the sensor can continuously compare the vapor pressure of the process material with the vapor pressure of the sealed-in reference fluid.

If the reference bulb is inserted into the process, there is no need for temperature compensation. This way the temperature of the reference fluid is always the same as that of the process. Consequently, if the two compositions are the same, their vapor pressures will also be identical. Therefore, the set point of the controller is zero with a small gap around it.

The operating principle of this analyzer makes it ideal for product quality control applications. For example, it can be used in distillation, dryer, evaporator, or similar applications where the product is a binary material and the goal is to continuously check the composition of the product.

As was shown in Figure 1.19c, in case of distillation column control applications, a reference fluid-filled bulb is connected to one side of a standard d/p transmitter, and the bulb is inserted into the liquid on the last tray of the column. The other side of the d/p cell is connected to the vapor space above that tray. This way, if the product being made and the reference fluid are the same, their vapor pressures will be identical and the d/p cell will read zero.

This is a very convenient analyzer for product composition control on optimized distillation columns, where the operating pressure is not constant but is "minimized" as a function of the available cooling tower water temperature. In such applications, a temperature measurement will not accurately reflect composition, because the boiling point varies with pressure.

On the other hand, a differential vapor pressure analyzer will correctly measure the binary composition under variable column pressure conditions, because as the column pressure drops, it also lowers the boiling point on the tray. This in turn will lower the reference bulb temperature and with it the vapor pressure inside the reference bulb. Therefore, as long as the tray composition is constant, the pressure differential will also be constant, even if the column pressure varies.

CONCLUSIONS

The limitations of this type of measurement includes the fact that it correctly reflects the composition of only binary materials and that it is also essential for the reference fluid to be stable. In addition, during upsets, the control loop can be effective only if the temperature of the reference fluid in the bulb will quickly follow the transient pressure or temperature changes in the column. In order to speed that response, the bulb and its heat transfer area should be maximized.

SPECIFICATION FORMS

When specifying differential pressure type vapor pressure analyzers, one is basically specifying a d/p detector or transmitter, plus a reference bulb. Therefore, the specification forms used fop d/p devices can also be used for these applications. To specify the application in general, the ISA Form 20P1001 can be used, while when a d/p transmitter is being specified, the ISA Form 20P2301 is applicable. With the approval on the International Society of Automation, these forms can be found on the next pages.

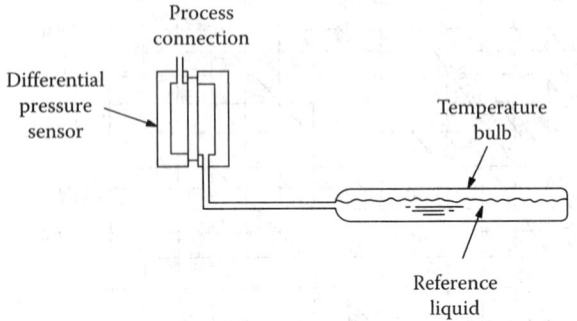

d/p cell measures the difference between the process and the reference vapor pressures.

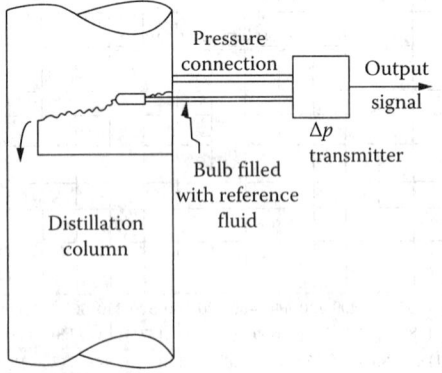

Installation of a d/p vapor pressure analyzer on a process where the bottom product of a distillation tower is controlled.

FIG. 1.19c
Differential vapor pressure transmitter design and installation.

1.19 Differential Vapor Pressure

DIFFERENTIAL PRESSURE DEVICE — Operating Parameters

RESPONSIBLE ORGANIZATION: ISA

#	Field		#	Field	
1			6	SPECIFICATION IDENTIFICATIONS	
2			7	Document no	
3			8	Latest revision	Date
4			9	Issue status	
5			10		

ADMINISTRATIVE IDENTIFICATIONS

12	Project number		Sub project no	
13	Project			
14	Enterprise			
15	Site			
16	Area		Cell	Unit
17				

SERVICE IDENTIFICATIONS

19	Tag no/Functional ident			
20	Related equipment			
21	Service			
22				
23	P&ID/Reference dwg number			
24	High side line/nozzle no			
25	High side conn pipe spec			
26	High side conn nom size		Rating	
27	High side conn termn type		Style	
28	High side conn schedule		Wall thickness	
29	High side conn length			
30	High side conn orient			
31	High side conn matl type			
32	Connection design code			
33	Low side line/nozzle no			
34	Low side conn pipe spec			
35	Low side conn nom size		Rating	
36	Low side conn termn type		Style	
37	Low side conn schedule no		Wall thickness	
38	Low side conn length			
39	Low side conn orientation			
40	Low side conn matl type			
41	Device insertion length			

SERVICE IDENTIFICATIONS Continued

43	Process conn diff length			
44	Zero datum conn length			
45	Inline hazardous area cl		Div/Zone	Group
46	Inline area min ign temp		Temp ident number	
47	Remote hazardous area cl		Div/Zone	Group
48	Remote area min ign temp		Temp ident number	

COMPONENT DESIGN CRITERIA

53	Component type		
54	Component style		
55	Output signal type		
56	Characteristic curve		
57	Type of protection		
58	Criticality code		
59	Max EMI susceptibility		Ref
60	Max temperature effect		Ref
61	Max response time		
62	Min required accuracy		Ref
63	Max dead band		
64	Avail nom power supply		Number wires
65	Min load capability		
66	Testing/Listing agency		
67	Test requirements		
68	Supply loss failure mode		
69	Signal loss failure mode		

PROCESS VARIABLES / MATERIAL FLOW CONDITIONS

#	Condition Identification				Units
75	High side pressure				
76	High side temperature				
77	High side matl phase type				
78	High side density				
79	High side specific gravity				
80	High side matl viscosity				
81	Low side pressure				
82	Low side temperature				
83	Low side matl phase type				
84	Low side density				
85	Low side specific gravity				
86	Low side matl viscosity				
87	Differential pressure				

CALCULATED VARIABLES

97	Calc zero el / suppr				
98	Calc measurement span				

PROCESS DESIGN CONDITIONS

#	Minimum	Maximum	Units
103			

MATERIAL PROPERTIES

132	Low side material name				
133	Low side composition				
134	Low side density at ref		At		
135	Low side NFPA health haz		Flammability	Reactivity	

MATERIAL PROPERTIES Continued

138	High side material name				
139	High side composition				
140	High side density at ref		At		
141	High side NFPA health haz		Flammability	Reactivity	

Rev	Date	Revision Description	By	Appv1	Appv2	Appv3	REMARKS

Form: 20P1001 Rev 1 © 2007 ISA

DIFFERENTIAL PRESSURE TRANSMITTER
Device Specification

1	RESPONSIBLE ORGANIZATION	
2		
3	(ISA)	
4		
5		

	SPECIFICATION IDENTIFICATIONS	
6		
7	Document no	
8	Latest revision	Date
9	Issue status	
10		

TRANSMITTER BODY

11			
12	Body/Flange type		
13	Process conn nominal size	Rating	
14	Process conn termn type	Style	
15	Vent/Drain location		
16	Mounting type		
17	Body/Flange material		
18	Vent/Drain material		
19	Bolting material		
20	Flange adapter material		
21	Gasket/O ring material		
22	Mounting kit material		
23			
24			
25			

SENSING ELEMENT

26		
27	Detector type	
28	Min diff press span	Max
29	Diaphragm/Wetted material	
30	Fill fluid material	
31		
32		
33		

TRANSMITTER

34	
35	Output signal type
36	Enclosure type no/class
37	Characteristic curve
38	Digital communication std
39	Signal power source
40	Transient protection
41	Integral indicator style
42	Signal termination type
43	Cert/Approval type
44	Span-Zero adjust lct
45	Failure/Diagnostic action
46	Enclosure material
47	
48	
49	
50	
51	
52	
53	
54	
55	
56	
57	
58	
59	

PERFORMANCE CHARACTERISTICS

60		
61	Max press at design temp	At
62	Min working temperature	Max
63	Accuracy rating	
64	Diff pressure LRL	URL
65	Min ambient working temp	Max
66–83		

ACCESSORIES

84	
85	Air set filter style
86	Air set gauges
87	Heating kit style
88	Remote indicator style
89	Manifold valve style

SPECIAL REQUIREMENTS

90	
91	
92	Custom tag
93	Reference specification
94	Special preparation
95	Compliance standard
96	Software configuration
97–99	

PHYSICAL DATA

100		
101	Estimated weight	
102	Overall height	
103	Removal clearance	
104	Signal con nominal size	Style
105	Mfr reference dwg	
106–108		

CALIBRATIONS AND TEST

			INPUT			OUTPUT OR SCALE	
110							
111	TAG NO/FUNCTIONAL IDENT	MEAS/SIGNAL/TEST	LRV	URV	ACTION	LRV	URV
112		Diff press-Analog output					
113		Diff pressure-Scale					
114		Diff press-Digital output					
115		Press-Digital output					
116		Temp-Digital output					
117							

COMPONENT IDENTIFICATIONS

	COMPONENT TYPE	MANUFACTURER	MODEL NUMBER
118			
119			
120–125			

Rev	Date	Revision Description	By	Appv1	Appv2	Appv3	REMARKS

Form: 20P2301 Rev 0 © 2001 ISA

Definitions

Reid vapor pressure is based on ASTM D323-99a Standard Test Method for vapor pressure of petroleum products. It is the vapor pressure of a chilled sample of gasoline or other fuel as measured in a test bomb at 100°F. The Reid vapor pressure is applicable only for gasoline, volatile crude oil, and other volatile petroleum products. It is not applicable for liquefied petroleum gases.

True vapor pressure is based on ASTM D2889-95a Standard Test Method for petroleum distillate fuels. It is the pressure of the vapor in equilibrium with the liquid at 100°F (the bubble point pressure at 100°F).

Abbreviation

RVP   Reid vapor pressure

Bibliography

Beaudouin, F., Intelligent transmitters for process control, (1995) http://www.sciencedirect.com/science/article/pii/001905789500012O.

Fielder, D., Vapor pressure with differential thermal measurement, (1990) http://www.dtic.mil/dtic/tr/fulltext/u2/a228699.pdf.

Harland, P. W., Pressure Gage Handbook, (1985) http://www.amazon.com/Pressure-Handbook-Mechanical-Engineering-Series/dp/0824774337.

IHS, Pressure sensors information, (2014) http://www.globalspec.com/learnmore/sensors_transducers_detectors/pressure_sensing/ pressure_sensors_instruments.

Lord, D. and Rudeen, D., Analysis of crude oil vapor pressures, (2005) http://citeseerx.ist.psu.edu/viewdoc/download?doi=10.1.1.473.9524&rep=rep1&type=pdf.

Miller, R. C., Accurate vapor pressure for refrigerants, (1992) http://docs.lib.purdue.edu/cgi/viewcontent.cgi?article=1186&context=iracc.

Mukhopadhyay, S. C., (2013) Intelligent sensing, instrumentation and measurements, http://www.amazon.com/Intelligent-Sensing-Instrumentation-Measurements-Measurement/dp/3642370268.

Pichler, H., Lutz, J., Why crude oil vapor pressure should be tested prior to rail transport, (2008) http://www.cscanada.net/index.php/aped/article/viewFile/5098/pdf_2.

Saylor, Pressure measurement, (2014) http://www.saylor.org/site/wp-content/uploads/2011/04/Pressure_measurement.pdf.

Tomczyk, J., Refrigerant pressures, states and conditions, (2005) http://www.achrnews.com/articles/refrigerant-pressures-states-and-conditions.

Wika Handbook, Pressure measurement, (2012) http://www.wika.us/upload/BR_RF_Handbook_en_us_18447.pdf.

1.20 Dioxin and Persistent Organic Pollutants Analyzers

D. H. F. LIU (1995)   **B. G. LIPTÁK** (2003, 2017)

Applications	Processes that inadvertently produce persistent organic pollutants (POP) such as industrial processes in which traces of chlorine are present, including combustion processes like waste incineration, chemical manufacturing, and paper bleaching
Type of sample	Gas sample containing particulates
Standard design pressure	Generally atmospheric or near atmospheric
Standard design temperature	–32°C to 815°C (–25°F to 1500°F)
Sample velocity	2–50 m/sec (7–167 ft/sec)
Materials of construction	316 or 304 stainless steel for pitot tubes; nickel, nickel-plated stainless, quartz, or borosilicate glass for nozzles
Cost	$8,000–$16,000 for a complete EPA particulate sampling system (EPA Reference Method 23); $75,000–$100,000 for laboratory-scale GC–MS unit
Partial list of suppliers	ALS Laboratory (http://www.alsglobal.eu/eng/category_dioxins.asp) EPA (http://water.epa.gov/scitech/methods/cwa/organics/dioxins/upload/2007_07_10_methods_method_dioxins_1613.pdf) MPU Gmb. (http://www.dioxin.de/) Pace Analytical (http://www.pacelabs.com/environmental-services/specialty-services/dioxins-furans/dioxin-technical-bulletins.htm) Sigma-Aldrich (http://www.sigmaaldrich.com/analytical-chromatography/sample-preparation/dioxin-prep-system.html) Supelco Analytical (http://www.sigmaaldrich.com/etc/medialib/docs/Supelco/General_Information/dioxin-brochure-jxb.Par.0001.File.tmp/dioxin-brochure-jxb.pd) Xenobiotic Detection Systems (http://www.epa.gov/wastes/hazard/wastemin/minimize/factshts/dioxfura.pdf)

INTRODUCTION

Dioxins is a general name for a large group of chemical compounds known as persistent organic pollutants (POPs). They are produced inadvertently by industrial processes in which traces of chlorine are present such as combustion processes like waste incineration, chemical manufacturing, and paper bleaching. Dioxins and furans can be found in the air, in water, and in contaminated soil. They are harmful to human health and and biologically are nearly nondegradable (persistent). As a result, their emission into the environment and food chain is strictly controlled. As a consequence, low levels of contamination have to be detected, providing a challenge to sample preparation and detection systems.

This chapter deals with the sampling and analysis of recalcitrant dioxin compounds. These measurements are required in ecological risk assessment and in the determination of the toxicity of various samples.

Dioxins and dioxin-like compounds (DLCs) are by-products of various industrial processes and are commonly regarded as highly toxic environmental pollutants and persistent organic pollutants (POPs). They include dioxins, which are polychlorinated dibenzo-*p*-dioxins (PCDDs). There are 75 PCDDs, and 7 of them are specifically toxic. In addition, there are polychlorinated biphenyls (PCBs), which also are not dioxins, but 12 of them have "dioxin-like" properties. Under certain conditions, PCBs may form dibenzofurans through partial oxidation.

The group of DLCs also includes furans, which are polychlorinated dibenzofurans (PCDFs). There are 135 congeners (derivatives differing only in the number and location of chlorine atoms). While, strictly speaking, they are not dioxins, 10 of them have "dioxin-like" properties.

The U.S. Environmental Protection Agency (EPA) regulates emissions from municipal waste combustors (MWCs) and sets emission limits for PCDDs and PCDFs. This chapter provides a brief summary of EPA Method 23, which is used in measuring the PCDD and PCDF emissions from municipal waste combustion (MWC) processes.

SAMPLING SYSTEMS

Figure 1.20a is a schematic of the sampling train that is used to measure the emission of PCDDs and PCDFs from MWCs.

A sample is withdrawn isokinetically from the stack through a probe, a filter, and a trap packed with Amberlite XAD-2 resin. Figure 1.20b shows the condenser and absorbent trap. The PCDDs and PCDFs are collected in the probe, on the filter, and on the solid adsorbent.

Chapter 1.3 provides a detailed discussion of the features and operation of stack sampling systems.

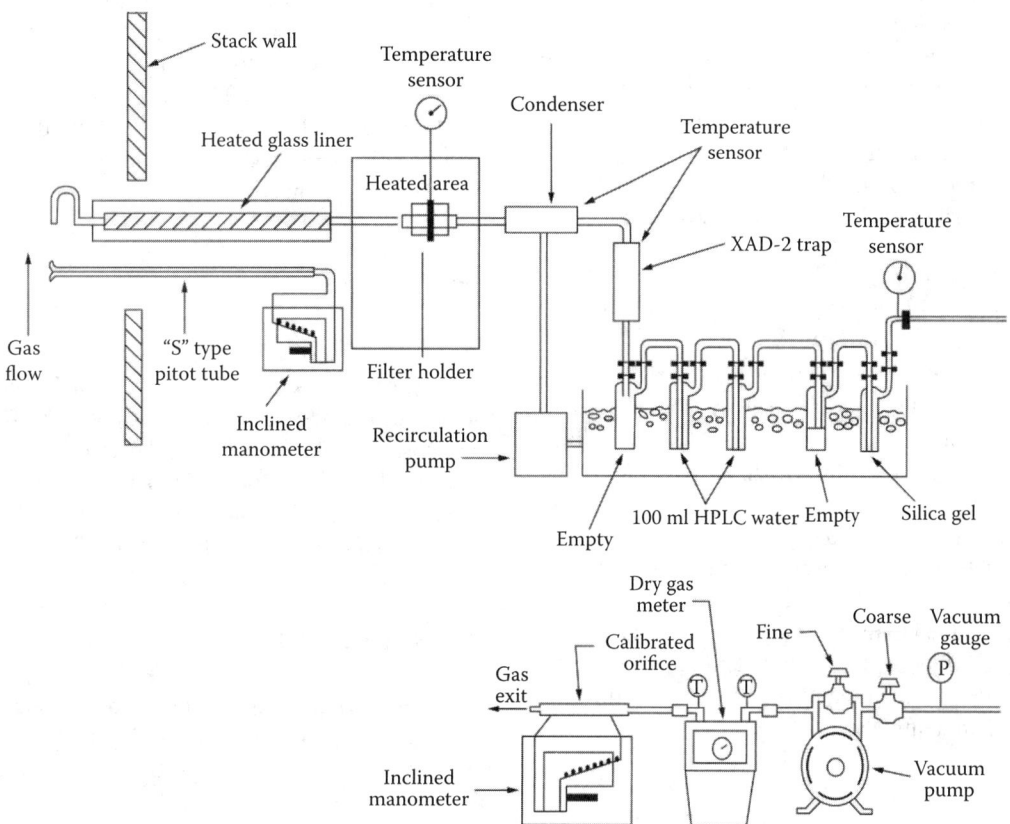

FIG. 1.20a
Sampling train.

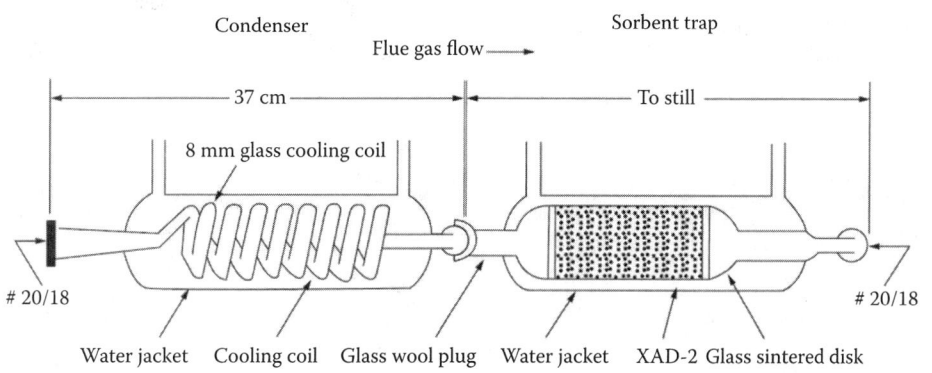

FIG. 1.20b
Condenser and absorbent trap.

Sample Recovery

At the end of the sampling period, the cleanup procedure begins as soon as the probe is removed from the stack. The procedure involves the collection of the deposited particulate matter in three containers (not shown in the figures) as follows. First, any particulate matter and fibers that adhere to the filter holder gasket are carefully transferred to a container (container 1). Next, the absorbent module (XAD-2 sorbent trap in Figure 1.20a) is removed from the train. After that, the material that is deposited in the nozzle, probe transfer lines, and the front half of the filter holder is quantitatively recovered first by brushing, while rinsing each with acetone. Then the probe is rinsed with methylene chloride and all the rinses are collected. The back half of the filter and the condenser are also rinsed with acetone, and the connecting line is soaked with methylene chloride. All the earlier rinses are collected in container 2. After that, the rinsing is repeated with toluene as the rinse solvent and the rinses are collected in container 3.

Sample Extraction

All the PCDDs and PCDFs are extracted from the particulate matter, the absorbent, and the rinses. The absorbent module (XAD-2) and the particulate cake from the filter are extracted with toluene in the Soxhlet apparatus.* The content of container 2 is evaporated at temperatures less that 37°C and is concentrated to dryness. This residue contains the particulate matter removed in the rinse of the train probe and nozzle.

The residue is added to the filter and the XAD-2 resin for extraction in the Soxhlet apparatus. In order to recover all the PCDDs and PCDFs, the material in container 3 goes through a series of multistep treatments.

Analysis

The extracted PCDDs and PCDFs are separated by high-resolution gas chromatography (Chapter 1.10), and each isomer is identified and measured with high-resolution mass spectrometry (Chapter 1.35). The total PCDDs and PCDFs are the sum of the individual isomers.

The gas chromatograph uses an oven that can maintain the separation column at the proper operating temperature and perform programmed increases in temperature at rates of at least 40°C/min (72°F/min). It uses a fused silica column with a length of 183 ft and an inside diameter (ID) of 0.01 in. (60 m × 0.25 mm ID), coated with DB-5, and a fused silica column that is 30 m × 0.25 mm ID (91 ft × 0.01 in.), coated with DB-225.

In operation, the oven temperature starts at 150°C (302°F) and is raised by at least 40°C/min (72°F/min) to 190°C (374°F), and then at 3°C/min (5.4°F/min) up to 300°C (572°F). The mass spectrometer is capable of routine operation at a resolution of 1:10,000 with a stability of ±5 ppm.

On ground glass connections of used glassware, any residual silicone grease sealant should be removed by soaking the glassware in chromic acid cleaning solution first.

CONCLUSIONS

The complexity of this method is such that to obtain reliable results, testers should be trained and experienced with the test procedures. All glass components of the sampling train upstream of and including the absorbing module should be cleaned. The method is described in Chapter 3A of the "Manual of Analytical Methods for the Analysis of Pesticides in Human and Environmental Samples."†

SPECIFICATION FORMS

When specifying dioxin analyzers only, please use the ISA form 20A1001, and when specifying both the analyzer and the composition or the properties of the process material of interest, use ISA form 20A1002. Both forms are reproduced with the permission of the International Society of Automation on the next pages.

---

* For the detailed operation of the Soxhlet extractor, refer to: https://en.wikipedia.org/wiki/Soxhlet_extractor.

† EPA, Manual of analytical methods for the analysis of pesticides in human and environmental samples, (1980) http://www3.epa.gov/ttnemc01/informd/tid-016.pdf.

1.20 Dioxin and Persistent Organic Pollutants Analyzers

1	RESPONSIBLE ORGANIZATION		ANALYSIS DEVICE	6	SPECIFICATION IDENTIFICATIONS			
2				7	Document no			
3	ISA		Operating Parameters	8	Latest revision		Date	
4				9	Issue status			
5				10				

11	ADMINISTRATIVE IDENTIFICATIONS			40	SERVICE IDENTIFICATIONS Continued			
12	Project number		Sub project no	41	Return conn matl type			
13	Project			42	Inline hazardous area cl		Div/Zon	Group
14	Enterprise			43	Inline area min ign temp		Temp ident number	
15	Site			44	Remote hazardous area cl		Div/Zon	Group
16	Area	Cell	Unit	45	Remote area min ign temp		Temp ident number	
17				46				
18	SERVICE IDENTIFICATIONS			47				
19	Tag no/Functional ident			48	COMPONENT DESIGN CRITERIA			
20	Related equipment			49	Component type			
21	Service			50	Component style			
22				51	Output signal type			
23	P&ID/Reference dwg			52	Characteristic curve			
24	Process line/nozzle no			53	Compensation style			
25	Process conn pipe spec			54	Type of protection			
26	Process conn nominal size	Rating		55	Criticality code			
27	Process conn termn type	Style		56	Max EMI susceptibility		Ref	
28	Process conn schedule no	Wall thickness		57	Max temperature effect		Ref	
29	Process connection length			58	Max sample time lag			
30	Process line matl type			59	Max response time			
31	Fast loop line number			60	Min required accuracy		Ref	
32	Fast loop pipe spec			61	Avail nom power supply			Number wires
33	Fast loop conn nom size	Rating		62	Calibration method			
34	Fast loop conn termn type	Style		63	Testing/Listing agency			
35	Fast loop schedule no	Wall thickness		64	Test requirements			
36	Fast loop estimated lg			65	Supply loss failure mode			
37	Fast loop material type			66	Signal loss failure mode			
38	Return conn nominal size	Rating		67				
39	Return conn termn type	Style		68				

69	PROCESS VARIABLES	MATERIAL FLOW CONDITIONS		101	PROCESS DESIGN CONDITIONS		
70	Flow Case Identification		Units	102	Minimum	Maximum	Units
71	Process pressure			103			
72	Process temperature			104			
73	Process phase type			105			
74	Process liquid actl flow			106			
75	Process vapor actl flow			107			
76	Process vapor std flow			108			
77	Process liquid density			109			
78	Process vapor density			110			
79	Process liquid viscosity			111			
80	Sample return pressure			112			
81	Sample vent/drain press			113			
82	Sample temperature			114			
83	Sample phase type			115			
84	Fast loop liq actl flow			116			
85	Fast loop vapor actl flow			117			
86	Fast loop vapor std flow			118			
87	Fast loop vapor density			119			
88	Conductivity/Resistivity			120			
89	pH/ORP			121			
90	RH/Dewpoint			122			
91	Turbidity/Opacity			123			
92	Dissolved oxygen			124			
93	Corrosivity			125			
94	Particle size			126			
95				127			
96	CALCULATED VARIABLES			128			
97	Sample lag time			129			
98	Process fluid velocity			130			
99	Wake/natural freq ratio			131			
100				132			

133	MATERIAL PROPERTIES			137	MATERIAL PROPERTIES Continued		
134	Name			138	NFPA health hazard		Flammability	Reactivity
135	Density at ref temp		At	139			
136				140			

Rev	Date	Revision Description	By	Appv1	Appv2	Appv3	REMARKS

Form: 20A1001 Rev 0

© 2004 ISA

	RESPONSIBLE ORGANIZATION	ANALYSIS DEVICE COMPOSITION OR PROPERTY Operating Parameters (Continued)		SPECIFICATION IDENTIFICATIONS	
1			6		
2			7	Document no	
3	ISA		8	Latest revision	Date
4			9	Issue status	
5			10		

	PROCESS COMPOSITION OR PROPERTY			MEASUREMENT DESIGN CONDITIONS				
11	Component/Property Name	Normal	Units	Minimum	Units	Maximum	Units	Repeatability
12								
13								
...								
37								

Rev	Date	Revision Description	By	Appv1	Appv2	Appv3	REMARKS

Form: 20A1002 Rev 0 © 2004 ISA

Definition

Dioxins is a general name for a large group of chemical compounds known as persistent organic pollutants (POPs). They are produced inadvertently by industrial processes in which traces of chlorine are present such as combustion processes like waste incineration, chemical manufacturing and paper bleaching. Dioxins and furans can be found in the air, in water, and in contaminated soil. They are harmful to human health and and biologically are nearly non-degradable (persistent). As a result, their emission into the environment and food chain is strictly controlled. As a consequence, low levels of contamination have to be detected, providing a challenge to sample preparation and detection systems.

Abbreviations

DLC	Dioxin-like compounds
GC	Gas chromatograph
MS	Mass spectrometry
MWC	Municipal waste combustors
PCB	Polychlorinated biphenyls
PCDD	Polychlorinated dibenzo-*p*-dioxins
PCDF	Polychlorinated dibenzofurans
POC	Persistent organic compounds
POP	Persistent organic pollutant
TCDD	Tetrachlorodibenzo-*p*-dioxin
TOC	Toxicity equivalence factor

Bibliography

AertoMet Engineering, Dioxin and furan testing, (2014) http://www.epastacktesting.com/dioxin-and-furan-stack-testing.php.

Buckley, T., Dioxin: EWG dioxin analysis, (2010) (http://www.ewg.org/dioxin/analysis).

Chlorene Detection Council, A comparison of dioxin risk characterizations, (2002) http://www.dioxinfacts.org/dioxin_health/public_policy/dr.pdf.

Danielson, C., Trace analysis of dioxin, (2007) (http://www.diva-portal.org/smash/get/diva2:145208/FULLTEXT01).

EPA, Dioxin treatment technologies, (1991) https://clu-in.org/contaminantfocus/default.focus/sec/Dioxins/cat/Treatment_Technologies/.

EPA, Method 23, Dioxins and furans, (2015) http://www3.epa.gov/ttnemc01/methods/method23.html.

EPA, Method 8290A, The analysis of polychlorinated dibenzo-p-dioxin and polychlorinated dibenzofurans by high resolution gas chromatography/high resolution mass spectrometry in test methods for evaluating solid waste, (2007) http://www3.epa.gov/epawaste/hazard/testmethods/sw846/pdfs/8290a.pdf.

EPA, Methods for the analysis of pesticides in human and environmental samples, (1980) http://www3.epa.gov/ttnemc01/informd/tid-016.pdf.

Eurofins, Dioxins and furans, (2010) http://www.eurofins.com/media/7656288/dioxine_ts_eng.pdf.

Supelco Analytical: Dioxin & PCB analysis, (2007) http://www.sigmaaldrich.com/etc/medialib/docs/Supelco/General_Information/dioxin-brochure-jxb.Par.0001.File.tmp/dioxin-brochure-jxb.pd.

Test America, Method for determining dioxins and furans in stack gases, (2015) http://www.testamericainc.com/services/source-air-testing/test-methods/method-0023a-dioxins-and-furans/method-0023a-sampling-analysis-procedures/.

1.21 Electrochemical Analyzers

R. A. GILBERT (1982)   **J. N. HARMAN III** (1995)
B. G. LIPTÁK (2003)   **S. VEDULA** (2017)

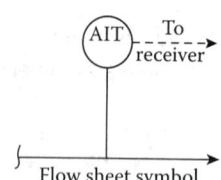

Flow sheet symbol

Design variations	Amperometric, coulometric, galvanometric, polarographic, and potentiometric principles.
Chapters discussing specific applications	1.8, carbon monoxide; 1.9, chlorine; 1.13, conductivity; 1.25, fluoride; 1.31, hydrogen sulfide; 1.33, ion-selective electrodes; 1.39, moisture in gas; 1.47, ORP probes; 1.49, oxygen; 1.51 and 1.52, ozone; 1.55, pH; 1.64, sulfur oxide; 1.26, toxic gases.
Inaccuracy	Generally 1% of full scale with absolute error values down to under 1.0 parts per billion (ppb).
Costs	Amperometric or polarographic residual chlorine transmitters cost about $6000, with the yearly cost of chemicals for the amperometric units adding another $1000 (somewhat less for polarographic).
	An electrochemical gas diffusion detector for hydrogen sulfide costs $2500.
	Potentiometric probes for pH, oxidation–reduction potential, and ion-selective measurements cost about $750, with associated transmitter about $2500.
	Electrochemical probes for nitrogen oxide detection in stacks cost $7000.
	The electrolytic dissolved oxygen probe and transmitter cost $3500.
	Costs for amperometric gas and dissolved ozone detector range from $4000 to $8000.
	An electrochemical membrane diffusion system to automatically monitor toxic gas concentration in 16 locations costs from $12,000 to $20,000.
	For more detailed cost information, refer to the Chapters listed earlier, where some of these specific analyzers are discussed.
Most popular suppliers	Emerson Process Management (www.emersonprocess.com)
	Endress + Hauser (www.endress.com)
	Hach (http://www.hach.com/gli)
	Mettler-Toledo (www.mt.com)
	Thermo Scientific (http://www.thermoscientific.com/en/home.html)
	Yokogawa (www.yokogawa.com/index.htm)
Other suppliers	ABB (www.abb.com/product/us/9aac100035.aspx)
	Advanced Sensor Technologies Inc. (www.astisensor.com/pH_ORP_Products.htm)
	Analytical technology Inc. (http://www.analyticaltechnology.com/public/product.aspx?ProductID = 1029)
	Bacharach Inc. (http://www.bacharach-inc.com/analyzers-monitors.htm)
	Bran & Luebbe Analyzing, Technicon Industrial Systems (http://www.spx.com/en/bran-luebbe/pd-mp-on-line-analyzers/)
	Broadley James (http://www.broadleyjames.co.uk/products_sensors_main.html)
	BW Technologies Ltd.(http://www.gasmonitors.com/main.cfm?cat = 60&page = singlegas)
	CEA Instruments Inc. (http://www.ceainstr.com/pdf_datasheets/txflamgardplus_Info.pdf)
	Cole-Parmer Instrument Co. (http://www.coleparmer.com/Category/pH_Meters_and_Water_Quality_Products/59358)
	Control Instruments Corp. (http://www.controlinstruments.com/products/gas-sensors)
	Custom Sensors and Technology (http://www.customsensors.com/index.php?q = prod/oxygen-sensors-and-tube-stubs)
	Detector Electronics Corporation (http://www.detronics.com/utcfs/ws-462/Assets/92–1054–3.1_GT3000.pdf)

(Continued)

Draeger Safety Inc. (http://www.draeger.com/Sites/enus_us/Pages/Oil-Gas-Industry/Sensor-EC-Electrochemical-sensors.aspx?navID = 2278)
Electro-Chemical Devices (http://www.ecdi.com/downloads/literature/do_sensor_series/Triton_DO9.PDF) (http://www.ecdi.com/products/steam-sterilizable-products.html)
Enmet Analytical Corp. (http://catalog.enmet.com/asset/SDS-97D-Literature.pdf)
GF Piping Systems (http://www.gfps.com/content/gfps/signet/en_US/sensors/phorp.html)
Honeywell (http://www.lesman.com/unleashd/catalog/analytical/Honeywell-APT4000/70–82–03–44.pdf) (http://www.lesman.com/unleashd/catalog/analytical/Honeywell-APT4000/70–82–03–45.pdf)
Horiba Instruments Inc. (http://www.horiba.com/process-environmental/products/water-quality-measurement/for-utility/)
International Sensor Technology (http://www.intlsensor.com/pdf/sm95manual.pdf)
Invensys (http://www.fielddevices.foxboro.com/en-gb/products/analytical-echem/)
Knick (http://www.knick-international.com/products/sensors/), Phoenix Electrode Co. (http://www.vl-pc.com/default/index.cfm/pH)
KWJ Engineering (http://www.kwjengineering.com/sensor-manufacturing)
Sensidyne Inc. (http://www.sensidynegasdetection.com/)
Sensorex (http://www.sensorex.com/products/categories/category/ph_sensors)
Sierra Monitor Corp. (http://www.sierramonitor.com/gas/about/Fire-Gas-Detection.php)
Teledyne Analytical Instruments (http://www.teledyne-ai.com/industrialsensors.asp)
Thermo Scientific (http://www.thermoscientific.com/en/products/ph-ise-conductivity-dissolved-oxygen-products.html)
Van London Co. (http://www.vl-pc.com/)

INTRODUCTION

The purpose of this chapter is to give an overall perspective of the many electrochemical analyzers used in the processing industries. These instruments will be discussed in more detail in later chapters of this volume. Electrochemical analyzers measure composition or concentration by detecting changes in voltage or current that occur between two electrodes in a solution (electrolyte) over time, because of the oxidation or reduction of that solution.

In galvanic and polarographic analyzers (such as CO, O_2, and toxic gas sensors discussed in Chapters 1.8, 1.49, and 1.26), the electrolyte is usually a gel. The gel is separated from the process stream by a membrane, and the electrochemical reaction in this electrolyte either occurs spontaneously (galvanic) or is caused by a polarizing voltage (polarographic).

In a cyclic voltammetry experiment, the working electrode potential is ramped linearly versus time. When cyclic voltammetry reaches a set potential, the working electrode's potential ramp is inverted. This inversion can happen multiple times during a single measurement. The current at the working electrode is plotted versus the applied voltage to give the cyclic voltammogram.

In squarewave voltammetry, a squarewave is superimposed on the potential staircase sweep. Oxidation or reduction of species is registered as a peak or trough in the current signal at the potential at which the species begins to be oxidized or reduced.

In potentiometric analyzers (such as Cl, conductivity, ion-selective, oxidation–reduction potential (ORP), and pH probes discussed in Chapters 1.9, 1.13, 1.33, 1.46, and 1.55), an electrical potential is detected, which is generated in response to the presence of dissolved ionized solids. The ORP probe detects the ratio of oxidizing to reducing agent. Conductivity sensors measure the solution's ability to conduct electricity, which depends on the concentration of dissolved ionized solids. It is a measurement of potential (E); there is no current (I) that passes between electrodes.

In amperometric analyzers (such as Cl, Fl, HS, moisture in gas, O_3, and toxic gases covered in Chapters 1.9, 1.25, 1.31, 1.39, 1.51, 1.52, and 158), if a strong oxidizing agent is present, the polarization of the measuring electrode is depolarized and a current flow is generated, which is proportional to the concentration of the oxidizing agent. It is the measurement of current (I) at fixed applied potential (E). Further developments to this application are pulsed amperometric detection (PAD). PAD is a technique used to detect certain classes of compounds, notably sugars and polyalcohols, among others. These compounds tend to foul the surface of an electrode, making ordinary constant-potential amperometric detection difficult. In PAD, cleaning potentials are applied to the electrode roughly once per second, interspersed with the detecting potential. Another new development in gas sensing technology is screen-printed electrochemical sensor (SPEC), a new-generation toxic gas technology that integrates amperometric gas sensor and printed electronics to achieve better monitoring performance. SPEC sensors can be fabricated in very small sizes and require only microwatts of power for operation, while meeting or exceeding the performance of conventional amperometric gas sensors. The sensors can be interfaced to wireless communication systems.

In coulometric analyzers (such as CO, oxygen in gas, and SO sensors discussed in Chapters 1.8, 1.49, and 1.64), the measured gas diffuses through the potassium hydroxide electrolyte to the cathode of the electrochemical cell, where it is reduced to hydroxyl ions. As an external voltage causes these ions to migrate to the anode, the resulting cell current is a measure of concentration.

VOLTAMMETRIC ANALYSIS

Voltammetric analyzers monitor the current, voltage, or charge that results from a specific oxidation–reduction reaction of the component of interest. During an oxidation–reduction reaction, one or more electrons are transferred from one atom or molecule to another.

This chapter presents a brief review of voltammetry, including a discussion of the concepts and properties of electrochemical probes, potentiometry, amperometry, polarography, and coulometry. The distinctions among voltammetric methods are summarized in Table 1.21a. This table lists the common voltammetric methods of analysis and their usual applications.

TABLE 1.21a
Summary of Voltammetric Methods

Method	Measured Variables	Online Instrument	Determinations[a]
Potentiometry	V	Yes	pH Anions (F–, S=, Cl–) Cations (NH_{4+}, Ca^{++}, Mg^{++})
Polarography	i, v	Yes	Pesticides Mercaptans Thiosulfates Chlorinated organics Toxic heavy metals
Stripping	i, v	No	Heavy metals Selenides Halogens Thioamides Trace metals
Amperometry	i, t, c	Yes	Sulfates Halides Phenols Aromatic amines Olefins
Controlled-potential coulometry	V i(t)	Yes	Precious metals Alloys Dissolved oxygen
Differential pulse polarography	i, v	No	Aromatic hydrocarbons Aromatic amines Phenol Ammonium salts Carboxylic anhydrides N-nitrosamines
Squarewave voltammetry	i	No	

Note: i, current; v, applied potential; c, concentration; i(t), current as a function of time.
[a] Partial list.

Current, Voltage, and Time

Successful application of voltammetry for monitoring and controlling process streams requires an understanding of the significant voltammetric variables. The fundamental requirements for voltammetry instrumentation are two electrodes, provisions for a current and voltage supply, and monitors for the respective voltage and current responses as a function of time. The characteristic response curves indicate the nature of the electrochemical analysis occurring in the solution between the electrodes. Figure 1.21a illustrates a surface that relates the three variables (current, voltage, and time), which are of interest in an electrochemical analysis.

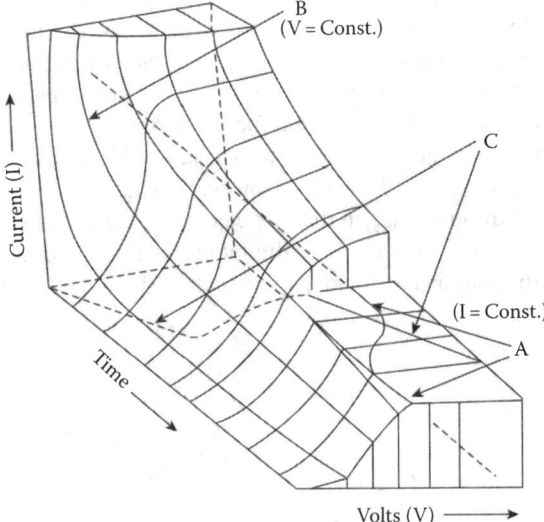

FIG. 1.21a
Response surface for voltammetry, showing curve A, the process of chronopotentiometry; curve B, amperometry; and curve C, the polarographic process.

The exact shape of a particular response curve in Figure 1.21a depends on the mass transport phenomena associated with the movement of the component of interest (analyte) to the working electrode. (The working electrode is the electrode in which the electrochemical process of interest occurs, while the reference electrode is passive. The usual working electrode is made from a metal, such as mercury, platinum, palladium, or gold.) Analyte diffusion to the working electrode is the usual transport mechanism required if the analytical expressions for the curves of Figure 1.21a are to be obtained. The control of diffusional mass transport is important for chronopotentiometry, amperometry, and polarography.

The electrochemical surface presented in Figure 1.21a is generated from the following equation*:

$$E = \left[(E° + (RT/nF) \ln \left(f_o/f_r \left(D_r/D_o \right)^{1/2} \right) \right] + (RT/nF) \ln \left[(K - it^{1/2})/(it^{1/2}) \right] \quad 1.21(1)$$

---
*Reimmuth, *Analytical Chemistry*, 32: 1509, 1960.

Equation 1.21(1) expresses the potential (E) of the working electrode in terms of E°, the standard thermodynamic potential; R, the gas constant; T, the absolute temperature; F, Faraday's constant; n, the number of electrons transferred for the specified oxidation–reduction reaction; f, the activities of oxidized and reduced analyte forms; i, the current; and K, a constant dependent upon particular analysis parameters other than voltage, current, or time.

Potentiometry

During a potentiometric analysis, such as in ORP, pH, or ion-selective measurements, no current is passing between electrodes. The potential difference (at zero current) is monitored between the measuring and reference electrodes. At the measuring electrode, an oxidation or reduction of a solution species is taking place. Such an analysis is called chronopotentiometry when the potential difference between a metallic measuring electrode and a reference electrode is monitored as a function of time.

Potentiometry may be done with a bare metal electron-conducting electrode and a reference electrode, in which case the technique responds to the oxidation–reduction couples present in the test solution. If a potentiometric determination is conducted using an ion-selective electrode and a reference electrode, one may determine solution species over a wide dynamic concentration range with good selectivity, and there is no need for oxidation–reduction couples in the solution. Examples of ion-selective electrodes are glass electrodes for pH (Figure 1.21b) and sodium ion determination; solid-state electrodes for fluoride, chloride, and sulfide determination;

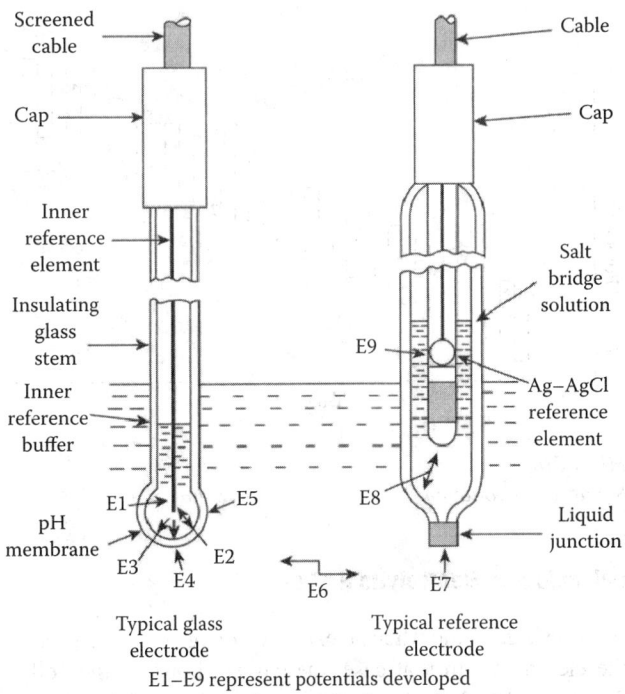

FIG. 1.21b
pH electrodes.

and liquid ion exchange electrodes for the determination of calcium, magnesium, and nitrate ions.

If a constant current is supplied to the electrodes, the progress of the electrochemical reaction is observed by monitoring the change in potential as a function of reaction time. This procedure is labeled chronopotentiometry, and a response curve would be similar to curve A in Figure 1.21a.

If the electrochemical process occurs at a constant potential, the reaction response is monitored as a changing current as a function of time. This type of analysis is called constant-potential voltammetry, or amperometry. A typical response curve is labeled B in Figure 1.21a.

Finally, monitoring reaction current as a function of applied potential is called polarography, and a typical polarographic curve is represented by C, which is a diagonal cut through the response surface, in Figure 1.21b.

Squarewave voltammetry: In squarewave voltammetry, a squarewave is superimposed on the potential staircase sweep. Oxidation or reduction of species is registered as a peak or trough in the current signal at the potential at which the species begins to be oxidized or reduced. In staircase voltammetry, the potential sweep is a series of stair steps. The current is measured at the end of each potential change, right before the next, so that the contribution to the current signal from the capacitive charging current is minimized. The differential current is then plotted as a function of potential, and the reduction or oxidation of species is measured as a peak or trough. Due to the lesser contribution of capacitive charging current, the detection limits for SWV are on the order of nanomolar concentrations (Figure 1.21c).

the electrodes are not consumed. Dissolved oxygen detection is a primary application of this type of probe.

Electrochemical probes are electrochemical cells that have been designed to respond to the presence of a particular analyte. Similar to electrochemical cells, electrochemical probes can be classified as either galvanic or electrolytic. Electrolytic probes require an external potential to be applied across the electrodes, whereas galvanic probes do not. Both types of probes require a pair of electrodes, usually metallic, which are immersed in an electrolyte so that current passes through the electrodes as the desired oxidation or reduction occurs at the surface of the working electrode.

The difference between an electrolytic and a galvanic probe is whether or not a potential from an external source is applied across the electrodes. Galvanic probes are constructed in such a manner that the presence of the analyte depolarizes the cell and allows a flow of electrons in the cell (Figure 1.21d). Because the electrochemical reaction occurs spontaneously, no external potential needs to be applied across the electrodes. However, galvanic probes have a major disadvantage in that the electrodes are consumed as the reaction continues. This limitation restricts their usefulness to monitoring situations in which periodic replacement of spent probes is acceptable.

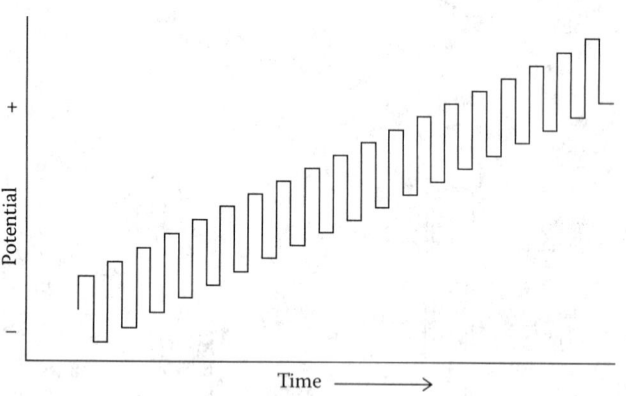

FIG. 1.21c
Squarewave voltammetry—increase of potential vs. time.

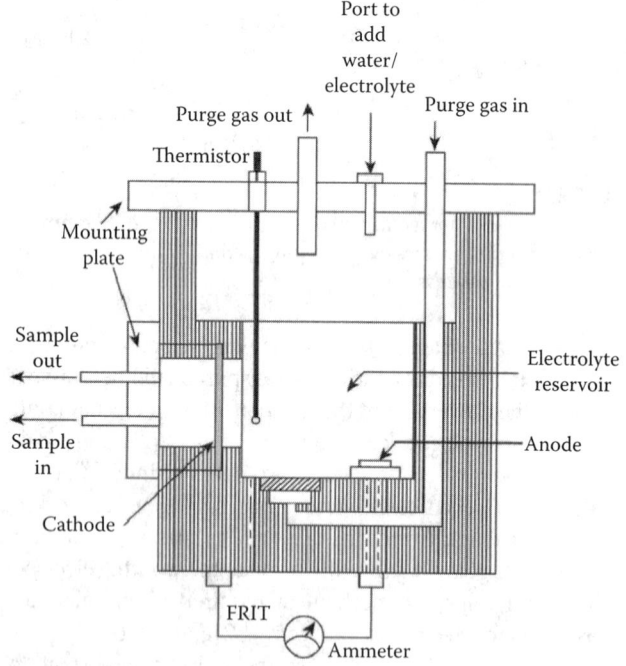

FIG. 1.21d
Galvanic oxygen detector cell. (Courtesy of Teledyne Analytical Instruments.)

Galvanic and Electrolytic Probes

A probe is galvanic when no external voltage is applied across the electrodes. In that case, the current flows as the cell is depolarized as diffusion of the analyte occurs. Electrodes are consumed during this operation and require periodic replacement. An electrolytic probe is similar to a galvanic one, except that a potential is applied across the electrodes and

Electrolytic probes are principally used as monitoring devices for dissolved oxygen in aqueous and nonaqueous media. Unlike with the galvanic probe, here a potential is applied from an external source across the electrodes. The choice of potentials is controlled by the analyte to be determined. The current from the probe is obtained at constant potential as a function of time, location of the probe within the process, or the process state.

Membrane-Covered Probes   To reduce the chemical interference from the process, a diffusion-limiting membrane can be placed over both electrodes. These membranes are permeable to the component of interest and are filled with the appropriate electrolyte. The thin analyte-permeable membrane separates the cell components and the electrolyte from the solution to be analyzed (Figure 1.21e).

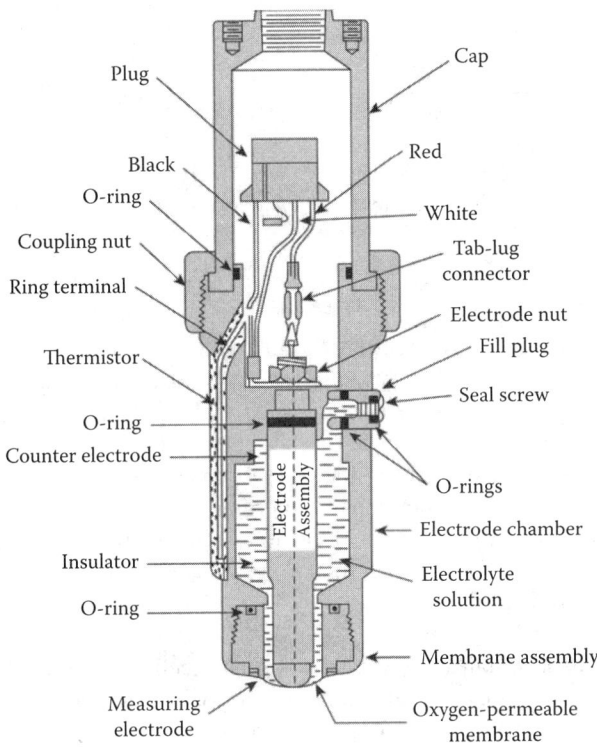

FIG. 1.21e
An oxygen-permeable membrane protects the working components of a galvanic dissolved oxygen probe. (Courtesy of ABB Inc., formerly Fischer & Porter.)

A deficiency of this technique is the dependence of the membrane permeability on the process temperature. This is usually corrected by measuring the process temperature and adjusting the gain of the analyzer unit to compensate for the permeability effects. Another problem is the interferences caused by dissolved gases other than oxygen, which permeate the membrane and cause interferences as they are oxidized or reduced at the working electrode. Fouling or coating of the probe necessitates frequent cleaning, although the equilibrium-dissolved oxygen sensor greatly reduces the fouling problems and also allows use in almost stagnant flows with very small face velocity at the probe.

Figure 1.21f shows a nominal configuration for an electrolytic probe. The probe consists of an inner and outer element. The inner element contains the cathode and anode with wires leading to the power supply and current monitoring circuit. The anode is positioned in a way that isolates the cathode from the anode. The outer element fits over the inner element and provides a space for the electrolyte. The membrane is mounted on the outside of this outer element and secured with an O-ring.

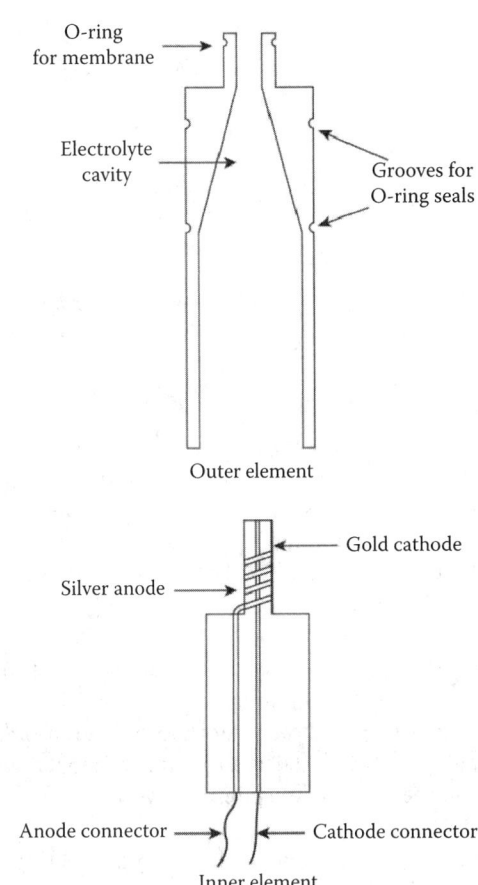

FIG. 1.21f
Configuration of an electrolytic probe. The inner element contains the anode and cathode, while the outer element isolates the electrolyte from the sample.

Sensors are available in disposable, nonrechargeable configurations, as well as in rechargeable configurations, in which access is provided to allow electrolyte change, electrode cleaning, and membrane replacement.

Amperometry

Amperometric titration is when the end point is determined by measuring the current (amperage) that passes through the solution at a constant voltage. Amperometry is the process of performing an amperometric titration. During such titration, the current flow is monitored as a function of time between

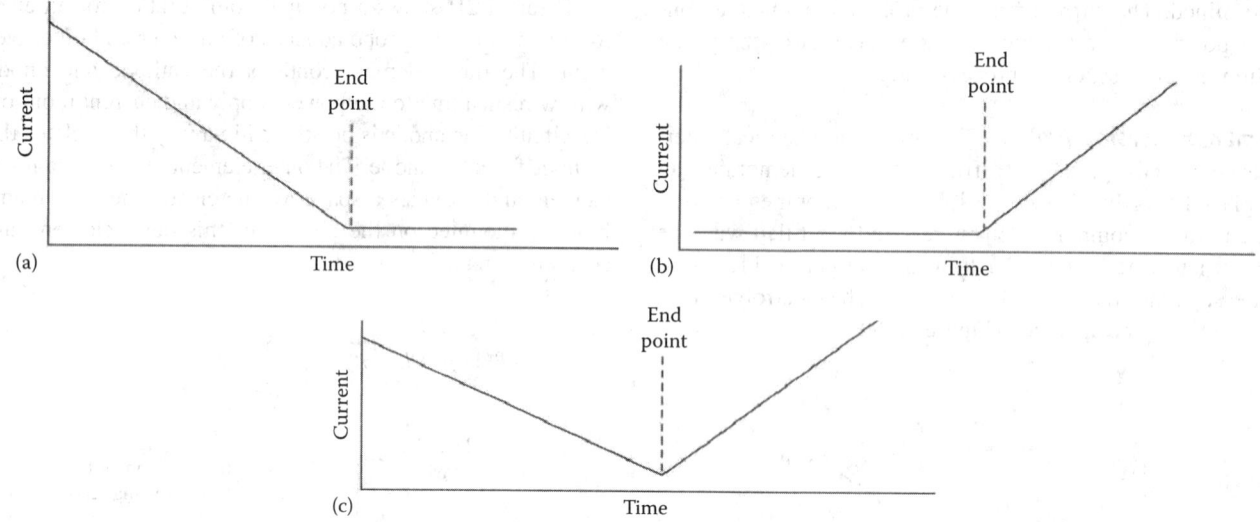

FIG. 1.21g
Response curves for polarographic cell system, showing curve (a), when analyte reacts at the electrode; curve (b), when reagent reacts at the electrode; and curve (c), when analyte and reagent react at the electrode.

working and auxiliary electrodes, while the voltage difference between them is held constant. Alternatively, the current can also be monitored as a function of the amount of reagent added to bring the titration of an analyte to the (stoichiometrically defined) end point. This is also called constant-potential voltammetry.

In amperometry, the titrating reagent is coulometrically generated, and the current that passes through a polarographic cell (containing either a solid or dropping mercury electrode at the appropriate oxidation or reduction potential) is measured as a function of reagent volume or time. Three types of response curves can be expected (Figure 1.21g).

Curve (a) is the result of the analyte reacting at the electrode, while the reagent does not. This curve indicates that the diffusion current is detected as long as the analyte remains. Once the reagent removes all the analyte, the current stops. This dead-stop end point technique is often used for amperometric determinations of dissolved oxygen.

In curve (b), the analyte is not reduced at the working electrode, but the reagent is. In this situation, there is no current flow until the analyte is completely destroyed and the reagent is allowed to react at the electrode surface.

Finally, curve (c) indicates the response when both analyte and reagent are electroactive. The increase after the equivalence point indicates that the current is the result of excess reagent reacting at the electrode surface.

Liquid chromatography followed by pulsed amperometric detection has become an important method for the separation and detection of polar aliphatic compounds. These compounds typically have poor detection properties, and their detection by optical methods requires derivatization. Also, dc amperometric detection, when possible, leads to diminishing response due to electrode fouling. PAD combines amperometric detection at a noble metal electrode with positive and negative potential excursions for "online" cleaning and reactivation of the electrode.

Solid Electrodes If the polarographic cell is replaced by a two-solid electrode system with a small applied potential (0.1–0.2 V), the response curves of interest are determined by a number of reversible reactions occurring in the cell at the voltage level impressed across the cell.

The set of reversible half-reactions of interest are symbolized as

$$R_r f\ R_o + e^- \qquad 1.21(2)$$

and

$$A_o + e^- f\ A_r \qquad 1.21(3)$$

where A_o and R_o are the oxidized forms and A_r and R_r are the reduced forms of the analyte and reagent, respectively. The fundamental concept of interest is the fact that the applied potential will support current flow only when both forms represented in Equations 21.2 or 21.3 are present in the

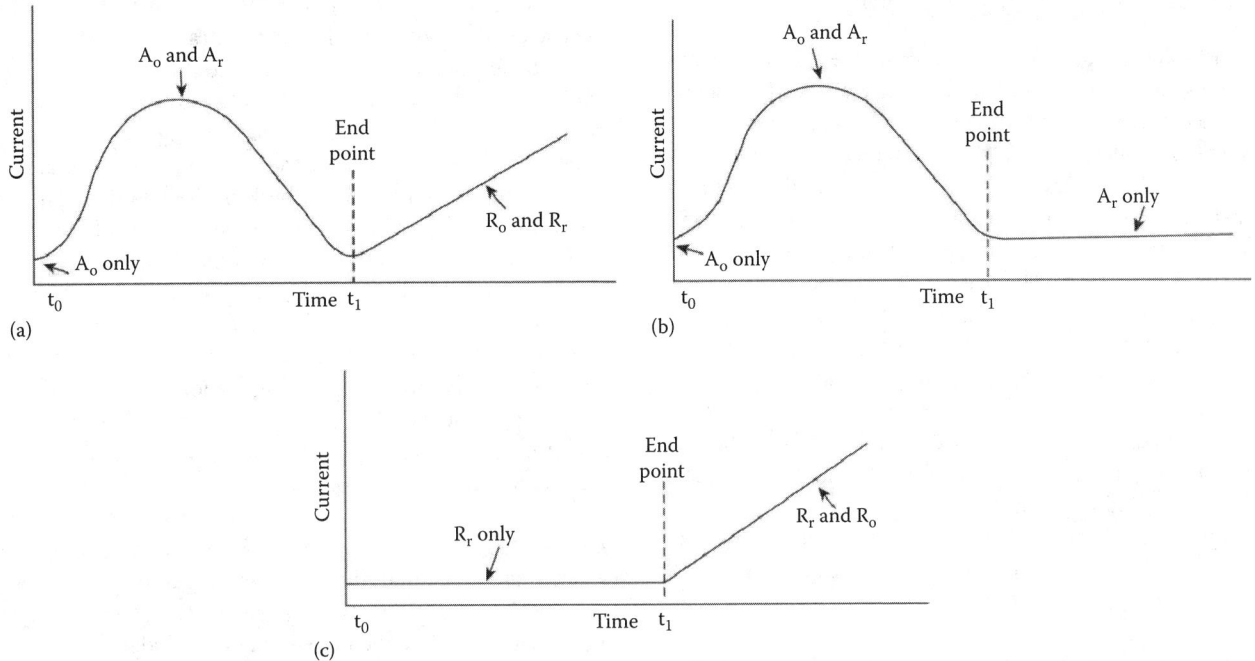

FIG. 1.21h
Response curves for two-solid electrode system, showing curve (a), when analyte and reagent are reversible; curve (b), when only the analyte is reversible; and curve (c), when only the reagent is reversible.

test solution. Figure 1.21h represents the possible response curves that illustrate this concept.

In curve (a) of Figure 1.21h, the absence of current at t_o is due to the lack of A_r. For values of t so that $t_o < t < t_1$, the current response is the result of varying concentrations of A_o and A_r as R_r is added. At $t = t_1$, the current returns to zero because of the absence of A_o and R_r. Finally, for $t > t_1$, the continuous addition of R_r, coupled with the R_o present from the redox reaction with A_o, results in a continuously increasing current flow.

Curve (b) indicates the response when only the analyte reaction is reversible. The curve from $t_o < t \leq t_1$ is identical to the same portion of curve (a) for the same reasons. However, for $t > t_1$, no current is detected because the reagent half-reaction is irreversible.

Curve (c) shows the results when the analyte reaction is irreversible but the reagent half-reaction is reversible. Current is not detected until R_r remains in the solution. This case is often called a dead-start titration because the current does not begin until all the analyte has been removed. Figure 1.21i illustrates the design of another membrane-type amperometric sensor.

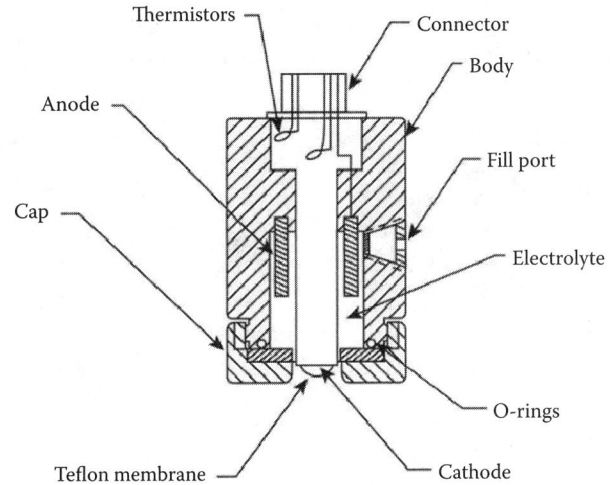

FIG. 1.21i
Membrane-type amperometric probe.

Polarography

Polarography is a process for monitoring the diffusion current flow between working and auxiliary electrodes as a function of applied voltage as it is varied systematically. The diffusion current that results is linearly dependent on the concentration of the analyte. Polarography can be applied using direct current, pulsed direct current, or alternating current (AC) voltage excitation waveforms. Dissolved oxygen determination is an example of an application for which polarography is used (Figure 1.21e).

Nobel Prize winner Jaroslav Heyrovsky developed direct current polarography in 1922. Polarography is performed by measuring the current flow, which is a function of the systematically varied electrode potential. Different types of polarographies can be performed as a function of the voltage variation patterns, voltage sweep rates, and the choice of working electrode materials.

Figure 1.21j shows a typical polarographic response curve. For this curve, the K term in Equation 21.1 is defined as

$$K = nFC°(7/3D_o)^{1/2} \qquad 1.21(4)$$

and

$$(nF/KT)(E - E° - \ln(f_o/f_r(D_o/D_r)^{1/2})) = \ln(i_d - i/i) \qquad 1.21(5)$$

where i_d is the diffusion current.

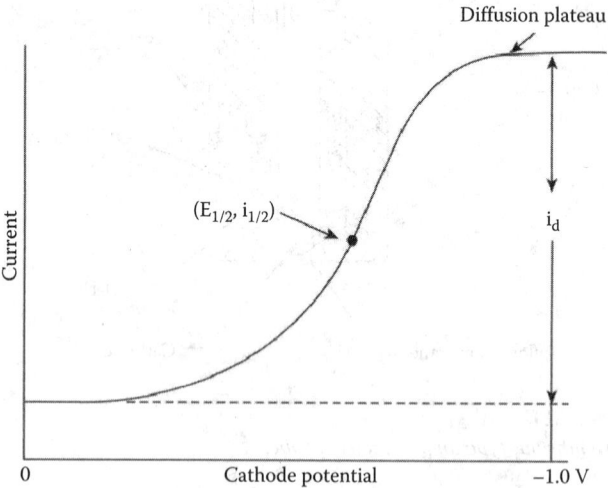

FIG. 1.21j
Polarogram showing typical current vs. cathode potential response. The diffusion current, i_d; the half-wave potential, $E_{1/2}$; and the half-wave current, $i_{1/2}$, are all illustrated.

The curve shows an S-shaped current response as the potential at the cathode, the working electrode where analyte reduction occurs, becomes more negative. Current limited by diffusion occurs at cathodic potentials beyond the potential required to initiate reduction. Thus, i_d is sufficient to reduce the diffusion-supplied analyte concentration to zero.

The significance of this is that the diffusion rate is controlled by the concentration gradient established between the concentration at the surface of the working electrode, which is held to zero as a result of the electrode reaction, and the concentration existing in the bulk of the solution. The current observed for the Faradic process of interest is directly proportional to the concentration of the desired analyte.

Subsequently, this technique was modified by changing the voltage excitation waveform to AC (AC polarography) or to pulsed direct current (pulse polarography or differential pulse polarography); these modifications resulted in increased sensitivity and better discrimination for the detection of the desired analyte in a background matrix of interferences.

Advantages One of the valuable aspects of a polarographic determination is that the potential observed at the point $i_{1/2} = i_d/2$ is unique to the analyte in the solution matrix. This occurs because the potential at $i_{1/2}$ is independent of the reactant concentrations but directly related to the standard potential for the analyte half-reaction. This inflection point potential is defined as the half-wave potential, $E_{1/2}$, and is a common parameter for qualitative identification of electroactive components in a solution.

When the reaction is fast and reversible and the diffusion coefficients of the oxidized and reduced forms of the analyte are virtually identical, $E_{1/2}$ becomes $E°$. Another value of polarography is that the diffusion current is linearly related to the concentration of the analyte.

Coulometry

Coulometry is the process of monitoring analyte concentration by detecting the total amount of electrical charge passed between two electrodes that are held at a constant potential or when constant current flow passes between them. Because the coulomb provides a direct connection through Faraday's law between reaction current and analyte concentration during a redox process, coulometry offers a direct method of determining analyte concentration.

Two types of coulometry are possible: constant current and controlled potential. Constant-current coulometry depends on analyte oxidation–reduction to support the specified current flow. When the supply of material is insufficient to carry this current, the electrode potential drifts until another reaction begins. The result of this process is a potential time relationship similar to the results obtained in chronopotentiometry, and the amount of charge passed is simply the product of the constant current and electrolysis time.

Controlled-Potential Coulometry By contrast, controlled-potential coulometry is conducted by maintaining a constant electrode potential while a current flow is measured. The number of coulombs is determined by integration of the reaction

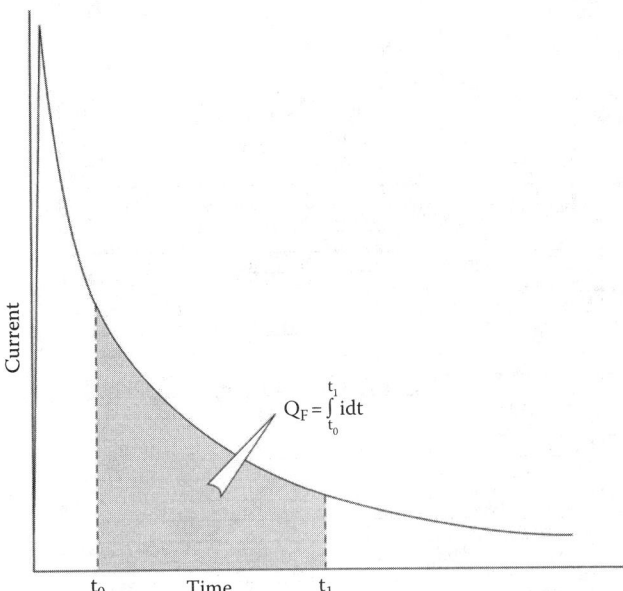

FIG. 1.21k
Response curve for controlled-potential coulometry, showing the relationship among current, i; time, t; and coulombs, Q_F, in the Faradic portion of the curve.

current as a function of electrolysis time. Figure 1.21k illustrates the current–time relationship for controlled-potential coulometry.

The choice of integration time is important because the current is the sum of the Faradic component and a capacitive component. As a result of this dual contribution, current integration is not started until the capacitive charging contribution is minimal.

CONCLUSIONS

Although instruments are available for all types of voltammetric analysis, currently only the potentiometric, polarographic, amperometric, and coulometric designs are available as online process analyzers. There are many techniques such as cyclic voltammetry, squarewave voltammetry, and PAD that are becoming popular methods.

Limitations

The limitations of this category of sensors are mostly related to the membranes. One major problem with the thin permeable membranes is the fact that the electrode response to the analyte is directly dependent on the condition of the membrane. Once the potential has been applied across the electrodes and the concentration depletion wave has reached the inner membrane wall, the amount of analyte that reaches the electrode depends on the diffusion of the analyte through the membrane. This situation is usually acceptable because the membrane condition is stable. The membrane itself is also subject to failure, and the surface of the membrane is exposed to the process fluid, which can alter its permeability to the analyte. These conditions make the long-term reliable use of electrolytic probes problematical, because they require constant maintenance and recalibration.

Amperometric end points pose another problem. To date, provisions for the various end points possible are not available, and care must be taken when determining analysis end points. Finally, instrumentation for coulometric methods requires that the working electrode be physically separated from the auxiliary electrode so that the reaction products do not migrate to the opposite electrode. Sintered glass disks are used to restrict the migration of reaction products between the electrodes, but they permit electrical conduction through the cell. Table 1.21a provides a summary of voltammetric methods and applications.

Advances

Significant advances have been made by the introduction of equilibrium-state voltammetric probes. These have the advantage of analyte sensitivities that are relatively independent of membrane coating and consume so little of the analyte as a result of the measurement process that they are able to work in processes where the sample flow rate past the face of the sensor is small.

During the past 10 years, there are developments in voltammetric sensors, with digital conversion of measurement signal. A microprocessor placed inside sensor head monitors the data coming from measurement and retains calibration information. An integrated chip converts measurement data to digital signal and exchanges with connected transmitter. The digital measurement signal is more accurate than analog signal, not affected by moisture and cable length. A precalibrated sensor can be installed and will be ready within a minute, which helps to avoid longer interruptions. The digital conversion of measurement signal provides many advanced diagnostics that were not possible with analog signals. Digital sensors provide information on remaining life of sensor by continuously analyzing sensor slope, impedance, the process conditions,

There are advancements in electrochemistry with the use of techniques such as cyclic voltammetry, squarewave voltammetry, and PAD in pharmaceutical industry. Another new development in amperometric sensors is SPEC, a new-generation toxic gas technology that integrates amperometric gas sensor and printed electronics to achieve better monitoring performance.

SPECIFICATION FORMS

When specifying electrochemical analyzers only, please use the ISA form 20A1001 and when specifying both the analyzer and the composition or the properties of the process material of interest, use ISA form 20A1002. Both forms are reproduced with the permission of the International Society of Automation on the next pages.

Analytical Measurement

1	RESPONSIBLE ORGANIZATION	ANALYSIS DEVICE	6	SPECIFICATION IDENTIFICATIONS	
2			7	Document no	
3	(ISA)	Operating Parameters	8	Latest revision	Date
4			9	Issue status	
5			10		

	ADMINISTRATIVE IDENTIFICATIONS				SERVICE IDENTIFICATIONS Continued		
11				40			
12	Project number		Sub project no	41	Return conn matl type		
13	Project			42	Inline hazardous area cl	Div/Zon	Group
14	Enterprise			43	Inline area min ign temp	Temp ident number	
15	Site			44	Remote hazardous area cl	Div/Zon	Group
16	Area	Cell	Unit	45	Remote area min ign temp	Temp ident number	
17				46			
18	SERVICE IDENTIFICATIONS			47			
19	Tag no/Functional ident			48	COMPONENT DESIGN CRITERIA		
20	Related equipment			49	Component type		
21	Service			50	Component style		
22				51	Output signal type		
23	P&ID/Reference dwg			52	Characteristic curve		
24	Process line/nozzle no			53	Compensation style		
25	Process conn pipe spec			54	Type of protection		
26	Process conn nominal size	Rating		55	Criticality code		
27	Process conn termn type	Style		56	Max EMI susceptibility	Ref	
28	Process conn schedule no	Wall thickness		57	Max temperature effect	Ref	
29	Process connection length			58	Max sample time lag		
30	Process line matl type			59	Max response time		
31	Fast loop line number			60	Min required accuracy	Ref	
32	Fast loop pipe spec			61	Avail nom power supply		Number wires
33	Fast loop conn nom size	Rating		62	Calibration method		
34	Fast loop conn termn type	Style		63	Testing/Listing agency		
35	Fast loop schedule no	Wall thickness		64	Test requirements		
36	Fast loop estimated lg			65	Supply loss failure mode		
37	Fast loop material type			66	Signal loss failure mode		
38	Return conn nominal size	Rating		67			
39	Return conn termn type	Style		68			

	PROCESS VARIABLES	MATERIAL FLOW CONDITIONS			PROCESS DESIGN CONDITIONS		
69				101			
70	Flow Case Identification		Units	102	Minimum	Maximum	Units
71	Process pressure			103			
72	Process temperature			104			
73	Process phase type			105			
74	Process liquid actl flow			106			
75	Process vapor actl flow			107			
76	Process vapor std flow			108			
77	Process liquid density			109			
78	Process vapor density			110			
79	Process liquid viscosity			111			
80	Sample return pressure			112			
81	Sample vent/drain press			113			
82	Sample temperature			114			
83	Sample phase type			115			
84	Fast loop liq actl flow			116			
85	Fast loop vapor actl flow			117			
86	Fast loop vapor std flow			118			
87	Fast loop vapor density			119			
88	Conductivity/Resistivity			120			
89	pH/ORP			121			
90	RH/Dewpoint			122			
91	Turbidity/Opacity			123			
92	Dissolved oxygen			124			
93	Corrosivity			125			
94	Particle size			126			
95				127			
96	CALCULATED VARIABLES			128			
97	Sample lag time			129			
98	Process fluid velocity			130			
99	Wake/natural freq ratio			131			
100				132			

	MATERIAL PROPERTIES				MATERIAL PROPERTIES Continued		
133				137			
134	Name			138	NFPA health hazard	Flammability	Reactivity
135	Density at ref temp	At		139			
136				140			

Rev	Date	Revision Description	By	Appv1	Appv2	Appv3	REMARKS

Form: 20A1001 Rev 0

© 2004 ISA

	RESPONSIBLE ORGANIZATION	ANALYSIS DEVICE COMPOSITION OR PROPERTY Operating Parameters (Continued)		SPECIFICATION IDENTIFICATIONS		
1			6			
2			7	Document no		
3	ISA		8	Latest revision		Date
4			9	Issue status		
5			10			

	PROCESS COMPOSITION OR PROPERTY			MEASUREMENT DESIGN CONDITIONS				
	Component/Property Name	Normal	Units	Minimum	Units	Maximum	Units	Repeatability
11								
12								
13								
14								
15								
16								
17								
18								
19								
20								
21								
22								
23								
24								
25								
26								
27								
28								
29								
30								
31								
32								
33								
34								
35								
36								
37								

Rev	Date	Revision Description	By	Appv1	Appv2	Appv3	REMARKS

Form: 20A1002 Rev 0
© 2004 ISA

Abbreviations

ORP Oxidation–reduction potential
PAD Pulsed amperometric detection
SPEC Screen-printed electrochemical sensor

Bibliography

Adams, V., *Water and Wastewater Examination Manual*, Chelsea, MI: Lewis, 1990.

Alegret and Merkoci, *Electrochemical Sensor Analysis*, Amsterdam, Netherlands: Elsevier, 2007.

Bard, A. J. and Faulkner, L. R., *Electrochemical Methods*, New York: John Wiley & Sons, 1980.

Bobacka, J., Ivaska, A., and Lewenstam, A., Potentiometric ion sensors, Chemical Reviews, 108: 329–351, 2008.

Bond, A. and Anterford, D., Comparative study of a wide variety of polarographic techniques with multifunctional instruments, *Analytical Chemistry*, 44: 721, 1972.

Bradley, H. J. and Tsai, L. S., *Amperometric Titration of Chlorine with a Modified pH Meter*, Washington, DC: U.S. Department of Agriculture, 1999.

Bratov, A., Abramova, N., and Ipatov, A., Recent trends in potentiometric sensor arrays—A review, Analytica Chimica Acta, 678: 149–159, 2010.

Bretherick, L., *Bretherick's Handbook of Reactive Chemical Hazards*, Stoneham, ME: Butterworth-Heinemann, 1999.

Buonauito, R. P., Oxygen analyzers, *Measurements and Control*, February 1990.

Cali, G. V., Improvements in pH control and dissolved oxygen instrumentation for industrial and municipal waste treatment, Conference on Application of U.S. Pollution Control Technology in Korea, Seoul, Korea, March 22, 1989.

Cardis, T. M., Managing data from continuous analyzers, 1992 ISA Conference, Houston, TX, October 1992.

Clark, K., Chlorine analyzers cut costs, improve performance, *InTech*, May 1998.

Evans, R., *Potentiometry and Ion Selective Electrodes*, New York: John Wiley & Sons, 1987.

Flato, J., The renaissance in polarographic and voltammetric analysis, *Analytical Chemistry*, 44: 75A, 1972.

Galster, H., *pH Measurement: Fundamentals, Methods, Applications, Instrumentation*, Weinheim, Germany: VCH, 1991.

Gray, J. R., Glass pH electrode aging characteristics, ISA/93 Technical Conference, Chicago, IL, September 1993.

Hitchman, M. L., *Measurement of Dissolved Oxygen*, New York: John Wiley & Sons, 1978.

Kimmel, D. W., LeBlanc, G., Meschievitz, M. E., and Cliffel, D., Electrochemical sensors and biosensors, Analytical Chemistry, 84: 685–707, 2012.

Kissinger, P. T. and Heinemann, W. R., *Laboratory Techniques in Electro-Analytical Chemistry*, New York: Marcel Dekker, 1984.

LaCourse, W. R., Pulsed Electrochemical Detection in High-Performance Liquid Chromatography, New York: Wiley, 1997, 324pp.

Lubert, K. H. and Kalcher, K., History of electroanalytical methods, Electroanalysis, 22: 1937–1946, 2010.

Mayrhofer, K. J. J., Wiberg, G. K. H., and Arenz, M., Impact of glass corrosion on the electrocatalysis on Pt electrodes in alkaline electrolyte, *Journal of the Electrochemical Society*, 155(1): P1, 2008.

NIOSH, *NIOSH Manual of Analytical Methods: Supplement*, 4th edn., Vols. 2, London, U.K.: BPI Information Services.

Osteryoung, J. and Osteryoung, R. A., Square Wave Voltammetry, Analytical Chemistry, 57: 101A, 1985.

Riley, T. et al., *Principles of Electroanalytical Methods*, John Wiley and Sons Inc., New York, NY, 1987.

Silvester, D. S., Recent advances in the use of ionic liquids for electrochemical, Sensing, Analyst, 136: 4871–4882, 2011.

Skoog, D. A., *Principles of Instrumental Analysis*, Pacific Grove, CA: Brooks/Cole Publishing, 1997.

Standard Methods for the Examination of Water and Wastewater, New York: APHA, AWWA, and WPCF, latest edition.

Wang, H. Y., Transient measurement of dissolved oxygen using membrane electrodes, *Biosensors*, 4(5): 273–285, 1989.

Weiss, M. D., Teaching old electrodes new tricks, *Control*, July 1991.

1.22 Elemental Analyzers

D. H. F. LIU (1995) **B. G. LIPTÁK** (2003, 2017)

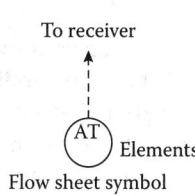

Flow sheet symbol

Type of instrument	A. Inductively coupled plasma (ICP) analytical atomic spectrometer
	A1. ICP-optical emission spectrometer (OES)
	A2. ICP-atomic emission spectrometer (AES)
	A3. ICP-mass spectrometer (MS)
	B. X-ray fluorescence (XRF) spectrometer
Element range	A. Virtually all elements
	B. Aluminium to Uranium
Applications	See Table 1.22a
Cost	A1. $50,000 to $200,000
	A3. $140,000 to $350,000
	B. $200,000 to $300,000 (online); $10,000 to $30,000 (portable)
Partial list of suppliers	Amptek (B) http://www.amptek.com/
	Bruker, (B) https://www.bruker.com/products/x-ray-diffraction-and-elemental-analysis/x-ray-fluorescence.html
	Horiba Scientific (B) (http://www.horiba.com/us/en/scientific/products/atomic-emission-spectroscopy/glow-discharge/)
	Jordan Valley (B) (www.forevision.com)
	Leco Corp. (http://www.laboratoryequipment.com/product-releases/2013/03/automated-shuttle-loaders-upgrade-elemental-analyzer)
	Leeman Labs Inc. Teledyne (A) (http://www.teledynewaterquality.com/pdfs/Water_Quality_Brochure.pdf)
	Olympus (http://www.olympus-ims.com/en/applications/x5000-petro-chemicals/)
	Oxford Instruments (B) (http://www.oxford-instruments.com/OxfordInstruments/media/x-ray-fluorescence/benchtop-xrf-brochures/x-supreme8000-brochure-june-2010.pdf)
	PerkinElmer Instruments (A) (http://www.perkinelmer.com/Catalog/Category/ID/Elemental%20Analysis)
	Seiko Instruments (http://www.seiko-instruments.de/files/brochure_sea_eu.pdf)
	Shimadzu Sci. Instruments (A) (www.shimadzu.com)
	SPECTRO Analytical Instruments (B) (http://www.spectrolive.com/condition-monitoring-the-role-of-elemental-analysis/)
	Thermo ARL (B) (http://www.thermoscientific.com/ecomm/servlet/productscatalog?storeId=11152&categoryId=82219&ca=xrf)
	Thermo Elemental (B) (http://www.thermoscientific.com/ecomm/servlet/productscatalog_11152__87466_-1_4)
	Varian Inc. (A) (www.varian.com)

INTRODUCTION

This chapter discusses the measurement of all elements. Of the 92 naturally occurring elements, approximately 30 metals and metalloids are potentially toxic to humans. They enter the environment through activities such as smelting, energy production, traffic, corrosion processes, landfills, and other activities. Heavy metals is the generic term for metallic elements having an atomic weight higher than 40.04 (the atomic mass of Ca). For most people, the main route of exposure to toxic elements is through the diet (food and water) and air. Therefore, there has been increasing concern, mainly in the developed world, about exposures, intakes and absorption of heavy metals by humans. Of particular concern are arsenic, barium, cadmium, chromium, lead, mercury, selenium, and silver.

Their allowable concentrations are very low and are set by various regulations. For example, for the cancer-causing arsenic, the Environmental Protection Agency (EPA) sets the limit of –0.01 parts per million (ppm) in drinking water, and the Occupational Safety and Health Administration (OSHA) sets the limit of 10 micrograms per cubic meter of workplace air (10 µg/m^3) for 8 hour shifts and 40 hour work weeks. Because of these low concentrations, very sensitive and accurate analyzers are needed for these measurement. Figure 1.22a provides some data for the average concentrations of the various elements in sea water. One such analyzer is the mass spectrometer, discussed in detail in Chapter 1.35.

Since its appearance in the 1960s, atomic absorption (AA) has become a widely used analytical tool. It is a popular analysis technique for elemental analysis in solids, liquids,

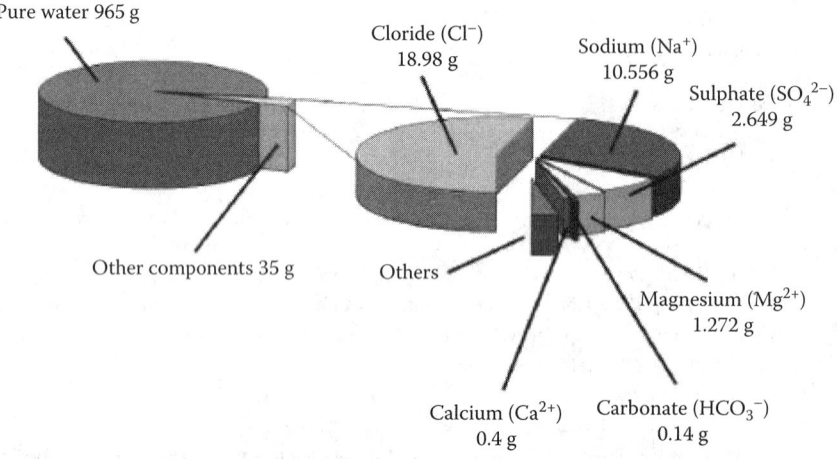

Element (ng/L)	North Pacific	North Atlantic
Mo	5000–10000	5000–10000
U	1800–2800	1800–2800
V	1630–1834	509–1172
As	1498–1798	1498–1573
Ni	117–704	117–352
Zn	7–536	7–105
Cu	32–286	64–127
Cr	156–260	14–165
Se	39–182	39–118
Mn	4–165	3–56
Cd	0,1–112	0,1–39
Fe	1–56	3–56
Pb	1–10	4–31
Co	0,8–3	1–2

FIG. 1.22a
The weights of various elements in a kilogram of sea water. (Courtesy of Marbef, Flanders Marine Institute (VLIZ).)

and loose powders of either organic or inorganic materials. Typical applications include the detection of sulfur (S) and lead (Pb) in refineries; the measurement of oxides in such raw materials as limestone, sand, bauxite, ceramics, slags, and sinters; and the sensing of major and minor elements (Cu, Fe, Ni, etc.) in food and chemical products. Compared to earlier techniques, AA is faster, more sensitive, and free from interference.

The inductively coupled plasma (ICP) atomic emission technique first appeared in the late 1970s and has been recognized as a technique with remarkable selectivity and sensitivity. It has simultaneous or rapid sequential multielement determination capability at the major, minor, trace, and ultratrace concentration levels without change of operating conditions.

X-ray fluorescence (XRF) is an analytical technique that is well suited to online applications. It has been used to measure the elemental composition of solids, liquids, and slurries. Solids analyzed have been in such diverse forms as powders, granules, sheets, or discrete parts. XRF is fast, reliable, and nondestructive (Figure 1.22b). It is also less expensive than other techniques, such as neutron activation, and introduces few health or safety hazards.

ATOMIC ABSORPTION SPECTROMETER

Figure 1.22c illustrates the basic operating principles of flame and furnace AA spectrometers. In flame AA, the sample is aspirated, nebulized, and passed into a flame that is burning at about 3000°K. The flame causes the sample compounds to dissociate. The sample atoms then can absorb radiation at wavelengths, called resonance lines, that are specific for the elements.

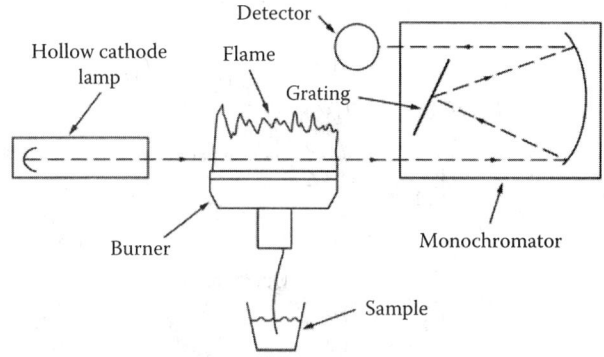

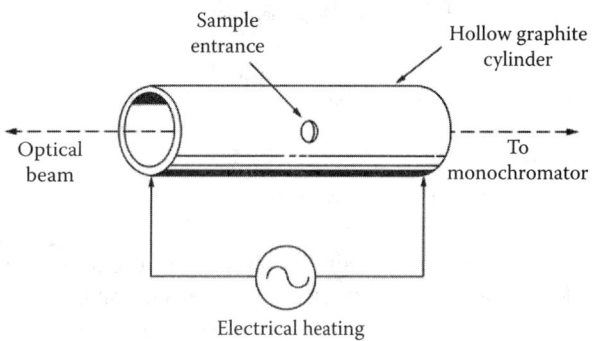

FIG. 1.22c
AA spectrometers.

The radiation source is a hollow cathode lamp in which the cathode contains the element being determined, and the monochromator is tuned to the appropriate resonance wavelength. Most commercial AA instruments can determine only one element at a time.

In furnace AA, a hollow graphite cylinder replaces the flame. The cylinder is heated electrically to a temperature high enough to atomize the sample. This unit is called a furnace atomizer, graphite furnace, or carbon rod. The device improves detection limits by 100- to 1000-fold, as compared to flame AA, but the furnace exhibits poorer precision and slower analysis speed, experiences more interference, and costs more.

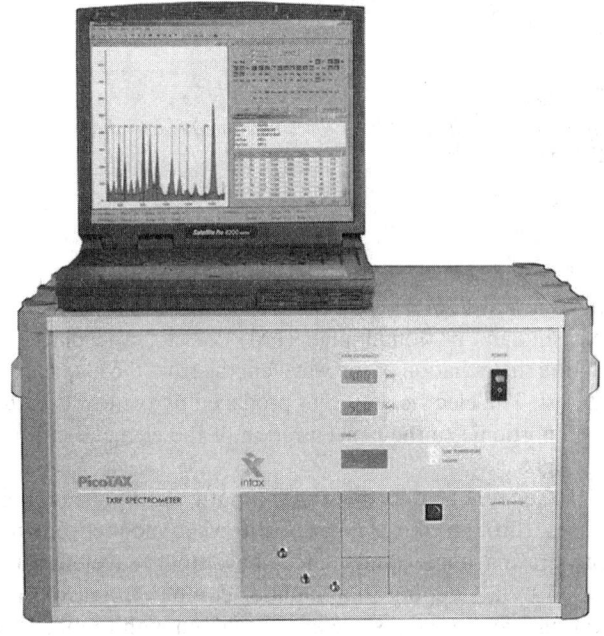

FIG. 1.22b
XRF spectrometer. (Courtesy of Roentec GmbH, Berlin, Germany.)

INDUCTIVELY COUPLED PLASMA DETECTOR

Figure 1.22d shows a schematic of an ICP-type detector. A stream of argon passes through a radio frequency (RF) induction coil, producing considerable heat. When "seed electrons" are introduced, an argon plasma is formed in the fireball region of the plasma from 8,000°K to 10,000°K.

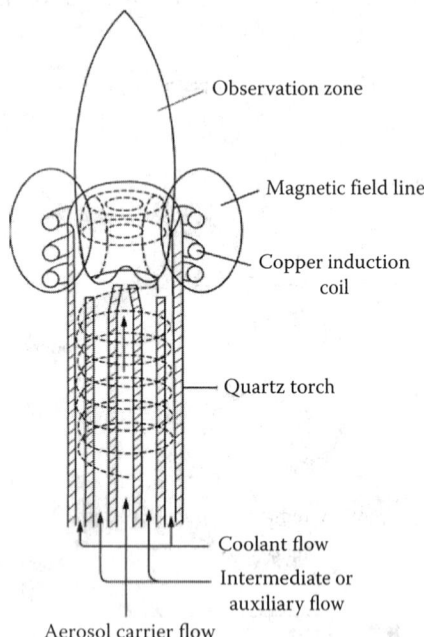

FIG. 1.22d
A schematic of an ICP-type detector.

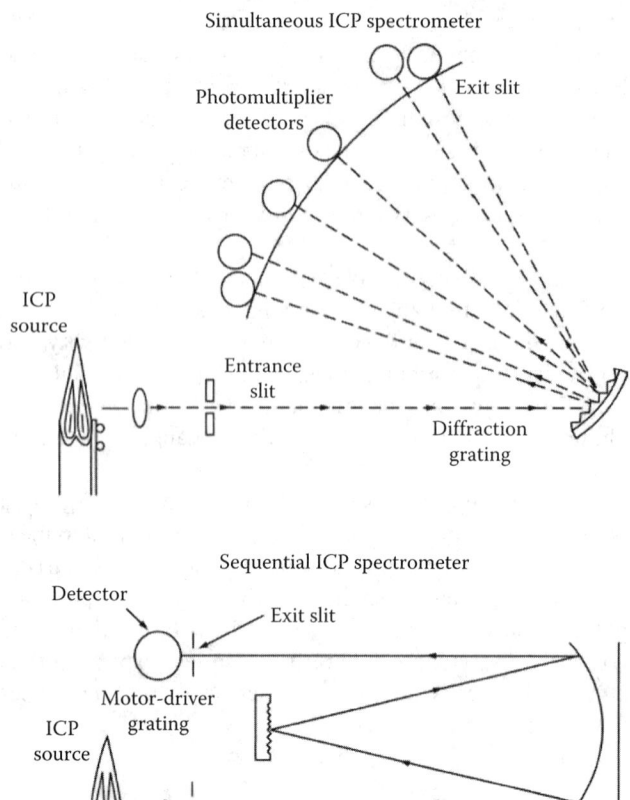

FIG. 1.22e
Operating principles of both the simultaneous and the sequential ICP spectrometers.

Argon is used to sustain the plasma because it requires less RF power than other gases. The outer flow supplies the plasma gas and cools the quartz torch. The inner flow carries the sample aerosol into the plasma. An optional middle flow helps to position the plasma and prevents carbon buildup on the tip of the central tube.

Operating Principle

Figure 1.22e illustrates the basic operating principles of both the simultaneous ICP spectrometer and the sequential ICP spectrometer. In either system, the sample is nebulized and introduced into the center of the plasma. The temperature causes the molecules to dissociate. At the same time, the atoms are excited and the excited atoms emit radiation. Every element emits radiation at its own characteristic wavelength.

In simultaneous ICP, the light passes through the entrance slit of a polychromator and is then diffracted by a grating. Exit slits and photomultiplier (PM) detectors are placed to capture the radiation of the wavelengths from the elements of interest. The electrical current produced by the PM detector is proportional to the concentration of the element emitting that light.

In sequential ICP, the light output from a plasma torch passes through the entrance slit of a monochromator. The grating angle (and hence the output wavelength) is varied by a stepping motor under computer control. The grating stays at each wavelength long enough to make the measurement. It then is driven rapidly to the next desired wavelength.

RELATIVE MERITS OF AA AND ICP

When deciding whether an AA instrument or an ICP instrument is the best option, the following should be considered:

1. *Speed of analysis*: The higher speed of ICP analysis stems from its capability to analyze several elements in a given sample. This produces a significant advantage in a sequential instrument, and an even greater advantage in a simultaneous instrument. In either case, speed in the two techniques differs by an order of magnitude.
2. *Flexibility*: The ICP technique gives a highly linear response, which enables simultaneous analysis of elements over a broad range of concentrations. This simplifies sample preparation and calibration. ICP requires no special cathode lamps for specific elements, so the sequential instruments become permanently available for analyzing any element. Also, the ICP technique enables the analysis of elements such as S and P that cannot be handled using the flame AA technique. It also provides increased capability for analyzing large series of elements such as B, W, lanthanides, and refractory in general.
3. *Matrix interference*: ICP spectroscopy is much more free from chemical and physical interference. An uncommonly high plasma temperature in an inert environment (Ar gas) reduces the problem of matrix interference.
4. *Detection limits*: The great sensitivity inherent in the technique and its improved signal-to-noise capacity give ICP better detection limits than those available with traditional flame AA. But the detection limits of furnace AA are better than those of a plasma for most elements. The exceptions are rare earth elements (B, U, Zr, etc.) plus nonmetals (S, Cl, I), which the AA furnace cannot measure.
5. *Precision*: Both systems use a pneumatic transfer system for samples and are very similar with respect to the degree of precision (i.e., repeatability) that can be obtained. There are no limitations to the use of ICP, as it can be used on gases, liquids, and solids (if spark ablation, laser ablation, or slurry introduction accessories are provided), so that vapors can be introduced into the plasma.

In spite of its advantages, the use of ICP has been limited by its relatively high cost, which can be justified only for large-scale analysis tasks or for solving problems that cannot be solved with flame AA. Because of increased productivity and lower prices, many laboratories have replaced AA spectrophotometers with ICP systems and left only the simpler tasks to be performed by flame AA spectrophotometers.

X-RAY FLUORESCENCE SPECTROMETER

XRF spectrometers irradiate the sample with a beam of high-energy x-rays that excite the elements present in the sample, so that they produce characteristic x-rays. This phenomenon is called x-ray fluorescence. XRF provides specific analyses of total elemental concentrations, without regard to chemical combinations.

The sample is excited by a source of electromagnetic radiation. This radiation excites the elements in the sample, causing them to fluoresce (i.e., give off their characteristic x-rays). The energy of these characteristic x-rays identifies the element, and the intensity is a measure of the element's concentration in the sample. The resulting signal is integrated over a period to give an average measure of concentration during this time.

Instrumentation

Online XRF analyzers can be configured in different ways to excite the process sample or to detect the characteristic fluorescence. The right combination of excitation and detection methods depends on the specific application. The x-ray source may be an x-ray tube or radioactive isotope (iron-55, curium-244, cadmium-109, or americium-241). The detection method may be either wavelength-dispersive (WDXRF) or energy-dispersive (EDXRF) XRF. When choosing either the anode used for the x-ray or a specific radioactive isotope, proper consideration should be given to the range of elements to be measured and the corresponding concentrations.

Figure 1.22f illustrates the principles of WDXRF and EDXRF instruments. With WDXRF, the x-ray energies are separated according to specific wavelengths by diffracting

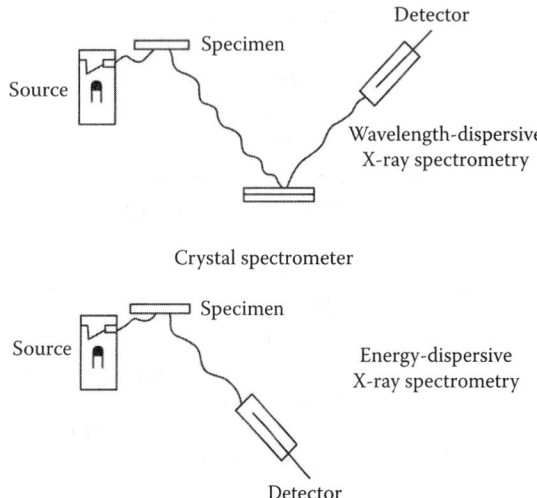

FIG. 1.22f
The detection methods used in XRF spectrometers.

crystals. The respective intensities are then measured by individual proportional counters. Sample excitation is by x-ray tube. With EDXRF, characteristic fluorescence is detected by a low-temperature solid-state detector or a gas-filled proportional counter and sorted electronically to produce an XRF spectrum of x-ray intensity energy. In EDXRF, sample excitation is by a low-level radioisotope source.

The WDXRF technique offers the best resolution, shortest analysis time, and highest sensitivity. It is well suited for the most demanding tasks, such as analyzing neighboring elements with large differences in concentrations. An EDXRF can be very compact and economical. It gives excellent sensitivity with good resolution, especially for light element analysis. The EDXRF systems are very useful in quality control, troubleshooting problems, etc.

Table 1.22a gives a partial list of proven and potential online XRF applications. A wide choice of sampling systems is now available for continuous monitoring of liquids in pipelines or tanks, granular solids in chutes, parts on conveyors, coatings on moving sheets of metals, or pigments in moving plastics or paper sheets. As online XRF technology advances, continued growth in the number of applications and installations can be expected.

TABLE 1.22a
List of Proven and Potential Online XRF Applications

Petroleum Refining
S in crude, refinery streams, gasoline, fuel oil, diesel, and MEA, DEA, and TEA stripper bottoms
V, Ni, and Fe in crude residues, cracking catalysts, coke, and bunker fuels
Ca, Zn, Cu, S, Cl, and Ba in lube-blending operations
S in sour water

Chemical and Petrochemical
Ti, Zn, Ca, and other fillers in polyolefins and other polymers
Sn, Br, Sb, P, Cl, or other fire retardants in fiber (nylon, Orlon, etc.) manufacture
S, Cl, and Br in elastomers
P in detergents
Monitoring catalysts—typically Co, Mo, Ni, Mn, Br, Ti, Pt, and Rh in processing
Blending additives manufacture (Ca, Zn, and Ba)

Inorganic Chemical Processing
Phosphoric acid manufacture
Metal treatment (plating) chemical manufacturing
Inks and paints (Ti, Zn, Pb, etc.); TiO_2 manufacture
China and other types of clay purification (Ti and Fe)
Photographic chemicals (As and Br)
Rare earths and rare earth chemical manufacture
Extractive hydrometallurgy (Cu, Zn, U, Hf, and Zr)
Glass manufacture—monitor Si, Fe, Ti, Ca, and K
Cement industry—monitor Fe, Ca, and Si

Pulp, Paper, and Rayon
Ca and Ti additives in pulp
Lignosulfates (S)
Wood preservatives (As, Cu, and Cr)
Viscose manufacture (Zn and S)

Catalyst Industry
Precious metal catalysts (Pt, Pd, and Rh)
Auto emission, reforming catalysts
Manufacture of Mo-, Co-, Ni-, and Zn-based catalysts
Catalyst recovery operations

Plating and Electroplating
Zn=Ni and Zn=Fe alloy coating of steel
Metals in pickling baths
Zn phosphating bath
Zn=Pb, Au=Ni=Br in plating baths for PC boards and electronic components

Metal Refining and Hydrometallurgical Operations
All base metals (Zn, Cu, Pb, Mo, Co, Ni, etc.)

SPECIFICATION FORMS

When specifying element analyzers only, please use the ISA form 20A1001, and when specifying both the analyzer and the composition or the properties of the process material of interest, use ISA form 20A1002. The following are the forms that are reproduced with the permission of the International Society of Automation.

1	RESPONSIBLE ORGANIZATION		ANALYSIS DEVICE		6		SPECIFICATION IDENTIFICATIONS		
2	ISA				7	Document no			
3			Operating Parameters		8	Latest revision		Date	
4					9	Issue status			
5					10				
11	ADMINISTRATIVE IDENTIFICATIONS				40	SERVICE IDENTIFICATIONS Continued			
12	Project number		Sub project no		41	Return conn matl type			
13	Project				42	Inline hazardous area cl		Div/Zon	Group
14	Enterprise				43	Inline area min ign temp		Temp ident number	
15	Site				44	Remote hazardous area cl		Div/Zon	Group
16	Area		Cell	Unit	45	Remote area min ign temp		Temp ident number	
17					46				
18	SERVICE IDENTIFICATIONS				47				
19	Tag no/Functional ident				48	COMPONENT DESIGN CRITERIA			
20	Related equipment				49	Component type			
21	Service				50	Component style			
22					51	Output signal type			
23	P&ID/Reference dwg				52	Characteristic curve			
24	Process line/nozzle no				53	Compensation style			
25	Process conn pipe spec				54	Type of protection			
26	Process conn nominal size		Rating		55	Criticality code			
27	Process conn termn type		Style		56	Max EMI susceptibility		Ref	
28	Process conn schedule no		Wall thickness		57	Max temperature effect		Ref	
29	Process connection length				58	Max sample time lag			
30	Process line matl type				59	Max response time			
31	Fast loop line number				60	Min required accuracy		Ref	
32	Fast loop pipe spec				61	Avail nom power supply			Number wires
33	Fast loop conn nom size		Rating		62	Calibration method			
34	Fast loop conn termn type		Style		63	Testing/Listing agency			
35	Fast loop schedule no		Wall thickness		64	Test requirements			
36	Fast loop estimated lg				65	Supply loss failure mode			
37	Fast loop material type				66	Signal loss failure mode			
38	Return conn nominal size		Rating		67				
39	Return conn termn type		Style		68				
69	PROCESS VARIABLES		MATERIAL FLOW CONDITIONS		101		PROCESS DESIGN CONDITIONS		
70	Flow Case Identification			Units	102	Minimum	Maximum	Units	
71	Process pressure				103				
72	Process temperature				104				
73	Process phase type				105				
74	Process liquid actl flow				106				
75	Process vapor actl flow				107				
76	Process vapor std flow				108				
77	Process liquid density				109				
78	Process vapor density				110				
79	Process liquid viscosity				111				
80	Sample return pressure				112				
81	Sample vent/drain press				113				
82	Sample temperature				114				
83	Sample phase type				115				
84	Fast loop liq actl flow				116				
85	Fast loop vapor actl flow				117				
86	Fast loop vapor std flow				118				
87	Fast loop vapor density				119				
88	Conductivity/Resistivity				120				
89	pH/ORP				121				
90	RH/Dewpoint				122				
91	Turbidity/Opacity				123				
92	Dissolved oxygen				124				
93	Corrosivity				125				
94	Particle size				126				
95					127				
96	CALCULATED VARIABLES				128				
97	Sample lag time				129				
98	Process fluid velocity				130				
99	Wake/natural freq ratio				131				
100					132				
133	MATERIAL PROPERTIES				137	MATERIAL PROPERTIES Continued			
134	Name				138	NFPA health hazard		Flammability	Reactivity
135	Density at ref temp		At		139				
136					140				
Rev	Date	Revision Description		By	Appv1	Appv2	Appv3	REMARKS	

Form: 20A1001 Rev 0 © 2004 ISA

Analytical Measurement

	RESPONSIBLE ORGANIZATION	ANALYSIS DEVICE COMPOSITION OR PROPERTY Operating Parameters (Continued)		SPECIFICATION IDENTIFICATIONS	
1			6		
2	ISA		7	Document no	
3			8	Latest revision	Date
4			9	Issue status	
5			10		

	PROCESS COMPOSITION OR PROPERTY			MEASUREMENT DESIGN CONDITIONS				
	Component/Property Name	Normal	Units	Minimum	Units	Maximum	Units	Repeatability
13								
14								
15								
...								
37								

(Rows 38–75: blank remarks/notes area)

Rev	Date	Revision Description	By	Appv1	Appv2	Appv3	REMARKS

Form: 20A1002 Rev 0 © 2004 ISA

Abbreviations

AA	Atomic absorption
AES	Atomic emission spectrometer
EDXRF	Energy-dispersive x-ray fluorescence spectrometer
ICP	Inductively coupled plasma
MS	Mass spectrometer
OES	Optical emission spectrometer
PM	Photomultiplier
RF	Radio frequency
WDXRF	Wavelength-dispersive x-ray fluorescence spectrometer
XRF	X-ray fluorescence spectrometer

Organization

ATSDR	Agency for Toxic Substance and Disease Registry

Bibliography

Agilent Technologies, Elemental impurity analysis, (2012) https://www.agilent.com/cs/library/primers/Public/5991-0436EN_Primer_Pharmaceuticals.pdf.

Amptek, What is XRF?, (2015) http://www.amptek.com/xrf/.

Archaerometry Laboratory, Overview of X-ray fluorescence, (2014) http://archaeometry.missouri.edu/xrf_overview.html.

Coastal Wiki, Elemental mass spectrometry trace element contaminants in the marine environment, (2013) http://www.coastalwiki.org/wiki/Elemental_mass_spectrometry_-_a_tool_for_monitoring_trace_element_contaminants_in_the_marine_environment.

InTech Europe, Heavy metals and human health, (2012) http://cdn.intechweb.org/pdfs/27687.pdf.

Liba, E., Elemental impurities, (2014) https://www.agilent.com/cs/library/whitepaper/Public/5990-9382EN_WhitePaper_ICP-MS_ICP-OES_Pharma.pdf.

Martin, S. and Griswold, W., Human health effects of heavy metals, (2009) http://www.engg.ksu.edu/chsr/files/chsr/outreach-resources/15HumanHealthEffectsofHeavyMetals.pdf.

Wang, J. and Balazs, M. K., Analytical techniques for trace elemental analysis, (2000) http://www-project.slac.stanford.edu/ssrltxrf/publications/balazs.pdf.

Wielopolski, L., Nuclear methodology for non-destructive multi-elemental analysis of large volumes of soil, (2011) http://cdn.intechopen.com/pdfs-wm/21129.pdf.

1.23 Fiber-Optic (FO) Probes and Cables

D. H. F. LIU (1995)   **M. W. REED** (2003)

B. G. LIPTÁK and R. S. HARNER (2017)

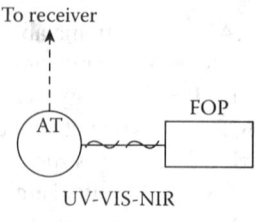

UV-VIS-NIR

Flow sheet symbol

Type of process fluid	Liquids, solids, gases, and multiphase.
Design pressure	Vacuum to 5000 psig (350 bar).
Design temperature	Extreme service, −100°C to +450°C.
Materials of construction	Stainless steel (316L) is most common, but high-nickel alloys and certain specialty metals are available.
Costs	Probes vary from $250–$1,500 for laboratory/feasibility versions to $2,000–$30,000 for process-capable units. Analyzer systems range from $2,000 for a basic photometer to over $100,000 for Raman and industrial FTNIR units. (For Raman analyzers, see Chapter 1.58, and for FTNIR, see Chapter 1.32.)
Advantages	Freedom from electromagnetic interference and noise, which in industrial plants can be caused by large equipment and separate power supply systems causing magnetic waves, ground surges, and ground potential differences. In addition, FO cables do not attract lightning strikes either.
Installation	For an initial fiber-optic network installation, it is advisable to obtain a full-system offering (one-stop shopping) for all the components with an end-to-end warranty for installation, operation, and maintenance.
Partial list of suppliers	ABB-Process Analytics Inc. (systems) (http://www.abb.com/product/us/9AAC100838.aspx)
	Applied Analytics (systems) (http://www.a-a-inc.com/spectrophotometer-principle/)
	Axiom Analytical, Inc. (probes) (http://www.goaxiom.com/process_products.html)
	Banner Engineering Inc. (optics) (http://www.bannerengineering.com/en-US/products/8/Sensors/26/Fiber-Optic-Sensors)
	Bran and Luebbe (systems) (http://www.spx.com/en/bran-luebbe/pd-mp-on-line-analyzers/)
	Brinkmann Instruments (systems) (http://www.metrohmusa.com/Products/Measurement/Physical-Measurements/Brinkmann-Probe-Colorimeter/index.html)
	B&W Tek, Inc. (systems) (http://bwtek.com/spectrometer-part-8-fiber-optic-probes/)
	Custom Sensors & Technology (systems) (http://www.customsensors.com/index.php?q=products/categories/Fiber%20Optic%20Probes)
	Dolan-Jenner Industries (optics) (http://www.dolan-jenner.com/Pro/Special_End&CFO.htm)
	Edmund Scientific (optics) (http://www.edmundoptics.com/illumination/fiber-optic-illumination/fiber-optic-light-guides/quartz-fiber-optic-light/1858)
	Guided-Wave, Inc. (systems) (http://www.guided-wave.com/products/probes)
	Hellma Analytics (probes) (http://www.mypatprobe.com/)
	InPhotonics, Inc. (systems) (http://www.inphotonics.com/probeselector.htm)
	Kaiser Optical Systems (systems) (http://www.kosi.com/na_en/products/raman-spectroscopy/raman-probes-sampling/raman-sampling-probe-heads-optics.php)
	Leoni Fiber Optics, Inc. (optics) (http://www.leonifo.com/leoni-fiber-optics-standard-products.php)

(Continued)

L.T. Industries, Inc. (systems) (http://www.ltindustries.com/prodprocess.htm)
Mettler-Toledo International, Inc. (systems) (http://us.mt.com/us/en/home/products/L1_AutochemProducts/ReactIR/ReactIR-45m-Reaction-Characterization.html)
Multimode Fiber Optics, Inc. (probes) (http://www.multimodefo.com/reflectance%20probes.htm)
Ocean Optics Inc. (systems) (http://www.oceanoptics.com/Products/fiberprobeoverview.asp)
OPSIS (http://www.cleanair.com/Equipment/OPSIS/opsis.htm)
PreSens GmbH (systems) (http://www.presens.de/products/category/category/sensor-probes.html)
Tec5USA. Inc. (probes) (http://www.tec5.com/text/107/en/fiber-optic-probes.html)
Thermo Scientific (probes) (http://www.thermoscientific.com/en/product/versa-fiber-optic-probe.html)

INTRODUCTION

This chapter discusses both fiber optic (FO) cables and FO probes but places the emphasis on the second, because in industrial applications, FO probes are more often used. The main difference between the two is the distance the information travels and the amount of information being carried. The extension cable of an FO probe is relatively short, and it carries a relatively small amount of information, while FO cables are long and carry large quantities of information. A single cable can be several thousand miles long and can transport information at a rate of billion book pages per second.

FIBER-OPTIC CABLES

The individual fibers are highly transparent, cylindrical conduits of light. When light enters a fiber that has a higher refractive index (RI) than the cladding surrounding it, the light stays inside that fiber, because the light is totally reflected by the cladding, and it just travels over long distances through the fiber as shown in Figure 1.23a.

The refractive index is defined as the speed of light (c) in vacuum divided by the speed of light through a particular material (v). For example, the refractive index of water is 1.33, meaning that light travels 1.33 times slower in water than it does in vacuum. The higher the RI index, the slower the light will travel through that material. In FO cables, the light in the fiber core travels slower (RI ~ 1.62) and in the cladding it travels faster (RI ~ 1.52). This combination creates a light conduit (waveguide) where total internal reflection occurs as the light travels through it.

Internet Supporting Cables

FO cables contain many optical fibers. In a "submarine" (under-sea) communication cable, the bundle of fibers is usually covered by seven layers of protection (Figure 1.23b).

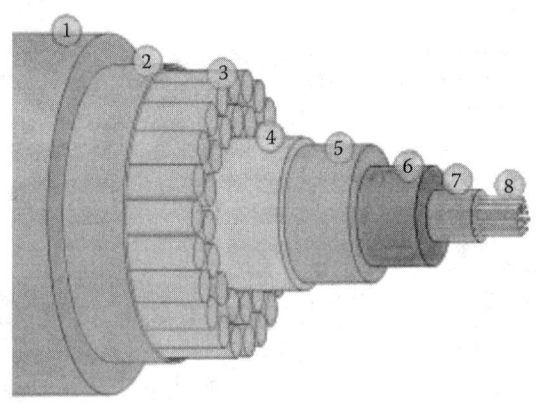

FIG. 1.23b

In a submarine (under-sea) communications cable, the optical fiber bundle is covered by seven layers of protection. (Courtesy of Wikipedia.)

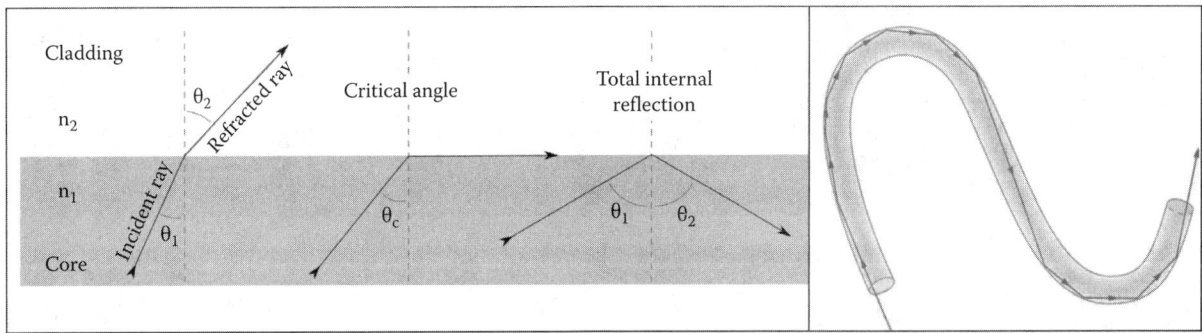

FIG 1.23a

The FO fibers are conduits of light providing total internal reflection, because the refractive index (RI) of their cladding is lower than that of their core.

The numbered layers in Figure 1.23a are 1-Polyethylene, 2-Mylar tape, 3-Stranded steel wires, 4-Aluminium water barrier, 5-Polycarbonate, 6-Copper or aluminium tube, and 7-Petroleum jelly. A modern cable is typically 69 millimeters (2.7 in.) in diameter and weighs around 10 kg per meter (7 lb/ft), although thinner and lighter cables are used for deep-water sections. The cables that connect the continents can deliver at a rate of several terabits (terabit = 10^{12} bits) per second. In order to give an idea of how much that capacity is, the Internet connection of a home usually ranges from one to 50 megabits (10^6 bits) per second.

When we send a photo or anything else through the Internet, we just click a button and so does the addressee when receiving it at the other end. We think little about the process how the photo gets there. We assume that it is by some magic that involves no material at all, or possibly some satellites, like does the GPS. Well, there is no magic here! That photo being sent and billions of others are all carried by FO cables.

We live in this strange digital age, when we use things without understanding them, and this is dangerous. This is similar to our cultural attitudes to computer viruses, but worst, because, while we do not understand how viruses work, we do know what they do. It can also be more than nuclear reactors, because, while we do not understand how cyber-attacks on nuclear reactors can be accomplished, we understand what meltdowns do. In the case of the Internet, we do not even know what the all the consequences would be of a meltdown. This "meltdown" is different from, say, the consequences of a power outage, because with electricity, we do have experience, we understand the consequences of losing it, and therefore, over the years, we have prepared for it by designing a variety of backup systems.

This is not the case with the Internet! On the one hand, bringing down the Internet is easy (all that is needed is to cut some cables), we not only have no backup, but we do not even know what the consequences will be if they are cut, and therefore, we do not really know what requires backup and what does not. Figure 1.23c shows the half million or so miles of cables that make the Internet work in this part of the world. Some have backup, others do not. Some are government owned; others are owned privately. They are of different ages and different constructions; their only common feature is that they all can be cut. The governments of some

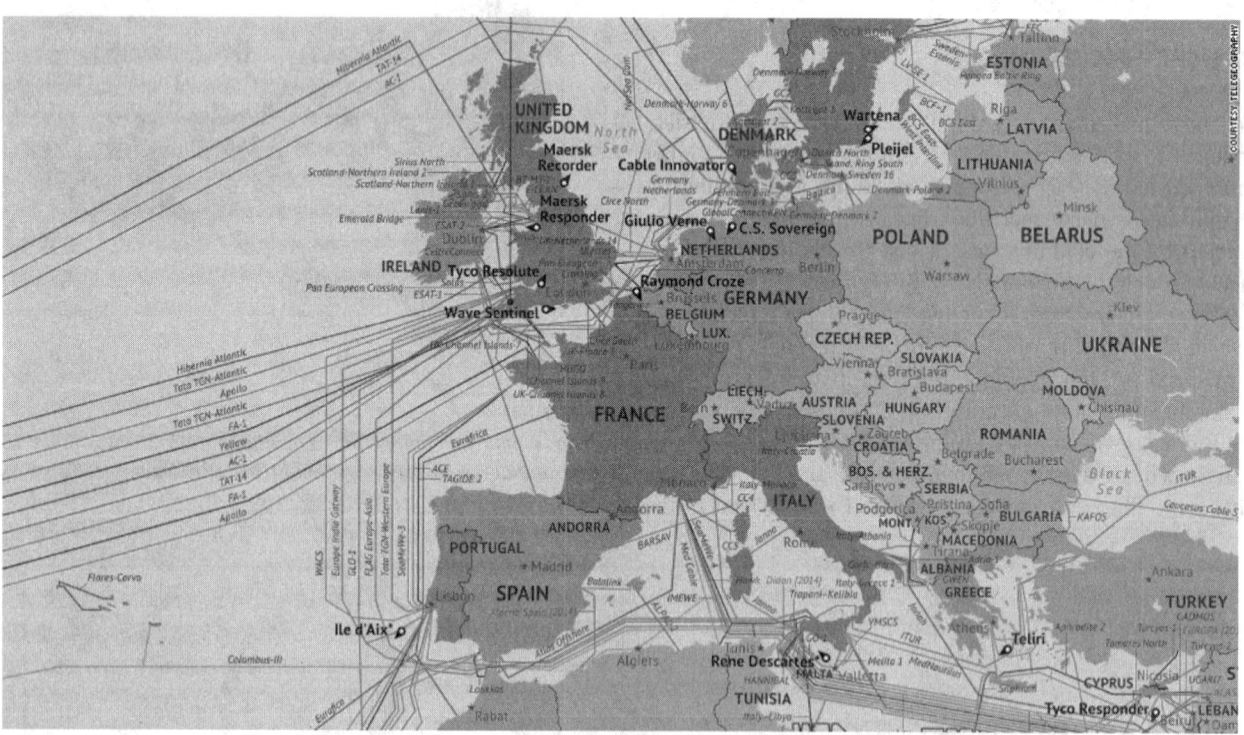

FIG. 1.23c
The network of "submarine" Internet cables laid between land-based stations. (Courtesy of CNN.)

closed societies are already doing that when they decide to deny Internet access to their citizens.

Actually, the Internet can be attacked even more efficiently than cutting cables, because it utilizes Internet exchange points (IXP). These exchange points reduce the per point delivery cost and improve routing efficiency, but they also increase the vulnerability, because many cables can be destroyed by destroying a single such junction point. In the United States, there are less than 100 such IXP junction points, where the cable networks come together, and these points are usually unprotected and are operated without backup.

This shows that the Internet industry could (and should) learn a lot from our industrial automation profession, before it is too late.

FIBER-OPTIC PROBES

Fiber-optic probes offer several advantages for chemical and optical measurements because the sensing area can be separated from the instrumentation by significant distance. Probes can be installed directly in a chemical process stream, while the spectrophotometer may be located hundreds of meters away in a protected area. A conventional sample conditioning system (SCS) is unnecessary, although certain aspects of the sample at the probe tip may need to be addressed. But the process sample need not be transported to a flow-through optical cell of a process spectrometer, so issues with sample degradation, loss of containment, and response delay are minimized. Compared to a laboratory instrument, the fiber-optic probe and extension fibers replace the sample cell and collimation optics of the conventional optical bench. Indeed, the optical fiber often serves as an extremely thick flexible lens exhibiting low loss over wide spectral ranges.

A well-designed probe can be successfully located in toxic, corrosive, explosive, or radioactive environments and withstand a wide range of temperature, pressure, and flow conditions. Fiber-optic cables usually are nonconductive and immune to electromagnetic interference. While exposed fiber can be broken easily, cables fabricated for industrial use contain the same strength members and jacketing used to protect copper in cable trays. Simplex optical fiber is often pulled through dedicated conduit to provide additional protection in rugged environments.

System Operation

Optical fibers transmit light very efficiently through a dielectric structure that supports total internal reflection (TIR) for certain electromagnetic wavelengths described as photons. While externally similar to a conventional *polymer fiber* with a circular cross section, a practical *optical fiber* consists of a circular core, a TIR cladding layer, and a protective coating. Most of the light injected into the core at one end emerges from the core at the other end. Losses are measured in fractions of a dB per kilometer at optimal wavelengths. Fibers intended for spectroscopy generally are constructed with larger-diameter cores (200–600 µm) of extremely pure silica to maintain low loss over a wide spectral range (ultraviolet–visible–near-infrared [UV–VIS–NIR]).

Measurement by fiber-optic probes falls into one of two categories depending on how the optical fibers interact with the process fluid. In the *intrinsic* mode, properties of the fiber such as microbending loss and attenuated total reflection are utilized to interrogate the environment in direct contact with the fiber. Sensitive interferometers and dispersive spectrometers are used to monitor the changes in phase, delay, and wavelength shifts in modified optical fibers, which are modulated by external stimuli. Single-mode and polarization-preserving optical fibers are used in long-distance communications where the temporal behavior of light pulses is modified. Intrinsic fiber-optic sensors are used in accelerometers, gyroscopes, hydrophones, and certain process parameters (temperature, pressure, and flow), but are outside the purview of this chapter.

In the *extrinsic* mode of measurement, optical fibers serve simply as light conduits from the tip of a probe to an analyzer location. Signal modulation during transit in the fiber cable should be minimized. The sensing tip could be an optical element, such as a window or prism, which is in contact with the sample. The tip might contain fluorescent material immobilized in a membrane that reemits photons at a different wavelength as the solution properties of the sample changes. But most common in industrial probes is transmission of light through a window that isolates the optics from the process and protects the optical fibers from contamination. The mechanical support and sealing mechanism for the optics constitute the practical definition of an industrial fiber-optic probe. If process sample breaches the seals, the measurement is likely invalidated. In this sense, the fiber-optic probe serves an analogous purpose for optics as a process thermowell serves for a resistance temperature device.

System Components

Fiber-optic probes are useless without dedicated light generation and detection apparatus. Commercial photometers and spectrometers accommodating optical fiber have been available since the 1980s. A schematic view of a legacy system

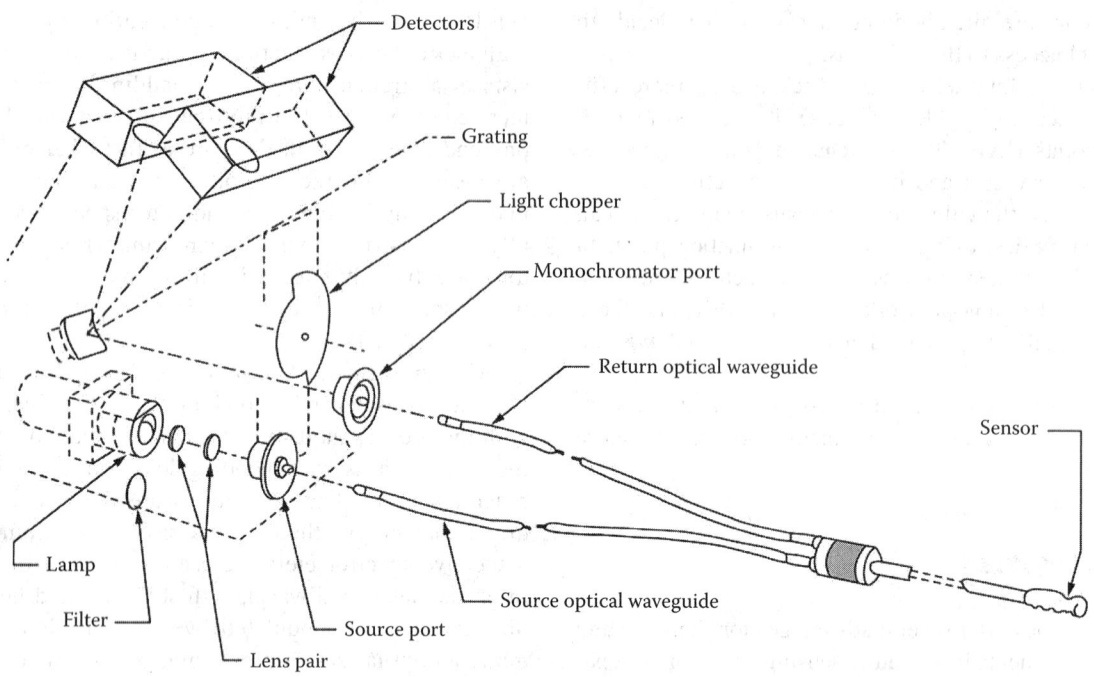

FIG. 1.23d
A schematic view of a legacy fiber-optic spectrophotometer. (Courtesy of Guided Wave, Inc.)

representing one of the first full-spectrum fiber-optic spectrophotometers is shown in Figure 1.23d.

This design was modular in that the grating and detectors could be exchanged and upgraded as necessary. By installing the proper components, the UV, VIS, and NIR regions could be accessed (200–2100 nm) with two grades of silica fiber. However, it also contained moving parts (grating servo and chopper) with limited lifetime and (ultimately) availability.

The fiber-optic probe depicted includes a source optical waveguide and a return waveguide that carry light to the probe sensor tip that interacts with the sample. In this case, the waveguides were single large-core optical fibers. The sensor tip contained a mirror to reflect light back, allowing it to pass through a sample gap in both directions. Full-spectrum measurement was accomplished with a slit/grating/detector combination. Other practical designs incorporated a simple band-pass filter* to monitor a single wavelength of light, a multiplicity of filters, and an optical train/mechanics to monitor several wavelengths simultaneously, or an array detector with a stationary dispersion element (Figure 1.23e). Larger-core fibers generally collect more light, but are less flexible on installation and cost more. Instruments designed for multiple fibers or bundles can increase the number of photons available for detection even further, but are limited to short runs due to manufacturing expense. Each waveguide option

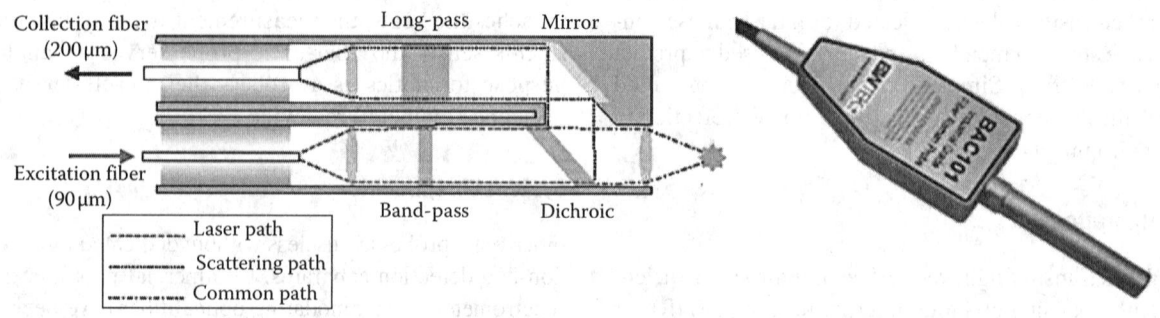

FIG. 1.23e
Fiber-optic sensor tip. (Courtesy of B & W Tec.)

---

* It is a filter that passes frequencies within a certain range and rejects (attenuates) frequencies outside that range. The bandwidth of the filter is the difference between the upper and lower cutoff frequencies.

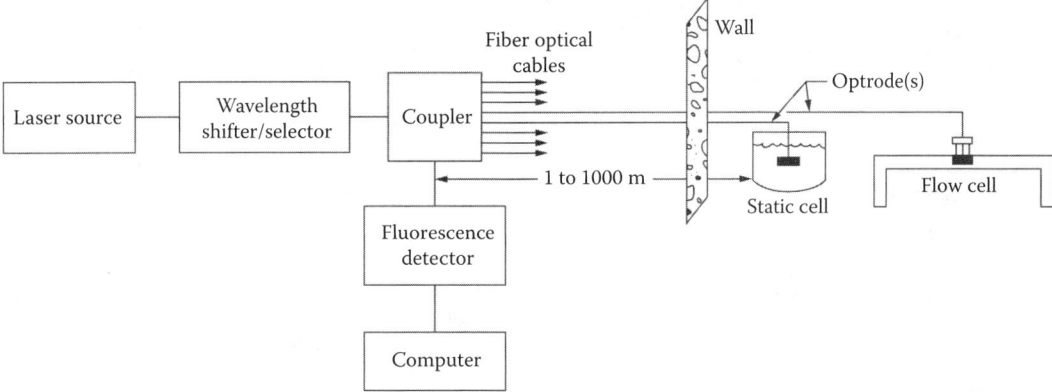

FIG. 1.23f
Analyzer schematic using an optical coupler to relay signals from remote sampling sites.

has specific interface requirements to maximize throughput (etendue*) and yield an effective light budget. A laboratory spectrophotometer can be converted to fiber optics, but it is seldom a perfect fit, because it has not been designed for use in an industrial environment.

For chemical processing applications, the analyzers must be rugged and have reliably established mean time between failure (MTBF) data. They typically must operate over a wide temperature range and sealed to prevent ingress of fugitive emissions, with self-diagnostics and auto restart capability. Fiber-optic instruments do have the luxury of operation in fully protected areas (analyzer house or control room), because the optical cable allows significant distance between the sensing point and the analyzer. Many times, this is the driving force for installation, that is, to remove the analyzer (and operator) from a hazardous area. This advantage may obviate a conventional sampling system, saving considerable expense in design of sample transportation and potential modification. In situ measurement in the pipe without sample removal can be real-time and maintenance free.

Because optical fibers are very small and relatively inexpensive, multipoint fiber-optic process measurements become practical. Figure 1.23f shows an optical coupler that allows a single detector or spectrometer to interrogate a multiplicity of fibers returning from probes.

A fiber-optic multiplexer coupled to multiple fiber-optic probes can increase the return on investment (ROI) for process installations where several related measurements are required. If a single analyzer is located at a central location and optical fibers extend to multiple probes, substantial cost savings can be realized over installing individual analyzers at each sample point. Multiplexers can be integrated into the analyzer, or stand alone with fiber-optic connector interfaces (Figure 1.23g). On the negative side, an analyzer failure causes the loss of data from all probes. Furthermore, many multiplexers have mechanical components and usually incur some light loss even under the best circumstances.

Fiber-optic distribution panel. (Courtesy of TE Connectivity.)

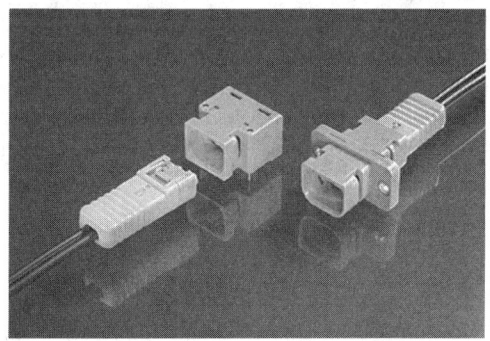

Fiber-optic connector. (Courtesy of Heilind.)

FIG. 1.23g
Fiber-optic connector interfaces.

Probe Designs

An industrial fiber-optic probe operates in the *extrinsic* mode where the fibers are protected from the process. This is usually accomplished by a window that is transparent to the photons of interest, or a sensing element that modifies the

---

* Geometric etendue (geometric extent), G, characterizes the ability of an optical system to accept light. It is a function of the area, S, of the emitting source and the solid angle, Q, into which it propagates. Etendue, therefore, is a limiting function of system throughput. Geometric etendue may be viewed as the maximum beam size the instrument can accept; therefore, it is necessary to start at the light source and ensure that the instrument, including ancillary optics, collects and propagates the maximum number of photons.

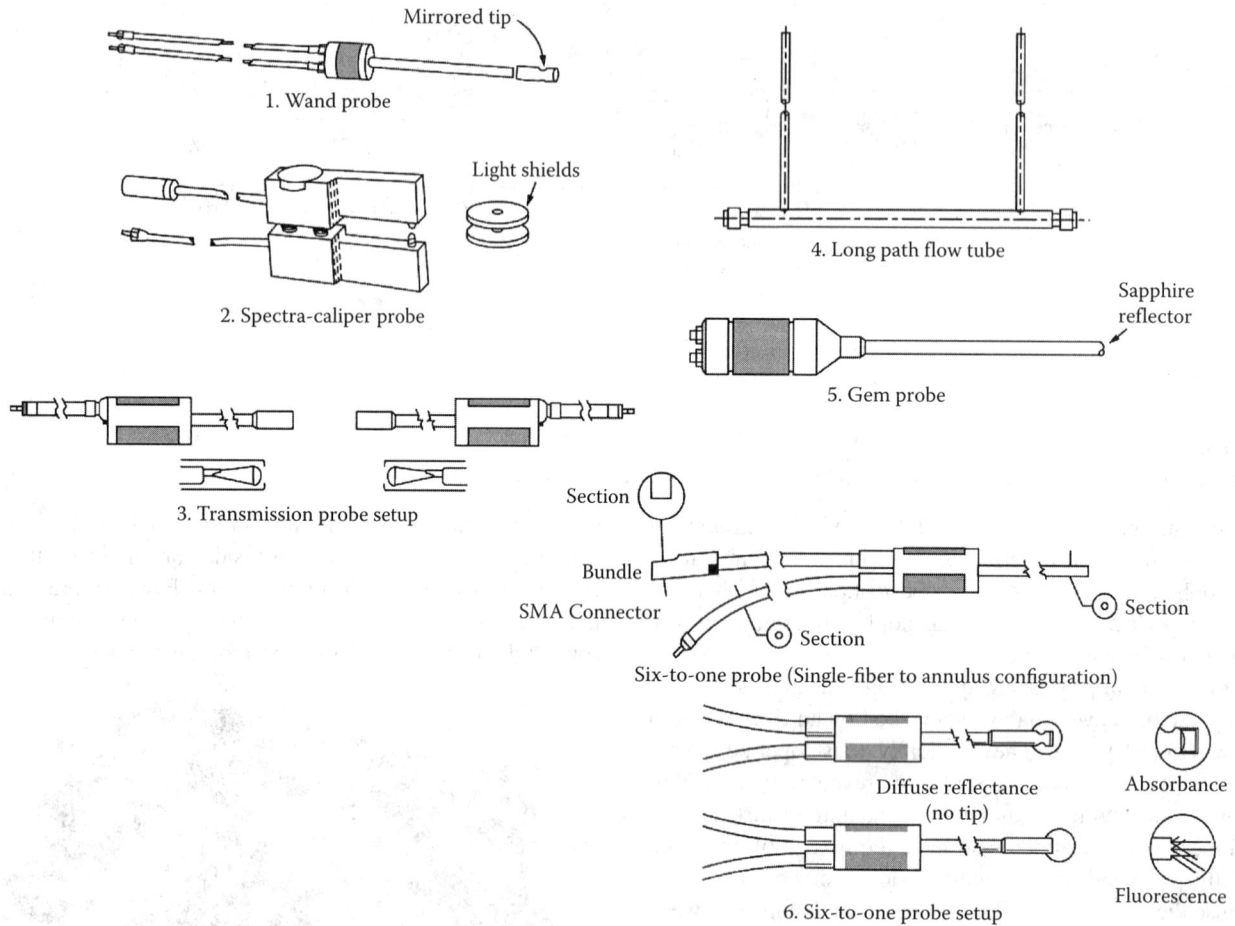

FIG. 1.23h
Early fiber-optic probe designs. (Courtesy of Guided Wave, Inc.)

photons in some reproducible manner. Fiber-optic probes are commercially available that do not protect the fibers adequately for industrial process use but may be applicable for general-purpose use. The flexibility gained with optical fibers can be useful even in lab probe designs and experimental setups. Figure 1.23h shows a variety of early probes for UV–VIS–NIR measurement.

Many of the probe designs utilize a cylindrical tube in which the optical fibers are protected. This shape is convenient for insertion through compression fittings or o-ring seals. The tube is typically machined from a metal compatible with the process. While stainless steel alloys are common, probes have been fabricated from high-nickel alloys (Monel™, Hastelloy™, Inconel™, etc.), titanium, and even tantalum for corrosion resistance. Glass-lined vessels and plastic line pipe present additional challenges when corrosion rate invalidates all metal choices. Probes coated with corrosion-resistant material can be successful in certain applications. Probes have been fabricated from ceramic or refractory material in extreme cases.

Sealing Techniques

Early probes used epoxy adhesive to seal optical fibers in direct contact with the sample or a glass window to separate the fibers from the process. Indeed, virtually all probes intended for laboratory use, that is, the general designation of "optrode" used in aqueous and biological media, do not utilize robust seals required for chemical processing. Several epoxies are formulated specifically for fiber-optic use but typically cannot handle the temperature, pressure, and corrosive nature of some process applications. Three general techniques have been developed for sealing a window at the end of an industrial fiber-optic probe.

Brazing Tapered sapphire windows can be press-fit into a corresponding metal fitting that has been heated and then cooled. These seals can be useful at moderate operating temperature. Unfortunately, the seal tends to deteriorate with temperature cycling and time. Catastrophic failure (window falls out) may occur at some elevated temperature when all compression is lost. Brazing is the introduction of the lower melting point metal/alloy between the body and window, which bonds to both surfaces. If brazing is done correctly, a reliable seal that has the advantage of high-temperature resistance and low permeability can be created. The braze material must wet both surfaces and withstand the stress generated in a body/window combination that is mismatched in thermal expansion coefficient.

While brazing is favored over press-fit, several disadvantages have been identified with this seal mechanism. Because the braze material is selected for physical characteristics (melting point, wetting capability, etc.), it may not have the same corrosion resistance of either the body metal or window. The stress of mismatched thermal expansion coefficient tends to flex the seal with temperature cycling, leading to fatigue. A probe may survive for a period of time and then suddenly fail (usually after the warranty period). While the most reliable brazed seals are made with engineered performs and multilayer metallizations, use of a program-controlled vacuum oven and machined fixtures are critical requirements for long-term success.

Elastomeric Gaskets When compared to brazed seals, gasket seals generally represent an intermediate temperature and pressure option, but excel in the corrosion resistance area. Compared to adhesive/epoxy seals used in biological sensors, the gasket can be specified to withstand the degree of chemical attack at the insertion point in the process. In chemical production, the seals used in existing process transmitters and pipe flange gaskets are usually known and well documented. Tables of chemical resistance can be consulted for new processes and extensions, but for critical applications, insult testing is always advisable.

The fiber-optic probe design shown in Figure 1.23i uses a pair of flat gaskets supporting a window in a spring-energized arrangement, which maintains a constant sealing force independent of large temperature changes of the environment.

As with o-ring sealed designs, expansion and contraction of dissimilar components (body, window, insert, etc.) is accommodated by design, not constrained or ignored. O-rings generally require more volume to implement the seal than gaskets, because initial compression of the o-ring at zero pressure differential must supply the sealing force, yet the thermal expansion of the o-ring must be accommodated in the gland dimensions.

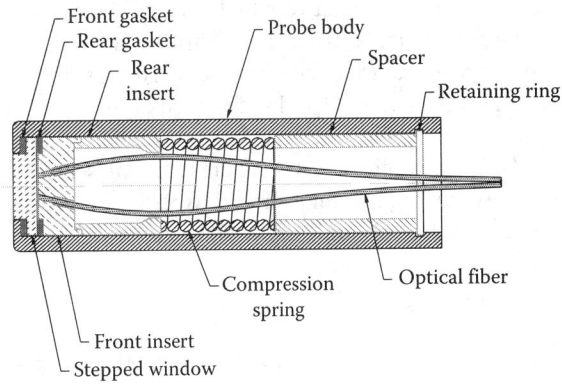

FIG. 1.23i

Spring-energized seal mechanism. (Courtesy of the Dow Chemical Company.)

Elastomeric seals are limited in temperature by the temperature rating of the polymer. Working above 200°C is problematic with all but the highest temperature rated materials (i.e., Kalrez™ compounds have excellent chemical resistance additionally). Pressure rating is also a function of the gland design or support structures. One seldom acknowledged advantage is that gaskets and o-rings can be replaced with relative ease, allowing the reuse of an expensive alloy body and/or window material in aggressive environments. Indeed, the internal optical fibers in Figure 1.23g might be exchanged with a single fiber and lens, converting a backscatter particle detection probe into a light collimation probe. The seal maintains a process-free internal environment for optical components. Elastomer permeation is a final concern, especially when a probe is subjected to alternating media, including solvents that tend to swell the polymer matrix. Even the best designed seal mechanism will not survive indefinitely if the gasket extrudes out of the gland under pressure.

Compression Rings Another alternative that combines the advantages of a spring-energized seal with the chemical resistance of a metal C-ring is shown in Figure 1.23j.

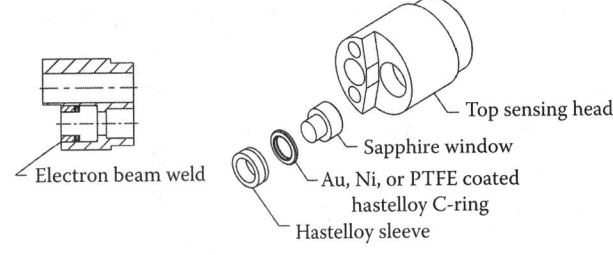

FIG. 1.23j

Welded metal seal assembly. (Courtesy of Axiom Analytical, Inc.)

This seal can operate at very high temperatures (>400°C) and pressures (up to 350 bar). In fact, structural ratings of the process flange and alloy, rather than the seal design, may limit the final process rating. One disadvantage is that the C-ring acts directly on the window, so surface preparation of the window and initial coating of the C-ring are critical to seal reliability. Furthermore, significant technical expertise is required in the e-beam welding of miniature components in tight spaces.

Process Interfaces

This chapter has focused on the use of fiber-optic probes in chemical process measurement, that is, in process vessels and pipes. Because one of the advantages of fiber-optic use is the ability to separate the sample point from the analyzer by significant distance, the robust nature of the probe has been emphasized. When a probe is inserted directly into the process, the tip of the probe replaces the conventional sampling system. Obviating the sample system does not eliminate all sampling issues. A probe in direct process contact is far less accessible and often subject to engineering extremes, corrosion mechanisms, and fouling. Even when a probe interface includes a double block and bleed bypass configuration or an automated extraction device, control of temperature, pressure, etc., may be inadequate for accurate analytical measurements. Local transmitters installed in the measurement leg may be tied to analyzer inputs to correct the response to standard conditions.

Single-Pass Probe The probes discussed in the paragraphs dealing with sealing were *single ended* in that a single penetration of the process is required. Figure 1.23k is an example of a commercial probe intended for installation in a process pipe or small vessel. The interface could be a standard pipe tee for large-diameter pipes, or extended for several feet if the nozzle is located on top of a vessel.

Virtually any process configuration can be accommodated with judicious use of probes, dip pipes, and support structures. The mechanical load on large units should be calculated based on fluid velocity, viscosity, harmonic resonance, and allowable deflection. Stress and strain due to normal operation as well as unusual events must be considered carefully. Maintenance procedures and operating discipline should include the possibility of a liquid-filled sample loop, so that the process probe does not become a very expensive pressure relief valve.

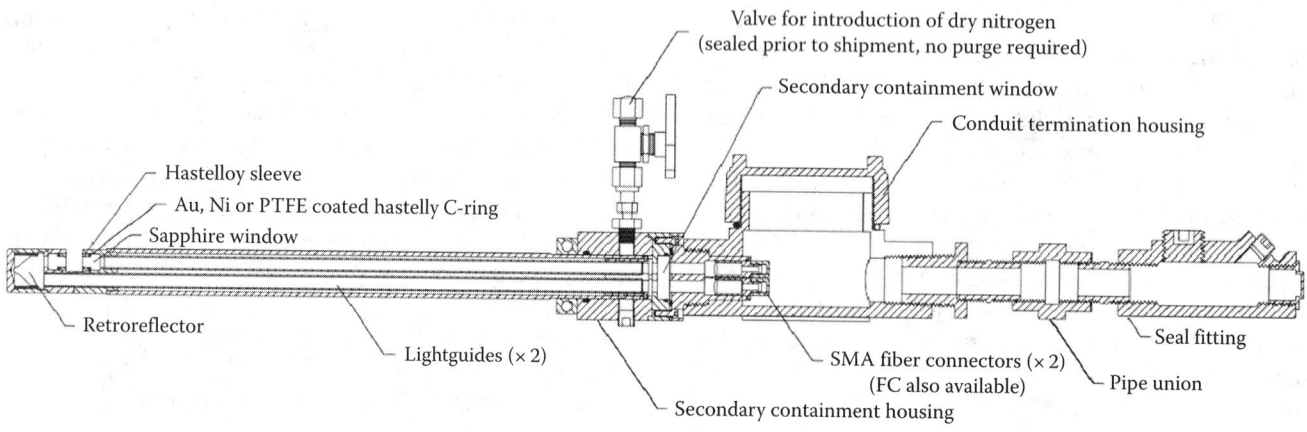

FIG. 1.23k
Single-pass transmission probe cross section. (Courtesy of Axiom Analytical, Inc.)

Cross-Pipe Dual Probe In some process configurations, the single-ended probe is not appropriate. Small-diameter pipes and tubing systems physically cannot accommodate some transmission probes, especially when the optimum optical path length approaches the inside diameter of the pipe. The emission and collection windows of a single-ended probe can be separated into two probes and pipe penetrations. Positioning is critical—geometrical alignment must be achieved and optical alignment maintained under process flow conditions. The separation distance of the probe tips creates an equivalent sample cell within the pipe. By incorporating sliding seals along the probe body and an external mechanism to move the probe(s), a variable path length can be achieved.

The analyzer flow-through cell, where the sample is extracted from the main process line, is really a miniature version of the cross-pipe configuration (Figure 1.23l).

Fiber-optic probes might still be preferred for convenience or to guarantee multiple levels of separation of the process and the analyzer. Some analyzers are designed with two enclosures and short fiber-optic connections in order to physically separate the sample system (wet box) from the analyzer and computer (dry box). Many permutations are possible, driven by the knowledge that a minor leak (single fitting or seal failure) in a sample line could destroy a very expensive analyzer system (replacement seldom covered by vendor warranty).

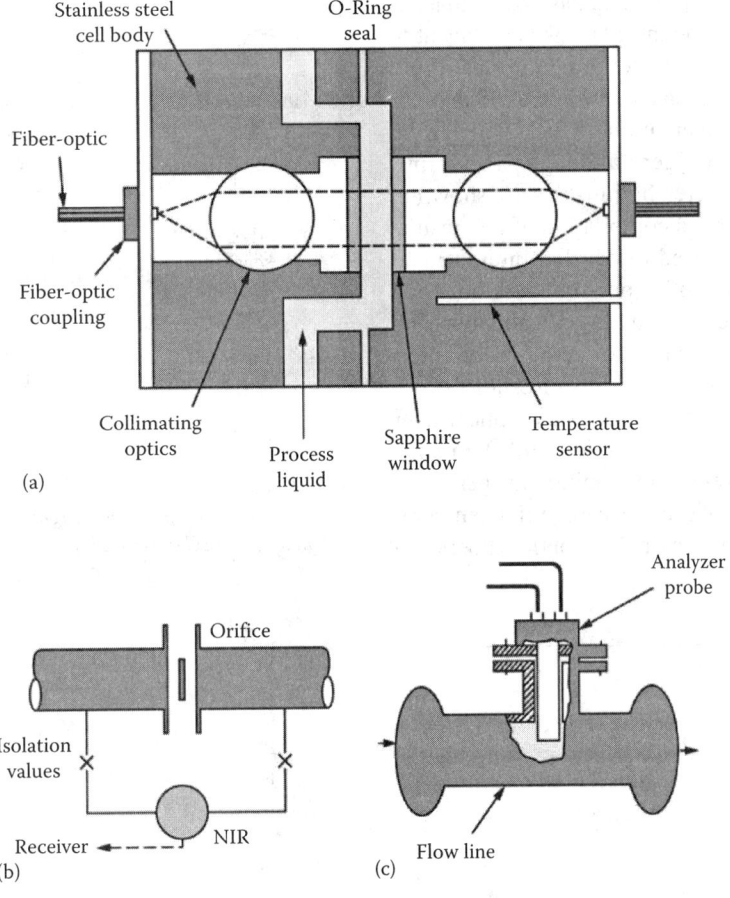

FIG. 1.23l
Fiber-optic interfaces: (a) miniaturized flow-through cell, (b) flow cell in pipe bypass configuration, and (c) single-pass probe installed in pipe tee.

SAFETY CONSIDERATIONS

Handling flammable or easily ignitable materials in chemical process environments is a known risk due to ignition by spark-producing equipment or by other means. In the United States, the National Electrical Code and National Fire Protection Association classify industrial areas based on the probability of a flammable mixture being present due to normal operations and abnormal events. Process equipment must be tested and labeled for use in hazardous service as specifically allowed by the standards.

Fiber-Optic Ignition

In the past, fiber-optic cables were considered "intrinsically safe" because none of the silica dielectric wavelengths conduct electrical energy. However, research has shown that the optical power in some fiber-optic installations is sufficient to cause ignition under a well-defined set of conditions. First, a minimum power level has been established at 35 mW for the most easily ignitable combination of fuel (i.e., carbon disulfide) and atmospheric oxygen, although other minimum ignition powers for selected flammables are shown in Figure 1.23m. Data shown are a compilation of test results from early publications and used to develop international standards. While most fiber-optic spectrometers intended for online measurement in classified areas do not approach the 35 mW optical power in the fiber or at the probe interface, laser-based designs may exceed this threshold.

Second, the optical power must escape the optical fiber (due to breakage or by design in a fiber-optic probe). Finally, a target particle must be present for a minimum period of time to raise its temperature above the autoignition temperature. Normally, the power levels carried by optical fibers and emitted via fiber-optic probes are lower than the minimum to cause ignition. One notable exception is a fiber-optic probe used in Raman measurement applications, where laser power can be hundreds of mW in current systems and the beam emitted from the probe is generally focused. Some Raman vendors have made design changes to accommodate the standards and mitigate the hazard during process operation in classified areas (see Chapter 1.58 for details).

The ANSI/ISA-60079-28 standard of 2013 specifies the requirements for the use of fiber optics in hazardous areas. A summary flowchart is shown in Figure 1.23n. While there are similarities to electrical standards for classified areas including mechanical protection schemes for cable and laser shutdown timing, an optical beam of photons is incapable of ignition at these low power levels (<2 W) without the presence of a target particle of the

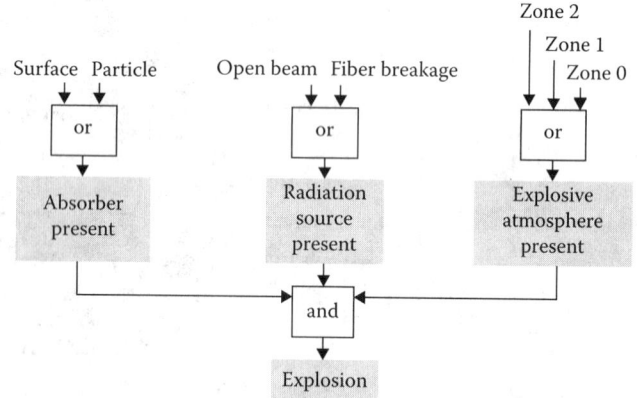

FIG. 1.23n
Requirements for fiber-optic ignition of a flammable atmosphere. (Courtesy of IEC 60079-28.)

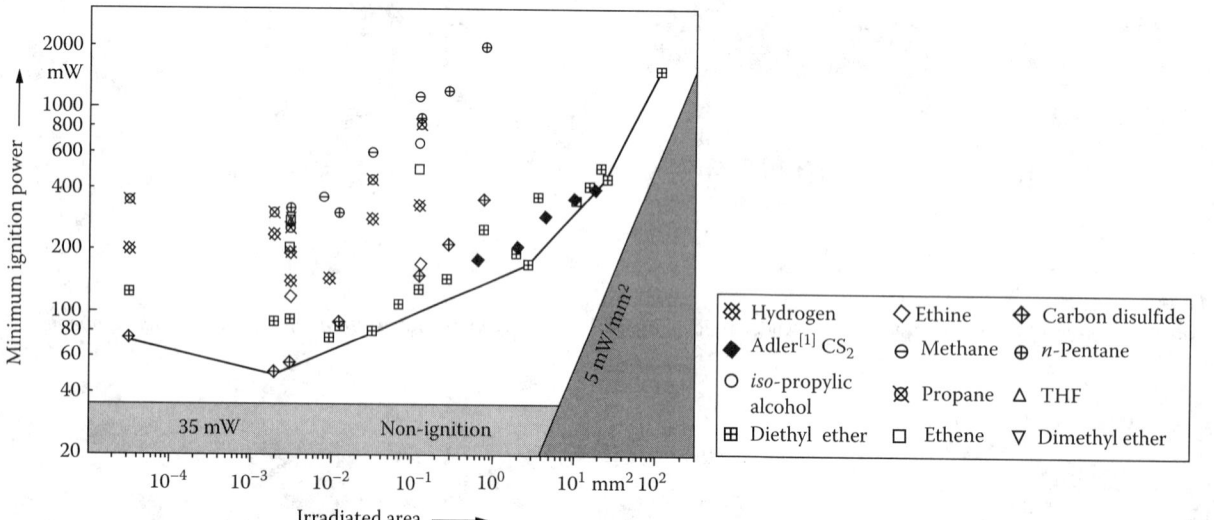

FIG. 1.23m
Minimum ignition power for selected flammables. (Courtesy of IEC 60079-28.)

optimum dimensions, absorptivity, and residence time. Furthermore, ignition in condensed phase systems (i.e., Raman probe in liquid sample) is impossible due to heat capacity considerations.

ENVIRONMENTAL EFFECTS

Laboratory-grade optical fiber is rarely used directly in industrial environments because of its fragile nature and susceptibility to environmental effects including sunlight, rain/snow, and temperature extremes. Bare optical fiber, when protected with various coatings, jackets, and strength members, is called a fiber-optic cable. Differential thermal expansion of these components within a poorly designed cable can induce excess stress in the optical fiber, leading to temperature-induced attenuation. The large-core fiber used for spectroscopy requires a different cable design when compared to smaller-diameter fibers in communications-grade cable.

While intrinsic sensors may utilize microbending loss to measure stress due to direct physical contact, stress-induced changes in fiber-optic transmission for analyzer systems must be avoided, whether in the probe or the extension fibers. Ruggedized cable laid in cable trays or pulled through conduit must be secured in a way to allow expansion and contraction due to ambient temperature changes. The outer jacket must be impervious to normal industrial environment as well as potential exposure to aggressive chemicals.

Care must be taken when attaching fiber-optic connectors to cabled fiber. A (SMA)* connectors are most often specified due to low cost and availability for large-core fiber. Alternate industrial connectors include FC and ST,[†] which use an integrated spring to keep the polished faces in physical contact (Figure 1.23o). The tolerance on outside diameter, the hole diameter, and centering is much tighter than with SMA. Much less variation in throughput is seen during frequent remating, assuming the surfaces are kept extremely clean. Environmental contaminants can be rejected by design so that the connector can be used outdoors or underground. Special connectors can be rated for military, avionic, or space applications. Most connectors use epoxy or other adhesive to retain the bare fiber for polishing. Thermally induced stress can cause connector failure due to epoxy delamination or fracture of the glass.

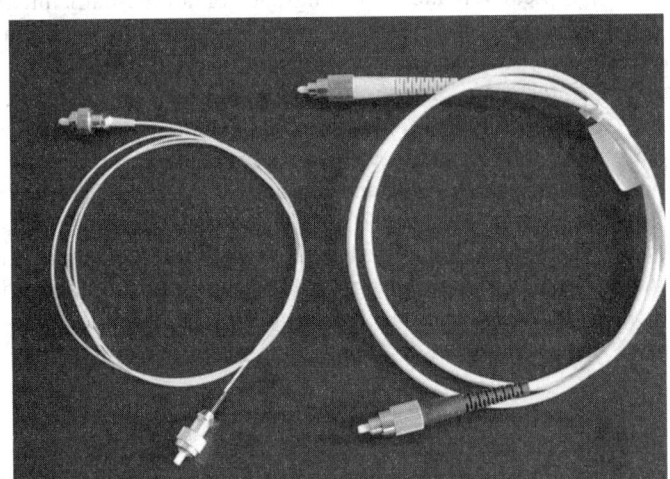

FC-APC/FC-PC patch cords

Standard male SMA connector: male body (inside threads) with male inner pin

FIG. 1.23o
Fiber-optic cable connectors.

---

* Optical SMA (SubMiniature version A) is a semi-precision threaded connector physically similar to the older RF coaxial connector by the same name.
† Connector design standards include FC, SC, ST, LC, MTRJ, MPO, MU, SMA, FDDI, E2000, DIN4, and D4.

APPLICATIONS

While it is beyond the scope of this chapter to list all the possible applications of industrial fiber-optic probes, a few categories will be discussed in the following text. Referring to silica optical fiber, the UV, VIS, and NIR spectral ranges can be covered with a minimum of two types of fiber-designated *low hydroxyl* and *high hydroxyl*. Extending into the deep UV (<230 nm) requires a solarization-resistant silica design. Use of silica beyond about 2200 nm is problematic due to scattering losses. Impurities and lattice dislocations in the core as well as absorption bands in the cladding are the primary loss mechanisms in silica fiber. Absorption bands are characterized by the vendor as losses in units of dB/km. Indeed, communication fiber has been optimized to reduce the losses at specific wavelengths, resulting in the original 850 nm, 1.31 μm, and 1.55 μm digital transmission bands (diode laser wavelengths) with manageable losses over many kilometers of fiber. However, photometric applications generally trade off loss in favor of wide/flat spectral range in order to avoid fiber-induced anomalies. Therefore, the highest-quality silica core and cladding are preferred in the manufacture of fiber intended for full-spectrum transmission.

Extending into the midinfrared region is possible using specialty fibers including chalcogenide glasses, polycrystalline silver halides, and hollow core structures with reflective silver and dielectric layers, although practical fiber length is limited by spectral losses and cost. Nevertheless, several vendors now utilize optical fibers that transmit in the IR and are integrated into a Fourier transform (FT) spectrophotometer. Applications are also being developed utilizing state-of-the-art quantum cascade lasers tuned to specific IR wavelengths for high-discrimination gas measurements. The flexible nature of the optical dielectric structure can be utilized to advantage in any instrument design requiring a few meters of fiber.

Absorption Measurement

The use of all-silica fiber and a stable optical path probe facilitates precision absorption measurement in the UV–VIS–NIR spectral regions in a process environment. Beam collimation through the optical path is preferred because process temperature changes affect the refractive index (RI) of the sample and therefore light throughput ($I°$) if the beam is focused. Temperature can also affect the effective path length in a probe design where the window separation and support are controlled by materials with high coefficient of thermal expansion. The same constraints on seals and optics inside the probe apply to the interface that supports the probes externally.

Typical industrial applications in the UV–VIS include chromophores, highly conjugated molecules and adducts, and binary mixtures. The lack of spectral features is offset by the high absorptivity in the UV allowing part per million (ppm) detection limits in some cases. ASTM Color and Yellowness Index analyzers in which the wavelengths are limited to the visible can benefit from the flexibility of fiber-optic transmission. Fibers with stable transmission characteristics in the blue/violet range are preferred because changes at the sub-milli-absorbance level can affect the predicted result. Unwanted physical modulation of the fiber transmission can be caused by environmental temperature fluctuations or other stresses in a fiber-optic cable.

NIR applications utilizing fiber-optic probes are numerous and best documented in publications and presentations. Some major areas include benzene and aromatics in fuel blends, gasoline octane number, diesel cetane number, hydroxyl number, epoxy number, water in organics, organics in water, and caustic and salt in water. While NIR development work was done in academic, industrial, and government laboratories more than 30 years ago, high-resolution FT-NIR benches were packaged for online use in the 1990s. Indeed, the nexus of rugged FT-NIR analyzers, industrial fiber-optic probes, and the field of chemometrics allowed modeling of NIR absorption for diverse chemical systems. Patent applications are still filed for novel methods and control algorithms.

While most gas-phase absorbance measurements are typically made with long-path or folded-path integrated cells to maximize detection limits, higher concentrations can be measured with fiber-optic probes in short-path interfaces. Direct in situ measurement of pressurized gaseous mixtures found in industrial plants may be advantageous, especially when the temperature and pressure are already measured and/or controlled. Precise pressure calibration is critical if results are referenced to laboratory methods at standard conditions. But the sample collection or conditioning system may be eliminated, allowing real-time concentration feedback (zero delay) for fast photometers.

Scattering

A version of the probe diagrammed in Figure 1.23h earlier was used for over 25 years to monitor process crystallizers for nucleation and solid growth using the simple backscatter (turbidity) signal. Forward scattering can be implemented effectively with optical fibers placed at small off-axis angles. Nephelometry* (scattering 90° off-axis) can also be designed into a fiber-optic probe with a slightly larger diameter. Indeed, combination systems where multiple measurements are made simultaneously through a single optical element can reduce the number of probes needed in industrial environments.

---

* A nephelometer is an instrument for measuring concentration of suspended particulates in a liquid or gas colloid. A nephelometer measures suspended particulates by sending a light beam through the fluid and measuring the particle density by detecting the light reflected from the particles.

Fluorescence Commercial dissolved oxygen probes are now readily available. The detection mechanism has not changed since the original implementation (Figure 1.23p), where oxygen quenches the fluorescence signal from indicator dye molecules held in a membrane at the fiber(s) tip.

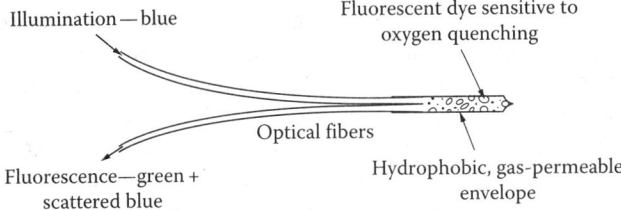

FIG. 1.23p
Historical fluorescence optrode.

Current designs immobilize the fluorophore in a solid matrix in direct contact with a gas or liquid sample, while a single fiber interrogates the matrix from a protected space within a probe body. Additional modifications, including time-resolved fluorescence detection, have elevated this optrode into practical industrial use.

Raman Raman analyzers (covered in full detail in Chapter 1.58) configured for online measurements, unlike other process photometers, are almost always implemented with a fiber-optic probe. Excitation laser wavelengths are usually in the VIS or NIR range and therefore compatible with silica fiber. Probes must have high collection efficiency due the low photon yield (often 10^{-6}) of the Raman scattering function in condensed phase. But focusing intense light at the tip of the probe can cause unwanted reactions, especially when adjacent to a window surface.

Gases can be measured, but a more efficient collection scheme is needed. Figure 1.23q shows an early design that attempts to increase Raman photon generation in a larger volume between mirrors.

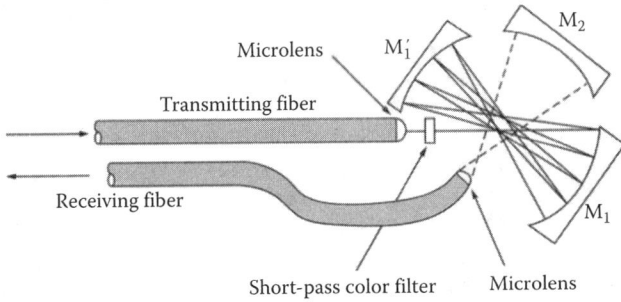

FIG. 1.23q
Legacy Raman gas sampling design.

Improved probe designs do not expose mirrors to the sample and integrate Raman scattering much more efficiently. The AirHead™ probe from Kaiser Optical Systems, Inc. incorporates the holographic filters and optics into a compact single-ended design that withstands 150°C and 1000 psig.

Critical Angle and Other Sensors

Process-capable sensors can be designed using light transmitted through crystals where certain angles are reflected and others are refracted. Again, the optical fiber may serve merely as a light conduit. However, more complex systems may use an optical fiber bundle as a flexible optical element with known aperture.

Level Sensors A simple case uses TIR in a polished glass hemisphere to create a level detector switch (Figure 1.23r). The RI difference between the glass and the material in contact with the glass changes drastically when the probe breaks the liquid surface.

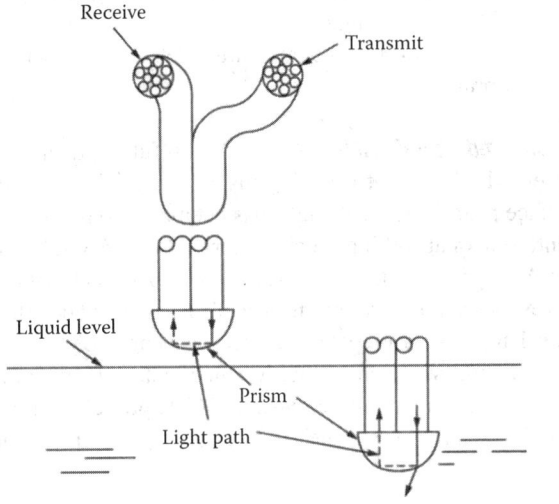

FIG. 1.23r
Optical level switch using hemispherical lens and optical fiber bundles.

When the glass (RI ~ 1.5) is in air (RI ~ 1.0), most of the light meets the critical angle requirement and stays within the window. But when the window is immersed in water (RI ~ 1.3), much of the light leaks out and less returns to the detector via the receive bundle. A simple threshold circuit is required to generate the on/off signal from the probe.

Refractive Index Measurement RI* can be measured directly by determining the angle at which light leaks out of a crystal with the correct geometry. Process refractometers utilizing fiber optics in the intrinsic mode (breakdown of the TIR within the fiber itself) have been proposed but suffer from a number of limitations including fragile nature of an unprotected fiber and multimode structure of light within the fiber. Process refractometers may utilize optical fiber bundles

---

* Refractive index of a substance is a function of the speed at which light, or any other form of radiation, propagates through that medium. It is defined as the speed of light (c) in vacuum divided by the speed of light through the substance (v). For example, the refractive index of water is 1.33, meaning that light travels 1.33 times slower in water than it does in vacuum.

to transmit light to a linear array detector where each fiber is allocated to a corresponding pixel. Significant interpolation is accomplished digitally to deduce the light/dark transition and correlate to RI.

Fabre Perot Interferometer   Sensors that are microfabricated to cause light to traverse a sensing gap, with dimensions on the order of the wavelength of light, cause optical interference that can be measured as a slight wavelength shift. RI is measured directly, but probes with sealed gap (microliter volume), containing a fluid with known temperature vs. RI curve, can be calibrated to display temperature. The optical fiber is used in the *extrinsic* mode, unlike *intrinsic* RI sensors where the fiber itself modulates the light. An example Fabre Perot system transmits excitation light from an light emitting diode (LED) in a single fiber and returns the wavelength-shifted light back to a beam splitter detector for evaluation and calibration.

Attenuated Total Reflectance   A useful optical effect, observed when light traveling inside a crystal is lost at the surface near the critical angle, has been used to measure concentration of absorbing species external to the crystal surface. An ATR probe monitors the loss of light due to absorption in the evanescent field at the surface of the crystal. The effective path length is on the order of the wavelength of light used, allowing measurement of highly concentrated absorbers without dilution. Furthermore, when an ATR probe is immersed in a slurry (solids suspended in a liquid), the continuous phase accounts for most of the evanescent wave absorption unless the solids attach to the crystal surface (fouling).

CONCLUSIONS

Fiber-optic probes are beneficial in industrial environments because they separate the sample/measurement zone from the optical analyzer/operator by distances from one meter to hundreds of meters or more. The difficulty of obtaining a process grab sample or constructing an automated SCS are prime drivers for use of an optical probe. While some probes may be appropriate for benign process conditions, more rugged designs are required for permanent installations and process environment extremes. Many fiber-optic systems can be considered intrinsically safe in hazardous environments, given the nonconducting cable construction and limited optical power levels. However, the user must be aware of the exceptions and consult the international standards when specifying a fiber-optic probe for use in classified locations.

SPECIFICATION FORM

When specifying Fiber-Optic probes and cables, the application can be described by using the ISA Form 20A1001. This form is provided on the next page with the permission of the International Society of America.

1.23 Fiber-Optic (FO) Probes and Cables

1	RESPONSIBLE ORGANIZATION	ANALYSIS DEVICE	6	SPECIFICATION IDENTIFICATIONS	
2			7	Document no	
3	(ISA logo)	Operating Parameters	8	Latest revision	Date
4			9	Issue status	
5			10		

	ADMINISTRATIVE IDENTIFICATIONS					
11						
12	Project number		Sub project no			
13	Project					
14	Enterprise					
15	Site					
16	Area		Cell		Unit	
17						

	SERVICE IDENTIFICATIONS		
18			
19	Tag no/Functional ident		
20	Related equipment		
21	Service		
22			
23	P&ID/Reference dwg		
24	Process line/nozzle no		
25	Process conn pipe spec		
26	Process conn nominal size	Rating	
27	Process conn termn type	Style	
28	Process conn schedule no	Wall thickness	
29	Process connection length		
30	Process line matl type		
31	Fast loop line number		
32	Fast loop pipe spec		
33	Fast loop conn nom size	Rating	
34	Fast loop conn termn type	Style	
35	Fast loop schedule no	Wall thickness	
36	Fast loop estimated lg		
37	Fast loop material type		
38	Return conn nominal size	Rating	
39	Return conn termn type	Style	

	SERVICE IDENTIFICATIONS Continued			
40				
41	Return conn matl type			
42	Inline hazardous area cl		Div/Zon	Group
43	Inline area min ign temp		Temp ident number	
44	Remote hazardous area cl		Div/Zon	Group
45	Remote area min ign temp		Temp ident number	
46				
47				

	COMPONENT DESIGN CRITERIA			
48				
49	Component type			
50	Component style			
51	Output signal type			
52	Characteristic curve			
53	Compensation style			
54	Type of protection			
55	Criticality code			
56	Max EMI susceptibility		Ref	
57	Max temperature effect		Ref	
58	Max sample time lag			
59	Max response time			
60	Min required accuracy		Ref	
61	Avail nom power supply			Number wires
62	Calibration method			
63	Testing/Listing agency			
64	Test requirements			
65	Supply loss failure mode			
66	Signal loss failure mode			
67				
68				

	PROCESS VARIABLES	MATERIAL FLOW CONDITIONS			PROCESS DESIGN CONDITIONS		
69				101			
70	Flow Case Identification		Units	102	Minimum	Maximum	Units
71	Process pressure			103			
72	Process temperature			104			
73	Process phase type			105			
74	Process liquid actl flow			106			
75	Process vapor actl flow			107			
76	Process vapor std flow			108			
77	Process liquid density			109			
78	Process vapor density			110			
79	Process liquid viscosity			111			
80	Sample return pressure			112			
81	Sample vent/drain press			113			
82	Sample temperature			114			
83	Sample phase type			115			
84	Fast loop liq actl flow			116			
85	Fast loop vapor actl flow			117			
86	Fast loop vapor std flow			118			
87	Fast loop vapor density			119			
88	Conductivity/Resistivity			120			
89	pH/ORP			121			
90	RH/Dewpoint			122			
91	Turbidity/Opacity			123			
92	Dissolved oxygen			124			
93	Corrosivity			125			
94	Particle size			126			
95				127			
96	CALCULATED VARIABLES			128			
97	Sample lag time			129			
98	Process fluid velocity			130			
99	Wake/natural freq ratio			131			
100				132			

	MATERIAL PROPERTIES				MATERIAL PROPERTIES Continued		
133				137			
134	Name			138	NFPA health hazard	Flammability	Reactivity
135	Density at ref temp	At		139			
136				140			

Rev	Date	Revision Description	By	Appv1	Appv2	Appv3	REMARKS

Form: 20A1001 Rev 0 © 2004 ISA

Definitions

Band-pass filter: It is a filter that passes frequencies within a certain range and rejects (attenuates) frequencies outside that range. The bandwidth of the filter is the difference between the upper and lower cutoff frequencies.

Critical angle is the minimum angle of incidence beyond which total internal reflection occurs, when light is traveling from a medium of higher to an other medium of lower index of refraction.

Entendue: Geometric etendue (geometric extent), G, characterizes the ability of an optical system to accept light. It is a function of the area, S, of the emitting source and the solid angle, Q, into which it propagates. Etendue, therefore, is a limiting function of system throughput. Geometric etendue may be viewed as the maximum beam size the instrument can accept; therefore, it is necessary to start at the light source and ensure that the instrument including ancillary optics collects and propagates the maximum number of photons.

Internet exchange point (IX or IXP) is a physical infrastructure through which Internet service providers (ISPs) and content delivery networks (CDNs) exchange Internet traffic. They reduce the per point delivery cost and improve routing efficiency. In other words, they enable the interconnection of a number of systems, primarily for the purpose of facilitating the exchange of Internet traffic.

Nephelometer: A nephelometer is an instrument for measuring concentration of suspended particulates in a liquidor gas colloid. A nephelometer measures suspended particulates by sending a light beam through the fluid and measuring the particle density by detecting the light reflected from the particles.

Refractive Index (RI): The refractive index of a substance is a function of the speed at which light, or any other form of radiation, propagates through that medium. It is defined as the speed of light (c) in vacuum divided by the speed of light through the substance (v). For example, the refractive index of water is 1.33, meaning that light travels 1.33 times slower in water than it does in vacuum.

SMA Connector: *Subminiature version A* (*SMA*) connectors are semiprecision coaxialRF connectors developed in the 1960s as a minimal connector interface for coaxial cable with a screw-type coupling mechanism. The connector has a 50 Ω impedance. It is designed for use from DCto 18 GHz.

Abbreviations

AIT	autoignition temperature
ATR	attenuated total reflection
CDN	content delivery network
FT	Fourier transform
IR	infrared
IXP	Internet exchange point
LED	light emitting diode
LTCO	long-term cost of ownership
μm	micrometers
MTBF	mean time between failure
NEC	National Electrical Code
NIR	near-infrared
Nm	nanometers
RI	refractive index
ROI	return on investment
RTD	resistance temperature device
SCS	sample conditioning system
SMA	subminiature (option A)
TIR	total internal reflection
UV	ultraviolet
VIS	visible

Organizations

IEC	International Electrotechnical Commission
ISA	International Society of Automation
NFPA	National Fire Protection Association
SPIE	International Society of Optics and Photonics

Bibliography

ANSI/ISA-60079-28 Explosive Atmospheres-Part 28: Protection of equipment and transmission systems using optical radiation, (2013) https://www.isa.org/store/products/product-detail/?productId=116786#sthash.aIcuOIfo.dpuf.

Azom, Identification and characterization of polymers using Raman spectroscopy, (2014) www.azom.com/article.aspx?ArticleID=10279.

Darginavicčius, J. et al., Generation of 30-fs ultraviolet pulses by four-wave optical parametric chirped pulse amplification, (2015) https://www.osapublishing.org/oe/fulltext.cfm?uri=oe-18-15-16096&id=203841.

Fiber Optics Association (FOA), Guide to telecommunications cable splicing, (2002) http://www.thefoa.org/splicebk.htm.

Fiber Optics Association (FOA), Reference guide to fiber optic cabling, (2009) http://www.thefoa.org/FOArgPC.html.

Fiber Optic Association (FOA), Guide to fiber optics, (2010) http://www.thefoa.org/tech/ref/OSP/install.html.

Fiber Optics Association (FOA), Reference guide to fiber optics, (2014) http://www.thefoa.org/FOArgfo.html.

Grattan, K. T. V., Fiber optic techniques for temperature measurement, (1995) http://link.springer.com/chapter/10.1007%2F978-94-011-1210-9_15.

Guided Wave, Guided wave fiber optic cable, (2013) http://www.guided-wave.com/fileadmin/media/infobase/product-data-sheets/1006_Fiber.pdf.

Hayes, J., Fiber optics technician's manual, (2006) http://www.amazon.com/Fiber-Optics-Technicians-Manual-Hayes/dp/1435499654.

Jackson, L. A., Submarine communication cable including optical fibres within an electrically conductive tube, (1978) http://patft.uspto.gov/netacgi/nph-Parser?Sect1=PTO1&Sect2=HITOFF&d=PALL&p=1&u=%2Fnetahtml%2FPTO%2Fsrchnum.htm&r=1&f=G&l=50&s1=4278835.PN.&OS=PN/4278835&RS=PN/4278835.

Jones, M. R., Use of blackbody optical fiber thermometers, (2001) http://tfaws.nasa.gov/TFAWS01/NASA/13Spacecr/kJONES.PDF.

LANshack, Fiber optic networks, (2015) http://www.lanshack.com/fiberoptic-tutorial-network.aspx.

NCBI Optical Fiber Networks for Remote Fiber Optic Sensors, (2012) http://www.ncbi.nlm.nih.gov/pmc/articles/PMC3355392/.

Nurmi, J., Shi, Z., and Hai, S., Optical fibers-Principle of operation, (2012) https://wiki.metropolia.fi/display/Physics/Optical+fibers+-+Principle+of+operation.

Omega, Introduction to fiber optics temperature measurement, (2013) http://www.omega.com/prodinfo/FiberOpticTemp.html.

OSHA, Fiber optic vibration sensor for remote monitoring in high power electric machines (2015) Giuliano Conforti, Massimo Brenci, Andrea Mencaglia, and Anna Grazia Mignani.

Ruiz-Perez, V. I., Optical fiber thermometer based on multimode interference effects, (2013) https://www.osapublishing.org/ViewMedia.cfm?uri=FiO-2013-FTh2B.5&seq=0.

SPIE, Fibers, (2014) http://spie.org/x32419.xml.

Yin, S., *Fiber Optic Sensors*, 2nd ed., (2011) http://books.google.com/books/about/Fiber_Optic_Sensors_Second_Edition.html?id=5KdCOguH9CsC.

Zang, Z. Y., Fiber optic fluorescence thermometry, (1996) http://link.springer.com/chapter/10.1007%2F0-306-47060-8_11.

1.24 Flame, Fire, and Smoke Detectors

R. NUSSBAUM (1969, 1982)　　**S. J. PATE** (2003)　　**B. G. LIPTÁK** (1995, 2017)

Flow sheet symbol: Furnace — XFE — To receiver

Types	A. Flame (pilot and main flame detectors in combustion processes).
	B. Fire (fire safety devices).
	C. Smoke (detect smoldering and the incipient of fires).
Smoke sensitivity	303 mm (2%/ft) obscuration ±0.6%/ft (Smoke sensitiviy requirement set be UL Standard UL217).
Costs	Battery-operated household smoke detectors are available from around $15. Addressable, intrinsically safe smoke detectors with software programmable test option and with both smoke and temperature sensors cost about $200. Flame rod or thermocouple-type pilot flame safeguard detectors cost about $500. Optical flame safeguards for burner management cost from $1000 to $3000, with the higher-priced units having explosion-proof packaging and intelligent electronics.
Further readings	For detailed discussions of IR detectors, refer to Chapter 1.32; for the coverage of spectrometry, read Chapter 1.62; and for an in-depth coverage of UV measurements, see Chapter 1.69.
Partial list of suppliers	Ansul Co. (B, C) https://www.ansul.com/en/us/pages/ProductDetail.aspx?productdetail = CHECKFIRE%20Detection%20and%20Actuation
	Apollo (A,B) http://www.apollo-fire.co.uk/search.aspx?search=fire+flame+sensors
	Autronica (A,B) http://www.autronicafire.no/Pages/Home.aspx
	Badger (B) (extinguishants) http://www.badgerfire.com/utcfs/Templates/Pages/Template-71/0,11119,pageId%3D6003%26siteId%3D603,00.html
	Chemetron (B) (extinguishants, alarms) http://www.chemetron.com/utcfs/Templates/Pages/Template-50/0,8061,pageId%3D62618%26siteId%3D5230,00.html
	Crowcon (B) (flame detectors) http://www.crowcon.com/uk/products/fixed-detectors/flame-detectors.html
	Detector Electronics Corp. (A, B) Systems Optical Flame and Gas Detection http://www.detronics.com/utcfs/Templates/Pages/Template-50/0,8061,pageId%3D2612%26siteId%3D462,00.html
	Eclipse (A) http://www.eclipsenet.com/products/t400/
	Fike Corp. Fike Metal Products Div. (B) http://www.fike.com/products/faccat.html
	Fire Lite, Honeywell (B,C) (fire panels, smoke alarm) http://www.firelite.com/products/ad355a.htm
	Fire Sentry, Honeywell (B, C) (optical fire detectors) http://www.firesentry.com/fs20x.php
	Fireye (A) (flame scanners) http://www.fireye.com/utcfs/Templates/Pages/Template-50/0,8061,pageId = 1232&siteId = 373,00.html http://www.fireye.com/utcfs/Templates/Pages/Template-50/0,8061,pageId%3D1241%26siteId%3D373,00.html
	Forney Corp. (A) http://www.forneycorp.com/Products/Pages/FlameDetectors.aspx
	General Monitors Inc. (A, B) (flame detectors) http://www.generalmonitors.com/Flame-Detectors/c/202;jsessionid=720C0EB38672CC8C18452AE78018B777.worker1?N = 10009&Ne = 10011
	Honeywell (B) (fire panels and smoke and heats) http://www.honeywelllifesafety.com/HLSP1business.html
	Intronics (A, B, C) (www.intronics.co.uk)
	Kidde-Fenwal Inc. (B, C) (fire panels smokes and heats, extinguishants) http://www.kidde-fenwal.com/utcfs/Templates/Pages/Template-50/0,8061,pageId%3D882%26siteId%3D386,00.html
	Life Safety Associates Inc. (B, C) (training organization) (www.lifesafety.com) Martin Control (A) http://martincontrol.com/flame-safeguards/
	Mickropack Ltd., (B) http://www.kellersinc.com/image/VIDEO%20FLAME%20DETECTION%20BROCHURE.pdf
	MSA Instrument Div. (B) (flame and gas detectors) http://us.msasafety.com/Fixed-Gas-%26-Flame-Detection/c/103?N=10136&Ne=10187&Nu=Product.Endeca_Rollup_Key&pageSize=100&view = grid
	Notifier, Honeywell (B) (fire panels) http://www.notifier.com/products/facp/Pages/NOTIFIER-fire-alarm-control-panels-overview.aspx
	Pyrotronics (B) (fire panels) http://www.buildingtechnologies.siemens.com/bt/global/en/firesafety/fire-detection/Pages/fire-detection.aspx

(Continued)

Reliable Fire Equipment (A, C) http://www.forneycorp.com/Products/Pages/FlameDetectors.aspx
Sierra Monitors (A, B, C) (optical fire detectors) (www.sierramonitor.com)
SimplexGrinnell Tyco (B, C) (fire panels) http://www.simplexgrinnell.com/ENUS/Solutions/FireDetectionAndAlarm/Pages/fire-detection-fire-alarm-simplexgrinnell.aspx
Talentum (B) http://www.talentum.co.uk/flame-detector-selection/uvir2 Tyco (A,B,C) http://www.tyco.com/solutions/fire-detection
Vinesys (B) http://www.vinesys.co.kr/main-eng.htm

INTRODUCTION

The sequence of events leading to an industrial explosion or fire accident includes the leaking and accumulation of a combustible material, the presence of oxygen, and an ignition source. The methods of preventing the evolution of such conditions have been discussed in several chapters. The goal of this chapter is to describe the sensors that can quickly detect the starting of a fire and to initiate both its suppression and the safe evacuation of the area.

This chapter covers both the types of instruments that guarantee the maintaining of combustion and the instruments that serve to warn the occupants of a building or industrial facility when a fire starts. The devices used for these purposes operate on similar principles and therefore can be discussed in the same chapter of the handbook.

First, the fire and smoke detectors will be discussed. This will be followed by a discussion of optical fire detectors, concluding with a description of fire safeguarding devices used in burner management.

A fire occurs in four distinct phases. In the first, or incipient, phase, warming causes the emission of invisible but detectable gases. In the second phase, smoldering, smoke is formed, so smoke detectors can be used. In the third phase, when the ignition temperature has been reached, flames are present and therefore their emitted radiation (infrared [IR] and ultraviolet [UV]) can be detected. In the fourth and last stage, heat is released; the temperature of the space starts to rise, and the use of thermal sensors becomes feasible. Obviously, the sooner the evolution of a fire is detected, the less damage it is likely to cause. Therefore, fire and smoke detectors are discussed here in the order of their applicability to the four stages of fires.

SMOKE DETECTORS

Most smoke sensors are either optical or ionization detectors, while others use both of these detection methods to increase their sensitivity. They can also detect and thus deter smoking in areas where it is banned. Smoke detectors in large commercial, industrial, and residential buildings are usually integrated into a central fire alarm system, while in family homes the smoke alarms are often powered only by a single disposable battery.

It is estimated that smoke detectors are installed in 93% of U.S. homes, but it is also estimated that about 30 % of these alarms are not working due to aging or due to the failure of the owners to replace dead batteries. The National Fire Protection Association (NFPA) estimates that nearly two-thirds of deaths from home fires occur in properties without working smoke detectors.

Ionization Chamber Sensors

In the early warming and incipient stage of fire, combustion products are emitted without visible smoke, flame, or heat release. Ionization chamber–type sensors are used to detect the presence of these gases by analyzing the composition of the atmosphere through the measurement of conductance. The ionization chamber contains two electrodes held at different potentials and a radioactive alpha particle source that ionizes the air in the chamber. The ionization current that results reflects the composition of the air and rises as the invisible combustion gas concentration rises.

The ionization chamber in a smoke detector consists of two metal plates at different voltages. Alpha particles from a source such as americium ionize the air molecules. The positive atoms flow toward the negative plate, as the negative electrons flow toward the positive plate. The movement of the electrons registers as a small but steady flow of current. When smoke enters the ionization chamber, the current is disrupted as the smoke particles attach to the charged ions and restore them to a neutral electrical state. This reduces the flow of electricity between the two plates in the ionization chamber. When the electric current drops below a certain threshold, the alarm is triggered (Figure 1.24a).

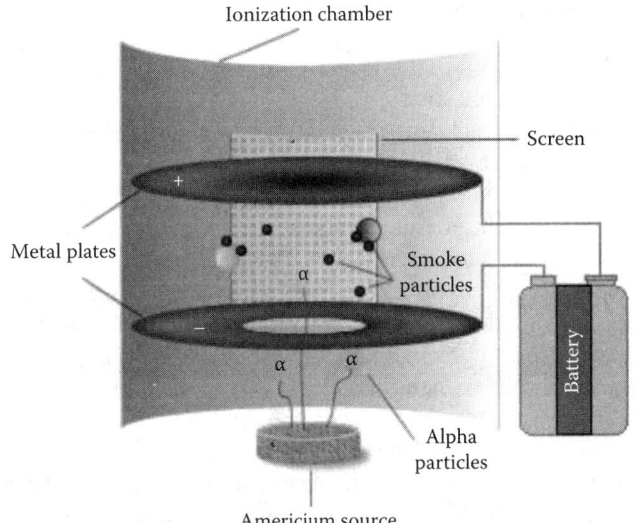

FIG. 1.24a

Illustration of the principle of an ionization chamber. (Courtesy of Fire Life Safety Consulting–FLS.)

Photoelectric Sensors

Once the fire starts to smolder and smoke is present, photoelectric sensors can be used to activate alarms. Most smoke detectors use a light beam and a photoelectric cell or transistor. As the smoke density rises, less light passes from the source to the receiver and an alarm is activated. Smoke detectors must be maintained so that dust and dirt accumulation will not cause false alarms. In most fires, the casualties are not caused by the heat of the fire but by the toxicity of the combustion gases and by asphyxiation from smoke. Therefore, early warning systems, such as photoelectric smoke detectors, are very important.

Figure 1.24b illustrates how the smoke particles deflect the light beam and trigger the smoke alarm.

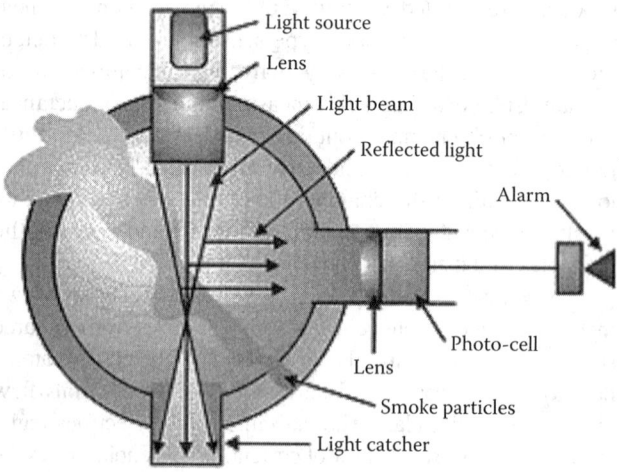

FIG. 1.24b
Sensing chamber in a photoelectric smoke detector. (Courtesy of D.E. Baker, Univerity of Missoury.)

FLAME AND FIRE DETECTORS

Thermal Sensors

There are two types of thermal sensors that are used in fire protection application. One is the rate-of-rise sensor, and the other is the absolute temperature sensor. Rate-of-rise alarms are usually set at 8.3°C–11.1°C (15°F–20°F) per minute. This rate of rise can be detected by either bimetallic pneumatic tube sensors or thermoelectric sensors. The fixed temperature sensors actuate an alarm when the space temperature reaches a present limit. They are usually either bimetallic or fusible link devices. In the fusible link devices, the melting of a low-melting-point solder activates the operation of sprinklers or other extinguishing devices.

Heat Detection Cables This type of cable is not used for temperature measurement, but for fire protection. These cables have both a digital and an analog design version and neither measures temperature nor signals the location of the fire, where along the length of the cable, the fire is located.

The digital version is more like a fuse that melts when a set temperature is reached. The analog version uses a thermistor or thermocouple. The thermistor version can only be used once, and after that, it has to be replaced. The thermocouple version can last through a fire and can be reused, but from then on, it is not as accurate as it was. The thermocouple version is like standard thermocouple sheath material, except that it has no hot junction. Instead, it has a carbon-based filler material that becomes conductive (making a semi junction) when exposed to extreme heat.

Combination Sensors Today, intelligent, addressable, and multisensing smoke detectors are also available. These are usually used in larger facilities where connection to addressable fire alarm control panels is needed. They can be provided with both photoelectric and thermal sensors to increase their immunity to false alarms (see Figure 1.24c).

FIG. 1.24c
Multisensor smoke alarm. (Courtesy of Fire-Lite Honeywell.)

Optical Flame Detectors

Once there is a flame, it emits a flickering radiation, which is mostly in the IR wavelength. Therefore, IR sensors can be used to detect the presence of flames. The flickering frequency of open flames (5–25 cps) allows the discrimination of flames from IR radiation generated by lightbulbs (at 120 cps) or from unmodulated ambient light sources. False alarms can still be caused by sunlight reflections from windows or rippling water surfaces, or by flickering neon signs. Therefore, the mounting location of the sensor should be carefully selected. Flame detectors should be used when combustible gases or flammable liquids are present, where the ignition is almost instantaneous and has practically no incipient or smoldering stage.

There are basically five types of optical detectors in addition to closed-circuit television (CCTV):

1. UV
2. IR
3. UV/IR
4. Dual IRs
5. Multi-IR

Flame detectors are specified by their cone of vision, which is defined by the field of view and the on-axis detection distance to the fire (Figure 1.24d). The angle of the field of view is defined by the 50% on-axis detection capability of the detector. Typically, these are about 90° (horizontally) although there are some detectors with fields of view of up to 120°, or with 100% off-axis detection capability. A distance to a specified size fuel source (typically 1 ft^2) is also specified. This fuel source is generally *n*-heptane although other fuels are often listed (e.g., gasoline and methane plume). The manufacturer's data sheet should indicate the off-axis detection distance, generally graphically (or as a percentage of the on-axis distance).

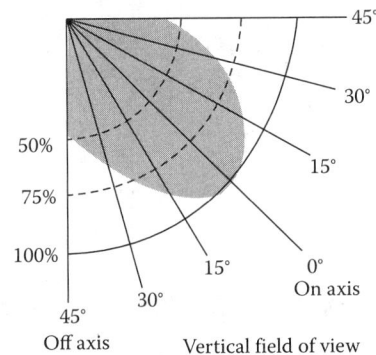

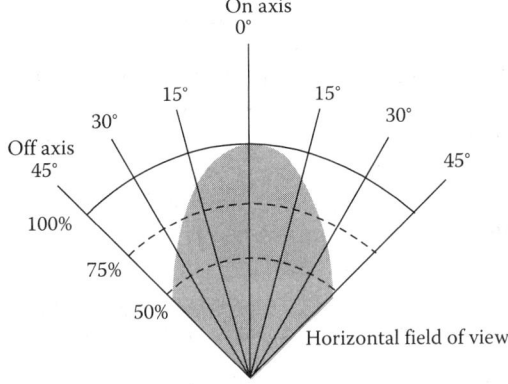

FIG. 1.24d
Cone of vision for a 90° with 50% off-axis sensitivity.

When selecting a flame detector, the specifying engineer must take into consideration the fuel source and the cone of vision of the selected detector to ensure adequate coverage of the potential hazard. Because these are optical devices, it is important to ensure that the detector has an unobstructed view to the hazard and that the selected detector automatically verifies the cleanliness of its lens. Several manufacturers offer an optical integrity check and therefore the ability of the detector to see a fire. The failure of the optical check should generate a fault alarm.

The flame detection algorithms within the flame detectors are highly specific, and often their response and capability to detect a fire will vary from manufacturer to manufacturer.

Tables 1.24a and 1.24b provide data on the flame detection distance and the false alarm sources, respectively, of the various optical fire sensors discussed later.

TABLE 1.24a
Typical Optical Flame Detection Distances (Feet)

Type	UV	IR	Dual IR	UV/IR	Multi-IR
Gasoline	90	85	100	100	210
Diesel	65	65	50	40	150
Methanol	50	50	20	55	150
Methane	80	45	25	90	100
Hydrogen	50	NR	15	NR	NR
Metal fires	15	NR	15	NR	NR
Black powder	15	40	40	NR	NR

Note: NR—no response.

TABLE 1.24b
False Alarm Source Impact

Type	UV	IR	Dual IR	UV/IR	Multi-IR
Arc welding	■	▲	▲	▲	▲
Modulated IR radiation	●	▲	●	●	●
Electrical arcs	■	▲	●	●	●
Radiation (nuclear)	■	●	●	●	●
Lightning	■	●	●	●	●
Grinding (metal)	■	●	●	●	●
Artificial lighting	■	●	●	●	●
Sunlight	▲	●	●	●	●

Note: ■, severe effect; ▲, some effect; ●, no effect.

Ultraviolet (UV) This optical technology was developed in the early 1970s using UV detection with Geiger Muller tubes. These were the first optical flame detectors and are based upon a Geiger Muller tube. The UV detectors count the number of pulses in the tube and give a total number of counts per second. The detector then alarms when a predetermined threshold has been exceeded. The UV detector is very good for flame detection in enclosed spaces. It is not so good for outdoor applications as there are too many potential sources for false alarms.

UV detectors are good for sensing hydrogen- and methanol-fueled fires (fires that emit predominantly in the UV spectrum) because these materials burn with a blue flame (i.e., strong UV source). These detectors are generally the fastest responding, typically 30 msec. The drawback is that they are prone to false alarming from strong UV sources such as arc welding and lightning. Even if not directly in the field of view, UV reflections will trigger an alarm. UV is attenuated by oil films and is obscured by smoke. Because of the sensitivity of UV detectors to lightning, it is generally not recommended for outdoor applications. Another potential cause of false alarms is static discharges that can be detected by some UV detectors.

FIG. 1.24e
UV-detecting fire scanner used for signaling fossil fuel fires. (Courtesy of Fireye.)

UV detectors are also used in self-checking scanners (Figure 1.24e), which can be used to detect UV emissions from fossil fuel flames such as natural gas, coke oven gas, propane, methane, butane, kerosene, light petroleum distillates, and diesel fuels. These scanners are available for use in Class I, Div. 2 Groups A, B, C, and D and Class II, Div. 2 Groups F and G hazardous locations.

For a detailed description of all ultraviolet analyzers, refer to Chapter 1.69.

Infrared (IR) In general, IR detectors are good for detecting hydrocarbon-based fires (i.e., fires that have strong IR emissions). IR detectors are generally not as fast as UV detectors to respond to a fire. A disadvantage of IR is that ice buildup can desensitize the detector (lessen its ability to detect a fire), but this can be overcome with heated optics. IR does not respond to electric arc welding unless the welding is very close to the detector, in which case the detector may alarm due to seeing the burning from the welding process. Some IR detectors have flicker and statistical analysis algorithms to minimize the effects of black body sources, a false alarm source.

For a detailed description of all ultraviolet analyzers, refer to Chapter 1.32.

UV/IR Detectors These detectors combine the best of both the UV and IR and result in fewer false alarms than UV or IR detectors alone. Depending on how the individual manufacturers use the two signals also affects their performance. Some UV/IR detectors "AND" both signals such that in order to generate an alarm, both UV "AND" IR sources must be present and exceeding their threshold levels. Other designs might "OR" the two sources such that only UV or IR will alarm. Yet, others may ratio the two signals; in order to generate an alarm, both sources of UV and IR must not only be present but also exceed a certain ratio.

One disadvantage is that the loss of one (either UV or IR) will prevent alarming except in the "OR" design. High background IR (engine body) may meet the IR condition in an "AND"-ed detector such that an arc or lightning would cause a false alarm. Also, it is possible to saturate the UV portion in a ratio-based detector, if it is so designed that the corresponding IR would have to be excessively large to signal a fire alarm.

Multispectrum (UV/dual IR/VIS) fire and flame detectors represent an advanced technology of electro-optical flame detectors. These designs are usually faster, and their false alarm is less likely to occur. They are operational over a wide temperature range and their detection range is usually longer than that of conventional UV/IR detectors (see Figure 1.24f).

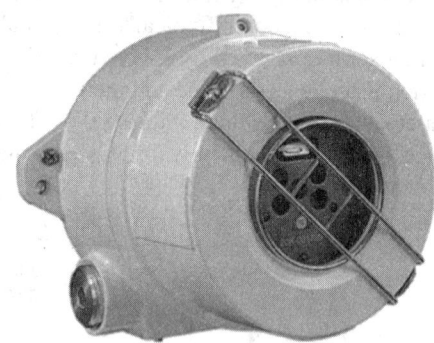

FIG. 1.24f
Multispectrum (UV/dual IR/VIS) fire and flame detector. (Courtesy of Honeywell Fire sentry.)

Dual and Multi-spectrum IR These detectors have two IR sensors. The dual IR sensors seem to be falling behind the more favored multispectrum (triple IR) detectors. These types of detectors generally have longer detection ranges than UV or IR sensors and are more fuel specific in their applications. Dual IRs can be desensitized by high background levels of IR, reducing their ability to detect a fire.

These detectors have three IR detectors. Each manufacturer's multispectrum detector is somewhat different. These differences are due to the differing (patented) flame detection algorithms used by the manufacturers (i.e., how the three optical IR signals are processed). In general, these detectors offer greater detection ranges and give fewer false alarms. They are currently the best performing flame detectors available.

Fire Detectors

Optical In selecting the right fire detection system, the user must analyze the potential hazard and false alarm sources and select the detector that can signal an identified fire with the highest immunity from false alarms (Tables 1.24a and 1.24b). The specific cone of vision of the selected flame detector should be evaluated to make sure that it will give adequate coverage to the particular fire hazard (Figure 1.24d). These detectors must also have an unobstructed view to the fire hazard area.

This is an emerging technology and there are only a few products to choose from. These types of detectors come either as black and white or as color images that are human viewable. As they are only sensitive to the red–green–blue spectrum, they are not suitable for blue/translucent flames from such fuels as hydrogen and methanol. The advantage of CCTV detectors is that the user is able to verify the presence of a fire before taking any action. Since this is an emerging technology, there is little information on their performance.

Some of these visual flame detectors can provide both flame detection and CCTV surveillance recording capabilities. They can also distinguish other radiation sources from radiation generated by fire. In the case of an alarm, the operators can study the live video of the CCTV images right in the control room (Figure 1.24g). The images before and after an incident can be used for post-incident analysis and can serve for future accident prevention.

FIG. 1.24g
Visual flame detectors can provide videos viewable in the control rooms. (Courtesy of Micropack Ltd.)

BURNER FLAME SAFEGUARDS

In this chapter, a brief description of the more widely used means of flame detection will be presented. After some basic theoretical discussions, the various devices will be described. The feature tabulation given in Table 1.24c will enable the reader to select the most desirable sensor for the application at hand. Functions of flame sensors as part of burner control systems will be discussed later.

Among the many characteristics of flame, the following have been successfully used to detect the present flame:

1. Heat generated
2. Ability to conduct electricity (ionization)
3. Radiation at various wavelengths, such as
   a. Visible
   b. IR
   c. UV

Methods of detecting these characteristics are described later.

TABLE 1.24c
Comparison of Flame Safeguards

Principle of Flame Detection	Rectification		Infrared	Visible Light	Ultraviolet
Type of Detector	Rectifying Flame Rod	Visible Light Rectifying Phototube	Lead Sulfide Photocell	Cadmium Sulfide Photocell	Ultraviolet Detector Tube
ADVANTAGES					
Same detector for gas or oil flame			✓		✓
Can pinpoint flame in three dimensions	✓				
Viewing angle can be orificed to pinpoint flame in two dimensions		✓	✓	✓	✓
Not affected by hot refractory	✓			✓	✓
Checks own components prior to each start	✓	✓	✓	✓	✓
Can use ordinary thermoplastic covered wire for general applications, no shielding needed	✓	✓			✓
No installation problem because of size				✓	✓
DISADVANTAGES					
Difficult to sight at best ignition point			✓		
Exposure to hot refractory may reduce sensitivity to flame flicker and require orificing			✓		
Flame rod subject to rapid deterioration and warping under high temperatures	✓				
Not sensitive to extremely hot premixed gas flame			✓	✓	
Temperature limit too low for some applications	✓	✓	✓	✓	
Shimmering of hot gases in front of hot refractory may simulate flame			✓		
Hot refractory background may cause flame simulation			✓		
Electric ignition spark may simulate flame					✓

Flame Rods

Besides heat detection, there are two basic principles of operation in flame rod detection of flames: conductivity and flame rectification. Conductivity systems are, for the most part, no longer used. Either type of system depends on the ability of the flame to conduct current when a voltage is applied across two electrodes in the flame. Heat from the flame causes molecules between the electrodes to collide with each other so forcibly as to knock some electrons out of the atoms, producing ions. This is called flame ionization.

Heat Sensor Rods The earliest flame sensors utilized the most obvious characteristics of the flame—namely, the heat generated. These devices were thermocouples, bimetallic elements, etc. For small installations and domestic burners, these devices were satisfactory and are still used. Their relatively slow response (2–3 min) renders them unsuitable, and indeed dangerous, for larger installations.

Let us take a large reforming furnace as an example. If it requires 1000 SCFM (0.47 m^3/sec) of fuel gas, and flame failure is detected only after 2–3 min, the 56–84 m^3 (2000–3000 ft^3) of unburned fuel gas admitted to the furnace will create an explosion hazard. A detector with a response of 4–6 sec or less will not permit sufficient amounts of gas to enter the furnace to cause an explosion. Another obvious disadvantage of these sensors is that since they only sense heat, they will be unable to distinguish between heat generated by flame and that radiated by the hot refractory.

Conduction Sensor Rods The first major breakthrough toward fast (but unreliable) detection of the flame was the discovery that flame is capable of conducting electricity. This is true because the flame, being a chemical reaction between a fuel and oxygen, liberates a large number of electrons. Because of ionization, the flame can conduct both direct (DC) and alternating (AC) currents, which are utilized to establish an electrical circuit. A rod immersed in the flame (flame rod) acts as one electrode, and the burner as the other. This proved to be fast but unreliable in that a high-resistance electrical short caused by, for example, faulty insulation could simulate the presence of flame.

Rectification Sensor Rods Any electrical device that offers a low resistance to the current in one direction but a high resistance to the current in the opposite direction is called a rectifier. An ideal rectifier is one with zero resistance in one direction and infinite resistance in the opposite direction.

When AC is passed through a rectifier, the current obtained will be rectified current. It will consist of only that portion of the input current to which the rectifier presented low resistance.

By making the area of one of the electrodes (the burner in this case) much larger than the other, conduction will essentially take place only in one direction (Figure 1.24h).

The explanation for the earlier phenomenon is that when ionization takes place, electrons are liberated from the gas molecules and are free to move about, constituting electric current. In addition to the freed electrons, the negative electrode contains many surplus electrons acquired through the negative side of the external circuit that makes the electrode negative. These surplus electrons repel each other and, given enough positive ions to attract them, will leave the negative electrode. The number of electrons leaving the negative electrode and entering the positive electrode determines the rate of current flow. It is apparent that the current flow depends on the number of positive ions that get near enough to the negative

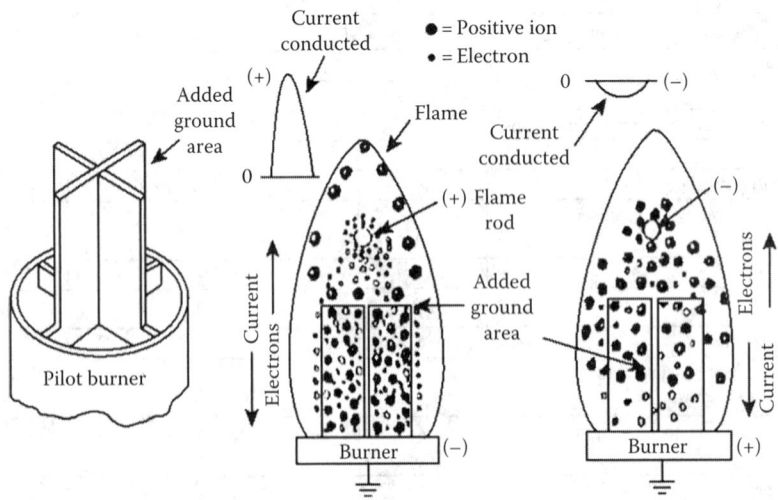

FIG. 1.24h
Rectification by flame.

electrode. Suppose the area of one electrode is made several times larger than the other and that electrode is negative. This will increase the flow of electrons to the other (positive) electrode.

If AC current is applied, the current through the flame will be rectified. Indeed, using this arrangement and thereby obtaining a *half-wave rectified* current, the lack of safety due to simulated flame was eliminated, as the circuitry associated responded only to a half-wave rectified signal. Unrectified AC or DC input to the detecting circuit will result in a safe shutdown. Because flame rods are in direct contact with the flame, they have to be cleaned or replaced often. Above approximately 1100°C (2000°F), very few metals can be used and even these become brittle. Fuels with high sulfur content burn with flames having low resistance, resulting in low flame rod output and leads to nuisance shutdowns.

To avoid direct contact with the flame and the maintenance problems arising from it, another obvious characteristic of the flame—its ability to radiate energy—was used.

Visible Radiation

Radiation emitted by the flame covers the energy spectrum for wavelengths corresponding to UV, visible, and IR ranges (Figure 1.24i).

Visible radiation occupies about 8% of the total bond of wavelengths radiated by the flame. To detect this position of the radiation, a rectifying phototube is used. It consists of a light-sensitive coated cathode of large surface and an anode encapsulated in a vacuum (Figure 1.24j).

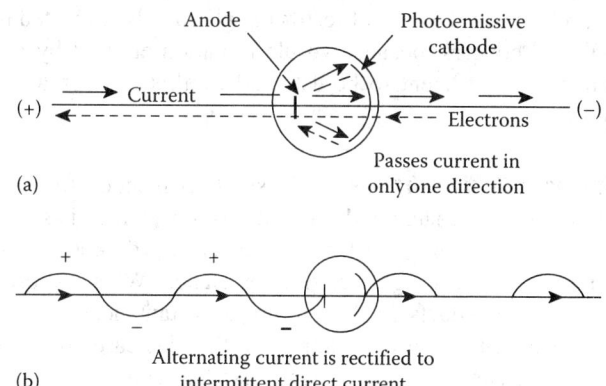

FIG. 1.24j
Rectifying phototube.

The number of electrons emitted by the cathode is approximately proportional to the light intensity. Conduction will take place only when the anode is positive.

If AC is applied, the phototube will act as a half-wave rectifier. The associated electronic circuit will respond only to such a rectified signal, and high-resistance shorts will not simulate flame. At high temperatures, the refractory (noncombustible insulator used to line interiors of furnaces)

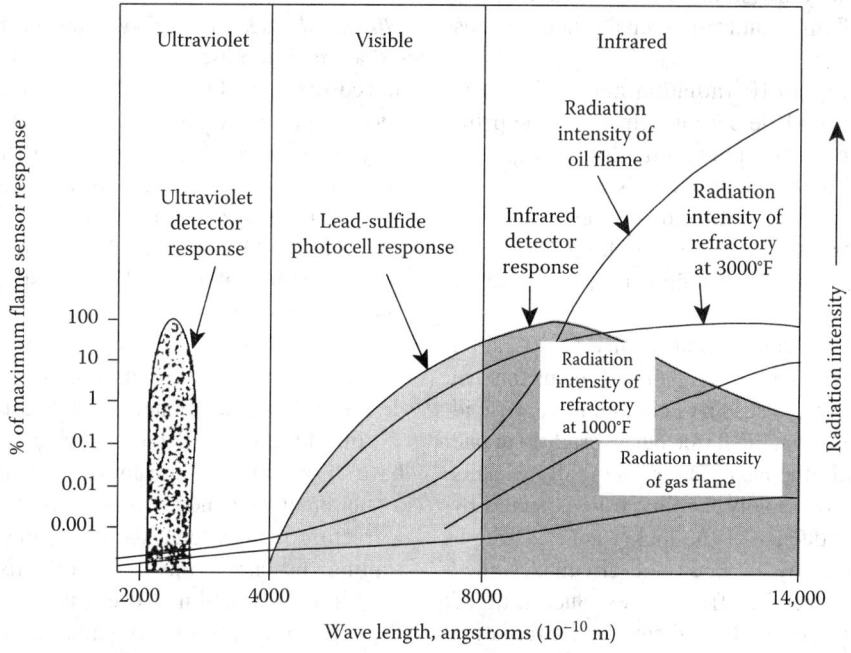

FIG. 1.24i
Applicable ranges of selected flame sensors and the range of hot refractor effect.

emits visible radiation. The detector might not be able to distinguish this radiation from the visible radiation emitted by flames; therefore, caution must be exercised in installing the detector in such a position that it will not be able to sight hot refractory. The use of rectifying photocells is limited to oil-fired burners, because visible radiation emitted by gas flame is insufficient to be detected by this type of flame sensor.

Cadmium Sulfide Photocell The single element of this type of photocell is coated with cadmium sulfide, which is sensitive to radiation in the visible spectrum only. In darkness, cadmium sulfide has high electrical resistance. When exposed to light, it conducts freely and almost instantaneously. The electrical resistance of cadmium sulfide decreases directly with increasing intensity of light. Like lead sulfide, it acts as a variable resistor and conducts current equally well in either direction.

This photocell is used in series with the coil of a low-voltage relay in an AC circuit. When the cell sees sufficient light to pass a given current, the relay "pulls in." The sensitive region of the cadmium sulfide cell is such that it will not respond to gas flames and can, therefore, be used with oil flames only. The performance of this cell will not be affected by the "shimmering" effects of hot refractory.

Infrared IR radiation covers the largest portion, approximately 90%, of the band of wavelengths emitted by the flame. Its intensity is by far the strongest of all radiation forms emitted. The IR radiation emitted is a more reliable means of detection than the visible radiation. It is emitted by gas as well as oil flames, and the intensity never drops to zero.

This cell is sensitive to IR radiation and is the most widely used device for such detection. Similar to the principle of the cadmium sulfide photocell, the principle of lead sulfide photocell is to change its resistance inversely to the IR radiation it is subjected to. It is a variable resistor and conducts electricity equally well in both directions without rectification. The current flow is a measure of flame strength.

The signal-sensing circuit responds only to so-called flame frequencies. When looking at a flame, it seems to burn steadily. But if the eyes were sensitive enough to very rapid changes, it could be seen that the flame burns brightly at one instant and less brightly the next. The flame really flickers or pulsates, but much too rapidly for this to be detected by the human eye. The frequency of the flicker is very irregular, but may be detected by an electronic circuit designed to accept only a proper band of frequencies. Such a tuned circuit is selective and will not be affected by IR radiation emitted by hot refractory, nor can a high-resistance short simulate a flame.

Lead sulfide photocells are often used on dual fuel applications for the detection of both pilot and main flames. The design shown in Figure 1.24k mounts on a 0.75 in. sighting pipe and is often used where rectifying photocells or flame rods are difficult to use.

FIG. 1.24k
Lead sulfide photocells detect infrared radiation emitted by the flames of natural gas, oil, and coal. (Courtesy of Honeywell Inc.)

Experiments show that a shimmering effect caused by movement of hot gases between the refractory and photocell can fool the circuit and flame simulation can occur. This so-called shimmering can be demonstrated by viewing objects through the heated air over a candle.

Ultraviolet Seemingly, UV is the least significant portion of radiation. It represents only about 1% of the total radiation and covers 10% of the emitted band of wavelengths. UV radiation is emitted by gas as well as oil flames. The device popularly known as a UV detector is a gas-filled tube with voltage applied between its two electrodes (anode and cathode).

The tube is conductive only if voltage is applied at its terminals, and it is subjected to UV radiation.

It is readily seen that a high-resistance short cannot simulate flame, nor can it be affected by the hot refractory. This is because no UV radiation will be emitted by refractories below approximately 1370°C (2500°F). Compared to the detectors so far mentioned, the UV cell is foolproof.

In addition to all these attributes, some UV detectors have an additional built-in safety feature. This consists of a shutter arrangement that blocks the view of the cell for a fraction of a second about 20 times each second, interrupting the circuit that can reestablish current flow again only if UV radiation still exists. This arrangement makes the detector responsive to flame failure within a fraction of a second, because it has to convince itself 20 times a second that flame exists.

Flame Safeguard Installation

Proper installation of flame sensors is essential to safe operation. Installations of these sensors fall into three categories:

1. Pilot and main flames supervised simultaneously—Checking the pilot and main flames simultaneously is the most desirable method of flame supervision. When this arrangement is used, the pilot is capable of safely igniting the main flame (Figure 1.24l). The flame rod in this case is located in the path of both the main flame and the pilot flame.

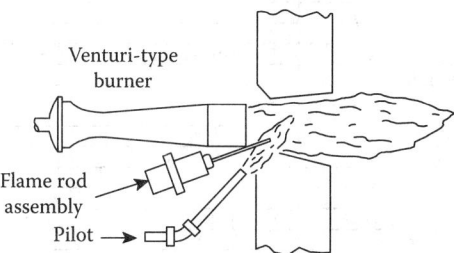

FIG. 1.24l
Pilot and main flame monitored by flame rod.

2. Only pilot and flame supervised—On some installations, it may be impossible to prove the pilot and main flames simultaneously because of the variations of the main flame envelope at different firing rates. This condition prevails, for example, in mechanical fan-type burners. The recommended practice of installation in this case is to supervise the pilot flame (Figure 1.24m).

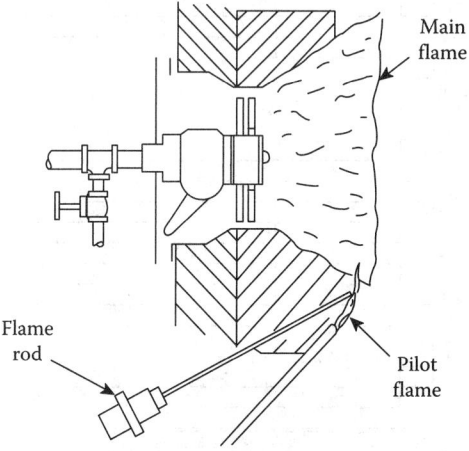

FIG. 1.24m
Mechanical fan-type burner. Pilot flame only is proved.

3. Only main flame supervised—In applications where the pilot flame is not lit continuously but only on lights off, the main flame is supervised. It is essential that the sensor be installed in such a way that it will be able to feel or see the flame during all variations in firing rates and draft adjustments.

The previously mentioned considerations are applicable for flame rods as well as for the different viewing devices. Additional care must be exercised in the installation of rectifying phototubes and lead sulfide photocells to ensure that incandescence of hot refractory or shimmering of hot air will not simulate the flame. This is accomplished by limiting the view of the devices through a restriction orifice.

Portholes for viewing devices must be cleaned regularly to ensure an unobstructed view and thus prevent nuisance shutdowns.

CONCLUSIONS

The order of presentation for the different types of flame sensors in this chapter was intended to show how, with the passage of time, the limitations of each design have been minimized or eliminated.

Starting with the heat sensors, it was seen that their response time rendered them unsuitable for industrial applications. The flame rods provided for fast response; however, their direct contact with the flame and the consequences, such as brittleness or oil and carbon deposits on the rod, limited their usefulness. Photocells utilizing the visible range of radiation were limited by their inability to reject radiation from the refractory, and their useful range covered only oil flames.

The IR detector eliminated most of the earlier shortcomings, but shimmering of hot air could still fool the detecting circuit. It is safe to say that UV detectors can be used reliably in any application. Their relatively high cost demands that the situation be carefully investigated for possible application of other sensors.

In an oil-fired furnace, the cadmium sulfide cell can be safely used if it is installed so that it will not see the hot refractory. For combination oil and gas furnaces, the lead sulfide cell will work satisfactorily if it is installed with its view pointed toward the flame through a restricting orifice so that it will not see shimmering hot air.

All devices except the flame rods share the common advantage of not being in contact with the fuel and the flame. The temperature limitations imposed on them are easily satisfied because the temperature outside of the furnace is unlikely to reach 66°C (150°F). The purpose of these installations is safety, and, therefore, cost consideration can be only secondary.

For a summary of flame guard sensor features, refer to Table 1.24c.

SPECIFICATION FORMS

When specifying flame, fire, or smoke detectors only, please use the ISA form 20A1001, and when specifying both the analyzer and the composition or the properties of the process material of interest, use ISA form 20A1002. The following are the forms that are reproduced with the permission of the International Society of Automation.

#				#				
1	RESPONSIBLE ORGANIZATION		ANALYSIS DEVICE	6	SPECIFICATION IDENTIFICATIONS			
2				7	Document no			
3	(ISA)		Operating Parameters	8	Latest revision		Date	
4				9	Issue status			
5				10				

#	ADMINISTRATIVE IDENTIFICATIONS			#	SERVICE IDENTIFICATIONS Continued			
11				40				
12	Project number		Sub project no	41	Return conn matl type			
13	Project			42	Inline hazardous area cl		Div/Zon	Group
14	Enterprise			43	Inline area min ign temp		Temp ident number	
15	Site			44	Remote hazardous area cl		Div/Zon	Group
16	Area	Cell	Unit	45	Remote area min ign temp		Temp ident number	
17				46				
18	SERVICE IDENTIFICATIONS			47	COMPONENT DESIGN CRITERIA			
19	Tag no/Functional ident			48				
20	Related equipment			49	Component type			
21	Service			50	Component style			
22				51	Output signal type			
23	P&ID/Reference dwg			52	Characteristic curve			
24	Process line/nozzle no			53	Compensation style			
25	Process conn pipe spec			54	Type of protection			
26	Process conn nominal size	Rating		55	Criticality code			
27	Process conn termn type	Style		56	Max EMI susceptibility		Ref	
28	Process conn schedule no	Wall thickness		57	Max temperature effect		Ref	
29	Process connection length			58	Max sample time lag			
30	Process line matl type			59	Max response time			
31	Fast loop line number			60	Min required accuracy		Ref	
32	Fast loop pipe spec			61	Avail nom power supply		Number wires	
33	Fast loop conn nom size	Rating		62	Calibration method			
34	Fast loop conn termn type	Style		63	Testing/Listing agency			
35	Fast loop schedule no	Wall thickness		64	Test requirements			
36	Fast loop estimated lg			65	Supply loss failure mode			
37	Fast loop material type			66	Signal loss failure mode			
38	Return conn nominal size	Rating		67				
39	Return conn termn type	Style		68				

#	PROCESS VARIABLES	MATERIAL FLOW CONDITIONS			#	PROCESS DESIGN CONDITIONS			
69					101				
70	Flow Case Identification			Units	102	Minimum	Maximum	Units	
71	Process pressure				103				
72	Process temperature				104				
73	Process phase type				105				
74	Process liquid actl flow				106				
75	Process vapor actl flow				107				
76	Process vapor std flow				108				
77	Process liquid density				109				
78	Process vapor density				110				
79	Process liquid viscosity				111				
80	Sample return pressure				112				
81	Sample vent/drain press				113				
82	Sample temperature				114				
83	Sample phase type				115				
84	Fast loop liq actl flow				116				
85	Fast loop vapor actl flow				117				
86	Fast loop vapor std flow				118				
87	Fast loop vapor density				119				
88	Conductivity/Resistivity				120				
89	pH/ORP				121				
90	RH/Dewpoint				122				
91	Turbidity/Opacity				123				
92	Dissolved oxygen				124				
93	Corrosivity				125				
94	Particle size				126				
95					127				
96	CALCULATED VARIABLES				128				
97	Sample lag time				129				
98	Process fluid velocity				130				
99	Wake/natural freq ratio				131				
100					132				

#	MATERIAL PROPERTIES			#	MATERIAL PROPERTIES Continued		
133				137			
134	Name			138	NFPA health hazard	Flammability	Reactivity
135	Density at ref temp		At	139			
136				140			

Rev	Date	Revision Description	By	Appv1	Appv2	Appv3	REMARKS

Form: 20A1001 Rev 0 © 2004 ISA

1	RESPONSIBLE ORGANIZATION	ANALYSIS DEVICE COMPOSITION OR PROPERTY Operating Parameters (Continued)	6	SPECIFICATION IDENTIFICATIONS
2			7	Document no
3			8	Latest revision — Date
4			9	Issue status
5			10	

	PROCESS COMPOSITION OR PROPERTY			MEASUREMENT DESIGN CONDITIONS				
	Component/Property Name	Normal	Units	Minimum	Units	Maximum	Units	Repeatability
13								
14								
15								
16								
17								
18								
19								
20								
21								
22								
23								
24								
25								
26								
27								
28								
29								
30								
31								
32								
33								
34								
35								
36								
37								

(Rows 38–75: blank remarks/notes area)

Rev	Date	Revision Description	By	Appv1	Appv2	Appv3	REMARKS

Form: 20A1002 Rev 0 © 2004 ISA

Definition

Flame rectification: The flame rectification is an electronic way to prove pilot flame before opening the main burner gas valve. It serves to prevent the main burner gas valve from opening if there is no pilot flame, or the temperature of the hot surface igniter has not yet been established. When in a combustion process the pilot flame is established, the flame acts as a path (conductor) that allows the DC current to flow from the flame rod to ground completing the circuit. This completed signal indicates that flame has been established and signals the combustion controls that the main gas valve can open and the hot surface igniter will start the main flame.

Abbreviations

CCTV   Closed-circuit television
IR   Infrared
UV   Ultraviolet

Organizations

NFPA   National Fire Protection Association
UL   Underwriters Laboratory

Bibliography

Baker, D. E., Residential fire detection, (1993) http://extension.missouri.edu/p/G1907.

Belanger, R., USNRC: Environmental assessment of ionization chamber smoke detectors, (2012) http://www.nrc.gov/reading-rm/doc-collections/nuregs/contract/cr1156/.

EPA, Smoke detectors and radiation, (2013) http://www.epa.gov/radiation/sources/smoke_alarm.html.

FEMA, Pre-fire protection engineering, (2008) http://www.usfa.fema.gov/downloads/pdf/nfa/higher_ed/pre_fire_protection_engineering.pdf.

Fike, Intelligent fire alarm control system, (2008) http://www.suppressionsystems.com/cybercat50.pdf.

Fireye, Flame safeguard controls in multi-burner environments, (1998) http://www.fireye.com/Literature/Documents/WV-96.pdf.

Flame Guard, Gas water heater with the flame guard® safety system installation, (2006) http://www.americanwaterheater.com/media/42179/bfgmanual.pdf.

FM Global, Approval standard for fire alarm signaling systems, (2014) http://www.fmglobal.com/assets/pdf/fmapprovals/3010.pdf.

Grainger, Types of smoke alarms and detectors, (2015) http://www.grainger.com/content/qt-types-smoke-alarms-detectors-366.

NFPA, Ionization vs. photoelectric smoke alarms, (2013) http://www.nfpa.org/safety-information/for-consumers/fire-and-safety-equipment/smoke-alarms/ionization-vs-photoelectric.

NFPA, Flammable and combustible liquids code, NFPA No. 30, (2015) http://www.nfpa.org/codes-and-standards/document-information-pages?mode=code&code=30.

NFPA, National fire alarm code No. 72, (2015) http://www.nfpa.org/codes-and-standards/document-information-pages?mode=code&code=72.

Nolan, D., *Handbook of Fire and Explosion Protection*, (2011) http://www.amazon.com/Handbook-Explosion-Protection-Engineering-Principles/dp/1437778577.

Saltsman, R., Ionization vs. photoelectric smoke alarms, (2013) http://www.startribune.com/local/yourvoices/146781575.html.

Sierra Monitor Corp., Flame detector selection guide, (2013), http://www.sierramonitor.com/docs/pdf/flame_detector_selection.pdf.

System Sensors, Smoke detectors, (2012) http://www.systemsensor.com/en-us/documents/system_smoke_detectors_appguide_spag91.pdf.

UL, Smoke alarms with sensitivity calibration provisions, (2009) http://ul.com/wp-content/uploads/2014/04/ul_SmokeAlarmCalibration.pdf.

UL, Standard for smoke alarm, UL217, (2015) http://ulstandards.ul.com/standard/?id=217_6.

Vandermeer, V., Fireeye, Flame safeguard controls, (2015) http://www.eclipsenet.com/products/t400/.

Vervalin, C. H., Fire protection manual, (1981) http://www.amazon.com/Protection-Manual-Hydrocarbon-Processing-Plants/dp/0872012883/ref=asap_bc?ie=UTF8.

1.25 Fluoride Analyzers

J. S. JACOBSON (1974, 1982)   **A. T. BACON** (1995)   **B. G. LIPTÁK** (2003, 2017)

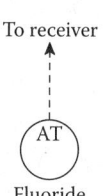

To receiver

Fluoride
Flow sheet symbol

Principles of operation	A. Detector tubes
	B. Electrochemical
	C. Paper tape photonics
	D. Ion mobility spectrometry
	E. Infrared spectrometry
	F. Ion-selective electrodes
	G. Silicon dioxide sensors
	H. Ion chromatography
	I. Titration
	J. Colorimetric
	K. Gas chromatography
Materials of construction	Surfaces of equipment contacting fluoride should be stainless steel, Teflon, epoxy, polyethylene, or polypropylene.
Concentrations measured	Monitoring for worker protection is usually in the 0–10 parts per million (ppm) range. Leak detection in the surroundings of hydrogen fluoride (HF) handling equipment requires sensitivity in the high ppm range. Monitoring for low-level environmental damage may require sensitivity in the parts per billion (ppb) range.
Sensitivity range	A. 0.1–20 ppm.
	B. Low ppm.
	C. 0.05–10 ppm.
	D. Low to mid ppm.
	E. Low ppb to high ppm, function of sampling time, volume.
	F. Qualitative response to high ppm levels.
	G–H. Measurement is usually in the ppm range; sensitivity depends on sampling parameters.
Calibration	Calibration of vapor analyzers usually is performed with permeation-type devices or with calibrated gas standards. Liquid-phase analyzers are calibrated using solutions of an appropriate fluoride compound such as sodium fluoride.
Accuracy	A. ±20% of actual measurement at low humidities, inaccurate at high humidities.
	B–D. ±10% of full scale.
	E. Varies with the type of instrument and humidity level.
	F, H–K. These laboratory and automated techniques are capable of providing high accuracy (±3% of actual measurement) under ideal conditions, but this is often compromised by inaccurate sample collection practices.
	G. Qualitative only.
Cost	A. $3/analysis, $50/sampling kit
	B. $1000–$3000 (portable monitor ~$1000)
	C. $3500/point
	D. $17,000
	E. $9000 (common IR) to $90,000 (FTIR)
	F. $500 for electrode; $1000 to $3000 for associated electronics (for water/lab analysis); $6,500 and up for automated system
	G. $1400/point
	H. $10,000–$20,000 (lab method)
	I. $500 (lab method)
	J. $1,000–$5,000 (lab method); $25,000 (automated system)
	K. $10,000 to $20,000 (lab method)

(Continued)

Partial list of suppliers ABB (F) (http://www.abb.us/product/seitp330/ea2c7179ae79ea98c12579cf00312cad.aspx), Emerson/Rosemount (F) (http://www2.emersonprocess.com/siteadmincenter/PM%20Rosemount%20Analytical%20Documents/Liq_ProdData_71–56.pdf)
Forbes Marshall (F) (http://www.pollardwater.com/pages_product/IEMPFC1000EaglePFC1000FluorideAnalyzer.asp)
Hach (F) (http://www.hach.com/ca610-fluoride-analyzer-with-reagents/product?id=7640284908)
Hanna Instruments (portable meter) (http://www.hannainst.com/usa/prods2.cfm?id=038007&ProdCode=HI%2098402)
Honeywell/Zellweger/MDA Scientific/Zellweger (C) (http://www.honeywellanalytics.com/Technical%20Library/EMEAI/ChemKey/Manuals/Chemkey%20TLD%20Operating%20Instructions%20Rev%201.2.pdf)
Los Gatos Research (E) (http://www.abb.us/product/seitp330/ea2c7179ae79ea98c12579cf00312cad.aspx)
Pollard Selective (F) (http://www.pollardwater.com/pages_product/IEMPFC1000EaglePFC1000FluorideAnalyzer.asp)
ProMinent (F) (http://www.prominent.ca/Products/Measuring-Control-and-Sensor-Technology/Parameter/Fluoride/Fluoride-Measurement.aspx)
Sensidyne (A, B) (http://www.sensidyne.com/colorimetric-gas-detector-tubes/detector-tubes/sensidyne_gas_detector_tube_number-156S_HYDROGEN_FLUORIDE0.17–30ppm.php)
Thermo Scientific/Dionex (H) (http://www.dionex.com/en-us/products/ion-chromatography/ic-rfic-systems/ic-titrator/lp-113695.html)
Thermo Scientific/Orion (F) (http://www.thermo.com/eThermo/CMA/PDFs/Product/productPDF_2330.pdf)

INTRODUCTION

Water fluoridation is the controlled addition of fluoride to a public water supply to reduce tooth decay. Fluoridated water has fluoride at a level that is effective for preventing cavities. Fluoridated water creates low levels of fluoride in saliva, which reduces the rate at which tooth enamel demineralizes and increases the rate at which it remineralizes in the early stages of cavities. Defluoridation is needed when the naturally occurring fluoride level exceeds recommended limits. A 1994 World Health Organization expert committee suggested a level of fluoride from 0.5 to 1.0 mg/l (milligrams per liter).

The number of analytical techniques available for the analysis of fluoride is fairly large. Yet, the most often used analyzers depend on ion-selective electrodes that are specific to fluoride. In the past, several of these techniques were used only in the laboratory. Table 1.25a lists a variety of automatic and manual methods for the detection of fluorides water, ambient air, stacks, and process streams. In order to select the most appropriate method for a given application, the physical state and properties of the fluoride compound must be taken into consideration.

TABLE 1.25a
Orientation Table for Fluoride Measurement Methods

	Automatic or Manual Method	Suitable for Water Analysis	Gas or Particulate Measurement	Suitable for Stack/ Process Monitoring	Suitable for Organic Fluoride Monitoring
Tubes	M	No	G	No	Yes[a]
Electrochemical	A	No	G	Some[b]	No
Paper tape	A	No	G	Some[b]	No
Ion mobility spectrometry	A	No	G	Yes	Yes[a]
Ion-specific electrodes	A/M[c]	Yes	G/P[d]	Some[b]	No
Silicon dioxide sensor	A	No	G	Leak detection	No
Ion chromatography	M	Yes	G/P[d]	Some[b]	No
Titration	M	Yes	G/P[d]	Some[b]	No
Colorimetric	A/M[c]	Yes	G/P[d]	Some[b]	No
Gas chromatography	A/M[c]	Yes	G	Some[b]	Yes
Infrared	A	No	G	Some[b]	Yes

[a] Required pyrolysis equipment available from manufacturer; other methods may be adapted with custom equipment.
[b] Suitability for stack monitoring depends on moisture present, interference, etc.
[c] Manual method can be performed in lab; automatic systems are available.
[d] Particulate measurement requires special collection techniques.

TYPES OF FLUORIDE COMPOUNDS

Fluorides can exist as gases, liquids, and solids. The most common form of fluoride is hydrogen fluoride (HF). This compound is very volatile. Its boiling point is 19.4°C (66.9°F), and therefore it is most often analyzed in the vapor state. All the methods that are described as gas or vapor fluoride analyzers detect this compound.

Other compounds, such as UF6 (uranium hexafluoride), which exist in the vapor phase, hydrolyze quickly in ambient air or solution to form HF, and thus can be analyzed indirectly by these methods. Other liquid and gaseous fluorides are much more stable and do not easily form HF. These compounds must be broken down prior to analysis. A common method of conversion is thermal decomposition, which is followed by analysis of the resulting HF.

Many solid fluorides are soluble in water and form fluoride ions. If fluoride is present in the air in the form of fine particles, these compounds can be collected on filters and analyzed by wet laboratory or automated methods. Since the automated wet methods operate by impinging an air sample into a wet collection stream, if total fluoride concentration measurement is required,* care must be taken to assure that the particles are freely admitted into the collection vessel.

Other fluoride-containing solids are not water soluble and require thermal decomposition at very high temperatures prior to analysis as HF.

DETECTOR TYPES

Water treatment facilities in most states are required to add hydrofluoric acid to drinking water. Liquid fluoride analyzers are used to control the amount of hydrofluoric acid sent out to the distribution system for public consumption. This is usually mandated by state Environmental Protection Division regulations.

In most cases, the goal of the analysis is to determine the quantitative amount of fluoride in a sample. In other cases, only the presence of HF needs to be determined. Qualitative sensors such as silicon dioxide are used to detect HF leakage from valve connections and other joints. These sensors do not provide quantitative information.

---

* ASTM D3267-12, Standard Test Method for Separation and Collection of Particulate and Water-Soluble Gaseous Fluorides in the Atmosphere (Filter and Impinger Method) 2012.

Most other devices discussed in this chapter provide some degree of quantitative information.

Analyzers are also grouped according to their operation, which can be automatic or manual. Purely manual devices such as detector tubes require an operator to collect and read the sample. Some of the manual laboratory techniques have been automated and depend on colorimetric, ion-specific electrodes and gas chromatography (GC) techniques for their analysis. Yet, other analyzers are fully automatic and do not require operator attention, except for maintenance.

Fluoride analyzers are also grouped on the basis of the sample phase into vapor and liquid analyzers. In the case of ion-specific electrodes, titration, and colorimetric methods, fluoride analysis requires an aqueous sample. On the other hand, the concentration of airborne fluoride is determined by using an impinged or absorbed sample. The remaining techniques are used for the analysis of the vapor phase.

The selection of an analyzer for a particular application will depend on the type of process sample, which is to be monitored, cost, accuracy, speed of response, range, and frequency of maintenance. These considerations are discussed later.

Detector Tubes

These devices utilize a hand-operated pump to draw a known volume of air through a calibrated tube (Figure 1.25a). The fluoride concentration in this case is indicated by the resulting color. The initial investment in equipment is very low, but this method is acceptable only if infrequent measurements are sufficient. Detector tubes are commonly used for spot monitoring in industrial hygiene applications.

FIG. 1.25a
Detector tubes are easy-to-use means of on-the-spot measurement of fluoride. (Courtesy of Sensidyne.)

Electrochemical Cells

An air sample is presented to the cell, either actively or passively (Figure 1.25b). The sample permeates into the cell, usually through a membrane, where an electrochemical reaction produces a current proportional to the concentration. The initial cost is relatively low. Some types require fairly high maintenance, although this is less of a concern on more modern designs. Cells are most often used in ambient air monitoring and are not easily adapted for process control. Some models are prone to freezing at low temperatures and do not operate well at humidity extremes.

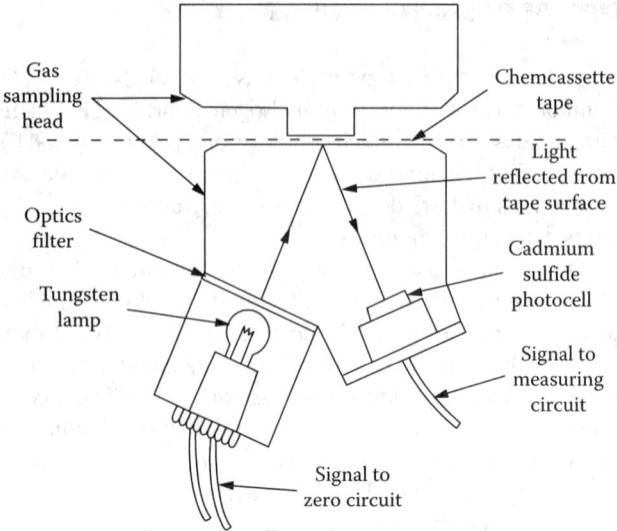

FIG. 1.25c
Fluoride concentration can be detected by monitoring the reflected light intensity from a chemically treated tape. (Courtesy of Honeywell/Zellweger Analytics/MDA Scientific Inc.)

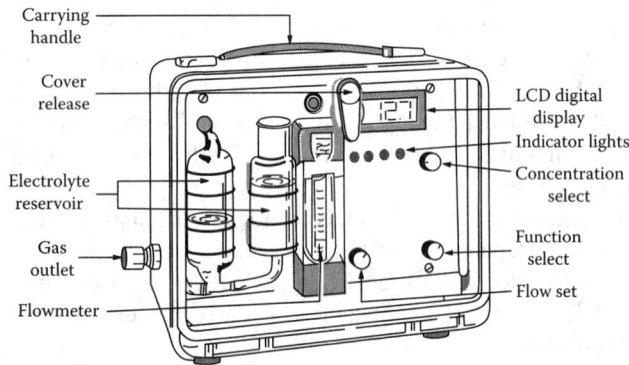

FIG. 1.25b
Portable gas monitor with electrochemical sensor. (Courtesy of Sensidyne Inc.)

Paper Tape

An air sample is drawn through a chemically treated tape, and the colorimetric change is read with a photometer (Figure 1.25c). The tape is automatically advanced as fluoride is detected. The initial cost is moderate, but expendables can be expensive if fluoride is constantly present. The tape has the advantage of not requiring an independent calibration, as each tape is calibrated at the factory. It is most often used for ambient air monitoring and is easily adapted for multipoint sampling. It is not commonly used for process control or outdoor applications.

Ion Mobility Spectrometry (IMS)

Air is drawn into the instrument and the sample permeates through a membrane into a cell, where it is ionized by a small radioactive source. The analysis technique is similar to that of time-of-flight mass spectrometry (MS) but is performed at

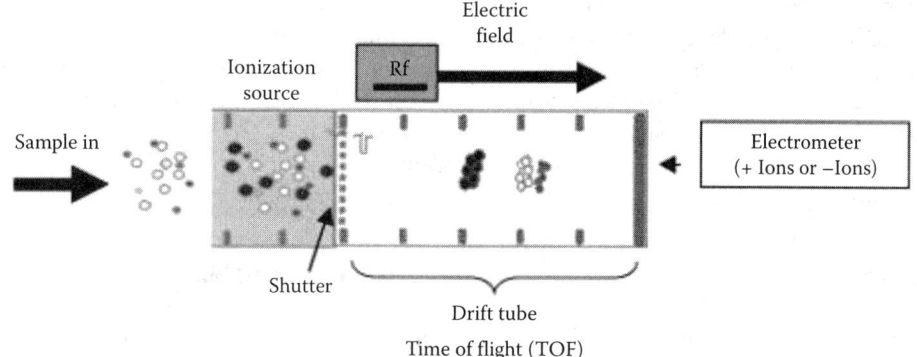

FIG. 1.25d
The principle of operation of ion mobility spectroscopy. (Courtesy of Eric Kirleis, Sionex Corp.)

atmospheric pressure. The initial cost is moderately high, but maintenance is minimal.

IMS identifies and detects chemicals based on time-of-flight (TOF) principles (Figure 1.25d). Essentially, IMS measures the time it takes a certain chemical ion to move through a uniform radio frequency (RF) electric field. Samples are ionized and the samples' ions are lined up using a shutter mechanism and then floated into a drift tube. The drift tube contains a homogeneous electric field, and this field moves the ions down the drift tube. The ions interact with neutral molecules within the drift tube, resulting in a TOF dependent on ion mass, size, and morphology. Once a target's TOF is established, the identification of the chemicals is based on their known TOF.

This technique has found wide acceptance in the petrochemical industry due to its high reliability, good interference rejection, and its ability to operate accurately at extremes of ambient temperature and humidity. This technique is most often used for outdoor ambient air monitoring and can be adapted for process control and stack monitoring. The chief disadvantage is the requirement for a constant supply of dry instrument air and purging in flammable environments.*

Infrared Spectroscopy

A gas sample is drawn into an optic cell where a characteristic infrared (IR) absorbance is used to determine concentration (Chapter 1.32). Several types are available, including nondispersive infrared and Fourier transform infrared (FTIR). Prices range from moderate to expensive, depending on analyzer type and features. Sensitivity is determined mainly by path length. In most cases, water vapor produces a strong interference, masking HF response.

The advantage of FTIR is the ability to monitor for several gases with the same analyzer. IR analyzers are typically used in applications such as stack monitoring, after sample cleanup. IR analyzers may also be used to monitor for certain fluorinated organic compounds.

Ion-Specific Electrodes (ISE)

The concentration of fluoride ions in solution is monitored by using an ion-selective electrode (Chapter 1.33), which provides a millivolt output signal in response to the activity of fluoride ions in the solution. The fluoride-selective ion electrode, in conjunction with a reference electrode, develops its potential across a doped fluoride crystal membrane. When the electrode is placed in a solution, the ions from the solution move to the surface of the membrane, and the electrical charge of the ions creates a potential difference across the membrane. At equilibrium, this potential difference is proportional to the activity of the fluoride ions in the solution.

These probes are inexpensive for laboratory applications but become expensive when used as automatic online monitors in continuous measurement applications. The advantages of selective ion detection of fluoride include their selectivity, large dynamic range, and good sensitivity.

Disadvantages include relatively high maintenance, which includes the need for buffering the solution sample using

---

* Bacon, T., Ion mobility spectroscopy applications for continuous emission monitoring, (1993) http://www.jusun.com.tw/IMS%20Application%20for%20CEM.pdf.

concentrated acetate or citrate for pH adjustment between 5.3 and 5.8. In addition, CDTA or ethylenediaminetetraacetic acid complexing agents are also used to prevent interference from metals. The electrodes are very sensitive to temperature effects and must be carefully controlled. This method is most often used for lab analysis and in applications that demand the selectivity and sensitivity provided.

Silicon Dioxide Sensors

Atmospheric HF causes etching of a thin layer of silicon dioxide. A light source and photo-optic receiver are used to monitor the rate of etching. Both initial and operating costs are low. Little maintenance is required other than periodic replacement of the disposable sensor tip. Because of the highly specific nature of the reaction and the ability to respond quickly to relatively high localized concentrations around flanges, etc., this method is most commonly used to monitor for leaks. The output is essentially qualitative.

Other Methods

The two most common methods of analyzing organic fluorides, such as Freons, are GC and pyrolysis followed by any of the conventional analysis procedures. GC methods may make use of a universal detector and simply rely on retention time to identify and quantitate the sample, or a halogen-specific detector such as a thermionic bead may be used. Pyrolysis kits are available for use with detector tubes.

Two other methods have been demonstrated as having potential for the detection of fluoride: MS and open-path laser-based systems. Commercial MS analyzers are available for lab and process use, and these instruments have the theoretical capability to identify both organic and inorganic fluorides in the gas phase, although their use would be very limited because of the relative complexity of these devices. Laser-based open-path systems have been demonstrated in the prototype stage but are not yet commercially available.

LABORATORY METHODS

The most common laboratory methods are ion-specific electrodes, ion chromatography, titration, and colorimetric methods. Ion-specific electrodes have been discussed earlier. Ion chromatography uses modified high-performance liquid chromatography equipment. Although expensive, this equipment can also determine a large number of other ions at the same time. Titration and colorimetric methods rely upon chemical reactions, which produce a color change. Both of these methods rely upon common lab equipment and procedures. All of these lab methods can be used in the analysis of water samples. Automated and semiautomated equipment exist for most of these methods.

SPECIFICATION FORMS

When specifying fluoride analyzers only, please use the ISA form 20A1001, and when specifying both the analyzer and the composition or the properties of the process material of interest, use ISA form 20A1002. Both forms are reproduced with the permission of the International Society of Automation on the next pages.

1.25 Fluoride Analyzers

1	RESPONSIBLE ORGANIZATION	ANALYSIS DEVICE	6	SPECIFICATION IDENTIFICATIONS	
2			7	Document no	
3	ISA	Operating Parameters	8	Latest revision	Date
4			9	Issue status	
5			10		

11	ADMINISTRATIVE IDENTIFICATIONS			40	SERVICE IDENTIFICATIONS Continued		
12	Project number		Sub project no	41	Return conn matl type		
13	Project			42	Inline hazardous area cl	Div/Zon	Group
14	Enterprise			43	Inline area min ign temp	Temp ident number	
15	Site			44	Remote hazardous area cl	Div/Zon	Group
16	Area	Cell	Unit	45	Remote area min ign temp	Temp ident number	
17				46			
18	SERVICE IDENTIFICATIONS			47			
19	Tag no/Functional ident			48	COMPONENT DESIGN CRITERIA		
20	Related equipment			49	Component type		
21	Service			50	Component style		
22				51	Output signal type		
23	P&ID/Reference dwg			52	Characteristic curve		
24	Process line/nozzle no			53	Compensation style		
25	Process conn pipe spec			54	Type of protection		
26	Process conn nominal size	Rating		55	Criticality code		
27	Process conn termn type	Style		56	Max EMI susceptibility	Ref	
28	Process conn schedule no	Wall thickness		57	Max temperature effect	Ref	
29	Process connection length			58	Max sample time lag		
30	Process line matl type			59	Max response time		
31	Fast loop line number			60	Min required accuracy	Ref	
32	Fast loop pipe spec			61	Avail nom power supply		Number wires
33	Fast loop conn nom size	Rating		62	Calibration method		
34	Fast loop conn termn type	Style		63	Testing/Listing agency		
35	Fast loop schedule no	Wall thickness		64	Test requirements		
36	Fast loop estimated lg			65	Supply loss failure mode		
37	Fast loop material type			66	Signal loss failure mode		
38	Return conn nominal size	Rating		67			
39	Return conn termn type	Style		68			

69	PROCESS VARIABLES	MATERIAL FLOW CONDITIONS			101	PROCESS DESIGN CONDITIONS	
70	Flow Case Identification			Units	102	Minimum	Maximum	Units
71	Process pressure				103			
72	Process temperature				104			
73	Process phase type				105			
74	Process liquid actl flow				106			
75	Process vapor actl flow				107			
76	Process vapor std flow				108			
77	Process liquid density				109			
78	Process vapor density				110			
79	Process liquid viscosity				111			
80	Sample return pressure				112			
81	Sample vent/drain press				113			
82	Sample temperature				114			
83	Sample phase type				115			
84	Fast loop liq actl flow				116			
85	Fast loop vapor actl flow				117			
86	Fast loop vapor std flow				118			
87	Fast loop vapor density				119			
88	Conductivity/Resistivity				120			
89	pH/ORP				121			
90	RH/Dewpoint				122			
91	Turbidity/Opacity				123			
92	Dissolved oxygen				124			
93	Corrosivity				125			
94	Particle size				126			
95					127			
96	CALCULATED VARIABLES				128			
97	Sample lag time				129			
98	Process fluid velocity				130			
99	Wake/natural freq ratio				131			
100					132			

133	MATERIAL PROPERTIES			137	MATERIAL PROPERTIES Continued	
134	Name			138	NFPA health hazard	Flammability	Reactivity
135	Density at ref temp		At	139			
136				140			

Rev	Date	Revision Description	By	Appv1	Appv2	Appv3	REMARKS

Form: 20A1001 Rev 0 © 2004 ISA

RESPONSIBLE ORGANIZATION	ANALYSIS DEVICE COMPOSITION OR PROPERTY Operating Parameters (Continued)	SPECIFICATION IDENTIFICATIONS			
ISA		Document no			
		Latest revision		Date	
		Issue status			

PROCESS COMPOSITION OR PROPERTY			MEASUREMENT DESIGN CONDITIONS				
Component/Property Name	Normal	Units	Minimum	Units	Maximum	Units	Repeatability

Rev	Date	Revision Description	By	Appv1	Appv2	Appv3	REMARKS

Form: 20A1002 Rev 0

Abbreviations

CDTA	Diaminocyclohexane-N,N-tetra-acetic acid
DTPA	Diethylenetriaminopentaacetic acid
EDTA	Ethylenediaminetetraacetic acid
FTIR	Fourier transform infrared
HF	Hydrogen fluoride
IMS	Ion mobility spectrometry
ISE	Ion-specific electrodes
NDIR	Nondispersive infrared
RF	Radio frequency
TOF	Time of flight
UF6	Uranium fexafluoride

Bibliography

ASTM D3267-12, Standard Test Method for Separation and Collection of Particulate and Water-Soluble Gaseous Fluorides in the Atmosphere (Filter and Impinger Method) 2012 http://www.astm.org/Standards/D3267.htm

Bacon, T., Ion Mobility Spectroscopy Applications For Continuous Emission Monitoring (1993) http://www.jusun.com.tw/IMS%20Application%20for%20CEM.pdf

Day, J. Choosing the right fluoride analyzer, WIOA Conference, (2009). http://www.wioa.org.au/conference_papers/09_vic/documents/JohnDay.pdf.

Emerson/Rosemount. Monitoring water quality (2011), http://www2.emersonprocess.com/siteadmincenter/PM%20Rosemount%20Analytical%20Documents/Liq_Brochure_91–6032.pdf.

Fluoridation (2015), http://www.prominentfluid.com.au/index.php?page=fluoridation.

Fresenius, W. Water Analysis, (1988). http://onlinelibrary.wiley.com/doi/10.1002/aheh.19890170211/abstract.

Omega, Fluoride Ion Selective Electrodes (1993) https://www.omega.com/manuals/manualpdf/M0780.pdf.

PMC, Determination of Fluoride in the Bottled Drinking Waters in Iran (2010) http://www.ncbi.nlm.nih.gov/pmc/articles/PMC3869560/ProMinent.

Thermo Scientific, Analysis of Fluoride in Drinking Water in the Presence of Interfering Metal Ions Such as Iron and Aluminum (2015) http://www.dionex.com/en-us/webdocs/81241-PO-IICS-Fluoride-16Sept2009-Lpn_2339-1.pdf.

Thermo Scientific/Orion. Fluoride analyzer (2013) http://www.wateronline.com/doc/thermo-scientific-orion-xp-fluoride-analyzer-0001

1.26 Hazardous and Toxic Gas Monitoring

E. L. SZONNTAGH (1995)   **J. M. JARVIS** (2003)
B. G. LIPTÁK (2017)

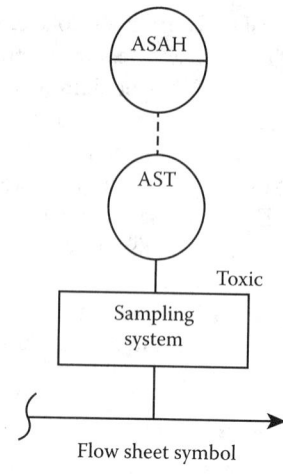

Flow sheet symbol

Applications   Safety-related monitoring of ambient air, enclosed spaces, and processes for hazardous gases and vapors, reduced oxygen; also used for the monitoring of fugitive emission sources in manufacturing or processing typically from stacks, valves, and fume hoods.
Types of Devices: (Most of these sensor designs are also discussed in other chapters as indicated, providing additional information.)
A. Continuous industrial monitors/alarms/transmitters
A1. Electrochemical cells (Chapter 1.21)
A2. Semiconductor sensors (Chapter 1.31)
A3. Catalytic bead sensors (Chapters 1.14, 1.8)
A4. Infrared (Chapters 1.32, 1.13)
A5. Colorimetric Chapter 1.31)
A6. Photoionization detectors (Chapters 1.14, 1.10)
A7. Flame ionization detectors Chapters 1.14, 1.10, 1.28)
A8. Open-path detectors (Chapter 1.62)
B. Personnel protection and dosage sensors
B1. Dosimeter and color-change badges for exposure monitoring
B2. Direct reading dosimeter (indicating) tubes
B3. Portable sample draw monitors
B4. Sorption tubes for thermal desorption
B5. Air sampling pumps
B6. Thermal desorption instruments
B7. Pocket-sized electrochemical monitors
C. Calibration equipment
C1. Calibrators for gas monitors

Costs   A1. $500–$2,200 per point of detection
A2. $200–$1,800 per point of detection
A3. $300–$1,600 per point of detection
A4. $1,400–$3,000 per point of detection
A5. $400–$4,600 per point of detection
A6. $2,800–$6,000 per detection point
A7. $4,000–$10,000 per detection point
A8. $4,000–$6,000 per detection point
B1. $1–$100 per badge
B2. $4–$1,100 per tube; $400 for hand pump; $800 for automated pump
B3. $45–$250 per point for testing service
B4. $35–$120 per tube
B5. $500–$2,000
B6. $1,700–$20,000
B7. $200–$1,800 for single gas monitors; $400–$5,000 for multiple gas monitors
C1. $200 for simple cylinder with regulator to several thousand dollars for gas generation equipment

INTRODUCTION

Normal atmospheric oxygen content is slightly in excess of 20%. If the oxygen level is reduced below 10% in an atmosphere, as a result of either oxygen displacement or removal, simple asphyxiation can occur, causing death. This simple asphyxiation can be caused by any gas, including even otherwise nontoxic ones. Most commonly, this hazard exists in enclosed spaces where there is an opportunity of a gas leak or buildup that dilutes or displaces the oxygen in an area. If a person enters the area, unconsciousness and death can occur within minutes.

Occupational Safety and Health Administration (OSHA) regulations require that oxygen concentration in air not fall below 19.5% during an 8 hr workday.

Hazardous gases are used in and produced by many industrial processes. Hazards include toxicity, carcinogenicity, flammability, and asphyxiation as primary concerns. To prevent worker exposures to these hazards, many electrical detection methods are employed. The electronic monitors are most often connected to alarms that alert workers to evacuate the area. There also may also be additional actions performed by the alarm such as logging exposure, summoning help, shutting down equipment, or providing additional ventilation.

In addition to electronic monitoring and control, other devices such as dosimeters and portable monitors are employed for personal protection. Other monitoring is done for industrial hygiene purposes such as quantifying exposures, determining fugitive emissions, and investigations (see Table 1.26a).

TABLE 1.26a
Partial List of Suppliers

Suppliers	Web Addresses	Electrochemical cells A1	Semiconductor sensors A2	Catalytic bead sensors A3	Infrared A4	Colorimetric A5	Photo ionization detectors A6	Flame ionization detectors A7	Open path A8	Exposure monitoring badges B1	Direct reading dosimeter tubes B2	Portable sample draw gas B3	Sorption tubes for thermal B4	Air sampling pumps B5	Thermal desorption instruments B6	Pocket-sized electrochemical B7	Calibration equipment C1
3M	www.3m.com									x		x				x	
Advanced Calibration Designs Inc.	http://goacd.com																x
Airwave Electronics Ltd.	www.gasdetect.com	x		x													
Analytical Systems International/Keco	www.liquidgasanalyzers.com			x	x	x		x									
Analytical Technology, Inc.	www.analyticaltechnology.com	x			x				x			x					
Avir Sensors	www.avirsensors.com				x												
Bacharach, Inc	www.bacharach-inc.com	x	x	x	x												
Bionics Instrument Co.	www.bionics-instruments.com	x		x	x							x					
Boreal Laser	www.boreal-laser.com				1				x								
BW Technologies	www.gasmonitors.com				x				x			x				x	x
CDS Analytical Inc.	www.cdsanalytical.com	x	x	x									x		x		
CEA Instruments, Inc.	www.ceainstr.com	x	x	x	x			x				x		x		x	
City Technology Ltd.	www.citytech.com	x	x	x												x	
Control Instruments Corporation	www.controlinstruments.com	x	x	x	x				x			x					
Crowcon Detection Instruments	www.crowcon.com	x	x	x												x	
Delphian Corporation	http://www.delphian.com	x	x	x	x							x					x
Detector Electronics Corporation	www.detronics.com	x			x												
DOD Technologies	www.dodtec.com	x	x	x	x	x	x				x	x		x		x	x
Draeger Safety Inc.	www.draeger.com	x	x	x	x	x	x			x	x	x				x	
Enmet	http://www.enmet.com	x	x	x	x							x				x	
Gas Measurement Instruments Ltd.	www.gmiuk.com	x	x	x	x		x		x			x				x	
General Monitors	www.generalmonitors.com																x
GfG Instrumentation Inc	www.gfg-inc.com	x												x		x	x
Honeywell Analytics.	www.honeywellanalytics.com	x	x	x	x	x	x					x		x		x	
IN USA, Inc.	www.inusacorp.com						x					x					
Industrial Scientific	www.indsci.com	x	x	x	x												x
International Sensor Technology	www.intlsensor.com	x	x	x	x		x					x				x	x
Interscan Corporation	www.gasdetection.com	x										x				x	x

(Continued)

TABLE 1.26a (Continued)
Partial List of Suppliers

Suppliers	Web Addresses	A1 Electrochemical cells	A2 Semiconductor sensors	A3 Catalytic bead sensors	A4 Infrared	A5 Colorimetric	A6 Photo ionization detectors	A7 Flame ionization detectors	A8 Open path	B1 Exposure monitoring badges	B2 Direct reading dosimeter tubes	B3 Portable sample draw gas	B4 Sorption tubes for thermal	B5 Air sampling pumps	B6 Thermal desorption instruments	B7 Pocket-sized electrochemical	C1 Calibration equipment
Ion Science Ltd.	http://www.ionscience.com						×									×	×
KEM Medical Products	www.kemmed.com									×							×
Kin-Tek	http://www.kin-tek.com																×
KWJ Engineering Inc.	http://www.kwjengineering.com/	×	×	×	×												
Mil-Ram Technology, Inc.	www.mil-ram.com	×	×	×	×		×		×			×					
MSA	http://www.msasafety.com/global/	×	×	×	×				×			×				×	
Neodym Technologies	www.neodymsystems.com	×	×	×													
Net Safety Monitoring Inc.	www.net-safety.com	×	×	×	×		×										
Pem-Tech	www.pem-tech.com	×	×	×								×					×
PureAire Monitoring Systems, Inc.	www.pureairemonitoring.com	×			×							×					×
RAE Systems Gas Detection	www.raesystems.com	×	×	×	×		×						×			×	×
RKI Instruments	www.rkiinstruments.com	×		×	×						×	×				×	
Scott Safety	www.scottsafety.com	×			×				×								
Senscient, Inc.	http://www.senscient.com								×								
Sensidyne, LP	www.sensidyne.com/	×									×			×		×	×
Servomex	www.servomex.com	×															
Sierra Monitor	www.sierramonitor.com			×			×					×		×			
SKC Inc.	www.skcinc.com									×	×	×	×			×	
Spectrex	http://spectrex-inc.com								×								
Synkera Technologies, Inc.	http://www.synkera.com/		×													×	×
Thermo Fisher Scientific Inc.	www.thermoscientific.com						×					×					
TQ Environmental	www.tqplc.com	×		×	×												
VICI Metronics	www.vici.com																×

1.26 Hazardous and Toxic Gas Monitoring

HAZARDOUS AREA CLASSIFICATION

OSHA classifies the hazardous areas according to Table 1.26b.

The International Standard IEC 60079-10 separates the hazardous areas into zones based upon the frequency of the occurrence and duration of an explosive gas atmosphere. The definitions of these zones are listed as follows:

Zone 0   An area in which an explosive gas atmosphere is present continuously or for long periods. (Rough Guide: More than 1000 hr/year).

Zone 1   An area in which an explosive gas atmosphere is likely to occur in normal operation. (Rough Guide: 10 hr or more/year but less than 1000 hr/year).

Zone 2   An area in which an explosive gas atmosphere is not likely to occur in normal operation and, if it does occur, is likely to do so only infrequently and will exist for a short period only. (Rough Guide: Less than 10 hr/year).

Zone 20   An area in which combustible dust, as a cloud, is present continuously or frequently, during normal operation, in sufficient quantity to be capable of producing an explosive concentration of combustible dust in a mixture with air.

Zone 21   An area in which combustible dust, as a cloud, is occasionally present during normal operation, in a sufficient quantity to be capable of producing an explosive concentration of combustible dust in a mixture with air.

Zone 22   An area in which combustible dust, as a cloud, may occur infrequently and persist for only a short period or in which accumulations of layers of combustible dust may give rise to an explosive concentration of combustible dust in a mixture with air.

TABLE 1.26b
Classification of Hazardous Areas According to OSHA

		Summary of Class I, II, III Hazardous Locations	
		Divisions	
Classes	Groups	1	2
I Gases, vapors, and liquids (Art. 501)	A: Acetylene B: Hydrogen, etc. C: Ether, etc. D: Hydrocarbons, fuels, solvents, etc.	Normally explosive and hazardous	Not normally present in an explosive concentration (but may accidentally exist).
II Dusts (Art. 502)	E: Metal dusts (conductive and explosive) F: Carbon dusts (some are conductive, and all are explosive) G: Flour, starch, grain, combustible plastic, or chemical dust (explosive)	Ignitable quantities of dust normally are or may be in suspension, or conductive dust may be present	Dust not normally suspended in an ignitable concentration (but may accidentally exist). Dust layers are present.
III Fibers and flyings (Art. 503)	Textiles, woodworking, etc. (easily ignitable, but not likely to be explosive)	Handled or used in manufacturing	Stored or handled in storage (exclusive of manufacturing).

Safety Instrument Performance Standards

The purpose of toxic gas detection is for personnel protection to warn against accidental exposure. Therefore, it is important that the instrumentation be reliable and perform accurately in the application. In the last decade, standards have been developed to ensure that detection instrumentation performs to a minimum set of specifications that will ensure adequate protection. These standards apply beyond those that might otherwise be applicable for hazardous area deployment.

International Society of Automation (ISA) has developed a set of standards that are applicable to electrical instruments for the determination of common toxic gases in the context of personnel protection. Among other gases, ANSI/ISA standards exist for hydrogen sulfide, carbon monoxide, ammonia, oxygen-deficient/enriched atmospheres and chlorine (see Bibliography).

In the European Union standard EN 45544-1 (see Bibliography) defines the set of minimum requirements for electrical apparatus used for direct detection and measurement of toxic gases and vapors. This standard, unlike the ISA standards, is generic for all toxic gas and vapor species. The standard is divided into four parts. Part 1 describes general requirements. Part 2 describes requirements for instruments that respond in the region of limit values. Part 3 describes requirements for apparatus designed for measuring concentrations well above limit values. Part 4 is a guide for selection, installation, use, and maintenance.

Factory Mutual Research Corporation (FMRC), similar to the EU standards, provides a generic set of performance requirements irrespective of the toxic gas species.

Levels of Toxicity

The definitions of terms related to flammable, explosive, and toxic gases are listed at the end of this chapter.

The mode of toxicity for some gases results from having a greater affinity for binding with blood hemoglobin than with oxygen. In this case, *chemical asphyxiation* occurs because of the displacement of the oxygen from the blood. Most notable in this category of toxic gases are carbon monoxide (CO) and hydrogen cyanide (HCN). In the case of CO, it is odorless and nonirritant. If the exposure is chronic, death can result even if CO is present in relatively low concentrations.

The modes of toxicity for other toxic gases are varied. The toxin can be dissolved in body fluids and can be transported within the body to injure biological tissues and organs. The toxicity of a gas or vapor depends on its specific toxicity; on the efficiency and effectiveness at which it is absorbed, transported, and metabolized; and on exposure time.

To describe the specific toxicity of each gas, threshold limit values (TLVs) have been established by the American Conference of Governmental Industrial Hygienists (ACGIH)* and permissible exposure levels (PELs) have been established by OSHA.† These values are generally measured as 8 hr time-weighted averages (TWAs), and the concentrations are expressed in parts per million (ppm).

It is customary to rate a substance with 100–1000 ppm exposure limits as moderately toxic, 10–100 ppm as highly toxic, and less than 10 ppm exposure limits as extremely toxic. The short-term exposure limit (STEL), which is generally 15 min, and the ceiling threshold limit value (TLV-C) of the concentration during any part of the working exposure have been established to limit exposures during short time periods. More details on the specific health effects of toxic substances can be found in the OSHA, EPA, etc., documents listed in the Bibliography.

Flammable and explosive gases are categorized as to their flammability or explosive hazard. Most safety systems for the detection of flammable or explosive gas are designed with ranges from 0% to 100% of the lower explosive limit (LEL).

Exposure Limits

Of greatest concern are extremely toxic gases and vapors. By convention, these are substances with TLV, PEL, STEL, or TLV-C values of 10 ppm or lower. A partial listing of gases and vapors with TLVs below 10 ppm is given in Table 1.26b.

In an ideal situation, there would never be any exposure to a toxic, carcinogenic, or oxygen-deficient atmosphere. However, in reality, there are potential trace amounts of these gases in the workplace. In acknowledgment of this guideline, safe exposure limits for each gas have been established. There are various organizations that have established limits. These guidelines are used by many to determine what alarm set points should be. OSHA and NIOSH substantially agree on the TWA values. Table 1.26c lists some of the common gases and vapors that are monitored for safety.

A couple of toxic gases deserve special mention because of their uniqueness. Mercury naturally occurs as the elemental metal. It is the only metal that is a liquid at room temperature and hence has significant vapor pressure. Mercury is widely used in industry as a catalyst and as an integral component in fluorescent lamps. Improper disposal of fluorescent lamps and inadvertent incineration provide an important pathway for mercury to enter the environment.

---

* ACGIH, Publications, (2015) https://www.acgih.org/forms/store/CommercePlusFormPublic/search?action=Feature.
† OSHA, Toxic and Hazardous Substances, Standards 1910-1000 (2006) https://www.osha.gov/pls/oshaweb/owadisp.show_document?p_table=STANDARDS&p_id=9991.

TABLE 1.26c
Recommend Exposure Limits of Some Common Gases and Vapors

	OSHA[a]	NIOSH[b] RELS[c]				
	PEL[d]	TWA[e]	IDLH[f]	STEL[g]	CEILING[h]	CA[i]
Acetic acid (vinegar)	10	10	50	15		
Acrylonitrile		1	85		10	85
Ammonia	50	25	300	35		
Aniline	5		100			100
Arsine	0.05		3			3
Benzene	1	0.1		1		500
Boron trifluoride	1	1	25		1	
Bromine	0.1	0.1	3	0.3		
Bromine pentafluoride		0.1	N.D.			
Bromoform	0.5	0.5	850			
1,3-Butadiene	1			5		2000
Carbon disulfide	20	1	500	10	30	
Carbon tetrabromide		0.1	N.D.	0.3		
Carbon tetrachloride	10		200	2		200
Carbonyl fluoride		2	N.D.	5		
Chlorine	1	0.5	10		0.5	
Chlorine dioxide	0.1	0.1	5	0.3		
Chlorine trifluoride	0.1	0.1	20		0.1	
Chloroform	50		500	2		500
Diborane	0.1	0.1	15			
1,1-Dimethyl hydrazine	0.5	0.06	15			15
Dimethyl sulfate	1	0.1	7			7
Diphenyl	0.2	0.2	15			15
Ethylene oxide	1	0.1	800		5	800
Fluorine	0.1	0.1	25			
Formaldehyde	0.75	0.016	20		0.1	20
Germane		0.2	N.D.			
Hydrazine	1	0.03	50		0.03	50
Hydrogen bromide	3	3	30		3	
Hydrogen chloride	5	5	50		5	
Hydrogen cyanide	10	4.7	50	4.7		
Hydrogen fluoride	3	3	30		6	
Hydrogen selenide	0.05	0.05	1			
Hydrogen sulfide	20	10	100		10	
Methyl isocyanate	0.02	0.02	3			
Methyl mercaptan	10	0.5	150			
Naphthalene	10	10	250	15		
Nickel carbonyl	0.001	0.001	2			2
Nitric acid	2	2	25	4		
Nitrobenzene	1	1	200			

[a] Occupational Safety and Health Administration.
[b] National Institute for Occupational Safety and Health.
[c] Recommended exposure limit of NIOSH.
[d] PEL: Permissible exposure limit, a legally enforceable occupational exposure limit established by OSHA.
[e] Time-weighted average.
[f] Immediately dangerous to life of health.
[g] Short-term exposure limit.
[h] Threshold limit value ceiling TLV.
[i] Substance that NIOSH considers to be a potential occupational carcinogen.

(Continued)

TABLE 1.26c (Continued)
Recommend Exposure Limits of Some Common Gases and Vapors

	OSHA[a]	NIOSH[b] RELS[c]				
	PEL[d]	TWA[e]	IDLH[f]	STEL[g]	CEILING[h]	CA[i]
Nitrogen dioxide	5	1	20			
Nitrogen trifluoride	10	10	1000			
2-Nitropropane	25		100			100
Oxygen difluoride	0.05	0.05	0.5		0.05	
Ozone	0.1	0.1	5		0.1	
Pentaborane	0.005	0.005	1	0.015		
Phenylhydrazine	5	0.14	15		0.14	15
Phosgene	0.1	0.1	2	0.2		
Phosphine	0.3	0.3	50	1		
Phosphorus oxychloride		0.1	N.D.	0.5		
Phosphorus trichloride	0.5	0.2	25	0.5		
Silane		5	N.D.			
Sodium azide		0.1	N.D.	0.3		
Stibine	0.1	0.1	5			
Sulfur dioxide	5	2	100	5		
Sulfur tetrafluoride		0.1	N.D.		0.1	
Tellurium hexafluoride	0.02	0.02	1			
Thionyl chloride		1	N.D.		1	
Vinyl chloride	1		N.D.			

Note: N.D.—Not Defined.

TABLE 1.26d
Recommended Exposure Limits for Mercury

	Mercury, Alkyl Compounds	Mercury, Aryl Compounds	Mercury, Inorganic Compounds
OSHA			
8 hr TWA	0.01 mg/m^3	0.1 mg/m^3	0.1 mg/m^3
Ceiling	0.04 mg/m^3	—	—
NIOSH			
8 hr TWA	0.01 mg/m^3, skin	0.05 mg/m^3, skin	0.05 mg/m^3, Skin
ST/ceiling	0.03 mg/m^3, (ST) skin	0.1 mg/m^3, (seiling) skin	0.1 mg/m^3, (Ceiling) Skin
IDLH	2 mg/m^3	10 mg/m^{3}s	10 mg/m^3
ACGIH			
8 hr TWA	0.01 mg/m^3, skin	0.1 mg/m^3, skin	0.025 mg/m^3, Skin
Short term	0.03 mg/m^3, skin	—	—

Mercury is a neurotoxin. The "mad hatters" of the nineteenth century suffered from mercury poisoning as a result of the mercury felting process that was prevalent at the time. In the United States, the disease became known as the "Danbury shakes" because of the large number of hatmakers in Danbury, Connecticut, who were afflicted with the disease. OSHA has established the PEL at 0.1 mg/m^3. ACGIH has established the recommended airborne exposure limit of mercury vapor at 0.05 mg/m^3 averaged over an 8 hr work shift.

Radon gas is a radioactive noble gas that occurs in nature through the radioactive decay of radium and uranium in soils. While it is chemically unreactive and hence nontoxic in that sense, it is radioactive and thought to be the leading cause of lung cancer after cigarette smoking (Table 1.26d).

Most emphasis on radon detection has been in the home and public building settings.* The generally accepted maximum permissible level of radon in these facilities is 4 pCi/l.

---
* EPA, Radon Measurement in Schools, (1993) http://www2.epa.gov/sites/production/files/2014-08/documents/radon_measurement_in_schools.pdf

DETECTOR TYPES

In the next paragraphs, a number of hazardous gas detector designs will be described, starting with the electrochemical ones that are one of the most common. These detectors are usually packaged in explosion-proof or intrinsically safe designs, are provided with both alarm contacts and analog or digital transmitter features, provide local displays, and are often self-calibrating. The appearance of a typical design is shown in Figure 1.26a, and its measurement capabilities are summed up in Table 1.26e.

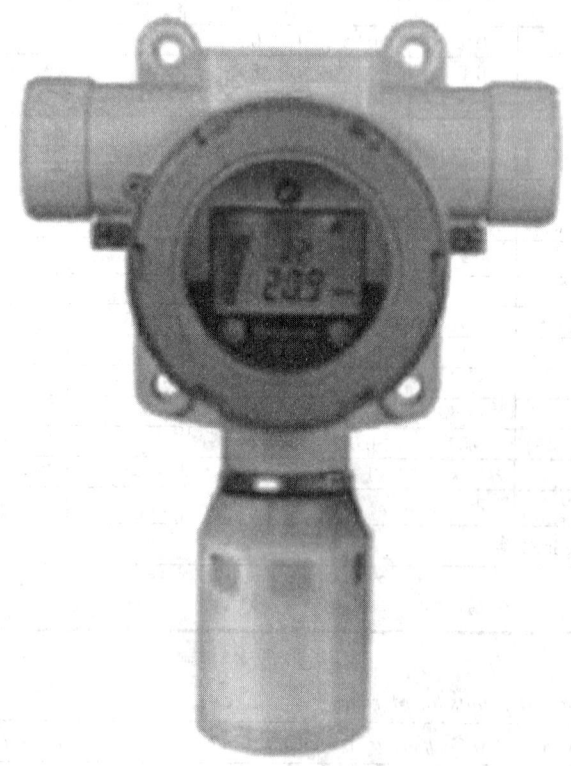

FIG. 1.26a
The packaging of a hazardous gas detector in explosion-proof housing. (Courtesy of Vulcain/Honeywell.)

TABLE 1.26e
Features of Oxygen, Toxic, and Combustible Gas Sensor Types

Sensor Family		Gas	User-Selectable Full-Scale Range	Default Range	Range Increments	Selectable Cal Gas Range	Default Cal Point
	mV	**Catalytic Bead Sensors**					
		Flammable 1–8•	20%–100% LEL	100% LEL	10% LEL		50% LEL
		Infrared Sensors					
		Methane	20%–100% LEL	100% LEL	10% LEL		50% LEL
		Propane	20%–100% LEL	100% LEL	10% LEL	30%–70% of selected full-scale range	50% LEL
		Carbon Dioxide	2% Vol.	2% Vol.	0.1% Vol.		1% Vol.
	EC	**Electrochemical Sensors**					
		Hydrogen Sulfide	10–100 ppm	50.0 ppm	0.1 ppm		10 ppm
		Carbon Monoxide	100–1000 ppm	300 ppm	100 ppm		100 ppm
		Hydrogen	1000 ppm only	1000 ppm	n/a		500 ppm
		Nitrogen Dioxide	10–50 ppm	10 ppm	5 ppm		5 ppm
	O_2	Oxygen	25% Vol. only	25% Vol.	n/a	20.9% Vol. (Fixed)	20.9% Vol.

Source: Courtesy of Vulcain/Honeywell.

Electrochemical Sensors

For more details of electrochemical detectors, see Chapter 1.21.

Electrochemical sensors have progressed much in the last 30 years. They have reduced the instability caused by excessive noise and drift, improved sensitivity and specificity, increased the working life span of these instruments, and simplified their calibration procedures. The latest electrochemical monitors are usually microprocessor controlled, stable, and easy to calibrate.

One of the most common types of detection is the electrochemical detector. This sensor has some form of electrolyte that will react with a target gas to generate a low-level electrical current. The amount of current generated is directly proportional to the amount of target gas sensed. A detector or transmitter evaluates the electrical signal, which is often in the nanoampere range and converts it into a usable form for connection to other equipment. Sensors have been developed that can detect a wide range of gas used or produced in industry. Most manufacturers provide both the sensor and the detector to match it. The output from the detector is then integrated into common alarm and monitoring systems. The chemistry, types of electrolyte, and alarm functions vary widely, but all electrochemical sensors have common attributes.

Electrolyte may be either a liquid, a gel, or a solid. The advantage of liquid electrolyte systems is that they can be recharged in the field by either adding or changing the electrolyte. The disadvantages include a propensity to dry out from high temperature or low humidity mandating more frequent service. Also liquid sensors must be handled and mounted so that the liquid does not leak. Gelled and solid electrolyte has a thicker consistency and can overcome the handling and mounting concerns of liquid electrolyte. It is also affected by temperature and humidity, but to a lesser degree.

Electrochemical sensors last from 1 to 2 years in most applications. Life is affected by exposure to gas, low humidity, high temperatures, and high air flows. These factors affect the electrolyte and hence the life of the sensor.

Some manufacturers have developed an automatic method to detect the useful life of the sensor and report it to the user. These sensors are typically used in high-tech and commercial applications where environmental conditions are favorable and the gases detected are toxic, oxygen, or hydrogen.

Oxygen Detectors

Oxygen measurement in gases (discussed in detail in Chapter 1.49) often use the amperometric detection method. Such a device is a metal/air battery in which the amount of current delivered is a function of how fast oxygen can diffuse to the cathode through a mass-flow-limiting capillary tube. A schematic view of a typical oxygen sensor is shown on the left of Figure 1.26b.

In these sensors, the current produced, and hence, the response to oxygen, is determined by the diffusion rate through the capillary. These sensors respond to the actual amount of oxygen available at the cathode and not the oxygen partial pressure. As a result, these sensors are sensitive to atmospheric pressure changes and can be adversely affected by pressure transients that occur as a matter of course in a sample system or when deployed in portable sensors.

Signal processing is employed in the sensor to minimize the effects of pressure transients. This same sensor can be used under a wider range of temperature variations, because the diffusion rate through the capillary is only weakly temperature dependent.[17]

Oxygen sensors can also use a Teflon-like membrane, similar to the technology used in dissolved oxygen electrodes for water analysis (Chapter 1.50). Sensors using this technology experience less output variation with pressure transients but are 10 times more sensitive to temperature changes. On the other hand, because diffusion through a polymer membrane is slower than through a capillary, the response time (t_{90}) of partial pressure sensors is typically 20–40 sec, whereas it is between 5 and 10 sec for mass-flow-controlled sensors.

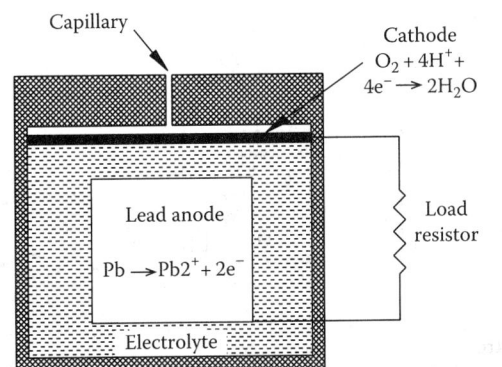 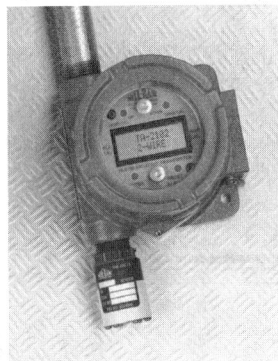

FIG. 1.26b

Amperometric oxygen detector operating principle (left) and on the right, a self-calibrating unit in explosion proof housing. (Courtesy of Mil-Ram.)

Amperometric Sensors

Sensors for toxic vapors differ structurally from units that detect oxygen, but they still are electrochemical cells that operate in the amperometric mode. These cells generate a current that is linearly proportional to the concentration of the gas present in the atmosphere.

The simplest form of an electrochemical cell is a two-electrode design, which consists of a semipermeable diffusion membrane, a reservoir of acid electrolyte, a sensing electrode, and a counter electrode. The toxic gas for which the cell was designed diffuses through the membrane and reacts at the working electrode, typically producing a number of H^+ ions as well as electrons. The H^+ ions travel through the electrolyte to the counter electrode, while the electrons travel through the external circuit and combine with the ions at the counter electrode. The current produced and measured in the external circuit is directly proportional to the concentration of the toxic gas present.

In many cases, the electrolyte is an aqueous acid. In some instances, it may be an organic gel or in a solid state. Formulations have been improved to reduce hygroscopic effects in low- and high-humidity environments that can have a negative effect on the working life of the electrolyte. Lifetimes in the latest models frequently exceed 2 years.

Electrodes and Electrolytes

The electrochemical sensor has electrodes that carry current from the electrolyte for connection to the detector. The electrodes are made from various metals but are often noble metals. Sensors have a minimum of two electrodes, but some models provide three or four. Two electrodes provide the electrical connection for sense and reference functions. The third is used as a reference or comparator. It is exposed to the same environmental conditions of the sense electrode but is not exposed to the gas sample. This permits the transmitter to compare the activity at the sense with the third. If both the reference and sense electrodes are generating current, the likely cause is environmental like a temperature, pressure, or humidity variation, and the detector can ignore the activity. If the sense electrode detects but not the reference, then the likely cause is a gas sense and is handled such. A four-electrode sensor provides for a second sense electrode that can be used in logic AND manner to provide additional alarm verification.

The membrane is a gas-permeable material that permits gas to enter the cell. The membrane also prevents the electrolyte from leaking from the cell. This is accomplished by designing the membrane pore size that is too small for the electrolyte to escape.

Two-electrode sensors are inexpensive and perform adequately on less-demanding applications. For industrial applications, the three-electrode designs are dominant. The third electrode permits the potential between the counter and the working electrodes to vary. This provides optimal catalytic efficiency with the aid of external circuitry and results in a greatly improved cell sensitivity. Additionally, the linearity of response and dynamic range of the cell are improved.

A diagram outlining the operation of both the two- and three-electrode sensor designs is shown in Figure 1.26c. A detailed discussion of their operation is provided in Chapter 1.21.

Toxic sensors respond to the partial pressure of the gas of interest, so the measured concentration will increase linearly with atmospheric pressure.

Toxic sensors also generally require a small amount of oxygen dissolved in the electrolyte for reactions at the counter electrode. As a consequence, the sensors should not be operated for long periods under anaerobic conditions. For short periods (10 min or less), the cell will operate without oxygen, because some dissolved oxygen will be present in the electrolyte. This situation is frequently exploited for sensor calibration, because compressed calibration gas with nitrogen can be used without fear of oxidation or reduction of shelf life.

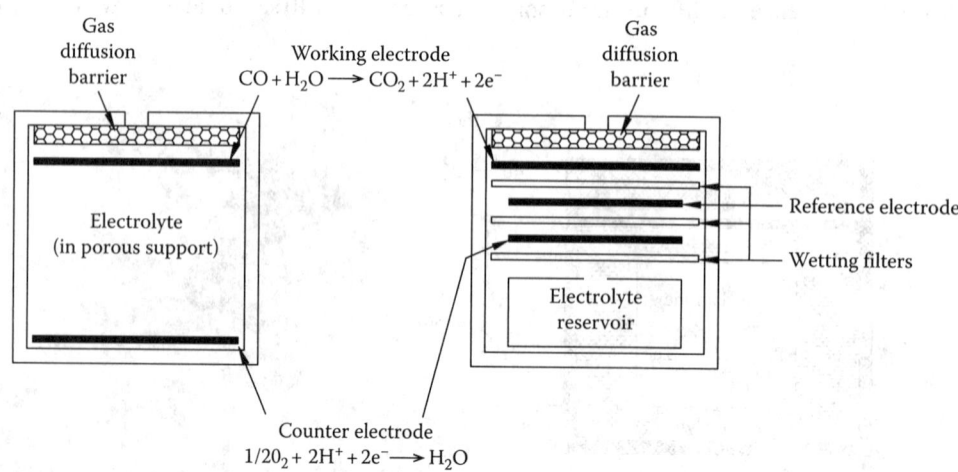

FIG. 1.26c

Diagram of the two- and three-electrode sensors.

All electrochemical sensors rely on catalytic reactions that are either accelerated or slowed as the operating temperature rises. This results in both response time and sensitivity changes. Generally, the sensitivity change with temperature variation is very repeatable and can be compensated with software or circuitry. At low temperatures, the speed of response can become intolerably slow. When the temperature becomes too low (around −35°C), the electrolyte can freeze, causing permanent damage to the cell.

The lifetime of the sensor is strongly dependent on relative humidity (RH) of the environment. The classic electrolyte in toxic sensors is generally 5 molar aqueous sulfuric acid. This concentration of acid generates 65% RH. For atmospheric RH above or below this level, the sensor will either absorb water or lose water to evaporation. Sensors can tolerate humidity variations between 15% and 90% RH. When the RH in the environment is over 90%, the cell will absorb moisture, and the electrolyte reservoir may eventually overflow, causing the cell to burst. Below 15% RH, the sensor will continue to lose moisture until there is not enough electrolyte for efficient ion transport, and the cell will fail for that reason.

Fixed Detectors

Electrochemical sensors are used in many permanent installations of toxic gas detectors. These sensors are designed for industrial use and frequently carry hazardous area approvals. Electrochemical sensors are also available with explosion-proof housings that are suitable for deployment in Class 1, Division 1, and CENELEC Zone 1 hazardous areas. Figure 1.26d shows a typical industrial toxic gas detector with an explosion-proof housing.

Because of the low-power requirements of the sensing element, these detectors are frequently designed to be intrinsically safe. Such designs operate on such low power as to be incapable of initiating an explosion. Intrinsically safe detectors tend to be less costly than the ones requiring explosion-proof housing, because they can utilize molded plastic housings rather than the bulky and expensive explosion-proof housings.

Most fixed detectors transmit their readings to the associated control and alarm system using 4–20 mA current loop. Usually, detectors that are designed for intrinsically safe operation are powered from a 4 to 20 mA current loop, which reduces the number of wires that must be installed. In the past decade, the low-cost, low-power, high-reliability, spread-spectrum radio telemetry systems found increasing acceptance for communications with central alarm systems.

Alternatively, if a central alarm system is not available, several detector models are available that provide self-contained alarming devices or provide strobe lights, sounders, and relays in addition to the 4–20 mA transmission signal loop. Figure 1.26e illustrates a toxic gas detector with built-in alarms.

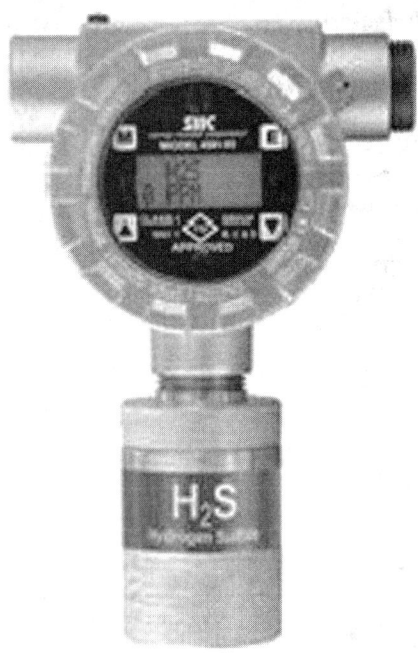

FIG. 1.26d
Physical appearance of a toxic gas detector module with display, alarm indicators, and 4–20 mA outputs in an explosion-proof housing. (Courtesy of Sierra Monitor Products.)

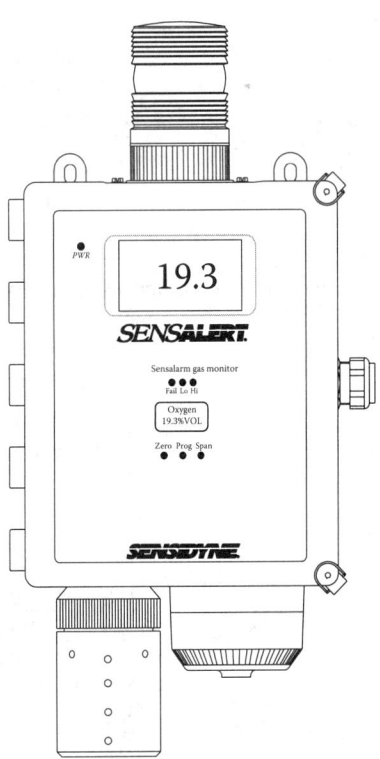

FIG. 1.26e
Illustration of an electrochemical sensor module with display, relay, sounder, and strobe to indicate alarms. (Courtesy of Sensidyne Inc.)

Portable Detectors

Electrochemical (and semiconductor) sensors are also available in pocket-sized, battery-powered personal monitors and in portable analyzers (Figure 1.26f). Pocket-sized, multichannel personal monitors are inexpensive and fast (t_{90} in around 20 sec). In addition to indicating toxic gas concentrations, they are provided with alarms and outputs set to actuate when OSHA limits are reached.

Some personal monitors can simultaneously detect several toxic gases. A typical sensor combination might consist of detectors for combustibles (typically, a low-power pellistor bead), for low oxygen, and for high CO and H_2S concentrations (electrochemical cells), for a total of four measurement channels. The sensors have a useful life of as long as 2 years and are inexpensive to replace. A partial list of available sensor types, their alarm set points, and inaccuracies is provided in Table 1.26f.

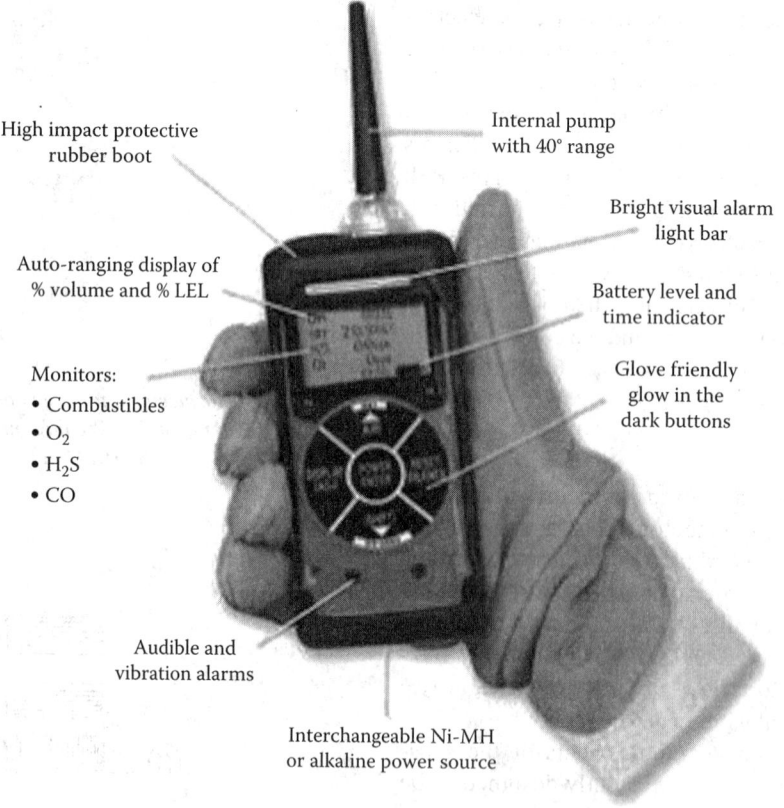

FIG. 1.26f
Typical pocket-sized portable electrochemical detector configured for four individual channels of detection. (Courtesy of RKI Instruments.)

TABLE 1.26f
Typical Range, Sensitivity, and Alarm Set Points of Portable Personal Protection Monitors[a]

Gas	Range	Resolution	Alarm Set Points (Low/High)
O_2	0%–30%	0.1%	19.5/23.5%
CO	0–500 ppm	1 ppm	35/200 ppm
H_2S	0–100 ppm	1 ppm	10/20 ppm
SO_2	0–20 ppm	1 ppm	2/10 ppm
NO	0–250 ppm	1 ppm	25/50 ppm
NO_2	0–20 ppm	0.1 ppm	1/10 ppm
NH_3	0–50 ppm	1 ppm	25/50 ppm
PH_3	0–5 ppm	0.1 ppm	1/2 ppm
Cl_2	0–10 ppm	0.1 ppm	0.5/5 ppm
HCN	0–100 ppm	1 ppm	4.7/50 ppm

[a] Abstracted from Cole-Parmer Catalog 2001/02, 2001.

Advantages and Limitations

Electrochemical cells have the advantage of simplicity, low cost, and reliability as compared to other technologies. Technologies such as flame ionization, mass spectrometry, fluorescence, ultraviolet (UV), and infrared (IR) analyzers provide better sensitivity and/or specificity than do electrochemical sensors.

It is rare for an electrochemical sensor to have sensitivity lower than 1 ppm. Additionally, an electrochemical sensor may have significant cross sensitivity to other substances. In most instances, environmental factors limit the set of substances that are likely to present interference for any given sensor installation, so the sensor provides utility in spite of cross sensitivities.

On the other hand, electrochemical sensors tend to be more stable, selective, and sensitive than sensors built on metal oxide semiconductor (MOS) technology. Because most electrochemical sensors use aqueous-based chemistries, MOS sensors have found utility in environments of either extremely high or low humidity. In these environments, electrochemical sensors tend to either dry out quickly or absorb moisture, resulting in a shortened operating life.

When the sensitivity or specificity of electrochemical sensors is inadequate, or the number of toxic vapors is high, other methodologies must be used.

Spectrometers

The preference is strong for toxic gas monitors that use solid-state technologies and have no moving mechanical parts. As a result, field-deployed instruments with warranties of up to 5 years are common.

Instruments based on nondispersive IR absorption, IR photoacoustic detection, pulsed UV fluorescence, and chemiluminescence measurements are all available. Fourier-transform IR (FTIR) and tunable diode laser spectrometers have also been utilized for high-sensitivity and high-specificity applications.

In the past decades, mass spectrometry has made tremendous strides (Chapter 1.35). Once thought of as being a maintenance-intensive technology, its reliability has been greatly improved. Advancements in miniaturization have also reduced their size such that some specially designed devices are field portable and smaller than a suitcase.

IR detectors detect based upon the principle that most gases absorb IR light. The absorption differs by gas and is relative to the concentration. These properties permit IR to be used to identify and quantify a target gas. The detectors have an IR-generating source, a gas cell, and a receiver. The source is chosen to generate a frequency of light that will be absorbed by the target gas. The light is passed through the gas sample in the cell, and the receiver interprets how much light is absorbed by the gas. Gas in the cell can be quantified by applying Beer–Lambert law, which defines a linear relationship between absorbance and concentration.

This type of IR detector is subject to the detection of gases similar in chemical makeup to the desired target gas. This is because absorbance is created by molecular bonds as they move. Then, compounds with similar bonds may be detected by this type of detector. To achieve specificity, a special type of IR detector, an FTIR, looks at a wide band of absorbance and develops spectra showing a complex representation of the absorbance of gas. This spectrum can then be used to precisely define the gas (s) in the sample. This technology is described in more detail elsewhere in this manual.

Applications

IR detectors are versatile and are available for many applications. Typical applications include combustible monitoring as well as toxic monitoring. Diatomic gases like hydrogen do not absorb IR well and therefore cannot be detected with IR sensors.

Several mass spectrometer packages have been designed for ambient air monitoring. Instruments are generally available with 32- or 64-channel multistream sampling capability, which lowers the unit cost per detection point. Published detection limits for some organic vapors are listed in Table 1.26g.

TABLE 1.26g
Mass Spectrometer Detection Limits with Typical Exposure Limits[a]

Compound	Detection Limit (ppm)	Typical Exposure Limit
Acrylonitrile	≤0.1	2 ppm OEL
Benzene	≤0.01	1 ppm OEL
Butadiene	≤0.1	10 ppm OEL
Carbon disulfide	≤0.01	10 ppm OEL
Chlorobenzene	≤0.01	50 ppm OEL
Epichlorohydrin	≤0.05	5 ppm OEL
Methyl iodide	≤0.01	5 ppm OEL
Trichloroethylene	≤0.05	100 ppm
Vinyl bromide	≤0.1	5 ppm
Vinyl chloride	≤0.1	5 ppm MEL

[a] Courtesy of Thermo ONIX.

Semiconductor Detectors

A semiconductor sensor is manufactured with a process that is similar to that used to manufacture integrated circuits. A metal oxide film is deposited over a silicon layer. This is bonded to a ceramic layer that is held at a constant temperature by an internal heater. Heat oxidizes the gas and changes the resistance of the oxide layer. The change is relative to the amount of target gas in the sample. The temperature is a design parameter chosen to respond to a certain family of gases.

Semiconductor detectors are very simple and fairly rugged. They are tolerant of high gas exposures and low humidity. However, they are not very specific as a single sensor will often respond to a wide variety of gases. These sensors are used in applications that do not need a high degree of accuracy or specificity. Applications include commercial and industrial flammable or toxic gas monitoring.

Catalytic Bead Detectors

The catalytic bead sensor is the oldest form of gas sensor having originally being used for mine safety. It is commonly used to detect combustible gases such as methane and hydrogen. The sensor features a bead composed of a fine wire coil with a heated outer covering of ceramic and a catalyst. When a combustible gas passes over this bead, it causes a small amount of combustion that in turn alters the temperature of the wire coil. The wire's resistance is raised by temperature variation and is relational to the amount of combustible gas in the sample. The resistance is read and is converted by a Wheatstone bridge in the detector or transmitter. Most designs use a second bead as a balance. This second bead is not exposed to the gas path and is used to filter out temperature fluctuations.

A catalytic bead detector can be expected to last from 2 to 5 years in most applications. This type of sensor is subject to damage (poisoning) from silicon-, lead-, and sulfur-containing compounds. Catalytic bead sensors are used exclusively to detect combustible gases. However, they are found in any application since they are little affected by hostile environments.

Photo and Flame Ionization Detectors

Photoionization detectors (PID) use a UV lamp to ionize the gas sample. The ionized gas produces charged particles. A detector converts this energy into an electrical current. The current is proportional to the amount of target gas in the sample. Different gases have significantly different ionization potentials. The UV source is chosen to provide the correct ionization potential for the target gas. The lamp life is extended on most detectors by creating a duty cycle with programmed on and off periods.

These detectors are most commonly used to detect volatile organic compounds (VOCs). VOCs are found in many manufacturing operations.

The flame ionization detector (FID) works much like the PID detector. Instead of having a UV lamp to ionize target gas in a sample, a flame is used for the same purpose. The flame is usually hydrogen because it burns without producing hydrocarbons.

These detectors are used to detect hydrocarbons. FIDs are used for combustion analysis as well as for gas detection applications.

Open-Path Detectors

These detectors are discussed in full detail in Chapter 1.62. Open-path detectors are capable of detection over a large area, usually outdoors. A laser source is aimed at a mating receiver. Any gas that is in the path of the laser will cause a drop in the energy received. Two separate readings are taken, one a reference and the other an active path. The ratio between the two determines the presence of gas. The gas concentration is actually the measurement of the amount of gas in the signal path. Due to this, it is expressed in PPM/meter or LEL/meter. Since the detector can neither pinpoint a sample nor give a simple concentration reading, a follow-up with a portable detector is often required.

These detectors are used to detect toxic and flammable gases in open areas. Open-path detectors are often used in petrochemical applications. Applications have to be carefully selected to eliminate false alarms from human or animal traffic. Additional concerns are fog and dust interrupting the sensing path.

DOSAGE SENSORS

The dosimeters in use include the various color-changing badges, direct reading tubes, and various sorption devices. Each is briefly described in the following paragraphs.

Color-Change Badges

These badges are worn by workers and serve to monitor their exposure to hazardous chemicals. Depending on the type of badge, the exposure period can be from 15 min to 8 hr. These badges register the TWA exposure level.

TABLE 1.26h
Personal Monitoring Toxic Gas Exposure Badges[a]

Analyte	Range (ppm x hr)	Minimum Detectable Concentration in 8 hr	Interferences
Acetone	20–24,000	2.5 ppm	NH_3, MEK, MIBK, MPK
Ammonia	4–300	0.5 ppm	RNH_2
Carbon disulfide	0.5–30	0.06 ppm	H_2S[b]
Carbon monoxide	10–525	1.25 ppm	Alkenes, H_2, H_2S
Chlorine	0.4–13	0.05 ppm	Br_2, I_2, HCl
Chlorine dioxide	0.1–1.4	0.013 ppm	NO_2
Ethanol	62–7,360	7.75 ppm	ROH, DMS
Formaldehyde	0.3–12	0.04 ppm	Acrolein
Glutaraldehyde	STEL (15 min)	0.04–0.95 ppm	None known
Hydrazine	0.01–0.8	0.002 ppm	Aromatic amines, MMH
Hydrogen sulfide	1–240	0.13 ppm	None known
Mercury	0.125–1.6 $mg/m^3 \times h$	0.02 mg/m^3	Strong oxidizers
Methanol	27–3200	3.38 ppm	ROH, DMS
Methyl ethyl ketone[c]	18–21,600	2.25 ppm	NH_3, acetone, MIBK, MPK
Methyl isobutyl ketone[c]	16–19,200	2.0 ppm	NH_3, MEK, Acetone, MPK
Nitrogen dioxide	0.5–13	0.06 ppm	O_3
Ozone	0.08–1.6	0.01 ppm	H_2O_2, NO_2[d]
Sulfur dioxide	0.1–16	0.013 ppm	None known

[a] Courtesy of K&M Environmental Inc.
[b] 3 ppm H_2S causes color development in cell 1, 10 ppm H_2S causes color development in cell 2.
[c] Coefficient must be applied to scale printed on badge.
[d] Ozone monitor is 10 times more sensitive to ozone than to nitrogen dioxide.

Table 1.26h lists some of the available badges and badge holders. The color change of an exposed badge can be self-developing, or the badge may need to be treated with a developing reagent. The color change can be interpreted by comparison to a color chart or by the use of electronic monitors.

Continuous Color-Change Monitor Colorimetric detectors use a paper substrate that is treated with a chemical reagent and then dried. This treated paper will then react and create a colored stain when a gas-bearing sample is passed through it. The detector has a mechanism that is like a reel-to-reel tape recorder that advances the tape at a very slow speed. As the tape is fed through the mechanism, a gas sample is drawn by a vacuum pump through the paper tape. If there is target gas in the sample, the gas will create a stain. The color and intensity define the gas and its concentration, respectively. The stain is read by a photodetector and then converted into a signal representing the concentration.

Colorimetric detectors are mostly used in high-tech applications. These detectors are capable of detecting most of the toxic gases used at very low levels. They are also well very suited for the clean environment with stable temperature and humidity in such facilities.

Continuous color-change concepts have also been employed to build continuous toxic gas monitors (Figure 1.26g) using a chemically treated tape that is incrementally exposed to the sample gas. If the gas of interest

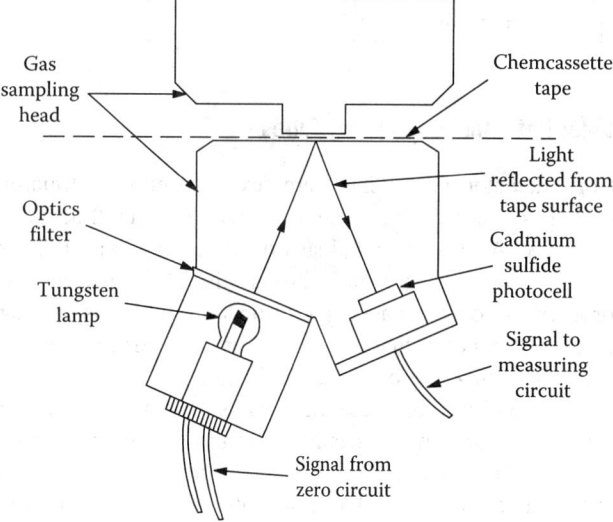

FIG. 1.26g
By monitoring the reflected light intensity from a chemically treated tape, toxic gas concentrations in the ppb range are detected. (Courtesy of MDA Scientific Inc.)

is present in the sample, the tape is *stained*, and the degree of staining is detected by measuring the reduction in the reflected light from the tape.

This spectrophotometric approach, coupled with microprocessor technology, is capable of detecting toxic gas concentrations in the ppb range and proving response times of under 30 sec. More than 50 toxic gases are detectable by this technique. A partial list of such detectable gases and their minimum concentrations is provided in Table 1.26i.

TABLE 1.26i
Substances Detectable by Continuous Toxic Gas Monitor[a]

Substance	Standard Range	Lower Detectable Limit
Ammonia	0–250 ppm	1 ppm
Chlorine	0–10 ppm	0.2 ppm
Diisocyanates	0–200 ppb	1 ppb
Hydrazines	0–1000 ppb	10 ppb
Hydrides	0–1000 ppb	3 ppb
Arsine	0–1000 ppb	2 ppb
Diborane	0–2000 ppb	50 ppb
Germane	0–1000 ppb	15 ppb
Hydrogen selenide	0–3000 ppb	5 ppb
PhosphineSilane	0–50 ppm	2 ppm
Hydrogen chloride	0–100 ppm	0.2 ppm
Hydrogen cyanide	0–100 ppm	0.8 ppm
Hydrogen fluoride	0–30 ppm	0.5 ppm
Hydrogen sulfide	0–50 ppm	0.5 ppm
Nitrogen dioxide	0–50 ppm	0.5 ppm
Phosgene	0–1000 ppb	12 ppb
p-Phenylenediamine	0–200 ppb	2 ppb
Sulfur dioxide	0–20 ppm	0.1 ppm

[a] Courtesy of MDA Scientific Inc.

TABLE 1.26j
Features and Capabilities of Color-Changing Dosimeter Tubes[a]

Description	Measuring Range (ppm)	Typical Applications	Shelf Life (yrs)
Ammonia (NH_3)	50–900	Process control	3
Ammonia (NH_3)	10–260	Industrial hygiene, leak detection	3
Benzene (C_6H_6)	5–200	Industrial hygiene	1
Carbon dioxide (CO_2)	0.05%–1.0%	Air contamination, concentration control	2
Carbon monoxide (CO)	10–250	Blast furnace, garage, combustion	1
Hydrogen sulfide (H_2S)	100–2000	Industrial raw gases, metallurgy	3
Hydrogen sulfide (H_2S)	3–150	Metal/oil refinery, chemical lab	3
Methyl bromide (CH_3Br)	5–80	Insert fumigation for mills, vaults, etc.	1
Nitrogen dioxide (NO_2)	0.5–30.0	Arc welding, acid dipping of metal products	1
Sulfur dioxide (SO_2)	20–300	Metal refining, waste gas analysis	2
Smoke tubes	—	Ventilation/air flow determination	—

[a] Courtesy of Cole-Parmer Instrument Co.

Color Detector (Dosimeter) Tubes

Color detector tubes (CDTs) are devices used for estimating toxic gas and vapor concentrations with an overall accuracy of ±25%. A hand-operated suction device pulls a predetermined amount of air sample through the CDT, and the color progression on the packed sorption bed (which is impregnated with a color-changing indicator material) registers the concentration of the particular toxic gas.

The purpose of these devices is to estimate concentrations before more accurate monitoring is performed. Concentrations by the color-band progression can be read within a few minutes. The disposable CDTs are relatively inexpensive, and a wide variety of some 250 different tubes are available. Table 1.26j lists some of the available tubes.

Dosimeter tubes are also available in diffusion-based designs (Figure 1.26h). They are used by snapping off the prescored end of the tube and a colored line of demarcation results in front of a scale that is labeled in units of ppm-hours of exposure.

For the measurement ranges, time periods, shelf lives, and other features of color-detecting dosimeter tubes, refer to Table 1.26k.

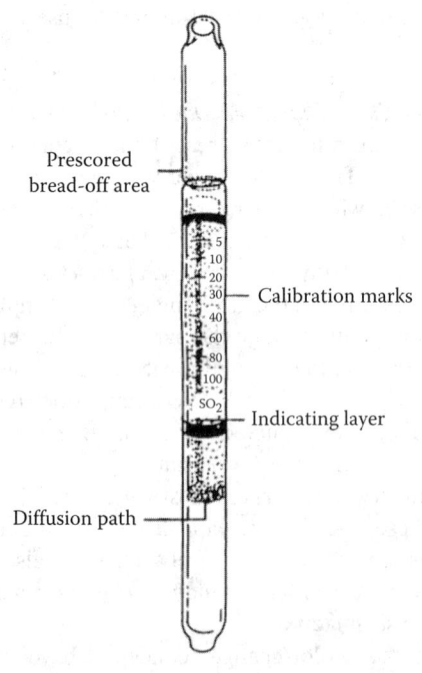

FIG. 1.26h
Diffusion-type dosimeter tube. (Courtesy of Sensidyne Inc.)

TABLE 1.26k
Passive, Diffusion-Type Dosimeter Tubes

Gas or Vapor to Be Measured Chemical Formula	Gastec Tube No. and Name	Measuring Range (ppm)	Measuring Time (h)	Shelf Life (Year)	Note	TLV-TWA,C (ACGIH) (ppm)
Acetaldehyde CH_3CHO	Formaldehyde	0.1–20	1–10	1*	—	C 25
	Acetone	4–1,200	1–10	2*	T	
	Methyl ethyl ketone	1.2–360	1–10	2*	T	
Acetic acid CH_3CO_2H	Acid gases	0.5–100	1–10	3	T	10
Acetic anhydride $(CH_3CO)_2O$	Acetic acid	0.45–90	1–10	3	T	5
Acetone CH_3COCH_3	Acetone	5–1,500	1–10	2*	T	500
	Methyl ethyl ketone	1.4–420	1–10	2*	T	
Ammonia NH_3	Ammonia	2.5–1,000	0.5–10	3	T	25
	Ammonia	0.1–10	1–10	2	T	
Benzene C_6H_6	Toluene	2.4–600	1–10	2	T	0.5
1,3-Butadiene $CH_2:CHCH:CH_2$	1,3-Butadiene	1.3–200	1–8	2	—	2
Carbon dioxide CO_2	Carbon dioxide	0.02–16%	0.5–10	2	T	5000
Carbon monoxide CO	Carbon monoxide	1.04–2,000	0.5–48	2	—	25
	Carbon monoxide	0.4–400	0.5–24	2*		
Chlorine Cl_2	Chlorine	0.08–100	0.5–24	2	—	0.5
	Trichloroethylene	2.4–240	1–8	1*	T	
Cumene $C_6H_5OH(CH_3)_2$	Toluene	3.4–850	1–10	2	T	50
trans-1,2-Dichloroethylene ClCH:CHCl	1,3-Butadiene	3.9–600	1–8	2	T	200
	Trichloroethylene	6–600	1–8	1*	T	
Dimethylamine $(CH_3)_2NH$	Ammonia	1.9–750	0.5–10	3	T	5
N,N-Dimethylethylamine $C_2H_5N(CH_3)_2$	Ammonia	4–1,600	0.5–10	3	T	—
Ethanol C_2H_5OH	Ethanol	100–25,000	1–10	3	—	1000
Ethyl benzene $C_6H_5C_2H_5$	Toluene	2.8–700	1–10	2	T	100
Ethylene $CH_2:CH_2$	1,3-Butadiene	1.56–240	1–8	2	T	—
Ethylene dichloride $ClCH_2CH_2Cl$	1,3-Butadiene	3.9–600	1–8	2	T	10
Formaldehyde HCHO	Formaldehyde	0.1–20	1–10	1*	—	C 0.3
Formic acid HCO_2H	Acetic acid	0.55–110	1–10	3	T	5
Furfural $C_5H_4O_2$	Formaldehyde	0.3–60	1–10	1*	—	2
Hydrazine N_2H_4	Ammonia	1.6–650	0.5–10	3	T	0.01
Hydrogen chloride HCl	Hydrogen chloride	1–100	1–10	3	TH	C2
	Trichloroethylene	1.8–180	1–8	1*	TH	
Hydrogen cyanide HCN	Hydrogen cyanide	1–200	1–10	2	—	C 4.7
Hydrogen fluoride HF	Hydrogen fluoride	1–100	1–10	3	H	C 3
	Hydrogen chloride	2.5–250	1–10	3	TH	
Hydrogen peroxide H_2O_2	Hydrogen peroxide	0.5–40	1–10	3	T	1
Hydrogen sulfide H_2S	Hydrogen sulfide	0.2–200	1–48	3	—	10
Isoprene $CH_2:C(CH_3)CH:CH_2$	1,3-Butadiene	2.6–400	1–8	2	TH	—
Methylamine CH_3NH_2	Ammonia	0.19–19	1–10	2	T	5
Methyl ethyl ketone $CH_3COC_2H_5$	Methyl ethyl ketone	2–600	1–10	2*	T	200
	Formaldehyde	0.125–25	1–10	1*	—	
	Acetone	6.5–1,950	1–10	2*	T	
Methyl isobutyl ketone $(CH_3)_2CHCH_2COCH_3$	Acetone	11.5–3,450	1–10	2*	T	50
	Methyl ethyl ketone	4–1,200	1–10	2*	T	
Nitric acid HNO_3	Hydrogen chloride	0.8–80	1–10	3	T	2
	Hydrogen fluoride	0.32–32	1–10	3	TH	
Nitrogen dioxide NO_2	Nitrogen dioxide	0.1–30	1–10	1*	—	3
	Nitrogen dioxide	0.01–3.0	1–24	1*	—	

(Continued)

TABLE 1.26k (Continued)
Passive, Diffusion-Type Dosimeter Tubes

Gas or Vapor to Be Measured Chemical Formula	Gastec Tube No. and Name	Measuring Range (ppm)	Measuring Time (h)	Shelf Life (Year)	Note	TLV-TWA,C (ACGIH) (ppm)
Styrene $C_6H_5CH{:}CH_2$	Toluene	26–6,500	1–10	2	T	20
Sulfur dioxide SO_2	Sulfur dioxide	10–600	1–5	3	T	2
	Sulfur dioxide	0.2–100	1–10	3	—	
Tetrachloroethane Cl_2CHCCl_3	Tetrachloroethylene	3–150	1–8	1*	T	25
	Trichloroethylene	1.5–150	1–8	1*	T	
Toluene $C_6H_5CH_3$	Toluene	2–500	1–10	2	T	50
Trichloroethylene $Cl_2C{:}CHCl$	Trichloroethylene	3–300	1–8	1*	T	50
Triethylamine $(C_2H_5)_3N$	Ammonia	5.3–2,100	0.5–10	3	T	1
Trimethylamine $(CH_3)_3N$	Ammonia	0.23–23	0.5–10	2	T	5
Vinyl chloride $CH_2{:}CHCl$	1,3-Butadiene	1.56–240	1–8	2	T	1
Xylene $C_6H_4(CH_3)_2$	Toluene	3.4–850	1–10	2	T	100

Source: Courtesy of Gastec.

Note: T, temp correction; H, humidity correction; *, refrigerator store; mesh, correction factor/chart.

Sorption-Type Dosimeters

These devices absorb toxic gases either passively (as a result of diffusion) or actively (when an air sample is pulled through them). After some period of exposure, the quantity of toxic gases that have been accumulated by them is measured. This is done by removing the devices and applying heat or solvent to desorb the toxic material.

Personnel dosimeters are either liquid filled or solid filled. The former usually requires wet chemical treatment before it can be analyzed for formaldehyde, ethylene oxide, and other substances, usually by a spectrophotometer. In this case, the time-consuming preparation requirement is a disadvantage (except for the direct-indicating types, which need no preparation or analyzer but are also less accurate). The time delay can be especially bothersome if the dosimeter has to be mailed to a laboratory for analysis.

Passive personal dosimetry (diffusion) is most valuable for double-checking monitoring data or for calibration. It is also useful when seeking concentration estimates before more elaborate instruments are procured. The solid-filled personnel dosimeters have to be desorbed (by solvents or, more and more frequently, by heat) for gas chromatographic or other instrumental analysis.

Their normal usage includes double-checking calibrations, determining sensitivity ranges of instruments, calculating useful flow rates and sampling-time duration for active sampling devices, and making approximate measurements before stationary instruments are obtained.

Active (also called dynamic) personnel sampling devices also need thermal desorption or wet chemical treatment after sample collection, but they do not rely on diffusion as do passive personnel dosimeters. In active dosimeters, an air-sampling pump continuously passes a known amount of air through the sorbent material. Components of the pure air pass through the sorbent while toxic gases and vapors are held back on the high surface area of the sorbent material. The sorbent material can be either liquid or solid.

Liquid-filled absorption devices are called *impingers*. They are fragile and therefore sensitive to mistreatment, and they are more troublesome to use than solid sorbent tubes. They can also cause injuries if the liquid leaks out.

Solid-filled active sampling devices (also called adsorptive, sorptive, or sorption tubes) are less sensitive to mistreatment than are impingers. The most frequent problem with these devices is breakthrough, which is a loss of the sorbed toxic gas (desorption), especially if flow rates are too high and/or sorption times are too long. These solid sorbent-filled trapping devices rely on surface adsorption rather than solution absorption.

TABLE 1.26l
Adsorbent Tubes for Trapping and Thermally Desorbing Organic Compounds[a]

Absorbent Tube	Application	Thermal Desorption Unit
Carbotrap 100	C_5–C_{12} Hydrocarbons	Dynatherm Tekmar CDS Analytical
Carbotrap 150	PCBs, PNAs, Alkyl benzenes, etc.	Dynatherm
Carbotrap 200	C2–C14 Hydrocarbons	Dynatherm
Carbotrap 300	EPA Methodologies TO-1, TO-2, TO-3 and other organics	Dynatherm Tekmar CDS Perkin Elmer
Carbotrap 370	C5–C30 Organics	Supelco

Notes: PCB, Polychlorinated biphenyls; PNA Polynuclear aromatic hydrocarbons; EPA Environmental Protection Agency.
[a] Abstracted from Supelco Chromatography Products, 1997.

Table 1.26l lists a number of sorption tubes with different packing solids. Although they can be desorbed by liquid solvents and injected as liquid samples into a gas chromatograph, thermal desorption can also be used by injecting the desorbed sample directly into a gas chromatograph as a vapor or gas. Sometimes, cold traps are also used before injecting, which can concentrate or focus the sample before thermal desorption and injection.

In addition to carbon-based sorption-tube fillings, alumina, silica gel, Tenax®, and many other gas–solid chromatography packing materials can also be used.

CALIBRATION

Toxic gas and toxic vapor monitors are only as good as is their calibration. Electrochemical sensors can be calibrated by prepared gas concentrations that are stored in pressurized gas cylinders, sampling bags, or calibration bottles (Figure 1.26i). Although high-pressure calibrating gas mixtures are commonly used, their reliability decreases with increasing shelf life and decreasing concentrations. Sorption, chemical reactions, diffusion, and heavy component settling by gravity can also reduce their reliability. Even stainless steel containers and periodic rolling cannot eliminate all the problems; therefore, alternative methodologies are preferred.

Low-pressure static gas mixtures in mixing bags or in rigid containers have the same problems in addition to having very limited capacities. They are still useful when used in syringe calibration for gas chromatography but are practically useless as continuous or semi-continuous calibration gas supplies.

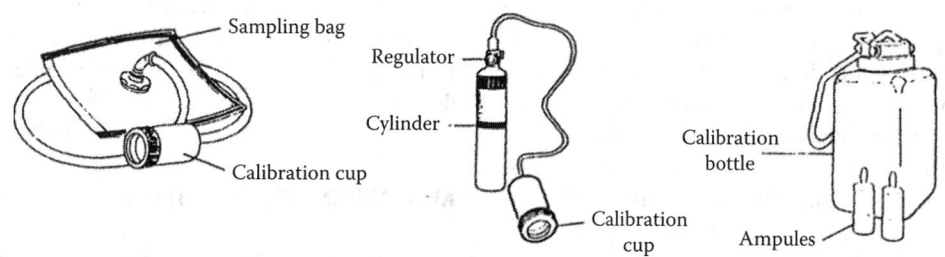

FIG. 1.26i
Accessories used in calibrating electrochemical toxic gas sensors. (Courtesy of Sensidyne Inc.)

Dynamic Calibrators

Dynamic calibrators operating on the permeation tube principle have been available for decades. The availability of a large number of gases and vapors (about 100 different toxic components) and the relatively high accuracy (±2%) of these systems place permeation systems among the most popular calibrating devices.

Another dynamic calibrator is the bubbler/dilution calibrator. In this design, instead of a permeation chamber, a bubbler or impinger is used. If the gas or vapor of interest is a liquid at 0°C (32°F), an ice/water bath can be used to provide temperature control.

The air is slowly bubbled through the liquid so that the air is saturated by the vapor of the liquid at the controlled temperature. The known flow rate of the saturated air can then be diluted by pure air. The bubbler liquid's vapor pressure can be calculated or can be taken from a handbook. The concentration of the yet undiluted stream can be calculated by Equation 26.1.

$$C\% = (100 \times \text{vapor pressure of the liquid}) / (\text{total atmospheric pressure}) \quad (26.1)$$

To increase accuracy, the same pressure sensor should be used to detect both the vapor and the atmospheric pressures. If the concentration is too high (it usually is), dilution by a large flow of pure air is necessary.

Calibrating Gas Generators A novel way of providing calibrating gases is to use electrochemical calibrating gas generators. They consist of electrochemical cells that provide coulometrically controlled quantities of specific gases that can be mixed into a fixed air flow. Accuracy is ±10%. Chlorine, chlorine dioxide, hydrogen, hydrogen cyanide, hydrogen sulfide, and ozone calibration gas generators are all available.

Some electrochemical detectors are also available with built-in gas generators so that bump tests can be performed automatically in the field to ensure detector operational integrity. Because, for the purposes of bump testing, absolute accuracy is not important, cross sensitivities can be utilized for testing cell integrity. Detectors with automatic sensor testing are available for bromine, chlorine, chlorine dioxide, fluorine, ozone, ammonia, carbon monoxide, hydrogen chloride, hydrogen cyanide, hydrogen fluoride, hydrogen sulfide, nitrogen dioxide, sulfur dioxide, and iodine.

Because calibration always involves toxic gases, all such procedures must be executed in well-ventilated areas, preferably inside high-efficiency exhaust hoods.

SELECTION

The following should be considered when selecting hazardous material monitoring system.

Target gas and detection range—While several detectors may detect a particular gas, it may not have suitable range. The detector's lower and higher detectable ranges need to be taken into account. The lower detectible limit obviously needs to be below the alarm set point. If the target gas will over-range the detector, it may be damaged or even fail to alarm.

Environmental conditions—The detector must be suitable for the application. If the detector is to be installed in a flammable or explosive atmosphere, it must be listed for such hazards. Furthermore, the temperature, humidity, and cleanliness of the application must be considered.

Interference gas—There may be other gases in the area that will cause false alarms. Other interferences can cause premature failures or even poison the sensor rendering it useless.

Maintenance requirements—The detector will require testing, calibration, and other maintenance. The detector should be chosen so that these activities can be done in a safe and effective manner. For instance, many detectors for use in hazardous environments can be calibrated with a magnetic wand without opening the detector. This allows calibration to be done without shutting down the process to safely open the detector.

Location—No matter what the application, there is one incontrovertible rule—the closer the detector is located to the source of a leak, the faster and more accurate the response will be.

Choice of diffusion vs. extractive detection—A diffusion detector simply exposes the detector to the ambient air. As gas diffuses into the area, it reaches the detector and it is processed. Most combustible detectors have diffusion style. Placement of the detector is critical to proper sampling. Extractive detectors rely on a pump or venturi to create a vacuum. The vacuum is connected to the gas detector. A sample line is installed to allow the gas to be drawn from a particular area. Colorimetric detectors are extractive detectors.

APPLICATION CONSIDERATIONS

Area Detection—Area detection is often the most difficult of applications to accomplish. When putting a detector in a room, several factors will need to be considered. If the gas being detected is heavier than air, the logical placement for the detector is close to the floor. This may be the best location, but consideration must be given to air flow. Even a modest amount of updraft will force a heavier-than-air gas upward, and the air in a high-tech clean room will certainly dissipate the sample. Thus, the gas weight relative to air is only a starting point to determining detector placement. A more realistic approach is to determine the likely source of a leak and consider which way the air flow will likely take the gas. Then, install the detector as close to the source of the leak as possible.

Duct Detection—A good detection method is to place the sensor in the exhaust duct of a piece of equipment handling toxic or flammable gas. By doing so, the vagaries of ambient air flow are overcome, and detection is more certain. When placing a sensor in a duct, the ideal placement is where airflow in the duct is laminar. To accomplish this, the detector is placed as far as practical from a bend, a wye, or a transition. When using an extractive detector in an exhaust duct, the vacuum on the sample line must be adequate to overcome the flow in the duct.

Stack Detection—To measure stack emissions, gas detectors are often placed in the exhaust stack. Most stacks are hostile environments, and care needs to be taken to prevent damage from temperature, moisture, acids, and cross interferences. The manufacturer should have technical resources to help with this type of application.

COMMUNICATION AND MAINTENANCE

All detectors discussed herein are connected to some form of control system. Most detectors have multiple relays that activate on different levels of detection. Almost all detectors can transmit analog values to a dedicated controller, a DCS, or a PLC system. Some also support HART and/or bus protocols too. Many controls are employed to ramp up exhaust fans, turn off equipment, and most importantly notify people in the area of an alarm.

All detectors require maintenance. Manufacturer recommendations vary from monthly to annual intervals. Basic maintenance would be to set the zero and calibrate the detector. Sometimes it is adequate to just "bump test" the detector at a regular interval. Whatever maintenance program is undertaken, it should be in accordance with the manufacturer's instructions.

SPECIFICATION FORM

When specifying hazardous or toxic gas monitors, the application can be described by using the ISA Form 20A1001. This form is provided on the next page with the permission of the International Society of America.

Analytical Measurement

1	RESPONSIBLE ORGANIZATION		ANALYSIS DEVICE	6	SPECIFICATION IDENTIFICATIONS		
2				7	Document no		
3	ISA		Operating Parameters	8	Latest revision		Date
4				9	Issue status		
5				10			

11	ADMINISTRATIVE IDENTIFICATIONS			40	SERVICE IDENTIFICATIONS Continued			
12	Project number		Sub project no		41	Return conn matl type			
13	Project				42	Inline hazardous area cl		Div/Zon	Group
14	Enterprise				43	Inline area min ign temp		Temp ident number	
15	Site				44	Remote hazardous area cl		Div/Zon	Group
16	Area		Cell	Unit	45	Remote area min ign temp		Temp ident number	
17					46				
18	SERVICE IDENTIFICATIONS				47				
19	Tag no/Functional ident				48	COMPONENT DESIGN CRITERIA			
20	Related equipment				49	Component type			
21	Service				50	Component style			
22					51	Output signal type			
23	P&ID/Reference dwg				52	Characteristic curve			
24	Process line/nozzle no				53	Compensation style			
25	Process conn pipe spec				54	Type of protection			
26	Process conn nominal size		Rating		55	Criticality code			
27	Process conn termn type		Style		56	Max EMI susceptibility		Ref	
28	Process conn schedule no		Wall thickness		57	Max temperature effect		Ref	
29	Process connection length				58	Max sample time lag			
30	Process line matl type				59	Max response time			
31	Fast loop line number				60	Min required accuracy		Ref	
32	Fast loop pipe spec				61	Avail nom power supply			Number wires
33	Fast loop conn nom size		Rating		62	Calibration method			
34	Fast loop conn termn type		Style		63	Testing/Listing agency			
35	Fast loop schedule no		Wall thickness		64	Test requirements			
36	Fast loop estimated lg				65	Supply loss failure mode			
37	Fast loop material type				66	Signal loss failure mode			
38	Return conn nominal size		Rating		67				
39	Return conn termn type		Style		68				

69	PROCESS VARIABLES	MATERIAL FLOW CONDITIONS			101	PROCESS DESIGN CONDITIONS		
70	Flow Case Identification			Units	102	Minimum	Maximum	Units
71	Process pressure				103			
72	Process temperature				104			
73	Process phase type				105			
74	Process liquid actl flow				106			
75	Process vapor actl flow				107			
76	Process vapor std flow				108			
77	Process liquid density				109			
78	Process vapor density				110			
79	Process liquid viscosity				111			
80	Sample return pressure				112			
81	Sample vent/drain press				113			
82	Sample temperature				114			
83	Sample phase type				115			
84	Fast loop liq actl flow				116			
85	Fast loop vapor actl flow				117			
86	Fast loop vapor std flow				118			
87	Fast loop vapor density				119			
88	Conductivity/Resistivity				120			
89	pH/ORP				121			
90	RH/Dewpoint				122			
91	Turbidity/Opacity				123			
92	Dissolved oxygen				124			
93	Corrosivity				125			
94	Particle size				126			
95					127			
96	CALCULATED VARIABLES				128			
97	Sample lag time				129			
98	Process fluid velocity				130			
99	Wake/natural freq ratio				131			
100					132			

133	MATERIAL PROPERTIES				137	MATERIAL PROPERTIES Continued		
134	Name				138	NFPA health hazard		Flammability	Reactivity
135	Density at ref temp		At		139				
136					140				

Rev	Date	Revision Description	By	Appv1	Appv2	Appv3	REMARKS

Form: 20A1001 Rev 0 © 2004 ISA

Definitions

Carcinogen (CA): Any substance that NIOSH considers to be a potential occupational carcinogen.

Ceiling limit (TLV-C): A concentration that should not be exceeded even instantaneously.

Highly toxic: A chemical that has a median lethal concentration (LC(50)) in air of 200 parts per million by volume or less of gas or vapor, or 2 mg/l or less of mist, fume, or dust, when administered by continuous inhalation for 1 hr (or less if death occurs within 1 hr) to albino rats weighing between 200 and 300 g each.

Immediately dangerous to life or health (IDLH): Limit concentrations used by the National Institute for Occupational Safety and Health (NIOSH) as respirator selection criteria.

Lower explosive limit/lower flammable limit (LEL/LFL): Lowest concentration of the substance in air (usually expressed in percent by volume) that will produce a flash or fire when an ignition source (heat, electric arc, or flame) is present. At concentrations lower than the LEL, propagation of a flame will not occur in the presence of an ignition source (OSHA 2012).

Permissible exposure limit (PEL): A legally enforceable occupational exposure limit established by OSHA, not only usually measured as an 8 hr TWA, but also may be expressed as a ceiling concentration exposure limit.

Recommended exposure limit (REL): The ceiling value that should not be exceeded at any time.

Short-term exposure limit (TLV-STEL): A maximum concentration for a continuous 15 min exposure period according to NIOSH or a maximum of four such periods per day, with at least 60 min between exposure periods, and provided the daily TLV-TWA is not exceeded, according to OSHA.

Threshold limit value (TLV): An occupational exposure value recommended by the *American Conference of Governmental Industrial Hygienists* (ACGIH), which it is believed that nearly all workers can be exposed day after day for a working lifetime without ill effect (ACGIH 2008).

Time-weighted average (TLV-TWA): A concentration limit for a normal 8 hr workday or 40 hr workweek, set by OSHA or TWA concentration for up to a 10-hour workday during a 40 hr workweek set by *NIOSH*.

Toxic: A chemical that has a median lethal concentration (LC(50)) in air of more than 200 parts per million but not more than 2000 parts per million by volume of gas or vapor, or more than 2 mg/l but not more than 20 mg/l of mist, fume, or dust, when administered by continuous inhalation for 1 hr (or less if death occurs within 1 hr) to albino rats weighing between 200 and 300 g each (OSHA 1994).

Upper explosive limit/upper flammable limit (UEL/UFL): The highest concentration of a vapor or gas (highest percentage of the substance in air) that will produce a flash of fire when an ignition source (e.g., heat, arc, or flame) is present. At higher concentrations, the mixture is too "rich" to burn; also see LEL.

Abbreviations

BEI	Biological Exposure Indices
CA	Carcinogen
CDT	Color-Detecting Tubes
DCS	Distributed Control System
HCN	Hydrogen Cyanide
IDLH	Immediately Dangerous to Life or Health
LC	Lethal Concentration
LEL	Lower Explosive Limit
LFL	Lower Flammable Limit
MOS	Metal Oxide Semiconductor
PEL	Permissible Exposure Levels
PLC	Programmable Logic Controller
ppm	Parts per million
REL	Recommended Exposure Limit
STEL	Short-Term Exposure Limit
TLV	Threshold Limit Value
TLV-C	Threshold Limit Value Ceiling TLV
TLV-STEL	Short-Term Exposure Limit TLV
TLV-TWA	Time-Weighted Average TLV
TWA	Time-Weighted Averages
UEL	Upper Explosive Limit
UFL	Upper Flammable Limit

Organizations

ACGIH	Conference of Governmental Industrial Hygienists
AIHA	American Industrial Hygiene Association
ANSI	American National Standards Institute
ASHRAE	American Society of Heating, Refrigerating, and Air-Conditioning Engineers
ATSDR	Agency for Toxic Substances and Disease Registry
BSI	British Standards Institute
CCOHS	The Canadian Centre for Occupational Health & Safety
CENELEC	European Committee for Electrotechnical Standardization
CSA	The Canadian Standards Association
EPA	Environmental Protection Agency
FMG	Factory Mutual Global
FMRC	Factory Mutual Research Corporation
GINC	Global Information Network on Chemicals
ILO	International Labor Organization
ISA	International Society of Automation
ISEA	International Safety Equipment Association
MSHA	The Mine Safety and Health Administration

NIJ	National Institute of Justice
NIOSH	National Institute for Occupational Safety and Health
NIST	National Institute of Standards and Technology
NSC	National Safety Council
OSHA	Occupational Safety and Health Administration
SEI	Safety Equipment Institute
SIRA	Society of Information Risk Analysts
USEPA	United States Environmental Protection Agency

Bibliography

ACGIH, Documentation of the threshold limit values and biological exposure indices, (2015) https://www.acgih.org/forms/store/ProductFormPublic/documentation-of-the-threshold-limit-values-and-biological-exposure-indices-7th-ed.

ACGIH, Publications, (2015) https://www.acgih.org/forms/store/CommercePlusFormPublic/search?action=Feature.

ACGIH, TLVs and BEIs, (2015) http://www.acgih.org/forms/store/ProductFormPublic/2015-tlvs-and-beis.

ANSI/ISA TR12.13.03-2009, Guide for combustible gas detection as a method of protection, (2009) https://www.isa.org/templates/one-column.aspx?pageid=111294&productId=116631.

ANSI/ISA 12.13.02-2012, Explosive atmospheres—Part 29-2: Gas detectors—Selection, installation, use and maintenance of detectors for flammable gases and oxygen, (2012) https://www.isa.org/templates/one-column.aspx?pageid=111294&productId=116358.

ANSI/ISA 12.13.01-2013, Explosive atmospheres—Part 29-1: Gas detectors—Performance requirements of detectors for flammable gases, (2013) http://webstore.ansi.org/RecordDetail.aspx?sku=ANSI%2FISA+12.13.01-2013.

BS EN 50402:2005+A1:2008, Electrical apparatus for the detection and measurement of combustible or toxic gases or vapors or of oxygen. Requirements on the functional safety of fixed gas detection systems, (2008) http://shop.bsigroup.com/ProductDetail/?pid=000000000030236469.

BS EN 45544-1:2015, Workplace atmospheres. Electrical apparatus used for the direct detection and direct concentration measurement of toxic gases and vapors, (2015) http://shop.bsigroup.com/ProductDetail/?pid=000000000030276471.

Calabrese, E. J. and Kenyon, E., (2011) Air Toxics and Risk Assessment, Lewis Publishers, http://www.belleonline.com/bios/calabrese.htm.

CDC, International Chemical Safety Cards, (2014) http://www.cdc.gov/niosh/ipcs/icstart.html.

CiTicesl, Toxic gas sensor, (1999) http://www.safetygas.com.tw/pdf/city%20technologies%20toxic%20gas%20sensors.pdf.

Draeger Tube and CMS Handbook, Leubeck, Germany: Draeger Gmbh, (2012) http://www.draeger.com/sites/assets/PublishingImages/Master/Oil_and_Gas/Upstream/9092086-Tubes-e-low.pdf.

Environment Agency, UK, A Review of the Toxicity and Environmental Behaviour of Hydrogen Bromide in Air, (2005) https://www.gov.uk/government/uploads/system/uploads/attachment_data/file/290736/scho0105bimw-e-e.pdf.

EPA, Radon measurement in schools, (1993) http://www2.epa.gov/sites/production/files/2014-08/documents/radon_measurement_in_schools.pdf.

EPA, Acute exposure guideline levels for airborne chemicals, (2015) http://www.epa.gov/oppt/aegl/index.htm.

FM, Approval standard for toxic gases, (2014) https://www.fmglobal.com/assets/pdf/fmapprovals/6340.pdf.

GFG, Photo and flame ionization detectors, (2013) http://goodforgas.com/wp-content/uploads/2013/12/AP1021_Using_PIDs_to_measure_toxic_VOCs_10_10_13.pdf.

Grzywacz, C. M., Monitoring for gaseous pollutants in museum environments, (2006) http://www.getty.edu/conservation/publications_resources/pdf_publications/pdf/monitoring.pdf.

Honeywell, Commercial gas detection, (2015) http://www.honeywellanalytics.com/~/media/honeywell-analytics/documents/english/2014commercialproductguidepg11008ncl.pdf?la=en.

InspectAPedia, Toxic gas exposure standards, (2015) http://inspectapedia.com/sickhouse/Gas_Exposure_Limits.php.

Lewis, R. J., Rapid guide to hazardous chemicals in the workplace, (2000) http://www.amazon.com/Rapid-Guide-Hazardous-Chemicals-Workplace/dp/0471355429.

Manes, G. H. et al., Real-time monitoring of volatile organic compounds in hazardous sites, (2011) http://cdn.intechopen.com/pdfs/22746.pdf.

McDermott, H. J., Air monitoring for toxic exposures, (2005) http://www.wiley.com/WileyCDA/WileyTitle/productCd-0471454354.html.

NIOSH, Electronic NIOSH pocket guide to chemical hazards, (2011) www.cdc.gov/niosh/npg/pgdstart.html.

NIOSH, *Pocket Guide to Chemical Hazards*, National Institute for Occupational Safety and Health, (2015) http://www.cdc.gov/niosh/npg/.

OSHA, Toxic and hazardous substances, standards 1910–1000, (2006) https://www.osha.gov/pls/oshaweb/owadisp.show_document?p_table=STANDARDS&p_id=9991.

OSHA, Toxic Industrial Chemicals (TICs) Guide, (2007) https://www.osha.gov/SLTC/emergencypreparedness/guides/chemical.html.

Sensidyne Detector Tube Selection Guide, (2006) http://www.raecorents.com/products/gasmonitoring/Sensidyne-AP-20S-Pump/Sensidyne_DetectorTube_Selector_0806_rM.pdf.

Urben, P. G., Bretherick-Handbook-Reactive-Chemical Hazards, (2006) http://www.amazon.com/Bretherick-Handbook-Reactive-Chemical-Edition-Two/dp/0123725631.

Web Links to Government Organizations

ACGIH (American Conference of Governmental Industrial Hygienists), www.acgih.org/.

AIHA (American Industrial Hygiene Association), Provides standards of ethical conduct for industrial hygienists, www.aiha.org/.

ATSDR (Agency for Toxic Substances and Disease Registry), An agency of the United States Department of Health and Human Services, www.atsdr.cdc.gov/atsdrhome.html.

BSI (British Standards Institute), www.bsi-global.com.

CCOHS (The Canadian Centre for Occupational Health & Safety), Canada's national center for health and safety information, www.ccohs.ca/.

CSA (The Canadian Standards Association), www.csa-international.org/.

FMG (Factory Mutual Global), A commercial insurance company with a research division that is an approval agency, www.ul.com/, Underwriters Laboratories, Inc., an approval agency, www.fmglobal.com.

GINC (Global Information Network on Chemicals), Useful information on hazardous substances, www.nihs.go.jp/GINC/.

ISA (International Society of Automation), www.isa.org/.

ISEA (International Safety Equipment Association), www.safety-central.org/isea.

MSHA (The Mine Safety and Health Administration), A United States Government agency regulating the mining industry, www.msha.gov/.

NIOSH (The National Institute for Occupational Safety and Health), www.cdc.gov/niosh/homepage.html.

NIST (National Institute of Standards and Technology), www.nist.gov/srm.

NSC (National Safety Council), www.nsc.org/.

OSHA (The Occupational Safety & Health Administration), A government agency that regulates to protect against occupational hazards, www.osha.gov/.

SEI (Safety Equipment Institute), www.seinet.org/.

SIRA (Society of Information Risk Analysts), Europe's leading independent research and technology organization in imaging and intelligent systems, www.sira.co.uk/.

USEPA (United States Environmental Protection Agency), www.epa.gov/.

1.27 Heating Value Calorimeters

A. P. FOUNDOS (1982) **D. LEWKO** (2003) **B. G. LIPTÁK** (1995, 2017)

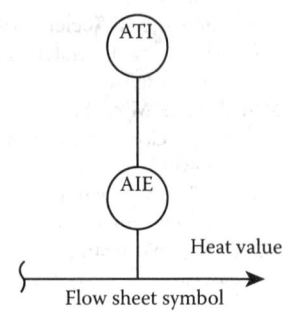

Flow sheet symbol

Type of designs	A. Direct measurement by burning of fuel gas
	B. Inferential by calculation from composition and physical analysis, including chromatography (Chapter 1.10), mass spectrometry (Chapter 1.35), and optical interferometry
	C. Special designs such as reaction calorimeters of Mettler-Toledo, designs for the measurement of partial molar heat capacities of biopolymers, isothermal titration of Calorimetry Sciences Corporation (TA Instruments), and total absorption calorimeters (Opal)
Applications	1. Custody transfer
	2. Process monitoring and control
	3. Blending and mixing of fuel gases
	4. Gas and liquefied natural gas (LNG) processing
	5. Compliance recording
Operation	a. Continuous
	b. Cyclic
	c. Portable
Performance	(1) Controlled environment
	(2) Varying ambient
	(3) High speed of response
	(4) Inaccuracy ±0.5% of full scale or better
	(5) Inaccuracy ±1.0% of full scale or better
	(6) Inaccuracy ±2.0% of full scale or better
Area classification	(gp) General purpose
	(ep) Explosion-proof
Cost	Under $10,000 [A, gp]
	$10,000–$15,000 [A, ep]
	$15,000–$30,000 [A/B, gp/ep]
Partial list of suppliers	AMCS-Chandler, (B, portable chromatograph) http://www.amcs.co.uk/index.php/products/item/9138_chandler_portable_natural_gas_chromatograph.html
	American Educational Products LLC, http://www.amep.com/standarddetail.asp?cid=3695
	Calorimetry Sciences Corporation (C) http://www.cityfos.com/company/CALORIMETRY-SCIENCES-in-LINDON-UT-143936.htm
	Chandler Engineering/Ranarex (B portable chromatograph) http://www.chandlerengineering.com/Products/LegacyProducts/model292.aspx
	Cosa Instrument Corp. (B, optical, portable) http://cosaxentaur.com/page/610/cosa-cv-pro-portable-calorific-value-measurement

(Continued)

Delta Instrument LLC - Union Instrument Gmbh. (A) http://www.deltainstrument.com/pdfs/cut_sheets/union/CWD2005_en_datasheet_2009_03A.pdf/DOWELLSCIENCEANDTECHNOLOGY(HK) (A, B, C) http://hkdowellltd.en.alibaba.com/search/product?SearchText=calorimeter

IKA (A bomb) http://www.ika.com/owa/ika/catalog.product_detail?iProduct=8802500

Mettler-Toledo Yamatake GC. (C) http://us.mt.com/us/en/home/products/L1_AutochemProducts/Reaction-Calorimeters-RC1-HFCal.html

Parr, (A bomb) http://www.scimed.co.uk/wp-content/uploads/2013/03/Introduction-to-bomb-calorimetry.pdf

Setarm Instrumentation, (C) http://www.setaram.com/

TA Instruments (C) http://www.tainstruments.com/product.aspx?siteid=11&id=260&n=1

INTRODUCTION

Heating Value or Calorific Value is the amount of heat produced by combustion of a unit quantity of a fuel. Common units used are Btu/lb, J/kg and kcal/kg. Therefore, calorific value is the energy contained in a fuel* or other materials, which can be determined by measuring the heat produced by the complete combustion of a specified quantity of it. This is usually expressed in joules per kilogram or, for gases, as mega joules per cubic meter. A calorimeter can be obtained by inserting a thermometer into a metal container filled with water and suspended above a combustion chamber. This design is called a bomb calorimeter, which is used to measure the heat of combustion of a particular reaction or the heat of combustion of a substance.

Calorimeters are analyzers that measure the heat value or energy content of fuels such as natural gas. There are two broad categories of this type of instrument: those that can be considered true calorimeters, because they are actually burning the gas and directly measuring its heating value, and inferential calorimeters, which analyze the composition of the gas or measure a physical parameter to determine the heating value.

Calorimeters can also measure the heat released by any other reaction. For example, to find the enthalpy change per mole of a substance A in a reaction between substances A and B, all that is needed is to insert them in a calorimeter and measure the initial and final temperatures as the reaction takes place. Multiplying this temperature change by the mass and specific heat capacities of the two substances gives the energy released by the reaction.

In this chapter, the emphasis is based on industrial calorimeters, and while microcalorimeters, titration calorimeters, reaction calorimeters, and partial molar heat capacities of biopolymers will be given less emphasis, some of their suppliers are also listed. Inferential calorimeters, which determine the heat content by calculation from the composition and physical analysis, including chromatography (Chapter 1.10) and mass spectrometry (Chapter 1.35), are not covered in detail either, because they are discussed in the earlier-noted chapters, but a few of their suppliers are also listed.

Table 1.27a provides a summary of the features of the more common calorimeter designs. For the definitions of the terms used in this chapter, refer to the end of the chapter.

---

* Gross or high heating value is the amount of heat produced by the complete combustion of a unit quantity of fuel. It is obtained when all products of the combustion are cooled down and when the water vapor formed during combustion is condensed.

  Net or lower heating value is the amount of heat produced by combustion of a unit quantity of a fuel minus the latent heat of vaporization of the water vapor formed by the combustion.

TABLE 1.27a
Summary of Calorimeter Features and Specifications

Type	Area Class: Direct	Area Class: Inferential	Area Class: General Purpose	Area Class: Ex-Proof	App 1	App 2	App 3	App 4	App 5	Continuously	Cyclic	Standard Sample	Empirical Calibration	Ambient Limits °F (°C)	Local Readout	Remote Transmitters	Range in Btu Full Scale	Accuracy ±% of Full Scale	Speed of Response (90%)
GROSS CALORIFIC VALUE																			
Water ΔT	✓		✓	✓	✓					✓	✓	✓	✓	72–77 (22–25)	✓	✓	130–3300	00.5	3 min
Air ΔT	✓		✓	✓	✓					✓	✓	✓		72–77 (22–25)	✓	✓	120–3600	00.5	15 min
Gas chromatograph		✓	✓	✓	✓	✓		✓			✓	✓		0–100 (–18 to 38)	✓	✓	Any	00.5	10 min
Adiabatic flame temperature	✓		✓	✓	✓					✓		✓		N/A	✓	✓	N/A	00.5	N/A
NET CALORIFIC VALUE																			
Airflow calorimeter	✓		✓	✓		✓		✓		✓		✓		50–90 (10–32)	✓	✓	130–3300	10.0	8 s
Gas chromatograph		✓	✓	✓			✓	✓			✓	✓		0–128 (–18 to 53)	✓	✓	Any	00.5	10 min
Expansion tube calorimeter	✓		✓				✓	✓		✓		✓		N/A	✓	✓	120–3300	10.0	3.5 min
Specific gravity		✓	✓			✓		✓		✓		✓		0–128 (–18 to 53)	✓	✓	Varies	20.0	N/A
Process chromatograph	✓		✓					✓	✓	✓			✓	60–90 (16–32)	✓	✓	150–3600	20.0	4.5 min
Thermopile calorimeter	✓	✓				✓		✓		✓		✓		N/A	✓	✓	150–3300	20.0	55 sec
WOBBE INDEX																			
Airflow calorimeter	✓		✓			✓	✓	✓		✓		✓		50–110 (10–43)	✓	✓	130–3300	00.75	8 sec
Gas chromatograph		✓	✓	✓	✓		✓	✓			✓	✓		0–120 (–18 to 49)	✓	✓	Any	00.5	10 min
Expansion tube calorimeter	✓		✓			✓	✓	✓		✓		✓		N/A		✓	120–3300	10.0	3.5 min
Thermopile calorimeter	✓	✓		✓		✓	✓	✓		✓		✓		N/A	✓	✓	150–3300	20.0	55 sec

[a] 1. Custody transfer, 2. process monitoring and control, 3. band mixing of fuel gases, 4. gas and liquefied natural gas (LNG) processing, 5. compliance recording.

UNITS AND WOBBE INDEX

When calorimeters are used for custody transfer of natural gas, the unit of measurement is often the BTU. One cubic foot (cf) of natural gas has an energy content of 1,027 BTUs. The Natural Gas Policy Act established the BTU as the unit of measurement of natural gas for pricing purposes, supplanting the traditional volume-based measurement. As a result, the market demand for custody transfer-type calorimeters markedly increased. In Europe and in other countries, natural gas calorimeters are usually calibrated in mega-Joule units.

The calorimeter's response time usually is the critical consideration for closed-loop control applications, while for custody transfer applications, it is accuracy. The calorimeter can measure either the gross calorific value (also called higher heating value) or net calorific value (also called lower heating value) or the Wobbe index (WI).

The WI is used to compare the combustion energy output of different composition fuel gases. If two fuels have identical WIs, then, at a given pressure and valve setting, the energy output will also be identical. Typically, variations of up to 5% are allowed. In essence, the WI is a measurement of the available potential heat, and it can be used in conjunction with the gas flow measurement to produce a measurement of heat flow rate.

DESIGN VARIATIONS

Bomb Calorimeters

A calorimeter can be obtained by inserting a thermometer into a container that is filled with water and is suspended above a combustion chamber. This design is called a bomb calorimeter, which is used to measure the heat of combustion of fuels. The main components of such a design are a small cup to contain the sample, oxygen, a stainless steel bomb, water, a stirrer, a thermometer, the dewar or insulating container (to prevent heat flow from the calorimeter to the surroundings), and ignition circuit connected to the bomb (Figure 1.27a).

In more recent calorimeter designs, the whole bomb is pressurized with pure oxygen (typically at about 30 bars) and a sample of about 1 g, and a fixed amount of water is placed in it. Once the charge is electrically ignited, the combustion heat passes through the stainless steel wall, thus raising the temperature of the steel bomb, its contents, and the surrounding water jacket. The temperature change in the water is then accurately measured with a thermometer. This reading, along with a bomb factor (which is dependent on the heat capacity of the metal bomb parts), is used to calculate the heat released. The bomb, with the known mass of the sample and oxygen, forms a closed system—no gases escape during the reaction.

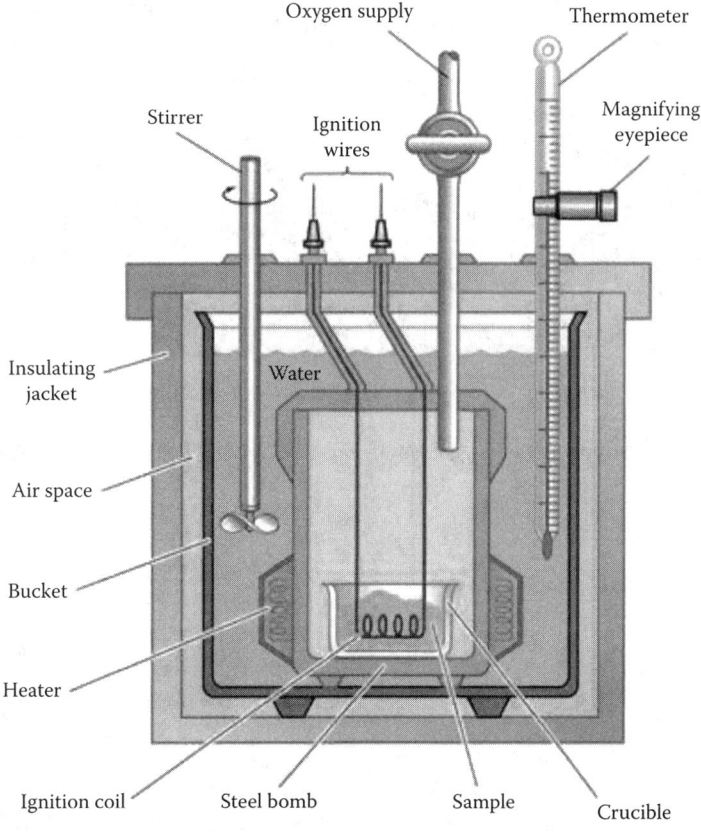

FIG. 1.27a
Components of a laboratory bomb calorimeter. (Courtesy of Helen Kollias.)

WATER TEMPERATURE RISE CALORIMETER

One variation of this design is illustrated in Figure 1.27b. Here, a constant (capillary-controlled) flow rate of the gas is mixed with a constant flow rate of oxygen (or air) and is burned. The resulting heat of combustion is removed by a constant flow rate of water. Both the flow rates and the temperatures of the entering streams are controlled at a constant value.

Thermistors measure the temperature of the water entering and leaving. The temperature difference or temperature rise is therefore a direct measure of the heating value of the burned gas. This calorimeter can be compensated for the variations in barometric pressure and can therefore measure the gross calorific value of fuel or waste gases.

Figure 1.27b also shows a calibration heater. During the calibration cycle, this electrical resistance heater is turned on while the gas flow is off. Thereby, a known accurately determined amount of heat is introduced, and the corresponding temperature rise on the water side is measured. When the temperature rise is the same as it was when gas was being burned, the amount of electric heat introduced matches the heating value of the gas. This type of calibration can be made any time and for any reading of the calorimeter.

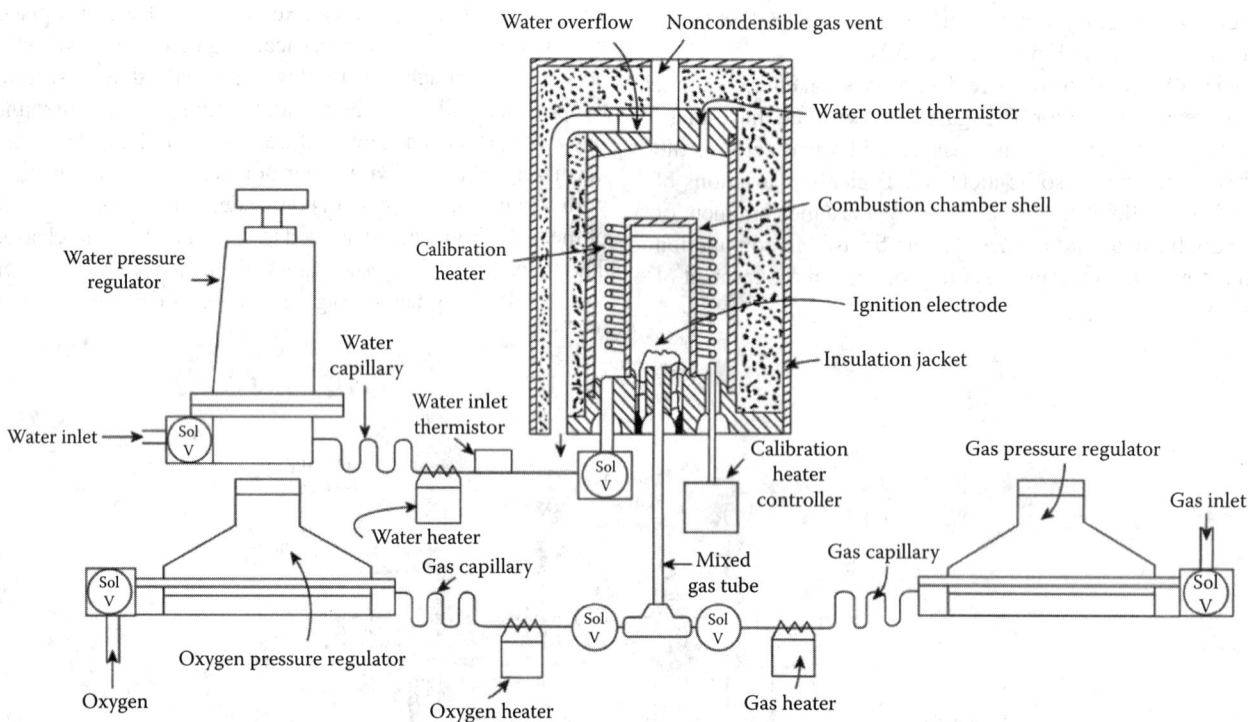

FIG. 1.27b
Water-temperature-rise-type calorimeter provided with electric heater for direct calibration. (From Christopher, D.E., Direct Energy Measurement, Measurements and Control, December 1991.)

Air Temperature Rise Calorimeter

The measurement is accomplished by continuously transferring all the combustion heat of a metered quantity of gas to a metered quantity of air (Figure 1.27c). The temperature rise of the air is measured and is related directly to gross calorific value of the gas. The unit can be modified so that the heat-absorbing air is not separated from the products of combustion, thus resulting in a more accurate measurement of the net calorific value.

Airflow Calorimeter

In this design, the variations in the heat released by the continuous burning of the fuel gas are offset by a continuous varying airflow that maintains the temperature of the products of combustion constant (Figure 1.27d). Thus, the airflow is correlated to the heating value of gas or to the WI. With the addition of a constant-volume gas-metering pump, and compensating for specific gravity variations, the instrument can be calibrated for the net calorific value.

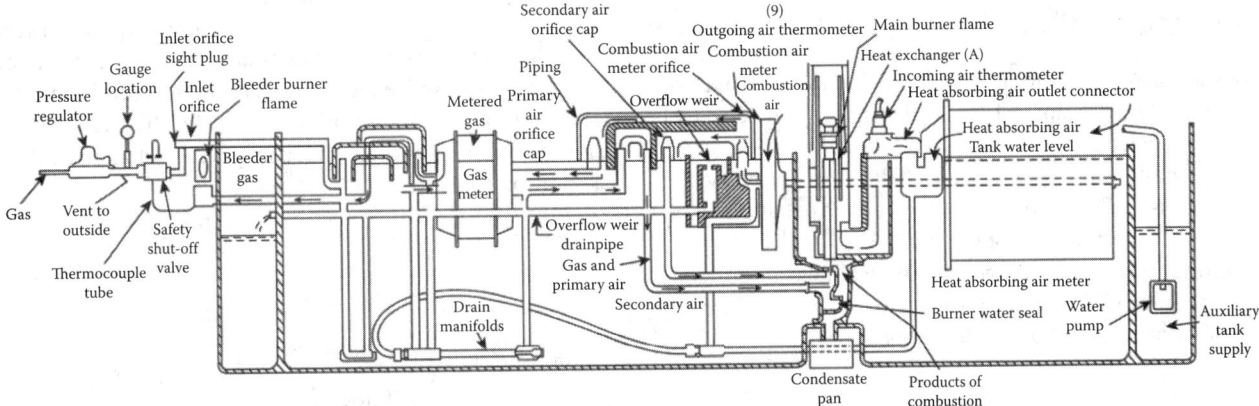

FIG. 1.27c
Air ΔT calorimeter.

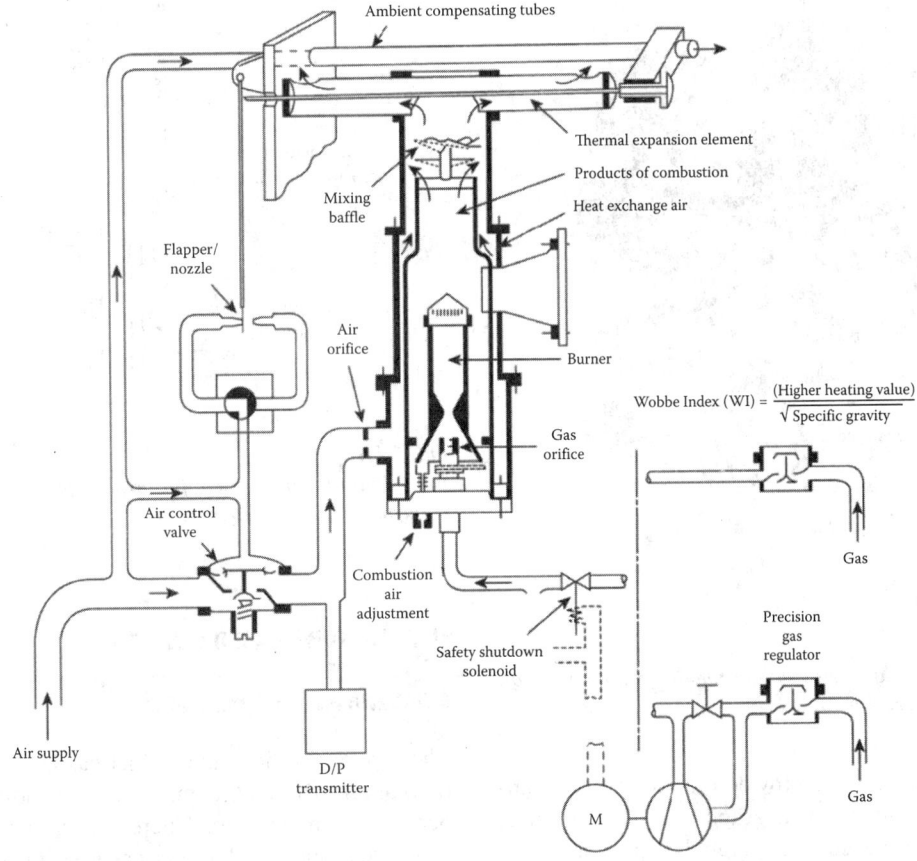

$$\text{Wobbe Index (WI)} = \frac{\text{(Higher heating value)}}{\sqrt{\text{Specific gravity}}}$$

FIG. 1.27d
Airflow calorimeter. CV, Calorific Value.

Residual Oxygen Calorimeter

In this design, a continuous gas sample is mixed with dry air at a precisely maintained ratio. The ratio is dependent on the BTU range of the gas to be measured. The fuel–air mixture is oxidized in a combustion furnace, and the oxygen concentration in the spent sample is measured using a zirconia oxide cell. With the addition of a precision-specific gravity cell, the analyzer can be calibrated to provide a measurement corresponding to the calorific value of the gas.

Chromatographic Calorimeter

A conventional chromatograph can also be used to analyze gas composition (Chapter 1.10), and a microprocessor can be used to calculate the heating value and specific gravity of the gas from empirical data held in memory by the microprocessor. This information can be used to calibrate the gross or net calorific value of the WI.

Chromatographic calorimeters are available in portable designs (Figure 1.27e), which are used to measure the heating value of natural gas in various locations.

relative density, composition, and compressibility using the AGA-8 method. The calorimeters come calibrated against a known-composition reference gas. Their memory can store 128 complete analyses and 32 calibrations.

Expansion Tube Calorimeter

In this design, the gas sample is delivered through a precision regulator system (Figure 1.27f), which is independent of specific gravity and atmospheric pressure. The gas is burned at the base of a differential expansion tube unit that responds to the temperature of the products of combustion and excess air. The differential signal is calibrated as the net calorific value. Modification of the regulator to respond to specific gravity changes allows calibration to the WI.

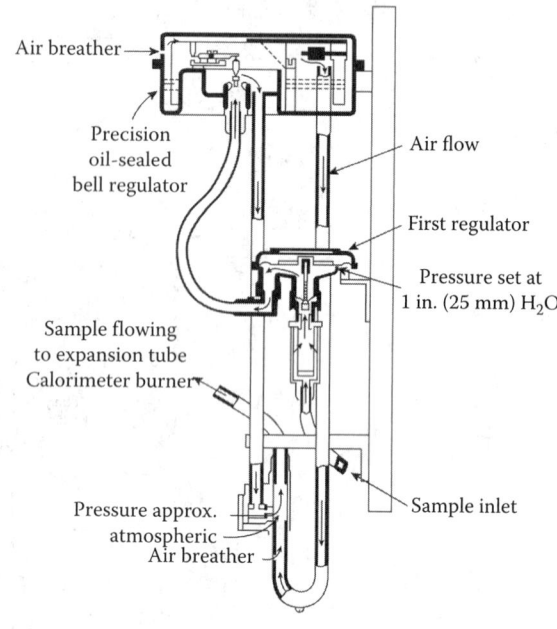

FIG. 1.27f
Precision regulator used for expansion tube-type calorimeter. (Courtesy of Cosa Instrument Corp.)

FIG. 1.27e
Portable natural gas calorimeter with chromatographic detector. (Courtesy of Chandler Engineering.)

In this design, software is provided to convert the composition readings of the chromatograph into heating value. The results can be displayed on the digital display or can be printed or transferred to a computer. The measurement takes about 12 min. These units can also calculate heating value,

TEMPERATURE-BASED DESIGNS

Adiabatic Flame Temperature

The gross calorific value of fuel gas is proportionate to the ratio of air to a fuel that maximizes the adiabatic flame temperature of the mixture. Therefore, in this design, flows are controlled, and two burners are used to obtain the mathematical derivative of the flame temperature composition that allows calibration for gross calorific value.

Thermopile

The thermopile calorimeter (Figure 1.27g) measures the temperature of the hot products of combustion mixed with a constant volume of air supplied by a fan. The sample to the burner is provided with an orifice bypass, which provides specific gravity compensation; thus, the resulting measurement is in terms of net calorific value. To measure the WI, the bleed is blocked, and the sample goes through the burner.

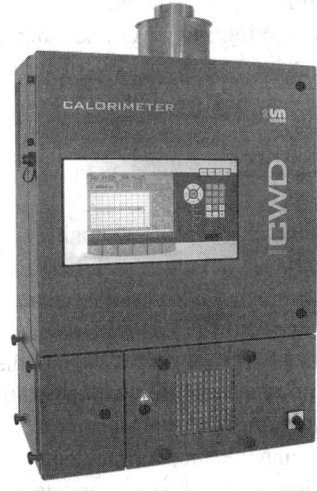

FIG. 1.27g
Thermopile calorimeters are also available for outdoor or hazardous area installations. (Courtesy of Delta Instrument LLC.)

Other Temperature-Based Designs

In heat flow calorimetry, either the temperature of the process or the temperature of the jacket is controlled. Heat is measured by monitoring the temperature difference between the heat transfer fluid and the process fluid. In addition, fill volumes (i.e., wetted area), specific heat, and heat transfer coefficient have to be determined to arrive at a correct value.

In heat balance calorimetry, the cooling/heating jacket controls the temperature of the process. Heat is measured by monitoring the heat gained or lost by the heat transfer fluid.

Adiabatic calorimeters are used to examine runaway reactions. Since the calorimeter runs in an adiabatic environment, any heat generated by the sample under test causes the sample temperature to increase, thus fueling the reaction. No adiabatic calorimeter is fully adiabatic, because some heat is always lost to the sample holder. A mathematical correction factor, known as the "phi-factor," can be used to correct the calorimetric result to account for these heat losses. The phi-factor is the ratio of the thermal mass of the sample and sample holder to the thermal mass of the sample alone.

Modulated-temperature differential scanning calorimeters (MTDSCs) are a type of differential scanning calorimeters, in which a small oscillation is imposed upon the otherwise linear heating rate.

MTDSC has a number of advantages. It facilitates the direct measurement of the heat capacity in one measurement, even in (quasi-)isothermal conditions. It permits the simultaneous measurement of heat effects that are reversible and not reversible at the timescale of the oscillation (reversing and nonreversing heat flows, respectively). It increases the sensitivity of the heat capacity measurement, allowing for scans at a slow underlying heating rate.

Optical Calorimeters

The optical gas calorimeters are new designs utilizing interferometry. Their main advantage is speed. In their portable design, they are capable of spot-checking measurements of natural gas or LPG at several locations, taking less than a minute at each. This innovative optical technology greatly improves the analysis speed while reducing the size and weight of the calorimeter.

Interferometry makes use of the principle of superposition to combine waves (light or electromagnetic) in a way that will cause the result to reflect the nature of the substance that the wave passes through. This works because when two waves with the same frequency combine, the resulting pattern is determined by the phase difference between the two waves—waves that are in phase will undergo constructive interference, while waves that are out of phase will undergo destructive interference. In analyzers, interferometers can be used in continuous-wave Fourier transform spectroscopy to analyze light containing features of absorption or emission associated with a substance or mixture.

Typically (see Figure 1.27h), a single incoming beam of coherent light will be split into two identical beams by a

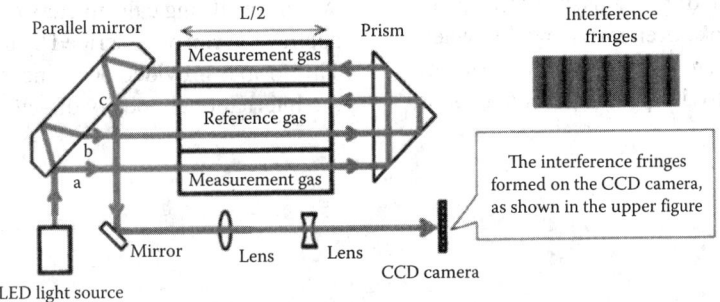

FIG. 1.27h
The operating principle of the calorimeter that utilizes the principles of interferometry. (Courtesy of Cosa Instrument Co.)

beamsplitter (parallel mirrors). Each of these beams travels a different route, called a path (a and b), and are recombined (c) before arriving at a detector. The path difference, the difference in the distance traveled by each beam, creates a phase difference between them. It is this introduced phase difference that creates the interference pattern between the initially identical waves. This phase difference is diagnostic of its cause along the paths, such as a difference in the refractive indexes of the gases along the paths.

This calorimeter therefore is an indirect one (like a chromatographic one) in so far that it determines the composition of the natural gas or other gas sample and, based on that composition, calculates the calorific value of the sample. The sensor in this case is an optic interferometer sensor (shown in Figure 1.27h), which measures the reflection of the measurement gas relative to the reflection of a reference gas and, based on this "reflection ratio" (using the Fourier interference fringes detected by a CCD camera), calculates the gas composition. Once the composition of the measurement gas is known, it calculates the heating value the same way as does a calorimeter using a chromatographic sensor.

APPLICATIONS

There are five general areas of application:

1. Custody transfer: Sale or purchase of fuel gas with accuracy being the primary consideration.
2. Process monitoring and control: To effect online manual or automatic control of process or efficient burning of the gas by using heat value measurement as one of the measured variables. Such applications include feed-forward control of fuel gas–fired heaters or boilers, stove-firing control, vaporizer control, and synthetic natural gas reactor control. In these applications, the measurement of heat value is important as a process measurement for efficient control of energy consumption or to limit the flow of heat to an energy-sensitive process. This application requires high speed of responses as well as reliability to achieve effective control.
3. Blending and mixing of fuel gas: To obtain a uniform quality gas or to utilize waste or by-product gases by blending them into the main fuel gas. Blending is often used to achieve a desired ratio of streams such as propane/air or blast furnace gas/coke oven gas. However, when by-product or waste gases are to be injected into a stream and they vary significantly in quantity, monitoring of the mixture is necessary to make proper use of the fuel. Speed of response and reliability are essential.
4. Processing of gas and liquefied natural gas (LNG) operations: LNG must be vaporized and conditioned for efficient consumption. Coke oven and blast furnace gases are the main fuels used by the steel industry, and refinery gas is a major by-product that is used in petroleum refining. All these fuel gases are produced in specific processing operations that require monitoring and conditioning for efficient use.
5. Compliance recording: For government-regulated energy transfer by pipeline and utility distributors to the consumers. Accuracy and traceability are essential criteria.

SAMPLE CONDITIONING

The requirements for the sample conditioning system are dependent on the limitations of the selected analyzer and the minimum, normal, and maximum limits of the process stream being monitored. In many industrial applications, fuel gases are generated as by-products of other processes, and these "off-gases" can be extremely dirty and require sample conditioning. In some cases, special consideration must be given to the dew point of the off-gas when designing the sample conditioning system. Refer to Chapters 1.2 for sample system options and design features.

CONCLUSION

Gaseous fuel energy is a costly commodity that is being consumed with much more care and efficiency than in the past. The key to its efficient consumption is measuring the available heat of the fuel gas. There are several calorimeters available for reliable online measurement of gas heating values. For closed-loop control applications, the designs that provide a speed of response of less than a minute (Table 1.27a) are preferred.

In addition, there are several manufacturers that offer calorimeters specifically for custody transfer applications. These analyzers offer improved accuracy at the expense of response time (Table 1.27a).

SPECIFICATION FORM

When specifying calorimeters or heating value detectors, the application can be described by using the ISA Form 20A1001. This form is provided on the next page with the permission of the International Society of Automation.

1.27 Heating Value Calorimeters

1	RESPONSIBLE ORGANIZATION		ANALYSIS DEVICE	6	SPECIFICATION IDENTIFICATIONS	
2				7	Document no	
3	ISA		Operating Parameters	8	Latest revision	Date
4				9	Issue status	
5				10		

11	ADMINISTRATIVE IDENTIFICATIONS			40	SERVICE IDENTIFICATIONS Continued		
12	Project number		Sub project no	41	Return conn matl type		
13	Project			42	Inline hazardous area cl	Div/Zon	Group
14	Enterprise			43	Inline area min ign temp	Temp ident number	
15	Site			44	Remote hazardous area cl	Div/Zon	Group
16	Area	Cell	Unit	45	Remote area min ign temp	Temp ident number	
17				46			
18	SERVICE IDENTIFICATIONS			47			
19	Tag no/Functional ident			48	COMPONENT DESIGN CRITERIA		
20	Related equipment			49	Component type		
21	Service			50	Component style		
22				51	Output signal type		
23	P&ID/Reference dwg			52	Characteristic curve		
24	Process line/nozzle no			53	Compensation style		
25	Process conn pipe spec			54	Type of protection		
26	Process conn nominal size		Rating	55	Criticality code		
27	Process conn termn type		Style	56	Max EMI susceptibility	Ref	
28	Process conn schedule no		Wall thickness	57	Max temperature effect	Ref	
29	Process connection length			58	Max sample time lag		
30	Process line matl type			59	Max response time		
31	Fast loop line number			60	Min required accuracy	Ref	
32	Fast loop pipe spec			61	Avail nom power supply		Number wires
33	Fast loop conn nom size		Rating	62	Calibration method		
34	Fast loop conn termn type		Style	63	Testing/Listing agency		
35	Fast loop schedule no		Wall thickness	64	Test requirements		
36	Fast loop estimated lg			65	Supply loss failure mode		
37	Fast loop material type			66	Signal loss failure mode		
38	Return conn nominal size		Rating	67			
39	Return conn termn type		Style	68			

69	PROCESS VARIABLES	MATERIAL FLOW CONDITIONS		101	PROCESS DESIGN CONDITIONS		
70	Flow Case Identification		Units	102	Minimum	Maximum	Units
71	Process pressure			103			
72	Process temperature			104			
73	Process phase type			105			
74	Process liquid actl flow			106			
75	Process vapor actl flow			107			
76	Process vapor std flow			108			
77	Process liquid density			109			
78	Process vapor density			110			
79	Process liquid viscosity			111			
80	Sample return pressure			112			
81	Sample vent/drain press			113			
82	Sample temperature			114			
83	Sample phase type			115			
84	Fast loop liq actl flow			116			
85	Fast loop vapor actl flow			117			
86	Fast loop vapor std flow			118			
87	Fast loop vapor density			119			
88	Conductivity/Resistivity			120			
89	pH/ORP			121			
90	RH/Dewpoint			122			
91	Turbidity/Opacity			123			
92	Dissolved oxygen			124			
93	Corrosivity			125			
94	Particle size			126			
95				127			
96	CALCULATED VARIABLES			128			
97	Sample lag time			129			
98	Process fluid velocity			130			
99	Wake/natural freq ratio			131			
100				132			

133	MATERIAL PROPERTIES			137	MATERIAL PROPERTIES Continued		
134	Name			138	NFPA health hazard	Flammability	Reactivity
135	Density at ref temp		At	139			
136				140			

Rev	Date	Revision Description	By	Appv1	Appv2	Appv3	REMARKS

Form: 20A1001 Rev 0 © 2004 ISA

Definitions

British Thermal Unit (BTU) is the amount of heat required to raise the temperature of 1 lb of water by 1°F at or near 60°F.

Calorific Value is the energy contained in a fuel or other materials, which can be determined by measuring the heat produced by the complete combustion of a specified quantity of it. This is usually expressed in joules per kilogram or for gases as mega joules per cubic meter.

Combustion Air Requirement Index (CARI) is a dimensionless number that indicates the amount of air required (stoichiometrically) to support the combustion of a fuel gas. Mathematically, the CARI is defined by Equation 1.27(1):

$$CARI = \frac{(Air-Fuel\ Ratio)}{\sqrt{(Sp.G)}} \qquad 1.27(1)$$

Dry BTU is the heating value of a gas, expressed on a dry basis. The common assumption is that pipeline gas contains 7 pounds (or less) of water vapor per million standard cubic feet (MSCF).

Gross calorific value is the heat value of energy per unit volume at standard conditions, expressed in terms of BTU per SCF or kilocalorie per cubic Newton meters (Kcal/N.m^3).

Gross or high heating value (HHV) is the amount of heat produced by the complete combustion of a unit quantity of fuel. It is obtained when all products of the combustion are cooled down and when the water vapor formed during combustion is condensed. Common units used are Btu/lb, J/kg and kcal/kg.

Heating value (HV) or calorific value (CV) is the amount of heat produced by combustion of a unit quantity of a fuel. Common units used are Btu/lb, J/kg and kcal/kg.

Net calorific value is the measurement of the actual available energy per unit volume at standard conditions, which is always less than the gross calorific value by an amount equal to the latent heat of vaporization of the water formed during combustion.

Net or lower heating value (LHV) is the amount of heat produced by combustion of a unit quantity of a fuel minus the latent heat of vaporization of the water vapor formed by the combustion. Common units used are Btu/lb, J/kg and kcal/kg.

Saturated BTU is the heating value that is expressed on the basis that the gas is saturated with water vapors. This state is defined as the condition when the gas contains the maximum amount of water vapors without condensation, when it is at base pressure and 60°F temperature.

Wobbe Index (WI) or Wobbe Number accounts for composition variations in terms of their effect on the heat value and specific gravity, which affect the flow rate through an orifice. In essence, the Wobbe Index is a measurement of the available potential heat, and it can be used in conjunction with the gas flow measurement to produce a measurement of heat flow rate. WI is an indicator of the interchangeability of fuel gases, such as natural gas, liquefied petroleum gas (LPG), and town gas. Equation 1.27(2) gives the definition of the Wobbe Index as a function of higher heating value (HHV) and specific gravity (SpG):

$$WI = HHV/\sqrt{SpG} \qquad 1.27(2)$$

Abbreviations

CARI	Combustion Air Requirement Index
CCD	Charge-coupled device
DSC	Differential scanning calorimeter
HHV	Higher heating value
LHV	Lower heating value
LNG	Liquefied natural gas
MSCF	Million standard cubic feet
MTDSC	Modulated-temperature differential scanning calorimeter
SNG	Synthetic natural gas
TM-DCS	Temperature-modulated differential scanning calorimetry
WI	Wobbe Index

Organization

AGA   American Gas Association

Bibliography

ASTM D1826 - 94, Standard test method for calorific (heating) value of gases in natural gas range by continuous recording calorimeter, (2010) http://www.astm.org/Standards/D1826.htm.

ASTM D7314, Standard practice for determination of the heating value of gaseous fuels using calorimetry and on-line/at-line sampling, (2015) http://www.astm.org/Standards/D7314.htm.

AZO Network, Temperature-modulated differential scanning calorimetry (TM-DSC), (2015) http://www.azom.com/article.aspx?ArticleID=4982.

Coleman, N. C., Craig, D. Q. M., Modulated temperature differential scanning calorimetry, (1996) http://www.sciencedirect.com/science/article/pii/0378517395044639.

Control Instruments, Selecting the Right Analyzer to Measure Heating Value in Flare Stacks (2015) http://www.controlinstruments.com/documents/selecting-right-analyzer-measure-heating-value-flare-stacks.

Couchman, J., The opal detector, (2002) http://www.hep.ucl.ac.uk/~jpc/all/ulthesis/node16.html.

Hohne, G. et al., Differential scanning calorimetry, (2003) http://www.springer.com/us/book/9783540004677.

Kaletunc, G., Calorimetry in food processing, (2009) http://www.wiley.com/WileyCDA/WileyTitle/productCd-0813814839.html.

Market Research, Global calorimeter industry report, 2015 http://www.marketresearch.com/QYResearch-Group-v3531/Global-Calorimeter-9121630/.

Parr, Introduction to bomb calorimetry, (2007) http://www.scimed.co.uk/wp-content/uploads/2013/03/Introduction-to-bomb-calorimetry.pdf.

Physics Classroom, Calorimeters and calorimetry, (2014) http://www.physicsclassroom.com/class/thermalP/Lesson-2/Calorimeters-and-Calorimetry.

Reed, B., Energy measurement using flow computers and chromatography, (2003) http://help.intellisitesuite.com/hydrocarbon/papers/5130.pdf.

Sokhansanj, S., The effect of moisture on heating values, (2011) http://cta.ornl.gov/bedb/appendix_a/The_Effect_of_Moisture_on_Heating_Values.pdf.

Sullivan, A. O., Oxygen bomb calorimeter experiment to find the calorific, (2013) http://www.academia.edu/5337823/Oxygen_Bomb_Calorimeter_Experiment_to_find_the_calorific.

University of Cambridge, The OPAL endcap electromagnetic calorimeter, (2015) http://www.hep.phy.cam.ac.uk/opal/opal_ee.html.

Wigmans, R., Calorimetry, (2000) http://ukcatalogue.oup.com/product/9780198502968.do.

1.28 Hydrocarbon Analyzers

R. J. GORDON (1974, 1982)   **B. G. LIPTÁK** (1995)   **I. VERHAPPEN** (2003)*
B. G. LIPTÁK and I. VERHAPPEN (2017)

To receiver

(AT) HC or combustibles

Flow sheet symbol

Type of measurements	A1. Total hydrocarbon by flame ionization
	B1. Methane by chromatography
	B2. Methane by flame ionization
	B3. Methane by mass spectrometry
	C1. Hydrocarbon classes by flame ionization
	C2. Hydrocarbon classes by mass spectrometry
	C3. Hydrocarbon classes by infrared
	D1. Individual hydrocarbons by chromatography
	E. Hydrocarbon dew point in natural gas (chilled mirror)
	F. Ion mobility spectroscopy
	G. Laser-induced Doppler Absorption Radar (LIDAR)
	H. Vertical Cavity Surface Emitting Laser (VCSEL)
Features (accuracy, etc.)	Note: For data concerning the features of the selected analyzer, refer to the following chapters:
	Chapter 1.10—Chromatography
	Chapter 1.14—Combustibles
	Chapter 1.32—Infrared
	Chapter 1.35—Mass Spectrometer
	Chapter 1.39—Chilled Mirror
	Chapter 1.58—Laser Absorption
	Chapter 1.62—Spectrometry
	Chapter 1.69—Ultraviolet
Ranges	From 0–10 ppm to 0%–100% lower explosive limit (LEL).
Reference method	Gas chromatography with flame ionization for nonmethanes.
Sensitivity	0.1 ppm (A1, B2, C1); chromatograph followed by electrochemical sensor can detect 50 ppb.
Costs	Infrared systems cost from $1,500 to $3,000 for a point system and $7,500 to $20,000 for an open-path system.
	Combustible gas battery-operated portable leak detectors $100–$1000 (see Chapter 1.14).
	Combustibles permanently installed multichannel system, $1000–$2500 per channel (see Chapter 1.14).
	Flame ionization detector for chromatographs, $3000–$5000.
	Flame or photoionization with sampled remote head installation, per-channel, $3000–$5000.
	Infrared. Infrared systems $1,500–$3,000 for a point and $7,500–$20,000 for an open-path system (see Chapter 1.62).
	Chromatographic units cost $25,000–$100,000 (see Chapter 1.10).
	Portable chromatograph with electrochemical detector $15,000 (see Chapter 1.10).
	Mass spectrometers start at $50,000 and can reach $250,000 (see Chapter 1.35).

(Continued)

---

* Reviewed by F. D. Martin.

Partial list of suppliers (For chromatograph, mass spectrometer, and infrared analyzer suppliers, also see Chapters 1.10, 1.32, and 1.35.)

ABB (C3 FT-NIR laboratory unit) http://new.abb.com/products/measurement-products/analytical/ft-ir-and-ft-nir-analyzers/laboratory-spectrometers/mb3600-hp10

Boreal Laser (G laser) http://www.boreal-laser.com/?s=monitoring&x=14&y=3

Buck Scientific (C3) infrared) http://www.bucksci.com/atomic-absorption-spectrophotometers/details/35/2/m530-infrared-spectrophotometer/hc-404-total-hydrocarbon-analyzer.html

Draeger (C3) http://www.draeger.com/sites/en_corp/Pages/Corporate/SearchResults.aspx?s=Corporate_EN_CORP&p=115&k=hydrocarbon

EcoTech (A1 flame ionization) http://ecotech.com/wp-content/uploads/2015/03/Baseline-D019-3-9000-Total-Carbon-Analyser.pdf

Emerson Process Management, Rosemount (A1 flame ionization) http://www2.emersonprocess.com/en-US/brands/rosemountanalytical/GasAnalysisSolutions/ContinuousGasAnalyzers/Hydrocarbon/Pages/index.aspx

General Monitors (combustibles) http://www.generalmonitors.com/Gas-Detectors/S4100C-Combustible-Gas-Detector/p/000140006500001031

Gow-Mac Instrument Co. (A1 flame ionization) http://www.gow-mac.com/Products/Instruments/GasAnalyzers/NEWTotalHydrocarbonGasAnalyzers.aspx

Honeywell, DMS (C3 combustibles, infrared) http:// http://www.detect-measure.com/zell.php

JUM Engineering, http://www.jum.com/info.htlm

Michell Instruments Ltd. (E hydrocarbon dew point) http://www.michell.com/us/news/condumax_II_hydrocarbon_dew-point_analyzer_result_driven_best_in_class_for_natural_gas_quality.htm

Mocon Inc. formerly Baseline Industries (A1flame ionization) http://products.baseline-mocon.com/category/hydrocarbon-analyzers

Particle Measuring Systems (I IMS) http://pmeasuring.com/products/airsentry-ii-mobile-amc-detection-system/

Sick (A1 flame ionization) http://www.sick.com/group/EN/home/products/product_news/analysis_and_process_measurement/Pages/gms810_fidor_total_hydrocarbon_analyzer.aspx

Siemens (A1 flame ionization) http://w3.siemens.com/mcms/sensor-systems/en/process-analytics/gas-analyzer-gas-analysis/extractive/hydrocarbons-total-content-measurement/pages/fidamat-6.aspx

Smiths Detection (B1, C2 chromatograph, mass spectrometer) http://www.smithsdetection.com/index.php/gaschromatography-mass-spectometry.html

Teledyne Analytical Instruments (C1) http://www.teledyne-ai.com/pdf/4000_Rev-.pdf

Thermoscientific (A1 flame ionization) http://www.thermoscientific.com/en/product/model-51i-total-hydrocarbon-analyzer.html

Vig Industries Inc. (A1 flame ionization) http://www.vigindustries.com/

Yokogawa, (H) http//www.yokogawa.com/us/technical-library/resources/white-papers/laser-spectroscopic-multi-component-hydrocarbon-analyzer/

INTRODUCTION

Hydrocarbon analysis is also discussed in several chapters listed in the following, and therefore it is advisable to also refer to these chapters when investigating the features and capabilities of a particular hydrocarbon analyzer type.

Chapter 1.14—Detection of combustibles
Chapter 1.10—Gas chromatographs
Chapter 1.26—Hazardous gas monitoring
Chapter 1.46—Oil-in-water measurement
Chapter 1.67—Total organic carbon analysis

The hydrocarbons comprise a large class of individual components (Table 1.28a). The most abundant hydrocarbon, methane, is nonreactive in photochemical reactions and is a less hazardous air pollutant. The remaining hydrocarbons vary widely in their complexity and reactivity. For this reason, the type of hydrocarbon analysis selected depends on the purpose for which data are needed.

To obtain highly detailed analyses for numerous hydrocarbons requires expensive and sophisticated equipment, whereas a single measurement for total hydrocarbons is simpler and somewhat less expensive. There are also methods of intermediate complexity, which permit the determination of properties without identification of the individual stream components.

TABLE 1.28a
Atmospheric Hydrocarbon Analyzers

Type	Hydrocarbon	
	Method	Limitations and Interferences
Total (as carbon)	FID	Some response to carbon-containing nonhydrocarbons
Methane	GC	Expensive equipment (can also be used for carbon monoxide)
Methane-only subtractive	Column preparation fussy column and FID	
	Mass spectrometry	Freeze-out required, expensive
Aromatics, olefins, paraffins	Subtractive columns and FID	Column preparation fussy
	Mass spectrometry	Freeze-out required, expensive, data reduction requirements large
	Infrared spectrometry	Freeze-out required, expensive, not total class coverage
	Ion-mobility spectrometry	Clean sample required, limited knowledge in industry
	Laser-induced absorption	Expensive, specialized support required
	Perimeter monitoring	Concentration/unit length rather than point value
Individuals	GC	Expensive, data reduction requirements large

ANALYZER TYPES

Flame Ionization Detectors

Flame ionization detectors (FIDs) are used to measure total hydrocarbons at low concentrations such as for pollution, leakage, or safety monitoring. In the FID analyzer, a flame of pure hydrogen, which contains almost no ions, is used (Figure 1.28a). When even traces of organic compounds are introduced by the sample, the hydrogen flame ionizes the carbon atoms, resulting in a large number of ions in the flame. The resulting ionic current is measured as an indicator of hydrocarbon concentration.

In dual-flame detectors, the concentration of total hydrocarbons and methane can be identified separately.

Flame ionization detection is the most suitable means of analysis for most hydrocarbons at the levels found in polluted air. It may be used alone for the measurement of total hydrocarbons or as a detector after separation by a column device such as a gas chromatograph. In FID analyzers, a sensitive electrometer detects the increase in ion intensity in a hydrogen flame when a sample containing organic compounds is introduced. The response is approximately in proportion to the number of organically bound carbon atoms in the sample, so the detector is basically a carbon atom counter. Carbon atoms bound to oxygen, nitrogen, or halogens, however, give a reduced response.

There is no response to nitrogen, carbon monoxide, carbon dioxide, or water vapor. The results are usually expressed in terms of the calibration gas used, for example, parts per million (ppm) carbon as methane. The response is rapid and can be as sensitive as 0.1 ppm. The response to various hydrocarbons is not perfectly uniform, and such variations should be taken into account in interpreting FID data. The FID analyzer is a reference method for total hydrocarbons in U.S. Air Quality Standards. Basically, it is an air-metering device attached to an FID (Figure 1.28b).

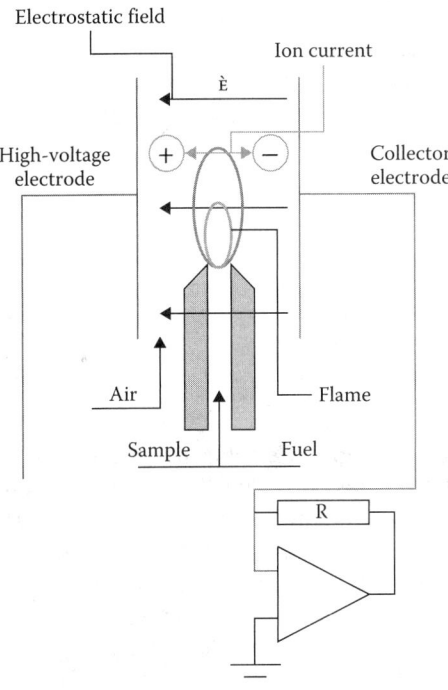

FIG. 1.28a
Flame ionization detector. (Courtesy of JUM Engineering.)

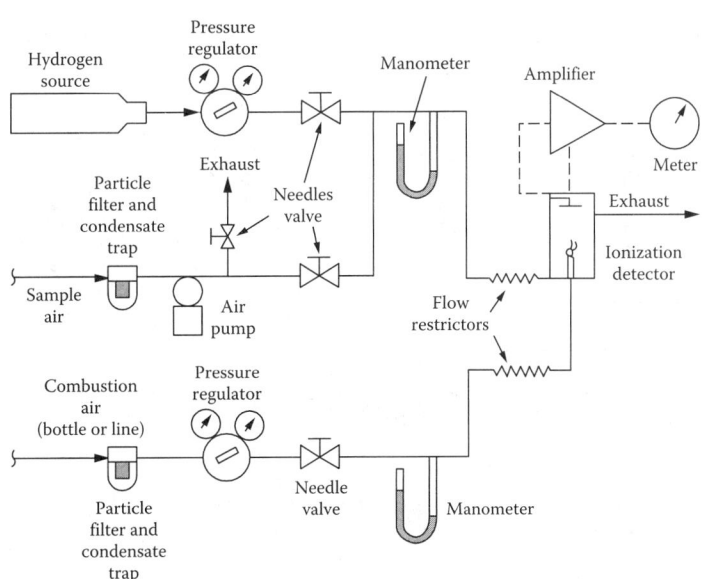

FIG. 1.28b
Hydrocarbon analyzer and hydrogen FID.

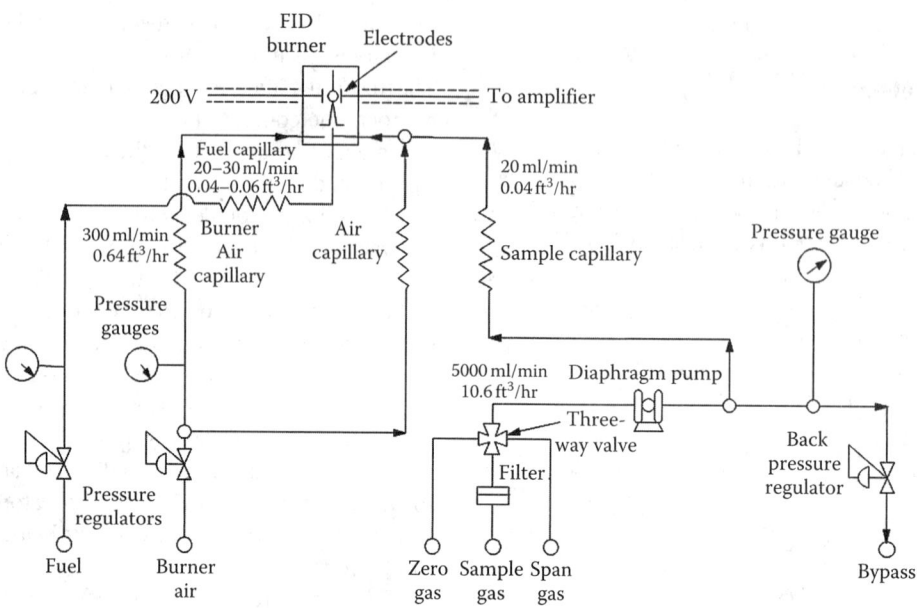

FIG. 1.28c
Sampling system for hydrocarbon analysis by an FID.

In microprocessor-controlled hydrocarbon analyzers, up to eight channels can be obtained for the simultaneous analysis of eight samples. Manufacturers also provide sampling systems (Figure 1.28c), which make remote calibration and automatic self-checking possible.

If discrete component analysis is not required, it is also possible to use liquid samples in conjunction with FID-type hydrocarbon analyzers if the hydrocarbons can be purged by nitrogen from the continuous liquid sample. Such a sampling system is shown in Figure 1.28d, where a nitrogen extraction column is used, serving to remove the hydrocarbon, which can be removed by purging. Here, the nitrogen is introduced at the bottom of the extraction column, and as it travels upward, it removes the purgeable hydrocarbons. The overflow level in the head tank, which seals the top of the column as the purge nitrogen leaves it, holds the extraction column pressure constant.

Benchtop microprocessor-controlled FID analyzers (Figure 1.28e) are often provided with graphic LCD touch screens, flameout indicators, and automatic fuel and sample line shutoff, ranges from 0 to 20,000 ppm, automatic zero, calibration, gas shutoff, programmable alarms, data logging, flow control for the gases, and both analog and digital outputs.

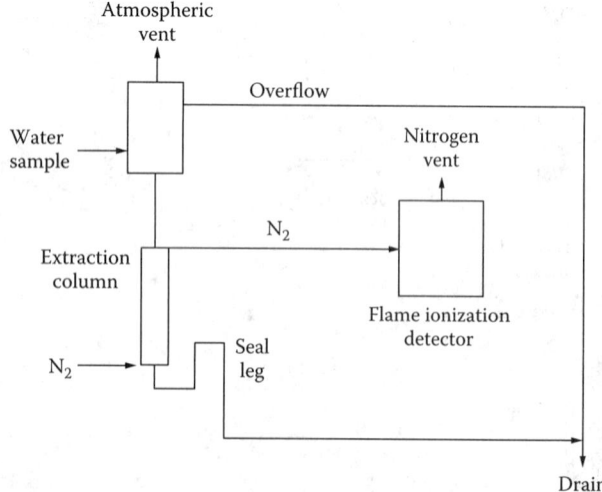

FIG. 1.28d
Detection of purgeable hydrocarbon concentration in water samples.

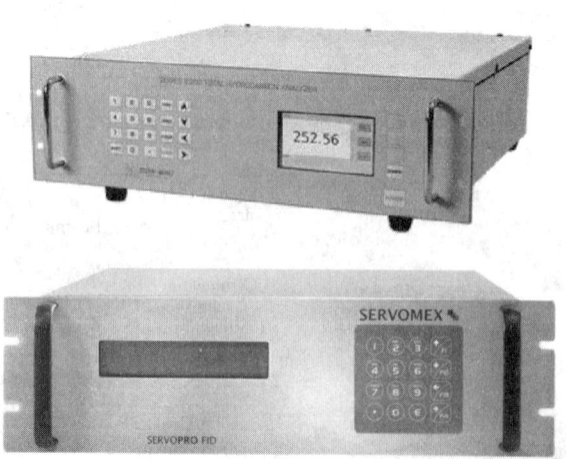

FIG. 1.28e
FID-based, smart, total hydrocarbon analyzer. (Courtesy of Gow Mac.)

Gas Chromatography

The only practical method for the measurement of specific hydrocarbons is gas chromatography (GC), which is discussed in full detail in Chapter 1.10. In GC, the basic apparatus (Figure 1.28f) is a carefully prepared tubular column of a finely divided solid with provisions for passing a steady flow of a carrier gas (often helium) through it. The column is temperature controlled, and the packing usually supports or is coated by a nonvolatile liquid phase.

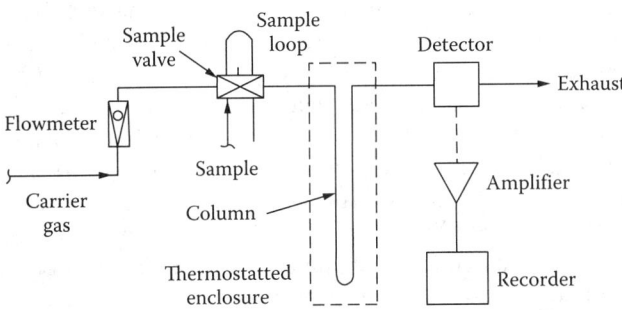

FIG. 1.28f
Gas chromatograph.

Preceding the column, there is a means for sample introduction (and sometimes sample splitting), and following the column is a detector. A high-sensitivity column usually employs an FID. There are a great many optional variations in the makeup of a GC instrument, involving sample valves, precolumns, columns, programmed temperature and pressure changes, stream splitters, and detectors.

The original GCs recorded the output on a strip chart (possibly with a peak integrator) or are converted to digital form for computer handling. In full-scale analyses for scores of hydrocarbons, data reduction from analog strip-chart recordings becomes very tedious and time consuming. Today's gas chromatographs use a computer interface to analyze and report the results of each analytical cycle as well as a wealth of related equipment health conditions.

Supplemental techniques increase the utility of GC for atmospheric hydrocarbons. A common method is to pass a sample of air through a small freeze-out trap, sweep out the air with helium, and then warm the trap and introduce the condensables to the GC column in one concentrated slug. This extends the lower limits of sensitivity to the parts-per-billion range. Subtractive columns may be in parallel or in series with conventional columns to help relate the various recorded peaks to specific hydrocarbon classes.

Calibration Complete GC calibration would require hundreds of different pure hydrocarbons and is never fully achieved. There are usually few unknowns left in careful work except in the more complex higher-molecular-weight ranges. Calibration is based on the fact that under carefully reproduced conditions, a given hydrocarbon will always require the same length of time to pass through the column to the detector. Ambiguities arising where components overlap may be resolved by change in column packing, increasing the analytical cycle time and column length, or sometimes by the use of subtractive columns. An elaborate but useful technique is to attach another analytical instrument such as an infrared (IR) or mass spectrometer to the outlet of the detector for peak identification.

In Figure 1.28g, an instrument is described for combined analysis of methane and carbon monoxide. This is a simple GC, set up to determine methane, with a programmed arrangement of valves and a catalytic carbon monoxide-to-methane converter. As a result, the readout has two peaks, one for methane directly and the other for methane formed by reduction of carbon monoxide.

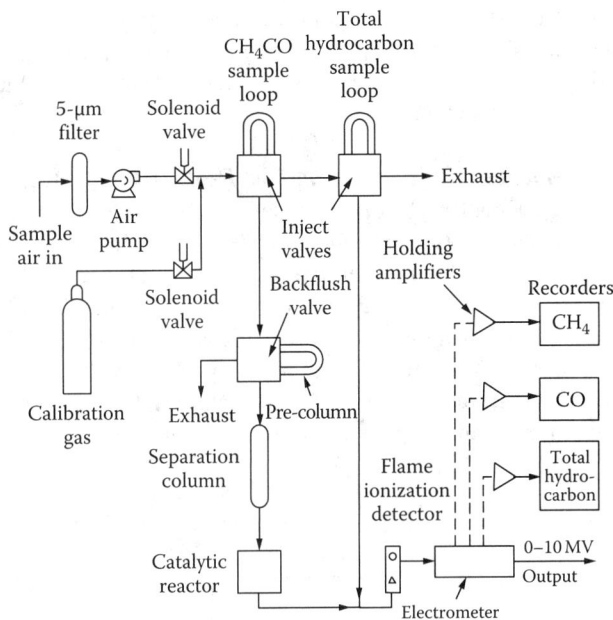

FIG. 1.28g
Methane and carbon monoxide analyzer.

Nonmethane Hydrocarbons A column of adsorptive charcoal may be treated with methane just until breakthrough occurs; that is, no more methane will be adsorbed, but other hydrocarbons are still retained. This column can then be used with an FID as an analyzer for methane only. Used in parallel or in switched alternation with the detector without any column, it allows the determination of nonmethane hydrocarbons by difference.

Reactive Hydrocarbons Certain specially prepared columns can be used to adsorb specific hydrocarbon classes. These are useful in GC and could be used with a simple FID analyzer alone to give analysis by broad classes. A column of crushed firebrick supporting mercuric sulfate will adsorb olefins and acetylenes, and one of palladium sulfate will then adsorb aromatic hydrocarbons (except benzene). A combination of these subtractive columns has been applied successfully to automobile exhaust analysis and for use in atmospheric work.

Spectrometric Methods

Either IR (Chapter 1.32) or mass spectrometry (Chapter 1.35) may be used with considerable advantage for individual hydrocarbon determination, although the sensitivity of both methods would require concentration such as by a freeze-out trap if used for atmospheric measurements. Since the calibration requirements are about equal to those needed for GC, the data interpretation is generally more complicated, and the extent of coverage is less; these methods have been widely supplanted by GC for atmospheric hydrocarbon detection work. These instruments may be useful for the analysis of a particular type or class of hydrocarbon and might make it unnecessary to set up a gas chromatographic capability in some cases.

Laser-Induced Doppler Absorption Radar (LIDAR)
Laser-induced Doppler absorption radar (LIDAR) can be used to remotely measure chemical concentrations in the atmosphere. Two different laser wavelengths are selected so that the molecule of interest absorbs one of the wavelengths while the other wavelength is selected to be in a region of minimal interference. The difference in intensity of the two returned signals is then used to determine the concentration of the gas being measured using the Lambert–Beer law.

By measuring the time taken for the light to travel and return, it is also possible to determine the distance from the base station to the measurement point. It is therefore possible to use LIDAR to "map" the chemical composition of an emission source in three dimensions and from a great distance.

Vertical Cavity Surface Emitting Laser (VCSEL)
Chromatographic natural gas analysis is intermittent, and it requires a carrier gas. Laser spectroscopic hydrocarbon analysis overcomes these limitations by providing continuous multicomponent analysis without the need for any utility. These instruments can also measure the calorific value of natural gas by irradiating the sample gas with a semiconductor laser beam. These tunable diode laser spectrometer (TDLS) units can measure absorption spectra of multi-component gases in a period of a few seconds by using the micro-electromechanical systems-vertical cavity surface emitting laser (MEMS- VCSEL) with a wavelength tuning range of approximately 50 nm, thereby obtaining the concentration of seven saturated hydrocarbon components with a carbon number of from one to five.

The surface-emitting laser design (Figure 1.28h) is suitable for multicomponent hydrocarbon analysis because it has a widely tunable wavelength range. As shown in Figure 1.28h, the light beam is split into three. The first beam passes through the gas cell containing the sample gas before reaching photodetector 1, the second traverses a methane gas cell for wavelength calibration and irradiates photodetector 2. The third beam goes directly into photodetector 3 and is used as the reference for the laser light intensity.

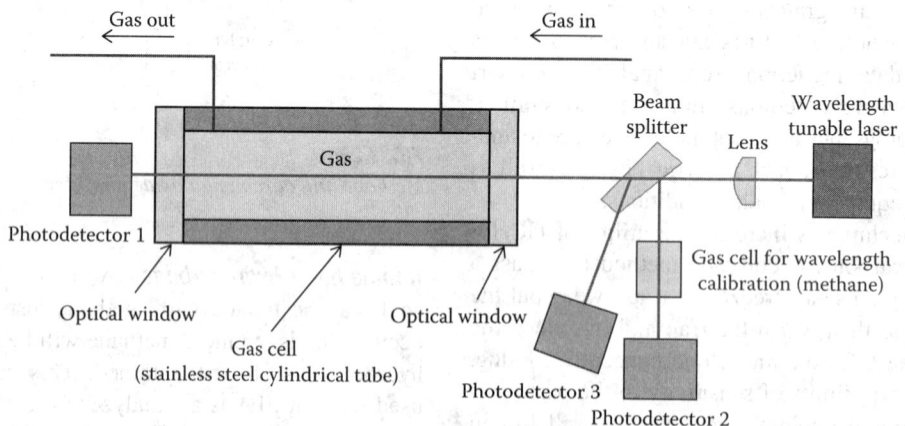

FIG. 1.28h
Tunable diode laser spectrometer (TDLS) for multicomponent hydrocarbon gas analysis. (Courtesy of Yokogawa.)

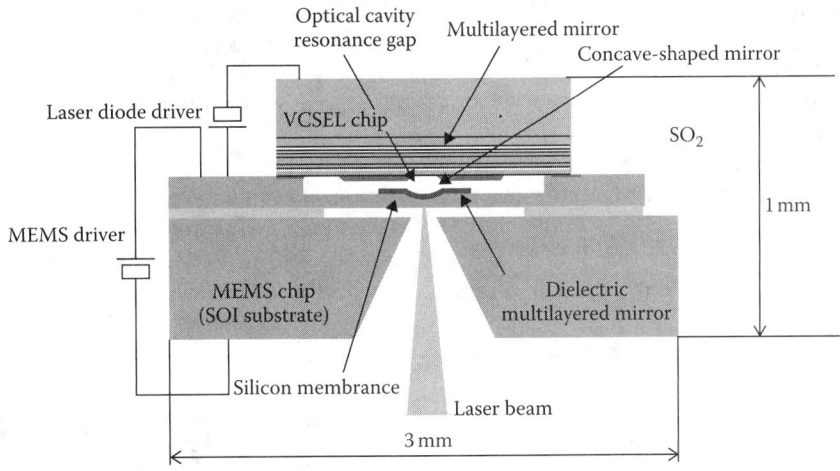

FIG. 1.28i
The design of the light source of the vertical-cavity surface-emitting laser (VCSEL). (Courtesy of Yokogawa.)

The laser light intensities are converted into electronic signals by the photodetector, and then the absorption spectrum is created by a computer, and finally the concentrations are calculated by using the multivariate analysis technique. Because the shape and the magnitude of an absorption spectrum are affected by gas temperature and pressure, the analyzer controls them to make them constant. Figure 1.28i shows this vertical-cavity surface-emitting laser with its widely tunable wavelength range, which is used for the light source of this analyzer.

Perimeter Monitoring Perimeter monitoring can be done with both ultraviolet (UV) and IR spectral equipment (Chapter 1.62). The selection between IR and UV is done by the absorption of the gases to be measured. Typical installations have a transceiver at one end and a retro-reflector at the far end. The light, typically at two wavelengths (reference and measurement), is transmitted to the mirror at the other end of the path of interest and then reflected back to a receiver in the same electronics enclosure. The resultant measurement is an average of the concentration of the gas of interest over the distance between the two units.

Newer models have separate transmitter and receiver assemblies, thus making it possible to measure around corners by using the appropriate reflectors. Path lengths of several hundred meters are typical, though distances of greater than 1 km are possible.

Portable IR and UV monitors and Fourier-transform near-infrared (FT-NIR) laboratory units are also popular for specific applications as described in Figure 1.28j.

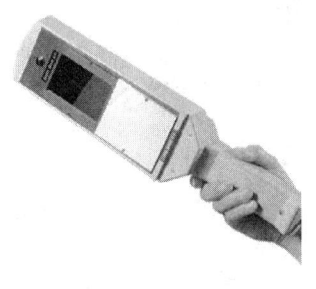

Handheld UV units are used as screening tools during spill cleanup or any application requiring the detection of hydrocarbons on the surface. (Courtesy of Thermo Scientific.)

Soil vapor monitors detect total hydrocarbons, including methane and distinguish between natural gases and other volatile compounds. (Courtesy of Thermo Scientific.)

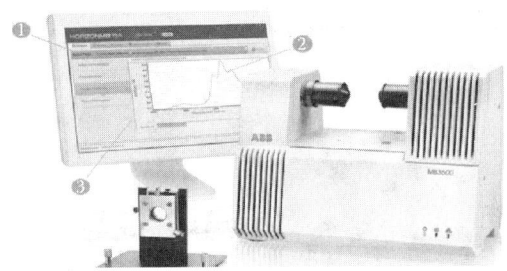

FT-NIR laboratory HC analyzer with menu selectable procedure (1), on-screen instructions (2), and report-generating capability (3). Preinstalled calibration for gasoline, diesel, etc., and available with flowing sample cell. (Courtesy of ABB.)

FIG. 1.28j
Examples of UV and IR analyzers for portable and laboratory applications.

Ion Mobility Spectroscopy

Ion-mobility spectrometry–mass spectrometry (IMS–MS), also known as ion-mobility separation–mass spectrometry, is an analytical chemistry method that separates gas-phase ions on a millisecond timescale using the IMS and uses mass spectrometry on a microsecond timescale to identify components in a sample.

Since this technique is not often used to monitor hydrocarbons, it is mentioned here only for the sake of completeness. While laser ionization almost selectively ionizes aromatic hydrocarbons, aliphatic hydrocarbons are only laser-ionized in case these contain conjugated double bonds.

This technique is best suited to detect elements in Group VIIA of the periodic table (Cl, Fl, I, etc.), though the range is being expanded and has also been applied to the detection of ammonia as well as other chemical agents, especially those associated with explosives. Laser ionization almost selectively ionizes aromatic hydrocarbons. The IMS is an analytical technique used to separate and identify ionized molecules in the gas phase based on their mobility in a carrier buffer gas.

Aspirating IMS is an ion-mobility spectrometry technology used to detect low or trace quantities of chemicals in the surrounding atmosphere. Sample flow is passed via ionization chamber and then enters to measurement area where the ions are deflected into one or more measuring electrodes by perpendicular electric field, which can be either static or varying (Figure 1.28k). The output of the sensor is characteristic of the ion-mobility distribution and can be used for detection and identification purposes.

Hydrocarbon Dew-Point Meter

The dew point of natural gas is that temperature at which significant condensation starts to be formed on a chilled surface. A chilled-mirror instrument of the dark spot design can be used for this measurement (Figure 1.28l). With this technique, almost invisible films, having sensitivities on the order of 1 ppm, become detectable. In this instrument, the optical surface is provided with a V-shaped depression.

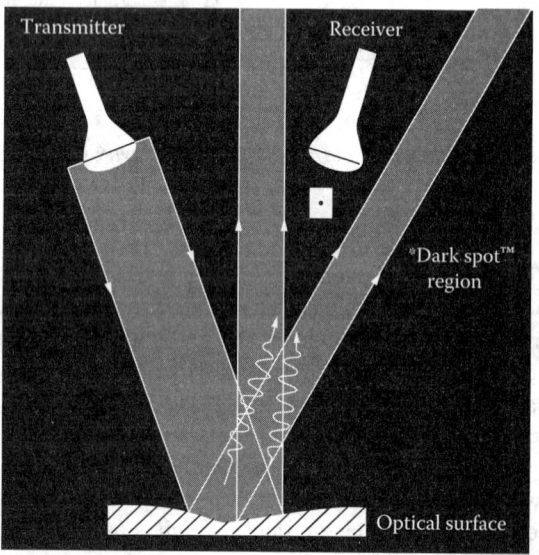

FIG. 1.28l
Measurement principle of the dark-spot-type chilled-mirror analyzer, which is used to measure the hydrocarbon dew point in natural gas. (Courtesy of Michell Instruments Ltd.)

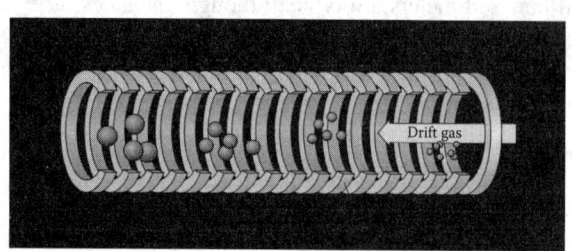

FIG. 1.28k
Ionization chamber of ion-mobility spectrometry–mass spectrometer.

When the surface is dry, the reflection from it results in a dark spot (as shown in Figure 1.28f). As hydrocarbon condenses on this surface, the scattered light intensity in the dark spot region is reduced. Optical fibers are used to detect this reduction of intensity, while miniature thermocouples measure the surface temperature of the mirror.

CALIBRATION

Only dynamic calibration is completely satisfactory. In this procedure, known concentrations of the gases to be measured are passed into the analytical system, preferably in exactly the same way as the unknown samples are collected. This may be difficult to accomplish because of low concentrations or because some components might condense or react with other components.

Standard calibration mixtures of one or several of the common hydrocarbons in carrier fluids are available commercially in high-pressure gas cylinders. These should always be checked against a reference standard or at least against a previously calibrated analyzer. If much calibration is necessary, it may be useful to set up a gas-handling and dilution system to prepare standard mixtures.

It is a fairly simple technique for hydrocarbons, such as butane (which can be liquefied at moderate pressure), to use permeation tubes for calibration. These are sealed tubes of a specific plastic, partially filled with liquid hydrocarbon. As long as some liquid remains, the hydrocarbon will effuse through the walls of a given tube at a rate depending only on temperature. To generate a known concentration, carrier gas is passed over the tube at a controlled rate in a temperature-controlled vessel. The tube may be calibrated gravimetrically.

In another technique, used to calibrate atmospheric hydrocarbon analyzers, a known amount of hydrocarbon is added by syringe or by crushing an ampoule, either in a large rigid vessel of known air volume or into the metered flow of air passing into a plastic bag. The bag must be of inert plastic; it has the advantage of collapsing as the sample is withdrawn, whereas a rigid vessel must be large in relation to the sample size in order to avoid substantial pressure differentials.

CONCLUSIONS

For total hydrocarbon measurement, the flame ionization analyzer is the most generally accepted detector. For proper operation, it requires the attention of operators, and it also consumes compressed gases, but it is reliable and accurate.

Nondispersive infrared (NDIR) sensors can identify only hydrocarbon classes, whereas two-wavelength IR detectors are able to identify individual hydrocarbon species. In this design, sensitivity is traded for specificity.

If individual hydrocarbon determination at low concentrations is required, chromatograph augmented with freeze-out is a common solution. When done manually, calibration is demanding, and the data reduction is time consuming, but computerized systems overcome that. The main limitation of chromatographic natural gas analysis is that the readings are intermittent and the analyzer requires a carrier gas.

Laser spectroscopic hydrocarbon analysis overcomes some of these limitations by providing continuous multi-component analysis without the need for utilities. These instruments operate by irradiating the sample gas with a semiconductor laser beam. These tunable diode laser spectrometers (TDLS) can measure the composition of multi-component gases in a few seconds, thereby obtaining the concentration of up to seven saturated hydrocarbon components that can have carbon number of from one to five.

Ion mobility spectroscopy and other spectrometric techniques are used only in special applications.

Analytical Measurement

SPECIFICATION FORMS

When specifying hydrocarbon analyzers only, please use the ISA form 20A1001, and when specifying both the analyzer and the composition or the properties of the process material of interest, use ISA form 20A1002. Both forms are reproduced with the permission of the International Society of Automation.

1	RESPONSIBLE ORGANIZATION		ANALYSIS DEVICE		6		SPECIFICATION IDENTIFICATIONS		
2	(ISA)				7	Document no			
3			Operating Parameters		8	Latest revision		Date	
4					9	Issue status			
5					10				

11	ADMINISTRATIVE IDENTIFICATIONS			40	SERVICE IDENTIFICATIONS Continued			
12	Project number		Sub project no	41	Return conn matl type			
13	Project			42	Inline hazardous area cl		Div/Zon	Group
14	Enterprise			43	Inline area min ign temp		Temp ident number	
15	Site			44	Remote hazardous area cl		Div/Zon	Group
16	Area	Cell	Unit	45	Remote area min ign temp		Temp ident number	
17				46				
18	SERVICE IDENTIFICATIONS			47				
19	Tag no/Functional ident			48	COMPONENT DESIGN CRITERIA			
20	Related equipment			49	Component type			
21	Service			50	Component style			
22				51	Output signal type			
23	P&ID/Reference dwg			52	Characteristic curve			
24	Process line/nozzle no			53	Compensation style			
25	Process conn pipe spec			54	Type of protection			
26	Process conn nominal size	Rating		55	Criticality code			
27	Process conn termn type	Style		56	Max EMI susceptibility		Ref	
28	Process conn schedule no	Wall thickness		57	Max temperature effect		Ref	
29	Process connection length			58	Max sample time lag			
30	Process line matl type			59	Max response time			
31	Fast loop line number			60	Min required accuracy		Ref	
32	Fast loop pipe spec			61	Avail nom power supply			Number wires
33	Fast loop conn nom size	Rating		62	Calibration method			
34	Fast loop conn termn type	Style		63	Testing/Listing agency			
35	Fast loop schedule no	Wall thickness		64	Test requirements			
36	Fast loop estimated lg			65	Supply loss failure mode			
37	Fast loop material type			66	Signal loss failure mode			
38	Return conn nominal size	Rating		67				
39	Return conn termn type	Style		68				

69	PROCESS VARIABLES	MATERIAL FLOW CONDITIONS		101	PROCESS DESIGN CONDITIONS		
70	Flow Case Identification		Units	102	Minimum	Maximum	Units
71	Process pressure			103			
72	Process temperature			104			
73	Process phase type			105			
74	Process liquid actl flow			106			
75	Process vapor actl flow			107			
76	Process vapor std flow			108			
77	Process liquid density			109			
78	Process vapor density			110			
79	Process liquid viscosity			111			
80	Sample return pressure			112			
81	Sample vent/drain press			113			
82	Sample temperature			114			
83	Sample phase type			115			
84	Fast loop liq actl flow			116			
85	Fast loop vapor actl flow			117			
86	Fast loop vapor std flow			118			
87	Fast loop vapor density			119			
88	Conductivity/Resistivity			120			
89	pH/ORP			121			
90	RH/Dewpoint			122			
91	Turbidity/Opacity			123			
92	Dissolved oxygen			124			
93	Corrosivity			125			
94	Particle size			126			
95				127			
96	CALCULATED VARIABLES			128			
97	Sample lag time			129			
98	Process fluid velocity			130			
99	Wake/natural freq ratio			131			
100				132			

133	MATERIAL PROPERTIES			137	MATERIAL PROPERTIES Continued		
134	Name			138	NFPA health hazard		Flammability	Reactivity
135	Density at ref temp		At	139			
136				140			

Rev	Date	Revision Description	By	Appv1	Appv2	Appv3	REMARKS

Form: 20A1001 Rev 0 © 2004 ISA

1.28 Hydrocarbon Analyzers

	RESPONSIBLE ORGANIZATION	ANALYSIS DEVICE COMPOSITION OR PROPERTY Operating Parameters (Continued)		SPECIFICATION IDENTIFICATIONS		
1			6			
2	ISA		7	Document no		
3			8	Latest revision		Date
4			9	Issue status		
5			10			

	PROCESS COMPOSITION OR PROPERTY			MEASUREMENT DESIGN CONDITIONS				
11	Component/Property Name	Normal	Units	Minimum	Units	Maximum	Units	Repeatability
12								
13–37								

Rev	Date	Revision Description	By	Appv1	Appv2	Appv3	REMARKS

Form: 20A1002 Rev 0 © 2004 ISA

Definition

Lambert–Beer law relates the attenuation of light, neutrons, etc., to the properties of the material through which they are passing.

Abbreviations

FID	Flame ionization detector
FTNIR	Fourier-transform near infrared
GC	Gas chromatograph
IMS-MS	Ion-mobility spectrometry–mass spectrometry
LCD	Liquid crystal display
LEL	Lower explosion limit
LIDAR	Laser-induced Doppler absorption radar
NDIR	Nondispersive infrared
TDLS	Tunable diode laser spectrometer
VCSEL	Vertical cavity surface emitting laser

Bibliography

ACGIH, Air sampling instruments for evaluation of atmospheric contaminants, (2009) http://www.acgih.org/store/ProductDetail.cfm?id=2100.

Emerson-Rosemount, Continuous emission monitoring, (2013) http://www2.emersonprocess.com/siteadmincenter/PM%20.

EPA Heath Consultants Inc., (2015) http://nlquery.epa.gov/epasearch/epasearch?typeofsearch=area&querytext=hydrocarbon&submit=Go&fld=outreach&areaname.

General Monitors, Fundamentals of combustible gas detection, (2010) http://s7d9.scene7.com/is/content/minesafetyappliances/Combustible%20Gas%20Detection%20White%20Paper.

Honeywell, Gas detection book, (2013) http://www.honeywellanalytics.com/~/media/honeywell-analytics/documents/english/11296_gas-book_v5_0413_lr_en.pdf?la=en.

International Sensor Technology, Sensor selection overview, (2015) http://www.intlsensor.com/pdf/sensorSelectionGuide.pdf.

JUM Engineering, What is a flame ionization detector?, (2000) http://www.jum.com/products/analyzers/fid.html.

McNair, H. M., Basic gas chromatography, (2009) http://www.amazon.com/Basic-Gas-Chromatography-Harold-McNair/dp/0470439548.

MSA, Gas detection handbook, (2007) http://www.gilsoneng.com/reference/gasdetectionhandbook.pdf.

OSHA, Use of combination oxygen and combustible gas detectors, (2013) https://www.osha.gov/dts/hib/hib_data/hib19900118.html.

Poole, C., Gas chromatography, (2011) http://www.amazon.com/Gas-Chromatography-Colin-Poole/dp/0123855403/ref=pd_sim_b_12.

Rando, R. J., Identity and analysis of total hydrocarbon, http://www.atsdr.cdc.gov/toxprofiles/tp123-c3.pdf.

Rood, D., The troubleshooting and maintenance guide for gas chromatographers, (2007) http://www.amazon.com/The-Troubleshooting-Maintenance-Guide-Chromatographers/dp/3527313737/ref=pd_sim_b_.Rosemount%20Analytical%20Documents/PGA_AN_42-PGA-AN-POWER-CEMS-Cogen.pdf.

Sherman, R. E., *Process Analyzer Sample-Conditioning System Technology*, New York: John Wiley & Sons, 2002. http://www.wiley.com/WileyCDA/WileyTitle/productCd-0471293644.html.

Shimadzu, The analysis of olefin and aromatic hydrocarbon in hydrocarbon mixture, (2010) http://www.shimadzu.com/an/literature/gcms/hko212180.html.

Sparkman, D., *Gas Chromatography and Mass Spectrometry: A Practical Guide*, 2nd edn., 2011. http://www.amazon.com/Gas-Chromatography-Mass-Spectrometry-Practical/dp/0123736285/ref=pd_sim_b_3.

Thermo Electron, Near infrared sensor for online measurement, (2005) http://www.thermo.com/eThermo/CMA/PDFs/Various/File_27448.pdf.

Yaws, C. L. and Baker, W., Matheson gas data book, (2001) https://books.google.com/books/about/Matheson_Gas_Data_Book.html?id=Sfvbgvu9OQMC.

Yokogawa, Laser spectroscopic multi-component hydrocarbon analyzer, (2015) http://www.yokogawa.com/us/technical-library/resources/white-papers/laser-spectroscopic-multi-component-hydrocarbon-analyzer/.

Zabetakis, M. G., Flammability characteristics of combustible gases and vapors, https://books.google.com/books/about/Flammability_Characteristics_of_Combusti.html?id=nolWGQAACAAJ.

Zemo Assoc., Petroleum hydrocarbon releases, (2007) http://www.swhydro.arizona.edu/archive/V6_N4/feature6.pdf.

1.29 Hydrogen Cyanide [HCN] Detectors

R. SREENEVASAN and B. G. LIPTÁK (2017)

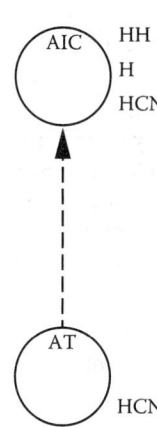

Applications	Cyanide destruction/cyanidation in gold processing plants
Range	0.2–20 ppm
Inaccuracy	± 0.1 ppm
Exposure consequences	
Concentration (ppm)	Limit and Consequences
AGEL values	See Table 1.29a
Up to 10	OSHA Limit for 8 hr average
10–50	Headache, dizziness, unsteadiness
50–100	Feeling of suffocation, nausea
100–200	Death in 30–60 min
Cost	Portable personal HCN monitor ~$500 Fixed-point HCN detector (electrochemical diffusion) ~$1,400 Stand-alone, Multichannel gas monitoring and alarm control system ~$5,000–$12,000
Partial list of suppliers	Air-Met Scientific (electro-chemical diffusion) https://www.airmet.com.au/assets/documents/product/catalogues/Industrial%20Scientific%20Catalogue%202014.pdf Altair (portable, electro-chemical) http://us.msasafety.com/Portable-Gas-Detection/Single-Gas/ALTAIR%26reg%3B-Pro-Single-Gas-Detector/p/000080000200001512 Calibrated Instruments Inc. (conductivity measurement based laboratory analyzer) http://www.calibrated.com/gasanalyzer.htm Crowcon (electro-chemical diffusion) http://www.crowcon.com/uk/products/fixed-detectors/xgard.html Detcon Inc. (electro-chemical diffusion) http://www.detcon.com/1-documents/data_sheets/1-sensors/model_100/Hydrogen%20Cyanide%20DM-100-HCN%20PDS.pdf Dräger Safety AG (electrochemical diffusion) http://www.draeger.com/sites/assets/PublishingImages/Products/cin_x-zone_5000/US/gas-detection-br-9041145-us.pdf Environmental Technology (Skalar UV) http://www.envirotech-online.com/news/water-wastewater/9/skalar_analytical/fully_automatic_cyanide_analyser/20240/ Gasmet (Fourier transform infrared spectroscopy for CEM applications) http:// http://www.gasmet.com/technology/spectra/250-hydrogen-cyanide Heinrich WösthoffMessetechnik GmbH (conductivity measurement-based laboratory analyzer) http://www.wösthoff.com/?content=products&pg=gasanalysensysteme&lang=EN Honeywell (electro-chemical diffusion) http://www.honeywellanalytics.com/~/media/honeywell-analytics/products/midas/documents/english/ss01115q_midas_s_hcn_e_hcn_spec-sheet_flr_3-27-12.pdf?la=en H. Wösthoff, GmbH, (conductivity) http://www.xn--wsthoff-90a.com/?content=products&pg=gasanalysensysteme&lang=EN Industrial Scientific Corporation (electrochemical diffusion) http://www.indsci.com/gas/hcn/ Sierra Monitor Corporation (electrochemical diffusion) http://www.sierramonitor.com/protect/all-fire-gas-products/electrochemical-toxic-gas-sensor-module Skalar Laboratory (Skalar UV) http://www.skalar.com/news/new-skalar-cyanide-analyzer

INTRODUCTION

Hydrogen cyanide is a colorless, rapidly acting, highly poisonous gas or liquid that has an odor of bitter almonds. Exposures to HCN may occur in industrial situations. Cyanide is clearly the lixiviant* of choice for gold leaching, with gold cyanidation being the most commonly used metallurgical technique throughout the world.

Gold cyanidation is a metallurgical technique for extracting gold from low-grade ore by converting the gold to a water-soluble coordination complex. It is the most commonly used process for gold extraction. Production of reagents for mineral processing to recover gold, copper, zinc, and silver represents approximately 13% of cyanide consumption globally, with the remaining 87% of cyanide used in other industrial processes such as plastics, adhesives, and pesticides. Due to the highly poisonous nature of cyanide, the process is controversial, and its usage is banned in a number of countries. Cyanide spills can have a devastating effect on rivers, sometimes killing everything for several miles downstream. However, the cyanide is soon washed out of river systems, and affected areas can soon be repopulated.

CYANIDE CHEMISTRY

The most toxic form of cyanide is free cyanide, which includes the cyanide anion itself [CN^-] and hydrogen cyanide. The cyanide ion [CN^-] is the ionic or alkaline form of hydrogen cyanide, displaying the typical properties of a weak acid (CyPlus 2001). As a weak acid, hydrogen cyanide incompletely dissociates in water.

The pH can significantly influence the equilibrium between the acid [HCN] and the alkaline form [CN^-]. At pH 9.36, CN^- and HCN are in equilibrium with each other. At a pH of 11, over 99% of cyanide remains in solution as CN^-, while at pH 7, over 99% of cyanide will exist as HCN.

For cyanidation, the pH region from 10 to 11 is of major importance, since gold dissolution with cyanide is usually performed in this pH domain (CyPlus 2001). Under equilibrium conditions, 81.3% of hydrogen cyanide is dissociated into the cyanide ion at pH 10, still leaving a significant amount as dissolved HCN (18.7%). Increasing the pH to 11 decreases the percentage of HCN to 2.3%. The pH dependence of the [HCN/CN^-] is of major importance for industrial applications.

Hydrogen cyanide has a relatively high vapor pressure (100 kPa at 26°C) and consequently volatilizes readily at the liquid surface under ambient conditions. The rate of volatilization depends on the hydrogen cyanide concentration in solution and, therefore, directly on the pH.

To completely avoid cyanide loss by volatilization, the cyanidation would have to be operated at pH > 12 with hydrogen cyanide dissociated to >99.8%. However, economic constraints restrict cyanidation operations at preferably pH 10.5–11. Similarly, cyanide destruction processes are also constrained to pH 10.5–11. Because of these constraints, gaseous hydrogen cyanide is likely to be present around the cyanidation and cyanide destruction areas of a gold processing plant.

EXPOSURE LIMITS

The U.S. OSHA specifies a permissible exposure limit of 10 ppm for hydrogen cyanide as an 8 hr time-weighted average concentration. The U.S. NIOSH has established a recommended exposure limit of 4.7 ppm for hydrogen cyanide.

Acute Exposure Guideline Levels (AEGLs) represent threshold exposure limits (exposure levels below which adverse health effects are not likely to occur) for the general public and are applicable to emergency exposures ranging from 10 min to 8 hr. Three levels of AEGLs have been developed—AEGL-1, AEGL-2, and AEGL-3. The three levels of AEGLs are defined as follows:

- AEGL-1 is the airborne concentration (expressed as ppm) of a substance above which it is predicted that the general population, including susceptible individuals, could experience notable discomfort, irritation, or nonsensory effects. However, the effects are not disabling and are transient and are reversible upon cessation of exposure.
- AEGL-2 is the airborne concentration (expressed as ppm) of a substance above which it is predicted that the general population, including susceptible individuals, could experience irreversible or other serious, long-lasting adverse health effects or an impaired ability to escape.
- AEGL-3 is the airborne concentration (expressed as ppm) of a substance above which it is predicted that the general population, including susceptible individuals, could experience life-threatening health effects or death (see Table 1.29a).

TABLE 1.29a

The U.S. EPA Summary Table of AEGL Values for Hydrogen Cyanide (ppm)

Classification	10 min	30 min	1 hr	4 hr	8 hr
AEGL-1 (nondisabling)	2.5	2.5	2.0	1.3	1.0
AEGL-2 (disabling)	17	10	7.1	3.5	2.5
AEGL-3 (lethal)	27	21	15	8.6	6.6

---

* See Definitions at the end of this chapter.

DETECTOR DESIGNS

Industrial, fixed-point HCN gas detectors, similar to other readily available toxic gas detectors (Figure 1.29a), operate based on electrochemical diffusion principle. Since this type of gas detector is described in Chapter 1.26, the principle will not be detailed here.

FIG. 1.29a
Portable multigas detector with LED display and both audible and visual ALARM. (Courtesy of Altair.)

Fourier-Transform Infrared (FTIR)

A Fourier-transform infrared spectroscopy (FTIR) gas analyzer (for details, see Chapter 1.32) detects gaseous compounds by their absorbance of infrared radiation. Because each molecular structure has a unique combination of atoms, each produces a unique infrared spectrum. From this, identification (qualitative analysis) and analysis (quantitative measurement) of the gaseous compounds is possible. An FTIR analyzer simultaneously measures multiple analytes in a complex gas matrix, detecting virtually all gas-phase species (both organic and inorganic).

Portable, continuous-flow analyzers can provide automatic determination and distillation, for different cyanide species such as total cyanide, free cyanide, weak acid, dissociable cyanide, and thiocyanate, applying EPA or ISO methodologies (Figure 1.29b).

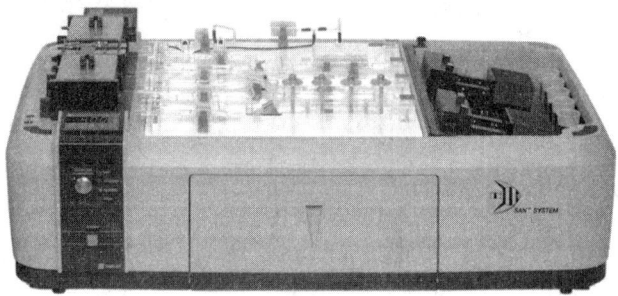

FIG. 1.29b
Automatic cyanide analyzer. (Environmental Technology.)

There are very few manufacturers who can supply FTIR gas analyzer. Gasmet™ FTIR gas analyzer collects a complete infrared spectrum (a measurement of the infrared light absorbed by molecules inside the sample gas cell) 10 times per second. Multiple spectra are co-added together according to selected measurement time (improving signal-to-noise ratio). The actual concentrations of gases are calculated from the resulting sample spectrum using patented modified classical least squares analysis algorithm.

The U.S. EPA draft standard OTM-29—"Sampling and Analysis for Hydrogen Cyanide Emissions from Stationary Sources," is applicable to the collection and analysis of gaseous cyanide [HCN] in the atmosphere and in suspended water droplets (specifically monitors emissions from power stations). FTIR gas analyzer is an ideal instrument for these CEM applications.

A special type of gas analyzer is used in laboratories to measure HCN gas concentration. The conductivity measurement principle utilized by H. Wösthoff, GmbH, introduces a previously metered sample gas into a suitable liquid reagent of measured electrical conductivity. The volumetrically proportioned streams of sample gas and liquid reagent combine, changing the conductivity of the reagent solution. The resulting difference in the conductivity of the reacted reagent solution is proportional to the concentration of the sample gas being measured.

Figure 1.29c shows an apparatus configuration in which liquid reagent conductivity is first measured at electrode 1 (before introduction of the sample gas) and then again at electrode 2 after the sample gas and reagent have thoroughly mixed in the reaction run and are again separated.

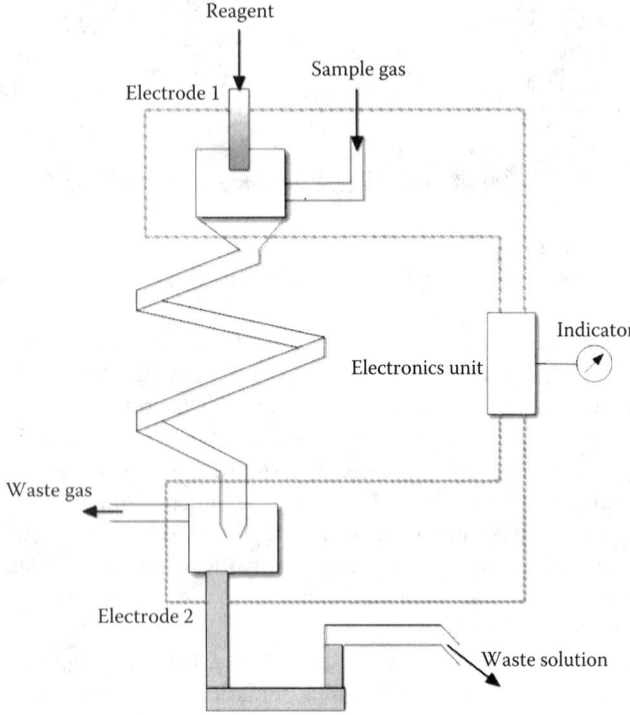

FIG. 1.29c
HCN gas analysis using conductivity measurement. (Courtesy of H. Wösthoff, GmbH.)

Hydrogen cyanide emissions from power station stacks or forest fires do not pose an immediate danger (exposure to lethal levels of HCN gas) to people. However, in industrial plants (gold processing plants), operators can be exposed to high levels of HCN gas. In these applications, the FTIR analyzers/conductivity measurement-based analyzers are not cost effective. As a consequence, the focus of this chapter is on electrochemical diffusion-type gas detectors and alarm systems.

Most common fixed-point hydrogen cyanide detectors (Figure 1.29d) are available in the range 0.2–20 ppm. Some manufacturers offer sensors with detection range 0.2–30 ppm. Very few manufacturers offer sensors with detection range 0.2–50 ppm (rate). Their advantage of being low-cost instruments is somewhat overshadowed by the fact that diffusion-type gas detectors can lose their detection efficiency over time.

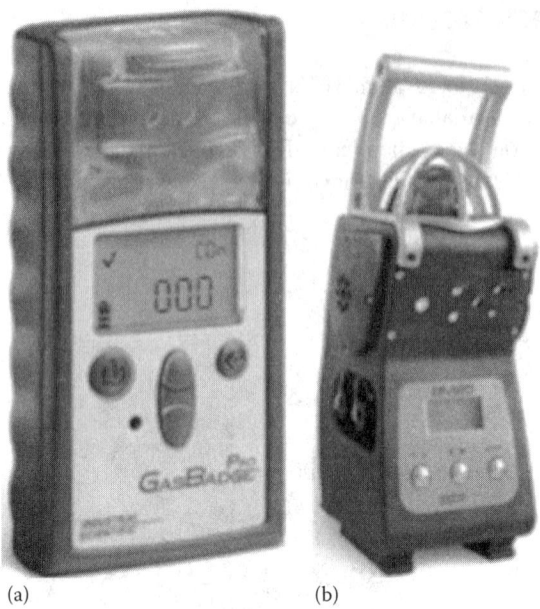

(a)　　　　　　　　(b)

FIG. 1.29d
Fixed-point hydrogen cyanide detector (a) and five-gas detector with sample draw capability (b). (Courtesy of Industrial Scientific.)

Typically, manufacturers recommend that the electrochemical sensors be replaced every 6 months.

Performance Standards

No known international performance standard exists for fixed-point hydrogen cyanide detectors (similar to ISA-92.0.01, Part 1-1998 Performance requirements for toxic gas detection instruments: Hydrogen sulfide [H_2S]). Most HCN detection systems (sensor + transmitter + alarm output) provide sensor fault alarm, high alarm (measured value exceeds alarm limit), plus 4–20 mA signal of the measured gas concentration. One large gold mining company decided to design and install hydrogen cyanide gas detection and alarming system with multiple alarms (4 off) and remote reset capabilities at its newest gold plant in Australia. In determining the warning alarm and evacuation alarm limits, the current U.S. Recommended Exposure Limits, NIOSH RELs (4.7 ppm averaged over 10 min, and 10 ppm averaged over 8 hr shift) have been considered.

Each HCN gas detector is required to provide the following alarms and local indications:

- Sensor fault alarm (white lamp indicator)—Alarm remains latched until the sensor fault is rectified.
- Warning alarm (amber lamp indicator)—Alarm is latched when the measured HCN concentration is >4.7 ppm. Alarm can be reset only when the measured

HCN concentration is <2 ppm. This alarm warns the operators to avoid the area (if possible).
- Evacuation alarm (red lamp indicator)—Alarm is latched when the measured concentration is >10 ppm. Alarm can be reset only when the measured HCN concentration is <7 ppm. This visual alarm requires the operators to evacuate the area.
- Audible alarm—Alarm is latched when the measured concentration is >10 ppm. Alarm can be reset only when the measured HCN concentration is <7 ppm. This audible alarm requires the operators to evacuate the area.

The earlier-mentioned hydrogen cyanide monitoring system (Figure 1.29e) was installed and commissioned in

FIG. 1.29e
Hydrogen cyanide detector + alarm + annunciators. (Courtesy of KD Fisher & Co Pty.)

TABLE 1.29b
Multiple Alarm Implementation—Hydrogen Cyanide Monitoring System

HCN Level (ppm)	Action
<2.0	Warning alarm is unlatched (AMBER light is de-energized); operators and visitors have unrestricted access to the area.
≥4.7	Warning alarm is latched (AMBER light is energized); operator and visitor access to the area is restricted.
<7.0	Evacuation alarm is unlatched (RED light and audible alarm are de-energized); operators and visitors are not allowed to work in the area unless equipped with full gas mask and respirators.
≥10.0	Evacuation alarm is latched (RED light and audible alarm are energized); operators and visitors have to evacuate the area immediately. They are not allowed to return unless it is safe to do so and permitted by the control room operators.

Australia; it is implementing the multiple alarms tabulated in Table 1.29b and has been operating satisfactorily since 2010.

CONCLUSION

Hydrogen cyanide monitoring practices are not uniform across the gold processing industry. It is hoped that the adoption of voluntary "Cyanide Code" will increase the awareness and implementation requirements for hydrogen cyanide monitoring.

SPECIFICATION FORMS

When specifying hydrogen cyanide [HCN] detectors only, one cam use the ISA form 20A1001, and when specifying both the analyzer and the composition or the properties of the process, which it will be monitoring, use the ISA form 20A1002. Both forms are reproduced with the permission of the International Society of Automation on the next pages.

Analytical Measurement

Operating Parameters

1	RESPONSIBLE ORGANIZATION		ANALYSIS DEVICE		6	SPECIFICATION IDENTIFICATIONS		
2					7	Document no		
3	(ISA logo)		Operating Parameters		8	Latest revision	Date	
4					9	Issue status		
5					10			

11	ADMINISTRATIVE IDENTIFICATIONS				40	SERVICE IDENTIFICATIONS Continued		
12	Project number		Sub project no		41	Return conn matl type		
13	Project				42	Inline hazardous area cl	Div/Zon	Group
14	Enterprise				43	Inline area min ign temp	Temp ident number	
15	Site				44	Remote hazardous area cl	Div/Zon	Group
16	Area		Cell	Unit	45	Remote area min ign temp	Temp ident number	
17					46			
18	SERVICE IDENTIFICATIONS				47			
19	Tag no/Functional ident				48	COMPONENT DESIGN CRITERIA		
20	Related equipment				49	Component type		
21	Service				50	Component style		
22					51	Output signal type		
23	P&ID/Reference dwg				52	Characteristic curve		
24	Process line/nozzle no				53	Compensation style		
25	Process conn pipe spec				54	Type of protection		
26	Process conn nominal size		Rating		55	Criticality code		
27	Process conn termn type		Style		56	Max EMI susceptibility	Ref	
28	Process conn schedule no		Wall thickness		57	Max temperature effect	Ref	
29	Process connection length				58	Max sample time lag		
30	Process line matl type				59	Max response time		
31	Fast loop line number				60	Min required accuracy	Ref	
32	Fast loop pipe spec				61	Avail nom power supply		Number wires
33	Fast loop conn nom size		Rating		62	Calibration method		
34	Fast loop conn termn type		Style		63	Testing/Listing agency		
35	Fast loop schedule no		Wall thickness		64	Test requirements		
36	Fast loop estimated lg				65	Supply loss failure mode		
37	Fast loop material type				66	Signal loss failure mode		
38	Return conn nominal size		Rating		67			
39	Return conn termn type		Style		68			

69	PROCESS VARIABLES	MATERIAL FLOW CONDITIONS			101	PROCESS DESIGN CONDITIONS		
70	Flow Case Identification			Units	102	Minimum	Maximum	Units
71	Process pressure				103			
72	Process temperature				104			
73	Process phase type				105			
74	Process liquid actl flow				106			
75	Process vapor actl flow				107			
76	Process vapor std flow				108			
77	Process liquid density				109			
78	Process vapor density				110			
79	Process liquid viscosity				111			
80	Sample return pressure				112			
81	Sample vent/drain press				113			
82	Sample temperature				114			
83	Sample phase type				115			
84	Fast loop liq actl flow				116			
85	Fast loop vapor actl flow				117			
86	Fast loop vapor std flow				118			
87	Fast loop vapor density				119			
88	Conductivity/Resistivity				120			
89	pH/ORP				121			
90	RH/Dewpoint				122			
91	Turbidity/Opacity				123			
92	Dissolved oxygen				124			
93	Corrosivity				125			
94	Particle size				126			
95					127			
96	CALCULATED VARIABLES				128			
97	Sample lag time				129			
98	Process fluid velocity				130			
99	Wake/natural freq ratio				131			
100					132			

133	MATERIAL PROPERTIES				137	MATERIAL PROPERTIES Continued	
134	Name				138	NFPA health hazard	Flammability	Reactivity
135	Density at ref temp		At		139			
136					140			

Rev	Date	Revision Description	By	Appv1	Appv2	Appv3	REMARKS

Form: 20A1001 Rev 0 © 2004 ISA

1.29 Hydrogen Cyanide [HCN] Detectors

	RESPONSIBLE ORGANIZATION	ANALYSIS DEVICE COMPOSITION OR PROPERTY Operating Parameters (Continued)		SPECIFICATION IDENTIFICATIONS			
1			6				
2			7	Document no			
3	ISA		8	Latest revision		Date	
4			9	Issue status			
5			10				

	PROCESS COMPOSITION OR PROPERTY			MEASUREMENT DESIGN CONDITIONS				
	Component/Property Name	Normal	Units	Minimum	Units	Maximum	Units	Repeatability
11								
12								
13								
...								
37								

(rows 38–75 blank)

Rev	Date	Revision Description	By	Appv1	Appv2	Appv3	REMARKS

Form: 20A1002 Rev 0 © 2004 ISA

Definition

Lixiviant is a liquid medium used in hydrometallurgy to selectively extract the desired metal from the ore or mineral. It assists in rapid and complete leaching. The metal can be recovered from it in a concentrated form after leaching.

Abbreviations

AEGL	Acute exposure guideline level
CEM	Continuous emission monitoring
CNF	Free cyanide
CNT	Total cyanide
FTIR	Fourier-transform infrared spectroscopy
HCN	Hydrogen cyanide
OTM	Other test method
PEL	Permissible exposure limit
ppm	Parts per million
REL	Recommended exposure limit

Organizations

AEG	American Environmental Group
ATSDR	Agency of Toxic Substances and Disease Registry
EPA	U.S. Environmental Protection Authority
ICME	International Council on Metals and the Environment
ICMI	International Cyanide Management Code
ISO	International Organization for Standardization
NIOSH	National Institute for Occupational Safety and Health
OSHA	Occupational Health and Safety Administration
UNEP	United Nations Environmental Program

Bibliography

ACGIH, TLVs and BEIs, (2015) http://www.acgih.org/forms/store/ProductFormPublic/2015-tlvs-and-beis.

Atmospheric Measurement Techniques, Measurements of hydrogen cyanide (HCN) and acetylene (C2H2), (2013) http://www.atmos-meas-tech.net/6/917/2013/amt-6-917-2013.pdf.

ATSDR, Toxic substances portal–hydrogen cyanide, (2014) http://www.atsdr.cdc.gov/MMG/MMG.asp?id=1141&tid=249.

Barrick, G., International cyanide management code, (2014) http://www.cyanidecode.org/signatory-company/barrick-gold-corporation.

Calabrese, E. J. and Kenyon, E., Air Toxics and Risk Assessment, Lewis Publishers, 2011, http://www.belleonline.com/bios/calabrese.htm.

EPA, US EPA draft standard OTM-29, Sampling and analysis of hydrogen cyanide emissions, (2011) http://www3.epa.gov/ttnemc01/prelim/otm29.pdf.

EPA, Acute exposure guideline levels for airborne chemicals, (2015) http://www.epa.gov/oppt/aegl/index.htm.

FM, Approval standard for toxic gases, (2014) https://www.fmglobal.com/assets/pdf/fmapprovals/6340.pdf.

Garcia, D. H., An introduction to the international cyanide code, (2010) http://www.br.srk.com/files/File/papers/Compliance-Cyanide-Code(1).pdf.

Honeywell, Commercial gas detection, (2015) http://www.honeywellanalytics.com/~/media/honeywell-analytics/documents/english/2014commercialproductguidepg11008ncl.pdf?la=en.

ICMI, International cyanide management code for the manufacture, transport and use of cyanide in the production of gold, (2012) http://www.cyanidecode.org/.

InspectAPedia, Toxic gas exposure standards, (2015) http://inspectapedia.com/sickhouse/Gas_Exposure_Limits.php.

Lewis, R. J., Rapid guide to hazardous chemicals in the workplace, (2000) http://www.amazon.com/Rapid-Guide-Hazardous-Chemicals-Workplace/dp/0471355429.

McDermott, H. J., Air monitoring for toxic exposures, (2005) http://www.wiley.com/WileyCDA/WileyTitle/productCd-0471454354.html.

NIOSH, Hydrogen cyanide, (2003) http://www.cdc.gov/niosh/docs/2003-154/pdfs/6017.pdf.

NIOSH, Pocket guide to chemical hazards, (2014) http://www.cdc.gov/niosh/npg/.

OSHA, Toxic and hazardous substances standards 1910–1000, (2006) https://www.osha.gov/pls/oshaweb/owadisp.show_document?p_table=STANDARDS&p_id=9991.

OSHA, Toxic Industrial Chemicals (TICs) Guide, (2007) https://www.osha.gov/SLTC/emergencypreparedness/guides/chemical.html.

OSHA, Hydrogen cyanide, (2011) http://www.osha.gov/dts/sltc/methods/validated/1015/1015.html.

Rubo, A., Kellens, R., Reddy, J., Steier, N., and Hasenpusch, W., Alkali metal cyanides, (2006) http://onlinelibrary.wiley.com/doi/10.1002/14356007.a08_159.pub3/abstract.

UNEP/OCHA Environment Unit, UN assessment mission-Cyanide Spillat Baia Mare, March 2000, http://reliefweb.int/report/hungary/cyanide-spill-baia-mare-romania-unepocha-assessment-mission-advance-copy.

Urben, P. G., Brethericks-Handbook-Reactive-Chemical Hazards, (2006) http://www.amazon.com/Brethericks-Handbook-Reactive-Chemical-Edition-Two/dp/0123725631.

Wikipedia, Gold cyanization, (2014) http://www.multimix.com.au/DOCUMENTS/Technical%20Bulletin1.pdf.

1.30 Hydrogen in Steam or Air Analyzers

B. G. LIPTÁK (2017)

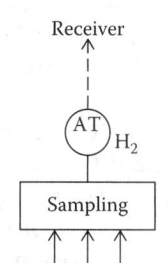

Types of designs	Thermal Conductivity, Chromatography, Solid State, Electrochemical, Palladium-based, Catalytic Bead, Mass Spectrometry, Laser
Explosive limits	LEL = 4% by volume, UEL = 75% by volume
Applications	Leak Detection, Explosion Protection, Hydrogen purity monitoring
Costs	Manual leak detectors: $100–$500 Continuous monitor: ~$ 1,000 Multiple stream detectors w/sampling: up to $10,000
Ranges	Alarm switch: Warning at 1%, alarm at 2%, leaking: 0–1000 ppm Handheld: 15ppm to 100%, Lithium Battery, Operating Time: 10 hr H_2-specific solid-state sensor for 0–2000 ppm H_2-specific catalytic sensor for 0%–100% of LEL Cerium oxide for 0.2%–4.0% of H_2 in air Semiconductor (1000 ppm—2%)
Partial list of suppliers	ABB, http://www.directindustry.com/prod/abb-measurement-analytics/product-56271-433046.html ATEQ (hydrogen leak detector) http://atequsa.com/product/hydrogen-portable-leak-tester/ BlueSens, http://www.bluesens.com/english/products/allgassensors/bcp-h2-sensor.html General Monitor, http://s7d9.scene7.com/is/content/minesafetyappliances/Catalytic%20Bead%20Sensors-Accessories%20Data%20Sheet HONEYWELL, http://www.lesman.com/unleashd/catalog/analytical/analyt_hwhydrogengas.htm H2 Scan, http://www.h2scan.com/pdfs/h2_monitoring.pdf INFICON (leak detector) http://products.inficon.com/en-us/nav-products/product/detail/sensistor%20sentrac%20hydrogen%20leak%20detector?path=Products%2FLeak-Detectors NexCeris, http://nexceris.com/sensors-alarm-systems/#sensors NOVA, http://catalog.nova-gas.com/item/continuous-multigas-analyzers-with-dew-point/ndustrial-process-gas-analyzer-systems-7900-series/model-7905a?&origin=keyword&by=prod&filter=0 Prism Gas Detection Pvt. Ltd. http://www.gasleakmonitor.com/xp-704.html RKI, http://www.rkiinstruments.com/pages/application_briefs/Hydrogen_Specific_Solutions.htm SBS, http://www.sbsbattery.com/products-services/by-product/battery-test-equipment-1/battery-monitoring-equipment/sbs-h2.html Synkera, http://www.synkerainc.com/products-services/chemical-sensing-analysis UST, http://www.synkerainc.com/products-services/chemical-sensing-analysis

INTRODUCTION

The concentration of hydrogen can be measured by a large variety of detectors. Most of these are described in detail in other chapters of this volume and therefore will only be briefly discussed here. These related chapters are

1.10 Chromatography
1.14 Combustibles
1.21 Electrochemical
1.34 Leak Detectors
1.35 Mass Spectrometry
1.62 Spectrometry
1.66 Thermal Conductivity

Hydrogen and Its Future Role

Some believe that in the twenty-second century and thereafter, hydrogen will be the key to the global energy economy, as it will serve to store, transport, and distribute solar energy.

Today, it is already the key energy source in space exploration, and it is also commonly used in many industrial processes. On the other hand, hydrogen is one of the most explosive gases, and leaking hydrogen cannot be seen or smelled. For this reason, it is essential that the concentration of hydrogen be continuously monitored, and if it approaches the lower explosive limit (LEL) of about 4% by volume in air or drops below its upper explosive limit (UEL) of 75% by volume, the appropriate safety interlocks be automatically triggered.

The nuclear power industry is particularly prone to hydrogen explosions when meltdowns occur. The first step in the sequence of events leading to a hydrogen explosion in that case is the formation of hydrogen, caused by the melting of the cladding of the fuel rods. This occurs at a temperature of around 1200°C, at which point the zirconium in the cladding oxidizes. The zirconium obtains the needed oxygen from the water in the reactor, and thereby, the splitting of water generates a large quantity of hydrogen:

$$Zr + 2H_2O = ZrO_2 + 2H_2 \qquad 1.30(1)$$

In order for an explosion to occur, the generated hydrogen has to travel to an area where oxygen is present, has to accumulate to a concentration exceeding its LEL of about 4%, and has to find an ignition source. Therefore, in a properly designed nuclear power plant, as soon as the presence of hydrogen is detected, venting must be automatically initiated.

PROPERTIES AND SAFETY

Hydrogen is very light (SpG = 0.07, air = 1), has an ignition temperature of 580°C–590°C (1076°F–1094°F), and a flammability range in air of 4%–75% by volume (Table 1.30a).

Hydrogen poses a number of hazards to human safety, from potential detonations and fires when mixed with air to leaking into external air where it can spontaneously ignite. Hydrogen detonation parameters, such as critical detonation pressure and temperature, strongly depend on the container geometry. Moreover, hydrogen fire, while being extremely hot, is almost invisible.

Because of the safety concerns, a number of safety guidelines have been issued. The U.S. Nuclear Regulatory Commission (USNRC) Regulatory Guide 1.97 covers hydrogen concentration within the containments. In 2015 at NREL, the Hydrogen Technologies Safety Guide has been issued. The current ANSI/AIAA standard for hydrogen safety guidelines is AIAA G-095-2004, Guide to Safety of Hydrogen and Hydrogen Systems. As NASA has been one of the world's largest users of hydrogen, this evolved from NASA's earlier guidelines, NSS 1740.16 (8719.16), which was retrieved in 2005. These documents cover both the risks posed by hydrogen in its different forms and ways to ameliorate them. There are international efforts in progress to harmonize codes and regulations through established organizations and frameworks such as IEA, IPHE, and ISO.

My recommendation concerning hydrogen monitoring is that if the presence of hydrogen within an enclosed space is detected (in any concentration), alarms should sound, and if the concentration reaches about 2.5% of LEL (about 1% in air), automatic safety interlocks should shut down all of all sparking devices and the space should be automatically vented without waiting for operator action or allowing the operators to override these automatic safety interlocks.

TABLE 1.30a
Hydrogen Weight and Volume Conversion

	Weight		Gas		Liquid	
	Pounds (lb)	Kilograms (kg)	Cubic feet (scf)	cu meters (Nm³)	Gallons (gal)	Liters (l)
1 pound	1.0	0.4536	192.0	5.047	1.6928	6.408
1 kilogram	2.205	1.0	423.3	11.126	3.377	14.128
1 scf gas	0.00521	0.00236	1.0	0.02628	0.00882	0.03339
1 Nm³ gas	0.19815	0.08988	38.04	1.0	0.3355	1.2699
1 gallon liquid	0.5906	0.2679	113.4	2.961	1.0	3.785
1 liter liquid	0.15604	0.07078	29.99	0.7881	0.2642	1.0

Source: Universal Gases Incorporated.
Scf (standard cubic foot) gas measured at 1 atmosphere and 70°F.
Nm³ (normal cubic meter) gas measured at 1 atmosphere and 0°C.
Liquid measured at 1 atmosphere and boiling temperature.

DETECTOR TYPES

Thermal Conductivity

Thermal conductivity detectors (TCDs) are the most widely used traditional hydrogen sensors. TCD measures the thermal conductivity of the sample taken from the monitored space and compares that to the thermal conductivity of a reference gas (often helium). The thermal conductivity of hydrogen is seven times higher than that of air. As a result, this measurement is sensitive, having a lower detection limit for hydrogen of about 50 ppm by volume.

Figure 1.30a at the top left illustrates a two-element TCD block assembly with the measuring filaments inserted into both the measuring and reference chambers. This is a flow-through design, using a single (replaceable) measuring and a single reference filament that can be configured into a classical Wheatstone bridge circuit with a high-quality regulated power supply, which guarantees stability. The cell is temperature controlled, while the thermistors are heated to elevated temperatures, and they lose their heat to the walls at a rate that is proportional to the thermal conductivity of the gas. The resulting temperature difference is related to the hydrogen concentration of the sample gas.

In addition to the two-element design, four-element thermal conductivity cell designs with recessed hot-wire filaments are also available, as shown at the top right of Figure 1.30a. Some cell designs include eight pairs to improve sensitivity. Vertical mounting is preferred to prevent sagging of the wire elements. The recessed elements generally provide an improved noise level, but poorer speed of response. The speed of response of the TCD is a function of the internal volume of the detector cell. The flow through the cell must be constant, usually around 500 ml/min.

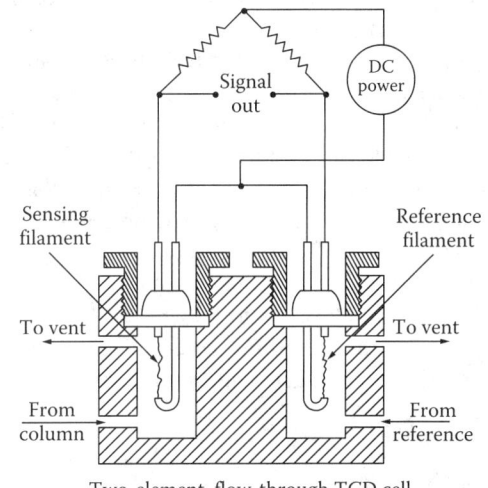

Two-element, flow-through TCD cell

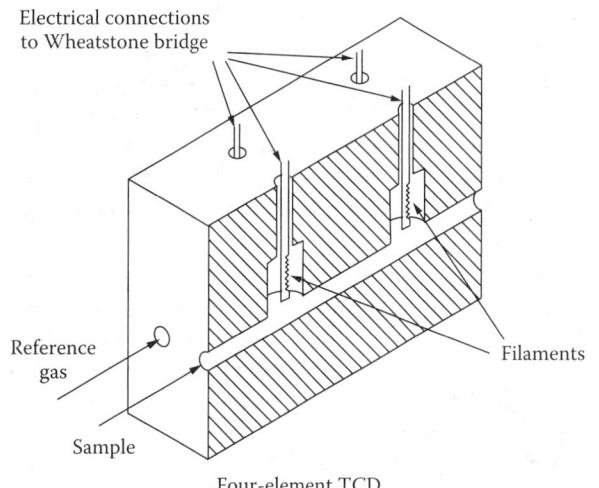

Four-element TCD

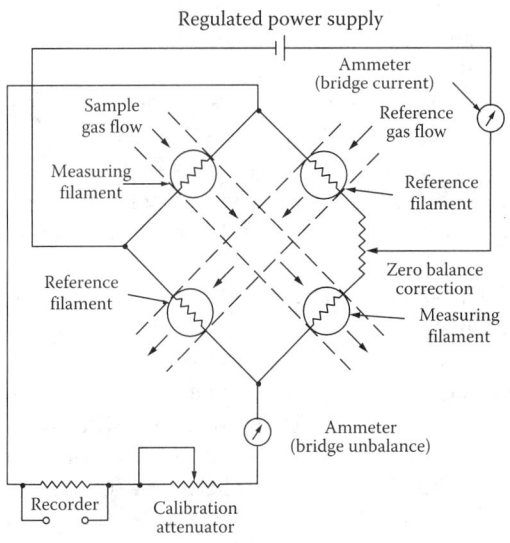

Wheatstone bridge circuit for TCD

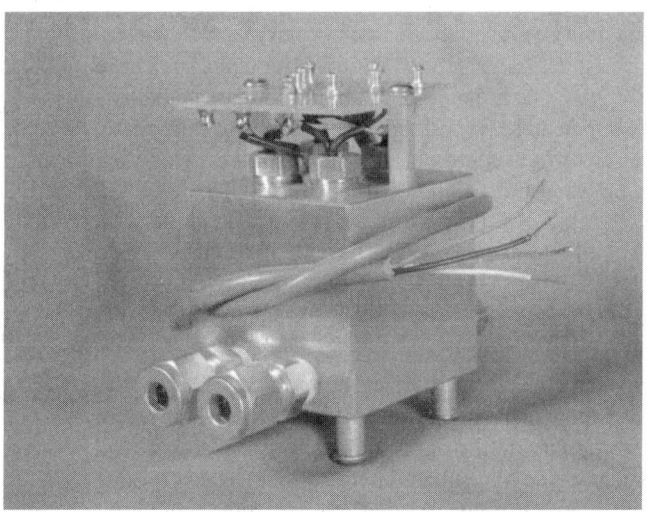

The physical appearance of a TDC cell

FIG. 1.30a
The design of the TCD cells.

Solid State

Driven particularly by the needs of the fuel cell industry, solid-state sensors are gaining popularity, due to their low cost, low maintenance, easy replacement, and general flexibility. Their designs include palladium, metal oxide semiconductor (MOS), catalytic bead (CB), and electrochemical designs.

In the diffusion head-type designs, the need for a sample pump or aspirator is eliminated. This eliminates moving parts and increases reliability. Therefore, diffusion head-type hydrogen detectors are commonly used when clean and dry samples are to be analyzed. One disadvantage of the diffusion head sensors is that plugging is difficult to detect because they cannot be furnished with low-sample-flow alarms as they generate sample movement by diffusion, density difference, convection, or similar effects.

Diffusion sensors usually operate in still air, but can also be provided with plant air aspirators or pumps to draw a sample flow over their sensors. Diffusion head designs are available with semiconductor sensors that respond to hydrogen as it is absorbed onto the doped surface of a MOS, by displaying a change in the resistance of the semiconductor surface. By varying the doping layer's thickness, the manufacturer can vary the responsiveness of the detector. Other designs use catalyst bead surface effect. There is a possibility of poisoning with both designs, so careful maintenance is required.

Both the diffusion-type electrochemical and the semiconductor sensors are available in the "hydrogen-specific" design that are not responsive to other gases such as hydrocarbons, and as such, they are suited for the task of continuous monitoring of hydrogen leaks.

Electrochemical

One of the most widely used hydrogen detectors are the electrochemical ones. They are well proved, simple, and have high sensitivity. Their operation is similar to that of a battery where an anode and a cathode are inserted into an electrolyte, and when hydrogen is present, it causes an electrochemical reaction between the anode and the cathode. This reaction generates a voltage or current, which is then measured as an indication of hydrogen concentration.

Palladium-Based

Palladium selectively absorbs hydrogen while forming palladium hydride, which can be measured in at least three ways. The most popular are the palladium resistor sensors. Second on popularity are the palladium field-effect transistors or capacitors. The third design variation is the palladium-coated optical sensor that converts hydrogen concentration into an optical signal.

Palladium-based thick film sensors using a ceramic base are popular. They are inexpensive, capable to operate at temperatures up to 200°C. Their response time is 20–30 sec, and their output signal (volts) are reasonably linear in the range of hydrogen concentrations of 1%–30%, but they are unreliable in areas where the air has high humidity or is saturated with steam.

Fiber-optic hydrogen sensors using a thin palladium film of 12 nm are one of the newer designs. Here the palladium film is deposited onto a glass substrate by thermal evaporation. The film is tested using 5% hydrogen in a nitrogen atmosphere. The absorption and desorption of hydrogen cause the optical properties of palladium to change depending on the concentration of hydrogen present in the atmosphere. Using a deuterium/halogen light source in conjunction with a UV/VIS spectrometer, the changes in the optical transmittance in the VIS spectra of the Pd film are monitored. This method is reported to be slow: the response and recovery times for the sensor are 60 sec and 300 sec, respectively. The sensor was reported to have good repeatability to continuous exposure cycles to 5% hydrogen with nitrogen being the carrier and recovery gas.

Hydrogen concentration and leak detectors are often palladium based, because palladium selectively absorbs hydrogen, forming palladium hydride. They should not be used if carbon monoxide, sulfur dioxide, or hydrogen sulfide is present in the monitored air. Their proper operation should be verifiable; they are able to detect hydrogen concentrations between 0.1% and 10% with a response time of under 1.0 sec in environments where the temperature is up to 80°C and the relative humidity is up to 98%. They have an accuracy of 5% FS, and their lifetime should preferably be 10 years but as a minimum longer than the time between scheduled maintenance.

Catalytic Bead (CB)

When mixtures of hydrogen in air come in contact with a heated and catalytically treated, fine, uniform, homogeneous platinum filament, combustion is induced at a temperature, which is considerably below the normal ignition temperature of hydrogen or other combustible gases. The heat generated by combustion can be measured by sensing the change of temperature of the filament by using thermocouples or by measuring the change of resistance of the filament.

CB sensors are a variation of the metal oxide sensors (MOS), having a sensing element coated by catalyst and a passive reference element for compensation.

In one design, half of a platinum element is covered by a thick film of nickel oxide. Here, when the cell is exposed to hydrogen, the catalyst layer converts (oxidizes) the hydrogen into water, while it itself is being heated by the heat of combustion, and the resulting temperature difference generates a voltage signal that is proportional to the H_2/air ratio.

In the catalytic combustion sensors, whether the measurement is based on the change of temperature or resistance, it is convenient to use two filaments. One filament is constantly exposed to the sample (detector filament). The other is hermetically sealed in an inert atmosphere (reference filament). If the measurement is based on temperature difference, the detection circuit is potentiometric, while if it is based on resistance, typically a Wheatstone bridge is used. One of the common limitations of catalytic combustion–type analyzers is the poisoning of the filament by silicon, sulfur, chlorinated compounds, or lead compounds; other limitations include relatively short useful life and slow response because of long sampling times.

These monitors are so maintenance prone because it is difficult to guarantee that the sample admitted into the analyzing cell is wholly representative of the hydrogen concentration in the monitored area. Also, the sample must also be free of particulate matter or moisture, and if the sample is at excessively high temperatures, it is advisable to use sample conditioners. Hydrogen being lighter than air, the samples naturally must be taken from near the ceiling of the monitored area. This may seem trivial, but is still worth mentioning.

Samples containing hydrogen also require flashback arrestors to dissipate the energy liberated by combustion. To ensure safe operation of the detectors, a variety of alarms are provided. These alarms can signal filament failure, power failure, alarm relay failure, and low sample flow rates. To ensure that an adequate amount of sample passes through the measuring chamber, flowmeters (rotameters) and needle valves can be provided for all except the diffusion head type of units.

Chromatography

Gas chromatography (GC) has the disadvantages of long response time (minutes), the need for carrier and calibration gases, which need to be resupplied periodically, and high maintenance. Therefore, using the earlier discussed dedicated hydrogen sensors is generally preferred. One exception to this is when it is necessary to quantify and identify several other gases besides hydrogen. Also, if the response time is of no concern, chromatographs can handle multiple samples.

Mass Spectrometry

Mass spectrometers, similarly to chromatographs, must be safety certified for operation in an industrial plant environment. They require air-conditioned shelters, safety monitors, calibration gases, and weekly maintenance by competent operators. They can identify many components in the sample stream. Similar to GCs, they require expensive sample systems and continuous bypass flows that add to their operating costs.

SAMPLING AND INSTALLATIONS

The selection of the sampling system design is usually made from among three basic configurations: (1) remote head (continuous measurement, continuous readout); (2) multiple head (continuous measurement, sequential readout); and (3) tube sampling system (sequential measurement, continuous readout).

The *remote head system* offers maximum flexibility, but it does that at the highest initial cost. This design typically consists of a number of locally mounted analyzer heads and an equal number of panel-mounted control and readout devices (Figure 1.30b). The maximum number of areas monitored from one central panel is a function of the capacity of the sample pumps (or aspirators) and of the physical size of the panel. Because the analyzer heads are located in the monitored areas, the speed of response is fast.

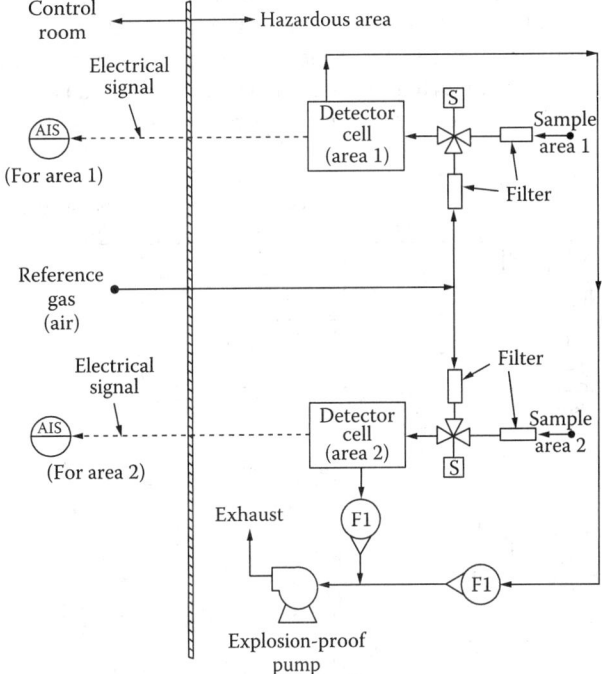

FIG. 1.30b
This figure illustrates the remote head system of combustible gas (hydrogen) monitoring.

In this configuration, samples are continuously drawn, and the electrical signal corresponding to the measured concentration is instantaneously transmitted to the control unit. The used sample is continuously withdrawn from the analyzer head through the tubing to the aspirator and is exhausted. This remote head system should be selected where fast response is essential, which justifies the cost.

In *multiple head systems*, the areas are monitored sequentially, which therefore results in time delays that in some applications can be unsafe. The multiple head system consists of a number of analyzer heads (one in each area to be monitored), one control unit with readout, and one or more sample pumps. The electrical circuit incorporates a single readout device common to all analyzing cells. In this configuration, the sample is drawn continuously to each sample chamber, and the pump continuously withdraws the expended sample.

The electrical output of each unit is transmitted to the panel, where sequential readout is provided. The dwell time for each area is typically 10 sec per area being monitored.

The tube sampling system consists of one analyzer head, one readout device, and a sample pump. This may be the least expensive arrangement, but sometimes the tubing cost can exceed the amount saved on the instrumentation. In this configuration, the samples from different areas are sequentially admitted to the common analyzer head, and the electrical signal is then transmitted to the readout device. A sample selector unit, consisting of time-sequenced solenoid valves, is arranged to admit one sample to the detector and connect all other sample lines to the sample pump. The sample is drawn continuously through each line. The sample selector is located at the analyzer head, so a fresh sample is always present at the sample selector. A clean gas purge can be provided after each analysis to prevent an erroneous reading caused by residual carryover in this type of system (see Figure 1.30c).

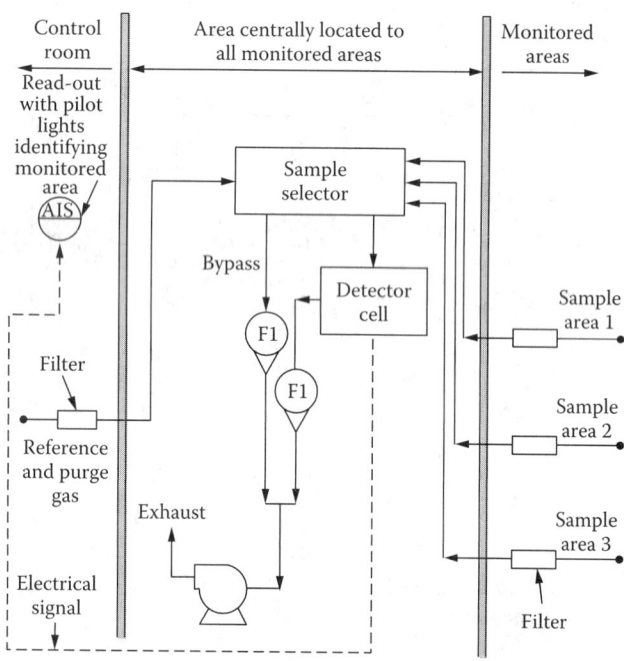

FIG. 1.30c
Tube sampling–based hydrogen monitoring is the slowest and usually the least expensive, if the sample tubes are short. It is slow, because sample has to travel through a sample tube before reaching the detector and because samples from several areas are analyzed in sequence.

As far as the detector cells are concerned, the following should be considered. While Wheatstone bridge cells use a fine helical filament, the thermocouple cell uses a heavy straight filament with a much longer useful life. In addition, the evaporation on the exposed filament results in a constant change of base resistance of the filament. In the Wheatstone bridge circuit, this change of base resistance produces a shifting of zero and requires frequent rebalancing of the bridge.

The temperature change measured by the thermocouple is independent of filament deterioration. Thus, for the thermocouple detector, the zero drift is reduced to a negligible amount even over long periods of time. Therefore, the thermocouple detector is often superior to the Wheatstone bridge–type detector.

Specialized Applications

Sensors are available for the monitoring hydrogen in such specific environments, such as where refrigerant gases are present (Figure 1.30d). One such ceramic sensor provides sensitive and rapid detection of the presence of hydrogen in the ambient air. This detector can be used stand-alone or be incorporated into the safety interlocks of equipment like those of fuel cell driven vehicles.

FIG. 1.30d
Hydrogen detector for fuel cell powered vehicles. (Courtesy of NexCeris.)

Another specialized application for a hydrogen sensor is for leak detection. Hydrogen leak detectors are available in portable designs (Figure 1.30e) and are popular, because hydrogen is less expensive than helium, when the gas is consumed for larger leak testing operations. While hydrogen is

FIG. 1.30e
Portable Hydrogen leak detector. (Courtesy of ATEQ.)

combustible in high concentrations, it is safe at the low concentrations of testing and it quickly and it is also nontoxic, nonpoisonous and is without any environmental impact. Refilling of the batteries is necessary every 8 hr.

CONCLUSIONS

For hydrogen leakage detection purposes, manual monitors are often sufficient. For explosion prevention purposes, continuous monitoring is recommended. This continuous monitoring should be integrated into an automatic interlock system, so that if dangerous levels of hydrogen concentration are detected, the associated process is automatically shut down, and the space is automatically vented.

Hydrogen monitors should be free from interference from other gases (should be specific to hydrogen), should be fast, should be of low cost, and should be of low maintenance. As was discussed in this chapter, the traditional sensor designs (thermal conductivity, solid state, electrochemical, and catalytic combustion) are widely used. The most promising solid-state design is using palladium, which is hydrogen specific and does not require oxygen for its operation.

SPECIFICATION FORMS

When specifying hydrogen in steam or air analyzers only, please use the ISA form 20A1001, and when specifying both the analyzer and the composition or the properties of the process material of interest, use ISA form 20A1002. Both forms are reproduced with the permission of the International Society of Automation on the next pages.

Analytical Measurement

ANALYSIS DEVICE — Operating Parameters

#	RESPONSIBLE ORGANIZATION		#	SPECIFICATION IDENTIFICATIONS		
1	ISA		6			
2			7	Document no		
3			8	Latest revision	Date	
4			9	Issue status		
5			10			

#	ADMINISTRATIVE IDENTIFICATIONS			#	SERVICE IDENTIFICATIONS Continued			
11				40				
12	Project number	Sub project no		41	Return conn matl type			
13	Project			42	Inline hazardous area cl		Div/Zon	Group
14	Enterprise			43	Inline area min ign temp		Temp ident number	
15	Site			44	Remote hazardous area cl		Div/Zon	Group
16	Area	Cell	Unit	45	Remote area min ign temp		Temp ident number	
17				46				
18	SERVICE IDENTIFICATIONS			47				
19	Tag no/Functional ident			48	COMPONENT DESIGN CRITERIA			
20	Related equipment			49	Component type			
21	Service			50	Component style			
22				51	Output signal type			
23	P&ID/Reference dwg			52	Characteristic curve			
24	Process line/nozzle no			53	Compensation style			
25	Process conn pipe spec			54	Type of protection			
26	Process conn nominal size	Rating		55	Criticality code			
27	Process conn termn type	Style		56	Max EMI susceptibility		Ref	
28	Process conn schedule no	Wall thickness		57	Max temperature effect		Ref	
29	Process connection length			58	Max sample time lag			
30	Process line matl type			59	Max response time			
31	Fast loop line number			60	Min required accuracy		Ref	
32	Fast loop pipe spec			61	Avail nom power supply			Number wires
33	Fast loop conn nom size	Rating		62	Calibration method			
34	Fast loop conn termn type	Style		63	Testing/Listing agency			
35	Fast loop schedule no	Wall thickness		64	Test requirements			
36	Fast loop estimated lg			65	Supply loss failure mode			
37	Fast loop material type			66	Signal loss failure mode			
38	Return conn nominal size	Rating		67				
39	Return conn termn type	Style		68				

#	PROCESS VARIABLES	MATERIAL FLOW CONDITIONS			#	PROCESS DESIGN CONDITIONS		
69					101			
70	Flow Case Identification			Units	102	Minimum	Maximum	Units
71	Process pressure				103			
72	Process temperature				104			
73	Process phase type				105			
74	Process liquid actl flow				106			
75	Process vapor actl flow				107			
76	Process vapor std flow				108			
77	Process liquid density				109			
78	Process vapor density				110			
79	Process liquid viscosity				111			
80	Sample return pressure				112			
81	Sample vent/drain press				113			
82	Sample temperature				114			
83	Sample phase type				115			
84	Fast loop liq actl flow				116			
85	Fast loop vapor actl flow				117			
86	Fast loop vapor std flow				118			
87	Fast loop vapor density				119			
88	Conductivity/Resistivity				120			
89	pH/ORP				121			
90	RH/Dewpoint				122			
91	Turbidity/Opacity				123			
92	Dissolved oxygen				124			
93	Corrosivity				125			
94	Particle size				126			
95					127			
96	CALCULATED VARIABLES				128			
97	Sample lag time				129			
98	Process fluid velocity				130			
99	Wake/natural freq ratio				131			
100					132			

#	MATERIAL PROPERTIES			#	MATERIAL PROPERTIES Continued		
133				137			
134	Name			138	NFPA health hazard	Flammability	Reactivity
135	Density at ref temp	At		139			
136				140			

Rev	Date	Revision Description	By	Appv1	Appv2	Appv3	REMARKS

Form: 20A1001 Rev 0   © 2004 ISA

	RESPONSIBLE ORGANIZATION	ANALYSIS DEVICE COMPOSITION OR PROPERTY Operating Parameters (Continued)		SPECIFICATION IDENTIFICATIONS			
1			6				
2			7	Document no			
3	ISA		8	Latest revision		Date	
4			9	Issue status			
5			10				

	PROCESS COMPOSITION OR PROPERTY			MEASUREMENT DESIGN CONDITIONS				
	Component/Property Name	Normal	Units	Minimum	Units	Maximum	Units	Repeatability
11								
12								

(rows 13–75)

Rev	Date	Revision Description	By	Appv1	Appv2	Appv3	REMARKS

Form: 20A1002 Rev 0

© 2004 ISA

Abbreviations

CB	Catalytic Bead
FET	Field-Effect Transistors
GC	Gas Chromatography
LEL	Lower Explosive Limit
MEMS	Microelectromechanical Systems
MOS	Metal Oxide Semiconductor
MS	Mass Spectrometry
SAW	Surface Acoustic Wave
TCD	Thermal Conductivity Detector
UEL	Upper Explosive Limit

Organizations

AIAA	American Institute of Aeronautics and Astronautics
ANSI	American National Standards Institute
IEA	International Energy Agency
IPHE	International Partnership for Hydrogen and Fuel Cells in the Economy
ISO	International Organization for Standardization
NASA	National Aeronautics and Space Administration
NREL	National Renewable Energy Laboratory
NSS	National Space Society
USNRC	U.S. Nuclear Regulatory Commission

Bibliography

AIAA Guide to Safety of Hydrogen and Hydrogen Systems (G-095-2004e), (2015) https://www.aiaa.org/StandardsDetail.aspx?id=3864.

DOE, Innovation in hydrogen monitoring and control systems, (2005) http://www.h2scan.com/pdfs/h2_monitoring.pdf.

General Monitors, Catalytic sensors and accessories, (2015) http://s7d9.scene7.com/is/content/minesafetyappliances/Catalytic%20Bead%20Sensors-Accessories%20Data%20Sheet.

General Monitor, Hydrogen detection in oil refineries, (2015) http://www.gmigasandflame.com/downloads/white-papers/Hydrogen-Gas-Detection.pdf.

IEEE, A review of palladium-based fiber-optic sensors for molecular hydrogen detection, (2012) http://ieeexplore.ieee.org/xpl/login.jsp?tp=&arnumber=5742670&url=http%3A%2F%2Fieeexplore.ieee.org%2Fiel5%2F7361%2F4427201%2F05742670.pdf%3Farnumber%3D5742670.

Jefferson Lab, The element hydrogen, (2011) http://education.jlab.org/itselemental/ele001.html.

Martin, L. P., Pham, A. Q., Glass, R. S., Electrochemical hydrogen sensor for safety monitoring (2004) http://www.sciencedirect.com/science/article/pii/S0167273804006344.

NASA, Safety standard for hydrogen and hydrogen systems, (2005) https://www.hq.nasa.gov/office/codeq/doctree/canceled/871916.pdf.

NASA, Glenn Safety Manual, Hydrogen, (2015) http://smad-ext.grc.nasa.gov/shed/pub/gsm/chapter_06.pdf.

Praxair, Liquid hydrogen, (2004) http://www.hydrogenandfuelcellsafety.info/resources/mdss/Praxair-LH2.pdf.

Rivkin, C., Burgess, R., and Buttner, W., Hydrogen technologies safety guide, (2015) http://www.nrel.gov/docs/fy15osti/60948.pdf.

Soundarrajan, P., and Schweighardt, F., Hydrogen sensing and detection, (2007) http://photonics.inescporto.pt/links/fct2013/metap28.pdf.

Universal Industrial Gases, Inc., Unit conversions for hydrogen, (2007) http://www.uigi.com/h2_conv.html.

University of Central Florida, Hydrogen basics, (2014) http://www.fsec.ucf.edu/en/consumer/hydrogen/basics/production-electrolysis.htm.

USNRC Regulatory Guide 1.97, Criteria for accident monitoring instrumentation for nuclear power plants, (2006) http://pbadupws.nrc.gov/docs/ML0615/ML061580448.pdf.

UST, Multi-range gas detector, (2005) http://www.synkerainc.com/products-services/chemical-sensing-analysis.

1.31 Hydrogen Sulfide Detectors

D. H. F. LIU (1995) **B. G. LIPTÁK** (2003, 2017)

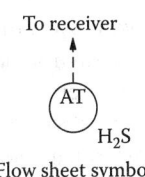

Flow sheet symbol

Analyzer types	Electrochemical gas diffusion, solid-state or gold-film sensors, tape staining, UV photometric, gas chromatography.
Ranges	0–50 ppm is typical for ambient air monitors, with maximum range up to 0–500 ppm. For process applications, ranges up to 0%–100% are available.
Inaccuracy	3%–5% of full scale for air monitors; 1% of full scale for process analyzers.
Costs	Pocket-size, battery-operated monitor costs $500–$1000; the cost of portable microprocessor-based diffusion-type units is about $2500; the cost of UV photometric designs is in the range of $10,000 and for chromatographic analyzers, prices are in excess of $20,000.
Partial list of suppliers	ABB Inc. (http://www02.abb.com/GLOBAL/FRABB/frabb028.nsf/viewunid/1FE27EEAFC83771EC1256B6E004C9525/$file/ppa+sulfide+gaz.pdf) AMETEK (http://www.ametekpi.com/products/WR-Model-880-NSL-Tail-Gas-Analyzer.aspx) Applied Analytics (http://www.a-a-inc.com/product.php?recordID=7) Arizona Instrument Co. (http://www.azic.com/products_jerome_631.aspx) Ashtead-Technology (http://ashtead-technology.com/productgroup/instruments/836/) Baseline-Mocon Inc. (http://www.thomasnet.com/profile/10067547/baselinemocon-inc.html?what=hydrogen+sulfide+monitors+and+analyzers&cov=NA&heading=1652288&searchpos=5&cid=10067547) EcoTech (http://www.thermoscientific.com/ecomm/servlet/productsdetail_11152___11961608_-1) EMMET Corporation (http://www.thomasnet.com/profile/00165231/enmet-corporation.html?what=hydrogen+sulfide+monitors+and+analyzers&cov=NA&heading=1652288&searchpos=1&cid=165231) Interscan Corporation (http://www.thomasnet.com/catnav/productdetail.html?cid=345230&cncat=1529&cnprod=3001014) Nova Analytical Systems (http://www.thomasnet.com/profile/10091207/nova-analytical-systems-inc.html?what=hydrogen+sulfide+monitors+and+analyzers&cov=NA&heading=1652288&searchpos=2&cid=10091207) Pem-Tech (http://www.pem-tech.com/on-stream-process-analyzers/pt605-hydrogen-sulfide-h2s.html) PemTech (http://www.pem-tech.com/gas-sensors/hydrogen-sulfide/pt295-solid-state-sensor-mos.html) Sierra Monitor Corporation (http://www.thomasnet.com/profile/00640726/sierra-monitor-corp.html?what=hydrogen+sulfide+monitors+and+analyzers&cov=NA&heading=1652288&searchpos=6&cid=640726) Sensidyne (http://sensidyne.com/uploads/7013951rK_Sensor_Man.pdf) Thermo Scientific (http://www.thermoscientific.com/ecomm/servlet/productsdetail_11152___11961608_-1)

INTRODUCTION

Hydrogen sulfide (H_2S) is a toxic gas found in many industrial environments. It is commonly monitored for personnel safety, environmental protection, and process control. This chapter will describe the more frequently used commercial instruments for ambient and online measurement of H_2S in process gases and liquids.

ANALYZER TYPES

The concentration of hydrogen sulfide can be measured by a large variety of detectors. Some are unique to H_2S measurement (such as the gold film, and lead acetate designs), but most are not unique and are also described in other chapters of this volume and therefore are only be briefly discussed here. These related chapters are

1.10 Chromatography
1.14 Combustibles
1.21 Electrochemical
1.26 Toxic Gas Sensors
1.29 Hydrogen Cyanide
1.34 Leak Detectors
1.35 Mass Spectroscopy
1.62 Spectrometry
1.66 Thermal Conductivity
1.69 UV Photometric

Electrochemical

For portable, battery-operated instrumentation for ambient air monitoring, the most common sensors are the fuel-cell-type electrochemical gas diffusion sensors.

Fuel cells convert the chemical energy of fuel and oxygen into electrical energy, while the electrode and the electrolyte remain unaltered. Fuel is converted at the anode into hydrogen ions, which travel through the electrolyte to the cathode, and electrons, which travel through an external circuit to the cathode. If oxygen is present at the cathode, it is reduced by these electrons, and the hydrogen and oxygen ions eventually react to form water.

The electrochemical sensors are available in pocket-size, battery-operated packaging, with replaceable "pop-out" sensors. The advantage of these units is that they do not require pumps or aspirators to pull in the sample, and they are unaffected by wind or variations in relative humidity.

These pocket-size units are usually configured for a number of channels, which might include H_2S, CO, O_2, and CH_4 (Figure 1.31a).

In the sensor, a pair of polarized electrodes is isolated from the ambient air by a gas-permeable membrane. As hydrogen sulfide diffuses through the membrane, an oxidation–reduction reaction occurs and the resulting

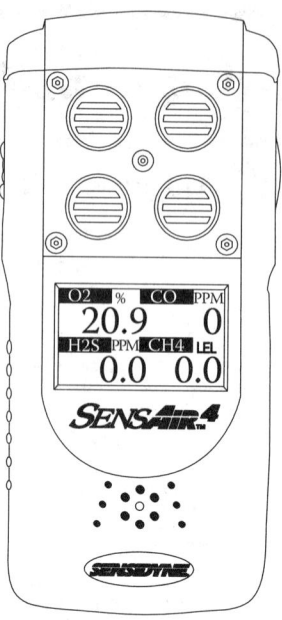

FIG. 1.31a
Pocket-size, battery-operated electrochemical monitor with four channels, including H_2S. (Courtesy of Sensidyne Inc.)

electrons cause a current flow, which is in proportion to the H_2S concentration in the air.

Some of the limitations of this design include interferences from background gases such as hydrogen, carbon monoxide, ethylene, sulfur dioxide, chlorine, methyl mercaptan, nitric oxide, and nitrogen dioxide. The units are also available as explosion-proof transmitters (Figure 1.31b) for operation at near-ambient temperatures and up to 10 psig (0.7 bar) services. The electrochemical sensors are preferred in pure oxygen or in benign atmosphere applications or where ruggedness is less important than accuracy.

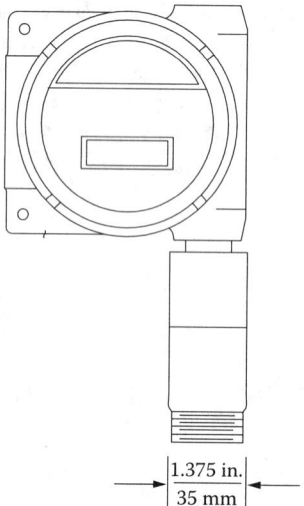

FIG. 1.31b
Diffusion-type electrochemical hydrogen sulfide detector.

Gold Film

Gold films absorb hydrogen sulfide and respond to its concentration by a proportional change in their electrical resistance. These analyzers usually include an internal pump that draws in the ambient air. One advantage of the gold-film sensor is that it is available in the parts per billion (ppb) range, and it is not sensitive to interference by SO_2, CO_2, or CO.

Solid State

Solid-state sensors are used in combustible gas detection equipment (see Chapter 1.14) for the measurements of H_2S in ambient air at the parts per million (ppm) level. The advantage of the solid-state approach is that there is no sampling system involved. The solid-state H_2S sensors are formed by depositing thin films on silicon chips, as shown in Figure 1.31c. Separate film layers serve as heaters, temperature-monitoring thermistors, and H_2S-sensitive metal oxide semiconductors. If the semiconductor is heated to a constant temperature in the 150°C–300°C range, a measurable decrease in electrical resistance occurs if exposed to H_2S. The resistance decrease is a logarithmic function of H_2S concentration.

They are microprocessor-based modular designs with both digital and analog outputs (4–20 mA), programmable alarms and with LED display providing continuous indication of the H_2S concentration in ppm. The calibration mode, alarm, and fault status of the sensor are also displayed on the LED (Figure 1.31d). Their recalibration and maintenance is convenient.

FIG. 1.31d
Microprocessor-based, solid-state H_2S transmitter. (Courtesy of Pemtech.)

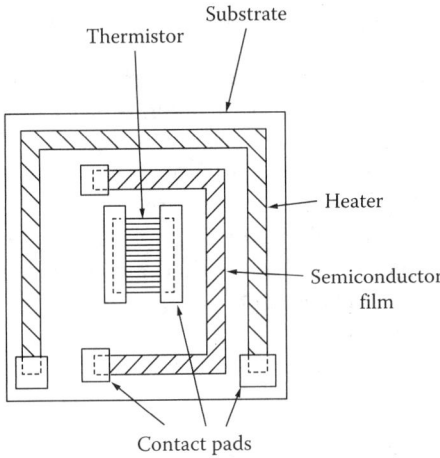

FIG. 1.31c
Schematic view of a solid-state H_2S sensor. (From Kaminski, C. and Poli, A., Electrochemical or solid state H_2S sensors: Which is right for you? InTech, June 1985, 55–61.)

In the case of the solid-state sensor, hydrogen, isopropanol, ethyl, and methyl mercaptan interfere with the measurement. In general, solid-state sensors are preferred to detect high concentrations or to operate under extreme temperatures or vibration or in corrosive atmospheres.

The solid-state metal oxide Semiconductor (MOS) sensors are suitable for explosion proof area applications.

Lead Acetate Tape

This type of analyzer system is used by industry for attaining compliance with the U.S. Environmental Protection Agency (EPA) Fuel Gas to Combustion Devices Regulations. It has a history of onstream field performance and reliability for measuring H_2S in gas streams such as coal gas, natural gas, and mixed propane–butane. It is specific to H_2S. The system's ability to report H_2S levels enables timely corrective actions to be taken to protect against upset, contamination, and corrosion buildup.

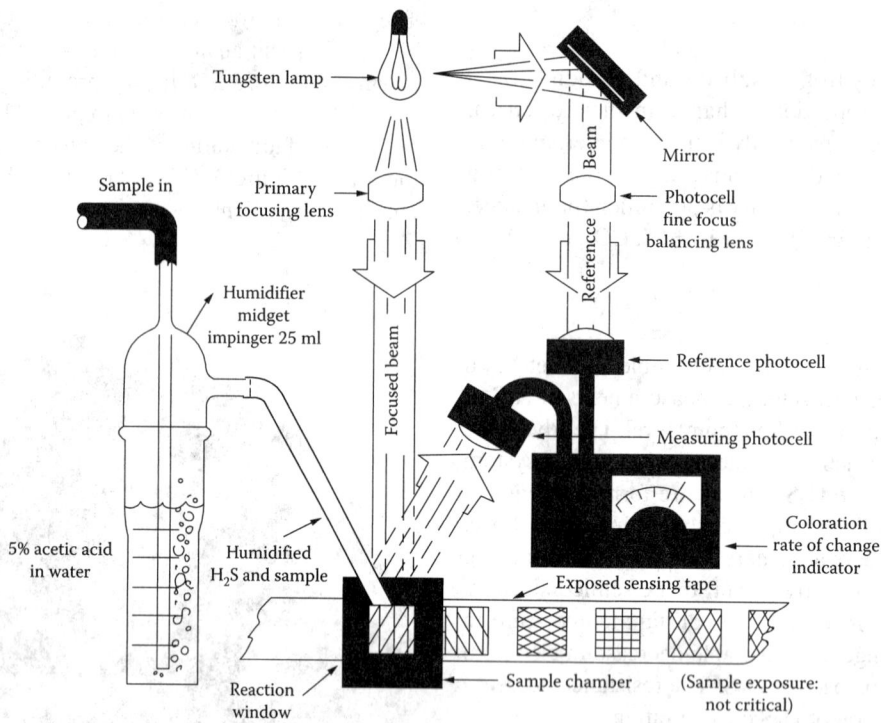

FIG. 1.31e
Rate of change of reflectance-type H_2S readout system.

Figure 1.31e shows the instrument and its operating principles. Sample at a constant flow rate enters a humidifier where it bubbles through a 5% acetic acid solution. The sample then flows into the reaction window of the sample chamber, where it passes over an exposed surface of paper sensing tape impregnated with lead acetate; the tape is automatically driven by a motor. Hydrogen sulfide reacts with lead acetate to form lead sulfide, causing a brown stain on the paper. The rate of reaction and resulting rate of color change is proportional to the concentration of H_2S in the sample.

An optical system, photocells, and first-rate derivative electronic processor provide an output voltage proportional to the rate of change of the photocell output voltage. This output is proportional to the concentration of H_2S in the sample. A reference photocell compensates for light intensity changes.

Ultraviolet

The measurement of the ultraviolet (UV) absorption of H_2S provides a sensitive and selective technique for monitoring H_2S concentrations in gas streams in which no other UV-absorbing compounds are present. Direct photometric analysis and the analysis systems described in the following paragraphs for monitoring low levels of H_2S are all operating successfully in gas processing plants and oil refineries.

Figure 1.31f shows a flow diagram of a direct photometric analyzer system for monitoring low levels of H_2S in interference-free gas streams. (See Chapter 1.69 for a detailed description of the split-beam UV analyzer.) The system uses a bypass stream to decrease sample lag and a self-cleaning filter to minimize particulate matter in the sample. The sample pressure is maintained within the sample cell with a backpressure regulator after the sample cell. The analyzer sample cell is purged with air or nitrogen when automatic zeroing is required for high-sensitivity measurements.

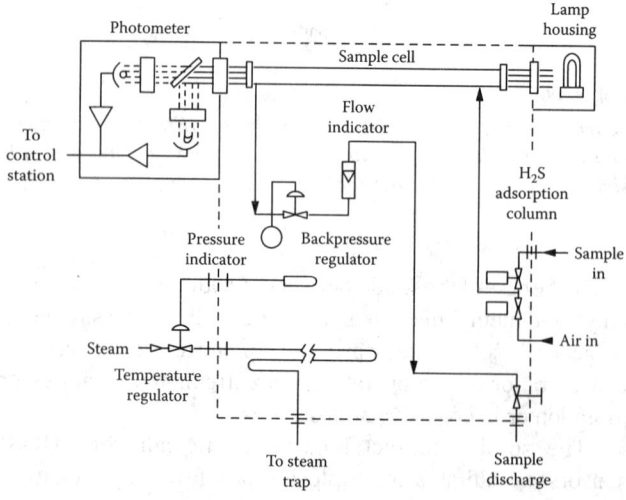

FIG. 1.31f
Direct low-level H_2S photometric analyzer.

Where the level of potentially interfering compounds is relatively low and not rapidly changing, the previous system has operated successfully using the sample gas with the H_2S selectively removed as the "zero" reference gas. Potentially interfering compounds are those with conjugated double bonds, such as 1,3-butadiene and aromatics and other sulfur compounds.

Where background absorbance is excessively high and changing rapidly, a special system has been developed for selective H_2S analysis. In this system, H_2S is extracted with a dilute ammonium hydroxide solution, and the strong UV absorption of the ammonium sulfide formed in solution is measured and calibrated for H_2S concentration in the gas stream.

Chromatograph

Process gas chromatographs have been designed for environmental monitoring of H_2S at the ppm level using the sulfur-specific flame photometric detector (FPD) (Figure 1.31g). See Chapter 227 for the basic principles of gas chromatography and a more detailed description of the FPD.

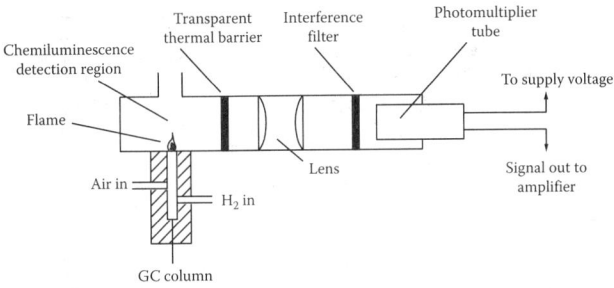

FIG. 1.31g
The flame photometric detector.

APPLICATIONS

For Claus sulfur recovery applications, a top of the pipe analyzer is available, which can measure both H_2S and SO_2. An installed unit is shown in Figure 1.31h.

FIG. 1.31h
Top of the pipe tail gas analyzer detects both H_2S and SO_2. (Courtesy of Ametek Process Instruments.)

SPECIFICATION FORMS

When specifying hydrogen sulfide analyzers only, one can use the ISA form 20A1001, and when specifying both the analyzer and the composition or properties of the process, which it will be monitoring, use the ISA form 20A1002. Both forms are reproduced with the permission of the International Society of Automation on the next pages.

Analysis Device — Operating Parameters

#	RESPONSIBLE ORGANIZATION		ANALYSIS DEVICE	#	SPECIFICATION IDENTIFICATIONS		
1	ISA			6			
2			Operating Parameters	7	Document no		
3				8	Latest revision	Date	
4				9	Issue status		
5				10			

#	ADMINISTRATIVE IDENTIFICATIONS			#	SERVICE IDENTIFICATIONS Continued		
11				40			
12	Project number	Sub project no		41	Return conn matl type		
13	Project			42	Inline hazardous area cl	Div/Zon	Group
14	Enterprise			43	Inline area min ign temp	Temp ident number	
15	Site			44	Remote hazardous area cl	Div/Zon	Group
16	Area	Cell	Unit	45	Remote area min ign temp	Temp ident number	
17				46			
18	SERVICE IDENTIFICATIONS			47			
19	Tag no/Functional ident			48	COMPONENT DESIGN CRITERIA		
20	Related equipment			49	Component type		
21	Service			50	Component style		
22				51	Output signal type		
23	P&ID/Reference dwg			52	Characteristic curve		
24	Process line/nozzle no			53	Compensation style		
25	Process conn pipe spec			54	Type of protection		
26	Process conn nominal size	Rating		55	Criticality code		
27	Process conn termn type	Style		56	Max EMI susceptibility	Ref	
28	Process conn schedule no	Wall thickness		57	Max temperature effect	Ref	
29	Process connection length			58	Max sample time lag		
30	Process line matl type			59	Max response time		
31	Fast loop line number			60	Min required accuracy	Ref	
32	Fast loop pipe spec			61	Avail nom power supply	Number wires	
33	Fast loop conn nom size	Rating		62	Calibration method		
34	Fast loop conn termn type	Style		63	Testing/Listing agency		
35	Fast loop schedule no	Wall thickness		64	Test requirements		
36	Fast loop estimated lg			65	Supply loss failure mode		
37	Fast loop material type			66	Signal loss failure mode		
38	Return conn nominal size	Rating		67			
39	Return conn termn type	Style		68			

#	PROCESS VARIABLES	MATERIAL FLOW CONDITIONS		#	PROCESS DESIGN CONDITIONS		
69				101			
70	Flow Case Identification		Units	102	Minimum	Maximum	Units
71	Process pressure			103			
72	Process temperature			104			
73	Process phase type			105			
74	Process liquid actl flow			106			
75	Process vapor actl flow			107			
76	Process vapor std flow			108			
77	Process liquid density			109			
78	Process vapor density			110			
79	Process liquid viscosity			111			
80	Sample return pressure			112			
81	Sample vent/drain press			113			
82	Sample temperature			114			
83	Sample phase type			115			
84	Fast loop liq actl flow			116			
85	Fast loop vapor actl flow			117			
86	Fast loop vapor std flow			118			
87	Fast loop vapor density			119			
88	Conductivity/Resistivity			120			
89	pH/ORP			121			
90	RH/Dewpoint			122			
91	Turbidity/Opacity			123			
92	Dissolved oxygen			124			
93	Corrosivity			125			
94	Particle size			126			
95				127			
96	CALCULATED VARIABLES			128			
97	Sample lag time			129			
98	Process fluid velocity			130			
99	Wake/natural freq ratio			131			
100				132			

#	MATERIAL PROPERTIES			#	MATERIAL PROPERTIES Continued		
133				137			
134	Name			138	NFPA health hazard	Flammability	Reactivity
135	Density at ref temp	At		139			
136				140			

Rev	Date	Revision Description	By	Appv1	Appv2	Appv3	REMARKS

Form: 20A1001 Rev 0 © 2004 ISA

1.31 Hydrogen Sulfide Detectors

	RESPONSIBLE ORGANIZATION		ANALYSIS DEVICE COMPOSITION OR PROPERTY Operating Parameters (Continued)		SPECIFICATION IDENTIFICATIONS		
1				6	Document no		
2				7	Latest revision		Date
3	(ISA logo)			8	Issue status		
4				9			
5				10			

	PROCESS COMPOSITION OR PROPERTY			MEASUREMENT DESIGN CONDITIONS				
11	Component/Property Name	Normal	Units	Minimum	Units	Maximum	Units	Repeatability
12								
13–37								

Rev	Date	Revision Description	By	Appv1	Appv2	Appv3	REMARKS

Form: 20A1002 Rev 0

© 2004 ISA

Abbreviations

FPD   Flame photometric detector
LED   Light emitting diode
MOS   Metal oxide semiconductor
UV   Ultraviolet

Bibliography

ACGIH, Documentation of the threshold limit values and biological exposure indices, (2015) https://www.acgih.org/forms/store/ProductFormPublic/documentation-of-the-threshold-limit-values-and-biological-exposure-indices-7th-ed.

ASTM D6021 Standard Test Method for Measurement of Total Hydrogen Sulfide in Residual Fuels by Multiple Headspace Extraction and Sulfur Specific Detection, (2012) http://www.astm.org/Standards/D6021.htm.

ASTM D7621-15 Standard Test Method for Determination of Hydrogen Sulfide in Fuel Oils by Rapid Liquid Phase Extraction, (2015) http://www.astm.org/Standards/D7621.htm.

ATSDR, Toxicological profile for hydrogen sulfide, (2006) http://www.atsdr.cdc.gov/toxprofiles/tp.asp?id=389&tid=67.

Calabrese, E. J. and Kenyon, E., (2011), Air Toxics and Risk Assessment, Lewis Publishers, http://www.belleonline.com/bios/calabrese.htm.

CiTicesl, Toxic gas sensor, (1999) http://www.safetygas.com.tw/pdf/city%20technologies%20toxic%20gas%20sensors.pdf.

EPA, Acute exposure guideline levels for airborne chemicals, (2015) http://www.epa.gov/oppt/aegl/index.htm.

FM, Approval standard for toxic gases, (2014) https://www.fmglobal.com/assets/pdf/fmapprovals/6340.pdf.

InspectAPedia, Toxic gas exposure standards, (2015) http://inspectapedia.com/sickhouse/Gas_Exposure_Limits.php.

Lewis, R. J., Rapid guide to hazardous chemicals in the workplace, (2000) http://www.amazon.com/Rapid-Guide-Hazardous-Chemicals-Workplace/dp/0471355429.

Lodge, J., Techniques for measurement of hydrogen sulfide and sulfur oxides, (2013) http://onlinelibrary.wiley.com/doi/10.1029/GM003p0024/summary.

McDermott, H. J., Air monitoring for toxic exposures, (2005) http://www.wiley.com/WileyCDA/WileyTitle/productCd-0471454354.html.

NIOSH, Pocket Guide to Chemical Hazards, National Institute for Occupational Safety and Health, (2015) http://www.cdc.gov/niosh/npg/.

OSHA, Toxic Industrial Chemicals (TICs) Guide, (2007) https://www.osha.gov/SLTC/emergencypreparedness/guides/chemical.html.

Petex, Hydrogen sulfide in production operations, (1996) http://www.amazon.com/Hydrogen-Sulfide-Production-Operations-Oil/dp/088698176X.

PetroWiki, Nanotechnology in hydrogen sulfide detection, (2014) http://petrowiki.org/Nanotechnology_in_hydrogen_sulfide_detection.

Scribd, Hydrogen sulfide training manual, (2006) http://www.scribd.com/doc/12756949/H2s-Manual.

1.32 Infrared and Near-Infrared Analyzers

J. E. BROWN (1969)   A. C. GILBY (1982)

B. G. LIPTÁK and T. M. CARDIS (1995)

E. H. BAUGHMAN (2003)   B. G. LIPTÁK (2017)

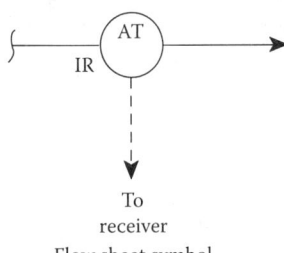

Flow sheet symbol

INFRARED ANALYZERS

Process streams	Gas or liquid, surface analysis of solids
Ranges and applications (See Table 1.32a)	Maximum range is usually 100%, with path length adjustment
	Ammonia—100 ppm
	Carbon monoxide—25 ppm
	Carbon dioxide—20 ppm
	Ethylene—100 ppm
	Hexane—100 ppm
	Methane—10 ppm
	Moisture (humidity)—50 ppm
	Nitrous oxide—10 ppm
	Propane—100 ppm
	Sulfur dioxide—100 ppm
	(*Notes*: 1. Some of these analytes can be measured by UV (Chapter 1.69), for example, sulfur dioxide.
	2. The minimum range is also a function of the matrix—the minimum for benzene in air is going to be much lower than that for benzene in gasoline.
	3. The normal range is a factor of 10, so ammonia could be 10–100 ppm or 1%–10%, but not 100 ppm to 10%.
	4. These are examples only, not an exclusive list.)
Operating pressure	Standard from atmospheric to 150 psig (10 bars); special up to 1000 psig (70 bars)
Operating Temperature	−40°C to 50°C (−40°F to 120°F) is standard; probe temperatures can be higher with special Arrangements
Humidity limitations	Up to 95% relative humidity (normally the instrument is purged, which negates the effect of humidity in the atmosphere).
Materials of construction	Cell bodies are available in all standard materials; windows can be made of sodium chloride, calcium fluoride, barium fluoride, sapphire, or zinc selenide
Cell lengths	For liquids, from 0.004 to 4 in. (0.1 to 100 mm); for gases, up to 40 m (130 ft) enclosed and any length for open-path monitoring
Warm-up time	15–20 min. (For most stable operation, allow 16 hr for warm-up.)
Repeatability	±1% of full scale
Linearity	±0.5 of full scale
Inaccuracy	±2% of span
Drift	±1% of full scale for zero and the same for span per day

(*Continued*)

Costs	Remember that the installation and upkeep costs are normally much larger than just the vendor costs given below. Single-beam portable or laboratory units cost $4000–$5000 Industrial nondispersive infrared analyzer with diaphragm capacitor costs $8,000 Multigas analyzer pulling in up to five gases from 50-m (150-ft) distances costs $25,000–$27,000 Microprocessor-based portable spectrometer with preprogrammed multicomponent identification capability for ambient air monitoring and with space for 10 user-defined standards for calibration, AC/DC converter, sample probe, and carrying case costs $20,000 Industrial FTIR costs $75,000–$125,000
Partial list of suppliers	ABB Process Analytics—Bomem http://new.abb.com/products/measurement-products/analytical Ametek http://www.ametekpi.com/products/Thermox-WDG-V-Combustion-Analyzer.aspx Aqua Measure http://www.aquameasure.com/methods.htm CAI http://www.gasanalyzers.com/products-ia.php Combustion http://www.cambustion.com/products/ndir500/operating-principle Control Instruments Corp. http://www.controlinstruments.com/technologies/infrared-analyzers Enviro-Analytical http://www.enviro-analytical.com/enviroproducts/ftir_analyzers.html Fuji Electric http://www.fujielectric.com/products/instruments/library/catalog/box/doc/ECNO325c.pdf Horiba Instruments http://www.horiba.com/us/en/automotive-test-systems/products/emission-measurement-systems/portable-emission-analyzers/details/mexa-584l-826/ Infrared Industries http://www.infraredindustries.com/product/ir-8400d-dual-stream-gas-analyzer/ International Sensor Technology http://www.intlsensor.com/pdf/products.pdf K2BW http://www.k2bw.com/5_c_18.htm Midac http://www.midac.com/i-series.html Moisture Register Products, of Aqua Measure Instruments, http://www.aquameasure.com/ NDC http://www.ndc.com/en/Products/At-Line-Near-Infrared-Analyzers/InfraLab-Meat-Analyzer.aspx Servomex http://www.servomex.com/servomex/web/web.nsf/en/servoflex-minimp-5200-multipurpose Shimadzu (A, B) http://www.shimadzu.com/an/spectro/uv/uv1800/uv2.html Siemens http://www.automation.siemens.com/mcms/sensor-systems/en/process-analytics/gas-analyzer-gas-analysis/extractive/ir-active-components/pages/ultramat-23.aspx Signal Instruments, http://www.k2bw.com/418.htm Teledyne http://www.teledyne-ai.com/products/7500.asp Thermo Scientific http://www.thermoscientific.com/en/search-results.html?keyword=Fourier+Transform+Infrared+Spectroscopy+%28FTIR%29&countryCode=US&matchDim=Y Unity Scientific, http://www.unityscientific.com/products/wet-chemistry/sample-preparation Wilks-Spectro Scientific Co. http://www.wilksir.com/products/infraran-specific-vapor-analyzers.html Yokogawa http://www.yokogawa.com/an/ir-gas/an-ir-gas-001en.htm Zeltex Inc. http://www.zeltex.com/portable/101.pdf Most popular: ABB, Servomex, and Siemens
Fiber-optics, sample systems, tools	Axiom, sample systems, fibers, both NIR and IR http://www.goaxiom.com/process_products_NIR-UV.html#process_nir_multiplexer Dave Mayes, a developer of spectroscopic tools (http://www.dsquared-dev.com) Equitech International Corp., fiber switches, fiber connections to the process, sampling systems http://www.equitechintl.com/Multiplexer.htm Fiber Tech Optica, fiber optics only http://www.fibertech.com/enterprise/fiber-optic-services/ Optec http://www.optek.com/Product_Detail.asp?ProductID=26 Remspec Corp. Fiber Optics http://ir-fiber.com/ Solutions Plus, Inc., makers of traceable standards, a division of Ricca Chemical http://www.riccachemical.com/

NEAR INFRARED ANALYZERS

Process fluids	Gas, liquid, or solid, but mostly liquid and solid
Some applications (See Table 1.32b)	Active ingredient in drugs Benzene in gasoline, 0.2%–1% Boiling points of gasoline, 50°C–200°C (122°F–392°F) Btu of natural gas (high pressure) Caustic in water 0.1%–10% Cetane of diesel fuel

(Continued)

	Molecular weight of small polymers
	Octanes of gasoline, 80–100
	Octanes of components of gasoline, 60–120
	Protein content of wheat
	p-Xylene concentration in mixture of aromatics
Operating pressure	150 PSI standard (10 bar)
	1000 PSI special (70 bar)
Ambient temperature	–40°C to 50°C (–40°F to 120°F) is standard. (Note: Since the ambient temperature changes, it will affect the spectrometer and it will require temperature stabilization.)
Stream temperature	This restricts cell material only; normally one keeps the temperature constant.
Humidity limitations	None—NIRs, like IRs, should be purged; this eliminates the humidity problem.
Materials of construction	Cell bodies in all standard materials; windows can be quartz (most common), sapphire, and others
Cell path lengths	For liquids, 0.04–4 in. (1–100 mm); for gas, long (unless sample is at high pressure so cell length becomes too long to be practical)
Warm-up time	Manufacturers normally quote minutes—recommend overnight for best stability
Repeatability	±0.01% of full scale
Linearity	±0.5% of full scale
Inaccuracy	±1% of span (depends on how well the "modeling" has been done; can be much better)
Drift	±0.01% of full scale and the same for span per day
Costs	$80,000–$180,000, depending on number of streams, distance between the analyzer and sample, and sample preparation required. (How can these costs be justified by the user? At one installation, the analyzer is determining 25 properties every 45 s. It is also possible to look at several streams and still update the control system as often as needed. At another installation, the plant estimated that the analyzer saved $15 million the first year it was in service.)
Partial list of suppliers	ABB Process Analytics—Bomen http://new.abb.com/products/measurement-products/analytical
	Bran and Luebbe—Technicon http://www.ebay.com/itm/Bran-Luebbe-IA450-Technicon-TechniServ-NIR-/221349142772
	Brimrose Corporation of America http://www.brimrose.com/products/nir_mir_spectrometers/sort_by_spectrometers.html
	Foss-NIR Systems http://www.foss.dk/
	Guided Wave http://www.guided-wave.com/products/spectrometers.html
	Hamilton Sundstrand (AIT Division—Analect) http://www.spectroscopyonline.com/spectroscopy//product/productDetail.jsp?id=48253
	Jasco Analytical Instruments http://www.jascoinc.com/products/spectroscopy/uv-visible-nir
	LTI http://www.LTIndustries.com
	Ocean Optics http://www.coleparmer.com/Product/Ocean_Optics_Chem_USB4_Visible_NIR_Spectrophotometer/WU-83500-10
	Rosemount Analytical, Inc.—Emerson http://www2.emersonprocess.com/siteadmincenter/PM%20Rosemount%20Analytical%20Documents/PGA_Manual_AOTF-NIR_200010.pdf
	Thermo Electron http://www.thermo.com/eThermo/CMA/PDFs/Various/File_27448.pdf
	Unity Scientific http://www.unityscientific.com/products/NIR/at-line/smartsampler.asp

INTRODUCTION

In the first part of this chapter, the infrared (IR) analyzers will be discussed, while the near-infrared (NIR) analyzers will be described in the second part of this chapter. This is not the only chapter where IR and NIR analyzers are discussed. As can be seen from the analyzer selection guide provided in Table 1.1a, IR and NIR analyzers are applicable to a wide range of analytical tasks.

It should also be noted that the boundaries between ultraviolet (UV), visible, NIR, and IR are slowly disappearing. Analyzers are evolving that are capable of operating in all of these spectra, and as the mathematical tools to handle full spectral ranges are becoming available. The addition of microprocessors or tabletop computers has enhanced the performance of these instruments by providing such features as self-calibration, self-diagnostics, and chemometric tools, for example, partial least squares (PLS) and principal component regression (PCR), to name two, while design modularity has contributed to simplifying maintenance.

For an overall view of where process analysis is going, the annual conference of the International Forum for Process Analytical Chemistry (IFPAC) can be recommended. It is held annually and normally contains sessions on process IR and NIR; for more information, see http://www.ifpac.com.

PRINCIPLES OF ANALYSIS

The composition of both gases and liquids can be analyzed by measuring their absorption or reflectance in the infrared (IR) or near-infrared (NIR) spectral regions. These analyzers can operate either in the photometric (absorption) or in the spectrophotometric (dispersion) mode, and some designs are capable to operate in both.

IR absorption (or reflection used with solid samples) is a technique that can be used successfully for continuous chemical analysis. The infrared region of the electromagnetic spectrum is generally considered to cover wavelengths from 0.8 to 20,000 μm. NIR normally covers 0.8 to 2500 μm, and classic IR covers the rest. For IR analysis, these limits are normally put in terms of frequency (cm^{-1}, wave numbers or the number of waves per cm): 4000–500 cm^{-1}, which corresponds to wavelengths of 2,500–20,000 μm.

Except for a small overlap region, sources and detectors that are needed in the NIR will not work in the IR, and vice versa. Some laboratory spectrometers have both sources and detectors so they can work in both areas. For the process, most gas analysis is done in the IR and most solid and liquid analysis is done in the NIR. The choice is based on workable path lengths.

Infrared radiation interacts with almost all molecules (except the homonuclear diatomics oxygen (O_2); nitrogen (N_2); hydrogen (H_2); chlorine (Cl_2); etc., and monatomics such as helium (He); neon (Ne); etc.) by exciting molecular vibrations and rotations that affect the dipole of the molecule (Figure 1.32a). The oscillating electric field of the IR wave interacts with the electric dipole of the molecule, and when the IR frequency matches the natural frequency of the molecule, some of the IR power is absorbed.

The pattern of wavelengths, or frequencies, absorbed identifies the molecule in the sample. The strength of absorption at particular frequencies is a measure of the concentration of the species. Analytical laboratory IR is largely concerned with identification, or qualitative analysis, while process IR is concerned with quantitative analysis. Some typical spectra are shown in Figure 1.32b. The NIR consists of overtones and combinations of these IR bands.

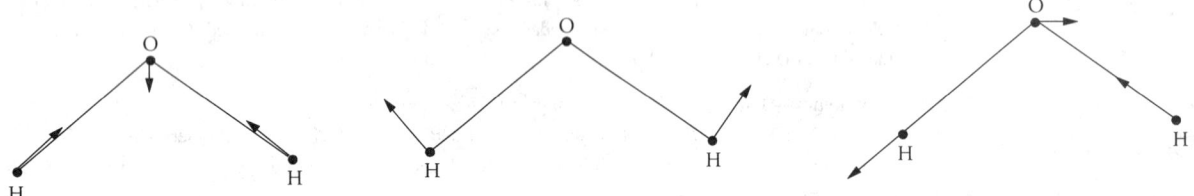

FIG. 1.32a
The three fundamental vibrations of the water molecule (left to right) are symmetric stretch, bend, and asymmetric stretch. The amplitudes of the vibrations have been exaggerated for clarity.

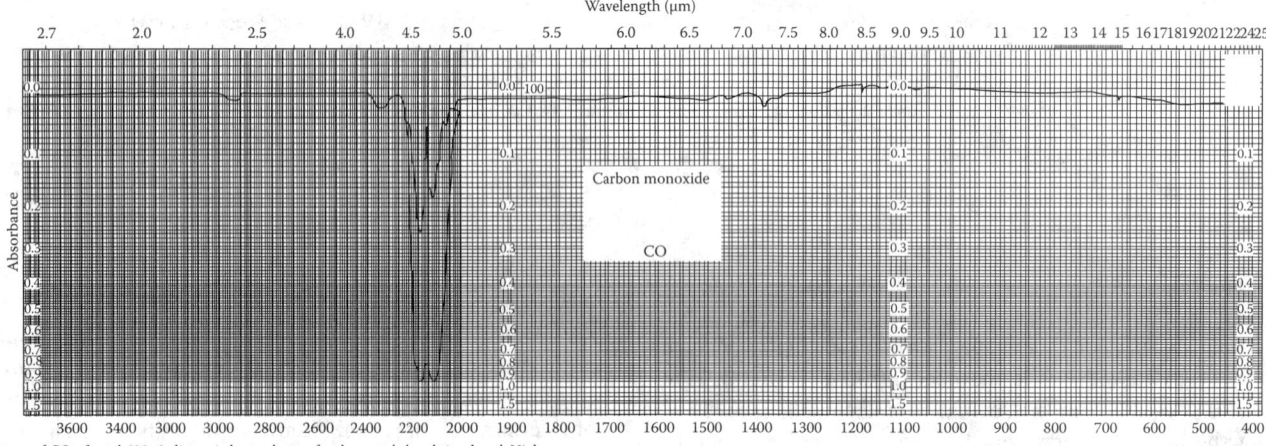

400 torr of CO of total 600. A distomic has only one fundamental absorbtion band. Higher resolution shows the band to consist of many sharp lines about 4 cm^{-1} aprt (rotational line structure).

FIG. 1.32b
Examples of IR spectra recorded using a laboratory double-beam spectrometer. All spectra are gas phase using a 2 in. (5-cm) cell with N_2 added to give a total pressure of 600 mmHg (torr). (Courtesy of Dow Chemical Co.) (Continued)

1.32 Infrared and Near-Infrared Analyzers   433

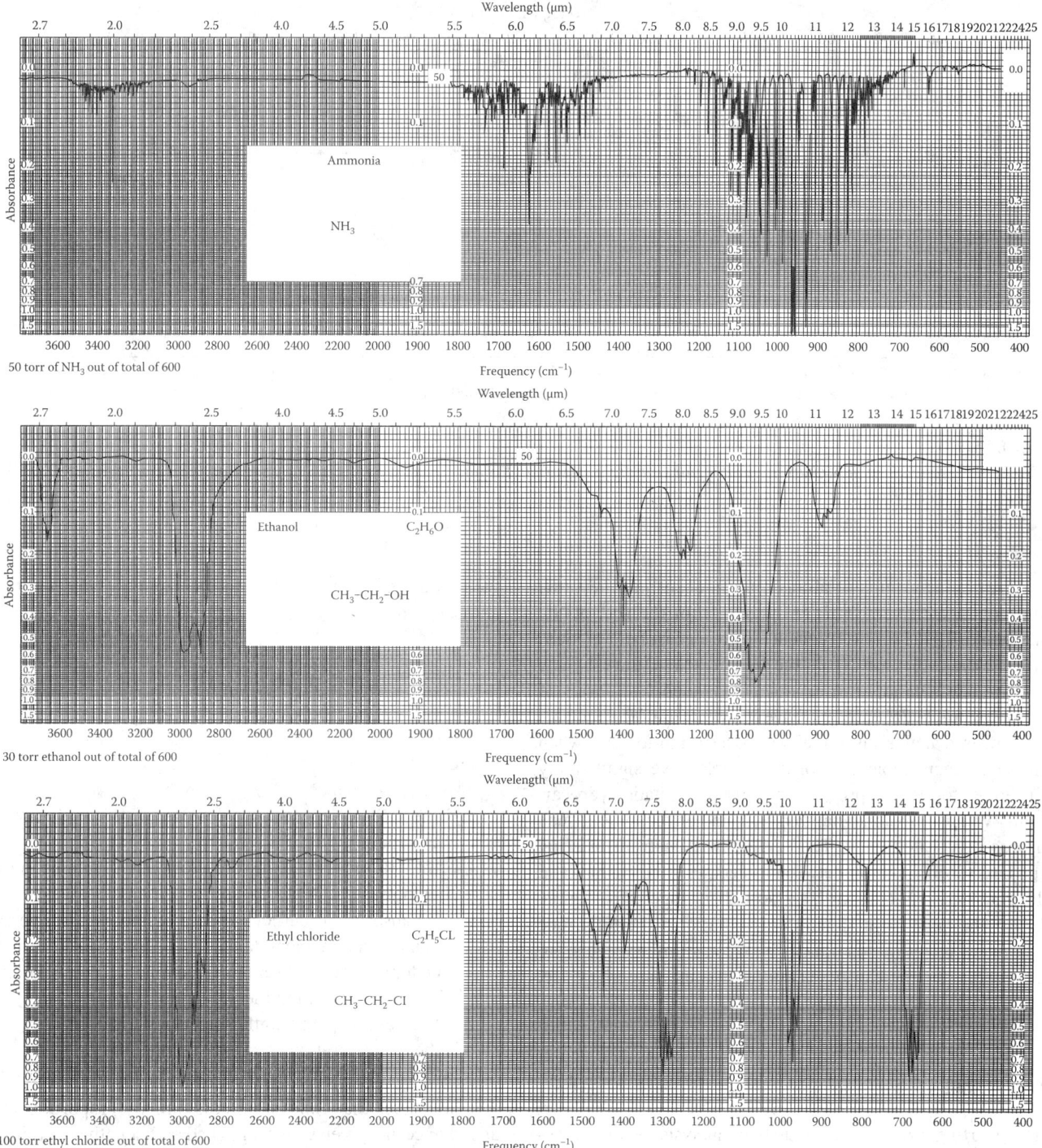

FIG. 1.32b (Continued)
Examples of IR spectra recorded using a laboratory double-beam spectrometer. All spectra are gas phase using a 2 in. (5-cm) cell with N_2 added to give a total pressure of 600 mmHg (torr). (Courtesy of Dow Chemical Co.)

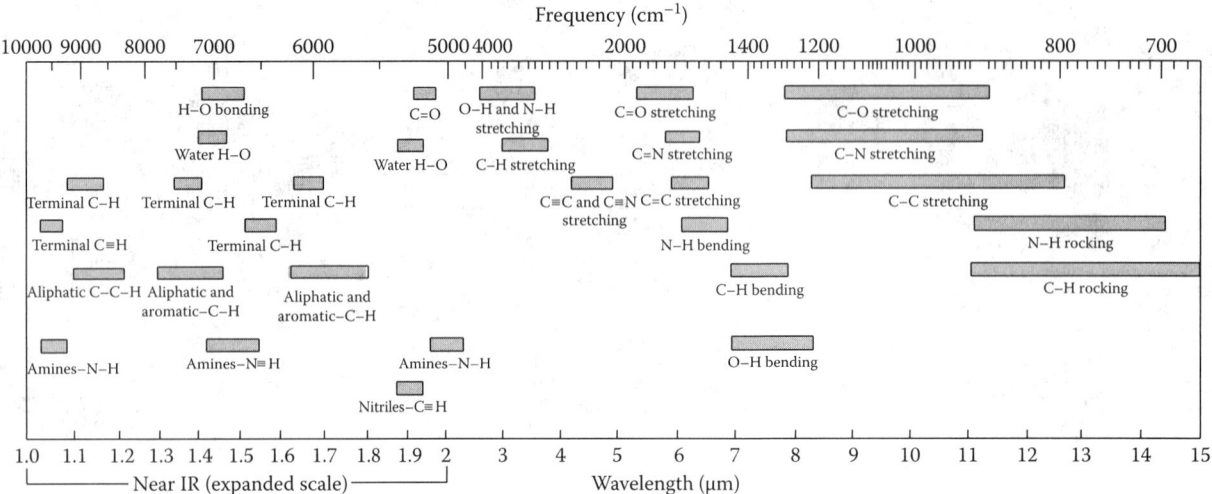

FIG. 1.32c
Functional group frequency chart. Fundamental vibrations absorb in the mid-IR; overtones and combination bond are 10–10,000 weaker and absorb in the NIR.

Particular groups of atoms tend to absorb at the same frequency with very little influence from the rest of the molecule. These group frequencies are a great help in identifying the molecules from the IR spectra (Figure 1.32c). On the other hand, similar molecules, such as a series of homologous hydrocarbons, have very similar IR spectra.

Infrared analysis is, therefore, most straightforward when the component molecules of the sample have significantly different atomic groupings. A mixture of aliphatic hydrocarbons would be better analyzed by another technique, such as gas chromatography. The part of the spectrum offering the best discrimination between molecules is between 7 and 15 μm, 1430 and 670 cm^{-1}, the so-called fingerprint region. Given the very large signal-to-noise ratio in the NIR, one can make very fine separations between similar species; for example, o-xylene can be measured in a mixture of xylene, ethylbenzene, and benzene.

Beer–Lambert Law

The starting point for quantitative analysis is the Beer–Lambert law, frequently just called Beer's law, which relates the amount of light absorbed to the sample's concentration and path length.

$$A = abc = \log_{10} \frac{I_0}{I} \qquad 1.32(1)$$

where
- A is the absorbance
- I is the IR power-reaching detector with sample in the beam path
- I_0 is the IR power-reaching detector with no sample in the beam path
- a is the absorption coefficient of pure component of interest at analytical wavelength; the units depend on those chosen for b and c; a newer term, ε, extinction coefficient, is the preferred term in the academic literature
- b is the sample path length, sometimes l is used
- c is the concentration of sample component

The law states that concentration is directly proportional to absorbance at a given wavelength and path length at specified temperature and pressure. Note that, however, the logarithm functions—frequently absorbances over 1—show nonlinear behavior, and sometimes problems start as low as 0.7.

Sometimes, isolated peaks in the IR can be found that correspond only to the item of interest, then calibration plots of A vs. c can be made up using known samples and used to analyze unknown ones (Figure 1.32d). Beer's law is also helpful in choosing the optimum path length for accurate analysis. (In some cases, this linear relationship is not observed. See the discussion of linearity later on in this chapter.)

There is also the case where overlapping spectral lines eliminates the simple application of Beer's law. Note: Not only concentration of molecular species can be identified, but also physical properties can frequently be measured. This is done more in the NIR than the IR, but can be done in both. A statistical approached called chemometrics, PLS and PCR, for example, is used to measure physical properties and solve the overlapping spectral lines problem referred to earlier. These very powerful programs require care in use and are more frequently used in NIR than IR.

Laboratory Spectrophotometers

Some of the laboratory spectrophotometers are of the dispersive design, meaning that a prism or grating is used to separate the spectral components in the IR radiation of the source. Most modern laboratory units are of the Fourier transform (FT) type, where a moving mirror generating an interference pattern of the wavelengths accomplishes dispersion (Figure 1.32e), and then the FT converts this detected signal into something useful. Fourier transformation (FT) is a mathematical process to convert raw data into the actual spectrum of the radiation absorption over a wide spectral range. This is superior to dispersive spectrometry, which measures IR radiation intensity over a narrow spectral range. It can be applied to optical spectroscopy, optical and infrared spectroscopy (FTIR, FTNIR), nuclear magnetic resonance (NMR), and magnetic resonance spectroscopic imaging (MRSI).

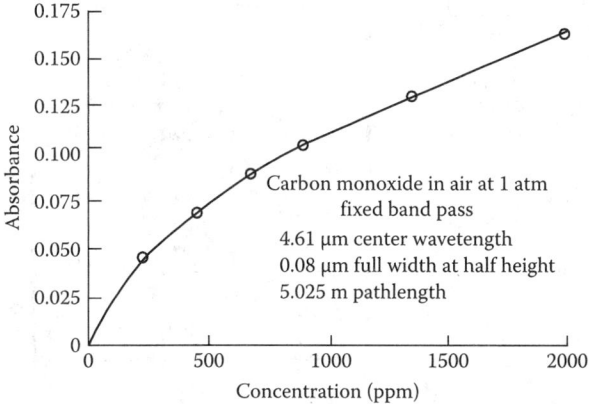

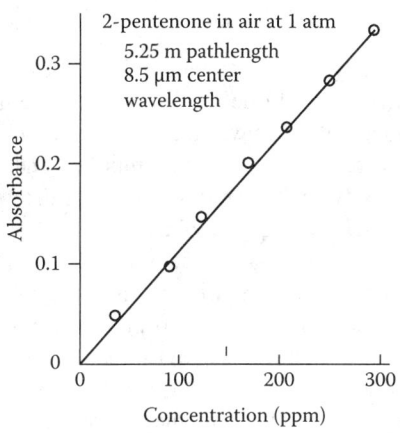

FIG. 1.32d
Calibration curves for a typical filter analyzer. Top: Carbon monoxide shows a large departure from the Beer–Lambert law. Bottom: 2-Pentenone gives a linear calibration plot.

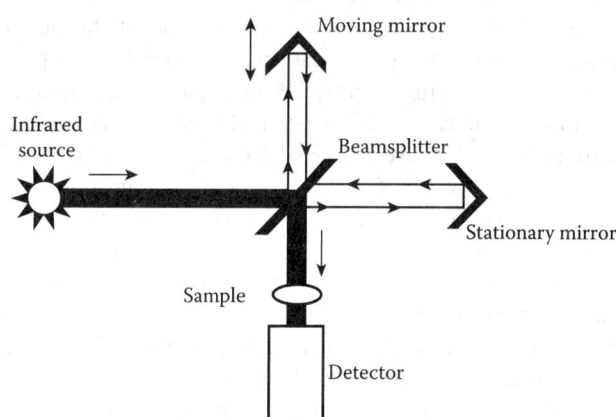

FIG. 1.32e
The operating principle of a laboratory spectrophotometer using interferograms in FTIR analysis.

436 *Analytical Measurement*

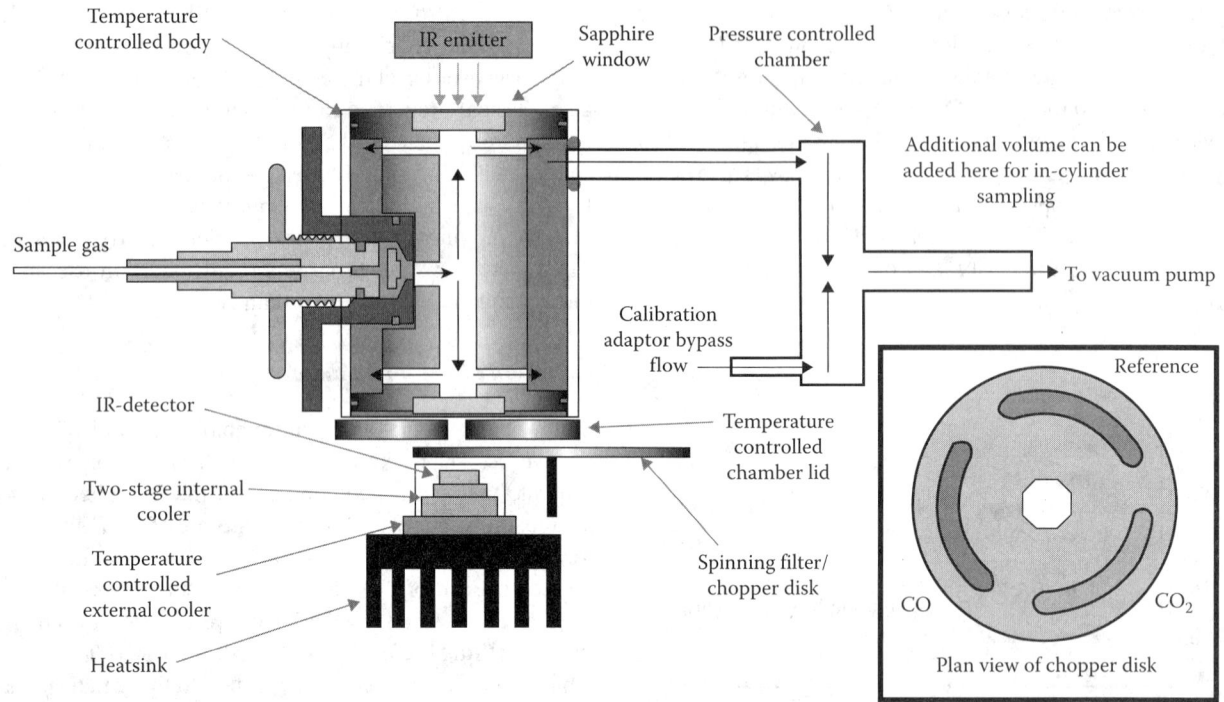

FIG. 1.32f
NDIR analyzers are often used to measure the carbon monoxide and carbon dioxide concentration of combustion emissions. (Courtesy of Combustion Ltd.)

In Figure 1.32e, the IR source is pointed at the chapter on Fourier transform spectrometers.

In this design, the IR source is pointed at the beam splitter, which divides the beam into two parts. One part is sent to the moving mirror, the other part to the stationary mirror. The two reflected beams are recombined at the beam splitter. Because the beam from the stationary mirror traveled a constant distance, while the one reflected from the moving mirror traveled a variable distance, when they are recombined, an interference pattern is created, because some of the wavelengths are strengthened, while others are weakened by the recombination. The result is an interferogram, which is sent through the sample, where some wavelengths are absorbed, while others are transmitted and reach the detector. The detector signal is sent to a computer, where an algorithm called a Fourier Transform interprets it. These Fourier Transform Infrared (FTIR) analyzers will be described later, in connection with Figure 1.32o.

Most industrial process analyzers are nondispersive infrared (NDIR) designs. However, there are some very good FTIRs on the market for process applications where several components need to be measured. Since oxygen and nitrogen do not absorb in the IR spectrum, dry air is frequently used as a zero reference gas; however, one needs to be aware of the very strong absorbance of CO_2 when one is doing this.

Detectors are mostly solid-state type, with some microphone types still in existence. The solid-state type, the most common today, functions by converting the incoming photons to an electric current. This current is amplified and then sent to the recording device. The microphone-type detectors are filled with an IR-absorbing gas, which is heated by the radiation it receives and expands as a consequence. It is this expansion that is measured. The microphone type usually lacks uniform sensitivity across the spectrum. The solid-state detectors are generally of lower sensitivity than the microphone type at the microphone type's peak sensitivity, but they have almost uniform sensitivity across the spectrum.

Figure 1.32f illustrates a nondispersive infrared (NDIR) analyzer, which is often used to measure the concentration of carbon oxides (CO and CO_2). Each of these gases absorb some infrared at their corresponding frequencies and based on that the NDIR detector is able to determine the volumetric concentration of CO or CO_2 in the sample. This type of design is also further discussed in the later paragraphs.

ANALYZER DESIGN CATEGORIES

Single-Beam Configuration

Designs of IR instruments can be separated into single- and dual-beam configurations. Single-beam analyzers are the main ones used in the process world (Figure 1.32g). They function by alternating filters in the beam path—the reference is chosen to not absorb the species of interest, but to offset any other species that would absorb at the measuring wavelength.

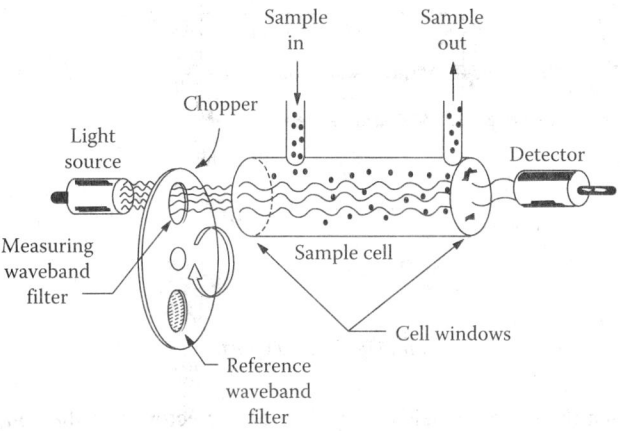

FIG. 1.32g
Single-beam IR analyzer provided with two filters, one for the sample measurement and the other for reference.

The proper choice of the reference wavelength is not a trivial task. The reference wavelength should be as close to the measuring wavelength as possible, and yet let the measurement be made. Drifts in the source or detector will be compensated best if the measuring and reference wavelengths are close together.

As has already been shown in Figure 1.32f for CO and CO_2 and is also shown in Figure 1.32g in a more generalized illustration, these filters are put on a chopper wheel in the light beam, and as the chopper alternatively spins one filter or the other into the optical path, the difference or ratio in the energies received at the detector will be a function of the concentration of the component of interest. Because filters normally change absorbance at a given wavelength with temperature, this approach must be temperature stabilized. Also note that dirt on the window will not affect the reading because it effects both the measuring and reference wavelength in the same way so that the ratio stays constant if the monitoring and reference wavelengths are close enough.

A less expensive design uses just the measuring wavelength, eliminating the filter wheel and reference filter. Problem: In that design if the source varies in intensity or temperature, they will both effect the measurement. Similarly, if the detector changes in sensitivity, so will the result. In other words, one needs to re-zero and re-span these instruments very frequently.

If no lenses or mirrors are used to direct the IR radiation from the source to the receiver, it is necessary to polish the interior walls of the sample cell to make them highly reflective. Sometimes gold foil is used to achieve this goal. The use of reflective walls can be very expensive because it can contribute to drift as contamination of the wall changes reflectivity.

Dual-Beam Configuration

In the dual-beam configuration, the IR radiation is allowed by the chopper to pass alternately through the sample or the reference tube (Figure 1.32h). The reference tube can provide a true zero when it is filled with a nonabsorbing gas, or it can act to balance out the gases which are not of interest, if filled with those gases. A narrow-band-pass optical filter is placed in front of the detector to limit the IR energy it receives to the wavelength that is characteristic of the component of interest. (Note: This severely limits the amount of energy getting to the detector; the narrower the band pass, the less energy, and the narrower the band pass, the more specific.)

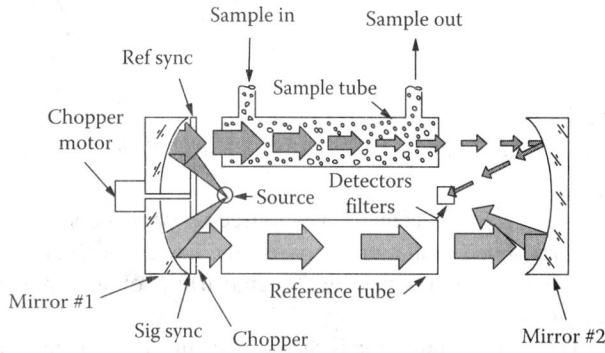

FIG. 1.32h
Dual-beam IR analyzer where the radiation alternatively passes through the sample tube, which contains the component of interest, or the zero-reference tube, which is free of absorbing gas.

If the sample contains the component of interest, this will attenuate the magnitude of the detected signal in the absorption band of the band-pass filter (Figure 1.32i). The use of the reference cell in the dual-beam configuration reduces the drift caused by power supply, detector changes, or some temperature fluctuations. The use of collimating optics also eliminates the need for internal reflection from the interior surfaces of the tubes, thus simplifying their construction and eliminating the associated drift due to dirt accumulating on the wall of the tubes.

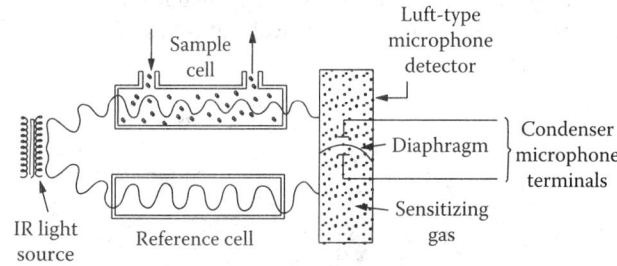

FIG. 1.32i
Dual-beam analyzer with internally reflective cell surfaces and Luft-type microphone detector. (For a discussion of Luft detectors, see DEFINITIONS at the end of this chapter.)

Dual-Beam Design for Stacks When the IR analyzer is used for in situ stack gas analysis, two different approaches can be used (1) the detectors and source on the same side of the

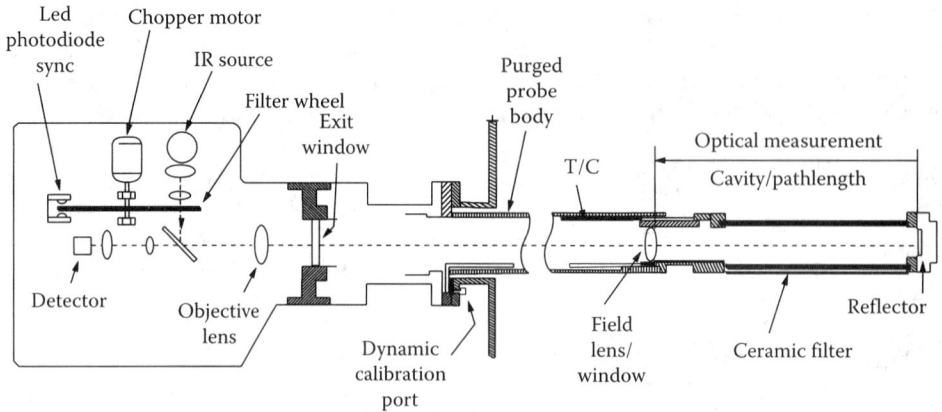

FIG. 1.32j
IR analyzer used for in situ stack gas analysis equipped with ceramic diffusion cell, reflector, and calibration port.

stack and (2) the detectors and source on opposite sides of the stack. For installation purposes, the first is much easier, but it requires an internal filter to remove any solids from the stream.

The first configuration is shown in Figure 1.32j. Note that only one opening is required in the stack, with no critical alignment across the stack required, as in the other option. It is very critical to keep the reflector mirror clean; some people use a ceramic filter for that purpose. This is the only type that the author has used, and it has worked well.

In the second system, shown in Figure 1.32k, both the reference and measuring wavelength are affected the same by scattering bodies in the stream; therefore, there is no need for a filter. This system is frequently used in pollution alarm situations. The reference cell is filled with somewhat less than the allowed emission of the gas of interest, and the IR beam leaving the stack is alternated between a neutral filter and the reference cell. When the stack gases are below the level of concern, the reference cell will remove the absorption lines that are characteristic of the gas of interest. Therefore, the total IR energy from the reference cell is at a reduced value. The analyzer is balanced by selecting a neutral filter, which reduces the energy from all the wavelengths to such an extent that the energy levels received by the detector from the neutral and reference gas filters will be the same.

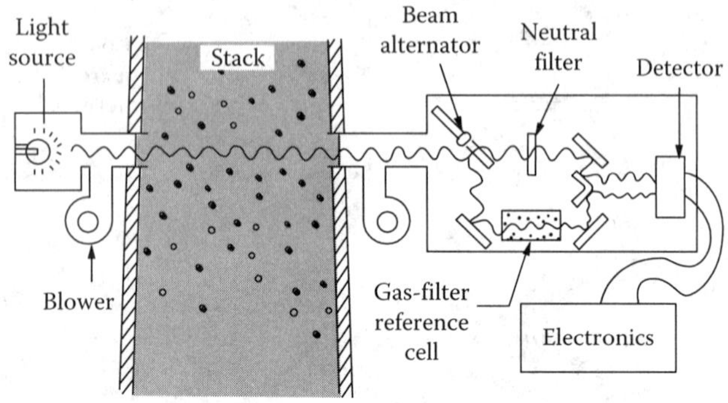

When the pollutant of interest is present in the stack, the energy content of the reference path is unaffected (because the absorption is already complete at the selected wavelength). On the other hand, the IR energy reaching the detector through the neutral filter is reduced (due to the absorption of the pollutant gas) and the ratio between the beams reflects the pollutant concentration at the level of concern in the stack. This measurement is unaffected by particulate concentration variations, within reason, as it affects both paths equally and their ratio is unaffected.

In these cross-stack, absorption spectroscopes, the sensor can be a very selective tunable diode laser (TDL). Such a design is shown in Figure 1.32l. This design is suitable for use in hazardous areas and includes both a transmitter and a receiver unit, which are connected to a junction box.

FIG. 1.32l
Cross-stack configuration of an absorption spectroscope. (Courtesy of ABB.)

LABORATORY ANALYZERS

Grating Spectrophotometers

The standard laboratory IR instrument was a double-beam, optical null spectrophotometer. Diffraction gratings have completely replaced prisms for separating the IR beam into its component wavelengths. Both approaches have been replaced by FTIR, discussed below. Most IRs, old and modern, use a heated *black-body* source that emits at all

FIG. 1.32k
Dual-beam IR analyzer for stack gas monitoring. (From Nelson, R.L., InTech, June 1987.)

IR wavelengths, and the thermal detector responds roughly equally at all wavelengths (Figure 1.32m).

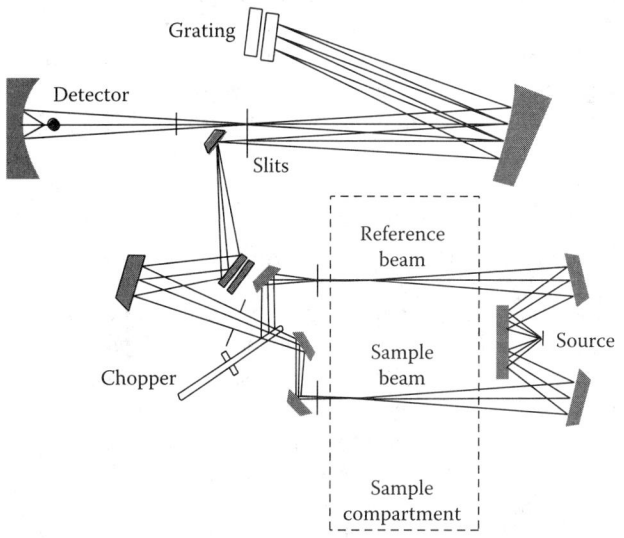

FIG. 1.32m
A widely used laboratory double-beam optical null spectrophotometer.

The complete spectrum is scanned by rotating the grating and measuring the light intensity passing through its exit slit. Sample cells are placed in the sample beam and, as the spectrum is scanned, the instrument drives an attenuator into the reference beam until the detector sees equal energy from both beams. These spectrophotometers are ideally suited for qualitative analysis.

There are computer libraries of spectra and search programs that allow the computer to find the major compound in the sample and sometimes some of the minor ones. In many cases, manufacturers have added computer data handling to make possible both qualitative and quantitative analyses.

Intelligent samplers are available to serve the automatic and unattended analysis of hundreds of samples in laboratories. One such design is shown in Figure 1.32n. In this design, a "ring cap" (sample cap) is loaded into the "towers." Each of the three towers contains 24 sample caps for a total of 72 samples. After loading, the sampler is programmer with the information concerning the contents and locations of the sample caps. The analysis of each sample takes only a minute or so and after that they are analyzed, the results are stored in the memory of the sampler.

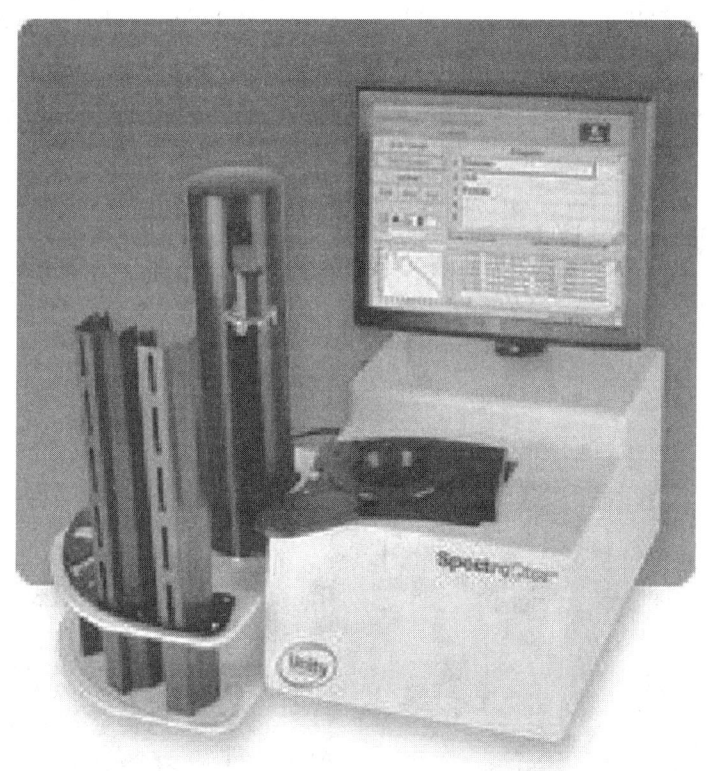

FIG. 1.32n
Intelligent sampler can automatically analyze up to 72 samples and place the results in memory. (Courtesy of Unity Scientific.)

Filter Spectrometers

A circular variable filter (CVF) selects the wavelength to be measured. Microcomputer-controlled, single-beam analyzers working from 4000 to 690 cm^{-1} have been designed for the quantitative analysis of mixtures. Measurements are made at each analytical wavelength in sequence, and the output is presented as a list of concentrations for mixtures of up to 10 components.

Narrow-IR-band-pass filters consist of multiple layers of dielectrics of alternating refractive index on a transparent substrate. They pass a band of wavelengths while rejecting all others. The width of the band pass is typically 1 to a few percent of the center wavelength of the item of interest. Spectral resolution is low when compared to that obtained with a grating instrument, but signal-to-noise ratios are higher. A CVF is made of dielectric layers of continuously varying thickness so that the wavelength selected depends on the angular position of the CVF wheel.

Fourier Transform Spectrometers

A quite different and very high-performance approach to analysis makes use of the Michelson interferometer or some modification thereof. Instead of separating the different wavelengths of light with a filter or a grating for measurement, the complete spectrum is encoded as an interferogram in a few seconds of measurement time, and the spectrum is computed by Fourier transform or fast Fourier transform.

Advantages include speed, full-spectrum method, and light throughput (one is not throwing away most of the light as one is looking at a different wavelength). The disadvantage is the moving parts. There has been a major effort to stabilize the moving parts, but they still require smooth movement for good spectra; there must be an attached computer to do the transform, but this can act as a storage device for some of the spectra.

A schematic of a typical FTIR or FTNIR spectrophotometer is shown in Figure 1.32o. Collimated light from the source is directed through the beam splitter. Approximately 50% of the light passes through the beam splitter to the fixed mirror, M1. The balance of the light is reflected to the moving mirror, M2. When these two beams are reflected off the mirror surfaces, they recombine at the beam splitter to give constructive and destructive interference as a function of the wavelength of light and the distance that M2 has moved.

The included helium–neon laser detection system in the FT analyzer monitors the position and velocity of the moving mirror. This results in excellent wavelength accuracy for the FT analyzers. The displacement of the moving mirror induces phase differences that result in an interferogram. In order to produce a spectrum of absorbance vs. wavelength, the interferogram must be transformed digitally from a plot of detector response vs. optical path difference. This calculation of the spectrum is carried out using Fourier transform or fast Fourier transform on the attached computer.

Since the FTIR or FTNIR analyzer measures all the wavelengths simultaneously on one detector, the entire spectrum can be used for quantitative analysis. FT spectrophotometers are used for multicomponent applications where the high optical resolution allows separation of interfering components. (Note: One should be careful with the term *optical resolution*. Some suppliers will talk about resolution as the number of data points per unit of the spectrum. This, in most cases, is not the same as their optical resolution. Without high optical resolution, you cannot separate some components, regardless of the data point resolution.)

FTIR and FTNIR spectrophotometers are being used online where the full spectrum is needed for analysis. One big advantage of the full-spectrum measurement is that the associated computer programs can be set up to detect unusual samples or impurities in the sample. These unusual samples are frequently referred to as outliers, but they are sometimes an early warning of an unwanted change in raw materials or in the process.

Tunable Lasers

On the face of it, this is the spectroscopist's dream come true. Laser diodes are available to cover the 4000 to 400 cm^{-1} region. A given diode may cover 200 cm^{-1} with small discontinuities about every cm^{-1} in its tuning. The devices (PbSSe or PbSnTe) have very high spectral resolution and good power output, but require liquid N$_2$ temperature or below, plus other complex support equipment. They are for the research spectroscopist, but they have been used in isolated cases of process analysis that could not be done by more standard methods.

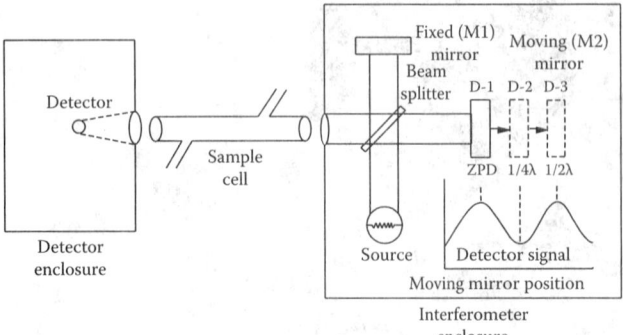

FIG. 1.32o
FTIR spectrophotometer.

Tunable laser spectrometers (TLS) are also used by NASA in analyzing the atmosphere on Mars. Figure 1.32p shows a demonstration of the unit in the visible spectrum, but the actual sensor works in the IR region. It shows how the laser beam bounces between the mirrors in the measurement chamber, which is filled by Martian air. By measuring the absorption at wavelengths specific to methane, carbon dioxide, or moisture, the Martian atmosphere is analyzed.

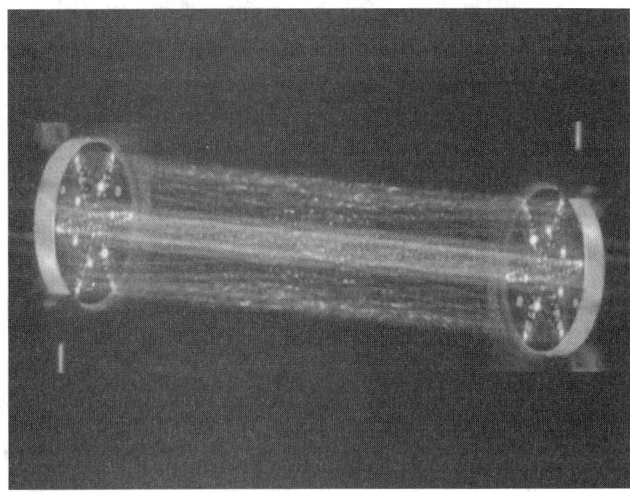

FIG. 1.32p
The measuring chamber of the tunable laser spectrometers that are used to analyze the composition of the Martian atmosphere. (Courtesy of NASA.)

Process Control Applications

Normally, laboratory instruments cannot be taken from the laboratory and put onto the process line because the laboratory instrument does not meet the fire code for the process area and it is not stable enough to run 7 days a week, 24 hr a day without the need for recalibration. Some companies, such as Dow Chemical, have done a good job in moving many laboratory instruments into the process arena, but with a high investment of manpower.

The Fourier transform spectrophotometer has successfully made the move from laboratory instrument to an online tool, as have diode array instruments. In many cases, however, the process instrument no longer looks like its laboratory cousin, and the cost is considerably higher by the time it is temperature stabilized and made explosion proof. (Process instruments should not need a safe, air-conditioned building to survive in the process world, but this can aid its stability and will certainly aid in its maintenance.)

ONLINE APPLICATIONS

Single-Component Analyzers

Single-component analyzers are nondispersive (absorption type) NDIR analyzers, used almost exclusively for gas analysis. These analyzers were invariably double beam and used gas-selective Luft detectors filled with the gas to be analyzed (Figure 1.32q), but today they use solid-state detectors. In the usual positive filtering mode, the IR beam from single or dual sources is chopped and passed in phase through the sample or reference cells to the detector. Sample molecules attenuate certain wavelengths, and the difference in power between sample and reference beams is sensed as a change in capacitance. This is amplified and displayed to give an output corresponding to concentration.

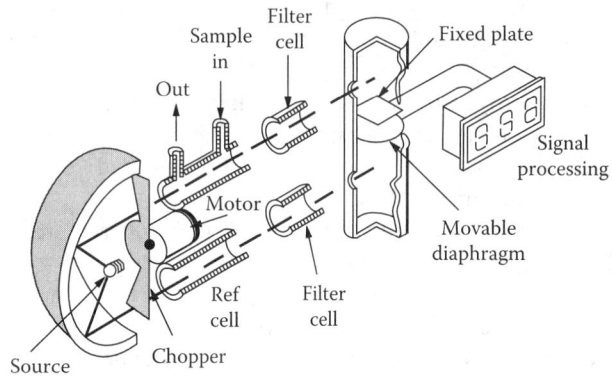

FIG. 1.32q
Typical NDIR analyzer using positive filtering. Both chambers of the detector are filled with a sample of the gas to be analyzed.

Selectivity is improved by adding a narrow-band-pass filter, which selects the wavelengths of interest, and filter cells filled with interfering species, which remove undesirable wavelengths from both beams. The NDIR technique is most sensitive and selective for small molecules whose spectral fine structure is resolved under ambient conditions. The spectral resolution of these instruments is effectively set by the width of the absorptions line and, in many applications, is much better than the typical laboratory grating spectrophotometer.

Luft Detector The Luft Detector has excellent sensitivity, but it must be temperature controlled and protected from external vibration. These analyzers need routine calibration for span and zero. (This is true for most process analyzers.)

Most modern instruments employ a solid-state, photoconductive detector, Luft-type and microflow detectors are still used.

The electrical conductance of solid-state detectors changes in response to the impact of an IR light beam and the resulting change in current flow is proportional to the IR energy. The solid-state detector has the advantage of being insensitive to external mechanical vibration.

The Luft-type detector is the oldest still in use. It is found in dual-beam analyzers and uses a pair of chambers that are filled with the adsorbing gas interest. One chamber receives light through the inert gas-filled reference cell, and the other receives it through the sample gas-filled cell. The chambers are separated by a capacitive diaphragm which distends in response to the difference in adsorption-induced gas expansion between the two detector chambers. This detector is extremely sensitive and selective, but is expensive to produce.

Over the years, various detectors have been developed for use with infrared gas analyzers. While most modern instruments employ a solid-state, photoconductive detector, Luft-type and microflow detectors are commonly encountered.

Generally, all detectors translate the difference in infrared energy into a sine wave that has a frequency determined by the chopping rate and a magnitude proportional to the energy difference.

The solid-state detector is constructed of a material that changes its electrical conductance in response to IR light energy falling on it. Changes in current flowing through the detection circuit are proportional to the difference in transmitted IR energy. The solid-state detector has the advantage of being insensitive to external mechanical vibration. In addition, careful selection of detector material allow for simplified optical filter design.

Dual-Chamber Luft Detector

The Luft-type detector is the oldest type found in common use. It is found in dual-beam analyzers and uses a pair of chambers that are filled with the adsorbing gas interest (Figure 1.32r). One chamber receives light through the inert gas-filled reference cell, and the other receives light through the sample gas-filled cell. The chambers are separated by a capacitive diaphragm which distends in response to the difference in adsorption-induced gas expansion between the two detector chambers. This detector is extremely sensitive and selective, but is expensive to produce.

The microflow detector is similar to the Luft detector, but employs a small orifice between the two detection chambers. Instead of differential pressure, the resulting flow is measured. A thermocouple arrangement in the orifice is used for this purpose. This detector is relatively immune to mechanical vibration, but is not as simple as the solid-state type.

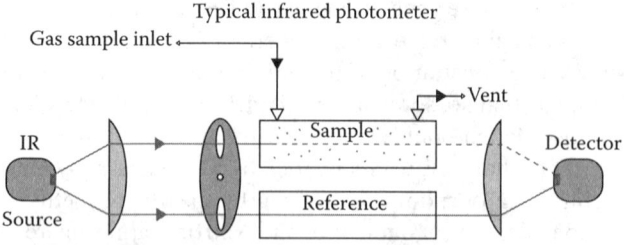

FIG. 1.32r
The dual-chamber, gas-filled Luft detector in a fully automated design. (Courtesy of Signal Instruments.)

Design Variations

Some important variants of this basic NDIR method with Luft detectors should be mentioned:

1. A second detector placed behind the first can be sensitized to measure and compensate for an interferent.
2. Detector chambers can be placed one behind the other with the chopper alternately delivering each beam to both. A different pressure of sensitizing gas in each detector chamber provides a built-in reference wavelength.
3. The combined beams, alternately admitted to the same chamber of the detector, cause heating and gas flow in and out of the chamber. A sensitive flowmeter gives an output proportional to the sample concentration.
4. In a technique called cross-flow modulation, the mechanical chopper is replaced by a valve that exchanges sample and reference gas between the two beams at a frequency of about 1 Hz. High sensitivity and zero stability are claimed for this technique, but reference gas must be supplied continuously, flow rates and temperatures must be carefully regulated, and response time is slow.

The solid-state detectors do not require the utilities that the Luft detector requires and are fast enough for the filter instruments. (Many of them do not keep up with a scanning grating, however.) For stability and reproducibility, it is hard to beat the filter instrument with a solid-state detector.

Gas Filter Correlation Gas filter correlation (GFC) spectrometers (Figure 1.32s) use a nonspecific thermal detector with a specifying reference gas cell. This is an example of negative filtering. The technique is most useful when high specificity is required and it is not practical to have a reference beam. The IR beam passes alternately through a cell containing a fixed quantity of the gas to be analyzed and through a similar cell filled with a zero gas, such as N_2. The combined beam passes through the sample, for example, across a smoke stack, and on to a thermal detector.

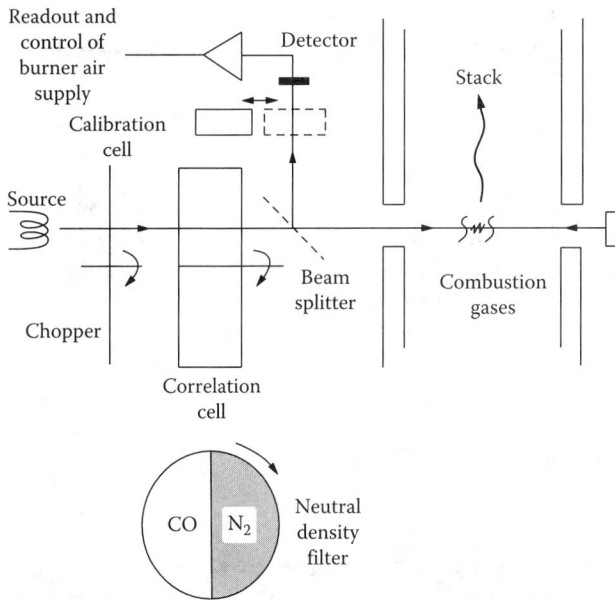

FIG. 1.32s
The gas filter correlation technique is an example of negative-filtering NDIR. The diagram shows how changes can be sensed in stack CO in the 0- to 1000-ppm range and used for burner control.

This system, which is similar to NDIR, works well for analyzing small gas phase molecules whose spectra have well resolved, rotational fine structure in the presence of strong interference from other species. One example is the analysis of carbon monoxide (CO) at the 100-ppm level in combustion gas-containing large concentrations of carbon dioxide (CO_2) and water (H_2O).

For accurate data, the sample must be removed from the optical path periodically to check for zero drift. A sealed calibration cell can be used to check span. (Note: Both of these require installation considerations.) Another configuration of the stack gas analyzer GFC has already been shown in Figure 1.32k.

It is possible to extend the GFC method to several gases with nonoverlapping spectra by placing a gas mixture in the correlation cell and inserting different narrow-band-pass filters sequentially into the beam.

Filter Analyzers Narrow-band-pass filters are used with nonspecific thermal detectors and are the most common IR units available today. Various configurations exist as follows:

1. Single-beam, dual-wavelength with analytical and reference (Figure 1.32g)
2. Single-beam, multiwavelength (one reference and up to 10 measuring wavelengths), same basic construction as shown in Figure 1.32t
3. Dual-beam, single-wavelength
4. Dual-beam, dual-wavelength

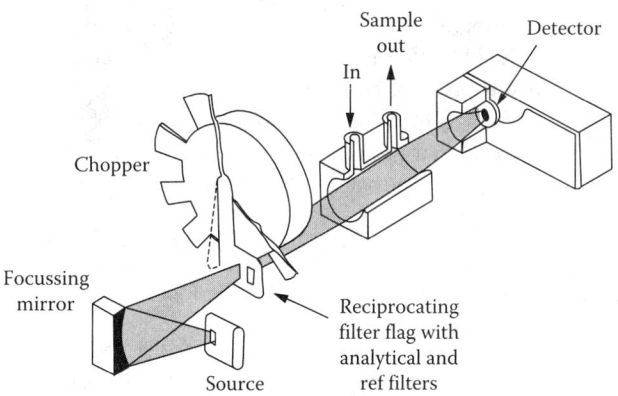

FIG. 1.32t
Optical schematic of a single-beam, dual-wavelength filter analyzer. In an alternate arrangement, larger filters can be mounted directly on the rotating chopper wheel. The principle has been extended to multiple wavelengths. If a microcomputer-controlled circular variable filter wheel replaces the reciprocating filter flag, the optical schematic becomes that of a programmable multicomponent analyzer.

The intent of the dual-beam design is to compensate for source output changes or detector or electronic gain changes. It cannot account for changes in sample cell transmission, such as dirt accumulation on windows or scattering particles in the sample stream.

The dual-wavelength design aims to use a reference wavelength not absorbed by the sample but affected by the earlier-mentioned sources of error to the same extent as the analytical wavelength. (Therefore, the reference wavelength should be as close to the analytical wavelength as possible.) The single-beam design is optically and mechanically simpler and lends itself easily to interfacing with a wide variety of sample cells.

Multiple-Component Fixed Filter Analyzer

When the components to be analyzed have absorption bands that do not overlap in the spectrum, the single-component analyzers can simply be expanded to measure multiple wavelengths. (Note: Having absorption bands that do not overlap does not imply that the absorption peaks are free from interference—only that some part of the band is free from other species.)

There are NDIR analyzers that use two detectors filled with two different gases, such as carbon monoxide and a hydrocarbon. The IR passes through first one detector and then the other. Filter analyzers are more easily expandable to multiple wavelengths, enabling a wide variety of process and quality control analyses to be made. Filters can be mounted on a rotating wheel or inserted sequentially into the beam by a cam mechanism. ABB can analyze up to 10 different gases with their multiwave units.

These instruments typically use analog signal processing. It is possible, but awkward, to compensate for spectral interference by cross-coupling between the different wavelength outputs. Digital signal processing simplifies this problem, but it is not widely used in multiple-wavelength fixed-filter analyzers. Note: Digital processing is very common with full-spectrum techniques.

Programmed Circular Variable Filter Analyzer

These microcomputer-controlled analyzers, derived directly from the CVF laboratory analyzers mentioned previously, are simple and rugged in optical design. The same basic analyzer can be programmed for any number of analyses up to a maximum of about 10 components. The programmable CVF offers a major advantage over using fixed filters.

It avoids the whole problem of filter selection and filter wavelength manufacturing tolerances and the prohibitive costs of obtaining fixed filters for one-of-a-kind analyses.

The built-in microcomputer makes interference compensation relatively straightforward. Instrument calibration coefficients and analytical wavelengths are obtained using known mixtures of the components of interest and placed in read-only memory (ROM).

The microcomputer enables self-diagnostic and self-checking features to give early warning to trouble and simplify the analyzer service.

This technology is most developed for workroom air quality monitoring.

PORTABLE ANALYZERS

Portable, single-beam IR photometers are also available as portable units (Figure 1.32u). They can serve the onsite measurement of a specific gas or gases that are present in the ambient air. They can also be used for leak detection, measurement of refrigerant gases, solvent vapors, and the

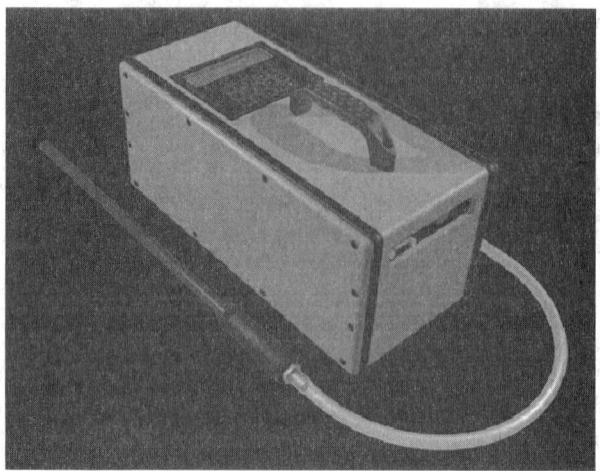

FIG. 1.32u
Portable, single-beam IR photometer. (Courtesy of Wilks-Spectro Scientific Co.)

measurement of other IR absorbing gases. This portable analyzer prompts the operator through the steps of the analysis and displays the concentration of the detected gases either in parts per million (ppm) or percent (%) units.

There are several portable IR analyzers. The one shown in Figure 1.32v is used for the measurement of octane. The unit can be calibrated to measure the octane of unleaded gasoline, ethanol blended fuel, leaded gasoline, and diesel fuel.

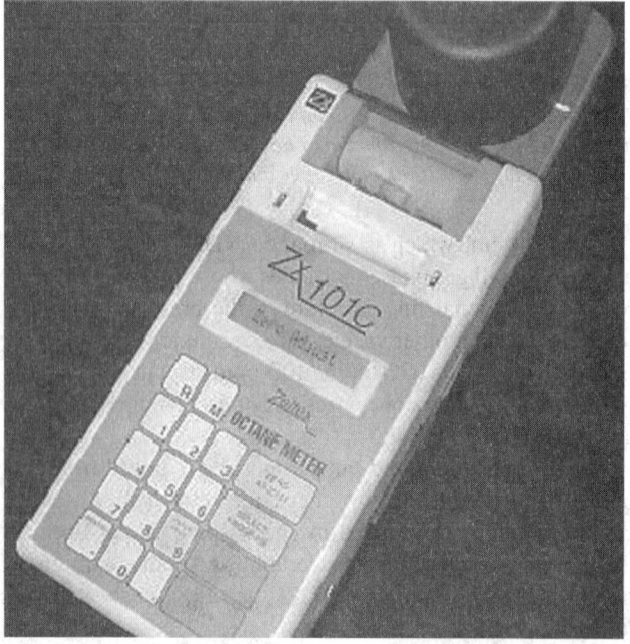

FIG. 1.32v
Portable octane analyzer. (Courtesy of Zeltex, Inc.)

INFRARED SOURCES

With the exception of the tunable diode laser, all the sources used in IR analyzers are of the black-body* type (Figure 1.32w). An element is heated to as high a temperature as is consistent with long operating life. The radiation varies as a function of wavelength as the source temperature increases.

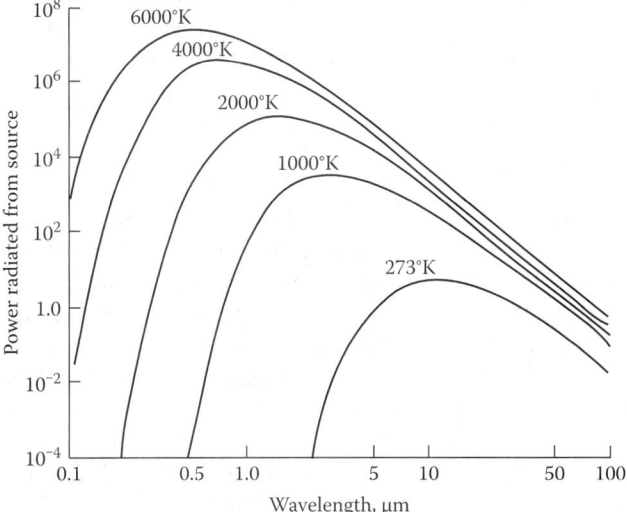

FIG. 1.32w
Radiated power in w./m²/str/μm from an ideal black body as a function of wavelength and temperature. Practical sources will have lower output for a given temperature to the extent that their emissivity is less than unity. The sun's radiation is close to a black body at 6000°K (5727°C) with peak output in the visible. A Nichrome wire source typically runs a little below 1000°K (727°C) with a peak output near 2.5 μm. Increasing the temperature increases output much more in the NIR than at longer wavelengths.

Sources used in process analyzers in the mid-IR depend on ohmic heating of an element, such as Nichrome wire, either exposed or embedded in a ceramic matrix. The metal oxide surface layer or ceramic has good emissivity in the mid-IR. The normal source temperature is in the 2000°C range. In NIR, the much hotter tungsten halogen lamp with a quartz envelope is an excellent source. (Remember, the shorter the wavelength, the higher the energy; therefore, the hotter source is required in NIR.) At wavelengths longer than 4 μm, the output drops due to envelope absorption, but even if it did not, the emissivity of tungsten decreases at longer wavelengths. (See the chapter on NIR for more comments on this source.)

---

* Black-body radiation is the electromagnetic radiation, having a specific spectrum and intensity that depends only on the temperature of the body (Figure 1.32w). A black-body at room temperature appears black, as most of the energy it radiates is infrared and cannot be seen by the human eye. When it becomes a little hotter, it appears dull red, and as its temperature increases further, it eventually becomes blindingly brilliant blue-white.

INFRARED DETECTORS

In all cases, IR energy is modulated so that an AC signal is detected and synchronous demodulation can be used to narrow the noise bandwidth. Beam chopping or modulation is best done after the sample. This avoids errors due to the detector sensing emission from hot samples. The error effect can be large, particularly at the longer wavelengths, beyond 5 μm.

NDIR Detectors

These are the gas-filled capacitive microphones already described (Figures 1.32g and 1.32h). They use low-modulation frequencies, 19 Hz or below, and are affected by external noise and vibration. They operate in the IR and are very sensitive. One can also use solid-state devices in this application.

Thermal Detectors

These are used in the mid-IR by filter analyzers. The pyroelectric detector is usually a wafer of $LiTaO_3$ about 20 μm thick, electroded on each side. It behaves electrically like a capacitor, and a current flows out of it if the temperature changes. It must be used in a chopped beam of IR, most commonly near 50 Hz. The minimum detectable signal is about 10^{-9} W. in a 1-Hz bandwidth, and it is somewhat less sensitive than a typical NDIR detector. Pyroelectric detectors must be mounted and sealed to avoid microphonic and acoustic interference. Thermopile detectors are available in the single-, dual-, and quad-element designs, where the term "element" refers to the size of the IR absorbing surface (Figure 1.32x).

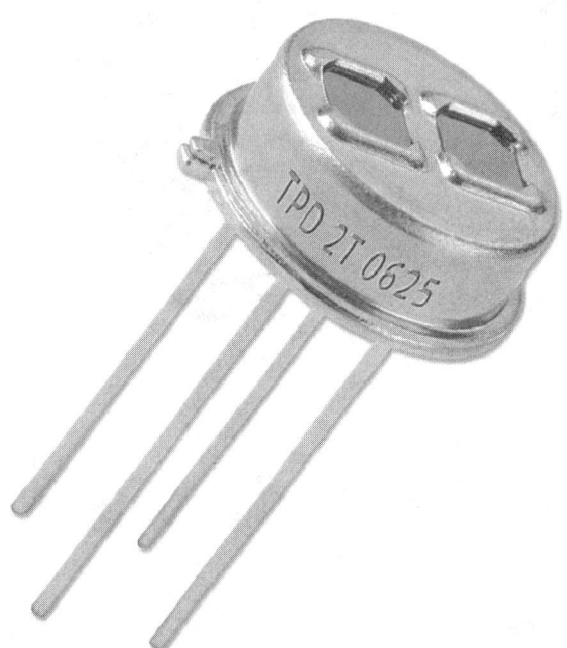

FIG. 1.32x
Photograph of a pyroelectric thermal detector; when in use, an IR-transmitting window, or lens, seals the air space in front of the detector. (Courtesy of Warsash Scientific.)

The sensitivity of the evaporated thermopile is best at lower-chopping frequencies, below 10 Hz. It is not only very rugged and less microphonic than the pyroelectric detector, but also less sensitive.

Photoconductive Detectors

In NIR, photon detectors such as PbS or PbSe can be used. They are two or three orders of magnitude more sensitive than the thermal detectors and operate best at higher chopping frequencies. The PbSe sensitivity extends almost to 5 µm, if thermoelectrically cooled, and can be used for such common analyses as $CH/CO/CO_2$. Response is strongly temperature dependent, in contrast with the earlier-mentioned thermal devices.

SELECTING THE CELL

Path Length Selection

There is an optimum cell path length for analyzing a particular sample. Too short a path length means a low sample absorbance, which gives a weak signal in comparison to instrument noise. Too long a path length results in very little energy reaching the detector and Beer's law often fails. Simple theory, based on the detector/preamplifier as the chief source of noise,[3] unaffected by the IR power reaching the detector, predicts that most accurate concentration measurements can be made when the sample transmits 1/e of the incident beam (e = 2.73).

This corresponds to a transmittance (T) of 36%, or an absorbance (A) of 0.43 A. However, any absorbance between 0.1 and 1.0 A, 80% to 10% T, will normally give good results. Measurements at higher absorbance tend to minimize the effect of certain kinds of electronic drift, while at lower absorbances, they minimize nonlinear ties in the analyzer's response. (see the paragraph titled CALIBRATION.)

Therefore, the cell path length should be chosen to put the sample absorbance in this desirable range. When analyzing trace gases, the absorbance is often near the lowest detection limit of the analyzer, even when the longest available path length is used, and that is the best that can be done.

Gas Cells

Figure 1.32y shows some gas cells used for process analysis. The common type transmits the IR beam straight through to the detector. The path lengths range from 0.1 mm to 50.0 cm (0.004 to 19.5 in.). Some of these long cells, especially those used on NDIR analyzers, are internally gold coated to act as a light pipe. The single-beam analyzers can interface with a wider range of cell type, such as multiple reflection cells with path lengths adjustable between 0.75 and 40 m (30.0 in. and 130.0 ft). These long path cells are especially valuable for analyzing trace contaminants in the air.

In choosing a gas cell, one should consider the volume if only small quantities of sample are available or if faster turnover is required. The most efficient cells have a high path length-to-volume ratio. Sample pressure and temperature must be controlled for accurate results because the instrument response depends on the number of sample molecules in the cell.

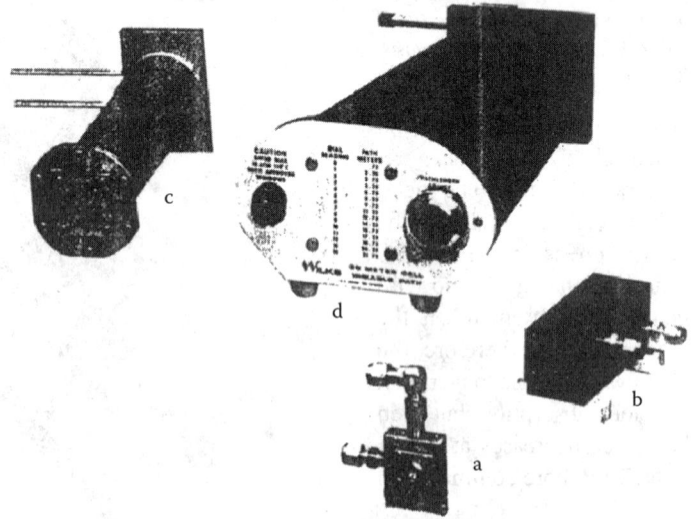

FIG. 1.32y
Cells for use with IR process analyzers. The longer-path-length cells are not conveniently interfaced with single-beam analyzers. (a) 2 mm (0.08 in.) cell, gas, or liquid; volume, 0.4 mL. (b) 10 cm (3.9 in.) gas cell (or liquid for NIR); volume, 50 mL. (c) 50 cm (19.5 in.) gas cell, two passes; volume, 600 mL. (d) 20 m (66 ft) variable-path gas cell; volume, 5.4 L.

Liquid Cells: Transmission Type

Liquid cells have much shorter path lengths to compensate for the higher-density samples. Thickness varies from 0.1 mm to 10.0 cm (0.004 to 4 in.), the longer cells being used in the NIR region, where sample absorption coefficients are lower. Sample streams must be carefully filtered to avoid cell plugging, particularly when very short (0.1-mm) IR cells are used.

Liquid Cells: Reflection Type

The multiple internal reflection (MIR) technique, also called attenuated total reflection (ATR), is a way of avoiding the problems of thin transmission cells. The IR beam makes multiple internal reflections at the surface of a high refractive index crystal wetted by the sample liquid of lower index of refraction (Figure 1.32z). The effective path length of the beam through the sample depends on the angle of incidence, the refractive indices of the sample and crystal, and the number of reflections.

The method is being used increasingly in mid-IR, with the sample being brought to the MIR cell via a sample loop from the process stream. Additionally, MIR cells have been put directly into the process stream with an optical connection to each end of the crystal. The big advantages of this approach are that no time is lost in the sample system and there is no need to remove solids, as they will not interfere with the measurement and can be used to follow batch reactions without withdrawing samples.

However, inline applications face a number of problems, including (1) mounting an analyzer directly in the process stream; (2) keeping the surface of the crystal, which is wetted by the process stream, free of contaminating films; (3) problems with running calibration checks on the systems; and (4) the difficulty of servicing the crystal without shutting down the process. (If the cell becomes coated with a film, all the spectrometer will see is that and will not see the changes in the process fluid.)

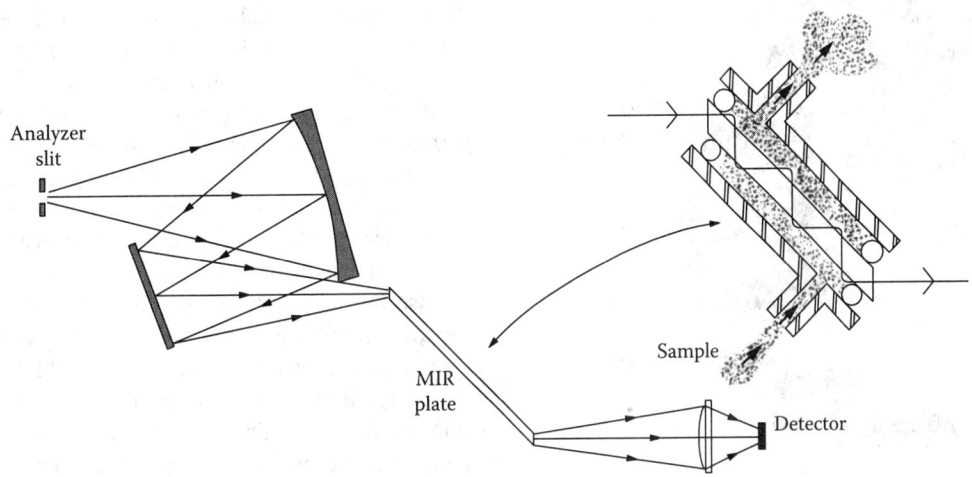

FIG. 1.32z
Optical schematic of an MIR cell suitable for use in a process sample loop interfaced with a single-beam filter analyzer. Materials used for the MIR crystal include sapphire, silicon, germanium, and zinc selenide.

Solid Samples

Sample composition can be determined by analyzing the spectra of IR diffusely reflected from the sample surface (Figure 1.32aa). Note that with these instruments, one is only analyzing the surface of the sample. The most common application is moisture measurement in, for example, paper as it is being made or feed stocks, such as wood chips (see Chapter 1.40). Filter analyzers utilizing this principle are used for analyzing grain and other food products for components such as protein, oil, and moisture. Most of this work is done in the NIR, where one sees not only the surface but also into the solid, because of the lower absorption coefficients (more detail is given later in the discussion of NIR).

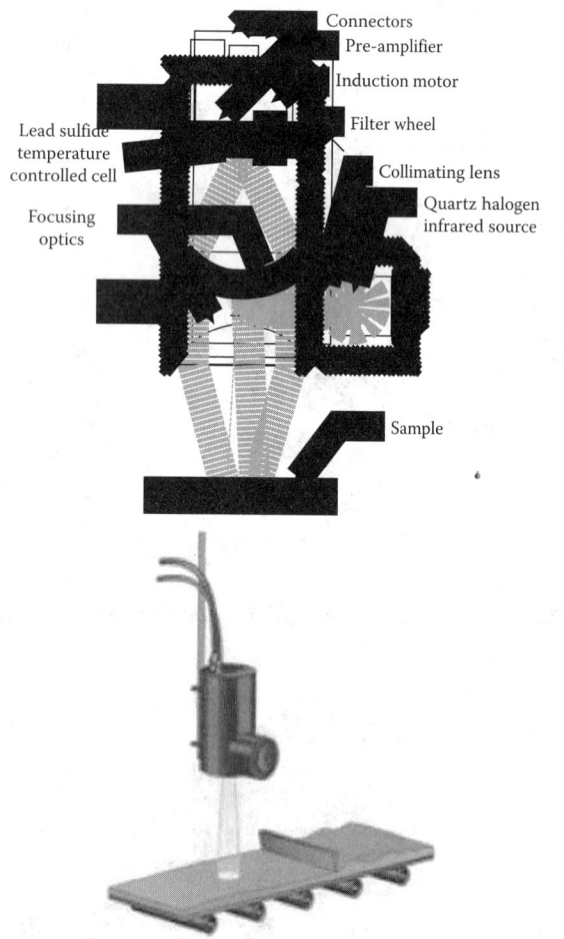

FIG. 1.32aa
Noncontact IR sensors are commonly used for moisture measurement of powders and bulk solids. (Courtesy of Moisture Register Products, a division of Aqua Measure Instrument Co.)

CALIBRATION

All IR analyzers require initial calibration with known samples. In general, the strategies of reference wavelengths and reference beams do not reduce sources of zero drift completely, and once in service, the analyzer zero and span must be checked and reset on a routine basis. The user's manual should give guidance on the steps involved.

With constant temperature and pressure samples, an IR analyzer is inherently span stable and, in principle, requires only that the cell path length be constant. However, depending on the analyzer design and operating environment, the span should be checked on a periodic basis. In general, the frequency of requiring span calibration is a function of the quality of temperature controls provided for the analyzer and the application. In NDIR and GFC analyzers, the characteristics of the gas-filled cells vary with temperature.

Filter analyzers as well as the NDIR and GFC types contain narrow-band-pass filters that change wavelength and absorbance with ambient temperature. For span stability, these filters must be temperature controlled. Long-term stability will be degraded if a high electronic scale expansion of gain is required, because the cell path length is not optimized for the job (as is frequently the case in the more sensitive analysis ranges).

Zero and span samples can be plumbed to a calibration port for use when required. Gas mixtures in cylinders are often used but should themselves be checked, as they are subject to change with age. (The gas can react with the cylinder wall and disappear from the mixture, or it can liquefy and be removed. The liquefaction can be compensated by heating the cylinder to raise the temperature and constantly stirring the sample; however, this also increases the probability of reaction.)

For gas analyses, the component of interest can sometimes be removed by filtering or catalytic oxidation to provide the zero gas. The zero check is frequently automated using a timer or is built into the memory of a microprocessor-controlled analyzer. Some instruments employ an automated span calibration analogous to the auto zero by inserting a sealed calibration cell or a secondary standard in the form of an attenuating filter into the analyzer beam.

Linearity

The detector response of an IR analyzer is not linearly related to concentration. The filter analyzers for many samples follow the Beer–Lambert law, and a logarithmic amplifier provides an acceptable linear output. However, if the analyzer cannot resolve the sample's absorption features, as is the case for most small gas-phased molecules and some liquids, the measured absorbance will fail to increase linearly at the high absorbances. This is the case with carbon monoxide (Figure 1.32d). These problems, and the nonlinear output of the NDIR and GFC analyzers, are frequently corrected by a linearizing circuit board.

PACKAGING

There is some uniformity of packaging among manufacturers of NDIR analyzers for panel mounting. At present, there are no intrinsically safe IR analyzers, and they must be packaged in a purged or explosion-proof housing to be acceptable in a hazardous area. Plumbing and sample cells containing flammable samples should be kept outside the enclosure containing the analyzer source and electronics unless flame arrestors are used in the sample lines.

It may be necessary to purge the analyzer head to prevent corrosion or to eliminate absorbing ambient gas molecules from the optical path outside the cell. Many analyzers have a remote readout and control panel option so that the analyzer head can be located close to the measurement site. The IR components are usually mounted on a vibration-isolated rigid structure, and the analyzer head is temperature controlled to some extent.

APPLICATIONS AND ADVANCES

Table 1.32a gives a summary of some common process applications for IR analyzers.

The advances in this technology involve several areas of development (Figure 1.32ab). The microprocessor has

TABLE 1.32a
IR Analyzer Applications Summary

Analyzer	Organic Vapors					Comments
	Carbon Monoxide	Carbon Dioxide	Simple Molecules	Complex Molecules	Organic Liquids	
Nondispersive infrared (NDIR)	♥	♥	♥			Single-component analysis: methane, ethylene, CO, CO_2, etc.
Mid-IR filter	♥	♥	♥	♥	♥	Same as above, including NH_3, vinyl chloride, methyl ethyl ketone, etc.
FTIR	♥	♥	♥	♥	♥	The advantage of the FTIR is that it can look at multiple species
Correlation spectrometer	♥					Stack analysis, single-component gas analysis

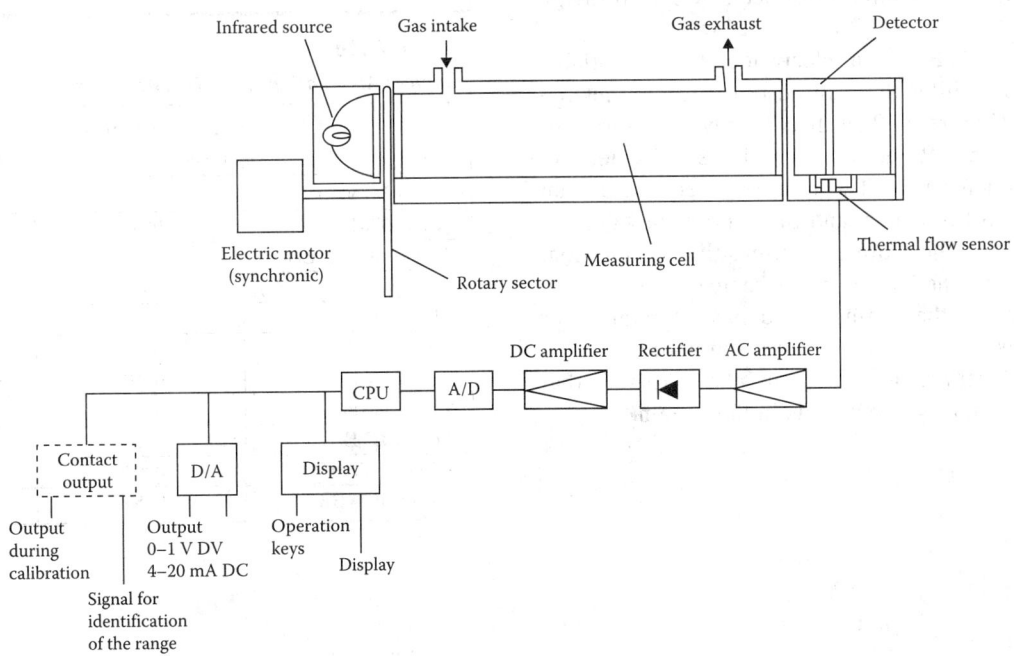

FIG. 1.32ab
Intelligent IR analyzer. (Courtesy of Yokogawa.)

contributed to the development of self-diagnostic and self-calibrating designs. Auto-calibration was also helped by the use of multiple reference cells when stable reference gases are available. Modular design in conjunction with self-diagnostics has simplified maintenance.

Intelligent units can have the following features and more:

- Self-diagnostics, notifying the operator of abnormalities
- Zero and span calibration requires only the pressing of a key while the reference gas is flowing.
- Can perform automatic calibration of zero and span at predefined intervals
- Remote switching of range for each gas

The growth of fiber-optic technology (Chapter 1.23) made the probe-type IR analyzer practical and extended the spectrum by allowing the same probe to use UV, visible, NIR, and IR forms of radiation. One should be careful with such extended spectrum applications, because windows that are transparent in one region of the spectrum may be opaque in another portion.

The online use of MIR or ATR crystal probes is also promising, although the task of keeping the probe surface clean is not easy to meet.

Another area of development is to minimize the number of moving parts, choppers, shutters, or beam alternators and eliminate the need for multiple paths in the IR analyzer. The goal can be met by the use of acousto-optic tunable filters (AOTFs), which are made of thallium–arsenic–selenide (TAS). This crystal can be tuned by a radio frequency oscillator, ultrasonic transducer, over a spectrum of 2–5.5 μm.

The TAS AOTF is an electronically controllable narrow-band-pass filter that can be tuned to any desired IR frequency (Figure 1.32ac). It can provide a series of chopped and tuned IR beams to simultaneously measure a variety of stack gases. The required frequencies can be selected in milliseconds, and the solid-state AOTF is small and rugged. Therefore, it is insensitive to vibration, but must be maintained at a constant temperature.

Table 1.32b lists the minimum and maximum ranges of some of the gases and vapors that are commonly detected by NDIR analyzers. Note: These numbers assume that there are no interfering compounds in the stream.

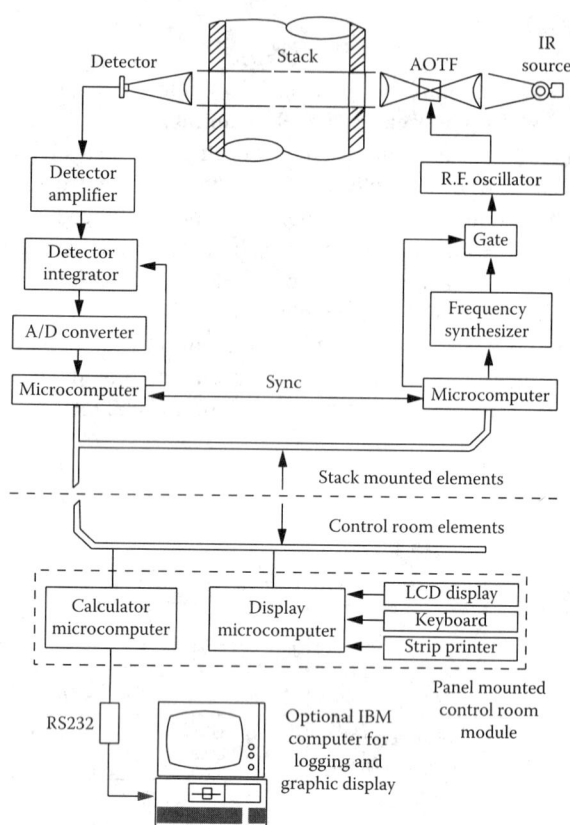

FIG. 1.32ac
The solid-state tunable crystal eliminates most of the moving parts from the IR analyzer and allows simultaneous measurement of many gases in the process stream. (From Nelson, R.L., InTech, June 1987.)

TABLE 1.32b
Typical Applications for NDIR Analyzers

Gas	Minimum Range (ppm) (for 10 m Cell)	Maximum Range (%)
Ammonia (NH_3)	0–100	0–10
Butane (C_4H_{10})	0–300	0–100
Carbon dioxide (CO_2)	0–20	0–100
Carbon monoxide (CO)	0–25	0–100
Ethylene (C_2H_4)	0–100	0–100
Hexane (C_6H_{12})	0–100	0–5
Methane (CH_4)	0–10	0–100
Nitrogen oxide (NO)	0–10	0–10
Propane (C_3H_8)	0–100	0–30
Sulfur dioxide (SO_2)	0–100	0–30
Water vapor (H_2O)	0–50	0–5

NEAR-INFRARED ANALYZERS

NIR absorptions are the overtones and combination bands of the IR, particularly involving a hydrogen atom. There are two major effects of this from the online analyzer point of view: (1) it is very difficult for a human being to say which band represents which compound and (2) the absorbances are much weaker than in the IR region (about 1/10 weaker per overtone).

This means that the first overtone is 1/10 weaker than the IR, and the second overtone 1/10 weaker than the first. At first glance, both effects appear to be undesirable. Yet, the second turns into a major advantage for analyzing either liquids or solids—the path length is much longer. For solids, this means that more than just the surface is seen and can be analyzed.

With hydrocarbon samples like gasoline, the path length in the IR region would be about 0.1 mm, while in the first overtone of the NIR, it can be about 1 mm, in the second overtone about 1 cm, and in the third overtone about 10 cm. For process applications, the longer path length has three advantages: (1) more uniform sampling, (2) fewer plugging problems, and (3) the window fouling is much less of a problem.

Short path lengths in the IR region can be handled with an ATR or MIR cell, but fouling remains to be a major problem. (Fouling is normally just a surface layer, and with a 10-cm cell, it is not seen.)

Interpreting the Absorption Bands

Returning to the first item, difficulty in the interpretation of the absorption bands. Normally in the NIR, the objective is not to do qualitative analysis, but quantitative analysis with a trained computer. The computer can distinguish quantities of individual chemical species, for example, p-xylene in a mixture of aromatics, including both o-xylene and m-xylene. However, the normal task is to do quantitative analysis of physical properties like hardness of wheat, octane number of gasoline, boiling points of gasoline, cetane number of diesel, etc.

How is the computer trained? A set of samples, normally 30–50, is prepared covering the range of interest of all the analytes and "I don't cares" in the system. The spectrum for each sample is determined and paired with the known analyte concentrations. These data are all put into a computer with a chemometric program that uses a system to fit the spectrum to the data. Once that fit has been determined, the generated model can be used for future samples within the range of the initial data.

It is important to remember the following:

1. Before using the model on real samples, it should be tested on some new samples that were not used to make the model. Doing this is desirable, even if time-consuming and inconvenient. Normally, one is using 30–50 samples, each having 700–1000 spectral data points attached; the computer can find a fit, but does it mean anything? The only way to find out is to test it.
2. Extrapolations from that initial data set are done at your risk. For gasoline, we used about 30 samples of each grade and time of year to build complete models to predict the various properties of the gasoline. Note: The 30 samples per grade per time of year are considered as a starting set, and the working models are based on 50–100 samples.

Sample Temperature Control

If the sample contains water and one is measuring the ionic solutes, then temperature control is absolutely required. Liquid water exists in different forms related to the number of hydrogen bonds, as a function of temperature, and the change of forms causes spectral shifts that will overcome any other measurements one wants to make concerning the ions dissolved in the water. In hydrocarbon streams, the jury is still out—some claim that sample temperature control within 1°C is required for good analysis, while others say they can compensate for changing temperatures.

Because, in any optical measurement, one is counting molecules in the optical path, temperature will affect the spectrum. Therefore, it is easiest to control the temperature rather than try to compensate for it when doing high-quality work.

Fiber-Optics

One of the big advantages of using the NIR range is the availability of optical fibers, but not all of them are suited for NIR applications. To be useful for spectroscopic applications, the fiber needs to be insensitive to outside light, temperature changes, and movement (vibrations). The outside light problem can be solved by proper choice of jacket or keeping the fiber encased in an opaque case. (Remember that just because the jacket is opaque in the visible spectrum does not mean it will be opaque in the NIR.)

Temperature changes can cause major problems. Vickers* shows that a 50°C temperature change on a 10 m fiber will cause a change in absorbance at over 1.0 A. K. The solution in such a case is to have a buffer that extracts the cladding modes and have the core, cladding, and buffer coiled inside the jacket to allow for thermal expansion and contraction without straining the fiber.

This system, in addition to adding expense to the fiber, lowers the total light throughput of the fiber, but makes it usable so that one is analyzing the light from the stream rather than from the fiber. Plastic-clad silica fiber is very inexpensive and therefore attractive, but it is totally unacceptable for hydrocarbon analysis; one sees the plastic cladding, and the depth of penetration is a function of temperature, which makes background correction very difficult.

A fiber-optic-based, semiportable (benchtop) NIR analyzer is shown in Figure 1.32ad. This unit is compatible with a variety of probes or sampling systems, is of the dual-beam design, can monitor up to 6 sample points and because of the fiber-optic design, allows safe remote measurement of hazardous samples.

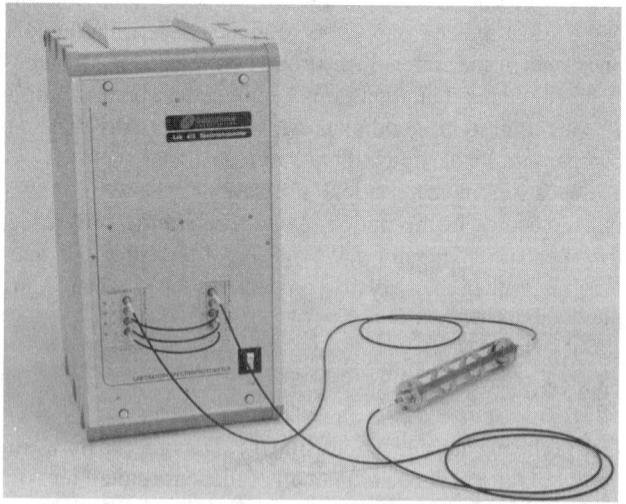

FIG. 1.32ad
Fiber-optic-based, semiportable (benchtop) NIR analyzer. (Courtesy of Guided Wave.)

---

* Vickers, G. H. et al., Influence of ambient temperature on the near-infrared transmission of optical fibers, Paper 178, Eastern Analytical Symposium, Somerset, NJ, 1990.

Types of NIRs

NIRs come in four basic types:

1. The filter instrument uses two or more filters to pick out reference and measuring wavelengths. These are the lowest-cost instruments and generally only measure one to a few analytes in the stream.
2. The dispersive units with single detectors and moving grating. The moving grating causes some wavelength instability that becomes a major problem with very sensitive measurements.
3. FTNIR is very similar to FTIR; however, since the frequency and wavelengths are inverse scales, at the shorter wavelengths in NIR there is a higher reproducibility factor required for the mirror movement. However, these have been successfully applied in some very rigorous applications, such as measuring the octane of gasoline.
4. The diode array instrument with no moving parts—note that the material of construction in the diode array sets the usable range of the instrument. The most common diodes are made from silicon that becomes transparent at wavelengths longer than 1100 nm; therefore, these diode arrays are only usable in the third overtone. [The diode array instrument has no moving parts but requires extreme temperature stabilization to prevent wavelength changes due to different distances between the grating and detector array as a function of temperature. One unit on the market controls the spectrograph temperature to ±0.1°C, while the outside temperature varies from −40°C to 50°C (−40°F to 120°F).]

Optical Sources

The optical source for any spectrophotometer is important, but it is even more critical with systems based on fiber-optics. The only light that is important in a fiber-optic instrument is the light that can be reproducibly launched onto the fiber. Therefore, big bulbs are not needed when trying to put light into a 200 μm inner-diameter fiber; the key is to have a short, *stable* filament.

Many filaments are long, compared to the fiber diameter, and the hot spot that is giving off the light moves along the filament. Therefore, sometimes the fiber is in focus of the hot spot and sometimes it is not. This is totally unsatisfactory. What one wants is a short filament, where the hot spot stays fixed and the filament does not sag, which requires a larger-diameter filament. This allows the light to be focused onto the fiber and gives a light bulb with a very long life, exactly what is needed for process analysis.

Gases

The only known NIR application on gas samples involves the Btu content of natural gas. In order to make this measurement, the natural gas line would have to be under high pressure, normally 1000 psi or so, in order to provide a sufficient number of molecules in the optical path to make the measurement.

Liquids

Under stable temperature conditions, ions in water can be measured. One analyzer measures the hydroxide ion concentration in a neutralization bath using a filter instrument with the reference chosen to eliminate effects of any other ions. Using third overtone diode array instruments, the octane numbers of gasoline can be determined and, at the same time, measure the MTBE concentration, the boiling points of the gasoline, the Reid Vapor Pressure (RVP), etc.

Crude oil properties have been measured using the second overtone region. The key is to properly model all these values when starting up the analyzer. (These same measurements have been made in the second overtone of the NIR using the FT approach.)

Solids

Modern NIR got its start with Dr. Karl Norris measuring such characteristics of agricultural products as the hardness of wheat, the protein content in wheat, the food value of alfalfa hay, etc. Because in the solids, the natural bandwidth was very wide, the original NIR instruments had poor optical resolution, but very high signal-to-noise ratios to see very small differences in the materials. When Dr. James Callis, University of Washington, Center for Process Analytical Chemistry (CPAC), used one of these instruments to determine the octane of gasoline, it became evident that a great deal of information was discarded due to the poor optical resolution.

The diode array instrument that was developed improved the optical resolution by a factor of 3 and actually also improved the signal-to-noise ratio, which made it a very usable instrument. This resolution is used, for example, to measure MTBE in gasoline by measuring the small shift in the methyl group attached to the oxygen of the MTBE vs. all the other methyl groups in the gasoline.

Calibration Transfer

One of the earlier problems associated with NIR analyzers was that each instrument was different. If a calibration model was build on instrument A, then it could not be used on B, even if B was from the same manufacturer and had the same model number. Dr. Bruce Kowalski and his group at the University of Washington, CPAC, developed methods to solve this problem that have now been incorporated into some of the commercial analyzers.

Since there is a significant amount of time put into developing a model, one should make sure that the model is transferable to a different analyzer or to the same analyzer after maintenance is done on it, for example, when the light bulb is changed.

SPECIFICATION FORMS

When specifying infrared analyzers only, one can use the ISA form 20A1001, and when specifying both the analyzer and the composition or the properties of the process, which it will be monitoring, use the ISA form 20A1002. Both forms are reproduced with the permission of the International Society of Automation on the next pages.

ANALYSIS DEVICE — Operating Parameters

RESPONSIBLE ORGANIZATION
1.
2.
3. ISA
4.
5.

SPECIFICATION IDENTIFICATIONS
6. Document no
7. Latest revision | Date
8. Issue status
9.
10.

ADMINISTRATIVE IDENTIFICATIONS
11.
12. Project number | Sub project no
13. Project
14. Enterprise
15. Site
16. Area | Cell | Unit
17.

SERVICE IDENTIFICATIONS
18.
19. Tag no/Functional ident
20. Related equipment
21. Service
22.
23. P&ID/Reference dwg
24. Process line/nozzle no
25. Process conn pipe spec
26. Process conn nominal size | Rating
27. Process conn termn type | Style
28. Process conn schedule no | Wall thickness
29. Process connection length
30. Process line matl type
31. Fast loop line number
32. Fast loop pipe spec
33. Fast loop conn nom size | Rating
34. Fast loop conn termn type | Style
35. Fast loop schedule no | Wall thickness
36. Fast loop estimated lg
37. Fast loop material type
38. Return conn nominal size | Rating
39. Return conn termn type | Style

SERVICE IDENTIFICATIONS Continued
40.
41. Return conn matl type
42. Inline hazardous area cl | Div/Zon | Group
43. Inline area min ign temp | Temp ident number
44. Remote hazardous area cl | Div/Zon | Group
45. Remote area min ign temp | Temp ident number
46.
47.

COMPONENT DESIGN CRITERIA
48.
49. Component type
50. Component style
51. Output signal type
52. Characteristic curve
53. Compensation style
54. Type of protection
55. Criticality code
56. Max EMI susceptibility | Ref
57. Max temperature effect | Ref
58. Max sample time lag
59. Max response time
60. Min required accuracy | Ref
61. Avail nom power supply | Number wires
62. Calibration method
63. Testing/Listing agency
64. Test requirements
65. Supply loss failure mode
66. Signal loss failure mode
67.
68.

PROCESS VARIABLES / MATERIAL FLOW CONDITIONS | PROCESS DESIGN CONDITIONS

#	Variable	Value	Units	#	Minimum	Maximum	Units
70	Flow Case Identification			102			
71	Process pressure			103			
72	Process temperature			104			
73	Process phase type			105			
74	Process liquid actl flow			106			
75	Process vapor actl flow			107			
76	Process vapor std flow			108			
77	Process liquid density			109			
78	Process vapor density			110			
79	Process liquid viscosity			111			
80	Sample return pressure			112			
81	Sample vent/drain press			113			
82	Sample temperature			114			
83	Sample phase type			115			
84	Fast loop liq actl flow			116			
85	Fast loop vapor actl flow			117			
86	Fast loop vapor std flow			118			
87	Fast loop vapor density			119			
88	Conductivity/Resistivity			120			
89	pH/ORP			121			
90	RH/Dewpoint			122			
91	Turbidity/Opacity			123			
92	Dissolved oxygen			124			
93	Corrosivity			125			
94	Particle size			126			
95				127			

CALCULATED VARIABLES
96.
97. Sample lag time
98. Process fluid velocity
99. Wake/natural freq ratio
100.

128.
129.
130.
131.
132.

MATERIAL PROPERTIES
133.
134. Name
135. Density at ref temp | At
136.

MATERIAL PROPERTIES Continued
137.
138. NFPA health hazard | Flammability | Reactivity
139.
140.

Rev	Date	Revision Description	By	Appv1	Appv2	Appv3	REMARKS

Form: 20A1001 Rev 0 © 2004 ISA

1.32 Infrared and Near-Infrared Analyzers

	RESPONSIBLE ORGANIZATION	ANALYSIS DEVICE COMPOSITION OR PROPERTY Operating Parameters (Continued)		SPECIFICATION IDENTIFICATIONS		
1			6	Document no		
2			7	Latest revision		Date
3	(ISA)		8	Issue status		
4			9			
5			10			

	PROCESS COMPOSITION OR PROPERTY			MEASUREMENT DESIGN CONDITIONS				
	Component/Property Name	Normal	Units	Minimum	Units	Maximum	Units	Repeatability
11								
12								
13								
...								
37								

(Rows 38–75: open notes area)

Rev	Date	Revision Description	By	Appv1	Appv2	Appv3	REMARKS

Form: 20A1002 Rev 0 © 2004 ISA

Definitions

Beer–Lambert law relates the absorption of light to the transmission properties of the material through which the light is traveling; in other words, it relates the amount of light absorbed to the concentration of the material of interest in the sample.

Black-body radiation is the electromagnetic radiation, having a specific spectrum and intensity that depends only on the temperature of the body (Figure 1.32w). A black-body at room temperature appears black, as most of the energy it radiates is infrared and that cannot be seen by the human eye. When it becomes a little hotter, it appears dull red, and as its temperature increases further, it eventually becomes blindingly brilliant blue-white.

Fourier transformation is a mathematical process to convert raw data into the actual spectrum of the radiation absorption over a wide spectral range. This is superior to dispersive spectrometry, which measures IR radiation intensity over a narrow spectral range. It can be applied to optical spectroscopy, optical and infrared spectroscopy (FTIR, FTNIR), nuclear magnetic resonance (NMR), and magnetic resonance spectroscopic imaging (MRSI).

Luft detectors consist of two chambers, divided by a diaphragm, and the same intensity of pulsed infrared radiation is received by both chambers. When the gas of interest flows through the sample cell, a reduction in radiation energy is received by the detector chamber, which causes the temperature and pressure to drop in that chamber. The amount of pressure drop is in proportion to the gas concentration, and this pressure difference between the two chambers causes a movement of the diaphragm, which is detected by capacitance.

Michelson interferometer splits a radiation source into two arms and reflects each back toward a beam splitter, which then combines their amplitudes interferometrically. The resulting pattern is directed to some type of photoelectric detector.

Monochromator A monochromator is an optical device that transmits a mechanically selectable narrow band of wavelengths of light or other radiation chosen from a wider range of wavelengths available at the input.

Photometer is an analyzer that measures the absorbance or transmittance at a single or at multiple specific wavelengths. In the multiple-wavelength configuration, calculations can be made to obtain the difference between, or ratio of, the measurements at two specific wavelengths.

Retroreflector is a device or surface that reflects light back to its source with a minimum of scattering. It is sometimes also called a retroreflector or cataphote.

Spectrophotometer is an analyzer that obtains the spectra of the sample through wavelength scanning. Continuous measurements can be obtained by periodically detecting the changes in the composition of the flowing sample. Software is available to enlarge or reduce the spectra obtained, detect peaks, or make various other calculations, based on the spectra.

Abbreviations

AOTF	Acousto-optic tunable filter
ATR	Attenuated total reflection
CEAS	Cavity enhanced (laser) absorption spectroscopy
CEM	Continuous emission monitor
CVF	Circular variable filter
FT	Fourier transform
FTIR	Fourier transform infrared analyzer
FTNIR	Fourier transform near-infrared analyzer
GFC	Gas filter correlation
MIR	Multiple internal reflection
MRSI	Magnetic resonance spectroscopic imaging
MTBE	Methyl t-butyl ether
NDIR	Nondispersive infrared
NIR	Near infrared
NMR	Nuclear magnetic resonance
PAS	Photoacoustic spectroscopy
PCR	Principal component regression
PDA	Photodiode arrays
PLS	Partial least squares
PMT	Photomultiplier tube
RVP	Reid vapor pressure
TAS	Thallium–arsenic–selenide
TDL	Tunable diode laser
TLS	Tunable laser spectrometer
UV/VIS	Ultraviolet–visible
VDT	Video display tube

Organizations

APHA	American Public Health Association
ASTM	American Society of Testing and Materials
CPAC	Center for Process Analytical Chemistry
IFPAC	International Forum for Process Analytical Chemistry

Bibliography

ABB, Cavity Enhanced (laser) Absorption Spectroscopy (CEAS) for continuous emission monitoring (2015) https://library.e.abb.com/public/0c18a972dad14bd1a8a598f5bd817c52/WP_ANALYTICAL_001-EN_A.pdf.

Bakeey, K. A., Spectroscopic Tools and Implementation Strategies (2010) http://www.wiley.com/WileyCDA/WileyTitle/productCd-047072207X.html.

Burns, D. A., *Handbook of Near-Infrared Analysis* (2001) http://books.google.com/books/about/Handbook_of_Near_Infrared_Analysis_Secon.html?id=XkALgZVXxQQC.

Coates Consulting, New Miniaturized Spectral Measurement Platforms Covering the Visible to the Mid-IR (2010) http://depts.washington.edu/cpac/Activities/Meetings/Fall/2010/documents/NeSSI_Workshop_November_2010_Coates.pdf.

Emerson-Rosemount: Continuous emission monitoring (2013) http://www2.emersonprocess.com/siteadmincenter/PM%20Rosemount%20Analytical%20Documents/PGA_AN_42-PGA-AN-POWER-CEMS-Cogen.pdf.

Harrison, K. M., *Grating Spectroscopes* (2012) http://www.amazon.com/Grating-Spectroscopes-Patrick-Practical-Astronomy/dp/1461413966/ref=sr_1_cc_1?s=aps&ie=UTF8&qid=1391295340&sr=1-1-catcorr&keywords=spectrometers+analyzers+books.

International Sensor Technology, Infrared Sensors (2011) http://search.aol.com/aol/search?q=International%20sensors%20Infrared%20gas%20sensors&s_it=keyword_rollover&ie=UTF-8&VR=3430.

Luma Sense Technologies, Photoacoustic Detection (2015) http://www.lumasenseinc.com/EN/solutions/techoverview/pas/.

Mark, H. and Griffiths, P., *Analysis of noise in Fourier transform infrared spectra* (2002) https://www.osapublishing.org/as/abstract.cfm?uri=as-56-5-633.

Mellon, M. G., *Analytical Absorption Spectroscopy* (2007) http://www.amazon.com/Analytical-Absorption-Spectroscopy-MG-Mellon/dp/1406751707/ref=sr_1_14?s=books&ie=UTF8&qid=1390740544&sr=1-14&keywords=industrial+spectrophotometer.

Mitchell, M. K. and Stapp, W., *Field Manual for Water Quality Monitoring*, 13th edn. http://www.amazon.com/Field-Manual-Water-Quality-Monitoring/dp/0757555462.

Sherman, R. E., *Process Analyzer Sample-Conditioning System Technology*, New York: John Wiley & Sons, 2002. http://www.wiley.com/WileyCDA/WileyTitle/productCd-0471293644.html.

Siesler, H. W., *Near-Infrared Spectroscopy: Principles, Instruments, Applications*, 2008. http://www.amazon.com/Near-Infrared-Spectroscopy-Principles-Instruments-Applications/dp/3527301496.

Smith, A. L., *Applied Infrared Spectroscopy*, http://books.google.com/books/about/Applied_infrared_spectroscopy.html?id=yiRRAAAAMAAJ.

Stellar Net, Inc., *Fiber Optic Spectrum Analyzers* (2013) http://www.stellarnet.us/popularconfigurations_fiberanlzr.htm.

Thermo Electron, *Near Infrared Sensor for Online Measurement* (2005) http://www.thermo.com/eThermo/CMA/PDFs/Various/File_27448.pdf.

Van der Maas, J. H., *Basic Infrared Spectroscopy*, London, UK: Heyden & Sons, Ltd., 1972. https://archive.org/details/BasicInfraredSpectroscopy.

1.33 Ion-Selective Electrodes (ISE)

R. T. OLIVER (1972, 1982) **S. S. LIGHT** (1995)
W. P. DURDEN (2003) **B. G. LIPTÁK** (2017)

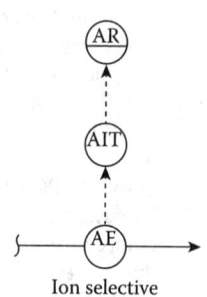

Ion selective
Flow sheet symbol

Types of electrode	Glass, solid state, solid matrix, liquid ion exchanger, gas sensing. For more data, see Tables 1.33a and 1.33m
Standard design pressure	Generally dictated by electrode holder; 0 psig for solid state and liquid ion exchanger; 0–100 psig (0–7 bars) for most electrode types; over 100 psig (over 7 bars) for solid-state designs
Standard design temperature	0°C–50°C (32°F–122°F) for solid matrix and liquid ion exchange; −5°C to 80°C (23°F–176°F) for most others, with 100°C (212°F) intermittent exposure being permissible
Range	From fractional parts per million (ppm) to concentrated solutions
Relative error	For laboratory devices ±0.1 mV. For direct measurements (process applications), an absolute error of ±1.0 mV is equivalent to a relative error of ±4% relative error in activity for monovalent ions and ±8% for divalent ions. For end-point detection or batch control, relative error ranges between ±0.1% and ±0.25%. For expanded-scale commercial amplifiers, error is better than ±1% of full scale.
Costs	Similar to those of pH electrodes (Chapter 1.55). Individual electrodes (measuring or reference only): $100–$500; Transmitters from $1000 up, depending on accessories
Partial list of suppliers	(Suppliers of pH probes and pH systems are listed in Chapter 1.55.) Advanced Sensor Technologies Inc. (http://www.astisensor.com/proise.htm) Advanced Sensor Technologies Inc. (http://www.astisensor.com/pH_ORP_Products.htm) Endress+Hauser (http://www.us.endress.com/#products/pH-transmitters) Fisher Scientific (http://www.fishersci.com/ecomm/servlet/Search?LBCID=98244456&keyWord=Ion+Selective+Electrodes&store=&nav=10397844&offSet=0&storeId=10652&langId=-1&fromSearchPage=1&searchType=PROD&typeAheadCat=prodCat) Hanna Instruments (http://www.hannainst.com/usa/prods2.cfm?id=042002) Honeywell (https://www.honeywellprocess.com/en-US/explore/products/instrumentation/analytical-instruments-and-sensors/ph-orp/Pages/default.aspx) Horiba Instruments Inc. (http://www.horiba.com/analysis-based-on/electrochemistry-reaction/potentiometry-ion-electrode-method/) Istek Inc. (http://www.istek.co.kr/shop/02_03_eng.php?part_idx=165) Laval Lab (http://www.lavallab.com/ion-selective-electrode/ion-selective-electrode-elit.htm) Metrohm USA (http://www.metrohmusa.com/Products/Titration/Electrodes/Ion-Electrodes/index.html) Mettler Toledo (http://us.mt.com/us/en/home/products/Laboratory_Analytics_Browse/Electrodes_family_page/Ion_electrodes_family_page.html) Nico 2000 Ltd (http://www.nico2000.net/datasheets/electrodes.html) Nico Scientific (http://www.nicosensors.com/Level_1/products.htm) Oakton Instruments (http://www.4oakton.com/proddetail.asp?prod=224&parent=144&value=detail) Omega (http://www.omega.com/subsection/ion-selective-electrodes.html) Pasco (http://www.pasco.com/family/ion-selective-electrodes/)

(Continued)

Sartorius (http://www.sartorius.com/en/search/?tx_solr%5Bq%5D=ion-selective+electrode&id=65&L=0&x=4&y=9)
Thermo Scientific (http://www.thermoscientific.com/en/products/ise-electrodes.html)
World Precision Instruments (http://www.wpiinc.com/index.php?xsearch%5B0%5D=ion+selective+electrodes&submit=go&xsearch_id=products_search_top&src=directory&srctype=lister&view=products)
WTW (http://www.wtw.de/en/products/lab/ise/electrodes.html)
Yokogawa (http://www.yokogawa.com/us/products/analytical-products/ph-orp/index.htm)

INTRODUCTION

Ion-selective electrodes (ISEs) comprise a class of primary elements used to obtain information related to the chemical composition of a process solution. They are electrochemical transducers that generate a millivolt potential when immersed in a conducting solution containing free or unassociated ions to which the electrodes are responsive. The magnitude of the potential is a function of the logarithm of the *activity* of the measured ion (not that of the total concentration of that ion) as expressed by the Nernst equation (Equation 1.33(1)).

The familiar pH electrode for measuring hydrogen ion activity is the best known of the ISEs and was the first one to be made commercially available (see Chapter 1.55). With few exceptions—notably the silver billet electrode for halide measurements and the sodium glass electrode—the pH electrode was the only satisfactory electrode available to the process industry prior to 1966.

Currently, more than two dozen electrodes are suitable for industrial use. Table 1.33a lists some of the commercially available electrodes for which industrial process applications have been reported. Additional electrodes are also used for research purposes.

TABLE 1.33a
Ion-Selective Electrodes

Ion/Species	Type of Electrode	Lower Detectable Limit, ppm	Principal Interferences
Ammonia	Gas sensing	0.009	Volatile amines
Bromide	Solid state	0.04	CN^-, I^-, $S^=$
Cadmium	Solid state	0.01	Ag^+, Hg^{++}, Cu^{++}, Fe^{++}, Pb^{++}
Calcium	Solid matrix/liquid membrane	0.2	Zn^{++}, Fe^{++}, Pb^{++}, Cu^{++}, Ni^{++}, Sr^{++}, Mg^{++}, Ba^{++}
Carbon dioxide	Gas sensing	0.4	Volatile weak acids
Chloride	Solid state	0.2	Br^-, CN^-, $S^=$, SCN^-, I^-
Chloride	Liquid membrane	0.2	ClO_4^-, Br^-, I^-, NO_3^-, OH^-, F^-, OAc^-, $SO_4^=$, HCO_3^-
Copper (II)	Solid state	0.006	Ag^+, Hg^{++}, Fe^{+++}
Cyanide	Solid state	0.01	$S^=$, I^-
Divalent cation[a]	Solid matrix/liquid membrane	0.001	
Fluoroborate (BF_4^-) (boron)	Liquid membrane	0.11	I^-, HCO_3^-, NO_3^-, F^-
Iodide	Solid state	0.006	$S^=$, CN^-
Lead	Solid state	0.2	Ag^+, Hg^{++}, Cd^{++}, Fe^{++}
Nitrate	Solid matrix/liquid membrane	0.3	Cl^-, ClO_4^-, I^-, Br^-
Nitrite	Gas sensing	0.002	CO_2, volatile weak acids
Perchlorate	Liquid membrane	0.7	Cl^-, ClO_3^-, I^-, Br^-, HCO_3^-, NO_3^-, etc.
Potassium	Liquid membrane	0.04	Cs^+, NH_4^+, H^+, Ag^+, $Tris^+$, Li^+, Na^+
Redox (platinum)	Solid state	Varies	All redox systems
Silver/sulfide	Solid state	0.01 Ag 0.003 S	Hg^{++}
Sodium	Glass	0.02	Ag^+, H^+, Li^+, Cs^+, K^+, Tl^+
Thiocyanate	Solid state	0.3	OH^-, Cl^-, Br^-, I^-, NH_3, $S_2O_3^=$, CN^-, $S^=$
Sulfur dioxide	Gas sensing	0.06	CO_2, NO_2, volatile organic acids

[a] Water hardness electrode is also known as the divalent cation electrode.

THEORY OF OPERATION

The potential developed across an ion-selective membrane is related to the ionic activity as shown by the Nernst equation:

$$E = \frac{2.3\,RT}{nF} \log \frac{a_1}{a_{int}} \quad\quad 1.33(1)$$

where

E is the potential developed across the membrane
a_1 is the activity of the measured ion in the process sample
a_{int} is the activity of the same ion in the internal reference solution
2.3 RT/nF is the Nernst slope, or slope of the calibration curve, and is a function of the absolute temperature T, the charge on the ion being measured n; R,* the gas law constant; and F,† the Faraday constant

Table 1.33b shows how the Nernst slope changes with temperature and with the charge on the ion. When the ratio of the two activities is unity, the potential across the membrane is zero.

TABLE 1.33b
Nernst Slopes (2.3 RT/nF)

Electrode Temperature		mV per Decade of Activity[a]	
°C	°F	n = ±1	n = ±2
0	32	54.19	27.10
10	50	56.17	28.08
20	68	58.17	29.08
25	77	59.16	29.58
30	86	60.15	29.58
40	104	62.15	29.58
50	122	64.12	32.03
60	140	66.10	33.05
70	158	68.09	34.04
80	176	70.07	35.04
90	194	72.05	36.02
100	212	74.04	37.02

[a] The slopes are positive for cations and negative for anions.

Equation 1.33(1) assumes that the membrane has identical selectivity properties on both sides. If for some reason this is not true, the equation is written as

$$E = E_{asy} + \frac{2.3\,RT}{nF} \log \frac{a_1}{a_{int}} \quad\quad 1.33(2)$$

---

* Gas Constant (R) is the amount of energy needed to change the temperature of a mole of gas by one degree. It is work per degree per mole. It may be expressed in any set of units representing work or energy (such as joules) and temperature (such as degrees Celsius, Kelvin, or Fahrenheit). When using Joules and K, the value of R is R = 8.314 Joules/(K mole).
† Faraday Constant (F) is the electric charge that is carried by one mole of electrons. It is expressed in Coulombs (C) per mole units and has a value of F = 96,500 C/mole.

where E_{asy} is the asymmetry potential and amounts to a few millivolts. This equation is simplified by the fact that a_{int} is fixed by the internal structure of the electrode, giving

$$E = E^{o\prime} + \frac{2.3\,RT}{nF} \log a_1 \quad\quad 1.33(3)$$

where $E^{o\prime}$ is a new constant.

A more complete equation takes also into account the electrode interferences and defines the selectivity coefficient. It is known as the Nikolsky equation:

$$E = E^{o\prime} + \frac{2.3\,RT}{nF} \log \left(a_i + k_{ij} a_j \right)^{z_i/z_j} \quad\quad 1.33(4)$$

where k_{ij} is the selectivity coefficient of interfering ion j with respect to measured ion i, each ion with charges z_i and z_j.

Reference Electrode

In actuality, the potential of a single electrode cannot be measured by itself. It can be measured only in conjunction with a reference electrode and a high-input impedance voltmeter (Figure 1.33a). The latter is necessary to prevent current from flowing through the electrode, an action that would tend to cause electrochemical reactions in the solution phase around the membrane.

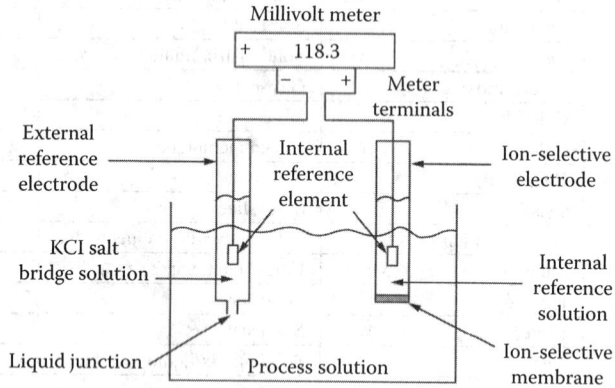

FIG. 1.33a
Ion-selective electrode measuring system.

The potential read on the voltmeter is equal to the algebraic sum of the potentials developed within the system. That is, the observed meter potential is the sum of the potentials developed by the measuring electrode, E; the reference electrode, E_{ref}; and a small but important liquid junction potential, E_j:

$$E_{meter} = E - E_{ref} + E_j \quad\quad 1.33(5)$$

Under normal operating conditions, the reference electrode is assumed to be constant, as is the liquid junction potential. However, this is not always the case. Substituting Equation 1.33(3) into Equation 1.33(5) and combining constant terms, including E_{ref} and E_j, gives the general form of the Nernst equation:

$$E_{meter} = E° + \frac{2.3\,RT}{nF} \log a_1 \quad \quad 1.33(6)$$

where $E°$ is a constant for a given electrode system at a specific temperature. It depends on the choice of reference electrode and includes the liquid junction potential.

The Nernst equation for the electrode pair can be written as an instrument input–output equation:

$$\text{Output} = A + B \log (\text{input}) \quad \quad 1.33(7)$$

where

The output is a millivolt signal to a meter
A is a zero adjustment
B is a span or slope adjustment around a temperature-independent, or isopotential, point

The input to the electrodes is the composition of the solution in terms of activity. Equation 1.33(6) states that the output of an electrode pair is linear with respect to the logarithm of the activity of the ion being measured (Figure 1.33b). The slope of the curve relating E_{meter} to $\log a_1$ is 59.16 mV (at 25°C for n = 1) or 29.58 mV (at 25°C for n = 2).

Concentration and Activity

Ignoring the effects of chemical reactions that would tie up ions, the activity of the ions is related to the analytical concentration, C, as shown in the following equation:

$$a = \gamma C \quad \quad 1.33(8)$$

where γ is the activity coefficient and is a measure of the interaction among ions in solution. It can be thought of as an empirical factor to explain the difference between the actual behavior of ions in solution and the ideal behavior. At zero ion concentration, that is, no ionic interaction, γ is taken as unity and the activity is equal to concentration. As the concentration increases, γ decreases at first, passes to a minimum value, and then rises, often to values greater than unity in very concentrated solutions.*

The activity coefficient is constant when the ionic composition of the solution is constant. Substituting Equation 1.33(8) into Equation 1.33(6) gives

$$E_{meter} = E + \frac{2.3\,RT}{nF} \log (\gamma C) \quad \quad 1.33(9)$$

or, at constant total ionic conditions, it gives

$$E_{meter} = E° + \text{constant} + \frac{2.3\,RT}{nF} \log C \quad \quad 1.33(10)$$

---

* Frankenthal, R. P., in *Handbook of Analytical Chemistry*, Meites, L., Ed., New York: McGraw-Hill, 1963, p. 1, Table 1-8.

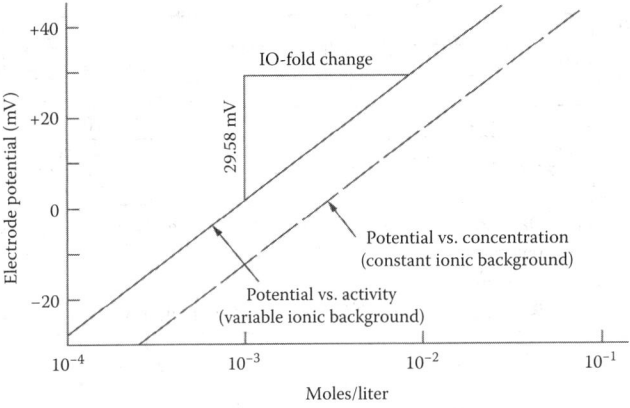

FIG. 1.33b
Electrode potential for calcium chloride solutions as a function of concentration and activity.

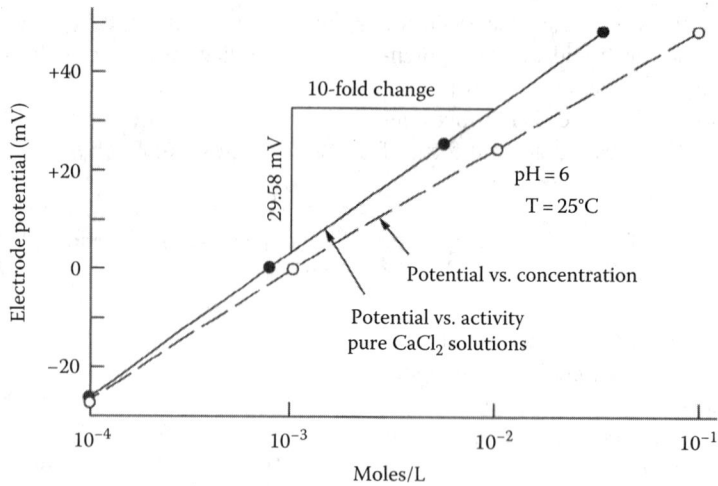

FIG. 1.33c
Concentration and ion activity vs. electrode potential.

where the constant term is $RT/nF \log \gamma$. Equation 1.33(10) is linear with respect to the concentration term (Figure 1.33b). However, if the ionic background of a solution varies, as in the preparation of a series of standards by dilution, the activity coefficient is no longer constant and Equation 1.33(9) is nonlinear (Figure 1.33c).

Equation 1.33(6) can predict the change in potential to be expected from a given change in activity or concentration (constant activity coefficient). For instance, if the activity changed twofold (100% change), then the potential developed is

$$E_{meter} = E° + \frac{2.3\,RT}{nF} \log 2a_1 \quad \quad 1.33(11)$$

Subtracting Equation 1.33(6) from Equation 1.33(11) gives

$$\Delta E = \frac{2.3\,RT}{nF} \log 2 \quad \quad 1.33(12)$$

TABLE 1.33c
Changes in Meter Potential for Changes of Activity[a]

σ_1/σ_2	ΔE to mV 25°C [77°F]		Equivalent pH Change (n = 1)
	n = 1°	n = 2°	
0.1	−60[b]	−30[b]	−1.0
0.25	−36	−18	−0.6
0.5	−18	−9	−0.3
0.79	−6	−3	−0.1
1.00	0	0	0.0
1.26	+6	+3	+0.1
2.00	+18	+9	+0.3
4.00	+36	+18	+0.6
10.00	+60[b]	+30[b]	+1.0

Note: Data are for positive ions. For negative ions, sign should be reversed.
[a] See Equation 1.33(12).
[b] Values rounded off from 59.16 and 29.58 (see Table 1.33b).

or 18 mV at 25°C for n = 1. A similar argument would show that for the same sample, an 18 mV decrease would be observed if the initial activity was cut in half (50% change).

This change is not dependent on the magnitude of a_1 and is the same whether measuring fluoride at the parts per million (ppm) level, chloride in 4% salt solutions, or the pH of a 15% sulfuric acid solution. Table 1.33c lists the changes in potential to be expected for up to a 10-fold change in activity. Column 1 shows the ratio of the original activity a_1 to the final activity a_2.

The last column shows the equivalent pH change if the [H⁺] were being measured. The data indicate that for precise measurements of small changes in activity, it is necessary to use an expanded-scale meter. For example, a span of 60 mV would allow a 10-fold change to be measured, using the full scale of the measuring instrument.

Ionic Strength Adjustment, Buffers

Introduction of the concept of the high-ionic-strength medium serves to remove or minimize two of the disadvantages of ISEs. The concentration, rather than the activity, may be interpreted directly from the observed electromotive force (emf) (see Equations 1.33(9) and 1.33(10)); some *chemical* and electrode interferences are removed.

The high-ionic-strength media are frequently designated by the acronym ISAB, which stands for ionic strength

TABLE 1.33d
Composition of Total Ionic Strength Adjustment Buffer, a High-Ionic-Strength Buffered Complexing Medium for Measuring Fluoride Ion Concentration

Sodium chloride	1.0 M
Acetic acid	0.25 M
Sodium acetate	0.75 M
Sodium citrate	0.001 M
Ionic strength	1.75 M
pH	5.0

Source: Frant, M.S. and Ross, J.S., Jr., *Anal. Chem.*, 40, 1169, 1968.

adjustment buffer. Table 1.33d illustrates the composition of total ionic strength adjustment buffer (TISAB). TISAB is used for rendering the fluoride ion electrode virtually specific for the measurement of the total concentration of fluoride ion in solution, even if the test solution has a pH outside of the acceptable range and fluoride-complexing ions such as iron or aluminum are present.

In spite of small contributions from the test solution, the high concentration of sodium chloride fixes the ionic strength at a virtually constant value. The acetate–acetic acid buffer holds the pH into the optimum range for the fluoride electrode, and the citrate complexes commonly interfering ions such as iron and aluminum more firmly than does the fluoride.

ISAB solutions for other ions have been described in the literature and are available commercially. To utilize ISAB solutions for continuous process measurements, reagent addition systems are employed that mix the test and ISAB solutions together, utilizing process pressure or peristaltic (or other types) pumps.

Temperature Effects*

There are three temperature effects on ion-selective measurements, including the T term in the Nernst equation (see Table 1.33b), the thermal characteristics of the electrodes, and the thermal characteristic of the solution. The T term in Equation 1.33(6) states that the potential produced by the electrode system is a function of temperature as well as of ion activity.

This effect can be compensated for, manually or automatically, by manipulating the input signal to the converter to indicate the true activity at the measured temperature. Temperature can also be compensated for by designing the electrode pair that there is a zero temperature error at a particular ion activity. Figure 1.33d shows this effect for the fluoride electrode used to control the fluoridation of public water supplies.

---

* Negus, L. E. and Light, T. S., Temperature coefficients and their compensation in ion-selective systems, *Instrumentation Technology*, 19, 23–26, 1972.

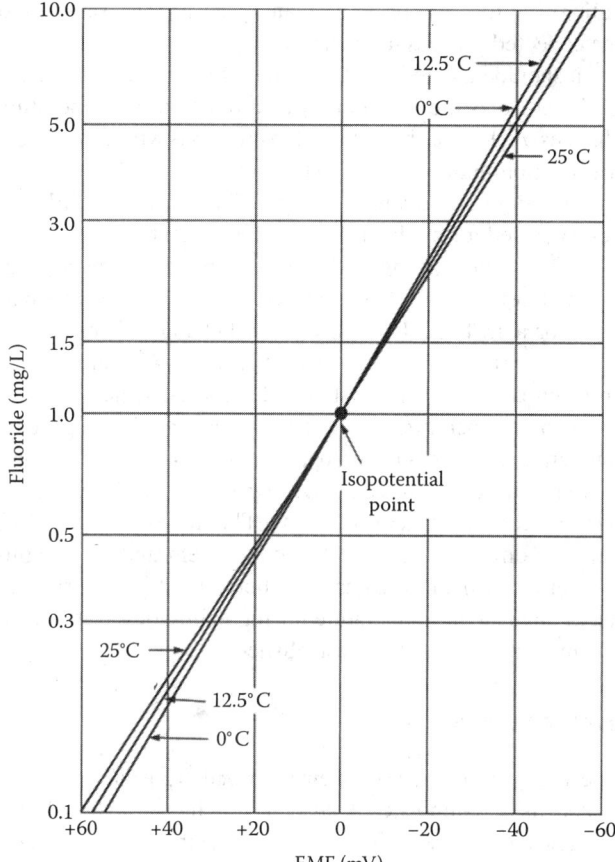

FIG. 1.33d
Isopotential point for fluoride electrode.

Isopotential Point The point at which the temperature curves intersect, 1 mg °F/l, is the control level for fluoridation. This point of intersection is called the isopotential point, and temperature effects are negligible on either side for a 10°C–15°C (18°F–27°F) change in temperature. The isopotential point for pH-measuring systems is normally about pH 7 and has a potential value of 0 mV.

The activity coordinate of the isopotential points of the solid-state electrodes is fixed during manufacture, whereas the millivolt coordinate is also dependent on the choice of reference electrode. However, due to the construction of the liquid-membrane electrode, the isopotential points can be changed to fit the process.

Electrode Internals and Calibration The second effect associated with temperature is created by the different internal thermal characteristics of the measuring and reference electrodes. This effect can be minimized if the internal elements of the measuring electrode are matched to the reference electrode. Most commercial pH systems employ matched internal elements for both electrodes.

The temperature effect on the chemistry of solutions is the third factor that can create an apparent error in measurement. This is difficult to quantify but can be offset by calibrating the system with preanalyzed process samples at the expected process temperature.

It should be noted that this is not a system error. The electrode indicates the true activity as a function of temperature changes. As long as the status of the process is what is required, the solution temperature effect is not important because the true activity is the quantity desired. This effect is usually not compensated for by the measuring instrumentation.

After sudden changes, time is required for a new state of thermal equilibrium to be reached (approximately one-half hour for a 18°F–27°F [10°C or 15°C] change). Therefore, it is important to remember that when electrodes are moved from process samples to standard samples, which are at a different temperature, or when sudden and large process temperature changes occur, one should wait until a new thermal equilibrium is reached. During this time, the potential of the electrode system will drift. The duration of the drift depends on the particular electrode system and the magnitude of the temperature change. Therefore, it is important to avoid changes in temperature during calibration or to allow thermal equilibrium to be established.

System Accuracy

The accuracy of measurements derived from an analytical system is a composite of all contributing variables. These variables for ion-selective measuring systems are the measuring electrode; the reference electrode, including the liquid-junction potential; the selective-ion potential converter; the recorder; and the temperature and solution errors.

The relationship between overall emf errors (ΔE) and ionic concentration (C) may be derived from the Nernst equation (1.33(6)) by substituting the values of the gas and Faraday constants R and F and assuming that the activity coefficient (γ) is unity in Equation 1.33(8) at 25°C:

$$\Delta E = \left(\frac{59.16}{n}\right) \log\left(\frac{1+RE}{100}\right) \quad \textbf{1.33(13)}$$

where RE represents the percent relative error in the concentration

$$RE = \frac{100 \Delta C}{C} \quad \textbf{1.33(14)}$$

A plot of Equation 1.33(13) is given in Figure 1.33e.

The relative error in measuring activity is dependent only on the *absolute* error in the emf and is independent of the activity range and of the size of the sample being measured. This is similar to an equal percentage valve in which equal incremental changes in valve opening (electrode potential) produce equal percentage changes in flow (RE in activity) for all valve openings—assuming constant differential pressure (constant temperature). Being a logarithmic device, an

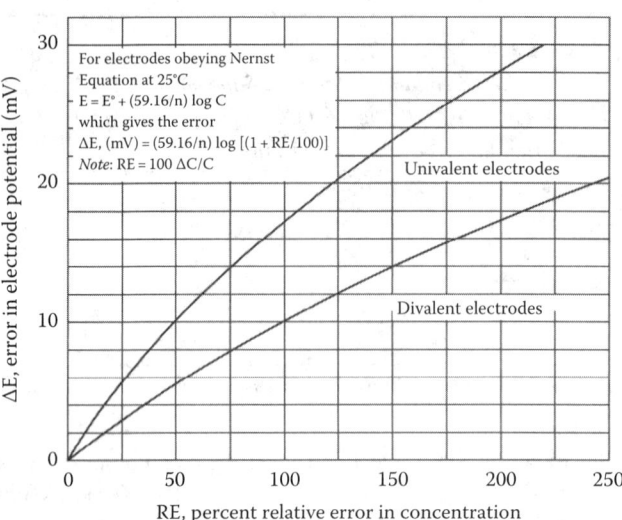

FIG. 1.33e
Relative error in concentration as a function of theoretical error in potential.

electrode gives a constant precision throughout its dynamic range. Concentrated solutions can be analyzed with the same accuracy as dilute solutions.

Laboratory Devices Laboratory measuring instruments for ISE measurements with an uncertainty of ± 0.1 mV are commercially available. It is possible to make laboratory pH measurements within an error of ± 0.002 pH units (equivalent to ± 0.12 mV). Similarly, under carefully controlled conditions, ISEs may be made repeatable within 0.1 mV. The accuracy attained to date in process instruments has been limited by the reference rather than by the ISEs.

Process Applications Ion-selective measurement systems for process applications are repeatable to ± 1 mV. For an electrode responding to univalent ions, an overall error of 1 mV corresponds to a 3.9% relative error in activity; for an electrode responding to divalent ions, the relative error is 7.8% per millivolt.

This means that they are roughly 5% within the value of activity, when measuring univalent ions, including H+, in acidic or basic solutions or 10% within that, when making divalent ion measurements. These figures apply only to direct electrode measurements. When electrodes are used as end point detectors in titrations or batch reactions or in differential systems, relative errors of 0.1% are possible.

ELECTRODE TYPES

Glass

ISEs are classified according to the type of sensing membrane employed. Glass electrodes are constructed from specially formulated glass and respond to ions by an ion exchange of mobile ions within the membrane structure. The membrane is fused to a glass body so that the outer surface makes contact with the sample or process stream, while the inner surface makes contact with an internal filling solution containing a constant activity of the ion for which the membrane is sensitive (Figure 1.33f).

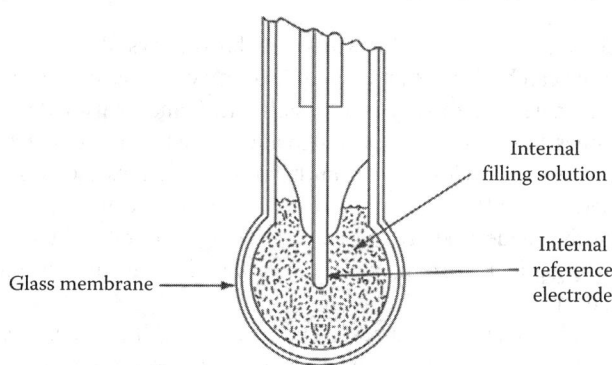

FIG. 1.33f
Conventional glass pH electrode.

A stable electrical contact is made with the internal solution by a silver wire coated with silver chloride. Other internal contacts have been used (mercury–mercurous chloride or thallium amalgam thallous chloride), but the silver–silver chloride is the most popular. The internal filling solution must contain a constant chloride ion activity and be saturated with silver chloride so that a stable potential is maintained at the metal salt–solution interface (Figure 1.33g).

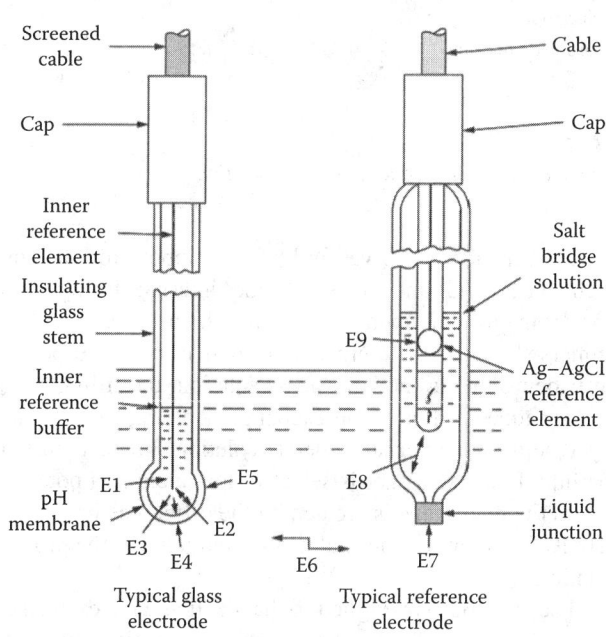

FIG. 1.33g
pH electrodes. Note: E1–E9 represent potentials developed between glass solution and reference electrode

In the conventional glass electrode, the internal solution is buffered at a pH of 7 and contains a chloride level similar to that in the external reference electrode (for more details on pH reference electrodes, see Chapter 1.55). Other glass electrodes in process use are the sodium, ammonium, and potassium ion electrodes.

In addition to the construction already described, the sodium electrode can also be prepared by slicing a thin section from a rod of sodium-sensitive glass and cementing it to an epoxy body (Figure 1.33h). This eliminates the familiar glass body of the pH electrode. Epoxy construction is not yet available for pH measurement due to difficulties inherent in cementing pH glasses to epoxy.

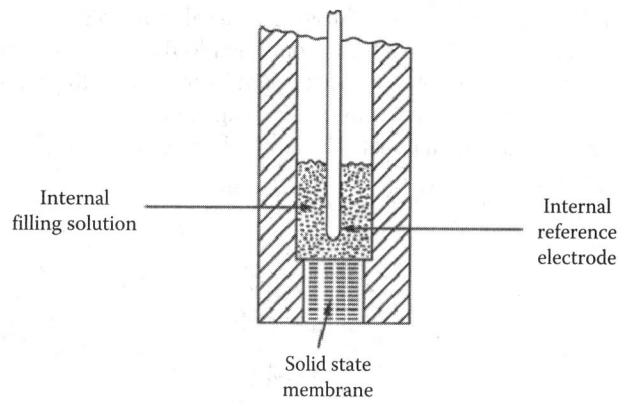

FIG. 1.33h
Solid-state sodium-sensitive membrane electrode.

A carbon dioxide and an ammonia electrode can be made from a pH electrode by covering the membrane with a permeable membrane sac filled with pH buffer. The respective gas in solution will selectively diffuse in or out of the permeable membrane, causing a pH change. The latter is dependent on the activity of the gas in the process solution.

Solid State

Solid-state electrodes are made of crystalline membranes, and there are scrupulous requirements for the size and charge of the mobile ions within the membrane. The composition of the membrane varies as a function of the required measurement. For instance, the fluoride electrode has a single crystal of doped lanthanum fluoride for a sensing membrane.

The silver and sulfide membranes are pressed pellets of insoluble silver sulfide. The small solubility of silver sulfide in solution prevents the coexistence of silver and sulfide ions, except in extremely small amounts, and the electrode can be used to measure either of these ions. Like the sodium electrode, these membranes are sealed in epoxy bodies (Figure 1.33h).

TABLE 1.33e
Solid-State Electrodes and Their Membrane Composition

Electrode	Membrane	Form
Fluoride	LaF_3	Single crystal
Silver/sulfide	Ag_2S	Pressed pellet
Chloride	AgX[a]	Single crystal
Bromide or iodide	$AgX–Ag_2S$	Pressed pellet
Cyanide	$AgI–Ag_2S$	Pressed pellet

[a] X = Cl, Br, or I.

Table 1.33e lists some of the commercially available solid-state electrodes and the composition of their sensing membranes.

Some pressed pellets and the single crystalline silver–salt membranes are capable of having a metal deposited on the surface and an electrical lead connected to the metal deposit (Figure 1.33i). A solid connector permits the use of the electrodes in any position without breaking electrical continuity. Also, there are no internal solutions to deteriorate with time or temperature.

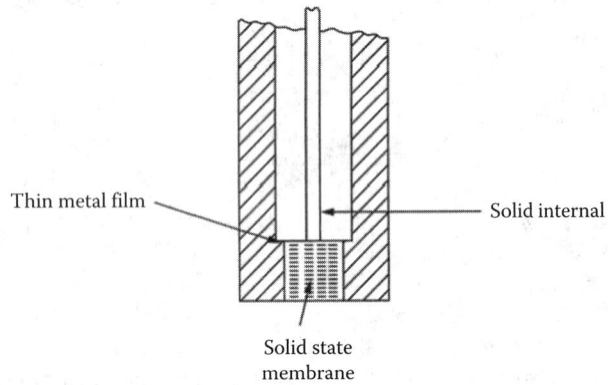

FIG. 1.33i
Solid-state membrane electrodes with solid internals.

Figure 1.33j shows a conventional silver wire or silver billet electrode that behaves identically to its corresponding solid-state electrode. However, small imperfections in the silver halide coating expose free silver metal to the process

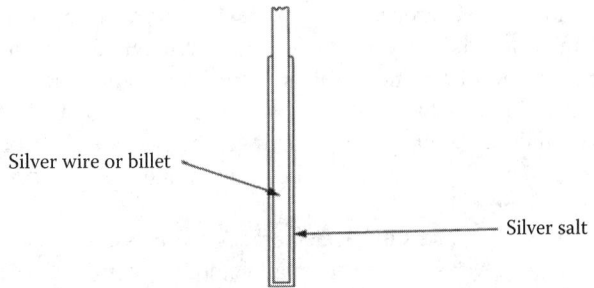

FIG. 1.33j
Conventional silver–silver electrode.

solution, thereby developing variable oxidation–reduction potentials, and these electrodes have not found wide use in industrial applications. When placed in clean, controlled environments, such as the inner-filling solution of ISEs, they produce stable reference potentials.

Liquid Ion Exchange

There are many ions for which no glass or crystalline membrane can be found that is suitable for process measurements. Fortunately, chemistry is a versatile field, and by using techniques familiar in ion exchange and solvent extraction technology, electrodes can be built for some of these ions. An inert hydrophobic membrane, such as a treated filter paper, can be made selective to certain ions by saturating it with an organic ion-exchange material dissolved in an organic solvent.

This feature requires a construction of the electrode, as shown in Figure 1.33k, which is a cross section of the tip of a liquid ion-exchange electrode. This electrode has two filling solutions, an internal aqueous filling solution in which the silver–silver chloride reference electrode is immersed and an ion-exchange reservoir of a nonaqueous water-immiscible solution, which *wicks* into the porous membrane.

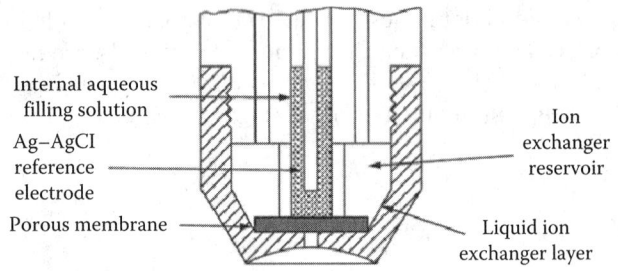

FIG. 1.33k
Divalent cation electrode tip in cross section.

The membrane serves only as a support for the ion-exchange liquid and separates the internal filling solution from the unknown solution in which the electrode is immersed. In effect, there is a sandwich, with the bottom layer being the unknown process solution, the filling being the nonaqueous liquid ion-exchange solution, and the top layer being the internal aqueous solution. For example, if the liquid ion exchanger is selective for calcium, a potential across the membrane is created by the difference in calcium activity between the internal filling solution and the process solution.

The electrode is designed so that the liquid ion exchanger, used as a sensing element, has a very small positive flow into the process stream. Therefore, liquid ion-exchange membrane electrodes require recharging with an ion exchanger.

Liquid ion-exchange membrane electrodes come in kit form. The kit contains an electrode body and sufficient ion exchanger, internal filling solution, and membranes to recharge the electrode many times. A single recharging should last several months in a properly designed system. Unlike the solid-state or glass electrodes, liquid-membrane electrodes cannot be used in nonaqueous solutions because they would dissolve the liquid ion exchanger. The body of the electrode is a chemical-resistant plastic.

Solid-state matrix electrodes are an improvement over liquid ion-exchange electrodes. The ion exchanger is permanently embedded in a plastic matrix, such as polyvinyl chloride, with a nonporous membrane. Rebuilding or solution replenishment is not required, and improved analytical performance and lower limits of detection are obtained.

MEASUREMENT RANGE

The upper limit of detection for ISEs is the saturated solution. However, due to the problems of making measurements with reference electrodes that have large liquid junction potentials (Chapter 1.55), the electrodes are specified as having an upper limit of 1 M. If the problems of large liquid junctions are brought under control, measurements can be made in saturated or nearly saturated solutions.

The lower limit of detection is usually determined by the solubility of the solid-state-sensing element or the liquid ion exchanger. The solution pH sometimes determines the lower limit of detection. Some dilute solutions are unstable, but activity measurements may be made if the solution is buffered with respect to the ion being measured, that is, if the free ion is in equilibrium with a relatively large excess of complexed ion. This is the case when free silver is measured in photographic emulsions or sulfide, cyanide, or fluoride in acid solutions.

INTERFERENCES

All ISEs are similar in principle of operation and use. They differ only in the details of the process by which the ion to be measured moves across the membrane and by which other ions are kept away. Therefore, a discussion of electrode interferences will have to be in terms of the membrane materials.

The glass electrodes and the solid matrix/liquid ion-exchange electrodes both function by an exchange of mobile ions within the membrane, and ion-exchange processes are not specific. Reactions will occur among many ions with similar chemical properties, such as the alkali metals, alkaline earths, or transition elements. Thus, a number of ions may produce a potential when a given ISE is immersed in a solution.

Even the pH glass electrode will respond to sodium ions at a very high pH (low hydrogen ion activity). Fortunately, an empirical relationship can predict electrode interferences, and a list of selectivity ratios for the interfering ions can be obtained by consulting the manufacturers' specifications or the chemical literature.

Solid–state matrix electrodes are made of crystalline materials, and interferences resulting from ions moving into the solid membrane are not to be expected. Interference is usually by a chemical reaction with the membrane. One, which is observed with the silver–halide membranes (for chloride, bromide, iodide, and cyanide activity measurements), involves reaction with an ion in the sample solution, such as sulfide, to form a more insoluble silver salt. As already mentioned, specific details of electrode side reactions can be found in the manufacturers' specifications and chemical literature.

Solution Interferences

A true interference is one that produces an electrode response that can be interpreted as a measure of the ion of interest. For example, the hydroxyl ion, OH^-, causes a response with the fluoride electrode at fluoride levels below 10 ppm. In addition, the hydrogen ion, H^+, creates a positive interference with the sodium ion electrode.

Often an ion will be regarded as interfering if it reduces the activity of the ion of interest through chemical reaction. It is true that this reaction (complexation, precipitation, oxidation–reduction, and hydrolysis) results in an activity of the ion that differs from the concentration of the ion by an amount greater than that caused by ionic interactions. However, the electrode is still measuring the true activity of the ion in the solution.

An example of solution interference will illustrate this point. Silver ion in the presence of ammonia forms a stable silver–ammonia complex that is not measured by the silver. Only the free, uncombined silver ion is measured. The total silver ion may be obtained from calculations involving the formation constant of the silver–ammonia complex and the fact that the total silver is equal to the free silver plus the combined silver. Alternately, a calibration curve can be drawn relating to the total silver (from analysis or sample preparation) to the measured activity. The ammonia is *not* an electrode interference.

Most of the confusion stems from the fact that analytical measurements have been in terms of concentration without regard to the actual form of the material in solution and electrode measurements often disagree with the laboratory analyst's results. However, the electrode reflects what is actually taking place in the solution at the time of measurement. This may be far more important in process applications than the more classic information. With some of the techniques suggested, the two measurements are often reconciled.

CALIBRATION

Calibration solutions for ion-selective systems are not normally buffered to resist changes, as are the standard solutions for pH systems. They can therefore be affected by dilution, evaporation, oxidation, or contamination by foreign matter in the process fluid. Thus, more care must be taken in preparing and handling these solutions than is generally needed in a typical pH application. Attention should be paid to eliminate carryover from one test solution to another from distilled water rinses.

Calibration solutions should be prepared in accordance with accepted principles of analytical chemistry. Many common chemical standards are available as stock solutions from laboratory supply houses. Generally, only solutions at a reasonably high concentration level (greater than 0.01 M or 100 ppm) should be made for storage. Serial dilutions of these stock solutions should be made at the time of use because very dilute solutions are particularly likely to lose some of their ions by absorption of the walls of the storage vessels. Use of high-grade plastic storage bottles is recommended.

Table 1.33f lists some solutions frequently used to check the performance of an ion-selective measuring system. When the ionic background is held constant (sulfide, chloride cyanide, and pH), the potential difference between two of these solutions is Nernstian (see Table 1.33b for data on the temperature sensitivity of Nerstian slopes). For others, the potential difference should be normalized to decide changes in activity.

To achieve the utmost in accuracy and meaningful measurements, the ion-selective measuring system should be standardized, and thus optimized, in a solution carefully chosen to be chemically similar to the process solution at the point of prime interest. This solution should be at a stable temperature near the actual process temperature ($\pm 3.6°F$ [$\pm 2°C$]). A grab sample of the process solution analyzed in the laboratory may be the best and most convenient standard to use.

TABLE 1.33f
Calibrating Solutions for Ion-Selective Electrodes

Electrode	Chemical Composition	Ionic Concentration	Approximate Ion Activity	Approximate emf vs. 1.0 M KCl, AgCl, Ag Reference at 77°F (25°C)
Hydrogen (pH)	0.05 M KH phthalate	4.008 pH buffer at 25°C	$10^{-4.008}$ M H$^+$/L	+143 (+178)[a]
	0.025 M KH$_2$PO$_4$ + 0.025 M + Na$_2$HPO$_4$	6.86 pH buffer at 25°C	$10^{-6.86}$ M H$^+$/L	−35 (0)[a]
	0.01 M borax	9.18 pH buffer at 25°C	$10^{-9.18}$ M H$^+$/L	−149 (−114)[a]
Fluoride	22.10 mg NaF/L	10.0 mg F$^-$/L	9.8 mg F$^-$/L	−59
	2.21 mg NaF/L	1.0 mg F$^-$/L	1.0 mg F$^-$/L	0.0
Chloride	1.00×10^{-2} M KCl in 1.00 M KNO$_3$	1.00×10^{-2} M Cl$^-$	0.61×10^{-2} M Cl$^-$	+118
	1.00×10^{-3} M KCl in 1.00 M KNO$_3$	1.00×10^{-3} M Cl$^-$	0.61×10^{-2} M Cl$^-$	+177
Silver	1.00×10^{-2} M AgNO$_3$	1.00×10^{-2} M Ag$^+$	0.90×10^{-2} M Ag$^+$	+443
	1.00×10^{-3} M AgNO$_3$	1.00×10^{-3} M Ag$^+$	0.96×10^{-3} M Ag$^+$	+385
Sulfide	1.00×10^{-1} M Na$_2$S in 1.00 M NaOH	1×10^{-1} M S$^=$	0.15×10^{-1} M S$^=$	−860
	1.00×10^{-3} M Na$_2$S in 1.00 M NaOH	1×10^{-3} M S$^=$	0.15×10^{-3} M S$^=$	−800
Cyanide	1.00×10^{-3} M NaCN in 1.00×10^{-1} M NaOH	1.00×10^{-3} M CN$^-$	0.76×10^{-3} M CN$^-$	−192
	1.00×10^{-4} M NaCN in 1.00×10^{-1} M NaOH	1.00×10^{-4} M CN$^-$	0.76×10^{-3} M CN$^-$	−133
Water hardness	1.00×10^{-2} M CaCl$_2$	1.00×10^{-2} M Ca^{++} or 1000 mg/L as CaCO$_3$	0.55×10^{-2} M Ca^{++} or 550 mg/L as CaCO$_3$	+34
	1.00×10^{-4} M CaCl$_2$	1.00×10^{-4} M Ca^{++} or 10 mg/L as CaCO$_3$	0.92×10^{-4} M Ca^{++} or 9.2 mg/L as CaCO$_3$	−18

[a] vs. 4 M KCL, AgCl, Ag reference electrode at 25°C.

Multiple Electrode Heads

Multiple electrode heads can accommodate up to six ion-selective measuring electrodes in combination with a single (common) reference electrode, for use in simultaneous multicomponent analysis and in combination with four- or eight-channel computer interfaces. The advantage of this design is operating cost reduction, because multiple measurements can be made while using a common head and connecting cables. In addition, because the reference electrode is more likely to fail and also because it is less expensive than the measuring electrodes, operating costs can be much reduced if the measuring electrodes are not replaced every time the reference electrode fails (Figure 1.33l).

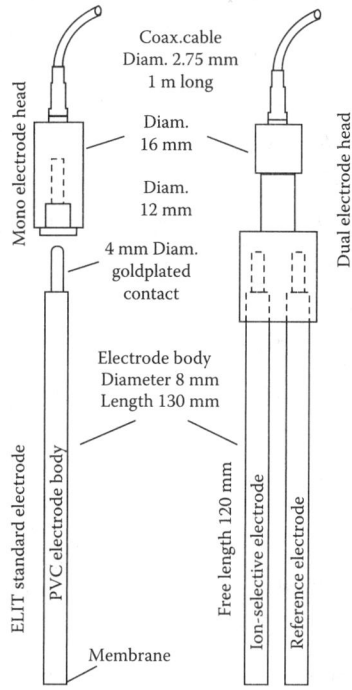

FIG. 1.33l
Multiple electrode heads are available for connecting up to six ISE electrodes with a single reference electrode, for use in simultaneous multicomponent analysis using four- or eight-channel computer interfaces. (Courtesy of Rundle, C. C. (2006), A Beginners Guide to Ion-Selective Electrode Measurements, http://www.nico2000.net/Book/Guide1.html.)

ADVANTAGES

Compared to other composition-measuring techniques, such as photometric, titrimetric, chromatographic, or automated-classic analysis, the ISE measurement has an impressive list of advantages. An electrode measurement is simple, rapid, nondestructive, direct, and continuous, and, therefore, easily applied to closed-loop process control. In this respect, it is similar to using a thermocouple for temperature control. Electrodes can also be used in opaque solutions and viscous slurries. In addition, the electrodes measure the free or active ionic species in the process, under process conditions, and consequently the status of a process reaction.

POTENTIAL DISADVANTAGES

However, there are several disadvantages. The specificity of ISEs is not quite as good as that of the glass pH electrode. Interferences vary from minor to major; the literature and manufacturers' data on limitations need to be consulted for each electrode. Also, the electrodes do not measure the total concentration of ions, the parameter that is often requested. The reason is that prior to the introduction of electrodes, concentration information was the only information available from the chemist due to his or her classic measurement techniques. Control laboratory chemists and process engineers are not accustomed to thinking in terms of activity, even when making pH measurements (Chapter 1.55). This habit may well disappear as this new ion-selective technique becomes more popular.

There are some manufacturers that have begun packing the electrodes separately and installing them in a junction head. This allows the ISEs to be independently replaced and stored dry. The reference electrodes are also independently replaced and stored wet. This allows the best of both worlds for storage life and cost of replacement.

However, there are times when concentration is a desirable measurement, for example, material balance calculations or pollution control. The knowledge of material balance allows a prediction as to where a process reaction will be in the future. This information is necessary if a process is to be controlled by introducing changes that will nullify those predicted.

In pollution control, it is generally accepted that many ions, even in the combined state, are detrimental to life forms. Fluorides, cyanides, and sulfides, to name a few, are harmful to fish and humans in many combined forms. Yet, their sum total is not measured, but they are detected individually by ISEs. Consequently, pollution control agencies usually require concentration information.

The electrodes can be used for this purpose if they are calibrated with solutions matching the process or with ISAB solutions (see later). If this is not satisfactory, an electrode can be used for online control, and separate grab samples can be analyzed by other procedures to obtain the information needed to comply with regulations.

Another disadvantage derives from a misunderstanding about precision and accuracy. Many classic analytical techniques name a relative error of $\pm 0.1\%$. ISEs name relative errors of $\pm 4\%$–8% (see "System Accuracy" section earlier in this chapter). In terms of pH, this is equivalent to a measurement of ±0.02 pH units—ordinarily regarded as a satisfactory measurement.

When used with some degree of understanding, ISEs can supply satisfactory composition information and afford closed-loop control that was previously unattainable. When in doubt, the user should consult with the electrode manufacturers or his or her own analytical chemists.

It is evident from the preceding discussion that ISE measurements and pH measurements using the glass electrode are not only identical in theory but also similar in practice. The electrodes are generally the same size and fit the pH holder assemblies.

Electrical insulation of the electrodes is as important as with the glass electrode. In addition, the electrodes are subject to the same fouling by oils and slimes as glass electrodes and can, in general, be cleaned using methods already proven for the pH electrodes. Because of these similarities, discussion of the application of ISEs is omitted here.

SPECIFICATION FORMS

When specifying Ion Selective Analyzers only, one can use the ISA form 20A1001 and when specifying both the analyzer and the composition or the properties of the process, which it will be monitoring, use the ISA form 20A1002. Both forms are reproduced with the permission of the International Society of Automation on the next pages. In case of specifying ORP pH detectors, refer to the specification forms at the end of Chapters 1.37 and 1.55.

ANALYSIS DEVICE — Operating Parameters

#	RESPONSIBLE ORGANIZATION		#	SPECIFICATION IDENTIFICATIONS	
1	ISA		6	Document no	
2			7	Latest revision	Date
3			8	Issue status	
4			9		
5			10		

#	ADMINISTRATIVE IDENTIFICATIONS			#	SERVICE IDENTIFICATIONS Continued			
11				40				
12	Project number		Sub project no	41	Return conn matl type			
13	Project			42	Inline hazardous area cl		Div/Zon	Group
14	Enterprise			43	Inline area min ign temp		Temp ident number	
15	Site			44	Remote hazardous area cl		Div/Zon	Group
16	Area	Cell	Unit	45	Remote area min ign temp		Temp ident number	
17				46				
18	SERVICE IDENTIFICATIONS			47				
19	Tag no/Functional ident			48	COMPONENT DESIGN CRITERIA			
20	Related equipment			49	Component type			
21	Service			50	Component style			
22				51	Output signal type			
23	P&ID/Reference dwg			52	Characteristic curve			
24	Process line/nozzle no			53	Compensation style			
25	Process conn pipe spec			54	Type of protection			
26	Process conn nominal size		Rating	55	Criticality code			
27	Process conn termn type		Style	56	Max EMI susceptibility		Ref	
28	Process conn schedule no		Wall thickness	57	Max temperature effect		Ref	
29	Process connection length			58	Max sample time lag			
30	Process line matl type			59	Max response time			
31	Fast loop line number			60	Min required accuracy		Ref	
32	Fast loop pipe spec			61	Avail nom power supply			Number wires
33	Fast loop conn nom size		Rating	62	Calibration method			
34	Fast loop conn termn type		Style	63	Testing/Listing agency			
35	Fast loop schedule no		Wall thickness	64	Test requirements			
36	Fast loop estimated lg			65	Supply loss failure mode			
37	Fast loop material type			66	Signal loss failure mode			
38	Return conn nominal size		Rating	67				
39	Return conn termn type		Style	68				

#	PROCESS VARIABLES	MATERIAL FLOW CONDITIONS			#	PROCESS DESIGN CONDITIONS		
69					101			
70	Flow Case Identification			Units	102	Minimum	Maximum	Units
71	Process pressure				103			
72	Process temperature				104			
73	Process phase type				105			
74	Process liquid actl flow				106			
75	Process vapor actl flow				107			
76	Process vapor std flow				108			
77	Process liquid density				109			
78	Process vapor density				110			
79	Process liquid viscosity				111			
80	Sample return pressure				112			
81	Sample vent/drain press				113			
82	Sample temperature				114			
83	Sample phase type				115			
84	Fast loop liq actl flow				116			
85	Fast loop vapor actl flow				117			
86	Fast loop vapor std flow				118			
87	Fast loop vapor density				119			
88	Conductivity/Resistivity				120			
89	pH/ORP				121			
90	RH/Dewpoint				122			
91	Turbidity/Opacity				123			
92	Dissolved oxygen				124			
93	Corrosivity				125			
94	Particle size				126			
95					127			
96	CALCULATED VARIABLES				128			
97	Sample lag time				129			
98	Process fluid velocity				130			
99	Wake/natural freq ratio				131			
100					132			

#	MATERIAL PROPERTIES			#	MATERIAL PROPERTIES Continued		
133				137			
134	Name			138	NFPA health hazard	Flammability	Reactivity
135	Density at ref temp		At	139			
136				140			

Rev	Date	Revision Description	By	Appv1	Appv2	Appv3	REMARKS

Form: 20A1001 Rev 0 © 2004 ISA

Analytical Measurement

	RESPONSIBLE ORGANIZATION	ANALYSIS DEVICE COMPOSITION OR PROPERTY Operating Parameters (Continued)		SPECIFICATION IDENTIFICATIONS		
1			6			
2	ISA		7	Document no		
3			8	Latest revision		Date
4			9	Issue status		
5			10			

	PROCESS COMPOSITION OR PROPERTY			MEASUREMENT DESIGN CONDITIONS				
	Component/Property Name	Normal	Units	Minimum	Units	Maximum	Units	Repeatability
11								
12								
13								
...								
37								

(rows 38–75 blank)

Rev	Date	Revision Description	By	Appv1	Appv2	Appv3	REMARKS

Form: 20A1002 Rev 0 © 2004 ISA

Definitions

Activity (a) a is the chemical activity for the relevant species, where a_{Red} is the reductant and a_{Ox} is the oxidant. $a_X = \gamma_X c_X$, where γ_X is the activity coefficient of species X. (Since activity coefficients tend to unity at low concentrations, activities in the Nernst equation are frequently replaced by simple concentrations.)

Crystalline membranes: Crystalline membranes have good selectivity, because only ions that can introduce themselves into the *crystal structure* can interfere with the *electrode* response. Selectivity of crystalline membranes can be for both *cation* and *anion* of the membrane-forming substance.

Faraday constant (F) is the electric charge that is carried by one mole of electrons. It is expressed in Coulombs (C) per mole units and has a value of F = 96,500 C/mole.

Gas constant (R) is the amount of energy needed to change the temperature of a mole of gas by one degree. It is denoted as Rg when referring to a particular gas and as Ra when referring to air. It is work per degree per mole. It may be expressed in any set of units representing work or energy (such as joules) and temperature (such as degrees Celsius, Kelvin or Fahrenheit). When using Joules and °K, the value of R is R = 8.314 Joules/(°K mole).

Glass membranes: *Glass* membranes are made from an ion-exchange type of glass, having good selectivity for H^+, Na^+, Ag^+, Pb^{2+}, and Cd^{2+}. The glass membrane has excellent chemical durability and can work in very aggressive media. A very common example of this type of electrode is the *pH glass electrode*.

Ion-exchange resin membranes: *Ion-exchange resins* are based on special organic *polymer* membranes that contain a specific ion-exchange substance (resin). This is the most widespread type of ion-specific electrode. They are also the most widespread electrodes with anionic selectivity. However, such electrodes have low chemical and physical durability as well as *survival time*.

Ion-selective electrode (ISE): It is a sensor that converts the activity of a specific ion dissolved in a solution into an electrical potential, which generates a voltage that is theoretically dependent on the logarithm of the ionic activity, according to the Nernst equation.

Abbreviations

ISAB   Ionic strength adjustment buffer
ISE   Ion-selective electrodes
SIE   Specific ion electrodes
TISAB   Total ionic strength adjustment buffer

Bibliography

Al Attas A. L., Construction and Analytical Application of Ion Selective Bromazepam Sensor (2009) http://www.electrochemsci.org/papers/vol4/4010020.pdf.

Applicon Analytical, Ion selective electrodes (2005) http://www.metrohm-applikon.com/Downloads/Support_BTL5.pdf.

Bailey, P. L., *Analysis with Ion-Selective Electrodes* (2003) http://onlinelibrary.wiley.com/doi/10.1002/anie.197708041/abstract.

Bergveld, P. and Sibbald, A., *Analytical and Biomedical Applications of Ion-Selective Field-Effect Transistors (ISFETS)*, Vol. 23, New York: Elsevier, http://www.amazon.com/Applications-Ion-Selective-Field-Effect-Transistors-Comprehensive/dp/044442976X.

Chemical Sensors Research Group (CSRG), Ion-selective electrodes (2006), http://csrg.ch.pw.edu.pl/tutorials/ise/.

Covington, A. K., Ion-Selective Electrodes Chemistry Chasette, http://www.rsc.org/images/21_Ion-selective_Electrodes_tcm18-29983.pdf.

Evans, A., *Potentiometry and Ion-Selective Electrodes*, New York: John Wiley & Sons, http://www.amazon.com/Potentiometry-Ion-selective-Electrodes-Analytical-Chemistry/dp/0471913928.

Freiser, H., Ed., *Ion-Selective Electrodes in Analytical Chemistry*, Vols. 1 and 2, New York: Plenum Press, 1978 and 1980.

Koryta, J., *Ion-Selective Electrodes*, London, U.K.: Cambridge University Press, http://www.annualreviews.org/doi/abs/10.1146/annurev.ms.16.080186.000305?journalCode=matsci.1.

Libert, A., Anion determination with ion selective electrodes using Gran's plots. Application to fluoride (1982) http://www.sciencedirect.com/science/article/pii/0165993682800629.

Linder, E., Toth, K., and Pungor, E., *Dynamic Characteristics of Ion-Selective Electrodes*, Boca Raton, FL: CRC Press, 1988.

Martin, C. R., Current trends in ion-selective electrodes (1982) http://www.sciencedirect.com/science/article/pii/0165993682800629.

New Mexico State University. Ion selective electrodes (2006) http://www.chemistry.nmsu.edu/Instrumentation/IS_Electrod.html.

Rundle, C. C. (2006), *A Beginners Guide to Ion-Selective Electrode Measurements*, http://www.nico2000.net/Book/Guide1.html.

YSI. Primer on ion selective measurement (2012) http://www.ysi.com/media/pdfs/ba76001-Online-ISE-Primer-e01.pdf.

1.34 Leak Detectors

R. K. KAMINSKI (1982)   **B. G. LIPTÁK** (1995, 2003, 2017)

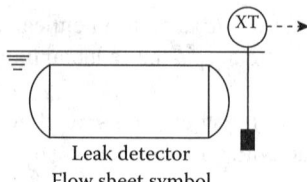

Leak detector
Flow sheet symbol

Types	1. Aboveground detectors
	1A. Pressurization
	1B. Bubbles, paints, etc.
	1C. Combustible and toxic
	1D. Ultrasonic
	1E. Thermal conductivity
	1F. Halogen leak detectors
	1G. Others (vacuum, thermograph, etc.)
	2. Underground leak detectors
Sensitivity	1A Variable
	1B 10^{-4} cm^3/s
	1C See chapters 1.14 and 1.26
	1D Variable
	1E 10^{-5} cm^3/s; see Chapter 1.66
	1F 10^{-7} cm^3/s
Costs	Case of bubbling foam bottles from $50.
	Portable, intrinsically safe combustible gas detector with 100 ppm sensitivity or portable toxic gas monitor: $250–$500.
	Ultrasonic leak detector $1500.
	Portable thermal conductivity leak detector with 10^{-5} cm^3/s sensitivity: $1600.
	Halogen leak detectors about $3000.
	Thermography sensors are around $10,000.
	Mass spectrometer is on average $25,000, but their cost can range from $3,000 to $150,000, depending on application complexity and accessories. Underground leak detector systems are available for single- and multiple-point gas monitoring applications (about $2,500 for single point and $10,000 for up to 12 points).
	Note: For a complete list of hazardous and toxic leak detector suppliers refer to Table 1.26a.
Partial list of suppliers	Acoustic Systems Inc. (1D) (http://www.wavealert.com/)
	American Gas and Chemical Co. (1B) http://www.amgas.com/ldpage.htm
	Barcharach (1C) (www.bacharach-inc.com) http://www.bacharach-inc.com/leak-detection.htm
	Cole-Parmer Instrument Co. (1B, 1C, 1E) (www.coleparmer.com) http://www.coleparmer.com/Category/Leak_Detectors/7740 http://www.coleparmer.com/buy/product/54068-uei-combustible-gas-leak-detector.html
	CTRL Systems Inc. (1D) http://www.ctrlsys.com/products/leak_detection.php
	Davis Instrument (1D) (www.davis.com) http://www.davis.com/Search/leak%20detectors
	Delphian Corp. (1C) (http://www.amgas.com/ldpage.htm)
	Detcon Inc. (1C) (http://www.detcon.com/3g-portable_gas_sensors.htm)
	FCI (1E) http://www.fluidcomponents.com/Industrial/Products/LevelSwitches/ProdLevelSwitches.asp
	Foxboro-Invensys Co. (1C,E) (http://resource.invensys.com/instrumentation/documentation/eib/pad/p4100021.pdf)
	Furness Controls (www.furness-controls.com) http://www.furness-controls.com/products/leak-detection/
	Gow-Mac Instrument Co. (www.gow-mac.com) http://www.gow-mac.com/products/prod_cat.cfm?cat_id = 7
	International Sensor Technology (1C) (http://www.intlsensor.com/portables.html)
	Laco Technologies (1F) (www.lacotech.com) http://www.lacotech.com/leaktesting/leaktestingsystems.aspx
	LDS Vacuum Products (1G) (www.lds-vacuum.com) http://vacuumshopper.stores.yahoo.net/leakdetectors.html
	Leybold Inficon Inc. (1F) (www.inficon.com) http://products.inficon.com/en-us/nav-products/Category/ProductGroup/pg_LeakDetectors?path = Products%2F

(Continued)

MKS Technology (1F) (www.mksinst.com) http://www.mksinst.com/product/Catalog.aspx?catalogID = 59
MSA Instrument Div. (1C, 1D) (http://us.msasafety.com/Fixed-Gas-%26-Flame-Detection/Gas-Detectors/ UltraSonic%26trade%3B-EX-5-Gas-Leak-Detector/p/000140008300001002)
Preferred Rimcor Instruments (www.preferred-mfg.com) http://www.preferred-mfg.com/products/pu/ HD-A2C-Oil-Water-Leak-Detector
Purpora Engineering (Pipe tester) http://purporaengineering.com/index.php?option=com_content&view=article&id=53&Itemid=86
RAE Systems (1C) http://www.raesystems.com/products/toxirae-pro
RKI Instruments (1C, 2) (http://www.rkiinstruments.com/pages/products.htm)
Scott Specialty Gases (1B) (www.scottgas.com) http://www.alspecialtygases.com/Prd_gas_detection.aspx
Severn Trent Services (1C) (www.severntrentservices.com) http://www.severntrentservices.com/Oil___Gas/Gas_Leak_Detectors_prodc_485.aspx
Sierra Monitor Corp. (1C) (www.sierramonitor.com) http://www.sierramonitor.com/gas/about/Fire-Gas-Detection.php http://www.sierramonitor.com/gas/IT/Toxic_gas_detector.php?gclid=CIuR0djr-rkCFewDOgodOCMA7w
SubSurface Leak Detection (1F) (http://www.subsurfaceleak.com/ld12_prod_pg.html)
Teledyne Analytical Instruments (1C) (www.teledyne-ai.com) http://www.teledyne-ai.com/searchresults.asp?cx = 011468841755678738939%3Aci-ajqagd-0&cof = FORID%3A9&q = leak+detectors&sa = Search
TM Electronics (1S, 1B, 1F) (www.tmelectronics.com) http://www.tmelectronics.com/product_list.htm#1;http://www.tmelectronics.com/product_list.htm#4;
U.E. Systems Inc. (1D) (www.uesystems.com) http://www.uesystems.com/new/?s = leak+detectors
Uson LP (1C) (www.uson.com) http://www.uson.com/leak-testers/

INTRODUCTION

This chapter describes a variety of sensors that are used in leak detection applications, including the measurement of leakage through valves and steam traps, tightness of aboveground tanks or pipes, and the measurement of tank leakage underground. Other chapters of this volume also discuss leakage detectors that serve specific tasks that are unique to specific industries, such as methane leakage in the process of drilling for oil or gas (Chapter 1.28) and hydrogen leakage in the nuclear industry (Chapter 1.30). The different sensors that are referred to in this chapter are also described in more detail elsewhere, such as

Chapter 1.14—Combustibles
Chapter 1.30—Hydrogen
Chapter 1.35—Mass Spectrometer
Chapter 1.45—Odor
Chapter 1.66—Thermal Conductivity
Chapter 1.69—Visible

When a particular chemical is leaking into the atmosphere in an industrial or laboratory environment, the sensor used to detect its ambient concentration can be a combustible or toxic gas analyzer or any of the long list of other analyzers described in this handbook. In this chapter, the emphasis is on those types of leak detectors that detect the actual leak as it occurs.

Chemical materials and fuels that leak into the atmosphere represent both a financial loss and a resource loss. Fluid leaks also waste energy because energy goes into the shipment as well as the processing of chemicals. In addition, the safety aspects of leak detection are extremely important because the escape of toxic, flammable, or explosive materials can cause accidents. See Table 1.34a for an orientation of a limited number of leak detection applications and sensor types that are suited for those applications.

TABLE 1.34a
Orientation Table for Leak Detectors

Application	Detector Options
Gas leaks, general	Handheld gas detectors
	Personnel protection indicators
	Wall-mounted gas detector
Water leaks	ColorMetrics
	Tracer dyes
Small parts	Immersit immersion fluid
	Tracer dyes
Unpressurized pipes and containers	ColorMetrics
	Ultrasonic leak detector
	Tracer dyes
Pressurized pipes and containers	Liquid leak detector
	Ultrasonic leak detector
Large parts	Liquid leak detector
Large containers	Hydrostatic leak testing
	Immersit immersion fluid
	Handheld gas detectors
	Ultrasonic leak detector
	ColorMetrics
	Tracer dyes
	Inspection penetrants
Oil/fuel leaks	ColorMetrics
	Tracer dyes
	Fuel Guard for AFVs

ABOVE GROUND DETECTION

Pressure- and Vacuum-Based Testing

One obvious method of leak detection is to evacuate the space inside the test object and measure if the vacuum leaks out (pressure rises) over time.

Pressurization by pneumatic or hydraulic means is probably the most widely used method in industry. The fall of a pointer on a pressure gauge can indicate a leak.

Various standards and codes explain specific procedures for piping, pneumatic instrumentation systems, and vessels. The recommended practice of the National Fire Protection Association goes into considerable detail on how to handle underground leaks. American Petroleum Institute Standard 527 describes a method for checking the leakage at the seats of safety relief valves.*

Test equipment is available for pressure testing, including special fittings for plugging off lines and testing joints.

Paints, Dyes, Bubbling Commercial formulations are available that will bubble or foam at the point of a leak. Safe materials are available for different chemical applications. Once the existence of a leak has been proved by pressurization, the bubble emission technique can help to isolate its location.

Aerosols, paints, and papers that change color due to chemical reaction are used in much the same way as the materials that bubble and foam. They are applied to the exact area that is being tested, and the color change spots the leak.

In detecting steam leakage from steam traps, a tape is fastened to the steam trap's outlet piping. If a spot on the tape remains silver, the trap is working. If live steam passes through the trap continuously, the spot turns black because of overheating.

Dyes can be used to label sewage steams to check flows and seepage areas. One company has biodegradable, fluorescent dyes that are still visible after dilution to 1 ppm in water.

Phosphorescent powders can be useful for identifying defective bags and other leaks in dust collectors. The powder is put into the dirty side of the collector, and it works its way through the system in a few minutes. After that time, the inside of the collector is checked with an ultraviolet light to find leakage points that are visible due to the glowing of the powder.

White smoke can be used to locate leaks in sewers, ducts, and other places that cannot be pressurized. Candles, bombs, or smoke generators with blowers produce the smoke that emerges at points of leakage.

Combustible or Toxic Leaks

If a liquid is toxic or flammable, a wide variety of analyzers can be considered to detect if it is leaking out into the atmosphere. These detectors have been described in detail in Chapters 1.14 and 1.26. These instruments can use infrared (IR), colorimetric, or a dozen other principles of operation. They range from handheld monitors (Figure 1.34a) to continuous, permanently installed sensors.

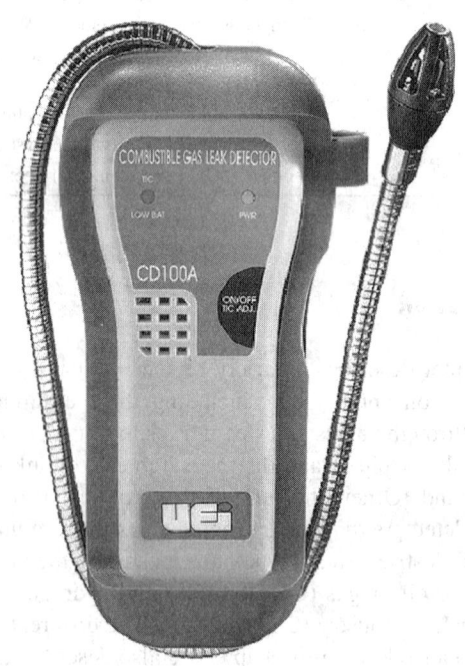

FIG. 1.34a
Combustible gas leak detector. (Courtesy of Cole-Parmer Instrument Co.)

Portable detector tube units provide measurements of low concentrations of gases and vapors. For a detailed list of the capabilities of a list of these tubes, refer to Table 1.26r. Most of the tubes have direct reading scales. The length of the discoloration in the tube provides the concentration reading. Some tubes contain chemicals that change color; these require a color comparison for the reading. Manufacturers have tubes for 100 or more specific analyses. Some of these tubes have been certified by the National Institute for Occupational Safety and Health.

For carbon monoxide, hydrogen sulfide, and a few other toxic materials, small badges and stickers (dosimeters) that change color as a function of total exposure are available.

---

* API Standard 527, Rev. 3, Seat tightness of pressure relief valves (2007). http://www.api.org/publications-standards-and-statistics/standards/whatsnew/publication-updates/new-refining-publications/api_std_527.

Personal Alarms

Personal alarm units can provide considerable protection against toxic materials (Figure 1.34b). These devices record or continuously integrate toxic levels to provide a time-weighted average exposure number. An audible alarm occurs if a preset threshold limit value is exceeded at any instant.

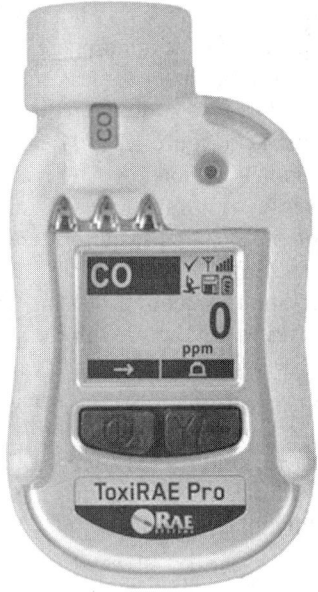

FIG. 1.34b
Personal wireless toxic gas monitor and alarm. (Courtesy of RAE Systems.)

Table 1.34b provides a summary of the typical alarm set points and inaccuracies of standard gas monitors.

TABLE 1.34b
Capabilities of Pocket-Sized, Portable Toxic Gas Monitors

	Alarm Set Point (ppm)	Inaccuracy
CO	200	±10 ppm FS; ±5 ppm in 0–200 range
H_2S	50	±10 ppm FS; ±3 ppm in 0–100 range
SO_2	20	±10 ppm FS; ±3 ppm in 0–100 range
NO_x	200	±10 ppm FS; ±5 ppm in 0–200 range
NO_2	20	±2 ppm FS

Source: Courtesy of Cole-Parmer Instrument Co.

The most common type of combustible gas detector utilizes the catalytic combustion principle. A platinum wire filament (in a Wheatstone bridge circuit) is usually mounted in a diffusion-sensing head. Sulfur compounds and halides can prevent proper operation of this sensor. In processes where flammable gases or vapors can escape the technique, serious consideration for intermittent and continuous monitoring must be warranted.

Table 1.34c lists the compounds that can be detected and the ppm sensitivity that can be expected of the measurement, using an intrinsically safe portable sensor, which is provided with a probe for focusing the detection to localized points of leaking. The unit is also suited for the detection of ammonia leaks.

TABLE 1.34c
Detection Limits of Portable Combustible Gas Detectors

Compounds	Formula	Sensitivity (ppm)
Acetylene	C_2H_2	50
Alcohols	R–OH	50
Hydrogen sulfide	H_2S	5
Gasoline	—	1
Sulfur dioxide	SO_2	5
Vinyl chloride	C_2H_3Cl	5
—	—	50

Source: Courtesy of Cole-Parmer Instrument Co.

DETECTOR TYPES

Ultrasonic Detectors

Fluids escaping from openings generate sonic and ultrasonic waves. With the acoustic emission technique, leaks can be detected from a distance. One vendor's explanation is that a 10 psig (68.9 kPa) gas leak might be detectable at 15 m (50 ft), but would be detected at only 3 m (10 ft) if the pressure were 1 psig (6.89 kPa). The method can be considered for the location of leakage from buried pipes and tanks. Although the technique has limitations for low-pressure gas problems, it can detect the gurgling of leaks from sewers and other low-pressure liquid lines.

One successful application of the ultrasonic leak detector is the detection of the leaking of steam traps from a distance. These portable probe-type units can also detect the inleakage of atmospheric air into vacuum equipment (Figure 1.34c).

Slash energy costs
Easily locate compressed air
and steam leaks

FIG. 1.34c
Ultrasonic leak detector. (Courtesy of CTRL Systems.)

Thermal Conductivity Detectors

Thermal conductivity leak detectors have cells that contain coils in bridge circuits. Heat dissipation increases with the concentration of the gas or gases in the sample, and the cooling effect changes coil resistance. As detailed in Chapter 1.66, these instruments will. The instrument will detect many different gases and has a wide dynamic range.

The portable, handheld leak detectors are provided with microprocessor-based intelligence. The unit is provided with a probe and a small fan that continuously draws in the gas that is present at the tip of the probe. The unit can be automatically zeroed (by pressing the rezero button) based on the ambient air sample around the probe. Next, the operator would identify the gas of interest and direct the probe to the probable point of leakage. The small fan then draws in a sample through the probe, and the sensor measures the resulting minute changes in thermal conductivity. Table 1.34d lists the detectable leak rates of some common gases.

TABLE 1.34d
Minimum Detectable Leak Rates of Some Common Gases Using Thermal Conductivity–Type Leak Detectors

Gas	Minimum Detectable Leak Rate	
	cc/s	ft^3/year
Helium	0.00001	0.11
Hydrogen	0.000005	0.06
R12	0.000006	0.07
R1301	0.000006	0.07
R134(A)	0.000004	0.04
SF6	0.00001	0.11
CO$_2$	0.00003	0.34
CH$_4$	0.00002	0.23
Argon	0.00002	0.23

Halogen Detectors

A halogen-bearing sample produces ionization current flow in a specially designed cell. The cell life of this design is shortened by high halogen concentrations. Although this instrument type is very valuable for checking refrigeration systems and for certain kinds of production testing, it is of less value in general industrial applications. One of its disadvantages is that the cell runs very hot, so these detectors cannot be used in a flammable atmosphere.

Electron-capture-type halogen leak detectors are the most sensitive (10^{-7} cm^3/s or even smaller) leak detectors available on the market (Figure 1.34d). Leak testing of reactors or other equipment is done by first evacuating the vessel and then filling it with a halogen gas, such as Freon-12 (dichlorodifluoromethane), and scanning the welds and other joints by a halogen sniffer probe.

When refrigeration systems are being leak-tested, the halogen is the working fluid inside the equipment. It should be noted that such halogens as Freon-12 (also called F-12, Refrigerant-12, or R-12) are not only destructive to the ozone layer round the Earth but can also be fatal if inhaled. When exposed to high temperature, R-12 breaks down into chlorine, hydrogen, chloride, hydrogen fluoride, and phosgene gas, which is highly toxic.

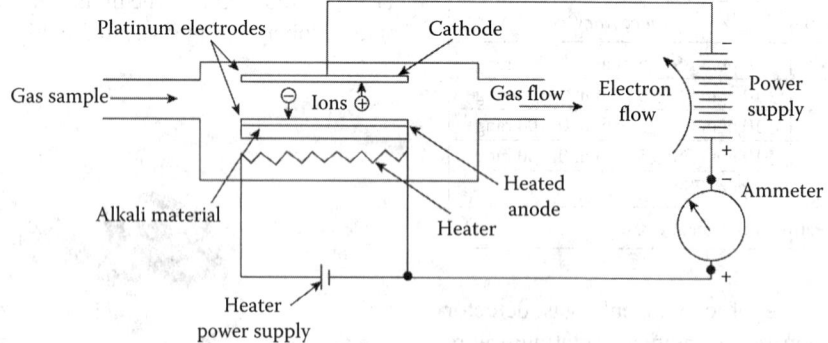

FIG. 1.34d
Schematic diagram of a heated anode halogen vapor sensor.

The latest edition of the *Non-Destructive Testing Handbook** of the American Society for Nondestructive Testing provides a large body of information on halogen tracer gases and leak testing. When vessels or heat exchangers are checked for leakage, they are usually filled with a pressurized mixture of air and R-12 gas. Table 1.34e provides information on the detectable leakage rates as a function of the richness of the air and R-12 gas mixture inside the equipment being tested.

TABLE 1.34e
Example of Percent R-12 Halogen Tracer Gas at 200 kPa Absolute (15 psig) to Detectable Leakage Rate[a]

% Tracer Gas	Leakage Rates			
	$Pa \cdot m^3/s$	$std \cdot cm^3/s$	oz/year	g/annum
100	9×10^{-7}	9×10^{-6}	0.05	1.5
50	1.8×10^{-6}	1.8×10^{-5}	0.1	3
25	3.6×10^{-6}	3.6×10^{-5}	0.2	6
10	9×10^{-6}	9×10^{-5}	0.5	15
5	1.8×10^{-5}	1.8×10^{-4}	1.0	30
1	9×10^{-5}	9×10^{-4}	5.0	150
0.5	1.8×10^{-4}	1.8×10^{-3}	10.0	300

Source: Courtesy of General Electric Co.
Note: $Pa \cdot m/s \; 2 \times 10 \; std \; cm^3/s$.
[a] Safety factor of 4.5 is included in these tracer gas concentrations. The assumed halogen leak detector sensitivity setting is 2×10^{-7}.

Figure 1.34e illustrates how the sniffer probe might be used on a vessel wall in an open atmosphere.

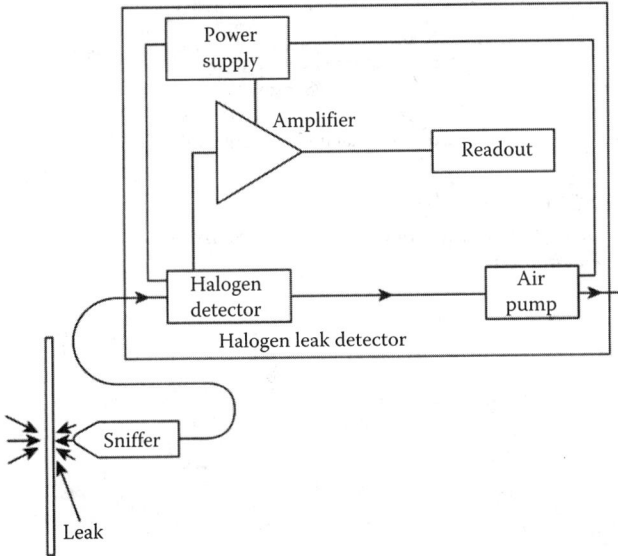

FIG. 1.34e
Standard sniffer probe and halogen leak detector.

---

* ASNT Level III Study Guide: Leak Testing Method (LT) (2004) https://www.asnt.org/Store/ProductDetail?productKey=42a1e772-ef52-464e-a20c-baded9ec19ad.

Figure 1.34f shows an adapter that can be used when testing lengths of tubing or piping for leakage.

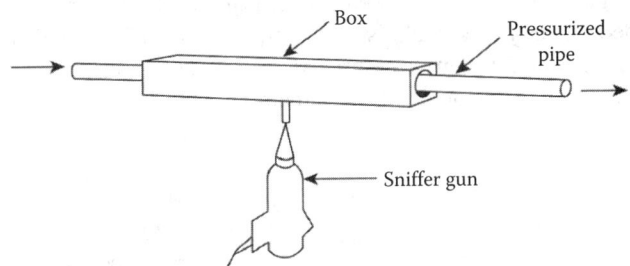

FIG. 1.34f
Special adapter for testing lengths of tubing or pipe.

Thermography

Thermography involves the detection of temperature differences by scanning in the IR region of the spectrum. The result is a thermal picture that is usually two colors, but sophisticated systems can convert the measurements into graded colors. If a hot fluid is escaping, it can be detected at a distance, above- or belowground, or under insulation. Leaking valves and steam traps are easily located through the use of this technique.

Thermography has already seen many other applications that only involve thermal energy. It is valuable for the inspection of insulation and refractory materials. As a maintenance tool, it can be used to check heater tubes and electrical equipment.

Mass Spectrometry

When used for leak detection, the mass spectrometer (see Chapter 1.35 for details) is usually made sensitive to helium. The helium acts as a tracer gas, inside or outside of the item that is being tested. These expensive instruments would be difficult to use in a chemical plant, but they have considerable value for production (assembly line) leak testing.

UNDERGROUND DETECTION

Underground leaks at refineries, tank farms, gas stations, and industrial waste sites require monitoring. The monitored materials can be flammable hydrocarbons or toxic substances.

Level Monitoring

The monitoring systems can be based on the measurement of either liquids or vapors. In installations where the tanks are double walled or where the tanks are installed in line sumps, the measurement of liquid levels can be sufficient (Figure 1.34g).

The level sensor can be any detector that is capable of distinguishing between water and hydrocarbons. One such level detector is the thermal dispersion type (Figure 1.34h), which utilizes the differences between the thermal conductivities of liquids to differentiate between them.

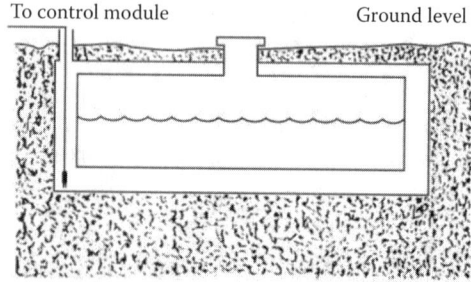

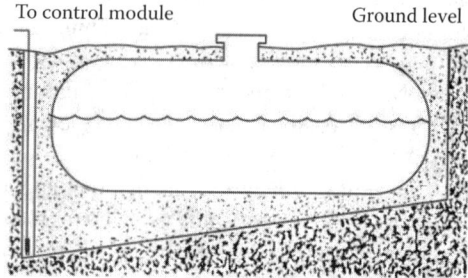

FIG. 1.34g
Thermal dispersion–type level switches can be used to detect leakage in double-walled underground tank installations.

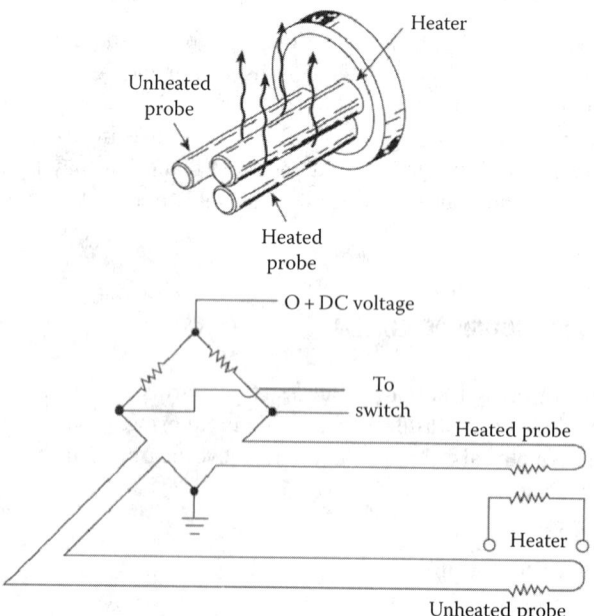

FIG. 1.34h
Thermal conductivity–type level switch. (Courtesy of Fluid Components International LLC.)

Soil Concentration Detection

When the tanks are not provided with double walls, it is common practice to detect the concentration of hydrocarbon vapors in the soil, instead of attempting to sense the presence of hydrocarbon liquids. This is because vapors migrate about 50 times faster than liquids. These sensors are usually set at gasoline concentrations of 0.5% by volume (or less) in air, which corresponds to 50% of the lower flammability limit for gasoline. Passive inground sensors can measure the concentration of hydrocarbon vapors in the soil (Figure 1.34i).

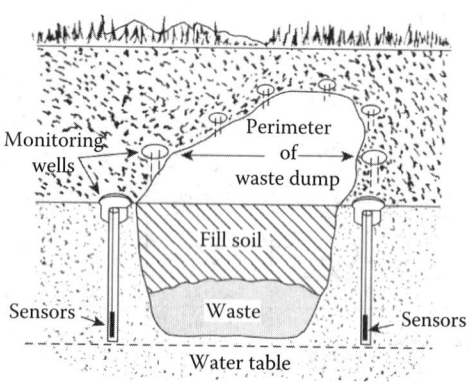

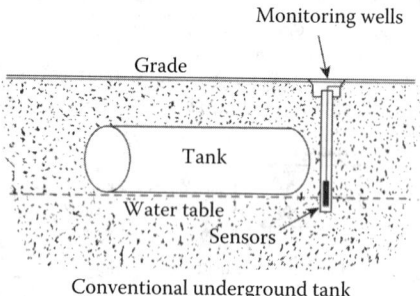

FIG. 1.34i
Passive, inground sensors detect the concentration of hydrocarbon vapors in the soil.

Aspirated Sensors

Aspirated underground leak sensors pull a vacuum inside an underground probe that is provided with slots to allow inleakage of vapors (Figure 1.34j). The slotted portion must never be fully covered by groundwater, and the slots must be protected from plugging. The aspirated vapor sample tends to speed up the measurement, because the hydrocarbon vapors will migrate toward the vacuum created by the slotted probe.

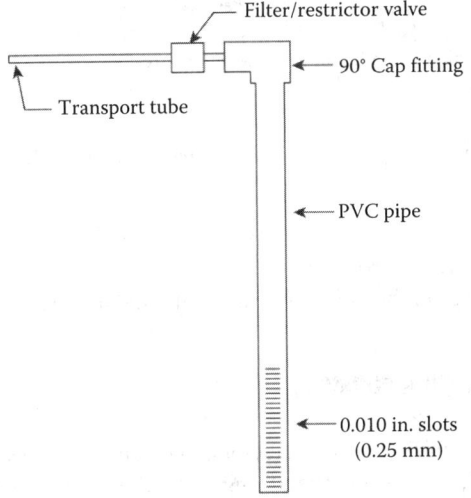

FIG. 1.34j
The typical underground test probe.

Another advantage of this design is that the sensors can be located in an accessible and clean area. With multiple probes in the ground, continuous concentration profile measurements are possible. The sensor can be a bulk semiconductor vapor analyzer, a combustibles detector, or any of the other leak detectors discussed earlier.

Standpipe Detector

Figure 1.34k shows the main components of a testing package that are capable of detecting minute leaks in underground tanks. Before the actual measurement is made, the groundwater level is measured, and the tank is pressurized to make it grow to reach its stable volume. Next, the tank contents are circulated, so that the temperature of the tank contents will be uniform within a fraction of a degree. Once the tank has been stabilized both in terms of its shape and the temperature

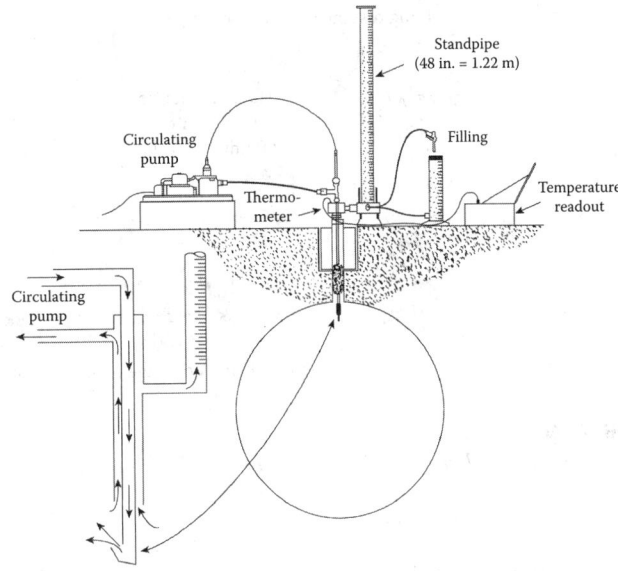

FIG. 1.34k
Petrotite test equipment. (Courtesy of Purpora Engineering.)

of its contents, then the level in the standpipe of Figure 1.34k can signal even a minute leakage.

NEW DEVELOPMENTS

Among the many leak detection applications, oil and natural gas pipeline leak testing is the most important. The methods of detection include hydrostatic testing after pipeline erection and leak detection during service. The leak detection systems (LDS) can use many techniques that can be divided into internal and external systems.* Internally based systems utilize field instrumentation such as flow, pressure, or fluid temperature sensors of conditions inside the pipes. Externally based systems usually use infrared radiometers or thermal cameras, vapor sensors, acoustic microphones, or fiber-optic cables to monitor the presence of hydrocarbons outside the pipelines.

One newer of the methods of leak detection is based on the fact that usually there is a difference between the temperature of the soil in which the hydrocarbon pipe is buried and the temperature of the flowing fluid. If leakage occurs at a particular location, such a difference will take place and

---

* API 1130, Computational Pipeline Monitoring for Liquid Pipelines (2002). https://law.resource.org/pub/us/cfr/ibr/002/api.1130.2002.pdf.

482 Analytical Measurement

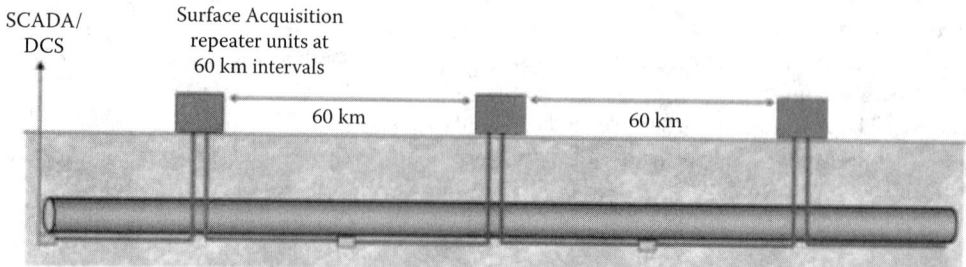

FIG. 1.34l
Fiber-optic leak detection of long pipelines. (Courtesy of R. Kluth and J. Worsley article published in WNPT magazine on 19 May, 2008.)

can be detected by laser-based fiber-optic Raman sensing (Figure 1.34l).

Another challanging application is to detect under-sea pipe leaks. This can be accomplished by accoustic techniques. In that case, the sensors are an array of hydrophones, which convert the acoustic pressure waves into electrical signals, configured to discriminate the noise of a leak from other sources of sound, including common ocean sounds (Figure 1.34m). The hydrophones obtain directional information, both horizontally and vertically, and can have a detection range of up to a 500 m radius from the sensor location. Data processing can be done subsea or topside, depending on the communication method and depth.

SPECIFICATION FORMS

When specifying Leak Detectors, one can use the ISA form 20A1001, and when specifying both the detector and the properties of the process fluid it will be monitoring, use the ISA form 20A1002. Both forms are reproduced with the permission of the International Society of Automation on the next pages.

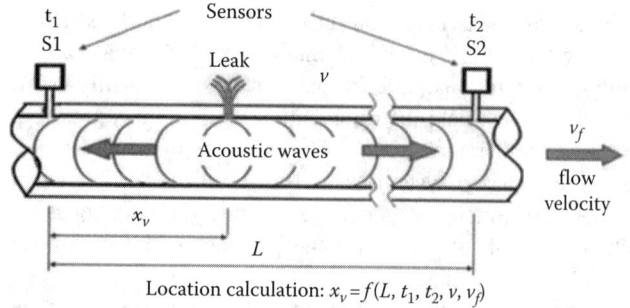

FIG. 1.34m
Sub-seas leakage detection based on the use of acoustic hydrophones. (Courtesy of C. Paschola's article in Marine Technology, September 25, 2014.)

1.34 Leak Detectors

ANALYSIS DEVICE — Operating Parameters

#	Field		#	Field	
1	RESPONSIBLE ORGANIZATION		6	SPECIFICATION IDENTIFICATIONS	
2			7	Document no	
3	(ISA)		8	Latest revision	Date
4			9	Issue status	
5			10		

#	ADMINISTRATIVE IDENTIFICATIONS		#	SERVICE IDENTIFICATIONS Continued	
11			40		
12	Project number	Sub project no	41	Return conn matl type	
13	Project		42	Inline hazardous area cl	Div/Zon Group
14	Enterprise		43	Inline area min ign temp	Temp ident number
15	Site		44	Remote hazardous area cl	Div/Zon Group
16	Area	Cell Unit	45	Remote area min ign temp	Temp ident number
17			46		
18	SERVICE IDENTIFICATIONS		47	COMPONENT DESIGN CRITERIA	
19	Tag no/Functional ident		48		
20	Related equipment		49	Component type	
21	Service		50	Component style	
22			51	Output signal type	
23	P&ID/Reference dwg		52	Characteristic curve	
24	Process line/nozzle no		53	Compensation style	
25	Process conn pipe spec		54	Type of protection	
26	Process conn nominal size	Rating	55	Criticality code	
27	Process conn termn type	Style	56	Max EMI susceptibility	Ref
28	Process conn schedule no	Wall thickness	57	Max temperature effect	Ref
29	Process connection length		58	Max sample time lag	
30	Process line matl type		59	Max response time	
31	Fast loop line number		60	Min required accuracy	Ref
32	Fast loop pipe spec		61	Avail nom power supply	Number wires
33	Fast loop conn nom size	Rating	62	Calibration method	
34	Fast loop conn termn type	Style	63	Testing/Listing agency	
35	Fast loop schedule no	Wall thickness	64	Test requirements	
36	Fast loop estimated lg		65	Supply loss failure mode	
37	Fast loop material type		66	Signal loss failure mode	
38	Return conn nominal size	Rating	67		
39	Return conn termn type	Style	68		

#	PROCESS VARIABLES	MATERIAL FLOW CONDITIONS			Units	#	PROCESS DESIGN CONDITIONS		
69						101	Minimum	Maximum	Units
70	Flow Case Identification					102			
71	Process pressure					103			
72	Process temperature					104			
73	Process phase type					105			
74	Process liquid actl flow					106			
75	Process vapor actl flow					107			
76	Process vapor std flow					108			
77	Process liquid density					109			
78	Process vapor density					110			
79	Process liquid viscosity					111			
80	Sample return pressure					112			
81	Sample vent/drain press					113			
82	Sample temperature					114			
83	Sample phase type					115			
84	Fast loop liq actl flow					116			
85	Fast loop vapor actl flow					117			
86	Fast loop vapor std flow					118			
87	Fast loop vapor density					119			
88	Conductivity/Resistivity					120			
89	pH/ORP					121			
90	RH/Dewpoint					122			
91	Turbidity/Opacity					123			
92	Dissolved oxygen					124			
93	Corrosivity					125			
94	Particle size					126			
95						127			
96	CALCULATED VARIABLES					128			
97	Sample lag time					129			
98	Process fluid velocity					130			
99	Wake/natural freq ratio					131			
100						132			

#	MATERIAL PROPERTIES			#	MATERIAL PROPERTIES Continued		
133				137			
134	Name			138	NFPA health hazard	Flammability	Reactivity
135	Density at ref temp	At		139			
136				140			

Rev	Date	Revision Description	By	Appv1	Appv2	Appv3	REMARKS

Form: 20A1001 Rev 0   © 2004 ISA

484 Analytical Measurement

	RESPONSIBLE ORGANIZATION	ANALYSIS DEVICE COMPOSITION OR PROPERTY Operating Parameters (Continued)		SPECIFICATION IDENTIFICATIONS	
1	ISA		6	Document no	
2			7	Latest revision	Date
3			8	Issue status	
4			9		
5			10		

	PROCESS COMPOSITION OR PROPERTY			MEASUREMENT DESIGN CONDITIONS				
	Component/Property Name	Normal	Units	Minimum	Units	Maximum	Units	Repeatability
11								
12								
...								
75								

Rev	Date	Revision Description	By	Appv1	Appv2	Appv3	REMARKS

Form: 20A1002 Rev 0 © 2004 ISA

Abbreviations

AFV	Alternate fuel vehicles
LDS	Leak detection system
TLV	Threshold limit value
TWA	Time-weighted average

Organizations

API	American Petroleum Institute
ASNT	American Society for Nondestructive Testing
ASTM	American Society for Testing And Materials
NFPA	National Fire Protection Association

Bibliography

ACGIH, Documentation of the Threshold Limit Values and Biological Exposure Indices (2015). https://www.acgih.org/forms/store/ProductFormPublic/documentation-of-the-threshold-limit-values-and-biological-exposure-indices-7th-ed.

ANSI/ISA TR12.13.03-Guide for Combustible Gas Detection as a Method of Protection (2009) https://www.isa.org/templates/one-column.aspx?pageid=111294&productId=116631.

API 1130, Computational Pipeline Monitoring for Liquid Pipelines (2002). https://law.resource.org/pub/us/cfr/ibr/002/api.1130.2002.pdf.

API Standard 527, Rev. 3, Seat tightness of pressure relief valves (2007). http://www.api.org/publications-standards-and-statistics/standards/whatsnew/publication-updates/new-refining-publications/api_std_527.

ASNT Level III Study Guide: Leak Testing Method (LT) (2004). https://www.asnt.org/Store/ProductDetail?productKey=42a1e772-ef52-464e-a20c-baded9ec19ad.

ASNT: Nondestructive Testing Handbook, Leak Testing (1997). https://www.asnt.org/Store/ProductDetail?productKey=f8086b38-a 950–4d5c-9d36-e40be4c663d5.

EPA: Automatic tank and line leak detection (2013). http://pubweb.epa.gov/region07/underground_storage_tanks/atgman.htm.

EPA, Acute Exposure Guideline Levels for Airborne Chemicals (2015). http://www.epa.gov/oppt/aegl/index.htm.

EPA, Underground Storage Tanks (USTs) Laws & Regulations (2015). http://www2.epa.gov/ust/underground-storage-tanks-usts-laws-regulations.

Geiger, G., Principles of Leak Detection (2008). http://malaysiaflow.com/pdf/KOG_PipePatrol/PipePatrol%20-%20Principles%20of%20Leak%20Detection%20EN.pdf.

General Monitors, Fundamentals of Combustible Gas Detection (2010). http://s7d9.scene7.com/is/content/minesafetyappliances/Combustible%20Gas%20Detection%20White%20Paper.

Kluth, R. and Worsley, J., Digital pipeline leak detection — using fibre-optic distributed temperature sensing (2008). http://www.processonline.com.au/content/instrumentation/article/digital-pipeline-leak-detection-using-fibre-optic-distributed-temperature-sensing-579438440.

McDermott, H. J. Air monitoring for toxic exposures (2005). http://www.wiley.com/WileyCDA/WileyTitle/productCd-0471454354.html.

NIOSH, Pocket Guide to Chemical Hazards, National Institute for Occupational Safety and Health (2015). http://www.cdc.gov/niosh/npg/.

OSHA, Toxic Industrial Chemicals (TICs) Guide (2007). https://www.osha.gov/SLTC/emergencypreparedness/guides/chemical.html.

OSHA, Use of combination oxygen and combustible gas detectors (2013). https://www.osha.gov/dts/hib/hib_data/hib19900118.html.

Paschola, C., Understanding Subsea Acoustic Leak Detection and Condition Monitoring (2014). http://www.marinetechnologynews.com/blogs/understanding-subsea-acoustic-leak-detection-and-condition-monitoring-e28093-part-1-700514i.

1.35 Mass Spectrometers

R. C. AHLSTROM (1982, 1995)   **R. A. GILBERT** (2003, 2017)

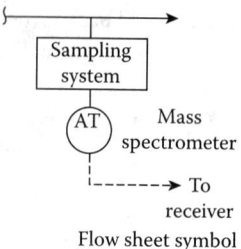

Flow sheet symbol

Type of sample	Vapor
Standard design	Atmospheric *pressure*
Sample temperature	50°C–200°C (122°F–424°F), vaporized sample with no condensation
Ambient temperature	15°C–35°C
Contacting materials	Inlet materials of construction designed to be compatible with sample
Auxiliary utilities required	Sometimes cooling unit
Cost	$3,000–$150,000, depending on analyzer section, application complexity, and data report and collection accessories Time-of-flight section: $1,25,000 Magnetic section (food processing industry): $100,000 RGA, $3000; multiple purchase, $2500 Quadrupole section: $30,000–$80,000 Ion-trapping section: $70,000–$120,000
Inaccuracy	±0.5% AR for most applications
Repeatability	±0.2% for most applications
Cycle time	0.5–4 sec per stream, depending upon applications
Special features	Database of mass spectra, programmable temperature control
Partial list of suppliers	Ametek Process Instruments (http://www.ametekpi.com/) Dycor-Dymaxion RGA (www.ametekpi.com/products/dycor-dymaxion-rga.aspx) Finnigan Mat Corp (www.biosurplus.com/store/products.) Jeol (http://www.jeol.co.jp/en/products/list_ms.html) Leyboldinficon (www.leyboldinficon.com MKS Instruments (www.mksinst.com) Perkin-Elmer Corp. (www.instruments.perkinelmer.com) Spectrum Automation (www.spectrumm-automation.com/mass-spectrometer.html) Waters Corp. (www.waters.com)

INTRODUCTION

Mass spectrometers (MSs) provide multicomponent analysis with the accuracy and reliability needed for process quality control at a cycle time necessary for closed-loop digital control. Instruments are available for sampling multiple streams for simultaneous concentrations of 10–15 components in less than 5 sec. Industry-based MSs provide a proper combination of the stream sampling, analysis, and data storage to meet the designated process needs.

OPERATION PRINCIPLES

Knowledge of the mass-to-charge (m/e) ratio for an atom or molecule in a sample is the basis of mass spectroscopy. For example, if a sample of diatomic nitrogen molecule with a mass of 28 was inserted into an MS, some of the sample molecules might become a single charged ion with an m/e of 28, while other sample molecules might become double charged ions with an m/e ratio of 14 and the spectrometer's detection system distinguishes these two groups of ions. This identification process is accomplished in a vacuum environment, typically in the order of 10^{-4}–10^{-6} Torr (mm Hg) in which the ions are created, separated, and ultimately detected. Consequently, for process stream applications, an MS configuration depends on the unit's online or off-line mission. In both cases, the system includes sampling ports that transition the sample from atmospheric to the lower operating pressure environment and the supporting vacuum pump subsystems that maintain the vacuum envelope the spectrometer requires. For the online spectrometer applications, this interface can be complicated.

Operationally, the small gas sample is introduced from atmospheric pressure through the inlet port into the ion generator section of the spectrometer. Sample ions are produced by the collision of rapidly moving electrons with the gas molecules to be analyzed. The electrons ionize the sample by removing one or more electron from an outer orbit of a neutral sample molecule. The resulting positive ions are extracted from the ion generator as fast as they are formed by means of electrostatic fields created by sets of electrodes that also accelerate and focus these ions into an ion beam. This ion beam is directed into the ion separation section of the instrument. Finally, these separate segregated ion groups are directed to an ion collector or detector, which produces an electric signal that is proportional to the number of ions in the detected group.

Sample Input

Preanalysis sample conditioning for an MS includes component separation and vaporization. It is very common to connect a gas chromatography system (Chapter 1.10) to the input port of an MS. After vaporization, the sample is drawn by suction into the MS through several possible types of inlet leaks. A sintered metal leak device is a commonly used input port because it allows a uniform molecular flow into the instrument based on the differential pressure across the sintered metal plate. Thin diaphragms made from Teflon may be used for corrosive service, and ultrafine metering valves can also be used. Also available are automatic metering valves that maintain a sample flow based on the pressure in the spectrometer's ionization chamber. While a variety of inlet leaks are available, one must be chosen that produces a stable flow of sample gas into the spectrometer and is compatible with the desired sample and the system's analyzer. Fortunately, most MS companies provide instruments with appropriate inlet port options and offer help in making the proper input port selection.

Sample Ionization

Several methods are available for producing ionic species from the sample gas. Electron bombardment is the most popular but requires periodic filament replacement. Spark generation and optical and chemical ionization methods are also used. Atmospheric ionization mass spectrometry is available and has been shown to be applicable to a wide range of compounds; however, most instruments used in process applications use vacuum environment–based systems with heated filaments. Since filament life and ease of replacement are of primary concern, most process MSs have dual-filament assemblies that allow switching from one to the other filament without any downtime. Filament replacement is then postponed until the next downtime occurrence or scheduled preventive maintenance event.

Figure 1.35a suggests the operation of a heated filament ion source. In this version, the molecules are shown entering the spectrometer from the top. After passing through a negatively charged repulsion screen, these neutral molecules interact with a focused electron beam to form a variety of intermediates, including an ion cloud that contains positive ions. These ions are pushed out of the ion generator section by the screen and then accelerated through an exit slit into the spectrometer's ion separation section. The sample ions are now z-axis aligned and directed into the magnet region of the ion separation section of the instrument. The vacuum system removes the bulk of the sample molecules and any other electron impact reaction products.

There are at least four options for the sample ionization section of an MS: (1) an open ion source for general high-vacuum applications, (2) a closed ion source for process and gas analysis, (3) a differential ion source for running both process pressure and high-vacuum applications simultaneously, and (4) a cross-beam ion source to enhance the minimum detection level for many different gas species. As with the case of the instrument's sample inlet port, most MS companies provide these options and offer help in making the appropriate ion generator selection.

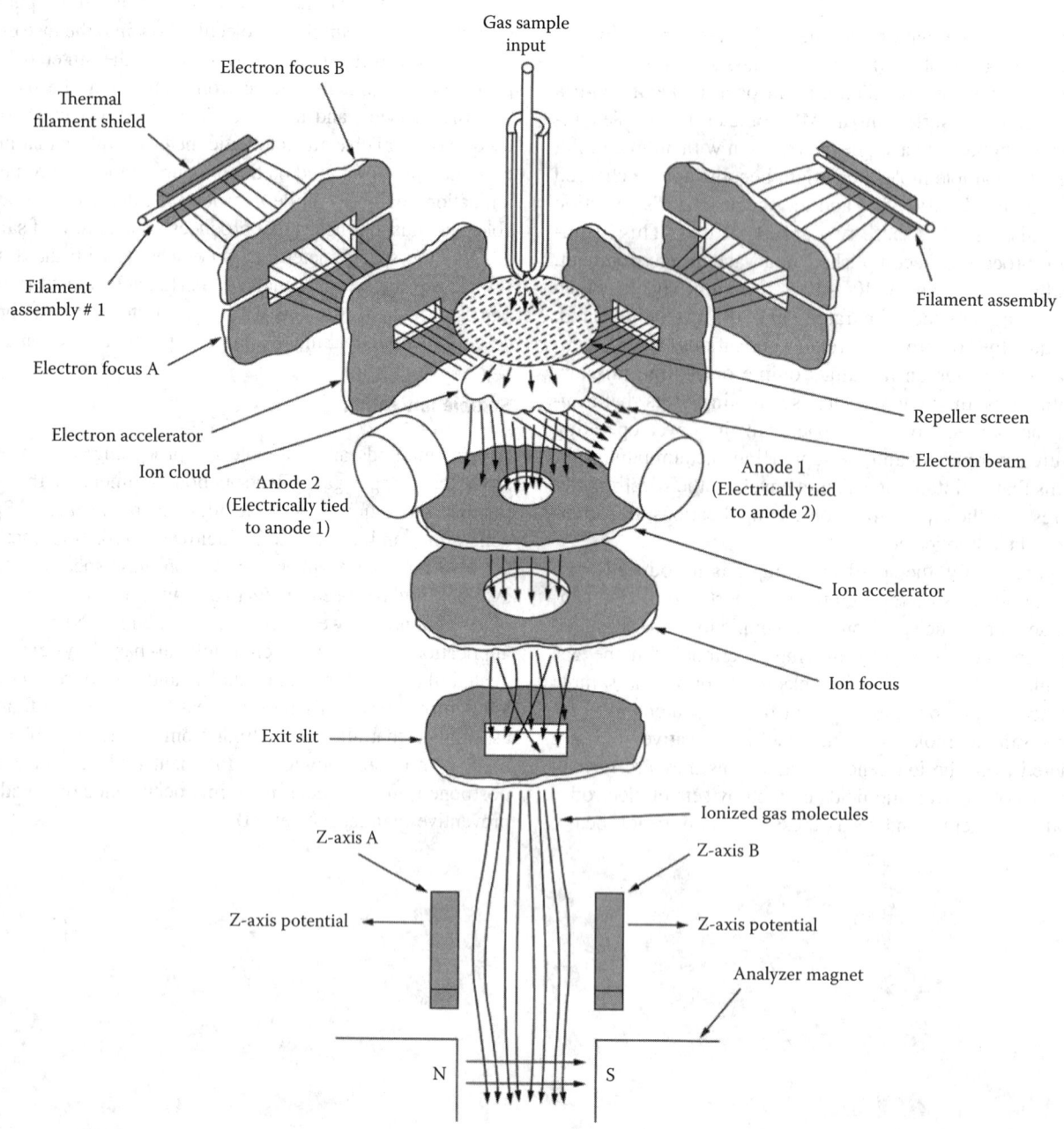

FIG. 1.35a
Ion source operation.

Ion Separation

Although the four major sections of an MS—the input port, the ion generator, the ion separation section, and the ion detector—are equally important, the behavior of the spectrometer is characterized by its ion separation section. There are several ion separator section options available. For instrumentation and process personnel, the selected instrument design must stress simplicity, automatic operation, and reliability; require a minimum of operator attention; and still meet the analysis needs. Further, it must be serviced and maintained by personnel with minimal MS technical expertise. The various separation options are discussed as follows.

Magnetic Sectors There are two types of magnetic sector instruments: fixed magnetic sector and electromagnetic-focusing sector. The fixed magnetic sector instrument utilizes a permanent magnet to produce a magnetic field at right angles to the direction of motion of the ion beam that exits the instrument's ion generation section. This normal interaction with the applied field forces the ions to bend in a circular trajectory proportional to their m/e ratios.

The concept of the fixed magnetic sector ion separation process is shown in Figure 1.35b. The top section of the illustration is the sample ionization section of the instrument which succinctly summarizes all of the details provided in Figure 1.35a. The center portion of the diagram illustrates the fixed magnetic sector ion separation portion of the instrument. In this case, eight multiple collectors are employed to obviate the need for scanning, vastly simplify the electronic system, and reduce the requirements associated with the selection of operational parameters. For each process application, the collectors are initially located at a predetermined point in the instrument focal plane. It is always desirable to measure the atmospheric composition of process stream components. This is possible when the ion collectors are placed along the magnetic sector focal plane to intercept the ion beams for nitrogen (m/e = 28), oxygen (m/e = 32), and argon (m/e = 40). With these three measurements, the output of the instrument could continuously monitor the atmosphere composition of the process stream and provide data for an appropriate control decision. In addition, the remaining ion output channels, five channels in this illustration, can be dedicated to monitor other components of the process stream.

The concept of the electromagnetic-focusing sector ion separation process is distinctly different from the multiple collector idea shown in Figure 1.35b. The electromagnetic-focusing sector utilizes changes in either the accelerating voltage or magnetic field to focus the desired ionic species onto a single collector. When accelerating voltage is used, the ions are focused and accelerated as a beam into a magnetic field. The acceleration voltage is varied to select the ions with a particular mass. For various ionic mass selections, the product of mass and acceleration voltage is a constant. Since the multiple mass component beam entering the magnetic field contains ions of essentially equal energy, the mass of each ion is distinguished by the ion's momentum. Because the radius of curvature of the ions in the magnetic field is different for ions of different momentum, only ions corresponding to one particular mass are focused on the collector. This signal is amplified and displayed as a signal proportional to the particular concentration (partial pressure) of the gas type with that mass.

Magnetic field variation is an alternate way to make a single mass ion beam exit an electromagnetic-focusing sector in an MS. The principal advantage of magnetic field scanning is the wider mass range achieved. The mass of the ion is not directly proportional to the magnetic field; however, circuits have been designed to provide linear spacing of masses exiting the focusing sector. When changing the magnetic field, the accelerating voltage is held constant at a value that corresponds to the higher mass of interest. When magnetic field variation is used to sort sample ions, the relative heights of different mass peaks do not occur in the same proportion as they do when the ion separation is accomplished by changing the accelerating voltage. In the latter case, the low-mass peaks are enhanced in both magnitude and resolution because of the more favorable accelerating voltage at which they appear. When the magnetic field is varied at an acceleration voltage corresponding to the highest mass of interest, this enhancement does not occur. Therefore, both the magnitude and resolution of mass peaks will be less.

Quadrupole Filter The use of a quadrupole mass filter to separate the sample ions that leave the ion generation

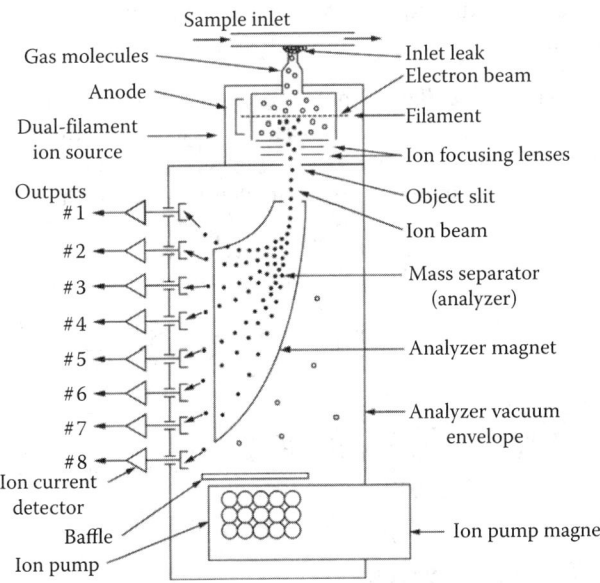

FIG. 1.35b
Fixed magnetic sector mass spectrometer operation.

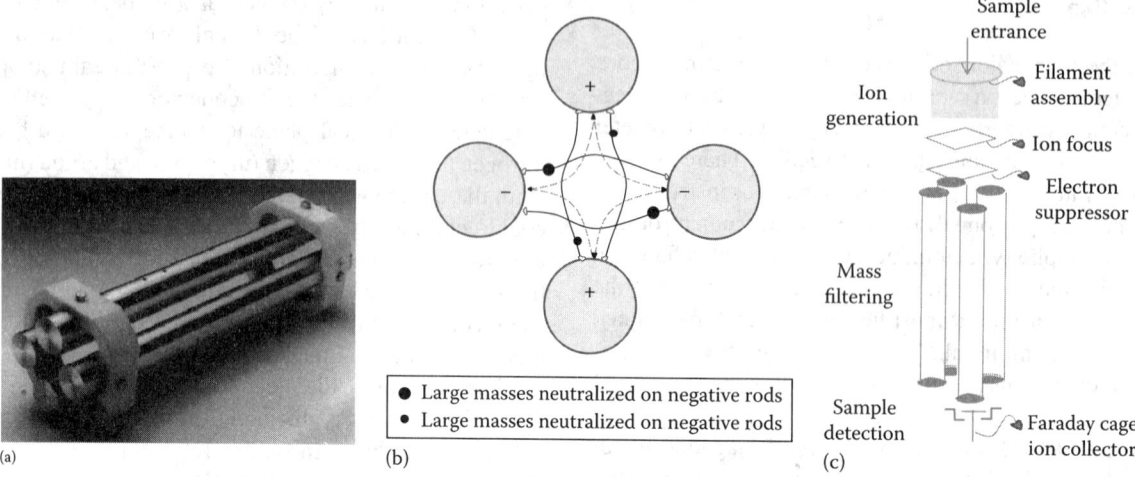

FIG. 1.35c
(a) Quadrupole rods extracted from small residual gas analyzer. (b) Top-down view of filtered mass pass through enclosed dotted region. (c) Side view illustration of quadrupole rods that includes ion generation (top) and ion detection (bottom) views.

section of an MS is also possible. Figure 1.35c outlines the operation of this type of ion separation section. The sample ions exit an ion source through the electron suppressor and then enter the quadrupole filter region of the instrument. Figure 1.35c(a) shows the four-pole set for a residual gas analyzer (RGA). This quadrupole m/e filter consists of four high-precision crafted cylindrical rods located precisely in an orthogonal array. The m/e-separated ion groups exit the quadrupole separation section and strike a single detector, which produces a signal proportional to the number of ions with the selected m/e ratio.

The two lower panels of Figure 1.35c provide a top-down view of the travel path, Panel (b), for ions entering the quadrupole ion separation portion of the instrument. As an ion moves at uniform speed among these rods, it undergoes complex oscillatory spiral motion transverse to its travel axis. The actual ion motion is considerably more complex than shown, and the path lines in the figure only represent two of the outer boundaries of particle motion. This motion pattern is created because diametrically opposing rod pairs are electronically connected to also form positive and negative rod pairs. The positive pair has a radio frequency (RF) voltage superimposed on a positive DC voltage, while the negative pair has a negative DC and RF voltage that is 180° out of phase with the positive pair. The varying electromagnetic field that results creates an environment that separates the ions by their m/e ratio.

For a given set of voltages, all ions below a given mass interact with the positive set of rods. These low-mass ions are neutralized and removed from the ion beam, while higher-mass ions remain in the ion beam. In turn, the negative set of rods interacts, neutralizes, and removes ions above a given mass value. The lower-mass ions remain in the ion beam and continue traversing the quadrupole section. By adjusting the RF-to-DC ratio, an ion passband can be created so that only ions with a specific narrow m/e ratio can exit the quadrupole ion separation section of the instrument. Under high-resolution conditions, the passed ions barely miss the rod pairs at the extremes of their transverse oscillatory motion, and the mass range that is allowed to exit the quadrupole filter is very narrow. A unique property of the quadrupole mass filter is the fact that the number of m/e ions that are passed is directly proportional to the voltage applied to the rods. Thus, by changing the voltages, various ionic masses of interest are allowed to pass through to the collector where they are measured.

Ion-Trapping Section Figure 1.35d illustrates an ion-trap MS that uses an interesting twist for its ion separation section. In this case, the geometry of the section plus the RF voltages applied combine to create an environment in which the

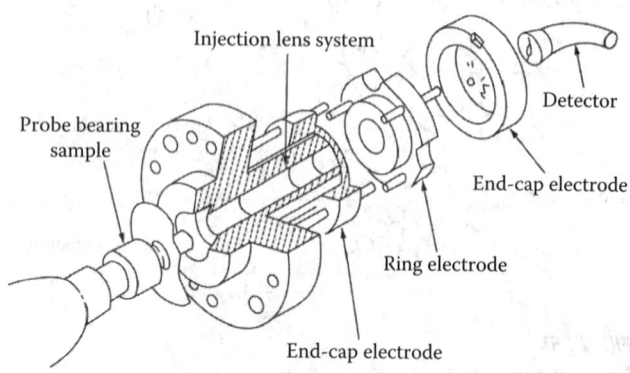

FIG. 1.35d
Ion-trap mass spectrometer.

ions undergo stable oscillations and remain trapped in that environment. Three electrodes, two end-cap electrodes that normally are at ground potential and between them a ring electrode to which an RF megahertz voltage is applied, generate the quadrupole electric field that can trap the sample ions. The ion separation is accomplished by introducing a change in operating voltage.

This alters the quadrupole electric field and allows trapped ions of a particular m/e ratio to adopt new but unstable trajectories. Thus, if the amplitude of the RF voltage applied to the ring electrode is systematically varied, ions of successively increasing m/e ratios are made to adopt unstable trajectories and to exit the ion trap, where they can be detected using an externally mounted electron multiplier.

Time-of-Flight Filter In time-of-flight MSs, the ionized sample is subjected to a negative-polarity-accelerating field, and then the ion beam is directed into a time-of-flight drift region. This drift region is the ion separation section of the spectrometer. The lighter ions travel faster through this region than the heavier ions, producing mass separation by the amount of time it takes for the ions to traverse this drift tube. To accomplish this type of mass separation, the ionized sample is introduced in discrete pulses of 20,000–35,000 pulses/s. On each cycle, the sample ions enter the flight tube (drift region) where the separation by mass values is accomplished. Upon exiting, the ions are then converted to electrons and amplified using an electron multiplier. The final amplified signal is detected by the detector's anode and fed to a suitable data acquisition system. Figure 1.35e illustrates the operational concepts for time-of-flight spectroscopy.

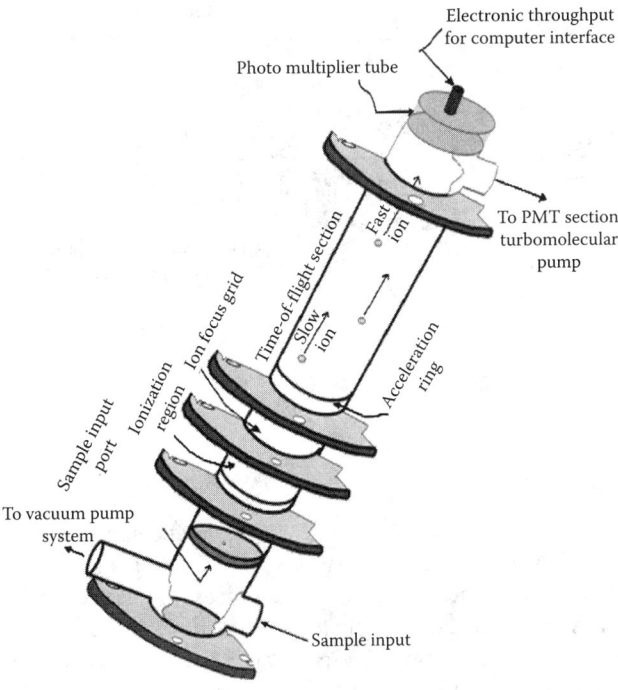

FIG. 1.35e
Time-of-flight mass spectrometer.

Ion Detection

It is important to understand that the ion collectors are not selective detectors. They will provide a signal if an ion with the appropriate m/e ratio emerges from the ion separator and strikes the detector surface. Thus, for this example, the conclusion that the atmospheric composition of the process stream is being monitored depends on the assumption that only ions created from the nitrogen, oxygen, and argon in air contribute to the collector signals at m/e ratios of 28, 32, and 40, respectively. If the process gas stream has other sources for these atoms, the conclusion is not immediately valid.

Ion detection errors develop for a variety of reasons. Errors that originate in the spectrometer's ion generation source or because of changes in conductance of the inlet leak are common. For the process stream atmospheric composition example, the three m/e ion collector outputs would be electronically scaled to create equal sensitivities for each measurement and then summed to create an analog of the sampled atmosphere. This error adjustment scheme is accomplished within the instrument with analog or digital circuitry. In either case, the sum is then compared to a fixed reference, and if it is properly calibrated, no error is created by this comparison. If a hypothetical drift is introduced, an error correction signal is created and is fed back to each channel through the instrument's gain adjusting elements. The gain gives an equal percentage change in each channel and drives the error to zero. The effectiveness of this technique for stabilizing calibration over extended periods has been proven in process applications. Error adjustments are accomplished by amplifier feedback circuits or digital correction techniques.

Vacuum Environment

To maintain the needed vacuum level within an MS, it is common to employ two vacuum pumps in series. With this pump arrangement, the removal of molecules from the instrument's vacuum environment becomes a two-stage pumping process. At the high-vacuum stage, the molecules enter the pumping system at a pressure of 1×10^{-5} or 1×10^{-6} Torr and are compressed to a pressure of 1×10^{-2} Torr. At this low-vacuum stage of the pumping process, the molecules enter the second pump and are compressed from 1×10^{-2} Torr to atmospheric pressure. The turbomolecular pump is a common selection for the high-vacuum pumping stage, while a root's blower or a rotary vane pump is a candidate for the low-vacuum-to-atmosphere pumping stage. Pump choice depends on required environmental conditions within the process being monitored. The pressure at both vacuum stages is monitored, and the vacuum sensor signals are used for diagnostic and control purposes. Vacuum sensors are discussed in Chapter 124.

For the spectrometer example presented within this chapter, an ion pump has been selected for the high-vacuum pumping stage. This choice is representative of a passive device that pumps by a combination of chemical gathering and physical burial. Like the cryopump, it does not require an additional pump connected in series; however, most systems include a parallel mechanical pumping system to lower the overall system-pumping burden. Figure 1.35f shows a typical ion pump. An electric field is established between the anode structure and the two cathodes, and a magnetic field is created by a permanent magnet. This creates a trapped electron cloud that continuously ionizes the neutral gas in the pump. The ions are accelerated out of the anode and impinge upon the cathode surface where they sputter cathode material. The tantalum and titanium cathode materials coat the pump structure. These two materials are very active getters and chemically combine with most of the ions present to remove them from the vacuum environment. Inerts are removed by actual burial under the sputtered material or by direct implantation.

RESIDUAL GAS ANALYZERS

The RGA represents the workhorse of mass spectroscopy with respect to industrial applications. This technology is a cost-effective way to meet the needs of several process applications. Figure 1.35g provides a conceptual view of the operation of an RGA. The technology is fundamentally a small low-resolution quadrupole MS that provides m/e resolutions that are suitable for certain types of process environments. In this illustration, the outer case that defines the vacuum environment envelope is not shown. The gas sample enters the ionization cage from the right and is ionized by electrons emitted from a hot filament. The ions are accelerated into the quadrupole section and then separated by their m/e ratio. Figure 1.35c provides more detail about a quadrupole mass filter. Similar m/e charge ratios enter the detector at the same time and are counted and displayed as an intensity signal at the specified m/e.

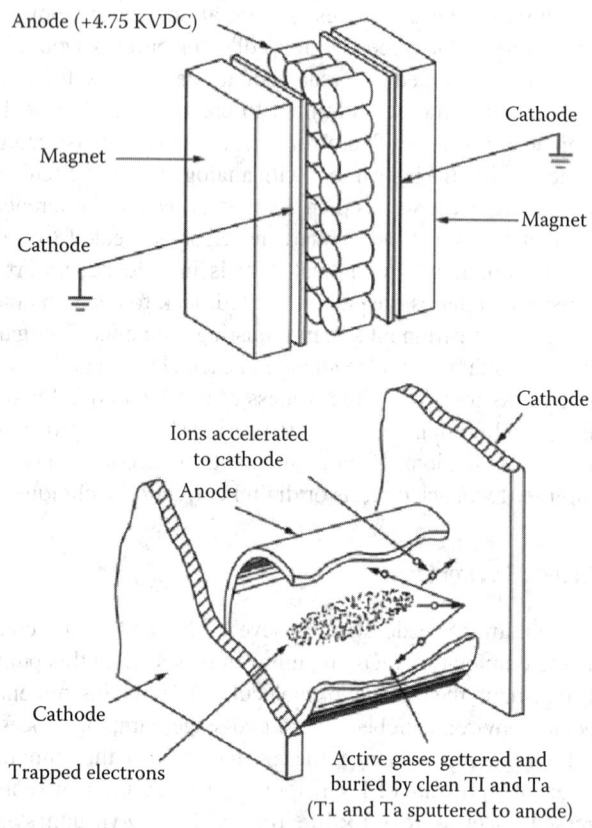

FIG. 1.35f
Ion pump operation.

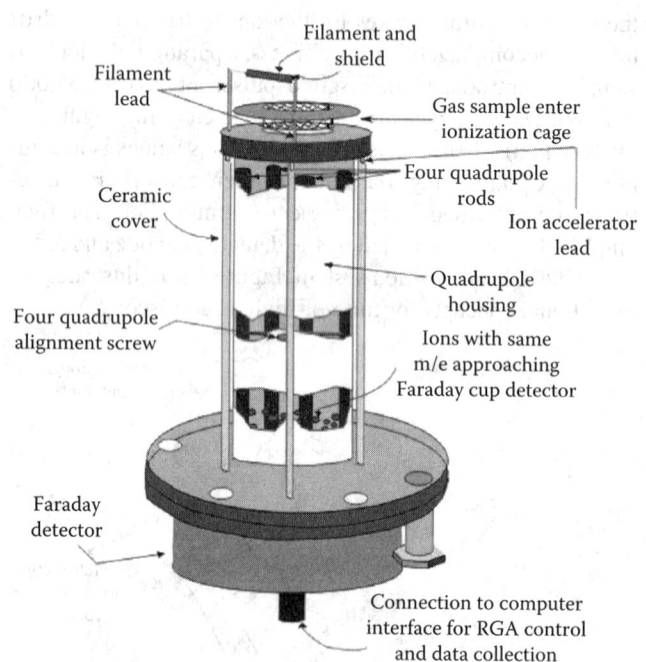

FIG. 1.35g
RGA operational concept.

Data Reduction and Presentation

It is to be expected that an MS system includes a complete computer-based data reduction system. In fact, for online industrial applications, there may be too many data manipulation options too easily available to shift operators. In any event, system manufacturers will help design the spectrometer output capabilities to meet your stated needs. In addition, there are an ample number of third-party vendors that can provide compatible software to meet the needs appropriate to any specific and special application.

The semiconductor, thin-film, MEMS and nanotechnologies depend on the RGA as online quality tools as well as vacuum leak detectors. In these process applications, the RGA serves as a diagnostic tool for monitoring the status of the process vacuum environment and the characterization of the process within that vacuum environment. The RGA is almost always added into sputter tool clusters and other unit operations within these industries.

CONCLUSIONS

The identification of a sample by the mass of its components is the appeal of mass spectroscopy and the driving force for its inclusion in other analytical techniques. The impact of increased genomic knowledge and research is now a major driver for these mass spectroscopy innovations. Natural product research as well as product production quality assurance protocols use high-resolution separation techniques, high-performance liquid chromatography, or capillary electrophoresis (HPLC or CE, respectively) combined with MSs. Application of these hybrid analysis platforms (CE-MS and HPLC-MS) is expanding to the quantification and distribution of constituents in biogenic raw materials, natural medicines, and biological materials obtained from model organisms, animals, and humans.

Innovations in ion production methods are also catapulting MSs into new applications. The introduction of an infrared laser to produce an electrospray ionizer has created new or faster protocols for off-line analysis in the fabric industry. In these dye identification and distribution in the fabric applications, the infrared matrix-assisted laser desorption electrospray ionization desorbs the dye from the sample that is then ionized by the electrospray. An electrospray MS ionizes the sample by forcing a solution with an organic solvent through a small heated capillary. On the very high extreme of the ionization energy range, accelerator mass spectroscopy (AMS), an ion accelerator, is combined with the low-energy analyzing magnet. This technology is directed toward nuclear physics applications that have a time reference that matches AMS requirements for long run times.

In summary, MSs are typically classified by the technology used to separate the ionic masses into clusters. MS examples include the quadrupole MS, the magnetic sector MS, the time-of-flight MS, and the ion-trap MS. (The RGA is an example of a quadrupole MS.)

For MS application in the process industry, the requirements of a dedicated, online analytical instrument might be stated as follows. The instrument is expected to perform a specific analytical task, on a continuous basis, for extended periods. It must exhibit long-term stability and accuracy while operating over its environmental range. An advantage of quadrupole instruments over electromagnetic-focusing ones is that the fields required to focus a particular mass can be changed very rapidly ($<10^{-3}$ s), which is especially valuable for computer-controlled measurements that require multiple-ion monitoring. The quadrupole ion trap is a mass spectrometer that has a broad mass range and provides molecular weight and structural information on biopolymers. It also has the greatest sensitivity of all mass spectrometers. The somewhat lower m/e resolution of the RGA is an acceptable trade-off when it is in situations where pressure, process stream disturbances because of contamination, and ease of assembly are the focus of attention.

SPECIFICATION FORMS

When specifying mass spectrometers, the ISA Form 20A1001 can be used to specify the detector device itself, and when specifying both the device and the nature of the application, including the composition and/or properties of the process sample, ISA Form 20A1002 can be used. Both forms are reproduced on the next pages with the permission of the International Society of Automation.

Analytical Measurement

ANALYSIS DEVICE — Operating Parameters

1	RESPONSIBLE ORGANIZATION		6	SPECIFICATION IDENTIFICATIONS		
2			7	Document no		
3	(ISA)		8	Latest revision	Date	
4			9	Issue status		
5			10			

	ADMINISTRATIVE IDENTIFICATIONS				
11					
12	Project number		Sub project no		
13	Project				
14	Enterprise				
15	Site				
16	Area		Cell	Unit	
17					

	SERVICE IDENTIFICATIONS			
18				
19	Tag no/Functional ident			
20	Related equipment			
21	Service			
22				
23	P&ID/Reference dwg			
24	Process line/nozzle no			
25	Process conn pipe spec			
26	Process conn nominal size		Rating	
27	Process conn termn type		Style	
28	Process conn schedule no		Wall thickness	
29	Process connection length			
30	Process line matl type			
31	Fast loop line number			
32	Fast loop pipe spec			
33	Fast loop conn nom size		Rating	
34	Fast loop conn termn type		Style	
35	Fast loop schedule no		Wall thickness	
36	Fast loop estimated lg			
37	Fast loop material type			
38	Return conn nominal size		Rating	
39	Return conn termn type		Style	

	SERVICE IDENTIFICATIONS Continued			
40				
41	Return conn matl type			
42	Inline hazardous area cl		Div/Zon	Group
43	Inline area min ign temp		Temp ident number	
44	Remote hazardous area cl		Div/Zon	Group
45	Remote area min ign temp		Temp ident number	
46				
47				

	COMPONENT DESIGN CRITERIA			
48				
49	Component type			
50	Component style			
51	Output signal type			
52	Characteristic curve			
53	Compensation style			
54	Type of protection			
55	Criticality code			
56	Max EMI susceptibility		Ref	
57	Max temperature effect		Ref	
58	Max sample time lag			
59	Max response time			
60	Min required accuracy		Ref	
61	Avail nom power supply			Number wires
62	Calibration method			
63	Testing/Listing agency			
64	Test requirements			
65	Supply loss failure mode			
66	Signal loss failure mode			

	PROCESS VARIABLES	MATERIAL FLOW CONDITIONS			Units
69					
70	Flow Case Identification				
71	Process pressure				
72	Process temperature				
73	Process phase type				
74	Process liquid actl flow				
75	Process vapor actl flow				
76	Process vapor std flow				
77	Process liquid density				
78	Process vapor density				
79	Process liquid viscosity				
80	Sample return pressure				
81	Sample vent/drain press				
82	Sample temperature				
83	Sample phase type				
84	Fast loop liq actl flow				
85	Fast loop vapor actl flow				
86	Fast loop vapor std flow				
87	Fast loop vapor density				
88	Conductivity/Resistivity				
89	pH/ORP				
90	RH/Dewpoint				
91	Turbidity/Opacity				
92	Dissolved oxygen				
93	Corrosivity				
94	Particle size				
95					
96	CALCULATED VARIABLES				
97	Sample lag time				
98	Process fluid velocity				
99	Wake/natural freq ratio				
100					

	PROCESS DESIGN CONDITIONS		
101	Minimum	Maximum	Units
102			
103			
104			
105			
106			
107			
108			
109			
110			
111			
112			
113			
114			
115			
116			
117			
118			
119			
120			
121			
122			
123			
124			
125			
126			
127			
128			
129			
130			
131			
132			

	MATERIAL PROPERTIES				MATERIAL PROPERTIES Continued		
133			137				
134	Name		138	NFPA health hazard		Flammability	Reactivity
135	Density at ref temp	At	139				
136			140				

Rev	Date	Revision Description	By	Appv1	Appv2	Appv3	REMARKS

Form: 20A1001 Rev 0 © 2004 ISA

1.35 Mass Spectrometers

		RESPONSIBLE ORGANIZATION		ANALYSIS DEVICE COMPOSITION OR PROPERTY Operating Parameters (Continued)				SPECIFICATION IDENTIFICATIONS			
1							6				
2							7	Document no			
3		ISA					8	Latest revision		Date	
4							9	Issue status			
5							10				

		PROCESS COMPOSITION OR PROPERTY			MEASUREMENT DESIGN CONDITIONS				
		Component/Property Name	Normal	Units	Minimum	Units	Maximum	Units	Repeatability
11									
12									
13									
14									
15									
16									
17									
18									
19									
20									
21									
22									
23									
24									
25									
26									
27									
28									
29									
30									
31									
32									
33									
34									
35									
36									
37									

Rev	Date	Revision Description	By	Appv1	Appv2	Appv3	REMARKS

Form: 20A1002 Rev 0 © 2004 ISA

Bibliography

Budzikiewicz, H., Selected reviews on mass spectrometric topics (online), *Mass Spectrometry Reviews*, 31(6), 2012.

Cook, R. G., McLuskey, S. A., and Kaiser, R. E., Ion trap mass spectroscopy, *Chemical and Engineering News*, American Institute of Engineers, New York, March 25, 1991.

Dearth, M. A., Evaluation of a commercial mass spectrometer for its potential to measure auto exhaust constituents in real time, *Industrial Engineering Chemical Research*, 38, 2002–2209, 1999.

Encyclopedia of Spectroscopy and Spectrometry, New York: Academic Press, 2000. http://booksite.elsevier.com/brochures/academicpress/.

Gas Chromatography and Mass Spectrometry: A Practical Guide, New York: Academic Press, 1996.

Hsu, Y., Lin, J., Chu, Q., Wang, Y., and Chen, C., Macromolecular ion accelerator mass spectrometer, *Analyst*, 138, 7384–7391, 2013.

http://pubs.rsc.org/en/content/ebook/978-1-84973-036-5#!divbookcontent.

Inorganic Mass Spectrometry: Fundamentals and Applications, New York: Marcel Dekker, 2000. http://www.dekker.com.

Johnson, R. C., Koch, K., and Cook, R. G., On-line monitoring of reactions of epichlorohydrin in water using liquid membrane introduction mass spectrometry, *Industrial Engineering Chemical Research*, 38, 343–351, 1999.

Lewis, G., Mass spectrometers move on-line, *InTech*, December 1989.

Li, M., Hu, B., Li, J., Zhang, X., and Chen, H., Extractive electrospray ionization mass spectrometry toward in situ analysis without sample pretreatment, *American Chemical Society*, 81, 7724–7731, 2009.

Liu, P., Lu, M., Zheng, Q., Zhang, Y., Dewald, H, and Chen, H., Recent advances of electrochemical mass spectrometry, *Analyst*, 138, 5519–5539, 2013.

Liu, S., Kleber, M., Takahashi, L., Nico, P., Keiluweit, M., and Ahmed, M., Synchrotron-based mass spectrometry to investigate the molecular properties of mineral-organic associations, *Analytical Chemistry*, 85, 6100–6106, 2013.

Mass Spectrometry and Nutrition Research, ISBN: 978-1-84973-036, Royal Society of Chemistry 2010. http://www.avs.org/.

Plasma Diagnostics, American Vacuum Society Classics, American Institute of Physics Press: New York, 1995. http://www.avs.org/.

Robertson, A. J. B., Mass spectrometry, *Rapid Communications in Mass Spectrometry*, 27(7), 2013.

Seger, C., Strum, S., and Stuppner, H., Mass spectrometer and NMR spectroscopy: Modern high-end detectors for high resolution separation techniques, *Natural Product Reports*, 30, 970–987, 2013.

Sparkman, O. D., *Mass Spectrometry Desk Reference*. Pittsburgh, PA: Global View Publication. ISBN 0-9660813-9-0. 2006.

Sparkman, O. D., *Mass Spectrometry Desk Reference*. Pittsburgh, PA: Global View Publishing. ISBN 0-9660813-2-3. Definition of spectrograph. *Merriam Webster*. 2008.

1.36 Mercury in Ambient Air

D. J. SIBBETT (1974, 1982) **R. S. SALTZMAN** (2003) **B. G. LIPTÁK** (1995, 2017)

Methods of detection	A. Ultraviolet, flameless atomic absorption
	B. Atomic absorption spectrophotometer
	C. Gold-film sensor
	D. Colorimetric titration
	E. Gas chromatography
	F. Atomic fluorescence
	G. Neutron activation
Inaccuracy	A. 2% on direct measurement
	B. 10% with concentrators
	C. 5% at 0.1 mg/m^3
	OSHA PEL: 0.1 mg/m^3 as total mercury (TWA)
Regulatory levels for emissions	100, 50, 10, and perhaps 5 µg/m^3
Range, in ambient air	1–200 ng/m^3
Ranges of analyzers	A. 0–400 volumetric ppb
	B. 0–45 or 0–75 µg/m^3
	C. 0–2 mg/m^3
Sensitivity	A. 1.0 ng/m^3 is about the best attainable
	B. 0.1 µg/m^3
	C. 0.003 mg/m^3
	F. 0.1 ppm in liquid impinger sample
Costs	A. Laboratory analyzer alone costs about $25,000; process analyzer alone costs about $30,000; an integrated sampler–detector system costs about $45,000 or more
	B. $100,000 and up
	C. Battery-operated, portable gold-film sensor with small flow pump costs $5,000
Partial list of suppliers	Also see the supplier lists of Chapters 1.22 and 1.26, for element and toxic substance detectors.
	A list of 49 mercury analyzer suppliers can be found at http://www.environmental-expert.com/companies/?keyword=mercury+analyzers
	ABB (A) (www.abb.com/analytical)
	Amptek http://www.amptek.com/
	AppliTek (http://www.applitek.com/en/offer/analyzers/air-quality-emission/stack-monitoring/online-mercury/)
	Aurora Instruments Ltd., http://www.aurorabiomed.com/atomic-fluorescence-spectroscopy/
	Bacharach Inc. (www.bacharach-inc.com)
	Bruker nanoanalytics, https://www.bruker.com/products/x-ray-diffraction-and-elemental-analysis/x-ray-fluorescence.html
	Buck Scientific (http://www.bucksci.com/catalogs/410%20Hg%20Analyzer.pdf)
	Cole-Parmer (http://www.coleparmer.com/Category/Cold_Vapor_Mercury_Analyzer/44404)
	EcoChem Analytics (http://www.ecochem.biz/Library/EcoChemCEMSAdvances.pdf)
	Fisher Scientific (www.fishersci.com)
	Gasmet (http://www.mercury2013.com/news/Gasmet-launches-new-Mercury-emissions-monitor/33/)
	Horiba Scientific, http://www.horiba.com/scientific/products/x-ray-fluorescence-analysis/tutorial/xrf-spectroscopy/
	LECO Corp. (www.leco.com)

(Continued)

Leeman Labs Inc. (www.leemanlabs.com)
Mercury Instr. GmbH (http://www.mercury-instrumentsusa.com/)
Ohio/Lumex (http://www.ohiolumex.com/products/ra82.php)
Perkin-Elmer Corp. (A, B) (www.instruments.perkinelmer.com)
P.S. Analytical (http://mercuryanalyser.com/mercury-analyzer/sir-galahad-ii-mercury-analyzer/)
Shimadzu http://www.shimadzu.com/an/elemental/wdxrf/xrf1800/xrf.html
Tekran Inc. (http://www.tekran.com/products/ambient-air/tekran-model-2537-cvafs-automated-mercury-analyzer/)
Teledyne (http://www.teledyneleemanlabs.com/products/mercury/index.asp)
Thermo Automation Syst. (www.thermo.com)

INTRODUCTION

Because the mercury concentration of interest in ambient air is in the range of nanograms per cubic meter (ng/m^3), no direct method of analysis exists and all measurements require the use of concentrators in sampling systems. Therefore, this chapter will start with descriptions of the various sampling and concentrator designs.

Instruments used in the measurement of mercury in air or gas samples are also discussed in other chapters of this handbook. These include the chapters on air monitoring (Chapter 1.4), chromatographs (Chapter 1.10), element detectors (Chapter 1.22), toxic detectors (Chapter 1.26), mass spectroscopy (Chapter 1.35), and ultraviolet analyzers (Chapter 1.69).

Air saturated with mercury will contain between 4 and 100 mg/m^3 of mercury in the temperature range between 7°C and 43°C (45°F and 110°F). Since saturation is not even approached under normal circumstances when the air is moving, sensitive analytical and sampling methods are needed when data are required on the actual ambient levels of mercury. For practical use, the methods employed must be relatively insensitive to interference and be otherwise appropriate for the collection of data under a wide range of operating conditions. For regulatory purposes, related to emissions, three or four levels of sensitivity are of importance: 100, 50, 10, and perhaps 5 μg/m^3 of mercury in the air. Accuracy within ±10% at these levels is required. However, ambient air monitoring requires the ability to determine levels in the 1–200 mg/m^3 range.

SAMPLING AND CONCENTRATION

Collection of mercury samples for accurate determination of concentrations found at ambient conditions or in industrial plants requires that

1. The size of the sample taken must be adequate to satisfy the sensitivity threshold requirements of the analytical technique utilized
2. The rate of sampling must be controlled so that efficient collection is achieved
3. The sampling interval selected must be appropriate for the required monitoring function
4. The collection device or procedure must be compatible in capacity and chemical reactivity of material components with the levels and properties of mercury and mercury compounds to be sampled
5. Effects of potential interference must be avoided
6. Elapsed time between collection and analysis must be minimized in order to avoid losses
7. It must be possible to evaluate all unknown factors in collection and analysis by using the appropriate calibration procedures

In the sampling of mercury and its compounds from the air, two basic procedures are available. These procedures utilize either the chemical reactivity of the sampled components in various aqueous media or the extraordinarily efficient extraction (amalgamation) of mercury from air by the noble metals.

In the latter case, reduction of mercury compounds to elemental mercury is required before the sampling process is carried out. Currently, a wet chemical procedure for sampling is recommended by the Environmental Protection Agency. This procedure requires considerable labor and upkeep to perform effectively.

Impinger Collection Methods

Impingers are specially designed bubble tubes used for collecting airborne chemicals into a liquid medium. With impinger sampling, a known volume of air is bubbled through the impinger containing a specified liquid. The liquid will chemically react with or physically dissolve the chemical of interest, which in this case is mercury.

In conjunction with the publication of the National Emission Standards for Hazardous Air Pollutants,[*] the Environmental Protection Agency has published two methods for sampling and analyzing mercury in gaseous and particulate emissions. The methods are similar with the exception that particulates must be sampled isokinetically, whereas gaseous samples are passed through filters to remove particles. This latter procedure requires

---

[*] EPA, *National Emission Standards for Hazardous Air Pollutants Compliance Monitoring*, 2015, http://www2.epa.gov/compliance/national-emission-standards-hazardous-air-pollutants-compliance-monitoring.

equilibration of the filter prior to generation of accurate data at low mercury levels.

In principle, samples are drawn through acidic aqueous iodine monochloride solutions utilizing stringent controls over flow rates, acidity, and other process variables. This procedure gives excellent recovery of organic mercury compounds and of mercury vapor. Samples obtained with this procedure may be stored.

Particulate Sampling Figure 1.36a shows the sampling train utilized for particulate and vapor samples. It consists of a Pyrex-lined, heated probe section containing a thermocouple for temperature measurement and a pair of pitot tubes and a manometer for gas flow measurement. This section is followed by five impingers of the Smith–Greenburg type placed in an ice bath and a filter for particle removal prior to passage of the gases through the two final absorber stations. These components are followed by miscellaneous components for control and calibration of the sample airflow.

For sampling particulates, the initial rate of gas flow is 2 scfh (57 slph). Flow should be maintained at a rate proportional to the flow through the stack sampled.

Vapor Sampling Figure 1.36b shows a simpler technique used for sampling of vapors without particulates. In this method, particulates are removed by a filter, at the inlet into the absorption train. Samples obtained by this procedure may be analyzed by a procedure utilizing a flameless atomic absorption spectrophotometer, which has been accepted by the Environmental Protection Agency for investigations.

A number of other collection procedures employing impingers have been examined. These utilize water, ethyl alcohol, isopropyl alcohol, potassium permanganate–sulfuric acid, potassium permanganate–nitric acid, or iodine–potassium iodine.

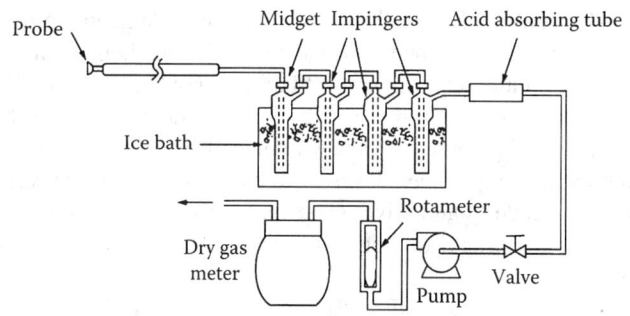

FIG. 1.36b
Vapor sampling train.

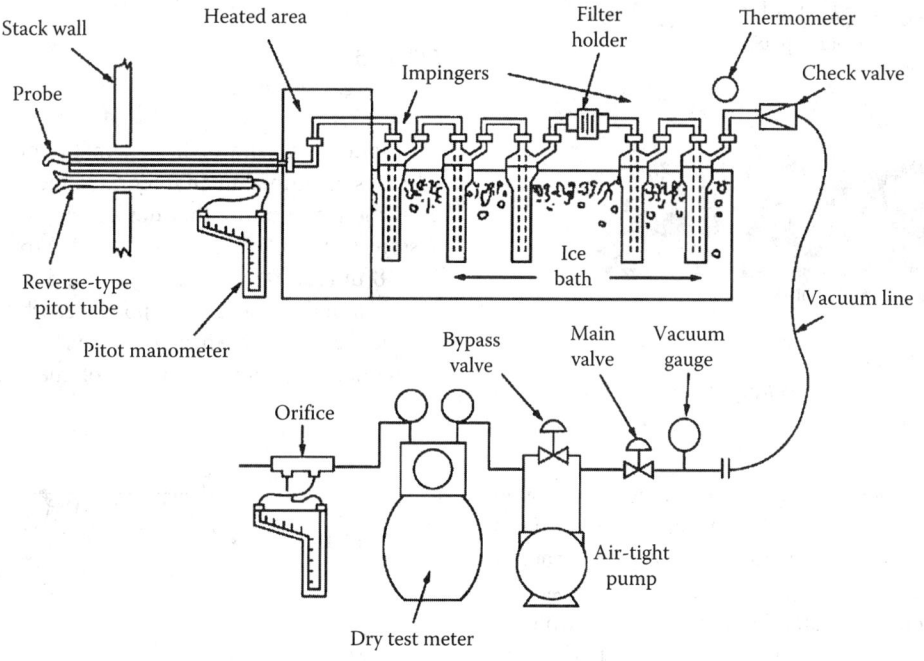

FIG. 1.36a
Particulate sampling train.

However, for effective collection of alkyl mercury compounds, acidic iodine monochloride solutions are preferable.

Amalgamation on Wettable Metals

A number of less laborious methods of sampling mercury vapor, compounds, and particulates have been developed that utilize absorption (amalgamation) of vapor on silver, gold, and alloys of these materials. In order to sample particulates and compounds of mercury, the trapping procedure must be preceded by a decomposition step. A catalytic process that achieves this objective is described later.

Examples of procedures for collecting elemental mercury vapor include the method shown in Figure 1.36c, which shows a simple absorption tube filled with gold or silver wool. The tube may be sized to satisfy any specific parameters of the sampled gas stream. Fine wire may also be used as packing for flow rates under 50 L/min, and mercury vapor removal can be made quantitative. Gold materials are required for gases containing hydrogen sulfide or sulfur dioxide.

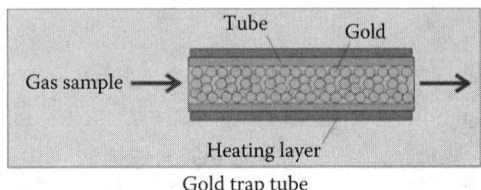
Gold trap tube

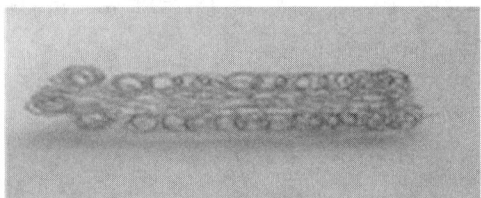
Gold wire filling

FIG. 1.36c
Absorption tube type mercury collection device. (Courtesy of Mercury Instr. GmbH.)

The gold trap selectively forms a mercury amalgam. Mercury is adsorbed and desorbed from the gold. Thus, interference from other compounds like SO2 is eliminated.

The gold trap drastically enhances detection sensitivity. If ultra-pure reagents and atomic absorption spectrometer (AAS) detectors are used, a measuring range from 1 ppt to 5000 ppt can be achieved, and with atomic fluorescence spectroscopy (AFS) detectors, measuring ranges of 0.05 ppt to 5000 ppt are possible.

When a simple absorption tube of this type is employed, the mercury vapor collected may be released for analysis in a flameless atomic absorption spectrometer (AAS) by utilizing an induction furnace (Chapter 1.22). In this procedure, extensively used by the U.S. Geological Survey, a continuously flowing airstream is used to pass the pulse of the mercury vapor into the AAS cell. This pulse is generated during the heating in the furnace.

Sampling Tubes The Canadian Department of the Environment employs a versatile method of sampling of gas streams. It uses a packing, which consists of finely woven silver mesh (Figure 1.36d). One version of the sampling tube is 5 in. (125 mm) high and utilizes a 1/4 in. (6 mm) inner diameter Pyrex tubing. Separation between the inner and outer tubes is 3/64 in. (0.1 mm). In order for this collection procedure to function effectively, relatively low airflow rates through the wire mesh must be used. Sampling rates of 1–5 L/min have been employed, but long sampling periods are required to make measurements at low air concentrations.

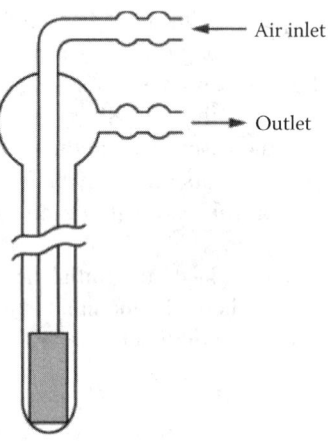

FIG. 1.36d
Mercury gas sampling tube.

Figure 1.36e shows the schematics of a collection device, which is contained in an integrated mercury analyzer. In the collection section of this device, air at rates from 0.03 to 7 scfm (1 to 200 L/min) is passed through a grid containing 48 m (160 ft) of 22-gauge wire arranged in a carefully spaced annulus. Mercury vapor is desorbed by direct passage of electrical current through the wire. Analysis is achieved by the flameless atomic absorption method in a photometer.

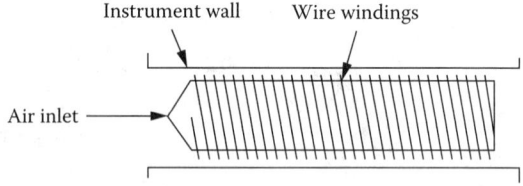

FIG. 1.36e
Mercury collection device.

Activated Absorption

The use of silver on a charcoal absorbent is highly efficient and may be used to remove mercury vapor from many gas streams. Extraction of the mercury vapor as a concentrated pulse for analysis, however, is relatively difficult. Induction heating is required to maximize the vapor concentration. Trapping is also possible with iodine-activated charcoal and mineral wool. Regeneration of these materials for repeated use presents difficulties not yet resolved.

Conversion of Mercury Vapors

In order to analyze for total mercury including particulates and vapor, reduction of compounds such as mercuric chloride, mercuric sulfide, dimethyl and diethyl mercury, pesticides, and fungicides to the elemental form must be achieved. Three techniques are in use: (1) thermal decomposition at 816°C–1093°C (1500°F–2000°F); (2) catalytic reduction or disproportionation, usually at 648°C–704°C (1200°F–1300°F) over materials such as cupric oxides; and (3) chemical conversion in solutions such as acidic iodine monochloride. Commercial equipment is available, which may be used in conjunction with sampling and analysis instrumentation.

DETECTION METHODS

No direct analytical method is capable of attaining accurate measurements in the range required for ambient air without the use of a concentrator, discussed earlier.

Ultraviolet Absorption

A wide variety of methods have been developed to measure airborne mercury-containing materials in the atmosphere based on the strong absorption of ultraviolet light at 253.7 nm by an elemental mercury vapor. (For a complete description of the various UV analyzers, see Chapter 1.69.) Since many other compounds absorb light in this range, although they do that at a much higher concentration than does mercury, some method of separating out these interferences is needed when low levels of mercury are to be measured.

Atomic Absorption Spectrophotometry Atomic absorption spectroscopy (AAS) is a procedure for the quantitative determination of chemical elements using the absorption of optical radiation by free atoms in the gaseous state. It was first used in the 19th century by Robert Wilhelm Bunsen in Germany. The atomizers most commonly used nowadays are flames and electrothermal (graphite tube) atomizers. The atoms are irradiated by an element specific radiation source or a monochromator is used to separate the element-specific radiation, which is finally measured by a detector (Figure 1.36f). Mercury is conveniently analyzed by atomic absorption techniques because the elemental vapor exists in atomic form under normal ambient conditions. Thus, when a cloud of mercury vapor is irradiated at an appropriate wavelength, mercury atoms absorb a portion of the energy of the beam.

In the standard atomic absorption technique, liquid samples are aspirated into an acetylene air flame where the atomic mercury is released and the absorption of light at 253.7 nm is measured by a spectrophotomultiplier (Figure 1.36g).

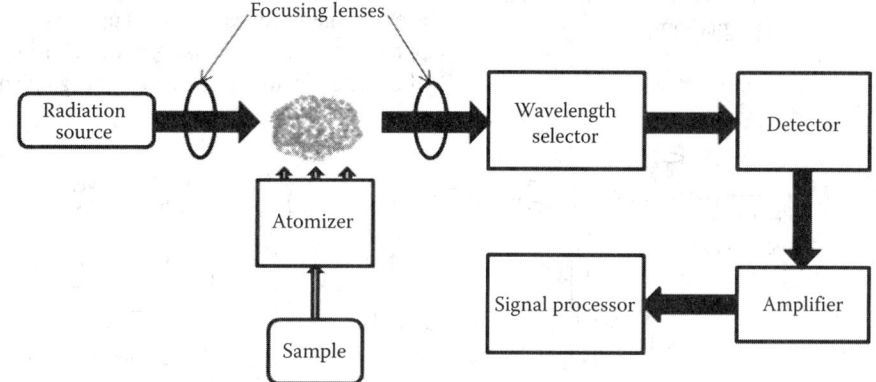

FIG. 1.36f
The main components of an atomic absorption spectroscope (AAS). (From https://en.wikipedia.org/wiki/Atomic_absorption_spectroscopy.)

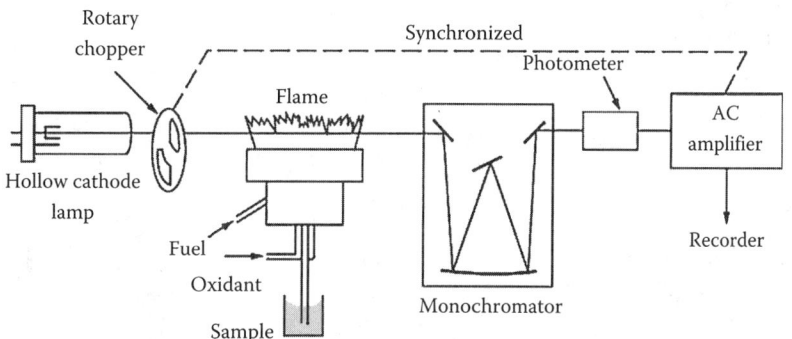

FIG. 1.36g
Atomic absorption spectrophotometer.

The sensitivity of 0.01 μg of mercury/mL of sample is readily attained by this procedure. A continuous series of calibration checks are required for maximal precision and accuracy.

Flameless Atomic Absorption Spectroscopy The burner–aspirator and flame indicated in the schematic of the conventional AAS (Figure 1.36g) may be replaced by a quartz-windowed glass cell, and the sample may be generated by reducing an aqueous sample of mercury compounds to elemental mercury.

Figure 1.36h shows the apparatus that may be used in conjunction with an AAS. In this procedure, a sample collected by utilizing a separate collection train is pipetted into the gas washing bottle. To this is added a solution of a reducing agent such as hydroxylamine sulfate or stannous chloride, and the bottle is closed.

With clean air supplied at a rate of 1.3 L/min, aeration is commenced and continued until a maximum peak height readout is obtained on the recorder. By calibration against acidic mercuric chloride standard solutions, the peak heights obtained with unknown samples can be compared with the calibration curve for direct determination of the mercuric ion concentration. This value may be converted to the gas-phase concentration of mercury.

A number of commercial units are available that use the principle of flameless atomic absorption spectroscopy for determination of mercury. Some designs require that samples be taken by use of a liquid sampling train and transported to the instrument, while others accept gas samples.

Colorimetric Methods

Dithizone In this method, air is drawn through an acidic solution (10% H_2SO_4) of potassium permanganate (4 g/100 mL), where the elemental mercury, inorganic mercury compounds, and the readily decomposable mercury–organic compounds are converted to mercuric ions.

Methyl mercury compounds require the use of permanganate at 100°C (212°F). The determination of mercury may be concluded colorimetrically whence the sample is titrated with dithizone and compared with the colors developed by standards. A spectrophotometer may also be used for quantitative comparison between the color developed by test samples and those of controlled standard quantities of mercury.

In the method developed by the American Conference of Government Industrial Hygienists, the sample is collected from the air by passing it through an aqueous solution of iodine (0.25%) and potassium iodine (3%). Mercury is extracted from a buffered solution (ammonium citrate, hydroxylamine hydrochloride, and phenol red at pH 8.5) with dithizone in chloroform (20 mg of dithizone/5 mL) until all color is removed. After washing the chloroform phase, the mercury is extracted into acidic (0.1 N HCl) potassium bromide (8%), shaken with a chloroform solution of 10 mg/L of dithizone, and filtered through cotton into a colorimeter cell. The optical density at 485 mm is used to determine the mercury. Calibration is carried out with mercuric chloride.

The dithizone method is relatively insensitive, requiring the use of large samples (about 10 μg are required) when the levels of mercury are low. It also requires considerable care on the part of the chemist to avoid the loss of organic mercury compounds.

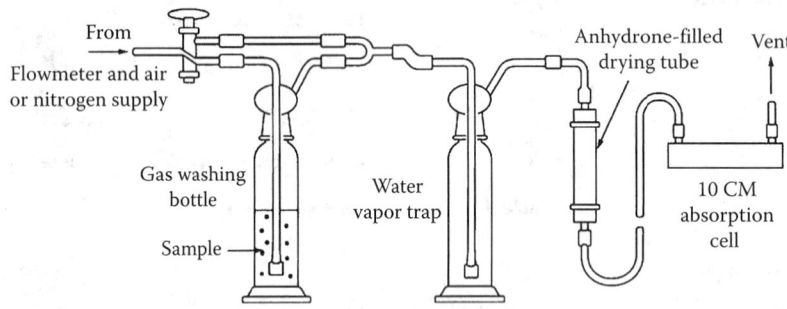

FIG. 1.36h
Apparatus for flameless atomic absorption.

Selenium Sulfide and Others Active selenium sulfide applied to paper as a coating may be used to detect the presence of elemental mercury vapor. The paper develops a black color on exposure to air containing mercury. The degree of blackening is a function of the concentration of mercury and the times of exposure. Quantitation may be made with a simple densitometer.

Mercury can also be detected with copper, iodine paper, selenium paper, and gold chloride on silica gel. These methods appear to have a limiting sensitivity in the 500–1000 µg/m³ range and a measurement error of ±5%.

Gas Chromatography for Mercury Organics

Organic mercurials can be determined at levels of approximately 1 ppb by gas chromatographic methods. Samples must be extracted into benzene and injected into a gas chromatographic column (Carbowax 20M on Chromosorb W) operating at 180°C (356°F), which separates methyl mercury, ethyl mercury, methoxyethyl mercury, phenyl mercury, and dimethyl mercury. Quantitation is obtained with an electron capture detector. Dimethyl mercury requires conversion into methyl mercury halide for analysis, however. Precision is reported to be about 12%, with 30% reliability.

Atomic Fluorescence Spectroscopy

Samples collected by any of the liquid impinger techniques (Figures 1.36a and 1.36b) may be analyzed by use of atomic fluorescence flame spectrophotometric methods. Figure 1.36i provides a schematic diagram of a typical fluorescence spectrophotometer and of some typical calibration data. Figure 1.36j shows an actual x-ray fluorescence spectrometer. A mercury concentration of 0.1 ppm in a collection fluid is the detection limit of this method.

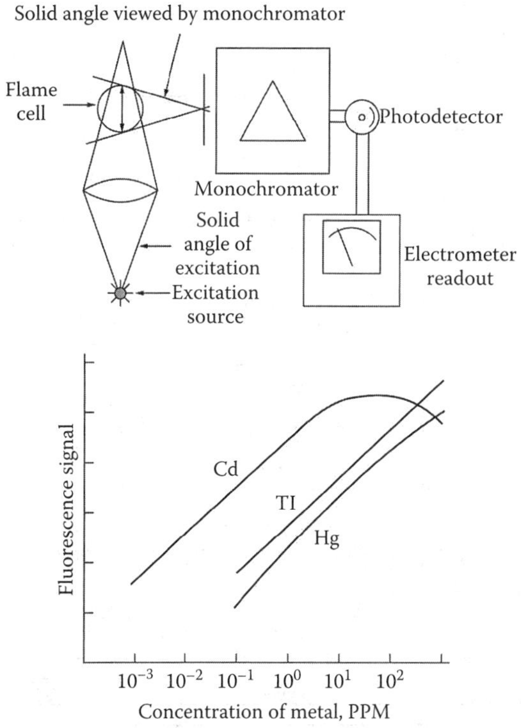

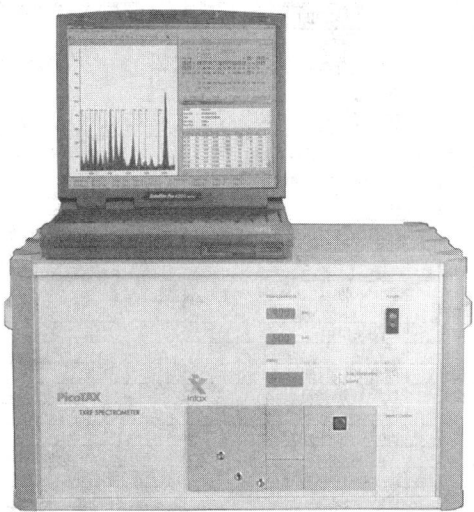

FIG. 1.36j
X-ray fluorescence spectrometer. (Courtesy of Rigaku Corporation.)

Other Analytical Procedures

A radiochemical procedure for the determination of mercury vapor has been developed based on the isotopic exchange that takes place when air samples containing mercury are passed through a solution of Hg^{203} mercuric acetate. The method potentially has good sensitivity but requires precautions, which are normal in handling radioisotopes.

Mercury that has been extracted from the air may be determined utilizing a neutron activation procedure. The measurement error and precision of this method is in the range of ±2%.

REGULATIONS

In the United States, the primary piece of federal legislation governing the airborne release of mercury in the Clean Air Act. Mercury is classified under the Act as a hazardous air pollutant and is thus subject to control under the National Emission Standards for Hazardous Air Pollutants (NESHAP) rather than the National Ambient Air Quality Standard (NAAQS). The key distinction is that the former is controlled by establishing performance standards under a program known as the maximum achievable control technology standards (MACT). However, as of early 2011, no federal limits of mercury from coal- and oil-fired electric utility steam generating units (EGUs) were on the books. The development of a regulatory policy framework to guide the emissions of mercury from power plants is ongoing, the major developments of which have occurred largely within the last decade.

FIG. 1.36i
Atomic fluorescence spectrophotometer.

SPECIFICATION FORMS

When specifying Mercury Analyzers one can use the ISA form 20A1001 and when specifying the detector and also describing the application ISA form 20A1002 can be used. The following are the forms that are reproduced with the permission of the International Society of Automation.

#	RESPONSIBLE ORGANIZATION	ANALYSIS DEVICE	#	SPECIFICATION IDENTIFICATIONS		
1			6			
2	(ISA)	Operating Parameters	7	Document no		
3			8	Latest revision	Date	
4			9	Issue status		
5			10			

#	ADMINISTRATIVE IDENTIFICATIONS			#	SERVICE IDENTIFICATIONS Continued		
11				40			
12	Project number	Sub project no		41	Return conn matl type		
13	Project			42	Inline hazardous area cl	Div/Zon	Group
14	Enterprise			43	Inline area min ign temp	Temp ident number	
15	Site			44	Remote hazardous area cl	Div/Zon	Group
16	Area	Cell	Unit	45	Remote area min ign temp	Temp ident number	
17				46			
18	SERVICE IDENTIFICATIONS			47			
19	Tag no/Functional ident			48	COMPONENT DESIGN CRITERIA		
20	Related equipment			49	Component type		
21	Service			50	Component style		
22				51	Output signal type		
23	P&ID/Reference dwg			52	Characteristic curve		
24	Process line/nozzle no			53	Compensation style		
25	Process conn pipe spec			54	Type of protection		
26	Process conn nominal size	Rating		55	Criticality code		
27	Process conn termn type	Style		56	Max EMI susceptibility	Ref	
28	Process conn schedule no	Wall thickness		57	Max temperature effect	Ref	
29	Process connection length			58	Max sample time lag		
30	Process line matl type			59	Max response time		
31	Fast loop line number			60	Min required accuracy	Ref	
32	Fast loop pipe spec			61	Avail nom power supply		Number wires
33	Fast loop conn nom size	Rating		62	Calibration method		
34	Fast loop conn termn type	Style		63	Testing/Listing agency		
35	Fast loop schedule no	Wall thickness		64	Test requirements		
36	Fast loop estimated lg			65	Supply loss failure mode		
37	Fast loop material type			66	Signal loss failure mode		
38	Return conn nominal size	Rating		67			
39	Return conn termn type	Style		68			

#	PROCESS VARIABLES	MATERIAL FLOW CONDITIONS			#	PROCESS DESIGN CONDITIONS		
69				Units	101	Minimum	Maximum	Units
70	Flow Case Identification				102			
71	Process pressure				103			
72	Process temperature				104			
73	Process phase type				105			
74	Process liquid actl flow				106			
75	Process vapor actl flow				107			
76	Process vapor std flow				108			
77	Process liquid density				109			
78	Process vapor density				110			
79	Process liquid viscosity				111			
80	Sample return pressure				112			
81	Sample vent/drain press				113			
82	Sample temperature				114			
83	Sample phase type				115			
84	Fast loop liq actl flow				116			
85	Fast loop vapor actl flow				117			
86	Fast loop vapor std flow				118			
87	Fast loop vapor density				119			
88	Conductivity/Resistivity				120			
89	pH/ORP				121			
90	RH/Dewpoint				122			
91	Turbidity/Opacity				123			
92	Dissolved oxygen				124			
93	Corrosivity				125			
94	Particle size				126			
95					127			
96	CALCULATED VARIABLES				128			
97	Sample lag time				129			
98	Process fluid velocity				130			
99	Wake/natural freq ratio				131			
100					132			

#	MATERIAL PROPERTIES			#	MATERIAL PROPERTIES Continued		
133				137			
134	Name			138	NFPA health hazard	Flammability	Reactivity
135	Density at ref temp	At		139			
136				140			

Rev	Date	Revision Description	By	Appv1	Appv2	Appv3	REMARKS

Form: 20A1001 Rev 0 © 2004 ISA

1.36 Mercury in Ambient Air

	RESPONSIBLE ORGANIZATION	ANALYSIS DEVICE COMPOSITION OR PROPERTY Operating Parameters (Continued)		SPECIFICATION IDENTIFICATIONS		
1			6			
2			7	Document no		
3			8	Latest revision		Date
4			9	Issue status		
5			10			

	PROCESS COMPOSITION OR PROPERTY			MEASUREMENT DESIGN CONDITIONS				
	Component/Property Name	Normal	Units	Minimum	Units	Maximum	Units	Repeatability
13								
...								
37								

Rev	Date	Revision Description	By	Appv1	Appv2	Appv3	REMARKS

Form: 20A1002 Rev 0 © 2004 ISA

Definitions

Atomic absorption spectroscopy (AAS) is a procedure for the quantitative determination of chemical elements using the absorption of optical radiation by free atoms in the gaseous state. It was first used in the 19th century by Robert Wilhelm Bunsen in Germany. The atomizers most commonly used nowadays are flames and electrothermal (graphite tube) atomizers. The atoms are irradiated by an element specific radiation source or a monochromator is used to separate the element-specific radiation, which is finally measured by a detector.

Impingers are specially designed bubble tubes used for collecting airborne chemicals into a liquid medium. With impinger sampling, a known volume of air is bubbled through the impinger containing a specified liquid. The liquid will chemically react with or physically dissolve the chemical of interest.

Abbreviations

AAS	Atomic absorption spectrometer
AFS	Atomic fluorescence spectroscopy
CFR	Code of federal regulations
FR	Federal register
GLI	Great Lakes Initiative
MACT	Maximum achievable control technology
MATS	Mercury and air toxics standards
NAAQS	National ambient air quality standards
NESHAP	National emission standards for hazardous air pollutants
PEL	Permissible exposure limits
ppt	Parts per trillion
scfh	Standard cubic foot per hour
slph	Standard liters per hour
TCLP	Toxicity characteristic leaching procedure
TWA	Time weighted average

Organizations

ACGIH	American Conference of Government Industrial Hygienists
FDA	Food and Drug Administration

Bibliography

Arizona Instrument Inc., *Mercury Vapor Analyzer*, 2011, http://www.azic.com/wp-content/uploads/2014/12/700-0046-Jerome-431-X-Mercury-Vapor-Analyzer-Users-Manual.pdf.

EPA, *Mercury and Air Toxics Standards (MATS)*, 2015, http://www3.epa.gov/mats/actions.html.

EPA, *National Emission Standards for Hazardous Air Pollutants Compliance Monitoring*, 2015, http://www2.epa.gov/compliance/national-emission-standards-hazardous-air-pollutants-compliance-monitoring.

EPA, *Sampling and Analysis for Atmospheric Mercury*, 1999, http://www.tekran.com/files/EPA_IO-5.pdf.

IKIMP Mercury Literature, 2013, http://www.mercurynetwork.org.uk/policylinks/.

Pirrone, N. and Mason, R., *Mercury Fate and Transport in the Global Atmosphere: Emissions, Measurements and Models*, Springer, 2010, http://www.springer.com/us/book/9780387939575.

SKC Sample Setup Guide, 2014, http://www.skcinc.com/catalog/pdf/instructions/1165.pdf.

UNEP, Global Mercury Assessment, 2013, http://www.unep.org/PDF/PressReleases/GlobalMercuryAssessment2013.pdf.

Welz, B., Becker-Ross, H., Florek, S., and Heitmann, U., *High-resolution Continuum Source AAS*, 2005, http://onlinelibrary.wiley.com/doi/10.1002/3527606513.fmatter/pdf.

Yang, Y., Chen, H., and Wang, D., *Spatial and Temporal Distribution of Gaseous Elemental Mercury in Chongqing, China*, 2009, http://link.springer.com/article/10.1007%2Fs10661-008-0499-8.

1.37 Mercury in Water

S. NISHI (1974, 1982)　　**B. G. LIPTÁK** (1995, 2003)　　**S. VEDULA** (2017)

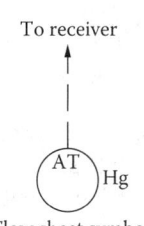

Flow sheet symbol

Pretreatment of sample	Wet oxidation is used for the detection of the total mercury level, and solvent extraction is used for the detection of organic mercury.
Method of detection	A. Colorimetric detection of total mercury level B. Atomic absorption or Atomic Fluorescence spectrophotometry of total mercury level C. Gas chromatograph with electron capture detector for Organic mercury D. Thin-layer chromatographic detection of organic mercury E. Ultraviolet detector preceded by wet chemistry package F. Thermal (Direct) decomposition
Ranges	The detection ranges for A and D are approximately 0.01–10 ppb of mercury in water, and for B and C, approximately 0.1 ppt–250 ppb of mercury in water. For E, the measuring ranges are from 0–1 μg/L to 0–100 ppm.
Costs	A. Colorimetric spectrophotometers for laboratory applications range from $4,000 to $10,000 B. About $100,000, depending on pretreatment of sample C. About $50,000 for laboratory application E. $50,000 or more F. $50,000
Partial list of suppliers	(For a more complete list, for chromatograph suppliers, see Chapter 1.10; for colorimeter suppliers, see Chapter 182; for atomic absorption analyzers, see Chapter 1.22; for chromatographs, see Chapter 1.10; for ultraviolet analyzer, see Chapter 1.69; and for wet chemistry analyzer, see Chapter 1.74.) Aurora Instruments (http://www.aurorabiomed.com/atomic-fluorescence-spectroscopy-3300.htm)—B Bacharach Inc. (http://www.bacharach-inc.com/PDF/MSDS/40991482.pdf)—Reagents Brooks Rand LLC (http://www.brooksrand.com/InstrumentManufacturingHome/MERXMMHgSystems)—B & C Buck Scientific (http://www.bucksci.com/catalogs/410%20Hg%20Analyzer.pdf?phpMyAdmin=97c9c4cc71cb0t6d5c9a3er5a21)—B Cetac (http://www.cetac.com/pdfs/Brochure_M-7600.pdf) and (http://www.cetac.com/pdfs/Brochure_M-8000.pdf)—B Cole-Parmer (A) (http://www.coleparmer.ca/Category/Cold_Vapor_Mercury_Analyzer/44404)—B LECO Corp. (http://www.leco.com/products/analytical-sciences/mercury/ama254-mercury-analyzer)—F Leeman Labs Inc. (http://www.teledyneleemanlabs.com/products/mercury/index.asp)—B & F Mercury Instr. GmbH-B (http://www.mercury-instruments.com/en-Mercury_Instruments_Products_PA_2.html) Milestone (http://milestonesci.com/products/direct-mercury-analysis.html)—B Perkin-Elmer Corp. (B, E) (http://www.perkinelmer.ca/CMSResources/Images/44–74467BRO_SMS100.pdf)—B P.S. Analytical (http://www.psanalytical.com/products/millenniummerlin.html)—B Shimadzu Scientific Instruments Tekran Inc. (http://www.tekran.com/products/laboratory/overview/)—B & C

INTRODUCTION

While Chapter 1.36 described the analyzers used in the monitoring of mercury concentration in ambient air samples, this chapter is devoted to the measurement of mercury in liquid or gas. There is unavoidably some overlap between the two chapters.

The analyzers used for the measurement of mercury are discussed in more detail in other chapters: chromatographs are covered in Chapter 1.10, colorimeters in Chapter 182, atomic absorption analyzers in Chapter 1.22, ultraviolet (UV) analyzers in Chapter 1.69, and wet chemistry analyzers in Chapter 1.74. Because these analyzers are mostly online, the emphasis in this chapter is placed on, but not limited to, the laboratory procedures, while the discussion of online units is not as detailed in this chapter.

The continuous online mercury monitors are used in many industrial processes, including both effluent and quality controls in chlorine–alkali plants, quality controls of drinking water and of sulfuric acid, control of industrial sewage and purification plants, monitoring the scrubber water in power plants and of waste incinerators, just to mention a few.

TOTAL MERCURY DETECTION

Sample Treatment

Mercury is present in the environment in organic and inorganic compounds, either dissolved in water or adsorbed on particulate matter or sediments. The total mercury level is the sum of organic mercury and inorganic mercury in a sample.

To analyze a variety of mercury compounds by the commonly available methods, it is necessary to digest the sample in order to effect decomposition of the accompanying substances, and thereby to convert the mercury present in the sample into inorganic mercury (Hg^{2+}).

There are four combinations of digestion methods used as detailed below:

1. Sample preoxidation with sulfuric acid + potassium permanganate
2. Sample preoxidation with hydrochloric acid + potassium permanganate
3. Sample preoxidation with hydrochloric acid + sodium chlorate
4. Sample preoxidation with Fenton's reagent

This digestion is carried out by mixing the sample with a mineral acid and a strong oxidizing agent and heating it. Out of the above methods, the preoxidation with sulfuric acid and potassium permanganate is most commonly used.

Heating is either by a water bath or by direct flame. The digestion must be carried out in the presence of excess $KMnO_4$ and, therefore, a fresh addition must be made whenever the color of the permanganate fades (the digestion being complete when the color of $KMnO_4$ no longer fades).

In case of biological materials where the samples are rich in organic matter, the samples can be best digested by heating with H_2SO_4 and HNO_3 until white fumes (sulfuric acid) evolve. The digestion is complete when the solution becomes clear. When carbonization of organic matters occurs, the digestion is continued with fresh addition of nitric acid. An apparatus such as that shown in Figure 1.37a prevents

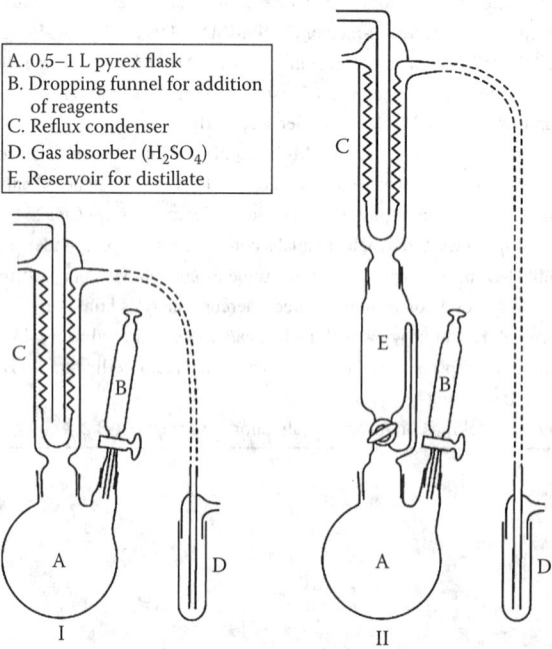

A. 0.5–1 L pyrex flask
B. Dropping funnel for addition of reagents
C. Reflux condenser
D. Gas absorber (H_2SO_4)
E. Reservoir for distillate

FIG. 1.37a
Digestion flasks for detection of mercury (I) in water and (II) in organic materials.

the small losses due to vaporization of mercury during the digestion.

Since upon completion of the digestion an excess of the oxidizing agent is present in the digested solution, it becomes necessary to reduce the remaining oxidizing agent. Hydroxylamine is best suited for the reduction of NO_2 or $KMnO_4$. Hydroxylamine alone, however, cannot reduce nitrous acid completely with nitric acid as a digesting liquor, and simultaneous addition of urea is advisable. The samples thus digested are then analyzed for mercury by one of the detectors described below.

Colorimetric Detection

Dithizone is widely used in the colorimetric analysis of mercury, where the absorption of visible light is detected as a measure of mercury concentration. When an aqueous solution of mercury with a wide pH range (0–13) is shaken with a solution of dithizone in chloroform, carbon tetrachloride, or benzene, a mercury complex is formed that dissolves in the organic layer. Dithizone is designated H2Dz, the mercuric (Hg^{2+}) dithizonate is represented as $Hg(HDz)_2$, and the mercurous (Hg^+) dithizonate as $Hg(HDz)$.

A solution of $Hg(HDz)_2$ in carbon tetrachloride shows a maximum absorption at the wavelength of 485–490 nm. Mercury forms $Hg(HDz)$ in the absence of water or a secondary dithizonate, HgDz, in aqueous alkaline solutions, which are deficient in dithizone. This compound gives a violet color.

Dithizone reacts with silver (Ag), copper (Cu), gold (Au), palladium (Pd), and platinum (Pt) in addition to Hg under acidic conditions, but it is generally sufficient to consider interference of Cu alone in environmental analysis. Metals, such as lead (Pb), cadmium (Cd), zinc (Zn), nickel (Ni), cobalt (Co), and iron (Fe), are not extracted from an acidic solution. Moreover, tin (Sn) and bismuth (Bi) are not extracted from a strongly acidic solution unless they are present in large quantities.

Interference by Copper Interference of Cu may be eliminated by one of the following methods:

1. When a small flow of a dithizone in carbon tetrachloride solution is added to a weakly acidic solution containing Hg and Cu, extraction of Hg is completed before that of Cu begins. Separation of Hg and Cu is thus possible.
2. Extraction of Hg and Cu can be carried out simultaneously with excess dithizone and then treated with a masking agent such as KBr, KI, and $Na_2S_2O_3$ to decompose only $Hg(HDz)_2$ and to transfer Hg^{2+} into the aqueous layer. Thereafter, Hg is again extracted with dithizone.
3. Ethylenediaminetetraacetic acid (EDTA) in the form of the disodium salt can be added as a masking agent of Cu, and Hg alone is extracted with dithizone. The extraction is carried out at a pH of 2.5 or higher, preferably 5.5, at which level Cu–EDTA is stable.

A relatively large quantity of Cl^{2-} also interferes with the extraction of Hg^{2+} under strongly acidic conditions. For instance, the extraction of Hg can be done without difficulty, when up to 1 M of Cl^- is in 50 mL of 1 N sulfuric acid. Extraction becomes incomplete as Cl^- exceeds 2.5 M. Under neutral to alkaline conditions, some interference is also observed when NH_4^+ is present in large quantities.

Analysis Procedure The procedure generally followed in the colorimetric determination of mercury by dithizone is as follows: The sample solution is extracted repeatedly with excess dithizone, and Hg^{2+} is captured in the solvent layer as $Hg(HDz)_2$. Contaminants are also removed in this step. The $Hg(HDz)_2$ is then decomposed by a suitable masking or oxidizing agent, and the Hg^{2+} is liberated and transferred into the aqueous layer. Here, it is reacted with a given excess if dithizone, under specified conditions, and its concentration is measured colorimetrically.

The extraction immediately preceding the colorimetric determination is carried out under acidic conditions. The color developed is a mixture of orange (due to $Hg(HDz)_2$) and green (due to the excess of dithizone). The absorbance of such mixed colors can be measured directly, or measurements may be made at 485–490 nm, which is the maximum absorption wavelength of $Hg(HDz)_2$. Higher sensitivity and accuracy can be obtained at 605–620 nm, which is the maximum absorption wavelength of dithizone.

With the single-color method, mercury is extracted with a solution of dithizone. The excess dithizone is removed by shaking it with dilute aqueous ammonia. The absorbance is measured at 485–490 nm. With this method, changes in the concentration of dithizone do not affect the results. $Hg(HDz)_2$ gradually fades in the light. This tendency becomes pronounced when impure dithizone is used, whereas a solution of high-grade dithizone remains stable for several hours in a lighted room.

Atomic Absorption Spectrophotometry

In atomic absorption spectrophotometry, a solution containing mercury is introduced directly into a flame. This technique does not give high sensitivity in the analysis of mercury, and it is difficult to detect mercury at a concentration below 0.2 ppm. Mercury can be analyzed with high sensitivity if the aqueous solution is first reduced to mercury vapor and then sent to the absorption cell by aeration, where it is analyzed by atomic absorption spectrophotometry without the use of flame. Stannous salts are best suited for reducing mercury in an aqueous solution. Mercuric (Hg^{2+}) ions are reduced by stannous salts to metallic mercury (Hg^0) according to Equation 1.37(1):

$$Hg^{2+} + Sn^{2+} \rightarrow Hg^0 + Sn^{4+} \qquad 1.37(1)$$

The metallic mercury thus formed is vaporized by aeration and is sent to the absorption cell (with quartz windows), where its absorbance is measured at a wavelength of 253.7 nm.

Figure 1.37b illustrates both the closed and open systems of measurement. Beer's law holds for the mercury levels of 0.01–10 µg/L at a cell length of approx. 200 mm involved in this method. There are also dedicated instruments for mercury measurement available, those have a fixed wavelength and typically reach lower detection limits than multielement atomic absorption spectrometers.

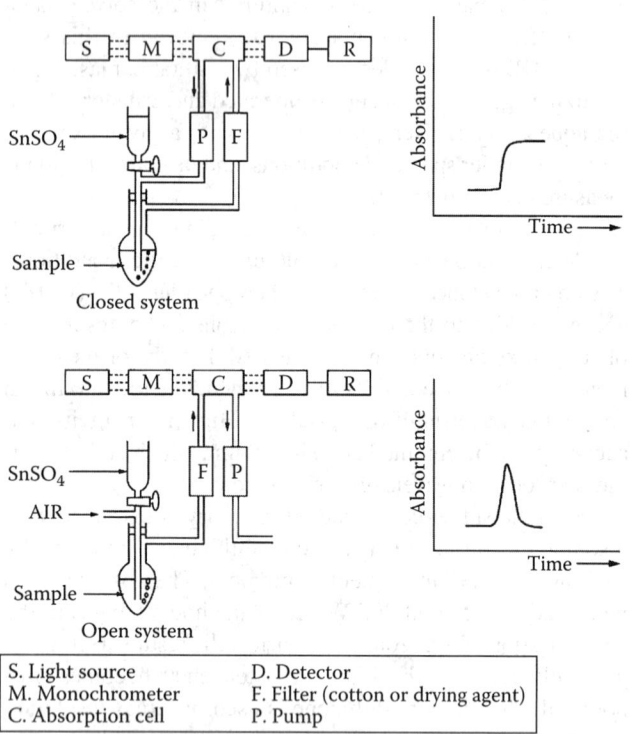

FIG. 1.37b
Flameless atomic absorption spectrophotometer with reduction–aeration hardware.

If large quantities of metal ions are present, it can be reduced by the addition of stannous salts. If these ions are not reduced, they may interfere with the reduction of mercury. Interfering elements, however, are rarely present in ordinary environmental samples in sufficient quantities to cause interference. If the aeration causes the vaporization of organic substances, which absorb UV rays, it will interfere with the analysis of mercury. Hence, it is necessary to decompose organic substances thoroughly.

Analysis Procedure A suitable amount of water sample is introduced into a 500 mL flask and 20 mL of sulfuric acid (1 + 1), plus 15–20 mL of $KMnO_4$ (6%) is added. After mixing, a reflux condenser is attached to the flask and the mixture is boiled. The $KMnO_4$ solution is replenished when it has been consumed. The flask is cooled when the color of $KMnO_4$ no longer disappears, and the excess $KMnO_4$ is reduced with a hydroxylamine sulfate solution.

A portion (containing 10 µg or less of mercury) is transferred into a flask such as the one shown in Figure 1.37b. Two milliliters of sulfuric acid (1 + 1) and 2 mL of stannous sulfate (10% in 2 N H_2SO_4 HCl) are added, and aeration is started at a rate of 2 L/min (stannous chloride is now more commonly used in place of Stannous sulfate, as it is more stable).

Closed-Circuit Method The vapor thus formed is driven into the vapor phase and circulated in a closed circuit. The UV absorption at 253.7 nm is detected in the optical cell of an atomic absorption spectrophotometer. A typical record of the UV absorbance is shown in Figure 1.37c. A calibration curve is prepared with standard solutions containing known amounts of mercury to assist in the quantitative analysis. The relative standard deviation for this method is reportedly between 2% and 10%.

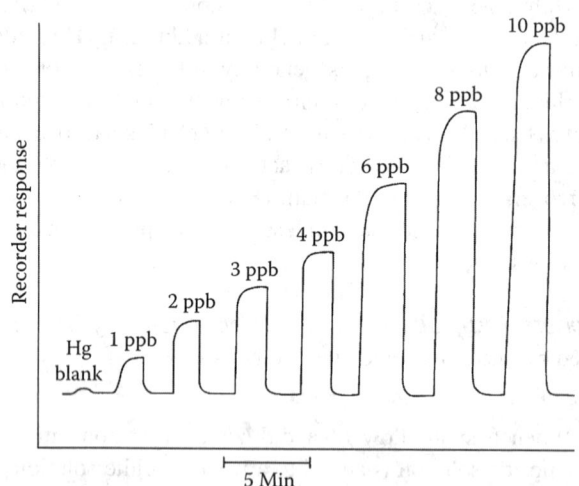

FIG. 1.37c
UV absorption record of vapor sample produced by aeration.

Open System In this method, the mercury is stripped from the reduced sample and flows through the measuring cell continuously without recirculation. A typical record of the UV absorbance is shown in Figure 1.37b.

Online Measurement

Figure 1.37d illustrates the components of an automatic mercury monitor with an UV detector. In the first step, the mercury is reduced to the elemental state by the addition of tin(II) chloride or $NaBH_4$. Subsequently, mercury is stripped from the aqueous phase by an airstream and is carried into an optical detector cell, which can be made of silica (Suprasil). Here, the mercury concentration is detected by UV absorption measurement at a wavelength of 253.7 nm.

This method of analysis is referred to as cold vapor atomic absorption spectroscopy (CVAAS). It is both sensitive (lowest range is 0–1 ppt) and selective for mercury. Measuring ranges are available from 1.0 ppt to 1.0 ppm. The reagent requirement is 3 L/week for each reagent. Transmitter outputs can be 4–20 mA analog or bidirectional RS 232. Some of these analyzers are fully automated.

Automatic operation can include unattended operation guaranteed by self-diagnostics for alarming when malfunctions are detected, for example, the loss of reagent, sample, or stripping airflows; leakage of fluids; photometer lamp burnout; or need for recalibration. The continuous-measurement mode can also be switched to a periodic one. Automatic calibration can also be performed. This is done by automatically switching from the sample stream to a calibration solution. Similarly, the zero can also be automatically reset, and rinsing steps can be automatically initiated for the purpose of providing automatic self-cleaning. This analyzer is illustrated in Figure 1.37e.

FIG. 1.37e
Mercury process analyzer. (Courtesy of Mercury Instruments GmbH.)

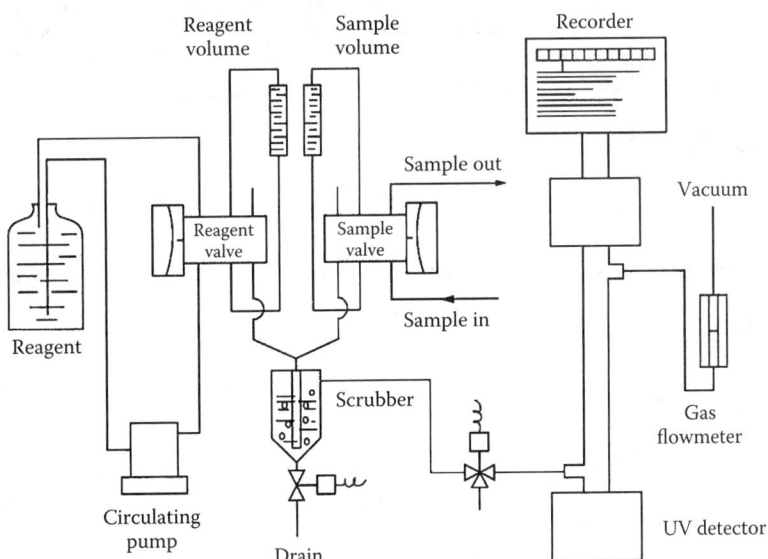

FIG. 1.37d
Continuous mercury monitor.

Online Atomic Fluorescence Spectrophotometry In contrast to Cold Vapor Atomic Absorption Spectrometer (CVAAS), the characteristics of Cold Vapor Atomic Fluorescence spectrometer (CVAFS)-based mercury analyzers include sub ppt detection limits and a much wider dynamic range than CVAAS. CVAFS instruments are usually available in two configurations; one which uses atomic fluorescence detection and one which employs a gold amalgamation system to preconcentrate mercury prior to atomic fluorescence detection. The detection limit for the simple fluorescence approach is about 0.2 ppt, whereas that for the preconcentration approach it can be as low as 0.02 ppt. The US EPA has method 245.7 for use without preconcentration and 1631 with preconcentration.

Figure 1.37f provides a schematic diagram of the CVAFS technique including the gold amalgamation system to preconcentrate the mercury prior to fluorescence detection.

CVAFS instruments typically utilize a peristaltic pump to introduce sample and stannous chloride along with a stream of pure, dry gas (typically argon) to a gas–liquid separator where it is bubbled through to release mercury vapor. The mercury is then transported by the carrier gas through a dryer and to an inert switching valve where it is either directed to a gold amalgamator trap for preconcentration prior to introduction to the fluorescence detector or directly to the detector bypassing the trap altogether. The drying stage is quite important in atomic fluorescence as water vapor and other molecular species can interfere with the measurement. Once in the detector, mercury vapor absorbs light at 254 nm creating fluorescent light at same wavelength. Fluorescence detection is usually made at 90° to the incident beam to minimize any incidental light scatter from the excitation source itself. The intensity of the fluorescent light is directly proportional to the concentration of mercury.

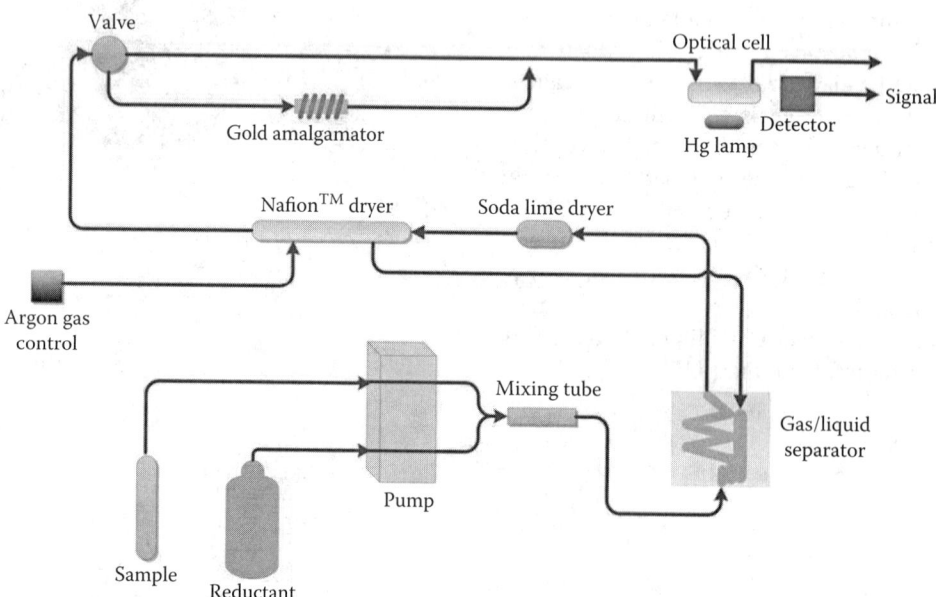

FIG. 1.37f
A schematic diagram of cold vapor atomic fluorescence analyzer with gold amalgamation. (Courtesy of Teledyne Leeman Labs.)

ORGANIC MERCURY DETECTION

Sample Treatment

In organomercury compounds of the RHgX type, R is an organic group, such as alkyl and phenyl, and X is an electronegative group, such as halogen and hydroxyl. Those carrying methyl, ethyl, and propyl groups are known to be the causes of Minamata disease (a severe neurological disorder resulting from poisoning by organic mercury and leading to severe permanent neurological and mental disabilities or death). When these compounds are present in water, it is first necessary to extract them with a suitable solvent.

The solubility of a compound of the RHgX type in organic solvents varies. If X is a halogen, the compound is soluble in aromatic hydrocarbons such as benzene and toluene. If X is an ion such as SO_4^{2-}, OH^- or $CH_3CO_2^-$, the solubility in hydrocarbons is extremely low. Therefore, it is first necessary to convert X to a halogen, and then the RHgX compound can be extracted with an organic solvent such as benzene and toluene. The aqueous solution is made acidic by addition of hydrochloric acid before extraction. The lower alkyl mercury compounds are moderately soluble in water, and a relatively large amount of solvent is necessary to effect quantitative extraction.

The organomercury compounds extracted in the organic solvent are generally contaminated and must be cleaned. The cleaning of an RHgX-type compound can be performed effectively by backextraction with an aqueous solution of a sulfur-containing compound such as cysteine. Organomercury compounds react with a sulfhydryl compound according to Equation 1.37(2) and move from the organic solvent phase to the aqueous phase.

$$RHgX + R'SH \rightarrow RHgSR' + HX \quad 1.37(2)$$

where
  Both R and R' are organic groups
  X is an electronegative group

Many of the organic compounds remain in the organic solvent phase, and the purpose of cleaning is thereby accomplished.

The transfer of the organomercury compounds from the organic solvent phase to the aqueous phase is quantitative. Therefore, when the backextraction is carried out by using a smaller volume of the aqueous solution against a known volume of the organic solvent, concentration of the organomercury compounds may be determined simultaneously with cleaning up.

The organomercury compounds extracted back into the aqueous layer are again liberated as RHgCl by the addition of hydrochloric acid to the aqueous solution according to Equation 1.37(3).

$$RHgSR' + HCl \rightarrow RHgCl + R'SH \quad 1.37(3)$$

The RHgCl liberated is extracted with a small amount of benzene and analyzed by gas chromatographic techniques.

Gas Chromatography

Lower alkylmercury and phenylmercury compounds vaporize upon heating and can be analyzed by gas chromatography. The use of an electron capture detector gives high sensitivity and is best suited for analysis of traces of organic mercury. For example, methylmercury can be detected to a level of 1×10^{-11} g.

Organomercury compounds are highly reactive with metals; consequently, it is not desirable to use metal tubing as column materials. A glass column is preferable. Also, a polar substance as a liquid phase in the column yields better results. The liquid phases most frequently used are polyethylene glycol, polydiethylene glycol succinate, and polybutanediol succinate.

The amount of the liquid phase to be coated on the support is preferably 5%–10% for analysis of alkylmercury and about 2% for analysis of phenylmercury. An increase in the coating amount causes an increase in bleeding gas and a decrease in the standing current of the detector, with resultant lower sensitivity when a high column temperature is used. The size of the liquid phase is reduced for high-sensitivity analysis.

The use of the subtractive technique is recommended for simplified identification of organomercury compounds by gas chromatography. When an organic solvent, such as benzene or toluene, which contains organomercury compounds, is mixed with the aqueous solution of a bivalent sulfur compound, such as $Na_2S_2O_3$ or cysteine, the organomercury compounds disappear from the organic solvent. If one compares the gas chromatograms before and after this treatment, one will be noticed that the peaks corresponding to the organomercury compounds disappeared or diminished markedly after treatment.

The analysis procedure is as follows: 10 mL of concentrated hydrochloric acid is added to 500 mL of the sample water containing methylmercury compounds. The resultant solution is mixed with 100 mL of benzene, and the mixture is allowed to settle. The aqueous layer is separated and is extracted again with 100 mL of fresh benzene. This is repeated three times, the aqueous layer is discarded, and the combined benzene layer is washed with distilled water.

The benzene layer is separated and is extracted back with 10 mL of a 0.1% aqueous solution of 1-cysteine. The aqueous layer is also separated, and 1 mL of concentrated hydrochloric acid and 2 mL of benzene are added. After mixing, the benzene layer is separated and dried over a small amount of anhydrous sodium sulfate. A 10 µL sample is analyzed by gas chromatography.

The column recommended contains 5% polydiethylene glycol succinate on Chromosorb W, 60–80 mesh, packed in glass tubing, 1 m long, and with a 3 mm inside diameter. Column temperature is set at 130°C (266°F), and the flow rate of carrier gas is maintained at 60 mL/min. Methylmercury chloride is eluted in 3–5 min under these conditions, and concentrations of 1 µg/L or less can be detected by this method.

Thin-Layer Chromatography

Thin-layer chromatography offers a simple and inexpensive method for analysis of organomercury compounds. Silica gel and alumina are mainly used as the adsorbent layer. The R_f values of organomercury compounds for a variety of developers are shown in Tables 1.37a and 1.37b. The mercury compounds are visualized by spraying the plate with a solution of dithizone. Ordinarily, mercury of the order of 0.5 µg can be identified visually in this manner. When developed as organomercury dithizonate, visualization becomes unnecessary and mercury on the order of 0.1 µg can be identified visually.

TABLE 1.37a
R_f Value of Organomercury Compounds

Compound	Developer[a]				
	A	B	C	D	E
Methylmercuric chloride, CH_3HgCl	0.35	0.59	0.29	—	0.42
Methylmercuric iodide, CH_3HgI	0.04	0.03	—	—	—
Ethylmercuric chloride, C_2H_5HgCl	0.41	0.74	0.50	0.72	0.46
Ethylmercuric phosphate, $(C_2H_5Hg)_2HPO_4$	0	0	0.49	0.67	—
Methoxyethylmercuric chloride, $CH_3OC_2H_4HgCl$	0.23	0.49	0.46	0.74	—
Phenylmercuric chloride, C_6H_5HgCl	0.51	0.73	0.56	0.76	—
Phenylmercuric iodide, C_6H_5HgI	0.74	0.84	0.92	0.86	—
Phenylmercuric acetate, $C_6H_5HgCH_3CO_2$	0.39	0.64	0.86	0.83	0.48

[a] Key: Adsorbent layer; silica gel, A. n-Hexane:acetone (85:15), B. n-Hexane:acetone (70:30), C. Butyl alcohol saturated with water, D. Isopropyl alcohol:water (90:10), and E. Chloroform.

TABLE 1.37b
R_f Value of Organomercury Dithizonates

Compound	Developer[a]					
	1	2	3	4	5	6
Methylmercury dithizonate, $CH_3Hg(HDz)$[b]	0.64	0.48	0.57	0.77	0.89	0.86
Ethylmercury dithizonate, $C_2H_5Hg(HDz)$	0.64	0.51	0.62	0.78	0.91	0.87
Methoxyethylmercury dithizonate, $CH_3OC_2H_4Hg(HDz)$	0.32	0.16	0.25	0.44	0.58	0.49
Ethoxyethylmercury dithizonate, $C_2H_5OC_2H_4Hg(HDz)$	0.44	0.23	0.34	0.55	0.71	0.67
Phenylmercury dithizonate, $C_6H_5Hg(HDz)$	0.48	0.34	0.46	0.62	0.72	0.69
Mercury dithizonate, $Hg(HDz)_2$	0.19	0.09	0.17	0.28	0.19	0.15

[a] Key: 1. Hexane:acetone (9:1), 2. Hexane:acetone (19:1), 3. Hexane:acetone (93:7), 4. Petroleum ether:acetone (9:1), 5. Hexane:acetone (19:1), 6. Petroleum ether:acetone (19:1), Adsorbent layer: 1–4 silica gel and 5, 6 alumina.
[b] HDz, dithizonate ligand.

DIRECT (THERMAL) DECOMPOSITION

The thermal decomposition technique with gold trap preconcentration is also called amalgam technique. This method of analysis consists of three phases: decomposition, collection, and detection.

The first stage of an analysis is known as the "decomposition phase." During this phase, a sample container is placed inside a prepacked combustion tube. The combustion tube, heated to approximately 750°C through an external coil, provides the necessary decomposition of the sample in a gaseous form. The evolved gases are then transported (via an oxygen carrier gas) through catalytic compounds, where all interfering impurities such as ash, moisture, halogens, and minerals are removed from the evolved gases.

Following decomposition, the cleaned, evolved gas is transported to an amalgamator for the "collection phase." The amalgamator, a small glass tube containing gold-plated ceramics, collects all of the mercury in the vapor. With a strong affinity for mercury and a significantly lower temperature than the decomposition phase, the amalgamator is heated to approximately 900°C, thereby releasing all mercury vapor to the detection system.

The released mercury vapor is transported to the final phase of the analysis, the "detection phase," where all vapors pass through atomic absorption spectrometer. The spectrometer uses light at a wavelength of 253.7 nm. This wavelength is specific to elemental mercury and will be absorbed by the mercury particles in the vapor for subsequent detection by a UV detector (Figure 1.37g).

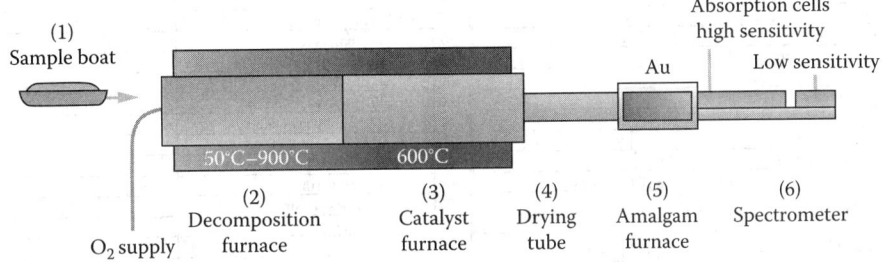

FIG. 1.37g
Direct mercury analysis by thermal decomposition. (Courtesy of Teledyne Leeman Labs.)

SPECIFICATION FORMS

When specifying mercury analyzers one can use the ISA form 20A1001 and when specifying the detector and also describing the application ISA form 20A1002 can be used. Both forms are reproduced on the following with the permission of the International Society of Automation.

#	RESPONSIBLE ORGANIZATION	ANALYSIS DEVICE	#	SPECIFICATION IDENTIFICATIONS	
1	(ISA logo)	Operating Parameters	6		
2			7	Document no	
3			8	Latest revision	Date
4			9	Issue status	
5			10		

#	ADMINISTRATIVE IDENTIFICATIONS		#	SERVICE IDENTIFICATIONS Continued		
11			40			
12	Project number	Sub project no	41	Return conn matl type		
13	Project		42	Inline hazardous area cl	Div/Zon	Group
14	Enterprise		43	Inline area min ign temp	Temp ident number	
15	Site		44	Remote hazardous area cl	Div/Zon	Group
16	Area	Cell Unit	45	Remote area min ign temp	Temp ident number	
17			46			
18	SERVICE IDENTIFICATIONS		47			
19	Tag no/Functional ident		48	COMPONENT DESIGN CRITERIA		
20	Related equipment		49	Component type		
21	Service		50	Component style		
22			51	Output signal type		
23	P&ID/Reference dwg		52	Characteristic curve		
24	Process line/nozzle no		53	Compensation style		
25	Process conn pipe spec		54	Type of protection		
26	Process conn nominal size	Rating	55	Criticality code		
27	Process conn termn type	Style	56	Max EMI susceptibility	Ref	
28	Process conn schedule no	Wall thickness	57	Max temperature effect	Ref	
29	Process connection length		58	Max sample time lag		
30	Process line matl type		59	Max response time		
31	Fast loop line number		60	Min required accuracy	Ref	
32	Fast loop pipe spec		61	Avail nom power supply		Number wires
33	Fast loop conn nom size	Rating	62	Calibration method		
34	Fast loop conn termn type	Style	63	Testing/Listing agency		
35	Fast loop schedule no	Wall thickness	64	Test requirements		
36	Fast loop estimated lg		65	Supply loss failure mode		
37	Fast loop material type		66	Signal loss failure mode		
38	Return conn nominal size	Rating	67			
39	Return conn termn type	Style	68			

#	PROCESS VARIABLES	MATERIAL FLOW CONDITIONS		#	PROCESS DESIGN CONDITIONS		
69				101			
70	Flow Case Identification		Units	102	Minimum	Maximum	Units
71	Process pressure			103			
72	Process temperature			104			
73	Process phase type			105			
74	Process liquid actl flow			106			
75	Process vapor actl flow			107			
76	Process vapor std flow			108			
77	Process liquid density			109			
78	Process vapor density			110			
79	Process liquid viscosity			111			
80	Sample return pressure			112			
81	Sample vent/drain press			113			
82	Sample temperature			114			
83	Sample phase type			115			
84	Fast loop liq actl flow			116			
85	Fast loop vapor actl flow			117			
86	Fast loop vapor std flow			118			
87	Fast loop vapor density			119			
88	Conductivity/Resistivity			120			
89	pH/ORP			121			
90	RH/Dewpoint			122			
91	Turbidity/Opacity			123			
92	Dissolved oxygen			124			
93	Corrosivity			125			
94	Particle size			126			
95				127			
96	CALCULATED VARIABLES			128			
97	Sample lag time			129			
98	Process fluid velocity			130			
99	Wake/natural freq ratio			131			
100				132			

#	MATERIAL PROPERTIES		#	MATERIAL PROPERTIES Continued		
133			137			
134	Name		138	NFPA health hazard	Flammability	Reactivity
135	Density at ref temp	At	139			
136			140			

Rev	Date	Revision Description	By	Appv1	Appv2	Appv3	REMARKS

Form: 20A1001 Rev 0

© 2004 ISA

1	RESPONSIBLE ORGANIZATION		ANALYSIS DEVICE COMPOSITION OR PROPERTY Operating Parameters (Continued)		6	SPECIFICATION IDENTIFICATIONS			
2					7	Document no			
3					8	Latest revision		Date	
4					9	Issue status			
5					10				

	PROCESS COMPOSITION OR PROPERTY			MEASUREMENT DESIGN CONDITIONS				
	Component/Property Name	Normal	Units	Minimum	Units	Maximum	Units	Repeatability
11								
12								
13								
...								
75								

Rev	Date	Revision Description	By	Appv1	Appv2	Appv3	REMARKS	

Form: 20A1002 Rev 0 © 2004 ISA

Abbreviations

CVAAS	Cold vapor atomic absorption spectroscopy
CVAFS	Cold vapor atomic fluorescence absorption spectroscopy
H_2O_2	Hydrogen peroxide
H2Dz	Dithizone
H_2SO_4	Sulfuric acid
HNO_3	Nitric acid
HCl	Hydrochloric acid
$KClO_3$	Potassium chlorate
$KMnO_4$	Potassium permanganate

Organization

EPA	Environmental Protection Agency

Bibliography

Control Staff, Mysteries of gold film sensors to be studied, *Control*, January 1989.

Denny, R. and Sinclair, R., *Visible and Ultraviolet Spectroscopy*, New York: John Wiley & Sons, 1987.

EPA—Analytical methods for mercury in national pollutant discharge elimination system (NPDES) permits, 2007, http://water.epa.gov/scitech/methods/cwa/metals/mercury/upload/2007_10_02_pubs_mercurymemo_analyticalmethods.pdf.

EPA—Analysis of mercury in water, 2012, http://www.amazon.com/Analysis-Mercury-Water-Preliminary-Methods/dp/1249430135/ref=sr_1_2?s=books&ie=UTF8&qid=1381863407&sr=1-2&keywords=mercury+analysis+in+water.

EPA—Guidance for implementation and use of EPA method 1631 for the determination of low-level mercury, 40 CFR Part 136, 2011, http://www.amazon.com/Guidance-Implementation-Determination-Low-Level-Mercury/dp/1244057282/ref=sr_1_fkmr0_1?s=books&ie=UTF8&qid=1386171525&sr=1-1-fkmr0&keywords=Method+EPA+245.1+Mercury+in+water.

EPA-Method 245.7, 2013, http://www.amazon.com/Method-245-7-Mercury-FluorescenceSpectrometry/dp/1289221383/ref=sr_1_1?s=books&ie=UTF8&qid=1386171199&sr=1-1&keywords=Mercury+in+water+by+Cold+Vapor+Atomic+Fluorescence+Spectrometry.

EPA-Method 7473, 2007, Mercury in solids and solutions by thermal decomposition, amalgamation and atomic absorption spectrometer, 2007, http://www.epa.gov/sam/pdfs/EPA-7473.pdf.

Ewing, G. W., ed., *Instrumental Methods of Chemical Analysis*, New York: McGraw-Hill, 1981.

Frei, R. W. and Hutzinger, O., *Analytical Aspects of Mercury and Other Heavy Metals in the Environment*, New York: Taylor & Francis, 1976.

Guilbault, G. G., ed., *Fluorescence*, New York: Marcel Dekker, latest edition.

Guilbault, G.G. 1990. *Practical Fluorescence*, 2nd edn., New York: Marcel Dekker, Inc.

Hughes, T. J., *The Finite Element Method*, Dover Publishers, Galveston, TX, 2000.

Kwon, Y. W., *The Finite Element Method Using MATLAB*, Boca Raton, FL: CRC Press, 2000.

Leithe, W., *The Analysis of Air Pollutants*, Ann Arbor, MI: Ann Arbor Science Publishers, latest edition.

Portable high-performance mercury analyzer for field use mercury sniffer, *The Rigaku Journal*, 15, 1998.

Sax, N. and Lewis, R., *Dangerous Properties of Industrial Materials*, New York: Van Nostrand Reinhold, 1988.

Seiler, H. and Sigel, H., *Handbook on Toxicity of Inorganic Compounds*, New York: Marcel Dekker, 1988.

Sharif, A., Monperrus, M., Tessier, E., and Amouroux, D., Determination of methyl mercury and inorganic mercury in natural waters at the pgL^{-1} level: Intercomparison between PT-GC-Pyr-AFS and GC-ICP-MS using ethylation or propylation derivatization, *Proceedings of the 16th International Conference on Heavy Metals in the Environment*, Vol. 1, 2013.

Smith, D. et al., *EPA's Sampling and Analysis Methods Database*, Chelsea, MI: Lewis Publishers, 1990.

Tatton, J. O. G. and Wagstaffe, P. J., Identification and determination of organomercurial fungicide residues by thin-layer and gas chromatography, *Journal of Chromatography*, 44, 284, 1969.

USEPA, Methyl mercury in water by distillation, aqueous ethylation, purge and trap and CVAFS, EPA Method 1630, January 2001.

USEPA-Method 245.7, Mercury in water by Cold Vapor Atomic Fluorescence Spectrometry, http://water.epa.gov/scitech/methods/cwa/metals/mercury/upload/2007_07_10_methods_method_mercury_1631.pdf.

USEPA-Method 7473, February 2007, http://www.epa.gov/osw/hazard/testmethods/sw846/pdfs/7473.pdf.

Vaughn, W. W., A Simple Mercury Vapor Detection for Geochemical Prospecting, Geological Survey Circular 540, Washington: U.S. Department of the Interior, 1967.

1.38 Moisture in Air
Humidity and Dew Point

A. BRODGESELL (1969, 1982)   **B. G. LIPTÁK** (1995, 2003)   **V. WEGELIN** (2017)

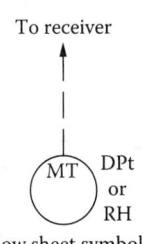

To receiver

Flow sheet symbol

Types of designs	A. Spectroscopic Absorption
	B. Electrolytic, solution resistance
	C. Thin-film capacitance (aluminum oxide or polymer)
	D. Surface condensation on chilled surface (D1, detected optically using chilled mirror; D2, detected electrically)
	*Legacy methods and designs**
	E. QCM: Quartz Crystal Microbalance
	F. Wet-dry bulb differential psychrometry
	G. Hair or synthetic fiber element
	H. Cellulose Element
Ranges	A. 0%–100% RH or −50°C to 50°C (−50°F to 120°F)
	B. 15%–95% RH,
	C. 0%–100% RH; when used as dew point sensor, −40°C to 10°C (−40°F to 50°F) is typical, with polymer sensors down to −60°C (96°F); 12%–99% RH or −64 to 79°C (−50 to 175°F) dew point.
	D. −70 to 85°C (−100 to 200°F) dew point or its equivalent (5% to 100% RH); high-temperature units available up to 177°C (350°F)
	E: 0.1 to 2500 ppmv
Inaccuracy	A. ±2% RH
	B. ±3% to 5% RH, ±3°C
	C. ±2% to 3% RH standard; 1% RH up to 90% RH and 2% RH over 90% RH is available (when used to detect dew point, an error of 4°F [2°C] is typical)
	D. 0.2°C–0.4°C (±0.3°F to 0.8°F) dew point or 1% RH between 20% RH and 90% RH
	E. ±5% AR
Costs	A. Primarily for weather stations, satellite installations. Building your own weather station can be costly. However, Internet apps are usually free, and can be downloaded and integrated into the control system if local ambient conditions are required.
	Laboratory instruments using these technologies are costly, $25K and up. There are currently limited choices for instruments of this type for industrial applications. For example, consider a TDLAS noncontact monitor for moisture in high temperature furnace applications.
	B, C: Most commercial (HVAC) and industrial products fall in this category:
	Home, small business unit: $50 to $150
	Sensor only, commercial grade: $150
	Sensor only, industrial grade: $1500.

* Legacy Methods and Designs relate to 4th Edition. Most of these are no longer commercially available, but may still be found in older installations. We provide this material to help those readers who work with older installations understand the design of these legacy designs.

(*Continued*)

Portable Hand Held unit: $300.

Display monitor: add $3000

Trending, data logging: add $3000

Sample System:add $3000

Special apps available, such as high temperature to 1400°C, Zirconium sensor.

D. Portable, battery-operated unit, $3,000; bench or meteorological unit with continuous balance and aspirator, $8,000 and up. Cycled chilled-mirror unit, from $3,000 to $5,000. NIST traceable test bench: $20,000.

E. Sensor only: $150. Handheld instrument: $300.

Partial list of suppliers

Alpha Moisture Systems (www.dew-point.com)

C: http://www.dew-point.com/dewpoint_transmitters/dewpoint_transmitter_AMT_Ex.html

Ametek Process Instruments (www.ametekpi.com)

A: http://www.ametekpi.com/products/WR-IPS-4.aspx

E: http://www.ametekpi.com/products/Model-5100.aspx

APT Instruments (www.aptinstruments.com)

C: http://www.aptinstruments.com/Merchant2/merchant.mvc?Screen=CTGY&Store_Code=AI&Category_Code=RH100

Bacharach, Inc. (http://www.bacharach-inc.com) http://www.bacharach-inc.com/th1800.htm

BartecGroup (www.bartec.de)

Chino Works America Inc. (www.chinoamerica.com)

C: http://www.chinoamerica.com/products/sensors/hnc/

CosaXentaur Instrument Corp. (www.cosaxentaur.com)

C, D: http://cosaxentaur.com/product_list.cxi?subsection=317

Davis Instruments (www.davisnet.com), http://www.davisnet.com/weather/products/weather_product.asp?pnum=06152

DIYTrade (cn.diytrade.com)

E: http://cn.diytrade.com/china/pd/8575596/%E7%9F%B3%E8%8B%B1%E6%99%B6%E4%BD%93%E5%BE%AE%E5%A4%A9%E5%B9%B3%E6%99%B6%E7%89%87.html

Dwyer Instruments (www.dwyer-inst.com)

B: http://www.dwyer-inst.com/Product/Humidity/HumidityTransmitters/Model657-1

C: http://www.dwyer-inst.com/Product/Humidity/HumidityTransmitters/SeriesRHP

EdgeTech Instruments. (www.EdgeTechInstruments.com)

C: http://www.edgetechinstruments.com/moisture-humidity/gallery/category/transmitters

D: http://www.edgetechinstruments.com/moisture-humidity/gallery/category/humidity-controllers

Emerson Process (www.emersonprocess.com) Formerly Bristol-Babcock, Rosemount Analytical Inc.

C: http://www2.emersonprocess.com/siteadmincenter/PM%20Rosemount%20Analytical%20Documents/PGA_AN_42-PGA-AN-GPD-XSTREAM-TM-AOX.pdf

ExTech-Online (www.extech-online.com)

C: http://www.extech-online.com/index.php?main_page=index&cPath=78_36_38

Fluke (en-us.fluke.com)

C: http://en-us.fluke.com/products/hvac-iaq-tools/fluke-971-hvac-iaq.html

General Electric Measurement and Control (www.ge-mcs,com) Formerly General-Eastern, Panametrics, Protimeter, Telaire

C, D: http://www.ge-mcs.com/en/moisture-and-humidity.html

D: http://store.ge-mcs.com/products/19793/ge-general-eastern-humilab

Honeywell Sensing and Controls (www.honeywell.com)

B, C: http://sensing.honeywell.com/products/humidity_sensors?Ne=2308&N=3217

Invensys (www.invensys.com). Formerly Foxboro

B: http://resource.invensys.com/instrumentation/documentation/eib/mi/mi_003-318.pdf

iSensors Corp. (www.isensors.com)

Jumo Process Controls (www.jumousa.com)

C: http://www.jumousa.com/products/humidity/4327/capacitive.html

Kahn Instruments Inc. (www.kahn.com)

C, D: http://www.kahn.com/hygrometers/

MEECO (www.meeco.com)

B: http://meeco.com/meeco/wp-content/themes/mecco/pdf/english/Accupoint_LP2.pdf

Mitchell Instrument Co, Inc. (www.mitchellinstrument.com)

(Continued)

E: http://www.mitchell.com/us/documents/QMA2030_97148_V3_US_1209.pdf
Nova Analytical (www.nova-gas.com)
C: http://catalog.nova-gas.com/category/dew-point-dp-?&plpver=10
Newport Electronics (www.newportus.com)
http://www.newportus.com/ppt/iTHX-SD.html
Ohmic Instrument Co. (www.ohmicinstruments.com)
C: http://www.ohmicinstruments.com/home/?page_id=40
Omega Engineering (www.omega.com)
C,D: http://www.omega.com/search/esearch.asp?start=0&perPage=10&summary=yes&sort=rank&search=Dewpoint&submit=Search&ori=Dewpoint
PhyMetrixInc, (C, D) (www.phymetrix.com), http://www.phymetrix.com/Prod_PPMa.htm, http://www.phymetrix.com/Prod_DMA.htm
Pro-Chem Analytic
C: http://www.instrumart.com/assets/Pro-Chem-DCM28-Data-Sheet.pdf
Rotronic Instrument Corp. (www. rotronic-usa.com)
B: http://www.rotronic-usa.com/index_products.html
Schneider Electric (www.Schneider-Electric.com). Formerly Square-D.
C: http://www.schneider-electric.com/products/ww/en/1200-building-management-system/1245-hvac-control-devices/7752-american-hvac-sensors/?xtmc=Humidity&xtcr=6
Shaw Meters (www.shawmeters.com)
D. http://www.shawmeters.com/sadp.html
Thermo Fisher Scientific (C) (www.thermoscientific.com) Formerly Barnant Co.
Tiger Optics: http://www.tigeroptics.com/app/tigeroptics/
A, E http://www.tigeroptics.com/app/tigeroptics/
Thunder Scientific (D) (www.thunderscientific.com)
D: http://www.thunderscientific.com/humidity_equipment/model_2500.html
VaisalaOyj. (C) (www.Vaisala.com)
A: http://www.vaisala.com/en/products/automaticweatherstations/Pages/MAWS201M.aspx
C: http://www.vaisala.com/en/products/dewpoint/Pages/default.aspx
Yankee Environmental Systems (A) (www.yesinc.com)
A: http://www.yesinc.com/products/raduv.html
C, D: http://www.yesinc.com/products/met-hyg.html
Yokagawa Corp. of America (B, C) (www.yca.com)
A: http://cdn2.us.yokogawa.com/Measurmentation_BU_02.pdf

Most popular   ABB, Emerson, Endress+Hauser, GE-Measurement and Control Solutions, Vaisala, Yokagawa

INTRODUCTION

Leonardo da Vinci was the first to attempt the measurement of humidity by weighing a ball of wool. Air is a mixture of oxygen, nitrogen, and water vapor. Water vapor follows the ideal gas laws in this mixture. The amount of water vapor per unit volume of air can vary by about 100,000–1 as one compares the water content of hot, saturated air to bone-dry arctic air. Although this specific measurement method is no longer used, it is important to note that the physics relating to the measurement has *not* changed (Figure 1.38i).

The most frequently used units in expressing the amount of water vapor in air include relative humidity (0%–100% RH); dew point (saturation) or wet bulb temperature (both expressed in °F or °C); and volume or mass ratio, expressed as parts per million volume (ppmv) or weight (ppmw).

OPERATING PRINCIPLES

Physics has not changed for centuries! Therefore, to understand measurement, it is important to understand the physics.

Humidity refers to the water vapor contained in the air at a particular temperature. Warm air has a greater capacity for water vapor than does cold air.

Relative humidity (RH) is the ratio of the actual partial pressure of the water vapor to the saturation vapor pressure at a particular temperature. The relationships of temperature,

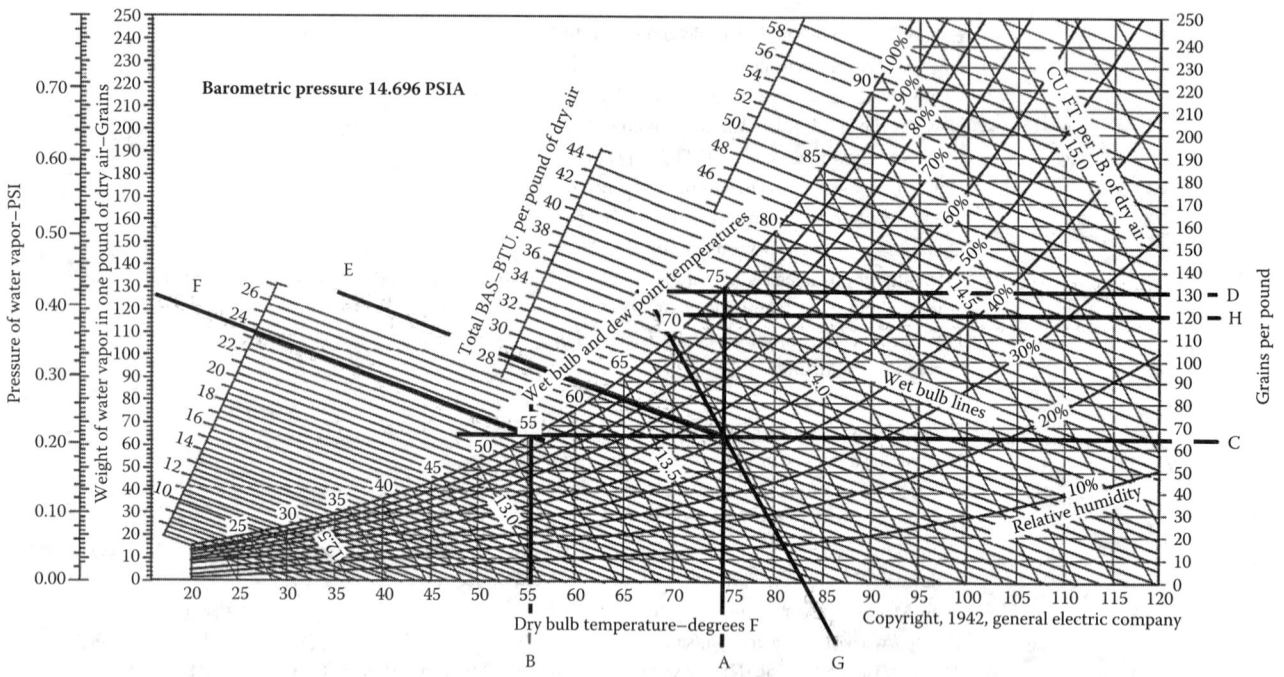

FIG. 1.38a
Psychrometric chart. See Chapter 2.2 for SI units. Note, this chart is based on one standard atmosphere, 14.696 psia, or 1 bar.

amount of water vapor, and associated energy in the air are given in the psychrometric chart. Knowing how to read the psychrometric chart not only helps us to understand the weather, but is essential in the engineering and design of many industrial and commercial control systems. These principles are illustrated in Figure 1.38a.

When the weatherman says the temperature is 75°F and 50% RH:

A. On the X scale find the dry bulb temperature = 75°F.
B. Draw a vertical line A to intersect the curved line 50% RH. Now, draw a horizontal line to the left. Where this line intersects the saturation line is the wet bulb temperature. This is where the atmosphere is saturated at 100% humidity, and water begins to condense. The vertical line B from this point is the wet bulb temperature, in this example 55°F. This is also known as the dew point.
At the saturation point, the atmosphere cannot contain any more water in vapor form, so it begins to condense.
C. Extend the horizontal line to the right. The Y scale represents the amount of moisture in the air, in grains per pound of dry air. From the chart, the amount of water in the air at 75°F and 50%RH is about 65 grains/lb of dry air.
D. Extend the vertical line A at 75°F until it intersects the saturation line. From this point, draw a horizontal line to the Y scale. This represents the maximum amount of water vapor the air at 75°F can contain, in this example 130 grains/lb of dry air.

Now, calculate the RH:

Observed water vapor at current conditions, where line A intersects line C: Read scale to the right.
Maximum water vapor at 75 F: where line A intersects line D: Read scale to the right.

Enthalpy in the Air

The atmosphere stores energy in the form of temperature and humidity. The measure of this stored energy is enthalpy, given as BTU/lb of dry air. During the day, enthalpy normally increases due to the radiant heating from the sun. During the evening, after the sun goes down, enthalpy decreases due to evaporation and radiation to upper layers of the atmosphere. The amount of stored energy can be calculated from the psychrometric chart (Figure 1.38a).

E. From point A, 75°F and 50% RH (the dry bulb condition), follow the slanted line to the left and slightly up to the scale that reads Total BAS. This is the enthalpy of the air, in BTU per pound of dry air. In this case, about 28.4 BTU/lb.
F. From Point B, 55°F and 100%RH (the wet bulb condition), follow the slanted line to the left and slightly up to the scale that reads Total BAS. This is the enthalpy of the air, in BTU per pound of dry air. In this case, about 23.6 BTU/lb. Thus, the total change in enthalpy is F − E, 23.6−28.4 BTU/lb = −4.8 BTU/lb of dry air.

G. From the chart, the weight of dry air at 75°F and 50% RH is 13.68 ft³/lb, or 0.073 lb/cf. 7000 grains = 1 lb. Therefore:
Weight of water vapor at 75°F, 50% RH = 65 grains = 0.009 lb/cf
Weight of water vapor at 55°F, 100% RH = 130 grains = 0.019 lb/cf

Note that the chart in Figure 1.38a is for atmospheric pressure of 1 standard atmosphere if the psia is 14.696. Changes in elevation and weather patterns need to be considered for a more accurate representation of pressure on humidity (the z-axis for a 3D diagram). For further details, refer to Psychrometric Tables for Obtaining the Vapor Pressure, Relative Humidity, and Temperature of the Dew Point by United States Weather Bureau (August 31, 2012).

APPLICATIONS

Using the Psychrometric Chart

Engineers use the psychrometric chart to determine how to size equipment that relies on atmospheric conditions to heat and cool industrial processes and buildings.

HVAC: Heating, Ventilating, and Air Conditioning

This regards the air quality specifications for occupied space, including humidity. ASHRAE, the American Society of Heating, Refrigerating and Air-Conditioning Engineers, sets the standards for occupied spaces, www.ashrae.org.

Cooling Note from the chart that as temperature drops, humidity increases. This is why water condenses on your glass of iced tea. Therefore, an easy way to reduce temperature is to add water. In hot, dry climates, atomizers spray water mist into the air to cool it down. This is the principle of a swamp cooler. This follows line E in Figure 1.38a. Enthalpy does not change. This is often referred to as free cooling, as no additional energy is required to cool the environment. Just the addition of water.

For process cooling towers, clean process water is pumped through the process to remove heat. Once pumped to the cooling tower, spraying the hot process water into the air inside the cooling tower causes some of the water to evaporate. Heat is exchanged from the process water to the atmosphere, which cools the process water. Process water is then returned back through the process.

H. However, note that in hot, humid climates, this method of cooling is not very effective. At 75°F and 90% RH, less than a 3°F reduction in temperature can be achieved simply by adding water (Line H).

Effect of Low Humidity As temperature rises, humidity decreases. Humidity also is a conductor of electricity. Thus in winter, the dryer air becomes an insulator, resulting in an electrostatic discharge:

- For electrical installations, the possibility of electrostatic discharge could damage electrical components.
- It can cause known health issues related to low humidity: flu, dehydration, and skin irritation.

Effect of High Humidity As temperature decreases, humidity increases. Thus, the outside of a glass of ice water on the table gets wet, as the moisture in the air condenses on it.

Now, it is necessary to cool electrical installations, to remove heat from the electronic components, and protect them from excessive heat. But in so doing, cooling the environment adds moisture:

- For electrical installations, cooling increases the possibility of condensation of the water vapor, potentially causing damage to electrical components.
- It can cause known health issues related to high humidity: mold growth, dust mites, pneumonia.

For these reasons, ASHRAE specifies a comfort zone of temperature and RH for office and residential environments (Figure 1.38b).

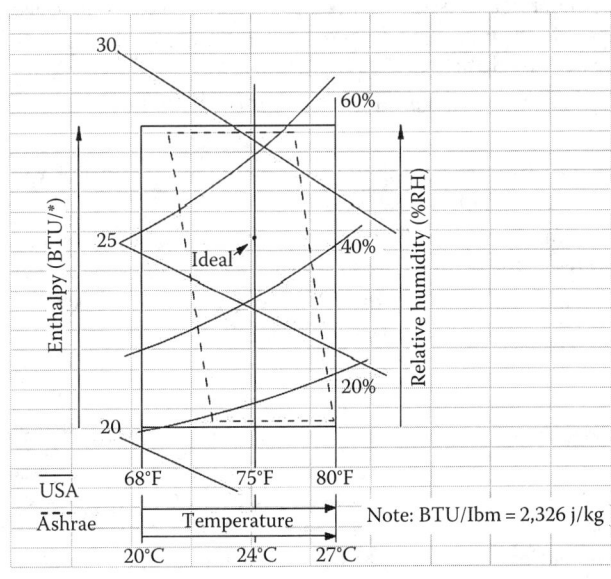

FIG. 1.38b
ASHRAE comfort zones.

A full discussion of air handler and building conditioning control strategies is found in Volume 2, Chapter 8.2, of the fourth edition of this handbook, including the management of humidity.

Clean Rooms, Incubators, Microenvironments, Respiration Therapy

Clean rooms are primary production facilities for pharmaceuticals or solid-state electronic assemblies. Issues with humidity in clean rooms include

- Low humidity, <30%—potential for electrostatic discharge
- High humidity, >50%—potential for bacteria and biological contaminant growth, corrosion.
- ISO Standard 14664-1—specifies air quality for clean rooms

Breathing Air

In numerous environments, supplied air is required. Examples include hospitals and breathing air packs. Issues with humidity in these environments include:

- Low humidity, <30%—growth of flu viruses, skin, eye irritation, dehydration.
- High humidity, >50%—potential for pneumonia, asthma, viruses, mites, fungi.

Cylinder Gas Association, CGA, www.cganet.com, develops and promotes safety standards and safe practices in the industrial gas industry.
Standards governing breathing air quality include:

- The limit of 67 ppmv, –50°F dew point, is the CGA, G-7.1 default
- American Society of Safety Engineers, (www.asse.org), ASSE 6000 Medical Gas Certification
- National Fire Protection Association, (www.nfpa.org), NFPA 99, Ch 5 Gas and Vacuum Systems

Airline Environment

You might be wondering why you experience jet lag. Here is one possible reason. Ambient air at –50°C is pulled from the ambient atmosphere at 35,000 ft, heated up to about 20°C, and introduced into the cabin. Result is humidity of less than 1%. Common issues are jet lag due to dehydration. New FAA (Federal Aviation Administration) rules now require humidifiers in cockpits.

Humidity Effect of Combustion

Combustion is the product of igniting a proper mixing of moles of oxygen with moles of fuel. For combustion of pure methane, for example, the stoichiometric balance is

$$4O_2 + 2CH_4 = 4H_2O + 2CO_2 + heat$$

Note that combustion adds moisture to the atmosphere. When you see a smokestack plume, it normally means that the moisture in the exhaust is condensing. As the plume cools, the moisture is recombined (as vapor) into the atmosphere, with a related increase of humidity.

In combustion control, we are interested in the molecular (mass) balance between oxygen and fuel. Oxygen only comprises 21% of air; most of the rest is nitrogen. From the psychrometric chart, higher temperature means lower density, thus lower mass flow. Therefore, even a small reduction in air temperature can lead to a significant increase in power output. An easy way to do this is adding moisture through a misting system to the air intake. A secondary benefit of misting is that the increased humidity of the air reduces the NOx emissions produced by the combustion process.

A practical limit to this strategy is to keep the air humidity below saturation levels. This will prevent condensation of water droplets and ice particles from entering the compressor and turbine and will avoid costly damage and erosion.

Calculation of Relative Humidity from First Thermodynamic Principles

Please keep in mind that the first principles of thermodynamics have not changed for centuries. However, our understanding of them has changed, and our abilities to capture these behaviors using thin-film and digital techniques allow us to deploy them in smaller, faster, and more efficient packages.

Therefore here, as an alternate to the psychrometric chart, let us calculate RH from the first principles of thermodynamics using steam tables (Volume 1, Chapter 10.6). These same relationships are ultimately embedded in the firmware of the thin-film designs we will discuss later. Remember also that these calculations are valid only for air at ambient conditions and should not be applied to other gases or if the temperature/pressure conditions differ significantly from the ambient. Specific humidity S, and relative humidity RH can be calculated using Equations 1.38(1) through 1.38(4). Equations 1.38(1) and 1.38(2) are derived from the definition of specific humidity assuming water vapor in an ideal gas. Equation 1.38(3) is derived from the first law of thermodynamics. Equation 1.38(4) is the definition of RH for an ideal gas.

$$S_2 = \frac{0.622 P v_2}{P - P v_2} \quad \quad 1.38(1)$$

$$S_1 = \frac{0.622 P_{v_1}}{P - P_{v_1}} \quad \mathbf{1.38(2)}$$

$$S_1 = \frac{0.24(t_2 - t_1) + S_2 h e_2}{h v_1 - h w_2} \quad \mathbf{1.38(3)}$$

$$\text{Relative humidity (\%)} = \frac{P_{v_1}}{P_{s_1}} \times 100 \quad \mathbf{1.38(4)}$$

where
- S_1, S_2 is the specific humidity
- t_1, t_2 is the temperature
- he_2 is the enthalpy of evaporation
- hv_1 is the enthalpy of saturated vapor
- hw_2 is the enthalpy of saturated water
- Pv_1, v_2 is the partial pressure of water vapor
- Ps_1 is the saturation pressure
- P is the total gas pressure

Subscripts 1 and 2 refer to dry and wet bulb conditions, respectively. Values of Ps_1, Pv_2, hv_1, hw_2, and he_2 are obtained from a steam table.

EXAMPLE

Determine the RH of air at atmospheric pressure when the dry and wet bulb thermometer readings are $t_1 = 80°F$ (27°C) and $t_2 = 60°F$ (16°C), respectively.

For t_1 and t_2 from the steam table,

- $Ps_1 = 0.5069$ psia
- $hv_1 = 1096.6$ BTU/lb mole
- $Pv_2 = 0.256$ psia
- $hw_2 = 28.06$ BTU/lb mole
- $he_2 = 1059.9$ BTU/lb mole

and, therefore, the specific humidity, using Equation 1.38(1), is

$$S_2 = 0.622 \frac{0.256}{14.7 - 0.256} = 0.011 \quad \mathbf{1.38(5)}$$

$$S_1 = \frac{0.24(60 - 80) + 0.011(1059.9)}{1096.6 - 28.06} = 0.00642 \quad \mathbf{1.38(6)}$$

$$0.00642 = 0.622 \frac{P_{v_1}}{14.7 - P_{v_1}} \quad \mathbf{1.38(7)}$$

solution $Pv_1 = 0.150$ psia

$$\text{Relative humidity} = \frac{0.150}{0.5069} \times 100 \quad \mathbf{1.38(8)}$$

solution = 29.6%

Internet Options for the Calculation of Relative Humidity

Numerous Internet options abound for software solutions to make RH calculations such as the ones shown earlier. Some sites also include applications that can be downloaded and run on a portable personal device (smart phone). Easier still, just go to www.weather.com, type in your location, and you will learn its RH.

Though such applications make the job easier, the user is cautioned to validate the accuracy of any of these applications before putting down money to construct a process to implement management and control of humidity.

SENSOR TECHNOLOGY

Sensor technology has evolved dramatically since this volume was last published. Most products in today's market for process control and building HVAC control are now of thin-film design. These are compact, rugged designs that tend to have better response, accuracy and repeatability of their earlier-generation counterparts. Most products include both humidity and temperature sensors in the same probe, so that all parameters necessary to calculate the enthalpy of a system are included. Many also include communication links for integration to other control system components. These links can vary from 4–20 mA outputs, to HART, and other industrial communication protocols.

Figure 1.38c gives the operating ranges of a number of humidity sensors. There is some overlap regarding the sensors available for the detection of moisture in air. HVAC and weather monitoring applications involve temperatures and pressures within normal ranges (−20°F to +140°F, atmospheric pressure, for example). Thus, these sensors tend to

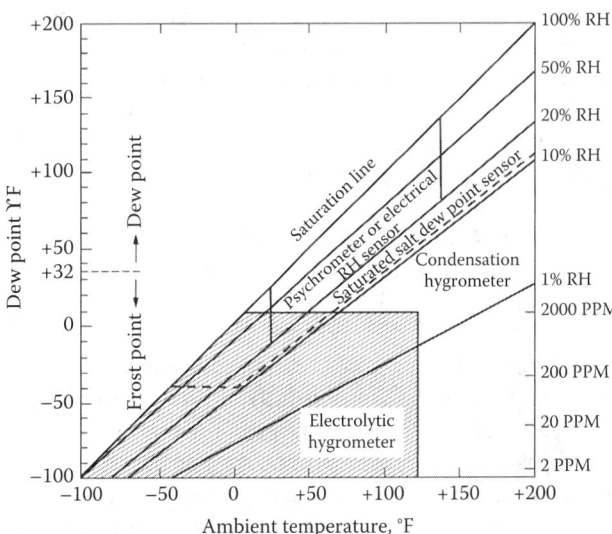

FIG. 1.38c
The operating ranges of different humidity sensors. (From Wiederhold, P.R., Instrum. Control. Syst., 1978.)

be of simpler design and of lower cost. In fact, a household temperature and humidity system can be purchased for about US$ 15 from the local hardware store.

Process applications normally involve much greater temperature, pressure, velocity, and humidity ranges. Also, for many processes, moisture content in parts per million is normally required. Industrial applications may also involve process contaminants that will attack the sensor substrate, vibration, and electrical hazard classification. For this reason, industrial sensors are normally much more rugged, and may require specialized sampling systems to assure reliable and repeatable results.

This chapter focuses on moisture in air. For moisture in other gases and liquids, please refer to Chapter 1.39. Moisture in solids is discussed in Chapter 1.40.

Today, most transmitters are equipped with digital communication links including 4–20 mA, Wireless HART.

Spectroscopic Absorption

Most national and international weather stations rely on spectroscopic absorption either from ground-based or, more commonly, from satellite-based instruments. The technologies used include

- FTIR: Fourier transform infrared
- NDIR: Non dispersive infrared
- TDLAS: Tunable diode laser absorption spectroscopy
- NMR: Nuclear magnetic resonance
- CRDS: Cavity ring down spectroscopy

Each of these methods operates in different frequency ranges, from microwave through infrared spectra, to ultraviolet. Each method discovers different parameters relating to all elements comprising the atmosphere. Integration of these patterns by complex simulation models provides comprehensive prediction of weather, molar concentrations, and other atmospheric conditions. Examples of such simulation models include:

- ISCCP, mathematical model from the National Environmental Satellite Data and Information Service: www.hitran.iao.ru.
- National Climatic Data Center (NCDC), National Oceanic and Atmospheric Administration (NOAA): www.modis-atmos.gsfc.nasa.gov

A detailed discussion of each of these methods is beyond the scope of this book. But hopefully inclusion of these terms will offer a link leading the reader to further research.

We enjoy the results of this technology on the nightly news. As an individual company, we could never hope to install or support such an infrastructure. However, as relates to our interest in humidity, these models provide useful information to us, in predicting and managing our process control environment.

Now, the great news is that we can now access this information instantly, at little or no cost. And if ambient conditions are an important factor for the proper monitoring and control of your plant process, it is quite a simple procedure to integrate this information into your control system. Please keep in mind that the same first principles previously discussed in the section "Calculation of Relative Humidity from First Thermodynamic Principles" apply to the models available online. (Figure 1.38d).

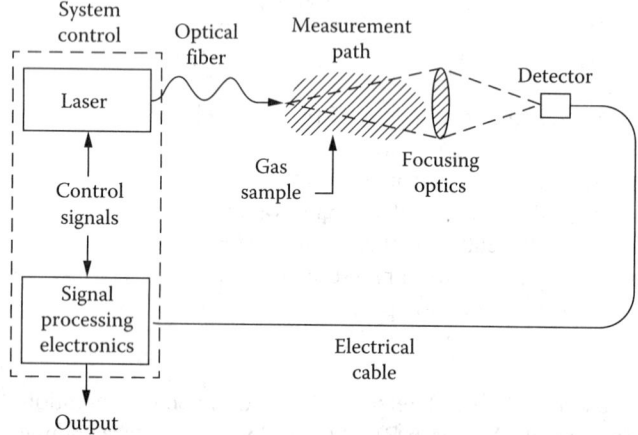

FIG. 1.38d
TDLAS spectroscopy principle of operation (Courtesy, http://www.tdlas.com/theory.shtml.)

Some laboratory-grade instruments offer spectroscopic analysis of air samples. Specifically, for this discussion, the absorption of electromagnetic energy by water vapor in the atmosphere predicts humidity in different ways. To date, only a few instruments using spectroscopic methods for industrial applications exist. Some progress has been made in creating instruments designed for industrial use with TDLAS for analysis of heat treat furnaces and NDIR for humidity monitoring of offices, retail stores, and warehouses. The benefit of these technologies is that the sensors do not need to contact the process and, thus, do not need to be designed to withstand temperature, pressure, or electrical characteristics of the process they are measuring. These instruments tend to be more expensive than those of other technologies. However, as the cost of technology falls, we could expect to find the benefits of spectroscopic adsorption to reach commercial and industrial workplaces.

CRDS is based on absorption spectroscopy. It works by attuning light rays to the unique molecular fingerprint of the sample species. By measuring the time it takes the light to fade or "ring down," you receive an accurate molecular count in milliseconds. The time of light decay, in essence, provides an exact, noninvasive, and rapid means to detect contaminants in the air, in gases, and even in the breath.

Electrolytic Cells

Based on Faraday's law of electrolysis, the electrolytic cell consists of a hollow glass tube with two spirally

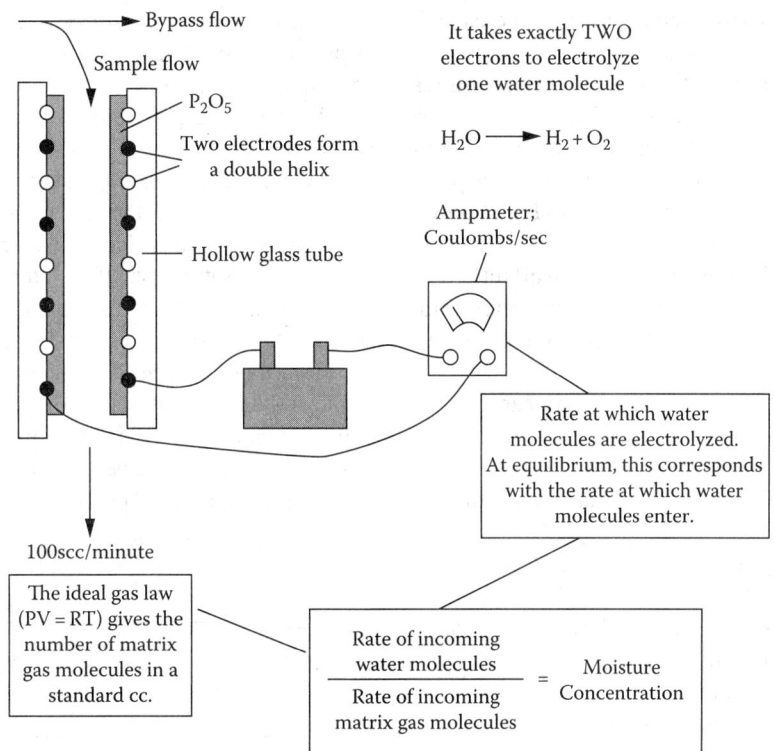

FIG. 1.38e
Electrolytic humidity probe. (Courtesy of MEECO, Inc.)

wound electrodes. Typically called a Dunsmore sensor, the electrodes are coated with a phosphorus pentoxide (P_2O_5) film. When the sample gas enters the cell at a known flow rate, the film absorbs all the moisture molecules present. By applying an electrical potential (voltage) to the electrodes, each absorbed water molecule is electrolyzed, generating a finite current. This current is precise and proportional to the amount of absorbed water. It is, therefore, an exact, direct measurement of the water vapor present in the sample gas. Figure 1.38e shows an example of this type of cell.

A variation of this technology is the Pope cell. The Pope cell is similar to the Dunmore element but, instead of lithium chloride solution, it uses polystyrene as a substrate for the conductive wire grid. In this design, polystyrene is treated with sulfuric acid, which produces a thin hygroscopic layer on its surface. Changes in humidity cause large changes in the impedance of this layer and, because the operating portion of the sensor is only its surface, its speed of response is fast—only a few seconds. An AC-excited Wheatstone bridge circuit can be used to measure the impedance.

THIN-FILM CAPACITANCE PROBES

In this design, a capacitor is formed by depositing a layer of porous aluminum oxide on a conductive aluminum substrate and coating the oxide with a thin film of condensation from evaporated gold. The aluminum substrate and the gold film serve as the electrodes of the capacitor (Figure 1.38f).

When exposed to air, the water vapor passes through the gold layer into the aluminum oxide dielectric and is absorbed

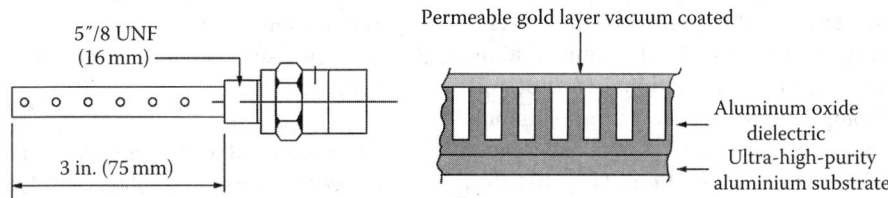

FIG. 1.38f
Thin-film aluminum oxide capacitance element, which can be packaged as a probe. (Courtesy of Mitchell Instruments Ltd.)

528 *Analytical Measurement*

by it. The amount of water absorbed determines the capacitance registered by the sensor.

To achieve accuracy specifications, each sensor requires a separate nonlinear calibration curve, and the unit must be periodically recalibrated to compensate for aging and contamination. These sensors are designed for low dew point measurements but can be disabled if exposed to high humidity or if wetted.

Materials for special application are available. For example, thin-film sensors constructed using zirconium rather than aluminum can withstand process temperatures up to 1400°C.

The advantages of both the electrolytic and capacitance type sensors are as follows:

- Ultra-small package size—surface-mount electronics are less than 0.5 mm.
- Probe-type packaging for direct insertion into process streams.
- Measurement range is wide and well suited for the detection of low dew points.
- High resolution, 14 bit and greater.
- Multiple alarm points, contact closures.
- Multiple communication protocols.
- Improved accuracy, repeatability, and stability.
- Most of these type sensors are also packaged with a thin film temperature probe to ease the conversion from dew point to relative humidity.
- Fast response.
- Autocalibration.

These characteristics make them suitable candidates for packaging for use in such applications as HVAC and many industrial plant processes, where cost is the primary concern.

Chilled Mirror

The optical chilled-mirror hygrometer is a primary method for dew point measurement. In this method, for a given mole fraction of water and gas pressure, the temperature of the mirror is cooled until its surface begins to fog. At this point, liquid water and water vapor pressures are in equilibrium. Once the water vapor partial pressures are determined, all definitions encountered in humidity as outlined earlier can be conveniently expressed.

These relationships have been experimentally and theoretically determined and tabulated by Goff and Gratch, Keenan, and Keyes,[3] and in the Smithsonian tables,[4] and they are constantly being refined. The Smithsonian tables, in particular, are used by the National Institute of Standards and Technology (NIST, formerly the National Bureau of Standards), www.nist.gov, as the reference documents for humidity sensor calibration. They list the saturation vapor pressures over plane surfaces of pure water and pure ice corresponding to the equivalent dew points or frost points.

Figure 1.38g illustrates that, at equilibrium partial pressure, when the gas mixture is saturated with ice water, the rate of water molecules leaving the atmosphere and condensing on the chilled surface is the same as the rate of water molecules leaving the chilled surface and reentering the atmosphere. At equilibrium saturation, the water vapor partial pressure of the condensate is equal to the water vapor partial pressure of the gas atmosphere. To establish this dynamic equilibrium at the mirror surface, it is necessary to precisely cool and control the mirror at the saturation temperature. A temperature element is then placed in thermal contact with the mirror, and the mirror temperature is utilized directly as the dew point or saturation temperature.

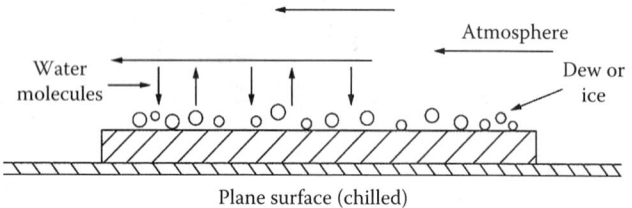

FIG. 1.38g
Equilibrium partial pressure illustration.

Cooling Methods

Historically, the cooling of the mirror surface has been accomplished with acetone and dry ice, liquid CO_2, mechanical refrigeration, and, more recently, thermoelectric heat pumps. Optical phototransistor detectors control automatically the surface at the dew point or frost point. Temperature instrumentation includes the entire spectrum from glass bulb thermometers to all types of electrical temperature elements.

The manually cooled, visually observed hygrometer is commonly known as the *dew cup*. It is a relatively inexpensive technique, and, when operated by an experienced and skilled technician, it is quite accurate. However, it does suffer from some limitations.

1. It is not a continuous measurement.
2. It is operator dependent, so the readings may vary from operator to operator.
3. Versions using expendable coolants require replacement supplies.

All of these difficulties are overcome by the thermoelectrically cooled, optically observed dew point hygrometer.

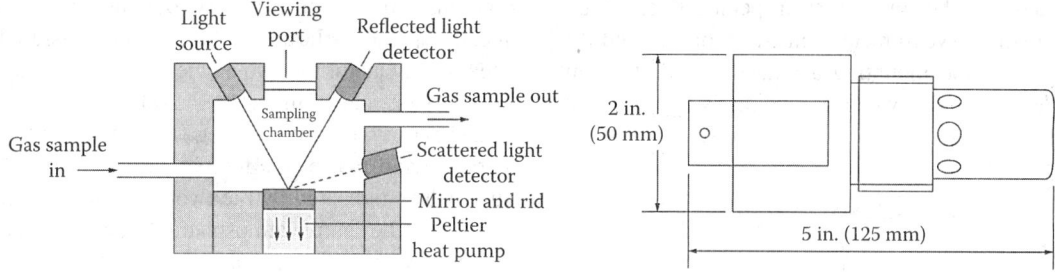

FIG. 1.38h
The cooled-mirror-type sensing element and the probe protector that houses it. (Courtesy of Mitchell Instruments Ltd.)

An instrument of this type is shown in Figure 1.38h. The mirror surface is chilled to the dew point by a thermoelectric cooler while a continuous sample of the atmosphere gas is passed over the mirror. The mirror is illuminated by a light source and observed by a photodetector bridge network. As condensate forms on the mirror, the change in reflectance is detected by a reduction in the direct reflected light level received by the photodetector because of the light-scattering effect of the individual dew molecules. This light reduction forces the optical bridge toward a balance point, reduces the input error signal to the amplifier, and proportionally controls the drive from the power supply to the thermoelectric cooler. This continuously maintains the mirror at a temperature at which a constant-thickness dew layer is retained. Independently, embedded within the mirror, a temperature-measuring element measures the dew point temperature directly.

QCM Quartz Crystal Microbalance In this design, the faces of a quartz crystal are covered with a hygroscopic polymer. Then a pair of electrodes are connected to this crystal. When voltage is applied to the sensor, a very stable oscillation occurs. As the amount of moisture adsorbed onto the polymer varies, the mass of the QCM changes producing a corresponding change in the frequency of oscillation. These easily measurable changes have a direct relation to the moisture concentration of the sample gas (Figure 1.38i).

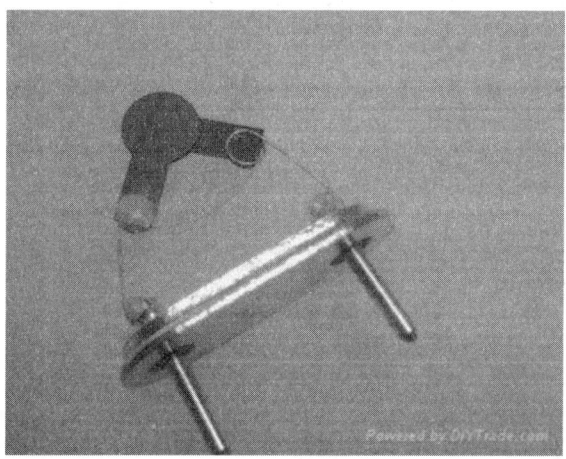

FIG. 1.38i
A quartz sensing element with electrodes. (Courtesy of DIYTrade.)

Legacy Designs: Historical Perspective

Legacy methods and designs relate to the 4th edition of this Handbook. Most of these designs are no longer commercially available, but may still be found in older installations. We provide this material to help those readers who work with older installations to help them understand the design of these legacy designs.

Hair or Synthetic Fiber Element This was the earliest design, Figure 1.38j. It has no practical applications today. However, some artistic and collectable designs may incorporate this technology.

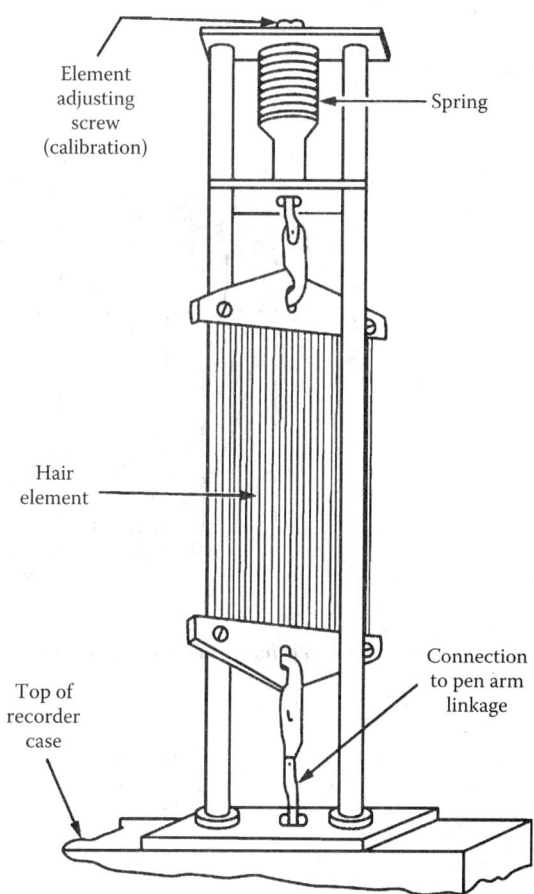

FIG. 1.38j
Hair hygrometer.

Sling Psychrometer For a historical perspective, Figure 1.38k shows a sling psychrometer consisting of wet and dry bulb thermometers and mounting arranged so that the unit can be whirled manually when a reading is to be made.

The instrument consists of two temperature-sensitive elements exposed to the atmosphere whose moisture level is to be measured. One of the two elements, the wet bulb, is wrapped with a wick soaked in water; the other element, the dry bulb, is left bare. Water evaporating from the wick lowers its temperature and that of the temperature element. The temperature values indicated by the two thermometers are related to the RH of the sample atmosphere. RH can be calculated from the equations presented earlier, using the wet and dry bulb temperature readings, or it can be read from a psychrometric chart such as the one shown in Figure 1.38a. For a reliable measurement, the sample velocity should be well in excess of 3 m/s (10 ft/s).

Cellulose Hygrometers Cellulose strips and other shapes are also used to measure humidity. Like human hair and synthetic materials, cellulose changes its dimensions as the water vapor concentration varies. This elongation has been used to build dial and digital indicators and also relatively inexpensive HVAC-type thermostats, transmitters, and recorders (Figure 1.38l).

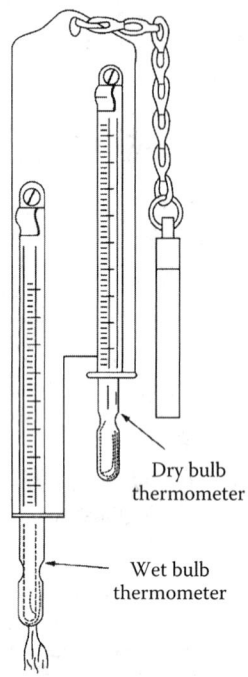

FIG. 1.38k
Sling psychrometer.

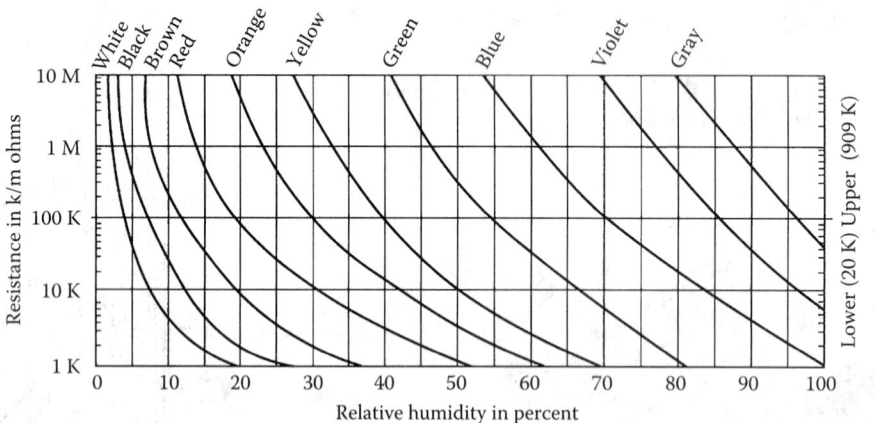

FIG. 1.38l
Ten-color-coded Dunsmore sensing element. (Courtesy Ohmic Instruments Co.)

ADDITIONAL CONSIDERATIONS

Installation

The instrument does not require a high-velocity sample to perform satisfactorily; in fact, sample velocity in excess of 0.3 m (1 ft) per second may result in a poor measurement and shortened element life. The reason is evident when one considers the measurement principle involved, namely, the change of solution vapor pressure and temperature on the sensing element by heating it to establish equilibrium with the sample. At high sample velocities, connective heat losses may swamp the measurement signal, and operation at higher power causes a reduction of element life.

The sensing element therefore must be located in a relatively quiescent zone or must be protected from direct impingement of the sample. In ducts, a sheet metal hood installed over the element and open on the downstream side is adequate. In pipelines, the element can be installed through the side outlet of a tee or can be mounted separately in a sampling chamber piped to the process line. The latter installation is preferred for samples under pressure, because the element can be serviced without shutting down the pipeline (Figure 1.38m).

Consideration needs to be given to the electrical and environmental conditions, with respect to industry codes such as the NFPA electrical classifications and the NEMA environmental classifications.

Condensing Atmospheres

The instruments discussed can be used on all non-corrosive gases in non-condensing atmosphere, if they do not react with the electrolytic substrate, the thin film, or water itself.

Sampling Systems

A sampling system is required when:

1. The sample temperature or pressure is above or below the operating limits of the sensing element
2. Flow rate needs to be maintained within the design limits of the sensor
3. The process line cannot be shut down for maintenance of the element, or entering the process environment is unsafe

When the process temperature is above the operating temperature limit of the sensor, a sample cooler must be installed upstream of the measuring element. The sampling system must be designed so that the sample will never be cooled to its dew point under any possible process or ambient conditions. Sample tubing should be small in diameter. The sensing element is to be mounted in a larger diameter sampling chamber, as shown in Figure 1.38n, to reduce sample velocity at the element.

Another need of a sampling system is to remove oils, particulate matter, and other contaminants that may attack the substrate. The design of the sample system must not change the process characteristics of the stream to be measured.

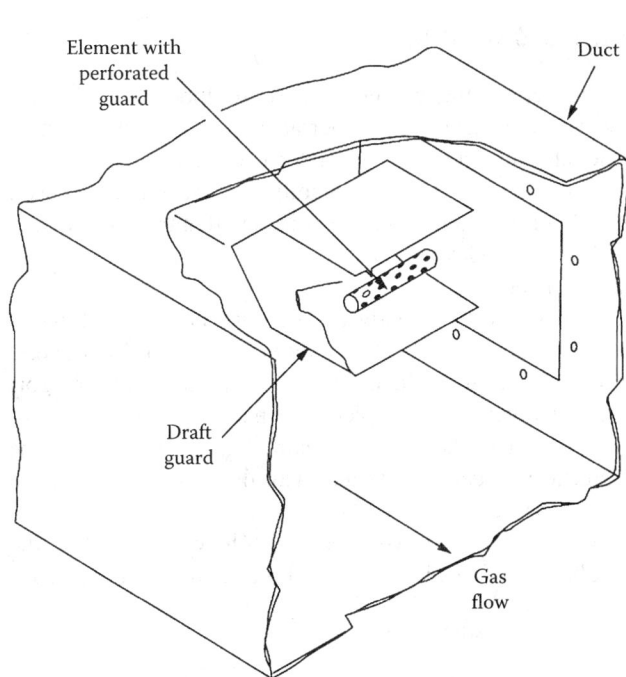

FIG. 1.38m
Duct installation with draft guard.

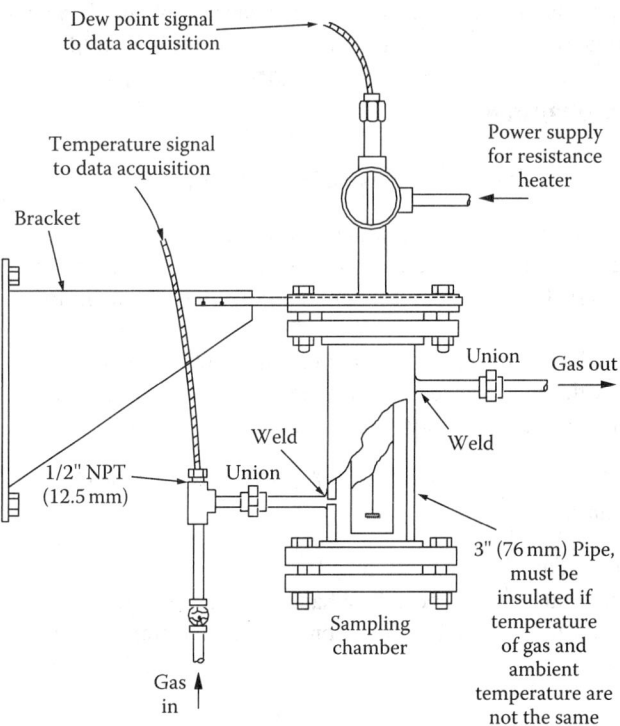

FIG. 1.38n
Probe installed in sampling chamber.

For high pressure systems, a pressure reducing or regulating valve may be necessary. For the measurement of stagnant or low-pressure systems, a pump may be required to draw the sample over the sensor and assure representative conditions.

The sensor requires periodic maintenance and must be removable from the line. Whenever such removal would result in a process shutdown, the sensor should be installed in a separate sampling chamber and shutoff valves provided in the sample piping.

In the sampling system, considerations of leakage, pressure and temperature gradients, and moisture absorption/desorption characteristics must be considered.

The problem of leakage is relative; that is, if the dew point being measured is close to the ambient dew point, leakage into the system may not bias the reading substantially. If the system is pressurized above the atmospheric pressure so as to create a leakage out of rather than into the system, the error introduced will be less.

The temperature stability of the sampling system components is also quite important. For any given equilibrium sampling condition, a specific amount of moisture will be absorbed onto the sampling system wetted surface, so control of the sample line temperature may be necessary to ensure an equilibrium condition.

Finally, the sample delivery components must be inert to components in the sample stream, and the material absorption/desorption characteristics cannot have an effect on overall system response. Stainless steel and nickel alloy tubing are normally the best nonhygroscopic materials, but copper, aluminum alloys, Teflon®, and polypropylene may also be suitable. Most other plastic and rubber tubing is unacceptable in all ranges. Process testing of sample system components is advised.

CALIBRATION

All sensors must be calibrated periodically:

1. To correct for sensor and electronics drift
2. To verify that no damage has been done to the sensor, such as attack or contamination from elements in the process
3. To remove oils and other contaminants in the process that may occlude the substrate and/or affect its accuracy.
4. To satisfy quality control procedures

There are three methods of calibration that need to be considered:

1. Autocalibration
2. Manual spot check by a handheld instrument
3. Reference and certification to a national standard

Autocalibration

During autocalibration, internal electronics of the sensor assembly heat the sensor, then, as it cools down to ambient temperature, both the humidity and temperature values are monitored. It thereby evaluates whether the humidity reading at 0% RH is correct. If it is not (Figure 1.38m), the drift is corrected by the microprocessor. In this figure, Pws is the temperature-dependent saturation pressure of the water vapor. This autocalibration technique allows the readings of low dew point values down to −60°C (−96°F) with errors under ±2°C. (Figure 1.38o).

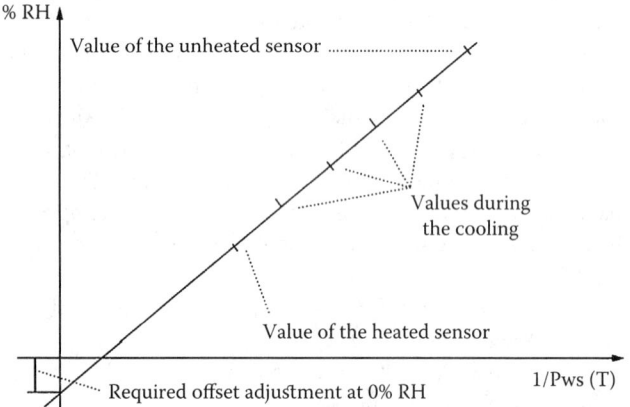

FIG. 1.38o
Autocalibration (Courtesy of VaisalaOyj.)

Handheld Calibration

In this method, the reading of the sensor mounted in the process is spot checked with a portable, handheld unit. Though handheld units normally do not possess the same speed of response, accuracy, or repeatability of the process sensor, they do provide a reasonable indication that the process sensor is in fact following the process and that no major damage has occurred.

For this to be effective, there must be a way of safely inserting the probe of the handheld unit into the process. In air-conditioning ducts, this could be a temporary port in the ductwork. In sampled systems, this could be a secondary port in the sample stream. The introduction of the handheld unit cannot interfere with the measurement of the primary sensor.

Corporate policy would establish how often to take the handheld readings, how to record them, and what corrective

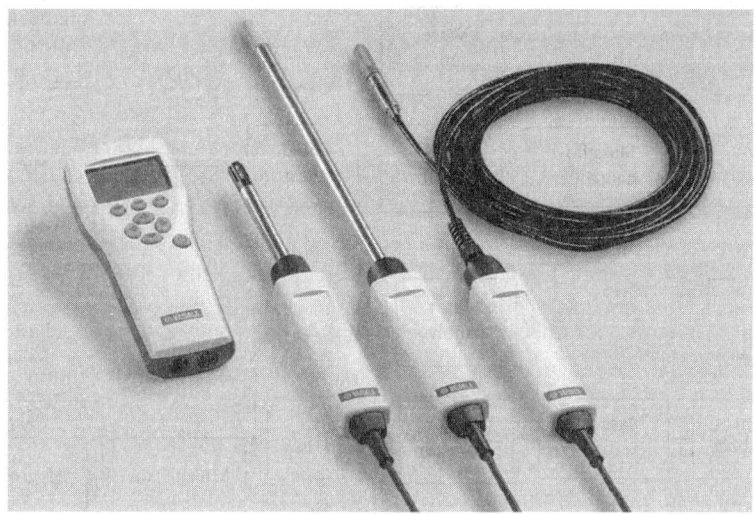

FIG. 1.38p
Handheld humidity and temperature meter for spot-checking applications. (Courtesy of VaisalaOyj.)

action to take if the primary humidity measurement differs significantly from the handheld unit (Figure 1.38p).

Reference to Standards

All statements related to the measurement accuracy and uncertainty of the sensor must be verified by testing to a standard. Standards for humidity measurement are established by the National Institute for Standards and Technology (NIST), Temperature and Humidity Group (www.nist.gov). This group also specifies how the testing is to be conducted. Standards from this group are globally coordinated with other standards setting bodies, such as the International Organization for Standardization (ISO) (www.iso.org).

For a manufacturer, company, or an agency to perform a reference calibration, and make claims of accuracy, it must show that the test equipment and test procedure used to make the calibration have been approved by NIST or ISO, and that an actual test has been performed on NIST equipment. Calibration to a reference standard includes a certificate that verifies that the testing agency has a proven chain of conformance to the NIST or ISO standard, range of the tests, and the degree of uncertainty of the measurement.

The frequency of calibration to a reference standard is a matter of agreement within specific industry groups. In some cases, a certificate of calibration may be a quality control requirement. The range of the tests is determined by the type of generator.

Two-Pressure Humidity Generator

The two-pressure humidity generator, Mark 2, includes a presaturator and a final saturator with a heatable tube connecting both. The presaturator generates high water mole fractions with minimal uncertainty. Here, clean, dry carrier gas is humidified to a dew point nearly equal to the final saturation temperature. The presaturator accomplishes nearly all the saturation. The final saturator performs small adjustments to ensure that the generated humidity is constant and determinable with minimal uncertainty, at the specified temperature and pressure required for the test.

An evolution of the Mark 2 is the hybrid humidity generator, or HHG, which adds an additional gas multiplexer, to improve uncertainty to 0.03%. This now introduces a new flow method: divided flow method.

At a given value of saturation temperature, two methods can now be used to lower the humidity while still knowing its value accurately: the two-pressure (2-P) technique and the divided flow method. The HHG is capable of using these two methods separately or together (hence the name *hybrid*). The two-pressure technique involves saturating the gas at an elevated pressure and afterward expanding the gas down to ambient pressure. The divided flow method involves diluting the saturated gas with dry gas using precisely metered streams of gas. Such a technique allows the generation of arbitrarily low humidity values while operating the saturator at convenient temperatures.

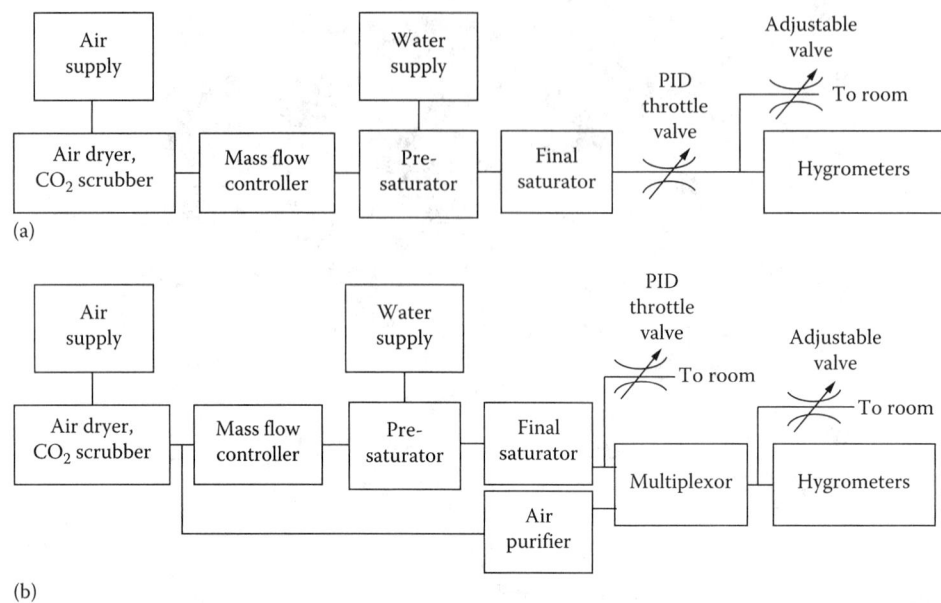

FIG. 1.38q
Schematic representation of the HHG in (a) two-pressure and (b) divided flow modes.

When performing hygrometer calibrations, the HHG operates in 1-P or 2-P mode for mole fractions greater than or equal to 1.6×10^{-3} (a frost point of $-15°C$). The HHG uses the divided flow method for mole fractions less than this value. A detailed explanation of this technology, along with the uncertainty calculations, is given in NIST standard SP250-83* (Figure 1.38q).

CALIBRATION METHOD

Calibrations are done at specified humidity points as requested by the end user, corresponding to the humidity range of the end user's process, and within the range of the sensor under test. Calibrations may be performed using humidity definitions of dew/frost point, RH, mole fraction, or mass ratio.

Before connecting the hygrometers to the hybrid generator, they are checked for cleanliness; any noticeable dirt inside a hygrometer or on its fittings is grounds for rejecting the instrument for calibration. In the case of chilled-mirror hygrometers, the hygrometer mirrors are cleaned with ethanol. Finally, tests are performed to ensure that calibration points requested for each hygrometer are within its measurement range. The pressure gauges of the hybrid generator are prepared for operation by correcting the zero offset for their readings.

The humidity test generator temperature and pressure parameters are then set to generate the desired humidity. For each nominal humidity point, a data acquisition application is used to sample and record temperature and pressure of the saturator, pressure inside the hygrometers, and wet-gas and dry-gas flows if the divided flow method is used, for a specified period of time. Typically, this would be for three to five test points, but may involve many more (Figure 1.38r).

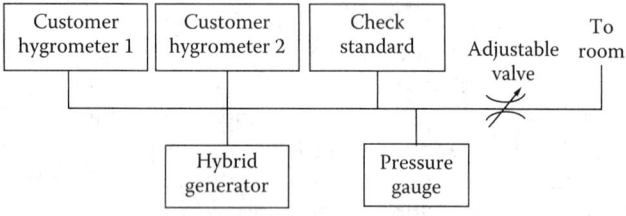

FIG. 1.38r
Connections of customer hygrometer and the check standard to hybrid generator.

Once the data are acquired for all requested humidity points, the results are assembled into a workbook. The new check standard data are compared to previous check standard data to validate the performance of the generator during this set of measurements. The final results comparing the customer hygrometer humidity to the HHG humidity are placed in a calibration report that proves the traceability of

---

* NIST Special Publication 250-83, Calibration of Hygrometers with the Hybrid Humidity Generator, NIST, December 2008, www.nist.gov.

the accuracy of this measurement from the sensor to international standards.

A commercially available dew point calibrator traceable to NIST standards is shown in Figure 1.38s.

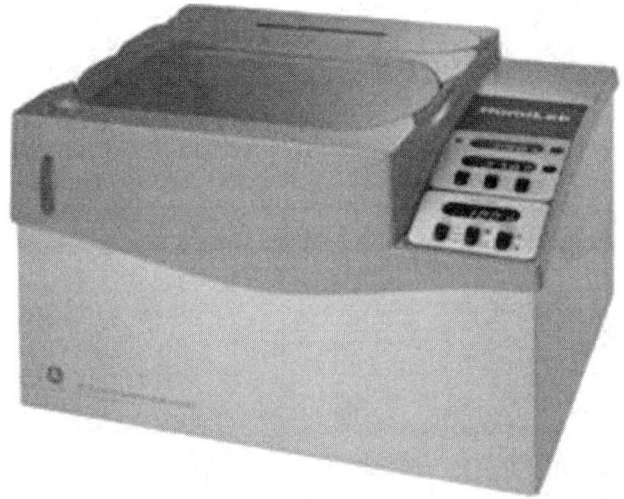

FIG. 1.38s
HumiLab chilled mirror calibrator. (Courtesy GE Infrastructure Testing.)

CONCLUSIONS

The instruments discussed here are generally applicable to atmospheric air. Operating temperature and pressure are usually limited to near-atmospheric conditions although, with some designs (such as capacitance), the operating limits can be extended. The principal feature common to all these instruments is that the output signal is not related directly to the moisture content but to a parameter that is related to the moisture content. Accuracy varies widely depending on design.

Sampling systems may be required to condition the sample and to protect the sensor. All sensors must be calibrated periodically.

SELECTION CRITERIA

When determining what technology to select, consider what is the purpose of this measurement:

- If curiosity, or monitor only, then low cost is best.
- If control, what accuracy levels are required for the process?
- If standards compliance is required, then select instruments that can be traceable to NIST standards.
- If you wish to become a standards compliance agency for calibration, select those instruments that perform calibrations referenced to national standards.
- Costs are proportional to performance.

Selection related to other environments may be found in other chapters:

- Chapter 1.39—Moisture in gases and liquids
- Chapter 1.22—Moisture in solids
- Chapter 112—High pressure applications
- Chapter 85—High temperature applications

SPECIFICATION FORMS

When specifying Mercury in Air analyzers, please use use the ISA form 20A2373 and when specifying the detector and also describing the application ISA form 20A1002 can be used. Both forms are reproduced on the following pages with the permission of the International Society of Automation.

1	RESPONSIBLE ORGANIZATION		HUMIDITY/DEWPOINT TRANSMITTER	6	SPECIFICATION IDENTIFICATIONS		
2			w/wo SWITCHES	7	Document no		
3	ISA		Device Specification	8	Latest revision		Date
4				9	Issue status		
5				10			

11	MOUNTING FITTING			60	PERFORMANCE CHARACTERISTICS		
12	Mounting/Fitting type			61	Max press at design temp		At
13	Process conn nominal size	Rating		62	Min working temperature		Max
14	Process conn termn type	Style		63	RH/Dewpoint accuracy		
15	Mounting/Fitting material			64	Temperature accuracy		
16				65	RH/Dewpoint repeatability		
17				66	Max response time		
18				67	Min ambient working temp		Max
19	SENSING ELEMENT			68	Contacts ac rating		At max
20	Sensor type			69	Contacts dc rating		At max
21	Probe style			70	Max sensor to receiver lg		
22	Temperature LRL	URL		71			
23	Humidity/Dewpoint LRL	URL		72			
24	Temperature sensor type			73			
25	Sensor protection style			74			
26	Probe length			75			
27	Sensor head material			76			
28				77			
29				78			
30				79			
31	LEAD WIRE AND EXTENSION			80			
32	Extension type			81			
33	RH/Dewpoint cable length			82			
34	Temperature cable length			83	ACCESSORIES		
35	Max cable operating temp			84	Calibrator		
36	Signal termination type			85	Humidity standards		
37				86	Replacement probe		
38				87			
39				88			
40	TRANSMITTER w/wo SWITCHES			89			
41	Housing type			90	SPECIAL REQUIREMENTS		
42	Output signal type			91	Custom tag		
43	Enclosure type no/class			92	Reference specification		
44	Local operator interface			93	Compliance standard		
45	Digital communication std			94	Calibration report		
46	Signal power source			95	Software configuration		
47	Qty of input sensors			96			
48	Contacts arrangement	Quantity		97			
49	Integral indicator style			98			
50	Signal termination type			99	PHYSICAL DATA		
51	Cert/Approval type			100	Estimated weight		
52	Mounting location/type			101	Overall width		
53	Calibration mode			102	Overall height		
54	Measurement compensation			103	Removal clearance		
55	Enclosure material			104	Signal conn nominal size		Style
56				105	Mfr reference dwg		
57				106			
58				107			
59				108			

110	CALIBRATIONS AND TEST		INPUT OR TEST			OUTPUT OR SCALE	
111	TAG NO/FUNCTIONAL IDENT	MEAS/SIGNAL/SCALE	LRV	URV	ACTION	LRV	URV
112		Humidity-Analog output					
113							
114		Humidity-Scale 1					
115		Other variable-Scale 2					
116		Temp-Analog output					
117		Setpoint 1-Output					
118		Setpoint 2-Output					
119							

120	COMPONENT IDENTIFICATIONS		
121	COMPONENT TYPE	MANUFACTURER	MODEL NUMBER
122			
123			
124			
125			

Rev	Date	Revision Description	By	Appv1	Appv2	Appv3	REMARKS

Form: 20A2373 Rev 0 © 2004 ISA

1.38 Moisture in Air

	RESPONSIBLE ORGANIZATION	ANALYSIS DEVICE COMPOSITION OR PROPERTY Operating Parameters (Continued)		SPECIFICATION IDENTIFICATIONS		
1			6	Document no		
2			7	Latest revision	Date	
3	(ISA logo)		8	Issue status		
4			9			
5			10			

	PROCESS COMPOSITION OR PROPERTY			MEASUREMENT DESIGN CONDITIONS				
	Component/Property Name	Normal	Units	Minimum	Units	Maximum	Units	Repeatability
13								
14								
15								
...								
37								

(Rows 38–75: open remarks/notes area)

Rev	Date	Revision Description	By	Appv1	Appv2	Appv3	REMARKS

Form: 20A1002 Rev 0   © 2004 ISA

Definitions

Absorption: In spectroscopy, the attenuation of electromagnetic radiation by a specific analyte. The amount of attenuation is directly proportional to each analyte's concentration.

Adsorption: The adhesion of a fluid in extremely thin layers to the surfaces of a solid.

Dew point: Temperature to which the air (or any gas) must be cooled in order that it becomes saturated with water.

Dry bulb: The temperature of air shielded from radiation and moisture. It is directly proportional to the mean kinetic energy of the air modules.

Enthalpy: A measure of the total energy of a thermodynamic system. It includes the internal energy, which is the energy required to create the system, and the amount of energy required to make room for it by displacing its environment and establishing its volume and pressure. Typical units: BTU/lb, kcal/kg.

Grains: A measurement of moisture. In the Bronze Age, the weight of a single grain of barley. Now specified as exactly 7000 grains = 1 lb.

Hygrometer: An apparatus that measures humidity.

Hygroscopic material: A material with great affinity for moisture.

Partial pressure: In a mixture of gases, the partial pressure of one component is the pressure of that component if it alone occupied the entire volume at the temperature of the mixture.

Relative humidity: The ratio of the mole fraction of moisture in a gas mixture to the mole fraction of moisture in a saturated mixture at the same temperature and pressure. Alternatively, the ratio of how much water vapor is in the air vs. the maximum it could contain at the particular temperature and pressure.

Saturated solution: A solution that has reached the limit of solubility.

Saturation pressure: The pressure of a fluid when condensation (or vaporization) takes place at a given temperature. (The temperature is the saturation temperature.)

Specific humidity: The ratio of the mass of water vapor to the mass of dry gas in a given volume.

Wet bulb: The temperature of air when cooled to saturation (100% RH) simply by evaporation of water into it.

Abbreviations

FTIR	Fourier transform infrared
g/m^3	grams of water per cubic meter; known as absolute humidity in moisture measurements.
lb/mmscf	Pounds of water per 1 million standard cubic feet
NDIR	Non dispersive infrared
NMR	Nuclear magnetic resonance
Ppmv	Parts per million by volume
Ppmw	Parts per million by weight
QCM	Quartz crystal microbalance
TDLAS	Tunable diode laser absorption spectroscopy

Standards Agencies

ASHRAE: American Society of Heating, Refrigerating, and Air-Conditioning Engineers, www.ashrae.org,

ASSE: American Society of Safety Engineers, www.asse.org

CGA: Cylinder Gas Association,, www.cganet.com. CGA G-7.1–2011 Grade D, Air/Gas Quality

ISCCP: International Satellite Cloud Climatology Project. The focus of the **ISCCP** is to collect weather satellite radiance measurements and to analyze them.www.isccp.giss.nasa.gov

ISO: International Organization of Standardization, www.iso.org

NCDC: National Climatic Data Center. **NCDC** is the world's largest active archive of weather data. www.ncdc.noaa.gov

NCSL: National Conference of Standards Laboratories International was formed in 1961 to promote cooperative efforts for solving the common problems faced by measurement laboratories. http://www.ncsli.org/

NFPA: National Fire Protection Association, www.nfpa.org

NIST: National Institute of Standards and Testing, www.nist.gov

NOAA: National Oceanic and Atmospheric Administration. NOAA is a federal agency focused on the condition of the oceans and the atmosphere.www.noaa.gov

SAE: Society of Automotive Engineers International.www.sae.org. ARP 5623, Aerospace Recommended Practice, Infrared Runway Surface, Ambient Air and Dew point Temperature.

Bibliography

ACGIH, *Air Sampling Instruments for Evaluation of Atmospheric Contaminants,* 9th ed., Cincinnati, OH: American Conference of Governmental Industrial Hygienists, 2001.

AIHA, *Workplace Environmental Exposure Level Guide Series, Full Set of 109 Individual Guides on Toxic Chemicals,* Akron, OH: American Industrial Hygiene Association, 1979–2002.

An Evaluation of Performance of Trace Moisture Measurement Methods, NPL, BOC Edwards, 2004 http://www.tigeroptics.com/TA/photo/view.php?gal=users;site,cms,files&s=orig&f=An+Evaluation+of+Performance+of+Trace+Moisture+Measurement+Methods.pdf

Cole, K. M. and Reger, J. A., Humidity calibration techniques, *Instrum. Contr. Syst.,* January 1970.

Cooper, F. G., Dew point vs. frost point measurements, *Sensors,* October 1991.

Cortina, V. B., Sampling systems for dew point hygrometers, *InTech,* March 1985.

Dadchauji, F. and Webster, S., Humidity instrumentation tubes or high temperatures, *Control,* January 1992.

Edge Tech Instruments, *Basic Humidity Definitions,* http://www.edgetechinstruments.com/docs/humdef.pdf.

Irving, C. L. and Higgins, C. T., HVAC System Design Narrows RH Control to ±1%, *Control Eng.,* 27(11), 150, 1980.

ISCCP, mathematical model from National Environmental Satellite Data and Information Service: www.hitran.iao.ru.

Kohler, H. M. and Mathew, A., Continuous *in-situ* and elevated temperature moisture measurement in high particulate reactive processes, *Proc. ISA 1996 Conference,* Paper 96–002.

Krigman, A., Moisture and humidity, *InTech,* March 1985.

McKinley, J. J., Using permeation tubes to create trace concentration moisture standards, National *Conference of Standards Laboratories Conference,* Atlanta, GA, July 1997.

Mettes, J. et al., Multipoint ppm moisture measurement using electrolytic cells, *ISA Conference,* Houston, TX, October 1992.

Meyers, R. A., Ed., *Encyclopedia of Analytical Chemistry: Instrumentation and Applications,* New York: John Wiley & Sons, 2000.

Moisture/humidity, *Meas. Contr.,* February 1992.

National Climatic Data Center (NCDC), National Oceanic and Atmospheric Administration (NOAA): www.modis-atmos.gsfc.nasa.gov.

Petersen, K. L., Portable dew-point generator, *Meas. Contr.,* February 1992.

Psychrometric Tables for Obtaining the Vapor Pressure, Relative Humidity, and Temperature of the Dew-Point by United States.

Schultz, G., Relative humidity measurement with an analog transmitter, *Meas. Contr.,* December 1992.

Sherman, L. H., Sensors and conditioning circuits simplify humidity measurement, *EDN,* May 16, 1985.

Shinskey, F. G., Humidity, dew point and wet-bulb temperature, *Instrum. Cont. Syst.,* 43, July 1975, and 15, August 1975.

Tannermann, J., Getting a grip on humidity, *InTech,* August 2002.

Understanding Air Density and its Effects, Jack Williams, USA Today, May 7, 2005, usatoday30.usatoday.com/weather/wdensity.htm

Weather Bureau(Aug 31, 2012)(http://www.amazon.com/s/ref=nb_sb_noss?url=search-alias%3Dstripbooks&field-keywords=humidity+and+dew+point)

Wiederhold, P. R., Humidity measurements, *Instrum. Technol.,* 45–50, August 1975.

Wiederhold, P. R., Cycling chilled-mirror hygrometer, *Meas. Contr.,* February 1993.

Yan, W. B. and Mallon, T., breakthrough technology measures parts-per-trillion moisture in gases, *Chem. Res.,* February/March 2002.

1.39 Moisture in Gases and Liquids

A. BRODGESELL (1969, 1982) **B. G. LIPTÁK** (1995)
J. F. TATERA (2003) **S. VEDULA** (2017)

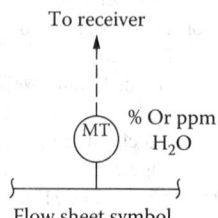

Flow sheet symbol

Types of designs	A. Electrolytic hygrometer
	B. Capacitance
	C. Impedance
	D. Piezoelectric
	E. Heat of adsorption
	F1. Infrared (IR)
	F2. NIR
	F3. Tunable diode Laser Absorption Spectroscopy
	G. Microwave
	H. Karl Fischer titrator (discussed under laboratory)
	I. Drying oven (laboratory)
	J. Dipole
	K. Cavity Ring Down
	L. Neutron
	M. Refractive Index
	N. Calibrators
Ranges	A. through G. see Table 1.39a
	H. 10 ppm to 100%
	I. 0.005%–100%
	J. 0%–100%
	K. ppt to ppm levels
	L. 0%–60%
	M. 2–12,225 ppm
	N. ppm to % ranges
Inaccuracy	A. ±5% of reading or 0.4 ppm for ppm ranges and ±4 ppb for ppb ranges
	B. and C. 3%
	D. from 1 to 2500 ppmv
	E. +/−3%
	F. 2% FS
	G. For a 1%–15% moisture range, error is within 0.5%
	H. and I. 0.5%–1%
	J. 0.25%
	K. +/−4%
	L. +/−0.5%
	M. ±2% of span

(Continued)

Costs   (For moisture in air and relative humidity sensors, also refer to Chapter 1.38 of this volume)
A. $10,000 to $20,000 with sample system
B. and C. $6,000–$15,000 for thin-film probe, $6,000–$250,000 for flow-through bypass analyzer
D. $45,000
F. and G. $30,000–$35,000
H. $15,000–$30,000
I. $15,000–$25,000
J. $25,000 to $35,000
K. $30,000–$60,000
L. $30,000–$40,000
M. $35,000–$45,000
N. $5,000 to $10,000

Partial list of suppliers   (For moisture measurement in air and solids, also refer to Sections 8.32 and 8.34)
Ametek Process Instruments (http://www.ametekpi.com/products/Model-3050.aspx) (D)
Ametek Process Instruments (http://www.ametekpi.com/products/WR-IPS-4.aspx) (F1)
Arizona Instrument Corp. (http://www.azic.com/products_computrac_max4000xl.aspx) (I)
(http://www.azic.com/products_computrac_vaporpro.aspx) (H)
Bartec Benke (http://www.bartec.de/homepage/deu/downloads/produkte/16_messtechnik/Hygrophil-F5673_e.pdf) (M)
CEM Corp. (http://www.cem.com) (G)
Cosa Xentaur (http://cosaxentaur.com/resources/article_level2a/516/DPT.D.BR.0001_B.pdf) (B)
http://cosaxentaur.com/resources/article_level2a/516/HDT_LQNewestUpdateVersion.pdf (H)
http://cosaxentaur.com/resources/article_level2a/594/XTDL-HT_Brochure.pdf (F3)
http://www.cscscientific.com/moisture/karl-fischer/ (H)
Dewcon Instruments Inc (http://www.pronamicscontrol.com/Dewcon_Humidity_Sensor.pdf) (J)
Emerson Process Management (G) (http://www2.emersonprocess.com/siteadmincenter/PM%20Roxar%20Documents/Flow%20Metering/Roxar%20Watercut%20meter%20Brochure.pdf)
Endress + Hauser Instruments (http://www.endress.com) (F3)
Foss NIR Systems Inc. (http://www.foss-nirsystems.com) (F2)
GE measurement and control (http://www.ge-mcs.com/en/moisture-and-humidity.html) (C, F3, N)
Illinois Instruments Inc. http://www.systechillinois.com/en/products/Gas/Moisture (A, C)
Kahn Instruments Inc. (http://www.kahn.com/hygrometers/index.html) (A, B)
Kin-Tek Laboratories Inc. (http://www.kin-tek.com/abgasst.html) (M)
MEECO Inc. (A) http://meeco.com/meeco/products/trace-ultra-trace-analyzers
Michell Instruments (http://www.michell-instruments.com) (B)
(http://www.michell.com/uk/category/dew-point-transmitters.htm)
Nova Analytical Systems Inc. (http://www.nova-gas.com) (B)
http://catalog.nova-gas.com/category/continuous-dew-point-only-analyzers
Panametrics Inc. (http://www.transcat.com/PDF/GEMIS1.pdf) (B, C, N)
Servomex https://www.servomex.com/80257230003F5977/0/580A98C5D5F3CC3B8025791A004723B4/$FILE/DF750_uk.pdf (N)
https://www.servomex.com/80257230003F5977/0/FA7621C1F9CF6C2C8025791A00470F1F/$FILE/DF745SG_uk.pdf (N)
Teledyne Analytical Instruments (B) http://www.teledyne-ai.com/pdf/8800.pdf
Thermo Electron Corp. (http://www.thermo.com) (F, L) https://static.thermoscientific.com/images/D01814~.pdf https://static.thermoscientific.com/images/D01815~.pdf
Tiger Optics LLC (http://www.tigeroptics.com) (K) http://tigeroptics.com/pdf/Brochures/TO_HALO_3_H2O_500_2.pdf
http://tigeroptics.com/pdf/Brochures/TO_HALO_KA_3.pdf
Vaisala (B) http://www.vaisala.com/en/products/moistureinoil/Pages/default.aspx

Most Popular:
Ametek, General Eastern, and Endress & Hauser

INTRODUCTION

There is a certain amount of overlap between the three sections that are devoted to the subject of moisture measurement in this handbook. (Chapter 1.38 describes the measurement of moisture in air; this chapter, in liquids and gases; and Chapter 1.40, in solids.) Some of the instruments, such as the chilled-mirror or the thin-film capacitance-type sensors discussed in Chapter 1.38, can operate not only as dew point or humidity sensors in air but also as moisture analyzers in other gases. Similarly, the electrolytic hygrometer, which is discussed in this chapter, could also be used to measure humidity or dew point in air. Similar overlaps exist between the coverage of this chapter and Chapter 1.40, dealing with moisture in solids, particularly with respect to microwave, neutron, and infrared analyzers. The reader is also reminded that separate chapters are devoted to some of the analyzers that can also be used for other analytical tasks besides moisture detection, such as infrared (Chapter 1.32) and autotitrators (Chapter 1.74). The reader is advised to refer to these chapters as well for additional information concerning their suppliers and features.

LABORATORY ANALYZERS

An old but still useful method of determining the moisture of a sample is to heat it and measure the resulting weight loss. Some drying ovens are provided with built-in weight scales and displays that indicate the sample moisture content directly in weight percentage. The reliability of this technique is primarily a function of the presence or absence of volatile materials other than water and the weighing procedure. Some built-in scale oven systems even have experience issues with the weighing procedure and air current effects on the sample holders.

Another widely used and accurate (sensitive to 5 ppm) method of moisture analysis in gas or liquid samples is by titration. The reagent is referred to as the Karl Fischer reagent, which consists of a solution of sulfur dioxide, iodine, and pyridine in methanol. The titration end points can be found automatically (see Chapter 1.74) by the measurement of the current flow through the solution. Recent concerns regarding the toxicity of pyridine have led to the use of substitute buffers like imidazole. The iodine water reaction is the same, and it is still classified as a Karl Fischer titration.

PROCESS ANALYZERS

Some of the industrial process analyzers are of the probe type (capacitance, fiber-optic infrared probes), others can look through the process stream in the pipe (microwave), and the majority require some form of a sampling system. Table 1.39a provides a summary of the ranges and other features of many of the process moisture analyzers. The subject of sampling (which has already been covered to some extent in Chapters 1.2, 1.3 and 1.10 of this volume) will also be addressed in the next chapter.

TABLE 1.39a
Summary of Moisture Analyzer Features

Type	Range	Sample Phase	Sample System Required	Remarks
Electrolytic hygrometer	0–500 to 0–2000 ppb and 1–1000 ppm	Clean gas Special sampling for liquids	Yes	Sample flow must be constant
Change of capacitance	0–10 to 0–1000 ppm	Clean gas or liquids	a	Sample temperature must be constant
Impedance type	0–20,000 ppm (−100 to +20°C dew point)	Clean gas or liquids	No, only for liquids	Sample temperature of liquids must be constant
Piezoelectric type	0.1–2500 ppm	Clean gas only	Yes	
Heat of absorption type	0–10 to 0–5000 ppm	Clean gas or liquid Special sampling for liquids	Yes	Sample flow must be constant
Infrared absorption Near Infrared Tunable Diode Laser Absorption spectroscopy	Gas; ppm-100%	Gas	b	
Microwave absorption or microwave frequency change due to load	0%–1% to 0%–70%	Liquids, slurries, and pastes only	No	Density and salinity of liquid affect measurement. Need to be compensated

a Available in probe form, but can be direct pipeline mounted only if flow velocity is under 1.6 fps (0.5 m/s).
b Fiber-optic probe (FOP) designs can be direct pipeline mounted without sampling.

Sampling Systems

Some sampling system features are peculiar to moisture analyzers, and these are covered below. The function of any sampling system is to deliver a clean, appropriately representative sample to the measuring element at the required pressure, temperature, and flow rate. The measurement of moisture, although simple in principle, is complicated by the fact that most materials adsorb and desorb moisture.

At moisture levels in the percent range, moisture absorption is not normally a serious problem; however, at moisture levels in the low parts per million (ppm) ranges, sampling system materials must be selected so as to minimize their effect on the moisture concentration in the sample. Depending on the levels being monitored, materials that are less likely to interfere with the moisture content in the sample are stainless steels, Teflon®, Viton®, Kel-F®, nickel and nickel-plated materials, and cadmium and cadmium-plated materials. Materials that should almost always be avoided are copper and its alloys, rubber, neoprene, and elastomers in general.

It is important to apply the material considerations to all parts of the sampling system, including the internals of valves and filters.

Electrolytic Hygrometer

The principle of measurement utilized involves the electrolysis of water into oxygen and hydrogen. Since two electrons are required for electrolysis of each water molecule, the electrolysis current is a measure of the water present in the sample. If the volumetric flow rate of the sample gas into the electrolysis cell is controlled at a fixed value, then the electrolysis current is a function of water concentration in the sample (Figure 1.39a). This relationship is illustrated in the following example.

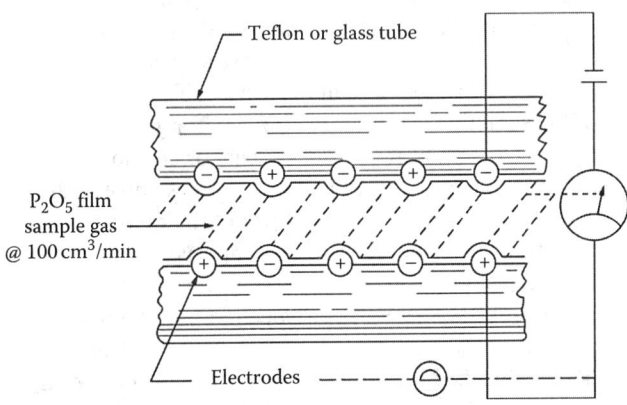

FIG. 1.39a
Operating principle of the electrolytic hygrometer.

Determine the water concentration in a sample when the sample flow rate is 100 cm³/min (0.212 ft³/h) at 37.8°C (100°F) and 10 psig (68.9 kPa). The electrolysis current is measured at 320 µA.

Electrolysis current = 320×10^{-6} C/s

$1\ C = 6.25 \times 10^{18}$ electrons

Therefore, the electrolysis current is equal to the electron flow of 2×10^{15} electrons/s. Since two electrons of charge are required to electrolyze one water molecule, the water flow into the cell is equal to 10^{15} molecules/s. If, for purposes of this example, the water present in the sample is considered an ideal gas, the volume of water entering the sample can be calculated using the ideal gas law.

$$V_2 = \frac{P_1}{P_2} \frac{T_2}{T_1} V_1 \qquad \mathbf{1.39(1)}$$

Avogadro's Law Avogadro's law states that equal volumes of different ideal gases at the same pressure and temperature contain the same number of molecules. The volumetric flow rate of water vapor can be calculated as

$$V_1 = 379\ N/A \qquad \mathbf{1.39(2)}$$

where
  N is the flow of water in molecules per second
  A is the Avogadro's number (A = 2.73283×10^{26} molecules/ [lb·mol])
  379 = volume of a pound-mole of ideal gas in cubic feet

Subscripts 1 and 2 refer to standard and sample conditions, respectively.

$$V_1 = \frac{379 \times 10^{15}}{2.73 \times 10^{26}} = 1.39 \times 10^{-9}\ \text{ft}^3/\text{sec} = 5.0 \times 10^{-6}\ \text{ft}^3/\text{h}$$
$$\mathbf{1.39(3)}$$

$$V_2 = 5 \times 10^{-6} \times \frac{560}{520} \times \frac{14.7}{24.7} = 3.2 \times 10^{-6}\ \text{ft}^3/\text{h} \qquad \mathbf{1.39(4)}$$

Since the sample flow is controlled at 0.212 ft³/h (100 cm³/min), the volume concentration of moisture is $3.2 \times 10^{-6}/0.212 = 15.1 \times 10^{-6}$. Expressed in parts per million (ppm), the moisture content is 15.1 ppm. The example above is intended to illustrate the fact that the hygrometer cell sees only a mass flow of water—number of water molecules electrolyzed per unit time. To obtain an output in the form of moisture content, the sample flow rate must be a known constant value. Furthermore, the accuracy of the output can never exceed the accuracy with which the sample flow is controlled.

The commercial electrolysis hygrometer cell consists of a small chamber containing two noble metal electrodes that support a thin layer of desiccant. Moisture in the sample is absorbed by the desiccant and electrolyzed by means of a voltage-regulated power supply connected to the electrodes.

Units are available for use in nonhazardous areas with the sample flow control, electrolysis cell, and electronics packaged as a single unit. When used in hazardous areas, the cell and flow controller are housed in an explosion-proof conduit, and the electronic circuitry is remote mounted (see Figure 1.39b).

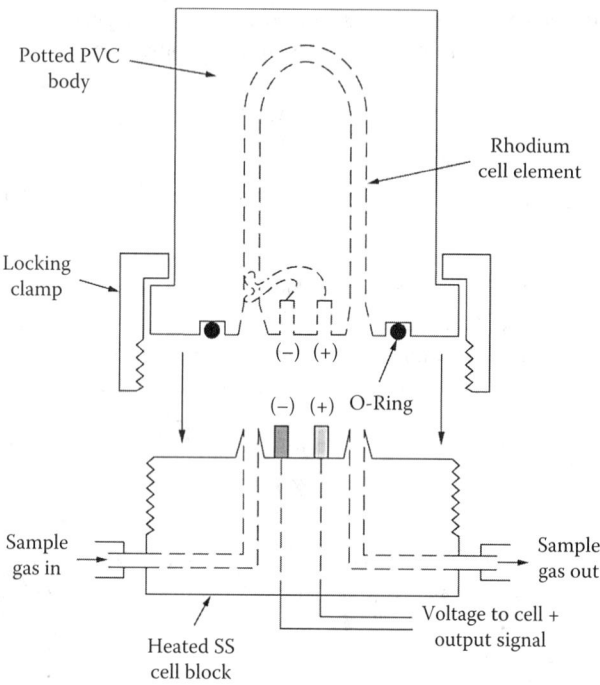

FIG. 1.39b
Electrolytic hygrometer cell.

Recombination Effect Although this instrument will operate satisfactorily with a variety of samples, a phenomenon called the *recombination effect* introduces large errors at low moisture levels in hydrogen-rich or oxygen-rich samples. Recombination is the reversion to water of the electrolysis products; it introduces an error into the measurement when the recombined oxygen and hydrogen are re-electrolyzed. Apparently, all electrodes catalyze this reaction, although some electrode materials do so more than others. The use of rhodium as the electrode material has been found to minimize recombination.

When monitoring very low moisture levels in oxygen-rich or hydrogen-rich atmospheres, the recombination produces a relatively large error even with the best choice of electrode materials. For such an application, two sensors are used; one measures at sample flow rate X and the other at sample flow rate 2X. Because the error due to recombination is a constant, subtraction of the two sensor outputs yields a signal that is independent of recombination.

Cell Limitations The cell will perform satisfactorily with a variety of samples; however, a number of sample materials will cause problems or should not be monitored with this instrument. Gases known to cause problems are hydrogen and oxygen. These gases have already been discussed in terms of the recombination effect, described earlier. Some other gases for which this instrument is unsuitable are unsaturated monomers, alcohols, amines, ammonia, hydrogen fluoride, and $CHClF_2$ (Freon®) refrigerant.

Alcohols are seen by the cell as water. Amines and ammonia usually react with the desiccant. Hydrogen fluoride can corrode the internals of the cell. The data collected on $CHClF_2$ refrigerant indicates an anomaly, although the reason for this is not fully understood. Unsaturated hydrocarbons such as butadiene or monomers with a strong tendency to polymerize cannot be monitored, as the cell will be quickly coated with polymer.

Generally, the instrument should not be used with samples whose components may deposit in the cell (condensable vapors). When a cell becomes contaminated, it will show a memory for polarity, that is, the outputs under forward and reverse flow through the cell will not be equal. The electrolytic hygrometer is suitable for most inert elemental gas applications and for other gases that do not react with P_2O_5.

A cell will lose sensitivity when exposed to moisture levels of a few parts per million over a period of weeks. This sensitivity loss is due to the elution of desiccant with the sample. However, this process occurs over a long period of time and the cell can be recoated fairly easily in the field during periodic maintenance.

Liquid Samples The electrolytic hygrometer cell can be exposed only to gases. When the process sample is in the liquid state, there are two ways of using the electrolytic hygrometer. Recall that this process must not result in transporting to the sensor vapors that will condense or be absorbed.

1. Sample vaporization: If the liquid sample has sufficiently high vapor pressure, the sample can be vaporized, and the measurement is then made as a gas sample. The sample is vaporized by means of a vaporizing regulator located as close as possible to the sample take-off point to minimize lag. The remainder of the sampling system is the same as for gas samples.
2. Sample stripping: If the sample cannot be vaporized, this method can be used to obtain a gas sample at the cell. The moisture is stripped from the liquid sample in a falling film column. Liquid sample is continuously metered into the top of the column, where it descends as a thin film. Dry nitrogen, metered into the bottom of the column, ascends and removes moisture from the descending film of liquid sample. A filter located at the top of the column removes any droplets entrained in the nitrogen before the gas is passed through the measuring cell.

A drain valve at the bottom of the column facilitates the removal of stripped sample. The nitrogen is dried before entering the column by passing through an electrolytic dryer, similar to the measuring cell, where moisture is removed

the capacitance would increase correspondingly from minimum to maximum. Water, with a dielectric constant of 80, would increase the capacitance to 80 times the value of the air capacitor.

The measuring cell of the hygrometer consists of two electrically insulated concentric metal cylinders to form the measuring capacitor plates (Figure 1.39e). The annulus between the cylinders is filled with alumina desiccant. Two porous metal discs support the cylinders and retain the desiccant in the annulus. The sample is allowed to flow through the annulus, and process water is absorbed or desorbed by the desiccant, which remains in equilibrium with the sample in terms of percent saturation. The desiccant thus amplifies the moisture content of the sample because the saturation level of the desiccant that is in equilibrium with the sample is very much higher than that of the sample.

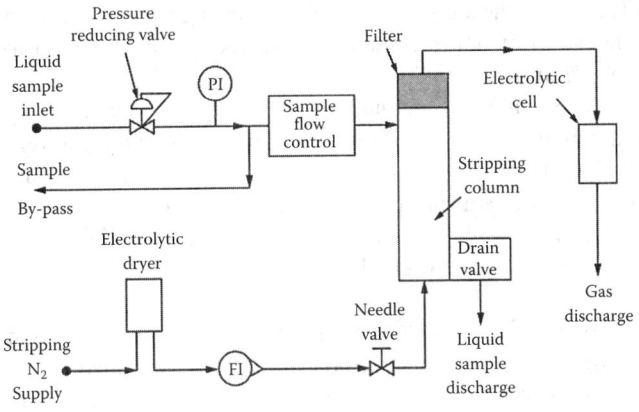

FIG. 1.39c
Liquid sample stripping.

by electrolysis. The supply nitrogen moisture level should be fairly low (less than 500 ppm) to avoid overloading the dryer (Figure 1.39c).

Sampling System As pointed out previously, the electrolytic hygrometer is sensitive only to the mass flow of water into the cell; the concentration of moisture is inferred from the known sample flow rate. Therefore, the accuracy of the readout is partially dependent on the accuracy of the sample flow control and the accuracy of the moisture measurement can never exceed that of the sample flow control. This fact should be kept in mind when selecting a sample flow controller.

Capacitance Hygrometer

The principle of measurement utilized is the change of capacitance associated with a change of the sample dielectric constant between capacitor plates. The value of capacitance is a function of plate area, plate spacing, and the dielectric constant of the material between the plates. The dielectric constant of a material has a unique value for each substance and is related to the polarity characteristics of the molecules. This property can, therefore, be used to detect the presence of a specific substance in a pipeline stream.

Figure 1.39d shows an air capacitor connected to a battery and the same capacitor with mica between the plates. The capacitance of mica is six times that of air because the dielectric constant, K, of mica is 6, and the K of air is 1. If the mica were inserted gradually between the plates, then

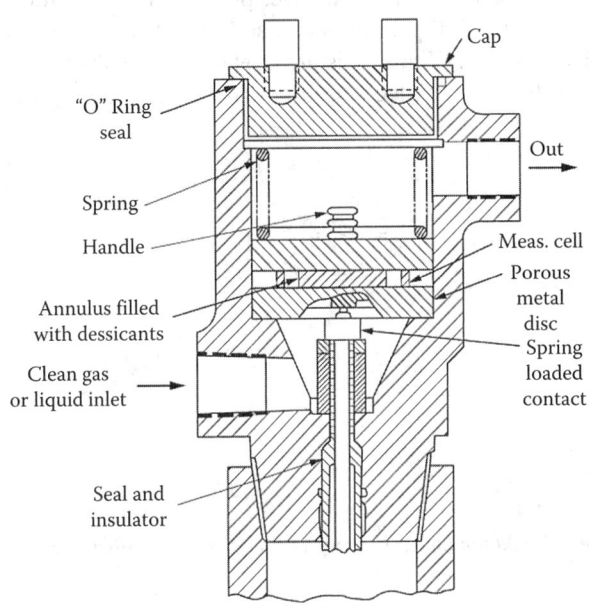

FIG. 1.39e
Capacitance-type measuring cell.

The measuring capacitor is part of an electrical circuit that includes a reference capacitor. This circuit is usually powered by a 15 kHz fixed-amplitude sine wave. The measuring and reference capacitors are switched alternately into the circuit such that its output voltage is a function of the connected capacitance. The output signal is a sine wave of

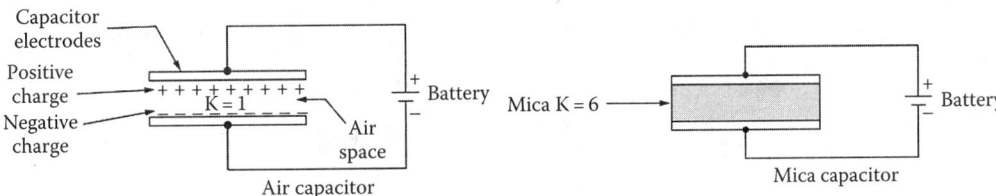

FIG. 1.39d
The capacitance is a function of the dielectric constant of the material filling the space between the capacitor plates.

the same frequency as the power signal and whose amplitude varies with the measuring and reference capacitors as they are switched into the circuit. This difference in amplitude is related to the measured capacitance, which, in turn, is a function of the moisture content of the sample.

The electronic chassis is normally mounted near the measuring cell and sampling system, which should be close to the sample take-off. In addition to an integral indicator and output signals of standard serial, milliampere, and millivolt ranges, the electronic chassis can be provided with multiple range switching and alarm contacts.

Sampling Systems It was pointed out previously that the desiccant in the measuring cell is the equilibrium with the sample in terms of percent saturation. For this reason, it is necessary to maintain the sample and measuring cell at constant temperatures. For example, propane at 38°C (100°F) will dissolve 300 ppm of water; at 21°C (70°F), it will dissolve 150 ppm. With a moisture level of 15 ppm, the propane would be 5% saturated at 100°F and 10% saturated at 70°F.

Therefore, the output signal would change 100% with no change in moisture content if temperature were allowed to vary from 100°F to 70°F.

The sample is maintained at constant temperature by passing it through a coil immersed in a constant temperature bath immediately upstream of the measuring cell. Additional temperature control of the sample is recommended even if the process stream is temperature controlled, because ambient conditions can affect the sample temperature.

Thin-Film Capacitance Probes The design and operation of thin-film capacitance probes have been described in the previous chapter in connection with Figures 1.38e and 1.38n. These units are also suitable for explosion-proof area classifications and can be used in measuring the moisture content of many gas and liquid samples (e.g., many liquid hydrocarbons). The probe units are suitable for installation of high-pressure processes and are provided with microprocessor-based electronics capable of self-diagnostics and of multiple measurements or unit conversions (Figure 1.39f). These probes are

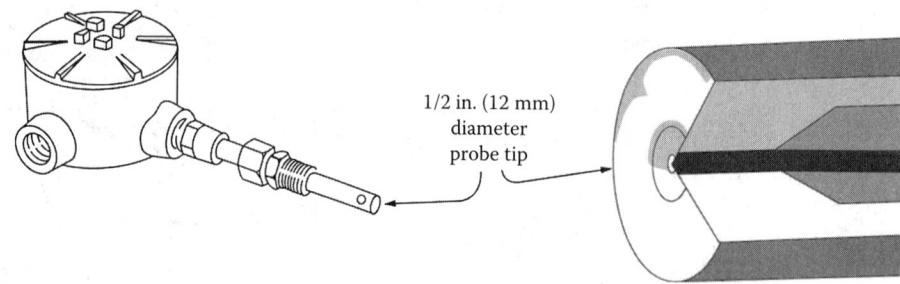

FIG. 1.39f
The gold/aluminum oxide probe can measure the moisture content of both gases and liquids. (Courtesy of Endress + Hauser Inc.)

available with a variety of covers/shields that can be used for a variety of sampling situations (particulate, bubbles, various viscosity liquids, etc.).

A unique configuration of a thin-film capacitance ceramic sensor is one with an integral calibration system. The *in situ* and automatic calibration is performed against a reference gas. The precision of this system is the result of the use of a chilled-mirror dew point hygrometer to measure the moisture content of the reference gas used in the calibration (Figure 1.39g).

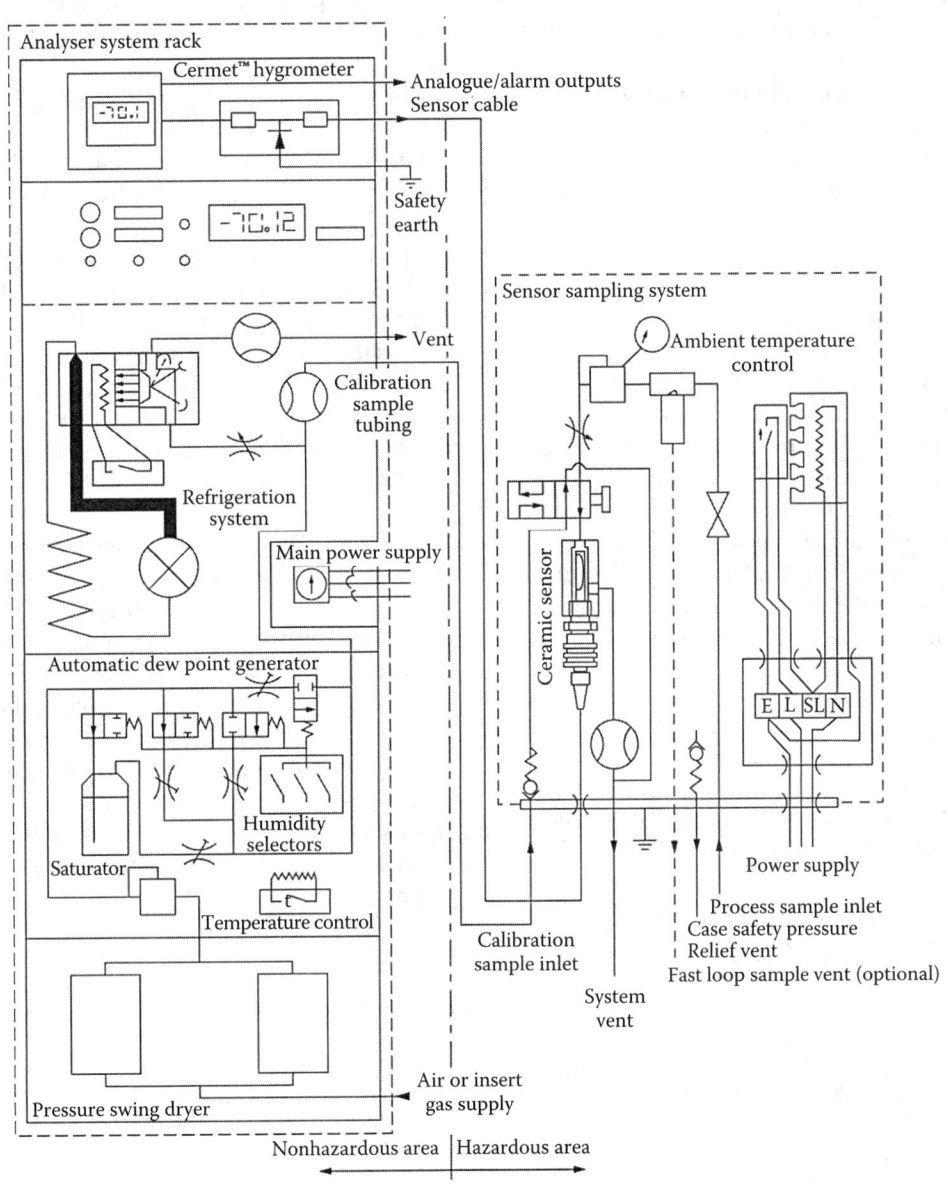

FIG. 1.39g
Thin-film ceramic sensor provided with integral and automatic means for recalibration. (Courtesy of Michell Instruments Ltd.)

Limitations The capacitance instruments are not suitable for polar materials such as alcohols, because these become conductive at the 15 kHz operating frequency and short the measuring capacitor. Data on instruments operating at higher frequencies were not available.

Free sulfur and iron oxide concentrate on the desiccant will short the element.

At sample viscosities above 500 Saybolt Seconds Universal (SSU), moisture should not contact the desiccant at reasonable flow rates, and the measurement becomes meaningless.

The life of the desiccant is partially a function of solids in the stream, which pass through the filter. While the desiccant can be replaced very readily, the short life of the desiccant excludes the use of this instrument on sample streams containing large quantities of fines.

Impedance Hygrometer

This instrument measures the water content of a sample by means of a probe whose electrical impedance is a function of the vapor pressure of moisture in the fluid. Impedance is the apparent opposition to the flow of alternating current. The probe consists of an aluminum strip that is anodized to form a porous layer of aluminum oxide. A thin coat of gold is applied over the aluminum oxide. Leads from the gold and aluminum electrodes of the probes connect the sensing element to the measuring circuitry (Figure 1.39h).

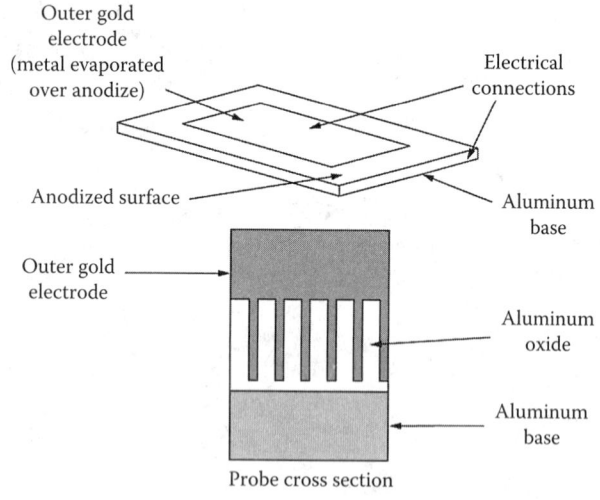

FIG. 1.39h
Impedance-measuring sensor.

Water vapor penetrates the gold layer and equilibrates on the aluminum oxide. The number of molecules adsorbed to the aluminum oxide is a function of the water vapor pressure in the sample. Each water molecule adsorbed contributes a distinct increment to the total conductivity of the aluminum oxide. The total probe impedance, the reciprocal of probe conductivity, is thus a measure of water vapor pressure in the sample. Water vapor pressure of a gas sample uniquely determines the dew point temperature and moisture content of the sample. The output is normally calibrated to these units, as they are more convenient to use than vapor pressure. In the case of certain liquid samples, the moisture content can be measured through the application of Henry's law.

In terms pertinent to this measurement, Henry's law states that, at constant temperature, the mass of water vapor dissolved in a given volume of liquid is in direct proportion to the partial pressure of water vapor in the sample. Henry's law can be restated in simpler form to read that the weight concentration of moisture in the sample is equal to the partial pressure of water vapor times a constant. However, Henry's law holds only for liquids with moderate solubility for water vapor, such as pure hydrocarbons, silicone oils, etc. The instrument cannot be used on liquid samples with a high solubility for water such as the alcohols, because the relationship expressed by Henry's law does not hold.

Installation The probe is designed for direct installation into the process stream. However, where danger of explosion exists, the probe is inserted into a small sample chamber connected to the process line with flashback arrestors located at the sample inlet and outlet. For liquid samples, temperature must be controlled and/or monitored and compensated for. Several probes can be connected to a single readout device and each probe can be monitored directly or by means of an automatic scanner. A serial, milliamp, and voltage output signals are available for recording or other functions; single or multiple range measurements are standard (Figure 1.39i or 1.39j).

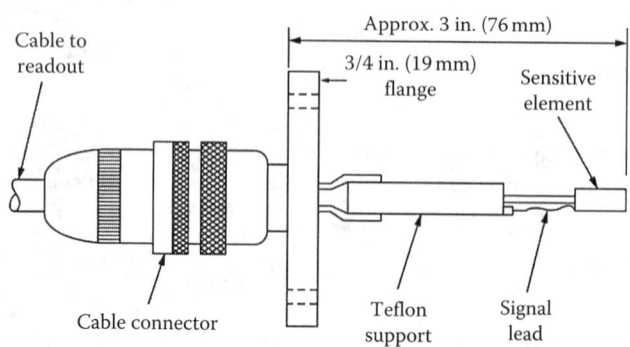

FIG. 1.39i
Flanged probe.

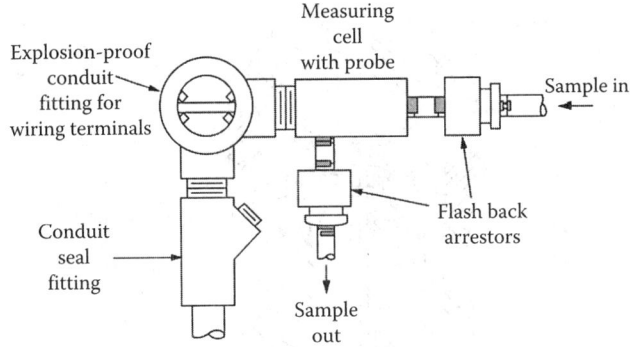

FIG. 1.39j
Impedance hygrometer installation for hazardous areas.

Sampling System Often a sampling system is not required on gas samples. On liquids, sample temperature must be held constant or monitored and compensated for. These probes are available with a variety of covers/shields that can be used for a variety of sampling situations (particulate, bubbles, various viscosity liquids, etc.). Often these shields can eliminate the need for, or simplify the design of a full-blown sampling system (Figure 1.39j).

Limitations The instrument can be used on all gases that are not corrosive to the probe and that will not spontaneously polymerize on contact with the probe materials. On liquids, the instrument is limited to those fluids with a moderate solubility for water vapor; thus, the instrument is not suitable for measurement in polar liquids such as the alcohols.

This sensor is suited for moisture measurement in hydrocarbons that consist of only hydrogen and carbon atoms or in halogenated hydrocarbons, amides, esters, many silicone oils, etc. The aluminum oxide probe is suited for moisture detection in cyclohexane, diethyl ether, liquefied natural gas, ethane, propane, butadiene, butane, styrene, and propylene. It is not recommended for applications where conductive solid particles, mercury, salts, or acids and bases are present. Hydrofluoric acid, for example, can damage the sensor, even in such low concentrations as 1 ppm.

The analyzer requires recalibration about twice a year and must be protected against excessive ambient temperatures (158°F or 70°C) and against excessive flow velocities. If the process stream flows at a velocity in excess of 1.7 fps (0.5 m/s), it is advisable to place the sensor in a slow bypass. The viscosity of the sample also needs to be taken into consideration.

Piezoelectric Hygrometer

Piezoelectric quartz crystals have a number of uses: in communications, to control frequencies; in industry to measure temperature and thickness of metal films; and to generate ultrasonic waves. In moisture measurement, advantage is taken of the oscillating crystal's sensitivity to deposits of foreign material on its surface. Commercially available crystals will show a frequency change of 2000 Hz/µg of materials deposited.

For moisture measurement, the quartz crystals are coated with a hygroscopic material and exposed to the sample. Water from the sample is absorbed by the coating, which increases the total mass and decreases the oscillating frequency of the crystals. To measure changes of decreasing moisture concentration and to simplify the frequency measurement, two crystals are used. One crystal is exposed to a wet sample and the other to a dry reference gas for a short period. Then, sample and reference gas flows are switched so that moisture is absorbed by one crystal while being desorbed by the other. A cycle timer controls this switching. The frequency changes between the two crystals are in proportion to their mass changes and the moisture content of the sample gas.

Sampling System Dry reference gas is obtained by passing part of the sample gas through an integral dryer, or an external source of dry nitrogen can be used. Two solenoid valves located upstream of the measuring cell switch the sample and reference gases from one crystal to the other. Sample gas flow to each crystal is regulated, although small variations of sample flow rate will not affect the measurement. For samples near atmospheric pressure, a vacuum pump may be used to draw sample through the measuring cell (Figure 1.39k). Some versions of this instrument design are being supplied with integral permeation tube calibration/benchmark systems to help ensure that the measurement is still accurate.

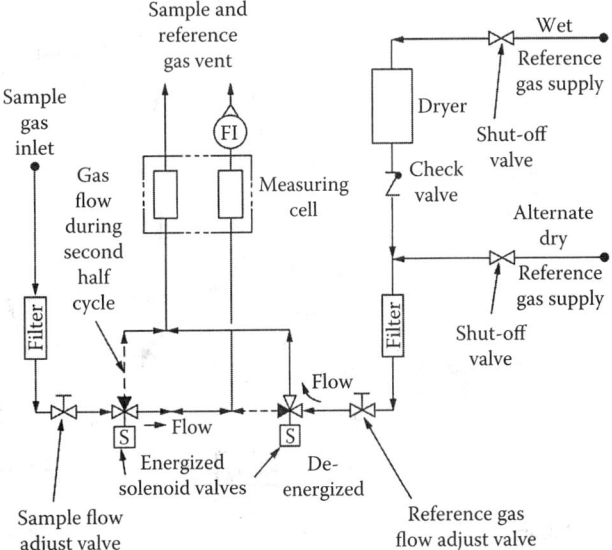

FIG. 1.39k
Piezoelectric hygrometer sampling system.

Limitations The instrument is not capable of measuring moisture in liquid phase samples. For this reason, it is applicable to only those liquid samples that can be stripped or completely vaporized within the operating limits of the instrument. In addition, monomers with a strong tendency to polymerize, such as butadiene and styrene, may coat the crystals and prevent proper operation.

Heat-of-Adsorption Hygrometer

The process of adsorption and desorption involves an exchange of energy. During adsorption, energy is released to the environment, and during desorption, energy is removed. When a wet gas is passed through a column of adsorbent, which selectively adsorbs moisture, the temperature rise due to heat liberation is in proportion to the amount of moisture being adsorbed. Other factors can affect the heat of adsorption. They include things like the nature of the adsorbent and operating temperature, and they are selected to maximize the heat of adsorption (Figure 1.39l).

The sensing element consists of two desiccant columns, each containing a number of thermocouples connected in series. This assembly is housed in a temperature-controlled, thermally insulated housing. In operation, wet sample gas is passed through one column and dry reference gas through the other. Since adsorption and desorption occur simultaneously, the net thermocouple output voltage represents the algebraic sum of the heat gained by one column and lost by the other. The reference gas and sample gas streams are

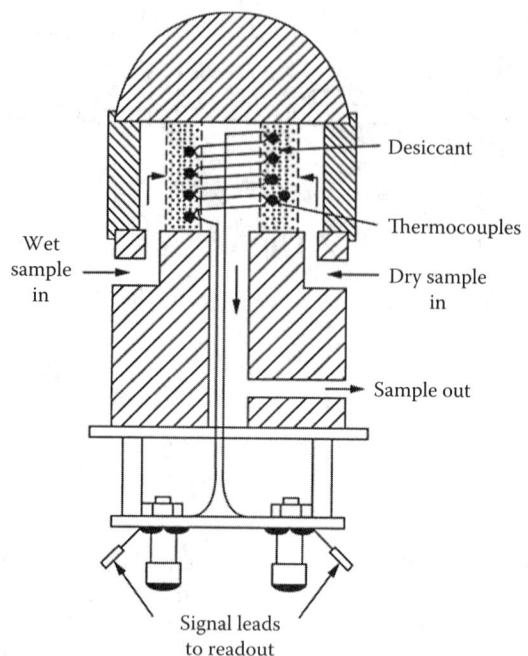

FIG. 1.39l
Measuring cell for the heat of an adsorption-type hygrometer.

switched on a time cycle basis to maintain dynamic conditions necessary for measurement in the sensing element. Sample flow is closely regulated, because moisture concentration is inferred from the known sample and reference flow rates (Figure 1.39m).

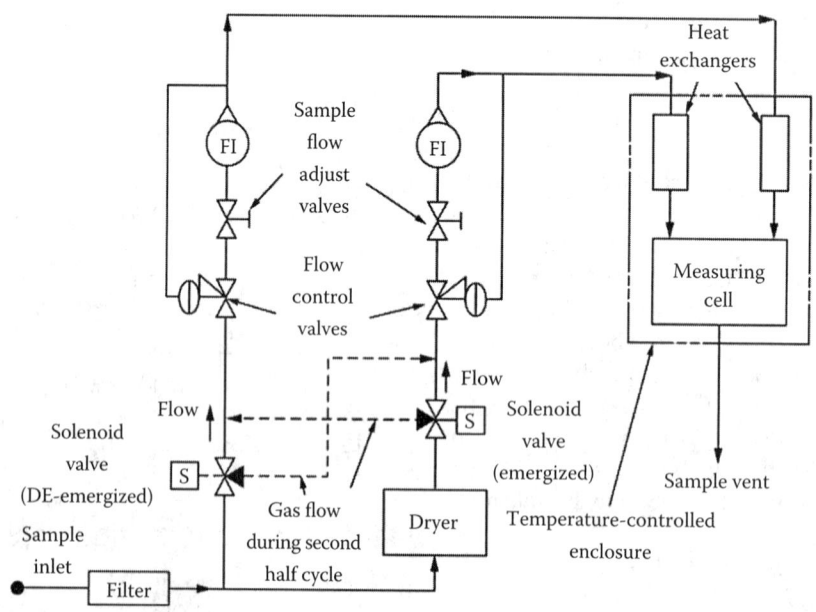

FIG. 1.39m
Sampling system for the heat of an adsorption hygrometer.

Sampling System The sampling system contains the following:

1. A dryer to remove all moisture from the sample gas so that the gas can be used as the reference.
2. Solenoid valves and cycle timer to switch reference and sample gas between the two desiccant columns.
3. Flow regulators to hold sample and reference gas flows constant.
4. Heat exchangers within the insulated enclosure to heat both reference and sample gas to the operating temperature of the sensing elements. The sample flow to the instrument is split into two equal flow rates at the sampling system; one half is the measured sample, and the other half is dried for use as the reference gas. The timer actuates solenoid valves in the sample and reference gas lines so that each desiccant column is exposed alternately to reference and to sample gas.

Limitations The sensing element cannot be exposed to liquid samples; however, the moisture content of liquid streams can be measured by one of two methods. Liquids with low boiling points can be vaporized at the sample take-off points by means of a vaporizing regulator, and the measurement is then performed on a gas sample. Or the moisture can be stripped from the liquid sample with dry gas, and from the moisture content of this gas the moisture concentration in the liquid sample can be inferred. Sample stripping is described in conjunction with electrolytic hygrometers.

Spectroscopy

Water absorbs electromagnetic radiation in the infrared (IR) region of the spectrum. Specifically, infrared radiation of 1.4 and 1.93 μm wavelength is absorbed strongly by water. By measuring the attenuation (decrease of light intensity) of a beam of this wavelength as it passes through a sample, the moisture content of the sample can be determined. However, other factors such as reflection and dispersion of the radiant energy will contribute to the attenuation. Therefore, it is necessary to either calibrate these factors out of the measurement or to use a reference wavelength that is not absorbed by moisture but is affected by all other factors to the same extent as the measuring wavelength. The difference in the attenuation of the measurement and reference wavelength is then a function of moisture content only.

The sensing element consists of three groups of components: an IR radiation source, sample cell, and radiation detector. The radiation source consists of a lamp, filters to pass the measuring and references wavelengths, and optics to direct the beam through the glass sample cell. The radiation pickup consists of optics to collect the transmitted radiation and a photocell to convert the electromagnetic energy to an electric current.

The measuring and reference wavelengths are allowed to impinge alternately on the photocell so that two sets of current pulses are produced. These pulses are converted into two direct current (DC) signal levels whose ratio represents the moisture content of the sample (Figure 1.39n). For more details on IR analyzers, refer to Chapters 1.23 and 1.32. Traditional spectroscopic techniques have not been successful at doing this in natural gas because methane absorbs light in the same wavelength regions as water. But if one uses a very high-resolution spectrometer, it is possible to find some water peaks that are not overlapped by other gas peaks.

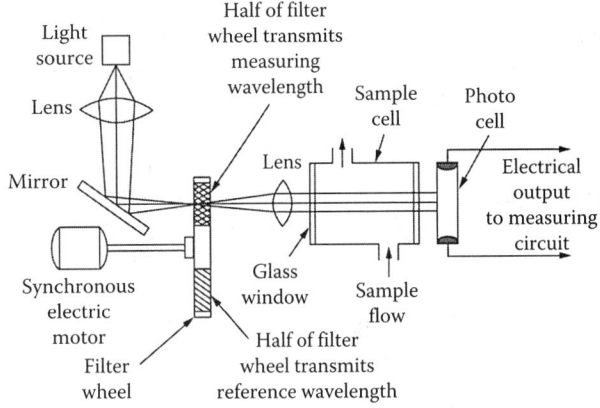

FIG. 1.39n
Schematic representation of the infrared moisture detector.

More complex spectrometers (IR, FTIR, NIR, and TDLAS) are also being used for moisture detection and monitoring applications when the application requires or justifies it (lower levels of detection, multiple simultaneous components [including moisture], complex sample matrices, etc.). One of the newer one is tunable diode laser adsorption spectroscopy (TDLAS), measures down to the 1 ppm with an appropriate sample. The tunable laser provides a narrow, tunable wavelength light source that can be used to analyze these small spectral features. According to the Beer–Lambert law, the amount of light absorbed by the gas is proportional to amount of the gas present in the light's path; therefore, this technique is a direct measurement of moisture. In order to achieve a long enough path length of light, a mirror is used in the instrument. The mirror may become partially blocked by liquid and solid contaminations, but since the measurement is a ratio of absorbed light over the total light detected, the calibration is unaffected by the partially blocked mirror (if the mirror is totally blocked, it must be cleaned).

The Tunable Diode Laser Absorption Spectroscopy (TDLAS) analyzer has a higher upfront cost compared to the analyzers above. However, the TDLAS technology is the only one that can meet any one of the following: the necessity for an analyzer that will not suffer from interference or damage from corrosive gases, liquids, or solids, or an analyzer that will react very quickly to drastic moisture changes, or an analyzer that will remain calibrated for very long periods of time.

Sampling System There are no special sampling system requirements for this instrument. Selection of hydrophobic (lacking affinity for water) materials for sampling system components is not critical, given that the instrument can measure only relatively high moisture content, where moisture contributed to the sample by the sampling system is insignificant in its effect on the measurement. When fiber-optic probes (FOPs) are used, the need for sampling systems is eliminated (Chapter 1.23).

Limitation There are several general limitations to the field of application of this instrument. The sample fluid must not be corrosive to the glass sample cell. It must have some minimum transparency for the measurement and reference wavelengths. Additionally, there cannot be any other components present in the sample that will selectively absorb either the measuring or reference wavelength.

Microwave Absorption Hygrometer

In the microwave frequency band of 20–22 GHz (K band), the wavelength is about 13–15 mm. In this frequency band, moisture (free water) responds uniquely. Only the water molecule produces molecular resonance. The microwave radiation can be guided by waveguides or transmitted through the process stream from its source to a detector.

The operating principle of this instrument is the same as that of the infrared absorption hygrometer, namely, selective absorption of electromagnetic energy by moisture in the sample. However, in this case, radiant energy in the microwave region is used. A transmitter provides a microwave beam that can either be transmitted through or reflected by the sample material. The receiving unit senses wave attenuation and phase shift and provides an output. The unit senses the mass of moisture in the beam path so that the readout is normally in terms of the mass of moisture per unit volume. However, with proper calibration, readout in weight percent can be achieved.

Several probe configurations are possible, depending on the application. Figure 1.39o shows the transmitting and receiving probes welded into a section of pipe. Internally, Teflon contacts the sample. This type of arrangement can be advantageous on slurries and pastes, because there are no obstructions to flow. The units described above measure the attenuation of a microwave beam in the sample by moisture.

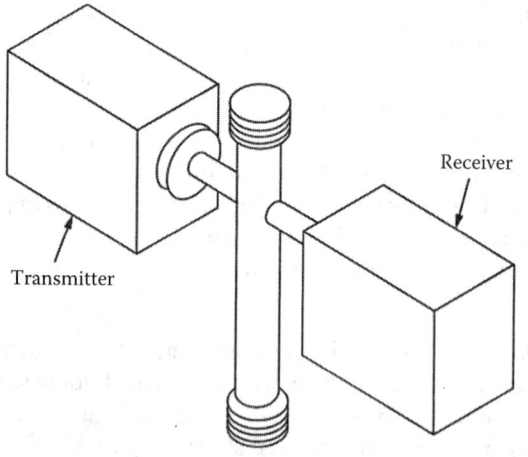

FIG. 1.39o
Microwave sensor installed in pipe.

The reflectance-type unit contains the source and detector in a single housing. Rather than detecting the transmitted microwave energy, this design measures the amount of energy reflected by the sample. The probe of the reflectance unit does, however, require contact with the process material. Selection of the type of probe to be used is determined by the reflective and absorptive properties of the sample at operating conditions and by the physical installation. An advantage of the microwave hygrometer over the infrared unit is that there are no optical windows that may coat and effect the measurement.

Advancements in microwave moisture analysis technology use an oscillator which is sensitive to changes in the phase of its load. As the load changes, the operating frequency changes accordingly. The frequency pulling of the oscillator is in response to the changing phase angle in measurement section due to change in dielectric constant of the fluids. Density and salinity variations of process fluid affect the measurement value.

Limitations The unit is not suitable for moisture measurement in gases.

Dipole Polarization Effect Moisture Sensor

The principle of this measurement relies on the fact that water molecules are formed by covalent bonds. These bonds share electrons unequally between the hydrogen and oxygen atomic nucleuses, and the result is an electrical asymmetry around the molecule (two slightly positively charged hydrogen atoms and a slightly negatively charged oxygen atom). When these molecules are subjected to an electric field, they tend to align with the electric field because of their dipole nature. Unfortunately, other influences such as surface polarization, ions, etc., can also create an electrical effect in the field.

By selecting an appropriate sensor field frequency (e.g., 20 kHz), the influence of these other influences can be reduced. Other polar molecules can interfere, but since water molecules are very specific, this is not considered a significant factor in most analysis. Ammonia (NH_3) is the closest to water, and its interference effect is listed as about 1:50 (± 50 g/m^3 NH_3 equals about ± 1 g/m^3 of water).

A dipole probe is depicted in Figure 1.39p. The sensor measurement chamber is an open capacitor that is protected from coating or attack by the process steam components by a membrane filter. The measurement output is directly proportional to the number of water molecules in the field and is reported as a mass per unit volume output. The probe can be made from several materials (stainless steel, inconel, etc.) and can even be coated with nonstick materials. The result is an available probe for most needs. The probe is usually heated to about 10°C above the highest process temperature to provide a constant measurement temperature and to minimize potential condensation and fouling. Versions of this probe design have been used to measure moisture in processes as hot as 430°C.

Because the measured signal is directly proportional to the number of water molecules in the chamber, calibration is very easy. A zero-reference capacitor is hermetically sealed in the probe and, once the zero reference is established and because the measurement is linear, you need only one other calibration point for the entire 0%–100% range. The calibration will not change for the life of the instrument as long as the physical integrity of the measuring and reference chambers is maintained.

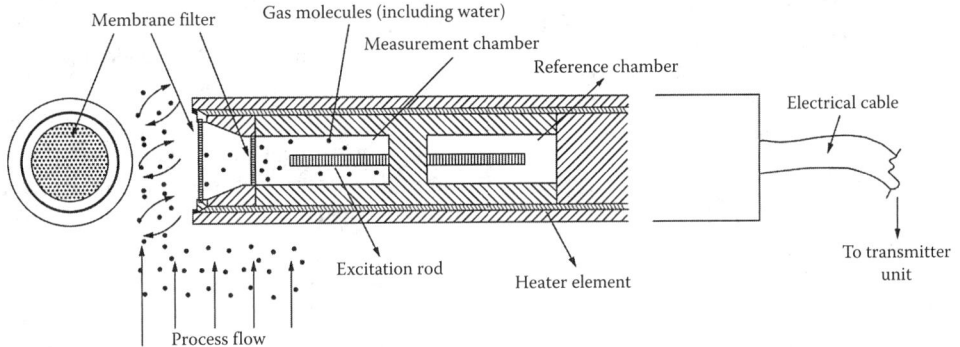

FIG. 1.39p
A dipole polarization effect moisture sensor. (Courtesy of Dewcon Instruments Inc.)

Sampling System There are no special sampling system requirements for this instrument. The probe is designed to insert directly into most processes. Extreme sample conditions may require some sample conditioning.

Limitations The unit is not suitable for moisture measurement in liquids.

Cavity Ring-Down Spectroscopy (CRDS): Moisture Analysis

This spectroscopic technique is different enough from those previously discussed that it will be discussed here on its own. The measurement principle is based on absorption spectroscopy. The difference between this and more traditional absorbance spectroscopy is that this measurement is based on time rather than just the magnitude of the absorbance. The actual measurement is based on a light intensity decay rate. By measuring the time, it takes for the appropriate molecular absorbance light fingerprint for particular sample species to fade or ring-down, you can measure extremely low levels of the absorbing contaminants.

A CRDS instrument is depicted in Figure 1.39q. A continuous-wave laser emits a beam of light energy through a highly reflective mirror into the absorbance cell (cavity). This cell is capable of reflecting the light back and forth many times and, by this multiple path design, will typically achieve total path lengths in the range of 100 km (vs. typically 100 m for typical absorbance multiple pass cells). Once the detector sees a preset level of light energy, the light source is diverted from the cavity. On successive passes of light, a small amount of light is sensed and, once the light *rings down*, the measurement is complete.

The sample concentration is determined from the shape of the ring-down decay curve. Figure 1.39r shows an example of a ring-down decay curve.

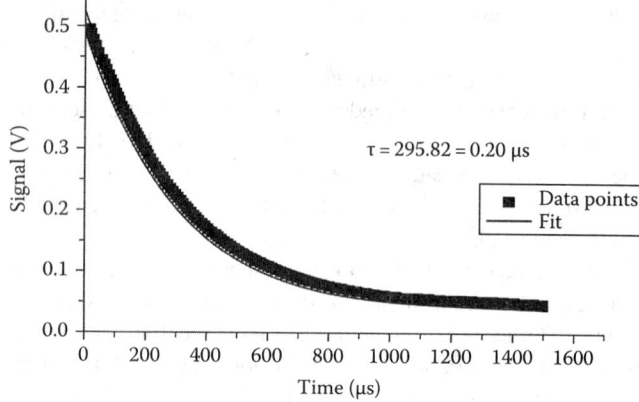

FIG. 1.39r
Typical ring-down decay curve. (Courtesy of Tiger Optics LLC.)

Because the measurement is time based, it is not subject to small variations in laser intensities. Because of its very long path length, where molecules in the sample are frequently exposed to the light, the sensitivity of the instrument is enhanced to a level of actually achieving part-per-trillion (ppt) levels of measurement on water in gases. This technology has been utilized primarily in the semiconductor industry, on ultrapure gases, and for biosafety. Robust process-hardened instruments for more typical process applications are under development.

Sampling System Special sampling systems are required for this instrument. The sample must be clean, temperature controlled, pressure controlled, and noncondensing in addition to being compatible with materials that it will contact.

Limitations The unit is not suitable for moisture measurement in liquids or condensing vapors. Gas samples should be adequately cleaned and/or conditioned in an appropriate sampling system.

Neutron Backscatter Moisture Online Analyzer

This measurement principle is essentially the same as that of the nuclear moisture gauge discussed in greater detail in Chapter 1.44. It is a neutron radiation source that is focused on a sample, and the neutron backscatter is measured and correlated to an appropriate moisture content. High-energy neutrons are focused on the sample. Backscattered low-energy neutrons that have interacted with a hydrogen nucleus are detected and correlated to the hydrogen/moisture density.

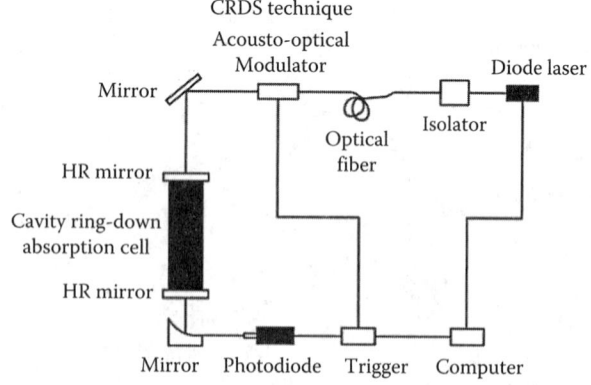

FIG. 1.39q
Cavity ring-down spectroscopy instrument. (Courtesy of Tiger Optics LLC.)

As a result of the advancements in this design, it is now also being considered for some nonsolid applications, although the majority of its applications is still on solid samples. Figure 1.39s depicts a typical installation.

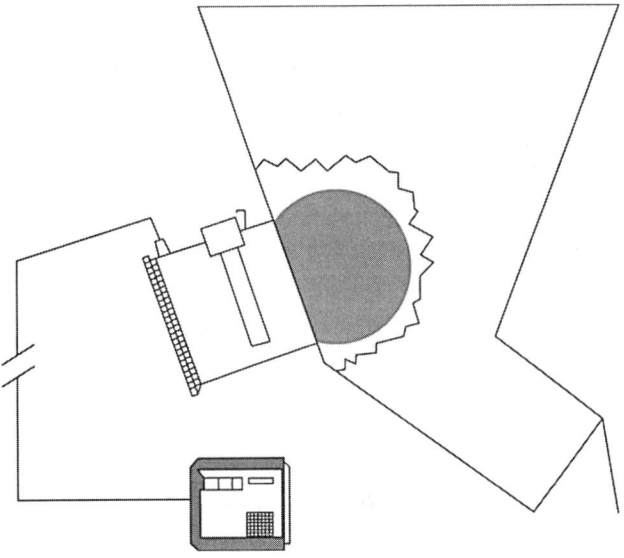

FIG. 1.39s
Example of a neutron backscatter moisture sensor installation. (Courtesy of Thermo Electron Corp.)

Because the measurement is so focused on the neutron and hydrogen nucleus interaction, it is not an appropriate measurement where hydrogen-containing species, other than water, can vary in concentration. This essentially eliminates the world of hydrocarbons, but it is still suitable for inorganic chemicals, salts, mineral applications, etc. Moisture in blast furnace coke is still the dominant application. Optional density compensation systems are available but are generally not required. This could be factor if you are encountering foam or slurries of significantly varying densities.

A sensor can typically measure a semisphere sample volume of approximately 18 in. (46 cm) radius. If one is monitoring a sample of varying consistency, the overall measurement accuracy can be increased by placing the sensor where it monitors a flowing sample. This effectively will give an integrated or averaged moisture value.

Sampling System There are usually no sampling system requirements for this instrument. The sensor is typically mounted directly on the process.

Limitations The unit is not suitable for moisture measurement in gases or samples with varying levels of hydrogen atoms (like most hydrocarbons). If the sample density varies significantly, a density measurement should be made for compensation.

Refraction-Based Moisture Analyzer

The sensor consists of a coated, very resistant and microporous substrate of quartz glass with a pore diameter in which only water molecules can enter (Figure 1.39t). This penetration happens because of the equilibrium (all gases balance automatically). Because of the different refraction index ($n_{air} = 1.0$; $n_{water} = 1.33$), the water molecules cause a shift in wavelength. With increasing moisture, the refraction spectrum is shifted toward larger light wavelength and readout by a polychromator. This shift (no measurement of intensity $\Rightarrow$ independence of fiber-optic cable length) is proportional to the dew point temperature of the gas. The sensor can be used in hazardous areas (Zone 0, or Class I/Div. 1). The sample temperature is monitored with an integrated temperature element in moisture sensor. Pressure can be monitored by an external pressure transmitter or entered as a constant value in the transmitter. Three different sensors can be connected to a single transmitter.

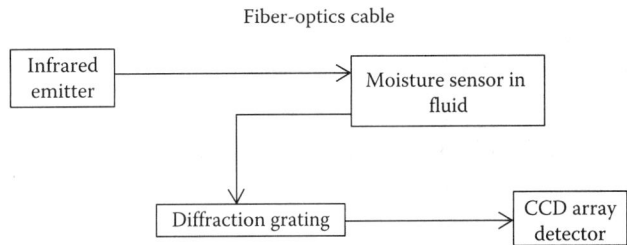

FIG. 1.39t
Refraction-based moisture analyzer.

Sampling System There are no special sampling system requirements for this instrument. The sensor is designed to insert directly into most processes through a retractor mechanism. The sensor can withstand maximum process pressure of 200 bar and maximum process temperature 60°C. If process conditions are different, sample conditioning may be required.

Calibration of Moisture Analyzers

Moisture calibrators are available in two conceptual designs. They are based on either saturated gas dilution or permeation tube designs. Some moisture analyzers include a calibration system integral to the instrument or as an available option. In addition to calibration, these can be helpful to test/ensure that the moisture sensor is still alive and responding.

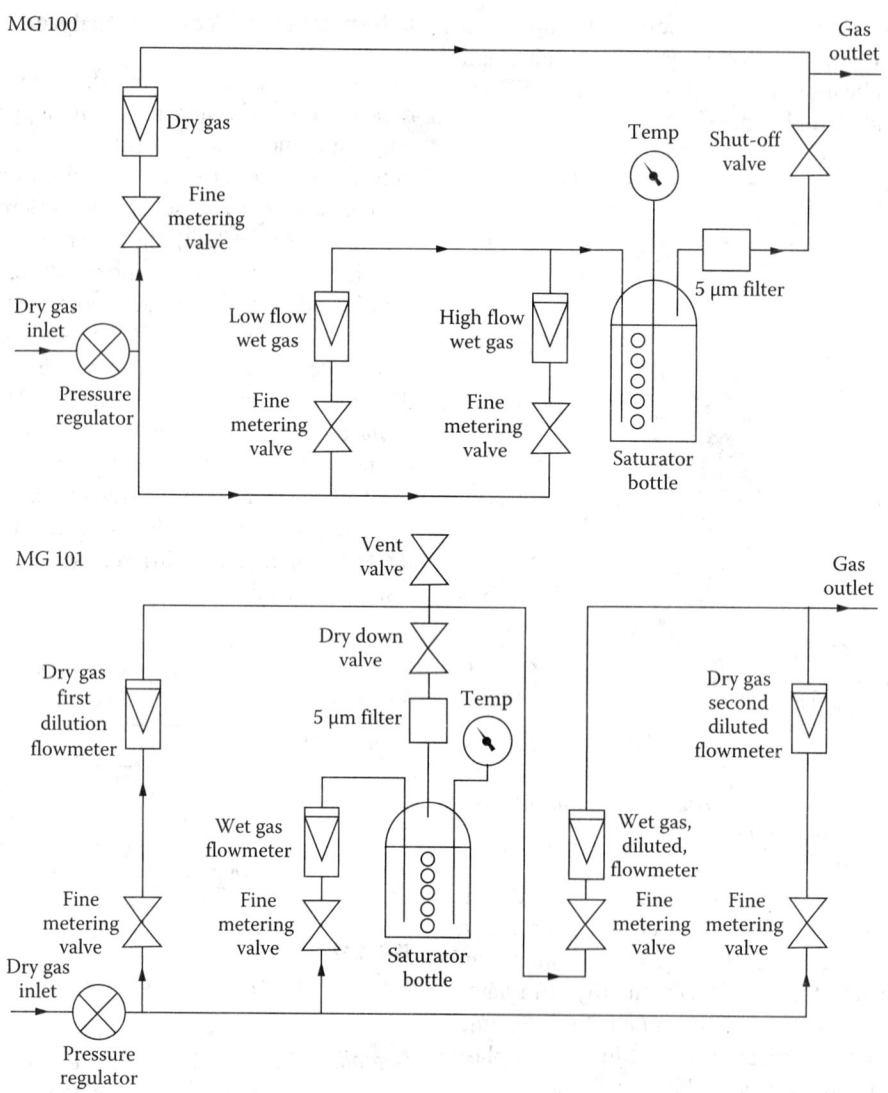

FIG. 1.39u
Example of gas dilution–based moisture calibration systems. (Courtesy of Panametrics Inc.)

The first design that will be reviewed is the saturated gas dilution system concept. Figure 1.39u depicts a standard range and a lower range design. Both start out with a water-saturated vapor stream at a known temperature and, consequently, known moisture content. This stream is then diluted in a controlled (ratio) fashion with a dry gas stream to yield the desired value. Obviously, the larger the single-stage dilution ratio, the more difficult it is to maintain precise control. The lower system in the figure offers multiple opportunities (stages) to dilute the saturated stream with dry gas and will allow you to achieve more precise control of calibration samples at lower values than a large-ratio single-stage dilution approach.

The second design concept is the permeation tube design. Figure 1.39v depicts a simple permeation tube moisture calibration design. Permeation tube designs can be used to generate known standards from dry gases or to

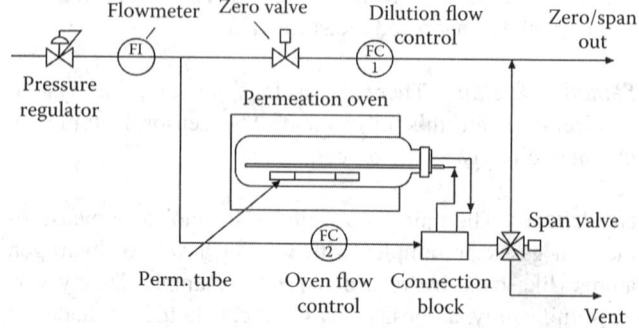

FIG. 1.39v
Example of a permeation tube–based moisture calibration system. (Courtesy of Kin-Tek Laboratories Inc.)

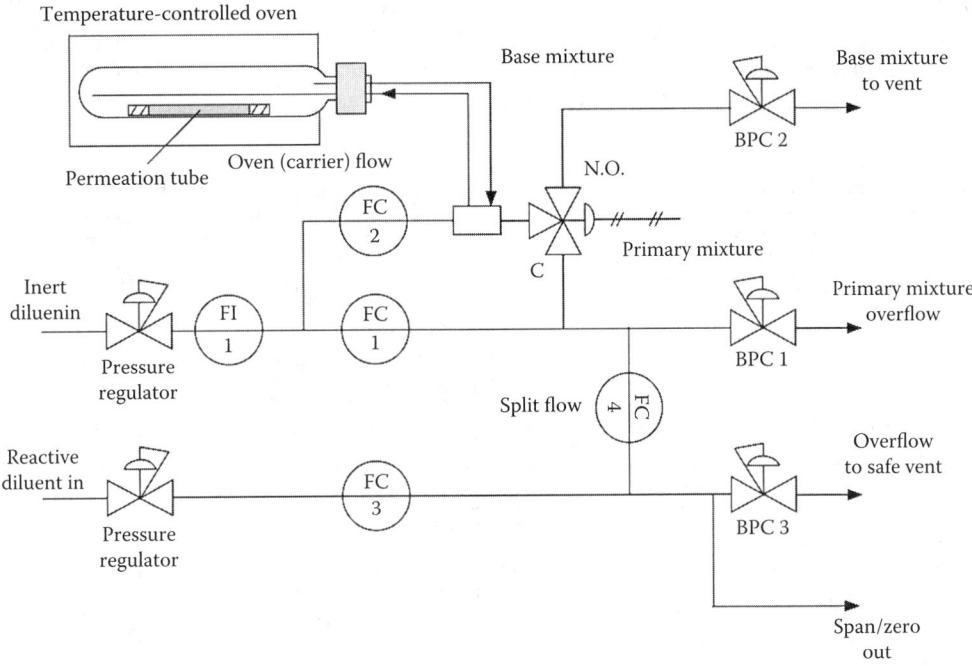

FIG. 1.39w
Example of a permeation tube–based moisture in reactive gases calibration system. (Courtesy of Kin-Tek Laboratories Inc.)

make known standard additions to samples or standards. They are based on the concept that, under controlled conditions, the rate/volume of moisture that will permeate through a permeation tube (its membrane wall) is constant. By also controlling these conditions and the dilution gas flow rates, one can at least generate a reproducible and predictable moisture standard. If one is adding it to a system, and if the dilution gas is truly dry or constant, it is possible to generate an actual moisture value. Typically, permeation systems operate at much lower flow rates than saturation dilution systems.

A recent advancement of the permeation tube design is depicted in Figure 1.39w. It depicts a permeation tube system that is designed to introduce moisture into gas streams that contain reactive components. Some examples of reactive gas components that were considered include ammonia, HCl, HBr, chlorine, silane, and others. The reactive component does not come in contact with the permeation tube, and the reactive matrix is formed by dilution after the water addition. These need to be evaluated on an individual application basis and surely will not work for all situations, but they are a nice addition to the analyst's toolbox.

Limitations Generally speaking, the saturated gas dilution design concept is better for generating higher concentration standards and higher standard flow rates. The permeation designs are better for lower concentration standards.

SPECIFICATION FORMS

When specifying analyzers for moisture in gases and liquids, one can use the ISA form 20A1001 and when specifying the detector and also describing the application ISA form 20A1002 can be used. The following are the forms that are reproduced with the permission of the International Society of Automation.

#	RESPONSIBLE ORGANIZATION	ANALYSIS DEVICE	#	SPECIFICATION IDENTIFICATIONS		
1			6			
2	(ISA logo)	Operating Parameters	7	Document no		
3			8	Latest revision	Date	
4			9	Issue status		
5			10			

#	ADMINISTRATIVE IDENTIFICATIONS			#	SERVICE IDENTIFICATIONS Continued		
11				40			
12	Project number	Sub project no		41	Return conn matl type		
13	Project			42	Inline hazardous area cl	Div/Zon	Group
14	Enterprise			43	Inline area min ign temp	Temp ident number	
15	Site			44	Remote hazardous area cl	Div/Zon	Group
16	Area	Cell	Unit	45	Remote area min ign temp	Temp ident number	
17				46			
18	SERVICE IDENTIFICATIONS			47			
19	Tag no/Functional ident			48	COMPONENT DESIGN CRITERIA		
20	Related equipment			49	Component type		
21	Service			50	Component style		
22				51	Output signal type		
23	P&ID/Reference dwg			52	Characteristic curve		
24	Process line/nozzle no			53	Compensation style		
25	Process conn pipe spec			54	Type of protection		
26	Process conn nominal size	Rating		55	Criticality code		
27	Process conn termn type	Style		56	Max EMI susceptibility	Ref	
28	Process conn schedule no	Wall thickness		57	Max temperature effect	Ref	
29	Process connection length			58	Max sample time lag		
30	Process line matl type			59	Max response time		
31	Fast loop line number			60	Min required accuracy	Ref	
32	Fast loop pipe spec			61	Avail nom power supply		Number wires
33	Fast loop conn nom size	Rating		62	Calibration method		
34	Fast loop conn termn type	Style		63	Testing/Listing agency		
35	Fast loop schedule no	Wall thickness		64	Test requirements		
36	Fast loop estimated lg			65	Supply loss failure mode		
37	Fast loop material type			66	Signal loss failure mode		
38	Return conn nominal size	Rating		67			
39	Return conn termn type	Style		68			

#	PROCESS VARIABLES	MATERIAL FLOW CONDITIONS		#	PROCESS DESIGN CONDITIONS		
69				101			
70	Flow Case Identification		Units	102	Minimum	Maximum	Units
71	Process pressure			103			
72	Process temperature			104			
73	Process phase type			105			
74	Process liquid actl flow			106			
75	Process vapor actl flow			107			
76	Process vapor std flow			108			
77	Process liquid density			109			
78	Process vapor density			110			
79	Process liquid viscosity			111			
80	Sample return pressure			112			
81	Sample vent/drain press			113			
82	Sample temperature			114			
83	Sample phase type			115			
84	Fast loop liq actl flow			116			
85	Fast loop vapor actl flow			117			
86	Fast loop vapor std flow			118			
87	Fast loop vapor density			119			
88	Conductivity/Resistivity			120			
89	pH/ORP			121			
90	RH/Dewpoint			122			
91	Turbidity/Opacity			123			
92	Dissolved oxygen			124			
93	Corrosivity			125			
94	Particle size			126			
95				127			
96	CALCULATED VARIABLES			128			
97	Sample lag time			129			
98	Process fluid velocity			130			
99	Wake/natural freq ratio			131			
100				132			

#	MATERIAL PROPERTIES			#	MATERIAL PROPERTIES Continued		
133				137			
134	Name			138	NFPA health hazard	Flammability	Reactivity
135	Density at ref temp		At	139			
136				140			

Rev	Date	Revision Description	By	Appv1	Appv2	Appv3	REMARKS

Form: 20A1001 Rev 0 © 2004 ISA

1.39 Moisture in Gases and Liquids

	RESPONSIBLE ORGANIZATION	ANALYSIS DEVICE COMPOSITION OR PROPERTY Operating Parameters (Continued)		SPECIFICATION IDENTIFICATIONS		
1-5	ISA		6	Document no		
			7	Latest revision	Date	
			8	Issue status		
			9-10			

	PROCESS COMPOSITION OR PROPERTY			MEASUREMENT DESIGN CONDITIONS				
Row	Component/Property Name	Normal	Units	Minimum	Units	Maximum	Units	Repeatability
11–37								

Rev	Date	Revision Description	By	Appv1	Appv2	Appv3	REMARKS

Form: 20A1002 Rev 0 © 2004 ISA

Abbreviations

CRDS   Cavity ring-down spectroscopy
FOP   Fiber-optic probe
IR   Infrared
NIR   Near infrared
PPM   Pats per million
TDLAS   Tunable diode laser absorption spectroscopy

Bibliography

Bailey, S. J., Moisture sensors, *Contr. Eng.*, September 1980.

Belkim, H. M., Factor affecting response characteristics of electrolytic instruments for detecting moisture, *AID/ISA Symposium*, Pittsburgh, PA, May 1970.

Cucchiara, O., The measurement of dissolved water in organic liquids using a hygrometer, *Anal. Instrum.*, 15, ISA, 1977.

Curl, R. F., Capasso, F., Gmachl, C., Kosterev, A. A., McManus, B., Lewicki, R., Pusharsky, M., Wysocki, G., and Tittel, F. K., Quantum cascade lasers in chemical physics, *Chem. Phys. Lett.*, 487(1–3), 1–18, 2010.

Dudek, J. et al., Semiconductor gases exact max, *InTech*, 32–33, July 2002.

Gumpert, R. and Pakulis, I. E., Three moisture measurement techniques, *InTech*, August 1981.

http://www.phasedynamics.com/index.php/products/ppmanalyzers (January 14, 2016).

Jutlia, J. M., Multicomponent on-stream analyzers for process monitoring and control, *InTech*, July 1979.

Kluczynski, P., Lundqvist, S., Westberg, J., and Axner, O. Faraday rotation spectrometer with sub-second response time for detection of nitric oxide using a cw DFB quantum cascade laser at 5.33 μm, *Appl. Phys. B*, 103(2), 451–459, 2010.

Kohler, H. M. and Mathew, A., Continuous in-situ and elevated temperature moisture measurement in high particulate reactive processes, *ISA1996 Conference Paper* #96-002.

Leading Edge Moisture measurements in natural gas, P&A Select Oil & Gas 2007 Markus Garber, Product specialist Moisture measurement.

Makeev, Y. V., Lifanov, A. P., and Sovlukov, A. S., Microwave measurement of water content in flowing crude oil, *Automat. Remote Contr.*, 74(1), 157–169, January 2013.

Mator, R. J., Trace moisture analyzers and their calibration, *Anal. Instrum.*, 12, ISA, 1975.

McKinley, J. J., Using permeation tubes to create trace concentration moisture standards, *National Conference of Standards Laboratories Conference*, Atlanta, GA, July 1997.

McKinley, J. J., Using permeation tubes to prepare trace moisture standards in reactive gases, *International Symposium on Humidity and Moisture*, Taipei, Taiwan (http://www.kin-tek.com/article/traceH2OinRgases4.pdf).

Mettes, J. and Beck, T., Manufacturers today require PPB moisture detection, *Meas. Contr.*, February 1992.

Mettes, J. et al., Multipoints PPM moisture measurement using electrolytic cells, *ISA Conference*, Houston, TX, October 1992.

Moisture analyzer helps extend period between dryer column recharging, *Contr. Eng.*, August 1980.

Polta, R. C., Anderson, C. T., and Stule, D. A., Monitoring moisture in dewatered sludge, *InTech*, June 1980.

Rothman, L. S., Gordon, I. E., Barber, R. J., Dothe, H., Gamache, R. R., Goldman, A., Perevalov, V., Tashkun, S. A., and Tennyson, J. HI-TEMP, the high-temperature molecular spectroscopic database, *J. Quant. Spectrosc. Radiat. Transf.*, 111, 2139–2150, 2010.

Scelzo, M. J., The trend is toward processing and measurement at "ultralow" moisture levels, *Meas. Contr.*, February 1992.

Sharpe, S. W., Johnson, T. J., Sams, R. L., Chu, P. M., Rhoderick, G. C., and Johnson, P.A., Gas-phase databases for quantitative infrared spectroscopy, *Appl. Spectrosc.*, 58(12), 10, 2004.

Studebaker, P., Moisture analyzers balance sensitivity with price, *Control*, November 1993.

Sutherland, D., Moisture analyzer calibration, *Meas. Contr.*, June 1993.

Wysocki, G., So, S. G., and Jeng, E., VCSEL based Faraday rotation spectroscopy with a modulated and static magnetic field for trace molecular oxygen detection, *Appl. Phys. B*, 2010.

Yan, W. B. and Mallon, T., Breakthrough technology measures parts-per-trillion moisture in gases, *Chem. Res.*, February/March 2002.

1.40 Moisture in Solids

A. BRODGESELL (1969, 1982)   **B. G. LIPTÁK** (1995, 2003)   **D. KOPECKÝ** (2017)

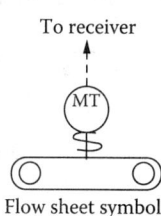

Flow sheet symbol

Types of designs	A. Fast neutron moderation
	B. Infrared
	C. Microwave
	D. Capacitance and dielectric constant sensors
	E. Impedance, resistance
	F. Nuclear magnetic resonance
	G. Radio frequency absorption
	H. Karl Fischer titrator (laboratory)
	I. Oven drying (loss on drying)
	J. Time domain reflectometry
Ranges	A. 2%–80%
	B. From 0% to 0.2% up to 1% to 90% maximum
	C. From 0%–1% to 0%–70%
	D. 4%–40%
	F. 0.01%–99.99%
	G. From 0%–35% to 0%–80%
	H. 10 ppm to 100% (from 2 ppm for absorption method)
	I. 0%–100%
	J. 0%–100%
Inaccuracy	For most, 0.5%–2% of full scale
	A. 0.1% if density is also measured and bound hydrogen is constant
	B. 0.5%–1% of calibrated range
	C. For a 0%–15% range, the error is within 0.5%; on installations where nuclear density gauging is included, the error can be within 0.1%
	F. 0.01% AR
	G. Error is under 0.05% moisture
	H. 5 ppm
	I. 0.2% AR
	J. 2% FS
Costs	A. About $25,000 if nuclear density gauge is also included
	B. From $10,000 to $15,000 for reflectance units; higher for transmission systems operating at three frequencies to obtain thickness and density compensation
	C. From $15,000 for microwave only, or from $25,000 if nuclear density gauge is included
	D. About $5000 and up for industrial units, $1000 for soil moisture detectors
	G. $7,000–$10,000; more if installation is included. Handheld meters are $1500–$2000; computerized units for the laboratory are $5000–$6000; online RF or IR systems cost from $7,000–$13,000; four-channel IR/RF costs range from $16,000–$36,000; online scanning, $40,000–$45,000
	H. About $10,000
	I. $5000 and up

(Continued)

Partial list of suppliers

Arizona Instruments LLC (I) (http://www.azic.com)
http://www.azic.com/products_computrac_lossondrying.aspx
http://www.azic.com/products_computrac_greenkfalternative.aspx
Automata, Inc. (J) (http://www.automata-inc.com)
http://www.automata-inc.com/Products/Sensors/All-Sensors/AQUA-TEL-TDR-T/
Berthold Technologies (A, C) (https://www.berthold.com)
https://www.berthold.com/en/pc/moisture-analyzer
https://www.berthold.com/en/pc/radiometric-moisture-measurement/moisture-measurement-system-lb-350
C.W. Brabender Instruments, Inc. (I) (http://www.cwbrabender.com)
http://www.cwbrabender.com/Moisture_Tester_MT-C.html
Campbell Scientific (J) (http://www.campbellsci.com)
http://www.campbellsci.com/tdr100
CEM Corp. (C, F) (http://www.cem.com)
http://www.cem.com/smart-trac-2.html
http://www.cem.com/smart-turbo.html
Dynamax (J) (http://www.dynamax.com)
http://www.dynamax.com/IrrigationControl.htm
Eb Research (F) (http://www.ebresearch.net)
http://www.ebresearch.net/pq_001.html
Enercorp Instruments, Ltd. (E) (http://www.enercorp.com)
http://www.enercorp.com/hum/kpm.htm
ESi Environmental Sensors, Inc. (J) (http://www.esica.com)
http://www.esica.com/products_moisturepoint.php
Global Water Instrumentation (J) (http://www.globalw.com)
http://www.globalw.com/products/at210.html
Imko Micromodultechnik, GMBH (J) (http://www.imko.de)
http://www.imko.de/en/products/industrial-moisture
James Instruments (C, D, E) (http://www.ndtjames.com)
http://www.ndtjames.com/Moisture-Testing-s/6.htm
Kett (B, E, I) (http://www.kett.com)
http://www.kett.com/productlines/prod03.aspx
http://www.kett.com/productlines/prod02.aspx
http://www.kett.com/productlines/prod06.aspx
http://www.kett.com/productlines/prod04.aspx
http://www.kett.com/productlines/prod07.aspx
Mitsubishi Chemical Analytech Co., Ltd. (H) (http://www.dins.jp)
http://www.dins.jp/dins_e/1prdcts/product.htm
Moisture Register Products (B, C, G, D) (http://www.aquameasure.com)
http://www.aquameasure.com/continuous.htm
http://www.aquameasure.com/rd.htm
http://www.aquameasure.com/portable.htm
NDC Infrared Engineering (B) (http://www.ndcinfrared.com)
http://www.ndc.com/en/Products/On-Line-NIR-Gauges.aspx
PNNL Sensors (http://www.technet.pnnl.gov)
Sensortech Systems, Inc. (B, G) (http://www.sensortech.com)
http://www.sensortech.com/moisture.aspx
Soilmoisture Equipment Corp. (J) (http://www.soilmoisture.com)
http://www.soilmoisture.com/subcategory.asp?cat_id=19
Spectrum Technologies, Inc. (J) (http://www.specmeters.com/)
http://www.specmeters.com/tdr300/
Thermo Scientific (A, C) (http://www.thermoscientific.com)
http://www.thermoscientific.com/ecomm/servlet/applicationscatalog_11152_L11761_82214_-1_1
Troxler (A) (http://www.troxlerlabs.com)
http://www.troxlerlabs.com/products/field_equipment.php

INTRODUCTION

Determination of the water quantity presented in solids is very important in many industrial and laboratory measurements. Selection of appropriate measurement methods and analyzers can significantly affect production rates or product quality. In specialized manufacturing processes, for example, based on weight transactions, very small errors in the determination of moisture can even lead to huge economic losses.

The next chapter deals with methods and analyzers for measurement of moisture in solids, which are mainly used in the industry. Some of them belong originally to laboratory methods. In the past, these methods required a long time to obtain accurate results, and often operators with great theoretical and practical experiences. With the gradual increase in integration and miniaturization of electronic circuits, the introduction of microprocessor technology, and improvements in computer technology and software, many of these methods were transformed into simple devices designed for a specific application, which do not require skilled operators. Thus, all presented methods are represented by industrially applicable analyzers. Only chemical methods such as Karl Fisher titration or gravimetric methods such as desiccation can still be considered as laboratory, since both methods require a higher degree of knowledge and precision.

The text is divided into subchapters devoted to individual measuring methods. For each of the methods, representative instruments, their function, and incorporation into the process are described.

In some cases, the moisture content of solids can be measured by the same types of analyzers that are used for the detection of moisture in liquid or gas samples. These devices, such as capacitance, infrared, and microwave sensors, have already been discussed in Chapters 1.38 and 1.39, and therefore will be treated only briefly in this chapter. The laboratory-type moisture analyzers, such as the Karl Fischer type units, can also be used on both liquid and solid samples.

It should also be noted that the main purpose of this chapter is to describe the available choices for solids moisture analyzers and is not to give detailed descriptions of the families of analyzers involved. These descriptions are given in separate chapters, such as Chapter 1.32 dealing with infrared analyzers, and Chapter 1.74 devoted to autotitrators, etc.

CHEMICAL METHODS

Among the number of existing (often experimental) chemical methods for determination of moisture in solids, only one should be considered as suitable for wider applications—the Karl Fisher titration.

The Karl Fisher autotitrators are accurate to 5 ppm, but they still require qualified and experienced operators. When solids are analyzed, the sample must first be dissolved in a solvent. If the solids are not soluble, a solvent must be used to leach the moisture out of the solids. For some autotitrators (e.g., for determination of moisture in minerals), high-temperature furnaces which transform moisture into gas phase are also available. The gas phase is transferred using dry inert gas like nitrogen into titration cell. Once the moisture is transferred into the titration cell, the titration is performed using the Karl Fischer reagent, which consists of a solution of sulfur dioxide, iodine, and pyridine (or better imidazole) in methanol. These reactions occur:

$$H_2O + I_2 + SO_2 + 3C_5H_5N = 2C_5H_5N \times HI + C_5H_5NSO_3 \quad \textbf{1.40(1)}$$

$$C_5H_5NSO_3 + CH_3OH = C_5H_5N \times HSO_4CH_3 \quad \textbf{1.40(2)}$$

The titration end point can be obtained volumetrically or coulometrically (see Chapter 1.74). Due to the nature of the Karl Fisher titration, where moisture is detected stoichiometrically, it should be used as calibration method.

GRAVIMETRIC METHODS

Gravimetric methods belong to the oldest techniques for measuring of moisture in solids. The principle of these methods is very simple; there are two basic approaches to measuring (1) the first one consists in weighing of a sample before and after its drying and (2) the second one uses weighting of the amount of condensed water or water absorbed by the desiccant, which was released from the sample by drying. Based on these two approaches, a number of procedures and instruments have been developed. They reflect the needs of moisture measurement in various types of materials, including those materials prone to thermal decomposition or organic matters. Today, gravimetric methods are the basic methods for determining the moisture content in solids; however, there are notable disadvantages of these methods for industrial use consisting in discontinuous nature of measurement, longer time required to achieve accurate results, and the need of staff with empirical experience with the particular type of samples.

Oven Drying (Loss on Drying)

The most widely used method for gravimetric determination of moisture in solids is oven drying followed by weighting. Preweighted sample is placed in a hot air oven. After several hours at temperatures from 105°C to 110°C, the weight loss of the sample is stabilized. Dry sample is then weighted again on external balance. The difference in weight ideally corresponds to the weight of the evaporated water. Thus, the concentration of moisture in the sample can be easily calculated. This method is commonly called oven drying, but in the literature can be found also as loss on drying (LOD). The advantage of oven drying is its simplicity and easily available devices, which usually consists of common laboratory equipment. The initial calibration is not required for oven drying, since it belongs to primary methods. On the contrary, the result of measurement may be loaded with number of errors that must be considered in the evaluation.

The basic problem of using common hot air oven is that over time, equilibrium between the pressure of water vapor above the sample and in space of oven is achieved. The moisture content of the sample is then determined by the sorption isotherm of the material. There is no further evaporation of water from the sample after equilibration gets established, which introduces a substantial error in the measurement. One of the ways to deal with this problem is to continuously wash the oven with dry air or to use one of drying agents.

Some measurement errors are also related to slow weighting of the dried sample on external balance, especially when hydrophilic materials that can absorb moisture from the air in a relatively short time are weighted. Manufacturers have developed a solution for this problem, so-called moisture balances, where balance is combined with oven.

Dependence showing weight loss of the sample during oven drying on time indicates that after some time weight loss of the sample gets stabilized. However, if the temperature in the oven increases, weight loss gets stabilized at a new, higher value. This behavior is observed even at temperatures well above 100°C and it is characteristic for complex natural materials, which contain free water on the surface or in capillaries as well as bound molecular water or water coming from decomposition of organic substances. Additionally, should other volatile substances be present in the sample, a significant distortion of the result due to their evaporation may occur. Temperature and time required for oven drying therefore cannot be defined explicitly, they are rather determined empirically.

Another problem associated with conventional hot air oven is formation of crusts on the surface of the dried organic matters (like when baking cakes), which prevents further evaporation of water from the sample. Additional drying of these samples leads to destruction of their original chemical structure. To avoid this phenomenon, a number of improvements were developed, especially regarding heat sources used to heat the sample. The sample can be heated by infrared or microwave radiation or by inductive heating.

Infrared lamps with temperature about 2000 K and with power up to several hundred watts are very cheap and can withstand hundreds of thousands of tests. Their radiation penetrates approximately 2–3 mm below the surface of the sample, thereby increasing the vaporization surface. In this case, the sample temperature often reaches around 70°C— that is significantly lower than that of conventional hot air dryers. Modern models of ovens use quartz heaters instead of infrared lamps. Quartz heaters operate at 2.6 µm and their lifetime is 5–10 times longer. Another type of heater is based on ceramic tubes with power of hundreds of watts.

Appropriate choice of the thermal radiator may also reduce the total drying time to several minutes. Therefore, conventional microwave ovens operating at 2.45 GHz offer good results as well. However, even here, accurate results are more dependent on operator's experience than on accuracy of the device. For some moisture balances, tables with temperatures and drying times required for particular materials are also available. These tables are made directly by the manufacturer and they are often included in the drying programs in the memory of the balances. With the birth of microprocessors, the temperature of drying has been also automated, and its waveform is often divided into several different modes suitable for different types of samples. Moisture balances are currently typically digital and fully automatic, but mechanical moisture balance for the environment with a high degree of interferences are still available. For all types of moisture balance, temperature compensation is necessary, because the hot air in the oven may affect measurement accuracy. Cross section of typical moisture balances is in Figure 1.40a.

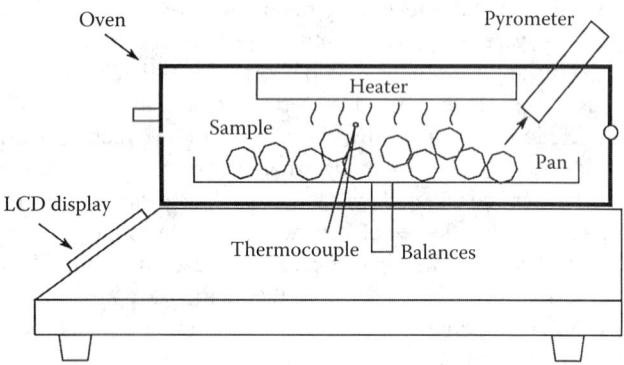

FIG. 1.40a
Moisture balances (lateral cross section, simplified).

Modern devices can be equipped with up to two thermometers (1) thermocouple, which allows monitoring of the temperature in the vicinity of the sample, and (2) infrared thermometer to monitor the temperature of the sample surface. These two temperatures are often different and their precise monitoring helps to significantly speed up drying of the sample and consequently also prevents damaging of the material.

Moisture balances available on the market are designed for most types of materials, from granular materials, fibers, for both organic and synthetic substances. Weight of the samples tends to reach up to hundreds of grams; some manufacturers provide equipment capable of measuring up to 10 samples at a time. By correctly choosing the time and drying temperature, residual moisture content of 0.01% can be achieved.

There are many modifications of the oven drying. One of them is drying under reduced pressure, which combines the previously described moisture balances with vacuum technology, and allows use of lower temperatures during the drying of the sample. Vacuum drying is thus particularly suitable where moisture is detected in temperature-prone materials, for example, pharmaceuticals. Other modifications are dryers that have built-in some type of hygrometer. Thus, moisture of the sample is determined from the vapor pressure of water inside the oven. The other oven drying methods should be associated with Karl Fisher titration.

Moisture balances are widely used in the chemical and pharmaceutical industry, paper industry, in food and beverage, pharmaceutical industry, the manufacture of plastics and composites, textiles, cosmetics, paints and coatings, and many others.

Desiccants

Determination of moisture in solids using desiccants is oven drying analogue. It uses high absorption ability of some substance, called desiccant, to absorb water vapors and thus effectively reduce the pressure of water vapors in an enclosed space. This is primarily a laboratory method that is used for thermally unstable substances, which decompose at temperatures above 60°C. This method is however characterized by lengthy drying process, which can take weeks or months.

A sample of the substance is placed in a desiccator containing a desiccant, which absorbs moisture from the air. During drying, pressure of water vapors in the vicinity of the sample is being reduced, which leads to gradual and slow drying. The desiccant must be periodically exchanged. The amount of water that this procedure removes from the sample is dependent on the efficiency of the desiccant. Desiccant efficiency is expressed by the minimum weight of water on a liter of dry gas obtained by using the desiccant. The effectiveness of chosen frequently used desiccants is given in Table 40.1a.

TABLE 1.40a
Effectiveness of Commercially Available Desiccants (They can be slightly different according to source)

Desiccant	Desiccant Efficiency ($mg \cdot dm^{-3}$)
P_2O_5	0.000025
$Mg(ClO_4)_2$	0.0005
BaO	0.00065
H_2SO_4 (100%)	0.003
Al_2O_3	0.005
KOH	0.014
Silica gel	0.03
CaO	0.2
$CaCl_2$	0.25
NaOH	0.8

Furthermore, speed of water absorption by the desiccant and possibility of its regeneration should be considered. From this perspective, barium oxide is the best. At the process of absorption, it is chemically changed to barium hydroxide with up to 16 water molecules; this process can be reversed by converting it to barium hydroxide monohydrate. A very sensitive desiccant—phosphor pentoxide, commonly used in instruments for measuring trace concentrations of moisture in gases, is less suitable, because a thin layer of phosphoric acid is formed on the surface, preventing further diffusion of water.

The method of using desiccants to determine moisture content in solids places great demands on time and service care; this method therefore remains only a complementary laboratory method.

Thermogravimetry

Thermogravimetry (TGA) is not the primary gravimetric method for the determination of moisture in solids, but it is usually used as a complementary analytical method for determining the behavior of substances at their gradual warming to their eventual final thermal decomposition. For this reason, the under-mentioned description is just a brief summary. For readers with deeper interest, specialized literature is listed at the end of the chapter.

TGA is a method, at which weight loss of the sample depending on temperature and time is measured. In practice, specialized device, called thermobalance, equipped with accurate balances and adjustable heating furnace is being used. A sample ranging from milligrams to tens of grams is placed on the pan of the balance in the furnace. The furnace is gradually heated to a defined temperature and the dependency of the weight loss of the sample on temperature is followed.

Measurements can be further extended to different types of working atmospheres (inert, oxidizing, reducing gases, corrosive gases and vapors, steam, liquids, etc.) with different pressure (from high vacuum, the atmospheric pressure, and over pressure). Apparatuses of some manufacturers are also equipped with one of the analytical techniques such as infrared or mass spectrometry, or gas chromatography, which allow studying of substances released during sample heating.

From the shape of the TGA curve, behavior of the sample at a given temperature or the processes of oxidation and corrosion, or dehydration can be evaluated. Dehydration studies are critical with regard to the measurement of moisture in solids, because they can measure the free and bound water in the sample.

Using data from the TGA in the industry is complicated. The reason is different behavior of standard samples with weight not exceeding 100 g and much heavier samples commonly found in industrial processes.

Lyophilization

Drying of samples such as tissues, cells, blood plasma, or microorganisms is often complicated because of their thermal instability. Water passing into gas phase changes its volume, which in turn also causes damage to the fragile structure of the dried samples.

Lyophilization (sometimes called freeze drying) is a technique that uses slow sublimation of liquids at low temperatures (to the temperature of liquid nitrogen −197°C) and generally under reduced pressure. Water from the lyophilized sample passes into the gas phase slowly, without melting, reducing the overall damage to the sample. Lyophilization also removes some other ailments of classical drying, for example, there is no formation of foam on the surface of freeze-dried samples, lyophilized sample does not create crust, sample remains sterile, because the low temperatures impede the growth of bacteria, oxidation of the sample at low temperatures is very slow, and finally there is not the above-mentioned degradation of thermally unstable samples.

Lyophilization is generally carried out in two steps. In the first step, most water is in the form of ice sublimated from the sample under high vacuum. Sample temperature is normally kept below 0°C, and although there are exceptions, it is assumed that up to −30°C most of the water in the sample freezes. After several days of continuous drying, only a few percent of water remain in the sample. In the following step, the temperature is increased, which causes disruption of the remaining bonds and molecules are released. This step causes that moisture in sample gets less than 1%.

Apparatuses for lyophilization of the samples are supplied with different temperature minimum as the desktop (tabletop) version as well as industrial systems. An accessory of lyophilizers is also a vacuum system.

Absorption and Condensation

In the introduction to gravimetric methods, there were mentioned two approaches to measurement of moisture in solids. The first approach, called Loss on Drying, was described in the paragraphs devoted to oven drying, desiccation, thermogravimetric analysis, and lyophilization. The second approach, called Gain on Drying, is based on direct weighting of the water which was evaporated from the sample. Two approaches can be used: (1) the absorption of the evaporated water by desiccant and subsequent weighting and (2) condensation of the evaporated water and subsequent weighting.

Commercially available devices use absorption methods for measuring of moisture in samples with very low water content (2 ppm or 1 μg ($3.527 \cdot 10^{-8}$ ounce) of water). The principle is based on the drying by stream of dry inert gas such as nitrogen N_2. Nitrogen stream with moisture and other volatile substances flow through carbon filters. Only water vapors are allowed to pass. Further, water vapors are absorbed by preweighted tube filled with desiccant (e.g., phosphorus(V) oxide). Here, the water is absorbed and the tube is weighted again.

The condensation method is used when the sample is composed from several volatiles. The sample is placed into a glass tube and heated. Water is evaporated from the sample by heating and condenses in cooler part of the tube. This part of the tube is removed after drying and weighed.

FAST NEUTRON MODERATION

Neutron Probe

In this instrument, fast neutrons (speed about 9600 km/s (6000 miles/s)) are moderated by the hydrogen atoms of measured solid. The amount of detected slow neutrons (speed about 2.7 km/s (1.7 miles/s)) is then proportional to the hydrogen content and it is used for moisture evaluation.

Neutrons are subatomic particles that are electrically neutral and have a mass approximately equal to the mass of a hydrogen atom. The source of fast neutrons is a mixture of isotope emitting alpha particles and beryllium. Alpha particles bombard beryllium atoms, resulting in neutron formation by the equation:

$$^{4}_{2}\alpha + ^{9}_{4}Be \rightarrow ^{1}_{0}n + ^{12}_{6}C + eV \quad \quad 1.40(3)$$

Isotope ^{241}Am (half-life 432.2 years) is commonly used as a source of alpha particles in commercially available devices. Alternatively, isotopes ^{226}Ra, ^{239}Pu, and ^{210}P are used, but they have a lot of disadvantages like smaller neutron yield or undesirable emission of gamma radiation.

Neutrons emitted from the source penetrate into material bulk and because they lack an electric charge, are not deflected by the negative and positive charges associated with atoms; they are deflected or reflected only by elastic collisions with the atomic nuclei. The slowdown is mainly by neutron collision with an atom of similar size.

Such an impact involves a change of momentum of the neutron and of the impacted nucleus. Because of the conservation of energy and momentum, after the impact, the energy of the neutron is decreased by an amount equal to the energy transferred to the impacted nucleus. The energy of the neutron after the impact is a function of the mass of the impacted atom.

This fact is explained by Equations 1.40(4) and 1.40(5) below. In these equations, m_1 and m_2, respectively, equal the mass of the neutron and the impacted atom. V_1 and V_3 are the neutron velocities before and after impact, and V_2 is the velocity of the impacted nucleus, which was initially at rest.

$$m_1 V_1 = m_2 V_2 + m_1 V_3 \text{ (conservation of momentum)}$$
$$1.40(4)$$

$$\frac{1}{2} m_1 V_1^2 = \frac{1}{2} m_2 V_2^2 + \frac{1}{2} m_1 V_3^2 \text{ (convervation of energy)}$$
$$1.40(5)$$

From Equation 1.40(4),

$$V_2 = \frac{m_1(V_1 - V_3)}{m_2} \quad \quad 1.40(6)$$

Substituting (1.40(6)) into Equation 1.40(5) and simplifying the results yields

$$V_3 = \frac{-V_1(m_2 - m_1)}{(m_1 + m_2)} \quad \quad 1.40(7)$$

If the mass of the impacted nucleus is equal to the mass of the neutron, all of the neutron's kinetic energy is transferred to that nucleus, and V_3 becomes zero.

The hydrogen atom, because it is most nearly equal in its mass to the neutron, is the most efficient energy absorber (or moderator) of neutrons. In the common inorganic substances, the greatest source of hydrogen nuclei is water; therefore, it can be considered that the rate of slow neutrons is proportional to the moisture. For comparison, 20 collisions are needed to slow neutron by hydrogen nucleus; by carbon it is 100 collisions and in the case of oxygen it is 150 collisions.

Slow neutrons gradually reach the energy around 0.025 eV, and they come into thermal equilibrium with the surrounding nuclei of atoms of the material. In some case, they are absorbed by the nuclei. Nuclei Cl, K, B, Li, Rh, Ag, Hg, Cd, In, Hf, Re, Ir, Au, and Br are very strong absorbers of slow neutrons and for this reason, they can distort measurements. It is essential to avoid measurement of moisture in solids with high concentration of strong absorbers.

There is a number of detectors for detection of slow neutrons; particularly scintillation detectors with photomultiplier, proportional counter tube filled with $^{10}BF3$, ^3He with amplitude discriminators which restrict detection of γ radiation, or Geiger–Müller tube shielded by cadmium. Detection by Geiger–Müller tube is based on cadmium isotope ^{113}Cd (12.3% in cadmium shielding), which releases gamma radiation by collision with slow neutrons. All detectors must be specific and sensitive to slow neutrons only.

Neutrons from a point source are spread in all directions. If the probe is surrounded by a material of constant moisture, slow neutrons wander until they are completely absorbed. The traveled distance depends on the moisture content and thanks to point source it creates a sphere with radius R. This sphere is called the sphere of influence of the neutron probe (for soil moisture probes reaches the size around 30–50 cm depending on humidity). The detector of slow neutrons is always placed at a distance less than the radius of the sphere of influence. Similarly, for the most accurate measurements, it is essential that the amount of the material fills more space than the volume of the sphere of influence.

The method has several limitations resulting from its essence. (1) The instrument is sensitive to all hydrogen atoms; chemically bonded hydrogen in the sample is also seen as water. This can be corrected for by calibration only if the quantity of bonded hydrogen in the sample is constant and the amount of hydrogen present as free water is of the same order of magnitude as the amount of bonded hydrogen. If they are not of the same magnitude, a change in moisture concentration might not result in a detectable signal change. This method is not accurate for measurement of substances containing a large amount of hydrocarbons (coal and soil with high humus content). (2) Measured materials must also contain low concentrations of strong absorbers of slow neutrons. (3) Due to the radioactivity of the neutron sources, it is necessary to comply with legal restrictions under the laws of the territory.

The advantage of this method is the possibility to measure granular materials, assuming no changes of moisture in the volume of sphere of influence. Measurement accuracy is not affected by temperature, pressure, conductivity, or varying particle sizes and even frozen water can be measured reliably.

Neutron probes are primarily used in the measurement of soil moisture for irrigation purposes (Figure 1.40b). The procedure of soil measurement is as follows: Firstly narrow inspection hole is dug into the ground, placed where the measurement is representative. This hole is loaded by aluminum tube with close contact with the soil. Aluminum is transparent to neutrons and does not slow them. The probe is lowered into the measurement of this case. Tube must be sealed between the measurements to avoid filling by surface water.

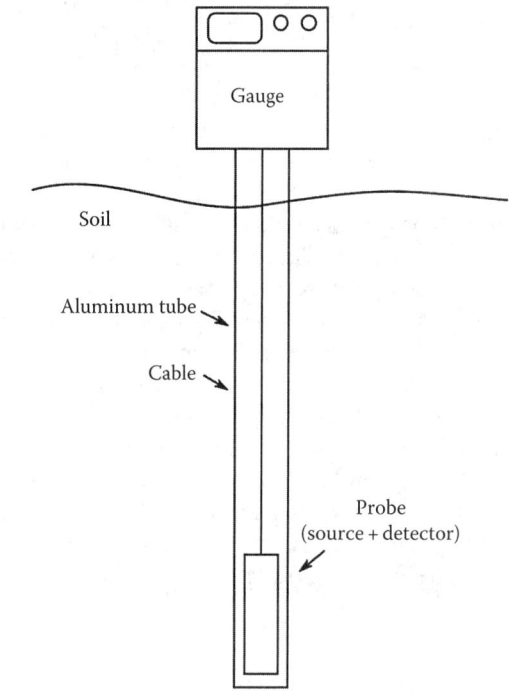

FIG. 1.40b
Neutron probe for measurement of moisture in soil.

COMBINED NEUTRON/GAMMA PROBES

Gamma radiation can be used to measure moisture in solids similarly as neutron probes. The main obstacle to high accuracy of this measurement is dependency of gamma absorption on density of the material. In the event that there is a change in density such as swelling, gamma probes are more suitable as a density meters. This feature can be used in combined neutron/gamma probes.

Neutron radiation source is ^{241}Am–Be, source of gamma radiation is usually ^{60}Co or ^{137}Cs. Neutron detector counter tube is filled with ^3He and for detection of gamma radiation scintillation detectors or Geiger–Müller tubes are used.

Combined neutron/gamma probes are used for moisture and density measurements of solid surfaces as asphalt,

568 Analytical Measurement

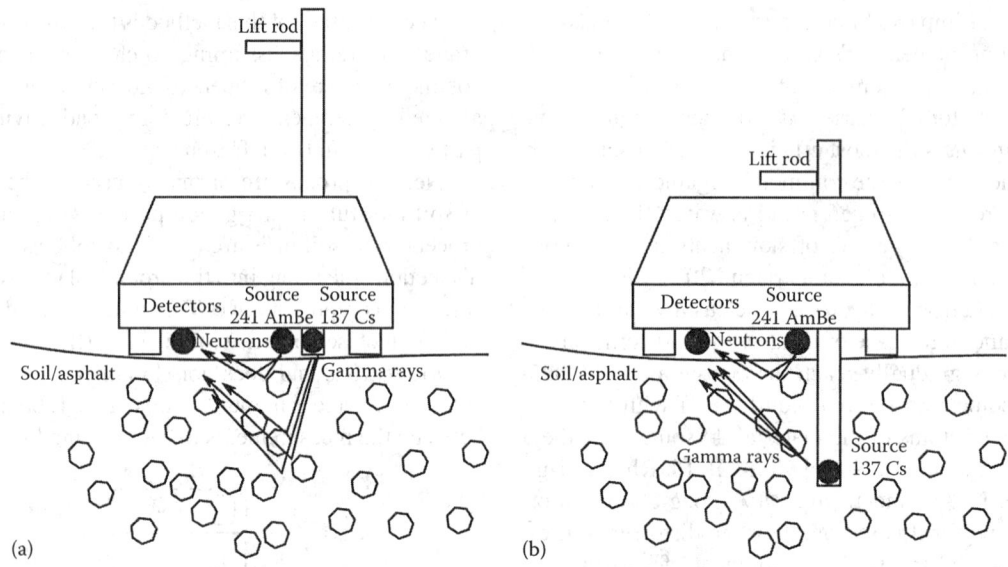

FIG. 1.40c
Measurement of moisture and density by combined neutron/gamma probe (a) backscattering (b) absorption.

concrete, soil, or in moisture and density measurements of materials on a conveyor belts. In case of soil, asphalt, or concrete measurement, they are used in two modes for (1) surface and (2) subsurface measurements (Figure 1.40c). For surface measurements, both neutrons and gamma sources are placed above the surface of the object and the density of the material is measured by backscattering. In the case of subsurface measurements, gamma radiation source is inserted into inspection hole and measure the attenuation rate of gamma radiation. Neutron source remains above the surface. If the density change is invoked only by moisture, gamma radiation can also be used for moisture measurement.

The sensing unit can be mounted over conveyors, as shown in Figure 1.40d, on bins, or on pipelines. The sample volume should be fairly large, 400 mm (16 in.) wide and a minimum of 50 mm (2 in.) thick for conveyors, or a minimum of 200 mm (8 in.) diameter pipe size.

On conveyors where the solids bed may be of nonuniform thickness, a scraper plow or a roller should be installed ahead of the measuring head. When measuring moisture in a bin or chute, care must be taken to prevent material buildup on the vessel walls. For applications involving solids with a tendency to cake, the vessel wall should be lined with a nonsticking plastic such as Teflon or be provided with a bin vibrator.

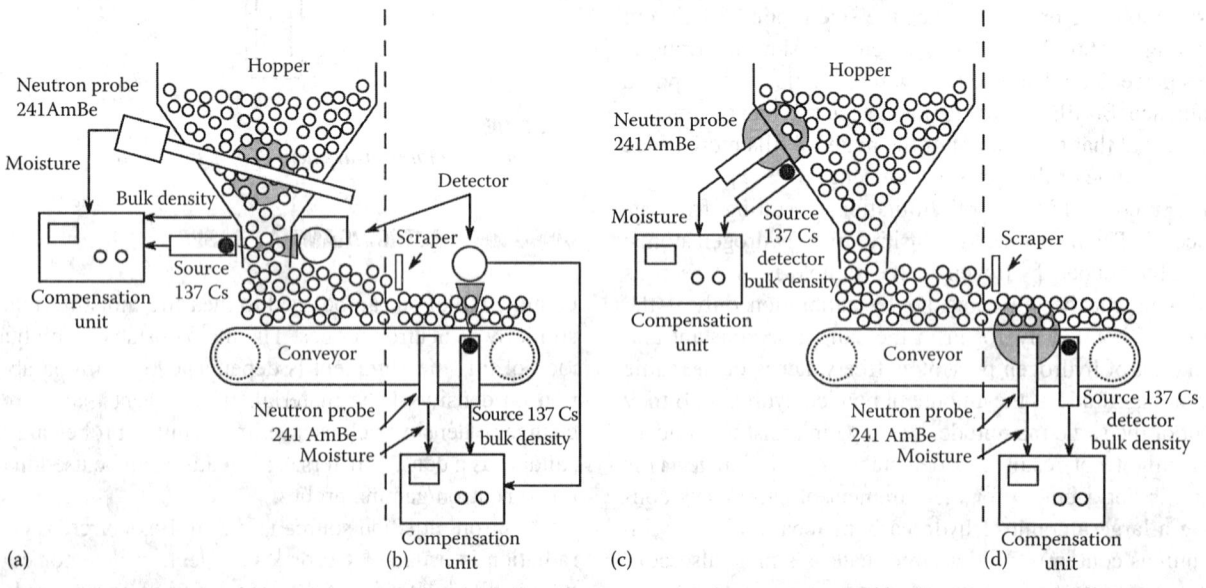

FIG. 1.40d
Measurement of moisture in solids with density compensation using gamma (a) hopper, absorption, (b) conveyor, absorption; (c) hopper, backscattering; (d) conveyor, backscattering.

INFRARED ABSORPTION AND REFLECTION

The water molecule has a permanent dipole moment, which is the cause of the water ability to absorb in the infrared spectrum of electromagnetic radiation. The water in the infrared spectrum shows characteristic absorption bands, which serve as an indicator in the presence of moisture in solids. The absorption bands are processed by nondispersive infrared moisture analyzers.

Infrared moisture analyzers usually process two characteristic absorption bands in the near infrared region—one in 1.9 or 1.4 μm and the second in 1.75, 1.6, or 1.3 μm. Since the solids absorb infrared radiation in a thin surface layer of material, measurement methods must be particularly reflective. Absorption is only exceptionally used for measuring of thin materials, such as paper.

The scheme of infrared moisture analyzer for measuring by reflection method is shown in Figure 1.40e. The sensing element consists of a light source, filters to pass the measuring and reference wavelengths, optics to direct the light beam onto the sample and to collect the reflected light, and a photocell to convert the reflected light into electrical current (Figures 1.40e and 1.40f). The sensor should be mounted so that the light aperture is a few inches from the sample.

The two current pulse outputs of the photocell, which correspond to the reflected measurement and reference wave-

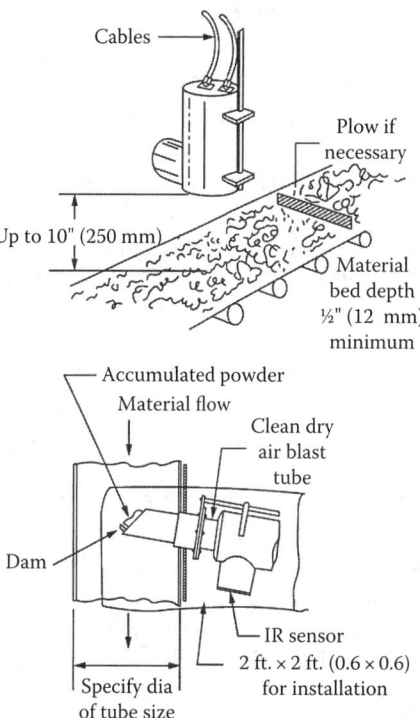

FIG. 1.40f
Installation of infrared moisture analyzers for belt and duct applications. (Courtesy of Aqua Measure Instrument Co., formerly Moisture Register Products.)

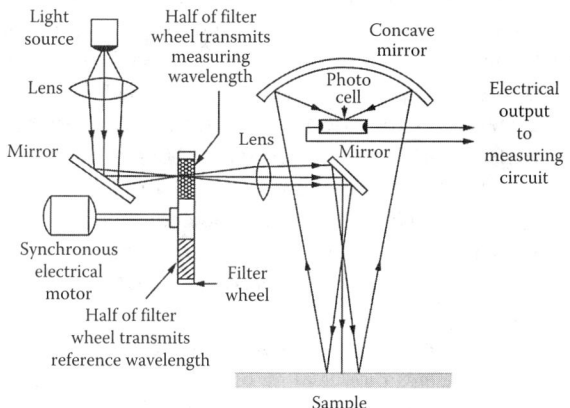

FIG. 1.40e
Infrared moisture measurement (reflectance). In microprocessor-based "quadra-beam" designs, four channels are used, and the measured reflections are rationed to reduce drift and stabilize the measurement.

lengths, are converted into two DC signals. The ratio of these signals is read out as moisture content.

The main advantage of infrared analyzers is the ability to measure very low levels of moisture (under 10%) without contact. If the reflectivity of the sample is not above a minimum value, the amount of light reflected (and the change due to absorption) will be too small to produce a useful output. Graphite, coal, metal powders, and inorganic pigments are all poor reflectors of infrared, so this instrument cannot be used to analyze their moisture content.

Because infrared waves do not penetrate very far below the surface of the sample, the moisture measurement obtained will be representative of the moisture content on the surface of the material. If the surface moisture content is not representative of the average moisture content of the sample, this sensor cannot be used. This is usually the case when the sample is exposed to the ambient atmosphere. Due to drift in the instrument, periodic recalibration is necessary.

TIME DOMAIN REFLECTOMETRY

Time domain reflectometry (TDR) is a method commonly used for fault detection of electric cables; however, it can be advantageously adapted for measurement of moisture in solids.

The principle of this method is as follows: Electromagnetic wave propagating through electric cable is partially or totally reflected back to the source on every impedance change. The speed of wave propagation and the elapsed time between the transmission and detection of wave reflection is used to easily calculate position of a place with reduced impedance (disorder). Reversely, if the distance of the place, where the change of impedance occurs, is known, it is easy to calculate the speed of wave propagation through material and thus its electrical properties.

Water greatly reduces the rate of electromagnetic wave propagation due to high permittivity and so TDR can be used to measure moisture content of solids. TDR apparatus for measurement of moisture in solids is basically composed of the following components (in commercially available devices they are all built in one electronic device) (1) a pulse generator, (2) a coaxial cable with the probe, (3) a sampler, and (4) an oscilloscope (Figure 1.40g).

The pulse generator generates electromagnetic waves in the form of rectangular pulses. They are sent to coaxial cable whose shield is connected to ground (electrical potential is 0 V). One end of the coaxial cable connects the pulse generator, the sampler, and the oscilloscope, and the second end is fitted with a probe (the open end of the cable). The probe usually consists of two wires with a length of several tens of centimeters. While the properties of the dielectric filling of coaxial cable are known, the dielectric properties of the probe are unknown. The probe is stuck inside the measured material which forms the dielectric. Reflected electromagnetic wave is registered by the sampler composed of a precise multimeter and timing device. The oscilloscope is used to display and evaluate waveform measured by the sampler.

The wave reflection occurs at the interface of (1) the coaxial cable and the probe, where there is an impedance change, and also at (2) the end of the probe, where the value of impedance approaches infinity. The voltmeter of the sampler measures the change in the amplitude of the reflected wave. Generally, there may be a decrease in the amplitude due to the lower impedance of the interface coaxial cable/probe, or to increase the amplitude, if the impedance is higher. Timing device of the sampler registers three times (1) the time at which the wave arrives from the pulse generator, (2) the time at which the reflected wave arrives from the interface coaxial cable/probe, and (3) the time at which the wave arrives from the open end of the probe. Because coaxial cable and probe lengths are well known, speed of wave propagation through the "dielectric" of the probe (measured solid with moisture) can be easily calculated. The evaluation of probe permittivity from the time of reflected wave at the interface coaxial cable/probe and reflected waves on the open end of the probe is called time domain analysis. Mathematical expression of permittivity is based on the Maxwell equations and it is beyond the scope of this text. Simplified relationship between permittivity ε_r of the material and the time needed to overcome the distance between interface coaxial cable/probe and the open end Δt is finally expressed by the following equation:

$$\varepsilon_r = \left(\frac{c_0 \Delta t}{2 l}\right)^2 \qquad 1.40(8)$$

where
c_0 is the speed of wave propagation in vacuum
l is the length of the probe

TDR probes are most frequently used to measure soil moisture. Their main advantage is the high precision and continuous measurement; the TDR method does not require calibration and partially it is not affected by presence of salts in the measured material. In contrast, the disadvantages can be seen in the complexity of the electronics and higher purchase costs.

Presence of salts may affect the accuracy of measurements at following circumstances. Their concentration in wet material may cause a short circuit between the two poles of the probe and prevents reflection of the wave. There could be two situations: (1) the concentration of salts in the soil is high, but the humidity is low, so there is no short circuit and (2) the salt concentration is high, but the soil is very wet, and measurement error may be affected.

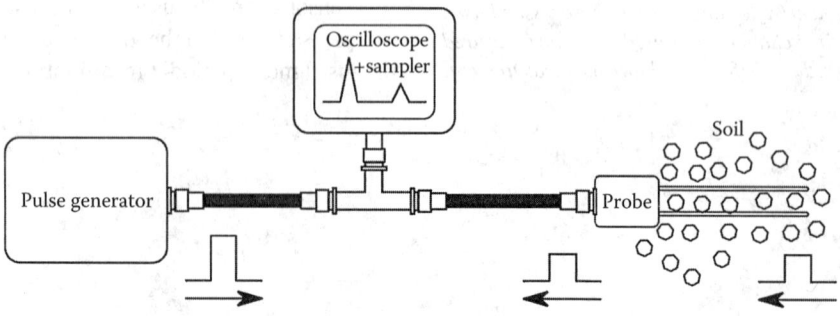

FIG. 1.40g
For explanation of basic principle of time domain reflectometry.

MICROWAVE ATTENUATION

The operation of the microwave-type moisture analyzer has already been described in Chapter 1.39, and that discussion will not be repeated here.

Microwave-type solids moisture analyzers have been used to measure the water content of plastic powders, chips and pellets, soaps, grains, powdered clay, paper, hardboard, and plastic products. To measure the moisture content of moving webs or other materials that are thinner than 0.5 mm (0.02 in.), multiple transmission is required.

Because the absorption of microwaves is affected by both bulk density and solids temperature, both of these variables must be measured and compensated for if precise readings are desired. Such a compensated unit is illustrated in a gravity flow (or pneumatically conveyed) bypass application as shown in Figure 1.40h.

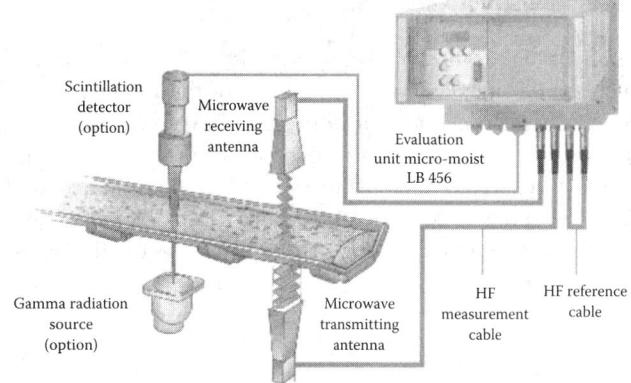

FIG. 1.40i
Noncontacting moisture analyzer can be provided with gamma radiation densitometer to compensate for variations in process density. (Courtesy of Berthold Technologies.)

Moisture in Coal

Microwave analyzers can also be used to measure the moisture content of coal as it is being conveyed on a belt, without sampling or crushing (Figure 1.40j). These units are also automatically corrected for density and temperature variations and can detect the moisture content within about 0.5% over a range of 0%–20%.

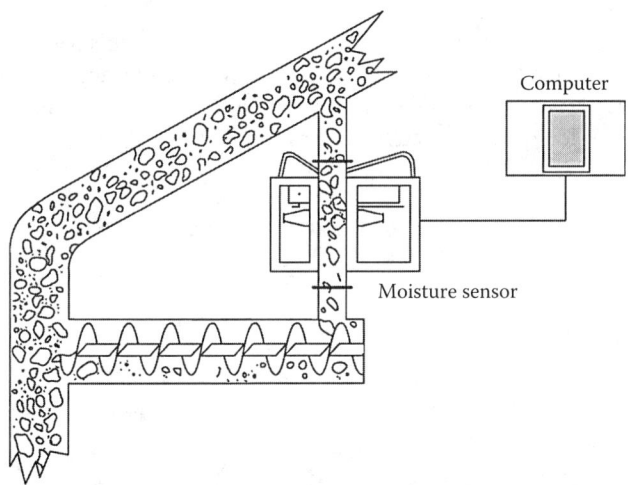

FIG. 1.40h
Microwave-type solids moisture analyzer with radiation densitometer correction. (Courtesy of ThermoMeasure Tech, formerly Kay-Ray/Sensall, Inc.)

Basis Weight Compensation

If contactless moisture detection is desired for a conveyor belt application where the product density varies, a microwave-type analyzer can be compensated. Basis weight compensation can be obtained by the addition of a low-activity gamma radiation source as shown in Figure 1.40i. Basis weight compensation is available not only on conveyor belt installations but also for measurements in chutes and in vertical ducts and pipes.

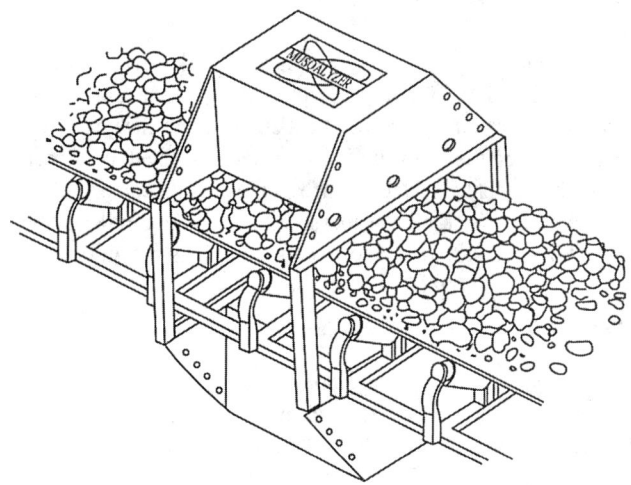

FIG. 1.40j
Continuous moisture-in-coal analyzer utilizing the microwave principle.

Sugar Industry Applications

Microwave moisture analyzers have also been developed for such applications as the measurement of the dry substance content in the process of sugar extraction. These sensors can measure the sugar concentration both in the pipelines and in the sugar tanks or crystallizers. The pipeline units are flanged at both ends, whereas the tank mounted ones can be inserted through the nozzles of the crystallizers or other containers (Figure 1.40k).

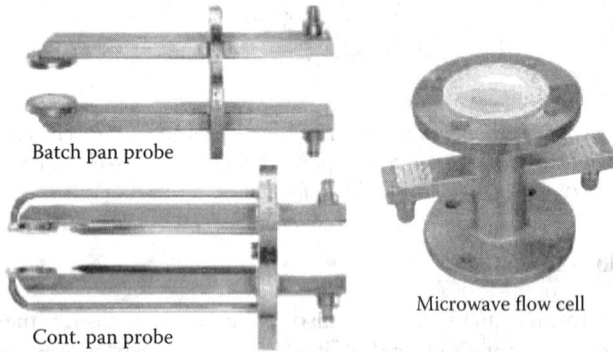

FIG. 1.40k
Sugar concentration measurement by the use of microwave analyzers. (Courtesy of Berthold Technologies.)

CAPACITANCE MOISTURE GAUGE

The capacitance-type hygrometer has been described in connection with Figure 1.39e in Chapter 1.39. In capacitance-type measurements of moisture, the great difference in dielectric constants between the dry sample and water is utilized to perform the measurement. As the changes in moisture content affect the dielectric constant, the measured capacitance reflects that change and the variations are read as changes in moisture level.

Sensor Designs

The measuring probe may consist of stainless steel electrodes embedded in plastic or in other insulators to form the measuring head of a capacitor (Figure 1.40l). The probe or measuring head lightly contacts the process material, which serves as the capacitor dielectric (the nonconducting material between the capacitor plates).

Another type of capacitive element is designed for moisture measurement in powders that are transported on conveyors. In that case, the capacitor plates are mounted at right angles to each other as shown in Figure 1.40m. In this design, the horizontal electrode rides on the powder surface while the vertical electrode is submerged. Other variations include force-fed measuring cells as shown in Figure 1.40n.

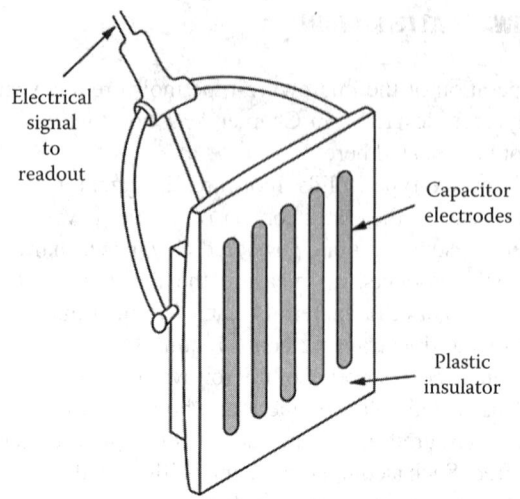

FIG. 1.40l
Capacitance-measuring head for moisture detection in solids.

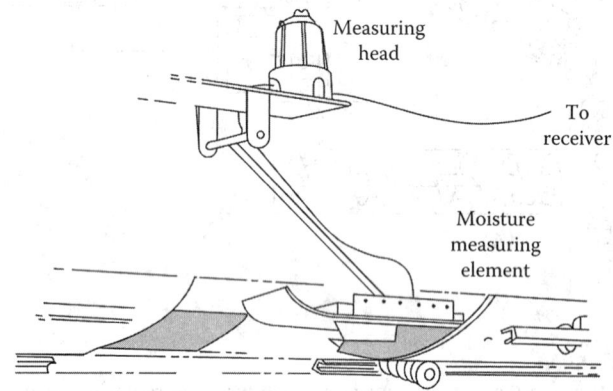

FIG. 1.40m
Ski-type probe for solids moisture detection.

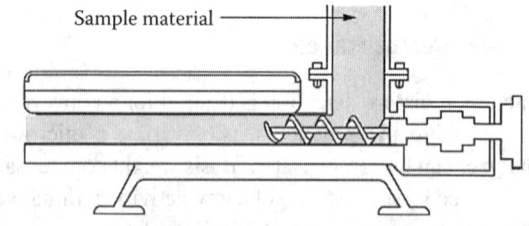

FIG. 1.40n
Force-fed measuring cell.

Capacitance is also used to detect the moisture content of agricultural products such as grains. The same principle can also be utilized to detect the moisture content of the soil by the insertion of a probe. Such a probe might consist of two electrodes with an insulating gap between them, and the capacitance circuit is used to measure the soil moisture as the dielectric constant of the insulator varies with the soil's moisture content.

The probe of the soil moisture monitor shown in Figure 1.40o is 3.4 in. in diameter and 27 in. long. It is normally installed vertically with intimate contact with the soil. The top 9 in. of the probe are passive, and only the bottom 18 in. section is used for measuring the soil moisture. The probe generates 0–5 V or 4–20 mA outputs, and several probes can be connected to a central computer, which can sound alarms or can automatically operate irrigation systems.

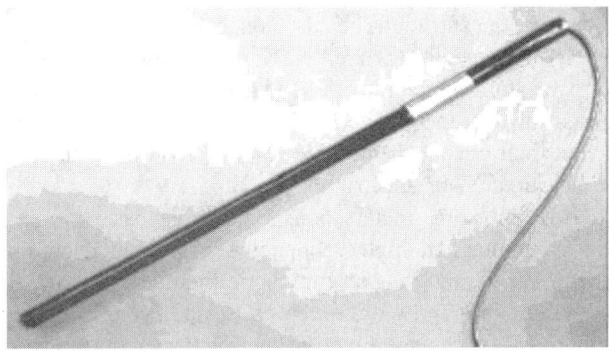

FIG. 1.40o
Dielectric constant monitoring probe for the remote measurement of soil moisture. (Courtesy of Automata, Inc.)

CAPACITANCE MEASUREMENT

Capacitance can be measured by two methods. One utilizes a capacitance bridge, which is analogous to the Wheatstone bridge, used for resistance measurement. Here, the bridge is excited by a radio frequency (RF) alternating current (AC) signal, while a servomotor continuously balances the capacitive reactance of the measuring capacitor with the reactance of a known capacitor. (Capacitive reactance of an AC circuit is the part of the impedance that is due to capacitance.)

Alternatively, the measuring capacitor can be connected in parallel with a known inductance to form a parallel resonant circuit. This circuit is powered by a constant frequency source. In this case, the output voltage becomes a function of connected capacitance. At resonance, this output is a maximum and decreases nonlinearly for larger and smaller values of capacitance. The nonlinearity can be minimized by an appropriate choice of circuit parameters and by narrowing the operation of the instrument to a limited portion of the resonance curve.

Limitations

In all capacitance measurements, the capacitance of the solids is a not a unique function of the moisture content of the process material; it is also affected by such factors as particle size, packing, and material density. Therefore, such moisture measurements will be accurate only if these variables are constant or can be compensated for.

RESISTANCE MOISTURE GAUGE

The electrical resistance of nonconducting solids is also a function of their moisture content. Instruments that take advantage of this fact use a balanced bridge circuit to detect these resistance changes (Figure 1.40p).

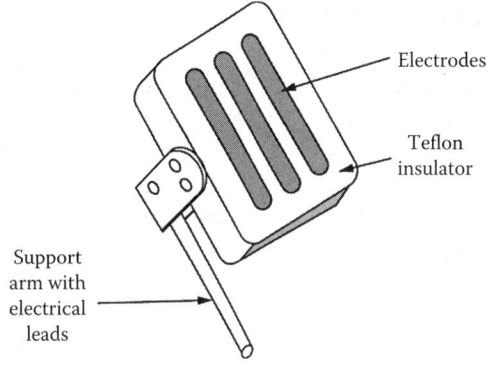

FIG. 1.40p
Resistance measuring head.

In this design, the measuring electrodes form a part of the bridge circuit as they contact the process solids. A regulated voltage is applied to the electrodes so that a small current flow is maintained through the solids. The changes in the moisture level result in a change of electrode current. Therefore, the amount of bridge imbalance is related to the moisture content. This probe also detects the capacitive reactance caused by the moisture level changes, but this portion of the measurement is continuously balanced by servomotor-driven reference capacitor.

Limitations

The accuracy of this measurement is strongly influenced by the contact pressure between sample and electrodes, the operating temperature, packing density, and particle size. In addition, the resistance values that are measured are very high, so good insulation is required at the electrodes to prevent leakage currents from introducing an error.

IMPEDANCE MOISTURE GAUGE

The moisture content solids can also be measured indirectly by detecting the moisture in the atmosphere above or near the process solids. The atmosphere near the solids is in equilibrium with the moisture content of the process materials. Therefore, several of the instruments discussed in Chapters 1.38 and 1.39 can be used in this type of measurement.

In particular, the change-of-impedance type of hygrometer is well suited, because its sensing element is small and can be easily mounted close to the solids surface (see Figure 1.39h). However, for this measurement to be effective, the element must be situated so that no drafts or convection currents occur, because even small air movement will influence the moisture measurement.

NUCLEAR MAGNETIC RESONANCE

Nuclear magnetic resonance (NMR) for the determination of moisture in solids is not limited only by laboratory use in large and complex experimental devices operated by skilled staff. On the market, there are compact, portable, fast, and accurate instruments designed specifically to measure the moisture in solids.

The following theoretical basis and the principles of NMR apparatuses for measuring moisture in solids are very simplified, but they are quite sufficient to understand the benefits of this method.

Hydrogen nucleus of water molecule rotates around its axis, creating a nuclear magnetic moment (for better imagination, the nucleus behaves like a bar magnet). If the nucleus is inserted into uniform magnetic field of permanent magnet with magnetic induction B_0, the magnetic moment of nucleus is oriented and the so-called precession motion around the axis of the magnetic field origins. If these atoms are further exposed to perpendicular alternating magnetic field with magnetic induction B_1, it changes the orientation of the magnetic moment associated with the absorption of energy. However, since the magnetic moment rotates against the magnetic field of permanent magnet, the position is thermodynamically unstable. Part of this energy is returned to the lattice and magnetic moment of nucleus returns to its original orientation. Absorption and return of high-frequency energy is a continuous process. Then the absorption characteristics of the molecule are dependent on the frequency of the alternating magnetic field. The biggest absorption is in the vicinity of resonance. The result of NMR measurement is the absorption spectrum dependent on the frequency.

Moisture meters based on NMR are dedicated only to the measurement of resonance of hydrogen nuclei (referred to as ^{1}H-NMR). For this reason, they do not measure directly the water concentration, but the concentration of hydrogen nuclei. Unlike neutron probes, NMR can distinguish between hydrogen atoms in the material and hydrogen atoms in water. NMR of hydrogen nuclei is directly dependent on the influence of their environment, which is reflected in the shape of the measured NMR spectrum.

NMR moisture meters consist of three basic parts: a permanent magnet, high-frequency source, and detection system. Permanent magnet (magnetic induction up to 2 T) is usually combined with sweep coils modifying magnetic induction B_0. This combination, permanent magnet/sweep coils, allows studying the signal in the vicinity of hydrogen resonance. The disadvantage of permanent magnets is their weight.

The sample is inserted into the high-frequency excitation coil, which also serves for detection. The coil is a part of the parallel resonant circuit, which changes its conductivity as a result of hydrogen nuclei resonance (Figure 1.40q).

NMR moisture meters are very accurate and currently also compact. They are used for the determination of

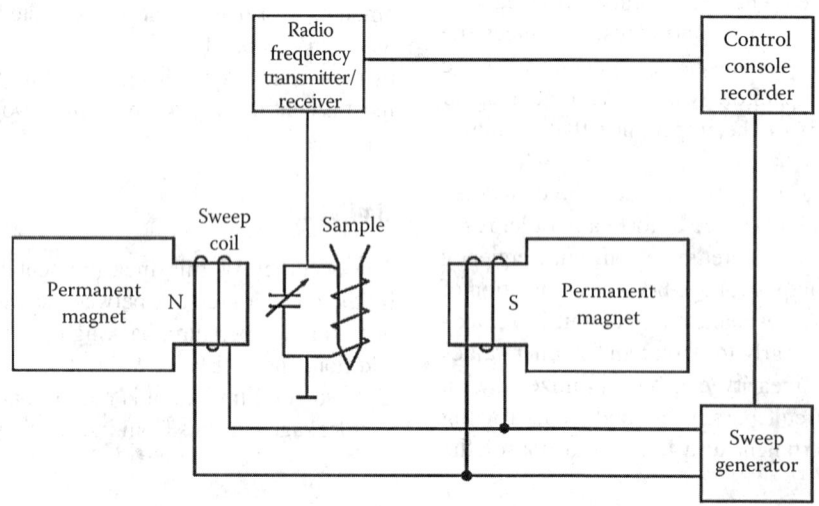

FIG. 1.40q
Simplified scheme of NMR moisture meter.

moisture in food such as cereals, flour, starch, dried fruit, etc. Some limitation is in the measurement of moisture in food with a high fat content, because fat hydrogen atoms are poorly distinguishable from the hydrogen atoms of water.

RADIO FREQUENCY ABSORPTION

Similar to microwaves, the attenuation of RF radiation by the process solids can also be used to detect moisture. The water molecules absorb the RF energy in the form of molecular motion. By means of special electrodes, portable units have been marketed for sheet-paper moisture measurements. Online, continuous moisture analyzers are also available for paper, food, chemical, gypsum wallboard, lumber, and other flat products. The analyzer must be calibrated to the particular product under test.

High-temperature RF moisture transmitters are also available for services up to 260°C (500°F). The analyzer can be installed inside dryers or kilns and is capable of continuous operation in contaminated atmospheres. All electronic components are mounted outside the kiln.

CONCLUSIONS

This text introduced a range of methods and analyzers for measuring of moisture in solids. It is not complete and comprehensive overview. Readers with deeper interest may find a number of other but less used or even bizarre methods and analyzers in the bibliography at the end of this chapter. The aim of this chapter, however, is to describe in particular those which are used in industrial practice, and are readily available in the form of commercial devices.

It can be concluded that none of the above-described analyzers can be universal and it is always necessary to consider what type of materials or products will be measured and under what conditions. Manufacturers make selection easier, because they produce usually highly specialized devices for a particular type of material or product. There are moisture balances for specific foods or paper, neutron probes and TDR probes to measure soil moisture, capacitance handhelds for paper and textiles, resistance handhelds for wood, etc.

In terms of accuracy and selectivity, NMR seems to be one of the best methods. This originally laboratory method has taken the form of relatively simple and affordable tabletop systems for direct measurement of moisture in solids. Similarly, neutron probes find application in continuous measurement of processes, although there are some legal limitations and sometimes problems with selectivity. On the other side, there are plenty of inexpensive simple handheld test instruments based on the measurement of electrical resistance or capacity or simple moisture balances with longer evaluation. Devices working with infrared radiation or microwave are ideal for belt conveyors and where it is necessary to evaluate the moisture in materials with complex texture.

The number of companies engaged in the production of analyzers for measuring of moisture in solids (see partial list at the beginning of chapter) shows that this is an important issue and therefore potential customer has a wide choice of models.

SPECIFICATION FORMS

When specifying moisture in solids detectors, one cam use the ISA form 20A1001 and when specifying both the analyzer and the requirements of the process application, use the ISA form 20A1002. Both forms are reproduced with the permission of the International Society of Automation on the next pages.

Analytical Measurement

Analysis Device — Operating Parameters

#	RESPONSIBLE ORGANIZATION		#	SPECIFICATION IDENTIFICATIONS	
1	ISA		6		
2			7	Document no	
3			8	Latest revision	Date
4			9	Issue status	
5			10		

ADMINISTRATIVE IDENTIFICATIONS

12	Project number		Sub project no	
13	Project			
14	Enterprise			
15	Site			
16	Area		Cell	Unit
17				

SERVICE IDENTIFICATIONS

19	Tag no/Functional ident		
20	Related equipment		
21	Service		
22			
23	P&ID/Reference dwg		
24	Process line/nozzle no		
25	Process conn pipe spec		
26	Process conn nominal size	Rating	
27	Process conn termn type	Style	
28	Process conn schedule no	Wall thickness	
29	Process connection length		
30	Process line matl type		
31	Fast loop line number		
32	Fast loop pipe spec		
33	Fast loop conn nom size	Rating	
34	Fast loop conn termn type	Style	
35	Fast loop schedule no	Wall thickness	
36	Fast loop estimated lg		
37	Fast loop material type		
38	Return conn nominal size	Rating	
39	Return conn termn type	Style	

SERVICE IDENTIFICATIONS Continued

41	Return conn matl type			
42	Inline hazardous area cl		Div/Zon	Group
43	Inline area min ign temp		Temp ident number	
44	Remote hazardous area cl		Div/Zon	Group
45	Remote area min ign temp		Temp ident number	

COMPONENT DESIGN CRITERIA

49	Component type		
50	Component style		
51	Output signal type		
52	Characteristic curve		
53	Compensation style		
54	Type of protection		
55	Criticality code		
56	Max EMI susceptibility		Ref
57	Max temperature effect		Ref
58	Max sample time lag		
59	Max response time		
60	Min required accuracy		Ref
61	Avail nom power supply		Number wires
62	Calibration method		
63	Testing/Listing agency		
64	Test requirements		
65	Supply loss failure mode		
66	Signal loss failure mode		

PROCESS VARIABLES — MATERIAL FLOW CONDITIONS

#	Variable				Units
70	Flow Case Identification				
71	Process pressure				
72	Process temperature				
73	Process phase type				
74	Process liquid actl flow				
75	Process vapor actl flow				
76	Process vapor std flow				
77	Process liquid density				
78	Process vapor density				
79	Process liquid viscosity				
80	Sample return pressure				
81	Sample vent/drain press				
82	Sample temperature				
83	Sample phase type				
84	Fast loop liq actl flow				
85	Fast loop vapor actl flow				
86	Fast loop vapor std flow				
87	Fast loop vapor density				
88	Conductivity/Resistivity				
89	pH/ORP				
90	RH/Dewpoint				
91	Turbidity/Opacity				
92	Dissolved oxygen				
93	Corrosivity				
94	Particle size				

CALCULATED VARIABLES

97	Sample lag time	
98	Process fluid velocity	
99	Wake/natural freq ratio	

PROCESS DESIGN CONDITIONS

#	Minimum	Maximum	Units
102			
103			
...			
132			

MATERIAL PROPERTIES

#	Field	Value	At
134	Name		
135	Density at ref temp		At
136			

MATERIAL PROPERTIES Continued

#	Field	Value	Flammability	Reactivity
138	NFPA health hazard			
139				
140				

Rev	Date	Revision Description	By	Appv1	Appv2	Appv3	REMARKS

Form: 20A1001 Rev 0 © 2004 ISA

	RESPONSIBLE ORGANIZATION	ANALYSIS DEVICE COMPOSITION OR PROPERTY Operating Parameters (Continued)		SPECIFICATION IDENTIFICATIONS		
1			6	Document no		
2	ISA		7	Latest revision		Date
3			8	Issue status		
4			9			
5			10			

	PROCESS COMPOSITION OR PROPERTY			MEASUREMENT DESIGN CONDITIONS				
	Component/Property Name	Normal	Units	Minimum	Units	Maximum	Units	Repeatability
11								
12								
13								
14								
15								
16								
17								
18								
19								
20								
21								
22								
23								
24								
25								
26								
27								
28								
29								
30								
31								
32								
33								
34								
35								
36								
37								

(Rows 38–75: blank)

Rev	Date	Revision Description	By	Appv1	Appv2	Appv3	REMARKS

Form: 20A1002 Rev 0

© 2004 ISA

Definitions

Desiccator usually glass sealable vessel containing desiccant

Dipole moment the measure of the electrical polarity of a system of charges

Gamma radiation electromagnetic radiation of high frequency ($>10^{19}$ Hz)

Geiger–Müller tube detector of ionizing radiation based on avalanche phenomenon

Impedance complex ratio of the voltage to the current in an alternating current

Infrared radiation electromagnetic radiation of frequency range of between 430 THz and 300 GHz

Microwave radiation electromagnetic radiation of frequency range between 300 MHz and 300 GHz

Neutron subatomic particle with no electric charge

Pyrometer noncontact thermometer

Reagent substance or compound that is added to a system in order to bring about a chemical reaction, or added to see if a reaction occurs

Scintillation detector detector of ionizing radiation based on luminescence

Sorption isotherm describes the equilibrium of the sorption of a material at a surface at constant temperature

Teflon brand name of polytetrafluoroethylene

Titration common laboratory method where known concentration and volume of titrant reacts with a solution of analyte to determine concentration

Abbreviations and Symbols

°C	Degrees of Celsius, unit of temperature
°F	Degrees of Fahrenheit, unit of temperature
µg	Microgram (= 10^{-6} kg), unit of mass
^{1}H-NMR	Proton nuclear magnetic resonance
AC	Alternating current
Ag	Silver, chemical element
Am	Americium, chemical element
Au	Gold, chemical element
B	Boron, chemical element; magnetic field
Be	Beryllium, chemical element
BF$_3$	Boron trifluoride
Br	Bromine, chemical element
c_0	Speed of wave propagation in vacuum
Cd	Cadmium, chemical element
Cl	Chlorine, chemical element
cm	Centimeter (= 0.01 m), unit of length
Co	Cobalt, chemical element
Cs	Cesium, chemical element
DC	Direct current
dm	Cubic decimeter, unit of volume
eV	Electron volt, unit of energy
G	Giga, prefix meaning 10^9
He	Helium, chemical element
Hf	Hafnium, chemical element
Hg	Mercury, chemical element
Hz	Hertz, unit of frequency
in.	Inch, unit of length
In	Indium, chemical element
Ir	Iridium, chemical element
K	Kelvin, unit of temperature; potassium, chemical element
l	Length
Li	Lithium, chemical element
LOD	Loss on Drying
m	Mass
mA	Milliampere (= 0.001 A), unit of electric current
mg	Milligrams (= 0.001 g), unit of mass
micron	Micrometer (= 10^{-6} m), unit of length
mm	Millimeter (= 0.001 m), unit of length
NMR	Nuclear Magnetic Resonance
P	Phosphorus, chemical element
ppm	Parts per million
Pu	Plutonium, chemical element
R	Radius
Ra	Radium, chemical element
Re	Rhenium, chemical element
RF	Radio Frequency
Rh	Rhodium, chemical element
t	time
T	Tesla, unit of magnetic induction
TDR	Time domain reflectometry
TGA	Thermogravimetry
V	velocity; volt, unit of electric voltage

Bibliography

Amoodeh, M. T., Khoshtaghaza, M. H., Minaei, S. 2006. Acoustic on-line grain moisture meter. *Computers and Electronics in Agriculture* 52(1–2): 71–78.

Austin, J., Rodriguez, S., Sung, P.-F. 2013. Utilizing microwaves for the determination of moisture content independent of density. *Journal Powder Technology*, 236(17–23).

Ayalew, G., Ward, S. M. 2000. Development of a prototype infrared reflectance moisture meter for milled peat. *Computers and Electronics in Agriculture* 28(1): 1–14.

Ballard, L. F. 1973. Instrumentation for Measurement of Moisture; Literature Review and Recommended Research, Highway Research Board, National Research Council.

Botsco, R. May 1970. *Microwave Moisture Measurement*, Instrumentation and Control Systems.

Bouraoui, M., Richard, P., Fichtali, J. 1993. A review of moisture content determination in foods using microwave oven drying. *Food Research International* 26(1): 49–57.

Cancilla, P., Barrette, P., Rosenblum, F. 2003. On-line moisture determination of ore concentrates: A review of traditional methods and introduction of a novel solution. *Minerals Engineering* 16(2): 151–163.

Černý, R. 2009. Time-domain reflectometry method and its application for measuring moisture content in porous materials: A review. *Measurement* 42(3): 329–336.

Dobriyal, P., Qureshi, A., Badola, R. 2012. A review of the methods available for estimating soil moisture and its implications for water resource management. *Journal of Hydrology* 458–459: 110–117.

Dubois, R., van Vuuren, P., Tatera, J. 2002. New sampling sensor initiative, An Enabling Technology, *47th Annual ISA Analysis Division Symposium*, Denver, CO, April 14–18, 2002.

Feber, G. Digital-to-analog converter with neutron probe or measuring moisture, Hutní Listy (Czechoslovakia), September 1982.

Fexa J., Široký K. 1983. *Měřenívlhkosti*. Praha, Czech Republic: SNTL.

Fussell, E. August 2001. Molding the future of process analytical sampling. *InTech*, p. 32.

Gardner, R. C. January 1968. Moisture/basis weight infrared gage for paper. *InTech*.

Garverick, L. and Senturia, S. D. January 1982. MOS device for AC measurement of surface impedance with application to moisture monitoring. *IEEE Transactions on Electron Devices* 29(1): 90–94.

Geary, P. J. *Measurement of Moisture in Solids: A Survey Based on Scientific and Technical Literature*. Chislehurst, U.K.: Sira Institute (ASIN 0900223006).

James, T. July 1982. Instrumentation in steelworks for coke moisture measurement. *Steel Times*.

Jin, X., Van der Sman, R. G. M., Gerkema, E. 2011. Moisture distribution in broccoli: Measurements by MRI hot air drying experiments. *Procedia Food Science* 1: 640–646.

Kizito, F., Campbell, C. S., Campbell, et al. 2008. Frequency, electrical conductivity and temperature analysis of a low-cost capacitance soil moisture sensor. *Journal of Hydrology* 352(3–4): 367–378.

Kohler, H. M. and Mathew, A. 1996. Continuous in-situ and elevated temperature moisture measurement in high particulate reactive processes, ISA 1996 Conference Paper #96–002.

Lee, L. Y. May 1984. On-line moisture analyzers. *InTech*.

McKinley, J. J. July 1997. Using permeation tubes to create trace concentration moisture standards. *National Conference of Standards Laboratories Conference*, Atlanta, GA.

McMahon, T. K. August, 2001. The new sampling/sensor initiative. *Control*.

Mettes, J. et al. October 1992. Multipoint PPM moisture measurement using electrolytic cells. *ISA Conference*, Houston, TX.

Mitchell, J., Milton, S. 1977. *Aquametry*, 2nd edn., Part 1, New York: John Wiley & Sons.

Nyström, J., Dahlquist, E. 2004. Methods for determination of moisture content in woodchips for power plants—A review. *Fuel* 83(7–8): 773–779.

Paper and Board Moisture Measurement. June 1983. Report PB83–866863, NTIS, Springfield, VA.

Pyper, J. W. 1985. The determination of moisture in solids. *Analyticachimicaacta* 170: 159–175.

Rezaei, M., Ebrahimi, E., Naseh, S. 2012. A new 1.4-GHz soil moisture sensor. *Measurement* 45(7): 1723–1728.

Salgado et al. 1982. *Neutron Gage for Coke Moisture Measurement in the Steel Industry*. Vienna, Austria: International Atomic Energy Agency.

Scelzo, M. J. February 1992. The trend is toward processing and measurement at ultralow moisture levels, *Measurements and Control*.

Sherman, R. E. 2002. *Process Analyzer Sample-Conditioning System Technology*. New York: John Wiley & Sons.

Sinibaldi, F. J. and Wenner, S. R. March 3, 1982. Can a TAPPI bore test measure moisture accurately? *TAPPI* 65: 103.

Staub, M. J., Gourc, J.-P., Laurent, J.-P. et al. 2010. Long-term moisture measurements in large-scale bioreactor cells using TDR and neutron probes. *Journal of Hazardous Materials* 180(1–3): 165–172.

Trabelsi, S., Nelson, S. O. 2006. Nondestructive sensing of bulk density and moisture content in shelled peanuts from microwave permittivity measurements. *Food Control* 17(4): 304–311.

Van den Berg, F. W. J., Hoefsloot, H. C. J., Smilde, A. K. 2002. Selection of optimal process analyzers for plant-wide monitoring. *Analytical Chemistry* 74(13): 3105–3111.

Van der Heijden, G. H., Huinink, H. P., Pel, L. 2011. One-dimensional scanning of moisture in heated porous building materials with NMR. *Journal of Magnetic Resonance* 208(2): 235–242.

Whiting, D., Nagi, M. March 1997. Evaluation of Troxler Model 4430 Water/Cement Gauge. Final report, American Society of Civil Engineers.

1.41 Molecular Weight of Liquids

A. BRODGESELL (1969, 1982) **B. G. LIPTÁK** (1995, 2003, 2017)

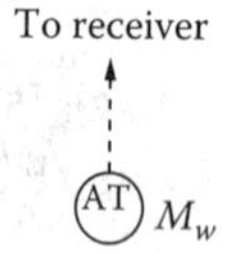

To receiver

(AT) M_w

Flow sheet symbol

Applications types	Measurement of the molecular weights of polymers and of other larger molecules
	A. Membrane osmometers
	B. Vapor pressure osmometers
	C. Light-scattering photometers
	D. Viscometers (also see Chapters 1.71 and 1.72)
	E. Liquid chromatographs (HPLC), gel permeation (also see Chapter 1.11)
	F. End group determination (nuclear magnetic resonance—NMR)
	G. Electron microscopy (also see Chapter 1.53)
	H. Ultracentrifuge sedimentation
Design pressure	Atmospheric
Design temperature	Usually up to 150°C (300°F)
Element material	Glass, Kel-F, gel–cellophane, stainless steel
Inaccuracy	5%–10% AR
Range	Molecular weights of 50 and higher
Costs	A, B, C. About $10,000.
	D. See Chapters 1.71 and 1.72. Mooney viscometers, processability testers, and rubber analyzers cost more, in the range of $25,000–$80,000.
	E. See Chapter 1.11, about $50,000.
	G. See also Chapter 1.53, about $100,000.
	H. Laboratory centrifuges, from $3,000 to $10,000.
Partial list of suppliers	For liquid chromatographs, see Chapter 1.11, and for viscometers, see Chapters 1.71 and 1.72.
	Advanced Instruments Inc. (A, B) http://www.aicompanies.com/index.cfm/products/?productId=1
	Beckman (H) (www.astbury.leeds.ac.uk)
	Beckman Coulter (E, H) (www.beckman.com) https://www.beckmancoulter.com/wsrportal/search/Liquid%20chromatographs/#2/10//0/25/1/0/asc/1/Liquidchromatographs///0/1//0/
	Bourevestnik Inc. (A, B) http://www.bourevestnik.ru/eng/Products/Analytical/mt5.html
	Brookfield Engineering Laboratories Inc. (D) (www.brookfieldengineering.com) http://www.brookfieldengineering.com/products/viscometers/index.asp
	Brookhaven Instruments (C) http://www.brookhaveninstruments.com/products/molecular_weight/p_MW_BI-MwA.html
	Cole-Parmer (E) http://www.coleparmer.com/Category/High_Performance_Liquid_Chromatography_HPLC/50089
	C.W. Brabender Instruments Inc. (D) (www.cwbrabender.com)
	Daniel (E) http://www.southeastern-automation.com/PDF/Emerson/Chromatograph%20Notes/NGL%20Fractionation%20Part%202%2027KB.pdf
	Dennis Kunkel Microscopy (G) (www.denniskunkel.com)
	Electron Microscope Unit (G) http://www.emunit.hku.hk/#/Home
	FEI Co. (G) (www.feicompany.com)
	Gonotec GmbH (freezing point osmometer) http://www.gonotec.com/products/osmomat-3000

(Continued)

Gynkotek HPLC Inc. (E) http://www.geminibv.nl/labware/gynkotek-hplc-pomp-p580a/gynkotek-hplc-pomp-p580-manual-eng.pdf
Hewlett-Packard (E) http://www.hpl.hp.com/hpjournal/94aug/aug94a4aa.pdf
Hitachi (E) http://www.hitachi-hta.com/products/life-sciences-chemical-analysis/liquid-chromatography?gclid=CJHuxYDZ67cCFRSi4AodeVsAeA
HPLC Technology Co. (E) (www.hplc.co.uk)
Knauer GmbH (A, B) http://www.knauer.net/en/products/product-types/osmometers.html
Labcompare (E) http://www.labcompare.com/Laboratory-Analytical-Instruments/35-Liquid-Chromatography-HPLC/
LEO Electron Microscopy (G) https://mai.ku.edu/Microscopy_and_Analytical_Imaging_Lab/LEO_SEM.html
Millipore S.A., EMD (E) http://www.millipore.com/catalogue/item/16–228
Pall Corp. (H) http://www.pall.com/main/laboratory/literature-library-details.page?id=40883
PerkinElmer Corp. (C) http://www.perkinelmer.com/technologies/molecular-spectroscopy/default.xhtml
Regis Technologies Inc. (E). (http://www.registech.com/chromatography)
Scanning Electron Microscope (G) (www.usyd.edu.au/sv/emu)
Siemens Energy & Automation (E) (http://news.usa.siemens.biz/press-release/gas-chromatograph/siemens-enhances-maxum-gas-chromatograph-color-touch-screen-display)
Sigma-Aldrich (F) http://www.sigmaaldrich.com/materials-science/polymer-science/polymer-analysis.html
Thermo Scientific (E, H) (http://www.dionex.com/en-us/products/liquid-chromatography/lp-71340.html)
Thomson Instrument (E) (http://www.htslabs.com/catalog/?d=flash) http://www.hplc.com/Shodex2/index_en.html
Waters Corp. (E) (http://www.waters.com/waters/en_HU/ACQUITY-Advanced-Polymer-Chromatography-System/nav.htm?cid=134724426&locale=en_HU)
Wyatt Technology Corp. (C & D) (http://www.wyatt.com/products/hardware/instruments.html) http://www.wyatt.com/products/hardware/viscostarii-viscometer.html
Yamazen (E) (http://yamazenusa.com/wp-content/themes/openair/images/green_flash_chromatography.pdf)

INTRODUCTION

The measurement of the molecular weight of gases and vapors has already been discussed in the chapters describing: Gas Chromatographs—Chapter 1.10), Hydrocarbon Analyzers—Chapte 1.28, IR, UV—Chapters 1.32 and 1.69 and several other chapters.

Therefore, this chapter will concentrate only on the molecular weight of liquids and particularly that of polymers. Some of the instruments used in the measurement of such larger molecules, are also discussed in chapters that are describing viscometers, mass spectrometers, and liquid chromatographs (Chapters 1.11, 1.35, 1.71, and 1.72).

Molecular weight measurement of gases and vapors has already been discussed in the chapters on density (Chapter 128), on gas chromatographs (Chapter 1.10), and on infrared (Chapter 1.10), UV (Chapter 1.69), spectrometers (Chapter 1.62), and several other chapters. Therefore, this section will concentrate only on the measurement of the molecular weight of liquids and particularly that of polymers. Some of the instruments used in the measurement of the larger molecules, such as viscometers, mass spectrometers, and liquid chromatographs, have already been covered in Chapters 1.11, 1.35, 1.71, and 1.72, so the reader is referred to those chapters.

POLYMERS

The word polymer has been derived from the Greek words 'poly' and 'meros', implying many parts, and refers to a characterizing feature of polymeric materials – namely their chain like structure. This structure is formed by developing chemical links between a number of monomers. For instance, polymerizing the monomer styrene, results in the polymer polystyrene (Figure 1.41a).

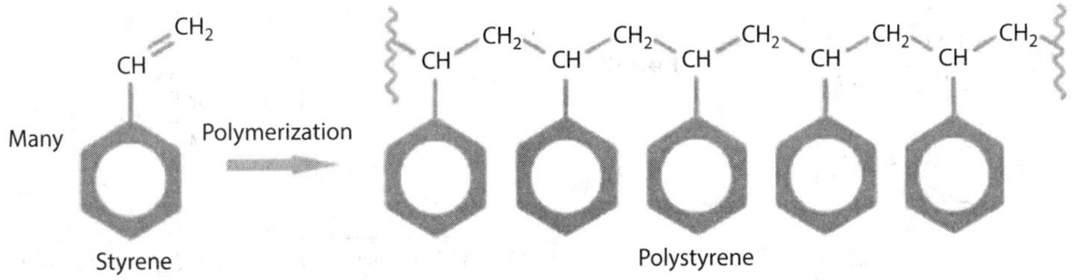

FIG. 1.41a
Polymers have a chain-like structure, made up of repeating monomers. (Courtesy of Malvern Instruments Ltd.)

Polymers consist of large molecules formed by the bonding of relatively simple and similar parts. The size and shape of the polymer molecules are determined by the number of basic building blocks and by the way these are linked together. For example, the basic blocks might be arranged in chains of various lengths, in chains with branches, or in two chains linked in several places. Thus, while the molecular weight of the basic building blocks might be known, polymers are a mixture of molecular species covering a wide range of sizes, and no definite molecular weight can be assigned to them.

However, an average molecular weight of this molecular weight distribution can be determined and is useful in predicting the physical properties of the polymer. Among the methods available for the determination of the average molecular weight are osmometry, light scattering, viscometry, gel-permeation chromatography (GPC), end group determination (nuclear magnetic resonance [NMR]), electron microscopy, ultracentrifuging, sedimentation, and diffusion.

Each of these methods yields an average that must be defined, since the averages obtained are not identical, as is illustrated by the following example.

EXAMPLE

Consider 10 lb of polymer consisting of the following:
 5 lb of molecular weight 10,000
 3 lb of molecular weight 20,000
 2 lb of molecular weight 120,000
The average of the weight fractions yields

$$10\bar{M} = 5 \times 10,000 + 3 \times 20,000 + 2 \times 120,000 \quad \bar{M} = 35,000 \quad 1.41(1)$$

The average molecular weight obtained is called the weight average and is designated by M_w.

The average of the mole fraction yields

$$\frac{10}{\bar{M}} = \frac{5}{10,000} + \frac{3}{20,000} + \frac{2}{120,000} \quad \bar{M} = 15,000 \quad 1.41(2)$$

This average is called the *number average molecular weight* and is designated by M_n.

The averages obtained are not identical. The weight average emphasizes the large molecules, and the number average the small ones. For this reason, an average molecular weight must always be defined. In general, M_w and M_n are defined as

$$M_w = \frac{\Sigma W_i M_i}{\Sigma W_i} = \frac{\Sigma n_i M_i^2}{\Sigma n_i M_i} \quad 1.41(3)$$

$$M_n = \frac{\Sigma W_i}{\Sigma n_i} = \frac{\Sigma n_i M_i}{\Sigma n_i} \quad 1.41(4)$$

where
 n_i represents the number of moles
 W_i is the weight of the fraction
 M_i is the molecular weight of molecules of size i

Additional averages can be defined as

$$M_z = \frac{\Sigma n_i M_i^3}{\Sigma n_i M_i^2} \quad 1.41(5)$$

and

$$M_v = \left(\frac{\Sigma n_i M_i^{a+1}}{\Sigma n_i M_i} \right)^{1/a} \quad 1.41(6)$$

where a is defined by the equation

$$[\eta] K (M_v)^a \quad 1.41(7)$$

where
 $[\eta]$ is the intrinsic viscosity of the polymer
 K is a constant

The definition of intrinsic viscosity is given under the discussion of viscometers by Equations 1.41(15) through 1.41(18).

ANALYZER DESIGNS

Osmometers

Osmotic pressure is defined as the pressure that must be applied to the solution to stop osmosis from the pure solvent. Osmotic pressure, π, is related to solution concentration and molecular weight of the solute as given by

$$\pi = \frac{RT}{M} c + Bc^2 + Cc^3 \quad 1.41(8)$$

where
 R is the ideal gas constant
 T is the absolute temperature
 c is the solution concentration
 B and C are the virial coefficients

The first term of the equation is the van't Hoff relationship* for ideal solutions. However, polymer solutions are anything but ideal, so at least the second term of Equation 1.41(8) can never be neglected. Because there are two unknowns, M and B, in the equation, at least two measurements are required. Equation 1.41(8) is therefore rewritten as

$$\frac{\pi}{c} = \frac{RT}{M} + Bc + Cc^2 \quad 1.41(9)$$

---

* van't Hoff factor is named after J.H. van't Hoff, is a measure of the effect of a solute upon properties such as osmotic pressure. The van't Hoff factor is the ratio between the actual concentration of particles produced when the substance is dissolved, and the concentration of a substance as calculated from its mass.

The measured values of π are plotted as π/c vs. c. The measurement points are linearly extrapolated to c = 0 where the terms containing the unknown virial coefficients go to 0. The value of M can then be calculated from the relationship given in

$$M = \frac{RT}{(p/C)_c} = 0 \qquad \textbf{1.41(10)}$$

The molecular weight M thus obtained is the number average molecular weight M_n.

Membrane Osmometer The membrane osmometer (Figure 1.41b) consists of two compartments separated by a semipermeable membrane. A pure solvent and a dilute polymer solution are introduced into the two compartments, respectively. The membrane acts as a filter to permit passage of solvent molecules but not that of the polymer. Migration of solvent molecules into the solution results in a change in pressure between the two compartments until osmotic pressure is reached. Osmotic pressure is defined as the pressure that must be applied to the solution to stop osmosis from the pure solvent. It is used to obtain the number average molecular weight.

Static osmometers are provided with a capillary connection to the solution compartment so that osmotic pressure can be read as hydrostatic head. The principal disadvantage of the static osmometer is the long time, which is required to reach equilibrium 1–2 hr and longer.

Automatic Osmometers Automatic osmometers (Figure 1.41c) are available, which determine the osmotic pressure dynamically. In these units, the rate of solvent flow through the membrane is measured as a function of externally applied pressure.

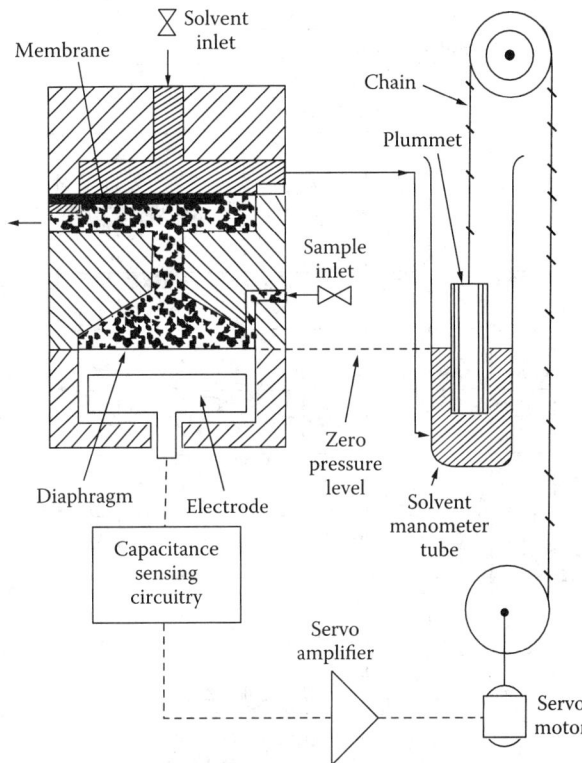

FIG. 1.41c
Automatic osmometer.

One type of automatic osmometer has a solvent reservoir mounted on a screw elevator and connected to the solvent side of the membrane by a capillary in which there is a small air bubble. A light source and photocell sense the position of the air bubble as it tends to move with any flow of solvent in the capillary. The photocell output activates a servomotor, which, in turn, positions the solvent reservoir to reduce pressure on the solvent compartment, thus stopping osmotic flow. As a result, the static head of the solvent is exactly opposed to the osmotic pressure of the solution.

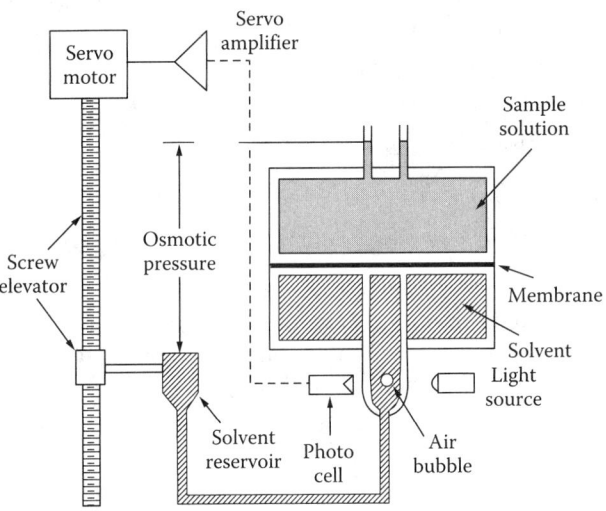

FIG. 1.41b
Membrane-type automatic osmometer.

The hydrostatic head on the solvent side can be continuously displayed in digital form, or the pressure can be recorded providing a permanent record of the measurement (Figure 1.41d).

Sample: Polystyrene	Temperature: 36°C		Solvent: Toluene
Density: 0.865 g/cm³ @27°C			
Concentration C/100 cr	0.5	1.5	2.0
Solvent pressure P_o cm	5.00	5.00	5.00
Sample pressure P cm	7.00	12.5	16.00
$\pi = P - P_o$	2.00	7.50	11.00
π/C	4.00	5.00	5.5

$$M_n = \frac{RT}{(\pi/c)\, c=0} = \frac{848 \times 309}{3.5 \times 0.865} = 86{,}000$$

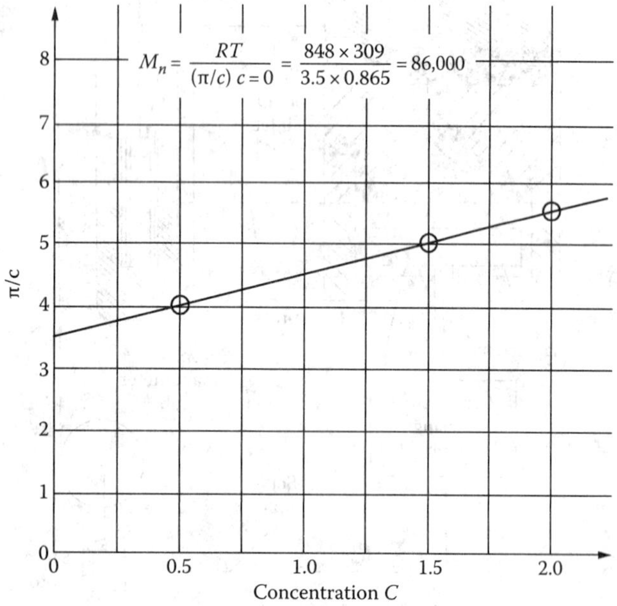

FIG. 1.41d
Membrane osmotogram.

Another automatic osmometer utilizes a diaphragm to sense an increased pressure in the solvent compartment, which results in a change of capacitance in an oscillator. A signal to a servomotor positions a plummet in the manometer to reduce the pressure on the solvent side of the membrane by an amount equal to the osmotic pressure.

Automatic osmometers are thermostatically controlled, because temperature enters into the measurement. With automatic osmometers, the operating temperature is a minimum of 6°C (10°F) above ambient to ensure good temperature control. However, for a given sample, the lowest possible operating temperature is selected to ensure longer membrane life and to inhibit possible corrosion.

Static osmometers are usually not provided with temperature control, so some means of maintaining a constant temperature must be provided. Usually, the measurement is made in a constant-temperature bath. The operating temperature depends on the melting point of the sample and the lowest temperature required to keep it in solution.

Vapor Pressure Osmometers The vapor pressure of any solvent is lowered by the addition of solute. If a pure solvent and a solution are placed in two containers and connected in a closed system, the pure solvent will evaporate and condense into the solution. The resultant temperature difference, due to the latent heat of evaporation of the solvent, can be detected as an indication of the molecular weight of the solute. In practice, the temperature difference is read as the output voltage of a resistance bridge using high-accuracy temperature sensors (Figure 1.41e).

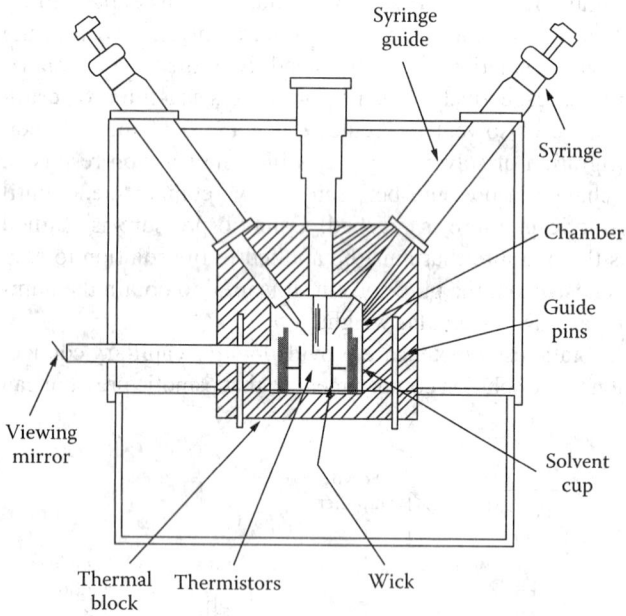

FIG. 1.41e
Vapor pressure osmometer.

The relationship between the bridge output voltage ΔV, the solution concentration c, and the average molecular weight M_n is given by

$$\left(\frac{\Delta V}{c}\right)_{c=0} = \frac{K}{M_n} \qquad 1.41(11)$$

where K is the calibration constant for the particular combination of solvent and type of thermistor used.

Although the amount of drop in vapor pressure is only a function of the amount of solute in solution and of the solute molecular weight, ΔV is a relative quantity depending on the solvent, operating temperature, and temperature-sensing element used. The instrument, therefore, must be calibrated with a known solution.

In practice, several values of ΔV are obtained at different concentrations, and these are plotted on a graph of $\Delta V/c$ vs. c (Figure 1.41f). The graph is then extrapolated linearly to $c = 0$ and the value of K determined. Once the value of K is known, M_n can be determined in the same manner for any polymer in the solution.

Sample: Fatty acid	Temperature: 35°C		Solvent: Toluene
Calibration factor $K = 350$			
Concentration C g/L	5	7.5	1.5
1. Output ΔV	1.40	2.15	4.35
2. Output ΔV	1.40	2.16	4.35
3. Output ΔV	1.41	2.14	4.35
ΔV average	1.40	2.15	4.35
$\Delta V/C$	0.280	0.287	0.290

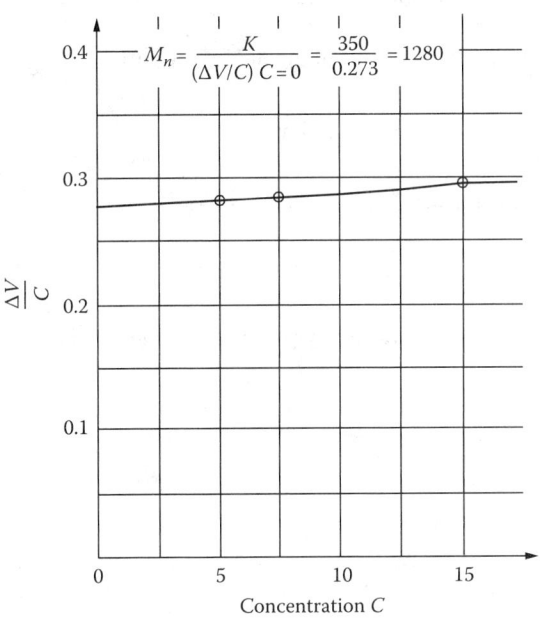

FIG. 1.41f
Vapor pressure osmotogram.

The vapor pressure osmometer consists of a thermostated chamber saturated with solvent vapor. Two thermistor beads are suspended in the chamber, and syringe guides are built into the chamber. Using the syringes, a drop of solvent and a drop of solution are placed on the reference and measuring thermistor, respectively.

As the solvent condenses, it warms the measuring thermistor. Therefore, its electrical resistance decreases. The thermistors are part of a resistance bridge whose output is in proportion to the temperature difference between measuring and reference thermistor. The resistance bridge output ΔV is plotted on a graph of $\Delta V/c$ vs. c.

The applicability of this instrument is limited because of the small difference in temperatures that need to be detected. Typically, for a 1% solution of 50,000 molecular weight polymer, the temperature difference is on the order of 0.0006°C (0.001°F).

Light-Scattering Photometers

Polymer molecules dissolved in a suitable solvent cause scattering of the incident light (Figure 1.41g). From the intensity and distribution angle of the scattered light, the weight average molecular weight, M_w, can be determined. When the molecules are small in comparison to the wavelength used, the scattered light has the same intensity for all angles of observation except for its polarization. For such molecules, a single observation at 90° to the incident light beam is sufficient.

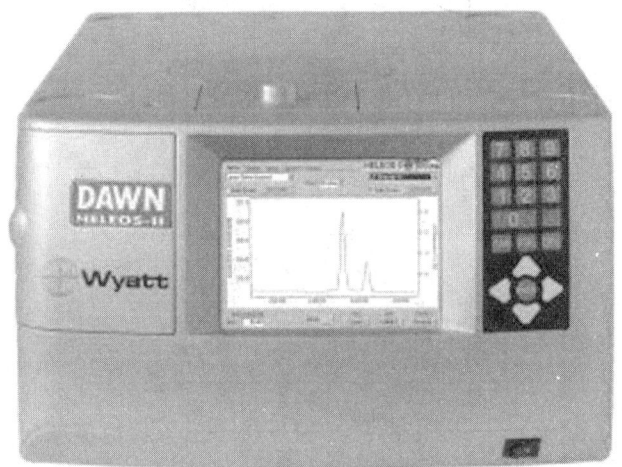

FIG. 1.41g
Light-scattering molecular weight detector used for characterization of proteins, conjugates, macromolecules, and nanoparticles. (Courtesy of Wyatt Technology Corp.)

If molecular size approaches the wavelength of light used, the light is scattered by portions of the molecule, which are widely separated. The light will undergo destructive interference, reducing the intensity of scattering. Destructive interference is a function of the light beam's angle. In other words, interference will be zero in the direction of the incident light and will increase with the angle of observation.

Zimm Plot

Zimm Plot The Zimm plot* (see Figure 1.41h) utilizes data taken over as large an angular range as possible and extrapolated to zero angle. The ordinate is intercepted at zero angle, when the concentration ($1/M_w$) is 0. The ordinate of the Zimm plot is K_c/R_θ and the abscissa is $\sin^2(\theta/2) + K_c$, where

$$K = \frac{2\pi^2 n^2 (dn/dc)}{A\lambda^4} \qquad 1.41(12)$$

where

- n is the index of refraction of the solution
- A is the Avogadro number
- dn/dc change of refractive index with concentration
- λ is the wavelength of light used

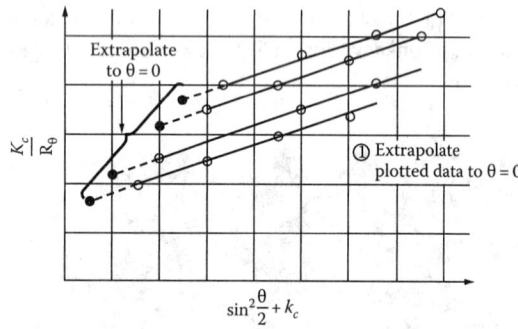

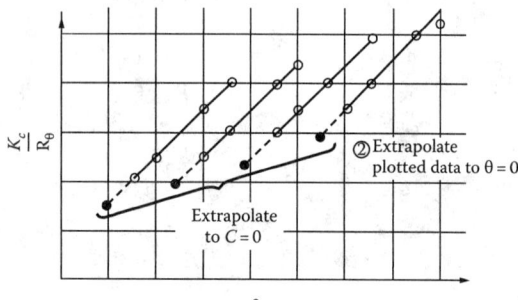

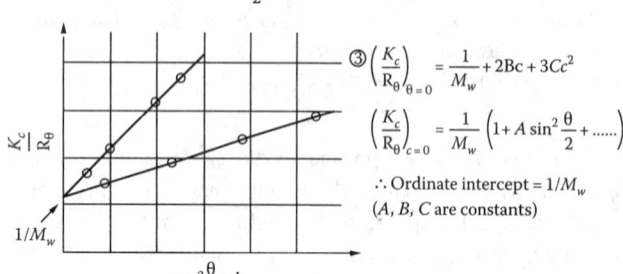

$$\left(\frac{K_c}{R_\theta}\right)_{\theta=0} = \frac{1}{M_w} + 2Bc + 3Cc^2$$

$$\left(\frac{K_c}{R_\theta}\right)_{c=0} = \frac{1}{M_w}\left(1 + A\sin^2\frac{\theta}{2} + \ldots\right)$$

∴ Ordinate intercept = $1/M_w$
(A, B, C are constants)

FIG. 1.41h
Zimm plot construction.

- θ is the angle of observation
- R_θ is the Rayleigh ratio† at the wavelength used for the fluid the sample cell is immersed in (ratio of intensities of scattered light to incident light)
- c is the solution concentration
- K is the arbitrary constant chosen to give a convenient spread between data points

To obtain the Zimm plot, four concentrations are usually prepared along with pure solvent, and the data are collected at a number of incident light angles (10 or more). The scattered intensity for a reference substance (I_s) is measured at 90°. The change in refractive index with concentration (dn/dc) is obtained with a differential refractometer.

Step 1: $I_\theta = I_s$ (calibration factor, which is supplied with instrument)

If the scattered light intensity (I_θ) of the material in which the cell is immersed is being measured, I_θ is obtained directly.

Step 2: Correct each measurement for solvent scattering,

$$I_\theta = I_\theta \text{ solution} - I_\theta \text{ solvent}$$

and correct for incident light polarization,

$$I \text{ corrected} - I_\theta \times \alpha$$

where $\alpha = \sin\theta$ for vertically polarized incident light and

$$\alpha = \frac{\sin\theta}{1 + \cos^2\theta}$$

for unpolarized incident light.
Step 3: Calculate K from Equation 1.41(12).
Step 4: Calculate the Zimm coordinates from the following equations:

$$\frac{K_c}{R_\theta} = \frac{K}{I \text{ corrected}} \times \frac{I_\theta}{R_\theta} \text{ (ordinate)} \qquad 1.41(13)$$

and

$$\sin^2\left(\frac{\theta}{2}\right) \times k_c \text{ (abscissa)} \qquad 1.41(14)$$

Photometer A narrow, high-intensity beam of light is generated by a light source, which usually is a mercury vapor lamp. It is provided with a filter to provide green light at 5461 Å or

---

* Zimm plot is a diagrammatic representation of data on light scattering by large particles. It is used for the simultaneous evaluation of the mass average molar mass, the second viral coefficient of the chemical potential, etc. Several modifications of the Zimm plot are in frequent use.

† Rayleigh ratio is the quantity used to characterize the scattered intensity at the scattering angle of the incident light or other forms of radiation.

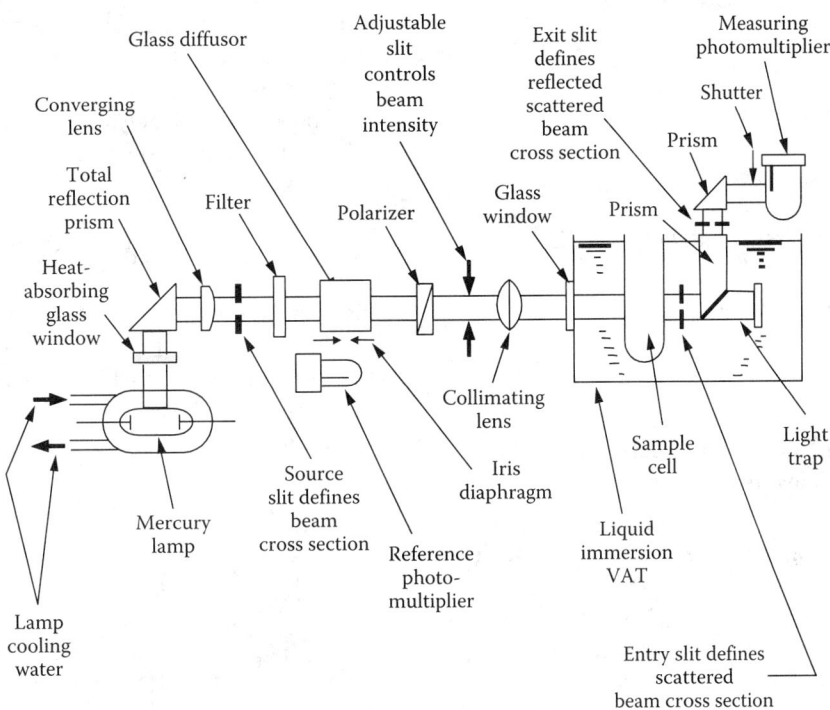

FIG. 1.41i
Light-scattering photometer.

blue light at 4358 Å (Figure 1.41i). This beam is focused on the sample cell, and light that is passing through the sample cell is absorbed in a light trap to avoid reflective interference with the measurement. The light, which is scattered by the sample, is converted by the measuring photomultiplier to an electrical current whose intensity is measured.

The photomultiplier is mounted on a platform, which can be rotated so that the viewing angle can be varied from 30° to 150° to the incident light beam. Since the light source intensity always varies somewhat, part of the generated light beam is permitted to impinge on a reference photomultiplier. The reference photomultiplier output can then be used to automatically compensate the measurement. Another way to perform the compensation is as follows: $I_\theta = i_\theta / i_s$, where I_θ represents the true measurement intensity, i_θ is the output of the measurement photomultiplier, and i_s is the output of the reference photomultiplier.

Design variations include the use of an immersion vat for the sample cell. This vat is filled with a fluid whose refractive index is the same as that of glass to eliminate refraction at the sample cell.

The sample for this instrument must be very carefully dedusted by filtering to remove all suspended particles. The use of a sample vat does not eliminate the requirement for removing all dust from the sample.

Viscometers

The various devices that are available for viscosity measurement are discussed in Chapters 1.70, 1.71, and 1.72.

To determine the molecular weight of a polymer from viscosity measurement, the polymer's intrinsic or limiting viscosity must first be determined (Figure 1.41j). Intrinsic viscosity $[\eta]$ is defined by the following relationships:

$$[\eta] = \lim_{c \to 0} \frac{\eta - \eta_0}{\eta_0 c} = \lim_{c \to 0} \frac{\eta_{sp}}{c} \qquad 1.41(15)$$

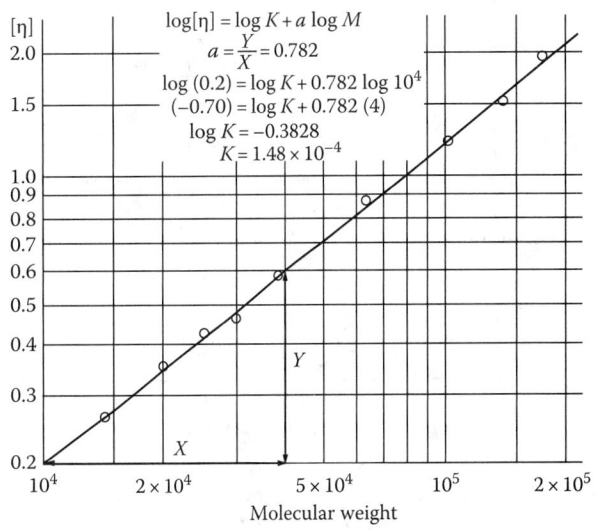

FIG. 1.41j
Viscometer calibration for K and a.

$$\eta_{sp} = \frac{\eta - \eta_0}{\eta_0} \qquad 1.41(16)$$

where

η is the viscosity of the solution
η_0 is that of the solvent
c is the solution concentration in g/mL or g/dL

Another definition of intrinsic viscosity is

$$[\eta] = \lim_{c \to 0} \frac{\ln \eta/\eta_0}{c} \qquad 1.41(17)$$

Intrinsic viscosity is related to the molecular weight in accordance with the Mark–Houwink equation

$$[\eta] = KM^a \quad \text{or} \quad \log[\eta] = \log K + a \log M \qquad 1.41(18)$$

where K and a are constants for a given polymer–solvent system at the temperature of the viscosity measurement.

Intrinsic Viscosity Viscosity measurements are made of a well-fractionated or monodisperse polymer whose molecular weight is known or has been measured by some other method (Figure 1.41k). The intrinsic viscosity is calculated by plotting $(\eta - \eta_0)/\eta_0$ vs. c and by extrapolating linearly to $c = 0$. If the plot of $(\eta - \eta_0)/\eta_0$ vs. c does not fall on a straight line, log η/η_0 can be plotted vs. c for better linearity.

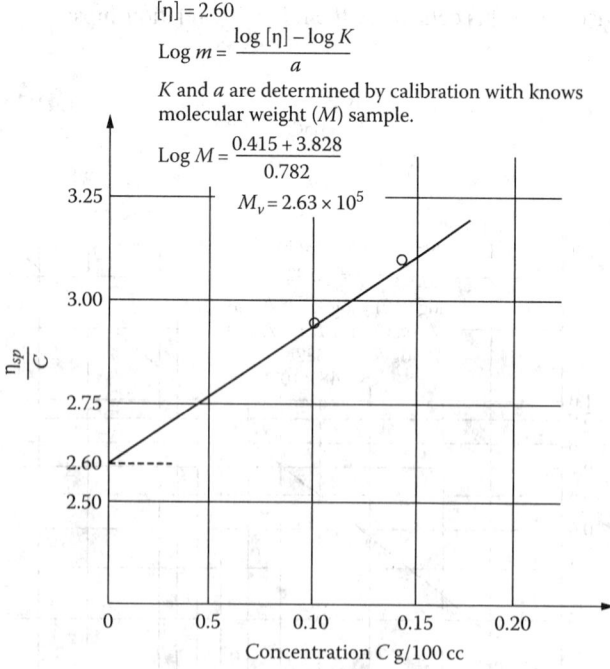

FIG. 1.41k
Determination of intrinsic viscosity and average molecular weight.

The values of intrinsic viscosity $[\eta]$ obtained are plotted as a function of the known molecular weight on log–log paper, and the constants a and K are evaluated. Once K and a are known, M_v for a polydisperse sample can be calculated.

The limitation of this method is that the empirical relationship $[\eta] = KM^a$ is valid only for linear polymers.

Gel-Permeation Chromatography

This technique is based on the chromatographic separation of molecules by size (Figure 1.41l). A solvent stream is split, and a polymer sample is added to one-half of the stream. The solution is directed into a column packed with a rigid, cross-linked styrene gel.

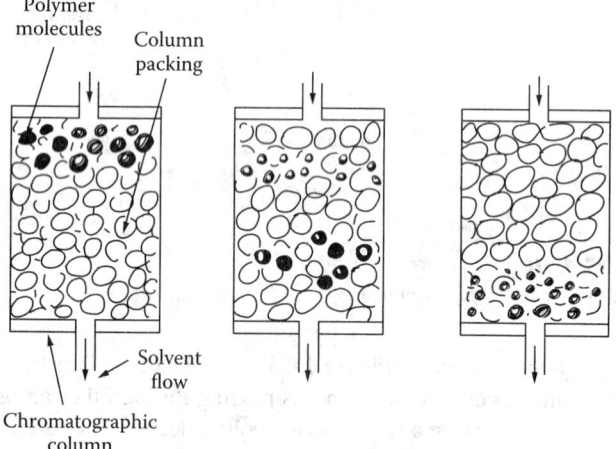

FIG. 1.41l
Three stages in the chromatographic separation of polymers: (left) at sample injection, (center) during separation, and (right) at sample elution.

As the polymer moves into the column, the smaller molecules diffuse into the gel pores, while the larger ones cannot penetrate and thus follow a shorter path. Molecules are therefore eluted from the column in the order of their sizes, with the smallest molecules eluting last (Figure 1.41m).

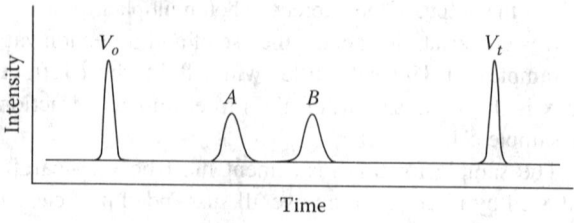

FIG. 1.41m
In a gel-permeation chromatography, the elution time relates to molecular weight, where V_o = no retention, V_t = complete retention, and A and B = partial retention.

1.41 Molecular Weight of Liquids

Differential Refractometer Sensor In the GPC, the solution and solvent are passed through the measuring and reference cells of a differential refractometer (see Chapter 1.59) in which the difference in refractive index between the sample and solvent is measured (Figure 1.41n). The output curve represents the relative abundance of molecules of a particular size from which a molecular weight distribution can be plotted and average molecular weights can be determined.

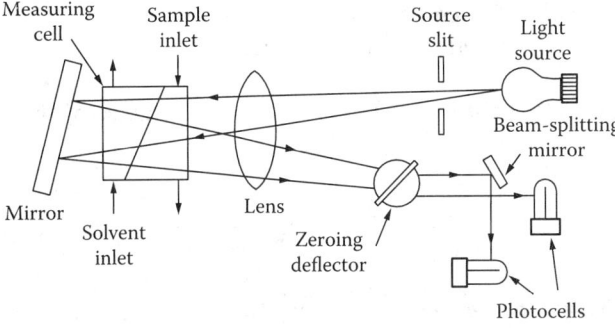

FIG. 1.41n
A schematic view of refractometer used with gel-permeation chromatography.

The differential refractometer utilizes a collimated light beam, which is passed through the reference and sample solutions and then is reflected back through both solutions by a mirror. The beam is split by another mirror, guaranteeing that equal amounts of light will fall on the two photosensors. The photosensor and two resistors form a resistance bridge whose output is proportional to the difference in the amounts of light falling on the photocells.

Changes in refractive index of the solution cause the light beam to shift so that unequal amounts of light will fall on the two photosensors. These refractive index variations are related directly to solution concentration. Thus, elution time and resistance bridge output will signal the relative abundance of molecules of a particular size.

Complete Instrument The instrument consists of a free-standing assembly containing the chromatograph portion refractometer, recorder, and control electronics.

The chromatograph portion consists of two sets of columns with column switching valves, sample injection valves, and solvent loop. Solvent from a reservoir is degassed in a heater, and flow is controlled by a positive displacement pump (Figure 1.11b).

The total solvent flow is split into two equal streams. One-half is passed through the sample loop, and the other half serves as the reference. Both the sample and solvent are passed through two column banks, respectively, before entering the refractometer cell. The solvent is then returned to the reservoir, and the sample is discharged to a sample collector. Each 5 mL increment of sample volume discharged is marked automatically on the recorder trace to indicate elution time.

The electronics consist of several subassemblies, including refractometer controls, automatic sample injection controls, and the main control assembly.

Molecular Weight from Chromatogram For the steps used in determining the molecular weight from the refractometer output curve, refer to Figures 1.41o through 1.41q.

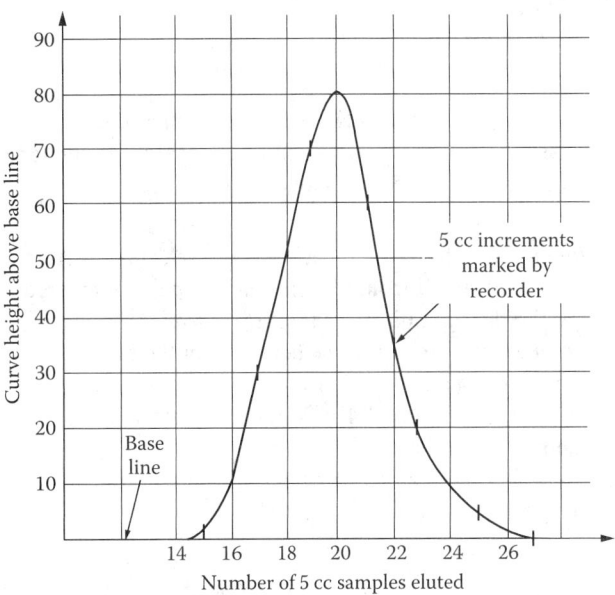

FIG. 1.41o
Chromatograph output distribution curve.

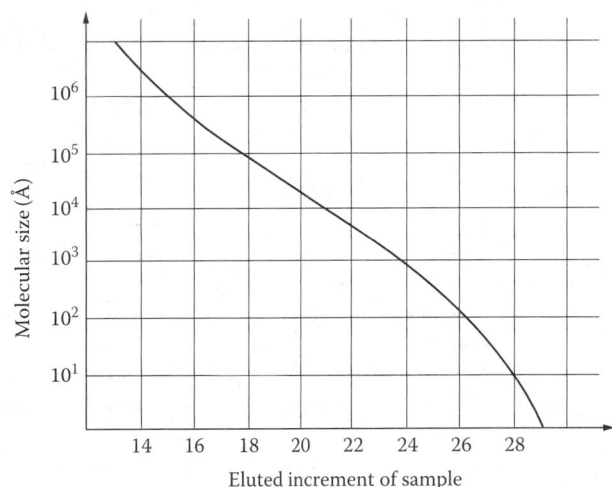

FIG. 1.41p
Calibration curve furnished with instrument.

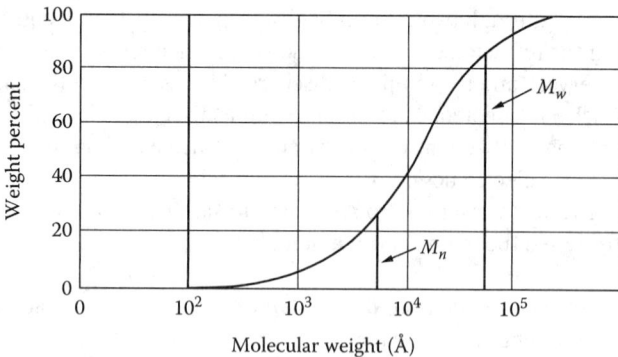

FIG. 1.41q
Weight percent cumulative plot.

To determine molecular weight from the GPC output, a table similar to Table 1.41a should be prepared. In the table, the columns are obtained as follows:

Column 1: In this column, each successive 5 mL fraction that is obtained from the chromatograph is numbered on the distribution curve (Figure 1.41o).
Column 2: Next, measure the height from the baseline to each numbered point on the distribution curve. The baseline in Figure 1.41o is drawn by the operator as shown in the diagram.
Column 3: Add successive heights to obtain cumulative height as shown.

Column 4: Normalize the data in Column 3 (0–100) by taking the individual cumulative heights and dividing them by the total cumulative height and multiplying the result by 100:

$$\left(\frac{\text{Col3}}{374} \times 100\right) \qquad 1.41(19)$$

Column 5: Tabulate the chain length from the calibration curve of count number vs. chain length in Figure 1.41p.
Column 6: Divide Column 2 by Column 5.
Column 7: Multiply Column 2 by Column 5.

The number average molecular weight is obtained by dividing Column 2 by Column 6 totals:

$$\frac{374}{0.107} = 3500 \text{ Å} \qquad 1.41(20)$$

The weight average molecular weight is obtained by dividing Column 7 by Column 2 totals:

$$\frac{19{,}565{,}810}{374} = 352{,}300 \text{ Å} \qquad 1.41(21)$$

Figure 1.41q is obtained by plotting Column 4 vs. Column 5. The total GPC system is shown in Figure 1.41r.

TABLE 1.41a
Determination of Molecular Weight from Chromatograph Output

1	2	3	4	5	6	7
Sample Count	Height Above Baseline	Cumulative Height	Cumulative Weight Percent	Chain Length (Å)	Number of Particles	Col. 2 × Col. 5
14	0	374	100.0	3,000,000	0	0
15	1	374	100.0	1,000,000	0.0000	1,000,000
16	10	373	99.8	450,000	0.0000	4,500,000
17	30	363	97.0	200,000	0.0002	6,000,000
18	50	333	89.0	80,000	0.0006	4,000,000
19	70	283	75.7	30,000	0.0023	2,100,000
20	80	213	57.0	15,000	0.0053	1,200,000
21	60	133	35.5	9,000	0.0067	540,000
22	35	73	19.5	5,000	0.007	175,000
23	20	38	10.2	2,000	0.010	40,000
24	10	18	2.7	900	0.011	9,000
25	5	8	1.3	300	0.017	1,500
26	2	3	0.5	140	0.024	280
27	1	1	0.3	30	0.033	30
Totals		374		1071		19,565,810

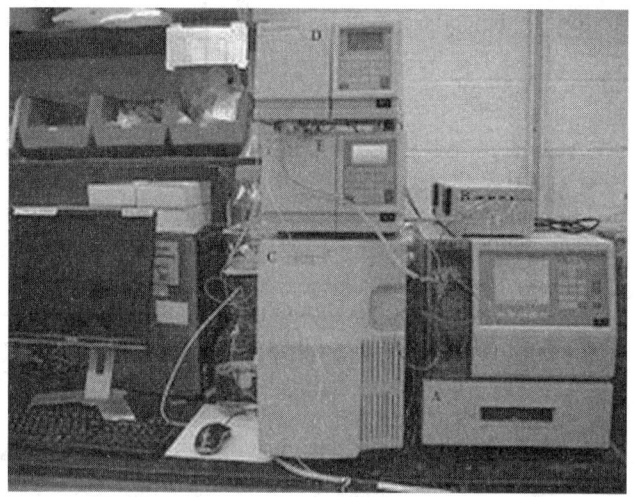

FIG. 1.41r
The total GPC system consists of the sample holder, the column itself, the pump, the refractive index detector, and/or the UV–visible detector. (Courtesy of Waters Corp.)

End Group Determination

Molecular weight determination by functional group analysis requires that the polymer molecule contain a known number of distinguishable groups. Such groups are carboxyl and hydroxyl groups or amino end groups. In linear polymers, the number average molecular weight is obtained as

$$M_n = \frac{2m}{x_e} \quad 1.41(22)$$

where

  m is the number of pounds of sample
  x_e the number of moles of end groups

However, this method can only be used where the number of end groups is known, as is the case with linear polymers. The number of end groups can be determined chemically or by means of an infrared spectrometer, which was calibrated with a known sample. This method fails at molecular weights above approximately 30,000 because the fraction of end groups becomes too small to be detected.

Electron Microscope

The optical microscope is limited in its power of resolution to the detection of objects that are larger than one-half of the wavelength of the shortest visible light used. This corresponds to approximately 8000 Å.

Electrons, however, do not behave only as particles but also as waves, with a wavelength of about 0.5 Å. Consequently, a beam of electrons can be used to detect particles much smaller than those visible by means of the optical microscope. In the electron microscope (Figure 1.41s), a heated filament provides a stream of electrons, which pass through a magnetic coil, acting as a condensing lens. Upon emerging from the object, the electrons pass through two additional magnetic coils, which act analogous to the objective and projector lenses of the optical microscope.

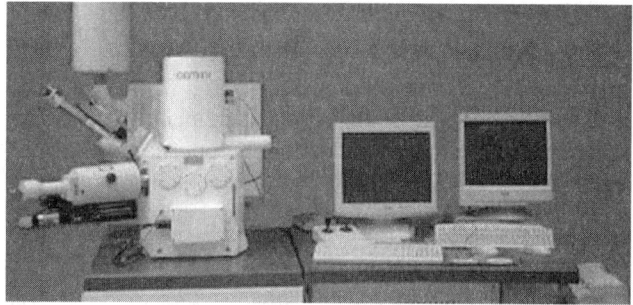

FIG. 1.41s
The electron microscope. (Courtesy of LEO Electron Microscopy.)

The focused electron beam impinges on a photographic film, where a much enlarged image is produced. The photographs can be further enlarged without distortion. Magnifications of 100,000:1 are possible. The lower limit of resolution obtainable is in the range of 15–20 Å.

Transmission Electron Microscope Transmission electron microscopes (TEMs) use electrons as *light source*, and their much lower wavelength makes it possible to get resolutions that are 1000 times better than those of a regular light microscope. The electron source at the top of the microscope emits electrons that travel through vacuum. The TEM uses electromagnetic lenses to focus the electrons into a thin beam, which travels through the process sample.

Depending on the density of the sample, some of the electrons will scatter and thereby disappear from the beam. At the bottom of the microscope, the unscattered electrons will hit a fluorescent screen and produce a *shadow image*. The image can be studied directly by the operator or can be photographed and transmitted for evaluation by others.

Advantages and Limitations The magnifying power of the electron microscope is sufficient to permit visual observation and measurement of polymer molecules. From a photograph of the polymer sample, the size of the molecules can be measured and plotted as molecular size distribution. Essentially, this distribution curve should be the same as the output curve of the GPC shown in Figure 1.41o. However, while the chromatograph automatically counts the number of molecules within each size range, the grouping and counting of molecules must be performed by the operator.

Since the microscope shows only molecular size, the makeup of the polymer molecule—the composition of its building blocks and distance between adjacent blocks—must be known before molecular weight can be calculated from the measured molecular size. One advantage of the electron microscope is that the molecular structure can be observed along with molecular size. In other types of measurements, the structure must be determined by independent measurements.

Ultracentrifuge

Sedimentation and diffusion measurements of polymer solutions both involve the frictional properties of the polymer molecules. Therefore, they are closely related, and both are needed to interpret the data obtained from the velocity ultracentrifuge.

Not all polymers are suitable for analysis by means of the ultracentrifuge. For the measurement to be successful, the polymer and solvent must differ in refractive index, the solvent must have a low viscosity, and the polymer must be soluble near room temperature. Finally, mixed solvents must be avoided because of complex corrections that must be made to the measurement.

Sedimentation Velocity There are two different types of ultracentrifuge measurements. In the sedimentation velocity method, the centrifuge is operated at from 20,000 to 60,000 rpm rotational speeds. These produce centrifugal fields that equal several hundred thousand times the acceleration of gravity. Under these conditions, the polymer molecules move under the influence of these centrifugal forces against the opposition of frictional forces.

The instrument determines the rate of movement and provides a rapid means of estimating the size, shape, and molecular weight of macrosolutes. It also allows the calculation of the coefficients of sedimentation and diffusion if the sedimenting components are well separated, and it allows the calculation of effective mass of solute components, particle asymmetry, and molecular weight.

The rate of sedimentation is related to the ratio of molecular weight to the frictional force. The frictional force is determined independently by diffusion or viscosity measurement. The fact that a second, independent measurement is required is one of the principal disadvantages of the sedimentation velocity method, because a viscosity measurement alone can also be used to determine the average molecular weight.

Sedimentation Equilibrium In the sedimentation equilibrium method, the centrifuge is operated at 10,000–30,000 rpm speeds for periods of several hours to days or weeks. Under constant conditions, the polymer will not pellet at the bottom of the cell but will be redistributed over time, with increasing concentration of solute as the distance from the center of rotation increases solely as a function of its molecular weight.

After a period of about a day, the effect of centrifugal forces on the solute particles will be balanced by diffusion, and the apparent concentration will no longer change. If the measurements are made with thermodynamically ideal solutions, the weight average molecular weight and higher averages can be obtained. In addition, homogeneity with respect to M_w, association constants, and aggregation states can also be calculated.

Instrument Construction The ultracentrifuge consists of an aluminum alloy rotor, which is several inches in diameter, and a solution cell mounted within the rotor near its periphery. The rotor is mounted in an evacuated chamber that is provided with windows for observation of the solution in the cell. The rotor may be driven by air or by an electric motor–operated oil turbine. Polymer concentration along the solution cell is measured by refractive index or light absorption detectors along the length of the cell.

CONCLUSIONS

Of the methods available for the determination of polymer molecular weight, none is ideally suited for in-line measurement, but the techniques of gel permeation and viscometry can be applied using automated sampling systems.

None of the methods produces a direct output in terms of molecular weight. Sedimentation and diffusion methods are not very useful for process applications because of the time required to obtain a measurement.

GPC, while its analysis time is in the range of 2–3 hr, provides a complete molecular weight distribution in addition to molecular weight averages.

Automatic membrane osmometers are relatively fast, but their useful range is between 10,000 and 300,000 molecular weight. At the lower end, special membranes can be used to extend the range to about 5000 molecular weight. Vapor pressure osmometers complement the useful range of the membrane osmometer at the lower end of the scale. The operating range of these instruments is up to approximately 20,000 molecular weight.

The analysis time and range of viscometers are comparable to those of the osmometers; however, their usefulness is limited to linear polymers. In terms of accuracy, range of application, and speed of analysis, the light-scattering photometer, and the electron microscope offer the best choice. These instruments, however, are much more costly than other types, and their advantages rarely can justify the expense.

SPECIFICATION FORMS

When specifying Molecular Weight Analyzers,, one cam use the ISA form 20A1001 and when specifying both the analyzer and the requirements of the process application, use the ISA form 20A1002. The following are the forms that are reproduced with the permission of the International Society of Automation.

#	RESPONSIBLE ORGANIZATION	ANALYSIS DEVICE	#	SPECIFICATION IDENTIFICATIONS		
1			6			
2	ISA		7	Document no		
3		Operating Parameters	8	Latest revision	Date	
4			9	Issue status		
5			10			
11	ADMINISTRATIVE IDENTIFICATIONS		40	SERVICE IDENTIFICATIONS Continued		
12	Project number	Sub project no	41	Return conn matl type		
13	Project		42	Inline hazardous area cl	Div/Zon	Group
14	Enterprise		43	Inline area min ign temp	Temp ident number	
15	Site		44	Remote hazardous area cl	Div/Zon	Group
16	Area	Cell	Unit	45	Remote area min ign temp	Temp ident number
17			46			
18	SERVICE IDENTIFICATIONS		47			
19	Tag no/Functional ident		48	COMPONENT DESIGN CRITERIA		
20	Related equipment		49	Component type		
21	Service		50	Component style		
22			51	Output signal type		
23	P&ID/Reference dwg		52	Characteristic curve		
24	Process line/nozzle no		53	Compensation style		
25	Process conn pipe spec		54	Type of protection		
26	Process conn nominal size	Rating	55	Criticality code		
27	Process conn termn type	Style	56	Max EMI susceptibility	Ref	
28	Process conn schedule no	Wall thickness	57	Max temperature effect	Ref	
29	Process connection length		58	Max sample time lag		
30	Process line matl type		59	Max response time		
31	Fast loop line number		60	Min required accuracy	Ref	
32	Fast loop pipe spec		61	Avail nom power supply		Number wires
33	Fast loop conn nom size	Rating	62	Calibration method		
34	Fast loop conn termn type	Style	63	Testing/Listing agency		
35	Fast loop schedule no	Wall thickness	64	Test requirements		
36	Fast loop estimated lg		65	Supply loss failure mode		
37	Fast loop material type		66	Signal loss failure mode		
38	Return conn nominal size	Rating	67			
39	Return conn termn type	Style	68			

#	PROCESS VARIABLES	MATERIAL FLOW CONDITIONS			#	PROCESS DESIGN CONDITIONS		
69					101			
70	Flow Case Identification			Units	102	Minimum	Maximum	Units
71	Process pressure				103			
72	Process temperature				104			
73	Process phase type				105			
74	Process liquid actl flow				106			
75	Process vapor actl flow				107			
76	Process vapor std flow				108			
77	Process liquid density				109			
78	Process vapor density				110			
79	Process liquid viscosity				111			
80	Sample return pressure				112			
81	Sample vent/drain press				113			
82	Sample temperature				114			
83	Sample phase type				115			
84	Fast loop liq actl flow				116			
85	Fast loop vapor actl flow				117			
86	Fast loop vapor std flow				118			
87	Fast loop vapor density				119			
88	Conductivity/Resistivity				120			
89	pH/ORP				121			
90	RH/Dewpoint				122			
91	Turbidity/Opacity				123			
92	Dissolved oxygen				124			
93	Corrosivity				125			
94	Particle size				126			
95					127			
96	CALCULATED VARIABLES				128			
97	Sample lag time				129			
98	Process fluid velocity				130			
99	Wake/natural freq ratio				131			
100					132			
133	MATERIAL PROPERTIES				137	MATERIAL PROPERTIES Continued		
134	Name				138	NFPA health hazard	Flammability	Reactivity
135	Density at ref temp		At		139			
136					140			

Rev	Date	Revision Description	By	Appv1	Appv2	Appv3	REMARKS

Form: 20A1001 Rev 0 © 2004 ISA

594 Analytical Measurement

	RESPONSIBLE ORGANIZATION	ANALYSIS DEVICE COMPOSITION OR PROPERTY Operating Parameters (Continued)		SPECIFICATION IDENTIFICATIONS		
1			6	Document no		
2	(ISA logo)		7	Latest revision		Date
3			8	Issue status		
4			9			
5			10			

	PROCESS COMPOSITION OR PROPERTY			MEASUREMENT DESIGN CONDITIONS				
11	Component/Property Name	Normal	Units	Minimum	Units	Maximum	Units	Repeatability
12								
13								
...								
75								

Rev	Date	Revision Description	By	Appv1	Appv2	Appv3	REMARKS

Form: 20A1002 Rev 0

© 2004 ISA

Definitions

Osmometry: Osmometry is the measurement of the osmotic strength of a substance. This is often used by chemists for the determination of average molecular weight.

Osmosis: The tendency of molecules of a solvent to pass through a semipermeable membrane from a less concentrated solution into a more concentrated one.

Osmotic pressure: It is the pressure that must be applied to the solution to stop osmosis from the pure solvent and it is related to solution concentration and molecular weight of the solute.

Rayleigh ratio is the quantity used to characterize the scattered intensity at the scattering angle of the incident light or other forms of radiation.

Rayleigh scattering is named after the British physicist Lord Rayleigh and is the scattering of light or other types of radiation by particles much smaller than the wavelength of the radiation. Rayleigh scattering does not change the state of the material, hence it is a paramagnetic process.

Refractive index (RI) or index of refraction of a material is a dimensionless number that describes how light propagates through that medium. It is defined as n = c/v, where c is the speed of light in vacuum and v is the speed of light in the medium. For example, the refractive index of water is 1.33, meaning that light travels 1.33 times faster in a vacuum than it does in water. The refractive index determines how much light is bent, or refracted, when entering a material.

van't Hoff factor is named after J.H. van't Hoff, is a measure of the effect of a solute upon properties such as osmotic pressure. The van't Hoff factor is the ratio between the actual concentration of particles produced when the substance is dissolved, and the concentration of a substance as calculated from its mass.

Vapor pressure osmometry: It is a technique to measure the number average molecular weightof a polymer. It is based uponRaoult's lawthat governs change in vapor pressureof a solution based on themole fractionof the solute.

Zimm plot is a diagrammatic representation of data on light scattering by large particles. It is used for the simultaneous evaluation of the mass average molar mass, the second viral coefficient of the chemical potential, etc. Several modifications of the Zimm plot are in frequent use.

Abbreviations

GPC   Gel-permeation chromatography
HPLC High-pressure liquid chromatography
NMR   Nuclear magnetic resonance
TEM   Transmission electron microscope

Bibliography

Ahuja, S. and Rasmussen, H.T. (eds.), *HPLC Method Development for Pharmaceuticals*, Academic Press, 2007. http://store.elsevier.com/HPLC-Method-Development-for-Pharmaceuticals/isbn-9780123705402/.

Allcock, H., Contemporary polymer chemistry, 2003. http://www.amazon.com/Contemporary-Polymer-Chemistry-3rd-Edition/dp/0130650560.

Annual Book of ASTM Standards, 2015. http://www.astm.org/BOOKSTORE/BOS/.

ASTM D4001-13, Standard test method for determination of weight-average molecular weight of polymers by light scattering, 2013. http://www.astm.org/Standards/D4001.htm.

Bender, T. A., Fast analytical separations with high-pressure liquid chromatography, 2013. http://pubs.acs.org/doi/abs/10.1021/ed300757a.

Brandrup, J., *Polymer Handbook*, 2003. http://www.wiley.com/WileyCDA/WileyTitle/productCd-0471479365.html.

Carraher, C. E., *Polymer Chemistry*, 2008. http://www.crcpress.com/product/isbn/9781420051025.

Cowie, J. M. G., *Polymers: Chemistry and Physics*, 2008. http://www.amazon.com/Polymers-Chemistry-Physics-Materials-Edition/dp/0849398134.

Denn, M. M., Polymer melt processing, 2008. http://www.amazon.com/Polymer-Melt-Processing-Foundations-Engineering/dp/0521899699/ref=sr_1_35?s=books&ie=UTF8&qid=1387658772&sr=1-35.

Dong, M. W., *Modern HPLC for Practicing Scientists*, Hoboken, NJ: John Wiley & Sons, 2006. http://www.amazon.com/Modern-HPLC-Practicing-Scientists-Michael/dp/047172789X/ref=pd_sim_b_4.

Fanali, S., Haddad, P. R., Poole, C., and Shoenmakers, P., *Liquid Chromatography: Fundamentals and Instrumentation*, Philadelphia, PA: Elsevier, 2013. http://www.amazon.com/Liquid-Chromatography-Instrumentation-Salvatore-Fanali/dp/0124158072.

Fried, J., Polymer science and technology, 2003. http://www.amazon.com/Polymer-Science-Technology-Joel-Fried/dp/0130181684/ref=sr_1_5?s=books&ie=UTF8&qid=1387658108&sr=1-5.

Majors, R., Fast and ultrafast HPLC on sub-2 µm porous particles, 2010. http://www.chromatographyonline.com/lcgc/article/articleDetail.jsp?id=333246&pageID=3#.

Malvern Instruments Ltd., Measuring Molecular Weight, Size and Branching of Polymers, 2015. http://www.azom.com/article.aspx?ArticleID=12255.

McMaster, M. C., *HPLC, A Practical User's Guide*, Wiley, 2007. http://onlinelibrary.wiley.com/doi/10.1002/jssc.200790030/abstract.

Menczel, J. D., Thermal analysis of polymers, 2009. http://www.amazon.com/Thermal-Analysis-Polymers-Fundamentals-Applications/dp/0471769177/ref=sr_1_22?s=books&ie=UTF8&qid=1387658556&sr=1-22.

Meyer, V. R., *Practical High-Performance Liquid Chromatography*, Hoboken, NJ: John Wiley & Sons, 2010. http://www.amazon.com/Introduction-Modern-Liquid-Chromatography-Snyder/dp/0470167548.

Pall Corp., Centrifugal devices, 2013. http://www.pall.com/main/laboratory/literature-library-details.page?id=40883.

Podzimek, S., The use of GPC coupled with a multi-angle laser light scattering photometer for the characterization of polymers, 2003. http://onlinelibrary.wiley.com/doi/10.1002/app.1994.070540110/abstract.

Rubinstein, M., Polymer physics, 2003. http://www.amazon.com/Polymer-Physics-Chemistry-M-Rubinstein/dp/019852059X/ref=sr_1_1?s=books&ie=UTF8&qid=1387657867&sr=1-1.

Rudin, A., The elements of polymer science, 2012. http://www.amazon.com/Elements-Polymer-Science-Engineering-Third/dp/0123821789/ref=sr_1_3?s=books&ie=UTF8&qid=1387657981&sr=1-3.

Schmitt, T. M., ASTM, methods for polymer molecular weight measurement, 2012. http://www.astm.org/DIGITAL_LIBRARY/MNL/PAGES/MNL12254M.htm.

Snyder, L. R., Kirkland, J. J., and Dolan, J. W., *Introduction to Modern Liquid Chromatography*, Hoboken, NJ: John Wiley & Sons, 2009. http://www.amazon.com/Introduction-Modern-Liquid-Chromatography-Snyder/dp/0470 167548.

Waksmudzka-Hajnos, M., High performance liquid chromatography in phytochemical analysis, 2010. http://www.amazon.com/Performance-Chromatography-Phytochemical-Analysis-Chromatographic/dp/142009260X/ref=pd_sim_b_43.

Waters, High Performance Liquid Chromatography, 2013. http://www.waters.com/waters/en_US/HPLC—High-Performance-Liquid-Chromatography/nav.htm?cid=10048919&locale=en_US.

Xiang, Y., Liu, Y., and Lee, M. L., Ultrahigh pressure liquid chromatography using elevated temperature. J. *Chromatogr*. A, 1104(1–2): 198–202, 2006. http://www.ncbi.nlm.nih.gov/pubmed/16376355.

1.42 Natural Gas Measurements

B. G. LIPTÁK and D. C. GALLEGO (2017)

INTRODUCTION

Natural gas (NG) is a hydrocarbon gas mixture. Its main component is methane which determines the market value of the natural gas. Other hydrocarbons, carbon dioxide, nitrogen, and hydrogen sulfide may be found in the composition of natural gas (Table 1.42a).

TABLE 1.42a
Geographic Variation in the Composition of Natural Gas

	LNG Composition (Mole Percent)				
Source	Methane	Ethane	Propane	Butane	Nitrogen
Alaska	99.72	0.06	0.0005	0.0005	0.20
Algeria	86.98	9.35	2.33	0.63	0.71
Baltimore gas and electric	93.32	4.65	0.84	0.18	1.01
New York City	98.00	1.40	0.40	0.10	0.10
Sand Diego gas and electric	92.00	6.00	1.00	—	1.00

Source: Liquid Methane Fuel Characterization and Safety Assessment Report, Cryogenic Fuels Inc. Report No. CFI-1600, December 1991.

It is formed when layers of decomposing plant and animal matter are exposed to intense heat and pressure over thousands of years. Most natural gas was created over time by two processes mechanisms: biogenic and thermogenic. Biogenic gas is created by organisms in marshes, bogs. landfills or shallow sediments. Deeper, Thermogenic gas is produced deeper in the earth, at greater temperatures and pressures, from buried organic material.

Natural gas is normally found in gaseous form, but it can also be liquefied or under special conditions it can also be in solid form (natural gas hydrates). Natural gas, like petroleum, is produced by the decomposition of dead marine microorganisms that are buried in low-oxygen environments (Biogenic mechanism). The accumulation of sediments on these microorganisms generates the high pressure and temperature conditions that form the petroleum and the natural gas (Thermogenic mechanism). The natural gas is contained in zones where the rocks are impermeable, normally gypsum formations deep underground.

Natural gas is used in many ways, such as:

- Electricity production (thermal power stations)
- Gas heating and domestic use
- Transportation (currently the natural gas may be used as fuel for cars and public transport)
- Plastic production (as raw material)
- Ammonia production (as raw material)
- Hydrogen production (as raw material)

NATURAL GAS INSTRUMENTATION

The key measurement devices that are used in the natural gas industry, serve to analyze its composition and to measure its flow rate. These and related devices will be discussed in this chapter.

When the natural gas is in liquid form (in order to ease its transportation), it is called liquified natural gas (LNG). For LNG applications, the instruments used should be impact tested at the design temperature (normally −320°F). The testing specimen should be handled according to ASTM E-23 and the tests should meet the requirements of ASTM A-370.

When the natural gas is in liquid form (in order to ease its transportation), it is called LNG. If the measurement device is for LNG applications, the materials of construction of the instruments should be subject to an impact test at the design temperature (normally −320°F). The testing specimen should be handled according to ASTM E-23 and the test requirements should meet ASTM A-370.

This chapter focuses on the measurement of the flow and on the composition of natural gas.

Flow Measurement

Section 2 of Volume 1 of this handbook provides a detailed discussion of the features and capabilities of all flow meter designs, including their accuracies, flow ranges, costs and manufacturers. In this chapter, only a few short paragraphs are provided to describe some of those flow detectors that are commonly used in the natural gas industry. While this information is useful, if the reader wants to gain a detailed understanding of the features and capabilities of these devices, it is recommended to refer to Volume 1 of the Instrument and Automation Engineers' Handbook (IAEH).

TABLE 1.42b
The Types of Flow Meters Used to Measure the Flow of Natural Gas and Their Main Features

Type	Orifice	Venturi	Turbine	Ultrasonic	Coriolis
Rangeability available	3:1	3:1	10:1	10:1	10:1
Accuracy at 100% flow	±1.5%FS ±0.75%FS[a]	±1.7%FS ±0.75%FS[a]	±0.5%AR ±0.25%AR[a]	±2%FS[a] ±0.5%FS[a]	±0.4%AR
Size range in inches	≥2 in.	≥3 in	¼–12 in.	≥1 in.	¼–14 in.
Estimated cost[b]	L	H	H	M	H
Measures mass or volume flow	Volume flow	Volume flow	Volume flow	Volume flow	Mass flow
Minimum Reynolds No.	8,000	200,000	Velocity 2–15 mm²/s	No limits	No limits
Maintenance frequency (years)	5 years	5 year	Periodic check	Periodic check	Periodic check
For detailed and reliable data, see chapters	1.38	1.52	1.48	1.49	1.34

[a] Calibrated.
[b] L, low; M, medium; H, high.

The information in Table 1.42b illustrates the prevailing views and practices in the natural gas industry. In general, one could comment that the capabilities of Coriolis, Turbine, and Venturi meters are understated while that of the ultrasonic flow meters are somewhat overstated in this table. It should also be noted that for subsea or multiphase measurements of natural gas flow, Section 2 of Volume 1 of the IAEH should be consulted.

Orifice Plates Orifice plates operate on a measurement principle based on the differential pressure generated across a restriction inserted into the pipe. Their pressure loss is relatively high as shown in Figure 1.42a. This type of flowmeter is discussed in detail in standards as ISO 5167 part 2, API 14.3.3, and AGA report No. 3 (AGA reports are the most used for guidance in natural gas measurements because they provide description and installation guidelines for orifices), Chapter 2.21 in Volume 1.

For natural gas measurements, the following should be considered when using orifice plates:

- Its use is not recommended where the ratio of full-scale flow to minimum flow exceeds 3:1, Table 2.21a in Volume 1.
- Pipe and plate size can be anywhere between 2 and 24 in, Table 2.21o in Volume 1.
- Without calibration, and with a turndown of 3:1, this industry expects the total error of the orifice-based flow measurement to be between ±1.4% and ±1.7% of full scale (FS). If a more accurate measurement is needed, the flowmeter should be calibrated in which case the error is expected to drop to ±0.75%FS. If the calibration is carried out, the orifice plate should be mounted in a calibrated meter run. The error (E) at any actual flow (AF) under 100% can be determined as $E = (\%FS)(FS/AF)^2$.
- Material of the element should be AISI 316 stainless steel as a minimum.

Orifice plates represent a cheap option for flow measurement and they are typically replaced every 5 or 6 years.

Venturi Tubes The venturi tube generates less pressure drop than do orifice plates (Figure 1.42a), but is still an energy-extractive flowmeter that measures flow as a function of the square root of the differential pressure across a contoured restriction. This flow sensor is used in processes where the permanent pressure drop must be minimized. The price of the venturi tube is higher than that of an orifice plate–based installation, but their accuracy is better, maintenance is less, and

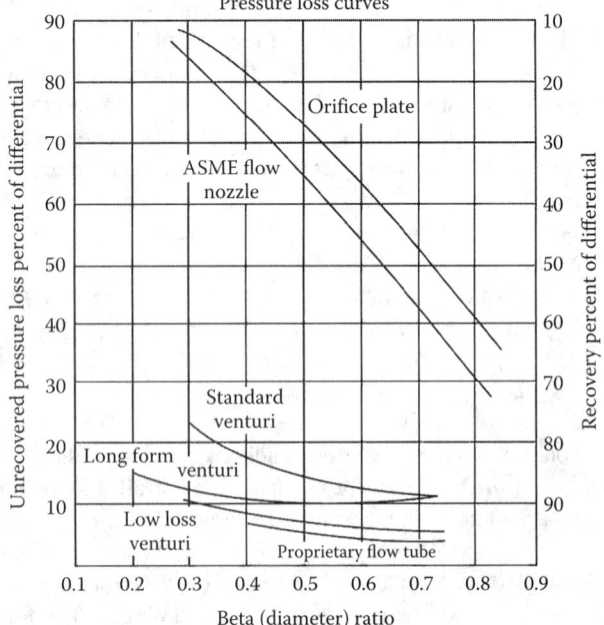

FIG. 1.42a
The pressure drop through a flow element is important when the flows are large, such as in natural gas flow measurement. In this respect, venture tubes are much superior to orifice plates or flow nozzles.

their life is longer. When the purpose of the flow measurement is custody transfer, the higher cost is well justified.

Turbine Flowmeters This flowmeter is an inferential device that measures the flow rate indirectly by using the natural kinetic energy of the flow as it passes through the blades of the turbine rotor. This causes the turbine to spin and as the blades move close to a magnetic coil, they interrupt the magnetic field produced by the coils so a pulse signal is generated. The frequency of this signal is directly proportional to the flow velocity and with the cross-sectional area of the pipe known it is possible to calculate the volumetric flow rate.

These meters require substantial maintenance (since they have moving parts and their maximum RPM must not be exceeded), their mounting is expensive, they require periodic calibration, and their price is higher than other types of flowmeters. Therefore, they are used only when high accuracy is required, such as in and for custody transfer applications.

Turbine meter performance and installation guidelines for natural gas measurement are included in AGA Report No. 7 "Measurement of Gas by Turbine Meters." This standard provides installation guidelines for various turbine meter body designs (inline, inline with a flow conditioning device, and angle bodied), inlet and outlet piping configurations (based on different meter body designs), and also provides data on the turbulence limits downstream. Turbine meter accuracy is generally ±1.0%AR although ±0.5%AR can be achieved with laboratory calibration. Turbine meter sizes available are typically between 1 and 8 in.

Ultrasonic Flow Meters As the name implies, these devices measure the process velocity by measuring the time taken for ultrasonic wave pulses to traverse an interior length of pipe section, both with and against the flow of the process. Time-of-Flight meters transmit ultrasonic signals (called chords) in fixed patterns (called paths) up and down or across the center of the pipe between two or more matching transducers located in or on the outside of the pipe.

Ultrasonic meter accuracy is affected by the flow conditions, excessive gas velocity, noise from control valves and regulators, or the presence of suspended particles of liquids or solids in the gas. Also, the fluid should not contain more than 18%–20% of CO_2 or a high (over 90%) H_2 concentration. (One of the advantages of these meters is that they do not create any pressure loss since the transducers are not invasive.)

Ultrasonic meter design, functionality, and installation standards for natural gas are provided in AGA Report No. 9 "Measurement of Gas by Multipath Ultrasonic Meters." The ultrasonic meter accuracy is about ±0.5%FS without any laboratory flow calibration and with laboratory calibration the accuracy obtained is improved to ±0.2%FS. (Other sources consider such as listed in Table 15.1 consider the accuracy to be less.) These meters also provide accurate bidirectional measurements if provided with proper flow conditioning devices. Ultrasonic meters for natural gas applications are available in sizes between 2 and 24 in.

Multipath ultrasonic flowmeters are normally used for LNG and natural gas measurements when required for custody transfer.

Ultrasonic flowmeters have also been introduced specifically for custody transfer measurement of cryogenic fluid processes in LNG (liquid natural gas) production. These meters are 8-path devices with an operating temperature range from −196°C to +60°C (−322°F to −140°F) (see Figure 1.42b).

FIG. 1.42b
Bidirectional, multipath, LNG flowmeter. (Courtesy of Emerson/ Daniel Measurement and Control, Inc.)

Coriolis Mass Flowmeters Coriolis mass flowmeters measure the force resulting from the acceleration caused by mass moving toward (or away from) a center of rotation. Similarly to flowing water in a loop of flexible hose that is "swung" back and forth in front of the body with both hands. Because the water is flowing toward and away from the hands, opposite forces are generated and cause the hose to twist. In a Coriolis mass flowmeter, the "swinging" is generated by vibrating the tube(s) in which the fluid flows.

FIG. 1.42c
Coriolis flowmeter for compressed natural gas service. (Courtesy of Siemens.)

The amount of twist is proportional to the mass flow rate of fluid passing through the tube(s). Sensors are used to measure the twist and generate a linear flow signal.

Coriolis flowmeters can be used on compressed natural gas service. Such small and efficient units provide relatively high capacity. For refueling applications, a flow rate of up to 500 kg/min ensures faster fueling. These units are designed to fit into integrated automation systems. This allows a broad range of measurement and control of compressed natural gas distribution, storage, and refueling (Figure 1.42c).

AGA Report No. 11 "Measurement of Natural Gas by Coriolis Meters" provides performance-based information and test methods for Coriolis meters.

Normal accuracy required by AGA is 1% FS, but most of flowmeters achieve ±0.5% Full Scale.

The advantages of Coriolis flowmeters are its ability to measure bidirectional flows; immunity to pulsation, unaffected by installation orientation; withstand service temperatures up to 400°F, not dependent on a velocity profile; and do not require flow conditioning devices in order to improve the accuracy. Normally, these flowmeters are supplied with both an indicating signal (4–20 mA) and a custody transfer signal (pulse signal). Also, other 4–20 mA signals for variables like density can be supplied.

Some disadvantages include their cost and that their size is limited to 16 in. (this can be solved by mounting several flowmeters in parallel). It can also be a limitation that the pressure drop across the meter should exceed 0.5 bars.

Analyzers

The natural gas must be analyzed in order to determine its properties and composition (Table 1.42a) that define the value of the gas on the market. Natural Gas (NG) and Liquefied Natural Gas (LNG) are usually analyzed for

- Composition: C_1, C_2, C_3, iC_4, nC_4, iC_5, nC_5, C_6+, C_6, and C_7+.
- Impurities: N_2, CO_2, CO, O_2, and H_2S.
- Physical properties: Specific gravity, compressibility, and the Wobbe index.

The sensors for the detection of some of these gases are covered in full detail in the following chapters:
Chapter 1.7: Carbon Dioxide
Chapter 1.28: Hydrocarbons
Chapter 1.30: Hydrogen
Chapter 1.31: Hydrogen Sulfide

Calorimeters (see Chapter 1.27) are used to calculate the heating value of the gas. Table 1.27a provides guidance for the selection of the appropriate calorimeter or Wobble Index detector for a particular application. If the heating value is not directly measured, indirect heating value determination is also used on the basis of gas composition.

Table 1.1a provides a listing of all the analyzers covered in this handbook and their suitability for the measurement of the concentration of a long list of substances. Table 1.1v identifies the main features of these analyzers.

Sampling Systems In order to obtain an accurate analysis of NG and LNG, it is essential to have the right sampling system, using the proper sample extractive method to obtain the sample. The sample should be maintained at the same pressure and temperature (Figure 1.42d) where the analyzer was calibrated, otherwise errors will result.

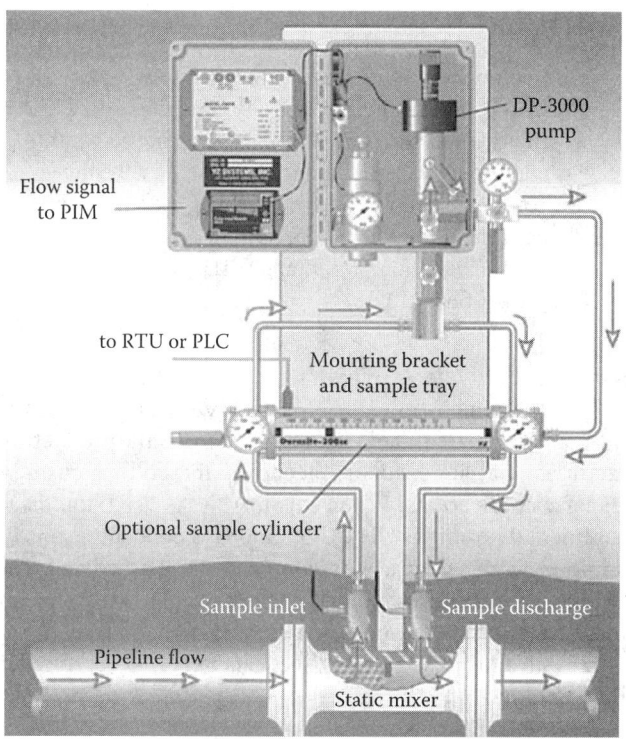

FIG. 1.42d
Calibrating Gas Analyzers using Constant Pressure Sample Cylinders. (Courtesy of Eastern Energy Services Pte Ltd.)

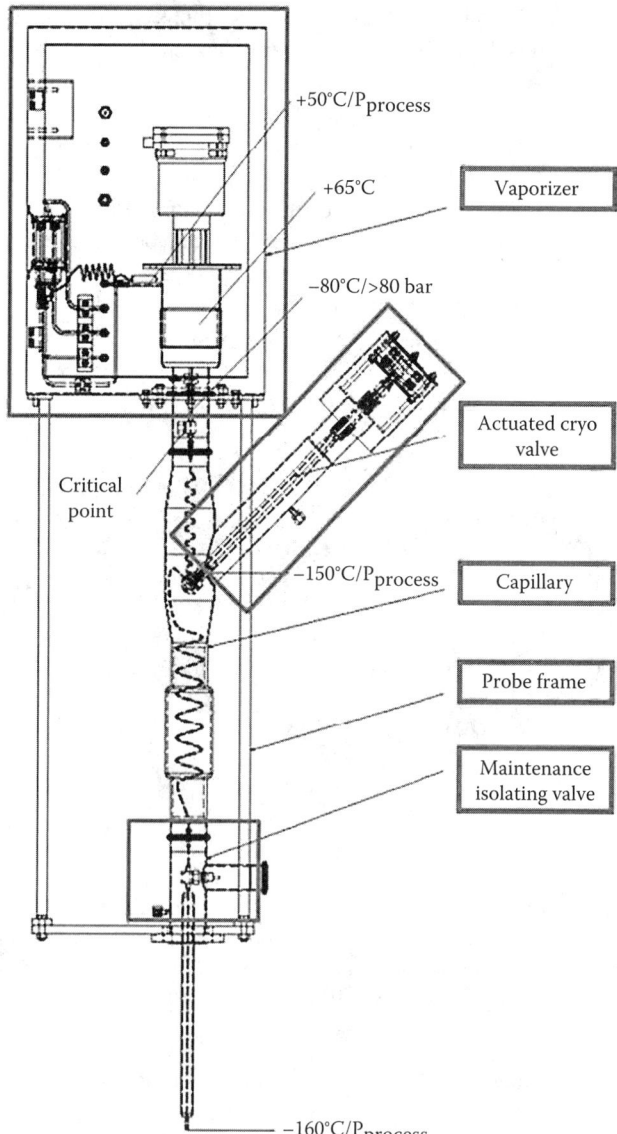

FIG. 1.42e
LNG sample probe and vaporizer system. (Courtesy of Topnir Systems.)

Descriptions of recommended sampling systems can be found in API MPMS Chapter 14 Section 1, GPA Standard 2166-05, ISO 10715, and LNG Custody Transfer Handbook (by GIIGNL).

Figure 1.42e illustrates an LNG sample probe and vaporizer system. In this system, the LNG enters through a capillary tube and its temperature rises from about −160°C to −150°C in the capillary tube. At the inlet of the vaporizer coil, a pressure reducing check valve is provided, which flashes the liquid sample. The vaporization of the approximately 0.5 cm^3 fluid at this "critical point" increases the temperature and the pressure and leads to pressures and temperatures of approximately 80 bars and −100°C. The temperature is controlled at +65°C in the vaporizer, in order to obtain approximately +55°C at the outlet of the vaporizer.

The analysis of NG and LNG can be performed using chromatography, mass spectrophotometry, and other methods. Currently, the most used method is gas chromatography.

Chromatographs This is the most common analysis method for NG and LNG applications, as it is accepted by all the standards. The general description and basic principle of operation of gas and liquid chromatographs are described in Chapters 1.10 and 1.11. The main characteristics of these analyzers are as follows:

- They require sampling, conditioning, and transport systems.
- They can be compact enough to be mounted near the sampling system (Figure 1.42f) or can be installed in analyzer shelters together with other analyzers. Currently, the compact designs are able to perform a complete analysis of the natural gas sample.

FIG. 1.42f
Compact natural gas chromatograph that measures hydrocarbon dew point and an expanded range of natural gas components (from C6+ to C9+). (Courtesy of ABB.)

- Chromatographs usually have two or more columns in different configurations. These configurations are described in ISO, ASTM, API, and GPA standards. A possible configuration is to have one column analyzing for N_2, C_1 to C_5 and another column for C_{6+}. Another configuration is to have one column for N_2, C_1, C_2, and CO_2 and another column for C_3, C_4, and C_5. Columns used in NG and LNG chromatographs are usually any of two types:
  - Packed Columns having lengths between 0.5 and 9 m with diameters between 2 and 6 mm. They are suitable for quantitative and qualitative analysis. These columns are described in standards such as GPA 2261 or ISO 6974 parts 3, 4, and 5.
  - Capillary Columns having lengths between 5 and 100 m with diameters between 0.1 and 0.5 mm. These columns provide very good separation of the different components. There are several sub-types of capillary columns:
    - WCOT: Wall-Coated Open Tubular columns.
    - SCOT: Support-Coated Open Tubular columns.
    - PLOT: Porous-Layer Open Tubular columns.
- Modern chromatographs are furnished with combinations of column types to improve the separation and the impurities removing process. For instance, ISO 6974 uses a combination of capillary columns (WCOT and PLOT), whereas GPA 2186 uses a combination of WCOT capillary columns and packed columns.
- The transport gas is usually Helium or Hydrogen with a high purity (higher than 99.995%).
- After the gas is separated by the chromatographs, each component is identified by a detector. They are either the Thermal Conductivity Detectors (TCD), which are sensitive to most components in the natural gas and provide a good linear response. For the detection of higher molecular weight (C_{6+}) components, Flame Ionization Detectors (FID) are often used. For full details on TCD and FIDs detectors, see Chapters 1.66 and 1.28.

Mass Spectrometers This method is starting to be used but not that much. It is used for the measurement of natural gas in gas treatment plants and ammonia production facilities, but it is not an accepted method in custody transfer applications. When this method is used, the measurement determines CH_4, CO_2, and H_2S concentrations in addition to the natural gas composition. The advantages of mass spectroscopy include that a single analyzer can measure several streams and can obtain the results faster than with chromatographs. See Chapter 1.35 for further information.

Moisture and H_2S Analyzers In order to determine the amount of impurities or composition of NG and LNG, the following analyzer designs are also used:

- Moisture concentration should be determined to avoid equipment damage or corrosion if other components like H_2S are also present and prevent a drop in market value of the gas, etc. Moisture and dew point measurements are discussed in detail in Chapter 1.39. H_2O contamination can be measured by using
  - TLS analyzers (Tunable Laser Spectroscopy).
  - Dew Point Analyzers.
  - Capacitance Cell Analyzers.
- H_2S can cause corrosion and therefore its concentration needs to be measured. H_2S can be measured by
  - Lead acetate method (Colorimetric method, Chapter 1.13).
  - FID detector on the chromatograph (Chapter 1.10).
  - UV-Spectroscopy (Chapter 1.69).

Mass spectroscopy method can be used in order to measure H_2S, although it is not used specifically for this application due to its price.

SPECIFICATION FORMS

When specifying Natural Gas Analyzers, use the ISA form 20A1001 and when specifying both the analyzer and the requirements of the process application, use the ISA form 20A1002. The following are the forms that are reproduced with the permission of the International Society of Automation.

#					#				
1	RESPONSIBLE ORGANIZATION		ANALYSIS DEVICE		6	SPECIFICATION IDENTIFICATIONS			
2					7	Document no			
3	ISA		Operating Parameters		8	Latest revision		Date	
4					9	Issue status			
5					10				
11	ADMINISTRATIVE IDENTIFICATIONS				40	SERVICE IDENTIFICATIONS Continued			
12	Project number		Sub project no		41	Return conn matl type			
13	Project				42	Inline hazardous area cl		Div/Zon	Group
14	Enterprise				43	Inline area min ign temp		Temp ident number	
15	Site				44	Remote hazardous area cl		Div/Zon	Group
16	Area		Cell	Unit	45	Remote area min ign temp		Temp ident number	
17					46				
18	SERVICE IDENTIFICATIONS				47				
19	Tag no/Functional ident				48	COMPONENT DESIGN CRITERIA			
20	Related equipment				49	Component type			
21	Service				50	Component style			
22					51	Output signal type			
23	P&ID/Reference dwg				52	Characteristic curve			
24	Process line/nozzle no				53	Compensation style			
25	Process conn pipe spec				54	Type of protection			
26	Process conn nominal size		Rating		55	Criticality code			
27	Process conn termn type		Style		56	Max EMI susceptibility		Ref	
28	Process conn schedule no		Wall thickness		57	Max temperature effect		Ref	
29	Process connection length				58	Max sample time lag			
30	Process line matl type				59	Max response time			
31	Fast loop line number				60	Min required accuracy		Ref	
32	Fast loop pipe spec				61	Avail nom power supply			Number wires
33	Fast loop conn nom size		Rating		62	Calibration method			
34	Fast loop conn termn type		Style		63	Testing/Listing agency			
35	Fast loop schedule no		Wall thickness		64	Test requirements			
36	Fast loop estimated lg				65	Supply loss failure mode			
37	Fast loop material type				66	Signal loss failure mode			
38	Return conn nominal size		Rating		67				
39	Return conn termn type		Style		68				
69	PROCESS VARIABLES	MATERIAL FLOW CONDITIONS			101	PROCESS DESIGN CONDITIONS			
70	Flow Case Identification			Units	102	Minimum	Maximum		Units
71	Process pressure				103				
72	Process temperature				104				
73	Process phase type				105				
74	Process liquid actl flow				106				
75	Process vapor actl flow				107				
76	Process vapor std flow				108				
77	Process liquid density				109				
78	Process vapor density				110				
79	Process liquid viscosity				111				
80	Sample return pressure				112				
81	Sample vent/drain press				113				
82	Sample temperature				114				
83	Sample phase type				115				
84	Fast loop liq actl flow				116				
85	Fast loop vapor actl flow				117				
86	Fast loop vapor std flow				118				
87	Fast loop vapor density				119				
88	Conductivity/Resistivity				120				
89	pH/ORP				121				
90	RH/Dewpoint				122				
91	Turbidity/Opacity				123				
92	Dissolved oxygen				124				
93	Corrosivity				125				
94	Particle size				126				
95					127				
96	CALCULATED VARIABLES				128				
97	Sample lag time				129				
98	Process fluid velocity				130				
99	Wake/natural freq ratio				131				
100					132				
133	MATERIAL PROPERTIES				137	MATERIAL PROPERTIES Continued			
134	Name				138	NFPA health hazard		Flammability	Reactivity
135	Density at ref temp		At		139				
136					140				

Rev	Date	Revision Description	By	Appv1	Appv2	Appv3	REMARKS

Form: 20A1001 Rev 0 © 2004 ISA

Analytical Measurement

1	RESPONSIBLE ORGANIZATION	ANALYSIS DEVICE COMPOSITION OR PROPERTY Operating Parameters (Continued)	6	SPECIFICATION IDENTIFICATIONS	
2			7	Document no	
3			8	Latest revision	Date
4			9	Issue status	
5			10		

	PROCESS COMPOSITION OR PROPERTY			MEASUREMENT DESIGN CONDITIONS				
	Component/Property Name	Normal	Units	Minimum	Units	Maximum	Units	Repeatability
13								
14								
15								
16								
17								
18								
19								
20								
21								
22								
23								
24								
25								
26								
27								
28								
29								
30								
31								
32								
33								
34								
35								
36								
37								

(Rows 38–75: blank remarks/notes area)

Rev	Date	Revision Description	By	Appv1	Appv2	Appv3	REMARKS

Form: 20A1002 Rev 0 © 2004 ISA

Abbreviations

FID	Flame Ionization Detector
PLOT	Porous-Layer Open Tubular columns
RPM	Revolution Per Minute
SCOT	Support-Coated Open Tubular columns
TCD	Thermal Conductivity Detector
TDLAS	Tunable Diode Laser Absorption Spectroscopy
TLS	Tunable Laser Spectroscopy
WCOT	Wall-Coated Open Tubular

Organizations

AGA	American Gas Association, http://www.aga.org/
API	American Petroleum Institute, http://www.api.org
ASTM	American Society of the International Association for Testing and Materials
GIIGNL	International Group of Liquefied Natural Gas Importers, http://www.giignl.org/
GPA	Gas Processors Association, http://www.gpaglobal.org
ISO	International Organization for Standardization, http://www.iso.org

Bibliography

AGA, Natural gas quality and gas interchangeability (2015), http://www.aga.org/our-issues/issuesummaries/Pages/NaturalGasQualityandGasInterchangeability.aspx.

AGA:3.1, Orifice metering of natural gas and other related hydrocarbon fluids (1990), https://archive.org/details/gov.law.aga. 3.1.1990.

AGA Report No. 4A, Natural gas contract measurement and quality clauses (2001), http://www.techstreet.com/products/1098305.

AGA Report No. 7, Turbine gas meter (1993), http://www.scribd.com/doc/56766106/AGA-Report-7-Turbine-Gas-Meter.

AGA Report No. 8, Compressibility factor of natural gas and related hydrocarbon gases (1994), http://www.techstreet.com/products/294.

AGA Report No. 9, Measurement of gas by multipath ultrasonic meters (2007), http://www.techstreet.com/products/1373965.

AGA Report No. 11, Measurement of natural gas by Coriolis meter (2003), http://www.techstreet.com/products/1206625.

API 2530, Orifice metering of natural gas (1985), http://www.amazon.com/Orifice-Metering-Natural-Report-2530/dp/9996820696.

API MPMS 14.1 Collecting and Handling of Natural Gas Samples for Custody Transfer, 7th edn. (2006), http://www.pngis.net/standards/21090/.

API MPMS, Chapter 14—Natural gas fluids measurement Section (2013), http://ballots.api.org/copm/cogfm/ballots/docs/ 14_3_2_%20ed%205cogfmballot.pdf.

API MPMS Chapter 14.3.3, Orifice Metering of Natural Gas (2015), http://www.api.org/publications-standards-and-statistics/standards/whatsnew/publication-updates/new-manual-of-petroleum-measurement-publications/api_mpms_14_3_3.

ASTM A370—15, Standard Test Methods and Definitions for Mechanical Testing of Steel Products (2015), http://www.astm.org/Standards/A370.htm.

ASTM E23—12c, Standard Test Methods for Notched Bar Impact Testing of Metallic Materials (2012), http://www.astm.org/Standards/E23.

Eastern Energy Services Pte Ltd. (2013), http://easternenergyservices.blogspot.com/2013/10/calibrating-gas-analysers-using.html.

GIIGNL, LNG Custody Transfer Handbook (2015), http://www.giignl.org/system/files/cth_version_4.00_-_february_2015.pdf.

GPA 2186, Tentative method for the extended analysis of hydrocarbon liquid (2002), http://www.techstreet.com/products/1374865.

GPA 2186-02, Mixtures containing nitrogen and carbon dioxide by temperature programmed gas chromatography (2002), http://www.techstreet.com/products/1374865.

GPA 2261, Analysis for natural gas and similar gaseous mixtures by gas chromatography (2000), http://www.techstreet.com/products/861546.

GPA Standard 2166-05, Obtaining natural gas samples for analysis by gas chromatography (2009), https://www.gpaglobal.org/publications/view/id/37/.

ISO 5167-2, Orifice plates (2003), https://www.iso.org/obp/ui/#iso:std:iso:5167:-2:ed-1:v1:en.

ISO 6974, Natural gas—Determination of composition with defined uncertainty by gas chromatography (2002), http://www.iso.org/iso/catalogue_detail.htm?csnumber=26238.

ISO 10715, Natural gas—Sampling guidelines (1997), http://www.iso.org/iso/catalogue_detail.htm?csnumber=18803.

LNG Custody Transfer Handbook by GIIGNL (2012), http://www.docstoc.com/docs/131252007/LNG-CUSTODY-TRANSFERHANDBOOK-Third-edition-GIIGNL.

Silva, J.E., Thermogenic mechanisms (2007), http://www.ncbi.nlm.nih.gov/pubmed/16601266.

Topnir, LNG Online Analysis (2013), http://www.topnir.com/product%20&%20services/services/lng.php#.

Turner, P. et al., Calibration effects during natural gas analysis using a quadrupole mass spectrometer (2004), http://www.sciencedirect.com/science/article/pii/S0165993604004030.

Voogd, J. et al., Fast quantitative analysis of natural gas by mass spectrometry using a new calibration procedure (1983), http://www.sciencedirect.com/science/article/pii/0020738183870156.

1.43 Nitrogen, Ammonia, Nitrite and Nitrate

C. E. HAMILTON (1974, 1982) **M. T. LEE-ALVAREZ** (2003)
B. G. LIPTÁK (1995, 2017)

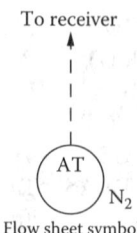

Flow sheet symbol

Nitrogen analysis method	1. Ammonia:
	  1A direct Nessler
	  1B direct phenate
	  1C distillation-titrimetric
	  1D distillation-Nessler
	  1E distillationphenate
	  1F microcoulometric
	  1G gas chromatographic
	  1H ion-selective electrodes
	  1I hypochlorite-chemiluminescence
	2. Nitrite:
	  2A sulfanilic acid
	  2B m-phenylenediamine
	3. Nitrate:
	  3A phenoldisulfonic acid
	  3B ultraviolet (UV) absorption
	  3C brucine
	  3D reduction
	  3E specific ion electrode
	  3F ion chromatography
	4. Total nitrogen:
	  4A Kjeldahl
	  4B combustion-chemiluminescence
	  4C gas chromatograph-chemiluminescence
	  4D combustion-electrochemical detector
	  4E combustion-thermal conductivity detector
	  4F persulfate digestion colorimeter

Costs	The cost of an autoanalyzer with colorimetric detection starts at about $35,000. The cost of specific ion electrodes (1H, 3E) alone for ammonia range from $500 to about $1,000. The cost of a chemiluminescence detector module alone, added to a TOC analyzer for simultaneous TN and TOC measurements starts at about $10,000, a laboratory TN only analyzer is about $35,000, while an online TN analyzers, depending on their sampling systems, costs about twice that. Chemiluminescence detectors for gas chromatographs are about $20,000 and thermal conductivity detection for TN detection about twice that.

(Continued)

Partial list of suppliers ABB, (1H, 3E) http://www.abb.com/product/us/9AAC100044.aspx?country=00
Antek Instruments (4C) (http://www.paclp.com/Lab_Instruments/Brand/Antek)
GE Power and Water (http://www.gewater.com) (4C)
Hach (1H) (http://www.hach.com/ammonia-ion-selective-electrode-ise-bnc-connector/product?id=7640488929)
Leco Corp. (http://www.leco.com) (4E)
Mitsubishi (4B, 4C) (http://cosaxentaur.com/page/561/total-nitrogen-analyzer-nsx-2100vn)
MultiTek (http://www.paclp.com/Lab_Instruments/MultiTek)
Ogranak-Krug (http://www.kruginter.com/product.php?id=91)
Shimadzu Corporation (http://www.shimadzu.com) (4B)
Signal (4B, 4C) http://www.k2bw.com/chemiluminescence.htm
Skalar (4C) http://www.skalar.com/news/new-model-nitrogen-detector-for-total-nitrogen-analyzer
Teledyne-Tekmar (4B) http://www.teledynetekmar.com/products/TOC/Apollo/documentation/TN%20Module.pdf
Thermo Scienticic, (4C) http://www.thermoscientific.com/en/product/model-17-i-i-i-ammonia-analyzer.html
Timberline (wet chemistry) http://www.timberlineinstruments.com/product/ammonia-analyzer/

For additional suppliers see the following chapters
Chapter 1.10: Gas Chromatographs
Chapter 1.13: Colorimeters
Chapter 1.33: Ion-Selective Electrodes
Chapter 1.69: UV Analyzers
Chapter 1.73: Water Quality Analyzers
Chapter 1.74: Wet Chemistry Analyzers

INTRODUCTION

The measurement of nitrogen in its different forms (ammonia, nitrite, nitrate, and total nitrogen) can involve a variety of laboratory operations or utilize a variety of process analyzers. The analyzers that can be used in the industrial process environment include chromatographs, ion-selective electrodes, ultraviolet analyzers, and wet chemistry analyzers using colorimetric end-point sensors. Each of these types of analyzers has already been discussed in other chapters and therefore, in-depth coverage of these analyzers will not be repeated here.

NITROGEN AND THE ENVIRONMENT

For many rivers in the proximity of humans, the largest input of nitrogen is from sewage whether treated or untreated. The nitrogen derives from breakdown products of proteins found in urine and feces. The differing forms of nitrogen are relatively stable in most river systems with nitrite slowly transforming into nitrate in well oxygenated rivers and ammonia transforming into nitrite/ nitrate. Elevated levels of nitrogen are often associated with overgrowths of plants or eutrophication.

Ammonium ions also have a toxic effect, especially on fish.. The toxicity of ammonia is dependent on both pH and temperature. The management of river chemistry to avoid ecological damage is particularly difficult in the case of ammonia as a wide range of potential scenarios of concentration, pH and temperature have to be considered and the diurnal pH fluctuation caused by photosynthesis considered. On warm summer days with high-bi-carbonate concentrations unexpectedly toxic conditions can be created.

Although the air is 79% nitrogen, the supply of nitrogen to plants and animals is limited by its availability in the usable chemical compounds. Atmospheric nitrogen is an inert gas that relatively few organisms can convert to usable forms such as ammonia, nitrate, and nitrite (Figure 1.43a).

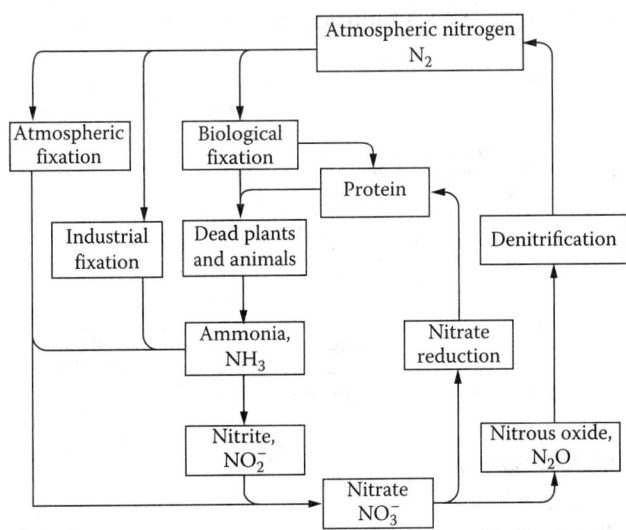

FIG. 1.43a
Nitrogen cycle.

Until the advent of large-scale industrial fixation processes for nitrogen, the major problem with nitrogen was its limited availability. The amount of nitrogen fertilizers produced since 1950 has increased many-fold, and it is estimated that its production may exceed 160 million tons in 2016.

Because not all of the nitrogen fertilizer is retained in the soil long enough for its intended use, several environmental problems have arisen in recent years. Nitrogen compounds in streams and lakes result in an increased growth of algae, which depletes the available oxygen for oxygen-dependent organisms. The rapid eutrophication of lakes is at least partly a result of excess nitrogen compounds in the runoff.

Nitrification and Denitrification

Another process, termed *nitrification*, in which ammonium ion is oxidized to nitrate or nitrite, can also create problems in the aquatic environment by depleting the available oxygen. Ammonium ion in normally alkaline waters is in equilibrium with free ammonia, which is toxic to fish and other aquatic forms. Microorganisms decompose organic nitrogen compounds by a process called *ammonification*, which converts the organic nitrogen compounds such as amino acids to ammonium ions. The ammonium ions created add to the problems described.

Another process that is part of the nitrogen cycle, *denitrification*, converts the nitrates or nitrites produced by nitrification into inert molecular nitrogen. Figure 1.43a shows the nitrogen cycle. Nitrate concentration, if higher than 45 mg/l in drinking water, causes an illness in infants called *methemoglobinemia*. The problems caused by nitrogen compounds in water have made it necessary to analyze for nitrogen in its many forms or oxidation states. The oxidation states of nitrogen are listed in Table 1.43a.

TABLE 1.43a
Oxidation States of Nitrogen

Oxidation State	Name	Symbol
+5	Nitrate ion	NO_3^{1-}
+3	Nitrite ion	NO_2^{1-}
0	Nitrogen	N_2
−1	Hydroxyl amine	NH_2OH
−3	Ammonia	NH_3

ANALYSIS GOALS

In nitrogen related analysis, the goal can be the determination of the quantity ammonia, nitrite, nitrate or total nitrogen. The types of analyzers used are also multiple and include chromatographs, wet chemistry, ion-selective detection, chemiluminescence, ultraviolet absorption, thermal conductivity, colorimetry, etc. The features and operation of all of these analyzers are described in other chapters, therefore here only their application and testing is discussed.

Ammonium Detectors

The direct Nessler* method is sensitive under optimal conditions to about 0.02 mg/l ammonia nitrogen. As a result of interferences such as hardness, iron, acetone, organic amines, and aldehydes, most water samples require a preliminary cleanup distillation. The Nessler reagent produces a yellow-to-brown color that can be compared visually to standards or can be read against a standard curve photometrically between the wavelengths of 400 and 500 nm (1 nm = 1 µm = 10 Å), depending on the photometer light pathlength chosen and on the concentration of the sample.

The direct phenate method, which uses the reaction of ammonia, hypochlorite, and phenol catalyzed by manganous ion to produce an intense blue color (indophenol), is subject to interferences from acidity and alkalinity. The blue color has a maximum absorbance at the wavelength of about 630 nm. Because of interferences, cleanup distillation is desirable if subsequent determinations are to be made for other forms of nitrogen.

The average relative standard deviation for the sample concentration range of 0.2–1.5 mg/l ammonia nitrogen by the above methods has been reported as 20.3% for the direct Nessler and 27.0% for the direct phenate methods. After cleanup distillation, both methods have an average standard deviation of 13%. If the sample is distilled from a phosphate buffered at a pH of 7.4 into a standard boric or sulfuric acid absorbent solution, the excess acid can be titrated with a standard base to determine the ammonia nitrogen. The relative standard deviation for this method, however, averages about 40%.

For certain water samples, it may be useful to distinguish the *free* ammonia nitrogen distilled from a neutral solution (pH 7.4), the *fixed* ammonia nitrogen distilled from solution at pH 10.5 or greater, and the *albuminoid*† nitrogen obtained by conversion with permanganate and alkaline distillation. Ammonia concentration can be rapidly determined by microcoulometry or by gas chromatography.

Probes are available to detect the ammonium ion concentration or total nitrogen (in the form of ammonium) in liquid streams. The probe measures the change in pH, which occurs in a thin film that is located between a permeable membrane and the pH glass electrode. The ammonia passes through the gas-permeable membrane in quantities proportional to its partial pressure (concentration) and causes a corresponding

---

* Nessler cylinders are used for colorimetric. The color of the sample in a Nessler cylinder is visually compared with a reference model. The tubes are often used to carry out a series of calibration solutions of increasing concentrations, which functions as a comparative scale.
† Albuminoid nitrogen is nitrogen which is still in the ammonium salts or still within the organic proteinoid molecules that are present in the water.

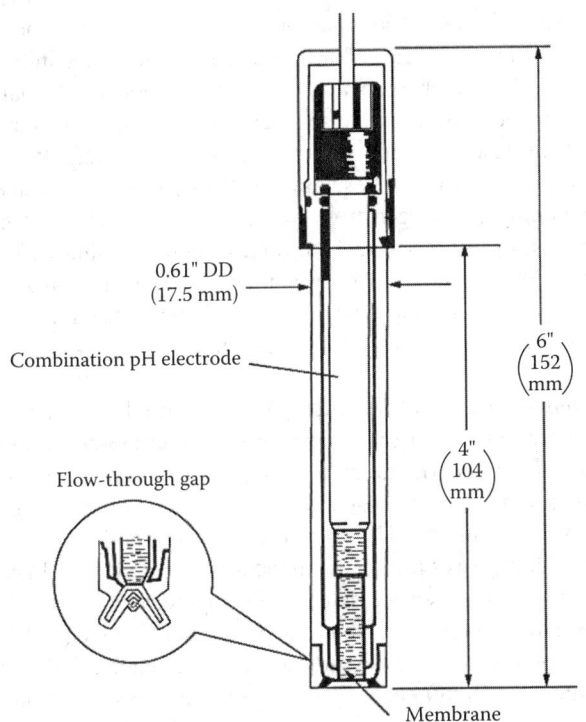

FIG. 1.43b
Probe for the direct determination of ammonia.

change in the film pH. Therefore, the measurement of pH reflects the concentration of ammonia. Figure 1.43b shows both the probe and the cap that can be added to its tip for use in continuous analysis in a closed-flow system.

Chemiluminescence Detector (CLD) A more sensitive method involves the indirect detection of ammonia using chemiluminescence. Chemiluminescence is produced when a chemical reaction produces an electronic excited species. As the excited species returns to the ground state, light is emitted. As ammonia consumes the hypochlorite, the chemiluminescence signal of the luminol–hypochlorite complex decreases leading to negative peaks. Another method involves the reaction of ammonia with hypochlorite, yielding a volatile compound that is combusted at 600°C. The NO produced reacts with ozone, and the resulting chemiluminescence is detected. The total time for the analysis is less than 3 min. Determination limit of as low as 1×10^{-8} M can be achieved using this method.

Nitrite Measurement

Nitrite, an intermediate stage in the nitrogen cycle, can result from the decomposition of natural protein. Another source of nitrite ion in water is its use as a corrosion inhibitor.

Nitrite analyses by standard methods are based on the diazotization of sulfanilic acid with nitrite in strong acid followed by coupling with alpha naphthyl amine hydrochloride. The purple color formed can be read on a photometer at a wavelength of 520 nm or compared visually to standards. The best concentration range for the use of this method is below 2 mg/l, but higher concentrations can also be determined by appropriate dilution with distilled water.

An alternative method for the higher concentration ranges uses the yellow–brown color of the reactant of nitrite with metaphenylenediamine in acid solution. Both these methods are pH dependent but otherwise suffer few significant interferences except for chlorides when present in concentrations greater than 15,000 mg/l. Many diazotization reactions have been proposed and used to eliminate or minimize these interferences.

Another way of measuring nitrite involves its chemical reduction to NO by reducing agents such as NaI, V(III), Ti(III), Cr(III), and Mo(VI) + Fe(II). The NO generated is detected using chemiluminescence.

Nitrate Measurement

Phenol disulfonic acid reacts with nitrate to produce a yellow color that can be measured at the maximum wavelength absorption between 410 and 480 nm, depending on sample concentration range and on colorimeter cell path length. Chlorides, even at 10 mg/l, represent a severe negative interference. Silver ion precipitation of chlorides has been used, but excess silver ion causes a brown precipitate, which interferes with the measurement. Nitrites at concentrations greater than 0.2 mg/l are a positive interference that can be removed with sodium azide.

The ultraviolet absorption at 220 nm wavelength can be used to measure nitrate. Chlorides represent no interference, but organic matter that absorbs at or near the same wavelength causes a positive variable interference. Hexavalent chromium and nitrite ions also interfere. Empirical correction for the normal interferences has been used successfully.

Nitrate can also be directly measured using ion-exchange chromatography (see Definitions at the end of the chapter for a discussion of IEC). Detection limits of 50 mg/l N have been reported. This method allows for sequential analysis of different anions. However, anions that have the same retention time as nitrates will interfere. Furthermore, an excessive amount of chlorides (>400 mg/l) results in decrease in signal for nitrates.

In the brucine method, nitrate reacts with brucine sulfate in a glacial acetic and dilute sulfuric acid mixture. The rate and intensity of color change with time and with temperature so that simultaneous development of the color to its maximum is also necessary for the calibration standards. At concentrations below 1 mg/l, results have significant negative bias. Chlorides above about 1000 mg/l and nitrites both interfere with the measurement.

Nitrate has been analyzed by reduction to ammonia or nitrite, which is then analyzed by one of the several methods previously described. Several different metals and alloys have also been used to determine nitrate and nitrite, but ferric and fluoride ions interfere with the analysis. An infrared method has also been reported for determining both nitrate and nitrite simultaneously, but it requires extensive concentration and cleanup procedures.

A specific ion electrode for nitrate is available that is sensitive to less than 1 mg/l nitrate ion, but interferences due to the normal ranges of chloride and bicarbonate ion activities make the results far from acceptable. However, for water systems in which these interferences are absent, or for higher nitrate (>10 mg/l) concentrations, the electrode is rapid and convenient. The electrode's liquid ion exchanger deteriorates with time, causing a significant error if it is not recalibrated frequently and rejuvenated at least monthly. Although Cl^{1-} and HCO_3^{1-} ions can be removed by precipitation, the procedure is difficult and time consuming. Automated systems have been developed for the several reductions and determinative colorimetric reactions.

Total Nitrogen

The increasing use of organic amines and other nitrogen compounds in manufacturing processes has led to increase in regulation in Europe for wastewater to be treated before being released in the environment. Standards such as ISO/TR 11905 Part 1 and 2, DIN 38 409, and DIN 38 406-E 5-1 include methods for the determination of nitrogen. The European Council requires that all wastewater treatment plants in major cities to measure total nitrogen (TN) by instrumental analysis.

Kjeldahl Method The total or Kjeldahl nitrogen standard method (for a discussion of the Kjeldahl Method, see Definitions at the end of this chapter) determines free ammonia and organically bound nitrogen in the −3 valence state but does not determine nitrites, azides, nitro, nitroso, oximes, or nitrates. Organic nitrogen is determined by subtracting the separately determined free ammonia nitrogen from the total nitrogen. The original analysis consists of several hours of digestion in boiling sulfuric acid, addition of toxic mercury compounds, then ammonia distillation and detection. The automated Kjeldahl has improved the inherent hazard of the test to the operator.

Chemiluminescence Detector (CLD) Figure 1.43c shows a flow diagram for a total nitrogen chemiluminescence analyzer. An aliquot of the sample is injected into a high-temperature furnace in an atmosphere of pure oxygen where nitrogen is converted to NO. The carrier gas transports the resulting steam and gases through the moisture control system. Inside the chemiluminescence detector, the NO formed reacts with ozone to produce an excited state of nitrogen dioxide (NO_2^*).

A photomultiplier tube adjacent to the reaction cell detects the photons (light) emitted as the NO_2^* returns to its ground state. The integrated signal is proportional to the amount of nitrogen in the sample. Concerns about recovery for different types of nitrogen species have been raised. Some manufacturers have employed more efficient catalysts, a NO_x converter, or swept the stream of gases through a reducing chamber containing a very strong reducing agent such as vanadium (III) chloride. Simultaneous measurement of carbon (see Chapter 1.67) can be conducted by allowing the gas stream to pass through an NDIR detector before the CLD detector.

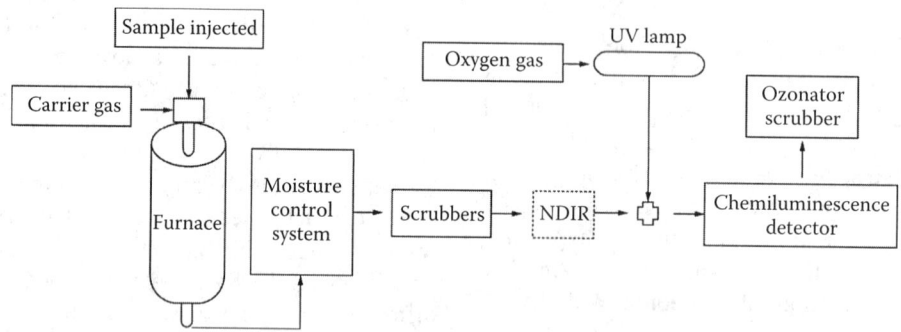

FIG. 1.43c
Chemiluminescence total nitrogen analyzer.

A typical range of measurement is 0.05–4000 mg/l N. Speciation of the different nitrogen species can be accomplished by using a gas chromatograph with the chemiluminescence detector. For higher levels of TN measurements, such as 30 mg/l to % levels, the NO_x produced during combustion is converted to N_2, which is detected with a thermal conductivity detector.

The NO produced from the combustion of a nitrogen-containing sample can also be analyzed by reacting with an electrolyte in an electrochemical cell. This reaction produces a measurable current, which is directly proportional to the amount of nitrogen in the original sample. The analytical range of measurement is 0.1–1000 mg N/l and precision of 1%–10% can be achieved, depending on the sample matrix.

SUMMARY

This chapter concentrated on the measurement of the concentration of nitrogen compounds in water. While these concentrations are important indicators of water quality, in this handbook the reader can find much more information on water quality determination. For example the detection of phosphate concentration is covered in Chapter 1.56, the measurement of pH in Chapter 1.55 and the overall subject of water quality analysis is covered in Chapter 1.73.

SPECIFICATION FORMS

When specifying analyzers for the measurement of nitrogen compounds in water (ammonia, nitrite, nitrate, etc.) use the ISA form 20A1001 and when the type of analyzer has already been selected and both the analyzer and the requirements of the process application are to be specified, use the ISA form 20A1002. Both forms are reproduced with the permission of the International Society of Automation on the next pages.

612 Analytical Measurement

ANALYSIS DEVICE — Operating Parameters

1	RESPONSIBLE ORGANIZATION		6	SPECIFICATION IDENTIFICATIONS	
2			7	Document no	
3	(ISA)		8	Latest revision	Date
4			9	Issue status	
5			10		

11	ADMINISTRATIVE IDENTIFICATIONS			40	SERVICE IDENTIFICATIONS Continued		
12	Project number		Sub project no	41	Return conn matl type		
13	Project			42	Inline hazardous area cl	Div/Zon	Group
14	Enterprise			43	Inline area min ign temp	Temp ident number	
15	Site			44	Remote hazardous area cl	Div/Zon	Group
16	Area	Cell	Unit	45	Remote area min ign temp	Temp ident number	
17				46			
18	SERVICE IDENTIFICATIONS			47			
19	Tag no/Functional ident			48	COMPONENT DESIGN CRITERIA		
20	Related equipment			49	Component type		
21	Service			50	Component style		
22				51	Output signal type		
23	P&ID/Reference dwg			52	Characteristic curve		
24	Process line/nozzle no			53	Compensation style		
25	Process conn pipe spec			54	Type of protection		
26	Process conn nominal size	Rating		55	Criticality code		
27	Process conn termn type	Style		56	Max EMI susceptibility	Ref	
28	Process conn schedule no	Wall thickness		57	Max temperature effect	Ref	
29	Process connection length			58	Max sample time lag		
30	Process line matl type			59	Max response time		
31	Fast loop line number			60	Min required accuracy	Ref	
32	Fast loop pipe spec			61	Avail nom power supply		Number wires
33	Fast loop conn nom size	Rating		62	Calibration method		
34	Fast loop conn termn type	Style		63	Testing/Listing agency		
35	Fast loop schedule no	Wall thickness		64	Test requirements		
36	Fast loop estimated lg			65	Supply loss failure mode		
37	Fast loop material type			66	Signal loss failure mode		
38	Return conn nominal size	Rating		67			
39	Return conn termn type	Style		68			

69	PROCESS VARIABLES	MATERIAL FLOW CONDITIONS				101	PROCESS DESIGN CONDITIONS		
70	Flow Case Identification				Units	102	Minimum	Maximum	Units
71	Process pressure					103			
72	Process temperature					104			
73	Process phase type					105			
74	Process liquid actl flow					106			
75	Process vapor actl flow					107			
76	Process vapor std flow					108			
77	Process liquid density					109			
78	Process vapor density					110			
79	Process liquid viscosity					111			
80	Sample return pressure					112			
81	Sample vent/drain press					113			
82	Sample temperature					114			
83	Sample phase type					115			
84	Fast loop liq actl flow					116			
85	Fast loop vapor actl flow					117			
86	Fast loop vapor std flow					118			
87	Fast loop vapor density					119			
88	Conductivity/Resistivity					120			
89	pH/ORP					121			
90	RH/Dewpoint					122			
91	Turbidity/Opacity					123			
92	Dissolved oxygen					124			
93	Corrosivity					125			
94	Particle size					126			
95						127			
96	CALCULATED VARIABLES					128			
97	Sample lag time					129			
98	Process fluid velocity					130			
99	Wake/natural freq ratio					131			
100						132			

133	MATERIAL PROPERTIES			137	MATERIAL PROPERTIES Continued		
134	Name			138	NFPA health hazard	Flammability	Reactivity
135	Density at ref temp		At	139			
136				140			

Rev	Date	Revision Description	By	Appv1	Appv2	Appv3	REMARKS

Form: 20A1001 Rev 0 © 2004 ISA

		RESPONSIBLE ORGANIZATION		ANALYSIS DEVICE COMPOSITION OR PROPERTY Operating Parameters (Continued)			SPECIFICATION IDENTIFICATIONS		
1					6				
2					7	Document no			
3					8	Latest revision		Date	
4					9	Issue status			
5					10				

	PROCESS COMPOSITION OR PROPERTY			MEASUREMENT DESIGN CONDITIONS				
	Component/Property Name	Normal	Units	Minimum	Units	Maximum	Units	Repeatability
12								
13								
...								
75								

Rev	Date	Revision Description	By	Appv1	Appv2	Appv3	REMARKS

Form: 20A1002 Rev 0 © 2004 ISA

Definitions

Albuminoid nitrogen is nitrogen which is still in the ammonium salts or still within the organic proteinoid molecules that are present in the water.

Chemiluminescence: Chemiluminescence is the emission of light resulting from a chemical reaction.

Denitrification: Denitrification is a process that involves nitrogen getting back into the atmosphere. It can be used when there is too much nitrogen in the soil because of the use of excessive amounts of fertilizers. The process is performed primarily by heterotrophic bacteria.

Eutrophication: The process by which a body of water acquires a high concentration of nutrients, especially phosphates and nitrates. These typically promote excessive growth of algae. As the algae die and decompose, high levels of organic matter and the decomposing organisms deplete the water of available oxygen, causing the death of other organisms, such as fish. Eutrophication is a natural, slow-aging process for a water body, but human activity greatly speeds up the process.

Ion-Exchange Chromatography (IEC) is a process that is able to separate polar molecules because of their affinity to the ion exchanger used. It works on charged molecules and is often used in water analysis and quality control.

Kjeldahl Method is an analytical chemistry method for measuring the quantity of nitrogen in chemical substances, developed by Johan Kjeldahl. It consists of heating the substance with sulphuric acid to liberate the reduced nitrogen as ammonium sulphate. After this potassium sulphate is added, followed by a number of other laboratory analysis steps which are concluded by back titration to determine the nitrogen content of the sample.

Nessler Cylinders are used for colorimetric. The color of the sample in a Nessler cylinder is visually compared with a reference model. The tubes are often used to carry out a series of calibration solutions of increasing concentrations, which functions as a comparative scale.

Nitrification: The process by which bacteria in soil and water oxidize ammonia and ammonium ions and form nitrites and nitrates. Because the nitrates can be absorbed by more complex organisms, as by the roots of green plants, nitrification is an important step in the nitrogen cycle.

Abbreviations

CLD	Chemiluminescence Detector
IEC	Ion Exchange Chromatography
NDIR	Nondispersive Infrared
TN	Total Nitrogen
TOC	Total Organic Carbon

Organizations

DIN	Deutsches Institut fur Normung
ISO	International Standards Organization

Bibliography

ASA Analytics, Total nitrogen in wastewater, 2013, http://www.asaanalytics.com/total-nitrogen.php.

Banica, F., Chemical Sensors and Biosensors: Fundamentals and Applications, 2012, http://www.wiley.com/WileyCDA/WileyTitle/productCd-0470710667.html.

Dominik, S., Process worldwide: Online analyzer for total nitrogen measurement, 2011, http://www.process-worldwide.com/analysis_lab_equipment/analytical_instruments/articles/314600/.

Ewing, G., Analytical Instrumentation Handbook, 2009, http://www.amazon.com/Ewings-Analytical-Instrumentation-Handbook-Edition-ebook/dp/B000SHLXRO.

Fresenius, W. et al., Water Analysis, 1988, http://www.springer.com/us/book/9783642726125.

GEC, Total Nitrogen Analyzer, 2015, http://www.gec.jp/CTT_DATA/WMON/CHAP_4/html/Wmon-087.html.

Golterman, H. L., Direct nessierization of ammonia and nitrate in fresh-water, 1991, http://www.limnology-journal.org/articles/limn/pdf/1991/01/limno19911p99.pdf.

ISO/TR 11905-1 and -2, Water quality—Determination of nitrogen, 1997, http://www.iso.org/iso/iso_catalogue/catalogue_tc/catalogue_detail.htm?csnumber=23630.

Miranda, K. et al., A rapid, simple spectrophotometric method for simultaneous detection of nitrate and nitrite, nitric oxide, 2001, https://www.researchgate.net/publication/12151822_A_Rapid_Simple_Spectrophotometric_Method_for_Simultaneous_Detection_of_Nitrate_and_Nitrite.

NICO, Method for determining the concentration of ammonium (NH_4^+) in aqueous solutions, 2015, http://www.nico2000.net/analytical/ammonium.htm.

Palmer, T. and Ross, M., Measuring Ammonia with Online Analyzers, 2001, http://www.wwdmag.com/analytical-instruments-spectrophotometric/measuring-ammonia-online-analyzers.

Paparone, F., What You Should Know About ISE Measurements of Ammonia, 2015, http://www.coleparmer.com/TechLibraryArticle/971.

Shugar, G. et al., Chemical Technicians' Ready Reference Handbook, 2012, http://www.amazon.com/Chemical-Technicians-Reference-Handbook-Edition/dp/0071745920.

Thermo Fischer Scientific, total nitrogen analysis, 2015, http://www.selectscience.net/products/total-nitrogen-and-sulfur-analyzer,-tn+ts-3000/?prodID=12619.

TOSOH, Ion Exchange Chromatography (IEC), 2015, http://www.separations.us.tosohbioscience.com/ServiceSupport/TechSupport/ResourceCenter/PrinciplesofChromatography/IonExchange.

Wang, X. J., et. al., Spatial and temporal variations in dissolved and particulate organic nitrogen in the equatorial pacific: biological and physical influences, 2008, http://www.biogeosciences.net/5/1705/2008/bg-5-1705-2008.pdf.

Wikipedia, Kjeldahl Method, http://en.wikipedia.org/wiki/Kjeldahl_method.

1.44 Nitrogen Oxide (NO$_x$) Analyzers

R. J. GORDON (1974, 1982)   **B. G. LIPTÁK** (1995, 2003, 2017)

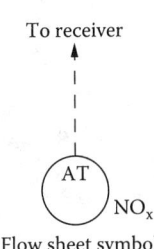

Flow sheet symbol

Analysis methods	A. Infrared
	B. Ultraviolet
	C. Chemiluminescent
	D. Colorimetric
	E. Electrochemical
	F. Gas chromatography
Applications	For industrial applications, the most often used analyzers are the nondispersive infrared, ultraviolet, and electrochemical types. The chemiluminescent and gas chromatographic techniques are less frequently used.
Reference method	Colorimetric, applied to integrated samples collected in alkaline solution
Ranges	A. 0–500 ppm to 0%–10%
	B. 0–100 ppm to 0%–100%
	C. 0–50 ppb to 0–10,000 ppm
	D. Down to ppb
	E. 0–500 to 0–2500 ppm
	F. Down to ppb
Inaccuracy	Generally 1%–2% of span, but some microprocessor-based electrochemical designs can be accurate within 2% of reading, and some chemiluminescent units are accurate to within 0.5% of span.
Costs	The installation and the upkeep costs are normally larger than the first costs of purchasing.
	A. Single-beam portable or laboratory units, $4,000–$5,000; an industrial NDIR, $8,000; a multiple-gas analyzer using up to five gases from 50 m (150 ft) distance, $25,000 to $27,000; FTIR, $75,000 to $125,000
	B. Laboratory spectrophotometers, from $2,500 to $5,000; industrial units, about $20,000
	C. About $10,000–$20,000, used laboratory units available at around $3,000
	D. About $50,000
	E. Pocket-sized personal monitors begin at $750; remote sensor heads begin at $1,500; onsite monitors with data loggers begin at $5,000; stack gas analyzer with printer, auto calibration, and probe costs $6,000–$8,000
	F. Installed cost in the range of $50,000–$100,000
Partial list of suppliers	For a more complete and global list of suppliers, refer to http://www.zhdanov.ru/classified-catalogue/manufacturers-and-suppliers/continuous-emission-monitoring-systems-cems-ie.htm and for additional
	Altech (A,C) http://www.altechusa.com/nitrogen.html
	Arrow Straight (E) (http://www.shelfscientific.com/cgi-bin/tame/newlaz/non.tam)
	Ashtead Technology (http://ashtead-technology.com/productgroup/instruments/904/)
	Chemiluminescent Wet Chemistry Analyzers (Chapter 1.74)
	Colorimetric Analyzers (Chapter 1.13)
	ECO Physics (C) (http://www.ecophysics-us.com/)
	EiUk (portable A, E) http://www.eurotron-uk.com/products/emission-analyser/11-rasi-900-1-emission-analyser.html
	Electrochemical Analyzers (Chapter 1.21)
	Environment SA (C) (http://www.environnement-sa.com/products-page/en/air-quality-monitoring-en/gas-analyzers-en/ac32m-analyzer-oxides-nitrogen-no-nox-no2-2/?cat=86)

(Continued)

Gas Chromatographs (Chapter 1.10)

GE Power&Water (C) (http://www.geinstruments.com/applications/medical-research)

Horbia Instruments (A, C) (http://www.horiba.com/us/en/automotive-test-systems/products/emission-measurementsystems/portable-emission-analyzers/details/mexa-720-nox-860/)

Infrared Analyzers (Chapter 1.32)

MSA Instruments (A) (http://ca.msasafety.com/search?text=nitrogen+oxide+analyzer&_requestConfirmationToken=0e758b2bd10ca18daebb7889b70e25eeaa606912)

Rosemont (A) (http://www2.emersonprocess.com/siteadmincenter/PM%20Rosemount%20Analytical%20Documents/PGA_PDS_MLT_2_103–6582_200612.pdf)

Siemens (C) (http://www.industry.usa.siemens.com/automation/us/en/process-instrumentation-and-analytics/processanalytics/pa-brochures/Documents/SIEMENS%20NOXMAT%20600%20Brochure.pdf)

Signal USA (C) (http://www.teledyne-api.com/manuals/04410D_200E.pdf)

Teledyne (A, C) (http://www.teledyne-ml.com/9841b.asp)

Tisch Environmental (http://tisch-env.com/products/25-Gas-Analyzers/39-TE-NA721/default.asp)

Ultraviolet Analyzers (Chapter 1.69)

INTRODUCTION

NO_x means the total concentration of NO and NO_2. Oxygen and nitrogen do not react at ambient temperatures. Therefore, nitrogen oxide sensors (NO_x) are usually high-temperature devices built to detect NO_x in combustion products, such as in smokestacks. In areas of high motor vehicle traffic, such as in large cities, the amount of nitrogen oxides emitted into the atmosphere can be significant because NO_x gases are formed whenever combustion occurs in the presence of nitrogen. It is also produced by lightening during thunderstorms, due to the extreme heat produced. The main cause of thermal NO_x formation is the combusting natural gas, but the burning. Fuel NO_x tends to dominate during the combustion of fuels, such coal can also have a significant nitrogen content and can generate NO_x, particularly when burned in combustors designed to minimize thermal NO_x.

During daylight, these concentrations are in equilibrium and the ratio NO/NO_2 is determined by both the intensity of the sunshine and the concentration of ozone which reacts with NO to again form NO_2. When NO_x and volatile organic compounds (VOCs) react in the presence of sunlight, they form ozone and photochemical smog and thereby cause air pollution. Ozone can cause adverse effects such as damage to lung tissue and reduction in lung function. Nitrogen oxides eventually form nitric acid when dissolved in atmospheric moisture, forming a component of acid rain. Small particles can penetrate deeply into sensitive lung tissue and damage it, causing premature death in extreme cases.

Many analyzers are capable of detecting the NO_x concentration, and their features, and capabilities are discussed in the following chapters:

Chapter 1.10   Gas Chromatographs
Chapter 1.13   Colorimeters
Chapter 1.21   Electrochemical
Chapter 1.32   Infrared
Chapter 1.62   Spectrometers
Chapter 1.69   Ultraviolet

The oxides of nitrogen are measured both in ambient air and in the gases emitted by industry. While the types of analyzers used for these two applications do overlap, here, an attempt will be made to separate them. Therefore, in this chapter, the devices more often used for industrial emission monitoring will be discussed first. These include infrared, ultraviolet, chemiluminescent, gas chromatographic, and electrochemical devices. Colorimetric and coulometric analysis is more often used for ambient air analysis and will be discussed later in this chapter.

DESIGNS FOR INDUSTRIAL EMISSION APPLICATIONS

Industrial monitoring usually involves the detection of NO_x concentration in hot combustion products. It will be noted in the following paragraphs, which of the NO_x detectors are suited for such applications, as not all of them are.

Paramagnetic Analyzers

Nitrogen oxide is attracted by the magnetic field and therefore is detectable by a paramagnetic analyzer. This instrument is described in this chapter under Oxygen Analyzers (Chapter 1.49) and is not widely used for the measurement of nitrogen oxide concentration.

Thermal Conductivity Analyzers

The thermal conductivity (see Chapter 1.66) of nitric oxide (NO) is slightly less than that of air—about 90%. Therefore, although thermal conductivity measurements have been attempted for NO analysis, it is neither a sensitive nor a selective means of measurement.

Nondispersive Infrared Analyzers

Nondispersive infrared (NDIR) analyzers are suitable for the determination of NO concentration. These devices are most often used in stack gas analyzer packages, which, in addition to NO_x, also detect the concentrations of carbon dioxide, carbon monoxide, sulfur dioxide, and opacity. Chapter 1.32 describes the NDIR analyzers, and Figures 1.32k and 1.32ad shows some of their stack-mounted variations.

Ultraviolet Analyzers

The absorbance of nitrogen dioxide (NO_2) in the ultraviolet range is shown in Figure 1.44a. Because, in most combustion processes, there is an interest in determining the concentrations of the both NO_x and SO_2, some of the suppliers of ultraviolet analyzers have combined the two tasks into a single analyzer whereby their concentrations are simultaneously measured (Figure 1.44b). Because NO is essentially

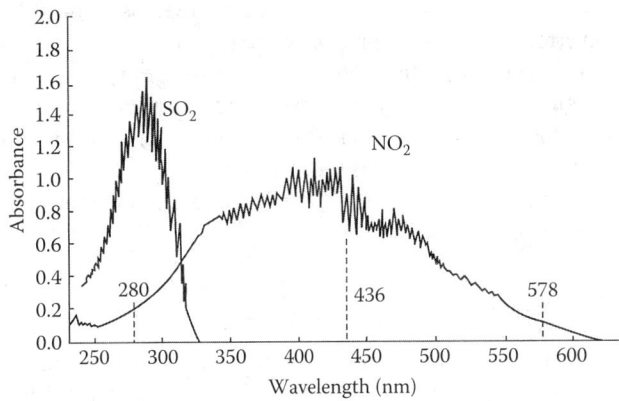

FIG. 1.44a
The absorbance of SO_2 and NO_2 in the ultraviolet range.

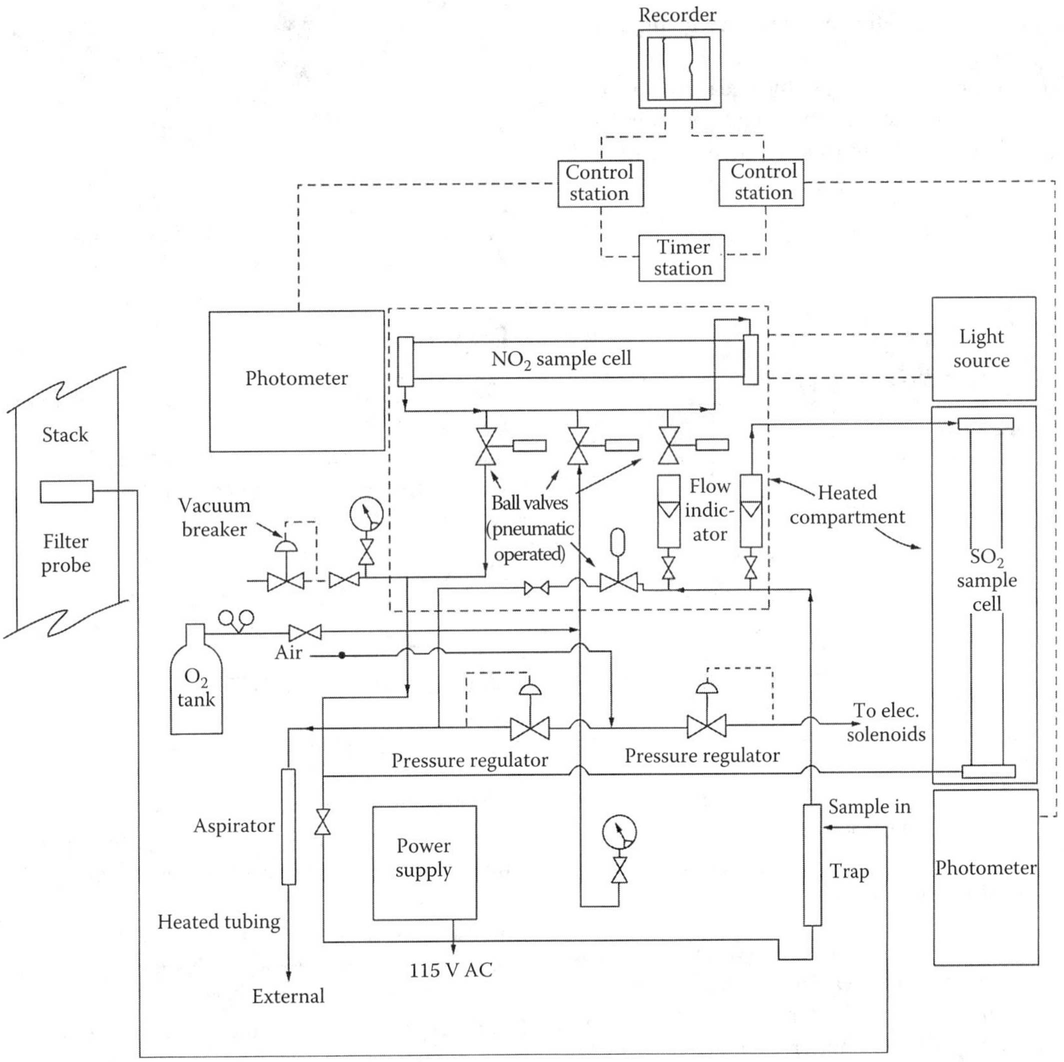

FIG. 1.44b
Ultraviolet analyzer used for the simultaneous measurement of NO_x and SO_2.

transparent in the visible and ultraviolet regions, it must be converted to NO_2 before it can be measured.

In Figure 1.44b, this conversion is achieved by contacting the sample gas with oxygen pressurized to five atmospheres. Once the NO is converted, the total NO_2 concentration is measured as NO_x. Ranges can be as narrow as 0–100 ppm or as wide as 0%–100% NO_x, and the measurement error is about 2% of full scale.

Chemiluminescent Analyzers

Another method of nitrogen oxide* determination makes use of the fact that NO reacts with ozone to form NO_2, and this reaction is accompanied with the release of light (chemiluminescence).

$$NO + O_3 \rightarrow NO_2 + h\nu \text{ (light } 0.6\text{–}3\mu) \quad \text{1.44(1)}$$

If excess ozone is present, the light emission is proportional to the amount of NO present in the sample. Instruments that operate on the basis of this principle are commercially available.

For the determination of NO_2 (which reacts with ozone rather slowly), some analyzers are provided with a catalytic converter to first reduce NO_2 to NO and then determine the sum of NO_2 and NO directly. The apparatus is schematically shown in Figure 1.44c.

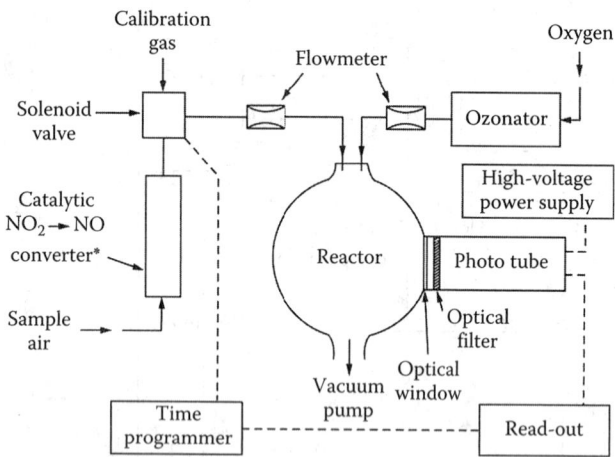

*Converter can be omitted if only NO is measured

FIG. 1.44c
Chemiluminescence nitric oxide analyzer.

Electrochemical Sensors

Portable or permanently installed probe-type electrochemical sensors are available for NO, NO_2, and NO_x measurement in stacks. These units are microprocessor based, are available with 4–20 mA transmitter outputs, and include

---
* Fontijn, A., Sabadell, A. J., and Ronco, R. J., *Anal. Chem.*, 42.

self-diagnostic and self-calibration features. The probe can be exposed to continuous temperatures of 850°C (1550°F), which for short periods can rise to 1200°C (2200°F).

As is discussed in detail in Chapter 1.14, state-of-the-art portable combustion analyzers (Figure 1.44d) combine non-dispersive infrared (NDIR) technology with electrochemical measurement cells, allowing the simultaneous measurement of O_2, CO, NO, NO_2, SO_2, H_2S, or H_2.

FIG. 1.44d
Portable combustion emission analyzer using both NDIR and electrochemical sensors to simultaneously measure the concentrations of seven gases, including NO and NO_2. (Courtesy of EiUk.)

Gas Chromatographs

There have been a limited number of reports of applications for analysis of oxides of nitrogen by gas chromatography. Such chromatographic columns are fairly short and are provided with electron capture detectors. These are the same types of chromatographs that are used for the analysis of peroxyacyl nitrates. For added details on gas chromatography, see Chapter 1.10.

AMBIENT AIR MONITORING

Calibration Methods

The reliability of dynamic calibration of gas analyzers is always superior to that of static calibration, but it is also much more difficult. This is because, in high concentrations, NO is rapidly oxidized to NO_2 by air, and NO_2 condenses and dimerizes (For a discussion of deminerization of NO_2, see *Definitions* at the end of this chapter.) at high concentrations. For these reasons, dynamic calibration for NO and NO_2 requires great care. In colorimetric or coulometric analysis, the NO in fact is not measured directly at all but is measured only after its oxidation to NO_2, so these analyzers are calibrated only for NO_2.

Dynamic Calibration Dynamic calibration for NO_2 requires preparation of a sample of inert gas containing a known concentration of NO_2. This sample may be obtained by gas-dilution techniques (with special precautions in handling NO_2), gravimetrically, electronically, or by use of a permeation tube. The permeation tube (as used with hydrocarbons) is described in Chapter 1.28. However, NO_2 permeation tubes are moisture sensitive. Therefore, during storage, these tubes should be protected from moisture, and only dry gases should be used for dilution. If these precautions are observed, the permeation tube is probably the most convenient method for the dynamic calibration of NO_2.

Static Calibration Static calibration is carried out with standard solutions of nitrite. Because this procedure does not detect the components of the gas in the sampling line, it is not as complete a test as is the dynamic calibration, but it is much simpler. The stoichiometric ratio of nitrite to NO_2 under controlled conditions is usually a constant value.

In the Griess–Saltzman method, the consensus (although there has been controversy) is that 0.72 mole of nitrite gives the color of 1 mole of NO_2. Using the Jacobs–Hochheiser method, this factor has been found to be 0.63. (These methods are further discussed in later paragraphs).

NO–NO₂ Combination Analysis

In continuous analysis, it is customary to determine both NO and NO_2 as NO_2 (Table 1.44a). The NO is oxidized to NO_2 by means of potassium permanganate or dichromate or by chromium trioxide, in various formulations. The efficiencies of conversion seem to depend on the length of service and, for the chromium oxidizers, they also depend on humidity. Aqueous permanganate seems to be the best choice, even though it may not be completely efficient (hence, NO may be underestimated).

In series analysis for NO and NO_2 (Figure 1.44e), the air passes through an NO_2 analyzer for measurement and removal of NO_2, then through an oxidizer to convert NO to NO_2, and finally through a second NO_2 analyzer. The second analyzer gives a measure of NO concentration. In one type of parallel analysis, two equal air streams are analyzed for NO_2, one of them after passage through an oxidizer. The latter gives total oxides of nitrogen (NO + NO_2) from which NO is found by difference from the other parallel analyzer. In a second type of parallel analysis, one stream is analyzed directly for NO_2, and the second is scrubbed free of NO_2 by passage through ascarite, followed by oxidation of NO to NO_2 and NO_2 analysis.

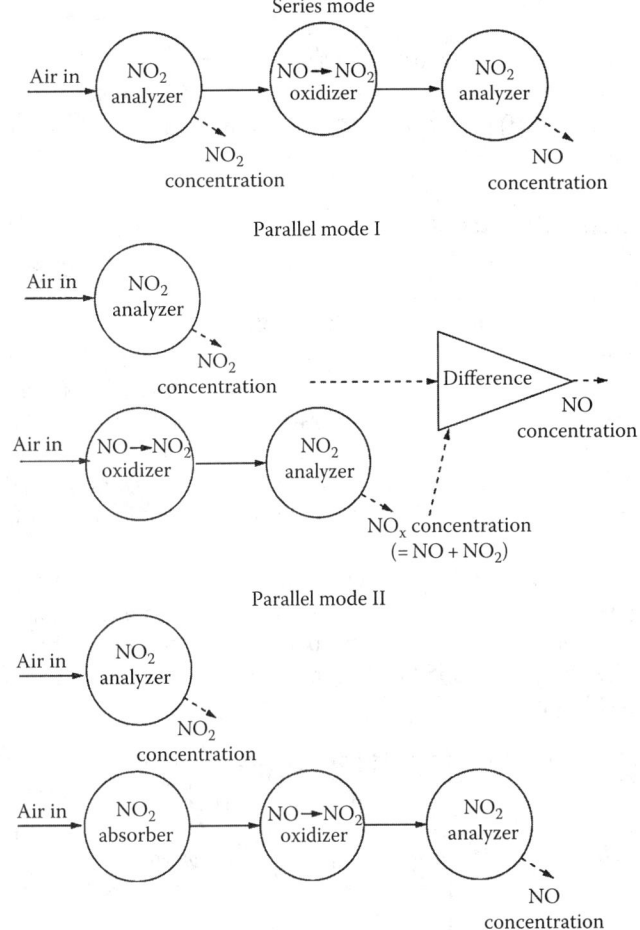

FIG. 1.44e
Nitrogen oxide analyzer modes.

TABLE 1.44a
Nitrogen Oxide Analyzers

General Method	Type	Advantages	Disadvantages
Colorimetric	Griess–Saltzman	Precise, thoroughly tested, widely used, continuous analysis	Short life of collected sample, sensitive reagents, NO oxidation required
	Jacobs–Hochheiser	Precise, stable after collection	Not adapted to continuous analysis, sensitive reagents, NO oxidation required
Coulometric		Simple apparatus, continuous analysis	Sensitive to other oxidants, NO oxidation required
Chemiluminescent		Dry gases only, sensitive photometry, continuous analysis	Requires ozone generator NO_2 catalytic reduction
Gas chromatography		Specific, frequent analysis	Not a developed instrument, expensive and complex
Electrochemical		Simple apparatus, continuous analysis	Sensitivity not high, NO oxidation required

Colorimetric Analyzers

There are two important colorimetric methods* EPA, Air Quality Criteria for Oxides of Nitrogen. (1993) for NO_2 determination. They are the Griess–Saltzman and the Jacobs–Hochheiser methods.

Griess–Saltzman Method The Griess–Saltzman method is used in most continuous colorimetric NO_2 analyzers. It is based on the reaction of NO_2 with sulfanilic acid to form a diazonium salt that couples with N-(1-napthyl)-ethylenediamine dihydrochloride to form a deeply colored azo dye. Air is passed into the reagent solution for not over 30 min. After that, time is allowed for development, and the color is measured at 550 nm.

The measurable range of concentrations is from 0.02 to 0.75 ppm. In manual use, the color is developed for 15 min and should be read within 1 hr (on a colorimeter or spectrophotometer). In a continuous analyzer (Figure 1.44f), the gas and liquid flow rates are adjusted for optimal response, and the developed color is potentiometrically read in a flow cell using a 550 nm filter. Response times are usually 5–15 min.

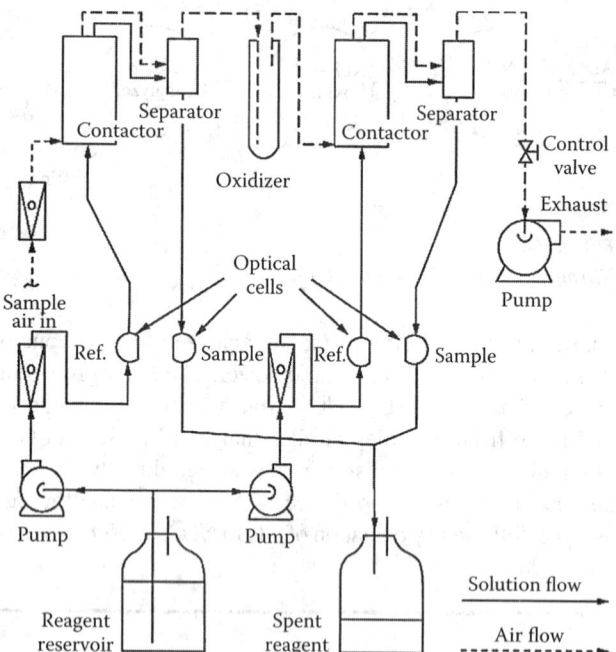

FIG. 1.44f
Colorimetric nitrogen oxide analyzer.

Jacobs–Hochheiser Method The Jacobs–Hochheiser method is the standard reference method used for the U.S. National Air Quality Standards. The reason for this is that the standards are based on the annual average, and this method allows collection of up to 24 h integrated samples and delays in analysis of at least 2 weeks. In contrast, with the Griess–Saltzman method, samples must be quickly analyzed.

In the Jacobs–Hochheiser method, the air is passed through aqueous sodium hydroxide so that the NO_2 is concerted to nitrite ion. Sulfur dioxide is removed from the solution by treatment with hydrogen peroxide and is acidified. The rest of the procedure is the same as for the Griess–Saltzman method except that sulfanilamide is used instead of sulfanilic acid. Efficiencies found with this procedure in the U.S. National Air Surveillance Network are approximately 35%.

PORTABLE MONITORS

For purposes of personal protection, battery-operated portable units are available (Figure 1.44d). These units are usually provided with one or two alarm set points and with memory for some thousands of data points along with their times and dates. Table 1.44b lists the ranges, resolutions, and alarm set points for a number of gases including NO and NO_2.

TABLE 1.44b
Typical Range, Sensitivity, and Alarm Set Points of Portable Personal Protection Monitors

Gas	Range	Resolution	Alarm Set Points (Low/High)
O_2	0%–30%	0.1%	19.5%/23.5%
CO	0–500 ppm	1 ppm	35/200 ppm
H_2S	0–100 ppm	1 ppm	10/20 ppm
SO_2	0–20 ppm	1 ppm	2/10 ppm
NO	0–250 ppm	1 ppm	25/50 ppm
NO_2	0–20 ppm	0.1 ppm	1/10 ppm
NH_3	0–50 ppm	1 ppm	25/50 ppm
PH_3	0–5 ppm	0.1 ppm	1/2 ppm
Cl_2	0–10 ppm	0.1 ppm	0.5/5 ppm
HCN	0–100 ppm	1 ppm	4.7/50 ppm

Source: Abstracted from Cole-Parmer Catalog 2001/2002, Vernon Hills, IL, 2001.

These pocket-sized, battery-operated, portable electrochemical detectors are usually provided with digital displays and audible alarms. They can be configured for one or more monitoring channels.

CONCLUSIONS

The most widely used conventional method of ambient air analysis is colorimetric. It is most often based on the use of the Griess–Saltzman reagent (a diazotization method). This is a fairly precise and dependable method but requires a great deal of attention.

The colorimetric method is specific for NO_2. To analyze for NO, an oxidation step is required. Of various oxidizer

---

* *Air Quality for Nitrogen Oxides*, U.S. Environmental Protection Agency, Air Pollution Control Office Publication No. AP-84, Chapter 5.

columns, the most commonly used is the permanganate one, but none is completely satisfactory. It is important to calibrate it over a range of NO concentrations and at various humidity levels.

The chemiluminescent method utilizes the reaction between ozone and NO. This is a dry gas method requiring an ozone generator and compressed gas. In the chemiluminescent method, a catalytic reduction of NO_2 to NO is required.

For industrial applications, the most often used analyzers are the NDIR, ultraviolet, and electrochemical types. The chemiluminescent and gas chromatographic techniques are less frequently used.

SPECIFICATION FORMS

When specifying analyzers for the measurement of nitrogen oxides (NO_x), use the ISA form 20A1001, and when the type of analyzer has already been selected and both the analyzer and the requirements of the process application are to be specified, use the ISA form 20A1002. Both forms are reproduced with the permission of the International Society of Automation on the next pages.

Analytical Measurement

ANALYSIS DEVICE — Operating Parameters

#	RESPONSIBLE ORGANIZATION		#	SPECIFICATION IDENTIFICATIONS		
1			6			
2			7	Document no		
3	ISA		8	Latest revision	Date	
4			9	Issue status		
5			10			

#	ADMINISTRATIVE IDENTIFICATIONS			#	SERVICE IDENTIFICATIONS Continued			
11				40				
12	Project number		Sub project no	41	Return conn matl type			
13	Project			42	Inline hazardous area cl		Div/Zon	Group
14	Enterprise			43	Inline area min ign temp		Temp ident number	
15	Site			44	Remote hazardous area cl		Div/Zon	Group
16	Area	Cell	Unit	45	Remote area min ign temp		Temp ident number	
17				46				
18	SERVICE IDENTIFICATIONS			47				
19	Tag no/Functional ident			48	COMPONENT DESIGN CRITERIA			
20	Related equipment			49	Component type			
21	Service			50	Component style			
22				51	Output signal type			
23	P&ID/Reference dwg			52	Characteristic curve			
24	Process line/nozzle no			53	Compensation style			
25	Process conn pipe spec			54	Type of protection			
26	Process conn nominal size		Rating	55	Criticality code			
27	Process conn termn type		Style	56	Max EMI susceptibility		Ref	
28	Process conn schedule no		Wall thickness	57	Max temperature effect		Ref	
29	Process connection length			58	Max sample time lag			
30	Process line matl type			59	Max response time			
31	Fast loop line number			60	Min required accuracy		Ref	
32	Fast loop pipe spec			61	Avail nom power supply			Number wires
33	Fast loop conn nom size		Rating	62	Calibration method			
34	Fast loop conn termn type		Style	63	Testing/Listing agency			
35	Fast loop schedule no		Wall thickness	64	Test requirements			
36	Fast loop estimated lg			65	Supply loss failure mode			
37	Fast loop material type			66	Signal loss failure mode			
38	Return conn nominal size		Rating	67				
39	Return conn termn type		Style	68				

#	PROCESS VARIABLES	MATERIAL FLOW CONDITIONS		#	PROCESS DESIGN CONDITIONS			
69				101				
70	Flow Case Identification		Units	102	Minimum	Maximum		Units
71	Process pressure			103				
72	Process temperature			104				
73	Process phase type			105				
74	Process liquid actl flow			106				
75	Process vapor actl flow			107				
76	Process vapor std flow			108				
77	Process liquid density			109				
78	Process vapor density			110				
79	Process liquid viscosity			111				
80	Sample return pressure			112				
81	Sample vent/drain press			113				
82	Sample temperature			114				
83	Sample phase type			115				
84	Fast loop liq actl flow			116				
85	Fast loop vapor actl flow			117				
86	Fast loop vapor std flow			118				
87	Fast loop vapor density			119				
88	Conductivity/Resistivity			120				
89	pH/ORP			121				
90	RH/Dewpoint			122				
91	Turbidity/Opacity			123				
92	Dissolved oxygen			124				
93	Corrosivity			125				
94	Particle size			126				
95				127				
96	CALCULATED VARIABLES			128				
97	Sample lag time			129				
98	Process fluid velocity			130				
99	Wake/natural freq ratio			131				
100				132				

#	MATERIAL PROPERTIES			#	MATERIAL PROPERTIES Continued		
133				137			
134	Name			138	NFPA health hazard	Flammability	Reactivity
135	Density at ref temp		At	139			
136				140			

Rev	Date	Revision Description	By	Appv1	Appv2	Appv3	REMARKS

Form: 20A1001 Rev 0 © 2004 ISA

1.44 Nitrogen Oxide (NO$_x$) Analyzers

	RESPONSIBLE ORGANIZATION	ANALYSIS DEVICE COMPOSITION OR PROPERTY Operating Parameters (Continued)		SPECIFICATION IDENTIFICATIONS		
1			6			
2	ISA		7	Document no		
3			8	Latest revision		Date
4			9	Issue status		
5			10			

	PROCESS COMPOSITION OR PROPERTY			MEASUREMENT DESIGN CONDITIONS				
	Component/Property Name	Normal	Units	Minimum	Units	Maximum	Units	Repeatability
13								
14								
...								
75								

Rev	Date	Revision Description	By	Appv1	Appv2	Appv3	REMARKS

Form: 20A1002 Rev 0

© 2004 ISA

Definitions

Demineralization of NO_2 is the joining of two NO_2 molecules. This pairing of odd electrons means that the molecule is no longer paramagnetic, but diamagnetic in nature. That is the reason why NO_2 exists as dimers, that is, N_2O_4.

Diamagnetic materials create induced magnetic fields in a direction opposite to an externally applied magnetic field and are repelled by the applied magnetic field.

Paramagnetic materials are attracted by an externally applied magnetic field and form internal induced magnetic fields in the direction of the applied magnetic field.

Abbreviations

NDIR   Nondispersive infrared
NO   Nitrogen oxide

Bibliography

ACGIH, Air Sampling Instruments for Evaluation of Atmospheric Contaminants, 2009, http://www.acgih.org/store/ProductDetail.cfm?id=2100.

AirNow, Air Quality Guide for Nitrogen Dioxide, 2011, http://airnow.gov/index.cfm?action=pubs.aqiguidenox.

BMC, Analysis of nitric oxides, 2011, http://bmcresnotes.biomedcentral.com/articles.

Brand-gaus chemiluminescence technology, 2004, http://www.brandgaus.com/NOxAnalyzerTechnology.htm.

California Environmental Protection Agency, Review of the Ambient Air Quality Standard for Nitrogen Dioxide, 2009, http://www.arb.ca.gov/research/aaqs/no2-rs/no2-rs.htm.

Clean Air Engineering, Stance and Source Emission Testing, 2014, http://www.cleanair.com/services/StackSourceTesting/index.htm.

Cooper, C. D., Alley, F. C., Air Pollution, 2002, http://www.amazon.com/Coopers-Pollution-Control-Edition-Hardcover/dp/B003WUI3I6.

EPA, Nitrogen Dioxide (NO2) Primary Standards, 2010, http://www3.epa.gov/ttn/naaqs/standards/nox/s_nox_index.html.

EPA, Nitrogen Dioxide, 2015, http://www3.epa.gov/airquality/nitrogenoxides/.

Federal Register, Standards of Performance for New Stationary Sources, 2011, https://www.federalregister.gov/articles/2011/03/21/2011-4495/standards-of-performance-for-new-stationary-sources-and-emission-guidelines-for-existing-sources.

Honeywell, Gas Detection Book, 2013, http://www.honeywellanalytics.com/~/media/honeywell-analytics/documents/english/11296_gas-book_v5_0413_lr_en.pdf?la=en.

ICS: Stack Sampler, 2015, http://www.indiamart.com/industrial-commercial-services/environment-equipments.html.

Lodge, J. P., ed., *Methods of Air Sampling and Analysis*, Lewis Publishers, Boca Raton, FL, http://www.amazon.com/Methods-Sampling-Analysis-James-Lodge/dp/0873711416.

MSA, Gas Detection Handbook, 2007, http://www.gilsoneng.com/reference/gasdetectionhandbook.pdf.

Purtz, E. P., Evaluation of electrochemical nitrogen oxide analyzers, 1997, http://www.ncbi.nlm.nih.gov/pubmed/9058250.

Sherman, R. E., Process Analyzer Sample-Conditioning System Technology, 2002, http://www.wiley.com/WileyCDA/WileyTitle/productCd-0471293644.html.

Teledyne Advanced Pollution Instrumentstion, MODEL 200E NITROGEN OXIDE ANALYZER, 2010, http://www.teledyne-api.com/manuals/04410d_200e.pdf.

1.45 Odor Detection

A. TURK (1974, 1982)   **W. P. DURDEN** (2003)   **B. G. LIPTÁK** (1995, 2017)

Flow sheet symbol: AT — Odor, To receiver

Methods of detection	A. Organoleptic, using humans as the sensors of odor
	B. Chromatograph with mass spectrometer, flame ionization, etc., sensors
	C. Electronic nose
Sensitivity of detection	A. About 0.2 ppb
	B. About 10–200 ppb
	C. In the range of ppt to ppb, depending on chemical
Costs	Chromatographs (Chapter 1.10): Online, usually with flame ionization sensor and accessories, about $100,000
	Portable chromatograph about $ 20,000
	Mass spectrometer (Chapter 1.35) about $100,000
	Handheld, portable, about $1,000
	Many gas chromatographs are sensitive enough to detect odors. For a list of gas chromatograph suppliers, refer to Chapter 1.10.
Partial list of suppliers	Alpha Mos (B—FID sensor) http://www.alpha-mos.com/analytical-instruments/heracles-electronic-nose.php
	Arjay Engineering, (UV fluorescence) http://www.arjayeng.com/hydrosense_2410.html.
	Brechbüehler AG (B—w/mass spectrometer) https://www.brechbuehler.ch/GC-O.126.0.html
	Cyranose (C) http://www.ideo.com/work/cyranose-320
	Electronic Products (C)
	Electronic Sensor Technology (B—portable with SAW sensor) http://www.estcal.com/products/model4200_portable_znose.html
	MOCON Inc. (B—olfactometer) http://www.mocon.com/aroma.php
	MSA Instrument Div. (C—portable, handheld) http://us.msasafety.com/Portable-Gas-Detection/c/114?N=10138&Ne=10187&isLanding=true
	Perkin-Elmer Corp. (B) http://www.perkinelmer.com/Catalog/Product/ID/N6590010
	Shinyei (handheld) http://www.sca-shinyei.com/odormeter

INTRODUCTION

Odor is a sensation associated with smell, which can be hard to quantify. The same quantities of different materials cause different odor intensities. The unit of odor intensity is based on the odor of tertiary butyl mercaptan (TBM; W = 1.0). Using that reference, H_2S, for example, has an odor intensity of 0.08 or 8% of TBMs. Most odorant substances contain sulfur. Table 1.45a lists a number of odorant substances and their relative odor intensities (W).

TABLE 1.45a
Relative Odor Intensity of Different Chemicals

Abbreviation	Name	Formula	Odor Intensity (W)
EM	Ethyl mercaptan ethanethiol	CH_3H	1.08
DMS	Dimethyl sulfide	$(CH_3)_2S$	1.0
IPM	Isopropyl mercaptan 20-propanethiol	$CH_3CHSHCH_3$	0.88
MES	Methyl ethyl sulfide	$(CH_3)_2CH_2S$	0.66
NPM	Normal propyl mercaptan 1-propanethiol	$(CH_3)CH_2CH_2SH$	0.85
TBM	Tertiary butyl mercaptan 2-methyl 2-propanethiol	$(CH_3)_3CSH$	1.00 (ref)
SBM	Secondary butyl mercaptan 1-methyl 1-propanethiol	$(CH_3)_2CH_2CHSH$	1.99
DES	Diethyl sulfide ethyl sulfide	$(C_2H_5)_2S$	0.22
Thiophane	Thiophane tetrahydrothiophene	C_4H_8S	1.63
EIS	Ethyl isopropyl sulfide		0.07

The human olfactory system is capable of detecting and identifying a wide variety of chemical structures and giving different responses to different materials. Commercially available instrumentation and chemical methods are generally restricted to particular chemical structures and give a similar response to all compounds with that structure.

ODOR MEASUREMENT

This chapter describes both organoleptic and chemical/instrumental methods of odor measurement. The organoleptic methods, which utilize the human olfactory system, are completely subjective. However, techniques are available that can convert subjective measurements into useful objective results.

Recent improvements in technology and increased research in the area of instrumentation have made dramatic improvements in the creation of instruments that are capable of surpassing the sensitivity of the human olfactory system. In fact, the instruments of today are approaching the sensitivity of the canine olfactory system, which is thought to be as much as a million times more sensitive than that of humans. These developments have overcome the two major shortcomings generally suffered by chemical/instrumental methods.

The human olfactory system is orders of magnitude more sensitive than currently available chemical/instrumental methods. Humans can detect and identify odors present in quantities to which commercially available instrumentation and chemical methods are insensitive.

Human Olfactory System

The human olfactory system actually involves more than the nose. As a stream of air is drawn in through the nostrils, it is warmed and filtered by passing over the three baffle-shaped turbinate bones in the upper part of the nose (Figure 1.45a). Some of the air swirls past the olfactory receptors located high up in the nasal passages just below the brain.

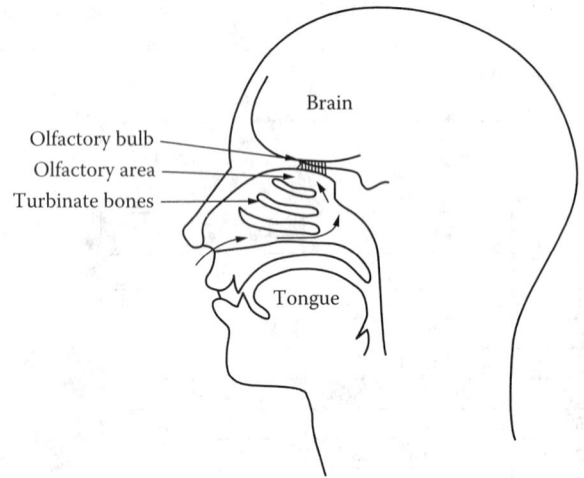

FIG. 1.45a
The human olfactory system.

These odor receptors consist of hairlike filaments attached to the end of the fibers of the olfactory nerve and the trigeminal nerve endings (Figure 1.45b). Upon being stimulated by odorous materials, these receptors send signals to the olfactory bulb, where they are relayed to higher centers of the brain. In these higher centers, the signals are integrated and interpreted in terms of the character and intensity of the odor.

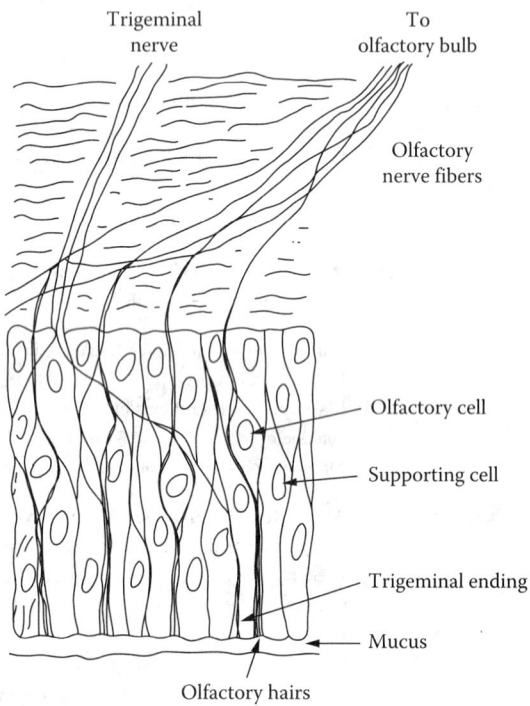

FIG. 1.45b
Section of an olfactory epithelium.

ODOR TESTING

Because of the extreme sensitivity of the human nose, the concentrations of odorants in air can be extremely low yet can produce a strong response. These low concentrations present a problem with regard to the manner by which samples are presented to the human sensor. Improper handling of an ambient air sample or improper preparation of an odorant standard can result in erroneous results because of adsorption of the odorants on the walls, incomplete mixing of an odorant with dilution air, or impure dilution air.

Errors can also result from the type of system used to bring the sample in contact with the olfactory sensors. The use of syringes or other means by which the nose is not immersed in the sample yields lower apparent odor intensities than methods whereby the nose is fully exposed to the sample.*

The use of an odor room can also yield misleading results because of natural odors generated by the body and adsorption

---

* Reckner, L. R. and Squires, R. E., Diesel Exhaust Odor Measurement Using Human Panels, 1968, SAE Paper 680444. http://papers.sae.org/680444/.

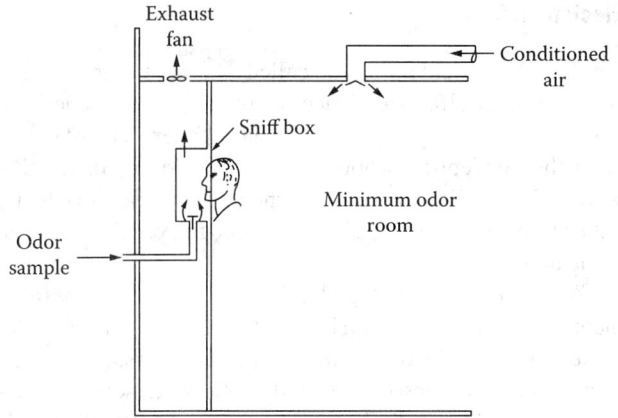

FIG. 1.45c
Diagram of sniff odor test system.

by clothing. The best approach appears to be the *sniff box* shown in Figure 1.45c. In this case, the human sensor is kept in an air-conditioned room into which charcoal-filtered air is blown to maintain a slight positive pressure. This prevents the sample in the sniff box and other contaminated air from entering the room. For measurement, the whole face is placed in the sample stream within the sniff box.

Odor Panels

The use of humans as measurement instruments introduces a variability that is difficult to control. This variability is due to the moods, biases, and other vagaries associated with the human sensor. Each human being is a unique creation, with differences in intelligence, persistence, sensitivity, experience, and interests. These differences make the response of an individual an unreliable indicator in itself. The use of a panel of individuals for the measurement of odor levels statistically eliminates the unreliability of the individual.

An odor panel generally consists of 5–10 people. The people who compose the odor panel need not have unusual olfactory abilities. However, they should be able to distinguish among odors of different intensities, discriminate among different odor qualities, and communicate the perceived sensations in terms of reference standards. They should also be emotionally receptive to making quantitative and discriminatory judgments without expressing their preference.

The primary tasks of an odor panel are the following:

1. Judge the relative intensity of an odor at different dilutions.
2. Discriminate among the different odor qualities.
3. Combine the intensity and quality of an odor to give a composite profile that can be communicated to the scientist in charge of the test.

The panelists are expected to follow instructions and also to render independent judgment reflecting their own sensations.

Tests In addition to exposing the panelists to various levels of odors similar to those to be measured, three tests can be used for training purposes. These are the triangle test, the intensity rating test, and the multicomponent identification test. In the triangle test, three samples are presented to the panelist at the same time. Two are identical, and the third is different in either intensity or quality. The panelist must select the odd sample.

The intensity rating test consists of a series of perhaps 20 dilutions of an odorant in an odorless medium. One sample is removed from the series, and the panelist is asked to determine the sample's proper place in the series according to its odor intensity.

The last test is the multicomponent odor identification test, in which three mixtures are presented to the panelist. These mixtures contain, in sequence, two, three, and four odors out of a possible total of eight known standards. The panelist is told how many components to look for and is asked to identify each of them.

This group of three tests develops a panelists' ability to distinguish between different odor intensities, discriminate among odor qualities, and communicate their sensations in terms of predetermined standards. After the group has some experience working together under a competent test director in actual measurement situations, highly accurate measurements of odor levels can be made.

ODOR DETECTORS

Gas Chromatography

The most successful instrumental method for the measurement of odor has been the gas chromatograph (see Chapter 1.10 for a detailed description of this analyzer). Gas chromatographs, after separating the components of a sample, can use a variety of detectors to identify the quantities of the individual components, including flame ionization, mass spectrometer (Figure 1.45d), and other sensors. In Kraft paper mills, for example, flame photometric detectors are used. These detectors can determine the concentrations of sulfur dioxide, hydrogen sulfide, and other

FIG. 1.45d
Odor-detecting chromatograph with mass spectrometer sensor. (Courtesy of Brechbüehler AG.)

odorous gases down to about 0.01 ppm (Chapters 1.31 ans 1.64). Success in detecting these sulfur-containing compounds using a coulometric cell detectors with platinum electrodes has also been reported (Chapters 1.64 and 1.65). Levels as low as 0.1 ppm were detected by this method.

Gravimetric sensors use gas chromatography (GC) coupled with surface acoustic wave and quartz crystal microbalance measurements. The GC separates the gas and deposits the chemical compound on a vibrating quartz crystal. This causes a change in the acoustic or sound propagation characteristics as a result of the mass change caused by the chemical absorbed. Response time is approximately 10 s.

The odor threshold value (OTV) of tetradecenal in water is 0.009 ppb.* Thus, the human nose, with its attendant sociological, psychological, and physiological complications, seems to be the only odor sensor available for fully successful odor measurement.

However, the detection of odor using chemical or chromatographic methods for determining measurable levels of individual chemicals in a vapor phase is not really odor detection but gas or chemical detection. The detected chemical(s) have an odor, and based on experience, the output of the chromatograph can be interpreted into an odor associated with, for example, the smell of a peach.

True odor analysis will not only determine that the odor is of a peach, but whether it is a green, ripe, or rotten peach. Today's instruments can now determine the *quality* of the odor in near real time. These odor detectors and analyzers are trained to recognize the odors and the levels of distinction by exposure to standards of odor. Once trained, these devices will provide a proportional reading of the odor, much the same as a human olfactory system but without the subjectivity.

Odor-detecting chromatographs can be permanently installed or can be portable (Figure 1.45e).

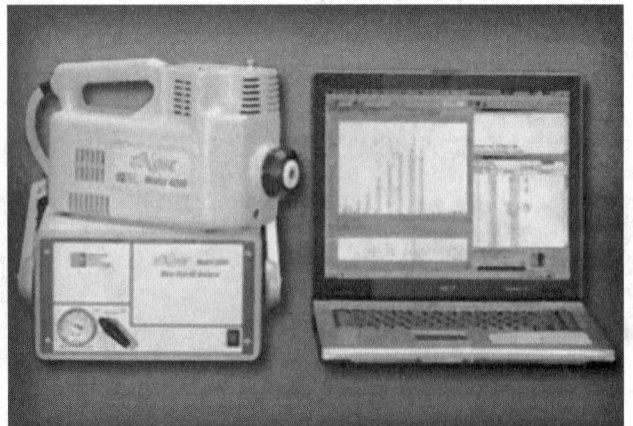

FIG. 1.45e
Portable chromatograph with SAW sensor. (Courtesy of Electronic Sensor Technology.)

---

* Widder, S. Symrise GmbH & Co. KG., 2008, http://www.google.com/patents/US7332468.

Electronic Nose

An electrochemical nose, also called an e-nose or micronose, is an artificial olfaction device having an array of chemical gas sensors, a sampling system, and a pattern classification algorithm to identify various gases, vapors, or odors. In other words, they utilize an array of chemical gas sensors and a pattern recognition algorithm to recognize, identify, and compare odors.

Advances in technology have led to significant improvements, both in the analysis capabilities and the sizes of these portable electronic nose-type detectors. The array of chemical gas sensors is usually solid-state sensors (chemoresistors, chemodiodes, or electrodes). All electronic noses use the same basic method of sensing and analysis (Figure 1.45f).

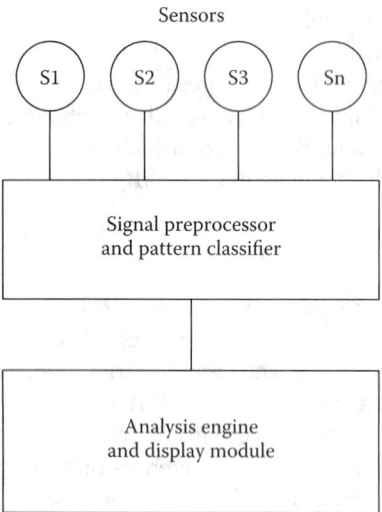

FIG. 1.45f
Typical schematic view of an electronic nose.

Primary odors are composed of polar molecules, organic vapors, and phthalocyanines. New sensors such as the chemosensors are devices that convert chemical composition into a quantifiable electrical format. Other new types include conductometric, optical, gravimetric, amperometric, calorimetric, potentiometric, and chemocapacitor sensors.

Training All electronic noses must be trained to recognize the odors as specific smells. Known samples are used to generate defined patterns, odor fingerprints, or (as they have been called) smell prints. These standards are used to define the parameters by which the training occurs. A major challenge is to interpret the odor signals into a specific reading of a specific odor with no false readings. The intelligence in the analyzer must emulate the human equivalent of subjective comparison. This is done through prediction algorithms and sophisticated pattern analysis such as multivariate, fuzzy logic, genetic algorithms, cluster analysis, discriminate function analysis, and adaptive models.

Polymeric Film Sensors

Conductometric sensors are made up of conductive polymers and metal oxide semiconductors. The first are composed of a polymeric film composite. This technology was developed concurrently in the United States and the United Kingdom. The work in the United States occurred at the California Institute of Technology in a joint effort with Jet Propulsion Laboratory. The developments in the United Kingdom occurred at the University of Manchester Institute of Science and Technology.

In both cases, the developed sensors consist of an array of pairs of electrical contacts that are electrically connected by a composite film. The composite film is composed of a special nonconductive polymer, which is mixed with carbon black in a homogeneous blend. This forms a conductive bridge between the electrical contacts.

The composite film is selected to absorb specific vapor analytes. As it does so, it swells, thereby increasing the distance between the carbon black particles. This first changes its resistance characteristics and then breaks the electrically conductive path in the film. The variance in the resistance between the electrical contacts is used as the output of the sensor. Because of the selected polymer, each sensor in the array is sensitive to a different level of concentration of a chemical. This allows relative concentrations to be determined.

Metal Oxide Sensors

The second type of sensor, metal oxide, employs a similar method of odor detection but uses metal oxide sensors in an array, reacting to the chemicals, changing the semiconductors' conductance, and producing the reading. The use of metal oxide sensors falls into two categories, thin and thick film. The thin film targets NO_x, H_2, and NH_3. The thick film is used for the detection of odors of alcohols, ketones, and combustible materials.

Other Sensors

Potentiometric sensors, using metal oxide semiconductor field effect transistor semiconductors, react to the chemicals in an odor, which change the resistance of the sensor to generate the reading. Chemocapacitors perform the same function by reacting to the chemicals in the odor by generating a change in the sensor capacitance.

Amperometric sensors use the catalytic oxidation of a chemical analyte to generate heat proportional to the amount of chemical present.

Optical sensors use specially coated optical fibers. The coating consists of specifically formulated fluorescent dyes with unique characteristics that cause the fibers to react to very specific chemicals, changing the optical characteristics in proportion to the amount of chemical present.

Calorimetric sensors are thermal sensors that measure the heat generated by the absorption of an analyte into the sensor coating. The more chemical present, the greater the heat generated.

Hydrocarbon odor can also be indirectly detected by on-line monitoring of ppm concentrations of oils in water (Chapter 1.46). In this design, noncontacting UV light is targeted at the flowing sample and the emulsified oils in the water fluoresces the light energy back out of the water at a signature wavelength. The intensity of the light at that wavelength is an indication of the ppm concentration (Figure 1.45g).

FIG. 1.45g
UV fluorescence detection of hydrocarbon concentration in water in the ppm range can be correlated to odor.

APPLICATIONS

This technology allows the analyzers to be small—at least desktop size and, in some cases, easily handheld (Figure 1.45h). There are indications that credit card–size devices will eventually be developed. A sensor can be designed to be as specific or as broad in application as desired. There are many possible applications in industrial and commercial safety, drug interdiction and enforcement, explosive detection, pharmaceuticals, medical, food and beverage, perfumes, breweries, wineries, etc.

E-Nose Applications

Anywhere an odor can be used to detect the presence or even absence of a substance, these electronic noses can be used. The cycle time is typically a few seconds but can be longer, depending on the chemical and the vapor state. Temperature dramatically affects the vaporization of most chemicals, thereby reducing the amount of chemicals to cause an odor. When a sensor/analyzer is used for batch applications, there is an additional time required to purge and clear the sensor array, allowing it to return to a baseline state.

SPECIFICATION FORMS

When specifying analyzers for the measurement of odor, use the ISA form 20A1001, and when the type of analyzer has already been selected and both the analyzer and the requirements of the process application are to be specified, use the ISA form 20A1002. Both forms are reproduced with the permission of the International Society of Automation on the next pages.

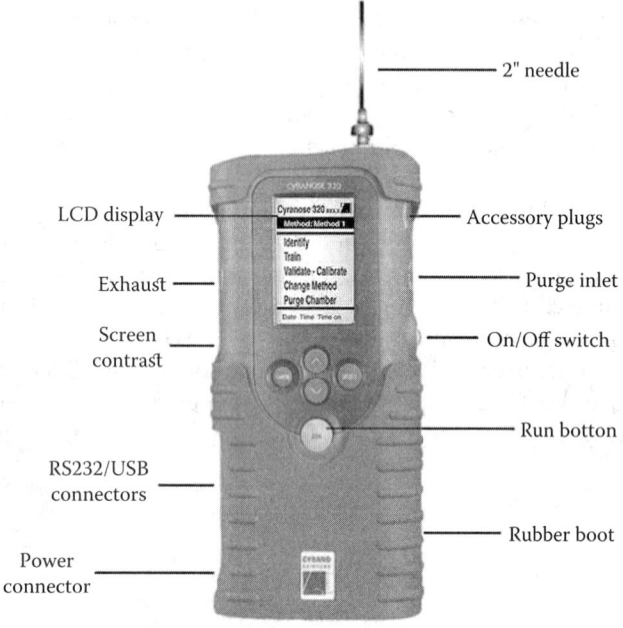

FIG. 1.45h
Handheld electronic nose or E-nose. (Courtesy of Electronic Products.)

1.45 Odor Detection

1	RESPONSIBLE ORGANIZATION		ANALYSIS DEVICE		6	SPECIFICATION IDENTIFICATIONS			
2					7	Document no			
3	(ISA)		Operating Parameters		8	Latest revision		Date	
4					9	Issue status			
5					10				

11	ADMINISTRATIVE IDENTIFICATIONS			40	SERVICE IDENTIFICATIONS Continued			
12	Project number	Sub project no		41	Return conn matl type			
13	Project			42	Inline hazardous area cl		Div/Zon	Group
14	Enterprise			43	Inline area min ign temp		Temp ident number	
15	Site			44	Remote hazardous area cl		Div/Zon	Group
16	Area	Cell	Unit	45	Remote area min ign temp		Temp ident number	
17				46				
18	SERVICE IDENTIFICATIONS			47				
19	Tag no/Functional ident			48	COMPONENT DESIGN CRITERIA			
20	Related equipment			49	Component type			
21	Service			50	Component style			
22				51	Output signal type			
23	P&ID/Reference dwg			52	Characteristic curve			
24	Process line/nozzle no			53	Compensation style			
25	Process conn pipe spec			54	Type of protection			
26	Process conn nominal size	Rating		55	Criticality code			
27	Process conn termn type	Style		56	Max EMI susceptibility		Ref	
28	Process conn schedule no	Wall thickness		57	Max temperature effect		Ref	
29	Process connection length			58	Max sample time lag			
30	Process line matl type			59	Max response time			
31	Fast loop line number			60	Min required accuracy		Ref	
32	Fast loop pipe spec			61	Avail nom power supply			Number wires
33	Fast loop conn nom size	Rating		62	Calibration method			
34	Fast loop conn termn type	Style		63	Testing/Listing agency			
35	Fast loop schedule no	Wall thickness		64	Test requirements			
36	Fast loop estimated lg			65	Supply loss failure mode			
37	Fast loop material type			66	Signal loss failure mode			
38	Return conn nominal size	Rating		67				
39	Return conn termn type	Style		68				

69	PROCESS VARIABLES	MATERIAL FLOW CONDITIONS				101	PROCESS DESIGN CONDITIONS			
70	Flow Case Identification				Units	102	Minimum		Maximum	Units
71	Process pressure					103				
72	Process temperature					104				
73	Process phase type					105				
74	Process liquid actl flow					106				
75	Process vapor actl flow					107				
76	Process vapor std flow					108				
77	Process liquid density					109				
78	Process vapor density					110				
79	Process liquid viscosity					111				
80	Sample return pressure					112				
81	Sample vent/drain press					113				
82	Sample temperature					114				
83	Sample phase type					115				
84	Fast loop liq actl flow					116				
85	Fast loop vapor actl flow					117				
86	Fast loop vapor std flow					118				
87	Fast loop vapor density					119				
88	Conductivity/Resistivity					120				
89	pH/ORP					121				
90	RH/Dewpoint					122				
91	Turbidity/Opacity					123				
92	Dissolved oxygen					124				
93	Corrosivity					125				
94	Particle size					126				
95						127				
96	CALCULATED VARIABLES					128				
97	Sample lag time					129				
98	Process fluid velocity					130				
99	Wake/natural freq ratio					131				
100						132				

133	MATERIAL PROPERTIES				137	MATERIAL PROPERTIES Continued		
134	Name				138	NFPA health hazard		Flammability	Reactivity
135	Density at ref temp		At		139				
136					140				

Rev	Date	Revision Description	By	Appv1	Appv2	Appv3	REMARKS

Form: 20A1001 Rev 0 © 2004 ISA

Analytical Measurement

	RESPONSIBLE ORGANIZATION	ANALYSIS DEVICE COMPOSITION OR PROPERTY Operating Parameters (Continued)		SPECIFICATION IDENTIFICATIONS	
1			6	Document no	
2	(ISA)		7	Latest revision	Date
3			8	Issue status	
4			9		
5			10		

	PROCESS COMPOSITION OR PROPERTY			MEASUREMENT DESIGN CONDITIONS				
11	Component/Property Name	Normal	Units	Minimum	Units	Maximum	Units	Repeatability
12								
13–37								

Rev	Date	Revision Description	By	Appv1	Appv2	Appv3	REMARKS

Form: 20A1002 Rev 0 © 2004 ISA

Definitions

Olfactometer is an apparatus for measuring the acuity of the sense of smell. It is used to detect and measure ambient odor dilution. Olfactometers are used to gauge the odor detection threshold of substances. To measure intensity, olfactometers introduce an odorous gas as a baseline against which other odors are compared.

Olfactory refers to systems relating to, or connected with, the sense of smell.

Organoleptic properties are the aspects of food or other substances as experienced by the senses, including taste, sight, smell, and touch, in cases where dryness, moisture, and stale-fresh factors are to be considered.

Surface acoustic wave (SAW) is an acoustic wave traveling along the surface of a material exhibiting elasticity, with an amplitude that typically decays exponentially with depth into the substrate.

Abbreviations

FID	Flame ionization detector
MOSFET	Metal oxide semiconductor field effect transistor
OTV	Odor threshold value
ppb	parts per billion
QCM	Quartz crystal microbalance
SAW	Surface acoustic wave
TBM	Tertiary butyl mercaptan

Organization

JPL	Jet Propulsion Laboratory

Bibliography

ASTM D6273, Standard Test Methods for Natural Gas Odor Intensity, 2015, http://www.astm.org/Standards/D6273.htm.

DeFrancesco, L., Cyrano Sciences, Portable electronic nose, 2000, http://www.the-scientist.com/?articles.view/articleNo/19507/title/The-Nose-Knows–Cyrano-Sciences–Electronic-Nose/.

Defranco, L., The nose knows: Cyrano Sciences' electronic nose, http://www.the-scientist.com/?articles.view/articleNo/19507/title/The-Nose-Knows–Cyrano-Sciences–Electronic-Nose/.

Gardner, J. W. and Bartlett, A brief history of electronic noses, 2002, http://www.sciencedirect.com/science/article/pii/0925400594870853.

Mathas, C., The Five Senses of Sensors—Smell, 2013, http://thifalthifal.blogspot.com/2013/11/e-nose.html.

NASA, Cyrano "Nose" the smell of success, 2001, https://spinoff.nasa.gov/spinoff2001/ps4.html.

Pearce, T. C., Schiffman, S. S., Nagle, H. T., and Gardner, J. W., Eds., Handbook of Machine Olfaction, Electronic Nose Technology, 2004, http://onlinelibrary.wiley.com/doi/10.1002/3527601597.ch4/summary.

PNAS, Trends in odor intensity for human and electronic noses, 1998, http://www.ncbi.nlm.nih.gov/pmc/articles/PMC20396/.

Ravah, S, Morgan-Sagastume, J. M., Methods of Odor and VOC Control, 2005, http://link.springer.com/chapter/10.1007%2F3-540-27007-8_3.

Reckner, L. R. and Squires, R. E., Diesel Exhaust Odor Measurement Using Human Panels, 1968, SAE Paper 680444. http://papers.sae.org/680444/.

Reeves, G., Understanding and Monitoring Hydrocarbons in Water, 2000, http://www.arjayeng.com/pdf_specs/Understanding_Hydrocarbons.pdf.

Röck, F., Barsan, N., and Weimar, U., Electronic Nose: Current Status and Future Trends, 2008, http://pubs.acs.org/doi/abs/10.1021/cr068121q.

Staples, E. J., Real time characterization of food & beverages using an electronic nose with 500 orthogonal sensors and VaporPrint imaging, 2000, http://www.estcal.com/tech_papers/papers/GeneralAnalysis/SenExpo2000C.pdf.

Stordeur, R.T., Stordeur, C.M., Levine, S.P., and Hoggatt, J.H., A New Microprocessor-Controlled Dynamic Olfactometer, 1981, http://www.tandfonline.com/doi/pdf/10.1080/00022470.1981.10465232.

Widder, S. Symrise GmbH & Co. KG. http://osdir.com/patents/Perfume-compositions/8-tetradecenal-fragrance-flavoring-substance-07332468.html, "8-tetradecenal as fragrance and flavoring substance", 2008, http://www.google.com/patents/US7332468.

Wilson, A. D. and Baietto, M., Applications and Advances in Electronic-Nose Technologies, 2009, http://www.mdpi.com/1424-8220/9/7/5099.

1.46 Oil in or on Water

C. P. BLAKELEY (1974, 1982)   **I. VERHAPPEN** (2003)   **B. G. LIPTÁK** (1995, 2017)

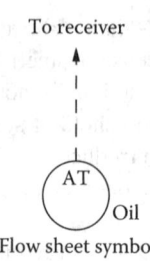

Flow sheet symbol

Types of designs	A. Reflected light oil slick detector (on–off), including laser
	B. Capacitance: available in probe form for interface detection or in a flow-through design or in a floating plate configuration for measuring the thickness of oil
	C. Ultraviolet (UV)
	D. Microwave, radio frequency, available as an interface probe, as a tape-operated tank profiler, or as an oil-in-water content detector
	E. Conductivity probes for interface detection
	F. Nuclear for interface detection
	G. Ultrasonic
	H. Near infrared
Related chapters	For details on the sensor categories, refer to
	Chapter 1.15 Conductivity (E)
	Chapter 1.28 Hydrocarbon Analyzers
	Chapter 1.32 Near Infrared (H)
	Chapter 1.69 Ultraviolet (C)
Inaccuracy	A. Generally from 1% to 5% of full scale
	B. The flow-through dual-concentric detector has a sensitivity of about 0.05%–0.1% FS water
	C. 0.1 ppm for a 0–10 ppm range
	D. Interface is detected to 5%; tank profile, to 1% or 3 cm, and water concentration, to 0.1%
Ranges	For the detector categories of conductivity, infrared, ultraviolet, etc., refer to the corresponding chapters.
	A. Generally from 0–50 ppm to 0%–100%
	B. Flow-through dual-concentric detector from 5% to 15% water in oil
	C. 0–10 ppm to 0–150 ppm oil in water
	D. Oil content detectable from 0% to 100%
Costs	In general, refer to the corresponding chapters.
	For example, a dual-wavelength ultraviolet unit for the measurement of oil in water in the 0–10 ppm to 0–150 ppm range and with auto-zero capability costs between $12,000 and $20,000, depending on the accessories required.
Partial list of suppliers	Agar Corp. (D—oil water mixture): http://www.agarcorp.com/literature/OW300.html, http://www.agarcorp.com/literature/ProductionSeparators.html
	Applied Analytics (C—oil in water): http://www.a-a-inc.com/product.php?recordID=13
	Cameron (G—ultrasonic pipeline interface): https://www.c-a-m.com/search?query=pipeline+interface
	Delta C Technologies (B—capacitance probe): http://www.delta-c.com/products.php?p=cat2
	Endress+Hauser Instruments (B—separation interface): http://www.us.endress.com/en/Endress-Hauser-group/industry-automation-expertise/industry-oil-gas/production-optimiz
	FMC Technologies (B—oil on water): http://www.fmctechnologies.com/en/FluidControl/Technologies/Invalco/AnalyticalInstruments/OilonWaterMonitors.aspx
	Foxboro, Invensys (Bouyancy): http://iom.invensys.com/EN/FoxboroEckardtDocs/EML1002_0BR_A_001_en.pdf
	GE Water & Power (oil on water slick detector): http://www.geinstruments.com/products-and-services/leakwise-oil-detectors
	Hach (A—nephelometer): http://www.hach.com/filtertrak-660-sc-laser-nephelometer-sensor-instrument-only/product-downloads?id=7640461566
	IMA (B—water in oil): www.ima.co.uk/products/by-measured-parameter/water-in-oil-meas…
	Interline (A—reflected laser light): http://www.interline.nl/media/1000219/odl_1600_091004_r3-interline.pdf
	KAM (D—oil in water): www.kam.com/kam-products/owd-oil-water-detector/

(Conitnued)

Marineinsight (E—oil on water): http://www.marineinsight.com/wp-content/uploads/2013/07/portable-gauge.jpg
Ohmart/Vega (F—oil–water separator interface): http://www.vega-americas.com/en/18832.htm
Orb Instruments Inc. (C—oil in water): http://www.orbinstruments.com/uv_oil.shtml
Phase Dynamics (D—water cut analyzer): http://www.phasedynamics.com/products.html
ProAnalysis (C—water in oil): http://www.oilinwater.com/technology/principle
(A—laser fluorescence): http://www.offshore-technology.com/contractors/environmental/proanalysis/
Teledyne Analytical Instruments (C—fiber optic): http://www.teledyne-ai.com/products/6650.asp
Tracerco (formerly Synetix) (F—water in oil online): http://www.tracerco.com/pdfs-uploaded/US%20Tracerco%20Density%20Gauge.pdf
Turner Designs (C—fluorescence design): http://www.turnerdesigns.com/customer-care/fluorometer-application-notes/oil-and-fuel-in-water-fluorometer-application-notes
Weatherford (H—NIR—water cut probe): http://www.weatherford.com/Products/Production/FlowMeasurement/RedEyeWater-CutMeter/index.htm

For additional suppliers, see
Chapter 1.15 Conductivity (E)
Chapter 1.28 Hydrocarbon Analyzers
Chapter 1.32 Near Infrared (H)
Chapter 1.69 Ultraviolet (C)

INTRODUCTION

The measurement of oil in or on water is an important requirement in applications related to both industrial and environmental pollution protection applications. Both types of applications are discussed in this chapter and are also covered in Chapter 1.28, in the chapter dealing with hydrocarbon analyzers. The most common applications of these sensors are used for interface measurements between the layers of oil and water in separation tanks and pipelines. Conductivity, capacitance, and ultrasonic-level probes are widely used for detecting such interfaces. In other applications, it is desirable to detect the amount of oil that is dispersed in a water stream. Ultraviolet analyzers are most often used for these measurements. These analyzers are also covered in more detail in Chapter 1.69, which is solely devoted to these types of sensors.

ENVIRONMENTAL POLLUTION SENSORS

Oil floating on water forms a mechanical barrier between the air and the water. It prevents oxygenation and kills oxygen-producing vegetation on the banks of streams. By coating the gills of fish, these materials also prevent breathing and cause the fish to suffocate. Outfalls from ships and municipal or industrial waste treatment plants must therefore be monitored for oil, and oil must be removed to prevent oil-bearing wastes from entering the receiving waters.

Continuous monitors are available to detect any hydrocarbons that are floating on the surface of the water. Hydrocarbon detectors are also discussed in detail in Chapter 1.28. Oil in the water is undesirable because it contributes to the biochemical oxygen demand (BOD)* and can also be toxic to aquatic biota, to the fish food in water, and possibly to the fish themselves.

Optical methods of detection require regular, conscientious maintenance for continuous and reliable performance. The capacitance approach used for the monitoring of oil film thickness on the water requires less maintenance but is limited to the detection of floating oil.

In this chapter, both water-in-oil and oil-in (or on)-water detectors will be discussed.

WATER-IN-OIL MEASUREMENTS

Capacitance

The capacitance of water is much higher (its dielectric constant is about 80) than that of oil (about 2), so measuring the dielectric constant is a convenient way to tell them apart. In addition to conventional capacitance probes, special dual-concentric

---

* BOD is the amount of oxygen required by aerobic microorganisms to decompose the organic matter in the water. It is used as a measure of the degree of water pollution because the oxygen required for the decomposition of the pollutants (the BOD) is the oxygen that is taken away from the amount that would otherwise be available to support fish and aquatic life. BOD is also called biological oxygen demand.

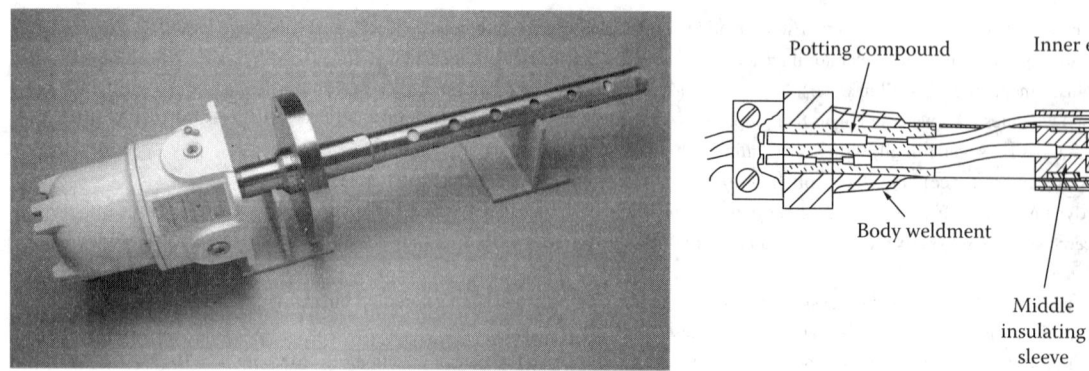

FIG. 1.46a
Dual-concentric capacitance probe, which is used to measure water in oil (water cut). (Courtesy of Delta C.)

designs (Figure 1.46a) are also available to detect the interface between water and oil in tanks. In addition, flow-through sensors are also available for in-pipeline applications.

The design of the flow-through version of the dual-concentric electrode is illustrated in Figure 1.46b. Here, the electrodes consist of two concentric pipes that are insulated from each other, thereby forming the capacitor through which flows the process stream. The flanged spool piece is available in sizes from 2 to 8 in. (50–200 mm) and is designed for operation at up to 150 psig (10.6 bar) and 100°C (212°F). The unit is available both for switching and for transmitting applications.

The water-in-oil sensor is most often applied for the purpose of setting the maximum amount of water that is allowed to be present in the oil. When that concentration is reached, the flow is diverted or other corrective action is taken.

Radio Frequency (Microwave)

When a cup containing water and oil is placed in a microwave oven, the water will heat up, but the oil will not. This is because shortwave radio frequency (RF) energy is absorbed much more efficiently by water than by oil.

In a radio-wave detector (Figure 1.46c), the transmitter produces waves that are of fixed frequency and contain a constant amount of energy. The more of this energy that is absorbed by the process fluid (the more water is in the mixture), the lower will be the voltage at the detector. The advantages of this design (relative to capacitance sensors) include wider range (0%–100%), lower sensitivity to buildup, insensitivity to temperature and salinity variations, and suitability for higher-temperature operations (up to 450°F, or 232°C).

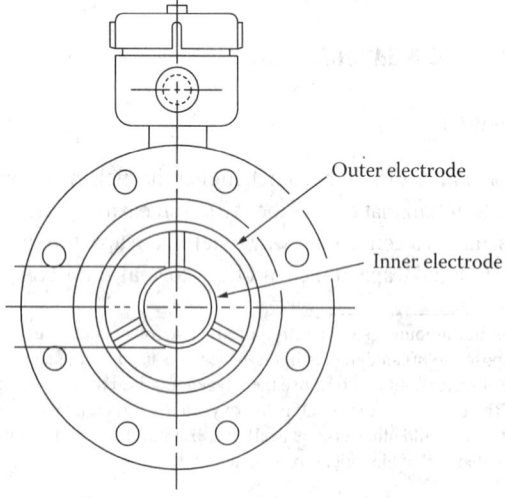

FIG. 1.46b
Flow-through water-in-oil detector utilizing two concentric capacitance electrodes.

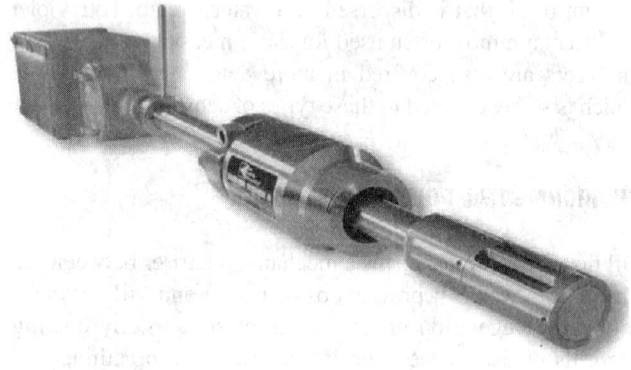

FIG. 1.46c
Microwave water-in-oil detector. (Courtesy of KAM.)

Water Dump Control RF probe-type water–oil interface detectors are used for "free water knockout" (Figure 1.46d) in which the probe is installed horizontally at an elevation of one-third of the diameter of the separator vessel and is set to

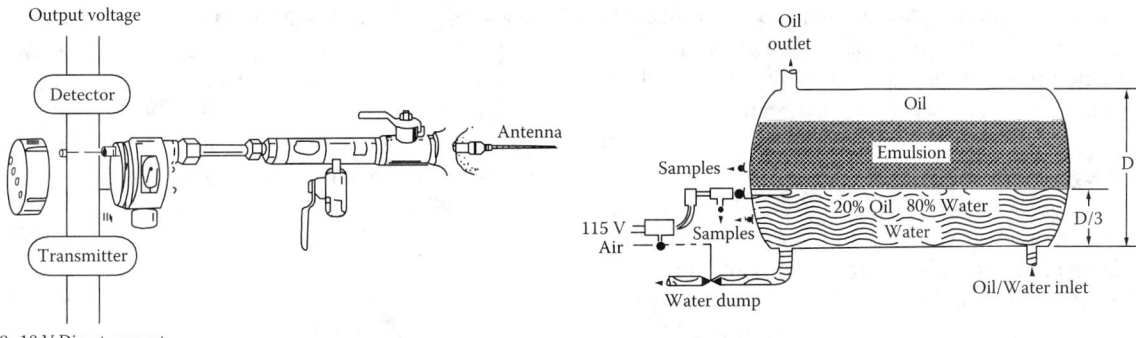

FIG. 1.46d
RF probe for the control of water dump valves on oil separators. (Courtesy of Agar Corp.)

open the water dump valve when the emulsion concentration drops below 20% oil (80% water). This way, the emulsion (rag layer) will build up above the probe. These instruments can also provide a 4–20 mA or other transmitted output signals and can detect water concentration within an error of about 5%.

Rag Layer Profiler Portable tank profilers are also available using RF principle of operation. Here, the tape-supported RF element is gradually lowered into a tank or a separator, which can be up to 30 m (100 ft) tall. As the sensor is lowered, it measures both the location of the interface (within an error of 0.12 in., or 3 mm) and the emulsion concentration throughout the tank height (from 0% to 100% within an error of 1% FS) (Figure 1.46e).

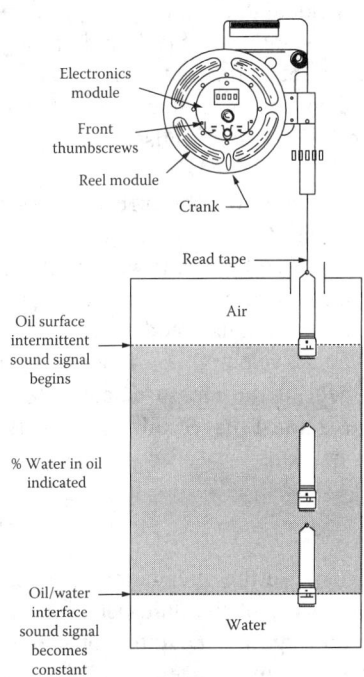

FIG. 1.46e
Tape-type tank profiler using radio frequency–based oil–water sensor. (Courtesy of Agar Corp.)

RF Probes A water-in-oil monitoring probe is also available that can detect the water concentration over a range of 0%–100% in tanks or pipelines (Figure 1.46f) within an error of 0.1% FS. All of these devices are available in explosion-proof construction and can be provided with digital displays.

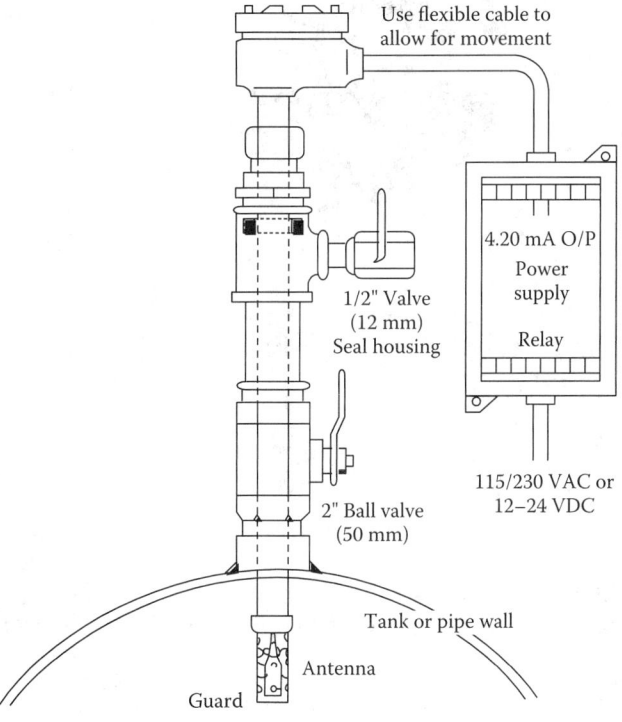

FIG.1.46f
RF-type water-in-oil concentration probes can detect the water concentration within an error of 0.1% FS. (Courtesy of Agar Corp.)

When applying this technology to measure oil in water, one must remember that there is a crossover point, at around 80% water concentration, at which the solution changes from being *water continuous* to *oil continuous*. This crossover results in a discontinuity in the analyzer output. Therefore, these devices suffer a significant loss in accuracy around this measurement point. In addition, many of these sensors are also sensitive to the salinity of the water phase. This can be corrected through the use of strapping tables, which should be selected by the user during calibration.

Conductivity-Based Interface

There is a substantial difference between the conductivities of water and hydrocarbons. This difference is often used as the basis for detecting the interface between oil and water (Figure 1.46g). Interface detectors are basically a kind of conductive level sensors that help in point level detection of conductive liquids. They use a low-voltage, current-limited power source applied across different electrodes. A conductive liquid that contacts both the longer and the shorter probes completes a conductive circuit. The change in conductivity is used to detect the oil/water interface.

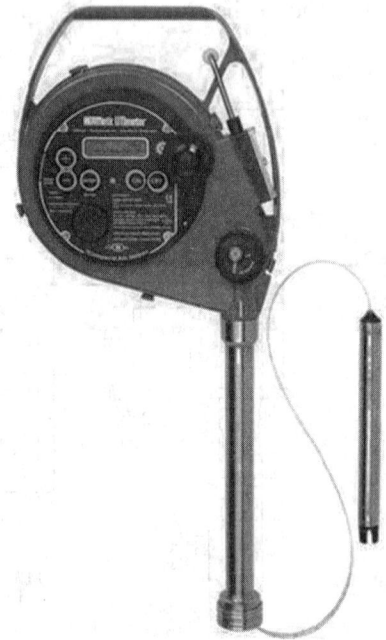

FIG. 1.46g
Conductivity-based interface detector. (Courtesy of Marineinsight.)

Vessel profiles can also be estimated by using an array of self-contained capacitance cells on a common mounting frame. If the appropriate software is available, the resultant capacitance profile of the vessel can be used to signal the multiple interfaces in that vessel to a resolution as good as 1 cm.

Ultrasonic Pipeline Interface

The time of passage of an ultrasonic pulse and the attenuation of that signal can be used to calculate both the density and the viscosity of the flowing fluid. Such instruments are used as pipeline oil–water interface detectors when they measure the densities and viscosities of the flowing fluids, which are substantially different from oil and water. These units mount between a pair of pipe flanges. They can measure the density and hence the amount of hydrocarbon in the flowing stream (Figure 1.46h).

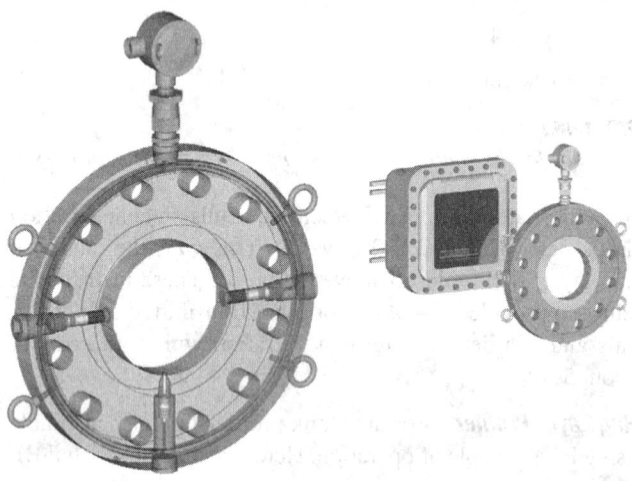

FIG. 1.46h
Wafer-type ultrasonic pipeline interface detector. (Courtesy of Cameron.)

Using these interface detectors, it is also possible to estimate fluid viscosity, which can also be used as interface detectors between batches of product in a pipeline. The viscosity is calculated from the slight rarefaction and compaction of molecules in the liquid, which cause viscous shearing of the fluid that absorbs acoustic energy. Kinematic fluid viscosity is determined from the energy loss of the meters' acoustic pulses as they pass through the fluid.

The analog outputs, which correspond to relative density and/or kinematic viscosity, can be used to detect the passing of an interface between two liquids in a pipeline. This information can be used to properly divert the flow into the correct tank. Relative density is determined by the change in sound velocity as a function of density and, because the object of the measurement is to compare the density of water against that of oil, an estimate of water or oil content can be made.

Nuclear

Several companies use the attenuation of gamma radiation to measure the density of the fluid between a source and a detector, thus allowing the preparation of a complete density profile for the vessel and hence an estimation of the *composition* inside. One manufacturer of this technology uses a strip source and detector cells (Figure 1.46i), and others use matched source and detector pairs that are moving up and down in a well to provide a density profile (Figure 1.46j).

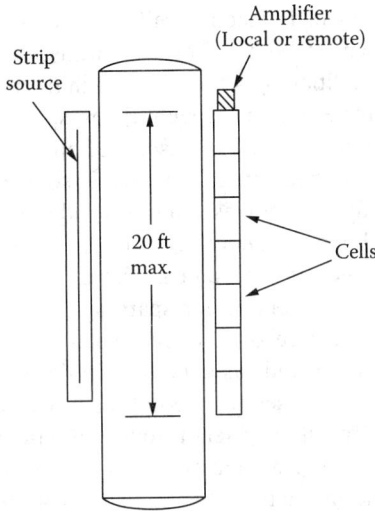

FIG. 1.46i
Strip-type nuclear source and cell receivers can be used for tank profiling.

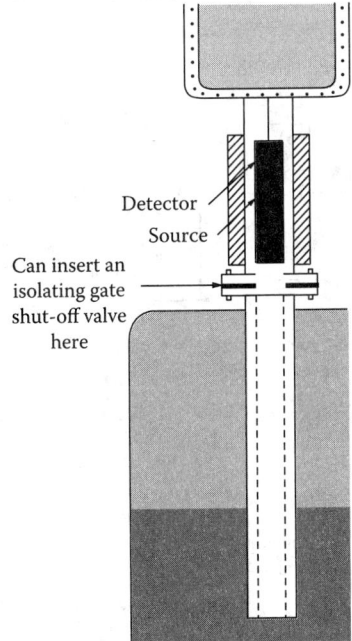

FIG. 1.46j
Backscatter-type traversing detector can detect interface levels while also drawing a specific gravity profile for the tank. (Courtesy of Ohmart Corp.)

As was the case for the conductivity array, software in this design is also used to determine the actual interfaces between the phases. This interface is normally at the inflection point between two layers of differing densities. The determination of the interface between oil and water can also be carried out

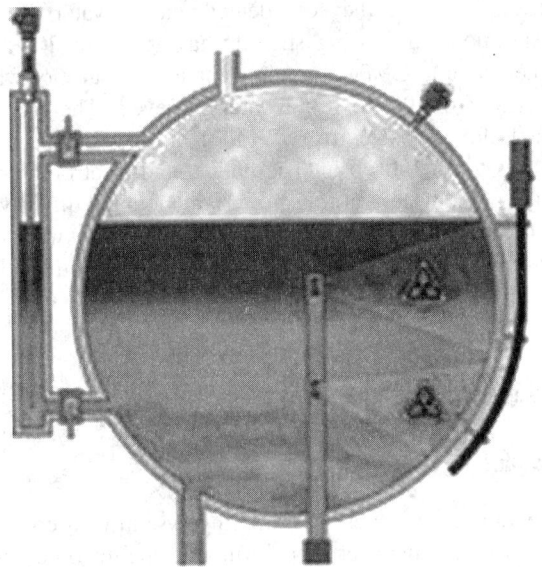

FIG. 1.46k
Oil separator provided with flexible radiation sensor. (Courtesy of Ohmart Corp.)

with a radiation-based flexible sensor (Figure 1.46k). The noncontact sensor measurement is not influenced by process pressure or process temperature. The unequal damping of the radioactive radiation in the different liquids enables a reliable detection of the interface. The placement of the radioactive sources directly in the container allows operation with low doses of radiation and achieves high measurement reliability.

Near Infrared

Near-infrared meters (Figure 1.46l) are available for water-cut measurement in three-phase flow (oil, water, and gas)

FIG. 1.46l
Multiphase water-cut detector. (Courtesy of Weatherford.)

applications, where they can detect relative water and oil concentrations in flowing streams having up to 20% gas volume fraction (GVF). It ignores changes in salinity while measuring from 0% to 100% water-cut levels. The probe is available for line sizes from 2 to 24 in. and for operating pressures up to 1500 psig. The optical and electronic components of the water-cut transmitter are mounted directly on top of the probe and can provide both real-time water-cut measurement and also average readings over specified time periods.

OIL-IN-WATER MEASUREMENTS

Ultraviolet

Figure 1.46m illustrates the sampling system of a continuous oil-in-water analyzer used for monitoring the oil content of steam condensate, recycled cooling water, and refinery or offshore drilling effluents.

The UV analyzer used in this system is a single-beam, dual-wavelength analyzer. This is superior to the single-wavelength designs because it is able to compensate for variations in sample sediment content, turbidity, and algae concentration, and also for window coatings. The cell operates according to Beer's law, which relates oil concentrations to UV energy absorption by the fixed length cell. The UV measuring band is centered at 254 nm, and the readings are sensitive to 0.1 ppm with a range of 0–10 ppm and provide a 90% response in 1 sec.

The instrument is automatically zeroed by the sample water being sent to both the measurement and the zeroing sides of the conditioning system. When in the measurement mode, the sample is sent through a high-speed, high-shear homogenizer that disperses all suspended oil droplets and the oil, which is adsorbed onto foreign matter so that the sample sent to the analyzer becomes a uniform and true solution.

Once per hour, the analyzer is automatically re-zeroed. In this mode, the sample water is sent through a filter that removes all the oil, and, after sparging,* it is sent to the analyzer. This oil-free zero-reference sample still contains all the other compounds as contained in the measurement. Therefore, it can be used to zero out this background.

When UV radiation is sent through an oil-contaminated water sample at a peak intensity of 365 nm, visible radiation is emitted. The intensity of this radiation can be measured by a photocell. The intensity of this radiation increases as the concentration of the fluorescent substance rises. At low concentrations (below 15×10^{-6}), the relationship between concentration and the visible radiation is essentially linear. In higher concentrations, some nonlinearity is experienced as a result of a saturation effect.

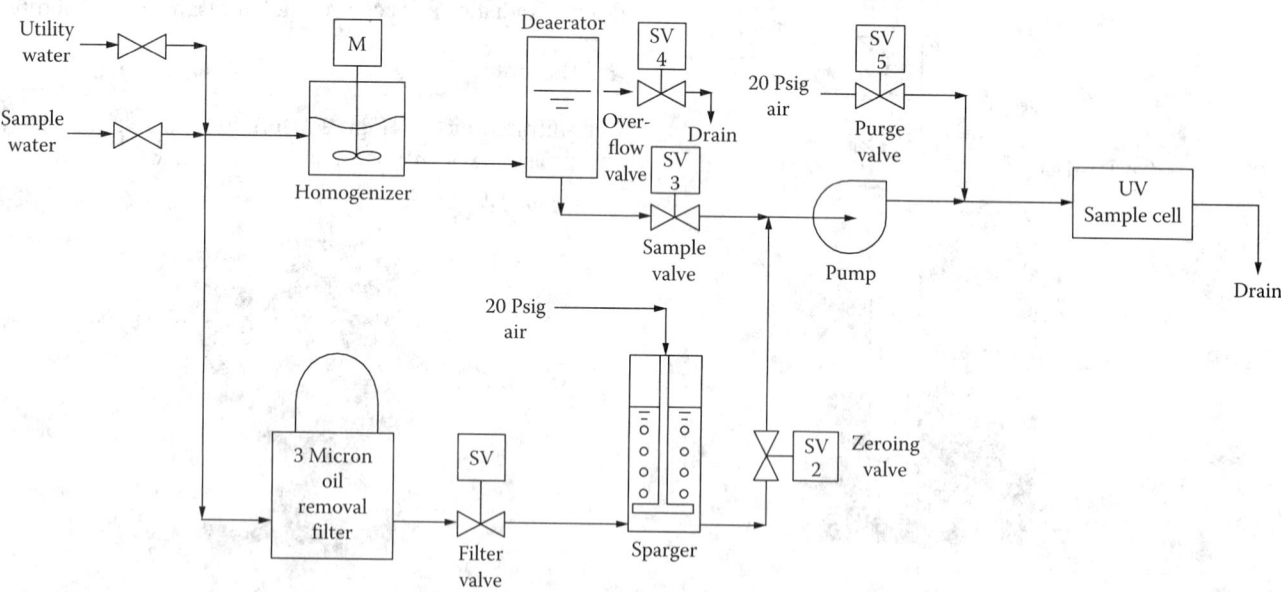

FIG. 1.46m
Sampling system of an ultraviolet oil-in-water analyzer with automatic-zero feature. (Courtesy of Teledyne Analytical Instruments.)

---

*Removal of dissolved gases.

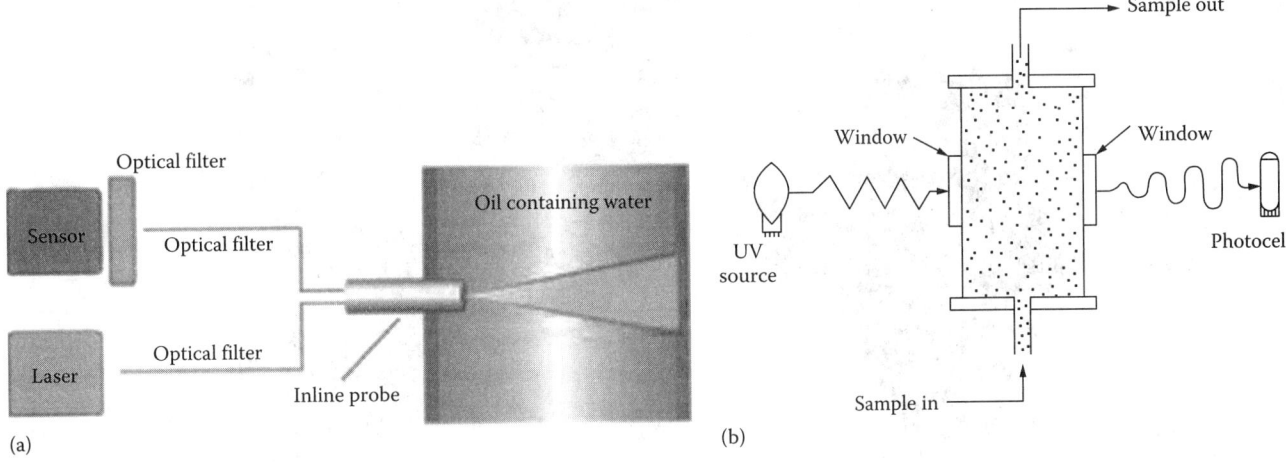

FIG. 1.46n
Oil-in-water measurement using ultraviolet light. (a) Reflected UV light through fiber-optic tubes (Courtesy of ProAnalysis). (b) Upflow sample through pipe with windows.

The most common configuration is to pass the process sample through the sensing head in an upflow direction (Figure 1.46n). The head is equipped with two windows that are set at right angles to each other so as to minimize the intensity of direct radiation from the source striking the photocell and also to reduce the effect of multiple scattering of the visible radiation. Optical filters at the incident and at the emergent windows are used (not shown) to reduce this effect to a negligible level.

For the detection of oil concentration in water, a falling-stream type of detector is also available. In this device, the sample stream is shaped into a rectangle (Figure 1.46o) as it falls through the viewing field of the UV beam and the photocell. Efficient optical filtration is important to overcome the unavoidable effects of direct reflection of incident radiation from the surface of the shaped stream.

OIL-ON-WATER MEASUREMENTS

Nephelometers Using Lasers

A nephelometer is an instrument used to measure the size and concentration of particles suspended in liquid or gas. This measurement is usually by the means of detecting the light these particles scatter. In oil-on-water applications, the nephelometer is intended to detect visible oil (hydrocarbon) slicks that are floating on fresh or salt water. It is usually mounted above a water surface as it detects the oil slicks providing immediate detection of oil leakage. It consists of two parts: a sensing head and a controller. The sensing head is inside an explosion-proof housing, which is supported on pontoons or float in the body of water. An S-shaped baffle can be used to direct the flowing water past the sensing head.

A laser beam is focused through a lens onto the surface of the water. A second lens refocuses the reflected light onto a photocell. When there is no oil on the water, only a minimum amount of reflection occurs. When floating oil is present, the reflected light intensity increases substantially because of the particles in the oil. The measurement is based on the differential between the outputs of the reflected light photocell and a reference photocell, which is measuring the output of the light source itself.

This instrument can provide an output signal that is proportional to the intensity of the reflected light, and it can

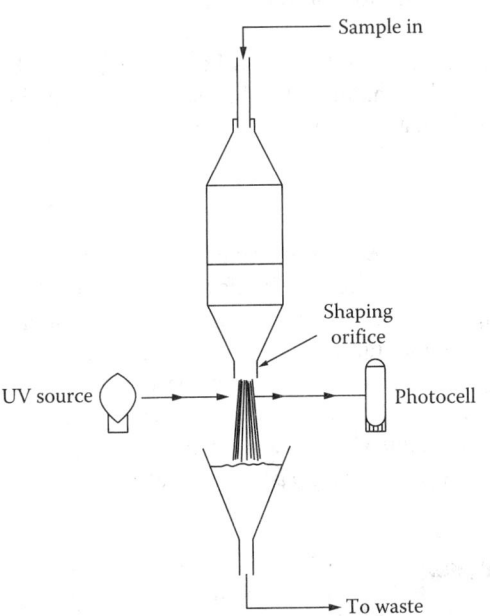

FIG. 1.46o
Falling-stream oil-in-water detector.

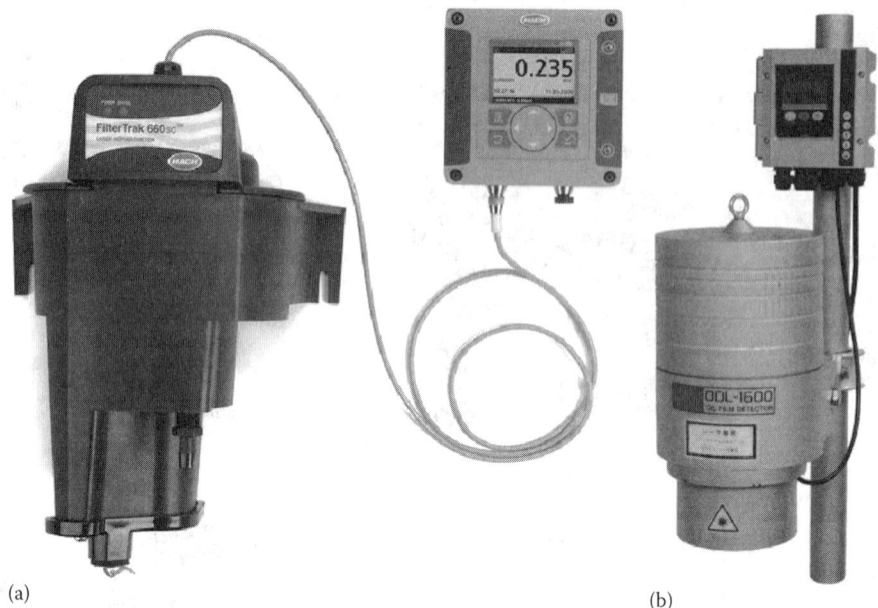

(a)　　　　　　　　　　　　　　(b)

FIG. 1.46p
Oil slicks can be detected by laser nephelometers, which are the change in the intensity of the reflection of a laser beam when an oil slick is present. (a) Laser nephelometer. (b) Oil slick detector. (a: Courtesy of Hach; b: Courtesy of Interline.)

also perform alarm functions (Figure 1.46p). The units are available both in intrinsically safe and in explosion-proof designs. They are provided with optical filters that eliminate the effects of sunlight and are also provided with microprocessor-based transmitters or controllers.

Care must be taken when using such reflected light measurements to block out sunlight and/or other stray light sources because, if they are allowed to be also reflected, they will introduce an error (typically on the high side).

Oil Slick Thickness Detection

When the amount of oil floating on the water is to be measured, the Lambert–Beer law [Equation 1.46(1)] is used to interpret the detected light intensity. The instrument measuring the thickness of the oil layer consists of a floating sensing head that can be connected by a shielded cable or by other means to a remote controller. The sensor detects the thickness of the oil layer on the water by capacitance measurement (Figure 1.46q).

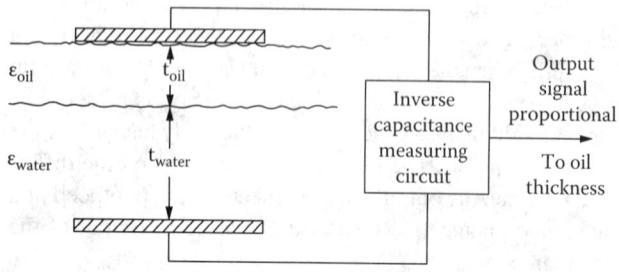

FIG. 1.46q
Parallel-plate capacitor used for the detection of the thickness of an oil slick layer on water.

The thickness of the oil layer is determined by solving Equation 1.46(1):

$$1/C = t_{oil}/\kappa_{oil}A + t_{water}/\kappa_{water}A \qquad \mathbf{1.46(1)}$$

where
  A is the effective area of one capacitor plate
  C is the capacitance
  t is the thickness
  κ is the dielectric constant

Because κ_{oil} = 1.9 to 2.1, whereas κ_{water} = 80, Equation 1.46(1) can be simplified by eliminating the second term, as shown in Equation 1.46(2):

$$1/C = t_{oil}/\kappa_{oil}A \qquad \mathbf{1.46(2)}$$

Thus, the inverse of capacitance is detected as a value that is proportional to the oil thickness. The circuit generates a DC signal in proportion to the inverse of capacitance and is used for remote transmission. The sensor takes advantage of the large differential in dielectric constants between oil and water. It is claimed that it is not confused by the presence of emulsified sludge (having a large dielectric constant) or by oily froth, which cannot pass under the float.

CONCLUSIONS

The on–off oil-on-water detector is capable of measuring as little as a few drops of petroleum floating on the surface of water, thus making it possible to detect those oil pollution levels that are visible to the human eye. It, therefore, serves a useful purpose as an alarm device downstream of plant outfalls and especially during and immediately after oil loading and unloading operations from tankers and tank trucks. This device presents the maintenance problems usually associated with optical measurements because the windows must be kept clean. The air column between the water surface and the window does reduces fouling due to splashing, but window cleanliness must be maintained for maximum sensitivity.

The oil thickness devices that are nonoptical require less maintenance. Because both devices can detect the absence or presence of oil slicks, they might also find application as oil spill monitors after oil transfer operations.

Floating on the surface of wastewater storage sumps or lagoons, the output of the oil thickness monitor can be used to start and stop oil reclamation equipment. These devices cannot be calibrated for a specific oil fraction, but just respond to any floating hydrocarbon.

The oil-in-water devices are optical, and even the falling-stream types require clean windows, although they are less subject to fouling than the sample chamber types. They must be calibrated for a specific type of oil, and other oil fractions will introduce errors. They were originally developed to monitor engine oil contamination in boiler feed water and condensate, which can be introduced by steam-driven feed water pumps.

These devices detect the presence of a specific hydrocarbon fraction in well-segregated waste streams. Where particle size is expected to exceed 5 μm, sample preparation prior to the UV analysis is necessary. Use of a high-shear mixer such as a blender has been found to produce a well-dispersed suspension suitable for measurement.

SPECIFICATION FORMS

There are some forms that can also be used for the specification of water and oil combination sensors. These include the following ISA forms reproduced on the next pages with the permission of the International Society of Automation:

Conductivity switches: Form 20A2381
Capacitance and RF switches and transmitters: Form 20L2101 and 20L2301
Nuclear switches and transmitters: 20L2141 and 20L2321
Ultrasonic switches and transmitters: 20L2171 and 20L2351

1	RESPONSIBLE ORGANIZATION	CONDUCTIVITY SENSOR w/wo INSERTION ASSEMBLY Device Specification	6	SPECIFICATION IDENTIFICATIONS	
2			7	Document no	
3	(ISA logo)		8	Latest revision	Date
4			9	Issue status	
5			10		
11	BODY HOLDER OR FITTING		60	INSERTION ASSEMBLY continued	
12	Body/Fitting type		61	Chamber wetted material	
13	Process conn nominal size	Rating	62	Compression ferrule matl	
14	Process conn termn type	Style	63		
15	Body/Fitting material		64		
16	Flange material		65	PERFORMANCE CHARACTERISTICS	
17	Seal/O ring material		66	Max press at design temp	At
18			67	Min working temperature	Max
19			68	Cell constant accuracy	
20			69	Measurement LRL	URL
21	SENSING ELEMENT		70	Temp compensation LRL	URL
22	Sensor type		71	Max temp response time	
23	Construction style		72	Max sensor to receiver lg	
24	Nominal cell constant		73		
25	Temperature sensor type		74		
26	Bore diameter		75		
27	Insertion/Immersion lg		76		
28	Temp sensor location		77		
29	Insulator material		78		
30	Electrode material		79		
31	Ext shaft/support matl		80		
32	Sheath material		81		
33			82		
34			83		
35			84		
36	LEAD WIRE AND EXTENSION		85	ACCESSORIES	
37	Extension type		86	Conductivity standard	
38	Cable length		87		
39	Max cable operating temp		88		
40	Signal termination type		89		
41	Cable jacket material		90		
42			91		
43	CONNECTION HEAD		92	SPECIAL REQUIREMENTS	
44	Housing type		93	Custom tag	
45	Enclosure type no/class		94	Reference specification	
46	Signal termination type		95	Compliance standard	
47	Cert/Approval type		96	Calibration report	
48	Mounting location/type		97		
49	Enclosure material		98		
50			99		
51	INSERTION ASSEMBLY		100	PHYSICAL DATA	
52	Assembly type		101	Estimated weight	
53	Isolation valve style		102	Overall height	
54	Process conn nominal size	Rating	103	Removal clearance	
55	Process conn termn type	Style	104	Signal conn nominal size	Style
56	Purge conn nominal size	Style	105	Mfr reference dwg	
57	Insertion/Immersion length		106		
58	Valve body material		107		
59	Valve seat material		108		

110	CALIBRATIONS AND TEST		INPUT OR TEST		OUTPUT	
111	TAG NO/FUNCTIONAL IDENT	MEAS/SIGNAL/SCALE	LRV	URV	LRV	URV
112		Measurement-Output signal				
113		Temp-Output signal				
114						
115						
116						
117						

118	COMPONENT IDENTIFICATIONS		
119	COMPONENT TYPE	MANUFACTURER	MODEL NUMBER
120			
121			
122			
123			
124			
125			

Rev	Date	Revision Description	By	Appv1	Appv2	Appv3	REMARKS

Form: 20A2381 Rev 0

1.46 Oil in or on Water

		CAPACITANCE OR RF ADMITTANCE LEVEL TRANSMITTER w/wo SWITCHES Device Specification			SPECIFICATION IDENTIFICATIONS	
1	RESPONSIBLE ORGANIZATION			6		
2				7	Document no	
3	ISA			8	Latest revision	Date
4				9	Issue status	
5				10		

	PROCESS CONNECTION				TRANSMITTER w/wo SWITCHES Continued	
11				60		
12	Body/Fitting type			61	Switch time delay	
13	Process conn nominal size	Rating		62	Enclosure material	
14	Process conn termn type	Style		63		
15	Wetted material			64		
16	Flange material			65		
17	Seal/O ring material			66	PERFORMANCE CHARACTERISTICS	
18				67	Max press at design temp	At
19				68	Min working temperature	Max
20	SENSING ELEMENT			69	Max discharge to sensor	
21	Configuration type			70	Accuracy rating	
22	Probe diameter			71	Min measurement span	Max
23	Insertion length			72	Min ambient working temp	Max
24	Inactive length			73	Contacts ac rating	At max
25	Integral cable length			74	Contacts dc rating	At max
26	Probe material			75	Max sensor to receiver lg	
27	Insulation/Sheath matl			76		
28	Weight material			77		
29				78		
30				79		
31	CONNECTION HEAD OR PREAMPLIFIER			80		
32	Type			81		
33	Measurement compensation			82		
34	Input signal URL			83		
35	Output signal type			84	ACCESSORIES	
36	Enclosure type no/class			85	Connecting cables length	
37	Power source			86	Power supply	
38	Cert/Approval type			87	Local indicator style	
39	Mounting location/type			88	Intrinsic safety barrier	
40	Enclosure material			89		
41				90		
42				91		
43	TRANSMITTER w/wo SWITCHES			92	SPECIAL REQUIREMENTS	
44	Housing type			93	Custom tag	
45	Measurement compensation			94	Reference specification	
46	Output signal type			95	Compliance standard	
47	Input signal URL			96	Calibration report	
48	Enclosure type no/class			97		
49	Characteristic curve			98		
50	Signal power source			99		
51	Contact arrangement	Quantity		100	PHYSICAL DATA	
52	Failsafe style			101	Estimated weight	
53	RFI protection			102	Overall height	
54	Integral indicator style			103	Removal clearance	
55	Cert/Approval type			104	Signal conn nominal size	Style
56	Mounting location/type			105	Mfr reference dwg	
57	Static discharge prot			106		
58	Failure/Diagnostic action			107		
59	Adjustable signal damping			108		

	CALIBRATIONS AND TEST		INPUT OR SETPOINT			OUTPUT OR SCALE	
110							
111	TAG NO/FUNCTIONAL IDENT	MEAS/SIGNAL/TEST	LRV	URV	ACTION	LRV	URV
112		Level-Analog output					
113		Level-Scale					
114		Level setpoint 1-Output					
115		Level setpoint 2-Output					
116		Level setpoint 3-Output					
117							

	COMPONENT IDENTIFICATIONS		
118			
119	COMPONENT TYPE	MANUFACTURER	MODEL NUMBER
120			
121			
122			
123			
124			
125			

Rev	Date	Revision Description	By	Appv1	Appv2	Appv3	REMARKS

Form: 20L2101 Rev 0

© 2001 ISA

CAPACITANCE OR RF ADMITTANCE LEVEL SWITCH
Device Specification

#	RESPONSIBLE ORGANIZATION		#	SPECIFICATION IDENTIFICATIONS	
1	ISA		6		
2			7	Document no	
3			8	Latest revision	Date
4			9	Issue status	
5			10		

PROCESS CONNECTION

11			
12	Body/Fitting type		
13	Process conn nominal size	Rating	
14	Process conn termn type	Style	
15	Wetted material		
16	Flange material		
17	Seal/O ring material		
18			
19			

SENSING ELEMENT

20	
21	Configuration type
22	Number of elements
23	Probe diameter
24	Insertion length
25	Inactive/Dead length
26	Integral cable length
27	Probe material
28	Insulation/Sheath matl
29	Weight material
30	
31	

CONNECTION HEAD OR PREAMPLIFIER

32	
33	Type
34	Measurement compensation
35	Input signal URL
36	Output signal type
37	Enclosure type no/class
38	Signal power source
39	Cert/Approval type
40	Mounting location/type
41	Enclosure material
42	
43	

SWITCH MECHANISM

44			
45	Housing type		
46	Element style		
47	Measurement compensation		
48	Output signal type		
49	Input signal LRL	URL	
50	Enclosure type no/class		
51	Signal power source		
52	Dead band type		
53	Contacts arrangement	Quantity	
54	Failsafe style		
55	Integral indicator style		
56	Cert/Approval type		
57	Mounting location/type		
58	Static discharge prot		
59	Failure/Diagnostic action		

SWITCH MECHANISM Continued

60	
61	RFI protection
62	Switch time delay
63	Enclosure material
64	
65	

PERFORMANCE CHARACTERISTICS

66		
67	Max press at design temp	At
68	Min working temperature	Max
69	Max discharge to sensor	
70	Repeatability/Sensitivity	
71	Min measurement span	Max
72	Min ambient working temp	Max
73	Contacts ac rating	At max
74	Contacts dc rating	At max
75	Max sensor to receiver lg	
76–84		

ACCESSORIES

85	
86	Connecting cables length
87	Mounting hardware
88	External relay
89	Intrinsic safety barrier
90–92	

SPECIAL REQUIREMENTS

93	
94	Custom tag
95	Reference specification
96	Compliance standard
97–99	

PHYSICAL DATA

100			
101	Estimated weight		
102	Overall height		
103	Removal clearance		
104	Signal conn nominal	size	Style
105	Mfr reference dwg		
106–108			

CALIBRATIONS AND TEST / SETPOINT / OUTPUT

#	TAG NO/FUNCTIONAL IDENT	MEAS/SIGNAL/TEST	LRV	URV	ACTION	LRV	URV
110							
111							
112		Level setpoint 1-Output					
113		Level setpoint 2-Output					
114		Level setpoint 3-Output					
115		Level setpoint 4-Output					
116							
117							

COMPONENT IDENTIFICATIONS

#	COMPONENT TYPE	MANUFACTURER	MODEL NUMBER
118			
119			
120–125			

Rev	Date	Revision Description	By	Appv1	Appv2	Appv3	REMARKS

Form: 20L2301 Rev 0 © 2001 ISA

NUCLEAR RADIATION LEVEL TRANSMITTER w/wo SWITCHES
Device Specification

#	RESPONSIBLE ORGANIZATION			#	SPECIFICATION IDENTIFICATIONS		
1				6	Document no		
2	(ISA logo)			7	Latest revision		
3				8		Date	
4				9	Issue status		
5				10			

#	RADIATION SOURCE		#	PERFORMANCE CHARACTERISTICS		
11			60			
12	Source type		61	Level accuracy rating		
13	Number of sources		62	Density accuracy rating		
14	Source size		63	Min measurement span	Max	
15	Fireproof design type		64	Response time		
16	Shutter style		65	Min ambient working temp	Max	
17	Position switch type		66	Contacts ac rating	At Max	
18	Mounting location/type		67	Contacts dc rating	At Max	
19	Collimation angle		68	Max sensor to receiver lg		
20	Source material		69			
21	Housing material		70			
22			71			
23			72			
24	SENSING ELEMENT		73			
25	Detector type		74			
26	Detector style		75			
27	Number of detectors		76			
28	Active length		77			
29			78			
30			79			
31	CONNECTION HEAD OR PREAMPLIFIER		80			
32	Type		81			
33	Housing type		82			
34	Measurement compensation		83			
35	Output signal type		84			
36	Enclosure type no/class		85	ACCESSORIES		
37	Signal power source		86	Connecting cables length		
38	Cert/Approval type		87	Mounting hardware		
39	Mounting location/type		88	Calibrator		
40	Enclosure material		89			
41			90			
42			91			
43	TRANSMITTER w/wo SWITCHES		92			
44	Housing type		93	SPECIAL REQUIREMENTS		
45	Measurement compensation		94	Custom tag		
46	Output signal type		95	Reference specification		
47	Enclosure type no/class		96	Compliance standard		
48	Characteristic curve		97			
49	Digital communication std		98			
50	Signal power source		99			
51	Contact arrangement	Quantity	100	PHYSICAL DATA		
52	Failsafe style		101	Estimated weight		
53	Integral indicator style		102	Overall length		
54	Cert/Approval type		103	Removal clearance		
55	Mounting location/type		104	Signal conn nominal size	Style	
56	Failure/Diagnostic action		105	Mfr reference dwg		
57	Enclosure material		106			
58			107			
59			108			

#	CALIBRATIONS AND TEST		INPUT OR SETPOINT			OUTPUT OR SCALE	
110							
111	TAG NO/FUNCTIONAL IDENT	MEAS/SIGNAL/TEST	LRV	URV	ACTION	LRV	URV
112		Level-Analog output					
113		Level-Digital output					
114		Level-Scale					
115		Density-Digital output					
116		Level setpoint 1-Output					
117		Level setpoint 2-Output					

#	COMPONENT IDENTIFICATIONS			
118				
119	COMPONENT TYPE	MANUFACTURER	MODEL NUMBER	
120				
121				
122				
123				
124				
125				

Rev	Date	Revision Description	By	Appv1	Appv2	Appv3	REMARKS

Form: 20L2141 Rev 0 © 2001 ISA

NUCLEAR RADIATION LEVEL SWITCH
Device Specification

#	RESPONSIBLE ORGANIZATION		#	SPECIFICATION IDENTIFICATIONS	
1	ISA		6		
2			7	Document no	
3			8	Latest revision	Date
4			9	Issue status	
5			10		

#	RADIATION SOURCE		#	PERFORMANCE CHARACTERISTICS	
11			60		
12	Shutter type		61	Repeatability/Sensitivity	
13	Number of sources		62	Response time	
14	Source size		63	Min ambient working temp	Max
15	Fireproof design type		64	Contacts ac rating	At max
16	Position switch type		65	Contacts dc rating	At max
17	Mounting location/type		66	Max sensor to receiver lg	
18	Source material		67		
19	Housing material		68		
20			69		
21			70		
22			71		
23	SENSING ELEMENT		72		
24	Detector type		73		
25	Number of detectors		74		
26	Active length		75		
27	Check source actuation		76		
28			77		
29			78		
30			79		
31	CONNECTION HEAD OR PREAMPLIFIER		80		
32	Type		81		
33	Housing type		82		
34	Output signal type		83		
35	Enclosure type no/class		84		
36	Signal power source		85		
37	Cert/approval type		86	ACCESSORIES	
38	Mounting location/type		87	Connecting cables length	
39	Enclosure material		88	Mounting hardware	
40			89		
41			90		
42			91		
43	SWITCH MECHANISM		92		
44	Housing type		93	SPECIAL REQUIREMENTS	
45	Output signal type		94	Custom tag	
46	Enclosure type no/class		95	Reference specification	
47	Signal power source		96	Compliance standard	
48	Contacts arrangement / Quantity		97		
49	Failsafe style		98		
50	Integral indicator style		99		
51	Cert/Approval type		100	PHYSICAL DATA	
52	Mounting location/type		101	Estimated weight	
53	Failure/Diagnostics action		102	Overall length	
54	Enclosure material		103	Removal clearance	
55			104	Signal conn nominal size	Style
56			105	Mfr reference dwg	
57			106		
58			107		
59			108		

#	CALIBRATIONS AND TEST		SETPOINT			OUTPUT	
110	TAG NO/FUNCTIONAL IDENT	MEAS/SIGNAL/TEST	LRV	URV	ACTION	LRV	URV
111							
112		Level setpoint 1-Output					
113		Level setpoint 2-Output					
114		Level setpoint 3-Output					
115		Level setpoint 4-Output					
116							
117							

#	COMPONENT IDENTIFICATIONS		
118			
119	COMPONENT TYPE	MANUFACTURER	MODEL NUMBER
120			
121			
122			
123			
124			
125			

Rev	Date	Revision Description	By	Appv1	Appv2	Appv3	REMARKS

Form: 20L2321 Rev 0 © 2001 ISA

1	RESPONSIBLE ORGANIZATION		NON-CONTACT ULTRASONIC LEVEL TRANSMITTER w/wo SWITCHES Device Specification		6	SPECIFICATION IDENTIFICATIONS	
2					7	Document no	
3					8	Latest revision	Date
4					9	Issue status	
5					10		

11	PROCESS CONNECTION			60	PERFORMANCE CHARACTERISTICS	
12	Body/Fitting type			61	Max press at design temp	At
13	Process conn nominal size		Rating	62	Min working temperature	Max
14	Process conn termn type		Style	63	Accuracy rating	
15	Wetted material			64	Level lower range limit	URL
16	Flange material			65	Min measurement span	Max
17	Seal/O ring material			66	Beam angle	
18				67	Min ambient working temp	Max
19				68	Contacts ac rating	At max
20				69	Contacts dc rating	At max
21	SENSING ELEMENT			70	Max sensor to receiver lg	
22	Detector type			71		
23	Detector style			72		
24	Insertion length			73		
25	Integral cable length			74		
26				75		
27				76		
28				77		
29	CONNECTION HEAD			78		
30	Type			79		
31	Enclosure type no/class			80		
32	Cert/Approval type			81		
33	Mounting location/type			82		
34	Enclosure material			83		
35				84	ACCESSORIES	
36				85	Connecting cables length	
37				86	Enclosure heater	
38	TRANSMITTER w/wo SWITCHES			87	Mounting hardware	
39	Housing type			88		
40	Measurement compensation			89		
41	Output signal type			90		
42	Enclosure type no/class			91		
43	Control sequence			92	SPECIAL REQUIREMENTS	
44	Characteristic curve			93	Custom tag	
45	Digital communication std			94	Reference specification	
46	Signal power source			95	Compliance standard	
47	Contact arrangement		Quantity	96	Calibration report	
48	Failsafe style			97		
49	Integral indicator style			98		
50	Cert/Approval type			99		
51	Mounting location/type			100	PHYSICAL DATA	
52	Failure/Diagnostic action			101	Estimated weight	
53	Signal processing			102	Overall height	
54	Enclosure material			103	Removal clearance	
55				104	Signal conn nominal size	Style
56				105	Mfr reference dwg	
57				106		
58				107		
59				108		

	CALIBRATIONS AND TEST		INPUT OR SETPOINT			OUTPUT OR SCALE	
110							
111	TAG NO/FUNCTIONAL IDENT	MEAS/SIGNAL/TEST	LRV	URV	ACTION	LRV	URV
112		Level–Analog output					
113		Level-Scale					
114		Level setpoint 1-Output					
115		Level setpoint 2-Output					
116		Level setpoint 3-Output					
117							

	COMPONENT IDENTIFICATIONS		
118			
119	COMPONENT TYPE	MANUFACTURER	MODEL NUMBER
120			
121			
122			
123			
124			
125			

Rev	Date	Revision Description	By	Appv1	Appv2	Appv3	REMARKS

Form: 20L2171 Rev 0

ULTRASONIC CONTACT-TYPE LEVEL SWITCH
Device Specification

#	RESPONSIBLE ORGANIZATION		#	SPECIFICATION IDENTIFICATIONS	
1			6		
2	(ISA)		7	Document no	
3			8	Latest revision	Date
4			9	Issue status	
5			10		

#	PROCESS CONNECTION		#	SWITCH MECHANISM Continued	
11			60		
12	Body/Fitting type		61	Enclosure material	
13	Process conn nominal size	Rating	62	Exterior coating material	
14	Process conn termn type	Style	63		
15	Wetted material		64		
16	Flange/Fitting material		65		
17	Seal/O ring material		66	PERFORMANCE CHARACTERISTICS	
18			67	Max press at design temp	At
19			68	Min working temperature	Max
20	SENSING ELEMENT		69	Repeatability/Sensitivity	
21	Configuration type		70	Min ambient working temp	Max
22	Nominal temp rating		71	Contacts ac rating	At Max
23	Number of sensors		72	Contacts dc rating	At Max
24	Probe diameter		73	Max sensor to receiver lg	
25	Insertion length		74		
26	Integral cable length		75		
27	Self test type		76		
28	Cert/Approval type		77		
29			78		
30			79		
31			80		
32	CONNECTION HEAD OR PREAMPLIFIER		81		
33	Type		82		
34	Output signal type		83		
35	Enclosure type no/class		84		
36	Signal power source		85	ACCESSORIES	
37	Gain type		86	Connecting cables length	
38	Cert/Approval type		87	Mounting hardware	
39	Mounting location/type		88	External relay	
40	Failure/Diagnostic action		89	Intrinsic safety barrier	
41	Enclosure material		90		
42			91		
43			92		
44	SWITCH MECHANISM		93	SPECIAL REQUIREMENTS	
45	Housing type		94	Custom tag	
46	Element style		95	Reference specification	
47	Input signal type		96	Compliance standard	
48	Output signal type		97		
49	Enclosure type no/class		98		
50	Control logic		99		
51	Gain type		100	PHYSICAL DATA	
52	Signal power source		101	Estimated weight	
53	Contacts arrangement	Quantity	102	Overall height	
54	Failsafe style		103	Removal clearance	
55	Integral indicator style		104	Signal conn nominal size	Style
56	Cert/Approval type		105	Mfr reference dwg	
57	Mounting location/type		106		
58	Failure/Diagnostic action		107		
59	Switch time delay		108		

#	CALIBRATIONS AND TEST		SETPOINT			OUTPUT	
110	TAG NO/FUNCTIONAL IDENT	MEAS/SIGNAL/TEST	LRV	URV	ACTION	LRV	URV
111							
112		Lg conn to sp 1-Output					
113		Lg conn to sp 2-Output					
114		Lg conn to sp 3-Output					
115		Lg conn to sp 4-Output					
116							
117							

#	COMPONENT IDENTIFICATIONS		
118			
119	COMPONENT TYPE	MANUFACTURER	MODEL NUMBER
120			
121			
122			
123			
124			
125			

Rev	Date	Revision Description	By	Appv1	Appv2	Appv3	REMARKS

Form: 20L2351 Rev 0 © 2001 ISA

Definitions

Beer–Lambert law relates the attenuation of light to the properties of the material through which the light is traveling.

Biochemical oxygen demand (BOD) is one of the most common measures of pollutant organic material in water. BOD indicates the amount of organic matter in the water that can be decomposed by microorganisms, which upon decomposition uses up the oxygen in the water. Therefore, a low BOD is an indicator of good quality water, while a high BOD indicates polluted water. Dissolved oxygen (DO) is the actual amount of oxygen available to fish and other aquatic life in the water. When the DO drops, aquatic life forms are negatively affected.

Capacitance is the ability of a body to store an electrical charge. Any object that can be electrically charged exhibits capacitance. A capacitor consists of two conductive parallel plates insulated from each other by a dielectric material between them. In such a parallel-plate capacitor, capacitance is directly proportional to the surface area of the conductor plates and inversely proportional to the separation distance between the plates. If the charges on the plates are $+q$ and $-q$, and the resulting voltage between the plates is V, then the capacitance C of that capacitor is $C = q/V$.

Dielectric constant is symbolized by the Greek letter kappa (κ) and is a property of an electrically insulating material (a dielectric), and its size is equal to the ratio of the capacitance of a capacitor filled with that material (C) divided by the capacitance of an identical capacitor that is placed in vacuum and has no dielectric material between its plates (C_0). In other words, $\kappa = C/C_0$. The insertion of a dielectric between the plates of, say, a parallel-plate capacitor always increases its capacitance or ability to store opposite charges on its plates. As such, it is a number without dimensions and is an expression of the extent to which a material concentrates electric flux.

Dielectric material is an electrical insulator that can be polarized by an applied electric field. This ability enables objects of a given size, such as sets of metal plates, to hold their electric charge for long periods of time and/or to hold large quantities of charge. When a dielectric is placed in an electric field, electric charges do not flow through the material the same way as they do in a conductor, but slightly shift from their average equilibrium positions causing dielectric polarization. Because of dielectric polarization, positive charges are displaced toward the field and negative charges shift in the opposite direction. This creates an internal electric field that reduces the overall field within the dielectric itself.

Nephelometer is an instrument used to measure the size and concentration of particles suspended in a liquid or gas. This measurement is usually by the means of detecting the light these particles scatter.

Rarefaction is the reduction of density, the opposite of compression. Like compression, which can travel in waves, rarefaction waves also exist in nature.

Sparger is a gas flushing device in which dissolved gases are removed from a liquid by bubbling inert gas through it.

Abbreviations

BOD   Biological oxygen demand (also called biochemical oxygen demand)
GVF   Gas volume fraction
RF   Radio frequency

Bibliography

Al-Azab, M., Oil pollution and its environmental impact (2011), http://www.amazon.com/Pollution-Environmental-Arabian-Developments-Sciences/dp/0444520600/ref=sr_1_3?s=books&ie=UTF8&qid=1411136766&sr=1-3&keywords=oil+measurement+in+or+on+water.

Burns, D.A., *Handbook of Near-Infrared Analysis* (2001), http://books.google.com/books/about/Handbook_of_Near_Infrared_Analysis_Secon.html?id=XkALgZVXxQQC.

Department on Energy and Climate Change, Guidance notes for petroleum measurements (2011), https://www.gov.uk/government/uploads/system/uploads/attachment_data/file/15563/DECC_Guidelines_-_Issue_8.pdf.

Duin, S., *Oil and Water* (2011), http://www.amazon.com/Oil-Water-Steve-Duin/dp/1606994921/ref=sr_1_2?s=books&ie=UTF8&qid=1411136551&sr=1-2&keywords=oil+in+or+on+water.

ISA, Measuring level interfaces, http://www.isa.org/InTechTemplate.cfm?template=/ContentManagement/ContentDisplay.cfm&ContentID=89173.

Mobrey ultrasonic sludge blanket (2012), http://www2.emersonprocess.com/siteadmincenter/PM%20Mobrey%20Documents/IP250.pdf.

Omega, Conductivity level switches (2014), http://www.omega.com/Green/pdf/LVCN_LVCF_LVCR_LVCP.pdf.

PLA Solutions, Slurry density measurement, http://plapl.com.au/slurry-density-measurement.html.

Schweinforth, R., Microwave measuring technology for sludge draining, http://www.berthold-us.com/industrial/productapplications/Municipa%20lWater%20Management/Microwave%20measurement%20for%20Sludge%20Draining.pdf.

Siesler, H.W., *Near-Infrared Spectroscopy: Principles, Instruments, Applications* (2008), http://www.amazon.com/Near-Infrared-Spectroscopy-Principles-Instruments-Applications/dp/3527301496.

Smith, A.L., Applied infrared spectroscopy, http://books.google.com/books/about/Applied_infrared_spectroscopy.html?id=yiRRAAAAMAAJ.

Sorrentino, R. and G. Bianchi, *Microwave and RF Engineering*, John Wiley & Sons (2010), http://www.google.com/search?hl=en&defl=en&q=define:microwave&ei=e6CMSsWUI5OHmQee2si1DQ&sa=X&oi=glossary_definition&ct=title.

Stellar Net Inc., Fiber optic spectrum analyzers (2013), http://www.stellarnet.us/popularconfigurations_fiberanlzr.htm.

Thermo Electron, Near infrared sensor for online measurement (2005), http://www.thermo.com/eThermo/CMA/PDFs/Various/File_27448.pdf.

Van de Kamp, W., The theory and practice of level measurement (2006), http://trove.nla.gov.au/work/33783362.

Webber, J., Properly measured liquid/liquid interfaces (2011), http://www.chemicalprocessing.com/articles/2011/properly-measure-liquid-liquid-interfaces/.

1.47 Oxidation–Reduction Potential (ORP)

R. T. OLIVER (1972) **R. H. JONES** (1982) **D. M. GRAY** (1995)
B. G. LIPTÁK (2003) **S. J. WEST** (2017)

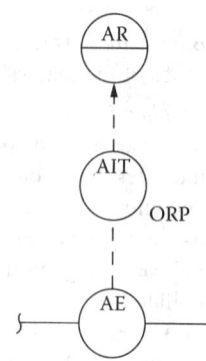

Flow sheet symbol

Design pressure	Vacuum to 150 psig (10.6 barg) is standard; special assemblies are available for up to 500 psig (35 barg)
Design temperatures	Generally from −5°C to 100°C (23°F to 212°F)
Materials of construction	Mounting hardware available in stainless steel, Hastelloy, titanium, PVC, CPVC, polyethylene, polypropylene, epoxy, polyphenylene sulfide, PTFE, and various elastomer materials; electrodes available in platinum or gold
Assemblies	Submersion, insertion, flow-through, and retractable
Cleaners	Ultrasonic, water, or chemical jet washer, brushes
Range	Any span between −2000 mV and +2000 mV
Inaccuracy	Typically ±10 mV and a function of the condition of the noble metal electrode and of reference electrode drift; repeatability about ±3 mV
Costs	Electrodes from $100 to $1000; portable or benchtop laboratory display and control units from $300 to $2500; transmitters from $500 to $2000; cleaners available from $500 to $7000
Partial list of suppliers	ABB (www.abb.us/product/us/9AAC100029.aspx)
	Advanced Sensor Technologies Inc. (www.astisensor.com/pH_ORP_Products.htm)
	Analytical Technology Inc. (www.analyticaltechnology.com/public/product.aspx?ProductID=1030)
	Broadley-James (www.broadleyjames.com/redox.html)
	EID Corp. (www.eidusa.com/ph_ORP.htm)
	Electro-Chemical Devices (www.ecdi.com/products/orp_series.html)
	Endress+Hauser (www.us.endress.com/eh/sc/america/us/en/home.nsf/#products/~liquid-analysis-instrument-ph-orp-measurement)
	Foxboro/Schneider Electric (www.fielddevices.foxboro.com/en-gb/products/analytical-echem/ph-and-orp/)
	GF Signet (www.gfps.com/content/gfps/signet/en_US/sensors/phorp/sensors/2734.html)
	Hach (www.hach.com/oxidation-reduction-potential-orp-/parameter-products?productType=instruments&parameterId=7639975457)
	Hamilton Co. (www.hamiltoncompany.com/products/sensors/c/13/)
	Hanna Instruments (www.hannainst.com/usa/subcat.cfm?id = 002)
	Honeywell (www.honeywellprocess.com/en-US/explore/products/instrumentation/analytical-instruments-and-sensors/ph-orp/Pages/default.aspx)
	Horiba Instruments, Inc. (www.horiba.com/application/material-property-characterization/water-analysis/water-quality-electrochemistry-instrumentation/electrodes-accessories/orp-electrodes/)
	Knick (www.knick-international.com/products/sensors/)

(Continued)

Lakewood Instruments (/lakewoodinstruments.com/530-series-orp-process-sensors/)
Mettler-Toledo (us.mt.com/us/en/home/products/Process-Analytics.html)
Nieuwkoop b.v. (www.nieuwkoopbv.nl/index.php/cPath/230_246_279)
Orion/ThermoScientific (www.thermoscientific.com/en/search-results.html?keyword=ORP+Electrodes&country
  Code=US&matchDim=Y)
Rosemount Analytical/Emerson (www2.emersonprocess.com/en-US/brands/rosemountanalytical/Liquid/Sensors/pH/
  Pages/index.aspx)
Sensorex (www.sensorex.com/products/categories/category/orp_sensors)
Swan Analytical Instruments (http://www.swan-analytical-usa.com/Catalog/en/ProductGroupOverview.aspx?
  subchapter=Chapter_feedwater&prdtGroup=Grp_Redox_Feed&prdtSubGroup=&prdtName=)
YSI/Xylem (www.ysi.com/search.php?q=orp)
Yokogawa (www.yokogawa.com/an/ph-orp/an-ph-orp-001en.htm)

INTRODUCTION

The measurement and control of oxidation–reduction potential (ORP or redox potential) are applied in many applications in both industrial processing and wastewater treatment. The more important applications include the following:

1. In ore leaching, metal is leached from the ore and converted to the desired state for further processing.
2. Toxic cyanides are oxidized to harmless reaction products in an oxidation–reduction reaction as part of the process to remove toxic heavy metals (Figure 1.47a).
3. In the pulp and paper industry, pulp is bleached with a variety of oxidants under ORP measurement and control.
4. Hexavalent chromium is reduced to the trivalent oxidation state as part of the process for removal of toxic chromium from metal finishing or from cooling tower blowdown wastewaters (Figure 1.47a).

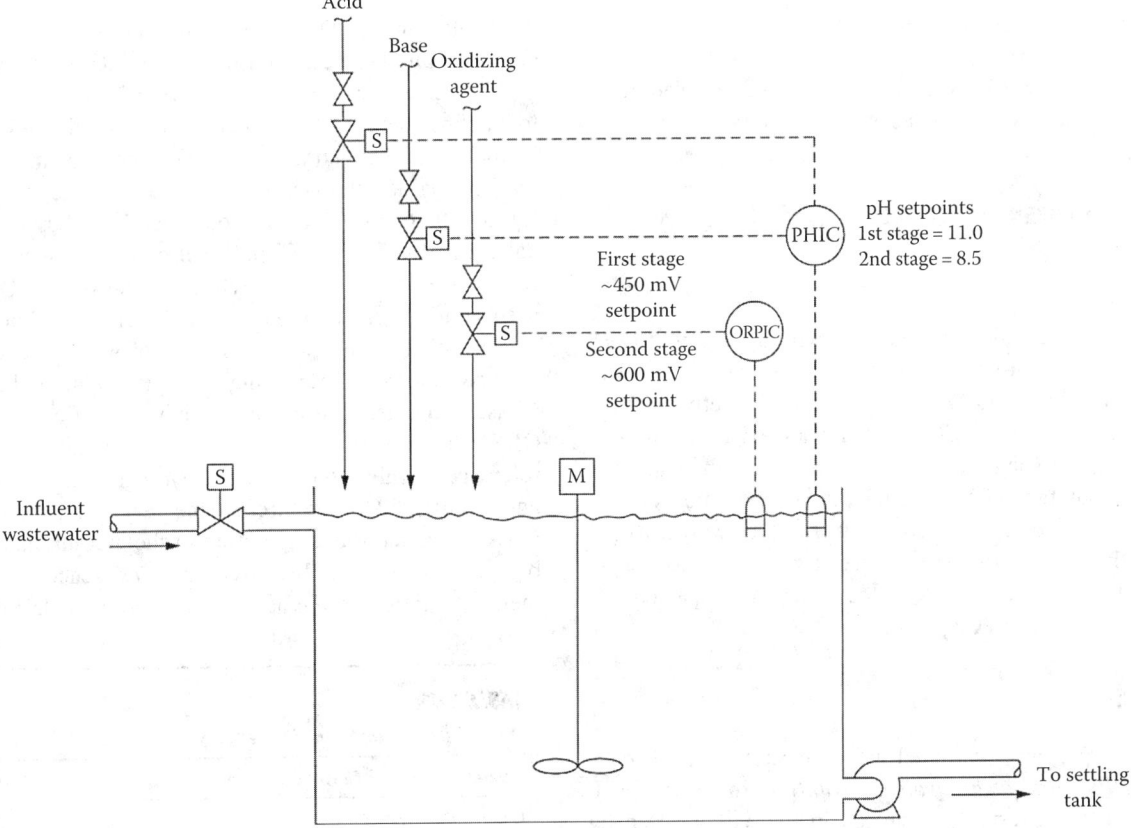

FIG. 1.47a
Batch oxidization of toxic cyanides.

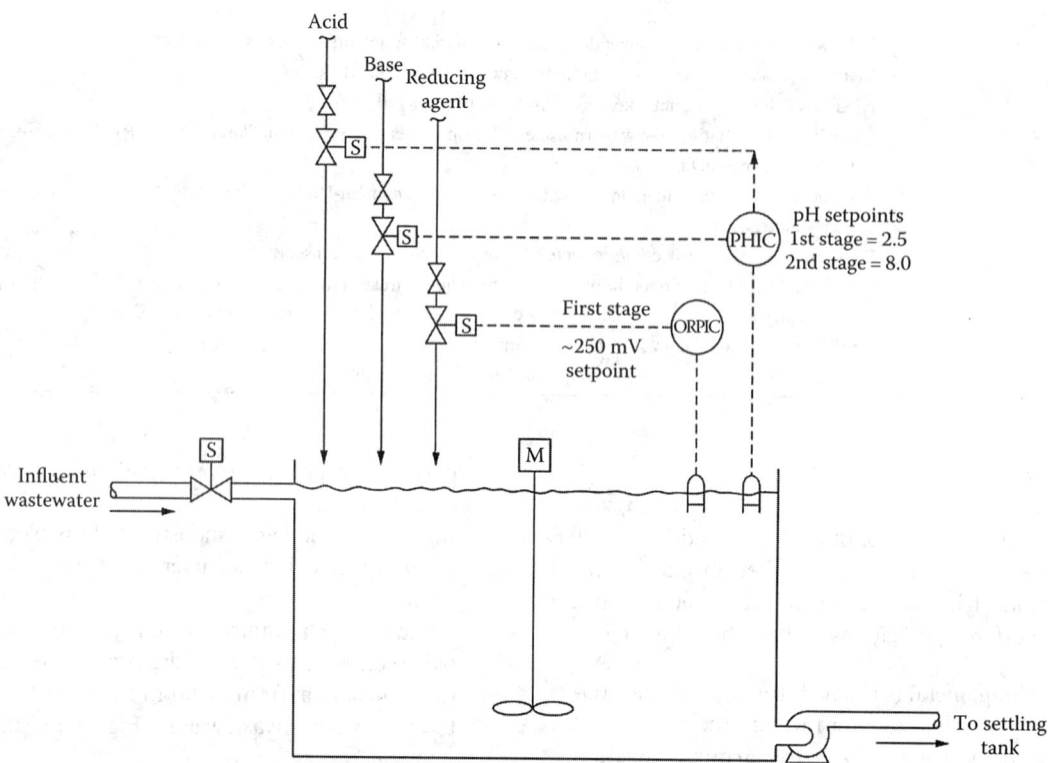

FIG. 1.47b
Batch reduction of toxic hexavalent chromium.

5. The manufacture of chlorine bleaches is controlled by ORP.
6. In sanitary wastewater treatment, ORP is used to control the addition of an oxidant for odor control (Figure 1.47b).

ORP MEASUREMENT PRINCIPLES

ORP

ORP is an electrochemical measurement analogous to pH. As pH is a measure of the activity of the hydrogen ion or proton in a solution, ORP is a measure of the activity of the electron in a solution. Electrons do not exist as independent species in solution (nor do protons), but their activity is reflected by the collective tendency of the components of a solution to donate or accept them. In other words, as pH reflects the tendency of the species in a solution to donate or accept protons, ORP reflects the tendency of species in a solution to donate or accept electrons.

ORP Scales

ORP and pH are measured using electrochemical cells (see *Cell Potential at Equivalence Point*). In the case of pH, the cell potential, usually expressed in mV, is converted by calibration to the pH scale. The pH scale is defined by standard solutions recognized by government institutions, for example, National Institute for Standards and Technology (NIST) in the United States, Deutsches Institutfür Normung (DIN) in Germany, and Japanese Industrial Standards (JIS) in Japan.

ORP is expressed simply in *raw* mV, without reference to a defined scale. This cannot be considered a best practice in analytical chemistry. The raw ORP cell potential depends on the type of reference half-cell used and the temperature. It is true that most ORP cells today use AgCl-based reference half-cells, but even AgCl half-cell potentials depend on the KCl concentration in the electrolyte. Many older ORP cells used saturated calomel reference half-cells, and though these have mostly been abandoned in favor of AgCl, not all operating procedures and reporting of ORP potentials have been adjusted to reflect this change. The best practice in reporting ORP values is to convert them to E_H, which indicates they are referenced against the standard hydrogen electrode (SHE), and report the temperature. At the very least, the reference ½-cell used and the temperature of the measurement should be specified. A short list of common reference ½-cells and their potentials vs. SHE at 25°C is found in Table 1.47a.

TABLE 1.47a
Potentials of Common Reference ½-Cells vs. SHE at 25°C

Reference ½-Cell/KCl Concentration	E vs. SHE (Volts)
Calomel/1.0 M KCl	+0.283
Calomel/sat'd KCl	+0.245
AgCl/1.0 M KCl	+0.222
AgCl/1.0 M KCl	+0.198

Oxidation and Reduction

The reacting molecules in the oxidation–reduction reactions either gain or lose electrons and therefore end up with different electron configurations after the reaction.

Oxidation in a chemical reaction is the overall process by which a molecule loses one or more electrons and increases its state of oxidation. An oxidant is a substance that is capable of oxidizing another chemical species in a process in which it acquires one or more electrons, which are lost by the species being oxidized. In this overall process, the oxidant itself is reduced.

Reduction is the overall process in which a species in a chemical reaction gains one or more electrons and decreases its state of oxidation. A reductant is a substance that is capable of reducing a chemical species. The reductant loses the electrons, which are gained by the species, and is itself oxidized in the overall process.

Redox Couples and Reversibility

A species that can undergo a redox reaction and the product of that reaction are called a redox couple. For example, in Equation 1.47(1), ferrous and ferric ions (Fe^{2+} and Fe^{3+}) comprise a redox couple. At thermodynamic equilibrium, only one redox couple will be present in a solution with significant concentrations of both its oxidized and reduced forms (unless there are other redox couples present with exactly the same potential—possible but improbable). However, in a solution that is not at equilibrium, more than one redox couple may be present. In this case, it is said that one or more of these couples are not *reversible* in solution. Couples that are reversible in solution are couples that react quickly with each other to achieve equilibrium. When equilibrium is achieved, only the couple with the highest concentration will remain.

The ORP is established at the surface of an inert metal electrode when members of a redox couple exchange electrons with the metal. The concept of reversibility described earlier for reactions in solution applies to these electron exchanges at the electrode surface as well. A stable or *poised* redox potential is established when the exchange current—the number of electrons exchanged per unit time—arising from a reversible redox couple is high compared to other electron exchanges. Electron exchanges may occur with other redox species that are not at equilibrium in the solution nor reversible at the electrode. These other exchanges are negligible in terms of establishing the ORP.

When the two conditions—equilibrium in the solution and reversibility at the electrode surface—are met, ORP can be related directly and precisely to the ratio of the oxidized to reduced form of a redox couple as described in detail in the following sections (Equation 1.47(9)). When these conditions are not met, ORP gives less exact but nevertheless qualitatively useful information about the redox state of a solution. For example, the redox couple formed between dissolved oxygen and water (third equation, Table 1.47b) does not achieve equilibrium in solution nor exchange electrons reversibly at an ORP electrode. Nevertheless, in the absence of some other dominating redox couples, oxygenated solutions will reliably exhibit a higher ORP than anoxic solutions and ORP can be useful for qualitatively assessing such a solution's state of oxygenation.

TABLE 1.47b
Reduction Potentials of Solution in ORP Measurement

Reduction	$E°$, V
$O_3 + 2H_3O^+ + 2e^- = O_2 + 3H_2O$	+2.070
$Cr_2O_7^{2-} + 14H_3O^+ + 6e^- = 2Cr^{3+} + 21H_2O$	+1.330
$O_2 + 4H_3O^+ + 4e^- = 2H_2O$	+1.229
$ClO^- + H_2O + 2e^- = Cl^- + 2OH^-$	+0.890
$Fe^{3+} + e^- = Fe^{2+}$	+0.770
Ag/AgCl electrode 4 M KCl	+0.199
$2H_3O^+ + 2e^- = H_2 + 2H_2O$	0.000
$Zn^{2+} + 2e^- = Zn$	0.763
$CNO^- + H_2O + 2e^- = CN^- + 2OH^-$	0.970
$Na^+ + e^- = Na$	−2.711

ORP Reaction

An ORP reaction involves an electron exchange that is capable of doing work. This capability is expressed in terms of potential for a half-cell, or electron, reaction. The potentials listed in Table 1.47b are for standard conditions, that is, where reactants and products are at unit activity. Voltages in this table are referenced to the standard SHE, which is assigned the value of 0.000 V.

Note that the reactions in Table 1.47b are written as reductions, which is the most often used convention. The term *ox/red* is used to indicate the oxidized form on the left side of the equation and the reduced form on the right. For example, the standard potential for ferric iron, Fe^{3+}, being reduced to ferrous, Fe^{2+}, is written as

$$E°_{Ox/Red} = +0.770 \text{ V} \qquad \textbf{1.47(1)}$$

It will be noted that $E_{Red/Ox} = -E_{Ox/Red}$, which simply means that polarity is reversed when the reaction is written as an oxidation reaction. For example,

$$Fe^{2+} = Fe^{3+} + e^-\; E°_{Red/Ox} = -0.770 \text{ V} \qquad \textbf{1.47(2)}$$

Half-Cell Reactions It is customary in dealing with oxidation–reduction reactions to write down separately the two half-cell reactions that make up the overall reaction. These are written so that known reactants are on the left and known products are on the right. The following is a general equation in which the hydronium ion participates in the reaction, and therefore the potentials are pH dependent.

Half Reaction

$$aOx_2 + yH_3O^+ + n_2e^- \rightleftarrows bRed_2 + wH_2O \quad E_{Ox_2/Red_2} \qquad 1.47(3)$$

Half Reaction

$$cRed_1 \rightleftarrows dOx_1 + n_1e^- \quad E_{Red_1/Ox_1} \qquad 1.47(4)$$

Overall Reaction

$$n_1aOx_2 + n_2cRed_1 + n_1yH_3O^+ \rightleftarrows n_1bRed_2 + n_2dOx_1 + n_1wH_2O \qquad 1.47(5)$$

Cell Potential

$$E_{overall} = E_{Ox_2/Red_2} + E_{Red_1/Ox_1} \qquad 1.47(6)$$

At any point in the reaction, the solution or cell potential is given by

$$E_{cell} = E_{Ox_2/Red_2} + E_{Ox_1/Red_1} \qquad 1.47(7)$$

which, at 25°C (77°F), is

$$E_{Ox_2/Red_2} = E°_{Ox_2/Red_2} + \frac{0.059}{n_2} \log \frac{[Ox_2]^a[H_3O^+]^y}{[Red_2]^b[H_2O]^w} \qquad 1.47(8)$$

and

$$E_{Ox_1/Red_1} = E°_{Ox_1/Red_1} + \frac{0.059}{n_1} \log \frac{[Ox_1]^d}{[Red_1]^c} \qquad 1.47(9)$$

Setting Equations 1.47(8) and 1.47(9) to equal each other and rearranging, we get

$$(n_1 + n_2)E_{cell} = n_2 E°Ox_2/Red + n_1 E°_{Ox_1/Red_1} + 0.059 \log$$
$$\times \frac{[Ox_2]^a[Ox_1]^d}{[Red_2]^b[Red_1]^e} + 0.059 \log H_3O^{+y} \qquad 1.47(10)$$

Two other relationships can also be given, one for any reaction point and the other for the equivalence point. At any point in the reaction after the start,

$$Red_2 = \frac{n_1 b}{n_2 d}[Ox_1] \qquad 1.47(11)$$

and at the equivalence point,

$$Ox_2 = \frac{n_1 a}{n_2 c}[Red_1] \qquad 1.47(12)$$

Cell Potential at Equivalence Point Substitution of the ratio relationship Ox_2^a/Red_2^b or Ox_1^d/Red_1^c, as derived from Equations 1.47(11) and 1.47(12), into Equation 1.47(10) will provide the cell potential for the equivalence point. It can be an interesting exercise to see how this equivalence point potential changes when differing values of H_3O^+ (i.e., pH) are substituted into Equation 1.47(10). By so doing, the profound effects that pH can have on equivalence and control point potentials will become apparent. This then shows that when hydronium, H_3O^+, or hydroxyl, OH^-, ions participate in ORP reactions, close control of pH may become just as important as close control of ORP.

Note again Equations 1.47(7) through 1.47(9), which state that

$$E°_{Ox_2/Red_2} + \frac{0.059}{n_2} \log \frac{[Ox_2]^a[H_3O]^y}{[Red_2]^b[H_2O]^w}$$
$$= E°_{Ox_2/Red_2} - E°_{Ox_1/Red_1} + \frac{0.059}{n_1} \log \frac{[Ox_1]^d}{[Red_1]^c} \qquad 1.47(13)$$

and

$$\frac{0.059}{n_1 n_2} \log \frac{[Red_2]^{n_1 b}[Ox_1]^{n_2 d}[H_2O]^{n_1 w}}{[Ox_2]^{n_1 a}[Red_1]^{n_2 c}[H_3O^+]^{n_1 y}} = E°_{Ox_2/Red_2} - E°_{Ox_1/Red_1} \qquad 1.47(14)$$

The equilibrium constant is given by

$$\frac{[Red_2]^{n_1 b}[Ox_1]^{n_1 d}}{[Ox_2]^{n_1 b}[Red_1]^{n_2 cIO}} = \frac{E°_{Ox_2/Red_2} - E°_{Ox_1/Red_2}}{0.059/n_1 n_2} \times [H_3O^+]^{O_1 y} \qquad 1.47(15)$$

Practical Measurements

Microprocessor-Based Units Microprocessor-based ORP analyzers generally provide wide rangeability with high-resolution digital displays, alarm/control set points, and output signal scaling limits.

Most chemical reactions that involve electron exchange are controlled near the equivalence point. A controlled excess of reagent is added to ensure that the reaction is driven to completion. Thus, most ORP reactions will be controlled just beyond the steep portion of the titration curve.

ORP SENSORS

The basic instrumentation used for ORP measurement closely parallels that used for pH measurement. In fact, many instrument suppliers use slightly modified pH analyzers for ORP detection, the main difference being changed sensitivity and a millivolt (mV) scale in place of a pH scale. The electrode hardware (i.e., the equipment used to install the electrodes in the process stream) is generally the same as that used in pH systems (Figures 1.47c and 1.47d).

There are two major differences between an ORP system and a pH system. One is the sensing electrode, which, in case of ORP measurement, is normally a noble metal (typically platinum or gold), although other metals and carbon have also been used on occasion. The second major difference is in temperature compensation. Process pH systems are typically temperature compensated, whereas ORP systems almost never are.

Nernst Equation

The basic thermodynamics apply to both pH and ORP, expressed by the classical Nernst equation. For oxidation–reduction half-cell reactions, this may be represented as follows:

$$E_{cell} = E°_{Ox/Red} + \frac{2.303\, RT}{nF} \log\left[\frac{Ox}{Red}\right] \quad 1.47(16)$$

where

E° is the potential under standard conditions of unit activity referred to the SHE
R is the gas constant, 1.986 cal per mol degree
F is the Faraday constant
T is the temperature in Kelvins
n is the number of electrons exchanged in the reaction

In the Nernstian representation of pH, n always equals 1. Note that even in the very abbreviated listing of Table 1.47b, the n values vary among reactions. This, plus the fact that a given ORP reaction may encompass side reactions, makes it quite clear why it is difficult, if not impossible, to temperature compensate an ORP reaction.

In Equation 1.47(19), the standard potential $E°_{Ox/Red}$ can be found in handbook tables, and values are usually given relative to the SHE. Therefore, Ecell in Equation 1.47(19) for the prevailing concentration is also relative to SHE.

E_{meas} is the value that is read on the meter. If E_{cell} is known from calculation or from actual measurement, its potential value for other systems can be readily converted by

$$E \text{ vs SHE} = E \text{ vs SCE} + E_{SCE/Red} \quad 1.47(17)$$

or

$$E \text{ vs SCE} = E \text{ vs SHE} - E_{SCE/Red} \quad 1.47(18)$$

or

$$E \text{ vs Ag/AgCl Ref} = E \text{ vs SHE} - E_{Ag/AgCl/Red} \quad 1.47(19)$$

Electrode Mounting

ORP and pH electrodes can often be mounted in the same tees or flow-through chambers and can have practically identical appearances. The electrode shown in

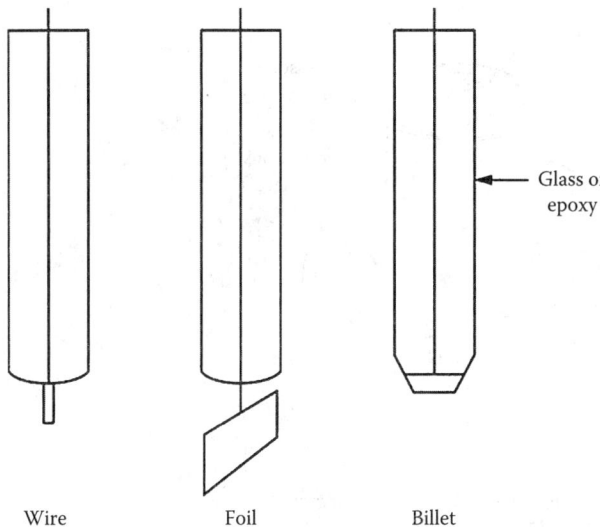

FIG. 1.47c
The various types of metallic ORP electrodes.

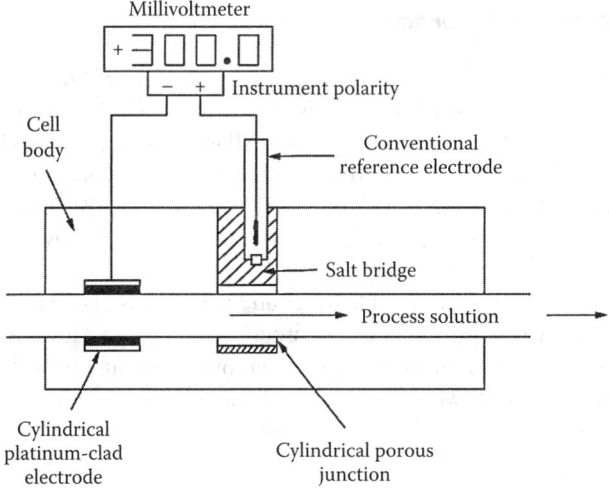

FIG. 1.47d
Cylindrical ORP electrode cell.

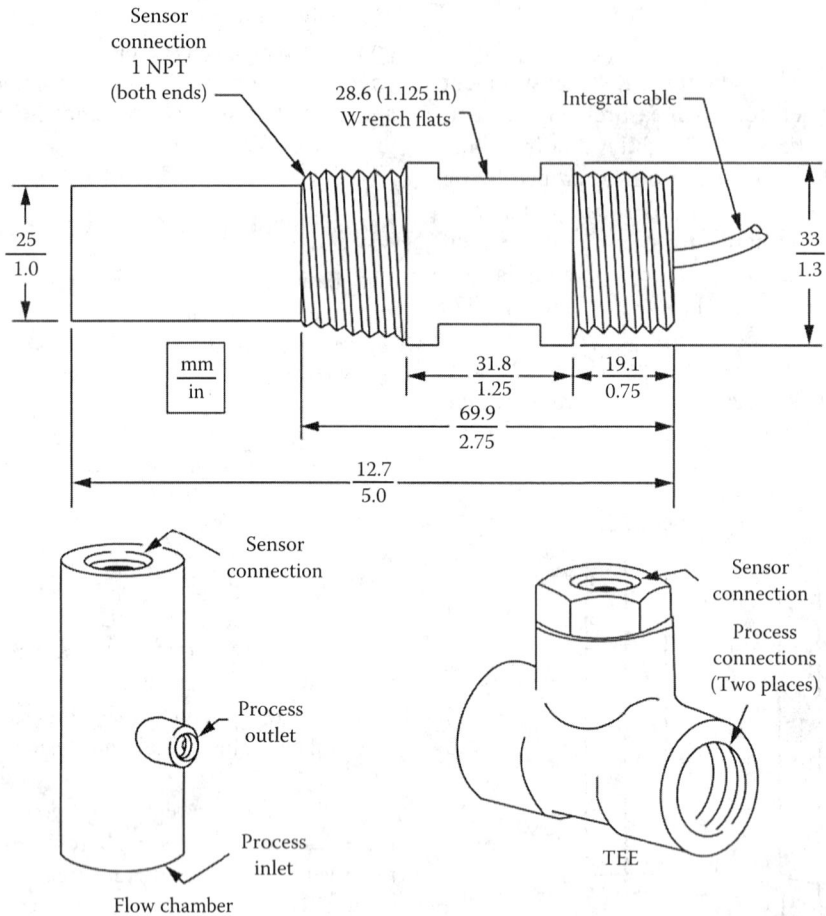

FIG. 1.47e
ORP and pH probes are packaged and mounted in the same way. (Courtesy of Foxboro/Schneider Electric.)

Figure 1.47e is a ruggedized, flat glass electrode having identical dimensions for both pH and ORP services. In case of the ORP probe, the only difference is the gold or platinum wire tip.

ORP APPLICATIONS

It is certainly possible by use of the equations developed earlier to determine the equivalence potential for an ORP control reaction. For a reaction involving H_3O^+ or OH^- ions, Equation 1.47(10) shows the rather profound effects of pH on the equivalence point potential. Equation 1.47(15) can be used to calculate the degree of completion of the reaction, and again, this relationship also shows the effects of pH. A reaction that pointedly illustrates these effects is the reduction of hexavalent chromium with sulfur dioxide or with bisulfite.

However, many prefer a more empirical approach, as exemplified by chromium reduction.

Chromium Reduction

The chromium reaction typically takes place at a pH of about 2.0–2.5 (Figure 1.47b). At this pH, there is a smell of sulfur dioxide when the sulfite ion is in slight excess. An experienced operator might adjust the control point potential to attain a slight odor of sulfur dioxide and then make further adjustments based on laboratory analysis for hexavalent chromium.

It is certainly possible to set up a system based on calculation when all reactants and products are known. However, this is seldom the case in industrial processes, and only the very innocent would proceed without analytical verification of the results.

The ORP responses in two of the most common ORP applications are illustrated by the titration curves in Figures 1.47f and 1.47g. These curves are only examples. Response can vary considerably from one specific installation and from one process composition to another. Actual control points must be finely adjusted after the start-up of a system. For details on how to configure and operate such control systems, refer to the volume on *process control* of this handbook.

Corrosion Protection

The following example illustrates how ORP can be used in conjunction with pH for corrosion control. It also introduces a tool known as the E_H–pH or Pourbaix diagram.*

The intimacy of the relationship between ORP and pH, illustrated in the previous example, is not surprising given the fact that they represent, respectively, the activities of the electron and proton in solution. The E_H–pH or Pourbaix diagram, where the composition of a solution, usually in contact with a metal, is mapped against ORP (E_H) on a vertical axis and pH on a horizontal axis, has proven itself a useful tool in fields including corrosion control, geochemistry, and hydrology. Figure 1.47h is a Pourbaix diagram for a solution containing chloride ions in contact with copper.† This diagram is one of a series created to guide control of corrosion in copper canisters used for the long-term storage of spent nuclear fuel in chloride-containing solutions. The dashed diagonal lines indicate the boundaries for the stability of water itself. Above the upper dashed line, water decomposes with the evolution

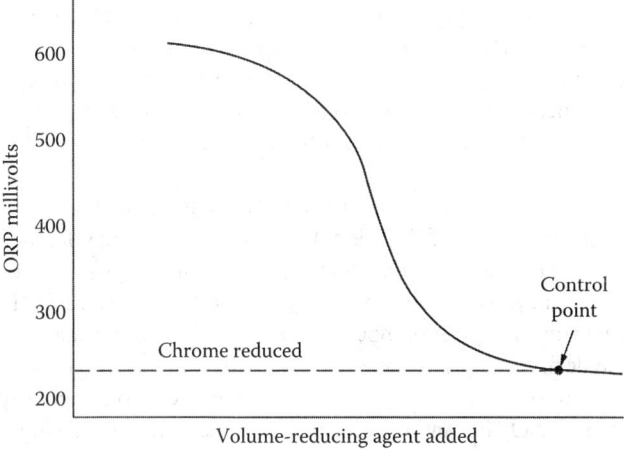

FIG. 1.47f
Chrome reduction titration curve.

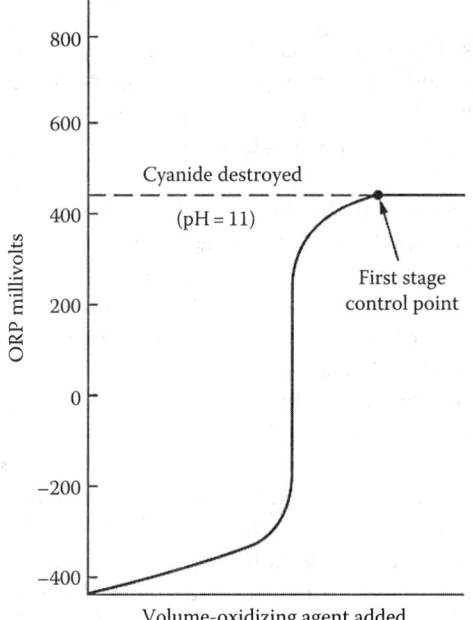

FIG. 1.47g
Cyanide oxidation titration curve.

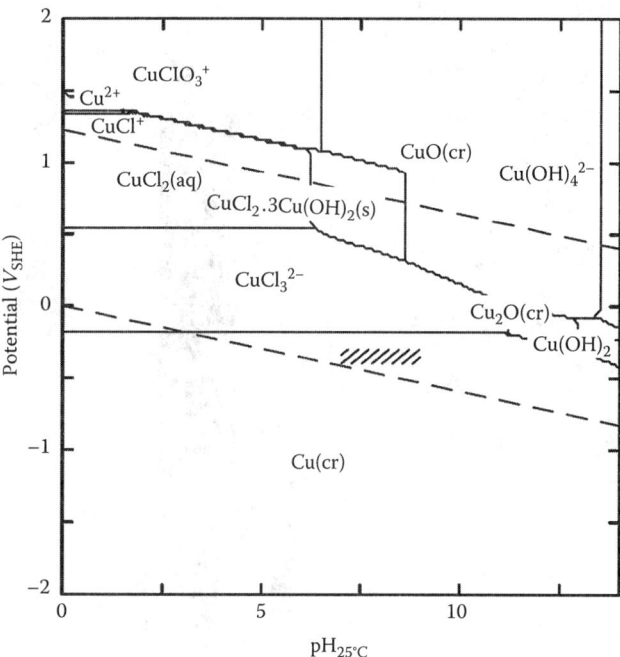

FIG. 1.47h
Pourbaix diagram for Cu in five molal chloride solution. (From Beverskog, B. and Petterson, S.-O., Pourbaix diagrams for copper in five m chloride solution, SKI Report 02:23, Swedish Nuclear Power Inspectorate, Klarabergviadukten 90, SE106 58, December 2002.)

---

* Delahay, P., Pourbaix, M., and Van Rysselberghe, P., Potential-pH diagrams, *J. Chem. Ed.*, December 1950.
† Beverskog, B. and Petterson, S-O., Pourbaix diagrams for copper in 5 m chloride solution, *SKI Report 02:23*, Swedish Nuclear Power Inspectorate, Klarabergviadukten 90, Stockholm, Sweden, SE106 58, December 2002.

of oxygen. Below the lower dashed line, water decomposes with evolution of hydrogen. Below the lowest solid line, metallic copper (Cu(cr)) is stable against corrosion. Above that line, the soluble copper species—corrosion products—are identified. The dashed box near the center of the diagram represents the range of ORP and pH values expected by the process operators.

ORP MAINTENANCE AND CALIBRATION

Maintenance

Maintenance of an ORP measuring and control system is generally comparable to that of a similar pH system. This includes replacement or refilling of the reference half-cell, in accordance with manufacturers' instructions, and cleaning of the ORP sensing surface. Cleaning of the sensing surface can be done manually with a mild abrasive (toothpaste and toothbrush have often served) or by means of an automated scrubber-type probe cleaner, like that shown in Figure 1.47i.

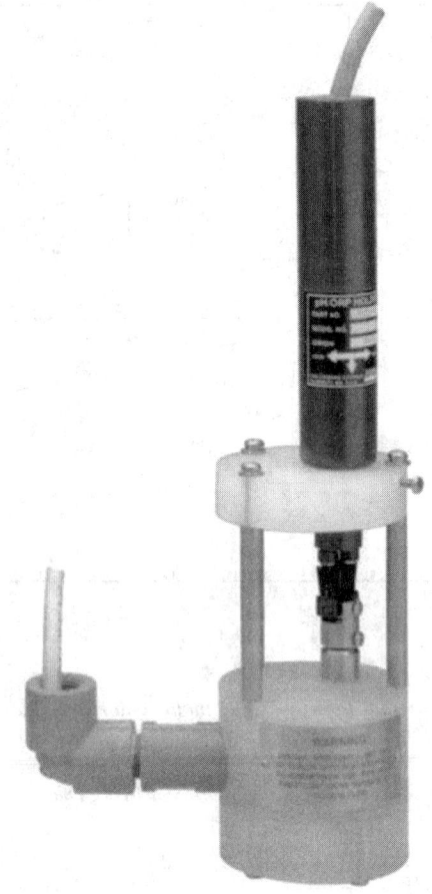

FIG. 1.47i
Scrubber-type probe cleaner. (Courtesy of Foxboro/Schneider Electric.)

Calibration

Calibration requirements for ORP systems have similarities to those of pH, but there are some differences:

1. The potentiometric accuracy that users generally require of ORP systems is not as great as for pH. Whereas 10 or 20 mV uncertainties are significant in many pH applications—corresponding to pH changes of 0.15–0.30 pH units—such uncertainties are often not significant in ORP control.
2. In pH measurement, a minimum of two calibration points is required to convert raw mV values to the pH scale. ORP, on the other hand, is expressed in mV, often without conversion to another scale. In cases where conversion to the SHE scale is desired (E_H), a one-point calibration is sufficient.

The most prevalent problems in ORP measurement are coating or poisoning of the metal ORP electrode or the reference electrode liquid junction that may result in sluggishness or inaccuracy. This measurement error, in turn, may result in charging the wrong amount of reagent and not maintaining the desired reagent excess.

Consult laboratory supply catalogs for commercially available ORP standards, as these are becoming increasingly more common. When either a change in span or a shift in potential is noted, the recommended corrective treatment is cleaning with aqua regia.

ORP CONTROL

ORP instruments are usually installed for control purposes. Because the measurement of ORP is quite similar to that of pH, it follows that many pH control considerations also hold for ORP. If, for example, an oxidizing solution such as dichromate in acid media is titrated with a reducing solution of ferrous ion, a titration curve that is quite similar to that of an acid–base titration can be generated. However, a more common practice is the reduction of hexavalent chromium with gaseous sulfur dioxide. Whereas it would be quite simple to catch samples of an acid waste and to titrate these samples with a standard sodium hydroxide solution, doing a titration of hexavalent chromium with sulfur dioxide would present some problems.

A more expedient way of determining reagent demand is by catching representative samples and having them analyzed chemically to determine hexavalent chromium. The reagent demand of sulfur dioxide can then be determined from the balanced oxidation–reduction overall reaction. From these data, a number can be assigned to the trivalent chromium concentration at the equivalence point and the operating pH level assigned. From Equations 1.47(10) and 1.47(15), equivalence point potential and unreacted hexavalent chromium at the equivalence point can be determined.

Residence Time

Unlike acid–base neutralizations in which the reaction is virtually instantaneous, ORP reactions are frequently time dependent. The reaction vessel might be sized accordingly to provide the required residence time. A case in point is the oxidation of cyanide with available chlorine.

The reaction between free cyanide, CN^-, with the hypochlorite ion, OCl^-, is quite rapid. However, in many cases where cyanide oxidation is employed, part of the cyanide may exist as metal–cyanide complex ions. The fraction of free cyanide ions is extremely small. For some metal–cyanide complexes, many hours of reaction time may be necessary to completely destroy the cyanide.

As was emphasized earlier, in ORP reactions involving hydronium, H_3O^+, or hydroxide, OH^-, ions, the measured potential at the equivalence point and the degree of completion of the reaction are both pH dependent. Therefore, pH control becomes as important as is ORP control (see Figures 1.47a and 1.47b). A constant potential is not indicative of an excess of reactant if pH is variable.

Most of the pH-related considerations that will be discussed in Chapter 1.55 also apply to ORP control. Of special importance is the provision of facilities to equalize the reagent demand by use of equalizing tanks or bleed-in tanks used to store and slowly emit strong solutions to the reaction vessel.

RECENT DEVELOPMENTS

As an analytical technique, ORP measurement is very mature, and as such, most recent technology advances are on the instrumentation side. As with other electrochemical techniques such as pH, dissolved oxygen, and conductivity, a clear trend is toward so-called smart or digital sensors. A smart sensor typically contains integral memory and analog-to-digital conversion electronics. Among the advantages of this approach are that every sensor can contain in its memory a unique identity or serial number, *date of birth*, calibration coefficients, and a history log that can be uploaded to a transmitter or control system at the time of connection. The digital communications link is capable of operating over longer cables than are possible with unamplified electrochemical signals. Most major ORP suppliers now offer digital ORP probes.

In spite of its maturity, ORP applications continue to expand at a steady pace. A survey or recent literature reveals new or refined applications in many fields including the following examples. Puig et al.* describe the use of ORP in the control of a biological batch reactor for the treatment of mixed municipal and industrial waste. Sistare et al.† apply ORP to the monitoring of reaction constituents in the synthesis of pharmaceuticals. Yu et al.‡ developed a process for control of a color removal process for textile industry wastewaters.

SPECIFICATION FORMS

When specifying ORP analyzers, the ISA Form 20A2341 can be used to specify the features required for the sensor itself, and when specifying both the sensor and the required transmitter and/or receiver, ISA Form 20A2342 can be used. These forms, with the permission of the International Society of Automation, are reproduced on the next pages.

---

* Puig, S., Corominas, L., Vives, M. T., Balaguer, M. D., Coloprim, J., and Colomer, J. *Ind. Eng. Chem. Res.*, 44(9), 3367–3373, March 2005.
† Sistare, F., St. Pierre Berry, L., and Mojica, C. A., Process analytical technology: An investment in process knowledge, *Org. Process Res. Dev.*, 9(3), 332–336, April 2005.
‡ Yu, R. F., Chen, H-W., Cheng, W-P., and Hsieh, P-H., Dosage control of the Fenton process for color removal of textile wastewater applying ORP monitoring and artificial neural networks, *J. Environ. Eng.*, 135(5), 325–332, 2009.

pH/ORP SENSOR w/wo INSERTION ASSEMBLY Device Specification

#	RESPONSIBLE ORGANIZATION		#	SPECIFICATION IDENTIFICATIONS	
1			6		
2	ISA		7	Document no	
3			8	Latest revision	Date
4			9	Issue status	
5			10		

#	BODY HOLDER OR FITTING			#	INSERTION ASSEMBLY continued	
11				60		
12	Body/Fitting type			61	Insertion/Immersion lg	
13	Process conn nominal size	Rating		62	Valve body material	
14	Process conn termn type	Style		63	Valve seat material	
15	Maximum probe diameter			64	Chamber wetted material	
16	Body/Fitting material			65	Compression ferrule matl	
17	Flange material			66		
18	Seal/O ring material			67		
19				68	PERFORMANCE CHARACTERISTICS	
20				69	Max press at design temp	At
21				70	Min working temperature	Max
22	SENSING ELEMENT			71	Min conductivity	
23	Sensor type			72	Max fluid velocity limit	
24	Construction style			73	Accuracy rating	
25	Measuring electrode shape			74	Measurement LRL	URL
26	Reference junction style			75	Temp compensation LRL	URL
27	Insertion/Immersion lg			76	Linearity	
28	Signal conn termn style			77	Max response time	
29	Temperature sensor type			78	Min ambient working temp	Max
30	Sensing surface matl			79	Max sensor to receiver lg	
31	Ref electrolyte material			80		
32	Ground pin material			81		
33	Reference element matl			82		
34	Primary junction matl			83	ACCESSORIES	
35				84	Electrolyte solution	
36				85	Buffer solution	
37				86	Cleaner type	
38	LEAD WIRE AND EXTENSION			87	Cleaner supply source	
39	Extension type			88	ORP std solution	
40	Cable length			89	Rotameter & valve	
41	Max cable operating temp			90		
42	Signal termination type			91		
43	Cable jacket material			92	SPECIAL REQUIREMENTS	
44				93	Custom tag	
45	CONNECTION HEAD OR PREAMPLIFIER			94	Reference specification	
46	Configuration type			95	Compliance standard	
47	Temperature compensation			96	Calibration report	
48	Enclosure type no/class			97		
49	Signal termination type			98		
50	Cert/Approval type			99		
51	Mounting location/type			100	PHYSICAL DATA	
52	Enclosure material			101	Estimated weight	
53				102	Overall height	
54				103	Removal clearance	
55	INSERTION ASSEMBLY			104	Signal conn nominal size	Style
56	Assembly type			105	Mfr reference dwg	
57	Isolation valve style			106		
58	Process conn nominal size	Rating		107		
59	Process conn termn type	Style		108		

#	CALIBRATIONS AND TEST		INPUT OR TEST		OUTPUT	
110						
111	TAG NO/FUNCTIONAL IDENT	MEAS/SIGNAL/SCALE	LRV	URV	LRV	URV
112		Measurement-Output signal				
113		Temp-Output signal				
114						
115						
116						
117						

#	COMPONENT IDENTIFICATIONS		
118			
119	COMPONENT TYPE	MANUFACTURER	MODEL NUMBER
120			
121			
122			
123			
124			
125			

Rev	Date	Revision Description	By	Appv1	Appv2	Appv3	REMARKS

Form: 20A2341 Rev 0 © 2004 ISA

pH/ORP/CONDUCTIVITY/RESISTIVITY TRANSMITTER/ANALYZER/MONITOR Device Specification

#			#		
1	RESPONSIBLE ORGANIZATION		6	SPECIFICATION IDENTIFICATIONS	
2			7	Document no	
3			8	Latest revision	Date
4			9	Issue status	
5			10		
11	TRANSMITTER OR ANALYZER		56	PERFORMANCE CHARACTERISTICS	
12	Housing type		57	Measurement accuracy	
13	Input sensor type		58	Temperature accuracy	
14	Input sensor style		59	Measurement LRL	URL
15	Output signal type		60	Temp compensation LRL	URL
16	Enclosure type no/class		61	Minimum span	Max
17	Control mode		62	Min ambient working temp	Max
18	Local operator interface		63	Contacts ac rating	At max
19	Characteristic curve		64	Contacts dc rating	At max
20	Digital communication std		65	Max sensor to receiver lg	
21	Signal power source		66		
22	Temperature sensor type		67		
23	Quantity of input sensors		68		
24	Preamplifier location		69		
25	Contacts arrangement	Quantity	70		
26	Integral indicator style		71		
27	Signal termination type		72		
28	Cert/Approval type		73		
29	Mounting location/type		74		
30	Failure/Diagnostic action		75		
31	Sensor diagnostics		76	ACCESSORIES	
32	Data history log		77	Remote indicator style	
33	Temperature compensation		78	Indicator enclosure	
34	Calibration function		79	Communicator style	
35	Enclosure material		80		
36			81		
37			82		
38			83	SPECIAL REQUIREMENTS	
39			84	Custom tag	
40			85	Reference specification	
41			86	Compliance standard	
42			87	Software configuration	
43			88	Software program	
44			89		
45			90		
46			91		
47			92	PHYSICAL DATA	
48			93	Estimated weight	
49			94	Overall height	
50			95	Removal clearance	
51			96	Signal conn nominal size	Style
52			97	Mfr reference dwg	
53			98		
54			99		
55			100		

#	CALIBRATIONS AND TEST		INPUT OR TEST			OUTPUT OR SCALE	
110							
111	TAG NO/FUNCTIONAL IDENT	MEAS/SIGNAL/SCALE	LRV	URV	ACTION	LRV	URV
112		Meas-Analog output 1					
113		Meas-Analog output 2					
114		Meas-Digital output					
115		Meas-Scale					
116		Temp-Scale					
117		Temp-Digital output					
118		Meas setpoint 1-Output					
119		Meas setpoint 2-Output					
120		Meas setpoint 3-Output					
121		Meas setpoint 4-Output					
122		Failure signal-Output					
123							

#	COMPONENT IDENTIFICATIONS		
124			
125	COMPONENT TYPE	MANUFACTURER	MODEL NUMBER
126			
127			
128			
129			

Rev	Date	Revision Description	By	Appv1	Appv2	Appv3	REMARKS

Form: 20A2342 Rev 0

Abbreviations

barg	Bar gauge
CPVC	Chlorinated (poly)vinyl chloride
ORP	Oxidation–reduction potential
psig	Pounds per square inch gauge
PVC	(Poly)vinyl chloride
SHE	Standard hydrogen electrode

Organizations

DIN	Deutsches Institutfür Normung
JIS	Japanese Industrial Standards
NIST	National Institute for Standards and Technology

Bibliography

Adams, V., *Water and Wastewater Examination Manual*, Lewis Publishers, Boca Raton, FL, 1990.

Copeland, A. and Lytle, D. A., Measuring the oxidation-reduction potential of important oxidants in drinking water, *J. Am. Water Works Assoc.*, 106, E10–E20, January 2014.

Dick, J. G., *Analytical Chemistry*, McGraw-Hill Professional, New York, 1973.

Flynn, D., *The NALCO Water Handbook*, 3rd ed., McGraw-Hill Professional, New York, July 2009. (http://www.amazon.com/gp/reader/0071548831/ref=sr_1_2?p=S0BV&keywords=oxidation-reduction+potential+control&ie=UTF8&qid=1391197393#reader_0071548831)

Fresenius, W., *Water Analysis*, Springer-Verlag, Berlin, Germany, 1988.

Latimer, W. M., *Oxidation Potentials*, Prentice Hall, Englewood Cliffs, NJ, 1952.

Light, T. S., Standard solution for redox potential measurements, *Anal. Chem.*, 44(6), 1038–1039, May 1972.

McNeil, B. M. and Harvey, L. M., *Fermentation: A Practical Approach*, Oxford University Press, New York, 1994.

Milazzo, G. and Caroli, S., *Tables of Standard Electrode Potentials*, John Wiley & Sons, New York, 1978.

Pauling, L., *General Chemistry*, W. H. Freeman, San Francisco, CA, 1970.

Rice, E. W., et al., (Eds.), Standard Methods for the Examination of Water and Wastewater, 22nd Edition, pp. 2–84 to 2–87, American Public Health Association, Washington, DC, 2012.

Ross, M., Treating water at ORP speed, *Water Technol.*, 19, 58–61, May 1996.

Shinskey, F. G., *pH and pIon Control in Process and Waste Streams*, John Wiley & Sons, New York, 1973.

Williams, D. S., *Online Monitoring of Wastewater Effluent Chlorination Using Oxidation Reduction Potential ORP vs Residual Chlorine Measurement: Werf Report Treatment Processes Project 99-wwf-6*, 412 pages, Int'l Water Association, January 30, 2005. (http://www.amazon.com/gp/product/184339717X)

Wu, H. and Wang, S., Impacts of operating parameters on oxidation-reduction and pretreatment efficacy in the pretreatment of printing and dyeing wastewater by Fenton process, *J. Hazard. Mater.*, 243, 86–94, December 2012.

1.48 Oxygen Demands (BOD, COD, TOD)

I. G. YOUNG (1974, 1982)   **J. F. TATERA** (2003)   **B. G. LIPTÁK** (1995, 2017)

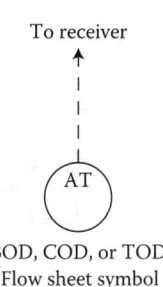

BOD, COD, or TOD
Flow sheet symbol

Types of measurements	A. Biological oxygen demand (BOD) B. Chemical oxygen demand (COD) C. Total oxygen demand (TOD) NOTE: The measurement of dissolved oxygen (DO) is covered in Chapter 1.50.
Sampling technique	Depending on the application, composite flow-averaged or time-averaged samples can automatically be collected or samples can also be obtained as grab samples for manual introduction into the analyzer. Instead of manual operation, continuous-flow automatic sampling is used, when the task is process monitoring and control.
Sample pressure	Typically collected at near atmospheric pressures
Sample temperature	Collected at process or ambient temperatures
Suspended solids	Varies with the application and the analyzer used. In many applications, both the liquid's and the solid's oxygen demand is of interest.
Materials of construction	Glass, quartz, Teflon, polyethylene, Tygon, polyvinyl chloride (PVC), stainless steel, ceramic
Ranges	A. 0.1–1500 mg/L is typical and higher with dilution B. 5–1500 mg/L is typical and higher with dilution or some methods C. 0–60,000 ppm
Inaccuracy	A. 3%–20% AR B. 2%–10% AR C. 2%–5% AR
Response time	A. 3 min. to 5 days B. 2–15 min. C. 3–10 min.
Costs	A. From $500 for a grab sample kit B. $500 for a test kit, from $8000 to $20,000 for a complete online automatic system C. $5000–$20,000
Partial list of suppliers	Applitech (A, B, C) http://www.applitechpharma.com/en/product.asp?id=155 Aqualytic (A) http://en.aqualytic.de/products/bod-measurement Camlab (B) http://www.camlab.co.uk/cod-chemical-oxygen-demand-c2865.aspx Challenge Environmental Systems Inc. (A, C) http://www.challenge-sys.com/ODM.html Endress and Hauser (B) http://www.endress.com/en/Tailor-made-field-instrumentation/liquid-analysis-product-overview/cod-analyzer-colorimetric-ca71cod GE Power and Water Co. (C), http://www.geinstruments.com/library/toc-analyzer-search-and-buyer-s-guide/toc-analyzer-sensor-search.html Global Water (A) http://www.globalw.com/products/oxitop.html Hach (A, B) http://www.hach.com/bodguide Hanna Instruments (B) http://www.hannainst.com/usa/subcat.cfm?id = 008 Liquid Analytical Resource (A, B, C) http://www.larllc.com/quicktocultra.html Mantec (A) http://www.mantech-inc.com/analytical/pc_bod.asp

(Continued)

Metex Corp. Ltd. (A) http://www.metexcorporation.com/Aqualytic/bod_measurement.htm
Orbeco-Hellige Co. (A) http://www.orbeco.com/download/Product%20Literature/Oxi700_BOD_System.pdf
Respirometry Plus LLC (A, B) http://www.respirometryplus.com/9.html
Scalar (A) http://www.skalar.com/news/the-sp1000-bod-analyzer-including-automated-sample-pipetting
ThermoScientific (B) http://www.thermoscientific.com/en/product/orion-3106-cod-analyzer.html
Toray Engineering Co. (C) http://www.toray-eng.com/measuring/water/oxygen-lineup/tod-810.html
Velp Scientific (A) http://www.velp.com/en/products/lines/2/family/31/
VWR (A, B) https://ni.vwr.com/app/Header?tmpl =/environment/bod_test.htm

INTRODUCTION

A significant portion of the total damage caused by the discharging of wastewater into lakes or rivers is because many of these discharges deplete the oxygen content of the receiving lake or river and the resulting impact that this depletion has on the lake's or river's ecosystem. This oxygen-depleting potential is usually expressed and quantified by biochemical oxygen demand (BOD—also called biological oxygen demand), chemical oxygen demand (COD), or total oxygen demand (TOD) measurements. These instruments measure the amount of oxygen that a liter of wastewater is expected to take from the receiving waters as its pollutants are degraded by oxygen-consuming (aerobic) bacteria.

BOD is the amount of oxygen needed by microorganisms to biologically degrade the organic compounds in a sample. Organics that are not biodegradable will report the BOD value as "zero." BOD analyzers utilize bacteria to oxidize the pollutants. In COD analyzers, the OD is usually measured through oxidation. TOD analyzers typically oxidize the sample in a catalyzed thermal combustion process and detect both the organic and inorganic impurities in a sample.

This section describes the various BOD, COD, and TOD analyzers. The main distinction between the various designs is in the speed at which the measurement is obtained and in the correlation of the resulting readings with such measurement results as the manometric BOD tests. Other measurements that are also of interest in connection with waste water quality measurements include dissolved oxygen analysis (Chapter 1.50) and waste water monitoring (Chapter 1.73).

OXYGEN DEMAND TESTING

The BOD of a sample of wastewater is the amount of elemental oxygen required to degrade the dissolved or suspended biodegradable organic material in the sample, using a population of bacteria to do the oxidation. This amount is in milligrams of oxygen per liter of waste water sample.

When the oxidation is carried out with a chemical reagent such as potassium dichromate, the oxygen equivalent of the amount of reagent that was required to degrade the waste in the sample is called the chemical oxygen demand (COD). If the waste can be oxidized both biologically and chemically, the BOD and COD readings will be the same. If this is the case, it depends on the nature of the waste. The subject of BOD/COD correlation will be discussed later.

Oxidation can also be caused by heating of the sample in a furnace in the presence of a catalyst and oxygen. In this case, the amount of oxygen that was required to degrade the waste in the sample is called total oxygen demand (TOD). If the heating in a furnace occurs in the presence of carbon dioxide, the result is called total carbon dioxide demand (TCO_2D).

The BOD test is perhaps the most important OD measurement for the analysis of effluent's oxygen demand from the receiving waters (streams, lakes, and rivers). Basically, the BOD test measures the amount of oxygen used by microorganisms that feed on organic pollutants in the water under aerobic conditions. In this test, a bacterial culture is added to the sample under well-defined conditions and oxygen utilization is measured.

Although test procedures are carefully defined,[*] it is difficult to obtain highly reproducible results, and the procedure is subject to the influence of many variables, particularly when the wastewater contains a variety of complex materials. Some of the factors that contribute to variations in BOD results are discussed below.

Seed

The seed is the bacterial culture that affects the oxidation of materials in the sample. If the biological seed is not acclimated to the particular wastewater, erroneous results are frequently obtained. Because different bacterial cultures are used in BOD measurements at different locations, it is not surprising that the results may be inconsistent. Also, recall the intent of the measurement. If the intent is to monitor plant feeds to help control the waste treatment operation, bacteria from the plant waste treatment process may be most useful, as it is probably acclimated to the plant's typical composition. If the intent is to predict the impact on receiving waters, a more random population may better predict this.

pH

The BOD results are also greatly affected by the pH of the sample, especially if it is lower than 6.5 or higher than 8.3.

---

[*] APHA, Standard Methods for the Examination of Water and Wastewater 22nd edn., 2015, http://www.amazon.com/Standard-Methods-Examination-Water-Wastewater/dp/0875530133.

Not only is oxidation of the material itself pH dependent, but so is that of the bacterial activity. In order to achieve uniform conditions, the sample should be buffered to a pH of about 7.

Temperature

Although the standard test condition calls for a temperature of 20°C (68°F), field tests often require operation at other temperatures and, consequently, the results tend to vary unless temperature corrections are applied (Figure 1.48a).

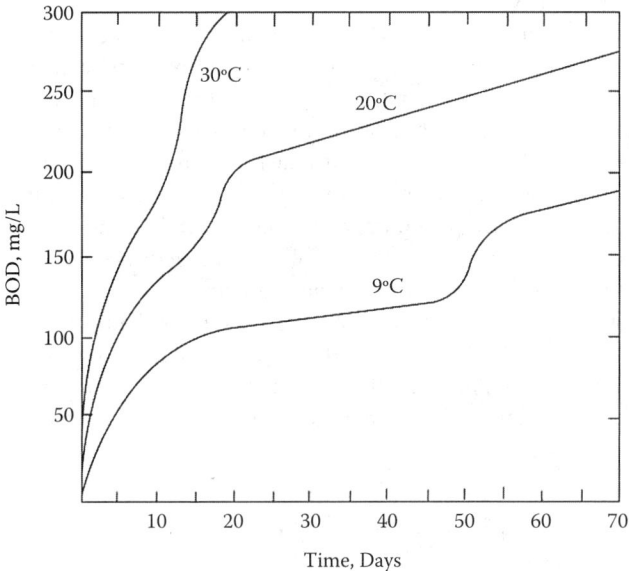

FIG. 1.48a
Progress of BOD at 9°C, 20°C, and 30°C (48°F, 68°F, and 86°F). The break in each curve corresponds to the onset of nitrification.

Toxicity

Toxic materials in the sample, although sample may be oxidizable or biodegradable, frequently have a biotoxic or biostatic effect on the biological seed. The presence of toxic materials of this type is indicated by an increase in the BOD value as a specific sample is diluted for the BOD test. Consistent values may be obtained either by removing the toxic materials from the sample or by developing a seed that is compatible with the toxic materials in the sample.

Incubation Time

The usual standard lab test incubation time is 5 days, although the time required for stabilization (complete biochemical degradation of materials in the water) may take as long as 20 or 30 days. The 5-day results may occur at a flat part of the OD vs. time curve, or they may occur at a steeply rising portion. Thus, depending on the type of seed and the type of oxidizable material, divergent results can be expected for this reason alone (Figure 1.48a).

On-line and more rapid BOD-based testing (like fast BOD and respirometry) typically reduce this time factor into the range of minute to hours. To accomplish this, they avoid long-term equilibrium requirements and usually optimize sample size to bacteria population and other relevant factors to give them a faster analysis with reasonable repeatability.

Nitrification

In the usual course of the BOD test, the oxygen consumption rises steeply at the beginning of the test owing to attack on carbohydrate materials. Another sharp increase in oxygen utilization occurs sometime during the 10th to 15th day in those samples containing nitrogenous materials (Figure 1.48a). Stated another way, the rate constant for attack on nitrogenous materials is much lower than that for attack on carbonaceous materials, and the demand due to nitrification is not appreciable until most of the carbonaceous material has been destroyed.

In view of difficulties and variability of the classic BOD determination, a rapid procedure that minimizes or eliminates these problems has been sought for many years. Although other procedures are used, the BOD continues to remain the universal standard method, supported by the force of tradition and the weight of legal authority in many jurisdictions.

Thus, those who are concerned with estimating the oxygen depletion load of effluent waters must be thoroughly acquainted with the BOD test and prepared to support other methods by suitable correlation to BOD results. Therefore, although OD may be measured in a number of ways, the 5-day BOD result is what is meant by OD in most cases.

BOD TESTS

Five-Day BOD Tests

If the BOD of a sample of water at 20°C (68°F) is measured as a function of time, a curve such as the one in Figure 1.48a is obtained. For the first 10–15 days, the curve is approximately exponential, but at about the 15th day a sharp increase is noted, which then falls off to a steady BOD rate.

Because of the length of time and because the curve does not flatten, a standard test period of 5 days has been adopted universally for the BOD procedure. This is a laboratory procedure requiring some skill and training to obtain concordant results. The procedure is described in greater detail in the literature; only a brief description is given here.

A measured portion of the sample to be analyzed is mixed with seeded dilution water so that after 5 days of incubation, the dissolved oxygen (DO) in the mixture is still sufficient for biological oxidation of materials in the sample. Of course, this cannot be known beforehand; consequently, a number of dilutions are run simultaneously for an unknown sample, or experience is used as a guide for well-defined samples.

The seeded dilution water contains phosphate buffer (including ammonium chloride), magnesium sulfate, calcium chloride, and ferric chloride, as well as a portion of seeding material. The former group of inorganic materials is frequently referred to as nutrients. The latter group is a suspension of bacteria in water, usually supernatant liquor from a domestic sewage plant.

Seeds may also be prepared from soil, developed from cultures in the laboratory, or obtained from receiving water 2–5 min downstream of the discharge. The DO content of the mixture is determined at the start of the test and again after 5 days of incubation at 20°C in a special BOD bottle.

DO Determination The DO may be determined by the Winkler titration method* or instrumentally with a DO membrane electrode. The difference in DO after 5 days is used to calculate the BOD of the original sample. Corrections must be applied for immediate OD (that due to inorganic reducing materials) and for the oxygen required by the bacteria themselves for sustaining life (endogenous metabolism). A blank sample is run to assist with this.

There is no standard against which the accuracy of the BOD test can be measured. The precision of the method is also difficult to ascertain because of the many variables. However, the single-operator precision of the method has been tested using a standard glucose–glutamic acid solution.

Using eight different types of seed materials, the single-operator precision was 11 mg/L at a level of 223 mg/L, or about 5%. It must be recognized that these results were obtained with highly skilled personnel under well-controlled laboratory conditions.

BOD Instruments A semiautomatic instrument has been designed to measure the BOD of as many as 11 samples. The samples have to be manually placed on the instrument turn table and the controls manually set. Means are provided for automatic re-aeration of those samples in which the DO has fallen to low values.

Measurements of the DO are made on a preset time schedule by the polarographic DO sensor. The capability of automatic re-aeration when the DO is low eliminates the need for dilution, leading to improved precision in the BOD results. The instrument consists of a measuring unit (DO probe, aerator, water-sealing mechanism, unplugging mechanism, sample bottle, and turntable) and a control unit, by which all of the operations are programmed. The measuring unit is housed in a chamber maintained at 20°C. Means for storing DO data on each sample are supplied, and the BOD is calculated from the DO values as already described. Figure 1.48b illustrates this instrument. For further information on dissolved oxygen measurement, refer to Chapter 1.50.

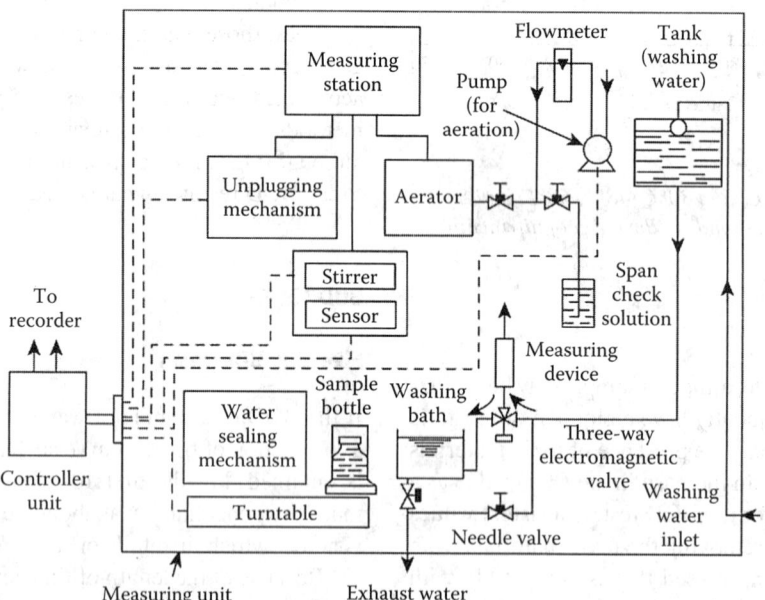

FIG. 1.48b
Semiautomatic BOD instrument.

---

* The Winkler method is a titration method to determine dissolved oxygen in the water sample. The steps of the Winkler method involve filling completely (no air space) a sample bottle with water. Next, the dissolved oxygen in the sample is "fixed" by adding a series of reagents that form an acid compound that is then titrated with a neutralizing compound that results in a color change. The point of color change is called the "endpoint," which corresponds to the dissolved oxygen concentration in the sample.

Extended BOD Test

As reflected in Figure 1.48a, continuation of the BOD test beyond five days shows a continuing rise in OD, with a sharp increase in BOD rate at the 10th day owing to nitrification. The latter process involves biological attack on nitrogenous organic material accompanied by an increase in BOD rate. The OD continues at a uniform rate for an extended time.

Knowledge of oxygen utilization of a polluted water supply is important because (1) it is a measure of the pollution load, relative to oxygen utilization by other life in the water; (2) it is a means for predicting progress of aerobic decomposition and the amount of self-purification taking place; and (3) it is a measure of the OD load removal efficiency by different treatment processes.

As a means for treatment plant control and setting the legal standards for wastewater effluents, the extended test is not used. However, it must be remembered that the five-day BOD does not represent the TOD load on the receiving water.

Manometric BOD Test

In the standard dilution method that was previously described, all the oxygen required must already be inside the BOD bottle, since it is sealed in a gas-tight manner at the initiation of the incubation period, and care is taken to prevent access of air into the sample. In the manometric procedure, the seeded sample is confined in a closed system that includes an appreciable amount of air.

As the oxygen in the water is depleted, it is replenished by the gas phase. A potassium hydroxide (KOH) absorber within the system removes any gaseous carbon dioxide generated by bacterial action. The oxygen removed from the air phase results in a drop in pressure that is removed with a manometer. This fall is then related to the BOD of the sample.

Thus, in the manometric method, the DO of the water remains at a moderately high level, close to saturation (9 mg/L at 20°C), whereas in the standard BOD, the DO falls continuously during the five-day incubation period to values near 1 mg/L. Despite this marked difference in conditions of DO during incubation, results in close agreement are obtained on many samples by the two different procedures.

An apparatus is commercially available in which the BODs of five samples can be determined simultaneously by the manometric method. A measured sample of the sewage or wastewater is placed in one of the bottles of the apparatus, and the bottle is connected to a closed-end mercury manometer (Figure 1.48c). Above the water sample a quantity of air is trapped. As bacteria in the sample utilize oxygen, it is replenished by oxygen from the air. The removal of oxygen from the air results in a lowering of the air pressure.

The fall in pressure is read on the mercury manometer directly in BOD units, assuming that the original air contained 21% oxygen. The preceding description assumes a sample that is already seeded. Of course, the method can

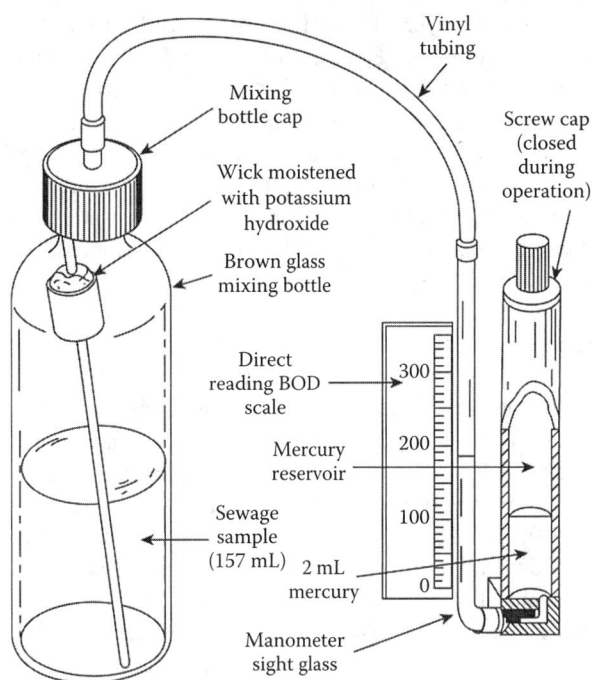

FIG. 1.48c
Manometric BOD apparatus.

be modified for those samples that require the addition of a bacterial culture. The procedure is carried out manually in the laboratory. In addition to the manipulations already described, it requires reading of the manometer by the laboratory technician.

Automatic Recording The manometric method also lends itself to automatic recording of the course of oxygen utilization, since it is possible to monitor the pressure continuously. This has been accomplished in an automatic respirometer now commercially available (Figure 1.48d).

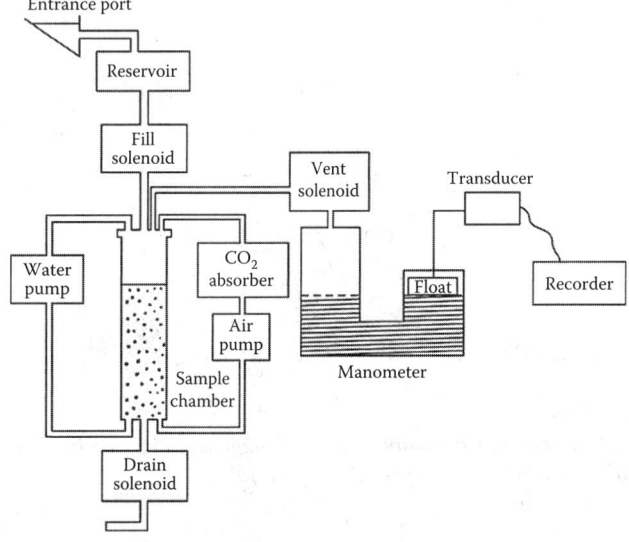

FIG. 1.48d
BOD determination by automatic respirometer.

The sample, from 1 to 4 L, is introduced into a closed system containing air. Countercurrent circulation of both air and water insures equilibrium between dissolved and gaseous oxygen. A carbon dioxide scrubber is provided in the gas-circulation line. A recording manometer detects the utilization of oxygen as the test is run for several hours. Published data indicate a correlation between the respirometer BOD and the standard BOD. Both laboratory and automated on-line versions of this instrument are available.

It must be recognized that BOD is inherently a time-consuming process and ill-suited to the requirements of process monitoring or control. The shortest period mentioned for the automatic respirometer is 2 hr, often too long for an effective control instrument. However, it is an excellent device for longer-term laboratory and process studies, since it can be made to simulate the activated sludge process.

Useful fast BOD and respirometric measurements have been reduced to times under an hour and have been useful in the control of several waste treatment operations, though they are more often utilized in a monitoring role.

BOD Determination in Minutes

When determining the BOD concentration in groundwaters, it might be acceptable to wait 5 days for the results, but in the control and operation of sewage treatment plants, it is not. The hold-up capacity of industrial and municipal wastewater treatment facilities and the desire for closed-loop control necessitates the use of a much faster sensor. One such analyzer is described in Figure 1.48e.

In this design, a bioreactor is filled with a large number of plastic rings, the interior of which are protected against mechanical abrasions and thereby providing a growth surface for organisms. A circulation pump serves to quickly distribute the sewage in the bioreactor and to keep the plastic rings in continuous motion.

The sewage concentration (nutrient level) in the reactor is maintained at a constant low value, which results in an OD of about 3 mg/L. This OD is measured and maintained constant by detecting the decrease in oxygen concentration at the points where the diluted sewage enters and leaves the reactor. The DO concentration at electrode 2 is kept at a constant value below that at electrode 1. If this difference drops, the sewage concentration is increased; if this difference rises, the concentration is reduced.

The sewage concentration (the nutrient concentration) in the bioreactor is adjusted by a computer. It varies the mixing ratio of sewage and dilution water. The total flow from the sewage and dilution water pumps is always 1 L/min, and it is the ratio of the two streams that is modulated. Therefore, this pumping ratio is an indication of the BOD concentration of the sewage sample.

Correlation between this fast BOD measurement and the five-day BOD obtained through conventional methods has been acceptable. Fouling of the piping was found to be minimal, and weekly recalibration of the oxygen electrodes was found to be satisfactory.

Advances in this field include the use of two cascade reactors that ensure complete biodegradation of the sample in the fastest time possible, for consistent and accurate correlation to BOD5. In addition to BOD, this BioMonitor of LAR simultaneously monitors the endogenous respiration of the activated sludge, for precise control of the treatment plant biomass. This enables plant operators to safeguard the biomass against sudden or long-term toxic shock, and to provide intelligent load equalization.

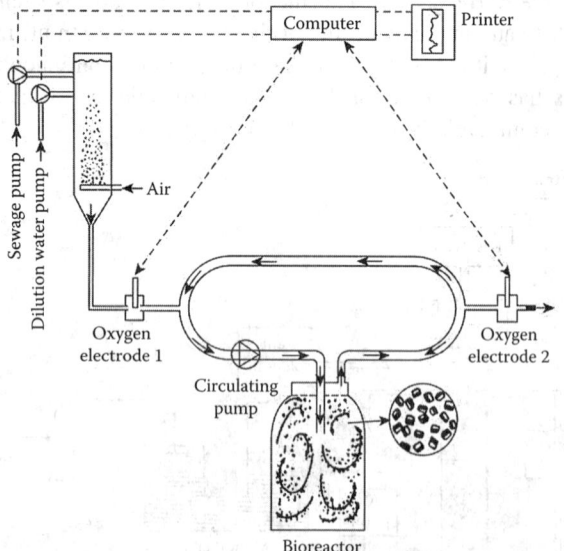

FIG. 1.48e
BOD assessment is obtained in a few minutes in a continuously circulated bioreactor where the oxygen take-up of the organisms controls dilution.

Electrolysis Instrument Electrolysis of water can supply oxygen to a closed system as incubation proceeds (Figure 1.48f). At constant current, the time during which electrolysis generates the oxygen to keep the system pressure constant is a direct measure of the OD (by Faraday's law*).

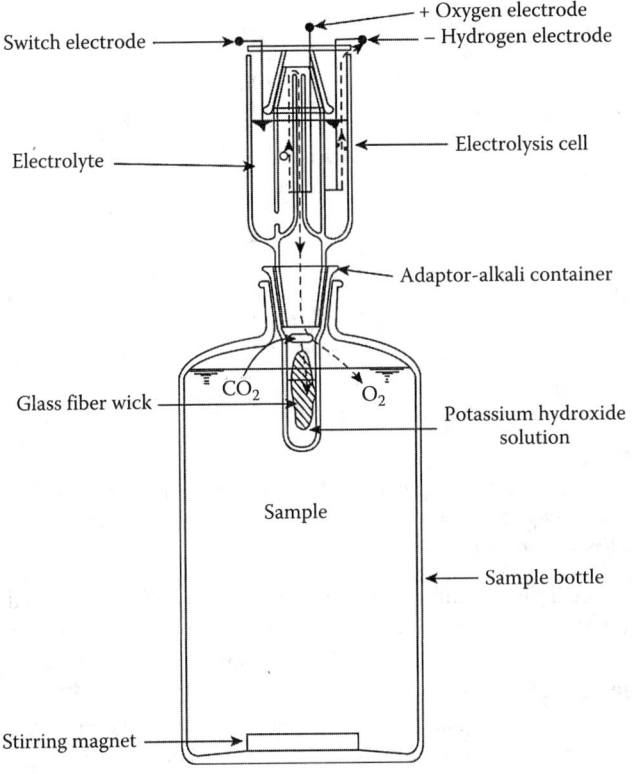

FIG. 1.48f
Electrolysis system for measuring BOD.

An instrument based on this principle permits the running of six samples simultaneously, and its readout gives BOD directly, in milligrams per liter for each sample. After starting the test run, operator attention is not required.

Robotic Analyzer The BOD robotic analyzer includes automated sample pipetting. By providing this further automation step in the BOD application, accurate, reproducible, and precise results can be obtained (Figure 1.48g), which eliminates the need for tedious manual handling procedures for this labor-intensive analysis method, which is carried out in accordance with EPA 405.1/ISO 5815- 1/ EN-1899-1/2.†

---

* Faraday's law of electrolysis states that the amount of substance produced at an electrode by electrolysis is proportional to the quantity of DC electricity transferred at that electrode.
† EPA Method 405.1: Biochemical Oxygen Demand, 1971. http://www.caslab.com/EPA-Method-405_1/.

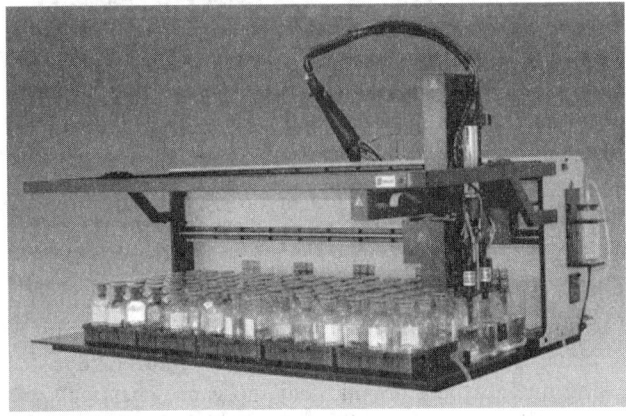

FIG. 1.48g
Robotic analyzer with automated sample pipetting. (Courtesy of Scalar.)

These units are compatible with all current models of oxygen sensors and probes (galvanic or optical DO probes) and can provide barcode identification. The robotic software package provided includes options for analysis scheduling, report printing, export options to LIMS/Excel, and a variety of quality control features.

COD MEASUREMENT

Dichromate COD Test

This laboratory method requires skill and training similar to that required for the BOD test. A sample is heated to its boiling point with known amounts of sulfuric acid and potassium dichromate. Loss of water is minimized by the use of a reflux condenser. After 2 h, the solution is cooled, and the amount of dichromate that reacted with oxidizable material in the water sample is determined by titrating the excess potassium dichromate with ferrous sulfate, using ferrous 1,10-phenanthraline (ferroin) as the indicator. The dichromate consumed is calculated as to oxygen equivalent for the sample and reported as milligrams of oxygen per liter of sample.

Interpretations of COD values are difficult, since this method of oxidation is markedly different from the BOD method. Although ultimate BOD values can be expected to agree with COD values, a number of factors may prevent this concordance. Among these, we may mention the following:

1. Many organic materials are oxidizable by dichromate but not biochemically oxidizable, and vice versa. For example, pyridine, benzene, and ammonia are not attacked by the dichromate procedure.
2. A number of inorganic substances such as sulfide, sulfites, thiosulfates, nitrites, and ferrous iron are oxidized by dichromate, creating an inorganic COD

that is misleading when estimating the organic content of wastewater. Although the factor of seed acclimation will give erroneously low results on the BOD tests, COD results are not dependent on acclimation.

3. Chlorides interfere with the COD analysis, and their effect must be minimized in order to obtain consistent results. The standard procedure provides for only a limited amount of chlorides in the sample. Despite these limitations, the dichromate COD has been useful in control of wastewater effluents from plants concerned with caustic and chlorine, dyeing and textiles, organic and inorganic chemicals, paper, paints, plating, plastics, steel, aluminum, and ammonia. This is usually accomplished by diluting the sample to achieve a lower chloride concentration and interference. This can be a problem for low COD concentration samples, as the dilution may dilute the COD concentration below the detection level or to the levels at which accuracy and repeatability are poor.

The term COD usually refers to the laboratory dichromate oxidation procedure, although it has also been applied to other procedures that differ greatly from the dichromate method but which do involve chemical reaction. These methods have been embodied in instruments both for manual operation in the laboratory and for automatic operation online. They have the distinct advantage of reducing analysis time from days (5-day BOD) and hours (dichromate, respirometer) to minutes.

Automatic On-Line Measurement

Figure 1.48h illustrates an on-line COD analyzer having ranges from 0 to 5000 ppm and with measurement cycle times that are adjustable from 10 min to 5 h. The sample flow can be continuous at rates up to 0.25 GPM (1.0 lpm) and can contain solid particles up to 100 μm in size. The automatic COD analyzer periodically takes a 5 cc sample from the flowing process stream and injects it into the reflux chamber after mixing it together with dilution water (if any) and with two reagents: dichromate solution and sulfuric acid.

The reagents also contain an oxidation catalyst (silver sulfate) and a chemical that complexes chlorides in the solution (mercuric sulfate). The mixture is boiled at 302°F (150°C) by the heater, and the vapors are recondensed by the cooling water in the reflux condenser. The solution is refluxed for a

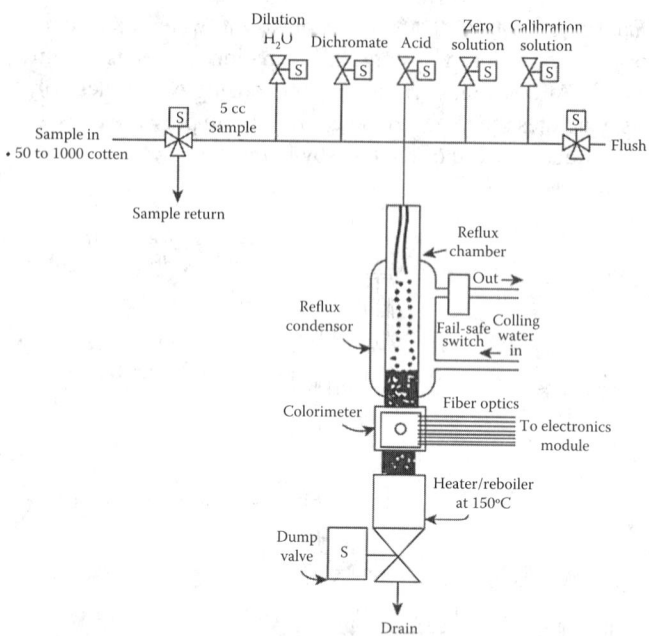

FIG. 1.48h

Automatic COD analyzer using dichromate reagent and fiber-optic colorimeter detector and providing features of adjustable reflux time and autocalibration.

preset time, during which the dichromate ions are reduced to trivalent chromic ions, as the oxygen-demanding organics are oxidized in the sample. The chromic ions give the solution a green color.

The COD concentration is measured by detecting the amount of dichromate converted to chromic ions by measuring the intensity of the green color through a fiber-optic detector. The microprocessor-controlled package is available with automatic zeroing, calibration, and flushing features. Analyzers utilizing this or other COD wet chemical reaction schemes are generally categorized as FIAs or CFAs, and some of the manufacturers of these types of analyzers do not advertise them as COD analyzers. To them, COD is just one application of their FIA or CFA capability.

In one instrument, a 20-μL water sample is manually injected into a carbon dioxide carrier stream and swept through a platinum catalyst combustion furnace. In that furnace, the pollutants are oxidized to carbon monoxide and water, and the water is removed from the stream by a drying tube, after which the reaction products receive a second platinum catalytic treatment.

A nondispersive infrared (NDIR) detector is used to measure the concentration of carbon monoxide, and a calibration chart is utilized to convert the readings into COD.

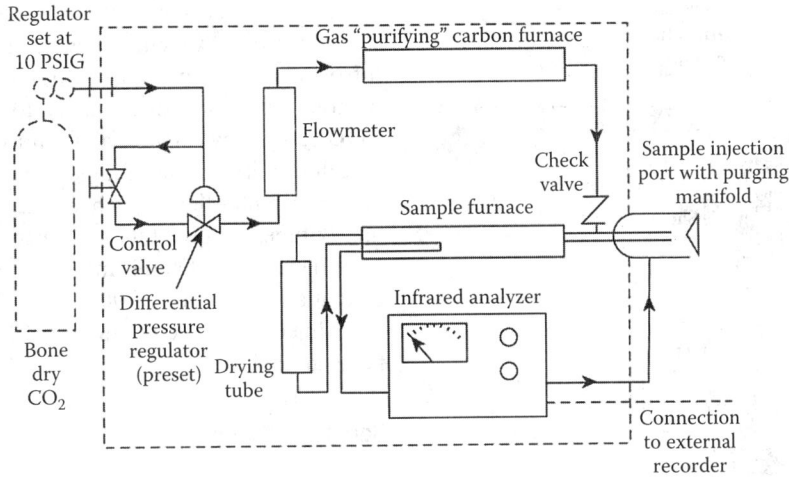

FIG. 1.48i
COD detection employing combustion in a carbon dioxide carrier and an NDIR sensor.

The analysis can be completed in 2 min. This instrument is available commercially for manual operation (Figure 1.48i). Data obtained on domestic sewage indicate excellent correlation between this method (frequently called CO_2D) and the standard COD.

Some other approaches to COD monitoring include methods that utilize ozone and OH–radicals as the oxidizing agents. One ozone-based scheme enriches dilution water with ozone and measures the dissolved ozone concentration of the dilution water and the residual dissolved ozone concentration downstream of the reaction chamber. It then calculates a COD value based on the difference in the two measurements and the ratios of dilution water and sample being delivered to the reaction chamber.

It has a response time of 3 min. Another OH–radical approach produces OH–radicals on an electrode by an electrical current. These OH–radicals are extremely strong oxidizing agents. An electrochemical measurement signal, based on the OH–radicals being converted and measured as OH–ions, results in a 30-s analysis with a range of 1–100,000 ppm that is said to correlate to the dichromate standard method.

The design in Figure 1.48j does not require the disposal of any analysis residues, cleaning solutions, or harmful reagents. It consists of a reactor vessel with magnetic stirrer with a lead dioxide, reference, and counter electrodes, three peristaltic pumps for defined supply of sample, the electrolyte concentrate, the regeneration solution and/or the calibration standard solution, magnetic pinch valves for the selection of sample, and reagent flows in the analytical part. It is provided with self-explaining software with integrated help files, splash-water-protected keypad, graphic-LCD-screen, presentation of the actual measurement values in the table form, displaying of the calibration values, daily protocols, and data storage for one year. It is also provided with software for automatic start up of the system in case of power interruption or in case of discontinuous usage.

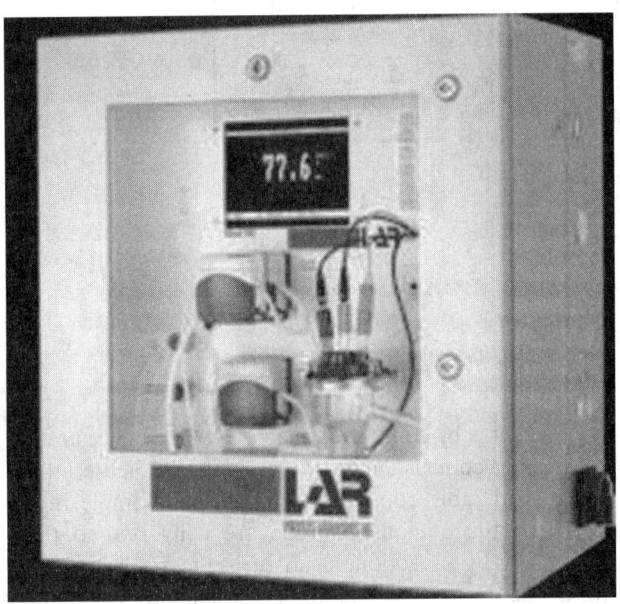

FIG. 1.48j
Automatic and continuous COD analyzer. (Courtesy of Liquid Analytical Resource.)

One thing to remember is that all of these techniques are based on a chemical oxidation reaction and that all reactions can experience interferences. The nice thing about having several options is that you can pick the one that is most suited to your sample matrix.

TOD MEASUREMENT

The TOD method is based on the quantitative measurement of the amount of oxygen used to burn the impurities in a liquid sample. Thus, it is a direct measure of the OD of the sample.

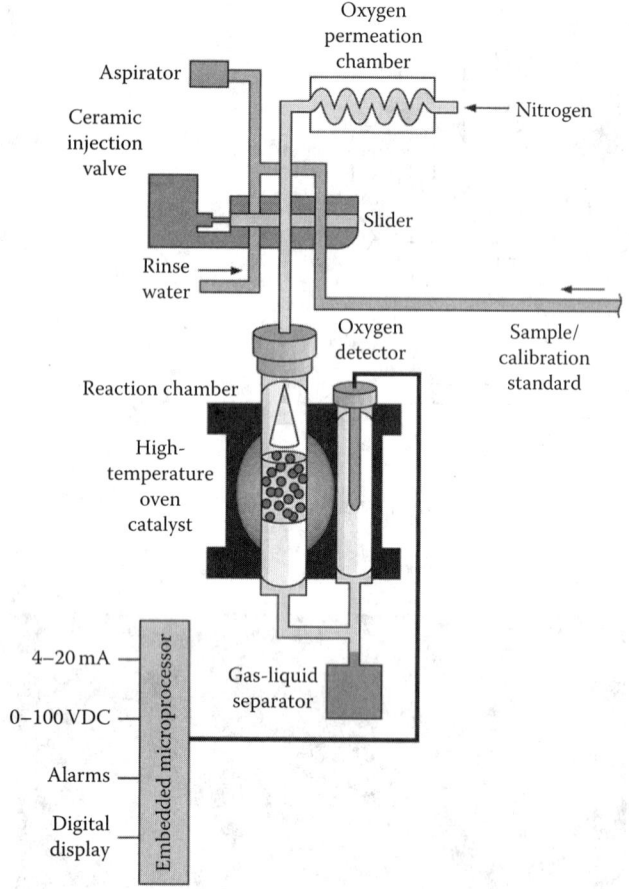

FIG. 1.48k
Basic components of an online TOD analyzer.

Measurement is by continuous analysis* of the concentration of oxygen in a combustion process gas effluent (Figure 1.48k).

The oxidizable components in a liquid sample introduced into the combustion tube are converted to their stable oxides by a reaction that disturbs the oxygen equilibrium in the carrier gas stream. The momentary depletion in the oxygen concentration in the carrier gas is detected by an oxygen detector and recorded as a negative oxygen peak. The TOD for the sample is obtained by comparing the recorded peak height or area to the peak sizes of the standard TOD calibration solutions, e.g., potassium acid phthalate (KHP).

Prepurified nitrogen from a cylinder passes through a fixed length of tube permeable to oxygen (usually silicone) into the combustion chamber packed with a solid catalyst, the gas scrubber, and then the oxygen detector. The baseline oxygen concentration is obtained as the nitrogen passes through the temperature-controlled permeation tube and can be varied to accommodate different TOD ranges by changing the nitrogen flow rate and the length of permeation tubing.

The combustion chamber is a length of Victor tubing, quartz tube, ceramic tube, or other metal tube that contains a platinum catalyst and is mounted in an electric furnace and held at a temperature of 900°C (1672°F). The aqueous sample is injected into this chamber, and the combustible components are oxidized.

Sampling Valves

Two basic types of injection valves are in use today, the sliding plate and the rotary sampling valve. Manual injection can be accomplished through a silicone rubber septum, if desired.

Figure 1.48l shows the main features of the sliding plate valve. Upon a signal from a cycle timer, the air actuator temporarily moves the valve to its "sample fill" position. At the same time, an air-operated actuator moves a 20-µL sample through the valve into the combustion tube. A stream of oxygen-enriched nitrogen carrier gas moves the slug of sample into the combustion tube.

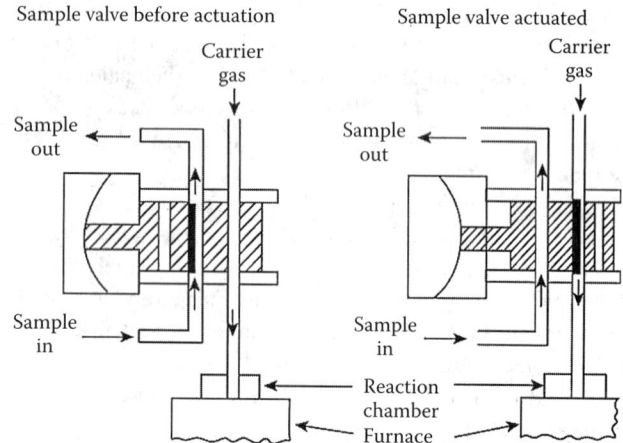

FIG. 1.48l
Automatic sliding plate liquid sample injection value.

Traditionally, sliding plate valves were manufactured from a variety of metals (stainless steel, Hastelloy, Monel, and others), but ceramic sliding plate sampling valves have been developed to provide reduced maintenance for this harsh application.

---

* Genthe, W. and Pliner, M., Total oxygen demand (TOD), 2015. http://www.wateronline.com/doc/total-oxygen-demand-tod-an-alternative-parameter-for-real-time-monitoring-of-wastewater-organics-0001.

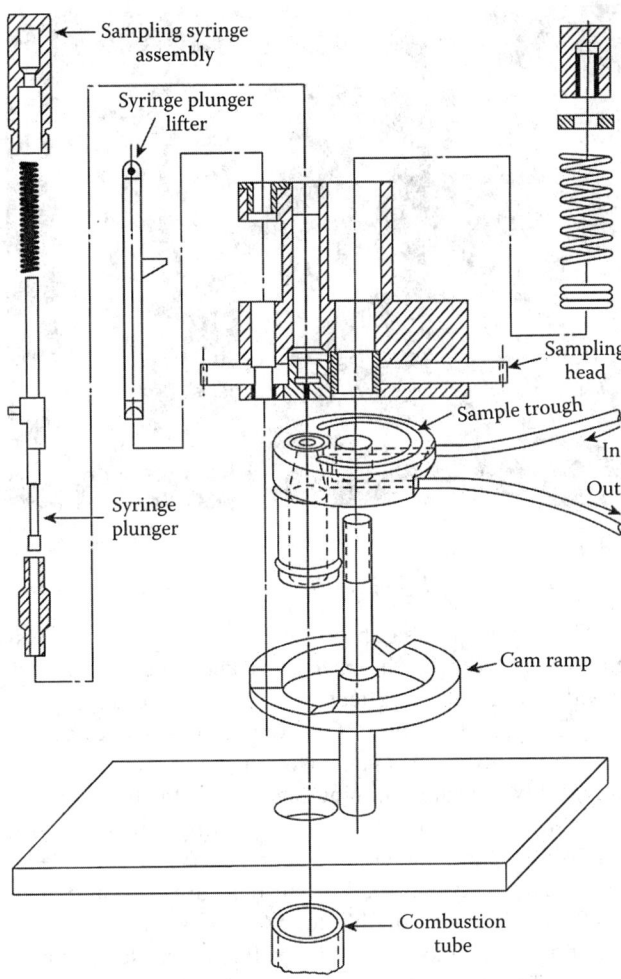

FIG. 1.48m
Rotary sample valve.

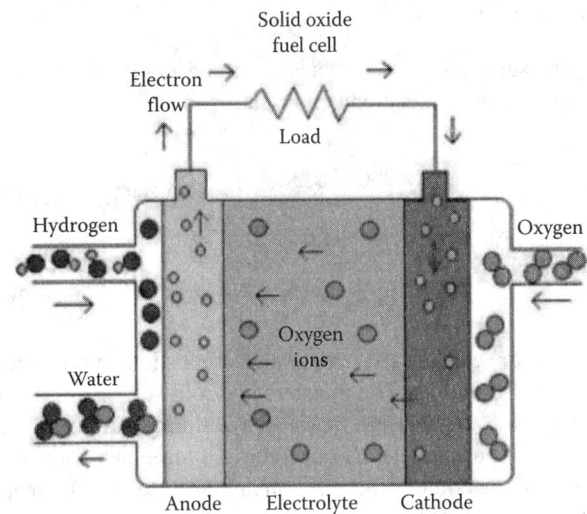

FIG. 1.48n
After the combustion of the impurities in the sample takes up some oxygen, the quantity of the oxygen remaining in the carrier gas is detected by measuring the electricity generated by the TOD's fuel cell, when it chemically oxidizes as much hydrogen into water, as there is oxygen available in the carrier gas. (Courtesy of Smithsonian Institution.)

The rotary sample valve is mainly used for on-stream TOD analyzers. Figure 1.48m shows the cross section of the valve, in which a motor continuously rotates a sampling head, which contains a built-in sampling syringe. For part of the time, the tip of the syringe is over a trough that contains the flowing sample. Two or more cam ramps along the rotational path cause the syringe plunger to rise and fall, thus rinsing the sample chamber. Just before the syringe reaches the combustion tube, it picks up a 20-μL sample. As it rotates over the combustion tube, it discharges the sample.

TOD Oxygen Detector

One type of TOD oxygen sensor is a platinum–lead fuel cell that generates a current in proportion to the oxygen content of the carrier gas passing through it. Before entering the cell, the gas is scrubbed in a KOH solution, both to remove acid gases and other harmful combustion products and to humidify the gas (Figure 1.48n).

The oxygen cell and the scrubber are located in a temperature-controlled compartment. The fuel cell output is monitored and zeroed to provide a constant baseline. The output peaks are linearly proportionate to the reduced concentration of oxygen in the carrier gas as a result of the sample's TOD.

Another type of TOD oxygen detector is a yttrium-doped zirconium oxide ceramic tube that has been coated on both sides with a porous layer of platinum. It is maintained at an elevated temperature and also provides an output that represents the reduction in oxygen concentration in the carrier gas as a result of the sample's TOD.

A more complete discussion of this detector technology can be found in Chapter 1.49 under "High-Temperature Zirconium Fuel Cells."

Calibration

Analysis is by comparison of peak heights or areas to a standard calibration curve. To prepare this curve, known TOD concentrations of a primary standard (KHP) are prepared in distilled and deionized water. Standard solutions are stable for several weeks at room temperature. Water solutions of other pure organic compounds can also be used as standards.

Several analyses can be made at each calibration concentration, and the resulting data are recorded as parts per million (ppm) TOD vs. peak height or area. Older instruments utilize a recorder, and within the normal instrument range, the plot is a straight line. Consequently, a single-concentration calibration of the recorder chart in milligrams per liter TOD can be made.

Newer instruments are microprocessor controlled, and the microprocessor stores the calibration results and outputs a result directly as ppm TOD.

Interferences

Nitrate salts and sulfuric acid will normally decompose under sample combustion conditions as follows:

$$2NaNO_3 \xrightarrow[Pt]{900°C} Na_2O + 1\tfrac{1}{2}O_2 + 2NO \quad \text{1.48(1)}$$

$$H_2SO_4 \xrightarrow[Pt]{900°C} H_2O + SO_2 + \tfrac{1}{2}O_2 \quad \text{1.48(2)}$$

This oxygen release results in a proportionate reduction in the TOD reading. The interference from sulfuric acid can be overcome by neutralizing it with sodium hydroxide. DO in the aqueous sample also becomes an unknown source of oxygen to the combustion reaction and lowers the TOD readings. But unless special precautions are taken, the standard solutions are also at or near saturation, thus automatically minimizing this potential interference. If absolute TOD levels are required, the oxygen equivalent of the interference should be added to the TOD reading.

Oxygen-saturated or air-saturated samples with very low TOD values pose a special problem, which can be circumvented by spiking the sample with a known concentration of a standard solution. The actual TOD will then be the analysis value minus the spiking concentration.

Heavy-metal ions give long-term interferences, usually by eventual reduction of the catalyst efficiency. Replacement of the combustion tube and a thorough cleaning of the catalyst is the only remedy. Materials like silanes or siloxanes also combust in the reactor and exhibit a positive TOD, but can leave a residue of silica in the reactor that can coat and desensitize the catalyst.

The catalyst and the reactor components can be coated when the sample is vaporized and combusted in a fashion that leaves some solids behind in the reactor. The total matrix of the sample should be evaluated before committing to a technique, not just the components being measured.

On-Line Monitoring

Continuous OD monitoring gives the operators real-time information on both the loading of the wastewater treatment plant and on toxic events. On the aeration basin, it can measure oxygen uptake data that are needed to control and optimize aeration and activated sludge recycling. These designs (Figure 1.48o) operate by measuring the amount of oxygen input needed to maintain a constant oxygen partial pressure that is in contact with a mixture of wastewater and microorganisms in the reactor vessel. Wastewater can be added from the influent or from the mixed liquor in the aeration basin. The effluent is removed from the reactor by a pump that has a rate setting equal to the sum of the wastewater and the mixed liquor input rates. The mass transfer rate to the reactor is kept high, thereby keeping the DO concentration in the reactor near saturation. This high mass transfer rate is accomplished by using an air diffuser and a headspace gas recycling pump.

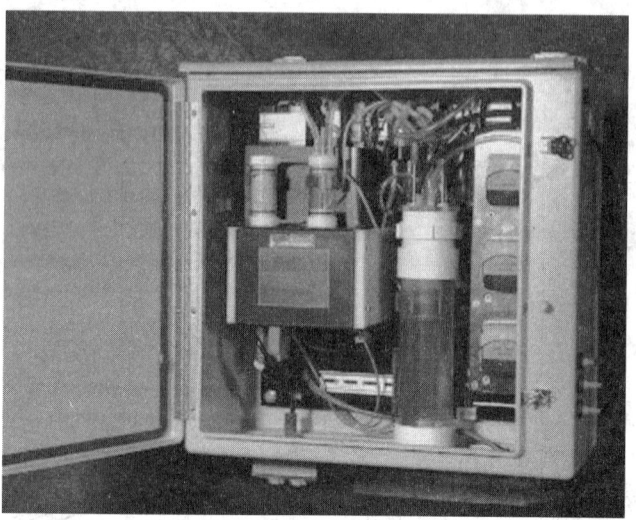

FIG. 1.48o
Online oxygen demand monitor. (Courtesy of Challenge Technology.)

The submersible sampling pump provided with the system is controlled by the package sequencing PLC and is provided with anti-clog features for a trouble-free operation. The system control PLC is connected to a touch screen monitor to allow user inputs involving changes in wastewater and mixed liquor pumping rates. The PLC touch screen displays the oxygen uptake rate in milligrams/liter/hour, and displays a graphic view of oxygen sensor data and level sensor data. Oxygen uptake rate data are transmitted in a 4–20 mA output, suitable for SCADA systems, and are also stored in memory for downloading for subsequent processing. The data file includes oxygen uptake rate in milligrams/liter/hour and the time and date of data recording.

Applications

Table 1.48a lists typical applications for the TOD analyzer. Instruments equipped with the rotary or sliding plate sampling

TABLE 1.48a
Application Suggestions for TOD Analyzers

Application	Purpose
Waste treatment plant influent	Determines loading of plant
Primary, secondary, and plant effluents	Determines efficiency of treatment and TOD load on the receiving waterway
Enforcement programs	Determines pollution levels
Industrial process control	Determines process efficiency and leaks by continuous monitoring
Research and development	Evaluates waste treatment processes
Stream surveillance	Monitors fresh, estuarine, and marine water quality
Municipal or industrial water treatment plants	Monitors quality of influent water prior to treatment
Boiler feed and high-purity water monitoring	Measures TOD with great sensitivity

valve lend themselves nicely to multistream analysis, since no carryover of the previous sample is present. If the streams have varying levels of TODs, each stream must be diluted to the basic range of the analyzer by appropriate automatic dilution techniques. On the laboratory analyzers, it is best to sample deionized water when the instrument is in a standby condition.

CORRELATION AMONG BOD, COD, AND TOD

Many regulatory agencies recognize as the basis for oxygen-depleting pollution control only the BOD or COD (preferably BOD) measurements of pollution load. The reason for this is that they are concerned with the oxygen-consuming pollution load on receiving waters, which is related to lowering of the DO due to bacterial activity.

Thus, if the other methods described are to be used to satisfy legal requirements of the oxygen-consuming pollution load in effluents or to measure BOD removal, it is necessary that a correlation be established between the other methods and BOD.

In order to summarize the features of various methods, it is assumed that the BOD is the standard reference method. The salient features of this method are: (1) a measurement of property (the BOD) of the sample, that is, the amount of oxygen required for bacterial oxidation of bacterial food in the water; (2) dependence of the OD on the nature of the food as well as on its quantity; and (3) dependence of the OD on the nature and the amount of bacteria.

Variations in OD due to variations in the amount (pounds per gallon) of food in the wastewater are expected; we are less able to deal with variations in OD when the amount of food is constant but changes occur in its BOD requirements. The same observations apply to the bacterial seed. Thus, variation in OD due to variation in the number of bacteria or their activity, as well as those due to the changes in the nature of their foodstuffs, leads to a systematic or bias error in BOD measurements, which cannot be predicted or corrected for.

The standard reference method is therefore inherently variable—one subject to analytical error. Researchers in an interlaboratory comparative study[14] employing a synthetic waste found standard deviations around the mean of ±20% for BOD and ±10% for COD.

BOD and COD Correlation

On the subject of correlation, some of the conclusions include:

1. A reliable statistical correlation between BOD and COD of a wastewater and its corresponding TOC or TOD can frequently be achieved, particularly when the organic strength is high and the diversity in dissolved organic constituents is low.
2. The relationship is best described by a least squares regression with the degree of fit expressed by the correlation coefficient—this applies to the characterization of individual chemical-processing and oil-refining wastewaters, not to all types of samples across the board.
3. The observed correspondence of COD–TOD was better than that of COD–BOD for the wastewaters. It was difficult to correlate BOD with TOD, particularly when the wastewater contained low concentrations of complex organic materials.
4. The BOD–COD or BOD–TOC ratios of an untreated wastewater are indicative of the biological treatment possible with the particular wastewater. As these ratios increased, higher treatment efficiencies by biological methods in terms of organic removal were noted.

Correlation between BOD and Other Methods Several papers have indicated high correlation between BOD and other methods. This can be achieved when the nature of the pollutant is constant and only its amount changes. For complex and varying mixtures, it is difficult or impossible to obtain good correlations.

The ability to truly correlate one measurement to another requires more than measurements that are measuring often-related properties. Correlation can also be highly dependent on the matrix of the sample being measured. Different measurement techniques often respond differently to different backgrounds and to different interferences.

For example, techniques like FID and ultraviolet (UV) absorption (Chapter 1.69) are being sold to make correlated COD measurements. Obviously, they are not measuring the ability to oxidize the sample. What they are measuring is a generalized organic content of a sample and trying to predict what a COD value of that sample would be, assuming the determined organic content represents it. While this is possible for some very well-characterized samples, this approach often fails to adequately handle surprises (process upsets or changes that result in significant sample matrix changes).

Many types of analyzers can measure various components that contribute to the BOD, COD, or TOD loading of a sample and, under the right circumstances, they can be reliably correlated to an appropriate BOD, COD, or TOD result. Many of these analyzers offer advantages like continuous or faster analysis and lower purchase, operating, and maintenance costs.

The key to instrument selection in this case is to insure that you have selected one that is appropriate for your process, sample, and measurement requirements. If in doubt, usually the safest path is to go with an instrument that is compatible with your sample and most directly measures the property of interest.

SPECIFICATION FORMS

When specifying BOD, COD, or TOD analyzers, the compositions and properties of the application can be specified by using the ISA Forms 20A1001 and 20A1002. If the measurement of dissolved oxygen (DO) is required, the specification form that can be used is ISA Form 20A2192. These forms are reproduced on the next pages with the permission of the International Society of America.

Analytical Measurement

1	RESPONSIBLE ORGANIZATION		ANALYSIS DEVICE	6	SPECIFICATION IDENTIFICATIONS	
2				7	Document no	
3	ISA		Operating Parameters	8	Latest revision	Date
4				9	Issue status	
5				10		

11	ADMINISTRATIVE IDENTIFICATIONS			40	SERVICE IDENTIFICATIONS Continued		
12	Project number		Sub project no	41	Return conn matl type		
13	Project			42	Inline hazardous area cl	Div/Zon	Group
14	Enterprise			43	Inline area min ign temp	Temp ident number	
15	Site			44	Remote hazardous area cl	Div/Zon	Group
16	Area	Cell	Unit	45	Remote area min ign temp	Temp ident number	
17				46			
18	SERVICE IDENTIFICATIONS			47			
19	Tag no/Functional ident			48	COMPONENT DESIGN CRITERIA		
20	Related equipment			49	Component type		
21	Service			50	Component style		
22				51	Output signal type		
23	P&ID/Reference dwg			52	Characteristic curve		
24	Process line/nozzle no			53	Compensation style		
25	Process conn pipe spec			54	Type of protection		
26	Process conn nominal size	Rating		55	Criticality code		
27	Process conn termn type	Style		56	Max EMI susceptibility	Ref	
28	Process conn schedule no	Wall thickness		57	Max temperature effect	Ref	
29	Process connection length			58	Max sample time lag		
30	Process line matl type			59	Max response time		
31	Fast loop line number			60	Min required accuracy	Ref	
32	Fast loop pipe spec			61	Avail nom power supply		Number wires
33	Fast loop conn nom size	Rating		62	Calibration method		
34	Fast loop conn termn type	Style		63	Testing/Listing agency		
35	Fast loop schedule no	Wall thickness		64	Test requirements		
36	Fast loop estimated lg			65	Supply loss failure mode		
37	Fast loop material type			66	Signal loss failure mode		
38	Return conn nominal size	Rating		67			
39	Return conn termn type	Style		68			

69	PROCESS VARIABLES	MATERIAL FLOW CONDITIONS			101	PROCESS DESIGN CONDITIONS	
70	Flow Case Identification			Units	102	Minimum	Maximum	Units
71	Process pressure				103			
72	Process temperature				104			
73	Process phase type				105			
74	Process liquid actl flow				106			
75	Process vapor actl flow				107			
76	Process vapor std flow				108			
77	Process liquid density				109			
78	Process vapor density				110			
79	Process liquid viscosity				111			
80	Sample return pressure				112			
81	Sample vent/drain press				113			
82	Sample temperature				114			
83	Sample phase type				115			
84	Fast loop liq actl flow				116			
85	Fast loop vapor actl flow				117			
86	Fast loop vapor std flow				118			
87	Fast loop vapor density				119			
88	Conductivity/Resistivity				120			
89	pH/ORP				121			
90	RH/Dewpoint				122			
91	Turbidity/Opacity				123			
92	Dissolved oxygen				124			
93	Corrosivity				125			
94	Particle size				126			
95					127			
96	CALCULATED VARIABLES				128			
97	Sample lag time				129			
98	Process fluid velocity				130			
99	Wake/natural freq ratio				131			
100					132			

133	MATERIAL PROPERTIES			137	MATERIAL PROPERTIES Continued		
134	Name			138	NFPA health hazard	Flammability	Reactivity
135	Density at ref temp		At	139			
136				140			

Rev	Date	Revision Description	By	Appv1	Appv2	Appv3	REMARKS

Form: 20A1001 Rev 0 © 2004 ISA

1.48 Oxygen Demands (BOD, COD, TOD)

	RESPONSIBLE ORGANIZATION	ANALYSIS DEVICE COMPOSITION OR PROPERTY Operating Parameters (Continued)		SPECIFICATION IDENTIFICATIONS	
1	ISA		6		
2			7	Document no	
3			8	Latest revision	Date
4			9	Issue status	
5			10		

	PROCESS COMPOSITION OR PROPERTY			MEASUREMENT DESIGN CONDITIONS				
	Component/Property Name	Normal	Units	Minimum	Units	Maximum	Units	Repeatability
11								
12								
13								
...								
75								

Rev	Date	Revision Description	By	Appv1	Appv2	Appv3	REMARKS

Form: 20A1002 Rev 0 © 2004 ISA

#	RESPONSIBLE ORGANIZATION	DISSOLVED OXYGEN/OZONE/CL2 TRANSMITTER/ANALYZER/MONITOR Device Specification	#	SPECIFICATION IDENTIFICATIONS	
1			6		
2	ISA		7	Document no	
3			8	Latest revision	Date
4			9	Issue status	
5			10		

#	TRANSMITTER OR ANALYZER		#	PERFORMANCE CHARACTERISTICS	
11			56		
12	Housing type		57	pH compensation LRL	URL
13	Input sensor type		58	Meas 1 accuracy rating	
14	Input sensor style		59	Meas 2 accuracy rating	
15	Output signal type		60	Temp accuracy rating	
16	Enclosure type no/class		61	Measurement LRL	URL
17	Control mode		62	Temp compensation LRL	URL
18	Local operator interface		63	Barometric pressure LRL	URL
19	Characteristic curve		64	Minimum span	Max
20	Digital communication std		65	Min ambient working temp	Max
21	Signal power source		66	Contacts ac rating	At max
22	Temperature sensor type		67	Contacts dc rating	At max
23	Quantity of input sensors		68	Max distance to sensor lg	
24	Preamplifier location		69		
25	Contacts arrangement	Quantity	70		
26	Integral indicator style		71		
27	Signal termination type		72		
28	Cert/Approval type		73		
29	Mounting location/type		74		
30	Failure/Diagnostic action		75		
31	Sensor diagnostics		76	ACCESSORIES	
32	Data history log		77	Remote indicator style	
33	Measurement compensation		78	Indicator enclosure	
34	Calibration function		79	Communicator style	
35	Enclosure material		80		
36			81		
37			82		
38			83	SPECIAL REQUIREMENTS	
39			84	Custom tag	
40			85	Reference specification	
41			86	Compliance standard	
42			87	Software configuration	
43			88	Software program	
44			89		
45			90		
46			91		
47			92	PHYSICAL DATA	
48			93	Estimated weight	
49			94	Overall height	
50			95	Removal clearance	
51			96	Signal conn nominal size	Style
52			97	Mfr reference dwg	
53			98		
54			99		
55			100		

#	CALIBRATIONS AND TEST		INPUT OR TEST			OUTPUT OR SCALE	
110	TAG NO/FUNCTIONAL IDENT	MEAS/SIGNAL/SCALE	LRV	URV	ACTION	LRV	URV
111							
112		Meas-Analog output 1					
113		Meas-Analog output 2					
114		Meas-Digital output					
115		Meas-Scale					
116		Temp-Scale					
117		Temp-Digital output					
118		Meas setpoint 1-Output					
119		Meas setpoint 2-Output					
120		Meas setpoint 3-Output					
121		Meas setpoint 4-Output					
122		Failure signal-Output					
123							

#	COMPONENT IDENTIFICATIONS			
124	COMPONENT TYPE	MANUFACTURER	MODEL NUMBER	
125				
126				
127				
128				
129				

Rev	Date	Revision Description	By	Appv1	Appv2	Appv3	REMARKS

Form: 20A2192 Rev 0 © 2004 ISA

Definitions

Electrolysis: Is electro-chemical decomposition process that occurs when direct current passes through a conductive liquid (an electrolyte, in other words a liquid that contains ions). When it passes between the two electrodes of an electrolytic cell, anions (positive ions) will deposit on the negative electrode (cathode) and cations (negative ions) will deposit on the positive electrode (anode). Therefore, in case of the electrolysis of water, hydrogen is formed at the cathode and oxygen at the anode.

Faraday's law of electrolysis: States that the amount of substance produced at an electrode by electrolysis is proportional to the quantity of DC electricity transferred at that electrode.

Faraday's law of induction: Is the principle on which the operation of electric motors, generators or solenoids are based. It states that a change in the magnetic environment of a wire coil will induce a voltage (emf - electromotive force) in it. It does not matter what caused the change in the magnetic environment (change in field strength, movement of a magnet, movement of the coil relative to the magnet, including rotation) a voltage will be induced.

Fuel cells: Are devices that convert the chemical energy of fuel into electricity through the chemical reaction of positively charged hydrogen ions combining with oxygen into water. Fuel cells are different from batteries in that they require a continuous source of fuel (hydrogen) and oxygen to sustain the chemical reaction. Fuel cells can produce electricity continuously for as long as these inputs are supplied.

Winkler method: Was discovered by Lajos Winkler (1863–1939) a Hungarian chemist. It is a titration method to determine dissolved oxygen in the water sample. The steps of the Winkler method involve the filling completely (no air space) a sample bottle with water. Next, the dissolved oxygen in the sample is "fixed" by adding a series of reagents that form an acid compound that is then titrated with a neutralizing compound that results in a color change. The point of color change is called the "endpoint," which corresponds to the dissolved oxygen concentration in the sample.

Abbreviations

BOD	Biochemical oxygen demand
CFA	Continuous flow analyzer
COD	Chemical oxygen demand
DO	Dissolved oxygen
EMF	Electromagnetic force
FIA	Flow injection analyzer
FID	Flame ionization detector
KHP	Potassium acid phosphate
KOH	Potassium hydroxide
LCD	Liquid crystal display
NDIR	Nondispersive infrared
OD	Oxygen demand
PEM	Proton exchange membrane
PVC	Polyvinyl chloride
TC	Total carbon
TIC	Total inorganic carbon
TOD	Total oxygen demand

Organizations

APHA	American Public Health Association
AWWA	American Water Works Association

Bibliography

Adams, V., *Water and Wastewater Examination Manual*, 1990. https://www.crcpress.com/Water-and-Wastewater-Examination-Manual/Adams/9780873711999.

ASTM D6238-98, Standard Test Method for Total Oxygen Demand in Water, 2011. http://www.astm.org/Standards/D6238.htm.

Bruckner, M., *The Winkler Method—Measuring Dissolved Oxygen*, 2015. http://serc.carleton.edu/microbelife/research_methods/environ_sampling/oxygen.html.

EPA Method 405.1: Biochemical Oxygen Demand, 1971. http://www.caslab.com/EPA-Method-405_1/

Fair, G., M., Geyer, J. C., Okun, D. A., *Water and Wastewater Engineering: Water Purification and Wastewater Treatment and Disposal*, 1966. http://www.abebooks.com/book-search/title/water-and-wastewater-engineering/author/fair-geyer-okun/.

Fresenius, W. et al., *Water Analysis*, 1988. http://www.springer.com/us/book/9783642726125.

Genthe, W. and Pliner, M., *Total Oxygen Demand (TOD)*, 2015. http://www.wateronline.com/doc/total-oxygen-demand-tod-an-alternative-parameter-for-real-time-monitoring-of-wastewater-organics-0001.

Ivandini, T. A., Saepudin, E., Wardah, H., Dewangga, N., and Einaga, Y., Development of a biochemical oxygen demand sensor using gold-modified boron doped diamond electrodes, 2012. https://scholar.google.co.id/citations?view_op=view_citation&hl=en&user=kC440cgAAAAJ&citation_for_view=kC440cgAAAAJ:IjCSPb-OGe4C.

Jouanneau, S., et al., Methods for assessing biochemical oxygen demand (BOD), 2013. http://people.ufpr.br/~heloise.dhs/TH058/2014_Jouanneau_metodos_DBO.pdf.keywords=oxygen+demand.

LAR, Fast, Precise Measurement of Biological Oxygen Demand, 2015. https://www.larllc.com/biomonitor.html.

Lin, L. T. (October 13, 2012), *Kinetic Study on Chemical Oxygen Demand Removal*, 2012. http://www.amazon.com/Kinetic-Chemical-Oxygen-Demand-removal/dp/3659249262/ref=sr_1_7?s=books&ie=UTF8&qid=1386012839&sr=1–7&

Rustum R., Adeloye A. J., and Scholz M. Applying Kohonen self-organizing map as a software sensor to predict the biochemical oxygen demand, 2008. http://www.ncbi.nlm.nih.gov/pubmed/18254396.

Smithsonian Institution, *Fuel Cell Basics*, 2008. http://americanhistory.si.edu/fuelcells/basics.htm.

WTW, *Determination of Biochemical Oxygen Demand (BOD)*, 2002. http://old.omnilab.de/hpb/export/2/BSB_E.PDF.

Yao, N. et al., A Novel Thermal Sensor for the Sensitive Measurement of Chemical Oxygen Demand, 2006. http://www.ncbi.nlm.nih.gov/pmc/articles/PMC4570432/.

1.49 Oxygen in Gases

R. K. KAMINSKI (1969, 1982)   **M. RAZAQ** (1995)
J. F. TATERA (2003)   **D. S. NYCE** (2017)

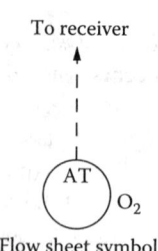

Flow sheet symbol

Design types	A. Deflection-type paramagnetic, dynamic dumbbell
	B. Thermal-type paramagnetic
	C. Dual-gas (differential pressure) paramagnetic
	D. Catalytic combustion
	E. Low-temperature electrochemical (galvanic, coulometric, polarographic)
	F. Zirconium oxide, voltage-mode high-temperature electrochemical
	G. Zirconium oxide, current-mode high-temperature electrochemical
	H. Alarm, pocket size, battery operated
	I. NIR spectroscopy
Sample requirement	Unless monitoring ambient air, all usually require a sampling system except F, G, and I, some of which can be inserted into the process as probes or utilize fiber-optics process interfaces.
Sample pressure and flow	Generally, near atmospheric pressures and low flow rates in the range of a few SCCM to a few SCFH. Most respond to partial pressure due to oxygen, and therefore will indicate higher with higher absolute pressure (except for those with a capillary-type diffusion barrier).
Materials of construction	Most are suitable for corrosive service, and types F, G, and I are also suitable for high-temperature service. Corrosive service may limit the life of electrochemical sensors.
Speeds of response	Most will reach 90% of full scale in <1 min; response times of types F and G are in ms, A: about 2 sec, B: about 3 sec, E: about 10 sec.
Operating temperatures	Most designs are suited for 93°C (200°F) service. Types F and G can operate at up to 621°C (1150°F) with stainless steel, up to 1204°C (2200°F) with mullite, and up to 1593°C (2900°F) with silicon carbide materials. Type E requires sample gas to be near ambient temperature as it reaches the sensing element.
Ranges	A. 0%–2%, 0%–25%, 0%–100%
	B. 0%–50%
	C. 0%–2%, 0%–25%, 0%–100%
	D. 0–2000 ppm
	E. From 0 to 1 ppm up to 0% to 100%
	F, G. Most often used for 0%–5%, 0%–10%, and 0%–25%, but can be obtained with ranges as wide as 1 ppm to 20% or 0.5% to 100%
	H. Usually set to alarm at about 18% oxygen concentration
	I. From 100 ppm, typically 0% to 2% or 0% to 100% range
Inaccuracy	Generally between 1% and 2% of full scale
	A. Error 0.02% oxygen between 98% and 100% oxygen
	D. 5% of span

(Continued)

Cost	(Does not include the cost of a sampling system that may be required in some cases. Sampling systems can have a wide range of cost, from $200 for a simple pump and rotameter, to over $10,000 for temperature control, particulate filters, and sometimes condensate removal).
	A. $3500
	B, C. $6,000 to $10,000
	D. $5000
	E. Portable monitoring units from $200; industrial analyzers with spans in the ppm range from $2000
	F. $1500 to $8000
	G. $600 to $3000
	H. Portable, battery-operated, pocket-size, low-oxygen alarms $130 to $500
	I. $20,000 to $25,000
Partial list of suppliers	ABB Inc. (B, C, E, F) (www.abb.com or www.abb.com/analytical)
	Ametek Inc. (D, F) (http://www.ametekpi.com/products/Thermox-WDG-V-Combustion-Analyzer.aspx)
	Bacharach Inc. (E) (www.bacharach-inc.com/analyzers-monitors.htm)
	BW Technologies (E, H) (www.gasmonitors.com)
	City Technology Ltd. (E) (www.citytech.co.uk)
	Cosa Instrument Corp. (E, F) (http://cosaxentaur.com/page/525/zirconia-oxygen-gas-analyzers)
	Enmet Corp. (H) (http://catalog.enmet.com/viewitems/sensor-transmitters/o-sub-2-sub-co-h-sub-2-sub-s-and-other-toxic-gases)
	Fuji Electric (F, G) (www.fujielectric.com/products/instruments/products/analyzer/top.html)
	GE (Panametrics) (B) (http://www.ge-mcs.com/en/gas/oxygen-analysis.html)
	Horiba Ltd. (E) (www.horiba.com/process-environmental/products/process/details/510-series-gas-analyzers-248/)
	Illinois Instruments Inc. (F) (http://www.systechillinois.com/en/intrinsically-safe-oxygen-analyzer-ec91_8.html)
	Industrial Scientific Corp. (E) (www.indsci.com/products/single-gas-detectors/GasBadgePro/)
	Land Combustion Inc. (F) (www.landinst.com/products/wdg-1200-1210-oxygen-analyser-systems)
	Michell Instruments (E) (www.michell.com/us/category/oxygen-analyzers.htm)
	MSA Instruments Div. (E, H) (us.msasafety.com/Fixed-Gas-%26-Flame-Detection/c/103;jsessionid=C330FD90100362E3796429A76AF78F79.worker1?N=10132+4294967095+4294966628429496 7067&Ne=10020&Nu=Product.Endeca_Rollup_Key&pageSize=100&view=grid)
	Neutronics Inc. (E) (www.ntroninc.com/products/index.html)
	NTRON (E, F) (http://www.ntron.com/Oxygen-analyzer-detector-monitor-systems.asp)
	Revolution Sensor Company (E, F, G) (www.rev.bz)
	RKI Instruments (E) (http://rkiinstruments.com/pages/01series.htm)
	Rosemount (Emerson) (A, E, F, G) (http://www2.emersonprocess.com/siteadmincenter/PM%20Rosemount%20Analytical%20Documents/PGA_PDS_X-STREAM_X2GP.pdf)
	Servomex (Delta-F) (A, B, E) (http://www.hummingbirdsensing.com/en/products/oxygen/paracube-micro)
	Sick Maihak Inc. (F) (http://www.sick.com/us1/en-us/home/products/product_portfolio/analyzers_systems/Pages/zirkor302.aspx)
	Siemens Energy & Automation (A) (www.sea.siemens.com)
	Systech Illinois (A) (http://www.systechillinois.com/en/paramagnetic-oxygen-analyzer-pm700_20.html)
	Teledyne Analytical Instruments (A, E, F) (www.teledyne-ai.com)
	Yokogawa Electric Corp. (F, G) (http://www.yokogawa.com/an/oxy-gas/an-zr22g_o2-001.htm)

INTRODUCTION

Oxygen is a diatomic molecule, meaning that one molecule of oxygen comprises two oxygen atoms (O_2). Oxygen has an atomic weight of 16, so its molecular weight is 32 (in atomic mass units, or AMU). Oxygen is vital to a large variety of industrial and biological oxidation and combustion processes. Many industries use pure oxygen, or conversely, use inert gases that are allowed to contain only a few parts per billion of oxygen as a contaminant, and both of these applications usually require analyzers to determine their oxygen concentrations. Dry atmospheric air contains approximately 20.95% of oxygen by volume, 78.09% nitrogen, 0.93% argon, plus trace amounts of other gases. The amount of water vapor in the atmospheric air varies and is usually represented in terms of relative humidity, which also depends on the temperature as well as the actual percentage of the air that is the water vapor. The fact that air can be relied upon to be about 20.9% oxygen is often utilized by using fresh air as a calibration source for oxygen monitors that operate in the percent oxygen range. Air can be used to calibrate some oxygen analyzers that operate in the PPM range, but then several hours of in-range gas must be sampled in order to rid the sampling and sensor system of the high oxygen levels before the analyzer can be returned to service at the low ppm level.

Specialized oxygen sensors are used for each type of application. Oxygen analyzers can depend on the paramagnetic or electrochemical properties of oxygen or can utilize catalytic combustion techniques. Some spectroscopic techniques are also in use, and are popular for monitoring O_2 while also monitoring other constituents in the process.

PARAMAGNETIC OXYGEN SENSORS

Oxygen is paramagnetic, meaning that it is strongly attracted to a concentrated magnetic field. This uncommon property, called *paramagnetism*, is shared by a few other gases such as nitric oxide, which has a susceptibility of about 43% of that of oxygen (Table 1.49a). Although other paramagnetic gases are not normally encountered in the processes where oxygen analysis is required, the table shows that some gases may have such a sufficient effect that the possibility of their presence should be evaluated. If there is an interfering background gas present, the measurement can be compensated accordingly, if needed, when the approximate amount of the interfering gas is known.

TABLE 1.49a
Magnetic Susceptibility of Various Interfering Process Gases Compared with Oxygen

Acetylene (C_2H_2)	−0.24
Carbon dioxide (CO_2)	−0.57
Carbon monoxide (CO)	+0.05
Chlorine	−0.77
Ethylene (C_2H_4)	−0.26
Helium	+0.29
Cyclo-hexane	−1.56
Hydrogen (H_2)	+0.24
Methane (CH_4)	−0.2
Nitric oxide (NO)	+42.6
Nitrogen (N_2)	−0.35
Nitrogen dioxide (NO_2)	+28.0
Oxygen (O_2)	+100.0

Note: +, Paramagnetic; −, Diamagnetic.

The atoms/molecules of some gases form a magnetic field in opposition to an externally applied magnetic field. This property is called diamagnetism, and such gases are repelled from the externally applied magnetic field. These diamagnetic gases include acetylene, methane, ethylene, carbon dioxide, and chlorine. In the table, paramagnetic gases are listed with a positive sign, whereas diamagnetic ones have a negative sign. In a paramagnetic oxygen analyzer, it is required that the paramagnetic property of the background gases be constant in order to permit the accurate measurement of oxygen concentration. The paramagnetic property of gaseous oxygen has been utilized in three basic types of industrial oxygen analyzer designs:

1. The *deflection-type* (dumbbell and coil) design suspends a dumbbell-shaped element, filled with nitrogen, within a magnetic field and surrounded by a motor coil that is used to keep the dumbbell in a constant position.
2. The *thermal* design (magnetic wind) depends on the flow of oxygen in the sample being directed by the magnetic field of a permanent magnet, and also on the decrease of the paramagnetic effect as the temperature of the paramagnetic gas (oxygen) increases in the vicinity of heating thermistors in the sensor.
3. In the *reference gas* design (dual-gas), two gases with different oxygen concentrations are combined in a magnetic field, and a differential pressure is generated.

Deflection Analyzer

In the deflection-type analyzer (Figure 1.49a), a dumbbell-shaped glass element is suspended within a concentrated magnetic field that is supplied by a permanent magnet assembly. The dumbbell is filled with nitrogen and is free to rotate along an axis perpendicular to the arm of the dumbbell. A motor coil (i.e., a wire loop) is positioned around the dumbbell and produces a magnetic field when a current is flowing in it. A small mirror is also mounted to the dumbbell. A beam of light is focused onto the mirror so that it is reflected to fall equally onto each of the two adjacent photo sensors when the dumbbell is in a balanced position (i.e., balancing between the force acting on the dumbbell due to the sample gas and permanent magnet,

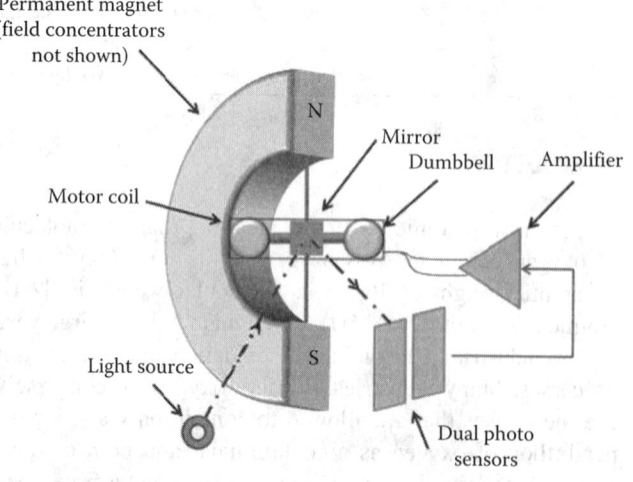

FIG. 1.49a
Deflection-type paramagnetic oxygen analyzer.

and the force on the dumbbell from the current flowing in the wire loop). When a test gas includes oxygen, the oxygen is attracted to the concentrated magnetic field, and is concentrated into that space, thus producing a pressure that is tending to rotate the dumbbell to push its ends out of that space. As the dumbbell rotates, light falls unequally upon the two photo sensors. This provides an error signal that is amplified and used to drive the motor coil, providing a "restoring force." The current applied to the motor coil is of a polarity and amplitude to exactly counteract the rotational force that was exerted on the dumbbell by the oxygen pressure. The restoring force returns the dumbbell to its original balanced position (this is called a force–balance system). The amount of current needed to keep the dumbbell in its original position is proportional to the oxygen concentration, and this current signal is conditioned and used to develop the output signal of the analyzer.

The sensing element is mounted on shock absorbers. Particulates and/or condensation in the sample gas are likely to cause major difficulties; therefore, sample filtering and conditioning should be used when necessary.

The most significant limitations of the dumbbell-type paramagnetic oxygen analyzer are its delicate nature and its sensitivity to shock and vibration. Variations in the magnetic susceptibility of the background gases in the sample or variations in the sample temperature (if not compensated) can also contribute to errors. The applications of this unit include the monitoring of combustion efficiency and the monitoring of purity of breathing air or of protective atmospheres. An advantage of this and other paramagnetic oxygen sensing technologies over electrochemical cells is that it does not wear out, or need replacement of the sensing element over time.

Thermal Analyzer

A flow-through type of *ring* element is illustrated in Figure 1.49b. In other designs, the sample gas diffuses into dead-ended cavities. In the case of the flow-through element illustrated, the paramagnetic oxygen content of the sample—after it enters the ring—is attracted by the magnetic field in the horizontal tube, where heating thermistors heat the sample gas. Oxygen loses its paramagnetic quality when it becomes hot. As the oxygen in the heated sample loses its paramagnetism, more oxygen is attracted to the magnetic field, increasing the gas pressure there and displacing the heated oxygen, causing it to flow toward a space of lower pressure, thus forming what is called a "magnetic wind." The magnetic wind flows out across the sensing thermistors. The heating and sensing

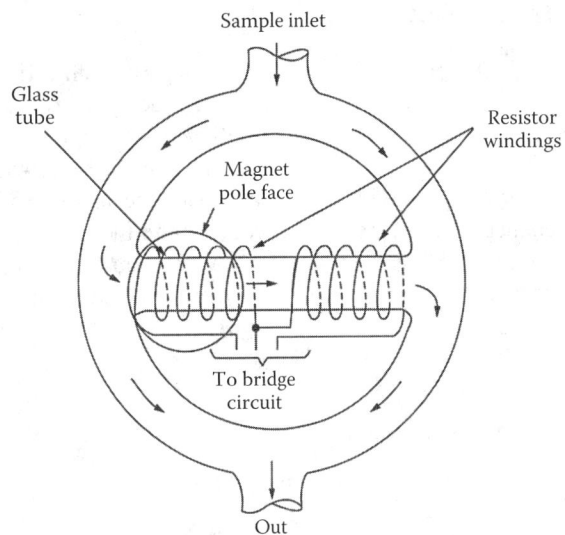

FIG. 1.49b
Measuring element in a thermal paramagnetic O_2 analyzer.

thermistors are connected in a Wheatstone bridge circuit to detect the resistance variations resulting from changes in the magnetic wind.

The flow rate of the magnetic wind and its effect on the temperatures of the thermistors are a function of oxygen concentration. Therefore, the Wheatstone bridge output signal is indicative of the oxygen concentration in the sample gas.

Because the heating and cooling of the thermistors is not only a function of the flow rate but also of the gas composition (affecting the thermal conductivity) and pressure, errors can be introduced by variations in these properties. The presence of diamagnetic materials (e.g., some other types of gases) can also introduce errors by affecting the magnetic wind. Errors can occur as a result of a change in sample pressure, because the magnetic susceptibility of oxygen varies as the square of the static pressure.

Some manufacturers offer pressure compensation in the form of special cells with compensating resistors that are also made part of the Wheatstone bridge circuit. Other suppliers require that precise pressure regulation be provided, within a few inches H_2O. Variations in barometric pressure can also cause measurement errors of up to ±2% on narrow ranges, because the effect of atmospheric pressure change is about 0.02% O_2 per inch H_2O (250 Pa).

In comparison with the dumbbell-type dynamic sensor, the advantage of the magnetic wind analyzer is its rugged design. Its disadvantages include the need for compensating for both temperature and for the thermal conductivity variations in the background gases of the sample. High temperatures can also cause degraded stability and reliability.

Dual-Gas Analyzer

Figure 1.49c shows a dual-gas cell. Two gases with different oxygen contents are brought together, producing a differential pressure. The reference gas can be 100% oxygen, nitrogen, or air. The reference gas passes through two ducts, one of which meets the sample gas in the magnetic field. Because both ducts are connected, the pressure, in proportion to the oxygen content of the gas sample, produces a flow that can be measured. All wetted parts can be made of stainless steel or tantalum.

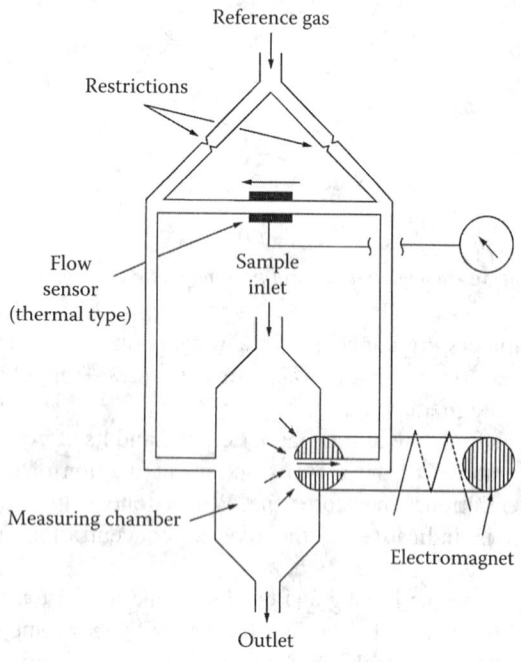

FIG. 1.49c
Measuring element in a dual-gas paramagnetic O_2 analyzer.

The dual-gas or differential-pressure-type oxygen analyzer is rugged, but it is sensitive to vibration, for which some manufacturers are able to compensate. As with all other paramagnetic analyzers, this one is also limited in its sensitivity to the percentage range and is not suited for trace oxygen measurements. This type of analyzer has largely lost favor due to its requirement for a reference gas, whereas other types do not require a continuous flow of reference gas.

CATALYTIC COMBUSTION OXYGEN SENSORS

This approach to oxygen analysis is very similar to techniques used for the measurement of combustible gases, and which are presented in Chapter 1.14 on combustible gas measurement in this handbook.

The analysis of oxygen concentration by catalytic combustion is accomplished by oxidizing a fuel and measuring the amount of heat generated. The sensor consists of a measuring cell and a reference cell, with a filament in each. The measuring filament is provided with a noble metal catalytic surface to oxidize the fuel, while the reference filament serves only to compensate for variations of temperature and thermal conductivity in the sample. In some designs, the unit is thermostatically maintained at a constant temperature (see Figure 1.49d). The filaments are connected in a bridge system.

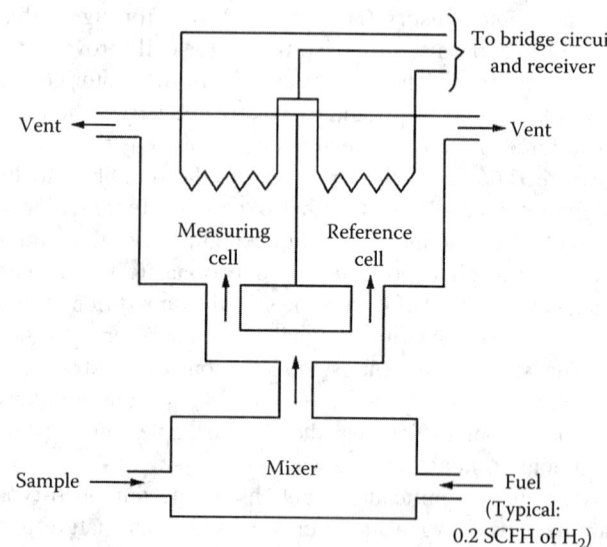

FIG. 1.49d
Catalytic combustion oxygen analyzer.

First, the sample and fuel are mixed, and then they enter both the measuring and reference cells. In the measuring cell, the fuel is burned in the presence of the filament having the catalyst. The resulting filament temperature is measured by detecting its resistance, which is a measure of the oxygen content of the gas sample.

In the reference cell, the temperature of the mixture is sensed, but no combustion takes place. The temperature difference between the two cells is attributed to the heat generated by combustion in the measuring cell and is used to calculate the amount of oxygen that was present to support the combustion.

ELECTROCHEMICAL OXYGEN SENSORS

Electrochemical oxygen sensors fall into three main categories, which are as follows:

1. High-temperature fuel cells, utilizing the conduction of oxygen ions (O^{2-}) from one electrode to another through a solid oxide electrolyte (e.g., zirconium dioxide, aka zirconia).
2. Ambient-temperature galvanic sensors, also called fuel cells, that operate by oxygen reduction at the cathode and dissolution of an active anode such as cadmium or lead in an electrolyte.
3. Polarographic sensors, consisting of three electrodes (cathode, anode, and a reference) and an electrolyte.

The operation of a polarographic sensor is similar to that of the galvanic sensors except that, here, an external potential is applied to the cathode to drive the oxygen reduction reaction.

The high-temperature fuel cell oxygen sensors are used for the measurements of gaseous oxygen, whereas the galvanic and polarographic oxygen sensors can be used for both gaseous and dissolved oxygen measurements in liquids.

All three types of electrochemical oxygen sensors typically measure the partial pressure due to oxygen and require either temperature control or temperature compensation. Some galvanic-type oxygen sensors also require the control of gas flow rate or compensation for its variation. Some fuel cell types measure percent oxygen, instead of partial pressure, utilizing a capillary instead of a membrane as the diffusion barrier. These are relatively temperature stable and do not normally require temperature compensation.

High-Temperature Zirconium Dioxide Fuel Cells

In these oxygen sensors, the oxygen is ionized in both a sample and a known reference gas stream. The sensing cell consists of a calcium-stabilized or yttria-stabilized zirconium dioxide solid electrolyte, which is provided with porous noble metal (typically platinum) electrodes on the inner and outer surfaces of the solid electrolyte (Figure 1.49e).

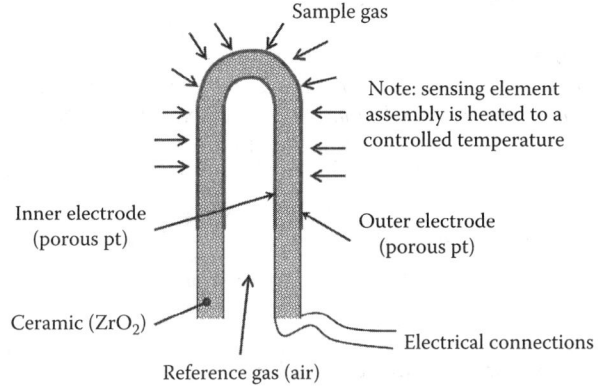

FIG. 1.49e
The flow of oxygen ions through a hot zirconium dioxide electrolyte causes a voltage difference across the thickness of the element.

The cell typically operates at a temperature of about 800°C (1472°F). When sample and reference gas streams come in contact with the electrode surfaces, the oxygen ionizes into O^{2-} ions. The oxygen ion concentration in each of the streams is a function of the partial pressure due to oxygen in the stream. The voltage potential difference developed between the electrodes depends on the difference in partial pressures due to oxygen between the sample gas and reference gas streams.

The electrode with more negative potential (higher oxygen concentration) will generate oxygen ions, whereas the electrode with the more positive potential (lower oxygen concentration) will convert oxygen ions into oxygen molecules. The cell reaction at the two electrodes can be expressed as

$$O_2 + 4e^- \rightarrow 2O^{2-} \quad \text{(at the cathode)} \quad \mathbf{1.49(1)}$$

$$2O^{2-} \rightarrow O_2 + 4e^- \quad \text{(at the anode)} \quad \mathbf{1.49(2)}$$

The open-circuit voltage relates to oxygen partial pressure by Equation 1.49(3), which is also referred to as the Nernst equation,

$$E = \frac{RT}{nF} * \ln\left(\frac{O_2 \text{ ref gas partial pressure}}{O_2 \text{ sample gas partial pressure}}\right) \quad \mathbf{1.49(3)}$$

where
  E is the open-circuit voltage developed
  R is the universal gas constant
  T is the temperature
  n is the number of electrons transferred per molecule of oxygen
  F is the Faraday's constant

From the above equation, it can be seen that the cell output signal changes logarithmically with the partial pressure due to oxygen in the sample stream when the partial pressure due to oxygen in the reference stream is constant.

The maximum detectable oxygen concentration in the sample stream may be greater than, less than, or equal to the oxygen concentration in the reference stream. When the oxygen concentration in the sample stream reaches that of the reference stream, the open-circuit voltage will be zero. If the oxygen concentration in the sample stream exceeds the oxygen concentration in the reference stream, the oxygen ions will move in the opposite direction, and the open-circuit voltage will be of the opposite polarity.

It should also be noted that the open-circuit potential is directly related to the temperature of the cell. Therefore, accurate control of the cell temperature or temperature compensation is required.

Cell Design and Limitations Figure 1.49f illustrates the design of a typical high-temperature zirconia cell. The temperature of the cell is maintained at about 800°C (1472°F). The sample and reference gas streams flow near the porous platinum electrodes, which are in contact with the solid oxide electrolyte. Oxygen ionizes at the electrode surfaces. Equilibrium between oxygen and oxygen ions is established at the electrode surfaces, and the concentration of oxygen in the sample stream is determined by measuring the open-circuit voltage.

Such oxygen sensors should be used only for those applications in which the sample contains no combustibles, because, at the elevated temperatures, the oxidizable material in the sample gas will stoichiometrically combine with oxygen. Consequently, if combustibles are present, the oxygen concentration in the sample gas will decrease, thereby causing an error in measurement. This limitation reduces the possibility of applying such oxygen sensors for the measurement of oxygen at sub-ppm levels.

Some versions of this design are available as transmitters, and others are marketed with integral indicators for both portable and permanent installations. These analyzers are available with very wide ranges. An analyzer scale might have graduations between 1 and 200,000 ppm or might cover a range of 0.5%–100%.

Self-Diagnostics and Self-Cleaning Zirconium dioxide stack probe analyzers are available with features of in-place calibration and self-diagnostics. Some designs can be serviced without removing the probe from the stack.

Probes are available in lengths of 0.5, 1, 2, 3, and 4 m (1.5, 3, 6, 9, and 12 ft) and should be inserted to 40%–50% of duct diameter to reach the most representative area. In some designs, the zirconium oxide sensing element is located at the tip of the probe (Figure 1.49g). In that case,

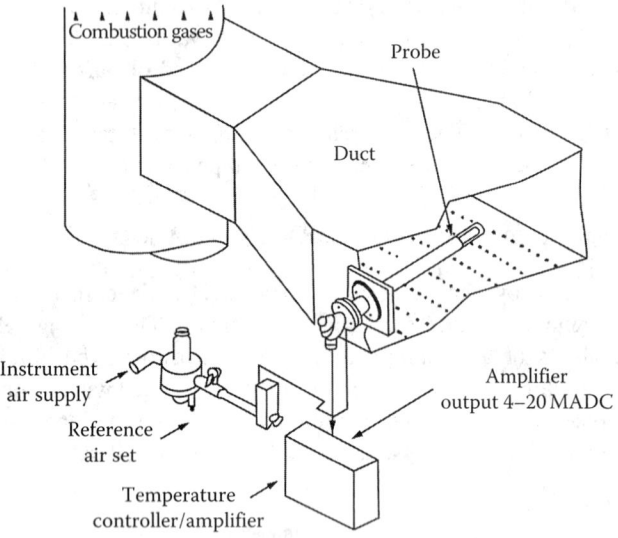

FIG. 1.49g
In-leakage through the duct joints and operating temperature must both be considered in selecting the location for installing the probe.

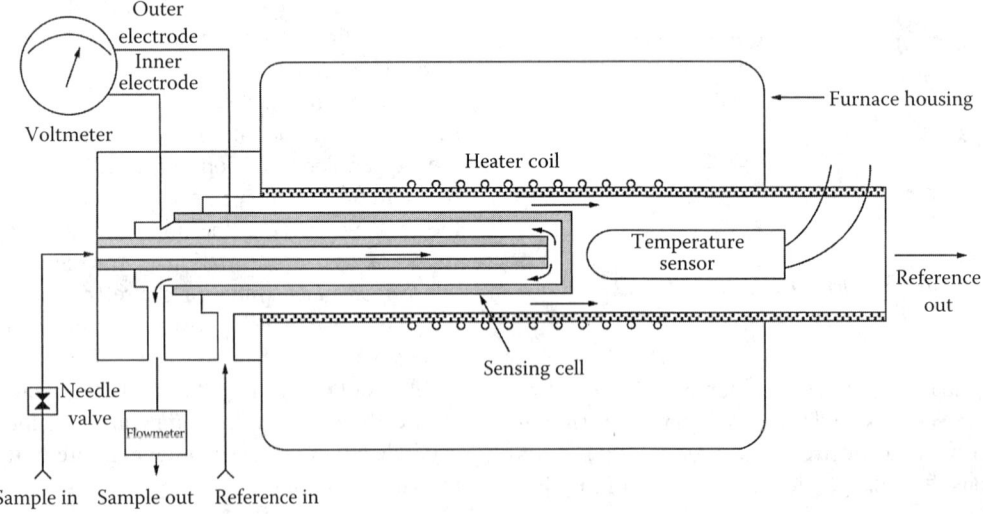

FIG. 1.49f
High-temperature zirconia electrochemical oxygen sensor.

the installation must be made at a location where the operating temperature is under 800°C (1472°F); otherwise, the sensor could not be maintained at its constant temperature of 800°C (1472°F).

In other designs, the sensing element is located outside the duct, and aspirators are used to bring the samples to it. These designs can be utilized for the samples with higher operating temperatures, because the gases can be cooled before they are brought to the sensor (Figure 1.49h).

During normal operation, cell activity causes the partial pressure due to oxygen within the cavity to be reduced to nearly zero. As oxygen from the sample gas diffuses into the cavity, it is pumped out by cell action, and this generates a current flow in the excitation circuit. The output is in direct proportion to the current flow, and to the oxygen concentration of the sample (Figure 1.49i).

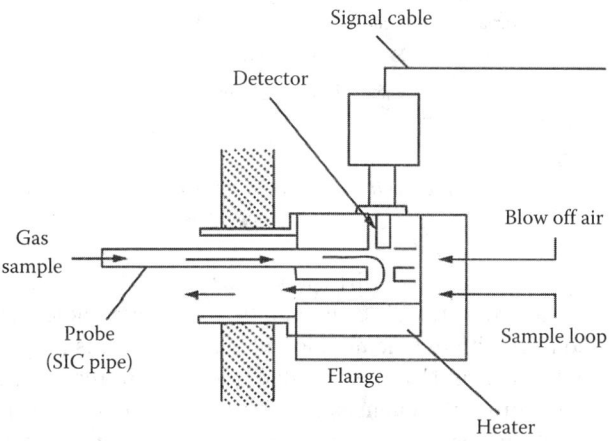

FIG. 1.49h
Self-cleaning high-temperature probe allows for use at operating temperatures up to 1593°C (2990°F), due to built-in temperature-controlled cooler.

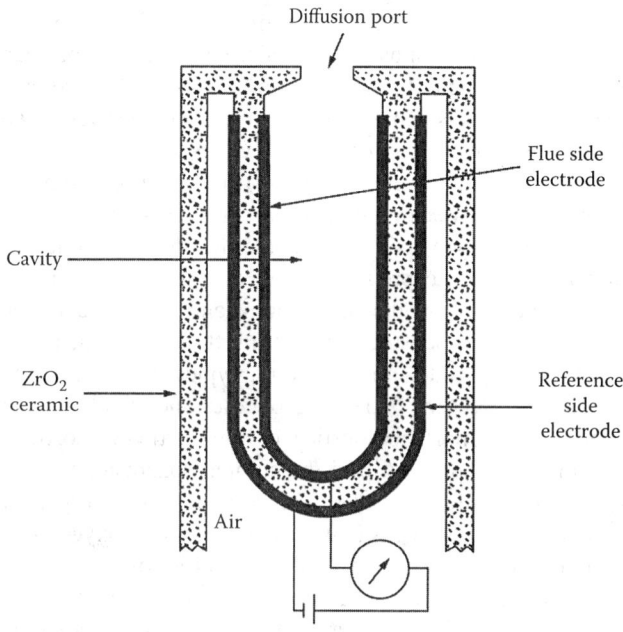

FIG. 1.49i
High-temperature diffusion-limited current-mode cell.

Because ductwork is likely to leak, and because combustion processes are operated under vacuum, the farther the sample point is from the burner, the larger is the potential for measurement error caused by the oxygen in-leakage. Therefore, zirconium oxide probes should be installed as close to the burner as their temperature limitation allows.

Other limitations of this sensor include its relatively short life (1–2 years) and its relatively high replacement cost. Sensor failure can be caused by shorting out the cell, which results in the diffusion of the platinum electrodes through the zirconium oxide layer. As the sensor ages, it also becomes more difficult to calibrate. These analyzers are therefore not recommended for trace oxygen measurement in the ppm range. Their primary application is in measuring the percentage of excess oxygen in combustion, or in steel-producing blast furnace processes. But this type of sensing element can be incorporated into a benchtop analyzer, with the sensing element having a heater and temperature sensor. If not exposed to corrosive or contaminating samples, this type of sensing element can last a long time.

High-Temperature Current-Mode Oxygen Sensors

The zirconia cell used in this sensor includes a cavity. A small excitation voltage has been applied across the sensing electrodes on the inside and outside surfaces of the cavity.

The cell can be used with samples of high sulfur dioxide content and can operate at temperatures of up to 1582°C (2880°F).

Inexpensive zirconium oxide oxygen sensors are also used in the fuel/air controls of internal combustion engines. These sensors typically operate at a temperature of around 316°C (600°F).

Galvanic Sensors (Fuel Cells)

Similar to the high-temperature solid metal oxide fuel cell O_2 sensors, galvanic-type O_2 sensors comprise a cathode, an anode, and an electrolyte.

A typical sensor may have a catalyst at the cathode that is made of fine particles or wire of silver or platinum. This metallic cathode also provides means of electrical connection at the cathode. A common material used as the anode is lead in a fine wire form such as lead wool. A metal housing can provide connection means for the anode, or a metallic electrode can be added to implement an anode connection when the housing material is nonconductive. The electrolyte may be potassium hydroxide, potassium acetate, or another polar solution. The lead anode material becomes oxidized or used up as oxygen enters the cell, with typical lifetimes in air of 6–24 months (and some newer types last longer than 24 months), depending on the sensor design and signal output level (i.e., a higher signal level would cause the anode material to be consumed more quickly). The sensitivity of these sensors is determined by their construction, by the choice of sensing cathodes, and the diffusion barriers. Analyzers utilizing these sensors are used for a variety of applications from ppm-level measurements to breathing air applications (20%–25%), and also up to 100% oxygen.

Material technology advances have enabled manufactures to develop fuel cells that are made of a variety of polymeric and composite materials. Galvanic sensors can be made very small and can operate at very low voltage and current levels. Consequently, they are extensively used for intrinsically safe and portable (even pocket-sized) applications. The author pioneered the use of fuel cell oxygen sensors for the inerting of industrial processes having flammable liquids or vapors. In this application, a sample gas is drawn from the process vessel, conditioned to remove particulates and condensing vapors, and presented to the fuel cell oxygen sensor. For inerting, the oxygen measurement is compared with a setpoint, typically in the range of 1%–4% oxygen, depending on the amount of oxygen needed to support combustion of the particular solvents and or vapors present, and the comparison result is used to control a solenoid valve. The valve admits nitrogen or another oxygen-free gas that is used to displace oxygen to a sufficiently low level to avoid the possibility of combustion to take place, should an ignition source become available. As an additional safety feature to detect a faulty solenoid or a fault in wiring to the solenoid, the control system also measures the current that is driving the solenoid valve. The author pioneered the use of miniature oxygen fuel cells in portable analyzers for personal safety, as well as the first digital portable combustion efficiency analyzer for measuring and adjusting oil burners used for home and industrial heating. In the combustion analyzer, oxygen concentration, temperature, and particulate matter are sampled in the flue gas of the oil burner. Leaks can be detected and repaired, the air/fuel mix can be adjusted on the burner appliance, as calculations based on the temperature and oxygen concentration are made within the analyzer to report the resulting burner efficiency. Fuel sensor type of oxygen sensing elements utilize a diffusion barrier that limits the rate of oxidation of the anode material. The diffusion barrier is typically either a membrane or a capillary, each having its own performance characteristics. But such sensors for low ppm measurements may have no diffusion barrier added.

Membrane Type In a membrane type of oxygen fuel cell sensor, the amount of oxygen entering the cell and oxidizing the anode is proportional to the partial pressure due to oxygen. At the cathode, oxygen is consumed as soon as it arrives, so the concentration of oxygen at the cathode is near zero. New oxygen comes in by diffusing into the membrane material, diffusing through the thickness of the membrane, and then exiting the other side of the membrane. A typical construction is shown in Figure 1.49j.

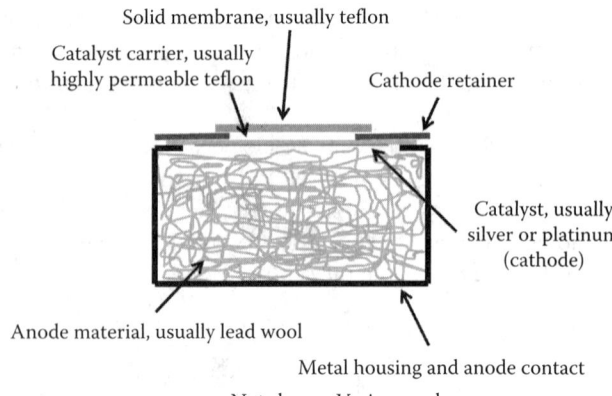

FIG. 1.49j
Membrane-type fuel cell responds to partial pressure due to oxygen.

Since the rate of diffusion of oxygen into the membrane is proportional to the number of oxygen molecules in contact with the membrane, the output signal is proportional to the partial pressure due to oxygen in the sample gas. Accordingly, changes in the absolute pressure of the sample gas result in proportional changes in the sensor output signal.

Capillary Type Some fuel cell designs utilize a capillary instead of a membrane to limit the flow of oxygen into the cell. This construction offers both advantages and disadvantages. For example, the capillary type is not sensitive to the sample gas pressure when the pressure is constant, but a sudden change in sample gas pressure will cause a momentary change in the reading until equalization occurs in a few seconds (Figure 1.49k).

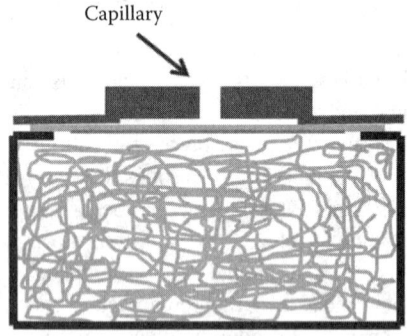

FIG. 1.49k
Capillary-type fuel cell responds to percent oxygen. Note: Housing, anode material, cathode material, catalyst carrier, cathode retainer, etc., may all be similar as with membrane type. Main difference is having capillary instead of solid membrane as diffusion barrier.

The membrane design is less responsive to fast concentration changes and is more prone to problems caused by coating with dusts and powders. The capillary design therefore offers an additional option to meet application requirements.

Several manufacturers offer both capillary and membrane systems. One major difference is that a sensor having a membrane diffusion barrier responds to the partial pressure due to oxygen, whereas one using a capillary responds to the percentage oxygen concentration. For example, a capillary version measuring air at one atmosphere and reading 20.9% will also indicate 20.9% with the same sample gas when the pressure is increased to two atmospheres. But a membrane-type sensor would indicate an oxygen concentration that is twice as high when the absolute pressure is doubled.

A capillary-type sensor employs a membrane to contain the electrolyte, and often also as a carrier for the cathode catalyst. A capillary is positioned on the outside of the membrane. The capillary and membrane are sized so that there is essentially near zero amount of oxygen in the space between the capillary and the membrane, because the oxygen is taken up quickly when it is near the membrane. If we assume that air is 20.9% oxygen and most of the remainder is nitrogen, then there remains stationary nitrogen molecules within the capillary, and oxygen molecules are moving through the capillary past the stationary nitrogen molecules. So the flow of oxygen into the sensor is mainly impeded by the stationary nitrogen molecules. If the air pressure were doubled, as posed in the example, there would be twice as many oxygen molecules available to flow into the sensor, but there would also be twice as many nitrogen molecules present to impede the flow. So the net result is that the amount of oxygen flowing into the sensor is relatively independent of the air pressure.

One consideration with the capillary design is a slight nonlinearity in the signal output vs. concentration. In a normal range of 0.5%–25% oxygen, the nonlinearity has little effect. But it must be considered with wider ranges. The output signal vs. oxygen concentration follows the relationship:

$$\text{Signal} = c * \ln\left(\frac{1}{1-C}\right) \qquad 1.49(4)$$

where
Signal is the output current of the sensing element (or its voltage across a load resistor, such as 50 Ω)
c is a constant that is calculated first with a known signal and concentration
C is the concentration of the sample gas (e.g., use 0.209 for 20.9% oxygen)
ln is the natural logarithmic function

Using this formula, one can calculate the output at another concentration when the output is known at a reference concentration, such as 20.9%.

Low ppm Level, Sensing Element Design O_2 sensors that employ smooth metal cathodes can achieve sensitivity for gaseous oxygen to sub-ppm levels with an excellent signal-to-noise ratio (S/N or SNR). However, sensitivity for gaseous oxygen in the sub-ppb level is achieved by using a high surface area metal-catalyzed gas diffusion electrode as the O_2-sensing cathode. In one product, a high surface area metal-catalyzed gas diffusion electrode consists of a porous hydrophobic electrically conductive Teflon™-carbon backing layer bonded to a thin layer of high surface area metal catalyst dispersed on a carbon support.

This type of electrode provides an effective surface area of up to 600 times the geometric area of the electrode, thus generating high signal output per unit O_2 concentration.

A typical galvanic O_2 sensor utilizing a high surface area metal-catalyzed gas diffusion electrode with a gaseous oxygen sensitivity of 1 ppb is illustrated in Figure 1.49l. The main body of the sensor is made of a polymeric material and generally has a U-shaped configuration that contains electrolyte, an anode, and a thermistor. The cathode is mounted on one side of the sensor wall.

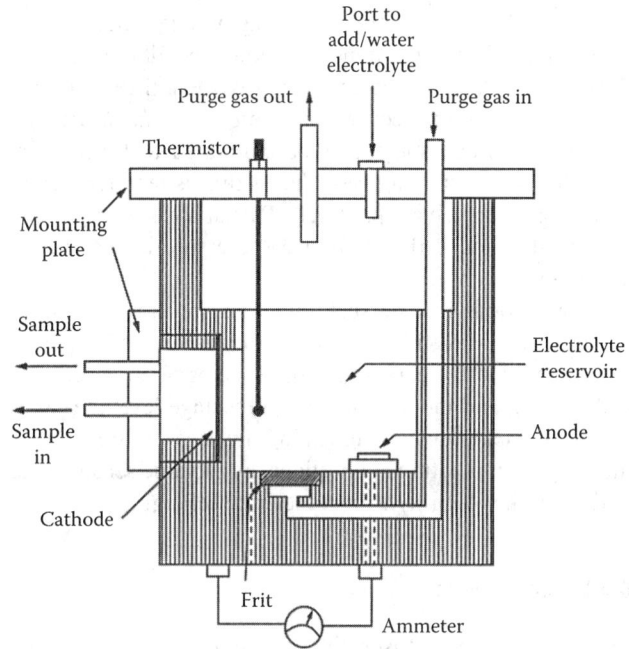

FIG. 1.49l
Electrochemical oxygen sensors utilizing high surface area metal-catalyzed gas diffusion electrode as the oxygen-sensing cathode. (Courtesy of Teledyne Analytical Instruments.)

A stainless steel plate with gas inlet and outlet tubes is attached to the same wall, with an O-ring between the plate and the sensor wall. The top of the sensor is sealed with a stainless steel plate and an O-ring to eliminate the exposure of inner components to ambient air. This plate also contains a port to fill the sensor with electrolyte or charge the sensor with water for continuous operation. The O_2 dissolved in the electrolyte is removed by bubbling oxygen-free gas through the porous frit at the bottom of the sensor to minimize interference during analysis.

The gas to be analyzed enters the cathode cavity via an inlet tube and then leaves via the outlet tube. During this process, it diffuses through the porous backing layer, reaches the catalyst surface exposed to the electrolyte, and immediately reacts, generating an electrical signal proportional to the partial pressure due to O_2 in the gas mixture. Such sensors generally require temperature and gas flow rate compensation.

Advantages and Limitations   A significant advantage of this type of O_2 sensor is the reduced susceptibility to contamination of the sensing element by particulates in the sample gas. This is because particulates cannot enter the main cell body. The sensing element allows the diffusion of gases only through the back surface of the electrode to the catalyst surface exposed to the electrolyte. Analyzers utilizing such O_2 sensors are useful for long-term online operations.

The galvanic sensor is relatively insensitive to shock and vibration as long as the shock level is not sufficient to deform the electrode conductor materials, but might read low as a result of a loss of sensitivity from aging and oxidizing the anode. Therefore, it might need periodic recalibration. Also, if exposed to large concentrations of oxygen, the sensor can take a long time to recover from the *oxygen shock*. Percent reading types can sometimes be calibrated to fresh air at 20.9% oxygen, but then several hours time is required before returning to service at ppm levels. Partial pressure types can be air calibrated if the atmospheric pressure is known, but also require recovery time.

Applications include such things as ppb trace measurements to validate semiconductor quality gases used in potentially hazardous or explosives atmospheres, food processing and storage, and more traditional percentage level measurements of stack gases and breathing air. It is not suitable for most high-temperature applications, as galvanic sensors tend to dry out the electrolyte and accelerate cell aging.

Coulometric Sensor

In the coulometric sensor, the sample gas diffuses through a diffusion barrier, such as a polytetrafluoroethylene membrane, to the cathode of the electrochemical cell, where it is reduced to hydroxyl ions. Because of the conductivity of the electrolyte, the ions migrate to the anode, where they are oxidized back to oxygen. An external voltage drives this reaction, which results in a cell current that is proportional to the oxygen concentration in the gas sample (Figure 1.49m).

A main advantage of this design is its nondepleting nature, as neither electrode undergoes a chemical change during measurement. On the other hand, the effective area of the electrodes can change with time as a result of contaminant deposition on the electrodes. Therefore, the sensor could require periodic recalibration and eventual replacement.

Furthermore, this sensor could also be subjected to oxygen shock similar to galvanic-type sensors, which use a gas diffusion barrier to control the amount of oxygen that enters the sensor. This sensor is suited for ppb, ppm, and percentage

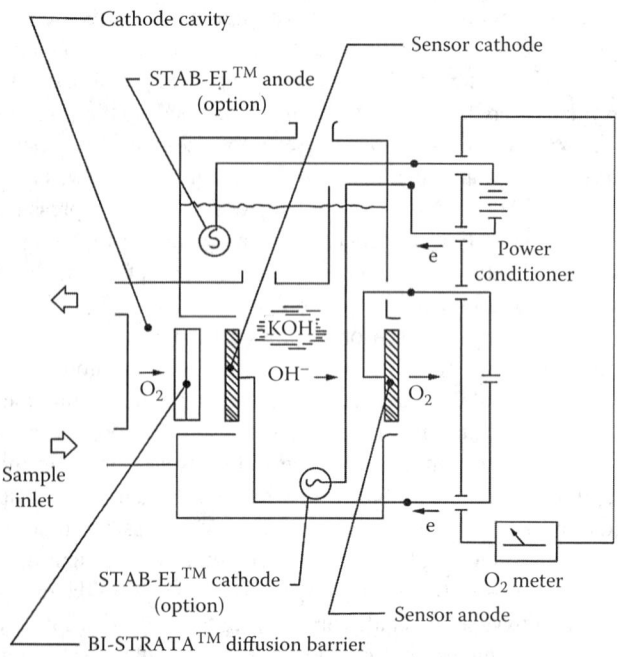

FIG. 1.49m
The coulometric electrochemical oxygen analyzer. (Courtesy of Delta F Corp.)

measurements up to 25%. Its areas of application are similar to those of the galvanic sensor.

Polarographic Sensor

The measuring cell of a polarographic sensor comprises a sensing electrode (cathode), a reference electrode (anode), and an electrolyte, which is usually potassium chloride. The cathode is separated from the sample gas by a permeable membrane (often PTFE) that permits the diffusion of oxygen through the electrolyte onto the sensing surface of the cathode. A voltage is applied between the two electrodes, causing the oxygen to be electrochemically reduced and producing an ionic current that is linear with the oxygen concentration of the sample.

This sensor is rugged and is insensitive to position and sample flow rate variations. On the other hand, it is slow to respond, because of the time required for the oxygen to diffuse through the membrane. It also requires pressure and temperature compensation and periodic (about yearly) replacement, as both the anode and the electrolyte are consumed during operation. This sensor is seldom used on gas services; its most common application is in dissolved oxygen measurement in liquids.

SPECTROSCOPIC OXYGEN DETECTION

In this handbook, the spectroscopic techniques are presented in depth in Chapters 1.32, 1.35, 1.62, and 1.69. They will be presented here only briefly.

It might seem like overkill to consider using a spectrophotometer to measure oxygen when there are so many simpler, less expensive, and effective ways to detect it. But one might need to monitor a process for more than just oxygen, and it might be an advantage to monitor multiple components of a gas simultaneously. But to the contrary, the use of the same analyzer for multiple-component measurements also increases the consequences of instrument failure. Therefore, critical measurements should always be made by dedicated instruments, and the reason that spectroscopy is mentioned here is that it is being used increasingly in applications where oxygen-specific analyzers were used in the past.

Mass Spectroscopy

Mass spectroscopy is used to monitor oxygen in a number of applications, including semiconductors, fermentation, pharmaceuticals, and other specialized areas. This instrument ionizes the sample and separates the ions according to their mass. In these applications, mass spectrometers are sometimes called *residual gas analyzers* because of the way they are used. Mass spectrometers are used when the determination (qualitative and quantitative) of more than one component is required in the sample. The application of this technique for the measurement of oxygen is in no way unique or special. The same techniques that are used for most other gases are also utilized for oxygen.

Near-Infrared Spectroscopy

Many gas concentrations can be measured by monitoring their ability to absorb light in the infrared region. NIR spectroscopy has been marketed for stack monitoring and other gas monitoring applications that include an oxygen measurement.

For a molecule to absorb infrared light and be measured in the infrared region, traditional thinking leads one to believe that it must have an electric dipole moment with which the IR radiation can interact. This typically requires a molecule that is made up of atoms in a structure that has generated an unbalanced electric field. Homonuclear molecules such as H_2, O_2, and N_2 contain two identical atoms and do not have a dipole moment. For this reason, they have long been thought of as being invisible in the IR spectrum.

For example, HCl has a hydrogen and a chlorine atom and is bound in a way that results in the chlorine being more electronegative than the hydrogen. The resulting molecule has an electrical dipole moment with which IR radiation can interact easily. By contrast, most homonuclear molecules absorb radiation only in the ultraviolet region of the spectrum. Molecular oxygen is one of the few exceptions.

The detection of molecular oxygen by the NIR spectroscopic technique involves a measurement that is often described as "a forbidden transition." Many molecules can absorb radiation (energy) and enter a temporary excited or radiative state. Usually, they quickly release this extra energy in the form of radiation. These levels of excitation typically occur within the bounds of traditional quantum rules, and the released energy (re-radiation) is typically described as fluorescence. The lifetime of these excited states is a small fraction of a second in duration.

Sometimes, when the re-radiation is not in compliance with the more traditional quantum rules, phosphorescence instead of florescence can be obtained, and the re-radiation can experience large time delays from the initial absorption. In the case of molecular oxygen, there is such a transition that can be monitored in the NIR region of the spectrum near 760 nm. It is described as a highly forbidden singlet delta state of oxygen. The fact that the state lies close to the oxygen molecular ground state means that oxygen can be measured in this "forbidden region" by NIR methods such as TDLAS.

TDLAS Designs More detailed information about TDLAS presented in Chapter 1.62. A TDLAS analyzer that, in the NIR region, is successfully used to make oxygen measurements is said to utilize *single-line absorption spectroscopy*, in which a diode laser is scanned across a chosen absorption line. The selection and application of the O_2 spectral line is proprietary to most vendors, but it is usually one of the oxygen lines around 760 nm (Figure 1.49n).

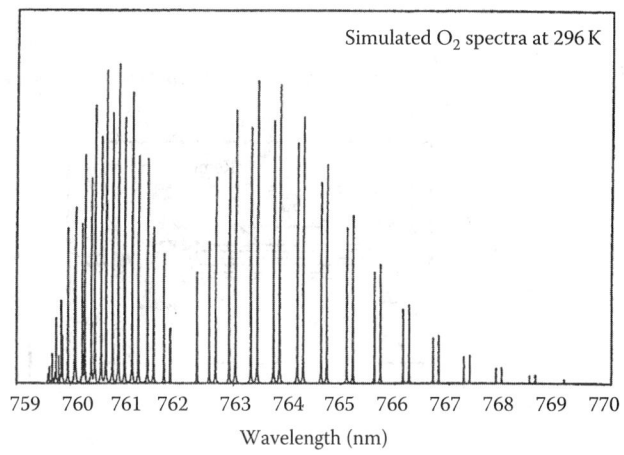

FIG. 1.49n
Simulated NIR O_2 spectra. (Courtesy of Unisearch Associates Inc.)

Suppliers employ wavelength modulation to enhance sensitivity. The laser source wavelength is modulated slightly as the absorption line is scanned. The resulting output signal is mathematically manipulated (i.e., decomposed into frequency components; usually, the second harmonic is used to generate the measurement). The result is an oxygen measurement by a spectroscopic technique that previously was not considered possible. It has been made possible only by advancements in laser technology, computing, and optics.

These analyzers are being used in incinerator stacks and other applications where samples are typically dirty or corrosive, and when high detection speeds are required. The incorporation of fiber-optic technology provided a high degree of installation flexibility. Figures 1.49o and 1.49p depict several of the installation layouts and sampling configurations that are in use.

Although neither NIR nor mass spectroscopic techniques will probably ever be used to measure oxygen only, they can be justified to simultaneously measure oxygen and a multiplicity of other components in a sample.

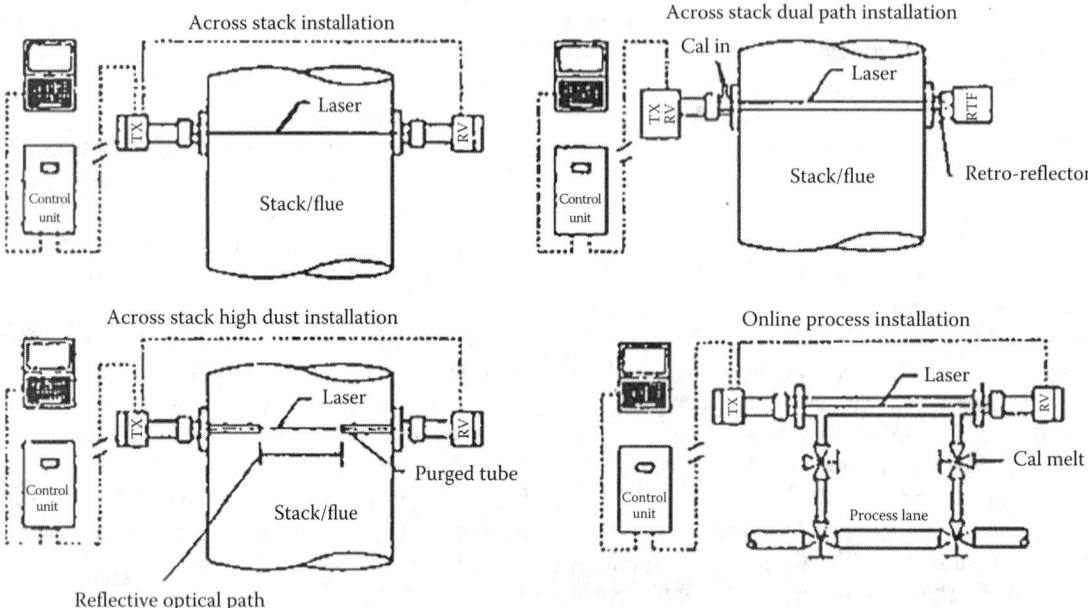

FIG. 1.49o
NIR tunable diode laser instrument layouts. (Courtesy of Analytical Specialties for Norsk Elektro Optikk A/S.)

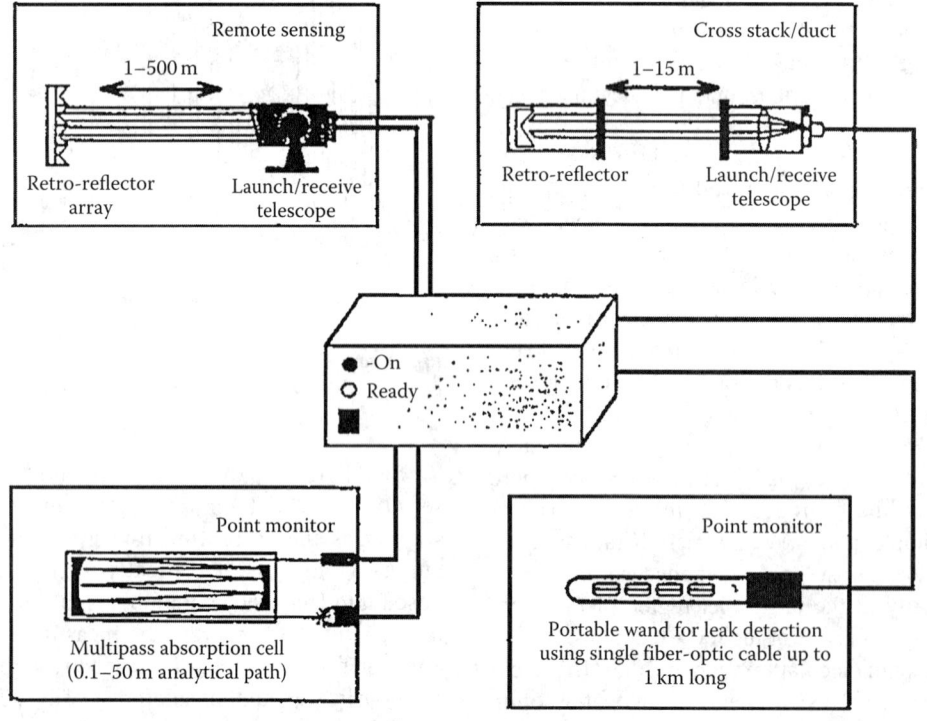

FIG. 1.49p
Examples of fiber-enabled NIR sampling configurations. (Courtesy of Unisearch Associates Inc.)

OTHER OXYGEN DETECTION METHODS

Several additional methods of oxygen detection can be considered, especially if sample components other than oxygen are of interest. For example, gas chromatography can be valuable for atmospheric analysis and to check the levels of oxygen, hydrogen, carbon monoxide and dioxide, organic solvents, Freon, methane, and others.

Chemical analyzers of the manual Orsat type are also available for a large variety of ranges from 0%–20% to 100% oxygen. Orsat gas analyzers have been designed for specific applications, such as combustion and blast furnace testing, and for checking the purity of medical and industrial oxygen. The Orsat method has also been traditionally used to measure the purity of carbon dioxide from fermentation processes (breweries, etc.), to ensure that the oxygen level was in the low ppm range before saving the carbon dioxide for use in carbonation or being sold for other processes. But it has always been difficult to obtain repeatable and reliable measurements in the low ppm oxygen range (less than 20 ppm) using this method. The author pioneered the use of electrochemical fuel cell oxygen sensors in a portable ppm analyzer for fermentation processes, obtaining repeatable and reliable results in the late 1980s. As a result, this measurement has been largely changed over to fuel cell oxygen sensors.

DEVELOPMENTS OVER THE LAST DECADE

Of the paramagnetic types, the deflection (dumbbell) technology is still the most popular. The sensing element does not wear out and can last indefinitely if not damaged by shock or contamination. Low-temperature electrochemical cells have been further refined to provide longer life and somewhat less sensitivity to interfering or contaminating gases.

Zirconia-type oxygen analyzers are robust and can last a long time. Voltage-mode zirconia sensing elements have been developed that combine the heating element with the sensing tube and temperature sensor. This enables better performance by allowing tight control of the sensing element temperature. A voltage-mode zirconia sensing element can measure from low ppm to high percent oxygen levels, and does not wear out unless contaminated or over-heated, but is more expensive than the current-mode type. Current-mode zirconia sensing elements are now more widely available. They have the advantages of being cheaper than the voltage mode, and have a higher signal output level. But the current-mode type must utilize a different sensing element for different ranges of oxygen concentration, and has a limited lifetime of about 5 years.

Portable and pocket oxygen analyzers typically utilize a fuel cell type of sensing element, and are also available as multi-gas monitors.

All technologies of oxygen analyzers have improved over the last decade due to further implementation of electronic capabilities. The greatest of which is communication. Analyzers may now communicate digitally over many protocols, including wireless transmission. The most important functions include calibration and data transfer. Various alarm functions are also available.

SPECIFICATION FORMS

When specifying analyzers of oxygen in gases, the compositions and properties of the application can be specified by using the ISA Forms 20A1001 and 20A1002. If the measurement of dissolved oxygen (DO) is required, the specification form that can be used is ISA Form 20A2192. These forms are reproduced on the next pages with the permission of the International Society of America.

Analytical Measurement

ANALYSIS DEVICE — Operating Parameters

#	RESPONSIBLE ORGANIZATION		#	SPECIFICATION IDENTIFICATIONS		
1	ISA		6			
2			7	Document no		
3			8	Latest revision	Date	
4			9	Issue status		
5			10			

#	ADMINISTRATIVE IDENTIFICATIONS			#	SERVICE IDENTIFICATIONS Continued			
11				40				
12	Project number	Sub project no		41	Return conn matl type			
13	Project			42	Inline hazardous area cl		Div/Zon	Group
14	Enterprise			43	Inline area min ign temp		Temp ident number	
15	Site			44	Remote hazardous area cl		Div/Zon	Group
16	Area	Cell	Unit	45	Remote area min ign temp		Temp ident number	
17				46				

#	SERVICE IDENTIFICATIONS			#	COMPONENT DESIGN CRITERIA		
18				47			
19	Tag no/Functional ident			48			
20	Related equipment			49	Component type		
21	Service			50	Component style		
22				51	Output signal type		
23	P&ID/Reference dwg			52	Characteristic curve		
24	Process line/nozzle no			53	Compensation style		
25	Process conn pipe spec			54	Type of protection		
26	Process conn nominal size	Rating		55	Criticality code		
27	Process conn termn type	Style		56	Max EMI susceptibility	Ref	
28	Process conn schedule no	Wall thickness		57	Max temperature effect	Ref	
29	Process connection length			58	Max sample time lag		
30	Process line matl type			59	Max response time		
31	Fast loop line number			60	Min required accuracy	Ref	
32	Fast loop pipe spec			61	Avail nom power supply		Number wires
33	Fast loop conn nom size	Rating		62	Calibration method		
34	Fast loop conn termn type	Style		63	Testing/Listing agency		
35	Fast loop schedule no	Wall thickness		64	Test requirements		
36	Fast loop estimated lg			65	Supply loss failure mode		
37	Fast loop material type			66	Signal loss failure mode		
38	Return conn nominal size	Rating		67			
39	Return conn termn type	Style		68			

#	PROCESS VARIABLES	MATERIAL FLOW CONDITIONS				#	PROCESS DESIGN CONDITIONS		
69						101			
70	Flow Case Identification				Units	102	Minimum	Maximum	Units
71	Process pressure					103			
72	Process temperature					104			
73	Process phase type					105			
74	Process liquid actl flow					106			
75	Process vapor actl flow					107			
76	Process vapor std flow					108			
77	Process liquid density					109			
78	Process vapor density					110			
79	Process liquid viscosity					111			
80	Sample return pressure					112			
81	Sample vent/drain press					113			
82	Sample temperature					114			
83	Sample phase type					115			
84	Fast loop liq actl flow					116			
85	Fast loop vapor actl flow					117			
86	Fast loop vapor std flow					118			
87	Fast loop vapor density					119			
88	Conductivity/Resistivity					120			
89	pH/ORP					121			
90	RH/Dewpoint					122			
91	Turbidity/Opacity					123			
92	Dissolved oxygen					124			
93	Corrosivity					125			
94	Particle size					126			
95						127			
96	CALCULATED VARIABLES					128			
97	Sample lag time					129			
98	Process fluid velocity					130			
99	Wake/natural freq ratio					131			
100						132			

#	MATERIAL PROPERTIES			#	MATERIAL PROPERTIES Continued		
133				137			
134	Name			138	NFPA health hazard	Flammability	Reactivity
135	Density at ref temp	At		139			
136				140			

Rev	Date	Revision Description	By	Appv1	Appv2	Appv3	REMARKS

Form: 20A1001 Rev 0

1.49 Oxygen in Gases

	RESPONSIBLE ORGANIZATION	ANALYSIS DEVICE COMPOSITION OR PROPERTY Operating Parameters (Continued)		SPECIFICATION IDENTIFICATIONS		
1			6			
2			7	Document no		
3	(ISA)		8	Latest revision		Date
4			9	Issue status		
5			10			

	PROCESS COMPOSITION OR PROPERTY			MEASUREMENT DESIGN CONDITIONS				
	Component/Property Name	Normal	Units	Minimum	Units	Maximum	Units	Repeatability
13								
14								
15								
16								
17								
18								
19								
20								
21								
22								
23								
24								
25								
26								
27								
28								
29								
30								
31								
32								
33								
34								
35								
36								
37								

Rev	Date	Revision Description	By	Appv1	Appv2	Appv3	REMARKS

Form: 20A1002 Rev 0

© 2004 ISA

1	RESPONSIBLE ORGANIZATION		DISSOLVED OXYGEN/OZONE/CL2		6	SPECIFICATION IDENTIFICATIONS	
2			TRANSMITTER/ANALYZER/MONITOR		7	Document no	
3	(ISA)		Device Specification		8	Latest revision	Date
4					9	Issue status	
5					10		

11	TRANSMITTER OR ANALYZER		56	PERFORMANCE CHARACTERISTICS	
12	Housing type		57	pH compensation LRL	URL
13	Input sensor type		58	Meas 1 accuracy rating	
14	Input sensor style		59	Meas 2 accuracy rating	
15	Output signal type		60	Temp accuracy rating	
16	Enclosure type no/class		61	Measurement LRL	URL
17	Control mode		62	Temp compensation LRL	URL
18	Local operator interface		63	Barometric pressure LRL	URL
19	Characteristic curve		64	Minimum span	Max
20	Digital communication std		65	Min ambient working temp	Max
21	Signal power source		66	Contacts ac rating	At max
22	Temperature sensor type		67	Contacts dc rating	At max
23	Quantity of input sensors		68	Max distance to sensor lg	
24	Preamplifier location		69		
25	Contacts arrangement	Quantity	70		
26	Integral indicator style		71		
27	Signal termination type		72		
28	Cert/Approval type		73		
29	Mounting location/type		74		
30	Failure/Diagnostic action		75		
31	Sensor diagnostics		76	ACCESSORIES	
32	Data history log		77	Remote indicator style	
33	Measurement compensation		78	Indicator enclosure	
34	Calibration function		79	Communicator style	
35	Enclosure material		80		
36			81		
37			82		
38			83	SPECIAL REQUIREMENTS	
39			84	Custom tag	
40			85	Reference specification	
41			86	Compliance standard	
42			87	Software configuration	
43			88	Software program	
44			89		
45			90		
46			91		
47			92	PHYSICAL DATA	
48			93	Estimated weight	
49			94	Overall height	
50			95	Removal clearance	
51			96	Signal conn nominal size	Style
52			97	Mfr reference dwg	
53			98		
54			99		
55			100		

110	CALIBRATIONS AND TEST		INPUT OR TEST		OUTPUT OR SCALE	
111	TAG NO/FUNCTIONAL IDENT	MEAS/SIGNAL/SCALE	LRV	URV	ACTION	LRV	URV
112		Meas-Analog output 1					
113		Meas-Analog output 2					
114		Meas-Digital output					
115		Meas-Scale					
116		Temp-Scale					
117		Temp-Digital output					
118		Meas setpoint 1-Output					
119		Meas setpoint 2-Output					
120		Meas setpoint 3-Output					
121		Meas setpoint 4-Output					
122		Failure signal-Output					
123							

124	COMPONENT IDENTIFICATIONS		
125	COMPONENT TYPE	MANUFACTURER	MODEL NUMBER
126			
127			
128			
129			

Rev	Date	Revision Description	By	Appv1	Appv2	Appv3	REMARKS

Form: 20A2192 Rev 0 © 2004 ISA

Abbreviations

AMU	Atomic mass unit. The standard unit that is used for indicating mass on an atomic or molecular scale, approximately equivalent to the mass of one neutron or one proton
HCl	Hydrogen chloride (a gas), or in aqueous solution, hydrochloric acid
IR	Infra red
NIR	Near infrared
Pa	Pascal, an SI unit of pressure, equal to 1 N/m^2
ppb	Parts per billion
ppm	Parts per million
PTFE	Polytetrafluoroethylene, also called by the trade name Teflon
S/N (or SNR)	Signal-to-noise ratio
TDLAS	Tunable diode laser absorption spectroscopy

Bibliography

Eggins, B. R., *Chemical Sensors & Biosensors*, John Wiley & Sons, Hoboken, NJ 2004.

Fusaro, D., Environment, energy concerns resuscitate oxygen analyzers, *Control*, February 1992, p. 48–50.

Harlow, V., *Oxygen Hacker's Companion*, Airspeed Press, Warner, NH, 2002.

IEC 61207-3, Gas analyzers—Expression of performance—Part 3: Paramagnetic oxygen analyzers, 2002.

Janata, J., *Principles of Chemical Sensors*, Plenum Press, New York, 1990.

Korotcenkov, G., *Handbook of Gas Sensor Materials: Properties, Advantages and Shortcomings*, Springer, New York, NY, 2013.

Rayburn, S., *The Foundations of Laboratory Safety*, Springer-Verlag, New York, 1990.

Ren, F., *Semiconductor Device-Based Sensors for Gas, Chemical, and Biomedical Applications*, CRC Press, Boca Raton, FL, 2011.

Servomex, Paramagnetic Oxygen Analysis, https://www.servomex.com/servomex/web/web.nsf/en/paramagnetic-oxygen-analysis accessed June 16, 2015.

Sherman, R., *Analytical Instrumentation: Practical Guides for Measurement & Control*, ISA, Research Triangle Park, NC, 1996.

Shermann, R., *Process Analyzer Sample Conditioning System Technology*, Wiley Interscience, Hoboken, NJ, 2001.

SST Sensing, Oxygen Sensor Operating Principle, www.sstsensing.com/download/file/fid/480, accessed June 6, 2015.

Toray Eng., Principle of Measurement, Zirconia Oxygen Analyzers, www.toray-eng.com/measuring/tec/zirconia.html, 2013, accessed June 16, 2015.

Webster, J., *Measurement, Instrumentation, and Sensors Handbook*, CRC Press, Boca Raton, FL, 2014.

West, H. and Zinn, C. D., Impact of OSHA/EPA process safety regulations on controls systems, *Proceedings of the 1992 ISA Conference*, Houston, October, 1992.

White, L. T., *Hazardous Gas Monitoring*, William Andrew Publishing, Norwich, NY, 2000.

Yokogawa, Oxygen Analyzers in General, www.yokogawa.com/an/faq/oxy/oxy_general.htm, 2011, accessed June 16, 2015.

1.50 Oxygen in Liquids (Dissolved Oxygen)

R. K. KAMINSKI (1969, 1982) **M. P. DZIEWATKOSKI** (2003)
B. G. LIPTÁK (1995, 2017)

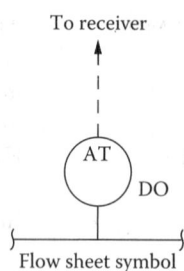

Flow sheet symbol

Types	A. Polarographic
	B. Galvanic
	C. Coulometric
	D. Multiple-anode
	E. Thallium
	F. Fluorescence
Operating pressure	Up to 50 psig (3.5 barg) or submersion depths of up to 8.3 m (25 ft)
Operating temperature range	0°C–50°C (32°F–122°F); special designs up to 80°C (175°F)
Flow velocity at sensing membrane	Preferably in excess of 1 ft/s (0.3 m/s); some can operate down to 0.2 ft/s (0.06 m/s)
Materials of construction	Typical material for sensor housing: PVC or stainless steel; possible electrode materials: platinum, gold, silver, zinc, lead, cadmium, and copper; for membrane assembly, ABS plastic or stainless steel mesh-reinforced silicone and/or Teflon® membrane
Speed of response	90% in 30 s; 98% in 60 s
Range	Common ranges are 0–5, 0–10, 0–15, and 0–20 ppm; special units are available with ranges up to 0–150 ppm or down to the 0–20 ppb range used on boiler feedwater applications. Systems can also be calibrated in partial pressure units.
Inaccuracy	Generally, ±1% to ±2% of span; industrial transmitter errors are generally within 0.02 ppm over a 0–20 ppm range. Thallium cells are available with a 0–10 ppb range and can read the DO within an error of 0.5 ppb.
Calibration	The three primary methods of calibration are: (1) Winkler titration, (2) Air-saturated water, and (3) Water-saturated air. Electrochemical sensors require more frequent calibrations than optical ones, which often can hold their calibration for months. The needed frequency of calibrations is dictated by experience and in some cases compensation for barometric pressure variations is needed. Temperature, barometric pressure, and salinity all affect calibration.
Costs	Portable, battery-operated, 1.5%–2% FS units that also read temperature cost from $500 to $2000; replacement probes range from $250 to $1000; 1% FS, microprocessor-based portable bench top units for laboratory or plant service cost from $1000 to $2000; industrial-quality (0.02 ppm error limit) DO probe and 4–20 mA transmitter range from $2000 to $3500; cleaning assemblies cost from $500 to $2000.
Partial list of suppliers	ABB, http://new.abb.com/products/measurement-products/analytical/continuous-liquid-analyzers/dissolved-oxygen/ads430
	ABB Inc. Instrumentation (http://www05.abb.com/global/scot/scot203.nsf/veritydisplay/61fe8c77c45d9654c1257815002e172d/$file/ds_9435-en_h.pdf)
	Analytical Technology, Inc. (http://www.analyticaltechnology.com/public/product.aspx?ProductID=1028)
	Broadley-James Corporation (http://www.broadleyjames.com/dir-sensorDO-main.html)
	Clean Instruments http://www.cleaninst.com/do500.htm
	Cole-Parmer http://www.coleparmer.com/Product/WTW_ProfiLine_Oxi_3205_Dissolved_Oxygen_Meter_Only/EW-53105-30
	Emerson Process Management, Rosemount Analytical http://www2.emersonprocess.com/en-us/brands/rosemountanalytical/liquid/sensors/do/pages/index.aspx
	Endress + Hauser Inc. (http://www.us.endress.com/eh/sc/america/us/en/home.nsf/#products/oxygen-sensors)

(Continued)

Global Water Instr. Inc. (http://www.globalw.com/products/fdo.html)
Great Lakes Instruments Inc. (http://www.hach.com/dissolved-oxygen/parameter-products?productType=instruments&usEpa=false&parameterId=7639974407&hideObsolete=false&pimContext=USen)
Greenspan Technology Pty Ltd. (www.greenspan.com.au)
Hach (http://www.wastewatercanada.com/Products/Hach/Parameters_Process/LDO-Spec.pdf)
Hanna Instruments (http://www.hannainst.com/usa/prods2.cfm?id=005002&ProdCode=HI%207609829-2)
HFScientific http://www.hfscientific.com/products/advantedge_dissolved_oxygen_electrode_sensor
Honeywell Inc. (http://honeywellphelectrodes.com/dissolvedoxygen.php)
Horiba Ltd. (http://www.horiba.com/es/scientific/products/water-quality/dissolved-oxygen-meters/)
IC Controls Ltd. (http://www.environmental-expert.com/products/ic-controls-model-855-ppm-dissolved-oxygen-analyzer-243143)
Invensys-Foxboro (http://iom.invensys.com/EN/Pages/Foxboro_MandI_Analytical_DissolvedOxygen.aspx)
Mettler Toledo Ingold Inc. (http://us.mt.com/us/en/home/products/ProcessAnalytics/DO_CO2_Sensor/level_2_DO_family_MARCH03_NEW.html)
OxySense (http://www.processinstruments.net/products/dissolved-oxygen-monitor.php)
Rosemount Analytical, Liquid Division (http://www2.emersonprocess.com/siteadmincenter/PM%20Rosemount%20Analytical%20Documents/Liq_ProdData_71-RDO.pdf)
Royce Technologies (http://www.royceinst.com/RefDocs/Royce_DissolvedOxygen.pdf)
Sensorex (http://www.sensorex.com/products/categories/category/do_process_sensors)
Teledyne Analytical Instruments (www.teledyne-ai.com)
ThermoScientific (http://www.thermoscientific.com/ecomm/servlet/productsdetail_11152___12769482_-1)
Van London Company Inc. (http://www.vl-pc.com/default/index.cfm/meters-and-accessories/knick-products/knick-sensors-including-memosens-and-fittings/knick-oxygen-sensors/se-303-dissolved-oxygen-sensor/)
Vernier, http://www.vernier.com/products/sensors/dissolved-oxygen-probes/do-bta/
Yokogawa Corp. of America (http://www.yokogawa.com/an/do402g/an-do-01en.htm)
YSI Inc. (https://www.ysi.com/parametersdetail.php?Dissolved-Oxygen-1)

INTRODUCTION

Limnology (the study of lakes) shows that dissolved oxygen is essential to support aquatic life. It is second in importance only to water itself. Dissolved oxygen is necessary to many forms of life including fish, invertebrates, bacteria, and plants. The amount of dissolved oxygen needed varies from creature to creature and range from 1 to 15 mg/L. Microbes such as bacteria and fungi also require dissolved oxygen as they decompose organic material.

Dissolved oxygen enters the waters through the air or is produced by aquatic plants such as phytoplankton, algae, and seaweed by photosynthesis. The aeration of water can be caused by waves, rapids, and waterfalls. As aquatic photosynthesis is light-dependent, the dissolved oxygen produced drops with the depth of the water, and also it peaks during daylight hours and declines at night.

Chapter 1.49 was devoted to the analysis of oxygen in gases. This chapter describes the sensors used to measure the concentration of oxygen in liquids. These sensors are most often used in the wastewater treatment industry, where the most commonly used dissolved oxygen (DO) detector designs are the polarographic, galvanic, and fluorescent designs. The different electrochemical cell-type sensors (e.g., polarographic, galvanic, coulometric) can be used in both gas and liquid applications.

Because dissolved oxygen concentration is most often measured by probe-type sensors and because it is essential to keep these probes clean, the importance of probe cleaners cannot be overemphasized. Probe cleaners are discussed in detail in Chapter 1.55, 1.33, 1.39, 1.47, and 229, and in most detail in Chapter 1.55 in connection with pH probes. Therefore, their treatment here is abbreviated and the reader is referred to those chapters.

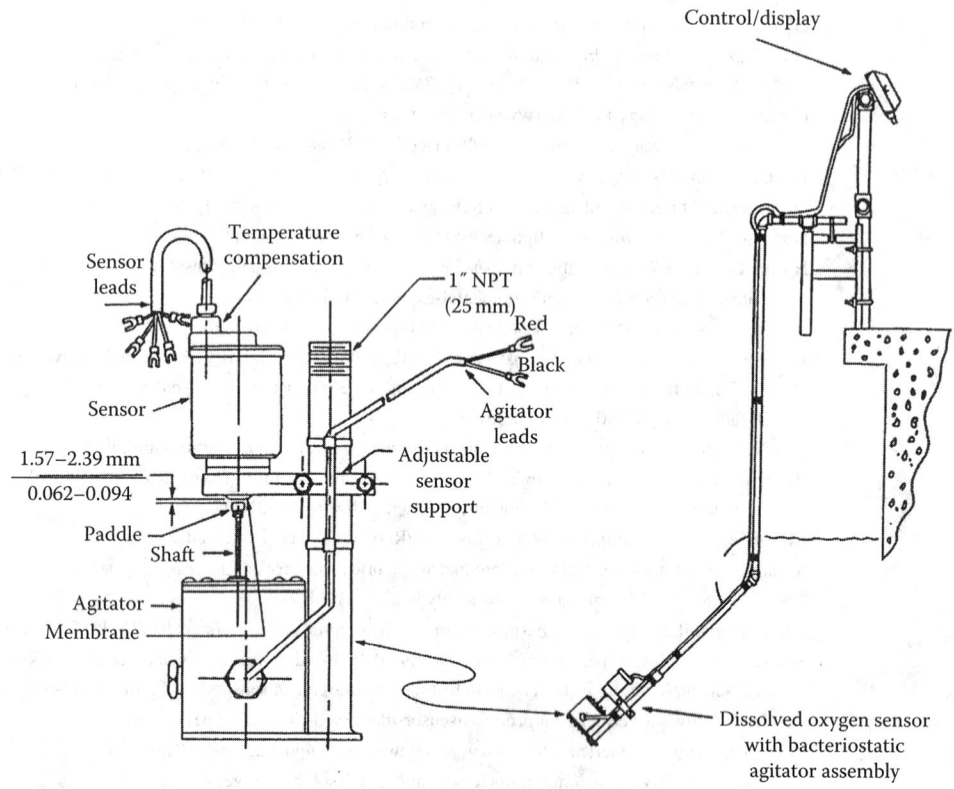

FIG. 1.50a
Cleaner assembly for keeping the membrane surface of a dissolved-oxygen probe free of buildup or biological growths.

Figure 1.50a shows a vibratory paddle-type probe cleaner. This unit cleans the membrane surface by a back-and-forth motion close to the surface. The agitator paddle is coated with a biological growth-inhibiting substance that minimizes the biological growth on the membrane surface.

Galvanic and polarographic sensors continue to dominate the dissolved oxygen measurement field for both online and off-line analysis. However, significant advances have been made with the introduction of equilibrium-state voltammetric and fluorescence-based probes. These have the advantage of analyte sensitivities, which are relatively independent of membrane coating, and, since they consume little or no analyte as a result of the measurement process, they are able to work in processes where there is little or no flow past the face of the sensor.

All oxygen detectors used for gaseous samples can also measure the oxygen concentration in liquid streams if the oxygen is first removed from the liquid. This approach has the major advantage that only a clean gas mixture contacts the oxygen detector.

ELECTROCHEMICAL DETECTORS

The galvanic and polarographic electrochemical analyzers most commonly consists of a probe that can be directly immersed in the process solution. They usually have six basic components:

1. A measuring cathode
2. A reference anode
3. A temperature sensor (thermistor)
4. An electronics module
5. Electrolyte solution (potassium chloride [KCl] or potassium hydroxide [KOH] are typically used)
6. An oxygen porous membrane

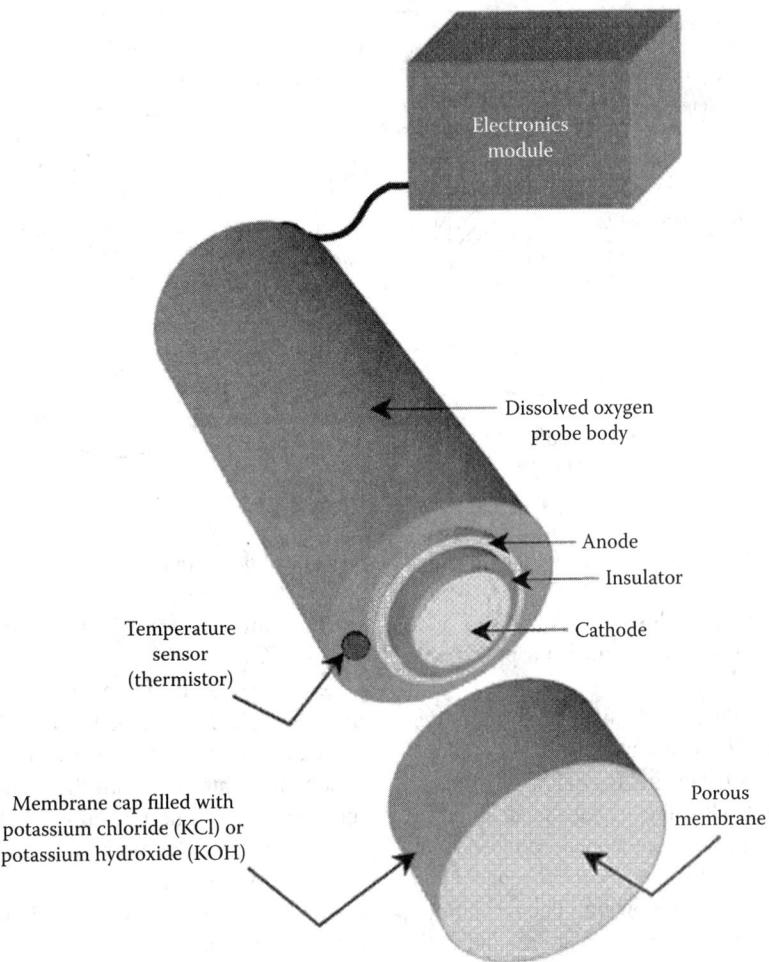

FIG. 1.50b
The main components of all electrochemical DO detector probes. The main difference between the designs is that the polarographic ones do, the galvanic ones do not use polarizing voltage. (Courtesy of ITA News.)

The measuring electrode sensor outputs a current to the electronics module in proportion to the concentration of dissolved oxygen in the process stream (Figure 1.50b).

Polarographic (Voltametric) Cells

In the polarographic cell, a polarizing voltage is applied to two electrodes that are in contact with an electrolyte, contained inside a porous membrane at the tip of the DO probe. When the oxygen molecules diffuse across the membrane and come into contact with the cathode, an electrical current flows from the cathode to the anode, which is in proportion to the amount of dissolved oxygen in the process sample. The analyzer electronics sense the flow of electrical current and produce an output reading of dissolved oxygen expressed in percent saturation or in milligrams per liter (mg/L).

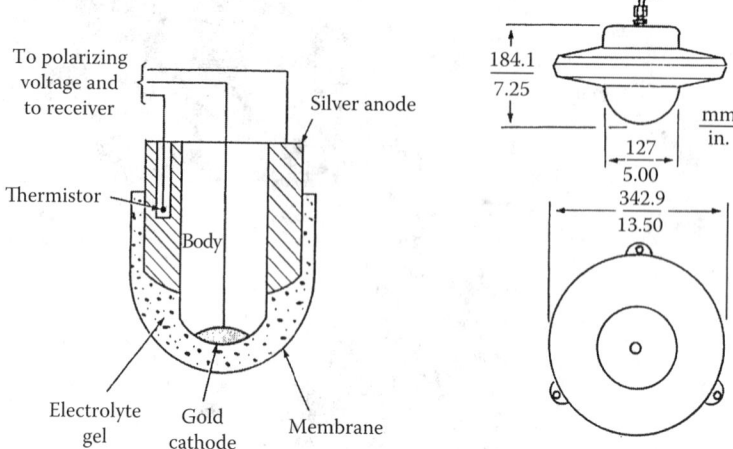

FIG. 1.50c
Probe-type polarographic cell oxygen detector and flotation collar in which it can be mounted.

The basic polarographic cell is illustrated in Figure 1.50c. It has *two* noble metal electrodes and requires a polarizing voltage to reduce the oxygen.

The dissolved oxygen in the sample diffuses through the membrane into the electrolyte, which usually is an aqueous KCl solution. If there is a constant polarizing voltage (usually 0.8 V) across the electrodes, the oxygen is reduced at the cathode, and the resulting current flow is proportional to the oxygen content of the electrolyte. This current flow is detected as an indication of oxygen content. The oxidation-reduction reactions, in the case of a gold–silver cell and with KCl electrolyte, are as follows:

At the gold cathode,

$$O_2 + 2H_2O + 4 \text{ electrons} \rightarrow 4(OH^-) \qquad 1.50(1)$$

and at the silver anode,

$$4Ag + 4(Cl^-) \rightarrow 4AgCl + 4 \text{ electrons} \qquad 1.50(2)$$

A variety of designs are available to extend the working life of the cell to several months. These include a number of membrane and electrolyte reservoir designs in addition to the use of a larger silver anode. The silver chloride at the anode can be converted back to silver by reversing the polarizing voltage.

Sample Temperature and Flow The polarographic cell, like galvanic cells, is affected by temperature. Therefore, it is necessary to either control the sample temperature or to provide temperature compensation to attain high-precision measurements. In case of DO measurement, high precision corresponds to ±1% to ±2% error. If the sample temperature is allowed to vary between 0°C and 43°C (32°F and 110°F), the measurement error in some designs will rise to approximately ±6%.

Both galvanic and polarographic cells require a minimum sample flow velocity. This is necessary to eliminate stagnant layers of sample over the membrane, which otherwise would interfere with the continuous transfer of oxygen into the cell. The higher sample velocities are also beneficial because of their scrubbing action. Some suppliers provide a combination cell that, for maximum cleaning effect, includes a sample pump that directs a 5 ft/s (1.5 m/s) flow velocity against the membrane.

Galvanic Cells

The main difference between the designs of polarographic and galvanic cells is that the polarographic cells do, the galvanic cells do not use polarizing voltage. Both cells use the same components. In a galvanic cell, the anode can be formed of lead and the cathode of silver, gold, or platinum. When the oxygen molecules diffuse across the membrane of the sensor, they react with the surface of the cathode, and an electrical current is produced, which the analyzer electronics compensate for temperature variations and produce a direct reading output.

The operating theory of the dissolved oxygen galvanic cell is the same as that of the galvanic cell used for gaseous samples. With the exception of the considerations related to the drying-out problem, all the design considerations are identical. The main difference is that most dissolved oxygen sensors are installed in dirty water, so they require special cleaners, agitators, and specialized sample systems.

The ranges of the galvanic cell dissolved oxygen analyzers can be as low as 0–20 ppb for applications such as the measurement of DO content of boiler feedwater.

All galvanic cells consist of an electrolyte and two electrodes (Figure 1.50d). The oxygen content of the electrolyte is equalized with that of the sample. The reaction is spontaneous; no external voltage is applied. In this reaction, the cathode reduces the oxygen into hydroxide, thus releasing four electrons for each molecule of oxygen (Equation 1.50(3)). These electrons cause a current flow through the electrolyte. The magnitude of the current flow is in proportion to the oxygen concentration in the electrolyte.

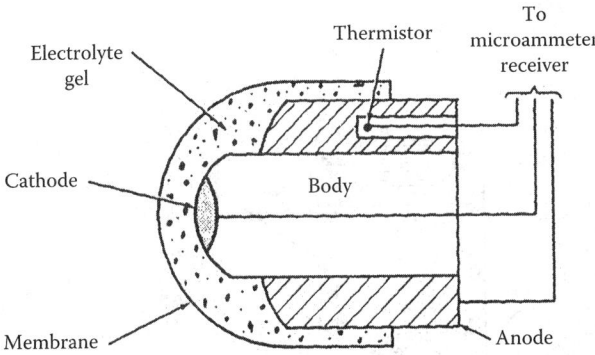

FIG. 1.50d
Probe-type galvanic cell oxygen detector.

The most common electrode materials are gold, silver, copper, and lead, and the most frequently used electrolyte is potassium hydroxide (KOH). The cathode must be a noble metal (silver or gold) for the cathode potential to reduce molecular oxygen when the cell circuit is closed. A base metal is selected for the anode (lead, cadmium, copper, zinc, or silver) with good stability and without any tendency toward passivation. The electrolyte (KOH, potassium chloride [KCl], or potassium bicarbonate [$KHCO_3$]) is selected to minimize its dissolution of the anode when the cell is open.

In the case of a lead anode, the cell reactions can be expressed as follows:

$$\text{Cathode: } O_2 + 2H_2O + 4 \text{ electrons} \rightarrow 4(OH^-) \quad \quad \mathbf{1.50(3)}$$

$$\text{Anode: } 2Pb \rightarrow 2(Pb^{++}) + 4 \text{ electrons} \quad \quad \mathbf{1.50(4)}$$

The galvanic cell designs are subject to various degrees of contamination by background gases in the process stream. As a very general rule, the following background gases can be considered harmless: argon, butane, carbon monoxide, ethane, ethylene, helium, hydrogen, methane, nitrogen, and propane. However, the following gases are likely to contaminate the cell: chlorine and other halogens, high concentrations of carbon dioxide, hydrogen sulfide, and sulfur dioxide.

Flow-Through Cells Special cells have been developed to minimize the effect of the background gases. When an acid gas (such as CO_2) is present that would neutralize a potassium hydroxide electrolyte in the background, a potassium bicarbonate electrolyte can be considered. Special cells are also available for the measurement of oxygen in acetylene and fuel gases.

In flow-through cell designs, sampling systems are usually required to bring the process stream to the analyzer, and to filter it, scrub it with caustic or treat it in other ways as preparation for the measurement. The probe-type membrane design does not require a sampling system if it can be located in a representative process area where the pressure, temperature, and velocity of the process stream are compatible with the cell's mechanical and chemical design.

Probe Design In this design (Figure 1.50d), the electrodes are wetted by an electrolytic solution that is retained by a membrane (usually Teflon) that acts as a selective diffusion layer, allowing oxygen to diffuse into the electrolyte solution while keeping foreign matter out. The sensor is usually mounted in a thermostatically controlled housing, and a thermistor is provided to compensate for minor temperature variations.

The ion current established in the electrolyte can be expressed by the following equation:

$$i_x = \frac{nFAP_mC_s}{L} \quad \quad \mathbf{1.50(5)}$$

where
- i_x is the ion current
- n is the number of electrons involved in the electrode reaction
- A is the area of the cathode surface
- F is the Faraday's constant
- P_m is the permeability coefficient of the membrane
- L is the thickness of the membrane
- C_s is the oxygen concentration

Equation 1.50(5) shows the relationship among the cell components. The Nernst equation* (Equation 1.50(1)) also applies to the galvanic cell and explains why electrode potentials are

---

*Nernst equation relates the concentration detected by a galvanic or polarographic cell and the electric signal they generate. In other words, it gives a formula that relates the numerical values of the concentration gradient, which causes ions to move through a permeable membrane and the corresponding values of the electrical gradient that opposes the movement of that charge.

a function of the absolute temperature. In addition, the ionic activity also varies with temperature, thus causing additional temperature sensitivity.

The characteristics of the membrane are critical to performance. The ideal membrane would be inert, stable, strong, permeable to oxygen, and impermeable to water molecules or other ions. In most cases, a compromise solution is accepted.

The design of a galvanic (amperometric) cell with gold–copper electrodes designed for rail mounting is shown in Figure 1.50e. The maintenance requirements of this design have been reduced by providing an electrolyte supply that lasts for 2–3 years and by making the membrane assembly easily replaceable. These analyzer systems are available in weatherproof housing, with 1% of span inaccuracy and with 4–20 mA transmitter output.

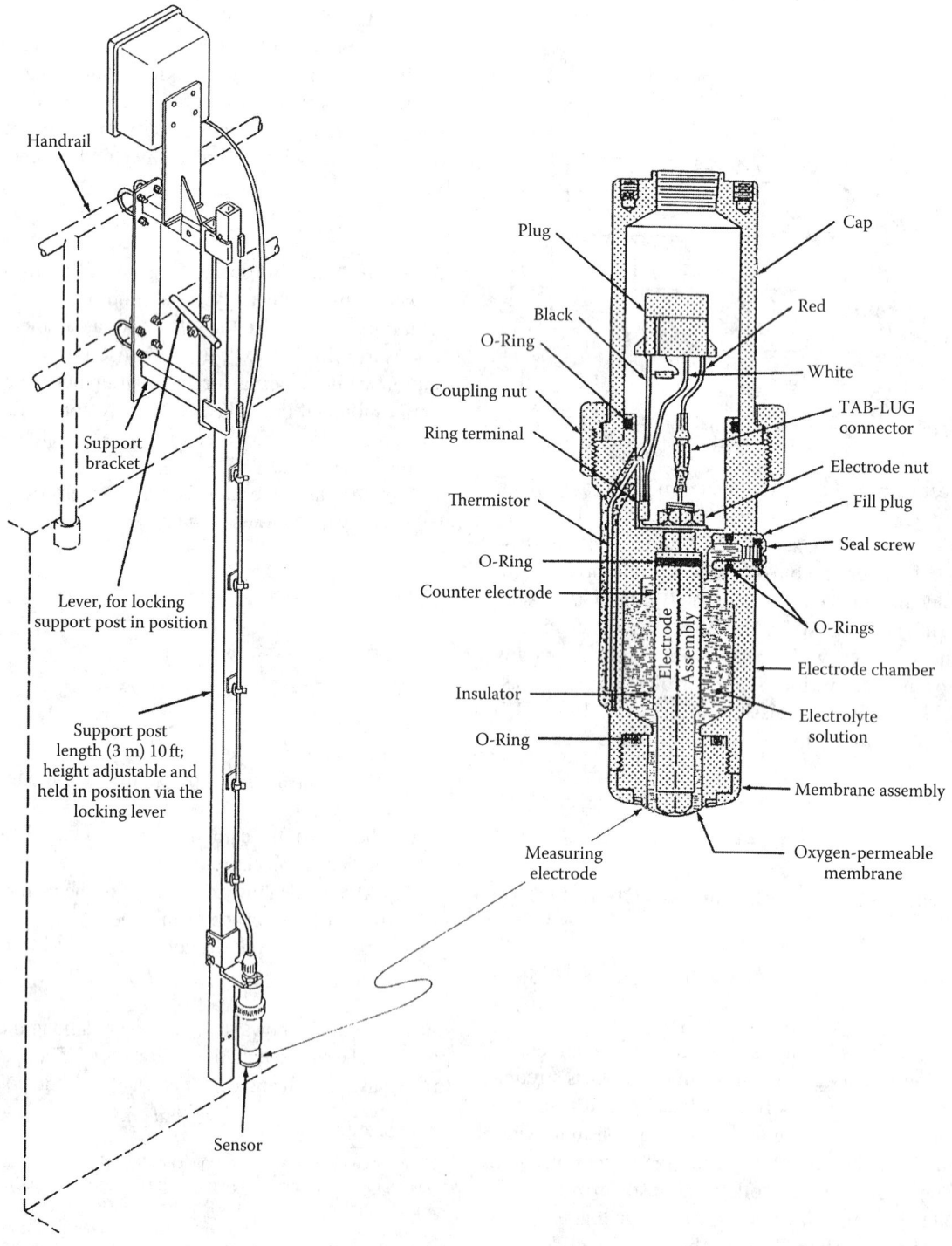

FIG. 1.50e
The design of a galvanic DO cell and its rail mounting installation. (Courtesy of ABB Inc., Zurich, Switzerland [formerly Fisher & Porter Co.])

Trace Oxygen Sensor In the flow-through cells, the process sample stream is bubbled through the electrolyte. The oxygen concentration of the electrolyte is therefore in equilibrium with the sample's oxygen content, and the resulting ion current between the electrodes is representative of this concentration.

In some trace analyzer designs, the cathode is made out of a porous metal, and the sample gas passes through this electrode, immersed in the electrolyte. Oxygen reduction tends to be complete within the pores of this electrode (Figure 1.50f).

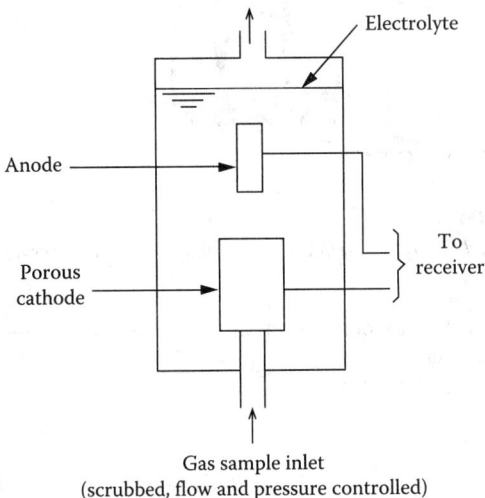

FIG. 1.50f
Flow-through trace oxygen analyzer cell.

These types of cells are usually provided with sampling consisting of (but not limited to) filtering and scrubbing components and flow, pressure, and temperature regulators.

Coulometric Sensor In the coulometric sensor, the sample gas diffuses through a diffusion barrier to the cathode of the electrochemical cell, where it is reduced to hydroxyl ions. Due to the conductivity of the potassium hydroxide electrolyte, the ions migrate to the anode, where they are oxidized back to oxygen. An external voltage drives this reaction, which results in a cell current that is proportional to the oxygen concentration in the gas sample.

The main advantage of this design is its nondepleting nature, as neither electrode undergoes a chemical change during measurement. Also, there is no need for periodic sensor replacement, and the unit is not subject to *oxygen shock*. This sensor is suited for ppb, ppm, and percentage measurements up to 25%. The areas of application of the coulometric analyzer are similar to those of the galvanic sensor.

Multiple-Anode Design The multiple-anode detector (Figure 1.50g) has three electrodes. Two are interspaced (noted by + and − in the figure) and covered with electrolyte. Oxygen is consumed at many cathodes but is generated at many anodes.

$$2H_2O \rightarrow O_2 + 4H^+ + 4e^- \quad \text{1.50(6)}$$

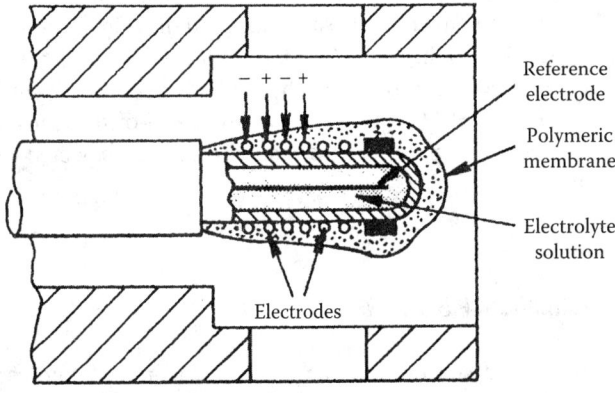

FIG. 1.50g
Multiple-anode oxygen detector.

Because a balance exists in this cell, there is no net generation of products. A number of advantages are claimed for this technique. These include that fouling affects only the response time, there is no deterioration of electrodes, and membrane replacement is not required.

Thallium Cells

Thallium cells are somewhat unique in their operating principle and cannot be classified into the category of either galvanic or polarographic cells. At the same time, they are of the electrochemical type. One thallium-electrode cell design is somewhat similar in appearance to the unit illustrated in Figure 1.50c except that it has no membrane or electrolyte. This cell has a thallium outer ring electrode and an inner reference electrode. When oxygen contacts the thallium, the following reaction takes place:

$$4Tl + O_2 + 2H_2O \rightarrow 4Tl(OH) \rightarrow 4(Tl^+) + 4(OH^-) \quad \text{1.50(7)}$$

The potential developed by the cell is a function of the thallous ion concentration at the face of the electrode, and the ion concentration is in proportion to the concentration of dissolved oxygen.

Differential Conductivity Detectors Another cell involving the reaction between thallium and oxygen is the thallium differential conductivity analyzer. This sensor detects the amount of thallous hydroxide formed by the measurement of conductivity. The sample conductivity is sensed both before and after the reaction, and the difference in conductivities is detected as a measure of dissolved oxygen content.

To eliminate temperature gradient problems, the actual instrument has both a sample and a reference flow stream. They are maintained at the same rate, and they flow through a similar flow path. Due to the high sensitivity of this measurement, accuracies of ±0.5 ppb are obtained over the minimum range of 0–10 ppb. The response speed of this sensor is relatively slow; 95% of the response can be expected in about 3 min.

FLUORESCENCE SENSORS

Fluorescence sensors operate on the principle of fluorescence quenching. Some chemical compounds have the ability to absorb light of a particular wavelength and then to re-emit light at a different (longer) wavelength. The intensity of the emitted light can be diminished when another chemical compound interacts with the compound responsible for the fluorescence. Because its intensity is lower than expected, the emitted fluorescence intensity is said to be *quenched*. This phenomenon can be used for the measurement of oxygen.

In this case, a compound containing ruthenium is immobilized in a gas-permeable matrix called a *sol-gel*. Sol-gels are very low-density, silica-based matrices suitable for immobilizing chemical compounds such as the ruthenium compound used in this measurement technique. Effectively, the sol-gel is equivalent to the membrane in a conventional DO sensor.

Using fiber-optics, light from a light-emitting diode is transferred to the backside of the sol-gel coating. The emitted fluorescence is collected from the backside of the sol-gel with another optical fiber and its intensity is detected by a photodiode. A simplified sensor design is shown in Figure 1.50h. If no oxygen is present, the intensity of the emitted light will be at its maximum value. If oxygen is present, the fluorescence will be quenched, and the emitted intensity will decrease.

The ratio of the emitted intensity when no oxygen is present (I_o) to the emitted intensity in the presence of oxygen (I) can be related to the oxygen partial pressure (which can

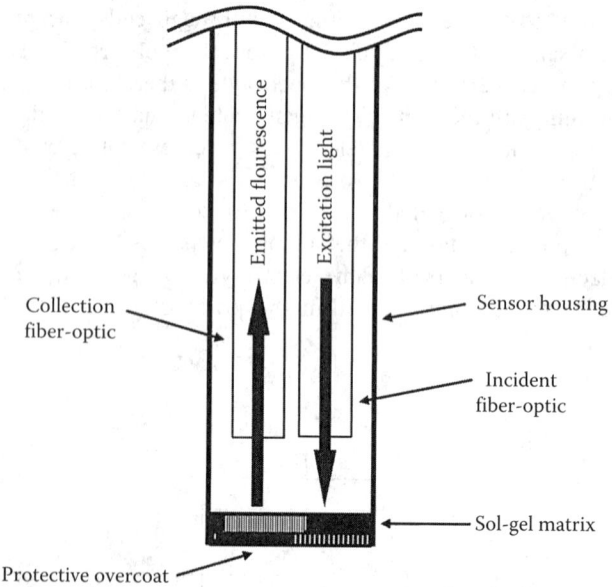

FIG. 1.50h
Simplified diagram of a fluorescence-based dissolved oxygen sensor.

be converted to concentration) through the Stern–Volmer equation below:

$$\frac{I_O}{I} = 1 + K(P_{O_2}) \qquad 1.50(8)$$

where

K is a temperature dependent constant
P_{O_2} is the partial pressure of oxygen

This equation shows that, in the absence of oxygen, the ratio I_O/I will be 1.0 and that it will increase linearly as the oxygen concentration increases.

To determine the oxygen content of a solution, it is first necessary to determine I_O when the oxygen concentration is zero. Hence, calibration consists of a minimum of two points: one measurement is made in a zero-oxygen environment, and the other is at near normal oxygen concentrations where measurements are to be made.

The dynamic range for this sensor type is from single-digit ppb through 40 ppm. Response times are approximately 30 s, and inaccuracy is typically 1% of the measurement range. The sensing element can be damaged by solutions having pH values greater than 10 and by solutions containing high levels of aggressive solvents such as aromatic compounds and ketones. The sol-gel membranes are available with a coating (such as silicone) to provide protection against abrasion.

WET CHEMISTRY ANALYZERS

Another method of dissolved oxygen detection involves wet-chemistry analyzers. These devices operate intermittently by taking a small sample and adding reagents to it. The reagents develop a color in the sample if a certain component is present. This technique can be applied to dissolved oxygen detection (in addition to many other applications) by colorimetrically determining the concentration of the unknown component.

CONSIDERATIONS APPLICABLE TO ALL SENSORS

For the purposes of the discussion that follows, it is assumed that oxygen is dissolved in ordinary water. For most applications, the sensors are actually inserted into more complex solutions. However, the major component is usually water and thus, for the sake of the discussion that follows, the physical and chemical properties of such a solution usually can be treated as pure water.

Dissolved oxygen sensors can also be used for measurements in solutions consisting primarily of organic solvents. In this case, one must be aware that the solubility characteristics of oxygen in such a solution can be very different from that in a solution consisting primarily of water. Therefore, the oxygen solubility algorithms preprogrammed into the transmitter for water may not accurately reflect the true oxygen concentration in a solvent.

Calibration Methods*

Calibration usually consists of exposing the sensor to a sample of known oxygen content and adjusting its output until it matches the known concentration in the sample. The three primary methods of calibration are as follows: (1) Winkler titration, (2) Air-saturated water, and (3) Water-saturated air. Electrochemical sensors require more frequent calibrations than optical ones, because this second category of sensors is more stable. They can often hold their calibration for months.

---

* YSI, Calibration Tips (2009) https://www.ysi.com/File%20Library/Documents/Tips/DO-6-Series-Sondes-DO-Calibration-Tips-Excerpt.pdf.

The needed frequency of calibrations is dictated by experience and in some cases compensation for barometric pressure variations is needed.

Many dissolved oxygen systems are designed to operate over a very wide concentration range—for example, from 0 to 100 mg/L or up to 500% saturation. Calibration can be based on either one or two points. Additionally, calibration can be performed in air- or oxygen-saturated media or in air (or oxygen) that is saturated with water vapor. The method of calibration will have an influence on accuracy and is often dictated by the desired measurement range.

Given this, manufacturers may suggest different calibration schemes based on the level at which the determination is to be made. For example, if all measurements are to be made at concentrations greater than 0.5 mg/L, then a one-point calibration will usually suffice. (This is not true for fluorescence-based sensors, as was discussed earlier.) In this case, the intercept of the calibration plot is set to equal zero current at zero concentration. This assumption is often valid, given that the background current for most sensors is usually very small in comparison to the values being measured.

On the other hand, if measurements are to be made at levels below 0.5 mg/L, then one will want to set the low calibration point (the zero signal) in an oxygen-free medium and then set the second calibration point in a medium of higher (and known) oxygen content. This procedure will ensure greater accuracy at lower concentrations.

Special care must be observed when calibrating membrane-covered sensors that consume oxygen (e.g., polarographic or galvanic) in a zero-oxygen environment. For this case, the sensor must be exposed to the oxygen-free medium (and polarized) for approximately 1 h prior to calibrating the zero point. Otherwise, residual oxygen in the electrolyte within the sensor will cause an erroneously high current measurement.

For an in-depth description of calibration methods, see YSI, Calibration Tips (2009) https://www.ysi.com/File%20Library/Documents/Tips/DO-6-Series-Sondes-DO-Calibration-Tips-Excerpt.pdf.

Temperature Compensation

Temperature is an important factor in dissolved oxygen measurements for two reasons. First, the solubility of oxygen in a liquid is temperature dependent and, second, the rate at which oxygen moves across the permeable membrane is also temperature dependent. The change in solubility is inversely proportional to the solution temperature, as shown in

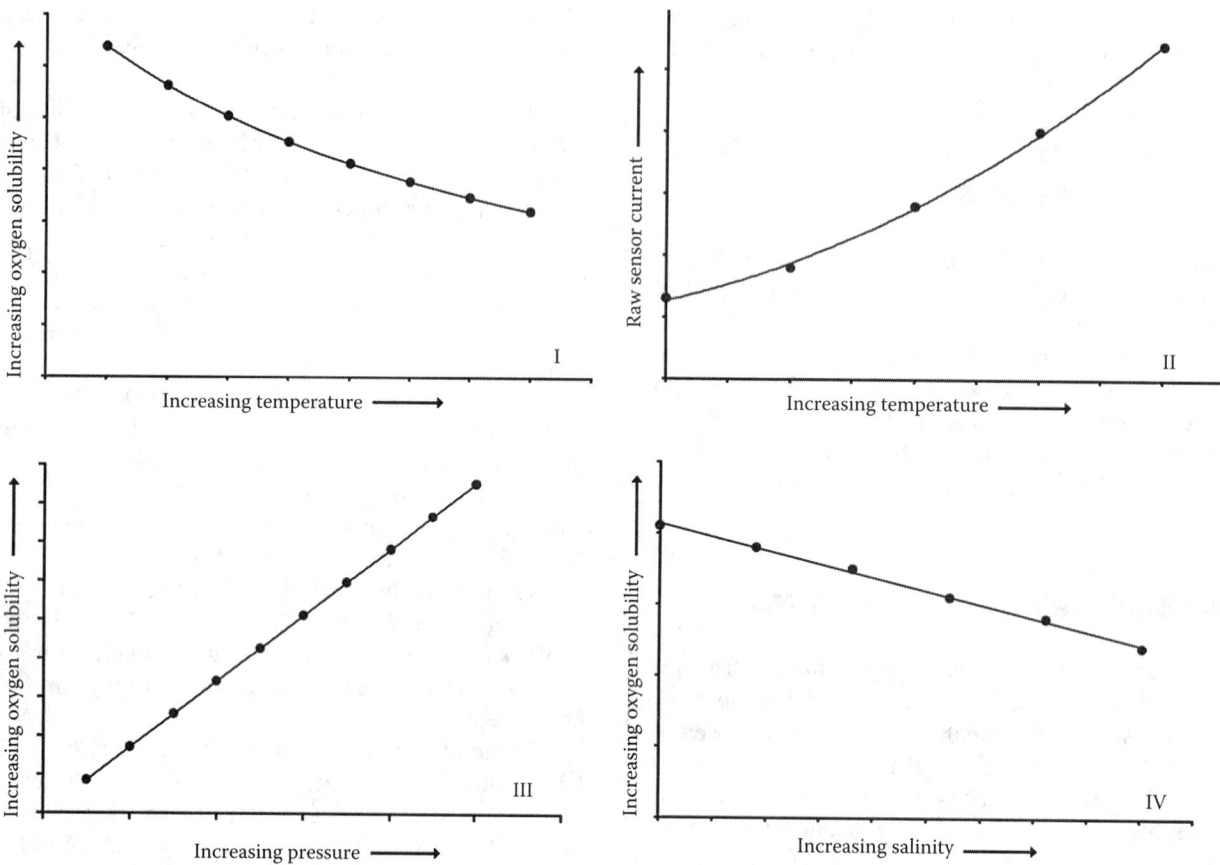

FIG. 1.50i
Variation of (I) dissolved oxygen concentration with liquid temperature, (II) sensor current with temperature, (III) dissolved oxygen concentration with pressure (total pressure or oxygen partial pressure), and (IV) dissolved oxygen concentration with salinity. (Graphs generated from data tabulated in Hitchamn, M. L., Measurement of Dissolved Oxygen (1978) http://www.alibris.com/Measurement-of-Dissolved-Oxygen-Michael-L-Hitchman/book/4255763.)

Figure 1.50i(I). If the user is interested in oxygen concentration values, then a solution temperature change must reflect a change in oxygen concentration on the transmitter display. Practically, all dissolved oxygen meters/transmitters employ an algorithm to determine oxygen concentration based on a solution temperature measurement.

Dissolved oxygen sensors are designed to operate under steady-state oxygen transport across the membrane. Ideally, the membrane permeability should never change after calibration. However, membrane permeability increases with increasing temperature. Hence, the sensor signal will increase with temperature, as shown in Figure 1.50i(II). For an uncompensated sensor, the change in signal with temperature can be 1%–6% per degree Celsius.

Erroneous results will be obtained if we do not account for this. Therefore, all dissolved oxygen sensors employ specific algorithms to correct for this effect. Algorithms are usually manufacturer specific, because membrane materials play an important role in determining the final mathematical form.

Both solubility and membrane permeability factors are accounted for by temperature compensation employed by the measurement loop. Dissolved oxygen sensors have built-in or adjacent temperature measuring devices that are actively monitored by the transmitter.

Pressure Effects

Dissolved oxygen sensors respond to the amount of oxygen that is present in a liquid. Because oxygen is a gas under normal conditions, the amount of oxygen in a liquid is usually expressed as the partial pressure of oxygen (P_{O_2}) or the mole fraction of oxygen (X_{O_2}, equal to 0.209 for air). The two quantities are related through the total gas pressure (P_T) of the system as follows:

$$P_{O_2} = X_{O_2} P_T \qquad \text{1.50(9)}$$

Furthermore, the solubility (concentration) of oxygen dissolved in a liquid at a given temperature and pressure follows a relatively simple relationship known as Henry's law,

$$S_{O_2} = K_H P_{O_2} = K_H X_{O_2} P_T \qquad \text{1.50(10)}$$

where K_H is a constant that depends on both temperature and liquid composition. This equation shows that solubility varies linearly with partial pressure and total pressure (Figure 1.50i(III)).

This is an important relationship, because dissolved oxygen transmitters require the input of a pressure value (barometric or system pressure) during calibration and operation. Some will even provide real-time pressure corrections using input from an external pressure sensor.

Some dissolved oxygen analyzers are suitable for measurements in both the gas- and liquid-phase applications. Additionally, calibration can be performed in either the gas or the liquid phase. In this case, the contribution of water vapor pressure to the total gas pressure (and hence the oxygen partial pressure) must be accounted for so as to achieve the highest possible accuracy. Therefore, some dissolved oxygen transmitters are designed so that the influence of water vapor can be entered as a relative humidity value during calibration and operation.

For example, oxygen sensors can be calibrated in water saturated with air, in air saturated with water vapor (in each case, the relative humidity being essentially 100%), or in ordinary air (in which case, the relative humidity is less than 100%). In each case, the appropriate water vapor (relative humidity) correction factor can be entered into the software.

Salinity Effects

The salinity (i.e., the amount of a salt dissolved in water) is usually expressed in terms of mass of salt per mass of solution or mass of salt per volume of solution (g/1000 g, ‰). In more general terms, this quantity can be expressed as the mass of chloride per volume of solution (g/L). A salinity value is directly obtainable from conductivity measurements.

The solubility of oxygen in water decreases nonlinearly with increasing salinity, as shown in Figure 1.50i(IV). The effect is most noticeable for salinity values greater than 2 g/1000 g. For comparison purposes, salt water has a salinity of approximately 35 g/1000 g. At 20°C and 1 atm of pressure, the solubility of oxygen in salt water is approximately 7.3 mg/L. The solubility of oxygen in pure water at the same temperature and pressure is 9.1 mg/L. Therefore, if salinity compensation were ignored, an error of almost 20% would result.

APPLICATIONS

The applications in which dissolved oxygen sensors are used include biopharmaceutical, water and wastewater, food and beverage, and chemical processing. Each application category dictates slightly different requirements for the sensor and transmitter. Biopharmaceutical applications require that the sensor body be of a hygienic design to allow for cleanability and to resist clean-in-place and sterilize-in-place conditions. Therefore, the sensor bodies are usually polished stainless steel and have no exposed grooves or crevices where matter can accumulate.

Additionally, membranes usually have an outer layer of Teflon to resist the buildup of coating. In contrast, in wastewater applications, the specifications are less stringent, so sensors are usually constructed from polymer materials, have lower temperature and pressure ratings, and are less resistant chemically.

Similar to biopharmaceutical applications, food and beverage usage usually requires that the sensor be of a hygienic design. However, it does not necessarily need to withstand sterilization conditions. Often, in food and beverage applications, the sensor is exposed only to temperatures that are high enough to pasteurize the process liquid.

Chemical process applications require sensor characteristics similar to those for biopharmaceutical and food and beverage applications. In these applications, too, the materials used must withstand higher temperatures and pressures than polymer materials can provide and must have good chemical resistance characteristics. For hazardous area applications, an intrinsically safe sensor must be considered, and manufacturers are usually able to supply intrinsic-safety-related information.

INSTALLATIONS

The specific application will determine whether a housing or other mounting structure is required. Dissolved oxygen sensors other than portable unit types are usually mounted in a support structure that offers a secure mounting point or protection from the process environment.

The mountings used for dissolved oxygen sensors are essentially the same as the ones used for pH sensors (Chapter 1.55). For simple wastewater applications, the sensors are usually mounted in a plastic pipe such that the electrical connections are protected from moisture and debris. In more demanding applications, sensors are mounted in a sturdier (usually stainless steel) housing that is resistant to high temperature, high pressure, and aggressive chemical environments. In these cases, the sensor/housing combinations are most often mounted in a weld-in socket on the process vessel or pipe.

Dissolved oxygen sensors can also be mounted in retractable housings; however, this is usually done only for applications in which the sensor can be cleaned, calibrated, and perhaps sterilized within the housing after being retracted from the process.

SPECIFICATION FORM

When specifying dissolved oxygen detectors or transmitters, the ISA Form 20A2192 can be used. That form is reproduced on the next page with the permission of the International Society of Automation.

Analytical Measurement

DISSOLVED OXYGEN/OZONE/CL2 TRANSMITTER/ANALYZER/MONITOR — Device Specification

RESPONSIBLE ORGANIZATION: ISA

SPECIFICATION IDENTIFICATIONS
7	Document no	
8	Latest revision	Date
9	Issue status	

TRANSMITTER OR ANALYZER
12	Housing type
13	Input sensor type
14	Input sensor style
15	Output signal type
16	Enclosure type no/class
17	Control mode
18	Local operator interface
19	Characteristic curve
20	Digital communication std
21	Signal power source
22	Temperature sensor type
23	Quantity of input sensors
24	Preamplifier location
25	Contacts arrangement — Quantity
26	Integral indicator style
27	Signal termination type
28	Cert/Approval type
29	Mounting location/type
30	Failure/Diagnostic action
31	Sensor diagnostics
32	Data history log
33	Measurement compensation
34	Calibration function
35	Enclosure material

PERFORMANCE CHARACTERISTICS
57	pH compensation LRL	URL
58	Meas 1 accuracy rating	
59	Meas 2 accuracy rating	
60	Temp accuracy rating	
61	Measurement LRL	URL
62	Temp compensation LRL	URL
63	Barometric pressure LRL	URL
64	Minimum span	Max
65	Min ambient working temp	Max
66	Contacts ac rating	At max
67	Contacts dc rating	At max
68	Max distance to sensor lg	

ACCESSORIES
77	Remote indicator style
78	Indicator enclosure
79	Communicator style

SPECIAL REQUIREMENTS
84	Custom tag
85	Reference specification
86	Compliance standard
87	Software configuration
88	Software program

PHYSICAL DATA
93	Estimated weight	
94	Overall height	
95	Removal clearance	
96	Signal conn nominal size	Style
97	Mfr reference dwg	

CALIBRATIONS AND TEST
TAG NO/FUNCTIONAL IDENT	MEAS/SIGNAL/SCALE	INPUT OR TEST LRV	URV	ACTION	OUTPUT OR SCALE LRV	URV
	Meas-Analog output 1					
	Meas-Analog output 2					
	Meas-Digital output					
	Meas-Scale					
	Temp-Scale					
	Temp-Digital output					
	Meas setpoint 1-Output					
	Meas setpoint 2-Output					
	Meas setpoint 3-Output					
	Meas setpoint 4-Output					
	Failure signal-Output					

COMPONENT IDENTIFICATIONS
COMPONENT TYPE	MANUFACTURER	MODEL NUMBER

Rev	Date	Revision Description	By	Appv1	Appv2	Appv3	REMARKS

Form: 20A2192 Rev 0 © 2004 ISA

Definitions

Galvanic (also called voltaic cell is named after Luigi Galvani or Alessandro Volta, respectively, and is an electrochemical cell that derives electrical energy from spontaneous oxidation-reduction (redox) reactions taking place within the cell. It generally consists of two different metals connected by an electrolyte (salt bridge) or individual half-cells separated by a porous membrane.

Nernst equation is named after Walther Nernst the German physical chemist who first formulated it. It relates the concentration detected by a galvanic or polarographic cell and the electric signal they generate. In other words, it gives a formula that relates the numerical values of the concentration gradient, which causes ions to move through a permeable membrane and the corresponding values of the electrical gradient that opposes the movement of that charge.

Permeability constant quantifies the diffusion of molecules, called the permeant, through a membrane or interface, where the permeant moves from high concentration to low concentration across the interface. Permeation can occur through most materials having a crystal structure including ceramics and polymers. Permeability depends on the temperature. The unit of permeation is the darcy (d or D), named after Henry Darcy, which is not an SI unit.

Polarography (also called voltametry) is an electrochemical method of analyzing solutions of reducible or oxidizable substances. It was invented by a Czech chemist, Jaroslav Heyrovský, in 1922. In the polarographic cell, a polarizing voltage is applied to two electrodes that are in contact with an electrolyte, which can be contained inside a porous membrane at the tip of the sensor probe. When the reducible or oxidizable substance molecules diffuse through the membrane and come into contact with one of the electrodes, an electrical current that is in proportion to the amount of the substance in the process stream flows from one electrode to the other. The analyzer output is expressed in percent saturation or in milligrams per liter (mg/L).

Abbreviations

F	Faraday's constant
Pm	Permeability coefficient
SI	Système International d'Unités

Bibliography

ASTM, D888–92 Standard test methods for dissolved oxygen in water, 1996. http://www.astm.org/DATABASE.CART/HISTORICAL/D888-92R96.htm.

CHEMetrics, Inc. Oxygen (dissolved)., 2010. http://www.chemetrics.com/Oxygen+(dissolved).

CHEMetrics, Inc. Oxygen (dissolved): Rhodazine D Method, 2013. http://www.chemetrics.com/products/pdf/oxygen_rhodazined.pdf.

Emerson Process Management. RDO optical dissolved oxygen sensor and analyzer, 2014. http://downloadily.com/docs/rosemount-dissolved-oxygen-sensor.html.

EPA. Dissolved oxygen and biochemical oxygen demand. In *Water Monitoring and Assessment*, 2012. http://water.epa.gov/type/rsl/monitoring/vms52.cfm.

Eutech Instruments. Dissolved oxygen electrodes, 1997. http://www.eutechinst.com/techtips/tech-tips16.htm.

Finesse, LLC. Dissolved oxygen sensor primer. In *Improved Process Measurement & Control*, no date. http://finesse.com/media/121732/Finesse.TruDO.primer.ApNote.pdf.

Hach. Luminescent dissolved oxygen, 2015. http://www.hach.com/PowerLDO.

Hach. Oxygen, dissolved: Indigo Carmine Method, 2013. http://www.hach.com/dissolved-oxygen-accuvac-ampules-lr-pk-25/product?id=7640193980.

Hargreaves, J. A. and Tucker, C. S., 2002. Measuring dissolved oxygen concentration, In *Aquaculture*. Southern Regional Aquaculture Center, 2002. https://srac.tamu.edu/index.cfm/event/getFactSheet/whichfactsheet/167/.

Hitchamn, M. L., Measurement of dissolved oxygen, 1978. http://www.alibris.com/Measurement-of-Dissolved-Oxygen-Michael-L-Hitchman/book/4255763.

ITA News, Dissolved oxygen, 2010. http://www.instrument.org/enews/2010/summer2010/itaenewssummer2010_files/Page435.htm.

Katznelson, R. Dissolved oxygen measurement principles and methods, 2004. http://www.waterboards.ca.gov/water_issues/programs/swamp/docs/cwt/guidance/311.pdf.

Lenntech, Why oxygen dissolved in water is important, 2015. http://www.lenntech.com/why_the_oxygen_dissolved_is_important.htm.

Mills, A. Optical oxygen sensors. Platinum metals, 1997. http://www.technology.matthey.com/article/41/3/115-127/.

Omega, Dissolved oxygen analyzer, 2015. http://www.omega.com/kwbld/dissolvedoxygenanalyzer.html.

Pasco. Dissolved oxygen probe care, 2014. http://www.pasco.com/support/technical-support/technote/techIDlookup.cfm?TechNoteID=497.

Thermo Orion. Polarographic dissolved oxygen probes, 2015. http://www.fondriest.com/thermo-orion-polarographic-dissolved-oxygen-probes.htm.

Wang, H. Y., Transient measurement of dissolved oxygen using membrane electrodes, 1989. http://deepblue.lib.umich.edu/handle/2027.42/28168.

White, A. F., Peterson, M. J., and Solbay, B. D., Measurement and interpretation of low levels of dissolved oxygen in ground water, 1990. http://onlinelibrary.wiley.com/doi/10.1111/j.1745-6584.1990.tb01715.x/abstract.

YSI, Calibration Tips, 2009. https://www.ysi.com/File%20Library/Documents/Tips/DO-6-Series-Sondes-DO-Calibration-Tips-Excerpt.pdf.

YSI. *Dissolved Oxygen Handbook*. 2009, http://cdn2.hubspot.net/hub/122959/file-16709566-pdf/docs/ysi_the_dissolved_oxygen_handbook_w39_0909_web.pdf.

1.51 Ozone in Gas

L. J. BOLLYKY (1974–2017)

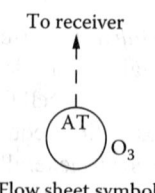

Flow sheet symbol

Types of designs	Photometric, UV, Amperometric, Electrochemical, Acoustic, Chemiluminescence
Applications	Ozone Output, Ozone Feed Gas, Ozone Off-Gas, Ozone Leak Detection, Ambient Ozone
Methods of detection	A. Ultraviolet light absorption A1. Single beam A2. Double beam B. Amperometric C. Thin-film semiconductor D. Acoustic E. Chemiluminescence
Sample flow rate	0.5–5 L/min (0.13–1.3 SCFM)
Ranges	0–100 ppb to 0–2000 ppb (ambient air) 0–2 ppm (leak detector) 0–10 ppm by volume (low concentration) 0%–5% by weight (medium–high concentration) 0%–25% by weight (high concentration)
Sensitivity of reading	0.001 ppm by volume (low concentrations) 0.001% by weight (medium and high concentrations)
Inaccuracy (% of reading)	A. 0.5%–1.0% B. 1%–5% C. 10% or more D. 0.5% E. 0.5%
Cost	A1. $4000–$7500 A2. $3,000–$15,000 B. $1200–$8000 C. $400–$1500 D. $1500 E. $12,500
Partial list of suppliers	Analytical Technology Inc. (ATI) (B) (http://www.analyticaltechnology.com; http://www.analyticaltechnology.com/public/product.aspx?productID = 1067) Anseros GmBH (A1), (A1) (http://www.anseros.de) BMT Messtechnik GmBH (A2) (http://www.bmt-berlin.de) Ebara Jitsugyo Co. (A1, C) (ej-kikaku@pop16.odn.ne.jp) Eco Sensors Inc. Div. of KWJ Eng. (C) (http://www.ecosensors.com)

(Continued)

In USA Inc., Northeast Controls Inc. (A1, C, D) (http://www.nciweb.net/in-usa)
Orbisphere Laboratories, Hach Co. (B) (http://www.orbisphere.com)
OSTI, Inc. distributor of BMT Messtechnik (A2) (http://www.osti-inc.com)
Rosemount Div. of Emerson Process Management (http://www.emersonprocess.com)
Teledyne Advanced Pollution Instrumentation Inc. (A1) (http://www.teledyne-api.com)
WEDECO of Xylem Inc. (A1) (http://www.xyleminc.com)
2B Technologies, An InDevR Co. (A) (http://www.twobtech.com)

New development — Improvements are being provided by the main suppliers (ATI, BMT, etc.) include Auto Calibration, Self-Diagnostics, Auto Sampling, and Auto Ranging.

INTRODUCTION

Ozone is an unstable gas that occurs naturally in the atmosphere. Its concentration is usually below 0.1 ppm at sea level but can be substantially higher concentrations at high altitudes* or in urban areas during the summer, where outdoor ozone can reach 0.2 ppm or more.[†]

At ground level, it is produced by nature through the interaction of sunlight with the oxygen in the air. In the upper atmosphere, it is produced by electrical storms, and in the lower atmosphere by lightning, or by the interaction of sunlight, oxygen, and organic pollutants (such as hydrocarbons) at ground level. The concentration of ozone at ground level (in smog) has been used as an indicator of the extent of pollution in the air.

Ozone is a powerful oxidizing agent and a very strong disinfectant. In industry, it is generated from air or oxygen by means of electrically powered devices called ozone generators. The major uses of ozone include the following:

1. Ozone can be used for the oxidation of pollutants (e.g., organic vapors and odorous materials) and also for disinfection.
2. It is also used for the oxidation of pollutants in water and wastewater treatment, for microflocculation, and for disinfection.
3. In the process industries, ozone is used for the bleaching of pulp paper, as a chemical oxidant, and for decolorization.

---

* Bollyky, L. J., *Ozone in Wastewater Treatment and Industrial Applications*, International Ozone Association, Zurich, Switzerland, 1989, 231–245.

† Bollyky, L. J., *Ozone in Wastewater Treatment and Industrial Applications*, International Ozone Association, Zurich, Switzerland, 1989, 562–592.

Ozone Monitoring

Gas phase ozone monitoring and analysis is most frequently done on air or oxygen streams, serving one of the following purposes:

1. Monitoring atmospheric ozone concentrations
2. Monitoring and controlling the ozone concentration in the output of ozone generators
3. Monitoring ozone concentrations in the off-gas streams of ozone treatment systems
4. Monitoring ambient ozone concentration in ozone generator rooms

For workspace safety, the Occupational Safety and Health Administration limits the allowable ozone concentration to 0.1 ppm. In areas where the general public is present, many guidelines and regulations set the limit at 0.05 or even 0.03 ppm.

Ozone monitoring devices can have "cross sensitivities" with oxides of nitrogen (NOx) and with chlorine compounds. For these reasons, it is recommended that the areas where ambient ozone is monitored not only be isolated from the outdoor air but also be separated from chlorine and hydrocarbon fumes.

The ambient ozone monitors should be installed at the height of the operators' head so as to detect the air that is breathed. The monitor should be located in still air, isolated from the outside air. It also should be kept out of sunlight, because sunlight can create interfering gases as it strikes plastics or hydrocarbons. If the ozone monitor is provided with a high alarm, it should be set just below 0.1 ppm, or the required limit.

After the ozone monitor is installed, its calibration should be checked on a quarterly basis. If the area of operation is dusty, wet, or potentially damaging in any other way, the calibration should be checked monthly.

Monitor Designs

Several methods have been used for the analysis and monitoring of ozone concentration in gas streams.*,† The most commonly used ozone monitors employ ultraviolet (UV) spectroscopy as the method of detection (see Chapter 1.69).

Other methods include the amperometric method, in which an electrochemical cell is isolated from the sample gas stream by a gas-permeable but liquid-impermeable membrane, similar to the ones used in dissolved oxygen measurement (Chapter 1.50). This method permits the use of the same monitor for both gas and liquid phase measurements.

The chemiluminescence method is used for ambient air ozone monitoring to eliminate potential interference by fine particles, mercury, organic compounds, and other substances that are known to affect the accuracy of photometric analyzers.

For ozone leak detection applications, the small and portable thin-film semiconductor-type monitor is well suited.

ULTRAVIOLET ANALYZER

Ozone is a strong UV absorber at the wavelength of 253.7 nm. It is used for UV ozone analyses. These instruments are available in a wide variety of complexity and sophistication, as is discussed in some detail in Chapter 1.69 (Figure 1.51a).

The generally accepted value of the ozone absorption coefficient is 3000 ± 30 M^{-1} cm^{-1} at 273 K and at 1 atm.‡ The UV measurement is an absolute method for the detection of ozone in the gas phase, because it does not need to be calibrated by using some other type of sensor.

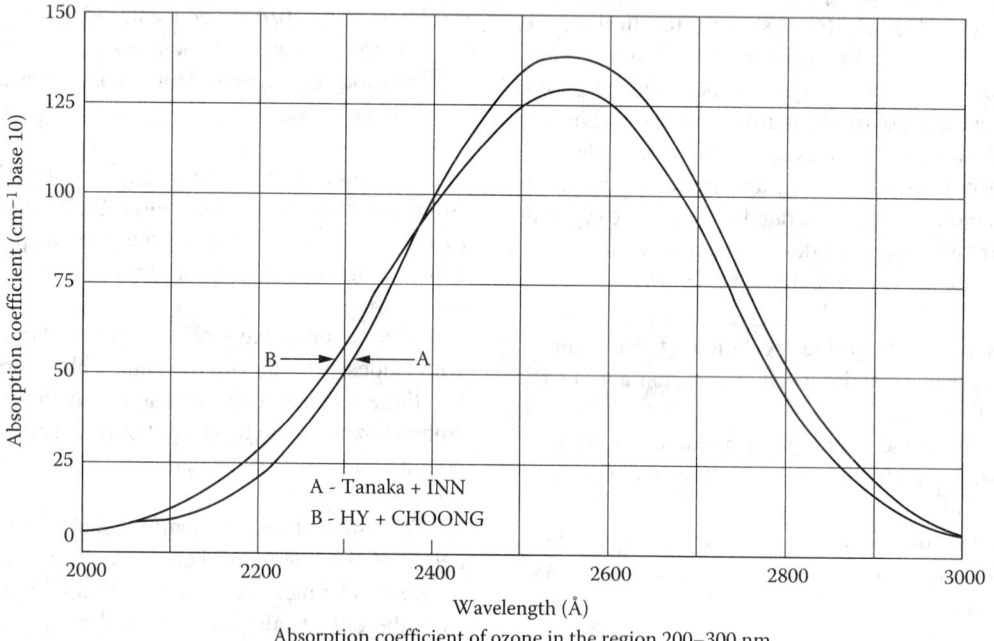

FIG. 1.51a
Absorption coefficient of ozone in the region 200–300 nm or 2000–3000 Angstrom (Å).

---

* Rice, R. G., Bollyky, L. J., and Lacy, W. J., *Analytical Aspects of Ozone Treatment of Water and Wastewater*, Lewis Publishers, Boca Raton, FL, 1986.
† Bollyky, L. J., *Ozone in Wastewater Treatment and Industrial Applications*, International Ozone Association, Zurich, Switzerland, 1989, pp. 562–592.

‡ American Water Works Association Research Foundation, Companie Generale des Eaux, *Ozone in Water Treatment Application and Engineering*, Lewis Publishers, Boca Raton, FL, 1991, 98–100.

Single-Beam Design

The single-beam design uses one UV light absorption cell. Figure 1.51b illustrates both a ppm and a percent concentration design. In these designs, every 5–20 sec, the sample gas stream and the ozone-free reference gas stream are alternately passed through the absorption cell. The UV light detector measures the intensity of the UV radiation as it leaves the sample cell.

The difference between the UV intensities that are detected when the sample is flowing and when the ozone-free reference sample is used is the basis for computing the ozone concentration in the sample. The computation is based on the Bee–Lambert law below:

$$A = abc = \log_{10} \frac{I_0}{I} \quad\quad 1.51(1)$$

where

A is the absorbance

I is the UV light intensity-reaching detector with sample in beam path

I_0 is the UV light intensity-reaching detector with no sample in beam path

a is the absorption coefficient of pure component of interest at analytical wavelength; the units depend on those chosen for b and c

b is the sample path length

c is the concentration of sample component

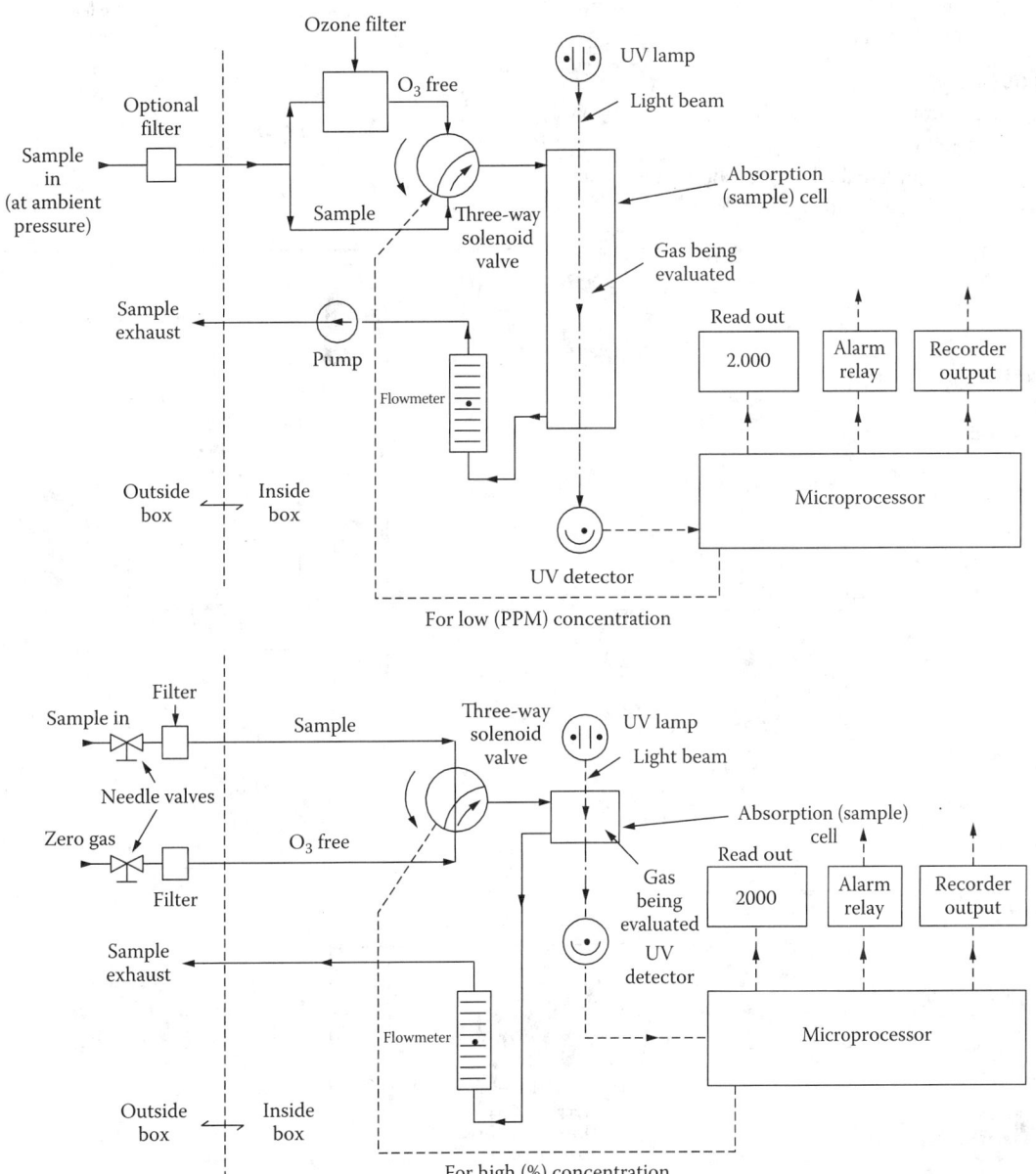

FIG. 1.51b
Single-beam ultraviolet analyzer can be used for both ppm and percent concentration detection of ozone.

Double-Beam Design

The double-beam monitor employs two UV light absorption cells. One monitors the sample stream and the other the ozone-free reference gas stream. Because this design does not require the frequent switching of gas streams, it requires less maintenance. It is also capable of correcting for monitor *cross sensitivity* to the presence of other gases in the sample gas streams and for dirt buildup on the windows.

Both types of UV monitors are available with pressure and temperature compensation. UV measurement is usually specific to ozone, because components normally present in the atmosphere or in the off-gases of ozone water treatment systems do not absorb UV at the 253.7 nm wavelength. Naturally, if ozone is being detected in an atmosphere that contains NOx or chlorine, the UV measurement will not be specific, and compensation is required.

Principal of Operation

Low-pressure Hg lamp is used as UV light source at 253.7 nm (Hartley Band).

The ozone concentration is calculated as shown in Figure 1.51c (also refer Figure 1.51d).

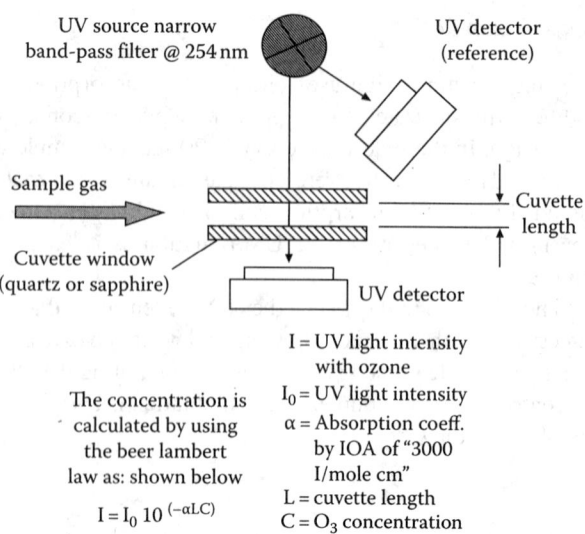

The concentration is calculated by using the beer lambert law as: shown below

$$I = I_0 \, 10^{(-\alpha L C)}$$

I = UV light intensity with ozone
I_0 = UV light intensity
α = Absorption coeff. by IOA of "3000 I/mole cm"
L = cuvette length
C = O_3 concentration

FIG. 1.51c
UV ozone monitor operation.

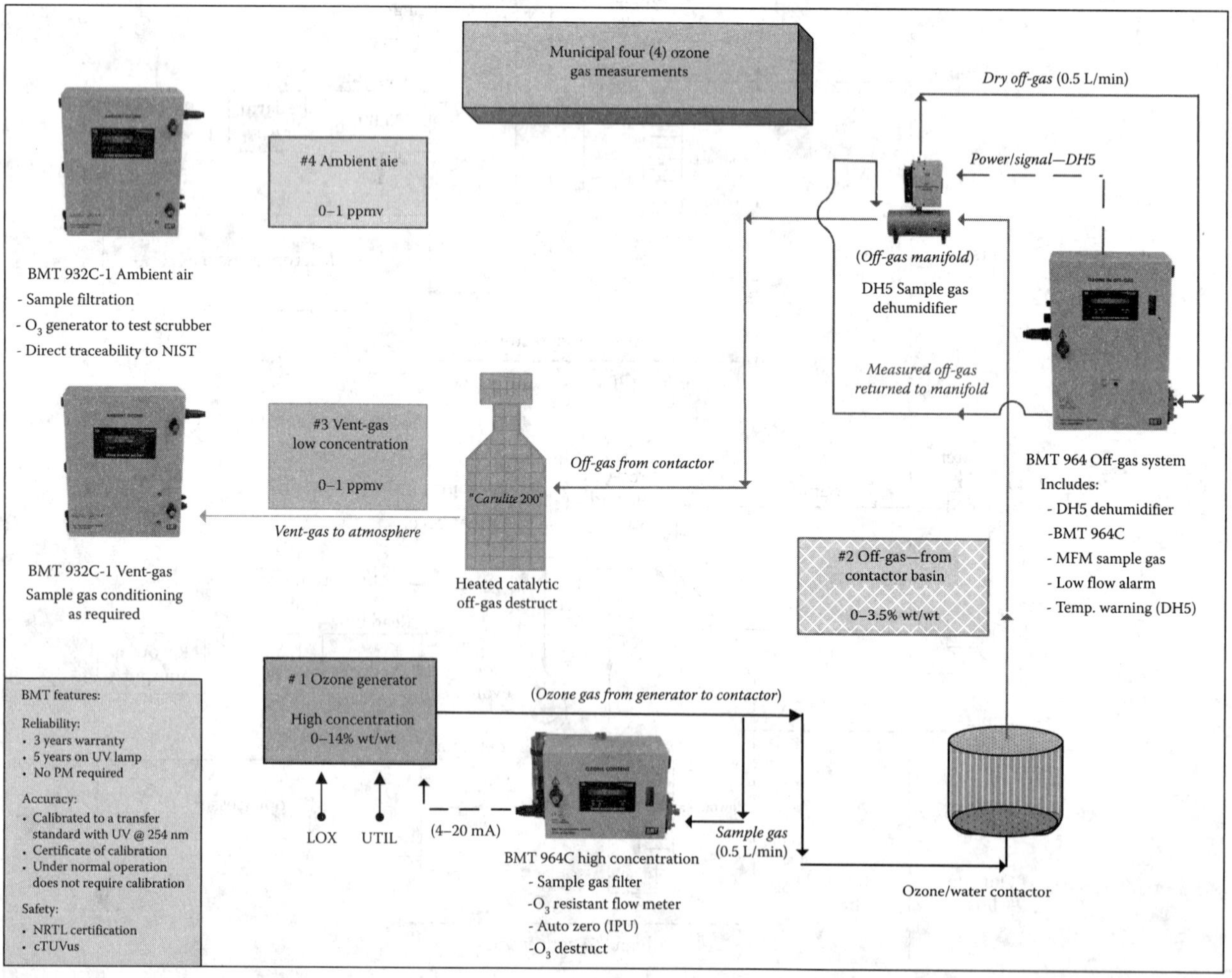

FIG. 1.51d
Four UV ozone analyzers in an ozone treatment municipal water plant, BMT Models 932C, 964, and 964C. (Courtesy of Osti, Inc.)

AMPEROMETRIC DESIGN

Figure 1.51e illustrates an amperometric electrochemical cell.* The cathode is gold, the anode is silver, and the electrolyte is aqueous potassium bromide (similar to Figure 1.51f). The cell is isolated from the sample by a gas-permeable, water-impermeable perfluoroalkoxy (PFA) membrane. The cathode and anode reactions are as follows:

$$\text{Cathode: } O_3 + H_2O + 2e^- = O_2 + 2OH^- \quad \mathbf{1.51(2)}$$

$$\text{Anode: } Ag + KBr = AgBr + K^+ + e^- \quad \mathbf{1.51(3)}$$

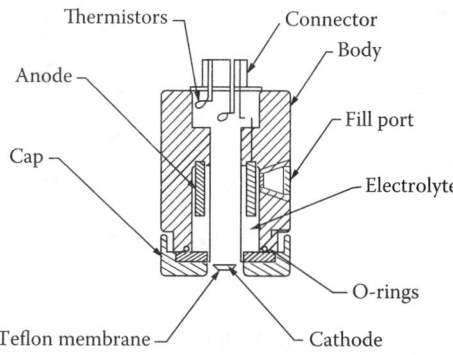

FIG. 1.51e
Membrane-type amperometric probe.

FIG. 1.51f
Electrochemical ozone sensor, ECO Model SM-EC. (Courtesy of KWJ Engineering, Inc.)

---

* Clark, L. C., Electrochemical Device for Chemical Analysis, U.S. Patent No. 2913386.

Calibration

The instrument can be calibrated by chemical methods,† against UV-type monitors, or by the air calibration method. The latter method is based on the fact that the ratio of permeation rates of any two gases is constant. Therefore, air calibration involves the determination of the current from the cell when the power supply is −0.7 V and the cell is in the air (which has a known oxygen concentration).

Once this "air current" is determined, the ozone concentration in the gas sample can be computed from the "ozone current" signal of the sample measured at 0.25 V and from the known O_2/O_3 permeation ratio. This air calibration method not only renders the monitor more convenient to use, but it also makes it less accurate relative to UV or chemical reference calibrations. The inaccuracy of the monitor is about 5% when air calibration is used.

THIN-FILM SEMICONDUCTOR DESIGN

This monitor uses a sensor consisting of an alumina substrate sandwiched between a thin-film platinum heater and a thin-film platinum electrode (Figure 1.51g). The electrode is coated with a thin film of semiconductor.

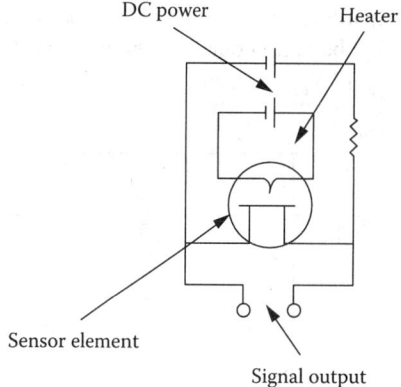

FIG. 1.51g
Thin-film semiconductor ozone monitor.

During measurement, the semiconductor is kept at an elevated temperature by the heater. Because of the adsorption and decomposition of ozone, the resistance of the semiconductor varies substantially in the presence of ozone. The change in resistance is measured, and a corresponding

---

† Rice, R. G., Bollyky, L. J., and Lacy, W. J., *Analytical Aspects of Ozone Treatment of Water and Wastewater*, Lewis Publishers, Boca Raton, FL, 1986.

FIG. 1.51h
Shirtpocket-size ozone sensor. (Courtesy of ECO Sensors Inc.)

voltage signal is generated that is linearly proportional to the ozone concentration.

The unit is available with a 0–10 ppm range and a 0.02 ppm sensitivity. It is a portable ozone monitor that is available with a belt clip pouch. It weighs about 6 oz and can be powered by 12 V_{dc} batteries (Figure 1.51h).

CHEMILUMINESCENCE DESIGN

This monitor is based on the measurement of chemiluminescent light emission from the reaction of ozone with nitrogen oxide, NO, reagent feed gas (Figure 1.51i).

$$NO + O_3 = NO_2 + h\nu + O_2$$

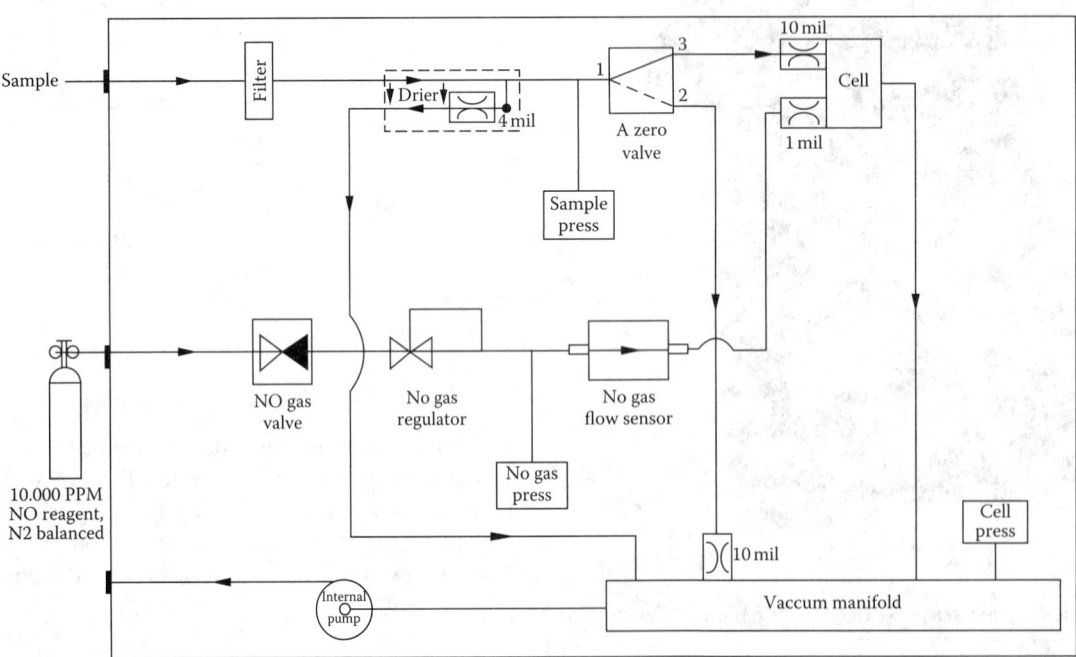

FIG. 1.51i
Chemiluminescence ozone monitor.

SPECIFICATION FORMS

When specifying analyzers for the measurement of ozone in gases, use the ISA form 20A1001, and when the type of analyzer has already been selected and both the analyzer and the requirements of the process application are to be specified, use the ISA form 20A1002. The following are the forms that are reproduced with the permission of the International Society of Automation.

1	RESPONSIBLE ORGANIZATION	ANALYSIS DEVICE	6	SPECIFICATION IDENTIFICATIONS		
2			7	Document no		
3		Operating Parameters	8	Latest revision	Date	
4			9	Issue status		
5			10			
11	ADMINISTRATIVE IDENTIFICATIONS		40	SERVICE IDENTIFICATIONS Continued		
12	Project number	Sub project no	41	Return conn matl type		
13	Project		42	Inline hazardous area cl	Div/Zon	Group
14	Enterprise		43	Inline area min ign temp	Temp ident number	
15	Site		44	Remote hazardous area cl	Div/Zon	Group
16	Area	Cell Unit	45	Remote area min ign temp	Temp ident number	
17			46			
18	SERVICE IDENTIFICATIONS		47			
19	Tag no/Functional ident		48	COMPONENT DESIGN CRITERIA		
20	Related equipment		49	Component type		
21	Service		50	Component style		
22			51	Output signal type		
23	P&ID/Reference dwg		52	Characteristic curve		
24	Process line/nozzle no		53	Compensation style		
25	Process conn pipe spec		54	Type of protection		
26	Process conn nominal size	Rating	55	Criticality code		
27	Process conn termn type	Style	56	Max EMI susceptibility	Ref	
28	Process conn schedule no	Wall thickness	57	Max temperature effect	Ref	
29	Process connection length		58	Max sample time lag		
30	Process line matl type		59	Max response time		
31	Fast loop line number		60	Min required accuracy	Ref	
32	Fast loop pipe spec		61	Avail nom power supply		Number wires
33	Fast loop conn nom size	Rating	62	Calibration method		
34	Fast loop conn termn type	Style	63	Testing/Listing agency		
35	Fast loop schedule no	Wall thickness	64	Test requirements		
36	Fast loop estimated lg		65	Supply loss failure mode		
37	Fast loop material type		66	Signal loss failure mode		
38	Return conn nominal size	Rating	67			
39	Return conn termn type	Style	68			
69	PROCESS VARIABLES	MATERIAL FLOW CONDITIONS	101	PROCESS DESIGN CONDITIONS		
70	Flow Case Identification	Units	102	Minimum	Maximum	Units
71	Process pressure		103			
72	Process temperature		104			
73	Process phase type		105			
74	Process liquid actl flow		106			
75	Process vapor actl flow		107			
76	Process vapor std flow		108			
77	Process liquid density		109			
78	Process vapor density		110			
79	Process liquid viscosity		111			
80	Sample return pressure		112			
81	Sample vent/drain press		113			
82	Sample temperature		114			
83	Sample phase type		115			
84	Fast loop liq actl flow		116			
85	Fast loop vapor actl flow		117			
86	Fast loop vapor std flow		118			
87	Fast loop vapor density		119			
88	Conductivity/Resistivity		120			
89	pH/ORP		121			
90	RH/Dewpoint		122			
91	Turbidity/Opacity		123			
92	Dissolved oxygen		124			
93	Corrosivity		125			
94	Particle size		126			
95			127			
96	CALCULATED VARIABLES		128			
97	Sample lag time		129			
98	Process fluid velocity		130			
99	Wake/natural freq ratio		131			
100			132			
133	MATERIAL PROPERTIES		137	MATERIAL PROPERTIES Continued		
134	Name		138	NFPA health hazard	Flammability	Reactivity
135	Density at ref temp	At	139			
136			140			
Rev	Date	Revision Description	By	Appv1 Appv2 Appv3	REMARKS	

Form: 20A1001 Rev 0

© 2004 ISA

1	RESPONSIBLE ORGANIZATION	ANALYSIS DEVICE COMPOSITION OR PROPERTY Operating Parameters (Continued)	6	SPECIFICATION IDENTIFICATIONS		
2	ISA		7	Document no		
3			8	Latest revision	Date	
4			9	Issue status		
5			10			

	PROCESS COMPOSITION OR PROPERTY			MEASUREMENT DESIGN CONDITIONS				
11								
12	Component/Property Name	Normal	Units	Minimum	Units	Maximum	Units	Repeatability
13								
14								
15								
16								
17								
18								
19								
20								
21								
22								
23								
24								
25								
26								
27								
28								
29								
30								
31								
32								
33								
34								
35								
36								
37								

38
39
40
41
42
43
44
45
46
47
48
49
50
51
52
53
54
55
56
57
58
59
60
61
62
63
64
65
66
67
68
69
70
71
72
73
74
75

Rev	Date	Revision Description	By	Appv1	Appv2	Appv3	REMARKS

Form: 20A1002 Rev 0

Abbreviations

nm   Nanometer
ppb   Parts per billion
ppm   Parts per million

Organizations

IOA   International Ozone Association
IUVA   International UV Association
OS&E   Ozone Science & Engineering

Bibliography

Christopher, R., Accuracy of high concentration ozone photometry, *IOA-UVA World Congress*, Paris, France, 2011.

Ciufia, V., An alternative method for the verification of high concentration ozone gas measurement (UV photometry vs KI), *Proceedings, IOA—IUVA World Congress*, Las Vegas, NV, 2013, pp. 148–149.

Hawe, E., Ozone detection using an integrating sphere as an optical absorption cell, *Journal of Physics: Conference Series*, 76, 012041, 2007.

O'Keeffe, S., Fitzpatrick, C., and Lewis, E., An optical fibre based UV and visible absorption spectroscopy system for ozone concentration monitoring, *Sensors and Actuators B*, 125, 372–378, 2007.

Ozone monitoring, Mapping and Public Outreach, Chapter 3 in *Air Now*, U.S. Environmental Protection Agency, Washington, DC, 2001. http://www.epa.gov/ttn.

TCE Marcus, Investigation of ozone concentration measurement by visible photo absorption method, *OS&E*, 35(3), 229–239, 2013.

Wallner, F., High concentration photometric ozone measurement primer, *Ozone News*, 35(1), 20–26, 2007.

1.52 Ozone in Water

L. J. BOLLYKY (1974–2017)

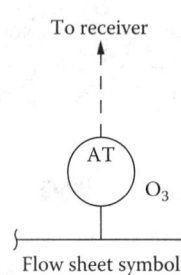

Flow sheet symbol

Type of design	Photometric, UV, Amperometric, Electrochemical
Applications	Ozone Treatment of Drinking Water, Wastewater, High Purity Water, Industrial Ozone Treatment Process Control, Ozone Contactor (Mass Transfer) Process Monitoring and Control
Methods of detection	A. Amperometric (Drinking Water, Wastewater, High Purity Water, Industrial Processes) A1. With bare electrodes A2. With membrane B. Stripping and gas phase detection (Drinking Water Contaminated with Fe and Mn) C. Ultraviolet absorption (High Purity Water, Clean Drinking Water) D. Colorimetric (Water Sample Testing)
Sample requirements	Flow: 100–500 cm^3/min for most, but type B requires 1.5 L/min Types C and D cannot tolerate the presence of suspended matter
Sensitivity (ppm = mg/L)	A1 and A2. 0.01 ppm B and C. 0.001 ppm D. 0.05 ppm
Ranges (ppm = mg/L)	A1 and B. 0–2 ppm A2. 0–200 ppm C. 0–5 ppm D. 0–1.5 ppm
Inaccuracy (ppb or percent of reading)	A1. 1 ppb A2. 5 ppb to 1% B. 1% C. 3% D. 1.5%
Costs	A1. $8000 A2. $2500–$8000 B. $5500–$8500 C. $3000–$5000 D. $300–$3000
Partial list of suppliers	Analytical Technology Inc. (ATI) (A2) (http://www.analyticaltechnology.com/public/productlist.aspx?CategoryID=1002; http://www.ozonesolutions.com/files/product/Q-45H/Q-45H_Manual.pdf) Anseros GmBH. (A1), (B) (http://www.anseros.de) BMT Messtechnik GmBH (A2) (http://www.bmt-berlin.de) Eco Sensors Inc. Div. of KWJ Eng. (B) (http://www.ecosensors.com) In USA Inc., Northeast Controls Inc. (B and C) (http://www.nciweb.net) OSTI, Inc. distributor of BMT Messtechnik (C) (http://www.osti-inc.com) Teledyne Advanced Pollution Instrumentation Inc. (C) (http://www.teledyne-api.com/Tseries_compound.asp)
New developments	Improvements are being provided by the main suppliers (ATI, BMT, etc.) include Auto Calibration, Self Diagnostics, Auto Sampling, and Auto Ranging.

INTRODUCTION

Dissolved ozone concentrations can be detected from as low as ppb to as high as 200 g/m^3. Ozone monitoring is used (1) for dissolved ozone measurement in water treatment, (2) in ozone treatment of wastewater, (3) for ozone contactor monitoring, and (4) for industrial ozone treatment process monitoring in general.

Ozone is primarily used for the treatment of water, wastewater, and water slurries such as paper pulp.*,†,‡ For these applications, ozone is generated from air or from oxygen by electrically powered ozone generators. The output of a typical ozone generator contains 1%–14% ozone by weight.

The ozone is transferred into the water in *ozone contactors*, which act as mass transfer devices as well as reactors for the treatment. The efficiency of the mass transfer in the contactor is monitored by measuring the ozone concentrations in the gas phase (see Chapter 1.51) at both the contactor's inlet and discharge. The dissolved ozone concentration in the ozone contactor is an indication of the progress of the treatment reaction or of disinfection.

In most water and wastewater treatment applications, when the ozone is generated from air, its solubility under ambient conditions is approximately 4 mg/L. When the ozone is generated from oxygen, its solubility is 15 mg/L.§

In these applications, the ozone is used for disinfection, chemical oxidation, or microflocculation. In paper pulp applications and in some wastewater treatment processes, it is also used for decolorization, bleaching, and disinfection.

Drinking Water Disinfection

The Surface Water Treatment Rules of the U.S. EPA¶ specify the use of CT values as the standard for drinking water disinfection using ozone or other chemical disinfectants. The CT value is the product of ozone (or disinfectant) concentration (C) in mg/L and the time of exposure of the water to the disinfectant in minutes (T). Therefore, to ascertain that the disinfection meets the U.S. EPA standard, the ozone concentration in the water (C) must be accurately monitored.

---

* Bollyky, L. J., *Ozone in Water Treatment*, International Ozone Association, Zurich, Switzerland, 1989.
† Bollyky, L. J., *Ozone in Wastewater Treatment and Industrial Applications*, International Ozone Association, Zurich, Switzerland, 1989.
‡ American Water Works Association, Research Foundation, Companie Generale des Eaux, *Ozone in Water Treatment, Application and Engineering*, Lewis Publishers, Boca Raton, FL, 1991.
§ Perry, R. H. and Green, D., *Perry's Chemical Engineer's Handbook*, 6th edn., McGraw-Hill, New York, 1984.
¶ Uman, M. F., National academy of engineering, EPA national primary drinking water regulations, *Federal Register*, 52(212), November 3, 1987.

The methods in use for the monitoring and analysis of ozone in water solutions** fall into two categories:

1. *Methods specific to ozone*. These methods are able to determine the ozone concentration with no or very little interference from the other chemicals. These methods are suitable for monitoring the progress of disinfection or of chemical oxidations. They include the indigo method, the amperometric-membrane method, the stripping and gas phase detection method, and the UV method (for deionized high-purity water).
2. *Methods that are not specific to ozone*. These methods determine the concentration of total oxidants and, if no other oxidant is present, they are suitable for controlling ozone generators or for controlling the rate of dosing ozone into the water. These methods include the amperometric-bare electrode method and the potassium iodide chemical method.

AMPEROMETRIC SENSORS

Amperometric analyzers utilize an electrochemical cell for the determination of ozone.†† Two types are in use. In the first design, the bare metal electrodes are in direct contact with the water, whereas in the second design the electrochemical cell is separated from the process water by a semipermeable membrane, as shown in Figure 1.51b.

Bare Metal Electrodes

In the bare-metal electrode design of the amperometric ozone monitor, the water, as it flows through the galvanic cell, serves as the electrolyte. The oxidizing agents present in the water are reduced at the cathode. The generated current is proportional to the concentration of all the oxidants. Therefore, ozone monitors of this type are not specific for ozone. Measurement errors caused by the presence of other oxidants and oxidation by-products can be mitigated by the proper choice of the galvanic electrode pairs (e.g., Au/Cu, Au/Ag, Pt/Cu, and Ni/Cu), and this choice depends on the other oxidants that are expected to be present.

In water treatment plants, when the predominant oxidant is ozone and the concentration of other oxidants is zero or is nearly constant, the rugged and reliable bare-metal electrode monitors can be used effectively. They can control

---

** Rice, G., Bollyky, L. J., and Lacy, W. J., *Analytical Aspects of Ozone Treatment of Water and Wastewater*, Lewis Publishers, Boca Raton, FL, 1986.
†† Clark, L. C., Electrochemical device for chemical analysis, U.S. Patent No. 2,913,386, 1956.

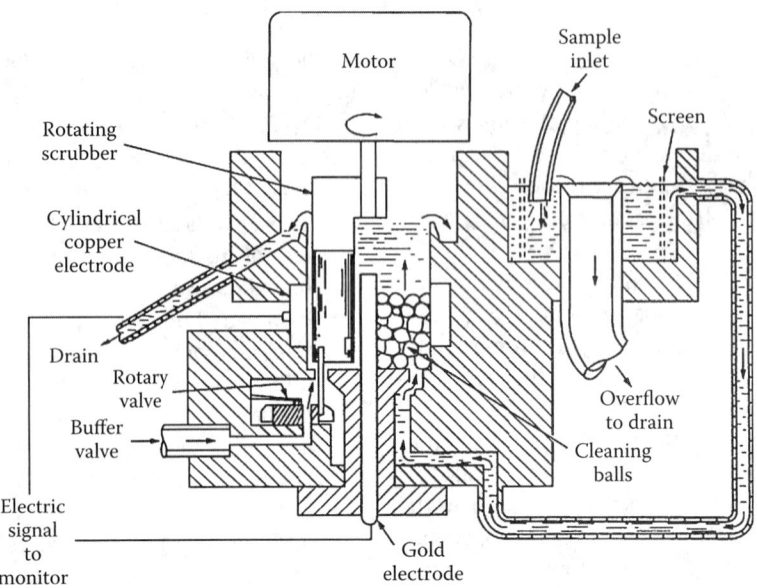

FIG. 1.52a
Amperometric ozone detector.

both the ozone generators and the settings of ozone dosage controllers (Figure 1.52a).

Membrane-Type Designs

Membrane-type amperometric ozone monitors are more selective to ozone, because the membrane keeps most of the other oxidants out of the electrochemical cell. The membrane is permeable only to gas—not to water. These monitors can be employed for the measurement of ozone in both water solutions and in the gas phase (Chapter 1.51). However, when the membrane is semipermeable to water, the monitor can measure ozone only in water solutions—not in the gas phase.

The unit illustrated in Figure 1.52b has a gold cathode, a silver anode, and a measuring cell that can be made out of PVC or 316L SS. It has a measuring range of 0–2 ppm and a repeatability that is the greater of 5 ppb or 5% of measurement. Its response time is 60 sec. It is available with two smart outputs, four dry-contact NO/NC relay outputs, and automatic recognition of its own status (Figure 1.52c).

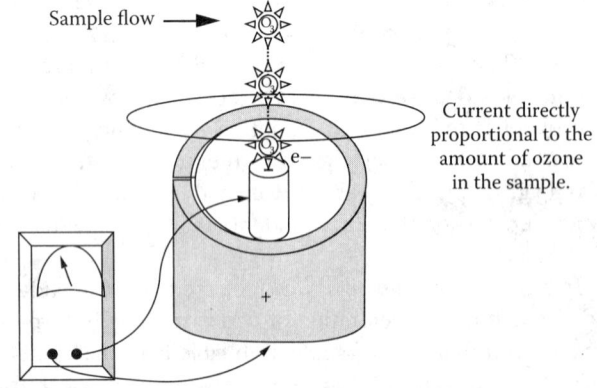

FIG. 1.52b
Polarographic sensor. Process in sensors: (1) Molecules of O_3 pass through the cell membrane. (2) O_3 reduced at the cathode to produce an electron. (3) The electron flows through the measurement cell as an electrical current. (4). The current is proportional to the number of ozone molecules which are directly proportional to the O_3 concentration of the sample. (5) Requires an electrolyte.

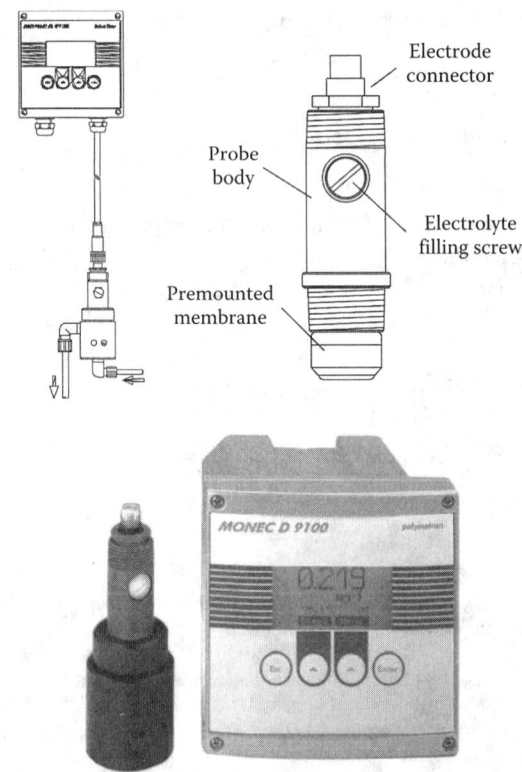

FIG. 1.52c
Online dissolved oxygen monitor for water treatment plants. (Courtesy of Polymerton SA.)

Amperometric ozone analyzers need to be calibrated every 1 or 2 months against some reference method such as the stripping, the UV, or the colorimetric indigo methods, which are described below.

STRIPPING AND GAS PHASE MONITORS

In this method, the dissolved ozone is stripped from the water by the use of inert gases such as nitrogen, oxygen, and air, which are bubbled through a packed-tower-type scrubber. UV monitors are used to analyze the gaseous ozone, which is stripped from the water. This method eliminates the potential interference by oxidants that are present in the water.

The stripping method is cumbersome and is subject to operational problems such as incomplete stripping, ozone decomposition during the stripping, and entrainment of liquid droplets into the gas sample. Therefore, calibration by a more reliable method is desirable. This can be obtained by determining the dissolved ozone concentration in the water prior to stripping by means of the indigo method.

ULTRAVIOLET ABSORPTION

The concentration of dissolved ozone in water can be measured by the absorption of ultraviolet (UV) light at wavelengths from 253.7 to 260 nm. However, there is an uncertainty concerning the molar absorption coefficient of aqueous ozone in natural waters. Values of 2600–3600 M^{-1} cm^{-1} are reported in the literature.*

Another complication is caused by the fact that water from natural sources, such as lakes, rivers, and wells, is likely to contain dissolved organic compounds that have strong UV light absorption at the wavelengths that are used for ozone detection. Therefore, the use of the UV absorption method is recommended only for the measurement of dissolved ozone in deionized, high-purity waters, clean drinking water, and clean process waters. UV absorbance of ozone in pure water is reported to be 2950 M^{-1} cm^{-1} at 258 nm.[†]

COLORIMETRIC METHOD

The best known, most frequently used colorimetric method is the indigo trisulfonate method for the measurement of ozone dissolved in water. This method was developed by Hoigne and Bader from 1979 to 1982.[†] It is considered to be the most selective, sensitive, precise, and fast method among all chemical methods. Interference by manganese ions, chlorine, hydrogen peroxide, and by the decomposition and oxidation products of ozone are all the lowest for this method. The price varies with the complexity and automation of the unit.

THE INDIGO METHOD

The indigo method utilizes the decolorization of the stock solution of blue potassium indigo trisulfonate by dissolved ozone. The disappearance of this blue color is monitored spectrophotometrically at the wavelength of 600 nm. The dissolved ozone concentration is calculated from the difference in light absorption of an untreated blank and of a treated sample, as follows:

$$\text{Dissolved } O_3 = \frac{(\Delta \text{absorption})(100)}{(f)(b)(v)} \quad 1.52(1)$$

where
- b is the path length of the cuvette in cm
- v is the volume of the ozone-containing sample added in mL (normally 90 mL) to raise the total volume of the sample plus the indigo solution to 100 mL
- f = 0.42 ± 0.1 cm^{-1} mg^{-1} L^{-1} indigo, based on molar absorptivity = 20,000 M^{-1} cm^{-1}

The grab sample analysis can be carried out with standard laboratory wares when the method is used for the calibration of other methods.

The indigo method is employed in the "AccuVac Ampul" ozone measurement. For this measurement, all the necessary chemicals are prepackaged under vacuum into a cuvette-ampul device. The sealed glass tip of this device is submerged in the ozone-containing water sample, and the tip is broken open to allow the water to enter the device and dissolve the chemicals therein. The cuvette-ampul is used directly for the spectrophotometric determination of the dissolved ozone.

The indigo method can also be automated by the use of gas-diffusion flow injection.[‡] In this procedure, the dissolved ozone in the sample stream is allowed to diffuse into an indigo solution stream through a 0.45 μm pore size Teflon® membrane. This causes the decolorization of the treated indigo solution stream, which is analyzed spectrophotometrically.

SPECIFICATION FORMS

When specifying analyzers for the measurement of ozone in water, use the ISA form 20A1001, and when the type of analyzer has already been selected and both the analyzer and the requirements of the process application are to be specified, use the ISA form 20A1002. Both forms are reproduced with the permission of the International Society of Automation on the next pages.

---

* American Water Works Association, Research Foundation, Companie Generale des Eaux, *Ozone in Water Treatment, Application and Engineering*, Lewis Publishers, Boca Raton, FL, 1991.
† Bader, H. and Hoigne, J., Determination of ozone in water by the indigo method: A submitted standard method, *Ozone Sci. Eng.*, 4, 169–176, 1982.
‡ Straka, M. R. et al., Residual aqueous ozone determination by gas diffusion flow injection analysis, *Anal. Chem.*, 57, 1799, 1985.

Analytical Measurement

1	RESPONSIBLE ORGANIZATION	ANALYSIS DEVICE	6	SPECIFICATION IDENTIFICATIONS
2			7	Document no
3	ISA	Operating Parameters	8	Latest revision — Date
4			9	Issue status
5			10	

	ADMINISTRATIVE IDENTIFICATIONS			SERVICE IDENTIFICATIONS Continued
11			40	
12	Project number	Sub project no	41	Return conn matl type
13	Project		42	Inline hazardous area cl — Div/Zon — Group
14	Enterprise		43	Inline area min ign temp — Temp ident number
15	Site		44	Remote hazardous area cl — Div/Zon — Group
16	Area	Cell — Unit	45	Remote area min ign temp — Temp ident number
17			46	
18	SERVICE IDENTIFICATIONS		47	
19	Tag no/Functional ident		48	COMPONENT DESIGN CRITERIA
20	Related equipment		49	Component type
21	Service		50	Component style
22			51	Output signal type
23	P&ID/Reference dwg		52	Characteristic curve
24	Process line/nozzle no		53	Compensation style
25	Process conn pipe spec		54	Type of protection
26	Process conn nominal size	Rating	55	Criticality code
27	Process conn termn type	Style	56	Max EMI susceptibility — Ref
28	Process conn schedule no	Wall thickness	57	Max temperature effect — Ref
29	Process connection length		58	Max sample time lag
30	Process line matl type		59	Max response time
31	Fast loop line number		60	Min required accuracy — Ref
32	Fast loop pipe spec		61	Avail nom power supply — Number wires
33	Fast loop conn nom size	Rating	62	Calibration method
34	Fast loop conn termn type	Style	63	Testing/Listing agency
35	Fast loop schedule no	Wall thickness	64	Test requirements
36	Fast loop estimated lg		65	Supply loss failure mode
37	Fast loop material type		66	Signal loss failure mode
38	Return conn nominal size	Rating	67	
39	Return conn termn type	Style	68	

	PROCESS VARIABLES	MATERIAL FLOW CONDITIONS				PROCESS DESIGN CONDITIONS		
69					101			
70	Flow Case Identification			Units	102	Minimum	Maximum	Units
71	Process pressure				103			
72	Process temperature				104			
73	Process phase type				105			
74	Process liquid actl flow				106			
75	Process vapor actl flow				107			
76	Process vapor std flow				108			
77	Process liquid density				109			
78	Process vapor density				110			
79	Process liquid viscosity				111			
80	Sample return pressure				112			
81	Sample vent/drain press				113			
82	Sample temperature				114			
83	Sample phase type				115			
84	Fast loop liq actl flow				116			
85	Fast loop vapor actl flow				117			
86	Fast loop vapor std flow				118			
87	Fast loop vapor density				119			
88	Conductivity/Resistivity				120			
89	pH/ORP				121			
90	RH/Dewpoint				122			
91	Turbidity/Opacity				123			
92	Dissolved oxygen				124			
93	Corrosivity				125			
94	Particle size				126			
95					127			
96	CALCULATED VARIABLES				128			
97	Sample lag time				129			
98	Process fluid velocity				130			
99	Wake/natural freq ratio				131			
100					132			

	MATERIAL PROPERTIES			MATERIAL PROPERTIES Continued
133			137	
134	Name		138	NFPA health hazard — Flammability — Reactivity
135	Density at ref temp	At	139	
136			140	

Rev	Date	Revision Description	By	Appv1	Appv2	Appv3	REMARKS

Form: 20A1001 Rev 0 © 2004 ISA

1.52 Ozone in Water

	RESPONSIBLE ORGANIZATION	ANALYSIS DEVICE COMPOSITION OR PROPERTY Operating Parameters (Continued)		SPECIFICATION IDENTIFICATIONS			
1			6				
2	ISA		7	Document no			
3			8	Latest revision		Date	
4			9	Issue status			
5			10				

	PROCESS COMPOSITION OR PROPERTY			MEASUREMENT DESIGN CONDITIONS					
	Component/Property Name	Normal	Units	Minimum	Units	Maximum	Units	Repeatability	
11									
12									
13									
...									
75									

Rev	Date	Revision Description	By	Appv1	Appv2	Appv3	REMARKS

Form: 20A1002 Rev 0   © 2004 ISA

Abbreviations

cm	Centimeter
cm^3	Cubic centimeter
ppb	Parts per billion
ppm	Part per million

Organizations

IOA	International Ozone Association
IUVA	International UV Association

Bibliography

ASTM, Standard Test Method for the Continuous Measurement of Dissolved Ozone in Low Conductivity Water, http://www.astm.org/Standards/D7677.htm

Binder, D. M. and Bollyky, L. J., Ozone removal of herbicides in drinking water, *Proceedings of 11th Ozone World Congress*, Vol. 1, 5-3-16, San Francisco, CA, 1993.

Bollyky, L. J., Ozone contactor design for drinking water treatment, emerging technology in practice, *Proceedings of AWWA Conference,* Cincinnati, OH, 1990.

Bollyky, L. J., Zebra mussel control with side stream ozonation: Design considerations, ozone for drinking water treatment, *Proceedings of the International Ozone Association Conference*, Vol. 183, Cambridge, MA, 1995.

Bollyky, L. J., Two stage AOP treatment of drinking water at Celina, OH, application and optimization of ozone for potable water treatment, *Proceedings of the International Ozone Association Conference*, Vol. 85, Ottawa, Ontario, Canada, 1996.

Bollyky, L. J., Benefits of ozone treatment for bottled water, *Proceedings of the International Ozone Association, 2002 Pan American Group Conference*, May 18–22, Raleigh-Durham, NC, 2002.

Bollyky, L. J. and Seiler, J., Removal of tributyl phosphate from water by advanced ozone oxidation methods, in *Ozone in Water Treatment*, L. J. Bollyky (ed.), Port City Press, NY, 1989.

Ciufia, V., An alternative method for the verification of high concentration ozone gas measurement (UV photometry vs. KI), *Proceedings IOA—IUVA World Congress*, Las Vegas, NV, pp. 148–149, 2013.

Ciufia, V., Innovations in dissolved O_3 measurement by UV photometry, *Proceedings, IOA—IUVA World Congress*, Las Vegas, NV, 2013.

Joost, R. and Bollyky, L. J., Optimization of ozone contactors for drinking water disinfection, in *Ozone in Water Treatment*, L. J. Bollyky (ed.), Port City Press, NY, 1989.

Mazzei, A. L. and Bollyky, L. J., Optimization mass transfer efficiency and ozone utilization with high efficiency venturi injectors, ozone in the Americas, *Proceedings of the International Ozone Association Conference*, Vol. 197, Toronto, Ontario, Canada, 1991.

Mazzei, A. L., Meyer, R. M., and Bollyky, L. J., Mass transfer of high concentration ozone with high efficiency injectors and degassing separators, ozone for drinking water treatment, *Proceedings of the International Ozone Association Conference*, Vol. 161, Cambridge, MA, 1995.

Mieog, J. et al., Verification of a sidestream ozone residual monitoring system to calculate ozone process CT and associated disinfection credits for a fullscale ozone contactor, *Proceedings, IOA—IUVA World Congress*, Las Vegas, NV, pp. 78–79, 2013.

Misco, T. J., Schiflacqua, M., and Bollyky, L. J., Ozone treatment for cryptosporidium control at Milwaukee, WI, USA, *Proceedings of the 13th Ozone World Congress*, Kyoto, Japan, Vol. 1, pp. 91–97, 1997.

Ozone measurement in potable water, *Water and Wastewater Ind.,* August 2002.

Ozone monitoring, Chapter 3 in *Air Now*, U. S. Environmental Protection Agency, Washington, DC, http://www.epa.gov/ttn.

1.53 Particle Size Distribution (PSD) Monitors

R. A. HERRICK (1974)   **A. WERTHEIMER** (1982)

B. G. LIPTÁK (1995, 2003, 2017)

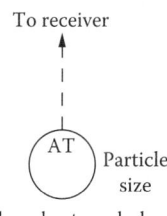

Flow sheet symbol

Measurement technique	A. Optical counting, particle sizing, photoanalysis
	B. Electron microscopy, imaging
	C. Image analysis (with A or B)
	D. Laser diffraction/scattering
	E. Electrical resistance and electrostatic
	F. Sieve analysis, air elutriation
	G. Sedimentation (photo or x-ray)
	H. Acoustic and Ultrasonic attenuation spectroscopy
	I. Bulk property—absorption, permeability
	J. Aerodynamic time of flight
	K. Sampling
Ranges by design	A. For particles suspended in air: over 0.2 μm
	B. 0.006–0.01 μm resolution
	D. Laser diffraction for flour milling applications: 10–5000 μm
	For air-suspended particle applications: 0.02–2000 μm
	Online optical (nonlaser): 2–200 μm
	Online laser diffraction: from nanometers to several millimeters
	E. 0.4–1200 μm (Coulter counter)
	F. Used for coarser particles, >75 μm but, with special designs, down to a few microns
	G. 1–100 μm for photo sedimentation, down to 0.2 μm with x-ray
	H. 25–600 μm
Ranges by application	*For continuous airborne particle counters*: 0–10 million or 0–1 billion
	Particles per cubic foot; ranges are selectable to count particles that are larger than 0.3, 0.5, 1, 3, 5, or 10 μm.
	For clean room atmospheres: down to 0.3 μm in size and down to 1 mg/m^3 in concentration
	For dry powder processes: 3–200 μm
Resolution	*Electron microscopy*: 0.006–0.01 μm
Inaccuracy	5% AR (particle size distribution)
Costs	A, E, G. $6,000–$18,000
	B. $25,000–$100,000
	C. $40,000–$80,000
	D. $15,000–$40,000
	F, I. $1,000–$8,000
	H. $6,000 for regular sludge densitometer, up to $50,000 for size distribution detection with referee method
Partial list of suppliers	Beckman Coulter, (E—Coulter particle counter) http://www.particle.com/wp-content/uploads/downloads/2013/11/Particle-Size-Analyzer-Brochure-Beckman-Coulter-LS-13-320.pdf
	Bühler AG, (D—Flour mill application) http://www.buhlergroup.com/global/en/products/online-psm-particle-size-measurement-myta.htm#.Ur7yuysmth0
	California Measurements Inc., (E, K) http://www.californiameasurements.com/html/product.html
	Climet Instruments Co., (D) http://www.climet.com/

(Continued)

Horiba Instruments, (D) http://www.horiba.com/us/en/semiconductor/products/processes/semiconductor-process/cmp-process/details/la-950-laser-particle-size-analyzer-108/

Loccioni, (H—powder coal application) http://environment.loccioni.com/wp-content/uploads/2013/01/POWdER.pdf

Malvern Instruments, (G, I) http://www.malvern.com/en/products/measurement-type/particle-size/default.aspx?utm_source=search_alliance&utm_medium=cpc&utm_campaign=Measurement%20Type

Norsk Elektro Optikk, (D) http://www.neo.no/products/

OPSIS, (D) http://www.opsis.se/Default.aspx?tabid=87&P_ItemId=21

Qutek, http://www.outotec.com/en/Products—services/Analyzers-and-automation/Particle-size-analyzers/#tabid-2

Shimadzu, (D—Laser diffraction) http://www.ssi.shimadzu.com/products/literature/testing/C060-E005A.pdf

Sympatec GmbH, (D) http://www.sympatec.com/

TSI Products, (A, D, E, J) http://www.tsi.com/Particle-Sizers/

INTRODUCTION

This chapter is devoted to the monitoring of the distribution of particle sizes in industries (such as coal powder or flour milling), in the indoor or outdoor air and/or in various emissions. It does not concentrate on the measurement of solid particles in liquids, because that subject is covered in Chapter 1.68. The coverage of this chapter is also limited to the measurement of solid particle *distribution* only, because the measurement of solid particle sizes is separately discussed in Chapter 1.54.

This chapter first introduces the reader to the field of particle size distribution measurement. After that, the designs of the various laboratory-type particle monitors are described. These units are classified according to sample size into small (a few 0.01 g in the sample), intermediate (a few grams), and large (a few 100 g) samples. After the description of the monitors designed to operate in the laboratory environment, the description of the continuous online particle distribution analyzers will conclude the chapter.

When the purpose of a measurement is to monitor the air in clean rooms, semiconductor production facilities, and other atmospheres, that are kept clean by High-efficiency particulate absorption (HEPA) filters (Figure 1.53a). The analyzers must be able to detect particles down to 0.3 μm in size

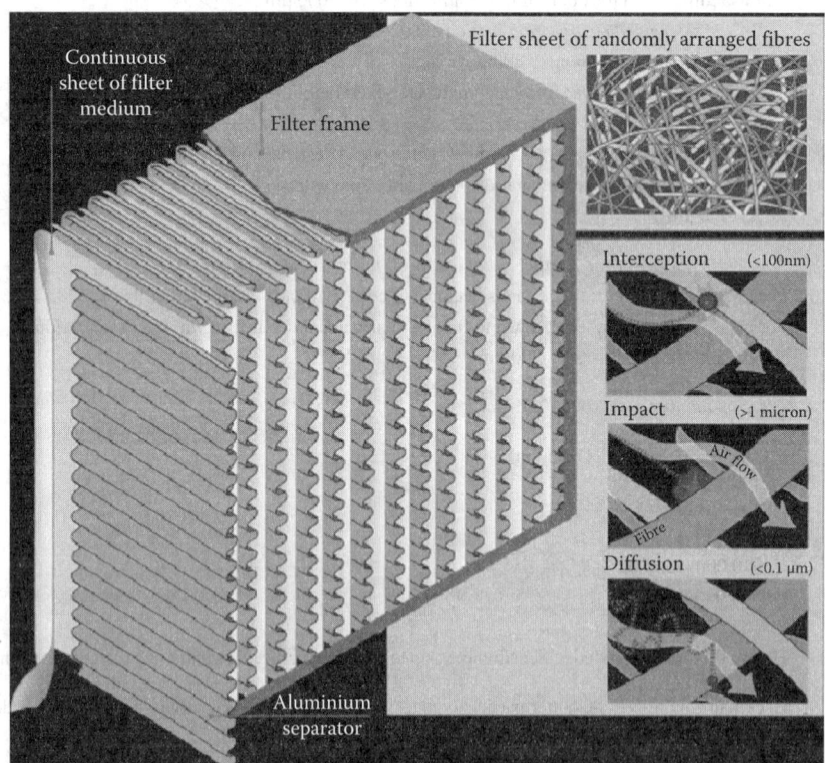

FIG. 1.53a
Functional description of a high absorption particulate arrestance (HEPA) filters.

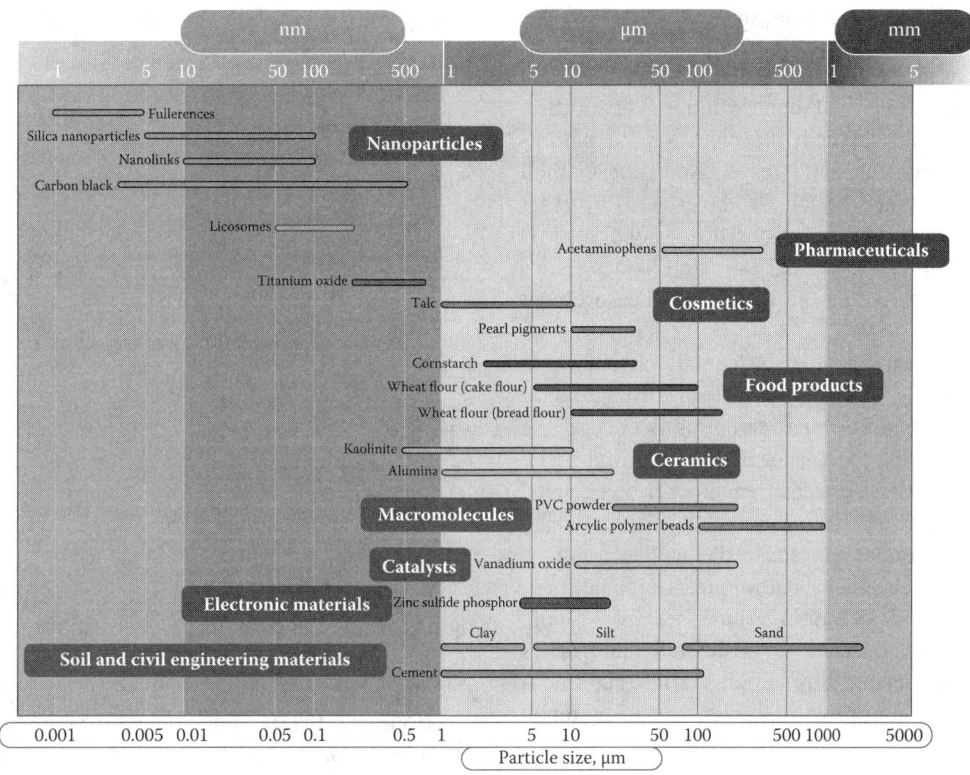

FIG. 1.53b
The range of particle size applications. (Courtesy of Shimadzu Corporation.)

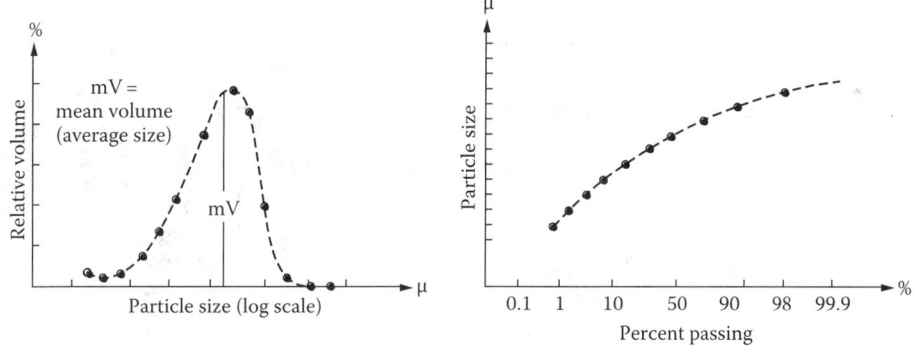

FIG. 1.53c
Typical displays of particle size distribution.

and must be sensitive down to the concentration of 1 mg/m³. The full range of applications is illustrated in Figure 1.53b.

When the purpose of the measurement is product quality control in dry powder processes (e.g., pigments, catalysts, clays, cements, and plastic powders), the analyzers provide data on both particle size distribution and cumulative percentage data in the range of 3–200 μm (Figure 1.53c).

APPLICATION OBJECTIVES

Accurate determination of particle size and distribution is critical in many industrial processes, such as grinding, agglomeration, crystallization, and emulsification. Objectives may include improving final product quality, as in ceramics, paints, and pigments; improving final product performance, as in abrasives and catalysts; improving process efficiency, as in crystallization and wastewater treatment; or minimizing energy consumption, as in grinding of ore for subsequent processing. In many processes, more than one of these objectives may be involved.

This chapter discusses the range of particle size measurement from approximately 0.01 to about 1000 μm. The goal of these measurements is determining the quality of material in a process rather than the detection of contaminants in an otherwise particle-free stream. The performance characteristics noted in this section are attainable from standard commercial instruments.

Particle Characteristics and Sampling

Particle size detectors differ in measurement principles and in size range, resolution, and concentration capabilities. Choosing the right measurement principle for the parameter of interest is important, but, irrespective of the method used, the reported particle size can also be affected by other factors such as shape or physical properties. Therefore, it is important to understand the critical particle characteristics for a particular process before the method of measurement is selected.

Accurate size measurement also depends on obtaining a representative sample of the bulk material. Bias toward smaller or larger sizes can occur if sampling is done at the wrong place or time in the process, or if a sample is extracted from a settled heap. The size measurement will be of value only if the sample represents the process.

The amount of material required for the analysis depends on the measurement technique. In some processes, it can be difficult to obtain a representative sample when only a small fraction of the collected material is inspected. One of the criteria by which particle size measuring techniques are classified is the amount of material inspected.

LABORATORY MEASUREMENTS

From the perspective of sample size, laboratory techniques can use small (a few hundredths of a gram), intermediate (a few grams), or large (hundreds of grams) amounts of sample material. These three categories will be separately discussed in the coming paragraphs.

Small Samples

Optical Microscopy Optical microscopy is often used as a standard method for particle size analysis to calibrate other particle sizing instruments. Size determination is based on the operator's criteria and judgment of parameters such as the longest dimension, the diameter of a sphere of equivalent cross section, and so on. When there is a large range of particle sizes, many particles must be counted to convert a number distribution to a weight distribution.

Size resolution is limited by the wavelength of light, which restricts the practical use to particles that are larger than about 1 μm. The depth of the field, or the region in which an object is in sharp focus, decreases as the magnification is increased. Size errors can arise from out-of-focus imagery, especially with spherical particles.

Electron Microscopy In this case, a fine electron beam is scanned across a sample, producing secondary electrons and other emissions from the material. The emissions are detected and displayed on a cathode ray tube to form an image, such as in a TV picture.

Operator interaction and judgment are required to determine the size. The size resolution is limited by the energy of the electron beam, and 0.006–0.01 μm resolution is typical for modestly priced equipment. Depth of field of the image is about 300 times greater than in the case of optical microscopy. However, a common source of error is in the evaluation of too few fields of particles.

Image Analyzers Automating the analysis of the image of a particle field reduces much of the tedium of counting large numbers of particles. Sophisticated electronic routines, augmented by an editing pointer, provide for rapid and error-free analysis of complex shapes. The input to an image analyzer is a high-resolution TV camera. Therefore, this technique must be combined with another device, such as an optical or electron microscope.

Coulter Reference Method The Coulter Principle is a "Reference Method" for particle size analysis. It is based on measuring the changes in electrical impedance produced by nonconductive particles suspended in an electrolyte. As shown in Figure 1.53d, a small opening (aperture) between electrodes is the sensing zone through which suspended particles pass. In the sensing zone, each particle displaces its own volume of electrolyte and this displaced volume is measured as a voltage pulse; the height of each pulse being proportional to the volume of the particle.

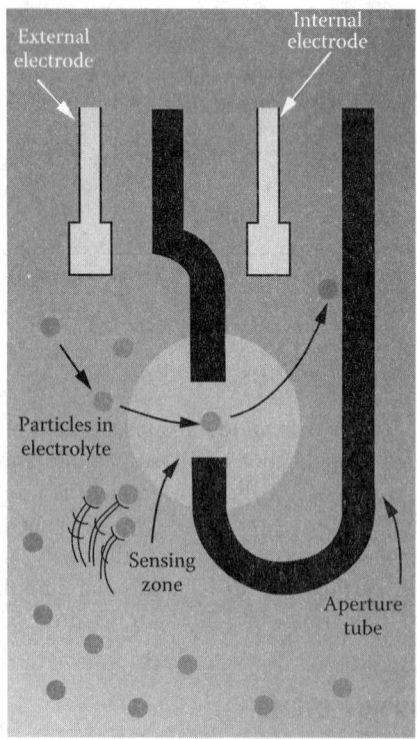

FIG. 1.53d
Illustration of the Beckman Coulter method of particle sizing and counting. (Courtesy of the Coulter Scientific Instrument Division.)

The quantity of suspension drawn through the aperture is precisely controlled to allow the system to count and size particles for an exact reproducible volume. Several thousand particles per second can be individually counted and sized. The results of this method of analysis are independent of particle shape, color, and density.

When this method is used, the sample is suspended in an electrolyte, and the particles are drawn through a small orifice that is immersed between the earlier-mentioned two electrodes and each particle, as it passes through this orifice, produces a momentary resistance change in proportion to its volume.

The pulses are then amplified, scaled, and separated into counting bins. Some level of training and skill is required to obtain good reproducibility, but, with such skill, high resolution can be achieved for narrow distributions. The size range is determined by the orifice diameter. For very broad distributions, blockage of the orifice can be troublesome, and different orifice sizes must be used to cover the full range. The practical lower particle size limit is around 0.5 µm.

Optical Scattering (Single Particle) When a light beam interacts with a particle, light is scattered or diffracted. The angular distribution and amount of scattered light depend on the particle size. The passage of individual particles generates scattered-light pulses, which are detected as they pass a probe beam. These pulses are assigned to counting bins according to their sizes.

The dynamic range of the electronics used will limit the size range, and the measurement accuracy is influenced by the composition of the particles, which are less than a few micrometers in size. The lowest size commonly measured is about 1 µm for a water-suspended material and a few tenths of 1 µm for an air-suspended material. Low signals for small particles and coincidence at high sample concentrations are the principal limitations.

Medium Size Samples

Light Scattering (Multiple Particle) When many particles are viewed by a light beam at the same time, a composite scattering pattern is generated. Several algorithms have been developed to extract the size distribution information from this scattering pattern.

By using low-angle scattering and optical analog filtering, the Microtrac approach generates a signal in proportion to the volume of material present in the light beam. The volume size histogram is calculated through a microprocessor program. By adding signals collected at high angles, which are used for measuring particle sizes less than 1 µm, the range obtained by this technique can cover from 0.1 to 1000 µm.

Resolution is lower than for single-particle analysis. Inaccuracy is a function of composition effects for particles below a few micrometers, but no special training or skill is needed to make the measurement. Materials can be suspended in either air or liquid. Other instruments also exist that utilize intermediate-size samples and operate on principles of forward scattering, but they are limited to particles larger than 1 µm.

Photo Sedimentation Photo sedimentation combines gravitational settling with optical attenuation. The sorting of the particles is based on their hydrodynamic properties, and sizing is based on the cumulative particulate cross section as a function of settling time. Care must be taken to avoid too rapid settling, convection currents, and improper or incomplete correction for optical scattering variations as a function of size and composition.

Photo sedimentation is limited to the size range of 1–100 µm. By replacing the optical beam with a well-collimated x-ray beam, this range can be extended to cover particles of a few tenths of 1 µm in size. Centrifugal sedimentation is also useful in this submicrometer size range.

Large Samples

Sieving This method is best suited to particle sizes that are predominantly coarser than 75 µm. In the sieving operation, the sample is shaken through a series of containers whose bases have regularly spaced, uniform-sized openings. The primary measure is the mass of material retained in each sieve, and the distribution is most often plotted as cumulative fraction vs. the nominal sieve aperture.

Several sieve series exist, including Tyler and the American Society for Testing and Materials (ASTM) in the United States (Table 1.53a) and British and German standards in Europe. Fixed ratios of sizes of square root or fourth root of 2 are common.

TABLE 1.53a
U.S. Sieve and Tyler Sieve Number Designations

U.S. Sieve No. Designation	Tyler Sieve Mesh Designation	Inches	mm
No. 40	4 Mesh	0.185	4.699
No. 8	8 Mesh	0.093	2.362
No. 10	9 Mesh	0.0787	2.0
No. 12	10 Mesh	0.065	1.651
No. 14	12 Mesh	0.0555	1.4
No. 16	14 Mesh	0.046	1.168
No. 20	20 Mesh	0.0328	0.833
No. 30	28 Mesh	0.0232	0.589
40	35 Mesh	0.0164	0.417

For particles smaller than about 75 µm, sieving becomes difficult. Problems include the increased cost of producing sieves with uniform apertures and the greater sophistication required of the mechanics of the sieving process. Wet, sonic, and air-jet sieving have all been used for this range, and, with care, particles down to a few micrometers can be handled.

Optical Methods Laboratories can also use transmissometry (light absorption across a sample), turbidity detection (light scattered at 90° to the light beam), and backscattering techniques, but because all three of these methods are also sensitive to particle size and composition changes, they are best suited for online monitoring of process streams.

Ultrasonic Attenuation When ultrasonic energy is transmitted through a slurry (Figure 1.53c), the amount of energy absorbed is dependent on the particle size, concentration, and ultrasonic frequency. In one approach, two pairs of sensors, each consisting of a transducer and a receiver, operate at two different frequencies.

For a given material, a correlation can be established between the size distribution as measured by a referee method and the attenuation of the higher-frequency beam. Corrections resulting from loading changes are based on the lower-frequency signal. This technique is restricted to particles between 25 and 600 μm, but high concentrations can be accommodated, which makes this approach attractive for online control. Entrained air bubbles can cause errors, so they must be avoided or eliminated.

ONLINE MEASUREMENT

Particle size analysis for industrial process control is especially valuable if the measurement can be rapid and continuous. The fully automated system can be complex, even if the size measurement technique is simple.

Optical Analyzer

As an example of a complete system, Figure 1.53e illustrates the package that includes components designed to extract a sample, measure the particle size through an optical multiple-particle analyzer, and report a specific measurement of the distribution for control.

As shown in Figure 1.53e, in the first step of an analysis cycle, water is charged into a mixing chamber through a debubbling circuit that prevents the air bubbles from the entering water. A circulating pump then agitates the water in the mixing chamber and circulates it through the sample cell, where the laser beam detector of the monitoring section is located.

When the slurry extraction by the sample extractor is actuated, a particulate sample is drawn and deposited into the mixing chamber. The circulating water carries the particles of the slurry through the sample cell. At the end of the measurement cycle, the mixing chamber is drained, rinsed, and refilled with water in preparation for the next sample.

The analyzer is capable of operating with wet slurry or with dry powder samples, and successive samples can be analyzed every few (2–15) min. The sample size is between 3 and 10 g, and the results are printed or displayed on particle

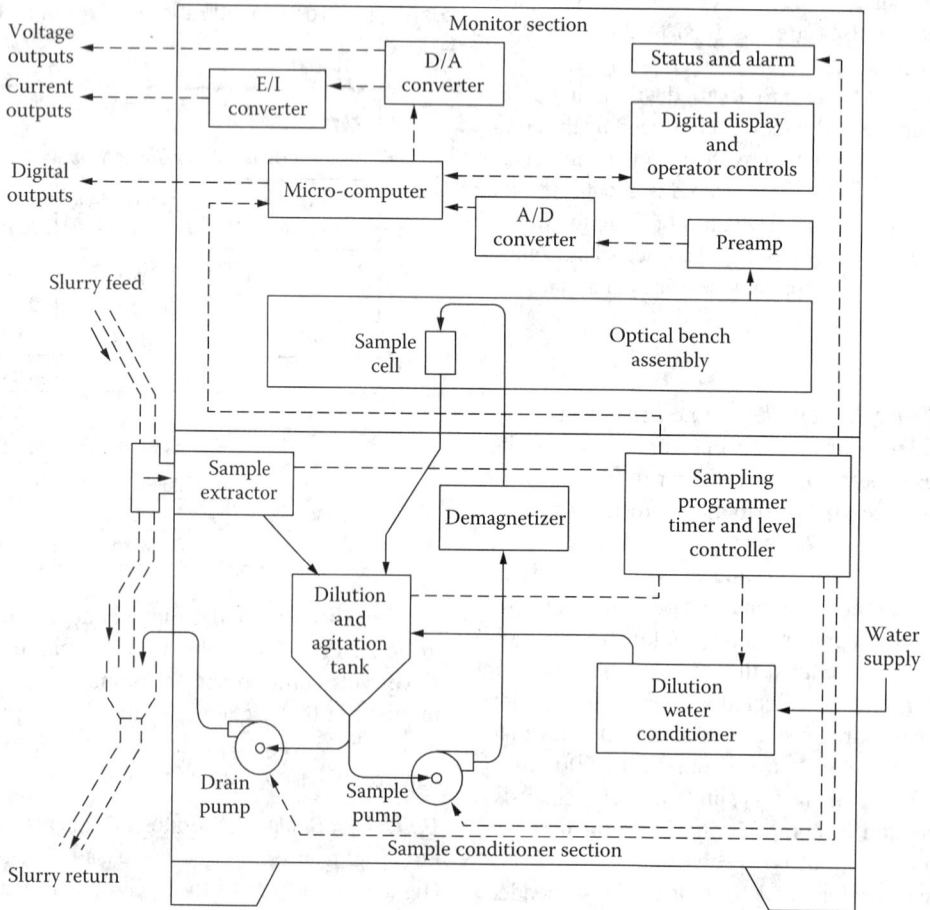

FIG. 1.53e
Block diagram of complete system for online process control using an optical particle analyzer.

size distribution curves (Figure 1.53c). The analyzer consumes about 1.5 gal (6 L) of water per cycle and 4.2 SCFH (120 L/hr) of air. The range of the analyzer is 0–200 μm, with 2.8 μm being the minimum reading.

Airborne-Particulate Counter

Figure 1.53f illustrates the elliptical-mirror-type airborne-particulate counter used in sizing and classifying powdered products or in inspecting the performance of HEPA filters, which are widely used in clean rooms.

The unit operates with a 0.25 SCFM (1 lpm) sample flow rate, which is directed to the focal point of its elliptical mirror, where the focused incident light is scattered by the particles in the sample. The scattered (Tyndall) light is collected by the elliptical mirror and is focused onto the photomultiplier detector.

Because the mirror completely surrounds the particles in the sample, it collects more light energy and therefore is capable of detecting smaller and fewer particles. The minimum particle size it is capable of detecting is about 0.3 μm, and the unit can operate at particle densities of up to 10 million particles/ft^3 (353 million particles/m^3).

The air sampling system includes a positive displacement pump and a pressure regulator (Figure 1.53d). The return air from the sensing chamber is filtered by an absolute filter and is reused as purge air. This purge air fills the sensor chamber and also surrounds (sheathes) the sample airstream to prevent the particles in the sample from entering the sensor chamber.

PARTICLE DISTRIBUTION MONITORS

Laser Diffraction

Today, laser diffraction is widely used in industrial online particle distribution monitoring. When this technique is used, the particle size is measured not by detecting the intensity of light but by evaluating the changes in the light pattern that varies according to particle size. Figure 1.53g illustrates how the pattern of the diffracted light intensity rises as the particle size drops.

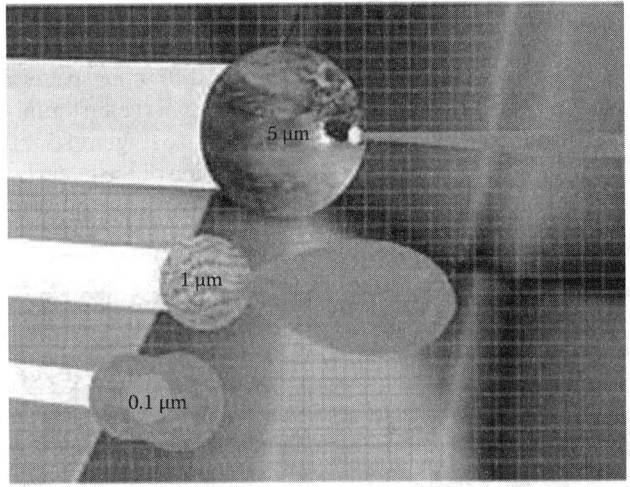

FIG. 1.53g
The diffracted light intensity drops as the particle size increases. (Courtesy of Schimadzu.)

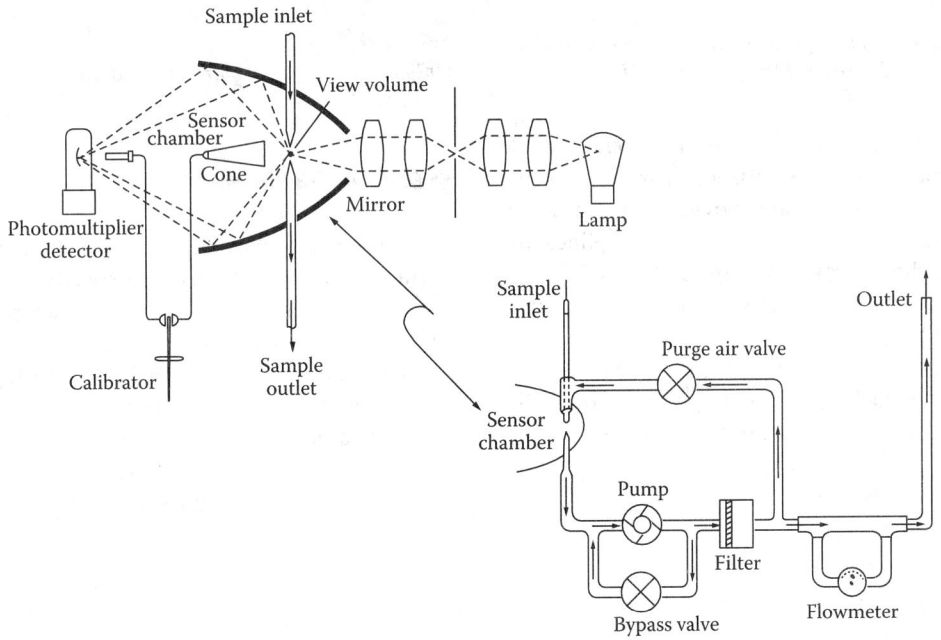

FIG. 1.53f
The elliptical-mirror-type airborne-particulate counter is used in clean rooms and is available with automatic sampling. (Courtesy of Climet Instrument Co.)

Because the light intensity distribution pattern of diffracted/scattered light changes according to the size of the particle, analyzers can tell the size of particles by detecting the diffracted light intensity distribution pattern. Laser diffraction measures the particle size distributions by measuring the angular variation in intensity of the scattered light as a laser beam passes through a dispersed particulate sample. Because large particles scatter light at small angles relative to the laser beam and small particles scatter light at large angles, the angular scattering intensity data can be analyzed to calculate the size of the particles responsible for creating the scattering pattern. The particle sizes are determined as the volume-equivalent sphere diameter and can range from nanometers to millimeters.

For a single spherical particle, the diffraction pattern shows a typical ring structure. The distance of the first minimum to the center depends on the particle size (Figure 1.53h). In particle sizing instruments, the acquisition of the intensity distribution of the diffracted light is usually performed with the help of a multielement photodetector.

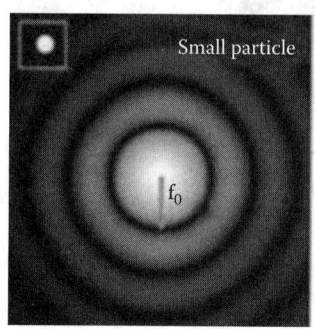
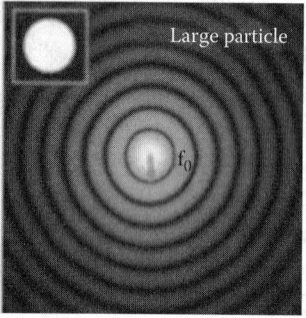

FIG. 1.53h
Diffraction patterns of laser light in forward direction for two different particle sizes. (Courtesy of Sympatec GmbH.)

Simultaneous diffraction on more than one particle results in a superposition of the diffraction patterns of the individual particles when they are moving and diffraction between the particles is averaged out. This simplifies the evaluation, so a mathematical algorithm can be used to convert the diffraction patterns into the *measurement of particle size distribution* (PSD).

More recently, low-angle laser light scattering has been combined with 90° or backscattering, the combination of different wavelengths, polarization ratio, white light scattering, etc. resulted in expanded measurement ranges from below 0.1 μm to about 1 cm. Laser diffraction currently is also the fastest method for particle sizing.

Particle size monitors (PSM) are, for example, used in the various grinding passages of flour mills. In these processes, the accurate knowledge of particle size distribution is essential to guarantee the quality and consistency of the end product. Online PSM can provide continuous monitoring of particle size in flour and/or semolina granulation processes, in the particle size range of 10–5000 μm (Figure 1.53i).

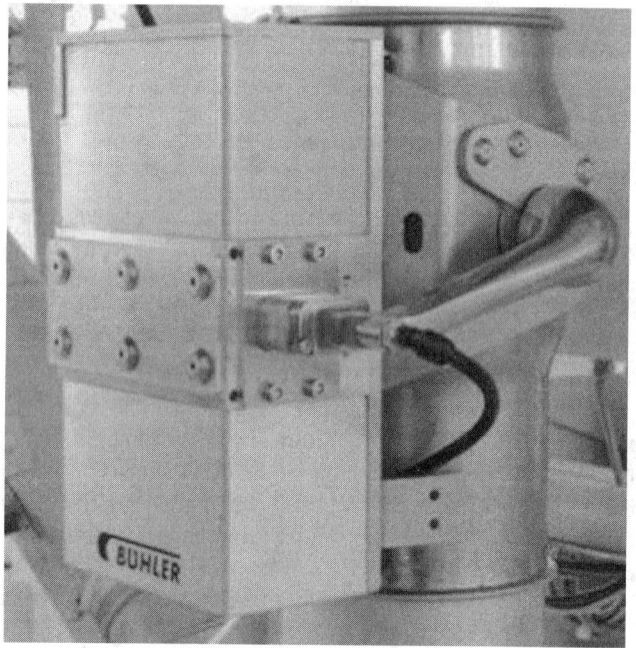

FIG. 1.53i
Online particle size monitor used in flour mills. (Courtesy of Bühler AG.)

SPECIFICATION FORMS

When specifying particle size distribution monitors, the ISA Form 20A1001 can be used to specify the analyzer device itself, and when specifying both the device and the composition and/or properties of the process sample, ISA Form 20A1002 can be used. Both forms are reproduced on the next pages with the permission of the International Society of Automation.

1.53 Particle Size Distribution (PSD) Monitors

#						#				
1	RESPONSIBLE ORGANIZATION		ANALYSIS DEVICE			6	SPECIFICATION IDENTIFICATIONS			
2						7	Document no			
3	ISA		Operating Parameters			8	Latest revision		Date	
4						9	Issue status			
5						10				
11	ADMINISTRATIVE IDENTIFICATIONS					40	SERVICE IDENTIFICATIONS Continued			
12	Project number		Sub project no			41	Return conn matl type			
13	Project					42	Inline hazardous area cl		Div/Zon	Group
14	Enterprise					43	Inline area min ign temp		Temp ident number	
15	Site					44	Remote hazardous area cl		Div/Zon	Group
16	Area		Cell	Unit		45	Remote area min ign temp		Temp ident number	
17						46				
18	SERVICE IDENTIFICATIONS					47				
19	Tag no/Functional ident					48	COMPONENT DESIGN CRITERIA			
20	Related equipment					49	Component type			
21	Service					50	Component style			
22						51	Output signal type			
23	P&ID/Reference dwg					52	Characteristic curve			
24	Process line/nozzle no					53	Compensation style			
25	Process conn pipe spec					54	Type of protection			
26	Process conn nominal size		Rating			55	Criticality code			
27	Process conn termn type		Style			56	Max EMI susceptibility		Ref	
28	Process conn schedule no		Wall thickness			57	Max temperature effect		Ref	
29	Process connection length					58	Max sample time lag			
30	Process line matl type					59	Max response time			
31	Fast loop line number					60	Min required accuracy		Ref	
32	Fast loop pipe spec					61	Avail nom power supply			Number wires
33	Fast loop conn nom size		Rating			62	Calibration method			
34	Fast loop conn termn type		Style			63	Testing/Listing agency			
35	Fast loop schedule no		Wall thickness			64	Test requirements			
36	Fast loop estimated lg					65	Supply loss failure mode			
37	Fast loop material type					66	Signal loss failure mode			
38	Return conn nominal size		Rating			67				
39	Return conn termn type		Style			68				
69	PROCESS VARIABLES		MATERIAL FLOW CONDITIONS			101	PROCESS DESIGN CONDITIONS			
70	Flow Case Identification				Units	102	Minimum	Maximum		Units
71	Process pressure					103				
72	Process temperature					104				
73	Process phase type					105				
74	Process liquid actl flow					106				
75	Process vapor actl flow					107				
76	Process vapor std flow					108				
77	Process liquid density					109				
78	Process vapor density					110				
79	Process liquid viscosity					111				
80	Sample return pressure					112				
81	Sample vent/drain press					113				
82	Sample temperature					114				
83	Sample phase type					115				
84	Fast loop liq actl flow					116				
85	Fast loop vapor actl flow					117				
86	Fast loop vapor std flow					118				
87	Fast loop vapor density					119				
88	Conductivity/Resistivity					120				
89	pH/ORP					121				
90	RH/Dewpoint					122				
91	Turbidity/Opacity					123				
92	Dissolved oxygen					124				
93	Corrosivity					125				
94	Particle size					126				
95						127				
96	CALCULATED VARIABLES					128				
97	Sample lag time					129				
98	Process fluid velocity					130				
99	Wake/natural freq ratio					131				
100						132				
133	MATERIAL PROPERTIES					137	MATERIAL PROPERTIES Continued			
134	Name					138	NFPA health hazard		Flammability	Reactivity
135	Density at ref temp		At			139				
136						140				

Rev	Date	Revision Description	By	Appv1	Appv2	Appv3	REMARKS

Form: 20A1001 Rev 0 © 2004 ISA

740 Analytical Measurement

	RESPONSIBLE ORGANIZATION	ANALYSIS DEVICE COMPOSITION OR PROPERTY Operating Parameters (Continued)		SPECIFICATION IDENTIFICATIONS		
1	ISA		6			
2			7	Document no		
3			8	Latest revision		Date
4			9	Issue status		
5			10			

	PROCESS COMPOSITION OR PROPERTY			MEASUREMENT DESIGN CONDITIONS				
	Component/Property Name	Normal	Units	Minimum	Units	Maximum	Units	Repeatability
11								
12								
13								
14								
15								
16								
17								
18								
19								
20								
21								
22								
23								
24								
25								
26								
27								
28								
29								
30								
31								
32								
33								
34								
35								
36								
37								

Lines 38–75 (blank)

Rev	Date	Revision Description	By	Appv1	Appv2	Appv3	REMARKS

Form: 20A1002 Rev 0 © 2004 ISA

Definitions

Air elutriation is a process for separating particles based on their size, shape, and density, using a stream of gas or liquid flowing in a direction usually opposite to the direction of sedimentation. This method is predominately used for particles with size >1 μm. The smaller or lighter particles rise to the top (overflow) because their terminal sedimentation velocities are lower than the velocity of the rising fluid.

Coulter principle is a "Reference Method" for particle size analysis. It is based on measuring the changes in electrical impedance produced by nonconductive particles suspended in an electrolyte. The sensing zone is a small opening (aperture) between electrodes through which suspended particles pass. In the sensing zone, each particle displaces its own volume of electrolyte and this displaced volume is measured as a voltage pulse; the height of each pulse being proportional to the volume of the particle.

HEPA (high-efficiency particulate arrestance) filters must satisfy certain standards of efficiency such as those set by the U.S. Department of Energy (DOE) to qualify as a HEPA filter. To meet these standards, an air filter must remove (from the air that passes through it) 99.97% of the particles that have a size of 0.3 μm.

Laser diffraction measures particle size distributions by measuring the angular variation in intensity of light scattered as a laser beam passes through a dispersed particulate sample. Large particles scatter light at small angles relative to the laser beam and small particles scatter light at large angles. The angular scattering intensity data can be analyzed to calculate the size of the particles responsible for creating the scattering pattern. The particle sizes are determined as the volume-equivalent sphere diameter. The size of the particle can range from nanometers to millimeters.

Tyndall effect also known as Tyndall scattering is light scattering by particles in a fine suspension. Tyndall states that the intensity of the scattered light depends on the fourth power of the frequency, so blue light is scattered much more strongly than red light. Under the Tyndall effect, the longer-wavelength light is more transmitted while the shorter-wavelength light is more reflected via scattering. An analogy to this wavelength dependency is that long-wave electromagnetic waves such as radio waves are able to pass through the walls of buildings, while short-wave electromagnetic waves such as light waves are stopped and reflected by the walls.

Abbreviations

HEPA   High-efficiency particulate absorption
PSD   Particle size distribution
PSM   Particle size monitor

Organizations

ASTM   American Society for Testing and Materials
DOE   United States Department of Energy

Bibliography

AIChE, Particulate Continuous Emission Monitoring (CD ROM). 1999. http://www.amazon.co.uk/Particulate-Continuous-Emissions-Monitoring-Proceedings/dp/0816908028.

Allen, T., Particle size measurement. 1975. http://onlinelibrary.wiley.com/doi/10.1002/aic.690220135/abstract.

ASTM, Particle size analysis. 2015. http://www.astm.org/search/fullsite-search.htm.

AZO Materials, Particle size—US Sieve Series and Tyler Mesh Size Equivalents. 2016. http://www.azom.com/article.aspx?ArticleID=1417.

Clarke, A. G., Industrial air pollution monitoring—gaseous and particulate emissions. 1997. http://www.amazon.co.uk/Industrial-Air-Pollution-Monitoring-Environmental/dp/0412633906.

EPA, Air pollution monitoring. 2015. http://www3.epa.gov/airquality/montring.html.

EPA, Particulate matter. 2015. http://www3.epa.gov/airquality/particlepollution/index.html.

EPA, Summary of the Clean Air Act. 1970. http://www.epa.gov/laws-regulations/summary-clean-air-act.

ETA, Visible emission observer. 2013. http://www.eta-is-opacity.com/ETA_VE_manual.pdf.

ISO 13320:2009, Particle size analysis—Laser diffraction methods. 2009. http://www.iso.org/iso/iso_http://www.iso.org/iso/iso_catalogue/catalogue_tc/catalogue_detail.htm?csnumber=44929.

Malvern, Laser diffraction. 2013. http://www.malvern.com/en/products/technology/laser-diffraction/default.aspx.

Ronaes, E. et al., Remote real-time monitoring of particle size distribution in drilling fluids during drilling of a depleted HTHP reservoir. 2009. http://www.onepetro.org/mslib/app/Preview.do?paperNumber=SPE-125708-MS&societyCode=SPE.

Schimadzu, Particle size distribution measurement by the laser diffraction/scattering method. 2013. http://www.shimadzu.com/an/powder/support/middle/m01.html.

Schimadzu, Paricle size analyzers. 2015. http://www.ssi.shimadzu.com/products/literature/testing/C060-E005A.pdf.

Shah, S. D., Monitoring transient particle size distributions. 2005. http://www.engr.ucr.edu/~dcocker/JA24.pdf.

Society of Automotive Engineers, Particle size distribution in the exhaust of diesel and gasoline engines. 2000. http://www.worldcat.org/title/particle-size-distribution-in-the-exhaust-of-diesel-and-gasoline-engines/oclc/44555320.

Sympatec, Laser diffraction methods. 2013. http://www.sympatec.com/Science/Characterisation/11_LaserDiffraction.html.

Sympatec GmbH, Particle size and shape analysis. 2015. http://www.sympatec.com/.

Wikipedia, Particulates. 2016. https://en.wikipedia.org/wiki/Particulates.

1.54 Particulate, Opacity, Air and Emission Monitoring

R. A. HERRICK (1974)   **G. F. McGOWAN** (1982)   **B. G. LIPTÁK** (1995, 2003, 2017)

To receiver

(AT) Opacity

Flow sheet symbol

Sampler and monitor types	A. High-volume air sampler
	B. Dichotomous sampler
	C. Tape sampler
	D. Manual sampler
	E. Piezoelectric crystal mass balance
	F. Impaction devices
	G. Radiometric devices, beta gauges
	H1. Charge transfer
	H2. Surface ionization
	I. Light, IR, etc. attenuation, transmission, opacity
	J. Light-scattering laser
	K. Visual observation
	L. Remote sensing
	M. Stack gas monitors
	M1. Sampling probes
	N. Smoke detector alarms
Potential applications	Visibility (I, J, K); fire/smoke detection (G, I); particle sizing (B, F, I, J)
Type of sample and installation	In situ (G, H, I, J, I, K, L); extractive (A, B, C, D, E, F, G); ambient air (A, B, C, D, E, F, G, I); flue/stack gas (D, F, G, H, I, J, K, L)
Standards, reference methods	For (A): EPA 40 CFR 50 Appendix B
	For (C): ANSI/ASTM D1704–78, EPA 40 CFR 60 Appendix A, Method 5
	For (D): ANSI/ASTM D3685–78
	For I: EPA 40 CFR 60 Appendix B, Perf Spec 1, EPA 40 CFR 60 Appendix A, Method 9
Dust measurement ranges	Light scattering mode: 0–200 mg/N m^3
Direct transmission mode	0.20–100 g/N m^3
Dust response time	0.05 sec
Dust measurement resolution	0.1 mg/N m^3 (scattered mode)
Standard smoke density ranges	In units of % opacity, 0%–18.7%, 0%–33.9%, 0%–64.5%, 0%–87.4%, and 0%–98.4%; in units of optical density, 0–0.09, 0–0.18, 0–0.45, 0–0.9, and 0–1.8
*Inaccuracy of smoke density transmissometers**	±3% of range
Costs	A and H1. <$2500
	B, C, D, E. $2500–$6500
	F. $1200–$5000
	G, H2, I. $5,000–$20,000
	J, L. $5,000–$100,000

* Transmissiometers measure the extinction coefficient of the atmosphere and serve to measure the visual range by sending a narrow laser beam through the air or other propagation medium. Atmospheric extinction is a wavelength-dependent phenomenon, but the most common wavelength in use for transmissiometers is 550 μm, which is in the middle of the visible waveband, and allows a good approximation of the visual range.

(Continued)

M. Around $5,000 for portable stack gas monitoring packages and about 40,000 for continuous stack gas composition and opacity monitoring

M1. The cost of the probes only in 1–3 m (3–10 ft) lengths with glass, quartz, or stainless steel lining is about $: from $2000 (The cost of a complete [EPA Reference Method 5] particulate sampling system, with both sampling and monitoring system is about $15,000.)

N. Simple smoke alarms start at $100.

Partial list of suppliers

For suppliers of stack particulate sampling trains (M, M1) see Chapter 1.4.

For suppliers of NIR, IR, visible, laser and UV scattering, and attenuation sensors (types I, J, and L), refer to the suppliers listed in Chapters 1.32, 1.62, and 1.69.

ABB Inc.—Analytical Products (I, J, L, M), http://www.abb.com/product/seitp330/09897ec98943f730c1257b880041c6d4.aspx?productLanguage=us&country=00

Ametek Thermox (A, I, J, M), http://www.ametekpi.com/products/FGA-900-Series-Stack-Gas-Emissions-Analyzer.aspx

Apollo Duct Smoke Detector (I—for HVAC), http://www.apollo-fire.co.uk/media/2864/pp2191_duct_smoke_detectors_issue_1.pdf

AUBURN (bag brake detector) http://auburnsys.com/content/triboguard-i-model-4001

AVL (A for particulates), https://www.avl.com/pss-i60-particulate-sampler

AWI (aviation related detectors) http://www.allweatherinc.com/wp-content/uploads/8365-C-041-Vis_Rev-c.pdf

Babbitt International Inc. (N—dust switch), http://www.babbittlevel.com/dem-products.html

Belfort Instrument (runway visual range detector) http://belfortinstrument.com/products/runway-visual-range/

Crypton (D, N diesel), http://www.crypton.co.za/Tto%20know/diesel_exhaust_emission_smoke_meter.html

Durag Inc. (I, M, M1—opacity, dust), http://www.durag.com/d_r_290.asp

Dwyer (I—opacity), http://www.dwyer-inst.com/articles/industry/PowderBulk/do-i-need-a-particulate-monitor/do-i-need-a-particulate-monitor.cfm

ENVCO—Environmental Equipment (A), http://www.envcoglobal.com/catalog/product/total-suspended-particulat…

Green Instruments (I—IR for oil mist), http://greeninstruments.com/web/G2000_brochure2012_WEB.pdf

HI-Q Environmental Product Co. (A), http://www.hi-q.net/products/outdoor-high-volume-air-samplers/3500-se

Horiba Instruments Inc. (I, M), http://www.horiba.com/fileadmin/uploads/Process-Environmental/Documents/ENDA-4000_Brochure.pdf

Lear Siegler Australia (A, M, M1), http://www.learsiegler.com.au/catalogue/25/cems_stack_emissions

Monitor (broken bag detector) http://www.monitortech.com/product_pe_broken.shtml

MSA Instrument Div. (C, F, M), http://www.msanet.com

Newterra (D), http://www.pacwill.ca

Norsk Elektro Optikk AS (J—laser for dust), http://www.neo.no/research/gasmonitors/dustmonitors.html

PCME Ltd. (B, D, M), http://www.pcme.com/product/pcme-qal-181

Preferred Instruments (I—smoke opacity), http://www.preferredinstruments.com/products/pi/JC-30D-Smoke-Opacity-Monitor

Protection 1 (L—smoke), http://www.protection1.com/home-security-systems/home-security-products/life-safety/wireless-smoke-detectors/

Rosemount Analytical Inc. Emerson (I, M), http://www2.emersonprocess.com/siteadmincenter/PM%20Rosemount%20Analytical%20Documents/Comb_Manual_OPM_4000_200810.pdf

Sensidyne, Inc. (D, M1), http://www.sensidyne.com

Sick Maihak Inc. (I, M), http://www.sick.com/us1/en-us/home/products/product_portfolio/analyzers_systems/Pages/dust_monitors.aspx

Siemens Applied Automation (M), http://www.sea.siemens.com

Sierra Monitor Corp. (A, B, F, M1), http://www.sierrainstruments.com/autotest/products/emissions-analytical

SKC Inc. (D, F—handheld dust monitor), http://www.skcinc.com/prod/770–300.asp

Thermo Fisher (B, D), http://www.fishersci.com/ecomm/servlet/Search?keyWord=Particulate+Monitors&store=Safety&nav=10396316&offSet=0&storeId=10652&langId=-1&fromSearchPage=1&searchType=PROD&typeAheadCat=prodCat

Thermo Scientific (G—Beta gauge) http://www.thermoscientific.com/en/product/beta-gauge-continuous-ambientparticulate-monitor-fh62c14.html

Thermo Scientific (handheld for miner protection), http://www.thermoscientific.com/en/product/pdm3600-personaldust-monitor.html

TSI Inc. (M) http://www.tsi.com/Aerosol-and-Dust-Monitors/

Unicontrol Inc. (I—smoke opacity), http://www.unicontrolinc.com/HC%20PDFs/A08740%20Universal%20Opacity%20Monitor%20Bulletin.pdf

Vayubodhan Upkaran Pvt. Ltd. (M1), http://www.tradeindia.com/fp197334/Handy-Stack-Sampler.html

R. A. Wagner (I—smoke opacity), http://www.keikaventures.com/s_smokemeters.php

Yokogawa Corp. of America (A, M, M1—particulates), http://www.yca.com; http://www.yokogawa.com/us/search/?q=particulate%20monitors

INTRODUCTION

This chapter is devoted to smoke and dust density measurement, which is performed most often by detecting opacity, or optical density. There are two major families of applications for these types of instruments. One is the measurement of the quality of ambient air, and the other is stack emission monitoring. The two applications overlap somewhat and some of the sensors discussed in this section can be used for both applications.

Other chapters in this volume that discuss subjects related to the detection of solid particle concentration include the following:

Chapter 1.62: On Spectrometry
Chapter 1.4:   On Stack Sampling
Chapter 1.3:   On Air Sampling and Monitoring

Sample collection is a major part of this measurement and therefore stack samplers and ambient dust collectors are also covered. The EPA plays a large role in determining if a plant must monitor their particulate emissions. The Clean Air Act* Amendments of 1990 require the U.S. EPA to regulate air pollutant emissions–based standards. These standards are known as the National Emissions Standards for Hazardous Air Pollutants (NESHAP) and the Maximum Achievable Control Technology (MACT) standards. MACT standards require facilities to meet specific emissions limits that are based on the emissions levels already achieved by the best-performing similar facilities.

This chapter is so organized that the definitions and theoretical concepts applicable to particulate measurement are discussed first. This is followed by the discussion of stack gas monitoring. The analyzers used for ambient air monitoring are discussed last.

THEORY

Suspended solids in gas will absorb, reflect, and scatter radiation (Figure 1.54a). These effects all attenuate the energy of radiation that is received at the detector. If the particle size is smaller than the wavelength of radiation, the attenuation is caused mostly by absorption. When the particle size is equal to or larger than the wavelength, the dominant form of attenuation is scattering.

Most particle sizes in industrial emissions fall between 0.1 and 50 μm, and the wavelength of light used for opacity measurement is in the range of 0.38–0.78 μm. Because light attenuation decreases with particle size when measuring submicron particles, it is desirable to use short wavelengths (0.38–0.78 μm) when measuring particulates in stacks.

---

* See details of the Clean Air Act at https://en.wikipedia.org/wiki/Clean_Air_Act_%28United_States%29

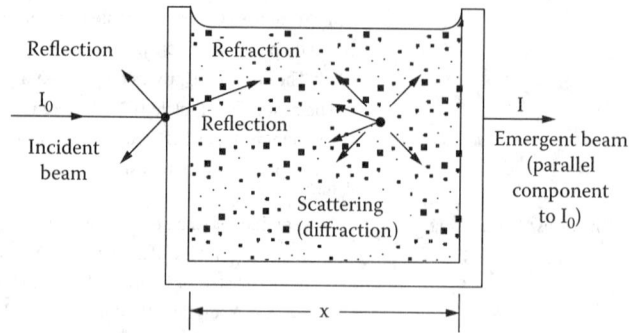

FIG. 1.54a
When parallel monochromatic (single-wavelength) light (I_0) enters a stack, parts of it are reflected, refracted, scattered, and absorbed, and only the remaining attenuated (I) light emerges from the stack. (From Particulate Continuous Emission Monitoring (CD ROM), American Institute of Chemical Engineers, New York, 1999.)

Opacity Measurement

The objective of opacity measurement is to observe visible emissions the same way they are observed by the human eye. The spectral sensitivity of the light-adapted human eye is *photopic*, having its maximum sensitivity at a wavelength of about 0.55 μm. To reproduce the human eye, the spectral distribution of the light source in an opacity meter should also be photopic.

Figure 1.54b shows the outputs of both a photopic and an incandescent light source. If an incandescent light source is used in an opacity meter, the readings will be in error. One component of the error will be the result of water absorption in

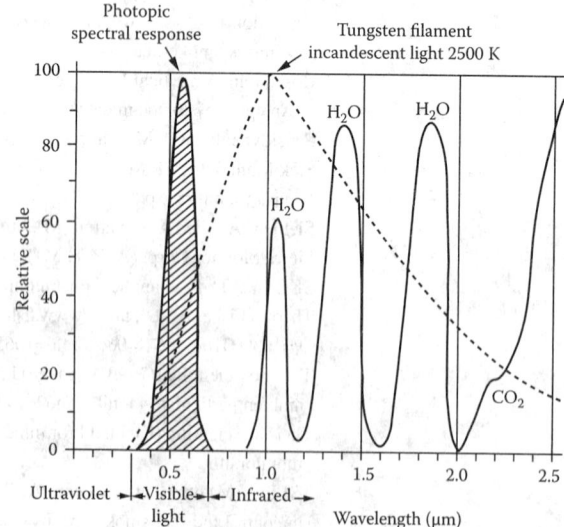

FIG. 1.54b
The spectral responses of "photopic" and "incandescent" light sources. With incandescent light, errors resulting from water absorption and low readings of submicron particulates will occur. (From Beutner, H.P., J. Air Pollution Control Assoc., September 1974.)

the near-infrared region, and the other error component, which will cause a low reading, will be caused by the effect of sub-micron particulates, to which the human eye is more sensitive. Therefore, transmissometers used for compliance with opacity regulations must operate in the visible photopic spectrum.

Units and Definitions

The ratio of I/I_0 (Figure 1.54a) is called transmittance (T). It is usually expressed as a percentage, with 100% corresponding to zero attenuation ($I = I_0$). Percent opacity (O) is the difference between 100% opacity and transmittance (T) ($O = 100 - T$). The Beer–Lambert relationship defines transmittance (T) as follows:

$$T = e^{-naql} \quad\quad 1.54(1)$$

where

T is the fraction of light transmitted (transmittance)
n is the number of particles per unit volume
a is the mean particle projected area
q is the particle extinction coefficient
l is the length of effluent path
e is the base of natural logarithm

The product of n, a, and q is often referred to as the *turbidity attenuation coefficient*.

Dust Loading The Lambert–Beer law can also be used to determine the dust loading in the stack (mg/m³), because dust loading can be determined for transmittance (T), as shown in Figure 1.54c. Optical density is a measure of light attenuation and is proportional to both path length and particulate concentration.

$$D = \text{opacity density} = \log_{10}\left[\frac{1}{(1-\text{opacity})}\right] \quad\quad 1.54(2)$$

Therefore, optical density can also be determined (Figure 1.54c) once the dust loading (mg/m³) is known. Numerically, one might say that transmittance becomes very small (<10%) or opacity very large (>90%) when optical density reaches 1.0; when it rises to 2.0, transmittance is almost 0%, and the substance is completely opaque. Some of the standard ranges for stack monitoring applications are given in Table 1.54a in both units of percent opacity and of optical density.

TABLE 1.54a
Standard Ranges of Smoke Density Monitor

Optical Density	% Opacity
0–0.09 and 0.18	0–18.7 and 33.9
0–0.18 and 0.45	0–33.9 and 64.5
0–0.45 and 0.90	0–64.5 and 87.4
0–0.90 and 1.80	0–87.4 and 98.4

Source: Courtesy of Lear Siegler Inc.

Ringelmann Card Numbers Another opacity measurement unit is based on Ringelmann card numbers. Ringelmann cards consist of a series of five cards that present graduated shades of gray from white to black for the evaluation of smoke density or opacity. The system utilizes a rectangular grill of black lines of definite width and spacing on a white background, which can be accurately reproduced (Figure 1.54d). The rule by which the chart may be reproduced is as follows:

Card 0. All white
Card 1. Black lines 1 mm thick, 10 mm apart, leaving white spaces 9 mm²

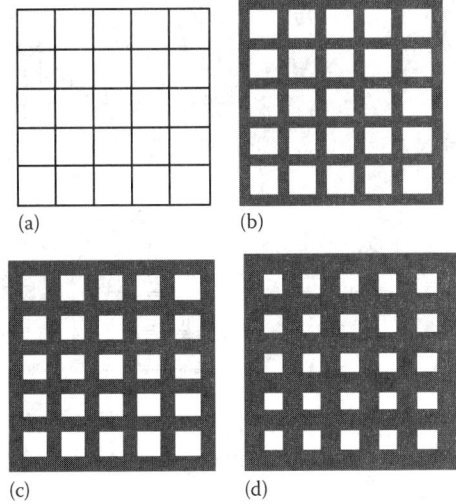

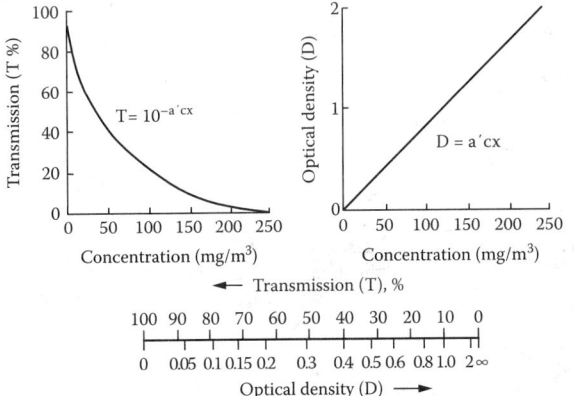

FIG. 1.54c
The relationship between transmission (T), concentration or dust loading, and optical density (D).

FIG. 1.54d
Ringelmann's scale for grading smoke density. (a), Card 1 (equivalent to 20% black) has black lines 1 mm thick, 10 mm apart, leaving white spaces 9 mm square; (b), Card 2 (equivalent to 40% black) has black lines 2.3 mm thick, 10 mm apart, leaving white space 7.7 mm square; (c), Card 3 (equivalent to 60% black) has black lines 3.7 mm thick, 10 mm apart, leaving white space 6.3 mm square; (d), Card 4 (equivalent to 80% black) has black lines 5.5 thick, 10 mm apart, leaving white spaces 4.5 mm square. Not shown: Card 0 (all white) and Card 5 (all black).

Card 2. Lines 2.3 mm thick, 10 mm apart, leaving white spaces 7.7 mm²
Card 3. Lines 3.7 mm thick, 10 mm apart, leaving white spaces 6.3 mm²
Card 4. Lines 5.5 mm thick, 10 mm apart, leaving white spaces 4.5 mm²
Card 5. All black

STACK EMISSION MONITORING

When all the components in the stack gas are to be simultaneously monitored, a combination system is required. This is because the absorption bands of the different components of interest are in different parts of the radiation spectrum.

Particulate Sampling

Particulate sampling has already been covered in Chapter 1.3 and therefore it is only briefly mentioned here. A complete Environmental Protection Agency (EPA) particulate sampling system (Reference Method 5) is composed of a pitot probe, a sampling unit, and a control unit with a vacuum pump. Figure 1.4a has shown this EPA particulate sampling train.

In some of these sampling packages, a microprocessor directs the automatic sampling (Figure 1.54e) method, which can be selected to follow U.S. EPA Method 5* or other international methods specified by VDI, BSI, or ISO. The microprocessor stores all measurements, reviews and diagnoses all inputs, controls the required

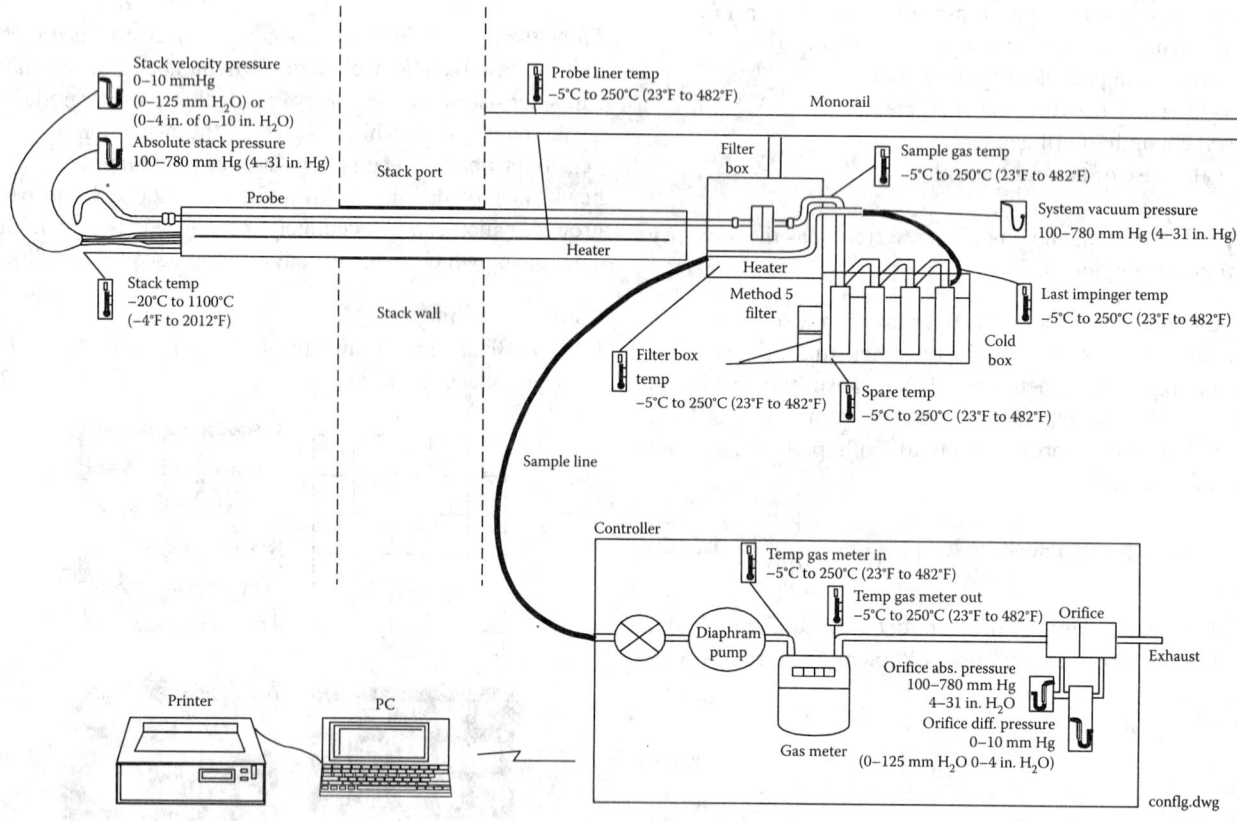

FIG. 1.54e
Components of an automatic sampling train (AST).

---

* EPA, Method 5–Determination of particulate matter emissions from stationary sources, 2011. http://www3.epa.gov/ttnemc01/promgate/m-05.pdf.

parameters, calculates isokinetic conditions, and reports the results either in a printed form or by transferring it to a floppy disk.

Besides the controller, such a package usually consists of a probe, a filter (hot) box, a cold box, a flexible sample line, glassware, a node box, and a monorail system. The probe is usually 0.9, 1.5, 2.1, or 3 m (3, 5, 7, or 10 ft) and made of stainless steel with a glass liner. Most probes are jacket heated and are provided with both a liner thermocouple and a stack temperature thermocouple.

Stack Gas Analysis and Sampling

In the "Introduction" section, I listed the chapters, where the sampling and monitoring of stack gases are covered. Therefore, the reader can refer to those chapters if his/her interest is in measuring the concentration of various gases or vapors in the stack emission stream. Therefore, here I will just mention that SO_2 and NO_2 absorption bands fall in the ultraviolet region (Figure 1.54f), whereas the absorption bands of CO, CO_2, and hydrocarbons fall in the infrared region (Figure 1.54g).

The stack monitoring sampling train packages have been discussed in Chapter 1.3 and therefore that coverage is not repeated here. The samples these packages (Figures 1.54h and 1.54i) provide are used to determine both the particulate concentration in the stack emission and the concentrations of the various vapors and gases of interest.

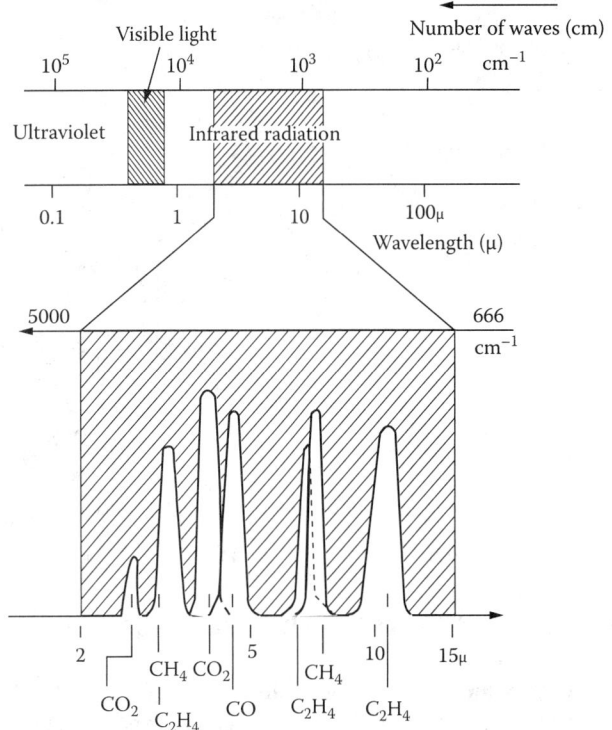

FIG. 1.54g
The absorption bands of CO_2, and CO_2 and hydrocarbons fall into the infrared region.

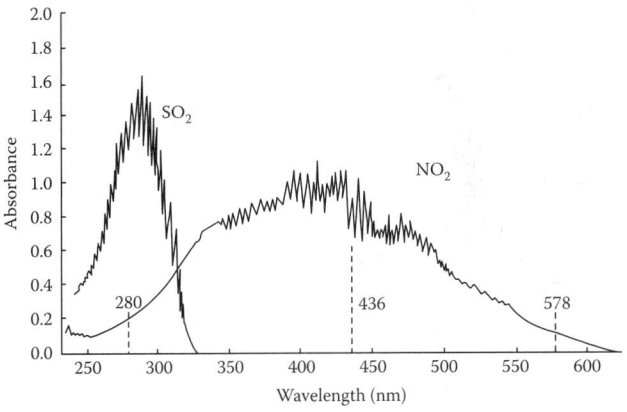

FIG. 1.54f
The absorbance of SO_2 and NO_2 in the ultraviolet range.

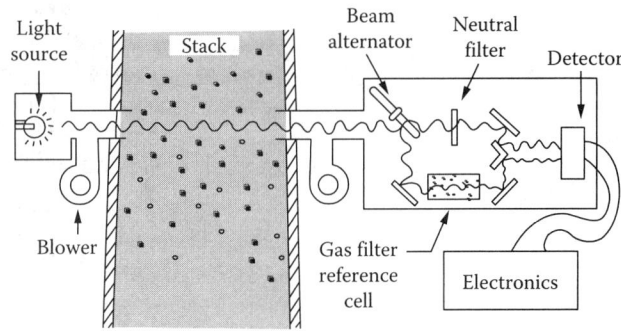

FIG. 1.54h
Dual-beam IR analyzer for multicomponent stack emission monitoring.

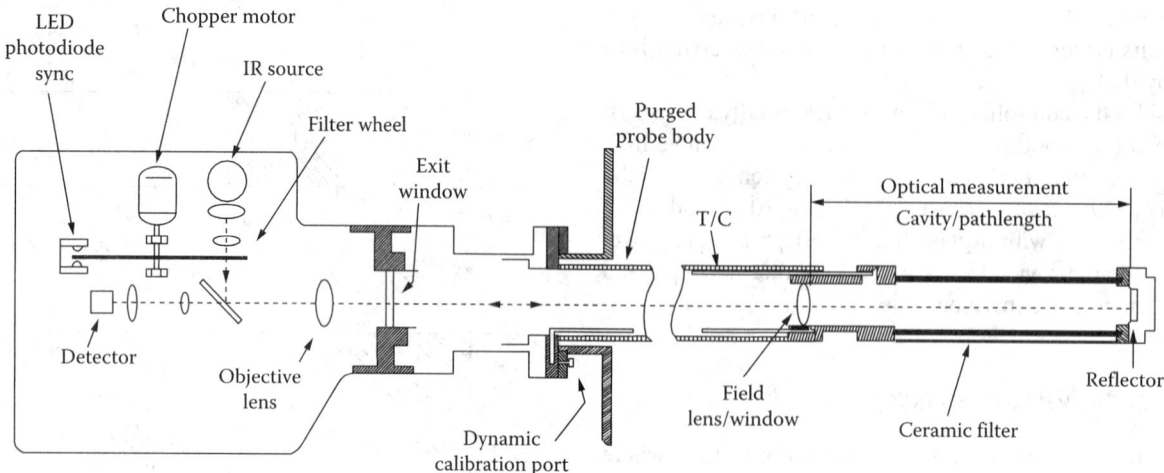

FIG. 1.54i
Stack gas monitor with ceramic diffusion cell.

LIGHT ATTENUATION AND TRANSMISSOMETERS

Opacity is measured in the visible region (Figure 1.54b). Transmissometers are the most accepted and frequently used means for the automatic and continuous monitoring of particulate in stack gases. In its most basic form, a transmissometer consists of a light source and a detector (Figure 1.54j). By referencing all measurements of the partially obstructed optical path to the clear path measurement, the transmissometer essentially provides a measurement of light attenuation, which can be expressed in units of optical density or opacity.

Optical density (also called extinction or absorbance). Optical density is related to transmittance (T), which is the ratio of the amount of radiation that passes through the sample (I), divided by the radiation that enters the sample (incident).

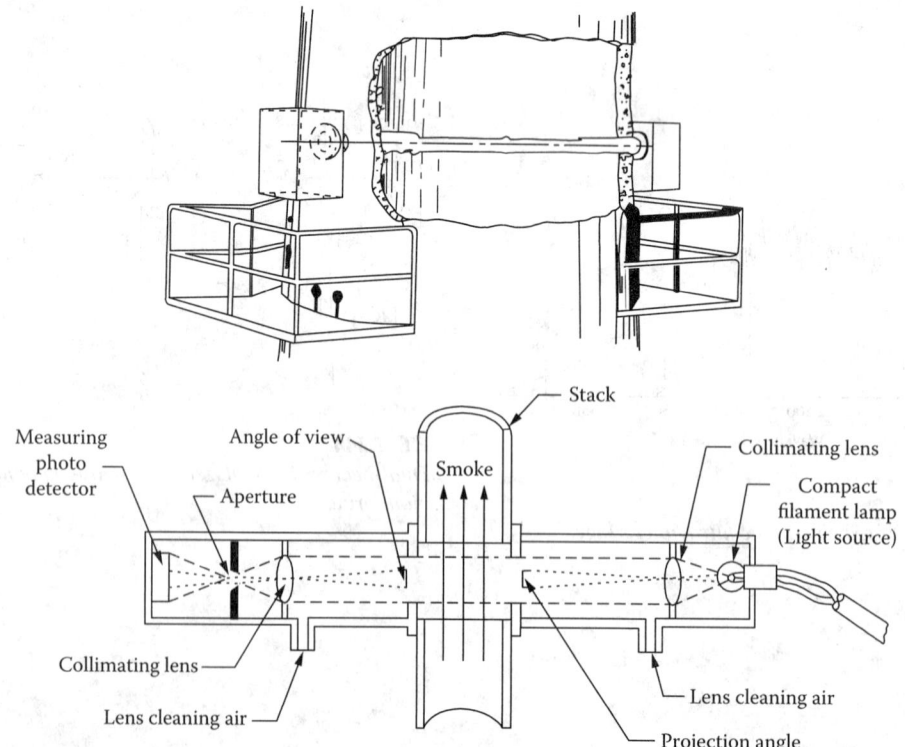

FIG. 1.54j
Single-pass transmissometer can detect smoke opacity. It is insensitive to minor alignment error or vibration but is affected by light or dirt again and by dirt building on the optics.

This ratio is called transmittance and is expressed as a percentage (%T = I/incident). Optical density (OD) is defined as OD = 2 − log10 %T. Therefore, the higher the transmittance, the lower the optical density, meaning that the sample is optically less dense and therefore less absorbing. For example, if 0.1% of the light is transmitted (%T = 0.1%), the optical density is 3, while if 10% of the light passes through (%T = 10%), the optical density is only 1.

Optical density, which is sometimes called *extinction*, varies linearly with particulate concentration if a relatively uniform distribution of particle size, composition, and path length exists. Opacity, on the other hand, varies from 0% to 100%—from clear to opaque conditions, respectively. Opacity can be related linearly to the Ringelmann numbers used by human observers. It is also the parameter typically regulated by the EPA to ensure proper operation and maintenance of particulate control equipment.

Transmissometers have become rather sophisticated, having been developed to achieve exceptional accuracy and reliability.

Transmissiometers measure the extinction coefficient of the atmosphere and serve to measure the visual range by sending a narrow laser beam through the air or other propagation medium. Atmospheric extinction is a wavelength-dependent phenomenon, but the most common wavelength in use for transmissometers is 550 nm (0.55 μm.) which is in the middle of the visible waveband, and allows a good approximation of the visual range.

Single-Pass Configuration

In a single-pass system, the light source is mounted on one side of a duct, and the detector is mounted on the other (Figure 1.54j). Consequently, the light passes through the gas or medium of interest only once. The major difficulty imposed by this configuration is its calibration. This is because the detector cannot be easily calibrated or exposed to a known light source without adding an additional light source specifically for that purpose, or by using other means that are not required in the normal measurement.

Double-Pass Configuration

A double-pass system (Figure 1.54k) incorporates both a light source and a detector on the transceiver side of the stack, and it provides a retroreflector on the opposite side. With this configuration, the light passes through the gas of interest twice, thereby increasing the sensitivity of the measurement. Furthermore, this configuration allows all the active components to be contained in the transceiver and permits a simple solenoid-mounted reflective surface to be incorporated into the transceiver to provide a simulation or calibration check of the zero-opacity condition. Internal filters can be activated sequentially to provide upscale checks of calibration.

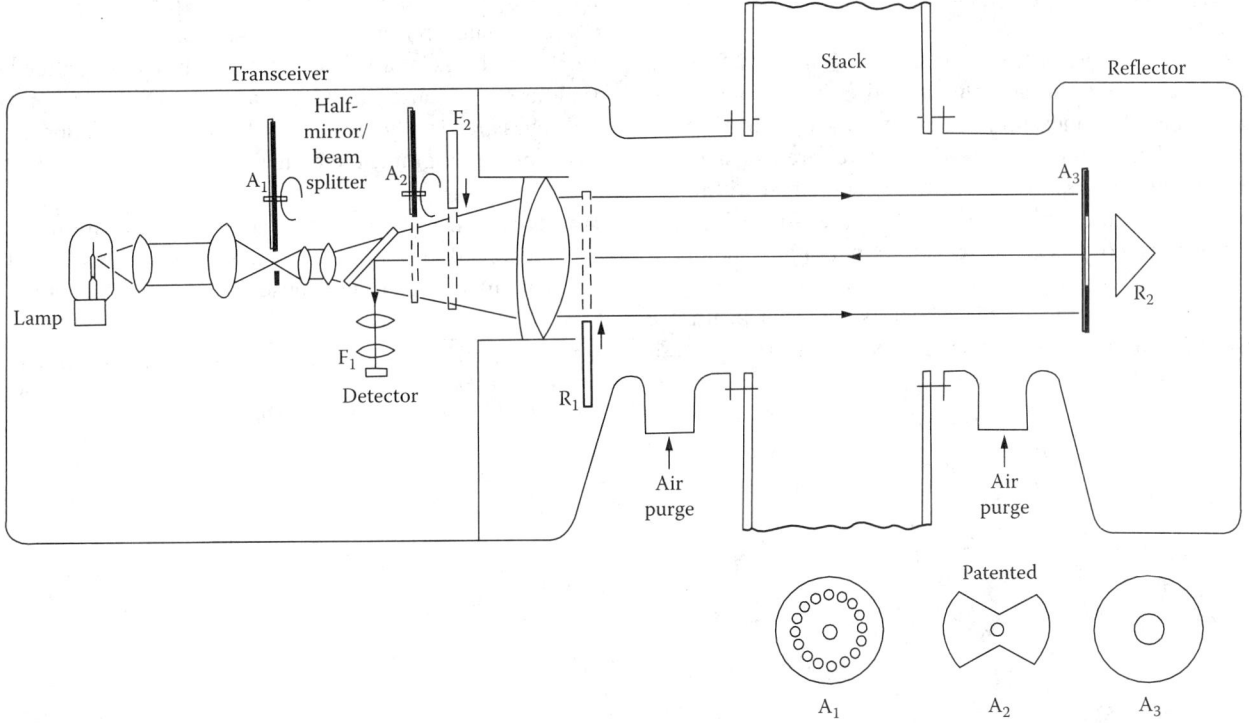

FIG. 1.54k
Double-beam, double-pass transmissometer. Legend: A_1, chopper wheel; A_2, Beam gating wheel; F_1, spectral filter; F_2, solenoid-activated neutral density filter; R_1, solenoid-activated zero cal reflector; R_2, retroreflector. Note: Allignment Bullseye not shown. (Courtesy of Lear Siegler Corp.)

The operation of the double-pass opacity analyzer shown in Figure 1.54k is as follows. The light from the lamp is filtered to generate a *photopic* light energy profile in accordance with Figure 1.54b. The chopper wheel A_1 (with holes in it) is located at the focal point of the condenser lenses. The half-mirror beam splitter splits the light beam into measuring and reference light beam segments. By the use of the beam gating wheel (A_2), the measurement single is periodically gated, making the analyzer insensitive to variations in ambient light.

The reference beam is established and used as a continuous indication of the intensity of the unattenuated light source. When the measuring beam is allowed to pass through the stack, it traverses the distance to the retroreflector (R_2), thereby increasing measurement sensitivity as it passes through the stack twice. The zero calibration reflector (R_1) is used to automatically rezero the instrument and thereby correct for dirt buildup on the windows; the neutral density filter (F_2) allows for span correction. The zero and span checks correct for lamp aging, dirt buildup, and drift. These checks can be performed automatically, under microprocessor control, once every hour.

Relative Performance Factors

Single-beam configurations (Figure 1.54i) essentially measure the detected light without compensation for changes in optics or in source or detector characteristics. Obviously, this is the simplest, most inexpensive, but least accurate technique.

Dual- or double-beam configurations (Figures 1.54h and 1.54k) internally split the light emitted from the source into two beams. The measurement beam is projected through the optical medium of interest and is referred to the second (reference) beam, which is totally contained within the instrument. If a separate detector is provided for each beam, as is done in so-called dual-beam configurations, compensation is continuously provided for light source variations. Furthermore, as long as the detectors for each beam are matched, only their differential errors appear in the overall measurement.

If the two beams can be modulated or time sequentially gated into a single detector, automatic compensation is provided for both light source and detector variations, which is typical of the more advanced forms of double-beam measurement systems (Figure 1.54k). Measurement accuracy can be further improved by the regulation of lamp voltage supplies and by automatic control of reference light levels.

The angles of both viewing and projection should be minimized to reduce the potential interference of scattered light. Most precision units exhibit less than 5° total angular divergence.

Spectral Characteristics The spectral response of the system determines its sensitivity to particulate size. Typically, the systems are matched to the photopic response of the light-adapted human eye so that the transmissometer sees exactly as does the human eye. This characteristic provides a peak response at 550 nm and provides a somewhat uniform response for particles as small as 0.3 μm. Longer wavelengths, such as 3.39 μm, can be used to provide more uniform response with respect to particle size as desired of a mass monitor.*
Multiwavelength devices have been designed that provide some indication of relative particle size distribution.

Optical Characteristics Typically, the more modern and precise instruments also incorporate a chopped light system, which makes the instrument insensitive to variations in ambient light. Also, by providing a uniformly illuminated beam much larger than the reflector or detector and by using an autocollimated retroreflector technique, the overall system can be made essentially insensitive to variations in alignment such as those caused by temperature and wind. Alignment bullseyes are usually provided to facilitate installation and to allow proper alignment to be maintained.

Air Purge Nearly all stack-mounted systems are provided with a forced-air purge system, which ensures a curtain of clean air between the exposed optical surface and stack

---

* Clarke, A. G., *Industrial Air Pollution Monitoring–Gaseous and Particulate Emissions*, 1997. http://www.amazon.co.uk/Industrial-Air-Pollution-Monitoring-Environmental/dp/toc/0412633906.

1.54 Particulate, Opacity, Air and Emission Monitoring

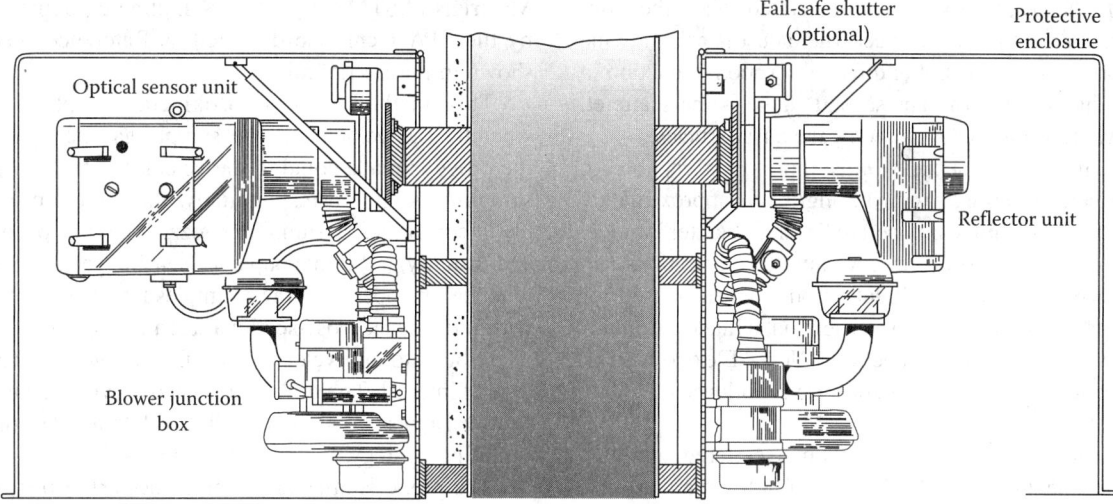

FIG. 1.54l
A 60 SCFM (1.7 m³/min) fan provides a curtain of clean and cold air between the instrument and the gases in the stack. (Courtesy of Lear Siegler Corp.)

gas. These units include both an air filtration system and a blower (Figure 1.54l).

LIGHT SCATTERING DESIGNS

When light strikes a particle, it scatters as a function of the particle's size, shape, and color. Light scatter can be measured as forward scatter, angular scatter, or backscatter.

Suspensions of particulate matter in the air scatter the light in all directions. Sophisticated devices are available for determining the concentration of particles in air and the size of these particles. Laser-based scattering devices measure particulate concentration and/or particulate size distribution. Such devices appear to be particularly useful in evaluating the behavior of particulate control equipment. For an in-depth discussion of open path laser spectrometry, refer to Chapter 1.62.

Laser-based Designs

This design exploits the phenomenon of light scattering, because the dust particles scatter the laser light the same way as particles appear to 'light up' when drifting through a ray of sunlight.

The laser-based stack monitors transmit light from a diode laser to two detectors located on the other side of the stack, diametrically opposite to the transmitter. By having both detectors on the other side of the stack, the interference from dust buildup on the optical windows and the maintenance need associated with their cleaning are both reduced or eliminated. As can be seen in Figure 1.54m, by this dual detector (direct and

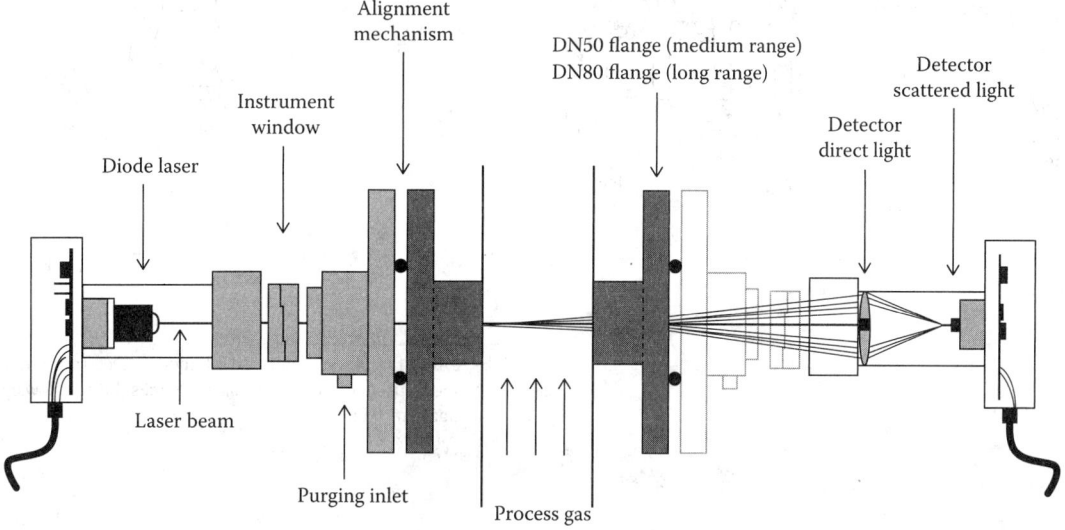

FIG. 1.54m
Laser light scattering stack emission dust monitor design. (Courtesy of Norsk Elektro Optikk AS.)

scattered) design, two signals are generated when the laser beam is sent across a stack or duct where dust is present: One signal represents the amount of direct absorption (reduction in intensity) due to the dust. The second signal is the scattered light, in which scattering is caused by the dust particles.

According to the supplier, this monitor is capable of detecting concentrations from 0.1 mg/m³ to approximately 200 mg/m³. The unit can operate in the "scatter" mode, which is normally chosen for low dust concentrations, up to 200 mg/N m³ or in the "direct" mode, which is used for higher concentrations. The mode selection is made automatically. According to the manufacturer, Norsk Elektro Optikk AS, the performance characteristics are as follows:

- Measurement range: 0–200 mg/N m³ (scattered mode)
- Measurement range: 0.20–100 g/N m³ (direct mode)
- Response time (pulse mode): 0.05 sec
- Resolution: 0.1 mg/N m³ (scattered mode)

STACK SAMPLERS

Manual Stack Samplers

Manual stack sampling techniques have been established by both the EPA and the American Society for Testing and Materials (ASTM). The stack sampling equipment required by the EPA technique designed as Reference Method 5 is shown in Figure 1.54n.*

This method uses an isokinetic sampling technique whereby the velocity of the sample gas being drawn into the probe is made equal to the velocity of the flue gas at the sampling point. Correspondingly, the sample probe assembly incorporates a sampling nozzle, an S-type pitot tube, and (usually) a thermocouple for sensing gas temperature. In practice, the operator draws samples from a variety of points within the stack cross-sectional area to minimize the effects of particulate and velocity stratification within the gas volume of interest. The probe is inserted or withdrawn manually from various access ports on the stack to achieve the desired sampling strategy.

The probe is attached to an adjacent hot box where the particulate is separated from the gas on a filter paper and/or cyclone separator. After removal of particulate, the gas proceeds to a cold box where the gas is cooled to 0°C (32°F) by a series of ice bath–cooled impingers. The gas then flows through a heated umbilical to the control unit where the gas flow is controlled and timed. The stack gas velocity and temperature are also read out from this unit.

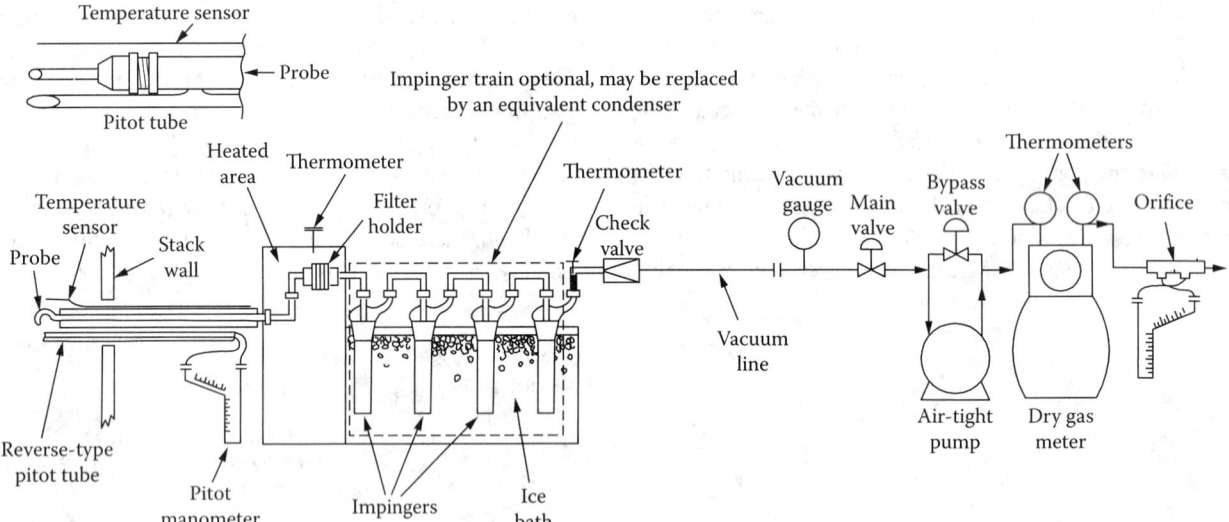

FIG. 1.54n
Manual stack sampler.

---

* EPA 40 CFR 60 Appendix A, Method 5; Determination of Particulate Matter Emissions from Stationary sources. http://www.epa.gov/ttn/emc/promgate/m-05.pdf.

Automatic Stack Samplers

As discussed earlier, a variety of options have been developed to make the system both easier to use and expandable to other gas measurements. Automatic microprocessor-controlled systems (Figure 1.54n) have been introduced that greatly reduce the required operator time and attention.

EPA Method 5 is one of the most widely used techniques available today and has been used to establish the data base for particulate-emitting stationary sources.

Broken Bag Monitors

Simplified transmissometer configurations are often used for detecting bag breaks in baghouse particulate control applications.* Transmissometers have also been widely used for highway visibility, runway visual range, highway tunnel, and roof vent monitoring. Exceptional accuracy and drift characteristics are required in such applications.

Typically, stack-mounted transmissometers are capable of up to 22.5 m (75 ft) separation, whereas visibility-monitoring units may have larger optics that provide path lengths of several hundred meters. For a more detailed discussion of open-path spectrometry, refer to Chapter 1.62.

Bag leak or bag break detectors are also available for the purpose when the goal is to measure the dust level in dust collector air exhausts and to signal if that level is abnormally high (Figure 1.54o).

The sensor

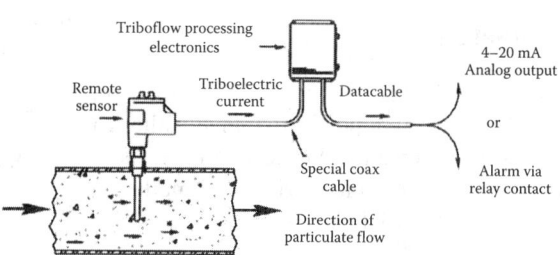

The installation

FIG. 1.54o
Dust collector installation of bag break detector. (Courtesy of Auborn.)

---

*AUBURN, Bag leak detector, 2014. http://auburnsys.com/content/triboguard-i-model-4001.

AMBIENT AIR OPACITY MONITORING

Particulate matter is the world's most ubiquitous air pollutant. Its natural sources include fires, volcanic action, vegetation, and the wind. Man-made sources include combustion, heating and abrasion processes, and material-handling activities, and these constitute another major source of the particulate burden on the atmosphere.

Airborne material can be measured as particulate matter in a wide range of sizes and characteristics, but larger sizes are limited as a result of the associated higher settling velocities. Small, inhalable particulate matter less than 15 µm in mass is the most damaging to human health. Consequently, techniques have also been developed to characterize and measure the size distribution of particulate matter (Chapter 1.53).

Particulate Concentration

Particulate material in ambient air is typically measured in terms of concentration, or weight per unit volume, although its presence may also be observed and measured in terms of visibility or light attenuation. The monitoring of air pollution has already been discussed in Chapter 1.4.

Visibility as a measured variable is a much more complex parameter than particulate concentration, as it involves the characteristics of the human eye. In addition to light attenuation, visibility is a function of contrast and background luminance, which are required to establish visual range measurements of runways for airport applications.

Particulate matter generated by combustion can also be measured in terms of concentration, but it is often specified according to optical density or opacity, which are light attenuation measurements. Opacity of smoke plumes is often related to the visual effects seen by a human observer.

Particulate concentrations range from that corresponding to a visibility of 50 miles (80 km) to the nearly opaque smoke plume arising from a coal-fired boiler without particulate emission controls. To measure such a range of concentrations, a wide variety of techniques have been developed for such monitoring.

In situ techniques generally mean those that can be used to measure particulate concentration in place. *Extractive* techniques include those that remove a sample from the medium to be measured and transport that sample to the measuring instruments.

Ambient Air Sampling

The air quality requirements are set by EPA as secondary standards and are tabulated in Table 1.54b. These include

TABLE 1.54b
Secondary Air Quality Standards

Pollutant	Long-Term Concentration Level ($\mu g/m^3$)	Short-Term Concentration Level[a] ($\mu g/m^3$)
Particulate	60—annual geometric mean	150—maximum 24 h concentration
Sulfur dioxide	60 (0.02 ppm)—annual arithmetic mean	260 (0.1 ppm)—maximum 24 h concentration
Carbon monoxide	—	10,000 (9 ppm)—maximum 8 h concentration or 40,000 (35 ppm)—maximum 1 h concentration
Photochemical oxidants	—	160 (0.08 ppm)—maximum 1 h concentration
Hydrocarbons	—	160 (0.24 ppm)—maximum 3 h concentration (6–9 AM)
Nitrogen dioxide	100 (0.05 ppm)—maximum arithmetic mean	

Note: Levels of air quality that are deemed adequate to protect the public welfare from any known or anticipated adverse effects.
[a] Stated concentration not to be exceeded more than once per year.

that the maximum 24 hr concentration of particulates must not exceed 150 µg/m³ more than once per year and that the annual geometric mean, as measured by the high-volume sampler, must not exceed 60 µg/m³.

High-Volume Sampler The basic instrument for determining the airborne concentration of particulate matter in the United States is the high-volume sampler (Figure 1.54p). The high-volume sampler provides data on the actual mass concentration of particulate matter in the air. A high-speed, multistage blower is used as the suction source. An adapter section allows the use of a flat 8 × 10 in. (203 × 254 mm) glass fiber filter. This filter has an efficiency of well over 99% for 0.3 µm particles.

An orifice plate is included on the back of the blower, which allows the monitoring of the airflow rate through the sampler. Twenty-four hours is a common sampling time period. During this period, the buildup of particles increases the filter resistance and decreases the airflow rate. When the decrease is small (e.g., less than 10%), the initial and final readings are averaged to determine the sampling rate.

If the decrease is significant, as might be expected in locations where the particulate concentration is high, it is desirable to use a recorder and determine the flow rate from that record. Alternatively, more precise measurements can be obtained by using instruments that include a self-regulating flow controller, which maintains a constant flow rate independent of filter loading.

The roof of the standardized sampling shelter has an annular opening sized so that particles larger than a nominal size of 100 µm are not drawn into the sampler. Because of the varying airflow rate and the effect of winds, this method provides only an approximation of a particle size. On the other hand, it can furnish a large body of ambient air quality data, and air quality standards have been promulgated based on the use of this standardized instrument.

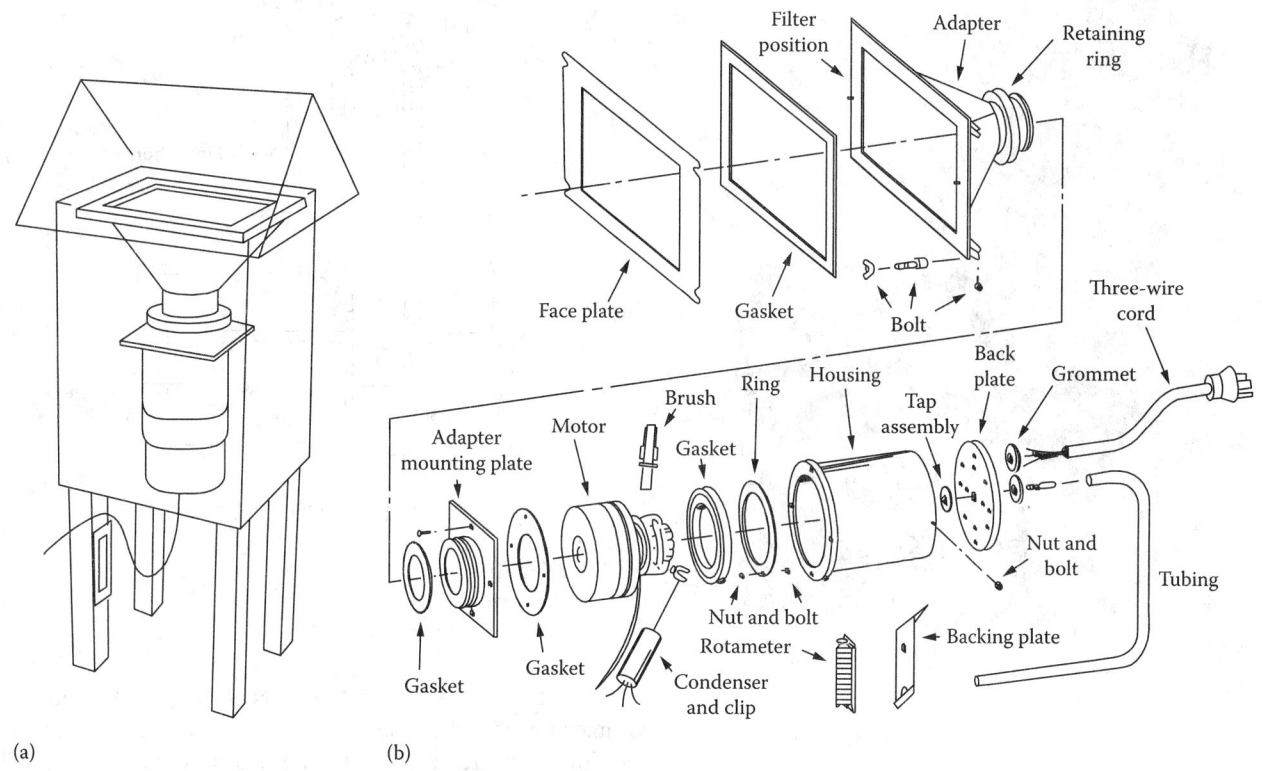

FIG. 1.54p
High-volume sampler. (a) Assembled sampler and shelter, (b) exploded view of typical high-volume air sampler.

The more automated high volume air sampler (Figure 1.54q) designs usually include two- or three-stage variable-speed centrifugal blowers, where the speed of the motor is controlled by a programmable logic controller (PLC) that accepts an input from an air mass flow meter. The PLC thereby maintains the preset flow rate. An illuminated graphic LCD display is provided which indicates the flow setpoint, actual flow, total volume of air that has been sampled, and the elapsed time. Networking and Communication option setups include two (selectable) RS232/RS485 ports, a 4–20 mA and/or 0–10 VDC analog output proportional to flow, the ability to send and receive SMS messages to/from any CDMA/GSM cellular phone for possibly alerting/reporting any predefined event via text message, remote, or local data acquisition, data logging in Microsoft Excel format, and a custom remote access utility that allows complete control of the unit from a remote location.

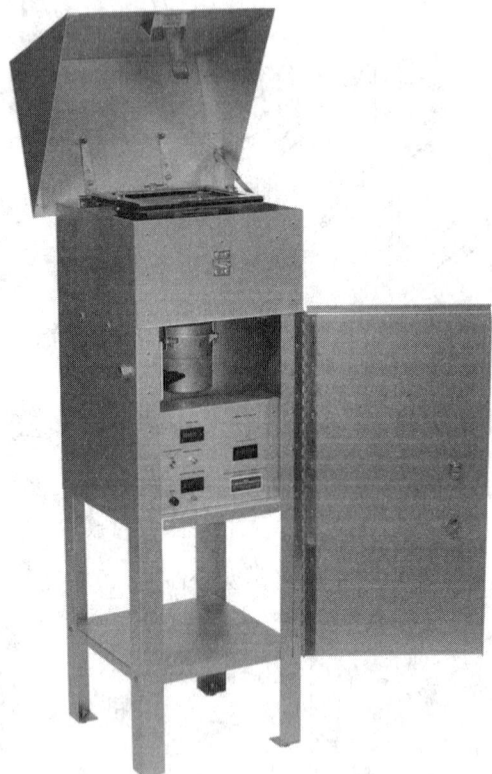

FIG. 1.54q
Automated high-volume air sampler. (Courtesy of HI-Q Environmental Product Co.)

Dichotomous Sampler Subsequent to the EPA's promulgation of the total suspended particulate standards, efforts have been made to establish the health effects of particulates in relation to their size. As a result of these studies, further data have been collected on the harmful effects of inhalable particulates less than 15 μm in size. In concert with this regulatory activity, the dichotomous sampler has been developed to meet the need for size-specific particulate collection.

The dichotomous sampler is generally designed to remove the particulates that are greater than 15 μm at the very the inlet sampling and transport mechanism. The remaining sample stream is then segregated into two size-specific regions, hence the name *dichotomous*.

A virtual impactor (shown in Figure 1.54r) is often used to draw off particulates that are smaller than 2.5 μm. As shown in Figure 1.54p, two sample streams are created. When these streams are collected on filters, one yields the particulate in the 15–2.5 μm size range and the other in the ≤2.5 μm size range.

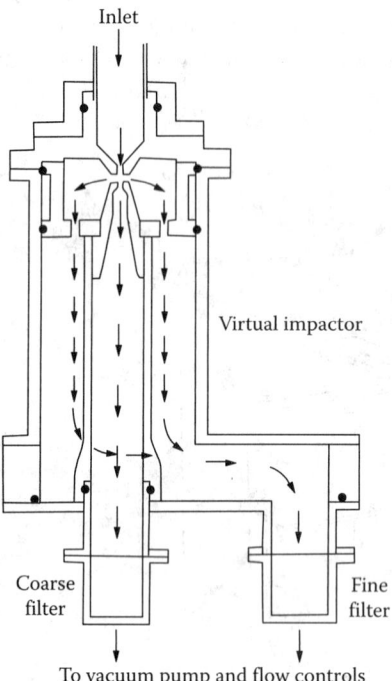

FIG. 1.54r
Virtual impactor.

Filters are typically 1.46 in. (37 mm) in diameter and may be of either glass fiber or membrane construction. Flow rates are accurately controlled to ensure the stability of the size cut points. Both automatic and manual models are available. Automatic units also provide for the automatic changing of the filters after a preselected sampling time interval has passed. Filters are also often analyzed using x-ray fluorescence to determine the elemental composition of the particulate.*

---

* Dzubay, T. G. and Stevens, R. K., Ambient analysis with dichotomous sampler and X-ray fluorescence spectrometer, *Environ. Sci. Technol.*, 9(7), July 1975.

Tape Sampler The tape sampler (Figure 1.54s) provides an indication of the soiling properties of the sampled air. It operates by drawing air through a paper tape at a rate of about 0.75 actual cubic feet per minute (ACFM) for periods of 1, 2, or more hours. Both filter paper and membrane tapes have been used.

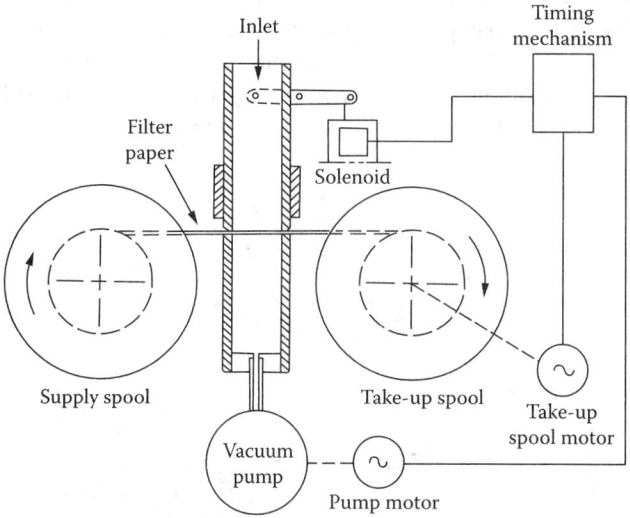

FIG. 1.54s
Tape sampler.

The soiling index (optical properties) of the filtered particulate is measured either by reflectance or transmittance. These optical measurements cannot be directly correlated with the mass concentration of particulate matter in the atmosphere except where particle size, shape, and color are constant and a series of comparative tests have been made using the high-volume sampler.

Soiling Index

Soiling index results are expressed in coefficients of haze per 1000 linear feet (COH/1000 LF) for transmittance measurements and in reflectance units of dirt shade (RUDS) for reflectance measurements. Coefficient of haze (COH) units are defined as 100 times the optical density determined by transmittance.

The reflectance unit of dirt shade is defined as 100 times the optical density determined by reflectance. RUDS are expressed either in terms of 10,000 LF of air (American Society for Testing and Materials unit) or 1,000 LF of air.

Some older tape samplers use an air pump, which is quite sensitive to system resistance. It is advisable to measure the airflow rate through both a clean spot and a dirty spot to determine whether an air volume correction to an average rate is advisable. The diameter of the spot is not a variable, because the results are expressed in terms of the length of the air column that is drawn through the spot.

Nephelometers

The nephelometer has had some use in air pollution studies. This device measures the angular scatter of light over a wide range in a flowing air sample. The instrument measures the scattering coefficient, which is in inverse proportion to the visual range. Thus, the nephelometer can be used to obtain a point measurement of visual range. The span of measured range with the nephelometer is from approximately 1 to 100 miles (1.6–160 km).

In industrial applications, instruments based on the Tyndall effect have been built. These instruments (nephelometers) utilize the phenomenon that particles that are invisible when viewed directly in the path of a strong light become visible when viewed from the side. The reflected Tyndall light intensity is proportional to the number of particles in the path of the light. This analyzer consists of a cylindrical sample cell with two Tyndall windows at each end, with their common axis perpendicular to the path of the entering light.

Piezoelectric Crystal Mass Balance

Some crystals, when excited by an alternating current, exhibit a resonant frequency that is proportional to the crystal mass. Quartz is commonly used and is readily available for this purpose. This principle is used to continuously monitor the mass concentration of particulate matter in air by trapping particulate from the impinging air stream on the sensing crystal while keeping the reference crystal clean.

Thereby, the resonant frequency of the sensing crystal changes, and a beat frequency develops. The particle mass is determined based on the mass vs. the frequency response characteristics of the crystal. Commercially available quartz crystals show a frequency shift of 2000 Hz/μg of material deposited.

A major limitation is the deposition and retention of particles on the sensing crystal. Two techniques, impaction and electrostatic precipitation, have been used commercially. Both approaches have weaknesses in collecting and retaining particles over the wide size range encountered in atmospheric aerosols.

Impaction Devices

Determination of the particle size distribution in aerosols is frequently desired. Although light-scattering devices can provide an indication of particle size distribution, most atmospheric measurements are made using impaction devices.

In principle, more inertial energy is required to impact small particles than large ones when they are impinged on a plane surface. This characteristic has been utilized through a series of impaction stages, with succeeding stages operated at higher velocities (Figure 1.54t).

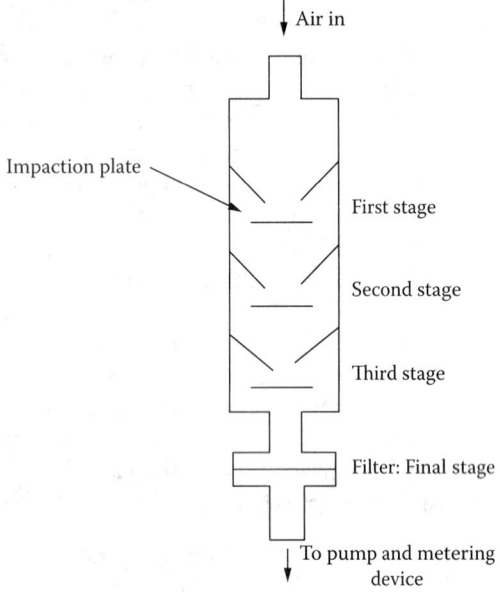

FIG. 1.54t
A schematic view of impaction device.

The upper practical size limit for impaction devices is about 10 μm. They are particularly useful in the size range of 0.5–7 μm, which corresponds to the approximate size range of most particulate matter in the ambient air.

A cascade impactor can be added to a manual stack sampling probe to evaluate stack emissions, and a fractionating sampler can be added to a high-volume sampler for ambient air analysis. In most cases, the impaction plate is removed from the sampling device and is weighed to determine the amount of sample. High plate tare weight is a potential source of inaccuracy.

Calibration Calibration of impaction devices is difficult. The normal practice involves the generation of a monodisperse aerosol of known particle size. Polystyrene spheres, such as manufactured for latex paint, have been used. Assuming that the dispersion technique generates a monodisperse aerosol (often the most difficult part of the procedure), the calibration is performed using a series of sizes. The calibration is in terms of aerodynamic size, that is, the inertial properties, which are a combination of particle size and specific gravity. The results from impaction sampling must state the specific gravity, whether measured or assumed, with the stated size distribution.

Radiometric Devices

Radiometric measurements of particulate are primarily accomplished using a radiation attention technique or by measuring the ionization effect of a radioactive source on ambient air. Typically, low-energy beta-type radioactive sources are used to minimize the human exposure considerations and associated regulatory problems.

In the typical beta gauge shown in Figure 1.54u, the air to be measured is drawn through a tape filter medium, where the particulate is collected. The radiometric transmission or attenuation of the filter medium is measured in the clean state (before collection) and in the dirty state (after collection). By appropriate processing of these two measurements, a measure is obtained that corresponds to the mass of collected particulate. The gas volume is usually accurately controlled so that the sample volume is precisely defined.

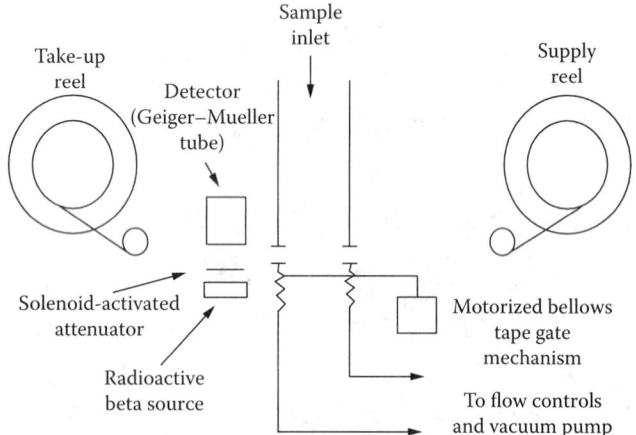

FIG. 1.54u
Beta-gauge particulate monitor.

The system is relatively insensitive to the normal chemical variation of particulate matter; however, it is especially responsive to hydrogen compounds. This technique has been widely applied to stack gas and ambient air measurements. The mechanical complexity and high costs, both the installed initial cost and maintenance costs, have limited the acceptance of these systems.

The ionizing technique has been applied to a variety of low-cost fire/smoke detectors. In this application, ambient air is exposed to a radiation source that partially ionizes the air molecules and allows a current flow through a pair of electrodes. As particulate concentrations increase, the radiation is absorbed as the particulate matter decreases the ionization and hence the current through the electrodes.

Charge Transfer (Triboelectricity)

The triboelectric effect (also known as *triboelectric charging*) is a type of contact electrification in which certain materials become electrically charged after they come into contact with another different material through friction. For example, rubbing a comb through the hair can build up triboelectricity. Most everyday static electricity is triboelectric. The polarity and strength of the charges produced differ according to the materials, surface roughness, temperature, strain, and other properties.

Both of these techniques utilize an in situ probe exposed, typically, to the stack or flue gas stream of interest. Using the charge transfer technique, a measurement is produced by detecting the charge transferred to an electrode surface on impaction with particulate matter.

The charge transfer mechanism requires that only dissimilar materials be involved. The same holds for the triboelectric charging phenomena. With appropriate electronic amplification and scaling, the output can be related to particulate concentration. This technique has not gained wide acceptance, presumably because of the difficulty in correlating charge transfer with particulate concentrations and in keeping the probe mechanism operable under the severe and diverse operating conditions experienced in stack applications.

Surface Ionization

In the surface ionization technique, airborne particulate decomposes when it impacts a hot wire, and there is a resultant release of ions that are then collected on an electrode. By detecting both the average ion current and analyzing the pulse height and rate, this instrument is able to provide an indication of total mass concentration and the relative distribution of particle sizes, respectively.

This technique has a high sensitivity to low particulate concentrations. Both techniques are essentially point measurements that provide no averaging effect including the overall stack or duct cross-sectional area.

Visual Observation

Human observation of smoke and dust was, of course, the initial stimulus to develop better techniques for control and monitoring. Human observation still remains one of the primary enforcement tools used by the Environmental Protection Agency (40 CFR 60 Appendix A, Reference Method 9) to ensure the proper operation and maintenance of particulate control equipment.

The EPA does provide training in smoke reading, and only certified observers are used for such enforcement activity. Studies have been undertaken to quantify the errors of such measurement techniques, and human frailties certainly are involved in addition to the angle of the sun, the sky cover, contrast, type of particulate, and condensables such as steam.

Remote Sensing

Several techniques that enable an observer to monitor smoke plume characteristics from a remote location were discussed in Chapter 1.62. Telephotometers have been used manually to establish smoke plume opacity based on the luminance of the smoke plume as measured with different background characteristics.

Light direction and ranging (LIDAR) has been developed as a research tool using a laser to quantify smoke plume characteristics. LIDAR is relatively expensive and complex but has proven useful in the research environment. It is normally contained in a mobile laboratory that can be driven to remote locations so as to provide good observations of a particular smoke plume. Aircraft have also been equipped with proper instrumentation to permit observation and measurement of smoke plume characteristics while they are flown through the plume.

CONCLUSIONS

Particulate-monitoring methods have been developed that achieve high standards of accuracy and reliability. Under the impetus of the EPA, such methods have become widely used in the United States. Future developments are anticipated in the particle sizing area as regulatory agencies become less concerned with the visual effects of smoke and more concerned with its health effects. Visual effects, highway and airport visibility in particular (Figure 1.54v), will continue to be of substantial interest as will the conditions of scenic vistas and parks.

Transmissiometer with revolving viewing window

Combination visibility sensor, utilizing both attenuation and forward scatter technologies

FIG. 1.54v
Advanced runway visual range (RVR) detectors are also capable of measuring the intensity of smoke, dust, or blowing sand. (Courtesy of AWI.)

SPECIFICATION FORMS

When specifying particulate or opacity monitors, the ISA Form 20A1001 can be used to specify the detector device itself, and when specifying both the device and the nature of the application, including the composition and/or properties of the process sample, ISA Form 20A1002 can be used. The following are the forms that are reproduced with the permission of the International Society of Automation.

#	RESPONSIBLE ORGANIZATION	ANALYSIS DEVICE	#	SPECIFICATION IDENTIFICATIONS		
1			6			
2	(ISA)	Operating Parameters	7	Document no		
3			8	Latest revision	Date	
4			9	Issue status		
5			10			

#	ADMINISTRATIVE IDENTIFICATIONS		#	SERVICE IDENTIFICATIONS Continued		
11			40			
12	Project number	Sub project no	41	Return conn matl type		
13	Project		42	Inline hazardous area cl	Div/Zon	Group
14	Enterprise		43	Inline area min ign temp	Temp ident number	
15	Site		44	Remote hazardous area cl	Div/Zon	Group
16	Area	Cell Unit	45	Remote area min ign temp	Temp ident number	
17			46			
18	SERVICE IDENTIFICATIONS		47			
19	Tag no/Functional ident		48	COMPONENT DESIGN CRITERIA		
20	Related equipment		49	Component type		
21	Service		50	Component style		
22			51	Output signal type		
23	P&ID/Reference dwg		52	Characteristic curve		
24	Process line/nozzle no		53	Compensation style		
25	Process conn pipe spec		54	Type of protection		
26	Process conn nominal size	Rating	55	Criticality code		
27	Process conn termn type	Style	56	Max EMI susceptibility	Ref	
28	Process conn schedule no	Wall thickness	57	Max temperature effect	Ref	
29	Process connection length		58	Max sample time lag		
30	Process line matl type		59	Max response time		
31	Fast loop line number		60	Min required accuracy	Ref	
32	Fast loop pipe spec		61	Avail nom power supply		Number wires
33	Fast loop conn nom size	Rating	62	Calibration method		
34	Fast loop conn termn type	Style	63	Testing/Listing agency		
35	Fast loop schedule no	Wall thickness	64	Test requirements		
36	Fast loop estimated lg		65	Supply loss failure mode		
37	Fast loop material type		66	Signal loss failure mode		
38	Return conn nominal size	Rating	67			
39	Return conn termn type	Style	68			

#	PROCESS VARIABLES	MATERIAL FLOW CONDITIONS		#	PROCESS DESIGN CONDITIONS		
69				101			
70	Flow Case Identification		Units	102	Minimum	Maximum	Units
71	Process pressure			103			
72	Process temperature			104			
73	Process phase type			105			
74	Process liquid actl flow			106			
75	Process vapor actl flow			107			
76	Process vapor std flow			108			
77	Process liquid density			109			
78	Process vapor density			110			
79	Process liquid viscosity			111			
80	Sample return pressure			112			
81	Sample vent/drain press			113			
82	Sample temperature			114			
83	Sample phase type			115			
84	Fast loop liq actl flow			116			
85	Fast loop vapor actl flow			117			
86	Fast loop vapor std flow			118			
87	Fast loop vapor density			119			
88	Conductivity/Resistivity			120			
89	pH/ORP			121			
90	RH/Dewpoint			122			
91	Turbidity/Opacity			123			
92	Dissolved oxygen			124			
93	Corrosivity			125			
94	Particle size			126			
95				127			
96	CALCULATED VARIABLES			128			
97	Sample lag time			129			
98	Process fluid velocity			130			
99	Wake/natural freq ratio			131			
100				132			

#	MATERIAL PROPERTIES		#	MATERIAL PROPERTIES Continued		
133			137			
134	Name		138	NFPA health hazard	Flammability	Reactivity
135	Density at ref temp	At	139			
136			140			

Rev	Date	Revision Description	By	Appv1	Appv2	Appv3	REMARKS

Form: 20A1001 Rev 0 © 2004 ISA

	RESPONSIBLE ORGANIZATION	ANALYSIS DEVICE COMPOSITION OR PROPERTY Operating Parameters (Continued)		SPECIFICATION IDENTIFICATIONS		
1			6			
2			7	Document no		
3	ISA		8	Latest revision		Date
4			9	Issue status		
5			10			

	PROCESS COMPOSITION OR PROPERTY			MEASUREMENT DESIGN CONDITIONS			
	Component/Property Name	Normal	Units	Minimum	Units	Maximum	Units	Repeatability

(rows 11–75)

Rev	Date	Revision Description	By	Appv1	Appv2	Appv3	REMARKS

Form: 20A1002 Rev 0 © 2004 ISA

Definitions

Coefficient of haze (also called smoke shade) is a measurement of visibility interference in the atmosphere. It is measured by drawing a 1000 ft air sample through an air filter and detecting the radiation intensity through the filter.

Nephelometer an instrument for measuring the extent or degree of cloudiness. It is used to determine the concentration or particle size of solid particles in suspensions by means of transmitted, scattered, or reflected light.

Optical density (also called extinction or absorbance): Optical density is related to transmittance (T), which is the ratio of the amount of radiation that passes through the sample (I), divided by the radiation that enters the sample ($I_{incident}$). This ratio is called transmittance and is expressed as a percentage (%T = $I/I_{incident}$). Optical density (OD) is defined as OD = $2 - \log_{10}$ %T. Therefore, the higher the transmittance, the lower the optical density, meaning that the sample is optically less dense and therefore less absorbing. For example, if 0.1% of the light is transmitted (%T = 0.1%), the optical density is 3, while if 10% of the light passes through (%T = 10%), the optical density is only 1.

Ringelmann scale is a scale for measuring the apparent density of Smoke. It has a five levels of density inferred from a grid of black lines on a white surface which, if viewed from a distance, merge into known shades of gray. There is no definitive chart, rather, Prof. Ringelmann provides a specification, where smoke level "0" is represented by white, levels "1" to "4" by 10 mm square grids drawn with 1, 2.3, 3.7, and 5.5 mm wide lines, and level "5" by all black.

Soiling index: It is a measure of the darkening potential of smoke and soot suspended in one cubic meter of air.

Transmissiometers: They measure the extinction coefficient of the atmosphere and serve to measure the visual range by sending a narrow laser beam through the air or other propagation medium. Atmospheric extinction is a wavelength-dependent phenomenon, but the most common wavelength in use for transmissometers is 550 nm (0.55 μm.) which is in the middle of the visible waveband, and allows a good approximation of the visual range.

Transmittance: In optics and spectroscopy, transmittance is the fraction of incident light (electromagnetic radiation) at a specified wavelength that passes through a sample. The terms visible transmittance (VT) and visible absorptance (VA), which are the respective fractions for the visible spectrum of light, are also used.

Triboelectric effect: (also known as *triboelectric charging*) is a type of contact electrification in which certain materials become electrically charged after they come into contact with another different material through friction. For example, rubbing a comb through the hair can build up triboelectricity. Most everyday static electricity is triboelectric. The polarity and strength of the charges produced differ according to the materials, surface roughness, temperature, strain, and other properties.

Visibility range is the distance at which a given standard object can be seen with the unaided eye.

Abbreviations

ACFM	Actual cubic feet per minute
AST	Automatic sampling train
CDMA	Code division multiple access
COH	Coefficient of haze
GSM	Global system mobile
LCD	Liquid crystal display
LIDAR	Light direction and ranging
LF	Linear Feet
MACT	Maximum achievable control technology
NESHAP	National Emission Standards for Hazardous Air Pollutants
RUDS	Reflectance units of dirt shade
RVR	Runway visual range
SMS	Short message service
TDLAS	Tunable diode laser absorption spectroscopy
VA	Visible absorbance
VT	Visible transmittance

Organizations

ASTM	American Society for Testing and Materials
BSI	British Standards Institution
EPA	Environmental Protection Agency
ISO	International Organization for Standardization
VDI	Verein Deutscher Ingenieure

Standards

ANSI/ASTM D3685–78; Standard Test Method for Sampling and Determination of Particular Matter in Stack Gases, http://www.astm.org/Standards/D3685.htm.

EPA, Standards of performance for new stationary sources, title 40, Part 60, Appendix A. 1991. http://www.deq.state.or.us/aq/forms/sourcetest/appendix_a1.pdf.

EPA 40, CRF Part 50: National Ambient Air Quality Standards for Particulate Matters, 2006, http://www.epa.gov/ttnamti1/files/ambient/pm25/pt5006.pdf.

EPA 40 CFR 60 Appendix A, Method 5; Determination of Particulate Matter Emissions from Stationary sources http://www.epa.gov/ttn/emc/promgate/m-05.pdf.

EPA 40 CFR 60 Appendix A, Method 9; Visual Determination of the Opacity of Emissions, http://www.epa.gov/ttnemc01/promgate/m-09.pdf.

EPA 40 CFR 60 Appendix B, Perf Spec 1, Standards of Performance for Stationary Sources http://www.deq.state.or.us/aq/forms/sourcetest/appendix_b.pdf.

Bibliography

American Institute of Chemical Engineers, Particulate continuous emission monitoring (CD ROM). 1999. http://www.amazon.co.uk/Particulate-Continuous-Emissions-Monitoring-Proceedings/dp/0816908028.

AMP, EPA Method sampling system. 2015. http://www.ampcherokee.com/v/vspfiles/photos/C002.0002-2T.jpg.

ASTM D3685/D3685M-13, *Standard Test Methods for Sampling and Determination of Particulate Matter in Stack Gases*, ASTM International. 2013. http://www.astm.org/Standards/D3685.htm.

ASTM D6331-14, *Standard Test Method for Determination of Mass Concentration of Particulate Matter from Stationary Sources at Low Concentrations* (*Manual Gravimetric Method*). 2014. http://www.astm.org/Standards/D6331.htm.

ASTM D6831-11, *Standard Test Method for Sampling and Determining Particulate Matter in Stack Gases Using an In-stack*, Inertial Microbalance. 2011. http://www.astm.org/Standards/D6831.htm.

AUBURN, *Bag Leak Detector*. 2014. http://auburnsys.com/content/triboguard-i-model-4001.

AVL. Smoke value measurement. 2008. https://www.avl.com/c/document_library/get_file?uuid=2b39210c-6937-43e4-b223-b5c2c11f91ac&groupId=10138.

Belfort Instrument, Runway visual range. 2015. http://belfortinstrument.com/products/runway-visual-range/.

Chang, D. P. Y., Transmissometer measurement of participate emissions from a jet engine test facility. 1977. http://www.tandfonline.com/doi/pdf/10.1080/00022470.1977.10470473.

Clarke, A. G., Industrial Air Pollution Monitoring—Gaseous and Particulate Emissions. 1997. http://www.amazon.co.uk/Industrial-Air-Pollution-Monitoring-Environmental/dp/toc/0412633906.

Clean Air Engineering, Stack and source emission testing. 2014. http://www.cleanair.com/services/StackSourceTesting/index.htm.

Conner, W. D. and Hodkinson, J. R., Optical Properties and Visual Effects of Smoke-Stack Plumes. 1972. http://search.aol.com/aol/search?enabled_terms=&s_it=wscreen50-bb&q=+Conner%2C+W.+D.+and+Hodkinson%2C+J.+R.+Optical+Properties+and+Visual+Effects+of+Smoke-Stack+Plumes%2C+.

Cooper, C. D. and Alley, F. C., Air pollution. 2002. http://www.amazon.com/Coopers-Pollution-Control-Edition-Hardcover/dp/B003WUI3I6.

EPA, Current knowledge of particulate matter (PM) continuous emission monitoring. 2002. http://www3.epa.gov/ttnemc01/cem/pmcemsknowfinalrep.pdf.

EPA, Method 5—Determination of particulate matter emissions from stationary sources. 2011. http://www3.epa.gov/ttnemc01/promgate/m-05.pdf.

EPA, National ambient air quality standards for particulate matter. 2013. https://www.gpo.gov/fdsys/pkg/FR-2013-01-15/pdf/2012-30946.pdf.

EPA, Opacity test methods. 2013. http://www.epa.gov/region9/air/phoenixpm/fip/method.html.

ETA, Visible emission observer. 2013. http://www.eta-is-opacity.com/ETA_VE_manual.pdf.

Harrison, R. M. et al. Air quality in urban environments. 2009. http://www.amazon.com/Quality-Environments-Environmental-Science-Technology/dp/1847559077/ref=pd_sim_sbs_b_1.

ICS: Stack Sampler. 2015. http://www.indiamart.com/industrial-commercial-services/environment-equipments.html.

Konstantinos, A., Smoke detection. 2008. http://www.demcare.eu/downloads/smokedet.pdf.

Majewski, W. A. and Hannu, J., Smoke opacity. 2013. http://www.dieselnet.com/tech/measure_opacity.php.

Monitor, Broken bag detector. 2015. Monitor (broken bag detector). http://www.monitortech.com/product_pe_broken.shtml.

Pedace, E. A. and Sansone, E. B., The relationship between "Soiling Index" and suspended particulate matter concentrations. 1972. http://www.tandfonline.com/doi/abs/10.1080/00022470.1972.10469644.

Preferred Instruments. Smoke opacity monitor installation. 2008. http://www.preferredinstruments.com/assets/documents/JC-30D,%20Smoke%20Opacity%20Monitor%20Installation%20Manual.pdf.

Teledyne Analytical Instruments, Particle measurement system. 2012. http://www.teledyne-api.com/manuals/07318b_602.pdf.

Thermo Scientific, Particulate monitoring. 2015. http://www.thermoscientific.com/en/products/particulate-monitoring.html.

TSI, Air quality monitoring. 2015. http://www.tsi.com/Air-Quality-Monitoring/.

1.55 pH Measurement

A. C. BLAKE (1969)   **T. S. LIGHT** (1972)

D. L. HOYLE and T. J. MYRON (1974, 1982)

G. K. McMILLAN (1995)   **J. R. GRAY** (2003, 2017)

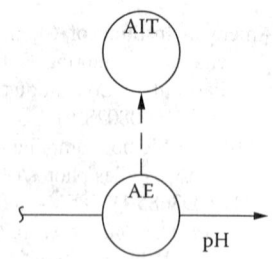

Flow sheet symbol

Standard design pressures	Vacuum to 100 psig (7 bars); special assemblies to 500 psig (35 bars)
Standard design Temperatures	Generally, −5°C–100°C (23°F–212°F); sterilizable, −30°C to 130°C (−22°F to 266°F); Glasteel®, (<5 pH) −5°C to 140°C (23°F–284°F)
Materials of construction	Electrode hardware: stainless steel, monel, Hastelloy®, titanium, epoxy, Kynar®, halar, polyvinyl chloride, chlorinated polyvinyl chloride, polyethylene, polypropylene, polyphenylene sulfide, Ryton®, Teflon®, various elastomer materials
Assemblies	Flow-through, submersion, insertion, and retractable
Cleaners	Ultrasonic, jet washer (chemical and water), and brush
Inaccuracy	Electrodes 0.02 pH; lab meters and displays 0.01 pH; transmitters 0.02 mA; installation effects 0.2 pH
Range	0–14 pH
Costs	pH sensors, $200 (general purpose) to $1300 (high purity water); lab meters, $200–$800; transmitters, $1000 (two-wire transmitters) to $2000 or more (line-powered transmitter); assemblies, $500–$2000; automatic cleaning systems, $8000 or more
Partial list of suppliers	ABB http://new.abb.com/products/measurement-products Advanced Sensor Technologies Inc. http://www.astisensor.com/pH_ORP_Products.htm Analytical Technology Inc. http://www.analyticaltechnology.com/public/product.aspx?ProductID=1071 Broadley-James http://www.broadleyjames.com/dir-sensorsPH.html Electro-Chemical Devices http://www.ecdi.com/products/t80.html Endress+Hauser http://www.us.endress.com/#products/pH-transmitters +GF+ Signet, http://www.gfps.com/content/gfps/signet/en_US/sensors/phorp/instruments.html GLI International http://www.hach.com/sc200-models/category-products?productCategoryId = 14371221969&secondPageNumber = 1&isNew = false&pimContext = USen Hach Co. http://www.hach.com/sc200-models/category-products?productCategoryId=14371221969&secondPageNumber=1&isNew=false&pimContext=USen Hamilton Co. http://www.hamiltoncompany.com/products/sensors/c/12/

(Continued)

Hanna Instruments http://www.hannainst.com/usa/prods2.cfm?id = 025004&ProdCode = HI 21
Honeywell https://www.honeywellprocess.com/en-US/explore/products/instrumentation/analytical-instruments-and-sensors/ph-orp/Pages/default.aspx
Horiba Instruments, Inc. http://www.horiba.com/us/en/process-environmental/products/water-treatment-environment/on-site-h-1-series/
Knick http://www.knick-nternational.com/products/measuring_systems/
Mettler-Toledo http://us.mt.com/us/en/home/products/Process-Analytics/transmitter.html
Pfaudler Inc. http://www.pfaudler.com/accessories_and_instruments.php
Rosemount Analytical http://www2.emersonprocess.com/en-US/brands/rosemountanalytical/Liquid/Instruments/Pages/index.aspx
Sensorex http://www.sensorex.com/products/categories/category/ph_sensors
Swan Analytical Instruments http://www.swan.ch/Catalog/en/ProductOverview.aspx
Teledyne Analytical Instruments http://www.teledyne-ai.com/products/lxt.asp
The Foxboro Company http://www.fielddevices.foxboro.com/en-gb/products/analytical-echem/ph-and-orp/
Thermo Scientific http://www.thermoscientific.com/en/products/ph-measurement.html
Van London—Phoenix Co. www.vl-pc.com
Yokogawa http://www.yokogawa.com/us/products/analytical-products/ph-orp/index.htm

INTRODUCTION

The pH detectors belong to the family of ion-selective electrodes, so the reader is advised to refer also to Chapter 1.33 in this chapter, which discusses the general topic of ion-selective electrodes.

One of the important considerations in the design of pH systems is the means by which the probes are kept clean. Probe cleaners are also discussed in other sections of this chapter, so the reader might also refer to Figures 1.55q, 1.55r, 1.55s, and 1.55t for the description of various probe-cleaning designs.

Measurement Error

The pH measurement error in process applications is an order of magnitude larger than the normally stated electrode error, which is a result of varying glass surface conditions, dissociation constants, streaming potentials, concentration gradients, and diffusion or liquid junction potentials of actual installations.

Optimistic users and suppliers who provide one electrode per point in the process may believe that they have achieved accuracies of 0.02 pH or better. Results of installations with three electrodes per point in the process show that the inaccuracy over a month is at best 0.25 pH and can be obtained only after frequent process calibration adjustments. In fact, if all three electrodes agree within a tenth of a pH for more than a few minutes in an industrial application, it indicates that the electrodes are probably coated, broken, or still have protective caps on them.*

In contrast, a bioreactor can have two pH electrodes consistently within 0.1 pH of one another, when initially standardized to a referee pH analyzer. This is because the reactor is well mixed avoiding concentration gradients and has a uniform controlled temperature.

---
* McMillan, G. K., Understand some basic truths of pH measurement, *Chem. Eng. Prog.*, 87(10), October 1991, pp. 30–37.

Measurement Range

The measurement of pH covers an incredibly wide range of extremely dilute acid and base concentrations (Figure 1.55a). For strong acids and bases, it can track changes from one to one millionth of a percent.* Thus, pH can be used as a sensitive indicator of deficient and excess acid and base reactant concentrations for chemical reactors and scrubbers.

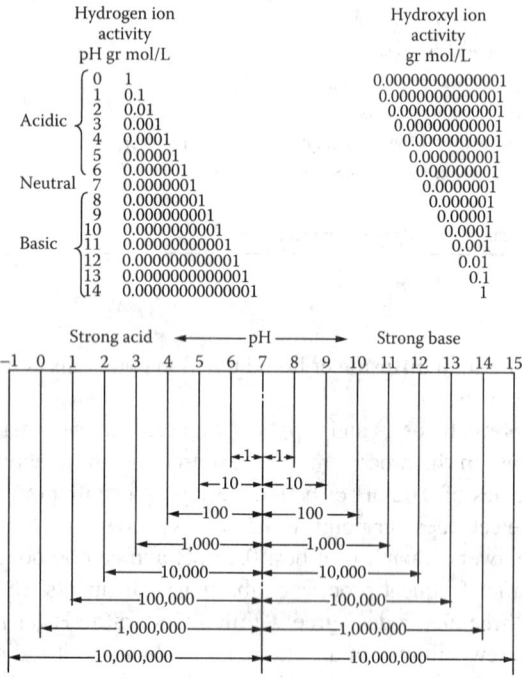

FIG. 1.55a
The logarithmic nature of pH.

For example, a few millionths of a percent of excess of sodium hydroxide (a strong base) is needed for chlorine destruction with sodium bisulfite. pH measurement is used to reduce sodium hydroxide addition to its bare minimum and still ensure complete use of sodium bisulfite.

Applications

In addition, pH measurement is used to correct the inference of hypochlorite concentration from an oxidation–reduction potential (ORP) measurement (Chapter 1.47). Since ORP curves drastically shift with pH, the use of ORP to infer the concentration of oxidizing or reducing species must be accompanied by an accurate pH measurement.[†]

Biological reactors use acids and bases to supply food to or neutralize the waste products of organisms. Cells are extremely sensitive to pH fluctuations. Genetically engineered bacteria tend to be weak and need particularly tight pH control. Thus, pH is critical to cell growth rate, enzyme reactions, and extraction of intercellular products.[‡]

The sensitivity of cells to pH has an even wider significance in that any food, drink, or drug ingested or injected, and any waste discharged to the environment, must have pH specifications to prevent damage to living matter and ecological systems. Stricter environmental regulations have increased the number and importance of pH measurements. Some environmental regulations have instantaneous limits on pH. An excursion outside the acceptable range for a fraction of a second can be a recordable violation.

The pH measurement system must be designed to prevent violation indications from spurious readings due to installation effects. For most discharges, the acceptable range lies between 6 and 9 pH. To ensure that a surface impoundment can be declassified (nonhazardous), the pH of all entering streams must always be between 2 and 12.5 pH.[§]

Whereas materials of construction are generally less sensitive to pH than are living cells, a range of pH must be maintained to prevent corrosion of metals or disintegration of plastics. Intermediate pH control is often needed to prevent damage to vessels, piping, and instrumentation.

THEORETICAL REVIEW

In pH measurement systems, a pH-responsive glass takes up hydrogen ions and establishes a potential at the glass surface with respect to the solution. This potential is related to the hydrogen ion activity of the solution by the Nernst relationship

$$E_g = E_g^o + \frac{2.303RT}{F} \log_{10} a \qquad 1.55(1)$$

where
- E_g is the sum of reference potentials and liquid junction potentials, which are constants (in millivolts)
- E_g^o is the potential when $a = 1$
- a is the hydrogen ion activity
- T is the absolute temperature in kelvins (°C + 273)
- R is the 1.986 calories per mol degree
- F is the Faraday (coulombs per mol)
- 2.303 is the logarithm conversion factor

The process variable pH is the negative logarithm of the hydrogen ion (i.e., proton) activity:

$$pH = -\log(a_H) \qquad 1.55(2)$$

If both sides of the equation are multiplied by −1, and the definition of an antilogarithm is used, the result shows that the hydrogen ion activity is equal to 10 raised to the negative

---

* McMillan, G. K., pH Control, Instrument Society of America, Research Triangle Park, NC, 1984.
† Shinskey, F. G., *pH and pION Control in Process and Waste Streams*, John Wiley & Sons, New York, 1973.
‡ McMillan, G. K., *Biochemical Measurement and Control*, Instrument Society of America, Research Triangle Park, NC, 1987.
§ McMillan, G. K., Woody's performance review–What's inline next for pH control, *InTech*, 35(1), January 1988.

power of pH. The lowercase p designates the mathematical relationship between the ion and the variable as a power function; the H denotes the ion is hydrogen:

$$a_H = 10^{-pH} \quad \quad 1.55(3)$$

For dilute aqueous (water) solutions, the activity coefficient is approximately unity, and the hydrogen ion concentration is essentially equal to the hydrogen ion activity. As the concentrations of acids, bases, and salts increase, the crowding effect of the ions reduces the hydrogen ion activity. Thus, an increase in salt concentration can cause an increase in pH reading, even though the hydrogen ion concentration is constant.

An acid is a proton donor, and a base is a proton acceptor. When an acid dissociates (breaks apart into its component ions), it yields a hydrogen ion and a negative acid ion. When a base dissociates, it gives a positive base ion and a hydroxyl ion that is a proton acceptor. Acids and bases that completely dissociate are called strong acids and bases. Weak acids and bases only partially dissociate. When water dissociates, the result is both a hydrogen ion (proton) and hydroxyl ion (proton acceptor).*

Thus, water acts as both an acid and a base. Neutralization is the association of hydrogen ions from acids and hydroxyl ion from bases to form water. The following equations show the dissociation (forward arrow) and association (backward arrow) of an acid, base, and water, respectively:

$$HA = H^- + A^- \quad \quad 1.55(4)$$

$$BOH = B^+ + OH^- \quad \quad 1.55(5)$$

$$H_2O = H^- + OH^- \quad \quad 1.55(6)$$

The product of the hydrogen and hydroxyl ion concentrations in water solutions is equal to 10 raised to the negative power of pK_w where K_w is the water dissociation constant:

$$[H^+][OH^-] = 10^{-pK_w} \quad \quad 1.55(7)$$

The neutral point is where the hydrogen ion concentration equals the hydroxyl ion concentration. At 25°C, this occurs at 7 pH. However, the pK_w decreases by about −0.35 pH per 10°C. A decrease in temperature will raise the neutral point and the pH reading of alkaline solutions, even though the concentration of acids and bases is constant. For example, a stream with strong electrolytes will change from 8 to 11 pH as it cools down from 90°C to 25°C.†

Temperature compensators correct for the change in millivolts per pH unit per the Nernst equation, but do not correct for this change in the actual pH of the solution with temperature. Since the change in pH readings with temperature of solution pH changes is often larger than that per the Nernst equation, microprocessor-based pH transmitters have added algorithms for solution pH correction.

Ion Concentrations

At 25°C, pK_w is 14, and the product of the hydrogen and hydroxyl ion concentrations is 10^{-14}. This relationship is shown in Table 1.55a for the pH range of 0–14. This means that a pH measurement can track 14 decades of hydrogen ion concentration and detect changes as small as 10^{-14} (at 14 pH). The concentration changes of strong acids and bases also follow the decade change per pH unit within this range. No other concentration measurement has such rangeability and sensitivity. This has profound implications for pH control.‡

TABLE 1.55a
Concentrations of Active Hydrogen and Hydroxyl Ions at 25°C at Different pH Values (in Gram-Moles/Liter) and Some Examples of Fluids That Have Corresponding pH Values

pH	Fluid Example	Hydrogen Ions	Hydroxyl Ions
0	4% sulfuric acid	1.0	0.00000000000001
1		0.1	0.0000000000001
2	Lemon juice	0.01	0.000000000001
3		0.001	0.00000000001
4	Orange juice	0.0001	0.0000000001
5	Cottage cheese	0.00001	0.000000001
6	Milk	0.000001	0.00000001
7	Pure water	0.0000001	0.0000001
8	Egg white	0.00000001	0.000001
9	Borax	0.000000001	0.00001
10	Milk of Magnesia	0.0000000001	0.0001
11		0.00000000001	0.001
12	Photo developer	0.000000000001	0.01
13	Lime	0.0000000000001	0.1
14	4% sodium hydroxide	0.00000000000001	1.0

Concentrated strong acids and bases have a pH that lies outside this range. For example, concentrated sulfuric acid has a pH of −10, and concentrated sodium hydroxide has a pH of 19 as measured by a hydrogen electrode. However, the set point of a pH loop is usually well within the 0–14 range.

Some feed-forward pH loops might need measurements outside this range, but the shortened life expectancy and increased error from the electrode at the extremes of the range make such measurements impractical. When the nominal pH value is expected to be near the extremes of the 0–14 pH range, conductivity (Chapter 1.15) is usually a better measurement choice for detecting acid and base concentrations.

pH MEASUREMENT

The hydrogen ion does not exist by itself in aqueous solutions. It is associated with a water molecule to form a hydronium

---

* Bates, R. G., *Determination of pH Theory and Practice*, John Wiley & Sons, New York, 1964, pp. 11–12.
† Kalis, G., How accurate is your on-line pH analyzer, *InTech*, 37(6), June 1990.

‡ McMillan, G. K., pH control—A magical mystery tour, *InTech*, 37(9), September 1984.

ion (H_3O^+). The glass measurement electrode (Figure 1.55b) develops a potential when hydronium ions get close enough to the glass surface for hydrogen to jump and become associated with hydronium ions in an outer layer of the glass surface as illustrated in Figure 1.55c.

This thin hydrated gel layer is essential for electrode response. The input to the pH measurement circuit is a potential difference between the external glass surface that is exposed to the process (E_1) and the internal glass surface that is wetted by a 7 pH solution (E_2). If the external glass

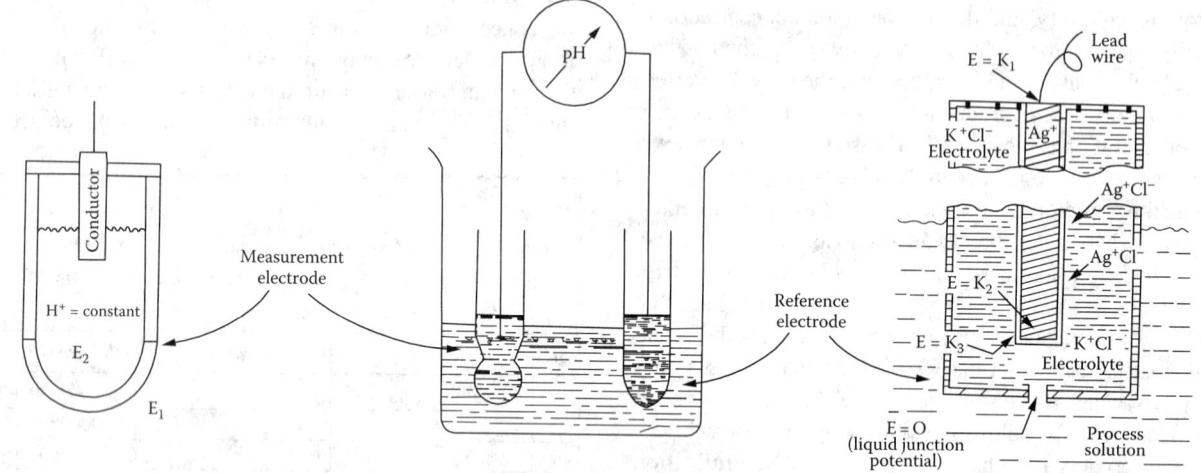

FIG. 1.55b
The traditional configuration of a glass measurement electrode and a flowing junction reference electrode.

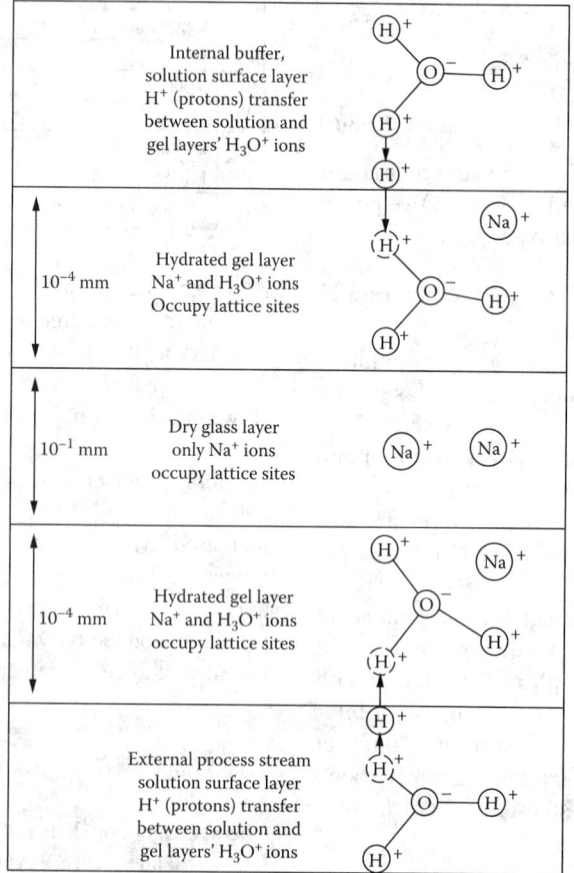

FIG. 1.55c
Functional layers of the glass membrane.

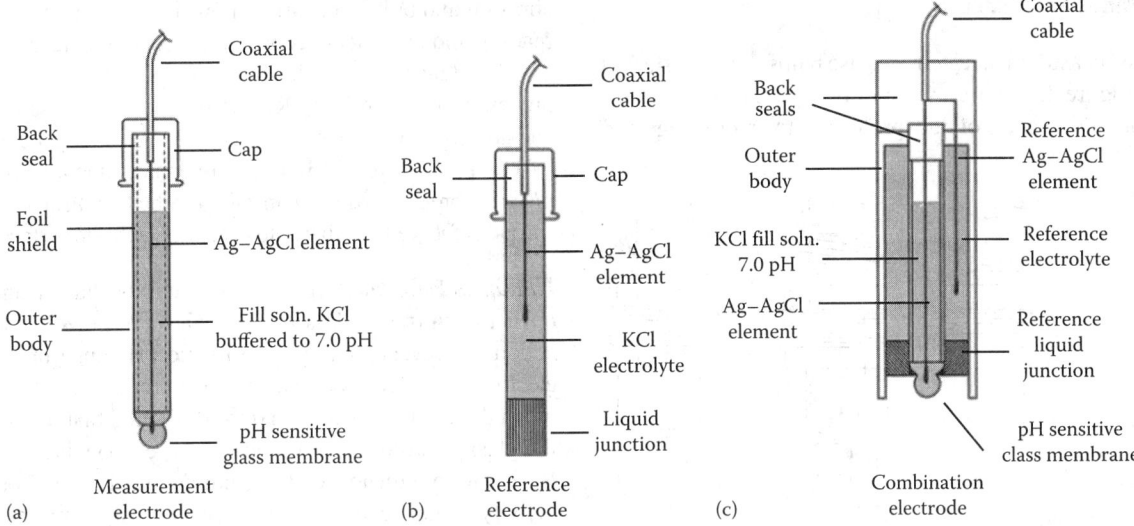

FIG. 1.55d
Basic electrode configurations. (a) Measurement electrode, (b) reference electrode, and (c) combination electrode.

surface is in exactly the same condition as the internal glass surface, the Nernst equations simplify to the following equation, where the potential difference in millivolts is proportional to the deviation of the process pH from 7 pH at 25°C:

$$E_1 - E_2 = 0.1984 * (T + 273.16) * (7 - pH) \qquad 1.55(8)$$

The basic components of measurement, reference, and combination electrodes are shown in Figure 1.55d.

Temperature Effects

The change in the process temperature in Kelvin changes the millivolts generated in an amount proportional to the deviations from 7 pH, as shown in Figure 1.55e. When the process is 7 pH, the millivolt potential is zero, and temperature has no effect on the number of millivolts.

The point of zero temperature effect is called the *isopotential point*, which depends not only on the temperature effect on the pH electrode but also on the temperature effect on the reference electrode. The actual isopotential point will change with temperature as a result of changes in the condition of the external gel layer from dehydration, abrasion, etching, and contamination.

Conventional temperature compensation circuits correct the millivolts only for the Nernst temperature effect and assume that the isopotential point is fixed at 7 pH. Some microprocessor-based transmitters provide an adjustable isopotential point. This is particularly important for nonglass measurement electrodes.

The isopotential point is obtained by differentiating the equation for the entire pH measurement cell, which includes both the pH electrode and the reference electrode, setting the result to zero, and solving for the isopotential pH. The $E^0(T)$ terms in this equation include the potential inside the

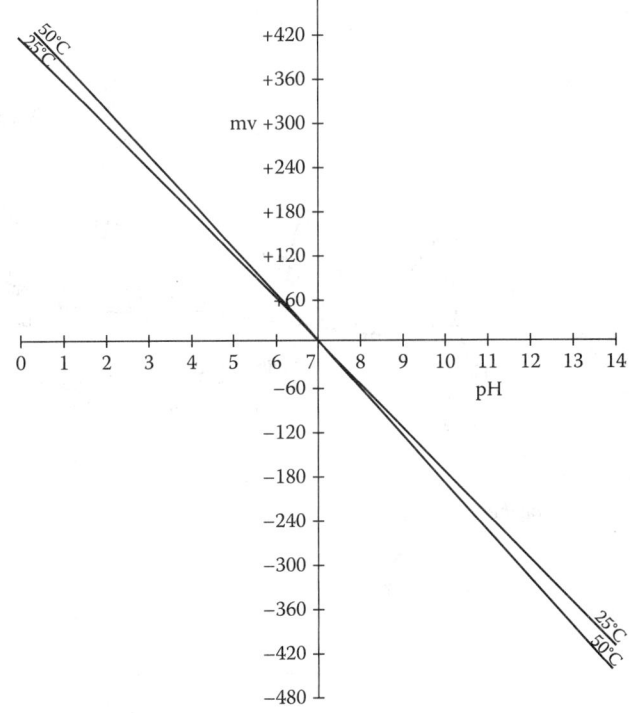

FIG. 1.55e
Nernst temperature effect on millivolts versus pH.

glass electrode, which consists of the inside surface of the glass electrode and its internal silver–silver chloride wire, and the reference electrode:

$$E(T) = E^0(T) + \left[\frac{dE^0(T)}{dT}\right] * (T - 298.15) - 0.1984 * (pH)$$

$$1.55(9)$$

pH Electrodes and Sensors

Most glass measurement electrodes use bulbs. Various shroud designs (Figure 1.55f) are used and represent a compromise for the conflicting goals of bulb protection from breakage and abrasion and bulb exposure to flow to ensure a representative reading and prevent areas of stagnation and material buildup.

The standard design has slots or holes to allow the flow of process around the bulb. Recessed shrouds are used for pulp and abrasive applications. The recessed area can get clogged or caked up and cover the bulb and reference junction. When the material breaks loose, it can take the bulb with it. A tapered shroud creates eddies that help keep the electrode surfaces clean.

Flat Glass Electrodes Flat glass electrodes have been developed to minimize glass damage and maximize a sweeping action to prevent fouling. The usable pH range of flat glass electrodes is often lower than bulb electrodes. The stated range of small button electrodes is 0 to 10 pH, and that of large flush flat glass electrodes, illustrated in Figure 1.55g, is 2–12 pH.

High sodium ion concentrations and low hydrogen ion activity (alkali error) can also have a larger effect on flat glasses, which lowers their upper pH range limit.

Some manufacturers offer several different glass formulations that are tailored to a process application. Figure 1.55h shows the temperature and pH range for four

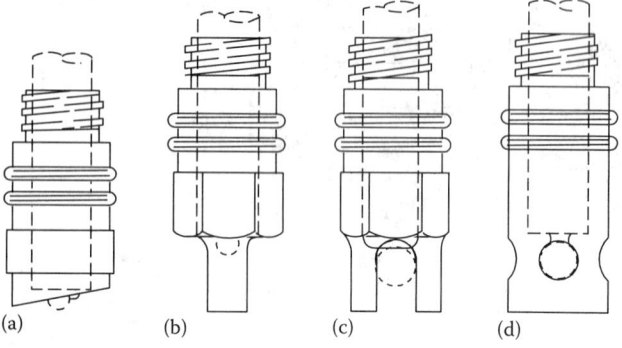

FIG. 1.55f
Various shroud designs. (a) Tapered, (b) recessed, (c) slotted, and (d) holes. (Courtesy of Electro-Chemical Devices.)

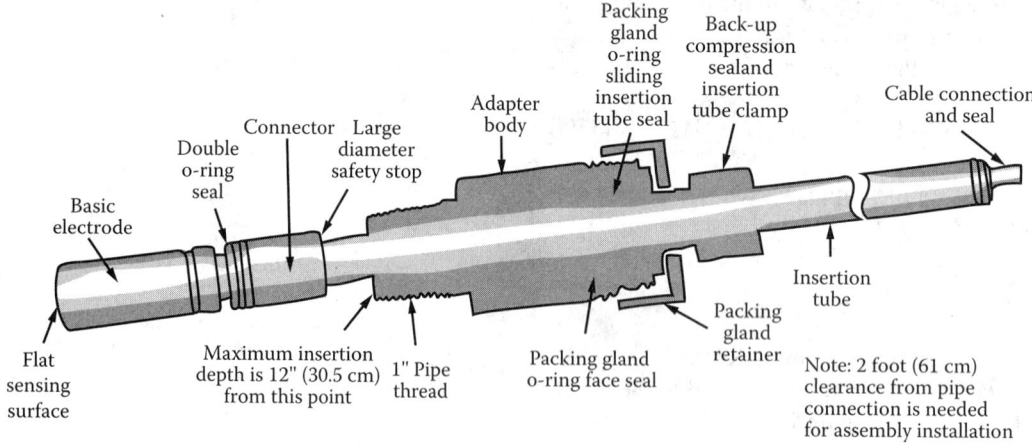

FIG. 1.55g
Flat glass probe. (Courtesy of Sensorex.)

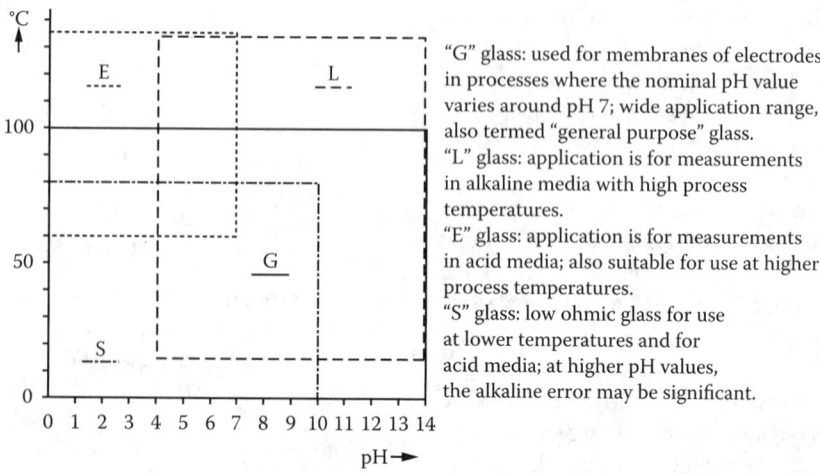

FIG. 1.55h
The temperature and pH ranges of four types of glass. (Courtesy of Yokogawa.)

different types of glass. Figure 1.55i illustrates how lower alkali error improves accuracy above 10 pH.

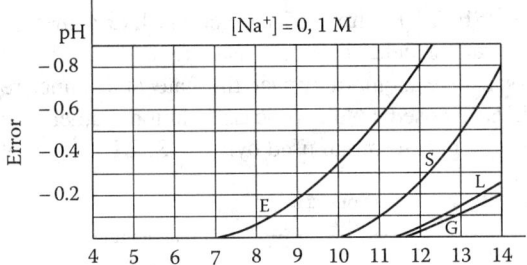

FIG. 1.55i
The alkaline errors of four species of glass. (Courtesy of Yokogawa.)

Higher glass impedance normally corresponds to higher resistance to chemical attack from strong bases (e.g., sodium hydroxide) and strong acids (e.g., hydrochloric acid). Thick and flat glasses are more abrasion resistant, but low-impedance glass must be used, which reduces its chemical resistance.

In the late 1980s and early 1990s, progress was made in reducing the impedance of sodium ion and chemical-resistant glass so it could be used at temperature as low as 5°C. In general, the specifications of a manufacturer's glass formulation should be checked against the needs of the pH application.

Glasteel Pfaudler has patented a measurement electrode made of Glasteel®, a unique material formed by the fusion of glass and steel (Figure 1.55j). The probe has greater structural

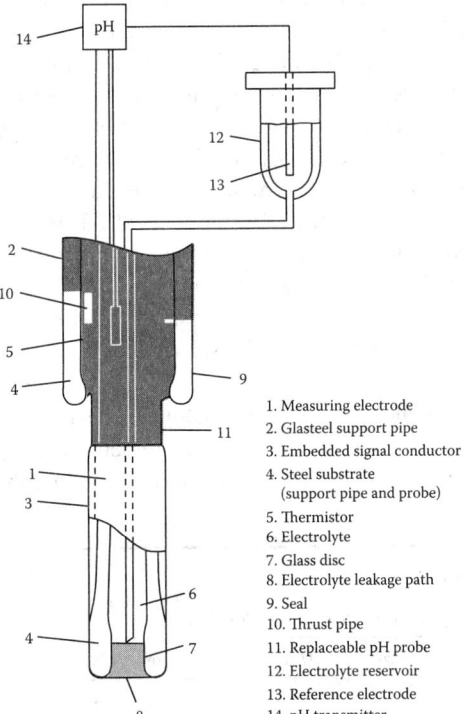

FIG. 1.55j
The construction of the replaceable Glasteel pH probe assembly. (Courtesy of Pfaudler.)

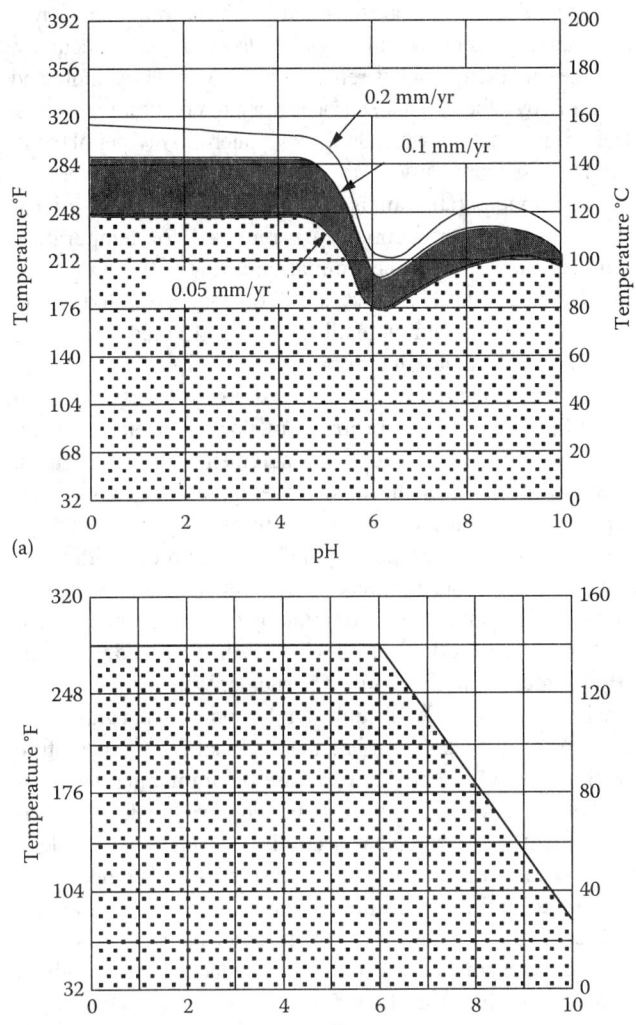

FIG. 1.55k
The Glasteel electrode's corrosion resistance and temperature rating are better than that of other electrodes. (a) Glass loss in mm/year and (b) shaded area is within allowable tolerance of 0.1 pH. (Courtesy of Pfaudler.)

endurance, greater corrosion resistance, and a higher temperature rating (140°C below 5 pH) as depicted in Figure 1.55k.

This probe is designed to withstand the rigors of heat, pressure, and dynamic agitation in reactors. It needs a modified or microprocessor-based transmitter, because its zero potential point lies between 1 and 3 pH. The cost of the Glasteel electrode is much higher than that of any other pH electrodes.

Metal Oxide pH Electrodes Before the glass measurement electrode was developed for industrial measurements, the antimony electrode was used. Antimony is a hard and brittle material. The electrode responds to pH by surface oxidation. Because oxidant accumulation on its surface deteriorates the accuracy and repeatability of the measurement, periodic cleaning is necessary.

The electrode's response is nonlinear and requires narrow calibration spans or special polynomials for adequate accuracy. The measurement range is reported to be 3–8 pH by some and 3–11 pH by others. This sensor is very temperature sensitive. The millivolt output, which is approximately 50 mV per pH unit, will increase from 1 mV/°C to 3 mV/°C as the pH increases.

The isopotential point is near 2 pH, so modified or microprocessor-based transmitters are required for temperature compensation. Yet another limitation is that antimony is also an ORP electrode. Therefore, it will respond to oxidizing and reducing species, and its reading will be affected by as little chlorine as a fraction of a ppm.

Antimony is usually used as a last resort. It has been used in applications where abrasion from slurries or etching from hydrofluoric acid causes glass measurement electrode failures. It cannot be used in the food, beverage, or pharmaceutical applications because of its toxicity. In recent years, work has been conducted toward developing glass electrodes with resistance to high concentrations of hydrofluoric acid, which, if fruitful, will eliminate many antimony electrode applications.

Other metal oxide electrodes have been evaluated for use as pH electrodes. Iridium oxide in particular was given considerable focus as a possible substitute for the conventional glass electrode in the early 1990s. It had the potential for a temperature rating of 200°C and held the promise of eliminating the problems of dehydration, etching, sodium ions, and breakage associated with glass measurement electrodes, as well as being able to make repeatable measurement outside the 0–14 pH range.

Unlike antimony, iridium oxide has excellent linearity and repeatability. However, as with other metal oxides, iridium oxide responds to oxidizing and reducing agents, which tend to severely offset the pH reading, even at low parts per million concentrations. Attempts were made to circumvent the effects of oxidizing and reducing agents through the use of polymer coatings, but these proved unsuccessful; therefore, the development efforts toward industrial iridium oxide sensors have largely been abandoned.

Ion-Selective Field Effect Transistors The most promising new pH sensor technology of the last three decades has been the development of the ion-selective field effect transistor (ISFET). Its greatest advantage is that it has a relatively small, integral pH-sensitive area instead of the glass bulb found in common pH electrodes, which can break and contaminate the process sample with broken glass. This makes the ISFET especially appealing for food applications, in which on-line pH measurements have been avoided in consideration of the risk of contamination by broken glass electrodes.

An ISFET (Figure 1.55l) functions like a metal oxide semiconductor field effect transistor (MOSFET), but it uses a pH-sensitive membrane over the gate or channel region, which is exposed to the process solution, rather than the conductive gate terminal used by a MOSFET.

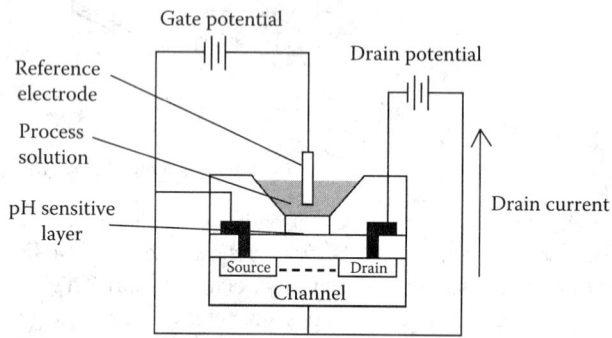

FIG. 1.55l
Ion-selective field effect transistor pH sensor.

When an ISFET is operated at a constant drain current, the measured gate-source voltage is a result of the potential of the conventional reference electrode and of the pH-sensitive membrane, which responds to pH in a Nernstian manner, similar to a conventional pH electrode. The special circuitry required to handle an ISFET measurement can be embedded in the pH analyzer or in an electronics module, which allows the ISFET sensor to be used with a conventional pH analyzer.

The major drawback of ISFET sensors relative to conventional glass electrodes had been their lack of robustness, and this has created problems in the food industry, which often requires exposure to strong alkaline solutions that are used for clean-in-place (CIP) sanitation, as well as steam sterilization. In recent years, the robustness of ISFET pH sensors has improved so that steam sterilization temperatures of up to 130°C or 135°C can be tolerated, but the ISFET sensors must still be withdrawn from the process during the CIP cycle to avoid shortening the life of the sensor.

Relative Performance of pH Sensors The relative performance of the different types of measurement electrodes under various operating conditions is summarized in Table 1.55b,

TABLE 1.55b
Relative Performance of Different pH Probe Designs Considering a Number of Performance Criteria

Type (Range)	Accuracy	Temperature	ORP	Sodium	Coating	Abrasion	Etching	Cost
Glass bulb (0–14 pH)	+	−	+	+	−	−	−	+
Flat glass (2–12 pH)	+	−	+	−	+	+	−	+
Glasteel (0–10 pH)	+	+	+	+	+	+	+	−
Antimony (3–11 pH)	−	−	−	−	−	+	+	+

Note: +, good; −, poor.

where the (+) sign refers to good, and the (−) sign indicates poor performance or rating.

Reference Electrodes

The reference electrode should provide a standard reference potential at a given temperature plus an electrical continuity between its internal electrode and the glass measurement electrode (Figure 1.55b). The standard reference potential is the result of a reproducible silver ion concentration maintained by a silver–silver chloride wire in contact with a potassium chloride solution.

The reference electrode has a junction where the internal electrolyte is in contact with the external process fluid. A liquid–liquid junction or diffusion potential develops as electrolyte ions and process ions migrate into the junction. When the charge accumulation becomes large enough to oppose further ion migration, the potential stops changing.

This potential introduces an error that is often overlooked in pH measurements. Figure 1.55m shows the error for various concentrations of potassium chloride liquid fills and different concentrations of hydrochloric acid and sodium hydroxide in the process fluid. The junction potential magnitude and time to equilibrium generally increase for gel-filled or solid-state reference electrodes.

silver ion in the reference electrolyte is disrupted, typically by precipitation or complexation of the dissolved silver, most notably by sulfide, iodide, and bromide in the process.

A flowing junction helps reduce such contamination and helps establish a quicker and more constant reference junction potential. However, flowing junctions require electrolyte pressurization above the process pressure and periodic refilling of the electrolyte. Figure 1.55j illustrates the use of pressurized remote-mounted reservoirs. The reference electrode is mounted in the external reservoir for additional protection against contamination and for improved temperature stability.

Flowing junctions are more common in applications in low-conductivity water and applications where there is a large concentration of contaminating ions. Unlike a flowing junction, which maintains electrolytic contact with the process by transport of the electrolyte solution, a nonflowing junction relies on the diffusion of reference electrolyte ions into the process and, as a result, will always be susceptible to the diffusion of process material through the liquid junction.

Combination Measurement/Reference Electrode The combination electrode is used mostly to reduce the maintenance and replacement cost of electrodes. It has a sealed reference electrode that surrounds the measurement electrodes and an annular reference junction (Figure 1.55n).

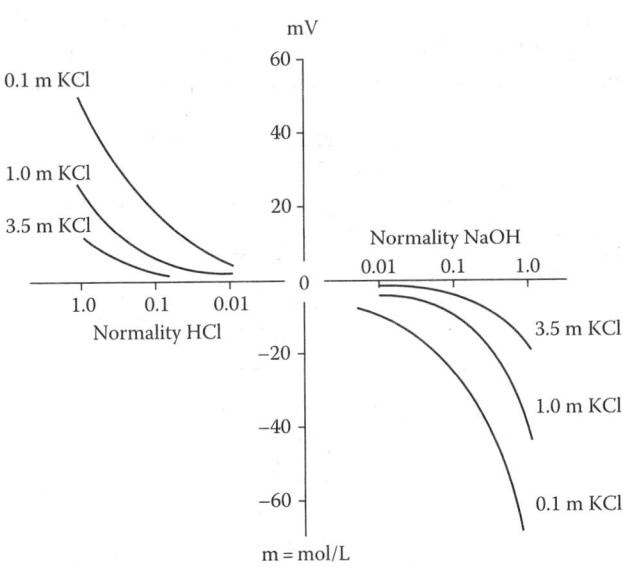

FIG. 1.55m
Charge accumulation at the liquid–liquid junction can oppose ion migration and introduce an error. The values of this error are shown for different concentrations of electrolyte (KCl) and different concentrations of acidic (HCl) or basic (NaOH) process fluids. (Courtesy of Yokogawa.)

Flowing and Double Junctions The reference junction is also an entry path of process material into the internals of the reference electrode. Electrolyte contamination can cause large shifts of the pH measurement. This happens when the equilibrium between the silver–silver chloride wire and the dissolved

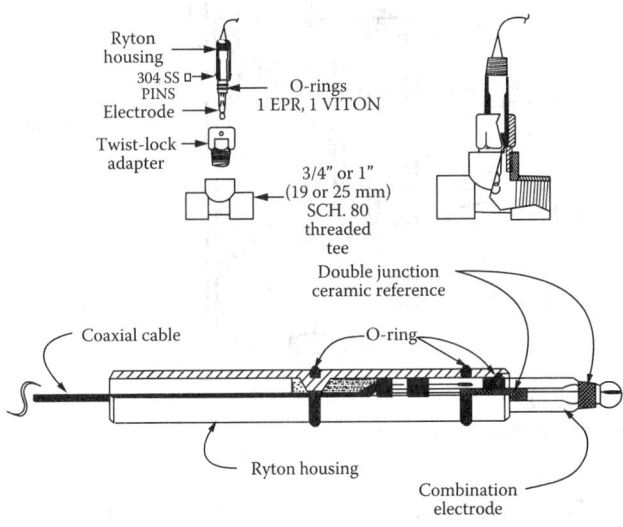

FIG. 1.55n
Combination electrode with double junction. (Courtesy of Van London.)

The construction features of a particular reference design in a combination electrode present a trade-off between (1) the effort to quickly establish and minimize a constant junction potential and (2) the goal of resisting contamination and coating. The user must decide which is more important for a given application.

A decrease in the area of the reference junction reduces contamination but increases coating problems. A thickening or solidification of the internal fill, the addition of junctions, and a decrease in the porosity of the junction reduce contamination problems but delay equilibrium and often increase the magnitude of the junction potential. Figure 1.55n shows the cross section of a combination electrode with a double-junction reference.

The potassium chloride solution is saturated and has a tendency to crystallize and reduce the diffusion rate and the associated potential at the junction. Also, silver from the silver–silver chloride internal element gets into the potassium chloride fill, reacts with sulfides and halides, and clogs the junction.

Multiple Junction References The most common approach to preventing contamination of nonflowing reference electrodes in combination pH sensors has been the use of multiple junction references (Figure 1.55o). These designs retard the diffusion of contaminants that reach the reference element by requiring them to sequentially diffuse through each junction of the reference.

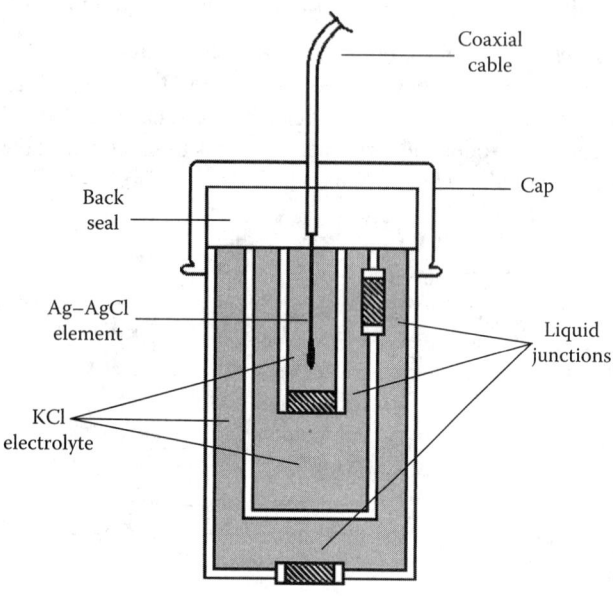

FIG. 1.55o
Triple junction reference electrode.

The electrolyte solutions in the reference are typically gelled to prevent transport of the contaminant to the next liquid junction by convection. The electrolyte solution in contact with the outermost liquid junction can be an electrolyte with a low silver ion concentration, such as potassium nitrate, to prevent precipitation of silver halides at the outermost liquid junction and precipitation of heavy metals in the process with chloride in the electrolyte solution. A number of reference fill solutions are offered, which have been formulated to improve performance in process solutions with moderate or high concentration of organic solvents, or solutions in which carbonate or sulfate scaling can be a problem.

Stiff Gel Reference and Ion Traps Mettler-Toledo has developed an Argenthal reference system that has a stiff gel with

silver ion traps to prevent silver ions from reaching contaminants that would precipitate them. It can use multiple liquid junctions or an aperture diaphragm to provide a direct contact between the process and the reference electrolyte. It provides performance advantages in streams with high salt concentrations, emulsions, suspensions, proteins, sulfides, hot alkalis, and pressure fluctuations.

Ion traps have become more common in reference electrodes. They isolate the silver/silver chloride wire with an ion exchange resin that minimizes the diffusion of silver ions into the outer fill solution. With silver ion concentration greatly reduced in the outer fill solution, the liquid junction is much less likely to be plugged by contaminants that precipitate silver ions.

The term "ion trap" has also been used to describe an outer reference electrode chamber designed with baffles and a gelled fill solution. The purpose of the baffles is to block poisoning contaminants from the process behind the baffles and prevent them from reaching the area of the reference containing the silver/silver chloride wire.

Differential Reference Another approach has been to dispense with the conventional silver–silver chloride reference altogether. Figure 1.55p describes Great Lakes Instruments' (now known as GLI International) patented differential reference electrode. It uses a second glass measurement electrode behind a double junction in a buffer solution for the reference electrode. The glass bulb isolates the silver–silver chloride element of the reference from sulfides and halides in the process. The buffer

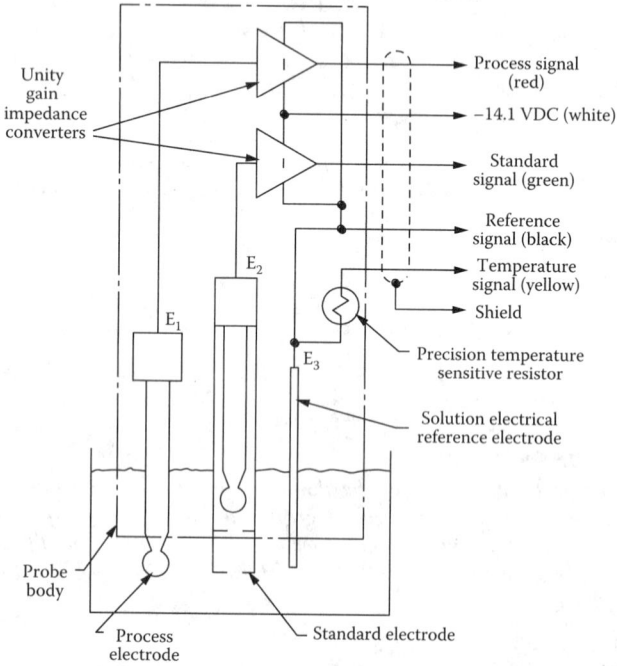

FIG. 1.55p
The differential electrode measures the electrical potential between the process and the standard electrode. (Courtesy of Great Lakes Instruments.)

solution is chosen to be compatible with the process fluids to ensure a constant buffer pH and constant reference potential.

A ground rod prevents errors and noise from ground potentials. The smaller positive ions in the buffer solution tend to move through the reference junction out into organic and pure water streams. This creates an offset in the pH reading. The differential electrode has an excellent performance record in wastewater streams with inorganic acids and bases.

Because of the lower thermal conductivity of most solid-state media, it is especially important for the temperature compensation sensor to be as close as possible to the tip to minimize thermal lag.

ELECTRODE CLEANERS

Significant natural self-cleaning by turbulent eddies requires a velocity of ≥5 ft/s past the electrode. A velocity of greater than 10 ft/s can cause excessive measurement noise and sensor wear. The area obstructed by the electrode must be subtracted from the total cross-sectional area when estimating the total area that is open to flow around the electrode.

Shrouds and Filters

The pressure drop at the restricted cross section should be calculated to ensure that there will be no cavitation. Some shroud designs such as the tapered shroud in Figure 1.55f create more effective eddy action. Flat surface electrodes get adequate cleaning action at velocities of 1–2 ft/s.

The addition of filters shifts the maintenance from the electrode to the filter. The filter usually needs to be changed more often than the electrode has to be cleaned. An extra filter is not recommended unless it is self-cleaning or can be automatically backwashed.

Automatic Cleaners

The four types of automatic cleaners are ultrasonic, brush, water jet, and chemical. The components of some of these assemblies are shown in Figure 1.55q, and their performance ratings for a variety of applications are given in Table 1.55c. These methods tend to concentrate on the removal of coatings from the sensing electrode. Particles caught and materials clogged in the porous reference junction are generally difficult to dislodge. The impedance of plugged reference junctions can get so high that it approaches an open circuit. The pH reading can drift.

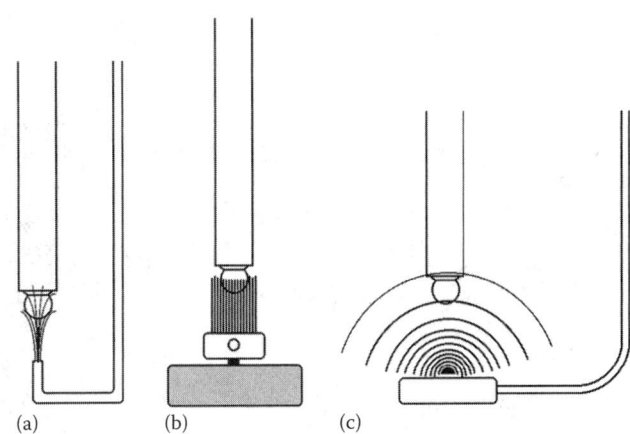

FIG. 1.55q
Design variations of on-line electrode cleaners. (a) Jet cleaner for water or chemical cleaning, (b) brush cleaner, and (c) ultrasonic cleaner.

TABLE 1.55c
Ratings for Various Types of Cleaners

	Application	Ultrasonic	Water Jet	Brushing	Chemical
Slime microorganism	Food, paper, pulp, aquatic weed	X	O	O	n
	Bacteria (activated sludge) whitewash	n	O	O	n
Oil	Tar, heavy oil	X	X	X	n
	Light oil	O	n	n	O
	Fatty acid, amine	X	O	X	O
Suspension	Sediment	O	X	X	O
	Metallic fines	n	X	X	O
	Clay, lime	n	O	X	O
Scale	Flocculating deposit neutralized effluent $CaCO_3$	n	n	n	n

Source: Courtesy of Horiba Instruments.
Note: O, recommend; n, applicable; X, not applicable.

776 *Analytical Measurement*

Ultrasonic Cleaners The ultrasonic cleaner uses ultrasonic waves to vibrate the liquid near electrode surfaces. Effectiveness depends on the vibration energy and velocity past the electrodes. Heavy-duty electrodes are needed to withstand the ultrasonic energy. The ultrasonic cleaner works well in processes where fine particles and easily supersaturated sediments are formed or in suspension. It can move loose and light particles and oil deposits. Most of the disappointments with ultrasonic cleaners come from applications in which coatings are difficult to remove. In general, their use has greatly declined in recent years in favor of water-jet and chemical cleaning, although they are still commercially available.

Brush Cleaners A brush cleaner removes coatings by rotating a soft brush around the measurement bulb. The brush does not reach the reference junction. It has an adjustable height and a replaceable brush, and it can be electrically or pneumatically driven. Soft brushes are used for glass and ceramic disks are used for antimony electrodes. Sticky materials can gunk up the brush and smear the bulb. As with ultrasonic cleaning, brush cleaner use has declined.

Water-Jet Cleaners The water-jet type of cleaner directs a high-velocity water jet to the measurement bulb. The reading of the loop will become erratic during washing. Therefore, the cycle timer that starts the jet should also freeze the pH reading and switch the pH controller to manual during the wash cycle and for at least 2 min after the wash period to provide for electrode recovery. The water jet works well in removing materials that are easily dissolved in water.

A high-pressure flanged assembly, which can be installed in a pipe tee, is shown in Figure 1.55r, and a

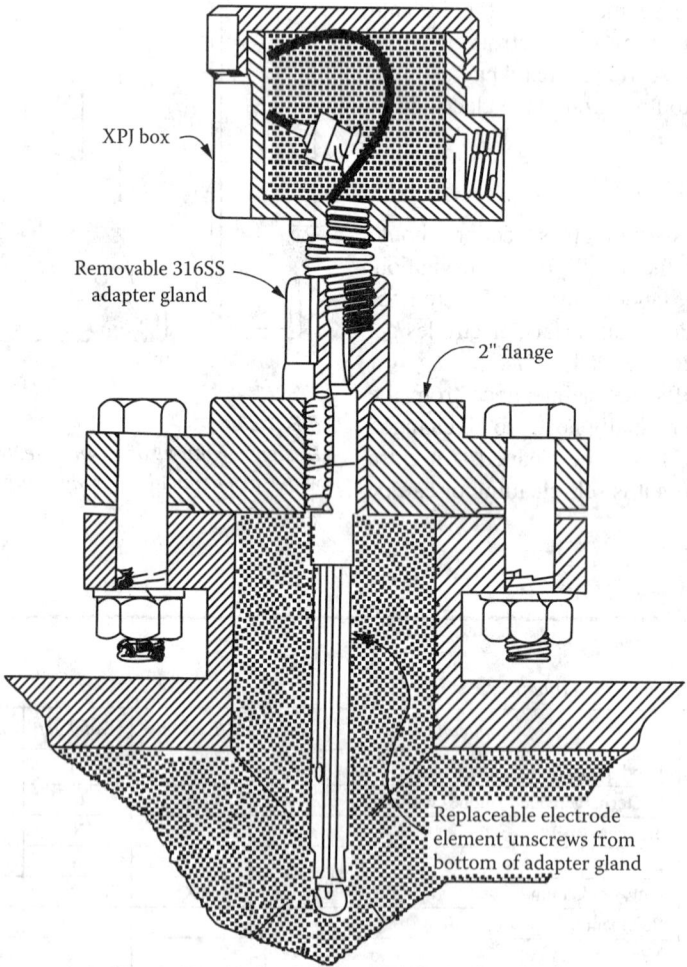

FIG. 1.55r
High-pressure electrode cleaning assembly. (Courtesy of Van London.)

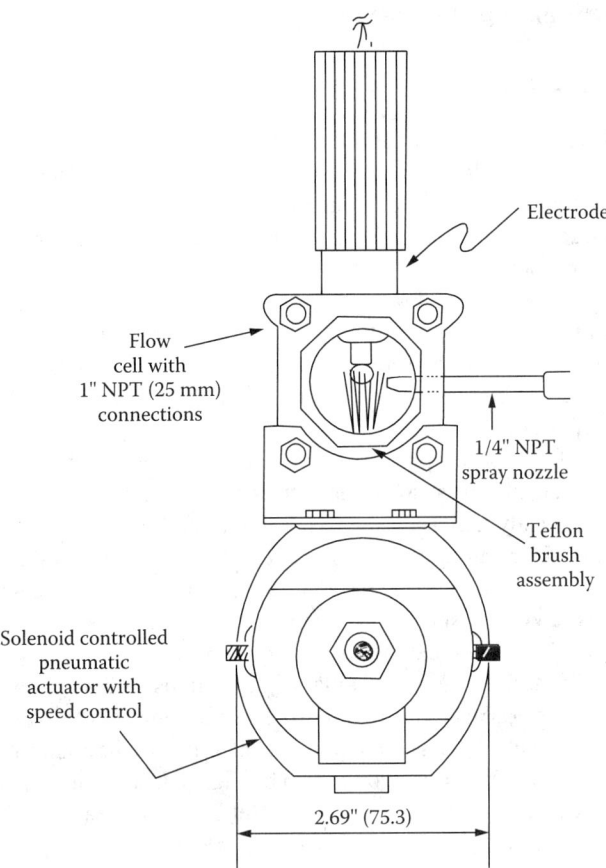

FIG. 1.55s
Brush cleaner with optional water jet.

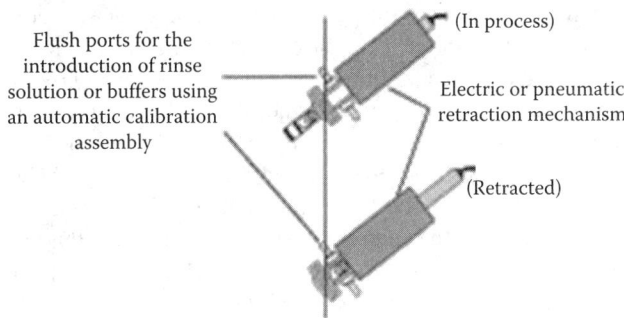

FIG. 1.55t
Retraction assembly.

flow-through brush cleaner and water-jet combination assembly is shown Figure 1.55s.

Chemical Cleaners Chemical probe cleaning uses a chemical jet such as a dilute acid or base that is compatible with the process. A base is typically used for resins and an acid for crystalline (carbonates) and amorphous precipitations (hydroxides). A dilute hydrochloric acid solution is frequently used.

Chemical cleaning tends to be the most effective of the available methods, but acid and base cleaners chemically attack glass and can contaminate the reference electrode. In addition, cleaning cycles that are too frequent or too long will cause premature failure of the glass electrode. As with the water jet, the cycle timer must hold the last pH reading and suspend control action during the wash cycle.

Redundant pH sensors can be installed in parallel so that while one electrode assembly is being reconditioned, the other is in control. In such installations, after the wash cycle, the reconditioned assembly is automatically returned to control.

Automatic Retraction An electrically or pneumatically retractable pH assembly can be used in automated on-line cleaning applications and for storage, regeneration, and calibration (Figure 1.55t). This design is particularly useful in applications where the duration of probe exposure must be short so as to protect it from glass surface deterioration and to avoid forming contamination of the reference fill. Such contamination can be caused by the hot caustic or by non-aqueous solutions.

In any mounting method, the electrode tips must be pointed down to prevent the air bubble inside the electrode fill from residing in the tip and drying out the inside surface. The air bubble is necessary to provide some compressibility to accommodate thermal expansion. An installation angle of 15° or more from the horizontal is sufficient to keep the bubble out of the tip. Some electrode designs eliminate the bubble and provide a flexible diaphragm for fill contraction and expansion.

Manual Cleaning Electrodes can be manually cleaned by soaking them for several hours in a dilute hydrochloric acid solution. In general, the solution used for cleaning should be chosen for its effectiveness in removing the contaminant while not attacking the wetted materials of the pH sensor. Exposure to cleaning solutions should be minimized, as it can cause contamination of the liquid junction.

Electrodes that are sluggish or have an insufficient span or efficiency can often be reactivated by soaking them for 1 min in a dilute solution of hydrofluoric acid in a nonglass container. The reactivation is created by the hydrofluoric acid dissolving part of the aged gel layer. The electrode should then be soaked overnight in its normal storage solution (typically 4 pH buffer).* The electrode's documentation should be checked for the recommended procedure and HF concentration before attempting reactivation.

On-Line Cleaning On-line electrode cleaning systems should be considered when manual cleaning becomes a maintenance burden. Selection of the cleaning method should be done in consultation with the manufacturer of the cleaning

---

* Ingold, W., pH electrodes storage, ageing, testing and regeneration, Ingold technical publication E-TH 7-2-CH.

system, given that several factors are peculiar to each process and cleaning method, and these can determine the difference between success and failure. Cleaning can be done with the sensor in the process or out of the process using automatic retraction and can be initiated manually by a timer or by a diagnostic alarm from the transmitter, typically high reference impedance, which gives an early indication of coating. During cleaning, the control system must be notified that the pH measurement is out of service or the pH output should be held at the last valid value or at a default value.

On-Line Cleaning Systems On-line cleaning systems have been developed, which go beyond simply flushing a retracted sensor with water or cleaner (Figure 1.55u). These systems can automatically retract the sensor, based on a timer, diagnostic alert, or manual initiation. The actual cleaning action of the sensor can use a preprogrammed sequence of cleaner, water or even air, or the sequence can be programmed by the user to provide effective cleaning for a particular application. In some cases, the cleaning medium can be applied in pulses rather than a continuous flow to more effectively dislodge coatings.

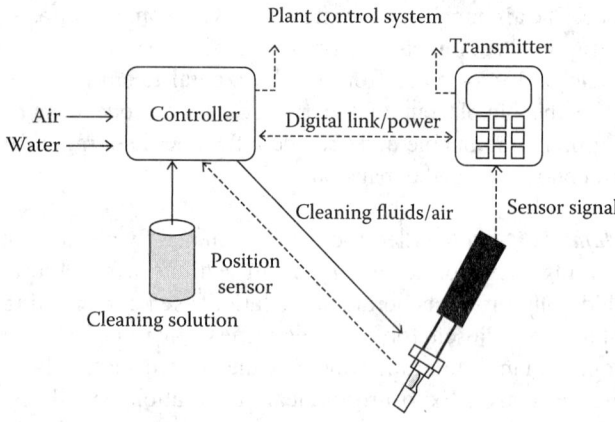

FIG. 1.55u
On-line sensor cleaning system.

The pH transmitters used with these systems provide the control for the system and the user interface for setting up the cleaning application. The cleaning system controller provides relays to supply air to retract the sensor and water for flushing, as well as the cleaning solution, which can be supplied to the controller using a pump or eductor. The controller also receives an input from the position sensor of the retraction assembly.

In applications using an analog signal from the transmitter, the cleaning system can notify the plant control system that the pH sensor is retracted and out of service through the use of discrete signals, or a cleaning cycle can be initiated by a discrete signal from the plant control system to the cleaning system. In installations using digital communications, such as Foundation Fieldbus or Profibus, the same information can be communicated to the plant control system over the bus by the transmitter.

APPLICATION PROBLEMS

High Salt Errors

Alkalinity error is caused by positively charged alkaline ions that move into the outer glass surface, displace the hydrogen ions, and decrease the activity of the hydrogen ion in the gel layer. The millivolts developed at the surface are proportional to the ratio of the activity of the hydrogen ion in the process fluid to that in the gel layer. Therefore, the decrease in activity in the gel layer increases the millivolts developed, which decreases the pH reading.* The result is a perceived increase in hydrogen activity of the process and a lowered pH reading.

The error disappears when the electrode is removed and inserted in buffer solutions. The smaller alkaline ions can more easily penetrate the glass matrix and cause a greater shift. Thus, lithium ions cause a larger error than sodium ions, and sodium ions cause a larger error than potassium ions. Because sodium ions are much more common, the effect is often called the *sodium ion error*.

Tight-matrix, high-pH glass formulations such as glass L in Figure 1.55h and general-purpose wide-range glass formulations such as glass G in Figure 1.55h minimize the alkalinity error. For alkalinity error caused by high salt concentrations, the decrease in pH readings is offset by the increase in pH caused by the decrease in the hydrogen ion activity.

High Salt Effect on Electrodes For an increase in sodium chloride concentration from 1 to 4 molarity at 25°C, the pH reading from a general-purpose glass dropped by only 0.2 pH in a test with Tris buffers. For lithium chloride, the pH reading decreased by 1.0 pH. For potassium chloride, the pH reading actually increased by ≈0.4 pH.

A separate test, documented in Figure 1.55v, with three electrodes installed in a circulation line showed that

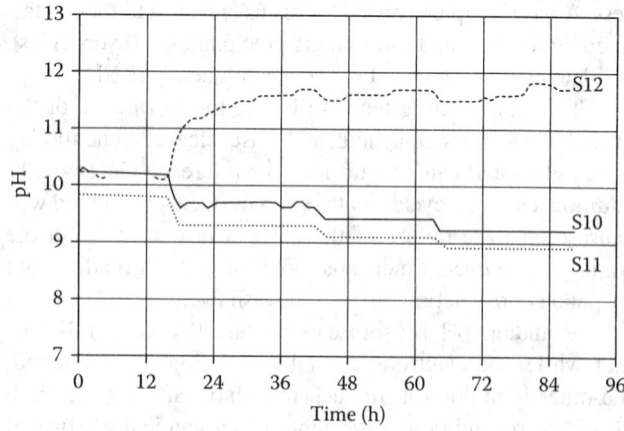

FIG. 1.55v
High salt (NaCl) effect on electrodes.

---
* Ingold, W., Principles and problems of pH measurement, Ingold technical publication E-TH 1-2-CH, pp. 17–18.

pH measurement decreased slightly for three additions of sodium chloride as expected for glass bulb measurement electrodes (S10 and S11) but went upscale by 1.5 pH for a flat glass measurement electrode (S12). Subsequent tests and calibration checks of the flat glass electrode confirmed that it was in good condition.

Sodium hydroxide solutions above 12 pH and hydrofluoric acid solutions below 3 pH will greatly shorten the life of the electrode. The chemical attack rapidly increases with temperature and may cause failure within a day. Sample cooling and dilution are often the only alternatives.

High Acid Errors

Acid error is attributed to the adsorption of acid molecules into the gel layer and the associated increase in the hydrogen ion activity of the gel layer.* Exposure to concentrated acids also dehydrates the gel layer and reduces the efficiency of the electrode. The net result is an increase in the pH reading. Acid error typically starts below 1 pH and becomes worse as the pH decreases and exposure time increases. A periodic water soaking or water-jet washing of the measurement electrode helps replenish the gel layer.

Solution Temperature Effects

The actual pH of the process stream changes with temperature due to changes in dissociating constants with temperature. Figure 1.55w shows the pH change with temperature for

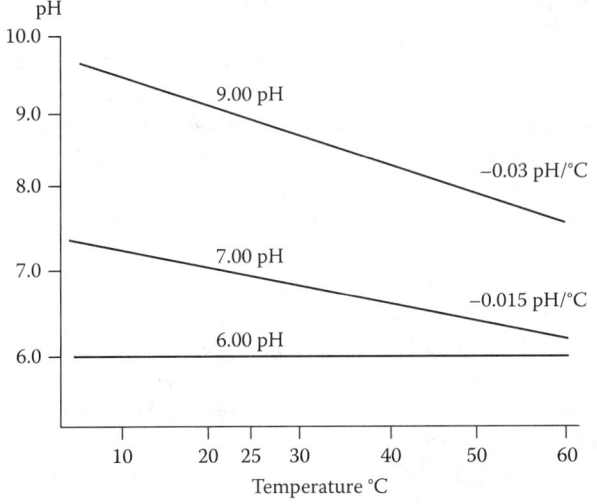

FIG. 1.55w
Temperature dependence of water and strong acid and base solutions.

---

* Ingold, W., Principles and problems of pH measurement, Ingold technical publication E-TH 1-2-CH, pp. 19–20.

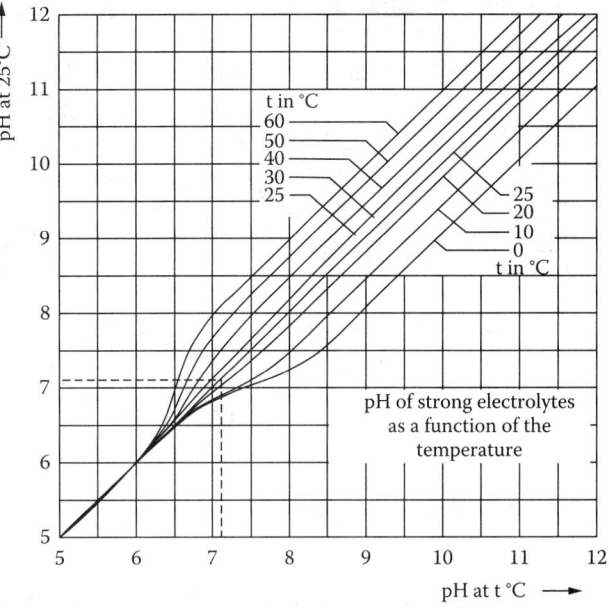

FIG. 1.55x
The pH of strong electrolytes changes with temperature. (Courtesy of Yokogawa.)

a neutral solution and 6.0 pH and 9.0 pH solutions with strong acids and bases. Figure 1.55x shows the temperature effect on the pH of strong electrolytes.

The dissociation constants of strong acids and bases are off the scale, and the effect shown in this figure is primarily the result of changes in the water dissociation constant. Strong acid solutions have virtually no temperature dependence, but neutral and strong base solutions will always exhibit temperature dependence. For weak acids and bases, the dissociation constants are on the scale, and their effect on pH is more difficult to predict. It usually requires a laboratory test.

Microprocessor-based transmitters offer polynomial or straight-line fits of the data to temperature-correct the measurement for actual pH changes. Most of these algorithms cannot correct for changes in the glass or in the internals of the electrode.

Water Concentration Errors

Water concentrations below 40% can also cause shifts in the pH measurement. When the addition of a solvent causes the water content to drop below 40%, the change in hydrogen ion activity can be significant enough to cause a noticeable change in pH. The addition of propylene glycol causes a decrease in the hydrogen ion activity and an increase in the pH reading.

For the same three electrodes (S10, S11, and S12) in the circulation line that were subjected to the high salt test shown

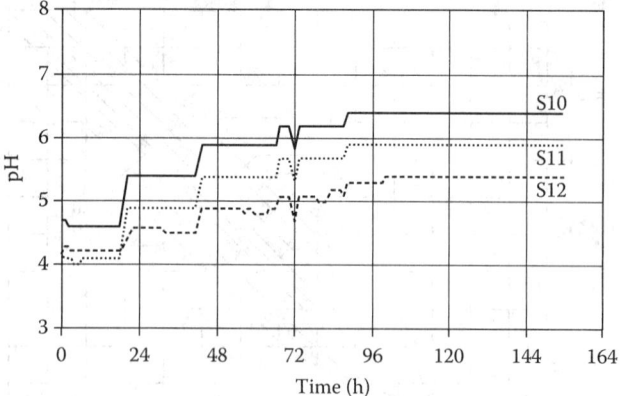

FIG. 1.55y
Low water effect on three electrodes, produced by the addition of propylene glycol.

in Figure 1.56v, four successive additions of propylene glycol were made to perform a low water test. The incremented pH measurements are shown in Figure 1.55y.

For applications with low water content, it helps to use a glass formulation with a large number of active sites and to automate the periodic water soaking or water-jet washing of the glass to hydrate the gel layer.

Nonaqueous Solutions

In nonaqueous solutions, the solvent can be an acid, base, alcohol, or hydrocarbon instead of water. An acid solvent acts as a proton donor, and the pH scale is shifted down. For example, the pH scale for an acetic acid solvent is −6 to −1 pH and for a formic solvent is −9 to −2 pH.

A base solvent acts as a proton acceptor, so the proton activity is decreased, and the scale is shifted up. An alcohol solvent acts as both a proton donor and acceptor like water so that changes in proton activity are moderated. The pH scale occupies about the same region as water but may extend slightly farther upscale and downscale. For example, the pH scale for an ethanol solvent is −4 to +16 pH.

A hydrocarbon solvent acts as neither a proton donor nor acceptor, so the solvent is passive to changes in proton activity. Consequently, the upper and lower limits of the pH scale are usually much farther apart than for water. For example, the pH scale for acetone solvent is −5 to +20 pH.*

If the process is an aqueous solution dispersed with an immiscible, nonaqueous solvent, much of the acid or base will be in the aqueous phase. It is better to make the pH measurement in a vessel that allows the phases to separate and confine the pH measurement to the aqueous phase.

---

* Bates, R. G., *Determination of pH Theory and Practice*, John Wiley & Sons, New York, 1964, 201–288.

When considering applications with a significant nonaqueous solvent content, it is important to note that most pH sensors are built with wetted materials designed for aqueous applications. In many cases, nonaqueous solvents can attack the o-rings, seals, and even body material of the sensor, causing the process to leak into the sensor and cause a short.

The presence of nonaqueous solvents can also cause unpredictable behavior at the liquid junction of a nonflowing reference, leading to measurement errors. Therefore, the use of a flowing reference should be considered.

Probe Coating and Low Conductivity

A slime coating of 1 mm thickness slows down the migration of hydronium ions to the glass surface enough to increase the pH measurement time constant from 10 sec to 7 min.[†] The slow electrode response greatly increases the period of oscillation of the control loop and causes poor control. The amplitude of the measurement oscillations is a much attenuated version of the real-world pH fluctuations. For thick coatings, a hydronium ion concentration is trapped, and the pH reading freezes.

Coatings also increase the impedance of the electrode and can cause a reading that steadily climbs up the scale as the coating thickens.

Conventional pH measurements of low-conductivity water (e.g., boiler feedwater and condensate) suffer from flow sensitivity, electrokinetic effects, sluggish response, changes in hydrogen activity from flow of reference electrolyte, actual pH changes from water dissociation constant changes with temperature, and isopotential point shifts with temperature.

Special systems have been designed to minimize these problems through the use of a constant head device to provide constant flow; stainless steel components to ensure that sensors are completely shielded and earthed; a low-resistance glass sensor in a high-velocity, low-volume cell with minimum distance between reference junction and glass to decrease system resistance; a low-leak reference junction to minimize hydrogen activity changes from electrolyte contamination of the sample; an in-line automated temperature-adjusted buffer facility to reduce thermal equilibrium and flow errors; and full temperature compensation to deal with Nernst slope, actual pH, and isopotential point changes with temperature.[‡]

Air bubbles that cling to or pass by the measurement bulb can cause noise in the reading and particles can have a

---

[†] Moore, R. L., Good pH measurement in bad process streams, *Instrum. Control Syst.*, 63(12), 39–43, 1990.

[‡] Rushton, C. and Bottom, A., Measuring pH in low conductivity waters, *Kent Tech. Rev.* (U.K.), 24, March 1979.

similar effect. Abrasion reduces the number of active sites on the glass surface and causes a shortened span and slow response.

INSTALLATION METHODS

There are two main principles for installation design. First, for pH control, the sensor location and assembly must be chosen to minimize transportation delays and sensor time constant. The additional dead time from a delayed and slow measurement increases the loop's period, control error, and sensitivity to nonlinearity.

Second, the installation must minimize how often the electrodes must be removed for maintenance (e.g., calibration and cleaning). Removal and manual handling increase error and reduce the life of the electrode. The fragile gel layer is altered by handling and the equilibrium achieved by the reference junction is upset.

Submersion Assemblies

Sample systems are undesirable, because they add transportation delay and increase cost, and there are problems associated with winterization and plugging. Therefore, it would seem that a submersion assembly would be best for control. However, velocities below 1 ft/s will dramatically slow the electrode measurement response because of the increased boundary layer near the glass surface[*,†,‡] and promote the formation of deposits that can further slow down the measurement.

The bulk velocity in even the most highly agitated vessels rarely exceeds 1 ft/s and is often much lower. This results in coating problems and a slow response. Removal of a submersion assembly is also time-consuming. The addition of various cleaners, such as those depicted in Figure 1.55z, helps reduce the number of times a submersion assembly must be removed.

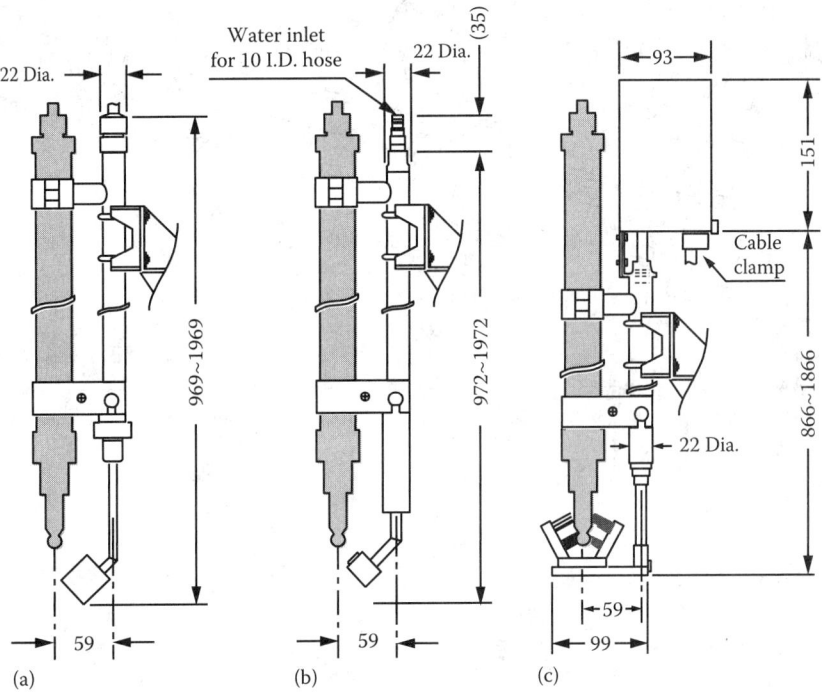

FIG. 1.55z
Submersion assemblies provided with various cleaners. (a) With ultrasonic cleaner, (b) with water-jet cleaner, and (c) with brushing cleaner. (Courtesy of Horiba Instrument.)

---

[*] Disteche, A. and Dubuisson, M., Transient response of the glass electrode to pH step variations, *Rev. Sci. Instrum.*, 25(9), 869–875, September 1954.

[†] Guisti, A. L. and Hougen, J. O., Dynamics of pH electrodes, *Control Eng.*, 8(4), 136–140, April 1961.

[‡] Hershkovitch, H. Z. and McAvoy, T. J., Dynamic modeling of pH electrodes, *Can. J. Chem. Eng.*, 56, 346–353, June 1978.

Retractable Units

The best location for most assembly probes, except in the most abrasive services, is in a recirculation line close to the vessel outlet. Installation downstream of the pump is preferred, because the strainer blocks, and the pump breaks up, clumps of materials that might otherwise break the electrodes. The retractable electrode provides the most straightforward and economical solution, as compared to simply inserting the sensor, because it allows the sensor to be removed for service without shutting down the process line or draining a vessel. However, accidents caused by removal of the restraining strap or omission of tubing ferrules have caused such assemblies to be banned from many plants.

Flow-through assemblies or direct probe insertions with block, drain, and bypass valves are often used. The flow is returned to the suction of the pump or the vessel. If the flow chamber has a cross-sectional area that is much larger than the process connections, the velocity will drop too low, response time will slow, and coating problems will increase.

Side entry into a vessel with a retractable probe is the standard installation for bioreactors, as illustrated in Figure 1.55aa.

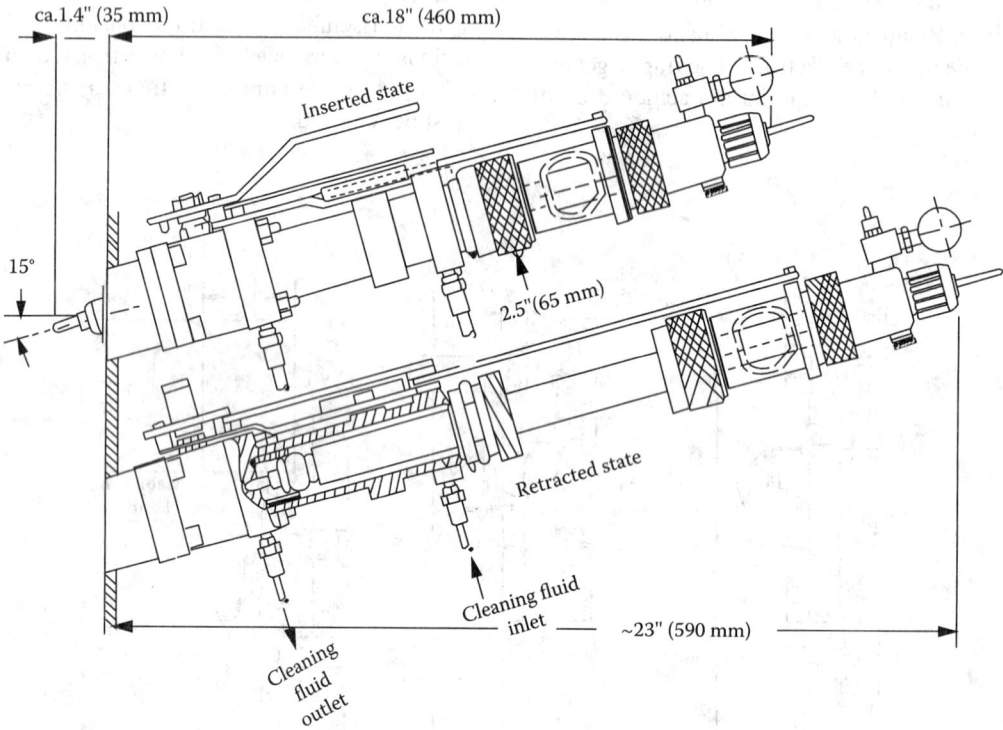

FIG. 1.55aa
Retractable, side entry probes, which are often used in fermentors.

Median Selector

A single pH measurement leads to a false sense of security. The use of two pH measurements creates endless questions as to why they do not agree. An alternative is to install three pH measurement systems and select the middle value (not the average). This method is preferred for pH control loops, especially where environmental limits exist, for the following reasons*:

1. It eliminates reaction to unreal spikes and fluctuations.
2. It reduces noise without introducing an additional lag.
3. It improves accuracy (reduces error band).
4. It ignores a single failure of any type.
5. It facilitates calibration while the loop is in automatic control.
6. It allows diagnosis of slow, dead, or inaccurate sensors.
7. It reduces unnecessary calibration requests and work.

A pH electrode is usually serviced only if it deviates from the median pH by more than one pH unit for more than several minutes. Median selection saves money by the elimination of unnecessary calibration and replacement. It can also increase on-stream time, reduce recordable violations, improve product quality, and reduce reagent consumption through improved sensor reliability and accuracy. However, triple redundant measurements do look expensive, and appearances sometimes become more important than performance.

Self-Diagnostics

An alternative to median selection is a self-maintaining installation of a single electrode. Some microprocessor-based transmitters have diagnostics based on electrode impedance measurements (Figure 1.55ab) and recovery time from a jet washing that can detect broken, coated, and unsubmerged sensors.

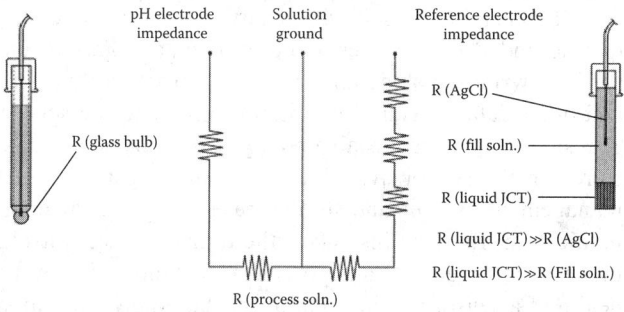

FIG. 1.55ab
pH sensor impedance diagnostic measurements.

---

*McMillan, G. K., *Tuning and Control Loop Performance*, 2nd ed., Instrument Society of America, Research Triangle Park, NC, 1990.

Glass Impedance Measurements Glass impedance measurements can be made between the glass electrode and a solution ground or the reference electrode, which has negligible impedance compared to the glass electrode. Glass electrode impedance is typically on the order of hundreds of megohms and is highly temperature dependent, decreasing by roughly one-half for every 8°C increase in temperature.

As a result, the impedance reading must be temperature compensated to avoid false low-impedance alarms. In some cases with low-impedance glass electrodes at high temperatures, the impedance drops below the measurable range of the diagnostic circuitry, and the analyzer shuts off the low-impedance alarm.

The glass impedance measurement is most useful in detecting glass electrode cracking or breakage, which results in a sudden drop in impedance. Low-impedance alarms are also useful in detecting shorts in the sensor wiring, whereas high-impedance alarms can detect an open circuit or severe coating of the pH electrode.

Reference Electrode Impedance Reference electrode impedance is necessarily measured between the reference electrode and a solution ground and is typically in the low kOhm range. The major source of reference electrode impedance is the liquid junction, and, thus, an increase in reference impedance can indicate that the liquid junction is plugged or coated or that the reference electrode is not in the process solution.

The reference electrode impedance is a much more sensitive indication of coating than glass electrode impedance. It should be noted that contamination of the reference electrode by process components does not result in an increase in reference electrode impedance unless it is accompanied by a plugging of the liquid junction.

Sensor Fault Signaling Sensor diagnostic errors are indicated on the analyzer display, and there is almost always a fault relay contact and fault current output value provided for remote indication of sensor faults. With the advent of pH analyzers with digital communications, the nature of the sensor fault can be identified remotely, enabling maintenance personnel to be better prepared to address the problem before visiting the installation.

CALIBRATION

The electrode should be washed with distilled water between buffer immersions. One should also remember that the actual pH of the buffer of process solutions can change with temperature† or with carbon dioxide absorption, and the reference junction can take a long time to equilibrate.

---

†Ingold, W., Calibration of pH electrodes, Ingold technical publication E-TH-5-2-CH, pp. 9–10.

A 10 pH buffer can drop 0.1 pH per day as a result of carbon dioxide absorption, and, in general, all buffers should be discarded after use and the stock buffer containers tightly sealed. The pH sensor should be allowed to come to the same temperature as the buffer solution. All buffers should be checked with an accurate laboratory electrode and for expiration date before use. For flowing junctions, care must be exercised to ensure that the reference electrolyte does not contaminate the buffer.

High-ionic-strength buffers should be employed for high-ionic-strength process applications. A process calibration should be performed 1–8 h after the electrode is commissioned. (The wait time depends on the degree of solidification of the reference fill.) A zero or standardization adjustment is used to make the pH sensor agree with a sample that was measured with fast and accurate laboratory electrodes immediately after sample withdrawal.

pH electrodes can have a wide range of time constants depending on the glass formulation, and the time constant tends to increase with the age of the electrode. The key to a good calibration is patience. The electrode must be allowed time to fully respond to the buffer or the slope determined by the calibration will be erroneously low. Response times much greater than normal are an indication that the electrode is at the end of its useful life.

Buffer Calibration Errors

Microprocessor-based pH analyzers often have features that help the user to avoid buffer calibration errors. Buffer values as a function of temperature are stored in memory for a two-point calibration, which allows the buffer value to be corrected for temperature during calibration. There is also a stabilization feature that prevents that analyzer from accepting a millivolt reading that has not fully responded to the buffer or millivolt changes due to warming or cooling of the sensor. Important diagnostic information is also provided after a buffer calibration in the form of the electrode slope, which is a measure of the sensitivity of the electrode and the zero offset, which indicates the reference electrode offset from an ideal value or asymmetry within the glass electrode.

Warnings are usually provided to prevent acceptance of a calibration that results in a slope and zero offset indicative of a bad sensor or a user error. Analyzers can also perform fully automatic calibrations if provided with the software and relays for controlling withdrawal of the sensor from the process through an insertion valve, rinsing, and introduction of the buffers.

pH electrodes should be stored in either a 4 pH buffer solution to help condition the glass surface or a 7 pH buffer solution to help prevent the loss of ions from the reference fill. Electrodes stored in distilled water for long periods of time will deteriorate as a result of the loss of ions from the measurement glass reference fill.

Cleaning and Calibration Systems

Automatic cleaning and calibration systems have been developed by augmenting a cleaning system with hardware, relays and pumps in controllers, and software in transmitters. The additional hardware and software support the transport of two or more buffer solutions and the calculation and control required for buffer calibration. Cleaning and calibration intervals can be programmed to occur at different intervals or initiated manually or by the plant control system. Cleaning can be triggered by a diagnostic alarm, and in some cases, the calibration interval can be adjusted based on process variable measurements.

New Developments

New pH Sensor Developments Among the sensor developments since the last publication of the *Instrument Engineers' Handbook* have been the combination pH and ORP electrodes, which, when used with the appropriate transmitter, can measure pH and ORP, as well as the redox value (rH), which gives the oxidizing or reducing power of a solution independent of pH. In many ORP applications, a pH measurement is important as pH usually affects the ORP of a solution. Using these sensors makes it possible to make both measurements with a single transmitter.

Smart pH Sensors Smart pH sensors have a memory that stores calibration, diagnostic, and process data and generally look no different than nonsmart pH sensors. The data gathered by smart sensors can include sensor identification information, such as the manufacturer, model code, serial number, and tag number of the installation; calibration data, including the slope and zero offset, and also the glass and reference electrode impedances at the time of calibration; and process data, such as the time in service, the minimum and maximum temperature exposure, the time the sensor was exposed to temperatures above certain limits, and the number of exposures to CIP and steam used for sterilization and SIP (steam in place).

pH transmitters used with smart sensors must be able to read and store the latest calibration data from the sensor, and write the slope and zero offset from calibrations, and other data measured by the transmitter to the sensor. Transmitters can make all or only a portion of the data stored in the sensor available to a control system's asset management system and its database using digital communication. In some instances, the transmitter can, itself, process the stored data and provide a local indication of the health of a sensor and its components based on calibration and process data and even give an estimate of the remaining life of the sensor.

One of the advantages of using smart pH sensors is the ability to calibrate the sensor in the laboratory or shop and then simply install the sensor in the field. This also makes it possible for an analytical specialist to do the calibration, examine the sensor, and check the data stored in the sensor, in a controlled environment, and a nonspecialist to do the final sensor installation.

Some smart sensors also have dedicated asset management software for use in the laboratory or shop. The sensor may be connected to a PC directly through an interface box or through a transmitter in the lab using a modem. This software can hold the complete calibration and process exposure history of all the smart pH sensors, analyze the health of various sensor components, and provide estimates of remaining sensor life. The slope and zero offset can be trended, and complete reports of pH sensor status can be generated.

Using the data gathered from pH sensors, which have been used at a particular sample point (tag number), the sample point can be characterized in terms of the effects of its temperature and composition on the life of a pH sensor and its components. Having this knowledge can allow the user to know when a sensor is likely to fail, based on the historical effects of the process on sensor components. Knowing which sensor component is most affected by the process can be used to select a sensor better suited to the demands of the process. This historical information can also be useful when contacting the pH sensor supplier regarding issues with the sensor (Figure 1.55ac).

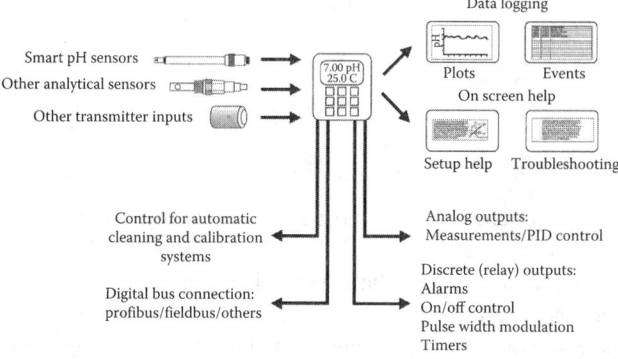

FIG. 1.55ac
Features available in line-powered transmitters.

Digital pH Sensors To function as a "smart" pH sensor, a sensor only needs to have a memory for storing calibration and other process data. The actual millivolt and RTD measurements can be connected to transmitter as raw or preamplified analog signals, as with any pH sensor. But in addition to having a memory, digital sensors also process the raw sensor signals in the head of the sensor and output a digital signal to the transmitter. This avoids problems that can sometimes occur with high impedance millivolt output of a pH sensor.

The Memosens® sensors developed by Endress+Hauser use inductive coupling to connect the sensor to its cable instead of a conventional metal to metal connector, which makes the sensor to cable connection waterproof and resistant to corrosion and increases the reliability of the connection over conventional connectors. This technology has been adopted by other pH sensor and transmitter manufacturers and is being offered as a de facto standard.

New pH Transmitter Developments pH transmitters have been developed with larger, often multilingual displays and more descriptive menu systems and keypads, which have made configuration and troubleshooting easier. Calibration at the local display is a stepwise process using prompts and usually includes buffer temperature correction and stabilization, which prevents the user from accepting a buffer value before the sensor has fully responded to the buffer solution.

Diagnostics in transmitters include glass and reference electrode impedance alarms, as well as alerts related to the temperature measurement and problems with the transmitter's electronics and memory. The use of a larger display makes it possible to more fully describe problems and can even offer corrective action.

These advances have been applied to two-wire transmitters and line power transmitters. In some two-wire transmitters, a second externally powered 4–20 mA output has been added to provide a second pH output with a different range or to output a secondary measurement such as temperature.

Line-powered transmitters have seen even more developments, such as the capability to control sensor cleaning and calibration systems. Data logging can also be available to track alarms and changes to the calibration constants and log runtime data including pH and secondary data such as temperature, sensor millivolts, RTD resistance, and analog outputs. In many cases, these data can be plotted on screen or downloaded to a portable memory device for a more detailed examination using a PC or laptop. Even though a plant or batch historian may be available, data logging provides a better tool for troubleshooting because it captures secondary measurements, which can show the root cause of a pH problem. When smart pH sensors are used, the data stored in the sensor can be viewed on the analyzer display.

Line-powered transmitters have become multiparameter and can include a one or more analytical measurements in addition to pH. In some cases, an additional measurement can be provided by 4–20 mA input to the transmitter, which allows it receive, scale, and even power pressure, temperature, flow, or any other transmitter with a 4–20 mA output. This allows the transmitter to provide a more complete solution beyond a simple pH measurement.

In addition to monitoring, analytical transmitters can provide control functionality. Multiparameter analyzers have more than one analog output, and one or more of these can be assigned to PID control for the measurements made. The user display and keypad are used for setup and tuning of the PID loop, and if a data logger is present, it can capture the PV of the control application and the control output.

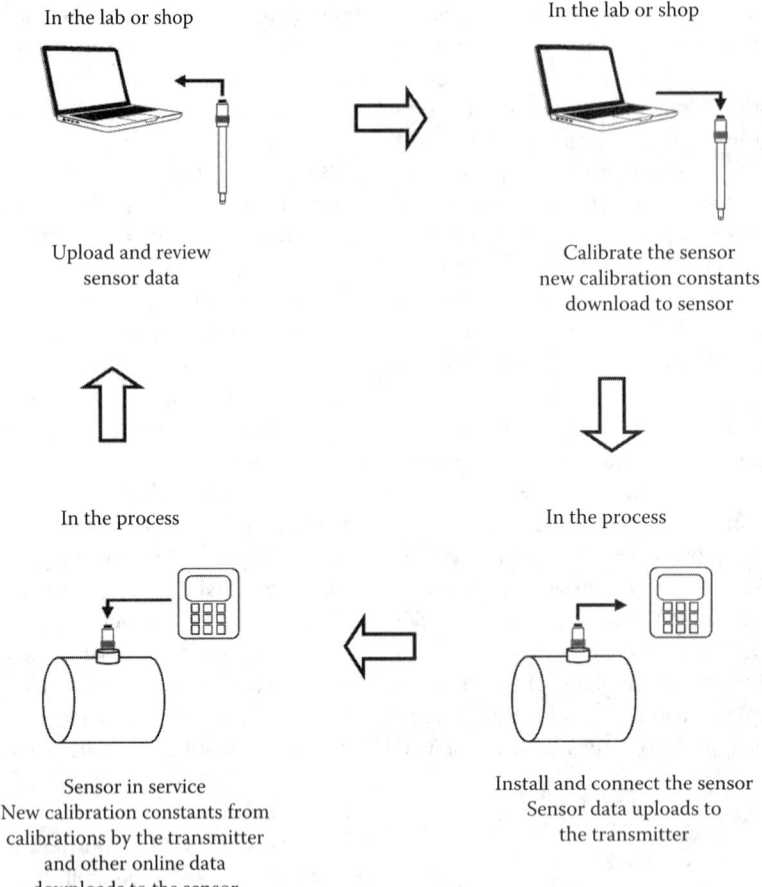

FIG. 1.55ad
How smart pH sensors are used.

If the transmitters have relays, discrete control functions can include simple high or low relay control, or possibly relay control with a delay timer, typically used to allow time chemical additions to mix. More sophisticated control can include pulse width modulation (Figure 1.55ad).

Smart pH Transmitters The term "smart" is used to describe a transmitter that has a digital output, in addition to or in place of an analog 4–20 mA output. This allows a user to have access to the transmitter's configuration parameters, diagnostic information, secondary measurements such as sensor millivolt input and temperature, and control parameters if they are present. The most common digital protocols used are Highway Addressable Remote Transducer (HART®), Profibus®, and Foundation® fieldbus. HART typically employs a digital signal superimposed on a 4–20 mA output, while Profibus and FOUNDATION fieldbus use digital physical layers.

Asset Management The advantage of smart transmitters to a pH user is the ability to access a transmitter remotely at a PC to check for diagnostic alarms, review configuration parameters, and even do single point calibrations. While PC software applications have been developed that simply list parameters, the preferred presentation of transmitter data is a Windows environment with the parameters organized according to function. The first of the two technologies used to create these windows is Electronic Device Descriptive Language (EDDL), which is a file created for a certain model of transmitter by its manufacturer that contains instructions for an asset management program to create windows. The second technology is the field device tool (FDT)/device type manager (DTM), where the FDT provides a frame application to display the information, and a DTM is an executable program for each type of transmitter that creates the windows for it.

Beyond the ease of accessing and quickly reviewing the status of a transmitter and its configuration, this technology can also provide a historian for changes to the transmitters parameters, including its calibration constants. It can also be used to create a configuration template and then download it to a transmitter, making the configuration of a number of

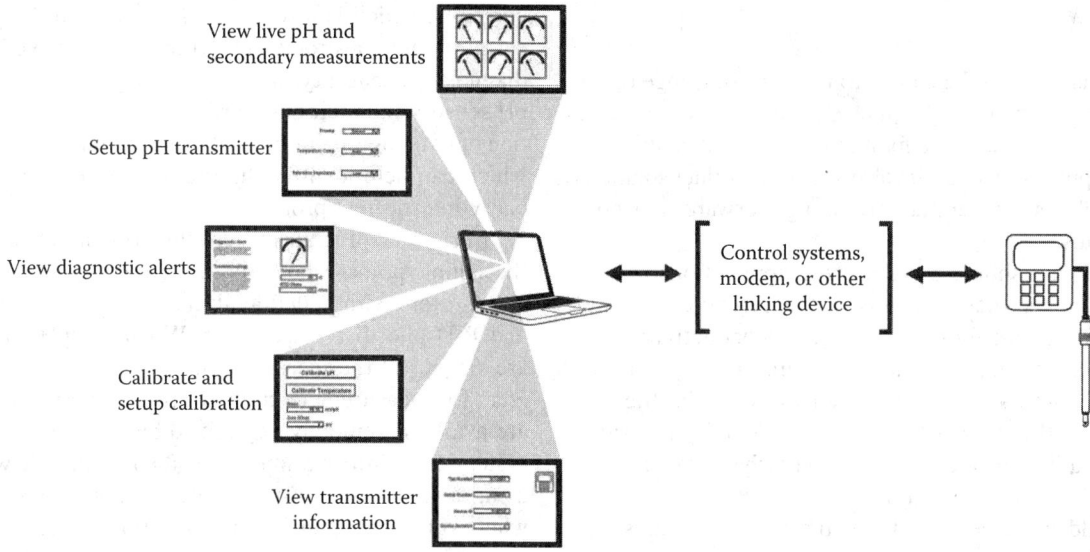

FIG. 1.55ae
Asset management.

transmitters faster and easier than using the local displays of each (Figure 1.55ae).

Wireless HART™ Transmitters Wireless HART is part of the new version 7 of HART that defines wireless communications using wireless transmitters in a self-organizing, mesh network configuration. This provides multiple paths for the wireless signal among the transmitters in the network and results in a more reliable network, less susceptible to a loss of the wireless signal due to the failure or blockage of any single transmitter. The network communicates to a gateway, which communicates to the control system using Modbus, Ethernet, or even wirelessly. A wireless pH transmitter is battery powered, and depending on the transmission rate, battery life can be as long as 4 years at a 1 min update rate. A wired transmitter can be used in the network using a Wireless HART adaptor, which can connect one to three wired transmitters to the network.

The advantage of wireless transmitters is that they can be installed without the time and expense of running power and signal wire to the sample point. Wireless pH transmitters have essentially the same feature set as a two-wire transmitter. In many cases, the sample point will have power available, and the use of a Wireless HART adaptor with a wired transmitter allows a multiparameter analyzer to be used, which can include other measurements in addition to pH, provide all the advanced features of a line-powered transmitter, and still be accessible from the control room wirelessly (Figure 1.55af).

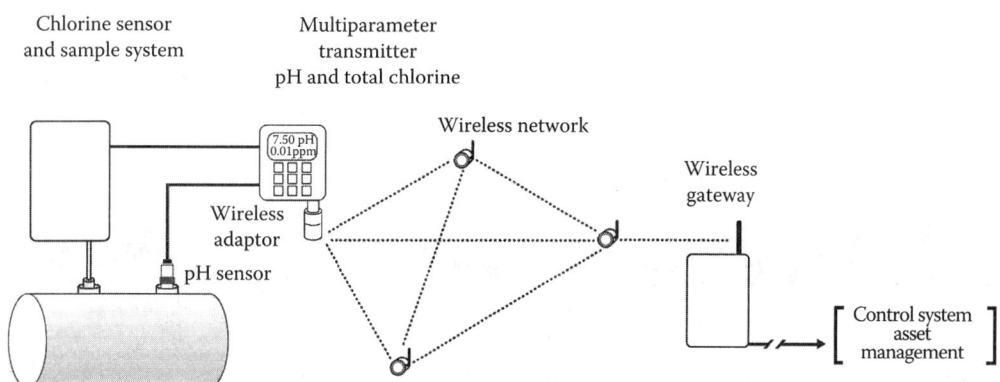

FIG. 1.55af
Monitoring pH and chlorine using a line-powered multiparameter transmitter and wireless adaptor.

CONCLUSION

The pH sensor should be chosen to meet the pH range of the application. The mounting method should be chosen to make the sensor easily accessible for maintenance and calibration. Most pH applications that involve water and dilute solutions at near ambient temperature are straightforward and pose few problems.

Special care should be taken with applications involving components that can poison the reference or foul the sensor, nonaqueous solvents, low conductivity, and extremes of temperature and pressure. In applications where the nominal pH is expected to be at the high or low limits of the 0–14 pH range, a conductivity measurement should be considered as an alternative measurement of acid and base concentrations.

It should be remembered that the pH of solutions can change with temperature. This is particularly the case with solutions that have a neutral or basic pH. It never hurts to consult several sensor suppliers as to the suitability of their various designs to a particular process application.

Smart pH sensor data can indicate how well the sensor performs in a particular application. Ideally, the calibration history will show how the sensor ages in the process but can also show that calibrations are not being performed correctly, which is also good information to have, though certainly not desirable. pH transmitters have increased diagnostic capability, not only through supporting smart pH sensors and their on-board diagnostics but also by addition of data logging of pH and its secondary measurements, which can help identify the root cause of problems, especially intermittent problems.

In general, pH transmitters can bring more capability to the sample point, especially in the form of multiparameter transmitters with added analytical or other measurements and PID and discrete control. When the pH transmitters are smart, all the measurements, diagnostics, and parameters for measurement configuration and controller tuning are available remotely, organized by an asset management system. This information can be made available wirelessly, using a wireless transmitter or wireless adaptor, in cases where running signal wire is cost prohibitive.

SPECIFICATION FORMS

When specifying pH measurements, ISA Forms 20A2341 and 20A2342 can be used. Both forms are reproduced on the next pages with the permission of the International Society of Automation.

#				#		
1	RESPONSIBLE ORGANIZATION		pH/ORP SENSOR w/wo INSERTION ASSEMBLY Device Specification	6	SPECIFICATION IDENTIFICATIONS	
2	(ISA)			7	Document no	
3				8	Latest revision	Date
4				9	Issue status	
5				10		
11	BODY HOLDER OR FITTING			60	INSERTION ASSEMBLY continued	
12	Body/Fitting type			61	Insertion/Immersion lg	
13	Process conn nominal size		Rating	62	Valve body material	
14	Process conn termn type		Style	63	Valve seat material	
15	Maximum probe diameter			64	Chamber wetted material	
16	Body/Fitting material			65	Compression ferrule matl	
17	Flange material			66		
18	Seal/O ring material			67		
19				68	PERFORMANCE CHARACTERISTICS	
20				69	Max press at design temp	At
21				70	Min working temperature	Max
22	SENSING ELEMENT			71	Min conductivity	
23	Sensor type			72	Max fluid velocity limit	
24	Construction style			73	Accuracy rating	
25	Measuring electrode shape			74	Measurement LRL	URL
26	Reference junction style			75	Temp compensation LRL	URL
27	Insertion/Immersion lg			76	Linearity	
28	Signal conn termn style			77	Max response time	
29	Temperature sensor type			78	Min ambient working temp	Max
30	Sensing surface matl			79	Max sensor to receiver lg	
31	Ref electrolyte material			80		
32	Ground pin material			81		
33	Reference element matl			82		
34	Primary junction matl			83	ACCESSORIES	
35				84	Electrolyte solution	
36				85	Buffer solution	
37				86	Cleaner type	
38	LEAD WIRE AND EXTENSION			87	Cleaner supply source	
39	Extension type			88	ORP std solution	
40	Cable length			89	Rotameter & valve	
41	Max cable operating temp			90		
42	Signal termination type			91		
43	Cable jacket material			92	SPECIAL REQUIREMENTS	
44				93	Custom tag	
45	CONNECTION HEAD OR PREAMPLIFIER			94	Reference specification	
46	Configuration type			95	Compliance standard	
47	Temperature compensation			96	Calibration report	
48	Enclosure type no/class			97		
49	Signal termination type			98		
50	Cert/Approval type			99		
51	Mounting location/type			100	PHYSICAL DATA	
52	Enclosure material			101	Estimated weight	
53				102	Overall height	
54				103	Removal clearance	
55	INSERTION ASSEMBLY			104	Signal conn nominal size	Style
56	Assembly type			105	Mfr reference dwg	
57	Isolation valve style			106		
58	Process conn nominal size		Rating	107		
59	Process conn termn type		Style	108		

#	CALIBRATIONS AND TEST		INPUT OR TEST		OUTPUT	
110						
111	TAG NO/FUNCTIONAL IDENT	MEAS/SIGNAL/SCALE	LRV	URV	LRV	URV
112		Measurement-Output signal				
113		Temp-Output signal				
114						
115						
116						
117						

#	COMPONENT IDENTIFICATIONS		
118			
119	COMPONENT TYPE	MANUFACTURER	MODEL NUMBER
120			
121			
122			
123			
124			
125			

Rev	Date	Revision Description	By	Appv1	Appv2	Appv3	REMARKS

Form: 20A2341 Rev 0 © 2004 ISA

pH/ORP/CONDUCTIVITY/RESISTIVITY TRANSMITTER/ANALYZER/MONITOR Device Specification

#	RESPONSIBLE ORGANIZATION		#	SPECIFICATION IDENTIFICATIONS	
1			6	Document no	
2			7	Latest revision	Date
3	ISA		8	Issue status	
4			9		
5			10		

#	TRANSMITTER OR ANALYZER		#	PERFORMANCE CHARACTERISTICS	
11			56		
12	Housing type		57	Measurement accuracy	
13	Input sensor type		58	Temperature accuracy	
14	Input sensor style		59	Measurement LRL	URL
15	Output signal type		60	Temp compensation LRL	URL
16	Enclosure type no/class		61	Minimum span	Max
17	Control mode		62	Min ambient working temp	Max
18	Local operator interface		63	Contacts ac rating	At max
19	Characteristic curve		64	Contacts dc rating	At max
20	Digital communication std		65	Max sensor to receiver lg	
21	Signal power source		66		
22	Temperature sensor type		67		
23	Quantity of input sensors		68		
24	Preamplifier location		69		
25	Contacts arrangement	Quantity	70		
26	Integral indicator style		71		
27	Signal termination type		72		
28	Cert/Approval type		73		
29	Mounting location/type		74		
30	Failure/Diagnostic action		75		
31	Sensor diagnostics		76	ACCESSORIES	
32	Data history log		77	Remote indicator style	
33	Temperature compensation		78	Indicator enclosure	
34	Calibration function		79	Communicator style	
35	Enclosure material		80		
36			81		
37			82		
38			83	SPECIAL REQUIREMENTS	
39			84	Custom tag	
40			85	Reference specification	
41			86	Compliance standard	
42			87	Software configuration	
43			88	Software program	
44			89		
45			90		
46			91		
47			92	PHYSICAL DATA	
48			93	Estimated weight	
49			94	Overall height	
50			95	Removal clearance	
51			96	Signal conn nominal size	Style
52			97	Mfr reference dwg	
53			98		
54			99		
55			100		

#	CALIBRATIONS AND TEST		INPUT OR TEST			OUTPUT OR SCALE	
110							
111	TAG NO/FUNCTIONAL IDENT	MEAS/SIGNAL/SCALE	LRV	URV	ACTION	LRV	URV
112		Meas-Analog output 1					
113		Meas-Analog output 2					
114		Meas-Digital output					
115		Meas-Scale					
116		Temp-Scale					
117		Temp-Digital output					
118		Meas setpoint 1-Output					
119		Meas setpoint 2-Output					
120		Meas setpoint 3-Output					
121		Meas setpoint 4-Output					
122		Failure signal-Output					
123							

#	COMPONENT IDENTIFICATIONS		
124			
125	COMPONENT TYPE	MANUFACTURER	MODEL NUMBER
126			
127			
128			
129			

Rev	Date	Revision Description	By	Appv1	Appv2	Appv3	REMARKS

Form: 20A2342 Rev 0 © 2004 ISA

Definitions

Activity is the effective concentration of an ion in solution, which takes into account the associations and interactions of the ion with other chemical species in solution. The activity of hydrogen ion in a solution will change if a salt is added, although the concentration of hydrogen ion remains constant. Since pH is a measure of hydrogen ion activity, the pH will change as well.

Isopotential point is the pH at which the millivolt output of a pH sensor does not change with temperature. For the vast majority of pH sensors, the isopotential point is 7.00 pH and 0.0 mV.

Nernst equation is an equation that relates the equilibrium reduction potential of a half-cell in an electrochemical cell to the standard electrode potential, temperature, activity, and reaction quotient of the underlying reactions and species used. It is named after the German physical chemist who first formulated it, Walther Nernst.

Abbreviations

CIP	Stands for clean in place. It is a cleaning process used in the food, beverage, and pharmaceutical industries to clean piping and vessels by alternately pumping cleaning solutions and rinse water through them.
HART	Stands for Highway Addressable Remote Transducer, which is the protocol of the HART Communications Foundation.
MOSFET	Is a metal oxide semiconductor field effect transistor, which is mainly used in microprocessors.
ORP	Is the oxidation–reduction potential of a solution, measured using a platinum or gold electrode and a reference electrode.
rH	Is the redox value that gives the oxidizing or reducing power of a solution independent of pH. It is calculated as follows: rH = (ORP/29) + 2pH.

Organizations

EDDL	www.eddl.org/Pages/default.aspx
FDT/DTM	www.fdtgroup.org/
Fieldbus FOUNDATION™	www.fieldbus.org/
HART Communication Foundation	www.hartcomm.org/
Profibus®	Profibus International www.profibus.com/

Bibliography

Bigalke, D., pH: Reducing error comparing in-line versus off-line techniques—Analysis 2005.

Butler, J. L., *Ionic Equilibrium—A Mathematical Approach*, Addison-Wesley, Boston, MA, 1964. http://www.amazon.com/Ionic-Equilibrium-Mathematical-James-Butler/dp/0201007304/ref=sr_1_fkmr0_1?s=books&ie=UTF8&qid=1391372829&sr=1-1-fkmr0&keywords=Ionic+Equilibrium%E2%80%94A+Mathematical+Approach.

Galster, H., *pH Measurement: Fundamentals, Methods, Applications, Instrumentation*, Wiley-VCH, New York, 1991. http://www.amazon.com/pH-Measurement-Fundamentals-Applications-Instrumentation/dp/3527282378/ref=sr_1_4?s=books&ie=UTF8&qid=1391367009&sr=1-4&keywords=Galster%2C+Helmuth.

McMillan, G., pH measurement and control opportunities—ISA automation week 2011: Analyzers, 2011.

McMillan, G. K., *Essentials of Modern Measurements and Final Elements in the Process*, International Society of Automation, Research Triangle Park, NC, 2010, 275–314. http://www.amazon.com/Essentials-Measurements-Elements-Process-Industry/dp/1936007231/ref=sr_1_9?s=books&ie=UTF8&qid=1391280359&sr=1-9&keywords=Gregory+McMillan.

McMillan, G. K. and Cameron, R. A., Advanced pH Measurement and Control, 3rd ed., The Instrumentation, Systems, and Automation Society (October 30, 2004), ISA, Research Triangle Park, NC, 2004. http://www.amazon.com/Advanced-pH-Measurement-Control-3rd/dp/1934394432/ref=sr_1_2?s=books&ie=UTF8&qid=1391280359&sr=1-2&keywords=Gregory+McMillan.

Merriman, D. C., Junction potential variations in pH, *Proceedings of the ISA/93 Technical Conference*, Chicago, IL, September 19–24, 1993.

Mooney, E. F., On-line photometric titrations for process control, *Proceedings of the 1992 ISA Conference*, Houston, TX, October 1992.

Moore, R. L., *Neutralization of Waste by pH Control*, Instrument Society of America, Research Triangle Park, NC, 1978. http://www.amazon.com/Neutralization-Waste-Water-Control-Monograph/dp/0876643837/ref=sr_1_1?s=books&ie=UTF8&qid=1391283014&sr=1-1&keywords=Neutralization+of+Waste+by+pH+Control.

Pence, G. and Baril, R., *Bringing pH Measurement Systems Up to Date*, InTech, ISA, Research Triangle Park, NC, February 2008.

Pfannenstiel, E., Process pH measurement continues evolution, *InTech*, ISA, Research Triangle Park, NC, October 2002.

Proudfoot, C. G. et al., Self-tuning PI control of a pH neutralization process, *Proc. Inst. Elec. Eng.*, 130(5), 1983, pp. 267–72. http://ieeexplore.ieee.org/xpl/login.jsp?tp=&arnumber=4642209&url=http%3A%2F%2Fieeexplore.ieee.org%2Fstamp%2Fstamp.jsp%3Ftp%3D%26arnumber%3D4642209.

Queeney, K.M., Why are there so many types of pH electrodes?—Analysis, ISA, Research Triangle Park, NC, 2003.

Queeney, K. M., Automated pH cleaning and calibration systems—A modular approach—ISA Expo 2004.

Skoog, D. A, Holler, F. J., and Crouch, S. R., *Principles of Instrumental Analysis*, 6th ed., Cengage Learning, ISA, Research Triangle Park, NC, 2006. http://www.amazon.com/Principles-Instrumental-Analysis-Douglas-Skoog/dp/0495012017/ref=sr_1_1?s=books&ie=UTF8&qid=1391370944&sr=1-1&keywords=Skoog%2C+D.+A.

Walker, T., *Difficulties Making Quality pH Measurements—POWID*, ISA, Research Triangle Park, NC, 2005.

Weiss, M. D., Teaching old electrodes new tricks, *Control*, July 1991.

Wescott, C. C., *pH Measurement*, Academic Press, New York, 1978, http://www.amazon.com/Ph-Measurements-C-Clark-Westcott/dp/0127451501/ref=sr_1_2?s=books&ie=UTF8&qid=1391281079&sr=1-2&keywords=pH+Measurement.

1.56 Phosphate Analyzer

W. H. PARTH (1974, 1982)   **B. G. LIPTÁK** (1995, 2003, 2017)

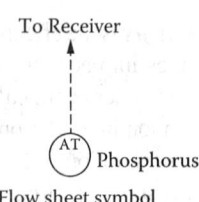

Flow sheet symbol

Methods of detection	A. Colorimetric B. Flame photometric C. Chromatographic
Operating pressure	Atmospheric
Materials of construction	Most analyzers can be obtained with wetted parts made out of stainless steel, glass, or Teflon
Inaccuracy	2%–3% of full scale
Analysis time	From a fraction of 1–15 min
Ranges	A. From 0–5 to 0–20 ppm B. From 1 to 100 ppm The overall capability of measurement for all types is from 0–10 ppb to 0–100 ppm
Costs	Laboratory units cost $10,000–$25,000. Industrial installations with sampling system included cost from $25,000–$100,000. Vane-type filters for sewage applications cost about $2500.
Partial list of suppliers	ABB (A): http://www05.abb.com/global/scot/scot203.nsf/veritydisplay/cd3cc50a3f00ee1dc12579bb005172cf/$file/ds_azt6p-en_d.pdf HACH Co. (A) http://www.hach.com/5500sc-phosphate-high-range-analyzer-2-channels-reagents-included/product?id=16215284756 Huakeyi-Digital (B) http://www.digitalcontrols.org/products/analyzers/phosphate-analyzer.html#HK-108W Indiamart (A,B,C) http://dir.indiamart.com/impcat/phosphate-analyzer.html Rosemount Analytical Inc. (A): http://www2.emersonprocess.com/siteadmincenter/PM%20Rosemount%20Analytical%20Documents/Liq_ProdData_71-CFA_Phosphate_3047.pdf Skalar (robotic): http://www.skalar.com/news/complete-automation-for-st-cod-and-total-phosphate-analysis SwanAnalyticalInstruments(A)http://www.swan-analytical-usa.com/Catalog/en/ProductGroupOverview.aspx?subchapter=Chapter_feedwater&prdtGroup=Grp_Phosphate_Feed&prdtSubGroup=&prdtName= Thermo Orion (A): http://www.le.com.my/wp-content/uploads/2013/02/19.-2095-Phosphate-Monitor.pdf Thermo Scientific (A,C) http://www.thermoscientific.com/content/tfs/en/product/orion-2095-phosphate-analyzer-accessories-orion-online-monitors.html Varian (C) http://users.stlcc.edu/Departments/fvbio/Chromatography_Gas_Varian_gc3800brochure18pgs.pdf Waltron (A): http://www.waltron.net/phosphate_analyzer_ai3042.asp

INTRODUCTION

This chapter discusses the measurement of both the organic and inorganic forms of phosphate. The organic form is called orthophosphate, which is an ester of phosphoric acid. It is a critical water pollutant. The inorganic form is phosphate (PO_4^{3-}), which is a salt of phosphoric acid. Inorganic phosphates are mined to obtain phosphorus, which is used in agriculture, and if it enters the waters, it can be lethal to fish. Total phosphate is the sum of both forms.

This chapter covers the various analyzer designs that are used to measure the concentration of phosphate. Each of these devices are fully described in the chapters listed below, and therefore, the detailed description of their designs are not repeated here.

Chapter 1.10   Chromatography
Chapter 1.13   Colorimetry
Chapter 1.73   Water Quality Monitoring
Chapter 1.74   Wet Chemistry

PHOSPHORUS IN WASTEWATER

The principal application of phosphorus analyzers is in the control of phosphate removal in sewage treatment plants. By knowing the flow rate and the phosphorus content of raw sewage, the required optimal quantities of chemical additives can be determined. In addition, a measurement of the phosphorus remaining after treatment indicates the quality of performance.

Phosphorus occurs in wastewater almost entirely in the form of phosphates, including orthophosphates, condensed phosphates (pyrophosphate, metaphosphate, and polyphosphate), and as organically bound phosphates. The various methods of phosphorus detection do not all respond to the total phosphorus present.

Other applications are those specific to the control of phosphate addition to high-pressure boiler water as a corrosion inhibitor, and to the measurement of elemental phosphorus in the elemental form from phosphorus extracting plants.

LABORATORY ANALYSIS

Colorimetric procedures are available to determine the concentration of soluble orthophosphate. In the commonly used aminonaphtholsulfonic acid method* of analysis, ammonium molybdate reacts with a dilute phosphorus solution to produce molybdophosphoric acid. This acid is then reduced to the intensely colored complex, molybdenum blue, by the combination of aminonaphtholsulfonic acid and sulfite reducing agents.

The stannous chloride method,* although slightly more sensitive, is similar to the method just described except for the substitution of stannous chloride for aminonaphtholsulfonic acid as the reducing agent.

CONTINUOUS COLORIMETER

A continuous phosphorus analyzer consists of a sample temperature controller, a multiple peristaltic pump for reagent and sample metering, a mixing and time-delay section, reagent storage, a colorimeter, and an electronic readout section.

---

* APHA/AWWA/WEF, Standard Methods for the Examination of Water and Wastewater. 2012. http://www.amazon.com/Standard-Methods-Examination-Water-Wastewater/dp/0875530133.

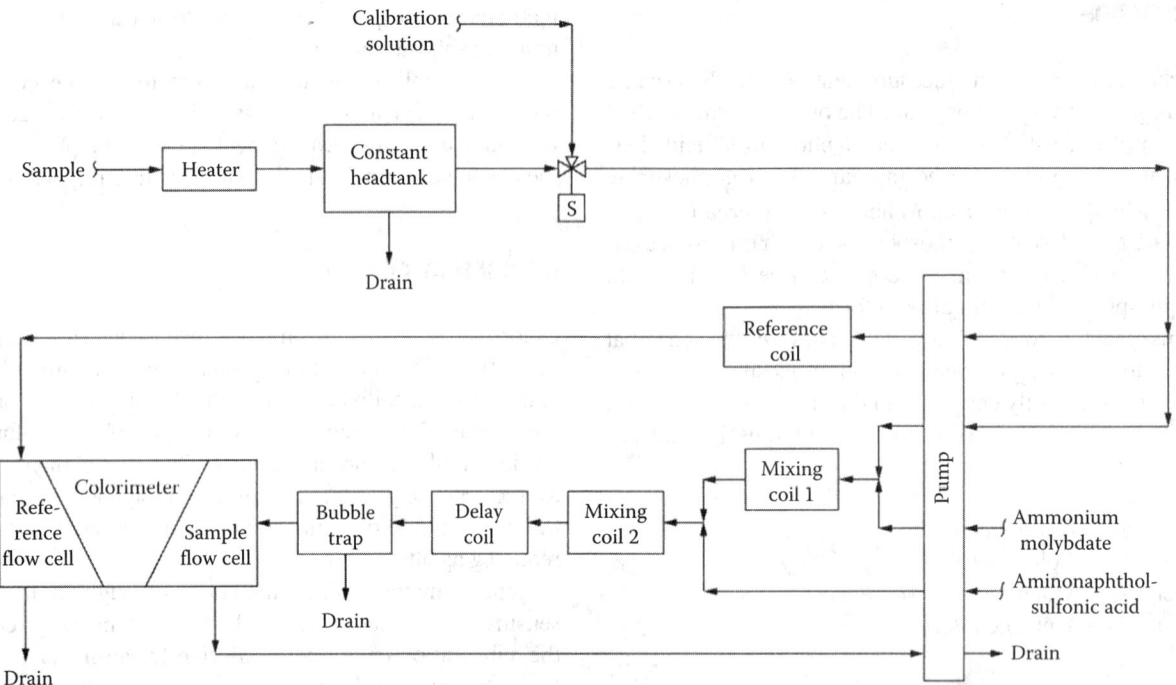

FIG. 1.56a
Orthophosphate analyzer.

Referring to Figure 1.56a, the water sample is brought to a constant temperature to ensure uniform sample reaction and rapid response time. The sample stream is degassed in the constant head tank and divided into two paths, one for the reference sample and the other for the reaction sample. A multiple peristaltic pump meters the sample streams as well as the addition of ammonium molybdate and aminonaphtholsulfonic acid solutions.

The reaction and reference samples pass through separate delay coils. In one coil, which gives about a 5-min delay, the reaction sample and the reagents complete the color reaction; in the other delay coil, the reference sample is experiencing the same delay, in addition to the time that the reaction sample spends in the mixing coils. The sample and reference streams from the delay coils are fed into the dual flow cells of a dual-beam colorimeter. A bubble trap ahead of the colorimeter removes any bubbles formed in the analyzer.

The photodetectors sense the difference in color intensity between the reaction and reference samples. In this case, the molybdenum blue complex is measured at a wavelength of 6900 Å. The electronic section amplifies the colorimeter output and, if necessary, linearizes it. The use of a dual-beam colorimeter automatically compensates for the variations in inherent color or turbidity of the sample.

Pumpless Design

Figure 1.56b shows a design where the use of peristaltic pumps has been eliminated by using pressurized reagent chambers from which the reagents are automatically supplied

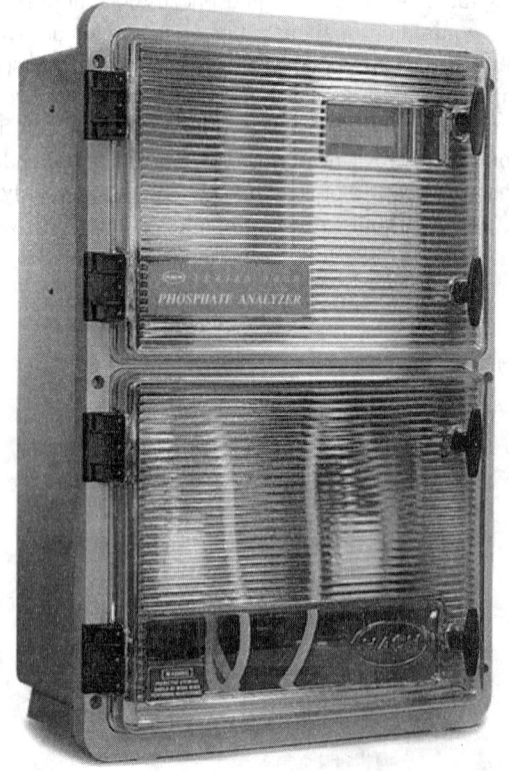

FIG. 1.56b
High-range phosphate analyzer. (Courtesy of Hach Co. http://www.hach.com/5500sc-phosphate-high-range-analyzer-2-channels-reagents-included/product?id=16215284756.)

to a set of microprocessor controller solenoid valves. In this design, the solenoid valves release the reagents in precisely controlled volumes, ensuring the accuracy of each test.

The more recent analyzer designs provide selectable outputs to drive recorders or other data acquisition equipment. The analyzer span can be programmed by the operator to cover any portion of the instrument's range. Usually, RS-232 Serial I/O ports provide two-way communication with external computers and/or printers. In most systems, concentration set-point alarms can also be programmed by the operator over the entire range, and each alarm can be programmed to signal high or low concentrations or high or low rates of concentration changes. Some of the better designs might include four alarm relays (two high voltage and two low voltage) each equipped with unpowered SPDT contacts. Usually, an interlock is also provided to automatically shut down the unit when the sample flow is interrupted and to restart the unit when flow is restored.

TOTAL PHOSPHATE ANALYSIS

Total inorganic phosphate can be measured by first hydrolyzing the sample with sulfuric acid at 95°C (203°F). This converts the phosphates in the meta-, pyro-, and polyforms to orthophosphates. Total phosphates (inorganic plus organic) can be determined by an additional step consisting of oxidizing organic compounds (e.g., by boiling in potassium persulfate) to split off the phosphate moiety, which is then available for reaction. Instruments that use a single-beam colorimeter in either a continuous or batch-type analyzer are also available.

FLAME PHOTOMETRIC ANALYSIS

This analyzer detects the photometric flame emission of phosphorous compounds in a hydrogen–air flame. This method was developed by Draegerwerk in Germany and was first applied to the detection of phosphorus compounds in air. As discussed in connection with Figure 1.10z, the method later was used as a detector for gas chromatography. Still later, the pulsed flame photometric detector (Figure 1.10ac) was developed, which uses the pulsed flame characteristics of phosphorus shown in Figure 1.56c.

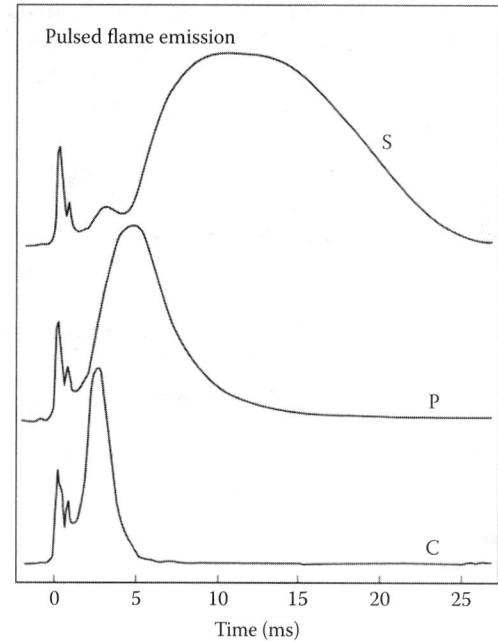

FIG. 1.56c
The separation in the time domain of phosphorus, sulfur, and carbon spectra. This PFPD analyzer provides infinite selectivity against hydrocarbon emission as well as unique heteroatom identification capability. (From Amirav, A. et al., Pulsed Flame Photometric Detector for Gas Chromatography, Tel Aviv University, Tel Aviv, Israel, 2001.)

Operation

The detector in Figure 1.56d contains a burner with separate delivery tubes for hydrogen and the air sample.

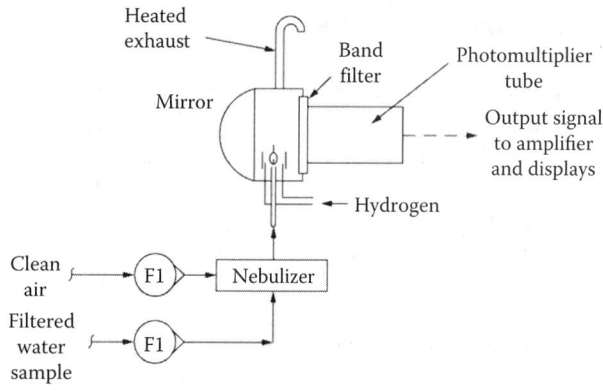

FIG. 1.56d
Flame photometric detector of phosphorus in water.

If phosphorus is present in the hydrogen-rich (reducing) flame, it will produce a strong luminescent emission between the wavelengths of 4850 and 5650 Å. This emission is at its maximum at 5260 Å and is isolated by a narrow bandpass interference filter.

The hydrogen and air are burned in a hollow tip that shields the flame from direct view of the mirror and photomultiplier tube. When phosphorus is present, the emission occurs above the shielded flame, and the light is transmitted directly and by way of the mirror through the filter to the photomultiplier tube. The shield offers specificity from carbon dioxide and hydrocarbons with flame emission at 5260 Å. The output from the photomultiplier tube is linear over several decades.

The detector can measure phosphorus in water by the addition of a nebulizer, which injects a mist of sample water into a clean air stream at a constant rate. The output is linear in the 1–100 ppm range.

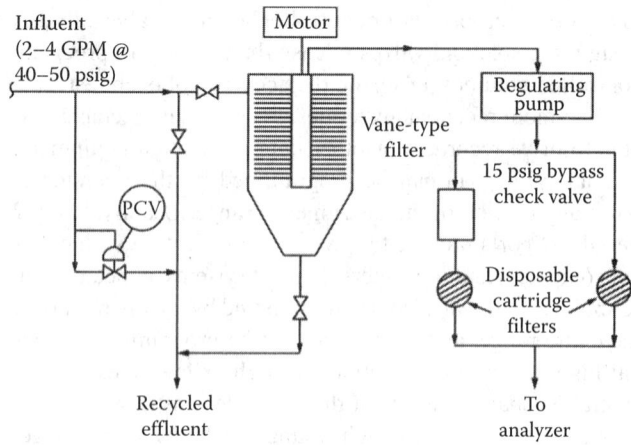

FIG. 1.56e
Raw sewage sample handling system.

CHROMATOGRAPHY

Where elemental phosphorus discharges into water, it has been found that a concentration of a few parts per billion is lethal to fish. Laboratory techniques* possessing this sensitivity have been developed.

The phosphorus is partially isolated by extraction in a suitable solvent such as benzene or isooctane. A portion of the extract is injected into a chromatograph utilizing a flame photometric detector. Mud and tissue samples can also be analyzed rapidly by this method.

SAMPLE HANDLING SYSTEMS

The successful application of the phosphorus analyzers just described depends on reliable delivery of a well filtered sample. Such systems are discussed in Chapters 1.3 and 1.73.

The sampling systems shown in Figures 1.56e and 1.56f are capable of handling raw sewage. The primary filter is of the motor-driven vane type with alternate stationary and rotating disks, and the clearance between the plates determines the degree of filtration.

The second stage of filtering consists of two disposable cartridge filters. A check valve diverts the flow to the second filter when the pressure drop across the first indicates that a change is necessary. The regulating pump between the filter stages ensures a constant flow rate in difficult determinations. Provision for backflushing the system is also advisable in some applications.

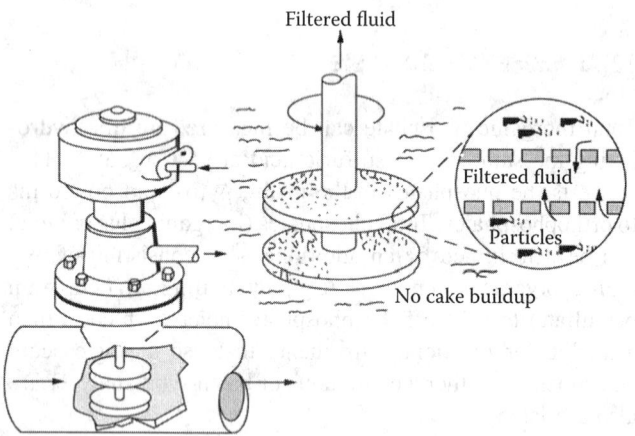

FIG. 1.56f
The rotary disc filter.

The selection of a good sample handling system from the options discussed in Chapter 1.3 is just as critical as the selection of the analyzer itself. Provisions for convenient calibration are also an essential part of a complete and successful installation.

SPECIFICATION FORMS

When specifying phosphate analyzers, the ISA Form 20A1001 can be used to specify the detector device itself, and when specifying both the device and the nature of the application, including the composition and/or properties of the process samples, ISA Form 20A1002 can be used. Both forms are reproduced on the next pages with the permission of the International Society of Automation.

---

* Zhenying, D. et al. Gas chromatographic analysis of trace amount of elemental phosphorus in wastewater, 1987. http://en.cnki.com.cn/Article_en/CJFDTOTAL-SPZZ198702019.htm.

1.56 Phosphate Analyzer

ANALYSIS DEVICE
Operating Parameters

1	RESPONSIBLE ORGANIZATION		6	SPECIFICATION IDENTIFICATIONS	
2			7	Document no	
3	ISA		8	Latest revision	Date
4			9	Issue status	
5			10		

	ADMINISTRATIVE IDENTIFICATIONS			SERVICE IDENTIFICATIONS Continued	
11			40		
12	Project number	Sub project no	41	Return conn matl type	
13	Project		42	Inline hazardous area cl	Div/Zon Group
14	Enterprise		43	Inline area min ign temp	Temp ident number
15	Site		44	Remote hazardous area cl	Div/Zon Group
16	Area	Cell Unit	45	Remote area min ign temp	Temp ident number
17			46		
18	SERVICE IDENTIFICATIONS		47		
19	Tag no/Functional ident		48	COMPONENT DESIGN CRITERIA	
20	Related equipment		49	Component type	
21	Service		50	Component style	
22			51	Output signal type	
23	P&ID/Reference dwg		52	Characteristic curve	
24	Process line/nozzle no		53	Compensation style	
25	Process conn pipe spec		54	Type of protection	
26	Process conn nominal size	Rating	55	Criticality code	
27	Process conn termn type	Style	56	Max EMI susceptibility	Ref
28	Process conn schedule no	Wall thickness	57	Max temperature effect	Ref
29	Process connection length		58	Max sample time lag	
30	Process line matl type		59	Max response time	
31	Fast loop line number		60	Min required accuracy	Ref
32	Fast loop pipe spec		61	Avail nom power supply	Number wires
33	Fast loop conn nom size	Rating	62	Calibration method	
34	Fast loop conn termn type	Style	63	Testing/Listing agency	
35	Fast loop schedule no	Wall thickness	64	Test requirements	
36	Fast loop estimated lg		65	Supply loss failure mode	
37	Fast loop material type		66	Signal loss failure mode	
38	Return conn nominal size	Rating	67		
39	Return conn termn type	Style	68		

	PROCESS VARIABLES	MATERIAL FLOW CONDITIONS				PROCESS DESIGN CONDITIONS		
69					101			
70	Flow Case Identification			Units	102	Minimum	Maximum	Units
71	Process pressure				103			
72	Process temperature				104			
73	Process phase type				105			
74	Process liquid actl flow				106			
75	Process vapor actl flow				107			
76	Process vapor std flow				108			
77	Process liquid density				109			
78	Process vapor density				110			
79	Process liquid viscosity				111			
80	Sample return pressure				112			
81	Sample vent/drain press				113			
82	Sample temperature				114			
83	Sample phase type				115			
84	Fast loop liq actl flow				116			
85	Fast loop vapor actl flow				117			
86	Fast loop vapor std flow				118			
87	Fast loop vapor density				119			
88	Conductivity/Resistivity				120			
89	pH/ORP				121			
90	RH/Dewpoint				122			
91	Turbidity/Opacity				123			
92	Dissolved oxygen				124			
93	Corrosivity				125			
94	Particle size				126			
95					127			
96	CALCULATED VARIABLES				128			
97	Sample lag time				129			
98	Process fluid velocity				130			
99	Wake/natural freq ratio				131			
100					132			

	MATERIAL PROPERTIES			MATERIAL PROPERTIES Continued		
133			137			
134	Name		138	NFPA health hazard	Flammability	Reactivity
135	Density at ref temp	At	139			
136			140			

Rev	Date	Revision Description	By	Appv1	Appv2	Appv3	REMARKS

Form: 20A1001 Rev 0 © 2004 ISA

Analytical Measurement

1	RESPONSIBLE ORGANIZATION	ANALYSIS DEVICE COMPOSITION OR PROPERTY Operating Parameters (Continued)	6	SPECIFICATION IDENTIFICATIONS		
2			7	Document no		
3	ISA		8	Latest revision		Date
4			9	Issue status		
5			10			

	PROCESS COMPOSITION OR PROPERTY			MEASUREMENT DESIGN CONDITIONS				
	Component/Property Name	Normal	Units	Minimum	Units	Maximum	Units	Repeatability
13								
14								
15								
16								
17								
18								
19								
20								
21								
22								
23								
24								
25								
26								
27								
28								
29								
30								
31								
32								
33								
34								
35								
36								
37								

Rev	Date	Revision Description	By	Appv1	Appv2	Appv3	REMARKS

Form: 20A1002 Rev 0 © 2004 ISA

Definitions

Orthophosphate is an organic chemical, an ester of phosphoric acid. This organic chemical is important in biochemistry and in bioechochemistry, because it is a critical water pollutant.

Phosphate (PO_4^{3-}) is an inorganic chemical, a salt of phosphoric acid. Inorganic phosphates are mined to obtain the phosphorus, which is used in agriculture and industry. If it enters the receiving waters, it is lethal to fish.

Abbreviation

SPDT   Single pole double throw

Organizations

APHA   American Public Health Association
AWWA   American Water Works Association
WEF   Water Environment Federation

Bibliography

Amirav, A. Pulsed flame photometric detector for gas chromatography, 2001. https://www.researchgate.net/publication/231170451_Pulsed-flame_photometer_A_novel_gas_chromatography_detector.

APHA/AWWA/WEF, Standard methods for the examination of water and wastewater, 2012. http://www.amazon.com/Standard-Methods-Examination-Water-Wastewater/dp/0875530133.

ASA Analztics, Total phosphorous in water or wastewater, 2016. http://www.asaanalytics.com/total-phosphorous.php.

Cleary, J. Analysis of phosphate in waste water, 2009. http://doras.dcu.ie/2969/2/v52–32.pdf.

EPA, Water resources, 2015. http://www.epa.gov/learn-issues/water-resources#quality.

Friends of Five Creeks, Total phosphate, 2010. http://www.fivecreeks.org/monitor/phosphate.shtml.

Lenntech, Phosphorus removal from wastewater, 2006. http://www.lenntech.com/phosphorous-removal.htm.

MIT, Running Fluorine or Phosphorus NMR on the Varian 300 (2007) Running Fluorine or Phosphorus NMR on the Varian 300

Strom, P. F. Technologies to remove phosphorus from wastewater, 2006. http://www.water.rutgers.edu/Projects/trading/p-trt-lit-rev-2a.pdf.

University of Toronto, Proton with phosphorus decoupling, 2006. http://www.chem.toronto.edu/facilities/nmr/NMRBlog/C1701201524/E20061219184514/index.html.

Wikipedia, Phosphate test. http://en.wikipedia.org/wiki/Phosphate_test.

Zhenying, D. et al. Gas chromatographic analysis of trace amount of elemental phosphorus in wastewater, 1987. http://en.cnki.com.cn/Article_en/CJFDTOTAL-SPZZ198702019.htm.

1.57 Physical Properties Analyzers for Petroleum Products

N. S. WANER (1972) **A. ALSTON** (1982) **D. E. PODKULSKI** (1995)
I. VERHAPPEN and B. G. LIPTÁK (2003, 2017)

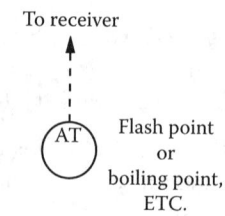

Flash point or boiling point, ETC.

Flow sheet symbol

Analyzer type	A. Distillation analyzer
	B. Vacuum distillation analyzer
	C. Simulated distillation by gas chromatography
	D. Air-saturated vapor pressure analyzer—continuous
	E. Air-saturated vapor pressure analyzer—cyclic
	F. Dynamic vapor pressure analyzer
	G. Continuous vapor—liquid ratio analyzer
	H. Differential-pressure pour pointer
	J. Viscous-drag pour pointer
	K. Optical-cloud point analyzer
	L. Freeze-point analyzer
	M. Low-temperature flash-pointer analyzer
	N. High-temperature flash-point analyzer
	O. Octane engine comparator analyzer
	P. Reactor-tube continuous octane analyzer
	Q. Near-infrared inferential measurements
Potential applications	Crude fractions (A, B, C, D, J, K, L, M); gasoline components/product (A, B, C, D, E, F, G, H, O, P); diesel (K, M); jet or kerosene (L, M); and lube oils (J, N)
Reference methods	ASTM D86 (A); ASTM D1160; (B); ASTM D2887/D3710 (C); ASTM D4953-90 (D, E, F); ASTM D1267 (F); ASTM D2533 (G); ASTM D97 (H, J); ASTM D2500 (K); ASTM D2386 (L); ASTM D56/D93 (M, N); and ASTM D2699/D2700 (O, P)
Costs	A. $40,000
	B. $56,000
	C. $40,000
	D. $38,000
	E. $28,000
	F. $26,000 to $35,000
	G. $85,000
	H. $37,000
	J. $41,000
	K. $28,000
	L. $33,000
	M. $44,000
	N. $41,000
	O. $220,000
	P. $86,000
	Q. $100,000

(Continued)

Partial list of suppliers ABB Process Analytics (C, E) http://www.abb.com/product/us/9AAC100838.aspx?country=CA
Applied Controls http://www.analyzer-systems.com/products.aspx
Applied Instrument Technologies (Q) http://www.aitanalyzers.com/process-analyzer-product.php?id=8
Bartec Benke (A, B, D, J, K, L, M, Q) http://www.energy-business-review.com/suppliers/bartec-benke-ebr-profile/products
Camlab (O) http://www.camlab.co.uk/portable-octane-analyzer-for-spark-ignition-engine-fuel-p32165.aspx
Core Labs (O) http://www.corelab.com/search.aspx?q=octane
Eravap http://www.eralytics.com/new/cms/prod.asp?id=194&mode=details
Grabner Instruments http://www.grabner-instruments.com/products/vaporpressure/vap-online.aspx
Hobré Instruments BV http://www.hobre.com/index.php/43-distributed-analyzers/81-icon-physical-property-analyzers.html
InfoMetrix (Chemometrics) http://infometrix.com/chemometrics/
Ircon http://www.iconscientific.com/index.php/solutions/vapourpressure.html
Koehlerin Instrument (J, K, M, N) http://www.koehlerinstrument.com/products/KLA-3TS.html
Modcon http://www.modcon-systems.com/project/process-nir-analyzer/
Ocean Optics (UOP) (P) http://oceanoptics.com/product/
ORB Instruments Inc. (A, F, H, K, L) (http://www.orbinstruments.com)
PAC incl. Precision Scientific (A, B, C, D, G, J, K, L) http://www.paclp.com/process_analytics/microdist/
Phase Technology (J, K, L) http://www.phase-technology.com/
Precision Scientific (A, B, C, E, G, H, K, L, M, N) http://www.paclp.com/process_analytics/brand/pspi
Siemens Applied Automation (C) https://www.industry.usa.siemens.com/automation/us/en/process-instrumentation-and-analytics/process-analytics/pa-brochures/Documents/MaxII_brochure_en.pdf
Thomas Scientific (M, N) http://www.thomassci.com/Instruments/Flash-Point-Testers/_/Manual-Flash-Point-Testers/

INTRODUCTION

This chapter focuses on the physical property analyzers that are widely used in the petroleum industry in refining, processing, blending, and pipelining operations, which assist in the evaluation of the properties of gasoline, diesel, kerosene, naphtha, and reformate, as well as optical, biomedical, and wire coatings. The analysis of the physical properties of other substances are not covered in this chapter, because the other chapters of this volume cover them.

Both the laboratory and the on-line, continuous analysis of petroleum products are covered, including the applicable ASTM Standard Test methods for refinery processes and products. In addition to the information on the front page of this chapter, Table 1.57a provides a more detailed overall orientation of the various physical property analyzers discussed in this chapter.

Prior to the introduction of on-stream analyzers, analyses were done in the laboratory on periodic grab samples, and the results were reported to the process unit operator at some later time, permitting set point adjustments of parameters such as flow, temperature, pressure, and level. Continuous on-stream plant analyzers offer many advantages over laboratory analyses, including the characteristics enumerated in the following.

TABLE 1.57a
Orientation Table for Petroleum Physical Properties Analyzers

Analyzer	Type	Range	Repeatability (±)	Flow Rate GPH (LPH)	Pressure psig (kPa)	Temperature °F (°C)	Cycle Time (min)	Cost (in $1000s)	Suppliers
Distillation	Distillation	5%–95% 38°C–343°C (100°F–650°F)	1% sample boiling range	1.2 (4.5)	20–150 (138–1035)	30 (17) below IBP	5	40	Precision Scientific
	Vacuum	343°C–538°C (650°F–1000°F)	1% sample boiling range	2.3 (8.7)	50–250 (345–1724)	180 (82) max	16	56	Precision Scientific
	Simulated distillation	2%–98% –17°C to 538°C (0°F–1000°F)	3°F–12°F (2°C–7°C)	4 (15) (sample inject)	5–150 (35–1035)	Below expected IBP	10–30	40	ABB - Asea Brown Boveri Siemens
Vapor pressure	Air-saturated continuous	2–19 psia	0.1 psia	1.6 (6)	10–100 (69–690)	50–110 (10–43)	2	38	Precision Scientific
	Air-saturated cyclic	0–20 psia	0.05 psia	Bypass flow			9	28	Asea Brown Boveri
	Dynamic	0–20 psia 0–200 psia	Equal to or better than ASTM	10–50 (38–190)	75–500 (520–3450)	70–120 (21–49)	0.75	26–35	Precision Scientific
Vapor/liquid	Continuous	10–30 V/L to 66°C (150°F)	0.5 V/L	2–4 (7.6–15.1)	50–150 (350–1050)	Normal blending range	3	85	Core Labs
Pour point	Differential pressure	–59°C to +10°C (–75°F to +50°F)	5°F (2°C)	2 (7.6)	<20 (138)	20 (11) above pour point	5–15	37	Precision Scientific Orb Instruments Inc.
Cloud point	Optical	–25°C to +15°C (–13°F to +59°F)	1°F (0.6°C)	3–5 (11.3–18.9)	250 (1724)	20 (11) above pour point	2	28	Precision Scientific Orb Instruments Inc.
Freeze point	Aviation fuels	–65°C to –10°C (–85°F to +14°F)	1°F (0.6°C)	3–5 (11.3–18.9)	250 (1724)	50–122 (10–50)	3	33	Precision Scientific Orb Instruments Inc.
Flash point	Low temperature	10°C–50°C (50°F–250°F)	2°F (1°C)	3.5 (13.2)	1000 (6895)	250 (139) max above flash point	1–7	44	Precision Scientific Orb Instruments Inc.
	High temperature	60°C–316°C (140°F–600°F)	3°F (1.5°C)	1 (3.8)	10–125 (69–863)	50 (28) below flash point	1–5	41	Precision Scientific
Octane	Comparator engine reactor tube	ASTM specs 2 octane numbers	ASTM specs 0.1 RON	ASTM specs 0.02 (0.06)	ASTM specs 50–500 (350–3500)	ASTM specs 100 (38)	ASTM specs 2	220 (86)	Core Labs

ADVANTAGES OF CONTINUOUS ANALYZERS

1. Continuous measurement of the stream, eliminating long time lags
2. Reduction of errors caused by unrepresentative samples or by changes in sample composition caused by sample handling
3. Elimination of human errors characteristic of nonautomated laboratory procedures
4. Ability to recognize process trends, thus permitting the automatic control of a given process variable by closed-loop control
5. Cost reductions resulting from minimization of laboratory analyses
6. Closer control resulting in smaller tolerances in final product specifications and reduction in quality "give-away"
7. Feasibility of implementing in-line blending systems, which result in economic benefits resulting from the elimination of tankage, and from increased system flexibility and better quality control
8. Ability to provide continuous inputs to computerized process control systems for plant optimization
9. Direct measurement of process variables rather than detection of properties by inference

DISTILLATION ANALYZERS

Laboratory Measurement Standards

Distillation analyzers (Figure 1.57a) were introduced to provide data on the volatility characteristics of process streams and separation efficiency of distillation units. ASTM Method D 86-IP-123 is the currently accepted laboratory standard for determining the boiling characteristics of petroleum products distilled at atmospheric pressure. The method employs a batch technique and approaches a single plate distillation process without reflux.

The petroleum products analyzed are complex mixtures of components, and a low level of fractionation is achieved. True boiling-point distillation, in columns with 15–100 theoretical plates and at reflux ratios of 5:1 or more, produces greater separation of components. The apparatus and procedures for true boiling-point determination are not standard, are complex, take longer to perform, and are not as widely used.

Distillation curves for a few hydrocarbons are shown in Figure 1.57b along with a comparison of curves generated by ASTM Method D 86-IP-123 and by true boiling-point determinations for kerosene.

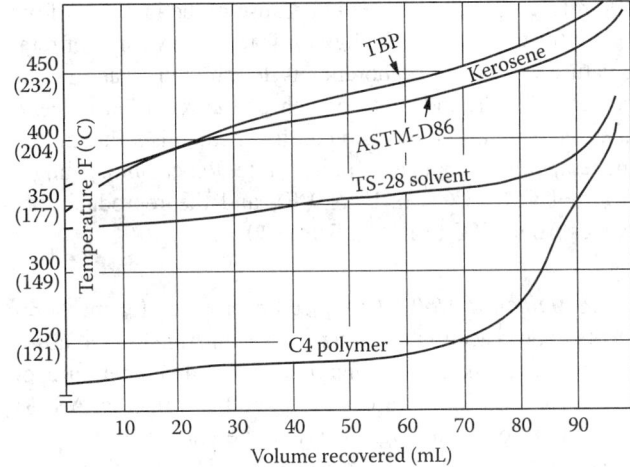

FIG. 1.57b
Distillation curves.

ASTM Method D 86-IP-123 A sample is heated in an Engler flask at a prescribed rate. Packing is not used, and some refluxing occurs as a result of condensation (Figure 1.57c).

FIG. 1.57a
Online distillation analyzer. (Courtesy of PAC.)

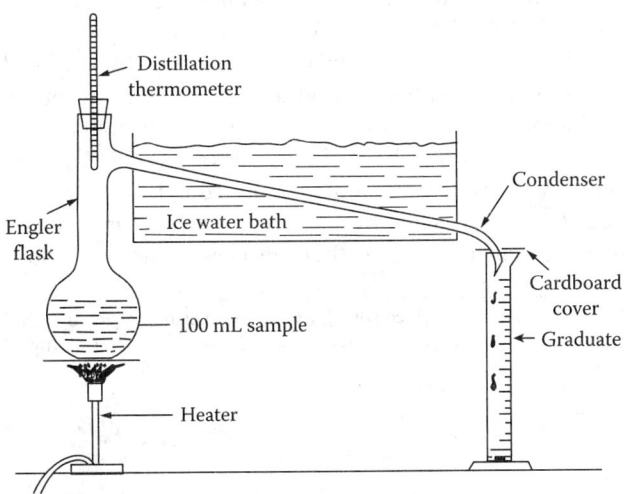

FIG. 1.57c
Apparatus for ASTM D 86 distillation test.

The vapors that are produced flow through a condenser immersed in an ice-water bath, and the distillate is collected in a graduated cylinder. The initial boiling-point temperature is defined as the reading of the thermometer at the instant the first drop of condensate falls from the lip of the condenser tube.

As the higher boiling fractions vaporize, condense, and collect in the graduate, corresponding temperature readings are recorded to permit the plot of a curve of temperature vs. percent of sample recovered. The end point or final boiling point is described as the maximum thermometer reading observed during the test; it usually occurs when all the liquid has been boiled off from the bottom of the flask.

Usually, the percentage recovered does not equal the 100 mL sample charge, partly because of the inability of the apparatus to condense the lightest fractions. A curve of temperature vs. percent evaporation is determined by adding the percent of light ends lost to each of the recorded percentages recovered. The precision of this method is a function of the temperature change vs. boiling rate. Repeatability ranges from ±1°C to 5°C (±2°F to 9°F), and the reproducibility ranges from ±3°C to 11°C (±5°F to 20°F).

ASTM Method D 1160

This method provides for measurements under vacuums, ranging from 1 mmHg (133 Pa) absolute to atmospheric to a maximum liquid temperature of 399°C (750°F). Results are not comparable with other ASTM distillation tests, although they may be converted to corresponding vapor temperature at 760 mmHg by reference to Maxwell and Bonnell vapor pressure charts.

The sample must be moisture free and is equivalent to a volume of 200 mL at the temperature of the receiver in which the condensed overheads are collected. The sample charge is boiled at a rate that produces a recovered distillate of approximately 4–8 mL/min. The overhead vapors are condensed in a jacketed condenser and receiver in which the circulating coolant is maintained within 3°C (±5°F) in the range of 32°C–77°C (90°F–170°F).

Measurement of the vapor temperature is by a special Kovar*-tipped thermocouple located at the side arm of the boiling flask leading to the condenser. Temperatures for the 5%, 10%, 20%, 30%, 40%, 50%, 60%, 70%, 80%, 90%, and 95% recovered volumes are reported unless the liquid's temperature, measured by a mercury in glass thermometer positioned in the boiling flask, reaches a value of 750°F

---

* Kovar is an iron–nickel–cobalt alloy with a coefficient of thermal expansion similar to that of hard (borosilicate) glass. It is used in the tip of thermocouples.

(399°C). In such cases, the test is terminated and the percentage recovered is also reported.

Depending on the percentage recovered and the operating pressure, repeatability varies from ±4.4°C to 5.6°C (±8°F to 10°F), and reproducibility is from 8.4°C to 17°C (15°F to 30°F). This method required considerably greater skill to obtain optimal results and is more complicated than ASTM Method D 86-IP-123.

ASTM Methods D 2887-89 and D 3710-88

These methods employ gas chromatographic analysis to determine boiling range distributions. D-2887 is for compounds that value an initial boiling point higher than 55°C (100°F) and a final boiling point less than 538°C (1000°F). Sample is injected into the analysis column, and the hydrocarbon temperature is raised at a repeatable rate during the analysis. The boiling-point temperature is related to time by injecting a calibration standard with known compounds and relating their boiling points to time.

The data gathered in this are not equivalent to data from ASTM D-86 or D-1160. Those are single-plate distillations. Results are close to those obtained by ASTM D 2892, true boiling-point chromatography. Repeatability as defined by ASTM is 2°C–7°C (3°F–12°F), depending on the value of the percent recovered. Reproducibility is 3°C–14°C (6°F–24°F) based on the value of percent recovered.

ASTM D-3710 is similar to D-2887 but is applicable to gasoline and gasoline components only. The maximum final boiling point is 260°C (500°F). Conditions are modified to allow discrete measurements of isopentane and lighter compounds. Temperature is related to time in the same manner. Repeatability and reproducibility are given as values based on the temperature/percent recovered slope and can range from 1°C to 6°C (2°F to 10°F).

Online Distillation Analyzers

Continuous plant distillation analyzers may be correlated with the results obtained by the ASTM laboratory methods, but they are not an exact duplication. In the laboratory (a batch technique), the temperature (of a given percentage that has been evaporated) is read off a rapidly rising vapor temperature as the sample is being evaporated, whereas in the plant the analyzer measures temperatures as a continuous process in which an equilibrium has been established for the given percentage evaporated to be monitored. This improves measurement precision but not necessarily accuracy, which depends on offline analyses.

End-Point Distillation Analyzers As shown in Figure 1.57d, the analyzer is a miniature process unit. There are two modes of operation based on the desired measurements of the percentage, which is boiled off. The following description applies to analyses of boiled-off percentages of 50% and greater. A conditioned and pressure-regulated sample is delivered at a rate of approximately 4 gal/h (15.2 L/h) to the top of a packed column through an inlet valve.

The inlet sample flow rate is governed by a float in a boiling pot below the column, which maintains an essentially constant level in the pot. A radiant heater boils the sample in the pot under an elevated pressure, which is held constant. The overheads are condensed at these pressures. The bottoms' flow from the boiler pot is metered by means of restriction orifice upstream of the outlet control valve.

Because the orifice and the packed columns are subject to the same differential pressure, the ratio of overheads to bottoms is fixed. The orifice size and heater wattage are selected for the particular sample and for the percentage evaporated point to be monitored. The bubble-point temperature of the bottoms can be correlated with the percent evaporated point to be monitored. As the sample's percentage evaporated temperatures change, the distillation process within the analyzers correspondingly adjusts the temperatures in the boiler pot.

For analyses of 50% boiled off and less, the analyzer operates in much the same manner. One difference is that the conditioned and pressure-regulated sample first enters a preheater coil at the top of the boiler pot and is then fed to the top of the packed column. However, a restricting orifice is located in the overhead line (at location 2) at the top of the packed column. Where required, a restrictor may also be used in the bottoms line.

A thermocouple in the overhead line (at location 1) at the top of the column measures the overhead vapor temperature and serves as the analyzer readout. By the suitable selection of restrictors and heater size, a given percentage of evaporated temperature between the 5% boiling point and the 50% point can be measured. Sample effluent may be returned to the process by an eductor or receiver and pump. A variation on this is a thin-film distillation analyzer that has the capability of measuring temperature at a set percentage boiled-off value or percentage off at a set temperature value.

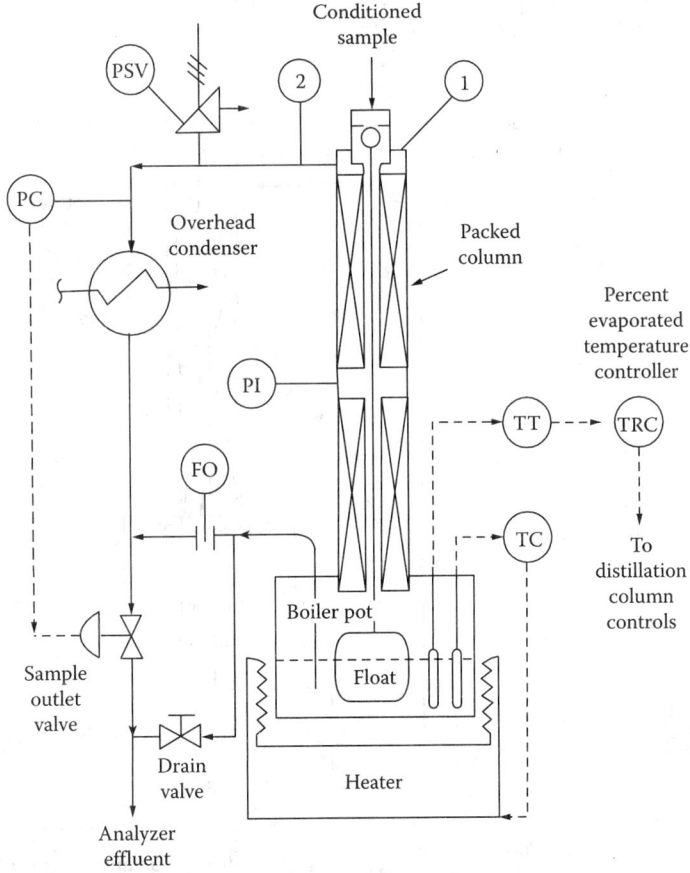

FIG. 1.57d

Packed column end-point analyzer. In an initial boiling-point analyzer, the temperature is detected in location 1 and a restriction orifice is added in location 2.

Vacuum Distillation Analyzer Some hydrocarbon feedstocks have high boiling points or may decompose if boiled at atmospheric pressure. To avoid decomposition and to reduce the boiling-point temperatures, the product may be distilled at a reduced pressure. Essentially, the same technique is used as shown in Figure 1.57d, except that the column and boiler pot are operated under a controlled vacuum.

A conditioned and pressure-regulated sample enters the top of the column through a metering valve controlled by a float in the reboiler pot so as to maintain a constant sample level in the pot. The sample is preheated in a heat-exchanger above the reboiler pot before entering at the top of the packed column (Figure 1.57e).

A fixed-wattage radiant heater boils the sample. Overheads from the column flow through a water-cooled condenser from which a metering pump at a constant rate withdraws some of the condensate, and the remaining distillate is refluxed to the top of the column through an overflow tube.

A second metering pump withdraws the bottoms fractions at a constant rate so that the ratio of overheads to bottoms flow fixes the percentage evaporated material to be measured. Both gear pumps are driven by the same motor and maintained at a constant temperature by immersion in individually heated oil baths. A thermocouple in the reboiler pot measures the bottoms' bubble-point temperature, which may be correlated with ASTM Method D 1160.

A vacuum pump removes the noncondensable vapors from the system, a pressure controller modulates vacuum pressures, and a vacuum surge tank stabilizes the system to avoid excessive pressure fluctuation. If wide variations in product end points are expected, the heat input to the reboiler pot may be regulated by an autoformer in the heater circuit to avoid column flooding due to excessive refluxing. A sight glass in the reflux line permits observation of the reflux rate so that optimal column loading conditions may be established.

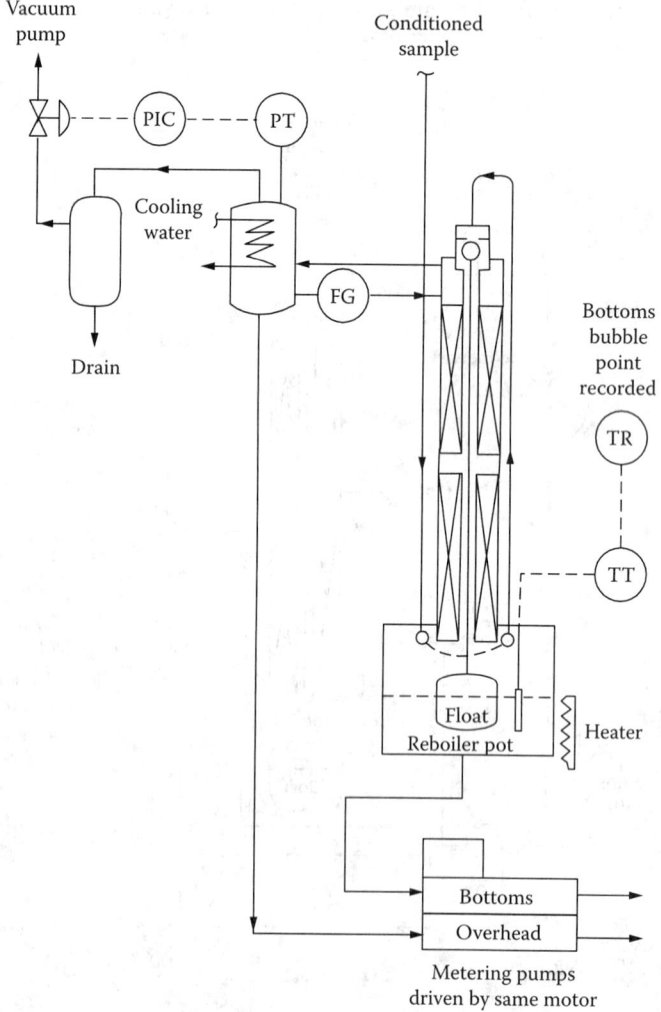

FIG. 1.57e
Packed-column vacuum distillation analyzer.

Horizontal Still Distillation Analyzer This analyzer, no longer in production, shown in Figure 1.57f, has two switch-selectable modes, the first of which permits the measurement of temperature at any preselected bottoms flow rate as a percentage of total sample flow rate through the analyzer. In the second mode, the percentage of bottoms to total sample flow at a specific temperature value is measured.

The distillation unit is essentially an horizontal still constructed with a series of baffles and weirs to promote separation of vapors and liquids at each compartment. The still is heated electrically. When the sample arrives at the last compartment, the vapor overheads and the liquid bottoms are at thermal equilibrium. A thermocouple in the bottoms section of the last compartment senses the temperature of distillation.

Vapor overheads leave the analyzer column at atmospheric pressure. They are condensed and pumped out of the analyzer system. The liquid bottoms are cooled, filtered, and then piped into a variable-speed pump system. Any escaping vapors from bottoms lines are condensed and returned to the bottoms pump. This prevents any significant losses in the volume of the bottoms product leaving the analyzer.

A differential pressure (dP) transducer continuously senses the level of the liquid bottoms at the suction side of the variable-speed pump. Heat exchangers ensure that the inlet pump and the variable-speed bottoms pump operate at the same sample temperature so as to minimize errors in flow measurement.

The voltage output of the dP transducer is generated by a voltage-to-frequency oscillator, which produces the pulse train that drives the stepper-motor-controlled variable-speed bottoms pump. In one control mode, the distillation temperature is preset, and the logic control varies the speed of the bottoms pump until a constant pump suction head is established. The capacity of the sample feed pump is fixed, and the pulse rate of the bottoms pump is converted into an equivalent flow volume, permitting the computation of the percentage distilled at temperature. In the second mode, the desired speed of the bottoms pump is preset, thus establishing a specific bottoms percentage. The logic control then varies the voltage to the heater until a constant pump suction head is established at the bottoms pump.

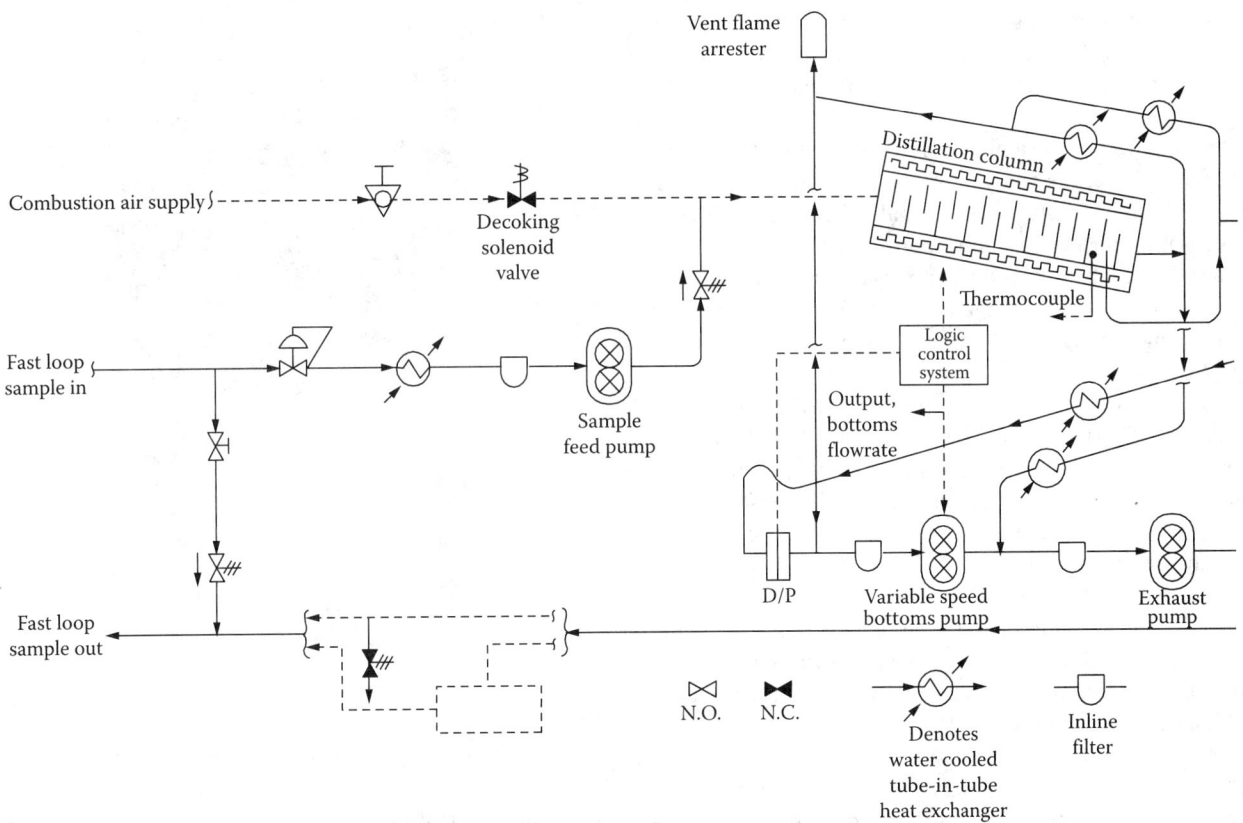

FIG. 1.57f
Horizontal still distillation analyzer.

Simulated Distillation by Gas Chromatography Temperature-programmed gas chromatography is discussed in Chapter 1.10. The analyzer injects a sample after the temperature-programmed oven has been cooled to ambient or subambient temperature, depending on the expected initial boiling point. The temperature is ramped at a set rate to a level necessary to elute all of the components present. The relatively nonpolar column elutes the sample to the detector in boiling-point order. The detector output is integrated with respect to a time–temperature relationship, and an output for the required temperature or percentage boiled off is calculated and provided as an output. Multiple outputs are available from this type of analyzer.

Calibration and Sampling Techniques employed in the calibration of distillation analyzers are the standard sample method and the spot (or grab) sample method. In the latter method, a sample is drawn off the process line, and its properties are determined (standardized) by replicate tests using ASTM Method D 86-IP-123. For ASTM Method D 86-IP-123, the standard sample is introduced into the analyzer, and the results are compared to the laboratory determinations.

Sampling systems include a reservoir tank for holding a standard sample and the means for its introduction into the analyzer. In addition to establishing a correlation between the analyzer readout and ASTM Method D 86-IP-123, it also serves to check analyzer performance in the event that its output is suspect. Care should be exercised to ensure that the standard sample remains stable and unaltered for the length of use.

The spot (or grab) sample technique involves the collection of samples at the analyzer without disturbing its operation and simultaneously notes the analyzer readouts with due compensation for the system response time. Analyzer readouts are then compared with laboratory determinations of the spot samples by ASTM Method D 86-IP-123. Again, care in collecting and handling the spot samples is of particular importance to ensure accurate results.

ASTM D-1160 calibrations are done in the same manner as for ASTM-86. ASTM D-2887 and D-3710 calibrations are done using a known mixture of straight-chain hydrocarbons as a way of setting the time vs. temperature relationship. The sample is injected and the area calculated for each time interval as is defined in ASTM D-2887. The cumulative area for each time interval is calculated and divided by the total area and multiplied by 100. This is the percentage boiled off and corresponds to the temperature established for that point by the time vs. temperature curve established for the analyzer.

Applications Distillation analyzers have significant use in the control of petroleum-refining processes. They have applications as varied as the production of fuel oils and the blending of gasolines, they are part of crude oil distillation units and alkylation units, and they are used for control of feedstocks and for control of the reflux ratio in fractionation towers.

Simulated distillation by gas chromatography (ASTM D-2887 and D-3710) is making large inroads into boiling-point measurement applications as it becomes more reliable and as refiners make the decision to use these measurements to control product quality instead of the laboratory-based D-86 method. They are primarily used on cut-point control and for blending component control.

VAPOR PRESSURE ANALYZERS

Reid Method (ASTM Method D 323-90)

Reid vapor pressure (RVP) is the measure of the volatility of gasoline. It is defined as the absolute vapor pressure of a liquid at 37.8°C (100°F) as determined by the test method ASTM-D-323.

The vapor pressure (Figure 1.57g) of petroleum products, except for liquefied petroleum gases and fuels containing oxygenates, is usually determined by the Reid method as given by ASTM Method D 323-90. Sample handling is especially critical with very volatile products because of the hazard of loss of light ends. The method provides four different procedures based on specific applications.

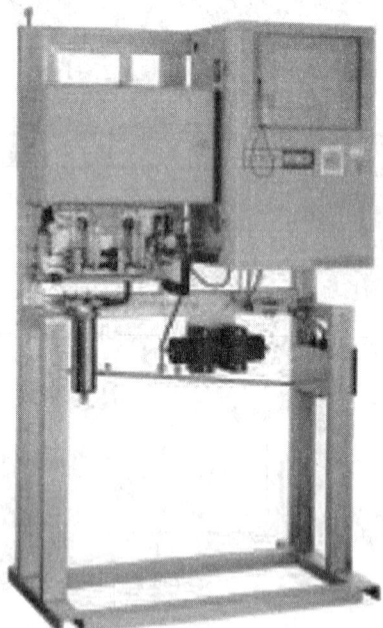

FIG. 1.57g
Vapor pressure analyzer. (Courtesy of Bartec Benke.)

Procedure A For products that have a vapor pressure less than 26 psia (179 kPa), a sample volume of a least 1 qt (0.95 L) but not greater than 2 gal (7.5 L) is collected and is placed as soon as possible in a bath maintained at 32°F–40°F (0°C–4°C).

The Reid vapor pressure (RVP) bomb consists of a liquid sample in a lower chamber that is coupled to an air chamber having a volume approximately four times that of the liquid and a Bourdon tube pressure gauge connected to the top of the air chamber. The liquid chamber is filled with a chilled sample and is quickly coupled to the air chamber with the attached pressure gauge.

The assembly is then immersed in a water bath maintained at 37.8°C ± 0.1°C (100°F ± 0.2°F). After a minimum of 5 min, the apparatus is removed and shaken vigorously, and the pressure is noted. It is then quickly reimmersed in the bath, and, at intervals of not less than 2 min, the procedure is repeated at least twice until two consecutive pressure readings are made that are within 0.05 psi (0.35 kPa) of each other. This reading is reported in pounds RVP (psia). Although the pressure is approximately the true vapor pressure in psia for some products, it should be noted that RVP is lower than the true vapor pressure because of the loss of some light ends during sample handling.

Repeatability varies from 0.1 psi (0.7 kPa) for products of 0–5 lb (0–22.5 kg) to 0.46 psia (3.2 kPa) for products having an RVP between 5 and 15 lb (22.5 and 67.5 kg). Corresponding reproducibility varies from 0.35 to 0.75 psi (2.4 to 5.0 kPa).

Dry Reid Method (ASTM Method D 4953-90)

This method is a modification of ASTM D 323-90, adding the requirement that the interior surfaces of the liquid and vapor chambers be free of water. Oxygenated compounds would tend to go into the water phase and reduce the measured vapor pressure. The method also describes a semiautomatic method that has improved precision. Repeatability of the manual method is 0.72 psi (5.0 kPa) and of the semiautomatic method is 0.26 psi (1.8 kPa). The corresponding reproducibility is 1.0 psi (7.3 kPa) and 0.94 psi (6.5 kPa).

Liquefied Petroleum Gases (ASTM Method D 1267-89)

ASTM Method D 1267-89 for the determination of the vapor pressure of liquefied petroleum gases (LPGs) is usually performed at 37.8°C ± 0.1°C (100°F ± 0.2°F) but may be run at temperatures as high as 70°C (158°F). The use of this method is limited to products whose vapor pressure does not exceed 225 psig (111.6 MPa). Essentially, the same apparatus is used as in the Reid method. The sample is not air saturated; however, 40% volume of the liquid sample is withdrawn from the completely filled bomb after a prescribed purging procedure to allow space for expansion of the product when it is immersed in the test bath.

The observed pressure is reported as LPG vapor pressure in pounds per square inch gauge pressure (psig) after corrections are made for gauge error and for standard barometric pressure at the test temperature. Repeatability is given as 1.8 psi (12 kPa). Reproducibility is given as 2.8 psi (19 kPa).

Air-Saturated Vapor Pressure Analyzer—Continuous

Sample is metered by a positive displacement pump at a rate of 100 cc/min through a heat exchanger immersed in a constant temperature bath at 37.8°C (100°F) (Figure 1.57h). The sample is then sprayed into an air-saturation chamber, also immersed in the bath, where it is saturated with air. A float-controlled needle valve discharges the air-saturated sample and vapors into the vaporizing chamber and also maintains a constant level of liquid in the chamber.

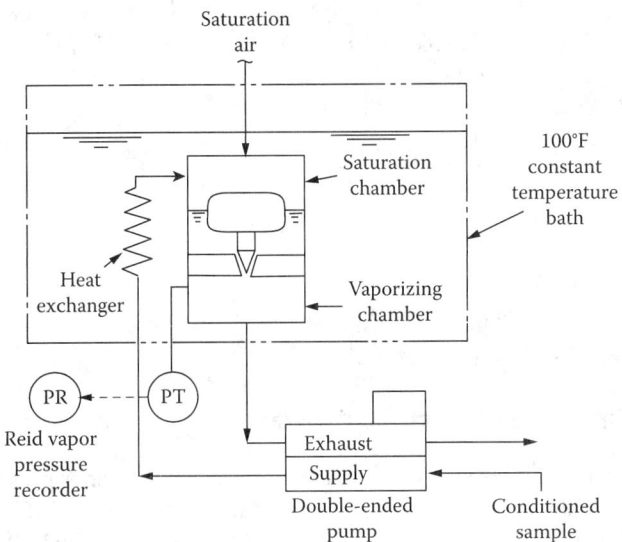

FIG. 1.57h
Air-saturated vapor pressure analyzer.

The sample supply pump is double ended and designed so that the exhaust end withdraws a flow of 500 cc/min of liquid and vapor mixture from the vaporizing chamber, thus establishing the 4:1 liquid–air ratio prescribed by the RVP (ASTM D323-90) method. The pressure in the vapor chamber is sensed by a pressure transmitter, which has a calibrated output signal representing the RVP. The exhaust pump jacket is also maintained at the 37.8°C (100°F) bath temperature.

Air-Saturated Vapor Pressure Analyzer—Cyclic

The analyzer uses a sample cell that is designed to meet the 4:1 vapor–liquid ratio. Temperature of the cell is controlled at 100°F (37.8°C) (Figure 1.57i). The analyzer first runs a fill cycle to flush out old sample. This is drained, and the cell is purged with air three times to dry it out. The air inside the cell is then allowed to come up to the cell temperature. Once again, the cell is filled with sample and allowed to come to equilibrium for 3 min. The pressure is measured and is compared to site elevation and a zero pressure measured when the cell was empty and an RVP value calculated. The cell is then drained for the start of the next cycle. The cycle time is about 9 min.

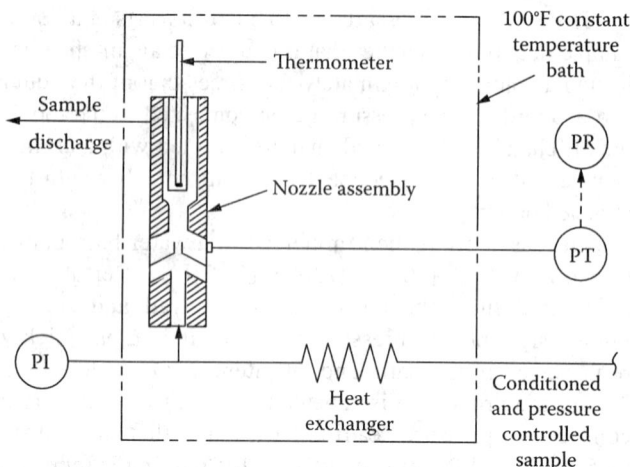

FIG. 1.57j
Dynamic vapor pressure analyzer.

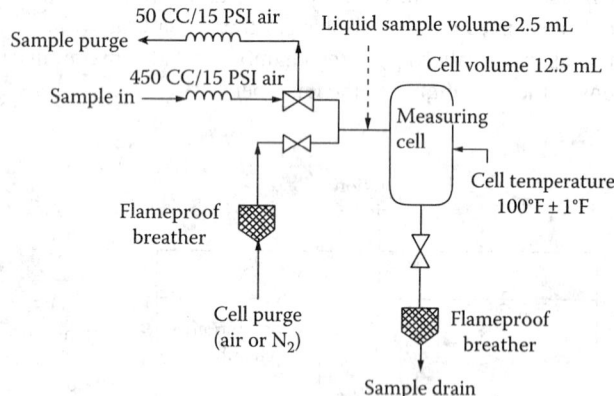

FIG. 1.57i
Cyclic, air-saturated vapor pressure analyzer.

Dynamic Vapor Pressure Analyzer

This analyzer may be calibrated to measure the vapor pressure of products covered by either ASTM Method D 1267-90, ASTM Method D 4953, or ASTM Method D 1267-90. The analysis is continuous, and the sample effluent may be returned either to a pressurized line or to an atmospheric receiver tank. Incoming sample is filtered and maintained at a constant pressure (Figure 1.57j). The sample is brought to a temperature of 37.8°C ± 0.06°C (100°F ± 0.1°F) in the heat exchanger, which is immersed in the constant temperature bath.

The sensing device is a modified jet pump element with the suction side dead-ended into the cell of an absolute pressure transmitter. The velocity of the sample in the small-diameter nozzle causes the pressure head to approach the vapor pressure of the least volatile component in the stream. This value is lower than the RVP, as determined by ASTM Method D 323-90. By reducing the efficiency, the system simulates the selective vaporization, which occurs in the 4:1 liquid volume (vapor–liquid ratio) in the RVP test apparatus.

One of the parameters affecting the efficiency of a jet pump, eductor, or aspirator is the location of the tip of the nozzle with respect to the throat of the downstream venturi. By adjusting the location of the nozzle, the operating efficiency may be adjusted so that the analyzer readout provides an essentially one-to-one correlation with the RVP. Figure 1.57k

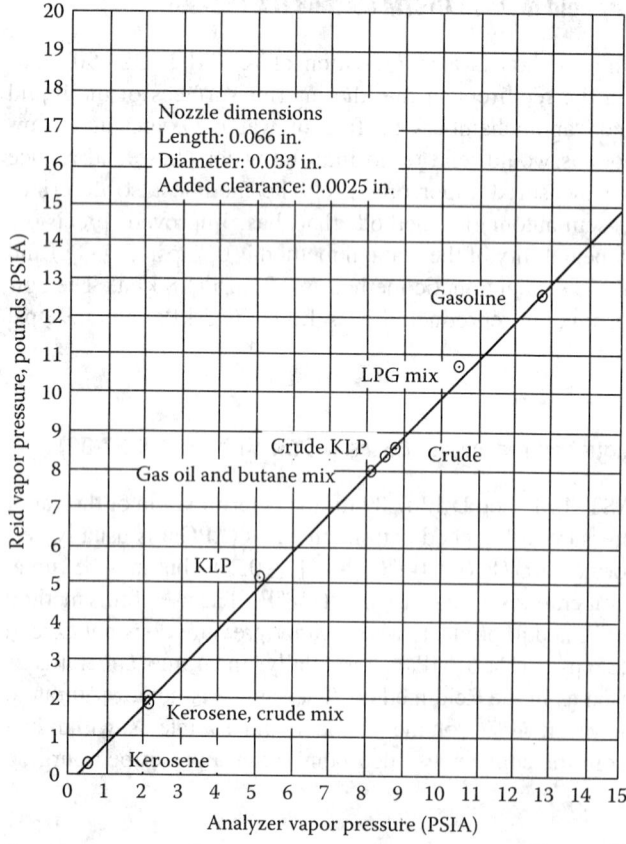

FIG. 1.57k
Dynamic vapor pressure analyzer results.

shows the relationship of the analyzer results against RVP for various hydrocarbons and blends.

When the analyzer has to return the sample into a pressurized line, a backpressure regulator is added to maintain a constant backpressure on the system and thereby eliminate the effects of varying return line pressure. An inlet pressure of roughly 45–50 psig (311–345 kPa) is required when the analyzer is discharging to atmosphere and the inlet pressure regulator is set at 40 psig (276 kPa). When the sample is returned to a pressurized system, the inlet pressure regulator is set at a value equal to 2.5 times the return line pressure plus 40.

Calibration

Calibration of the cyclic air-saturated analyzer is done using a special syringe and pure components. The pressure reading is compared to known values at the 38°C (100°F) cell temperature. The other analyzers are calibrated using the standard sample method and the spot (or grab) sample method, which are equally reliable if carefully performed.

Sample collection and handling are critical in both cases, as is the manner in which a stable and uniform standard sample is preserved during delivery to the analyzer. Standards used for the online analyzers should be pressured into the analyzers using a water displacement system to prevent a vapor space from forming over the standard, thus reducing the standard accuracy.

Applications

Recent environmental regulations are putting pressure on refiners to reduce the vapor pressure of gasoline components and therefore the vapor pressure of the product gasoline. This means that more of these analyzers will be installed at the crude, reformer, alkylation, and cracking units to control vapor pressure at the source. They have been used with LPG, but they have been used more extensively to monitor the vapor pressure of pipeline-transported products, thereby avoiding vapor locking the pumps and minimizing safety hazards during tanker loading operations. They may also be used to detect product interfaces at pipeline receiver stations.

VAPOR–LIQUID RATIO ANALYZERS

Relative volatility is a measure of the tendency to vaporize. The higher the volatility, the higher the vapor pressure of the

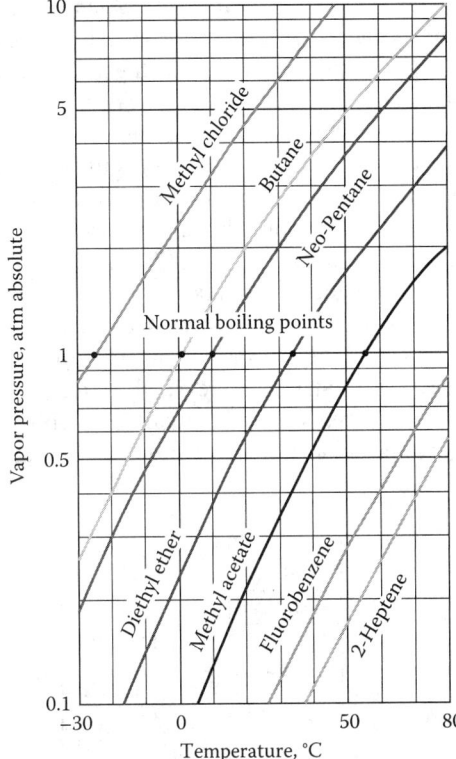

FIG. 1.57l

The higher the vapor pressure of a liquid at a particular temperature, the mode volatile that liquid is (the higher it's V–L ratio). (Courtesy of Mbeychok.)

liquid is at the same temperature (Figure 1.57l). In distillation processes, the higher the volatility, the easier it is to separate the component from the less volatile ones. In the case of automotive fuels, it is desirable to use higher volatility gasoline in colder regions, but not be high enough to cause vapor lock. The proportion of butane increases the volatility of the blended product and because butane is relatively inexpensive, cost is reduced by increasing its proportion. One measure of volatility is the vapor–liquid ratio (abbreviated as or V/L or V–L).

Front-end volatility must be closely controlled in a gasoline blend to permit the greatest use of light blending components without incurring vapor lock during operation of a gasoline engine.

Vapor–Liquid ratio (or V/L or V–L) is a measure of volatility, the tendency of a mixture of liquids to vaporize. It expresses the volume ratio of vapor to liquid that will exist at a given temperature and atmospheric pressure. Automotive fuel specifications usually state the temperature at which V–L is 20 ($T_{(V/L=20)}$). This temperature is where vapor lock can begin. A fuel has high volatility when it's $T_{(V/L=20)}$ is low. Volatility should be relatively high in low ambient temperature regions but should not be high enough to cause vapor lock.

The vaporization of a fuel in the carburetor of an engine is predicable neither from distillation temperatures nor from pressure tests. The curves of Figure 1.57m show how four different fuels (with some similar properties) exhibit different volatility characteristics as determined by measuring their vapor–liquid (V–L) ratios at various temperatures. In the past, front-end volatility was determined by employing indirect measurements such as the RVP; the 10%, 20%, and 50% evaporated temperatures; and the temperature corresponding to a given V–L ratio (usually 20). This procedure and the required computations are time-consuming, cumbersome, and inaccurate.

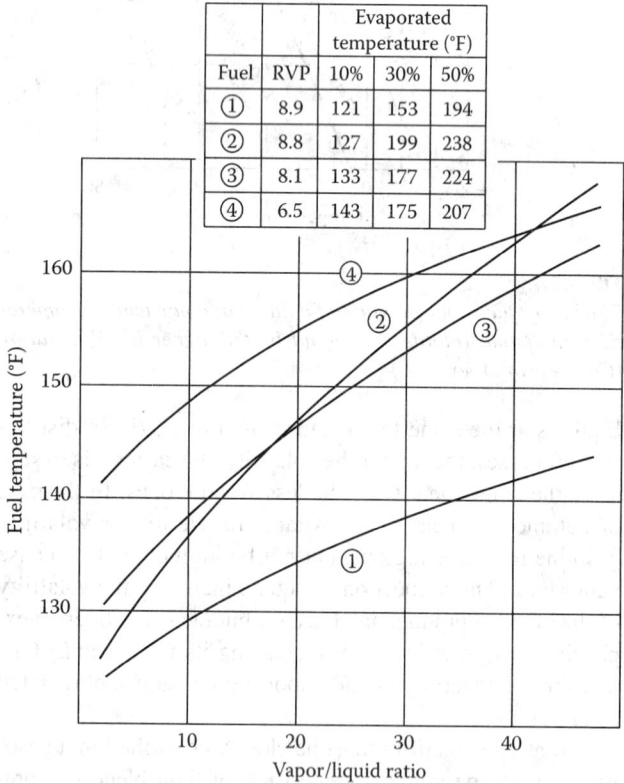

FIG. 1.57m
Volatility effects on gasoline vaporization.

Volatility Test (ASTM Method D 2533-90)

The laboratory technique determines the V–L ratio by direct measurement for a given reference temperature. ASTM Method D2533-90 uses a special burette constructed with a stopcock at the top, 0.5 cc graduations, a short bottom arm fitted with a rubber septum, and a long bottom arm connected to a 250 mL leveling bulb by rubber tubing. The burette is filled with pure dry glycerin, and a sample of 1 cc or less is injected into the burette through the rubber septum by a hypodermic syringe.

The burette is then placed in a controlled temperature bath. As the sample vaporizes, the 250 cc leveling bulb, referenced to atmospheric pressure through the drying tube, is raised or lowered to maintain absolute, which compensates for the prevailing barometric pressure. The volume of vapor is indicated on the burette's graduations, and, because the liquid volume injected is known, the V–L ratio can be calculated. Repeatability for a V–L ratio of 20 using glycerol as the burette fluid is 2.7°C (±4.8°F).

Continuous Vapor–Liquid Ratio Analyzer

Figure 1.57n shows a block diagram of the analyzer. A slipstream from a sample loop is conditioned and cooled to approximately 2°C–4°C (35°F–40°F) before it enters a metering gear pump. The differential pressure across the pump is controlled to guarantee the constant delivery of liquid sample to a vapor–liquid separator at a rate of 25 cc/min. The vapor–liquid separator is enclosed in a constant temperature bath controlled at the desired test temperature.

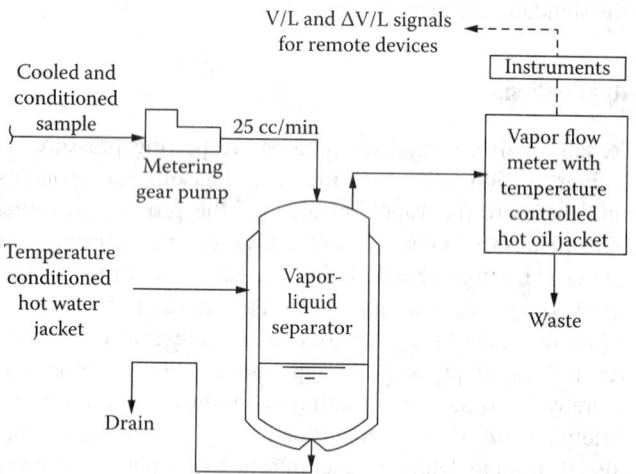

FIG. 1.57n
Continuous vapor–liquid ratio analyzer.

Vaporization occurs at or near atmospheric pressure, with the residual liquid phase being discharged through a liquid seal, and the vapors are measured by a low-pressure drop flowmeter. The meter is maintained at 93°C (200°F) to prevent condensation and to establish a temperature reference for vapor flow measurements. With the inlet liquid flow rate fixed, the V–L ratio may be computed after compensations are made for the vapor temperature and the barometric pressure. In addition to computing the V–L ratio, the deviation of V–L from a preselected V–L set point is also displayed and provided as an output signal.

Calibration and Applications

A V–L analyzer may be calibrated by the introduction of a sample of known volatility or by the grab sample technique.

These analyzers are primarily applicable to in-line gasoline blending operations. An in-line blending system offers both economic advantages and greater flexibility, speed, and ease in switching blend formulas to meet production requirements. Although precise metering systems provide the means for implementing an in-line blending system, the final blend properties are still only implied rather than precisely known. Vapor pressure alone or in combination with distillation characteristics has been shown to be insufficient to provide efficient control of front-end volatility, because of inaccuracies and lags.

The ability to control light components during blending to meet environmental requirements for volatility provides a very significant economic incentive. A V–L analyzer, in an in-line gasoline blending system, may be used to reset the set point of a butane flow control system.

POUR-POINT TEMPERATURE ANALYZERS

Pour-Point Test (ASTM Method D 97-87)

The standard laboratory procedure for measuring the flow characteristics of petroleum oils is given in ASTM Method D 97-87. The sample must be heated without stirring to 46°C (115°F) or 83°C (15°F) above the expected pour-point temperature before starting the test. A thermometer immersed in a jacketed sample test jar and a cooling bath is used.

The cooling rate of the sample is fixed as it is examined at 3°C (5°F) intervals to ascertain whether it will flow when the test jar is tilted. When no sample movement is detected in the tilted jar after a five-second interval, the pour point is reported as 3°C (5°F) above the indicated temperature. All pour-point values are reported in multiples of 3°C (5°F). This point corresponds to a viscosity of about 500,000 cSt. Repeatability is given as 3°C (5°F) and reproducibility as 6°C (10°F).

Pour-point analyzers were developed in an attempt not only to automate a laboratory procedure for process control but also to improve the accuracy of such measurements.

Pour Point by Differential Pressure

This analyzer (Figure 1.57o) is cyclic in operation, with one cycle consisting of five basic sequences. The analyzer is pro-

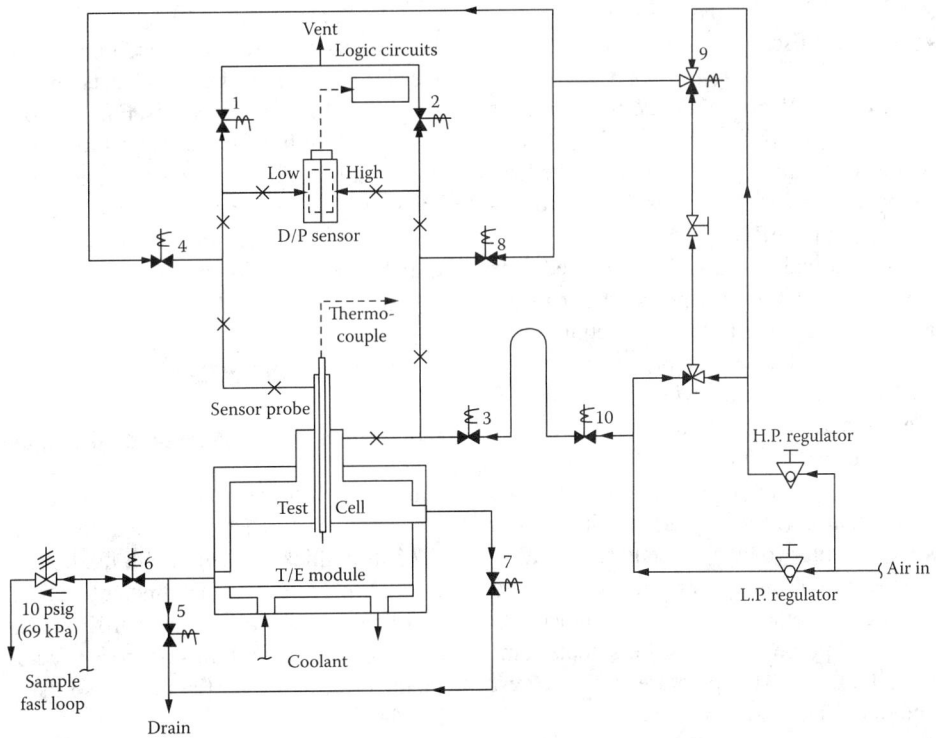

FIG. 1.57o
Pour Point measurement by detecting differential pressure (D/P).

grammed so that it resets to sequence no. 1 on startup or on interruption of any kind. The sequences are as follows:

1. *Forced drain sequence*: Power to the thermoelectric cooler (T/E) module is reversed, warming the cell and draining all of the residue.
2. *Fill sequence*: The T/E module continues to heat the cell. Depending on product viscosity, either low- or high-pressure air is applied to the test cell while sample flows into the cell and out to drain by way of the overflow. This air pressure prevents the sample from entering the sensing lines of the analyzer.
3. *Level sequence*: Air is used to apply pressure to the surface of the sample and force surplus produce out to drain. Power to the T/E module is applied at normal polarity to initiate sample cooling. At termination of this sequence, all valves deenergize, and the sensing probe is submerged below the surface of the sample.
4. *Equalizing sequence*: The test cell, the differential pressure (dP) sensor, and all lines connecting these two components are vented to atmosphere, establishing a condition of zero differential pressure in the sensing system. At the end of this sequence, all valves close. Low-pressure air is held between two of the valves.
5. *Pulse sequence*: The low-pressure air trapped between valves expands into the test cell, producing a momentary pressure pulse against the surface of the liquid outside the sensor probe and against the high-pressure diaphragm of the dP sensor. The output voltage of the dP sensor rises immediately, reflecting this pressure pulse.

The pressure pulse on the surface of the liquid also tends to push liquid into the sensor probe, compressing the air inside the probe. As the sensor probe is connected to the low-pressure diaphragm of the dP sensor, this results in a reduction of the output voltage of the dP sensor.

The control logic of the analyzer has a built-in fixed time delay that gives the sample time to compress the air inside the sensor probe after each pressure pulse. The output of the dP sensor is then compared to an adjustable set-point voltage. If the output of the dP sensor is below set point, another equalizing sequence and another pulse sequence occur.

As the sample cools, its resistance to flow increases, and its ability to compress the air inside the sensor probe decreases. This results in an increase of output voltage from the dP sensor. Eventually, this voltage exceeds set point. A peak-holding relay trips, and the analyzer resets to the forced drain sequence to begin another cycle. A thermocouple installed inside the sensor probe measures the sample temperature at all times. Pour-point temperature is the lowest temperature attained in each measurement cycle.

Viscous-Drag Pour Pointer

Sample flows into a 23 mL cup in which an alloy ball is suspended. A thermocouple is attached to the surface of the ball to sense sample temperature. A thermoelectric cooler removes heat from the sample as the cup turns. When the sample reaches the pour point, the viscosity is great enough to turn the ball, deflecting a light beam that is aimed at the ball and reflected to a detector. The analyzer notes the temperature and then adds 3°C (5°F) to the value to conform to the ASTM requirements. After the pour point is sensed, power to the thermoelectric module is reversed. The sample is heated and flushed out prior to initiation of the next cycle.

Calibration

Either the standard or the grab sample method is suitable. The convenience of a locally available standard sample provides a rapid check on the analyzer performance as well as serving as a means of calibration. The greater initial cost for this feature should be weighed against the delay incurred in obtaining a laboratory analysis of the spot sample.

The repeatability of ASTM Method D 97-87 for pour point is only about 5°F (2.8°C), and its reproducibility is only 10°F (5.6°C). To improve the accuracy of the ASTM results, a sufficient number of determinations are advisable, because the precision of the process analyzer usually exceeds the accuracy of the laboratory method.

Applications

Pour-point temperature measurements are used to characterize the heavy end of fuel oil cuts from vacuum and atmospheric distillation processes. Fuel oil grades 1–4 are affected by these specifications. When a product is sufficiently free of wax that a cloud-point determination becomes meaningless, the pour point may be used as an index of the temperature at which flow will be impeded due to semisolidification rather than by the formation of wax crystals.

CLOUD-POINT ANALYZERS

Cloud-Point Tests (ASTM Method D 2500-88)

This method is applicable to products with a cloud point below 49°C (120°F) that are transparent in a layer 1.5 in. (37.5 mm) thick. A cylindrical, flat-bottomed, clear glass test jar 1.25 in. (31.3 mm) in diameter by 4.75 in. (118.8 mm) long, and having a scribed sample fill line 2.125 in. (53.13 mm) above the inside bottom surface is used. A cork holds a thermometer coaxially in the test jar so that its bulb rests at the bottom.

A watertight jacket with an 0.25 in. (6.25 mm) thick cork or felt pad at the bottom holds the test jar when it is immersed in a cooling water bath. The sample must be dried at a temperature at least 13.9°C (25°F) above the approximate cloud point so as to remove any moisture and minimize trace water-haze formation. The test jar is fitted with a cork or felt ring approximate 0.1875 in. (4.69 mm) thick, which is positioned 1 in. (25.4 mm) from the bottom to keep it centered in the jacket.

The test is begun by placing the jar vertically in a −1.1°C to 1.7°C (30°F–35°F) water bath so that the jacket projects no more than 1 in. (25.4 mm) from the bath liquid. At intervals of 1.1°C (2°F), the jar is quickly but gently removed and examined for formation of a wax haze. (This examination should take 3 sec at most.) A water haze is generally uniform throughout the sample, whereas a wax crystal haze always appears first at the bottom of the jar.

If the cloud point is not detected when the sample reaches 10°C (50°F), the sample is transferred to a second bath, which also contains a test jacket, and is maintained at −17.8°C to −15°C (0°F–5°F), and the test is continued. Successively lower temperature baths are used as required for low-temperature cloud-point products. The temperature at which a distinct wax haze is first observed, expressed in increments of 1°C (2°F), is reported as the cloud point. For gas oils, repeatability is 2°C (4°F) and reproducibility is 4°C (8°F). For other oils, both repeatability and reproducibility are 5.6°C (10°F).

Cloud and pour point can be detected by the same analyzer (Figure 1.57p).

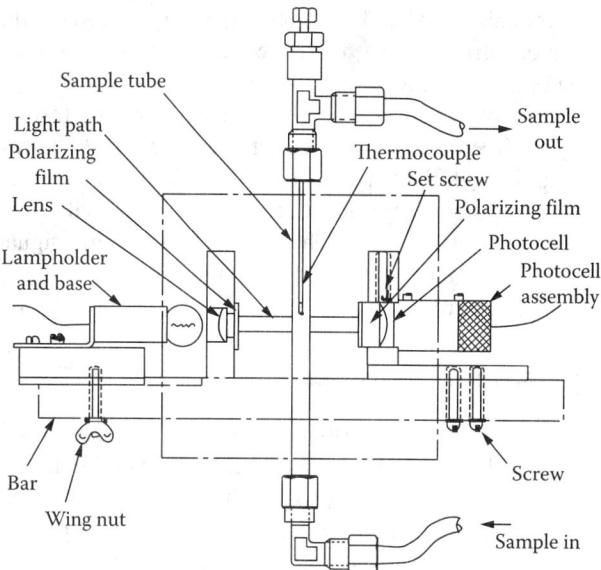

FIG. 1.57q
Optical cloud-point analyzer.

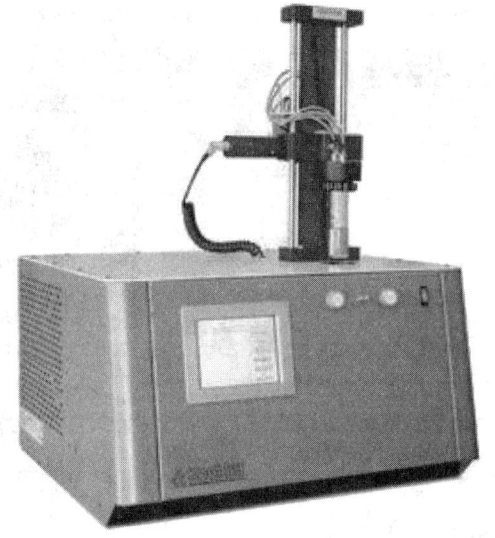

FIG. 1.57p
Automatic cloud and pour-point analyzer. (Courtesy of Koehlerin Instrument.)

Optical Cloud-Point Analyzer

The cloud-point analyzer (Figure 1.57q) detects the temperature at which wax crystals first appear as a liquid hydrocarbon sample is cooled. A measuring cycle begins when a three-way solenoid valve, which normally circulates a slipstream sample, is energized and introduces sample into the test cell. After the cell is flushed clear of all residue from a previously tested sample, the solenoid valve de-energizes, and a thermoelectric module reduces sample temperature until the analyzer detects the presence of wax crystals. At that point, the cloud-point temperature is represented as the lowest temperature the sample attains, as measured by a thermocouple, and the next measuring cycle is initiated.

The detection system is based on a light beam that is directed through a polarizing lens, through the sample cell and a second polarizing lens, and into a photocell. The electrical resistance of the photocell varies with the amount of light to which it is exposed.

Polarizing lenses permit only light vibrating in a single plane to pass through them. For example, light rays may be controlled so they vibrate only in a vertical plane if the transmission axis of the lens is vertical. If two lenses are installed so that the axis of light transmission of the second lens is perpendicular to that of the first, no light rays will pass through the second lens. The test cell lenses are assembled in this manner, and the photocell senses the presence of light.

As the sample is cooled, it eventually reaches the temperature at which crystals of ice and/or wax appear. These crystals reflect some light rays and bend others so that the plane of vibration of some light rays changes as they pass through the crystals. Some of this reoriented light then passes through the second polarizing lens and to that photocell.

Ice crystals are doubly refractive. As light rays pass through them, ice crystals will change the path of the light rays several times. However, ice crystals have very low refractive angles, so they do not seriously affect the critical direction of the light rays as they approach the second lens.

Wax crystals are also doubly refractive but have very high angles of refraction and, therefore, have a very pronounced effect of the light rays passing through them.

The photocell, which senses little light passing through a crystal-free sample, will detect quantities of light rays passing through the lenses as a result of the presence of ice crystals and very large quantities of light as a result of the presence of wax crystals. The photocell, therefore, signals the first appearance of wax crystals by a significant change in its electrical resistance.

Calibration and Applications

The technique used to calibrate pour-point analyzers is also used to calibrate cloud-point analyzers. The measurement of the cloud point applies only to petroleum oils that are transparent in layers 1.2 in. (37.5 mm) thick and contain paraffin waxes or other compounds capable of forming crystals prior to total solidification. Its major application has been to gas oils and cycle oils not only to meet specifications but also to facilitate product transport and to prevent filter clogging during cold weather.

FREEZING-POINT ANALYZER

Aviation Fuel Tests (ASTM Method D 2386-88)

The freezing point as defined by this method is the temperature at which the last hydrocarbon crystal (formed during cooling) melts after the sample temperature is allowed to rise. This temperature must be within 3°C (5.4°F) of the temperature at which the appearance of hydrocarbon crystals is first observed.

A jacketed clear-glass tube is filled with either dry air or nitrogen at atmospheric pressure. The tube's outer diameter (OD) is 30 mm, and the tube's inner diameter (ID) is 18 mm, with a 2 mm space between tubes. The overall length is 237 mm. The plug in the sample tube holds a total-immersion thermometer and a stirring rod formed with three spiral loops at the bottom and positioned slightly below the thermometer bulb. The tube is filled with 25 cc of fuel and placed in a clear vacuum flask (70 mm ID and 280 mm long) containing a coolant such as alcohol or solid carbon dioxide.

The sample is stirred vigorously during cooling, and the temperature at which hydrocarbon crystals first appear is noted, ignoring any haze that may form at about −10°C (14°F) as a result of dissolved water in the sample. The sample tube is then removed and allowed to warm up slowly while the sample is continuously stirred. The temperature at which the last crystal disappears is reported as the freezing-point temperature, provided it is within 3°C of the crystal formation point temperature. Otherwise, the test must be repeated. Results are reported in increments of 0.5°C (0.9°F). Repeatability is 1°C (1.8°F) and reproducibility is 2.5°C (4.5°F).

Freezing-Point Analyzer for Aviation Fuel

The principle of measurement is exactly the same as that of the cloud-point analyzer. The sample cell is cooled by a mechanical refrigeration unit, which is supplied with the analyzer. At detection of the cloud point, the power to the thermoelectric module is removed, and the sample is allowed to warm up. As the crystals disappear, the light falling on the photocell is reduced. Once the light falls below a threshold level, the cycle is terminated, and the freeze-point value is updated.

Calibration and Applications

Calibration of the fuel analyzer variation should be based on the comparison of analyzer readouts with multiple laboratory test results. Its principal application is in the processing of aviation fuels such as JP-4, kerosenes, and similar products. Determination of purity is needed in the production of benzene, toluene, ethylbenzene, o-xylene, p-xylene, and phthalic anhydride. These materials must be able to be supercooled and have a specified freezing-point temperature.

Calibration of hydrocarbon purity analyzers is best achieved by using a sample of known purity. A certified standard thermistor may also be used if a sample of known purity is unavailable. Comparison with the ASTM methods can also establish the purity of a sample, which can then be used for analyzer calibration (Figure 1.57r).

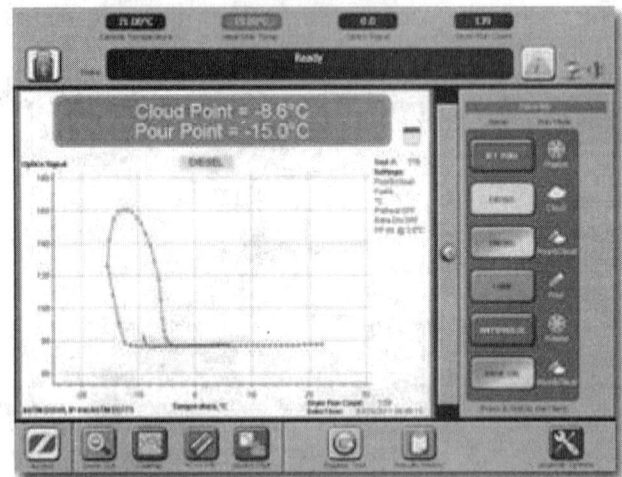

FIG. 1.57r

Analyzer to detect freezing, pour, and cloud points. (Courtesy of Phase Technology.)

FLASH-POINT ANALYZER

Flash point of a flammable liquid is the lowest temperature at which it can form an ignitable mixture with oxygen. At this temperature, the vapor may cease to burn when the source of ignition is removed. The flash point is not related to the temperatures of the ignition source or of the burning liquid, which are much higher. The flash point is often used as one descriptive characteristic of liquid fuel.

Flash-Point Tests (ASTM Methods D 56-90 and D 93-90)

Plant analyzers are intended to correlate with ASTM Method D 56-90 (Tag closed tester) and ASTM Method D 93-90-IP-34/85 (Pensky–Martens closed tester). ASTM Method D 56-90 is for materials with a viscosity less than 5 cSt at 40°C (104°F) and a flash point below 93°C (299°F). A sample of 50 cc is used at a temperature of at least 11°C (20°F) below the expected flash point.

The sample cup is immersed in a bath whose temperature may be raised at a prescribed rate. Thermometers measure the bath and sample temperature. The sample lid prevents loss of sample vapors and directs a small flame into the cup periodically. The flash point is defined as the lowest sample temperature to cause ignition of the vapor above the sample at 1 atm absolute.

Repeatability is 1.1°C (2°F) for flash points below 60°C (140°F) and 1.7°C (3°F) for flash points between 60°C and 93°C (140°F and 190°F). Reproducibility is 3.3°C (6°F) for flash points below 13°C (55°F), 2.2°C (4°F) for flash points between 13°C and 59°C (55°F and 139°F), and 3.3°C (6°F) for flash points from 60°C to 93°C (140°F to 190°F).

Pensky–Martens test is a closed-cup test, used in the determination of the flash-point temperature of lubricating oils, liquids containing suspended solids and liquids that tend to form a surface film during testing. It is performed by heating and stirring the test specimen while at regular intervals, the stirring is stopped and heat from an ignition source is applied until, in each cycle, and ignition of the vapors (flashing) occurs. The peak temperature reached after this occurs in each cycle is the flash point temperature.

The Pensky–Martens closed tester is for materials with an indicated flash-point temperature as high as 370°C (698°F). Approximately 4.2 in.^{3} (6.7×10^{-5} m^3) of sample is used, and the sample cup is heated directly by either a gas or electric heater at a prescribed rate. The sample cup lid is designed to support a sample stirrer, a mercury-in-glass thermometer, and an apparatus for periodically exposing the vapor above the sample to a test flame. The repeatability is 5.5°C (10°F) for materials with a flash point above 104°C (220°F) and the reproducibility is 8.5°C (15°F).

Flash-point temperature analyzers are available for online applications with microprocessor-based intelligence and in explosion-proof designs (Figure 1.57s).

FIG. 1.57s
Intelligent flash-point analyzer. (Courtesy of ORB Instruments.)

Low-Temperature Flash-Point Analyzer

Sample is fed to the analyzer's heating chamber at a constant rate, and air is added at a rate of 600 cc/min (Figure 1.57t). The air sample mixture is heated at a controlled rate before it enters the flash cup and overflows to maintain a constant level. The vapor rises into the vapor space and is periodically

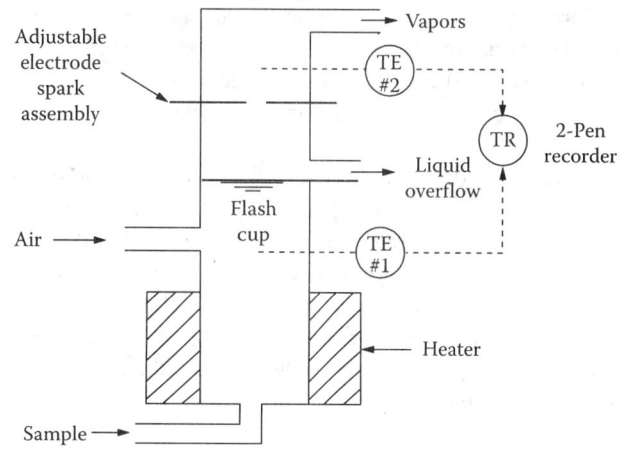

FIG. 1.57t
Low-range flash-point analyzer.

exposed to a high-voltage spark. Thermocouple 1 in the flash cup measures the temperature of the air-sample mixture, and thermocouple 2 in the vapor space responds to the temperature rise when vapor ignition occurs.

When the sample is heated to its flash point and its vapor ignited, the temperature in the vapor space increases. This shuts the heater off and causes recorder pen 1 (to which thermocouple 1 is connected) to be driven downscale. The peaks of the resulting sawtoothed record indicate the flash-point temperature for each analysis cycle.

This design is mounted on a frame with all accessory components piped and wired for field installation. The accessories include (1) a sample conditioning system to filter and coalesce free water from the sample, regulate sample pressure, and indicate coalescer bypass flow; (2) a mechanical refrigeration system and temperature controller to cool the sample below its flash point; (3) an air compressor, filter, and flowmeter to supply combustible air; (4) a duplex positive-displacement pump to provide a constant rate of sample flow to the analyzer and return analyzer effluent and coalescer bypass to the pressurized process line; and (5) block, check, relief, and backpressure valves for isolation and the ability to withdraw sample for calibration.

High-Temperature Flash-Point Analyzer

Sample from a slipstream is metered to the system at a constant rate by one head of a duplex positive-displacement pump and is preheated to a fixed temperature below the flash point as determined from the preceding analysis. It is mixed with air at a rate of 1500 cc/min. A final heater provides the additional heat required to bring the air-sample mixture to the flash-point temperature. The liquid entering the flash chamber is returned to process by the second head of the duplex pump, and the rising vapors are exposed to a high-voltage spark every 10 s.

Ignition is detected by the deflection of a diaphragm caused by the combustion pressure pulse. The control circuit increases or decreases the final heater output depending on whether ignition has occurred. At the same time, the preheater controls are also adjusted to maintain the desired temperature differential. Flash-point temperature is sensed by a thermocouple in the flash chamber liquid and displayed on a recorder chart.

Calibration and Applications

Either the spot sample or standard sample method may be used. In either case, care must be exercised when a low flash-point sample is used so as to prevent loss of light ends. The analyzer can be applied to the control of crude and vacuum distillation, dewaxing, solvent extraction and stripping, deasphalting, blending, residual fuel oil processing, and pipeline interface detection.

OCTANE ANALYZERS

Octane Tests (ASTM Methods D 2699 and D 2700)

Treatment of this subject will be brief because of the vast complexity involved. Basically, two methods are employed in which in a standard engine with an unknown fuel is compared with a standard or reference fuel. One method (ASTM Method D 2699) yields a motor octane number (MON) in which the engine is run at 900 rpm, and the second method (ASTM Method D 2700) provides a research octane number (RON) in which the engine is run at 600 rpm.

The difference in octane number obtained by these methods (the spread) is indicative of city driving at low speeds as compared to highway engine performance. Standard fuels are based on normal heptane (zero rating) blended with isooctane (100 rating), with the octane number equal to the percentage of isooctane in the blend. When the unknown fuel produces the same knocking as a standard fuel blend, it is rated equal to the octane rating of the standard blend. The RON is higher than the MON, with the spread increasing with increasing octane numbers.

Octane Comparator Analyzer (ASTM Method D-2885)

These analyzers are based on ASTM Method D 2885, an online comparator system, using a standard engine with a modified carburetor fuel delivery system and standard detonation pickup and knockmeter. Comparison is made between the process stream and a prototype fuel, which serves as the standard or octane number reference point from which an octane number difference is determined as readout. This is accomplished by locating two thermocouples in the tube 1 in. (2.54 cm) apart and equidistant from the temperature peak. Any movement of the peak due to a change in fuel octane rating is sensed by a differential temperature controller and causes a compensating change in reactor tube pressure to restore the temperature peak location.

Calibration and Applications

The online comparator engine automatically calibrates itself by running the proto fuel during its operation. The reactor tube analyzer requires a similar source of calibration gasoline of known octane.

These analyzers can trim octane *give-away* during blending to approximately 0.1 octane above specification requirements. A more important application is to measure the octane of blending components such as reformate, alkylate, and catalytic cracked gasolines so that these units optimize their production.

Octane analyzers are also available in inexpensive portable designs for spark ignition testing (Figure 1.57v).

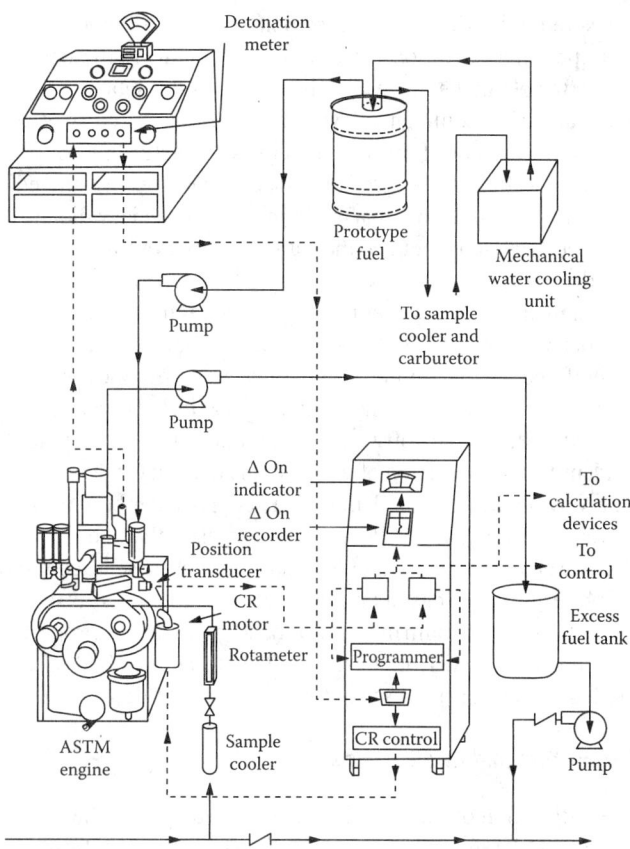

FIG. 1.57u
Octane analyzer equipment.

the analyzer readout. Figure 1.57u illustrates the equipment diagram for such an analyzer.

System accuracy depends on the performance of the standard engine, which must be properly maintained for optimal system operation. Also, the prototype fuel octane number should be determined to within ±0.1 ON (octane number) or better, because it serves as the reference for stream comparison measurements.

Reactor-Tube Continuous Octane Analyzer

This analyzer monitors the reactions that precede engine knocking, the parameters of which may be controlled and correlated with octane number. A fuel and controlled air volume mixture is delivered at a rate of 1 cc/min to a reactor tube maintained at an elevated temperature. Partial oxidation reactions in the tube produce a peak temperature, the location of which is related to octane number.

Higher-octane fuels cause the peak temperature to move away from the tube inlet, whereas increasing the reactor tube pressure moves the peak closer to the tube inlet. Consequently, if the temperature peak location is fixed by varying the reactor tube pressure as fuel octane number varies, the pressure may be correlated with octane number and used as the analyzer

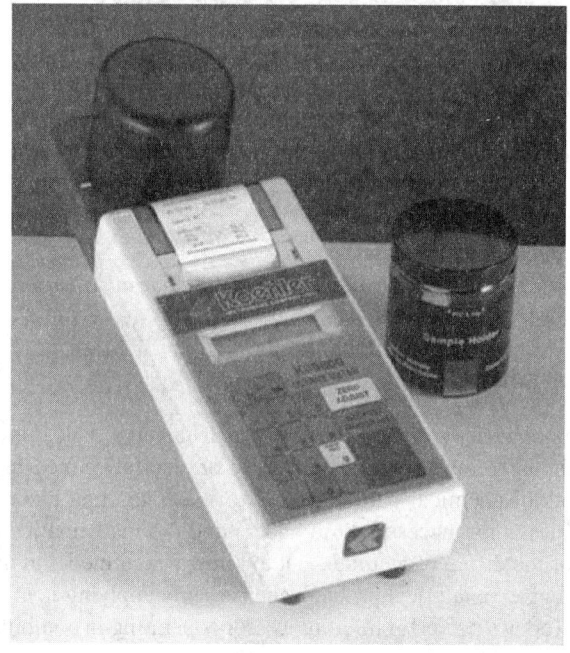

FIG. 1.57v
Portable octane analyzer. (Courtesy of Camlab.)

NEAR-INFRARED ANALYZERS

A detailed description of this category of analyzers is presented in Chapter 1.32. Near-infrared (NIR) analyzers do not directly measure the physical properties, such as boiling, pour, or flash points. Instead, a model is utilized that provides a relationship between the property of interest and the absorption of near-infrared radiation at one or more wavelengths in the spectrum, which is measured by the analyzer. Consequently, the NIR analyzers represent a secondary technique of measurement.

An online design can provide real-time monitoring of chemical and physical properties (Figure 1.57w).

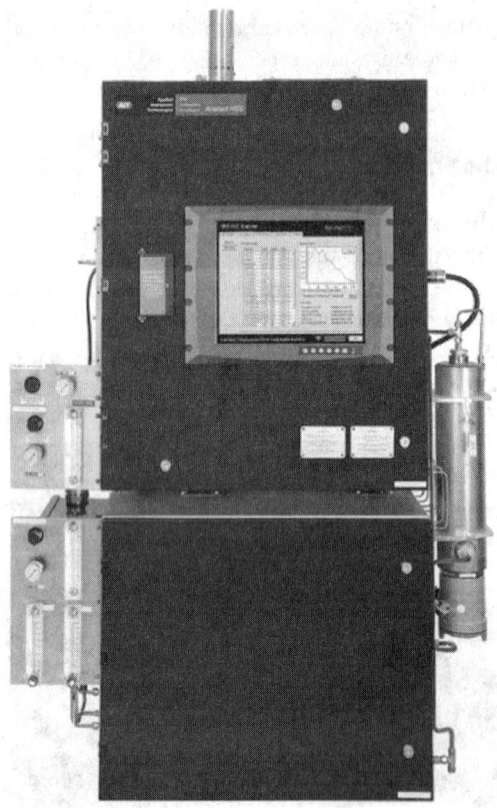

FIG. 1.57w
Online monitor of the physical properties of crude oil. (Courtesy of Applied Instrument Technology.)

Chemometrics

Chemometrics refers to using linear algebra calculation methods, to make either quantitative or qualitative measurements of chemical data. The key to understanding chemometrics is not necessarily to understand the mathematics of all of the different methods, but to know which model to use for a given analytical problem and properly applying it.

Today's powerful microprocessors operating in combination with CCD detector arrays make it possible to capture a full spectra of light vs. one wavelength. In addition, rugged fiber-optic probes (Chapter 1.23) make it possible to measure composition and other process properties by detecting the changes in spectra. The approach used in modeling these relationships is called *chemometrics* and is now an entire field of study for chemists and statisticians. Chemometrics uses mathematical tools and algorithms such as partial least squares (PLS) to assist in identifying correlation between the often small changes in the received spectra and the property of interest. The most commonly used part of the spectrum used for these measurements is the near infrared.

The process of finding such correlation involves the comparison of a large number of process samples that are measured by conventional means while their NIR spectrum is also measured. It is critical that the sample set cover all representative operating conditions that are expected during the operation of the plant; the model is only as good as the data on which it is built. The confidence level of predicting the correlation outside the bounds of the sample set is still relatively low.

Once sufficient samples and spectra have been gathered, this data set is divided into a modeling data set and a validation data set. (One must not use the same data to prove a model as one did to create it, or it is not really a test.) The modeling set is then mathematically manipulated, often taking first and second derivatives of the spectra as a form of *normalization* and to reinforce the magnitude of the changes between the separate wavelengths measured, and compared against the change in the actual data. The software and operator are normally able to identify a combination of wavelengths that vary in a predictable way against the modeling data set. This can be that one wavelength increases while another decreases, that the ratio of two wavelengths changes, or a combination of these affects.

After a model has been identified, it is tested against the validation set to confirm that it indeed predicts the property of interest reliably. If it does, it is ready to be installed in the field and used for process control.

Calibration and Applications

The calibration of these analyzers is valid only over the range of the model for which the relationship was established. Therefore, to avoid errors, one should not extrapolate beyond this range. Because of the flexibility of this technique, it has been applied to most of the physical property measurements discussed in this chapter.

NEW DEVELOPMENTS

As with most measurement technologies, though the standards on which they are based have changed little, the past decade has seen an increased use of microprocessors in the analyzers themselves. In many cases, the physical property analyzers are being replaced with spectroscopic and other techniques that are less "mechanical" in nature and though a secondary measurement technique the analyzer itself is inherently more reliable.

CONCLUSIONS

ASTM-based Physical Property analyzers will continue to play an important role in hydrocarbon analysis because they represent the conditions in which the products are being processed or used. The advent of microprocessors is also seeing the use of the analyzers in wide scale use—as, for example, with the knock measurement being incorporated in automobile engine blocks and integrated into vehicle fuel injection controls. The trend of on-line physical property analyzers becoming more and more available and of the inclusion of microprocessor-based intelligence makes this field of analysis even more valuable than in the past.

SPECIFICATION FORMS

When specifying physical property analyzers, the ISA Form 20A1001 can be used to specify the detector device itself, and when specifying both the device and the nature of the application, including the composition and/or properties of the process samples, ISA Form 20A1002 can be used. The following are the forms that are reproduced with the permission of the International Society of Automation.

#	RESPONSIBLE ORGANIZATION		ANALYSIS DEVICE	#	SPECIFICATION IDENTIFICATIONS		
1				6			
2	(ISA)		Operating Parameters	7	Document no		
3				8	Latest revision	Date	
4				9	Issue status		
5				10			
11	ADMINISTRATIVE IDENTIFICATIONS			40	SERVICE IDENTIFICATIONS Continued		
12	Project number	Sub project no		41	Return conn matl type		
13	Project			42	Inline hazardous area cl	Div/Zon	Group
14	Enterprise			43	Inline area min ign temp	Temp ident number	
15	Site			44	Remote hazardous area cl	Div/Zon	Group
16	Area	Cell	Unit	45	Remote area min ign temp	Temp ident number	
17				46			
18	SERVICE IDENTIFICATIONS			47			
19	Tag no/Functional ident			48	COMPONENT DESIGN CRITERIA		
20	Related equipment			49	Component type		
21	Service			50	Component style		
22				51	Output signal type		
23	P&ID/Reference dwg			52	Characteristic curve		
24	Process line/nozzle no			53	Compensation style		
25	Process conn pipe spec			54	Type of protection		
26	Process conn nominal size	Rating		55	Criticality code		
27	Process conn termn type	Style		56	Max EMI susceptibility	Ref	
28	Process conn schedule no	Wall thickness		57	Max temperature effect	Ref	
29	Process connection length			58	Max sample time lag		
30	Process line matl type			59	Max response time		
31	Fast loop line number			60	Min required accuracy	Ref	
32	Fast loop pipe spec			61	Avail nom power supply		Number wires
33	Fast loop conn nom size	Rating		62	Calibration method		
34	Fast loop conn termn type	Style		63	Testing/Listing agency		
35	Fast loop schedule no	Wall thickness		64	Test requirements		
36	Fast loop estimated lg			65	Supply loss failure mode		
37	Fast loop material type			66	Signal loss failure mode		
38	Return conn nominal size	Rating		67			
39	Return conn termn type	Style		68			

#	PROCESS VARIABLES	MATERIAL FLOW CONDITIONS		#	PROCESS DESIGN CONDITIONS		
69				101			
70	Flow Case Identification		Units	102	Minimum	Maximum	Units
71	Process pressure			103			
72	Process temperature			104			
73	Process phase type			105			
74	Process liquid actl flow			106			
75	Process vapor actl flow			107			
76	Process vapor std flow			108			
77	Process liquid density			109			
78	Process vapor density			110			
79	Process liquid viscosity			111			
80	Sample return pressure			112			
81	Sample vent/drain press			113			
82	Sample temperature			114			
83	Sample phase type			115			
84	Fast loop liq actl flow			116			
85	Fast loop vapor actl flow			117			
86	Fast loop vapor std flow			118			
87	Fast loop vapor density			119			
88	Conductivity/Resistivity			120			
89	pH/ORP			121			
90	RH/Dewpoint			122			
91	Turbidity/Opacity			123			
92	Dissolved oxygen			124			
93	Corrosivity			125			
94	Particle size			126			
95				127			
96	CALCULATED VARIABLES			128			
97	Sample lag time			129			
98	Process fluid velocity			130			
99	Wake/natural freq ratio			131			
100				132			

#	MATERIAL PROPERTIES			#	MATERIAL PROPERTIES Continued		
133				137			
134	Name			138	NFPA health hazard	Flammability	Reactivity
135	Density at ref temp		At	139			
136				140			

Rev	Date	Revision Description	By	Appv1	Appv2	Appv3	REMARKS

Form: 20A1001 Rev 0 © 2004 ISA

Analytical Measurement

		RESPONSIBLE ORGANIZATION	ANALYSIS DEVICE COMPOSITION OR PROPERTY Operating Parameters (Continued)		SPECIFICATION IDENTIFICATIONS		
1				6			
2		ISA		7	Document no		
3				8	Latest revision		Date
4				9	Issue status		
5				10			

		PROCESS COMPOSITION OR PROPERTY			MEASUREMENT DESIGN CONDITIONS				
		Component/Property Name	Normal	Units	Minimum	Units	Maximum	Units	Repeatability
11									
12									
13									
14									
15									
16									
17									
18									
19									
20									
21									
22									
23									
24									
25									
26									
27									
28									
29									
30									
31									
32									
33									
34									
35									
36									
37									

Line	Notes
38	
39	
40	
41	
42	
43	
44	
45	
46	
47	
48	
49	
50	
51	
52	
53	
54	
55	
56	
57	
58	
59	
60	
61	
62	
63	
64	
65	
66	
67	
68	
69	
70	
71	
72	
73	
74	
75	

Rev	Date	Revision Description	By	Appv1	Appv2	Appv3	REMARKS

Form: 20A1002 Rev 0 © 2004 ISA

Definitions

Autoformer: Has only one coil which performs the functions of both the primary and secondary windings of a conventional transformer.

Chemometrics refers to using linear algebra calculation methods to make either quantitative or qualitative measurements of chemical data. The key to understanding chemometrics is not necessarily to understand the mathematics of all of the different methods, but to know which model to use for a given analytical problem and properly applying it.

Engler flask: Standard distilling flask usually of 100 mL capacity used to determine the volatility characteristics of petroleum products (such as gasoline, naphtha, or kerosene).

Flash point of a flammable liquid is the lowest temperature at which it can form an ignitable mixture with oxygen. At this temperature, the vapor may cease to burn when the source of ignition is removed. The flash point is not related to the temperatures of the ignition source or of the burning liquid, which are much higher. The flash point is often used as one descriptive characteristic of liquid fuel.

Knockmeter: Ultrasonic sensor to detect when a spark from the spark plug does not ignite the fuel properly leading to incomplete combustion.

Kovar is an iron-nickel-cobalt alloy with a coefficient of thermal expansion similar to that of hard (borosilicate) glass. It is used in the tip of thermocouples.

Maxwell and Bonnell chart: Correlations for converting actual distillation temperatures to atmospheric distillation temperature equivalents. ASTM D 1160.

Pensky–Martens test is a closed-cup test, used in the determination of the flash-point temperature of lubricating oils, liquids containing suspended solids, and liquids that tend to form a surface film during testing. It is performed by heating and stirring the test specimen, while at regular intervals, the stirring is stopped and heat from an ignition source is applied until, in each cycle, an ignition of the vapors (flashing) occurs. The peak temperature reached after this occurs in each cycle is the flash point temperature.

Reid vapor pressure (RVP) is a measure of the volatility of gasoline. It is defined as the absolute vapor pressure of a liquid at 37.8°C (100°F) as determined by the test method ASTM-D-323.

Relative volatility is a measure of the tendency to vaporize. The higher the volatility, the higher the vapor pressure of the liquid is at the same temperature. In distillation processes, the higher the volatility, the easier it is to separate the component from the less volatile ones. In the case of automotive fuels, it is desirable to use higher volatility gasoline in colder regions, but not be high enough to cause vapor lock. The proportion of butane increases the volatility of the blended product, and because butane is relatively inexpensive, cost is reduced by increasing its proportion. One measure of volatility is the vapor–liquid ratio (abbreviated as or V/L or V–L).

Vapor–Liquid ratio (or V/L or V–L) is a measure of volatility, the tendency of a mixture of liquids to vaporize. It expresses the volume ratio of vapor to liquid that will exist at a given temperature and atmospheric pressure. Automotive fuel specifications usually state the temperature at which V–L is 20 ($T_{(V/L=20)}$). This temperature is where vapor lock can begin. A fuel has high volatility when it's $T_{(V/L=20)}$ is low. Volatility should be relatively high in low ambient temperature regions but should not be high enough to cause vapor lock.

Abbreviations

CCD	Charge-Coupled Device
cSt	centistoke
JP-4	Jet Propellant-4
LPG	Liquefied petroleum gas
MON	Motor Octane Number
NIR	Near-Infrared
ON	Octane Number
PLS	Partial Least Squares
RON	Research Octane Number
RVP	Reid Vapor Pressure
T/E	Thermo Electric
V-L	Vapor–Liquid diagram

Organizations

ASTM	American Society for Testing and Materials
ISA	International Society of Automation

Standards

ASTM D56-05, Standard test method for flash point by tag closed cup tester, (2010) http://www.astm.org/search/fullsite-search.html?query=ASTM%20D56&.ASTM D86-12, Standard test method for distillation of petroleum products at atmospheric pressure, (2014) http://www.astm.org/Standards/D86.

ASTM D97-12, Standard test method for pour point of petroleum products, (2014) http://www.astm.org/search/fullsite-search.html?query=ASTM%20D97&.ASTM D1160-13. Standard test method for distillation of petroleum products at reduced pressure, (2014) http://www.astm.org/search/fullsite-search.html?query=ASTM%20D1160&.

ASTM D1267-12, Standard test method for gage vapor pressure of liquefied petroleum (LP) gases (LP-gas method), (2014) http://www.astm.org/search/fullsite-search.html?query=ASTM%20D1267%20&.

ASTM D1500-12, Standard test method for ASTM color of petroleum products (ASTM color scale), (2012) http://www.astm.org/search/fullsite-search.html?query=ASTM%20D2500%20&.

ASTM D2386-06, Standard test method for freezing point of aviation fuels, (2012) http://www.astm.org/search/fullsite-search.html?query=ASTM%20D2386%20&.

ASTM D2699-13b, Standard Test method for research octane number of spark-ignition engine fuel, http://www.astm.org/search/fullsite-search.html?query=ASTM%20D2699&resStart=0&resLength=10&.

ASTM D2887-13, Standard test method for boiling range distribution of petroleum fractions by gas chromatography 1, 2, http://www.astm.org/search/fullsite-search.html?query=ASTM%20D2887&.

ASTM D4953-06, Standard test method for vapor pressure of gasoline and gasoline-oxygenate blends (dry method), (2012) http://www.astm.org/search/fullsite-search.html?query=ASTM%20D4953&resStart=0&resLength=10&.

Bibliography

Brereton, R.G., *Chemometrics: Data Analysis for the Laboratory and Chemical Plant*. John Wiley & Sons, 2003. http://www.wiley.com/WileyCDA/WileyTitle/productCd-0471489786.html.

California Air Resources Board, Vapor to liquid volume ratio, 2011. http://www.arb.ca.gov/vapor/eos/eo-vr201/eo-vr201j/vr201j_e05.pdf.

Clevett, K.J., Process analyzer technology, 1986. https://books.google.com/books/about/Process_analyzer_technology.html?id=usBTAAAAMAAJ.

InfoMetrix, Chemometrix, 2014. http://infometrix.com/chemometrics/.

Intertec, Physical properties testing, 2014. http://www.intertek.com/analysis/materials/physical-properties-testing/.

Kowalyk, D., Physical properties analysis of petroleum products using FT-NIR, 2000. http://www.cbeng.com/resources/whitepaper/PhysicalPropertiesAnalysis.pdf.

Parkash, S., Refining Processes Handbook, Gulf Professional Publication, 2003. http://www.amazon.com/Refining-Processes-Handbook-Surinder-Parkash/dp/075067721X/ref=sr_1_fkmr0_1?ie=UTF8&qid=1452432119&sr=8-1-fkmr0&keywords=Parkash%2C+S.%2C+Refining+Processes+Handbook%2C+Gulf.

Shanovsky, G. et al., Integrated monitoring for optimizing crude distillation, 2012. http://www.modcon-systems.com/wp-content/uploads/2014/04/PTQ-CDU-Control.pdf.

Sherman, R.E., Process analyzer sample-conditioning system technology, 2002. http://www.wiley.com/WileyCDA/WileyTitle/productCd-0471293644.html.

1.58 Raman Analyzers

R. MANOHARAN and N. K. SETHI (2003)

B. G. LIPTÁK and A. V. DESHMUKH (2017)

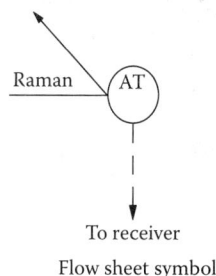

Flow sheet symbol

Process fluids	Liquids, powders, slurries, emulsions (gases), particularly suited for polymers
Operating pressure	Analyzer (ambient), sample interface 3000 psig (206 bars)
Operating temperature	Analyzer 0°C–50°C (32°F–122°F), sample interface 350°C (662°F)
Humidity limitations	20%–100% noncondensing
Calibration transfer	Intra-analyzer transfer after maintenance and transfer between analyzers
Sample interface	Invasive, noninvasive
Wetted material	Stainless steel, titanium, Hastelloy® and Kynar® probe body, sapphire and quartz windows or lens
Fiber-optics	Low-hydroxy silica, multimode fibers, hundreds of meters
Update time	A few seconds to minutes
Online analysis	Yes
Warm-up time	60 min
Measurement	Continuous
Repeatability	0.1% or less
Linearity	Over the entire measurement range
Laser safety	Required
Costs	Analyzers for process applications cost $60,000–$150,000, depending on the packaging, ruggedness, availability of analyzer standardization, and availability of simultaneous multichannel capability
	A benchtop, low-resolution Raman instrument costs $20,000 with probes and other accessories
	Laboratory Raman instruments cost $50,000–$100,000, depending on the technology (dispersive or FT-based) and also available sampling accessories
	Optical fibers for process installation cost $15–$30/ft/channel (consisting of three cables), depending on the packaging options.
	The fiber-optics probe costs $2,000–15,000, depending on the application requirements and required ruggedness
Partial list of suppliers	AIT, http://www.aitanalyzers.com/raman-spectroscopy-rpm
	ARI, www.atmrcv.com/technology.html
	Bruker Optics, Inc., http://www.bruker.com/products/infrared-near-infrared-and-raman-spectroscopy.html
	BWTek, http://bwtek.com/products/bac101/
	CIC Photonics (fiber optic probe) http://www.globalspec.com/local/4776/NM
	Direct Industry, http://www.directindustry.com/industrial-manufacturer/raman-spectrometer-80689.html

(Continued)

FPO, http://www.freepatentsonline.com/7102746.html
Horiba Group, http://www.horiba.com/scientific/products/raman-spectroscopy/
InPhotonics, http://www.inphotonics.com/Raman-spectrometer.htm
Jasco Inc., http://www.jascoinc.com/material-science/raman-spectrometers
Kaiser Optical Systems, Inc. (optical probes), http://www.kosi.com/raman-spectroscopy/raman-sampling-probe-heads-optics.php
Oxford Instruments, http://www.andor.com/learning-academy/raman-spectroscopy-an-introduction-to-raman-spectroscopy
Process Instruments, Inc., http://www.process-instruments-inc.com/pages/petroleum.html
Raman Systems, Inc., http://www.ramansystems.com/
Real-Time Analyzers, http://www.rta.biz/content/ramanid_portable_raman_analyzer.asp
Renishaw, plc, http://www.renishaw.com/en/raman-spectroscopy—6150
Rigaku, http://www.rigakuraman.com/products/
Rosemount Analytical, Emerson, http://www2.emersonprocess.com/siteadmincenter/PM%20Rosemount%20Analytical%20Documents/PGA_Manual_Raman_200009.pdf
Semrock (laser optics) http://www.semrock.com/raman-spectroscopy.aspx
Thermo Nicolet, http://www.thermoscientific.com/en/products/fourier-transform-infrared-spectroscopy-ftir.html
TSI, http://www.tsi.com/Raman-Spectrometers/

INTRODUCTION

Raman scattering was discovered in 1928, by Chandrasekhara Venkata Raman. It can be used as a tool to characterize any nonmetallic material. It is the shift in wavelength of the inelastically scattered radiation that provides the chemical and structural information contained in the scattered light.

Raman scattering is similar to infrared (IR) absorption (Chapter 1.32) in that they are both vibrational spectroscopies that probe the discrete vibrations of molecules and molecular bonds. The Raman effect relies on the scattering of the incident photons, whereas IR occurs through light absorption. These vibrational techniques have different selection rules for the types of molecular vibrations that can be observed, and so they are often considered complementary.[*,†,‡]

In 1930, the Raman effect became the analytical tool for nondestructive analysis. In 1960, the technique found more interest due to the advent of light amplification by stimulated emission of radiations (LASERs). The Fourier transform (FT) Raman instrumentation appeared in 1986. The developments in instrumentation were mainly due to the advancements in sources, detectors, optics, computing, newer electronics, and digital signal processing methods.

Raman analyzers can be considered for applications in which IR or near-infrared (NIR) detectors are suited for making the measurement. The cost of a Raman process analyzer typically exceeds that of other analyzers. Therefore, gas chromatography (GC) is preferred to Raman spectroscopy except in cases where continuous and fast readings are more important than cost and precision. Raman technology is distinctly advantageous for the analysis of viscous liquids such as polymer melts.

This chapter describes Raman spectroscopy for process control measurement and analysis. The first part of this chapter deals with the principles of Raman spectroscopy. Next, the technological developments that transformed Raman spectroscopy from an esoteric laboratory technique into an industrial process analytical tool are presented, and the components of the Raman process analyzer are described in detail.

Following this discussion, the benefits of Raman spectroscopy are compared with NIR and mid-IR absorption spectroscopy methods. Finally, the industrial process applications of Raman analyzer technology and its future directions are discussed.

Raman spectroscopy finds a wide number of applications ranging from laboratory analysis to field process analyzers and portable monitors. Raman spectroscopy has been a well-established technique in the field of analytical chemistry, biomedical, semiconductors, polymer, pharmaceutical, and other process industries. Raman analyzers can work in harsh environments even under high pressures and temperatures. As compared to Fourier transform infrared (FTIR), aqueous solutions could easily be handled using Raman instrumentation.

Almost no sample preparation is required for Raman analyzers. Multiple probes could be used with a single unit taking advantage of multiplexing. It is also flexibility available in many modern Raman instruments to choose among the excitation wavelengths to match the various sample properties.

---

[*] Long, D. A., *The Raman Effect: A Unified Treatment of the Theory of Raman Scattering by Molecules*, John Wiley & Sons, New York, 2002. http://www.wiley.com/WileyCDA/WileyTitle/productCd-0471490288.html.

[†] Silva, E., Raman spectroscopy: A comprehensive review, 2013. http://www.academia.edu/1131363/Raman_Spectroscopy_A_Comprehensive_Review.

[‡] Campbell, R., A spectroscopic probes of protein structure, 1984. http://link.springer.com/chapter/10.1007/978-1-4757-1505-7_9.

A typical online Raman spectrum is shown in Figure 1.58a with the different peaks from many of its major chemical components. The peak frequency shift (i.e., the location along the x-axis) yields the sample composition, and the peak intensity yields the concentration of that particular component. A key advantage with the Raman effect is that it is a linear process.

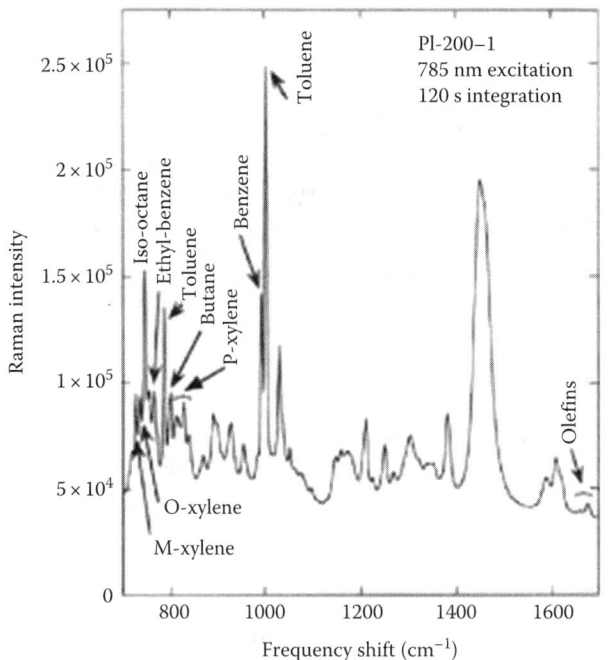

FIG. 1.58a
The Raman spectrum of gasoline yields both the composition and concentration of each component. (Courtesy of Process Instruments.)

PRINCIPLES OF RAMAN SPECTROSCOPY

Light scattering is the deflection of a light ray from a straight path. Most objects that one sees are visible due to light scattering from their surfaces. The scattering of light depends on the wavelength of the light being scattered and can be elastic or inelastic. Elastic scattering of light is also called Raleigh scattering and is a scattering mechanism that results in the uniform scattering of the incoming light by atoms or molecules in all directions.

Raman scattering is the shift in wavelength of the inelastically scattered monochromatic radiation that provides the chemical and structural information contained in the scattered light. When the sample is irradiated by a monochromatic laser light, having a wavelength from 532 to 785 nm, the resulting weak Raman signal can be used to detect the composition of the sample.

Optical Spectroscopy

In optical spectroscopy, light from a source is allowed to impinge on the sample to be examined. Molecules in the sample respond to the incident light by absorbing, scattering, or changing the energy of photons. Optical spectroscopy is the analysis of change between the incident and return light. Optical spectroscopy can be used to detect, identify, characterize, and quantify chemical components in the sample.

Vibrational spectroscopy, where the light characteristics change as a result of various modes of molecular vibrations, is the most powerful form of optical spectroscopy methods for providing quantitative chemical information. NIR absorption, mid-IR absorption, and Raman scattering are three major forms of vibrational spectroscopy methods. For an in-depth discussion of NIR and IR spectroscopy methods, the reader is referred to Chapter 1.32 of these analyzers. Although reference to NIR and IR methods is essential when discussing the merits of Raman spectroscopy, the focus of this chapter is on the use of Raman scattering for chemical analysis and process applications.[*,†,‡]

Raman Scattering

In Raman spectroscopy, a beam of monochromatic light is used to irradiate the sample and the scattered radiations are observed. Monochromatic beams are obtained from a LASER source. The Raman spectrum consists of sharp narrow peaks, so-called as Stokes and anti-Stokes, located on either side of a strong Rayleigh peak.

When a beam of monochromatic light is incident on matter, it could be reflected, transmitted, and scattered. The frequency of almost 99% of the scattered beam is the same as that of the incident radiations. Rayleigh scattering is also called elastic scattering and is a scattering mechanism that results in the uniform scattering of the incoming light by atoms or molecules in all directions. In Raleigh scattering, there is no change in the energy of photons in the scattering process. This occurs when electrons in the molecule vibrate according to the applied electric field of incident beam. So there is a resonance. However, it was observed by C. V. Raman in 1928 that a small portion (1 in 10^6–10^9 photons) of the scattered secondary radiations was of different frequencies than the incident beam. There is a change in the energy of the photon, and this is inelastic scattering. There is

---

[*] Long, D. A., *The Raman Effect: A Unified Treatment of the Theory of Raman Scattering by Molecules*, John Wiley & Sons, New York, 2002. http://www.wiley.com/WileyCDA/WileyTitle/productCd-0471490288.html.
[†] Silva, E. Raman spectroscopy: a comprehensive review, 2013. http://www.academia.edu/1131363/Raman_Spectroscopy_A_Comprehensive_Review.
[‡] Campbell, R. A spectroscopic probes of protein structure, 1984. http://link.springer.com/chapter/10.1007/978-1-4757-1505-7_9.

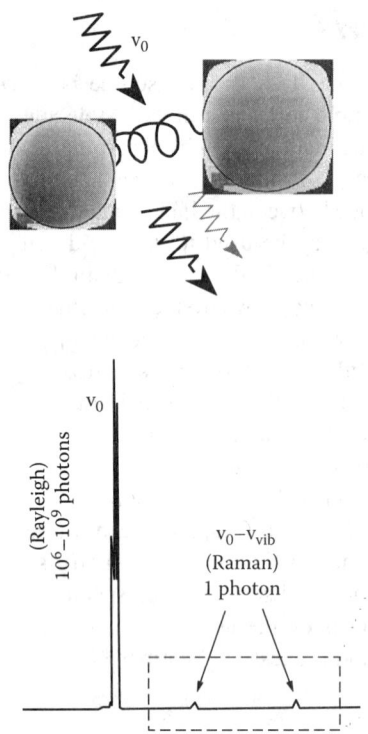
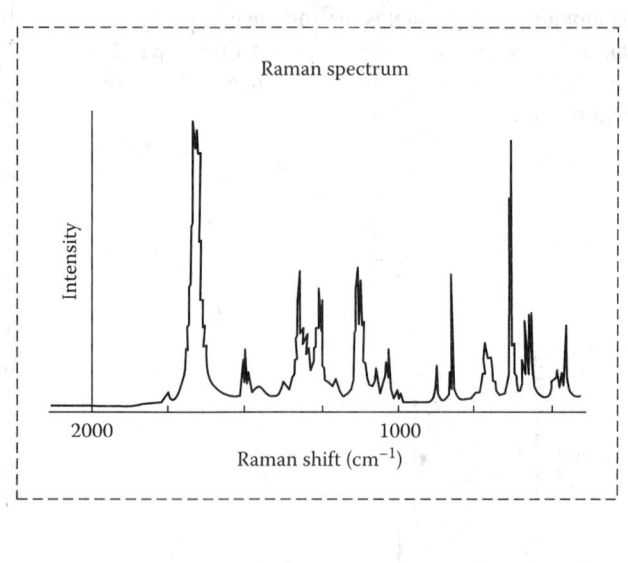

FIG. 1.58b
Pictorial representation of Raman scattering.

no resonance in this process. This type of scattering is called as Raman scattering.

A pictorial description of the Raman scattering process is shown in Figure 1.58b. When monochromatic light interacts with molecules, most of the photons are scattered without any change in energy.* This process is called elastic or Rayleigh scattering, and it occurs when the electrons in a molecule oscillate in resonance with the applied electric field of the incident light. However, a small number of photons (1 out of 10^6–10^9) are inelastically scattered and undergo a change in energy.

The inelastic scattering process is called *Raman scattering*. The change in photon energy occurs because during the time the electrons oscillate in resonance with the applied electric field, a molecule may vibrate. A vibrational mode that changes molecular polarizability (dipole moment induced by electric field) results in a change of incident photon energy. The difference in energy between the incident photons and inelastically scattered photons is called *Raman shift*. A plot of the intensity of the inelastically scattered light as a function of the energy change is called *Raman spectrum* (Figure 1.58b).

Figure 1.58c shows a schematic of the Raman and Rayleigh scattering processes. There are two kinds of Raman scattering events. In event (a), incident photons lose energy to the molecule, causing it to go to an excited, vibrational state ($v_0 \rightarrow v_1$). As a result, scattered photons will have less energy (Stokes Raman) compared to the incident photon.

In event (b), the incident photons gain energy from the molecule, because the molecules lose energy by going from a higher to lower vibrational state ($v_1 \rightarrow v_0$). In this case, the Raman scattered photon will have higher energy than the incident photons (anti-Stokes Raman). At room temperature, most molecules are in the vibrational ground state (v_0), so there is a greater probability that incident photons will lose energy to the molecules during the interaction, that is, Stokes Raman. Generally, Raman spectroscopy refers to Stokes Raman unless specified otherwise. Thus, Raman scattering is observed at lower energy or longer wavelength compared to that of the incident light.

---

* The relationships between energy, frequency, wave number, and the wavelength are as follows:
  $E = hv$ (h is Planck's constant, and v is frequency in s^{-1}).
  $E = hc\upsilon$ (c is the speed of light in cm/s, and υ is wave number in cm^{-1}).
  $v = 1/\lambda$ (λ is wavelength in nm).

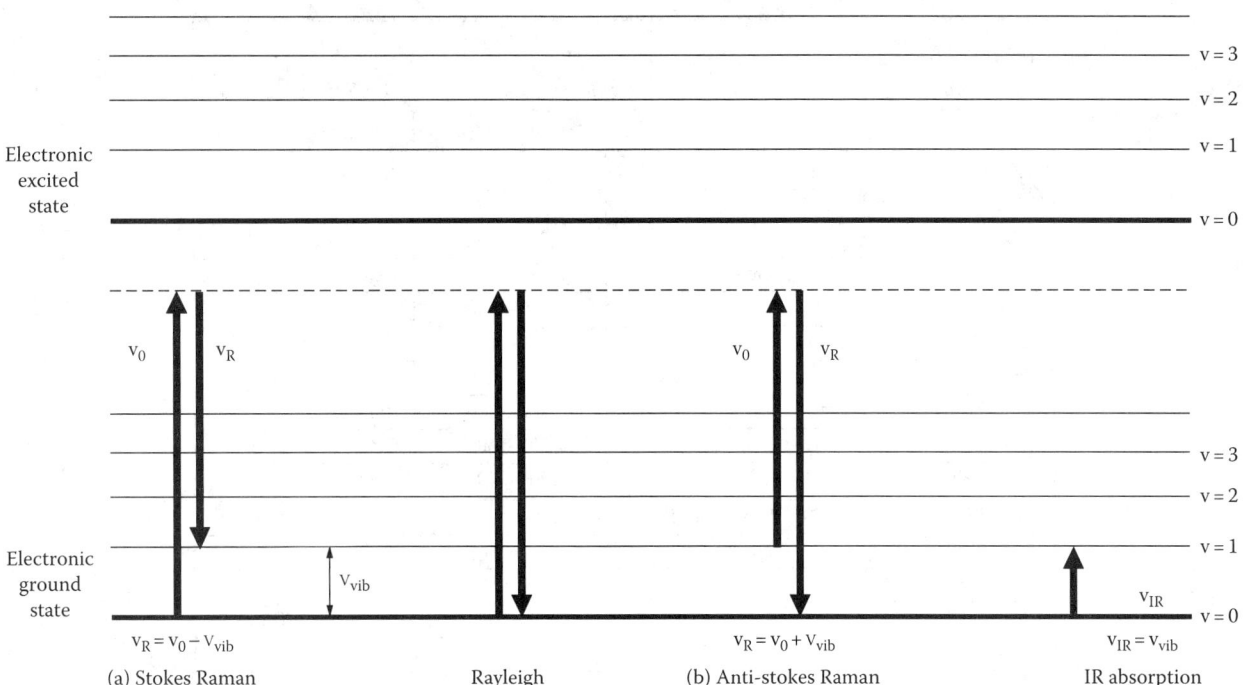

FIG. 1.58c
Schematic of energy-level diagram for Raman and Rayleigh scattering processes.

The difference between the energy of incident photons and Raman scattered photons is called the Raman shift.

Spectral Information

Raman scattering intensity is related to the number of molecules in the sample according to the following equation:

$$I_R = I_0 \cdot \sigma_R \cdot N \cdot (v_0 - v_R)^4 \qquad 1.58(1)$$

where
- I_R is the Raman signal intensity
- I_0 is the incident light intensity
- σ_R is the Raman scattering cross section that is related to molecular polarizability
- N is the number of molecules in the sample volume probed by the light beam
- v_0 is the frequency of the excitation light
- v_R is the Raman scattering frequency

This relationship permits the use of Raman for quantitative analysis.

As in mid-IR absorption, Raman scattering also probes vibrations of the molecular bonds. The Raman shift frequency (v_R; see Figure 1.58c) is equivalent to the energy of mid-IR absorption for vibrational modes of a given group in the molecule (e.g., C–H stretch in CH_4 methane). In a classical model of molecular vibration, the Raman shift (expressed in energy, wave number, wavelength, or frequency [see footnote on page 828]) is related to the molecular properties using an analogy of mass attached to a spring (Figure 1.58b) as follows:

$$\Delta v_{vib} = \frac{1}{2\pi C} \sqrt{\frac{k}{\mu} \frac{1}{2}} \qquad 1.58(2)$$

where
- Δv_{vib} is the Raman shift or absorption energy in the mid-IR region (expressed in cm^{-1} units)
- C is the speed of light
- k is the strength of the bond (or force constant of the spring, in N/m)
- μ is the reduced mass given as $m_1 m_2/(m_1 + m_2)$, where m_1 and m_2 are atomic mass attached to the spring or bond

As the relationship shows, the Raman shift is related to the strength of the bond and the atomic masses of the molecule. For instance, for the carbon–carbon vibrations in functional groups with increasing bond strength, C–C, C=C, and C≡C, the Raman shift occurs at increasing frequencies, ~1000, ~1650, and ~2300 cm^{-1}, respectively.

Although Raman and mid-IR provide similar information, the selection rules for observations are different. Selection rules determine which vibrational modes are observed in the spectrum. Raman scattering arises due to a change in polarizability during vibration, whereas IR absorption occurs due to a change in dipole moment. Thus, for instance, homonuclear diatomic molecules (N_2, O_2, H_2, etc.) show Raman bands but not mid-IR absorption bands,

TABLE 1.58a
Molecular Functional Groups Observable by Raman and Their Raman Shift Positions

Raman Shift (cm^{-1})	Functional Groups	Raman Shift (cm^{-1})	Functional Groups
3400	(O–H, N–H)	1440–1400	N=N (aromatic)
3300	≡C–H	1420–1370	Polyaromatic ring
3100–3000	=C–H	1380–1330	C–H deform, C–(NO$_2$)
3000–2800	(C–H) stretch	1310–1290	CH$_2$ twist
2600–2550	(S–H)	1310–1250	Amide II
2260–2220	C≡N	1300–600	C–C (chain)
2200	S–H	1270–1250	C–H deform
2250–2100	C≡C	1260–1200	Ring breath
2050	C=C=O (aliphatic)	1250–1000	C=S
1970–1900	C=C=C (aliphatic)	1200–800	C–O–C
1830–1700	C–O (carbonyl, ester)	1100–1050	C–S (aromatic)
1750–1630	C=C (aliphatic)	900–850	O–O
1680–1630	Amide I band	800–650	C–S (aliphatic)
1670–1660	C=N stretch	800–500	C–X (X=Cl, Br, I)
1670–1630	Amide I band	800–400	Si–Cl
1670–1610	C=N stretch	650–200	P–Cl
1650–1590	N–H (amines)	600–350	P–P
1640	Water (weak band)	550–450	Si–O–Si
1630–1540	C=C (aromatics, polyenes)	550–430	S–S
1580–1550	N=N (aliphatic)	450–150	Metal–O
1560–1530	C–NO$_2$	400–250	C–C (aliphatics)
1480–1400	C–H bend		

because their only vibrational mode does not change the dipole moment of the molecule.

Table 1.58a lists various molecular functional groups that can be observed and quantitatively analyzed using Raman spectroscopy.

INSTRUMENTATION

Raman scattering is inherently weak and is difficult to observe without intense monochromatic excitation and a sensitive detector. Modern Raman spectrometer systems use lasers as excitation sources. Visible laser light (400–700 nm) from an argon ion, helium–neon, or krypton ion laser is commonly used in laboratory Raman instruments. One of the drawbacks of visible laser excitation is fluorescence interference.

Almost all complex samples, especially biological systems, exhibit fluorescence. Because the fluorescence signal is three to five orders of magnitude stronger than the Raman signal, even trace fluorescence from impurities can mask weak Raman signals. For this reason, NIR laser-based Raman instruments have superseded visible laser instruments. Fluorescence decreases rapidly at longer excitation wavelengths, and most materials exhibit little or no fluorescence emission when excited with NIR lasers.

In order to enhance Raman scattering (surface-enhanced Raman scattering [SERS]), a silver-doped solgel film can be used. The technique involves first doping silver ions as a silver ammine complex [Ag(NH$_3$)$^{2+}$] into a solgel matrix, followed by chemical reduction of the doped silver ions into the corresponding silver metal particles. The silver-doped solgel films coated on a glass substrate exhibit strong enhancement (24,000-fold) of Raman scattering from molecules adsorbed on the surface Figure 1.58d).

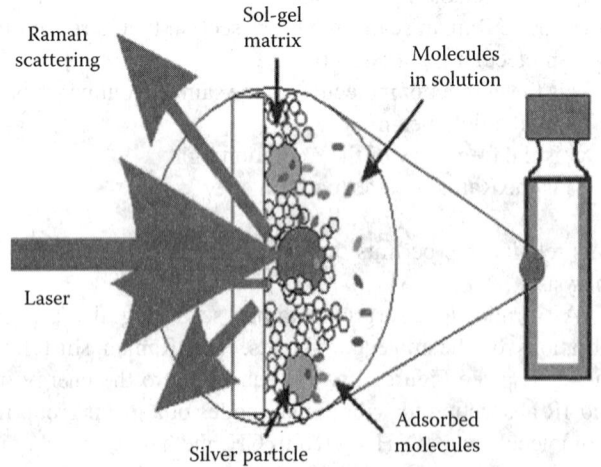

FIG. 1.58d

Enhancing Raman scattering by the use of solgel matrix. (Courtesy of NASA.)

Fourier Transform Raman Instrumentation

FT Raman is the first technology to use NIR excitation to collect fluorescence-free Raman signals.* The heart of this system is a wavelength-stabilized Michelson interferometer, one arm of which is scanned in length to produce a time-varying interference signal. The Raman spectrum is computed from the FT of this signal. FT Raman setup employs continuous-wave Nd:YAG laser ($\lambda = 1.064$ μm) and a cryogenically cooled indium gallium arsenide (InGaAs) detector.

Three advantages of an FT spectrometer are (1) Conne's advantage, excellent wavenumber repeatability; (2) Fellgett's multiplex advantage, information in all wavelengths collected at the same time; and (3) Jacquinot's advantage, increased throughput.

The main disadvantage of FT-based analysis is its lower signal-to-noise ratio (S/N); it is limited by both reduced signal strength at the longer NIR wavelength or lower frequency (signal strength is proportional to v_0^4) and the relatively high noise levels associated with InGaAs detectors. Furthermore, for remote spectroscopic applications, optical fibers are used for light delivery and signal collection. In such instances, the small numerical aperture (NA) of fibers (0.22) is not readily matched to a Michelson interferometer, thus losing the throughput advantage of the interferometer.

Michelson interferometer was invented by Albert Abraham Michelson. It uses a beamsplitter in which a light is split into two arms. Each of those is reflected back toward the beamsplitter which then combines their amplitudes interferometrically. The resulting interference pattern is typically directed to some type of photoelectric detector or camera.

Fluorescence was a source of problems when attempting to observe the Raman spectra, at the time when the Raman Effect was discovered, because the whole Raman spectrum was masked by fluorescence. The major difference between fluorescence and the Raman effect is that the Raman effect is not a resonant effect, it takes place at any frequency of incident light and maintains a constant separation from the excitation frequency.

FT Raman was introduced in 1964 by Chantry and Gebbie. However, it took more than 20 years to make it viable due to the lack of technology available in those days. The problem of fluorescence was reduced (almost eliminated) by measuring the Raman spectra in the NIR region. Thus, FT Raman is the first technology to use NIR excitation to collect fluorescence-free Raman signals.*

Conventional Raman spectroscopy lacks in frequency precision; therefore, high-precision subtractions are not possible. Furthermore, high-resolution Raman experiments are not possible with conventional method.

It is also important to optically, or otherwise, filter out or attenuate the Rayleigh line. The Rayleigh line is 10^6 times stronger than Stokes-shifted Raman lines. Stokes shift, named after the Irish physicist George G. Stokes, is the difference in frequency between maximum positions of the absorption (fluorescence) and the emission (Raman) spectra of the same electronic transition. When an atom absorbs energy, it enters an excited state, and one way for it to relax is to lose energy by emitting a photon. When the emitted photon has less energy than the absorbed photon, this energy difference is called the Stokes shift. If the emitted photon has more energy, the energy difference is called an anti-Stokes shift.

In order to have observable Raman spectrum, this Rayleigh line must be attenuated to the extent that is comparable with the strongest Raman lines. Modern digital filtering methods could also help in extracting the Rayleigh line. There are other approaches in digital signal processing like windowing, baseline corrections, signal analysis, and estimation that also find important place in FT Raman spectrum estimation. Next, stray light due to laser excitation must be eliminated/minimized; otherwise, it might saturate the detector electronics.

Certain important aspects of FT Raman are that some Raman instruments can also provide IR and Raman capabilities in the same instrument. It must be noted that the Raman Stokes and anti-Stokes are collected simultaneously by an FT Raman instrument and the frequency accuracy in FT Raman is much better as compared to conventional Raman instrument.

Raman versus IR Spectroscopy The diameter of a LASER beam that is used in Raman spectroscopy is typically 1–2 mm, and a very small sample area is required for obtaining the Raman spectrum. In many cases, a very small quantity of sample is available and this is the major advantage of Raman spectroscopy over IR spectroscopy.

Water does affect Raman spectrum being a weak Raman scatterer. Therefore, Raman spectra in aqueous solutions could be easily obtained. Many biological compounds in aqueous solutions could therefore be studied using Raman spectroscopy. This is not, however, true about the IR spectrum, as IR radiations are strongly absorbed in aqueous solutions.

Raman spectra could be obtained by placing a sample in a glass tube. However, IR radiations are absorbed by glass tubes, and therefore quartz sample holders are required for IR.

---

* Chase, B., A new generation of Raman Instrumentation, *Appl. Spectrosc.*, 48(7), 14A–19A, 1994, and references cited therein. https://www.researchgate.net/publication/238187065_A_New_Generation_of_Raman_Instrumentation.

Raman spectroscopy essentially requires a LASER source. The sample might be heated or there could be photodecomposition.

A common advantage of vibrational spectroscopy, both Raman and IR spectroscopies, is that it is applicable to solids, gases, or liquid solutions.

Dispersive Raman Instrumentation

Technological advances have resulted in dispersive Raman instruments that employ imaging spectrographs and charge-coupled device (CCD) detectors. A schematic of a dispersive NIR Raman system is shown in Figure 1.58e. It shows the basic components of an FT Raman spectrometer. The heart of this system is a wavelength-stabilized Michelson interferometer. It consists of one movable mirror, a fixed mirror, and a beam splitter. A laser source is used to excite the sample. The movable mirror arm is scanned in length to produce a time-varying interference signal, so-called as interferogram. The Raman spectrum is computed from the FT of this signal. FT Raman setup employs continuous-wave Nd:YAG laser ($\lambda = 1.064$ μm) and a cryogenically cooled indium gallium arsenide (InGaAs) detector. The power of laser used in FT Raman ranges from 1 to 10 W.

The NIR laser is filtered to remove unwanted sidebands from the laser and is focused on the sample. The scattering signal is collected by a second lens, filtered by a Raman edge filter to reject intense Rayleigh scattering, and focused onto an imaging spectrograph equipped with a grating. A CCD array detector mounted at the output focal plane of the spectrograph detects the Raman signal. Commercial systems also use optical fibers instead of lenses to deliver and collect light. Various fiber-optic sampling arrangements have been reviewed in detail.*

The primary gain in S/N of a CCD dispersive system as compared to that of an FT system results from the low noise level of a cooled CCD array detector. However, there are two factors to be optimized in selecting the appropriate NIR excitation laser. First, in the visible wavelength, a CCD has very high efficiency, but fluorescence strongly interferes with the Raman spectra.

In the NIR region, although fluorescence interference is reduced or eliminated, the CCD detector quantum efficiency drops considerably as a result of silicon absorption. In selecting the optimal excitation wavelength, a trade-off must be made between these two factors. For most complex samples, the excitation wavelength range generally falls between 750 and 830 nm. Comparative S/N studies have shown that CCD/dispersive system can be several orders of magnitude more sensitive than FT Raman operation.[†,‡]

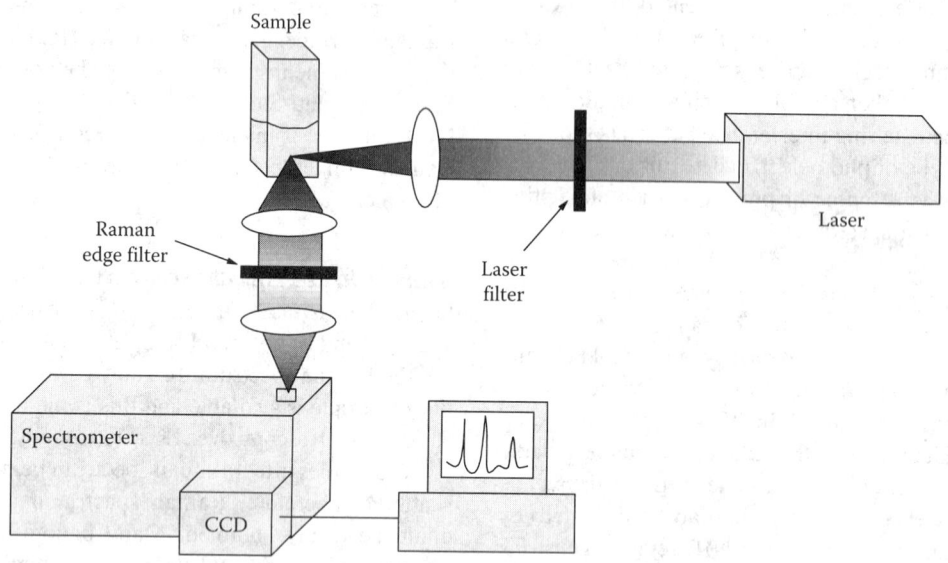

FIG. 1.58e

Schematic of a dispersive Raman instrument.

---

* Lewis, I. R. and Griffiths, P. R., Raman spectrometry with fiber optic sampling, *Appl. Spectrosc.*, 50(10), 12A–30A, 1996. https://www.osapublishing.org/as/abstract.cfm?uri=as-50-10-12A.
† Chase, B., A new generation of Raman Instrumentation, *Appl. Spectrosc.*, 48(7), 14A–19A, 1994, and references cited therein. https://www.researchgate.net/publication/238187065_A_New_Generation_of_Raman_Instrumentation.
‡ McCreery, R. L., Modern techniques in Raman spectroscopy, in *Instrumentation for Dispersive Raman Spectroscopy*, Laserna, J. J. (Ed.), John Wiley & Sons, New York, 1996, pp. 41–72. http://www.wiley-vch.de/publish/dt/AreaOfInterestCH00/availableTitles/0-471-95774-7/.

Raman scattering is inherently weak and is difficult to observe without intense monochromatic excitation and a sensitive detector. Modern Raman spectrometer systems use lasers as excitation sources. Visible laser light (400–700 nm) from an argon ion, helium–neon, or krypton ion laser is commonly used in laboratory Raman instruments. One of the drawbacks of visible laser excitation is fluorescence interference.

Almost all complex samples, especially biological systems, exhibit fluorescence. Because the fluorescence signal is three to five orders of magnitude stronger than the Raman signal, even trace fluorescence from impurities can mask weak Raman signals. For this reason, NIR laser-based Raman instruments have superseded visible laser instruments. Fluorescence decreases rapidly at longer excitation wavelengths, and most materials exhibit little or no fluorescence emission when excited with NIR lasers.

RAMAN PROCESS ANALYZERS

The Raman process analyzer is designed to produce continuous data that are reliable and reproducible for process analysis and control. Some of the key requirements for Raman process analyzers include (1) ruggedness of analyzer, probes, and optical filters; (2) data accuracy and repeatability; (3) long-term stability; (4) operating environment-independent performance; (5) standardized, analyzer-independent data to allow for calibration transfer between analyzers; (6) smart diagnostics for analyzer performance and predictive maintenance; (6) industry-standard communication and outputs for control; (7) availability of outputs, diagnostic parameters, and alarms for operation and maintenance; (8) laser safety for personnel and the process; (9) protection against failure; (10) packaging and certification for field applications; (11) ease of operation and maintenance; (12) multilevel passwords and security features; (13) high uptime (high mean time to failure and low mean time to repair); and (14) low total cost of ownership.

Components

Figure 1.58f shows the layout of a typical Raman analyzer that uses fiber-optics for process application. In a Raman process system, light is filtered and delivered to the sample via excitation fiber. Raman scattered light is collected by collection fiber(s) in the fiber-optic probe, filtered, and sent to the spectrometer via return fiber-optical cable(s). The CCD camera detects the signal to provide the Raman spectrum. In a Raman process analyzer, the components and their implementation will be different from those used in laboratory instruments.

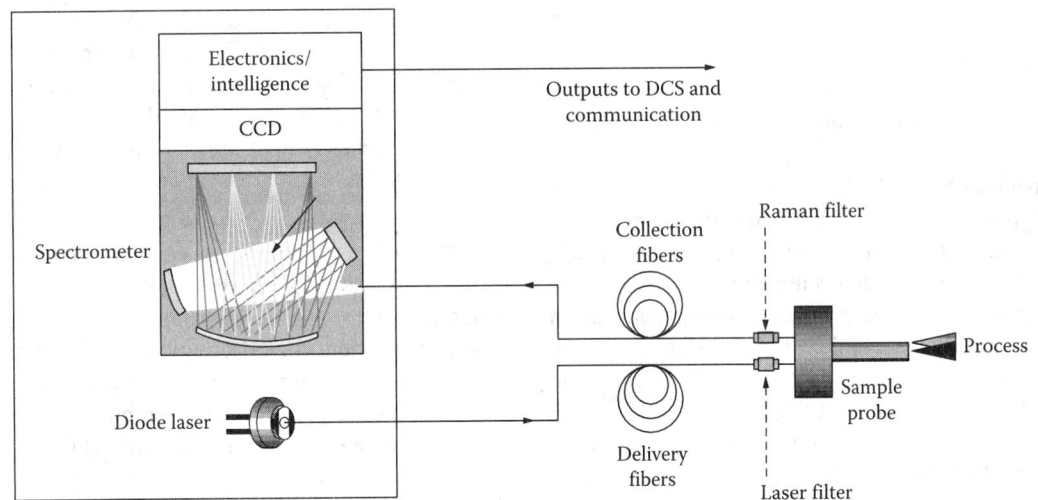

FIG. 1.58f
Schematic of fiber-optic Raman analyzer for process measurements.

Laser Excitation Source To take advantage of low-noise CCD cameras and to minimize fluorescence interference, NIR diode lasers are used in process instruments. Advances in telecommunications, where NIR diode lasers are used as a laser pump source, have significantly contributed to the development of NIR diodes.*

Available diodes have broad line width, are very compact (as small as 1 × 2 in. [25 × 51 mm] with integrated Peltier cooler, thermistor temperature sensor, and photomonitor), offer tens of watts of power, and are relatively inexpensive. However, these multimode devices also exhibit line shape instability, a phenomenon known as *mode hopping* (Figure 1.58g).

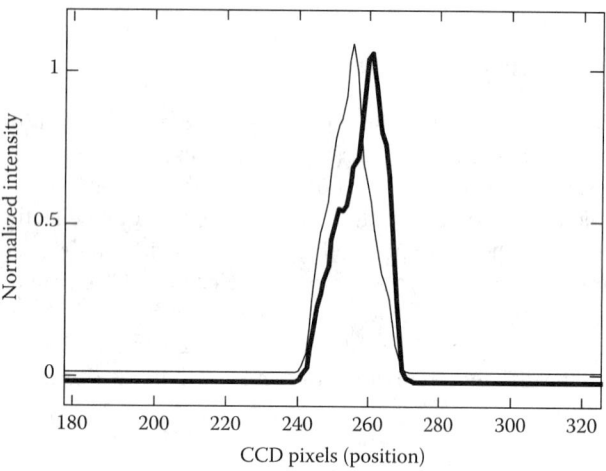

FIG. 1.58g
Instability of multimode diode laser caused by mode hopping, illustrating the need for compensation.

Raman spectra obtained with a multimode laser exhibit lower resolution resulting from the broad spectra line shape, and they also show instability. In most applications, lower spectral resolution can be circumvented with the use of multivariate analysis for extracting concentration, much as in NIR. The more significant problem is the instability of the Raman spectrum due to laser mode hopping. To compensate for this, two approaches have been employed.

The first approach uses a Raman reference standard. A small portion of the laser beam is split and sent to a reference channel containing a small sample of a chemically stable material, for example, an industrial diamond. The Raman output of the reference diamond is sent to the same CCD camera. Thus, for every process sample spectrum, there is a reference spectrum of the diamond. A transfer function (convolution function) is calculated to match the spectrum of the reference material to its true spectrum. The calculated transfer function is then applied on the sample spectrum to produce a true sample spectrum that is independent of laser mode hopping.

In the second approach, a small portion of the laser beam is directly monitored for mode hopping, and this information is used to compensate for the laser instability.

A multimode diode can be made into a narrow-line-width, single-mode laser by forcing one of the modes to lase and providing stabilization using a diffraction grating.† Diffraction Gratings are optical components used to separate light into its component wavelengths. When there is a need to separate light of different wavelengths with high resolution, a diffraction grating is the most often used tool of choice. A diffraction grating is made by making many (up to 6,000 lines/cm) parallel scratches on the surface of a flat piece of transparent material. The scratches are opaque, but the areas between the scratches transmit light. A single-mode laser has narrow line width, resulting in high-resolution Raman spectra. However, single-mode lasers are inefficient, relatively large in size, and more expensive compared to multimode devices. The trade-off between resolution and total cost of ownership should be considered for each application.

Some analyzers use visible laser excitation (e.g., green) for high-sensitivity applications where the process sample is free of fluorescence emission. The use of the green laser offers high sensitivity due to v^4 factor (see Equation 1.58(1)) and allows detection of C–H and N–H stretching modes in the Raman shift region, 2800–3400 cm^{-1} (see Table 1.58a). Again, the total cost of ownership, fluorescence interference, required sensitivity, and spectral range desired should be taken into consideration in deciding on a visible laser-based Raman analyzer.

Spectrometer A typical spectrograph consists of mirrors and a grating for dispersion. Ruled diffraction grating spectrographs use aluminum-coated optical surfaces and are not very efficient for NIR region. Further improvement in throughput is achieved by using gold-coated optical components. The development of holographic optics offers highly efficient holographic gratings with additional advantages such as fewer optical elements and compact size, and the light beam acceptance angle can be matched with that of the optical fibers for increased system throughput.† The critical

---

* Komachi, Y. et al. Improvement and analysis of a micro Raman probe, 2016. https://www.osapublishing.org/ao/abstract.cfm?uri=ao-48-9-1683#figanchor12.

† Pan, M.-W. et al. Fiber-coupled high-power external-cavity semiconductor lasers for real-time Raman sensing, 1998. https://www.osapublishing.org/ao/abstract.cfm?uri=ao-37-24-5755.

‡ Chase, B., A new generation of Raman Instrumentation, *Appl. Spectrosc.*, 48(7), 14A–19A, 1994, and references cited therein. https://www.researchgate.net/publication/238187065_A_New_Generation_of_Raman_Instrumentation.

requirement of all spectrographs is temperature stability. The analyzer must be very well temperature controlled or located in a temperature-controlled environment.

Detector A charge coupled detector (CCD) is the primary choice for dispersive instruments because of its extremely low dark current at low temperatures (as low as 0.05 electrons/s at −40°C). Raman process instruments use detectors that are cooled by two- or three-stage Peltier thermoelectric coolers.

NIR Raman instruments use back-thinned CCD cameras that have higher quantum efficiency in the NIR range as compared to front-illuminated devices. A CCD camera is a 2D array device and is available in a variety of pixel formats (512 × 256, 512 × 512, 1024 × 512, 1024 × 1024, etc.). The x-axis is the dispersion axis, and the y-axis is the intensity axis. The number of x-axis pixels determines spectral resolution (if applicable) and spectral coverage. The 2D array allows simultaneous measurement of multiple channels by arranging the light inputs vertically along the y-axis. Thus, the pixel format of the y-axis is chosen to accommodate multiple optical inputs into the spectrometer.

Raman instruments with multiple inputs are available for a variety of reasons:

1. Multiple optical fibers from a single stream can be coupled into the spectrometer for increased optical throughput and sensitivity.
2. The dual-fiber optic input beam, one for referencing and one for the process stream, can be used for dual-channel referenced Raman spectroscopy.
3. A split architecture where the top and bottom portions of the CCD measure two different spectral ranges increases spectral resolution and spectral coverage.
4. In one design, up to six channels can be accommodated for simultaneous measurement of two references and four process streams.

Unlike scanning devices with which one wavelength unit is measured at a time, the 2D array format of the CCD, combined with its very high read rate on the order of megahertz, allows simultaneous measurement of all channels in real time. The advantage of simultaneous measurement is that any transient drift in laser power does not affect the spectral line shape but only overall amplitude, which can be compensated for by normalizing the spectrum.

Fiber-Optic Light Delivery and Collection Optical fibers are used to deliver light and collect Raman signals remotely. Silica fibers with a core diameter ranging from 50 to 500 μm are used. The larger the core diameter, the less the bending radius (flexibility). Raman spectroscopy uses low-hydroxy fibers to minimize fiber background interference in Raman data and also to minimize light attenuation caused by fiber absorption.

Different fiber-optic configurations are employed in Raman spectroscopy (Figure 1.58h), depending on the required sensitivity. These include one excitation and six surrounding collection fibers (circular geometry), one excitation and three collection fibers (triangle geometry), and one excitation and one collection fiber.

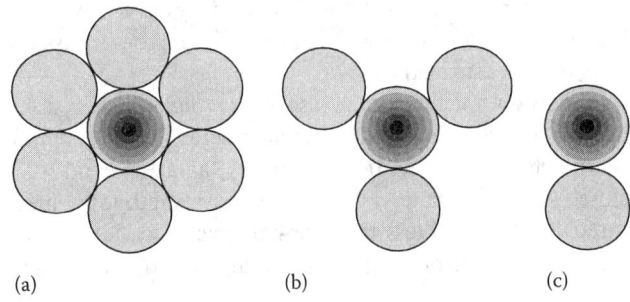

FIG. 1.58h
Various optical fiber geometries for laser excitation (dark circle) and Raman collection (gray circles). The configurations can be (a) circular, (b) triangular and (c) single collector types.

Although possessing lower sensitivity as compared to other geometries, most analyzers nonetheless incorporate one-excitation, one-collection fiber geometry. This configuration is simple to implement, is less expensive, requires no alignment in fiber-to-fiber coupling, and is easy to install in the field. With low-hydroxy fibers, light can be transmitted over hundreds of meters without a significant decrease in throughput. The main issue is that as the laser traverses through optical fibers, optical fibers generate silica Raman bands and/or a broad fluorescence background.

Proper filtering is therefore needed (see next paragraph). For field installation, optical fibers are protected with PVC or metal jackets. Fibers are terminated with SMA 905, FC, ST connectors. Reproducible fiber-to-fiber coupling needs to be considered in selecting connectors for the fiber-optic cables.

Laser and Raman Filters In order to observe inherently weak Raman signals, the laser needs to be monochromatic, that is, free of any laser sideband interference. In addition, the intense Rayleigh scattering should be completely prevented from reaching the detector. Laser and Raman filters, respectively, serve this purpose. Laser filters must be bandpass filters of very high attenuation (10^7 or more) so as to completely eliminate fiber background and Rayleigh scattering. The filter must also have very high transmission (80% or more) at the laser wavelength to pass laser light efficiently (Figure 1.58i).

Raman filters, on the other hand, are long-pass (or notch) filters. Raman filters have very high attenuation at the laser wavelength to prevent laser radiation from entering the collection fiber and the detector. Raman filters also have very high transmission at the Raman frequency region to pass the Raman light efficiently.

Analyzer manufacturers use either dielectric or holographic filters. Holographic filters are more efficient and have sharp cutoff characteristics that allow Raman spectral lines close (~100 cm^{-1} or less) to the laser wavelength to be measured. Dielectric filters that use hard metal oxide coatings are less efficient but exhibit less temperature dependence.

Selection of appropriate filter technology depends on the location of the analyzer (open field, shelter, control room), location of the filters (protected or unprotected), required sensitivity, and how close to Rayleigh scattering the Raman bands need to be measured.

Sample Interface Sample interface is critical in all process applications, because analyzer performance and reliability are highly dependent on the sampling system. (For a detailed discussion of the features and capabilities of fiber-optic probes, refer to Chapter 1.23.)

Typically, there is no need for a sample extraction, handling, and conditioning system for Raman analyzers, because an optic probe can be directly interfaced to the process stream. The probe consists of optical fibers for laser excitation and Raman signal collection. There are several excitation fiber and collection fiber geometries for use in Raman spectroscopy.* Most common for process application is one-excitation, one-collection fiber geometry, because of its simplicity and ruggedness.

Raman Probes The use of fiber-optic probes not only lowers the initial installation project costs, it requires less maintenance over the analyzer lifetime. Generally, a sampling system can

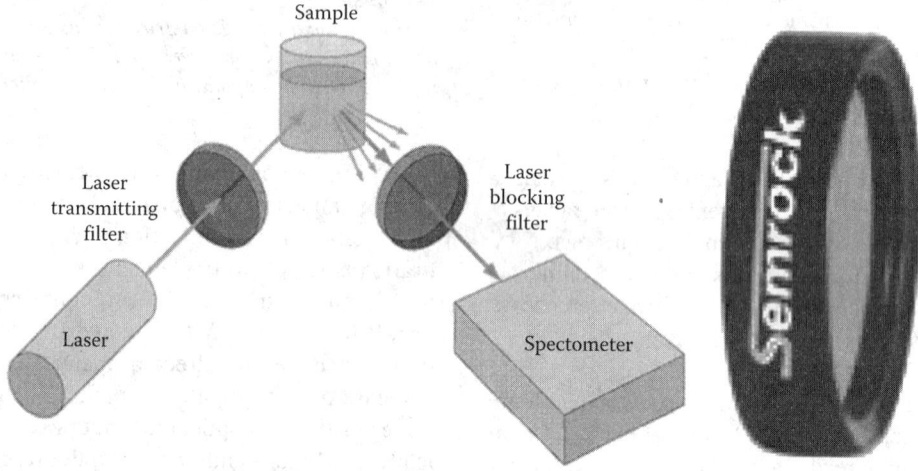

FIG. 1.58i
Raman filtering. (Courtesy of Semrock.)

---

* Lewis, I. R. and Griffiths, P. R., Raman spectrometry with fiber optic sampling, *Appl. Spectrosc.*, 50(10), 12A–30A, 1996. https://www.osapublishing.org/as/abstract.cfm?uri=as-50-10-12A.

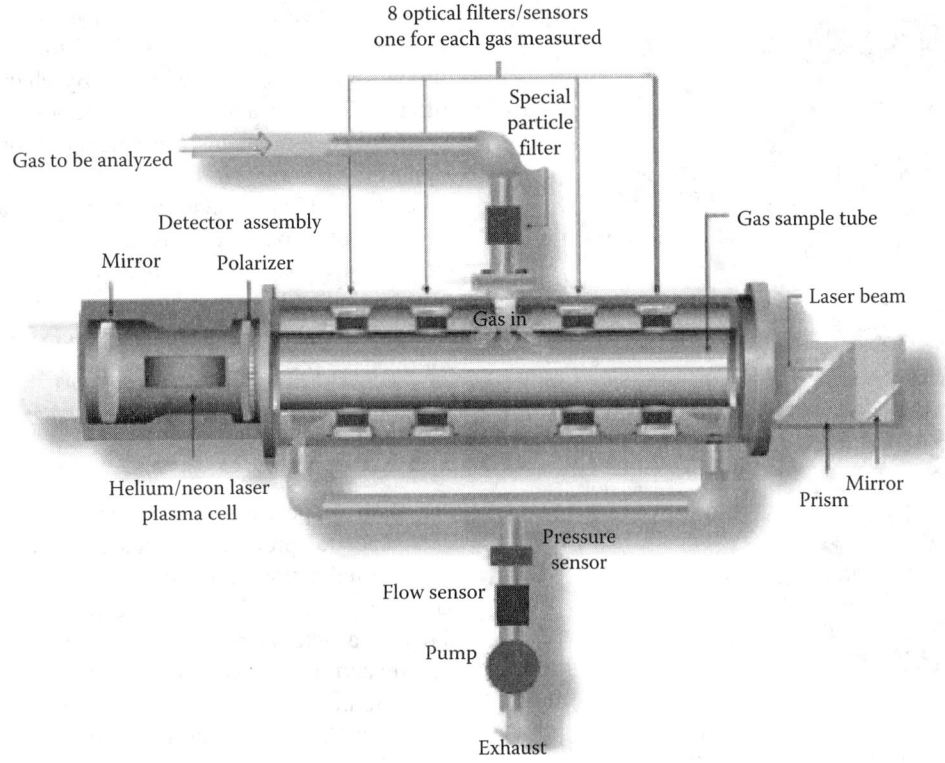

FIG. 1.58j
Raman analyzer sampling system. (Courtesy of ARI.)

require more maintenance (Figure 1.58j) than the analyzer itself. By eliminating the sampling system, a major need for continuous maintenance of the analyzer is also eliminated.

The two types of Raman probes are invasive and noninvasive. The requirements for an invasive or insertion probe are that it should be industrially hardened, have a high degree of chemical/corrosion resistance, and be able to withstand very high process temperatures and pressures. Developments in glass–metal welding and welded metal pressure seals have made available robust, invasive Raman probes that can withstand up to 350°C (662°F) and 3000 psig (206 bars) and operate under highly corrosive conditions. Different types of sleeves are available; stainless steel, titanium, Hastelloy, and Kynar can be used in corrosive applications.

Probes are available with the laser and filters either external to the probe body or integrated within the probe body. Figure 1.58k shows invasive probes with different sleeve constructions.

An insertion probe may be used in conjunction with retraction mechanism for safely retracting the probe from the

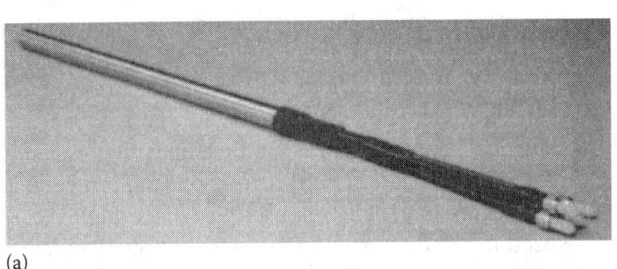
(a)

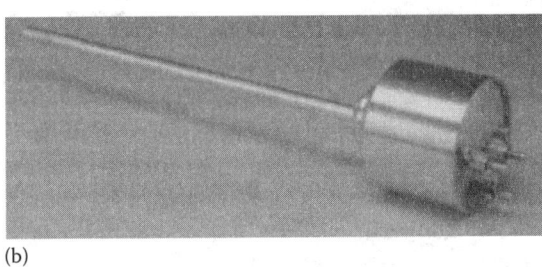
(b)

FIG. 1.58k
Raman process probes. (a) Invasive titanium sleeve probe (Courtesy of CIC Photonics, Inc.) and (b) invasive probe with stainless steel construction and gold brazing of fibers to the body (Courtesy of Rosemount Analytical, Inc.).

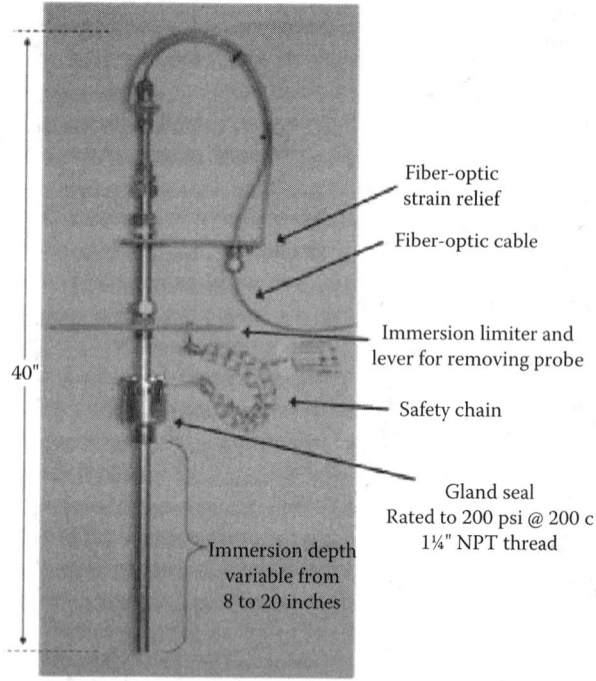

FIG. 1.58l
Sampling system for Raman analyzer. (Courtesy of Process Instruments.)

process for maintenance. Such probes are also available in the intrinsically safe design (Figure 1.58l).

In applications where the process integrity cannot be breached for hazardous or sterility reasons and where the processes are operating at extreme pressure, temperatures, or corrosive conditions, a noninvasive probe should be considered.

Typically, in noninvasive probe implementation, the probe is placed outside the process container, and the laser excitation and Raman collection are done through optically transparent windows such as quartz or sapphire. Figure 1.58m

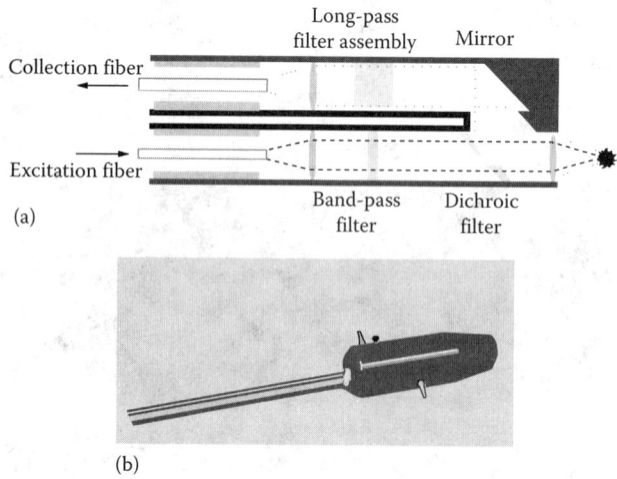

FIG. 1.58m
Focused probe with integrated laser and Raman filters. (a) Schematic of the beam path and (b) process probe for invasive and noninvasive applications. (Courtesy of InPhotonics, Inc.)

shows one such design, where a lens at the tip focuses the laser beam and also collects the Raman scattering signal through the transparent window. By changing lenses with different focal distances or by using a variable focal length assembly, measurements can be done through 12 mm thick sapphire windows.

Chlorate bleach reactions in the pulp and paper industry and nitration reactions in the chemical processing industry are examples of hazardous and corrosive processes. Cell culture monitoring and fermentation monitoring are examples of sterile processes that would use noninvasive probes.

Data Analysis

The Raman spectrum must be processed before it can be used. Spectral processing includes noise filtering, smoothing, normalization, reference compensation, fluorescence background removal, and correction for scattering, especially in a heterogeneous media. The Raman spectrum is then related to a concentration of chemical components via calibration. A simple calibration is the intensity or band area ratio method, which is internally normalized.

Polymerization Application   For instance, in a polymerization reaction, the monomer concentration would decrease as the polymer concentration increases. Thus, the ratio of bands associated with monomer and polymer molecules can be used to easily estimate the degree of polymerization. In cases where the spectrum of the mixture can be expressed as a linear superposition of the spectra of individual component line shapes and their Raman cross sections, relative concentration of various components can be easily estimated.*

Also, where the Raman signal from the component of interest can be isolated, the signal strength can be related to its concentration by measuring Raman spectra of several known standards that span the range of concentration of interest.

Multivariable Predictive Analysis   For most applications, however, a simple linear equation (e.g., Equation 1.58(1)) cannot be used because of noise, variability in measurement, spectral overlap of various bands, internal absorption, diffuse scattering, etc. For such applications, multivariate predictive statistical analysis (such as multiple linear regression [MLR], partial least squares [PLS] regression, and principal component regression [PCR]) is used to reduce highly correlated spectral information into a few orthogonal (uncorrelated) spectral vectors.

---

* Manoharan, R., Wang, T., and Feld, M. S., Histochemical analysis of biological tissues using Raman spectroscopy, *Spectrochim. Acta A*, 52, 215–249, 1996. http://www.academia.edu/19768231/Histochemical_analysis_of_biological_tissues_using_Raman_spectroscopy.

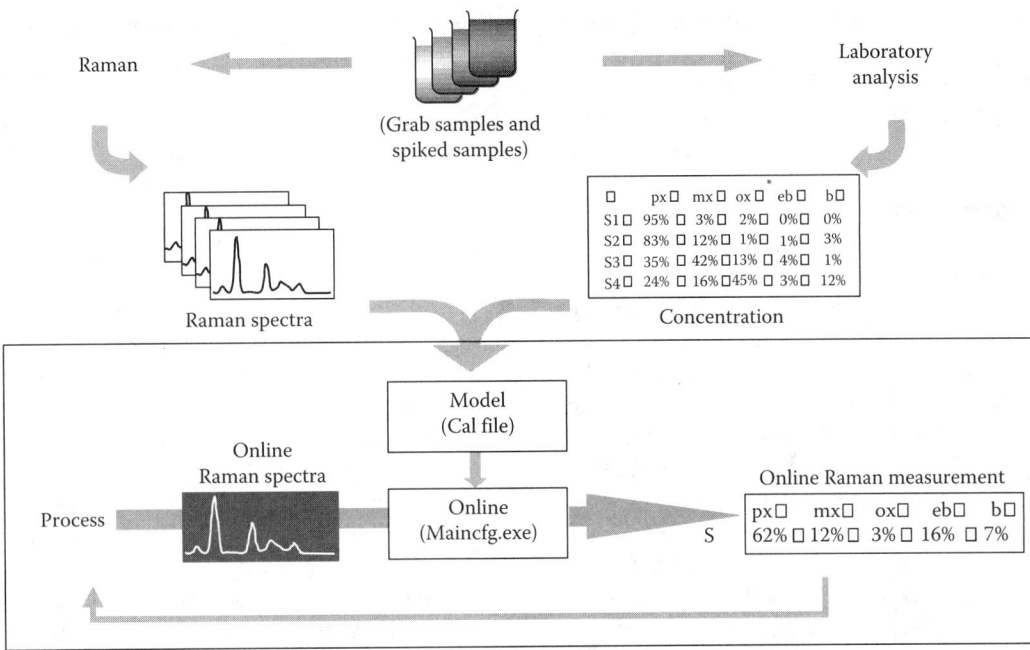

FIG. 1.58n
Analyzer calibration scheme using laboratory analysis as the standard. Calibration model is verified for accuracy and robustness and downloaded to the analyzer for online prediction.

The scaling factors of the spectral vectors are then related to concentration via multidimensional curve fitting or regression analysis.* Nonlinear analysis using an artificial neural network is also increasingly being used in Raman spectroscopy.* In the multivariate analysis, the analyzer is trained first, using a set of standard samples of known concentration.

Figure 1.58n illustrates such a calibration scheme. Samples are collected from the process (sometimes spiked with components to get the span range desired), and their Raman spectra are obtained. The same samples are then analyzed via standard laboratory chemical analysis. Spectral information and known chemical composition from the laboratory are related using a multivariate model. The model validity is then checked using unknown samples to get an estimate of prediction error. Naturally, the calibration accuracy cannot exceed that of the laboratory analysis used to build the model, although repeatability is usually better.

Calibration Model A calibration scheme built outside of the process environment should be considered preliminary only and must be fine-tuned when the analyzer is operating in an actual process. An effective way to fine-tune an existing calibration scheme, or to build one from scratch, is by periodically collecting samples for laboratory analysis while the analyzer is operational.

Correlating Raman spectral response in actual process conditions with laboratory analysis is the best way to build accurate, process-specific calibration schemes, whether based on peak integration or statistical modeling. This, however, takes time and patience to do well. Also, a good calibration model must include the extremes of the process range. For a stable manufacturing process, it can be frustrating to wait for the process to swing from one extreme to another and, perhaps more significantly, to be able to collect lab samples precisely when the process conditions are at the extremes.

It is highly desirable (if possible) to install an automatic lab sampler that can be triggered by the analyzer to collect a sample for lab analysis when the analyzer detects a value in the extreme of its detection range or detects unusual process conditions that are not built into the preliminary model. Such a trigger can be based on predicted data limits or on modeling parameters such as spectral residuals (a measure of how well the model-calculated spectrum matches with the measured spectrum) and/or Mahalanobis distances (a measure of how close the measured spectrum is from the mean of all the data used in the calibration, measured in standard deviation). Clearly, whenever a sample is collected for laboratory analysis, the concurrent Raman spectrum should be saved for diagnostics and for inclusion in future calibration updates.

When the analyzer is fully operational with a satisfactory calibration scheme, it is good practice to continue to routinely check its calibration status. Again, this can be done best with collected samples and laboratory analysis. Use of an automated sampler is preferred to avoid human errors in sample collection. The calibration check frequency can be determined based on process stability, calibration scheme stability, and failure rate. Clearly, access to a well-equipped and efficient laboratory is essential for building and maintaining good calibration schemes for Raman analyzers.

---

* Martens, H. and Naes, T., *Multivariate Calibration*, John Wiley & Sons, New York, 1992. http://www.wiley.com/WileyCDA/WileyTitle/productCd-0471930474.html.

Calibration Transfer Generally, a calibration model is valid as long as the analyzer components are not modified. Only a few available analyzers are capable of calibration transfer, after major analyzer maintenance, between two streams on the same analyzer or between two different analyzers made by the same vendor.

Two major approaches to calibration transfer are (1) multivariate standardization, where the spectral domain of one analyzer is mapped onto that of the other analyzer by modeling spectral differences and correcting for those differences, and (2) analyzer standardization, where all analyzers are made to yield identical spectra from a given sample using standards.

Raman analyzer standardization itself is a multistep process.* First, laser instability is compensated using, for example, diamond as Raman reference. Second, the CCD pixel in the x-axis is converted to Raman shift frequency by calibrating it with known emission lines of argon, neon, or one of the recent American Society for Testing and Materials (ASTM) standards (sulfur, cyclohexane, polystyrene, etc.) and with a known Raman line of diamond. Third, intensity response of the analyzer is corrected over the entire Raman shift range using a white light source or fluorescence color glasses.[†]

Properly done analyzer standardization ensures spectral reproducibility and facilitates calibration transfer. Calibration transfer reduces redundancy by eliminating the need to develop an entirely new calibration model where one already exists for a similar analyzer and process. Calibration transfer saves tremendous amounts of time and effort in developing a new calibration model for each occasion when analyzer maintenance is performed or a new channel or a new analyzer is installed for the same process application.

Diagnostics and Maintenance

An analyzer that has good diagnostic tools allows for safe operation and reduces analyzer downtime. Two areas of diagnostics are calibration model performance and monitoring the mechanical/electrical health of the analyzer.

For multivariate calibration models, spectral residuals and Mahalanobis distance are indicative of the spectral quality and data reliability. Spectral shape changes can provide a variety of diagnostic information such as process change, insufficient process flow, laser performance, and probe fouling.

Analyzer health parameters that should be automated include laser life (based on output power/current), fan life (if present), laser current, laser operating temperature, analyzer internal temperature, CCD operating temperature, and loss of purge air. Availability of these parameters to DCS and asset management systems can automatically trigger analyzer maintenance and repair. A Raman analyzer maintenance scheme should include scheduled preventive maintenance to replace analyzer parts that have known average operational life and rigorous data quality monitoring protocols that can identify problems.

Laser Safety

Raman analyzers use Class IIIB to Class IV lasers. Personnel laser safety is of paramount importance in process implementation. Raman analyzers are available with safety features that warn of laser status and can initiate automatic shutdown. In process applications, the analyzer must be capable of shutting off the laser automatically or reducing the energy to nonhazardous level when internal analyzer diagnostics indicate a potential disruption of the laser loop or when the possibility of personnel exposure exists.

In one scheme, a photodiode in the analyzer continuously monitors the Rayleigh scattering light from the process picked up by a dedicated optical fiber in the probe. When the probe is withdrawn or the laser cable is broken, the absence of Rayleigh scattering immediately triggers safe-mode operation. When the fault condition is cleared, the laser automatically returns to normal mode of operation.

Another consideration is the possibility of the laser inducing hot spots in the process line or the sampling cell. Hot spots are an important safety consideration if the process can potentially contain solid particulates or explosive gas mixtures that absorb at the laser frequency. The possibility then exists that particulates or gas mixtures may be heated to high temperatures by the laser and become an explosion hazard in the presence of hydrocarbons.

Although the relative mass of process flow provides a sufficiently large heat sink to eliminate the risk of explosion, caution is prudent if the nature of the process stream warrants it. It is generally best to do a thorough safety review before installing a Raman analyzer and to develop safety guidelines for each installation.

Outputs and Communication

Raman analyzers provide industry-standard communication protocols. Generally, a Modbus or Ethernet protocol is preferred over analog (4–20 mA) connection, because a large amount of information can be transmitted and monitored remotely, including a number of safety and instrument health factors. Ethernet capability also allows for remote access of the analyzer for technical experts to diagnose and rectify problems both within the plant and from the vendor site.

Packaging

Because Raman analyzer technology is still relatively new compared to, for example, NIR and GC technologies, standards

---

* Shaheen, M. and Manoharan, R., Raman process analyzer calibration transfer through analyzer standardization, *SPIE Proc.* 3859, 24–28, 1999. http://proceedings.spiedigitallibrary.org/proceeding.aspx?articleid=909764.

[†] McCreery, R. L., Photometric standards for Raman spectroscopy, in *The Handbook of Vibrational Spectroscopy*, Chalmers, J. M. (Ed.), 2002. http://www.wiley.com/WileyCDA/WileyTitle/productCd-0471988472.html.

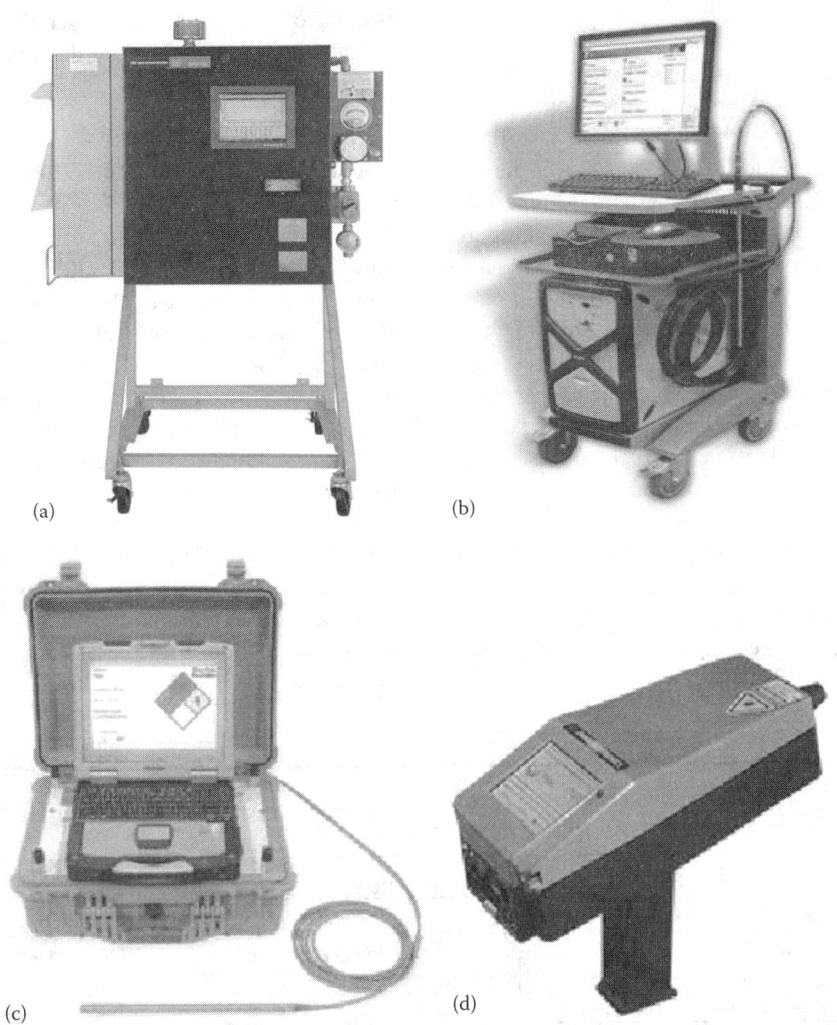

FIG. 1.58o
Packaging options of Raman analyzers: (a) multichannel analyzer (Courtesy of AIT), (b) portable benchtop analyzer (Courtesy of Kaiser Optical Systems), (c) portable Raman analyzer with 5 m optic probe (Courtesy of Real-Time Analyzers), and (d) handheld analyzer (Courtesy of TSI).

for analyzer packaging are still evolving. Since Raman analyzers are compatible with fiber-optics for remote analysis, some vendors offer benchtop analyzers that can be located in a control room. However, if the analyzer is located very far from the process stream, then the cost of packaged fiber-optics can be significant, especially if multifiber geometry is used. The various forms of available packaging are shown in Figure 1.58o.

A few Raman analyzers are available with NEMA12 and NEMA 4X ratings. Z-purged analyzers for hazardous applications are also available.

Installation and Maintenance

How a Raman process analyzer is installed and calibrated to a large extent defines its future maintenance requirements. While it is true that there is a correlation between installation and future maintenance for any process analyzer, this is particularly true for Raman analyzers because of the range of installation options.

Whether one chooses to install a complicated sampling system, using an invasive or noninvasive probe, has significant impact on future maintenance needs of a Raman analyzer. Other factors, such as whether the analyzer hardware is located in a hazardous or nonhazardous area and whether the analyzer is housed in an analyzer shelter or in a freestanding purged cabinet, also affect the analyzer's maintenance needs.

Beyond the specific maintenance requirements arising from how the analyzer was installed, the other maintenance requirements for the Raman analyzer tend to be minimal and pertain only to replacing consumable items such as filter and laser on a scheduled basis or when they break down.

Probe Designs

The most cost-effective way to install and operate a Raman analyzer is by using an invasive or noninvasive probe. The use of fiber-optic probes not only lowers the initial installation project costs; it also requires less maintenance over the analyzer lifetime. Generally, a sampling system requires more maintenance than the analyzer itself. By eliminating the sampling system, a major need for continuous maintenance of the analyzer is also eliminated.

When using a fiber-optic probe, the other analyzer hardware (e.g., electronics, detector, laser, computer) can be located in a nonhazardous area by using fiber-optic cable to connect the probe. Such an installation scheme eliminates the need for an explosion-proof purged cabinet or an expensive customized analyzer shelter house. Analyzer installation project costs are always a significant factor for a manufacturing site.

All Raman analyzer installations should include an easily accessible sampling point to collect process samples for laboratory analysis. Laboratory analysis is generally required for verifying and updating analyzer calibration. A good practice is to have a sampling feature that allows automatic sampling for laboratory analysis when the analyzer indicates unusual process conditions.

ADVANTAGES AND DISADVANTAGES

A decision to use a Raman analyzer depends on the availability and practicality of alternative analyzer technologies. For example, it is clearly not desirable to employ a Raman analyzer where a GC can perform the same analysis. This is because a GC is more sensitive, more familiar to the average user, and more widely accepted. Therefore, unless a good case can be made based on speed of analysis or low-maintenance requirements of the Raman analyzer, a GC would be preferred.

Generally, a Raman analyzer works where NIR and mid-IR analyzers are applicable. The Raman analyzer, however, combines the best aspects of NIR and IR absorption spectroscopy methods with some of its unique advantages. Table 1.58b lists the advantages and disadvantages

TABLE 1.58b
Advantages and Disadvantages of Mid-IR, Near-IR, and Raman Methods

	NIR Absorption	Mid-IR Absorption	Raman Scattering
Selection rule	Change in dipole moment	Change in dipole moment	Change in polarizability
Spectral band profile	Overtone/combination bands, broad (nonspecific)	Fundamental band, narrow (high specificity)	Fundamental bands, narrow (high specificity)
Sensitivity	Moderate (ppm, low percentage)	Very high (ppb, ppm)	Low (low percentage, ppm in favorable cases)
Qualitative identification	Difficult	Yes. Extensive spectral library available	Yes. Comprehensive library not available
Sample	Transparent, bubble, and particulate-free liquids. Powder samples that require diffuse reflectance probe	Transparent liquid, thin solid pellets, and ideal for gases. Special ATR probes needed	Same probe used for all samples: liquids, slurries, emulsions, powders, solids, and samples with particulates and bubbles
Signal-to-concentration relationship	Logarithmic	Logarithmic	Linear
Quantitative analysis	Mathematical/statistical modeling	Simple arithmetic possible	Simple arithmetic possible
Transfer of calibrations	Difficult	Difficult	Relatively easy
Sample handling	Yes	Yes	No, noninvasive or in situ
Water interference	Good sensitivity to and interference by water	Very high sensitivity to and strong interference by water	No or minimal water interference
Temperature dependence	Sensitive to process temperature variation	Sensitive to process temperature	Wide temperature variation tolerated
Sampling probe	Double sided or single sided when used with reflectance	Double sided or special ATR probes	Single ended
Probe fouling	More likely	More likely	Less likely or not likely with noninvasive
Remote via fiber-optics	Long fibers, low-hydroxy silica fibers	Short fibers, expensive chalcogenide fibers	Long fibers, low-hydroxy silica fibers
Wetted optical window	Inert (sapphire, silica, quartz)	Fragile (ZnSe, CaF2, KBr)	Inert (sapphire, silica, quartz)

of Raman spectroscopy compared to mid-IR and NIR absorption spectroscopies.

Advantages

Like mid-IR, Raman bands represent fundamental modes of vibration. Bands are narrow, are molecule specific, and provide quantitative chemical analysis. Like NIR absorption, Raman spectroscopy permits the use of optical fibers for remote analysis. Additionally, the following unique advantages of Raman spectroscopy can be noted:

1. It involves minimal or no water interference, an important feature because there is an increasing trend to move away from organic-based to water-based processes due to environmental considerations.
2. The technology is equally amenable to a variety of samples ranging from gases, liquids, slurries, emulsions, powders, and solids to samples with particulates and bubbles. This enables Raman use in heterogeneous systems such as catalytic process, emulsion polymerization, fermentation monitoring, and cell culture monitoring.
3. Noninvasive probes can be used and are a key feature in many applications where the process must not be breached.

Limitations

The limitations of Raman spectroscopy are its low sensitivity as compared to IR absorption and fluorescence interference from analytes or impurities in the sample. By optimizing sample and collection geometry, sensitivity can be improved, and by utilizing NIR radiation as the excitation source, fluorescence can be minimized.

APPLICATIONS

Raman analyzer technology can be applicable for chemical composition analysis in a range of manufacturing industries (e.g., specialty chemicals, petrochemicals, petroleum refining, pharmaceutical, agriculture, and pulp and paper). Raman spectroscopy is a developing technology, and a good amount of research and planning is necessary before deciding on employing Raman technology for a process analyzer. In this regard, it is best for industrial technical experts and analyzer technology providers to work together to develop new applications for Raman analyzers.

One application area in which Raman technology is distinctly advantageous is for the analysis of viscous liquids, for example, polymer melts. Use of a noninvasive probe precludes the necessity of building a sample-handling system. Sampling system designs for handling solids or highly viscous liquids tend to be very expansive and costly, and they have lower reliability than those for free-flowing liquids and vapors.

For polymers and, in general, for macromolecules, it is typically not the chemical composition but the relationship of the chemical composition to the physical properties of bulk density, viscosity, etc., that is more valuable from a process viewpoint. Raman analysis of the chemical signature of the process stream generally can be correlated to the physical properties using multivariate statistical models. This is one area where the speed, stability, and reliability of Raman analysis offer an attractive means for real-time process control.

The range of applications in which Raman analyzer technology can be employed to provide valuable real-time process information is extensive. A summary of typical Raman applications is provided in Table 1.58c.

TABLE 1.58c
Typical Process Applications for Raman Spectroscopy

Industry	Process Applications
Chemical processing	Endpoint determination
	Reaction monitoring
	Hydrocarbon analysis
	Emulsion polymerization
	Copolymer reactions
	Heterogeneous catalytic reactions
	Nitration reactions
	Esterification reactions
	Batch and continuous distillations
	Chemical reactor safety and efficiency monitoring
	Polymer crystallinity
	Inorganic analysis
Petrochemical	Monitoring xylene separation process
	Paraxylene purity analysis
	Styrene monomer/polymer reaction
	Endpoint determination
	Reaction monitoring
Pharmaceutical	Fermentation
	Cell culture monitoring
	Bromination reaction
	Raw material verification and screening
	Intermediate analysis
Pulp and paper	Pulp bleaching
	Chlorine dioxide production
Food and beverage	Modified starch
	Hydrogenation of oils and fats
	Sugars
	Endpoint determination
	Fermentation

CONCLUSION

The cost of a Raman process analyzer typically exceeds that of other analyzers used in the process industry.

The CCD camera and lasers are the most expensive components of the Raman analyzer. Indeed, to reduce cost on a per-process-stream basis, Raman analyzers often include multichannel capability. Up to four process streams can be analyzed with a single CCD camera by splitting the lasers. Advances in CCD and laser technologies continue to lower the cost of these components, and that will help to reduce overall analyzer cost.

Raman calibration transfer is an important technical area that needs to be improved. Currently, there is no traceability of standards used in x- and y-axis calibration of Raman spectrum. Progress in this direction is being made as ASTM has adapted eight standards for x-axis calibration of Raman spectrum.

The availability of inexpensive, rugged process probes that can work reliably within a high-pressure, high-temperature, and corrosive process environment will enable the Raman analyzers to be used in a wider range of process applications.

There is a need to improve the sensitivity of the technique. Efficient light collection by probes and effective coupling of the Raman signal to the spectrograph, among other things, would enhance Raman sensitivity. Multipass sample cell design should increase Raman sensitivity, especially for gases. Experimental optical probe designs have already been demonstrated for gas sampling using Raman analyzers.

Another inherent problem with Raman spectroscopy is fluorescence interference. As described earlier, the choice of CCDs as low-noise detectors does not allow excitation wavelengths above 850 nm because of the lack of sensitivity in the corresponding Raman region. For complete avoidance of fluorescence, excitation wavelengths above 1000 nm are needed. A number of applications will become germane for Raman analysis by exciting and collecting light beyond 1000 nm and using new generation of CCD cameras.

Continuing efforts to improve detection limits, fluorescence elimination, traceability of calibration standards, calibration capability, mechanical reliability, and lower cost, coupled with its inherent technical advantages, portend a bright future for Raman technology for online process analysis.

SPECIFICATION FORMS

When specifying Raman analyzers, the ISA Form 20A1001 can be used to specify the features required for the device itself, and when specifying both the device and the nature of the application, including the composition and/or properties of the process samples, ISA Form 20A1002 can be used. With the permission of the International Society of Automation, both forms are reproduced on the next pages.

1	RESPONSIBLE ORGANIZATION		ANALYSIS DEVICE		6	SPECIFICATION IDENTIFICATIONS	
2					7	Document no	
3	ISA		Operating Parameters		8	Latest revision	Date
4					9	Issue status	
5					10		

	ADMINISTRATIVE IDENTIFICATIONS					SERVICE IDENTIFICATIONS Continued		
11					40			
12	Project number		Sub project no		41	Return conn matl type		
13	Project				42	Inline hazardous area cl	Div/Zon	Group
14	Enterprise				43	Inline area min ign temp	Temp ident number	
15	Site				44	Remote hazardous area cl	Div/Zon	Group
16	Area		Cell	Unit	45	Remote area min ign temp	Temp ident number	
17					46			

	SERVICE IDENTIFICATIONS				COMPONENT DESIGN CRITERIA	
18				47		
19	Tag no/Functional ident			48		
20	Related equipment			49	Component type	
21	Service			50	Component style	
22				51	Output signal type	
23	P&ID/Reference dwg			52	Characteristic curve	
24	Process line/nozzle no			53	Compensation style	
25	Process conn pipe spec			54	Type of protection	
26	Process conn nominal size	Rating		55	Criticality code	
27	Process conn termn type	Style		56	Max EMI susceptibility	Ref
28	Process conn schedule no	Wall thickness		57	Max temperature effect	Ref
29	Process connection length			58	Max sample time lag	
30	Process line matl type			59	Max response time	
31	Fast loop line number			60	Min required accuracy	Ref
32	Fast loop pipe spec			61	Avail nom power supply	Number wires
33	Fast loop conn nom size	Rating		62	Calibration method	
34	Fast loop conn termn type	Style		63	Testing/Listing agency	
35	Fast loop schedule no	Wall thickness		64	Test requirements	
36	Fast loop estimated lg			65	Supply loss failure mode	
37	Fast loop material type			66	Signal loss failure mode	
38	Return conn nominal size	Rating		67		
39	Return conn termn type	Style		68		

	PROCESS VARIABLES	MATERIAL FLOW CONDITIONS					PROCESS DESIGN CONDITIONS			
69						101				
70	Flow Case Identification				Units	102	Minimum	Maximum		Units
71	Process pressure					103				
72	Process temperature					104				
73	Process phase type					105				
74	Process liquid actl flow					106				
75	Process vapor actl flow					107				
76	Process vapor std flow					108				
77	Process liquid density					109				
78	Process vapor density					110				
79	Process liquid viscosity					111				
80	Sample return pressure					112				
81	Sample vent/drain press					113				
82	Sample temperature					114				
83	Sample phase type					115				
84	Fast loop liq actl flow					116				
85	Fast loop vapor actl flow					117				
86	Fast loop vapor std flow					118				
87	Fast loop vapor density					119				
88	Conductivity/Resistivity					120				
89	pH/ORP					121				
90	RH/Dewpoint					122				
91	Turbidity/Opacity					123				
92	Dissolved oxygen					124				
93	Corrosivity					125				
94	Particle size					126				
95						127				
96	CALCULATED VARIABLES					128				
97	Sample lag time					129				
98	Process fluid velocity					130				
99	Wake/natural freq ratio					131				
100						132				

	MATERIAL PROPERTIES				MATERIAL PROPERTIES Continued		
133				137			
134	Name			138	NFPA health hazard	Flammability	Reactivity
135	Density at ref temp	At		139			
136				140			

Rev	Date	Revision Description	By	Appv1	Appv2	Appv3	REMARKS

Form: 20A1001 Rev 0 © 2004 ISA

Analytical Measurement

	RESPONSIBLE ORGANIZATION	ANALYSIS DEVICE COMPOSITION OR PROPERTY Operating Parameters (Continued)		SPECIFICATION IDENTIFICATIONS	
1			6		
2	ISA		7	Document no	
3			8	Latest revision	Date
4			9	Issue status	
5			10		

	PROCESS COMPOSITION OR PROPERTY			MEASUREMENT DESIGN CONDITIONS				
	Component/Property Name	Normal	Units	Minimum	Units	Maximum	Units	Repeatability
11								
12								
13								
14								
15								
16								
17								
18								
19								
20								
21								
22								
23								
24								
25								
26								
27								
28								
29								
30								
31								
32								
33								
34								
35								
36								
37								

(Rows 38–75: blank remarks/notes area)

Rev	Date	Revision Description	By	Appv1	Appv2	Appv3	REMARKS

Form: 20A1002 Rev 0

© 2004 ISA

Definitions

Diffraction gratings are optical components used to separate light into its component wavelengths. When there is a need to separate light of different wavelengths with high resolution, a diffraction grating is the most often used tool of choice. A diffraction grating is made by making many (up to 6000 lines/cm) parallel scratches on the surface of a flat piece of transparent material. The scratches are opaque, but the areas between the scratches transmit light.

Elastic scattering of light is also called Rayleigh scattering and is a scattering mechanism that results in the uniform scattering of the incoming light by atoms or molecules in all directions.

Light scattering is the deflection of a light ray from a straight path. Most objects that one sees are visible due to light scattering from their surfaces. The scattering of light depends on the wavelength of the light being scattered.

Michelson interferometer was invented by Albert Abraham Michelson. It uses a beamsplitter in which a light is split into two arms. Each of those is reflected back toward the beamsplitter which then combines their amplitudes interferometrically. The resulting interference pattern is typically directed to some type of photoelectric detector or camera.

Monochromatic light is light that of one color; in other words it is light having a single wavelength.

Raman scattering: Raman scattering or the Raman effect is the inelastic scattering of a photon. It was discovered by C. V. Raman and K. S. Krishnan in liquids and by G. Landsberg and L. I. Mandelstam in crystals. The effect had been predicted theoretically by A. Smekal in 1923.

Raman scattering was discovered in 1928, by Chandrasekhara Venkata Raman. It can be used as a tool to characterize any nonmetallic material. It is the shift in wavelength of the inelastically scattered radiation that provides the chemical and structural information contained in the scattered light. The Raman shifted photons can have higher or lower energy, depending upon the vibrational state of the molecule under study. When the sample is irradiated by a monochromatic laser light, having a wavelength from 532 to 785 nm, the resulting Raman signal is weak and require sensitive detectors to be measured.

Raman spectroscopy: It is a spectroscopic technique used to observe vibrational, rotational, and other low-frequency modes in a system. It relies on inelastic scattering, or Raman scattering, of monochromatic light, usually from a laser in the visible, near-infrared, or near-ultraviolet range. The laser light interacts with molecular vibrations, phonons, or other excitations in the system, resulting in the energy of the laser photons being shifted up or down. The shift in energy gives information about the vibrational modes in the system.

Raman spectrum: It is the change in wavelength of light scattered while passing through a transparent medium, the collection of new wavelengths being characteristic of the scattering medium and differing from the fluorescent spectrum in being much less intense and in being unrelated to the absorption of the medium.

Stokes shift named after the Irish physicist George G. Stokes, is the difference in frequency between maximum positions of the absorption (fluorescence) and the emission (Raman) spectra of the same electronic transition. When an atom absorbs energy, it enters an excited state, and one way for it to relax is to lose energy by emitting a photon. When the emitted photon has less energy than the absorbed photon, this energy difference is called the Stokes shift. If the emitted photon has more energy, the energy difference is called an anti-Stokes shift.

Abbreviations

CCD	Charge-coupled device
FT	Fourier transform
FTIR	Fourier transform infrared
GC	Gas chromatography
MLR	Multiple linear regression
NA	Numerical aperture
NIR	Near infrared
PCR	Principal component regression
PLS	Partial squares regression
SERS	Surface-enhanced Raman scattering
S/N	Signal-to-noise ratio

Organization

ASTM American Society for Testing and Materials

Bibliography

Azom. Identification and characterization of polymers using Raman spectroscopy, 2014. www.azom.com/article.aspx?ArticleID=10279.

Chalmers, J. M., *Handbook of Vibrational Spectroscopy*, 2002. http://www.wiley.com/WileyCDA/WileyTitle/productCd-0471988472.html.

Chase, B., A new generation of Raman instrumentation, 1994, and references cited therein. https://www.researchgate.net/publication/238187065_A_New_Generation_of_Raman_Instrumentation.

Clarke, R. H., Low-resolution Raman Spectroscopy as an Analytical Tool for Organic Liquids, http://52ebad10ee97eea25d5e-d7d40819259e7d3022d9ad53e3694148.r84.cf3.rackcdn.com/UK_OCE_Raman_Systems_Article_DS.pdf.

Da Silva, S. W., Raman spectroscopy of cobalt ferrite nanocomposite in silica matrix prepared by sol–gel method, 2006. http://www.sciencedirect.com/science/article/pii/S002230930600281X.

Horiba, Process Raman systems, 2015. https://www.horiba.com/fileadmin/uploads/Scientific/Documents/Raman/processbro.pdf.

Kempfert, K. D., Jiang, E. Y., Oas, S., Coffin, J., Detectors for Fourier transform spectroscopy, 2001. http://mmrc.caltech.edu/FTIR/Nicolet/DetectorsforFTIR1204.pdf.

Komachi, Y. et al., Improvement and analysis of a micro Raman probe, 2016. https://www.osapublishing.org/ao/abstract.cfm?uri=ao-48-9-1683#figanchor12.

Lenain, B., Raman spectroscopy, a valuable tool, 2010. http://depts.washington.edu/cpac/Activities/Meetings/Satellite/2010/Tuesday/Lenain_CPAC%20ROME2010R.pdf.

Lewis, I. R. and Edwards, G. M., Eds., *Handbook of Raman Spectroscopy: From the Research Laboratory to the Process Line*, Marcel Dekker, New York, 2001. http://onlinelibrary.wiley.com/doi/10.1002/jrs.1117/abstract.

Lewis, I. R. and Griffiths, P. R., Raman spectrometry with fiber optic sampling, *Appl. Spectrosc.*, 50(10), 12A–30A, 1996. https://www.osapublishing.org/as/abstract.cfm?uri=as-50-10-12A.

Long, D. A., *The Raman Effect: A Unified Treatment of the Theory of Raman Scattering by Molecules*, John Wiley & Sons, New York, 2002. http://www.wiley.com/WileyCDA/WileyTitle/productCd-0471490288.html.

Maruszewski, K., Raman Spectra of Molecules Adsorbed on Ag Centers in Sol-Gel Matrices, 2003. http://kmim.wm.pwr.edu.pl/blog/wp-publications/maruszewski2003/.

McCreery, R. L., *Raman Spectroscopy for Chemical Analysis*, Wiley-Interscience, New York, 2001. http://www.amazon.com/Spectroscopy-Chemical-Analysis-Richard-McCreery/dp/0471252875.

Oxford Instruments, Raman spectroscopy, 2013. http://www.andor.com/learning-academy/raman-spectroscopy-an-introduction-to-raman-spectroscopy.

Pan, Ming-Wei, et al., Fiber-coupled high-power external-cavity semiconductor lasers for real-time Raman sensing, 1998. https://www.osapublishing.org/ao/abstract.cfm?uri=ao-37-24-5755.

Schulte, A. and Guo, Y., Laser Raman spectroscopy, 2006. http://link.springer.com/chapter/10.1007/0-387-37590-2_15.

Semrock, Filter types for Raman spectroscopy applications, 2014. http://www.semrock.com/filter-types-for-raman-spectroscopy-applications.aspx.

Silva, E., Raman spectroscopy: a comprehensive review, 2013. http://www.academia.edu/1131363/Raman_Spectroscopy_A_Comprehensive_Review.

1.59 Refractometers

J. E. BROWN (1969, 1982)   **B. G. LIPTÁK** (1995, 2003, 2017)

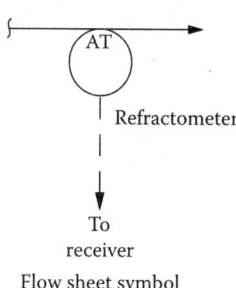

Refractometer

To receiver

Flow sheet symbol

Applications	Liquids and slurries. Measures the concentrations of dissolved solids or water-soluble liquids in the Food/Beverage, Chemical, Pharmaceutical, Sugar, Textile, Paper, Semiconductor, Coolant and Lubricant industries. The differential refractometer is more sensitive than the critical angle ones, but they require a sample-handling system and are limited to applications in which the sample is clean. The critical angle refractometer can monitor dirtier streams and can be installed in line.
Materials of construction	Stainless steel and glass are standard, with Teflon®? gaskets
Operating pressure	Up to 1000 psig (70 barg)
Operating temperature	Up to 1371°C (2500°F)
Recommended flow velocity	For in-line units in excess of 5 fps (1.5 m/sec)
Ranges	*Spans of industrial in-line units*: from 1.25 to 1.65 RI or 0–100% Brix (1.33–1.54 RI) *Battery Acid application*: 1.00–1.30 RI (battery acid) *Urine*: 1.333–1.360 RI *Other*: 1.333–1.520, 1.435–1.520, 1.300–1.700 RI In percent salinity, 0%–10% and 0%–28% In % Brix units, 0%–10%, 0%–20%, 0%–32%, 28%–62%, 58%–90%. 0%–90% Brix for detecting freezing point of coolants, −51°C to 0°C (−60°F to 32°F) In window-washer-fluid concentration, 0%–100% Brix
Speed of response	3–30 sec
Drift	From 0.5% per month to 1% per 24 hr.
Inaccuracy	The inaccuracy of a good industrial in-line unit, having a span of 0.08 RI, is about 0.0002 RI or 0.25% FS. The least accurate units have an error of 5% of span. For hand-held laboratory units in % Brix units, for a range of 0%–95%, the error is 0.1% FS (0.05% FS in high-precision refractometers). The corresponding errors in RI units are 0.0002 and 0.0001, respectively.
Costs	Portable, hand-held, battery-operated refractometer, $100–$600 average $200 Specialized, digital, hand-held units for battery acid, antifreeze, and window washer applications cost about $1000 High-precision benchtop laboratory units cost about $3000, (refurbished $2000) Industrial in-line transmitters cost about $7500 In-line sensors in sizes up to 3 in. (75 mm) cost $8,000 and 4 in. (100 mm) cost $10,000 A beverage analysis system with ±0.02 Brix inaccuracy, cost about $25,000

(Continued)

Partial list of suppliers

AFAB Enterprises (% solids) http://www.refractometer.com/index.php/pr-111 Anton Paar (smart, critical angle) http://www.anton-paar.com/corp-en/footer/search/?tx_solr%5Bq%5D=refractometers&store_context=
APT Instruments (sugar, protein, glycol) http://aptinstruments.com/Merchant2/merchant.mvc?Screen=CTGY&Store_Code=AI&Category_Code=RI
Bellingham+Stanley (smart, juice) http://www.bellinghamandstanley.com/ltd/refractometers.html
Camlab Ltd. (automatic, digital) http://www.camlab.co.uk/automatic-digital-refractometer-rfm300-series-p14615.aspx
Cole-Parmer Instrument Co (Digital, bench) http://www.coleparmer.com/Product/ATAGO_Digital_Pocket_Refractometer_0_to_93_Brix/EW-02941-53
Electron Machine Corp. (paper liquor, acid, glucose, binary mixtures) http://www.electronmachine.com/products_e-scan-refractometer.php
Extech Instrument Corp. (sucrose, salinity, battery) http://www.extech.com/instruments/categories.asp?catid=17
Hanna Instrument Inc. (sugar, glycol, salt water) http://hannainst.com/usa/search_results.cfm?q=refractometers
K-Patents Inc. (on-line, digital, paper pulp) http://www.kpatents.com/dd23_pulp.php
Kruess Optronic (process, online, digital) http://www.kruess.com/laboratory/products/refractometers/process-refractometers/
Kyoto Electronics Mfg. (portable, Brix) http://www.kyoto-kem.com/en/products/refract/index.html
Mettler-Toledo Ltd. (Brix, Baume, SpG) http://us.mt.com/us/en/home/products/Laboratory_Analytics_Browse/Refractometry_Family_Browse_main.html
Misco (smart, inline) http://www.misco.com/
NSG Precision Cells (handheld, benchtop, digital) http://www.precisioncells.com/search/results/refractometer/1
Reichert Technologies (Benchtop, automatic, digital) http://www.reichert.com/products.cfm?pcId=315
Rudolph Research (Food, Beverage, Chemical, Petroleum, and Pharmaceutical) http://rudolphresearch.com/products/refractometers/
Schmidt & Haensch (digital, inline) http://www.schmidt-haensch.com/en/products/prod/process-instruments//ipr-compact-1/
Vee Gee Scientific Inc.—Smith & Haensch (benchtop) http://www.veegee.com/refractometers.html

INTRODUCTION

Refraction is the change in the direction of light (or other waves) at the interface of two transmission mediums. When an incident wave passes from one medium to another at any incident angle (α) other than perpendicular, its phase velocity changes, while its frequency remains constant, This results is a change in direction, measured as the angle of refraction (β).

Phase velocity of a wave is the velocity at which a phase point on it (such as the crest) travels. The phase velocity (v_p) is the ratio of the wavelength lambda (λ) and the period of oscillation (T), so that $v_p = \lambda / T$.

A pencil standing in water appears to be broken at the water line. If sugar is added to the water, the pencil will appear to bend even more. This is because light travels slower in water than through air, and even slower when materials are dissolved in water. This phenomenon makes refractometers widely useable in many industries, including the Food/Beverage, Chemical, Pharmaceutical, Sugar, Textile, Paper, Semiconductor, Coolant and Lubricant industries.

The refractive index is the ratio of the velocity of light in a vacuum to its velocity in a particular substance (Figure 1.59a). The refractive index (RI) of a substance is a dimensionless number that describes how fast light, propagates through that medium. It is defined as $RI = N' = C/V$, where C is the speed of light in vacuum and V is the speed

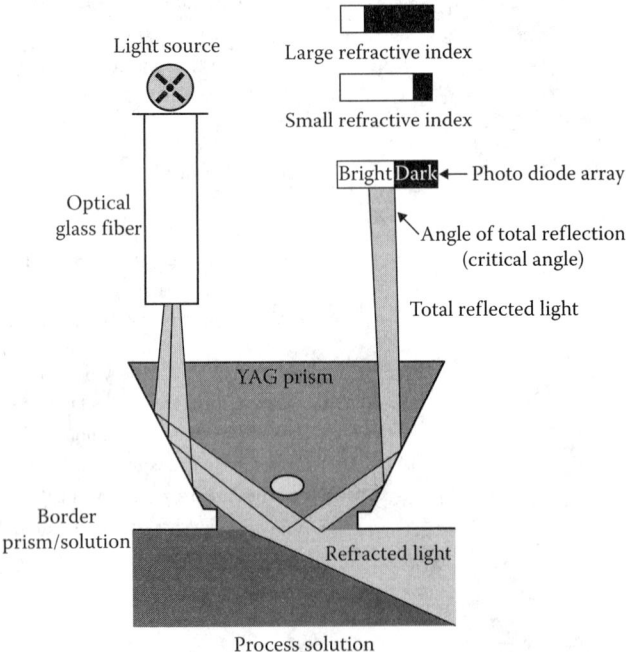

FIG. 1.59a

The operating principle of a refractometer. (Courtesy of Schmidt & Haensch.)

of light in the substance. For example, the refractive index of water is 1.33, meaning that light travels 1.33 times slower in water than it does in vacuum.

UNITS (RI and BRIX)

The index of refraction, or refractive index (RI), is the ratio between the speed of light in vacuum and the speed of light in the substance of interest. RI therefore is unity when the light travels in vacuum at a speed of 0.98 billion fps or 186,300 miles/sec (0.3 billion m/sec). The RI of air is 1.0003, and the RI of most gases, liquids, and solids is between 1 and 2. See Table 1.59a for the refractive index values of a variety of substances.

TABLE 1.59a
Refractive Index Table

Acetic acid	1.3718	Formic acid	1.3714
Acetone	1.3588	Glycerol	1.4729
Acrylic acid	1.4224	Glycol	1.4318
Amyl acetate	1.4012	Heptane	1.3876
Benzene	1.5011	Hexane	1.3749
Butyl acetate	1.3951	Hexanol	1.4135
Butyl alcohol	1.3993	Hydrazine	1.470
Butylene	1.3962	Hydrogen chloride	1.256
Carbon disulfide	1.6295	Lead tetraethyl	1.5198
Carbon tetrachloride	1.4631	Menthol	1.458
Chlorobenzene	1.5248	Methyl alcohol	1.3288
Chloroform	1.4464	Methyl ethyl ketone	1.3807
Cycloheptane	1.4440	Nitric acid	1.397
Cyclohexane	1.4262	Nonane	1.4055
Cyclohexanone	1.4503	Octane	1.3975
Cyclopentane	1.4065	Pentane	1.3575
Decane	1.41203	Perchloroethylene	1.5053
Di-ethyl benzene	1.4955	Phenol	1.5425
Di-methyl benzene	1.4972	Propanol(n)	1.3851
Di-ethyl ether	1.3497	Pronanol(iso)	1.3776
Ethyl acetate	1.3722	Styrene	1.5434
Ethyl alcohol	1.3624	Toluene	1.4969
Ethylbenzene	1.4952	Water	1.3330

Note: All data based on 20°C (68°F).

Some industries use their own units rather than the index of refraction. Two examples are the sugar and citrus juice industries. They prefer to use the % Brix scale, which refers to the weight-percent of sugar concentration, corresponding to the number of grams of sugar contained in 100 g of solution. Figure 1.59b provides the conversion between RI and the % Brix units.

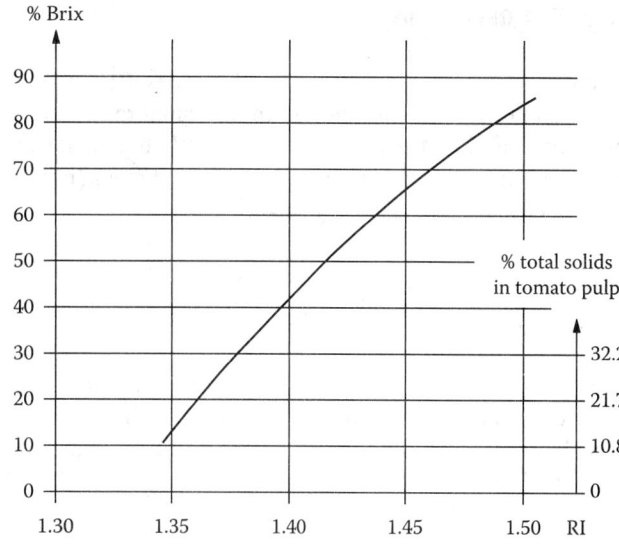

FIG. 1.59b
Refractive index correlations in food industries.

Table 1.59b gives the typical % Brix readings for a number of food items and the errors with which refractometers can measure them.

TABLE 1.59b
The % Brix Values of Different Food Industry Products and the Errors within which Refractometers Can Read These Values

Food Industry Products	Range of Typical % Brix Values	Inaccuracy (%)
Juices: Concentrated orange	12–72	0.25
Concentrated lemon	12–72	0.25
Concentrated apple	12–72	0.25
Concentrated grape	12–72	0.25
Single-strength juices	0–25	0.1
Strawberry jam	36–82	0.25
Grape preserves	12–72	0.25
Apple sauce	0–25	0.12
Tomato juice	0–25	0.25
Tomato paste (below 15%)	0–25	0.25
Tomato paste (above 15%)	12–72	0.5
Catsup	12–72	0.25
Beef extract	0–25	0.1
and Beef "stock"	12–72	0.25
Condensed skim milk	12–72	0.25
Sweetened condensed milk	12–72	0.5
Lactose	12–72	0.25
Lactose	0–25	0.1
Condensed whey	12–72	0.5
Chocolate malt syrup	12–72	0.25
Chocolate syrup	12–72	0.5
Maple syrup	12–72	0.25
Molasses	12–72	0.25
Blackstrap	12–72	0.25

THEORY OF OPERATION

A refractometer measures the refractive index (RI), which is the ratio of the velocity of light in a vacuum to its velocity in the material of interest. Refraction of light (angularity change) occurs at the interface of two different media unless the incidence of light is perpendicular.

$$RI = n' = \frac{V_{(vacuum)}}{V_{(material)}} \qquad 1.59(1)$$

Snell's law expresses the relationship between the angle of incidence and angle of refraction when light is passed through the interface of two different materials (Figure 1.59c) as

$$n' = \frac{\sin \alpha}{\sin \beta} \qquad 1.59(2)$$

The refractive index of a material is usually expressed in terms of air as a standard.

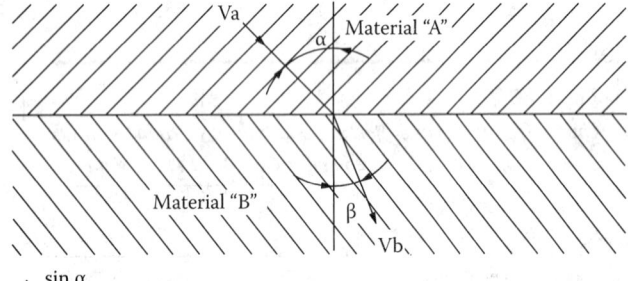

$n' = \dfrac{\sin \alpha}{\sin \beta}$
α is the angle of incidence
β is the angle of refraction
Va is the light velocity in meterial "A"
Vb is the light velocity in meterial "B"
n' is the refractive index of meterial "A" to material "B".

FIG. 1.59c
Illustrations of refraction terms.

Critical Angle of Refraction

A second phenomenon is observed with materials of different indices of refraction: when the angle of incidence exceeds a certain angle, the light ceases to be refracted and is totally reflected (Figure 1.59d). This angle is called the *critical angle* and is defined as

$$\phi_c = \arcsin \frac{n'}{n} \qquad 1.59(3)$$

The process refractometers in use today measure either the refraction angle or the critical angle of refraction.

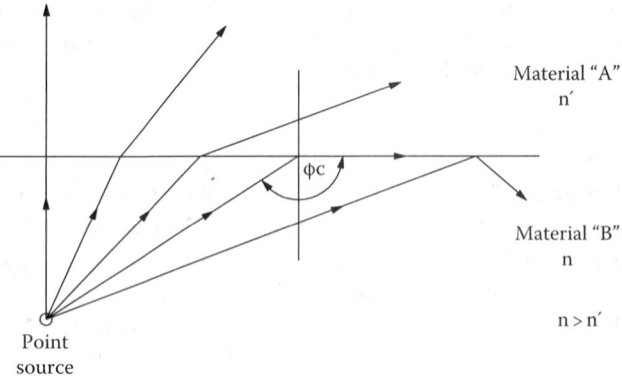

FIG. 1.59d
Illustration of refraction and reflectance.

Table 1.59c provides a summary of the features and capabilities of both of these refractometer designs.

TABLE 1.59c
Refractometer Comparison Summary

Features	Differential Refractometers	Critical Angle Refractometer
Measurement principle	Changes in angle of refraction	Changes in angle of total reflectance (critical angle)
Type sample	Clean translucent liquid	Liquids—clean or turbid, slurries
RI ranges (full-scale)	0.1–0.00005 RI unit	0.1–0.005 RI
Sample flow rate	2–10 cc/min	Depends on type installation (unlimited)
Type of installation	Requires sample system	Can be used with or without sample system; can be installed in pipeline or vessel
Pressure rating	40 psig (276 kPa)	250 psig (1.7 MPa)
Ambient temperature	−1°C to 49°C (30°F–120°F)	−1°C to 149°C (30°F–300°F)

The differential refractometer measures the refraction angle changes as a function of sample RI. This is generally done by holding the incident light angle constant. In that case, Snell's law becomes

$$n = \left(\frac{1}{\sin \beta} \right) \qquad 1.59(4)$$

The refractive indices of all materials vary with temperature, so the temperature must be compensated for unless the measurement is made at the standard reference temperature of 20°C (68°F). Because temperature affects RI, it must be (1) held constant or (2) measured and compensated for in the process analyzer.

REFRACTOMETER DESIGNS

Refractometers measure either the refraction angle or the critical angle of refraction. The angle of refraction detecting designs can have single or dual pass configurations. Their packaging can be for hand-held, bench-top or for in-line use, and their features can include probe and fiber-optic versions and a variety of cleaning and automatic intelligence capabilities.

Differential

Single-Pass Figure 1.59e illustrates the single-pass differential refractometer, which consists of a tungsten filament source, mask, sample flow cell with a sealed reference inner cell, beam splitter, and two opposed phototubes. The reference cell is filled with a solution having an RI value that corresponds to approximately the midscale of the measurement range.

output is used to relocate the beam splitter, thereby rebalancing the light beam equally on the phototubes. The position of the beam splitter is then tracked as the measurement of process RI.

The reference cell angle and the distance between the cell and detector system determine the sensitivity and range of the design. A folded light path using mirrors helps maintain the unit in a reasonable package, even for a 40 in. (1 m) path length.

Two-Pass The two-pass differential refractometer uses a triangular-shaped reference cell (Figure 1.59f) and a reflecting mirror. The reference cell shape presents two nonnormal surfaces that interface with the sample to create two refraction angles for the forward pass and two for the return pass.

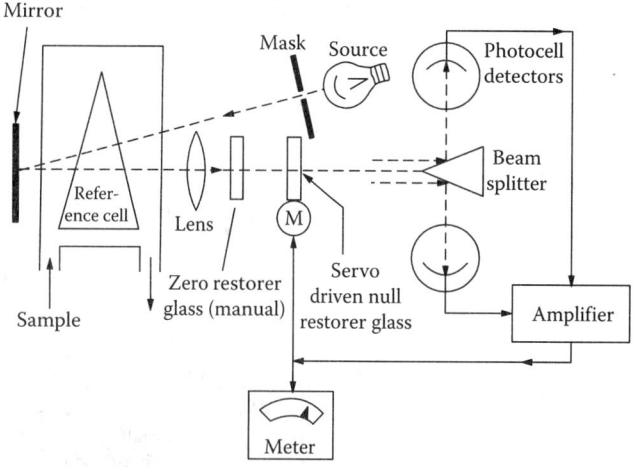

FIG. 1.59f
Two-pass refractometer block-diagram.

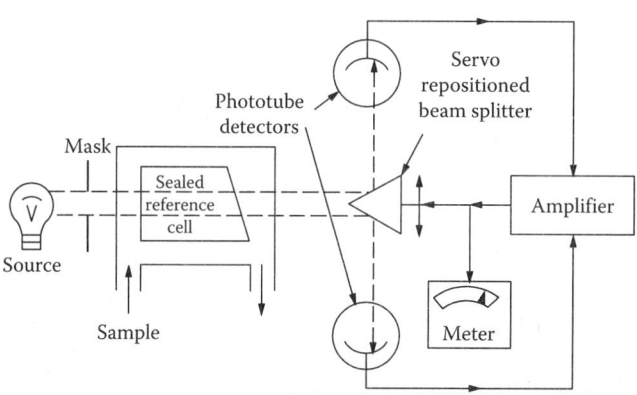

FIG. 1.59e
Single-pass differential refractometer design.

The light beam passes unaltered through the perpendicular cell windows, creating a refraction angle only at the nonnormal window interface. The measured magnitude of the refraction angle is proportional to the difference between the RIs of the sample and of the reference solution.

The detector consists of a beam splitter and two opposed phototubes or photocells. The detector circuit can be designed so that the amplifier reads the ratio of the outputs from the two phototubes as a measurement of the displacement, or as the refraction of the light beam. Phototube irregularities, such as unequal surface response and unequal aging, limit the sensitivity, linearity, and range of this design. Therefore, a null-balance system is usually employed in which the ratio

The reference cell is filled with a solution that has an RI value corresponding to about mid-scale of the measured range. The optical system is zeroed using a sample solution with an RI of the zero value and by adjusting the zero meter reading. The displaced light beam (resulting from changes in sample RI) is detected by a change in the light intensities received by the two photocells, or phototubes, which result in more light impinging on one than on the other.

The amplifier senses the unbalance and changes the position of the null restorer glass via a servo drive to rebalance the light on the photocells. The servomotor also drives a helipot that tracks the position of the null glass as the RI measurement varies.

Temperature Effects Sample temperature is critical in obtaining accurate refractive index measurements. The differential refractometer generally employs a sample temperature controller to maintain the sample at a constant temperature. In some designs, the entire prism assembly is also controlled for maximum stability.

The range of this refractometer is determined by the shape of the reference cell (the angle between two refracting surfaces), the thickness of the zero restorer glass, and the span attenuation in the meter circuit. This design is very sensitive, compact, and flexible.

Flowing Reference Cell The sealed reference cell can be replaced by a flowing cell to allow a differential measurement between two process streams. This approach has been used in blending operations to continuously monitor the difference in RI before and after blending. Extraction and filtration processes can use this system to monitor the difference in RI continuously before and after the extraction step.

Critical-Angle

The *critical angle* is the angle of incident at which light is totally reflected (see Figure 1.59d). The critical-angle refractometer receives the reflected light from a prism interface with the process sample (Figure 1.59g).

The light beam is focused on the prism–sample interface with different incident angles progressing across the width of the beam. Some of this light is transmitted into the sample medium, but reflection will occur only when the incident angles are larger than the critical angle. As the RI of the sample changes, the portion of the light beam that is reflected will also change, so the width of the reflected beam will change.

The point at which the change from refraction to reflection occurs is followed by the detector photocells. As the light beam width changes, more or less light falls on the measuring detector, causing an imbalance between the photocells. The amplifier senses this imbalance and repositions a restorer glass using a servo-balance system until the

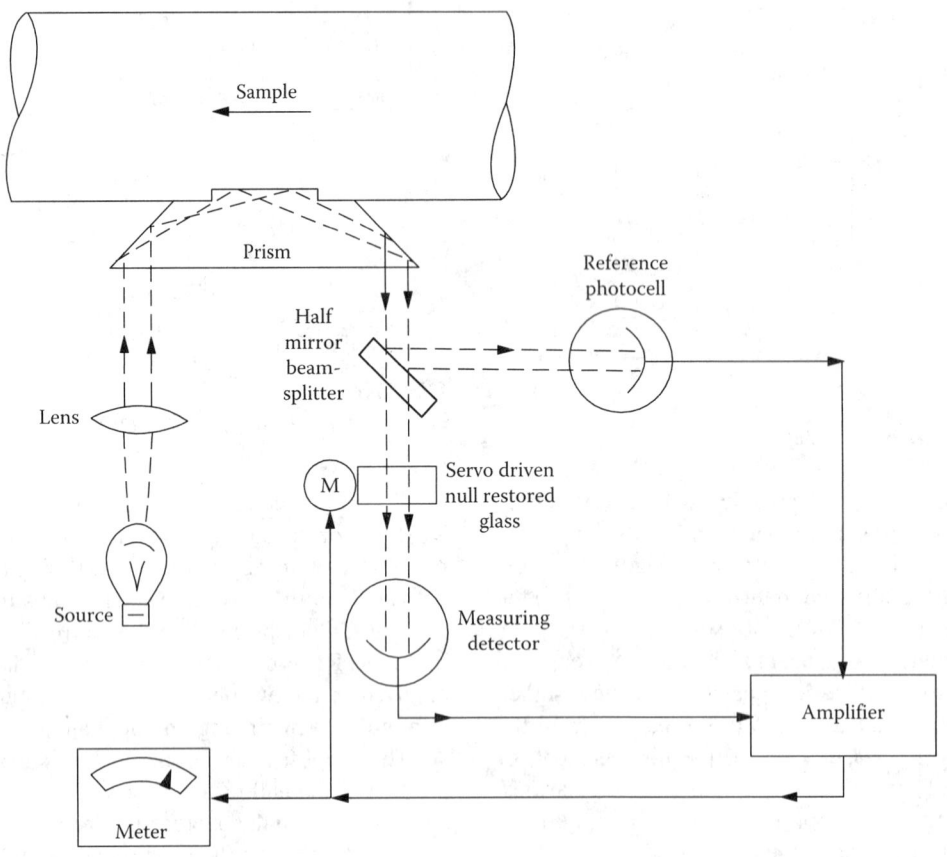

FIG. 1.59g
Schematic representation for the critical angle refractometer.

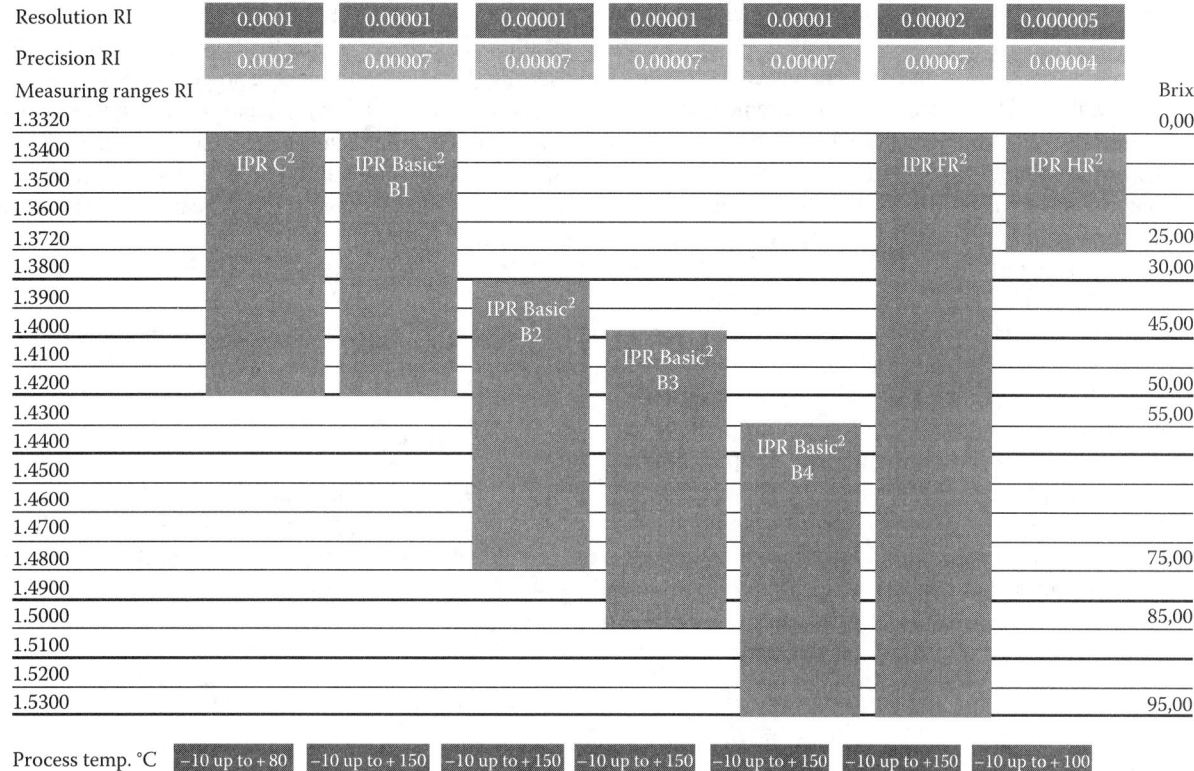

FIG. 1.59h
The measurement ranges (RI and Brix) and other characteristics of different models of critical angle refractometers. (Courtesy of Schmidt + Haensch.)

photocells are again in balance. A helipot in tandem with the null glass drive senses the position of the null glass as a measure of the critical angle.

The range of this instrument is determined by selection of prism glass material and restorer glass thickness. The measurement ranges (RI and Brix), resolutions, precisions and operating temperature ranges of one manufacturer's products are shown in Figure 1.59h.

Automatic

Figure 1.59i is a schematic of an automatic refractometer. In this design a light source is directed under a wide range of angles onto a prism surface which is in contact with a sample. Depending on the difference in the refractive index between prism material and sample, the light is partly transmitted or totally reflected. The critical angle of total reflection is

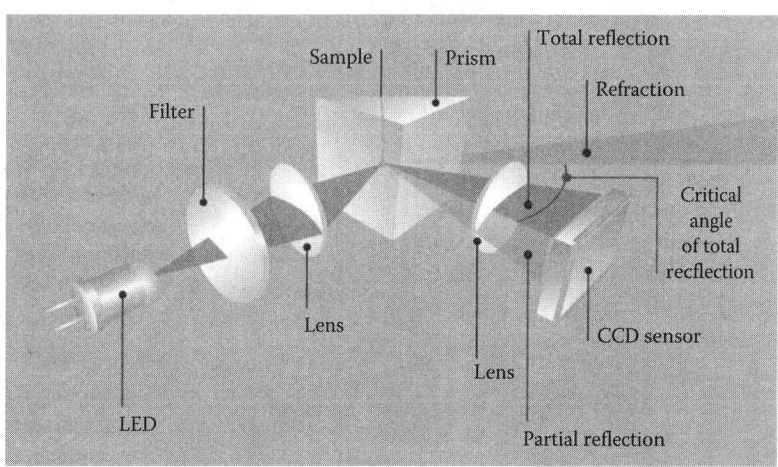

FIG. 1.59i
Illustration of the operation of a critical angle refractometer. (Courtesy of Anton Paar GmbH.)

determined by measuring the reflected light intensity as a function of the incident angle.

The refractive index of a sample can be measured manually or automatically. The automatic measurement of the refractive index of the sample is based on the determination of the critical angle of total reflection. In this design a light source, usually a long-life LED, is focused onto a prism surface. An interference filter guarantees the specified wavelength. Due to focusing the light onto a spot on the prism surface, light is reflected over a wide range of angles. Depending on the refractive index of the sample, which is in direct contact with the prism, the incoming light below the critical angle of total reflection is partly transmitted into the sample, whereas, as the angles of incidence rises, the light is totally reflected. This reflected light intensity is a function of the incident angle, which is measured with a high-resolution sensor array and from it the refractive index of the sample is calculated. This determination of the refraction angle is independent of vibrations and other environmental disturbances.

Mounting and Compensation This unit is frequently installed in line by mounting the prism assembly in a pipe and inserting this pipe section in the process line. The design also allows vessel mounting with the prism assembly inserted in a flange that can be attached to a vessel (storage tank, mixing tank, etc.). Such units can be used on slurry streams, opaque samples, and viscous samples that are not suited for measurement by differential refractometers.

If the temperature varies, temperature compensation can be provided by including a resistance thermometer or a thermistor-type sensor in the analyzer probe.

Design Variations

Figure 1.59j illustrates a probe-type refractometer design that is used in the food, paper, and chemical industries. This critical angle refractometer utilizes fiber optics and is microprocessor controlled.

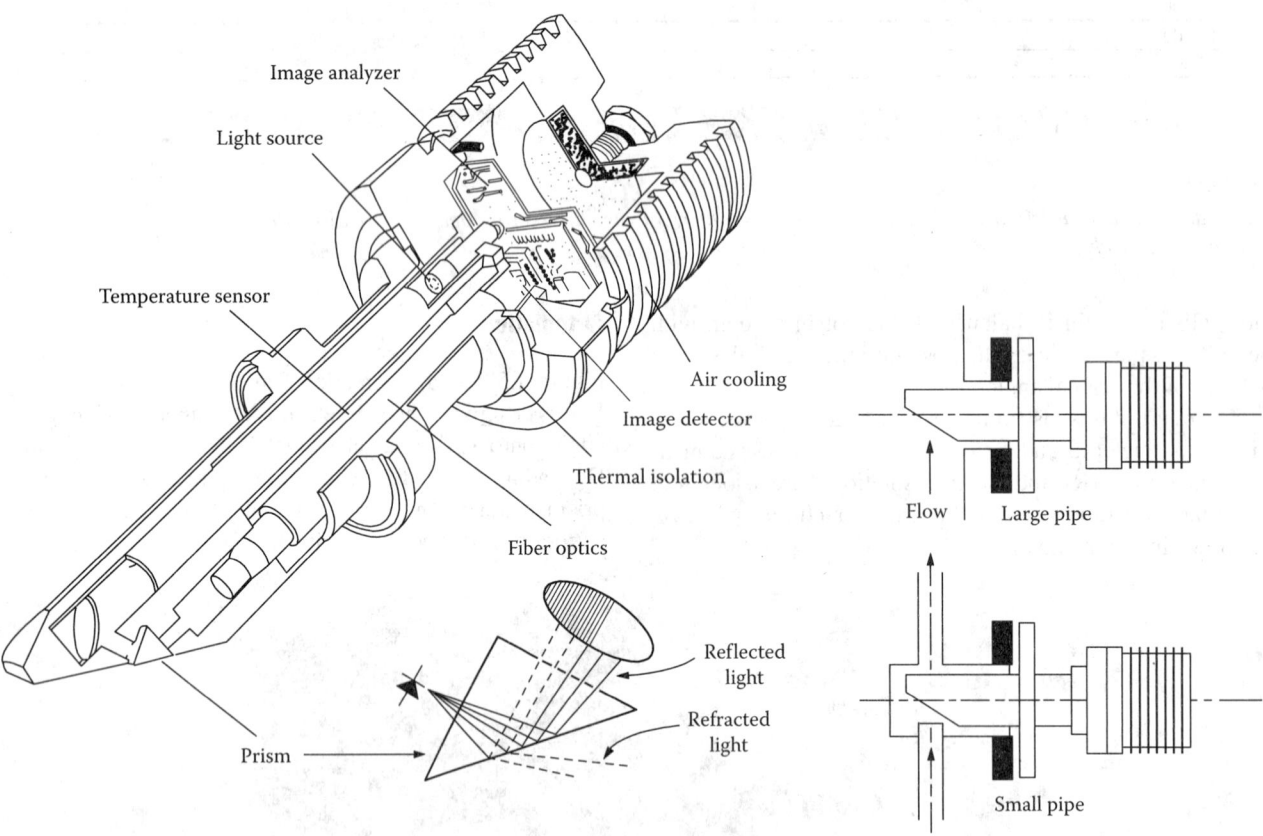

FIG. 1.59j
Probe-type critical angle refractometer. (Courtesy of K-Patents.)

The prism surface is wetted by the process fluid and, when light rays at different angles are directed at that surface, the rays arriving at a steep angle are mostly refracted, so the corresponding image sector is dark. On the other hand, the rays arriving at a flat angle are mostly reflected, so the corresponding image sector is light. The critical angle of reflection corresponds to the line where the light and dark regions meet. Hundreds of photocells are employed to accurately measure the position of the shadow edge and thereby to detect concentration of the process fluid, which rises as the dark area increases.

Mounting and Cleaning Options Inline, industrial process refractometers can be installed directly in pipe lines or can measure the RI in a bypass arrangement. They can be of the flow-through variety or can be flange mounted on tanks and pipes (Figure 1.59k). These units can also be provided by a variety cleaning systems including the use of cleaning fluids and mechanical or ultrasonic cleaners.

Mounted in 90° elbow hose connections

Mounted in flange for tank or vessel with cleaning nozzle opposite the prism face

FIG. 1.59k
Online refractometer installation and cleaning options. (Courtesy of Schmidt + Haensch.)

858 Analytical Measurement

Handheld and Benchtop Designs Handheld refractometers (Figure 1.59l) require only a few droplets of sample from juice, juice concentrate, beverage, soft drink, wine, beer, coffee, milk products, jam, jelly, molasses, liquid sugar, fructose, glucose, sucrose, acids, bases, fats, oils, solvents, pulp, paper, paint, glue, paraffin, wax, resin and other products to measure their RI, Brix, specific gravity or other concentration and purity related characteristics. These instruments are also available in intelligent designs, capable of converting the readings for both loval display purposes or for transmitting them in analog or digital forms.

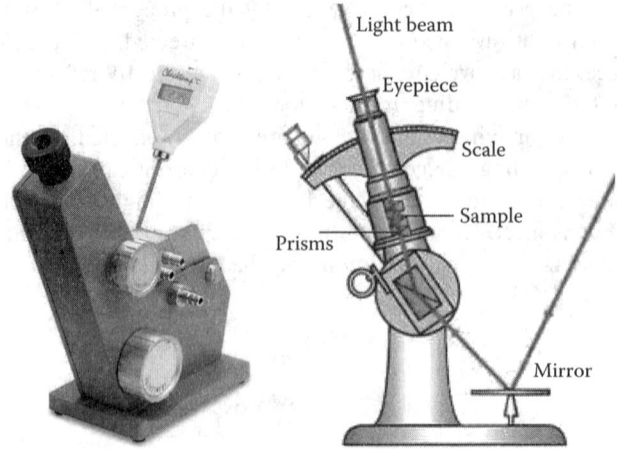

(a) Manual, benchtop (Courtesy of VEE GEE Scientific, Inc.)

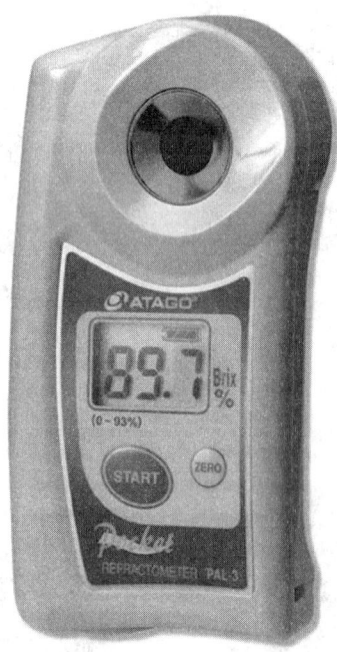

FIG. 1.59l
Handheld digital refractometer. (Courtesy of Cole-Parmer-Atago.)

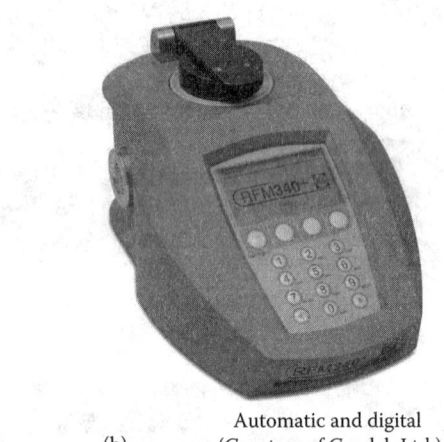

(b) Automatic and digital (Courtesy of Camlab Ltd.)

FIG. 1.59m
Optical and digital refractometers for laboratory use.

Automatic, digital benchtop units (Figure 1.59m) used for laboratory analysis of grab samples are usually provided with even more intelligence than what was listed above for the handheld units. These often include self-diagnostics, conversion, transmission, multiple data manipulation, historical display generation and many others.

Reflected Light Measurement

Some refractometer designs measure the total amount of reflected light, which also varies with the refractive index of the process fluid. One instrument (Figure 1.59n) uses a sapphire rod as a light guide, and it measures both the intensity of the light entering the rod and also the intensity of the reflected light. The percentage of reflected light varies with the refractive index and therefore with concentration.

Fiber-optic Probe   Figure 1.59o describes a reflective fiber optic probe design. Here, a small fiber optic probe (0.25 or 0.125 in. [6 or 3 mm] diameter) is used. The light that is transmitted to the prism is reflected into the fiber optic probe. When the prism is immersed in a process fluid that has a higher refractive index than that of the prism, the reflected light will significantly decrease.

Such detectors can measure concentration, because refractive index changes with concentration. One advantage of these temperature-compensated fiber optic probes is that a single readout instrument can monitor up to four probes inserted in four different locations in the process. This lowers the unit cost of measurement.

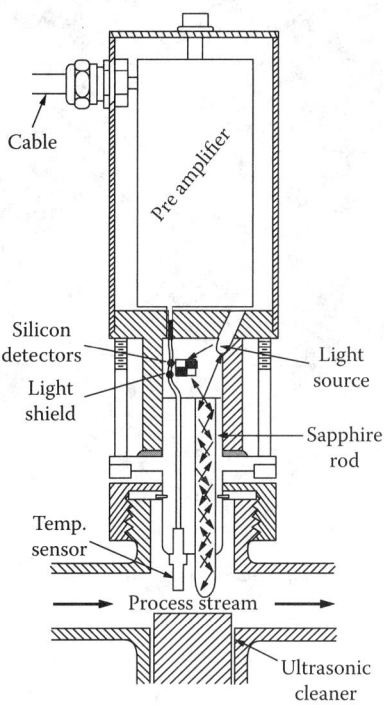

FIG. 1.59n
In-line, reflected light detecting refractometer with ultrasonic cleaner.

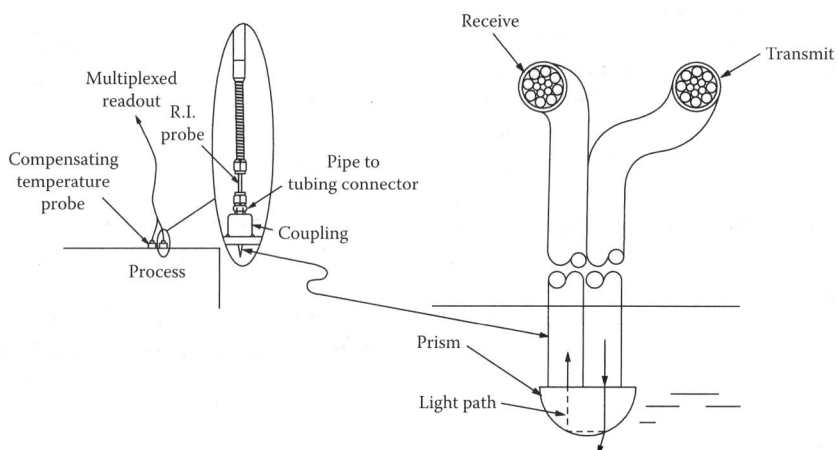

FIG. 1.59o
Four temperature compensated fiber-optic RI probes can share the same readout instrument.

CONCLUSIONS

The refractometer is a nonselective instrument that measures the RI of the entire sample. The RI of a mixture follows the simple mixture law,

$$N_{mixture} = C_a N_a + C_b N_b + C_c N_c + \cdots \qquad 1.59(5)$$

where
$C_a, C_b, \ldots$ are the molecular concentrations of components a, b, c, …
$N_a, N_b, \ldots$ are the respective RI of components a, b, …

Thus, when used as a composition analyzer, the refractometer can monitor only binary mixtures.

Table 1.59c lists the RI ranges that the differential and the critical angle refractometers can handle, and Equation 1.59(5) can be used to determine the RI range required for an expected concentration change.

Because these are optical devices, it is necessary to keep their prism surfaces clean. The types of probe cleaners discussed in Chapters 1.2 and 1.55 are also applicable to these instruments. The standard options offered by most suppliers include chemical or water washers, among other options.

Another limitation of this analyzer is a result of the temperature sensitivity of the angle of refraction. All refractometers must be temperature compensated unless used in a constant-temperature process.

Refractometers are available in handheld, benchtop or industrial in-pipe designs. Their automatic digital versions are very versatile and are available with state-of-the-art software based measurement manipulation and display abilities. Their applications are very wide, they can measure the concentrations of juices, juice concentrates, beverages, soft drinks, wine, beer, coffee, milk products, jams jellies, tomato pastes, molasses, liquid sugar content, fructose, glucose, sucrose, acids, bases, fats, oils, solvents, pulp, paper, paint, glue, paraffin's, waxes, resins, etc.. When used for binary mixtures, the refractometer can be a very accurate analytical tool (Figure 1.59p), but in other designs it is usually unable to detect trace impurities.

Mixtures with trace components having RI values close to one of the major components may act as binary samples and can give satisfactory results. The differential refractometer is more sensitive than the critical angle refractometer, but it requires a sample-handling system and is limited to applications in which the sample is clean. The critical angle refractometer can monitor dirtier streams and can be installed in line.

FIG. 1.59p
On-line refractometer for measuring the concentration of binary mixtures. (Courtesy of Electron Machine Corp.)

SPECIFICATION FORMS

When specifying Refractometers, the ISA Form 20A1001 can be used to specify the features required for the device itself, and when specifying both the device and the nature of the application, including the composition and/or properties of the process materials, ISA Form 20A1002 can be used. These forms, with the permission of the International Society of Automation, are reproduced on the next pages.

1	RESPONSIBLE ORGANIZATION		ANALYSIS DEVICE	6	SPECIFICATION IDENTIFICATIONS		
2	(ISA)			7	Document no		
3			Operating Parameters	8	Latest revision	Date	
4				9	Issue status		
5				10			

11	ADMINISTRATIVE IDENTIFICATIONS				40	SERVICE IDENTIFICATIONS Continued			
12	Project number		Sub project no		41	Return conn matl type			
13	Project				42	Inline hazardous area cl		Div/Zon	Group
14	Enterprise				43	Inline area min ign temp		Temp ident number	
15	Site				44	Remote hazardous area cl		Div/Zon	Group
16	Area		Cell	Unit	45	Remote area min ign temp		Temp ident number	
17					46				
18	SERVICE IDENTIFICATIONS				47				
19	Tag no/Functional ident				48	COMPONENT DESIGN CRITERIA			
20	Related equipment				49	Component type			
21	Service				50	Component style			
22					51	Output signal type			
23	P&ID/Reference dwg				52	Characteristic curve			
24	Process line/nozzle no				53	Compensation style			
25	Process conn pipe spec				54	Type of protection			
26	Process conn nominal size		Rating		55	Criticality code			
27	Process conn termn type		Style		56	Max EMI susceptibility		Ref	
28	Process conn schedule no		Wall thickness		57	Max temperature effect		Ref	
29	Process connection length				58	Max sample time lag			
30	Process line matl type				59	Max response time			
31	Fast loop line number				60	Min required accuracy		Ref	
32	Fast loop pipe spec				61	Avail nom power supply			Number wires
33	Fast loop conn nom size		Rating		62	Calibration method			
34	Fast loop conn termn type		Style		63	Testing/Listing agency			
35	Fast loop schedule no		Wall thickness		64	Test requirements			
36	Fast loop estimated lg				65	Supply loss failure mode			
37	Fast loop material type				66	Signal loss failure mode			
38	Return conn nominal size		Rating		67				
39	Return conn termn type		Style		68				

69	PROCESS VARIABLES	MATERIAL FLOW CONDITIONS			101	PROCESS DESIGN CONDITIONS	
70	Flow Case Identification			Units	102	Minimum	Maximum	Units
71	Process pressure				103			
72	Process temperature				104			
73	Process phase type				105			
74	Process liquid actl flow				106			
75	Process vapor actl flow				107			
76	Process vapor std flow				108			
77	Process liquid density				109			
78	Process vapor density				110			
79	Process liquid viscosity				111			
80	Sample return pressure				112			
81	Sample vent/drain press				113			
82	Sample temperature				114			
83	Sample phase type				115			
84	Fast loop liq actl flow				116			
85	Fast loop vapor actl flow				117			
86	Fast loop vapor std flow				118			
87	Fast loop vapor density				119			
88	Conductivity/Resistivity				120			
89	pH/ORP				121			
90	RH/Dewpoint				122			
91	Turbidity/Opacity				123			
92	Dissolved oxygen				124			
93	Corrosivity				125			
94	Particle size				126			
95					127			
96	CALCULATED VARIABLES				128			
97	Sample lag time				129			
98	Process fluid velocity				130			
99	Wake/natural freq ratio				131			
100					132			

133	MATERIAL PROPERTIES				137	MATERIAL PROPERTIES Continued		
134	Name				138	NFPA health hazard	Flammability	Reactivity
135	Density at ref temp		At		139			
136					140			

Rev	Date	Revision Description	By	Appv1	Appv2	Appv3	REMARKS

Form: 20A1001 Rev 0

862 Analytical Measurement

	RESPONSIBLE ORGANIZATION	ANALYSIS DEVICE COMPOSITION OR PROPERTY Operating Parameters (Continued)		SPECIFICATION IDENTIFICATIONS	
1	ISA		6	Document no	
2			7	Latest revision	Date
3			8	Issue status	
4			9		
5			10		

	PROCESS COMPOSITION OR PROPERTY			MEASUREMENT DESIGN CONDITIONS				
11	Component/Property Name	Normal	Units	Minimum	Units	Maximum	Units	Repeatability
12								
13–37								

Rev	Date	Revision Description	By	Appv1	Appv2	Appv3	REMARKS

Form: 20A1002 Rev 0 © 2004 ISA

Definitions

Baumé degrees were introduced by the French pharmacist Antoine Baumé in 1768 as liquid density units. The Baumé degrees are variously noted as degrees Baumé, B°, Bé° or simply Baumé. On this scale, the density of distilled water is zero, and there are separate scales for liquids heavier and for liquids lighter than water.

Brix degree (°Bx) or percentage (%Bx) gives the sugar content of aqueous solutions. It is the weight-percent of sugar concentration, corresponding to the number of grams of sugar contained in 100 g of solution. If the solution contains dissolved solids other than pure sucrose, then the °Bx only approximates the dissolved solid content.

Brix scale is calibrated to show the weight-percent of sugar, corresponding to the number of grams of sugar contained in 100 g of solution.

Phase velocity of a wave is the velocity at which a phase point on it (such as the crest) travels. The phase velocity (v_p) is the ratio of the wavelength lambda (λ) and the period of oscillation (T), so that $v_p = \lambda / T$.

Refraction is essentially a surface phenomenon. When light is incident at a transparent surface, the transmitted component of the light (that which goes through the interface) changes direction at the interface. Another component of the light is reflected at the surface. The refracted beam changes direction at the interface and deviates from a straight continuation of the incident light ray. Refraction of light is the most commonly observed phenomenon, but any type of wave can refract when it interacts with a medium, for example, when sound waves pass from one medium into another or when water waves move into water of a different depth.

Refractive Index is the ratio of the velocity of light in a vacuum to its velocity in a specified medium. The refractive index (RI) of a substance is a dimensionless number that describes how fast light, propagates through that medium. It is defined as $RI = N' = C/V$, where C is the speed of light in vacuum and V is the speed of light in the substance. For example, the refractive index of water is 1.33, meaning that light travels 1.33 times slower in water than it does in vacuum.

Snell's law (also known as the Snell–Descartes law and the law of refraction) is a formula used to describe the relationship between the angles of light incidence and light refraction, when referring to light passing through a boundary between two different substances, such as water, glass and air. Snell's law therefore states that for a given pair of media and for a wave of a single frequency, the ratio of the sines of the angle of incidence (α) and the angle of refraction (β) is equivalent to the ratio of its phase velocities (v_1/v_2) in the two media.

Abbreviation

RI   Refractive Index

Bibliography

AFAB Enterprises, About refractometers, (2013) http://www.refractometer.com/.

Cornell Edu, Using refractometer to measure the sugar concentration, (2004) http://maple.dnr.cornell.edu/kids/refractometer.htm.

Hanson, J., Refractometry, (2003) http://www2.ups.edu/faculty/hanson/labtechniques/refractometry/interpret.htm.

IHS, In-line refractometer, (2016) http://www.globalspec.com/industrial-directory/inline_refractometer.

ISO, Ophthalmic instruments—Eye refractometers, (2003) http://www.iso.org/iso/catalogue_detail.htm?csnumber=34276.

Marcus, S., Refractometers revisited, (2008) http://www.pfonline.com/articles/refractometers-revisited.

MEL, Refractometers, (2013) http://encyclopedia.che.engin.umich.edu/Pages/ProcessParameters/Refractometers/Refractometers.html.

Mettler Toledo, Brix—Sugar determination, (2013) http://us.mt.com/dam/Analytical/Density/DE-PDF/BRIX-Sugar_Determination.pdf.

MISCO, Refractometer applications, (2013) http://www.misco.com/refractometer-support/refractometer-forum/refractometer-applications.

Mital, K. S., Measuring the effect of sugar concentration on the refractive index, (2010) http://www.usc.edu/CSSF/History/2010/Projects/J1913.pdf.

Pastorek, C., Refractometry, (2002) http://chemistry.oregonstate.edu/courses/ch361-464/ch362/refract02.htm.

Rudolph Research Analytical, Refractometer selection guide, (2013) http://rudolphresearch.com/products/refractometers/selection-guide/.

1.60 Rheometers

J. J. MAGDA (2003)   **B. G. LIPTÁK** (2017)

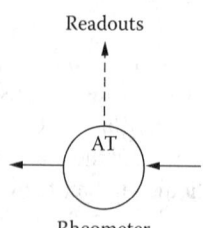

Types	A. Cone-and-plate
	B. Parallel disk
	C. Coaxial cylinder
	D. Rectangular torsion
	E. Bending and tension/compression
	F. Extensional flow rheometer
	G. Extensional capillary rheometer
Linear properties detected	A. Elastic storage modulus versus frequency
	B. Viscous loss modulus versus frequency
Nonlinear properties detected	A. Viscosity versus shear rate
	B. First normal stress difference versus shear rate
	C. Second normal stress difference versus shear rate
	D. Extensional viscosity versus rate of extension
Frequency range	0.01–200 Hz
Pressure	Atmospheric
Temperature range	−150°C to +600°C
	Generally in the $3,000–$10,000 range.
	For the costs of some of the specific design, see http://www.alibaba.com/showroom/rheometer-price.html
Cost	$30,000–$200,000
Partial list of suppliers	Anton Paar (http://www.anton-paar.com/Rheometer/59_USA_en?productgroup_id=23&Rheometer)
	Brabender (extrusion) http://www.brabender.com/english/plastics/products/extruders/extrusion-rheometers.html
	Brookfield http://www.brookfieldengineering.com/products/rheometers/laboratory-yr-1-yield.asp
	Dynisco (online) http://www.dynisco.com/online-rheometer-viscosensor
	Fann (High temperature) http://www.halliburton.com/fann/products/oil-well-cement-testing/viscosity/rheometer-50sl.page?node-id=hlz0hxqc
	Göttefert (capillary) http://www.capillary-rheometer.com/index.html
	Haake, Caber (extensional) http://www.eu-softcomp.net/FILES/caber.pdf
	Instron (smart and capillary) http://www.instron.us/wa/product/Smart-Rheo.aspx
	KSV Nima (Interfacial) http://www.ksvnima.com/products/ksv-nima-isr
	Malvern http://www.malvern.com/en/products/product-range/kinexus-range/kinexus-ultra-plus/default.aspx
	ORL (online) http://www.onlinerheometer.com/about-the-olr/product-brochure/
	Pavement Interactive "Shear Rheometer" http://www.pavementinteractive.org/article/dynamic-shear-rheometer/
	Qingdao Guangyue Rubber Machinery Manufacturing Co., Ltd. http://qdguangyue.en.alibaba.com/company_profile.html
	Qualiest International (capillary) http://www.worldoftest.com/capillary-rheometer
	RheoSense http://www.rheosense.com
	Rosand (capillary) http://www.bioprocessonline.com/doc/rosand-capillary-rheometers-0001
	TA Instruments http://www.tainstruments.com/lpage.aspx?id=290&n=1&siteid=11
	Thermo Scientific (extensional) http://www.thermoscientific.com/en/product/haake-caber-1-capillary-breakup-extensional-rheometer.html, http://www.thermoscientific.com/en/products/rheometers.html

INTRODUCTION

Rheometers characterize the viscoelastic properties of polymer melts and of other materials. They are used for those fluid flows that cannot be defined by viscosity alone (see Chapters 1.70, 1.71 and 1.72 for a detailed discussion of the measurement of viscosity) and require more parameters to be measured and controlled in order to define the polymer's flow characteristics. Rheometers are quality control tools and serve to assess the processability of resins, are used by plastics compounders, and also serve as R&D tools to help determine which resin best fits a particular process or application.

These capabilities of rheometers can justify their consideration, in spite of their relatively high cost, when high-quality polymeric (thermoplastic) materials are to be manufactured. The cost of rheometers has been dropping over the years, and today, it is down to about $40,000. There are two distinctively different types of rheometers. The ones that measure the applied shear stress or shear strain are called rotational or shear rheometers, whereas rheometers that apply extensional stress or extensional strain are called extensional rheometers.

Rheometers facilitate the conversion of measured quantities, such as force, torque, pressure, and angular velocity, into stress and strain. These two tensorial properties allow the derivation of all other rheological properties. Due to the importance of these values, a number of geometries have been developed to allow their accurate measurement based specifically on the material system under consideration, several of the most important of which are shown in Figure 1.60a.

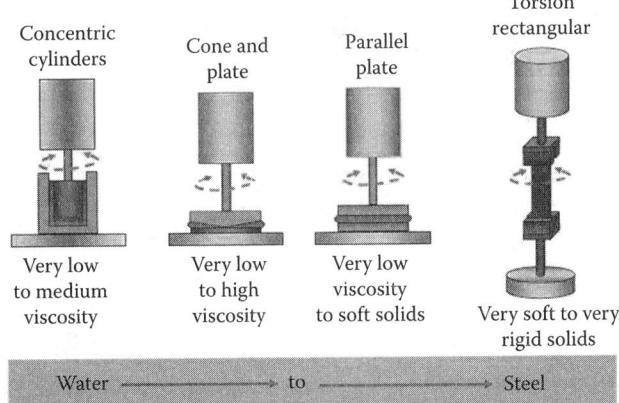

FIG. 1.60a
Rheometer designs. (Courtesy of Society of Plastics Engineers, Texas A&M University.)

POLYMER CHARACTERIZATION

Some of the justifications for performing rheological analysis on molten thermoplastics prior to molding or extrusion operations are as follows:

1. *Solvent-free characterization of the molecules present in the polymer*: Thermoplastic raw materials are a mixture of molecular species that vary widely in size and degree of molecular branching (see Chapter 1.41 on molecular weight measurement). Rheological analysis of a small amount of the molten polymer without solvent can be used to detect batch-to-batch variations in the polymeric raw material that may be deleterious to the properties of the final product. An industrial viscometer such as a melt flow indexer will suffice to detect large variations in the average molecular weight of the polymer. However, the entire molecular weight distribution (MWD) of the polymer sample can only be determined with the extra capabilities of a rheometer.

2. *Measurement of polymer flow properties that are key to the stability of polymer processing flow*: When attempting to switch the raw material used in an existing manufacturing process, it is important to maintain the values of key polymer flow properties, or significant modifications of the manufacturing process will be necessary. Many of these key nonlinear flow properties (viscosity, extensional viscosity, and normal stress differences) can be measured using a rheometer at high shear rates or high extension rates under conditions that mimic the conditions present in the actual polymer processing flow.

Solid Polymer Measurements

Rheometers are also useful for analyzing solids manufactured from polymers. Polymer solids, of course, cannot flow in a rheometer, but they nonetheless exhibit both viscous and elastic behavior when undergoing oscillatory deformations and hence are said to be *viscoelastic materials*. Some of the reasons for using oscillatory rheometers to perform dynamic mechanical analysis (DMA) on viscoelastic solids are as follows:

1. *Determination of the glass transition temperature*: The temperature range over which a solid part can be used depends to a large extent on the glass transition temperature (T_g) of the polymer from which it is fabricated. DMA is a very sensitive technique for measuring T_g. For an amorphous polymer, the value of the dynamic shear modulus measured in DMA experiments at fixed frequency drops by three orders of magnitude when the temperature exceeds T_g.

2. *Performance prediction*: Polymeric solids differ from other materials in that the mechanical properties depend strongly on temperature and time as well as on the polymer molecular weight distribution, the presence of additives or fillers, and the process conditions used to fabricate the part. Hence, tabulated mechanical properties for different polymers are of limited accuracy, and DMA experiments used with time–temperature superpositioning (discussed later) provide a better method for estimating the long-term mechanical behavior of a given polymeric solid part.

TESTS TO DISTINGUISH LIQUIDS FROM SOLIDS

The goal of scientific rheometry is to study the mechanical forces in a liquid or a solid sample subjected to a known and controlled velocity field or strain field.

Shear-Strain Tests

A liquid, by definition, cannot resist a shear stress at rest, which means that the shear stress must eventually decay to zero in the absence of flow. Hence, the rheological experiment known as the *shear stress relaxation* shown schematically in Figure 1.60b is an excellent method for determining whether a given polymer sample is a viscoelastic liquid or a viscoelastic solid.

In this experiment, one imposes a chosen value of the shear strain on the sample and measures the resulting shear stress as a function of time. Initially, the sample is in a stress-free rest state between two parallel semi-infinite solid plates separated by a gap L. At time equal zero, the top plate is suddenly moved to the right by a chosen distance Δz, while the bottom plate is held stationary. Assuming good adhesion with no slip at the sample/plate interface, this imposes a time-independent shear strain γ on the sample equal to $\Delta z/L$.

The horizontal force F exerted to keep the bottom plate stationary is measured as a function of time. The value of this force divided by the sample-contacting area A of the plate is equal to the shear stress τ. If τ eventually decays to zero, the sample is a liquid; otherwise, the sample is a solid. In a linear stress relaxation experiment, the sample is subjected to a small strain that only slightly perturbs the sample on a molecular level from its equilibrium state. For sufficiently small strains, the linear shear stress relaxation modulus G is given by

$$G = \frac{\tau}{\gamma} \qquad 1.60(1)$$

The key test for a liquid is, "Does G approach zero after the passage of a long time period (t)?" A more practical way to answer this question is by the use of oscillatory shear experiments in dynamic mechanical analysis (DMA), a technique described in more detail later. The influence of time t in a linear stress relaxation experiment is equivalent to the influence of the period of oscillation in DMA. Hence, the key test for a liquid can be rephrased as, "Does the elastic storage modulus G′ (defined later) approach zero at low frequencies?"

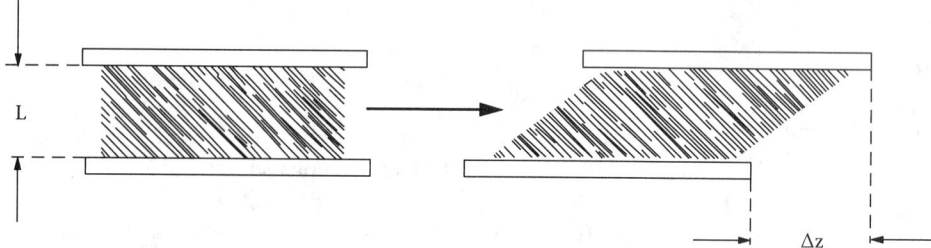

FIG. 1.60b
Stress relaxation test at fixed shear strain.

Shear Creep Tests

The *shear creep test* is complementary to the stress relaxation experiment. In shear creep, one imposes a chosen value of the shear stress and measures the resulting value of the strain as a function of time. Initially, the sample is in a stress-free rest state between two parallel semi-infinite solid plates separated by a gap L. At time zero, the top plate is suddenly pulled to the right with a constant force per unit area ($\tau = F/A$), while the bottom plate is held stationary. This imposes a time-independent uniform shear stress τ on the sample, assuming good adhesion at both sample–plate interfaces.

If, after the passage of a long time period, the top plate is observed to move a certain distance Δz to the right and stop, then the sample is a solid. If, on the other hand, the top plate does not stop moving but reaches a terminal velocity V, then the sample is a liquid, and the velocity field within the liquid is linear and called *rectilinear simple shear flow*. In this case, the shear rate or rate of strain κ is calculated as V/L, and the shear viscosity μ is defined as the shear stress divided by the shear rate.

Unless κ is very small, simple shear flow is a so-called nonlinear flow, because the alignment of the polymer molecules in the sample will be significantly perturbed by the flow. Because the degree of perturbation increases with shear rate κ, it is not surprising that the shear viscosity value μ also varies with κ, which is called *non-Newtonian behavior*.

Nonlinear Shear Flow

A non-Newtonian liquid in shear flow has three independent nonlinear flow properties: viscosity (μ), first normal stress difference (N_1), and second normal stress difference (N_2).

Shear viscosity μ is a measure of the liquid's resistance to flow, and N_1 and N_2 are manifestations of liquid elasticity. For Newtonian liquids such as water, N_1 and N_2 are both zero. However, polymer molecules in steady shear flow are stretched along the flow direction and compressed along the direction perpendicular to the plates in Figure 1.60c, which gives rise to nonzero N_1 and N_2 values. Although N_1 and N_2 are difficult to measure, there is a growing consensus that their values are important to the behavior of thermoplastics in practical polymer processing flows.

In extrusion, when a thermoplastic exits the die into air or another cooling fluid, the amount of extrudate die swell depends strongly on the value of N_1 for the thermoplastic. The value of N_2 is thought to be critical to the stability of multilayer coextrusion, which is a manufacturing process in which two immiscible thermoplastics are extruded through the same die so as to produce a layered product. Even if the two thermoplastics have identical viscosity curves, a difference in N_2 values may result in a flow disturbance that disrupts the layered structure.

SELECTION

In selecting a rheometer design to analyze a polymeric sample, the following questions should be considered:

1. What class of deformation or flow is most relevant, shear or extension (elongation) or a combination of these two basic types, such as bending?
2. Should the sample deformation rate be constant or oscillatory? A constant rate of deformation is impossible for a viscoelastic solid. For constant deformation rate experiments, what is the necessary range of shear rates or extension rates? For oscillatory deformations, what is the necessary range of frequencies?
3. Will the rheometer control the amount of stress applied to the sample (i.e., stress-controlled rheometer) or the amount of strain applied to the sample (i.e., strain/strain-rate-controlled rheometer). Strain/strain-rate-controlled rheometers cannot be used to measure the true value of the yield stress for weak solids.
4. Will the sample adhere to the stainless steel fixtures of the rheometer so that the no-slip boundary condition is satisfied, or will it be necessary to clamp the sample? Are large solid particles suspended in the sample, comparable in size to the gap of the proposed rheometer fixture? If so, a fixture with a larger gap must be used.
5. Is the sample corrosive, volatile, combustible, or otherwise destructive to stainless steel rheometer fixtures? To study the gelation of network-forming materials like epoxies, disposable rheometer fixtures will probably be required.
6. What is the required temperature range for testing? Liquid nitrogen cooling may be required for subambient testing temperatures.
7. What is the sensitivity required for stress or strain measurement? What is the maximum stress value to be measured or applied to the sample?

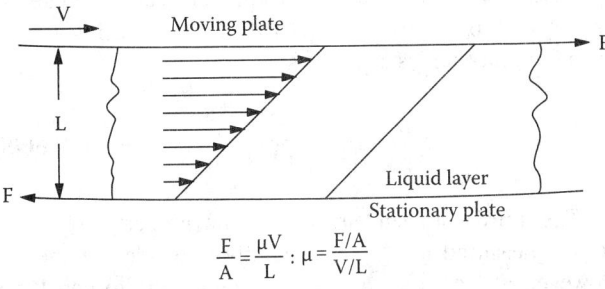

FIG. 1.60c
Variables in a rectilinear simple shear flow test, the variables are F—constant force; A—unit area; μ—viscosity; V—velocity; L—gap between plates.

DETECTOR DESIGNS

Cone and Plate

In shear creep experiments, all viscoelastic liquids are expected to exhibit the velocity field known as simple shear flow at steady state (Figure 1.60c), barring the occurrence of certain flow disturbances that will be discussed later. For liquids with long relaxation times, the use of rotational shear flow geometries is the most practical method for achieving this steady-state condition in the laboratory.

The cone-and-plate flow geometry (Figure 1.60d) consists of a stationary flat plate (really, a disk) and a rotating cone with a very shallow cone angle. The tip of the cone is truncated by a small amount, on the order of 50–100 μm. During testing, the cone is lowered so that its tip, if it were present, would just touch the center of the plate. The liquid sample fills the narrow gap between the cone and the plate and is held in place during testing by surface tension forces, provided that the inertia is not too large.

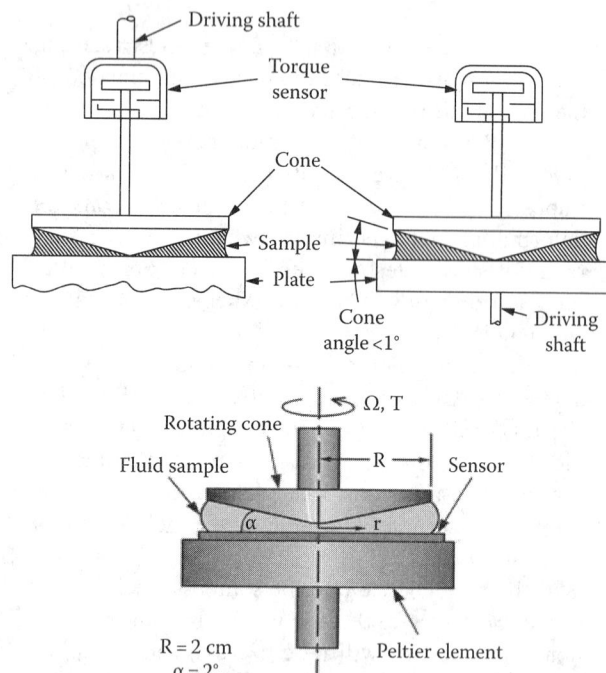

FIG. 1.60d
Cone-and-plate rheometer sensing element design.

The temperature of the sample is often monitored with thermocouples embedded in the rheometer plate. In a strain-rate-controlled cone-and-plate rheometer, a variable-speed motor drives the cone at a precise rotation speed while the torque required to keep the rheometer plate stationary is calculated from the measured deflection of a torsion spring. Hence, the strain-controlled rheometer can be operated in a mode very similar to the idealized stress relaxation experiment (Figure 1.60b).

In a stress-controlled cone-and-plate rheometer, a constant torque is applied to the drive shaft of the cone, and the rotation of the cone that results is measured with an optical encoder. Hence, the stress-controlled rheometer can be operated in a mode very similar to the idealized shear creep test already discussed. For both types of rheometers, the cone rotation speed Ω and the torque T will have chosen or measured values at steady state, assuming that the sample is a liquid. The velocity field in the liquid is expected to be identical on a local level to rectilinear simple shear flow. The cone-and-plate flow geometry is widely used for testing non-Newtonian liquids, because the shear rate value κ is the same everywhere within the sample and is given by

$$\kappa = \frac{\Omega}{\alpha} \qquad 1.60(2)$$

where
  κ is the shear rate value
  Ω is the rotation speed of the cone
  α is the cone angle

Assuming that the velocity field in the sample is simple shear flow as expected, the torque T on the rotating cone is related to the shear stress τ and cone radius R by fluid mechanics:

$$\tau = \frac{3T}{(2\pi R^3)} \qquad 1.60(3)$$

Shear Flow Properties

In Equation 1.60(3), R is the radius of the cone and the plate. The non-Newtonian shear viscosity is defined as shear stress τ divided by shear rate κ:

$$\mu = \frac{\tau}{\kappa} \qquad 1.60(4)$$

Equations 1.60(2) through 1.60(4) are solved simultaneously to give the viscosity value for both stress-controlled and strain-rate-controlled cone-and-plate rheometers. Due to elasticity, a polymeric liquid will tend to push the cone and plate apart during flow. By measuring the total vertical force F_N exerted on the plate during testing, one can calculate the first normal stress difference (N_1) of the sample using a result from fluid mechanics as follows:

$$N_1 = \frac{2F_N}{(\pi R2)} \qquad 1.60(5)$$

The third independent shear flow property, N_2, cannot be measured using commercially available rheometers. However, at least one company is attempting to develop a new cone-and-plate design in which N_2 can be calculated from measurements of the difference in pressure between the center and the rim of the rheometer plate.

Torque Ranges The most important cone-and-plate rheometer design feature is the method by which the shear stress is measured (or the strain, in a stress-controlled rheometer).

In a strain-rate-controlled rheometer, the motor that drives the cone has a typical rotation speed range of 0.001–100 rad/s. Typical cone angles range from 0.02 to 0.10 rad. Hence, according to Equation 1.60(2), the theoretical shear rate available with the rheometer is 0.01–5000 s^{-1}. However, the actual available shear range is usually much smaller as a result of torque transducer sensitivity limitations at low shear rates and flow disturbances at high shear rates (see the following).

Because the onset of flow disturbances varies considerably with the material being tested, no shear rate range is given in the feature summary at the beginning of this section. In a rate-controlled rheometer, both the torque T and the total vertical force F_N are usually calculated from the deflection of a spring measured using a displacement transducer of a differential transformer type. Care must be taken that the vertical deflection of the spring used to measure F_N does not cause a significant change in the gap between the cone and the plate. Because rotational friction from the instrument itself is detrimental to sensitivity, top-of-the-line rheometers employ air bearings on the cone drive shaft.

A given rheometer is sometimes said to cover a certain measurable range of fluid viscosities, but it is really more accurate to say that the rheometer covers a certain measurable range of torques. A typical rheometer transducer can be used accurately over a torque range covering perhaps four orders of magnitude, from 0.00001 to 0.1 N m. Alternative transducers covering both higher and lower torque ranges are also available.

For a given rate-controlled rheometer, the minimum viscosity value that can be measured can be calculated from the minimum measurable torque value, the maximum possible shear rate, and the maximum available cone-and-plate radius using Equations 1.60(2) through 1.60(4). These equations show that, for low-viscosity fluids, the torque value T, and hence measurement accuracy, can be enhanced by increasing the shear rate κ or by increasing the cone-and-plate radius R.

Selection The choice between a strain/strain-rate-controlled rheometer and a stress-controlled rheometer often depends on whether stress relaxation data or shear creep data are more desirable. However, the stress-controlled rheometer can be operated in strain-controlled mode through use of a feedback control loop that manipulates the value of the imposed torque; however, time to set strain is longer than for a strain-controlled rheometer. In addition, strain-rate-controlled rheometers should not be used to study the flow properties of a sample that might be a weak solid, because the driven rotation of the cone will damage the sample structure.

Observation of hysteresis in the viscosity curve (μ vs. κ) indicates that the structure of the sample has been damaged by flow. The properties of a fragile solid can be measured by using a strain-controlled rheometer in low-amplitude oscillatory measurements (discussed in a later section) or by using a stress-controlled rheometer. In a stress-controlled rheometer, a strain transducer is used to measure sample strain or rate of strain in response to a chosen shear stress value τ. If the sample has a yield stress value that exceeds τ, then the measured steady-state rotational speed of the cone will be zero, a result that indicates that the sample is a weak solid.

Flow Field Disturbances The goal of scientific rheometry as applied to liquids is to study the mechanical response of the liquid when subjected to a known and controlled velocity field. Unfortunately, there is no way to know beforehand whether a given rheological sample will exhibit the expected velocity field in the cone-and-plate device or any other rheometer. If the actual velocity field within the rheometer is disturbed from the expected flow pattern, then the rheometer software will probably calculate incorrect values for the viscosity and other rheological properties.

Because of flow field disturbances, it is virtually impossible to measure μ, N_1, and N_2 values with a cone-and-plate flow geometry at the high shear rate values typical in manufacturing operations (κ = 10–1000 s^{-1} for extrusion, κ = 1,000–50,000 s^{-1} for injection molding). Figure 1.60e shows a sketch of one such flow disturbance called *edge fracture*. These flow disturbances are thought to be driven by elasticity or inertia and can be delayed to higher shear rates by reducing the cone angle or the cone-and-plate radius. They do not occur when the cone-and-plate geometry is used for linear rheological measurements. This is a principal reason why linear properties G′ and G″ are more commonly measured in industrial laboratories than N_1 and N_2, even though linear rheological properties have no direct relevance to the flow conditions actually present in thermoplastic processing flows, which are usually very nonlinear.

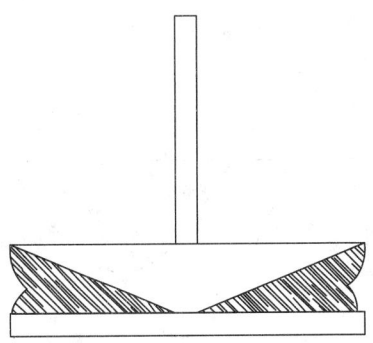

FIG. 1.60e
Flow disturbance due to edge fracture of sample in cone-and-plate instrument.

Dynamic Mechanical Analysis Linear rheological properties are measured when the strain field or velocity field only slightly perturbs the sample on a molecular level from its equilibrium state. For molten thermoplastics, linear rheological measurements have no direct relevance to polymer processing flows, because most practical manufacturing flows are highly nonlinear. However, linear rheological measurements are still very useful for characterizing the polymer molecules or making a rheological *fingerprint* of the raw material.

Linear rheological properties are most commonly measured in oscillatory experiments called dynamic mechanical analysis (DMA). For a liquid in oscillatory shear, the typical experimental setup is identical to that for simple shear flow (Figure 1.60d), with the sole exception being that the cone does not rotate continuously in one direction but, rather, alternates between clockwise and counterclockwise rotations periodically at an angular frequency ω.

The rheometer plate is held stationary, which imposes a sinusoidal shear strain on the sample, as sketched in Figure 1.60f. If $\Delta\Phi$ is the maximum angular displacement of the cone, then the amplitude of the strain curve is given by

$$\gamma_{max} = \frac{\Delta\Phi}{\alpha} \qquad 1.60(6)$$

where α is the cone angle. The torque necessary to keep the plate from rotating is measured and used to calculate the shear stress τ via Equation 1.60(3). Typical behavior for the shear stress τ is also plotted against time in Figure 1.60f. Note that the shear stress and the shear strain are both sinusoidal functions of time, and they both have the same angular frequency ω. However, there is a phase shift between the two curves—the strain curve lags the stress curve by a constant phase angle denoted δ.

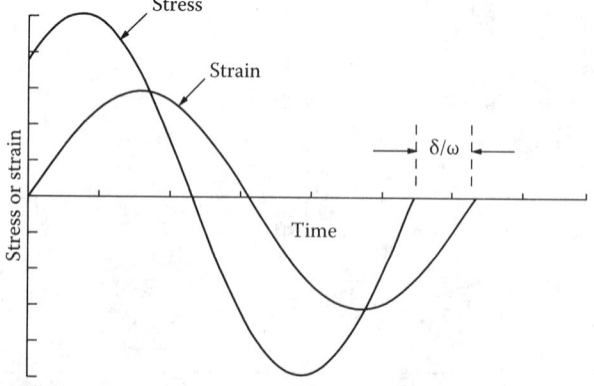

FIG. 1.60f
Typical stress and strain curves in DMA.

Viscoelastic Shear Moduli The amplitude of the shear stress curve divided by the amplitude of the shear strain curve is denoted G*, the dynamic shear modulus. Thus, the oscillatory shear experiment yields two independent rheological properties: G* and δ. To obtain linear rheological properties, one needs to check that the measured values of G* and δ are independent of γ_{max}. An alternative set of properties are the elastic storage modulus (G′) and the viscous loss modulus (G″), given by G′ = G* cos δ, G″ = G* sin δ.

For a cyclical shear deformation process at frequency ω, G′ is a measure of the sample's ability to store elastic strain energy, and G″ is a measure of viscous energy dissipation. The relative importance of energy dissipation and energy storage is given by the loss tangent tan δ

$$\tan\delta = \frac{G''}{G'} \qquad 1.60(7)$$

The value of tan δ is infinite for a purely viscous liquid, zero for a perfectly elastic solid, and intermediate between these two extremes for a real viscoelastic material. However, the mechanical behavior of most viscoelastic materials (solid or liquid) becomes more like the perfectly elastic solid as the frequency is increased, which may be explained as follows.

When a polymeric sample is subjected to a shear strain, the polymer molecules have many different modes for relaxing the stress with widely varying relaxation times. At very low oscillation frequencies, almost all of these relaxation modes are fast enough to be operative, so energy storage is minimal. Conversely, at very high frequencies, most of the polymer molecules do not have sufficient time to relax during the period of oscillation ($2\pi/\omega$). Thus, the polymer molecules remain stretched throughout the entire oscillation cycle, giving rise to increased elastic energy storage and increased shear modulus G′.

Molecular Weight Distribution Polymeric materials have a wide spectrum of relaxation times, with each relaxation time corresponding to a different possible mode of mechanical relaxation. As explained in the preceding section, one can probe the mechanical relaxation processes that occur at different time scales by measuring oscillatory shear properties at various frequencies. The largest molecules in the sample contribute to relaxation processes occurring at long times or low frequencies, whereas small molecules contribute to relaxation at high frequencies.

In recent years, rheometer companies have introduced software that can be used to invert curves of G′ and G″ versus ω measured for a molten thermoplastic to give the molecular weight distribution (MWD) of the sample (Chapter 1.41). This method of estimating the MWD is more sensitive to the presence of polymer branches and high-molecular-weight *tails* in the sample than traditional solution-based methods such as gel permeation chromatography (Chapter 1.41).

However, the theory on which the inversion software is based is still evolving and thus may not work for all types of polymers. Dynamic curves measured on samples in the solid state cannot be inverted to give the MWD.

Time–Temperature Superposition Time–temperature superposition is an approximate method used by experimental rheologists to extend the effective frequency range of a rheometer in DMA experiments. As an example, suppose one is interested in predicting the in-use mechanical performance of a polymeric solid subjected to a constant stress for 1 week at 25°C. The appropriate frequency range for DMA of the solid at 25°C would be (1 week)$^{-1}$ ≈ 10^{-6} s^{-1}, which might be too low to access with a given rheometer. However, if the empirical time–temperature shift factor for the sample (a_T) is known to be 1×10^{-3} for shifting between 25°C and 90°C, then the appropriate frequency range for DMA at 90°C is 10^{-6} s^{-1}/a_T = 10^{-3} s^{-1}, which may be accessible to the rheometer motor.

If the principle of time–temperature equivalency is valid for this sample, then the value of G* measured at 90°C and a frequency of ω = 10^{-3} s^{-1} is approximately equal to the value of G* for the same sample at 25°C and ω = 10^{-6} s^{-1}. Unfortunately, this principle fails for many complex polymer materials. In particular, the principle fails if the sample has a phase transition in the temperature range between the use temperature and the testing temperature.

Conversion among Rheological Properties Linear rheological properties measured in oscillatory shear can be rigorously converted to linear rheological properties defined for stress relaxation tests, shear creep tests, or any other linear experiment. The software needed for the conversion is often supplied with the rheometer. However, the software cannot be used to calculate nonlinear rheological properties such as the non-Newtonian viscosity versus shear rate.

Parallel Disk

The principal advantage of the cone-and-plate instrument is that it imposes a uniform value of the strain rate κ on the sample in flow experiments and a uniform value of the shear strain γ in DMA experiments. This is particularly useful when analyzing liquids such as molten thermoplastics having properties that strongly depend on κ.

Unfortunately, the cone-and-plate geometry cannot be used to analyze solids. Recall that the cone, which has a truncated tip, should be slowly lowered into the sample until the truncated region is only 50–100 μm above the plate, depending on the cone. This cannot be done with solid samples. Probably, it should not be done either with liquid samples containing particles larger than a few micrometers in size, because particles of this size are large enough to disturb the flow field in the narrow gap region of the cone-and-plate instrument.

Geometry The parallel-disc rheometer geometry (Figure 1.60g) is obtained by replacing the cone in a cone-and-plate instrument with a stainless steel disc of the same diameter. All other features of the rheometer, including transducers and drive motors, remain the same. The sample must be cut or molded into a disc of uniform thickness and radius equal to the rheometer disc radius. The sample is sheared in between the two parallel stainless steel discs or plates, with the same rheometer gap at all radial positions. The parallel-disc geometry can be used to study a solid sample of any thickness, provided that the sample adheres to the surfaces of the discs. For analysis of suspensions, one can minimize wall-slip errors by choosing the gap to be much larger than the suspended particle size.

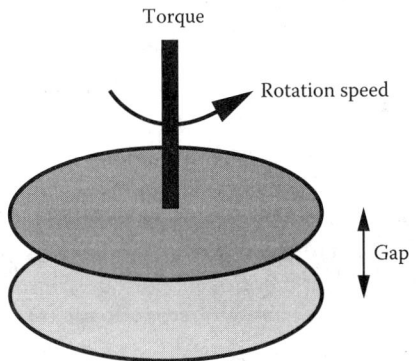

FIG. 1.60g
Parallel-disk instrument, where the "gap" = L.

The parallel-disc instrument, like the cone-and-plate, can be used to make either stress-controlled or strain-controlled measurements on the sample in steady-shear flow or oscillatory shear strain. The flow field in the instrument is expected to be equivalent to simple shear flow (Figure 1.60c) on a local level, but the local value of the shear rate κ increases with radial coordinate. The maximum value of the shear rate occurs in the sample located at the rim of the parallel discs, and is equal to

$$\kappa_{max} = \frac{R\Omega}{L} \qquad 1.60(8)$$

where
- R is the radius of the discs
- Ω is the angular speed of the rotating disc
- L is the gap between the discs

Viscosity μ® N_1, G', or G''. What is this®, it should be,

Limitation The principal disadvantage of the instrument arises from the variation of the shear rate value within the sample. As a consequence, the parallel-disc instrument cannot be used to make a single-point measurement of the viscosity of a non-Newtonian liquid. This may be seen from the fluid mechanics equation used to calculate the viscosity. If the shear rate at the rim of the discs is κ_{max}, and the torque on the discs is T, then the sample viscosity value at shear rate equal to κ_{max} is given by

$$\mu = T(2\pi R 3 \kappa max)^{-1} \left[\frac{3 + d\ln(T/(2\pi R3))}{d\ln(\kappa max)} \right] \qquad 1.60(9)$$

Equation 1.60(9) contains the derivative of the torque with respect to the shear rate at the rim, in contrast to the analogous equation (Equation 1.60(3)) used with the cone-and-plate instrument. Hence, to determine the viscosity at a single value of the shear rate, one has to measure the torque at several different rim shear rate values and then numerically evaluate the derivative in Equation 1.60(9).

Rectangular Torsion

The parallel disc instrument can be used only with solid samples that adhere to the stainless steel discs so that the *no slip* boundary condition is satisfied. *Slip* refers to the occurrence of relative motion between the steel and the sample located at the steel/sample interface. When slip occurs during DMA experiments, the computer software will use a value for the strain that is larger than the true value imposed on the sample.

One way to detect the occurrence of slip is to see if slight compression of the sample between the parallel plates has a drastic effect on the DMA results. When slip is a problem, the rectangular torsion instrument (Figure 1.60h) offers a viable alternative to the parallel-disc instrument for solid samples. The sample must be cut or molded into a thin rectangular specimen of uniform thickness.

The specimen is clamped into the instrument at both ends, and the clamping force is adjusted so that no slip can occur. During testing, one clamp undergoes low amplitude oscillatory rotations at frequency ω while the torque required to keep the other clamp stationary is measured with a torque transducer. The experimentally determined properties are the dynamic shear moduli of the sample, G' and G'', as a function of frequency.

Torsional DMA can be used to identify the specimen glass transition temperature as the temperature at which G' drops precipitously at fixed frequency. However, torsional DMA cannot be used to determine the MWD of the specimen.

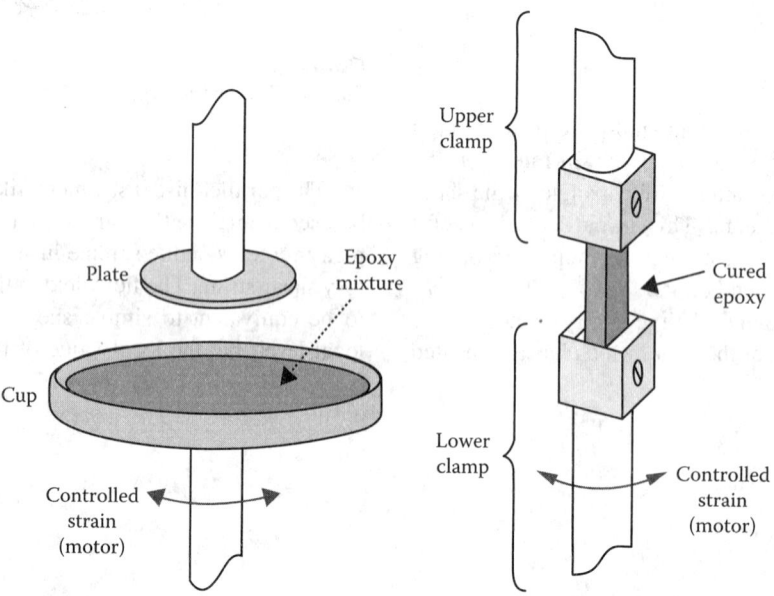

FIG. 1.60h
Rectangular torsion rheometer. (Courtesy IOP Science.)

Coaxial Cylinder

The principal advantage of the coaxial-cylinder instrument (Figure 1.60i) is a large surface area for contact between the sample and the instrument, which increases torque levels and hence is advantageous for measuring low viscosity values. However, the instrument cannot be used to measure normal stress differences and cannot easily be used with solid samples. The liquid sample completely fills the narrow gap or annulus between two concentric cylinders.

During testing, the outer cylinder (called the *cup*) is rotated while the inner cylinder (the *bob*) is held stationary. When the cup is rotated continuously, the flow field in the sample looks like simple shear flow (Figure 1.60c) on a local level. However, as with the parallel-disc instrument, the shear rate value varies with location within the sample. The sample viscosity is calculated from the measured torque on the bob. Linear shear moduli G′ and G″ can also be measured in experiments in which the cup undergoes oscillatory rotations.

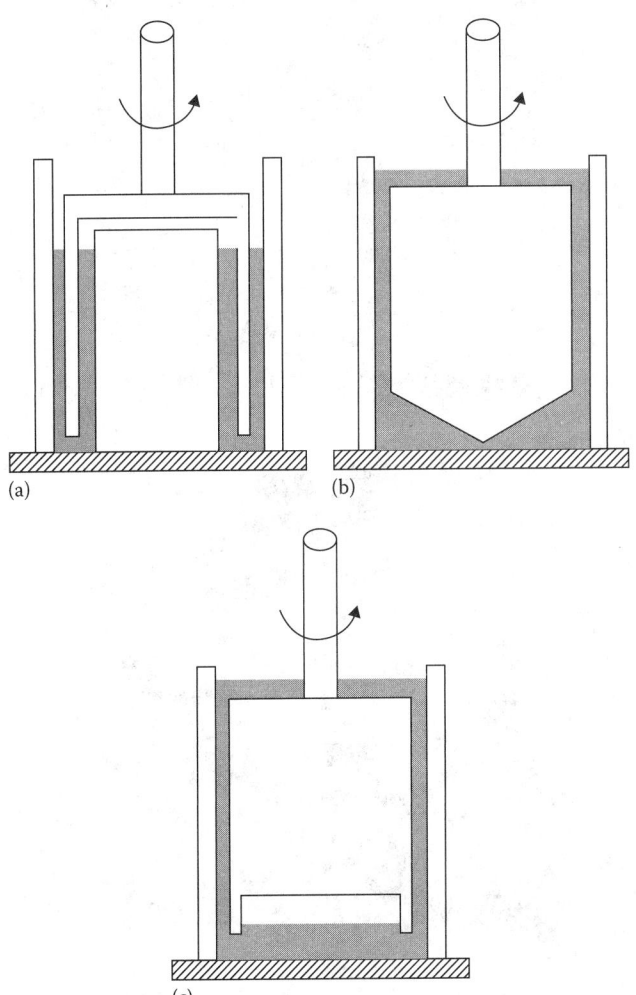

FIG. 1.60i

Three alternative cylindrical designs. (a) is the double gap, (b) the one with cone and plate at the bottom, and (c) is the hollow cavity at the bottom configuration, which serves to trap the air. (Courtesy of National Institute of Standards and Technology.)

Tension/Compression and Bending

All of the instruments considered thus far impose a shear deformation (Figure 1.60b) on the sample. However, the behavior of a solid sample during tension, compression, or bending may also be of interest, and instruments are also available for these types of deformations.

In fact, the principal difference between a so-called *rheometer* and a *dynamic mechanical analyzer* is the number of instruments available with the latter for applying various bending and tensile deformations on solid samples. The bending and tensile/compression deformations are usually applied to the specimen as an oscillatory function of time at angular frequency ω. When a small time-independent uniaxial tensile deformation is applied to a specimen, the ratio of the tensile stress to the tensile strain is defined as the linear elastic tensile modulus E.

In oscillatory experiments, one measures at each frequency the tensile storage modulus (E′) and tensile loss modulus (E″), properties which are analogous to G′ and G″, defined earlier for shear. In fact, for a material that does not undergo volume changes during deformation, the dynamic tensile moduli are exactly three times larger than the corresponding dynamic shear moduli.

Extensional Flow

The value of the extensional viscosity μ_{ex} is a measure of a liquid's resistance to stretching during flow. For commercial thermoplastics, the value of μ_{ex} is important in plastics-forming operations such as fiber spinning and blown-film extrusion.

Figure 1.60j shows an idealized representation of uniaxial extensional flow. Here, the cylinder of liquid is stretching along its longitudinal axis while contracting in the radial dimension so that the volume remains constant. To obtain a constant extension rate, the stretching force F is adjusted in such a way that the length z of the liquid cylinder increases exponentially with time t.

$$z \propto \exp(\varepsilon t) \qquad 1.60(10)$$

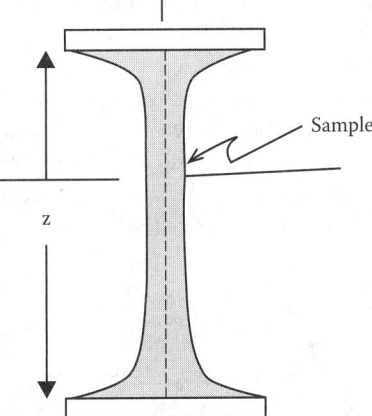

FIG. 1.60j

Uniaxial extensional flow rheometer sensor.

Here, ε is the rate of extension, analogous to the rate of strain κ in shear flow. The stretching force F and the cylinder cross-sectional area A are both functions of time. The extensional viscosity μ_{ex} is a material property defined as

$$\mu_{ex} = F/(A\varepsilon) \qquad 1.60(11)$$

Thus, for a given rate of elongation ε, the necessary stretching force increases with the elongational viscosity of the liquid μ_{ex}. Two thermoplastic samples may have identical shear viscosity curves (μ vs. κ) and completely different extensional viscosity curves (μ_{ex} vs. ε).

For some thermoplastics, the measured value of μ_{ex} calculated with Equation 1.60(11) never reaches a steady value, even though the extension rate is constant. There is little doubt that μ_{ex} is a key property for polymer processing flows that has been comparatively neglected in rheometry.

Extensional Viscosity Detectors Rheometers that have been built for measuring the extensional viscosity have used different techniques. These included the capillary breakup, rotary clamp stretching, stagnation point flow, and opposing jets techniques. However, it has proven difficult to build a robust and accurate extensional flow rheometer for industrial use.

Between 1987 and 1990, the International Committee on Rheology organized a round-robin test of extensional flow rheometers. In this test, the value of μ_{ex} was measured for an international standard fluid in various laboratories around the world. The test results were disappointing, with values measured in different laboratories varying by as much as several orders of magnitude. More recently, the filament stretching technique has found favor among academic laboratories for measurements of the extensional viscosity.

Capillary

The capillary rheometer is used to study the pressure-driven flow of a molten thermoplastic through a capillary. The viscosity of the process sample can be calculated from measurements of volumetric flow rate versus pressure drop along the capillary axis. Ordinarily, capillary rheometers are used only to calculate the viscosity; hence, these instruments are covered under the discussion of industrial viscometers (Chapter 1.71).

However, it is possible to use capillary rheometry to measure N_1 as well as the viscosity. This can be done by varying the length of the capillary (at fixed diameter) and noting the effect on the pressure drop within the entrance region of the capillary.

CONCLUSIONS

Molten thermoplastics have several elastic properties in addition to the viscosity that can be measured by using rheometers. True material properties can be measured only if the rheometer subjects the sample to a controlled velocity field or to a controlled deformation. Hence, the popular melt flow indexer, while useful, cannot be used to measure a true material property such as viscosity.

Probably the most significant innovation of recent years was the development of an online process rheometer for use in controlling polymer processing and similar operations, which can continuously measure the true elastic properties of a thermoplastic during processing (Figure 1.60k).

The online rheometer (OLR) continuously measures the flow properties of chemical or polymer products in the flowing pipeline, allowing the quality control of their flow properties through rheology. The flow properties of the flowing product are directly related to its pourability, viscosity, spreadability, tack, elasticity, and other textural or flow properties that can thereby be controlled quickly and online, without the need for taking grab samples for laboratory analysis.

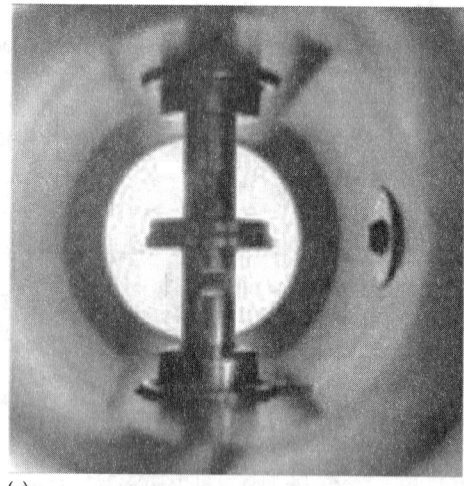

(a)

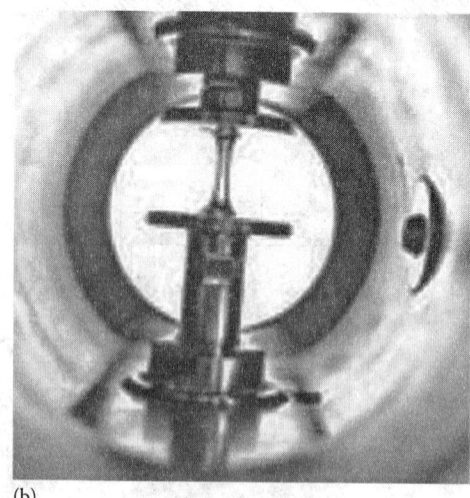

(b)

FIG. 1.60k
The operation of the online rheometer—OLR. (a) Inside the OLR with the oscillatory squeeze-flow plates closed and (b) Inside the OLR with the oscillatory squeeze-flow plates open. (Courtesy of OnLine Rheometer Group of Rheology Solutions Pty Ltd, Bacchus Marsh, Victoria, Australia.)

SPECIFICATION FORMS

When specifying rheometers, the ISA Form 20A1001 can be used to specify the features required for the device itself, and when specifying both the device and the nature of the application, including the composition and/or properties of the process materials, ISA Form 20A1002 can be used. The following are the forms that are reproduced with the permission of the International Society of Automation.

#	RESPONSIBLE ORGANIZATION	ANALYSIS DEVICE — Operating Parameters	#	SPECIFICATION IDENTIFICATIONS	
1			6		
2	ISA		7	Document no	
3			8	Latest revision	Date
4			9	Issue status	
5			10		

#	ADMINISTRATIVE IDENTIFICATIONS		
11			
12	Project number	Sub project no	
13	Project		
14	Enterprise		
15	Site		
16	Area	Cell	Unit
17			
18	SERVICE IDENTIFICATIONS		
19	Tag no/Functional ident		
20	Related equipment		
21	Service		
22			
23	P&ID/Reference dwg		
24	Process line/nozzle no		
25	Process conn pipe spec		
26	Process conn nominal size	Rating	
27	Process conn termn type	Style	
28	Process conn schedule no	Wall thickness	
29	Process connection length		
30	Process line matl type		
31	Fast loop line number		
32	Fast loop pipe spec		
33	Fast loop conn nom size	Rating	
34	Fast loop conn termn type	Style	
35	Fast loop schedule no	Wall thickness	
36	Fast loop estimated lg		
37	Fast loop material type		
38	Return conn nominal size	Rating	
39	Return conn termn type	Style	

#	SERVICE IDENTIFICATIONS Continued		
40			
41	Return conn matl type		
42	Inline hazardous area cl	Div/Zon	Group
43	Inline area min ign temp	Temp ident number	
44	Remote hazardous area cl	Div/Zon	Group
45	Remote area min ign temp	Temp ident number	
46			
47			
48	COMPONENT DESIGN CRITERIA		
49	Component type		
50	Component style		
51	Output signal type		
52	Characteristic curve		
53	Compensation style		
54	Type of protection		
55	Criticality code		
56	Max EMI susceptibility	Ref	
57	Max temperature effect	Ref	
58	Max sample time lag		
59	Max response time		
60	Min required accuracy	Ref	
61	Avail nom power supply		Number wires
62	Calibration method		
63	Testing/Listing agency		
64	Test requirements		
65	Supply loss failure mode		
66	Signal loss failure mode		
67			
68			

#	PROCESS VARIABLES	MATERIAL FLOW CONDITIONS		Units
69				
70	Flow Case Identification			
71	Process pressure			
72	Process temperature			
73	Process phase type			
74	Process liquid actl flow			
75	Process vapor actl flow			
76	Process vapor std flow			
77	Process liquid density			
78	Process vapor density			
79	Process liquid viscosity			
80	Sample return pressure			
81	Sample vent/drain press			
82	Sample temperature			
83	Sample phase type			
84	Fast loop liq actl flow			
85	Fast loop vapor actl flow			
86	Fast loop vapor std flow			
87	Fast loop vapor density			
88	Conductivity/Resistivity			
89	pH/ORP			
90	RH/Dewpoint			
91	Turbidity/Opacity			
92	Dissolved oxygen			
93	Corrosivity			
94	Particle size			
95				
96	CALCULATED VARIABLES			
97	Sample lag time			
98	Process fluid velocity			
99	Wake/natural freq ratio			
100				

#	PROCESS DESIGN CONDITIONS		
101			
102	Minimum	Maximum	Units
103 – 132			

#	MATERIAL PROPERTIES			MATERIAL PROPERTIES Continued		
133 / 137						
134 / 138	Name			NFPA health hazard	Flammability	Reactivity
135 / 139	Density at ref temp	At				
136 / 140						

Rev	Date	Revision Description	By	Appv1	Appv2	Appv3	REMARKS

Form: 20A1001 Rev 0 © 2004 ISA

Analytical Measurement

	RESPONSIBLE ORGANIZATION	ANALYSIS DEVICE COMPOSITION OR PROPERTY Operating Parameters (Continued)		SPECIFICATION IDENTIFICATIONS	
1	ISA		6	Document no	
2			7	Latest revision	Date
3			8		
4			9	Issue status	
5			10		

	PROCESS COMPOSITION OR PROPERTY			MEASUREMENT DESIGN CONDITIONS				
	Component/Property Name	Normal	Units	Minimum	Units	Maximum	Units	Repeatability
11								
12								
13								
...								
75								

Rev	Date	Revision Description	By	Appv1	Appv2	Appv3	REMARKS

Form: 20A1002 Rev 0 © 2004 ISA

Definitions

Newtonian fluid: In a Newtonian fluid, the relationship between the shear stress and the shear rate is linear and this proportionality is the coefficient of viscosity.

Non-Newtonian fluid: In a non-Newtonian fluid, the relationship between the shear stress and the shear rate is not a constant (viscosity) and can even be time dependent. Therefore, non-Newtonian is a fluid whose flow properties differ in any way from those of Newtonian ones. Most commonly the viscosity (the measure of a fluid's ability to resist gradual deformation by shear or tensile stresses) of non-Newtonian fluids is dependent on shear rate or shear rate history. Many salt solutions and molten polymers are non-Newtonian fluids, as are many commonly found substances such as ketchup, custard, toothpaste, starch suspensions, paint, blood, and shampoo.

Peltier effect is whereby heat is given out or absorbed when an electric current passes across a junction between two materials.

Rheometer: It characterize the viscoelastic properties of polymer melts and other materials. They are laboratory instruments that are used to measure the nature of the flow of polymers and other a liquids, suspensions, or slurries when a force causes them to move. They are used for those fluid flows that cannot be defined by viscosity alone and require more parameters to be measured and controlled in order to define the polymer's flow characteristics. Rheometers are quality control tools and serve to assess the processability of resins, are used by plastics compounders, and also serve as R&D tools to help determine which resin best fits a particular process or application.

Viscosity is the fluid's resistance to gradual deformation caused by shear or tensile stresses.

Abbreviations

DMA   Dynamic Mechanical Analysis
MWD   Molecular Weight Distribution
OLR   Online Rheometer

Organization

NIST   National Institute of Standards and Technology

Bibliography

Baird, D. G., First normal stress difference measurements for polymer melts at high shear rates in a slit-die using hole and exit pressure data (2008) http://www.che.vt.edu/Faculty/Baird/Papers/2008_1.pdf.

Barnes, H. A., *A Handbook of Elementary Rheology* (2000) http://citeseerx.ist.psu.edu/showciting?cid=24183441.

Ceramic Industry, Using rheometry to improve manufacturing (2002) http://www.ceramicindustry.com/articles/print/using-rheology-to-improve-manufacturing.

Chhabra, R. P. and Richardson, J. F., Non-Newtonian flow in the process industries (1999). http://www.sciencedirect.com/science/article/pii/B9780750637701500038.

Clasen, C., Capillary breakup extension rheometry, 2010. http://cheric.org/PDF/KARJ/KR22/KR22-4-0331.pdf.

Dealy, J. M. and Wissbrun, K. F., *Melt Rheology and Its Role in Plastics Processing*, Van Nostrand Reinhold, New York, 1999. http://www.abebooks.com/book-search/isbn/0792358864/.

Gotro, J., Rheometers (2014) http://polymerinnovationblog.com/rheology-thermosets-part-2-rheometers/.

Harward Edu., Rheometry (2011) http://soft-matter.seas.harvard.edu/index.php/Rheometry.

Hassager, O., Danish Polymer Centre, Extensional rheometry (2013) http://www.dpc.kt.dtu.dk/Instrumentation/Extensional_Rheometry.aspx.

Macosko, C. W., *Rheology: Principles, Measurements, and Applications*, Wiley-VCH, New York, 1994. http://www.wiley.com/WileyCDA/WileyTitle/productCd-0471185752.html.

Malvern, Rotational Rheometry (2012) http://www.anamet.cz/sites/all/storage/school_of_rheology_part_1_rotational.pdf.

Rheology Solutions Pty Ltd (ORL), Rheology solutions (2015) http://www.onlinerheometer.com/downloads/olr004.pdf.

Schweitzer, T., A cone-partitioned plate rheometer cell with three partitions (CPP3) to determine shear stress and both normal stress differences for small quantities of polymeric fluids (2013) http://scitation.aip.org/content/sor/journal/jor2/57/3/10.1122/1.4797458.

Xu, J. et al., Use of a sliding plate rheometer to measure the first normal stress difference at high shear rates (2007) http://link.springer.com/article/10.1007/s00397-006-0156-5?no-access=true.

1.61 Sand Concentration and Subsea Pipeline Erosion Detectors

H. S. GAMBHIR and S. YADAV (2017)

Instrument types	Passive acoustic sand detector
	Intrusive and erosion based sand detector
Applications	Subsea or onshore pipeline monitoring
Method of detection	A. Acoustics
	B. Intrusive
Design pressure	Up to 9,775 psig (674 barg) for passive acoustic sensors
	Up to 10,790 psig (744 barg) for intrusive erosion based sensors
Design temperature	Temperature limit at pipe surface is from −40°F to +392°F (−40°C to +200°C); for instrument electronics, the limit is +10°F to +140°F (−12°C to +60°C)
Design water depth	Up to 3000 m
Unit of measurement	Micrometer (μm)
Accuracy	0.1% of full scale or better
Range	0–4 mm; metal loss resolution 5 nm
Wetted materials of construction	One of the manufacturers provides probe material AISL 316L as standard (other material on request); Erosion element material is Monel® 400
Cost	$40,000–$80,000 for passive acoustic type
	$90,000–$120,000 for intrusive erosion based
Partial list of suppliers	Aquip Systems PTY Ltd (A & B), http://www.aquip.com.au/products/acoustic-sand-detectors/
	Clampon (A), http://www.clampon.com/
	Emerson Process Management—ROXAR (A & B), http://www2.emersonprocess.com/en-US/brands/roxar/sanderosion/subseasanderosion/Pages/SubseaSand.aspx
	Teledyne Cormon (A & B), http://www.cormon.com/products/sand_erosion.aspx

INTRODUCTION

This chapter describes the technology behind the detection of sand and erosion in most of the oil and gas development fields. It also identifies the type of measurements, its principle of operation, and the installation and retrieval mechanism.

BACKGROUND[*,†,‡]

The production of sand particles along with the process fluid in the oil and gas subsea can lead to erosion and damage of the pipeline, process systems and clogging of filters, etc.

---

[*] Bill Hedges and Andy Bodington, A comparison of monitoring techniques for improved erosion control: A field study.
[†] Birkeland R., S.E. Lilleland, R. Johnsen, N.A. Braaten, Erosion monitoring manages sand production, *Oil & Gas Journal*, September 28, 1998, http://www.ogj.com/articles/print/volume-96/issue-39/in-this-issue/general-interest/erosion-monitoring-manages-sand-production.html.
[‡] Earles, D.M., C.W. Stoesz, A.S. Amaral, B. Hughes, J.G. Pearce, H.A. DeJongh, and F.H.K. Rambow, Real-time monitoring of sand control completions (SPE 134555), Shell Exploration and Production, September 19–22, 2010.

The sand production indicates the oil and gas reservoir degradation or, in the worst case, a possible collapse of the reservoir. The production of sand could also fill up the process equipment due to the settling of sand. This would necessitate the repair/replacement of process equipment, pipeline, and result in production and revenue loss. Therefore, an effective sand and erosion monitoring system is essential for any subsea/onshore field development that have the potential to produce sand.

Apart from being an important safety device, continuous sand monitoring system can help in optimizing well productivity by determining the maximum sand-free rate for a well. With the increase in the production from the well, generally it is followed by an increase in sand production. With a reliable sand monitoring system, the production from the well can be reduced as soon the sand is detected. However, with a continuous sand monitoring system, one can check whether sand production has a tendency to reduce over time due to consolidation of the producing reservoir. The production from any of the wells can be further optimized and the productivity of the well can be increased.

PASSIVE ACOUSTIC SAND DETECTOR*,†,‡,§

Passive acoustic sand detector consists of an acoustic sensor that is clamped onto the pipe wall (Figure 1.61a). This acoustic sensor can be installed on each subsea well or the place where the sand detection is required. The acoustic sensor picks up the amount of particles colliding with the pipe wall by registering the generated series of ultrasonic pulses. Therefore, it has to be installed after a 90° bend. This type of sand detector is not intrusive and thus can be retrieved very easily.

Principle of Operation

The solid particles drifting in the gas collides with each other and against the pipe inner wall, this collision creates the ultrasonic acoustic pulse sequence like a noise. Frequency of this noise spectrum is in MHz magnitude.

The ultrasonic signal generated due to the collision of the sand particles is transmitted through the pipe wall and picked up by the acoustic subsea sensor. A preamplifier

---

* Clampon SandQ™, *Monitor Brochure*, Clampon, Bergen, Norway, June 2011.
† Goodman, H.E., Wellbore integrity, sand management and frac pack, *Journal of Petroleum Technology*, 61, 74, September 2009.
‡ ClampOn Acoustic Sand Detectors, http://www.aquip.com.au/products/acoustic-sand-detectors/.
§ Roxar Sand monitor, http://www2.emersonprocess.com/en-us/brands/roxar/sanderosion/topsidesanderosion/pages/roxarsandmonitor.aspx.

FIG. 1.61a
Acoustic sensor mounted on a pipeline using clamps. (From Andrews, J. et al., Production enhancement from sand management philosophy (SPE94511)—A case study, SPE Statoil ASA.)

is connected to the acoustic transducer by a miniature coaxial cable. The preamplifier is connected to a signal processing board that outputs a 4–20 mA current signal and also a Modbus signal in logarithmic proportion to the measured acoustic energy. The sand detector can be installed on the pipelines at water depth up to 3000 m below the mean sea level.

Advantages

- Noninvasive
- Easy to install and relocate
- Contains no moving parts
- Can be mounted in hostile environments
- Incorporates sensitive, intelligent, and intrinsically safe sensors
- Same sensor can perform PIG detection, spectrum analysis, and more

Limitations

A problem frequently encountered in passive acoustic sand monitoring is interference from sources other than solid particles in the flow stream such as noise from liquid/gas mixtures, droplets in high velocity gas wells, mechanical/structural noise, and electrical interference. With new technologies developed by some of the manufacturers, good signal-to-noise ratio that is vital for quality measurement is filtered to some extent.

INTRUSIVE AND EROSION BASED SAND DETECTOR*,†,‡,§

The second type of sand detector is based on measuring the change in electrical resistance of sensing elements that are eroded by the sand. This is an intrusive sensor and can be located on the subsea wells or any location where sand/erosion detection is envisaged. It cannot be retrieved by itself; however, it can be designed to be installed on the structural module that can be retrieved subsea.

Principle of Operation

According to the Ohm's law

$$\Delta E = R \cdot I$$

where
  ΔE is the potential drop across the element
  R is the resistance of the element
  I is the current

Also,

$$R = \frac{\rho \cdot L}{T \cdot W}$$

where (as per Figure 1.61b),
  ρ is the resistivity of the element (property of the material)
  L is the length of the element
  T is the thickness of the element
  W is the width of the element

---

* Birkeland, R., S. E. Lilleland, R. Johnsen, and N. A. Braaten, Erosion monitoring manages sand production, *Oil & Gas Journal*, September 28, 1998, http://www.ogj.com/articles/print/volume-96/issue-39/in-this-issue/general-interest/erosion-monitoring-manages-sand-production.html.
† McLaury, B.S. and S.A. Shiraz, Generalization of API RP 14E for erosive service in multiphase production, Paper SPE 56812, *SPE Annual Technical Conference and Exhibition, Houston*, TX, October 3–6, 1999.
‡ Goodman, H.E., Wellbore integrity, sand management and frac pack, *Journal of Petroleum Technology*, 61, 74, September 2009.
§ Subsea Sand-Erosion Sensors, http://www2.emersonprocess.com/en-US/brands/roxar/sanderosion/subseasanderosion/Pages/SubseaSand.aspx.

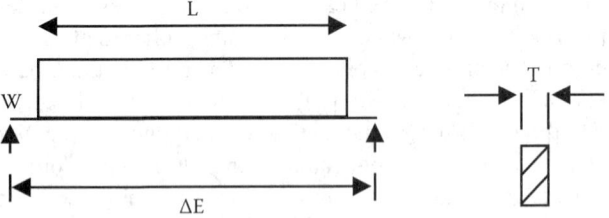

FIG. 1.61b
Resistance element.

Each element is connected to electrical wires and fed with a constant current (I). As thickness (T) of the element reduces due to erosion, its resistance (R) will increase. By measuring the potential drop (ΔE) across the element, the change in resistance of each element can be continuously monitored and the thickness reduction of the element can be quantified down in the nanometer (nm) range.

The elements are redundantly coupled, and up to four individual elements can be used on one probe (Figure 1.61c). A reference element is included at the back of the probe body to compensate for temperature changes that influence the resistance of the element material. This reference element is exposed to the same temperature but is not exposed to the erosive sand particles.

The length of the sand probe is adjusted to cover the internal pipe diameter. The element thickness is determined based on the information of expected production rates, maximum tolerable continuous sand production, and average reservoir grain size. Based on the required lifetime of the sensor and a certain safety margin, the total element erosion of the element can be calculated to determine the required thickness of the element.

Sample specification of sand detector is given in Table 1.61a.

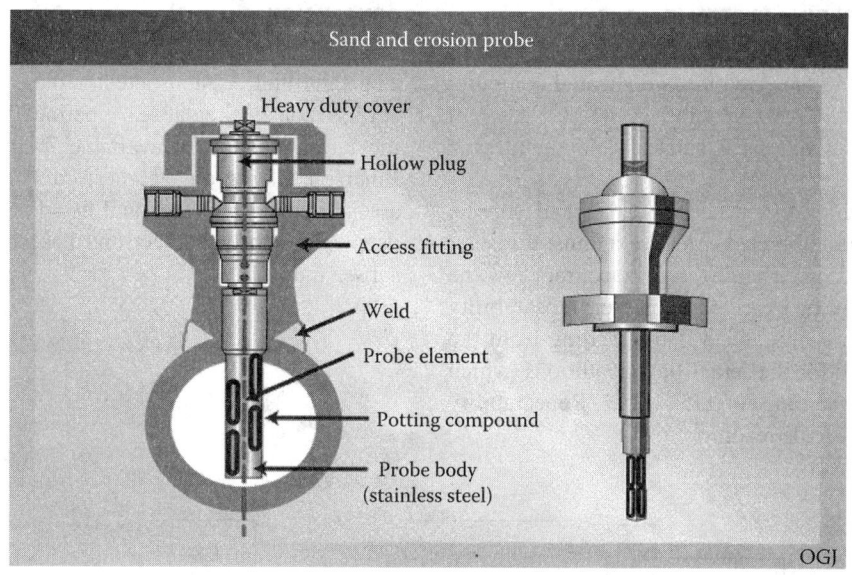

FIG. 1.61c
Intrusive and erosion based sand detector. (From Hank Rawlins, C., Sand management methodologies for sustained facilities operations (SPE 64645-MS), SPE, eProcess Technologies.)

TABLE 1.61a
Example of Detailed Specification of Sand Detector (A)

Category	Comparison Criteria	Sand Detector (A) and (B)
Operating envelope	Particle size	15–2,000 μm
	Particle concentrations	≥1 ppm
	Flow velocities	1–80 m/s
	Time base	Selectable multiple of minutes
	Pressure	Pipe line pressure rating (up to 15,000 psi)
	Response time	<1 s
Accuracy and repeatability	Accuracy in measuring quantity of particles greater than 325 mesh at 95%–100% GVF	±10%–25% Absolute for subsea product without sand injections
	Repeatability	1%
Maintenance	Calibration	In-house (during fabrication)
	Sensitivity to required input	*Trending*: N/A
		Quantified: Accuracy of input important
	Recommended maintenance	No need
Intrusive erosion based sensor	Sensor material	Inconel, Carbon steel
	Flange/hub mounting	Usually ANSI or API
	Metal loss resolution	10 nm to 0.5 μm dependent on element thickness
Interface	Pipe dimensions	Any
	Piping requirements	Typically, sensors are installed downstream of 90 degree bend for the acoustic type
	Power consumption	Approx. 60–100 mA at 24 VDC
	Communication methods	Digital RS 485/422 (ASCII, Binary, Modbus) 4–20 mA output, Relay output, Canbus, and Profibus
	Approx. weight (dry)	Approx. 10–25 kg
	MTBF	25 years

TYPICAL SAND MONITORING SYSTEM/DATA FLOW

Several sensors can be connected to its dedicated computer or directly to the plant SCADA system.

A schematic shows a typical subsea sand monitoring system (Figure 1.61d).

No sand monitoring system can be calibrated subsea. Sensors are tested and calibrated in-house against a master sensor before installation. The deviation (accuracy) in the range of 25%–50% can be expected. However, repeatability is ±1%. Accuracy can be improved if the reading from the sand sensor is adjusted for the actual flow conditions, which will give accuracy in the range of ±15%–25%. Repeatability remains at ±1% of full-scale reading.

CONCLUSION

The sand and erosion detector measures the sand particles and erosion in the subsea pipeline. The subsea devices are the latest technology available in the market supplied by Emerson, Clamp-on Weatherford, Aquip, etc. These devices are proven design and qualified for the installation in the deepwater subsea project and has been widely used in the subsea industry.

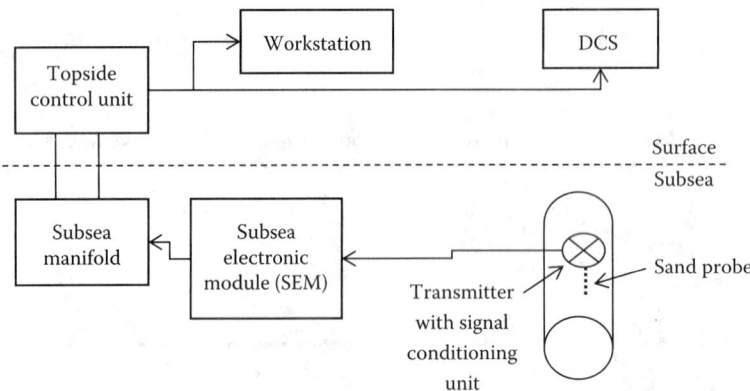

FIG. 1.61d
Typical sand monitoring system.

SPECIFICATION FORMS

For sand and erosion detectors, the ISA Form 20A1001 can be used to specify the features required for the device itself, and when specifying both the device and the nature of the application, including the composition and/or properties of the process materials, ISA Form 20A1002 can be used. The following are the forms that are reproduced with the permission of the International Society of Automation.

1	RESPONSIBLE ORGANIZATION		ANALYSIS DEVICE		6	SPECIFICATION IDENTIFICATIONS			
2					7	Document no			
3	(ISA)		Operating Parameters		8	Latest revision		Date	
4					9	Issue status			
5					10				
11	ADMINISTRATIVE IDENTIFICATIONS				40	SERVICE IDENTIFICATIONS Continued			
12	Project number		Sub project no		41	Return conn matl type			
13	Project				42	Inline hazardous area cl		Div/Zon	Group
14	Enterprise				43	Inline area min ign temp		Temp ident number	
15	Site				44	Remote hazardous area cl		Div/Zon	Group
16	Area		Cell	Unit	45	Remote area min ign temp		Temp ident number	
17					46				
18	SERVICE IDENTIFICATIONS				47				
19	Tag no/Functional ident				48	COMPONENT DESIGN CRITERIA			
20	Related equipment				49	Component type			
21	Service				50	Component style			
22					51	Output signal type			
23	P&ID/Reference dwg				52	Characteristic curve			
24	Process line/nozzle no				53	Compensation style			
25	Process conn pipe spec				54	Type of protection			
26	Process conn nominal size		Rating		55	Criticality code			
27	Process conn termn type		Style		56	Max EMI susceptibility		Ref	
28	Process conn schedule no		Wall thickness		57	Max temperature effect		Ref	
29	Process connection length				58	Max sample time lag			
30	Process line matl type				59	Max response time			
31	Fast loop line number				60	Min required accuracy		Ref	
32	Fast loop pipe spec				61	Avail nom power supply			Number wires
33	Fast loop conn nom size		Rating		62	Calibration method			
34	Fast loop conn termn type		Style		63	Testing/Listing agency			
35	Fast loop schedule no		Wall thickness		64	Test requirements			
36	Fast loop estimated lg				65	Supply loss failure mode			
37	Fast loop material type				66	Signal loss failure mode			
38	Return conn nominal size		Rating		67				
39	Return conn termn type		Style		68				
69	PROCESS VARIABLES		MATERIAL FLOW CONDITIONS		101	PROCESS DESIGN CONDITIONS			
70	Flow Case Identification			Units	102	Minimum	Maximum	Units	
71	Process pressure				103				
72	Process temperature				104				
73	Process phase type				105				
74	Process liquid actl flow				106				
75	Process vapor actl flow				107				
76	Process vapor std flow				108				
77	Process liquid density				109				
78	Process vapor density				110				
79	Process liquid viscosity				111				
80	Sample return pressure				112				
81	Sample vent/drain press				113				
82	Sample temperature				114				
83	Sample phase type				115				
84	Fast loop liq actl flow				116				
85	Fast loop vapor actl flow				117				
86	Fast loop vapor std flow				118				
87	Fast loop vapor density				119				
88	Conductivity/Resistivity				120				
89	pH/ORP				121				
90	RH/Dewpoint				122				
91	Turbidity/Opacity				123				
92	Dissolved oxygen				124				
93	Corrosivity				125				
94	Particle size				126				
95					127				
96	CALCULATED VARIABLES				128				
97	Sample lag time				129				
98	Process fluid velocity				130				
99	Wake/natural freq ratio				131				
100					132				
133	MATERIAL PROPERTIES				137	MATERIAL PROPERTIES Continued			
134	Name				138	NFPA health hazard		Flammability	Reactivity
135	Density at ref temp		At		139				
136					140				
Rev	Date	Revision Description		By	Appv1	Appv2	Appv3	REMARKS	

Form: 20A1001 Rev 0 © 2004 ISA

1	RESPONSIBLE ORGANIZATION		ANALYSIS DEVICE COMPOSITION OR PROPERTY Operating Parameters (Continued)		6		SPECIFICATION IDENTIFICATIONS		
2					7	Document no			
3					8	Latest revision		Date	
4					9	Issue status			
5					10				
11		PROCESS COMPOSITION OR PROPERTY				MEASUREMENT DESIGN CONDITIONS			
12		Component/Property Name	Normal	Units	Minimum	Units	Maximum	Units	Repeatability
13									
14									
15									
16									
17									
18									
19									
20									
21									
22									
23									
24									
25									
26									
27									
28									
29									
30									
31									
32									
33									
34									
35									
36									
37									

Rev	Date	Revision Description	By	Appv1	Appv2	Appv3	REMARKS

Form: 20A1002 Rev 0

Definitions

Gas Volume Fraction (GVF): GVF, or gas volume fraction, is defined as the ratio of the gas volumetric flow rate to the total volumetric flow rate. The total volumetric flow rate is the sum of the liquid volumetric rate and the gas volumetric flow rate. These volumetric flows are usually expressed in actual (not standardized) volumetric terms.

PIG Detection: Pipeline Inspection Gauges (PIG) is run inside the pipeline for the inspection, cleaning, and gauging of the internal diameter. PIG detectors are devices with some type of indicator to detect the passage of a pig in the pipeline during a pig run.

PIG detectors are usually located near the launcher and receiver ends to confirm successful launch and receipt of the pigs or any trouble area of the pipeline, such as at a tee, elbow, wye, etc. They can be equipped with electrical switches and connected to a PLC to be used to operate piping equipment, such as pumps, compressors, valves, mixing tanks, etc., when strategically located along the pipeline or plant.

Wet Gas: The "wet gas" term is used to denote a natural gas flow containing a relatively small amount of free liquid by volume; usually, this may be limited up to about 10%. The mass ratio of gas to liquid varies significantly with pressure for a constant Gas Volume Fraction.

Abbreviations

DCS	Distributed control system
GVF	Gas volume fraction
MTBF	Mean time between failures
PIG	Pipeline inspection gauges
PPM	Parts per million
PSIG	Pounds per square inch gauge
SCADA	Supervisory control and data acquisition
SEM	Subsea electronic module

Organizations

ANSI	American National Standards Institute
API	American Petroleum Institute
ASCII	American Standard Code for Information Interchange

1.62 Spectrometers, Open Path (OP)

J. M. JARVIS (2003)   **B. G. LIPTÁK** (2017)

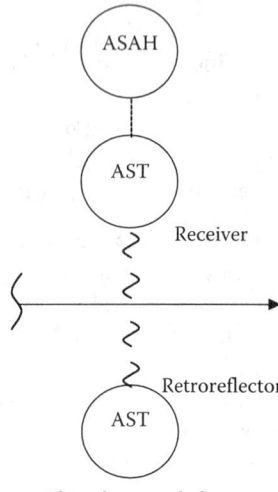

Flow sheet symbol

Applications	Ambient air quality (or fence-line) monitoring for the detection of toxic or hazardous vapors, for emission monitoring, as well as combustible vapors in personnel safety applications. They are used for safety monitoring, security monitoring, stack emissions monitoring, process monitoring, and environmental monitoring. For the suitable *regimes of operation in the electromagnetic spectrum for various path sensors*. See Figure 1.62d.
Types of devices	A. FTIR (Fourier-Transform Infrared) including multiple gas
	A1. Ambient air quality monitor
	A2. Process monitor including combustibles
	A3. Stack monitor
	B. CRDS (Cavity Ring-Down Spectroscopy)
	B1. Ambient air quality monitor
	B2. Process monitor including combustibles
	B3. Stack monitor
	C. TDL (Tunable Diode Laser)
	C1. Ambient air quality monitor
	C2. Process monitor including combustibles
	C3. Stack monitor
	D. OPUV (Open-Path Ultraviolet), DOAS (Differential optical absorption spectroscopy)
	D1. Ambient air quality monitor
	D2. Process monitor including combustibles
	D3. Stack monitor
Costs	A. $50,000 for minimal instrument configuration; $150,000 typical installed cost
	$5,000 to $10,000 and up
	B. $75,000 and up
	C. $35,000–75,000, depending on configuration
	$25,000 and up
	D. $50,000 and up
Partial list of vendors	ABB (Fixed A2); http://www.abb.us/product/us/9AAC202331.aspx
	Boreal Laser Inc. (C1, C2); http://www.boreal-laser.com/externalContent/GasFinder2_brochure.pdf
	Bruker (A); http://www.bruker.com/news-records/single-view/article/bruker-expands-infrared-remote-sensing-product-line-with-open-path-air-monitoring-system-ops.html
	Cascade Technologies (C2, C3); http://www.cascade-technologies.com/CT3400— Extractive-Multigas-Analyser/
	CIC Photonics (Fixed A1, A2, A3); http://www.irgas.com/IRGAS/#
	Det-Tronics (A,B,C) http://www.det-tronics.com/AboutUs/Pages/AboutUs.aspx
	Drager Safety Inc. (D2); http://www.draeger.com/sites/en_aunz/Pages/Chemical-Industry/Draeger-Polytron-7000.aspx

(Continued)

Ecotech (A, C, D); http://www.ecotech.com.au/gas-analyzers/open-path-aqms-systems
Gasmet Technologies Inc. (Fixed and portable A1, A2, A3); http://www.gasmet.com/na
Honeywell Analytics (Fixed A1, C1); http://www.honeywellanalytics.com/en-US/products/HighTechGovernmentSystems/ACM150/Pages/ACM150.aspx
Horiba (Fixed A2); http://www.horiba.com/us/en/semiconductor/products-old/dry-process-monitoring-control/details/ftir-gas-analyzer-402/
IMACC. (A1, A2); http://imacc-instruments.com/open-path-monitoring/
Midac Corporation (Fixed A2, A3) http://www.midac.com/titan-ol.html
MKS Instruments (Fixed A1, A2, A3); http://www.mksinst.com/product/category.aspx?CategoryID=89
Norsk Elektro Optikk (C2); http://www.neo.no/products/
OP-FTIR (A); http://www.optra.com/opftir.html
Optra Inc. (A) http://www.optra.com/images/PS-OP_FTIR.pdf
Picarro, Inc. (Fixed B2); http://www.picarro.com/
Siemens Environmental Systems (C3); http://www.industry.usa.siemens.com/automation/us/en/process-instrumentation-and-analytics/process-analytics/pa-brochures/Documents/LDS6%20Catalog%20Extract.pdf
Thermo Fisher Scientific Inc. (A); http://www.thermoscientific.com/ecomm/servlet/productsdetail?productId=16271471&groupType=PRODUCT&searchType=0&storeId=11152&from=search
Tiger Optics LLC (C1, C2); http://www.tigeroptics.com/app/tigeroptics/products/index.php?category_type=markets&category_id=10
Unisearch Associates (A, C); http://www.unisearch-associates.com/instruments.html
Unisearch Associates, (C) http://www.unisearch-associates.com/instruments/lasir.html

INTRODUCTION

Open-path monitoring is used for analytical measurements in remote and inaccessible locations. It can utilize infrared (IR), visible, and laser or ultraviolet (UV) forms of radiation. The technique is also used for perimeter monitoring of structures and facilities. Such monitoring is either to detect the release of low levels of toxic and hazardous vapors or to signal the release of combustible hydrocarbons at relatively much higher levels. The use of open-path monitoring is driven by either the need for probing an area without physical intrusion or the need to monitor an area that is larger than one that can be cost effectively monitored with a requisite number of point detectors.

Open-path monitoring is a subset of techniques collectively called long-path monitoring. Long-path monitoring includes the use of multi-pass reflection cells to cover distances as great as hundreds of meters and achieve high detection sensitivity.

This chapter confines itself to the discussion of instruments based on absorption. It omits discussion of LIDAR (acronym for light detection and ranging, an abbreviation formed in analogy with RADAR) techniques because of their very specialized nature. Also omitted is the discussion of gas cloud imaging technology, which is based on backscatter absorption. This technology is very new and has not yet been widely adopted in practical industrial situations.

TYPES OF DESIGNS

The four measurement technologies, open-path Fourier transform infrared (OP-FTIR), open-path ultraviolet (OP-UV), open-path tunable diode laser (OP-TDLAS), and cavity ringdown spectrometry (CRDS), all basically employ spectrophotometers designed for operation in a particular part of the spectrum. All spectrophotometers contain a light source, some transfer optics, a spectral dispersion element or optical filters, and optical detectors.

Multiple gas FTIR monitors and spectrometers are analyzers used to detect multiple gases simultaneously. These systems are commonly used for safety monitoring, security monitoring, stack emissions monitoring, process monitoring, and environmental monitoring. FTIR systems are available in both stationary and portable versions, capable of monitoring for several gases in one sample. Some systems share a single FTIR analyzer among multiple sampling points through the use of a manifold and solenoids that rotate through several monitoring points.

A relatively new technology is the CRDS. CRDS systems have several unique applications for the detection of contaminants in high-purity gas, environmental monitoring, food safety, and open-area natural gas detection. Both systems also have laboratory variants as well as units designed for industrial monitoring. Both systems use sophisticated software to analyze results and provide user control.

IR, UV, and TDLAS

Figure 1.62a shows the primary components of the IR, UV, and TDLAS sensors. All these designs have a light source that illuminates the gas in the open path. The light source is collimated with transfer optics and is directed through the open-path length to the receiving optics. The light is transferred through dispersing and filtering elements and is finally received by the optical detector. Note in Figure 1.62a that the dispersion elements can be utilized to pre-disperse the light prior to its passing through the open-path sample, or it can be dispersed after passing through the sample. In OP-FTIR, the interferometer dispersing element is most often configured to modulate and pre-disperse the light prior to passing through the sample. This configuration prevents thermal (blackbody) radiation from being added to the spectrum of the IR source.

Table 1.62a shows how the IR, UV, and TDLAS elements differ, and Figure 1.62b illustrates the operation of the TDLAS sensor.

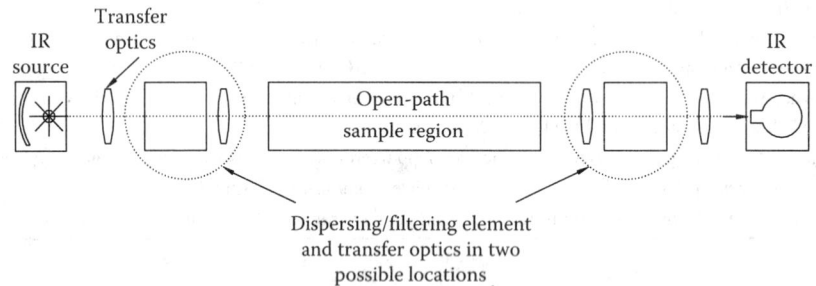

FIG. 1.62a
A schematic of a generalized open-path sensor.

TABLE 1.62a
The Main Features and Characteristics of the Infrared (FTIR), Visible (TDLAS) and Ultraviolet (UV) Radiation Based Designs

	OP-FTIR	OP-TDLAS	OP-UV
Wavelength range (µm)	Atmospheric transmission windows between 2.94–4.16 and 8.0–14.3	Single wavelengths between 0.8 and 1.7	Atmospheric transmission window starting at 0.260 extending to visible
Light source	Heated globar	Solid-state diode laser	High-pressure xenon arc
Dispersion element	Michelson interferometer[a]	None required except for electronics to scan laser wavelength	Grating spectrometer
Transfer optics	Gold-coated mirrors	Glass lenses or mirrors	Silver mirrors
Optical detector(s)	Liquid-nitrogen-cooled mercury–cadmium–telluride photodiode	Indium–gallium–arsenide photodiode	Linear CCD array

[a] These are not dispersion elements but still serve the purpose of isolating wavelengths for subsequent radiometric analysis by the optical detectors.

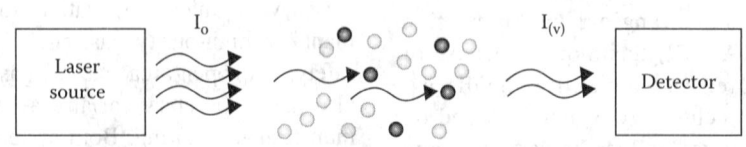

FIG. 1.62b
The operation of the TDLAS sensor. (Adapted from Sorokina, I.T. and Vodopyanov, K.L., Solid-State Mid-Infrared Laser Sources [2003].)

Blackbody Radiation Interference

As shown in Figure 1.62c, there is significant IR emission in the ambient (27°C) blackbody curve that lies within the usable wavelength range of OP-FTIR systems. The effect of uncompensated thermal background radiation is to introduce a bias and/or variability in the measured gas concentrations that have absorption bands that lie in this region.

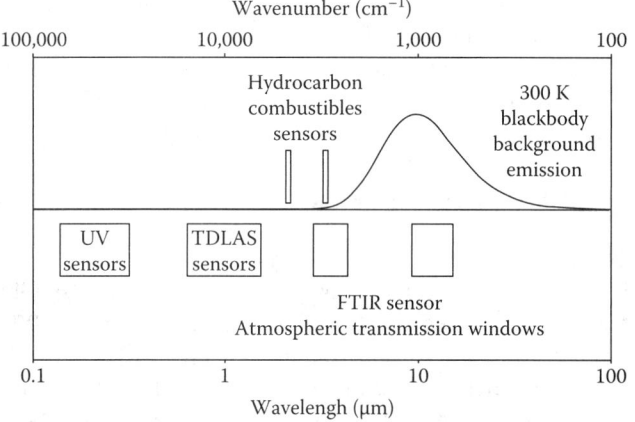

FIG. 1.62c
The regimes of operation in the electromagnetic spectrum for various path sensors. A 300°K (27°C) blackbody emission spectrum is superimposed on the diagram. Note that atmospheric absorption limits the range of FTIR measurements to two atmospheric transmission windows.

The amount of ambient blackbody radiation is generally insignificant at shorter wavelengths and has little effect on OP-HC, TDLAS, and OP-UV measurements. For these instruments, the dispersing element generally resides with the detector. This architecture provides a convenient, compact, and simple instrument implementation.

There can be extenuating circumstances, however, if the detector occasionally is pointed in a direction that has reflected sunlight. Sunlight approximates a 6000°K blackbody and subsequently has significant radiation through the visible portion of the spectrum into the UV. If proper consideration is not given to this potential interference in the design of the instrument, incorrect concentration measurements will be made.

Interferometers

Interferometers are widely used in science and industry for the measurement of small displacements, refractive index changes and surface irregularities. In analytical science, interferometers are used in continuous wave Fourier transform spectroscopy to analyze the absorption or emission associated with a substance or mixture. In an interferometer, two light rays from a common source combine at a half-silvered mirror to reach detector. They may either interfere constructively (strengthening in intensity) if their light waves arrive in phase, or interfere destructively (weakening in intensity) if they arrive out of phase, depending on the exact distances between the three mirrors (Figure 1.62d).

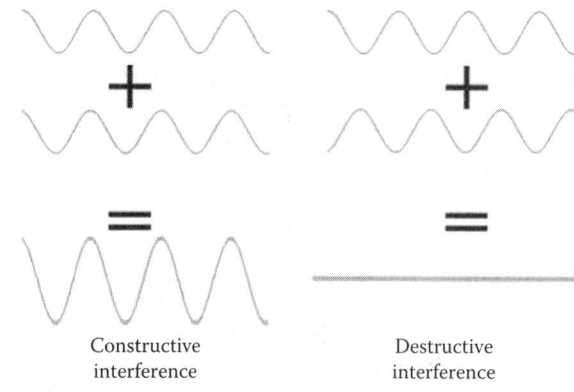

FIG. 1.62d
In the interferometer, the two light rays from the same source can recombine constructively or destructively as a function of their phases upon arrival.

Before the 1980s, IR measurements were made using dispersive spectrometers employing diffraction gratings. Today, as a result of low-cost computing, improved optical designs, and optical detectors, Fourier transform interferometers have almost totally replaced dispersive spectrometers for this task.

Interferometers have much increased optical efficiency and, as a result, yield readings faster and with higher signal-to-noise ratios, and offer higher spectral resolution than corresponding dispersive instruments. Additionally, as a fundamental consequence of interferometry, wavelength measurement precision is vastly improved over dispersive spectrometers.

However, as advantageous as interferometry is in the IR portion of the spectrum, the advantages quickly disappear at shorter wavelengths. At visible and UV wavelengths, the optical efficiency advantage is nullified by vastly more efficient optical detectors and light sources. (UV spectrometry in general is discussed in more detail in Chapter 1.69.) The mechanical tolerances for interferometry at short wavelengths become prohibitive as well. Therefore, dispersive spectrometry is still widely used for short wavelengths.

OPEN-PATH ABSORPTION

Open-path absorption spectrophotometry takes advantage of the fact that a beam of photons can interact with molecules in the atmosphere, depending on the photons' wavelength. As the photons interact with and are absorbed by the molecules, they are removed from the beam. The reduction in the number of photons in the beam is a measure of the number (or concentration) of molecules in their path.

Beer's Law

Beer's law relates the absorption of light to the transmission properties of the material through which the light is traveling. In other words, it relates the amount of light absorbed to the concentration of the material of interest in a sample. For light at a given wavelength, $I(\lambda)$, and a gas concentration profile that varies as a function of position, $c(l)$, over the optical path, L, this phenomenon can be expressed in the following differential form:

$$\frac{dI(\lambda)}{I(\lambda)} = \varepsilon(\lambda) \cdot c(l) \cdot dl \qquad 1.62(1)$$

The term $\varepsilon(\lambda)$ is called the absorptivity coefficient. This coefficient is determined by the characteristics of the absorbing molecule and depends strongly on the wavelength and more weakly on the temperature and pressure.

Generally, the concentration profile, $c(l)$, is not known. If the concentration is assumed to be a constant and uniform average value over the optical path, $\bar{c}$, referred to as the path-average concentration (PAC), then the equation can be integrated to a more familiar form of the Beer's law:

$$\int_{Io}^{I} \frac{dI(\lambda)}{I(\lambda)} = \varepsilon(\lambda) \cdot \int_{0}^{L} c(l) \cdot dl \Rightarrow -\ln\left[\frac{I(\lambda)}{Io(\lambda)}\right] = \varepsilon(\lambda) \cdot \bar{c} \cdot L \qquad 1.62(2)$$

where
  $Io(\lambda)$ is the original intensity of the optical beam at a given wavelength
  $I(\lambda)$ is the intensity of the beam after absorption by the molecules in the atmosphere

The fraction $I(\lambda)/Io(\lambda)$ is termed the "transmittance," as it denotes the fraction of transmitted light through the atmosphere. The negative logarithm of the transmittance is termed "absorbance."

Once a sample has been qualified, it needs to be quantified. FTIR relies on the Beer–Lambert law. This law defines the relationship between absorbance and concentration:

$$A = \varepsilon\, b\, c \qquad 1.62(3)$$

where
  A is the absorbance
  ε is the molar absorptivity with units of L mol^{-1} cm^{-1}
  b is the path length of the sample cell
  c is the concentration of the compound in solution, expressed in mol L^{-1}

Absorbance is the extent to which the intensity of a beam of IR radiation is decreased on passing through the gas sample in the gas cell. Transmittance is an expression of the intensity of the beam of IR radiation passing through the same sample of air. If no gases are present at a given wave number, the transmittance is 1.0 or 100%, and the absorbance is 0. The relationship between absorbance (A) and transmittance (T) is expressed as

$$A = -\log T \qquad 1.62(4)$$

Absorbance, as defined by the Beer–Lambert law, is directly proportional to the concentration of the sample gas when other key variables are held constant, but transmittance is not directly proportional. Therefore, absorbance is used as a basis of measuring the concentration of gases in the FTIR analysis.

Path-Integrated Concentration

The concentration profile integrated over the path length is termed path-integrated concentration (PIC). It is simply the PAC multiplied by the measurement path length. PIC has units of mass per unit volume (often μg m^{-3}) times path length in meters. Using the ideal gas law, the temperature and pressure at which the measurements were made, and the molecular weight of the species being measured, these units can be converted to the more familiar volumetric units of ppm*m or ppb*m in the case of toxic vapor determination. For combustible gas determination, the more appropriate units are in terms of the lower explosive limit, LEL*m.

An open-path instrument natively measures PIC. This is frequently misunderstood by newcomers to the field. An open-path instrument operating over a 100 m path length will produce the same PIC reading for a narrow, concentrated 1 m thick plume having a vapor concentration of 300 ppm as it will for a 3 ppm vapor cloud dispersed across the entire 100 m path length. In each of these situations, the PAC would be 3 ppm. This is illustrated in Figure 1.62e.

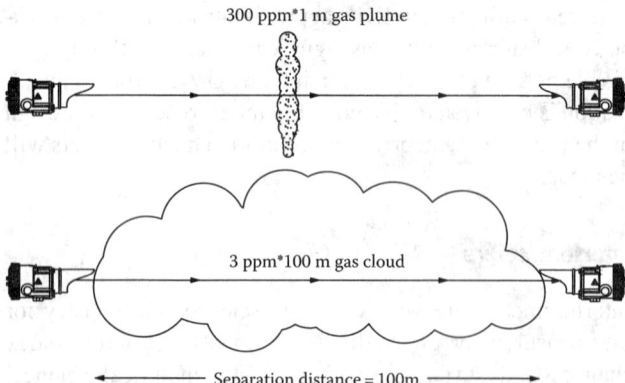

FIG. 1.62e

Illustration of path-integrated concentration and relationship to path-average concentration for narrow plumes and widely dispersed vapor clouds.

OP-FTIR SPECTROMETRY

OP-FTIR instrumentation consists of five major components: an IR light source, interferometer, transmitter and receiver transfer optics, IR detector, and data system/controller. The IR light source is generally a glow bar with high emissivity across the IR spectrum. The detector can either be located at the other end of the separation distance or a reflector can be located there and the detector will sense the reflected signal at the same location as where the light source is (Figure 1.62f). In case of the reflected signal configuration, the system sends a modulated IR beam from the base unit to a reflector at a typical distance of 50–250 m. The IR light that is returned from the reflector to the base is analyzed for the presence of the gases of interest. With simultaneous characterization using the complete IR spectrum, a wide range of gases can be detected automatically.

FIG. 1.62f
Open-path air monitor. (Courtesy of Bruker, Billerica, MA.)

Interferometer

The interferometer is generally one of the Michelson variants and is used to modulate the light as a function of wavelength. The interferometer utilizes a moving mirror to generate an optical path difference between two beams of light that are allowed to interfere with each other. The position of the moving mirror must be measured to a tolerance of a fraction of the wavelength of light. This is frequently accomplished by using a highly stable tracking laser and counting interference fringes.

The moving mirror and the requirement for high-precision measurement of the optical path difference are two of the vulnerabilities of the interferometer mechanism. They limit reliability over time, operating temperature, and vibration. Much effort has been put into the design of interferometer systems to increase their reliability. The Michelson interferometer serves to sample gases and develop an interferogram. The interferogram is then converted mathematically to an IR spectrum. The basics of the interferometer are shown in Figure 1.62g.

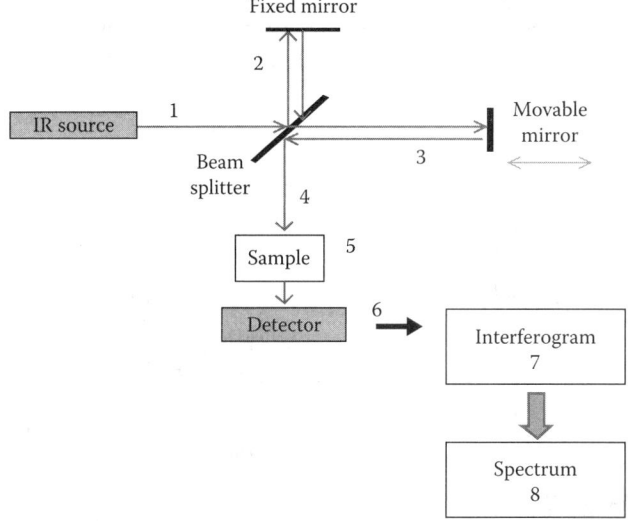

FIG. 1.62g
The interferometer.

The measurement consists of the following steps:

1. The IR source generates IR over a multifrequency range that shines on the beam splitter. The beam splitter divides the light into two paths.
2. A fixed mirror bounces the IR back toward the sample.
3. A movable mirror is adjusted to recombine the refracted light to obtain the maximum signal.
4. The recombined light is sent through the sample cell that contains the gas to be analyzed. IR will be absorbed as discussed previously based upon the species and the concentration of gas.
5. Sample is contained in a gas cell that permits the IR to pass through.
6. A detector converts the remaining IR into an electrical signal.
7. This signal is then converted to an interferogram.
8. The interferogram is manipulated mathematically into a spectrum that then is used for analysis.

Sample Cell

The sample cell is an important component of the FTIR system. First and foremost, the cell must contain the gas sample. For durability, it must be made of a material that is compatible with the gases that will be sampled. Furthermore, it cannot react with the gases sampled, or the results will be affected. Most cells are made from stainless steel or glass.

The results obtained from an FTIR sample are better as the optical path is increased. Many of the cells in use today provide for a path of 3–10 m in length. Obviously, it is impractical to have a cylinder of that length as part of the FTIR instrument. The solution is a folded path cell. In the folded path cell (Figure 1.62h), there are a series of mirrors and two windows are provided to permit the IR beam to enter to permit the IR to enter, then bounce back and forth several times before exiting the cell. The total of all the paths added up give the total path length of the cell.

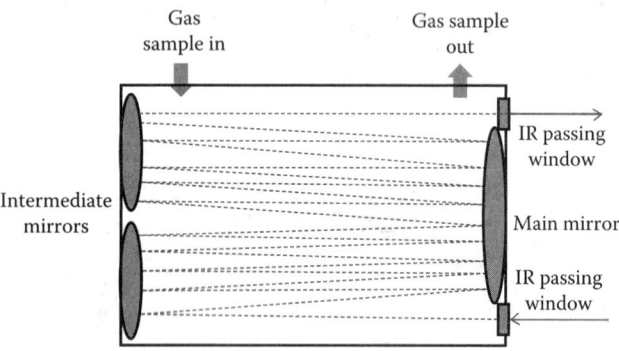

FIG. 1.62h
Folded path sample cell.

In the folded path cell, another consideration is the makeup of the IR passing window. Like the cell, these windows must stand up the rigors of the gases to be sampled. Some materials include potassium bromide (KBr), potassium chloride (KCl), and calcium bromide (CaBr). All pass IR well but have varying degrees of resistance to moisture and chemicals in a sample.

As mentioned before, the total IR path through the sample cell length affects its sensitivity. Shorter path cells are used for higher detection limits in the high PPM or % v/v level. If the user wants low PPM or even PPT level resolutions, then a longer path cell will be required. As cell path gets longer, the volume of the cell increases. Larger volumes require longer sample fill times and purge times after the analysis has occurred. This time constraint may be a factor in applying an FTIR in a continuous monitoring application, especially if multiple points are sequentially sampled. Also as the IR path gets longer, there is a consideration for the amount of IR energy that eventually reaches the detector, as the energy is lower over the longer path, the signal-to-noise ratio decreases.

Transfer Optics and the Detector Types

Transfer optics are generally telescopes of the Newtonian or Cassegrainian configuration. Reflective optics are required, because inexpensive lenses that can cover the full useful range of wavelengths are not available.

The optical detector receives the modulated light. A cooled mercury cadmium telluride detector is most often used to obtain the required sensitivity for the measurement. The detectors are cooled using liquid nitrogen (LN_2) or by a Sterling cycle heat pump.

The LN_2 Dewar must be refilled every few days, which adds to operating maintenance costs. Sterling cycle coolers reduce the daily maintenance burden but are relatively expensive and have relatively short life spans. In some applications, in which a relatively low-resolution measurement can be tolerated, detectors operating at room temperature can be used to address this maintenance issue.

The more commonly used detector is the deuterated triglycine sulfate (DTGS) detector. The DTGS detector is stable at normal room temperature and provides a stable and linear response over a wide range of frequencies making it a good choice for industrial monitoring applications.

Software and Computer Equipment Once the detector *reads* the gas sample, the next component in the sampling system is the software and the computer that it runs on. This is where there is a wide divergence between different systems. The number of gases that can be detected, the accuracy of detection, and the speed of detection are all affected by the software and the computer.

The first step in identifying (qualifying) and measuring (quantifying) a gas sample is to convert the data points received from the detector from an interferogram into a spectrum. The interferogram contains the information of IR absorption from many frequencies simultaneously. To qualify and quantify the gas in the sample, it needs to be converted to a plot of the absorbance or transmittance at each unique frequency of interest, which is the spectrum. The means of doing this conversion is the mathematical technique known as Fourier transformation. The transformed data can then be compared to a known reference spectrum for identification.

As mentioned before, the frequencies at which each gas absorbs IR are unique and consistent. This makes it easy to compare a sample to a known sample. The known sample can either be in the form of a library, a matrix, or a method. The form differs from unit to unit and computer resources available. The larger number of comparisons to be made will affect the speed of detection and requires more computing resources. To this end, some manufacturers limit the number of gases that are detectable to speed up the process. This is done with a method or a matrix that limits the number of gases and/or the number of frequencies for comparison, which results in faster qualification.

Calibration and Reference Zero To get an accurate measurement, the system needs a reference to compare the absorbance against. This is accomplished by taking a *background* reading before the start of the analysis. This background is usually an inert gas such as nitrogen, which is both clean and dry. As mentioned previously, N_2 is diatomic that does not absorb IR and can thus provide a zero reference. The background will be used as a reference ratio in the software to negate the effects from factors that could cause errors such as

- Detector drift
- Dirty or fogged mirrors
- Degradation of the optical path
- Degradation of the IR source

Other variables need to be controlled to get accurate FTIR readings. These include the need to verify that the wavelength of the path is accurately represented. If there were a wave number offset, the gas would not be either accurately qualified or quantified. This is done by using a laser to provide a reference through the same optical path that the interferometer uses. A laser is a reliable source of monochromatic (single-frequency) light. Commonly, a He–Ne laser is used as it provides light in the visible light range. Since this light is at much higher wavelength than the IR range used for detection, it provides a good reference source and alleviates errors in resolving wavelengths of absorption when taking a sample.

When choosing an FTIR to monitor multiple gases, there are several considerations that need to be made to assure a successful application. These variables need to be controlled to get successful results.

Sample Cell

The sample needs to be chosen to match the gas(es) of interest. The longer the total IR path length, the better the resolution will be. If the target detection range is in the parts-per-billion (e.g., a toxic gas) range, a longer path length (e.g., 10 m) will be appropriate. On the other hand, if the range for detection is in the percent volume area, then a short cell (e.g., 10 cm) will be more appropriate. The small cell that has a smaller volume will fill faster and provide good resolution for the detection level desired.

Sample Temperature Temperature is often a concern with applications like stack monitoring. The temperature that the analyzer system can handle is largely controlled by its materials of construction. As with any instrument, this temperature must be taken into account.

Another concern is the temperature differential between the sample point and the analyzer location. If a warm humid sample is drawn into a cold sample cell, there will be condensation formed. This moisture can foul the cell mirrors and cause faults. Worse, if the moisture is caustic, it can permanently damage the cell or windows. It is possible in some applications to overcome these problems by heating the sample, sample line, or the sample cell.

Sample Pressure Most FTIR analyzers are designed to operate at atmospheric pressure. If a sample is at a higher pressure than the design pressure, the quantification will be inaccurate. As demonstrated before, the Beer–Lambert law defines the relationship between absorbance and concentration. If the cell has a sample with a higher-than-normal pressure, there will be more molecules of the target gas in the cell. Consequently, the absorbance will be higher, and the calculated concentration inaccurate. This demonstrates why it is important to maintain a constant pressure in the analyzer from sample to sample. Most manufacturers have an ability to correct for minor deviations in pressure.

Sample Concentration The list of gases to be monitored and the range of concentration become the matrix or the method used for identifying and quantifying gas samples. This is normally provided by the system manufacturer. Information is provided as to which gases are to be monitored and in what range. When the method is written and tested, peaks are chosen for each gas. The peaks must be unique within the group of gases, and multiple peaks are used to help qualify the gas sample. In addition, the selected peaks must be used to quantify the sample using partial least squares or classical least squares to solve the area under the peak and thus quantify the sample.

Data System/Controller

The data system/controller performs a myriad of functions. It monitors the operation of the interferometer and synchronizes the collection of spectral data from the IR detector. It demodulates the spectral data using a fast Fourier transform algorithm and then does spectral manipulation to ratio the background and to compensate for background interferences and variation. As a final step, the corrected spectrum is processed by a pattern recognition and/or quantification algorithm that calculates PAC (e.g., average ppm) or PICs (e.g., ppm*m).

Instrument Configurations

There are a number of instrumental configurations for OP-FTIR instruments. The simplest OP-FTIR systems are the so-called bistatic configurations. In these configurations, either a light source illuminates the open-path region and is post-modulated with the interferometer, or the light source is pre-modulated before illuminating the open-path region. The arrangement of the components of this design is shown in Figure 1.62i. This configuration derives its name from the fact that both the transmitter and the receiver must be fixed in a static position and precisely aimed at each other.

The first real-time OP-FTIR systems had a direct-source, post-modulated bistatic configuration. This configuration has the advantage of simplicity and hence is the least expensive design, but it also has a major disadvantage. This is a result of the varying thermal emission of near-field objects that produce a varying spectral background at frequencies below 2000 cm^{-1}. If uncompensated, this background will produce a negative bias on the measured concentration.

Most of this varying background emission and its effects can be compensated. This is done by turning off the source and by subtracting the spectrum while it is off. The direct-source configuration has the advantage of being easily adaptable to a passive monitoring configuration where the thermal emission from the background or hot gas plumes acts as the IR source.

The pre-modulated bistatic configuration solves the background emission problem, because the source modulation occurs much faster than do changes in background emissions. However, since the detector and interferometer are separated by some distance, an additional cable is required to communicate between them.

Bistatic configurations in general have the requirement of supplying power at both the receiver and the transmitter, which can be a disadvantage in some locations. Additionally, there is a requirement for alignment at both receiver and transmitter, which can be time consuming for mobile systems. It may be less of an issue for permanent fixed systems.

The so-called monostatic configurations were developed to address issues raised with bistatic designs. In a monostatic

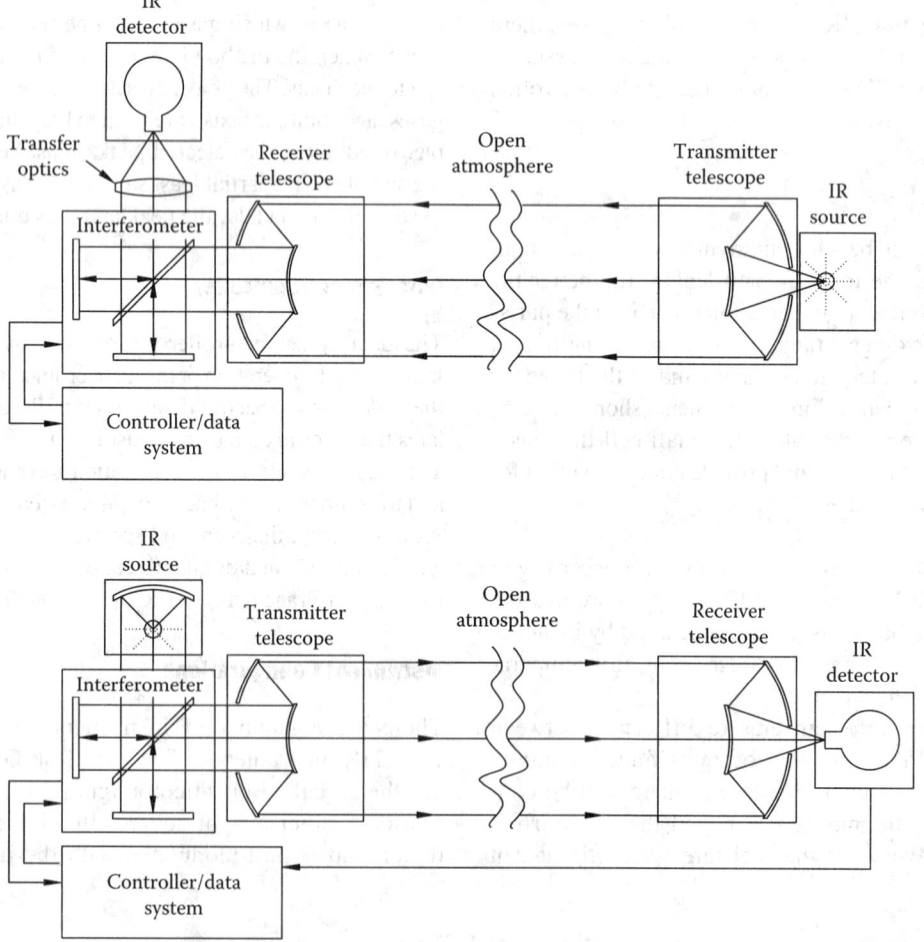

FIG. 1.62i
Bistatic configurations for OP-FTIR. Post-modulated configuration (top) uses interferometer to modulate the light after it passes through the open path. The premodulated configuration (bottom) uses the interferometer to modulate the light prior to passing through the open path.

configuration, all of the optical components of the transmitter and receiver are in the same location, and a retroreflector is used to return the light from the transmitter to the receiver. This configuration derives its name from the fact that only the transceiver portion of the instrument needs to be precisely pointed, as the retroreflector returns light to its source regardless of orientation. A diagram of two monostatic configurations is shown in Figure 1.62j.

The single-telescope monostatic configuration is relatively simple to implement and is inexpensive. The use of a single telescope makes alignment simpler and lowers costs. The corner-cube array returns the light to the direction from which it came. This property reduces the divergence of the beam on its return path back to the detector compared to the divergence that would result from a flat mirror. Also, the retroreflector array can be very large so as to capture and return essentially all of the divergent signals from the telescope.

However, this design requires a beam splitter in the optical path that removes 50% of the light from the outgoing beam and 50% of the light from the return beam for an overall loss of 75% of the total light intensity.

The dual-telescope monostatic configuration has greater optical efficiency, as it does not utilize a beam splitter in the optical path. It utilizes a translating retroreflector, which is essentially a portion of a very large cube. This single large retroreflector does not have the divergence reversal properties of the corner cube array. The second telescope adds cost and complexity to the system.

OP-FTIR sensitivities for various molecular species are listed in Table 1.62b.

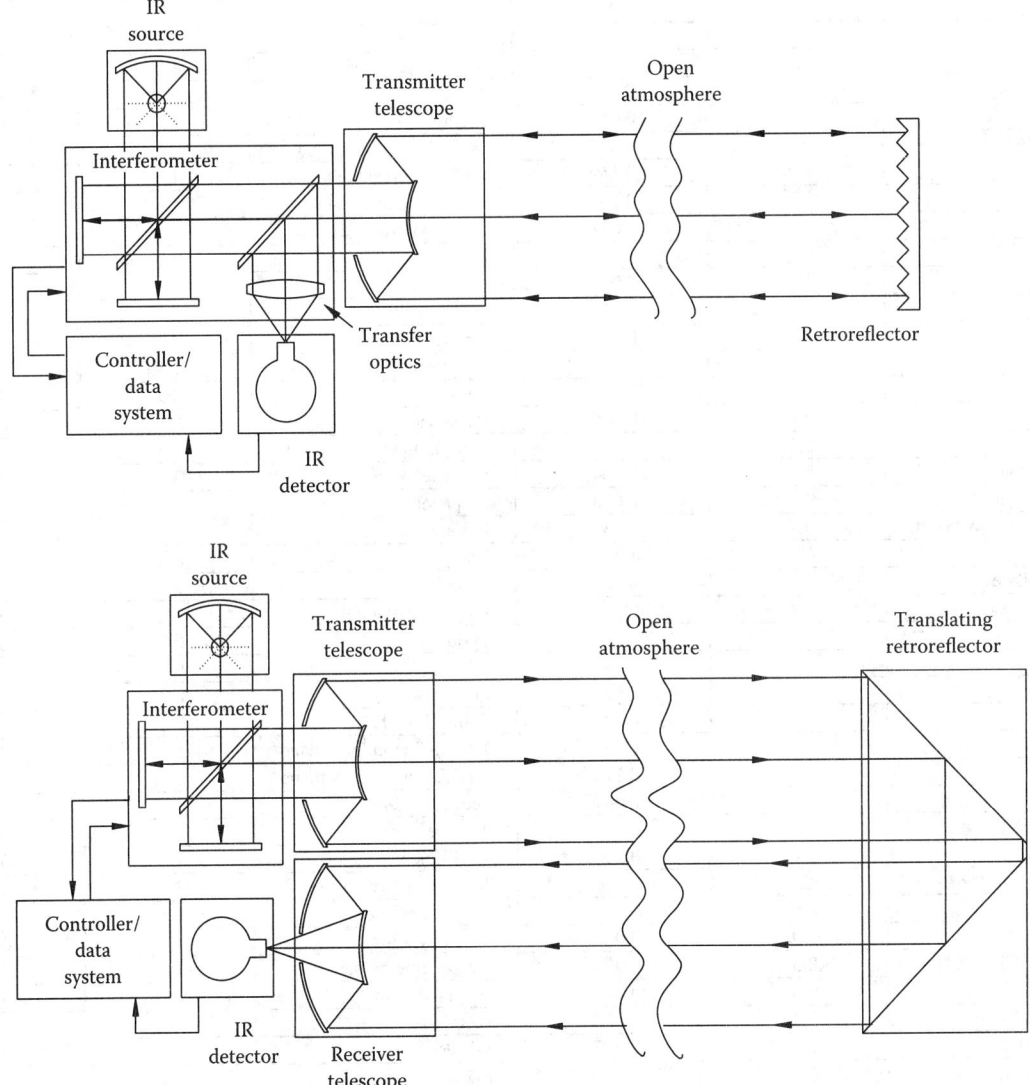

FIG. 1.62j
Two monostatic configurations for OP-FTIR both using pre-modulated configurations. The single-telescope configuration (top) uses a corner-cube retroreflector array and is less optically efficient. The dual-telescope configuration (bottom) is more optically efficient, but requires more telescope complexity.

TABLE 1.62b
Typical Detection Limits for Open-Path FTIR

Species	PI-MDC Field Data[a] (ppm*m)	PI-MDC Noise Limited (ppm*m)	Species	PI-MDC Field Data (ppm*m)	PI-MDC Noise Limited (ppm*m)
Acetaldehyde	6	0.081	Hydrocarbon continuum	3	0.06
Acetic acid	1.5	0.088	Hydrogen chloride	0.6	0.071
Acetone	9	1.3	Hydrogen cyanide	1.5	0.016
Acetonitrile	15	0.13	Hydrogen sulfide	90	51.2
Acetylene	0.6	0.006	Isobutane	0.6	0.027
Acrolein	1.5	0.15	Isobutanol	1.2	0.071
Acrylic acid	3	0.088	Isobutyl acetate	1.5	0.024
Acrylonitrile	1.8	0.179	Isobutylene	1.2	0.064
Ammonia	0.6	0.057	Isoprene	1.2	0.06
Benzene	7.5	0.6	Isopropanol	3	0.084
1,3-Butadiene	0.6	0.054	Isopropyl ether	3	0.038
Butanol	4.5	0.114	Methanol	1.2	0.077
1-Butene	3	0.145	Methylamine	6	0.137
cis-2-Butene	7.5	0.276	Methyl benzoate	6	0.038
trans-2-Butene	3	0.145	Methyl chloride	18	0.289
Butyl acetate	1.5	0.027	Methylene chloride	1.5	0.068
Carbon monoxide	0.3	0.086	Methyl ether	3	0.089
Carbon tetrachloride	0.6	0.01	Methyl ethyl ketone	12	0.197
Carbonyl sulfide	0.6	0.014	Methyl isobutyl ketone	4.5	0.072
Chlorobenzene	3	0.159	Methyl mercaptan	12	0.462
Chloroethane	3	0.16	Methyl methacrylate	1.5	0.06
Chloroform	0.6	0.021	2-Methyl propene	0.6	0.03
m-Cresol	6	0.132	Morpholine	0.6	0.016
o-Cresol	1.2	0.063	Nitric acid	0.3	0.037
p-Cresol	3	0.077	Nitric oxide	7.5	0.26
Cyclohexane	0.9	0.01	Nitrogen dioxide	15	0.045
1,1-Bromoethane	1.5	0.101	Nitrous acid	1.5	0.013
m-Dichlorobenzene	0.9	0.027	Ozone	0.9	0.104
o-Dichlorobenzene	0.9	0.025	Phosgene	0.3	0.015
p-Dichlorobenzene	0.6	0.05	Phosphine	0.6	0.109
1,1-Dichloroethane	3	0.104	Propane	3	0.057
1,2-Dichloroethane	9	0.095	Propanol	6	0.137
1,1-Dichloroethylene	0.6	0.03	Propionaldehyde	3	0.186
Dimethylamine	6	1.72	Propylene	1.2	0.107
Dimethyl disulfide	3	0.101	Propylene dichloride	3	0.247
1,4-Dimethyl piperazine	0.9	0.034	Propylene oxide	3	0.157
1,4-Dioxane	0.6	0.037	Pyridine	6	0.057
Ethane	3	0.094	Silane	0.3	
Ethanol	3	0.158	Styrene	0.3	0.138
Ethyl acetate	1.2	0.025	Sulfur dioxide	9	0.046
Ethylamine	6	0.131	1,1,1,2-Tetrachloroethane	1.2	0.059
Ethylbenzene	6	0.119	1,1,2,2-Tetrachloroethane	6	0.174
Ethylene	0.3	0.041	Tetrachloroethylene	0.6	0.059
Ethylene oxide	3	0.025	Toluene	7.5	0.073
Ethyl mercaptan	15	0.14	1,1,1-Trichloroethane	1.2	0.031
Formaldehyde	1.5	0.139	1,1,2-Trichloroethane	3	0.067
Formic acid	0.6	0.034	Trichloroethylene	0.6	0.06
Furan	0.9	0.02	Trimethylamine	3	0.038
Halocarb-11 (CCl_3F)	0.3	0.012	1,2,4-Trimethylbenzene	1.5	0.098
Halocarb-12 (CCl_2F_2)	0.3	0.016	Vinyl chloride	1.2	0.097
Halocarb-22 ($CHClF_2$)	0.3	0.017	m-Xylene	3	0.111
Halocarb-113 ($CFCl_2CF_2Cl$)	0.6	0.033	o-Xylene	6	0.044
Hexafluoropropene	0.3	0.019	p-Xylene	6	0.102

Source: Courtesy of Industrial Monitoring and Control Corporation, Round Rock, TX.
Notes: Field data typical for atmospheric path with 20,000 ppm H_2O, 360 ppm CO_2, and other atmospheric gases present.
Noise-limited levels assume an RMS noise of 5×10^{-5} using reasonably accessible bands.
[a] PI-MDC, Path-Integrated-Minimum Detectable Concentration.

OP-UV SPECTROMETRY

OP-UV spectrometry can be used to measure vapors or gases that have weak absorption characteristics and therefore low sensitivities in the IR spectrum. These include such compounds as nitrogen oxides, sulfur dioxide, benzene, and also homonuclear diatomic molecules, such as chlorine, that have no IR absorption spectra but acceptably strong UV absorption.

The UV spectra are much less specific than the IR spectra and do not have well-defined and separated absorption features. The numbers of compounds that can be determined by UV are much fewer than ones that are absorbing in the IR spectra. Still, the earlier-listed and environmentally very important compounds make OP-UV an important complementary methodology.

The Spectrometer

A schematic of an OP-UV spectrometer with a monostatic configuration is shown in Figure 1.62k. Bistatic configurations are available as well and are generally provided with fixed source.

The light source of the open-path UV-DOAS generally is either a tungsten halogen or xenon arc lamp, although deuterium lamps can also be used. The light is collimated before being transmitted through a telescope to a receiving unit (Figure 1.62l). Unless operated in the passive mode,

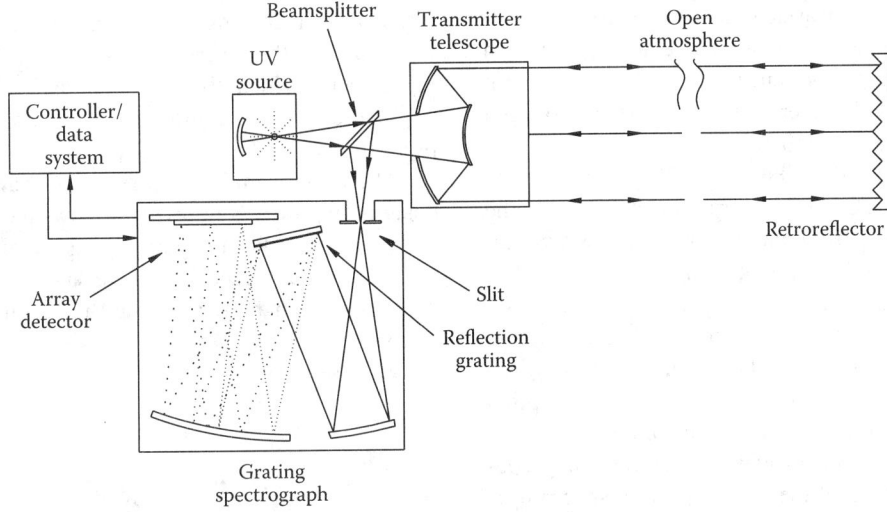

FIG. 1.62k
Schematic representation of an OP-UV spectrometer.

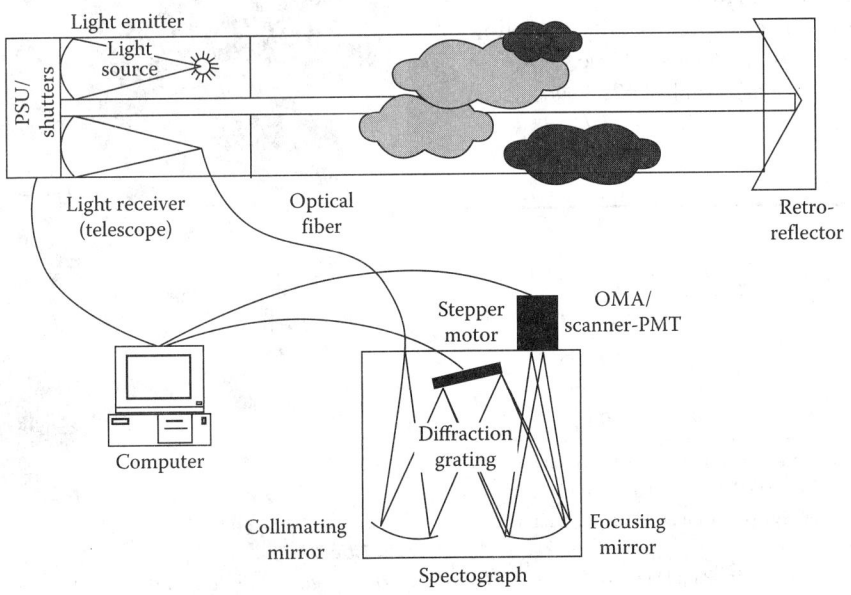

FIG. 1.62l
Monostatic configuration of a UV-DOAS system.

the system requires a sending-and-receiving telescope. Incoming light can be focused either directly into the spectrometer or onto a fiber-optics cable where it is transmitted to the spectrometer. There are some commercial systems that use one telescope for both transmitting and receiving and are restricted to a monostatic configuration.

Sources are generally high-pressure xenon arc lamps that emit a continuum from the UV to the visible spectra. The spatial extent of the arc can be very small, on the order of 250 µm, which lends itself to collimation with very low divergence. As a result, UV instruments can operate with path lengths up to several kilometers.

The long distances are facilitated by the low optical divergence and by the fact that water vapor and carbon dioxide do not absorb in the UV spectra as they do in the IR. On the other hand, the short UV wavelengths can be strongly scattered by aerosols such as fog, snow, smoke, and dust, which greatly degrade their performance. The short wavelength limit is determined by the interference of molecular oxygen at wavelengths shorter than 260 nm.

The detector depicted in Figure 1.62l is a solid-state array detector such as a CCD. The array detector permits the acquisition of intensities from multiple wavelengths simultaneously and without mechanically moving parts.

The spectra of many molecules in the UV, notably SO_2 and NO, in most cases have periodic absorption patterns resulting from the interaction of the vibrational energy levels with the electronics absorption. The overall absorption band can be quite wide and broadly overlapping. However, the periodicity can be used to identify the absorbing species. A Fourier transform of the UV absorption spectrum is often used to pretreat the data to isolate the absorbing species based on the spatial periodicity. The peaks thus produced can be further treated with spectral quantification algorithms such as the classical least squares method.

OP-UV has been used to detect and quantify the following gases: NO, NO_2, NO_3, formaldehyde, ozone, SO_2, benzene, toluene, and o, m, p-xylenes. Table 1.62c lists selected gases and detection limits available with OP-UV.

TABLE 1.62c
Minimum Detection Limits for Selected Gases by OP-UV

Species	Detection Limit (ppb*m)[a]
Nitrogen oxide	100–225
Benzene	42–150
Ammonia	310–5870

Source: Myers, J. et al., Environmental Technology Verification Report, Opsis Inc. AR-500 Ultraviolet Open-Path Monitor, ETV Advanced Monitoring Systems Center, Battelle, Columbus, OH, September 2000.

[a] Integration times vary from 1 to 5 min; Path length varied from 100 to 250 m.

OP-TDLAS SPECTROMETRY

OP-TDLAS is a relatively recent technology that has been applied commercially only to air monitoring within the past decade. Recently, these instruments have been made sufficiently rugged and reasonably simple to use in other applications.

Previously, lead salt diodes were used, emitting laser radiation in the 3–20 µm mid-IR portion of the IR spectrum. Instruments based on these lasers were capable of making very sensitive measurements, sometimes in the parts-per-trillion range. They were also very fast, having measurement times as low as 1/10 sec.

On the other hand, complexity and cost limited their commercial acceptance. The lead salt lasers required liquid nitrogen cooling for operation, and IR absorptions accessible to the lasers were subject to pressure-broadening effects, which limited their use to point monitoring applications using multi-pass cells operating at reduced pressure.

These gas analyzers are also available in portable designs (Figure 1.62m), which allows movement from location to location for analyzing the gas emissions in a variety of places. The collected data are saved in onboard memory, which can later be downloaded into computers for further analysis.

FIG. 1.62m
Portable gas emission analyzer. (Courtesy of Unisearch Assoc http://www.unisearch-associates.com/company.html.)

Diode Lasers

Diode lasers originally developed and manufactured in large volumes for telecommunications applications have been adopted to OP-TDLAS applications.

Various diodes operate at or near room temperature and emit radiation between 0.6 and 2.5 μm in the near-IR portion of the spectrum. Wavelength tuning over relatively short wavelength ranges is achieved by temperature tuning as well as by varying the injection current into the diode itself. Temperature tuning is most often performed using a thermostat, which controls a thermoelectric cooler. Current tuning is performed using a programmable current source and is done at relatively high frequencies in the kilohertz range.

The absorption ranges that are accessible with these near-IR wavelengths are less sensitive than those in the mid-IR, but they are less susceptible to pressure-broadening effects. This permits their use in the open-path monitoring applications (Figure 1.62n). The reduction in detection sensitivity is partially recovered by the use of high-frequency wavelength modulation and by improved signal-detection techniques. Detection limits in the range of low ppm*m of target molecules are generally achievable.

Because these diode lasers are used in large volumes for telecommunications applications, their costs are an order of magnitude or two lower than those of lead salt diodes. Additionally, with near-IR wavelengths, traditional and low-cost glass optical materials can be used to fabricate the supporting optical assemblies instead of exotic and expensive mid-IR optical materials.

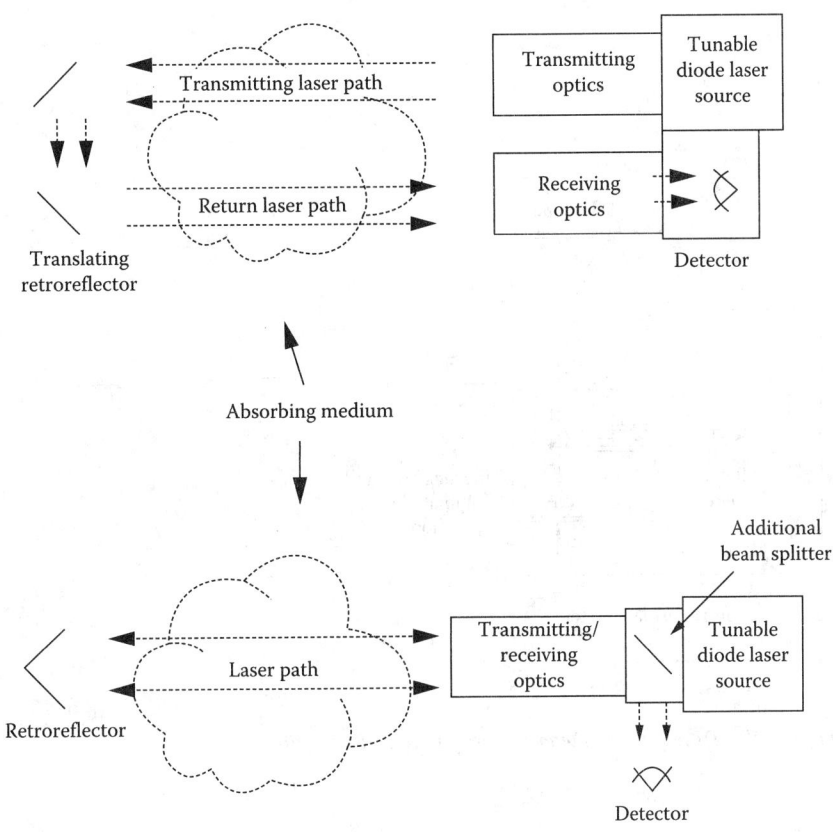

FIG. 1.62n
Deployment configuration of a monostatic tunable diode-laser unit. (Courtesy of EPA, Washington, DC.)

Applications

Molecules having absorption spectra that are accessible with diode lasers include NO, NO_2, HF, HCl, HCN, HI, NH_3, C_2H_2, CO, CO_2, H_2S, and CH_4. In some fortuitous instances, diode lasers have wide enough tunability to simultaneously measure multiple gases that have closely spaced absorption features. This is the case with HF and CH_4 and with CO and CO_2. Generally, however, a separate laser is required for each gas analyte of interest. Some commercial devices permit the fitting of more than one laser into a device so that multiple gases can be measured.

Principle of Operation

A schematic diagram illustrating the principle of operation of an OP-TDLAS is shown in Figure 1.62o. The monostatic configuration using a retroreflector is representative of commercial architectures.

In this configuration, laser light from a diode laser is directed to a beam splitter and then onto a steering mirror in the telescope, which directs the light out into the open atmosphere. The light interacts with the target gas molecules over a one-way optical path of up to 1 km and eventually falls onto a retroreflector array. The retroreflector array returns the light back to the telescope, which serves to focus it onto the optical detector.

The small portion of the light intercepted by the beam splitter is directed through a reference cell where a sample of the target analyte gas is present. The absorption of light by the gas in the reference cell is used to generate a feedback signal to the diode-laser thermostat to keep the laser wavelength accurately tuned to the center of a gas absorption line.

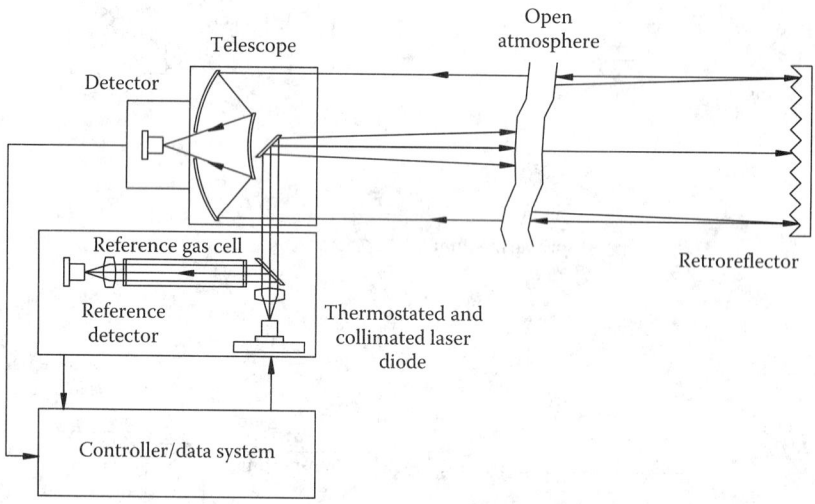

FIG. 1.62o
Schematic representation of an OP-TDLAS spectrometer in monostatic configuration.

Wavelength Modulation Spectrometry All commercial OP-TDLAS devices use some form of wavelength modulation to perform the measurement. The principle of wavelength modulation spectrometry is shown in Figure 1.62p. With this methodology, the laser, which is tuned to the center of an absorption line, is modulated at high frequency about the absorption maximum.

Temperature tuning is used to adjust the diode-laser output to the center of the absorption line. Current tuning using a programmable current source is used to modulate the laser wavelength at kilohertz frequencies. As the laser wavelength passes through the peak absorption wavelength, the detected laser intensity drops and then increases as the laser moves to one side of the line or the other.

A laser wavelength modulation frequency of f produces an intensity modulation of the laser at frequency $2f$ because of absorption of the analyte gas. The $2f$ intensity modulation is detected using lock-in detection. Lock-in detection at high frequency provides signal-to-noise enhancement that somewhat negates the general low absorptivity of the absorption lines. Measurements are typically made at a rate of one sample per second.

Table 1.62d lists some representative gases of interest for air monitoring that can be measured using TDLAS along

TABLE 1.62d
Some Representative Gases and Approximate PIC Detection Limits for OP-TDLAS Assuming Ability to Measure Absorbance to 1 Part in 10^5

Species	Detection Limit (ppm*m)
HF	0.2
H_2S	20
NH_3	5.0
CH_4	1.0
HCl	0.15
HCN	1.0
CO	40
NO	30
NO_2	0.2

Source: Frish, M.B. et al., Handheld laser-based sensor for remote detection of toxic and hazardous gases, SPIE Paper No. 4199-05 in *Water, Ground and Air Pollution Monitoring*, Boston, MA, November 5, 2000.

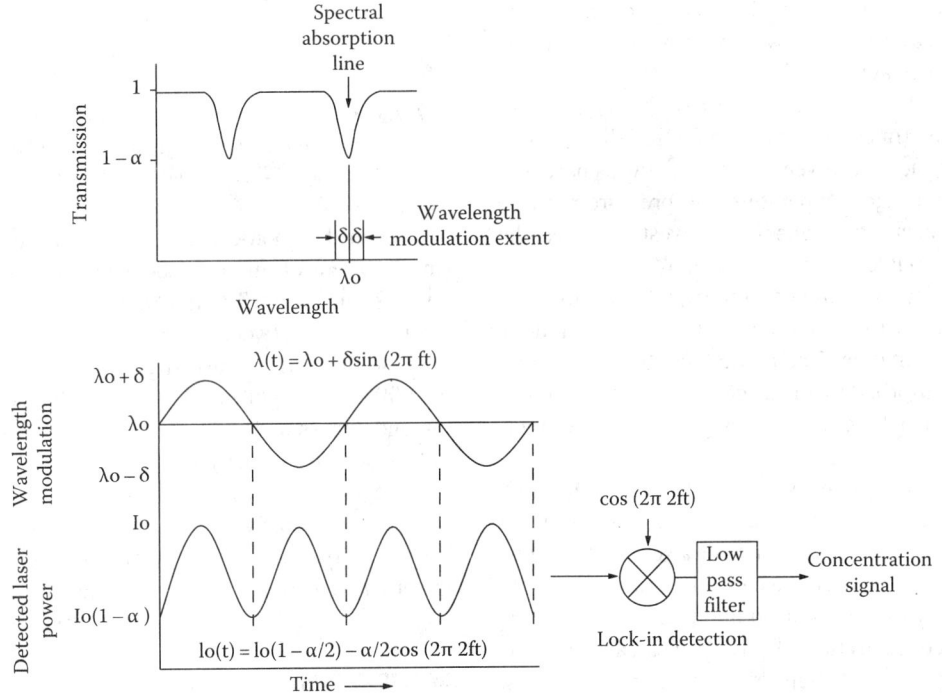

FIG. 1.62p
Principle of wavelength modulation spectrometry—the laser wavelength is tuned to an absorption line and modulated at a frequency f. The laser intensity as it passes through the sample is modulated at a frequency 2f. The concentration signal is recovered using lock-in detection techniques.

with their approximate detection limits. For industrial applications, hydrogen fluoride (HF) monitoring by OP-TDLAS has proven to be very useful because of the low detection limits achievable and because of the lack of interference.

OP-CRDC Spectroscopy

A relatively recent development in spectroscopy is the CRDS. It provides significant enhancement in resolution over FTIR. CRDS systems can regularly detect trace gases at levels in the parts-per-billion (ppb) or even parts-per-trillion (ppt) range. The unique absorption characteristics of each gas as described for FTIR provide the qualification and quantification in CRDS. Applications for CRDS include stack monitoring, environmental monitoring, trace gas analysis, and high-tech facility monitoring.

CRDS uses a laser as an IR source. A sample cell (cavity) with two (or more) high-quality mirrors holds the sample. The laser fills the cavity with light that bounces between the mirrors. A photodiode detector detects the light passing through the cell. At a predetermined level, the laser is shut off or shuttered. The light left in the cell will then leak out (ring-down). The time required to leak to zero is measured by the photodiode. Any gas in the cell will affect the ring-down time. This can be compared to a known reference with no gas present.

To identify the gas, the laser is tuned to different frequencies that are known to be absorbed by the gas or gases of concern. After the gas is identified, it can be quantified by measuring the decay time that will be proportional to the amount of gas in the cavity.

Due to the light scattering in the cavity, the effective path length is many times more than an FTIR cell. This fact makes CRDS very accurate even at levels as low as parts per trillion (ppt). Like FTIR, temperature and pressure must be controlled. To get accurate results in the lowest of ranges, the control of these variables is most critical. While the CRSD can provide precision, it is limited in the number of different compounds it can test for. The range of the tunable diode is the cause of this limitation. The diode must be tunable to a frequency that is absorbed by the target gas. CRDS is mostly used in applications where one to four gases must be monitored at a time.

Nearly every small gas-phase molecule (e.g., CO_2, H_2O, H_2S, NH_3) has a unique near-IR absorption spectrum. At subatmospheric pressure, this consists of a series of narrow, well-resolved sharp lines, each at a characteristic wavelength. Because these lines are well spaced and their wavelength is well known, the concentration of any species can be determined by measuring the strength of this absorption, that is, the height of a specific absorption peak. But in conventional IR spectrometers, trace gases provide far too little absorption to measure, typically limiting sensitivity to the parts per million at best. CRDS avoids this sensitivity limitation by using an effective path length of many kilometers. It enables gases to be monitored in seconds or less at the parts-per-billion level and some gases at the parts-per-trillion level.

In CRDS, the beam from a single-frequency laser diode enters a cavity defined by two or more high-reflectivity mirrors. The analyzer in Figure 1.62q uses a three-mirror cavity to support a continuous-traveling light wave. This provides superior signal to noise compared to a two-mirror cavity that supports a standing wave. When the laser is on, the cavity quickly fills with circulating laser light. A fast photodetector senses the small amount of light leaking through one of the mirrors to produce a signal that is directly proportional to the intensity in the cavity.

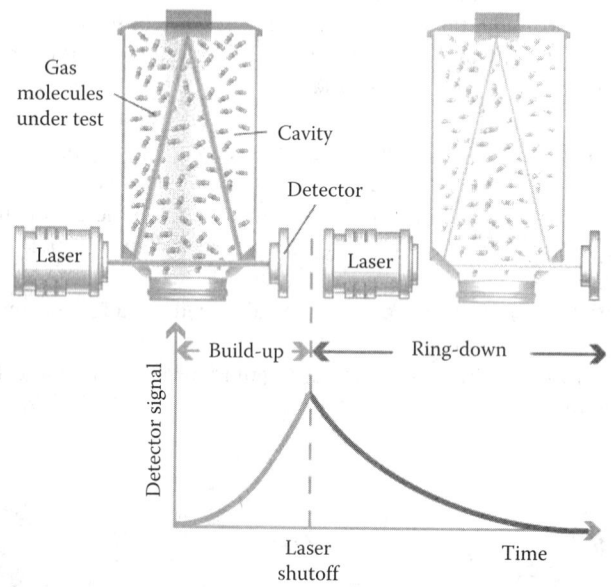

FIG. 1.62q
Illustration of the buildup and ring-down phases in a CRDS analyzer. (Courtesy of Picarro Inc., Sunnyvale, CA.)

When the photodetector signal reaches a threshold level (in a few tens of microseconds), the continuous-wave laser is abruptly turned off. The light already within the cavity continues to bounce between the mirrors (about 100,000 times), but because the mirrors have slightly less than 100% reflectivity (99.999%), the light intensity inside the cavity steadily leaks out and decays to zero in an exponential fashion. This decay, or "ring down," is measured in real time by the photodetector, and the amount of time it takes for the ring down to happen is determined solely by the reflectivity of the mirrors (for an empty cavity). In the illustrated cavity of only 25 cm in length, the effective path length within the cavity can be over 20 km.

APPLICATIONS

The two most common uses of open-path vapor detection are the detection of toxic and combustible vapors. These needs have resulted in instruments that have evolved specifically for these applications. Open-path toxic gas detection is generally used for very low-level detection—often in fence-line monitoring roles for the estimation of emissions from a facility.

There are many more open-path combustible hydrocarbon (OP-HC) detectors in use than there are toxic detectors. In a petrochemical plant, there can be as many as 150 OP-HC detectors, but only about half a dozen or fewer open-path toxic detectors. Combustible detection requires much less instrument sensitivity, but instruments are highly engineered to provide much more protection against false alarms and to deliver high availability in excess of 99.9%. Immunity to false alarms is very important, because, when an alarm is actuated, executive action is frequently taken. Nondispersive infrared spectroscopy is most often used in OP-HC detectors.

Toxic Gas Measurement

The types of open-path toxic gas detector designs include OP-FTIR spectroscopy. This is a very commonly used technique that utilizes a number of very specific absorption bands in the IR spectrum. A single instrument can be configured for the measurement of a wide variety of multiple components, and these sensors are highly adaptable for survey work.

In applications in which FTIR does not have sufficient sensitivity, OP-UV spectroscopy is frequently employed. This methodology can be used for applications involving the detection of homonuclear diatomic molecules (chlorine, bromine, etc.), which have no IR absorption, or of molecules that absorb only weakly in the IR region, such as benzene, sulfur dioxide, and nitrogen oxides.

Because low-cost, highly reliable, solid-state diode lasers have been developed and became available for high-volume telecommunications applications, a new class of open-path detectors has been developed and applied to a subset of toxic measurement applications. Instruments in this category utilize the ability of the diode laser to scan over very short wavelength intervals. This method of measurement is referred to as open-path tunable diode-laser absorption spectroscopy (OP-TDLAS).

Combustible Vapor Measurement

A very important use of open-path gas detection is to measure the concentration of combustible vapors. These vapors typically are hydrocarbons such as methane and ethylene, so the abbreviation OP-HC is frequently used. Detection is not strictly limited to pure hydrocarbons but can also be used to detect a number of organic vapors having a near-IR absorption spectrum due to CH bonds.

In terms of numbers of detectors sold, this application far dominates the market. OP-HC detectors can be found in hundreds of oil and gas production facilities and, in lesser numbers, in downstream transportation and distribution facilities as well as petrochemical facilities. In recent years, OP-HC detectors have seen increased usage to augment detection by traditional point hydrocarbon detectors, especially in situations where the probability of detecting a leak with a point detector is low.

OP-HC detection is especially useful for perimeter monitoring of tank farms, process areas, and other places where combustible vapor leaks can happen over a widely dispersed area. Because the emphasis is on safety applications, these detectors are optimized for low maintenance, avoidance of false alarms, and low costs.

These detectors typically are housed in flameproof enclosures and are suitable for deployment into Class 1, Division 1, or European Committee for Electrotechnical (CENELEC) Zone 1 hazardous areas. These devices typically have onboard heaters to melt snow and ice so that they can operate unattended and uninterrupted in inclement weather.

Figure 1.62r shows a simplified schematic representation of an OP-HC detector. The bistatic configuration is most commonly used. Monostatic configurations using a retroreflector require electrical power for heaters and subsequent flameproofed enclosure, which somewhat reduces the advantage of that configuration for this application.

The detection principle relies on a two-channel nondispersive photometer. The active channel is equipped with an optical filter that limits light to the detector to the hydrocarbon

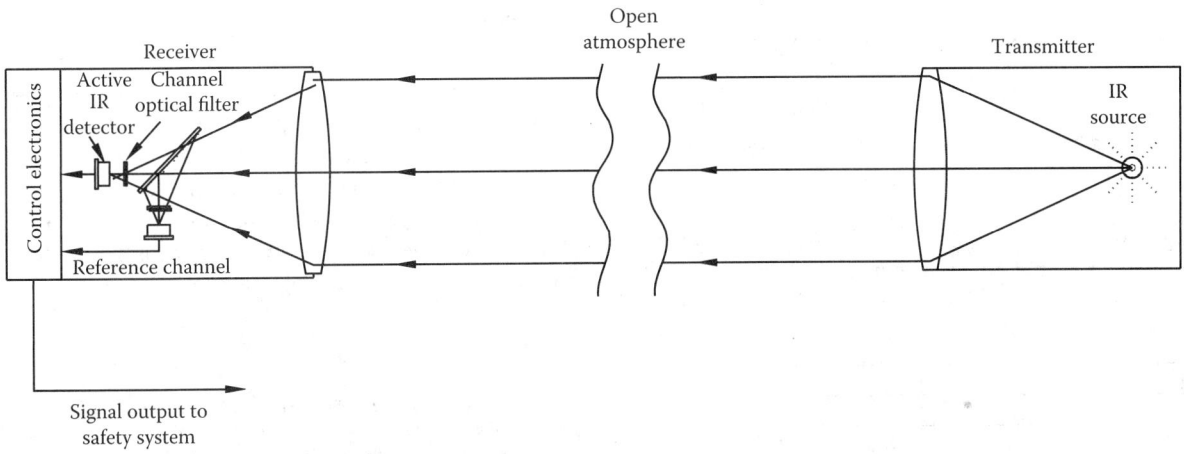

FIG. 1.62r
Schematic representation of an NDIR-based OP-HC detector.

absorption region of the spectrum. The reference channel observes the intensity of light in an adjacent portion of the spectrum that is free from hydrocarbon absorption. The optical filters are carefully specified to minimize false gas signals from differential absorption resulting from moisture as well as changes in the spectral output characteristics of the source over time.

Interferences These devices utilize bright, modulated sources with receiver detection electronics tuned to the modulation frequency. Typically, a xenon flash lamp is used; less typically, a modulated tungsten filament or micro-fabricated IR source is used. The source is modulated to mitigate against solar interference, either directly as the sun enters the field of view of the detector at sunrise or sunset or indirectly as reflections from waves in offshore installations or other background objects.

Solar radiation that falls within the field of view of the receiver and that has intensity modulated frequency components within the acceptance bandwidth of the receiver electronics can result in a false gas signal. Source modulation is performed at frequencies where there is minimal solar modulation. The short pulse duration of the xenon lamp permits the utilization of a very narrow acceptance bandwidth filter in the receiver electronics.

Transmission Distance Distances between the transmitter and the receiver vary typically between 10 and >120 m. Most vendors supply at least two models: a short-range unit for operation from 10 to 60 m and a longer-range unit operating to the >120 m distance. The primary difference between the units is the efficiency of the collimation and condensing optics. In most cases, the worst-case visibility, resulting from fog and other atmospheric phenomena, presents the practical limit to transmitter–receiver separation.

Table 1.62e illustrates this point by showing how the maximum acceptable transmitter–receiver separation varies with visibility in fog. Additionally, for very short-range applications <10 m, such as in ventilation ducts and small enclosures, the units can be equipped with optical attenuators fitted to the transmitter to keep the intensity of light within acceptable levels for the receiver.

The PIC range of the devices are usually set at around 0–5 LEL*m. This corresponds to a range of from 0% to 25% * m for methane (where 1 LEL of methane is 5%). Low and high alarms are generally set for 2 and 3 LEL*m.

OP-HC Advantages and Limitations The use of PIC for open-path monitoring has been the source of much debate and confusion, especially when attempts are made to correlate the range with point hydrocarbon detectors that are generally ranged from 0 to 1 LEL (0%–100% LEL). Using PIC, it is clearly possible that the concentration of gas will exceed, in some location, in the beam's path, the LEL. On the other hand, many point sensors would be required to be able to place a point detector at the precise location where such localized concentrations occur.

Much discussion and research has been done to arrive at a rational range setting strategy for OP-HC devices. There has been considerable work done at the Christian Michelson Institute in Norway on vapor cloud explosions. It has been found that, for oil and gas facilities, where largely unconfined explosions can be expected, flame front velocities below 100 m s^{-1} will not produce significant damage to personnel or facilities. Experiments have also demonstrated that flames need a run-up distance of approximately 5.5 m. Therefore, vapor clouds with dimensions less than this may not cause substantial damage.*,†

The primary signal output from an OP-HC detector is almost always a 4–20 mA current loop. The output is generally tied into a dedicated fire and gas safety system that can trigger some form of executive action in the facility. Secondary outputs are frequently available as well. Digital protocols such as HART and Modbus are common and permit the devices to be tied to PLCs and asset management computer systems.

SPECIFICATION FORMS

When specifying open-path spectrometers, the ISA Form 20A1001 can be used to specify the features required for the device itself, and when specifying both the device and the nature of the application, including the composition and/or properties of the process materials, ISA Form 20A1002 can be used. These forms, with the permission of the International Society of Automation, are reproduced on the next pages.

TABLE 1.62e
Typical Maximum Distance between Transmitter and Receiver over Which Proper OP-HC Operation Can Be Maintained at a Worst-Case Fog Visibility

Worst-Case Visibility in Fog[a] (m)	Transmitter/Receiver Separation Distance (m)
6	10
15	25
18	30
24	40
30	50
36	60

Source: Courtesy of Detector Electronics, Bloomington, MN.
[a] As measured by Meteorological Optical Range.

---

* Nolan, D. P., *Handbook of Fire & Explosion Protection Engineering Principles for Oil, Gas, Chemical, & Related Facilities*, 2013. https://books.google.com/books/about/Handbook_of_Fire_Explosion_Protection_En.html?id=tzR_rfTxx0YC.
† Bjerketvedt, D. et al., *Gas Explosion Handbook*, 1993. http://www.worldcat.org/title/gas-explosion-handbook/oclc/66939498.

1.62 Spectrometers, Open Path (OP)

1	RESPONSIBLE ORGANIZATION		ANALYSIS DEVICE	6	SPECIFICATION IDENTIFICATIONS		
2	(ISA logo)			7	Document no		
3			Operating Parameters	8	Latest revision	Date	
4				9	Issue status		
5				10			

	ADMINISTRATIVE IDENTIFICATIONS				40	SERVICE IDENTIFICATIONS Continued		
11					41	Return conn matl type		
12	Project number		Sub project no		42	Inline hazardous area cl	Div/Zon	Group
13	Project				43	Inline area min ign temp	Temp ident number	
14	Enterprise				44	Remote hazardous area cl	Div/Zon	Group
15	Site				45	Remote area min ign temp	Temp ident number	
16	Area		Cell	Unit	46			
17					47			
18	SERVICE IDENTIFICATIONS				48	COMPONENT DESIGN CRITERIA		
19	Tag no/Functional ident				49	Component type		
20	Related equipment				50	Component style		
21	Service				51	Output signal type		
22					52	Characteristic curve		
23	P&ID/Reference dwg				53	Compensation style		
24	Process line/nozzle no				54	Type of protection		
25	Process conn pipe spec				55	Criticality code		
26	Process conn nominal size		Rating		56	Max EMI susceptibility	Ref	
27	Process conn termn type		Style		57	Max temperature effect	Ref	
28	Process conn schedule no		Wall thickness		58	Max sample time lag		
29	Process connection length				59	Max response time		
30	Process line matl type				60	Min required accuracy	Ref	
31	Fast loop line number				61	Avail nom power supply		Number wires
32	Fast loop pipe spec				62	Calibration method		
33	Fast loop conn nom size		Rating		63	Testing/Listing agency		
34	Fast loop conn termn type		Style		64	Test requirements		
35	Fast loop schedule no		Wall thickness		65	Supply loss failure mode		
36	Fast loop estimated lg				66	Signal loss failure mode		
37	Fast loop material type				67			
38	Return conn nominal size		Rating		68			
39	Return conn termn type		Style					

	PROCESS VARIABLES	MATERIAL FLOW CONDITIONS			PROCESS DESIGN CONDITIONS		
69							
70	Flow Case Identification		Units	102	Minimum	Maximum	Units
71	Process pressure			103			
72	Process temperature			104			
73	Process phase type			105			
74	Process liquid actl flow			106			
75	Process vapor actl flow			107			
76	Process vapor std flow			108			
77	Process liquid density			109			
78	Process vapor density			110			
79	Process liquid viscosity			111			
80	Sample return pressure			112			
81	Sample vent/drain press			113			
82	Sample temperature			114			
83	Sample phase type			115			
84	Fast loop liq actl flow			116			
85	Fast loop vapor actl flow			117			
86	Fast loop vapor std flow			118			
87	Fast loop vapor density			119			
88	Conductivity/Resistivity			120			
89	pH/ORP			121			
90	RH/Dewpoint			122			
91	Turbidity/Opacity			123			
92	Dissolved oxygen			124			
93	Corrosivity			125			
94	Particle size			126			
95				127			
96	CALCULATED VARIABLES			128			
97	Sample lag time			129			
98	Process fluid velocity			130			
99	Wake/natural freq ratio			131			
100				132			

	MATERIAL PROPERTIES				MATERIAL PROPERTIES Continued		
133				137			
134	Name			138	NFPA health hazard	Flammability	Reactivity
135	Density at ref temp		At	139			
136				140			

Rev	Date	Revision Description	By	Appv1	Appv2	Appv3	REMARKS

Form: 20A1001 Rev 0

© 2004 ISA

	RESPONSIBLE ORGANIZATION	ANALYSIS DEVICE COMPOSITION OR PROPERTY Operating Parameters (Continued)		SPECIFICATION IDENTIFICATIONS	
1			6	Document no	
2	ISA		7	Latest revision	Date
3			8		
4			9	Issue status	
5			10		

	PROCESS COMPOSITION OR PROPERTY			MEASUREMENT DESIGN CONDITIONS				
	Component/Property Name	Normal	Units	Minimum	Units	Maximum	Units	Repeatability
11								
12								
13								
14								
15								
16								
17								
18								
19								
20								
21								
22								
23								
24								
25								
26								
27								
28								
29								
30								
31								
32								
33								
34								
35								
36								
37								

Rows 38–75: (blank notes area)

Rev	Date	Revision Description	By	Appv1	Appv2	Appv3	REMARKS

Form: 20A1002 Rev 0

© 2004 ISA

Definitions

Beer's law relates the absorption of light to the transmission properties of the material through which the light is traveling. In other words, it relates the amount of light absorbed to the concentration of the material of interest in a sample.

Black body radiation is radiation emitted by a black body, which is held at constant, uniform temperature. A black body is a theoretical object that absorbs 100% of the radiation that hits it and emits all of it. The black body reflects no radiation and appears perfectly black. In practice, no material has been found to absorb all incoming radiation; even graphite absorbs only about 97% of the radiation received. The theoretical black body is also a perfect emitter of radiation. All objects emit radiation when their temperature is above absolute zero. This radiation has a specific spectrum and intensity that depends only on the temperature of the body.

Interferometry is an important investigative technique. Interferometers are widely used in science and industry for the measurement of small displacements, refractive index changes and surface irregularities. In analytical science, interferometers are used in continuous wave Fourier transform spectroscopy to analyze light containing features of absorption or emission associated with a substance or mixture. In an interferometer, two light rays from a common source combine at the half-silvered mirror to reach a detector. They may either interfere constructively (strengthening in intensity) if their light waves arrive in phase, or interfere destructively (weakening in intensity) if they arrive out of phase, depending on the exact distances between the three mirrors.

Abbreviations

CCD	Charge-Coupled Device
CRDS	Cavity Ring-Down Spectrometry
DOAS	Differential Optical Absorption Spectroscopy
DTGS	Deuterated Triglycine Sulfate
FTIR	Fourier Transform Infrared
HF	Hydrogen fluoride
LEL	Lower Explosion Limit
LIDAR	Light Detection and Ranging
MCT	Mercury Cadmium Telluride
NDIR	Nondispersive Infrared
OP-CRDS	Open-path cavity ring-down spectrometry
OP-FTIR	Open-path Fourier transform infrared
OP-HC	Open-path hydrocarbon
OP-TDLAS	Open-path tunable diode laser
OP-UV	Open-path ultraviolet
PAC	Path-Average Concentration
PIC	Path-Integrated Concentration
PI-MDC	Path-Integrated-Minimum Detectable Concentration
RADAR	Radio Detection and Ranging
TDL or TDLAS	Tunable Diode Laser

Organizations

ACGIH	American Conference of Governmental Industrial Hygienists
AIHA	American Industrial Hygiene Association
ATSDR	Agency for Toxic Substances and Disease Registry
BSI	British Standards Institute
CCOHS	Canadian Centre for Occupational Health & Safety
CENELEC	European Committee for Electrotechnical Standardization
CSA	Canadian Standards Association
GINC	Global Information Network on Chemicals
ISEA	International Safety Equipment Association
MSHA	Mine Safety and Health Administration
NIOSH	National Institute for Occupational Safety and Health
NIST	National Institute of Standards and Technology
NSC	National Safety Council
OSHA	Occupational Safety & Health Administration
SEI	Safety Equipment Institute
UL	Underwriters Laboratories
USEPA	United States Environmental Protection Agency

Government Agencies

Agency for Toxic Substances and Disease Registry. An agency of the United States Department of Health and Human Services. www.atsdr.cdc.gov/atsdrhome.html.

American Conference of Governmental Industrial Hygienists for over 60 years, ACGIH has been a well-respected organization for individuals in the industrial hygiene and occupational health and safety industry. www.acgih.org/.

American Industrial Hygiene Association. Provides standards of ethical conduct for industrial hygienists www.aiha.org/.

British Standards Institute. www.bsi-global.com.

Canadian Centre for Occupational Health & Safety. Canada's National Center for Health and Safety Information. www.ccohs.ca/.

Canadian Standards Association. www.csa-international.org/.

Electronic NIOSH Pocket Guide to Chemical Hazards. www.cdc.gov/niosh/npg/pgdstart.html.

Europe's leading independent research and technology organization in imaging and intelligent systems. www.sira.co.uk/.

Factory Mutual Global. A commercial insurance company with a research division that is an approval agency. www.fmglobal.com.

Global Information Network on Chemicals. Useful information on hazardous substances. www.nihs.go.jp/GINC/.

International Safety Equipment Association. www.safetycentral. org/isea.
ISA—International Society of Automation www.isa.org/.
ISA—The Instrumentation, Systems and Automation Society. www. isa.org/.
Mine Safety and Health Administration. A United States Government agency regulating the mining industry. www.msha.gov/.
National Institute for Occupational Safety and Health. www.cdc. gov/niosh/homepage.html.
National Institute of Standards and Technology. Standard reference materials. www.nist.gov/srm.
National Safety Council. www.nsc.org/.
Occupational Safety & Health Administration. A government agency that regulates to protect against occupational hazards. www.osha.gov/.
Safety Equipment Institute. www.seinet.org/.
Underwriters Laboratories, Inc., an approval agency. www.ul.com/.
United States Environmental Protection Agency. www.epa.gov/.

Bibliography

ACGIH, *Air Sampling Instruments for Evaluation of Atmospheric Contaminants*, 9th edn., American Conference of Governmental Industrial Hygienists, Cincinnati, OH, 2001. http://www.alibris.com/Air-Sampling-Instruments-for-Evaluation-of-Atmospheric-Contaminants-Acgih/book/9453180.

AIHA, *Workplace Environmental Exposure Level Guide Series, Full Set of 109 Individual Guides on Toxic Chemicals*, American Industrial Hygiene Association, Akron, OH, 1998. http://www.amazon.co.uk/361-ea-99-Workplace-Environmental-Exposure-Series/dp/093262703X.

ASHRAE, Standard 110–1995, *Method of Testing Performance of Fume Hoods*, American Society of Heating, Refrigerating, and Air-Conditioning Engineers, Atlanta, GA, 1995. http://webstore.ansi.org/RecordDetail.aspx?sku=ANSI%2fASHRAE+110-1995.

Berden, G. and Engeln, R., *Cavity Ring-Down Spectroscopy: Techniques and Applications*, Wiley-Blackwell Publishing Ltd., 2009. http://onlinelibrary.wiley.com/doi/10.1002/9781444308259.ch1/summary.

Bjerketvedt, D. et al., *Gas Explosion Handbook*, 1993. http://www.worldcat.org/title/gas-explosion-handbook/oclc/66939498.

Bretherick, L., *Bretherick's Handbook of Reactive Chemical Hazards*, Butterworth-Heinemann, Burlington, MA, 1999. http://eng.monash.edu/materials/assets/documents/resources/ohs/bretherick-vol1.pdf.

Busch, K. W. and Busch, M. A., *Cavity-Ringdown Spectroscopy*, American Chemical Society, 1999. https://global.oup.com/academic/product/cavity-ringdown-spectroscopy-9780841236004?cc=us&lang=en&.

Calabrese, E. J. and Kenyon, E., *Air Toxics and Risk Assessment*, Lewis Publishers, Boca Raton, FL, 1991. https://www.bookdepository.com/Air-Toxics-Risk-Assessment-Edward-Calabrese/9780873711654.

Cullis, C. F. et al., *Detection and Measurement of Hazardous Gases*, Butterworth-Heinemann, Burlington, MA, 1981. http://www.worldcat.org/title/detection-and-measurement-of-hazardous-gases/oclc/7328268.

EPA, Open path technologies, 2013. http://www.clu-in.org/programs/21m2/openpath/tdl/.

Frish, N. B., et al. The next generation of TDLAS analyzers, 2007. http://proceedings.spiedigitallibrary.org/proceeding.aspx?articleid=825042.

Lide, D. R. et al. (eds.), *CRC Handbook of Chemistry and Physics*, 83rd edn., CRC Press, Boca Raton, FL, 2002. https://bookmaderligh.files.wordpress.com/2015/10/crc-handbook-93rd-edition-pdf.pdf.

Ness, S. A., *Air Monitoring for Toxic Exposures: An Integrated Approach*, John Wiley & Sons, New York, 1991. https://books.google.com/books/about/Air_monitoring_for_toxic_exposures.html?id=FVpCAQAAIAAJ.

NIOSH, *Manual of Analytical Methods*, 4th edn., Vols. I and II, DIANE Publishing, Washington, DC, 2014, http://www.cdc.gov/niosh/docs/2003-154/default.html.

Nolan, D. P., *Handbook of Fire & Explosion Protection Engineering Principles for Oil, Gas, Chemical, & Related Facilities*, 2013. https://books.google.com/books/about/Handbook_of_Fire_Explosion_Protection_En.html?id=tzR_rfTxx0YC.

Phillips, D. and Griffith, D., Open-path FTIR spectrometer, 2014. http://smah.uow.edu.au/cac/openpath/index.html.

Science Direct, Vibrational spectroscopy, 2007. http://www.sciencedirect.com/science/article/pii/S092420310600141X.

Shao, L., Griffiths, P. R., and Levtem, A. B., Advances in data processing for open-path Fourier transform infrared spectrometry, 2010. http://www.ncbi.nlm.nih.gov/pubmed/20879801.

Sorokina, I. T. and Vodopyanov, K. L., *Solid-State Mid-Infrared Laser Sources*, 2003. http://www.springer.com/us/book/9783540006213.

Unisearch Associates, Instruments—LasIR, 2016. http://www.unisearch-associates.com/instruments/lasir.html.

Woodford, C. Interferometers, 2015. http://www.explainthatstuff.com/howinterferometerswork.html.

1.63 Streaming Current Particle Charge Analyzer

W. F. GERDES (1974, 1982)   **B. G. LIPTÁK** (1995, 2003, 2017)

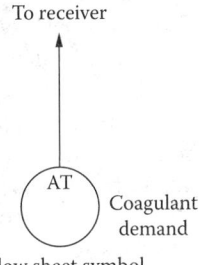

Flow sheet symbol

Applications	Continuous control of streaming current can guarantee proper clarification of beverages, dewatering, water treatment, thickening of suspensions, etc. by adjusting the addition of coagulant chemicals, based on the measurement of the surface charge on particles
Materials of construction	Stainless steel, silver, and Teflon®
Sample size required	In laboratory grab samples: about 10 cc In online bypass flow based installations
Approximate cost	$10,000–$20,000
Partial list of suppliers*	Accufloc (monitor) http://www.accufloc.com/ BTG Instruments (laboratory analyzer) http://www.btg.com/products.asp?langage=15&appli=5&numProd=357&cat=prod Chemtrac Inc. (online and multiple) http://beabout.com/resources/clientfiles/CHEM-Water%20Treatment%20Innovations.pdf Colloidal Dynamics (monitor) http://www.colloidal-dynamics.com/zetacad.php Dispersion Technology (Zeta potential and laboratory) http://www.dispersion.com/products Hach (online monitor) http://www.hach.com/af7000-streaming-current-monitor-230v/product?id=10452743002 HF Scientific (laboratory) http://www.hfscientific.com/products/microtscm_streaming_current_monitor Kenelec Scientific (laboratory and controller) http://www.kenelec.com.au/products/cid/168/category/Streaming%20Current%20Monitors/parent/0/t/products Micrometrix (portable laboratory) http://micrometrix.com/wp-content/uploads/2013/11/SCM1-Data-Sheet-110813m.pdf Milton Roy (online monitor and control) http://www.miltonroy.com/MiltonRoy/en/Products/Controllers/Streaming-Current-Detectors Pi (laboratory and online) http://www.processinstruments.net/products/water-treatment-controllers/online-charge-analyzer/ Sentrol Systems (online) http://www.sentrolsystems.com/ XPRT Environmental (monitor) http://www.environmental-expert.com/products/accufloc-streaming-current-monitor-120345

* Suppliers' List for the fifth edition was updated with the help of Francisco Alcala, Automation Specialist, CDM Smith.

INTRODUCTION

The treatment of waste water requires the removal of very small, finely divided particles, having surface electric charges (colloids) that are repelled by one another and unless removed, would remain dispersed in a liquid for a long time. Coagulation is the process of forming semisolid lumps from these particles by adding coagulants that remove this surface electric charge and cause the small colloidal particles to coagulate and sediment out from the stream. The amount of colloid particles in the wastewater can be detected by measuring the streaming current or potential of the water. Therefore, the dosage of coagulant addition can be set to bring the streaming current value (SCV) of the treated water to zero.

The Streaming Current Detector (SCD) is a valuable tool to monitor and control the required rate at which coagulants need to be added in water or wastewater treatment plants. The SCD is available as an in-line instrument, which can continuously control water quality. In addition to water and wastewater treatment applications, it can serve the same functions in the paper, food, chemical, petroleum, and other industries where the optimized addition of sometimes expensive charge altering chemicals is needed.

The streaming current monitor (SCM) is a fundamental tool for controlling the rate of addition of coagulants during wastewater treatment. These control loops increase the rate of coagulant addition, as the streaming current of the wastewater increases. The increased concentrations of coagulants will cause the small colloidal particles to coagulate and settle out from the stream and as the concentration of colloid particles is reduced, the streaming potential also drops. As this occurs, the SCM controller reduces the rate of coagulant addition and thereby, while keeping the concentration of colloidal particles low, also conserves the use of coagulants.

Consequently, the accurate measurement of surface charges on the suspended particles is important when studying the adsorption characteristics of colloidal suspensions because the streaming current detector (SCD) can estimate the amount of treatment material required to continuously control the addition of coagulation chemicals.

Less directly, SCD can compare the ability of alternative treating chemicals to influence the electric charge of particles, and it can determine the effect of pH on a chemical's ability to promote or enhance coagulation. The most effective use of chemicals requires a recognition of the role of electrical charges in stabilizing suspensions.

OPERATING PRINCIPLE

In ionic liquids, any interface with a solid or a second liquid will have an electrical charge that is caused by the preferential adsorption or positioning of ions. The liquid adjacent to the interface surface will contain excess charges of the opposite sign, called counterions.

If charged particles are attached to a filter or capillary wall, these counterions can be swept downstream by a stream of water. This flow of charges of predominantly one sign constitutes a current called the *streaming current*. In the case of an insulating capillary, the return path is by ionic conduction through the liquid in the stream. With suitable electrodes, the return path can be arranged so as to measure this streaming current.

The Streaming Current

The van der Waals force causes particles that carry high charges to be preferentially adsorbed to the surfaces of a cylinder and piston as shown in Figure 1.63a. When such a particle is moved upward by the piston movement, its counterions

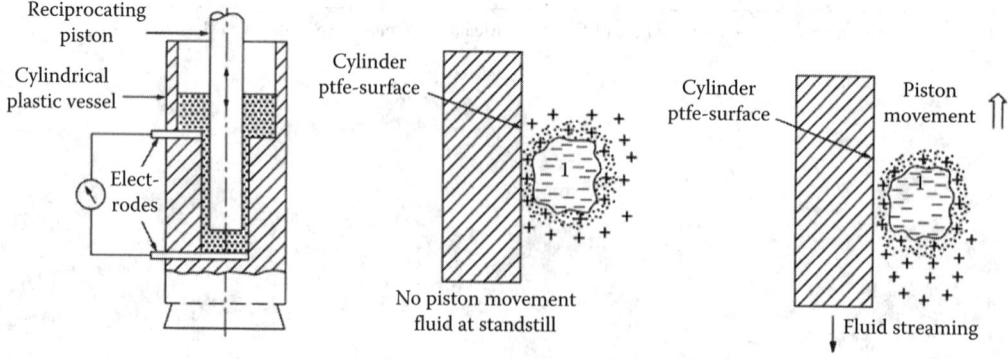

FIG. 1.63a
In the boundary of the diffuse layer, at the cylinder surface, a cloud of counterions is sheared off by the streaming fluid. (Courtesy of BTG Instruments.)

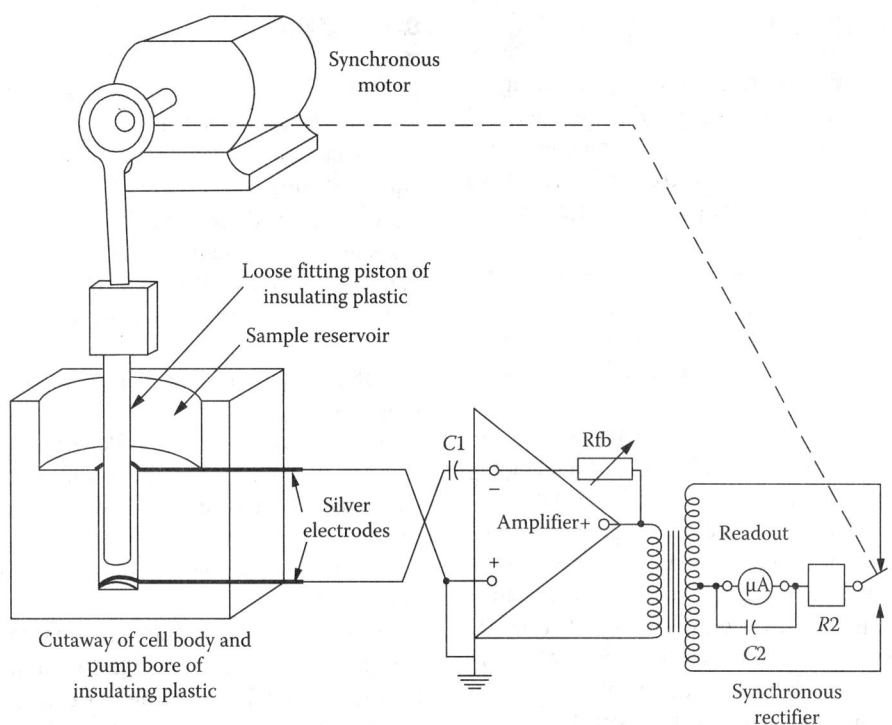

FIG. 1.63b
Streaming current detector.

are sheared off as the fluid moves in the opposite direction relative to the piston. The totality of these ions is the streaming potential detected by this analyzer.

Because the energy level of the streaming current is small as compared to the random signals from electrodes and from inadvertent thermocouples, it is advantageous to develop a reversing or alternating signal. This signal of known frequency can be more readily distinguished from noise that occurs at other frequencies.

Determination of polarity of the charge is by phase-sensitive rectification through the action of a switch that is synchronized with the instantaneous flow of the liquid. This alternating design permits a pump and capillary to be combined into a single unit with a loosely fitted piston reciprocating in a dead-ended bore (Figure 1.63b). Cleaning is easy, because the piston and bore can be separated by disassembly.

In case of a laboratory analyzer (Figure 1.63c), the sample is placed in the measuring cell and a pH probe is inserted. As the piston in the cell oscillates causing a high flow rate, the charged ions are adsorbed to the cell wall and are separated from their counterions creating a streaming current. The two electrodes shown in Figure 1.63b pick up this current and display it, usually next to the measured pH value. A polyelectrolyte of opposite charge is added until reaching the charge zero point. The digital display indicates the polarity and charge value (streaming current) of the sample water, waste water, or other chemical.

FIG. 1.63c
Laboratory streaming current analyzer. (Courtesy of Micrometrix.)

Treatment Chemical Selection

To estimate the amount of treatment that a process liquid requires, a volumetric sample is titrated with the chemical that is intended for use for treatment at a known concentration until a zero signal is obtained. To compare alternative treating chemicals, identical samples can be titrated with the various chemicals.

When it is desired to study the effect of a change in pH on treating requirements, identical samples are titrated at various pH levels. The pH effect can be significant, because the tolerance of chemicals can be considerably different at low or high pH values.

In the usual treatment plant, the SCD may continuously control the feed of cationic chemicals. The need for changing the charging rate of the chemical additives arises from variations in the flow rate of the process fluid, changes in the loading of suspended solids in the process fluid, variations in the unit demand of the solids, or any combination of these factors.

The main advantage of SCD control is its fast response to changes in the operating conditions. Charge neutralization occurs almost as soon as the treating chemical is dispersed in the stream; therefore, samples may be taken 1 or 2 min after addition of the chemical.

CALIBRATION

Although readout is in arbitrary units, reproducibility can be demonstrated by standard samples. Sensitivity is constant, but zero shift can occur as a result of inadequate cleaning or the use of strong surface-active agents. Unfortunately, there is no absolute zero standard. Colloidal suspensions are useless as routine standards, because they become unstable with time. Suspensions made stable by excess surface-active materials leave a stubborn residue on the SCD.

Often, the most useful standards are two buffer solutions chosen so that their pHs are on either side of the isoelectric point of the capillary material. The isoelectric point is defined as the pH at which the material exhibits zero charge. Upscale or downscale shifts of buffer readings will indicate the bias of the instrument caused by contamination of the surfaces. Buffers near pH 4 and 5 are useful.

Dielectric constant, pH, and total ionic content all exert legitimate influences on the streaming current reading. Lowering the pH of a sample shifts its SCD reading upscale, and increasing the salt content by adding a balanced, neutral salt shifts the reading toward zero (Figure 1.63d).

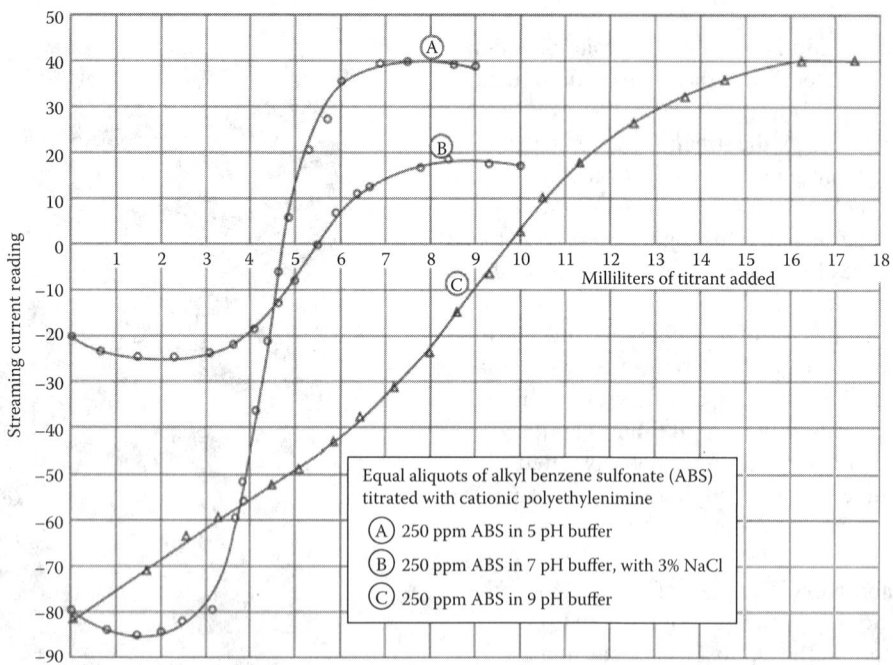

FIG. 1.63d
Titration curves indicating effect of pH and salt on SCD reading.

APPLICATIONS

Most applications involve either titration of the sample or prior treatment in the plant, because a single reading on untreated material provides little information. This is because SCD readings are almost independent of the concentration of suspended solids. Titration can be performed with as little sample as will submerge the active part of the instrument.

Sampling

If batch samples are used, they should be large enough to permit rinsing the SCD apparatus several times. Skimming or decanting can be used to remove sand, larger solid particles, and oil globules. Little information is lost by the removal of larger particles, because charge is a surface phenomenon, and the fines provide most of the surface.

When a continuous sample is required, a self-cleaning bypass filter should be used (see Figure 1.63e). Periodic back-flushing or cleaning of the sampling line is also recommended.

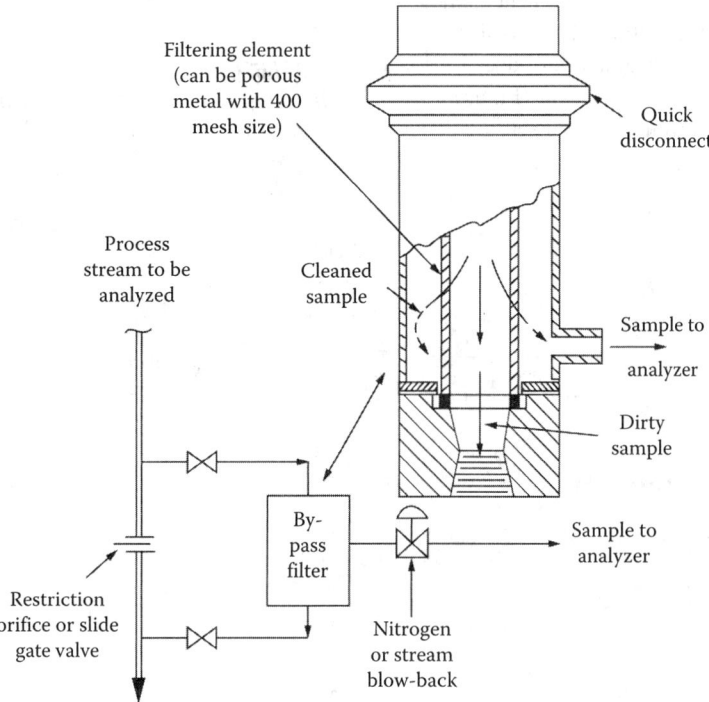

FIG. 1.63e
Self-cleaning bypass filter design and installation.

Control System

For continuous control, the SCD measurement signal is fed to a two-mode controller that modulates the chemical feed pump or control valve (Figure 1.63f). The transmitter signal can be sent through a 0.1–0.2 min filter to remove noise. The two-mode feedback controller itself is usually set for a PB = 500% and I = 2 repeats/min.

The maximum and minimum opening of the additive control valve can be limited so that a loss of sample will not cause an open loop. In the old pneumatic systems, pressure regulators provided the limits for valve openings, but today they are all electronic.

When automatic control of the rate of coagulant charging is being installed on a water or wastewater treatment application, the SCM is installed before the sedimentation tank (settling) and is controlling the dosage of coagulant introduced prior to the settling process in water treatment plant (Figure 1.63g). The ideal level of streaming current for good settling depends

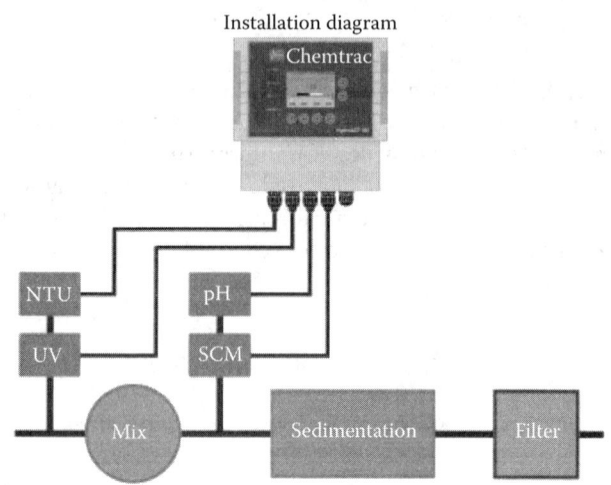

FIG. 1.63g
Installation of the streaming current meter (SCM) in a waste treatment plant. (Courtesy of Chemtrac Inc.)

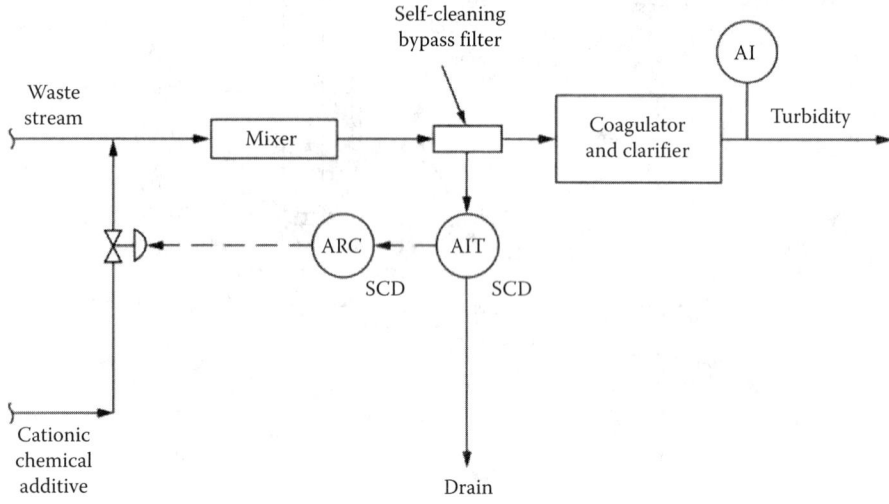

FIG. 1.63f
Chemical addition control utilizing a streaming current detector.

on the particular plant design. This is determined by either jar tests or by trial and error. In any case, once it is known, a Streaming Current Monitor can be used to automate the control of coagulant dosing even at times when the concentration of the incoming water changes.

In case of an online SCM monitor, the sensor usually is a teflon piston reciprocating into a stainless steel, teflon, and epoxy sensor chamber, oscillating at a rate of about 200 strokes/min. The design is usually such that both the sensor chamber and the piston can be removed if maintenance is needed. The unit shown in Figure 1.63h has two LED displays, the lower one displays the streaming current reading and the upper one displays the streaming current reading's offset from the set point.

This unit is available with two 4–20 mA outputs and 2 alarm relay outputs in addition to RS485/Modbus digital communications. Push-button calibrations (both zero and span) can be performed on any samples, and the zero offset can be manually adjusted at any time. A built-in PID controller can automatically control the coagulant dosing rate as shown in Figure 1.63f. An automatic liquid cleaning system is also available, which flushes out the deposited flock from the sensor under the control of an adjustable timer.

The SCD controller set points can be determined on the basis of turbidity measurement downstream of the coagulation equipment. The set point of the SCD controller can be directly adjusted by this turbidity controller in a cascade arrangement. In some processes, more than one chemical additive is used and therefore a sequence of additions and the attendant interactions must be considered in the control configuration.

SPECIFICATION FORMS

When specifying streaming current analyzers, the ISA Form 20A1001 can be used to specify the features required for the device itself and when specifying both the device and the nature of the application, including the composition and/or properties of the waste water or other process materials, ISA Form 20A1002 can be used. These forms, with the permission of the International Society of Automation, are reproduced on the next pages.

FIG. 1.63h
Online streaming current monitor used in automatic control of coagulant charging. (Courtesy of Accufloc.)

Analytical Measurement

ANALYSIS DEVICE — Operating Parameters

#	RESPONSIBLE ORGANIZATION			#	SPECIFICATION IDENTIFICATIONS		
1				6			
2				7	Document no		
3				8	Latest revision	Date	
4				9	Issue status		
5				10			

#	ADMINISTRATIVE IDENTIFICATIONS				#	SERVICE IDENTIFICATIONS Continued		
11					40			
12	Project number		Sub project no		41	Return conn matl type		
13	Project				42	Inline hazardous area cl	Div/Zon	Group
14	Enterprise				43	Inline area min ign temp	Temp ident number	
15	Site				44	Remote hazardous area cl	Div/Zon	Group
16	Area	Cell		Unit	45	Remote area min ign temp	Temp ident number	
17					46			
18	SERVICE IDENTIFICATIONS				47			
19	Tag no/Functional ident				48	COMPONENT DESIGN CRITERIA		
20	Related equipment				49	Component type		
21	Service				50	Component style		
22					51	Output signal type		
23	P&ID/Reference dwg				52	Characteristic curve		
24	Process line/nozzle no				53	Compensation style		
25	Process conn pipe spec				54	Type of protection		
26	Process conn nominal size		Rating		55	Criticality code		
27	Process conn termn type		Style		56	Max EMI susceptibility	Ref	
28	Process conn schedule no		Wall thickness		57	Max temperature effect	Ref	
29	Process connection length				58	Max sample time lag		
30	Process line matl type				59	Max response time		
31	Fast loop line number				60	Min required accuracy	Ref	
32	Fast loop pipe spec				61	Avail nom power supply		Number wires
33	Fast loop conn nom size		Rating		62	Calibration method		
34	Fast loop conn termn type		Style		63	Testing/Listing agency		
35	Fast loop schedule no		Wall thickness		64	Test requirements		
36	Fast loop estimated lg				65	Supply loss failure mode		
37	Fast loop material type				66	Signal loss failure mode		
38	Return conn nominal size		Rating		67			
39	Return conn termn type		Style		68			

#	PROCESS VARIABLES	MATERIAL FLOW CONDITIONS				#	PROCESS DESIGN CONDITIONS		
69						101			
70	Flow Case Identification				Units	102	Minimum	Maximum	Units
71	Process pressure					103			
72	Process temperature					104			
73	Process phase type					105			
74	Process liquid actl flow					106			
75	Process vapor actl flow					107			
76	Process vapor std flow					108			
77	Process liquid density					109			
78	Process vapor density					110			
79	Process liquid viscosity					111			
80	Sample return pressure					112			
81	Sample vent/drain press					113			
82	Sample temperature					114			
83	Sample phase type					115			
84	Fast loop liq actl flow					116			
85	Fast loop vapor actl flow					117			
86	Fast loop vapor std flow					118			
87	Fast loop vapor density					119			
88	Conductivity/Resistivity					120			
89	pH/ORP					121			
90	RH/Dewpoint					122			
91	Turbidity/Opacity					123			
92	Dissolved oxygen					124			
93	Corrosivity					125			
94	Particle size					126			
95						127			
96	CALCULATED VARIABLES					128			
97	Sample lag time					129			
98	Process fluid velocity					130			
99	Wake/natural freq ratio					131			
100						132			

#	MATERIAL PROPERTIES				#	MATERIAL PROPERTIES Continued		
133					137			
134	Name				138	NFPA health hazard	Flammability	Reactivity
135	Density at ref temp		At		139			
136					140			

Rev	Date	Revision Description	By	Appv1	Appv2	Appv3	REMARKS

Form: 20A1001 Rev 0 © 2004 ISA

1	RESPONSIBLE ORGANIZATION	ANALYSIS DEVICE COMPOSITION OR PROPERTY Operating Parameters (Continued)	6	SPECIFICATION IDENTIFICATIONS		
2			7	Document no		
3			8	Latest revision		Date
4			9	Issue status		
5			10			

	PROCESS COMPOSITION OR PROPERTY			MEASUREMENT DESIGN CONDITIONS				
	Component/Property Name	Normal	Units	Minimum	Units	Maximum	Units	Repeatability
13								
14								
15								
16								
17								
18								
19								
20								
21								
22								
23								
24								
25								
26								
27								
28								
29								
30								
31								
32								
33								
34								
35								
36								
37								

Rev	Date	Revision Description	By	Appv1	Appv2	Appv3	REMARKS

Form: 20A1002 Rev 0

© 2004 ISA

Definitions

Coagulation is the process of forming semisolid lumps in a liquid. Coagulant is the material that causes the transformation of some of the materials in the liquid to turn into a soft, semisolid, or solid mass.

Colloids are very small, finely divided solids (particles that do not dissolve) that remain dispersed in a liquid for a long time due to their small size and electrical charge. One property of colloid systems that distinguishes them from true solutions is that colloidal particles scatter light. The particles of a colloid selectively absorb ions and acquire an electric charge. All of the particles of a given colloid take on the same charge (either positive or negative) and thus are repelled by one another.

Electrolyte an electrolyte is any substance containing free ions that make the substance electrically conductive. Therefore, an electrolyte is a compound that ionizes when dissolved in water or any other ionizing solvents. Electrolytes commonly exist as solutions of acids, bases, or salts. Electrolyte solutions are normally formed when a salt is placed into a solvent such as water and the salt (a solid) dissolves into its components.

Flocculation suspended particles are brought together through collisions to form bigger particles called floc that may be settled or filtered out of the water.

Isoelectric point is the pH at which the material exhibits zero charge.

Streaming current: Streaming current and streaming potential are two interrelated electrokinetic phenomena of surface chemistry and electrochemistry. They are the electric current or potential that are developed when an electrolyte is forced to flow through a channel or porous plug having electrically charged walls.

van der Waals force is a weak attractive force between atoms or nonpolar molecules caused by a temporary change in dipole moment arising from a brief shift of orbital electrons to one side of one atom or molecule, creating a similar shift in adjacent atoms or molecules.

Zeta potential is the electrostatic potential generated by the accumulation of ions at the surface of the colloidal particle. Particle charge can be controlled by modifying the suspending liquid. Modifications include changing the liquid's pH or changing the ionic species in solution.

Abbreviations

ABS	Alkyl benzene sulfonate
I	Integral
PB	Proportional band
PID	Proportional-integral-derivative
SCD	Streaming current detector
SCM	Streaming current monitor
SCV	Streaming current value

Organization

AWWA   American Water Works Association

Bibliography

Adams, V., *Water and Wastewater Examination Manual*, Lewis Publishers, Boca Raton, FL, 1990. http://www.downloadbooks-forfree.net/epubpdf/water-and-wastewater-examination-manual.

AWWA, *Coagulation, Flocculation, and Sedimentation*, Denver, CO, 2005. http://test75.awwa.org/store/productdetail.aspx?productid=8091.

AWWA, M37 *Operational Control of Coagulation and Filtration Processes*, 3rd edn., Denver, CO, 2011. http://www.awwa.org/store/productdetail.aspx?ProductId=6726.

AWWA, *Water Treatment*, Denver, CO, 2003. https://books.google.co.za/books/about/Water_Treatment.html?id=WO6A_4JAdVsC.

Benjamin, M.M., *Water Quality Engineering: Physical/Chemical Treatment Process*, John Wiley & Sons, Hoboken, NJ, 2013. http://www.amazon.com/Water-Quality-Engineering-%20Treatment-Processes/dp/1118169654/ref=sr_1_12?s=books&%20ie=UTF8&qid=1390080053&sr=1-12&keywords=coagulation%20+and+flocculation.

Bratby, J., *Coagulation and Flocculation in Water and Wastewater Treatment*, IWA Publishing, London, U.K., 2008. http://www.amazon.com/Coagulation-Flocculation-Water-Wastewater-Treatment/dp/1843391066/ref=sr_1_2?s=books&ie=UTF8&qid=1390079414&sr=1-2&keywords=coagulation+and+flocculation.

CMC, What is zeta potential, (2013) http://www.colloidmeasurements.com/zeta.html.

Crittenden, J.C. et al., *MWH's Water Treatment: Principles and Design*, John Wiley & Sons, Hoboken, NJ, 2012. http://www.amazon.com/MWHs-Water-Treatment-Principles-Design/dp/0470405392/ref=sr_1_7?s=books&ie=UTF8&qid=1390079723&sr=1-7&keywords=coagulation+and+flocculation.

Csuros, M., *Microbiological Examination of Water and Wastewater*, Lewis Publishers, Boca Raton, FL, 1999. https://books.google.com/books/about/Microbiological_Examination_of_Water_and.html?id=AtjDUn5KfG0C.

Dobias, B., *Coagulation and Flocculation*, 2nd edn., Taylor & Francis, Boca Raton, FL, 2005. http://www.amazon.com/Coagulation-Flocculation-Second-Surfactant-Science/dp/1574444557/ref=sr_1_1?s=books&ie=UTF8&qid=1390079253&sr=1-1&keywords=coagulation+and+flocculation.

Engelhardt, T.L., Coagulation, flocculation and clarification of drinking water, (2010) http://www.hach.com/cms-portals/hach_com/cms/documents/pdf/applicationseminars/Coagulation-Flocculation-and-Clarification.pdf.

EScubed Ltd., Zeta potential—Streaming potential, (2011) http://www.escubed.co.uk/sites/default/files/zeta_potential_(an013)_streaming_potential.pdf.

Fairhurst, D., An overview of the zeta potential, (2013) http://www.americanpharmaceuticalreview.com/Featured-Articles/133232-An-Overview-of-the-Zeta-Potential-Part-1-The-Concept/.

Hendricks, D.W., *Water Treatment Unit Processes: Physical and Chemical*, CRC Press, Boca Raton, FL, 2006. http://www.crcpress.com/product/isbn/9780824706951.

Jones, E., Coagulation, flocculation, sedimentation, and filtration, (2011) http://www.amazon.com/COAGULATIONFLOCCULATION-SEDIMENTATION-FILTRATIONJONES/dp/B006ZZ7ZLG/ref=sr_1_9?s=books&ie=UTF8&qid=1390079884&sr=1-9&keywords=coagulation+and+flocculation.

Kramer, L. and Horger, J., Streaming current monitor used to optimize coagulant dosages, (2015) http://www.waterworld.com/articles/print/volume-17/issue-3/editorial-focus/streaming-current-monitor-used-to-optimize-coagulant-dosages.html.

Lyklema, J., Fundamentals of interface and colloid science, (2014) http://www.sciencedirect.com/science/bookseries/18745679.

Micrometrics, Optimizing coagulation with the streaming current meter, (2014) http://info.ncsafewater.org/Shared%20Documents/Web%20Site%20Documents/Spring%20Conference/SC13_Presentations/OM_T_AM_08.00_Veal.pdf.

Sibiya, S.M., Evaluation of the streaming current detector (SCD) for coagulation control, (2014) http://www.sciencedirect.com/science/article/pii/S1877705814001362.

Walker, G., Colloids, (1999) http://sst-web.tees.ac.uk/external/U0000504/Notes/colloids/collintro.html.

Zeta-Meter, Everything you want to know about coagulation & flocculation, (1993) http://www.zeta-meter.com/coag.pdf.

1.64 Sulfur Dioxide and Trioxide

R. A. HERRICK (1974, 1982)   **B. G. LIPTÁK** (1995, 2003, 2017)

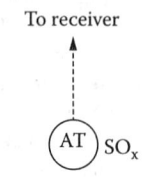

Flow sheet symbol

Possible methods of detection	A. Colorimetric (Chapter 1.13)
	B. Conductimetric (Chapter 1.15)
	C. Correlation spectrometry (Chapter 1.62)
	D. Coulometric (Chapter 1.21)
	E. Electrochemical (Chapter 1.21)
	F. Flame photometric (Chapter 1.54)
	G. Infrared (Chapter 1.32)
	H. Thermal conductivity (Chapter 1.66)
	I. Ultraviolet (pulsed fluorescence) (Chapter 1.69)
	J. Chromatographic (Chapter 1.10)
Most often used method	G. Infrared and I. Ultraviolet
Applications	Types A, B, D, E, F, and H are mostly used for ambient air monitoring (see Chapter 1.4 for details), whereas types C, G, and I are for stack gas monitoring (see Chapters 1.32, 1.62 and 1.69).
Reference method	A. Pararosaniline; an EPA study (CPA 70–101) found coulometry to be equivalent to this reference method.
Sample pressure	Where sampling is involved, the sample pressure is usually near atmospheric.
Sample temperature	In ambient air analysis, types A, B, D, E, F, and H are generally limited to about 49°C (120°F). Stack gas monitors (C, G, and I) are not so limited, as they usually are not in direct contact with the sample but are protected by air cooling.
Sample flow rate	Usually between 0.1 and 1.0 ACFM (2.8 and 28 alpm)
Materials of construction	Because of the reactivity of sulfur oxide gases, glass, Teflon®, and stainless steel are preferred; PVC should be avoided.
Speed of response	Types C, E, F, G, and I are nearly instantaneous; types A, B, and D depend on the liquid capacity of the automated wet chemistry system.
Ranges	Concentration in ambient air should stay below 0.1 ppm in the short range and below 0.02 ppm in the long range; therefore, the sensor range should not exceed 0–1 ppm.
	Stack monitors require 0–1000 ppm ranges.
	A, B, D, and F. From 0.01 to 5 ppm, but with dilution, systems can measure up to 5000 ppm
	E. 0–15 ppm
	C, G, and I. From 0 to 100 ppm up to 0%–100%, but most applications are from 500 to 10,000 ppm (type C can detect down to 5 ppm)
	H. 0%–20%

(Continued)

Inaccuracy	Generally, 1%–3% of full scale; types G and I can have errors as low as 0.5% of full scale. In ambient air measurement, the absolute limit on precision is 0.005 ppm.
Costs	Pocket-sized gas detection tubes and personal monitors $50–$500. Ultraviolet laboratory spectrometers cost from $2500 to $5000. Portable ambient monitors for multiple gas detection cost $4,000–$7,000, and permanently installed ambient air quality monitors configured for multiple-sample or multiple-gas monitoring is $10,000–$20,000. For details on open path spectrometry units see Chapter 1.62.
	Stack gas monitor costs range from $25,000 to $35,000, depending on materials of construction, accessories, and the number of components analyzed, which can include opacity. Passive, remote-sensing correlation spectrometers start at $35,000. Thermal conductivity analyzers cost about $10,000.
Partial list of suppliers	ABB, (Flame photometric) http://www.abb.com/product/seitp330/512997fa1e40403585257881005db3b2.aspx
	ACMAS Technologies (Coulometric) http://www.measuring-meters.com/pdf/coulometric-sulphur-analyzer.pdf
	Aeroqual, (UV) http://www.aeroqual.com/product/sulfur-dioxide-analyzer-module
	Airway Electronics (Total sulfur) http://gasdetect.com/products/gas-analyzers/ae2430-sulfur-analyzer/
	AMETEK, (I, Stack Gas) http://www.ametekpi.com/searchresults.aspx?Keywords=sulfur%20dioxide%20analyzer
	Analytical Systems (Total Sulfur analyzer) http://liquidgasanalyzers.com/product/total-sulfur-analyzer/
	EcoTech, (I, Ambient) bhttp://www.ecotech.com/gas-analyzers/so2-analyzer
	GDW, (detection tubes), http://www.gasdetectionwarehouse.com/rae-systems-colorimetic-tubes-10-per-box-sulfur-dioxide-so2/
	Horiba, (I, Ambient) (http://www.horiba.com/us/en/process-environmental/products/ambient/details/apsa-370-ambient-sulfur-dioxide-monitor-272)
	Nova, (IR) http://catalog.nova-gas.com/Asset/460%20Process%20SO2%20Analyzer.pdf
	Sensidyne, (Flue gas) http://www.sensidyne.com/colorimetric-gas-detector-tubes/detector-tubes/103sf-sulphur-dioxide-in-flue-gas.php
	Shimadzu, (G, I, Flue Gas) (http://www.shimadzu.com/an/industry/environment/e8o1ci0000000bep.htm, http://www.shimadzu.com/an/monitoring/continuous/soa307.html)
	Signal (IR, UV) http://www.signalinstruments.com/Gas-SO2.htm
	Teledyne, (I, Ambient) (http://www.teledyne-ai.com/pdf/6200t_Rev-A.pdf, http://www.teledyne-api.com/products/101_102e.asp)
	Thermo Scientific, (I, Ambient) (http://www.thermoscientific.com/ecomm/servlet/productsdetail_11152___11961389_-1)
	Tisch Environmental (UV) https://tisch-env.com/continuous-analyzers/sulfur-dioxide-analyzer
	Yokogawa, (Flue Gas) http://www.yokogawa.com/an/download/manual/IM11G04G01-01E.pdf

INTRODUCTION

Sulfur dioxide (SO_2) is a colorless gas and is toxic in large amounts. It smells like burnt matches. When oxidized to sulfur trioxide, it is readily transformed to sulphuric acid mist. Acid rain and snow causes the pH to drop to about 4 and can kill both plants on land and fish in the waters. SO_2 is also a precursor to sulphates, which are one of the main components of respirable particles in the atmosphere.

Most sulfur dioxide is produced by the combustion of elemental sulfur and therefore the flue gas from coal-fired power plants and similar facilities require desulfurization. To control the concentration of SO_2 emissions, accurate analyzers are required.

The predominant sulfur oxide in the atmosphere is sulfur dioxide (SO_2). Some sulfur trioxide (SO_3) is also formed in combustion processes, but it rapidly hydrolyzes to sulfuric acid, which is considered to be a particulate matter.

TABLE 1.64a
History of the National Ambient Air Quality Standards (NAAQS) for Oxides of Sulfur during the Period 1971–2012

Final Rule/Decision	Primary/Secondary	Indicator[a]	Averaging Time	Level[b]	Form
1971 36 FR 8186 April 30, 1971	Primary	SO_2	24 hr	0.14 ppm	Not to be exceeded more than once per year
			Annual	0.03 ppm	Annual arithmetic average
	Secondary		3 hr	0.5 ppm	Not to be exceeded more than once per year
			Annual[c]	0.02 ppm	Annual arithmetic average
1973 38 FR 25678 September 14, 1973	Secondary	Secondary 3 hr SO_2 standard retained, without revision; secondary annual SO_2 standard revoked.			
1996 61 FR 25566 May 22, 1996	Primary	Existing primary SO_2 standards retained, without revision.			
2010 75 FR 35520 June 22, 2010[d]	Primary	SO_2	1 hr	75 ppb	99th percentile, averaged over 3 years[e]
		Primary annual and 24 hr SO_2 standards revoked.			
2012 77 FR 20218 April 3, 2012	Secondary	Existing secondary SO_2 standard (3 hr average) retained, without revision.			

[a] SO_2 = sulfur dioxide.
[b] Units of measure are in parts per million (ppm) and parts per billion (ppb).
[c] The 1971 final rule also included a secondary 24 hr SO_2 standard of 0.1 ppm, maximum 24 hr concentration not to be exceeded more than once per year, as a guide to be used in assessing implementation plans to achieve the annual standard.
[d] The 1 hr SO_2 standard added in 2010 is a primary standard.
[e] The form of the 1 hr standard is the 3-year average of the 99th percentile of the yearly distribution of 1 hr daily maximum SO_2 concentrations.

Therefore, this chapter concentrates on the monitoring of sulfur dioxide. Table 1.64a gives the history of the National Ambient Air Quality Standards (NAAQS) over the last four decades. From this table, it can be seen that the EPA limits on the allowable SO_2 concentration became more demanding with the passage of time. As of this writing, the secondary air quality standard for sulfur dioxide is 60 μg/m³ (0.02 ppm), but it is advisable to be conservative and design the process equipment for half of that. The 1 hr primary standard is 75 ppb, and the form of the 1 hr standard is the 3-year average of the 99th percentile of the yearly distribution of 1 hr daily maximum SO_2 concentrations.

Sulfur dioxide (SO_2) is the product of the combustion of sulfur compounds and causes significant environmental pollution. Main sources of SO_2 in the environment are various industrial processes such as the burning of coal in power stations, the extraction of metals from ore, and the combustion of fuel within automobiles.

Sulfur dioxide is a noxious gas that can cause respiratory damage as well as impair visibility when in high concentrations. Sulfur dioxide also has the potential to form acid rain (H_2SO_4), which causes health, environmental, and infrastructural problems to human society.

In most ambient air monitoring projects, SO_2 is one of six pollutants that EPA requires to be measured.

Applications

The oxides of sulfur are measured both in ambient air, where their concentration is usually a small fraction of 1 ppm, and in stack flue gas and other industrial emissions, where their concentrations are in hundreds of ppm (see Figure 1.64a).

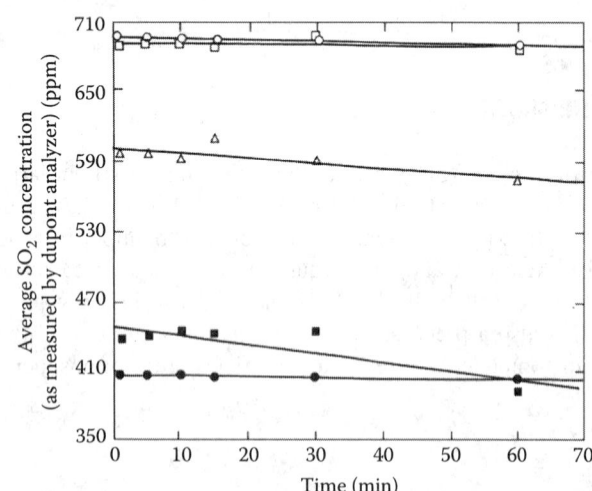

FIG. 1.64a

Sulfur dioxide readings (on different dates) obtained by monitoring the stack of a power-generating boiler burning pulverized coal. (From Homolya, J.B., Monitoring SO_2 emissions from stationary sources, Proceedings of the ISA Conference, October 1973.)

Because the analyzers used for these two types of application do overlap, they are discussed separately here.

While in the first page of this chapter some 10 methods of SO_2 measurements were listed, the fluorescence-type ultraviolet (UV) analyzers are most often used. Here, the monitoring of sulfur oxides in the ambient air will be discussed after the coverage of industrial analyzers.

INDUSTRIAL ANALYZER TYPES

Flue gas in the stack effluents contains orders of magnitude higher concentrations of sulfur dioxide than does ambient air. Consequently, few of the techniques used to detect the trace amounts of sulfur dioxide in ambient air are applicable to source monitoring, except after massive dilution.

The colorimetric procedure is not applicable to source monitoring at all, mainly because of the interference of nitrogen oxides. The analyzers that are generally favored for source monitoring include infrared (IR), UV, and correlation spectrometry, but the most often used is UV.

Nondispersive Infrared (NDIR)

Sulfur dioxide absorbs radiation over a broad range of wavelengths, which include both the IR and UV regions of the spectrum. Chapter 1.32 describes the nondispersive infrared (NDIR) analyzers in some detail. Figure 1.64b shows one of their stack-mounted variations.

These devices are most often used in flue gas analyzer packages, where several stack gases and opacities are

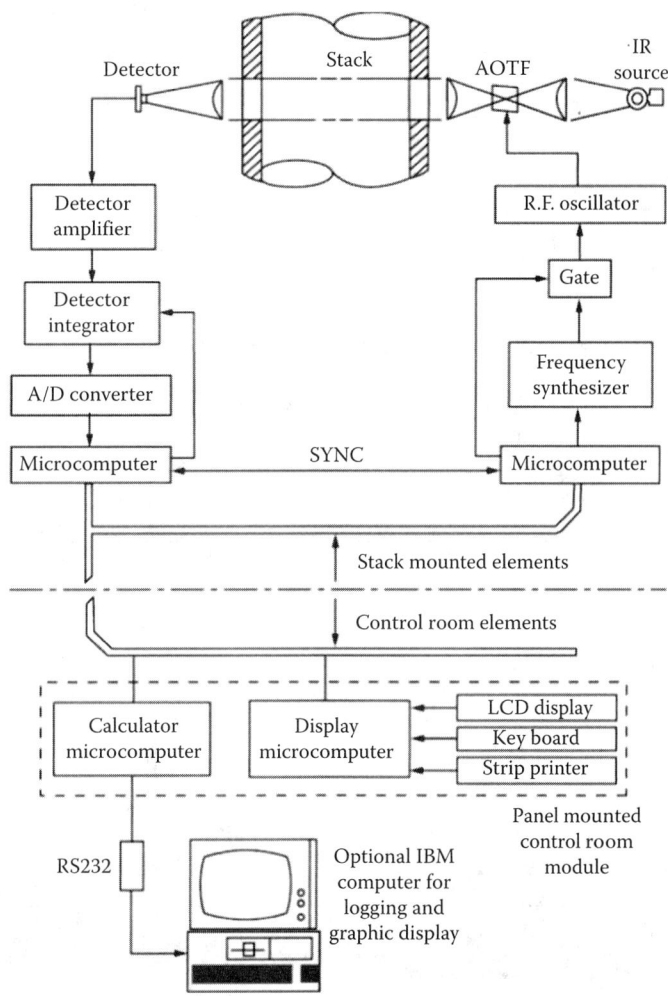

FIG. 1.64b
IR analyzer with solid-state tunable crystal, which allows for the simultaneous measurement of many gases in the stack. (From Nelson, R.L., InTech, 1987.)

measured together. Chapter 1.32 discusses the NDIR analyzer using a positive filtering scheme, illustrated in Figure 1.64c.

Ultraviolet

UV analyzers are discussed in detail in Chapter 1.69. They are frequently used in sulfur dioxide applications. The absorbance of SO_2 in the UV range is shown in Figure 1.64d.

Because, in most stack-monitoring applications, there is an interest in measuring the concentrations of both NO_x and SO_2, some suppliers offer a single analyzer for the simultaneous monitoring of both gases (Figure 1.44b). This analyzer is not inserted into the process, but samples are taken from the process into the analyzer, and the sample gases must be filtered before they reach the detector cell (sample chamber) to prevent a buildup of particulates in the cell. Similarly, the moisture content of the sample must also be controlled to prevent condensation.

UV analyzers of the probe design can be inserted into the stack. Figure 1.64e illustrates a self-cleaning, self-calibrating, microprocessor-operated probe design, capable of measuring both NO and SO_2 simultaneously.

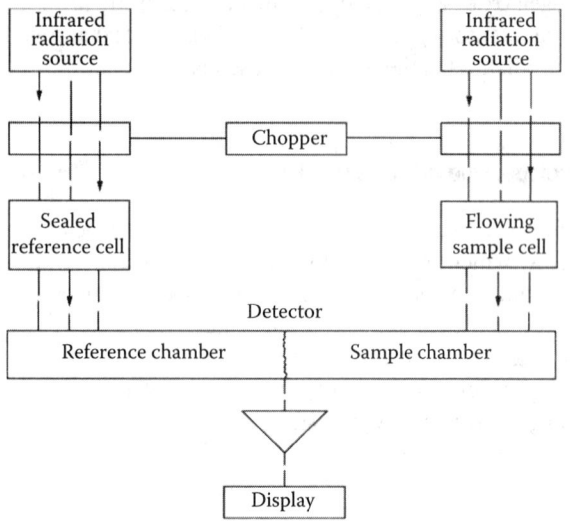

FIG. 1.64c
Nondispersive infrared analyzer.

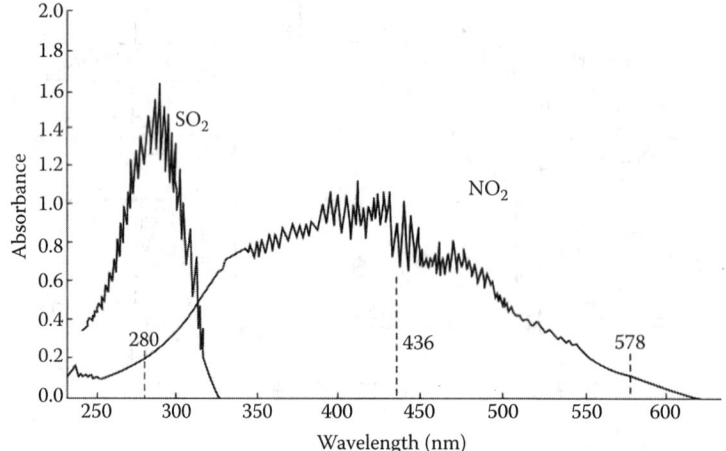

FIG. 1.64d
The absorbance of SO_2 and NO_2 in the ultraviolet range.

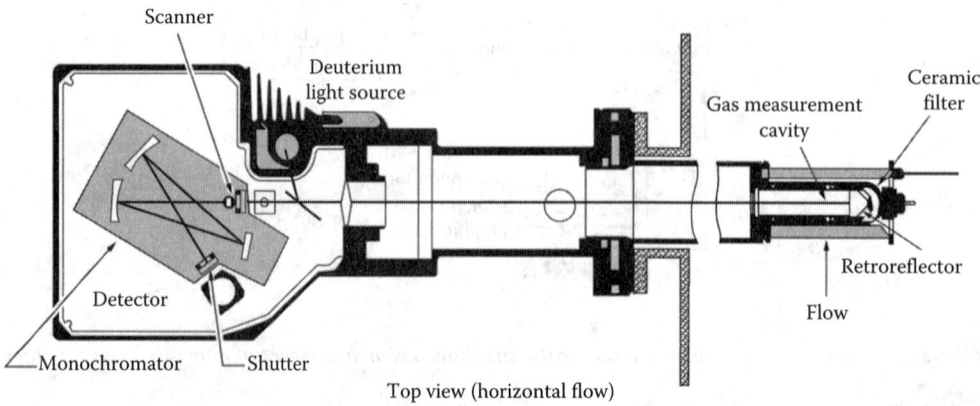

FIG. 1.64e
Self-calibrating, self-cleaning, self-drying probe-type analyzer that simultaneously detects the concentrations of SO_2 and NO_2 at temperatures up to 800°F.

Correlation Spectrometry

As discussed in detail in Chapter 1.62, the correlation spectrometer can be used at a location remote from the source to determine the sulfur dioxide content of stack gases. The instrument shown in Figure 1.64f is specific for SO_2 because of the photographic optical mask of the SO_2 spectrum.

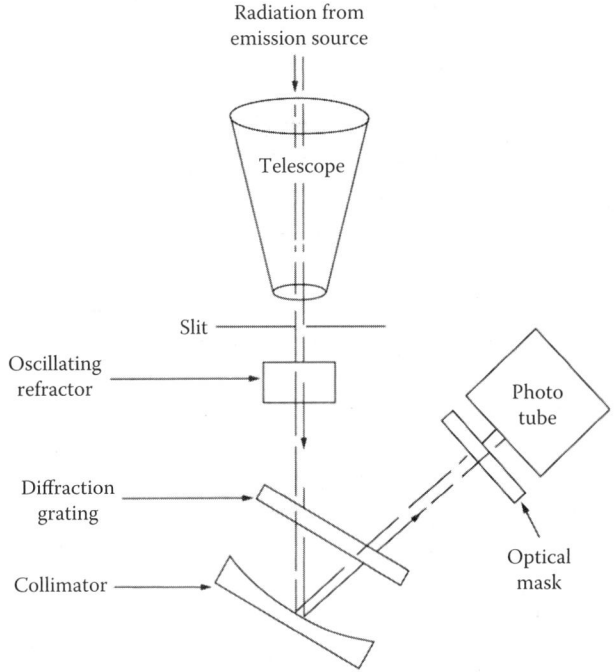

FIG. 1.64f
Remote sensing correlation spectrometer.

In this design, the natural background level of UV radiation is used to zero the instrument. When the optics are pointed across a stack plume containing sulfur dioxide, the resulting change in the UV radiation in the wavelength pattern of SO_2 is detected at the instrument. If the stack plume diameter and stack gas density are both known, the sulfur dioxide concentration can be determined.

Thermal Conductivity

Thermal conductivity analyzers are discussed in detail in Chapter 1.66 and therefore will be described only briefly here.

If the thermal conductivity of air is taken as 1.0, the thermal conductivity of SO_2 is 0.344. This difference is substantial enough to allow the measurement of sulfur dioxide in air at higher concentrations. The minimum range of these detectors is about 0%–3% SO_2 in air and 0%–1% air in SO_2 full scale. Thermal conductivity analyzers can also be used for measuring sulfur dioxide in nitrogen.

AMBIENT AIR ANALYZERS

The measurement of sulfur dioxide in air has been performed for many years because of the simplicity of manual analysis based on the ability of sulfur dioxide to reduce a starch–iodine solution. The earliest sulfur dioxide detectors were automated wet chemical devices that measured the increase in the conductivity of a solution after air had been intimately mixed in it by some contacting procedure.

The U.S. Environmental Protection Agency (EPA) has deleted conductivity-type analyses from their approved list of ambient air sampling instruments, because this technique is nonspecific for SO_2 but measures all acid gases. The EPA has designated the colorimetric technique using pararosaniline as the reference procedure, with flame photometric and coulometric detection as acceptable alternative procedures.

Calibration

The standard procedure for calibrating sulfur dioxide instruments involves the preparation of known concentrations of SO_2 in air. This can be accomplished with high accuracy when permeation tubes, made of Teflon tubing and containing liquid sulfur dioxide, are maintained at constant temperature while a constant flow of air is passed over the tube. The most desirable procedure is a system calibration rather than an instrument calibration, with the known concentration introduced at the system inlet.

Colorimetric Analyzers

The reference technique for determining the sulfur dioxide content of atmospheric air is based on the absorption of SO_2 from an air sample by a solution of sodium tetrachloromercurate, which, upon the addition of formaldehyde and a pararosaniline dye, forms a strong purple

926 Analytical Measurement

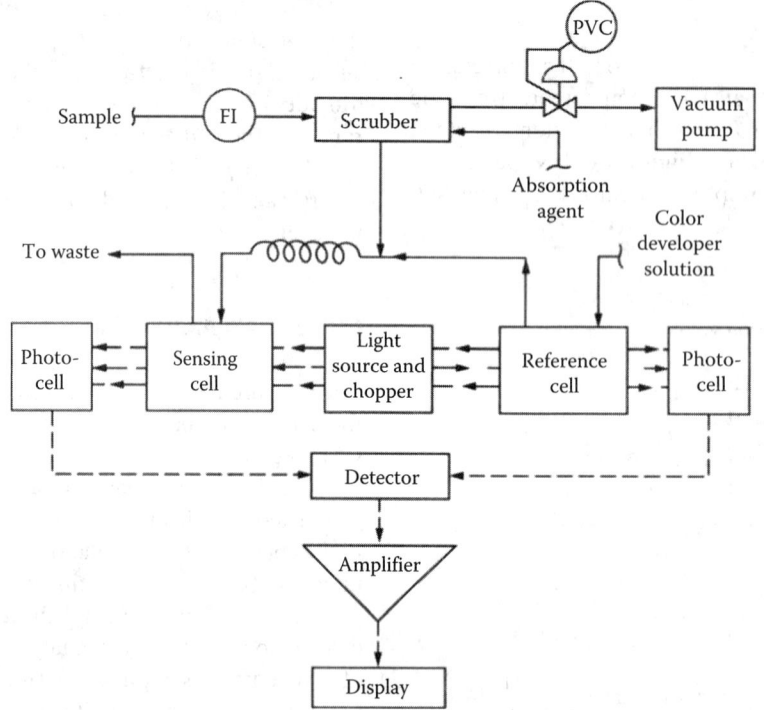

FIG. 1.64g
Automatic colorimetric analyzer for the detection of oxides of sulfur in air.

dye complex. Several manufacturers have automated this procedure and offer packages such as the one shown in Figure 1.64g.

This family of analyzers requires frequent maintenance because of the complexity of the plumbing and the tendency of the dye complex to plate out on the cell windows, thus reducing sensitivity. Properly maintained, they provide an excellent record of sulfur dioxide concentrations in the air, because they operate on the basis of a chemical reaction, which is specific for sulfur dioxide.

Conductimetric Analyzers

There is more experience with the operation of this family of analyzers than with any other in sulfur dioxide detection applications. The instruments are fairly simple electronically, but the associated piping does create some maintenance problems. The basic electric circuit diagram is shown in Figure 1.64h.

To cause a change in the resistance of the measurement cell (Rm) as the sulfur dioxide concentration of the sample changes, some of the available instruments use a weak solution of sulfuric acid and hydrogen peroxide as an absorbent, whereas others use distilled water. These instruments must be temperature compensated, because the change in resistance (conductance) over the normal ambient temperature range is nearly as large as the conductance changes that can

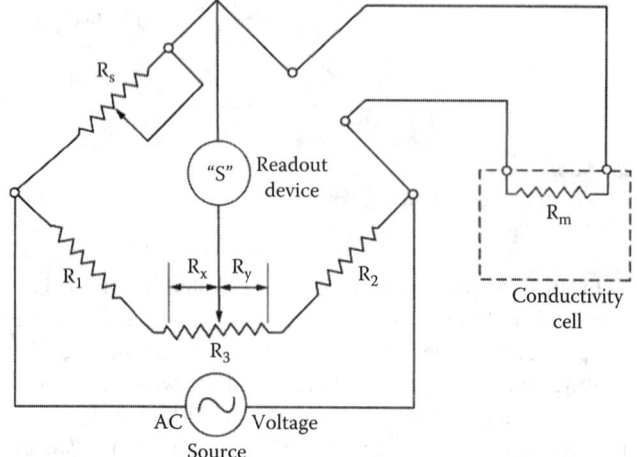

FIG. 1.64h
The electric bridge circuitry for the measurement of conductivity.

be expected as a result of the variation in sulfur dioxide concentration of ambient air.

Conductimetric analysis is not specific to sulfur dioxide. The presence of acidic or basic gases must be compensated for; otherwise, they will interfere with the accuracy of the reading. Gases such as hydrogen chloride will cause positive interference, and basic gases such as ammonia will introduce a negative interference.

Coulometric Analyzers

From a solution of bromine or iodine, the reducing action of sulfur dioxide will electrogenerate free bromine or iodine. These elements can then be detected to give an indication of the SO_2 content of the sample air stream (Figure 1.64i).

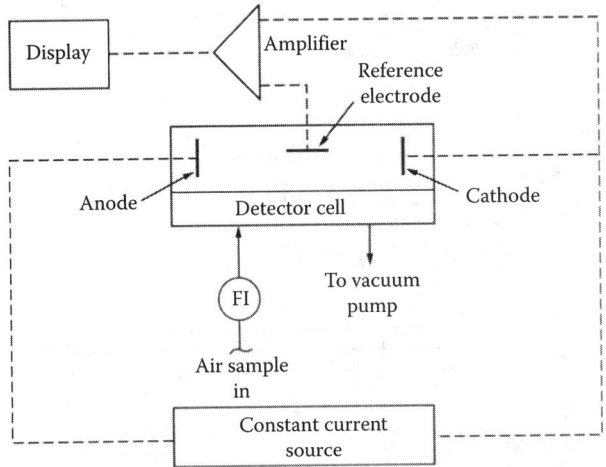

FIG. 1.64i
Coulometric analyzer for oxides of sulfur.

The required supply of reagents for coulometric analyzers is usually less than the reagent requirements of either colorimetric or conductimetric analyzers. At least one commercially available coulometric analyzer is capable of measuring ozone (Chapter 1.51) concurrently with sulfur dioxide.

Flame Photometric Analyzers

Flame photometric analyzers are discussed in detail in Chapters 1.10 and 1.56. These analyzers utilize the phenomenon that, when an air stream containing sulfur is burned in a hydrogen-rich flame, radiation is generated in a wavelength band centered at 394 μm. Flame photometric detectors can measure sulfur concentrations in air down to levels less than 0.01 ppm. The electrical output signal of this detector (Figure 1.64j) is logarithmic. Typical instrument ranges can cover two decades of concentration, for example, 0.01–1.0 ppm, with the capability to switch to a higher range.

This instrument measures total sulfur in the sample stream. To provide discrimination between various sulfur compounds, additional hardware is available from some manufacturers. An all-Teflon gas chromatographic column preceding the flame photometric analyzer can be used to separate the sulfur compounds quantitatively. This combination (Chapter 1.10) has proved extremely useful for the measurement of reduced sulfur compounds, for example, hydrogen sulfide and dimethyl disulfide in the air.

Electrochemical Analyzers

When air passes over an element consisting of a semipermeable membrane, an electrolyte, and a voltage-sensitive pickup, some gases will selectively migrate through the membrane and generate a signal in the electrolyte (Figure 1.64k). This phenomenon has been used to measure the concentration of many gases in ambient air, including sulfur dioxide.

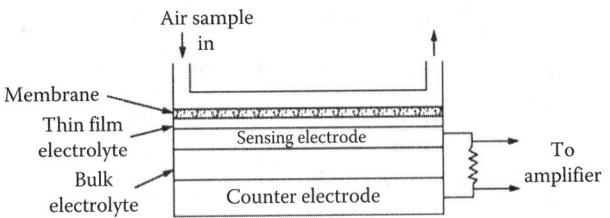

FIG. 1.64k
Electrochemical gas analyzer.

The electrochemical analyzers have had limited use in the measurement of air pollutants but are used as the sensor in pocket-sized portable indicators and alarms (see Figure 1.64l). Operational problems can develop in maintaining the proper moisture content of the membrane and the proper electrolyte strength because of the migration of water vapor.

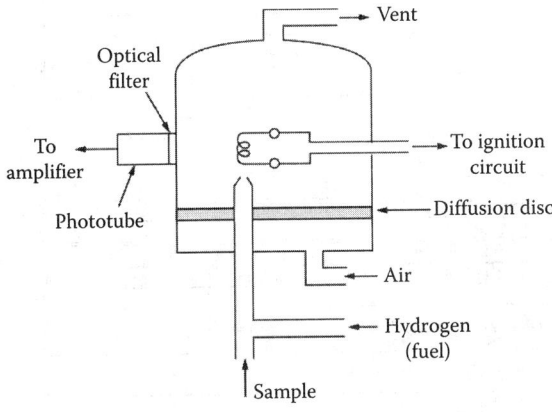

FIG. 1.64j
Flame photometric analyzer.

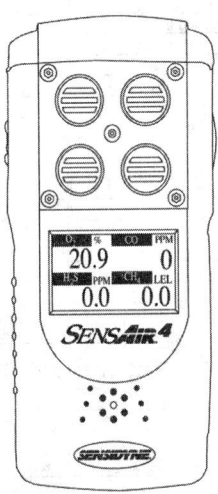

FIG. 1.64l
Portable electrochemical detectors are available for the measurement of up to four gases. (Courtesy of Teledyne API.)

928   Analytical Measurement

SPECIFICATION FORMS

When specifying sulfur dioxide analyzers, the ISA Form 20A1001 can be used to specify the features required for the device itself, and when specifying both the device and the nature of the application, ISA Form 20A1002 can be used. The following are the forms that are reproduced with the permission of the International Society of Automation.

#	RESPONSIBLE ORGANIZATION	ANALYSIS DEVICE Operating Parameters	#	SPECIFICATION IDENTIFICATIONS		
1			6			
2	(ISA logo)		7	Document no		
3			8	Latest revision	Date	
4			9	Issue status		
5			10			

#	ADMINISTRATIVE IDENTIFICATIONS			#	SERVICE IDENTIFICATIONS Continued			
11				40				
12	Project number	Sub project no		41	Return conn matl type			
13	Project			42	Inline hazardous area cl	Div/Zon		Group
14	Enterprise			43	Inline area min ign temp	Temp ident number		
15	Site			44	Remote hazardous area cl	Div/Zon		Group
16	Area	Cell	Unit	45	Remote area min ign temp	Temp ident number		
17				46				
18	SERVICE IDENTIFICATIONS			47				
19	Tag no/Functional ident			48	COMPONENT DESIGN CRITERIA			
20	Related equipment			49	Component type			
21	Service			50	Component style			
22				51	Output signal type			
23	P&ID/Reference dwg			52	Characteristic curve			
24	Process line/nozzle no			53	Compensation style			
25	Process conn pipe spec			54	Type of protection			
26	Process conn nominal size	Rating		55	Criticality code			
27	Process conn termn type	Style		56	Max EMI susceptibility	Ref		
28	Process conn schedule no	Wall thickness		57	Max temperature effect	Ref		
29	Process connection length			58	Max sample time lag			
30	Process line matl type			59	Max response time			
31	Fast loop line number			60	Min required accuracy	Ref		
32	Fast loop pipe spec			61	Avail nom power supply			Number wires
33	Fast loop conn nom size	Rating		62	Calibration method			
34	Fast loop conn termn type	Style		63	Testing/Listing agency			
35	Fast loop schedule no	Wall thickness		64	Test requirements			
36	Fast loop estimated lg			65	Supply loss failure mode			
37	Fast loop material type			66	Signal loss failure mode			
38	Return conn nominal size	Rating		67				
39	Return conn termn type	Style		68				

#	PROCESS VARIABLES	MATERIAL FLOW CONDITIONS			#	PROCESS DESIGN CONDITIONS		
69					101			
70	Flow Case Identification			Units	102	Minimum	Maximum	Units
71	Process pressure				103			
72	Process temperature				104			
73	Process phase type				105			
74	Process liquid actl flow				106			
75	Process vapor actl flow				107			
76	Process vapor std flow				108			
77	Process liquid density				109			
78	Process vapor density				110			
79	Process liquid viscosity				111			
80	Sample return pressure				112			
81	Sample vent/drain press				113			
82	Sample temperature				114			
83	Sample phase type				115			
84	Fast loop liq actl flow				116			
85	Fast loop vapor actl flow				117			
86	Fast loop vapor std flow				118			
87	Fast loop vapor density				119			
88	Conductivity/Resistivity				120			
89	pH/ORP				121			
90	RH/Dewpoint				122			
91	Turbidity/Opacity				123			
92	Dissolved oxygen				124			
93	Corrosivity				125			
94	Particle size				126			
95					127			
96	CALCULATED VARIABLES				128			
97	Sample lag time				129			
98	Process fluid velocity				130			
99	Wake/natural freq ratio				131			
100					132			

#	MATERIAL PROPERTIES				#	MATERIAL PROPERTIES Continued		
133					137			
134	Name				138	NFPA health hazard	Flammability	Reactivity
135	Density at ref temp		At		139			
136					140			

Rev	Date	Revision Description	By	Appv1	Appv2	Appv3	REMARKS

Form: 20A1001 Rev 0   © 2004 ISA

	RESPONSIBLE ORGANIZATION	ANALYSIS DEVICE COMPOSITION OR PROPERTY Operating Parameters (Continued)		SPECIFICATION IDENTIFICATIONS		
1			6			
2	ISA		7	Document no		
3			8	Latest revision	Date	
4			9	Issue status		
5			10			

	PROCESS COMPOSITION OR PROPERTY			MEASUREMENT DESIGN CONDITIONS				
	Component/Property Name	Normal	Units	Minimum	Units	Maximum	Units	Repeatability
11								
12								
...								
75								

Rev	Date	Revision Description	By	Appv1	Appv2	Appv3	REMARKS

Form: 20A1002 Rev 0

© 2004 ISA

Abbreviations

NAAQS   National Ambient Air Quality Standards
NDIR   Nondispersive Infrared

Bibliography

ASTM, Standard test methods for sulfur dioxide content of the atmosphere (West-Gaeke method), (2015) http://www.astm.org/Standards/D2914.htm.

Chatterjee, S. et al., Colorimetric determination of sulphur dioxide and sulphites in various environmental samples, (2013) http://onlinelibrary.wiley.com/doi/10.1002/jccs.200300038/abstract.

Eaton, W.C., *Performance of Automated Ambient Sulfur Dioxide Analyzers*, Taylor & Francis, Boca Raton, FL, 1993. https://www.researchgate.net/publication/254278347_Performance_of_Automated_Ambient_Sulfur_Dioxide_Analyzers_with_Respect_to_a_Proposed_5Minute_Ambient_Air_Quality_Standard.

EPA, Sulfur dioxide, (2015) http://www3.epa.gov/airquality/sulfurdioxide/health.html, http://www3.epa.gov/airtrends/sulfur.html.

EPA, U.S. EPA: Federal Register 62(133), 40 CFR Part 80 Part II, p. 37337 (July 11, 1997); 64(92), 40 CFR Parts 80, 85 and 86, p. 26055 (May 13, 1999); 65(28), 40 CFR Part 80, pp. 6752–6774 (February 10, 2000); and 66 (12), 40 CFR Part 80, pp. 5002–5141 (January 18, 2001).

Nadkari, R.A.K., The challenge of sulfur analysis in the fuels of the future, (2004) http://www.astm.org/SNEWS/JUNE_2004/nadkarni_jun04.html.

National Atmospheric Emissions Inventory, Sulphur dioxide, (2014) http://naei.defra.gov.uk/overview/pollutants?pollutant_id=8.

Ontario, Ministry of the Environment and Climate Change, (2010) http://www.airqualityontario.com/science/pollutants/sulphur.php.

Rolan, J., Intercomparison of measurements of sulfur dioxide in ambient air, (2012) http://onlinelibrary.wiley.com/doi/10.1029/96JD03587/abstract.

SciTech, Costs and benefits of sulfur oxide control, (1983) http://www.osti.gov/scitech/biblio/5435389.

Siemens, Total sulfur analyzer 2, (2011) http://www.industry.usa.siemens.com/automation/us/en/process-instrumentation-and-analytics/solutions-for-industry/environmental-monitoring/Documents/MaxumFlareTS_PIABR-0010–1010.pdf.

State of Alaska, Department of Environmental Conservation: Operating procedure for SO2 monitoring, (2012) http://dec.alaska.gov/air/am/SO2_SOP_13feb12.pdf.

1.65 Sulfur in Oil and Gas

B. G. LIPTÁK (2003)   **A. F. BUENO** (2017)

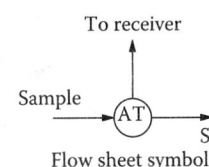

Flow sheet symbol

Type of sample	Crude oil, diesel, gasoline, middle distillates, kerosene, gas oil, jet fuel, lubricating oils, fuel oil, natural gas, and refinery gas
Method of analysis	A. Gas Chromatography with Flame Photometry B. Hydrogenolysis and Rateometric Colorimetry C. Sulfur Chemiluminescence D. Pyrolysis and Ultraviolet Fluorescence E. X-Ray Fluorescence
Reference methods	A. ASTM D7041 B. ASTM D4045 (liquid) and D4468 (gas) C. ASTM D5623 (liquid) and D5504 (gas) D. ASTM D5453 (liquid) and D6667 (gas and LPG) E. ASTM D4294 (EDXRF) and D7039 (MWDXRF)
Range	A. 0.5–100 mg/kg B. 0.02–10 mg/kg (or higher, by dilution) C. 0.1–100 mg/kg D. 1–8000 mg/kg E. 2–500 mg/kg
Sensitivity	A. 0.5 mg/kg (ppm) B. 10 µg/kg (ppb) C. 1 mg/kg (ppm) D. 250 µg/kg (ppb) E. 0.6 mg/kg (ppm)
Cost	A. U.S. $75,000 (process unit); U.S. $65,000 (lab unit) B. U.S. $60,000 (process unit) C. U.S. $100,000 (process or lab unit) D. U.S. $100,000 (process unit); U.S. $65,000 (lab unit) E. U.S. $100,000 (process unit); U.S. $75,000 (lab unit)
Partial list of suppliers	ABB: A (http://www.abb.com/product/seitp330/512997fa1e40403585257881005db3b2.aspx) Agilent Technologies: C (https://www.chem.agilent.com/en-US/Newsletters/accessagilent/archive/pages/sulfur_detector.aspx?cid = 7964) Analytical Systems International: B (http://liquidgasanalyzers.com/product/total-sulfur-analyzer/) Galvanic Applied Sciences Inc.: (http://www.galvanic.com/products/protech), C (http://www.galvanic.com/products/sulfurchrome-gc) HobréInstruments: E (http://hobre.com/products/metorex-c100-on-line-elemental-analysis/) Horiba: E (http://www.horiba.com/us/en/scientific/products/elemental-analyzers/sulfur-and-chlorine-in-oil/) O I Analytical: A (http://www.oico.com/default.aspx?id=product&productID=38) PAC: D (http://www.paclp.com/Process_Analytics/6000 and http://www.paclp.com/Process_Analytics/6200) Perkin Elmer: C (http://www.perkinelmer.com/Catalog/Product/ID/NARL8300)

(Continued)

Rigaku: E (http://www.rigaku.com/products/xrf/minizsulfur)
ShimadzuCorp.: E (http://www.ssi.shimadzu.com/products/productgroup.cfm?subcatlink=xrayinstr)
Teledyne Analytical Instruments: D (http://www.teledyne-ai.com/products/6400-tsg.asp)
Thermo Scientific: D (http://www.thermoscientific.com/ecomm/servlet/productsdetail_11152_L11016_80562_11962694_-1)
XOS: E (https://xos.com/products/sindie-online/)

Applications   The choice of method of analysis depends on factors such as sample physical state, concentration range, and the need for speciation of sulfur compounds. Liquid or gaseous samples can be analyzed by methods A, B, C, and D. Method E can be used only for liquid samples. Samples of high volatility could also impair the results of that method due to evaporation by heating during the analysis. If it is necessary to quantify sulfur compounds individually, method A is the choice, because it allows speciation. Alternatively, method C can also be used, when attached to a gas chromatograph. For the determination of H_2S, method B is the most suitable because it is based on a highly selective reaction to that analyte. This is also the method of choice for low ppb levels.

INTRODUCTION

Sulfur is a common constituent in petroleum products. It is usually present in the form of elemental sulfur, hydrogen sulfide, mercaptan, and thiophene. Sulfur dioxide, produced by burning fossil fuels, is an important air pollutant. Sulfur also inhibits the performance of automotive catalytic converters designed to reduce nitrogen oxide emissions. Environmental authorities are tightening the limits on sulfur levels in petroleum products to a few milligrams per kilogram in gasoline and diesel. Online analysis of sulfur is relevant to control desulfurization processes and to ensure product specifications in blending systems.

There are several ways to determine the concentration of sulfur. Some methods are specific for liquid samples, others for gases. Some of them report the total sulfur only, and others provide speciation of sulfur compounds.

GAS CHROMATOGRAPHY WITH FLAME PHOTOMETRIC DETECTION

This is an old and accurate method for the quantification of sulfur in liquid and gases samples. Four steps are involved in the analysis: sample injection, oxidation, separation, and detection. A fixed volume (0.1–1 μL) of sample is injected into the analyzer using a sample injection valve. The sample is vaporized and transported by a carrier gas (air) into a high-temperature (>900°C) furnace, where it is burned with air. The compounds present in the sample are oxidized, and sulfur species are converted to sulfur dioxide. The gases are then split, and about 80% of the flow is vented through a vent valve. The remainder of the flow is transferred by the carrier gas to a chromatographic column where the combustion products are separated. The packed column, usually made of steel covered with a passivation layer, shall provide a complete separation of sulfur dioxide and carbon monoxide. This operation is necessary because carbon monoxide acts as an interferent since it quenches fluorescence produced by sulfur dioxide.

A flame photometric detector (FPD) is used to determine the concentration of sulfur dioxide, which is proportional to the sulfur amount in the sample. The components are burned in a hydrogen-rich flame, and sulfur dioxide is reduced to disulfur and excited. In a cooler zone of the detector, it decays and releases energy, emitting light at a specific wavelength of 300–425 nm, centered at 394 nm. A narrow bandpass filter selects radiation around that wavelength to the photomultiplier tube (PMT). There, light strikes a photosensitive surface, and a photon knocks an electron loose. The electron is amplified inside the PMT for an overall gain of up to a million. Figure 1.65a shows a diagram of a chromatograph

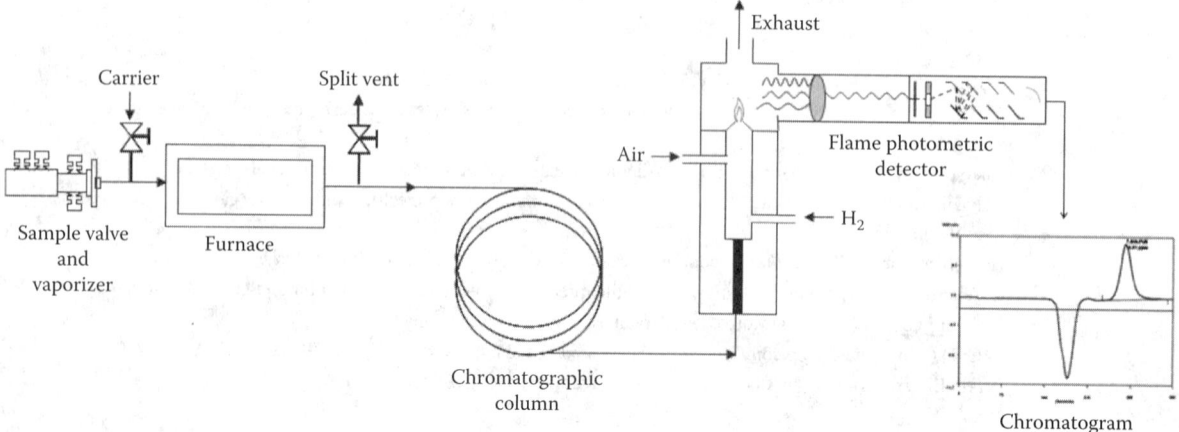

FIG. 1.65a
Gas chromatograph with flame photometric detector.

with an FPD. The negative peak in the chromatogram is caused by the carbon monoxide quenching effect. The peak of sulfur dioxide appears around 210 sec.

A more recent detector, called pulsed flame photometric detector (PFPD), utilizes a propagating flame. The combustor is filled with the combustible gas mixture. The flame is ignited, propagates through the combustor, and burns out when all the fuel is consumed. This process is repeated three to four times a second. The reactions promoted by the propagating flame produce element-specific light emissions at determined wavelengths and lifetimes. The associated information of emission lifetime and wavelengths brings more sensitivity, selectivity, and better signal to noise ratio compared to traditional FPD. This technology allows for the detection of very low sulfur concentrations.

Gas chromatography with flame photometric detection is a matrix-independent, interference-free method. The drawbacks include nonlinear response with sulfur concentration, relatively high response time (typical analysis time is 5–6 min), and the need for utilities such as zero-grade air (carrier gas) and chromatographic-grade hydrogen (flame fuel).

HYDROGENOLYSIS AND RATEOMETRIC COLORIMETRY

Originally developed for the determination of hydrogen sulfide (H_2S) and also known as the lead acetate tape method, this technique has been modified to permit the quantification of total sulfur in liquid and gases.

The gas (or vaporized liquid) sample is mixed with hydrogen and then transferred to a pyrolyzer furnace made of a quartz or ceramic tube at a temperature of 900°C or higher. At that temperature, hydrocarbons are cracked to form methane, and sulfur compounds are converted to H_2S (Equation 1.65(1)):

$$R - HS + H_2 \rightarrow H_2S + RH \qquad 1.65(1)$$

The gas containing H_2S is humidified by bubbling through a solution of acetic acid in water. The humidified gas passes through a paper tape that has been impregnated with lead acetate. H_2S reacts with lead acetate according to the following simplified reaction:

$$H_2S + Pb \rightarrow PbS + H_2O \qquad 1.65(2)$$

Lead sulfide produces a brown stain on the tape. The H_2S amount, which is proportional to the concentration of sulfur in the sample, is determined by measuring the rate of change of reflectance caused by darkening when lead sulfide is formed. A red light–emitting diode illuminates the tape, and a solid-state silicon photodiode detector is used to determine the intensity of the stain (Figure 1.65b). The tape is kept in a cassette and moves between samples, so that each sample utilizes a new section of paper.

Results are given in volumetric basis, so in order to obtain the concentration in mg/kg (ppmw), a density correction is necessary. The reaction of sulfur is very selective, which makes this technique almost interference free. It is also very sensitive and able to detect very low concentrations (mg/kg or even μg/kg).

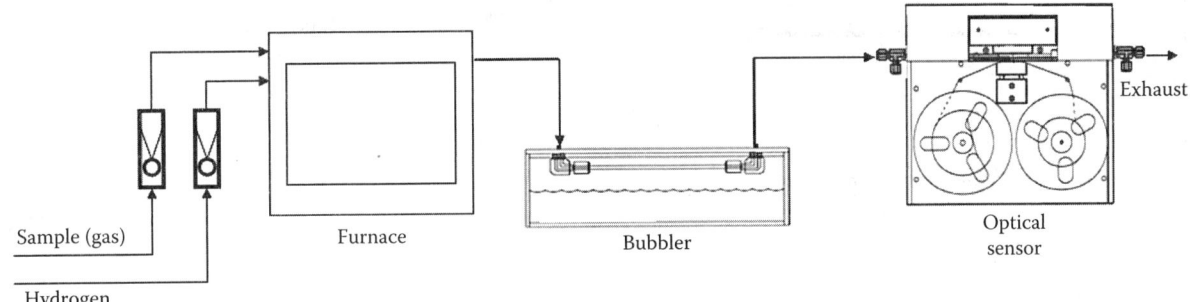

FIG. 1.65b
Diagram of a lead acetate tape analyzer.

Response time is about 3 min. The drawback is that the analyzer has moving parts, and optical alignment is critical. Hydrogen is necessary for the reaction, and the lead acetate–impregnated paper tape is a consumable. Lead is a cumulative poison and represents a dangerous waste.

SULFUR CHEMILUMINESCENCE

Three steps are involved is this method: combustion, reaction with ozone, and photometric detection. The combustion furnace is made of two externally heated ceramic tubes, kept at 750°C. The sample is introduced with air into the furnace through a larger tube under vacuum and mixed with hydrogen flowing countercurrent in a small tube. The compounds occurs by means of a hydrogen flame. Sulfur components are transformed into sulfur monoxide, according to Equations 1.65(3) and 1.65(4):

$$R - S + O_2 \rightarrow SO_2 + H_2O + CO_2 \quad 1.65(3)$$

$$SO_2 + H_2 \rightarrow SO + H_2O \quad 1.65(4)$$

Sulfur monoxide is transferred under vacuum to the chemiluminescence chamber, where it reacts with ozone (generated by the ozone generator, by flowing air through a steel tube at 7500 VCA), producing an electronically excited state of sulfur dioxide that releases light at about 350 nm upon relaxation (Equations 1.65(5) and 1.65(6)):

$$SO + O_3 \rightarrow SO_2^* + O_2 \quad 1.65(5)$$

$$SO_2^* \rightarrow SO_2 + h\upsilon \quad 1.65(6)$$

The light is detected by a PMT. Figure 1.65c shows a diagram of a sulfur chemiluminescence analyzer.

This method has high sensitivity, selectivity, and linear response with sulfur concentration. When coupled with gas chromatography, it is able to provide speciation of sulfur compounds. It requires continuous flow of air and hydrogen for the reaction as well as helium, used as a carrier gas.

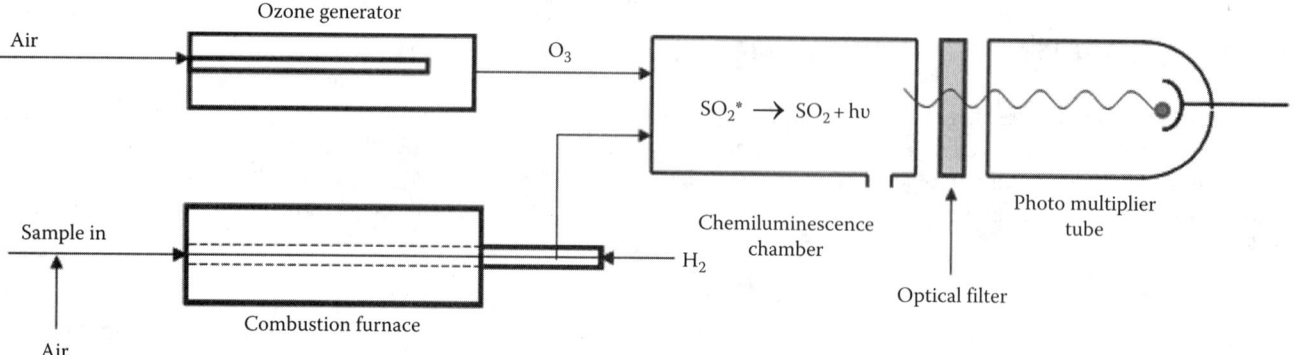

FIG. 1.65c
Diagram of a sulfur chemiluminescence analyzer.

PYROLYSIS AND ULTRAVIOLET FLUORESCENCE

This method involves heating the sample with an excess of oxygen and then exposing the combusted material to UV light at 210 nm for detection. Sulfur dioxide, produced by combustion, absorbs UV light and emits radiation at a specific wavelength (about 350 nm) that is detected by the PMT inside the sulfur detector. The signal produced by the PMT is sent to the controller for processing and calculation of the concentration. The reactions described by Equations 1.65(7) through 1.65(9) occur:

$$R\text{--}S + O_2 \rightarrow SO_2 + \text{other combustion products (1000°C)} \quad \mathbf{1.65(7)}$$

$$SO_2 + h\upsilon \rightarrow SO_2^* \quad \mathbf{1.65(8)}$$

$$SO_2^* \rightarrow SO_2 + h\upsilon' \quad \mathbf{1.65(9)}$$

The analysis process begins after the controller positions the air-actuated sample block valve, which allows sample containing sulfur to flow to the six-port rotary injector valve. The sample injection valve introduces a fixed amount of sample into the carrier stream. The sample is injected into the pyrolysis tube where it is combusted with an excess of oxygen. Oxygen is supplied to the inlet tube inside the pyrolysis tube to assist the flow of the sample. Oxygen is supplied again to the pyrolysis tube at a higher flow rate to provide an oxygen-rich environment for combustion and conversion of the sample at temperatures from 1000°C to 1100°C.

Sample products after pyrolysis pass through a membrane dryer to remove water (because it quenches the fluorescence signal) and then enter the sulfur fluorescence chamber inside the detector where it is exposed to an ultraviolet light source. Because a filter lens allows only the appropriate wavelength of light to reach the PMT, the current produced by the PMT is proportional to the amount of SO_2 being fluoresced. The effluent is then vented.

Figure 1.65d shows a diagram of an analyzer based on pyrolysis and ultraviolet fluorescence detection.

The method provides a very linear response with sulfur concentration.

The analyzer requires continuous flow of inert gas (argon or helium) and oxygen. The sample must be dry and free of dust. Since the sample is introduced as a fixed volume, the result is expressed in a volumetric basis. If the result in a mass basis, such as mg/kg, is required, use sample density to convert it.

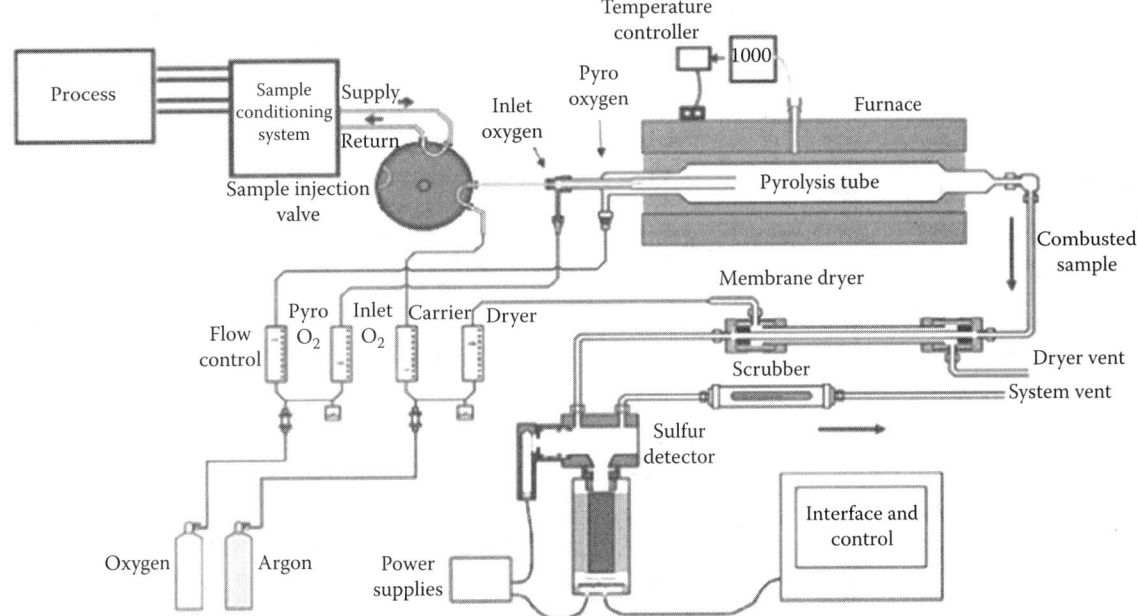

FIG. 1.65d
Diagram of an analyzer based on pyrolysis and ultraviolet fluorescence detection. (Courtesy of PAC.)

X-RAY FLUORESCENCE

When a beam of x-ray hits an atom, it promotes ionization by expelling electrons of its inner shells. In that situation, the electronic structure of the atom becomes unstable, and electrons in more energetic shells move to the inner shells to fill the holes. As the electron drops to the lower energy state, energy is released in the form of a photon, which is equal to the energy difference of the two shells involved. The transition of an electron from shell L to shell K is called Kα. From shell M to K, Kβ. From shell M to L, Lα, and so on.

There are different types of analytical methods based on x-ray fluorescence spectrometry. The most relevant for sulfur online analyzers are the energy-dispersive (EDXRF) and monochromatic wavelength-dispersive (MWDXRF) instruments.

In EDXRF instruments, the sample flows through a cell with a thin beryllium window, while it is irradiated by a polychromatic beam from an x-ray tube. Elements in the sample are excited by the x-ray and release energy in the form of fluorescence. A detector receives radiation from the sample and counts the pulses at different energy levels. Several elements can be determined simultaneously that way.

In a sulfur-specific analyzer, filters are used to discriminate between sulfur Kα from other elements' x-ray radiations.

Figure 1.65e shows a diagram of a typical EDXRF analyzer.

In the MWDXRF analyzer, the polychromatic beam from the x-ray tube hits a monochromator that separates radiation according to its wavelength. A specific wavelength, suitable to excite the K-shell electrons of sulfur, is focused onto the sample. Again, the radiation emitted by the sample is separated by a second monochromator that sends the fluorescent sulfur Kα radiation at 0.5373 nm to the detector.

A diagram of a WDXRF analyzer is shown in Figure 1.65f.

When working on low-sulfur samples, care must be taken with the cleanness of the window. Any residual sample stick to the window may interfere with the result and cause "memory effect," that is, contamination from previous samples. In order to avoid this, one manufacturer may add a window cleaning device (generally using an organic solvent) in the analyzer configuration, while the other may provide a system to replace the window using a beryllium tape cassette.

It is a very accurate and fast method. Typical analysis time ranges from 1 to 5 min. No consumable gases are necessary. Results are given on a weight basis, with no need for conversions.

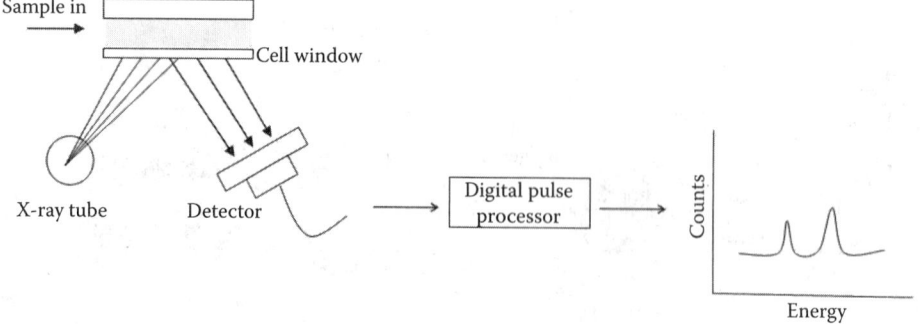

FIG. 1.65e
Diagram of an EDXRF analyzer.

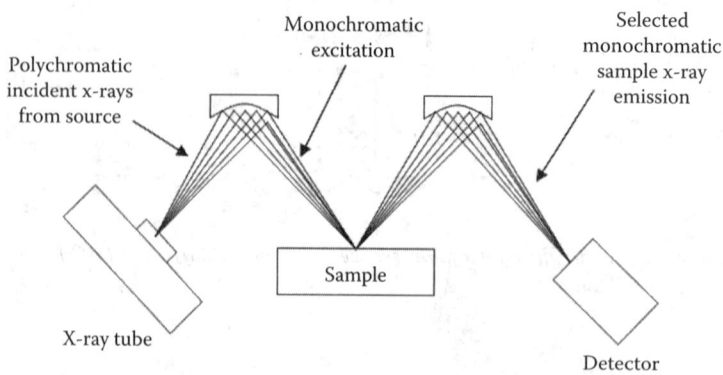

FIG. 1.65f
Diagram of an MWDXRF analyzer. (Courtesy of XOS.)

Sampling and Calibration

Analysis of gaseous sulfur compounds is challenging due to the reactivity of these substances. Ideally, analysis is performed on-site to eliminate loss of sulfur or contamination. Sampling must be performed using nonreactive containers, such as inert material–coated vessels, Tedlar® bags, and polypropylene fittings. Plastic containers shall not be used because volatile materials may be lost by diffusion through the walls of container. Laboratory equipment must be inert or passivated as well to ensure reliable results.

Accurate sulfur standards are required for calibration. Permeation and compressed gas standards should be stable, of high purity, and of the highest available accuracy. Gas standards may include hydrogen sulfide, carbonyl sulfide, or methyl mercaptan in a matrix of nitrogen, helium, or methane.

For liquid samples, calibration may be performed using standards such as dibenzothiophene, butyl sulfide, or thianaphthene in organic solvents, usually toluene, isooctane, or xylene.

New Developments

With stricter environmental requirements and demand for more efficient catalysts, the sulfur content of products has reduced considerably, generating the so-called ultralow fuels.

The analytical challenge is to determine increasingly lower concentrations, with simpler, faster, and more robust instruments. In this sense, x-ray fluorescence stands out, with its significant advances in the last decade. EDXRF has long been recognized as a good method for the measurement of sulfur in the range of 100–500 mg/kg. By reducing the background and increasing the count rate, using a new dual chamber proportional counter detector, modern analyzers achieve a detection limit of 1 mg/kg.

The new WDXRF technique uses x-ray fluorescence monochromatic excitation, which can provide much better signal to background ratio than the traditional polychromatic excitation used by EDXRF. Detection limits of 0.15 mg/kg can now be achieved by WDXRF.

Another area that showed progress is related to the sample collection and transportation. Dealing with ultralow sulfur products requires special care to avoid losing components by reaction, adsorption, or absorption. New materials, such as amorphous silicon-treated electropolished stainless steel tubing, are now available for the efficient preservation of samples containing sulfur as low as µg/kg levels, ensuring reliable analytical results.

CONCLUSIONS

There are several methods for the quantification of sulfur in oil and gas, and each one has its advantages and limitations.

When there is a need for sulfur compound speciation, method A or C (coupled with a gas chromatograph) must be used. Gas chromatography analyzers require the continual use of gases such as air and hydrogen, which leads to increased cost of operation and safety concerns. Analysis times may be increased when heavier sulfur compounds are present, such as methylbenzothiophene or diphenyl sulfide, which has a retention time of 28 min.

Method B is very good for total sulfur and H_2S quantifications. It also has excellent sensitivity and has been the choice for low ppb levels. The analyzer is robust, but needs hydrogen and lead acetate tape. Disposal of lead-impregnated paper may be a problem.

Method D has very good sensitivity and also requires gases such as argon or helium and oxygen. The method is free of interferences. Heavy oils may lead to coke formation, and the removal and cleanup of the pyrolysis tube would be required.

Method E is very robust, being almost maintenance free. It does not require the use of gases. Interference is the matter for this method, and differences between the elemental composition of samples and the calibration standards can result in biased sulfur determinations. Only liquid samples can be analyzed by method E.

SPECIFICATION FORMS

When specifying sulfur in oil or gas analyzers, the ISA Form 20A1001 can be used to specify the features required for the device itself and when specifying both the device and the nature of the application, including the composition and/or properties of the process materials, ISA Form 20A1002 can be used. These forms, with the permission of the International Society of Automation, are reproduced on the next pages.

Analytical Measurement

1	RESPONSIBLE ORGANIZATION		ANALYSIS DEVICE	6	SPECIFICATION IDENTIFICATIONS	
2				7	Document no	
3	(ISA)		Operating Parameters	8	Latest revision	Date
4				9	Issue status	
5				10		

11	ADMINISTRATIVE IDENTIFICATIONS			40	SERVICE IDENTIFICATIONS Continued		
12	Project number		Sub project no	41	Return conn matl type		
13	Project			42	Inline hazardous area cl	Div/Zon	Group
14	Enterprise			43	Inline area min ign temp	Temp ident number	
15	Site			44	Remote hazardous area cl	Div/Zon	Group
16	Area	Cell	Unit	45	Remote area min ign temp	Temp ident number	
17				46			
18	SERVICE IDENTIFICATIONS			47			
19	Tag no/Functional ident			48	COMPONENT DESIGN CRITERIA		
20	Related equipment			49	Component type		
21	Service			50	Component style		
22				51	Output signal type		
23	P&ID/Reference dwg			52	Characteristic curve		
24	Process line/nozzle no			53	Compensation style		
25	Process conn pipe spec			54	Type of protection		
26	Process conn nominal size	Rating		55	Criticality code		
27	Process conn termn type	Style		56	Max EMI susceptibility	Ref	
28	Process conn schedule no	Wall thickness		57	Max temperature effect	Ref	
29	Process connection length			58	Max sample time lag		
30	Process line matl type			59	Max response time		
31	Fast loop line number			60	Min required accuracy	Ref	
32	Fast loop pipe spec			61	Avail nom power supply		Number wires
33	Fast loop conn nom size	Rating		62	Calibration method		
34	Fast loop conn termn type	Style		63	Testing/Listing agency		
35	Fast loop schedule no	Wall thickness		64	Test requirements		
36	Fast loop estimated lg			65	Supply loss failure mode		
37	Fast loop material type			66	Signal loss failure mode		
38	Return conn nominal size	Rating		67			
39	Return conn termn type	Style		68			

69	PROCESS VARIABLES	MATERIAL FLOW CONDITIONS		101	PROCESS DESIGN CONDITIONS		
70	Flow Case Identification		Units	102	Minimum	Maximum	Units
71	Process pressure			103			
72	Process temperature			104			
73	Process phase type			105			
74	Process liquid actl flow			106			
75	Process vapor actl flow			107			
76	Process vapor std flow			108			
77	Process liquid density			109			
78	Process vapor density			110			
79	Process liquid viscosity			111			
80	Sample return pressure			112			
81	Sample vent/drain press			113			
82	Sample temperature			114			
83	Sample phase type			115			
84	Fast loop liq actl flow			116			
85	Fast loop vapor actl flow			117			
86	Fast loop vapor std flow			118			
87	Fast loop vapor density			119			
88	Conductivity/Resistivity			120			
89	pH/ORP			121			
90	RH/Dewpoint			122			
91	Turbidity/Opacity			123			
92	Dissolved oxygen			124			
93	Corrosivity			125			
94	Particle size			126			
95				127			
96	CALCULATED VARIABLES			128			
97	Sample lag time			129			
98	Process fluid velocity			130			
99	Wake/natural freq ratio			131			
100				132			

133	MATERIAL PROPERTIES			137	MATERIAL PROPERTIES Continued		
134	Name			138	NFPA health hazard	Flammability	Reactivity
135	Density at ref temp	At		139			
136				140			

Rev	Date	Revision Description	By	Appv1	Appv2	Appv3	REMARKS

Form: 20A1001 Rev 0 © 2004 ISA

1.65 Sulfur in Oil and Gas

Abbreviations

EDXRF	Energy-dispersive x-ray fluorescence
FPD	Flame photometric detector
MWDXRF	Monochromatic wavelength dispersive X-ray fluorescence
PFPD	Pulsed flame photometric detector
PMT	Photomultiplier tube
UV	Ultraviolet

Organizations

ASTM	International
ISA	International Society of Automation

Bibliography

ASTM Standard D4045, 2004, *Standard Test Method for Sulfur in Petroleum Products by Hydrogenolysis and Rateometric Colorimetry*, ASTM International, West Conshohocken, PA, 2015. www.astm.org.

ASTM Standard D4294, *Standard Test Method for Sulfur in Petroleum and Petroleum Products by Energy Dispersive X-Ray Fluorescence Spectrometry*, ASTM International, West Conshohocken, PA, 2010a. www.astm.org.

ASTM Standard D4468, 1985, *Standard Test Method for Total Sulfur in Gaseous Fuels by Hydrogenolysis and Rateometric Colorimetry*, ASTM International, West Conshohocken, PA, 2011. www.astm.org.

ASTM Standard D5453, *Standard Test Method for Determination of Total Sulfur in Light Hydrocarbons, Spark Ignition Engine Fuel, Diesel Engine Fuel, and Engine Oil by Ultraviolet Fluorescence*, ASTM International, West Conshohocken, PA, 2012a. www.astm.org.

ASTM Standard D5504, *Standard Test Method for Determination of Sulfur Compounds in Natural Gas and Gaseous Fuels by Gas Chromatography and Chemiluminescence*, ASTM International, West Conshohocken, PA, 2012b. www.astm.org.

ASTM Standard D5623, 1994, *Standard Test Method for Sulfur Compounds in Light Petroleum Liquids by Gas Chromatography and Sulfur Selective Detection*, ASTM International, West Conshohocken, PA, 2014. www.astm.org.

ASTM Standard D6667, *Standard Test Method for Determination of Total Volatile Sulfur in Gaseous Hydrocarbons and Liquefied Petroleum Gases by Ultraviolet Fluorescence*, ASTM International, West Conshohocken, PA, 2014. www.astm.org.

ASTM Standard D7039, *Standard Test Method for Sulfur in Gasoline, Diesel Fuel, Jet Fuel, Kerosine, Biodiesel, Biodiesel Blends, and Gasoline-Ethanol Blends by Monochromatic Wavelength Dispersive X-Ray Fluorescence Spectrometry*, ASTM International, West Conshohocken, PA, 2015. www.astm.org.

ASTM Standard D7041 2010e1, *Standard Test Method for Determination of Total Sulfur in Light Hydrocarbons, Motor Fuels, and Oils by Online Gas Chromatography with Flame Photometric Detection*, ASTM International, West Conshohocken, PA, 2010b. www.astm.org.

Barone, G. et al., Impact of sampling and transfer component surface roughness and composition on the analysis of low-level and mercury containing streams, *Annual ISA Analysis Division Symposium—Proceedings*, Houston, TX, 2005.

Chen, Z.W., Wei, F., Radley, I., and Beumer, B., Low-level sulfur in fuel determination using monochromatic WD XRF—ASTM D 7039–04, *J. ASTM Int.*, 2(8), 49–53, September 2005. http://www.xos.com/wp-content/uploads/Chen%202005_ASTM.pdf (accessed May 17, 2015).

Clemons, J. and Johnson, B., Trace total sulfur measurement with a flame photometric detector, (2003) http://www.norskanalyse.no/files/52125.pdf (accessed May 17, 2015).

Clevett, K.S., *Process Analyzer Technology*, John Wiley & Sons, New York, pp. 314–321, 1986.

Greaves, D., Rolley, S., Burgess, A., and Murphy, D., Integration of total sulfur and Wobbe index analysis using UV spectroscopy and TDL oxygen measurement, *Annual ISA Analysis Division Symposium—Proceedings*, League City, TX, 2011.

Greyson, J.C., *Carbon, Nitrogen and Sulfur Pollutants*, Marcel Dekker, New York, 1990.

Hidy, G.M., *Atmospheric Sulfur*, Academic Press, New York, 1994.

Higgins, M., Barone, G., Smith, D., and Neeme, T., A comparison of surface adsorption effects in mercury and sulfur analyzer systems, *ISA Expo 2007—Proceedings*, Houston, TX, 2007.

Melda, K.J., Total sulfur in fuels: A proven process gas chromatographic technique to measure the reduced limits, *Annual ISA Analysis Division Symposium—Proceedings*, 2000.

Spitler, R.W., Pasquale, A., and Friudenberg, R., Using pulsed UV fluorescence, *Hydrocarbon Engineering*, February 2003. http://www.docstoc.com/docs/130700260/Using-pulsed-UV-fluoresence-Thermo (accessed May 17, 2015).

Tarkanic, S. and Crnko, J., Rapid determination of sulfur in liquid hydrocarbons for at-line process applications using combustion/oxidation and UV-fluorescence detection, *Journal of ASTM International*, 2(9), 12980, October 2005.

Thind, S., Evaluation of sulfur measurement analytical techniques for sulfur in gases and fuel, (June/July 2004). http://www.cianalytics.com/Portals/0/Articles/eval_sulphur_measurement_analytical_techniques_gases_fuel.pdf (accessed May 17, 2015).

1.66 Thermal Conductivity Detectors

J. E. BROWN (1969, 1982)   **B. G. LIPTÁK** (1995, 2003, 2017)

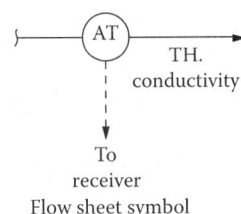

Flow sheet symbol

Applications	Best suited for the detection of gases or vapors in binary mixtures, for leakage detection or as detectors on chromatographs. Particularly suited for the measurement of high thermal conductivity gas analysis, such as hydrogen and helium.
Design pressure	Near atmospheric
Sample temperature	2°C–43°C (35°F–110°F)
Cell materials of construction	Brass, stainless steel, Monel
Range	The full span of the analyzer should correspond to a minimum of a 2% change in the thermal conductivity of the gas mixture.
Inaccuracy	Inaccuracy is 1%–2% of full scale for binary samples when the thermal conductivity of each constituent is accurately known. Published thermal conductivity data can be in error by as much as 5%. Interpretation of readings on multicomponent mixtures requires additional measurements and analysis.
Cost	A portable leak detector with 10^{-5} cc/s sensitivity costs about $1,600. Industrial analyzers for binary mixtures range from $2,500 to $6,000. Analyzers with higher sensitivity and/or in corrosion-resistant materials cost from $6,000 to $12,000.
Partial list of suppliers	ABB Process Analytics www.abb.com/analytical Buck Scientific (for gas chromatograph) http://www.bucksci.com/atomic-absorption-spectrophotometers/details/3119/394/gas-chromatographs/gc-detectors/thermal-conductivity-detector-tcd.html Cole-Parmer Instrument Co. (for gas chromatograph) http://www.coleparmer.com/Search/thermal%20conductivity%20detectors ESS (continuous) http://essme.com/thermal-analyzer/?op=m7 GE XMTX (hydrogen in gas transmitter) http://www.ge-mcs.com/en/gas/hydrogen-analysis-and-thermal-conductivity/xmtc-transmitter.html Gow-Mac Instrument Co. (for gas chromatograph) http://www.gow-mac.com/products/prod_cat.cfm?prod_id=29&cat_id=10 Honeywell (inline transmitter) https://www.honeywellprocess.com/en-US/explore/products/instrumentation/analytical-instruments-and-sensors/thermal-conductivity/Pages/7866-thermal-conductivity-analyzer.aspx MSA Instrument Div. (combustible gas) http://site.msagasmonitors.com/MSA/pdf/tankscope-combustible-gas-indicator_manual.pdf Nova (Portable or continuous H2) http://catalog.nova-gas.com/item/all-categories/ontinuous-process-hydrogen-gas-analyzer-430-series/series-430n4?&plpver=10&origin=keyword&filter=&by=prod Perkin Elmer (refinery gas) http://www.perkinelmer.com/Catalog/Product/ID/NARL8105 PID Analyzer LLC (analyzer and controller) http://www.hnu.com/secure/pdf/204TCAnalyze.pdf Siemens Applied Automation (for gas chromatograph, etc.) http://www.automation.siemens.com/mcms/sensor-systems/en/process-analytics/process-gas-chromatograph/Pages/microsam.aspx Teledyne Analytical Instruments (explosion proof, trace, portable) http://www.teledyne-ai.com/industrial/cat_specialapps.asp#tcd Thermco Instrument Corp. (panel or wall mounted) http://www.thermco.com/ss119.pdf Vici Valco Instrument Co. (microvolume) http://www.vici.com/instr/tcd.php

INTRODUCTION

Thermal conductivity (k, λ, or κ) is the property of a material to conduct heat from a warmer surface through the material to a colder surface. It has the units of heat energy transferred per unit of time and per unit area, divided by the temperature gradient, which is the temperature difference between the two surfaces divided by the distance between the plates. Thermal conductivity can have the units of Btu/(ft·sec·°F) that equals 0.893 kcal/(cm·min·°C). In everyday usage, the thermal conductivity (TC) is expressed as the ratio of the thermal conductivity of the material of interest (T) divided by that of some reference material such as air (TC_{air}).

Composition measurement by detecting the thermal conductivity of gases is one of the simplest and oldest methods of analyzing process streams. Early developments by the British resulted in an instrument of this type, which was called a *katharometer* or *catharometer*. The name still persists in Europe.

This technique takes advantage of the facts that different substances have a varying capacity to conduct heat energy from a heat source. This ability differs for each gas. It is called *thermal conductivity* and can be expressed in various unit systems such as BTU/h/ft²/°F/in., W/s/cm²/°C/cm, kerg/s/cm²/°C/cm, and so on.

This is a simple, rugged, inexpensive, reliable, and easily maintained, but nonspecific, analyzer that can determine the composition of only binary mixtures. It is neither very sensitive, nor very fast, but it is well suited for many chromatographic and leak detection applications.

Chapters that also discuss the use of thermal conductivity detectors (TCDs) include Chapter 1.10 on Chromatography, Chapter 1.34 on Leak detection, and Chapter 1.14 on Combustible Detection.

THERMAL CONDUCTIVITY

Thermal conductivity is often expressed as a factor relating the ability of a particular gas to conduct heat to that of air at various temperatures (Table 1.66a). In practice, continuous thermal conductivity analyzers measure a *change* in heat dissipation by comparing the change with a *reference* condition.

TABLE 1.66a
Thermal Conductivity Factors

	R_0[a]	R_{100}[a]
Acetone	0.406	0.546
Acetylene	0.776	0.900
Air	1.000	1.000
Ammonia	0.897	1.086
Argon	0.709	0.725
Benzene	0.370	0.573
Carbon dioxide	0.614	0.690
Carbon monoxide	0.964	0.962
Chlorine	0.322	0.381
Ethylene	0.735	0.919
Ethane	0.807	0.970
Helium	6.230	5.840
Hydrogen	7.130	6.990
Methane	1.318	1.450
Nitrogen	0.996	0.996
Oxygen	1.043	1.052
Pentane(n)	0.520	0.702
Refrigerant 12	0.354	0.356
Sulfur dioxide	0.344	0.377

[a] R_0, R_{100} = Thermal conductivity of gas/thermal conductivity of air at 0°C and 100°C, respectively.

Measurement Ranges

It has been known for more than a century that the heat-conducting ability of various gases differs considerably. Therefore, by measuring the thermal conductivity of a binary mixture, one can determine the composition of the mixture. The accuracy of the measurement is a function of the reliability of the thermal conductivity data used for the gases making up the mixture, and those data are not always accurate.

Some of the common thermal conductivity analyzer applications include the measurement of *hydrogen* in air, nitrogen, carbon dioxide, carbon monoxide, argon, blast furnace gases, and reformer gases; *helium* in air or nitrogen; *methane* in air; *propane* in air; *Freon*® in air; and *carbon dioxide* in air, nitrogen, or flue gases.

A general rule is that the full span of the analyzer should correspond to a minimum 2% change in the thermal conductivity of the gas mixture. Table 1.66b gives the full-scale ranges of some binary gas mixtures that will result in at least a 2% change in the thermal conductivity of the mixture. It can be noted that, as the thermal conductivity of a gas (helium, hydrogen) increases, so does the sensitivity of measurement; therefore, the range can be narrower.

TABLE 1.66b
Ranges of Gas Mixture Compositions Suitable for Measurement by Thermal Conductivity

Gases in the Mixture	Range of Concentrations of the First Gas in the Second (%)	An Error of 1% of Full Scale Corresponds to (%)
Air in carbon dioxide	0–5	0.05
Air in helium	0–2.5	0.025
Air in oxygen	0–0	0.4
Air in sulfur dioxide	0–1	0.01
Argon in nitrogen	0–7	0.07
Carbon dioxide in air	0–7	0.07
Carbon dioxide in nitrogen	0–7	0.07
Carbon dioxide in oxygen	0–6.5	0.065
Helium in air	0–0.5	0.005 (50 ppm)
Helium in hydrogen	0–12	0.12
Hydrogen in helium	0–10	0.1
Hydrogen in nitrogen	0–0.3	0.003 (30 ppm)
Nitrogen in argon	0–5	0.05
Nitrogen in carbon dioxide	0–5	0.05
Nitrogen in hydrogen	0–2.5	0.025
Nitrogen in oxygen	0–55	0.55
Oxygen in air	0–38	0.38
Oxygen in carbon dioxide	0–4.5	0.045
Oxygen in nitrogen	0–52	0.52
Sulfur dioxide in air	0–3	0.03

TCD ANALYZER

The TCD can measure the composition of binary mixtures at moderate sensitivity.

The main components of a thermal conductivity analyzer are the measuring cell, regulated power supply, Wheatstone bridge, and case temperature control.

Detectors

The measuring cell consists of a relatively large mass of metal to provide a stable heat sink. Through the metal block, flow passages are drilled or formed, and a recessed cavity is machined for inserting a heat source and sensing element, such as a hot-wire filament. The cell material must be compatible with the process gas sample and must also have a high thermal conductivity coefficient. Stainless steel is generally employed.

The TCD is one of the more widely used detectors for gas chromatography because it is simple in construction, rugged,

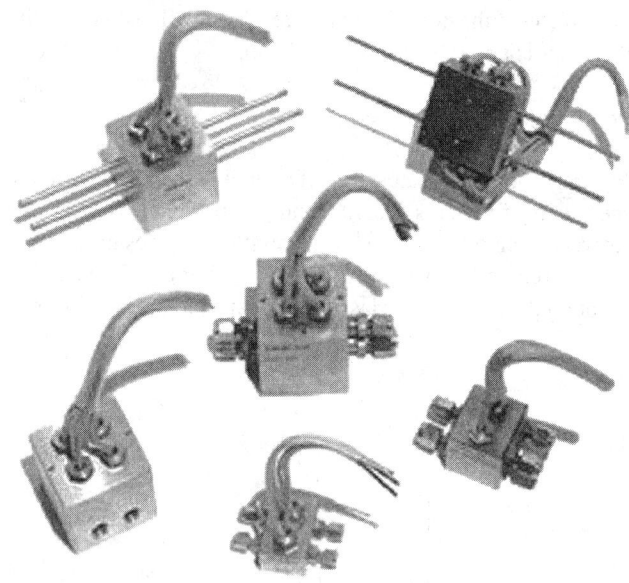

FIG. 1.66a
Design variations of thermal conductivity detectors used for chromatography applications. (Courtesy of Gow-Mac.)

versatile, sensitive, relatively linear over a wide range, and inexpensive (Figure 1.66a).

The detector transducers can be resistance wires (filaments) or thermistors (beads of metallic oxide). The operation of these heated elements is similar except that filaments have positive, and thermistors have negative, resistance coefficients. This means that, as the temperature rises, the resistance of filaments also rises, but the resistance of thermistors drops.

The choice between them is based on sensitivity and temperature considerations. In general, thermistors are used for ambient and sub-ambient applications, whereas resistance wire filaments are used at higher temperatures. In terms of sensitivity, the thermistors are superior.

Hot-wire filaments are now in prevalent use as a result of improved filament designs. Up to 1965, thermal conductivity analyzers frequently used thermistors to achieve the desired sensitivity. Thermistors are beads of metallic oxides with a thin coating (typically glass) over the surface. This coating tends to crack with excess heat, and when this element is used in gases with high hydrogen content, the oxides are reduced by the hydrogen, and drift is experienced. Glass bead thermistors develop frequent failures, particularly when used in high-hydrogen-containing gases. The hot-wire filaments develop surface temperatures between 204°C and 400°C (400°F and 750°F) and are sometimes used with a catalyst coating to promote catalytic cracking of hydrocarbons to further increase system sensitivity.

Once the selection is made, the next step is to pick the right material to withstand the corrosiveness of the sample gas. The filaments are usually made from tungsten or platinum alloy materials, but nickel, rhenium–tungsten, and gold-plated tungsten filaments are also available. The wire diameter is about 0.03 mm (0.001 in.).

TCD Cells

Figure 1.66b illustrates a TCD block assembly with the measuring filaments inserted into both the measuring and reference chambers. This is a flow-through design using a single (replaceable) measuring and a single reference filament configured into a classical Wheatstone bridge circuit.

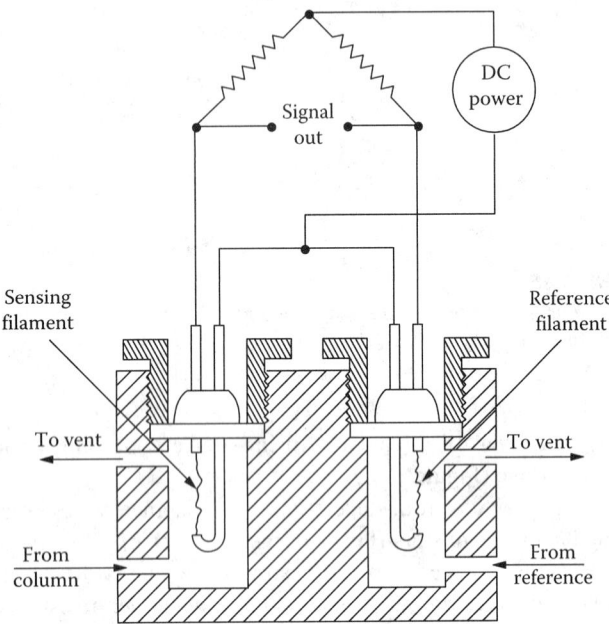

FIG. 1.66b
The design of a two-element, flow-through TCD cell.

In addition to the two-element design, four-element thermal conductivity cell designs with recessed hot-wire filaments are also available, as shown in Figure 1.66c.

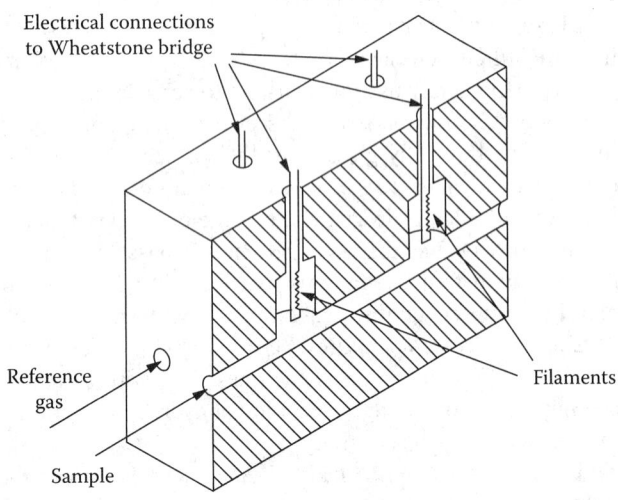

FIG. 1.66c
Four-element thermal conductivity cell.

The elements are used in pairs, one in the sample stream and one in the reference gas. One or two pairs are normally used, but some cell designs include eight pairs to improve sensitivity. Vertical mounting is preferred to prevent sagging of the wire elements. The recessed elements generally provide an improved noise level but poorer response speed.

The response speed of the TCD is a function of the internal volume of the detector cell. The flow through the cell must be constant (usually in the range of 15–100 mL/min for chromatographic and 500 mL/min for process analyzers).

Bridge Circuits

The sensing system can be a basic Wheatstone bridge that uses a high-quality regulated power supply (Figure 1.66d) rated between 100 and 300 mA. Analyzer stability is primarily a function of power supply voltage regulation. In addition to the means of improving sensitivity, it is more practical to use low-noise operational amplifiers on the bridge output. However, a low signal-to-noise ratio is required of the basic bridge output.

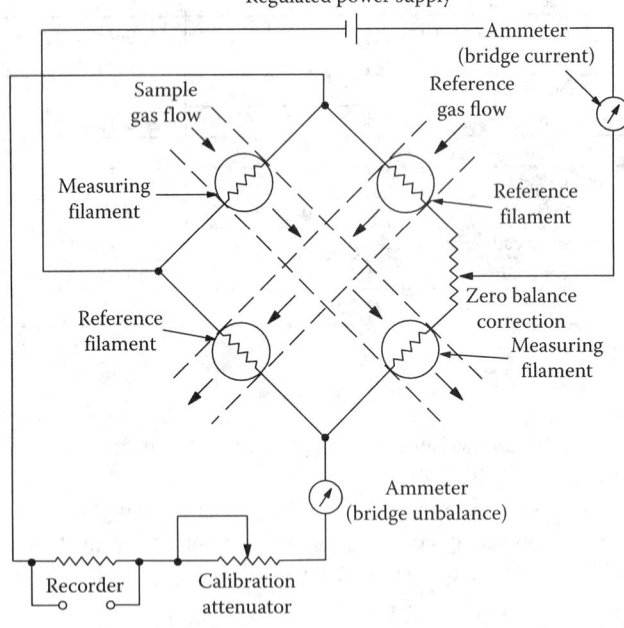

FIG. 1.66d
Typical Wheatstone bridge measuring circuit for thermal conductivity analyzer.

To provide faster response and to protect the filaments if the flow of carrier gas fails, the latest TCD designs favor constant current or constant resistance bridges (see Figures 1.66e and 1.66f).

Case temperature control provides a constant-temperature environment for the measuring cell to enhance stability. Various temperature control systems are used, ranging from off/on thermal switches with bare strip heaters to the more

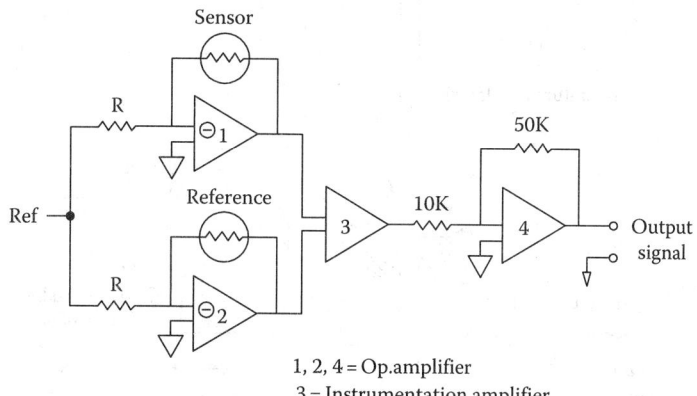

FIG. 1.66e
The electric circuit for providing constant current to a TCD utilizing thermistor bead elements. (From Annino, R. and Villabolos, R., Process Gas Chromatography: Fundamentals and Applications, ISA, Copyright 1992 ISA. Used with permission. All rights reserved.)

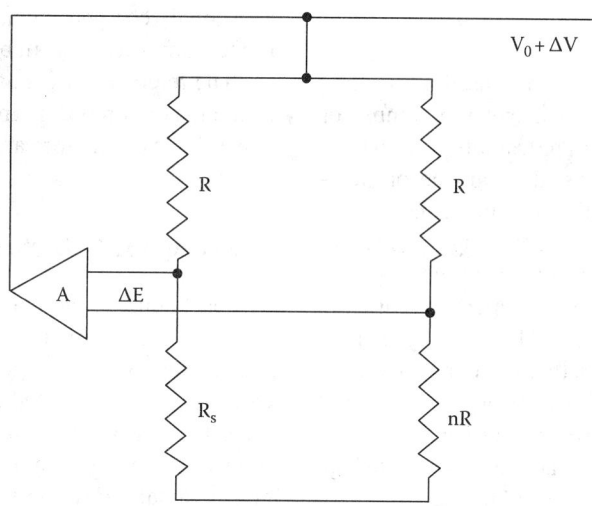

FIG. 1.66f
The electric circuit for providing a constant resistance/temperature circuit for a TCD. (From Annino, R. and Villabolos, R., Process Gas Chromatography: Fundamentals and Applications, ISA, Copyright 1992 ISA. Used with permission. All rights reserved.)

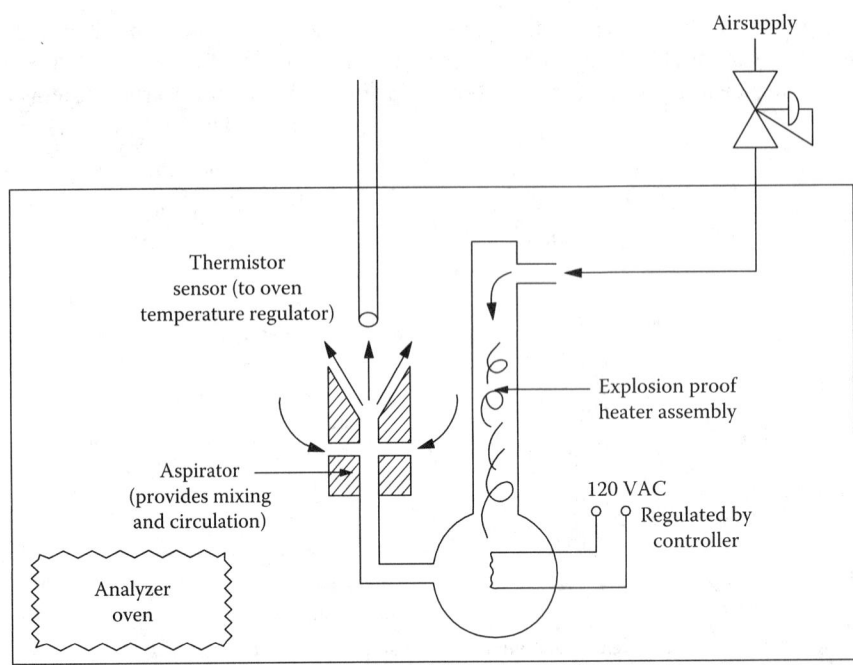

FIG. 1.66g
Oven temperature regulator using thermistor-type sensor provides sensitive temperature control.

refined ones for chromatographic ovens (see Figure 1.66g). Analyzer stability can be adversely affected by inadequate temperature control.

Operation

The thermal conductivity analyzer, when used as a chromatographic detector, operates with a metered sample of 50–200 cc/min (flow controlled). The sample flow passes through the measuring cell and across the filaments or glass-bead thermistors, which are "hot" from being heated by the Wheatstone bridge. This resistance heating provides an elevated temperature on the surface of the filament.

Heat energy is conducted from the filament, through the flowing gas, to the walls of the cell. The quantity of heat thus conducted is a function of the thermal conductivity of the flowing gas. When a sample stream of lower thermal conductivity than that of the zero standard gas is introduced, less heat is conducted away, and the filament surface temperature (and its resistance, if it is a hot-wire filament) increases. This causes an unbalance in the Wheatstone bridge. The degree of unbalance can be calibrated in terms of composition.

If a thermistor-type detector is used, it has a negative temperature coefficient, so it will unbalance the bridge in the opposite direction, thus requiring a polarity reversal for the ammeter connections. Otherwise, the operation is identical.

Reference Filament   Reference filaments are used to provide compensation for temperature and barometric pressure variations. Because the reference filaments are in opposite legs of the bridge relative to the measuring filaments, small temperature variations in the cell should affect both filaments equally and therefore cancel out. The reference gas can be sealed for this purpose, but a *flowing reference* provides additional compensation, because the venting pressure reflects any variations in barometric pressure.

The reference gas is usually a single gas that is the same as the major component in the sample gas. Reference flow is generally less than the sample flow, 40–100 cc/min.

Packaging and Calibration

The packaging of thermal conductivity analyzers varies with suppliers, but most manufacturers separate the electronics (power supply and bridge) from the temperature-controlled case. In some cases, the two parts can be physically separated by a distance of up to 100 ft (30 m). In most designs, the cabinets require little more panel or wall space than do conventional transmitters.

Calibration is accomplished using known samples to establish an empirical calibration reference.

One notable exception to the earlier-mentioned general rule is the analysis of hydrogen in hydrocarbons. From Table 1.66a, it can be seen that hydrogen has a very high R_0 value relative to most hydrocarbon gases. In such cases, the sample often can be considered to be a hydrocarbon–hydrogen binary mixture and calibrated using an "average" background of non-hydrogen components. The error in the measurement will depend on the true concentrations of the other components, but in many applications, ±5% full-scale inaccuracy is easily attainable.

Hydrogen detector transmitters are available for online applications (see Figure 1.66h). The accuracy of these units is usually 2% FS. Their linearity and repeatability are between 0.5% FS and 1% FS. Their span and zero drift are about 0.5% per week, and their response to a 90% of span change is about 20 sec.

FIG. 1.66h
XMTC Thermal Conductivity Binary Gas Transmitter which can be used for on-line application. (Courtesy of GE Panametrics.)

CONCLUSIONS

Although simple in design, this analyzer has a major limitation: only binary mixtures can be accurately measured by it. The analyzer is nonspecific; because it measures the total thermal conductivity of the process sample, it cannot distinguish or identify the component that causes the conductivity change in multicomponent mixtures. Therefore, its applications are limited to binary or binary-like mixtures. Some industrial gas streams are binary mixtures and do require analysis.

The advantages of the thermal conductivity analyzer include its low cost, simplicity, reliability, and reasonable speed of response. Its main limitation is that it can measure only binary mixtures. In addition to its nonspecific nature, the need for empirical calibration further restricts its use.

It is also recommended that all water vapors be removed from the measurement sample by drying. Because of the earlier-listed limitations, its applications are few and usually involve binary mixtures for applications such as leakage measurement and chromatography or the detection of hydrogen or helium in applications where the thermal conductivity of the "background" gases is low and relatively constant.

Thermal conductivity detectors are best suited for applications in which the conductivity of the gas of interest is much higher or much lower than that of air. For this reason, some of the recommended binary mixture measurement applications include the measurement of the concentrations of the following gases, which have the following thermal conductivities relative to air at 0°C: acetylene 0.78, ammonia 0.90, butane 0.68, carbon dioxide 0.55, chlorine 0.32, ethane 0.75, helium 5.97, hydrogen 7.15, methane 1.25, sulfur dioxide 0.35 xenon 0.21.

SPECIFICATION FORMS

When specifying thermal conductivity detectors, the ISA Form 20A1001 can be used to specify the features required for the device itself and when specifying both the device and the nature of the application, including the composition and/or properties of the process materials, ISA Form 20A1002 can be used. These forms, with the permission of the International Society of Automation, are reproduced on the next pages.

1	RESPONSIBLE ORGANIZATION		ANALYSIS DEVICE	6	SPECIFICATION IDENTIFICATIONS		
2				7	Document no		
3	ISA		Operating Parameters	8	Latest revision		Date
4				9	Issue status		
5				10			

11	ADMINISTRATIVE IDENTIFICATIONS				40	SERVICE IDENTIFICATIONS Continued			
12	Project number		Sub project no		41	Return conn matl type			
13	Project				42	Inline hazardous area cl		Div/Zon	Group
14	Enterprise				43	Inline area min ign temp		Temp ident number	
15	Site				44	Remote hazardous area cl		Div/Zon	Group
16	Area		Cell	Unit	45	Remote area min ign temp		Temp ident number	
17					46				
18	SERVICE IDENTIFICATIONS				47				
19	Tag no/Functional ident				48	COMPONENT DESIGN CRITERIA			
20	Related equipment				49	Component type			
21	Service				50	Component style			
22					51	Output signal type			
23	P&ID/Reference dwg				52	Characteristic curve			
24	Process line/nozzle no				53	Compensation style			
25	Process conn pipe spec				54	Type of protection			
26	Process conn nominal size		Rating		55	Criticality code			
27	Process conn termn type		Style		56	Max EMI susceptibility		Ref	
28	Process conn schedule no		Wall thickness		57	Max temperature effect		Ref	
29	Process connection length				58	Max sample time lag			
30	Process line matl type				59	Max response time			
31	Fast loop line number				60	Min required accuracy		Ref	
32	Fast loop pipe spec				61	Avail nom power supply			Number wires
33	Fast loop conn nom size		Rating		62	Calibration method			
34	Fast loop conn termn type		Style		63	Testing/Listing agency			
35	Fast loop schedule no		Wall thickness		64	Test requirements			
36	Fast loop estimated lg				65	Supply loss failure mode			
37	Fast loop material type				66	Signal loss failure mode			
38	Return conn nominal size		Rating		67				
39	Return conn termn type		Style		68				

69	PROCESS VARIABLES	MATERIAL FLOW CONDITIONS				101	PROCESS DESIGN CONDITIONS		
70	Flow Case Identification				Units	102	Minimum	Maximum	Units
71	Process pressure					103			
72	Process temperature					104			
73	Process phase type					105			
74	Process liquid actl flow					106			
75	Process vapor actl flow					107			
76	Process vapor std flow					108			
77	Process liquid density					109			
78	Process vapor density					110			
79	Process liquid viscosity					111			
80	Sample return pressure					112			
81	Sample vent/drain press					113			
82	Sample temperature					114			
83	Sample phase type					115			
84	Fast loop liq actl flow					116			
85	Fast loop vapor actl flow					117			
86	Fast loop vapor std flow					118			
87	Fast loop vapor density					119			
88	Conductivity/Resistivity					120			
89	pH/ORP					121			
90	RH/Dewpoint					122			
91	Turbidity/Opacity					123			
92	Dissolved oxygen					124			
93	Corrosivity					125			
94	Particle size					126			
95						127			
96	CALCULATED VARIABLES					128			
97	Sample lag time					129			
98	Process fluid velocity					130			
99	Wake/natural freq ratio					131			
100						132			

133	MATERIAL PROPERTIES				137	MATERIAL PROPERTIES Continued		
134	Name				138	NFPA health hazard		Flammability	Reactivity
135	Density at ref temp		At		139				
136					140				

Rev	Date	Revision Description	By	Appv1	Appv2	Appv3	REMARKS

Form: 20A1001 Rev 0 © 2004 ISA

		RESPONSIBLE ORGANIZATION	ANALYSIS DEVICE COMPOSITION OR PROPERTY Operating Parameters (Continued)		SPECIFICATION IDENTIFICATIONS	
1				6		
2		ISA		7	Document no	
3				8	Latest revision	Date
4				9	Issue status	
5				10		

		PROCESS COMPOSITION OR PROPERTY			MEASUREMENT DESIGN CONDITIONS				
		Component/Property Name	Normal	Units	Minimum	Units	Maximum	Units	Repeatability
11									
12									
13									
...									
75									

Rev	Date	Revision Description	By	Appv1	Appv2	Appv3	REMARKS

Form: 20A1002 Rev 0 © 2004 ISA

Definitions

Katharometer is the older name used for the TCD, still sometimes used in Europe.

Thermal conductivity: Thermal conductivity is the ability of a gas to conduct heat. It is often expressed as a ratio relative to the thermal conductivity of air.

Abbreviations

CG	Gas chromatography
ECD	Electron capture detector
FID	Flame ionization detector
ICP	Inductively coupled plasma
TCD	Thermal conductivity detector

Bibliography

Agilent Technology, *Thermal conductivity detector, troubleshooting tips*, (2013) http://www.chem.agilent.com/Library/Support/Documents/a16093.pdf.

Bakir, E., Thermal conductivity of cross-linked polymers, (2011) http://www.amazon.com/s/ref=nb_sb_noss_1?url=search-alias%3Dstripbooks&field-keywords=thermal%20conductivity&sprefix=thermal+c%2Cstripbooks&rh=i%3Astripbooks%2Ck%3Athermal%20conductivity.

Emerson Rosemount, *Measuring hydrogen sulfide in refinery gas*, (2013) http://www2.emersonprocess.com/siteadmincenter/PM%20Rosemount%20Analytical%20Documents/GC_AN_42-PGC-AN-REFINING-H2S-TCD.pdf.

Hinshaw, J.V., *Thermal conductivity*, (2006) http://www.chromatographyonline.com/lcgc/Column%3A+GC+Connections/The-Thermal-Conductivity-Detector/ArticleStandard/Article/detail/283449.

Hussain, W., *Thermal conductivity and dielectric properties of composites*, (2012) http://www.amazon.com/Thermal-Conductivity-Dielectric-Properties-composites/dp/3659145580/ref=sr_1_7?s=books&ie=UTF8&qid=1390396804&sr=1-7&keywords=thermal+conductivity.

Kiss, L., *Thermal Conductivity31/Thermal Expansion19: Proceedings of 31st Int'l Thermal Conductivity Conference* 2013. http://www.amazon.com/Thermal-Conductivity-Expansion-Proceedings-Symposium/dp/1605950556/ref=sr_1_4?s=books&ie=UTF8&qid=1390396614&sr=1-4&keywords=thermal+conductivity.

MSA, *Gas Detection*, 5th edn., 2009. http://www.gilsoneng.com/reference/gasdetectionhandbook.pdf.

Royal Society of Chemistry, *CHNS elemental analysis*, (2008) http://www.rsc.org/images/CHNS-elemental-analysers-technical-brief-29_tcm18-214833.pdf.

Tritt, T.M., *Thermal conductivity: Theory, properties, and applications*, (2005) http://www.amazon.com/Thermal-Conductivity-Properties-Applications-Physics/dp/0306483270

Vajpai, A., *Thermal conductivity of nanofluids*, (2012) http://www.amazon.com/s/ref=nb_sb_noss_1?url=search-alias%3Dstripbooks&field-keywords=thermal%20conductivity&sprefix=thermal+c%2Cstripbooks&rh=i%3Astripbooks%2Ck%3Athermal%20conductivity.

Valco Instruments, *Microvolume thermal conductivity detector*, http://www.vici.com/support/manuals/tcd_man.pdf.

Yaws, C.L., *Handbook of Transport Property Data, Viscosity, Thermal Conductivity*, Gulf Professional Publishing, Houston, TX, 1995. http://books.google.com/books/about/Handbook_of_transport_property_data.html?id=4fVUAAAAMAAJ.

1.67 Total Carbon and Total Organic Carbon (TOC) Analyzers

I. G. YOUNG (1974, 1982) **B. G. LIPTÁK** (1995)
M. T. LEE-ALVAREZ (2003) **S. VEDULA** (2017)

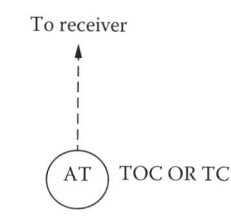

To receiver

AT TOC OR TC

Flow sheet symbol

Oxidation methods for TC and TOC measurement	A. Catalyzed Ozone Hydroxyl Radical Oxidation with NDIR Detection
	B. Combustion Oxidation with NDIR Detection
	C. UV/Heated Persulfate Oxidation with NDIR detection
	D. UV Light Oxidation with NDIR detection or differential conductivity
	E. Boron Doped Diamond (BDD) with NDIR detection
Samples	Laboratory sample sizes range from 0.01 to 40 cc. For in-line applications, the sample flow rates range from 0.25 to 30 cc/min
Flowing sample solids	Up to 2 mm in diameter soft particulates in Catalyzed Ozone Hydroxyl Radical Oxidation (Two Stage Advanced Oxidation) method.
	Size of particles up to <0.5 mm in diameter for other oxidation methods
Materials of construction	Glass, quartz, Teflon®, stainless steel, Hastelloy, polyethylene, and PVC
Measurement cycle time	A. 6.5 min
	B. 6.4 min including sampling and oxidation
	C. 5–7 min including sample pretreatment
	D. 3 hr for 1–25 samples
	E. 5–10 min interval
Utilities or reagents	Inorganic carbon is removed by acidification followed by sparging. 10% (v/v) H_3PO_4 (phosphoric acid) or 5% HNO_3 (nitric acid), or 1.8% sulfuric acid each in deionized water
	Carrier gas: CO_2 free air, nitrogen, or oxygen
	UV persulfate method: 1.5 M $Na_2S_2O_8$ (sodium persulfate),
	Oxidation with hydroxyl radicals method: 1.2 N Sodium Hydroxide (NaOH) Reagent
Ranges	See Table 1.67a below for further details.
	A. 0–10 to 0–20,000 mg/L
	B. 0–10 to 30,000 mg/L
	C. 0–1 to 0–1000 mg/L
	D. 0.5 μg/L to 2000 μg/L
	E. 0.1–25 ppm Carbon
Sensitivity	A. 0.6 mg/L
	B. 1.0 mg/L
	C. 1.0% of smallest measuring range
	D. 300 ppb C
	E. 0.5 ppb
Inaccuracy	A. ±2% of span
	B. ±2% of span
	C. ±2% of span
	D. ±1 ppb or n5%, whichever is greater
	E. ±10%

(Continued)

Costs	A, B. $40,000–$50,000 including PC and software, $30,000–$40,000 with auto sampler
	C. $30,000–$35,000 for a manual analyzer, $40,000–$45,000 with autosampler and software
	D. $5,000–$8,000 for portable colorimeters, $15,000 for spectrophotometers, $1,500 COD reactor, $1,200 reagent kit
	E. $30,000–$40,000
Partial list of suppliers	Anatel (B) (http://www.hach.com/anatel-toc600-toc-analyzer/product-downloads?id = 7640480108)
	BioTector Analytical Systems Ltd. (A) (http://www.hach.com/b7000)
	Endress +Hauser (A) (https://portal.endress.com/wa001/dla/5000276/3462/000/00/TI448CEN_1009.pdf) (https://portal.endress.com/wa001/dla/5000260/7579/000/00/TI424CEN_0609.pdf)
	GE Instruments (B) (http://www.geinstruments.com/products-and-services/toc-analyzers-and-sensors/previous-generation-toc-models)
	Hach (A, B) (http://www.hach.com/anateltoc) (http://www.hach.com/wdm-astrotoc-uv-on-line-total-organic-carbon-analyzer/product?id = 10741523775)
	Mettler Toledo (B) (http://us.mt.com/us/en/home/products/ProcessAnalytics/Thornton_TOC_Family.html)
	OI Analytical (A, E) (http://www.oico.com/default.aspx?id = product&productid = 121)
	Rosemount Analytical Inc. (A) (http://www.southeastern-automation.com/PDF/Emerson/PAD/Gas_PDS_2100C_103-2100C_200403.pdf)
	Shimadzu Corporation (A) (http://www.shimadzu.com/an/toc/online/toc-4200.html)
	Sick Limited (http://www.sick.com/us1/en-us/home/products/product_portfolio/analyzers_systems/Pages/tocor700.aspx)
	Teledyne Tekmar (A, D) (http://www.teledynetekmar.com/products/TOC/Torch/documentation/Torch_Product_Brochure.pdf) (http://www.teledynetekmar.com/products/TOC/Fusion/documentation/Fusion_Product_Brochure.pdf
	UIC Inc. (C) (http://www.uicinc.com/SystemSheets/CM135.pdf)

INTRODUCTION

A variety of wastewater parameters can be monitored to assess the pollutant load that these wastes represent. The total organic carbon (TOC) concentration of the waters and wastewaters is often used as a process control monitoring tool and as a quality indicator. The classical biochemical oxygen demand (BOD) and the chemical oxygen demand (COD) have long been employed for similar purposes. All three readings have their strengths and weaknesses. Monitoring of Total Carbon and Total Organic Carbon in water is a two-step process involving an oxidation process followed by the detection of the resulting carbon dioxide/carbonate.

Advantages and Limitations

A TOC analysis is very rapid and accurate, but it measures only the organic carbon content and it therefore does not detect the pollutant load that is represented by inorganic nitrogen-based or phosphorus-based molecules. TOC analyzers with Catalyzed Ozone Hydroxyl Radical Oxidation (Two Stage Advanced Oxidation) technology can be built with TN (Total Nitrogen) and/or TP (Total Phosphorus) modules for the measurement of nitrogen and phosphorus-based compounds. A BOD analysis is slow, but it measures the biologically degradable molecules that exert an oxygen demand on the receiving waters, and its readings will vary if the bioassay used is changed. Thus, the BOD, in addition to being a lengthy 5- or 7-day analysis, gives significantly different readings as a function of the bioassay used.

The COD analysis suffers from shortcomings in oxidation efficiencies and interferences (e.g., chlorides), although its analysis time is reduced (e.g., 2 hr).

Direct correlation between TOC, BOD, and COD usually is possible. In-line TOC measurement is used as an alternative to laboratory COD and BOD analysis across the globe and TOC is recognized by many licensing authorities. With proper interpretation, the TOC can represent a rapid and frequently accurate method of assessing the pollution load levels of municipal and industrial wastes.

Carbon Measurement Techniques

To determine the TOC content of a sample, first its total inorganic carbon (TIC) content is typically eliminated. The TIC present in a water sample is usually in the form of inorganic bicarbonates and carbonates. One of two techniques can be used to remove TIC.

In the first technique, the inorganic carbon components are analyzed independently and then subtracted from the total carbon (TC). In this case, the TOC is determined by the difference between TC and TIC. The TOC result obtained from this technique includes the possible volatile (purgeable) organic carbon (VOC) compounds present in the sample.

$$TC - TIC = TOC$$

The other technique is to acidify the sample down to a pH of typically 2 or less, followed by a brief gas sparging to drive off the inorganic carbon dioxide formed by the acidification. Any carbon remaining after this sparging represents TOC when there is no significant concentration of volatile organic compounds in the sample. If the concentration of volatiles in the sample is high, the TOC result obtained after the TIC sparging represent Non-Purgeable Organic Carbon (NPOC), because some or all of the volatile organics will also be removed during the TIC sparging process. Therefore, when the TIC and VOC concentration in a sample is negligible or very low, the TC found in the sparged sample is considered to be equal to the TOC content of the sample.

Analyzer Development

As a result of the rapid acceptance and usefulness of TOC analysis as a laboratory method, online TOC analyzers became available in the late 1960s. The TOC analyzer was first introduced in 1964 as a single-channel TC analyzer using a catalytic oxidation combustion step followed by the analysis of the resulting carbon dioxide. In this analyzer, the inorganic carbon (IC) was either removed by acid sparging or was determined by titration. A few years later, a second channel was added to this analyzer, which permitted parallel determination of the IC in a second heated reaction chamber.

Since the earlier period, several other techniques such as UV/Heated/Persulfate oxidation have appeared with various changes in the methodology of detection. Their success has been limited by the relative complexity of these continuous analyzers. By 1980, there were at least four distinctly different methods in use to accomplish TOC analysis. Among these detection methods are considerable variations, depending on the instrument manufacturer using them and on the level of carbon concentration in the analyzed process sample. However, when applied to in-line analysis, these methods suffer from a number of limitations. These limitations include the inability to handle dirty samples, sample containing salts, oils, fats, greases and particulates, etc., often resulting in frequent cleaning and requirement for sample filtration.

Catalyzed Ozone Hydroxyl Radical Oxidation (Two Stage Advanced Oxidation) technology has been developed in 1995 as an alternative for the combustion and UV/Heated/Persulfate oxidation technologies for in-line analysis due to the limitations of these traditional oxidation technologies.

A new reagentless oxidation method developed in 2006 used to continuously monitor the total organic carbon level in process water streams. The samples are drawn into the analyzer at 5–10 min intervals from a fill and spill sampling system. Phosphoric acid is introduced into the syringe to sparge and remove the inorganic (TIC) content. The TIC-free sample is then transferred into the reaction chamber and oxidized by hydroxyl radicals, peroxides, and ozone formed by special boron doped diamond (BDD) electrodes.* Organic compounds are oxidized and converted to CO_2, which is measured by a solid state nondispersive infrared (SSNDIR) detector to calculate the TOC content. The technique is somewhat constrained in the range of organics it can effectively oxidize, so applications are limited to situations where TOC response is attributable to a single known analyte such as in condensate return or cooling water monitoring. It may not perform well in cases where TOC response is attributable to multiple unknown organic compounds such as waste water. It is also not recommended for regulatory compliance applications.

---

* Kisacik, L., Stefanova, A., Earnst, S., and Baltruschat, H., Oxidation of carbon monoxide, hydrogen peroxide and water at boron doped diamond electrode: The competition for hydroxyl radicals, Institute for Physics and Theoretical Chemistry, Bonn, Germany, p. 761, April 7, 2013.

TABLE 1.67a
Official Methods of TOC Determination

Oxidation Method	Detector	Analytical Range	Official Methods
Combustion	NDIR	0.004–25,000 mg/L	EPA 415.1, 9060A Standard Method 5310B ASTM D2579 ASTM D5173 DIN 38 409 H3 ISO 8245 AOAC 973.47 USP 643
Combustion	Coulometric	2–50,000 mg/L	ASTM D4129 ASTM D513
UV	NDIR or conductivity	0.0005–0.5 mg/L	USP 643 ASTM D4839 ASTM D4779
UV/persulfate	Membrane/conductivity	0.0005–50 mg/L	USP 643 Standard Method 5310C ASTM 5904
UV/persulfate	NDIR	0.002–10,000 mg/L	EPA 415.1, 9060A Standard Method 5310C ASTM D2579 ASTM D4839 ASTM D4779 ASTM D5173 ISO (Draft) 8245 AOAC 973.47 USP 643
Heated persulfate	NDIR	0.002–1,000 mg/L	EPA 415.1, 9060A Standard Method 5310C ASTM D2579 ASTM D5173 ISO (Draft) 8245 AOAC 973.47 USP 643
Catalyzed ozone hydroxyl radical oxidation (two stage advanced oxidation)	NDIR	0.006–40,000 mg/L	EPA 415.1 ASTM D5173 ISO 8245 ISO EN 1484 DIN 38409-H3
Boron doped diamond	NDIR	0.1–25 mg/L	ASTM or USEPA Approvals not available

Official Methods of TOC Determination

A user's choice may also be limited by industry and government standards that apply to a particular analysis. Table 1.67a provides a list of different oxidation/detection techniques and their corresponding U.S. and EU methods.* Most TOC analyzers used in the laboratory include an autosampler for higher sample throughput and a PC, which allows easier operation of the instrument. A separate module for the analysis of solids is also available from most manufacturers.

DETECTOR TYPES

Nondispersive Infrared Analyzers

The TOC analyzers first oxidize and convert all organic matter to CO_2. After this conversion, a continuous flow of oxygen or air carrier gas transports the resulting cloud of steam and CO_2 through a condenser/cooler and/or water trap into a

---

* Wallace, B., Purcell, M., and Furlong, J., Total organic carbon analysis as a precursor to disinfection byproducts in potable water: Oxidation technique considerations, *J. Environ. Monit.*, 4, 35–42, 2002.

nondispersive infrared (NDIR) analyzer, where the CO_2 concentration is measured. The detector compares the peak area of the sample to the peak area of a calibration standard and calculates the carbon concentration value.

Differential Conductivity Conductivity of the sample is monitored before and after oxidation to accurately predict Total organic carbons.

Oxidation Methods

Catalyzed Ozone Hydroxyl Radical Oxidation (Two Stage Advanced Oxidation) The Catalyzed Ozone Hydroxyl Radical Oxidation,* which was developed as an alternative to the standard oxidation methods of combustion and UV/Heated/Persulfate oxidation, overcome many of the recognized barriers associated with in-line TOC measurement. The sample is oxidized by hydroxyl radicals, which are created by exposing high pH reagents to ozone. Therefore, a large volume of sample, containing high concentrations of suspended solids, salts up to 30% w/w and calcium up to 12% w/w can be injected into the reaction chamber and oxidized with ease. The oxidation process itself is a self-cleaning technology. It is possible to measure fatty and fibrous materials without any risk of clogging or contamination. The volume of the sample injected into the reactor of the analyzer is automatically adjusted for the optimum measuring range.

A volume of unfiltered sample liquid is injected into the reactor. A base reagent is added and the sample is oxidized with hydroxyl radicals that are generated by exposing high pH reagents to ozone. The oxidation of organic compounds present in the sample takes place and both the major compound "carbonates" and the minor compound "oxalates" are formed. An acid reagent, which contains a catalyst, such as manganese, is added into the reactor. In the presence of the catalyst and ozone, oxalates are broken down to carbon dioxide and all carbonates and carbon dioxide are sparged at pH near 1. Carbon dioxide gas is carried by the carrier gas though a cooler and measured using a nondispersive infrared (NDIR) sensor. The result is displayed as TC (total carbon). The TC result represents the sum of TOC (which includes any volatile organics) and TIC (total inorganic carbon) present in the sample. The IC in the sample is sparged in the form of carbon dioxide gas by the addition of the acid reagent and measured by an NDIR sensor. The result is displayed as TIC. TOC result, including the volatiles, is then calculated from the difference between the TC and TIC measurements, as TOC = TC − TIC. Figure 1.67a illustrates a Two-Stage Advanced Oxidation analyzer using an NDIR sensor.

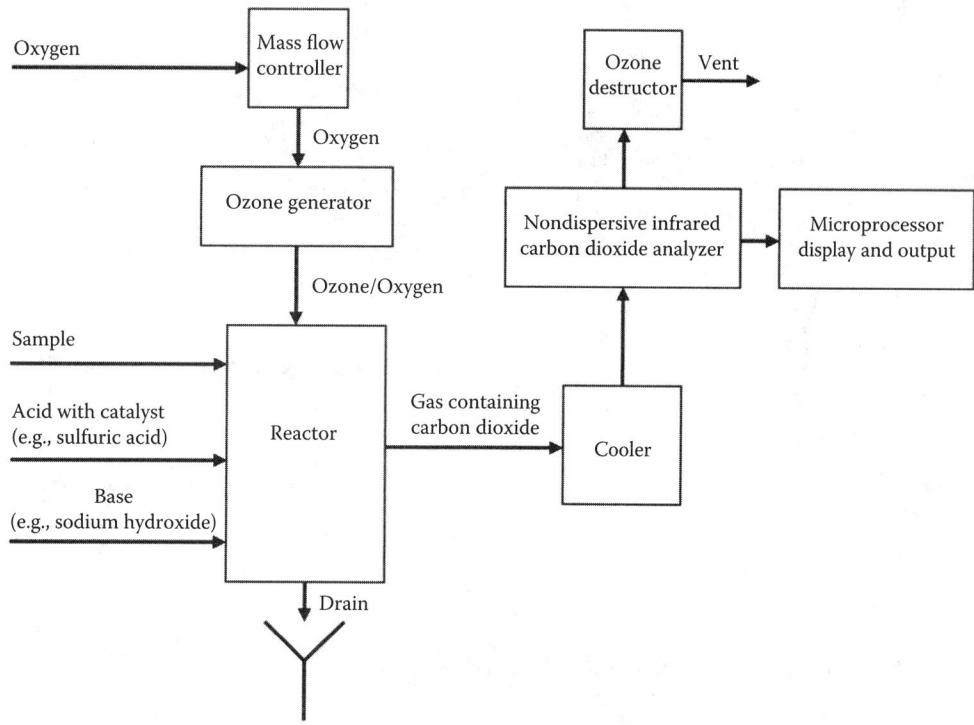

FIG. 1.67a
Catalyzed ozone hydroxyl radical oxidation. (Two stage advanced oxidation) analyzer with NDIR sensor. (Courtesy of BioTector Analytical System Ltd.)

---

* *Annual Book of ASTM Standards*, American Society for Testing and Materials, D5173, Philadelphia, PA, February 15, 2015.

The Catalyzed Ozone Hydroxyl Radical Oxidation analyzers can be configured to measure TOC as TC-TIC, TIC, and TOC (NPOC), TC and VOC without any mechanical changes in the system. The analyzer is supplied typically with three TOC analysis ranges and has automatic range change capability where the analyzer selects the optimum operating range depending on the TOC concentration in the sample.

The Two-Stage Advanced Oxidation technology using catalyzed ozone hydroxyl radical oxidation has proven its ability to handle samples with high concentration of salts in the industry, together with the fact that it can handle samples containing high levels of fats, oils, greases, and particulates. It has gained recognition as an extremely reliable technique for in-line TOC monitoring. The Two-Stage Advanced Oxidation method is also available for the analysis of Total Nitrogen (TN) and Total Phosphorous (TP), providing the same unrivaled reliability in a wide range of wastewater and industrial process water applications including both dirty and clean applications, such as the municipal/wastewater treatment plants, refineries, food, pharmaceutical, chemical industries, airport deicing, pulp and paper plants, and many other customized applications.

High-Temperature Combustion The original and the expanded high-temperature methods*,† both contained a high-temperature furnace and a combustion tube to catalytically oxidize all carbonaceous species (TC) into carbon dioxide (CO_2). Figure 1.67b illustrates a high-temperature combustion analyzer using an NDIR sensor. Sample is delivered into the furnace, which contains a combustion tube. The temperature within the combustion tube is maintained at about

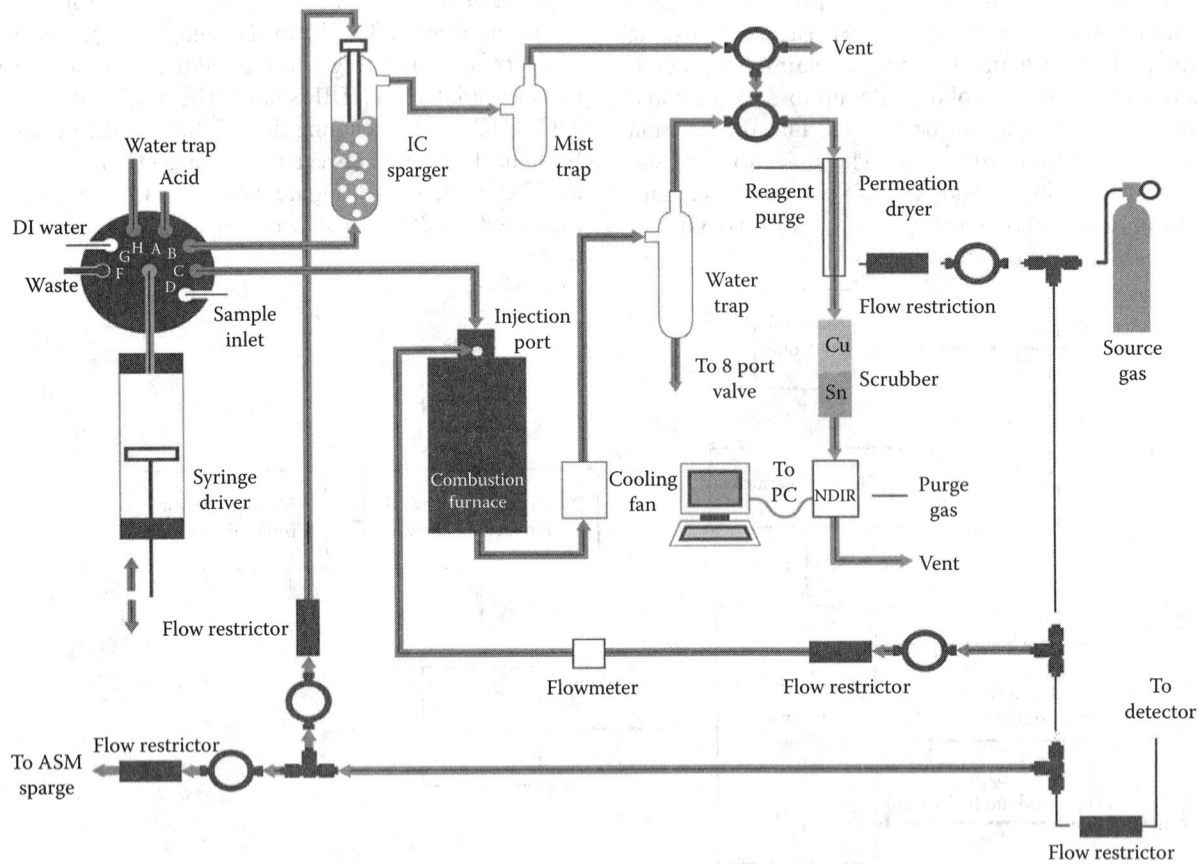

FIG. 1.67b
Catalytic combustion analyzer with NDIR sensor. (Courtesy of Tekmar-Dohrmann.)

---

* Van Hall, C.E., Saranco, J., and Stenger, V.A., Rapid combustion method for the determination of organic substances in aqueous solutions, *Anal. Chem.*, 35(3), 315–318, 1963.
† Van Hall, C.E. and Stenger, V.A., An instrumental method for rapid determination of carbonate and total carbon in solutions, *Anal. Chem.*, 39(4), 503–507, 1967.

680°C–1000°C. Catalysts such as platinum or cobalt oxide, which are placed on a suitable substrate, promote the conversion of carbon in the sample to CO_2. The stream of gas passes through a halide scrubber and then enters the NDIR detector, where CO_2 is measured.

The high-temperature combustion method has found some utility in monitoring wastewaters, salty waters,[*] and particulate matrices.[†] However, as discussed in Standard Method 5310B, the formation and fusion of dissolved salts in the combustion furnace and thus requirement for frequent maintenance and furnace replacement is a well-known phenomenon, which is the core disadvantage of the combustion oxidation technology. The next challenge for the high-temperature techniques is the magnitude and variability of the instrument blank.[‡] Instrument manufacturers have developed better catalysts and improved their methods to minimize their effects. They have also lowered the detection limit, making these analyzers suitable for drinking water and some pharmaceutical applications.

Wet Chemical Oxidation Wet chemical methods have the advantage of better sensitivity, because they analyze a large volume of sample (up to 20 mL) and thereby increase the intensity of the resulting NDIR signal for a given concentration.[§] In this analysis, the sample is also delivered to an oxidation chamber, where the addition of a persulfate reagent and acid in the presence of UV light or heat (~100°C) results in the formation of mainly sulfate radicals with minute amount of hydroxyl radicals that converts carbon to CO_2. The ratio of generated sulfate radicals to hydroxyl radicals is typically 99:1 in UV/Heated/Persulfate systems. The CO_2 produced is continuously sparged and is transported to the detector by the carrier gas.

The main disadvantages of the UV/Heated/Persulfate systems is that large organic particulates and very large organic molecules such as tannins, lignins, and humic acid are not oxidized fully or oxidized very slowly, because persulfate oxidation is rate limited. Highly turbid samples and turbidity is a major interference factor as the intensity of the UV light reaching the sample matrix is reduced resulting in sluggish or incomplete oxidation. The oxidation of the organic matter is inhibited by chloride concentrations above 0.05% (500 ppm).

For trace amounts of carbon (less than 1 mg/L), the UV light by itself is sufficient for converting the organic carbon to CO_2. Ultraviolet light, with or without the persulfate reagent, has been widely used in low-level TOC applications in the semiconductor and the pharmaceutical industries. Other wet chemical methods include photocatalytic oxidation, which makes use of a 400 nm near-UV energy source, water, and a titanium dioxide slurry, which are used to produce the hydroxyl radicals.

Inorganic Carbon Introducing an aliquot of the sample into an inorganic carbon (IC) chamber and adding some acid results in the IC being converted to CO_2, this can then be measured with an NDIR detector. When the IC concentration is very large as compared with the TOC, such as in drinking waters, the difference method may not be as accurate as the acid sparge technique. For such applications, the IC must be manually or automatically acidified and sparged from the sample before TOC analysis. This method of analysis is available for both laboratory and online applications.

Automatic On-Line Design Figure 1.67c illustrates the catalytic-oxidation-type TOC analyzer, which has been adopted for continuous online operation. This design incorporates an acid injection system that converts the IC into carbon dioxide. The generated CO_2 is removed in a sparging chamber before the sample reaches the analyzer. If the sparging portion of the sampling system is left off, the analyzer becomes a TC analyzer.

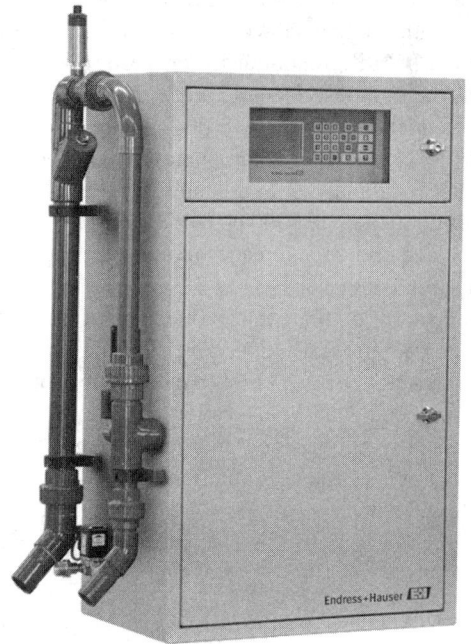

FIG. 1.67c
TOC analyzer using high temperature thermal reactor. (Courtesy Endress+Hauser AG.)

---

[*] Spyres, G., Nimmo, M., Worsfold, P.J., Achterberg, E.P., and Miller, A.E., Determination of dissolved organic carbon in seawater using high temperature catalytic oxidation techniques, *Trends Anal. Chem.*, 19(8), 498–506, 2000.

[†] Lee-Alvarez, M., Total organic carbon analysis of particulated samples, *Tekmar Dohrmann Appl. Note*, 9(22), 1999.

[‡] Urbansky, E.T., Total organic carbon analyzers as tools for measuring carbonaceous matter in natural waters, *J. Environ. Monit.*, 3, 102–112, 2001.

[§] Wallace, B., Purcell, M., and Furlong, J., Total organic carbon analysis as a precursor to disinfection byproducts in potable water: Oxidation technique considerations, *J. Environ. Monit.*, 4, 35–42, 2002.

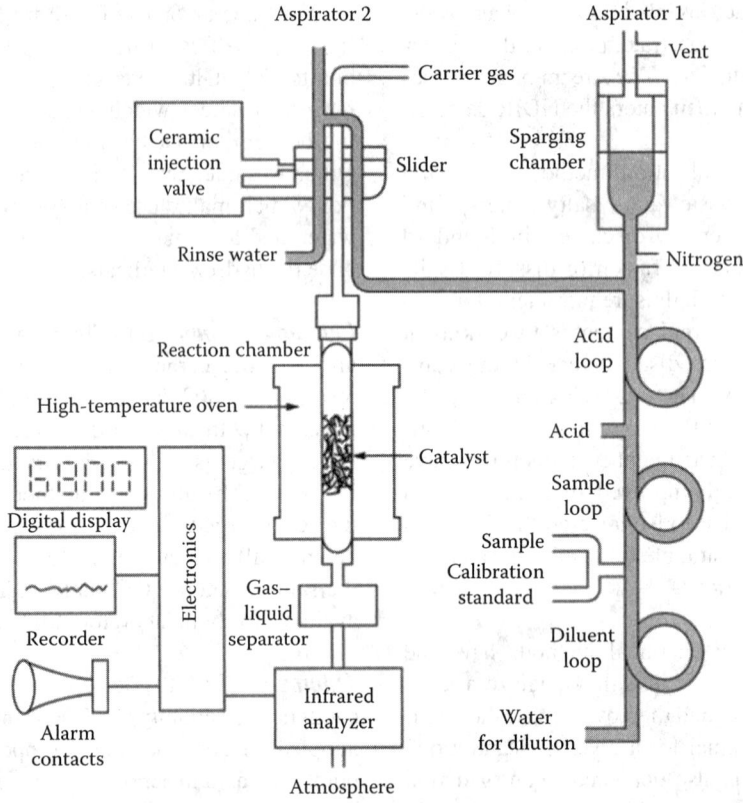

FIG. 1.67d
Online, nondispersive infrared-type total organic carbon (TOC) analyzer.

If a second, low-temperature reaction chamber is added in parallel with the one shown in Figure 1.67d, the analyzer becomes a TOC differential analyzer, performing the same manner as the unit described in Figure 1.67c. The cycle time of this instrument is 2.5 min in the TC mode of operation and 6 min in the TOC mode. It can operate unattended indoors or outdoors in an analyzer house, and it can handle samples with solids content up to 1000 ppm and with particle sizes up to 200 μm, because an automatic water rinse is applied after each measurement cycle.

The accuracy of the sample size is guaranteed by the design of the ceramic injection valve. The calibration of the instrument is automatically checked every time a known standard is introduced. During autocalibration, the analyzer runs three consecutive calibration standards, averages the results, and adjusts the instrument calibration to within preset limits or activates an alarm.

The analyzer is also provided with dilution and automatic range change capability for conditions when the concentration exceeds the operating range. The carrier gas (nitrogen) consumption is 100 cc/min at 50 psig (3.5 barg), and the acid consumption in the TOC mode is about 1.0 gal (3.8 L) per month.

Aqueous Conductivity*

This method of analysis employs the ultraviolet-persulfate oxidation of organic carbon coupled with the measurement of TOC using conductivity. Most systems utilizing the conductivity detection method can be used for both laboratory and in-line applications.

---

* *Annual Book of ASTM Standards*, American Society for Testing and Materials, D5997-96 and D4519, Philadelphia, PA, February 15, 2015.

Samples are introduced using an autosampler or through online sample tubing. Figure 1.67e illustrates a high-sensitivity TOC analyzer used in semiconductor and pharmaceutical applications, both in the laboratory and online. The TOC analysis is performed by conversion of IC to CO_2 after addition of an acid. The CO_2 formed may be removed from the system by degassing, or it can stay in the stream and be measured as IC. The flow stream is divided into two channels for the separate measurement of TC/TOC and IC.

In the TC/TOC channel, the carbon in the sample is oxidized to CO_2 by the action of the persulfate reagent and the UV lamp. The CO_2 in the liquid stream passes through a semipermeable membrane and then dissolves in high-purity water. A conductivity

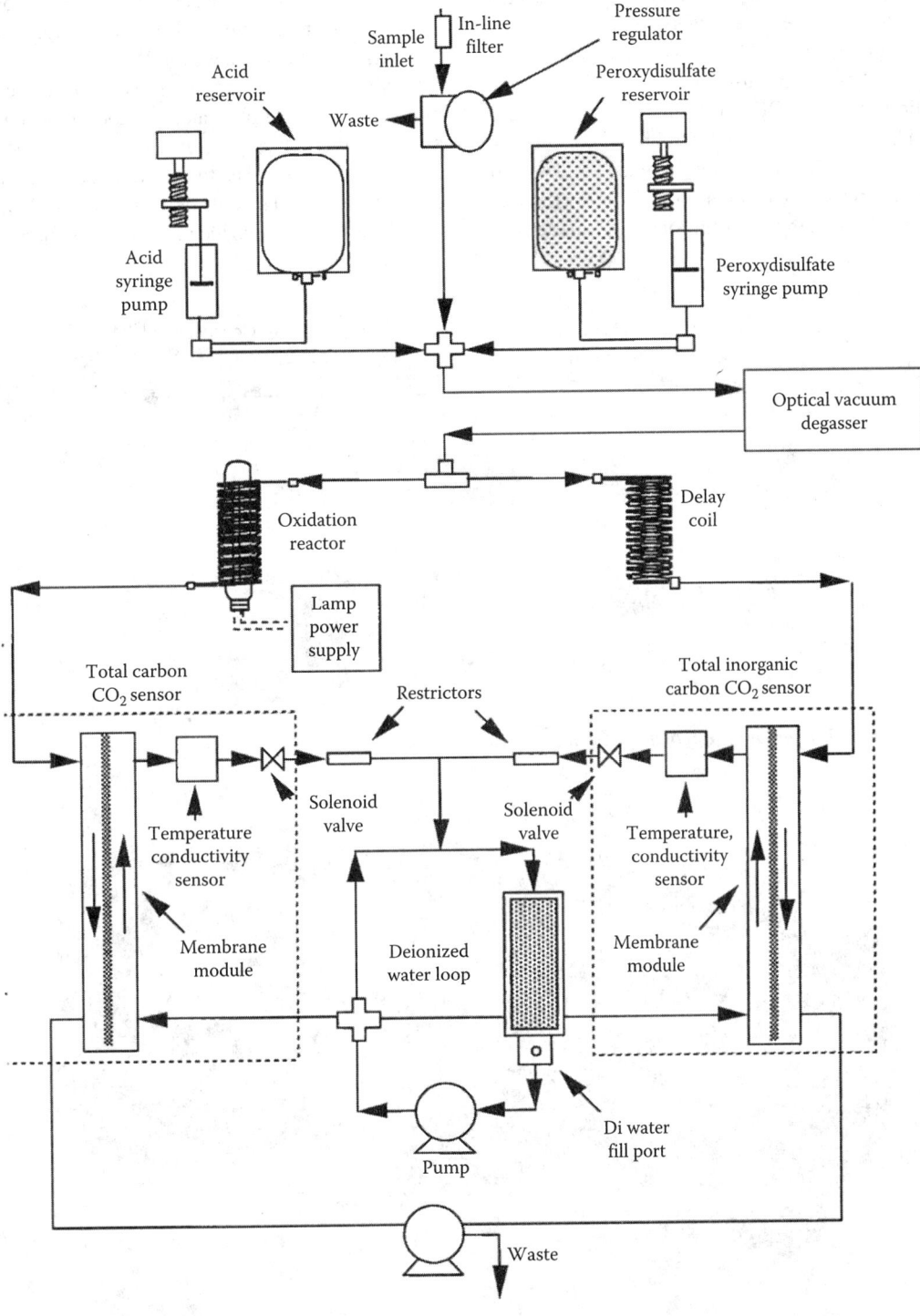

FIG. 1.67e
Online conductivity TOC analyzer. (Courtesy of GE Analytical Instruments.)

cell located in the stream measures the increase in conductivity caused by the CO_2 and relates this to the calibration data.

A chemometric equation is used to determine the relationship between equilibrium concentrations of H^+, CO_2, HCO_3^-, and OH^- and conductivity. The temperature of the conductivity cell is monitored so that readings can be adjusted to compensate for changes in temperature. Continuous pumping through a mixed bed of ion-exchange resin purifies the water on the conductivity side of the membrane. This method has the advantage of being able to measure very low concentrations of carbon while using relatively small volumes of sample.

This analyzer has found use in semiconductor and power applications where TOC levels of less than 1 µg/L are required. The semipermeable hydrophobic membrane protects against and minimizes the interference from higher concentrations of chloride and other ions. This method of analysis is limited to cases in which carbonaceous matter can be introduced into the reaction zone, because the inlet system limits the size of the particles that can be introduced. In addition, the wet oxidation technique has very low recovery if the samples contain particulate materials.

Online Conductivity Another method that has found use specifically in online semiconductor reclaiming applications involves differential conductivity measurements, before and after oxidation of the water stream, as it moves through a quartz coil— refer Figures 1.67f and 1.67g.

This instrument allows for TOC updates once per second. A parallel stream goes through a stopped-flow cell, which captures a 0.3 mL sample-water aliquot. Total organic carbon is measured through a dynamic end-point detection method. The stopped-flow cell serves to calibrate the continuous cell every 2–3 min. A typical sample flow rate of 30–300 mL/min can be used. The analyzer also includes an alarm for high TOC values.

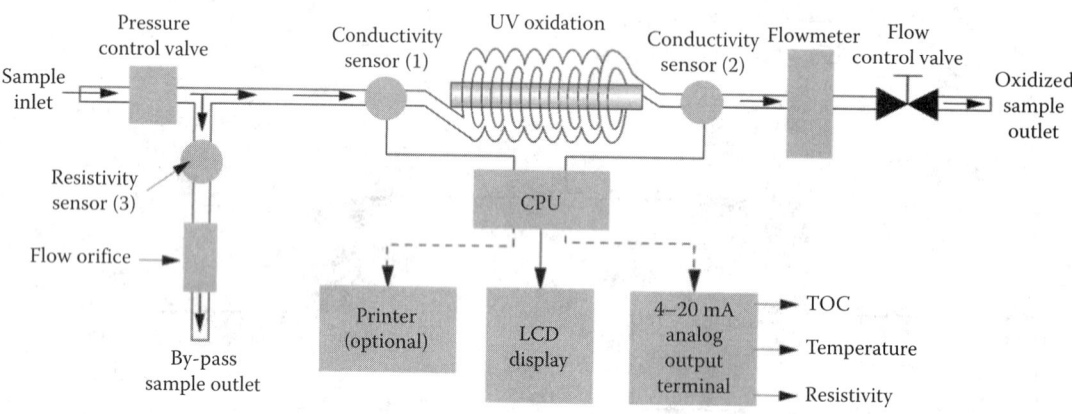

FIG. 1.67f
TOC measurement using differential conductivity. (Courtesy Mettler Toledo.)

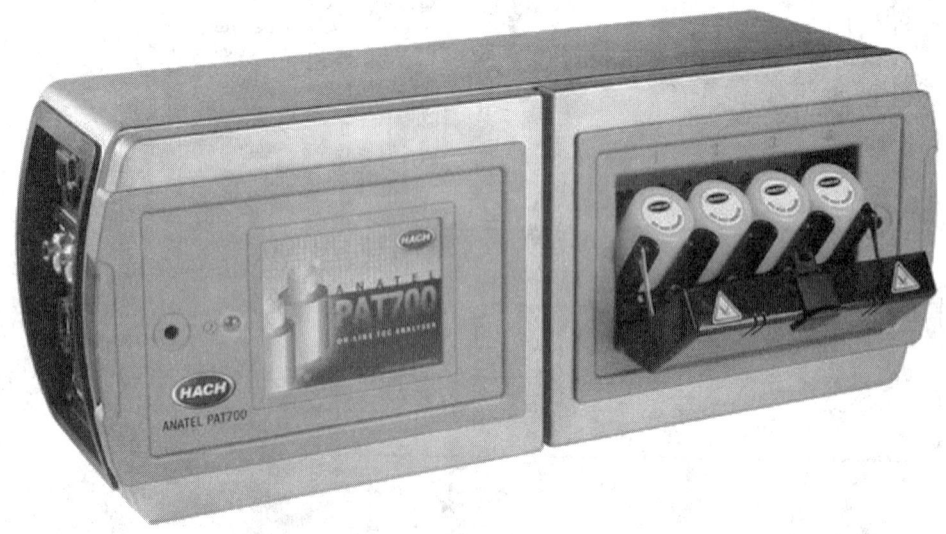

FIG. 1.67g
TOC analyzer using UV light for oxidation with conductivity detector. (Courtesy Hach.)

Coulometric Analysis

In coulometric detectors, the carbon dioxide is indirectly titrated. In this analyzer, the carrier gas with the CO_2 reaches the electrochemical cell containing ethanolamine and an acid–base indicator dye. As the CO_2 reacts with the ethanolamine, a strong titratable acid is formed, and the strong blue color of the solution fades. This change is detected by a photometer.

At the cathode, an equivalent amount of hydroxide is electrochemically generated that neutralizes the acid that has formed and returns the cell to its original color. The current required to produce the hydroxide ions is directly proportional to the concentration of the carbon dioxide from the sample.

This detector neither needs calibration standards nor is affected by water from the sample. It also gives a linear signal in its entire measuring range.

Colorimetric Analysis

This method provides an inexpensive TOC monitoring method for optimizing treatment processes. However, the results cannot be used to report the TOC values, which are required by the D/DBP rule for drinking water.

In this analyzer, the sample is first acidified and then sparged of TIC. In the outside vial, the organic carbon reacts with the persulfate and acid. The vial is placed for 2 hr in a COD reactor at 40.5°C (105°F) to allow for digestion of the organic compounds. The CO_2 generated diffuses through the inner ampule containing a pH indicator reagent. Carbonic acid is formed, which changes the pH of the solution. The degree of color change associated with the change in pH of the solution is related to the amount of organic carbon present in the sample.

Flame Ionization Detector

Flame ionization detector (FID) analyzers are no longer manufactured for TOC analysis. However, they are still in use for VOC analysis.

In the FID-type analyzer (Figure 1.67h), a small acidified sample is transported in the presence of an oxidizer through a heated vaporization zone. Here, the organic carbon in the form of CO_2, plus any volatile organic carbon (VOC), is driven off. The residual sample is sent through a pyrolysis zone to convert the remaining TOC to CO_2. The CO_2 subsequently is converted to methane in a nickel-reduction step, and the resulting methane is measured by an FID detector. The VOC is separated from the CO_2 in a bypass column, reduced to methane, and routed to the same FID for an additional VOC analysis to be added to the dissolved organic carbon value.

Another method that also employs the FID to directly analyze the VOC after TIC (CO_2) as removed is shown in Figure 1.67i. In this method, the catalytic oxidation combustion

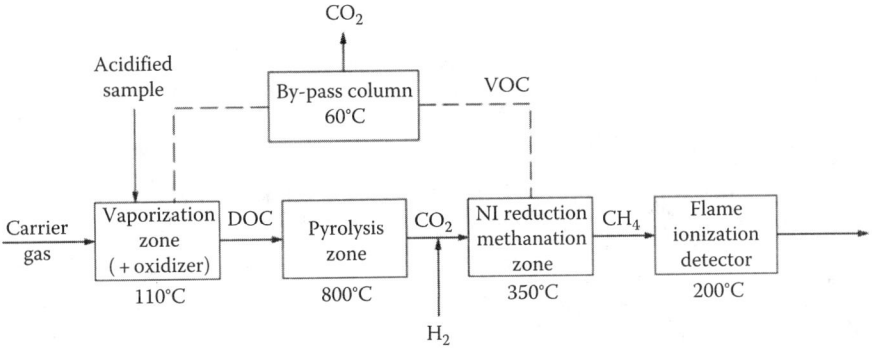

FIG. 1.67h
Flame ionization detector.

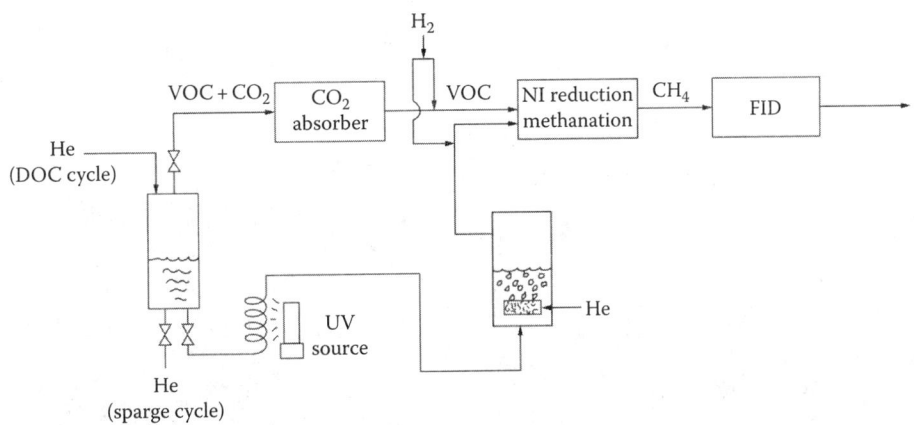

FIG. 1.67i
FID analyzer with wet oxidation.

is replaced by a wet oxidation step. Persulfate is added to the sample, and the solution is then exposed to ultraviolet (UV) radiation to enhance the efficiency of oxidation. The resulting CO_2 is sparged and converted in the nickel-reduction methanator, and its concentration is measured in the FID analyzer.

CONCLUSION

There are TOC sensors and TOC analyzers available in the market. TOC sensors are small, portable, fast, and less expensive than analyzers. There are battery operated TOC measuring device available from GE. TOC sensors are less accurate than TOC analyzers. If the intended TOC device use is for regulatory reporting, managing an important process control variable, real-time release, or other critical-to-quality product attributes, accuracy is essential and a TOC analyzer is appropriate. On the other hand, if the intended use is for general TOC monitoring—not for making critical quality decisions—then, other characteristics may be more important than accuracy and a TOC sensor may be appropriate. Sensors are typically used to *monitor* a process while analyzers are more suited to manage a process. Data from sensors are used for information only, and data from analyzers are used to make critical quality decisions.

TOC can be measured both online and offline. Off-line measurements (lab methods) are typically used for high concentrations (>1 ppm). Online measurements are typically used for sub-ppm (<1000 ppb) detection and quicker response than the lab methods. Most of the industries Thornton services use online measurements for the faster response, which is required for process control. Laboratory TOC analyzers are the least susceptible to variations in water sample conditions, but they require frequent maintenance requirement.

The TOC analyzers based on catalyzed Ozone Hydroxyl Radical Oxidation have reported to be highly successful due to the larger diameter internal sample tubes. These types of analyzers need maintenance once in 6 months. The TOC analyzer with UV/persulfate-based method also reported to be successful in special applications.

SPECIFICATION FORMS

When specifying total carbon or total organic carbon analyzer, the ISA Form 20A1001 can be used to specify the features required for the device itself and when specifying both the device and the nature of the application, ISA Form 20A1002 can be used. These forms, with the permission of the International Society of Automation, are reproduced on the next pages.

1.67 Total Carbon and Total Organic Carbon (TOC) Analyzers

1	RESPONSIBLE ORGANIZATION		ANALYSIS DEVICE	6	SPECIFICATION IDENTIFICATIONS		
2				7	Document no		
3	ISA		Operating Parameters	8	Latest revision	Date	
4				9	Issue status		
5				10			

11	ADMINISTRATIVE IDENTIFICATIONS			40	SERVICE IDENTIFICATIONS Continued		
12	Project number		Sub project no	41	Return conn matl type		
13	Project			42	Inline hazardous area cl	Div/Zon	Group
14	Enterprise			43	Inline area min ign temp	Temp ident number	
15	Site			44	Remote hazardous area cl	Div/Zon	Group
16	Area	Cell	Unit	45	Remote area min ign temp	Temp ident number	
17				46			
18	SERVICE IDENTIFICATIONS			47			
19	Tag no/Functional ident			48	COMPONENT DESIGN CRITERIA		
20	Related equipment			49	Component type		
21	Service			50	Component style		
22				51	Output signal type		
23	P&ID/Reference dwg			52	Characteristic curve		
24	Process line/nozzle no			53	Compensation style		
25	Process conn pipe spec			54	Type of protection		
26	Process conn nominal size	Rating		55	Criticality code		
27	Process conn termn type	Style		56	Max EMI susceptibility	Ref	
28	Process conn schedule no	Wall thickness		57	Max temperature effect	Ref	
29	Process connection length			58	Max sample time lag		
30	Process line matl type			59	Max response time		
31	Fast loop line number			60	Min required accuracy	Ref	
32	Fast loop pipe spec			61	Avail nom power supply		Number wires
33	Fast loop conn nom size	Rating		62	Calibration method		
34	Fast loop conn termn type	Style		63	Testing/Listing agency		
35	Fast loop schedule no	Wall thickness		64	Test requirements		
36	Fast loop estimated lg			65	Supply loss failure mode		
37	Fast loop material type			66	Signal loss failure mode		
38	Return conn nominal size	Rating		67			
39	Return conn termn type	Style		68			

69	PROCESS VARIABLES	MATERIAL FLOW CONDITIONS				101	PROCESS DESIGN CONDITIONS		
70	Flow Case Identification				Units	102	Minimum	Maximum	Units
71	Process pressure					103			
72	Process temperature					104			
73	Process phase type					105			
74	Process liquid actl flow					106			
75	Process vapor actl flow					107			
76	Process vapor std flow					108			
77	Process liquid density					109			
78	Process vapor density					110			
79	Process liquid viscosity					111			
80	Sample return pressure					112			
81	Sample vent/drain press					113			
82	Sample temperature					114			
83	Sample phase type					115			
84	Fast loop liq actl flow					116			
85	Fast loop vapor actl flow					117			
86	Fast loop vapor std flow					118			
87	Fast loop vapor density					119			
88	Conductivity/Resistivity					120			
89	pH/ORP					121			
90	RH/Dewpoint					122			
91	Turbidity/Opacity					123			
92	Dissolved oxygen					124			
93	Corrosivity					125			
94	Particle size					126			
95						127			
96	CALCULATED VARIABLES					128			
97	Sample lag time					129			
98	Process fluid velocity					130			
99	Wake/natural freq ratio					131			
100						132			

133	MATERIAL PROPERTIES					137	MATERIAL PROPERTIES Continued		
134	Name					138	NFPA health hazard	Flammability	Reactivity
135	Density at ref temp			At		139			
136						140			

Rev	Date	Revision Description	By	Appv1	Appv2	Appv3	REMARKS

Form: 20A1001 Rev 0

© 2004 ISA

1	RESPONSIBLE ORGANIZATION		ANALYSIS DEVICE COMPOSITION OR PROPERTY Operating Parameters (Continued)	6	SPECIFICATION IDENTIFICATIONS		
2				7	Document no		
3				8	Latest revision	Date	
4				9	Issue status		
5				10			

	PROCESS COMPOSITION OR PROPERTY			MEASUREMENT DESIGN CONDITIONS				
	Component/Property Name	Normal	Units	Minimum	Units	Maximum	Units	Repeatability
11								
12								
13								
14								
15								
16								
17								
18								
19								
20								
21								
22								
23								
24								
25								
26								
27								
28								
29								
30								
31								
32								
33								
34								
35								
36								
37								

(Rows 38–75: blank)

Rev	Date	Revision Description	By	Appv1	Appv2	Appv3	REMARKS

Form: 20A1002 Rev 0

© 2004 ISA

Abbreviations

BDD	Boron Doped Diamond
BOD	Biochemical Oxygen Demand
COD	Chemical Oxygen Demand
CO_2	Carbon Dioxide
D/DBP	Disinfectants/Disinfection Byproducts
FID	Flame Ionization Detection
ISO	International Organization for Standardization
NDIR	NonDispersive Infrared
NPOC	Non Purgeable Organic Carbon
SSNDIR	Solid State NonDispersive Infrared
TC	Total Carbon
TIC	Total Inorganic Carbon
TN	Total Nitrogen
TOC	Total Organic Carbon
TP	Total Phosphates
VOC	Volatile Organic Carbon

Organizations

AOAC	Association of Analytical Communities
ASTM	American Society for Testing and Materials
DIN	Deutsches Institut für Normung e.V
EPA	Environmental Protection Agency
USP	U.S. Pharmacopeial

Bibliography

Adams, V.D., *Water and Wastewater Examination Manual*, Lewis Publishers, Boca Raton, FL, 1989.

Annual Book of ASTM Standards, Part 31, American Society for Testing and Materials, Philadelphia, PA, pp. 467–470, D 2579, 1974.

D'Alessandro, P.L. and Characklis, W.G., Simple measurement technique for soluble BOD progression, *Water Sewage Works*, 119(9), 106107, September 1972.

Davis, E.M., BOD vs. COD vs. TOC vs. TOD, *Water Wastes Eng.*, 32–38, February 1971.

Dawson, R., *Data for Biochemical Research*, Oxford University Press, Oxford, U.K., p. 568, 1990.

Furlong, J., Booth, B., and Wallace, B., Selection of TOC analayzer-Analytical considerations, (2000) http://www.teledynetekmar.com/resources/documents/TOC/TOC_AP-010.pdf.

Geisler, C., Andrews, J.F., and Schierjott, G., New COD analysis arrives, *Water Wastes Eng.*, 11, 26–29, April 1974.

Handbook for Monitoring Industrial Wastewater, U.S. Environmental Protection Agency, Cincinnati, OH, Technology Transfer, pp. 5–7 to 5–11, August 1973.

Helms, J.W., *Rapid measurement of organic pollution by total organic carbon and comparison with other techniques*, U.S. Department of Interior, Geological Survey, Water Resources Division, Openfile Report, Menlo Park, CA, May 21, 1970.

Horan, M., A discussion on significance of complete oxidation of all oxidizable compounds present in liquid sample where target component to be analyzed requires oxidation prior to measurement, *54th ISA AD Symposium*, Houston, TX, 2009.

Horan, M., On-line TOC Analysis by two stage advanced oxidation process for applications containing high levels of salt, *55th ISA AD Symposium*, New Orleans, LA, 2010.

Jenkins, K., Vanderwielen, A., Armstrong, J., Leonard, L., Murphy, G., and Piros, N., Application of total organic carbon analysis to cleaning validation, *J. Pharm. Sci. Technol.*, 50(1), 6–15, 1996.

Jones, R.H. and Degaforde, A.F., Application of a high-sensitivity total organic carbon analyzer, *ISA Trans.*, 7(4), 267–272, 1968.

Kehoe, T.J., Determining TOC in waters, *Environ. Sci. Technol.*, 11(2), 137–139, February 1977.

McNeil, B. and Harvey, L., *Fermentation: A Practical Approach*, IRL Press, New York, 1990.

Ratliff, T.A., *The Laboratory Quality Assurance System*, Van Nostrand Reinhold, New York, 1989.

Roesler, J.F. and Wise, R.H., Variables to be measured in wastewater treatment plant monitoring and control, *J. Water Pollut. Contr. Fed.*, 46(7), 1769–1775, July 1974.

Siepmann, F. and Teutscher, M., Abschlussberichtzum F + E Vorhaben 102-WA 161 Bau und Erprobung eines Mess, BMFT Abt Umweltforschung, 1984.

Tool, H.R., Manometric measurement of the biochemical oxygen demand, *Water Sewage Works*, June 1967.

Van Hall, C.E., Barth, D., and Stenger, V.A., Elimination of carbonates from aqueous solutions prior to organic carbon determination, *Anal. Chem.*, 37(6), 769–771, May 1965.

Williams, R.T., The carbonaceous analyzer as a water pollution research tool, *Proceedings of 21st Annual ISA Conference and Exhibit*, New York, 1966.

1.68 Turbidity, Sludge and Suspended Solids

A. BRODGESELL (1969, 1982)   **B. G. LIPTÁK** (1995, 2003)
R. SREENEVASAN (2017)

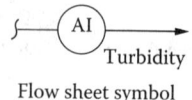

Flow sheet symbol

Types	Laboratory units can be manual or flow-through types; industrial turbidity transmitters in the process industry are available in both probe and flow-through designs
Design pressures	Up to 250 psig (17 barg)
Design temperatures	120°C (250°F) standard; 232°C (450°F) special
Construction materials	Stainless steel, glass, and plastics
Ranges	In ppm silica units from 0–0.5 to 0–1000 Nephelometric, nonratio in NTU units from 0 to 40 Ratio white light turbidity meters in NTRU units from 0 to 10,000 Nephelometric near IR, nonratio in FNU units from 0 to 1,000 Nephelometric near IR, ratio in FNRU units from 0 to 10,000 Surface scatter turbidity meters in NTU units from 0.5 to 10,000 Formazin backscattering in FBU units from 100 to 10,000+ Backscattering in BU units from 100 to 10,000+ Formazin attenuation in FAU units from 20 to 1000 Light Attenuation in AU units from 20 to 1000 Nephelometric, Multibeam in FTMU units from 0.02 to 4000 Laser turbidity in mNTU units from 0.007 to 5 Sludge density probe with reciprocating piston has a range from 0%–0.1% to 0%–10% of suspended solids
Inaccuracy	0.5%–2% of full scale for most designs; 5% of full scale for the reciprocating piston probe type suspended solids sensor, which is used on sludge applications
Costs	Standard reference solutions for calibration cost $90 per bottle; laboratory turbidity meters cost from $675 to $1800; laboratory nephelometers with continuous-flow attachment cost $1750; industrial turbidity transmitters cost from $2250 to $4500, depending on features and materials of construction. Sludge-density-detecting, self-cleaning probe with internal reciprocating piston and indicating transmitter costs $6800; ultrasonic suspended solids transmitters are $4500–$5400.
Partial list of suppliers	For additional manufacturers, refer to Section 6 of Volume 1, on density measurements, and Chapter 1.13, on colorimeters. ABB Process Automation (http://www.abbaustralia.com.au/product/us/9AAC100082.aspx) BTG Instruments (http://www.btg.com/products.asp?langage=1&appli=7&numProd=582&cat=prod) Cole-Parmer (http://www.coleparmer.com/Category/Turbidity_Meters_and_Turbidity_Testing_Products/7360) Emerson Process (http://www2.emersonprocess.com/en-US/brands/rosemountanalytical/Liquid/Systems/T1056/Pages/index.aspx) Endress + Hauser (http://www.au.endress.com/eh/sc/asia/au/en/home.nsf/#products/~liquid-analysis-instrument-turbidity-measurement) Galvanic Applied Sciences Inc. (http://www.galvanic.com/monitek.htm) Hach (http://www.hach.com/turbidityguide) HF Scientific Inc. (http://www.hfscientific.com/products/c/online_turbidity) Markland Specialty Engineering Ltd. (http://www.sludgecontrols.com/our-products/automatic-sludge-level-detector/) McNab (http://www.themcnab.com/Products/HSB/Model_HSB_high_sensitivity_clarity_analyzer.htm) Mettler Toledo (http://au.mt.com/au/en/home/products/Process-Analytics/turbidity-meter.html) Sigrist Process-Photometer (http://www.photometer.com/en/products/bycategory.html?categoryid=4) Turner Designs (http://www.turnerdesigns.com/products/online-and-inline-fluorometers)

INTRODUCTION

In the area of turbidity measurement, there is some overlap between the coverage of this chapter and of Chapter 1.13, which covers colorimeters. Similarly, the discussion of sludge and suspended solids measurements in this chapter does have some overlap with the discussion of the various densitometer designs in Section 6 of Volume 1.

Turbidity meters (similarly to the nephelometers discussed in Chapter 1.54) measure the cloudiness of a fluid by detecting the intensity of transmitted or reflected light. The cloudiness detected by turbidity meters is caused by finely dispersed suspended particles that, when exposed to a visible or infrared (IR) light, will scatter it. The cloudier the process fluid (the higher its turbidity), the more scattering will occur and therefore the less light will be transmitted through a sample. If the sensor photocell is placed at a 90° angle to the light path, the cloudier the process fluid, the more scattered light it will detect.

Light Absorption and Scattering

The turbidity meter detects light scattering, but the colorimeter measures light absorption (Figure 1.68a). If a liquid has no color at all, and the attenuation of light intensity is the result only of the scattering effect of solid particles, a colorimeter (Chapter 1.13) can also be used to detect turbidity. All that is needed is to convert from the standard units of colorimetry (0%–100% transmittance or 0–2 absorbance) into turbidity units.

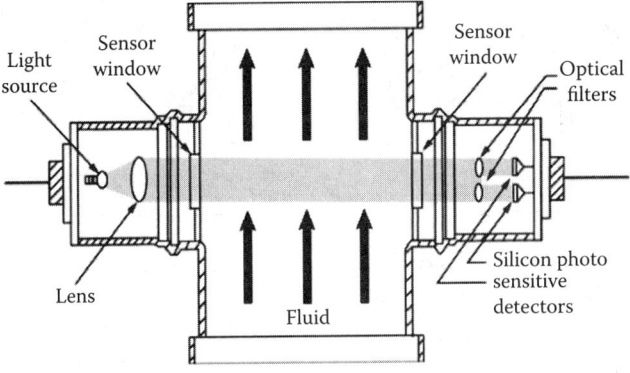

FIG. 1.68a
The design of an in-line color monitor. (Courtesy of Monitek.)

Other analyzers can also be used to detect turbidity. For example, sludge density measurements (Chapter 6.8 in Volume 1) can also be converted into turbidity units. Similarly, consistency readings (Chapter 1.16), which are normally in units of weight percent of solids, can also be expressed in turbidity units. Figure 1.68b illustrates such a consistency meter, which operates with reflected IR light transmitted over optical fibers.

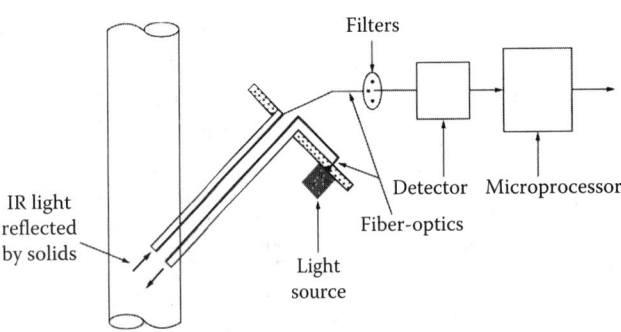

FIG. 1.68b
Probe-type, self-cleaning fiber-optic consistency detector. (Courtesy of KajaaniProcess Measurements LTD, a member of the ABB Group.)

Units of Turbidity

The three main methods of light intensity detection are perpendicular scattering (nephelometry), backscattering, and forward scattering. Different turbidity units have evolved in connection with these different designs.

The Jackson turbidity unit (JTU), for example, is a purely optical scale that correlates well with forward-scattering measurements. The value of one JTU corresponds to the turbidity of a liter of distilled water when 1 mg (1 ppm) of diatomaceous fullers earth (an inert material) is suspended in it.

Nephelometric turbidity units (NTUs) are based on U.S. EPA-approved and stable polymeric suspension standards and correlate well with perpendicular scattering designs. Formazin turbidity units (FTUs) are based on a Formazin polymer standard and also correlate well with perpendicular scattering designs.

Conversion among Turbidity Units The correlation between different turbidity units is given in Table 1.68a. However, because different turbidity instruments respond differently to variations in particle size, color, index of refraction, etc., one cannot obtain accurate correlation between them. These different scales include two Formazin scales in use. The error in turbidity measurements cannot be less than the precision at which the calibrating reference standard solution is available. In the case of Formazin standards, this variation can approach 1%. The NTU reference standards are more stable and last longer than the FTU ones.

The various turbidity units all measure the amount of solid particles in suspension. When parts per million (ppm) units are used, they can refer to the weight of the particular solids in suspension but, because this would require individual calibration using the particular solids, they usually refer instead to ppm of silica (silicon dioxide). Therefore, if the cloudiness (turbidity) of the process sample is the same as the turbidity that would result when 1 mg of silica is mixed in a liter of distilled water, the turbidity reading is called to be 1 ppm on the silica scale.

Interferences in Turbidity Measurements Turbidity measurement is affected by a range of interferences. The most prevalent and common interferences are listed in Table 1.68b. Turbidity interferences will either cause positive or negative bias. Negative biases result in measured values being below the true values and can become more significant as the value increases. Positive interferences are typically associated with

TABLE 1.68a
Conversion between Different Units of Turbidity

Turbidity Unit	Kieselgur Units (SiO_2)	Absolute Units A.E.	Formazin Turbidity Units E.B.C.	Formazin Turbidity Units A.S.B.C.	Jackson Turbidity Units	Helm Units ($BaSO_4$)	Mastic Units	Langrohr Units Reciprocal
Kieselgur units according to German Standards method (1 mg SiO_2 per liter dist. H_2O = 1 ppm)	1	0.000445	0.1	6.9	1	4	8	0.000465
Absolute units A.E. (Zeiss–Pulfrich turbidity unit)	2250	1						
Formazin turbidity units E.B.C.	10	0.00445	1	69	10	40	80	0.00465
Formazin turbidity units A.S.B.C.	0.145	0.000065	0.0145	1	0.145	0.58	1.16	0.0000675
Jackson turbidity units	1	0.000445						
Helm units ($BaSO_4$–Suspension)	0.25	0.00011	0.025	1.72	0.25	1	2	0.000116
Mastic units (1 drop mastic solution per 50 mL dist. H_2O; 50 drops ~1 mL)	0.125	0.000056	0.0125	0.86	0.125	0.5	1	0.0000582
Langrohr units according to German Standards method, reciprocal (cm light path at 25 mm dia.)	2150	0.956						1

Source: Courtesy of Sigrist Process Photometer.

TABLE 1.68b
Typical Interferences Associated with Turbidity Measurement

Interference	Effect on the Measurement
Absorbing particles (colored)	Negative bias (reported measurement is lower than the actual turbidity)
Color in the matrix	Negative bias if the incident light wavelengths overlap the absorptive spectra within the sample matrix
Particle size	Either positive or negative bias (wavelength dependent)
	Large particles scatter long wavelengths of light more readily than will small particles
	Small particles scatter short wavelengths of light more efficiently than long wavelengths
Bubbles	Positive bias and can impact measurement accuracy at all turbidity levels
Sample cell variations	Either positive or negative bias. This can be minimized through the use of matching and indexing techniques and the application of silicone oils to reduce reflections due to scratches. The impact of this interference is most severe at low turbidity values
Stray light	Positive bias (reported measurement is slightly higher than the actual turbidity)
Particle density	Negative bias (reported measurement is lower than the actual turbidity)
Particle settling	Positive or negative bias can result from due to the rapid settling of particles and depending on the length of time to perform a measurement. This is typically associated with grab sample, and laboratory/portable benchtop measurements
Instrument based	Degradation of instrument optical components can have both positive and negative impacts on measurement, but bias is usually negative
Contamination	Positive bias (reported measurement is higher than the actual turbidity)

extremely low measured turbidity values (value <0.1 NTU) and are significant in highly pure waters (e.g., filtered drinking water).

TURBIDITY ANALYZER DESIGNS

Forward Scattering or Transmission Type

Similar to the design of color monitors, in the forward scattering or transmission-type turbidity sensor design (Figure 1.68c), the light source is on one side of the process sample, and the detector is on the other. Therefore, what is detected in this case is the total attenuation. When this attenuation is the result of absorption by color, the unit is called a *colorimeter* (Figure 1.68a); when the attenuation is caused by light being scattered by solid particles, the unit is called a *turbidity meter*.

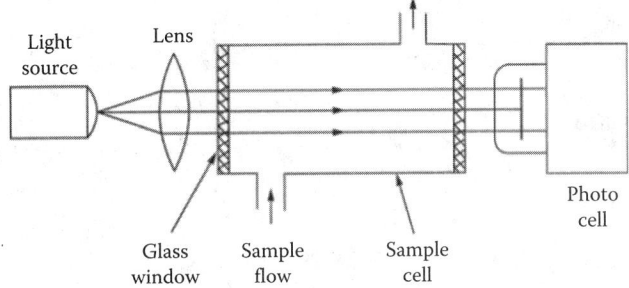

FIG. 1.68c
Schematic of transmission-type turbidity meter.

Dual-Beam Design When both color and solids are present, the total attenuation is the sum of the effects of absorption and scattering. Consequently, the single-beam turbidity analyzer can be used only if there is no color or if the color is constant and its effect can be zeroed out. When the background absorbance or color varies, a dual-beam or split-beam analyzer is needed.

Such a unit is described in Figure 1.68d. It uses two light paths, with one passing through the unfiltered process sample cell and the other through a reference sample cell containing the filtered process fluid. The output of the analyzer is proportional to the difference of optical

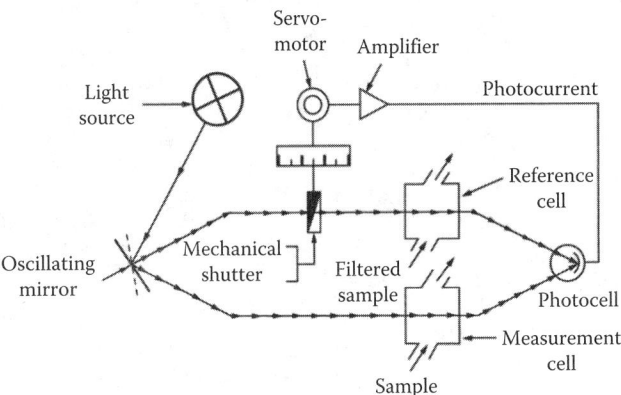

FIG. 1.68d
Oscillating dual-beam, forward-scattering turbidity analyzer. (Courtesy of Sigrist Process Photometer.)

absorbance between the two cells, which corresponds to the solid particles that are present in the sample but absent in the reference cell.

The particular design illustrated in Figure 1.68d has an oscillating mirror, which alternately directs the light beam (alternating 600 times/s) to the measuring and reference cells. The photocell detector converts the intensity differential of the two beams into a photocurrent, which modulates the opening of a mechanical shutter so that the differential between them will be zero. Therefore, the more solid particles present in the sample, the farther the shutter needs to be closed; therefore, the position of the shutter can be read as turbidity.

If, instead of using the filtered sample, the reference cell is filled with other reference materials, this same instrument can be used to measure properties such as color, fluorescence, and so on.

Ratioing This technology involves the use of two or more detectors to determine the turbidity value within a sample. A second technology uses the combination of two incident light sources and two detectors.

Multiple Detection Angles In this design, one primary detector is used, which is typically oriented at a 90° angle relative to the incident light beam (often referred to as the primary nephelometric detector). Other detectors will be at various angles including an attenuated, backscatter, and forward scatter angles. A software algorithm is often used to produce the turbidity

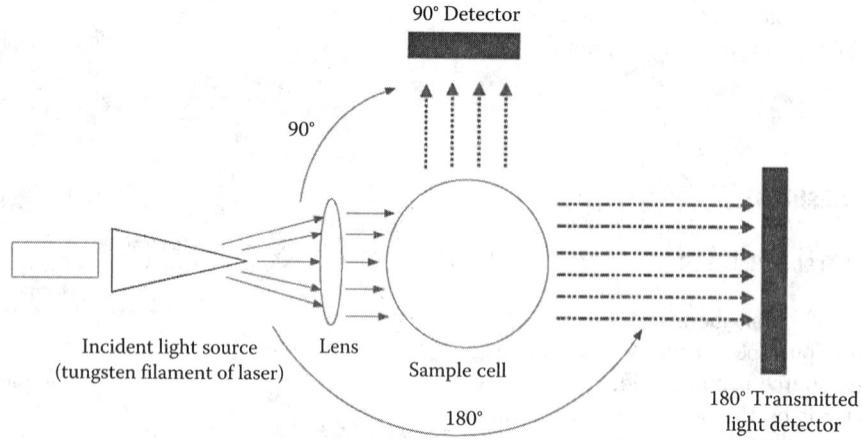

FIG. 1.68e
Geometric arrangement of detectors for ratio measurement.

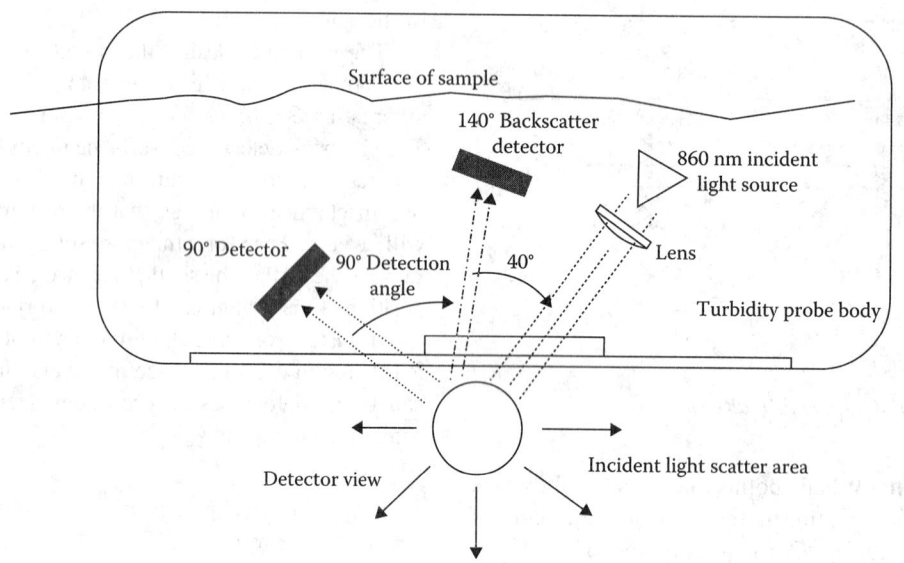

FIG. 1.68f
Ratio technique as applied to in situ measurement.

measurements from the combination of detectors. These detectors can help compensate for color interference and in optical changes such as lamp degradation (Figures 1.68e and 1.68f).

Dual Light Source and Dual Detectors   In this unique approach a combination of light sources that are geometrically oriented at 90° angle to each other are used. The detectors are also oriented at 90° angle to each other and at 90° and 180° angle to each of the light sources. In one phase of measurement, a detector will be the nephelometric (90°) detector and the other detector will be at 180° to the light source that is powered. In the second phase of the measurement, the second light source will be powered and the detector positions from phase one are reversed. A software algorithm is then used to generate turbidity value from different measurement phases. The combination of the two phases provides a turbidity

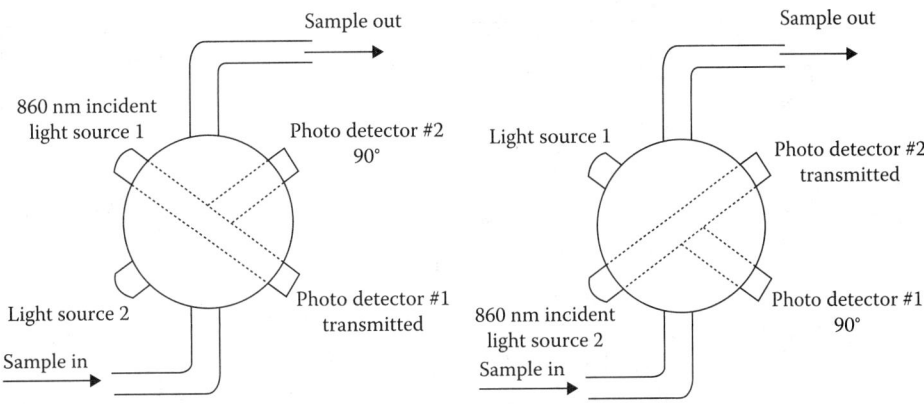

FIG. 1.68g
Dual-light source, dual-detector arrangement for turbidity measurement.

measurement that is corrected for color adsorption, fouling of optics, and any optical changes that can occur (Figure 1.68g).

Laser-Type Meter In the in-line laser-type turbidity meter (Figure 1.68h), a thin ribbon of light is transmitted across the process stream and, after it is attenuated by the process fluid, it falls on detector no. 1. If solids are present in the process fluid, some of the light will be scattered. This scattered light is collected and falls on detector no. 2. The ratio of the two signals from the detectors is an indication of the amount of solids in the process stream (turbidity). Because it is a ratio signal, it is unaffected by light source aging, line voltage variations, or background by light intensity variations.

The laser-type detector is less sensitive to interference by gas bubbles than are other turbidity meters, because the laser-based light ribbon is very thin (about 2 mm). Therefore, when a bubble passes through this ribbon, it causes a pulse that can be filtered out.

In general, in-line turbidity meters are less subject to bubble interference than are the turbidity analyzers, which require sampling. This is because the in-line units do not lower the operating pressure of the stream, so dissolved gases are not encouraged to come out of solution.

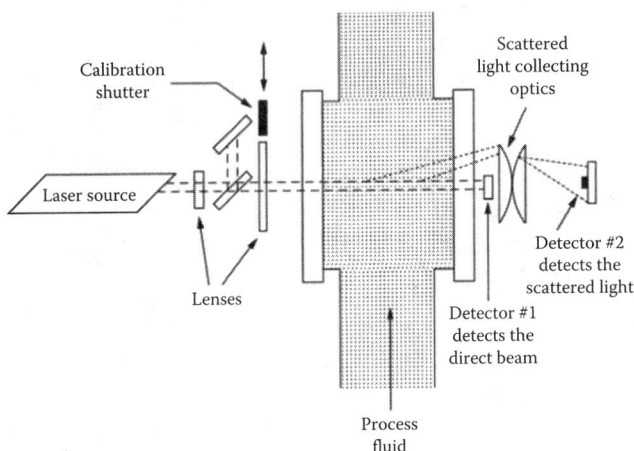

FIG. 1.68h
Laser-type in-line turbidity meter detects separately the total attenuation and the amount of scattering.

Summary of Known Turbidity Instrument Designs ASTM and United States Geological Survey (USGS) have recognized the lack of traceability of turbidity measurements in historical databases. In an attempt to improve data quality and collection, distinct turbidity reporting units were developed that are now based on the instrument terminology. Each technology is traced to a unique turbidity unit. Table 1.68c provides a summary of known turbidity technologies that are available and their respective reporting units.

TABLE 1.68c
Summary of Known Instrument Designs, Applications, Ranges, and Reporting Units

Design and Proposed Reporting Unit	Prominent Application and Major Interference Concerns	Key Design Features	Suggested Application and Operating Ranges
Nephelometric non-ratio (NTU)	White light turbidity meters. These designs comply with U.S. EPA 180.1 for low level turbidity monitoring. Color is major negative interference and optical variations cannot be compensated with this technique.	The incident light source is a tungsten filament lamp that is operated at a color temperature between 2200 and 3000 K. The detector is centered at 90° relative to the incident light beam.	Regulatory for drinking water. The optimal operating range is 0.0–40 units.
Ratio white light turbidity meters (NTRU)	Complies with U.S. EPA Interim Enhanced Surface Water Treatment Rule Regulations and Standard Methods 2130B. Can be used for both low level and high level measurements. Color interference (negative) is reduced and lamp variations are compensated for with this technique.	The incident light source is a tungsten filament lamp that is operated at a color temperature between 2200 and 3000 K. Several detectors are used in the measurement. A primary detector is located at 90° to the incident light beam. Other detectors are located at other angles. An instrument algorithm uses combination of detector readings to generate the turbidity value.	Regulatory for drinking water and wastewater (0–40 units). The technology can potentially measure up to 10,000 units.
Nephelometric near-IR turbidity meters, nonratiometric (FNU)	The instrument design is compliant with ISO 7027. The wavelength is less susceptible to color interferences. The light source is very stable over time because its output can be highly controlled. This technique is applicable for samples with color and for low level monitoring. Only samples that absorb light above 800 nm can result in negative interference.	This technology uses a light source in the near IR range (830–890 nm). The detector is centered at 90° relative to the incident light beam.	Regulatory compliance in Europe for drinking water and wastewater (0–40 units). The technology can potentially measure up to 1000 units.
Nephelometric near IR turbidity meters, ratiometric (FNRU)	Complies with ISO 7027. This technique is applicable for samples with high levels of color and for monitoring to high turbidity levels. Samples that absorb light above 800 nm will result in some negative interference.	This technology uses a light source in the near IR range (830–890 nm). The design uses several detectors in the measurement. A primary detector is located at 90° to the incident light beam. Other detectors are located at other angles. An instrument algorithm uses combination of detector readings to generate the turbidity value.	Regulatory compliance monitoring in Europe for drinking water and wastewater (0–40 units). The technology can potentially measure up to 10,000 units.
Surface scatter turbidity meters (NTU)	Turbidity is determined through light scatter at or near the surface of a sample. Negative color interferences are reduced when compared to the nonratio nephelometric method.	The incident light source is a tungsten filament lamp that is operated at a color temperature between 2200 and 3000 K. The detector is centered at 90° relative to the incident light beam. Both the detector and light source are mounted in a defined position immediately above the sample.	Sample flows through the instrument. This is good watershed monitoring instrument and can measure from 0.5–10,000 units.
Formazin backscatter (FBU)	This technology is not applicable for most regulatory purposes. It is best applied to samples with high turbidity and is commonly used in trending applications. Absorbance and color above 800 nm will result in negative interference.	This design applies a near IR monochromatic light source in the 780–900 nm range as the incident light source. The scattered light detector is positioned at 30 ± 15° relative to the incident light beam.	This technology is best suited for in situ measurement, in which a probe is placed in a sample for continuous monitoring purposes. It is best applied to turbidities in the 100–10,000+ unit range.

(Continued)

TABLE 1.68c (Continued)
Summary of Known Instrument Designs, Applications, Ranges, and Reporting Units

Design and Proposed Reporting Unit	Prominent Application and Major Interference Concerns	Key Design Features	Suggested Application and Operating Ranges
Backscatter unit (BU)	This technology is not applicable for most regulatory purposes. It is best applied to samples with high turbidity. The measurement will be susceptible to any visible color and particle absorbance that will result in a negative interference.	The design applies a white spectral source (400–680 nm range). The detector geometry is at $30° \pm 15°$ relative to the incident light beam.	This technology is best suited for in situ measurements in which sample color is part of turbidity measurement. It is best applied to turbidities in the 100–10,000+ unit range.
Formazin attenuation unit (FAU)	This technology may be applicable for some regulatory applications. The measurement is commonly performed with spectrophotometers. It is best suited for samples with high level turbidity. Particle absorption is a prominent interference.	The incident light beam is at a wavelength 860 ± 30 nm. The detector is geometrically centered at $180°$ relative to incident beam. This is typically an attenuation measurement.	This measurement is part of ISO 7027 regulation. The optimal turbidity measurement range is between 20 and 1000 units.
Light attenuation unit (LAU)	This technology is not applicable for some regulatory applications. This is commonly performed with spectrophotometers. Color and absorption are prominent interferences if their respective absorption spectrum is the same as the output spectrum of the incident light.	The wavelength of the incident light is in the 400–680 nm range. The light scatter detector is geometrically centered at $180°$ relative to incident beam. This is an attenuation measurement.	This is best applied to samples in which color is part of the turbidity measurement. Best for samples with turbidity in the range of 20–1000 units.
Nephelometric turbidity multi-beam unit (FTMU)	This technology is compliant to ISO 7027. It is applicable to regulatory monitoring for drinking water, wastewater, and industrial monitoring applications. This technology is very stable and will be immune to color absorbance below 800 nm. Above 800 nm, color and particle absorbance interferences will be reduced.	The technology consists of two light sources and detectors. The light sources comply with ISO 7027. The detectors are geometrically centered at $90°$ relative to each incident light beam. The instrument measures in two phases in which the detectors are at either $90°$ or $180°$ relative to the incident light beam, depending on the phase. An instrument algorithm uses a combination of detector readings to calculate the reported value.	Regulatory monitoring at low turbidity levels (0.02–40 units). The technology can measure up to 4000 units.
Laser turbidity unit (mNTU)	The application is for the monitoring of filter performance and breakthrough. Color interference can occur, and it absorbs the same wavelength of light that is emitted by the light source. However, color is typically significant in filtered samples.	The technology consists of an incident laser light source at 660 nm and a detector that is a high-sensitivity PMT design. The detector is centered at $90°$ relative to the incident light beam.	Regulatory monitoring of drinking water effluent and membrane systems. Range is 0.007–5 units. Reports in mNTU where 1 NTU = 1000 mNTU.

Suspended Solids and Sludge Density Sensors Figure 1.68i illustrates a probe-type suspended solids detector, widely used in biological sludge applications. The probe contains a reciprocating piston that, during its forward stroke, expels a sample (every 15–40 sec) while wiping clean the optical glass of its internal measurement chamber. After that, it pulls in a fresh sample during its return stroke. This device measures the total attenuation of the light, which, in the case of biological sludge applications, is mostly due to suspended solids. Because of its self-cleaning nature, the need for cleaning and maintenance is minimized.

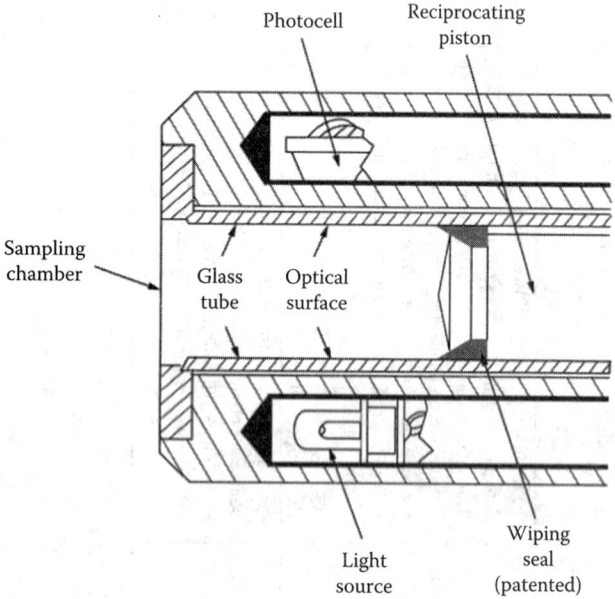

FIG. 1.68i
Probe-type suspended solids detector used on biological sludge applications. (Courtesy of Galvanic Applied Sciences Inc.)

This unit is available with a 4–20 mA transmitter output, which is updated every 15–40 sec and has a full-scale inaccuracy of 5% over the suspended solids ranges between 0%–0.1% and 0%–10%.

Scattered Light Detectors (Nephelometers)

Turbidity instruments utilize a light beam projected into the sample fluid to effect a measurement. The light beam is scattered by solids in suspension, and the degree of light attenuation and the amount of scattered light produced can both be related to turbidity. The light scattering is called the *Tyndall effect*, and the scattered light itself is called the *Tyndall light*.

A constant-candle-power lamp provides a light beam for measurement and, if one or more photosensors are appropriately placed, they can convert the measured light intensity into an electrical signal for readout. Usually, the photosensor is provided with a thermostat-controlled heater that maintains a constant temperature, because this measurement is temperature sensitive.

The supply voltage of the lamp must be kept constant within at least 0.5% to eliminate errors resulting from source intensity variations, because the measured light is referenced to the source. Deposits formed on the flow chamber windows interfere with accurate measurement and, thus, the windows require frequent cleaning or automatic compensation.

Instruments for measuring scattered light vary in design. One type uses a flow chamber similar to the one in Figure 1.68c, except that the window for the measured light is located at 90° to the window for the incident light (Figure 1.68j). One window transmits light beams into the measuring chamber, and the other at right angles to the first transmits scattered light to the photomultiplier sensor.

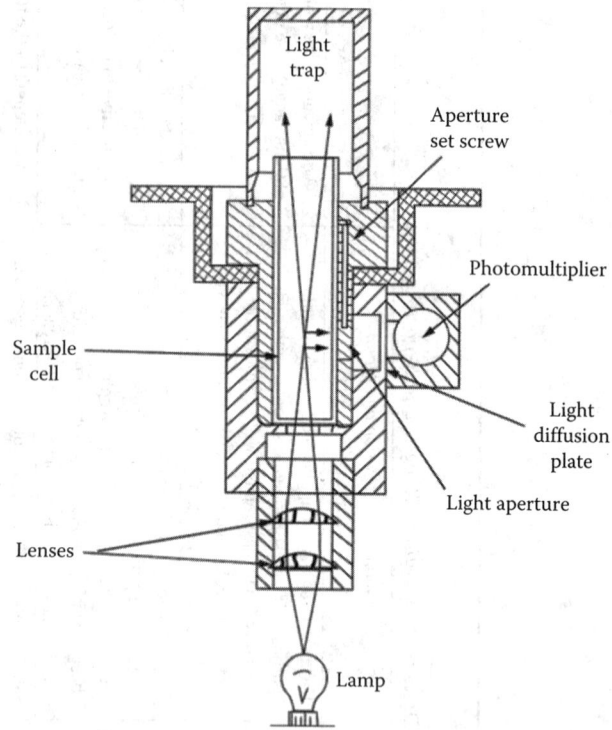

FIG. 1.68j
Light-scattering turbidity meter.

A light trap is located opposite the incident light window to eliminate reflection. With this arrangement, dissolved colors do not affect the measurement; however, instrument sensitivity is decreased if color is present because some light is absorbed by it. Variations of the basic unit include designs provided with two source beams and with two photosensors in conjunction with two pairs of opposed windows as shown in Figure 1.68k.

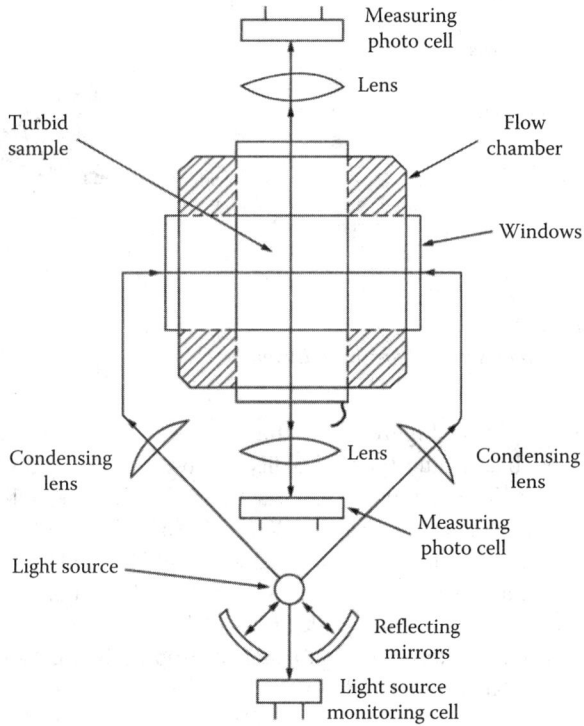

FIG. 1.68k
Turbidity meter with two light beams, two windows, and two measuring photocells.

Some designs utilize a separate photosensor to monitor the lamp output and to adjust the lamp supply voltage through a feedback circuit, which maintains the light intensity at a constant value.

Probe Design For wastewater and biological sludge applications, probe-type turbidity transmitters are preferred. One design (Figure 1.68l) uses an infrared light source and measures the intensity of the 90° scattered light that results.

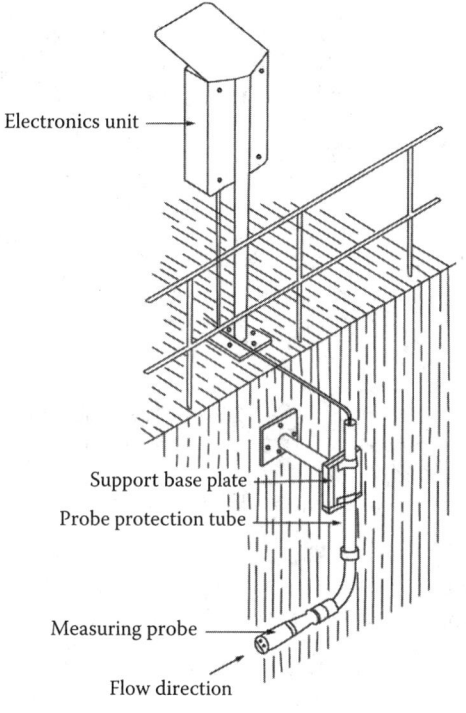

FIG. 1.68l
Automatically cleaned 90° scattered light-detecting turbidity transmitter. (Courtesy of BTG Instruments.)

These units are microprocessor based and are provided with built-in compensation for ambient light variations. They are also provided with wipers for automatic cleaning of the dip or insertion probe. The cleaning frequency is adjustable between 1 and 6 hr and, while the probe is being cleaned by a wiper, the output signal of the transmitter is held constant at its last value.

Backscatter Turbidity Sensors

The turbidity meters that do not have optical windows are not plagued by the problem of coating and deposit buildup on these windows. Figures 1.68m and 1.68n show two such windowless designs.

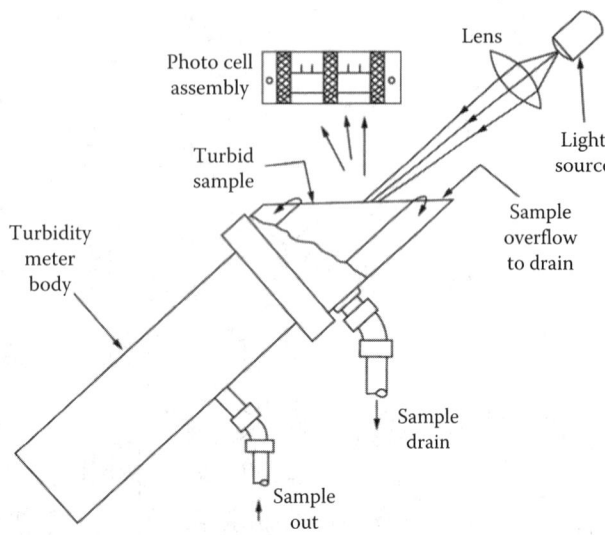

FIG. 1.68m
Surface scatter turbidity meter.

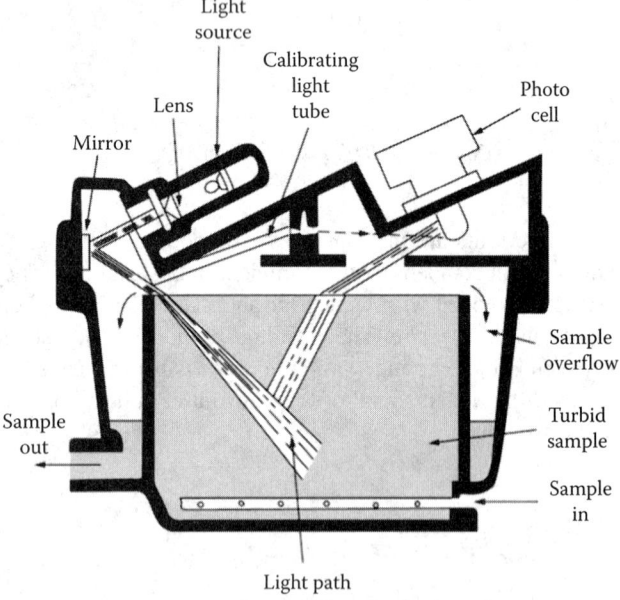

FIG. 1.68n
Atmospheric backscatter-type turbidity meter.

The unit shown in Figure 1.68j measures the surface scatter of a light beam. Particles on the surface of the fluid scatter the light in the direction of the photocell sensor. The design in Figure 1.68k projects the light beam into the sample so that, in this design, all particles in the path of the light beam contribute to the measurement.

The principal disadvantage of these designs is that they must operate at atmospheric pressure and at low sample flow rates (Figure 1.68o).

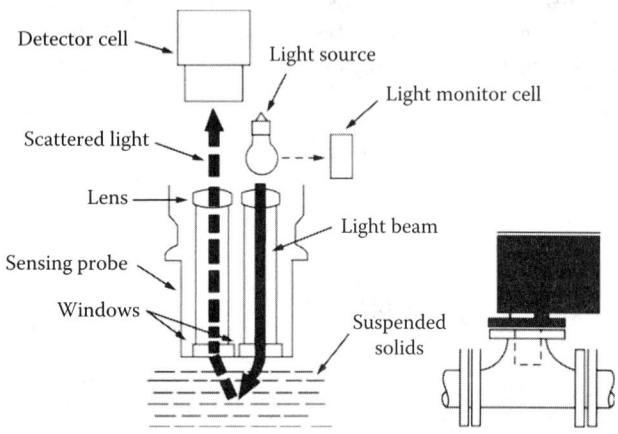

FIG. 1.68o
In-line backscatter-type turbidity analyzer.

In-Line Units Figure 1.68l illustrates the in-line version of the backscatter-type turbidity analyzer, which can be installed in either pipes or vessels. In this design, a 180° backscatter effect is measured. The units are suited for higher temperature applications (up to 232°C, or 450°F) and for high concentrations of solids. Ranges from 10 to 5000 ppm up to 5% to 15% on the silica scale are available.

A backscattering design using fiber-optic light cables is also available (Figure 68.2b).

Density-Based Sensors

If the concentration of suspended solids can be correlated to density, a variety of densitometers can be considered for coating- or sludge-type applications. The common advantage of these designs is that they do not depend on optical sensors and therefore do not require windows, which require cleaning.

A variety of such densitometers are described in some detail in Section 6 of Volume 1, so only a few references will be made here. The designs that are best suited for sludge-type applications are the ones that consist of only open pipes without any restrictions. Of these, Figure 6.6b illustrates the Coriolis type, Figure 6.7f the radiation type, Figures 6.8b through 6.8d the sonic and ultrasonic sludge densitometers, Figure 6.9e the vibrating twin-tube design, and Figure 6.10c the weighed U-tube densitometer.

CONCLUSIONS

Turbidity measurement is fairly simple in theory. The practical problems with these instruments include the problems posed by light source intensity changes, deposits on optical windows, and the presence of dissolved colors in the sample.

Units are now available that automatically correct not only for these effects but also for the variations in the ambient light intensity and even for gas bubbles. Self-cleaning probe designs are also available.

Multiple detection angles/dual light sources with dual detectors are some of the recent improvements, which aim to overcome the measurement limitations imposed by light source intensity changes and/or variations in ambient light intensity.

The other recent innovation (popular in water and wastewater industries) is the communication capabilities embedded in the sensor electronics. A single turbidity controller can now monitor up to a maximum of 8 turbidity sensors.

Additional 24 networked sensors can be monitored by installing three more expansion boards to the controller.

(Hach—http://www.hach.com/sc1000-multi-parameteruniversal-controller-display-module-without-gsm-gprs-/product?id=7640165862.)

The turbidity detector should be selected on the basis of the type of information needed (transmission, 90°, or 180° scatter), the nature and concentration of the solids that are to be detected, and the materials of construction required for the application.

SPECIFICATION FORMS

When specifying turbidity, sludge or suspended solids analyzers, the ISA Forms 20A2351 and 20A2352 can be used. These forms, with the permission of the International Society of Automation, are reproduced on the next pages.

TURBIDITY OR SUSPENDED SOLIDS SENSOR w/wo INTEGRAL TRANSMITTER
Device Specification

#	Field		#	Field		
1	RESPONSIBLE ORGANIZATION		6	SPECIFICATION IDENTIFICATIONS		
2			7	Document no		
3			8	Latest revision	Date	
4			9	Issue status		
5			10			

BODY HOLDER OR FITTING
11	BODY HOLDER OR FITTING				
12	Body/Fitting type				
13	Process conn nominal size		Rating		
14	Process conn termn type		Style		
15	Maximum probe diameter				
16	Body/Fitting material				
17	Seal/O ring material				

SENSING ELEMENT
22	Sensor type
23	Construction style
24	Measurement wavelength
25	Cleaning type
26	Light source style
27	Insertion/Immersion lg
28	Temperature sensor type
29	Measuring window matl

SIGNAL OR INTEGRAL TRANSMITTER
34	Configuration type
35	Output signal type
36	Enclosure type no/class
37	Digital communication std
38	Signal power source
39	Integral indicator style
40	Signal termination type
41	Cert/Approval type
42	Mounting location/type
43	Calibration function
44	Enclosure material

LEAD WIRE AND EXTENSION
48	Extension type
49	Cable length
50	Signal termination type
51	Cable jacket material

INSERTION ASSEMBLY
55	Assembly type			
56	Isolation valve style			
57	Process conn nominal size		Rating	
58	Process conn termn type		Style	
59	Insertion/Immersion lg			

INSERTION ASSEMBLY continued
61	Valve body material
62	Seal/O ring material
63	Chamber wetted material

PERFORMANCE CHARACTERISTICS
67	Max press at design temp		At
68	Min working temperature		Max
69	Min flow rate limit		Max
70	Max submersion depth		
71	Meas accuracy rating		Ref
72	Measurement repeatability		Ref
73	Measurement LRL		URL
74	Temp compensation LRL		URL
75	Max sensor to receiver lg		

ACCESSORIES
86	Calibration kit

SPECIAL REQUIREMENTS
92	Custom tag
93	Reference specification
94	Compliance standard
95	Calibration report

PHYSICAL DATA
100	Estimated weight		
101	Face-to-face dimension		
102	Overall length/height		
103	Removal clearance		
104	Signal conn nominal size		Style
105	Mfr reference dwg		

CALIBRATIONS AND TEST
#	TAG NO/FUNCTIONAL IDENT	MEAS/SIGNAL/SCALE	INPUT OR TEST LRV	URV	ACTION	OUTPUT LRV	URV
111							
112		Meas-Analog Output 1					
113		Meas-Analog Output 2					
114		Temp-Output signal					
115							
116							
117							

COMPONENT IDENTIFICATIONS
#	COMPONENT TYPE	MANUFACTURER	MODEL NUMBER
120			
121			
122			
123			
124			
125			

Rev	Date	Revision Description	By	Appv1	Appv2	Appv3	REMARKS

Form: 20A2351 Rev 0 © 2005 ISA

1.68 Turbidity, Sludge and Suspended Solids

1	RESPONSIBLE ORGANIZATION		TURBIDITY OR SUSPENDED SOLIDS TRANSMITTER OR CONTROLLER Device Specification			6	SPECIFICATION IDENTIFICATIONS		
2	(ISA)					7	Document no		
3						8	Latest revision		Date
4						9	Issue status		
5						10			
11	OPERATING PARAMETERS					53	PERFORMANCE CHARACTERISTICS		
12	Project number			Sub project no		54	Measurement accuracy		Ref
13	Project					55	Temperature accuracy		
14	Enterprise					56	Measurement repeatability		Ref
15	Site					57	Measurement LRL		URL
16	Area		Cell	Unit		58	Min ambient working temp		Max
17	Related equipment					59	Contacts ac rating		At max
18	Service					60	Contacts dc rating		At max
19						61			
20	P&ID/Reference dwg Number					62			
21	Remote hazardous area cl		Div/Zone	Group		63			
22	Remote area min ign temp			Temp ident number		64			
23						65			
24						66			
25	TRANSMITTER OR INDICATOR OR CONTROLLER					67			
26	Housing type					68			
27	Input sensor/signal type					69	ACCESSORIES		
28	Range selection style					70	Mounting hardware		
29	Output signal type					71	Mating connector		
30	Enclosure type no/class					72			
31	Control mode					73			
32	Local operator interface					74			
33	Characteristic curve					75			
34	Digital communication std					76	SPECIAL REQUIREMENTS		
35	Signal power source					77	Custom tag		
36	Temperature sensor type					78	Reference specification		
37	Quantity of meas signals					79	Compliance standard		
38	Aux input signal type					80	Software configuration		
39	Contacts arrangement			Quantity		81	Software program		
40	Integral indicator style					82			
41	Signal termination type					83			
42	Cert/Approval type					84			
43	Mounting location/type					85	PHYSICAL DATA		
44	Failure/Diagnostic action					86	Estimated weight		
45	Sensor diagnostics					87	Overall width		
46	Cleaning control type					88	Overall height		
47	Measurement compensation					89	Overall depth		
48	Calibration function					90	Signal conn nominal size		Style
49	Enclosure material					91	Mfr reference dwg		
50						92			
51						93			
52						94			

110	CALIBRATIONS AND TEST		INPUT OR TEST			OUTPUT OR SCALE	
111	TAG NO/FUNCTIONAL IDENT	MEAS/SIGNAL/SCALE	LRV	URV	ACTION	LRV	URV
112		Meas-Analog output 1					
113		Meas-Analog output 2					
114		Meas-Analog output 3					
115		Temp-Output signal					
116		Measurement-Scale					
117		Temperature-Scale					
118		Meas setpoint 1-Output					
119		Meas setpoint 2-Output					
120		Meas setpoint 3-Output					
121		Meas setpoint 4-Output					
122		Meas setpoint 5-Output					
123		Meas setpoint 6-Output					
124		Failure signal-Output					
125							
126	COMPONENT IDENTIFICATIONS						
127	COMPONENT TYPE	MANUFACTURER		MODEL NUMBER			
128							
129							
130							
131							
132							

Rev	Date	Revision Description	By	Appv1	Appv2	Appv3	REMARKS

Form: 20A2352 Rev 0 © 2005 ISA

Definitions

JTU: The Jackson turbidity unit (JTU) corresponds to the turbidity of a liter of distilled water when 1 mg (1 ppm) of diatomaceous fullers earth (an inert material) is suspended in it.

NTU: Nephelometric turbidity units (NTUs) are based on U.S. EPA-approved polymeric suspension standard (http://water.epa.gov/scitech/methods/cwa/bioindicators/upload/2007_07_10_methods_method_180_1.pdf).

Tyndall effect: A light beam projected into a sample stream is scattered by solids in suspension, and the degree of light attenuation and the amount of scattered light produced can both be related to turbidity. The light scattering is called the *Tyndall effect*, and the scattered light itself is called the *Tyndall light*.

Abbreviations

AU	Attenuation unit
BRU	Backscatter ratio unit
BU	Backscatter unit
FAU	Formazin attenuation unit
FBRU	Formazin backscatter ratio unit
FBU	Formazin backscatter unit
FNMU	Formazin nephelometric multibeam unit
FNRU	Formazin nephelometric ratio unit
FNU	Formazin nephelometric unit
NTMU	Nephelometric turbidity multibeam unit
NTRU	Nephelometric turbidity ratio unit
NTU	Nephelometric turbidity unit

Organizations

ASTM	American Society for Testing and Materials
EPA	Environment Protection Authority (Australia)
ISO	International Organization for Standardization
US EPA	U.S. Environmental Protection Agency

Bibliography

ASTM, *Annual Book of ASTM Standards*, American Society for Testing and Materials, West Conshohocken, PA, 2002. www.normas.com/ASTM.

Becker, W. et al., *Optimizing Ozonation for Turbidity and Organic Removal*, American Water Works Association, Denver, CO, 1996.

Buffle, J., ed., *In situ Monitoring of Aquatic Systems*, John Wiley & Sons, New York, 2000.

Capano, D., The ways and means of density, *InTech*, November 2000.

Dubois, R., van Vuuren, P., and Tatera, J., New sampling sensor initiative: An enabling technology, *47th Annual ISA Analysis Division Symposium*, Denver, CO, April 14–18, 2002.

Ewing, G.W., *Analytical Instrumentation Handbook*, Marcel Dekker, New York, 1997.

Extance, P. et al., Intelligent turbidity monitoring using fiber optics, *Proceedings of First International Conference Optical Fiber Sensors*, IEE, London, U.K., 1983.

Frenkel, M., Schwartz, R., and Garti, N., Turbidity measurements as a technique for evaluation of water-in-oil emulsion stability, *J. Disper. Sci. Technol.*, 3(2), 195–207, 1982.

Fussell, E., Molding the future of process analytical sampling, *InTech*, p. 32, August 2001. http://www.isa.org/Content/ContentGroups/InTech2/Departments/Standards_Update/200133/Molding_the_future_of_process_analytical_sampling.htm.

Gordon, I., Vibrating element technology for measuring liquid density in process applications, *Proceedings of ACHEMA*, Frankfurt, Germany, 2000.

Hashimoto, M. et al., An automated method for bacterial test by simultaneous measurement of electrical impedance and turbidity, *Jpn. J. Med. Eng.*, 19(1), 23–29, February 1981.

Hilliand, C.L., Design Solutions for Typical Problems of Colorimetric Process Analyzers, October 1992.

ISO, *ISO 7027 Water Quality—Determination of Turbidity*, International Standardization Organization, Geneva, Switzerland, 1999.

Kaiser, J. et al., *Bioindicators and Biomarkers of Environmental Pollution*, Science Publishers, Enfield, NJ, 2001.

Kotnik, P. et al., Accurate density measurement using oscillating density meters, *Proceedings of ACHEMA*, Frankfurt, Germany, 2000.

Krigman, A., Turbidity instrumentation: Clearing muddy waters, *InTech*, February 1984.

Letterman, R.D., *A Study of Low-Level Turbidity Measurements*, AWWA Research Foundation, Denver, CO, 2002.

McMahon, T.K., The new sampling/sensor initiative, *Control*, August 2001.

Mitchell, M.K. and Stapp, W., *Field Manual for Water Quality Monitoring*, 12th edn., LaMotte Company, Chestertown, MD, 2000.

Sadar, M., *Making Sense of Turbidity Measurements—Advantages in Establishing Traceability between Measurements and Technology*, Hach Company, Loveland, CO.

Sadar, M., Turbidity instrumentation—An overview of today's available technology, *FISC Turbidity Workshop Sponsored by USGS*, Reno, NV, April 2002.

Sherman, R.E., *Process Analyzer Sample-Conditioning System Technology*, John Wiley & Sons, New York, 2002.

The basics of photoelectric sensors, *Control*, April 1989.

Thomson, M., Interfacing sample handling systems for on-line process analyzers, Yokogawa Corporation of America, Sugar Land, TX, 2002.

Van den Berg, F.W.J., Hoefsloot, H.C.J., and Smilde, A.K., Selection of optimal process analyzers for plant-wide monitoring, *Anal. Chem.*, 74(13), 3105–3111, 2002.

Waller, M.H., Measurement and control of paper stock consistency, *Proceedings ISA Conference*, Houston, TX, October 1992.

Weiss, M.D., Particle size analysis goes on line, *Control*, August 1990.

1.69 Ultraviolet and Visible Analyzers

J. E. BROWN (1969, 1982)   **T. M. CARDIS** (1995)   **B. G. LIPTÁK** (2003, 2017)

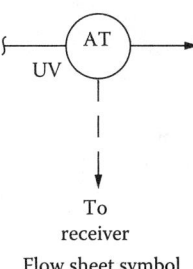

Flow sheet symbol

Configurations	UV and visible analyzers can operate either in the photometric (absorption) or in the spectrum (dispersive) mode. A. Photometers are nondispersive analyzers in which the source radiates over its full ultraviolet (UV) spectrum, and discrete wavelengths are separated by narrow (2–10 nm) bandpass filters. UV photometers are normally used for the measurement of a key single component in a process stream. Most process analyzers are nondispersive. B. Spectrophotometers are dispersive analyzers in which a prism is used to separate the spectral components of the UV spectrum. Most laboratory analyzers are dispersive.
Types of designs	Optics can be configured as single-beam, split-beam, dual-beam, or flicker photometers. The process stream can pass through a measurement chamber, or probe-type analyzers can be inserted into the process stream using a retroreflector configuration. Most detectors are capable of measuring the intensity of one wavelength at a time, whereas photodiode arrays can detect all wavelengths simultaneously.
Type of sample	Gas, vapor, and liquid
Design pressure	Normally atmospheric, but it can also be pressurized up to 150 psig (10.7 barg) to enhance sensitivity when analyzing gas samples. The fiber-optic, diode-array type of in-line analyzer can operate at up to 730 psig (50 barg).
Sample temperature	Standard units operate from 0°C to150°C (32°F–302°F); stack gas analyzer can operate at up to 427°C (800°F).
Materials of construction	Quartz and sapphire windows; other parts available in stainless steel, Hastelloy®, Monel®, titanium, Teflon®, Kynar®, and all other conventional materials
Wavelength ranges	Standard ranges include 200–800 nm or 400–1100 nm.
Cell lengths	Standard units are from 1 to 1000 mm (0.039–39.4 in.); for fiber-optic diode-array analyzer, 0.3–10 mm.
Repeatability	±1% of full scale
Inaccuracy	±2% of full scale for most; ±1% of full scale for the fiber-optic diode-array type of UV–visible analyzer
Cost	Portable battery-operated or benchtop laboratory UV/visible (200–1000 nm) photometers, spectrophotometers, and scanning spectrophotometer range from $1500 to $3000. Portable UV stack gas monitors range from $5000 to $7000. Cost of permanently installed stack monitoring packages range from $25,000 to $35,000 depending on materials of construction, accessories, and number of components analyzed. Industrial UV photometers cost in the without sampling systems cost in the range of $20,000–$25,000. Industrial photodiode array spectrophotometers cost from $50,000 to $70,000. Industrial fiber-optic spectrophotometers cost from $40,000 to $60,000.
Partial list of suppliers	ABB Analytical (multiwave) http://www.abb.com/product/us/9AAC128814.aspx Amatrac (Multiwave) http://www.anatrac.com/eng/item/PRO30061.html Ametek Process Instruments (A, B) http://www.ametekpi.com/searchresults.aspx?Keywords=UV/VIS%20analyzers Cole-Parmer Instrument Co. (A, B) http://www.coleparmer.com/Category/Visible_UV_Visible_Spectroscopy/5469 Emerson Process Analytics: Rosemount (A, B) http:// http://www2.emersonprocess.com/siteadmincenter/PM%20Rosemount%20Analytical%20Documents/PGA_PDS_X-STREAM_X2FD.pdf Horiba Scientific (Monochromators) http://www.horiba.com/us/en/scientific/products/monochromators/ Jasco Analytical Instruments (A, Laboratory) http://www.jascoinc.com/spectroscopy/uv-visible-nir

(Continued)

OPSIS (Laser diode) http://www.opsis.se/Products/ProductsCEMProcess/LD500LaserDiodeGasAnalyser/tabid/1071/Default.aspx
Optec (Fiber-optic) http://www.optek.com/en/af45-af46-uv-sensors.asp
Shimadzu (A, B) http://www.shimadzu.com/an/spectro/uv/uv1800/uv2.html
Teledyne Analytical Instr. (A) http://www.teledyne-ai.com/pdf/6000.pdf
Thermo Scientific (Multiple) http://www.thermoscientific.com/en/products/ultraviolet-visible-visible-spectrometry-uv-vis-vis.html
Toray Engineering Ltd. (A, water analyzers) http://www.toray-eng.com/measuring/water/uv-lineup/uvt-300.html
Vernier (Portable, A) http://www.vernier.com/products/sensors/spectrometers/ultraviolet-range/vsp-uv/
Water Online, (A,B) http://www.wateronline.com/doc/ets-uv-for-drinking-water-disinfection-0001

INTRODUCTION

The composition of both gases and liquids can be analyzed by measuring their absorption or reflectance in the ultraviolet (UV)–visible (VIS) spectral region (Figure 1.69a). These analyzers can operate either in the photometric (absorption) or in the spectrophotometric (dispersion) mode, and some designs are capable to operate in both.

A photometer measures the absorbance or transmittance at a single or at multiple specific wavelengths. In the multiple-wavelength configuration, calculations can be made to obtain the difference between, or ratio of, the measurements at two specific wavelengths.

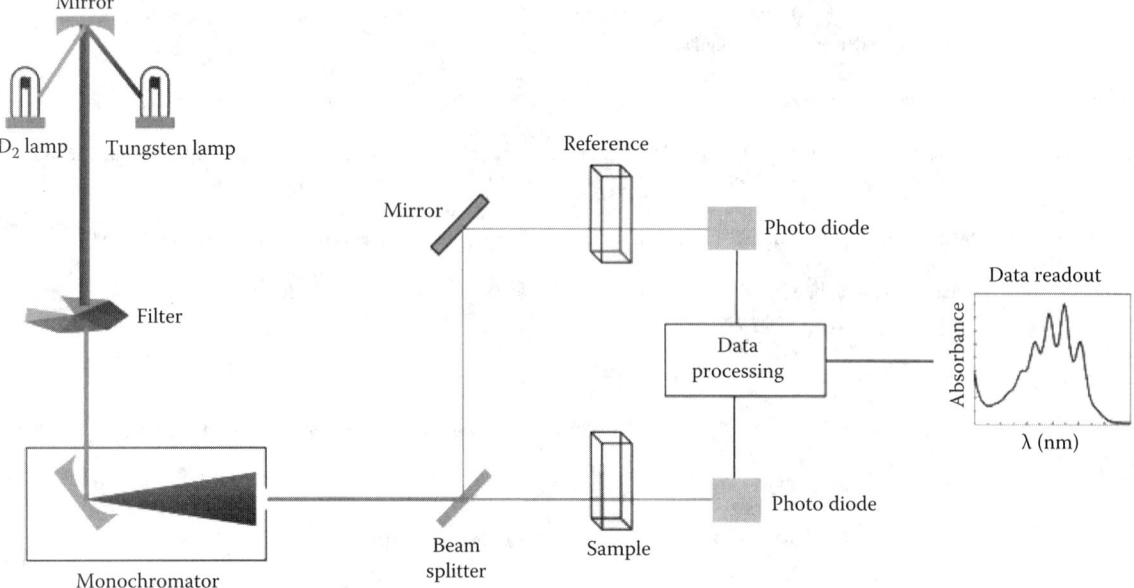

FIG. 1.69a
The main components of a UV/VIS analyzer.

A spectrophotometer obtains the spectra of the sample (Figure 1.69b) through wavelength scanning. Continuous measurements can be obtained by periodically detecting the changes in the composition of the flowing sample. Software is available to enlarge or reduce the spectra obtained, detect peaks, or make various other calculations, based on the spectra.

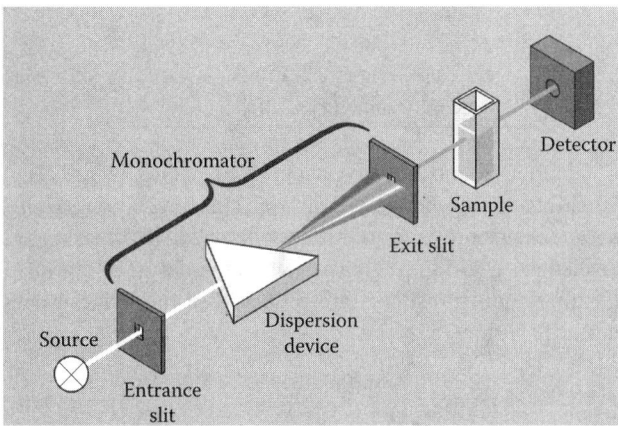

FIG. 1.69b
In the visible range, the color red has the lowest dispersion angle and the color violet has the highest.

In addition to the coverage of this chapter, UV/VIS analyzer designs and applications are discussed in other chapters, including the following:

Chapter 1.13: Colorimeters
Chapter 1.23: Fiber-optic sensors
Chapter 1.62: Spectrometry

UV analyzers are widely used in the determination of the concentrations of many materials, including but not limited to:

Chapter 1.31: Hydrogen sulfide
Chapter 1.32: Hydrocarbons
Chapter 1.44: Nitrogen oxide
Chapter 1.51: Ozone
Chapter 1.64: Sulfur dioxide, etc., applications

For a more complete list of UV absorbing materials refer to Table 1.69b.

UV THEORY

Ultraviolet (UV) radiation occurs at wavelengths between the range of 100–400 nm. This wavelength is shorter than that of visible light, but longer than x-rays. Though usually invisible to humans, under some conditions children and young adults can see ultraviolet down to a wavelengths of about 310 nm. UV, VIS, and near-infrared (NIR) photometers and spectrophotometers are widely used in the process industries. The use of UV analyzers started out in laboratories, but since 1940 they have also been used on-line in industrial applications.

UV Absorption

Fewer compounds absorb in the UV region than in the IR region, and the UV absorption pattern of a compound is not as distinctive (not as narrow) as its IR *fingerprint*. On the other hand, UV analyzers provide better selectivity in applications in which the sample contains air and humidity, because these materials do not absorb in the UV region.

UV analyzers are also more sensitive than IR detectors. Figure 1.69c shows the absorbance of sulfur and nitrogen

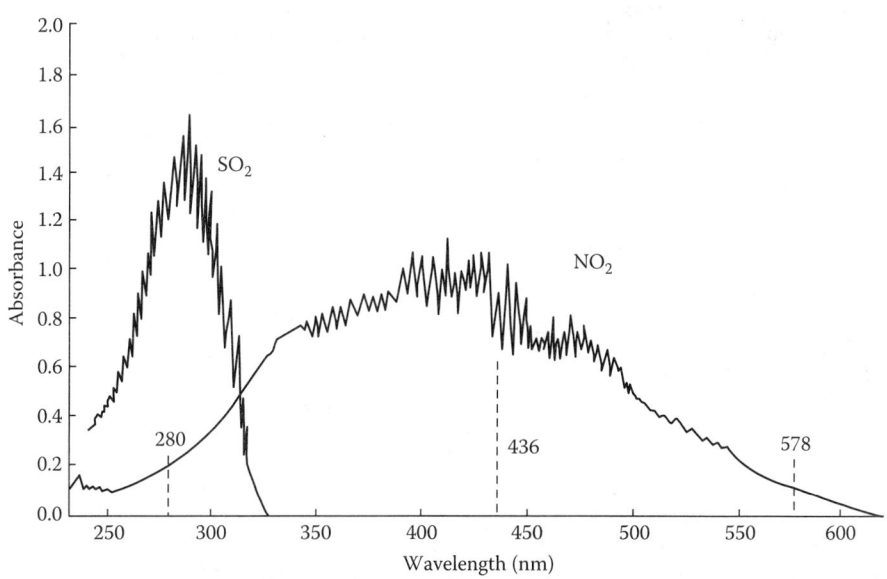

FIG. 1.69c
The absorbance characteristics of both SO_2 and NO_2 in the ultraviolet range.

dioxides in the UV region. On an equal path length basis, the UV absorbance of liquids is stronger than that of vapors in proportion to their densities.

This chapter starts with some theoretical discussions of the principles of UV radiation. This is followed by an explanation of different analyzer configurations and their main components.

Radiation Spectrum

The UV, VIS, and IR regions of the electromagnetic radiation spectrum are shown in Figure 1.69d. The UV region covers the wavelengths from 200 to 400 nm.* The VIS region extends from 400 to 800 nm, and the NIR region covers from 0.8 to 2.50 μm.

Nanometer units are commonly used in the UV/VIS region, and micrometers (or microns) are normally used in the NIR region (Table 1.69a). The UV–VIS–NIR is a relatively small part of the electromagnetic radiation spectrum, and the shorter the wavelength (the higher the frequency), the more penetrating the radiation becomes.

TABLE 1.69a
Relationship of Wavelength Terms

Angstrom = Å = 10^{-10} m
Micrometer = μm = 10,000 Å = 10^{-6} m
Nanometer = nm = 10 Å = 10^{-9} m
Meter = 10^3 nm = 10^6 μm = 10^9 nm = 10^{10} Å

Compounds absorb at various frequencies of radiation depending on the energy of their molecular transitions. High-energy electronic transitions are observed in the shorter-wavelength UV–VIS regions. Moderate-energy (vibrational and rotational) transitions are observed in the longer-wavelength IR region.

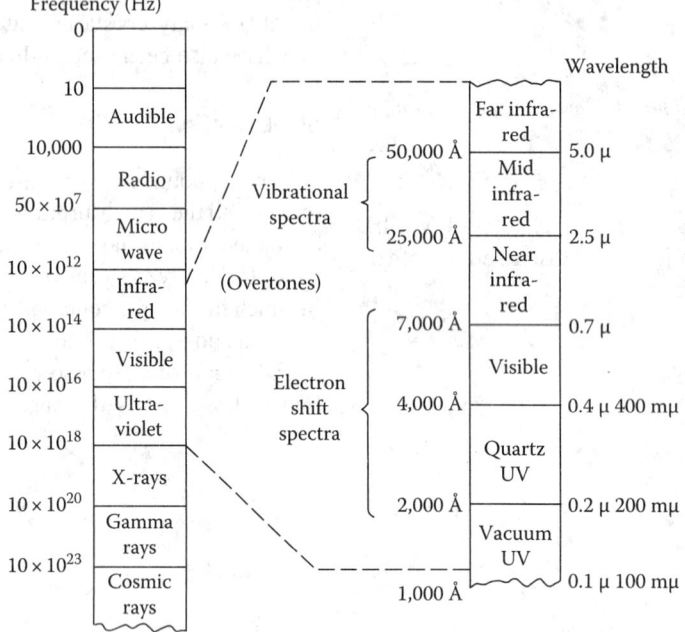

FIG. 1.69d
Electromagnetic radiation spectrum from the audible frequency up to that of the cosmic rays.

---

* Nanometer (nm) is equal to one billionth of a meter (0.000000001 m).

Beer–Lambert Law

Absorption spectroscopy is the measurement of radiation intensity at spectral wavelengths that are characteristic of the compound being monitored. When radiation enters a process sample, only part of the light is transmitted through (Figure 1.69e) it. The remainder of the radiation is absorbed or reflected, depending on the concentration of the sample and on the wavelength being measured. If particles are present in the sample, part of the radiation will be scattered.

The Beer–Lambert law is the fundamental law that governs quantitative analysis by absorption spectroscopy. This law states

$$A = abc \qquad \textbf{1.69(1)}$$

where
- A is the absorbance
- a is the molar absorptivity
- b is the sample path length
- c is the concentration of absorbing species

The molar absorptivity (a) is a constant that is characteristic of the chemical species being measured, and the path length of the sample cell is fixed. Therefore, the only variable that causes the absorbance (A) to vary is the concentration of the component being measured.

The absorbance of a sample is measured indirectly by measuring the transmitted radiation. The relationship between absorbance and transmittance is

$$A = \log\left(\frac{1}{T}\right) \qquad \textbf{1.69(2)}$$

where
- A is the absorbance
- T is the transmittance

Calibration methods for photometers involve the measurement of absorbance for several concentrations of the component being measured. When a plot of absorbance versus concentration gives a linear response, the method obeys the Beer–Lambert law. When the photometer exhibits a nonlinear response, a linearizer circuit is used to compensate for the nonlinearity.

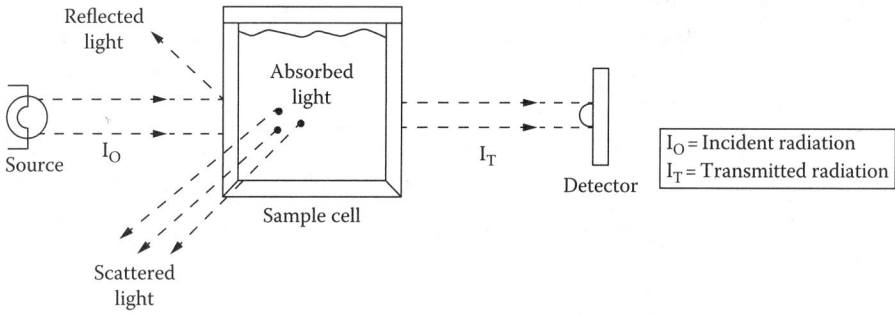

FIG. 1.69e
Absorption of light.

UV-Absorbing Compounds

The UV spectra of some typical absorbers are shown in Figure 1.69f. The characteristics of these spectra are notably different from the spectra of absorbers in the IR region (Figure 1.32b). For this reason, the UV analyzer can be expected to be less specific than an IR instrument.

If other UV-absorbing compounds are also present in the sample, more interference can be expected because of the broad characteristics of their spectra. On the other hand, UV analyzers are generally capable of greater sensitivity than their IR counterparts, so trace analyses are common for UV. This is due in part to the broad energy absorption and greater absorptivity of some compounds and to the availability of stronger sources that permit the use of longer cells.

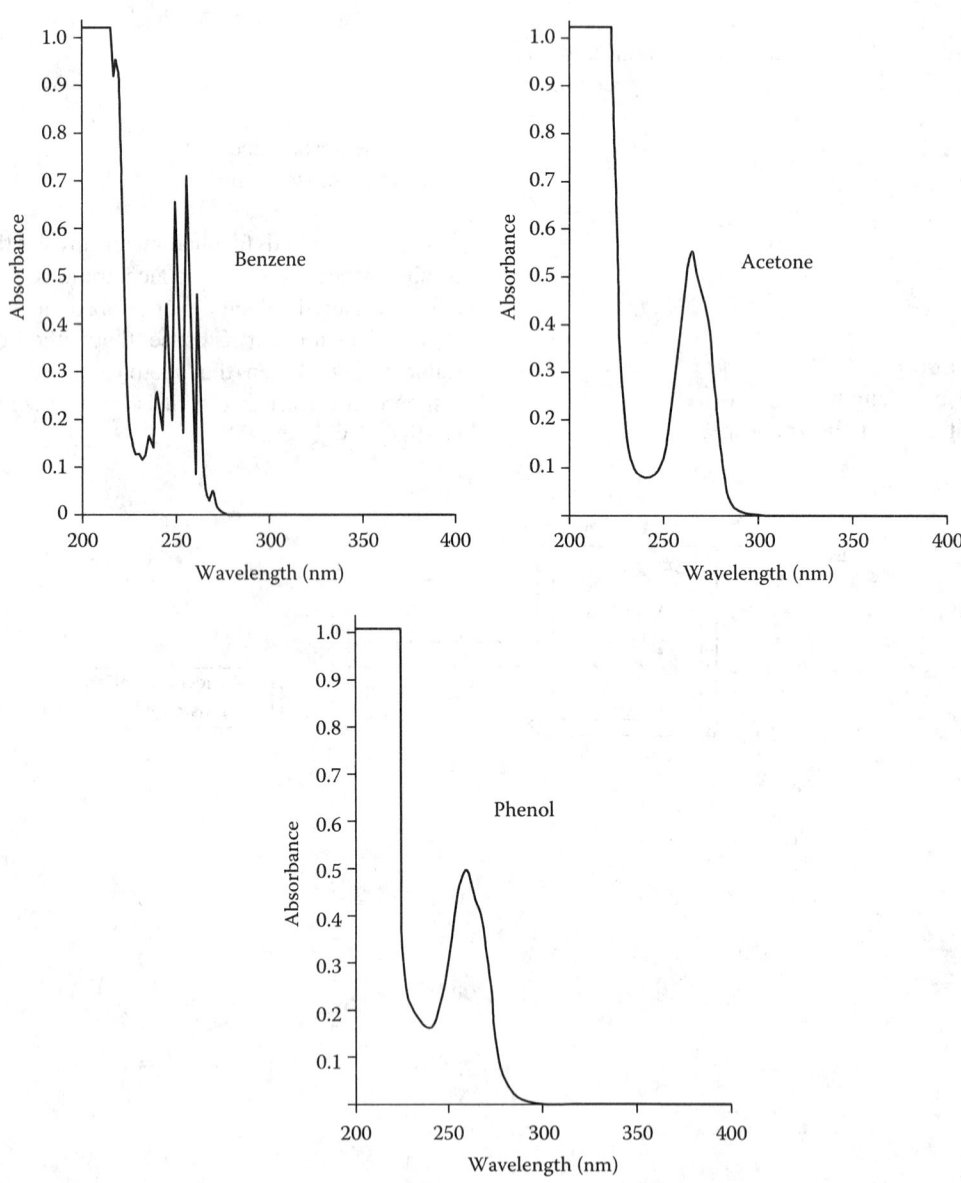

FIG. 1.69f
Ultraviolet spectra of benzene, acetone, and phenol.

UV Absorption Spectrum

An absorption spectrum of a compound is obtained by plotting the intensity of transmitted light or absorbance as a function of wavelength. UV spectra for three typical UV absorbers are presented in Figure 1.69f.

The benzene spectrum was obtained from a vapor phase sample. The acetone and phenol spectra were run on liquid samples dissolved in water. An important observation in these spectra is the high degree of overlap for the absorption bands in the 240–280 nm region. Care therefore must be exercised in UV applications to ensure that other compounds present in the process stream do not interfere with the measurement of the component of interest.

A useful feature of the UV region is that water does not absorb UV radiation. This allows for the measurement of trace levels of UV absorbers in aqueous samples. Generally, UV photometers have very good sensitivity.

The UV is a high-energy region of the electromagnetic spectrum where electronic transitions are observed. Absorption of radiation occurs in the UV region (200–400 nm) when an electron is excited from a ground energy state to an excited energy state. UV photometers are used principally to measure aromatic compounds, unsaturated compounds, and other compounds that contain pi electrons or unshared electron pairs. Another group of compounds that absorb UV are chromophores such as carbonyls, nitrates, and sulfoxides. A partial list of UV-absorbing compounds is given in Table 1.69b.

TABLE 1.69b
Partial List of UV-Absorbing Compounds

Acetic acid	1,3-Cyclopentadiene	Nitrogen dioxide
Acetone	Dimethylformamide	Ozone
Aldehydes	Elemental halogens	Perchloroethane
Ammonia	Ethylbenzene	Phenol
Aromatic compounds	Ferric chloride	Phosgene
	Fluorine	Potassium permanganate
Benzene	Formaldehyde	Proteins
Bromine	Formic acid	Salts of transition metals
Butadiene (1,3)	Furfural	Sodium hypochlorite
Caffeine	Hydrogen peroxide	Styrene
Carbon disulfide	Hydrogen sulfide	Sulfur
Carbon tetrachloride	Iodine	Sulfur dioxide
Chlorine	Isoprene	Sulfuric acid
Chlorine dioxide	Ketones	Toluene
Chlorobenzene	Mercury	Trichlorobenzene
Crotonaldehyde	Naphthalene	Uranium hexafluoride
Cumene	Nitric acid	Xylene (*ortho*, *meta*, and *para*)
Cylohexanol	Nitrobenzene	

Applications

UV photometers are normally used for the measurement of a key single component in a process stream. The following applications are examples of typical UV photometer measurements:

1. Residual chlorine in water in pulp and paper process
2. Total aromatics in wastewater
3. Chlorine in dichloroethane in vinyl chloride processes
4. Hydrogen sulfide and sulfur dioxide in the Claus sulfur recovery process in the petroleum refining industry
5. Sulfur dioxide in incinerator stack emissions

UV ANALYZER COMPONENTS

Photometers and spectrophotometers can be used for online monitoring of process streams. A diagram of the five main components that make up a spectrophotometer is shown in Figure 1.69g.* These components are enumerated as follows:

1. Source (provides radiation for the spectral region being measured)
2. Monochromator (a device used to select narrow bands of wavelengths)
3. Sample cell (contains the sample and provides an appropriate path length)
4. Detector (a device that measures transmitted energy and converts it into an electrical signal)
5. Readout device (provides a means of displaying or recording the measurement results)

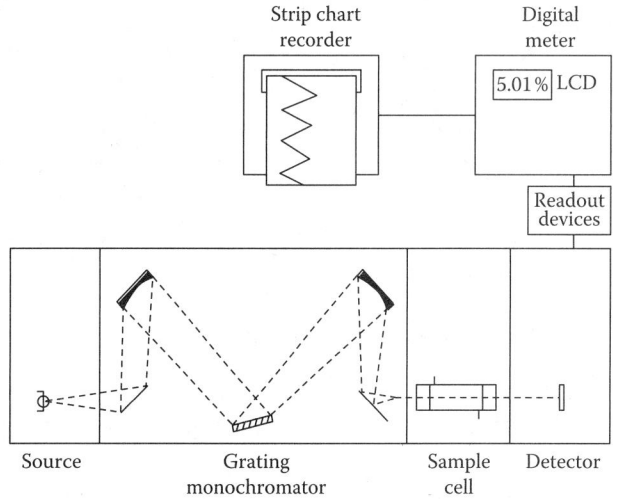

FIG. 1.69g
Block diagram showing the main components of a spectrophotometer.

---

* Thomas, M.J.K. et al., *Ultraviolet and Visible Spectroscopy*, John Wiley & Sons, New York, 1996.

Radiation Sources

The purpose of the radiation source is to provide sufficient energy of radiation to make the desired measurement in the region of spectral interest. Table 1.69c lists some of the sources commonly used in process photometers. The two sources used in the VIS and NIR regions are tungsten filaments and quartz-halide lamps.

The cadmium, mercury, and zinc vapor sources that are used in the UV region are *emission line* sources. The output of these sources provides radiation as narrow discrete emission lines at a high-energy level. Mercury vapor lamps are often used because of their long service life. Deuterium arc sources provide a broad range of UV radiation at all of the wavelengths in the UV region. The energy of the deuterium source is relatively lower than the energy of the mercury source.

Emission Types

Two types of UV energy sources are used: broad and discrete line emission sources. In open-path UV spectrometers (Chapter 1.62), the most commonly used sources are high-pressure xenon arc lamps, which emit a continuum from the UV to the VIS spectra (Figure 1.69h).

The broad emission sources provide energy in a broad wavelength band, and narrow band filters can be used to isolate the wavelengths of interest. These sources provide all wavelengths in the region, but some have a low-emission or low-energy level at a particular wavelength. Such low-energy sources include hydrogen or deuterium discharge lamps, tungsten lamps, and tungsten–iodine lamps.

Discrete line sources use gas discharge lamps with narrow lines of emission. These sources emit radiation energy

TABLE 1.69c
Typical Sources and Detectors

	Ultraviolet	Visible	Near-Infrared
Sources	Cadmium vapor	Tungsten filament	Tungsten filament
	Mercury vapor	Quartz-halide	Quartz-halide
	Zinc vapor		
	Deuterium arc		
	Tungsten–iodine		
Detectors	Photomultiplier tube	Photomultiplier tube	Lead sulfide
	Silicon photodiode	Silicon photodiode	Germanium photodiode
	Diode array	Diode array	Diode array
			Pyroelectric

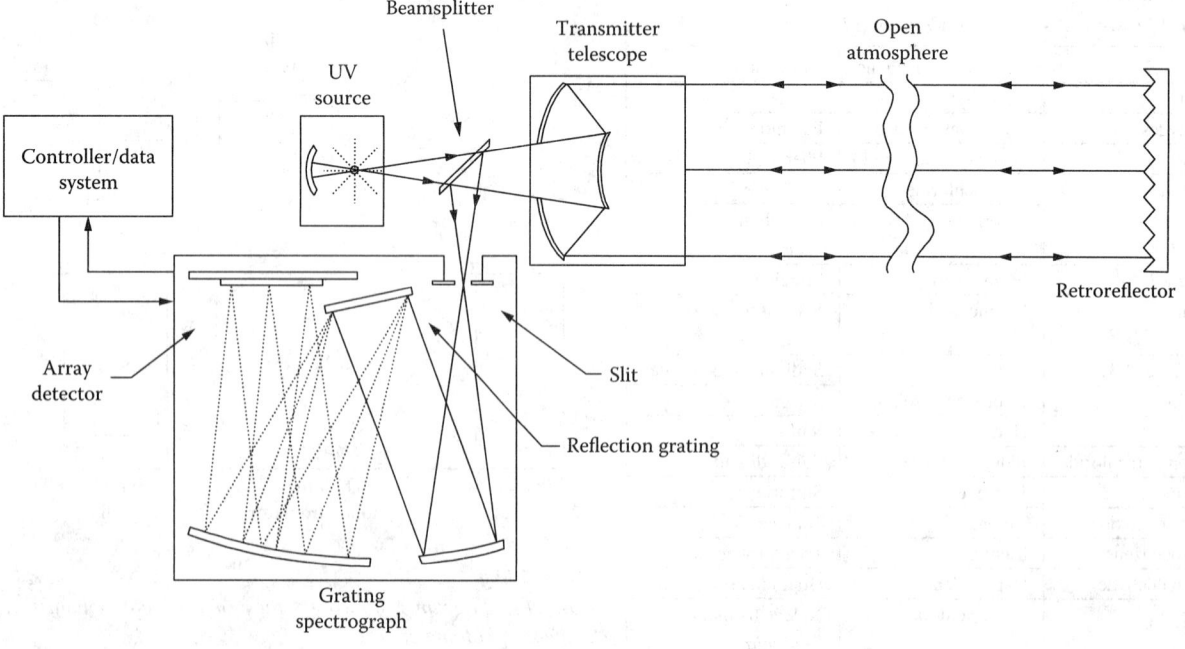

FIG. 1.69h
How the UV source is used in an open-path spectrometer.

at various discrete wavelengths at a high-energy level. The undesired are filtered, leaving only the wavelength of interest. However, the gas discharge lamps are limited to the spectral emissions of mercury, cadmium, zinc, and thallium. Therefore, not all wavelengths are available.

Selecting the Wavelength

Selection of measuring and reference wavelengths is generally a compromise between the maximum absorbance wavelength of the component of interest and the wavelengths available from the gas discharge lamps. This usually leads to the selection of a wavelength on the side of the absorption peak.

Of the radiation sources used in the UV region, the tungsten-filament-type incandescent lamps are the least expensive. These continuous-spectrum sources can be used in UV photometers, but they have sufficient energy only at wavelengths exceeding 350 nm. Tungsten–iodine cycle lamps can be used down to 300 nm. Mercury vapor lamps are the most useful UV sources because of their high intensity and long life (up to 5 years).

Medium-pressure mercury lamps can operate down to 300 nm, and low-pressure ones down to 254 nm. Zinc discharge lamps are useful because of their 214 nm emission line, as are chromium ones for their 228 nm emission line, but they are not as stable as mercury lamps. Hydrogen and deuterium lamps are delicate and expensive, and their lives seldom exceed 3 months.

Monochromator

Monochromators are instruments that transmit mechanically selectable narrow wavelength bands (λ) of UV or other types of radiation chosen from a wider range of wavelengths from the entering radiation. Glass is not transparent to UV radiation, so quartz prisms are used to disperse it. The monochromator can use the phenomenon of optical dispersion by prisms or can use diffraction grating (Figure 1.69l) to separate the wavelength bands (λ). The gratings operate in the reflective mode and they can be triangular mirrored half-prisms, producing linear dispersion or holographic gratings having sinusoidal grove profiles (cross-section).

Both dispersive and nondispersive monochromators are used in photometric analysis. Spectrophotometers are dispersive instruments, and photometers are nondispersive devices. The function of the monochromator is to disperse light from a source and selectively send a narrow spectral band to the sample and the detector (Figure 1.69i).

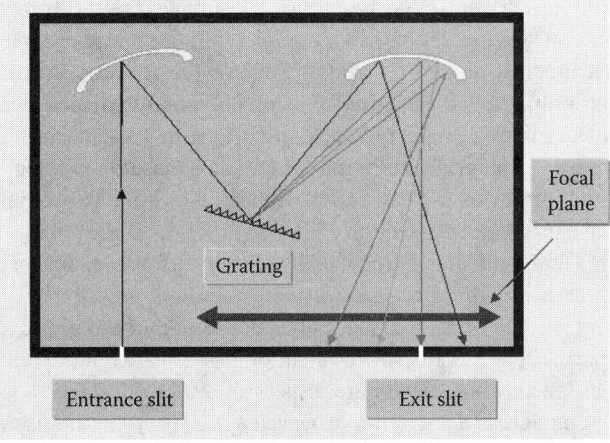

Czerny–Turner grating monochromator

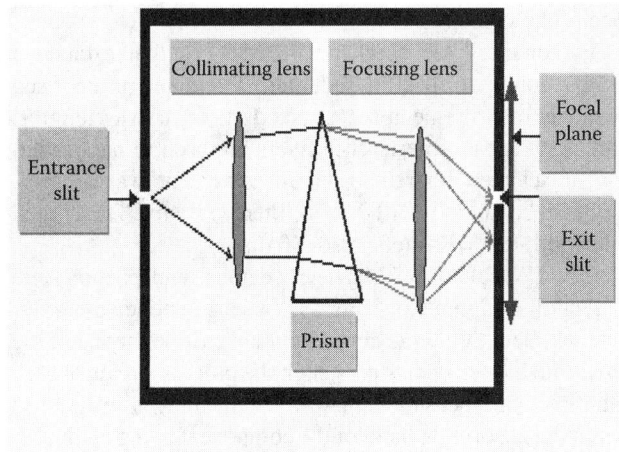

Bunsen prism monochromators

FIG. 1.69i
Monochromator designs.

Optical Dispersion

Optical dispersion is the separation of incoming radiation into wavelength bands (λ). When a narrow wavelength band (λ) of UV, or other type of radiation is to be selected, that can be done by dispersive or non-dispersive selectors. Filters are called non-dispersive (λ) selectors while prisms and gratings are called dispersive ones. A linear grating is an optically flat polished surface having dense and parallel groves (1000 or more per mm). The two types of gratings are the transmission and reflection (diffraction) types. In spectroscopic instruments, almost all gratings are of the reflection type. The gratings can produce linear or holographic dispersion. Linear dispersion is produced by triangular flat surfaced gratings. Holographic gratings are made by laser technology which produces a pattern of straight lines with a sinusoidal grove profile (cross-section).

The dispersing element is usually a diffractive grating, which is a highly polished mirror with a number of parallel lines scribed on its surface. For each position of the grating, a narrow band of dispersed radiation passes through the exit slit. Spectrophotometers are dispersive devices that are used to scan across a spectrum of wavelengths; they can be used to make measurements at several wavelengths. This capability allows for the analysis of multiple components with a spectrophotometer.

Photometers are nondispersive devices that exclude a large amount of spectral radiation. Photometers are used to make measurements at selected discrete wavelengths. Narrow bandpass interference filters are used to pass radiation at selected reference and measurement wavelengths. A typical bandwidth for UV/VIS filters is 10 nm. The typical bandwidth for NIR filters is 20–80 nm.

The UV spectra of benzene, acetone, and phenol were presented in Figure 1.69f. For such measurements, the reference wavelength filter is normally selected where none of the background components present in the process stream absorb radiation. The measurement wavelength filter is selected to match the absorption band of the component being analyzed. The ratio of the transmitted light at the reference and measured wavelengths is measured by the photometer. Normally, photometers are used to measure a single component in a process stream.

Measuring Cell

The purpose of the measuring cell is to contain a representative sample from the process steam. The major components of the cell are the cell body, windows, and O-rings. The proper selection of cell materials is very important for the successful application of a process photometer. Stainless steel is the material most commonly used for cell bodies. Other metals such as Monel, Hastelloy, and titanium are also used. Plastic cell bodies made of Teflon or Kynar are used in some applications. Quartz, sapphire, and glass cell windows are used in the UV–VIS–NIR spectral regions.

The sealing of the sample in the cell is accomplished with O-ring gaskets. Viton®, ethylene–propylene, and Kalrez® O-rings are commonly used in sample cells. An important parameter for the cell is the selection of an appropriate path length. The distance between the windows determines the path length of the cell. Path lengths from 1 mm to 1 m (0.039–39.4 in.) are used in process photometers. The sensitivity of a photometric measurement is dependent on the path length. For a particular measurement, a long path length will provide more sensitivity than a short one.

Detectors

The detectors used in process UV analyzers include phototubes, photomultiplier tubes (PMTs), and photocells. The photoelectric effect is used in the vacuum phototube to produce a current proportional to the energy striking the tube cathode. A phototube with UV response has a long life and a low temperature coefficient.

The PMT offers very sensitive detection of UV and VIS light, but large radiation energy levels will damage its light-sensitive surface. This detector has a high temperature coefficient.

The photocell (photovoltaic) is a semiconductor light detector of the barrier-layer type. It develops a current that is proportional to the light intensity, but this current output is not linear with the radiation energy level reaching the detector. This may not be detrimental when used in a null-balance detection system, and the relatively low cost of this device (because a voltage supply is not required) can make it attractive.

Table 1.69c lists the commonly used detectors in the UV–VIS–NIR regions. PMTs have traditionally been used in UV/VIS instruments. In the PMT, the photoelectric effect is used to produce a current proportional to the radiation that is striking the cathode of the tube. The gain voltage of the photocathode can be adjusted to obtain the desired sensitivity from the PMT.

Photodiodes are solid-state detectors that are smaller in size and lower in cost than PMT detectors. Silicon photodiodes are semiconductor detectors that are used in the UV/VIS region. Germanium photodiodes are used in the NIR region. A more recent development in photometric analyzers is the use of photodiode arrays (PDAs). The PDA detectors are used throughout UV–VIS–NIR regions. Many discrete detectors can be located in a very close space in the PDA.

This array of diode detectors allows for all of the wavelengths to be measured simultaneously. The PDA detectors can be used for multicomponent applications. Lead sulfide, germanium photodiodes, and pyroelectric detectors are used in the NIR photometers.

Readouts

Analog meters, digital meters, strip chart recorders, and video display tubes are examples of readout devices used in photometers and spectrophotometers. Online photometers are usually microprocessor operated and usually have both a 4–20 mA analog output and a digital output, which is sent to the process control computer. Spectrophotometers can incorporate a strip chart recorder to record a spectrum of the analyzed sample. The spectrum is a plot of percentage transmission or absorbance readings as a function of wavelength.

UV ANALYZER DESIGNS

The proven industrial designs used for process UV analyzers include the following:

1. Single beam
2. Split beam
3. Dual beam, single detector
4. Dual beam, dual detector
5. Flicker photometer
6. Photodiode
7. Retroreflector

Single-Beam

This simple design is limited to easy applications. The optical system (Figure 1.69j) consists of a source and two phototube detectors. The light source and detectors are so aligned that both detectors receive radiation from the same portion of the source. A sample cell and an interference filter are positioned between one detector and the source. The filter is selected to isolate the wavebands in which only the component of interest will absorb. The cell length is sized to give a sufficiently long absorbing path to provide good sensitivity (i.e., high absorbers require shorter paths). A filter usually identical to the measuring filter is positioned between the source and the reference detector.

The amplifier, or control circuit, compares the outputs of the two phototubes, and the difference in their outputs is related to the UV energy absorbed by the sample. The reference detector is used to provide compensation for changes in line voltage and source decay. In the past, some process analyzers were built without this feature. They detected only the energy change at the measuring detector, but these analyzers were found to be subject to drift and unreliable.

The opposed-beam design is a simple, low-cost, moderate-accuracy instrument that is used for simple, low-sensitivity measurements. The output of this instrument is affected by dirt and bubbles in the sample and by drift in the detector circuit.

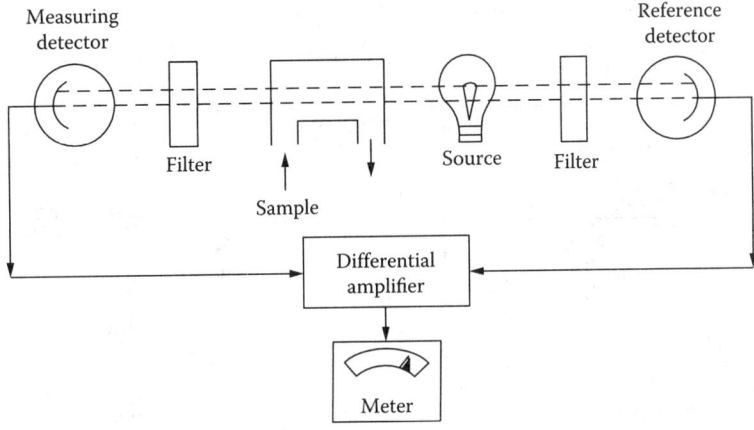

FIG. 1.69j
Opposed-beam UV analyzer.

Split-Beam

The split-beam UV analyzers use a single-beam optical system (Figure 1.69k). A single sample cell is located between the source and the detectors. A beam splitter (semi-transparent mirror) is placed at the outlet of the sample to create two paths for the radiation energy. There is one beam for the measuring wavelength and one beam for the reference wavelength.

Interference filters and/or broadband filters are used to isolate the desired wavelengths. The reference wavelength is selected in a region where the component of interest absorbs weakly or, preferably, not at all.

One beam is directed to the measuring PMT through an optical filter, which passes light at the measuring wavelength. The other beam is directed to the reference PMT through an optical filter, which passes light at the reference wavelength. The phototubes convert both light signals to electric currents proportional to the light intensity in each beam. The outputs of the phototubes are compared by the amplifier, which provides an output signal that is proportional to the concentration of the measured component. This output signal is reported on the output meter of the control section of the analyzer.

The use of the split-beam design with a reference detector offers the advantage of minimizing or eliminating the effect of other weak UV absorbers in the sample. In addition, changes in source intensity, sample turbidity (dirt or bubbles in the sample), and dirt buildup on the cell windows are seen equally in both wavelengths and do not affect the measurement as long as the absolute energy intensity does not drop below the sensitivity of the detector.

The three housings of the analyzer are easily separated. For most applications, the split-beam system offers high sensitivity and accuracy, low drift, and moderate cost.

Dual-Beam, Single-Detector

The dual-beam, single-detector UV analyzer uses two optical paths, a single sample cell, and a single photomultiplier detector (Figure 1.69l). One path includes the sample cell, and the other path is used for reference, as it does not pass through the sample. The paths are recombined, and both pass through an interference filter that isolates the wavelengths selected for the measurement.

An interrupter or *chopper* is used to alternately block the measuring and reference beams. This creates pulses of energy

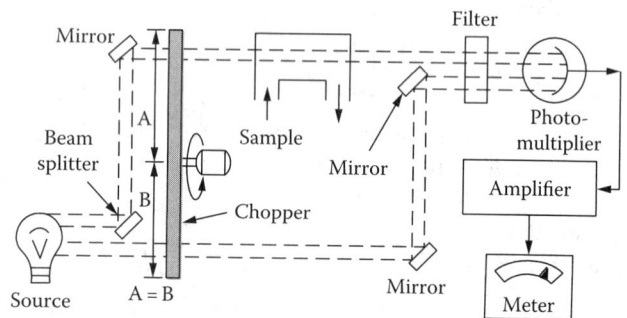

FIG. 1.69l
Dual-beam, single-detector analyzer.

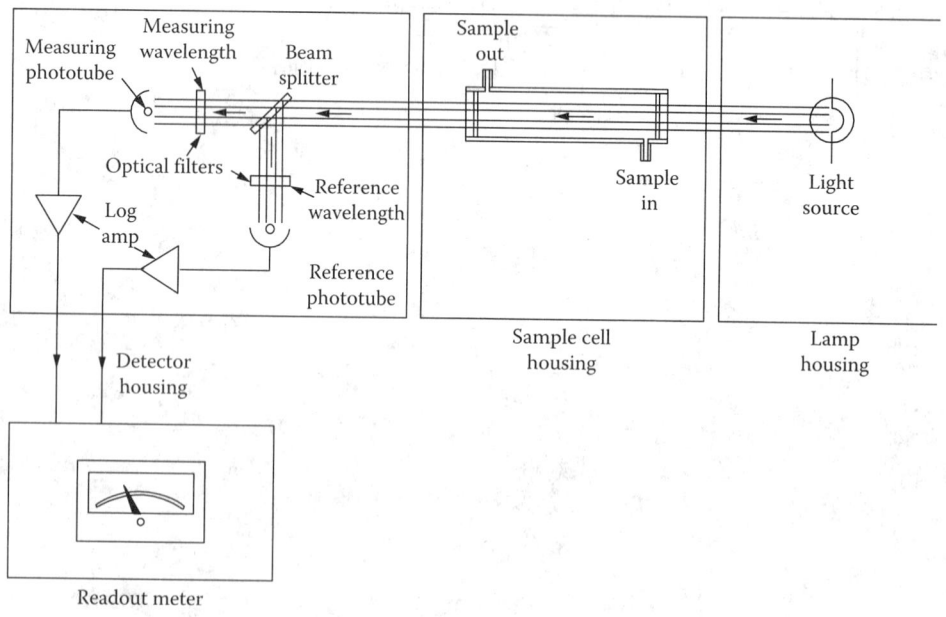

FIG. 1.69k
Split-beam photometer.

through the sample cell that are 180° out of phase with the energy pulses through the reference beam. These pulses are received consecutively by the photomultiplier, and if the sample cell is filled with a nonabsorber, they are equal in magnitude.

When the sample contains the component of interest, the energy pulse intensity and the output of the detector for the measuring pulse are both reduced. The control amplifier receives these pulse outputs and demodulates, or converts, their ratio to a usable analog signal.

This design gives good compensation for source intensity changes and detector drift, but turbid samples and samples leaving dirt deposits will cause drift. Samples of this nature sometimes require the use of two sample cells and two separate filters—one in each path.

The filters create a measuring and a reference wavelength as in the split-beam, dual-detector system. The sample cells may or may not experience window coating at the same rate, which only partially solves the drift problem. Use of a chopper introduces a moving mechanical member that increases maintenance.

The dual-beam, single-detector system offers high sensitivity and accuracy at moderate cost, but zero drift may occur if the process sample is dirty.

Dual-Beam, Dual-Detector

The dual-beam, dual-detector design isolates the wavelength used for the measurement before the beam splitter (Figure 1.69m) and uses separate phototube detectors for the measuring and the reference signals. This design is a combination of the optics of the dual-beam, single-detector analyzer (see Figure 1.69l), and the detector of the split-beam analyzer (see Figure 1.69k).

This design provides performance similar to that of the dual-beam, single-detector design. It provides high sensitivity and good accuracy at a moderate cost, but dirt in the samples can create drift errors.

Flicker Photometer

The flicker photometer uses a single-cell optical bench (Figure 1.69n) and a rotating disk with two interference filters. The radiation transmitted by the sample cell alternately passes through these filters, which are selected to produce the desired measuring and reference wavelengths. The detector receives these energy pulses, and the amplifier calculates their ratio, which changes as the concentration of the component of interest varies.

A schematic of a rotating fixed filter photometer is shown in Figure 1.69n. The light from the source passes through lenses L1 and L2 and is focused through the reference and the measurement filters on the rotating-filter wheel. A chopper motor is used to rotate the filter wheel. The light is chopped into reference and measurement signals. The chopped light passes through lens L3, where the light is collimated before entering the sample cell. The collimated light passes through the sample cell and onto lens L4, where the light is focused onto the detector.

The electronics of the analyzer calculate the log of the ratio of the reference and measurement signals. This calculated output is proportional to the concentration of the measured component. Rotating-filter photometers compensate for source decay, for dirt or bubbles in the sample, for detector drift, and for cell window obscuration, because these changes equally affect the reference and measurement signals.

The use of a single detector eliminates drift in analyzer response, which can result from uneven aging of multiple detectors. The use of a rotating filter introduces moving parts that require maintenance. Accuracy and stability are good, and cost is moderate.

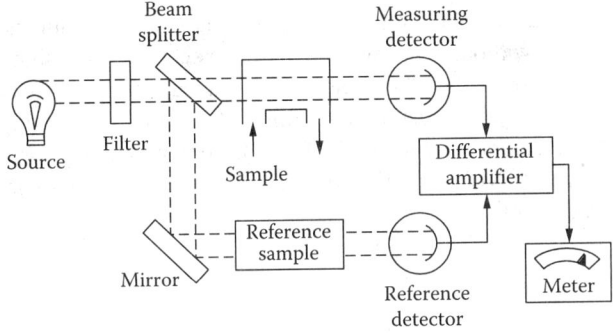

FIG. 1.69m
Dual-beam, dual-detector analyzer.

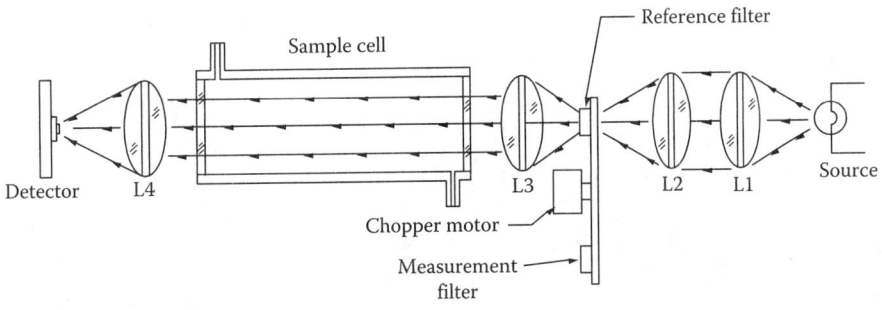

FIG. 1.69n
Multiwave flicker photometer design.

These types of analyzers can be provided with both UV and IR sources and detectors. In this type of a multichannel photometer (Figure 1.69o), the reference wavelength is chosen to fall in a location where the components of interest have little or no absorbance. The role of the software in the microprocessor is to eliminate interferences by applying the right response factor to each filter. This way, the absorbance readings are then converted into composition measurements. One of the advantages of this design is its ability to compensate for both the source and detector aging and for cell window obstruction. The design of these multicomponent analyzers also allows the sample cell to be isolated from the electronics.

Photodiode Array (PDA) Spectrophotometers

PDA spectrophotometers monitor all wavelengths in the spectrum simultaneously. A schematic of a diode array instrument is presented in Figure 1.69p.

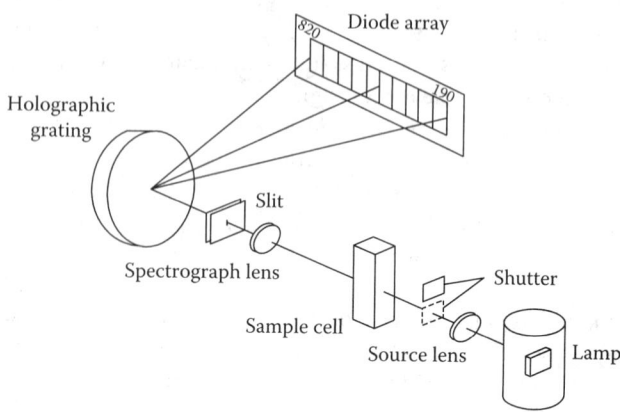

FIG. 1.69p
Photodiode array spectrophotometer components.

The light from the source passes through a lens and a shutter and through the sample cell. The light passes into the detector assembly, where it is focused onto a slit and onto a holographic grating, which disperses the light that strikes the PDA detector. The diode array detector is a series of linearly spaced silicon photodiode detectors that measure absorbance at a specific spectral bandwidth.

The advantage of this design is that the spectrum is scanned without any moving parts except for the shutter. Data from the PDA detectors are acquired in parallel, which results in a fast analysis time. Most of the work with PDA detectors has been in the UV/VIS region, but the short-wavelength NIR region (800–1100 nm) is utilized in some PDA devices.

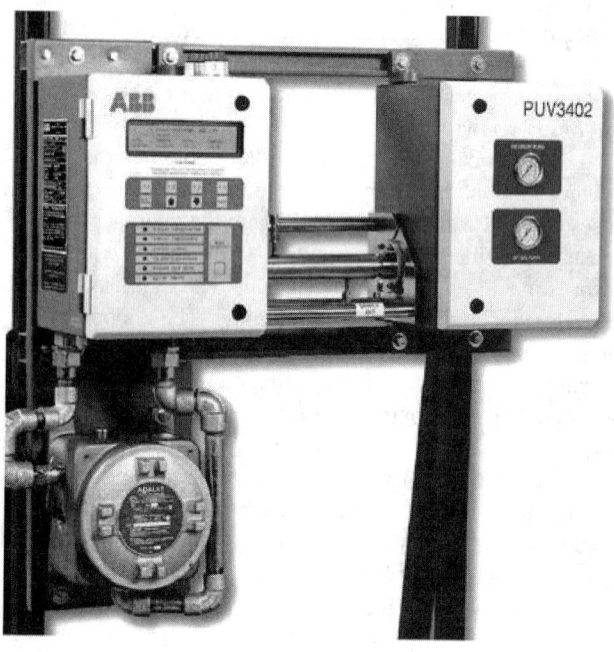

FIG. 1.69o
This multichannel process photometer can be provided with either UV or IR source and detectors, and it can analyze both liquid and gas samples. (Courtesy of ABB, Anatrac.)

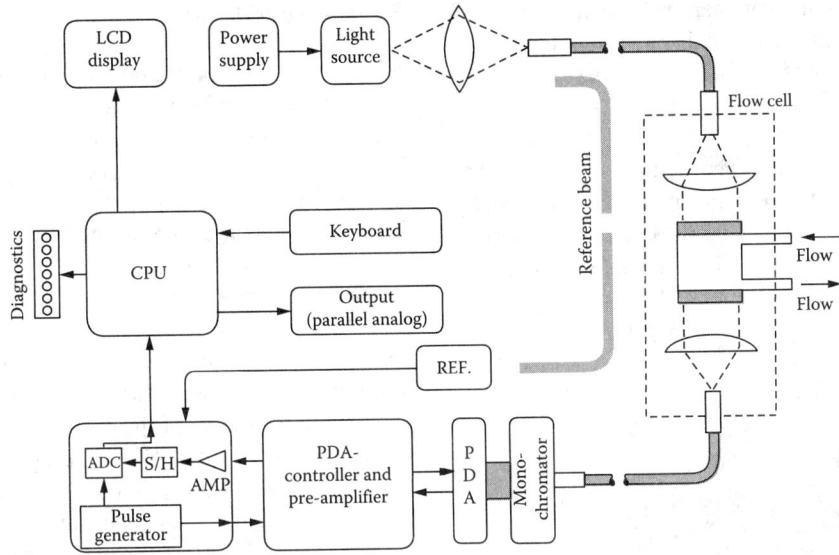

FIG. 1.69q
Photodiode array-type UV analyzer with fiber-optic cables.

Analyzers are also available that couple fiber-optic cables (Chapter 1.23) with PDA devices (Figure 1.69q). Analysis of hydrogen sulfide and sulfur dioxide in sulfur recovery and the determination of octane number are two examples of PDA applications.

Scanning Spectrophotometers

Scanning spectrophotometers are dispersive devices that normally utilize diffraction gratings to scan across a spectral region. Scanning devices can be used for multiple-component applications. Scanning spectrophotometers can be used in the UV, VIS, and NIR regions.

The measurement of moisture, protein, and carbohydrate in wheat and soybeans by Karl Norris of the USDA was one of the prominent early applications of an NIR scanning device. An important development has been the interfacing of fiber-optic wavelengths with conventional scanning spectrophotometers (Figure 1.69r).

The optical waveguide is usually a single fiber-optic cable. An advantage of using fiber-optics is the elimination of the sample-handling system. In the fiber-optic design, the polychromatic light from the source passes through lenses and filters and onto the fiber-optic cable. The light is transferred along the cable to the sample probe. The sample-modified light is then collected by a second fiber cable and transferred to the monochromator, where it is diffracted into individual wavelengths and measured by the detectors.

The data from the analyzer are processed by a personal computer (PC). The PC allows for the use of multivariate calibration techniques such as partial least squares.

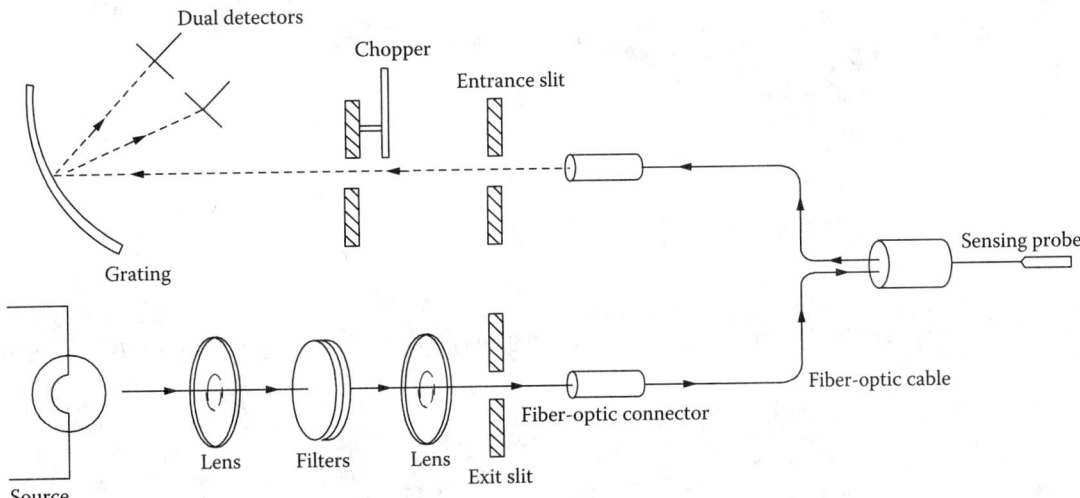

FIG. 1.69r
Scanning spectrophotometer with fiber-optic interfacing sensing probe.

The measurement of multicomponent solvent mixtures and the determination of octane number for gasoline samples are two typical applications performed on fiber-optic scanning spectrometers.

In addition to fiber-optic probes, in-line sensors are also available (Figure 1.69s). These units eliminate the need for sampling as the optical path passes directly through the process stream. The fiber-optic connector can be used from NIR to UV spectral ranges and is suited for operating temperatures of up to 240°C (464°F), operating pressures of up to 325 bars (4713 psig), and for path lengths from 1 mm to 1.0 m.

FIG. 1.69s
Fiber-optic transmission sensor. (Courtesy of Optek.)

Retroreflector Probes

Figure 1.69t illustrates a microprocessor-based UV analyzer probe, which can be used for monitoring both sulfur dioxide and nitrogen oxide in stack gas. In this design, the UV light from a deuterium source is projected through the gas measurement cavity inside the filter at the end of the probe. The retroreflector at the tip of the measurement cell returns the measurement beam to a point at which a monochromator separates the light into discrete spectral bands.

The monochromator has two exit slits, one for sulfur dioxide and the other for nitrogen oxide, that allow their corresponding wavelengths to impinge on the detector. By using the second derivative of the absorption spectra, the measurement is compensated for source aging, line voltage variations, broadband absorption, dirt buildup on the optics, and optical misalignment. A small shutter in the monochromator moves back and forth between the NO and the SO_2 beams, providing sequential reading of both at 1 min intervals.

The analyzer is provided with self-calibration capability, including auto-zero correction, auto-span check, and correction. The ceramic filter at the probe tip is self-cleaning and self-drying; at preset intervals, a blast of hot, pressurized instrument air is forced into the measurement cavity, which forces the entrained water or tar out of the probe and dries the filter. The analyzer is suitable for stack gas applications where the particulate loading, water vapor content, and temperature are all high (up to 427°C or 800°F).

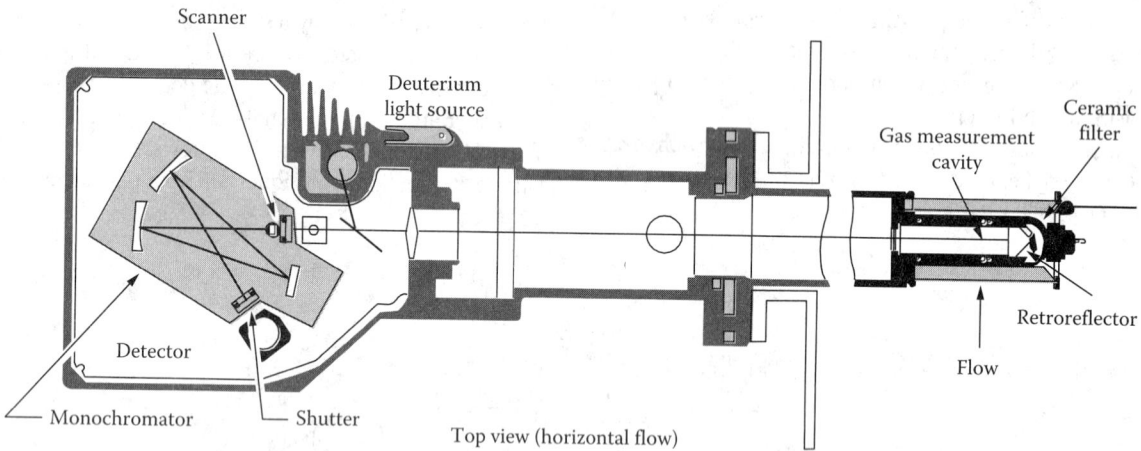

FIG. 1.69t
Self-calibrating, self-cleaning, and self-drying probe-type analyzer simultaneously detects the concentrations of SO_2 and NO at temperatures up to 800°F.

The analyzer measures only NO, so, if evaluations of NO_2 or NO_x are needed, additional measurements must be provided. The life of the deuterium lamp can be as short as 3 months.

VISIBLE AND NIR ANALYZERS

The absorption band and hardware used in VIS and NIR photometers are very similar to those used for UV photometers. See Chapter 1.32 for a full discussion of IR and NIR analyzers.

Visible Light Photometers

Absorption of light in the VIS region is also a result of electronic transitions. The VIS region of the electromagnetic spectrum (400–800 nm) is the region to which the human eye responds. A familiar example of VIS light is the dispersion of white light through a prism, which results in a rainbow of colors.

The observation of color arises from the absorption and emission of light by a sample. The intensity of the absorption of light at a characteristic wavelength is proportional to the intensity of the color. Photometers that operate in the VIS region are referred to as colorimeters (Chapter 1.13).

Split-beam and rotating fixed filter designs are also used in VIS photometers. The major difference with UV photometers is that a tungsten filament or quartz-halide source might be used. VIS photometers are frequently calibrated to correlate with accepted laboratory color units such as Saybolt, ASTM, and American Public Health Association (APHA). For example, the APHA units are used to measure the degree of yellowness of a sample. The APHA standards are based on known concentrations of platinum–cobalt solutions.

The following color measurements are examples of VIS photometer applications:

1. Yellow color of diesel oil—ASTM units, APHA units
2. Food color—APHA units
3. Color of purified organics—APHA units
4. Process water in papermaking

Near-Infrared Photometers

As detailed in Chapter 1.32, NIR photometers have been used for online process applications since the 1950s. The absorption bands observed in the NIR are overtone and combination bands of the rotation and vibration bands that occur in the fundamental IR region. Most of the useful bands in the NIR region involve hydrogenic functional groups such as O–H, C–H, and N–H.

The intensity of the overtone and combination bands is much weaker than the fundamental bands. The low molar absorptivity of the NIR region can be used to advantage in process applications, because longer path length sample cells can be used. The longer cells are less susceptible to plugging by particulate material, and the effects of coatings on the cell windows can also be minimized in the longer cells.

The two methods of sampling used in the NIR are transmission and diffuse reflectance. Transmission measurements are made on clear homogeneous liquid and gaseous samples. Solid samples that are optically opaque cannot be measured in the transmission mode. Solids and slurries are usually measured by a diffuse reflectance technique in which a portion of the light is transmitted through the sample and a portion of the light is returned to the surface. The reflected light is measured at the detector.

The measurement of water in organic liquids is a common application for NIR analyzers. Diffuse reflectance measurements of moisture in slurries and solids (Chapters 1.39 and 1.40) are also important applications. The following are examples of NIR applications:

1. Water in ethylene dichloride in vinyl chloride processes
2. Water in dimethylacetamide in nylon processes
3. Alcohol in hydrocarbons in the petrochemical industry
4. Moisture in paper slurries to check dryer efficiency
5. Moisture in grain in the food industry

CONCLUSIONS

The packaging of UV and VIS analyzers usually separates the amplifier/controller from the optical system (See Figure 1.69r). Some designs provide complete housing separation of the sample cell to isolate the electrical components from the flowing sample. The sample cell can then be temperature controlled when necessary, independent of the remaining optical system.

Vapor bubbles cannot be tolerated in the sample, because they will generate noise in the optical reading. If a gas sample pressurized cell is used, the cell pressure must be kept constant.

The UV analyzer has a unique and convenient feature that it offers a simple means of checking the calibration by use of a selected interference or broadband filter. Once the calibration (using known samples) is completed, a filter that absorbs at the wavelengths of interest can be used thereafter. The filter is selected to give equivalent absorbance of some percentage of full scale when placed in the measuring beam with a nonabsorber in the sample cell. Nitrogen or water is commonly used to purge the cell during this operation.

Online photometers and spectrophotometers are widely used for composition measurements in the UV–VIS–NIR spectral regions. Photometers are nondispersive devices that are used to make measurements at selected discrete wavelengths. Photometers are normally used for single-component measurements in process control applications. Spectrophotometers are dispersive devices that scan across a spectrum of wavelengths.

Spectrophotometers are often used for multiple-component measurements. The UV analyzer can be used to measure certain compounds having UV absorption characteristics if other UV-absorbing compounds are not present in the sample or if their presence can be compensated for.

Samples may be gas or liquid, but the sample must be relatively free of dirt and must be in a single phase. Trace measurements are also possible. Fiber-optic diode array designs can provide multicomponent liquid composition analysis, and retroreflector probes can provide self-cleaning and self-calibrating stack gas composition measurement.

SPECIFICATION FORMS

When specifying ultraviolet or visible analyzers, the ISA Form 20A1001 can be used to specify the features required for the device itself and when specifying both the device and the nature of the application, including the composition and/or properties of the process materials, ISA Form 20A1002 can be used. These forms, with the permission of the International Society of Automation, are reproduced on the next pages.

1.69 Ultraviolet and Visible Analyzers

#	RESPONSIBLE ORGANIZATION		ANALYSIS DEVICE	#	SPECIFICATION IDENTIFICATIONS			
1				6				
2	(ISA logo)		Operating Parameters	7	Document no			
3				8	Latest revision		Date	
4				9	Issue status			
5				10				

#	ADMINISTRATIVE IDENTIFICATIONS			#	SERVICE IDENTIFICATIONS Continued			
11				40				
12	Project number		Sub project no	41	Return conn matl type			
13	Project			42	Inline hazardous area cl		Div/Zon	Group
14	Enterprise			43	Inline area min ign temp		Temp ident number	
15	Site			44	Remote hazardous area cl		Div/Zon	Group
16	Area	Cell	Unit	45	Remote area min ign temp		Temp ident number	
17				46				
18	SERVICE IDENTIFICATIONS			47	COMPONENT DESIGN CRITERIA			
19	Tag no/Functional ident			48				
20	Related equipment			49	Component type			
21	Service			50	Component style			
22				51	Output signal type			
23	P&ID/Reference dwg			52	Characteristic curve			
24	Process line/nozzle no			53	Compensation style			
25	Process conn pipe spec			54	Type of protection			
26	Process conn nominal size		Rating	55	Criticality code			
27	Process conn termn type		Style	56	Max EMI susceptibility		Ref	
28	Process conn schedule no		Wall thickness	57	Max temperature effect		Ref	
29	Process connection length			58	Max sample time lag			
30	Process line matl type			59	Max response time			
31	Fast loop line number			60	Min required accuracy		Ref	
32	Fast loop pipe spec			61	Avail nom power supply			Number wires
33	Fast loop conn nom size		Rating	62	Calibration method			
34	Fast loop conn termn type		Style	63	Testing/Listing agency			
35	Fast loop schedule no		Wall thickness	64	Test requirements			
36	Fast loop estimated lg			65	Supply loss failure mode			
37	Fast loop material type			66	Signal loss failure mode			
38	Return conn nominal size		Rating	67				
39	Return conn termn type		Style	68				

#	PROCESS VARIABLES	MATERIAL FLOW CONDITIONS			#	PROCESS DESIGN CONDITIONS		
69					101			
70	Flow Case Identification			Units	102	Minimum	Maximum	Units
71	Process pressure				103			
72	Process temperature				104			
73	Process phase type				105			
74	Process liquid actl flow				106			
75	Process vapor actl flow				107			
76	Process vapor std flow				108			
77	Process liquid density				109			
78	Process vapor density				110			
79	Process liquid viscosity				111			
80	Sample return pressure				112			
81	Sample vent/drain press				113			
82	Sample temperature				114			
83	Sample phase type				115			
84	Fast loop liq actl flow				116			
85	Fast loop vapor actl flow				117			
86	Fast loop vapor std flow				118			
87	Fast loop vapor density				119			
88	Conductivity/Resistivity				120			
89	pH/ORP				121			
90	RH/Dewpoint				122			
91	Turbidity/Opacity				123			
92	Dissolved oxygen				124			
93	Corrosivity				125			
94	Particle size				126			
95					127			
96	CALCULATED VARIABLES				128			
97	Sample lag time				129			
98	Process fluid velocity				130			
99	Wake/natural freq ratio				131			
100					132			

#	MATERIAL PROPERTIES			#	MATERIAL PROPERTIES Continued		
133				137			
134	Name			138	NFPA health hazard	Flammability	Reactivity
135	Density at ref temp	At		139			
136				140			

Rev	Date	Revision Description	By	Appv1	Appv2	Appv3	REMARKS

Form: 20A1001 Rev 0

© 2004 ISA

1000 Analytical Measurement

	RESPONSIBLE ORGANIZATION	ANALYSIS DEVICE COMPOSITION OR PROPERTY Operating Parameters (Continued)		SPECIFICATION IDENTIFICATIONS
1	ISA		6	
2			7	Document no
3			8	Latest revision — Date
4			9	Issue status
5			10	

	PROCESS COMPOSITION OR PROPERTY			MEASUREMENT DESIGN CONDITIONS				
	Component/Property Name	Normal	Units	Minimum	Units	Maximum	Units	Repeatability
13								
14								
15								
16								
17								
18								
19								
20								
21								
22								
23								
24								
25								
26								
27								
28								
29								
30								
31								
32								
33								
34								
35								
36								
37								

(Rows 38–75: blank notes area)

Rev	Date	Revision Description	By	Appv1	Appv2	Appv3	REMARKS

Form: 20A1002 Rev 0   © 2004 ISA

Definitions

Beer–Lambert law relates the absorption of light to the transmission properties of the material through which the light is traveling. In other words, it relates the amount of light absorbed to the concentration of the material of interest in the sample.

Monochromators are instruments that transmit mechanically selectable narrow wavelength bands (λ) of UV or other types of radiation chosen from a wider range of wavelengths from the entering radiation. Glass is not transparent to UV radiation, so quartz prisms are used to disperse it. The monochromator can use the phenomenon of optical dispersion by prisms or can use diffraction grating to separate the wavelength bands (λ). The gratings operate in the reflective mode and they can be triangular mirrored half-prisms, producing linear dispersion or holographic gratings having sinusoidal grove profiles (cross-section).

Nanometer is equal to one billionth of a meter (0.000000001 m).

Optical dispersion is the separation of incoming radiation into wavelength bands (λ). When a narrow wavelength band (λ) of UV or other type of radiation is to be selected, that can be done by dispersive or non-dispersive selectors. Filters are called non-dispersive (λ) selectors while prisms and gratings are called dispersive ones. A linear grating is an optically flat polished surface having dense and parallel groves (about 1000 per mm). The two types of gratings are the transmission and reflection (diffraction) types. In spectroscopic instruments, almost all gratings are of the reflection type. The gratings can produce linear or holographic dispersion. Linear dispersion is produced by triangular flat surfaced gratings. Holographic gratings are made by laser technology which produces a pattern of straight lines with a sinusoidal grove profile (cross-section).

Abbreviations

CEM	Continuous-Emission Monitor
NIR	Near-infrared
nm	Nanometer (one billionth of meter)
PDA	Photodiode arrays
PLS	Partial least squares
PMT	Photomultiplier tube
UV/VIS	Ultraviolet–visible
VDT	Video display tube

Organizations

APHA	American Public Health Association
ASTM	American Society of Testing and Materials

Bibliography

Callis, J.B. and Christian, G.D., *Trace Analysis: Spectroscopic Methods for Molecules*, John Wiley & Sons, New York, 1986. http://onlinelibrary.wiley.com/doi/10.1002/ange.19870991046/abstract.

Emerson-Rosemount, Continuous emission monitoring, (2013) http://www.altechusa.com/upload/file/MIR-HCL%202013.pdf.

Gorog, S., *Ultraviolet-Visible Spectrometry in Pharmaceutical Analysis*, CRC Press, Boca Raton, FL, 1995. http://www.sigmaaldrich.com/catalog/product/sigma/z369306?lang=en&region=US.

Lang, L., Absorption spectra in the ultraviolet and visible region, (2000) http://www.amazon.com/Absorption-Spectra-Ultraviolet-Visible-Region/dp/0124363202.

Mellon, M.G., Analytical absorption spectroscopy, (2007) http://www.amazon.com/Analytical-Absorption-Spectroscopy-M-G-Mellon/dp/1406751707.

Misra, P., Ultraviolet spectroscopy and UV lasers, (2002) http://www.amazon.com/Ultraviolet-Spectroscopy-And-Lasers-Practical/dp/0824706684.

Palmer, C., *Diffraction Grating Handbook*, 6th edn., Newport Corporation, Rochester, NY, 2005. http://www.amazon.com/Diffraction-Grating-Handbook-Christopher-Palmer/dp/B00122ZEJC.

Perkampus, H.-H., Analytical applications of UV-VIS spectroscopy, (1992) http://link.springer.com/chapter/10.1007/978-3-642-77477-5_4.

Perkin Elmer, Validating UV/visible spectrophotometers, (2012) http://www.perkinelmer.com/CMSResources/Images/44-136839TCH_Validating_UV_Visible.pdf.

Pharma Change, Ultraviolet-visible (UV-Vis) spectroscopy—Derivation of Beer–Lambert law, (2012) http://pharmaxchange.info/press/2012/04/ultraviolet-visible-uv-vis-spectroscopy-%e2%80%93-derivation-of-beer-lambert-law/.

Rao, C.N.R., *Ultra-Violet and Visible Spectroscopy*, 3rd edn., Butterworth-Heinemann, London, U.K., 1967. http://www.alibris.com/Ultra-violet-and-visible-spectroscopy-chemical-applications-C-N-R-Rao/book/6888030.

Rosemount, UV analyzer, (1998) http://www2.emersonprocess.com/siteadmincenter/PM%20Rosemount%20Analytical%20Documents/PGA_Manual_890_199801.pdf.

Samson, J.A., Vacuum ultraviolet spectroscopy, (2000) http://www.sciencedirect.com/science/book/9780126175608.

Sherman, R.E., *Process Analyzer Sample-Conditioning System Technology*, John Wiley & Sons, New York, 2002. http://www.wiley.com/WileyCDA/WileyTitle/productCd-0471293644.html.

Soovali, L. et al., Uncertainty sources in UV-Vis spectrophotometric measurement, (2006) http://link.springer.com/article/10.1007/s00769-006-0124-x?no-access=true.

Stellar Net Inc. Fiber optic spectrum analyzers, (2013) http://www.stellarnet.us/popularconfigurations_fiberanlzr.htm.

Thomas, M.J.K. et al., *Ultraviolet and Visible Spectroscopy*, John Wiley & Sons, New York, 1996. http://www.amazon.com/Ultraviolet-Visible-Spectroscopy-Analytical-Chemistry/dp/0471967432/ref=sr_1_sc_2?s=books&ie=UTF8&qid=1390739266&sr=1-2-spell&keywords=ultraviolet+masurements.

Water Online, (A,B) ETS-UV for drinking water disinfection, (2016) http://www.wateronline.com/doc/ets-uv-for-drinking-water-disinfection-0001.

Workman, J., *Applied Spectroscopy: A Compact Reference for Practitioners*, Academic Press, San Diego, CA, 1998. http://www.amazon.com/Applied-Spectroscopy-Compact-Reference-Practitioners/dp/0127640703.

WOW, Ultraviolet–visible spectroscopy, (2015) http://us.wow.com/wiki/Ultraviolet-visible_spectroscopy.

1.70 Viscometers
Application and Selection

C. H. KIM (1969, 1982)   **B. G. LIPTÁK** (1995, 2003)   **M. ZAVRŠNIK** (2017)

Definition of viscosity	Viscosity is a fundamental characteristic of all liquids and provides a measure of a fluid's resistance to flow or shear. It describes the internal friction of a moving fluid. It is expressed as dynamic or absolute viscosity η (Eta) and as kinematic viscosity ν (Nu), where $\nu = \eta/\rho$ and ρ (Rho) is the fluid's density.
Viscosity units	Dynamic viscosity–SI Unit—Pa·s (Pascal-second) 1 Pa·s = 1 Ns/m^2 = 1 kg/s m (other commonly used unit is mPa·s); 1 mPa·s = 1 cP (Centipoise)
	Kinematic viscosity–SI Unit—m^2/s (square meters per second) (the most commonly used unit is mm^2/s); 1 mm^2/s = 1 cSt (Centistokes)
Types of viscous behavior	Newtonian (or ideally viscous fluids—independent of the applied shearing) and non-Newtonian (influenced by the amount of shearing)

INTRODUCTION

In industrial plants, viscosity measurements serve to determine the resistance of fluids to flow, define the behavior of various concentrations of slurries, or measure the molecular weight of polymers. For ideal viscous fluids measured at constant temperature, the value of the ratio of the shear stress τ (Tao) to the corresponding shear rate $\dot{\gamma}$ (Gamma dot) is dynamic (absolute) viscosity η (Eta).

Chapters 1.70 through 1.72 deal with viscosity measurement. This chapter provides some general orientation on viscometer selection and application, Chapter 1.72 discusses laboratory units, and Chapter 1.71 covers industrial viscometers. In addition, the reader is referred to the related detectors that measure consistency (Chapter 1.16) and molecular weight (Chapter 1.41).

This chapter begins with some background on viscous behavior, followed by an orientation table that provides guidelines to assist the reader in the selection and application of viscometers, in the form of a listing of the features and capabilities of both laboratory and industrial viscometers. Finally, the chapter is concluded with some definitions of terms and units that are used in connection with viscometry and with definitions of the different types of viscous behavior exhibited by industrial fluids.

THEORY OF VISCOUS BEHAVIOR

Viscosity is a fluid property that defines the fluid's resistance to flow. In order to provide some mathematical formalism, a simple two-plate model will be used. Consider a liquid between two closely spaced parallel plates as shown in Figure 1.70a.

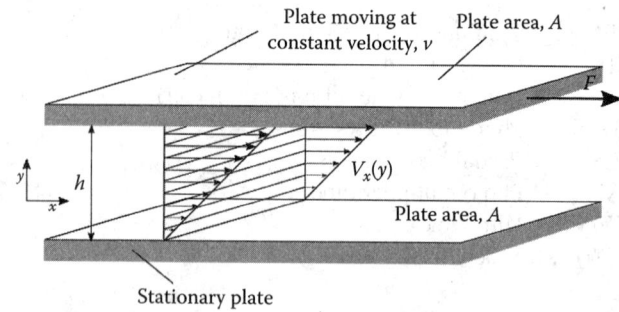

FIG. 1.70a
Two-plate model—laminar flow.

A force, F, applied to the top plate causes the fluid adjacent to the upper plate to be dragged in the direction of F. The applied force is communicated to neighboring layers of fluid below, each coupled to the driving layer above, but with diminishing magnitude. This results in the progressive decrease in the velocity of each fluid layer, as shown by the decreasing velocity vector away from the upper plate. In this system, the applied force is called a *shear* (when applied over an area, it is called a *shear stress* τ—Tau)

$$\tau = \frac{F}{A}\left[\frac{N}{m^2}\right] = [Pa] \qquad 1.70(1)$$

and the resulting deformation rate of the fluid, as illustrated by the velocity gradient $v_x(y)$, is called the *shear rate* $\dot{\gamma}$—Gamma dot (in old literature, the shear rate is sometimes given the variable D):

$$\dot{\gamma} = \frac{v}{h}\left[\frac{N}{s\,m}\right] = \left[\frac{1}{s}\right] = [s^{-1}] \qquad 1.70(2)$$

The mathematical expression describing the viscous response of the system to the shear stress is simply

$$\underset{\text{Shear stress}}{\tau} = \underset{\text{(Shear rate)}}{\dot{\gamma}} * \underset{\text{(Viscosity)}}{\eta} \qquad 1.70(3)$$

where

- τ is the shear stress, which is the force per unit area exerted on the upper plate in the *x*-direction (and hence is equal to the force per unit area exerted by the fluid on the upper plate in the *x*-direction under the assumption of a no-slip boundary layer at the fluid–upper plate interface)
- $\dot{\gamma}$ is the gradient of the *x*-velocity in the *y*-direction in the fluid
- η is the dynamic viscosity

$$\eta = \frac{\tau}{\dot{\gamma}} \qquad 1.70(4)$$

In fluid mechanics, diffusion of momentum is a more useful description of viscosity where the motion of a fluid is considered without reference to force. This type of viscosity is called the *kinematic viscosity* ν—nu—and is derived by dividing dynamic viscosity by mass density:

$$\nu = \frac{\eta}{\rho} \qquad 1.70(5)$$

Stokes' Law

Stokes' falling ball principle, published in 1851, was based on his investigations of spheres falling through liquids. The theory is based on the simplest measuring scheme: a solid sphere is used as a moving body. The sphere falls along the axis of a cylindrical tube filled with liquid being tested, and the velocity of the sphere is measured. It is assumed that the tube is vertical. The theory of motion of a solid sphere is valid under the following assumptions:

- The motion is steady, that is, the rate of sphere descent is constant.
- The inertia effects are considered negligible.
- Liquid is Newtonian.
- The wall effect on a sphere motion is neglected, that is, the sphere radius is much smaller than the tube radius.
- The sphere moves strictly in a vertical direction.

The measured viscosity is inversely proportional to the steady-state velocity of sphere descent and linearly proportional to the difference in densities of sphere and fluid:

$$\eta = \frac{r^2\left(\rho_{sphere} - \rho_{fluid}\right)g}{18v} \qquad 1.70(6)$$

where

- η is the viscosity
- r is the radius of the sphere
- v is the steady-state velocity of sphere descent
- ρ is the density (sphere and fluid)
- g is the gravitational constant

Hagen–Poiseuille Law

Capillary viscometers measure viscosity by detecting the flow or the pressure drop of Newtonian process liquid through a capillary under isothermal laminar flow conditions. According to the Hagen–Poiseuille law, the pressure drop of a Newtonian liquid passing through a capillary tube is directly proportional to its viscosity if the fluid's temperature and flow rate are kept constant. The following equation describes the Hagen–Poiseuille law, which governs the flow of fluids through capillaries:

$$\eta = \frac{\Delta p}{L}\frac{\pi R^4}{8Q} \qquad 1.70(7)$$

where

- η is the viscosity
- R is the radius of the capillary
- L is the length of the capillary
- Q is the volume flow rate
- Δp is the pressure drop across the length of the capillary
- g is the gravitational constant
- π is the mathematical constant (3.14159)

If we know the density ρ, kinematic viscosity can be calculated using Equation 1.70(5).

Intrinsic Viscosity

To determine the molecular weight of a polymer, the intrinsic viscosity or limiting viscosity must be determined (Figure 8.62b). Intrinsic viscosity $[\eta]$ is defined by the relationship

$$[\eta] = \lim_{c \to 0} \frac{\eta - \eta_0}{\eta_0 c} = \lim_{c \to 0} \frac{\eta_{sp}}{c} \qquad 1.70(8)$$

$$\eta_{sp} = \frac{\eta - \eta_0}{\eta_0} \qquad 1.70(9)$$

where
- η is the viscosity of the solution
- η_0 is the viscosity of the solvent
- R is the radius of the capillary
- c is the solution concentration in grams per milliliter or grams per deciliter (Figure 1.70b)

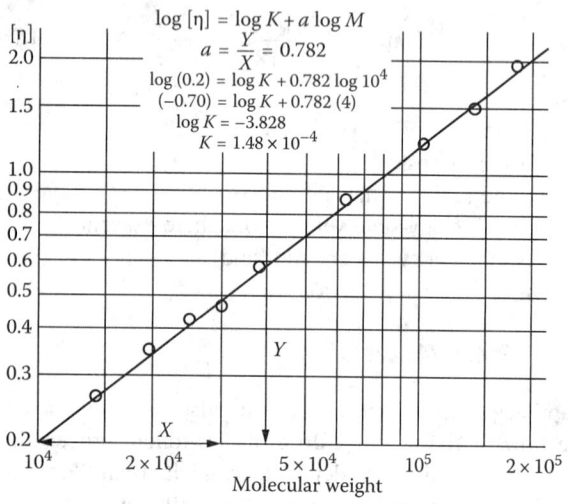

FIG. 1.70b
Relationship between intrinsic viscosity and molecular weight.

Another definition of intrinsic viscosity is

$$[\eta] = \lim_{c \to 0} \frac{\ln \eta / \eta_0}{c} \qquad 1.70(10)$$

Intrinsic viscosity is related to molecular weight as expressed by the Mark–Houwink equation:

$$[\eta] = K M^a \quad \text{or} \quad \log[\eta] = \log K + a \log M \qquad 1.70(11)$$

where K and a are constants for a given polymer–solvent system at the temperature of the viscosity measurement.

Newtonian and Non-Newtonian Fluids

Isaac Newton is credited with first suggesting a model for the viscous property of fluids in *Principia* published in 1687. Newton proposed that the resistance to flow caused by viscosity is proportional to the velocity at which the parts of the fluid are being separated from one another because of the flow; hence, the viscosity of a Newtonian or ideal liquid is one where the viscosity is constant as the rate of shear increases. Fluids that obey Newton's law of viscosity are called Newtonian fluids.

Fluids that do not obey Newton's law of viscosity are called non-Newtonian fluids. Therefore, one cannot speak of the viscosity of a non-Newtonian fluid without specifying the shear stress or the velocity gradient (shear rate) at which the resistance to deformation is of interest. Consequently, the viscometers used to measure non-Newtonian substances must be provided with accurate means of detecting the velocity gradient. Generally, non-Newtonian fluids are complex mixtures: slurries, pastes, gels, polymer solutions, etc. (Figure 1.70c).

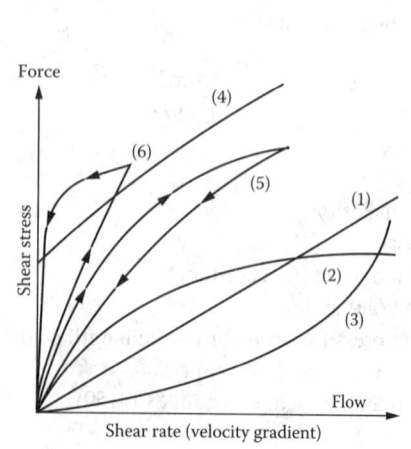
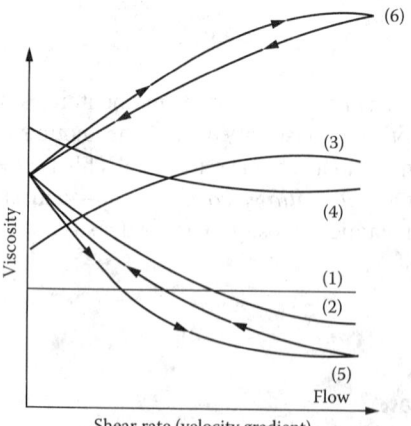

FIG. 1.70c
The viscosity and shear stress of Newtonian and non-Newtonian fluids when these fluids are deformed at various shear rates.

Newtonian Fluids Figure 1.70c illustrates the behavior of Newtonian and non-Newtonian fluids. When a fluid is Newtonian (curve 1 in Figure 1.70c), its viscosity is unaffected by shear rate, and the relationship between force (stress) and resulting flow (velocity) is linear. Some of the Newtonian fluids include gasoline, kerosene, mineral oils, water, and salt solutions in water.

Pseudoplastics Pseudoplastics (curve 2) are shear-thinning materials whose apparent viscosity drops as shear rate rises. Some such substances exhibit a yield stress above which the apparent viscosity drops, so that a unit increase in driving force results in more and more flow. Pseudoplastic materials include ketchup/tomato sauce, modern paints, paper pulp solutions, and polymer solutions.

Dilatant Fluids Dilatant (curve 3) materials are shear-thickening substances. Their apparent viscosity increases as the shear rate rises, and more and more stress is required to obtain the same increase in flow. Dilatant materials include quicksand, starch, peanut butter, and many candy compounds.

Plastic Solids Plastic solids (curve 4) are true plastics in the sense that they normally behave like solids, but when the shear stress (force) reaches their yield point, they start to behave as viscous fluids and start to *cold flow*. Most plastics, chewing gum, tar, and some oils exhibit this behavior.

Thixotropic Materials A material that has a viscosity that decreases with time under constant shear stress and takes a fixed time to return to a more viscous state after the stress has been removed is said to be thixotropic (curve 5). Thixotropic substances include coatings, sealants, creams, and suspensions.

Rheopectic Substances A material that has a viscosity that increases with time under constant shear stress and takes a fixed time to return to a less viscous state after the stress has been removed is said to be rheopectic (curve 6). Rheopectic substances include latex dispersions and clay solutions in water.

Apparent Viscosity Readings Apparent viscosity is the viscosity of a non-Newtonian fluid measured at specified shear conditions. It is important to understand that the apparent viscosity depends on the design of the viscometer that is measuring it. Each viscometer exposes the process sample to a different experience in terms of shear stress and shear rate, so non-Newtonian substances will register different apparent viscosity readings (Table 1.70a).

TABLE 1.70a
Viscosities of Different Materials in Different Units of Viscosity, Measured at Constant Temperature of 23°C (70°F)

Saybolt Universal (SSU)	Centistokes (cSt)	Centipoise[a] (cp)	Typical Liquid at 70°F[b]
31	1.00	1.0	Water
35	2.56	2.05	Kerosene
50	7.40	5.92	No. 2 fuel oil
80	15.7	12.6	No. 4 fuel oil
100	20.6	16.5	Transformer oil
200	43.2	34.6	Hydraulic oil
300	65.0	52.3	SAE 10W oil
500	110	88.0	SAE 10 oil
1,000	220	176	SAE 20 oil
2,000	440	352	SAE 30 oil
5,000	1,100	880	SAE 50 oil
10,000	2,200	1,760	SAE 60–70 oil
50,000	10,800	8,640	Molasses B

Source: Courtesy of Cole-Parmer Instrument Co.
[a] Centistokes × specific gravity = centipoise. Specific gravity is assumed to be 0.8 except in the case of water.
[b] Use actual specific gravity for liquid in question.

Typical Viscosities and Conversion among Units

Newton's hypothesis defines absolute viscosity of fluid as

$$\text{Absolute viscosity} = \frac{\text{shear stress}}{\text{shear rate}} = \frac{\text{shear stress}}{\text{velocity gradient}}$$

$$= \frac{(F/A)g_c}{(u/l)}, \text{poise} \qquad 1.70(12)$$

For the viscosity of common liquids and for conversion factors between SI units and some common non-SI units, see Tables 1.70b and 1.70c. There are countless other non-SI units of viscosity associated with different industries, and unfortunately, there is often no standard method for converting these into dimensionally correct SI viscosity units. An approximate conversion chart is provided in Table 1.70d.

TABLE 1.70b
Typical Viscosities of Materials Measured at 20°C

Material	Dynamic Viscosity (mPa·s)
Air	0.015
Gasoline, petrol (octane)	0.538
Water	1.0
Ethanol	1.2
Mercury	1.55
Blood plasma	1.70
Light oils	10
Glycol	20
Sulfuric acid	25
Motor oils	50–1,000
Olive oils	Approx. 1,000
Glycerin	1480
Honey	Approx. 10,000
Bitumen	20 kPa·s

TABLE 1.70c
Definition of Units and Their Equivalences

Unit System	Derived Viscosity Unit	Unit Name	Equivalence
Units for dynamic viscosity			
SI	kg/s·m or Pa·s	Pascal-second	1 Pa·s = 10 Poise
CGS	g/cm·s or dyne·s/cm²	Poise	1 Poise = 0.1 Pa·s
FPS	lb$_F$·s/in.²	Reyn	1 Reyn = 6894.8 Pa·s
Units for kinematic viscosity			
SI	m²/s	Square meters per second	1 m²/s = 10,000 Stokes
CGS	cm²/s	Stokes	1 Stoke = 10^{-4} m²/s
FPS	in.²/s, ft²/s	/	1 in.²/s = 645.16 mm²/s

TABLE 1.70d
Approximate Viscosity Conversion Chart

Note: Scales in lower section read directly in centipoises. Those above read in centistokes (to covert into centipoises, multiply by liquid specific gravity).

TABLE 1.70e
Conversion between mPa·s (Centipoise—cP) and Other Units of Absolute Viscosity

Name (Definition)	Abbreviation	Value Equivalent to 1 mPa·s
kgf-s/m^2	—	0.00010197
kgm/m-h	—	3.6
lbf-s/ft^2	—	0.00002088
lbf-s/in.^{2}	—	0.000000175
lbm/ft-s	—	0.000672
lbm/ft-h	—	2.42
lbm/in.-s	—	0.000056

Numerous converters or conversion tables can be found by searching the World Wide Web. A short summary is given in Table 1.70e.

Kinematic Viscosity The value of the kinematic viscosity ν (in cm^2/s units) can be obtained approximately from the indications of the following viscometers (which all give their readings in seconds) by the associated equations:

Saybolt Universal
  when $32 < t < 100$, $\nu = 0.00226t - 1.95/t$
  when $t < 100$, $\nu = 0.00220t - 1.35/t$

Saybolt Furol
  when $25 < t < 40$, $\nu = 0.00224t - 1.84/t$
  when $t > 40$, $\nu = 0.216t - 0.60/t$

Redwood No. 1 (English)
  when $34 < t < 100$, $\nu = 0.00260t - 1.79/t$
  when $t > 100$, $\nu = 0.00247t - 0.50/t$

Redwood Admiralty
  (English), $\nu = 0.027t - 20/t$
  Engler (German), $\nu = 0.00147t - 3.74/t$

VISCOMETER SELECTION AND APPLICATION

Table 1.70f, which is an orientation table, lists a range of viscometers and compares the features of the various designs to assist the reader in selecting the right one for the application at hand. When several choices appear to be acceptable for a particular application, the reader is advised to read about each in the following chapters before making the final selection. Chapter 1.72 covers the laboratory-type viscometer designs, and Chapter 1.71 describes the industrial in-line viscometers.

TABLE 1.70f
Orientation Table for Industrial Viscometers

Category	Types of Design		Provides Continuous Signal	In-Line Device	Laboratory Device	Local Readout	Remote Readout Trans.	Temp. Compensation	Gas	Newtonian	Non-Newtonian	Maximum Design Pressure (×10⁵ Pa)	Maximum Design Temperature (°C)	Inaccuracy (±%) (1) Based on Full Scale (2) Based on Measurement	Minimum Samples Size of Flow Rate (dm³·hr)	Applicable Viscosity Ranges in mPa·s
Laboratory	Bubble time	Manual			✓	✓				✓		1	25	2–10(2)	13 cm³	
	Capillary	Manual timing			✓	✓				✓		1	150	0.35(2)	20 cm³	
		Auto timing			✓	✓	✓			✓		1	150	0.35(2)	0.4 cm³	
		Extrusion		✓	✓	✓	✓			✓	✓	35	180	2(2)	30 cm³	
	Falling/rolling element	Ball			✓	✓	✓		✓	✓	✓	1000	180	0.5(2)	0.1 cm³	
		Needle													1	
					✓	✓		✓		✓		1	150	0.5(2)	10 cm³	
	Orifice/efflux cup	Saybolt	✓		✓	✓				✓		1	120	0.5(2)	60 cm³	
		Ford	✓		✓	✓				✓		1	27	2(2)	150 cm³	
		Zahn	✓		✓	✓				✓		1	27	2(2)	44 cm³	
		Auto			✓	✓	✓			✓		1	27	5(2)	44 cm³	
	Oscillating element	Piston			✓	✓	✓	✓		✓	✓	1	180	1(1)	2 cm³	
	Rotational	Coaxial cylinder	✓		✓	✓	✓			✓	✓	1	1600	0.5(2)	2 cm³	
		Cone and plate	✓		✓	✓	✓			✓	✓	1000	1000	0.5(2)	0.1 cm³	
		Parallel Plate	✓		✓	✓	✓			✓	✓	1000	1000	0.5(2)	0.2 cm³	
	Stabinger	Stabinger	✓		✓	✓	✓			✓		1	105	0.35(2)	2.5 cm³	
	Vibrating	Damping	✓		✓	✓	✓	✓		✓		1	100	1(1)	10 cm³	
Industrial	Capillary	Differential pressure	✓	✓	✓	✓	✓			✓		47	480	1–2(1)	4–15	
		Backpressure	✓	✓	✓	✓	✓			✓		35	100	1(1)	4	
	Coriolis	Torsional movement	✓	✓		✓	✓	✓		✓	✓	100	150	5(2)	—	
	Falling element	Falling ball		✓	✓	✓	✓	✓		✓		22	180	1(1)	—	
		Falling piston		✓	✓	✓	✓	✓		✓		35	350	1(1)	—	
		Falling needle		✓	✓	✓	✓	✓		✓		135	350	1(1)	—	
	Float	Single float	✓	✓		✓	✓			✓	✓	45	230	4(2)	180–480	
		Two-float	✓	✓	✓	✓	✓	✓		✓	✓	22	230	2–4(2)	60–540	
		Concentric	✓	✓	✓	✓	✓			✓		45	230	2–4(2)	480	
	Plastometers	Cone and plate		✓	✓	✓	✓				✓	8	200	0.5(1)	25 cm³	
		Kneader	✓	✓	✓	✓	✓	✓			✓	1	300	1(1)	80 cm³	Arbitary units are used
		Capillary	✓	✓	✓	✓	✓	✓			✓	345	300	2(1)	0.6 #/h	MU (money units)
	Rotational element	Rotating spindle	✓	✓	✓	✓	✓	✓		✓	✓	70	1300	1(1)	—	
		Gyrating element	✓	✓		✓	✓	✓		✓	✓	70	300	1(2)	—	
		Agitator Power	✓	✓		✓	✓			✓	✓	10	100	~5(1)	—	
		Dynamic liquid pressure	✓	✓		✓	✓	✓		✓	✓	25	200	1(2)	—	
	Oscillating element	Blade		✓	✓	✓	✓			✓	✓	16	60	1(1)		
		Piston	✓	✓	✓	✓	✓	✓		✓	✓	690	370	1(2)		
	Vibrating element	Torsional	✓	✓	✓	✓	✓	✓		✓	✓	350	450	2(2)		
		Reed rod	✓	✓	✓	✓	✓	✓		✓	✓	500	300	1(1)	—	

Viscosity range columns: 10^{-1}, 1, 10, 10^2, 10^3, 10^4, 10^5, 10^6, 10^7, 10^8, 10^9

——— Normal range   - - - - With special modifications

Selection

In selecting a viscometer for a specific task, the following should be considered:

1. Is this instrument for laboratory use or for continuous measurement in the plant for control?
2. What type of materials will this viscometer handle?
   a. Highly volatile? Closed system needed?
   b. Newtonian fluids, non-Newtonian fluids, or both?
   c. Rheological characteristics of the material—pseudoplastic, thixotropic, dilatant, etc.
   d. Corrosiveness of the fluids.
   e. Does the fluid contain solids? What are the special characteristics of this slurry or emulsion?
   f. What are the operating temperatures and pressures of the fluids?
   g. Do the sample composition and/or viscosity (due to reaction or time lag) change with time? Is a low lag time for manual sampling and testing sufficient, or is on-stream measurement essential?
   h. What is the relationship between viscosity and operating temperature?
3. Area classification—does the viscometer need to be explosion proof?
4. What is the viscosity ranges to be measured, and at which shear rates the measurement should be performed?
5. What levels of accuracy (maximum error allowable), sensitivity, and repeatability (for continuous process viscometer) are required?
6. What special features are needed?
   a. Remote indication or recording
   b. Automatic operation
   c. Automatic closed-loop control
   d. Temperature compensating system
7. What is the viscometer response time requirement?
8. What are the flow conditions—laminar or turbulent?

Applications

A viscosity measurement can be of value for one of the following two reasons:

1. It is very difficult to size a pump, pipeline, orifice meter, or agitator without knowing the viscosity of the process fluid. In any operation where liquids are used (spraying, coating, or dipping processes), the viscosity of the fluid determines the effectiveness of the process and the quality of the finished product. In short, viscosity is one of the most important process properties.
2. Viscosity readings can vary as a function of other process variables. These include molecular weight and its distribution in polymers, lubricating oils, and other substances, as well as the concentration, specific gravity, color, size, shape, and distribution of solids in a slurry or in an emulsion. All of these can cause viscosity variations.

Viscometers can be used for several purposes, primarily (1) to ensure that the finished product meets specifications, (2) to perform routine laboratory testing, (3) for scientific research, and (4) for in-line process control. Each is described briefly in the following.

Finished Product Specification For such applications, the appropriate type of viscometer has been specified by industry standards for product testing. Test procedures should be carefully followed, and test results correctly reported.

Routine Laboratory Testing Simple-to-operate and easy-to-clean viscometers should be considered for this purpose. Generally, the coaxial-cylinder type viscometer is well suited. More robust viscometers that do not require special training are generally selected for field laboratory testing, like efflux cups or similar. For small sample sizes, cone-and-plate rotating viscometers should be considered.

Scientific Research Study For scientific research purposes, accuracy and versatility should be the main selection considerations. If extreme accuracy is desired, consider the automatic capillary-tube viscometer or Stabinger viscometer (see Figure 1.70d). Cone-and-plate rotational viscometers are the most versatile units and are the only one that enable special rheological analysis for viscoelastic materials. If it is important to record the results to maintain a permanent record, both of the previously mentioned viscometers have the appropriate capability. For the measurement of gas and vapor viscosity, the falling ball viscometer is the best option.

FIG. 1.70d
A single process viscosity system can measure kinematic or dynamic viscosity, density as well as viscosity index. (Courtesy of Anton Paar GmbH.)

In-Line Process Control There are several options for continuously monitoring the viscosity of an industrial process stream, and they will be presented in Chapter 1.71. Once the unit is installed, very little operator labor will be required, and there will be no need to remove and handle samples from the process stream. Measurements can be taken at much shorter time intervals, and the viscosity will be known at exactly the conditions (temperature, pressure, flow, etc.) associated with the process. In choosing the in-process viscometer, same selection criteria as for laboratory units should be considered. Each of the possible designs may be suitable for some materials and applications, but may not cover all. Whether or not one can live with the associated compromises depends on the nature of the liquids being measured and the goals set. Characterization of Non-Newtonian's is difficult even in the best laboratory viscometer under ideal laboratory circumstances.

Since viscosity is a strong function of temperature, viscometer designs that provide automatic temperature compensation and do not require density corrections are likely to provide more accurate readings.

Besides the measurement values, further information about the status of the devices is useful and can help the plant operator, DCS system, or maintenance staff to take actions as required. Besides the standard 4–20 mA signaling, modern process viscometers therefore enable a wide variety of fieldbus connectivity and employ some level of self-diagnosis, like the NAMUR recommendation NE 107 (self-monitoring and diagnosis).

SPECIFICATION FORMS

When specifying viscometers, use the ISA form 20A1001 and when the type of analyzer has already been selected and both the analyzer and the requirements of the process application are to be specified, use the ISA form 20A1002. Both forms are reproduced with the permission of the International Society of Automation on the next pages.

ANALYSIS DEVICE
Operating Parameters

#	RESPONSIBLE ORGANIZATION		#	SPECIFICATION IDENTIFICATIONS		
1	(ISA logo)		6			
2			7	Document no		
3			8	Latest revision	Date	
4			9	Issue status		
5			10			

#	ADMINISTRATIVE IDENTIFICATIONS			#	SERVICE IDENTIFICATIONS Continued			
11				40				
12	Project number	Sub project no		41	Return conn matl type			
13	Project			42	Inline hazardous area cl		Div/Zon	Group
14	Enterprise			43	Inline area min ign temp		Temp ident number	
15	Site			44	Remote hazardous area cl		Div/Zon	Group
16	Area	Cell	Unit	45	Remote area min ign temp		Temp ident number	
17				46				
18	SERVICE IDENTIFICATIONS			47				
19	Tag no/Functional ident			48	COMPONENT DESIGN CRITERIA			
20	Related equipment			49	Component type			
21	Service			50	Component style			
22				51	Output signal type			
23	P&ID/Reference dwg			52	Characteristic curve			
24	Process line/nozzle no			53	Compensation style			
25	Process conn pipe spec			54	Type of protection			
26	Process conn nominal size	Rating		55	Criticality code			
27	Process conn termn type	Style		56	Max EMI susceptibility		Ref	
28	Process conn schedule no	Wall thickness		57	Max temperature effect		Ref	
29	Process connection length			58	Max sample time lag			
30	Process line matl type			59	Max response time			
31	Fast loop line number			60	Min required accuracy		Ref	
32	Fast loop pipe spec			61	Avail nom power supply			Number wires
33	Fast loop conn nom size	Rating		62	Calibration method			
34	Fast loop conn termn type	Style		63	Testing/Listing agency			
35	Fast loop schedule no	Wall thickness		64	Test requirements			
36	Fast loop estimated lg			65	Supply loss failure mode			
37	Fast loop material type			66	Signal loss failure mode			
38	Return conn nominal size	Rating		67				
39	Return conn termn type	Style		68				

#	PROCESS VARIABLES	MATERIAL FLOW CONDITIONS				#	PROCESS DESIGN CONDITIONS		
69						101			
70	Flow Case Identification				Units	102	Minimum	Maximum	Units
71	Process pressure					103			
72	Process temperature					104			
73	Process phase type					105			
74	Process liquid actl flow					106			
75	Process vapor actl flow					107			
76	Process vapor std flow					108			
77	Process liquid density					109			
78	Process vapor density					110			
79	Process liquid viscosity					111			
80	Sample return pressure					112			
81	Sample vent/drain press					113			
82	Sample temperature					114			
83	Sample phase type					115			
84	Fast loop liq actl flow					116			
85	Fast loop vapor actl flow					117			
86	Fast loop vapor std flow					118			
87	Fast loop vapor density					119			
88	Conductivity/Resistivity					120			
89	pH/ORP					121			
90	RH/Dewpoint					122			
91	Turbidity/Opacity					123			
92	Dissolved oxygen					124			
93	Corrosivity					125			
94	Particle size					126			
95						127			
96	CALCULATED VARIABLES					128			
97	Sample lag time					129			
98	Process fluid velocity					130			
99	Wake/natural freq ratio					131			
100						132			

#	MATERIAL PROPERTIES			#	MATERIAL PROPERTIES Continued		
133				137			
134	Name			138	NFPA health hazard	Flammability	Reactivity
135	Density at ref temp	At		139			
136				140			

Rev	Date	Revision Description	By	Appv1	Appv2	Appv3	REMARKS

Form: 20A1001 Rev 0 © 2004 ISA

	RESPONSIBLE ORGANIZATION	ANALYSIS DEVICE COMPOSITION OR PROPERTY Operating Parameters (Continued)		SPECIFICATION IDENTIFICATIONS		
	ISA			Document no		
				Latest revision	Date	
				Issue status		

	PROCESS COMPOSITION OR PROPERTY			MEASUREMENT DESIGN CONDITIONS				
	Component/Property Name	Normal	Units	Minimum	Units	Maximum	Units	Repeatability

Rev	Date	Revision Description	By	Appv1	Appv2	Appv3	REMARKS

Form: 20A1002 Rev 0 © 2004 ISA

Definitions

Absolute (dynamic) viscosity η *(Eta)*: Constant of proportionality between applied stress and resulting shear velocity (Newton's hypothesis).

Apparent viscosity: Viscosity of a non-Newtonian fluid under given conditions. Same as *consistency*.

Consistency: Resistance of a substance to deformation. It is the same as viscosity for a Newtonian fluid and the same as apparent viscosity for a non-Newtonian fluid.

Fluidity: Reciprocal of absolute viscosity; the unit in the cgs system is the rhe, which equals 1/poise.

Hagen–Poiseuille law (flow through a capillary)

$$Q = \frac{\pi R^4}{8 \eta L}(p_1 - p_2) \qquad 1.70(13)$$

Kinematic viscosity v *(Nu)*: Dynamic viscosity/density = $v = \eta/\rho$.

Pascal-second (Pa·s): Internationally accepted unit of absolute (dynamic) viscosity. Pa·s = newton-s/m² = 10 poise = 1000 centipoise.

Poise (μ): Unit of dynamic or absolute viscosity (dyne-s/cm²).

Poiseuille (Pi): Suggested name for the new international standard unit of viscosity, the pascal-second.

Relative viscosity: It is the viscosity of the solution related to the viscosity of the solvent.

$$\eta_r = \frac{\eta}{\eta_s} \qquad 1.70(14)$$

Saybolt Furol seconds (SFS): Time units referring to the Saybolt viscometer with a Furol capillary, which is larger than a universal capillary.

Saybolt universal seconds (SUS): Time units referring to the Saybolt viscometer.

Saybolt viscometer (universal, Furol): Measures time for a given volume of fluid to flow through standard orifice; units are seconds.

Shear rate: Defined as

$$\dot{\gamma} = \frac{v}{h} \qquad 1.70(15)$$

Pronounced "gamma-dot" with velocity v[m/s] and the distance h[m] between the plates (for linear velocity profiles). The unit of shear rate is [1/s] ("reciprocal second"). Sometimes following terms are used as synonymous: shear gradient, velocity gradient, strain rate, rate of deformation. In case of nonlinear velocity profiles, the differential notation of 1.70(14) is used.

Shear viscometer: Viscometer that measures viscosity of a non-Newtonian fluid at several different shear rates.

Specific viscosity (or *"viscosity relative increment"*): DIN 1342-2 no longer recommends the term "specific viscosity"; therefore, "relative viscosity change" is expressed as

$$\frac{(\eta - \eta_s)}{\eta_s} \qquad 1.70(16)$$

Stoke: Unit of kinematic viscosity v (cm²/s).
Stress: Force/area (*F/A*).

Abbreviations

CGS   Centimeter Gram Second
FPS   Foot–pound–second
SI   Système international d'unités

Organizations

DIN   DeutschesInstitutfürNormung—German Institute for Standardization
ISA   International Society of Automation

Bibliography

Agoston, A. et al., Viscosity sensors for engine oil condition monitoring—Application and interpretation of results, *Sensor Actuator A*, 121, 327–332, 2005.

Ancey, C., *Notebook—Introduction to Fluid Rheology*, Laboratoire hydraulique environnementale (LHE), ÉcolePolytechnique Fédérale de Lausanne, Écublens, Switzerland, Version 1.0, July 4, 2005.

Anderson, M.A., Bruno, B.A., and Smith, L.S., *Viscosity Measurement—Draft*, Department of Mechanical Engineering, Union College, Schenectady, NY, 2012.

Bandrup, J. and Immergut, E., *Polymer Handbook,* 3rd edn., John Wiley & Sons, New York, 1989.

Barnes, H.A., *A Review of the Rheology of Filled Viscoelastic Systems, Rheology Reviews*, The British Society of Rheology, London, U.K., 2003.

Basker, V.R. et al., Evaluation of an online torsional oscillatory viscometer for kraft black fluid, *PaperijaPuu/Pulp and Timber,* 82(7), October 2000.

Bhuyan, M., *Measurement and Control in Food Processing*, Taylor & Francis Group, Boca Raton, FL, 2007. http://www.taylorandfrancis.com/books/details/9780849372445/.

Bourne, M.C., *Food Texture and Viscosity Concept and Measurement,* Academic Press, New York, 1997.

Chhabra, R.P., *Non-Newtonian Flow and Applied Rheology*, Butterworth-Heinemann/IChemE, Amsterdam, the Netherlands, 2008. http://www.icheme.org/shop/books/elsevier%20books/non-newtonian%20flow%20and%20applied%20rheology.aspx?sc_trk=follow%20hit,{CB04A670-D58F-44E5-8F79-5354C478472D},Non%20Newtonian%20flow.

Dealy, J.M., Viscometers for online measurement and control, *Chemical Engineering,* October 1, 1984.

Deshpande, A.P. et al., *Rheology of Complex Fluids*, Springer Science+Business Media, New York, 2010. http://link.springer.com/book/10.1007%2F978-1-4419-6494-6.

Dutka, A.P. et al., Evaluation of a capillary-coriolis instrument for online viscosity and density measurement, *Proceedings of TAPPI Process Control, Electrical and Instrumentation Conference (ISA)*, March 1997.

Goodwin, J.W. et al., *Rheology for Chemists*, The Royal Society of Chemistry, Cambridge, U.K., 2008, http://pubs.rsc.org/en/Content/eBook/978-0-85404-839-7#!divbookcontent.

Hallikainen, K.E., Viscometry, *Instrumentation and Control System*, November 1972.

Han, C.D., *Rheology and Processing of Polymeric Materials*, Volume 1 Polymer Rheology, Oxford University Press, Inc., Oxford, U.K., 2007. http://global.oup.com/academic/product/rheology-and-processing-of-polymeric-materials-9780195187823;jsessionid=40C304B0312269960B4D964340B88A10?cc=us&lang=en&.

Helle, H. et al., Comparing a 10 MHz thickness-shear mode quartz resonator with a commercial process viscometer, *Sensor Actuator B*, B81(2–3), January 2002.

Krigman, A., Viscosity measurement: Still sticky, but stepping ahead steadily, *InTech*, November 1985.

Langer, G. and Werner, U., Measurements of viscosity of suspensions in different viscometer flows and stirring systems, *German Chemical Engineering*, August 1981.

Malkin, A.Y. and Isayev, A.I., *Rheology—Concepts, Methods, and Applications*, ChemTec Publishing, Toronto, Ontario, Canada, 2012. http://www.chemtec.org/proddetail.php?prod=1-895198-33-X.

Mansion, D., State of the art in transducers viscometer, *Nouvel Automatisme*, June 1984.

Matuski, F.J. and Scarna, P.C., Instrument makes on-line viscosity control of slurries possible, *Control Eng.*, 28(13), 1981.

Mezger, T.G., *The Rheology Handbook*, 3rd edn., Vincent Network GmbH & Co., Hannover, CO, 2011. http://www.european-coatings.com/silver.econtent/catalog/vincentz/european_coatings/buecher/european_coatings_tech_files/the_rheology_handbook.

Mizier, M.O., The measurement of the viscosity of liquids, *Mesures*, March 1984.

Rabinovich, V.A. et al., *Viscosity and Thermal Conductivity of Individual Substances*, Begell House, New York, 1997.

Roussel, G. and du Parquet, J., *Development of a Fully Automatic Viscometer*, Society of Automatic Engineers, Paper #82149, Warrendale, PA, October 1982.

Sheble, N., How do you like your mashed potatoes? *InTech*, June 2002.

Skeist, I., *Handbook of Adhesives*, 3rd edn., Van Nostrand Reinhold, New York, 1989.

Spearot, J.A. ed., *Oil Viscosity: Measurement and Relationship to Engine Operation*, ASTM, 1989.

Steltzer, W.D. and Schulz, B., Theory and measurement of the viscosity of suspensions, *High Temp. High Press.*, 15(3), 289–298, 1983.

Viscometers, *Measurement Control*, June 1993.

Walsh, L., *Quality Management Handbook*, Marcel Dekker, New York, 1986.

Wunderlich, T., Ultrasound pulse doppler method as a viscometer for process monitoring, *Flow Meas. Instrum.*, 10(4), 1999.

Zhang, Z. et al., Viscosities of lead silicate slags, *Miner. Metall. Process.*, 19(1), February 2002.

1.71 Viscometers
Industrial

C. H. KIM (1969, 1982) **B. G. LIPTÁK** (1995)
J. E. JAMISON (2003) **M. ZAVRŠNIK** (2017)

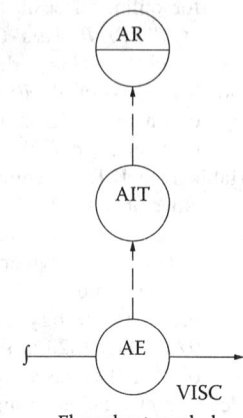

Flow sheet symbol

Types	A. Capillary
	A1. Differential pressure
	A2. Backpressure
	B. Falling element
	B1. Falling ball
	B2. Falling piston
	B3. Falling needle
	C. Float in variable-area flowmeter
	C1. Single-float
	C2. Two-float
	C3. Concentric
	D. Plastometers
	D1. Cone and plate
	D2. Kneader
	D3. Capillary extrusion
	E. Rotational element
	E1. Rotating cone
	E2. Agitation power
	E3. Dynamic liquid pressure
	F. Oscillating element
	F1. Oscillating blade
	F2. Oscillating piston
	G. Vibrating
	G1. Torsional oscillating element
	G2. Reed/Rod
	H. Coriolis mass flowmeter
	H1. Measuring tube torsional movement
Design pressures	Temperatures, viscosity
Ranges, errors	See orientation Table 1.71a.

(Continued)

Materials of construction	A, B. Hardened stainless steel or other corrosion-resistant metals
	C. Same choices as for variable-area flowmeters
	D, E. Hardened stainless steel, sealing—diamond like carbon
	F. Hardened stainless steel and Teflon®, sealing ring—silicone rubber
	G, H. Wide selection of corrosion-resistant materials and coatings
Speed of response	A. From 1 sec to about 5 min
	B2. Cycle time from 30 sec to 3 min
	C, D, E. A few seconds
	F1. Strokes every 2 sec, readings averaged over an adjustable number of strokes
	F2. About 30 sec
	H. Less than 1 sec
Costs	A1. From $1000 using basic designs implementing just additional differential pressure transmitters, up to $75,000 for complete systems including ASTM D 445 probe conditioning
	B1. Falling ball viscometer, from $7,000 to $16,000; high-pressure design with digital readout of ball roll time corrected for density, $23,000
	B2. From $2500 to $8000; a driven-piston element, $4000
	C1, C2. Two-float indicators with brass or stainless steel fittings in sizes from 12 to 38 mm (0.5–1.5 in.), from $1500 to $3000
	C3. 38 mm (1.5 in.) transmitter in stainless steel with 2 MPa flanges, $8000
	D. $10,000–$60,000, depending on features and on degree of automation
	E. From $10,000 to $25,000, explosion-proof up to $35,000
	F. From $8,000 to $25,000
	G. From $10,000 to $16,000, depending on materials of construction and transmitter features, explosion-proof up to $35,000
	H. $6,000–$20,000, depending on line size for the in-line treatment
Partial list of suppliers	Anton Paar GmbH (E) (http://www.anton-paar.com/corp-en/products/group/inline-viscometer/)
	ATAC Ltd. (A) (http://atacgroup.com/products/viscometers/)
	Automation Products (G) (http://www.dynatrolusa.com/Products/tabid/89/agentType/ViewType/TypeID/1/Viscosity-System.aspx)
	BartecBenke GmbH (A) (http://www.bartec.de/homepage/eng/20_produkte/190_analysengs/s_20_190_20.asp?ProdID=607)
	Brabender Messtechnik GmbH & Co. KG (E) (http://www.brabender-mt.de/index.php?id=173)
	Brookfield Engineering Laboratories Inc. (E, F) (http://www.brookfieldengineering.com/products/viscometers/process.asp)
	BTG Eclépens S.A. (F) (http://www.btg.com/products.asp?langage=1&appli=7&numProd=523&cat=prod)
	Edwin H. Benz Co., Inc.(D), (http://www.benztesters.com/plast.htm)
	Emerson Electric Co. (G) (http://www2.emersonprocess.com/en-US/brands/micromotion/density-viscosity/viscosity/Pages/index.aspx)
	Endress + Hauser Consult AG(H) (http://www.us.endress.com/#product/83I)
	Fisher and Porter a unit of Asea Brown Boveri Ltd. (C) (http://www.abb.com/product/us/9AAC100410.aspx)
	Galvanic Applied Sciences USA Inc. (G) (http://www.galvanic.com/nametre.htm)
	Hydramotion Ltd. (G) (http://www.hydramotion.com/)
	Lamy Rheology (E) (http://www.lamyrheology.com/English/Produits/194_RM-100L.html)
	Marimex Industries GmbH & Co. KG (G) (http://www.marimex.de/eng-products.php)
	Norcross Corporation (B) (http://www.viscosity.com/p_sensors_il.asp)
	ORB Instruments, Inc. (A) (http://www.orbinstruments.com/viscosity.shtml)
	PAC, LP (A, F) (http://www.paclp.com/Process_Analytics/Brand/Cambridge%20Viscosity)
	Sofraser (G) (http://www.sofraser.com/products/process-viscometer)
	Stony Brook Scientific Ltd. (B) (http://www.stonybrooksci.com/Products/ProductsFallingControlledNeedleViscometer.htm)

INTRODUCTION

Viscosity is an important parameter for quality control in a wide range of industrial processes and is one of the most important fluid properties for the production of many significant industrial substances. In particular, non-Newtonian viscosity behavior of heterogeneous mixtures (i.e., dispersions and suspensions) and polymeric solutions is an important material property that contributes to a fluid's performance and is often the main source of problems in handling, processing, and application.

This chapter covers industrial viscometers that can be installed directly in process line, a bypass loop or immersed in process vessels. The portable and benchtop laboratory viscometers will be discussed in the next chapter. The distinction between the two groups is not a sharp one, and some of the rotary, capillary, cone-and-plate, and piston designs are actually used in both services. Fluid viscosity can change significantly depending on the process conditions. Factors that affect viscosity include temperature, shear rate, time, and pressure. One must consider these effects when choosing and using a viscometer in the process line.

In this chapter, SI units Pascal seconds (Pa·s) or milli-Pascal second (mPa·s) for the dynamic viscosity and meter square per second (m^2/s) or millimeter square per second (mm^2/s) for kinematic viscosity will be used. Fairly common non-SI unit for dynamic viscosity are Poise (P) or centi-Poise (cP) and Stokes (St) or centi-Stokes (cSt) for kinematic viscosity. For conversion between these and other units, refer to Tables 1.70b and 1.70c and the Appendix A2.2.

There are four common industrial methods for the determination of viscosity: pressure drop or flow rate in conduit (capillary), drag (on rotational geometries or objects in the flow), time-of-flight (falling or pulled objects), and energy absorption (vibrating spheres, rods, blades, etc.). A summary of industrial viscometer features and capabilities is given in the Orientation Table 1.71a. In the following paragraphs, a general representation of individual methods with some illustrative examples will be given, the various viscometer designs are discussed as listed in the *orientation table* in Table 1.71a.

TABLE 1.71a
Orientation Table for Industrial Viscometers

Types of Design		Provides Continuous Signal	In-Line Device	Laboratory Device	Local Readout	Remote Readout Trans.	Temp. Compensation	Gas	Newtonian	Non-Newtonian	Maximum Design Pressure ($\times 10^5$ Pa)	Maximum Design Temperature (°C)	Inaccuracy (±%) (1) Based on Full Scale (2) Based on Measurement	Minimum Samples Size of Flow Rate ($dm^3 \cdot h$)
Capillary	Differential pressure	✓	✓	✓	✓	✓			✓	✓	47	480	1–2(1)	4–15
	Backpressure	✓	✓	✓	✓	✓			✓		35	100	1(1)	4
Coriolis	Torsional movement	✓	✓		✓	✓	✓		✓	✓	100	150	5(2)	—
Falling element	Falling ball			✓	✓	✓	✓		✓	✓	22	180	1(1)	—
	Falling piston			✓	✓	✓	✓		✓	✓	35	350	1(1)	—
	Falling needle			✓	✓	✓	✓		✓	✓	135	350	1(1)	—
Float	Single float	✓	✓	✓	✓	✓	✓		✓	✓	45	230	4(2)	180–480
	Two-float	✓	✓	✓	✓	✓			✓	✓	22	230	2–4(2)	60–540
	Concentric	✓	✓	✓	✓	✓				✓	45	230	2–4(2)	480
Plastometers	Cone and plate			✓	✓	✓				✓	8	200	0.5(1)	25 cm³
	Kneader			✓	✓	✓				✓	1	300	1(1)	80 cm³
	Capillary			✓	✓	✓				✓	345	300	2(1)	0.6 #/h
Rotational element	Rotating spindle	✓	✓	✓	✓	✓	✓		✓	✓	70	1300	1(1)	—
	Gyrating element	✓	✓		✓	✓	✓		✓	✓	70	300	1(2)	—
	Agitator	✓	✓		✓	✓	✓			✓	10	100	~5(1)	—
	Power Dynamic liquid pressure	✓	✓		✓	✓	✓		✓	✓	25	200	1(2)	—
Oscillating element	Blade		✓	✓	✓	✓	✓		✓	✓	16	60	1(1)	
	Piston	✓	✓	✓	✓	✓	✓		✓	✓	690	370	1(2)	
Vibrating element	Torsional	✓	✓	✓	✓	✓	✓		✓	✓	350	450	2(2)	
	Reed rod	✓	✓	✓	✓	✓	✓		✓	✓	500	300	1(1)	—

Applicable Viscosity Ranges in mPa·s: 10^{-1}, 1, 10, 10^2, 10^3, 10^4, 10^5, 10^6, 10^7, 10^8, 10^9

Arbitrary units are used MU (money units)

CAPILLARY VISCOMETERS

Continuous capillary viscometers are successfully used in oil refineries to control various products such as fuel oils, hydraulic oils, lubricating oils, fuels, and various grades of asphalts. The simplicity of the capillary viscometer has also resulted in a wide usage within some other industries including glycerol, foams, ice cream, milk, and cream (Rogers and Brimelow, 2000).

As shown in Chapter 1.70, the capillary viscometers measure viscosity by measurement of the differential pressure at a constant volume flow of the sample through the capillary or measurement of the volume flow through the capillary at a given differential pressure. According to the Hagen–Poiseuille law and Equation 1.70(7), the pressure drop under idealized flow conditions (laminar, isothermal, stationary flow conditions, Newtonian flow behavior, etc.) is linearly proportional to viscosity. Therefore, at constant flow rate, the differential pressure between the ends of the capillary can be used as a measure of viscosity.

Continuous capillary viscometers are primarily designed to measure the viscosity of Newtonian liquids. The Newton's hypothesis assumed that the viscosity of a fluid is independent of the rate of shear or of the shearing force of deformation, as long as the flowing temperature and pressure are fixed. Fluids that behave in this manner are called Newtonian. Because differential pressure transmitters are used to measure the viscosity, they are also readily adaptable to the automatic control of processes.

In most capillary-type viscometers, the temperature effect on the measurement is eliminated not by temperature compensation but by temperature controls. When the capillary is contained in a temperature-controlled environment and a constant velocity pump is used to produce a constant liquid flow rate, very high accuracies can be achieved. The measurement span of capillary viscometers can be influenced by capillary bore diameter and length. To minimize the end effects (the nozzle effect at the tip of the capillary), the use of large-diameter bores and long capillary tubes is recommended.

Limitations

Although this type of continuous viscometer is simple enough to be field fabricated (rather than purchased), one of its chief disadvantages is that, for accurate and reliable measurements, the capillary tube must be kept absolutely clean. However, fouling is likely to occur, because the capillary tube diameter is very small—usually in the range of 1.25–5 mm.

Because the sample flow rates must be low, this requirement limits the continuous capillary viscometer to bypass installations, where the goal is to automatically control the processes with minimal time lag.

Other potential error sources include (1) fluctuating flow rates through the capillary tube, (2) a dirty or plugged capillary, (3) leakage in the viscometer, (4) incorrectly set or drifting transducer zero and span, (5) fluctuating fluid pressure, (6) insufficient sample supply pressure, and (7) fluid temperature fluctuations due to thermostat malfunctions or to variation in sample supply temperature.

Calibration

The continuous capillary viscometer can be calibrated using Equation 1.70(7). Either the diameter of the capillary is calculated when the viscosity of the process fluid is known, or one can prepare the calibration curve of viscosity versus pressure drop using several fluids of known viscosity. Such a calibration should produce a straight line that goes through the zero point of the viscosity and pressure drop coordinates.

There are basically two types of viscometers that utilize the continuous capillary principle. One type measures the pressure drop across the capillary tube, and the other measures the upstream pressure as the sample flows through the capillary tube.

Differential Pressure Type

Figure 1.71a is a schematic flow diagram of a differential pressure viscometer, which requires a bypass installation. An external strainer is provided to remove any solids from the sample before it enters the viscometer. A constant sample flow rate (at about 3.8 dm^3/h or 1 GPH) is maintained by a precision metering pump, which is driven by a synchronous motor.

Two heat exchangers, before and after the metering pump, are used to keep the sample fluid in thermal equilibrium with the thermostatic bath. A relief valve protects against damage caused by excessive pressures that may occur in a blocked capillary.

The pressure drop across the capillary is measured with a differential pressure transducer, which is connected to the inlet and outlet sides of the capillary. The differential pressure transmitter output is a linear indication of the process viscosity and is used for indicating, recording, or controlling the process.

This type of viscometer can measure viscosities up to 2500 mPa·s and can operate at temperatures up to 116°C (240°F). If high-pressure capillaries and high-pressure metering pumps are used, the viscosity measurement can go up to 15,000 mPa·s and the line pressures can reach 46 bar (670 psi) at 480°C (900°F).

The range and span of the differential pressure transmitter determine the range of a single capillary tube. Overall inaccuracy is about ±1% of full scale, with repeatability being about the same. The average response time is about 2 min, but it varies with the length of the sample loop.

In-Line Design To reduce the response time of the viscometer to less than 1 sec, the capillary tube can be inserted into the main process stream, and sample flow rate can be increased to about 15 dm^3/h as shown in Figure 1.71b. In this

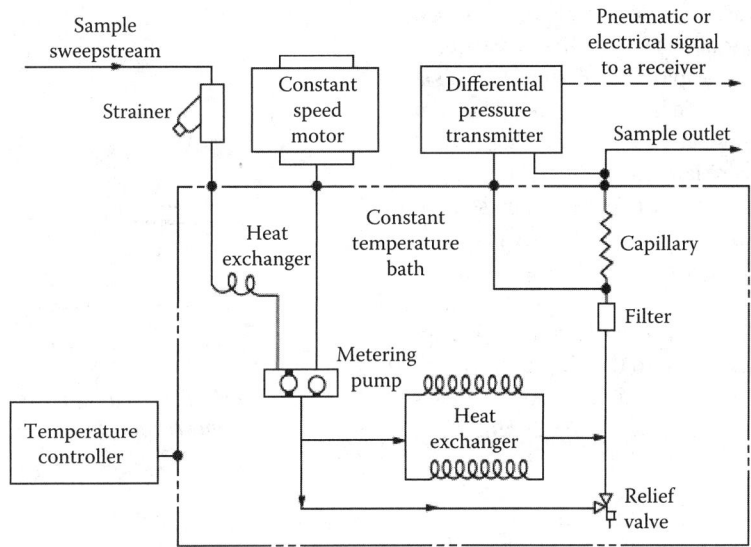

FIG. 1.71a
Schematic flow diagram of a differential-pressure-type continuous capillary viscometer.

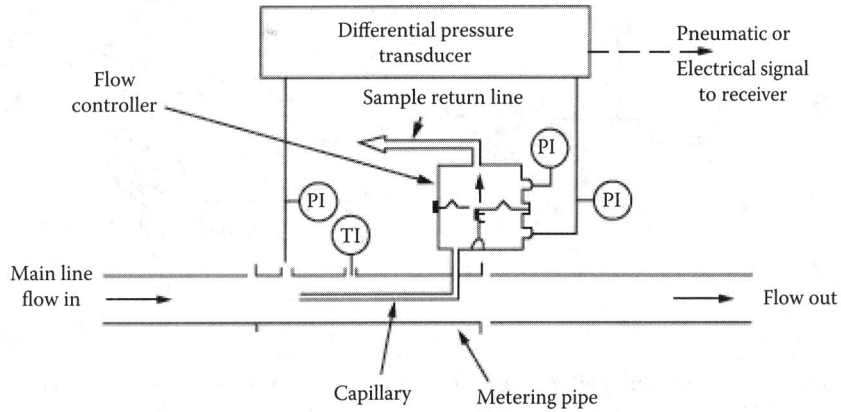

FIG. 1.71b
Schematic diagram of an in-line, differential-pressure-type continuous capillary viscometer.

design, the constant flow rate through the capillary tube is maintained by a pressure-regulated, diaphragm-type flow controller.

The error in this measurement is higher than with the external sampling design, and inaccuracy is about ±2.0% of full scale when measuring viscosity in the range of 5–30 mPa·s.

One advantage of this design is its increased speed of response. Another advantage is that, because the viscosity measurements are made at the process stream temperature, this viscometer can be used as a viscosity controller that maintains constant fluid viscosity by adjusting temperature. Applications include controlling fuel oil viscosity to maintaining optimal atomization patterns in industrial furnaces, heating plants, steam power stations, and marine boilers.

Dual-Capillary Design Figure 1.71c illustrates a balanced dual-capillary viscometer that is used for controlling the viscosity of fuel oil to boilers and other burners. The dual-capillary design is superior to the single-capillary units, because it is unaffected by variations in fuel line pressure.

An electric motor-driven gear pump continuously pulls a small amount of fluid, flowing in the housing, through an inlet capillary tube at a constant flow rate. This flow is then forced through an identical outlet capillary tube. Since both capillary tubes are identical, the difference between pump suction and discharge pressures is a linear function of viscosity. This design assures that rapid changes in line pressures (like in large diesel engines) do not affect the viscosity measurement. The pressure difference is measured by a d/p cell and is kept constant by adjusting the heat input into the oil preheater.

This in-line viscometer is provided with 50 mm DIN flanges, and the flow through it can be up to 380 dm³/min.

The fuel oil pressure can be up to 35 bar (500 psi), and the temperature can be up to 149°C (300°F). The differential pressure range of the d/p cell is 0–3.8 m of water, which in units of viscosity can correspond to 0–70 mPa·s.

Pipe Section as Viscometer If the flow is constant or can be measured, and therefore it is possible to compensate for the variations in flow rate, the pressure drop through any pipe section (Figure 1.71d) can be used to detect viscosity. As long as the inner walls of the pipe section are clean, and as long as the temperature and flow variations are compensated for, the detected pressure drop will be linearly proportional to viscosity. To enhance the reliability of the measurement while establishing laminar flow conditions these types of viscometers must have sufficiently long entrance and exit sections.

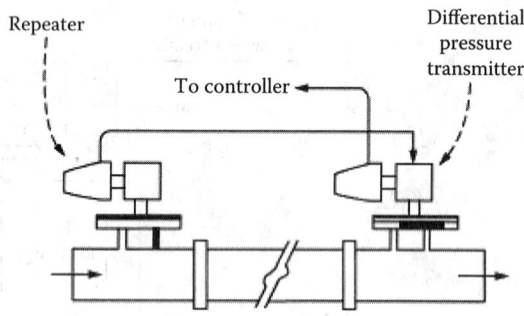

FIG. 1.71d
Differential pressure measurement, using extended diaphragm sensors across any section of process pipe to detect viscosity.

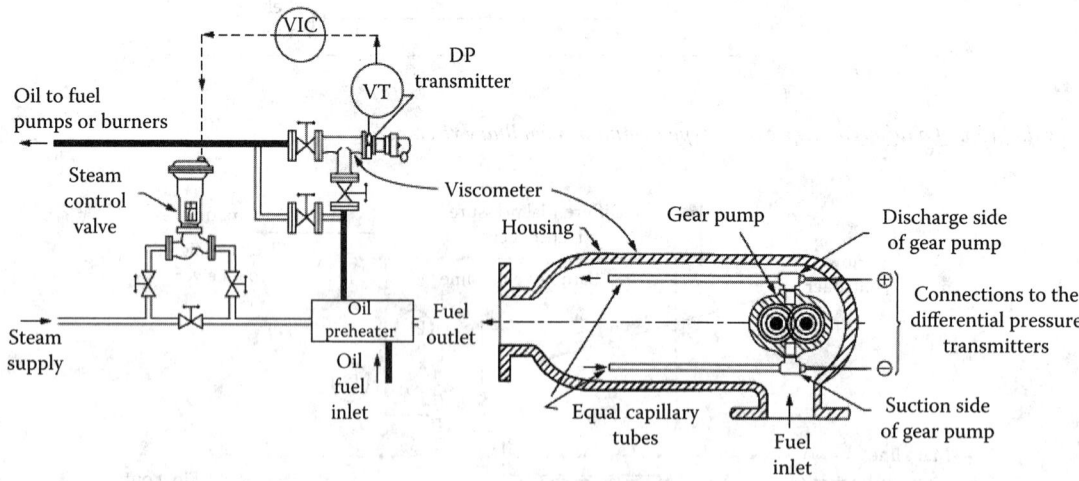

FIG. 1.71c
Balanced dual-capillary viscometer used to control oil viscosity, which guarantees proper atomization and therefore improved combustion efficiency. (Courtesy of Conameter Corporation.)

Backpressure Type

The operation of this viscometer is quite similar to the differential pressure type except that it measures only the upstream pressure to a capillary tube that discharges to the atmosphere or returns the sample to a pressure-regulated process line as shown in Figure 1.71e.

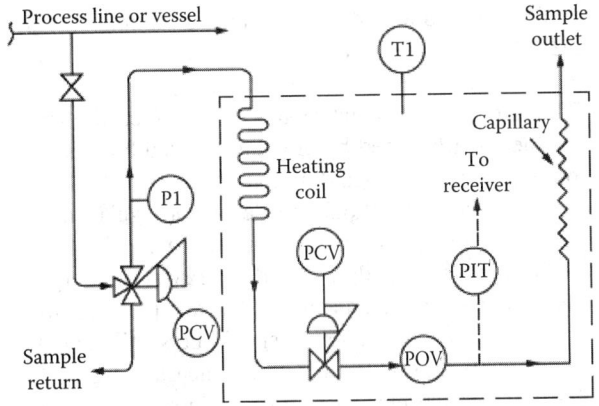

FIG. 1.71e
Schematic system diagram of a backpressure-type continuous capillary viscometer.

The sample fluid is continuously fed to the instrument from the process line or from a vessel and the temperature is maintained by flowing through a heat exchanger, which is immersed in a constant temperature bath. The sample then passes through a pressure regulator followed by a flow controller. Under these conditions of constant flow rate and constant temperature, the sample pressure at the entrance to the measuring capillary tube is linearly proportional to the viscosity of the liquid.

The inlet pressure to the capillary is sensed by a strain gauge (or any other pressure measuring principle) and converted to the desired units of viscosity. Because this viscometer measures only the inlet pressure to the capillary tube, it is extremely important to maintain the outlet side at constant pressure by discharging the capillary to atmosphere or to a pressure-regulated vessel or pipeline.

The backpressure-type viscometer can measure viscosities within the range of 5–500 mPa·s at temperatures up to 100°C (210°F). Its overall inaccuracy and repeatability are ±1% of full scale, and its response time is 3–6 min.

FALLING ELEMENT VISCOMETERS

Falling element viscometers determine viscosity by measuring the drag force acting on a falling element under specific flow conditions. There are a variety of falling elements that can be used such as balls, pistons, needles, cylinders, etc. There are three forces acting on the object; the buoyancy force and the drag force act upward while the gravitation force (weight of the element) acts down. The drag force is composed of a shearing force (due to the fluid) and a pressure force (due to flow separation). Falling element viscometers are generally designed to operate in the Stokes (creeping) flow regime which is characterized by a lack of flow separation and occurs for very low Reynolds numbers (Re < 0.1). The element will fall or slide down the tube and the instrument will measure the time it takes for the element to travel between two timing points. Using a combination of earlier assumptions, the viscosity can then be calculated by terms of time, geometry, and density difference.

It is a pulsed measurement and is principally designed for clean liquids. It has the potential to be very accurate −1% based on full scale; however, its dependence upon a fixed element/wall gap can limit the range over which viscosity can be measured. This principle is inherently less tolerant of particles and product buildup compared to other techniques.

Falling-Piston Viscometer

The falling-piston viscometer has been used in paper sizing, printing, coating, polymerization, starch conversion, textile sizing, and blending process applications. The working principle of the falling-piston viscometer is quite similar to the falling-ball viscometer, discussed in Chapter 1.72 in connection with Figure 1.71c.

The measuring element consists of a piston inside a measuring tube as shown in Figure 1.71f. The measuring element can be installed in a tank (open or closed) or in a liquid-filled pipeline, as long as the measuring tube is completely immersed in the fluid.

During the filling phase, the piston, which is resting at the bottom of the tube, is automatically raised by an air lifting mechanism or by a motor-cam mechanism. As the piston is raised, a sample of the liquid is drawn in through the openings in the sides of the tube and fills the measuring tube as the piston is withdrawn.

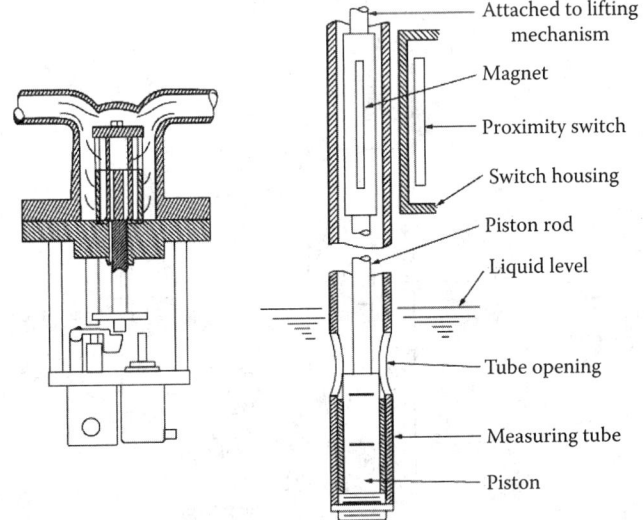

FIG. 1.71f
The falling-piston viscometer.

Process Pressure-Operated Design In some in-line units, the piston can be raised by the process fluid itself. In such designs, the measuring tube is filled through an opening at the bottom of the tube, and the filling and measuring cycles are controlled by the periodic opening and closing of a valve. The pressure drop across the measuring tube is about 0.35 bar.

During the measuring phase, the piston assembly is allowed to fall by gravity, expelling the sample out of the tube through the same route that it entered. The time of fall is a measure of viscosity, using the clearance between the piston and the inside wall of the measuring tube as the measuring orifice. The timed interval is then displayed on an indicator or recorded.

When high-viscosity materials are to be detected, and it is desired to make the measurement within a reasonable cycle, a two-way air cylinder is used to lift and force down the piston. This design also accommodates the use of any mounting position that is desired.

The viscosity range of this sensor is from 0.1 to 10^6 m Pa·s, whereby each piston has a 100:1 range. The inaccuracy of this viscometer is 1% of full scale, and the reproducibility and sensitivity are 0.25% and 0.1% of full scale, respectively.

Because the cycle frequency of once every 2 min up to two cycles per minute, it is applicable to both batch and continuous process applications, with or without automatic process-control capabilities. Pressure and temperature ratings of the standard in-line unit are from full vacuum to 20 bar (300 psi) and for up to 340°C (650°F). A special high-pressure unit is available for operating pressures up to 35 bar (500 psi).

Precautions As in the case of all other viscometers, operating temperature and pressure should be specified and kept constant when using this sensor. The error caused by small variations in the process fluid temperature can be substantial, so the use of either a temperature-controlled sampling system or a temperature-compensated design should be used.

Naturally, if compensation is the choice, the viscosity versus temperature relationship must be accurately known.

Because this viscometer operates in a batch manner, it should not be used where fast (less than 1 min) response time is required. Where a sampling loop is used, the rate of sample pumping should be set to minimize the lag time. To obtain reasonable reproducibility, the following should be observed:

1. Avoid any vapor entrainment in the sample liquid caused by agitation or boiling.
2. Avoid turbulence.
3. Calibrate the instrument regularly to correct for measurement drift caused by gradual material buildup or for wear of the piston and tube.
4. The measuring unit should be cleaned at regular intervals—frequency depends on the rate of material buildup.
5. An in-line filter should be used to remove any larger sized solids or foreign materials. The maximum allowable size of solids remaining should be small enough that they will not interfere with the measurement.
6. Sensitivity requirements should be known for blending process applications.
7. Do not use this viscometer if the process liquids have poor flow characteristics.
8. Erratic readings may result if this detector is subjected to severe vibration.

Falling-Slug or Falling-Ball Viscometers

This instrument automatically measures the time required for a cylindrical slug of a specific density to fall a given distance in a vertical tube, which is filled with the process liquid at a constant temperature.

As shown in Figure 1.71g, this viscometer operates by the sample pump first purging the system of the previous sample

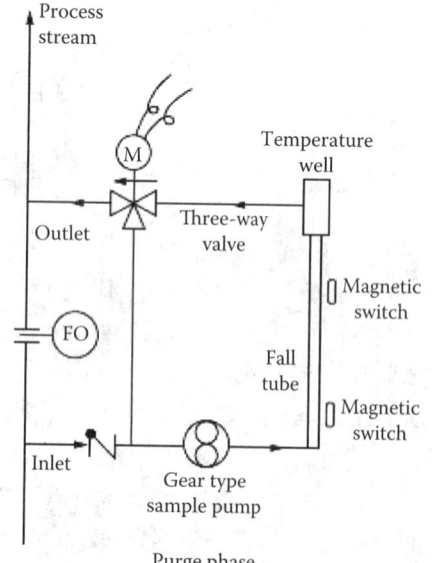
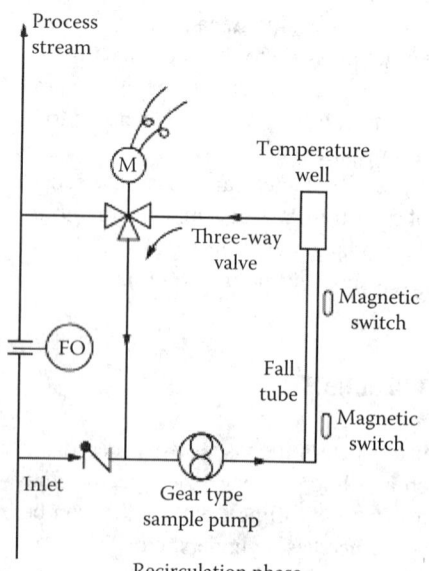

FIG. 1.71g
The purge and the recirculating (measuring) cycle of the falling-slug viscometer.

and then introducing a fresh one. Two separate thermostats control the purge and recirculation cycles by changing the direction of flow through a three-way valve.

In the recirculation phase, the sample pump flow raises the slug to the top of the fall tube. When the sample temperature has stabilized and matches the thermostat set point, the pump and the sample flow stops, thereby permitting the slug to fall. As it does, it actuates two magnetic switches that are attached to the side of the fall tube, separated by 25–500 mm. The resultant time measurement is directly proportional to the viscosity of the sample.

The viscosity range of this detector is 10–1,000,000 mPa·s. The specific ranges for particular applications are field selectable by adjusting the distance between the two magnetic switches on the fall tube. Full scale indication can be selected from 10 to 250 sec in five steps. The inaccuracy and the reproducibility of this viscometer are both about ±1% full scale, depending on the precision of the thermostat and of the time recorder. Errors caused by small variations in the process fluid temperature can be substantial.

The viscometer is designed to operate at temperatures up to 150°C (300°F) and at pressures up to 14 bar (200 psi). This detector is suited for continuous viscosity measurement applications in which it is sufficient to make a reading once every 3 min. The falling-slug viscometer is recommended for use only on clean process fluids that are not shear sensitive.

FLOAT VISCOMETERS

The float viscometer has been successfully used to maximize combustion efficiency by controlling the viscosity of fuel oils in marine and stationery boilers. Other areas of application have been in measuring the viscosity of cement slurry, starch, glue, and petroleum products (motor oils).

The operating principle is similar to that of the variable area flowmeter (Chapter 2.50 in Volume 1 of this Handbook) where the viscous drag force on a float is proportional to the orifice opening required (between float and tapered tube) to move the fluid through that orifice at a constant flow rate.

In a rotameter-type flowmeter, the forces acting on the float are affected by the flow rate, by the float and liquid-specific gravity, and by the viscosity of the fluid being metered. For flow-metering applications, the floats are designed so that the viscous drag area is relatively small, so the float is relatively insensitive to viscosity while being sensitive to flow rate and density changes.

In the viscometer version of this design, the flow rate through the variable area meter is held constant. Therefore, if the position of the float changes, that change is an indication of a change in fluid viscosity and density or in kinematic viscosity. To increase its sensitivity, the viscometer float is designed with a large viscous drag area. To obtain accurate viscosity measurements with float-type viscometers, the flow rate must be constant. Because this required flow control can be obtained in three different ways, there are three different designs: single-float, two-float, and concentric.

Precautions

When using float-type viscometers, the following recommendations should be observed:

1. The float viscometer must be installed vertically, with the outlet at the top.
2. The instrument should be installed in a vibration-free location.
3. The process fluid temperature should be carefully controlled, or temperature compensation should be employed. If no automatic compensation is provided, viscosity versus temperature curves should be available for use by operating personnel.
4. The flow rate and the pressure of the process fluid should both be constant and smooth, pulsating metering pumps should not be used.
5. The process fluid should be free of foreign material to prevent plugging of the small orifice (2.5 mm, diameter minimum) inside the viscometer. The use of in-line filters is recommended to remove all foreign materials.
6. Install the viscometer in a bypass line to permit the flushing of the main pipeline and to facilitate viscometer maintenance and service.
7. Make certain that the specified operating conditions are not exceeded.
8. The viscometer should be periodically recalibrated using a known viscosity fluid. At the time of calibration, the tube and float should also be cleaned.
9. The sample fluid flow rate through the instrument should be sufficient to give good speed of response and sensitivity, but the rate of flow should not cause turbulence within the viscometer, which would result in erratic readings.
10. All air and vapor entrainment should be removed from the sample liquid.
11. When the process temperature is other than ambient, make sure the viscometer is insulated or stream traced, as required.

Single-Float Design

The single-float viscometer is a continuous and direct reading viscosity instrument. As illustrated in Figure 1.71h, a positive displacement pump (other flow control devices can also be used) provides the constant sample flow rate through the instrument. The recommended flow rate is between 2.9 and 7.6 dm^3/min. Automatic temperature compensation is not available. If a metering pump is used to generate the constant flow rate, the temperature rise through the pump should be measured and corrected for by the use of a viscosity versus temperature curve.

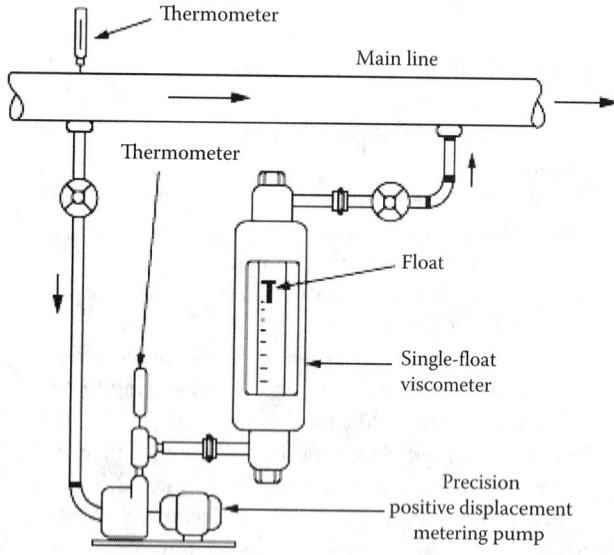

FIG. 1.71h
Typical piping arrangement of a single-float viscometer.

The single-float viscometer is also available as a transmitter that is used when remote displays or controls are required. In this case, the float position is detected by the use of an armature attached to the float extension rod with a magnetic sensing device around its outer periphery.

The glass tube viscometer is rated for 232°C (450°F) temperature and for 6.2 bar (90 psi) pressure. With steel-tube viscometers, the pressure rating depends on the operating temperature, such as a unit for 45 bar (650 psi) service at 232°C (450°F).

The single-float viscometer can be used to measure the viscosity of non-Newtonian fluids at less than 400 mPa·s and can handle Newtonian fluids up to 10,000 mPa·s. The rangeability of this viscometer is between 3:1 and 6:1. The inaccuracy of this sensor is ±4% of actual indication, and the reproducibility is ±4% of indication.

Two-Float Design

This is a relatively low-cost viscometer designed to provide intermittent viscosity local indication (not for transmission). The design incorporates two floats as illustrated in Figure 1.71i. The upper float is sensitive to fluid flow rate, and the lower is sensitive to viscosity. When making a measurement, the fluid flow rate is first manually adjusted to a constant value as indicated by the position of the upper float. Under such conditions, but only while maintaining the volumetric flow constant, the position of the other float indicates the viscosity of the fluid on a direct reading scale.

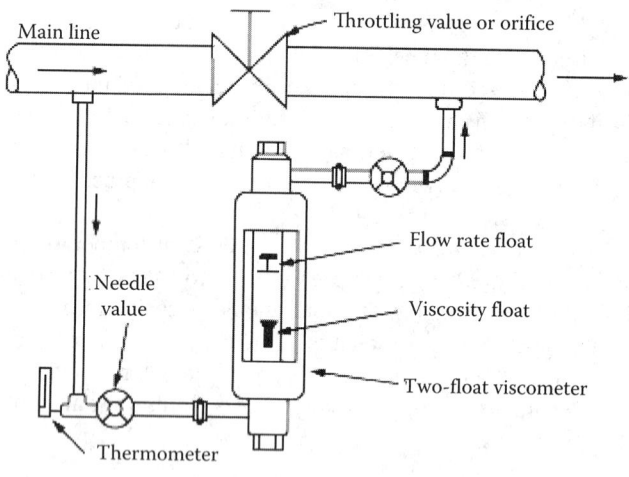

FIG. 1.71i
Typical bypass piping arrangement of a two-float viscometer.

In the main pipeline, a throttling valve or an orifice plate can be used between the viscometer inlet and outlet connections to produce the required pressure differential to provide sufficient flow through the instrument. Depending on size, the required flow rate is 0.95–9.5 dm^3/min.

The throttling valve or orifice plate is not required if the outlet can be discharged to a location where the pressure is lower than that of the main pipe. A needle valve should be used on the viscometer inlet to allow sensitive and accurate flow rate adjustment. As a function of size, the unit is rated for pressures up to 20 bar (300 psi) at 232°C (450°F), depending on size.

The two-float viscometer can measure the viscosity of Newtonian fluids from 0.3 to 250 mPa·s. It has a rangeability of 10:1 and an inaccuracy of ±4% of indication when the viscosity is higher than 35 mPa·s. When measuring lower viscosities, the error is ±2%.

Concentric Design

The concentric viscometer (Figure 1.71j) consists of a differential pressure regulator that maintains a constant pressure drop across the meter, and a variable area flowmeter, which is provided with a viscosity-sensitive float. As the fluid enters the instrument, it splits into two streams.

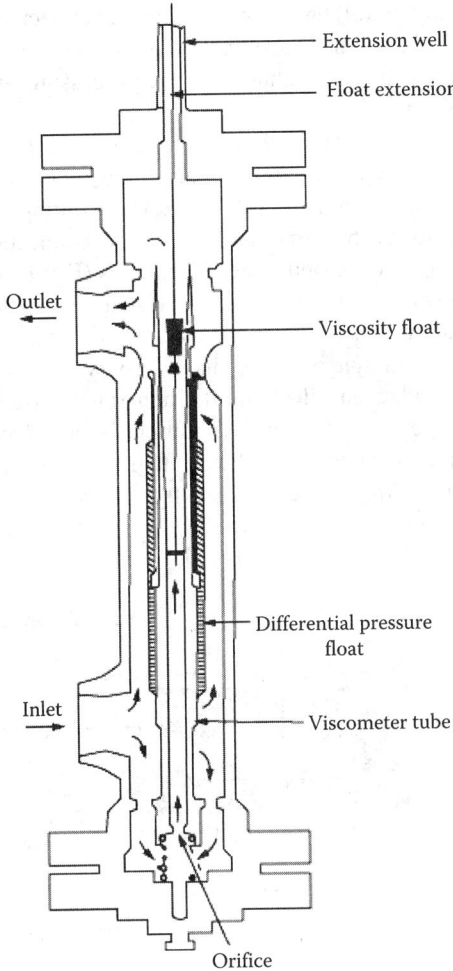

FIG. 1.71j
Cross-sectional view of a concentric viscometer.

The portion of the fluid that flows upward on the outside, around the differential pressure float, is used to control the pressure drop across the meter. The upper end of the differential pressure float acts as a control valve so that, when the flow rate changes, it throttles the flow to maintain a constant pressure drop. The pressure drop across this portion of the meter is determined by the weight of the differential pressure float.

The portion of the fluid that flows downward enters the viscometer tube through an orifice and then passes the viscosity-sensitive float. The flow rate through the viscometer tube is maintained constant, because the fluid passes through the orifice at a fixed pressure drop. Thus, constant flow rate is maintained, which is a requirement for the measurement of viscosity.

An extension attached to the viscosity float transmits its movement through a magnetic coupling to a receiver for display, recording, or automatic process control, as desired.

The unit may be placed in either the main process line or in a sample line, depending on the process flow rates. This viscometer has a flow-insensitive range, so the flow rate through it should not exceed 114 dm^3/min (30 GPM) (see Figure 1.71k).

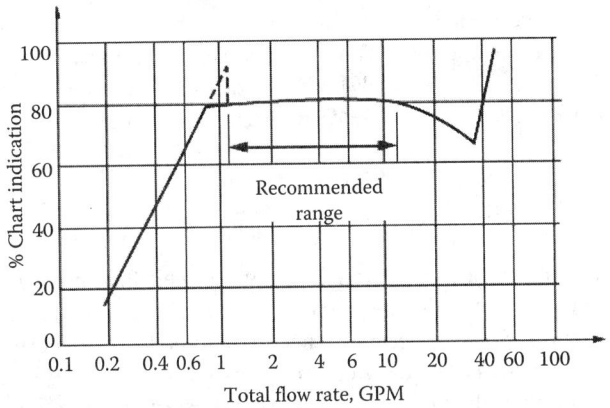

FIG. 1.71k
The characteristics viscosity reading of a concentric viscometer as a function of flow rate through the meter.

The recommended piping configuration for bypass installation is shown in Figure 1.71l. The restriction orifice shown in this figure should be sized for approximately 0.28 bar (4 psi) pressure drop. If a valve is used in place of an orifice, the correct valve position can be determined by slowly opening the valve until the viscometer reading levels off. Further opening will cause the reading to start to fall off and then, when turbulent conditions evolve, rise sharply, as illustrated in Figure 1.71k.

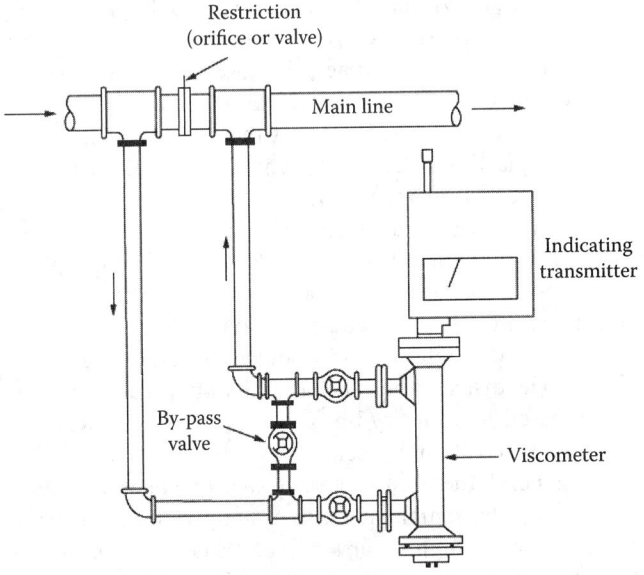

FIG. 1.71l
Bypass installation of a concentric viscometer.

The valve openings that produce steady viscosity readings are in the operating range. This procedure should be carried out when the main line flow is at a maximum; after that, the valve should be locked in place to prevent unauthorized adjustments. If the main line flow rate is lower than 7.6 dm^3/min (2 GPM), a booster pump should be used to provide adequate flow rate through the meter. The unit is rated for 45 bar (650 psi) pressure at a temperature of 232°C (450°F).

The concentric viscometer can measure viscosities in the range of 0.5–550 mPa·s and is available with a span range of 10:1. Its inaccuracy is ±2% of indication for viscosities lower than 35 mPa·s, and it is ±4% for higher viscosities.

PLASTOMETERS

Plastometers are employed to study the melt flow behavior of plastic materials or the molecular weight distribution of polymers. Some materials do not begin to flow unless a critical shear stress, termed yield stress, is exceeded. A simplistic definition of the yield stress of a solid is essentially the point at which, when increasing the applied stress, the solid first shows liquid-like behavior, that is, continual deformation. At stresses smaller than this characteristics value, the system behaves almost like solid and deforms elastically. As shear rate increases, the developing stresses become greater than the yield stress and the material begins to flow. This phenomenon is called *plastic behavior*.

Plastic Behavior

Prof. Eugene Bingham assumed that the flow rate for such materials was zero when they experienced stresses below a critical value and was linear with stresses above this limit value. A body that behaves in such a manner is called a *Bingham solid* or an *ideal plastic*. Extremely few solids behave in this manner. For most plastic materials, the yield value is not sharp and shows a nonlinear behavior with a delayed and imperfect recovery. The different types of plastics behave differently, and their behavior also changes with temperature.

For this reason, a plastometer gives only a relative indication of polymer behavior, thus characterizing a plastic material. Because the shear stress/shear rate relationship is based on the molecular behavior of materials, the plasticity can be related to the molecular weight of a material or to its distribution. The plasticity of plastic materials is expressed in arbitrary units such as Melt Flow Rate or Mooney Viscosity or percent of full scale.

In general, the plastometer consists of a heating chamber to keep the sample at constant temperature, a mechanism to apply a high shear force, and a device to measure torque or flow rate of the sample material and thereby detect the plasticity of the sample. Plastometers can be used in a laboratory (Chapter 1.72) to study polymeric behavior or used online to control processes manually or automatically. This type of measuring device is an invaluable tool in manufacturing plastics and synthetic rubbers.

Cone-and-Plate Plastometer

The cone-and-plate plastometer is used to conduct tests in evaluating crude rubber, rubber compounds, or reclaims; in the control of mill breakdown of polymer molecules; in the determination of time to scorch; in the calculation of optimal cure time; and for the evaluation of the processing characteristics of plastics.

This plastometer incorporates the features of the Mooney plastometer and is designed to meet the requirements of ASTM standard test method D-1646. The working principle of this sensor is the same as discussed in connection with laboratory-type cone-and-plate viscometers (Figure 1.72n).

As shown in Figure 1.71m, the instrument is designed to eliminate polymer "ball-up" and slippage tendencies by confining the sample in a disc-shaped cavity. Machined serrations provided on all of the die and rotor surfaces prevent slippage. In operation, the sample is introduced into a cylindrical test chamber where integral heaters are used to bring it to the predetermined test temperature (up to 200°C/400°F).

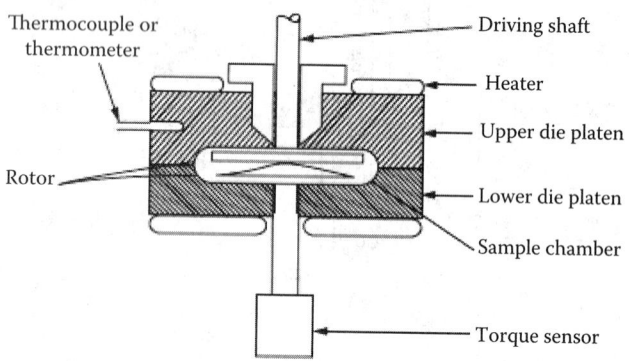

FIG. 1.71m
Cone-and-plate plastometer.

After the warm-up period has been completed, the plastic behavior of the sample is tested by driving the rotor at either a constant speed (normally at 2 rpm) or at various speeds between 0.05 and 2 rpm to test for relative molecular weight. The shearing action that takes place between the rotor and the die cavity is measured by some torque sensing mechanism and related to specimen viscosity. This type of plastometer typically operates within a range of 0–200 (Mooney) units has a repeatability of ±0.5 units.

The Unit of Mooney The Mooney unit is an arbitrary unit used to measure the plasticity of raw, or unvulcanized rubber; the plasticity in Mooney units is equal to the torque, measured on an arbitrary scale, on a disk in a vessel that contains rubber at a temperature of 100°C and rotates at 2 rpm.

Mooney gives the following equation to determine the average viscosity $\bar{\eta}$ of solids:

$$\bar{\eta} = \frac{G\,188.44\,g\,\theta}{2\pi\omega R^3 \left(\dfrac{R}{2a} + \dfrac{h}{b}\right)} \qquad 1.71(1)$$

where
  G is the gauge reading
  g is the acceleration from gravity
  θ is the pitch radius of worn gear
  ω is the angular velocity of rotor, rad/s
  R is the radius of rotor, in.
  a is the vertical clearance between rotor and stator above or below the rotor, in.
  h is the thickness of rotor, in.
  b is the effective radial clearance between rotor and stator

Kneader Plastometer

This instrument is widely used in measuring the viscosity of highly viscous materials like plastics and rubber. It measures the behavior of melt viscosity under conditions that are very similar to those that exist in the processing equipment.

Many different shapes of measuring heads are available (see Figure 1.71n) to accommodate a wide range of viscosities. In this plastometer, the jacket heater keeps the sample at a constant temperature (up 300°C/570°F). The kneader is driven by a dynamometer (available with variable-speed and fixed-speed drives) and the appropriate resistance encountered by the mixing blades is detected and the transmitted.

The torque is transmitted for recording or to automatically control processes with medium-viscosity, free-flowing fluids. The readings and recordings are on a 1000-unit division arbitrary scale, used for indicating shearing torque of the specimen being tested. Repeatability is ±1% of full scale. The measurement range can be adjusted.

Capillary Plastometer

This plastometer is designed for use in polymer manufacturing plants and is based on the capillary-tube viscometer principle.

The capillary plastometer shown in Figure 1.71o is calibrated to record either melt flow rate (flow rate of polyethylene through an open-ended capillary at 190°C (374°F) and 2.9 bar (43.2 psi) pressure. It can also record CIL flow index (190°C/374°F and 103 bar/1500 psi for polypropylene) or can alternately record both melt flow rate and CIL melt indexes.

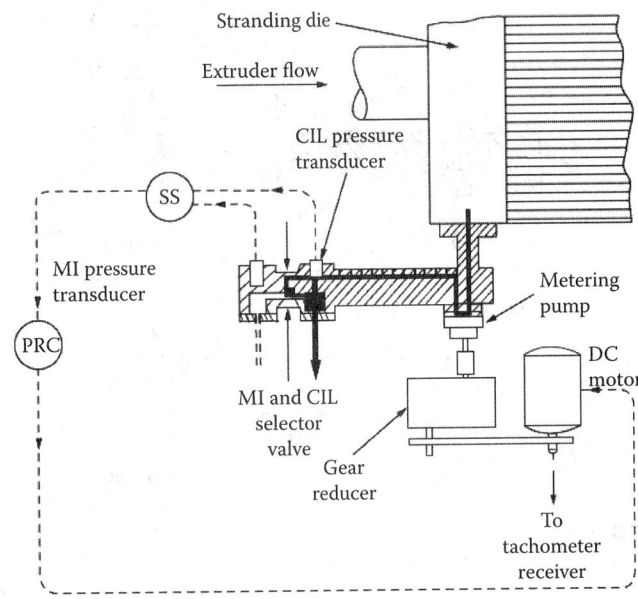

FIG. 1.71o
Pressure-controlled capillary plastometer.

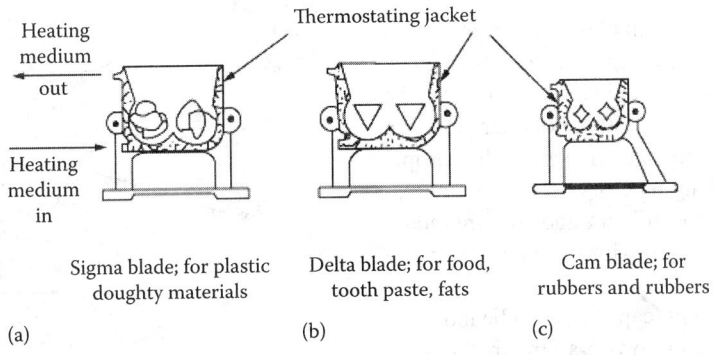

FIG. 1.71n
Measuring heads of kneader plastometer.

These two measurements for a particular polymer can be interpreted as both a molecular weight parameter and as a molecular weight distribution parameter. The capillary-type plastometer is not as versatile as the previously described plastometers.

In feeding the capillary-type plastometer, the polymer melt is first conditioned to be at some specific temperature and pressure. It is then extruded through a capillary of suitable dimensions. The rate of flow of the polymer through the capillary is measured by detecting the output of the tachometer on the metering pump and is recorded in units appropriate for the test (Figure 1.71p).

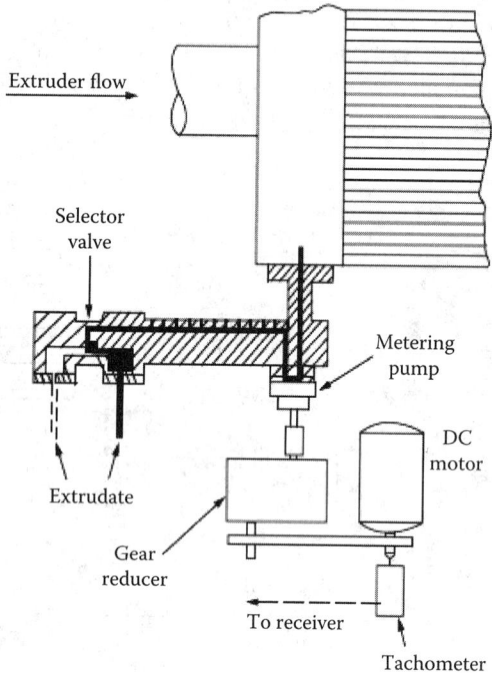

FIG. 1.71p
Flow-sensing capillary plastometer.

The operating range of this plastometer is 0–200 MI and 0–100 CIL units, with a repeatability within ±2% of full scale. The unit is designed to operate at up to 300°C (572°F) and up to 345 bar (5000 psi) in pressure.

Advantages One advantage of this instrument is that it can be used as an automatic online process control device for both viscous fluids and plastic solids. Especially for the solution polymerization processes, the polymer solution can be directly analyzed in this plastometer as the sample is received from the processing reactors through an autosampling device that flashes the solvent and the unreacted monomers, and melts the polymer prior to sending it to the plastometer.

For plastic solids, the die unit (capillary) can be mounted to the pelletizing extruder of the process stream for measurement. The capillary-type plastometer is ideally suited for study of plastic behavior of materials that are processed through injection molding or finishing operations.

Limitations The following conditions might degrade the accuracy of the measurements obtained using capillary-extrusion viscometers:

1. Nonuniformity of sample shear rate
2. Entrance effects—energy losses at the capillary entrance and discharge points
3. Compressibility of fluids
4. Pressure loss produced by the flow in the sample chamber
5. Temperature gradient created by shear-induced heat

When performing scientific research, great care should be taken to correct for these errors. If that is done, capillary-extrusion viscometers have excellent reproducibility and are well suited for routine industrial and scientific work.

ROTATIONAL ELEMENT VISCOMETERS

The principle of operation for rotational viscometers relies upon the liquid to be measured being sheared between two surfaces—one being stationary and the other rotating. In general, the principle of measuring the rate of rotation of a solid shape (cone, disk, cylinder, or spindle) in a viscous medium upon application of a known force or torque required to rotate the solid shape at a definite angular velocity is implemented. The main advantage of the rotational element viscometers as compared to many other viscometers is its ability to continuously replace the sample while operating at a given shear rate; thus, the measurement of non-Newtonian apparent viscosity is possible, as well as dynamic viscosity measurement of Newtonian fluids. Assuming constant geometry, the shear rate can be changed by varying the rotation speed of the rotational element.

Rotary Spindle Design

In the basic design illustrated in Figure 1.71q, a motor drives a cage coupled through a calibrated spring to a spindle arm that supports the spindle or cylinder in the fluid being measured.

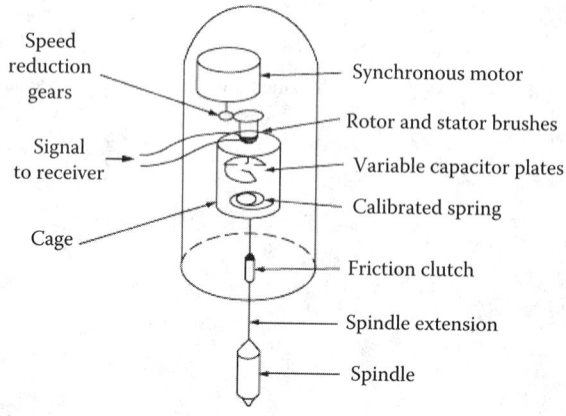

FIG. 1.71q
The operation of the rotating-spindle-type industrial viscometer.

During measurement, the spring tends to wind up until the force it generates equals the viscous drag on the spindle; hence, an increase in viscosity will be indicated by an increase in the deflection of the spring. At this point, the cage and spindle both rotate at the same speed but with a definite angular relationship to each other. This angular relationship is proportional to the torque on the spring.

Depending on the manufacturer, several methods can be used to convert the angular relationship into a viscosity reading. One of the methods is shown in Figure 1.71r, where one side of a variable capacitor is attached to the cage and the other to the spindle arm. A capacitance is thus made proportional to measured viscosity. Another option is to use a potentiometer mounted in the cage with its free member connected to the spindle arm. In this case, resistance is made directly proportional to the viscosity being measured. Regardless of the torque measurement method used, torque together with the rotational speed and measurement geometry is used to provide the viscosity reading.

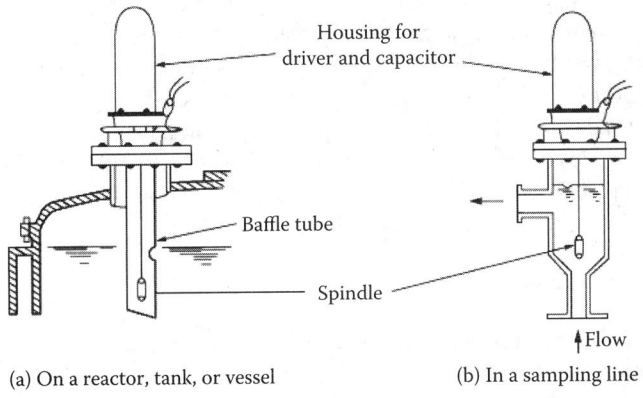

FIG. 1.71r
Tank and pipeline installation of the rotating spindle viscometer.

When measuring low viscosities, the variable capacitor-type design is preferred. This type of viscometer can measure viscosity ranges from 0–10 to 0–50,000 mPa·s. Its inaccuracy is ±1% of full scale between the ranges of 0–100 and 0–50,000 mPa·s. Its repeatability is ±0.3% of full scale for all ranges.

The response time of this type of viscometer is less than 30 sec, and the measurement range can be varied over by changing the size or shape of the spindle. The measurement range of this viscometer can be varied over a 30:1 ratio by changing the size or shape of the spindle. The standard spindle speed is 50 rpm. In some installations, the maximum range can be increased to 0–100,000 mPa·s.

Installation This type of viscometer may be installed on a tank, an open vessel, or in a sampling line (Figure 1.71r). For in-line installations, the flow rate of the process liquid should be constant, nonpulsating, and laminar (less than 11.4 dm³/min or 3 GPM for a standard 100 mm/4 in. pipe). Typically, the flow rate of the fluid through the system is neglected for process measurements; however, doing this can cause problems if measurements are made at low shear rates, or the liquid is very shear sensitive. If either of these process conditions exists, the fluid flow rate may contribute significantly to the overall shear rate and thus considerably affect the measured value. This technology is particularly prevalent in applications where lower viscosity liquids are being used. Swirling motion in the measurement chamber may be reduced by installation of a spider-type deflector in the inlet port.

The viscometer head is designed to operate at temperatures up to 71°C (160°F) and at pressure ranges of 8 mmHg absolute to 8.9 bar (100 psi). The head should be purged with dry air or other inert gases at a rate of 2.4×10^{-4} m³/s (0.5 ft³/min) or greater. A positive pressure differential of 0.2 bar (3 psi) should be maintained between the viscometer and the process vessel to prevent liquid or vapor penetration into the head.

The temperature of the mounting flange on which the viscometer is placed should not exceed 316°C (600°F). Above 104°C (220°F), it is necessary to use a cooling pad and place it between the flange and the viscometer. The process liquid temperature can be up to 1370°C (2500°F).

Performance The rotating spindle viscometer is not affected by normal industrial vibration. It is provided with over range protection through a friction clutch on the spindle extension. It is simple to clean the spindle. The sample fluid should be maintained at a constant temperature, or the viscosity measurement should be compensated for temperature variations. Logarithmic temperature-compensating units are available and are matched to the viscosity–temperature relationship shown by the particular process fluid. Logarithmic compensation to within 1% error over a span of 6°C (10°F) can be provided.

A baffle tube should always be installed when this viscometer is used in an agitated tank or reactor to obtain laminar flow condition. However, if very violent agitation or bubbling conditions exist, this viscometer should not be used at all. The spindle should always be immersed, and the liquid level should be kept constant to obtain reproducible readings.

The frequency of the power supply should be checked. Wide deviations from the specified frequency will cause spindle speed changes and can introduce errors in the viscosity measurement of non-Newtonian fluids, because their apparent viscosity is related to shear rate.

The rotating-cone viscometer is an electromechanical device, so a monthly preventive maintenance and cleaning schedule is recommended. If the unit is operated continuously, the speed reduction gears will need to be replaced on a yearly basis.

Because of its simple design, ease of cleaning, and non-clogging features, this viscometer has been successfully used in monitoring solid and liquid blending processes. Rotating-cone viscometers are quite versatile in their application over wide ranges of viscosities and are equally applicable for measuring the viscosities of Newtonian and non-Newtonian fluids.

Magnetically Coupled Design Magnetic couplings are used in the industry to transmit torque through a gap. This can be done through a nonmagnetic containment barrier such as stainless steel, allowing complete isolation of the inner magnetic hub from the outer magnetic hub. There are no contacting parts. Couplings may be hermetically sealed so that they may work in harsh environments, such as chemical hazardous applications, high temperature, and high pressure environments. There are various designs possible how to use the magnetic coupling for rotation element viscometers. One of the options is shown in Figure 1.71s. Here, a magnetic coupling is provided between the electronic detector (which is at atmospheric pressure) and the rotating sensor in the measuring cell (which is operated at the process pressure).

With this separation between atmospheric and pressurized areas, the requirement for purging is eliminated and, in the measuring cell, operating conditions can be up to 196 bar (2850 psi) at 20°C/68°F or 111bar (1620 psi) at 300°C/572°F.

The magnetically coupled viscometer should not be used to measure fluids that contain fiber, ferrite, or abrasive materials, because they interfere with the operation of the magnetic coupling or can damage the stainless steel–sapphire bearings.

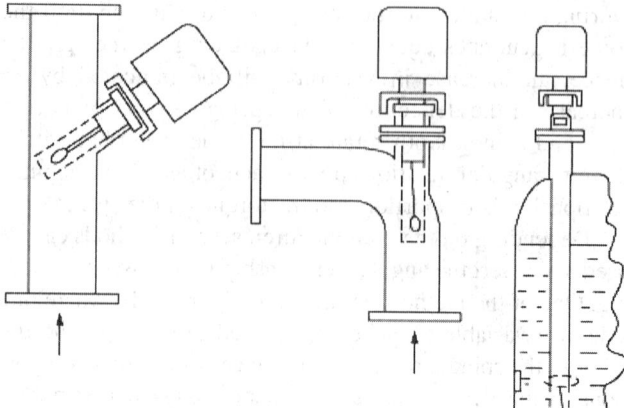

FIG. 1.71t
Pipeline and vessel installations of the gyrating element viscometer. (Courtesy of Brabender Messtechnik GmbH &Co. KG.)

substance. A differential gear with a metering spring and an inductive transducer detects the torque.

The drive and the torque-sensing mechanism are isolated from the process side by stainless steel bellows that are capable of operating at up to 70 bar (1000 psi) operating pressures and 300°C (572°F) operating temperatures. The drive motor speed is adjustable from 15 to 120 rpm, and the unit can detect viscosities between 10 and 10 million mPa·s (the overall range is subdivided into several working ranges). For pipeline installation, the minimum pipe size is 100 mm (4 in.).

Agitator Power

This type of instrument is widely used in the paper industry to control and measure the consistency of paper pulp slurries. In this viscometer, instead of measuring the torque exerted on a rotating cone, a transmitting current-meter is used to detect the torque exerted on the process agitator. It measures the current consumed as the agitator is being driven in a mixing tank.

As shown in Figure 1.71u, because most of the industrial agitator motors are oversized, the torque response of this

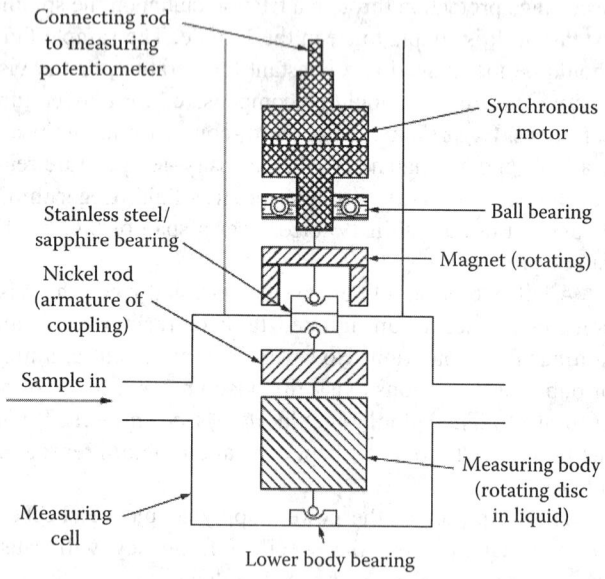

FIG. 1.71s
Magnetically coupled rotational industrial viscometer.

Gyrating-Element Viscometer

The viscometer illustrated in Figure 1.71t differs from the previously described. A sensing device moved by a drive shaft moves on a circular route within the substance being monitored. Because of the inclined position of the device it may be called a gyratory motion. The torque transmitted by the drive shaft is proportional to the viscosity of the

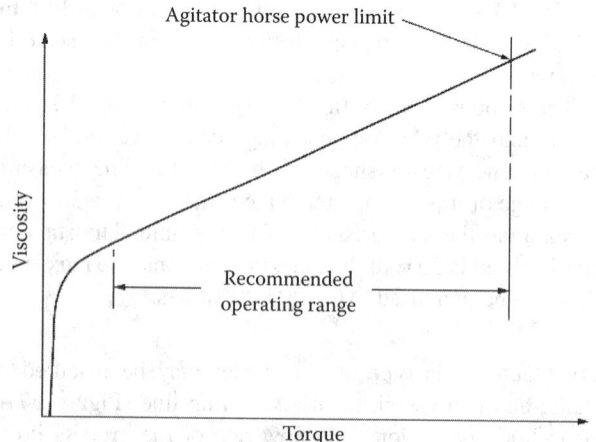

FIG. 1.71u
Relationship between viscosity and torque in agitator power viscometer.

sensor is poor at low viscosities. The size of the motor and impeller should be selected so that the agitator will operate in a region where the viscosity versus torque relationship is linear. The motor, reducer, bearing assembly, and pressure seal are all part of the viscometer, and any change in characteristics of these components would also affect the power consumption of the motor and, in turn, the viscosity reading.

This instrument is simple and easy to install, but it has low sensitivity and a low ratio between agitator power current change and changes in viscosity of the fluid. Many different impeller designs are available to improve the sensitivity and rangeability. This type of instrument should not be used with fluids that are thixotropic or rheopectic (materials whose viscosity changes with duration of agitation or shear).

The accuracy and sensitivity of this method of viscosity detection are both poor, but its repeatability is reasonable at about ±1% of full scale.

Dynamic Liquid Pressure

The process viscometer is based on a similar hydrodynamic effect which makes slide bearings work. In such bearings, the relative motion between two surfaces induces shear stress in a lubricant and leads to the formation of a lubricant film separating the sliding surfaces. The pressure necessary to carry the load can only develop if the film is wedge shaped, so that the variable surface will be slightly inclined. The pressure distribution in the wedge-shaped film depends on the surface velocity and the dynamic viscosity of the lubricant.

This viscosity dependence is utilized to measure the viscosity given a fixed geometrical configuration of the sliding surfaces. The functional principle is shown in Figure 1.71v. A rotor and a static outer surface define the wedge-shaped gap of the viscometer configuration. The rotor can be powered directly from a rigid shaft or using a magnetic coupling. The fluid under test is conveyed into the gap by the action of the rotor. The opposite static outer surface is rigidly fixed on the entry side and open-ended at the outlet. Due to the pressure rise in the gap, the outer surface is slightly displaced, acting as a spring. The displacement between the fixed and flexible parts of the spring element is proportional to the fluids viscosity and measured with an inductive sensor. The sensor coil is mounted on the fixed part of the spring while the counter piece is attached to the flexible part.

Besides the mechanical components outlined, the viscometer incorporates an electric drive with closed-loop control to achieve stable flow conditions and shear rate in the wedge-shaped gap. An inductive displacement sensor measures the deflection of the spring, offering the advantage of being independent of the process fluid and decoupled from the rotating motion and hence not influenced by bearing friction changes (Figure 1.71w). A temperature sensor immersed

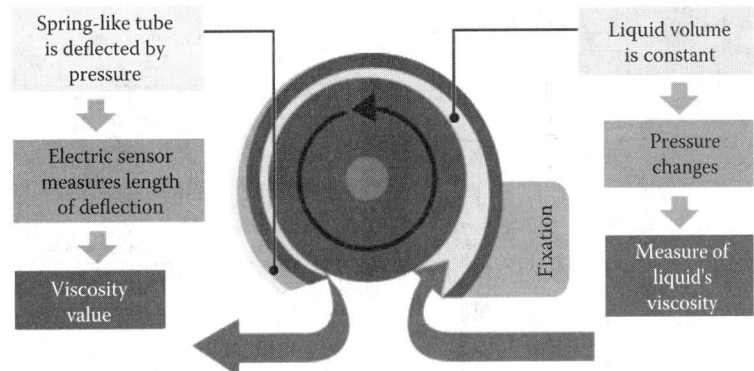

FIG. 1.71v
Partially open tube with a rotating cylindrical shaft. (Courtesy of Anton Paar GmbH.)

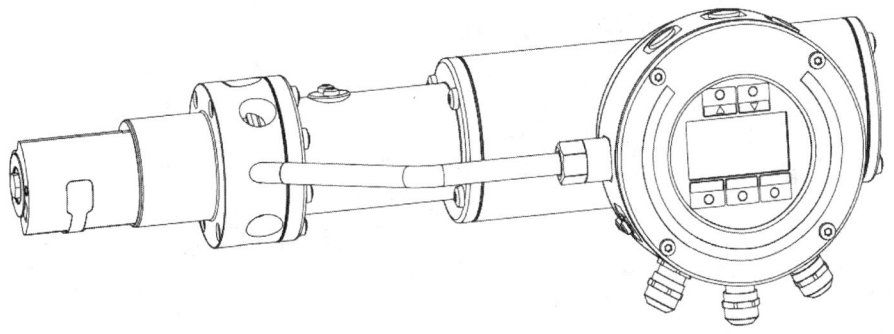

FIG. 1.71w
Dynamic liquid-pressure-type viscometer. (Courtesy of Anton Paar GmbH.)

in the process liquid is used to compensate for temperature variations of the sensor front-end and to provide process temperature information.

By design, new sample is continuously drawn into the gap. The transducer signal is acquired with a robust contactless measurement, allowing operating at up to 25 bar (360 psi) operating pressures and 200°C (392°F) operating temperatures. This type of viscometer can measure viscosities up to 50,000 mPa·s. Its inaccuracy is ±1% of adjusted range and its repeatability is ±0.5% of adjusted range. The viscometer can be setup with a local operating terminal or the display is easily removed from the electronic housing and setup as a remote operating terminal at eye level. A transmitter version in standard and explosion-proof construction is also available.

OSCILLATING VISCOMETERS

Oscillating Blade

Figure 1.71x illustrates the oscillating blade viscometer in which, as the blade moves between the fixed side walls, it discharges the process fluid through the openings of one wall of the measuring chamber while pulling in a fresh sample through the other. The blade is rotated around its fulcrum by a plunger/solenoid mechanism as its tip travels the distance between points A and B. When the stroke is completed, the polarity of the current to the plunger coil is reversed, and the blade is returned to its original position.

Adjustable mechanical stops are provided to limit the stroke. The time of travel between points A and B is detected optically as, upon the completion of the stroke, a beam interrupter blocks a light beam. The total stroke time is about 0.2 sec, and the blade makes a stroke about every 2 sec. The average time of several strokes (adjustable) is measured as an indication of viscosity.

The indicating in-line transmitter can detect viscosities between 10 and 100,000 mPa·s. It has a minimum span of 50 mPa·s and a maximum span of 100,000 mPa·s. The maximum allowable liquid velocity in the process pipe can be up to 2 m/s (6.5 ft/s), but the fluid should not contain particles that are larger than 6 mm (0.2 in.) in diameter.

The maximum allowable liquid velocity in the process pipe can be up to 2 m/s, but the fluid should not contain particles that are larger than 1 mm in diameter.

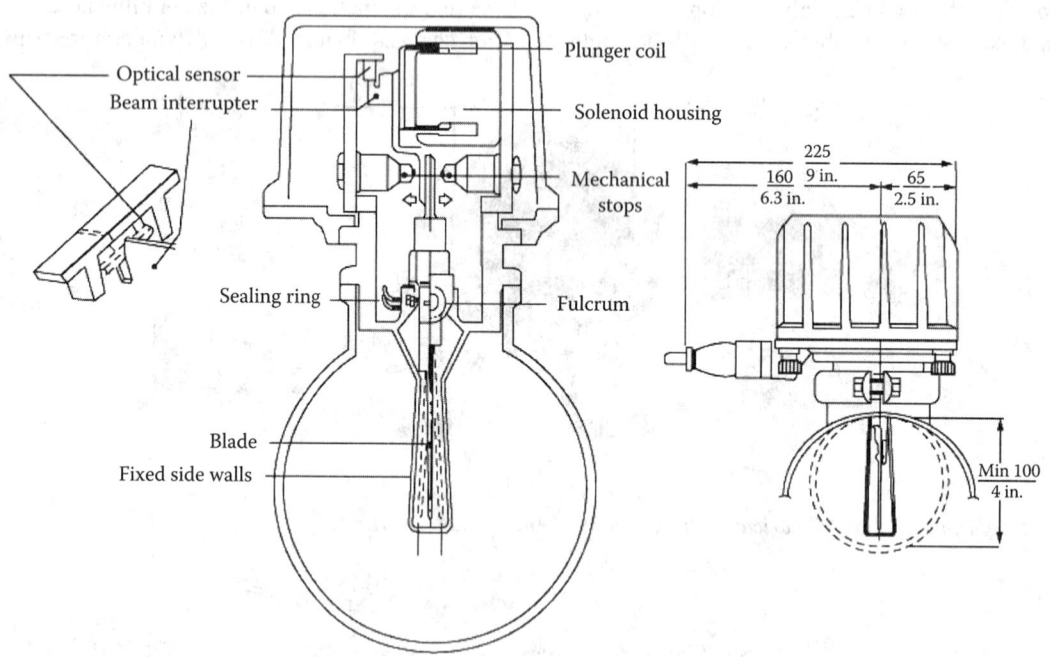

FIG. 1.71x
Oscillating-blade-type viscosity transmitter. (Courtesy of BTG Eclépens S.A.)

Oscillating Piston

The oscillating piston viscometer is shown in Figure 1.71y. With the sensor inserted into the pipeline, the magnetic piston is surrounded by the fluid sample deflected into the measurement chamber. Two coils (A and B) inside the sensor body are used to magnetically force the piston back and forth a predetermined distance (about 5 mm). The viscosity of the process fluid resists the movement of the piston. By alternatively powering the coils with a constant force, the piston's round trip travel time is measured. An increase in viscosity is sensed as a slowed piston travel time. The time required for the piston to complete a two-way cycle is an accurate measure of viscosity. The deflecting fence acts to continuously deflect fresh sample into the measurement chamber. About 30 sec is needed to make a measurement. The viscometer is provided with an integral resistance temperature detector that provides temperature compensation.

The oscillating piston viscometer can operate at high pressures (up 690 bar/10,000 psi) and at high temperatures (315°C/600°F). The wetted parts of the sensor are made of corrosion-resistant alloys (Inconel 718, Hastelloy C276, and stainless steel 17-4PH) and Teflon.

Using diverse pistons, different viscosity ranges can be covered. Each piston provides a useful range of about one order of magnitude of viscosity measurement covering viscosity ranges from 0.2 to 2 mPa·s up to 2,000–20,000 mPa·s. For low viscosity measurements, large pistons with smaller clearance between the piston and the chamber wall are used, and for high viscosity measurements, smaller pistons with greater annular clearance are implemented.

To use this sensor, the process fluid should be free of solid particles that are larger than about 25 μm in size. The inaccuracy claimed is 2% of actual reading or better.

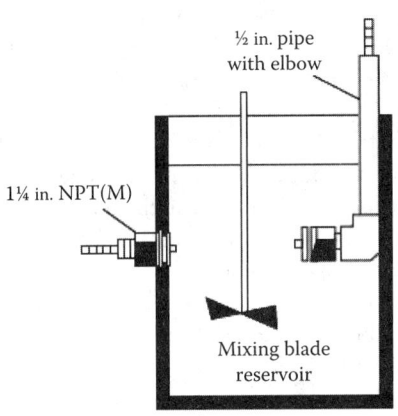
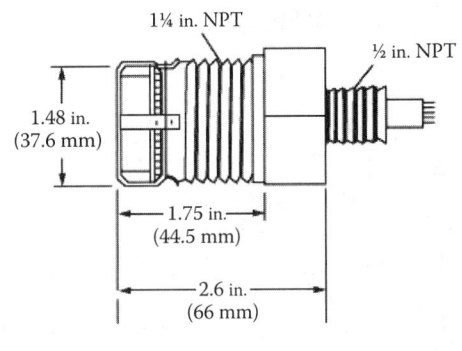
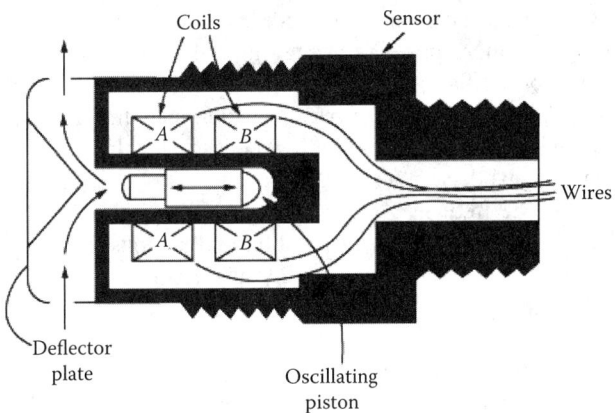

FIG. 1.71y
As the piston is made to oscillate by coils A and B, which are alternately energized, the time of piston travel is registered as viscosity. (Courtesy of PAC, LP.)

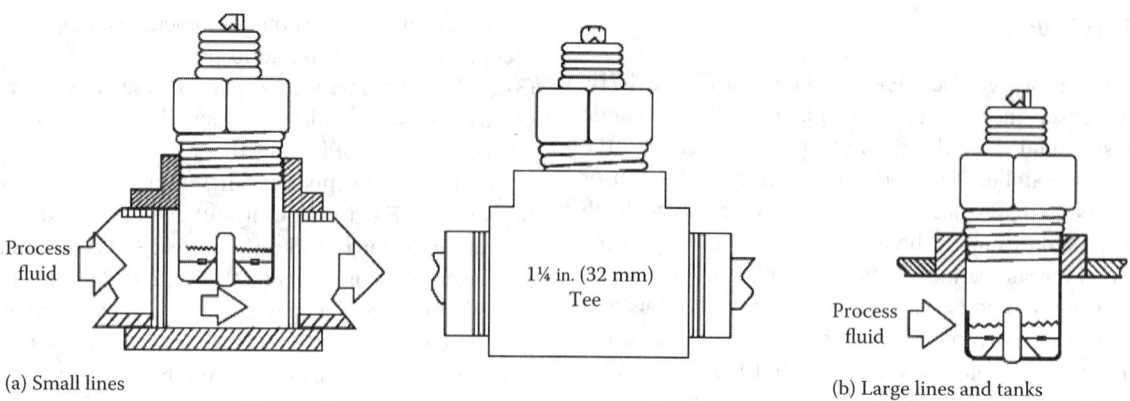

FIG. 1.71z
Pipeline installation of the oscillating piston viscometer. (Courtesy of PAC, LP.)

The electronics are available in explosion-proof construction and can be either the local indicator or the transmitter type. The pipeline installation of this detector is illustrated in Figure 1.71z.

VIBRATIONAL VISCOMETERS

Vibrational viscometers determine viscosity by forcing a resonator to vibrate in the fluid of interest and then measuring the damping that is associated with fluid viscosity.

The resonator's damping may be measured by one of several methods:

1. Measuring the power input necessary to keep the oscillator vibrating at a constant amplitude. The higher the viscosity, the more power is needed to maintain the amplitude of oscillation.
2. Measuring the decay time of the oscillation once the excitation is switched off. The higher the viscosity, the faster the signal decay.
3. Measuring the frequency of the resonator as a function of phase angle between excitation and response waveforms. The higher the viscosity, the larger the frequency change for a given phase change.

The most common vibrational viscometers are torsional oscillation designs and vibrating reeds and roods.

Torsional Oscillation Design

Torsional oscillation viscometers are considered surface-loaded systems because they respond to a thin layer of fluid at the surface of the sensor. The higher the viscosity, the larger the damping imposed on the resonator. The sensor can either be spherical, cylindrical, or rod shaped. The spherical elements can be used between 10 and 100,000 mPa·s, the cylindrical tips from 0.1 to 10,000, and the rod-shaped ones from 100 to 1,000,000 mPa·s. Most types can be exposed to high pressures (350 bar/5000 psi) and high temperatures (454°C/850°F) with specific modifications.

An example of the torsional oscillation viscometer using the first sensing method is shown in Figure 1.71aa. The sensor is electromechanically oscillated in a torsional motion by an inner shaft around its central axis. The amplitude of oscillation is only 1 μm, and the power required to maintain that amplitude is proportional to the material's viscosity. This change in power requirement is converted to the viscosity reading. To provide temperature compensation, the temperature of the process fluid is detected by a internal thermocouple which is inserted through a central hole in the inner drive shaft.

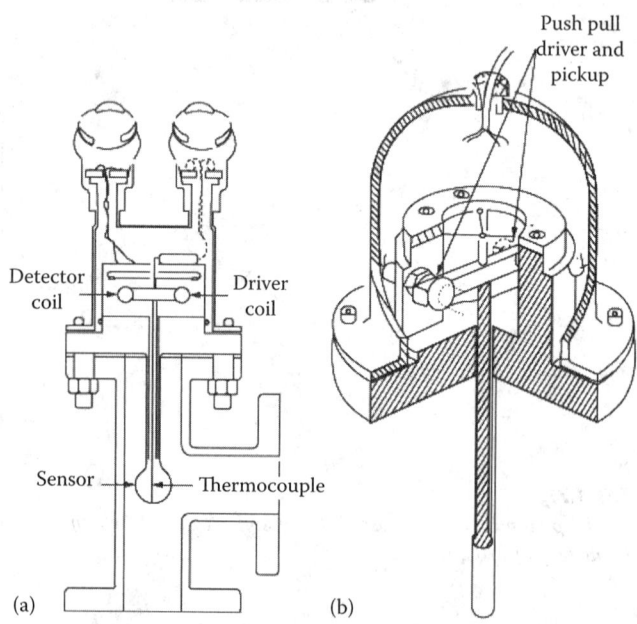

FIG. 1.71aa
(a) Torsional oscillation-type viscometers with spherical and (b) rod-type probe sensors. (Courtesy of Galvanic Applied Sciences USA Inc.)

This type of sensor design is especially suitable for fluids which may clog tubes or rotating chambers. The mounting base is typically a standard ANSI piping flange or sanitary clamp connections for the Pharmaceutical and Food Process Industries.

To minimize inaccuracy (error), automatic range switching is available. By always selecting the smallest span to measure the prevailing viscosity, the error can be reduced to 2% of actual reading. In addition, a compensating filter is provided to minimize the effect of pipeline vibrations.

The viscometer transmitter indicates the output reading as dynamic viscosity at a specific density. To correct for density variations, the density of the process fluid can be manually dialed in or automatically compensated with external signal input.

Vibrating-Reed/Rod Viscometer

The vibrating-reed viscometer consists of a frequency generator, a vibrating spring rod immersed in the probe, and a pickup unit and can be installed directly into the process vessels or pipeline (Figure 1.71ab).

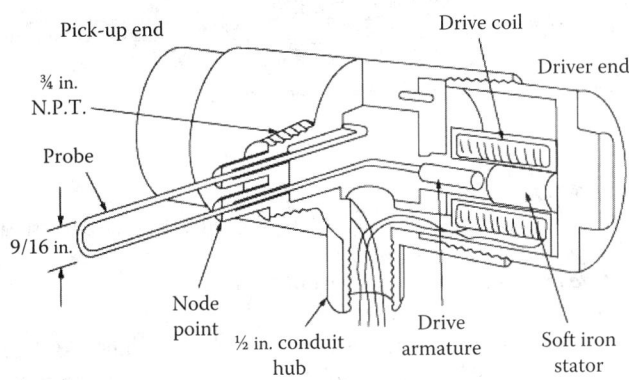

FIG. 1.71ab
Vibrating-reed viscometer.

The drive coil is excited at a certain input frequency. This produces a pulsating magnetic field that causes the drive armature to vibrate at the same frequency. The mechanical vibration of the drive armature is transmitted to the probe through an all-welded, pressure-sealed node that is located where the amplitude of vibration is zero.

The principle of operation is based on the fact that the amplitude of probe vibration depends on the viscosity of the process media. The resistance to the shearing action caused by the probe's vibration increases with an increase in the viscosity of the process media. The amplitude of probe vibration is sent through a second welded node point by the upper spring rod to the pickup end of the detector.

The pickup end is similar to that of the driver end, except that a permanent magnet is used to induce an oscillating voltage in the pickup coil. The magnitude of the voltage generated is a measure of the process viscosity. For accurate measurements, *laminar* flow conditions and complete immersion of the probe should be maintained. Therefore, the installation of this type of viscometer directly in an agitated vessel is not recommended. Material buildup on the probe can occur if the process material has a tendency to adhere or if it contains long fibrous materials. Slight or loose buildup can be removed by periodic in-line purging. If the buildup can be severe, this type of viscometer should not be used.

Performance Inaccuracy of the instrument is about ±1% of full scale, and reproducibility is a little better. There are five different detector units to cover the viscosity ranges from 10^{-1} to 10^5 mPa·s at relatively low shear rate. Typical response curves of four different detector units are shown in Figure 1.71ac.

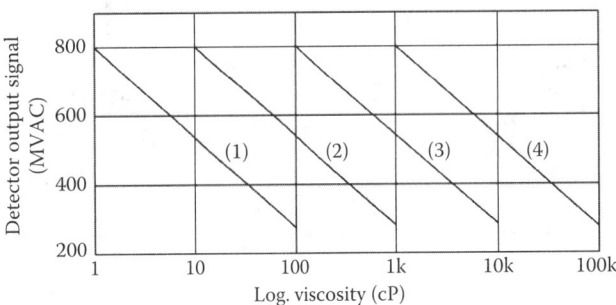

FIG. 1.71ac
The response curves of four vibrating-reed viscometers.

Because the viscosity of fluids depends on their temperature and pressure, it is mandatory to maintain the process pressure and temperature constant. As shown in Figure 1.71ad, if the fluid temperature must be allowed to fluctuate, temperature compensation is required. Before ordering a viscometer for in-process use, the fluid's viscosity must be measured as a function of temperature so that this characteristic curve can be used in the converter.

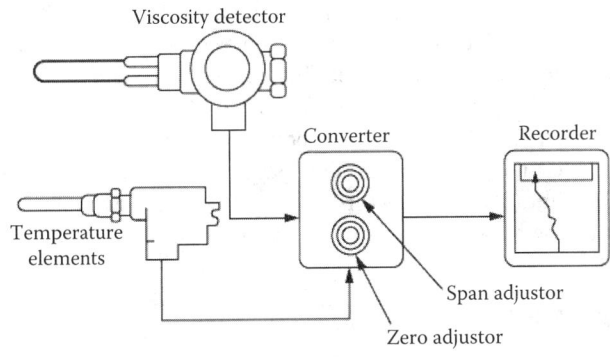

FIG. 1.71ad
Complete loop diagram of a temperature-compensated vibrating-reed-type viscosity recorder.

For accurate measurements, *laminar* flow conditions should be maintained. Therefore, the installation of this type of viscometer directly in an agitated vessel is not recommended. If it must be done, the viscosity reading should be taken only when the agitator is off.

Installation and Calibration Complete immersion of the probe in the process fluid is essential. Figure 1.71ae illustrates the recommended pipeline installation of this viscometer probe in vertical upward flow. Complete immersion of the probe is required at all times. To improve the accuracy of measurement, the recommended viscometer installation is in a temperature, pressure, and flow rate controlled sample loop.

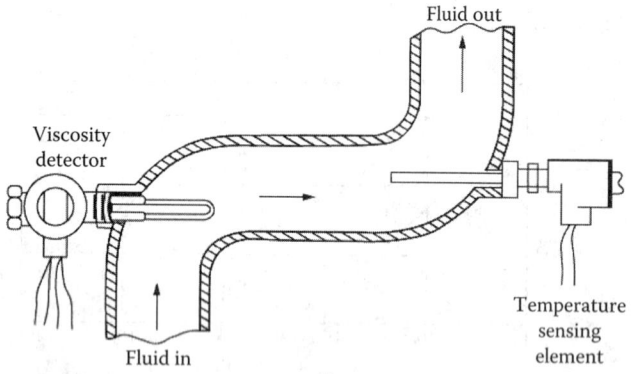

FIG. 1.71ae
Installation recommendation for an in-line vibrating-reed viscometer.

The amplitude of vibration will change with condition and age of the probe. It is good practice to calibrate the instrument regularly when it is used for critical operations. Material buildup on the probe can occur if the process material has a tendency to adhere or if it contains long fibrous materials. Slight or loose buildup can be removed by periodic in-line purging. If the buildup can be severe, this type of viscometer should not be used.

Because the detector unit transmits a low-voltage signal, the use of shielded wires is recommended. The wires should be grounded at both ends to minimize noise pickup. Supply voltage variations between 105 and 125 V will have no harmful effects. The *frequency of supply power should be checked*, because the probe vibration frequency is controlled by the line frequency. If the vessel or pipeline vibrates, and this vibration frequency is close to 120 Hz, it can interfere with the reliability of viscosity measurement by this sensor.

CORIOLIS MASS FLOWMETER

Since many decades, instrument designers have used the principle of Coriolis force to measure mass flow and density with Coriolis flowmeters. Most of these meters also include temperature measurement and are thus true multivariable instruments.

A traditional method for measuring viscosity with mass flowmeters is to add a differential pressure meter around the mass flowmeter, whereby the viscosity is measured by determining the pressure drop through the mass flowmeter. In this case, the capillary design is used and the viscosity calculation is based on the Hagen–Poiseuille equation.

Measuring Tube Torsional Movement

A more recent method is to enhance the measurement of the rotational force exerted by fluids flowing through an oscillating measurement tube which is traditionally used to calculate mass flow and density (Figure 1.71af).

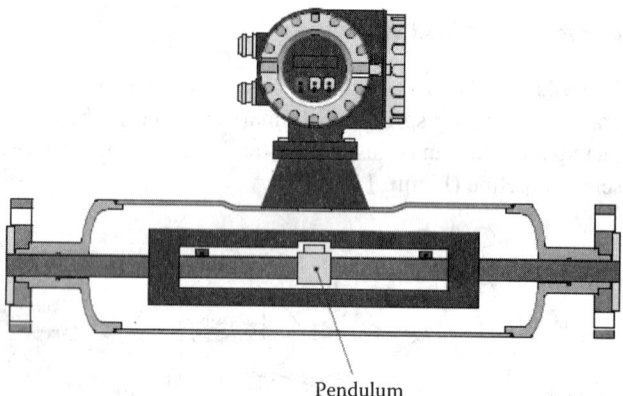

FIG. 1.71af
Mass flowmeter/measuring tube torsional movement type (pendulum on the measuring tube balancing the system for immunity to external forces). (Courtesy of Endress + Hauser.)

Coriolis flowmeters use a torsion mode balance system for optimum balance and isolation from external influences. A pendulum is attached to the middle of the oscillating measuring tube to provide a balancing force (Figure 1.71ag). This pendulum oscillates counterphase to the tube, thus compensating for the momentum of the measuring tube.

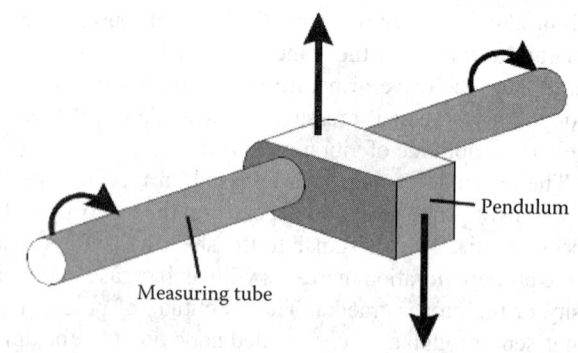

FIG. 1.71ag
Mass flowmeter/measuring tube torsional movement type (oscillation movement of the measuring tube and pendulum). (Courtesy of Endress + Hauser.)

Due to the rotational motion of the tube, the fluid is forced to a rotational motion. Depending on the fluid's viscosity, different velocity profiles of the fluid are generated (Figure 1.71ah). The gradient of the velocity profile induces shear forces in the fluid, which dampen the oscillation of the measuring tube and can be measured via the excitation current necessary to maintain tube oscillation. The excitation current can then be used to calculate the viscosity of the fluid.

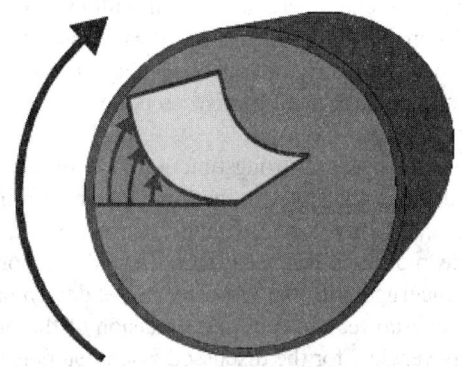

FIG. 1.71ah
Mass flowmeter/measuring tube torsional movement type (cross section of the tube, showing the velocity profile of the fluid). (Courtesy of Endress + Hauser.)

Measurement inaccuracy is ±5% of actual reading and a repeatability is ±0.5% of actual reading. The in-line meter can measure viscosities up to 20,000 mPa·s for sample temperatures up to 150°C (302°F) and line pressured up to 100 bar (1450 psi).

NEW DEVELOPMENTS

Trends

The accelerated growth of smart instrumentation in the process industry over the past 20 years has also affected process viscometers.

With the development of microelectronics, communication standardization, and diagnostic protocols, the process viscometer of today and tomorrow is becoming part of the smart manufacturing processes, optimized to enable cost-effective manufacturing of the customized products of the future. The following distinctive trends can be observed:

Display and human machine interface—process viscometers are able to display the measured viscosity and process temperature. The display layout, display units, calibrations, and adjustments are done locally or using the provided field communication. One can optionally display the process viscosity at a reference temperature by selecting the proper calculation method, whereby the internal process temperature and process viscosity data are used for calculation. The devices combine human-centered design which is intuitive to use with multilevel access control.

Communication—The standard analog output for process viscosity and process temperature is clearly enhanced and integrates into data networks using different fieldbus technologies (Foundation fieldbus, Profinet, Modbus, and EtherNet/IP, to name just a few). This allows remote users to access internal values and reconfigure the functions of the process viscometer.

Self-diagnostics—smart process viscometers perform self-diagnosis and monitor internal signals for faults. One of the recommendations used is the NAMUR NE107 for device diagnostics. The NE107 recommendations provide specific advice on how diagnostic information from intelligent devices should be presented to those that need it, from operators to maintenance technicians and engineers, in a standard format. This cuts complexity and reduces the training required for operators and technicians, contributing to improved safety and instrument availability.

Integrating smart process viscometers together with proper process automation systems will play a key role in modern plant management. This can significantly reduce operating costs and help achieve operational excellence. Many of the benefits are still being discovered as users gain more experience with these technologies in real-world plant settings.

Upcoming Technologies Some additional technologies which could be implemented for process viscosity measurement are briefly described in the paragraphs in the following.

Surface acoustic wave sensors are a class of microelectromechanical systems (MEMS) which rely on the modulation of surface acoustic waves to sense a physical phenomenon. The sensor transduces an input electrical signal into a mechanical wave which, unlike an electrical signal, can be easily influenced by physical phenomena. The device then transduces this wave back into an electrical signal. Changes in amplitude, phase, frequency, or time delay between the input and output electrical signals can be used to measure the presence of the desired phenomenon like density and viscosity.

Measurement of velocity profile within the pipe (either by Nuclear Magnetic Resonance, laser Doppler velocimetry, or acoustics) and calculation of viscosity. Here, the approach is to recognize that a fluid undergoing tube flow experiences shear rates ranging from zero at the tube center to a maximum at the wall. Independent of the constitutive relation, the shear stress distribution can be calculated from the pressure drop. The shear rate is determined by using a spatially resolved velocimetric technique to measure the radial dependence of the fluid velocity, from which a range of shear rates in the tube can, in principle, be obtained by differentiation. Dividing the shear stress at each radial position by the shear rate gives the shear viscosity over shear rates that theoretically range from zero at the tube center to a maximum at the tube wall.

Squeeze Flow. A common definition of "squeezing" phenomenon is the large deformation (or a flow) of a soft material (or a viscous fluid) between two nearly solid surfaces approaching each other. Consequently, the gap between surfaces is changing in time and the sample is ejected from the gap. In many applications, the gap is small in comparison to the other dimensions of the surfaces, so squeezing flow is mainly associated to thin film hydrodynamics, lubrication, and rheology of complex fluids. The squeezing flow has both extensional and shears components, but in the limit of small gap and low relative velocity the shear is considered dominant. Generally, one is measuring the normal force on the top or bottom plate as a function of time whilst it is moving with a prescribed sinusoidal motion. Fourier transforms of the applied and transmitted signals are calculated and from this analysis the values of viscosity are calculated.

CONCLUSIONS

Rheological properties of industrial materials are one of the most important indicators of their quality. Furthermore, rheological or flow properties of process streams directly affect the design and operation of processes. Therefore, rheological characterization of samples, taken in some critical steps in production, provides crucial insight into product quality control and process economics. However, there are some difficulties in taking samples one by one out of the process and examining them. For example, viscosity, which is the most important rheological property, must be determined in a range of shear rate as most of the fluids in industry are non-Newtonian. Moreover, rheological properties of many samples are closely related to the flow conditions in the process. That is why the measurements in laboratory conditions may lead to inaccurate results and process measurements are necessary.

All presented industrial viscometers show some element of novelty in terms of the way they measure viscosity or how they ensure that liquid is replenished in the measuring region. Some viscometer designs provide intermittent readings or require bypass or sampling system, others provide continuous readings or/and can be inserted directly into the process pipe or vessel.

Whether or not one can live with the associated compromises depends on the nature of the liquids being measured. Non-Newtonian characterization is difficult even in the best offline viscometer under ideal laboratory circumstances. To measure such liquids in the process is very difficult, as most instruments are calibrated to give an equivalent Newtonian viscosity.

Since viscosity is a strong function of temperature, viscometer designs that provide automatic temperature compensation and do not need any density corrections are likely to provide more accurate readings. In addition, one must also be concerned with the viscosity range, design pressure, design temperature, and required precision of the measurement. An overview for the discussed viscometers is given in Table 1.71a.

SPECIFICATION FORMS

When specifying industrial viscometers, use the ISA form 20A1001 and when the type of analyzer has already been selected and both the analyzer and the requirements of the process application are to be specified, use the ISA form 20A1002. Both forms are reproduced with the permission of the International Society of Automation on the next pages.

1.71 Viscometers: Industrial

1	RESPONSIBLE ORGANIZATION		ANALYSIS DEVICE	6	SPECIFICATION IDENTIFICATIONS		
2	(ISA)			7	Document no		
3			Operating Parameters	8	Latest revision		Date
4				9	Issue status		
5				10			

	ADMINISTRATIVE IDENTIFICATIONS				SERVICE IDENTIFICATIONS Continued			
11				40				
12	Project number		Sub project no	41	Return conn matl type			
13	Project			42	Inline hazardous area cl		Div/Zon	Group
14	Enterprise			43	Inline area min ign temp		Temp ident number	
15	Site			44	Remote hazardous area cl		Div/Zon	Group
16	Area	Cell	Unit	45	Remote area min ign temp		Temp ident number	
17				46				
18	SERVICE IDENTIFICATIONS			47				
19	Tag no/Functional ident			48	COMPONENT DESIGN CRITERIA			
20	Related equipment			49	Component type			
21	Service			50	Component style			
22				51	Output signal type			
23	P&ID/Reference dwg			52	Characteristic curve			
24	Process line/nozzle no			53	Compensation style			
25	Process conn pipe spec			54	Type of protection			
26	Process conn nominal size	Rating		55	Criticality code			
27	Process conn termn type	Style		56	Max EMI susceptibility		Ref	
28	Process conn schedule no	Wall thickness		57	Max temperature effect		Ref	
29	Process connection length			58	Max sample time lag			
30	Process line matl type			59	Max response time			
31	Fast loop line number			60	Min required accuracy		Ref	
32	Fast loop pipe spec			61	Avail nom power supply			Number wires
33	Fast loop conn nom size	Rating		62	Calibration method			
34	Fast loop conn termn type	Style		63	Testing/Listing agency			
35	Fast loop schedule no	Wall thickness		64	Test requirements			
36	Fast loop estimated lg			65	Supply loss failure mode			
37	Fast loop material type			66	Signal loss failure mode			
38	Return conn nominal size	Rating		67				
39	Return conn termn type	Style		68				

	PROCESS VARIABLES	MATERIAL FLOW CONDITIONS			PROCESS DESIGN CONDITIONS		
69				101			
70	Flow Case Identification		Units	102	Minimum	Maximum	Units
71	Process pressure			103			
72	Process temperature			104			
73	Process phase type			105			
74	Process liquid actl flow			106			
75	Process vapor actl flow			107			
76	Process vapor std flow			108			
77	Process liquid density			109			
78	Process vapor density			110			
79	Process liquid viscosity			111			
80	Sample return pressure			112			
81	Sample vent/drain press			113			
82	Sample temperature			114			
83	Sample phase type			115			
84	Fast loop liq actl flow			116			
85	Fast loop vapor actl flow			117			
86	Fast loop vapor std flow			118			
87	Fast loop vapor density			119			
88	Conductivity/Resistivity			120			
89	pH/ORP			121			
90	RH/Dewpoint			122			
91	Turbidity/Opacity			123			
92	Dissolved oxygen			124			
93	Corrosivity			125			
94	Particle size			126			
95				127			
96	CALCULATED VARIABLES			128			
97	Sample lag time			129			
98	Process fluid velocity			130			
99	Wake/natural freq ratio			131			
100				132			

	MATERIAL PROPERTIES				MATERIAL PROPERTIES Continued		
133				137			
134	Name			138	NFPA health hazard		Flammability	Reactivity
135	Density at ref temp		At	139			
136				140			

Rev	Date	Revision Description	By	Appv1	Appv2	Appv3	REMARKS

Form: 20A1001 Rev 0 © 2004 ISA

	RESPONSIBLE ORGANIZATION	ANALYSIS DEVICE COMPOSITION OR PROPERTY Operating Parameters (Continued)		SPECIFICATION IDENTIFICATIONS		
1			6			
2			7	Document no		
3			8	Latest revision	Date	
4			9	Issue status		
5			10			

	PROCESS COMPOSITION OR PROPERTY			MEASUREMENT DESIGN CONDITIONS				
	Component/Property Name	Normal	Units	Minimum	Units	Maximum	Units	Repeatability
13								
14								
15								
16								
17								
18								
19								
20								
21								
22								
23								
24								
25								
26								
27								
28								
29								
30								
31								
32								
33								
34								
35								
36								
37								

Rev	Date	Revision Description	By	Appv1	Appv2	Appv3	REMARKS

Form: 20A1002 Rev 0 © 2004 ISA

Definitions

Accuracy: Degree of conformity of an indicated value to a recognized accepted standard value, or ideal value.

Bar: Bar is a non-SI unit of pressure, defined as exactly equal to 100,000 Pa and is accepted for use with the SI.

Bingham plastic: A Bingham plastic is a viscoplastic material that behaves as a rigid body at low stresses but flows as a viscous fluid at high stress.

Bypass: Means that only part of the liquid is taken and passed through the viscometer—for instance, in a bypass loop.

Hastelloy: Hastelloy is the registered trademark name of Haynes International, Inc. The trademark is applied as the prefix name of a range of different highly corrosion-resistant metal alloys, loosely grouped by the metallurgical industry under the material term "superalloys" or "high-performance alloys."

Inconel: A family of austenitic nickel–chromium-based superalloys. The name is a trademark of Special Metals Corporation.

In-Line: Is used to describe the situation where all the liquid whose viscosity is being measured passes through or around the process viscometer.

Newtonian: Flow model of fluids in which a linear relationship exists between shear stress and shear rate, where the coefficient of viscosity is the constant of proportionality.

Non-Newtonian: Any laminar flow that is not characterized by a linear relationship between shear stress and shear rate.

Online: Online measurement is any measurement which provides (quasi) continuous process viscosity data (could be an in-line or bypass installation).

psi: Pound per square inch or, more accurately, pound-force per square inch. Unit of pressure based on avoirdupois units. It has been largely replaced by the Pascal. 1 psi = 6894.75 Pa.

Rheopecty: A reversible time-dependent increase in viscosity at a particular shear rate; the longer the fluid undergoes shearing, the higher its viscosity.

Teflon: DuPont brand name for Polytetrafluoroethylene (PTFE)—a synthetic fluoropolymer of tetrafluoroethylene.

Thixotropy: A reversible time-dependent decrease in viscosity at a particular shear rate. Shearing causes a gradual breakdown in structure over time.

Abbreviations

GPM   Gallons per minute
MI   Melt index
psi   Pound per square inch
SI   Système international d'unités

Organizations

ANSI   American National Standards Institute
CIL   Canadian Industries Limited
DIN   Deutsches Institut für Normung—German Institute for Standardization

Bibliography

Abnett, A. and Garvey, R., New orbital ball type of viscometer, *Proceedings of the Joint Conference on Integrated Monitoring, Diagnostics, and Failure Prevention*, Mobile, AL, NTIS Accession Number AD-P010 179/0/XAB, April 1996.

Ahmed, J. et al., *Novel Food Processing—Effects on Rheological and Functional Properties*, Taylor & Francis Group, LLC, Boca Raton, FL, 2010. http://www.crcpress.com/product/isbn/9781420071191.

Antlinger, H. et al., Utilizing pressure waves for sensing the properties of liquids, *Proceedings of the GMe Forum*, Johannes Kepler Universität Linz, Linz, Austria, 2011.

Arola, D.F. et al., Pointwise observations for rheological characterization using nuclear magnetic resonance imaging, *J. Rheol.*, 43, 9–30, 1999.

ASTM Standard, ASTM-D-445-97/XAB, *Kinematic Viscosity of Transparent and Opaque Liquids (the Calculation of Dynamic Viscosity)*, September 1998.

Bandrup, J., Immergut, E.H., Grulke, E.A., and Bloch, D., eds., *Polymer Handbook*, 4th edn., John Wiley & Sons, New York, 1999.

Barnes, H.A., Thixotropy: A review, *J. Non-Newtonian Fluid Mech.*, 1997.

Barnes, H.A., *A Handbook of Elementary Rheology*, University of Wales, 2000.

Basker, V.R., Dutka, A.P., Crisalle, O.D., and Fricke, A.L., Evaluation of an online torsional oscillatory viscometer for kraft black liquor, *PaperijaPuu/Pulp and Timber*, 82(7), October 2000.

Batra, R.L., *Falling Cylinder Viscometer for Casson Fluids*, Jamshedpur Regional Institute of Technology, Jamshedpur, India, 1981.

Birkhofer, B.H., Ultrasonic in-line characterization of suspensions, PhD thesis, ETH Zurich, Zurich, Switzerland, 2007.

Black, S.S., Automated differential viscometry leads to 300 percent increase in analysis rate, *Sci. Comput. Autom.*, 5, April 1996.

Boetcher, U., Ultraschallspektroskopische und rheologische Charakterisierung vonkonzentrierten o/w-Emulsionen mit Stabilisatoren, PhD thesis, Technischen Universität Dortmund, Dortmund, Germany, 2009.

Bullon, H.H., Shear rate determination in a concentric cylinder viscometer, Defence Science and Technology Organization, NTIS accession number AD-A315 366/5, July 1996.

Carr, K.R., Rotational viscometer for high-pressure, high-temperature fluids, Patent application 6-501-313, June 6, 1983.

Chang, V., Zambrano, A., Mena, M., and Millan, A., A sensor for online measurement of the viscosity of non-Newtonian fluids using a neural network approach, *Sensor Actuator A*, 47(1–3), March–April 1995.

Chu, B. and Wang, J., Magnet enhanced optical falling needle/sphere rheometer, *Rev. Sci. Instrum.*, 63(4), April 1992.

Clubb, D.O. et al., Quartz tuning fork viscometers for helium liquids, *J. Low Temp. Phys.*, July 2004.

Coblas, D. et al., Rheological caracterisation of viscous fluids in oscillatory squeezing flows, *UPB Sci. Bull.*, D72(3), 2010.

Cox, C.J., Park, M.J., and Wilson, C.A., Use of the capillary tube viscometer for the determination of the flow behavior of tall oil soap, *Trans. New Zeal. Inst. Prof. Eng.*, March 1983.

Cutrone, L., Use of the Brookfield viscometer to predict rheological performance of coatings, *J. Coatings Tech.*, January 1984.

D'Avila, M.A. et al., Magnetic resonance imaging (MRI): A technique to study flow anmicrostructure of concentratedemulsions, *Braz. J. Chem. Eng.*, 2005.

Dutka, A.P., Crisalle, O.D., Fricke, A.L., and Kalotay, P., Evaluation of a capillary-coriolis instrument for online viscosity and density measurements of kraft black liquor, *Proceedings of the TAPPI Process Control, Electrical and Instrumentation Conference (ISA)*, March 1997.

Ferry, J.D., Oscillation viscometry—Effects of shear rate and frequency, *Meas. Control*, September–October 1977.

Fitzgerald, J.V., Matusik, F.J., and Walsh, T.M., Inline viscometry, *Meas. Control*, December 1987.

Fournier-Villalon, P., Bertrand, J., and Couderc, J.P., Using a Brookfield viscometer for determining rheological properties of newtonian, plastic and pseudoplastic media, *Entropie*, 20(120), 1984.

Frisman, M.L., Rotation viscometer for readily unmixing suspensions, *Indust. Lab.*, October 1984.

Galvin, G.D., Hultin, J.F., and Jones, B., Development of a high-pressure, high-shear rate capillary viscometer, *J. Non-Newtonian Fluid Mech.*, January 1981.

Ghosh, T.K. et al., *Viscosity of Liquids*, Springer, Dordrecht, the Netherlands, 2007. http://www.springer.com/materials/mechanics/book/978-1-4020-5481-5.

Giap, S.G.E., The hidden property of Arrhenius-type relationship: Viscosity as a function of temperature, *J. Phys. Sci.*, 2010.

Gillis, K.A., Mehl, J.B., and Moldover, M.R., Greenspan acoustic viscometer for gases, *Rev. Sci. Inst.*, 67(5), May 1996.

Gimera, M., Software quality assurance plan for viscometer, Department of Energy Report, NTIS accession number DE95002145, October 1994.

Glicerina, V. et al., Rheological, textural and calorimetric modifications of darkchocolate during process, *J. Food Eng.*, 2013.

Goodwin, J.W. et al., *Rheology for Chemists*, Royal Society of Chemistry, Cambridge, U.K., 2008. http://pubs.rsc.org/en/Content/eBook/978-0-85404-839-7#!divbookcontent.

Grupp, J., Torsion viscometer for liquid crystals, *Rev. Sci. Inst.*, June 1983.

Helle, H., Valimaki, H., and Lekkala, J., Comparing a 10 MHz thickness-shear mode quartz resonator with a commercial process viscometer in monitoring resol manufacturing process, *Sensor Actuator B*, B81(2–3), January 2002.

Jethra, R., Viscosity measurement, *ISA Trans.*, 33(3), September 1994.

Kalotay, P., Online viscosity measurement using coriolis mass flowmeters, *Flow Meas. Instrum.*, 5(4), October 1994.

Khachatryan, G.M. et al., Device for the automatic continuous measurement of the dynamic viscosity of polymers, *Fibre Chem.*, May–June 1983.

Kim, J.H. et al., Rheological characteristics of poly(ethylene oxide) aqueous solutionsunder large amplitude oscillatory squeeze flow, *Korea Aust. Rheol. J.*, 24(4), 257–266, 2012.

Koseli, V., Zeybek, S., and Uludag Y., Online viscosity measurement of complex solutions using ultrasound Doppler velocimetry, *Turk. J. Chem.*, 2006.

Kurano, Y., Accurate measurement of viscosity under high pressure with a falling-sphere viscometer, *High Temp. High Press.*, 26(1), 1994.

Lucklum, F. et al., Miniature flow-through resonator cellfor density and viscosity sensing, *Proceedings of Eurosensors XXIV*, September 2010.

McCarthy, K.L. et al., Relationship between in-line viscosity and Bostwick measurement during ketchup production, *J. Food Sci.*, 2009.

Mia, S., Prediction of tribological and rheological properties of lubricating oils by sound velocity, PhD thesis, Graduate School of Science and Engineering, Saga University, Saga, Japan, 2010.

Nielsen, S., *Food Analysis*, Springer, New York, 2010. http://www.springer.com/food+science/book/978-1-4419-1477-4.

Nindo, C.I. et al., Viscosity of blueberry and raspberry juices for processing applications, *J. Food Eng.*, 2005.

Ohene, F., Livingston, C., Matthews, C., and Rhone, Y., Study of pressure drop in a capillary tube viscometer for a two-phase flow, Department of Energy Report #DOE/PC/91292-T17, NTIS accession number DE96000151, 1995.

Oppliger, H.R., Matusik, F.J., and Fitzgerald, J.V., New technique accurately measures low viscosity on-line, *Control Eng.*, July 1985.

Roger, U., Hessenkemper, H., and Roth, P., Glass conditioning by viscosity control, *Glass Sci. Tech.*, 69(8), August 1996.

Rogers, E.K. and Brimelow, C.J.B., *Instrumentation and Sensors for the Food Industry*, Woodhead Publishing Limited, Cambridge, U.K., 2001.

Sawhney, I.K. et al., Development of capillary tube viscometer for measuring flow properties of khoa, *J. Institute Eng.*, August 1984.

Sexton, W.C., Rotating spindle viscometer, *International and 42nd Annual Symposium and Exposition on Scientific Glassblowing*, NTIS accession number DE970610182/XAB, June 1997.

Sheen, S.H., Chien, H.T., and Raptis, A.C., In-line ultrasonic viscometer, *1994 Review of Progress in Quantitative Nondestructive Evaluation Conference*, Sponsored by Department of Energy, NTIS accession number DE95002982, October 1994.

Smith, R.E., Brookfield viscometers for determination of low-shear viscosity and leveling behavior, *J. Coatings Tech.*, January 1984.

Smith, R.S. et al., Measurements of the rheological properties of standard reference material2490 using an in-line micro-Fourier rheometer, *Korea Aust. Rheol. J.*, 16(4), December 2004.

Steiner, G., Gautsch, J., Breidler, R., and Plank, F., A novel fluid dynamic inline viscometer suitable for harsh process conditions, *Proceedings of Eurosensors XXIV*, September 2010.

Villemaire, J.P. and Agassant, J.F., Apparent viscosity measurements using a capillary viscometer, *Polymer Process Eng.*, 1(3), 221, 1983–1984.

Viscometers, *Meas. Control*, June 1993.

Vrentas, J.S., Mechanical energy balances for a capillary viscometer, *J. Non-Newtonian Fluid Mech.*, March 1983.

Walsh, T.M., Continuous viscosity control improves quality, *Adhesives Age*, December 1989.

Webber, P.J. and Savage, J.A., Measurement of the viscosity of chalcogenide glasses by parallel plate technique, *J. Mater. Sci.*, March 1981.

Wunderlich, T. and Brunn, P.O., Ultrasound pulse Doppler method as a viscometer for process monitoring, *Flow Meas. Instrum.*, 10(4), December, 1999.

Zhang, Z. and Reddy, R.G., Viscosities of lead silicate slags, *Miner. Metall. Process.*, 19(1), February 2002.

1.72 Viscometers
Laboratory

C. H. KIM (1969, 1982)   **B. G. LIPTÁK** (1995, 2003)   **M. ZAVRŠNIK** (2017)

Types	A. Bubble time
	B. Capillary
	B1. Manual capillary tube
	B2. Auto-timing capillary tube
	B3. Capillary extrusion (influx)
	B4. Capillary extrusion (efflux)
	C. Falling/Rolling element
	C1. Falling/Rolling ball
	C2. Falling needle
	D. Orifice/Efflux cup
	D1. Saybolt
	D2. Ford
	D3. Zahn
	D4. Auto timing
	E. Oscillating element
	E1. Piston type
	F. Rotational
	F1. Coaxial cylinder
	F2. Cone and plate
	F3. Parallel plate
	G. Stabinger
	H. Vibrating
Operating pressures, temperatures, ranges, and inaccuracies	See Table 1.72a
Materials of construction	A., B1., B2. Glass
	B3, B4. Hardened stainless steel
	C. Aluminum, brass, stainless steel
	D. Glass and corrosion-resistant alloys
	E. Wetted parts are quartz, borosilicate; needles are glass
	F. Stainless steel, nickel-plated brass, plastic, ceramic, platinum
	G. Copper, PTFE, PEEK
	H. Wide selection of corrosion-resistant materials and coatings
Costs	A. $100–$250
	B1. $75–$300 for capillary tubes; $500–$3000 for thermostatic baths
	B2. $5,000–$8,000 (with auto sampler up to $100,000)
	B3. $1500–$3000
	B4. $5,000–$15,000; computerized processability tester, $70,000

(Continued)

C1. From $1000
C2. $21,000 for research unit used in petroleum industry
D1-3. $50–$300 for cup; $250 for calibration
D4. $2000 and up
E. From $10,000
F. From $4000
G. From $20,000–$50,000
H. From $3500

Partial list of suppliers

Anton Paar GmbH (C, F, G) (http://www.anton-paar.com/corp-en/products/group/viscometer/; http://www.anton-paar.com/corp-en/products/group/rheometer/)
Brookfield Engineering Laboratories Inc. (F) (http://www.brookfieldengineering.com/products/viscometers/laboratory.asp; http://www.brookfieldengineering.com/products/rheometers/laboratory.asp)
Cannon Instruments Inc. (B, F) (http://www.cannoninstrument.com/)
Cargille Laboratories Inc. (A) (http://www.cargille.com/vistube.shtml)
Cole-Parmer Instrument Co. (B, C, D, F) (http://www.coleparmer.com/Search/viscosity)
GoettfertWerkstoff Preufmaschinen GmbH (B) (http://www.goettfert.de/index.php/produkte/kapillarrheometer)
KaltecScientific Inc. (F) (http://www.kaltecsci.com/dv10.html)
Lamy Rheology (F) (http://www.lamyrheology.com/Francais/Categorie/1_Viscosimetres.html)
Lauda-Brinkmann, LP. (B) (http://www.lauda-brinkmann.com/glasscapillary.html)
Malvern Instruments Ltd (B, F) (http://www.malvern.com/en/products/measurement-type/rheology-viscosity/default.aspx)
Norcross Corp. (D) (http://www.viscosity.com/p_ec_sc.asp)
OFI Testing Equipment, Inc. (D, F) (http://www.ofite.com/products.asp?category=Viscometers)
PAC, LP (B, E) (http://www.paclp.com/Lab_Instruments/Application/Physical%20Properties/Viscosity)
Ravenfield Designs (F) (http://www.ravenfield.com/)
SelectraSrl (D) (http://www.selectrasrl.it/tazza_gb.htm)
SI Analytics GmbH (B) (http://www.si-analytics.com/produkte/kapillarviskosimetrie.html)
TA Instruments (F) (http://www.tainstruments.com/lpage.aspx?id=290&n=1&siteid=11)
Thermo Scientific Inc. (C, F) (http://www.thermoscientific.com/en/products/viscometers.htmlhttp://www.thermoscientific.com/en/products/rheometers.html)
Theta Industries Inc. (F) (http://www.theta-us.com/viscometer/viscometer.html)
Vindum Engineering (H) (http://vindum.com/products/viscolite-700/)

INTRODUCTION

The distinction between laboratory and industrial viscosity analyzers is not a sharp one. In this handbook, most of the inline, automatic viscometers are discussed in Chapter 1.71, which covers industrial viscometers, whereas most units that are of portable or bench-top design are considered to be of the laboratory type and are discussed in this chapter.

A number of designs (capillary, rotational, cone and plate, piston) are available for both laboratory and inline applications and therefore are included in both chapters. An orientation table providing a summary of the features and capabilities of laboratory viscometers is given in Table 1.72a.

TABLE 1.72a
Orientation Table for Laboratory Viscometers

Types of Design		Provides Continuous Signal	In-Line Device	Laboratory Device	Local Readout	Remote Readout Trans.	Temp. Compensation	Fluids: Gas	Fluids: Newtonian	Fluids: Non-Newtonian	Maximum Design Pressure (× 10⁵ Pa)	Maximum Design Temperature (°C)	Inaccuracy (±%) (1) Based on Full Scale (2) Based on Measurement	Minimum Samples Size of Flow Rate (dm³ per hour)	Applicable Viscosity Ranges in mPa·s
Bubble time	Manual			✓	✓				✓		1	25	2–10(2)	13 cm³	1 – 10⁴
Capillary	Manual timing			✓	✓				✓		1	150	0.35(2)	20 cm³	1 – 10⁵
	Auto timing	✓		✓	✓	✓			✓		1	150	0.35(2)	0.4 cm³	10⁻¹ – 10⁵
	Extrusion		✓		✓	✓			✓	✓	35	180	2(2)	30 cm³	10² – 10⁸
Falling rolling element	Ball	✓		✓	✓	✓		✓	✓	✓	1000	180	0.5(2)	0.1 cm³	10⁻¹ – 10⁵
	Needle			✓	✓	✓	✓		✓		1	150	0.5(2)	10 cm³	10² – 10⁷
Orifice/efflux cup	Saybolt			✓	✓				✓		1	120	0.5(2)	60 cm³	10 – 10⁴
	Ford			✓	✓				✓		1	27	2(2)	150 cm³	10 – 10³
	Zahn			✓	✓				✓		1	27	2(2)	44 cm³	10 – 10³
	Auto	✓		✓	✓	✓			✓	✓	1	27	5(2)	44 cm³	10 – 10³
Oscillating element	Piston	✓		✓	✓	✓	✓		✓	✓	1	180	1(1)	2 cm³	1 – 10⁴
Rotational	Coaxial cylinder			✓	✓	✓			✓	✓	1	1600	0.5(2)	2 cm³	10⁻¹ – 10⁹
	Cone and plate			✓	✓	✓			✓	✓	1000	1000	0.5(2)	0.1 cm³	10⁻¹ – 10⁷
	Parallel Plate	✓		✓	✓	✓	✓		✓	✓	1000	1000	0.5(2)	0.2 cm³	10⁻¹ – 10⁹
Stabinger	Stabinger	✓		✓	✓	✓	✓		✓		1	105	0.35(2)	2.5 cm³	10⁻¹ – 10⁴
Vibrating	Damping	✓		✓	✓	✓	✓		✓		1	100	1(1)	10 cm³	1 – 10⁶

The viscosity ranges in this chapter are given in Pascal seconds (Pa·s) or in square meters per second (m²/s) units. For conversions among them and other units, refer to Table 1.72b.

TABLE 1.72b
Definition of Units and Their Equivalences

Unit System	Derived Viscosity Unit	Unit Name	Equivalence
Units for dynamic viscosity			
SI	kg/s·m or Pa·s	Pascal second	1 Pa·s = 10 Poise
CGS	g/cm·s or dyne·s/cm²	Poise	1 Poise = 0.1 Pa·s
FPS	lbF·s/in.²	Reyn	1 Reyn = 6894.8 Pa·s
Units for kinematic viscosity			
SI	m²/s	Square meters per second	1 m²/s = 10,000 Stokes
CGS	cm²/s	Stokes	1 Stoke = 10^{-4} m²/s
FPS	in.²/s, ft²/s	/	1 in.²/s = 645.16 mm²/s

LABORATORY VISCOMETER DESIGNS

In this chapter, the designs of laboratory viscometers are discussed in alphabetical order, starting with the bubble-time, capillary, falling/rolling element, orifice and oscillating-piston types, and concluding with the rotational and vibrational designs.

Bubble-Time Viscometers

In a bubble-time viscometer, a liquid streams downward in the annular zone between the glass wall of a sealed tube and the perimeter of a rising air bubble (rising "object"). The viscometer measures the time for the bubble to rise a specific distance which is proportional to the difference between the bubble and measured liquids density. For a rising gas bubble, one can neglect the air density and consequently the bubble viscometer determines the kinematic viscosity. The faster the bubble rises, the lower the viscosity and vice versa. Typical only transparent liquids are being measured.

In practical applications, bubble viscometers use carefully manufactured precision tubes that are calibrated using known viscosity standards. The kinematic viscosity is calculated by measuring the time and multiplying it with the calibration constant or by comparison. When using the comparison method, the laboratory technician determines the viscosity of a fluid by comparing the speed of bubble rise in a tube having the same dimensions as the standard tube, which is filled with a fluid of known viscosity. Using a set of standard tubes, a viscosity range from 22 to 100,000 mm²/s, in uniform, logarithmically even increments of about 26% can be covered. The viscosity of the fluid under test is assumed to be equal to the viscosity of that fluid in the standard tube that had the nearest bubble speed.

Another type, rather than requiring sets of standard tubes, consists of only one standard tube, and it determines the viscosity of the fluid by reading a precalibrated scale in Saybolt seconds. The distance the bubble travels during a fixed time period is related by the scale to viscosity.

The direct timing of the bubble is based on the observation that the nominal viscosity (in cm²/s) of the liquid in the bubble tube (10.65 ± 0.025 mm inside diameter) is numerically equal to the time (in seconds) required for the bubble to travel a distance of 73 mm.

Accuracy and Limitations Because of the influence of specific gravity and surface tension, the accuracy of this detector is low and depends on the scale range being used. For best results, when the timing method is used to obtain kinematic viscosity, precautions must be taken to ensure that the glass tube is completely vertical, that it has a standard bore, and that good temperature control is provided. A tube tipped by only one radius off the vertical axis will give an error of approximately 10% in the time of bubble travel. A temperature variation of only 0.6°C will cause a 5% variation in the timed bubble travel.

Care should be exercised to ensure that the bubbles are of uniform size. For low-viscosity liquids with a bubble speed below 5 sec, it is advised to make comparisons against predetermined standard values.

One advantage of this type of viscometer is that it requires no calibration or recalibration. The bubble-time viscometer is well suited for applications where evaporation losses must be avoided. It is not suitable for liquids containing crystal, fiber, or gel particles. Because of its simplicity and low cost, this method of measurement is suited for routine industrial use by operators without special training.

Capillary Viscometers

Capillary viscometers are most widely used for measuring viscosity of Newtonian liquids. They are inexpensive, easy to operate, and require a simple temperature control. The physical basis of the capillary viscometer is the Hagen–Poiseuille law. According to that, the pressure drop Δp of a Newtonian liquid passing through a capillary tube of inside radius R and length L is directly proportional to its viscosity if the fluid's temperature and flow rate Q are kept constant

$$\eta = \frac{\Delta p}{L} \frac{\pi R^4}{8Q} \qquad 1.72(1)$$

This results in two different fundamental measurement principles: (1) measurement of the differential pressure at a constant volume flow or (2) measurement of the volume flow through the capillary at a given differential pressure, usually by noting the time required for a known volume of liquid to pass two measurement marks on the glass body or accurately defined fixed sensors. Their geometry resembles a U-tube with at least two reservoir bulbs connected to a capillary tube passage, whereby four primary design types can be distinguished: original Ostwald, the modified Ostwald, the suspended level (Ubbelohde), and the reverse flow capillary viscometers.

The Ostwald viscometer is one of the simplest capillary tube viscometer. This manual viscometer consists of a sample reservoir and a capillary tube. As shown in Figure 1.72a, the hydrostatic head of the fluid causes the sample liquid to flow through the capillary. A clock for measuring the efflux time of the fixed liquid volume and a thermostatic bath with temperature controls completes the apparatus.

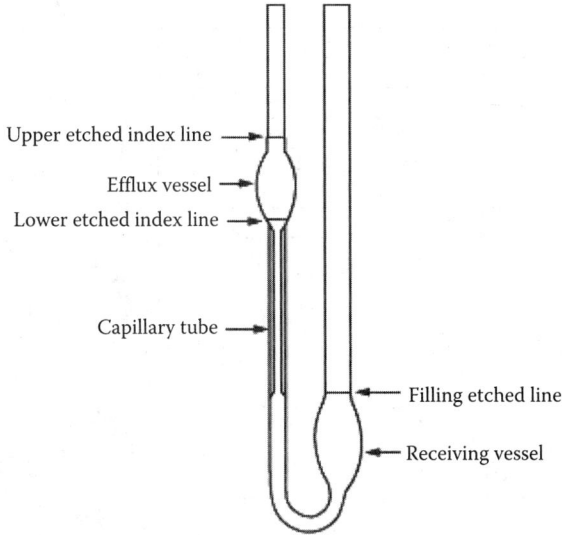

FIG. 1.72a
Ostwald manual capillary viscometer.

Using the Ostwald Viscometer The method of measuring kinematic viscosity is detailed in the ASTM D-445 standard. In practice, a sample liquid of fixed volume is charged to the lower receiving vessel, and the viscometer is placed in a thermostatic bath. After time is allowed for the sample liquid to reach thermal equilibrium (about 5–30 min), the sample is drawn up into the efflux vessel by suction until the level is above the upper etched index line.

The fluid is then permitted to flow down through the capillary by releasing the suction. The laboratory technician starts a stopwatch when the fluid surface passes the upper etched index line. The watch is stopped when the surface passes the lower etched index line of the efflux vessel. This efflux time (t) is multiplied by the viscometer calibration constant to obtain the kinematic viscosity of the sample fluid.

Accuracy and Limitations Because of the small driving force provided by the hydrostatic head of the fluid, and because of the change in hydrostatic head with time, the suitability of capillary tube viscometer is restricted to low-viscosity Newtonian fluids. On the other hand, it is a simple, accurate, and convenient detector for the kinematic viscosity range of 0.2–20,000 mm^2/s.

If the ASTM test procedure is followed carefully, a repeatability of ±0.11% and a reproducibility of about ±0.65% of reading may be achieved. To achieve this type of accuracy and repeatability, (1) the constant-temperature bath should be maintained with a uniformity of ±0.01 K, (2) efflux times of 200–1000 sec are required, and (3) precise calibration should be performed at the same temperature at which the device will operate and exact volume for the reference liquid and the test liquid should be used.

Ostwald-type viscometers can cause significant error in the measurement if the viscometer is not vertical in alignment, whereby 1° deviation from vertical axis will introduce 1% error in the head. The accurate knowledge of density is necessary to adjust the volume at different test temperatures.

In addition to measuring kinematic viscosity, capillary tube viscometers are used for intrinsic viscosity determinations, for molecular weight measurements (see Chapter 1.41), and for the study of molecular shapes of natural and synthetic polymers.

Because the Ostwald viscometer is operated under vented (atmospheric) conditions, and because the time lag between taking the sample and making the measurement is large, it should not be used on samples that evaporate or deteriorate when exposed to atmospheric humidity. This type of viscometer is recommended for use under static and stable conditions. Also, the sample liquid should be filtered before making the measurement to prevent clogging of the capillary tube by solids.

The capillary tube–type viscometer, if the temperature of the thermostatic bath is well controlled, is capable of very accurate measurements, is inexpensive and easy to operate, and needs little or no maintenance aside from cleaning.

Calibration The viscometer constant K is determined individually for each glass capillary viscometer by way of calibration. The determination of the constants is done by a simultaneous flow-time measurement in the viscometers to be calibrated (test specimens) and in the reference standard. During this comparative measurement, the viscometers that should be inspected and the reference viscometer must be placed simultaneously in the same thermostat bath. The test liquid, where the viscosity must not be known exactly, is filled into both viscometers, tempered, and the flowthrough time then measured. The constants of the viscometers to be inspected are then calculated according to the following equation:

$$K = \frac{(K_{standard})(t_{standard})}{t} \qquad 1.72(2)$$

where

K is the constant of the viscometer that is tested
t is the flow time of the viscometer to be tested
$K_{standard}$ is the constant of the reference viscometer
$t_{standard}$ is the flow time of the reference viscometer

To ensure a high statistical certainty, two measurement cycles involving seven flow-time measurements each are run, with the first measurement of the respective measurement cycle being considered as preliminary. In order to verify the determination result, the inspected viscometer is checked with a second reference. If the determined constants from the first and second cycle correlate within 0.1%, the constant is accepted (Tables 1.72c through 1.72e).

TABLE 1.72c
Approximate Viscosity of ASTM Viscosity Standard in mm²/s

ASTM Viscosity Standard	Approximate Kinematic Viscosity in mm²/s,[a] at							
	−53.88°C[b]	−40°C	20°C	25°C	30°C	37.77°C	50°C	98.88°C
S-3	300	80	4.6	4.0	—	3.0	—	1.2
S-6	—	—	10	9.0	—	6.0	—	1.8
S-20	—	—	44	35	—	20	—	3.9
S-60	—	—	160	120	—	60	—	7.7
S-200	—	—	700	480	—	200	—	16
S-600	—	—	2,500	1,600	—	600	280	32
S-2,000	—	—	9,000	5,700	—	2,000	—	76
S-8,000	—	—	38,000	22,000	—	8,000	—	—
S-30,000	—	—	—	—	50,000	27,000	11,000	—

[a] 1 mm²/s = 1 CentiStoke.
[b] °F = 32 + 1.8*°C.

TABLE 1.72d
Approximate Viscosity of ASTM Viscosity Standards in mPa·s

ASTM Viscosity Standard	Approximate Viscosity in mPa·s,[a] at					
	18.3°C[b]	25°C	30°C	37.77°C	50°C	98.88°C
S-3	3.8	3.3	—	2.5	—	0.9
S-6	8.6	7.7	—	5.1	—	1.5
S-20	38	30	—	17	—	3.2
S-60	140	100	—	51	—	6.3
S-200	620	430	—	180	—	14
S-600	2,200	1,400	—	530	250	32
S-2,000	7,900	5,000	—	1,700	—	63
S-8,000	34,000	19,000	—	7,000	—	—
S-30,000	—	—	46,000	24,000	9,500	—

[a] 1 mPa·s = 1 Centipoise.
[b] °F = 32 + 1.8*°C.

TABLE 1.72e
Approximate Viscosity of ASTM Viscosity Standards in Saybolt Universal Seconds

ASTM Viscosity Standard	Approximate Viscosity in SUS, at	
	37.77°C[a]	98.88°C
S-3	36	—
S-6	46	—
S-20	100	—
S-60	290	—
S-200	930	—
S-600	—	150

[a] °F = 32 + 1.8*°C.

Automatic Capillary Viscometer Automatic capillary tube viscometers provide automatic measurements while using conventional glassware. They consist of a microprocessor-based control package that provides temperature controls, automates the influx and efflux operations, performs the measurement of the efflux time, and generates automatic data display or printout.

Precise results within ±0.44%–1.61% reproducibility and ±0.11%–0.61% repeatability may be obtained while the temperature is maintained within ±0.01°C. Automatic influx controls (same level every time) eliminate drainage errors, while the electronic timing unit provides 0.01 sec or better resolution.

Once the viscometers have been loaded with samples, the controls start the pump and valves to influx the sample into the measuring bulb. Here, a predetermined time (about 3 min) is provided for the sample to reach bath temperature. At the end of this equilibrium period, the discharge of the sample from the efflux vessel is started. When fluid passes the upper detector, the timing counter starts and it continues until all the fluid passes the lower detector. Detecting mechanism includes optical, thermal conductivity, and ultrasonic detection. Highly viscous liquids and staining liquids cannot be tested if they tend to cling to or stain the capillary tubes.

The timing counter utilizes a display with readings. It is preferred that the photocells be actuated by light refraction rather than changes in the intensity of transmitted light because this eliminates the effects of sample color variation.

The utility requirements usually include an external cooling medium, a nitrogen gas supply, a vacuum source, and electric power supply. The viscometer might contain several capillary tube assemblies to permit parallel operation.

Highly viscous liquids and staining liquids cannot be tested if they tend to cling to or stain the capillary tubes. Because of the high accuracy obtainable, this viscometer is well suited for research work, even though it is more expensive than others. This automatic capillary tube viscometer can be converted for use as a process viscometer for Newtonian fluids by providing an automatic sampling system with adjustable influx pressure regulation. Because such a viscometer will give only intermittent viscosity measurements, it is not suitable for monitoring processes that can experience rapid changes.

Intrinsic Viscosity and Molecular Weight Intrinsic viscosity (also called *limiting viscosity number* (*LVN*)) measurement is useful for determining the molecular weight and/or the shape of polymer molecules in a solution. The intrinsic viscosity of dilute polymer solutions can be detected by efflux time measurements if the polymer solution is dilute enough to allow the sample to have Newtonian characteristics. The steps in the measurement are as follows. First, a value for the viscosity ratio (η_r = relative viscosity) is calculated.

$$\eta_r = \frac{t}{t_0} \qquad 1.72(3)$$

where

t is the efflux time of the solution
t_0 is the efflux time of the solvent

After this step, the specific viscosity (η_{sp}) is derived as follows:

$$\eta_{sp} = \eta_r - 1 \qquad 1.72(4)$$

Then, intrinsic viscosity [η_I] is calculated using either one of the following formulas:

$$[\eta_I] = \left(\frac{\eta_{sp}}{C}\right)_{C \to 0} \qquad 1.72(5)$$

$$[\eta_I] = \left(\frac{\log_e \eta_r}{C}\right)_{C \to 0} \qquad 1.72(6)$$

where C is the concentration

Either or both of the bracketed quantities can be plotted against concentration and can be extrapolated to infinite dilution to obtain a reading of intrinsic viscosity as illustrated in Figure 1.72b.

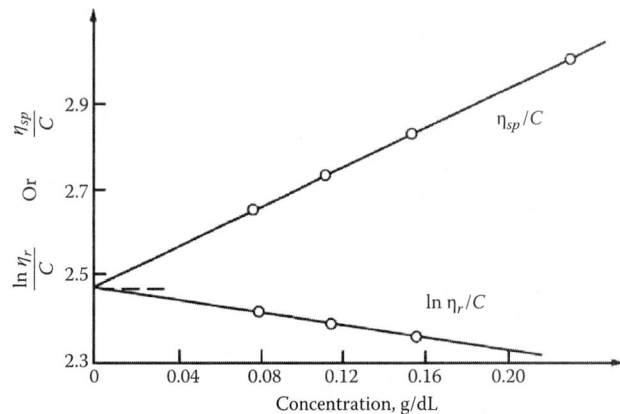

FIG. 1.72b
Intrinsic viscosity determination from specific or relative viscosity.

Mark–Houwink Relation The Mark–Houwink relation expresses the relationship between intrinsic viscosity and the (viscosity average) molar mass M.

$$[\eta_I] = KM^a \quad 1.72(7)$$

or, in logarithmic form

$$\log[\eta_I] = \log K + a \cdot \log M \quad 1.72(8)$$

where

M is the molar mass

K and a are the constants for a given polymer-solvent system that have to be determined by experiments

Efflux time measurements are made of well fractionated or monodisperse systems, and intrinsic viscosity readings $[\eta_I]$ are calculated. The molecular weight (M) is measured osmotically or by light scattering, and the values of $[\eta_I]$ are plotted against M using double log scale as shown in Figure 1.72c. From this plot, the value of "a" is calculated as the slope, and the value of K is determined from the slope "a" and from a single point on the line. Once K and a have been determined on the basis of measurement, M can be calculated easily from Equation 1.72(7).

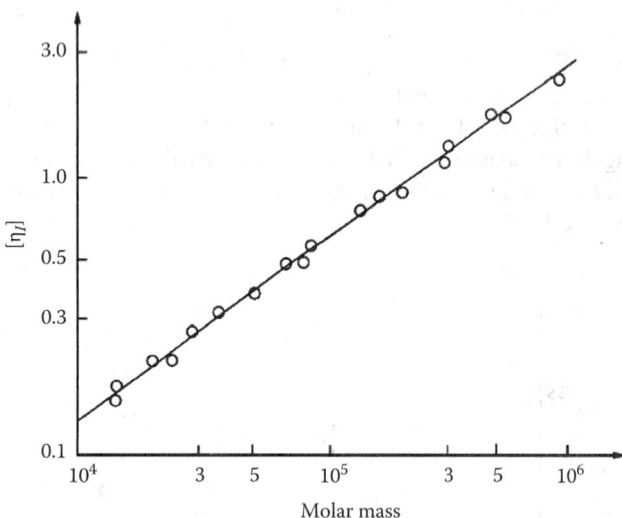

FIG. 1.72c
Molar mass determination from intrinsic viscosity.

Capillary Extrusion Viscometer The use of capillary tube viscometers is often limited to the measurements of low-viscosity Newtonian fluids because of the small driving force available to initiate the liquid flow. For high viscous liquids, the capillary extrusion viscometer is introduced. It is essentially a ram extruder, whereby a piston is accurately driven down a high precision, heated bore. Inside the bore is the liquid to be measured, and under the action of the driven piston, the sample is extruded through a die of known geometry.

One of the possible designs is described in Figure 1.72d. It utilizes compressed nitrogen gas from a commercial cylinder to drive the sample through a capillary. The process sample is charged to a thermostatically controlled steel reservoir, and a free-floating follower plug rides on top of the sample to distribute the pressure evenly over the surface. A series of extrusions may be made through an orifice at varying pressures. The rate of flow, in cubic centimeters per second, is calculated from the weight of the extruded sample and from the elapsed time during each extrusion. The driving force for the extrusion is measured by a pressure gauge. From these readings, the viscosity, the rate of shear, and the stress are all calculated.

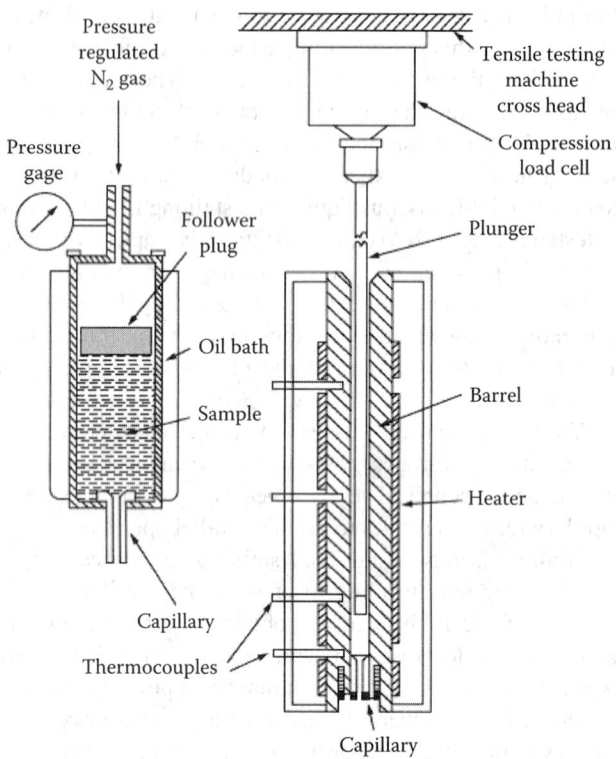

FIG. 1.72d
Schematic diagram of capillary extrusion viscometer.

Another type utilizes an automated tensile tester with a constant speed to extrude materials through a capillary. A sample is placed in a thermostatically controlled extrusion chamber, and a piston is attached to the moving cross head to force the material through a capillary at a fixed rate. The force required to drive the piston is detected as the compression of load cell attached to the cross head.

The rate of flow can be determined by the piston speed or by measuring the weight of the extruded material and the elapsed time of the test. The piston speed can be changed easily, and a series of tests can be performed at different shear rates on the same sample. The recorded load history makes it possible to study dilatant, thixotropic, or other non-Newtonian materials. The flow properties (viscosity, shear stress, and shear rate) are calculated from the measured data and from the constants of the viscometer geometry.

Still another type of viscometer measures the influx time of the material through a capillary. The liquid sample is forced through an open-ended capillary by a constant air pressure, which is generated by a vibrating vertical piston. The influx time (in poise) of the sample liquid to travel a marked distance along the capillary tube is measured as viscosity.

For these viscometers, a variety of capillary lengths and diameters are available. The capillary extrusion viscometer can serve to characterize non-Newtonian fluids having viscosities up to 10^5 Pa·s and over a shear rate range of 1–10^4 sec^{-1}. The inaccuracy of these measurements is within ±2% of reading. These viscometers are valuable for the study of materials that are processed in injection molding operations because the tests themselves are carried out under closely simulated processing conditions.

Although the technique of capillary extrusion viscometry appears simple, the following factors influence the measurement:

1. Thermal degradation
2. Chemical degradation
3. Compressibility/density effects of fluids
4. Shear heating
5. Pressure loss produced by the flow in the sample chamber

These factors can result in errors because they conflict with the assumptions that were made in defining Equation 1.72(1) for Newtonian fluid behavior in a capillary tube. When performing scientific research, great care should be taken to correct for these errors. If that is done, the capillary extrusion viscometers have excellent reproducibility and are well suited for routine industrial and scientific work.

Falling/Rolling Element Viscometers

Falling/rolling viscometers determine the viscosity by measuring the drag force acting on a falling/rolling element, which is allowed to fall or roll under gravity through a viscous medium. Typically, a spherical element is used for falling or rolling; however, there are a variety of falling elements that can be used such as needles, cylinders, etc. Use of falling/rolling elements requires a separate measurement of density to calculate kinematic viscosity.

When the gravitational force is balanced by the viscous resistance of the fluid, the element attains a uniform velocity. By measuring this velocity, the viscosity can be determined.

If we consider a homogeneous liquid with infinite extension, the viscous resistance of the sphere moving with a velocity v is equal to the force due to the difference in density between sphere and liquid (Stokes law). The viscosity can thus be calculated as

$$\eta = \frac{r^2(\rho_{sphere} - \rho_{fluid})g}{18v} \qquad 1.72(9)$$

where
- η is the viscosity
- r is the radius of the sphere
- v is the steady-state velocity of sphere descent
- ρ is the density (sphere and fluid)
- g is the gravitational constant

Manual Falling Ball Unit Typically, the falling ball viscometer consists of a precision bore glass tube and calibrated glass or special steel balls of optical precision, as shown in Figure 1.72e. The tube has a length of about 200 mm and a bore of about 16 mm, and it is mounted in a glass water jacket that is held at a specified angle to the vertical.

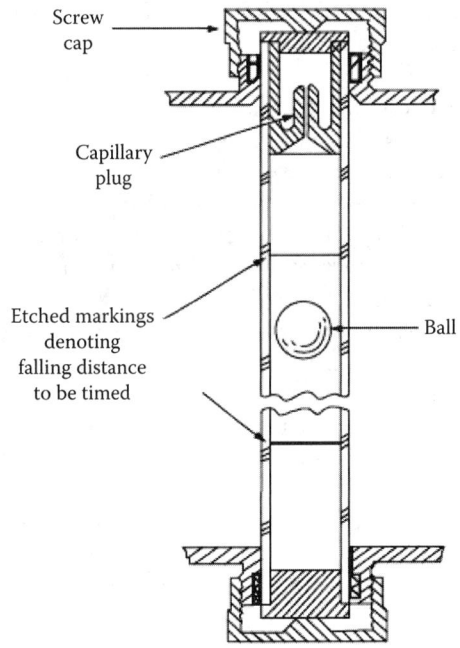

FIG. 1.72e
Falling ball viscometer.

A special capillary plug provides a positive seal and prevents introduction of air into the column during the tube inversion process. This permits as many readings to be taken as are desired, using a single sample, when measurements need to be double-checked or when the goal is to determine the rate of variation of viscosity with temperature. The closed system allows operation without the effects of skin formation, surface tension, evaporation, and other losses of material.

The measurement is made by detecting the time it takes for the ball to fall through an accurately calibrated distance, determined by two marks on the glass tube. The absolute viscosity of the fluid can be found by multiplying that time interval by a factor that reflects the effects of temperature, specific gravity of the fluid, and the characteristics of the ball used, in accordance with Equation 1.72(3).

The use of several calibrated balls of different sizes in the same glass tube covers the viscosity ranges from 0.01 to 1,000,000 mPa·s. This range covers from various gases (hydrogen, air, carbon dioxide, etc.) to such viscous liquids as will hardly pour through a 16 mm opening. If the temperature of the liquids in the glass jacket is precisely maintained (20°C), for the usual range of viscosities (10–600 mPa·s), the viscosity can be determined within an uncertainty of ±0.3%. Outside this viscosity range, a ±0.5% uncertainty can be expected. To obtain this level of accuracy, accurate knowledge of liquid density and accurate correction for ball density both become very critical. The variation of ball density with temperature is usually too small to have a measurable effect on the ball constant; however, for extremely high-accuracy measurements, a correction factor should be used as recommended by the manufacturer.

Care should be taken to minimize subjective errors. Insufficient accurate time measurement is often the cause of error. Lack of attention to temperature is another potential error source because viscosity at room temperature may vary from 2.5% to 16% for each degree change in temperature, depending on the type of liquid to be tested. Controlling temperature within hundredths of a degree is absolutely essential for accurate results.

Other possible sources of error are vibration during the time of the fall, inaccurate leveling of the viscometer, foreign matter or gas bubbles in the sample liquid, thermometer errors, and internal heating of the liquid by absorption of infrared radiation from direct sunlight or other sources.

The advantages of this viscometer include its operational simplicity, and the virtual elimination of subjective variables. It is well suited for routine industrial use as well as for precise scientific measurements of gases and Newtonian liquids, including dark- or opaque-colored fluids as long as they are free from crystal and gel particles (Figure 1.72f).

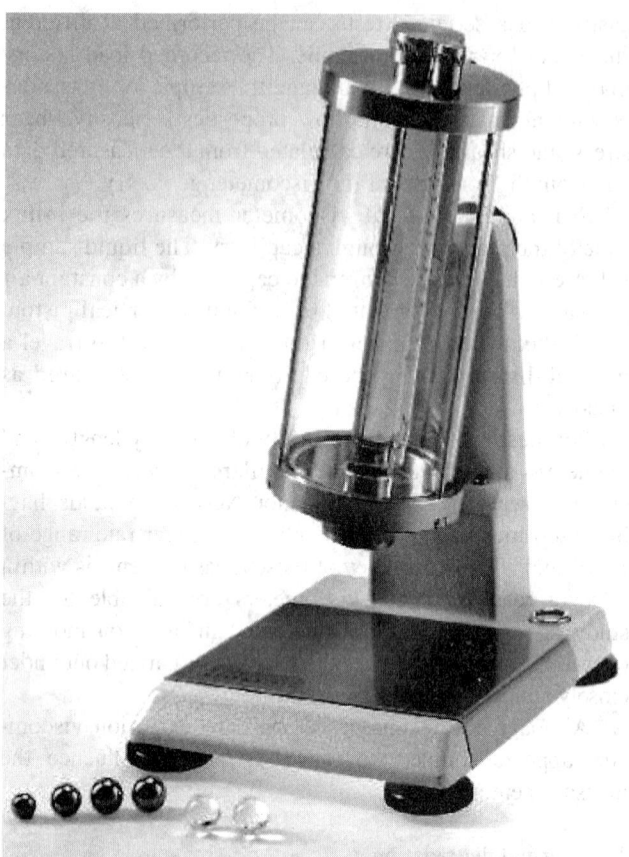

FIG. 1.72f
Manual falling ball viscometer. (Courtesy of Thermo Fisher Scientific Inc.)

Automatic Falling Ball Viscometer In order to simplify the viscosity measurement procedure, the falling ball viscometer is equipped with controls and timers. Generally, the viscometer consists of two units: the viscosity test assembly (equipped with a precision bore tube, heating and temperature control components, and the ball position detector) and the auxiliary control unit. The auxiliary control unit contains the temperature controller, an electrical timer with reset, and the electrical circuits. The viscosity test chamber, which is mounted on a trunnion, can be rotated to various angular positions. The electrical timer is activated by the ball release signal and, when the ball closes the contact at the lower end of the tube, the timer stops. The timer is accurate to within 0.001 sec. The fall time of the ball is proportional to the sample's viscosity, and the readings are interpreted by calibration curves.

The automatic falling ball viscometer is simple and safe to operate, and it can measure fluid viscosity under simulated process conditions. High accuracy (±0.5%) is obtained by judicious selection of the proper calibrated ball size and by adjusting the angular position of the precision bore tube.

Automatic Rolling Ball Viscometer with Variable Inclination Angle Similar to the falling ball method, the viscosity can also be measured using a sphere rolling down an inclined cylindrical tube filled with test liquid. In case of the automatic versions, the viscometers are equipped with thermoelectric temperature control and with an actuating drive that allows to vary inclinations by 15°–80° to both sides. Thereby automatic repetitions of runtimes and a shear rate variation by a factor of 4 is possible. Additionally, possible leveling errors are eliminated by averaging the forth and back runtimes.

Small size steel balls are running in glass capillaries with various sizes to fit different viscosity ranges (more than 100-fold per capillary). Runtimes (starting at 2.5 sec) are measured to a determinability of 0.05% using inductive sensors, which additionally allow for measurements of opaque samples. Sample volume is from 100 to 800 μL depending on capillary size.

Dynamic and kinematic viscosity is calculated by entering the sample's density or in combination with an automatic densitometer. Full automation by samples changers and the combination with other measuring devices is possible.

Relative and intrinsic viscosity of polymer solutions can be measured under zero shear conditions by extrapolating over the variable shear rate (Figure 1.72g).

FIG. 1.72g
Automatic rolling ball viscometer with variable inclination angle together with a densitometer and sample changer. (Courtesy of Anton Paar GmbH.)

Falling Needle Viscometer In addition to spheres, other geometries like cylinders and needles have also been introduced to improve the accuracy of the measurement. The replacement of falling balls with falling needles reduces wall correction effects and at the same time it is much simpler to vary the densities of the needles. Consequently, this results in an extremely stable falling motion. The working principle is basically the same as for falling ball viscometers and can be used for Newtonian and non-Newtonian liquids. A needle with hemispherical ends and a weighted tip or a controlled needle with hemispherical ends and extension bar plus external weights falls through the test liquid with the longitudinal axes of the needle and the measuring cylinder parallel to the gravity vector. The needle quickly reaches its terminal velocity, which is then measured either visually or electronically. A knowledge of the terminal velocity, the fluid and needle densities, and the needle geometry is sufficient to determine the viscous properties of the fluid.

Orifice/Efflux Cup Viscometers

The principle of these viscometers is similar to glass capillary viscometers, except that the flow through a short capillary does not satisfy or even approximate the Hagen–Poiseuille, fully developed, pipe flow. Orifice viscometers measure viscosity by comparing the time it takes a known volume of liquid to pass through an orifice with the time it takes the same amount of fluid with known viscosity to pass through the same orifice. The efflux time reading represents relative viscosity for comparison purposes and is expressed as "viscometer seconds" (like Saybolt seconds or Engler seconds). Empirical conversion tables or formulas specific to each instrument must be used to determine the kinematic viscosity.

Cup viscometers are relatively cheap, general not very accurate, and labor intensive. To obtain high accuracy in an efflux cup viscometer, the liquid-holding vessel and orifice should be temperature controlled. There are 40 or more design variations of these viscometers, some of which are described in the following chapters.

Saybolt Viscometer Saybolt viscometer is used for testing petroleum products and covers the measurement of viscosities at temperature between 21°C and 99°C (ASTM D88) and between 121°C and 232°C (ASTM E 102). For accurate viscosity measurements, the temperature of a test sample in the viscometer should not vary more than ±0.03 K after reaching the selected test temperature.

Each orifice and cup assembly should be calibrated in periodic intervals by measuring the efflux time at 37.8°C of an appropriate oil standard. Three types of orifices are available: for universal, furol, and asphalt applications. Respectively, the furol and asphalt orifices have an efflux time of approximately 1/10th and 1/100th that of the universal orifice. The cup–orifice combination should always be selected to provide an efflux time within the range of 20–100 sec.

Of these three types of efflux cups, the universal orifice (Saybolt universal viscometer) is the most commonly used, and its efflux time is designated as Saybolt universal seconds (SUS). For conversions between SUS and other viscosity units, refer to ASTM D2161-05e1. The Saybolt universal

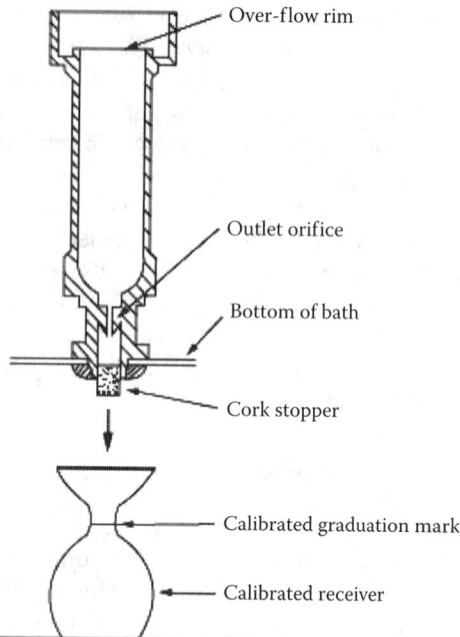

FIG. 1.72h
Saybolt viscometer.

TABLE 1.72f
Conversion of Ford Cup Seconds

Cup Number	Equation (Where t = Time of Flow [s] and v = Kinematic Viscosity [mm²/s])	Drain Time Seconds		Kinematic Viscosity (mm²/s)	
		Min	Max	Min	Max
1	$v = 0.49\,(t - 35.0)$	55	100	10	35
2	$v = 1.44\,(t - 18.0)$	40	100	25	120
3	$v = 2.31\,(t - 6.58)$	20	100	49	220
4	$v = 3.85\,(t - 4.49)$	20	100	70	370
5	$v = 12.1\,(t - 2.00)$	20	100	200	1200

viscometer, as shown in Figure 1.72h, measures the time required for 60 cc of sample fluid to flow out through an orifice having the dimensions of 0.176 cm in diameter and 1.225 cm in length. Saybolt universal seconds (t) can be converted to kinematic viscosity (v) by the following equations:

$$\text{When } t < 100 \text{ sec}, \quad v = 0.226t - \frac{195}{t} \text{ mm/s}^2 \quad \mathbf{1.72(10)}$$

$$\text{When } t > 100 \text{ sec}, \quad v = 0.220t - \frac{135}{t} \text{ mm/s}^2 \quad \mathbf{1.72(11)}$$

Ford Cups Ford cup viscometers are used for the measurement of low-viscosity liquids that do not deviate much from the ideal (Newtonian) liquid behavior. ASTM D-1200-58 standard method defines the procedure for determining the viscosity of paints, varnishes, lacquers, and related liquid materials using the Ford cup viscometers. Ford cups are provided with a conical bottom fitted with a standardized orifice. The measurement results are expressed as efflux time, in seconds, and they measure the time it takes for the liquid to flow through the orifice until the first break occurs in the stream. Efflux time may be converted to kinematic viscosity by the use of the equations provided in Table 1.72f.

The particular cup-and-orifice combination should be so selected as to provide an efflux time within the range of 20–100 sec. The test should be performed at a controlled temperature of 25°C, and the temperature drift during the test should not exceed 0.28°C as determined by a thermometer in the efflux stream. A wide range of construction materials are available for the test components, but the body usually is made of aluminum, and the orifices are made of brass. The cups and orifices are interchangeable, but the whole assembly must be calibrated each time the orifice is changed. The Ford cups should be recalibrated on a regular schedule to correct for the errors caused by orifice enlargement due to cleaning of material deposits in the orifice.

Zahn Cups The Zahn cup viscometer also known as dip-type viscometer is widely used by paint manufacturers to check their products during the manufacturing stages. It is also utilized for measuring the viscosity of many coatings, such as varnishes and lacquers, because of its convenience and speed. It is essentially a bullet-shaped container having a defined volume from 43 to 49 cm³ and an orifice in the bottom. To ensure that it is properly leveled, it is suspended from a ring and ball.

Zahn seconds represent the efflux time required for the volume of liquid to flow through the orifice. The test is made by filling the stainless steel cup with the sample liquid and then measuring the time period during which a steady stream is flowing through the bottom opening of the cup. The measurement is stopped when the cup first begins to drip. For best results, liquids should flow through the calibrated orifice in the bottom of the cup in approximately 20–80 sec. Efflux time may be converted to kinematic viscosity by the use of the equations provided in Table 1.72g.

TABLE 1.72g
Conversion of Zahn Cup Seconds (ASTM D 4212)

Cup Number	Equation (Where t = Time of Flow [s] and v = Kinematic Viscosity [mm²/s])	Drain Time Seconds		Kinematic Viscosity (mm²/s)	
		Min	Max	Min	Max
1	$v = 1.1\,(t - 29)$	35	80	5	60
2	$v = 3.5\,(t - 14)$	20	80	20	250
3	$v = 11.7\,(t - 7.5)$	20	80	100	800
4	$v = 14.8\,(t - 5)$	20	80	200	1200
5	$v = 23t$	20	80	400	1800

Automatic Efflux Cup The automatic efflux cup viscometer is essentially an automated Zahn cup viscometer, with the steps of filling, efflux timing, and solvent washing operations being controlled (Figure 1.72i). The programmer allows the liquid to enter the cup until the overflow detector actuates. Drainage of fluid from the bottom of the cup and from the overflow is sensed by deflection of torsion wire-mounted vanes, which operate sensitive switches that control the efflux timer.

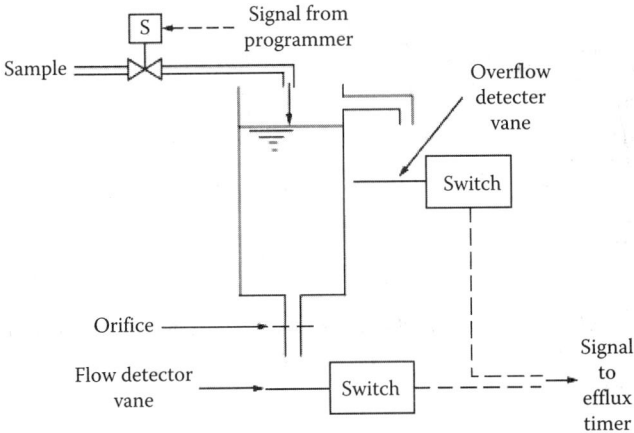

FIG. 1.72i
Automatic efflux cup viscometer.

The automatic efflux cup viscometer is a low-cost online viscometer. It is recommended for applications in which accuracy is not critical and intermittent measurements with substantial time lags between measurements can be tolerated.

Oscillating Piston Viscometer

The unit illustrated in Figure 1.71z can be used in the laboratory or as an online industrial viscometer. The measurement involves alternately energizing two coils (A and B). The coils cause a piston, located in the inner cavity of the probe, to move back and forth within the cavity. The piston is held within the cavity by the deflector plate, which also serves to direct the process fluid into the measurement cavity. Because movement of the piston is resisted by the viscosity of the process fluid that fills the cavity, its travel time increases as the process viscosity rises. The travel time of the piston is measured as an indication of viscosity. The piston is lightweight, to minimize gravity and vibration effects, and the inward and outward travel times are averaged to minimize flow force effects. About 30 sec is needed to make a measurement.

This viscometer is provided with an internal resistance temperature detector for temperature compensation.

This viscometer is provided with an internal resistance temperature detector for temperature compensation. It can operate at high pressures (up to 700 bar/10,000 psi) and high temperatures (315°C or 600°F). The wetted parts of the sensor are made of stainless steel and Teflon. It can be provided with a choice of pistons to cover viscosity ranges from 0.1 to 2 mPa·s up to 1,000 to 20,000 mPa·s. The process fluid should be free of solid particles that exceed about 25 μm in size. The accuracy claimed is 2% of actual reading or better. The electronics are available in explosion-proof construction and can serve as both display and transmitter units.

Rotational Viscometers

The rotational viscometer is probably the most widely used rheometer (see also Chapter 1.60). The operating principle is to measure the resistance on a solid shape rotating in the liquid of interest. A sample is put in between a fixed "holding" part (usually a cup or a lower plate) and a moving "measuring system" that is fixed to the rheometer motor. The moving part is influenced by the sample, and we can measure the sample influence on the rheometer motor. In other words, how much the motor is slowed down by the sample. The measuring unit can either preset a certain torque and measure the resulting speed, or preset a certain speed and measure the resulting torque. The torque is related to the shear stress, and the speed is related to the shear rate. If we know these two values, the viscosity can be calculated according to Equation 1.70(4).

The main advantage of the rotational viscometer is the ability to continuously operate under steady shear rate or shear stress conditions, so that other steady-state measurements can be performed. That way, time or temperature dependencies can be detected and determined. For these reasons, rotational viscometers are among the most widely used class of instruments for rheological measurements.

A number of rotational viscometers with different designs are available. The most important geometries are the concentric cylinder, cone-and-plate, and parallel-plate system. The choice of the measuring system depends on applicative, practical thoughts. As a very rough rule, we can say that concentric cylinders are used for low-viscosity liquids, and for fast drying samples; parallel plates for samples containing particles larger than 5 μm, and for highly viscous and viscoelastic materials such as polymer melts; cone and plate for all other samples and for all kind of viscoelastic materials without particles.

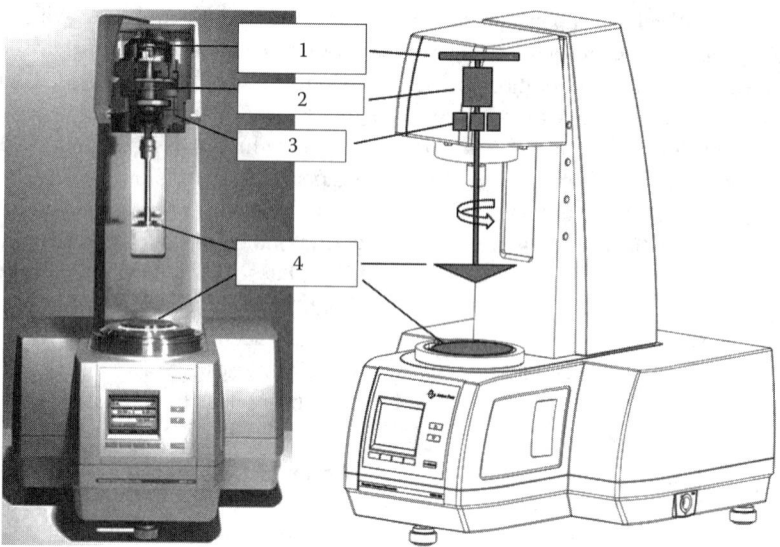

FIG. 1.72j
Rotational viscometer; position sensor (1), motor (2), air bearing (3) sustaining the motor and the upper measuring system, measurement system (4). (Courtesy of Anton Paar GmbH.)

Figure 1.72j illustrates a rotational viscometer.

In general, all rotational viscometers have five features in common: a torque or speed measuring system (could be mechanical, optical, inductive, etc.), a motor, a bearing system between motor and measuring system, a measuring system comprised of a two coaxial cylinders, a cone and a plate, two plates or some other configuration and an electronic controller. Often, additional temperature control of the sample is integrated into the unit by means of Peltier elements or external thermostat to provide assigned temperature conditions for the test.

Coaxial Cylinder Viscometer Concentric cylinders are widely employed to measure shear viscosities of "complex" liquids like polymer solutions, certain surfactant solutions, concentrated suspensions of colloidal particles, and many composite colloidal and polymeric materials in chemical, food, and other technologies.

Liquid contained in a narrow annulus between two coaxial cylinders, one or both of which rotate, experiences a nearly uniform shear rate. In the simplest situation, one cylinder is stationary and the other is set in motion with either a constant velocity or constant torque. The inner cylinder is often referred as bob and the external one as cup (see Figure 1.72k). For rotational tests there are two modes of operating: (1) the Searle method—here the bob is set in motion and the cup is stationary and (2) the Couette method—here the cup is set in motion and the bob is stationary. In industrial laboratories, almost all coaxial cylinder viscometers are working using the Searle method.

To determine the viscosity, the sample is put in between a fixed "holding" part and a moving part "measuring system" that is fixed to the coaxial cylinder viscometer motor. This moving part touches the sample and so we can measure which influence the sample has on the movement of the

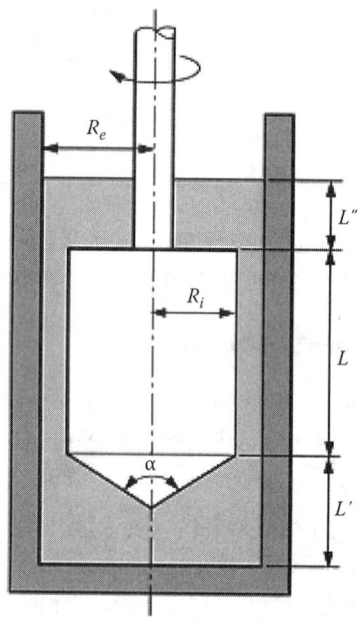

FIG. 1.72k
Coaxial cylinder measuring system according to DIN 53019/ISO 3219. (From Mezger, T.G., The Rheology Handbook, 3rd edn., Vincent Network GmbH & Co. KG, 2011, http://www.european-coatings.com/silver.econtent/catalog/vincentz/european_coatings/buecher/european_coatings_tech_files/the_rheology_handbook.)

coaxial cylinder viscometer motor; in other words, how much the motor is slowed down by our sample.

A coaxial cylinder viscometer can either preset a certain torque and measure the resulting speed, or preset a certain speed and measure the resulting torque. When tests are

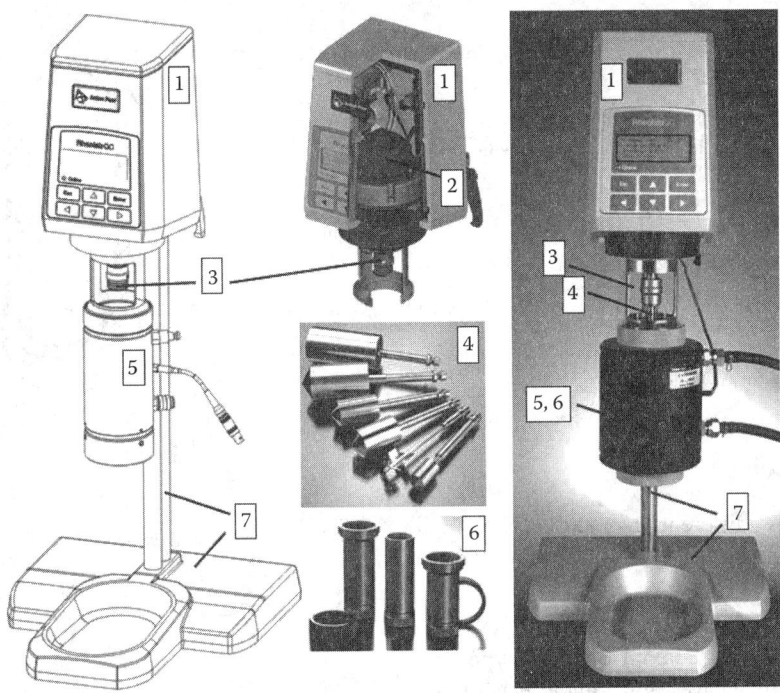

FIG. 1.72l
Coaxial cylinder viscometer; head (1) enclosing a high resolution optical encoder and the synchronous EC motor (2), coupling for connecting measuring systems (3), measuring systems—upper part (bob, 4), temperature control system for the measuring cell (5), cups for the measuring systems (6), stand (7). (Courtesy of Anton Paar GmbH.)

performed at controlled shear rate (CSR mode), the rotational speed is preset and the shear rate is calculated based on the gap $R_e - R_i$ and the circumferential speed of the shearing area. The flow resistance moment of the breaking is measured. The torque is converted into shear stress using the shear area of the measuring system, which is defined by geometry. The dynamic viscosity is calculated from shear stress and shear rate according to Equation 1.70(3). In controlled shear stress tests (CSS mode), the torque is preset and the shear stress is calculated from the torque and shear area. The achieved rotational speed due to the applied torque is measured. The rotational speed is converted to shear rate by appropriate geometry factors. The dynamic viscosity is again calculated using Equation 1.70(3) (Figure 1.72l).

Non-Newtonian Fluids In the design of the coaxial cylinder viscometer, many assumptions were made to give the appearance of Newtonian fluid behavior; that is, steady laminar flow, isothermal flow, no slippage at the wall, constant temperature, viscosity unaffected by shear rate, etc. Thus, the coaxial cylinder viscometer provides a rapid means of obtaining reproducible absolute viscosity readings of Newtonian fluids by subjecting the sample to a shear stress.

Uncertainty does arise when the process sample is non-Newtonian. The fundamental problem is that the very property that we are interested in measuring is affected by the shear applied by the viscometer. Therefore, it is important to accurately define the limits of the shearing conditions that the viscometer can generate. The major sources of errors in coaxial-cylinder–type viscometers when used on non-Newtonian fluids are as follows:

1. The shear stress, shear rate, and (therefore) the viscosity vary across the gap between the sample cup and the bob used to measure the viscosity.
2. The end effect (i.e., the contribution to the torque that arises in the fluid between the end of the bob and the bottom of the cup) is another error source. The end effect can be minimized by specifying the immersion depth, by using bottomless cups, or by trapping air beneath the bob.
3. The stress-induced heat generation within the fluid at high shear rates can also be problematic.

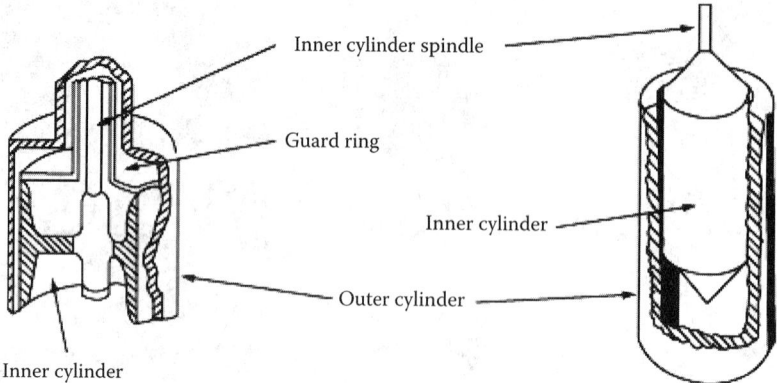

FIG. 1.72m
Coaxial cylinder viscometers.

For these reasons, many different geometrical bob and sample holder configurations have been developed (see Figure 1.72m).

Performance When using the coaxial cylinder–type viscometers, several factors may have significant effects on the overall performance. For that reason, the following operating conditions should be specified:

1. Speed of rotation
2. Spindle size and shape
3. Temperature of liquid
4. Size and shape of sample container
5. Depth of liquid from bottom of the container
6. Bob immersion depth
7. Elapsed time to rotation before reading is taken

The factors and conditions that are undesirable and therefore should be guarded against include the following:

1. Dirt on the torque measuring device
2. Dirty, corroded, pitted, or deformed spindle
3. Off-center placement of the spindle in the container
4. Lack of regular recalibration
5. Turbulent flow (low-viscosity fluid with high speed)
6. Large air bubbles in the sample

Therefore, when measuring the viscosity of non-Newtonian fluids, (1) obtain a viscometer that provides the most uniform shear rate throughout the measured sample, and (2) use a consistent experimental procedure.

Cone-and-Plate Viscometer The common feature of a cone-and-plate viscometer is that the fluid is sheared between a flat plate and a cone with a low angle, separated by a small gap; see Figure 1.72n. The particles within the sample should be smaller than one tenth of the smallest gap between cone and plate.

The geometric configuration of a cone and a plate provides uniform shear rate and stress throughout the fluid

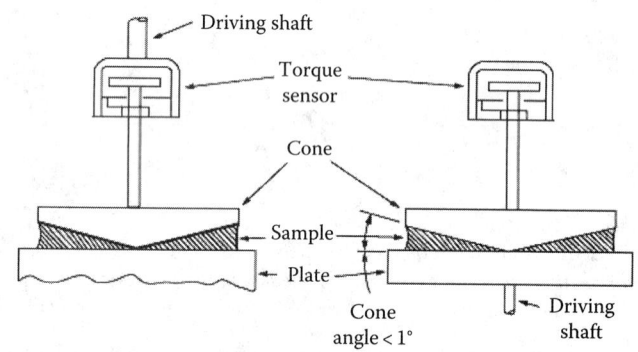

FIG. 1.72n
Cone-and-plate viscometers.

sample at a given angular velocity as a result of the linear increase in both the sample thickness and the tangential velocity as the distance from the center increases. In addition, the influence of shear-induced heat within the fluid at high shear rates is substantially reduced because the sample is a very thin layer.

As shown in Figure 1.72n, the cone-and-plate viscometer consists of a flat plate and a cone with a very small angle (less than 1°). The apex of the cone almost touches the flat plate surface, and the fluid sample fills the narrow gap between the cone and the plate. Capillary action keeps the sample materials in place during operation.

A continuously variable-speed motor drives the rotating plate. The rotary speed is precisely maintained by an electronic velocity control. In some cone-and-plate viscometer designs, the flat plate is rotated while the conical disc is held by a torsion measuring head; in others, the conical disc is rotated while the bottom flat plate is stationary. Torsion bar movement can be measured by different methods as already noted in the previous chapter.

In general, temperature of the sample is measured by temperature sensors embedded in the plate, and precise temperature control is maintained. The gap between cone and plate is kept constant by a servo system. The relative position of plate

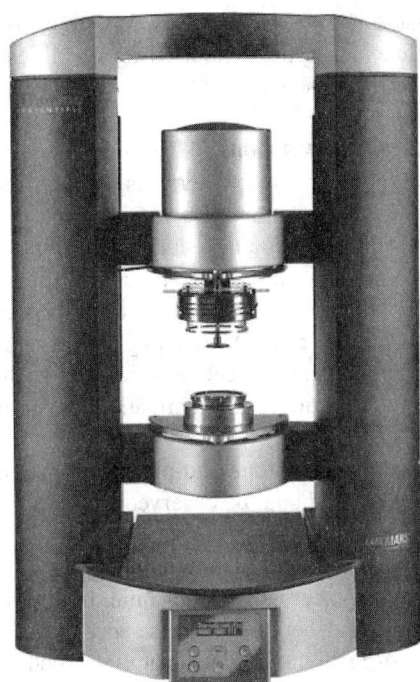

FIG. 1.72o
Cone- and plate-viscometer. (Courtesy of Thermo Fisher Scientific Inc.)

to cone can be detected by magnetic induction principle, a ceramic proximity gauge or other techniques (Figure 1.72o).

Parallel Plates Viscometer Parallel plate viscometers consists of two parallel plates separated by a gap filled with the liquid to be measured. Unlike the cone and plate and concentric cylinder geometries where the gap separating the surfaces is fixed, the parallel plate system has the advantage of flexible gaps. In addition, the shear rate is no longer uniform, but depends on radial distance from the axis of rotation and on the gap. The viscometer is useful forgels, pastes, soft solids, polymer melts, and other materials that are intolerant to narrow gaps.

Relative Measuring Systems Calculation of rheological parameters into absolute units on the basis of raw data is only possible if standardized geometries of absolute measuring systems are used. When using relative measuring systems, the preset conditions are not as required for an accurate rheological evaluation. In most cases, it is impossible to calculate shear rate values from raw data, since the geometry of the rotating bob (spindle) and the dimensions of the shear gap do not meet the required conditions (according to ISO).

Figure 1.72p illustrates a spindle viscometer. All these viscometers have three features in common: a mechanical means of driving a spindle at a constant speed, a torque-measuring device, and a means of correlating shear rate (spindle diameter and speed of rotation).

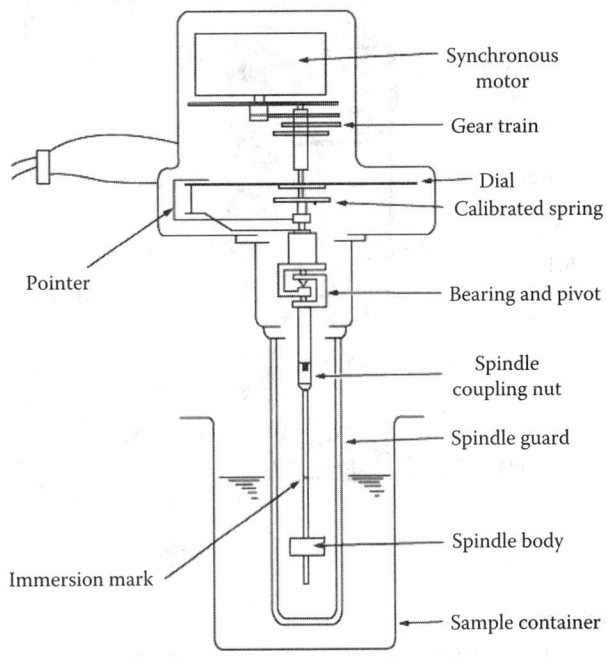

FIG. 1.72p
Spindel viscometer.

Using weights, one of these viscometer designs measures the force required to rotate a spindle or bob in the sample at a specified speed. The weights are suspended at the free end of a cable while the other end is wound around a pulley, which is geared to the shaft of the spindle. A standard one-pint friction-top can is used as a container. The weights are a measure of torque. This unit is inexpensive and is used for viscosity ranges from 10 to 500,000 mPa·s.

Another spindle viscometer rotates a flat, disc-shaped stirrer at a constant speed in a half-pint size sample container, which is firmly placed on a turntable equipped with torque-measuring capability. A spring is attached to the spindle of the turntable spindle and is used to counteract the torque that is generated by the sample. The angular deflection of the turntable is measured as the viscosity of the sample fluid. This type of viscometer is used for viscosity measurements, in the range from 10 to 1500 mPa·s.

In another spindle viscometer design, the sample container is rotated at a constant speed, and the torque required to restrain the freely suspended spindle is measured by detecting the spring-controlled angular deflection of the supporting torque wire. In some cases, the spring movement is converted to a current signal by use of the poles of a magnet, and the current required to restrain the spring is detected as an indication of viscosity. This instrument is good for low- to medium-viscosity fluids and operates at relatively low shear rates.

Still another type of spindle viscometer drives the spindle at a desired rotary speed through a calibrated spring, and the

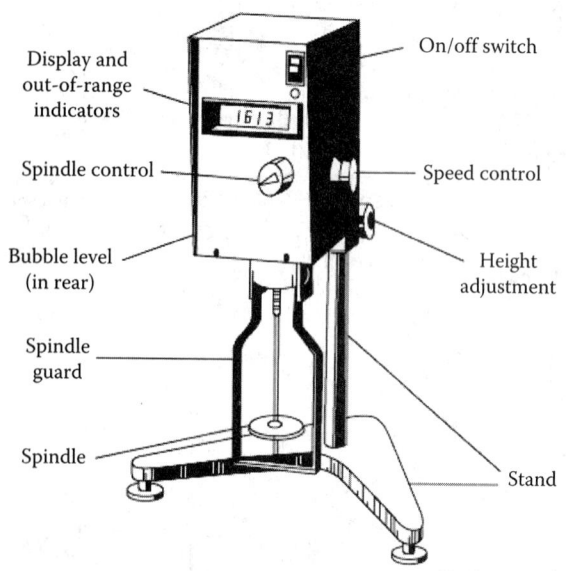

FIG. 1.72q
Spindle viscometer measures the torque required to rotate various spindles when immersed in the test fluid. It can detect viscosities from 10 to 8,000,000 mPa·s with error of 1%–2% of full scale. (Courtesy of Cole-Parmer Instrument Co.)

viscosity of the fluid is measured by the degree to which the spring is wound (Figure 1.72q). Constant speed is maintained by a synchronous motor, and speeds of rotation are changed by a gear train (up to eight speeds). The spring torque can be used for remote transmission of measured data.

For most of the spindles, the shear stress might be calculated from the measured torque since the geometry of the shear area is known. The shear rate, however, is not defined since there is no narrow concentric gap like for coaxial cylinder viscometers. The shear gap is therefore termed "infinitely wide," that is, the cup radius us assumed to be $Re = \infty$. For this reason, homogeneous shear conditions throughout the entire sample can-not be guaranteed. Therefore, the viscosity values cannot be determined since a well-defined shear rate value is required in order to calculate the corresponding viscosity value.

Since it is not possible to calculate a viscosity value without a defined shear rate (according to Newton's viscosity law), viscosity specifications that are based on tests performed with this kind of spindles therefore are relative viscosity values and should be termed as such or as instrument dependent values. Many users are speaking of Brookfield units or BU, after the name of a widespread viscometer type. Comparison of "Brookfield viscosities," which are also specified in mPa·s or Pa·s to absolute viscosity values that are measured when using absolute measuring systems always, will result in deviations due to the these reasons.

Advantages and Precautions The chief advantage of the commercial rotational viscometer is its ability to record rheograms automatically for non-Newtonian fluids that exhibit shear-dependent and time-dependent behavior over a wide range of shear stress and rates. The control unit gives uniform acceleration up to a selected maximum speed at a preselected acceleration rate. On reaching the preset maximum, the platen automatically decelerates to zero at the selected rate. Quick application and termination of strain or hold speed are also possible.

In connection with cone-and-plate viscometers, the following precautions should be observed:

1. Avoid any air bubbles in the sample
2. Avoid use of excess amount of fluid around the cone periphery to further minimize the edge effect
3. Avoid testing high-viscosity materials such as polymers, which tend to ball up and leave a gap between the cone and plate unless the processing temperature is high enough to melt the materials and ensure proper contact

The cone-and-plate viscometer is the most versatile of all currently available designs and is an excellent rheometer. It is capable of measuring not only absolute viscosity of Newtonian fluids but also elasticity and all the other flow properties such as dilatancy, thixotropy, yield value, shear stress/shear rate, and apparent viscosity/shear rate. It can make these measurements over wide and precisely controlled ranges of temperatures and shear rates. The main limitation of this type of viscometer compared to other laboratory units is its high cost.

Stabinger Viscometer The Stabinger viscometer is a miniaturized Couette type rotational viscometer. It combines the accuracy of kinematic glass capillary viscometers with the wide range and continuous measurement of rotational viscometers. The standard test method for the Stabinger viscometer is described in ASTM D7042.

A lightweight hollow rotor floats in a rotating tube filled with the sample liquid. The centrifugal forces hold the rotor in the axis of rotation as its density is lower than the liquid's density. A magnet inside the rotor induces eddy currents in the cell block, thereby retarding the rotors speed. The equilibrium speed of the

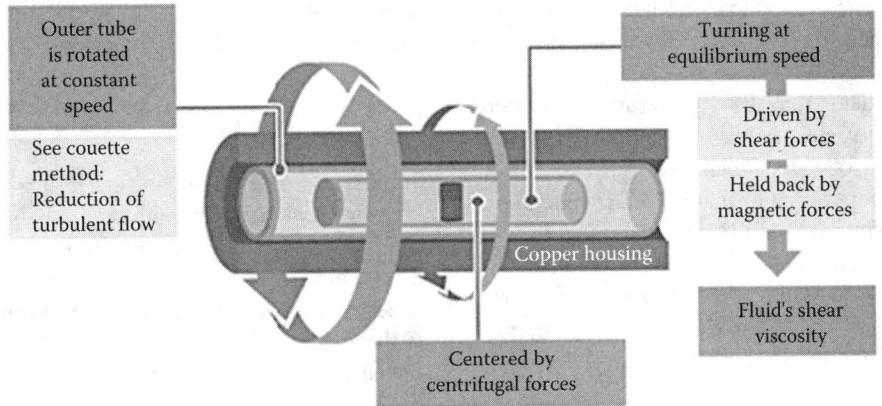

FIG. 1.72r
Working principle of Stabinger Viscometer. (Courtesy of Anton Paar GmbH.)

rotor is a measure for the dynamic viscosity. It is measured by counting the frequency of the magnetic field (see Figure 1.72r).

An integrated oscillating U-tube simultaneously measures the density of the sample, allowing to calculate the kinematic viscosity.

These viscometers are equipped with a fast thermoelectric temperature control, allowing also the determination of viscosity index or programmable temperature scans within a fraction of an hour. Optional external liquid cooling enables measurements at very low temperatures for jet fuels, brake fluids, and hydraulic oils.

Sample volume is approximately 3 mL, solvent consumption typically 10 mL. The measuring range is from 0.2 to 20,000 mPa·s, 0.6 to 3 g/cm³ and +5°C to 105°C (with counter cooling −56°C). Dynamic and kinematic viscosity, density, shear rate, and shear stress are calculated by the instrument. Full automation by samples changers and the combination with other measuring devices is also possible.

Vibrational Viscometers

Vibrational viscometer is an instrument that operates by measuring the damping of an oscillating electromechanical resonator immersed in a fluid whose viscosity is to be determined. The resonator generally oscillates as a cantilever beam or tuning fork. The higher the viscosity, the larger the damping imposed on the resonator. Several methods can be employed to determine the resonator's damping. Please see Chapter 1.71 for further details.

TRENDS

The application of technology in today's viscosity labs is inevitable to achieve timely progress and remain competitive. Apart from the fact that the modern laboratory should be designed in a way to optimize its physical layout, one can detect several trends with regard to instrumentation use or properties.

Automation

In general, laboratory automation refers to the use of technology to streamline or substitute manual manipulation of equipment and processes. Applications for lab automation range from the use of sample exchange stations to fully automated robotic stations for a high-throughput operation. The amount of automation that any lab requires depends on its situation. While an academic research lab may choose to use only some instruments to increase productivity and eliminate a tedious task, a chemical company will probably want to automate all phases of its research.

Fully and semiautomated work stations can prepare or manipulate hundreds of samples in a single day for high-throughput work. Work stations are usually dedicated to perform a fixed set of tasks and require the operator to refill them with new samples when needed to remain in active operation.

Robotic systems are even more sophisticated as they can perform many of the tasks that would normally be left for a human being. These tasks might involve gripping a tube or other object and emptying its contents or moving the object from one environment to another. The total cost per analysis is decreased and reproducibility of results is increased since the very same procedure is applied for all samples to be analyzed. Robotic automation also enables handling of hazardous materials that could harm the operating personnel.

Lab Management Systems

The increased use of high-throughput systems means an extraordinary increase in the amount of data generated in laboratories. In those circumstances, application of information technology to the handling of laboratory data and information is important. That type of handling often requires laboratory information management systems (LIMS). A typical automated system can log in samples; create batches and work lists, and group samples by test or other common attribute. A LIMS can generate schedules for testing and tracking samples in a lab as they move from one department to the next. It can also log data from testing instruments, thereby increasing productivity and helping to eliminate errors. Modern laboratory viscometers therefore must be able to interconnect with laboratory information management systems and/or SAP or similar enterprise software that is managing business operations. For special pharmaceutical and medical device-related activities, the laboratory viscometer should also be able to comply with 21 CFR Part 11 function (electronic signature, audit trail, and data archive).

Adaptability

The traditional boundaries between different scientific disciplines are being blurred. The resulting laboratories that support research or quality control activities must be highly flexible and adaptable, changing along with laboratory operations. Similar to requirements for laboratory environment (layout, furniture, supports, etc.), there is a prerequisite that one should be able to adapt laboratory equipment when necessary. This could mean adding some kind of automation with regard to cleaning or sample handling, or enhancing the primary viscosity measurement with other physical or chemical properties as needed. Additional properties modules should ideally reuse the sample without operator intervention.

Smart User Interfaces

Recent improvements in display technology and embedded operating systems allow the implementation of touch-activated screens for simplified, intuitive, and responsive user-interface. No user manual is required. Although complex behind the scenes, the user experience is easy. This relieves restrains of skilled operators' availability and further reduced the cost and maximizes productivity.

Smart Accessories

Typical accessory one can find in laboratories for viscosity measurements include temperature baths or Peltier temperature devices, additional temperature probes, different bobs or spindles, different cylinders, pressure cells, cameras to observe samples during shearing, sample exchangers, etc. If those accessories are equipped with some kind of "intelligence," one could call them smart. Modern smart accessories together with a proper laboratory viscometer enable automatic component recognition and system configuration. All measuring geometries, thermostatic devices, additional temperature probes, sample exchangers, and others are automatically recognized by the laboratory instrument. This information is taken into consideration when carrying out manual or (semi) automatic measurements. With computer-controlled measurements, the information is read into the software as soon as the accessory is connected to the instrument, providing a plug-and-play functionality. This rules out errors that occur when using incorrect geometries and centralizes the control.

SPECIFICATION FORMS

When specifying laboratory viscometers, use the ISA form 20A1001 and when the type of analyzer has already been selected and both the analyzer and the requirements of the process application are to be specified, use the ISA form 20A1002. Both forms are reproduced with the permission of the International Society of Automation on the next pages.

1.72 Viscometers: Laboratory

	RESPONSIBLE ORGANIZATION	ANALYSIS DEVICE		SPECIFICATION IDENTIFICATIONS		
1			6			
2			7	Document no		
3	(ISA logo)	Operating Parameters	8	Latest revision	Date	
4			9	Issue status		
5			10			

	ADMINISTRATIVE IDENTIFICATIONS				SERVICE IDENTIFICATIONS Continued		
11				40			
12	Project number	Sub project no		41	Return conn matl type		
13	Project			42	Inline hazardous area cl	Div/Zon	Group
14	Enterprise			43	Inline area min ign temp	Temp ident number	
15	Site			44	Remote hazardous area cl	Div/Zon	Group
16	Area	Cell	Unit	45	Remote area min ign temp	Temp ident number	
17				46			

	SERVICE IDENTIFICATIONS				COMPONENT DESIGN CRITERIA		
18				47			
19	Tag no/Functional ident			48			
20	Related equipment			49	Component type		
21	Service			50	Component style		
22				51	Output signal type		
23	P&ID/Reference dwg			52	Characteristic curve		
24	Process line/nozzle no			53	Compensation style		
25	Process conn pipe spec			54	Type of protection		
26	Process conn nominal size	Rating		55	Criticality code		
27	Process conn termn type	Style		56	Max EMI susceptibility	Ref	
28	Process conn schedule no	Wall thickness		57	Max temperature effect	Ref	
29	Process connection length			58	Max sample time lag		
30	Process line matl type			59	Max response time		
31	Fast loop line number			60	Min required accuracy	Ref	
32	Fast loop pipe spec			61	Avail nom power supply	Number wires	
33	Fast loop conn nom size	Rating		62	Calibration method		
34	Fast loop conn termn type	Style		63	Testing/Listing agency		
35	Fast loop schedule no	Wall thickness		64	Test requirements		
36	Fast loop estimated lg			65	Supply loss failure mode		
37	Fast loop material type			66	Signal loss failure mode		
38	Return conn nominal size	Rating		67			
39	Return conn termn type	Style		68			

	PROCESS VARIABLES	MATERIAL FLOW CONDITIONS			PROCESS DESIGN CONDITIONS		
69				101			
70	Flow Case Identification		Units	102	Minimum	Maximum	Units
71	Process pressure			103			
72	Process temperature			104			
73	Process phase type			105			
74	Process liquid actl flow			106			
75	Process vapor actl flow			107			
76	Process vapor std flow			108			
77	Process liquid density			109			
78	Process vapor density			110			
79	Process liquid viscosity			111			
80	Sample return pressure			112			
81	Sample vent/drain press			113			
82	Sample temperature			114			
83	Sample phase type			115			
84	Fast loop liq actl flow			116			
85	Fast loop vapor actl flow			117			
86	Fast loop vapor std flow			118			
87	Fast loop vapor density			119			
88	Conductivity/Resistivity			120			
89	pH/ORP			121			
90	RH/Dewpoint			122			
91	Turbidity/Opacity			123			
92	Dissolved oxygen			124			
93	Corrosivity			125			
94	Particle size			126			
95				127			
96	CALCULATED VARIABLES			128			
97	Sample lag time			129			
98	Process fluid velocity			130			
99	Wake/natural freq ratio			131			
100				132			

	MATERIAL PROPERTIES				MATERIAL PROPERTIES Continued		
133				137			
134	Name			138	NFPA health hazard	Flammability	Reactivity
135	Density at ref temp	At		139			
136				140			

Rev	Date	Revision Description	By	Appv1	Appv2	Appv3	REMARKS

Form: 20A1001 Rev 0 © 2004 ISA

Analytical Measurement

	RESPONSIBLE ORGANIZATION	ANALYSIS DEVICE COMPOSITION OR PROPERTY Operating Parameters (Continued)		SPECIFICATION IDENTIFICATIONS	
1			6		
2			7	Document no	
3	ISA		8	Latest revision	Date
4			9	Issue status	
5			10		

	PROCESS COMPOSITION OR PROPERTY			MEASUREMENT DESIGN CONDITIONS				
	Component/Property Name	Normal	Units	Minimum	Units	Maximum	Units	Repeatability
11								
12								
...								
37								

Rev	Date	Revision Description	By	Appv1	Appv2	Appv3	REMARKS

Form: 20A1002 Rev 0   © 2004 ISA

Definitions

Absolute (dynamic) viscosity η (*Eta*): Constant of proportionality between applied stress and resulting shear velocity (Newton's hypothesis) under simple steady shear.

Apparent viscosity: The value of viscosity evaluated at some nominal average value of the shear rate. The apparent viscosity applies, for instance, in the capillary method, where a range of shear rates are employed.

Flow curve: A graphical representation of the behavior of flowing materials in which shear stress is related to shear rate.

Newtonian: Flow model of fluids in which a linear relationship exists between shear stress and shear rate, where the coefficient of viscosity is the constant of proportionality.

Non-Newtonian: Any laminar flow that is not characterized by a linear relationship between shear stress and shear rate.

Rheology: The science of the deformation and flow of matter.

Shear: The relative movement of parallel adjacent layers.

Shear rate: The rate of change of shear with time.

Shear stress: The component of stress that causes successive parallel layers of a material body to move, in their own planes (i.e., the plane of shear), relative to each other.

Shear-thickening: An increase in viscosity with increasing shear rate during steady shear flow.

Shear-thinning [pseudoplastic]: A decrease in viscosity with increasing shear rate during steady shear flow.

Thixotropy: A reversible time-dependent decrease in viscosity at a particular shear rate. Shearing causes a gradual breakdown in structure over time.

Abbreviation

SI   Système international d'unités

Organizations

ASTM	ASTM International, formerly known as the American Society for Testing and Materials
CFR	Code of Federal Regulations
DIN	Deutsches Institutfür Normung–German Institute for Standardization
ISA	International Society of Automation
ISO	International Organization for Standardization
LIMS	Laboratory Information Management System
SAP	A German multinational software corporation that makes enterprise software to manage business operations and customer relations

Bibliography

Baird, D.G. et al., *Polymer Processing Principles and Design*, John Wiley & Sons, New York, 1998.

Bandrup, J. and Immergut, E., *Polymer Handbook*, 3rd edn., John Wiley & Sons, New York, 1989.

Barnes, H. A., The yield stress—A review or 'pantarei'—Everything flows? *J. Non-Newtonian Fluid Mech.*, 81, 133–178, 1999.

Bourne, M.C., *Food Texture and Viscosity Concept and Measurement*, Academic Press, New York, 1997.

Cho, Y.I., Hartnett, J., and Lee, W.Y., Non-Newtonian viscosity measurements in the intermediate shear rate range with falling-ball viscometer, *J. Non-Newtonian Fluid Mech.*, April 1984.

Dearly, J.M., Official nomenclature, *J. Rheol.*, 28(3), 181–195, 1984.

Dontula, P., Macosko, C.W., and Scriven, L.-E., Origins of concentric cylinders viscometry, *J. Rheol.* 49(4), 807–818, July/August 2005.

Dutka, A.P. et al., Evaluation of a capillary-coriolis instrument for online viscosity and density measurement, *Proceedings of the TAPPI Process Control, Electrical and Instrumentation Conference* (*ISA*), March, 1997.

Fitzgerald, J.V. and Matusik, F.J., A viscometer for many purposes, *Meas. Control*, June 1986.

Ghosh, T.K. et al., *Viscosity of Liquids*, Springer, Dordrecht, the Netherlands, 2007. http://www.springer.com/gp/book/9781402054815.

Giap, S.G.E., The hidden property of Arrhenius-type relationship: Viscosity as a function of temperature, *J. Phys. Sci.*, 21(1), 29–39, 2010.

Goodwin, J.W. et al., *Rheology for Chemists*, Royal Society of Chemistry, Cambridge, U.K., 2008. http://pubs.rsc.org/en/Content/eBook/978-0-85404-839-7#!divbookcontent.

Hackley, V.A. et al., *Guide to Rheological Nomenclature: Measurements in Ceramic Particulate System*, Institute of Standards and Technology Special Publication 945 Nat. Inst. Stand. Technol. Spec. Publ. 946, 31pp., January 2001. http://fire.nist.gov/bfrlpubs/build01/PDF/b01014.pdf.

Han, C.D., *Rheology and Processing of Polymeric Materials*, Vol. 1: Polymer Rheology, Oxford University Press, Inc., Oxford, U.K., 2007. http://global.oup.com/academic/product/rheology-and-processing-of-polymeric-materials-9780195187823;jsessionid=40C304B0312269960B4D964340B88A10?cc=us&lang=en&.

Harsveldt, A., Capillary viscometer to characterize the rheology behaviour of blade coating colours, *Paper Technol. Indust.*, January/February 1981.

Jung, F. et al., The capillary tube-plasma viscometer, *Biomedizinische Technik*, November 1983.

Kim, H.K., Co, A., and Ficke, A.L., Viscosity of black liquors by capillary measurements, Symposium Series, No. 207, Vol. 77, AICHE, 1981.

Kohini, J.L. and Plutchok, G.J., Predicting steady and oscillatory shear rheological properties of CMC/guar blend using the bird-carreanconstitutine model, *J. Texture Studies*, 18, 31–42, 1987.

Melekhin, A.N. and Sokolov, L.K., Investigation of static and dynamic characteristics of an automatic ball-type viscometer, *Teploenergetika*, April 1981.

Mezger, T.G., *The Rheology Handbook*, 3rd edn., Vincent Network GmbH & Co., Hannover, Germany, 2011. http://www.european-coatings.com/silver.econtent/catalog/vincentz/european_coatings/buecher/european_coatings_tech_files/the_rheology_handbook.

Rabinovich, V.A. et al., *Viscosity and Thermal Conductivity of Individual Substances*, Begell House, New York, 1997.

Schramm, G., An introduction to rheology and viscometry: Rotary viscometers, *Regulacion y MandoAutomatico*, November 1984.

Selby, T.W., Use of scanning Brookfield technique to study the critical degree of gelation, SAE paper #910746, Society of Automotive Engineers, Warrendale, PA, February 1991.

Sheble, N., How do you like your mashed potatoes? *InTech*, June 2002.

Skeist, I., *Handbook of Adhesives*, 3rd edn., Van Nostrand Reinhold, New York, 1989.

Viscometers, *Meas. Control*, June 1993.

Walsh, L., *Quality Management Handbook*, Marcel Dekker, New York, 1986.

Zhang, Z. et al., Viscosities of lead silicate slags, *Miner. Metall. Process.*, 19(1), February 2002.

1.73 Water Quality Monitoring

C. P. BLAKELEY (1974)   **G. L. COMBS** (1995)   **B. G. LIPTÁK** (2003, 2017)

Measured water quality parameters	Ammonia, biochemical oxygen demand (BOD), chemical oxygen demand (COD), *Parameters* chlorine, color, conductivity, dissolved oxygen (DO), fluoride, mercury, nitrates, odor, oil, ORP, ozone, pH, phosphorus, temperature, total carbon (TC), total inorganic carbon (TIC), total organic carbon (TOC), total oxygen demand (TOD), turbidity, volatile organic compounds (VOC). For a list of volatile organic compounds that are detected in waste water, refer to Table 1.73e.
Physical variables	Air temperature, rainfall, river flow or stage (level), sunlight, wind direction, and speed
Cost	Costs vary widely, depending on system requirements and components to be measured. For example, a single-parameter measuring system could cost as little as $3000, whereas a completely integrated monitoring system capable of simultaneous measurement of multiple volatile organic components at the ppb level and equipped with a full complement of user-specified data transmission options may cost up to $100,000. Sample system, installation, and the cost of building a suitable shelter can further increase the total cost. For the costs of individual sensors, refer to the individual chapters listed in the INTRODUCTION of this chapter.
Partial list of suppliers	ABB Inc. (B, C) "Continuous On-line Water Monitoring" https://library.e.abb.com/public/1facb219029da0b3c12578140039de5f/DS_7976-EN_E.pdf
Bran and Luebbe Inc. (automated, online B,C) http://www.fluidquip.com.au/uv-technologies/products/branluebbe/
Environmental Analytical Systems Inc. (automated, B) http://www.enviro-analytical.com/enviroproducts/aqualab_automated_water_quality_analyzer.html
Foxboro-Schneider (B) http://www.fielddevices.foxboro.com/en-gb/products/analytical-echem/
GE Power and Water "Water Quality Monitoring" (C) http://www.gewater.com/water-quality-monitoring.html
Hach Co. (B,C) http://www.hach.com/smartprobes, and http://www.hachppa.com/
Hydrolab Corp. (A, B) http://www.hachhydromet.com/web/ott_hach.nsf/id/pa_hydrolab_water_quality_sensors.html
LaMotte (A, portable smart units), http://www.lamotte.com/en/water-wastewater/instrumentation/2000-01.html
Libelium-Schneider (C) http://www.libelium.com/smart-water-sensors-monitor-water-quality-leakages-wastes-in-rivers-lakes-sea/
Markland Specialty Engineering Co. (sampler) http://www.sludgecontrols.com/our-products/automatic-duckbill-sampler/
OTT Hydromet (B) http://www.ott.com/en-us/products/water-quality-1/
Rosemount, Emerson (B, C) http://www2.emersonprocess.com/siteadmincenter/PM%20Rosemount%20Analytical%20Documents/Liq_Brochure_91-6032.pdf
Siemens Energy and Automation Inc. (C) http://w3.siemens.com/mcms/water-industry/en/process-products-systems/Pages/products-systems-and-solutions.aspx
Skalar Analytical BV (C) http://www.skalar.com/analyzers/on-line-analyzers-complete-on-line-water-monitoring
SMG Interlink (B) http://smglink.com/product-category/water/
Teledyne Analytical Instruments (B) http://www.teledyne-ai.com/appbinder/ab0003.pdf, http://www.teledyne-ai.com/pdf/lxt-800.pdf, http://www.teledyne-ai.com/products/lxt220do8.asp, etc.
Thermo Scientific, Dionex Corp. (B): http://www.dionex.com/en-us/markets/environmental/water-analysis/lp-71597.html and http://www.thermoscientific.com/en/community/water-analysis.html
Waltron Ltd. (B) http://www.waltronltd.com/dissolved_oxygen_analyzer_9062.asp and http://www.waltronltd.com/products.asp
Yokogawa (remote monitoring) http://www.yokogawa.com/product/appnotes/app-remotemonitor.htm
YSI (B,C) https://www.ysi.com/search?k=water+AND+quality+AND+monitor, and http://blog.ysi.com/blog/?Tag=wastewater+monitoring

INTRODUCTION

The rapid growth of the human population and of industry has focused attention on the quality and condition of our rivers, lakes, and drinking water supplies. Stringent laws have been passed to limit the concentration of elements that are hazardous to health as well as to provide a mechanism to monitor the general quality of the environment.

Declining water quality has become a global issue of concern as human populations grow, industrial and agricultural activities expand, and climate change threatens to cause major alterations to the hydrological cycle. Globally, the most prevalent water quality problem is eutrophication, a result of high-nutrient loads (mainly phosphorus and nitrogen), which substantially impairs beneficial uses of water. Major nutrient sources include agricultural runoff, domestic sewage (also a source of microbial pollution), and industrial effluents and atmospheric inputs from fossil fuel burning and bush fires. Lakes and reservoirs are particularly susceptible to the negative impacts of eutrophication because of their complex dynamics, relatively longer water residence times, and their role as an integrating sink for pollutants from their drainage basins.

Nitrogen concentrations exceeding 5 mg per liter of water often indicate pollution from human and animal waste or fertilizer runoff from agricultural areas. Figure 1.73a shows global trends in nitrate concentration in the water bodies.

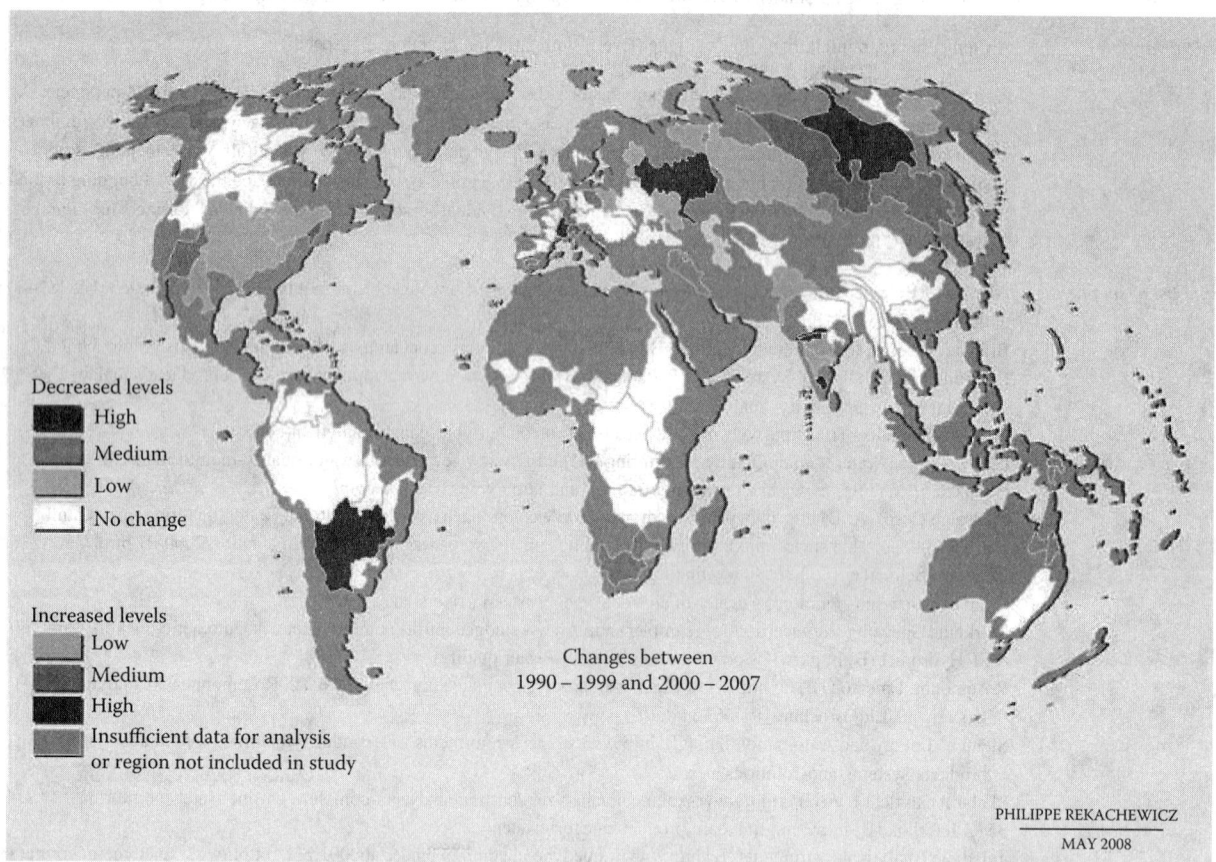

FIG. 1.73a
Global trends in the nitrate levels show that the worst rate of increase is occurring in Asia.

Related Information Sources

Water quality monitoring systems provide a means of measuring the condition of water. These systems range from single-parameter measurements to complex systems integrating multiple sensors with data manipulation and recording capabilities and, in many cases, with means for remote communications and data transfer. The measurement of individual water quality parameters is not going to be discussed in this chapter, because they are covered in detail in the following chapters:

Chapter 1.5—Ammonia
Chapter 1.74—Autotitrator, wet chemistry
Chapter 1.6—Biometers
Chapter 1.48—BOD (biochemical oxygen demand)
Chapter 1.9—Chlorine
Chapter 1.48—COD (chemical oxygen demand)
Chapter 1.13—Color
Chapter 1.15—Conductivity
Chapter 1.18—Cyanide
Chapter 1.50—DO (dissolved oxygen)
Chapter 1.21—Electrochemical
Chapter 1.25—Fluoride
Chapter 1.28—Hydrocarbons
Chapter 1.33—ISE (ion selective electrode)
Chapter 1.37—Mercury
Chapter 1.43—Nitrates
Chapter 1.45—Odor
Chapter 1.46—Oil
Chapter 1.47—ORP (oxidation–reduction potential)
Chapter 1.52—Ozone
Chapter 1.55—pH
Chapter 1.56—Phosphorus
Chapter 1.67—TOC (total organic carbon)
Chapter 1.48—TOD (total oxygen demand)
Chapter 1.68—Turbidity

Regulations

Water quality monitors can be categorized by purpose as described later.

General use systems are usually permanent installations used primarily for data collection for public information, notification, and similar purposes.

Regulatory systems serve to confirm compliance with a particular regulation or water quality standard that was imposed by a regulating agency, either on the local, state, or federal level. These packages are characterized by the stringent requirements that are implemented in standardizing equipment and techniques as well as implementing formal procedures for operation, calibration, and quality assurance monitoring.

Some of the key national and international regulations and standards include:

EPA, Federally promulgated water quality standards for specific states, territories, and tribes (2015) http://www.epa.gov/wqs-tech/federally-promulgated-water-quality-standards-specific-states-territories-and-tribes.

EPA, Regulatory information by topic: Water (2016) http://www.epa.gov/regulatory-information-topic/regulatory-information-topic-water.

UNEP (United Nations Environmental Programme (2006) http://www.un.org/waterforlifedecade/pdf/global_drinking_water_quality_index.pdf.

WHO, Guidelines for drinking-water quality (2010) http://www.who.int/water_sanitation_health/dwq/guidelines/en/.

Industrial systems serve a wide range of uses, including regulatory compliance monitoring, incoming water quality monitoring, aqueous process reaction monitoring, process leak detection monitoring, and biological/aeration pond monitoring. These systems are often typified by the use of online instrumentation that, in some cases, is suitable for operation in hazardous plant locations in addition to offering continuous operation with minimal attention.

MONITORING SYSTEM COMPONENTS

Regulatory agencies have conducted numerous studies to evaluate and determine the efficiency of a variety of types of systems and technologies. These evaluations have resulted in increasing the required quality and the number of suppliers and monitoring techniques that are available. The reliability of such systems has also improved over the years, as has their flexibility, which was required to meet unique demands of individual installations.

A water quality analyzer may be as simple as a probe inserted into water and connected to a conveniently located recorder or as complex as a data collection system with multiple sensors distributed in several locations and connected to a centralized data recording and compiling system.

A water quality monitoring system usually consists of a sampling system, a group of sensors, and a data transmission/logging section. For the package to function properly, these components must be fully integrated and coordinated in their automatic operation.

Many water quality parameters are temperature dependent, so their measurement requires temperature compensation. For this reason, water temperature is always measured by a separate temperature transmitter or by sensors that are included in the analyzers. Other parameters monitored while evaluating the quality of open water include the flow velocity of a river or the level of a water body, the wet and dry bulb temperature of the air, wind speed and direction, and solar radiation intensity (see Chapter 215 for weather stations).

Sampling Systems

A key element in any continuous analysis is the sample transport system. It is essential to provide a reliable and reproducible means of bringing a sample to a sensor or analyzer.

Figure 1.73b depicts a method of placing the sensor directly into the monitored water body for continuous sensing. This technique is used to measure variables such as pH, conductivity, dissolved oxygen (DO), and water temperature. The submersible assembly is anchored directly in the stream and is ideal for monitoring water quality parameters in protected areas.

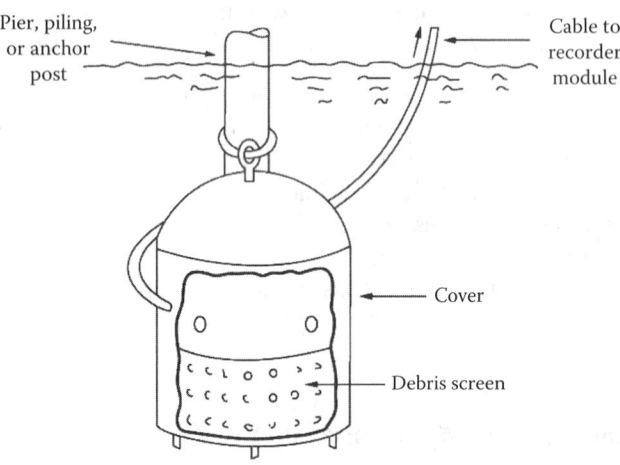

FIG. 1.73b
Submersible sensor housing that can accommodate a number of local sensors. (pH, conductivity, temperature, etc.)

It is possible in some cases to place the sensor directly into water, but most installations require a water sample to be transferred to a permanently installed analyzer. Figure 1.73c depicts a permanent water quality monitoring installation. The sampling system in such an installation might include filters, pumps, gauges, valves, and transport lines.

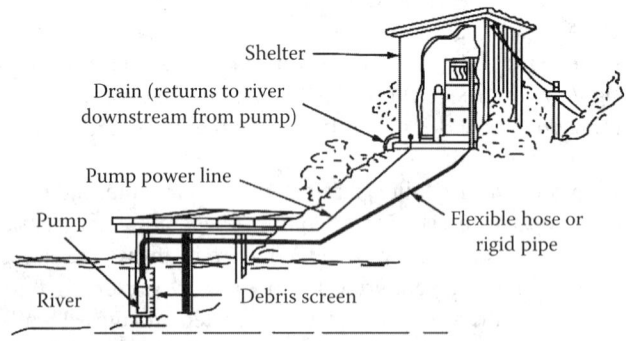

FIG. 1.73c
Permanent installation of a water quality monitoring analyzer package inside an analyzer shelter.

Duckbill Samplers Duckbill samplers should be considered when samples are to be collected at remote locations, from below tanks, sewers, channels, sumps, lakes, or rivers.

As shown in Figure 1.73d, this device has no moving mechanical components—only a rubber (EPT, Buna-N®,

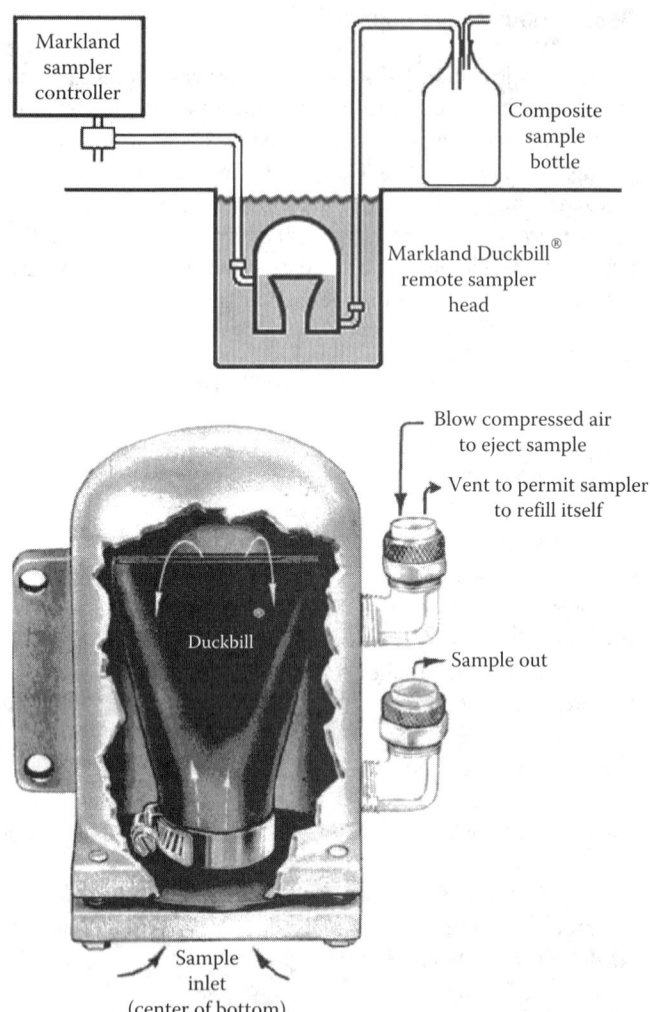

FIG. 1.73d
Remote sampler head for composite sample collection at remote locations. (Courtesy of Markland Specialty Engineering Ltd.)

or Viton®) bucket-shaped *duckbill* inside a housing made of PVC, aluminum, or stainless steel. This rubber insert closes around fibers and particulate matter without jamming and is operated by compressed air. An automatic controller is provided with which the user can adjust the frequency at which samples are taken into a composite sample collection bottle.

The sample enters through the bottom of the sampler by gravity. The sampler is installed below the monitored water or process liquid (or sludge) level, and it traps some air at the top of the chamber as it fills. When a sample is required, compressed air is introduced, which closes the duckbill inlet and discharges the sample through the bottom of the chamber. When a new sample is to be drawn into the sample chamber, the compressed air is vented, and a new fill cycle is initiated.

Sample Transport In all applications, care should be taken to ensure that a representative sample is sent to the analyzer. Sampling systems in general, and the components of liquid sampling systems, are discussed in Chapter 1.2.

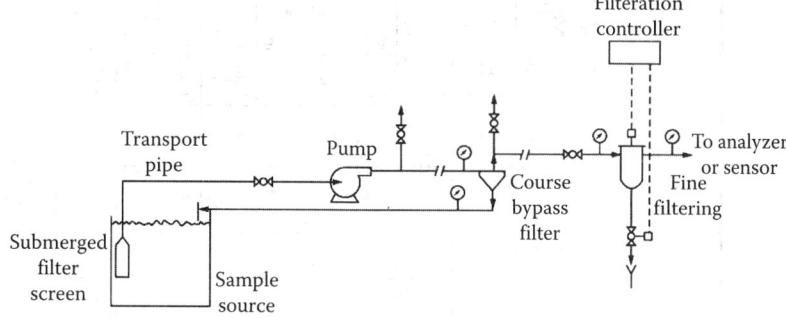

FIG. 1.73e
Schematic of a typical sample transport system.

Figure 1.73e illustrates a typical sample transport system. Pre-filtering may be necessary when the presence of particulate or algae might interfere with the operation of pumps or valves. Screens, Y-type strainers, or cyclone-type filters may be appropriate in other applications. However, for some measurements, filtering of a sample may change the concentration to be measured or change some other characteristics of the sample. Therefore, the effect on the integrity of the sample should be considered before selecting the method of filtering.

Pumps can be submersible. The exact choice of pump should consider the type of analysis being performed. For example, DO and volatile organic compound (VOC) analyses may limit the choice of pumps, because their concentrations can be altered by changing pressures or by any cavitation that occurs in the pump. In industrial applications, the pump must be rated for the correct hazardous location. Care should also be taken to ensure that the pump can handle particulate loading that might be present in the sample on a continuous basis or during excessive runoff (as, e.g., might occur during rain).

Transport lines may require heat tracing or other forms of heating to protect the sample from freezing during cold seasons. In addition, the presence of corrosive materials such as chlorides, strong acids, and bases in the sample may also limit the materials of construction that can be used in the sample transport system.

Analyzer Types

Table 1.73a lists a number of components that are typically measured in a water sample, plus other water quality characteristics, along with typical concentration ranges and their available methods of analysis. In some instances, there are several suitable methods of analysis. Trace concentrations of components are monitored in drinking, industrial, and ultrapure water. Other applications include the measurement of specific pollutants in municipal and industrial wastewater streams.

TABLE 1.73a
Partial List of Measured Components in Water

	Typical Instrument Measuring Range	Analysis Method
Acidity	1.0–3.0 pH	Titration
Alkalinity	0–300 ppm	Titration
Aluminum	0–50 ppb	Colorimetry
Ammonia	0–10 ppm	Colorimetry, ISE, chromatography
Boron fluoride	0–500 ppm	ISE
Bromide	0–2.0 ppm	ISE
Cadmium	0–20 ppm	ISE, chromatography
Calcium	0–300 ppm	Titration, ISE
Chloride	0.1–200 ppm	ISE, colorimetry
Chromate	0–100 ppb	Colorimetry
Copper	0–20 ppb	Colorimetry, ISE, chromatography
Cyanide	0–100 ppb	Colorimetry, titration, ISE, chromatography
Fluoride	0–2.0 ppm	ISE
Hardness (Ca + Mg)	0–2.5 ppm	Titration
Hydrazine	0–50 ppb	Colorimetry
Hydrogen sulfide/sulfide	0–500 ppb	Colorimetry/ISE, chromatography
Iron	0–20 ppb	Colorimetry
Lead	0–1 ppm	ISE, chromatography
Manganese	0–200 pm	Colorimetry
Nitrate/nitrite	0–1 ppm	Colorimetry
Phenols	0–5 ppm	Colorimetry
Phosphate	0–1 ppm	Colorimetry
Silica	0–10 ppb	Colorimetry
Sodium	0–10 ppb	ISE
Urea	0–100 ppm	Colorimetry

Note: ISE, ion-selective electrode.

Ion-Selective Electrodes

As shown in Table 1.73b, a wide variety of electrode designs are available for measuring the presence of different ions (see Chapter 1.33 for more details). These electrodes develop an

TABLE 1.73b
Specific Ion Electrodes

Ion	Electrode pH Range	Anions										Cations										Electrode Type
		ClO^-	I^-	ClO_3^-	Br^-	$S^=$	CN^-	$S_2O_3^-$	NH_3	OH^-	Cl^-	NO_3^-	$HPO_4^=$	HCO_3^-	Cu^{++}	Mg^{++}	Ba^{++}	H^+	Fe^{++}	Ni^{++}	Zn^{++}	Na^+	K^+	Ca^{++}	Sr^{++}	
Chloride	0–14		×		×	×	×	×	×																	Solid state
Sulfide	0–14		None													None										Solid state
Nitrate	2–12	×	×	×	×																					Liquid membrane
Fluoride	1–8.5		×		×					×	×	×	×	×												Solid state
Divalent cation	7–8																	Sensitive to most divalent actions								Liquid membrane
Calcium	5.5–11.0															×	×									Liquid membrane
Copper^{++}	2.5															×	×		×	×	×	×	×	×	×	Liquid membrane
Lead	4–7														×				×	×	×	×	×	×		Liquid membrane
Cyanide	0–14		×		×					×	×	×	×	×								×				Solid state
Hardness	5.5–11														×				×	×	×					Liquid membrane

Interfering Ions

electrical potential reflecting the ionic activity in the process stream. It is important to realize that the activity, not the concentration, of an ion is actually being measured. In dilute solutions, however, the activity may approach the concentration.

Conductivity Sensors   Specific conductance or conductivity provides an estimate of the total dissolved ionized solids in water (see Chapter 1.15 for details). It is a nonspecific method of measurement, as it is sensitive to all ions in the sample. Changes in the conductivity of water are usually indicative of a discharge or runoff of strong acids, bases, or other highly ionizable material. Instruments are typically capable of measuring from 0 to 1000 mS per cm. Conductance of water can vary widely between various bodies and is highly temperature dependent.

pH Probes   pH is a measurement of the hydrogen ion activity in a sample (see Chapter 1.55 for details). It is defined as the negative logarithm of the concentration of the active hydrogen ions, expressed in gram moles per liter. Typical pH values range from 0 (high acidity) to 14 (high alkalinity).

A pH of 7 indicates a neutral solution. Rapid changes in pH can signal a discharge or spill, because the pH of open bodies of water would normally change only gradually.

ORP Detectors   Oxidation–reduction potential (ORP) is a measurement of the oxidizing or reducing materials in water (see Chapter 1.47). It is an electrochemical measurement of the potential developed by these materials when present in water. Under normal circumstances, a body of water will have both oxidizing and reducing components in nearly equal concentrations. A sudden discharge can therefore be detected by detecting the changes in ORP. As are most others, ORP measurement is temperature dependent.

Oxygen Demand Detectors   Biochemical oxygen demand (BOD), chemical oxygen demand (COD), and total oxygen demand (TOD) analyzers (see Chapter 1.48) consist of a sampling and conditioning section that is followed by the analyzer itself. Analysis typically involves quantifying the amount of oxygen uptake by the sample.

The DO concentration of water is a function of the solubility of oxygen at a particular temperature (Table 1.73c)

TABLE 1.73c
Water Solubility of Oxygen

Temp. °C	Chloride Ion Concentration in Water, mg per L					Difference per 100 mg per L Chloride
	0	5,000	10,000	15,000	20,000	
0	14.6	13.8	13.0	12.1	11.3	0.017
1	14.2	13.4	12.6	11.8	11.0	0.016
2	13.8	13.1	12.3	11.5	10.8	0.015
3	13.5	12.7	12.0	11.2	10.5	0.015
4	13.1	12.4	11.7	11.0	10.3	0.014
5	12.8	12.1	11.4	10.7	10.0	0.014
6	12.5	11.8	11.1	10.5	9.8	0.014
7	12.2	11.5	10.9	10.2	9.6	0.013
8	11.9	11.2	10.6	10.0	9.4	0.013
9	11.6	11.0	10.4	9.8	9.2	0.012
10	11.3	10.7	10.1	9.6	9.0	0.012
11	11.1	10.5	9.9	9.4	8.8	0.011
12	10.8	10.3	9.7	9.2	8.6	0.011
13	10.6	10.1	9.5	9.0	8.5	0.011
14	10.4	9.9	9.3	8.8	8.3	0.010
15	10.2	9.7	9.1	8.6	8.1	0.010
16	10.0	9.5	9.0	8.5	8.0	0.010
17	9.7	9.3	8.8	8.3	7.8	0.010
18	9.5	9.1	8.6	8.2	7.7	0.009
19	9.4	8.9	8.5	8.0	7.6	0.009
2	13.8	13.1	12.3	11.5	10.8	0.015
20	9.2	8.7	8.3	7.9	7.4	0.009
21	9.0	8.6	8.1	7.7	7.3	0.009
22	8.8	8.4	8.0	7.6	7.1	0.008
23	8.7	8.3	7.9	7.4	7.0	0.008
24	8.5	8.1	7.7	7.3	6.9	0.008
25	8.4	8.0	7.6	7.2	6.7	0.008
26	8.2	7.8	7.4	7.0	6.6	0.008
27	8.1	7.7	7.3	6.9	6.5	0.008
28	7.9	7.5	7.1	6.8	6.4	0.008
29	7.8	7.4	7.0	6.6	6.3	0.008
3	13.5	12.7	12.0	11.2	10.5	0.015
30	7.6	7.3	6.9	6.5	6.1	0.008
31	7.5					
32	7.4					
33	7.3					
34	7.2					
35	7.1					

Note: Dissolved oxygen concentration in water, mg per liter (as a function of chloride ion concentration and temperature).

TABLE 1.73d
Variations of Dissolved Oxygen with Below-Atmospheric Pressures

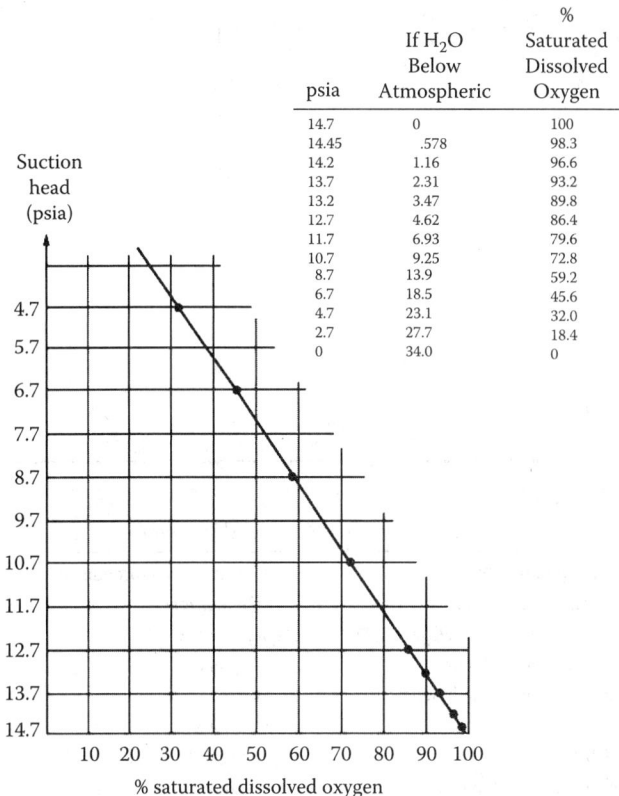

psia	If H_2O Below Atmospheric	% Saturated Dissolved Oxygen
14.7	0	100
14.45	.578	98.3
14.2	1.16	96.6
13.7	2.31	93.2
13.2	3.47	89.8
12.7	4.62	86.4
11.7	6.93	79.6
10.7	9.25	72.8
8.7	13.9	59.2
6.7	18.5	45.6
4.7	23.1	32.0
2.7	27.7	18.4
0	34.0	0

and barometric pressure (Table 1.73d). DO content usually describes the condition of the receiving waters (Chapter 1.50), whereas BOD, COD, and TOD usually describe the pollutant concentration in the discharged wastewaters.

BOD measurements provide information on the amount of oxygen required for the biochemical degradation of the organic matter in water. BOD measurement is rather slow, because it requires a minimum 5-day incubation period for analysis.

COD measures the quantity of oxygen that organics and certain oxidizable inorganic compounds that are present in water will consume. In COD measurements, oxygen is obtained from the reduction of a dichromate solution. The amount of oxygen required to oxidize the organics in the sample provides information on the COD concentrations in water.

TOD measures the oxygen demands of both the organic and inorganic compounds that react with and consume oxygen. The readings reflect the consumption of oxygen in units of mg O_2 per liter of sample or in ppm.

Dissolved Oxygen Sensors DO concentration indicates the amount of oxygen present in water and therefore the ability of receiving waters to support life. Electrochemical probes (see Chapter 1.50) are often used to measure the amount of DO in water. In these probes, the DO passes through a membrane and is reduced at the cathode. The amount of current produced is an indication of the DO concentration.

TC and TOC Analyzers Total organic carbon and total carbon (TC) analyzers (see Chapter 1.67) convert the carbonaceous molecules to CO_2, which is then measured by infrared or other detectors. The carbon in the sample can be converted to CO_2 catalytically or by means of high-temperature (750°C) oxidation.

Total organic carbon measurements are often performed by the addition of an acid followed by sparging, which serves to remove the inorganic carbon from the sample. In the TC mode, no sparging takes place, because organic and inorganic carbons are measured together.

Turbidity Meters Turbidity meters typically consist of a sample cell that is provided with a light source, lens, and photocells for detection (see Chapter 1.68). The water sample flows through the cell, and the cloudiness, caused by suspended solids, is measured and recorded. Turbidity analyzers detect the amount of light that is either absorbed or scattered by suspended solids before reaching the detector, which is placed 180° to the incident light path.

Nephelometers Nephelometry measures the light scatter caused only by suspended solids in the sample. Nephelometric measurements are therefore made with their detectors at a 90° angle to the incident light path.

The unit of measurement most often used is the nephelometric turbidity unit (NTU). As the amount of suspended particles in the sample increases, the amount of light scatter also increases, yielding a higher NTU value.

Wet Chemistry Analyzers Sophisticated hardware such as autotitrators (see Chapter 1.74) and flow injection analyzers

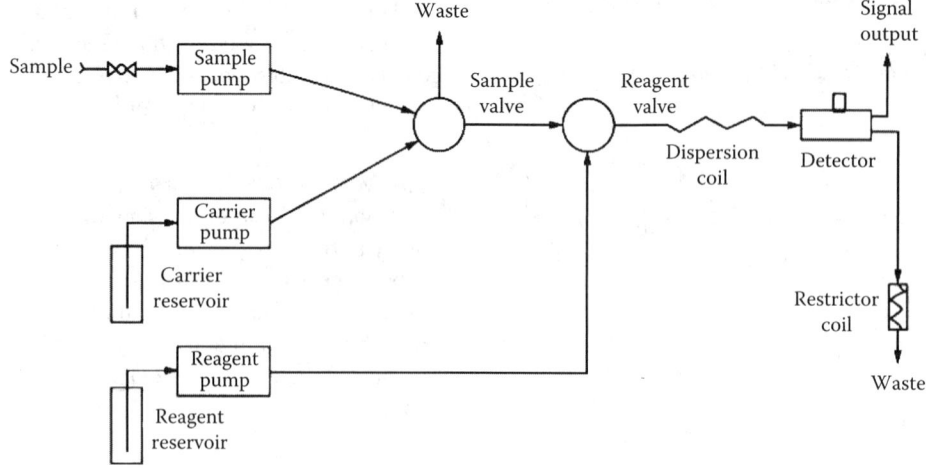

FIG. 1.73f
Simplified schematic of the configuration of a flow injection analyzer (FIA) package.

(FIAs) are also used in water quality monitoring. Figure 1.73f gives a schematic of an FIA. These analyzers use wet chemical methods to quantify a wide range of specific ions and other components found in water. These instruments incorporate sophisticated liquid-handling sections.

These analyzers usually include a sampling section, which is followed by the conditioning section that performs the analysis by mixing reagents with the sample. The actual measurement of concentration is done either by end-point titration or by colorimetric evaluation, using ultraviolet or other detectors.

VOC Detection The concentration of specific VOCs in a water stream can provide the plant operator with information about the condition of a process or of a piece of equipment. The reasons for monitoring VOC are varied. The goal can be (1) process monitoring of an aqueous reaction; (2) spill and/or leak detection, allowing the plant operator time to divert the spilled chemical, thus avoiding contamination of a waterway or biological pond; or (3) monitoring effluent discharges to the environment for regulatory compliance purposes.

While TOC and TC are important water quality parameters (see Chapter 1.67), particularly in the petroleum and chemicals industries, their values often are limited by a lack of sensitivity or by a lack of specific information about a particular compound of interest.

In the past, monitoring for specific organic compounds has been limited to tedious laboratory techniques requiring highly trained personnel and elaborate equipment. Later a variety of online process monitors have been introduced that have automated these techniques and analyses.

Table 1.73e lists a number of specific VOCs that are frequently analyzed in wastewater. Under normal conditions, there might be only a very low ppb level of a particular VOC

TABLE 1.73e
Volatile Organic Compounds Monitored in Water

1,1-Dichloroethylene	1,1,1,2-Tetrachloroethane
Methylene chloride	Ethylbenzene
trans-1,2-Dichloroethylene	m,p-Xylene
1,1-Dichloroethylene	o-Xylene
cis-1,2-Dichloroethylene	Styrene
2,2-Dichloropropane	Bromoform
Chloroform	Isopropylbenzene
Bromochloromethane	Bromobenzene
1,1,1-Trichloroethane	1,2,3-Trichloropropane
Carbon tetrachloride	1,1,2,2-Tetrachloroethane
1,1-Dichloropropene	n-Propylbenzene
Benzene	2-Chlorotoluene
1,2-Dichloroethane	4-Chlorotoluene
Trichloroethylene	1,3,5-Trimethylbenzene
1,2-Dichloropropane	tert-Butylbenzene
Dibromomethane	1,2,4-Trimethylbenzene
Bromomethane	sec-Butylbenzene
Bromodichloromethane	1,3-Dichlorobenzene
cis-1,3-Dichloropropene	1,4-Dichlorobenzene
Toluene	p-Isopropyltoluene
trans-1,3-Dichloropropene	1,2-Dichlorobenzene
1,1,2-Trichloroethane	n-Butylbenzene
Tetrachloroethene	1,2-Dibromo-3-chloropropane
1,3-Dichloropropane	1,2,4-Trichlorobenzene
Chlorodibromomethane	Hexachlorobutadiene
1,2-Dibromoethane	Naphthalene
Chlorobenzene	1,2,3-Trichlorobenzene

in a wastewater stream. However, during a plant upset or due to other abnormal conditions, concentrations may quickly rise to high levels.

Knowledge of the concentration of a particular compound can provide the operator with precise information as to the location of a particular leak or equipment failure. In addition, because there are regulations that limit the allowable concentrations of particular compound classes or particular compounds, information about a compound's precise concentration is required.

Chromatographs

Generally, chromatographs are the most sophisticated semi-continuous monitors utilized in water analysis. They are often used when the concentration of specific organics in a water stream needs to be determined (see Chapters 1.10 and 1.11). More recently, chromatographs have also been used to measure the concentration of specific ions (hence, ion chromatography) in water.

When used for water analysis, the gas chromatograph requires the addition of specialized sample conditioning components. Most chromatographs are provided with either sparging or purging and trapping devices to separate the components of interest from the water sample.

Sample Obtained by Sparging
Sparging is also known as "gas flushing." It is a technique of bubbling an inert gas, such as nitrogen, argon, or helium through a liquid, in order to remove dissolved gases from it. Sparging is often used when gas chromatographs measure the concentration of volatile organic compounds in wastewater. Figure 1.73g shows a sparger-type sample conditioning system in front of a chromatograph. The water sample is introduced through a pressure regulator (PCV) to establish a constant pressure. The sample is then heated to a specified temperature to enhance the sparging process, and its flow rate is regulated.

After the flow regulator, the sample is continuously introduced into the sparging vessel, where helium or some other purge gas is sparged through the sample. As the purge gas

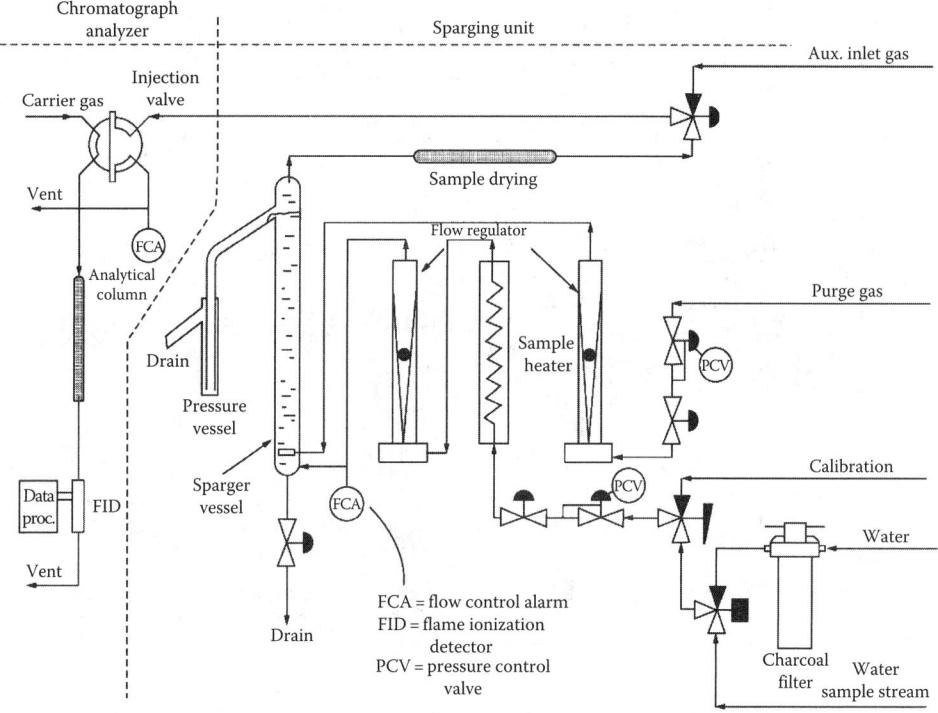

FIG. 1.73g
Volatile organic concentration in a water sample can be detected by gas chromatography where the gas sample for the chromatograph is obtained by sparging.

disperses, it rises through water and effectively extracts a portion of the VOCs. This purge gas, which now contains some volatiles, is either sent to the chromatograph for analysis or passed through a trap (hence the name "purge and trap") for further concentration before being introduced into the chromatograph.

The chromatograph separates the sample into its individual components and thus provides specific concentration readings for each. VOC concentrations as low as 0.1 ppb can be detected using these techniques.

Multivariable, Unattended Systems Multiparameter packages are available for the monitoring of surface waters, reservoirs, and water intake sites. The package shown in Figure 1.73h automatically and continuously monitors pH, conductivity, DO, electrolytic conductivity, turbidity, and temperature. For each of the parameters, a transmitter is provided with local indication and alarm, 4–20 mA analogue, and RS485 outputs. The package is pre-piped and pre-wired. Automatic recalibration of pH and DO, plus self-diagnostics for electrode status checks, is also provided. Against algal growth and plugging, automatic biocide cleaning is provided with programmable cleaning frequency. Air cleaning is also available to protect against sensor contamination.

When the purpose of an installation is to monitor the water quality of lakes and rivers, the monitoring package must be designed to operate unattended for 30–60 days

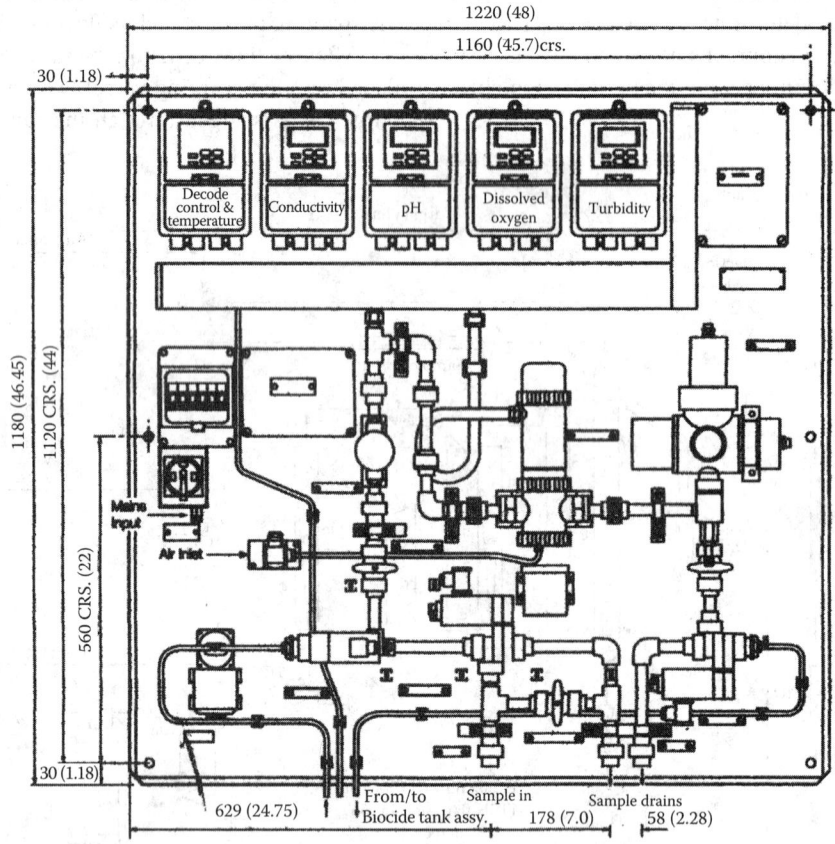

FIG. 1.73h
The design of a pre-piped and pre-wired, five-parameter water quality monitoring package, which is capable of unattended operation in remote locations. (Courtesy of ABB Inc.)

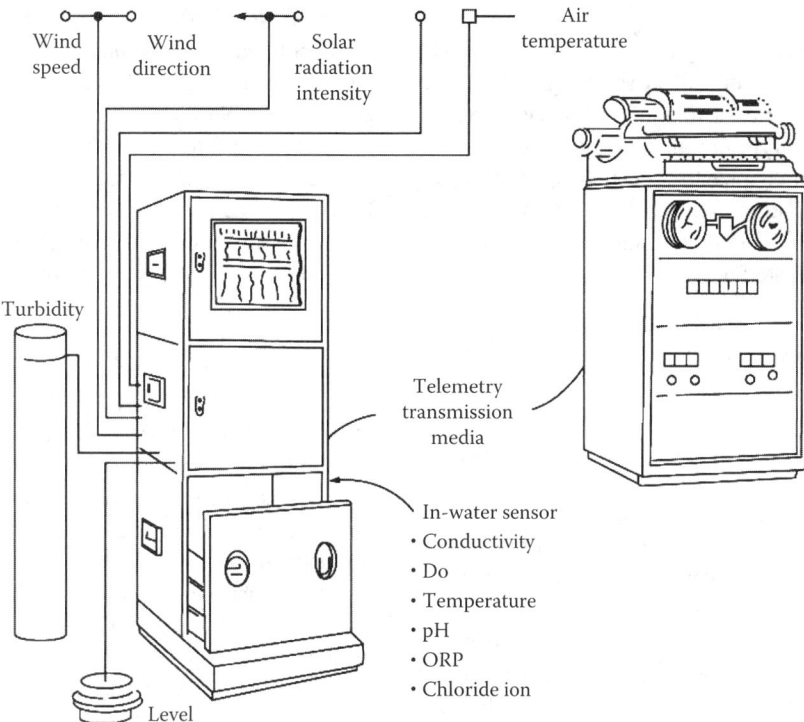

FIG. 1.73i
A complex package of an automatic water quality data measurement, collection, and transmission system.

at a time. Such a monitoring station is illustrated in Figure 1.73i.

The unattended monitoring station usually consists of sensors for weather-related data, plus some of the water quality sensors discussed earlier. The measurements are transmitted by telephone lines or through radio telemetry to centralized data collection systems, where the collected data are analyzed and stored.

In industrial applications, the water quality data usually are not transmitted by telemetry but by hard-wired analog transmission or over digital networks. When the water quality analyzer is located in an industrial plant, its housing is usually designed for its intended environment (including explosion proofing if required), without the need for additional analyzer houses.

On the other hand, when the water quality monitoring station is located in a remote area, it is common practice to place the system inside a shelter, near the open water body being monitored (Figure 1.73c).

Portable Spectrophotometer

During the last decades, substantial advances occurred in the field of manual water quality monitoring also. Figure 1.73j shows such a smart spectrophotometer, which is capable of memorizing up to 25 calibrations and over 40 preprogrammed tests. This microprocessor-based unit also provides automatic wavelength selection; it can be carried in a small suitcase, where it is also supplied with sample tubes, sample cell holders, and battery chargers.

FIG. 1.73j
Microprocessor-based, portable water quality monitor. (Courtesy of LaMotte, SMART Spectrophotometer.)

CONCLUSIONS

A short chapter like this cannot fully cover the gigantic topic of water and wastewater monitoring. It can serve only to give an overview of the many types of analyzers available and to give direction on which chapter to read if further information is needed on a particular analyzer design. Water quality is regulated both nationally and internationally and some of these key standards are listed below:

EPA, Regulatory information by topic: Water, (2016) http://www.epa.gov/regulatory-information-topic/regulatory-information-topic-water.

EPA, Federally promulgated water quality standards for specific states, Territories, and Tribes (2015) http://www.epa.gov/wqs-tech/federally-promulgated-water-quality-standards-specific-states-territories-and-tribes.

UNEP (United Nations Environmental Programme (2006) http://www.un.org/waterforlifedecade/pdf/global_drinking_water_quality_index.pdf.

WHO, Guidelines for drinking-water quality (2010) http://www.who.int/water_sanitation_health/dwq/guidelines/en/.

Wikipedia, Drinking water quality standards (2015) https://en.wikipedia.org/wiki/Drinking_water_quality_standards.

In Table 1.73f, a summary is provided of the limits set or recommended by some national and international organizations for a number of materials and water quality parameters. The table lists the concentration limits set by WHO, the European Union, EPA, and Ministry of Environmental Protection of China. In the table, if a nation has no standard for the particular substance, that is noted by ns and if standards possibly exist, but could not be found when the table was prepared, that was noted by ″. The units used in the table are either micrograms per liter (µg/L = ppb) or miligrams per liter (mg/L = ppm).

Unfortunately, the numbers in this table border on "wishful thinking," because the goals listed are not being met in 50% of the water supplies of the planet. This is not only the case in the third world. For example, the limit for lead concentrations given in the table range from 10 to 15 ppb. Yet, just recently (January, 2016), in the United States, after an accident in the city of Flint, Michigan, the lead concentration of the drinking water was well above this range and in a few homes has reached 13,000 ppb. The main cause of this accident was the lead piping, some of which is still in use in some 20% of the American communities.

TABLE 1.73f
Comparision of Water Quality Standards

Parameter	World Health Organization	European Union	United States	China
Temp. °C	5,000	10,000	15,000	20,000
Acrylamide	″	0.10 µg/L	″	″
Arsenic	10 µg/L	10 µg/L	10 µg/L	50 µg/L
Antimony	ns	5.0 µg/L	6.0 µg/L	″
Barium	700 µg/L	ns	2 mg/L	″
Benzene	10 µg/L	1.0 µg/L	5 µg/L	″
Benzo(a)pyrene	″	0.010 µg/L	0.2 µg/L	0.0028 µg/L
Boron	2.4 mg/L	1.0 mg/L	″	″
Bromate	″	10 µg/L	10 µg/L	″
Cadmium	3 µg/L	5 µg/L	5 µg/L	5 µg/L
Chromium	50 µg/L	50 µg/L	0.1 mg/L	50 µg/L (Cr6)
Copper	″	2.0 mg/L	TT	1 mg/L
Cyanide	″	50 µg/L	0.2 mg/L	50 µg/L
1,2-Dichloroethane	″	3.0 µg/L	5 µg/L	″
Epichlorohydrin	″	0.10 µg/L	″	″
Fluoride	1.5 mg/L	1.5 mg/L	4 mg/L	1 mg/L
Lead	″	10 µg/L	15 µg/L	10 µg/L
Mercury	6 µg/L	1 µg/L	2 µg/L	0.05 µg/L
Nickel	″	20 µg/L	″	″
Nitrate	50 mg/L	50 mg/L	10 mg/L (as N)	10 mg/L (as N)
Nitrite	″	0.50 mg/L	1 mg/L (as N)	″
Pesticides (individual)	″	0.10 µg/L	″	″
Pesticides—Total	″	0.50 µg/L	″	″
Polycyclic aromatic hydrocarbons	″	0.10 µg/L	″	″
Selenium	40 µg/L	10 µg/L	50 µg/L	10 µg/L
Tetrachloroethene and Trichloroethene	40 µg/L	10 µg/L	″	″

Note: Courtesy of Wikimedia Foundation, Inc., https://en.wikipedia.org/wiki/Drinking_water_quality_standards.

SPECIFICATION FORMS

When specifying water quality monitors, the ISA Form 20A1001 and 20A1002 can be used to specify the features required for the total system and when specific individual parameters, such as pH, ORP, turbidity, or DO are to be monitored, the ISA Forms 20A2341, 20A2351, or 20A2192 can be used. The following are the forms that are reproduced with the permission of the International Society of Automation.

1	RESPONSIBLE ORGANIZATION		ANALYSIS DEVICE	6	SPECIFICATION IDENTIFICATIONS	
2				7	Document no	
3	(ISA)		Operating Parameters	8	Latest revision	Date
4				9	Issue status	
5				10		
11	ADMINISTRATIVE IDENTIFICATIONS			40	SERVICE IDENTIFICATIONS Continued	
12	Project number		Sub project no	41	Return conn matl type	
13	Project			42	Inline hazardous area cl	Div/Zon / Group
14	Enterprise			43	Inline area min ign temp	Temp ident number
15	Site			44	Remote hazardous area cl	Div/Zon / Group
16	Area	Cell	Unit	45	Remote area min ign temp	Temp ident number
17				46		
18	SERVICE IDENTIFICATIONS			47		
19	Tag no/Functional ident			48	COMPONENT DESIGN CRITERIA	
20	Related equipment			49	Component type	
21	Service			50	Component style	
22				51	Output signal type	
23	P&ID/Reference dwg			52	Characteristic curve	
24	Process line/nozzle no			53	Compensation style	
25	Process conn pipe spec			54	Type of protection	
26	Process conn nominal size	Rating		55	Criticality code	
27	Process conn termn type	Style		56	Max EMI susceptibility	Ref
28	Process conn schedule no	Wall thickness		57	Max temperature effect	Ref
29	Process connection length			58	Max sample time lag	
30	Process line matl type			59	Max response time	
31	Fast loop line number			60	Min required accuracy	Ref
32	Fast loop pipe spec			61	Avail nom power supply	Number wires
33	Fast loop conn nom size	Rating		62	Calibration method	
34	Fast loop conn termn type	Style		63	Testing/Listing agency	
35	Fast loop schedule no	Wall thickness		64	Test requirements	
36	Fast loop estimated lg			65	Supply loss failure mode	
37	Fast loop material type			66	Signal loss failure mode	
38	Return conn nominal size	Rating		67		
39	Return conn termn type	Style		68		
69	PROCESS VARIABLES	MATERIAL FLOW CONDITIONS		101	PROCESS DESIGN CONDITIONS	
70	Flow Case Identification		Units	102	Minimum / Maximum / Units	
71	Process pressure			103		
72	Process temperature			104		
73	Process phase type			105		
74	Process liquid actl flow			106		
75	Process vapor actl flow			107		
76	Process vapor std flow			108		
77	Process liquid density			109		
78	Process vapor density			110		
79	Process liquid viscosity			111		
80	Sample return pressure			112		
81	Sample vent/drain press			113		
82	Sample temperature			114		
83	Sample phase type			115		
84	Fast loop liq actl flow			116		
85	Fast loop vapor actl flow			117		
86	Fast loop vapor std flow			118		
87	Fast loop vapor density			119		
88	Conductivity/Resistivity			120		
89	pH/ORP			121		
90	RH/Dewpoint			122		
91	Turbidity/Opacity			123		
92	Dissolved oxygen			124		
93	Corrosivity			125		
94	Particle size			126		
95				127		
96	CALCULATED VARIABLES			128		
97	Sample lag time			129		
98	Process fluid velocity			130		
99	Wake/natural freq ratio			131		
100				132		
133	MATERIAL PROPERTIES			137	MATERIAL PROPERTIES Continued	
134	Name			138	NFPA health hazard	Flammability / Reactivity
135	Density at ref temp	At		139		
136				140		
Rev	Date	Revision Description	By	Appv1 Appv2 Appv3	REMARKS	

Form: 20A1001 Rev 0 © 2004 ISA

		RESPONSIBLE ORGANIZATION	ANALYSIS DEVICE COMPOSITION OR PROPERTY Operating Parameters (Continued)			SPECIFICATION IDENTIFICATIONS	
1				6			
2		ISA		7	Document no		
3				8	Latest revision		Date
4				9	Issue status		
5				10			

		PROCESS COMPOSITION OR PROPERTY			MEASUREMENT DESIGN CONDITIONS				
11		Component/Property Name	Normal	Units	Minimum	Units	Maximum	Units	Repeatability
12									
13									
14									
15									
16									
17									
18									
19									
20									
21									
22									
23									
24									
25									
26									
27									
28									
29									
30									
31									
32									
33									
34									
35									
36									
37									

Rows 38–75: (blank)

Rev	Date	Revision Description	By	Appv1	Appv2	Appv3	REMARKS

Form: 20A1002 Rev 0 © 2004 ISA

1	RESPONSIBLE ORGANIZATION		DISSOLVED OXYGEN/OZONE/CL2 TRANSMITTER/ANALYZER/MONITOR Device Specification	6	SPECIFICATION IDENTIFICATIONS		
2				7	Document no		
3				8	Latest revision		Date
4				9	Issue status		
5				10			

11	TRANSMITTER OR ANALYZER		56	PERFORMANCE CHARACTERISTICS	
12	Housing type		57	pH compensation LRL	URL
13	Input sensor type		58	Meas 1 accuracy rating	
14	Input sensor style		59	Meas 2 accuracy rating	
15	Output signal type		60	Temp accuracy rating	
16	Enclosure type no/class		61	Measurement LRL	URL
17	Control mode		62	Temp compensation LRL	URL
18	Local operator interface		63	Barometric pressure LRL	URL
19	Characteristic curve		64	Minimum span	Max
20	Digital communication std		65	Min ambient working temp	Max
21	Signal power source		66	Contacts ac rating	At max
22	Temperature sensor type		67	Contacts dc rating	At max
23	Quantity of input sensors		68	Max distance to sensor lg	
24	Preamplifier location		69		
25	Contacts arrangement	Quantity	70		
26	Integral indicator style		71		
27	Signal termination type		72		
28	Cert/Approval type		73		
29	Mounting location/type		74		
30	Failure/Diagnostic action		75		
31	Sensor diagnostics		76	ACCESSORIES	
32	Data history log		77	Remote indicator style	
33	Measurement compensation		78	Indicator enclosure	
34	Calibration function		79	Communicator style	
35	Enclosure material		80		
36			81		
37			82		
38			83	SPECIAL REQUIREMENTS	
39			84	Custom tag	
40			85	Reference specification	
41			86	Compliance standard	
42			87	Software configuration	
43			88	Software program	
44			89		
45			90		
46			91		
47			92	PHYSICAL DATA	
48			93	Estimated weight	
49			94	Overall height	
50			95	Removal clearance	
51			96	Signal conn nominal size	Style
52			97	Mfr reference dwg	
53			98		
54			99		
55			100		

110	CALIBRATIONS AND TEST			INPUT OR TEST		OUTPUT OR SCALE	
111	TAG NO/FUNCTIONAL IDENT		MEAS/SIGNAL/SCALE	LRV	URV	ACTION	LRV	URV
112			Meas-Analog output 1					
113			Meas-Analog output 2					
114			Meas-Digital output					
115			Meas-Scale					
116			Temp-Scale					
117			Temp-Digital output					
118			Meas setpoint 1-Output					
119			Meas setpoint 2-Output					
120			Meas setpoint 3-Output					
121			Meas setpoint 4-Output					
122			Failure signal-Output					
123								

124		COMPONENT IDENTIFICATIONS		
125	COMPONENT TYPE	MANUFACTURER		MODEL NUMBER
126				
127				
128				
129				

Rev	Date	Revision Description	By	Appv1	Appv2	Appv3	REMARKS

Form: 20A2192 Rev 0

© 2004 ISA

Definitions

Nephelometer is an instrument used for the measurement of the concentration of suspended solid particulates by the use of a light beam. The light detector is located 90° to the side of the source beam. This way, the amount of light received by the detector is a function of particle density in the process stream. Nephelometers are calibrated using the particulate of interest, so that particle color, shape, and reflectivity will automatically be compensated for.

Sparging is also known as "gas flushing." It is a technique of bubbling an inert gas, such as nitrogen, argon, or helium through the sample liquid and thereby remove the dissolved gases from it. Sparging is often used when gas chromatographs are used to measure the concentration of volatile organic compounds in wastewater.

Abbreviations

BOD	Biochemical oxygen demand
COD	Chemical oxygen demand
DO	Dissolved oxygen
EPT	Ethylene-propylene-diene terpolymer
FIAs	Flow injection analyzers
ISE	Ion-selective electrode
NTU	Nephelometric turbidity unit
ORP	Oxidation–reduction potential
ppb	Parts per billion
ppm	Parts per million
TC	Total carbon
TIC	Total inorganic carbon
TOC	Total organic carbon
TOD	Total oxygen demand
VOC	Volatile organic compound

Organizations

ASTM	American Society for Testing and Materials
EPA	Environmental Protection Authority
UNEP	United Nations Environmental Programme
WHO	World Health Organization

Bibliography

ABB, Multi-parameter water quality monitor, (2013) https://library.e.abb.com/public/1facb219029da0b3c12578140039de5f/DS_7976-EN_E.pdf.

ASTM, Water testing standards, (2016) http://www.astm.org/Standards/water-testing-standards.html.

Buffle, J., ed., *In situ Monitoring of Aquatic Systems*, John Wiley & Sons, New York, 2000. http://www.sensorsportal.com/HTML/BOOKSTORE/aquatic_systems.htm.

Diersing, N., *Water Quality: Frequently Asked Questions*, Florida Brooks National Marine Sanctuary, Key West, FL, 2009. http://www.scirp.org/reference/ReferencesPapers.aspx?ReferenceID=1458406.

Duffy, M., Analysis of ppb levels of organics in water by means of purge-and-trap, capillary gas chromatography and selective detectors, (1988) http://www.sciencedirect.com/science/article/pii/S0021967301846550.

EPA, Federally promulgated water quality standards for specific states, territories, and tribes, (2015) http://www.epa.gov/wqstech/federally-promulgated-water-quality-standards-specificstates-territories-and-tribes.

EPA, Regulatory information by topic: Water, (2016) http://www.epa.gov/regulatory-information-topic/regulatory-information-topic-water.

EPA, Water quality criteria, (2015) http://www.epa.gov/wqc.

EPA, Water resources, (2012) http://www.epa.gov/learn-issues/water-resources#our-waters.

Franson, M.A., *Standard Methods for the Examination of Water and Wastewater*, American Public Health Association, Washington, DC, 2005. https://books.google.com/books/about/Standard_Methods_for_the_Examination_of.html?id=buTn1rmfSI4C.

Fresenius, W. et al., *Water Analysis*, Springer-Verlag, Berlin, Germany, 1988. http://www.springer.com/us/book/9783642726125.

Rosemount, Emerson, (2013) http://www2.emersonprocess.com/siteadmincenter/PM%20Rosemount%20Analytical%20Documents/Liq_Brochure_91–6032.pdf.

Schieck, D. and Brown, F., On-line analysis of volatile organics in water using head space gas chromatography, (1996) http://www.sciencedirect.com/science/article/pii/0019057896000079" \l "COR1" \o.

UN, Cleaning the waters, (2010) http://www.unep.org/PDF/Clearing_the_Waters.pdf.

UN, Water quality, (2013) http://www.un.org/waterforlifedecade/quality.shtml.

United Nations Environmental Programme (UNEP), (2006) http://www.un.org/waterforlifedecade/pdf/global_drinking_water_quality_index.pdf.

Water UK, Water quality standards, (2015) http://www.water.org.uk/policy/drinking-water-quality/water-quality-standards.

World Health Organization (WHO), Guidelines for drinking-water quality, (2010) http://www.who.int/water_sanitation_health/dwq/guidelines/en/.

WHO, *Guidelines for Drinking-Water Quality*, 4th edn., Geneva, Switzerland, Retrieved April 2, 2013. http://www.who.int/water_sanitation_health/dwq/gdwq3rev/en/

WOW, Water quality, (2010) http://us.wow.com/wiki/Water_quality.

1.74 Wet Chemistry and Autotitrator Analyzers

C. P. BLAKELEY (1974, 1982)   **E. H. BAUGHMAN** (2003)   **B. G. LIPTÁK** (1995, 2017)

Types of designs	A. Autotitrators are volumetric devices operated by the detection of pH, color, RedOx or zeta potential end points. B. Colorimetric wet chemistry analyzers (see Chapter 1.13). For a list of colorimetric applications see Table 1.74j. C. Flow injection analyzers.
Samples	No suspended solids are allowed when colorimetric units are used, and minimal solids are allowed with the others.
Inaccuracy	1%–3% full scale; accuracy is much affected by calibration.
Costs	Automatic laboratory units which can automatically and sequentially analyze up to 100 samples, with operations including the dispensing of diluents and titrants into each, stirring, measuring, and recording their pH values, and then automatically moving sample tubes to the rinsing station, cost about $15,000. Karl Fischer and potentiometric autotitrators start at around $6,000. The cost of online industrial units ranges from $10,000 to $50,000.
Partial list of suppliers	Analikticon (A multi titration) http://www.analyticon.com/products/laboratory/AUT-701-Automatic-Titrators.php Applitek (A) http://www.applitek.com/en/offer/analyzers/water-quality/single-parameter/titrilyzer/ Cole-Parmer Instrument Co. (A) http://www.coleparmer.com/Search/auto%20titrator Emerson Process, Rosemount Analytical (A, B) http://www2.emersonprocess.com/en-US/news/pr/Pages/911-wetchemistry.aspx Galbraith (testing laboratory) http://www.galbraith.com/wet-chemistry.htm Galvanic Applied Sciences (A, B) http://www.galvanic.com/Sentinel.pdf Global FIA (C) http://www.globalfia.com/ Hanna Instruments (A) http://hannainst.com/products/titrators.html JM Science Inc. (A) http://www.jmscience.com/ Metrohm (A,B) http://surface.metrohm.com/Atline_Online/Online/index.html Mettler Toledo (A) http://us.mt.com/us/en/home/products/Laboratory_Analytics_Browse/Product_Family_Browse_titrators_main/Product_Family_Automated_Systems_main.html MSI (C) http://www.msi.ucsb.edu/services/analytical-lab/instruments/flow-injection-analyzer Scientific Gear (A) http://www.scientificgear.com/Automatic-Titrators Skalar (A,B, robotic) http://www.skalar.com/analyzers/robotic-analyzers-custom-made-automation-solutions Spectralab Instruments (A) http://www.spectralabinstruments.com/automatic_potentiometric_titrators_main.php Thermo Fischer (A replacement parts)) http://corporate.thermofisher.com/content/tf/en/search-results.html?q=autotitrator Tytronics, a Metrisa Company (A) http://www.pollutiononline.com/companyprofile/dacab2d9-f73a-43a8-b4f3-3df52ea7f879 Unity Scientific (A) http://www.unityscientific.com/products/chem-analyzers-and-sample-prep/ http://www.unityscientific.com/products/chem-analyzers-and-sample-prep/smartchem-200.asp?src=westco

INTRODUCTION

Wet chemistry uses observation to analyze materials in the liquid phase. Wet chemistry is also called bench chemistry since many tests are performed at lab benches. Advances in this technology resulted in the automation of the steps of sample preparation, dilution, additions, mixing, heating, dialysis, extractions, distillation, digestion, phase-separation, hydrolysis, ion-exchange/reduction and more.

Wet chemistry analyzers and autotitrators are jokingly called "chemist in the box" devices because they automatically and accurately perform the operations that a laboratory chemist would perform manually to gain the same end result. Because of their complexity, one might think of them as the forerunners of robotics in the field of process control.

From the hardware point of view, flow injection analysis (FIA) methods eliminate much of the complexity but add the requirement of precise timing. The laboratory variations of these units actually use robotic arms to automatically and sequentially analyze up to 100 samples while recording the data, rinsing the sample tubes, etc.

Readers who are interested in maintaining contact and staying up to date with the most recent developments in the field of wet chemistry analyzers should participate in the conferences on process analyzers organized by IFPAC.*

TITRATION END POINTS

Titration is a volume measurement process in which an unknown concentration of a substance is determined by detecting the volume of a standard reagent that is needed to complete a reaction (to reach a reaction end point). The *point of equivalence* represents the perfect stoichiometric balance between the unknown substance and the reagent. The error in titration—the difference between the equivalence point and the actual end point—is called the *indicator blank*.

An acid–base titration curve is shown in Figure 1.74a where the x-coordinate is the volume of titrant added since the beginning of the titration, and whose y-coordinate is the concentration of the analyte at the corresponding stage of the titration. In an acid–base titration, the y-coordinate is usually the pH of the solution. The titration curve reflects the strength of the corresponding acid and base. For a strong acid and a strong base titration, the curve will be relatively smooth and very steep near the equivalence point. Because of this, a small change in titrant volume near the equivalence point results in a large pH change.

There are other types of titrations. Redox titrations are based on a reduction–oxidation reaction between an oxidizing agent and a reducing agent. A potentiometer or a redox indicator is usually used to determine the end point of this titration, which is often indicated by color change of the solution. For example, for the analysis of wines for sulfur dioxide requires iodine as an oxidizing agent. In this case, starch is used as an indicator; a blue starch–iodine complex is formed in the presence of excess iodine, signaling the end point. Other redox titrations do not require an indicator due to the intense color of the constituents.

Titration endpoint can also be detected by monitoring the zeta potential (ζ), which is a key indicator of the stability of colloidal dispersions, in which particles of a substance are evenly dispersed in a continuous phase of another substance. The magnitude of the zeta potential indicates the degree of electrostatic repulsion between adjacent, similarly charged particles in the dispersion. If a colloidal suspension is stable, its zeta potential is high, i.e., the solution or dispersion will resist aggregation. When the potential is small, attractive forces may exceed this repulsion and the dispersion may break and flocculate (coagulate).

Zeta potential titrations are titrations in which the completion is monitored by the zeta potential, rather than by an indicator, in order to characterize heterogeneous systems, such as colloids. One of the uses is to determine the isoelectric point when surface charge becomes zero, achieved by changing the pH or adding surfactant. Another use is to determine the optimum dose for flocculation or stabilization. A zeta potential titration system is shown in Figure 1.74b.

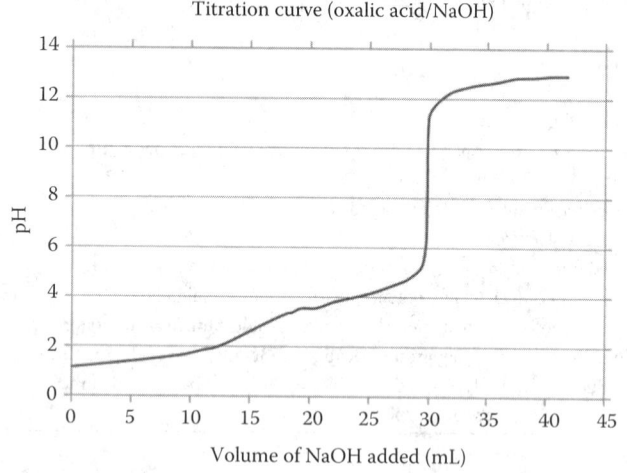

FIG. 1.74a
A typical titration curve of a strong acid titrated with a strong base.

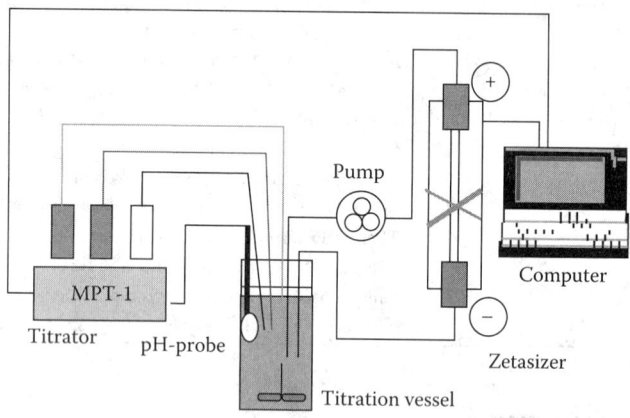

FIG. 1.74b
Zeta potential titration.

---

* International Forum for Process Analytical Chemistry, www.ifpac.com, (847) 548–1800.

End Point Detectors

Titrations can be grouped by the *type of reaction*, which can be acid–base, oxidation–reduction (ORP), complexation, or precipitation. Titrations can also be grouped by the *method of end point determination*, which (1) can be colorimetric, using pH-sensitive dyes, (2) can involve electrochemical electrodes, which can detect voltage (potentiometric, such as pH, specific ions, etc.), current (amperometric), resistance (conductivity), or total charge (coulometric), or (3) can be spectrophotometric or photometric titrations, which use the changes in radiation absorption to detect the end point (a form of the colorimetric detector but no added reagent is required).

When titration is done automatically, the sample is automatically delivered to the analyzer, the delivery of the titrant is automatically terminated when it reaches the end point, and the results are automatically recorded or transmitted for control purposes. Autotitrators are usually capable of targeting end points within a 0–14 pH or 0–2000 mV range.

Continuous and Batch Designs

When used for process control, the autotitrator can be continuous or batch type. For closed-loop control, the continuous titrators are preferred because their dead time and sample transportation lag times are shorter. They detect the titrant flow rate required to bring the sample (at a constant flow rate) to its end point. Autotitrator-based, closed-loop control systems are used for neutralization applications in which the sample (e.g., wastewater) is buffered, so pH measurement alone is insufficient to determine the amount of neutralization agent requirement.

The operation of a simple, batch-type, automatic titrator is shown in Figure 1.74c. Here, the sample is continuously received and is returned to the process through a three-way sample valve. When the field-adjustable program calls for a new sample, the three-way valve opens, and the reaction cell is filled with a new sample. The sample is magnetically stirred, and a titrant is added by an accurate piston pump. A less accurate peristaltic pump can be used to add another reagent, such as an indicator.

Temperature-compensated electrodes automatically detect the end points, and, when the data has been registered, the reaction cell is drained and vented. Depending on the programming, the measurement cycle can be followed by an automatic recalibration cycle using a standard reference solution, a rinse cycle, or another sample.

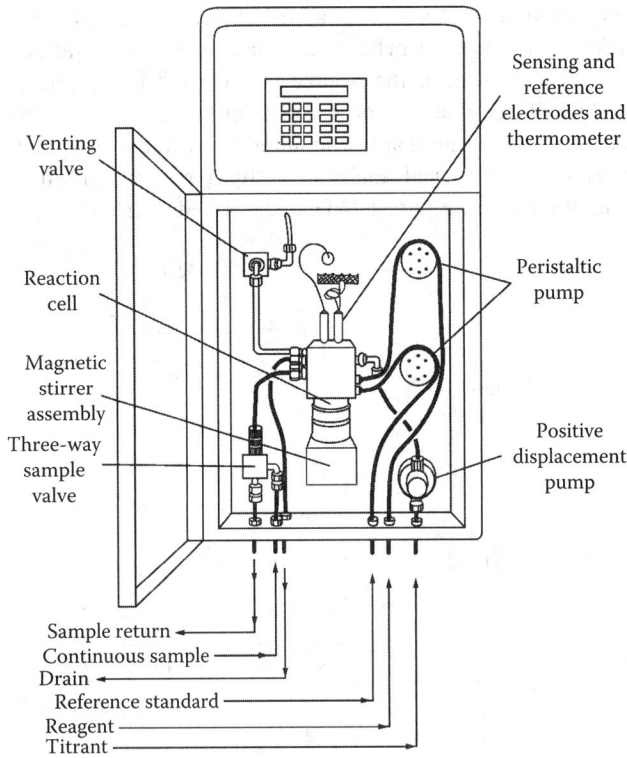

FIG. 1.74c
Field-programmable autotitrator.

VOLUMETRIC TITRATORS

Volumetric analyzers are devices that automatically perform titrations either to electrometric end points or to their colorimetric equivalents. High-precision instruments are available as well as simpler standard devices that perform valuable functions with adequate accuracy and repeatability.

Two distinct types of titrators are available (pH and colorimetric end point) for pollution control work, and these are generally limited to the conventional acid–base tritrations. The development of specific ion activity electrodes and of miniaturized submersible color probes using fiber-optics for light transmission and collection are beginning to open the doors to automatic titrations of a wider variety.

High Precision Designs

The high-precision automatic titrator performs in the exact same manner as a device used by a chemist in the laboratory. The titration beaker is flushed out with the sample, and then a known volume of sample is added. An electrode assembly measures the sample pH. Under constant, gentle agitation, titrant is added until the desired, preselected end point is reached. When that occurs, the titration is stopped and the volume of the titrant that was required to reach the end point is recorded. A typical analyzer of this type, in simplified form, is shown in Figure 1.74d and described later.

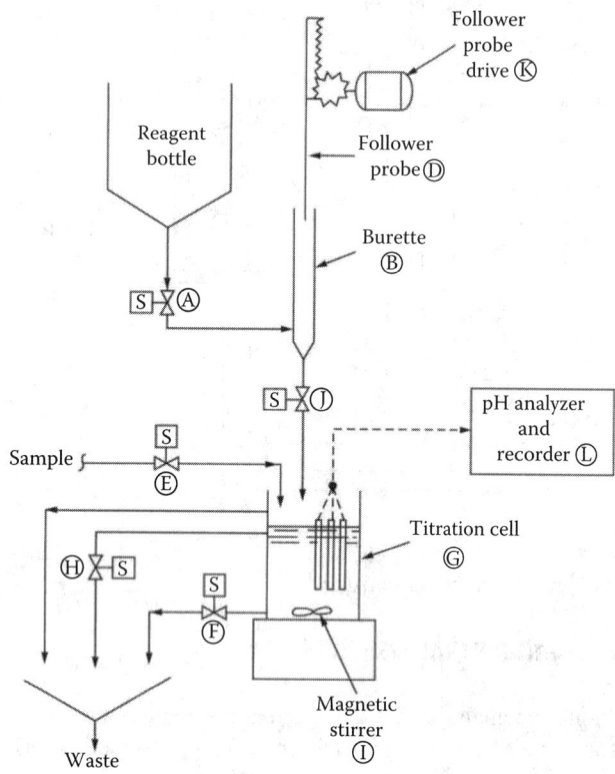

FIG. 1.74d
The main components of a volumetric analyzer (autotitrator).

Operating Sequence A programmer signals the fill valve (A) to open and to fill the burette (B) with reagent from the storage bottle until the level in the burette reaches the full mark. At this time, the conductivity follower probe (D) is at maximum height. Next, the programmer opens the sample fill valve (E) and drain valve (F) and, after a few minutes of flushing, the sample drain valve is closed, and the cell (G) is allowed to fill and overflow to waste. Then, the sample fill valve and the constant-level drain valve (H) are both closed.

At that time, the magnetic stirrer (I) is started and the regent valve (J) is opened. The pH of the solution in the titration vessel (G) is continuously measured and recorded. As the titrant level in the burette drops below the tip of the follower probe, the conductivity bridge is broken, and the drive motor (K) drives the probe down until the tip of the probe is again in contact with the solution. When the preset end point has been reached, the pH analyzer (L) closes the titrant valve (J), and the distance traveled by the follower probe is an indication of the volume of titrant used to complete the titration.

Multiple End Points Titrations to two end points can also be made. This permits automation of the classic phenolphthalein–methyl-orange titration, for example, of a dibasic molecule. The first end point is represented by the first plateau and may be stored in the electronic memory of the device. Most units do not use recorder pens anymore but compare the readings against electronically determined plateaus that would be seen if a pen were present.

The titration then proceeds to the second predetermined end point. The second plateau of the recording pen corresponds to the milliliters of reagent used for the complete titration. Control can be based on the first end point, the second end point, or a computed value using both end points.

In the classic phenolphthalein–methyl-orange titration with acid being the titrant, the final solution pH is always acidic. It is also possible to titrate acidic samples to one or two basic end points. In these applications, it may be desirable to interpose into the flush-and-fill cycle (before titrations for electrode maintenance) an occasional acid rinse. Such an occasional acid rinse will keep the system operational; without it, the metal hydroxides or carbonates might precipitate out on the windows or electrodes and eventually make them nonfunctional.

This type of volumetric analyzer is not as sensitive to environmental conditions as its colorimetric counterpart, and it does not require as clean a sample. However, it is a relatively sophisticated electronic instrument, so it should be accorded an operating environment quality similar to that of a control room.

Simpler Designs

A simple, continuous automatic titrator is illustrated in Figure 1.74e. It utilizes the classic method of titrating to obtain color change end points. In operation, an indicator solution is added to a known volume of sample, and the titrant is then admitted until the color change indicates that the desired end point has been reached. At this point, the volume of titrant used is read as an indication of the concentration of the active species that were present in the sample.

This instrument can also operate as a continuous analyzer. In that configuration, the various flows are detected by measuring the pressure drop (d/p) through capillary tubes that are directly proportional to the laminar flows in those tubes. (The assumption that d/p and flow are linearly related is correct only if the fluid viscosity is constant. This might not be the case if the flowing fluid is a hydrocarbon or if the fluid temperature is not accurately controlled.)

In the titrator, the sample flow is maintained constant by a fixed float valve. An automatic raising or lowering of the titrant float valve can vary the titrant flow. A colorimeter is adjusted to sense the end point color, and it is set to cause the float valve to open or close (drive up or down) so as to maintain this color. In that case, the float valve elevation (or position) is an indication of the prevailing concentration and can also be recorded.

Because the effectiveness of this device depends on color development and on maintaining the end point color, it is imperative that the samples be free of suspended matter. To perform a two-end-point titration, two titrators are needed: one for the equivalent of the phenolphthalein end point detection, and a second for the equivalent of the methyl-orange end point measurement, which was cited in the classic titration example earlier. It is then necessary to use the outputs from each titrator and to use external electronics for the calculations.

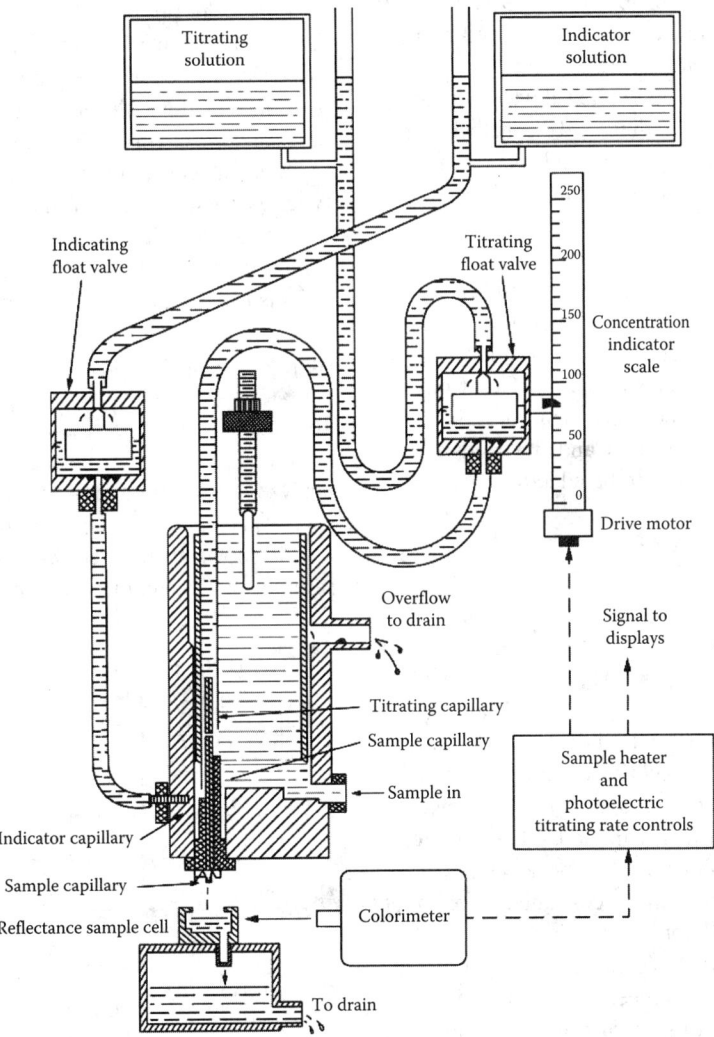

FIG. 1.74e
The operation of a continuous automatic titrator that is using color change as a means of determining the end point.

Applications

One widespread problem facing the pollution control engineer is the precipitation of heavy metals and similar applications that too often are lumped in with pH control. Consequently, conventional pH instrumentation is frequently used when it has little chance of success or is likely to give constant trouble because of electrode fouling. To overcome these problems, engineers frequently choose highly sophisticated control systems, which might include both feed-forward and feedback control loops, bias and variable ratio stations, and a variety of electrode cleaning devices ranging from mechanical wipers to ultrasonic cleaners.

Most heavy-metal precipitation problems are similar to the hot or cold lime soda softening techniques used in the preparation of water for boiler use and to the preparation of hard water for municipal drinking water suppliers. Cold or hot process softeners are readily controlled by titration. Under manual conditions, a flowmeter on the influent flow transmits pulses to a counter for every 379 L (100 gal) metered. The counter accumulates a predetermined number of pulses and activates a repeat cycle timer.

An operator collects a sample of softened water, titrates to the phenolphthalein end point (P reading) and the methyl-orange end point (M reading), and then makes the calculation 2P − M. He then adjusts the timer to feed sufficient lime and soda ash to maintain the desired value, which is close to zero. An automatic titrator can also perform this function, and electronic circuitry can make the computation and adjust the timer settings.

Another simple pH control problem is in the operation of cooling towers where hot water from distillation units is cooled but, because the water tends to pick up CO_2 from the atmosphere, it forms an acidic, corrosive solution. The addition of a base can easily counteract this, and one needs a titrator to determine how much base is to be added.

COLORIMETRIC ANALYZERS

For a more detailed discussion of colorimeters, refer to Chapter 1.13.

A chemist who is well trained, experienced, and working unhurriedly in a well equipped laboratory can analyze samples very accurately. Such conditions seldom if ever exist in a plant where the engineers or operators want answers immediately after they send their sample to the laboratory. In a way, their attitude is understandable, because they want to find out what is going on in their process and answers that arrive several hours later will not help much.

High-precision, automatic colorimetric analyzers perform in the manner that plant engineers would like the laboratory chemist to perform. In this instrument, the sample and reagents are accurately metered and proportioned. Reagents are added in identical sequences, with mixing and time delay steps between additions, as is required by the chemistry and the laboratory procedure. Where reactions are temperature dependent (and most are), a constant-temperature bath is included in the analysis hardware. Sample preparation by distillation or dialysis can also become part of the automatic analytical procedure if required. The analysis results provide about the same precision that can be obtained by an experienced chemist, and they demonstrate better repeatability.

These analyzers contain delicate, high-precision electronic components that emit low-level signals requiring enormous amplification for recording and control. Therefore, it is generally advisable to make some investment in analyzer site selection and preparation. Vibration, dirt, and dust should be eliminated, and sudden ambient temperature variations should be avoided. An air-conditioned environment is ideal for the analyzer, and proper maintenance is also essential. It should also be realized that wet chemical procedures frequently liberate corrosive or deleterious gases. These must be vented from the analyzer cubicle to prevent corrosion of the electronic components and must be expelled from the temperature-conditioned room.

As colorimetry is a measure of light absorption resulting from the color produced by the chemical reactions, it is essential that the sample fed to the automatic analyzer be free from suspended matter. This requirement can be relaxed somewhat by using a dual-wavelength analyzer to detect the color change, but dual-wavelength colorimeters are uncommon and expensive.

Reagent consumption for colorimetric analyzers of all types varies with the analysis to be performed. Generally, the high-pressure units use proportionately less reagent than the standard devices, but the reagent used by the high-pressure units is usually costlier and harder to prepare.

High Precision Designs

To accurately ratio the flow streams of a sample to the flows of dilution solvents and reagents, most high-precision laboratory analyzers use a multiple-head peristaltic pump that receives several plastic tubes and pumps all these liquids simultaneously.

As shown in Figure 1.74f, the ratio control between the sample and reagent fluids is achieved by the careful selection

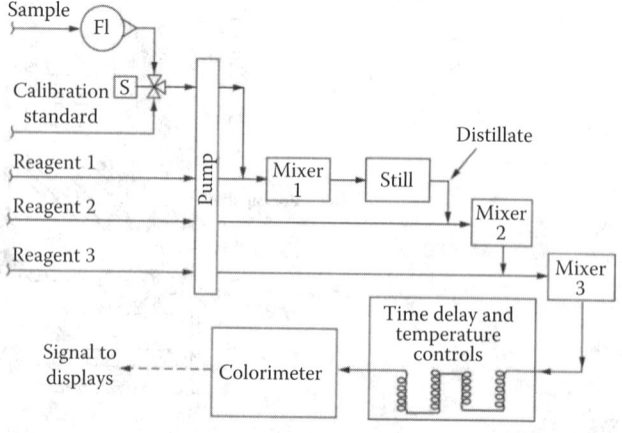

FIG. 1.74f

Precision colorimetric analyzer configuration, which includes a sample distillation step, as might be used for phenol or cyanide detection.

and control of the internal diameters of the tubes. Mixing is achieved by changing the tube diameters repeatedly downstream of the pump (thereby changing velocities), by changing flow directions, or both. It is difficult to convert this technology to automatic, online operation because of the high maintenance requirements of the pump and tubes.

Segmented Flow Design One analyzer, called the *segmented flow analyzer*, introduces air through one of the pump tubes to create discrete air gaps between segments of fluid. It then passes these liquid segments through horizontally mounted glass or plastic coils. The air gaps, while keeping the samples separate, permit free-fall of the liquid segments, thus achieving mixing of sample and reagent. Coils provide retention time. Where temperature control is required, these retention time coils can be made part of a constant-temperature bath that is thermostatically controlled and provided with an adjustable set point.

The reacted sample then passes through a "debubbler" and on to the analyzing section that frequently is a colorimeter. The colorimeter establishes a baseline when there is no sample in the stream and compares that to the reading when a sample is present. The advantage of this approach is that it compensates for the drift in light source intensity, electronic drift, and window fouling.

Other approaches use two beams from the same light source and split the beams, but this does not compensate for any window fouling. One needs to calibrate the difference in readings between the baseline and sample concentration, and this is done with standards.

Split Beam Design Figure 1.74g describes an older design, the split-beam colorimeter. Here, a single beam from light source (A) passes through an optical wavelength filter (B) and is split into two beams by the beam splitter (C). Next, the sample beam passes through the sample flow cell (E) and through an optical filter (F), which diffuses the light beam over a wide surface area of the measuring photocell detector (G_1). The other beam passes through a movable polarizing lens (H), a stationary Polaroid lens (J), and an optical filter (F) before it strikes an identical, matched photocell (G_2).

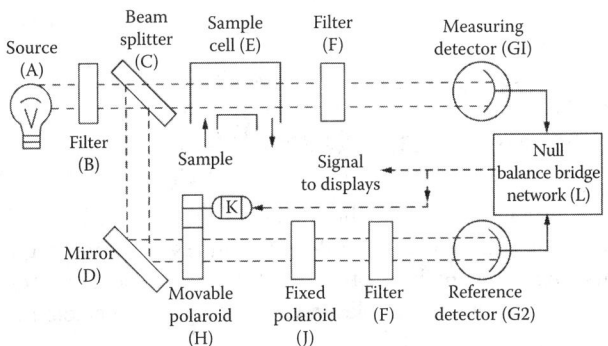

FIG. 1.74g
The main components of a dual-beam colorimeter.

With distilled water or a pure solvent in the sample cell, a maximum amount of light is transmitted to cell G_1. With the polarizing lenses aligned in parallel, maximum light is also received by photocell G_2. Therefore, the outputs from the two photocells are equal, and the bridge is in balance.

With a colored solution in the sample cell, the light intensity striking photocell G_1 is attenuated. Output G_1 is now less than output G_2, and the error signal drives the movable polarizing lens through drive motor (K) until the polarizing system has decreased the light intensity striking G_2 so that the outputs of the two photocells are again equal. The distance traveled by the movable polarizing system is a function of color intensity of the sample and can be recorded as the concentration of the constituent analyzed.

Most colorimetric analyses obey Beer's law. This law relates the absorption of light to the transmission properties of the material through which the light is traveling. In other words, it relates the amount of light absorbed to the concentration of the material of interest in the sample. The colorimetric analyzers tend to be nonlinear. Usually, however, the range requirement for a specific application is usually narrow enough that the pertinent portion of the curve may be considered linear or can be "linearized" with a log amplifier.

Calibration Colorimetric analyzers are generally equipped with automatic zero and standardization features. At specific intervals, the reagent feed is stopped, and the sample is introduced only into the sample flow cell. After a predetermined period of flushing with pure, unreacted sample, the analyzer reading should be zero. Therefore, under these conditions, the reading is reset to zero, thus compensating for sample or flow cell discoloration and for electronic drift.

Next, the sample is diverted to waste, the reagent system is reactivated, and, instead of a sample, a laboratory-prepared standard solution is introduced. After an interval sufficient to replace all liquids in the sample flow cell with this freshly reacted standard solution, the output signal is automatically compared to the signal level expected for the standard solution strength, and adjustments are made if needed.

The length of the optical path (i.e., the depth of reacted sample penetrated by the light beam) is critical. In an analysis for trace amounts of contaminants in which the color developed by the chemical reaction is expected to be weak, a long optical path is essential. The longer the light path, the less signal amplification is needed and, therefore, the greater will be accuracy of the analysis.

Conversely, if the color intensity is high, a short optical path is preferred so that the high-color intensity does not attenuate the light beam excessively. If the absorbance is too high (greater than ≈ 1), Beer's law will fail, and the signal will no longer be linear with the log amplifier.

High-precision analyzers generally have interchangeable flow cells that may range from 50 to 250 mm (2 to 10 in.) in optical path length. The longer the path length, the more critical is the requirement to remove solids or light scattering particles from the sample.

Simpler Design

The simple, continuous colorimetric analyzers are generally limited to analyses where no more than two or three reagents need to be added simultaneously. Although some of them can proportion the reagents to the sample by individual head vessels and capillaries, other designs employ multiple solution pumping heads coupled to a common drive motor. Each pumping unit operates in the same phase and is individually adjustable.

Mixing of sample and reagent is accomplished in many ways, from mechanical stirring in the sample cell to changes in velocities and direction, to free-fall into a head vessel that, in turn, discharges into the colorimeter flow cell.

Figure 1.74h illustrates such an unsophisticated colorimeter. The light from a single light source (A) is collimated by a lens (B) before it passes through the sample cell (C), in a single or double pass as illustrated. The light then passes through the color filter (E) on its way to the measuring photocell (F_1).

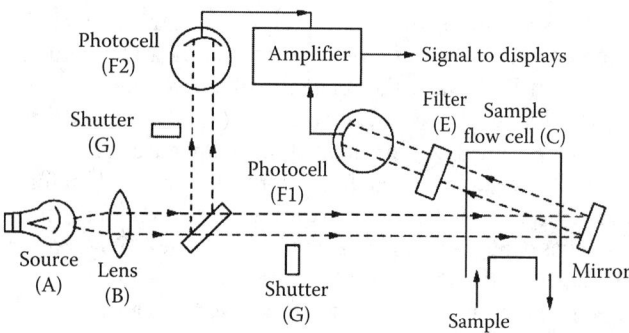

FIG. 1.74h
The components of an unsophisticated, standard colorimeter.

A second light beam passes directly to a second reference photocell (F_2). Both beams can be attenuated by a shutter (G). These shutters are used for zero and full-scale adjustment. The output differential is generally displayed in terms of constituent concentration on a nonlinear meter scale, or the output can be run through a log amplifier and reported on a linear scale. Outputs for recording or other displays are also available.

Calibration Ordinarily, these simple analyzers are manually standardized against calibrated standard slides that are provided by the manufacturers. These may be in the form of one or more calibrated orifices that limit the amount of light striking the measuring photocell, or in the form of optical filters serving the same purpose.

To overcome the effect of ambient temperature changes on photocell output, colorimeters may be equipped with thermostatically controlled heating elements that make the system relatively free from drift. It should be emphasized that if the operator checks only the detector of the colorimeter for drift, he will not know if the complete analyzer is in calibration. The only way to check the whole system is send a standard sample through the sample-conditioning system and into the analyzer.

On-Off Batch Design

Continuous analysis is not necessary if only an alarm or on/off control is required on process streams that are relatively stable. For these applications, automatic batch analyzers have been developed. These are usually single reagent devices in which the set point for alarm or control is fixed by the reagent.

Hardness analysis is an example. Based on the Versene method,* reagents have been developed that undergo a dramatic color change from green to red when hardness exceeds the chelating properties of the reagent. Reagents exhibit this vivid color change for a wide range of water hardness values, such as 0.75, 1.5, 5.0, 9.0, and so on up to 50 ppm or more of calcium carbonate.

This design is shown in Figure 1.74i, where a photosensitive device is employed that sees only green light, a circumstance denoting that the concentration in the sample being analyzed is below the reagent set point. At the color change from green to red, the wavelength filter screens out all light in the red wavelength. The photocell then receives no light, and an alarm signal is generated.

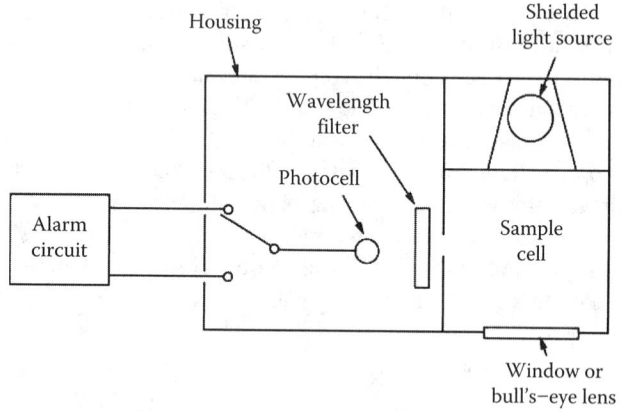

FIG. 1.74i
The configuration of an on/off batch colorimeter.

The sample cell is illuminated from the back and is visible to the operator. Fail-safe features are inherent in this approach. Failure of the light source, the photocell, the reagent, or the sample flow, if not visually detected by

---
* ASTM D1126-12, Standard Test Method for Hardness in Water, ASTM International, (2012) http://www.astm.org/Standards/D1126.htm.

the operator, will be noted as a result of energizing the alarm circuit. With this type of analyzer, it is obviously not possible to record analytical values. However, it is possible to use a simple voltmeter-type event recorder that is actuated by the alarm circuitry and will record the time of day and duration of the alarm condition.

Applications

In the control of water pollution, automatic colorimetric analyzers are used in the monitoring of influents into a treatment facility, in the monitoring of effluents from a treatment facility, and in the control of the actual treatment or abatement process.

Table 1.74a lists some of the more common analyses required by pollution abatement legislation, along with the minimum concentration or ranges that are also commercially available. To meet specific monitoring requirements, one can (with the assistance of the instrument manufacturers) develop or adapt chemistry to provide measurements that will satisfy the specific needs.

If, for example, a manufacturing process generates a waste stream containing hexavalent chromium, Cr^{6+}, and it is to be treated in a batch process, it may be sufficient to perform the following steps. Fill the treatment tank with the waste, and make a manual analysis to determine the amount of treatment chemicals that need to be added. After the treatment is complete, and prior to discharge, just manually analyze the result.

If this same waste is to be treated by a continuous process, it is very likely that conductivity measurements can be used more economically than colorimetry for the control of chemical additions. The effluent from continuous treatment, however, should be monitored for residual hexavalent chromium content after dilution with other wastes.

If, on the other hand, the alkaline chlorination process treats cyanide waste continuously, it may be more economical to monitor the effluent indirectly for residual chlorine rather than for cyanide by colorimetric means. The presence of free available chlorine in excess of 5 ppm would denote the absence of cyanide, because these two chemicals cannot coexist.

Where colorimetric analysis is warranted, efficient sample preparation and careful analyzer site selection are essential.

FLOW INJECTION ANALYZERS (FIA)

Flow injection analysis (FIA) is based on the injection of a liquid sample into a moving, nonsegmented continuous carrier stream of a suitable liquid. The injected sample forms a zone, which is then transported toward a detector that continuously records the changes in absorbance, electrode potential, or other physical parameter resulting from the passage of the sample material through the flow cell. Flow Injection evolved into Sequential Injection and Bead Injection which are novel techniques based on flow programming.

FIA has been pioneered by Jarda Ruizcka at the Center for Process Analytical Chemistry (CPAC), at the University of Washington in Seattle. FIA looks at the change in detector reading caused by the interaction of the sample and reagent. Detectors can be electrochemical devices, optical devices, and so on, and the advantage of looking at the change in signal rather than the absolute signal exists because this approach provides a built-in drift correction.

Much has been published about FIA, but its online process applications have been few. One such application was

TABLE 1.74a
Available Automatic Colorimetric Analyses

Types of Analysis	Minimum Concentration (ppm)	High Precision Analyzer	Standard Analyzer
Aluminum	10	♥	
Ammonia	10	♥	
Borate	10	♥	
Carbonate	4000	♥	
Chlorate	12%	♥	
Chloride	10	♥	♥
Chlorine	5	♥	♥
Chromium Cr^{6+}	5	♥	♥
Chromium total	5	♥	
Copper	10	♥	
Cyanide	5	♥	♥
Fluoride	5	♥	♥
Hardness total	500	♥	♥
Iron	10	♥	
Magnesium	150	♥	
Manganese	2%	♥	♥
Phenol	5	♥	
Phosphate total	100	♥	♥
Silica	15	♥	♥
Sulfate	500	♥	
Surfactants (anionic only)		♥	
Chemical oxygen demand (COD)		♥	

described in Chapter 1.73, where it is used in water quality monitoring applications. Described next are two very different and successful online applications of the technology. The laboratory application of the technique will be discussed first.

Laboratory Design

Classic laboratory FIA consists of a peristaltic pump moving a reagent stream past a detector (Figure 1.73j). The period between samples gives an excellent baseline for the detector. Periodically, sample is injected into the moving stream of reagent, reaction takes place, and the detector responds. For example, if the sample were an acid, the reagent would be a base, which would partially neutralize the sample. The detector, a pH meter, would note the drop in the pH of the stream as the sample passed. The method is extremely reproducible.

In automated wet chemistry, analyzers up to 16 analytical measurements can be made on a single sample simultaneously and start-up, shut-down, dilutions, re-runs, clean-up, and raw data storage can all be automated. The automatic processing steps can include dilutions, additions, mixing, heating, dialysis, extractions, distillation, digestion, phase-separation, hydrolysis, ion-exchange/reduction, and more.

The samplers serve to minimize both operators' time and errors. They can handle high workloads automatically and outside working hours, without any operator supervision. These samplers are basically robotic devices able to read barcodes, handle both pre- and post-dilution runs, prepare standards, use random access, multiple ranges, perform rinsing, handle customized containers, in short they can offer a very high degree of automation in the laboratory.

Industrial Designs

The factors that hinder FIA's acceptance by the processing industries include two basic needs.

1. There exists a need to conserve the amount of reagent consumed because, in the process world, there is a lot of sample, and the only way to keep it fresh is to let it flow continuously. For this reason, the roles of the reagent and the sample should be reversed.
2. It is desirable to eliminate the peristaltic pump because of the short lifetime of the pump tubes and because of its lack of long-term reproducibility.

The peristaltic pump is capable of generating low flows. These low flows reduce reagent consumption and allow sufficient time between the sample injection and detection points for the reaction to take place. Another big advantage of the peristaltic pump, when used for FIA, stems from the pump pulses, which promote mixing of the sample with the reagent.

One can provide low flow without a peristaltic pump by using pressurized reservoirs for reagents. If the reservoir is at constant pressure, which is easy to maintain with an air or nitrogen gas pressure, constant flow is guaranteed, because the pressure drop through the piping is also constant. To substitute for the mixing action of the pulses from the peristaltic pump, one can install a coiled piece of tubing or a static mixer (Figure 1.74j).

FIG. 1.74j
Random access, computer controlled sampler. (Courtesy of Skalar.)

Application Example 1

To illustrate the problems and advantages associated with the reversing of the sample and reagent, let us look at the problem of detecting 100–1000 ppm of polymer in a butane stream. The polymer is polyisobutylene, and its molecular weight is 1000–1500.

It was noted that if the butane sample was put into isopropyl alcohol (IPA), the solution became cloudy. It was also shown that the amount of "cloudy" formation was directly related to the amount of polymer in the butane solution. The problem was that the "cloud" caused by the undissolved polymer would coat the windows of the detector cells and would distort the detector readings.

The solution was found by injecting IPA into the butane stream, which does not result in detector drift and causes no cloud formation, even when polymer is present. This was the case because the sample had more butane in it than required to cause precipitation when the IPA was injected.

The analyzer devised for this application had a 10-port valve to trap a small sample of butane-polymer between two larger samples of IPA, which are then trapped in the butane solution. In this way, a detectable turbidity from the sample is trapped in the IPA, and the normally flowing stream of butane cleans the turbidity cell. The IPA requirements for this analysis were very small, because only the two "traps" had to be refilled between samples.

Application Example 2

In this application, the problem was to measure the hydrogen sulfide (H_2S) and carbon dioxide (CO_2) in an aqueous amine stream. This process can be found in many natural gas plants and in refineries where the amine streams are used to extract H_2S and CO_2 from a hydrocarbon vapor stream. In this process, the pressure on the amine is reduced after picking up the H_2S and CO_2, and the temperature is raised, which releases the H_2S and CO_2 from the amine so it can return for another load.

From the control point of view, the following two unresolved questions were costing the plant money.

1. How hard do we need to "strip" the amine? If too much stripping is applied, the required heat is wasted, and amine degradation results, which also costs money.
2. What flow of amine is necessary to meet product specifications? If the flow is excessive, it wastes energy, because it has to be heated; in addition, it extracts some of the product from the hydrocarbon stream. If too little flow (or too little heat) is used and the product does not meet specifications, the product is wasted to flare. (One gas plant achieved energy savings of $366,000 per year by optimizing flows, and identified more than $1,000,000 per year in extra gas that had been extracted with their previous control system.)

So the question is, how can we measure H_2S and CO_2 in an aqueous amine stream? Both are acids, and the CO_2 becomes H_2CO_3 when it dissolves in water. Therefore, simple acid–base titration will give only the sum of the two. The laboratory method is to precipitate the H_2S as PbS and then titrate the H_2CO_3. But this approach is not attractive for online use.

UV spectroscopy of the aqueous amine solutions "sees" only the H_2S and ignores the CO_2. In addition, the CO_2 amine stream can become very corrosive and destroy the unit.

If the aqueous amine is put into a strong acid solution, the H_2S and CO_2 go into the gas phase. The H_2S has a strong UV absorbance with almost no IR absorption, and the CO_2 has a strong IR absorbance with none in the UV. Therefore, the goal is to inject a known amount of the amine solution into an acid bath and strip the H_2S and CO_2, which, when sent to a UV and IR detector, will provide the required information.

This was achieved by sending the amine stream through a 10-port valve filling the sample loop. In fact, two loops were used, one for the amine coming from the contactor and one for the amine from the stripper. The loops were sized so that each would contain about the same amount of H_2S under normal operation. Periodically, the valve turned, allowing water to carry one of the loops of the aqueous amine to the acid bath. The bath was stripped with nitrogen to mix the solutions and force the gases past the UV and IR detectors.

To avoid excess corrosion while obtaining a fast reaction, 3N H_2SO_4 was used as the acid. While this product was developed at Amoco and they hold the patent on it, Hamilton Sundstrand (Orbital) now has the rights to make and sell it. It should be noted that a small amount of H_2S is released in this analysis and should be vented to a safe place.

Calibration

One of the problems with online analyzers is that they need calibration. With the H_2S/CO_2 amine analyzer, we chose Solutions Plus Inc.* to make us basic NaOH solutions of Na_2S and Na_2CO_3. When these solutions hit the acid, H_2S and CO_2 are released to calibrate the system, yet the method provides an aqueous system that is stable.

Unfortunately, there are not many process FIAs on the market.[†] Most process FIAs are now built to solve a problem that needs to be solved by whatever company has the problem.

CONCLUSIONS

Wet chemistry uses observation to analyze materials in the liquid phase. It can involve as many types of analysis as the number of tests that are performed on laboratory benches.

---

* Solutions Plus Inc., St. Louis, MO, (636) 349–4922, solnplus@accessus.net.
† Global FIA, (253) 549–2223, info@globalfia.com (company produces only process FIAs).

Advances in this technology resulted in the automation of such testing steps as sample preparation, dilution, additions, mixing, heating, dialysis, extractions, distillation, digestion, phase-separation, hydrolysis, ion-exchange/reduction and more.

In this chapter the discussion focused on automatic systems and on end point detection techniques including the volumentric autotitrators, colorimetric endpoint sensors and the various flow injection systems. This coverage represents only a small segment of the wet chemistry analyzers that are available to detect such materials as protein, fat, ammonia, calcium, phosphate, total nitrogen, total phosphorous and urea, just to mention a few. Wet chemistry can also detect the concentrations of various elements (Table 1.74a), environmental conditions like euthropication or BOD and COD plus just about any analytical variable from acidity and alcatinity to salinity and turbidity. To assist the reader with further information on this large range of applications, further reading material is provided under the BIBLIOGRAPHY listing at the end of the chapter.

SPECIFICATION FORM

When specifying wet chemistry analyzers, both the detector to be used and the details of the application can be specified by using the ISA form 20A1002. This form is reproduced with the permission of the International Society of Automation on the next page.

	RESPONSIBLE ORGANIZATION	ANALYSIS DEVICE COMPOSITION OR PROPERTY Operating Parameters (Continued)		SPECIFICATION IDENTIFICATIONS		
	ISA			Document no		
				Latest revision		Date
				Issue status		

PROCESS COMPOSITION OR PROPERTY			MEASUREMENT DESIGN CONDITIONS				
Component/Property Name	Normal	Units	Minimum	Units	Maximum	Units	Repeatability

Rev	Date	Revision Description	By	Appv1	Appv2	Appv3	REMARKS

Form: 20A1002 Rev 0 © 2004 ISA

Definitions

Beer–Lambert law relates the absorption of light to the transmission properties of the material through which the light is traveling. In other words, it relates the amount of light absorbed to the concentration of the material of interest in the sample.

Diprotic acid is an acid such as H_2SO_4 that contains within its molecular structure two hydrogen atoms per molecule capable of dissociating (i.e., is ionizable) in water. The first dissociation will, in the case of sulfuric acid, occur completely, but the second one will not. Diprotic acids are of particular note in regards to titration experiments, where a pH versus titrant volume curve will clearly show two equivalence points for the acid. This occurs because the two ionization capable hydrogen atoms on the acid molecule do not leave the acid at the same time.

Equivalence point is where the perfect stoichiometric balance exists between an unknown substance and the reagent. The error in titration—the difference between the equivalence point and the actual end point—is called the indicator blank.

FIA is an approach to chemical analysis that is accomplished by injecting a plug of sample into a flowing carrier stream. The principle is similar to that of segmented flow analysis (SFA) but no air is injected into the sample or reagent streams.

Zeta potential (ζ) is the electrokinetic potential in colloidal dispersions. It is a key indicator of the stability of colloidal dispersions, in which particles of a substance are evenly diapersed in a continuous phase of another substance. The magnitude of the zeta potential indicates the degree of electrostatic repulsion between adjacent, similarly charged particles in the dispersion. If a colloidal suspension is stable, its zeta potential is high, i.e., the solution or dispersion will resist aggregation. When the potential is small, attractive forces may exceed this repulsion and the dispersion may break and flocculate. So, colloids with high zeta potential (negative or positive) are electrically stabilized while colloids with low zeta potentials tend to coagulate.

Abbreviations

FIA   Flow injection analysis
IPA   Isopropyl alcohol
ORP   Oxidation reduction potential
SFA   Segmented flow analysis

Organizations

CPAC   Center for Process Analytical Chemistry
IFPAC   International Forum for Process Analytical Chemistry

Bibliography

Analyticon, Automatic Titration (2013) http://www.analyticon.com/products/laboratory/AUT-701-Automatic-Titrators.php.

ASTM D1126-12, Standard Test Method for Hardness in Water, ASTM International, (2012) http://www.astm.org/Standards/D1126.htm.

Beran, J.O., *Laboratory Manual for Principles of General Chemistry* (2010) http://www.alibris.com/Laboratory-Manual-for-Principles-of-General-Chemistry-J-A-Beran/book/3707125.

Harris, D.C., *Quantitative Chemical Analysis* (2010) https://books.google.com/books/about/Quantitative_Chemical_Analysis.html?id=tHR4nsLDFAwC.

Jeffrey, G.H. et al., *Vogel's Textbook of Quantitative Chemical Analysis*, 5th edn. (1989) http://www.amazon.com/Vogels-Textbook-Quantitative-Chemical-Analysis/dp/0582446937.

Kalafatis, A.D. et al., Identification of time-varying pH processes using sinusoidal signals (2005) http://www.sciencedirect.com/science/article/pii/S0005109804003139.

Lipps, W., Choosing a Wet Chemistry Analyzer (2007) http://ezinearticles.com/?Choosing-a-Wet-Chemistry-Analyzer&id=757772.

Meyers, R.A., Encyclopedia of Analytical *Chemistry: Applications, Theory, and Instrumentation* (2000) http://www.abebooks.com/book-search/isbn/9781119991205/.

Ruzicka, J., Flow Injection Analysis (2013) http://www.flowinjectiontutorial.com/index.html.

Scott, W.E., *Principles of Wet End Chemistry* (1996) http://www.tappi.org/content/pdf/member_groups/Paper/0101R241.pdf.

Skoog, D.A., *Principles of Instrumental Analysis* (2007) https://books.google.com/books?id=NCl6PwAACAAJ&source=gbs_book_other_versions.

Appendix

A.1 DEFINITIONS 1103

A.2 ABBREVIATIONS, ACRONYMS, AND SYMBOLS 1150

Symbols 1169

A.3 ORGANIZATIONS 1170

A.4 FLOWSHEET AND FUNCTIONAL DIAGRAMS SYMBOLS* 1173

Introduction 1173
Documentation 1173
   Flow Sheets 1174
   Instrument Symbols 1174
   Code Letters 1174
   Instrument Index 1174
Lettering and Symbols 1175
   Measurement Symbols 1178
   Signal and Connection Symbols 1184
   Final Control Element Symbols 1186
   Function and Calculation Symbols 1193
   Logic Function Symbols 1200
   Electrical Schematics Symbols 1209

Loop Diagram 1211
   Terminal, Power Supply and Action Symbols 1212
   Graphic Symbols 1213
   Examples of Graphic Symbols 1213
   Graphic Symbol Applications 1213
Device and Function Symbols 1213
   Examples 1214
   Process Variable Measurements 1214
   Instrument Signal Connections 1216
   Alarm Indicators 1217
   Functional Diagrams 1217
   Multipoint and Multifunction Instruments 1218
   Process Control Description 1218
Conclusions 1219
Abbreviations 1219

A.5 CONVERSION AMONG ENGINEERING UNITS 1220

Introduction 1220
History 1220
Modern and International Units 1220
Tables in This Chapter 1221
   Engineering Conversion Factors 1223
   Area Unit Conversion Factors 1238
   Density Conversion Factors 1239
   Energy and Power Conversion Factors 1239
   Flow Unit Conversion Factors 1241
   Length Unit Conversion Factors 1242

Power Unit Conversion Factors 1242
Pressure Unit Conversion Factors 1243
Specific Enthalpy, Entropy, Volume Conversion 1247
Temperature Conversion 1248
Time Unit Conversion 1248
Velocity Unit Conversion 1248
Viscosity Unit Conversion 1250
Volume Unit Conversion 1253
Weight Unit Conversion 1255

A.6
CHEMICAL RESISTANCE OF MATERIALS 1257

A.7
COMPOSITION AND PROPERTIES OF METALLIC AND OTHER MATERIALS 1280

Introduction 1280

A.8
STEAM AND WATER TABLES 1287

Definitions 1287

A.1 Definitions

This chapter provides a complete list of all definitions that are needed to understand the terms used in this volume. In addition, Chapter 1.9 provides additional definitions of terms used in the field of analog process control and in instrumentation testing.

1001 Architecture: refers to a control loop (SIF) configuration where the loop component has no backup and therefore when that component fails, the loop fails.

1002 Architecture: refers to a control loop (SIF) configuration where the loop component does have one backup and therefore when one of the two components fail, the loop keeps functioning.

2003 Architecture: refers to a control loop (SIF) configuration where the loop component has two backups and therefore the loop keeps functioning if one or two of the three instruments fail.

A

Absolute accuracy: The closeness of a measurement to a traceable standard, such as the National Institute of Standards and Technology (NIST) or National Bureau of Standards.

Absolute (dynamic) viscosity (η, eta): Constant of proportionality between the applied stress and the resulting shear velocity (Newton's hypothesis) under simple steady shear.

Absorbance (A): Ratio of the radiant energy absorbed by a body to the corresponding absorption of a blackbody at the same temperature. Absorbance equals emittance on bodies whose temperature is not changing. ($A = 1 - R - T$, where R is the reflectance and T is the transmittance.)

Absorption: The taking in of a fluid to fill the cavities in a solid.

Accumulation: Accumulation is the pressure increase over the maximum allowable working pressure of a tank or vessel during discharge through the pressure relief valve (PRV). It is given either in percentage of the maximum allowable working pressure or in pressure units such as or in bars or in pounds per square inch.

Accuracy (International Society of Automation, ISA): In process instrumentation, the degree of conformity of an indicated value to a recognized accepted standard value, or ideal value.

Accuracy, measured (ISA): The maximum positive and negative deviations observed in testing a device under specified conditions and by a specified procedure.

Accuracy of measurement (NIST): Closeness of the agreement between the result of a measurement and the value of the measurand. Because accuracy is a qualitative concept, one should not use it quantitatively or associate numbers with it. (NIST also advises that neither precision nor inaccuracy should be used in place of accuracy.)

Accuracy (Webster): Freedom from error or the absence of error. Synonyms: precision, correctness, exactness. (In this sense, the term is a qualitative, not quantitative, concept.)

Acoustic couplant: An acoustically conductive gel intended to link or acoustically couple two solid materials ensuring minimum attenuation of sound waves traveling through each material.

Acoustic emission: Acoustic emission is the sound waves produced when a material is internally changed as it undergoes stress caused by an external force. This occurrence is the result of a small surface displacement of a material produced due to stress waves generated when the energy in a material or on its surface is released rapidly. The stress waves generated can be used for inspection, quality control, and process monitoring.

Active ultrasonics: When an ultrasonic signal is sent into the process and the flow or other process variable is measured by analyzing the signal that returns.

Activity (a): The chemical activity for the relevant species, where a_{Red} is the reductant and a_{Ox} is the oxidant. $a_X = \gamma_X c_X$, where γ_X is the activity coefficient of species X. (Since activity coefficients tend to unity at low concentrations, activities in the Nernst equation are frequently replaced by simple concentrations.)

Actual reading error (%AR) (ISA): In process instrumentation, the algebraic difference between indication and the ideal value of the measured signal. It is the quantity that, algebraically subtracted from the indication, gives the ideal value.

Admittance (A): The reciprocal of the impedance of a circuit. Admittance of an AC circuit is analogous to the conductance of a DC circuit. Expressed in units of siemens (S).

Adsorption: The adhesion of a fluid in extremely thin layers to the surfaces of a solid.

Air elutriation: A process for separating particles based on their size, shape, and density, using a stream of gas or liquid flowing in a direction usually opposite to the direction of sedimentation. This method is predominately used for particles with size greater than 1 µm. The smaller or lighter particles rise to the top (overflow) because their terminal sedimentation velocities are lower than the velocity of the rising fluid.

Albuminoid nitrogen: Is nitrogen that is still in the ammonium salts or still within the organic proteinoid molecules that are present in water.

Alpha Curve (or zero-power temperature coefficient of resistance): In case of resistance bulbs, it is the relationship between the resistance change of a resistance temperature detector (RTD) vs. temperature historically taken over the temperature range of 0°C–100°C (32°F–212°F). For platinum RTDs, the alpha parameter value is 0.00385 unit resistance change per degree C. The difference between this straight-line approximation and the exact curve is about 0.4°C, with the maximum error occurring at roughly 50°C. If this error is acceptable, there is no point in doing any additional calibration.

Alpha (α) radiation: Consists of fast-moving helium atoms, positively charged particles having two neutrons and two protons. They have high energy, typically in the MeV range, but due to their large mass, they are stopped by just a few inches of air or a piece of paper.

Ambient pressure: Standard ambient pressure is defined by NIST as the absolute pressure of 101.325 kPa (14.696 psi, 1 atm). An unofficial but commonly used standard for absolute pressure is 100 kPa (14.504 psi, 0.986 atm).

Ambient temperature: Standard temperature is defined by NIST as 20°C (293.15°K, 68°F). An unofficial but commonly used definition of standard ambient temperature is 298.15°K (25°C, 77°F).

Ampacity: The current (amperes), a conducting system can support without exceeding the temperature rating assigned to its configuration and application.

Ampere (A): Is the SI unit of the flow of electric current, named after André-Marie Ampére (1775–1836). One ampere is equivalent to the flow of one Culomb (C) of electric charge—which is the electric charge of 6.241×10^{18} protons—per sec.

Amperometric titration: Titration in which the end point is determined by measuring the current (amperage) that passes through the solution at a constant voltage.

Amperometry: The process of performing an amperometric titration. The current flow is monitored as a function of time between working and auxiliary electrodes while the voltage difference between them is held constant; in other designs, the current is monitored as a function of the amount of reagent added to bring about titration of an analyte to the stoichiometrically defined end point. Also called constant potential voltammetry.

Amplifier: A device that enables an input signal to control power from a source independent of the signal and thus is capable of delivering an output that bears some relationship to and is generally greater than the input signal.

Anion: A negatively charged ion such as chloride, carbonate, or sulfate.

Apparent viscosity: The value of viscosity evaluated at some nominal average value of the shear rate. The apparent viscosity applies, for instance, in the capillary method where a range of shear rates are employed.

Aron method: Refers to a wattmeter design that contains two pendulum clocks with coils around their pendulum bobs. One is accelerated and the other is slowed in proportion to the current flow. It was patented by Hermann Aron in 1883.

Atomic absorption spectroscopy (AAS): Is a procedure for the quantitative determination of chemical elements using the absorption of optical radiation by free atoms in the gaseous state. It was first used in the nineteenth century by Robert Wilhelm Bunsen in Germany. The atomizers most commonly used nowadays are flames and electrothermal (graphite tube) atomizers. The atoms are irradiated by an element-specific radiation source, or a monochromator is used to separate the element-specific radiation, which is finally measured by a detector.

Attenuation backplane: Loss of communication signal strength. Physical connection between individual components and the data and power distribution buses inside a chassis.

Autoformer: Has only one coil, which performs the functions of both the primary and the secondary windings of a conventional transformer.

Automatic/manual station: A device that enables an operator to select an automatic signal or a manual signal as the input to a control valve or other controlling element. The automatic signal is normally the output of a controller, while the manual signal is the output of a manually operated device.

Availability: The probability that a device is successful at a time t when needed (c.f. safety availability = $1 - PFD_{avg}$).

β Ratio: The diameter ratio of the orifice bore diameter to the pipe diameter.

B

Backlash: In process instrumentation, a relative movement between interacting mechanical parts, resulting from looseness when motion is reversed.

Backpressure: The pressure on the discharge side of a PRV. This pressure is the sum of the superimposed and the built-up backpressures. The superimposed backpressure is the pressure that exists in the discharge piping of the relief valve when the valve is closed.

Balanced safety relief valve: A safety relief valve with the bonnet vented to atmosphere. The effect of backpressure on the performance characteristics of the valve (set pressure, blowdown, and capacity) is much less than on the conventional valve. The balanced safety relief valve is made in three designs: (1) with a balancing piston, (2) with a balancing bellows, and (3) with a balancing bellows and an auxiliary balancing piston.

Balling degrees: Is a standard for the measurement of the density of sugar solutions expressed as a percentage of the weight of the solute on (grams of sucrose per 100 g of solution). Degrees Balling (°B) is sometimes used as a substitute for degrees Plato (°P). However, the two are not equivalent since the Balling scale was calibrated at 17.5°C and not at 20°C as is the Plato scale.

Balun (balanced/unbalanced): A device used for matching characteristics between a balanced and an unbalanced medium.

Band-pass filter: An optical or detector filter that permits the passage of a narrow band of the total spectrum. It excludes or is opaque to all other wavelengths.

Bandwidth: Data-carrying capacity; the range of frequencies available for signals. The term is also used to describe the rated throughput capacity of a given network medium or protocol.

Bar: A non-SI unit of pressure, defined as exactly equal to 100,000 Pa and is accepted for use with the SI.

Barkometer degrees: Unit of specific gravity used in the tanning industry.

Barrier cable glands and poured seals: The means to "seal" air gaps within cabling to prevent a pressure pulse traveling outside an enclosure (Figure A.1a).

Baseband: A communication technique whereby only one carrier frequency is used to send one signal at a time. Ethernet is an example of a baseband network; also called *narrowband*, contrast with *broadband*.

Batching pig: A utility pig that forms a moving seal in a pipeline to separate liquid from gas media, or to separate two different products being transported in the pipeline.

Form of equipment protection/cable type	Nonarmoured cables	Armoured cables
Increased safety-Ex e (IEC 60079-7)		
Flameproof-Ex d (IEC 60079-1)		
Flameproof compound barrier -Ex d (IEC 60079-1)		

FIG. A.1a
Cable glands for hazardous areas. (Courtesy of CMP Products, http://www.cmp-products.com/cable-glands-for-hazardous-areas.)

The most common configurations of batching pigs are cup pigs and sphere pigs.

Baud rate: The speed of signaling or the rate of communication between two instruments connected by an RS-232 interface. One baud (or bit) is one pulse or one code element per second. The more common rates range from 300 to 9600.

Baumé degrees: Were introduced by the French pharmacist Antoine Baumé in 1768 as liquid density units. The Baumé degrees are variously noted as degrees Baumé, B°, Bé°, or simply Baumé. On this scale, the density of distilled water is zero, and there are separate scales for liquids heavier and for liquids lighter than water.

Becquerel (Bq): Unit used to quantify the activity of any radioactive material in disintegrations per second. The rate of activity of 1 g of radium (Ra226) is 3.7×10^{10} Bq.

Beer–Lambert law: Relates the light to the transmission properties of the material through which the light is traveling. In other words, it relates the amount of light absorbed to the concentration of the material of interest in a sample.

Bell hole: An excavation in a local area to permit a survey, inspection, maintenance, repair, or replacement of pipe section.

Bell nipple: A section of a large-diameter pipe fitted to the top of the blowout preventers that the flow line attaches to via a side outlet, to allow the drilling mud to flow back to the mud tanks.

Beta (β): *Radiation* consists of electrons.

Beta (β) radiation: Consists of beta particles from the nucleus, the term "beta particle" being an historical term used in the early description of radioactivity. These high-energy electrons have a greater range of penetration than alpha particles but much less than gamma particles. Beta radiation hazard is greatest if ingested.

Beta ratio (β): For circular restriction, beta (β) is the ratio of the diameter of the restriction divided by the inside diameter of the pipe. The ratio of the velocity in the pipe and the velocity in the restriction is beta square (β^2). The beta (β) of noncircular cross sections is the square root of the ratio of the area of the restriction divided by the area of the noncircular conduit.

Bifilar winding: Is a coil that contains two closely spaced, parallel windings. Wire can be purchased in bifilar form, usually as different colored enameled wire bonded together.

Bingham plastic: A viscoplastic material that behaves as a rigid body at low stresses but flows as a viscous fluid at high stress.

Biochemical oxygen demand (BOD): One of the most common measures of pollutant organic material in water. BOD indicates the amount of organic matter in the water that can be decomposed by microorganisms, which upon decomposition uses up the oxygen in the water. Therefore, a low BOD is an indicator of good quality water, while a high BOD indicates polluted water. Dissolved oxygen (DO) is the actual amount of oxygen available to fish and other aquatic life in the water. When the DO drops, aquatic life forms are negatively affected.

Biot number (modulus): A dimensionless quantity used in heat transfer calculations. Named after the French physicist Jean-Baptiste Biot (1774–1862), it gives a simple index of the ratio of the heat transfer resistances *inside of* and *at the surface of* a body.

Blackbody: The perfect absorber of all radiant energy that strikes it. The blackbody is also a perfect emitter. Therefore, both its absorbance (A) and emissivity (E) are unity. The blackbody radiates energy in predictable spectral distributions and intensities, which are a function of the blackbody's absolute temperature (Figure 4.11a).

Black-body radiation: Is the electromagnetic radiation emitted by a black body, which is held at constant, uniform temperature. A black body is a theoretical object that absorbs 100% of the radiation that hits it and emits all of it. The black body reflects no radiation and appears perfectly black. In practice, no material has been found to absorb all incoming radiation; even graphite absorbs only about 97% of the radiation received. The theoretical black body is also a perfect emitter of radiation. All objects emit radiation when their temperature is above absolute zero. This radiation has a specific spectrum and intensity that depends only on the temperature of the body.

Blowdown (blowback): The difference between the set pressure and the reseating (closing) pressure of a PRV, expressed in percent of the set pressure or in bars or in pounds per square inch.

Blowout preventers: Devices installed at the wellhead to prevent fluids and gases from unintentionally escaping from the wellbore.

Bode diagram: In process instrumentation, a plot of log gain (magnitude ratio) and phase angle values on a log frequency base for a transfer function. See Figure A.1b.

Bolometer: Thermal detector that changes its electrical resistance as a function of the radiant energy striking it.

Bonding: The practice of creating safe, high-capacity, reliable electrical connectivity between associated metallic parts, machines, and other conductive equipment.

Boroscope: An optical device consisting of a rigid or flexible tube with an eyepiece on one end and an objective lens on the other, linked together by an optical system in between. The optical system is usually surrounded by optical fibers used for illumination of the remote object. An internal image of the illuminated object is formed by the objective lens and magnified by the eyepiece, which presents the image to the viewer's eye.

Boyle's Law: States that the absolute pressure of a given mass of ideal gas is inversely proportional to its volume if its temperature is constant. Mathematically, Boyle's law can be stated as $P_1V_1 = P_2V_2$, because the product of the pressure and volume of a given mass of ideal gas is constant as long as its temperature is constant. Boyle's law was experimentally proved by Robert Boyle and his colleagues.

Bragg gratings: A fiber Bragg grating (FBG) is a type of distributed Bragg reflector constructed in a short segment of optical fiber that reflects particular wavelengths of light and transmits all others. This is achieved by creating a periodic variation in the refractive index of the fiber core, which generates a wavelength-specific dielectric mirror. An FBG can, therefore, be used as an in-line optical filter to block certain wavelengths or as a wavelength-specific reflector.

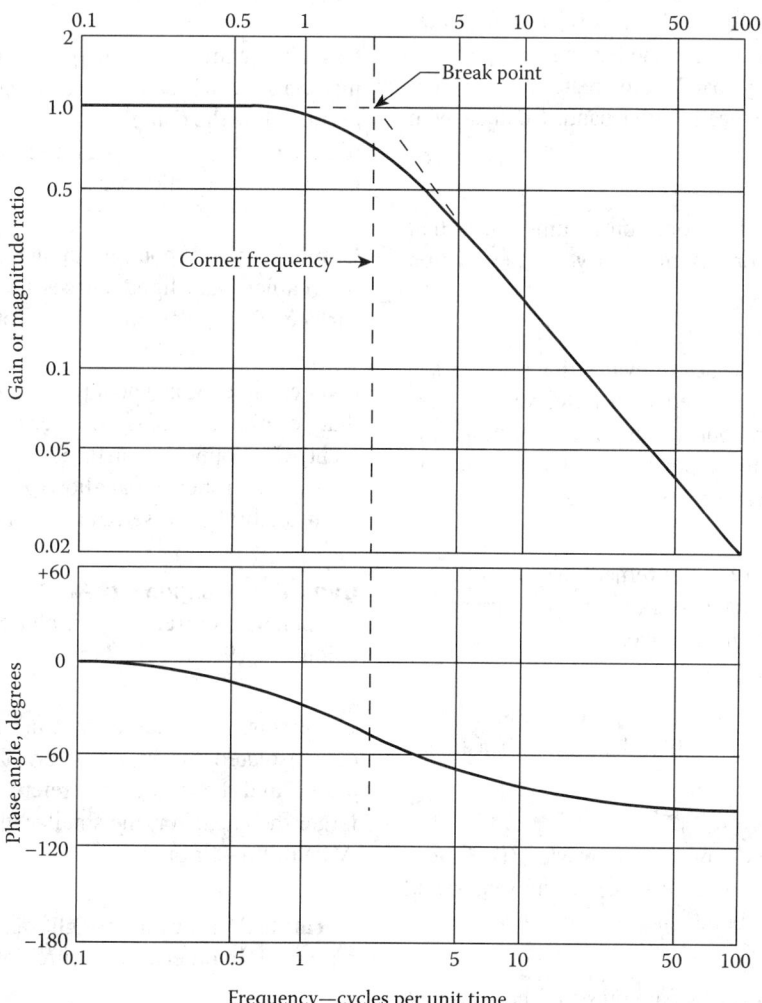

FIG. A.1b
Illustration of a typical Bode diagram.

Brightness pyrometer: Uses the radiant energy on each side of a fixed wavelength of the spectrum. This band is quite narrow and usually centered at 0.65 μm in the orange–red area of the visible spectrum.

British Thermal Unit (BTU): The amount of heat required to raise the temperature of 1 lb of water by 1°F at or near 60°F.

Brix degree (°Bx): Or percentage (%Bx) gives the sugar content of aqueous solutions. It is the weight-percent of sugar concentration, corresponding to the number of grams of sugar contained in 100 g of solution. If the solution contains dissolved solids other than pure sucrose, then the °Bx only approximates the dissolved solid content.

Brix scale: Calibrated to show the weight percent of sugar, corresponding to the number of grams of sugar contained in 100 g of solution.

Broadband: A communication technique that multiplexes multiple independent signals simultaneously, using several distinct carriers. A common term in the telecomm industry to describe any channel having a bandwidth greater than a voice-grade channel (4 kHz). Also called wideband. Contrast with baseband.

BTU "dry": "Dry basis". The common assumption is that pipeline gas contains 7 lb (or less) of water vapor per million standard cubic feet.

BTU "saturated": This is the heating value that is expressed on the basis of gas being saturated with water vapors. This state is defined as the condition when the gas contains the maximum amount of water vapors without condensation, when it is at base pressure and 60°F.

Bubbler: A level or density measurement system utilizing flow of a gas into pipe or tubing to transfer a pressure at one place to another more convenient place.

Buckle: A partial collapse of the pipe due to excessive bending associated with soil instability, landslides, washouts, frost heaves, earthquakes, etc.

Building automation and control networks (BACNet): A data communication protocol for transmitting sensor and control information within buildings.

Built-up backpressure: Variable backpressure that develops as a result of flow through the PRV after it opens. This is an increase in pressure in the relief valve's outlet line caused by the pressure drop through the discharge headers.

Buoyancy force: The force on an object caused by the surrounding fluid. Used by Archimedes to determine the exact volume of a valuable object.

Burning: A form of fire where the flame does not spread or diffuse but remains at an interface where fuel and oxidant are supplied in proper proportions.

Bypass: Only that part of the liquid is taken and passed through the viscometer—for instance in a bypass loop.

C

Calibration digs: Exploratory excavations, or bell holes, of portions of the pipeline in which an in-line inspection tool has recorded an indication.

Calibration error: The difference between the size of a dead weight corresponding to near the full capacity of the scale and the actual reading on the digital or analog readout.

Calorific value: The energy contained in a fuel or other materials, which can be determined by measuring the heat produced by the complete combustion of a specified quantity of it. This is usually expressed in joules per kilogram or, for gases, as mega joules per cubic meter.

Camera pig: A configuration pig that carries a video or film camera and light sources for photographing the inside surface of a pipe on an intermittent or continuous basis.

Capacitance (C): The property of an electrical insulator that permits the storage of energy through electric displacement when opposite surfaces are held at different voltages. This is the property that allows AC to appear to flow through an insulator. Expressed in units of farads.

Capacitive component: An AC can be separated into two components; current with a phase relation to the excitation voltage of +90° is called the *capacitive component*.

Capacitor: A device consisting of two conductors electrically isolated by an insulator. The conductors are called *plates*, and the insulator is referred to as the *dielectric*. The larger the capacitor, the smaller its impedance and the more AC will flow through it.

Carat: Is 24 times the weight ratio of the jewelry piece and pure gold. In other words, pure gold is 24 carat.

Carcinogen (CA): Any substance that the National Institute for Occupational Safety and Health (NIOSH) considers to be a potential occupational carcinogen.

Casing head: A large metal flange welded or screwed onto the top of the conductor pipe (also known as drive pipe) or the casing and is used to bolt the surface equipment such as the blowout preventers (for well drilling).

Cation: A positively charged ion such as sodium, potassium, or hydrogen.

Cation conductivity: The conductivity of a water sample that has been passed through a column containing a cation-exchange resin that replaces cations with hydrogen ions. It is used in steam plant water analysis.

Ceiling limit (TLV-C): A concentration that should not be exceeded even instantaneously.

Ceilometer: Is light or laser source instrument used to determine the height of the base of the clouds or the transmissivity (opacity) of the air, to measure the concentration of volcanic ash or aerosols within the atmosphere.

Celsius: Is a thermodynamic scale introduced by Andrew Celsius (1701–1744. Until 1954, 0°C was defined as the freezing, and 100°C was defined as the boiling point of water. In 1954, the SI units were introduced resulting in a slight change, because both the degree Celsius and the Kelvin units were redefined as precisely 1 part in 273.16 (approximately 0.00366) of the difference between absolute zero and the triple point of water. Additionally, the null point was redefined as precisely 273.15°C (−273.15°C = 0 K and 0°C = 273.15 K).

Charge-Coupled Device (CCD): Serves digital imaging by converting photons into electron charges. They are widely used where high-quality image data are needed.

Characteristic impedance: The impedance that, when connected to the output terminals of a transmission line, appears to be infinitely long, for there are no standing waves on the line, and the ratio of voltage to current is the same for each point of the line (nominal impedance of waveguide).

Chatter: Rapid, abnormal reciprocating variations in lift during which the disc contacts the seat.

Chemiluminescence: Chemiluminescence is the emission of light resulting from a chemical reaction.

Chemometrics: Refers to using linear algebra calculation methods to make either quantitative or qualitative measurements of chemical data. The key to understanding chemometrics is not necessarily to understand the mathematics of all of the different methods, but to know which model to use for a given analytical problem and properly applying it.

Chimney effect: Chimney effect is the drawing in of the cold outside air at the ground level of tall buildings, because the weight of the column of cold air outside the building is higher than the weight of the warm air column inside the building.

Chlorinated polyvinyl chloride: A low-cost, reasonably inert polymer used for some noninsertion sensors. It is easily solvent welded. Its maximum temperature range is about 225°F.

Chronopotentiometry: Process in which the potential difference between a metallic measuring electrode and a reference electrode is monitored as a function of time. At the measuring electrode, an oxidation or reduction of a solution species takes place.

Cladding: Thin-walled metal tube that forms the outer jacket of a nuclear fuel rod.

Cleaning pig: A utility pig that uses cups, scrapers, or brushes to remove dirt, rust, mill scale, or other foreign matter from the pipeline. Cleaning pigs are run to increase the operating efficiency of a pipeline or to facilitate inspection of the pipeline.

Closing pressure (reseat pressure): The pressure (measured at the valve inlet) at which the valve closes, flow is substantially shut off, and there is no measurable lift.

Coagulation: The process of forming semisolid lumps in a liquid. Coagulant is the material that causes the transformation of some of the materials in the liquid to turn into a soft, semisolid, or solid mass.

Coax: Jargon meaning *coaxial cable*, which consists of a center wire surrounded by low-K insulation, surrounded by a second shield conductor. It has the characteristics of low capacitance and inductance for transmission of high-frequency current.

Cold differential test pressure (CDTP): The pressure at which the PRV is adjusted to open during testing. The CDTP setting includes the corrections required to consider the expected service temperature and backpressure.

Colloids: Very small, finely divided solids (particles that do not dissolve) that remain dispersed in a liquid for a long time due to their small size and electrical charge. One property of colloid systems that distinguishes them from true solutions is that colloidal particles scatter light.

The particles of a colloid selectively absorb ions and acquire an electric charge. All particles of a given colloid take on the same charge (either positive or negative) and thus are repelled by one another.

Combined error: The error that also reflects nonlinearity and hysteresis. It is defined as the maximum deviation from a straight line, which is drawn between the no-load point and the full rated output. It is expressed as % FS and is measured when the load is increasing and again when the load is dropping.

Combustible flyings and fibers: Combustible materials that include cotton and wood shavings.

Combustible material: Any solid, dust, flyings, fibers, etc., that can ignite under certain conditions.

Combustion air requirement index (CARI): A dimensionless number that indicates the amount of air required (stoichiometrically) to support the combustion of a fuel gas. Mathematically, CARI is defined by Equation A.1(1)

$$\text{CARI} = \frac{\text{air} - \text{fuel ratio}}{\sqrt{(\text{Sp.G})}} \qquad \text{A.1(1)}$$

Common cause failure: A failure of two or more systems due to a single cause.

Common mode: When two or more systems fail in the same way due to a common cause.

Complementary metal oxide semiconductor (CMOS—pronounced "see-moss"): A low-power semiconductor chip using both negative and positive polarity circuits.

Compressibility factor (Z): The ratio of the molar volume of a gas to the molar volume of an ideal gas at the same temperature and pressure. It is a useful thermodynamic property for modifying the ideal gas law to account for the real gas behavior. In general, deviation from ideal behavior becomes more significant; the closer a gas to a phase change, the lower the temperature or the higher the pressure.

Computer worm: A worm is a self-replicating virus that does not alter files but resides in active memory and duplicates itself there. Worms use parts of an operating system that are automatic and usually invisible to the user. It is common for worms to be noticed only when their replication has already consumed system resources, slowing or halting the operation of the system. They differ from viruses as viruses need a host file, make their presence known by presenting messages, and take up computer memory leading to system crashes. Worms, on the other hand, exist inside other files, often in Word or Excel documents, and cause the entire document to travel from computer to computer as a worm.

Condensate pot, condensate chamber: A small vessel in steam metering service installed in the pressure sensing lines, which maintains a constant level of condensate with steam vapor in the upper part and condensate in the lower. See text for discussion.

Conductance (G): Is the degree to which an object conducts electricity, calculated as the ratio of the current that flows to the potential difference present. It is the reciprocal of resistance, expressed in units of Siemens (formerly called *mhos*).

Conductivity (g): The reciprocal of resistivity. All solids and liquids have some degree of conductivity. For the purpose of this section, any material above 1 µS/cm will be considered conductive (for example, most metals, water with any ions present, coke, carbon black, soaking-wet insulators, and aqueous external emulsions). Expressed in units of siemens/cm.

Cone of acceptance: Is also called aperture and is the maximum angle of incident light that can actually make to the fiber core without being bounced off or diffracted (bent) out of the cladding.

Configuration pig: An instrumented pig that collects data relating to the inner contour of a pipe wall or of the pipeline. Geometry pigs, camera pigs, and mapping pigs are types of configuration pigs.

Consistency: Is the term used to describe the solid content of a solid–liquid mixture. It is calculated as the ratio of the dry weight of the solids, divided by the total weight of the mixture. Mixtures having under 5% consistency are considered to be low, between 5% and 15%, medium, and over 15% as high consistency mixtures. Knowing the specific gravity of the solids and of the liquid phase, consistency can be determined (or approximated) on the basis of density measurement.

Constant backpressure: Backpressure that does not change under any condition of operation whether or not the PRV is closed or open.

Constant failure rate: The most common way to model industrial failures means that mean time to failure can be modeled as the inverse of failure rate.

Consultative Committee for Space Data Systems: Telemetry and telecom software library that provides reference implementation of the international protocol standard for transmission and reception of data in radio frequency communications with spacecraft.

Controller Area Network (CAN, CANbus): A multimaster broadcast serial-message-based digital communication format having bitwise arbitration to identify communication priority, originally used for vehicle data communication but now the CANopen version is used for industrial automation, with specifications controlled by CANcia (www.cancia.org).

Conventional safety relief valve: A safety relief valve with the bonnet vented either to atmosphere or internally to the discharge side of the valve. The performance characteristics (set pressure, blowdown, and capacity) are directly affected by changes of the backpressure on the valve.

Cooperative target: A target that is designed to reflect light to the detector of a sensor. Targets include glass corner cube retroreflectors and retroreflective tape made by several manufacturers. In some applications, mirrors may also be used as cooperative targets.

Coriolis force: Is an effect whereby a mass moving in a rotating system experiences a force (the Coriolis force) acting perpendicular to the direction of motion and to the axis of rotation. On the earth, the effect tends to deflect moving objects to the right in the northern hemisphere and to the left in the southern and is important in the formation of cyclonic weather systems.

Corrosion, external: Metal loss due to electrochemical, galvanic, microbiological, or other attack on the pipe due to environmental conditions surrounding the pipe.

Corrosion, internal: Metal loss due to chemical or other attack on the steel from liquids on the inside of the pipe.

Corrosion, pit: Local concentrated cell corrosion on the external or internal surfaces that results from the generation of a potential (voltage) difference set up by variations in oxygen concentrations within or outside the pit. The oxygen-starved pit acts as the anode, and the pipe surface acts as the cathode.

Corrosion, selective: A localized corrosion attack along the bond line of electric-resistance welding and flash welding, which leads to the development of a wedge-shaped groove that is often filled with corrosion products.

Corrosion, stress cracking: Stress-corrosion cracking is a progressive intergranular and/or transgranular cracking that results from a combination of applied tensile stress, cathodic protection currents, and a corrosive environment.

Coulomb (C): Was introduced by Charles-Augustin de Coulomb (June 14, 1736–August 23, 1806). It is the electric charge of one ampere-second. Conversely, an electric current of one ampere (A) represents the flow of one coulomb of electric charge past a specific point in one second. 6.24×10^{18} electrons have one coulomb electric charge. This is also true of 6.24×10^{18} positrons or 6.24×10^{18} protons, although these two types of particles carry a charge of opposite polarity to that of the electron.

Coulometry: Process of monitoring analyte concentration by detecting the total amount of electrical charge passed between two electrodes that are held at constant potential or when constant current flow passes between them.

Coulter principle: A "reference method" for particle size analysis. It is based on measuring the changes in electrical impedance produced by nonconductive particles suspended in an electrolyte. The sensing zone is a small opening (aperture) between electrodes through which suspended particles pass. In the sensing zone, each particle displaces its own volume of electrolyte and this displaced volume is measured as a voltage pulse; the height of each pulse being proportional to the volume of the particle.

Count: The smallest increment that the digital display registers.

Cracking, fatigue: Progressive cracking in the base material, weld, or weld zone, caused by pressure cycling or oscillatory stresses associated with the operation of the system.

Cracking, grid weld: Cracks in the weld or weld zone of the butt weld that connect sections of pipe.

Cracking, seam weld: Cracks in the weld or weld zone of the longitudinal seam weld of the pipe.

Critical angle: Is the minimum angle of incidence beyond which total internal reflection occurs, when light is traveling from a medium of higher to another medium of lower index of refraction.

Criticality: The relationship between the production and the consumption of neutrons. When a reactor's neutron population is steady, the fission chain reaction is self-sustaining and the reactor's condition is called "critical." When the power output is increased, the reactor's neutron production exceeds losses and this condition is called "supercritical"; however, when power production is reduced, losses dominate and the condition is called "subcritical."

Cryoelectronics: The field of research regarding superconductivity at low temperatures.

Cryogenics: The branches of physics and engineering that involve the study of very low temperatures, how to produce them, and how materials behave at those temperatures.

Crystalline membranes: Crystalline membranes have good selectivity because only ions that can introduce themselves into the crystal structure can interfere with the electrode response. Selectivity of crystalline membranes can be for both cation and anion of the membrane-forming substance.

Crystallography: How atoms are arranged in an object; the direct relationship between these arrangements and material properties (conductivity, electrical properties, strength, etc.).

CT scan: Computed tomography scan. Detailed images of internal organs are obtained by this type of sophisticated x-ray device.

Cup pig: A utility pig supported and driven by cups made of a resilient material such as neoprene or polyurethane. At least one of the cups forms a piston-like seal inside the pipe.

Curie (Ci): The unit used to quantify the activity of any radioactive material in disintegrations per second. One curie corresponds to the activity of 1 g of Ra226 ($Ci = 3.7 \times 10^{10}$ disintegrations per second), whether it is produced by Ra or some other material.

Curie temperature (T_c): Is named after Pierre Curie who showed that magnetism was lost at a critical temperature. It is the temperature where a material's "permanent" magnetism changes to "induced" magnetism. Higher temperatures make magnets weaker as spontaneous (permanent) magnetism only occurs below the Curie temperature. In a paramagnetic material the magnetization is directly proportional to an applied magnetic field and if heated, the magnetization is inversely proportional to temperature.

Current transformers (CT): Are used for the measurement of alternating current. When current in a circuit is too high to apply directly to measuring instruments, a current transformer produces a reduced current proportional to the current in the circuit. A current transformer isolates the measuring instruments from what may be very high voltage in the monitored circuit.

Cycling life: The specified minimum number of full-scale excursions or specified partial range excursions over which a device will operate as specified without changing its performance beyond specified tolerances.

D

Darcy–Weisbach equation: It relates the pressure drop due to friction in a pipe to the average velocity of the flow of an incompressible fluid. The equation is named after Henry Darcy and Julius Weisbach.

Damping: (1) (noun) The progressive reduction or suppression of oscillation in a device or system. (2) (adjective) Pertaining to or the product of damping. *Note 1*: The response to an abrupt stimulus is said to be "critically damped" when the time response is as fast as possible without overshoot; "under-damped" when overshoot occurs; or "overdamped" when response is slower than critical. *Note 2*: Viscous damping uses the viscosity of fluids (liquids or gases) to effect damping. *Note 3*: Magnetic damping uses the current induced in electrical conductors by changes in magnetic flux to effect damping.

Data servers: A standard interface to provide data exchange between field devices and data clients.

Datum line: The center of the differential pressure (d/p) transmitter measuring element is often called the datum line. All liquid-level measurements are referenced from this datum line, or point.

Dead band: The range through which an input can be varied without causing a change in the output.

Dead load: The static load (weight) that does not vary and is registered at zero applied load.

Decibel (dB): Decibel is commonly used in acoustics as a unit of sound pressure level. The reference pressure in air is set at the typical threshold of perception of an average human. Decibel expresses the ratio of two amounts of acoustic signal power equal to 10 times the common logarithm of this ratio on a scale from zero for the average least perceptible sound to about 130 for the average pain level.

$$L_p = 20 \log_{10}\left(\frac{p_{rms}}{p_{ref}}\right) dB$$

where p_{ref} is equal to the standard reference sound pressure level of 20 µPa in air or 1 µPa in water.

Deflagration or explosion: A form of fire where the flame front advances through a gaseous mixture at subsonic speeds.

Degasser: A device that separates air and/or gas from the drilling fluid. It is typically mounted on top of mud tanks.

Deionized: Refers to water of extremely high purity, with few ions to carry current. If exposed to air for any significant period, it will have a conductivity of about 5 µS/cm because of dissolved CO_2.

Deminerization of NO_2: Is the joining of two NO_2 molecules. This pairing of odd electrons means that the molecule is no longer paramagnetic, but diamagnetic in nature. That is the reason why NO_2 exists as a dimer, that is, N_2O_4.

Demultiplexing: Separation of multiple input streams that were muliplexed into a common physical signal back into multiple output streams.

Denitrification: Denitrification is a process that involves nitrogen getting back into the atmosphere. It can be used when there is too much nitrogen in the soil because of the use of excessive amounts of fertilizers. The process is performed primarily by heterotrophic bacteria.

Density: Mass per unit volume.

Dent: A local depression in the pipe surface caused by mechanical damage that produces a gross disturbance in the curvature of the pipe without reducing the pipe wall thickness.

Depth of field: The span of distances over which a rangefinder can accurately measure distance. This may be limited by the focus of the light collection optics as well as the maximum distance at which enough light is reflected to the sensor. These two factors will determine how the sensor's sensitivity changes with distance.

Derrick: The support structure for the equipment used to lower and raise the drill string into and out of the wellbore.

Desander/desilter: Contains a set of hydrocyclones that separate sand and silt from the drilling fluid. Typically mounted on top of mud tanks.

Desiccator: Usually glass sealable vessel containing dessicant.

Design pressure: This pressure is equal or less than the maximum allowable working pressure. It is used to define the upper limit of the normal operating pressure range.

Detector: A device that measures the amount of energy radiated by an object. It can be a thermal detector or a photodetector. Thermal detectors respond to radiation by changing their volume, capacitance, or generation of millivoltages; they can be thermocouples, thermopiles, pneumatic detectors, or bolometers. Their common feature is their relatively slow response. Photodetectors are semiconductors that produce a signal in proportion to the photon flux that strikes them.

Detonation: A form of fire where the advancement of the flame front occurs at supersonic speeds.

Device description: A clear, unambiguous, structured text description that allows full utilization/operation of a field device by a host/master without any prior knowledge of the field device.

Dew point: Saturation temperature of a gas–water vapor mixture (the temperature at which water condensation occurs as a gas is cooled).

Diagnostics: The means of detecting potential failures in a system by automatic tests.

Diamagnetic materials: Create an induced magnetic field in a direction opposite to an externally applied magnetic field and are repelled by the applied magnetic field.

Dielectric: A material which is an electrical insulator or in which an electric field can be sustained with a minimum of dissipation of power.

Dielectric compensation: A scheme by which changes in insulating liquid composition or temperature can be prevented from causing any output error. It requires a second sensor and homogeneous liquid. A dielectric is a material that is an electrical insulator or in which an electric field can be sustained with a minimum of dissipation of power.

Dielectric constant: A material characteristic expressed as the capacitance between two plates when the intervening space is filled with a given insulating material divided by the capacitance of the same plate arrangement when the space is filled with air or is evacuated. Therefore, it is the ratio of the permittivity of a substance to the permittivity of free space and as such, an expression of the extent to which a material concentrates electric flux and is the electrical equivalent of relative magnetic permeability.

Dielectric strength of insulation: The ability to withstand electrical breakdown upon application of higher voltages (e.g., instrument cable insulation will break down when subject to voltages found in power circuits).

Diffraction gratings: Are optical components used to separate light into its component wavelengths. When there is a need to separate light of different wavelengths with high resolution, then a diffraction grating is the most often used tool of choice. A diffraction grating is made by making many (up to 6,000 lines/cm) parallel scratches on the surface of a flat piece of transparent material. The scratches are opaque, but the areas between the scratches transmit light.

Diffuse reflection: Occurs when light strikes a target and is scattered over a wide angle. Plain white paper or flat (not glossy) wall paint are good diffuse materials. Diffuse targets are the best uncooperative targets and may be measured to over a wide range of incident angles (up to 80° for some materials).

Digit: The smallest increment of weight that the digital display resolves.

Diodes: Two-terminal electronic (usually semiconductor) devices that permit current flow predominantly in only one direction.

Dioxins: Is a general name for a large group of chemical compounds known as persistent organic pollutants (POPs). They are produced inadvertently by industrial processes in which traces of chlorine are present such as combustion processes like waste incineration, chemical manufacturing and paper bleaching, Dioxins and furans can be found in air, water, and contaminated soil. They are harmful to human health and biologically are nearly nondegradable (persistent). As a result, their emission into the environment and food chain is strictly controlled. As a consequence, low levels of contamination have to be detected, providing a challenge to sample preparation and detection systems.

Dipole moment: The measure of the electrical polarity of a system of charges.

Diprotic acid: An acid such as H_2SO_4 that contains within its molecular structure two hydrogen atoms per molecule capable of dissociating (i.e., ionizable) in water. The first dissociation will, in the case of sulfuric acid, occur completely, but the second one will not. Diprotic acids are of particular note in regards to titration experiments, where a pH versus titrant volume curve will clearly show two equivalence points for the acid. This occurs because the two hydrogen atoms on the acid molecule, which are capable of ionization, do not leave the acid at the same time.

Disbonding: Any loss of bond between the protective coating and steel pipe as a result of coating adhesion failure, chemical attack, mechanical damage, hydrogen concentrations, etc.

Discontinuity: An abrupt change in the shape (or impedance) of a waveguide (creating a reflection of energy).

Displacer: A small tank, typically cylindrical and often made of metal, which is suspended in a fluid for the purpose of measuring level, interface, or density and which does not move with level or density. The float moves with level and density.

Dither: A useful oscillation of small magnitude, introduced to overcome the effect of friction, hysteresis, or recorder pen clogging.

Doppler effect: Is the change in frequency of sound, light or the waves as the distance between the observer and the source changes. If the distance drops, the frequency rises and if it increases, the frequency drops.

Double block and bleed: A statement widely used for manual valve. It has a compact design that consists of two valves on the process line and two valves on the venting line.

Dribble flow: The last increment of charge in batch processes, which is added at a much reduced rate.

Drift: An undesired change in output over a period of time, which is not caused by a change in the input, environment, or load.

Drill bit: A device attached to the end of the drill string that breaks apart the rock being drilled. It contains jets through which the drilling fluid exits. A tricone bit comprises three conical rollers with teeth made of a hard material, such as tungsten carbide. The teeth break rock by crushing as the rollers move around the bottom of the borehole. A polycrystalline diamond compact bit has no moving parts and works by scraping the rock surface with disk-shaped teeth made of a hard material, such as tungsten carbide or diamond.

Drill line: Thick, stranded metal cable threaded through the two blocks (traveling and crown) to raise and lower the drill string.

Drill pipe: A joint of hollow tubing used to connect the surface equipment to the bottom hole assembly and acts as a conduit for the drilling fluid.

Drip leg: A section of pipe or tubing in gas service arranged to capture liquids for eventual draining.

Dry BTU: The heating value of a gas, expressed on a dry basis. The common assumption is that pipeline gas contains 7 lb (or less) of water vapor per million standard cubic feet.

Dry bulb: The temperature of air shielded from radiation and moisture. It is directly proportional to the mean kinetic energy of the air modules.

Dry well: See *primary containment*

Dust-ignition-proof: Enclosed in a manner to exclude ignitable amounts of dust or amounts that might affect performance. Enclosed so that arcs, sparks, and heat otherwise generated or liberated inside of an enclosure will not cause ignition of exterior accumulations or atmospheric suspensions of dust.

Dynamometers: Is a device used for the measurement of force, moment of force (torque), or power. For example, the power produced by an engine, motor, or other rotating prime mover can be calculated by simultaneously measuring torque and rotational speed (rpm). A dynamometer can also be used to determine the torque and power required to operate a driven machine such as a pump. In that case, a driving dynamometer is used. A dynamometer that is designed to be driven is called an absorption or passive dynamometer. A dynamometer that can either drive or absorb is called a universal or active dynamometer.

Dynamometers: Force-measuring instruments.

E

Eccentric loading test: Also called shift test or off-center loading test. It is tested by weighing in the center of the four quadrants of the weighing platform.

EEPROM: Stands for Electrically Erasable Programmable Read-Only Memory and is a type of nonvolatile memory used in computers and other electronic devices to store small amounts of data that must be saved when power is removed, for example, calibration tables or device configuration.

Effective coefficient of discharge: A coefficient used to calculate the minimum required discharge area of the PRV.

Elastic scattering of light: Is also called Rayleigh scattering and is a scattering mechanism that results in the uniform scattering of the incoming light by atoms or molecules in all directions.

Electrochemical process: The changes in voltage or current flow that occur between two electrodes in a solution (electrolyte) over time. The oxidation or reduction of the analyte provides data related to concentration.

Electrolysis: Is an electrochemical decomposition process that occurs when direct current passes through a conductive liquid (an electrolyte, in other words a liquid that contains ions). When it passes between the two electrodes of an electrolytic cell, anions (positive ions) will deposit on the negative electrode (cathode) and cations (negative ions) will deposit on the positive electrode (anode). Therefore, in the case of the electrolysis of water, hydrogen is formed at the cathode and oxygen at the anode.

Electrolyte: Any substance containing free ions that make the substance electrically conductive. Therefore, an electrolyte is a compound that ionizes when dissolved in water or any other ionizing solvents. Electrolytes commonly exist as solutions of acids, bases, or salts. Electrolyte solutions are normally formed when a salt is placed into a solvent such as water and the salt (a solid) dissolves into its components.

Electrolytic probe: A probe that is similar to a galvanic probe, except that a potential is applied across the electrodes, and the electrodes are not consumed. Dissolved oxygen detection is a primary application of this type of probe.

Electromagnet: Is a type of magnet in which the magnetic field is produced by an electric current. The magnetic field disappears when the current is turned off. Electromagnets usually consist of a large number of closely spaced turns of wire that create the magnetic field.

Electromagnetic wave (energy): A disturbance that propagates outward from any electric charge, which oscillates or is accelerated; far from the charge, it consists of vibrating electric and magnetic fields, which move at the speed of light and are at right angles to each other and to the direction of motion.

Electron microscope: Is a scientific instrument that uses a beam of highly energetic electrons to examine objects on a very fine scale.

Electronegativity: A relative indication of the ability of an atom to attract the electrons of a nearby molecule.

Electronvolt (eV): Is a unit of energy equal to approximately 160 zeptojoules (symbol zJ) or 1.6×10^{-19}. By definition, it is the amount of energy gained (or lost) by the charge of a single electron moved across an electric potential difference of one volt.

Electrophoresis: Is the motion of particles in a fluid under the influence of an electric field. This phenomenon was observed for the first time in 1807 by Ferdinand Frederic Reuss who noticed that the application of a constant electric field caused clay particles in water to migrate. This is the basis for a number of analytical techniques for separating molecules by size, charge, or binding affinity.

Elementary charge (or elementary positive charge) (e): Is the charge carried by a single proton, which is of the opposite sign of the electric charge carried by a single electron. This charge has a measured value of approximately $1.602176565(35) \times 10^{-19}$ coulombs. The magnitude of the elementary charge was first measured by Robert A. Millikan in 1909.

Emissivity or emittance (E): The ratio of the radiant energy emitted by an object to the radiant energy that a blackbody would emit at that same temperature. If the emittance is same at all wavelengths, the object is called a gray body. Some industrial materials change their emissivity with temperature and sometimes with other variables also. Emissivity always equals absorption, and it also equals 1 minus the sum of reflectance and transmittance ($E = A\ 1 - T - R$).

Encoders (rotary): Converts the angular position or motion of a shaft or axle to an analog or digital code. The encoder can be absolute or relative (incremental). The output of absolute encoders indicates the position of the shaft while the output of incremental encoders provide information about the motion (rotary speed) of the shaft. Encoder detectors can be mechanical, capacitive, optical, or magnetic.

Engler flask: Standard distilling flask, usually of 100 mL capacity, used to determine the volatility characteristics of petroleum products (such as gasoline, naphtha, or kerosene).

Enthalpy: A measure of the total energy of a thermodynamic system. It includes the internal energy, which is the energy required to create the system, and the amount of energy required to make room for it by displacing its environment and establishing its volume and pressure. Typical units are BTU/lb, kcal/kg.

Enthtalpy (h): Specific enthalpy is a thermodynamic quantity equivalent to the total energy content of the unit mass of a substance relative to a reference condition. It is equal to the internal energy (u) of the system plus the product of pressure (p) and specific volume (v), in other words, $h = u + pv$. In the English System, the usual unit of enthalpy is BTU/lbm while in the SI (metric) system, the usual unit is kilojoules per kilogram, where BTU = 1054.8 J = 1.055 kJ and lbm = 0.4536 kgm and, therefore, BTU/lbm = 2.324 kJ/kgm. In addition to joules and BTUs, other energy units such as ft-lbf and calories are also used. Enthalpy cannot be measured; only the difference between enthalpies can be determined if referred to a reference state at which it is given an arbitrary value, usually zero. In steam tables, the reference value ($h = 0$) is the enthalpy of saturated water at 32°F. For other substances, the reference values are different. For example, for Freon-12, it is the enthalpy of saturated liquid at −4.44°C (−40°F).

Entropy (s): Entropy and energy are both abstract concepts, which human beings are devised to aid in describing thermodynamic observations. Specific entropy (entropy per unit mass) is a thermodynamic quantity representing the unavailability of the thermal energy of a substance for conversion into mechanical work (it is often interpreted as the degree of disorder or randomness in the system). Entropy is a property of a substance, and it approaches zero at the absolute zero temperature. It is most often used quantified relative to a reference state at which it is given an arbitrary value, usually zero. In steam tables, the reference value ($s = 0$) is the enthalpy of saturated water at 32°F. For other substances, the reference values are different. For example, for Freon-12, it is the entropy of saturated liquid at −4.44°C (−40°F). In the English System, the most often used entropy unit is BTU/(lbm°R), which in metric units equals 4186.8 J/(kg °K) or 4.187.8 kJ/(kg °K).

Eps: Span shift of the transmitter, caused by the difference between the pressure at which it was calibrated and at which it is being operated.

Epz: Zero shift of the transmitter, caused by the difference between the pressure at which it was calibrated and at which it is being operated.

Equivalence point: Where the perfect stoichiometric balance exists between an unknown substance and the reagent. The error in titration—the difference between the equivalence point and the actual end point—is called the *indicator blank*.

Equivalent time sampling: The process that captures high-speed electromagnetic events in real time (nanoseconds) and reconstructs them into an equivalent time (milliseconds), which allows easier measurement with present electronic circuitry.

Erosion: Destruction or removal of material by abrasive action of moving fluids (or gases), usually accelerated by the presence of solid particles or matter in suspension.

Error: Most often defined as the difference between the measured value of one data point and the true value of the measurand.

Error (ISA): In process instrumentation, the algebraic difference between the indication and the ideal value of the measured signal. It is the quantity that, algebraically subtracted from the indication, gives the ideal value.

Ethernet: A family of computer networking protocols that serve local area networks. It was standardized in 1983 as IEEE802.3. The original 10BASE5 Ethernet used coaxial cable as a shared medium. Later the coaxial cables were replaced with twisted pair and fiber-optic links in conjunction with hubs or switches. Data rates were periodically increased from the original 10 MB/s to 100 GB/s.

Ets: Span shift of the transmitter, caused by the difference between the temperature at which it was calibrated and at which it is being operated.

Etz: Zero shift of the transmitter, caused by the difference between the temperature at which it was calibrated and at which it is being operated.

Eutrophication: The process by which a body of water acquires a high concentration of nutrients, especially phosphates and nitrates. These typically promote excessive growth of algae. As the algae die and decompose, high levels of organic matter and the decomposing organisms deplete the water of available oxygen, causing the death of other organisms, such as fish. Eutrophication is a natural, slow-aging process for a water body, but human activity greatly speeds up the process.

Excitation voltage: The voltage applied to the input terminals of an electrical bridge circuit.

Explosion proofing: The design means to prevent potential electrical sparking/heating in a device from igniting a flammable or combustible material.

Explosive range for an explosive mixture of vapor or gas: That range of gas concentration between the lower explosive limit and the upper explosive limit. Outside these ranges, the gas will not ignite.

F

Fahrenheit: Is a thermodynamic scale introduced by Daniel Gabriel Fahrenheit (1686–1724). On Fahrenheit's original scale the lower defining point was the lowest temperature to which he could reproducibly cool brine (defining 0°) while the highest was that of the average human core body temperature (defining 96°). Today, the scale is usually defined by two fixed points: the freezing point of water at as 32°, and the boiling point of water is defined to be 212°, and a 180° separation, as defined at sea level and standard atmospheric pressure. Absolute temperature in Fahrenheit units is −459.67°F or [°R] = [°F] + 459.67.

Fail dangerous: Failures of a safety instrumented function (SIF) that lead to a dangerous condition.

Fail spurious: Failures when there are not actual hazards.

Fan laws: The relationships between pressure, impeller diameter and speed, required power and flow in a centrifugal pump or fan.

Farad (F): Is the SI unit of capacitance, named after the English physicist Michael Farad. A 1.0 Farad capacitor, when charged with 1.0 coulomb of electrical charge, will have a potential difference of 1.0 V between its plates. The farad is an inconveniently large unit, both electrically and physically and therefore the more convenient and commonly used units are microfarad, nanofarad, and picofarad.

Faraday Constant (F): Is the electric charge that is carried by one mole of electrons. It is expressed in Coulombs (C) per mole units and has a value of F = 96,500 C/mole.

Faraday's law of electrolysis: States that the amount of substance produced at an electrode by electrolysis is proportional to the quantity of DC electricity transferred at that electrode.

Faraday's law of induction: Is the principle on which the operation of electric motors, generators, or solenoids are based. It states that a change in the magnetic environment of a wire coil will induce a voltage (emf—electromotive force) in it. It does not matter what caused the change in the magnetic environment (change in field strength, movement of a magnet, and movement of the coil relative to the magnet, including rotation); a voltage will be induced.

Fast Fourier Transforms (FFTs): Algorithms convert signals from the time or space domains to a representation in the frequency domain and thereby manage to reduce the complexity of the calculations.

Fault tolerance: The ability of a system to operate to acceptable specifications when some of its parts are not operational.

Feed-in tariff: A policy mechanism to support and accelerate investment in renewable energy technologies. It achieves this by offering long-term contracts to renewable energy producers, typically based on the cost of generation of each technology. Technologies such as wind power are awarded a lower per kilowatt-hour price, while technologies such as solar PV and tidal power are offered a higher price, reflecting higher costs. Usually, the price paid is reduced by time (tariff degression).

Fiber Bragg grating: A type of distributed Bragg reflector constructed in a short segment of optical fiber that reflects particular wavelengths of light and transmits all others. This is achieved by creating a periodic variation in the refractive index of the fiber core, which generates a wavelength-specific dielectric mirror. An FBG can, therefore, be used as an in-line optical filter to block certain wavelengths, or as a wavelength-specific reflector.

Fiber optic thermometer: Its operating principles include phase interference, fiber deformation, noncontacting open fiber with single or dual wavelength IR, blackbody radiation, and fluoroscopic types. It is used in electromagnetically strongly influenced environment, in microwave fields, power plants, or explosion-proof areas and wherever measurement with other temperature sensors is not possible. The fiber optical sensor is completely non-conductive and offers complete immunity to RFI, EMI, NMR, and microwave radiation with high temperature operating capability, intrinsic safety, and noninvasive use.

Fieldbus: The name of a family of industrial computer network protocols used for real-time distributed control and standardized as IEC 61158. In the early hierarchy of these protocols, the human–machine interface was at the top and it was linked to a middle layer (consisting of a simple programmable logic controller [PLC]) via a non-time-critical communications system (e.g., ethernet). At the bottom of this chain were the measurement sensors and switches sending information to the PLC and the manipulated devices (actuators, electric motors, console lights, valves, and alarms) receiving control instructions from the PLC.

Firewall: A router or access server designated as a buffer between any public networks and private network.

Fission power: One uranium 235 atom, caused by the impact of a neutron can produce, in addition to the fission products, three neutrons and release 202.5 MeV = 3.244×10^{-11} J of energy. This total fission power is hard to measure accurately but can be estimated on the basis of neutron flux measurement, if that is periodically compensated for effects such as uranium consumption in the fuel rods and for other effects. Also, once the heat balance based "thermal power" is determined, the actual "fission power" can be inferred from it, if such smaller thermal components as the heat introduced by the pumps, piping, and purification losses are also incorporated.

Flame path: The parallel metal surfaces found in flameproof and explosion-proof enclosures that cool hot gases as they exit that enclosure into a potentially explosive environment.

Flame rectification: Flame rectification is an electronic way to prove pilot flame before opening the main burner gas valve. It serves to prevent the main burner gas valve from opening if there is no pilot flame, or the temperature of the hot surface ignitor has not yet been established. When the pilot flame is established in a combustion process, the flame acts as a path (conductor) that allows the DC current to flow from the flame rod to the ground, completing the circuit. This completed signal indicates that flame has been established and signals the combustion controls that the main gas valve can open and the hot surface igniter will start the main flame.

Flammable gases: Flammable gases include hydrogen, propane, and ethylene.

Flammable material: A fluid, gas, vapor, or mist that can ignite under certain conditions.

Flammable vapors and mists: Flammable materials that include petrol/gasoline, benzene, and solvents under the right pressure and temperature conditions.

Flash point: The lowest temperature at which a flammable liquid gives off enough vapors to form a flammable or ignitable mixture with air near the surface of the liquid or within the container used. Many hazardous liquids have flash points at or below room temperatures. They are normally covered by a layer of flammable vapors that will ignite in the presence of a source of ignition.

Flexible bearings: The component in the load cell that allows the force coil to move without friction.

Flocculation: Suspended particles are brought together through collisions to form bigger particles called floc, which may be settled or filtered out of the water.

Flow curve: A graphical representation of the behavior of flowing materials in which shear stress is related to shear rate.

Flow injection analysis: An approach to chemical analysis that is accomplished by injecting a plug of sample into a flowing carrier stream. The principle is similar to that of segmented flow analysis, but no air is injected into the sample or reagent streams.

Flowmeter primary element: The device mounted internally or externally to the fluid conduit, which produces a signal with a defined relationship to the fluid flow in accordance with known physical laws relating the interaction of the fluid to the presence of the primary device. *Note*: The primary device may consist of one or more elements necessary to produce the primary device signal.

Flowmeter secondary device: The device that responds to the signal generated by the primary element and converts it to a display or to an output signal that can be displayed or transmitted and interpreted as flow rate or quantity. *Note*: The secondary device may consist of one or more elements as needed to translate the primary device signal into standardized or nonstandardized display or transmitted units.

Flow nozzles: Nozzles that produce overall lower permanent pressure loss as compared with orifice plates. Also, for the same generated pressure differential, flow nozzles will provide nearly twice the maximum flow rate. Flow nozzles have no sharp edges in the flow stream to erode and cause calibration shift. Flow nozzles behave similarly to Venturi flowmeters, except they have no pressure recovery exit cone. The generated differential pressure is the total pressure drop.

Fluidity: Reciprocal of absolute viscosity; the unit in the CGS system is rhe, which equals 1/poise.

Fluorinated ethylene propylene (FEP): A fluorocarbon that is extremely chemically inert. It melts at a reasonable temperature and can be plastic welded fairly easily. It is difficult to bond with adhesives. Maximum temperature range is limited to the 150°C (300°F) area.

Flutter: Rapid, abnormal reciprocating variations in lift during which the disc does not contact the seat.

Flux power: The total amount of neutrons generated by the fission of fuel in the reactor.

Fluxgate compass: Is an electromagnetic device which senses the direction of the horizontal component of the earth's magnetic field. The advantage of this compass over a magnetic compass is that the measurement is in the electronic form and therefore can be easily digitized and transmitted.

Formazin nephelometric unit (FNU): Formazin turbidity units are based on ISO-approved formazin polymer standard (http://www.iso.org/iso/catalogue_detail?csnumber=30123).

Foundation fieldbus: A bidirectional communications protocol used for communications among field devices and to the remotely located control system. It utilizes either twisted pair or fiber media to communicate between multiple nodes (devices) and the remote controller. The controller requires only one communication point to communicate with up to 32 nodes; this is a drastic improvement over the standard 4–20 mA communication method, which requires a separate communication point for each communication.

Fourier transformation: Is a mathematical process to convert raw data into the actual spectrum of the radiation absorption over a wide spectral range. This is superior to dispersive spectrometry, which measures IR radiation intensity over a narrow spectral range. It can be applied to optical spectroscopy, optical and infrared spectroscopy (FTIR, FTNIR), nuclear magnetic resonance (NMR), and magnetic resonance spectroscopic imaging (MRSI).

Free chlorine: The amount of chlorine that exists in the form of HOCl in the water device on the control system.

Freeness: Is the ability of pulp and water mixture to release or retain water on drainage.

Frequency shift keying: A frequency modulation method in which a carrier wave is switched between two discrete frequencies to indicate a binary "1" or "0," also called "mark" and "space," respectively.

Fuel cells: Cells that convert the chemical energy of fuel and oxygen into electrical energy while the electrode and the electrolyte remain unaltered. Fuel is converted at the anode into hydrogen ions which travel through the electrolyte to the cathode, and electrons, which travel through an external circuit to the cathode. If oxygen is present at the cathode, it is reduced by these electrons, and the hydrogen and oxygen ions eventually react to form water.

Fuel rods: Zirconium-clad uranium oxide rods in which the concentration of U235 can be 2% or higher.

Full-range output: The range of output over which the output responds to changes in the input.

Full scale: The maximum measurand value within the calibrated range of a transmitter (or other sensor). The minimum value is called "zero," and the difference between zero and full scale is called "span."

Full-scale range (ISA): The region between the limits within which a quantity is measured, received, or transmitted, expressed by stating the lower and upper range values.

G

Galvanic probe: A probe for which no external voltage is applied across electrodes; current flows as the cell is depolarized when diffusion of the analyte occurs. Electrodes are consumed during this operation and require periodic replacement.

Gamma (γ) radiation: This form of radiation has no mass and consists of electromagnetic waves that are comparable to x-rays. The distinction between γ-rays and x-rays is that γ-rays are produced within the nucleus, while x-rays are produced—by the transition of an electron—from the outer to the inner orbit. Gamma rays are denoted as γ and are electromagnetic radiations of high frequency and, therefore, high energy. Gamma rays are ionizing radiations and are thus biologically hazardous. They are classically produced by the decay from high-energy states of atomic nuclei (gamma decay). They have an electromagnetic radiation of high frequency greater than 10^{19} Hz.

Gas constant: A physical constant expressed in units of energy (i.e., the pressure–volume product) per temperature increment per mole. It is specific to a particular gas. It is denoted as R_g when referring to a particular gas and as R_a when referring to air.

Gas constant (R): Is the amount of energy needed to change the temperature of a mole of gas by one degree. It is denoted as Rg when referring to a particular gas and as Ra when referring to air. It is work per degree per mole. It may be expressed in any set of units representing work or energy (such as joules) and temperature (such as degree Celsius, Kelvin, or Fahrenheit). When using Joules and °K, the value of R is R = 8.314 Joules/(°K mole).

Gas groups: Groupings of gases (shown as follows) that have similar properties with respect to ignition energies.

Group I—Coal mining (methane)
Group II—Gases in other industries
Group IIA—Propane as representative gas
Group IIB—Ethylene as representative gas
Group IIC—Hydrogen as representative gas

Gas–liquid ratio: The gas volume flow rate relative to the total liquid volume flow rate (oil and water); all volumes converted to volumes at standard pressure and temperature.

Gas volume fraction: The ratio of the gas volumetric flow rate to the total volumetric flow rate. The total volumetric flow rate is the sum of the liquid volumetric rate and the gas volumetric flow rate. These volumetric flows are usually expressed in actual (not standardized) volumetric terms.

Gauging pig: A utility pig that is permanently deformable by obstructions in the pipeline and thus, upon retrieval from the line, provides evidence of the worst-case obstruction in a given pipeline segment.

Geiger–Mueller tube: A Geiger–Müller tube (or GM tube) is the sensing element of a Geiger counter instrument that can detect a single particle of ionizing radiation. It was named after Hans Geiger, who invented the principle in 1908, and Walther Müller, who collaborated with Geiger in developing it further in 1928. It is a type of gaseous ionization detector.

Detector of ionizing radiation based on avalanche phenomenon. It measures radiation through the ionization of a halogen gas at about 500 V DC potential.

Gel pig: A utility pig composed of a highly viscous gelled liquid. These pigs are often used for pipeline cleaning and are sometimes called gelly pigs.

Generatix: A geometric element that generates a geometric figure, especially a straight line that generates a surface by moving in a specified fashion. Also called *generator*.

Geometry pig: A configuration pig designed to record conditions, such as dents, wrinkles, ovality, bend radius, and angle, and occasionally indications of significant internal corrosion, by making measurements of the inside surface of the pipe.

Glass membranes: Glass membranes are made from an ion-exchange type of glass having good selectivity for H^+, Na^+, Ag^+, Pb^{2+}, and Cd^{2+}. The glass membrane has excellent chemical durability and can work in very aggressive media. A very common example of this type of electrode is the pH glass electrode.

Gooseneck: A thick metal elbow connected to the swivel and standpipe that supports the weight of and provides a downward angle for the kelly hose to hang from.

Gouges: Mechanical or forceful removal of metal from a local area of the surface on the pipe that may work to harden the pipe and make it more susceptible to cracking.

Grains: A measurement of moisture. In the Bronze Age, the weight of a single grain of barley. Now specified as exactly 7000 grains = 1 lb.

Gravity of Earth (denoted by g): Refers to the acceleration that the Earth imparts to objects on or near its surface. In SI units, this acceleration is measured in newtons per kilogram (N/kg or $N \cdot kg^{-1}$), and it has an approximate value of 9.81 m/s^2 (about 32.1740 ft/s^2). This means if the effects of air resistance are ignored, the speed of an object falling freely near the Earth's surface will increase by about 9.81 m (32.2 ft) per second every second. The precise strength of Earth's gravity varies with location. The nominal "average" value at the Earth's surface, by definition, is 9.80665 m/s^2 (about 32.1740 ft/s^2).

Gray (Gy): Gray expresses *absorbed* radiation, and sievert expresses *equivalent* radiation. It is the SI unit of radiation dose in units of joules (J) per kilogram. One gray equals 1 J of energy received per kilogram of mass (1 J/kg). For example, if a human organ weighing 2 kg is exposed to radiations that corresponds to energy of 6×10^{-5} J, then the radiation dose would be 6×10^{-5} J/2 kg = 3×10^{-5} J/kg or simply 3×10^{-5} Gy. Gray is used to quantify for high levels of radiation doses that cause tissue damage, while sievert is used for quantifying low-level doses.

Gray body: An object having an emittance of less than unity, but this emittance is constant at all wavelengths (over that part of the spectrum where the measurement takes place). This means that gray body radiation curves are identical to the ones shown in Figure 4.11a, except that they are dropped down on the radiated power density scale.

Gross calorific value: Is the amount of heat liberated by the complete combustion of unit weight of coal (or other fuel) under specified conditions; the water vapor produced during combustion is assumed to be completely condensed. The heat value of energy per unit volume at standard conditions, expressed in terms of BTU per SCF or kilocalorie per cubic Newton meters (kcal/N m^3).

Gross or high heating value: Is the amount of heat produced by the complete combustion of a unit quantity of fuel. It is obtained when all products of the combustion are cooled down and when water vapor formed during combustion is condensed.

Gross weight: The total weight of the material of interest plus the container and wrapping or the weight of a vehicle or container when it is loaded with goods.

Gross weight-in: The total weight set point while charging material into a tank or vessel.

Gross weight-out: The total weight set point while discharging material from a tank or vessel.

Ground: A conducting connection, whether intentional or accidental, between an electrical circuit or equipment and the Earth, or to some conducting body that serves in place of Earth. (See NFPA 70–100.)

Ground fault protector: A device used to open ungrounded conductors when high currents, especially those resulting from line-to-ground fault currents, are encountered.

Guard: The *electronic guard* (called a *shield* in some RF-level literature) consists of a concentric metallic element with an applied voltage identical to the voltage on the conductor it is "guarding." This negates the capacitance between the guarded conductor and the outside world.

Guided wave radar: A contact radar technology where time domain reflectometry (TDR) has been developed into an industrial-level measurement system where a probe immersed in the medium acts as the waveguide.

H

Hagen–Poiseuille law: Flow through a capillary

$$Q = \frac{\pi R^4}{8\eta L}(p_1 - p_2)$$

Half-life: Is the amount of time required to describe how quickly unstable atoms undergo radioactive decay. The term was introduced by Ernest Rutherford in 1907.

Hall effect: The production of a voltage difference (the Hall voltage) across an electrical conductor, transverse to an electric current in the conductor and a magnetic field perpendicular to the current. The Hall coefficient is defined as the ratio of the induced electric field to the product of the current density and the applied magnetic field.

HART: The HART Communications Protocol (Highway Addressable Remote Transducer Protocol) is an early implementation of fieldbus, a digital industrial automation protocol. Its most notable advantage is that it can communicate over analog instrumentation wiring, sharing the pair of wires used by the older system.

Hard spots: Local changes in the hardness of the steel in the pipe resulting from nonuniform quenching procedures during manufacture or changes in the chemistry of the steel. Hard spots, when stressed, are subject to failure from mechanisms such as hydrogen stress cracking.

Hastelloy: The registered trademark name of Haynes International, Inc. The trademark is applied as the prefix for a range of different highly corrosion-resistant metal alloys, loosely grouped by the metallurgical industry under the material term "superalloys" or "high-performance alloys."

Hazardous area: A location in which there is a likelihood of the presence of flammable gases, powders, dust, fibers, and other ignitable materials. Industrial equipment in such locations is normally protected by implementing explosion-proof, intrinsically safe, or purged equipment and practices. Such equipment is approved for use in hazardous locations by independent approval bodies, including UL, FM, CSA, and others around the world.

Hazen–Williams equation: Is an empirical relationship which relates the flow of water in a pipe with the physical properties of the pipe and the pressure drop caused by friction. It is named after Allen Hazen and Gardner Stewart Williams.

Heel: The material that remains in the vessel after the dump or empty cycle is completed.

Heel limit: The maximum allowable weight setting for the heel, which usually is internally programmable.

Hertz (Hz): In SI units, it is the unit of frequency (named after the German physicist Heinrich Hertz). One hertz means that an event repeats once per second.

High pressure liquid chromatography: Distinguished from traditional (low pressure) liquid chromatography because operational pressures are significantly higher (50–350 bar), while ordinary liquid chromatography typically relies on the force of gravity to pass the mobile phase through the column.

Highly toxic: A chemical that has a median lethal concentration (LC[50]) in air of 200 parts per million by volume or less of gas or vapor, or 2 mg/l or less of mist, fume, or dust, when administered by continuous inhalation for 1 hr (or less if death occurs within 1 hr) to albino rats weighing between 200 and 300 g each.

Holidays: Discontinuities in a coating, such as pinhole cracks, gaps, or other flaws, that allow areas of the base metal to be exposed to any corrosive environment that contacts the coating surface.

Home run wiring: Wire between the cabinet where the Fieldbus hosts or centralized control system resides and the first field junction box or device.

Hooke's law: Is the of elasticity, discovered by the English scientist Robert Hooke in 1660, which states that for relatively small deformations of an object is directly proportional to the deforming force or load. The deforming force may be applied to a solid by stretching, compressing, squeezing, bending, or twisting. Thus, a metal wire exhibits elastic behavior according to Hooke's law because the small increase in its length when stretched by an applied force doubles each time the force is doubled. Mathematically, Hooke's law states that the applied force F equals a constant k times the displacement or change in length x, or $F = kx$. The value of k depends not only on the kind of elastic material under consideration but also on its dimensions and shape.

Hot tap: The functionality to install and remove a measuring instrument in an active pipe while it is pressurized.

Hub (shared): Multiport repeater joining segments into a network.

Hydrostatic head: The pressure created by the depth (height) of liquid.

Hygrometer: An apparatus that measures humidity.

Hygroscopic material: A material with great affinity for moisture.

Hysteresis: The property of an element that its output, responding to a given excursion of the input, depends on the history of prior excursions and on the direction of the current. *Note 1*: It is usually determined by subtracting the value of dead band from the maximum measured separation between upscale going and downscale going indications of the measured variable (during a full-range traverse, unless otherwise specified) after transients have decayed. This measurement is sometimes called hysteresis error or hysteretic error. *Note 2*: Some reversal of output may be expected for any small reversal of input; this distinguishes hysteresis from dead band.

Hysteresis error: The variation in the readings that are registered when repeatedly weighing the same object or quantity.

I

Ignition energy: The energy needed to cause ignition of a material.

Ignition temperature: The surface temperature of exposed equipment needed to cause ignition of a material.

Immediately dangerous to life or health: Limit concentrations used by the NIOSH as respirator selection criteria.

Immiscible: Liquids that do not dissolve in each other, or nonmixing liquids.

Impedance: Maximum voltage divided by maximum current in an alternating current circuit. Impedance is composed of resistive, inductive, and capacitive components. As in DC circuits, the quantity of voltage divided by current is expressed in ohms (Ω).

Impedance (Z): Opposition to the flow of AC. Impedance in AC circuits is analogous to resistance in DC circuits. Expressed in units of ohms (Ω).

Impingers: Are specially designed bubble tubes used for collecting airborne chemicals into a liquid medium. With impinger sampling, a known volume of air is bubbled through the impinger containing a specified liquid. The liquid will chemically react with or physically dissolve the chemical of interest.

Impulse line: The tubing between tank connection and the d/p transmitter to transfer the process pressure.

Inaccuracy: A quantitative presentation of the difference between the actual value (measured) and the true value, expressed as a percentage of either span or full-scale value.

Incident ray: The ray of light approaching the object.

Inclusions: Foreign materials or particles in a metal matrix. These are usually compounds such as oxides, sulfides, or silicates, but may be any substance that is foreign to the matrix, whether it is soluble or insoluble.

Inconnel: A family of austenitic nickel-chromium-based superalloys. The name is a trademark of Special Metals Corporation.

Inferred pH: A calculation that uses the standard and cation conductivity of a water sample in a steam plant that provides an estimate of the pH of the water based on models developed from the type of water treatment used.

In-flight quantity: The material that enters a receiving vessel after the feed valve is closed.

Infrared radiation: Electromagnetic radiation of frequency range of between 430 THz and 300 GHz. That portion of the spectrum whose wavelength is longer than that of red light. Only the portion between 0.7 and 20 μm gives usable energy for radiation detectors.

In-line: The situation where all the liquid whose viscosity is being measured passes through or around the process viscometer.

In-line inspection (ILI): The inspection of a pipeline from the interior of the pipe using an in-line inspection tool.

In-line inspection tool: The device or vehicle, also known as an "intelligent" or "smart" (ILI tool) pig, that uses a nondestructive testing technique to inspect the wall of a pipe. An in-line inspection tool is a type of instrumented tool.

Insolation (incoming solar radiation): A measure of solar radiation energy received on a given surface area during a given time. It is also called *solar irradiation* and expressed as "hourly or daily irradiation." The unit recommended by the World Meteorological Organization is megajoules per square meter (MJ/m^2). An alternative unit of measurement is the langley (1 thermochemical calorie per square centimeter or 41,840 J/m^2). Practitioners in the business often use watt-hours per square meter (Wh/m^2). In the United States, the yearly insolation ranges from 800 to 2600 kWh per year.

Integrated circuitry: Also referred to as an IC, a chip, or a microchip, it is a set of electronic circuits on one small plate ("chip") of semiconductor material, normally silicon. This can be made much smaller than a discrete circuit made from independent components.

Integrated circuitry (IC) temperature sensor: Is a two terminal integrated circuit temperature transducer that produces an output current proportional to absolute temperature. The sensor package is small with a low thermal mass and a fast response time. The most common temperature range is 55°C–150°C (–58°F to 302°F). The solid state sensor output can be analog or digital.

Intelligent pig or tool: See *In-line inspection tool*.

Interoperability: A marketing term with a blurred meaning. One possible definition is the ability for like devices from different manufacturers to work together in a system and be substituted one for another without loss of functionality at the host level (HART).

Interface: The boundary between two fluids of differing densities.

Interference, common mode: A form of interference that appears between measuring circuit terminals and ground.

Interference, electromagnetic: Any spurious effect produced in the circuits or elements of a device by external electromagnetic fields. *Note*: A special case of interference from radio transmitters is known as "radio frequency interference."

Interferometry: Is an important investigative technique. Interferometers are widely used in science and industry for the measurement of small displacements, refractive index changes, and surface irregularities. In analytical science, interferometers are used in continuous wave Fourier transform spectroscopy to analyze light containing features of absorption or emission associated with a substance or mixture. In an interferometer, two light rays from a common source combine at the half-silvered mirror to reach a detector. They may either interfere constructively (strengthening in intensity) if their light waves arrive in phase, or interfere destructively (weakening in intensity) if they arrive out of phase, depending on the exact distances between the three mirrors.

Internet exchange point (IX or IXP): Is a physical infrastructure through which Internet service providers (ISPs) and content delivery networks (CDNs) exchange Internet traffic. They reduce the per point delivery cost and improve routing efficiency. In other words, they enable the interconnection of a number of systems, primarily for the purpose of facilitating the exchange of Internet traffic.

Intrinsic safety: Available energy is limited under all conditions to levels too low to ignite the hazardous atmosphere. This method is useful only for low-power equipment such as instrumentation, communication, and remote control circuits.

Instrumented pig: A vehicle or device used for internal inspection of a pipe, which contains sensors, electronics, and recording or output functions integral to the system. Instrumented pigs are divided into two types:

1. Configuration pigs, which measure the pipeline geometry or the conditions of the inside surface of the pipe
2. In-line inspection tools that use nondestructive testing techniques to inspect the wall of the pipe for corrosion, cracks, or other types of anomalies

Ion-exchange chromatography (IEC): Is a process that is able to separate polar molecules because of their affinity to the ion exchanger used. It works on charged molecules and is often used in water analysis and quality control.

Ion-exchange resin membranes: Ion-exchange resins are based on special organic polymer membranes that contain a specific ion-exchange substance (resin). This is the most widespread type of ion-specific electrodes. They are also the most widespread electrodes with anionic selectivity. However, such electrodes have low chemical and physical durability as well as "survival time."

Ion-selective electrode: A sensor that converts the activity of a specific ion dissolved in a solution into an electrical potential, which generates a voltage that is theoretically dependent on.

Isoelectric point: Is the pH at which the material exhibits zero charge.

Isopotential point: The pH at which the millivolt output of a pH sensor does not change with temperature. For the vast majority of pH sensors, the isopotential point is 7.00 pH and 0.0 mV the logarithm of the ionic activity, according to the Nernst equation.

ITS-90: The International Temperature Scale of 1990, an equipment calibration standard for making measurements on the Kelvin and Celsius temperature scales. It is an approximation of the thermodynamic temperature scale that facilitates the comparability and compatibility of temperature measurements internationally. ITS-90 offers defined calibration points ranging from 0.65°K to approximately 1358°K (−272.5°C to 1085°C) and is subdivided into multiple temperature ranges, which overlap in some instances.

J

Joule (J): Is the SI unit of work (or energy). It is named after James Joule (December 24, 1818–October 11, 1889). It equals the work which is done by the force of one Newton when its point of application moves one meter in the direction of the force's action (1 joule = $2.77777778 \times 10^{-7}$ kWh).

K

Katharometer: The older name used for the thermal conductivity detector, still sometimes used in Europe.

Kellydrive: A square, hexagonal, or octagonal tubing inserted through and is an integral part of the rotary table that moves freely vertically while the rotary table turns it.

Kellyhose: A flexible, high-pressure hose that connects the standpipe to the kelly (or more specifically to the gooseneck on the swivel above the kelly) and allows free vertical movement of the kelly, while facilitating the flow of the drilling fluid through the system and down the drill string.

Kelvin: Is a thermodynamic scale introduced by William Lord Kelvin (1824–1907). Kelvin is not referred to as a degree, but is often used in conjunction with the degree Celsius, which has the same magnitude. Absolute zero (the zero point on the Kelvin scale) is obtained by subtracting 273.16 K from the temperature of the triple point of water (0.01°C), which is equivalent to −273.15°C (−459.67°F).

Kick: Is when oil, gas, or water enters into the wellbore. When the bottom-hole pressure becomes less than the formation pressure and the permeability is great enough, formation fluid will enter the wellbore causing a kick.

Kinematic viscosity, ν(Nu): Dynamic viscosity/density = $\nu = \eta/\rho$.

Kjeldahl method: Is an analytical chemistry method for measuring the quantity of nitrogen in chemical substances, developed by Johan Kjeldahl. It consists of heating the substance with sulfuric acid to liberate the reduced nitrogen as ammonium sulfate. After this, potassium sulfate is added, followed by a number of other laboratory analysis steps that are concluded by back titration to determine the nitrogen content of the sample.

Knife edge: An assembly consisting of a sharp-edged support that allows a second device to pivot on the sharp edge. Most often used in mechanical scales and allows some rotation of second device with very little friction.

Knockmeter: Ultrasonic sensor to detect when a spark from the spark plug does not ignite the fuel properly, leading to incomplete combustion.

Kovar: Is an iron-nickel-cobalt alloy with a coefficient of thermal expansion similar to that of hard (borosilicate) glass. It is used in the tip of thermocouples.

Kovar tip: Thermocouple alloy.

L

Lack of fusion: In a weld, any area or zone that lacks complete melting and coalescence (fusion) of a portion of the weld.

Lack of penetration: In the welding process, failure to achieve fusion of the base metal to the desired or planned depth.

Lambda: The common term used for the failure rate λ.

Lambert–Beer law: Relates the attenuation of light, neutrons, etc., to the properties of the material through which they are passing.

Lamination: A type of imperfection or discontinuity with separation or weakness, usually aligned parallel to the worked surface of a metal.

Laser diffraction: Measures particle size distributions by measuring the angular variation in the intensity of light scattered, as a laser beam passes through a dispersed particulate sample. Large particles scatter light at small angles relative to the laser beam and small particles scatter light at large angles. The angular scattering intensity data can be analyzed to calculate the size of the particles responsible for creating the scattering pattern. The particle sizes are determined as the volume equivalent sphere diameter. The size of the particle can range from nanometers to millimeters.

Laser power: The optical power level emitted by the laser in a sensor. If the power is quadrupled, the maximum range will be doubled. Laser power is expressed in milliwatts (mW) or watts.

Laser target: The surface where a laser spot hits, from which light is reflected to the detector in an optical sensor. Target reflectance is the most important factor in determining the maximum range of a sensor.

Latency: Measures the worst-case maximum time between the start of a transaction and the completion of that transaction.

Launcher: A pipeline facility used for inserting a pig into a pressurized pipeline.

LC circuit: Is an inductor (L) and capacitor (C) circuit connected together. It can act as a resonator, storing energy which is oscillating at the circuit's resonant frequency.

Lift: The rise of the disc in a PRV.

Light scattering: Is the deflection of a light ray from a straight path. Most objects that one sees are visible due to light scattering from their surfaces. The scattering of light depends on the wavelength of the light being scattered.

Limit switch: Is a switch preventing the travel of an object in a mechanism past some predetermined point, mechanically operated by the motion of the object itself.

Line driver: Inexpensive amplifier and signal converter that conditions digital signals to ensure reliable transmissions over extended distances without the use of modems.

Linearity: The closeness to which a curve approximates a straight line. *Note 1*: It is usually measured as a nonlinearity and expressed as linearity; for example, a maximum deviation between an average curve and a straight line. The average curve is determined after making two or more full-range traverses in each direction. The value of linearity is referred to the output unless otherwise stated. *Note 2*: As a performance specification, linearity should be expressed as independent linearity, terminal-based linearity, or zero-based linearity. When expressed simply, linearity is assumed to be independent linearity.

See *conformity*.

Linear variable differential transformer: A type of position sensor having a primary excited by an AC carrier wave and two or more secondaries magnetically coupled with the primary by a movable core of magnetically permeable material.

Lixiviant: A liquid medium used in hydrometallurgy to selectively extract the desired metal from the ore or mineral. It assists in rapid and complete leaching. The metal can be recovered from it in a concentrated form after leaching.

Load cell creep: Change in load cell output signal that occurs with the passage of time, while no change occurs in either the load or the environmental conditions and other variables. It is usually measured while full scale (rated) load is applied and is expressed as a percent of full scale (rated output) over a particular period of time (% FS/hr, etc.).

Load cell deflection: The displacement along the primary axis of the load cell as the load is increased from no load to full (rated) load.

Load cell measuring cell: The part of the load cell that measures the load (weight or force) and generates an electronic signal related to that measurement. Also called *mechanical system or force motor*.

Load cell shunt calibration: Electrical simulation of load cell output by insertion of known shunt resistors between appropriate points within the circuitry.

Load cell temperature compensation: The automatic correction of any errors that might be caused by a change in the temperature. On less accurate scales, this compensation is done within the load cells, while on more accurate weighing systems it is done in software.

Load cell zero balance: The output signal of a load cell with excitation, but with no load applied, expressed in percent of rated output.

Load—safe maximum: The maximum percent of rated capacity that can be applied without causing a permanent shift in the load cell performance, which would exceed the specified allowable amount.

Load—ultimate maximum: The maximum percent of rated capacity, which can be applied without causing structural failure.

Lower explosive limit : The lowest concentration of gas or vapor in air where, once ignition occurs, the gas or vapor will continue to burn after the source of ignition has been removed.

Lorenz force: Is the force on a charged particle moving through a region containing both electric and magnetic fields. It is named after H. A. Lorentz who discovered it in 1965.

Lower range limit: Lowest value of the measured variable, which the device is capable of measuring.

Lower range value: Lowest value of the measured variable that the device was adjusted to measure.

Lowpass filters: Are used to remove high-frequency interference or noise from low-frequency signals.

Luft detectors: Consist of two chambers, divided by a diaphragm, and the same intensity of pulsed infrared radiation is received by both chambers. When the gas of interest flows through the sample cell, a reduction in radiation energy is received by the detector chamber, which causes the temperature and pressure to drop in that chamber. The amount of pressure drop is in proportion to the gas concentration, and this pressure difference between the two chambers causes a movement of the diaphragm, which is detected by capacitance.

M

Magnetic Flux Leakage (MFL): Is a nondestructive testing method often used to detect pitting or corrosion in pipelines. MFL uses a powerful magnet to magnetize the steel pipe wall and detect if the magnetic field "leaks," which occurs if corrosion or missing metal is present. In an MFL tool, a magnetic detector is placed between the poles of the magnet to detect the leakage field. Analysts interpret the chart recording of the leakage field to identify damaged areas and to estimate the depth of metal loss.

Magnetoresistance (magnetoresistive): The change in electrical resistance of a current-carrying conductor when a magnetic field is applied.

Magnetostriction (magnetostrictive): A technology used in linear position sensors, in which the position of a movable permanent magnet is detected by measuring the time period between application of a current pulse to a linear waveguide, and detection of a signal pulse near one end of the waveguide.

Manchester: A digital signaling technique that contains a signal transition at the center of every bit cell.

Manifold: The assembly of connections and valves between the impulse lines and the secondary to allow convenient calibration and replacement of the sensor.

Manning formula: Is also known as the Gauckler–Manning formula, or Gauckler–Manning–Strickler formula in Europe. In the United States, it is very frequently called Manning's Equation. The Manning formula is an empirical equation estimating the average velocity of a liquid flowing in a conduit that does not completely enclose the liquid, that is, in an open channel.

Manufacturing range: A range around the specified burst pressure within which the marked or rated burst pressure must fall. Manufacturing range is not used in ISO standards.

Mapping pig: A configuration pig that uses inertial sensing or some other technology to collect the data that can be analyzed to produce an elevation and plan view of the pipeline route.

Mass: The amount of a material defined by a resistance to acceleration, not to be confused with weight.

Maximum allowable operating pressure: The maximum pressure expected during normal operation.

Maximum allowable working pressure: This is the maximum pressure allowed for continuous operation. As defined in the construction codes (ASME B31.3) for unfired pressure vessels, it equals the design pressure for the same design temperature. The maximum allowable working pressure depends on the type of material, its thickness, and the service conditions set as the basis for design. The vessel may not be operated above this pressure or its equivalent at any metal temperature other than that used in its design; consequently for that metal temperature, it is the highest pressure at which the primary PRV can be set to open.

Maximum operating thermal limit: The license of all nuclear power plants states that the maximum operating thermal power (MOTP) of the plant cannot approach the thermal power limit (TPL) closer than the uncertainty margin (UM) (MOTP < TLP – UM). That UM is a function of the precision at which the various components of the heat balance calculation could be measured at the time.

MDQ: In chromatography, MDQ is the minimum amount of solute that will produce a detector signal twice the peak-to-peak noise of the detector. The dynamic range of the detector is the range of concentration of the test solute over which a change in concentration results in a measurable change in detector signal.

Measurand: The physical quantity that is measured by a sensor.

Measurement uncertainty: The limits to which a specific value may extend with some confidence. The most commonly used confidence in uncertainty analysis is 95%, but other confidences may be employed where appropriate.

Mechanical emissivity enhancement: Mechanically increasing the emissivity of a surface to near-blackbody conditions (using multiple reflection).

Metal loss: Any of a number of types of anomalies in pipe in which metal has been removed from the pipe surface, usually due to corrosion or gouging.

Micellar liquid chromatography: A form of reversed-phase liquid chromatography that uses an aqueous micellar solutions as the mobile phase.

Michelson interferometer: Was invented by Albert Abraham Michelson. It uses a beam splitter in which light is split into two arms. Each of those is reflected back toward the beam splitter, which then combines their amplitudes interferometrically. The resulting interference pattern is typically directed to some type of photoelectric detector or camera.

Micron: 0.001 mm. 10,000 Å. A unit used to measure wavelengths of radiant energy.

Micron or micrometer (μ or μm) I: Is an SI unit of length equaling one-millionth of a meter or one-thousandth of a millimeter.

Microwave and radar: The microwave spectrum is usually defined as electromagnetic energy ranging from approximately 1 to 100 GHz in frequency, but older usage includes lower frequencies. Most common applications are within the 1–40 GHz range. The frequency range of radar (8.0–12.0 GHz) is part of that range. Radar is a trade name that refers to instruments that operate in the above frequency range.

Microwave radiation: Electromagnetic radiation of frequency range between 300 MHz and 300 GHz.

MKS system: A system of units that expresses any given measurement using the fundamental metric units of the meter, kilogram, and/or second (MKS) units. It was in 1901 when Giovanni Giorgi proposed to the Associazione Elettrotecnica Italiana that this system be used as an international system.

Modal dispersion: Is a distortion mechanism occurring in multimode fibers, in which the signal is spread in time because the propagation velocity of the optical signal is not the same for all modes. Other names for this phenomenon include multimode distortion, modal distortion, intermodal distortion, intermodal dispersion, and intermodal delay distortion.

Modbus: A data communication messaging protocol used to establish a master/slave communication between intelligent devices.

Modem: Modulator–demodulator; a device that converts digital and analog signals. At the source, a modem converts digital signals to a form suitable for transmission over analog communication facilities. At the destination, the analog signals are returned to their digital form. Modems allow data to be transmitted over voice-grade telephone lines.

Monochromatic light: Is light that of one color; in other words, it is light having a single wavelength.

Monochromator: Is an instrument that transmits mechanically selectable narrow wavelength bands (λ) of UV or other types of radiation chosen from a wider range of wavelengths from the entering radiation. Glass is not transparent to UV radiation, so quartz prisms are used to disperse it. The monochromator can use the phenomenon of optical dispersion by prisms or can use diffraction grating to separate the wavelength bands (λ). The gratings operate in the reflective mode, and they can be triangular mirrored half-prisms, producing linear dispersion or holographic gratings having sinusoidal grove profiles (cross section).

Moody chart: Or Moody diagram is a nondimensional graph that relates the Darcy–Weisbach friction factor, the Reynolds number, and the relative roughness in a circular pipe. It can be used determining the pressure drop or flow rates in pipes.

MooN: A means of describing a redundancy scheme for an SIF where M paths of N total redundant paths are required for successful operation.

Morphology: The shape and size of the particles making up the object; the direct relationship between these structures and their material properties (ductility, strength, reactivity, etc.).

Mud pump: A reciprocal type of pump used to circulate drilling fluid through the system.

Mud tank: Often called mud pits and stores drilling fluid until it is required down the wellbore.

Multiphase flow: Two or more phases flowing simultaneously in a conduit. For example, this document deals in particular with multiphase flows of oil, gas, and water.

Multiphase flow rate: The total amount of the two or three phases of a multiphase flow flowing through the cross section of a conduit in unit time. The multiphase flow rate should be specified as multiphase volume flow rate or multiphase mass flow rate.

Multiplexing: A scheme that allows multiple logical signals to be transmitted simultaneously across a single physical channel. Compare with demultiplexing.

N

N-type semiconductor materials (negative): Have more electrons than "holes" and carry current mainly in the form of negatively charged electrons ("opants") in a manner similar to the conduction of current in a wire. They have more electrons than holes and these extra electrons do not fit into the regular crystal pattern of pure silicon.

Narrowband pyrometer: A radiation pyrometer that is sensitive to only a narrow segment of wavelengths within the total radiation spectrum. Optical pyrometers are one of the devices in this category.

National Electrical Code (NEC or NFPA 70): A standard for the safe installation of electrical wiring and equipment in the United States.

NEMA enclosures: NEMA defines standards for various grades of electrical enclosures typically used in industrial applications. Each is rated to protect against designated environmental conditions. A typical NEMA enclosure might be rated to provide protection against environmental hazards such as water, dust, oil, or atmospheres containing corrosive agents.

Nephelometer: An instrument for measuring the extent or degree of cloudiness. It is used to determine the concentration or particle size of solid particles in suspensions by means of transmitted, scattered, or reflected light.

Nernst equation: An equation that relates the equilibrium reduction potential of a half-cell in an electrochemical cell to the standard electrode potential, temperature, activity, and reaction quotient of the underlying reactions and species used. It is named after the German physical chemist who first formulated it, Walther Nernst.

Nessler cylinders: Are used for colorimetric. The color of the sample in a Nessler cylinder is visually compared with a reference model. The tubes are often used to carry out a series of calibration solutions of increasing concentrations, which functions as a comparative scale.

Net batch-out: A weighing mode of operation where the set point determines the amount to be withdrawn from the vessel.

Net calorific value: The measurement of the actual available energy per unit volume at standard conditions, which is always less than the gross calorific value by an amount equal to the latent heat of vaporization of the water formed during combustion.

Net metering: When a building receives its electricity from the grid through bidirectional electric meters, the quantity of net metered electricity is the difference between what the household or building received from the grid and the amount it has sent to the grid over some time period.

Net or lower heating value: Is the amount of heat produced by the combustion of a unit quantity of a fuel minus the latent heat of vaporization of the water vapor formed by the combustion.

Net weigh: The weight of material excluding its container or wrapping.

Net weigh-in: A mode of weighing where the set point for the total material to be added is based on zero, which corresponds to the weight that is already in the vessel.

Network: All of the media, connectors, and associated communication elements by which a communication system operates.

Newton: Is the standard unit of force in the metric (MKS) system and corresponds to the mount of force needed to accelerate 1 kg of mass at the rate of 1 m per second squared. The name honors the English physicist and mathematician Isaac Newton, who laid down the foundations of classical mechanics.

Neutron: Subatomic particle with no electric charge.

Neutron fluence: Neutron fluence is the time-integrated neutron flux in units of neutrons/cm^2.

Neutron flux (φ): Neutron flux (φ) is the total number of neutrons that are passing through a 1 cm^3 volume in a second. (Mather equation $\varphi = nv$). Neutron flux detectors provide an important measure of the fission power being released in a nuclear reactor.

Newtonian fluid: In a Newtonian fluid, the relationship between the shear stress and the shear rate is linear and this proportionality is the coefficient of viscosity.

Nitrification: The process by which bacteria in soil and water oxidize ammonia and ammonium ions and form nitrites and nitrates. Because the nitrates can be absorbed by more complex organisms, as by the roots of green plants, nitrification is an important step in the nitrogen cycle.

Non-fragmenting disc: A rupture disc design that when burst does not eject fragments that could interfere with the operation of downstream equipment (i.e., relief valves).

Non-incendiary equipment: Equipment that in its normal operating condition would not ignite a specific hazardous atmosphere in its most easily ignited concentration. *Note*: The electrical circuits may include sliding or make-and-break contacts releasing insufficient energy to cause ignition. Wiring that, under normal conditions, cannot release sufficient energy to ignite a specific hazardous atmospheric mixture by opening, shorting, or grounding shall be permitted using any of the methods suitable for wiring in ordinary locations.

Non-Newtonian fluid: In a non-Newtonian fluid, the relationship between the shear stress and the shear rate is not a constant (viscosity) and can even be time dependent. Therefore, non-Newtonian is a fluid whose flow properties differ in any way from those of Newtonians. Most commonly, the viscosity (the measure of a fluid's ability to resist gradual deformation by shear or tensile stresses) of non-Newtonian fluids is dependent on shear rate or shear rate history. Many salt solutions and molten polymers are non-Newtonian fluids, as are many commonly found substances such as ketchup, custard, toothpaste, starch suspensions, paint, blood, and shampoo.

Normal mode rejection: The ability of a circuit to discriminate against a normal mode voltage. *Note*: It may be expressed as a dimensionless ratio, a scalar ratio, or in decibels as 20 times the $\log_{10}$ of that ratio.

NTU: Nephelometric turbidity units (NTUs) are based on U.S. EPA-approved polymeric suspension standard.

Numeric aperture (N): Describes the range of angles within which light that is incident on the fiber will be fully reflected and therefore transmitted along it. If this angle is exceeded, some of the light will be lost.

O

Obstructions: Any restriction or foreign object that reduces or modifies the cross section of the pipe to the extent that flow is affected or in-line inspection pigs can become stuck (ovality, collapse, dents, undersized valves, wrinkles, bends, weld drop through). Also any foreign object in the pipeline.

Off-center loading: See *Eccentric Loading Test*, also called shift test.

Ohm (Ω): Was named after German physicist Georg Simon Ohm. It is the SI unit of electrical resistance. One ohm is the resistance between two pints of a conductor if it transmits one ampere of current when subjected to a potential difference of 1 volt. The definition of the "ohm" unit was revised several times.

Oil immersion: Concept in which equipment is submerged in oil to a depth sufficient to quench any sparks that may be produced. This technique is commonly used for switchgears, but it is not utilized in connection with instruments.

OIML R60, 2000: A recommendation prepared by the Organisation Internationale de Métrologie Légale (OIML), an intergovernmental organization created to promote the global harmonization of the legal metrology procedures that underpin and facilitate international trade. It prescribes the principal metrological static characteristics and static evaluation procedures for load cells used in the measurement of mass. It provides the authorities with uniform means for determining the metrological characteristics of load cells used in measuring instruments that are subjected to metrological controls.

Olfactometer: An apparatus for measuring the acuity of the sense of smell. It is used to detect and measure ambient odor dilution. Olfactometers are used to gauge the odor detection threshold of substances. To measure intensity, olfactometers introduce an odorous gas as a baseline against which other odors are compared.

Olfactory: Refers to systems relating to, or connected with, the sense of smell.

Online: Any measurement that provides (quasi) continuous process viscosity data (could be an in-line or bypass installation).

OPC (Open Platfrom Communication): The interoperability standard for the secure and reliable exchange of data in the industrial automation, space and in other industries. It is a platform that ensures the seamless flow of information among devices from multiple vendors. The OPC Foundation is responsible for the development and maintenance of standards. The OPC standards are a series of specifications developed by industry vendors, end-users, and software developers. These specifications define the interface between Clients and Servers, as well as Servers and Servers, including access to real-time data, monitoring of alarms and events, access to historical data and other applications.

Operating conditions: Conditions to which a device is subjected, not including the variable measured by the device. Examples of operating conditions include ambient pressure, ambient temperature, electromagnetic fields, gravitational force, inclination, power supply variation (voltage, frequency, harmonics), radiation, shock, and vibration. Both static and dynamic variations in these conditions should be considered.

Operating pressure: The operating pressure of a vessel is the pressure, in pounds per square inch gauge, to which the vessel is usually subjected in service. A processing vessel is usually designed for a maximum allowable working pressure (in pounds per square inch gauge) that will provide a suitable margin above the operating pressure in order to prevent any undesirable operation of the relief device. It is suggested that this margin be approximately 10%, or 25 PSI (173 kPa), whichever is greater. Such margin will be adequate to prevent the undesirable opening and operation of the PRV caused by minor fluctuations in the operating pressure.

Operating pressure margin: The margin between the maximum operating pressure and the set pressure of the PRV.

Operating pressure ratio: The ratio of the maximum operating pressure to the marked burst pressure expressed as a percentage (common U.S. definition). The ratio of the maximum operating pressure to the minimum of the performance tolerance expressed as a percentage (common ISO definition).

Operating ratio of a ruptured disc: (1) The ratio of the maximum operating pressure to the marked burst pressure expressed as a percentage (common U.S. definition). (2) The ratio of the maximum operating pressure to the minimum of the performance tolerance expressed as a percentage (common ISO definition).

Operating reactivity margin: The operating reactivity margin (ORM) number expresses the "degree of control," in units of the equivalent number, of control rods remaining in the reactor. In other words, ORM is the ratio of the extra reactivity obtained if all control rods are withdrawn divided by the effect if only one rod is withdrawn.

Optical dispersion: Is the separation of incoming radiation into wavelength bands (λ). When a narrow wavelength band (λ) of UV or other type of radiation is to be selected, that can be done by dispersive or nondispersive selectors. Filters are called nondispersive (λ) selectors while prisms and gratings are called dispersive ones. A linear grating is an optically flat polished surface having dense and parallel groves (about 1000 per mm). The two types of gratings are transmission and reflection (diffraction) types. In spectroscopic instruments, almost all gratings are of the reflection type. The gratings can produce linear or holographic dispersion. Linear dispersion is produced by triangular flat-surfaced gratings. Holographic gratings are made by laser technology that produces a pattern of straight lines with a sinusoidal grove profile (cross section).

Optical fiber: Is a flexible transparent fiber of extremely pure glass or plastic, generally between 10 and 200 microns in diameter, through which light can be transmitted by successive internal reflections.

Optical pyrometer: Also called a brightness pyrometer, it uses a narrow band of radiation within the visible range (0.4–0.7 μm) to measure temperature by color matching and other techniques.

Organic dusts: Carbon-based compounds (e.g., flour, coal, polyethylene, etc.).

Organoleptic properties: The aspects of food or other substances as experienced by the senses, including taste, sight, smell, and touch, in cases where dryness, moisture, and stale-fresh factors are to be considered.

Orifice fittings: The devices in the pipeline containing the orifice plate. See text for more details.

Orifice union: A pipe union with a flow restricting orifice plate installed.

Orthogonal coordinate system: Is a curvilinear coordinated system where each family of surfaces intersects the others at right angles. One of the orthogonal coordinate systems is the three-dimensional Cartesian system (x, y, z), because its coordinate surfaces x = constant, y = constant, and z = constant are planes that meet at right angles to one another, that is, are perpendicular.

Orthophosphate: Is an organic chemical, an ester of phosphoric acid. This organic chemical is important in biochemistry and in bioechochemistry because it is a critical water pollutant.

Ovality: A condition in which a circular pipe forms into an ellipse, usually as the result of external forces.

Overpressure: The pressure increase over the set pressure of the primary relief device. When the set pressure is the same as the maximum allowable operating pressure, the accumulation is the same as the overpressure. Pressure increase over the set pressure of the primary relieving device is overpressure. *Note*: From this definition, it will be observed that, when the set pressure of the first (primary) safety or relief valve is less than the maximum allowable working pressure of the vessel, the overpressure may be greater that 10% of set pressure.

Override safety control: A safety strategy that, upon the detection of critical safety conditions, takes over control by overriding all other controls, no matter if they are initiated by other control systems or by operators and brings the process into a "safe state," thereby preventing accidents.

P

P-type semiconductor materials (positive): Have more "holes" (electron deficiencies) than electrons. A hole has a positive electric charge, equal and opposite to the charge on an electron. In a semiconductor material, the flow of holes occurs in a direction opposite to the flow of electrons. P-type materials have an absence of electrons to complete the regular crystalline structure of pure silicon.

Paramagnetic materials: Are attracted by an externally applied magnetic field and form internal induced magnetic fields in the direction of the applied magnetic field.

Partial pressure: In a mixture of gases, the partial pressure of one component is the pressure of that component if it alone occupied the entire volume at the temperature of the mixture.

Particle tracking velocimetry: A term commonly used to refer to analysis of multiphase mixtures using equations of state and standard laboratory analysis methods. The particle tracking velocimetry signature helps the development of a correlation algorithm to improve the accuracy of the in situ multiphase measurement.

Particulate air pollution: Is defined by the U.S. EPA as an air-suspended mixture of both solid and liquid particles. They are often separated into three classifications: coarse, fine, and ultra-fine particles. Coarse particles have a diameter of between 10 µm and 2.5 µm and settle relatively quickly whereas fine (0.1–2.5 µm in diameter) and ultra-fine (<0.1µm in diameter) particles remain in suspension for longer. To put these sizes into perspective, human hair has a diameter of 50–70 µm and a grain of sand has a diameter of 90 µm.

Pascal-second (Pas): Internationally accepted unit of absolute (dynamic) viscosity. Pas = N-s/m^2 = 10 P = 1000 cP.

Passive ultrasonics (or sonics): When the process itself is generating sound or ultrasound, which is then analyzed and interpreted as indication of process variables (flow, leakage, etc.). The source of the sound can be the turbulence of the flowing fluid, the impact of solid particles or bubbles on the pipe walls, or any other noise-generating property.

PDVF (Polvinylidene fluoride): This fluorocarbon has substantially lower temperature limits than the others (250°F or 120°C) and is less inert chemically. It is dissolved by the ketones (acetone, MEK, MIBK) and attached by benzene and high concentrations of sulfuric acid. The most insidious enemy is caustic, which causes brittleness and cracking. It has much better toughness and abrasion resistance that the other fluorocarbons, as well as unique electrical properties ($K = 8$).

PE (Polyethylene): A low-temperature insulation that is compatible with a wide range of corrosives but is attacked by most petroleum products. Generally limited to situations in which fluoro- and chlorocarbons are not allowed, such as the tobacco and nuclear power industries. Maximum allowable temperature is in the 80°C (180°F) area.

PEEK (Polyether etherketone): A high-temperature, injection molded polymer that is chemically quite inert. This material has wide chemical application. Temperature capability is high at 225°C–260°C (450°F–500°F). Avoid any liquids with "phenol" in their names. Adhesive bonding to the molded parts would be difficult.

Peltier effect: Whereby heat is given out or absorbed when an electric current passes across a junction between two materials.

Pelton wheel: A water impulse turbine invented by Lester Allan Pelton in the 1870s. The Pelton wheel extracts energy from the impulse of moving water, as opposed to its weight like traditional overshot waterwheel. Pelton's paddle geometry was designed so that when the rim runs at half the speed of the water jet, the water leaves the wheel with very little speed, extracting almost all of its energy.

Pensky–Martens test: Is a closed-cup test, used in the determination of the flash-point temperature of lubricating oils, liquids containing suspended solids and liquids that tend to form a surface film during testing. It is performed by heating and stirring the test specimen, while at regular intervals, the stirring is stopped and heat from an ignition source is applied until, in each cycle, an ignition of the vapors (flashing) occurs. The peak temperature reached after this occurs in each cycle is the flash point temperature.

Percent passage: Percent passage = 100 − percent rejection; it is a measure of how much ionic contamination passes through the membrane with the permeate.

Percent rejection: Percent rejection = 100 * [C(feed water) − C(permeate)]/C(feed water), where $C(\cdot)$ is conductivity; it is a measure of how much ionic contamination is being removed by reverse osmosis.

Permissible exposure limit: A legally enforceable occupational exposure limit established by the Occupational Safety and Health Administration (OSHA), usually measured as an 8 hr time-weighted average, but also may be expressed as a ceiling concentration exposure limit.

Personal area network: A small network comprising personal devices, focused on a single person (i.e., cell phone, Bluetooth device, etc.).

PFA: Perfluoroalkoxy, a fluorocarbon that is quite inert chemically, melts at a fairly high temperature, and is easily plastic welded. It can be used up to 290°C (550°F), but as a probe insulation, it is generally limited to 175°C (350°F) because of bonding limitations with the metal rod.

PFA (Teflon): Perfluoroalkoxy, a fluorocarbon that is quite inert chemically, melts at a fairly high temperature, and is easily plastic welded. It can be used up to 290°C (550°F), but, as a probe insulation, it is generally limited to 350°F (175°C) because of bonding limitations with the metal rod. It is available in pellet or powder form. PFA combines the processing ease of conventional thermoplastic resins with the excellent properties of polytetrafluoroethylene (PTFE).

Pharmacopeia (USP): Is a nongovernment organization that endorses public health by establishing up to the minute standards to safeguard the quality of medicines and other health care technologies.

Phase difference sensor: A contact radar technology; unlike TDR-based systems, which measure using sub-nanosecond time intervals, derives level information from the changes in phase angle.

Phase shift: Of a transfer function, a change of phase angle with test frequency, as between points on a loop

phase characteristic: Of a signal, a change of phase angle with transmission.

Phase velocity of a wave: Is the velocity at which a phase point on it (such as the crest) travels. The phase velocity (v_p) is the ratio of the wavelength lambda (λ) and the period of oscillation (T) so that $v_p = \lambda / T$.

Phosphate (PO_4^{3-}): Is an inorganic chemical, a salt of phosphoric acid. Inorganic phosphates are mined to obtain the phosphorus, which is used in agriculture and industry. If it enters the receiving waters, it is lethal to fish.

Photocells: Are photoelectric, photovoltaic, or photoconductivity devices, which produce a current or voltage when exposed to light or other types of electromagnetic radiation.

Photodetector: Measures thermal radiation by producing an output through release of electrical changes within its body. They are small flakes of crystalline materials such as CdS or InSb, which respond to different portions of the spectrum, consequently showing great selectivity in the wavelengths at which they operate.

Photometer: An analyzer that measures the absorbance or transmittance at a single or at multiple specific wavelengths. In the multiple-wavelength configuration, calculations can be made to obtain the difference between, or ratio of, the measurements at two specific wavelengths.

Pi electron (π electron): Is an electron that resides in the Pi bond of a double or triple bond.

Piezoelectric effect: The production of electricity or electric polarity by applying a mechanical stress to certain crystals.

Piezoelectric sensor: Is a device that measures changes in dynamic pressure. The prefix piezo is Greek for "press" or "squeeze." They are typically not suited for static pressure measurements. Dynamic pressure measurement applications include turbulence, blast, ballistics, and engine combustion monitoring. Their capabilities include fast response, ruggedness, high stiffness, extended ranges, and the ability to measure quasi-static pressures.

PIG (pipeline inspection gauges) detection: A run inside the pipeline for the inspection, cleaning, and gauging of the internal diameter. Pig detectors are devices with some type of indicator to detect the passage of a pig in the pipeline during a pig run. Pig detectors are usually located near the launcher and receiver ends to confirm successful launch and receipt of the pigs or any trouble area of the pipeline, such as at a tee, elbow, wye, etc. They can be equipped with electrical switches and connected to a PLC to be used to operate piping equipment, such as pumps, compressors, valves, mixing tanks, etc., when strategically located along the pipeline or plant.

Pixel: A physical point or the smallest addressable element in an all-points-addressable display device; so it is the smallest controllable element of a picture represented on a screen. LCD pixels are manufactured in a 2D grid, using dots or squares, and sweep rates. Each pixel is an element of an original image, so more pixels provide more accurate representations of the original. In color image systems, a color is typically represented by three or four component intensities such as red, green, and blue, or cyan, magenta, yellow, and black.

Pixel (picture element): A dot that represents the smallest graphic unit of display in a digital image, used in machine vision and camera technology.

Planar process: Is one by which integrated circuits (IC) are built without manual wiring. The process uses photographic techniques to expose a silicon substrate to create silicon oxide insulators and doped regions acting as conductors. Joining the integrated circuits and using p-n junction, silicon wafers were created in 1960 by using visible or UV light to etch channels into the back of wafers and filling them with epoxy.

Plank's theory: Planck's law of radiation predicts the level of radiation emitted per unit surface area of a blackbody at each specific wavelength. According to Plank's law, the radiation emission peaks at shorter wavelengths as the temperature rises.

Plenum: Air distribution ducting, changer, or compartment.

PM10: Refers to particles smaller than 10 µm. They include dust, pollen, and mold spores. The World Health Organization (WHO) recommends that the PM10 exposure should not exceed 50 µg/m³ mean in any 24 hr period and 20 µg/m³ mean annually.

PM2.5: Refers to particles smaller than 2.5 µm. They include combustion particles, organic compounds, and metals. The World Health Organization (WHO) recommends that the PM2.5 exposure should not exceed 25 µg/m³ mean in any 24 hr period and 10 µg/m³ mean annually.

Poise (μ): Unit of dynamic or absolute viscosity (dyn-s/cm²).

Poiseuille (Pi): Suggested name for the new international standard unit of viscosity, the pascal-second.

Polarity (electric): Is present in all electric circuits and can be negative or positive. Electrons always flow from the negative to the positive pole. In direct current (DC) electric circuits, electrons flow in one direction only. In alternating current

(AC) circuits, the two poles alternate between negative and positive, and the direction of the electron flow reverses back and forth. The pole with relatively more electrons is said to have negative polarity; the other is assigned positive polarity. If the two poles are connected by a conductive path such as a wire, electrons flow from the negative pole toward the positive pole. In physics, the theoretical direction of current flow is considered to be from positive to negative by convention, opposite to the flow of electrons.

Polarography: Process for monitoring the diffusion current flow between working and auxiliary electrodes as a function of applied voltage as it is systematically varied. The concentration of analyte allows for the flow of the diffusion current, which linearly dependent on the analyte concentration. Polarography can be applied using direct current, pulsed direct current, and alternating current voltage excitation waveforms. Dissolved oxygen determination is an example of an application for which polarography is used.

Polytetrafluoroethylene (PTFE): Is a synthetic fluoropolymer of fluoroethylene that has numerous applications. The best-known brand name of PTFE-based formulas is Teflon by the DuPont Company, which discovered the compound. It has one of the lowest coefficients of friction against any solid.

Porosity: Small voids or pores, usually gas filled, in the weld metal.

Positive void coefficient: Increased replacement of water by steam increases reactivity, which increases temperature that further increases boiling.

Potentiometer: A resistance element having first and second terminals, and a third terminal connected with a wiper. The wiper moves along the resistance element, making electrical contact with it. A small current is passed through the resistance element, causing a voltage to develop across it. The voltage at the wiper terminal is a proportion of the difference between the two resistance element terminals, depending on the wiper position.

Potting: Refers to the use of a potting compound to completely surround all live parts, thereby, excluding the hazardous atmosphere; has been proposed as a method of protection. There is no known usage except in combination with other means.

Poundal: It is the foot-pound-second (FPS) unit of force, defined as the force producing an acceleration of 1 ft/s^2 on a mass of 1 lb.

Power factor: The ratio of total watts to the total root mean square (rms) volt-amperes.

$$F_p = \frac{\sum \text{watts per phase}}{\sum \text{rms volt-amperes per phase}} = \frac{\text{active power}}{\text{apparent power}}$$

Note: If the voltages have the same waveform as the corresponding currents, power factor becomes the same as phasor power factor.

PP: Polypropylene, similar to PE. Used for low cost and where fluorocarbons and chlorocarbons are excluded. Maximum temperature is in the area of 200°F.

Precision: A measure of repeatability indicates the ability to repeat measurements within narrow limits.

Precision or random error: The error that is a function of the instrument or system's overall hardware design and reaffirmed as such by the manufacturer in the carefully controlled environment of their facility or flow lab. This is the error listed on the manufacturer's spec sheet.

Pressure, maximum working: The maximum total pressure permissible in a device under any circumstances during operation, at a specified temperature. It is the highest pressure to which it will be subjected in the process. It is a designed safe limit for regular use. *Note*: Maximum working pressure can be arrived at by two methods: (1) designed: by adequate design analysis, with a safety factor; (2) tested: by rupture testing of typical samples.

Pressure regulator: A valve-type device designed to control a reduced pressure at its outlet.

Pressure relief valve: This is a generic term that might refer to relief valves, safety valves, and pilot-operated valves. The purpose of a PRV is to automatically open and to relieve the excess system pressure by sending the process gases or fluids to a safe location when its pressure setting is reached.

Pressure, surge: Operating pressure plus the increment above operating pressure to which a device may be subjected for a very short time during pump starts, valve closings, etc.

Pressure transient: A sudden increase in the steam pressure in the reactor will result in a sudden decrease in the steam-to-water ratio. The increase in the ratio of water to steam will lead to an increase in neutron moderation (because water is a better moderator than steam), which in turn will cause an increase in the power output of the reactor. In short, an increase in steam pressure increases power generation and a decrease in steam pressure reduces the power generation of the reactor. This sequence is called a "pressure transient."

Primary axis: The axis along which the load cell is designed to be loaded. It normally is its geometric center line.

Primary containment: The reinforced, "reverse light bulb" shaped containment of the reactor and its associated equipment. It is also referred to as dry well, because no water is supposed to be accumulated in it.

Primary element (in flow service): Typically an orifice plate, insert flow tube, venturi tube, cone meter, pitot tube, averaging pitot tube, which has a predictable relationship between flow and pressure difference.

Primary standard: A measuring instrument calibrated at a national standard laboratory such as NIST and used to calibrate other sensors.

Probability of failure on demand: A term that is much used, but poorly defined. Probability of failure on demand (PFD) is the sum of spurious trips and covert or hidden failures of an instrument, and the PFD for the safety instrumented system (SIS) is the sum of PFDs for each element in the system. PFD is used in the self-certification of its products by the manufacturer (self-certification), or other independent agency.

Process fieldbus (PROFIbus): A fieldbus-based automation standard that links a controller or control systems with decentralized field devices (sensors and actuators) using application-independent device profiles, with specifications controlled by Profibus International (www.profibus.com).

Process hazard analysis: An OSHA directive that identifies safety problems and risks within a process, develops corrective actions to respond to safety issues, and preplans alternative emergency actions to be triggered if safety systems fail. The process hazard analysis (PHA) must be conducted by a diverse team that has specific expertise in the process being analyzed. Many consulting and engineering firms also provide PHA services. PHA methodologies can include a "what-if analysis," hazard and operability study, failure mode and effects analysis, and fault tree analysis.

Profibus (process fieldbus): A data communication protocol for fieldbus communication in automation technology.

Prompt-gamma neutron activation analysis (PGAA): Is a technique to simultaneously measure the amount of elements in small samples by irradiating it with a beam of neutrons. The elements in the sample absorb some of these neutrons and emit prompt gamma rays. The energies of these rays identify the elements that have captured the neutrons, while the intensities of the peaks reveal their concentrations.

Proof: A unit of specific gravity used in the alchol industry.

Protocol: Formal description of a set of rules and conventions that govern how devices on a network exchange informtion.

Proximity sensors: Are sensors which convert information on the movement or presence of an object into an electrical signal. They are sensors which detect the presence of objects within some distance from it without making physical contact. Because this is a very large unit, a unit equal to one-trillionth of a farad (the picofarad, pF) is commonly used in RF circuits.

PTV: A term commonly used to refer to analysis of multiphase mixtures using equations of state (EOS) and standard laboratory analysis methods. The PVT signature helps the development of a correlation algorithm to improve the accuracy of the in-situ multiphase measurement.

Pulse width modulation (PWM): Typically, a square wave in which the on time to off time duty cycle varies as a percentage, such as from 0 to 100% duty cycle, to represent a signal.

Purge: Flowing a fluid into a system often for measurement reasons.

Purging, pressurization, ventilation: This refers to the maintenance of a slight positive pressure of air or inert gas within an enclosure so that the hazardous atmosphere cannot enter. Relatively recent in general application, it is applicable to any size or type of equipment.

PVDF: Polyvinylidene fluoride, a fluorocarbon that has substantially lower temperature limits than others (250°F or 120°C) and is less inert chemically. It is dissolved by the ketones (acetone, MEK, MIBK) and attacked by benzene and high concentrations of sulfuric acid. The most insidious enemy is caustic, which causes brittleness and cracking. It has much better toughness and abrasion resistance than other fluorocarbons as well as unique electrical properties ($K = 8$).

Pyranometer: Used to measure broadband solar irradiance on a planar surface and is a sensor that is designed to measure the solar radiation flux density (in watts per meter square) from a field of view of 180°.

Pyrheliometer: Can be any of a variety of devices that measure all the intensity of solar radiation received at the Earth. In one design, sunlight enters the instrument through a window and is directed onto a thermopile that converts heat to an electrical signal that can be recorded.

Pyrometer: Noncontact thermometer.

Q

Quevenne degree: A specific gravity unit used in expressing the fat content of milk.

R

Raceway: General term for enclosed channels, conduit, and tubing designed for holding wires and cables.

Rad: Expresses *absorbed* radiation, just as doe gray. (Gray = 100 rad)

Radar: Radio detection and ranging; a system using beamed and reflected radio frequency energy for detecting and locating objects, measuring distance or altitude, navigating, homing, bombing, and other purposes; in detecting and ranging, the time interval between transmission of the energy and reception of the reflected energy establishes the range of an object in the beam's path.

Radiation Quality Factor (Q): Is the factor by which the absorbed dose (red or gray) of radiation must be multiplied to obtain a quantity that expresses, the biological damage (rem or sievert) to the exposed tissue. The term, quality factor, has now been mostly replaced by "radiation weighting factor (WR)".

Radiation weighting factor (WR): Is the term that formerly was called radiation quality factor (Q).

Radioactive isotopes: Atoms with a different number of neutrons than a usual atom, with an unstable nucleus that decays, emitting alpha, beta, and gamma rays until the isotope reaches stability. Once it is stable, the isotope becomes another element entirely. Radioactive decay is spontaneous, so it is often hard to know when it will take place or what sort of rays it will emit during decay.

Radio frequency: A frequency that is higher than sonic but less than infrared. The low end of the RF range is 20 kHz, and its high end is around 100,000 MHz.

Radio frequency interference: A phenomenon in which electromagnetic waves from one source interfere with the performance of another electrical device.

Radius bends: The radius of the bend in the pipe as related to the pipe diameter (D). For example, a 3D bend would have radius bends.

Rag layer: The colorful name for a liquid layer of an undesired quantity of liquid floating above the heavier liquid phase and below the lighter layer in a tank. A radius of three times the diameter of the pipe measured to the centerline of the pipe.

RAM: Random access memory (RAM) is a form of computer data storage. A random access device allows stored data to be accessed directly in any random order. In contrast, other data storage media such as hard disks, CDs, DVDs and magnetic tape, as well as early primary memory types such as drum memory, read and write data only in a predetermined order, consecutively, because of mechanical design limitations.

Raman scattering: Raman scattering or the Raman effect is the inelastic scattering of a photon. It was discovered by C. V. Raman and K. S. Krishnan in liquids, and by G. Landsberg and L. I. Mandelstam in crystals. It can be used as a tool to characterize any nonmetallic material. It is the shift in wavelength of the inelastically scattered radiation that provides chemical and structural information contained in the scattered light. The Raman shifted photons can have higher or lower energy, depending upon the vibrational state of the molecule under study. When the sample is irradiated by a monochromatic laser light, having a wavelength from 532 to 785 nm, the resulting Raman signal is weak and requires sensitive detectors to be measured.

Raman spectroscopy: A spectroscopic technique used to observe vibrational, rotational, and other low-frequency modes in a system. It relies on inelastic scattering, or Raman scattering, of monochromatic light, usually from a laser in the visible, near-infrared, or near-ultraviolet range. The laser light interacts with molecular vibrations, phonons, or other excitations in the system, resulting in the energy of the laser photons being shifted up or down. The shift in energy gives information about the vibrational modes in the system.

Raman spectrum I: The change in wavelength of light scattered while passing through a transparent medium; the collection of new wavelengths is characteristic of the scattering medium and differing from the fluorescent spectrum in being much less intense and in being unrelated to the absorption of the medium.

Random failures: Physical failures due to "random" stresses.

Range (ISA): The region between the limits within which a quantity is measured, received, or transmitted, expressed by stating the lower and upper range values.

Range, elevated zero: A range in which the zero value of the measured variable, measured signal, etc., is greater than the lower range value. *Note 1*: The zero may be between the lower and upper range values, at the upper range value, or above the upper range value. *Note 2*: Terms suppression, suppressed range, or suppressed span are frequently used to express the condition in which the zero of the measured variable is greater than the lower range value. The term "elevated zero range" is preferred.

Range limit, lower: The lowest value of the measured variable that a device can be adjusted to measure. *Note*: The following compound terms are used with suitable modifications to the units: measured variable lower range limit, measured signal lower range limit, etc.

Range limit, upper: The highest value of the measured variable that a device can be adjusted to measure. *Note*: The following compound terms are used with suitable modifications to the units: measured variable upper range limit, measured signal upper range limit, etc.

Range, suppressed zero: A range in which the zero value of the measured variable is less than the lower range value. (Zero does not appear on the scale.) *Note 1*: For example, 20–100. *Note 2*: Terms elevation, elevated range, or elevated span are frequently used to express the condition in which the zero of the measured variable is less than the lower range value. The term "suppressed zero range" is preferred.

Range value, lower: The lowest value of the measured variable that a device is adjusted to measure. *Note*: The following compound terms are used with suitable modifications to the units: measured variable lower range value, measured signal lower range value, etc.

Range value, upper: The highest value of the measured variable that a device is adjusted to measure. *Note*: The following compound terms are used with suitable modifications to the units: measured variable upper range value, measured signal upper range value, etc.

Rangeability (recommended by IEH): Rangeability of a sensor is the measurement range over which the error statement, in the units of a percentage of actual reading, is guaranteed. Rangeability (R) of a flowmeter is the flow range over which the error does not exceed the guaranteed specified value.

Rankine: Is a thermodynamic scale introduced by Willian John Macquon Rankine in 1859. It is the absolute temperature scale in Fahrenheit units. The zero on the Rankine scale corresponds to −459.67°F and the size of each Rankine degree is the same as one Fahrenheit degree. Therefore, °R = °F + 459.67. Conversion from Rankine to other units is: °R =°F + 459.67 = (°C + 273.15) × 9⁄5 = K × 9⁄5.

Rated capacity: The maximum load that the load cell is designed to measure.

Rated output: The algebraic difference between the outputs at no load and at rated load.

Rated relieving capacity: The maximum relieving capacity of the PRV. This data is normally provided on the nameplate of the PRV. The rated relieving capacity of the PRV exceeds the required relieving capacity and is the basis for sizing the vent header system.

Ratio pyrometer: See two-color pyrometer.

Raw conductivity: A measured conductivity that is not compensated for temperature changes.

Rayleigh disk: An acoustic radiometer used to measure particle velocity, consisting of a thin disk set at an angle of 45° to a sound beam; the particle velocity is calculated from the resulting torque on the disk.

Rayleigh ratio: Is the quantity used to characterize the scattered intensity at the scattering angle of the incident light or other forms of radiation.

Rayleigh scattering: Is named after the British physicist Lord Rayleigh and is the scattering of light or other types of radiation by particles much smaller than the wavelength of the radiation. Rayleigh scattering does not change the state of the material; hence, it is a paramagnetic process.

Reactance (X): That part of the impedance of a circuit that is caused by either capacitance, inductance, or both. Expressed in units of ohms (Ω).

Reactivity: The rate of heat release in a nuclear reactor. This is the portion of nuclear fission energy that is available to generate steam.

Readability: The value of the finest division on an analog scale or the smallest increment of weight that a digital display resolves.

Reagent: Substance or compound that is added to a system in order to bring about a chemical reaction, or added to see if a reaction occurs.

Rear mount: A technique for making long inactive sections by mounting the probe on the end of a pipe, with its coax cable running through the pipe to the top of the tank. The coax must survive the process temperature, so it is often of high-temperature construction.

Recommended exposure limit: The ceiling value that should not be exceeded at any time.

Rectifier: An electronic device such as a semiconductor diode or valve that converts an AC into a DC by suppression or inversion of the alternate half cycles.

Redundancy: A design by which additional paths are in place in case one path fails.

Reference accuracy: Usually defined as the accuracy that includes only the effects of hysteresis, linearity, and repeatability. Modern field electronic instrumentation reference accuracy is typically in the 0.1% FS range, and some of the latest transmitters coming on the market are even more accurate in the 0.03%–0.08% FS accuracy range. Field calibrators have accuracy ratings in the 0.02%–0.08% FS of reading range and deadweight testers are in the range of 0.01%–0.05% FS of reading.

Reference junction: That thermocouple junction which is at a known or reference temperature. *Note*: The reference junction is physically that point at which the thermocouple or thermocouple extension wires are connected to a device or where the thermocouple is connected to a pair of lead wires, usually copper.

Reference junction compensation: A means of counteracting the effect of temperature variations of the reference junction, when allowed to vary within specified limits.

Reflectance or reflectivity (R): The amount of light a target reflects, expressed as a percentage of the incident light. Diffuse reflectance refers to the amount of light scattered in all directions by a diffuse target. Specular reflectance refers to the amount of light reflected by a mirror. Reflectance will depend on target color and composition and on the frequency of the light being reflected. Diffuse surfaces typically vary from 3% to 95% reflectance.

Reflected ray: The ray of light that leaves the object.

Reflection angle: The angle between the reflected ray and the normal.

Refraction: Refraction is essentially a surface phenomenon. When light is incident at a transparent surface, the transmitted component of the light (which goes through the interface) changes direction at the interface. Another component of the light is reflected at the surface. The refracted beam changes direction at the interface and deviates from a straight continuation of the incident light ray. Refraction of light is the most commonly observed phenomenon, but any type of wave can refract when it interacts with a medium, for example, when sound waves pass from one medium into another or when water waves move into water of a different depth.

Refractive index (RI): Or index of refraction of a material is a dimensionless number that describes how light propagates through that medium. It is defined as n = c/v, where c is the speed of light in vacuum and v is the speed of light in the medium. For example, the refractive index of water is 1.33, meaning that light travels 1.33 times faster in a vacuum than it does in water. The refractive index determines how much light is bent, or refracted, when entering a material.

Refractive index (n): Is the ratio of the velocity of light in a vacuum to its velocity in a specified medium. It determines how much light is bent, or refracted, when entering a material. It is defined as the ratio of the speed of light in vacuum, c = 299,792,458 m/s, and the phase velocity (v) of light in the medium. Historically, air at a standardized pressure and temperature has been used as a reference medium, having a refraction index of unity.

Reid vapor pressure: Is based on ASTM D323-99a Standard Test Method for vapor pressure of petroleum products. It is the vapor pressure of a chilled sample of gasoline or other fuel as measured in a test bomb at 100 F. The Reid vapor pressure is applicable only for gasoline, volatile crude oil, and other volatile petroleum products. It is not applicable for liquefied petroleum gases.

Relative humidity: The ratio of the mole fraction of moisture in a gas mixture to the mole fraction of moisture in a saturated mixture at the same temperature and pressure. Alternatively, the ratio of the amount of moisture in a gas mixture to the amount of moisture in a saturated mixture at equal volume, temperature, and pressure (the ratio of how much water vapor is in the air versus the maximum it could contain at the particular temperature).

Relative viscosity: The viscosity of the solution related to the viscosity of the solvent.

$$\eta_r = \frac{\eta}{\eta_s}$$

Relative volatility: Is a measure of the tendency to vaporize. The higher the volatility, the higher the vapor pressure of the liquid is at the same temperature. In distillation processes, the higher the volatility, the easier it is to separate the component from the less volatile ones. In the case of automotive fuels, it is desirable to use higher volatility gasoline in colder regions, but not be high enough to cause vapor lock. The proportion of butane increases the volatility of the blended product, and because butane is relatively inexpensive, cost is reduced by increasing its proportion. One measure of volatility is the vapor–liquid ratio (abbreviated as or V/L or V-L).

Reliability: That a system be successful for an interval of time.

Relief valve: An automatic pressure-relieving device actuated by the static pressure upstream of the valve, which opens in proportion to the increase in pressure over the operating pressure. It is used primarily for liquid service.

Relieving pressure: The sum of opening pressure plus overpressure. The pressure, measured at the valves inlet, at which the relieving capacity is determined.

REM: The unit of exposure to external radiation (roentgen + equivalent + man). A rem is a measure of the dose to body tissue in terms of its estimated biological effect relative to a dose of 1 R of x-ray. A person receives the dose of 1 rem when exposed to 1 R of radiation in any time period. A person should not receive more than 250 rems over an entire lifetime.

Reopening pressure: The opening pressure when the pressure is raised as soon as practicable after the valve has reseated or closed from a previous discharge.

Repeatability (ISA): The closeness of agreement among a number of consecutive measurements of the output for the same value of the input under the same operating conditions, approaching from the same direction, for full-range traverses.

Repeatability (NIST): Closeness of agreement between the results of successive measurements of the same measurand carried out under the same conditions of measurement. Repeatability may be expressed quantitatively in terms of the dispersion characteristics of the results.

Reproducibility (ISA): In process instrumentation, the closeness of agreement among repeated measurements of the output for the same value of input made under the same operating conditions over a period of time, approaching from both directions.

Reproducibility (NIST): Closeness of agreement between the results of measurements of the same measurand, carried out under changed conditions of measurement.

Residual chlorine: The difference between total and free chlorine.

Resistance (R): The opposition offered by an electrical conductor to the flow of direct electric current. Expressed in units of ohms (Ω).

Resistive component: An AC can be separated into two components. The portion that is in phase with the excitation voltage is called the resistive component.

Resistivity (ρ): The property of a conductive material that determines how much resistance a unit cube will produce. Expressed in units of ohm-centimeters (Ω-cm).

Resistor–capacitor circuit (RC): Is also called an RC filter and it is an electric circuit composed of resistors and capacitors driven by a voltage or current source. RC circuits can be used to filter a signal by blocking certain frequencies and passing others. The two most common RC filters are the high-pass, the low-pass, and the band-stop filters.

Resolution: The smallest amount of input signal change that the instrument can detect reliably. This term is determined by the instrument noise (either circuit or quantization noise).

Resonance: Of a system or element, a condition evidenced by large oscillatory amplitude, which results when a small amplitude of periodic input has a frequency approaching one of the natural frequencies of the driven system.

Retroflector: A device or surface that reflects light back to its source with a minimum of scattering. It is sometimes also called a retroflector or cataphote.

Retroreflection: The reflection of light off a target back in the direction from which it came, for a wide range of angles of incidence. Retroreflection is achieved through multiple reflections within a retroreflector.

Reverse osmosis: A technique for water purification that forces the feed water at elevated pressure through a semipermeable membrane, which holds back ions and other molecules, resulting in permeate that is purified and less conductive than the feed water.

Reynolds number (Re): Reynolds number expresses the ratio of inertial forces to viscous forces. At a very low Reynolds number, viscous forces predominate and inertial forces have little effect. The Reynolds number reflects the degree of turbulence of the flowing fluid. At low Reynolds numbers, the flow is laminar; at around 10,000, it gradually becomes turbulent. The Reynolds number is calculated as follows: $Re = 3160(SG)(Q)/(ID)\mu$.

Rheology: The science of the deformation and flow of matter.

Rheometers: Characterize the viscoelastic properties of polymer melts and other materials. They are laboratory instruments that are used to measure the nature of the flow of polymers and other liquids, suspensions, or slurries when a force causes them to move. They are used for those fluid flows that cannot be defined by viscosity alone and require more parameters to be measured and controlled in order to define the polymer's flow characteristics. Rheometers are quality control tools and serve to assess the processability of resins, are used by plastics compounders, and also serve as R&D tools to help determine which resin best fits a particular process or application.

Rheopecty: A reversible time-dependent increase in viscosity at a particular shear rate; the longer the fluid undergoes shearing, the higher its viscosity.

Rheostat: Constructed in a similar way as a potentiometer but with contact to only one end of the resistance element. Moving the wiper provides a variable resistance depending on the wiper position.

Richter degrees: A unit of specific gravity used in their alcohol industry.

Ringelmann scale: A scale for measuring the apparent density of smoke. It has five levels of density inferred from a grid of black lines on a white surface which, if viewed from a distance, merge into known shades of gray. There is no definitive chart, rather Prof. Ringelmann provides a specification, where smoke level 0 is represented by white; levels 1–4 by 10 mm square grids drawn with 1, 2.3, 3.7, and 5.5 mm wide lines, and level 5 by all black.

RMS–surface roughness: Surface roughness expressed in root mean square of the roughness in micro-inch units.

Rockwell scale: Is a hardness scale based on the indentation of materials. The Rockwell test determines the hardness by measuring the depth of penetration of an indenter under a large load compared to the penetration made by a preload. There are different scales, denoted by single letters (Rockwell A, B, C, etc.), corresponding to different loads or indenters. The result is a dimensionless number noted as HRA, HRB, HRC, etc., where the last letter refers to the respective Rockwell scale.

Rodding out: Using a rod or poker to open up a clogged connection.

Roentgen (R): The dose of radiation exposure defined as the quantity of radiation that will produce one electrostatic unit of ionization to one cubic centimeter of dry air. A 1 Ci source will produce a dose of 1 R at a receiver placed 1 m (3 ft) away from the source for 1 hr. The dose rate unit is the roentgen/hour (R/hr). NIST in 1998 defined it as 2.58×10^{-4} Ci/kg (1 Ci/kg = 3876 R).

Root mean square: A value representing the amount of DC signal that is required to generate the same amount of heat in the same load as does the particular AC signal.

Root sum square: Method to calculate the error as the square root of the sum of the squares of the individual error components.

Root valve: The first valve of the process.

Resistance temperature detector: A component having a resistance that varies with temperature, usually constructed as a platinum conductor (wire or circuit trace) on a substrate material such as ceramic. Most commonly the resistance at 0°C is 100 or 1000 Ω.

Reversed-phase chromatography: Also called hydrophobic chromatography, it includes any liquid chromatographic method that uses a hydrophobic stationary phase. In the 1970s, most liquid chromatography was performed using a solid support stationary phase (also called a "column") containing unmodified silica or alumina resins.

RS232: Short for recommended standard-232C, a standard interface approved by the Electronic Industries Alliance (EIA) for connecting serial devices. In 1987, the EIA released a new version of the standard and changed the name to EIA-232-D. And in 1991, the EIA teamed up with Telecommunications Industry Association (TIA) and issued a new version of the standard called EIA/TIA-232-E. Many people, however, still refer to the standard as RS-232C, or just as RS-232.

Rupture tolerance: The tolerance range on either side of the marked or rated burst pressure within which the rupture disc is expected to burst. Rupture tolerance may also be represented as a minimum–maximum pressure range. Also referred to as performance tolerance in ISO standards.

RYTON: Is "Solvay Specialty Polymers" trade name for polyphenylene sulfide.

S

Safety integrity level: A relative level of risk reduction provided by a safety function, or to specify a target level of risk reduction. In simple terms, SIL is a measurement of performance required for a SIF. In the European functional safety standards based on the IEC61508 standard, four SILs are defined, with SIL 4 being the most dependable and SIL 1 being the least. A SIL is determined based on a number of quantitative factors in combination with qualitative factors such as development process and safety life cycle management.

Safety instrumented function: Designed to prevent or mitigate a hazardous event by taking a process to a tolerable risk level. An SIF is composed of a combination of logic solver(s), sensor(s), and final element(s) and is assigned a safety integrity level (SIL) depending on the degree of risk that can be tolerated. One or more SIFs comprise an SIS.

Safety instrumented system: An SIS is designed to prevent or mitigate hazardous events by taking the process to a safe state when predetermined safe condition limits are violated. An SIS is composed of a combination of logic solver(s), sensor(s), and final control element(s). Other common terms for SISs are safety interlock systems, emergency shutdown systems, and safety shutdown systems. An SIS can consist of one or more SIFs.

Safety relief valve: An automatic pressure-actuated relieving device suitable for use as either a safety or relief valve.

Safety valve: An automatic pressure-relieving device actuated by the static pressure upstream of the valve and characterized by rapid and full opening or pop action. It is used for steam, gas, or vapor service.

Sampling period: The time interval between observations in a periodic sampling control system.

Sand filling: All potential sources of ignition are buried in a granular solid, such as sand. The sand acts partly to keep the hazardous atmosphere away from the sources of ignition and partly as an arc quencher and flame arrester. It is used in Europe for heavy equipment, it is not used in instruments.

Saturated BTU: The heating value that is expressed on the basis that the gas is saturated with water vapors. This state is defined as the condition when the gas contains the maximum amount of water vapors without condensation, when it is at base pressure and 60°F temperature.

Saturated solution: A solution that has reached the limit of solubility.

Saturation: A condition in which probe-to-ground RF current is determined solely by the impedance of the probe insulation. Increased conductivity in the saturating medium, even to infinity, will not cause a noticeable change in that current or in the transmitter output.

Saturation pressure: The pressure of a fluid when condensation (or vaporization) takes place at a given temperature. (The temperature is the saturation temperature.)

Saybolt Furol seconds: Time units referring to the Saybolt viscometer with a Furol capillary, which is larger than a universal capillary.

Saybolt universal seconds: Time units referring to the Saybolt viscometer.

Saybolt viscometer (universal, Furol): Measures time for a given volume of fluid to flow through standard orifice; units are seconds.

Scale: A multiplication factor to be applied to the 0%–100% transmitter output to obtain the measurement physical value quantitatively.

Scintillation: Generates a flash of light when certain materials absorb ionizing radiation.

Scintillation counter: A device that measures ionizing radiation.

Scintillation detectors: Scintillation detectors are gamma radiation detectors that utilize inorganic crystals that emit light when exposed to gamma-radiation. They sense the light photons resulting from gamma rays incident on certain crystal materials. This is the most sensitive but least stable detector. It is particularly affected by variations in environmental conditions such as temperature.

Scramming the reactor: Automatically shoving all control rods into the core (using gravity, hydraulics, or a mechanical spring force) to stop the chain reaction.

SD Cards (or Secure Digital Cards): Are ultra flash memory cards designed to provide high-capacity memory for cameras in a small size.

Sealing: Excluding the atmosphere from potential sources of ignition by sealing such sources in airtight containers. This method is used for components such as relays, not for complete instruments.

Seal-off pressure: The pressure, measured at the valve inlet after closing, at which no further liquid, steam, or gas is detected at the downstream side of the seat.

Secondary element, or secondary: The indicating gauge, manometer, or transmitter displaying or transmitting a signal derived from the sensed pressure or pressure difference.

Seebeck coefficient: This coefficient gives the amount of voltage generated (in microvolts) by a temperature change of 1°C in a particular thermocouple.

Seebeck effect: The conversion of temperature differences directly into electricity when two different metals are joined in two places, with a temperature difference between the junctions. This is because two metals respond differently to the temperature difference and create a current loop and a magnetic field.

Segment: The section of a network that is terminated in its characteristic impedance. Segments are linked by repeaters to form a complete network.

Semiconductors: Are materials which can conduct electricity only if they receive energy from increasing temperature or from doping. They are solid chemical elements or compounds, that can conduct electricity under some conditions but not others, making it a good medium for the control of electrical current. Their conductance varies depending on the current or voltage applied to a control electrode, or on the intensity of irradiation by infrared, visible light, ultraviolet (UV), or x-ray radiation. The specific properties of a semiconductor depend on the impurities, or "dopants". A single integrated circuit (IC), such as a microprocessor chip, can do the work of a set of vacuum tubes that would fill a large building and require its own electric generating plant.

Sensitivity: The ratio of the change in output magnitude to the change in the input, which causes it after the steady state has been reached. *Note 1*: It is expressed as a ratio with the units of measurement of the two quantities stated. (The ratio is constant over the range of a linear device. For a nonlinear device, the applicable input level must be stated.) *Note 2*: Sensitivity has frequently been used to denote the dead band. However, its usage in this sense is deprecated since it is not in accordance with the accepted standard definitions of the term.

Service: Term used by NFPA-70 (NEC) to demarcate the point at which utility electrical codes published by IEEE (NESC) take over. Includes conductors and equipment that deliver electricity from utilities.

Set pressure (opening pressure): The pressure at which the relief valve is set to open. It is the pressure measured at the valve inlet of the PRV at which there is a measurable lift, or at which discharge becomes continuous as determined by seeing, feeling, or hearing. In the pop-type safety valve, it is the pressure at which the valve moves more in the opening direction compared to corresponding movements at higher or lower pressures. A safety valve or a safety relief valve is not considered to be open when it is simmering at a pressure just below the popping point, even though the simmering may be audible.

Shale shaker: Separates drill cuttings from the drilling fluid before it is pumped back down the wellbore.

Shear: The relative movement of parallel adjacent layers.

Shear rate: The rate of change of shear with time.

Shear stress: The component of stress that causes successive parallel layers of a material body to move, in their own planes (i.e., the plane of shear), relative to each other.

Shear thickening: An increase in viscosity with increasing shear rate during steady shear flow.

Shear thinning (pseudoplastic): A decrease in viscosity with increasing shear rate during steady shear flow.

Shear viscometer: Viscometer that measures viscosity of a non-Newtonian fluid at several different shear rates.

Short-term exposure limit: A maximum concentration for a continuous 15 min exposure period according to NIOSH or a maximum of four such periods per day, with at least 60 min between exposure periods, and provided the daily threshold limit value–time-weighted average (TLV–TWA) is not exceeded, according to OSHA.

Shunt: Is an electronic component which allows electric current to pass around a point in the circuit by creating a low-resistance bypass path. The origin of the term is in the verb "to shunt" meaning to follow a different path.

Side loading: Any load acting 90° to the primary axis of loading on a weight scale.

Sievert (Sv): Sievert expresses *equivalent* radiation, while gray expresses *absorbed* radiation. Sievert is a measure of the health effect of low levels of ionizing radiation on the human body as it expresses radiation dose quantities. Quantities that are measured in sieverts are intended to quantify the health risk in terms of cancer induction, genetic damage, etc. The Sv unit is used in most nations, except for the United States, where rems are still used. The Sv equals 100 rems. Conventionally, sievert is not used for high levels of radiation, which produce deterministic effects. These effects are compared to the physical quantity absorbing the dose and are measured by the unit gray (Gy).

Sikes degree: A unit of specific gravity used in the alcohol industry.

Silistors: Are thermistors with positive resistance/temperature coefficient and are made of silicon or barium, lead and strontium titanates.

Simmer (WARN): The condition just prior to opening at which a spring-loaded relief valve is at the point of having zero or negative forces holding the valve closed. Under these conditions, as soon as the valve disc attempts to rise, the spring constant develops enough force to close the valve again.

Slammer worm: The slammer worm, also known as Sapphire and SQL Hell, is one of the fastest computer worms yet. In the case of a cyber attack on a nuclear power plant, it infected more than 90% of vulnerable hosts within 10 min and reached its full scanning rate (of more than 55 million scans/second) in 3 min.

Slug: The FPS unit of mass, defined as the mass that will be accelerated by 1 ft/s^2, when a 1 lb$_f$ acts upon it. It is equivalent to approximately 32.2 lb mass.

Smart electric meter: Communicates the electric utility and enables two-way communication between the meter and the central system. It is this two-way communication capability that differentiates it from remote meter readers systems, which operate only in one direction.

Smart field service: Is a microprocessor-based process transmitter or actuator that supports two-way communications with a host, digitizes the transducer signals, and digitally corrects its process variable values to improve system performance. The value of a smart field device lies in the quality of date it provides.

Snell's law: Also known as the Snell–Descartes law and the law of refraction, it is a formula used to describe the relationship between the angles of light incidence and light refraction, when referring to light passing through a boundary between two different substances, such as water, glass, and air.

Sorption isotherm: Describes the equilibrium of the sorption of a material at a surface at constant temperature.

Span: The algebraic difference between the upper and lower range values.

Note 1: For example:
Range 0°F to 150°F, Span 150°F
Range −20°F to 200°F, Span 220°F
Range 20°F to 150°C, Span 130°C

Note 2: The following compound terms are used with suitable modifications to the units: measured variable span, measured signal span, etc.

Note 3: For multirange devices, this definition applies to the particular range that the device is set to measure.

Sparger: In chemistry, sparging, also known as gas flushing in metallurgy, is a technique that involves bubbling a chemically inert gas, such as nitrogen, argon, or helium, through a liquid. This can be used to remove dissolved gases from the liquid.

Specific gravity: The ratio of a fluid density to a reference density. For liquids the density of a "standard" water is the reference; for gases the density of air at a specified pressure and temperature is used.

Specific humidity: The ratio of the mass of water vapor to the mass of dry gas in a given volume.

Specific viscosity (or viscosity relative increment): DIN 1342–2 no longer recommends the term "specific viscosity"; therefore, "relative viscosity change" is expressed as

$$\frac{(\eta - \eta_s)}{\eta_s}$$

Specific volume (v): Is property of materials defined as its volume per unit mass. Specific volume is inversely proportional to density. If the density of a substance doubles its specific volume, as expressed in the same base units, is cut in half. In the International System of Units (also referred to as SI or metric units), it is usually given as the number of cubic meters occupied by 1 kg of the particular substance (m^3/kg) or in terms of the number of cubic centimeters occupied by one gram of a substance (cm^3/g). To convert m^3/kg to cm^3/g, multiply by 1000; conversely, multiply by 0.001.

Spectral emissivity: The ratio of emittance at a specific wavelength or very narrow band to that of a blackbody at the same temperature.

Spectrophotometer: An analyzer that obtains the spectra of the sample through wavelength scanning. Continuous measurements can be obtained by periodically detecting the changes in the composition of the flowing sample. Software is available to enlarge or reduce the spectra obtained, detect peaks, or make various other calculations, based on the spectra.

Specular reflection: Specular reflection occurs when light strikes a shiny or mirror-like surface and is reflected away at one angle. Glass, liquid surfaces, and polished metals are specular and generally require a sensor configured specifically for specular surfaces. Acuity makes versions of the 4000 that can be used on specular surfaces, including liquids.

Sphere pig: Spherical utility pig made of rubber or urethane. The sphere may be solid or hollow, filled with air or liquid. The most common use of sphere pigs is as a batching pig.

Stand: A section of two or three joints of drill pipe connected together and stood upright in the derrick.

Standard deviation (σ): This is a quantitative value expressing the variation (dispersion) from the mean value of test results of a number of instruments.

Standard temperature and pressure: NIST uses a temperature of 20°C (293.15°K, 68°F) and an absolute pressure of 101.325 kPa (14.696 psi, 1 atm). In chemistry, IUPAC established the standard temperature and pressure (informally abbreviated as STP) as a temperature of 273.15°K (0°C, 32°F) and an absolute pressure of 100 kPa (14.504 psi, 0.986 atm, 1 bar). An unofficial but commonly used standard is the standard ambient temperature and pressure (SATP) as a temperature of 298.15°K (25°C, 77°F) and an absolute pressure of 100 kPa (14.504 psi, 0.986 atm). The International Standard Metric Conditions for natural gas and similar fluids are 288.15 K (59.00°F; 15.00°C) and 101.325 kPa.

Start-to-leak pressure: The pressure at the valve inlet at which the relieved fluid is first detected on the downstream side of the seat before normal relieving action takes place.

Static error band: A representative indication of a combination of errors at room temperature, usually including nonlinearity, hysteresis, and repeatability.

Steam–water ratio: The volume of steam compared to the volume of water in boiling water.

Stiction: Combination of sticking and slipping when stroking a control valve.

Stiction (static friction): Resistance to the start of motion, usually measured as the difference between the driving values required to overcome static friction upscale and downscale.

Stiffness: In process control, the ratio of change of force (or torque) to the resulting change in deflection of a spring-like element. *Note*: Stiffness is the opposite of compliance.

Stoke: Unit of kinematic viscosity $\nu/(cm^2/s)$.

Stokes shift: Is named after the Irish physicist George G. Stokes and is the difference in frequency between maximum positions of the absorption (fluorescence) and the emission (Raman) spectra of the same electronic transition. When an atom absorbs energy, it enters an excited state, and one way for it to relax is to lose energy by emitting a photon. When the emitted photon has less energy than the absorbed photon, this energy difference is called the Stokes shift. If the emitted photon has more energy, the energy difference is called an anti-Stokes shift.

Streaming current: Streaming current and streaming potential are two interrelated electrokinetic phenomena of surface chemistry and electrochemistry. They are the electric current or potential developed, when an electrolyte is forced to flow through a channel or porous plug having electrically charged walls.

Stroboscope: An instrument for determining the speed of cyclic motion (as rotation or vibration) that causes the motion to appear slowed or stopped. This is achieved by (1) a revolving disk with holes around the edge through which an object is viewed, (2) a device that uses a flashlight that is intermittently illuminate a moving object, or (3) a disk with marks that is viewed under intermittent light.

Stress: Force/area (F/A).

Subchannel: In broadband terminology, a frequency-based subdivision creating a separate communication channel.

Superimposed backpressure: Variable backpressure that is present in the discharge header before the PRV starts to open. It can be constant or variable, depending on the status of the other PRVs in the system.

Supervisory control and data acquisition: A telemetry system comprising a main computer station that allows an operator to monitor data and control I/O.

Suppression pool: A body of water usually located below the primary containment (also called the dry well), which serves to condense steam either escaping through leakage or intentionally relieved into it by relief valves. This pool is also called wet well or, if doughnut shaped (as is the case in older plants), the torus.

Suppression ratio (of a suppressed zero range): The ratio of the lower range value to the span.
Note: For example:
Range 20–100,
Suppression ratio = 20/80 = 0.25

Surface acoustic wave: An acoustic wave traveling along the surface of a material exhibiting elasticity, with an amplitude that typically decays exponentially with depth into the substrate.

Surveyor's theodolite: A precision instrument for measuring angles in the horizontal and vertical planes. Theodolites have been adapted for specialized purposes in fields like metrology and engineering technology. A modern theodolite consists of a movable telescope mounted within two perpendicular axes: the horizontal or trunnion axis and the vertical axis. When the telescope is pointed at a target object, the angle of each of these axes can be measured with great precision, typically to seconds of arc.

Switched hub: A multiport bridge joining networks into a larger network.

Swivel: The top end of the kelly that allows the rotation of the drill string without twisting the block.

Synchronous serial interface: A binary digital communication protocol common for use with encoders and position sensors that clocks out data from a field device into a receiving device synchronous with a clock signal provided by the receiving device.

Systematic error: The error resulting from the meter's actual field installation.

Systematic failures: Sometimes called functional failures and are almost always attributable to design or specification faults.

Systematic or bias error: The error resulting from the meter's actual field installation.

T

Tachometers: An instrument measuring the rotation speed of a shaft or disk, as in a motor or other machine. The device usually displays revolutions per minute on a calibrated analog dial or on a digital display. The word comes from the Greek words tachos (meaning speed) and metron (meaning measure).

Tap: The take-off point in the wall of the tank or vessel.

Tare: The weight of an empty container or the action of deducting the weight of the container from the total weight, so that the indicator directly reads the net weight only. For example, if the capacity of the scale is 500 and 200 lb is tared, the remaining net capacity of the scale is 300 lb.

TCP/IP (Transmission Control Protocol/Internet Protocol): The basic communication language or protocol of the Internet. When one has direct access to the Internet, the computer is provided with a copy of the TCP/IP program just as every other computer that one might send messages to or get information from.

Teflon: Brand name of polytetrafluoroethylene.

Temperature, ambient: The temperature of the medium surrounding a device. *Note 1*: For devices that do not generate heat, this temperature is the same as the temperature of the medium at the point of device location when the device is not present. *Note 2*: For devices that generate heat, this temperature is the temperature of the medium surrounding the device when it is present and dissipating heat. *Note 3*: Allowable ambient temperature limits are based on the assumption that the device in question is not exposed to significant radiant energy sources.

Temperature effect on rated output: The amount of change in the rated output that is caused by a change in ambient temperature. It is expressed as percentage change in rated output per 100°F change in ambient temperature.

Temperature effect on zero balance: The change in zero balance caused by a change in ambient temperature. It is expressed as change in zero balance in percent of rated output per 100°F change in ambient temperature.

Temperature range, compensated: The range over which the load cell is compensated to maintain the rated output and the zero balance within specific limits.

Temperature range, safe: The extremes of temperature within which the load cell can operate without permanent adverse effect on any of its performance characteristics.

(P)TFE*: Tetrafluoroethylene (*P* stands for *polymerized*, which is understood). The oldest, highest temperature and most inert fluorocarbon probe insulation. Extremely difficult to adhesive bond, it is usable up to 290°C (550°F). On probes, its temperature limit is determined by the type of bonding to the probe rod (300°F, 450°F, or 550°F). This is the most common probe insulation in the industry. Because it never melts (but disintegrates, producing HF, at >600°F), it is difficult to fabricate, impossible to plastic weld, and exhibits a high degree of microporosity. Can be destroyed by butadiene and styrene monomer.

---

* Most people interchange the name Teflon® with TFE. This is *completely* incorrect but understandable. TFE was the first fluorocarbon polymer to carry the trade name Teflon at E. I. DuPont. Dupont chose to use the Teflon trade name for a whole *family* of fluorocarbon resins, so FEP and PFA made by Dupont are also called Teflon. To complicate the matter, other companies now manufacture TFE, FEP, and PFA, which legally cannot be Teflon, since that name applies only to DuPont-made polymers.

Thermal conductivity: The ability of a gas to conduct heat. It is often expressed as a ratio relative to the thermal conductivity of air.

Thermal megawatt (MWt): The thermal energy generated by a nuclear reactor. This energy is about three times the electric energy that is generated, because some two-thirds of it is wasted during conversion and due to the discharging of the waste heat into the environment.

Thermal power: The total amount of heat generated by the reactor. This quantity of energy is less than the "flux power" because not all the neutrons impact a uranium atom in the fuel and this quantity of energy is more than the energy represented by the generated electric power because of the heat losses and equipment efficiencies involved.

Thermistor: A resistor having a resistance that varies significantly with temperature than do standard resistors. Thermistors are widely used as inrush current limiters, temperature sensors, self-resetting overcurrent protectors, and self-regulating heating elements. Materials used in a thermistor are usually a ceramic or polymer, while resistant temperature detectors use pure metals.

Thermocouple: A combination of two different conductors, usually metal alloy wires, producing a voltage difference in proportion to temperature. Usually comprising a first alloy connected to a second alloy and forming a "hot junction"; the other end of the second alloy also connected to another conductor of the first alloy, forming a "cold junction." The difference in temperatures between the hot and cold junctions producing a measured voltage that indicates the temperature difference.

Thermopile: Measures thermal radiation by absorption to become hotter than its surroundings. It is a number of small thermocouples arranged like the spokes of a wheel with the hot junction at the hub. The thermocouples are connected in series, and the output is based on the difference between the hot and cold junctions.

Thixotropy: A reversible time-dependent decrease in viscosity at a particular shear rate. Shearing causes a gradual breakdown in structure over time.

Threshold limit value: An occupational exposure value recommended by the American Conference of Governmental Industrial Hygienists (ACGIH) to which nearly all workers can be exposed day after day for a working lifetime without ill effect (ACGIH 2008).

Throughput: The maximum number of transactions per second that can be communicated by the system.

Thyristors: Are solid-state semiconductors with four layers of alternating N- and P-type material. They are bi-stable switches conducting when their gate receives a current trigger, and continue to conduct while they are forward-biased (that is, while the voltage across the device is not reversed). A three-lead thyristor is designed to control the larger current of its two leads by combining that current with the smaller current or voltage of its other lead, known as its control lead. In contrast, a two-lead thyristor is designed to "switch on" if the potential difference between its leads is sufficiently large, a value representing its breakdown voltage.

Time, dead: The interval of time between initiation of an input change or stimulus and the start of the resulting observable response.

Time domain reflectometry: An instrument that measures the electrical characteristics of wideband transmission systems, subassemblies, components, and lines by feeding in a voltage step and displaying the superimposed reflected signals on an oscilloscope equipped with a suitable time–base sweep.

Timeout: An event that occurs when one network device expects to hear from other network devices within a specified period of time but does not. The resulting timeout usually results in a retransmission of information or the dissolving of the session between the two devices.

Time-weighted average: A concentration limit for a normal 8 hr workday or 40 hr workweek, set by OSHA or TWA concentration for up to a 10 hr workday during a 40 hr workweek set by the NIOSH.

Titration: Common laboratory method where known concentration and volume of titrant reacts with a solution of analyte to determine concentration.

Topology: (1) Physical arrangement of network nodes and media within an enterprise networking structure. (2) The surface features of an object (how it looks) or its texture; a direct relation between these features and the material's properties (hardness, reflectivity, etc.).

Torr: Is a pressure unit that is named after Evangelista Torricelli, who discovered the principle of the barometer in 1644. It is defined as the standard atmospheric pressure divided by 760. Thus, 1 Torr is about 133.3 Pascals (Pa). Historically, 1 Torr was intended to equal the hydrostatic head pressure of one millimeter of mercury, but subsequent redefinitions made them slightly different (by less than 0.000015%). The Torr is not part of the International System of Units (SI), but it is often combined with the metric prefix of "milli", such as millitorr (mTorr).

Torus: See *Suppression pool*.

Total available chlorine: The chlorine that exists in water in any of the following forms: $HOCl$, NH_2Cl, NCl_3, or $NHCl_3$.

Total emissivity: The ratio of the integrated value of all spectral emittances to that of a blackbody.

Total probable error: Total error calculated by the root sum square method.

Total Suspended Particles (TSP): Is an archaic regulatory measure of the mass concentration of particulate matter (PM) in air. It was affected by the size selectivity of the inlet to the filter that collected the particles. The size cut varied with wind speed and direction and was from 20 to 50 μm (microns) in aerodynamic diameter. Under windy conditions, the mass tended to be dominated by large wind-blown soil particles of relatively low toxicity.

Townsend avalanche: Is a gas ionization process where free electrons are accelerated by a strong electric field results in electrical conduction through the gas "avalanche" multiplication caused by the ionization of molecules by ion impact.

Toxic: A chemical that has a median lethal concentration (LC[50]) in air of more than 200 parts per million but not more than 2000 parts per million by volume of gas or vapor, or more than 2 mg/l but not more than 20 mg/l of mist, fume, or dust, when administered by continuous inhalation for 1 hr (or less if death occurs within 1 hr) to albino rats weighing between 200 and 300 g each (OSHA 1994).

Tralles degree: A unit of specific gravity used in the alcohol industry.

Transducer: Is a device that converts variations in a physical quantity, such as pressure into an electrical signal, or vice versa. In a broader sense, the transducer is a device that converts any form of energy into another form of energy, such as a loudspeaker converts electric into sound energy.

Transient faults: Faults that are intermittent and may be caused by environmental conditions (e.g., temperature and EMF interference).

Transient overshoot: The maximum excursion beyond the final steady-state value of output as a result of an input change.

Transient overvoltage: A momentary excursion in voltage occurring in a signal or supply line of a device, which exceeds the maximum rated conditions specified for that device.

Transistors: Three-terminal, solid-state electronic devices. They are usually made of silicone, gallium-arsenide, or germanium and used for amplification and switching in integrated circuits.

Transitional flow: The flow rate between laminar flow and turbulent flow. It is characterized by a Reynolds number between ~2500 and >8000 dependent on flow rates, specific gravity, viscosity, and pipe size.

Translucency: Is a characteristic of a substance which allows light to pass through, but at interfaces with substances of a different index of refraction, they scatter the light. This is in contrast with a transparent medium which not only allows the transport of light but also allows for image formation.

Transmittance or transmissivity (T): In optics and spectroscopy, transmittance is the fraction of incident light (electromagnetic radiation) at a specified wavelength that passes through a sample. The terms visible transmittance and visible absorptance, which are the respective fractions for the visible spectrum of light, are also used.

Transmitter: A transducer that responds to a measured variable by means of a sensing element and converts it to a standardized transmission signal, which is a function only of the measured variable.

Trap: Pipeline facility for launching and receiving tools and pigs.

Triboelectric effect (also known as triboelectric charging): A type of contact electrification in which certain materials become electrically charged after they come into contact with another different material through friction. For example, rubbing a comb on hair can build up triboelectricity. Most everyday static electricity is triboelectric. The polarity and strength of the charges produced differ according to the materials' surface roughness.

True vapor pressure: Is based on ASTM D2889-95a Standard Test Method for petroleum distillate fuels. It is the pressure of the vapor in equilibrium with the liquid at 100 °F (the bubble point pressure at 100 °F).

Try valve: Valves commonly found on tanks at specified levels used to determine the presence or character of liquids.

Turndown, or turndown ratio: The ratio of the highest full scale to the least full scale that can still deliver the rated accuracy is called the turndown ratio, also called the turndown of the transmitter.

Twaddell degree: A unit of specific gravity used in the sugar, tanning, and acid industries.

Two-color pyrometer: Measures temperature as a function of the radiation ratio emitted around two narrow wavelength bands. Also called ratio pyrometer.

Tyndall effect (also known as Tyndall scattering): Light scattering by particles in a fine suspension. Tyndall states that the intensity of the scattered light depends on the fourth power of the frequency, so blue light is scattered much more strongly than red light. Under the Tyndall effect, the longer-wavelength light is more transmitted while the shorter-wavelength light is more reflected via scattering. An analogy to this wavelength dependency is that longwave electromagnetic waves such as radio waves are able to pass through the walls of buildings, while shortwave electromagnetic waves such as light waves are stopped and reflected by the walls, temperature, strain, and other properties.

U

Ultrasound: Is an oscillating pressure wave with a frequency greater than the upper limit of the human hearing range. Ultrasound has the same properties as "normal" (audible) sound, except that humans cannot hear it. Although this limit varies from person to person, it is approximately 20,000 Hz in healthy, young adults. Ultrasound devices operate at frequencies over 20 kHz up to several gigahertz.

Uncertainty (IAEH): Measurement uncertainty is expressed to a confidence level of 95%, and it is the limit to which an error may extend.

Uncertainty (Webster): A feeling of unsureness about something.

Uncertainty margin: The uncertainty at which the thermal power of the plant can be measured. It is included in the license of the plant, based on the precision of the measurements at the time of start-up.

Unclassified or general-purpose area: A plant area where standard electrical equipment may be installed.

Universal gas constant: The universal gas constant (R) is a physical constant. It is also called the ideal gas constant or molar gas constant. It is part of the ideal gas law, which defines the behavior of ideal gases as $PV = nRT$. The value of R in SI units is 8.314 and has the units of $J/K \cdot mol$ or $(kPa*m^3)/(kgm*mol*°K)$. In English units, $R = 10.73$ $(psia*ft^3)/(lbm*mol*°R)$.

Upper explosive limit: The highest concentration of gas or vapor in air in which a flame will continue to burn after the source of ignition has been removed.

Upper range limit: Highest value of the measured variable, which the device is capable of measuring.

Upper range value: Highest value of the measured variable that the device was adjusted to measure.

Utility pig: Pig that performs relatively simple mechanical functions, such as cleaning the pipeline.

V

van der Waals force: A weak attractive force between atoms or nonpolar molecules caused by a temporary change in dipole moment arising from a brief shift of orbital electrons to one side of one atom or molecule, creating a similar shift in adjacent atoms or molecules.

van't Hoff factor: Is named after J.H. van't Hoff, is a measure of the effect of a solute upon properties such as osmotic pressure. The van't Hoff factor is the ratio between the actual concentration of particles produced when the substance is dissolved and the concentration of a substance as calculated from its mass.

Vapor-Liquid ratio (or V/L or V-L): Is a measure of volatility, the tendency of a mixture of liquids to vaporize. It expresses the volume ratio of vapor to liquid that will exist at a given temperature and atmospheric pressure. Automotive fuel specifications usually state the temperature at which V-L is 20 ($T_{(V/L=20)}$). This temperature is where vapor lock can begin. A fuel has high volatility when it's $T_{(V/L=20)}$ is low. Volatility should be relatively high in low ambient temperature regions but should not be high enough to cause vapor lock.

Varactors: Are voltage-controlled capacitors used in voltage-controlled oscillators, parametric amplifiers, and frequency multipliers.

Variable backpressure: Backpressure that varies due to changes in operation of one or more PRVs connected into a common discharge header.

Velocity gradient (shear): Rate for change of liquid velocity across the stream: V/L for linear velocity profile, dV/dL for nonlinear velocity profile. Units are $V - L = ft/sec/ft = sec^{-1}$.

Velocity head: The velocity head is calculated as $v^2/2\,g$, where v is the flowing velocity and g is the gravitational acceleration (9.819 m/s² or 32.215 ft/s² at 60° latitude).

Vena contracta: A fluid dynamics term dating from the very early days of fluid research. It refers to the point in the restricted flow stream of smallest size, thus the point of maximum velocity and lowest pressure.

Venturi: A tube carrying a fluid flow; the tube narrows to a smaller diameter, causing a pressure increase at the inlet and an increase in the speed of fluid flow in the narrow portion. A venturi flow transmitter uses the change in pressure and flow speed to indicate the volumetric flow rate of the fluid.

Venturi effect: Is the reduction in the pressure of the flowing fluid that results when a fluid flows through a constricted section of pipe. The Venturi effect is named after Giovanni Battista Venturi (1746–1822), an Italian physicist.

Vernier scale: A device that lets the user measure more precisely than could be done by reading a uniformly divided straight or circular measurement scale. It is scale that indicates where the measurement lies in between two of the marks on the main scale.

Vibrating hose: A flexible, high pressure hose (similar to the kelly hose) that connects the mud pump to the stand pipe. It is called the vibrating hose because it tends to vibrate and shake (sometimes violently) due to its close proximity to the mud pumps.

Vibration amplitude unit (g): Is equal to the Earth's gravitational acceleration, which is g = 9.8 m/s² = 32.3 ft/s². What we feel as vibrations is that the object is being repeatedly displaced at a very high frequency. The reason the vibration amplitude is expressed in terms of acceleration (g) instead of a force (N) or the displacement (mm) is that at high frequency the system has less time to be displaced and because heavier objects are displaced less with the same force.

Viscosity: The fluid's resistance to gradual deformation caused by shear or tensile stresses.

Void coefficient: The measure of the influence of voiding on reactor power generation. Positive void coefficient (VC) means that an increase in core temperature (more boiling) increases power generation. Most nuclear reactors are designed to have a negative VC, so that the reactor reduces power generation if core temperature rises, such as if cooling is lost. Chernobyl had a positive VC.

Void fraction: The portion of the coolant volume that is made up by steam bubbles. An increase in void fraction can either increase or decrease core reactivity (steam generation), depending on the design.

Voiding: The proportion of steam bubbles in the cooling water.

Volt (V): Is named after the Italian physicist Alessandro Volta (1745–1827). It is the Standard International (SI) unit of electric potential or electromotive force, which is a potential that appears across a resistance of one ohm when a current of one ampere flows through that resistance. Average and instantaneous voltages can be assigned with a negative (–) or positive (+) polarity with respect to a zero, or ground reference potential.

Voltage transformers: Are used to increase or decrease the voltages of alternating current. This is done by varying the current in the transformer's primary winding, which creates a varying magnetic flux in the transformer core and a varying magnetic field impinging on the transformer's secondary winding. This varying magnetic field at the secondary winding induces a varying voltage in the secondary winding. Transformers can thus be designed to efficiently change AC voltages from one voltage level to another within power networks.

Voting: The means to decide which of the redundant paths are operating correctly.

W

Water cut: The water volume flow rate, relative to the total liquid volume flow rate (oil and water), both converted to volumes at standard pressure and temperature. The water cut is normally expressed as a percentage.

Water-in-liquid ratio: The water volume flow rate relative to the total liquid volume flow rate (oil and water) at the pressure and temperature prevailing in that section.

Watt (W): Is named after the Scottish inventor James Watt (January 19, 1736–August 25, 1819). It is the SI unit of power, equivalent to one joule per second, corresponding to the power in an electric circuit in which the potential difference is one volt and the current flow is one ampere.

Watt peak capacity (Wp): In case of photovoltaic (PV) solar cells, this is the nominal capacity or the capacity which is realized under certain test conditions. Wp is not the regular power output but instead the maximum capacity of a module under optimal conditions. These Standard Test Conditions (STC) are specified in standards such as IEC 61215, IEC 61646 and UL 1703; specifically the light intensity is 1000 W/m^2, with a spectrum similar to sunlight hitting the earth's surface at latitude 35°N in the summer at air mass 1.5 and with the temperature of the cells being 25°C.

Waveguide: A device that constrains or guides the propagation of electromagnetic waves along a path defined by the physical construction of the waveguide; includes ducts, a pair of parallel wires, and a coaxial cable.

Weighing accuracy: The intended meaning of the term is inaccuracy, and it relates to the size of the error. It is usually expressed as a percentage, and it is important to determine if the percentage is that of full scale (% FS) or that of actual reading (% AR). If the measured weight is in the lower part of the detector's range, the % AR error will much exceed % FS. Historically, a balance was a device that determined mass by balancing an unknown mass against a known mass as with a 2 pan balance. In modern weighing machines, balances are usually of the design that uses a force restoration mechanism to create a force to balance the force generated by the unknown mass.

Wet bulb: The temperature of air when cooled to saturation (100% relative humidity) simply by evaporation of water into it.

Wet gas: Denotes a natural gas flow containing a relatively small amount of free liquid by volume; usually this may be limited up to about 10%. The mass ratio of gas to liquid varies significantly with pressure for a constant GVF.

Wet leg: In certain level installations that use differential pressure sensors, there is a vertical pipe or tube filled with liquid connected at or near the top of the tank to provide a constant reference pressure at the low pressure tap on the sensor. The high side is connected to the bottom of the tank; this has the effect of having the indicated pressure increase when vessel level increases. The sensor zero output signal is adjusted to account for the high pressure in the wet leg.

Wet well: See *Suppression pool*.

Wetted part: Element in direct contact with the process medium.

Wheatstone bridge: A reference to an electric measuring scheme where the voltages across two resistances are adjusted to have a minimum voltage difference by adjusting the bridge.

Wideband (total) pyrometer: A radiation thermometer that measures the total power density emitted by the material of interest over a wide range of wavelengths.

Wiedemann effect: The twisting of a ferromagnetic rod through which an electric current is flowing when the rod is placed in a longitudinal magnetic field. It was discovered by the German physicist Gustav Wiedemann in 1858. It is one of the manifestations of magnetostriction in a field formed by the combination of a longitudinal magnetic field and a circular magnetic field created by an electric current.

Winkler method: Was discovered by Lajos Winkler (1863–1939), a Hungarian chemist. It is a titration method to determine dissolved oxygen in the water sample. The steps of the Winkler method involve the filling completely (no air space) a sample bottle with water. Next, the dissolved oxygen in the sample is "fixed" by adding a series of reagents that form an acid compound that is then titrated with a neutralizing compound that results in a color change. The point of color change is called the "endpoint," which corresponds to the dissolved oxygen concentration in the sample.

Wireless local area network: A localized network connecting computers in a limited space, such as a school or an office.

Wobbe Index (WI) or Wobbe Number: Accounts for composition variations in terms of their effect on the heat value and specific gravity, which affect the flow rate through an orifice. In essence, the Wobbe Index is a measurement of the available potential heat, and it can be used in conjunction with the gas flow measurement to produce a measurement of heat flow rate. WI is an indicator of the interchangeability of fuel gases, such as natural gas, liquefied petroleum gas (LPG), and town gas. The equation below gives the definition of the Wobbe Index as a function of higher heating value (HHV) and specific gravity (SpG):

$$WI = HHV/\sqrt{SpG}$$

Worm, computer: See Computer worm.

Worm, slammer: See *Slammer worm*.

Worst-case error: The sum of the maximum errors contributed by each error component.

Wrinkles: Ripples that occur on the inner radius of a pipe when the pipe is cold bent.

X

X and Y Purge: X purge allows a general-purpose instrument to be placed in a Division 1 area, while Y purge allows for a Division 2-rated instrument to be used in a Division 1 area. These are Instrument Society of America definitions of purge systems.

Xenon burnout: In nuclear power generation, xenon poisoning occurs at low power output when xenon135 formation inhibits the fission reaction.

X-rays: Forms of electromagnetic radiation having a wavelength in the range of 0.01–10 nm, corresponding to frequencies in the range 30 PHz to 30 EHz (3×10^{16} Hz to 3×10^{19} Hz) and energies in the range 100 eV to 100 keV.

The distinction between x-rays and gamma rays is a function of their origin: x-rays are emitted by electrons, while gamma rays are emitted by the atomic nucleus. An alternative method for distinguishing between x-rays and gamma radiation is on the basis of wavelength, with radiation shorter than some arbitrary wavelength, such as 10^{-11} m, defined as gamma rays. These definitions usually coincide since the electromagnetic radiation emitted by x-ray tubes generally has a longer wavelength and lower photon energy than the radiation emitted by radioactive nuclei.

Y

Young's modulus: Also known as the tensile modulus and expresses the stiffness of elastic solid materials. It measures the force (per unit area) that is needed to stretch (or compress) a material sample. It is named after the 19th-century British scientist. The term modulus is the diminutive of the Latin term modus which means measure.

Z

Zero: The minimum measurand value within the calibrated range of a transmitter (or other sensor). Whereas the maximum value is called "span," and the difference between zero and full scale is called "span."

Zero elevation: When zero is elevated, a bias is added to the reading of the transmitter. This elevation calibration is used when the LO side senses higher pressure than the HI side and the reading DP will be negative, which disables proper output reading.

Zero float: The zero shift in load cell caused by one complete cycle of successive rated tensile and compressive loading.

Zero return: The difference in zero balance measured immediately before rated load application of a specific duration and when the output has stabilized after the removal of the load.

Zero shift: Any parallel shift of the input–output curve.

Zero suppression: For a suppressed zero range, the measured variable zero is below the lower range value. It may be expressed either in units of the measured variable or in percent of span.

Zeta potential: The electrostatic potential generated by the accumulation of ions at the surface of the colloidal particle. Particle charge can be controlled by modifying the suspending liquid. Modifications include changing the liquid's pH or changing the ionic species in solution.

Zimm plot: Is a diagrammatic representation of data on light scattering by large particles. It is used for the simultaneous evaluation of the mass average molar mass, the second viral coefficient of the chemical potential, etc. Several modifications of the Zimm plot are in frequent use.

Zone 0, or continuous grade of release: Where there is continuous release source or frequent release source of flammable gases.

Zone 1, or primary grade of release: Where there is periodic or occasional release during normal operation.

Zone 2, or secondary grade of release: Where there is no release of flammable gases under normal operation (or only infrequently and for short periods).

A.2 Abbreviations, Acronyms, and Symbols

1H-NMR	Proton nuclear magnetic resonance
2D	Two dimensional
3D	Three dimensional

A

a	(1) Acceleration; (2) model coefficient
A	(1) Area; (2) ampere, symbol for basic SI unit of electric current; (3) admittance
Å	Angstrom ($=10^{-10}$ m)
AA	Atomic absorption
AAS	Atomic absorption spectrometer
ABS	Alkyl benzene sulfonate
abs	Absolute (e.g., value)
ABWR	Advanced Boiling Water Reactor
ac	Alternating current
ACFM	Actual cubic feet per minute; volumetric flow at actual conditions in cubic feet per minute (=28.32 alpm)
Ack	Acknowledge or acknowledged
ACL	Asynchronous connectionless
ACMH	Actual cubic meters per hour
ACMM	Actual cubic meters per minute
ACS	Analyzer control system
ACSL	Advanced continuous simulation language
A/D	Analog to digital, also analog-to-digital converter
AD	Actuation depth
ADC	Analog digital converter
ADIS	Approved for draft international standard circulation
ADS	Automatic depressurization system
ADV	Atmospheric dump valve
A&E	Alarm and event
AE	Acoustic emission
AED	Atomic emission detector
AEGL	Acute exposure guideline level
AES	Atomic emission spectrometer
AET	Acoustic emission testing
AF, a-f	Audio frequency
AFD	Adjustable frequency drive
AFS	Atomic fluorescence spectroscopy
AFV	Alternate fuel vehicles
AFW	Auxiliary feedwater system
Ag	Silver, chemical element
AH	Ampere hour
AI	Analog input
AIRAPT	Advancement of high-pressure science and technology
AIT	Autoignition temperature
a(k)	White noise
ALARA	As low as reasonably achievable
ALARP	As low as reasonably practicable
alpm	Actual liters per minute
AM	(1) Alarm management; (2) amplitude modulated
Am	Americium, chemical element
AMI	Advanced metering infrastructure
amp	Ampere; also A
AMPS	Advanced mobile phone system (cellular system)
AMR	Automated meter reading
AMS	Asset management systems or analyzer maintenance station
amu	Atomic mass unit
ANN	Artificial neural networks
ANPR	Automatic number plate reader
AO	Analog output
AOC	Automatic operation controls
AOPS	Automatic overfill prevention system
AOTF	Acousto-optic tunable filters
AP	Access point
APC	Automated process control
APDU	Application (layer) protocol data unit
API	Application programming interface or absolute performance index
°API	API degrees of liquid density
APM	Applicable pulse modulation
APTD	Air pollution technical data
%AR	Percent actual reading
AR	Actual reading
ARA	Alarm response analyses
ARIMA	Autoregressive integrated moving average
ARP	Address resolution protocol
ARQ	Automatic repeat request—an error checking routine/protocol
ART	Adaptive resonance theory
AS	Air source
ASCII	American standard code for information interchange
AS-i	Actuator sensor interface

ASIC	Application-specific integrated circuit	BJT	Bipolar transistor
ASK	Amplitude shift keying	°Bk	Barkometer degrees of liquid density
ASOS	Automated surface observing system	blk	Black (wiring code color for ac "hot" conductor)
ASP	Accident sequence precursor		
ASRS	Alarm system requirement specification	BMS	Burner management system
AST	Automatic sampling train	BO	Black oil
asym	Asymmetrical; not symmetrical	BOD	Biochemical oxygen demand
ATC	(1) Amplitude threshold curve; (2) automatic temperature correction	BOM	Black oil model
		BOP	Blowout Protector
ATEX	Atmosphères Explosibles, European directive 94/9/ec "centipoises explosibles"	bp, b.p.	(1) Backpropagation; (2) boiling point
		BPCS	Basic process control system
ATG(S)	Automatic tank gauge (system)	bpd	Barrel per day
atm	Atmosphere (= 14.7 psi)	Bps, b/sec	Bits per second—a measure of data speed
ATP	Adenosine triphosphate	BPSK	Binary phase shift keying
ATR	Attenuated total reflection	Bq	5 centipois, symbol for derived SI unit of radioactivity, joules per kilogram, J/kg
AU	Attenuation unit		
Au	Gold, chemical element	Br	Bromine, chemical element
AUI	Attachment unit interface	°Br	Brix degrees of liquid density
aux	Auxiliary	BRU	Backscatter ratio unit
AVSS	Audiovisual surveillance system	BS	British standard
AWG	American wire gauge	BSP	British Standard Product
AWPAG	All weather precipitation accumulation gauge	BSR	Blind shear ram
		BSW	Basic sediment and water
		bt	Billion tons
B		BTL	Bottom tangent line
B	Boron, chemical element; magnetic field	BTU	British thermal unit
b	Dead time	BU	Backscatter unit
°Ba	Balling degrees of liquid density	BWG	Birmingham wire gauge
BACNet	Building automation & control networks	BWR	Boiling water reactor
bar	(1) Barometer; (2) unit of atmospheric pressure measurement (−1 kPa)	B2B	Business to business
		C	
barg	Bar gauge	c	(1) Velocity of light in vacuum (3×10^8 m/s); (2) centi, prefix meaning 0.01
BARS	Non-SI unit of pressure		
bbl	Barrels (−0.1589 m^3)	C	Coulombs, symbol for discharge coefficient, capacitance
BC	Ballast control		
BCD	Binary coded decimal	°C	Celsius degree of temperature
BCS	Batch control system	ca.	*Circa* (about, approximately)
BDD	Boron doped diamond	CA	Carcinogen
BDP	Boiler design pressure	CAC	Channel access code
Be	Beryllium, chemical element	CAD	Computer-aided design
°Bé	Baumé degrees of liquid density	Cal	Calorie (gram = 4.184 J); also g-cal
BEI	Biological exposure indices	CAN (CANbus)	Controller area network
BEMS	Building energy management systems	CAPEX	Capital expenditure
BF3	Boron trifluoride	CARI	Combustion air requirement index
BFO	Beat frequency oscillator	CASE	Common application service elements
BFW	Boiler feedwater	CATV	Community antenna television (cable)
BHA	Bottom-hole assembly	CB	Catalytic bead
BHIIF	Baseline risk index of initiating effects	cc	Cubic centimeter
BHP	Bottom-hole pressure	cc/min, ccm	A flow unit, cubic centimeters per minute
bhp	Break horse power (= 746 W)	CCD	Charge coupled device
BiCMOS	Bipolar complementary metal oxide semiconductor – see CMOS	CCF	(1) Common cause failure; (2) combination capacity factor

CCR	Central control room	CL	Corrected level
CCS	Computer control system	CL 1	Electrically hazardous, Class 1, Division 1, Groups C or D
Ccs	Constant current source		
CCSDS	Consultative committee for space data systems—telemetry and telecom software library that provides reference implementation of the international protocol standard for transmission and reception of data in RF communications with spacecraft	CLD	Chemiluminescence detector
		CLP	Closed-loop potential factor
		CM	Condition monitoring or communication (interface) module
		cm	Centimeter (= 0.01 m), unit of length
		CMF	Coriolis mass flowmeter
CCTV	Closed circuit television	CMMS	Computer maintenance management systems
CCW	Counterclockwise		
CD	(1) Dangerous coverage factor; (2) compact disk; (3) collision detector; (4) conductivity detector	CMOS	Complementary metal oxide semiconductor (pronounced "see-moss")—a low-power semiconductor chip using both negative and positive polarity circuits
Cd	Cadmium, chemical element		
cd	Candela, symbol for basic SI unit of luminous intensity	CMPC	Constrained multivariable predictive control
CDDP	Cellular digital data packet	cmph, m^3/hr	Cubic meter per hour
cdf	Cumulative distribution function	CMR	Colossal magnetoresistance, having greater magnetoresistance than a typical GMR element
CDMA	Code division multiple access – an efficient protocol used in North America and a few other places for wireless telephones		
CDN	Content delivery network	CN_F	Free cyanide
CDPD	Cellular digital packet data	CNI	ControlNet International
CDT	Color detecting tubes	CNID	Constrained noninformative distribution
CDTA	Diaminocyclohexane-N,N-tetra-acetic acid	CN_T	Total cyanide
		CNWAD	Weak acid dissociable cyanide
CDTP	Cold differential test pressure	CO	(1) Controller output; (2) carbon monoxide
CEAS	Cavity enhanced (laser) absorption spectroscopy	Co	Cobalt, chemical element
		Co/Nox	Carbon monoxide/ nitrogen oxides
CEM	Continuous emission monitoring	CO_2	Carbon dioxide
CEMS	Continuous emission monitoring system	CO_2 D	Carbon dioxide demand
CENP	Combustion engineering nuclear power	Co-60	Cobalt 60
CFA	Continuous flow analyzer	COD	Chemical oxygen demand
CFM, cfm	Coriolis flowmeter or cubic foot per minute	COH	Coefficient of haze
		COM	Component object model
CFR	Code of federal regulations	cos	Cosign (trigonometric function)
CF/yr	Cubic foot per year	COT	Coke oven temperature
CG	Gas chromatography	COTS	Commercially off the shelf
cGMP	Current good manufacturing practice	cp, c.p.	(1) Candle power; (2) circular pitch; (3) center of pressure (cp and ctp sometimes are used to centipoise
CGR	Condensate to gas ratio		
CGS	Centimeter gram second		
CH	Critical high level	CPA	Coal particle analyzer
CI	Cast iron	cpm	Cycles per minute; counts per minute
Ci	Curie, the unit of the activity of radioactive materials	CPP	Central processing platform
		CPS	Computerized procedure system
CIM	Computer integrated manufacturing	cps	(1) Cycles per second (hertz, Hz); (2) counts per second; (3) centipoises (=0.001 Pa·S)
CIP	(1) Clean in place; (2) computer-aided production or control and information protocol (an application layer protocol supported by DeviceNet, Control-Net, and Ethernet/IP)		
		CPU	Central processing unit
		CPVC	Chlorinated (poly)vinyl chloride
CJ	Common junction	CR	Corrosion rate
Cl	Chlorine, chemical element	CRC	Cyclic redundancy check

CRDS	Cavity ring-down spectroscopy	DCV	Direct current volts
CRLF	Carriage return-line fed	DD	Device description
CRT	Cathode ray tube	D/DBP	Disinfectants/disinfection by products
CS	Carbon steel	DDC	Direct digital control
Cs	Cesium, chemical element	DDE	Dynamic data exchange
Cs-137	Cesium 137	dE	Change in polarization voltage
CSL	Car seal locked	DENSD	Density at actual conditions
CSMA/CD	Carrier sense multiple access with collision detection (csma/cd)	DENSTD	Density at standard conditions
		DES	Data encryption standard
CSO	Car seal open	DETF	Double end tuning fork
CSS	Central supervisory station	DFB	Distributed feedback
CSSB	Compatible single side band	DFIR	Diffused infrared
cSt	Centistokes	DFR	Digital fiber-optic refractometer
CSTR	Continuous-stirred tank reactor	DFT	Digital Fourier transform
CT	(1) Computed tomography; (2) cooling tower or the product of C for disinfectant concentration and T for time of contact in minutes; (3) current transformers	DH	Data highway
		DI	Digital input
		dI	Change in current flow
		dia	Diameter, also d and ϕ
CTDMA	Concurrent time domain multiple access	DIAC	Dedicated inquiry access code
CTMP	Chemithermomechanical pump	DIN	Deutsch industrie norm
CV	Calorific value	DIP	Dust ignition proof
CVAAS	Cold vapor atomic absorption spectroscopy	DIR	Diffused infrared
CVAFS	Cold vapor atomic fluorescence absorption spectroscopy	DIS	Draft international standard
		D(k)	Measured disturbance
		d(k)	Unmeasured disturbance
CVF	Circular variable filter	DLC	Dioxin-like compounds
cvs	Comma-separated variables	DLE	Data link escape
CW	(1) Continuous wave; (2) clockwise	DLL	Dynamic link library
CWOP	Citizen weather observer program	dm	Cubic decimeter, unit of volume
D		DM	Delta modulation
D	Diameter; also dia and ϕ derivative or derivative time of controller	DMA	Dynamic mechanical analysis
		DMM	Digital multimeters
d	(1) Derivative; (2) differential as in dx/dt; (3) deci, prefix meaning 0.1, (4) depth; (5) day	DMR	Digital microwave radio system
		DMS	Dead man switch
		DN	Nominal diameter
DA	Data access	DNA	Deoxyribonucleic acid
D/A	Digital to analog	DO	Dissolved oxygen
DAC	Device access code	DOAS	(1) Dedicated outdoor air system; (2) differential optical absorption spectroscopy
DACU	Data acquisition and control unit		
DAE	Differential algebraic equation	DP, d/p	(1) Differential pressure; (2) dynamic positioning
DAMPS	Digital advanced mobile phone system		
DART	Dynamically Arc Recognition and Termination	d/p cell	Differential pressure transmitter (a Foxboro-Emerson trademark)
DASH7	A radio network defined for "smart tags" specified by ISO/IEC 18000-7	DPCM	Differential pulse code modulation
		DPD	N,N-diethyl-p-phenylenediamine
dB	Decibels	DPDT	Double pole double throw (switch)
DBB	Double block and bleed	dpi	Dots per inch
DBPSK	Differential binary phase shift keying	DPV	Depressurization valve
DC	Dielectric constant	DPVS	Depressurization valve system
dc	Direct current	DQPSK	Differential quadrature phase shift keying
DCE	Data communications equipment	DRESOR	Distribution of ratio of energy scattered by the medium or reflected by the boundary surface
DCMS	Downhole corrosion-monitoring system		
DCS	Distributed control system		

DS	Secure digital	ECTFE/PFA	Ethylene chloro-tetra-fluoro-ethylene (Halar)–copolymer coated
DSB	Double side band	EDD	Electronic device description (file)
DSC	Differential scanning calorimeter	EDDL	Electronic device description language
DSL	Digital subscriber line	EDS	Electronic data sheet (DeviceNet)
DSM	Demand side management—electrical demand (KWD); heating, ventilation, and air conditioning (HVAC)	EDTA	Ethylenediaminotetra acetic acid
		EDXRF	Energy-dispersive x-ray fluorescence spectrometer
DSN	Distributed sensor networks	E/E/PE	Electrical/electronic/programmable electronic
DSP	Digital signal processing		
DSR	Direct screen reference	E/E/PES	Electrical/electronic/programmable electronic system
DSS	Drilling sensor sub		
DSSS	Direct sequence spread spectrum	EEPROM	Electrically erasable programmable read-only memory
DT	(1) Dead time; (2) delay termination of output		
DTC	Digital temperature compensation	EFD	Engineering flow diagram
DTE	Data terminal equipment	EFV	Excess flow valve
DTGS	Deuterated triglycine sulfate	EGA	Enhanced graphics adapter
DTM	Device type manager (an active X-component for configuring an industrial network component, a DTM "plugs into" an FDT)	EHC	Electrohydraulic control
		EHM	Equipment health management
		eHz	10^{18} (exa) Hz
		EIV	Explosion isolation valve
DTMF	Dual tone multifrequency	e(k)	Feedback error
DTPA	Diethylene triaminopenta acetic acid	EL	Elastic limit
DVM	Digital voltmeter	ELSD	Evaporating light scattering detector
DU	Dangerous component failure occurred in leg, but undetected	EMAT	Electromagnetic acoustic transducer
		EMF, Emf	Electromotive force
dx	Infinitesimal change in x	EMI	Electromagnetic interference
dy	Infinitesimal change in y	EMI/RFI	Electromagnetic and radio-frequency interference
E			
E	(1) Electric potential in volts; (2) scientific notation as in $1.5E-03 = 1.5 \times 10^{-3}$; (3) modulus of elasticity, pound-force/in.^{2} (kPa); (4) error, deviating from the truth; (5) Exa (10^{18})	em(k)	Process/model error
		EMS	Electronic metering system
		EN	European norm
		$E°_{cell}$	Standard cell potential
		$E°_{red}$	Standard Half-Cell Reduction Potential
		EOS	Equations of state
e	(1) Error; (2) base of natural (Naperian) logarithm, (3) exponential function; also exp $(-x)$ as in e^{-x}	EPA	Enhanced performance architecture; Environmental Protection Agency
		EPC	Engineering-procurement-construction
E{.}	Expected value operator	EPCM	Engineering-procurement-construction management (companies)
EA	Exhaust air		
EAI	Enterprise application integration	EPDM	Ethylene propylene diene terpolymer
EAM	Enterprise asset management	EPL	Equipment protection level
EB	Empirical Bayes	EPS	(1) Emergency protection system; (2) encapsulated postscript file; (3) electronic pressure scanner; (4) emergency power supply
EBCDIC	Extended binary code for information interchange		
EBR	Electronic batch records		
ECCS	Emergency core cooling system	Eps	Pressure caused error in span
ECD	(1) Electron capture detector; (2) equivalent circulating density	EPT	Ethylene-propylene-diene terpolymer
		Epz	Pressure caused error in zero
E_{cell}	Cell potential (electromotive force) at the temperature of interest	E_{red}	Half-cell reduction potential at the temperature of interest
ECKO	Eddy current killed oscillator	ERM	Enterprise resource manufacturing
ECL	Emitter coupled logic	ERP	(1) Enterprise resource planning; (2) effective radiated power
ECN	Effective carbon number		

ERW	Electric-resistance welds	FDM	Frequency division multiplexing
(E)SBWR	Economic and simplified boiling water reactor	FDMA	Frequency division multiplex access
		FDS	Flame-detection system
ESD	(1) Emergency shutdown; (2) electrostatic discharge	FDT	Field device technology
		FE	Final elements
ESN	Electronic serial number	FEED	Front end engineering and design
ET	(1) Electromagnetic testing; (2) Eddy current testing	FEGT	Furnace exit gas temperature
		FEMA	Failure mode analysis
ETFE	Ethylene tetrafluoroethylene	FEP	Fluorinated ethylene propylene
ETS	Equivalent time sampling	FES	Fixed end system
Ets	Temperature caused error in span	FET	Field effect transistors
Etz	Temperature caused error in zero	FF	(1) Foundation fieldbus; (2) feedforward; (3) flat faced
EUI-64	Extended unique identifier; a 64-bit numeric "name" assigned to a device on a network; assigned by the IEEE	FF-HSE	Foundation fieldbus, high-speed internet
		FFIC	Flow ratio–indicating controller
		FFT	Fast Fourier transform
eV	Electron volt, unit of energy	FG	Flow glass (condensate pot)
EWS	Engineering workstation	FGS	Fire and gas system
Exd	Explosion proof protection	FH	Frequency hopping
Exe	Increased safety protection	Fhp	Fraction horsepower (e.g., ¼-hp motor)
Exi	Intrinsically safe protection	FHSS	Frequency hopped spread spectrum
Exm	Encapsulated protection	FI	Flow indicator
Exn	Nonincendive protection	FIA	Flow injection analysis
Exo	Oil-filled protection	FIC	Flow indicator controller
Exp	Pressurization protection	FID	Flame ionization detector
Exq	Sand- or quartz-filled protection	FIE	Flame ionization element
Exs	Special protection	FIFO	First in, first out
		FISCO	Fieldbus intrinsically safe concept
F		FIT or FiT	Feed-in tariff
F	Faraday Constant, the number of coulombs per mole of electrons, symbol for derived SI unit of capacitance, ampere-second per volt $A \cdot s/V$	fl.	Fluid
		fl.oz.	Fluid ounce (−29.57 cc)
		FM	(1) Factory mutual; (2) fade margin; (3) frequency modulated
f, freq.	(1) Oscillation frequency, cycles/s; (2) some equations, the y vs. x relation	FMCW	Frequency modulated carrier (continuous) wave
FA	Frame application (for fdt/dtm)	FMEA	Failure mode and effects analysis
FARM	Frame acceptance and reporting mechanism	FMECA	Failure modes, effects, and criticality analysis
FAT	Factory acceptance test	FMEDA	Failure modes, effects and diagnostic analysis
FAU	Formazin attenuation unit		
FBAP	Function block application process (FF)	FMS	Fieldbus message specification or fieldbus messaging services/system
FBD	Function block diagram		
FBG	Fiber Bragg grating	FNICO	Fieldbus nonincendive concept
FBRU	Formazin backscatter ratio unit	FNMU	Formazinnephelometric multibeam unit
FBU	Formazin backscatter unit	FNRU	Formazinnephelometric ratio unit
FC	Flow controllers	FNU	Formazinnephelometric unit
FCE	Final control element	FO	(1) Flow open; (2) fiber-optic; (3) fail open
FCOR	Filtering and correlation (method)		
FCP	Final circulating pressure	FOC	Fiber-optic cable
FCS	Frame check sequence	FOD	Failure on demand (pfd)
FD	Fluorescence detector	FOP	Fiber-optic probe
FDE	Fault disconnection electronics	FOPP	Fiber-optic patch panel
FDI	Field device integration	FOSV	Full opening safety valve
FDL	Fieldbus data link		

fp, f.p.	Freezing point	GLI	Great Lakes initiative
FPC	Fine particle content	GLR	Gas liquid ratio
FPD	Flame photometric detector	G-M	Geiger–Mueller tube, for radiation monitoring
fpm, ft/min	Feet per minute (−0.3048 m/m)		
fps, Ft/s	Feet per second (−0.3048 m/s)	G_m	Model transfer function
FPS	Foot–pound–second	g/m^3	Grams of water per cubic meter, for moisture measurements, also known as absolute humidity.
Fps	Frames per second		
FR	Federal register		
FRM	Frequency response method	GMDSS	Global maritime distress and safety system
FRO	Full range output		
FS, fs	Full scale (also called URV)	GMM	Gradient magnetometer method
FSC	Fail safe controller	GMR	Giant magnetoresistance, having greater magnetoresistance than a typical MR element
FSD	Full-scale deflection		
FSK	Frequency shift keying		
FSL	Low flow switch	GOR	Gas/oil ratio
FSU	Floating storage unit	GOV	Gross observed volume
FT	(1) Flow transmitter; (2) Fourier transform	GP	Gaussian process
		G_p	Process transfer function
FTA	Fault tree analysis	GPA	General platform alarm
FTAM	File transfer and management	gph, gal/hr	Gallons per hour (=3.785 l/hr)
ftH$_2$O	Feet of water	gpm, gal/m	Gallons per minute (=3.785 l/m)
FTIR	Fourier transform infrared spectroscopy	GPRS	General packet radio service
FTNIR	Fourier transform near infrared analyzer	GPS	Global positioning system
FTP	File transfer protocol	grn	Green (wiring code color for grounded conductor)
FTS	Fault-tolerant system		
FTU	Formazin turbidity unit	GSC	Gas-solid chromatography
FW	Flash welds	GSD	(1) General station description; (2) Profibus version of an electronic data sheet

G

G	Giga, prefix meaning 10^9; (2) process gain; (3) conductance	GSM	Global system mobile—the world's most used protocol for wireless telephones
g	(1) Gram; (2) acceleration resulting from gravity (=9.806 m/s^2), unit of vibration amplitude; (3) conductivity	GSV	Gross standard volume
		Gt/yr	Giga tons per year
		GUI	Graphical user interface
		GUM	ISO guide for the expression of uncertainty in measurements.
GaAs	Gallium arsenide		
gal	Gallon (−3.785 liters)	GVF	Gas volume fraction
GB	Gigabyte, 1,000,000,000 bytes	GVO	Gas volume
GbE	Gigabit Ethernet	GWP	Gross world product
gbps, GBPS	Gigabits per second	GWR	Guided wave radar
G_c	Feedback control transfer function	Gy	Gray, symbol for derived SI unit of absorbed dose, joules per kilogram, J/kg
GC	Gas chromatograph		
g-cal	Gramcalorie, see also cal		
G_D	Unmeasured disturbance transfer function		
GD	(1) Measured disturbance transfer function; (2) approximate feedforward transfer function model		

H

		°H	Degrees Fahrenheit [t°F − 32)/1.8]
GDCS	Gravity-driven cooling system	H	(1) Humidity expressed as pounds of moisture per pound of dry air; (2) henry, symbol of derived SI unit of inductance, volt-second per ampere, V·s/A
GEN	Generation		
GEOS	Geosynchronous Earth orbit satellite		
GFC	Gas filter correlation		
G_{ff}	Feedforward transfer function	h	(1) Height; (2) hour
gHz	10^9 (giga) Hz	H1	Field-level fieldbus; also refers to the 31.25 kbps intrinsically safe SP-50, IEC61158-2 physical layer
GIAC	General inquiry access code		
		H$_2$Dz	Dithizone

H_2O_2	Hydrogen peroxide		HTTP, http	Hypertext transfer protocol
H_2SO_4	Sulfuric acid		HVAC	Heating, ventilation and air conditioning
HAD	Historical data access		HVD	High-voltage differential
HART	Highway addressable remote transducer		hwd	Height, width, depth
HAZOP	Hazard and operability studies hazard		HWO	Hydraulic workover unit
HC	Horizontal cross-connect		Hz	Hertz, unit of frequency (cycles/second)
HCDP	Hydrocarbon dew point			
HCF	HART communication foundation		**I**	
HCl	Hydrochloric acid		I	Human–system interface
HCN	Hydrogen cyanide		I	(1) Integral time of a controller in unit of time/repeat; (2) moment of inertia, in.^{4} (cm^4)
He	Helium, chemical element			
HEC	Header error check		i	Representing the identifier for the data point, variable, or coefficient
HEPA	High-efficiency particulate absorption			
HETP	Height equivalent to a theoretical plate		IA	Instrument air
HF	Hydrogen fluoride or hydrofluoric acid		IAC	Inquiry access code
Hf	Hafnium, chemical element		IAE	Integral of absolute error
HFE	Human factor engineering		IAQ	Indoor air quality
HFT	Hardware fault tolerance		IBC	Induction balanced coils
Hg	Mercury, chemical element		IBOP	Inside blowout preventer
HH	High-high tank level		I&C	Instrumentation and control or information and control
HHU	Hand-held unit			
hhv	Higher heating value		IC	(1) Integrated circuit; (2) isolation condenser; (3) intermediate cross-connect; (4) inorganic carbon
HI	High			
HID	Helium ionization detector			
HIPPS	High integrity pressure protection system		ICA	Independent component analysis
HIPS	High-integrity protective system		ICCMS	Inadequate core cooling monitoring system
HIST	Host interface system test			
HIU	Hydrostatic interface units		ICFD	In-core flux detector
HLCS	High-level control system		ICMP	Internet control message protocol
HLW	High-level nuclear waste		ICP	Initial circulating pressure
HMA	Hybrid multispectral analysis		ID	Inner diameter
HMDS	Hexamethyldisiloxane		IDC	Initial demonstration of capability
HMI	Human–machine interface		IDLH	Immediately dangerous to life or health
HNO_3	Nitric acid		I&E	Instrument and electrical
hor.	Horizontal		IE	(1) Ignition energy; (2) instrument earth
HOV	High-occupancy vehicle		IEC	Ion exchange chromatography
HP, hp	(1) High pressure; (2) horse power (U.S. equivalent is 746 W)		IEEE 802.15.4	Describes the physical and mac layers of this personal area network standard
HPCI	High-pressure core injection system or high pressure coolant injection system		IEPE	Integrated electric piezoelectric
			IETF	Internet engineering task force
HPCS	High-pressure core spray system		IFD	Intelligent field devices
HPIC	High-pressure injection cooling		IIS	Internet information server
HPLC	High-pressure (or high precision) liquid chromatography or high-performance liquid chromatograph		IL	Instruction list
			ILD	Instrument loop diagrams
			ILI	In-line inspection instruments
H&RA	Hazard and risk analysis		ILW	Intermediate level nuclear waste
HS	Hand select (switch)		IMC	Internet model control
HSE	High-speed Ethernet		iMEMS	Integrated microelectromechanical system
HSP	Hydrostatic pressure			
HSTG	Hydrostatic tank gauging		IMS-MS	Ion-mobility spectrometry–mass spectrometry
HTG	Hydrostatic tank gauging			
HTML	Hypertext markup language		IMU	Inertial measurement unit

In	Indium, chemical element	JFET	Junction field effect transistors
in.	Inch (= 25.4 mm)	JIT	Just-in-time (manufacturing)
InGaAs	Iridium gallium arsenide	JNID	Jeffrey's noninformative prior distribution
in. H_2O, or	A pressure unit, expressed as inches of water column	JNT	Johnson noise thermometer
inHg	Inches mercury	JP-4	Jet propellant-4 (50)
in-lb	Inch-pound (=0.113 N × m)	**K**	
INWC	Inches of water column, a pressure created by a level of water at a standard temperature	K	(1) Kelvin, unit of temperature, not used with degree symbol; potassium, (2) chemical element; (3) flow coefficient
I/O	Input/output	k	Kilo, prefix meaning 1000
I/P	Current to pneumatic converter	kbs, kbps, kb/sec	Kilobits per second
IP	(1) Ingress protection; (2) internet protocol, the basic addressing standard of the internet; (3) ionization potential	kBps, kB/sec	Kilobytes per second
		k-cal	Kilogram calories (= 4184 J)
I-P	Current-to-pressure conversion	Kc	Proportional sensitivity
IPA	Isopropyl alcohol	$KCLO_3$	Potassium chlorate
IPF	Instrument protective functions	kg	Kilogram
IPL	Independent protection layer	Kg	Kilograms mass
IpoS	Incorrect position switch	kg/hr	Kilograms per hour
IPPS	International practical pressure scale	kg-m	Kilogram-meter (torque = 7.233 foot-pounds)
IPTS	International practical temperature scale	kg/m^2	Kilograms per square meter (pressure unit)
Ipv6	Internet protocol version 6; latest internet addressing standard	kg/m^3	Kilograms mass per cubic meter, density
IR	Infrared	KHP	Potassium acid Phosphate
Ir	Iridium, chemical element	KHz	Kilo hertz
IRT	Infrared Testing	kip	1000 pounds (=453.6 kg)
IS	Intermediate system	km	Kilometer
ISA100 Wireless	A standard for process data acquisition and control; ANSI/ISA is a 100.11a and IEC 62734	$KMnO_4$	Potassium permanganate
		KOH	Potassium hydroxide
		K_p	Proportional gain of a PID controller
ISAB	Ionic strength adjustment buffer	kPa	Kilopascals, a pressure unit
ISE	(1) Ion-selective electrodes; (2) integral of squared error	KPLS	Kernel partial least square
		kVA	Kilovolt-amperes
ISFET	Ion-sensitive field effect transistors	kW	Kilowatts
ISH	Current sense (switch) high	kWD	Kilowatt demand
ISM	Industrial, scientific, medical; a set of frequencies allocated by governments for shared and unlicensed uses	kWh	Kilowatt-hours (=3.6 × 10^6)
		KWM	Kill weight mud
		L	
ISP	Internet service provider or interoperable system provider	L	(1) Length; (2) inductive, expressed in henrys
IT	Information technology	l	Liter (=0.001 m^3 = 0.2642 gal)
ITAE	Integral of absolute error multiplied by time	L2F	Laser two-focus anemometer
ITS	International temperature scale	LA	Level alarm
ITSE	Integral of squared error multiplied by time	Lab	CIE color space (rectangular coordinates) for human color recognition
ITT	Intelligent temperature transmitters	LACE	Least accurate component error
ITV	Industrial television	LACT	Lease automatic custody transfer
IVDS	Input variable and delay selection	LAHH	High-high tank alarm
IXP	Internet exchange point	LAN	Local area network
J		LAS	Link active schedule (FF)
J	Joule, symbol for derived SI unit of energy, het or work, Newton-meter, N·m	lat	Latitude

lb	Pounds	LO	(1) Low; (2) lock open
lb/min	Pounds per minute	LOC	Limiting oxygen concentration
lb/mmscf	Pounds of water per 1 million standard cubic feet	LOCs	Levels of concern
		LOD	Loss on drying
Lbf	Pounds force	LOF	Lack of fusion
lbm	Pounds mass	log, $\log_{10}$	Logarithm to base 10; common logarithm
lbm/ft^3	Pounds mass per cubic foot, density	LOI	Local operation interface
lbm/gal	Pounds mass per gallon	long.	Longitude
lbm/mol	Pounds mass per molecule	LOPA	Layers of protection analysis
LC	(1) Lethal concentration; (2) level controller; (3) liquid chromatography	LOS	Loss of sight
		LP	Low pressure
LCD	Liquid crystal display	LPC	Large particle content
Lch	CIE color space (cylindrical coordinates for lightless, chroma and hue	LPCI	Low-pressure injection system
		LPCS	Low-pressure core spray system
LCL	Lower confidence limit	LPG	Liquefied petroleum gas
LCM	Life cycle management	lph	Liters per hour (0.2642 g/hr)
LCSR	Loop current step response	lpm	Liters per minute (0.2642 g/m)
LD	Ladder diagram	LPR	Linear polarization resistance (lpr)
LDA	Laser Doppler anemometer	LPSA	Laser particle size analyzer
LDP	Large display panel	LQ	Living quarter
LDS	Leak detection system	LQG	Linear quadratic Gaussian
LEC	(1) Local exchange carrier; (2) lower explosive limit	LR	Linear regression
		LRC	Longitudinal redundancy check
LED	Light-emitting diode	LRL	Lower range limit
LEL	Lower explosive limit	LRV	Lower range value
LEOS	Low Earth orbit satellites	LS	Logic series latching device
LER	Licensee event report	LSH	High-level switch
lf	Linear feet	LSHH	High-high tank level switch
LFE	Linear flow element	LSL	Low-level switch
LFL	Lower flammable limit	LT	(1) Leak testing; (2) level transmitter
LG	Level gauge	LTCO	Long term cost of ownership
LGN	Liquefied natural gas	LTE	Long term evolution – used worldwide for 4g digital data communications
LGR	Liquid gas ratio		
l/hr	Liters per hour	LTI	Lineal time-invariant
LI	Level indicator	LVDT	Linear variable differential transformer
Li	Lithium, chemical element	LVN	Limiting viscosity number
LIBS	Laser induced breakdown spectroscopy	lx	Lux, symbol for derived SI unit of illuminance, lumen per square meter, lm/m^2
LIC	Level indicator controller		
LIDAR	Light direction and ranging	LY	1:1 repeater
LiFePO$_4$	Lithium iron phosphate	**M**	
lim	Limit	M	(1) Motor; (2) 1000 (commerce only); (3) Mach number; (4) molecular weight-mole; (5) mega prefix meaning 10^6; (6) module constant
lin	Linear		
liq	Liquid		
LIW	Loss in weight		
LLC	Logical link control	m	(1) Mass, lb (kg); (2) meter, symbol for basic SI unit of length; (3) milli, prefix meaning 10^{-3}; (4) minute (temporal)
LLW	Low-level nuclear waste		
lm	Lumen, symbol for derived SI unit of luminous flux, candela-steradian, cd·sr		
		MA	(1) Moisture analyzer; (2) motor actuator
		mA	Milliampere(= 0.001 a), unit of electric current
ln	Naperian (natural) logarithm to base e		
LNG	Liquefied natural gas	MAC	Media access control; part of the data link layer protocol—the highest level of protocol defined for any IEEE 802 standard
LNT	Linear no-threshold		

MACID	Medium access control identifier	MIB	Management information bass
MACT	Maximum achievable control technology	MIC	Minimum igniting current
mADC	Milliampere, direct current	micro	Prefix meaning 10^{-9}; (2) μ (mu), sometimes (incorrectly) u, as in ug to mean μg [both meaning microgram (=10^{-9} kg)]
MAE	Minimum absolute error		
MAOP	Maximum allowable operating pressure		
MAP	Manufacturing automation (access) protocol	μg	Microgram (= 10^{-6} kg), unit of mass
MATS	Mercury and air toxics standards	micron	Micrometer (= 10^{-6} m), unit of length (a term now considered obsolete)
MAU	Media access unit		
MAWP	Maximum allowable working pressure	MIE	Minimum ignition energy
max.	Maximum	MIMO	Multiple input multiple output
MB	Megabyte, 1,000,000 bytes	MIMOSA	Machinery information management open system alliance
Mb	Megabit, 1,000,000 bits		
Mb/d	Million barrels per day	min.	(1) Minute (temporal), also m. (2) minimum; (3) mobile identification number
Mbps. Mb/sec	Million bits per second		
MC	Main cross connect	MINE	Maximal information nonparametric exploration
mCi, mC	Millicurie (=0.001 Ci)		
MCM	Monte–Carlo method	MIR	Multiple internal reflection
m.c.p.	Mean candle power	MIS	Management information system
MCR	Main control panel	MISO	Multiple input single output
MCT	Mercury cadmium telluride	ml	Milliliter (= 0.001 cc)
MD	MIRA detector	MLC	Micellar liquid chromatography
MDBS	Mobile data base station	MLP	Multilayer perceptron
MDIS	Mobile data intermediate system	MLR	Multiple linear regression
MDQ	Minimum detectable quantity	mm	Millimeter (= 0.001 m), unit of length
m/e	Mass-to-energy ratio	mmf	Magnetomotive force in amperes
med.	Medium or median	mmHg	Millimeters mercury
MEDS	Medium Earth orbit satellite	MMI	Man–machine interface
MEMS	Microelectromechanical systems	mmpy	Millimeters per year
m.e.p.	Mean effective pressure	MMS	(1) Machine monitoring system; (2) mineral management service; (3) manufacturing message services
MES	Manufacturing execution system		
MESG	Maximum experimental safe gap		
Mesh	A layout or arrangement in which messages are relayed by neighbors toward their final destination	mmWC	Millimeters of water column
		MOC	Material of construction
		MODEM	Modulator/demodulator
MeV	megaelectron volt	MOF	Maximum operating flow
MFC	mass flow controllers	mol	Mole, symbol for basic SI unit for amount of substance
MFD	mechanical flow diagram		
MFE	magnetic flux exclusion	mol.	Molecule
mfg	manufacturer or manufacturing	mol%	Percent by moles of a process stream
MFL	magnetic flux leakage	MON	Motor octane number
mg	milligrams (= 0.001 g), unit of mass	MOON	M out of N voting system
MGD	million gallons per day	MOPS	Manual overfill prevention system
MGHr	Million gallons per hour	MOS	Metal oxide semiconductor
MGMR	Multi-gas/multi-range	MOSFET	Metal oxide semiconductor field effect transistor
mH$_2$O	Meters of water		
mho	Outdated unit of conductance, replaced by Siemens (S)	MOTL	Maximum operating thermal limit
		MOTP	Maximum operating thermal power
mHz	Hz 106 (mega)	mp, m.p.	Melting point
MHz	Mega hertz	MPa	Mega pascal (10^6 pascals = 10 bars = 145 psi)
MI	(1) Methane ice; (2) melt index; (3) mutual information		
		MPC	Model predictive control
mi	Mile (−1.609 km)	MPCA	Multivariate principle component analysis

MPD	Managed pressure drilling	MWC	Municipal waste combustors
MPE	Mean power emitted	MWD	Molecular weight distribution
MPFM	Multiphase flowmeter	Mwe	Megawatt electrical
MPH, mph, mi/hr	Miles per hour (= 1.609 km/hr)	MWL	Minimum working level
		MWP	Maximum working pressure
mps, m/s	Meters per second	MWth	Megawatt thermal
MPMS	Manual of petroleum measurement standards, issued by API	**N**	
MPS	Manufacturing periodic/aperiodic services	N	(1) Newton, symbol for derived SI unit of force, kilogram-meter per second squared, $kg \cdot m/s^2$; (2) neutron; 3) number of data in a sample
MPT	Minimum pressurization temperature		
mpy	Mill per year		
MR	Magnetoresistance, change in resistivity of a conductor due to the presence of a magnetic field	n	(1) Nano, prefix meaning 10^{-6}; (2) refractive index
		N_0	Avogadro's number (= $6.0^{23} \times 10^{23}$ mol^{-1})
mR	(1) Milliroentgen (=0.001 R); (2) millirem	N-16	Nitrogen-16
mrem	Millirem	NA	Numerical aperture
MRP	Material requirement planning or manufacturing resource planning	NAAQS	National ambient air quality standards
		NaI	Sodium iodide (crystal)
MRSI	Magnetic resonance spectroscopic imaging	NAND	Not and
		NaOH	Sodium hydroxide
MS	Mass spectrometry	NAT	Network address translation
ms, msec	Millisecond (=0.001 sec)	NB	Nominal bore, internal diameter of a pipe in inches
MSA	Metropolitan statistical areas		
MSB	Most significant bit	NC	Numeric controller
MSCF	Million standard cubic feet	NC, N/C, N.C.	Normally closed
MSD	Most significant digit	NCAP	Network capable application processors
MSDS	Material safety data sheet	NDE	Nondestructive evaluation
MSE	Mean squared error	NDIR	Nondispersive infrared
MSIV	Main steam isolation valve	NDM	Normal disconnect mode
MSPCA	Multiscale principle component analysis	NDT	Nondestructive testing
MSPI	Mitigating systems performance index	NEC	National electric code
		NESC	National Electric Safety Code
MSS	Main steam system	NESHAP	National emission standards for hazardous air pollutants
mSv	Millisievert		
μSv	Microsievert	NeSSI	New sample system initiative
MT	(1) Measurement test; (2) magnetic particle testing	NEXT	Near-end crosstalk
		NFL	Normal fill level or maximum working level
MTBE	Methyl *t*-butyl ether		
MTBF	Mean time between failure	NH_3	Ammonia
MTDSC	Modulated temperature differential scanning calorimeter	NH_4Cl	Ammonium chloride
		NI	Noninformative
MTPL	Maximum thermal power level	NIC	Network interface card
MTSO	Mobile telephone switching office	NIP	Normal incidence pyrheliometer
MTTF	Mean time to failure	NIR	Near infrared
MTTFD	Mean time to fail dangerously	nm	Nanometer, equals 1 billionth of a meter (0.000000001 m)
MTTFS	Mean time to spurious failure		
MTTR	Mean time to repair	NMR	Nuclear magnetic resonance
MTU	Master terminal unit	NO, N/O, N.O.	(1) Normally open; (2) nitrogen oxide
MUR	Measurement uncertainty recapture	NOP	Normal operating pressure
MUX	Multiplexer	NORM	Naturally occurring radioactive material
mV	Millivolt		
MVC	Minimum variance controller	NPD	Nitrogen phosphorous detector
MW	(1) Megawatt; (2) molecular weight	NPN	Negative–positive–negative transistor

NPOC	Nonpurgeable organic carbon	OP-FRIR	Open path Fourier-transform infrared
NPS	(1) National pipe standard; (2) nominal pipe size	OP-FTIR	open-path Fourier transform infrared
NPSC	American standard straight coupling pipe thread	OP-HC	Open-path hydrocarbon
NPSL	American standard straight locknut pipe thread	OP-TDLAS	Open-path tunable diode-laser absorption spectroscopy
NPSM	American standard straight mechanical pipe thread	OP-UV	Open-path ultraviolet
NPT	American standard pipe taper thread (also known as ANSI/ASME B1.20.1)	OPP	Overfill prevention process
		OPS	Overfill prevention system
		or	Orange (typical wiring code color)
NPTR	American standard taper railing pipe thread	ORM	Operating reactivity margin
		ORP	Oxidation–reduction potential
NQA	Erosol-based detector	OS	Operator station or operating system
NRM	Normal response mode	OSC	Override safety controls
NRZ	Nonreturn zero (refers to a digital signaling technique)	OSFP	Open shortest path first
		OSGP	Open smart grid protocol
NS	Nominal pipe size, the internal diameter of a pipe in inches	OSI	Open system interconnect (model)
		OSI/RM	Open system interconnect/reference model
NSSS	Nuclear steam supply system	OT	(1) Operator terminal; (2) open tubular
NTC	Negative resistance/temperature coefficient	OTDR	Optical time domain
		OTM	Other test method
NTMU	Nephelometric turbidity multibeam unit	OTV	odor threshold value
NTP	Normal pressure and temperature conditions	oz	Ounces mass

O

OAD	Outside air damper
OCB	Oxygen combustion bomb
OCD	Orifice-capillary detector
OCR	Optical character recognition
OD	(1) Outer diameter; (2) oxygen demand
ODBC	Open database connectivity or communication
OESOFT	Optical emission spectrometer
oft, OFT	Optical fiber thermometry
ohm	Unit of electrical resistance
OHVD	Overheight vehicle detection
OJT	On-the-job training
OLE	Object linking and embedding
OLE_DB	Object linking and embedding data base
OLR	Online rheometer
OLS	Ordinary least squares
OMMS	Optical micrometer for micromachine
ON	Octane number
OPC	Object link embedding (OLE) for process control
OPEX	Operating expenditure
OP-CRDS	open-path cavity ring-down spectrometry

NTRU	Nephelometric turbidity ratio unit
NTSC	National television standard code
NTU	Nephelometric turbidity unit
NUT	Network update time

P

P	(1) Phosphorus, chemical element; (2) pressure; (3) proportional; (4) pico, prefix meaning 10^{-12}; (5) peta (10^{15}); (6) variable for resistivity
PA	(1) Public address; (2) plant air
Pa	Pascal, symbol for derived SI unit of stress and pressure, Newtons per square meter, N/m^2
P&ID	Piping and instrumentation diagram
P.B.	Push button
PABS	Absolute pressure
PAC	Path average concentration
PAD	Pulsed amperometric detection
PAL	Phase alteration line
PAM	Pulse amplitude modulation
PAN	Personal area network
PAS	(1) Process automation system (successor to DCS); (2) photoacoustic spectroscopy
Pas, Pa·s	Pascal-second, a viscosity unit
PATM	Atmospheric pressure
PB	(1) Profibus; (2) proportional band
PBX	Private branch exchange
PC	(1) Pressure control; (2) personal computer
PCA	Principle component analysis
PCB	Polychlorinated biphenyls
PCCS	Passive containment cooling system
PCD	Polycrystalline diamond compact
PCDD	Polychlorinated dibenzo-p-dioxines

PCDF	Polychlorinated dibenzofurans	P/I	Pneumatic to current (conversion)
PCM	Pulse code modulation	PIC	(1) Pressure indicating controller; (2) path integrated concentration
PCR	(1) Principal component regression; (2) polymerase chain reaction	P&ID	Piping and instrumentation diagram
PCS	Process control system	PID	(1) Photoionization detector; (2) proportional-integral-derivative control
PCCS	Personal computer control system		
pct	Percent; also %	PIG	Pipeline inspection (integrity) gauge
PCTFE	Polychlorotrifluoroethylene	PI-MDC	Path integrated minimum detectable concentration
PCV	(1) Primary containment vessel; (2) pressure control valve		
		PIMFC	Pressure-insensitive mass flow controller
PD	(1) Proportional and derivative; (2) positive displacement	PIMS	Process information management system
		PIP	Process industry practices
PDA	(1) Personal digital assistant; (2) photodiode array	PIR	Precision infrared radiometer
		PIV	Particle image velocimetry
PDD	Pulsed discharge detector	PLC	Programmable logic controller
PDDC	Power density distribution control	PLL	Phase-locked loop
PDF	Probability of failure on demand	PLOT	Porous-layer open tubular columns
pdf	(1) Probability distribution function; (2) portable document format	PLS	Partial least square
		PM	(1) Proton monitors; (2) photomultiplier
PDM	Positive displacement meter	PMA	Physical medium attachment
PdM	Predictive maintenance	PMBC	Process model based control
PDS	Phase difference sensor	PMD	Photomultiplier detector
PDU	Protocol data unit	PMF	Probability mass function
PDVF	Polyvinylidene fluoride	PMMA	Para-methoxymethamphetamine
PE	(1) Power earth (2) polyethylene	PMMC	Permanent magnet moving coil
PECD	Pulsed electron capture detector	PMT	Photomultiplier tube
PED	Pressure equipment directive	PMUD	Mud pressure
PEEK	Polyether ether ketone	PNO	Profibus nutzerorganisation
PEL	Permissible exposure limit	PNP	Positive–negative–positive transistor
PEM	Proton exchange membrane	PO	Pulse output
PES	Programmable electronic systems	POC	Persistent organic compounds
PF, p.f.	Power factor	POP	Persistent organic pollutant
pF	Picofarad (10^{-12} F)	POPRV	Pilot-operated pressure relief valve
PFA	Per-fluoro-alkoxy copolymer	PORV	Power-operated relief valve or pilot-operated relief valve
PFC	Procedure functional chart		
PFD	(1) Probability fail dangerous; (2) process flow diagram	PP	(1) Polypropylene; (2) pyrophosphate
		PPA	Power purchase agreement
PFDavg	Average probability of failure on demand	PPB, ppb	Parts per billion
		PPC	Plant process computer
PFPD	Pulsed flame photometric detector	PPE	Personal protective equipment
PGAA	Prompt-gamma neutron activation analysis	PPG	Pounds per gallon
		ppm	Parts per million
PGC	Process gas chromatograph	PPM	Pulse position modulation
PGNAA	Prompt/pulsed gamma neutron activation analyzers	ppmv	Parts per million by volume
		ppmw	Parts per million by weight
pH	Acidity or alkalinity index (logarithm of hydrogen ion concentration)	PPP	Point-to-point protocol
		PPS	Polyphenylene sulfide
		ppt	Parts per trillion
PHA	Process hazard analysis	PRA	Probabilistic risk assessment
PHID	Pulsed helium ionization detector	PRC	Pressure-recording controller
PhT	Phase transition	PRD	Pressure relief valve
pHz	10^{15} (peta) Hz	precip	Precipitate or precipitated
PI	(1) Pressure indicator; (2) proportional-integral; (3) pulse induction	PROFIbus	Process Field Bus
pi, pI	Poiseuille, a viscosity unit		

PROM	Programmable read-only memory	°Q	Quevenne degrees of liquid density
PRV	Pressure relief valve	q^{-1}	Backward shift operator
PS	(1) Power supply (module); (2) pressure switch	QA	Quality assurance
		QAM	Quadrature amplitude modulation
PSAT	Prestartup acceptance test	QCM	Quartz crystal microbalance
PSD	Photosensitive device	QPLS	Quadratic partial least square
PSG	Phosphosilicate glass	QPSK	Quadrature phase shift keying
PSH	Pressure sense (switch) high	qt	Quart (−0.9463 liter)
PSI	Prestartup inspection	QV	Quaternary variable
psi, lb/in.²	Pounds per square inch	QVNS	Quantized voltage noise source
psia	Pounds per square inch, absolute		
psid	Pounds per square inch, differential	**R**	
psig	Pounds per square inch gauge (pressure)	R, r	(1) Resistance, electrical in ohms; (2) resistance, thermal, meter-Kelvin per watt, m·K/W; (3) roentgen, the unit of radiation dose, exposure to X and gamma radiation (= 2.58×10^{-4} C/kg); (4) gas constant (= 8.317×10^7 erg·mol^{-1}, °C^{-1})
psir	Pound per square inch relative		
PSK	Phase shift keying		
PSL	Low-pressure switch		
PSM	(1) Plant (process) safety management; (2) particle size monitor		
PSSR	Pre-startup safety review	r, rad	Radius
PSTN	Public switched telephone network	r^2	Multiple regression coefficient
PSU	Poststartup	%RA	Reference accuracy
PSV	Pressure safety valve	RA	Return air
PT	(1) Dye penetrant testing; (2) pressure transmitter: (3) liquid penetrant testing	R/A	Reverse action
		Ra	Radium, chemical element
		Ra-226	Radium 226
pt	Point, part, or pint (=0.4732 liter)	RACON	Marine radar beacon
PTC	(1) Positive resistance/temperature coefficient; (2) performance test code	RAD	Return air damper
		rad	Radian, symbol for SI unit of plane angle measurement or symbol for accepted SI unit of absorbed radiation dose (−0.01 Gy)
PTFE	Polytetrafluoroethylene (Teflon)		
PTP	Precision time protocol		
PTZ	Pan tilt zoom		
Pu	Plutonium, chemical element	RADAR	Radio detection and ranging
PUVF	Pulsed ultraviolet fluorescence	RAID	Redundant array of inexpensive disks
PV	(1) Photo voltaic; (2) process variable (measurement); (3) HART primary variable	RAM	Random access memory
		RASCI	Responsible for, approves, supports, consults, informed
PVC	(1) Positive void coefficient; (2) polyvinyl chloride	RBMK	Reactor Bolohoj Moshosztyl Kanalnyj
		RC	Resistance-capacitor circuit
PVDF	Polyvinylidene fluoride	RCIC	Reactor core isolation cooling system (HPCI, HPCS is also used)
PVHI	Process variable high (reading or measurement)		
		RCS	Reactor control system
PVLO	Process variable low (reading or measurement)	RCU	Remote control unit
		RD	Rupture disk
PVT	(1) Pressure, volume, temperature; (2) polyvinyl tubing	RDG, rdg	Reading
		RDP	Remote desktop protocol
PWM	Pulse width modulation	Re	(1) Rhenium, chemical element; (2) Reynolds number
PWR	Pressurized water reactor		
PX	Paraxylene	REL	Recommended exposure limit
PZT	Lead-zirconate-titanate ceramic	rem	Roentgen equivalent man (measure of absorbed radiation dose by living tissue)
Q			
Q	(1) Quantity of heat in joules (J); (2) reaction quotient; (3) electric charge	Re_D	Reynolds number corresponding to a particular pipe diameter
q	(1) Rate of flow; (2) electric charge in coulombs ©	ReSSA	Recursive soft sensing algorithm

Reyn	British unit of dynamic viscosity named in honor of Osborne Reynolds, (1 reyn = 1 $lb_f \cdot sec \cdot inches^{-s}$)	RS232	Recommended standard 232 for voltage level serial data
rev	Revolution, cycle	RS422	Recommended standard 422 for current loop serial data
RF, rf	(1) Radiofrequency; (2) raised face	RS485	Recommended standard 485 for multi-drop communications
RFC	Reversible fuel cell	RSA	Required safety availability
RFCMOS	Radio-frequency complementary metal oxide semiconductor	RSD	Relative standard deviation
RFD	Radio flow detector	RSL	Received signal level
RFF	Remote fiber fluorimetry	RSS	Root sum squared
RFI	Radio-frequency interference	RT	Radiographic testing
RFQ	Request for quote	R(T)	Resistance to temperature characteristics
RGA	Residual gas analyzer	RTD	Resistance temperature detector
RGB	Red-green-blue	RTE	Radiative transfer equation
rgm	Reactive gaseous mercury	RTJ	Ring-type joint
RH	Relative humidity	RTN	Return to normal
Rh	Rhodium, chemical element	RTO	Real-time optimization or operation
rH	Redox value that gives the oxidizing or reducing power of a solution independent of ph	RTOS	Real-time operating system
		RTR	Remote transmission request
		RTS	Ready (or request) to send
RHC	Reheat coil	RTS/CTS	Request to send/clear to send
RHR	Residual heat removal	RTU	Remote terminal unit
RI	Refractive index	RTV	Room temperature vulcanizing silicone
RID	Refractive index detector	RUDS	Reflectance units of dirt shade
RIP	Routing information protocol	RV	Relief valve
RIPS	Rotary inductive position sensor, having an inductance that changes as a core rotates within one or more sensing coils	RVDT	Rotary variable differential transformer, similar to an LVDT, but with the core moving in a rotary fashion
R(k)	Set point	RVI	Remote virtual interface
RM	Radiation monitor	RVLC	Low-capacity relief valve
RMI	Remote meter reader	RVP	Reid vapor pressure
RMM	Removable memory unit	RVR	Runway visual range
RMS, rms	(1) Root mean square; (2) rotary mirror sleeves	RVTP	Relevance variance tracking precision
		RWL	Reference wet leg
RMSE	Relative mean squared error	RWS	Remote workstation
RoHS	Restriction of hazardous substances	**S**	
ROI	return on investment	S	Siemens (Siemens/cm), symbol for unit of conductance, amperes per volt, A/V
ROM	(1) Remote operations manager; (2) read-only memory		
		s	(1) Second (also sec), symbol for basic SI unit of time; (2) Laplace variable; (3) sample-based estimate of σ
RON	Research octane number		
ROP	Rate of penetration		
ROV	Remote operated vehicles	s^2y	Sample variance of output y
RPC	(1) Reverse phase chromatography; (2) remote procedure call (RFC1831)	SA	Supply air
		SAP	(1) Service access point; (2) standard ambient pressure
RPLS	Recursive partial least squares		
RPM, rpm, r/min	Revolution per minute	SAT	(1) Site acceptance test; (2) supervisory audio tone
RPS	Reactor protection system	sat.	Saturated
RPV	Reactor pressure vessel	SATP	Standard ambient temperature and pressure
RRF	Risk reduction factor		
RRT	Relative response time (time required to remove most of the disturbance)	SAW	Surface acoustic wave
		SBO	Station blackout
RS	(1) Recommended standard; (2) reset/set	SC	System codes

SCADA	Supervisory control and data acquisition	SIL	Security (safety) integrity level
SCC	Stress corrosion cracking	sin	Sine, trigonometric function
SCCM	Standard cubic centimeter per minute	SIPD	Self-indicating pocket dosimeter
SCD	(1) Streaming current detector; (2) sulfur chemiluminescence detector	SIR	Statistical inventory reconciliation
		SIS	Safety instrumented system
SCE	Saturated calomel electrode	SIT	System integration testing
SCFH	Standard cubic feet per hour	SKU	Stock keeping units
SCFM	Standard cubic feet per minute (air flow at 1.0 atm and 70°F)	SLAMS	State and local air monitoring stations
		SLC	Safety life cycle
SCM	Streaming current monitor	SLCS	Standby liquid control system
SCMH	Standard cubic meter per hour	slph	Standard liters per hour
SCMM	Standard cubic meters per minute	slpm	Standard liter per minute
SCO	Synchronous connection oriented	SMA	Scale manufacturers' association
SCOT	Support-coated open tubular	SMPT	Simple mail transfer (management) protocol
SCR	Silicon-controlled rectifier	SMR	Specialized mobile radio
SCS	Sample conditioning system	SMS	Short message service
SCV	Stream current value	SMSD	Summon detector
SD	Component in leg has failed safe and failure has been detected	SMYS	Specified minimum yield stress
		S/N (or SNR)	Signal-to-noise ratio
SDI-12	Serial/digital interface	SNCR	Selective noncatalytic reductions
SD Card	Secure digital card	SNG	Synthetic natural gas
SDIU	Scanivalve digital interface unit	SNM	Special nuclear materials
SDN	Send data with no acknowledgement	SNMP	Simple network management protocol
SDND	Self-powered neutron detector	SNR	Signal-to-noise ratio
SDNN	Supervised distributed neural networks	SOAP	Simple object access protocol (an Internet protocol that provides a reliable stream-oriented connection for data transfer)
SDP	Significance determination process		
SDS	Smart distribution system		
SEA	Spokesman election algorithm	SOC	State of charge
SEM	Subsea electronic module	SOESOLAS	Sequence of events
SER	Sequence of event recorder	SONAR	Sound navigation and ranging
SERS	Enhanced Raman scattering	SOP	Standard operating procedure
SFA	Segmented flow analysis	SORV	Stuck open relief valve
SFC	Sequential function chart	sos	Speed of sound
SFD	(1) Scintillation fiber detectors; (2) system flow diagram; (3) start of frame delimiter	SP	(1) Spontaneous potential; (2) set point
		SPAR	Standardized plant analysis risk
SFF	Safe failure fraction	SPC	Statistical process control
SFI	Sight flow indicator	SPD	Self-powered (neutron) detector
SFR	Spurious failure rate	SPDS	Safety parameter display system
SG, S.G.	Specific gravity	SPDT	Single pole double throw
SGC	Smart gas chromatograph	SPEC	Screen printed electrochemical sensor
SHE	Standard hydrogen electrode	SPL	Sound pressure (power) level
SH&E	Safety, health and environmental	SPND	Self-powered neutron detector
SHM	Structural health monitoring	SPRT	standard platinum resistance thermometer
SHS	Sample handling system	SPST	Single pole single throw
SI	Système international d'unités (international units—metric)	SPVD	Self-powered vehicle detector
		sq	Square, squared
SICP	Shut-in casing pressure	SQC	Statistical quality control
SID	System identification digit (number)	SQL	Structured (or standard) query language
SIDPP	Shut-in drill pipe pressure	Sr	Steradian, symbol for SI unit of solid angle measurement
SIE	Specific ion electrodes		
SIF	Safety instrument function	SRD	Send and request data with reply
SIG	Special interest group	SREM	Standard radiation environment monitors

SRPD	Self reading pocket dosimeter
SRS	Safety requirements specification
SRV or SVR	Safety relief valve
SSB	Single side band
SSI	Synchronous serial interface
SSL	Secure socket layers
SSNDIR	Solid-state nondispersive infrared
SSU	Saybolt seconds universal
ST	Structural text
std.	Standard
STEL	Short-term exposure limit
STEP	Standard for exchange of product model data
STP	(1) Standard temperature and pressure corresponding to 70°F (21.1°C) and 14.7 psia (1 atm abs) and; (2) shield twisted pair
STR	Spurious trip rates
SU	Security unit or component in leg has failed safe and failure has not been detected
SUS	Sybold universal seconds
SV	(1) Secondary variable; (2) safety valve
Sv	Sievert
SVV	Code safety valve
S/W	Steam to water ratio
T	
T½	Half-life
T	(1) Tesla, symbol for derived SI unit of magnetic flux density, webers per square meter, Wb/m^2; (2) absolute temperature; (3) ton (metric = 1000 kg); (4) tube period, s (time for one cycle of oscillation); (5) tera, prefix meaning 10^{-12}
t	(1) Time; (2) thickness
tan	Tangent, trigonometric function
TAS	Thallium-arsenic-selenide
tau, τ	Process time constant (seconds)
tb	Trillion barrels
TBM	Tertiary butyl mercaptan
t/c	Thermal coefficient (also sometimes used for thermocouple)
TC	(1) Thermocouple; (2) temperature controller; (3) total carbon
TCD	Thermal conductivity detector
TCDD	Tetrachlorodibenzo-p-dioxin
tcf	Trillion cubic feet
TCLP	Toxicity characteristic leaching procedure
TCP	Transmission control protocol
TDLAS	tunable diode laser absorption spectroscopy
TDLS	Tunable diode laser spectroscopy
TCP/IP	Transmission central protocol/internet protocol
TCT	Tank capacity table
TCV	Temperature control valve
T_d	Derivative time (in seconds) of a PID controller
td	Process dead time (seconds)
TDC	Tilting disk check
TDL	Tunable diode laser
TDLAS	Tunable diode laser absorption spectroscopy
TDM	Time division multiplexing
TDMA	Time division multiple access
TDR	Time domain reflectometry
T/E	Thermoelectric
TEM	Transmission electron microscope
TFE	Tetrafluoroethylene
TFT	Thin film transistor
TG	Thermogravimetry
tHz	1912 (tera) Hz
TI	(1) Time interval between proof tests (test interval); (2) temperature indicator
Ti	Integral time (in seconds) of a PID controller
TIC	(1) Theils inequality coefficient; (2) temperature indicating controller; (3) total inorganic carbon
TIFF	Tagged image file format
TIR	Total internal reflection
TISAB	Total ionic strength adjustment buffer
TIW	Texas iron works valve
TLC	Top-of-the-line corrosion
TLD	Thermoluminescent dosimeters
TLS	Tunable laser spectrometer
TLV	Threshold limit value
TLV-STEL	Short-term exposure limit TLV
TLV-TWA	Time-weighted average TLV
TMP	Thermochemical pulp
TMR	Triple modular redundancy
TN	Total nitrogen
TOC	(1) Total organic carbon; (2) toxicity equivalence factor
TOD	Total oxygen demand
TOF	Time of flight
TOP	Technical and office protocol
TOU	Time of use monitoring
TP	Total phosphates
TPE	Total probable error
TPL	Thermal power limit
TQM	Total quality management
TR	Temperature recorder
T/R	Transmit-receive design

T/R	Transmit/recorder	UT	(1) Ultrasonic testing; (2) ultrasonic thickness
TRC	Temperature recording controller		
TRP	Thermal reactor power	UTG	Ultrasonic thickness gauge
TRUW	Transuranic nuclear waste	UTGM	Multiple echo UTG
T.S.	Tensile strength	UTP	Unshielded twisted pair
TS	Total sulfur analysis	UTS	Ultimate tensile stress
TSF	Tailings storage facility	UUP	Unshielded untwisted pair
TSH	Temperature sense (switch) high	UV	Ultraviolet
TTFM	Transit-time flow measurement	UV/DAD	Ultraviolet diode array detector
TTP	Through the probe	UV/VIS	Ultraviolet/visible
TUR	Test uncertainty ratio	UV/VIS/NIR	Ultraviolet/visible/near infrared
TV	Tertiary variable	UXO	Unexploded ordinance
TVD	True vertical depth		
°Tw	Twaddell degrees of liquid density	**V**	
TWA	Time weighted average	V	Volt, symbol for derived SI units of voltage, electric potential difference and electromotive force, watts per ampere, W/A
U			
u	Used incorrectly as prefix = 10^{-6} when the Greek letter μ is not available	v	Velocity
		VA	Visible absorbance
UA	Unified architecture (of OPC)	Vac	Voltage, alternating current
UART	Universal asynchronous receiver transmitter	VARACTOR	Voltage-sensitive capacitors
		VAV	Variable air volume (damper)
UBD	Underbalanced drilling	VBA	Visual Basic for Applications
UBET	Unbiased estimation	VBR	Variable bore ram
UCL	Upper confidence limit	VC	Void coefficient
UCMM	Unconnected message manager	VCE	Vapor cloud explosion
UDP	Universal datagram protocol, a basic message format standard for the internet	VCF	Volume correction factor
		VCSEL	Vertical cavity surface emitting laser
		VDF	Vacuum fluorescent display
UEL	Upper explosive limit	VDT	Video display tube
$u_{fb}(k)$	Feedback controller output	VDU	Video display unit
UF6	Uranium hexafluoride	vert.	Vertical
UFD	Utility flow diagram	VF	Void fraction
UFL	Upper flammable limit	VFD	Variable frequency drive
$u_{fk}(k)$	Feedforward controller output	VFIR	Very fast infrared
UHF	Ultrahigh frequency	VHF	Very high frequency
UHSDS	Ultrahigh-speed deluge system	VI	Visual inspection
UID	Ultrasonic interface detector	VIP	Video image processor
UIP	User interface plug-in	VIS	Visible
u(k)	Controller output	V-L	Vapor–liquid ratio
μm	Micrometers	VLF	Very low frequency
UM	Uncertainty margin	VLSI	Very-large-scale integration
UML	Universal modeling language	V/M	Voltmeter
UNACK	Unacknowledged	VME	Virsa Module Europa (IEEE 1014-1987)
UNZ	Upper null zone	VMS	Vibration monitoring system
UPS	Uninterruptible power supply	VOC	Volatile organic carbon
UPV	Unfired pressure vessel	VR	Virtual reality
URL	Upper range limit	VRML	Virtual reality modeling language
URV	Upper range value	vs.	Versus
USB	Universal serial bus	VT	(1) Visual testing; (2) visible transmittance
USCG	US Coast Guard		
USP	US Pharmacopeia	V&V	Verification and validation
UST	Ultrasonic thermometer	VVER	Vodo-Vodyanoi Energetichesky Reactor

W	
W	(1) Watt, symbol for derived SI unit of power, joules per second, J/s; (2) weight (also wt)
w	(1) Width; (2) mass flow rate
w.	Water
WAN	Wide area network
WAT	Water system
WATD	Dirty water system
Wb	Weber, symbol for derived SI unit of magnetic lux, volt-seconds, V·s
WBC	White balance control
WC	Water column
WCE	Worst-case error
WCOT	Wall-coated open tubular
WDXRF	Wavelength dispersion x-ray fluorescence
WG	(1) Wave guide; (2) standard British wire gauge
WGR	Water-to-gas ratio
wh	White, (wiring code color for ac neutral conductor)
WI	Wobble index
WIA-PA	Chinese wireless communications standard for process automation; IEC 62601
Wi-Fi	Wireless fidelity; the common name for the IEEE 802.11 family of wireless standards; the Wi-Fi alliance is the organization responsible for certifying devices conforming to the ieee 802.11 set of standards
WirelessHART	A standard for process data acquisition; IEC 62591
WLAN	Wireless local area network
WLPI	White-light polarization interferometry
WLR	Water-in-liquid ratio
WMS-2f	Wavelength-modulation-spectroscopy with second-harmonic detection
WOB	Weight of bit
Wp	Watts-peak capacity
WPAN	Wireless personal area network—(see PAN)
WS	Workstation
WT	Water tank
wt	Weight, also W
Wt%	Percent by weight of a process stream
WYSIWYG	What you see is what you get
X	
X	Reactance in ohms
x	(1) Measured value; (2) model input variable
XFI	Purge flow indicator
XL/A	Pilot lamp A
XML	Extensible markup language
x-ray	Electromagnetic radiation with a wavelength <100 Å
XRF	X-ray fluorescence
XYZ	Tristimulus function
Y	
Y	Expansion factor
y	Response variable
y(k)	Process output
yr	Year
yd	Yard (=0.914 m)
Z	
Z	(1) Atomic number; (2) electrical impedance (complex), expressed in ohms
z	The number of moles of electrons transferred in the cell reaction or half-reaction
ZEB	Zero energy band
ZigBee	A series of specifications from the ZigBee alliance defining additional communications layers above the MAC layer.
ZJ	Zeta joules (10^{21} joules).
ZSC	Position detector switch

SYMBOLS

ϕ	Radius of laser ray point
ΔP	Pressure difference
ΔPA	Differential pressure alarm
ΔPRC	Differential pressure recording controller
ΔPSL	Low-low differential pressure switch
ΔPT	Differential pressure transmitter
ΔV	Voltage difference
$\bar{X}$	The average, a sample-based estimate of μ
k	Spring constant, pound-force/in. or lb/s^2 (N/m or kg/s^2)
$\frac{\partial y}{\partial x_i}$	Rate of change of y with respect to the ith x, keeping all other values constant
$\in$	Finite, but small, change representing the error range
Σ	Sum
μ	(1) Micron; (2) mean or true value
$t\nu,\alpha$	Critical value of the t-statistic
α	Level of significance = 1 − (desired confidence limit)/100%
α	Symbol for alpha-radiation
$\beta-$	Symbol for beta-radiation
γ	Symbol for gamma-radiation and photons
ν	$N - 1$ = statistical degrees of freedom
ρ	Density
ρf	Fluid density, lb/in.^{3} (g/cc)
σ	Standard deviation
ω_n	Natural frequency, rad/s
Ω	Ohm

A.3 Organizations

A

AATCC	American Association of Textile Chemists and Colorist
ABNT	Associação Brasileira de Normas Técnicas
ACGIH	American Conference of Government Industrial Hygienists
AEG	American Environmental Group
AGA	American Gas Association
AGMA	American Gear Manufacturers Association
AIAA	American Institute of Aeronautics and Astronautics
AIHA	American Industrial Hygiene Association
AISC	American Institute of Steel Construction
ANSI	American National Standards Institute
AOAC	Association of Analytical Communities
AOCS	American Oil Chemists Society
APHA	American Public Health Association
API	American Petroleum Institute
ASCE	American Society of Civil Engineers
ASHRAE	American Society of Heating, Refrigerating and Air Conditioning Engineers
ASM	(1) American Society of Microbiology; (2) Abnormal Situation Management (Consortium)
ASME	American Society of Mechanical Engineers
ASNT	American Society of Non-Destructive Testing
ASPE	American Society of Plumbing Engineers
ASPO	Association for the Study of Peak Oil and Gas
ASSE	American Society of Safety Engineers
ASTM	American Society for Testing and Materials
ATEX	Atmosphere Explosive
ATSDR	Agency of Toxic Substances and Disease Registry
AWS	American Welding Society
AWWA	American Water Works Association

B

BASEEFA	British Approvals Service for Electrical Equipment in Flammable Atmospheres
BPCS	Business planning and control system
BSI	British Standards Institution

C

CARB	California Air Resources Board
CCOHS	The Canadian Centre for Occupational Health & Safety
CE	European Standards Commission
CEESI	Colorado Engineering Experiment Station Inc.
CEN	European Committee for Standardization
CENELEC	European Committee for Electrotechnical Standardization
CGA	Cylinder Gas Association
CIE	Commission International del'Eclairage
CIL	Canadian Industries Limited
CPAC	Center for Process Analytical Chemistry
CSA	Canadian Standards Association
Cyanide Code	International Cyanide Management Code

D

DEMKO	Denmark's Elektriske Material Kontrol
DICA	Direction Des Carburants (France)
DIN	Deutsches Institut für Normung
DIN	German Standard
DNV	Det Norske Veritas (Norway)
DOE	United States Department of Energy
DOT	U.S. Department of Transportation

E

EDDL	Electronic Device Descriptive Language
EDP	Environmental Protection Division
EEMUA	Engineering Equipment and Materials Users' Association
EFNDT	European Federation for Non-Destructive Testing
EFTA	European Free Trade Association
EI	Evaluation International
EIA	Energy Information Administration
EN	European Standards
EOTC	European Organization for Conformity Assessment
EPA	Environmental Protection Authority
EPIX	Equipment Performance and Information Exchange
ETACS	European Testing and Assessment of Comparability of Online Sensors/Analytics

A.3 Organizations 1171

ETSI	European Telecommunications Standards Institute	IFPAC	International Forum for Process Analytical Chemistry
EURAMET	European Association of National Metrology Institutes	IFRF	International Flame Research Foundation
		IIEAG	International Instrument Evaluation Agreement Group
exida	A leading authority in the field of functional safety. They serve to meet the growing need for companies to become more knowledgeable about the requirements for safety related applications. exida has evolved from two offices in the US & Germany to a global support network. They provide consulting, product testing, certification, assessment, cybersecurity, and alarm management services.	ILO	International Labor Organization
		IMO	International Maritime Organization
		INPO	Institute of Nuclear Power Operations
		IOA	International Ozone Association
		IP	Institute of Petroleum (UK)
		IPHE	International Partnership for the Hydrogen Economy
		ISA	International Society of Automation
		ISCCP	International Satellite Cloud Climatology Project

F

FDA	Food and Drug Administration
FDT/DTM	Field device tool/device type manager
FHWA	Federal Highway Administration
	Fieldbus Foundation™
FM	Factory Mutual
FMG	Factory Mutual Global
FMRC	Factory Mutual Research Corporation
FOA	Fiber Optic Association

ISEA	International Safety Equipment Association
ISMA	International Satellite Monitoring Agency
ISO	International Organization for Standardization
ISWM	International Society of Weighing and Measurement
ITA	Instrumentation Testing Organization
ITS-90	International Temperature Scale of 1990
ITU	International Telecommunication Union
IUPAC	International Union of Pure and Applied Chemistry
IUVA	International UV Association

G

GAWP	Georgia Association of Water Professionals
GIIGNL	International Group of Liquefied Natural Gas Importers
GINC	Global Information Network on Chemicals
GOST	Russian Standards Gosudarstvennyy Standart
GPA	Gas Processors Association
GRI	Gas Research Institute (U.S.)

J

JIS	Japanese Industrial Standards
JPL	Jet Propulsion Laboratory

K

KEMA	Keuring van Elektrotechnische Materialen te Arnhem
KOSHA	Certification for Safety Components

H

HART	Communication Foundation, www.hartcomm.org/
HI	Hydraulic Institute

L

LIPA	Long Island Power Authority

M

MMS	Mineral Management Service
MSHA	The Mine Safety and Health Administration
MSS	Manufacturers Standardization Society

I

IAEA	International Atomic Energy Agency
ICAO	International Civil Aviation Organization
ICME	International Council on Metals and the Environment
ICMI	International Cyanide Management Institute
ICNDT	International Committee for Non-Destructive Testing
ICRP	International Commission on Radiological Protection
IEA	International Energy Agency
IEC	International Electrotechnical Commission
IECEx	Certification of Personnel Competencies Scheme
IEEE	Institute of Electrical and Electronics Engineers

N

NACE	National Association of Corrosion Engineers
NAMUR	User Association of Automation Technology in Process Industries
NASA	National Aeronautics and Space Administration

NBS	National Bureau of Standards
NCDC	National Climatic Data Center
NCSL	National Conference of Standards Laboratories International
NCWM	National Conference on Weights and Measures
NDP	Norwegian Petroleum Directorate
NEC	National Electrical Code
NEMA	National Electrical Manufacturers Association
NEPSI	National Supervision and Inspection Center for Explosion Protection and Safety
NFPA	National Fire Protection Association
NIJ	National Institute of Justice
NIOSH	National Institute for Occupational Safety and Health
NISO	National Information Standards Organization
NIST	National Institute of Standards and Technology
NOAA	National Oceanic and Atmospheric Administration
NRC	Nuclear Regulatory Commission
NREL	National Renewable Energy Laboratory
NSC	National Safety Council
NSS	Network Security Services
NVLAP	National Voluntary Laboratory Accreditation Program

O

OECD	Organization for Economic Cooperation and Development
OIML	International Organization of Legal Metrology (French: Organisation Internationale de Métrologie Légale)
OS&E	Ozone Science & Engineering
OSHA	Occupational Safety and Health Administration

P

PESO	Petroleum and Explosives Safety Organisation
PHMSA	Pipeline and Hazardous Material Safety Administration
PPSA	Pigging Products and Services Association
Profibus®	Profibus International
PSF	Pipeline Simulation Facility
PTB	Physikalische-Technische Bundesanstalt

R

RIIS*	Research Institute of Industrial Safety (of Japan)

S

SAA	Australian Safety Administration
SAE	Society of Automotive Engineers International
SAIC	Science Applications International Corp.
SAMA	Scientific Apparatus Makers Association
SASO	Saudi Arabian Standard Organization
SEI	Safety Equipment Institute
SIRA	Society of Information Risk Analysts
SIREP	Sira Instrumentation panel
SPE	Society of Petroleum Engineers
SPIE	International Society of Optics and Photonics
SSPC	The Society for Protective Coatings
SWE	SIREP-WIB-EXERA

T

TAPPI	Technical Association of the Pulp and Paper Industry
TEPCO	Tokyo Electric Power Company
TIIS	Technology Institution of Industrial Safety
TNO	Toegepast Natuurwetenschappelijk Onderzoek (Netherlands Organisation for Applied Scientific Research)
TUV	Technischer Uberwachungs Verein (Germany)

U

UL	Underwriters Laboratories (an independent, not-for-profit, product safety and certification organization)
UNEP	United Nations Environmental Program
USCG	U.S. Coast Guard
USDA	United States Department of Agriculture
USEPA	United States Environmental Protection Agency
USGS	United States Geological Survey
USNRS	U.S. Nuclear Regulatory Commission
USP	US Pharmacopeia

V

VDE	Verband der Elektrotechnik
VDI	Verein Deutscher Ingenieure (Commission of Air Pollution Prevention)
VIM	International Vocabulary of Basic and General Terms in Metrology

W

WEF	Water Environment Federation
WIB	Werkgroup voor Instrument Beoordeling (Process Automation Users' Association or International Instrument Users' Association)
WMO	World Meteorological Organization

---

* Name was changed to National Institute of Industrial Safety, NIIS. Was then amalgamated with the National Institute of Industrial Health to be the National Institute of Occupational Safety & Health, JNIOSH.

A.4 Flowsheet and Functional Diagrams Symbols*

B. G. LIPTÁK (2017)

INTRODUCTION

Engineers are visual people, and the world is in the process of globalization. One of the consequences of these facts is the need for accurate and *standardized visual communication* among automation professionals. The international standardization of symbols and of letter abbreviations serves this purpose within our profession. This standardization is necessary to allow for the accurate and detailed transfer of technology among engineers of the various nations that are speaking different languages. Internationally standardized symbols allow all engineers around the world to accurately understand any and all automation, measurement or control designs.

The International Society of Automation (ISA) made a major contribution by developing the set of symbols that are presented in this chapter, which today are allowing the members of the automation profession to communicate around the world.

This chapter excerpts ISA's Standard 5.1 and provides recommendations for the standardized use of identification letters (abbreviations), graphic, binary logic, device, function, primary element, transmitter, electrical, and final element symbols. This information is described by text and is also tabulated for easier access. The index of these tables is given as follows:

Table A.4a (page 1176)—Identification Code Letters
Table A.4b (page 1177)—Instrumentation Device and Function Symbols
Table A.4c (page 1177)—Miscellaneous Device and Function Symbols
Table A.4d (page 1178)—Primary Element and Transmitter Symbols
Table A.4e (page 1179)—Measurement Related Abbreviations
Table A.4f (page 1180)—Primary Element Symbols
Table A.4g (page 1182)—Secondary Instrument Symbols
Table A.4h (page 1183)—Auxiliary and Accessory Device Symbols
Table A.4i (page 1184)—Instrument to Process Connection Symbols
Table A.4j (page 1185)—Instrument-to-Instrument Connection Symbols
Table A.4k (page 1187)— Final Control Element Symbols
Table A.4l (page 1188)—Final Control Element Actuator Symbols
Table A.4m (page 1190)—Regulator Relief Device Symbols
Table A.4n (page 1192)—Control Valve Failure Positions
Table A.4o (page 1193)— Functional Diagram Symbols
Table A.4p (page 1194)—Function Block Symbols
Table A.4q (page 1201)—Binary Logic Symbols
Table A.4r (page 1209)—Electrical Schematic Symbols

This standard establishes a uniform means of depicting and identifying instruments or other devices and their inherent functions and application software functions used for measurement, monitoring, and control, by presenting a designation system that includes identification schemes and graphic symbols. A significant part of this chapter has been extracted from the 2009 revision of the standard: ANSI/ISA-5.1-2009 and ISA ANSI/ISA-5.4-1991 (ANSI—American National Standards Institute).

The purpose of this chapter is to provide the reader with an internationally understood uniform means of depicting and identifying all instruments, control systems, and functions used for the measurement, monitoring, and control of automated systems. These symbols and identification codes are used in the preparation of engineering flow sheets and piping and instrument diagram (P&ID).

For the definitions of the various terms used in this chapter, refer to table Chapter A.1 in Appendix of this volume.

DOCUMENTATION

In designing a control system for a plant, the key documents used include, (1) the P&ID (piping and instrument diagram, also called flow sheet), which describes all key equipment in the plant, (2) the loop diagrams, which describe each control loop, including communications in full detail, (3) the specifications of each instrument and (4) the instrument index, which lists each instrument in the plant, identifies their related documents and status.

---

* The material in this chapter was mostly excerpted from the International Society of Automation (ISA) standard ANSI/ISA-5.1-2009, titled Instrumentation Symbols and Identification, with the permission of ISA.

Flow Sheets

The design documents that are prepared to show all the equipment, piping, and instrumentation on a single drawing are referred to as "flow sheets" or as piping and instrumentation drawings (P&IDs). An example of how a part of such a flow sheet would appear and what would be shown on it is illustrated in Figure A.4a. This figure describes the piping and control instrumentation of a distillation column.

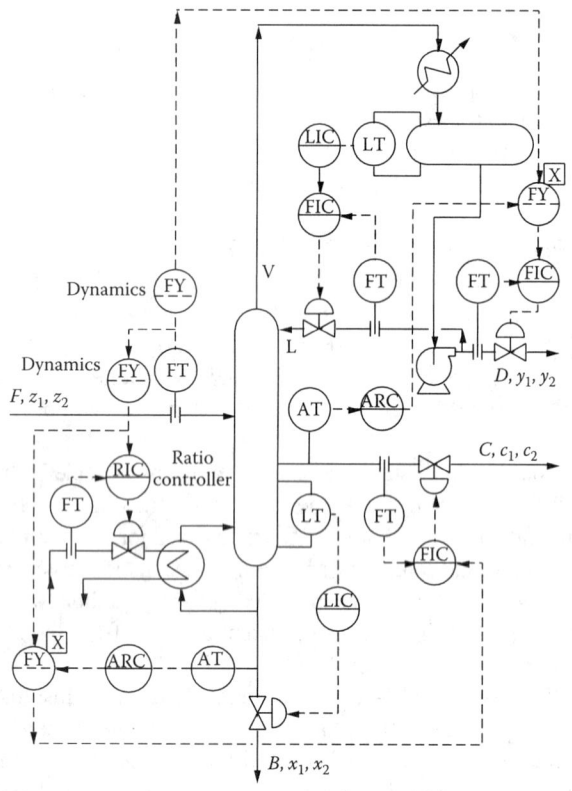

FIG. A.4a

A section of a P&ID diagram, showing the equipment, piping, and instrumentation of a distillation column.

On the P&ID diagram shown in Figure A.4a, there are a number of control loops, serving to control flows (FIC), levels (LIC), compositions (ARC), and ratios (RIC). Each of these loops consists of at least three components: the sensor, the controller, and the final control element (control valve). On a P&ID, these loop components are shown only schematically, without describing their power supplies, accessories, or the way how their signals are passed from one to another. When working on the detailed design, construction, checkout, start-up, operation, maintenance, rearrangement, or reconstruction of industrial plants, more detail is needed. Such details are provided by the individual control *loop diagrams*.

Instrument Symbols

Each main component of a control loop on the flow sheet is represented by a *bubble* (a circle) or other shapes (diamond, square, etc.), identifying the type of device (such as analog, digital, and logic). In case of an analog control system, a horizontal line is drawn, if the component is located remotely, or no line is drawn if it is not. In the upper half of the bubble, code letters are placed to identify the type of device. The identification code letters in the bubble vary widely, but the most common combination is to have three letters as follows:

Code Letters (Table A.4a)

First: Identifies the process variable (the measurement, such as F for flow or T for temperature).

Second: Identifies further aspects of the measurement (such as D for difference or R for ratio) or the type of display or purpose (such as I for indicator, R for recorder, T for transmitter, A for alarm, S for switch, Q for totalizer).

Third: Usually identifies if the unit is also a controller by the letter C.

Within the lower half of the bubble, below the letter combination above combination, a number is inserted that is unique to that device in the whole plant. The first digit of this number is usually the flow sheet number on which it is located, while the following digits are consecutive starting with one. In case it is felt that there is further information, which is important enough to appear on the flow sheet, outside (usually below) the bubble, one would add it. For example, in case of analyzers, the measurement identification letter of "A" could be considered insufficient, so one might add IR for infrared or UV for ultraviolet, etc., under the bubble.

Instrument Index

One of the documents used on larger projects is the instrument index, which serves to collect all the information related to each instrument. On a larger project there, it might be a document of a hundred or more pages, with the blank form of each page containing columns for the information shown in Figure A.4b. Each row in that index is dedicated to a particular instrument on the job, and as the design progresses, more and more columns get filled out on it.

In the first column goes the "Tag Number" of the instrument, which in case of a temperature transmitter on flow sheet 17 might be TT-17xx. In the second column goes the location, which might be a pipe or a vessel, say V-9 for the ninth vessel on that flow sheet. The type of transmitter is given in the next column, while the fourth column gives the corresponding installation detail drawing number. As the project progresses, the remaining columns also get filled in with information on the pipe or vessel specification (materials of construction, design pressure, etc.), the need for air supply if any, special supports, heat, or other type of tracing, electrical aspects, including power supply.

Once the instrument has been fully specified, the number of the requisition in which the specification can

Instr. No.	Location	Type	Detail DWG No.	Pipe spec.	Air sup-ply	Sup-port dot.	Tree-ing	Elec. req.	REQ. No.	P.O. and vendor	Remarks

DR | CHK | APP | Scale

For

Instrument index

Contract | DWG. No. | REV.

BÉLA G. LIPTÀK Associates

FIG. A.4b
A page of a typical Instrument Index. (Note: In the column titles, "Elec. req." indicates if electric power supply is needed, "REQ. NO." is the requisition number of the specification of the instrument and "P.O." is the purchase order number.)

be found is filled in, and once they have been sent out for bids, the quotations are analyzed and the vendor are selected, and the purchase order number and the vendor's name are added.

LETTERING AND SYMBOLS

Table A.4a provides a more detailed set of information concerning the meanings of the earlier-described code letters and their combinations, serving to form the alphabetic building blocks of the Instrument and Function Identification System. Table A.4a defines and explains the meanings of the individual letters when used to identify loop and device functions.

Naturally, the user is free to modify or deviate from the lettering recommended in Table A.4a, but if the user does, it is recommended to describe these differences on the various job documents, including the user's engineering and design standards, guidelines, and the legends on the flow sheets and installation drawings.

The instrument identification letters tabulated in Table A.4a are placed into the upper half-function symbol "bubbles"; the shapes and the lines across these "bubbles" describe the nature and location of the instrument as explained in Tables A.4b and A.4c.

Table A.4c provides a listing of additional, less frequently used flow sheet function symbols.

TABLE A.4a
Identification Code Letters

	First Letters (1)		Succeeding Letters (15)		
	Column 1	Column 2	Column 3	Column 4	Column 5
	Measured/Initiating Variable	Variable Modifier	Readout/Passive Function	Output/Active Function	Function Modifier
A	Analysis		Alarm		
B	Burner, combustion		User's choice	User's choice (5)	User's choice (5)
C	User's choice (3a)(5)			Control (23a)(23e)	Close (27b)
D	User's choice (3a)(5)	Difference, differential, (11a)(12a)			Deviation (28)
E	Voltage (2)		Sensor, primary element		
F	Flow, flow rate (2)	Ratio (12b)			
G	User's choice		Glass, gage, viewing device (16)		
H	Hand (2)				High (27a)(28a)(29)
I	Current (2)		Indicate (17)		
J	Power (2)		Scan (18)		
K	Time, schedule (2)	Time rate of change (12c)(13)		Control station (24)	
L	Level (2)		Light (19)		Low (27b)(28)(29)
M	User's choice (3a)(5)				Middle, intermediate (27c)(28)(29)
N	User's choice (5)		User's choice (5)	User's choice (5)	User's choice (5)
O	User's choice (5)		Orifice, restriction		Open (27a)
P	Pressure (2)		Point (test connection)		
Q	Quantity (2)	Integrate, totalize (11b)	Integrate, totalize		
R	Radiation (2)		Record (20)		Run
S	Speed, frequency (2)	Safety (14)		Switch (23b)	Stop
T	Temperature (2)			Transmit	
U	Multivariable (2)(6)		Multifunction (21)	Multifunction (21)	
V	Vibration, mechanical analysis (2)(4)(7)			Valve, damper, louver (23c)(23e)	
W	Weight, force (2)		Well, probe		
X	Unclassified (8)	X-axis (11c)	Accessory devices (22), Unclassified (8)	Unclassified (8)	Unclassified (8)
Y	Event, state, presence (2)(9)	Y-axis (11c)		Auxiliary devices (23d)(25)(26)	
Z	Position, dimension (2)	Z-axis (11c), safety instrumented system (30)		Driver, actuator, unclassified final control element	

Note: Due to space limitations, the explanatory notes in parentheses are not included, but can be found in ANSI/ISA-5.1-2009.

TABLE A.4b
Instrumentation Device and Function Symbols

No.	Shared Display, Shared Control (1)		C	D	Location and Accessibility (7)
	A — Primary Choice or Basic Process Control System (2)	B — Alternate Choice or Safety Instrumented System (3)	Computer Systems and Software (4)	Discrete (6)	
1	○ in □	◇ in □	⬡	○	• Located in field. • Not panel, cabinet, or console mounted. • Visible at field location. • Normally operator accessible.
2	○ in □ with line	◇ in □ with line	⬡ with line	○ with line	• Located in or on front of central or main panel or console. • Visible in front of panel or on video display. • Normally operator accessible at panel front or console.
3	○ in □ with dashed line	◇ in □ with dashed line	⬡ with dashed line	○ with dashed line	• Located in rear of central or main panel. • Located in cabinet behind panel. • Not visible in front of panel or on video display. • Not normally operator accessible at panel or console.
4	○ in □ with line (double)	◇ in □ with line (double)	⬡ with line	○ with line	• Located in or on front of secondary or local panel or console. • Visible in front of panel or on video display. • Normally operator accessible at panel front or console.
5	○ in □ with dashed line (double)	◇ in □ with dashed line (double)	⬡ with dashed line	○ with dashed line	• Located in rear of secondary or local panel. • Located in field cabinet. • Not visible in front of panel or on video display. • Not normally operator accessible at panel or console.

Note: Due to space limitations, the explanatory notes in parentheses are not included, but can be found in ANSI/ISA-5.1-2009.

TABLE A.4c
Miscellaneous Device and Function Symbols

No	Symbol	Description

1	□	• Signal processing function:
		  • Locate in upper right or left quadrant of symbols above.
		  • Attach to symbols above where affected signals are connected.
		  • Insert signal processing symbol from Table A.4r
		  • Expand symbol by 50% increments for larger function symbols.

2	◇ with C	• Panel-mounted patchboard plug-in point. • Console matrix point. • C-12 equals patchboard column and row respectively, as an example.
3 (7)(8)	◇ with I	• Generic interlock logic function. • Undefined interlock logic function.
4 (7)(8)	◇ with I	• 'AND' interlock logic function.
5 (7)(8)	◇ with I	• 'OR' interlock logic function.
6	○○	• Instruments or functions sharing a common housing. • It is not mandatory to show a common housing. • Notes shall be used to identify instruments in common housings not using this symbol.
7	○ with X	• Pilot light. • Circle shall be replaced with any symbol from column D in Table A.4b if location and accessibility needs to be shown.

Note: Due to space limitations the explanatory notes in parentheses are not included, but can be found in ANSI/ISA-5.1-2009.

Measurement Symbols

The type of detector or sensing element that is used in making a particular measurement can also be described on the flow sheet in various levels of detail. Tables A.4d through A.4h describe most of the standardized symbols that serve that purpose.

TABLE A.4d
Primary Element and Transmitter Symbols

No	Symbol	Description
1	(1a) (2) — ?E (*)	• Generic primary element, bubble format. • Notation (*) from Table A.4e should be used to identify type of element. • Connect to process or other instruments by symbols from Tables A.4i and A.4j • Insert in or on process flow line, vessel, or equipment.
2	(1a) (2) (3) — ?T (*)	• Transmitter with integral primary element, bubble format. • Notation (*) from Table A.4e should be used to identify type of element. • Connect to process or other instruments by symbols from Tables A.4i and A.4j. • Insert in or on process flow line, vessel, or equipment.
3	(1a) (2) (3) — ?T / ?E (*)	• Transmitter with close coupled primary element, bubble format • Notation (*) from Table A.4e should be used to identify type of element. • Connecting line shall be equal to or less than 0.25 in. (6 mm). • Connect to process or other instruments by symbols from Tables A.4i and A.4j. • Insert element in or on process flow line, vessel, or equipment.
4	(1a) (3) — ?E (*) — ?T	• Transmitter with remote primary element, bubble format. • Notation (*) from Table A.4e should be used to identify type of element. • Connecting line shall be equal to or greater than 0.5 in. (12 mm). • Connect to process or other instruments by symbols from Tables A.4i and A.4j. • Insert element in or on process flow line, vessel, or equipment.
5	(1b) (3) — ?T / #	• Transmitter with integral primary element inserted in or on process flow line, vessel, or equipment, bubble/graphic format. • Insert primary element symbol from Table A.4f at #. • Connect to other instruments by symbols from Table A.4f.
6	(1b) (3) — ?T / #	• Transmitter with close-coupled primary element inserted in or on process flow line, vessel, or equipment, bubble/graphic format. • Insert primary element symbol from Table A.4f at #. • Connecting line shall be equal to or less than 0.25 in. (6 mm). • Connect to other instruments by symbols from Table A.4f.
7	(1b) (3) — # — ?E	• Transmitter with remote primary element inserted in or on process flow line, vessel, or equipment, bubble/graphic format. • Insert primary element symbol from Table A.4f at #. • Connecting line may be any signal line from Table 5.2c. • Connecting line shall be equal to or greater than 0.5 in. (12 mm). • Connect to other instruments by symbols from Table A.4f.

Note: Due to space limitations, the explanatory notes in parentheses are not included, but can be found in ANSI/ISA-5.1-2009.

TABLE A.4e
Measurement Related Abbreviations

Analysis			
AIR = excess air	H_2O = water	O_2 = oxygen	UV = ultraviolet
CO = carbon monoxide	H_2S = hydrogen sulfide	OP = opacity	VIS = visible light
CO_2 = carbon dioxide	HUM = humidity	ORP = oxidation–reduction	VISC = viscosity
COL = color	IR = infrared	pH = hydrogen ion	
COMB = combustibles	LC = liquid chromatograph	REF = refractometer	
COND = electrical conductivity	MOIST = moisture	RI = refractive index	
DEN = density	MS = mass spectrometer	TC = thermal conductivity	
GC = gas chromatograph	NIR = near infrared	TDL = tunable diode laser spectrometer	

Flow			
CFR = constant flow regulator	OP = orifice plate	PT = pitot tube	VENT = Venturi tube
CONE = cone	OP-CT = corner taps	PV = pitot venturi	VOR = vortex Shedding
COR = Coriolis	OP-CQ = circle quadrant	SNR = sonar	WDG = wedge
DOP = Doppler	OP-E = eccentric	SON = sonic	
DSON = Doppler sonic	OP-FT = flange taps	TAR = target	
FLN = flow nozzle	OP-MH = multihole	THER = thermal	
FLT = flow tube	OP-P = pipe taps	TTS = transit time sonic	
LAM = laminar	OP-VC = vena contracta taps	TUR = turbine	
MAG = magnetic	PD = positive displacement	US = ultrasonic	

Level			
CAP = capacitance	GWR = guided wave radar	NUC = nuclear	US = ultrasonic
d/p = differential pressure	LSR = laser	RAD = radar	
DI = dielectric constant	MAG = magnetic	RES = resistance	
DP = differential pressure	MS = magnetostrictive	SON = sonic	

Pressure			
ABS = absolute	MAN = manometer	VAC = vacuum	
AVG = average	P-V = pressure vacuum		
DRF = draft	SG = strain gage		

Temperature			
BM = bimetallic	RTD = resistance temp detector	TCK = thermocouple type K	TRAN = transistor
IR = infrared	TC = thermocouple	TCT = thermocouple type T	
RAD = radiation	TCE = thermocouple type E	THRM = thermistor	
RP = radiation pyrometer	TCJ = thermocouple type J	TMP = thermopile	

Miscellaneous			
Burner, combustion	*Position*	*Quantity*	*Radiation*
FR = flame rod	CAP = capacitance	PE = photoelectric	α = alpha radiation
IGN = igniter	EC = Eddy current	TOG = toggle	β = beta radiation
IR = infrared	IND = inductive		γ radiation = gamma radiation
TV = television	LAS = laser		n = neutron radiation
UV = ultraviolet	MAG = magnetic		
	MECH = mechanical		
	OPT = optical		
	RAD = radar		
Speed	*Weight, force*		
ACC = acceleration	LC = load cell		
EC = eddy current	SG = strain gage		
PROX = proximity	WS = weigh scale		
VEL = velocity			

TABLE A.4f
Primary Element Symbols

	No	Symbol (4)	Description
Analysis	1		• Conductivity, moisture, etc. • Single element sensing probe.
Analysis	2		• pH, ORP, etc. • Dual element sensing probe.
Analysis	3		• Fiber-optic sensing probe.
Burner	4		• Ultraviolet flame detector. • Television flame monitor.
Burner	5		• Flame rod flame detector.
Flow	6		• Generic orifice plate. • Restriction orifice.
Flow	7		• Orifice plate in quick-change fitting.
Flow	8		• Concentric circle orifice plate. • Restriction orifice.
Flow	9		• Eccentric circle orifice plate.
Flow	10		• Circle quadrant orifice plate.
Flow	11		• Multihole orifice plate
Flow	12	(*)	• Generic venturi tube, flow nozzle, or flow tube. • Notation from Table A.4e required at (*) if used for more than one type.
Flow	13		• Venturi tube.
Flow	14		• Flow nozzle.
Flow	15		• Flow tube.
Flow	16		• Integral orifice plate.
Flow	17		• Standard pitot tube.
Flow	18		• Averaging pitot tube.
Flow	19		• Turbine flowmeter. • Propeller flowmeter.
Flow	20		• Vortex shedding flowmeter
Flow	21		• Target flowmeter.
Flow	22	(4) M ((((a) (b)	• Magnetic flowmeter.

(Continued)

TABLE A.4f (Continued)
Primary Element Symbols

	No	Symbol (4)	Description
Flow	23	(4) ΔT (a) (b)	• Thermal mass flowmeter.
Flow	24		• Positive displacement flowmeter.
Flow	25		• Cone meter. • Annular orifice meter.
Flow	26		• Wedge meter.
Flow	27		• Coriollis flowmeter.
Flow	28		• Sonic flowmeter. • Ultrasonic flowmeter.
Flow	29		• Variable-area flowmeter.
Flow	30		• Open channel weir plate.
Flow	31		• Open channel flume.
Level	32		• Displacer internally mounted in vessel.
Level	33		• Ball float internally mounted in vessel. • May be installed through top of vessel.
Level	34		• Radiation, single point. • Sonic.
Level	35		• Radiation, multi-point, or continuous.
Level	36		• Dip tube or other primary element and stilling well. • May be installed through side of vessel. • May be installed without stilling well.
Level	37		• Float with guide wires. • Location of readout should be noted, at grade, at top, or accessible from a ladder. • Guide wires may be omitted.

(Continued)

TABLE A.4f (Continued)
Primary Element Symbols

	No	Symbol (4)	Description
Level	38		• Insert probe. • May be through top of vessel.
Level	39		• Radar.
Pressure	40	PE (*)	• Strain gage or other electronic type sensor. • Notation (*) from Table A.4e should be used to identify type of element. • Connection symbols 6, 7, 8, or 9 in Table A.4i are used if connection type is to be shown. • Bubble may be omitted if connected to another instrument.
Temperature	41	TE (*)	• Generic element without thermowell. • Notation (*) should be used to identify type of element, see Table A.4e • Connection symbols 6, 7, 8, or 9 in Table A.4i are used if connection type is to be shown. • Bubble may be omitted if connected to another instrument.

Note: Due to space limitations, the explanatory notes in parentheses are not included, but can be found in ANSI/ISA-5.1-2009.

TABLE A.4g
Secondary Instrument Symbols

	No	Symbol (4)	Description
Flow	1	FG	• Sight glass.
Level	2	LG	• Gage integrally mounted on vessel. • Sight glass.
Level	3	LG	• Gage glass externally mounted on vessel or standpipe. • Multiple gages may be shown as one bubble or one bubble for each section. • Use connection 6, 7, 8, or 9 in Table A.4i if connection type is to be shown.
Pressure	4	PG	• Pressure gage. • Use connection 6, 7, 8, or 9 in Table A.4i if connection type is to be shown.
Temperature	5	TG	• Thermometer. • Use connection 6, 7, 8, or 9 in Table A.4i if connection type is to be shown.

Note: Due to space limitations, the explanatory notes in parentheses are not included, but can be found in ANSI/ISA-5.1-2009.

TABLE A.4h
Auxiliary and Accessory Device Symbols

	No	Symbol (4)	Description
Analysis	1	AW	• Sample insert probe, flanged. • Sample well, flanged. • Use connection 7, 8, or 9 in Table A.4i if flange is not used.
Analysis	2	AX	• Sample conditioner or other analysis accessory, flanged. • Represents single or multiple devices. • Use connection 7, 8, or 9 in Table A.4i if flange is not used.
Flow	3	FX	• Flow straightening vanes. • Flow conditioning element.
Flow	4	P	• Instrument purge or flushing fluid. • Instrument purge or flushing device or devices. • Show assembly details on drawing legend sheet.
Pressure	5		• Diaphragm pressure seal, flanged, threaded, socket welded, or welded. • Diaphragm chemical seal, flanged, threaded, socket welded, or welded. • Use connection 6, 7, 8, or 9 in Table A.4i if connection type is to be shown.
Pressure	6		• Diaphragm pressure seal, welded. • Diaphragm chemical seal, welded.
Temperature	7	TW	• Thermowell, flanged. • Test well, flanged. • Bubble may be omitted if connected to another instrument. • Use connection 7, 8, or 9 in Table A.4i if flange is not used.

Note: Due to space limitations, the explanatory notes in parentheses are not included, but can be found in ANSI/ISA-5.1-2009.

Signal and Connection Symbols

The flow sheet also serves to show where the output signal of an instrument is directed, how many other instruments it is sent to, and what type of signal or connection it is. It can be a tube, such as sampling, a capillary, or a fiber-optic tube, it can be an analog or on–off pneumatic or electronic output signal, or it can also be wired or wireless digital signal. The symbols used to distinguish these types of connections are tabulated in Tables A.4i and A.4j.

TABLE A.4i
Instrument to Process Connection Symbols

No	Symbol	Application
1	————————	• Instrument connections to process and equipment. • Process impulse lines. • Analyzer sample lines.
2	— — — (ST) — — —	• Heat [cool] traced impulse or sample line from process. • Type of tracing indicated by: [ET] electrical, [ST] steam, [CW] chilled water, etc.
3	⊥	• Generic instrument connection to process line. • Generic instrument connection to equipment.
4	⊥⊥	• Heat [cool] traced generic instrument impulse line. • Process line or equipment may or may not be traced.
5	◎	• Heat [cool] traced instrument. • Instrument impulse line may or may not be traced.
6	T	• Flanged instrument connection to process line. • Flanged instrument connection to equipment.
7	○	• Threaded instrument connection to process line. • Threaded instrument connection to equipment.
8	□	• Socket welded instrument connection to process line. • Socket welded instrument connection to equipment.
9	■	• Welded instrument connection to process line. • Welded instrument connection to equipment.

Note: Due to space limitations, the explanatory notes in parentheses are not included, but can be found in ANSI/ISA-5.1-2009.

TABLE A.4j
Instrument-to-Instrument Connection Symbols

No	Symbol	Application
1	(1) IA ───────	• IA may be replaced by PA [plant air], NS [nitrogen], or GS [any gas supply]. • Indicate supply pressure as required, for example, PA-70 kPa, NS-150 psig, etc.
2	(1) ES ───────	• Instrument electric power supply. • Indicate voltage and type as required, for example, ES-220 Vac. • ES may be replaced by 24 Vdc, 120 Vac, etc.
3	(1) HS ───────	• Instrument hydraulic power supply. • Indicate pressure as required, for example, HS-70 psig.
4	(2) ──/──/──	• Undefined signal. • Used for Process Flow Diagrams. • Used for discussions or diagrams where type of signal is not of concern.
5	(2) ──//──//──	• Pneumatic signal, continuously variable or binary.
6	(2) ── ── ── ──	• Electronic or electrical continuously variable or binary signal. • Functional diagram binary signal.
7	(2) ───────	• Functional diagram continuously variable signal. • Electrical schematic ladder diagram signal and power rails.
8	(2) ──┴──┴──	• Hydraulic signal.
9	(2) ──✕──✕──	• Filled thermal element capillary tube. • Filled sensing line between pressure seal and instrument.
10	(2) ──∩──∩──	• Guided electromagnetic signal. • Guided sonic signal. • Fiber-optic cable.
11	(3) (a) ∩ ∩ (b) ⋀ ⋀	• Unguided electromagnetic signals, light, radiation, radio, sound, wireless, etc. • Wireless instrumentation signal. • Wireless communication link.
12	(4) ──○──○──	• Communication link and system bus, between devices and functions of a shared display, shared control system. • DCS, PLC, or PC communication link and system bus.
13	(5) ──●──●──	• Communication link or bus connecting two or more independent microprocessor or computer-based systems. • DCS-to-DCS, DCS-to-PLC, PLC-to-PC, DCS-to-Fieldbus, etc. connections.
14	(6) ──◇──◇──	• Communication link and system bus, between devices and functions of a fieldbus system. • Link from and to "intelligent" devices.
15	(7) ──○────○──	• Communication link between a device and a remote calibration adjustment device or system. • Link from and to "smart" devices.
16	──⊙──⊙──	• Mechanical link or connection.
17	(3) (a) [(#)/(##)] rounded (a) [(#)/(##)] rounded (b) [(#)/(##)] arrow right (b) [(#)/(##)] arrow left	• Drawing-to-drawing signal connector, signal flow from left to right. • (#) = Instrument tag number sending or receiving signal. • (##) = Drawing or sheet number receiving or sending signal.

(Continued)

TABLE A.4j (Continued)
Instrument-to-Instrument Connection Symbols

No	Symbol	Application
18	(*)⊢──	• Signal input to logic diagram. • (*) = Input description, source, or instrument tag number.
19	──⊣(*)	• Signal output from logic diagram. • (*) = Output description, destination, or instrument tag number.
20	──[(*)⟩	• Internal functional, logic, or ladder diagram signal connector. • Signal source to one or more signal receivers. • (*) = Connection identifier A, B, C, etc.
21	[(*)⟩──	• Internal functional, logic, or ladder diagram signal connector. • Signal receiver, one or more from a single source. • (*) = Connection identifier A, B, C, etc.

Note: Due to space limitations, the explanatory notes in parentheses are not included, but can be found in ANSI/ISA-5.1-2009.

Final Control Element Symbols

In addition to the measurement and the receiver device, the flow sheet also shows the location and type of the final control element. This can be a variable-speed pump, a positioned control rod, or a control valve. In a control valve, the flow sheet can show what type it is (globe, butterfly, etc.); what are its size, characteristics, and failure position; what type of actuator it has (pneumatic, hydraulic, electric, etc.); and what accessories it is provided with (handwheel, positioner, converter, booster, solenoid, etc.). Tables A.4k through A.4n tabulate some of the recommended symbols to be used to depict these features and components.

TABLE A.4k
Final Control Element Symbols

No	Symbol	Description
1	(1) (2) (a), (b)	• Generic two-way valve. • Straight globe valve. • Gate valve.
2	(2) (3)	• Generic two-way angle valve. • Angle globe valve. • Safety angle valve.
3	(2)	• Generic three-way valve. • Three-way globe valve. • Arrow indicates failure or unactuated flow path.
4	(2)	• Generic four-way valve. • Four-way four-ported plug or ball valve. • Arrows indicates failure or un-actuated flow paths.
5	(2)	• Butterfly valve.
6	(2)	• Ball valve.
7	(2)	• Plug valve
8	(2)	• Eccentric rotary disc valve.
9	(1) (2) (a), (b)	• Diaphragm valve.
10	(2)	• Pinch valve.
11	(2)	• Bellows sealed valve.
12	(2)	• Generic damper. • Generic louver.
13	(2)	• Parallel blade damper. • Parallel blade louver.
14	(2)	• Opposed blade damper. • Opposed blade louver.
15	(4)	• Two-way on–off solenoid valve.
16	(4)	• Angle on–off solenoid valve.
17	(4)	• Three-way on–off solenoid valve. • Arrow indicates de-energized flow path.
18	(4)	• Four-way plug or ball on–off solenoid valve. • Arrows indicates de-energized flow paths.
19	(4)	• Four-way five-ported on–off solenoid valve. • Arrows indicates de-energized flow paths.
20	(5)	• Permanent magnet variable speed coupling.
21	(6)	• Electric motor.

Note: Due to space limitations the explanatory notes in parentheses are not included, but can be found in ANSI/ISA-5.1-2009.

TABLE A.41

Final Control Element Actuator Symbols

No	Symbol	Description
1	(7) 	• Generic actuator. • Spring-diaphragm actuator.
2	(7) 	• Spring-diaphragm actuator with positioner.
3	(7) 	• Pressure-balanced diaphragm actuator.
4	(7) 	• Linear piston actuator. • Single acting spring opposed • Double acting.
5	(7) 	• Linear piston actuator with positioner.
6	(7) 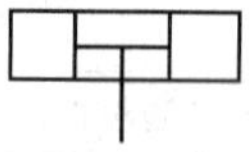	• Rotary piston actuator. • May be single acting spring opposed or double acting.
7	(7) 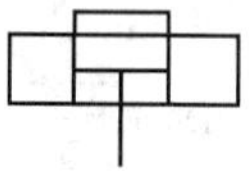	• Rotary piston actuator with positioner.
8	(7) 	• Bellows spring opposed actuator.
9	(7) 	• Rotary motor–operated actuator. • Electric, pneumatic, or hydraulic. • Linear or rotary action.
10	(7) 	• Modulating solenoid actuator. • Solenoid actuator for process on–off valve.
11	(7) 	• Actuator with side-mounted hand wheel.
12	(7) 	• Actuator with top-mounted handwheel.
13	(7)	• Manual actuator. • Hand actuator.
14	(7) 	• Electrohydraulic linear or rotary actuator.
15	(7) 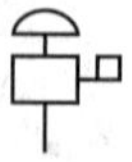	• Actuator with manual actuated partial stroke test device.

(Continued)

TABLE A.4l (Continued)
Final Control Element Actuator Symbols

No	Symbol	Description
16	(7)	• Actuator with remote-actuated partial stroke test device.
17	(8)	• Automatic reset on–off solenoid actuator. • Nonlatching on–off solenoid actuator.
18	(8)	• Manual or remote reset on–off solenoid actuator. • Latching on–off solenoid actuator.
19	(8)	• Manual and remote reset on–off solenoid actuator. • Latching on–off solenoid actuator.
20	(9)	• Spring or weight-actuated relief or safety valve actuator.
21	(9)	• Pilot-actuated relief or safety valve actuator. • Pilot pressure sensing line deleted if sensing is internal.

Note: Due to space limitations, the explanatory notes in parentheses are not included, but can be found in ANSI/ISA-5.1-2009.

TABLE A.4m
Regulator Relief Device Symbols

No	Symbol	Description
1	XXX	• Automatic flow regulator. • XXX = FCV without indicator. • XXX = FICV with integral indicator.
2	(1) (2) FICV (a) / (b)	• Variable-area flowmeter with integral manual adjusting valve. • Instrument tag bubble required with (b).
3	FICV	• Constant flow regulator.
4	FG	• Flow sight glass. • Type shall be noted if more than one type used.
5	FO	• Generic flow restriction. • Single stage orifice plate as shown. • Note required for multistage or capillary tube types.
6	FO	• Restriction orifice hole drilled in valve plug. • Tag number shall be omitted if valve is otherwise identified.
7	TANK	• Level regulator. • Ball float and mechanical linkage.
8		• Backpressure regulator. • Internal pressure tap.
9		• Backpressure regulator. • External pressure tap.
10		• Pressure-reducing regulator. • Internal pressure tap.
11		• Pressure-reducing regulator. • External pressure tap.
12		• Differential pressure regulator. • External pressure taps.
13		• Differential pressure regulator. • Internal pressure taps.

(Continued)

TABLE A.4m (Continued)
Regulator Relief Device Symbols

No	Symbol	Description
14		• Pressure-reducing regulator with integral outlet pressure relief and pressure gauge.
15		• Generic pressure safety valve. • Pressure relief valve.
16		• Generic vacuum safety valve. • Vacuum relief valve.
17		• Generic pressure–vacuum relief valve. • Tank pressure–vacuum relief valve.
18		• Pressure safety element. • Pressure rupture disk. • Pressure relief.
19		• Pressure safety element. • Vacuum rupture disk. • Vacuum relief.
20		• Temperature regulator. • Filled thermal system.
21		• Thermal safety element. • Fusible plug or disk.
22		• Generic moisture trap. • Steam trap. • Note required for other trap types.
23		• Moisture trap with equalizing line.

Note: Due to space limitations, the explanatory notes in parentheses are not included, but can be found in ANSI/ISA-5.1-2009.

TABLE A.4n

Control Valve Failure Positions

No	Method A (1) (10)	Method B (1) (10)	Definition
1	FO		• Fail to open position.
2	FC		• Fail to closed position.
3	FL		• Fail locked in last position.
4	FL/DO		• Fail at last position. • Drift open.
5	FL/DC		• Fail at last position. • Drift closed.

Note: Due to space limitations, the explanatory notes in parentheses are not included, but can be found in ANSI/ISA-5.1-2009.

Function and Calculation Symbols

In addition to the three main components of the control loop (measurement, control, and final control element), there can be a large number of other components. These can include summers, integrators, linearizers, square root extractors, and a wide range of converters, calculators, and function generators. Tables A.4o and A.4p tabulate some of the recommended symbols to be used to depict these.

TABLE A.4o
Functional Diagram Symbols

No	Symbol (1) (2)	Description
1	○ [*]	• Measuring, input, or readout device. • [*] = Instrument tag number. • Symbols from Table A.4 may be used.
2	(3) (4) — stacked rectangles (*)/(*)	• Automatic single-mode controller.
3	(3) (4) — rectangle (*) over two rectangles (*) (*)	• Automatic two-mode controller.
4	(3) (4) — rectangle (*) over three rectangles (*) (*) (*)	• Automatic three-mode controller.
5	(3) (4) — rectangle (*)	• Automatic signal processor.
6	(4) — diamond (*)	• Manual signal processor.
7	(3) (4) — trapezoid (*)	• Final control element. • Control valve.
8	(3) (4) — trapezoid with positioner bar (*)	• Final control element with positioner. • Control valve with positioner.

Note: Due to space limitations, the explanatory notes in parentheses are not included, but can be found in ANSI/ISA-5.1-2009.

TABLE A.4p
Function Block Symbols

No	Function Symbol (1) (2)	Equation / Graph	Definition
1	Summation □ Σ □ Σ □	$M = X_1 + X_2 \ldots + X_n$	• Output equals algebraic sum of inputs.
2	Average □ Σ/n □ Σ/n □	$M = X_1 + X_2 \ldots + X_n/n$	• Output equals algebraic sum of inputs divided by number of inputs.
3	Difference □ Δ □ Δ □	$M = X_1 - X_2$	• Output equals algebraic difference of two inputs.
4	Multiplication □ × □ × □	$M = X_1 \times X_2$	• Output equals product of two inputs.
5	Division □ ÷ □ ÷ □	$M = X_1 \div X_2$	• Output equals quotient of two inputs.

(Continued)

TABLE A.4p (Continued)
Function Block Symbols

No	Function / Symbol (1) (2)	Equation / Graph	Definition
6	Exponential $\boxed{X^n}$ $\boxed{\quad X^n \quad}$	$M = X^n$ (graph of X vs t linear; graph of M vs t exponential)	• Output equals nth power of input.
7	Root extraction $\boxed{\sqrt[n]{}}$ $\boxed{\quad \sqrt[n]{} \quad}$	$M = \sqrt[n]{X}$ (graph of X vs t linear; graph of M vs t root curve)	• Output equals nth root of input. • If 'n' omitted, square root is assumed.
8	Proportion (3) (a) $\boxed{K}$ (b) $\boxed{P}$ (3) (a) $\boxed{\quad K \quad}$ (b) $\boxed{\quad P \quad}$	$M = KX$ or $M = PX$ (graphs with break at t_1)	• Output proportional to input. • Replace 'K' or 'P' with '1:1' for volume boosters. • Replace 'K' or 'P' with '2:1', '3:1', etc., for integer gains.
9	Reverse proportion (3) (a) $\boxed{-K}$ (b) $\boxed{-P}$ (3) (a) $\boxed{\quad -K \quad}$ (b) $\boxed{\quad -P \quad}$	$M = -KX$ or $M = -PX$ (graphs with break at t_1)	• Output inversely proportional to input. • Replace '−K' or '−P' with '−1:1' for volume boosters. • Replace '−K' or '−P' with '−2:1', '−3:1', etc., for integer gains.

(Continued)

TABLE A.4p (Continued)
Function Block Symbols

No	Function / Symbol (1) (2)	Equation / Graph	Definition
10	Integral (3) (a) [∫] (b) [I] (3) (a) [∫] (b) [I]	$M = (1/T_I)\int X dt$ (Graph: X step input from t_1 to t_2; M ramps up from t_1 to t_2)	• Output varies with magnitude and time duration of input. • Output proportional to time integral of input. • T_I = Integral time constant.
11	Derivative (3) (a) [d/d] (b) [D] (3) (a) [d/dt] (b) [D]	$M = T_D (dx/dt)$ (Graph: X ramp; M step at t_1)	• Output proportional to time rate of change of input. • T_D = derivative time constant.
12	Unspecified function [$f_{(x)}$] [$f_{(x)}$]	$M = f(x)$ (Graph: X linear ramp; M bell-shaped curve)	• Output is a nonlinear or unspecified function of the input. • Function defined in note or other text.
13	Time function [$f_{(t)}$] [$f_{(t)}$]	$M = Xf(t)$ (Graph: X step at t_1; M s-curve rising)	• Output equals a nonlinear or unspecified time function times the input. • Output is a nonlinear or unspecified time function. • Function defined in note or other text.
14	Conversion [I/P] [I/P]	$I = P, P = I$, etc (Graph: X linear ramp; M linear ramp)	• Output signal type different from that of input signal. • Input signal is on the left and output signal is on the right. • Substitute any of the following signal types for 'P' or 'I': • A = Analog H = Hydraulic • B = Binary I = Current • D = Digital O = Electromagnetic • E = Voltage P = Pneumatic • F = Frequency R = Resistance

(Continued)

TABLE A.4p (Continued)
Function Block Symbols

No	Function Symbol (1) (2)	Equation Graph	Definition
15	High signal select `[ > ]` `[ > ]`	$M = X_1$ for $X_1 > X_2$ $M = X_2$ for $X_1 \leq X_2$	• Output equals greater of 2 or more inputs.
16	Middle signal select `[ M ]` `[ M ]`	$M = X_1$ for $X_2 > X_1 > X_3$ or $X_3 > X_1 > X_2$ $M = X_2$ for $X_1 > X_2 > X_3$ or $X_3 > X_2 > X_1$ $M = X_3$ for $X_1 > X_3 > X_2$ or $X_2 > X_3 > X_1$	• Output equals middle value of three or more inputs.
17	Low signal select `[ < ]` `[ < ]`	$M = X_1$ for $X_1 \leq X_2$ $M = X_2$ for $X_1 \geq X_2$	• Output equals lesser of 2 or more inputs.
18	High limit	$M = X$ for $X \leq H$ $M = H$ for $X \geq H$	• Output equals the lower of the input or high limit values.

(Continued)

TABLE A.4p (Continued)
Function Block Symbols

No	Function Symbol (1) (2)	Equation / Graph	Definition
19	Low limit	$M = X$ for $X \geq L$ $M = L$ for $X \leq L$	• Output equals the higher of the input or low limit values.
20	Positive bias	$M = X_1 + b$ $M = [-]X_2 + b$	• Output equal to input plus an arbitrary value.
21	Negative Bias	$M = X_1 - b$ $M = [-]X_2 - b$	• Output equal to input minus an arbitrary value.
22	Velocity limiter (3) (a) (b) (3) (a) (b)	$dM/dt = dX/dt$ for $dX/dt \leq H$, $M = X$ $dM/dt = H$ for $dX/dt \geq H$, $M \neq X$	• Output equals input as long as the input rate of change does not exceed the limit value that establishes the output rate of change until the output again equals the input.

(Continued)

TABLE A.4p (Continued)
Function Block Symbols

No	Function Symbol (1) (2)	Equation / Graph	Definition
23	High signal monitor H H	(State 1) $M = 0 @ X < H$ (State 2) $M = 1 @ X \geq H$	• Output state is dependent on value of input. • Output changes state when input is equal to or higher than an arbitrary high limit.
24	Low signal monitor L L	(State 1) $M = 1 @ X \leq L$ (State 2) $M = 0 @ X > L$	• Output state is dependent on value of input. • Output changes state when input is equal to or lower than an arbitrary low limit.
25	High/low signal monitor HL HL	(State 1) $M = 1 @ X \leq L$ (State 2) $M = 0 @ L < X < H$ (State 3) $M = 1 @ X \geq H$	• Output states are dependent on value of input. • Output changes state when input is equal to or lower than an arbitrary low limit or equal to or higher than an arbitrary high limit.
26	Analog signal generator A A	No equation No graph	• Output equals a variable analog signal that is generated: a. Automatically and is not adjustable by operator. b. Manually and is adjustable by operator.
27	Binary signal generator B B	No equation No graph	• Output equals an on–off binary signal that is generated: a. Automatically and is not adjustable by operator. b. Manually and is adjustable by operator.

(Continued)

1200 Appendix

TABLE A.4p (Continued)
Function Block Symbols

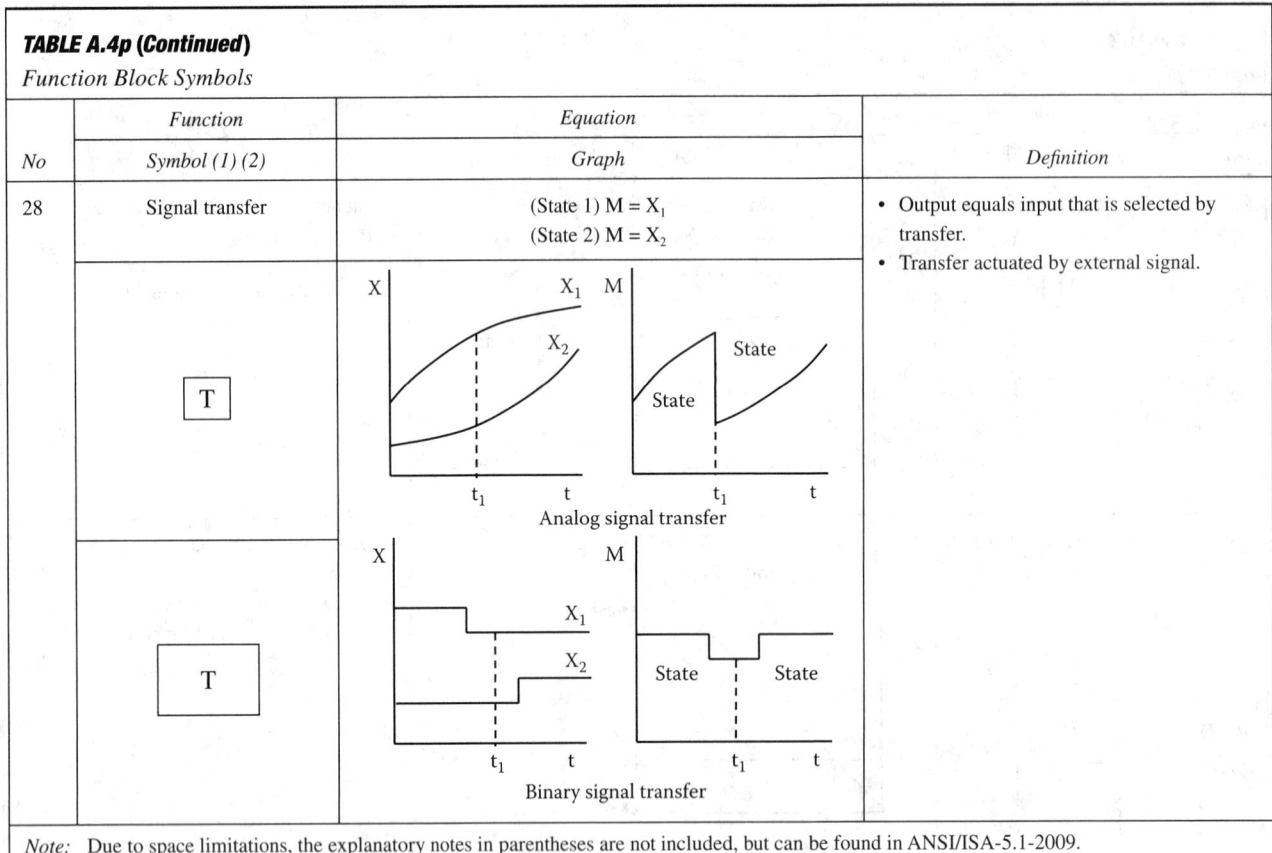

No	Function Symbol (1) (2)	Equation / Graph	Definition
28	Signal transfer	(State 1) $M = X_1$ (State 2) $M = X_2$ Analog signal transfer Binary signal transfer	• Output equals input that is selected by transfer. • Transfer actuated by external signal.

Note: Due to space limitations, the explanatory notes in parentheses are not included, but can be found in ANSI/ISA-5.1-2009.

Logic Function Symbols

In many cases, the control loop performs some logic function either in addition to or instead of its continuous control or monitoring and display functions. Table A.4q shows the symbols used to describe some of these logic functions.

TABLE A.4q
Binary Logic Symbols

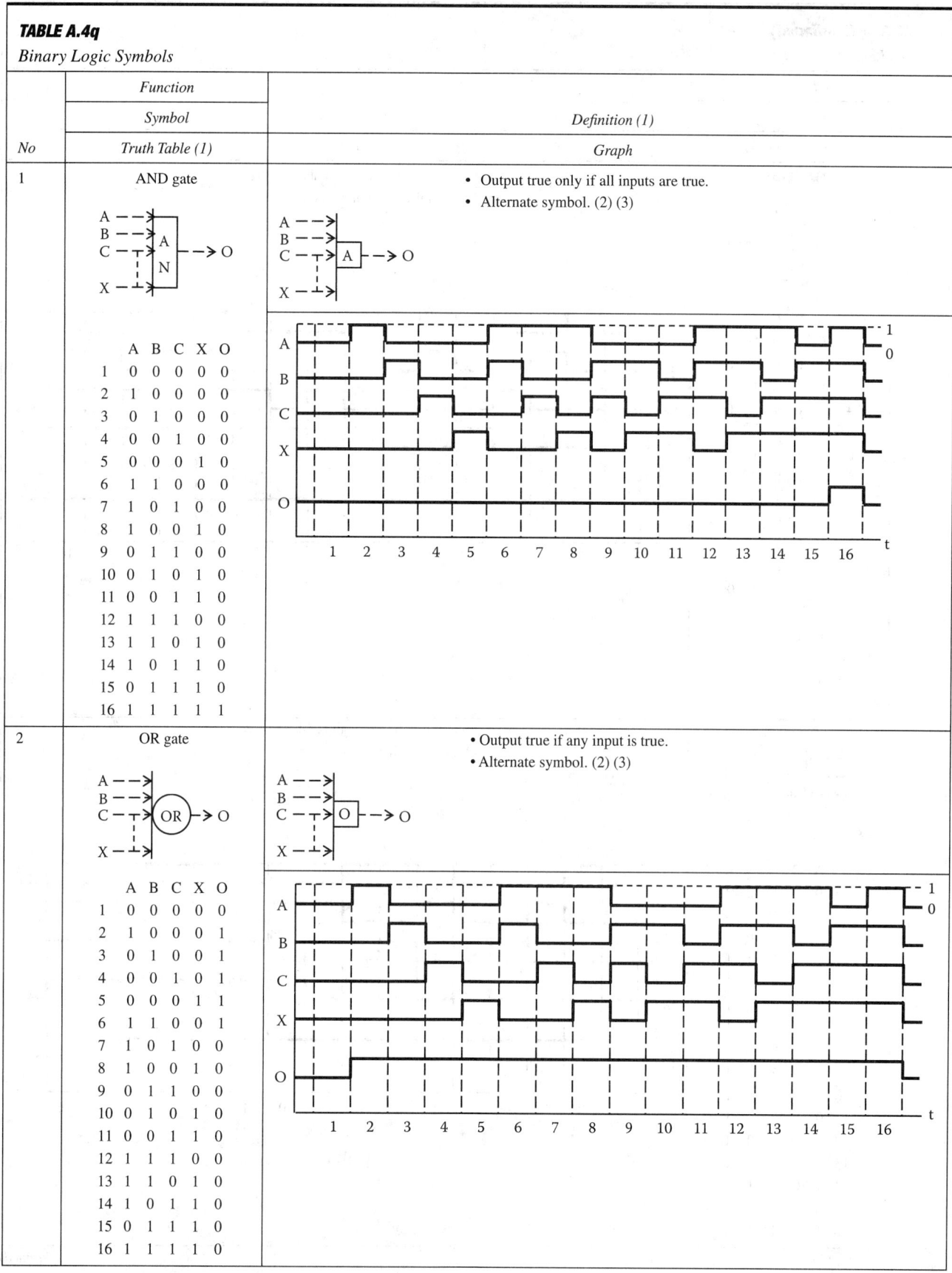

(Continued)

TABLE A.4q (Continued)
Binary Logic Symbols

No	Function / Symbol / Truth Table (1)	Definition (1) / Graph
3	NAND gate A, B, ..., X → [NAND] → O 　　A B C X O 1　0 0 0 0 1 2　1 0 0 0 0 3　0 1 0 0 0 4　0 0 1 0 0 5　0 0 0 1 0 6　1 1 0 0 0 6　1 1 0 0 0 7　1 0 1 0 0 8　1 0 0 1 0 9　0 1 1 0 0 10　0 1 0 1 0 11　0 0 1 1 0 12　1 1 1 0 0 13　1 1 0 1 0 14　1 0 1 1 0 15　0 1 1 1 0 16　1 1 1 1 0	• Output true only if all inputs are false. • Output false if any input is true.
4	NOR gate A, B, C, ..., X → (NOR) → O 　　A B C X O 1　0 0 0 0 1 2　1 0 0 0 1 3　0 1 0 0 1 4　0 0 1 0 1 5　0 0 0 1 1 6　1 1 0 0 1 7　1 0 1 0 1 8　1 0 0 1 1 9　0 1 1 0 1 10　0 1 0 1 1 11　0 0 1 1 1 12　1 1 1 0 1 13　1 1 0 1 1 14　1 0 1 1 1 15　0 1 1 1 1 16　1 1 1 1 0	• Output true if any input is false. • Output false if any input is true.

(Continued)

TABLE A.4q (Continued)
Binary Logic Symbols

No	Function / Symbol / Truth Table (1)	Definition (1) / Graph
5	Qualified OR gate Greater or equal to 'n' A ---≫ B ---≫ C --⊤≫ (≥n) ---→ O X --⊥≫ A B C X O 1 0 0 0 0 1 2 1 0 0 0 0 3 0 1 0 0 0 4 0 0 1 0 0 5 0 0 0 1 0 6 1 1 0 0 1 7 1 0 1 0 1 8 1 0 0 1 1 9 0 1 1 0 1 10 0 1 0 1 1 11 0 0 1 1 1 12 1 1 1 0 1 13 1 1 0 1 1 14 1 0 1 1 1 15 0 1 1 1 1 16 1 1 1 1 1	• Output true if number of true inputs is greater than or equal to 'n'. • Truth table and graph are for n = 2.
6	Qualified OR gate Greater than 'n' A ---≫ B ---≫ C --⊤≫ (>n) ---→ O X --⊥≫ A B C X O 1 0 0 0 0 0 2 1 0 0 0 0 3 0 1 0 0 0 4 0 0 1 0 0 5 0 0 0 1 0 6 1 1 0 0 0 7 1 0 1 0 0 8 1 0 0 1 0 9 0 1 1 0 0 10 0 1 0 1 0 11 0 0 1 1 0 12 1 1 1 0 1 13 1 1 0 1 1 14 1 0 1 1 1 15 0 1 1 1 1 16 1 1 1 1 1	• Output true if number of true inputs is greater than 'n'. • Truth table and graph are for n = 2.

(Continued)

TABLE A.4q (Continued)
Binary Logic Symbols

No	Function / Symbol / Truth Table (1)	Definition (1) / Graph
7	**Qualified OR gate Less or equal to 'n'** A ──▷ B ──▷ C ──▷ ≤n ──▷ O X ──▷ 　　A B C X O 1　0 0 0 0 1 2　1 0 0 0 1 3　0 1 0 0 1 4　0 0 1 0 1 5　0 0 0 1 1 6　1 1 0 0 1 7　1 0 1 0 1 8　1 0 0 1 1 9　0 1 1 0 1 10　0 1 0 1 1 11　0 0 1 1 1 12　1 1 1 0 0 13　1 1 0 1 0 14　1 0 1 1 0 15　0 1 1 1 0 16　1 1 1 1 0	• Output true if number of true inputs is less than or equal to 'n'. • Truth table and graph are for n = 2.
8	**Qualified OR gate Less than 'n'** A ──▷ B ──▷ C ──▷ <n ──▷ O X ──▷ 　　A B C X O 1　0 0 0 0 1 2　1 0 0 0 1 3　0 1 0 0 1 4　0 0 1 0 1 5　0 0 0 1 1 6　1 1 0 0 0 7　1 0 1 0 0 8　1 0 0 1 0 9　0 1 1 0 0 10　0 1 0 1 0 11　0 0 1 1 0 12　1 1 1 0 0 13　1 1 0 1 0 14　1 0 1 1 0 15　0 1 1 1 0 16　1 1 1 1 0	• Output true if number of true inputs is less than 'n'. • Truth table and graph are for n = 2.

(Continued)

TABLE A.4q (Continued)
Binary Logic Symbols

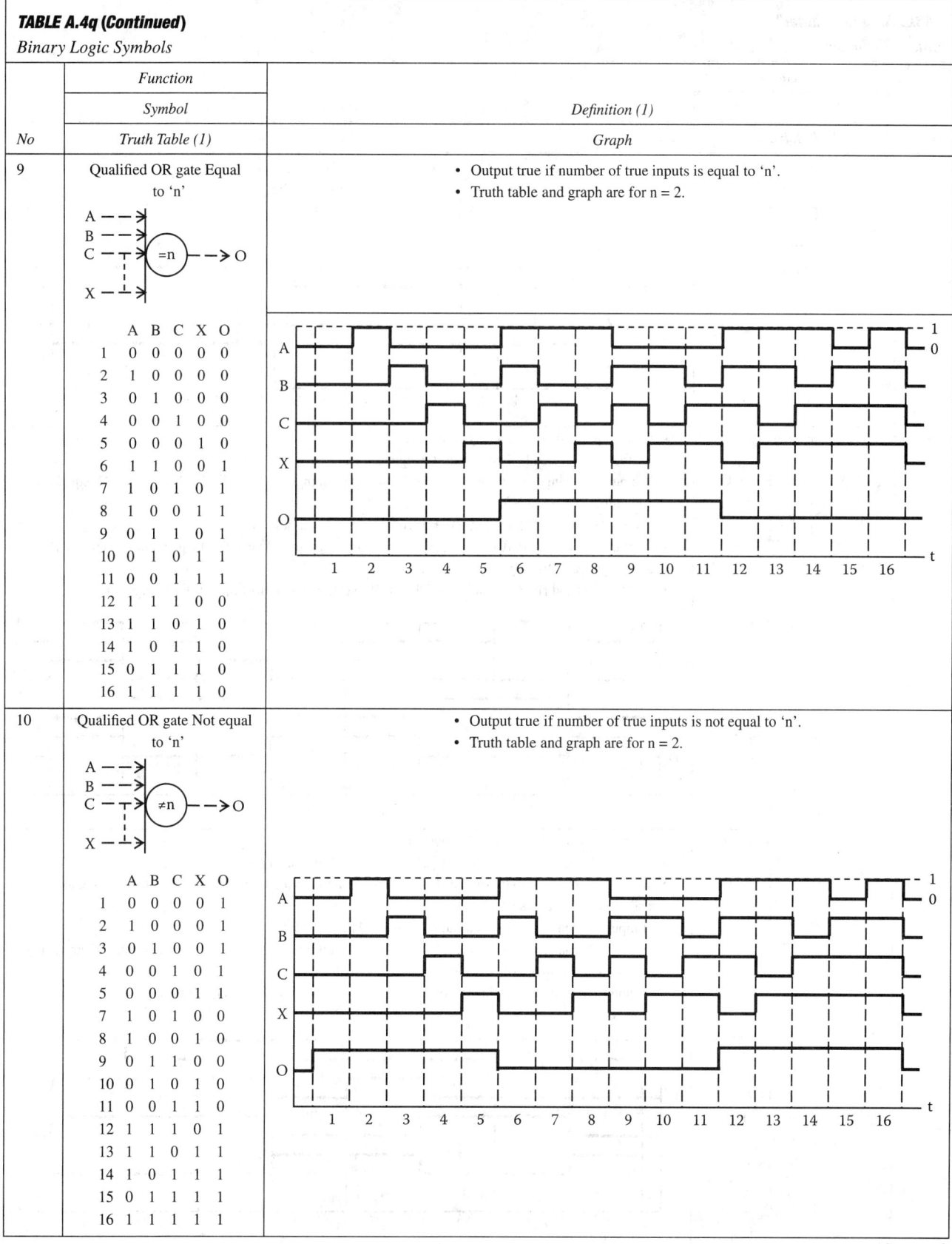

No	Function / Symbol / Truth Table (1)	Definition (1) / Graph
9	Qualified OR gate Equal to 'n' A, B, C, X → =n → O A B C X O 1 0 0 0 0 0 2 1 0 0 0 0 3 0 1 0 0 0 4 0 0 1 0 0 5 0 0 0 1 0 6 1 1 0 0 1 7 1 0 1 0 1 8 1 0 0 1 1 9 0 1 1 0 1 10 0 1 0 1 1 11 0 0 1 1 1 12 1 1 1 0 0 13 1 1 0 1 0 14 1 0 1 1 0 15 0 1 1 1 0 16 1 1 1 1 0	• Output true if number of true inputs is equal to 'n'. • Truth table and graph are for n = 2.
10	Qualified OR gate Not equal to 'n' A, B, C, X → ≠n → O A B C X O 1 0 0 0 0 1 2 1 0 0 0 1 3 0 1 0 0 1 4 0 0 1 0 1 5 0 0 0 1 1 7 1 0 1 0 0 8 1 0 0 1 0 9 0 1 1 0 0 10 0 1 0 1 0 11 0 0 1 1 0 12 1 1 1 0 1 13 1 1 0 1 1 14 1 0 1 1 1 15 0 1 1 1 1 16 1 1 1 1 1	• Output true if number of true inputs is not equal to 'n'. • Truth table and graph are for n = 2.

(Continued)

TABLE A.4q (Continued)
Binary Logic Symbols

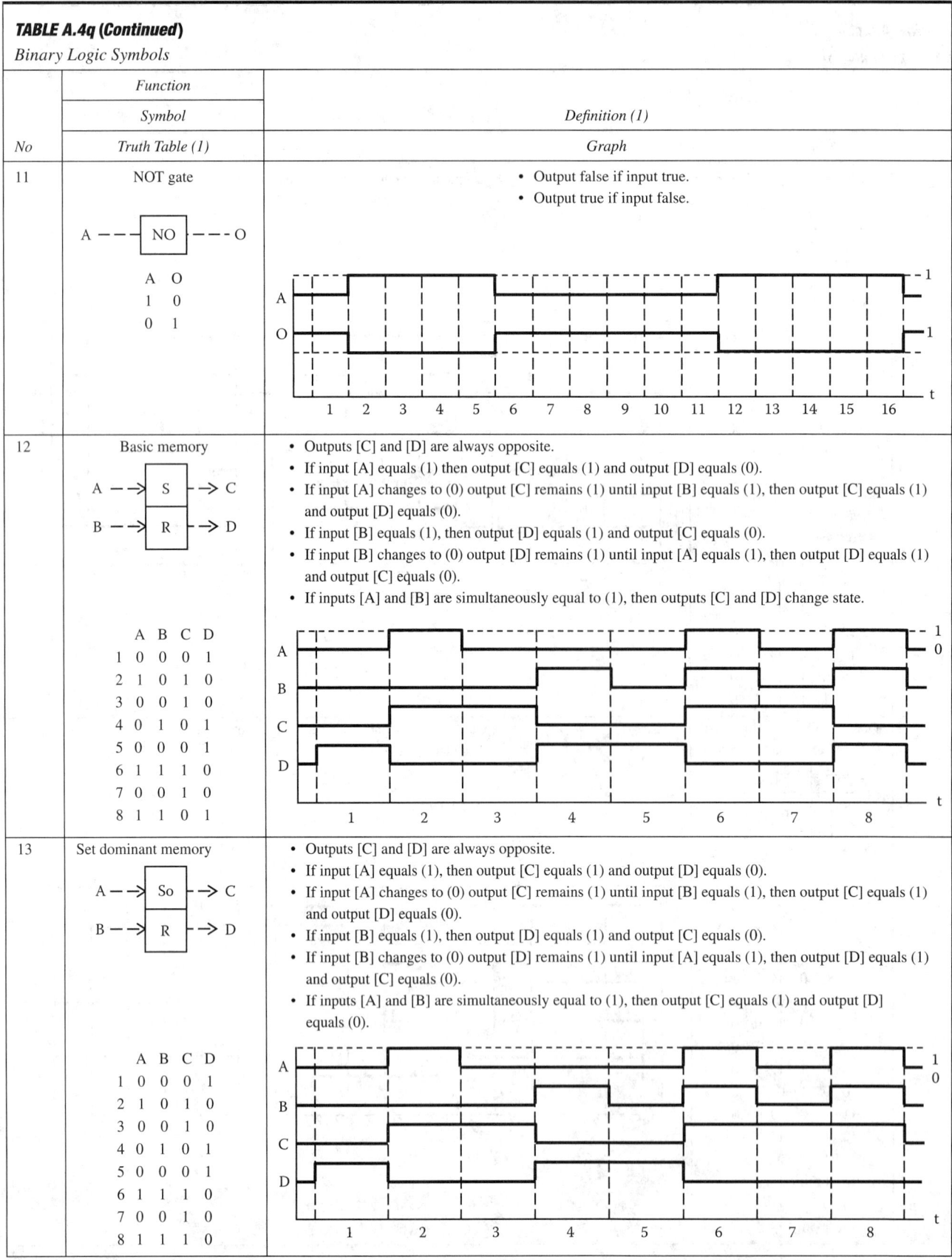

TABLE A.4q (Continued)
Binary Logic Symbols

No	Function / Symbol / Truth Table (1)	Definition (1) / Graph
14	Reset dominant memory A --> [S] --> C B --> [Ro] --> D A B C D 1 0 0 0 1 2 1 0 1 0 3 0 0 1 0 4 0 1 0 1 5 0 0 0 1 6 1 1 0 1 7 0 0 0 1 8 1 1 0 1	• Outputs [C] and [D] are always opposite. • If input [A] equals (1), then output [C] equals (1) and output [D] equals (0). • If input [A] changes to (0) output [C] remains (1) until input [B] equals (1), then output [C] equals (1) and output [D] equals (0). • If input [B] equals (1), then output [D] equals (1) and output [C] equals (0). • If input [B] changes to (0) and output [D] remains (1) until input [A] equals (1), then output [D] equals (1) and output [C] equals (0). • If inputs [A] and [B] are simultaneously equal to (1), then output [C] equals (0) and output [D] equals (1).
15	Pulse duration–fixed I --> [t PD] --> O NONE	• Output [O] changes from (0) to (1) and remains (1) for prescribed time duration (t) when input [I] changes from (0) to (1).
16	Time delay–off I --> [t DT] --> O NONE	• Output [O] changes from (0) to (1) when input [I] changes from (0) to (1). • Output [O] changes from (1) to (0) after input [I] changes from (1) to (0) and has been equal to (0) for time duration ('t).

(Continued)

TABLE A.4q (Continued)
Binary Logic Symbols

No	Function / Symbol / Truth Table (1)	Definition (1) / Graph
17	Time delay–on I →[t GT]→ O R ↗ NONE	• Output [O] changes from (0) to (1) after input [I] changes from (0) to (1) and [I] remains (1) for prescribed time duration (t). • Output [O] remains (1) until Input [I] changes to (0) or optional Reset [R] changes to (1).
18	Pulse duration–variable I →[t LT]→ O R ↗ NONE	• Output [O] changes from (0) to (1) when input [I] changes from (0) to (1). • Output [O] changes from (1) to (0) when Input [I] has been equal to (1) for time duration (t), Input [I] changes from (1) to (0), or optional Reset [R] changes to (1).

Note: Due to space limitations, the explanatory notes in parentheses are not included, but can be found in ANSI/ISA-5.1-2009.

Electrical Schematics Symbols

The details of electric wiring and of the associated components usually do not appear on the flow sheet, but separate electric drawings are prepared for them. The symbols used in these drawings are shown in Table A.4r.

TABLE A.4r
Electrical Schematic Symbols

No	Symbol (1)	Description
1	○	• Device wiring point. • Device wiring terminal.
2	(2)	• Normally open single circuit momentary pushbutton switch. • Form A switch contact. • Stack symbols to form multipole switches. • Combine with symbols 5 or 6 to form toggle or rotary actuated switches.
3	(2)	• Normally closed single circuit momentary pushbutton switch. • Form B switch contact. • Stack symbols to form multipole switches. • Combine with symbols 5 or 6 to form toggle or rotary actuated switches.
4	(2)	• Normally closed/normally open double circuit momentary pushbutton switch. • Form C switch contact. • Stack symbols to form multipole switches. • Combine with symbols 5 or 6 to form toggle or rotary actuated switches.
5	(3)	• Two-position toggle or rotary maintained position pushbutton switch actuator. • Combine with symbols 2, 3, and 4 to form single or multipole switches.
6	(3)	• Three-position toggle or rotary maintained position pushbutton switch actuator. • Combine with symbols 2, 3, and 4 to form single or multipole switches.
7	(4)	• Single-pole normally open toggle switch. • Form A switch contact. • Combine with symbols 10–15.
8	(4)	• Single-pole normally closed toggle switch. • Form B switch contact. • Combine with symbols 10–15.
9	(4)	• Double pole normally closed /normally open toggle switch. • Form C switch contact. • Combine with symbols 10–15.
10		• Rotary selector switch.
11	(5)	• Pressure switch actuator.
12	(5)	• Differential pressure switch actuator.
13	(5)	• Liquid level switch actuator.
14	(5)	• Temperature switch actuator.
15	(5)	• Flow switch actuator.
16	(5)	• Foot switch actuator.
17	(*)	• Relay operating coil. • (*) = Relay designator, such as: • a. Instrument tag number if assigned. • b. RO1, RO2, R4, R5, MR10, etc.

(Continued)

TABLE A.4r (Continued)
Electrical Schematic Symbols

No	Symbol (1)	Description
18		• Normally open relay contact. • Form A contact.
19		• Normally closed relay contact. • Form B contact.
20		• Normally open, normally closed relay contact. • Form C contact.
21	(*)	• On time delay. • Moves after relay coil is energized and set time has elapsed. • (*) = Set time.
22	(*)	• Off time delay. • Moves after relay coil de-energizes and set time has elapsed. • (*) = Set time.
23	(*)	• Transformer. • (*) = Rating, 220/120 Vac or Vdc, etc.
24	(6) (a) (*) (b) (*)	• Fuse, nonresettable. • (*) = Rating, 2 A, 5 A, etc.
25		• Thermal overload.
26	(*)	• Circuit interrupter, single-pole, manual reset. • (*) = Rating, 10 A, 15 A, etc.
27	(*)	• Circuit interrupter, three-pole, manual reset. • (*) = Rating, 15 A, 20 A, etc.
28	(*)	• Circuit breaker, single-pole, manual reset. • (*) = Rating, 20A, 30A, etc.
29	(*)	• Circuit breaker, three-pole, manual reset. • (*) = Rating, 20 A, 25 A, etc.
30		• Bell.
31		• Horn or siren.
32		• Buzzer.
33		• Solenoid coil.

(Continued)

TABLE A.4r (Continued)
Electrical Schematic Symbols

No	Symbol (1)	Description
34		• Pilot light.
35		• Battery
36		• Ground
37	(6) (a) (b)	• Connection conventions a) and b): • Left = Not connected. • Right = Connected.

Note: Due to space limitations, the explanatory notes in parentheses are not included, but can be found in ANSI/ISA-5.1-2009.

LOOP DIAGRAM

A loop diagram describes only a single instrument loop, but does that in full detail. The ISA provided some guidelines for the preparation of loop diagrams in their standard ISA-5.4-1991 "Instrument Loop Diagrams" approved on September 9, 1991. The examples used in ISA-5.4 are somewhat outdated as they reflect the state of the art of automation prior to the digital age, when analog and sometimes pneumatic control loops were used. Yet, they correctly show the type of information, which some of the traditional loop diagrams should include and how such data should be presented. For that reason, they are included here.

The instrument loop diagram contains all associated electrical and piping connections needed to accommodate the intended uses. As a minimum, an instrument loop diagram should contain the following information:

- Identification of the loop and loop components shown on the P&IDs.
- Word description of loop functions.
- Any special features or functions of shutdown and safety circuits.
- Indication of the interrelation to other instrumentation loops, including overrides.
- interlocks, cascaded set points, shutdowns, and safety circuits.
- All point-to-point interconnections with identifying numbers or colors of electrical cables, conductors, pneumatic multitubes, and individual pneumatic and hydraulic tubing.
- Identification of interconnections including junction boxes, terminals, bulkheads, ports, and grounding connections.
- General location of devices such as field, panel, auxiliary equipment, rack, termination cabinet, cable spreading room, I/O cabinet, etc.
- Energy sources of devices, such as electrical power, air supply, and hydraulic fluid supply and identification of voltage, pressure, and other applicable requirements.
- For electrical systems, identification of circuit or disconnect numbers.
- For process lines and equipment, sufficient information to describe the process side of the loop and to provide clarity of control action, including information on what is being measured and what is being controlled.
- Actions or fail-safe positions (electronic, pneumatic, or both) of control devices such as controllers, switches, control valves, solenoid valves, and transmitters (if reverse acting).

Some of the additional types of information that might be included at the user's discretion include the following:

- Process equipment, lines, and their identification numbers, source, designation, or flow direction.
- Reference to supplementary records and drawings, such as installation details, P&IDs, location drawings, wiring diagrams or drawings, and instrument specifications.
- Specific location of each device, such as elevation, area, panel subdivision, rack or cabinet number and location, and I/O location.

- Cross-reference between loops that share a common discrete component, such as multipen recorders and dual indicators.
- References to equipment descriptions, manufacturers, model numbers, hardware types, specifications or data sheets, purchase order numbers, etc.
- Signal ranges and calibration information, including set point values for switches, and alarm and shutdown devices.
- Software reference numbers, such as I/O addresses, control block types and names, network interfaces, and point names.
- Engraving or legend information that helps identify the instrument or accessory.
- Accessories, tagged or otherwise identified, such as regulators, filters, purge meters, manifold valves, and root valves.
- References to manufacturer's documentation such as schematics, connection details, and operating instructions.
- Color code identification for conductors or tubes that use numbers for differentiation.

The minimum size for the original drawing should be 11 in. by 17 in. The instrument loop diagram should typically contain only one loop, and that one loop should not be shown on multiple pages or sheets where practical. An attempt should be made that where loops share common components, they be completely shown on a single diagram. It is also desirable to provide space for future additions of components and loop data.

Terminal, Power Supply and Action Symbols

The flow sheet symbols discussed earlier in this chapter also apply for the instrument loop diagrams. However, those symbols are expanded to include information concerning connection points, energy source (electrical, air, hydraulic), and instrument action.

The method of describing terminals or bulkheads is shown in Figure A.4c.

The method of describing instrument terminals or ports is shown in Figure A.4d.

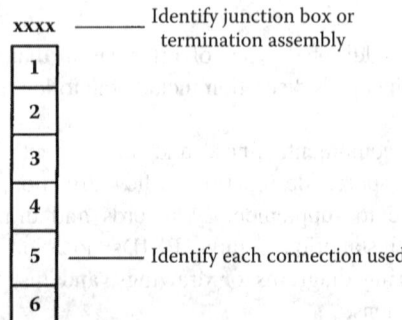

FIG. A.4c
Terminal and bulkhead symbol.

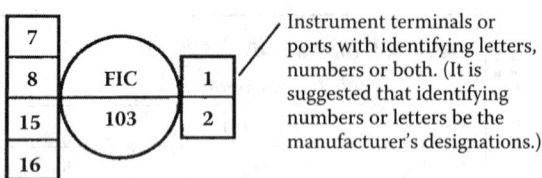

FIG. A.4d
Terminal or port designation.

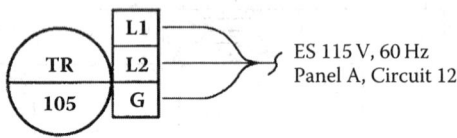

FIG. A.4e
Electrical power supply identifications.

In the area of energy supply, Figure A.4e illustrates the way electrical power supply should be shown, including the supply voltage level, the circuit number, or disconnect identification.

Figure A.4f shows the method for identifying air supplies, including the air supply pressure.

Figure A.4g describes the method for identifying hydraulic fluid supplies, including the hydraulic fluid supply pressure.

Figure A.4h shows the recommended method of describing instrument action. The figure shows the direction of the instrument signal by placing appropriate letters close to the

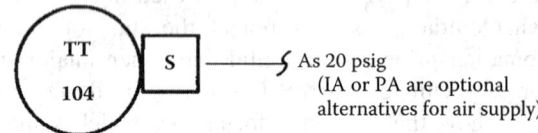

FIG. A.4f
Air supply identification.

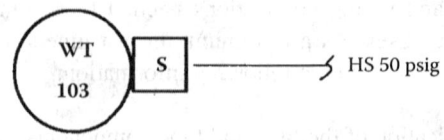

FIG. A.4g
Hydraulic fluid supply identification.

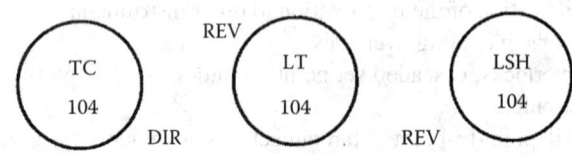

FIG. A.4h
The action designations for various instruments.

instrument bubble. It also uses the letters "DIR" if the output signal increases as input (measured variable) increases and used the letters "REV" if the output signal decreases as the value of the input (measured variable) increases. However, since most transmitters are direct acting, the designation DIR is optional for them.

GRAPHIC SYMBOLS

This informative annex to the standard describes the use of graphic symbols that are used to depict instrument loop devices and functions, application software functions, and the interconnections between them that is logical, unique, and consistent in application with a minimum of exceptions, special uses, or requirements.

Graphic symbols, when used with identification letters and numbers constructed as describes, should as a minimum describe the functionality of, and if assigned a loop number provide a unique identity for, each device and function shown.

Instrument identification applied to graphic symbols should include, as a minimum, an alphabetic functional identification to identify the functionality of devices and functions shown in the diagrams.

Brief explanatory notes or other text may be added adjacent to a symbol or in the note section of a drawing or sketch to clarify the meaning or purpose of a device or function.

Loop Identification Number numerals complete the identity of the loop being shown.

Lettering fonts should be similar to Arial Narrow and be a minimum of 3/32 in. (1.125 mm) high and a maximum of 13 characters per inch wide.

A Loop Number prefix should be used as follows:
Not be used with bubbles on drawings but indicated in the note section.
Be used with bubbles if more than one prefix is being used.
Be used in text.

Examples of Graphic Symbols

Instrument bubble symbols should use the upper half of each symbol for Functional Identification Letters and the lower half of each symbol for Loop Numbers:
Five characters or less.

Six characters or more relieve sides of bubble or enlarge bubble as required.

  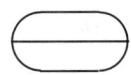 

Functional diagramming, binary logic, and electrical schematic symbols should be tagged at either (A), (B), (C), or (D).

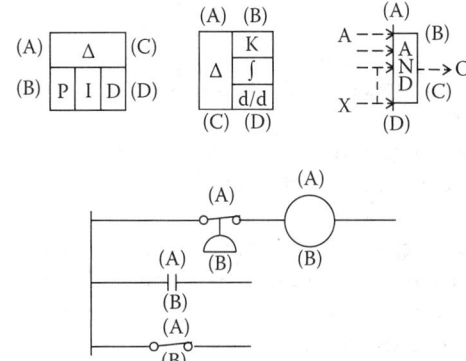

Graphic Symbol Applications

Graphic symbols provide representations of the instrumentation and functions required for process, machine, or equipment measurement, indication, control, modulation, and switching of variables by any or all of the following applications:

- Instrument diagrams
- Functional diagrams
- Binary logic diagrams
- Electrical schematics

The most common uses for the following:

Instrument diagrams are process flow diagrams (PFDs), piping and instrumentation diagrams (P&IDs), engineering flow diagrams (EFDs), and mechanical flow diagrams (MFDs).

Functional diagrams are instrument loop device and function details and application software details for microprocessor-based control and monitoring systems.

Binary logic diagrams are complex, interlocking, and stepwise logic programming and application software for microprocessor-based binary logic systems.

Electrical schematics are electrical diagrams for motor and other on–off control.

All of the applications may be used to prepare sketches and drawings for books, magazines, journals, and instruction and maintenance manuals.

DEVICE AND FUNCTION SYMBOLS

Instrumentation devices and functions are constructed for sketches, drawings, and diagrams by the use of the generic bubble and other geometric symbols and specific graphic symbols found in Clause 5.

It is not necessary to show a symbol or a bubble for every device or function required by a loop if the need for the device or function or its tag number is clearly understood, for example,

Symbols are not required, but may be used, for control valve positioners and stream sample conditioner components.
Bubbles are not required, but may be used, for orifice plate, thermocouple, and control valve graphic symbols.

When *smart* drawings, such as computer-generated P&IDs, that are linked to instrument indexes or data sheets are used, a bubble or graphic symbol to which an instrument tag number is attached should be used for all devices and functions that are to be indexed or require data sheets.

Examples

PFDs are developed by process engineers to provide basic process data and to describe process operation. Simple instrument diagrams are used to indicate the primary process control measurements and controlled streams required to operate the process. Process monitoring and alarm points and secondary and auxiliary controls and monitors are not shown but are added during the detailed process design and P&ID development.

A simple flow control requirement should be shown on a PFD as follows:

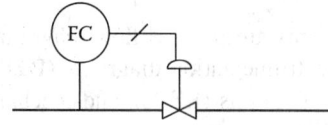

Instrument tag numbers should not be assigned on PFDs. A typical instrument diagram developed from the PFD diagram:

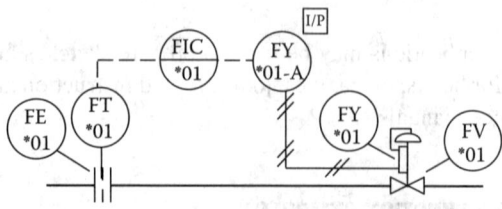

Bubbles [FE-*01] and [FY-01-B] are optional and not recommended.
Bubble [FV-*01] is optional but is recommended.

Typical equipment and function-oriented functional diagrams developed from the PFD diagram:

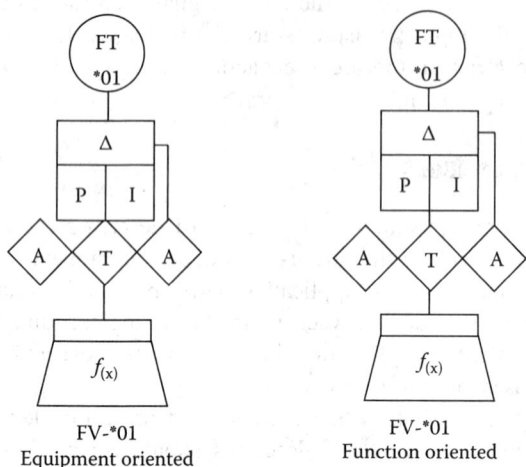

Process Variable Measurements

Process variable measurement devices are inserted in or mounted on pipelines and equipment to measure a physical property or to analyze a chemical composition and include but are not limited to the following:

Primary elements, such as orifice plates and thermocouples, that generate analog signals, position mechanical devices, or are used by transmitters to generate signals compatible with the control system.
Transmitters with integral primary elements, such as vortex shedding flowmeters and filled-capillary temperature devices that generate signals compatible with the control system.

Process measurements are indicated by
Bubbles as shown in Table A.4d for
Generic primary elements
Primary elements that do not have a graphic symbol in Table A.4f
Users who elect not to use graphic symbols from Table A.4f
Graphic symbols from Table A.4j
Analyzer primary element located in a process slip stream or in a process stream or equipment with or without accessory devices, such as sample conditioners that contain components that are not normally shown, and with the type of analyzer and the component of interest noted at (**) and (***), respectively:
With sample conditioners

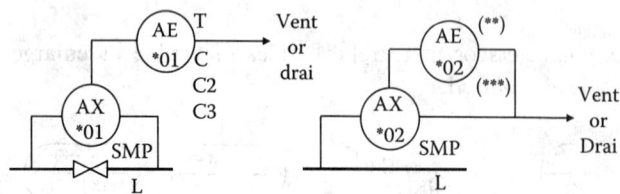

Without sample conditioners

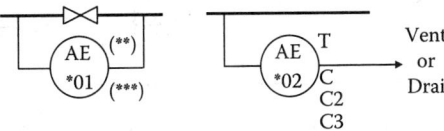

Analyzer primary element or transmitter inserted in process stream or equipment

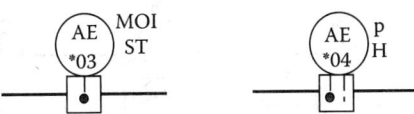

Orifice plate primary elements, with or without optional flow arrow, use generic orifice plate symbol with transmitter bubble connected to indicate orifice tap location for flange taps, corner taps, pipe taps, and vena contracta taps, respectively.
Single process connection, corner taps, pipe taps, and vena contracta taps are indicated by notation.

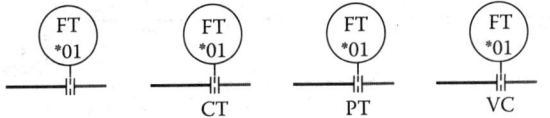

Double process connection, pipe taps, and vena contracta taps are indicated by notation.

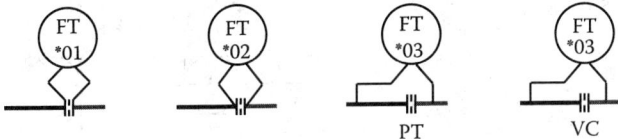

Process root block valves should be shown as required by the piping engineering group.
Orifice meter tubes or runs that are specified and requisitioned by the instrument group should be shown on drawings and sketches by the following:

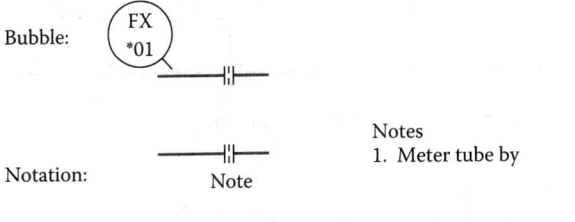

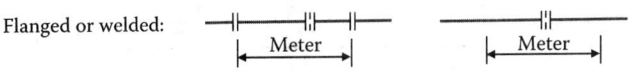

A.4 Flowsheet and Functional Diagrams Symbols 1215

Final Control Elements Final control elements installed in pipelines and equipment modulate or manipulate the process stream or equipment to affect the loop measured variable.

Final control elements include, but are not limited to, control valves, solenoid valves, louvers, dampers, motors, variable speed drives, and machine components.

Control valves are generally pneumatically operated and furnished with positioners that may

Be actuated by a pneumatic or an electronic signal
Not be shown if all control valves are furnished with positioners

Control valves with pneumatic or electronic signal:
Without positioner

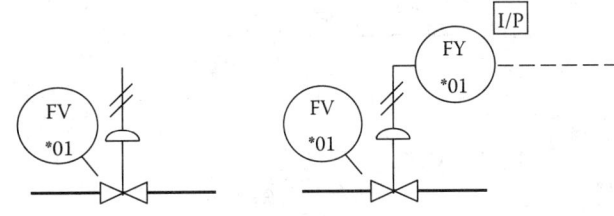

With positioner

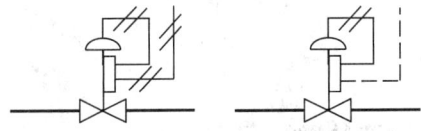

Cross-hatches from positioner to actuator are optional:
With tripping solenoid, with and without positioner:

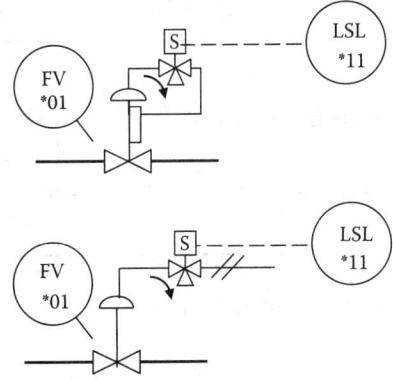

Instrumentation with integral components that measure process variables and transmit control and other functions as an integral part of a transmitter

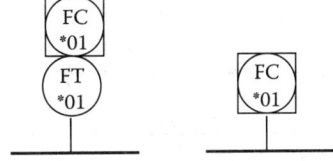

Manipulates control valves as an integral part of a control valve positioner

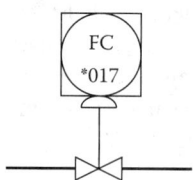

Is an integral assembly that contains a transmitter, a controller, and a control valve

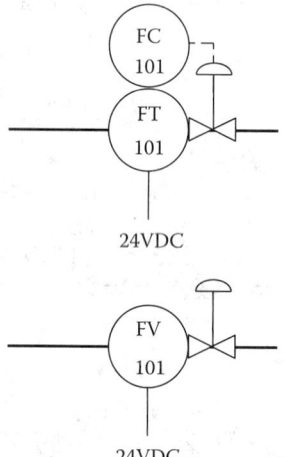

Instrument Signal Connections

Pneumatic discrete instrumentation

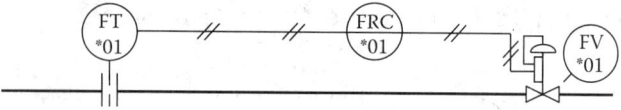

Electronic discrete instrumentation

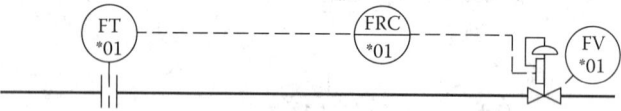

Shared display, shared control instrumentation

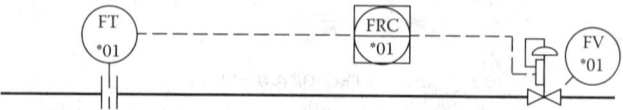

Shared display, shared control instrumentation, with diagnostic and calibration bus on field wiring

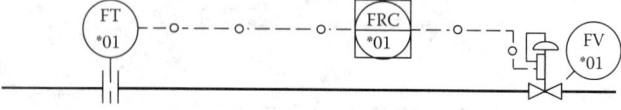

Shared display, shared control, and wireless instrumentation

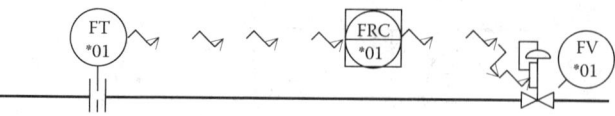

Shared display, shared control instrumentation, primary and alternate systems, no inter-bus communication

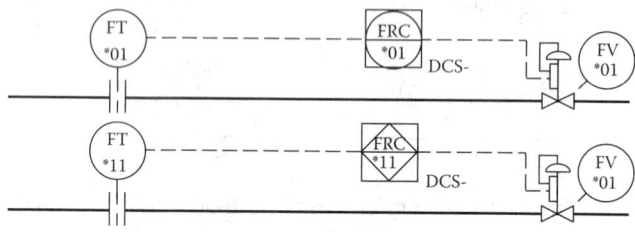

Shared display, shared control, primary, and alternate systems, with inter-bus communication

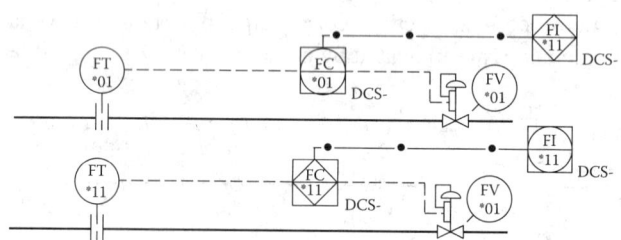

Shared display, shared control and fieldbus instrumentation, inter-bus communication
Fieldbus transmitter/controller and electronic valve positioner

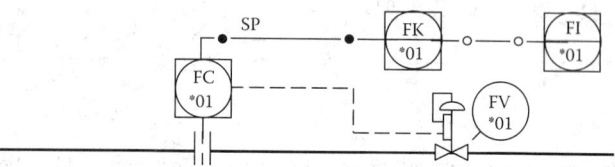

Fieldbus valve positioner/controller and electronic transmitter

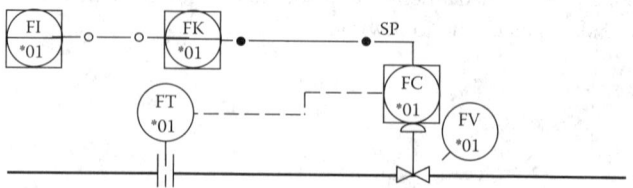

Fieldbus transmitter and valve positioner/controller

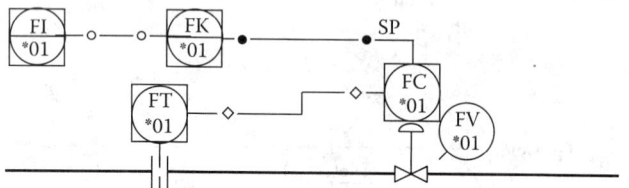

Fieldbus valve positioner/controller, transmitter, and indicator

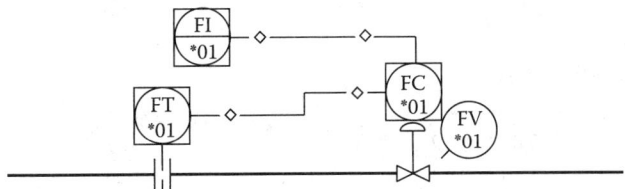

Separate devices or application software functions that do not require separate bubbles or tag numbers for each function

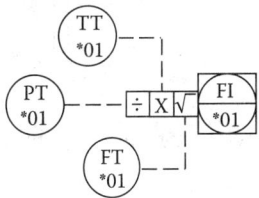

Fieldbus integral transmitter, controller, and valve positioner

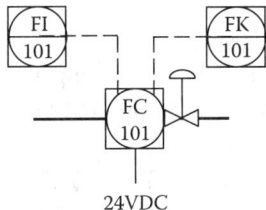

Integral devices or application software functions that do not require separate bubbles or tag numbers for each function

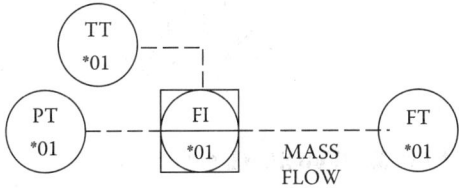

Instrument and functional diagrams should not be used to specifically identify signal tubing, wiring, and bus construction methods used to implement a monitoring and control system.
Function block symbols
Signal processing functions should be identified by
Appended to a bubble if an Instrument/Tag Number is required

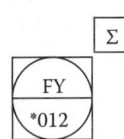

Attached tangentially to the affected bubble and in line with the signal if the function is an integral part of the affected bubble
An example of a common application is the calculation of mass flow with an orifice plate primary element

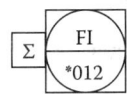

Separate devices or functions that require separate bubbles and tag numbers

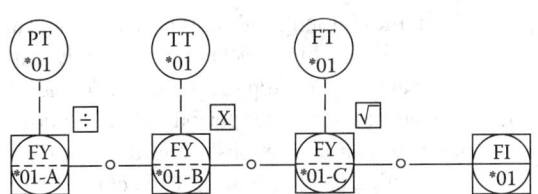

Alarm Indicators

Shared control and monitor functions generally allow the indication of four configurable alarms for process variables and set point deviations.
Only alarms that are to be configured are shown.

Instrument diagramming
Process variable alarms

```
FT      FC   HH
*01 ----*01  H
             L
```

Process variable deviation from set point alarms

```
FT      FC   DH
*01 ----*01
```

Process variable deviation from set point and process variable alarms

```
FT      FC   DH
*01 ----*01  H
             L
```

Functional Diagrams

Process variable alarms

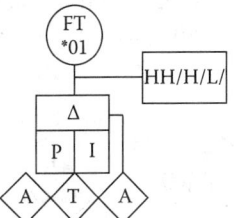

Process variable deviation from set point alarms

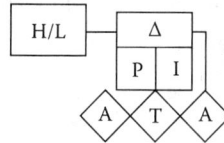

Process variable deviation from set point and process variable alarms

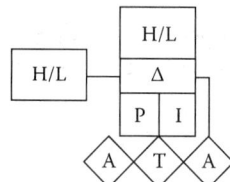

Multipoint and Multifunction Instruments

Multipoint instruments are single or multivariable indicators or recorders that receive input signals from two or more primary elements or transmitters.

Multivariable instruments are controllers that receive input signals from two or more primary elements or transmitters and control one manipulated variable.

Multifunction instruments are controllers that receive input signals from two or more primary elements or transmitters and control two or more manipulated variables.

Single variable or multivariable multipoint recorders for two or three points are drawn with bubbles either
Tangent to each other in the same order, left to right, as the pen or pointer assignments

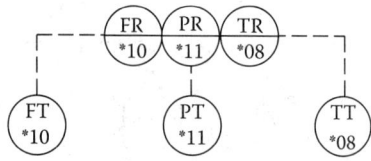

Separate from each other with pen number indicated and a note defining the multipoint instrument

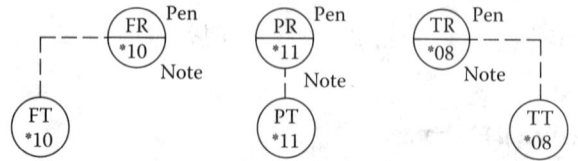

Multipoint indicators and recorders for four or more points are drawn with bubbles separate from each other, with point number indicated by adding a suffix to the tag numbers:

Single variable

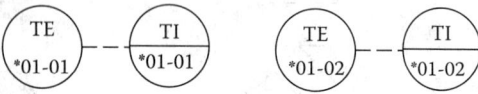

Multivariable

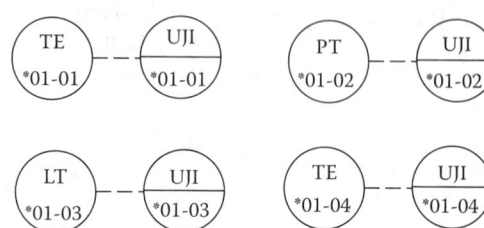

A multivariable controller example drawn with bubbles for each measured variable input, the output to the final control element, and measured variable indicators

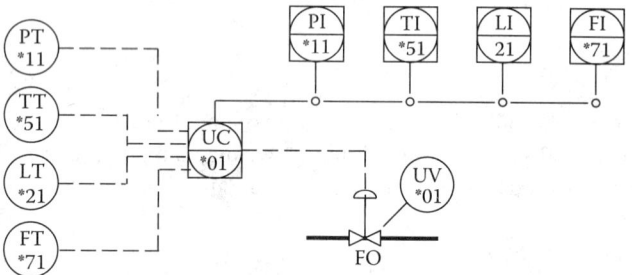

Variable multifunction controller example drawn with bubbles for measured variable inputs, controller and indicator functions, and final control elements

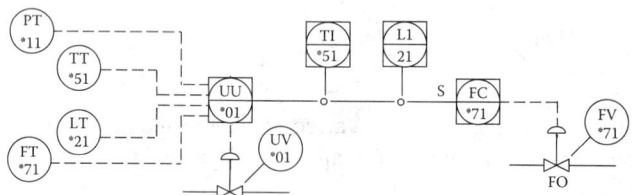

An example of instrument, functional, and electrical diagrams for a simple process

Process Control Description

Process description
    Tank periodically fills with a liquid, in small and large volumes over long and short time periods.
Control description
    Control system design for the following:
        Small volumes for long and short periods should allow tank to fill to a high level to automatically start the pump and then to stop the pump at a low level.
        Large volumes for long periods should allow the pump to run continuously and maintain a fixed level with a level to flow cascade control loop.

Pump control is selected by a three-position Hand-Off-Auto (H-O-A) selector switch:
  Method (a) selector switch is in "HAND" position.
  Method (b) selector switch is in "AUTO" position.
Pump should be stopped at any time:
  Automatically if low level is exceeded
  By operating the stop push button
  Switching the H-O-A selector to "OFF" position

Instrument diagram

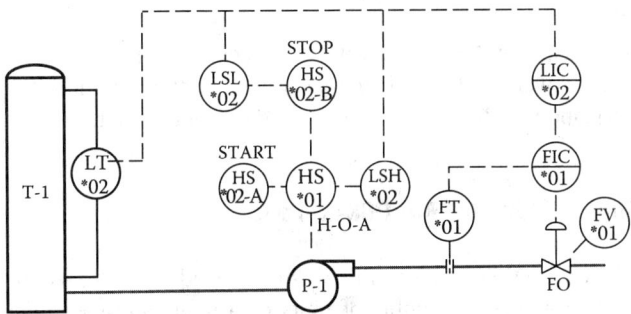

Functional diagram

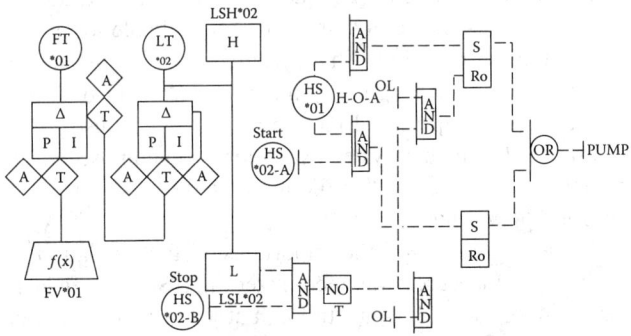

Electrical schematic diagram

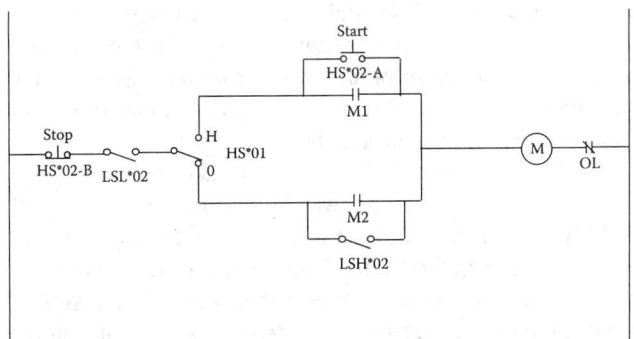

CONCLUSIONS

The material in this chapter was mostly excerpted from the International Society of Automation (ISA) standard ANSI/ISA-5.1-2009, titled Instrumentation Symbols and Identification, with the permission of ISA.

In addition to the material contained in this chapter, symbols may also be developed to show devices and functions that are not covered here by this standard. If new or revised symbols are developed, it is recommended to submit them to the ISA-5.1 committee for inclusion in the next revision of that standard.

Abbreviations

ACS   Analyzer control system
BMS   Burner management system
CCS   Computer control system
CEMS   Continuous emissions monitoring system
DCS   Distributed control system
MMS   Machine monitoring system
PCCS   Personal computer control system
P&ID   Piping and Instrumentation Diagram
PLC   Programmable logic controller
SIS   Safety instrumented system
VMS   Vibration monitoring system

For a more detailed list of all instrumentation and control related abbreviations, refer to Table A.4g.

A.5 Conversion among Engineering Units

B. G. LIPTÁK (2017)

INTRODUCTION

Converting a quantity from one unit to another is both time consuming and a potential source of errors. This chapter provides a number of conversion tables that can be used for that purpose. In addition, one can also use the Internet where software tools are available to have the conversions done automatically.

For engineering units one might use

1. http://www.engineeringtoolbox.com/unit-converter-d_185.html
2. http://www.gordonengland.co.uk/conversion/specific_heat.htm

For other more general conversions among units, one can use

1. http://www.onlineconversion.com/
2. http://en.wikipedia.org/wiki/Conversion_of_units#Area

HISTORY

Length units were used as early as the third millennium BC both in Egypt and in the Indus Valley. To our knowledge, the first units to measure length were the Mesopotamian *cubit* and Indian *dhanus*. The cubit was defined as the distance from the elbow to the tip of the middle finger. The foot evolved from the cubit through a complicated transformation is not yet fully understood. The Greeks and Romans inherited the foot from the Egyptians and the Roman mile was first defined as 1480 m (5000 ft). It was also used in England during the occupation but was later changed to ~1609 m (5280 ft) or 8 furlongs (one furlong was defined as 40 rods), etc. The introduction of the yard (0.9144 m) as a unit of length came later and possibly corresponded to two cubits and was divided into 2, 4, 8, and 16 parts called the half-yard, span, finger, and nail.

Units of mass were based on the *grain* that was first used to measure noble metals and corresponded to the weight of a grin of wheat. Goods in commerce were originally traded by number or volume. When weighing of goods began, units of mass were first based on the weight of a volume of water that corresponded to a volume of grain. Other weight units were also based on the volumes of water, such as the *talent*, which was the weight of a cubic *foot* of water, and the Egyptian *hon* corresponded to the weight of a *cubic palm* of water.

MODERN AND INTERNATIONAL UNITS

The decimal system of units was conceived in the sixteenth century when the jumble of units of weights and measures used by the different nations and trades started to interfere with commerce. It was not until 1790, however, that the French National Assembly requested the French Academy of Sciences to work out a system of units suitable for adoption by the entire world. This system, based on the meter (metre) as a unit of length and the gram as a unit of mass, was adopted as a practical measure to benefit industry and commerce. The metric system was soon adopted in scientific and technical circles, but in the more conservative areas of the world, it was not.

In the United States, the importance of the regulation of weights and measures was recognized in 1787 in Article 1, Section 8, of the U.S. Constitution, but it referred to the old *English* units and it took another century for the metric system to be even *legalized* in 1866.

In 1893, the international units of meter and kilogram also became the fundamental standards of length and mass in the United States, but that only meant that the foot and pound were *defined in relation to metric units*. Some states were more progressive. For example, Utah mandated in 1895 that "The Metric System shall be taught in the public schools of the State," but even that mandate was later repealed.

After the middle of the twentieth century, both England and Canada converted to the metric system, and practically all nations, except the United States, completed their *metrification*. Near the end of the twentieth century, Congress designated the metric system as the *preferred* system of weights and measures, but left this conversion voluntary.

As of this writing, the use of the metric system in the United States is still not mandatory although industries that are heavily involved in international trade, such as the military, have converted, while others, particularly the ones whose activity is mostly domestic, such as the general construction and the highway construction industries, are still using the so-called *imperial* or *English* system of units.

A.5 Conversion among Engineering Units

TABLES IN THIS CHAPTER

This handbook was first published for American readers only, but with each edition, it became more international, and in its fourth edition, about half of the volumes were already sold overseas. For this reason, I have encouraged the authors of this fifth edition to use both the metric and the English units, and in most cases, they have done so.

The conversion tables provided in this chapter will help the readers to convert among the two systems of units and also among the many units that are unique to the various industries. The symbols used for the various units are also given in the tables, while the symbols used for their multipliers are as follows:

1. k for 10^3 (kilo)
2. M for 10^6 (mega)
3. G for 10^9 (giga)
4. T for 19^{12} (tera)
5. P for 10^{15} (peta)
6. E for 10^{18} (exa)

Table A.5a provides the definitions and symbols used in the International System (SI) of Units.

Table A.5b provides two alphabetical listings of multipliers to be used to convert among the various units. Some of the coverage of these tables is redundant but their combination covers just about all engineering units that are in wider use.

The tables that follow are specific to individual variables such as flow, pressure, and temperature, etc.

Table A.5a (page 1222)—International Unit Symbols
Table A.5b (page 1223)—Multipliers to Convert Engineering Units
Table A.5c (page 1238) —Units of Area
Table A.5d (page 1238)—Area Unit Definitions
Table A.5e (page 1239)—Units of Density
Table A.5f (page 1239)—Density Unit Definitions
Table A.5g (page 1239)—Energy and Heat Unit Conversion
Table A.5h (page 1240)—Energy Unit Definitions
Table A.5i (page 1241)—Flow Unit Conversion (Mass)
Table A.5j (page 1241)—Flow Unit Conversion (Volumetric)
Table A.5k (page 1242)—Flow Unit Definitions (Volumetric)
Table A.5l (page 1242)—Length Unit Conversion
Table A.5m (page 1242)—Power Unit Conversion
Table A.5n (page 1243)—Power or Heat Flow Unit Definitions
Table A.5o (page 1243)—Pressure Unit Conversion (Courtesy of DGSI)
Table A.5p (page 1244)—Pressure Unit Conversion
Table A.5q (page 1245)—Pressure Conversion (psi to kg/cm^2)
Table A.5r (page 1246)—Pressure Conversion (head pressure)
Table A.5s (page 1247)—Pressure Unit Definitions
Table A.5t (page 1247)—Specific Enthalpy and Latent Heat Conversion
Table A.5u (page 1247)—Specific Heat and Entropy Unit Conversion
Table A.5v (page 1248)—Specific Volume Conversion
Table A.5w (page 1248)—Temperature Conversion
Table A.5x (page 1248)—Time Unit Conversion
Table A.5y (page 1248)—Velocity Unit Conversion
Table A.5z (page 1249)—Velocity Unit Definitions
Table A.5aa (page 1250)—Viscosity Unit Conversion
Table A.5ab (page 1253)—Volume Unit Conversion
Table A.5ac (page 1253)—Volume Unit Definitions
Table A.5ad (page 1255)—Weight Unit Conversion
Table A.5ae (page 1256)—Weight Unit Definitions

TABLE A.5a
International Unit Symbols

Quantity	Unit	SI Symbol	Formula
\multicolumn{4}{c}{*Base Units*}			
Length	Meter	m	—
Mass	Kilogram	kg	—
Time	Second	s	—
Electric current	Ampere	A	—
Thermodynamic temperature	Kelvin	K	—
Amount of substance	Mole	mol	—
Luminous intensity	Candela	cd	—
Supplementary Units			
Plane angle	Radian	rad	—
Solid angle	Steradian	sr	—
Derived units			
Acceleration	Meter per second squared	—	m/s^2
Activity (of a radioactive source)	Disintegration per second	—	(disintegration)/s
Angular acceleration	Radian per second squared	—	rad/s^2
Angular velocity	Radian per second	—	rad/s
Area	Square meter	—	m^2
Density	Kilogram per cubic meter	—	kg/m^3
Electric capacitance	Farad	F	$A \cdot s/V$
Electrical conductance	Siemens	S	A/V
Electric field strength	Volt per meter	—	V/m
Electric inductance	Henry	H	$V \cdot s/A$
Electric potential difference	Volt	V	W/A
Electric resistance	Ohm	Ω	V/A
Electromotive force	Volt	V	W/A
Energy	Joule	J	$N \cdot m$
Entropy	Joule per Kelvin	—	J/K
Force	Newton	N	$kg \cdot m/s^2$
Frequency	Hertz	Hz	(cycle)/s
Illuminance	Lux	lx	lm/m^2
Luminance	Candela per square meter	—	cd/m^2
Luminous flux	Lumen	lm	$cd \cdot sr$
Magnetic field strength	Ampere per meter	—	A/m
Magnetic flux	Weber	Wb	$V \cdot s$
Magnetic flux density	Tesla	T	Wb/m^2
Magnetomotive force	Ampere	A	—
Power	Watt	W	J/s
Pressure	Pascal	Pa	N/m^2
Quantity of electricity	Coulomb	C	$A \cdot s$
Quantity of heat	Joule	J	$N \cdot m$
Radiant intensity	Watt per steradian	—	W/sr
Specific heat	Joule per Kilogram-Kelvin	—	$J/kg \cdot K$
Stress	Pascal	Pa	N/m^2
Thermal conductivity	Watt per meter-Kelvin	—	$W/m \cdot K$
Velocity	Meter per second	—	m/s
Viscosity, dynamic	Pascal-second	—	$Pa \cdot s$
Viscosity, kinematic	Square meter per second	—	m^2/s
Voltage	Volt	V	W/A
Volume	Cubic meter	—	m^3

Engineering Conversion Factors

TABLE A.5b
*Multipliers to Convert Engineering Units**

To Convert	Into	Multiply by	To Convert	Into	Multiply by
A			Ampere-turns/in.	amp-turns/meter	39.37
Abampere	ampere (A)	1.000 000*E+01	Ampere-turns/in.	gilberts/cm	0.495
Abcoulomb	statcoulombs	2.998×10^{10}	Ampere-turns/meter	amp/turns/cm	0.01
Abcoulomb	coulomb (C)	1.000 000*E+01	Ampere-turns/meter	amp-turns/in.	0.0254
Abfarad	farad (F)	1.000 000*E+09	Ampere-turns/meter	gilberts/cm	0.01257
Abhenry	henry (H)	1.000 000*E+09	Angstrom	meter (m)	1.000 000*E−10
Abmho	siemens (S)	1.000 000*E+09	Angstrom unit	in.	3.937×10^{-9}
Abohm	ohm (Ω)	1.000 000*E+09	Angstrom unit	meter	1×10^{-10}
Abvolt	volt (V)	1.000 000*E+08	Angstrom unit	micron or μm	1×10^{-4}
Acre	sq chain (Gunters)	10	Are	acre (U.S.)	0.02471
Acre	rods	160	Are	meter^2 (m^2)	1.000 000*E+02
Acre	sq links (Gunters)	1×10^5	Ares	sq yards	119.6
Acre	hectare or sq hectometer	0.4047	Ares	acres	0.02471
			Ares	sq meters	100.0
Acre (U.S. survey)[a]	meter^2 (m^2)	4.046 873E+03	Astronomical unit	kilometers	1.495×10^8
Acre-feet	cubic ft	43,560.00	Astronomical unit	meter (m)	1.495 979E+11
Acre-feet	gallons	3.259×10^5	Atmosphere (standard)	pascal (Pa)	1.013 250*E+05
Acre-foot (U.S. survey)[a]	meter^3 (m^3)	1.233 489E+03	Atmosphere (technical = 1 kgf/cm^2)	pascal (Pa)	9.806 650*E+04
Acres	sq ft	43,560.00	Atmospheres	ton/sq in.	0.007348
Acres	sq meters	4,047.00	Atmospheres	cm of mercury	76.0
Acres	sq miles	1.562×10^{-3}	Atmospheres	ft of water (at 4°C)	33.9
Acres	sq yards	4,840.00	Atmospheres	in. of mercury (at 0°C)	29.92
Ampere hour	coulomb (C)	3.600 000*E+03	Atmospheres	kg/sq cm	1.0333
Ampere-hours	coulombs	3,600.00	Atmospheres	kg/sq meter	10,332.00
Ampere-hours	farads	0.03731	Atmospheres	pounds/sq in.	14.7
Amperes/sq cm	amps/sq in.	6.452	Atmospheres	tons/sq ft	1.058
Amperes/sq cm	amps/sq meter	10^4	**B**		
Amperes/sq in.	amps/sq cm	0.155	Bar	pascal (Pa)	1.000 000*E+05
Amperes/sq in.	amps/sq meter	1,550.00	Barn	meter^2 (m^2)	1.000 000*E−28
Amperes/sq meter	amps/sq cm	10^{-4}	Barrel (for petroleum, 42 gal)	meter^3 (m^3)	1.589 873E−01
Amperes/sq meter	amps/sq in.	6.452×10^{-4}	Barrels (oil)	gallons (oil)	42.0
Ampere-turns	gilberts	1.257	Barrels (U.S., dry)	cubic in.	7056.0
Ampere-turns/cm	amp-turns/in.	2.54	Barrels (U.S., dry)	quarts (dry)	105.0
Ampere-turns/cm	amp-turns/meter	100	Barrels (U.S., liquid)	gallons	31.5
Ampere-turns/cm	gilberts/cm	1.257	Bars	atmospheres	0.9869
Ampere-turns/in.	amp-turns/cm	0.3937	Bars	dynes/sq cm	10^6

* Where fewer than six decimal places are shown, more precision is not warranted. When the first digit discarded is less than 5, the last digit retained should not be changed. For example, 3.463 25, if rounded to four digits, would be 3.463; if rounded to three digits, 3.46. When the first digit discarded is greater than or if it is a 5, followed by at least one digit other than 0, the last figure retained should be increased by one unit. For example, 8.376 52, if rounded to four digits, would be 8.377; if rounded to three digits, 8.38. When the first digit discarded is exactly 5, followed only by zeros, the last digit retained should be rounded upward if it is an odd number, but no adjustment made if it is an even number. For example, 4.365, when rounded to three digits, would become 4.36. The number 4.355 would also round to the same value, 4.36, if rounded to three digits.

[a] Since 1893, the U.S. basis of length measurement has been derived from metric standards. In 1959, a small refinement was made in the definition of the yard to resolve discrepancies both in this country and abroad, which changed its length from 3600/3937 m to 0.9144 m exactly. This resulted in the new value being shorter by two parts in a million.

(Continued)

TABLE A.5b (Continued)
Multipliers to Convert Engineering Units

To Convert	Into	Multiply by	To Convert	Into	Multiply by
Bars	kg/sq meter	1.020×10^4	Btu (International Table)/s · ft² · °F	watt per meter²-Kelvin (W/m² · K)	2.044 175E+04
Bars	pounds/sq ft	2,089.00			
Bars	pounds/sq in.	14.5	Btu (thermochemical) · ft/ h · ft² · °F (k, thermal conductivity)	watt per meter-Kelvin (W/m · K)	1.729 577E+00
Baryl	dyne/sq cm	1.000			
Board foot	meter³ (m³)	2.359 737E–03			
Bolt (U.S. cloth)	meters	36.576	Btu (thermochemical) · in./ h · ft² · °F (k, thermal conductivity)	watt per meter-Kelvin (W/m · K)	1.441 314E–01
British thermal unit (39°F)	joule (J)	1.059 67E+03			
British thermal unit (59°F)	joule (J)	1.054 80E+03			
British thermal unit (60°F)	joule (J)	1.054 68E+03	Btu (thermochemical) · in./ s · ft² · °F (k, thermal conductivity)	watt per meter-Kelvin (W/m · K)	5.188 732E+02
British thermal unit (International Table)[b]	joule (J)	1.055 056E+03			
British thermal unit (mean)	joule (J)	1.055 87E+03	Btu (thermochemical)/ s · ft² · °F	watt per meter²-Kelvin (W/m² · K)	2.042 808E+04
British thermal unit (thermochemical)	joule (J)	1.054 350E+03	Btu (thermochemical)/ h · ft² · °F (C, thermal conductance)	watt per meter²-Kelvin (W/m² · K)	5.674 466E+00
Btu	liter-atmosphere	10.409			
Btu	ergs	1.0550×10^{10}			
Btu	foot-lb	778.3	Btu (thermochemical)/ft²	Joule per meter² (J/m²)	1.134 893E+04
Btu	gram-calories	252.0	Btu (thermochemical)/ft² · h	watt per meter² (W/m²)	3.152 481E+00
Btu	horsepower-hr	3.931×10^{-4}	Btu (thermochemical)/ ft² · min	watt per meter² (W/m²)	1.891 489E+02
Btu	joules	1,054.80			
Btu	kilogram-calories	0.252	Btu (thermochemical)/ft² · s	watt per meter² (W/m²)	1.134 893E+04
Btu	kilogram-meters	107.5	Btu (thermochemical)/h	watt (W)	2.928 751E+01
Btu	kilowatt-hr	2.928×10^{-4}	Btu (thermochemical)/in.² · s	watt per meter² (W/m)²	1.634 246E+06
Btu (International Table) · in./h · ft² · °F (k, thermal conductivity)	watt per meter-Kelvin (W/m · K)	1.442 279E–01	Btu (thermochemical)/lb	joule per kilogram (J/kg)	2.324 444E+03
			Btu (thermochemical)/ lb · °F (c, heat capacity)	joule per kilogram-Kelvin (J/kg · K)	4.184 000E+03
Btu (International Table) · in./s · ft² · °F (k, thermal conductivity)	watt per meter-Kelvin (W/m · K)	5.192 204E+02	Btu (thermochemical)/min	watt (W)	1.757 250E+01
			Btu (thermochemical)/s	watt (W)	1.054 350E+03
Btu (International Table) · ft/h · ft² · °F (k, thermal conductivity)	watt per meter-Kelvin (W/m · K)	1.730 735E+00	Btu/hr	foot-pounds/sec	0.2162
			Btu/hr	gram-cal/sec	0.07
			Btu/hr	horsepower-hr	3.929×10^{-4}
			Btu/hr	watts	0.2931
Btu (International Table)/ft²	joule per meter² (J/m²)	1.135 653E+04	Btu/min	foot-lb/sec	12.96
Btu (International Table)/h	watt (W)	2.930 711E–01	Btu/min	horsepower	0.02356
Btu (International Table)/h · ft² · °F (C, thermal conductance)	watt per meter²-Kelvin (W/m² · K)	5.678 263E+00	Btu/min	kilowatts	0.01757
			Btu/min	watts	17.57
			Btu/sq ft/min	watts/sq in.	0.1221
Btu (International Table)/lb	joule per kilogram (J/kg)	2.326 000*E+03	Bucket (Br. Dry)	cubic cm	1.818×10^4
Btu (International Table)/ lb · °F (c, heat capacity)	joule per kilogram-Kelvin (J/kg · K)	4.186 800*E+03	Bushel (U.S.)	meter³ (m³)	3.523 907E–02
			Bushels	cubic ft	1.2445
Btu (International Table)/s	watt (W)	1.055 056E+03	Bushels	cubic in.	2,150.40

[b] This value was adopted in 1956. Some of the older International Tables use the value 1.055 04 E+03. The exact conversion factor is 1.055 055 852 62*E+03. At the same time, it was decided that any data in feet derived from and published as a result of geodetic surveys within the United States would remain with the old standard (1 ft = 1200/3937 m) until further decision. This foot is named the U.S. survey foot.

As a result, all U.S. land measurements in U.S. customary units will relate to the meter by the old standard. All the conversion factors in these tables for units referenced to this footnote are based on the U.S. survey foot, rather than the international foot.

Conversion factors for the land measures given below may be determined from the following relationships:

1 league = 3 miles (exactly)
1 rod = 16 feet (exactly)
1 section = 1 square mile (exactly)
1 township = 36 square miles (exactly)
1 chain = 66 feet (exactly)

(Continued)

TABLE A.5b (Continued)
Multipliers to Convert Engineering Units

To Convert	Into	Multiply by
Bushels	cubic meters	0.03524
Bushels	liters	35.24
Bushels	pecks	4.0
Bushels	pints (dry)	64.0
Bushels	quarts (dry)	32.0

C

To Convert	Into	Multiply by
Cal (International Table)/g	joule per kilogram (J/kg)	4.186 800*E+03
Cal (International Table)/g·°C	joule per kilogram-Kelvin (J/kg·K)	4.186 800*E+03
Cal (thermochemical)/g·°C	joule per kilogram-Kelvin (J/kg·K)	4.184 000*E+08
Cal (thermochemical)/cm·s·°C	Watt per meter-Kelvin (W/m·K)	4.184 000*E+02
Cal (thermochemical)/cm²	joule per meter² (J/m²)	4.184 000*E+04
Cal (thermochemical)/cm²·min	Watt per meter² (W/m²)	6.973 333E+02
Cal (thermochemical)/cm²·s	watt per meter² (W/m²)	4.184 000*E+04
Cal (thermochemical)/g	joule per kilogram (J/kg)	4.184 000*E+03
Cal (thermochemical)/min	watt (W)	6.973 333E–02
Cal (thermochemical)/s	watt (W)	4.184 000*E+00
Caliber (inch)	meter (m)	2.540 000*E–02
Calorie (15°C)	joule (J)	4.185 80E+00
Calorie (20°C)	joule (J)	4.181 90E+00
Calorie (international table)	joule (J)	4.186 800*E+00
Calorie (kilogram, International Table)	joule (J)	4.186 800*E+03
Calorie (kilogram, mean)	joule (J)	4.190 02E+03
Calorie (kilogram, thermochemical)	joule (J)	4.184 000*E+03
Calorie (mean)	joule (J)	4.190 02E+00
Calorie (thermochemical)	joule (J)	4.184 000*E+00
Calories, gram (mean)	Btu (mean)	3.9685×10^{-3}
Candle/sq cm	lamberts	3.142
Candle/sq in.	lamberts	0.487
Carat (metric)	kilogram (kg)	2.000 000*E–04
Centares (centiares)	sq meters	1.0
Centigrade	Fahrenheit	(C° × 9/5) + 32
Centigrams	grams	0.01
Centiliter	ounce fluid (U.S.)	0.3382
Centiliter	cubic in.	0.6103
Centiliter	drams	2.705
Centiliters	liters	0.01
Centimeter of mercury (0°C)	pascal (Pa)	1.333 22E+03
Centimeter of water (4°C)	pascal (Pa)	9.806 38E+01
Centimeter-dynes	cm-grams	1.020×10^{-3}
Centimeter-dynes	meter-kg	1.020×10^{-8}
Centimeter-dynes	pound-feet	7.376×10^{-8}
Centimeter-grams	cm-dynes	980.7
Centimeter-grams	meter-kg	10^{-5}
Centimeter-grams	pound-feet	7.233×10^{-5}
Centimeters	feet	3.281×10^{-2}
Centimeters	inches	0.3937
Centimeters	kilometers	10^{-5}
Centimeters	meters	0.01
Centimeters	miles	6.214×10^{-6}
Centimeters	millimeters	10.0
Centimeters	miles	393.7
Centimeters	yards	1.094×10^{-2}
Centimeters of mercury	atmospheres	0.01316
Centimeters of mercury	feet of water	0.4461
Centimeters of mercury	kg/sq meter	136.0
Centimeters of mercury	pounds/sq ft	27.85
Centimeters of mercury	pounds/sq in.	0.1934
Centimeters/sec	feet/min	1.1969
Centimeters/sec	feet/sec	0.03281
Centimeters/sec	kilometers/hr	0.036
Centimeters/sec	knots	0.1943
Centimeters/sec	meters/min	0.6
Centimeters/sec	miles/hr	0.02237
Centimeters/sec	miles/min	3.728×10^{-4}
Centimeters/sec/sec	feet/sec/sec	0.03281
Centimeters/sec/sec	km/hr/sec	0.036
Centimeters/sec/sec	meters/sec/sec	0.01
Centimeters/sec/sec	miles/hr/sec	0.02237
Centipoise	pascal-second (Pa s)	1.000 000*E–03
Centistokes	meter² per second (m²/s)	1.000 000*E–06
Chain	inches	792.00
Chain	meters	20.12
Chains (surveyors' or Gunter's)	yards	22.00
Circular mil	meter² (m²)	5.067 075E–10
Circular mils	sq cms	5.067×10^{-6}
Circular mils	sq mils	0.7854
Circular mils	sq inches	7.854×10^{-7}
Circumference	radians	6.283
Clo	Kelvin meter² per watt (K·m²/W)	2.003 712E–01
Cord feet	cubic feet	16
Cords	cord feet	8
Coulomb	statcoulombs	2.998×10^{9}
Coulombs	farads	1.036×10^{-5}
Coulombs/sq cm	coulombs/sq in.	64.52
Coulombs/sq cm	coulombs/sq meter	10^{4}
Coulombs/sq in.	coulombs/sq cm	0.155
Coulombs/sq in.	coulombs/sq meter	1,550.00
Coulombs/sq meter	coulombs/sq cm	10^{-4}
Coulombs/sq meter	coulombs/sq in.	6.452×10^{-4}
Cubic centimeters	cubic ft	3.531×10^{-5}
Cubic centimeters	cubic in.	0.06102
Cubic centimeters	cubic meters	10^{-6}
Cubic centimeters	cubic yards	1.308×10^{-6}
Cubic centimeters	gallons (U.S. liq.)	2.642×10^{-4}
Cubic centimeters	liters	0.001

(Continued)

TABLE A.5b (Continued)
Multipliers to Convert Engineering Units

To Convert	Into	Multiply by
Cubic centimeters	pints (U.S. liq.)	2.113×10^{-3}
Cubic centimeters	quarts (U.S. liq.)	1.057×10^{-3}
Cubic feet	bushels (dry)	0.8036
Cubic feet	cubic cm	28,320.00
Cubic feet	cubic in.	1,728.00
Cubic feet	cubic meters	0.02832
Cubic feet	cubic yards	0.03704
Cubic feet	gallons (U.S. liq.)	7.48052
Cubic feet	liters	28.32
Cubic feet	pints (U.S. liq.)	59.84
Cubic feet	quarts (U.S. liq.)	29.92
Cubic feet/min	cubic cm/sec	472.0
Cubic feet/min	gallons/sec	0.1247
Cubic feet/min	liters/sec	0.472
Cubic feet/min	pounds of water/min	62.43
Cubic feet/sec	million gallons/day	0.646317
Cubic feet/sec	gallons/min	448.831
Cubic inches	cubic cm	16.39
Cubic inches	cubic feet	5.787×10^{-4}
Cubic inches	cubic meters	1.639×10^{-5}
Cubic inches	cubic yards	2.143×10^{-5}
Cubic inches	gallons	4.329×10^{-3}
Cubic inches	liters	0.01639
Cubic inches	mil-feet	1.061×10^{5}
Cubic inches	pints (U.S. liq.)	0.03463
Cubic inches	quarts (U.S. liq.)	0.01732
Cubic meters	bushels (dry)	28.38
Cubic meters	cubic cm	10^{6}
Cubic meters	cubic ft	35.31
Cubic meters	cubic in.	61,023.00
Cubic meters	cubic yards	1.308
Cubic meters	gallons (U.S. liq.)	264.2
Cubic meters	liters	1,000.00
Cubic meters	pints (U.S. liq.)	2,113.00
Cubic meters	quarts (U.S. liq.)	1,057.00
Cubic yards	cubic cms	7.646×10^{5}
Cubic yards	cubic ft	27.0
Cubic yards	cubic in.	46,656.00
Cubic yards	cubic meters	0.7646
Cubic yards	gallons (U.S. liq.)	202.0
Cubic yards	liters	764.6
Cubic yards	pints (U.S. liq.)	1,615.90
Cubic yards	quarts (U.S. liq.)	807.9
Cubic yards/min	cubic ft/sec	0.45
Cubic yards/min	gallons/sec	3.367
Cubic yards/min	liters/sec	12.74
Cup	meter^3 (m^3)	2.365 882E−04
Curie	Becquerel (Bq)	3.700 000*E+10
D		
°F·h·ft^2/Btu (International Table) (R, thermal resistance)	Kelvin meter^2 per watt (K·m^2/W)	1.761 102E−01
°F·h·ft^2/Btu (thermochemical) (R, thermal resistance)	Kelvin meter^2 per watt (K·m^2/W)	1.762 280E−01
Dalton	gram	1.650×10^{-24}
Day (mean solar)	second (s)	8.640 000E+04
Days	seconds	86,400.00
Decigrams	grams	0.1
Deciliters	liters	0.1
Decimeters	meters	0.1
Degree (angle)	radian (rad)	1.745 329E−02
Degree Celsius day (sidereal)	Kelvin (K) second (s)	8.616 409E+04
Degree centigrade	(See footnote c.)	$t_K = t_{°C} + 273.15$
Degree fahrenheit	degree Celsius	$t_{°C} = (t_{°F} − 32)/1.8$
Degree fahrenheit	Kelvin (K)	$t_K = (t_{°F} + 459.67)/1.8$
Degree rankine	Kelvin (K)	$t_K = t_{°R}/1.8$
Degrees (angle)	quadrants	0.01111
Degrees (angle)	radians	0.01745
Degrees (angle)	seconds	3,600.00
Degrees/sec	radians/sec	0.01745
Degrees/sec	revolutions/min	0.1667
Degrees/sec	revolutions/sec	2.778×10^{-3}
Dekagrams	grams	10.0
Dekaliters	liters	10.0
Dekameters	meters	10.0
Denier	kilogram per meter (kg/m)	1.111 111E−07
Drams	grams	1.7718
Drams	grains	27.3437
Drams	ounces	0.0625
Drams (apothecaries' or troy)	ounces (avoirdupois)	0.1371429
Drams (apothecaries' or troy)	ounces (troy)	0.125
Drams (U.S., fluid or apothecaries)	cubic cm	3.697
Dyne		1.000 000*E−05
Dyne/cm	erg/sq millimeter	0.01
Dyne/cm	newton (N)	1.000 000*E−07
Dyne/cm^2	newton-meter (N·m) pascal (Pa)	1.000 000*E−01
Dyne/sq cm	atmospheres	9.869×10^{-7}
Dyne/sq cm	inches of mercury at 0°C	2.953×10^{-5}
Dyne/sq cm	inches of water at 4°C	4.015×10^{-4}
Dynes	grams	1.020×10^{-3}
Dynes	joules/cm	10^{-7}

c The SI unit of thermodynamic temperature is the kelvin (K), and this unit is properly used for expressing thermodynamic temperature and temperature intervals. Wide use is also made of the degree Celsius (°C), which is the SI unit for expressing Celsius temperature and temperature intervals. The Celsius scale (formerly called centigrade) is related directly to thermodynamic temperature (kelvins) as follows:

1. The temperature interval one degree Celsius equals one kelvin exactly.
2. Celsius temperature (t) is related to thermodynamic temperature (T) by the equation $t = T − T_0$, where $T_0 = 273.15$ K by definition.

(Continued)

TABLE A.5b (Continued)
Multipliers to Convert Engineering Units

To Convert	Into	Multiply by
Dynes	joules/meter (newtons)	10^{-5}
Dynes	kilograms	1.020×10^{-6}
Dynes	poundals	7.233×10^{-5}
Dynes	pounds	2.248×10^{-6}
Dynes/sq cm	bars	10^{-6}
E		
Electronvolt	joule (J)	1.602 19E–19
Ell	cm	114.3
Ell	in.	45
Em, pica	in.	0.167
Em, pica	cm	0.4233
EMU of capacitance	farad (F)	1.000 000*E+09
EMU of current	ampere (A)	1.000 000*E+01
EMU of electric potential	volt (V)	1.000 000*E–08
EMU of inductance	henry (H)	1.000 000*E–09
EMU of resistance	ohm (Ω)	1.000 000*E–09
Erg	joule (J)	1.000 000*E–07
Erg	joule (J)	1.000 000*E–07
Erg/(cm² · s)	watt per meter² (W/m²)	1.000 000*E–03
Erg/(cm² · s)	watt per meter² (W/m²)	1.000 000*E–03
Erg/s	watt (W)	1.000 000*E–07
Erg/s	watt (W)	1.000 000*E–07
Erg/sec	dyne-cm/sec	1.000
Ergs	Btu	9.480×10^{-11}
Ergs	dyne-centimeters	1.0
Ergs	foot-pounds	7.367×10^{-8}
Ergs	gram-calories	0.2389×10^{-7}
Ergs	gram-cm	1.020×10^{-3}
Ergs	horsepower-hr	3.7250×10^{-14}
Ergs	joules	10^{-7}
Ergs	kg-calories	2.389×10^{-11}
Ergs	kg-meters	1.020×10^{-8}
Ergs	kilowatt-hr	0.2778×10^{-13}
Ergs	watt-hours	0.2778×10^{-10}
Ergs/sec	Btu/min	$5,688 \times 10^{-9}$
Ergs/sec	ft-lb/min	4.427×10^{-6}
Ergs/sec	ft-lb/sec	7.3756×10^{-8}
Ergs/sec	horsepower	1.341×10^{-10}
Ergs/sec	kg-calories/min	1.433×10^{-9}
Ergs/sec	kilowatts	10^{-10}
ESU of capacitance	farad (F)	1.112 650E–12
ESU of current	ampere (A)	3.335 6E–10
ESU of electric potential	volt (V)	2.997 9E+02
ESU of inductance	henry (H)	8.987 554E+11
ESU of resistance	ohm (Ω)	8.987 554E+11
F		
Farad/sec	ampere (absolute)	9.6500×10^4
Faraday (based on carbon-12)	coulomb (C)	9.648 70E+04
Faraday (chemical)	coulomb (C)	9.649 57E+04

To Convert	Into	Multiply by
Faraday (physical)	coulomb (C)	9.652 19E+04
Farads	microfarads	10^6
Farads	ampere-hours	26.8
Farads	coulombs	9.649×10^4
Fathom	meter	1.828804
Fathom	meter (m)	1.828 8E+00
Fathoms	feet	6.0
Feet	centimeters	30.48
Feet	kilometers	3.048×10^{-4}
Feet	meters	0.3048
Feet	miles (naut.)	1.645×10^{-4}
Feet	miles (stat.)	1.894×10^{-4}
Feet	millimeters	304.8
Feet	mils	1.2×10^4
Feet of water	atmospheres	0.0295
Feet of water	in. of mercury	0.8826
Feet of water	kg/sq cm	0.03048
Feet of water	kg/sq meter	304.8
Feet of water	pounds/sq ft	62.43
Feet of water	pounds/sq in.	0.4335
Feet/100 feet	percent grade	1.0
Feet/min	cm/sec	0.508
Feet/min	ft/sec	0.01667
Feet/min	km/hr	0.01829
Feet/min	meters/min	0.3048
Feet/min	miles/hr	0.01136
Feet/sec	cm/sec	30.48
Feet/sec	km/hr	1.097
Feet/sec	knots	0.5921
Feet/sec	meters/min	18.29
Feet/sec	miles/hr	0.6818
Feet/sec	miles/min	0.01136
Feet/sec/sec	cm/sec/sec	30.48
Feet/sec/sec	km/hr/sec	1.097
Feet/sec/sec	meters/sec/sec	0.3048
Feet/sec/sec	miles/hr/sec	0.6818
Fermi (femtometer)	meter (m)	1.000 000*E–15
Fluid ounce (U.S.)	meter³ (m³)	2.957 353E–05
Foot	meter (m)	3.048 000*E–01
Foot (U.S. survey)[a]	meter (m)	3.048 006E–01
Foot of water (39.2°F)	pascal (Pa)	2.988 98E+03
Foot of water (39.2°F)	pascal (Pa)	2.988 98E+03
Foot/pounds/sec	Btu/min	0.07717
Footcandle	lux (lx)	1.076 391E+01
Foot-candle	lumen/sq meter	10.764
Footlambert	candela per meter² (cd/m²)	3.426 259E+00
Foot-pounds	Btu	1.286×10^{-3}
Foot-pounds	ergs	1.356×10^7

[a] Since 1893, the U.S. basis of length measurement has been derived from metric standards. In 1959, a small refinement was made in the definition of the yard to resolve discrepancies both in this country and abroad, which changed its length from 3600/3937 m to 0.9144 m exactly. This resulted in the new value being shorter by two parts in a million.

(Continued)

TABLE A.5b (Continued)
Multipliers to Convert Engineering Units

To Convert	Into	Multiply by	To Convert	Into	Multiply by
Foot-pounds	gram-calories	0.3238	Gallon (U.S. liquid) per hp·h(SFC, specific fuel consumption)	meter^3 per joul.e (m^3/J)	1.410 089E–09
Foot-pounds	hp-hr	5.050 × 10^{-7}			
Foot-pounds	joules	1.356			
Foot-pounds	kg-calories	3.24 × 10^{-4}	Gallon (U.S. liquid) per minute	meter^3 per second (m^3/s)	6.309 020E–05
Foot-pounds	kg-meters	0.1383			
Foot-pounds	kilowatt-hr	3.766 × 10^{-7}	Gallons	cubic cm	3,785.00
Foot-pounds/min	Btu/min	1.286 × 10^{-3}	Gallons	cubic ft	0.1337
Foot-pounds/min	foot-pounds/sec	0.01667	Gallons	cubic in.	231.0
Foot-pounds/min	horsepower	3.030 × 10^{-5}	Gallons	cubic meters	3.785 × 10^{-3}
Foot-pounds/min	kg-calories/min	3.24 × 10^{-4}	Gallons	cubic yards	4.951 × 10^{-3}
Foot-pounds/min	kilowatt	2.260 × 10^{-5}	Gallons	liters	3.785
Foot-pounds/sec	Btu/hr	4.6263	Gallons (liq. Br. Imp.)	gallons (U.S. liq.)	1.20095
Foot-pounds/sec	horsepower	1.818 × 10^{-3}	Gallons (U.S.)	gallons (Imp.)	0.83267
Foot-pounds/sec	kg-calories/min	0.01945	Gallons of water	pounds of water	8.3453
Foot-pounds/sec	kilowatts	1.356 × 10^{-3}	Gallons/min	cubic ft/sec	2.228 × 10^{-3}
Free fall, standard (g)	meter per second^2 (m/s^2)	9.806 650*E+00	Gallons/min	liters/sec	0.06308
Ft·lbf	joule (J)	1.355 818E+00	Gallons/min	cubic ft/hr	8.0208
Ft·lbf/h	watt (W)	3.766 161E–04	Gamma	tesla (T)	1.000 000*E–09
Ft·lbf/min	watt (W)	2.259 697E–02	Gauss	lines/sq in.	6.452
Ft·lbf/s	watt (W)	1.355 818E+00	Gauss	webers/sq cm	10^{-8}
Ft·poundal	joule (J)	4.214 011E–02	Gauss	webers/sq in.	6.452 × 10^{-8}
Ft/h	meter per second (m/s)	8.466 667E–05	Gauss	webers/sq meter	10^{-4}
Ft/min	meter per second (m/s)	5.080 000*E–03	Gauss	tesla (T)	1.000 000*E–04
Ft/s	meter per second (m/s)	3.048 000*E–01	Gilbert	ampere (A)	7.957 747E–01
Ft/s^2	meter per second^2 (m/s^2)	3.048 000*E–01	Gilberts	ampere-turns	0.7958
Ft^2	meter^2 (m^2)	9.290 304*E–02	Gilberts/cm	amp-turns/cm	0.7958
Ft^2	meter^2 (m^2)	9.290 304*E–02	Gilberts/cm	amp-turns/in.	2.021
Ft^2/h (thermal diffusivity)	meter^2 per second (m^2/s)	2.580 640*E–05	Gilberts/cm	amp-turns/meter	79.58
Ft^2/h (thermal diffusivity)	meter^2 per second (m^2/s)	2.580 640*E–05	Gill (U.K.)	meter^3 (m^3)	1.420 654E–04
Ft^2/s	meter^2 per second (m^2/s)	9.290 304*E–02	Gill (U.S.)	meter^3 (m^3)	1.182 941E–04
Ft^2/s	meter^2 per second (m^2/s)	9.290 304*E–02	Gills	liters	0.1183
Ft^3 (volume; section modulus)	meter^3 (m^3)	2.831 685E–02	Gills	pints (liq.)	0.25
			Gills (British)	cubic cm	142.07
Ft^3 (volume; section modulus)	meter^3 (m^3)	2.831 685E–02	Grad	degree (angular)	9.000 000*E–01
			Grad	radian (rad)	1.570 796E–02
Ft^3/min	meter^3 per second (m^3/s)	4.719 474E–04	Grade	radian	0.01571
Ft^3/s	meter^3 per second (m^3/s)	2.831 685E–02	Grain (1/7000 lb avoirdupois)	kilogram (kg)	6.479 891*E–05
Ft^4 (moment of section)^d	meter^4 (m^4)	8.630 975E–03			
Furlongs	miles (U.S.)	0.125	Grain (lb avoirdupois/7000)/gal (U.S. liquid)	kilogram per meter^3 (kg/m^3)	1.711 806E–02
Furlongs	rods	40.0			
Furlongs	feet	660.0			
G			Grains	drams (avoirdupois)	0.03657143
G/cm^3	kilogram per meter^3 (kg/m^3)	1.000 000*E+03	Grains (troy)	grains (avdp.)	1.0
			Grains (troy)	grams	0.0648
Gal	meter per second^2 (m/s^2)	1.000 000*E–02	Grains (troy)	ounces (avdp.)	2.0833 × 10^{-3}
Gallon (Canadian liquid)	meter^3 (m^3)	4.546 090E–03	Grains (troy)	pennyweight (troy)	0.04167
Gallon (U.K. liquid)	meter^3 (m^3)	4.546 092E–03	Grains/Imp. Gal	parts/million	14.286
Gallon (U.S. dry)	meter^3 (m^3)	4.404 884E–03	Grains/U.S. gal	parts/million	17.118
Gallon (U.S. liquid)	meter^3 (m^3)	3.785 412E–03	Grains/U.S. gal	pounds/million gal	142.86
Gallon (U.S. liquid) per day	meter^3 per second (m^3/s)	4.381 264E–08	Gram	kilogram (kg)	1.000 000*E–03

^{d} This is sometimes called the moment of inertia of a plane section about a specified axis.

(Continued)

TABLE A.5b (Continued)
Multipliers to Convert Engineering Units

To Convert	Into	Multiply by	To Convert	Into	Multiply by
Gram-calories	Btu	3.9683×10^{-3}	Horsepower	kilowatts	0.7457
Gram-calories	ergs	4.1868×10^{7}	Horsepower	watts	745.7
Gram-calories	foot-pounds	3.088	Horsepower (550 ft·lbf/s)	watt (W)	7.456 999E+02
Gram-calories	horsepower-hr	1.5596×10^{-6}	Horsepower (550 ft lb/sec)	horsepower (metric) (542.5 ft-lb/sec)	1.014
Gram-calories	kilowatt-hr	1.1630×10^{-6}	Horsepower (boiler)	Btu/hr	33.479
Gram-calories	watt-hr	1.1630×10^{-3}	Horsepower (boiler)	kilowatts	9.803
Gram-calories/sec	Btu/hr	14.286	Horsepower (boiler)	watt (W)	9.809 50E+03
Gram-centimeters	Btu	9.297×10^{-8}	Horsepower (electric)	watt (W)	7.460 000*E+02
Gram-centimeters	ergs	980.7	Horsepower (metric)	watt (W)	7.354 99E+02
Gram-centimeters	joules	9.807×10^{-5}	Horsepower (metric) (542.5 ft lb/sec)	horsepower (550 ft-lb/sec)	0.9863
Gram-centimeters	kg-cal	2.343×10^{-8}	Horsepower (U.K.)	Watt (W)	7.457 0E+02
Gram-centimeters	kg-meters	10^{-5}	Horsepower (water)	Watt (W)	7.460 43E+02
Gram-force/cm^2	pascal (Pa)	9.806 650*E+01	Horsepower-hr	Btu	2,547.00
Grams	dynes	980.7	Horsepower-hr	ergs	2.6845×10^{13}
Grams	grains	15.43	Horsepower-hr	ft-lb	1.98×10^{6}
Grams	joules/cm	9.807×10^{-5}	Horsepower-hr	gram-calories	641,190.00
Grams	joules/meter (newtons)	9.807×10^{-3}	Horsepower-hr	joules	2.684×10^{6}
Grams	kilograms	0.001	Horsepower-hr	kg-calories	641.1
Grams	milligrams	1,000.00	Horsepower-hr	kg-meters	2.737×10^{5}
Grams	ounces (advp.)	0.03527	Horsepower-hr	kilowatt-hr	0.7457
Grams	ounces (troy)	0.03215	Hour (mean solar)	second (s)	3.600 000E+03
Grams	poundals	0.07093	Hour (sidereal)	second (s)	3.590 170E+03
Grams	pounds	2.205×10^{-3}	Hours	days	4.167×10^{-2}
Grams/cm	pounds/inch	5.600×10^{-3}	Hours	weeks	5.952×10^{-3}
Grams/cubic cm	pounds/ cubic ft	62.43	Hundredweight (long)	kilogram (kg)	5.080 235E+01
Grams/cubic cm	pounds/cubic in.	0.03613	Hundredweight (short)	kilogram (kg)	4.535 924E+01
Grams/cubic cm	pounds/mil-foot	3.405×10^{-7}	Hundredweights (long)	pounds	112
Grams/liter	grains/gal	58.417	Hundredweights (long)	tons (long)	0.05
Grams/liter	pounds/1,000 gal	8.345	Hundredweights (short)	ounces (avoirdupois)	1,600
Grams/liter	pounds/cubic ft	0.062427	Hundredweights (short)	pounds	100
Grams/liter	parts/million	1,000.00	Hundredweights (short)	tons (metric)	0.0453592
Grams/sq cm	pounds/sq ft	2.0481	Hundredweights (short)	tons (long)	0.0446429
H			**I**		
Hands	centimeters	10.16			
Hectare	meter^2 (m^2)	1.000 000*E+04	In./s	meter per second (m/s)	2.540 000*E–02
Hectares	acres	2.471	In./s^2	meter per second^2 (m/s^2)	2.540 000*E–02
Hectares	sq feet	1.076×10^{5}	In.^{2}	meter^2 (m^2)	6.451 600*E–04
Hectograms	grams	100.0	In.^{3} (volume; section modulus)^e	meter^3 (m^3)	1.638 706 E–05
Hectoliters	liters	100.0			
Hectometers	meters	100.0	In.^{3}/min	meter^3 per second (m^3/s)	2.731 177E–07
Hectowatts	watts	100.0	In.^{4} (moment of section)^4	meter^4 (m^4)	4.162 314E–07
Henries	millihenries	1,000.00	Inch	meter (m)	2.540 000*E–02
Hogsheads (British)	cubic ft	10.114	Inch of mercury (32°F)	pascal (Pa)	3.386 38E+03
Hogsheads (U.S.)	cubic ft	8.42184	Inch of mercury (60°F)	pascal (Pa)	3.376 85 E+03
Hogsheads (U.S.)	gallons (U.S.)	63	Inch of water (39.2°F)	pascal (Pa)	2.490 82 E+03
Horsepower	Btu/min	42.44	Inch of water (60°F)	pascal (Pa)	2.488 4 E+02
Horsepower	ft-lb/min	33,000.00	Inches	centimeters	2.540
Horsepower	ft-lb/sec	550	Inches	meters	2.540×10^{-2}
Horsepower	kg-calories/min	10.68	Inches	miles	1.578×10^{-5}

^{e} The exact conversion factor is 1.638 706 4*E–05.

(Continued)

TABLE A.5b (Continued)

Multipliers to Convert Engineering Units

To Convert	Into	Multiply by
Inches	millimeters	25.40
Inches	mils	1,000.00
Inches	yards	2.778×10^{-2}
Inches of mercury	atmospheres	0.03342
Inches of mercury	feet of water	1.133
Inches of mercury	kg/sq cm	0.03453
Inches of mercury	kg/sq meter	345.3
Inches of mercury	pounds/sq ft	70.73
Inches of mercury	pounds/sq in.	0.4912
Inches of water (at 4°C)	atmospheres	2.458×10^{-3}
Inches of water (at 4°C)	inches of mercury	0.07355
Inches of water (at 4°C)	kg/sq cm	2.540×10^{-3}
Inches of water (at 4°C)	ounces/sq in.	0.5781
Inches of water (at 4°C)	pounds/sq ft	5.204
Inches of water (at 4°C)	pounds/sq in.	0.03613
International ampere	ampere (absolute)	0.9998
International volt	volts (absolute)	1.0003

J

To Convert	Into	Multiply by
Joules	Btu	9.480×10^{-4}
Joules	ergs	10^7
Joules	foot-pounds	0.7376
Joules	kg-calories	2.389×10^{-4}
Joules	kg-meters	0.102
Joules	watt-hr	2.778×10^{-4}
Joules	Btu	9.480×10^{-4}
Joules	ergs	10^7
Joules	foot-pounds	0.7376
Joules	kg-calories	2.389×10^{-4}
Joules	kg-meters	0.102
Joules	watt-hr	2.778×10^{-4}
Joules/cm	grams	1.020×10^4
Joules/cm	dynes	10^7
Joules/cm	joules/meter (newtons)	100.0
Joules/cm	poundals	723.3
Joules/cm	pounds	22.48
Joules/cm	grams	1.020×10^4
Joules/cm	dynes	10^7
Joules/cm	joules/meter (newtons)	100.0

K

To Convert	Into	Multiply by
Kayser	1 per meter (1/m)	1.000 000*E+02
Kelvin	degree Celsius	$t°_C \cdot t_K - 273.15$
Kgf·m	newton-meter (N·m)	9.806 650*E+00
Kgf·s²/m (mass)	kilogram (kg)	9.806 650*E+00
Kgf/cm²	pascal (Pa)	9.806 650*E+04
Kgf/m²	pascal (Pa)	9.806 650*E+00
Kgf/mm²	pascal (Pa)	9.806 650*E+06
Kilocalorie (international table)	joule (J)	4.186 800*E+03
Kilocalorie (mean)	joule (J)	4.190 02E+03
Kilocalorie (thermochemical)	joule (J)	4.184 000*E+03
Kilocalorie (thermochemical)/min	watt (W)	6.973 333E+01
Kilocalorie (thermochemical)/s	watt (W)	4.184 000*E+03
Kilogram meters	Btu	9.294×10^{-3}
Kilogram meters	ergs	9.804×10^7
Kilogram meters	foot-pounds	7.233
Kilogram meters	joules	9.804
Kilogram meters	kg-calories	2.342×10^{-3}
Kilogram meters	kilowatt-hr	2.723×10^{-6}
Kilogram-calories	Btu	3.968
Kilogram-calories	foot-pounds	3,088.00
Kilogram-calories	hp-hr	1.560×10^{-3}
Kilogram-calories	joules	4,186.00
Kilogram-calories	kg-meters	426.9
Kilogram-calories	kilojoules	4.186
Kilogram-calories	kilowatt-hr	1.163×10^{-3}
Kilogram-force (kgf)	newton (N)	9.806 650*E+00
Kilograms	dynes	980,665.00
Kilograms	grams	1,000.00
Kilograms	joules/cm	0.09807
Kilograms	joules/meter (newtons)	9.807
Kilograms	poundals	70.93
Kilograms	pounds	2.205
Kilograms	tons (long)	9.842×10^{-4}
Kilograms	tons (short)	1.102×10^{-3}
Kilograms/cubic meter	grams/cubic cm	0.001
Kilograms/cubic meter	pounds/cubic ft	0.06243
Kilograms/cubic meter	pounds/cubic in.	3.613×10^{-5}
Kilograms/cubic meter	pounds/mil-foot	3.405×10^{-10}
Kilograms/meter	pounds/ft	0.672
Kilograms/sq cm	dynes	980,665
Kilograms/sq cm	atmospheres	0.9678
Kilograms/sq cm	feet of water	32.81
Kilograms/sq cm	inches of mercury	28.96
Kilograms/sq cm	pounds/sq ft	2,048.00
Kilograms/sq cm	pounds/sq in.	14.22
Kilograms/sq meter	atmospheres	9.678×10^{-5}
Kilograms/sq meter	bars	98.07×10^{-6}
Kilograms/sq meter	feet of water	3.281×10^{-3}
Kilograms/sq meter	inches of mercury	2.896×10^{-3}
Kilograms/sq meter	pounds/sq ft	0.2048
Kilograms/sq meter	pounds/sq in.	1.422×10^{-3}
Kilograms/sq mm	kg/sq meter	10^6
Kilolines	maxwells	1,000.00
Kiloliters	liters	1,000.00
Kilometers	centimeters	10^5
Kilometers	feet	3,281.00
Kilometers	inches	3.937×10^4
Kilometers	meters	1,000.00
Kilometers	miles	0.6214
Kilometers	millimeters	10^6
Kilometers	yards	1,094.00

(Continued)

TABLE A.5b (Continued)
Multipliers to Convert Engineering Units

To Convert	Into	Multiply by	To Convert	Into	Multiply by
Kilometers/hr	cm/sec	27.78	Lb/ft·h	pascal-second (Pa·s)	4.133 789E–04
Kilometers/hr	ft/min	54.68	Lb/ft·s	pascal-second (Pa·s)	1.488 164E+00
Kilometers/hr	ft/sec	0.9113	Lb/ft^2	kilogram per meter^2 (kg/m^2)	4.882 428E+00
Kilometers/hr	knots	0.5396			
Kilometers/hr	meters/min	16.67	Lb/ft^3	kilogram per meter^3 (kg/m^3)	1.601 846E+01
Kilometers/hr	miles/hr	0.6214			
Kilometers/hr/sec	cm/sec/sec	27.78	Lb/gal (U.K. liquid)	kilogram per meter^3 (kg/m^3)	9.977 633E+01
Kilometers/hr/sec	ft/sec/sec	0.9113			
Kilometers/hr/sec	meters/sec/sec	0.2778	Lb/gal (U.S. liquid)	kilogram per meter^3 (kg/m^3)	1.198 264E+02
Kilometers/hr/sec	miles/hr/sec	0.6214			
Kilopond	Newton (N)	9.806 650*E+00	Lb/h	kilogram per second (kg/s)	1.259 979E–04
Kilowatt-hr	Btu	3,413.00			
Kilowatt-hr	ergs	3.600 × 10^{13}	Lb/hp·h (SFC, specific fuel consumption)	kilogram per joule (kg/J)	1.689 659E–07
Kilowatt-hr	ft-lb	2.655 × 10^6			
Kilowatt-hr	gram-calories	859,850.00	Lb/in.^{3}	kilogram per meter^3 (kg/m^3)	2.767 990E+04
Kilowatt-hr	horsepower-hr	1,341			
Kilowatt-hr	joules	3.6 × 10^6	Lb/min	kilogram per second (kg/s)	7.559 873E–03
Kilowatt-hr	kg-calories	860.5			
Kilowatt-hr	kg-meters	3.671 × 10^5	Lb/s	kilogram per second (kg/s)	4.535 924E–01
Kilowatt-hr	pounds of water evaporated from and at 212°F	3.5.3	Lb/yd^3	kilogram per meter^3 (kg/m^3)	5.932 764E–01
Kilowatt-hr	pounds of water raised from 62° to 212°F	22.75	Lbf·ft	newton-meter (N·m)	1.355 818E+00
			Lbf·in.	newton-meter (N·m)	1.129 848 E–01
Kilowatts	Btu/min	56.92	Lbf·in./in.	newton-meter per meter (N·m/m)	4.448 222 E+01
Kilowatts	ft-lb/min	4.426 × 10^4			
Kilowatts	ft-lb/sec	737.6	Lbf·s/ft^2	pascal-second (Pa·s)	4.788 026E+01
Kilowatts	horsepower	1.341	Lbf·s/in.^{2}	pascal-second (Pa·s)	6.894 757E+03
Kilowatts	kg-calories/min	14.34	Lbf/ft	newton per meter (N/m)	1.459 390E+01
Kilowatts	watts	1,000.00	Lbf/ft^2	pascal (Pa)	4.788 026E+01
Kip (1000 lbf)	newton (N)	4.448 222E+03	Lbf/in.	newton per meter (N/m)	1.751 268E+02
Kip/in^2 (ksi)	pascal (Pa)	6.894 757E+06	Lbf/in.^{2} (psi)	pascal (Pa)	6.894 757E+03
Km/h	meter per second (m/s)	2.777 778E–01	Lbf/lb (thrust/weight [mass] ratio)	newton per kilogram (N/kg)	9.806 650E+00
Knot (international)	meter per second (m/s)	5.144 444E–01			
Knots	ft/hr	6,080.00	Lbf·ft/in.	newton-meter per meter (N·m/m)	5.337 866 E+01
Knots	kilometers/hr	1.8532			
Knots	nautical miles/hr	1.0	League	miles (approx.)	3.0
Knots	statute miles/hr	1.151	League	meter (m)	(See footnote a.)
Knots	yards/hr	2,027.00	Light year	miles	5.9 × 10^{12}
Knots	ft/sec	1.689	Light year	kilometers	9.46091 × 10^{12}
Kw·h	joule (J)	3.600 000*E+06	Light-year	meter (m)	9.460 55E+15
	L		Lines/sq cm	gauss	1.0
			Lines/sq in.	gauss	0.155
Lambert	candela per meter^2 (cd/m^2)	1/π*E+04	Lines/sq in.	webers/sq cm	1.550 × 10^{-9}
Lambert	candela per meter^2 (cd/m^2)	3.183 099E+03	Lines/sq in.	webers/sq in.	10^{-8}
Langley	joule per meter^2 (J/m^2)	4.184 000*E+04	Lines/sq in.	webers/sq meter	1.550 × 10^{-5}
Lb·ft^2 (moment of inertia)	kilogram meter^2 (kg·m^2)	4.214 011E–02	Links (engineer's)	inches	12.0
Lb·in.^{2} (moment of inertia)	kilogram meter^2 (kg·m^2)	2.926 397E–04	Links (surveyor's)	inches	7.92

[a] Since 1893, the U.S. basis of length measurement has been derived from metric standards. In 1959, a small refinement was made in the definition of the yard to resolve discrepancies both in this country and abroad, which changed its length from 3600/3937 m to 0.9144 m exactly. This resulted in the new value being shorter by two parts in a million.

(Continued)

TABLE A.5b (Continued)
Multipliers to Convert Engineering Units

To Convert	Into	Multiply by	To Convert	Into	Multiply by
Liter[f]	meter^3 (m^3)	1.000 000*E−03	Meters/sec/sec	cm/sec/sec	100.0
Liters	bushels (U.S. dry)	0.02833	Meters/sec/sec	ft/sec/sec	3.281
Liters	cubic cm	1,000.00	Meters/sec/sec	km/hr/sec	3.6
Liters	cubic ft	0.03531	Meters/sec/sec	miles/hr/sec	2.237
Liters	cubic in.	61.02	Mho	Siemens (S)	1.000 000*E+00
Liters	cubic meters	0.001	Mi/h (international)	meter per second (m/s)	4.470 400*E−01
Liters	cubic yards	1.308×10^{-3}	Mi/h (international)	kilometer per hour (km/h)	1.609 344*E+01
Liters	gallons (U.S. liq.)	0.2642			
Liters	pints (U.S. liq.)	2.113	Mi/min (international)	meter per second (m/s)	2.682 240*E+01
Liters	quarts (U.S. liq.)	1.057	Mi/s (international)	meter per second (m/s)	1.609 344*E+03
Liters/min	cubic ft/sec	5.886×10^{-4}	Mi^2 (international)	meter^2 (m^2)	2.589 988 E+06
Liters/min	gals/sec	4.403×10^{-3}	Mi^2 (U.S. survey)[a]	meter^2 (m^2)	2.589 998 E+06
Lumen	spherical candle power	0.07958	Microfarad	farads	10^{-6}
Lumen	watt	0.001496	Micrograms	grams	10^{-6}
Lumen/sq ft	lumen/sq meter	10.76	Microhms	megohms	10^{-12}
Lumens/sq ft	foot-candles	1.0	Microhms	ohms	10^{-6}
Lux	foot-candles	0.0929	Microinch	meter (m)	2.540 000*E−08
M			Microliters	liters	10^{-6}
Maxwell	Weber (Wb)	1.000 000*E−08	Micron	meter (m)	1.000 000*E−06
Maxwells	kilolines	0.001	Microns	meters	1×10^{-6}
Maxwells	webers	10^{-8}	Mil	meter (m)	2.540 000*E−05
Megalines	maxwells	10^6	Mile (international nautical)	meter (m)	1.852 000*E+03
Megohms	microhms	10^{12}	Mile (international)	meter (m)	1.609 344*E+03
Megohms	ohms	10^6	Mile (statute)	meter (m)	1.609 3E+03
Meter-kilograms	cm-dynes	9.807×10^7	Mile (U.K. nautical)	meter (m)	1.853 184*E+03
Meter-kilograms	cm-grams	10^5	Mile (U.S. nautical)	meter (m)	1.852 000*E+03
Meter-kilograms	pound-feet	7.233	Mile (U.S. survey)[a]	meter (m)	1.609 347 E+03
Meters	centimeters	100.0	Miles (naut.)	miles (statute)	1.1516
Meters	feet	3.281	Miles (naut.)	yards	2,027.00
Meters	inches	39.37	Miles (naut.)	feet	6,080.27
Meters	kilometers	0.001	Miles (naut.)	kilometers	1.853
Meters	miles (naut.)	5.396×10^{-4}	Miles (naut.)	meters	1,853.00
Meters	miles (stat.)	6.214×10^{-4}	Miles (statute)	centimeters	1.609×10^5
Meters	millimeters	1,000.00	Miles (statute)	feet	5,280.00
Meters	yards	1.094	Miles (statute)	inches	6.336×10^4
Meters	varas	1.179	Miles (statute)	kilometers	1.609
Meters/min	cm/sec	1.667	Miles (statute)	meters	1,609.00
Meters/min	ft/min	3.281	Miles (statute)	miles (naut.)	0.8684
Meters/min	ft/sec	0.05468	Miles (statute)	yards	1,760.00
Meters/min	km/hr	0.06	Miles/hr	cm/sec	44.7
Meters/min	knots	0.03238	Miles/hr	ft/min	88.0
Meters/min	miles/hr	0.03728	Miles/hr	ft/sec	1.467
Meters/sec	ft/min	196.8	Miles/hr	km/hr	1.609
Meters/sec	ft/sec	3.281	Miles/hr	km/min	0.02682
Meters/sec	kilometers/hr	3.6	Miles/hr	knots	0.8684
Meters/sec	kilometers/min	0.06	Miles/hr	meters/min	26.82
Meters/sec	miles/hr	2.237	Miles/hr	miles/min	0.1667
Meters/sec	miles/min	0.03728			

[a] Since 1893, the U.S. basis of length measurement has been derived from metric standards. In 1959, a small refinement was made in the definition of the yard to resolve discrepancies both in this country and abroad, which changed its length from 3600/3937 m to 0.9144 m exactly. This resulted in the new value being shorter by two parts in a million.

[f] In 1964, the General Conference on Weights and Measures adopted the name liter as a special name for decimeter. Prior to this decision, the liter differed slightly (previous value, 1.000028 dm^3), and in expression of precision volume measurement, this fact must be kept in mind.

(Continued)

TABLE A.5b (Continued)
Multipliers to Convert Engineering Units

To Convert	Into	Multiply by	To Convert	Into	Multiply by
Miles/hr/sec	cm/sec/sec	44.7	\multicolumn{3}{c}{*O*}	
Miles/hr/sec	ft/sec/sec	1.467	Oersted	ampere per meter (A/m)	7.957 747E+01
Miles/hr/sec	km/hr/sec	1.609	Ohm (International)	ohm (absolute)	1.0005
Miles/hr/sec	meters/sec/sec	0.447	Ohm centimeter	ohm-meter ($\Omega \cdot m$)	1.000 000*E−02
Miles/min	cm/sec	2,682.00	Ohm circular-mill per foot	ohm-millimeter^2 per meter ($\Omega \cdot mm^2/m$)	1.662 426E−03
Miles/min	ft/sec	88.0			
Miles/min	km/min	1.609	Ohms	megohms	10^{-6}
Miles/min	knots/min	0.8684	Ohms	microhms	10^6
Miles/min	miles/hr	60	Ounce (avoirdupois)	kilogram (kg)	2.834 952E−02
Mil-feet	cubic in.	9.425×10^{-6}	Ounce (troy or apothecary)	kilogram (kg)	3.110 348E−02
Millibar	pascal (Pa)	1.000 000*E+02	Ounce (U.K. fluid)	meter^3 (m^3)	2.841 307E−05
Milliers	kilograms	1,000.00	Ounce (U.S. fluid)	meter^3 (m^3)	2.957 353E−05
Milligrams	grains	0.01543236	Ounce/sq in.	dynes/sq cm	4309
Milligrams	grams	0.001	Ounce-force	newton (N)	2.780 139E−01
Milligrams/liter	part/million	1.0	Ounces	drams	16.0
Millihenries	henries	0.001	Ounces	grains	437.5
Milliliters	liters	0.001	Ounces	grams	28.349527
Millimeter of mercury (0°C)	pascal (Pa)	1.333 22E+02	Ounces	pounds	0.0625
Millimeters	centimeters	0.1	Ounces	ounces (troy)	0.9115
Millimeters	feet	3.281×10^{-3}	Ounces	tons (long)	2.790×10^{-5}
Millimeters	inches	0.03937	Ounces	tons (metric)	2.835×10^{-5}
Millimeters	kilometers	10^{-6}	Ounces (fluid)	cubic in.	1.805
Millimeters	meters	0.001	Ounces (fluid)	liters	0.02957
Millimeters	miles	6.214×10^{-7}	Ounces (troy)	grains	480.0
Millimeters	mils	39.37	Ounces (troy)	grams	31.103481
Millimeters	yards	1.094×10^{-3}	Ounces (troy)	ounces (avdp.)	1.09714
Millimicrons	meters	1×10^{-9}	Ounces (troy)	pennyweights (troy)	20.0
Million gals/day	cubic ft/sec	1.54723	Ounces (troy)	pounds (troy)	0.08333
Mils	centimeters	2.540×10^{-3}	Ounces/sq in.	pounds/sq in.	0.0625
Mils	feet	8.333×10^{-5}	Oz (avoirdupois)/ft^2	kilogram per meter^3 (kg/m^3)	3.051 517E−01
Mils	inches	0.001			
Mils	kilometers	2.540×10^{-8}	Oz (avoirdupois)/gal (U.K. liquid)	kilogram per meter^3 (kg/m^3)	6.236 021E+00
Mils	yards	2.778×10^{-5}			
Miner's inches	cubic ft/min	1.5	Oz (avoirdupois)/gal (U.S. liquid)	kilogram per meter^3 (kg/m^3)	7.489 152E+00
Minims (U.K.)	cubic cm	0.059192			
Minims (U.S., fluid)	cubic cm	0.061612	Oz (avoirdupois)/in^3	kilogram per meter^3 (kg/m^3)	1.729 994E+03
Minute (angle)	radian (rad)	2.908 882E−04			
Minute (mean solar)	second (s)	6.000 000E+01	Oz (avoirdupois)/yd^2	kilogram per meter^2 (kg/m^2)	3.390 575E−02
Minute (sidereal)	second (s)	5.983 617E+01			
Minutes (angles)	degrees	0.01667	Ozf · in.	Newton-meter (N · m)	7.061 552E−03
Minutes (angles)	quadrants	1.852×10^{-4}	\multicolumn{3}{c}{*P*}	
Minutes (angles)	radians	2.909×10^{-4}			
Minutes (angles)	seconds	60.0	Parsec	miles	19×10^{12}
Month (mean calendar)	second (s)	2.628 000E+06	Parsec	kilometers	3.084×10^{13}
Myriagrams	kilograms	10.0	Parsec	meter (m)	3.085 678E+16
Myriameters	kilometers	10.0	Parts/million	grains/U.S. gal	0.0584
Myriawatts	kilowatts	10.0	Parts/million	grains/Imp. gal	0.07016
\multicolumn{3}{c}{*N*}		Parts/million	pounds/million gal	8.345
Nepers	decibels	8.686	Peck (U.S.)	meter^3 (m^3)	8.809 768E−03
Newton	dynes	1×10^5	Pecks (U.K.)	cubic in.	554.6

(Continued)

TABLE A.5b (Continued)
Multipliers to Convert Engineering Units

To Convert	Into	Multiply by	To Convert	Into	Multiply by
Pecks (U.K.)	liters	9.091901	Poundals	grams	14.1
Pecks (U.S.)	bushels	0.25	Poundals	joules/cm	1.383×10^{-3}
Pecks (U.S.)	cubic in.	537.605	Poundals	joules/meter (newtons)	0.1383
Pecks (U.S.)	liters	8.809582	Poundals	kilograms	0.0141
Pecks (U.S.)	quarts (dry)	8	Poundals	pounds	0.03108
Pennyweight	kilogram (kg)	1.555 174E–03	Pound-feet	cm-dynes	1.356×10^7
Pennyweights (troy)	grains	24.0	Pound-feet	cm-grams	13,825.00
Pennyweights (troy)	ounces (troy)	0.05	Pound-feet	meter-kg	0.1383
Pennyweights (troy)	grams	1.55517	Pound-force (in.f)[h]	Newton (N)	4.448 222E+00
Pennyweights (troy)	pounds (troy)	4.1667×10^{-3}	Pounds	drams	256.0
Perm (0°C)	kilogram per pascal-second-meter^2 ($kg/Pa \cdot s \cdot m^2$)	5.721 35E–11	Pounds	dynes	44.4823×10^4
			Pounds	grains	7,000.00
			Pounds	grams	453.5924
Perm (23°C)	kilogram per pascal-second-meter^2 ($kg/Pa \cdot s \cdot m^2$)	5.745 25E–11	Pounds	joules/cm	0.04448
			Pounds	joules/meter (newtons)	4.448
			Pounds	kilograms	0.4536
	kilogram per pascal-second-meter ($kg/Pa \cdot s \cdot m$)	1.453 22E–12	Pounds	ounces	16.0
			Pounds	ounces (troy)	14.5833
			Pounds	poundals	32.17
Perm · in. (0°C)	kilogram per pascal-second-meter ($kg/Pa \cdot s \cdot m$)	1.453 22E–12	Pounds	pounds (troy)	1.21528
			Pounds	tons (short)	0.0005
			Pounds (avoirdupois)	ounces (troy)	14.5833
Perm · in. (23°C)	kilogram per pascal-second-meter ($kg/Pa \cdot s \cdot m$)	1.459 29E–12	Pounds (troy)	grains	5,760.00
			Pounds (troy)	grams	373.24177
			Pounds (troy)	ounces (avdp.)	13.1657
Phot	lumen per meter^2 (lm/m^2)	1.000 000*E+04	Pounds (troy)	ounces (troy)	12.0
			Pounds (troy)	pennyweights (troy)	240.0
Pica (printer's)	meter (m)	4.217 518E–03	Pounds (troy)	pounds (avdp.)	0.822857
Pint (U.S. dry)	meter^3 (m^3)	5.506 105E–04	Pounds (troy)	tons (long)	3.6735×10^{-4}
Pint (U.S. liquid)	meter^3 (m^3)	4.731 765E–04	Pounds (troy)	tons (metric)	3.7324×10^{-4}
Pints (dry)	cubic in.	33.6	Pounds (troy)	tons (short)	4.1143×10^{-4}
Pints (liq.)	cubic cm	473.2	Pounds of water	cubic ft	0.01602
Pints (liq.)	cubic ft	0.01671	Pounds of water	cubic in.	27.68
Pints (liq.)	cubic in.	28.87	Pounds of water	gallons	0.1198
Pints (liq.)	cubic meters	4.732×10^{-4}	Pounds of water/min	cubic ft/sec	2.670×10^{-4}
Pints (liq.)	cubic yards	6.189×10^{-4}	Pounds/cubic ft	grams/cubic cm	0.01602
Pints (liq.)	gallons	0.125	Pounds/cubic ft	kg/cubic meter	16.02
Pints (liq.)	liters	0.4732	Pounds/cubic ft	pounds/cubic in.	5.787×10^{-4}
Pints (liq.)	quarts (liq.)	0.5	Pounds/cubic ft	pounds/mil-foot	5.456×10^{-9}
Planck's quantum	erg/sec	6.624×10^{-27}	Pounds/cubic in.	gm/cubic cm	27.68
Point (printer's)	meter (m)	3.514 598*E–04	Pounds/cubic in.	kg/cubic meter	2.768×10^4
Poise	gram/cm sec	1.00	Pounds/cubic in.	pounds/cubic ft	1,728.00
Poise (absolute viscosity)	Pascal-second (Pa · s)	1.000 000*E–01	Pounds/cubic in.	pounds/mil-foot	9.425×10^{-6}
Pound (lb avoirdupois)[g]	kilogram (kg)	4.535 924E–01	Pounds/ft	kg/meter	1.488
Pound (troy or apothecary)	kilogram (kg)	3.732 417E–01	Pounds/in.	gm/cm	178.6
Poundal	newton (N)	1.382 550E–01	Pounds/mil-foot	gm/cubic cm	2.306×10^6
Poundal/ft^2	pascal (Pa)	1.488 164E+00	Pounds/sq ft	atmospheres	4.725×10^{-4}
Poundal · s/ft^2	pascal-second (Pa · s)	1.488 164E+00	Pounds/sq ft	feet of water	0.01602
Poundals	dynes	13,826.00			

[g] The exact conversion factor is 4.535 923 7*E–01.
[h] The exact conversion factor is 4.448 221 615 260 5*E+00.

(Continued)

TABLE A.5b (Continued)
Multipliers to Convert Engineering Units

To Convert	Into	Multiply by
Pounds/sq ft	inches of mercury	0.01414
Pounds/sq ft	kg/sq meter	4.882
Pounds/sq ft	pounds/sq in.	6.944×10^{-3}
Pounds/sq in	atmospheres	0.06804
Pounds/sq in.	feet of water	2.307
Pounds/sq in.	inches of mercury	2.036
Pounds/sq in.	kg/sq meter	703.1
Pounds/sq in.	pounds/sq ft	144.0

Q

To Convert	Into	Multiply by
Quadrants (angle)	degrees	90.0
Quadrants (angle)	minutes	5,400.00
Quadrants (angle)	radians	1.571
Quadrants (angle)	seconds	3.24×10^5
Quart (U.S. dry)	meter^3 (m^3)	1.101 221E–03
Quart (U.S. liquid)	meter^3 (m^3)	9.463 529E–04
Quarts (dry)	cubic in.	67.2
Quarts (liq.)	cubic cm	946.4
Quarts (liq.)	cubic ft	0.03342
Quarts (liq.)	cubic in.	57.75
Quarts (liq.)	cubic meters	9.464×10^{-4}
Quarts (liq.)	cubic yards	1.238×10^{-3}
Quarts (liq.)	gallons	0.25
Quarts (liq.)	liters	0.9463

R

To Convert	Into	Multiply by
Rad (radiation dose absorbed)	gray (Gy)	1.000 000*E–02
Radians	degrees	57.30
Radians	minutes	3,438.00
Radians	quadrants	0.6366
Radians	seconds	2.063×10^5
Radians/sec	degrees/sec	57.3
Radians/sec	rev/min	9.549
Radians/sec	rev/sec	0.1592
Radians/sec/sec	rev/min/min	573.0
Radians/sec/sec	rev/min/sec	9.549
Radians/sec/sec	rev/sec/sec	0.1592
Revolutions	degrees	360.0
Revolutions	quadrants	4.0
Revolutions	radians	6.283
Revolutions/min	degrees/sec	6.0
Revolutions/min	radians/sec	0.1047
Revolutions/min	rev/sec	0.01667
Revolutions/min/min	radians/sec/sec	1.745×10^{-3}
Revolutions/min/min	rev/min/sec	0.01667
Revolutions/min/min	rev/sec/sec	2.778×10^{-4}
Revolutions/sec	degrees/sec	360.0
Revolutions/sec	radians/sec	6.283
Revolutions/sec	rev/min	60.0
Revolutions/sec/sec	radians/sec/sec	6.283
Revolutions/sec/sec	rev/min/min	3,600.00
Revolutions/sec/sec	rev/min/sec	60.0
Rhe	1 per pascal-second (1/Pa·s)	1.000 000*E+01
Rod	chain (Gunters)	0.25
Rod	meters	5.029
Rod	meter (m)	(See footnote a.)
Rods	feet	16.5
Rods (surveyors' meas.)	yards	5.5
Roentgen	coulomb per kilogram (C/kg)	2.58E–04

S

To Convert	Into	Multiply by
Scruples	grains	20
Second (angle)	radian (rad)	4.848 137E–06
Second (sidereal)	second (s)	9.972 696E–01
Seconds (angle)	degrees	2.778×10^{-4}
Seconds (angle)	minutes	0.01667
Seconds (angle)	quadrants	3.087×10^{-6}
Seconds (angle)	radians	4.848×10^{-6}
Section	meter^2 (m^2)	(See footnote a.)
Shake	second (s)	1.000 000*E–08
Slug	kilogram	14.5
Slug	pounds	32.17
Slug	kilogram (kg)	1.459 390E+01
Slug/ft·s	Pascal-second (Pa·s)	4.788 026E+01
Slug/ft^3	kilogram per meter^3 (kg/m^3)	5.155 788E+02
Sphere	steradians	12.57
Square centimeters	circular mils	1.973×10^5
Square centimeters	sq ft	1.076×10^{-3}
Square centimeters	sq in.	0.155
Square centimeters	sq meters	0.0001
Square centimeters	sq miles	3.861×10^{-11}
Square centimeters	sq millimeters	100.0
Square centimeters	sq yards	1.196×10^{-4}
Square feet	acres	2.296×10^{-5}
Square feet	circular mils	1.833×10^8
Square feet	sq cm	929.0
Square feet	sq in.	144.0
Square feet	sq meters	0.0929
Square feet	sq miles	3.587×10^{-8}
Square feet	sq millimeters	9.290×10^4
Square feet	sq yards	0.1111
Square inches	circular mils	1.273×10^6
Square inches	sq cm	6.452
Square inches	sq ft	6.944×10^{-4}
Square inches	sq millimeters	645.2

---

[a] Since 1893, the U.S. basis of length measurement has been derived from metric standards. In 1959, a small refinement was made in the definition of the yard to resolve discrepancies both in this country and abroad, which changed its length from 3600/3937 m to 0.9144 m exactly. This resulted in the new value being shorter by two parts in a million.

(Continued)

TABLE A.5b (Continued)
Multipliers to Convert Engineering Units

To Convert	Into	Multiply by	To Convert	Into	Multiply by
Square inches	sq mils	10^6	\multicolumn{3}{c}{**T**}	
Square inches	sq yards	7.716×10^{-4}	Tablespoon	meter^3 (m^3)	1.478 676E–05
Square kilometers	acres	247.1	Teaspoon	meter^3 (m^3)	4.928 922E–06
Square kilometers	sq cm	10^{10}	Temperature (°C) + 17.78	temperature (°F)	1.8
Square kilometers	sq ft	10.76×10^6	Temperature (°C) + 273	absolute temperature (°K)	1.0
Square kilometers	sq in.	1.550×10^9	Temperature (°F) – 32	temperature (°C)	5/9
Square kilometers	sq meters	10^6	Temperature (°F) + 460	absolute temperature (°R)	1.0
Square kilometers	sq miles	0.3861	Tex	kilogram per meter (kg/m)	1.000 000*E–06
Square kilometers	sq yards	1.196×10^6			
Square meters	acres	2.471×10^{-4}	Therm	joule (J)	1.055 056E+08
Square meters	sq cm	10^4	Ton (assay)	kilogram (kg)	2.916 667E–02
Square meters	sq cm	10^4	Ton (long)/yd^3	kilogram per meter^3 (kg/m^3)	1.328 939E+03
Square meters	sq in.	1,550.00			
Square meters	sq miles	3.861×10^{-7}	Ton (long, 2240 lb)	kilogram (kg)	1.016 047E+03
Square meters	sq millimeters	10^6	Ton (metric)	kilogram (kg)	1.000 000*E+03
Square meters	sq yards	1.196	Ton (nuclear equivalent of TNT)	joule (J)	4.184E+09^i
Square miles	acres	640.0			
Square miles	sq ft	27.88×10^6	Ton (refrigeration)	watt (W)	3.516 800E+03
Square miles	sq km	2.59	Ton (register)	meter^3 (m^3)	2.831 685E+00
Square miles	sq meters	2.590×10^6	Ton (short)/h	kilogram per second (kg/s)	2.519 958E–01
Square miles	sq yards	3.098×10^6			
Square millimeters	circular mils	1,973.00	Ton (short)/yd^3	kilogram per meter^3 (kg/m^3)	1.186 553E+03
Square millimeters	sq cm	0.01			
Square millimeters	sq ft	1.076×10^{-5}	Ton (short, 2000 lb)	kilogram (kg)	9.071 847E+02
Square millimeters	sq in.	1.550×10^{-3}	Ton-force (2000 lbf)	Newton (N)	8.896 444E+03
Square mils	circular mils	1.273	Tonne	kilogram (kg)	1.000 000*E+03
Square mils	sq cm	6.452×10^{-6}	Tons (long)	kilograms	1,016.00
Square mils	sq in.	10^{-6}	Tons (long)	pounds	2,240.00
Square yards	acres	2.066×10^{-4}	Tons (long)	tons (short)	1.12
Square yards	sq cm	8,361.00	Tons (metric)	kilograms	1,000.00
Square yards	sq ft	9.0	Tons (metric)	pounds	2,205
Square yards	sq in.	1,296.00	Tons (short)	kilograms	907.1848
Square yards	sq meters	0.8361	Tons (short)	ounces	32,000.00
Square yards	sq miles	3.228×10^{-7}	Tons (short)	ounces (troy)	29,166.66
Square yards	sq millimeters	8.361×10^5	Tons (short)	pounds	2,000.00
Statampere	ampere (A)	3.335 640E–10	Tons (short)	pounds (troy)	2,430.56
Statcoulomb	coulomb (C)	3.335 640E–10	Tons (short)	tons (long)	0.89287
Statfarad	farad (F)	1.112 650E–12	Tons (short)	tons (metric)	0.9078
Stathenry	henry (H)	8.987 554E+11	Tons (short)/sq ft	kg/sq meter	9,765.00
Statmho	siemens (S)	1.112 650E–12	Tons (short)/sq ft	pounds/sq in.	2,000.00
Statohm	ohm (Ω)	8.987 665E+11	Tons of water/24 hr	pounds of water/hr	83.333
Statvolt	volt (V)	2.997 925E+02	Tons of water/24 hr	gallons/min	0.16643
Stere	meter^3 (m^3)	1.000 000*E–04	Tons of water/24 hr	cubic ft/hr	1.3349
Stilb	candela per meter^2 (cd/m^2)	1.000 000*E+04	Torr (mmhg, 0°C)	pascal (Pa)	1.333 22E+02
Stokes (kinematic viscosity)	meter^2 per second (m^2/s)	1.000 000*E–04	Township	meter^2 (m^2)	(See footnote a.)

^{a} Since 1893, the U.S. basis of length measurement has been derived from metric standards. In 1959, a small refinement was made in the definition of the yard to resolve discrepancies both in this country and abroad, which changed its length from 3600/3937 m to 0.9144 m exactly. This resulted in the new value being shorter by two parts in a million.

^{i} Defined (not measured) value.

TABLE A.5b (Continued)
*Multipliers to Convert Engineering Units**

To Convert	Into	Multiply by	To Convert	Into	Multiply by
U			Watts	horsepower	1.341×10^{-3}
Unit pole	weber (Wb)	1.256 637E–07	Watts	horsepower (metric)	1.360×10^{-3}
V			Watts	kg-calories/min	0.01433
Volt (absolute)	statvolts	0.003336	Watts	kilowatts	0.001
Volt/inch	volt/cm	0.3937	Watts (abs.)	Btu (mean)/min	0.056884
W			Watts (abs.)	joules/sec	1.0
W·h	joule (J)	3.600 000*E+03	Webers	maxwells	10^8
W·s	joule (J)	1.000 000*E+00	Webers	kilolines	10^5
W/cm^2	watt per meter^2 (W/m^2)	1.000 000*E+04	Webers/sq in.	gauss	1.550×10^7
W/in.^{2}	watt per meter^2 (W/m^2)	1.550 003E+03	Webers/sq in.	lines/sq in.	10^8
Watt (International)	watt (absolute)	1.0002	Webers/sq in.	webers/sq cm	0.155
Watt-hours	Btu	3.413	Webers/sq in.	webers/sq meter	1,550.00
Watt-hours	ergs	3.60×10^{10}	Webers/sq meter	gauss	10^4
Watt-hours	foot-pounds	2,656.00	Webers/sq meter	lines/sq in.	6.452×10^4
Watt-hours	gram-calories	859.85	Webers/sq meter	webers/sq cm	10^{-4}
Watt-hours	horsepower-hr	1.341×10^{-3}	Webers/sq meter	webers/sq in.	6.452×10^{-4}
Watt-hours	kilogram-calories	0.8605	**Y**		
Watt-hours	kilogram-meters	367.2	Yard	meter (m)	9.144 000*E–01
Watt-hours	kilowatt-hr	0.001	Yards	centimeters	91.44
Watts	Btu/hr	3.4129	Yards	kilometers	9.144×10^{-4}
Watts	Btu/min	0.05688	Yards	meters	0.9144
Watts	ergs/sec	107.0	Yards	miles (naut.)	4.934×10^{-4}
Watts	ft-lb/min	44.27	Yards	miles (stat.)	5.682×10^{-4}
Watts	ft-lb/sec	0.7378	Yards	millimeters	914.4
Watts	horsepower	1.341×10^{-3}	Yd^2	meter^2 (m^2)	8.361 274E–01
Watts	Btu/hr	3.4129	Yd^3	meter^3 (m^3)	7.645 549E–01
Watts	Btu/min	0.05688	Yd^3/min	meter^3 per second (m^3/s)	1.274 258E–02
Watts	ergs/sec	107.0	Year (365 days)	second (s)	3.153 600E+07
Watts	ft-lb/min	44.27	Year (sidereal)	second (s)	3.155 815E+07
Watts	ft-lb/sec	0.7378	Year (tropical)	second (s)	3.155 693E+07

Note: When a figure is to be rounded to fewer digits than the total number available, the procedure should be as follows: An asterisk (*) after the sixth decimal place indicates that the conversion factor is exact and that all subsequent digits are zero.

Area Unit Conversion Factors (Tables A.5c and A.5d)

TABLE A.5c
Units of Area

1 cir mil[a]
 = 0.000000785 sq in.

1 sq in.
 = 1,273,200 cir mils
 = 0.00694 sq ft
 = 0.000772 sq yd

1 sq ft
 = 144 sq in.
 = 0.01111 sq yd
 = 0.00002296 acres

1 sq yd
 = 1296 sq in.
 = 9 sq ft
 = 0.0002066 acres

1 acre
 = 43,560 sq ft
 = 4840 sq yd
 = 4096 sq m

[a] A cir (circular) mil is the area of a circle of 1/1000 in. dia. Thus, a round rod of 1 in. dia. has an area of 1,000,000 cir miles.

TABLE A.5d
Area Unit Definitions

Name of Unit	Symbol	Definition	Area in SI Units
Acre (international)	ac	$\equiv$ 1 ch $\times$ 10 ch = 4840 sq yd	$\equiv$ 4 046.856 4224 m^2
Acre (U.S. survey)	ac	$\equiv$ 10 sq ch = 4840 sq yd, also 43,560 sq ft	$\approx$ 4 046.873 m^2 [15]
Are	a	$\equiv$ 100 m^2	= 100 m^2
Barn	b	$\equiv$ 10^{-28} m^2	= 10^{-28} m^2
Barony		$\equiv$ 4000 ac	$\approx$ 1.618 742 $\times$ 10^7 m^2
Board	bd	$\equiv$ 1 in. $\times$ 1 ft	= 7.741 92 $\times$ 10^{-3} m^2
Boiler horsepower equivalent direct radiation	bhp EDR	$\equiv$ (1 ft^2) (1 bhp) / (240 BTU$_{IT}$/h)	$\approx$ 12.958 174 m^2
Circular inch	circ in.	$\equiv \pi/4$ sq in.	$\approx$ 5.067 075 $\times$ 10^{-4} m^2
Circular mil; circular thou	circ mil	$\equiv \pi/4$ mil^2	$\approx$ 5.067 075 $\times$ 10^{-10} m^2
Cord		$\equiv$ 192 bd	= 1.486 448 64 m^2
Dunam		$\equiv$ 1 000 m^2	= 1 000 m^2
Guntha		$\equiv$ 121 sq yd	$\approx$ 101.17 m^2
Hectare	ha	$\equiv$ 10 000 m^2	$\equiv$ 10 000 m^2
Hide		$\approx$ 120 ac (variable)	$\approx$ 5 $\times$ 10^5 m^2
Rood	ro	$\equiv$ ¼ ac	= 1 011.714 1056 m^2
Section		$\equiv$ 1 mi $\times$ 1 mi	= 2.589 988 110 336 $\times$ 10^6 m^2
Shed		$\equiv$ 10^{-52} m^2	= 10^{-52} m^2
Square (roofing)		$\equiv$ 10 ft $\times$ 10 ft	= 9.290 304 m^2
Square chain (international)	sq ch	$\equiv$ 66 ft $\times$ 66 ft = 1/10 ac	$\equiv$ 404.685 642 24 m^2
Square chain (U.S. survey)	sq ch	$\equiv$ 66 ft (U.S.) $\times$ 66 ft (U.S.) = 1/10 ac	$\approx$ 404.687 3 m^2
Square foot	sq ft	$\equiv$ 1 ft $\times$ 1 ft	$\equiv$ 9.290 304 $\times$ 10^{-2} m^2
Square foot (U.S. survey)	sq ft	$\equiv$ 1 ft (U.S.) $\times$ 1 ft (U.S.)	$\approx$ 9.290 341 161 327 49 $\times$ 10^{-2} m^2
Square inch	sq in.	$\equiv$ 1 in. $\times$ 1 in.	$\equiv$ 6.4516 $\times$ 10^{-4} m^2
Square kilometer	km^2	$\equiv$ 1 km $\times$ 1 km	= 10^6 m^2
Square link (Gunter's) (international)	sq lnk	$\equiv$ 1 lnk $\times$ 1 lnk $\equiv$ 0.66 ft $\times$ 0.66 ft	= 4.046 856 4224 $\times$ 10^{-2} m^2
Square link (Gunter's) (U.S. survey)	sq lnk	$\equiv$ 1 lnk $\times$ 1 lnk $\equiv$ 0.66 ft (U.S.) $\times$ 0.66 ft (U.S.)	$\approx$ 4.046 872 $\times$ 10^{-2} m^2
Square link (Ramsden's)	sq lnk	$\equiv$ 1 lnk $\times$ 1 lnk $\equiv$ 1 ft $\times$ 1 ft	= 0.09290304 m^2
Square meter (SI unit)	m^2	$\equiv$ 1 m $\times$ 1 m	= 1 m^2
Square mil; square thou	sq mil	$\equiv$ 1 mil $\times$ 1 mil	= 6.4516 $\times$ 10^{-10} m^2
Square mile	sq mi	$\equiv$ 1 mi $\times$ 1 mi	= 2.589 988 110 336 $\times$ 10^6 m^2
Square mile (U.S. survey)	sq mi	$\equiv$ 1 mi (U.S.) $\times$ 1 mi (U.S.)	$\approx$ 2.589 998 47 $\times$ 10^6 m^2
Square rod/pole/perch	sq rd	$\equiv$ 1 rd $\times$ 1 rd	= 25.292 852 64 m^2
Square yard (international)	sq yd	$\equiv$ 1 yd $\times$ 1 yd	$\equiv$ 0.836 127 36 m^2
Stremma		$\equiv$ 1 000 m^2	= 1 000 m^2
Township		$\equiv$ 36 sq mi (U.S.)	$\approx$ 9.323 994 $\times$ 10^7 m^2
Yardland		$\approx$ 30 ac	$\approx$ 1.2 $\times$ 10^5 m^2

Density Conversion Factors (Tables A.5e and A.5f)

TABLE A.5e
Units of Density

1 lb/cu in.
 = 1728 lb/cu ft
 = 0.864 tons[a]/cu ft
 = 23.3 tons/cu yd
 = 231 lb/gal

1 ton/cu yd
 = 0.0429 lb/cu in.
 = 74.1 lb/cu ft
 = 0.0370 tons/cu ft
 = 9.90 lb/gal

1 lb/cu ft
 = 0.000579 lb/cu in.
 = 0.000500 tons/cu ft
 = 0.0135 tons/cu yd
 = 0.1337 lb/gal

1 lb/gal
 = 0.00433 lb/cu in.
 = 7.48 lb/cu ft
 = 0.00374 tons/cu ft
 = 0.1010 tons/cu yd

1 ton/cu ft
 = 1.157 lb/cu in.
 = 2000 lb/cu ft
 = 27 tons/cu yd
 = 267 lb/gal

[a] Tons are short = 2000 lb.

Energy and Power Conversion Factors (Tables A.5g and A.5h)

TABLE A.5g
Energy and Heat Unit Conversion

1 Btu
 = 9340 in.-lb
 = 778.3 ft-lb
 = 0.0002938 kW h[a]
 = 0.0003931 hp h

1 kW h
 = 3,413 Btu
 = 31,873,000 in.-lb
 = 2,656,100 ft-lb
 = 1.342 hp h

1 in.-lb
 = 0.0001070 Btu
 = 0.0833 ft-lb
 = 0.00000003137 kW h
 = 0.0000000421 hp h

1 hp h
 = 2,544 Btu
 = 23,760,000 in.-lb
 = 1,980,000 ft-lb
 = 0.7455 kW h

1 ft-lb
 = 0.001284 Btu
 = 12 in.-lb
 = 0.000000376 kW h
 = 0.000000505 hp h

[a] 1 kW h = 3413 Btu, and 1 Btu = 778.3 ft-lb.

TABLE A.5f
Density Unit Definitions

Name of Unit	Symbol	Definition	Density in SI Units
Gram per milliliter	g/mL	≡ g/mL	= 1000 kg/m^3
Kilogram per cubic meter (SI unit)	kg/m^3	≡ kg/m^3	= 1 kg/m^3
Kilogram per liter	kg/L	≡ kg/L	= 1000 kg/m^3
Ounce (avoirdupois) per cubic foot	oz/ft^3	≡ oz/ft^3	≈ 1.001153961 kg/m^3
Ounce (avoirdupois) per cubic inch	oz/in^3	≡ oz/in^3	≈ 1.729994044 × 10^3 kg/m^3
Ounce (avoirdupois) per gallon (imperial)	oz/gal	≡ oz/gal	≈ 6.236023291 kg/m^3
Ounce (avoirdupois) per gallon (U.S. fluid)	oz/gal	≡ oz/gal	≈ 7.489151707 kg/m^3
Pound (avoirdupois) per cubic foot	lb/ft^3	≡ lb/ft^3	≈ 16.01846337 kg/m^3
Pound (avoirdupois) per cubic inch	lb/in^3	≡ lb/in^3	≈ 2.767990471 × 10^4 kg/m^3
Pound (avoirdupois) per gallon (imperial)	lb/gal	≡ lb/gal	≈ 99.77637266 kg/m^3
Pound (avoirdupois) per gallon (U.S. fluid)	lb/gal	≡ lb/gal	≈ 119.8264273 kg/m^3
Slug per cubic foot	slug/ft^3	≡ slug/ft^3	≈ 515.3788184 kg/m^3

TABLE A.5h
Energy Unit Definitions

Name of Unit	Symbol	Definition	Energy in SI Units
Barrel of oil equivalent	boe	$\approx 5.8 \times 10^6$ $BTU_{59°F}$	$\approx 6.12 \times 10^9$ J
British thermal unit (ISO)	BTU_{ISO}	$\equiv 1.0545 \times 10^3$ J	$= 1.0545 \times 10^3$ J
British thermal unit (International Table)	BTU_{IT}		$= 1.055\ 055\ 852\ 62 \times 10^3$ J
British thermal unit (mean)	BTU_{mean}		$\approx 1.055\ 87 \times 10^3$ J
British thermal unit (thermochemical)	BTU_{th}		$\approx 1.054\ 350 \times 10^3$ J
British thermal unit (39°F)	$BTU_{39°F}$		$\approx 1.059\ 67 \times 10^3$ J
British thermal unit (59°F)	$BTU_{59°F}$	$\equiv 1.054\ 804 \times 10^3$ J	$= 1.054\ 804 \times 10^3$ J
British thermal unit (60°F)	$BTU_{60°F}$		$\approx 1.054\ 68 \times 10^3$ J
British thermal unit (63°F)	$BTU_{63°F}$		$\approx 1.0546 \times 10^3$ J
Calorie (International Table)	cal_{IT}	$\equiv 4.1868$ J	$= 4.1868$ J
Calorie (mean)	cal_{mean}	$1/100$ of the energy required to warm one gram of air-free water from 0°C to 100°C @ 1 atm	$\approx 4.190\ 02$ J
Calorie (thermochemical)	cal_{th}	$\equiv 4.184$ J	$= 4.184$ J
Calorie (United States; FDA)	cal	$\equiv 1$ kcal	$= 1000$ cal $= 4184$ J
Calorie (3.98°C)	$cal_{3.98°C}$		≈ 4.2045 J
Calorie (15°C)	$cal_{15°C}$	$\equiv 4.1855$ J	$= 4.1855$ J
Calorie (20°C)	$cal_{20°C}$		≈ 4.1819 J
Celsius heat unit (International Table)	CHU_{IT}	$\equiv 1\ BTU_{IT} \times 1$ K/°R	$= 1.899\ 100\ 534\ 716 \times 10^3$ J
Cubic centimeter of atmosphere; standard cubic centimeter	cc atm; scc	$\equiv 1$ atm $\times 1$ cm^3	$= 0.101\ 325$ J
Cubic foot of atmosphere; standard cubic foot	cu ft atm; scf	$\equiv 1$ atm $\times 1$ ft^3	$= 2.869\ 204\ 480\ 9344 \times 10^3$ J
Cubic foot of natural gas		$\equiv 1\ 000\ BTU_{IT}$	$= 1.055\ 055\ 852\ 62 \times 10^6$ J
Cubic yard of atmosphere; standard cubic yard	cu yd atm; scy	$\equiv 1$ atm $\times 1$ yd^3	$= 77.468\ 520\ 985\ 2288 \times 10^3$ J
Electronvolt	eV	$\equiv e \times 1$ V	$\approx 1.602\ 177\ 33 \times 10^{-19} \pm 4.9 \times 10^{-26}$ J
erg (cgs unit)	erg	$\equiv 1$ g$\cdot$cm^2/s^2	$= 10^{-7}$ J
Foot-pound-force	ft lbf	$\equiv g \times 1$ lb $\times 1$ ft	$= 1.355\ 817\ 948\ 331\ 4004$ J
Foot-poundal	ft pdl	$\equiv 1$ lb$\cdot$ft^2/s^2	$= 4.214\ 011\ 009\ 380\ 48 \times 10^{-2}$ J
Gallon-atmosphere (imperial)	imp gal atm	$\equiv 1$ atm $\times 1$ gal (imp)	$= 460.632\ 569\ 25$ J
Gallon-atmosphere (U.S.)	U.S. gal atm	$\equiv 1$ atm $\times 1$ gal (U.S.)	$= 383.556\ 849\ 0138$ J
Hartree, atomic unit of energy	E_h	$\equiv m_e \cdot \alpha^2 \cdot c^2\ (= 2\ Ry)$	$\approx 4.359\ 744 \times 10^{-18}$ J
Horsepower-hour	hp$\cdot$h	$\equiv 1$ hp $\times 1$ h	$= 2.684\ 519\ 537\ 696\ 172\ 792 \times 10^6$ J
Inch-pound-force	in. lbf	$\equiv g \times 1$ lb $\times 1$ in.	$= 0.112\ 984\ 829\ 027\ 6167$ J
Joule (SI unit)	J	The work done when a force of one newton moves the point of its application a distance of one meter in the direction of the force[27]	$= 1$ J $= 1$ m$\cdot$N $= 1$ kg$\cdot$m^2/s^2 $= 1$ C$\cdot$V $= 1$ W$\cdot$s
Kilocalorie; large calorie	kcal; Cal	$\equiv 1\ 000\ cal_{IT}$	$= 4.1868 \times 10^3$ J
Kilowatt-hour; Board of Trade Unit	kW$\cdot$h; B.O.T.U.	$\equiv 1$ kW $\times 1$ h	$= 3.6 \times 10^6$ J
Liter-atmosphere	l atm; sl	$\equiv 1$ atm $\times 1$ L	$= 101.325$ J
Quad		$\equiv 10^{15}\ BTU_{IT}$	$= 1.055\ 055\ 852\ 62 \times 10^{18}$ J
Rydberg	Ry	$\equiv R_\infty \cdot h \cdot c$	$\approx 2.179\ 872 \times 10^{-18}$ J
Therm (E.C.)		$\equiv 100\ 000\ BTU_{IT}$	$= 105.505\ 585\ 262 \times 10^6$ J
Therm (U.S.)		$\equiv 100\ 000\ BTU_{59°F}$	$= 105.4804 \times 10^6$ J
Thermie	th	$\equiv 1\ Mcal_{IT}$	$= 4.1868 \times 10^6$ J
Ton of coal equivalent	TCE	$\equiv 7\ Gcal_{th}$	$= 29.288 \times 10^9$ J
Ton of oil equivalent	TOE	$\equiv 10\ Gcal_{th}$	$= 41.84 \times 10^9$ J
Ton of TNT	tTNT	$\equiv 1\ Gcal_{th}$	$= 4.184 \times 10^9$ J

Flow Unit Conversion Factors (Tables A.5i, A.5j, and A.5k)

TABLE A.5i
Flow Unit Conversion (Mass)

1 lb/s	1 lb/day
= 60 lb/min	= 0.00001157 lb/s
= 3600 lb/h	= 0.000694 lb/min
= 86,400 lb/day	= 0.0417 lb/h
= 2,628,000 lb/month[a]	= 30.4 lb/month
= 31,536,000 lb/year	= 365 lb/year

1 lb/min	1 lb/month
= 0.01667 lb/s	= 0.000000381 lb/s
= 60 lb/h	= 0.0000228 lb/min
= 1440 lb/day	= 0.001370 lb/h
= 43,800 lb/month	= 0.0329 lb/day
= 525,600 lb/year	= 12 lb/year

1 lb/h	1 lb/year
= 0.0002778 lb/s	= 0.0000000317 lb/s
= 0.01667 lb/min	= 0.000001903 lb/min
= 24 lb/day	= 0.0001142 lb/h
= 730 lb/month	= 0.002740 lb/day
= 8760 lb/year	= 0.0833 lb/month

[a] Month used is exactly 1/12 year = 30.4 days.

TABLE A.5j
Flow Unit Conversion (Volumetric)

1 cu ft/s	1 gal/s
= 60 cu ft/min	= 0.1337 cu ft/s
= 3600 cu ft/h	= 8.02 cu ft/min
= 7.48 gal/s	= 481 cu ft/h
= 448.8 gal/min	= 60 gal/min
= 26,930 gal/h	= 3600 gal/h

1 cu ft/min	1 gal/min
= 0.01667 cu ft/s	= 0.002228 cu ft/s
= 60 cu ft/h	= 0.1337 cu ft/min
= 0.1247 gal/s	= 8.02 cu ft/h
= 7.48 gal/min	= 0.01667 gal/s
= 448.8 gal/h	= 60 gal/h

1 cu ft/h	1 gal/h
= 0.0002778 cu ft/s	= 0.0000371 cu ft/s
= 0.01667 cu ft/min	= 0.002228 cu ft/min
= 0.002078 gal/s	= 0.1337 cu ft/h
= 0.1247 gal/min	= 0.0002778 gal/s
= 7.48 gal/h	= 0.01667 gal/min

TABLE A.5k
Flow Unit Definitions (Volumetric)

Name of Unit	Symbol	Definition	Volumetric Flow in SI Units
Cubic foot per minute	CFM[citation needed]	$\equiv 1$ ft^3/min	$= 4.719474432 \times 10^{-4}$ m^3/s
Cubic foot per second	ft^3/s	$\equiv 1$ ft^3/s	$= 0.028316846592$ m^3/s
Cubic inch per minute	in^3/min	$\equiv 1$ in^3/min	$= 2.7311773 \times 10^{-7}$ m^3/s
Cubic inch per second	in^3/s	$\equiv 1$ in^3/s	$= 1.6387064 \times 10^{-5}$ m^3/s
Cubic meter per second (SI unit)	m^3/s	$\equiv 1$ m^3/s	$= 1$ m^3/s
Gallon (U.S. fluid) per day	GPD[citation needed]	$\equiv 1$ gal/d	$= 4.381263638 \times 10^{-8}$ m^3/s
Gallon (U.S. fluid) per hour	GPH[citation needed]	$\equiv 1$ gal/h	$= 1.051503273 \times 10^{-6}$ m^3/s
Gallon (U.S. fluid) per minute	GPM[citation needed]	$\equiv 1$ gal/min	$= 6.30901964 \times 10^{-5}$ m^3/s
Liter per minute	LPM[citation needed]	$\equiv 1$ L/min	$= 1.6 \times 10^{-5}$ m^3/s

Length Unit Conversion Factors (Table A.5l)

TABLE A.5l
Length Unit Conversion

1 in.
= 0.0833 ft
= 0.0277 yd
= 0.0000158 mi

1 ft
= 12 in.
= 0.333 yd
= 0.000189 mi

1 yd
= 36 in.
= 3 ft
= 0.000568 mi

1 mi
= 63360 in.
= 5280 ft
= 1760 yd

Power Unit Conversion Factors (Tables A.5m and A.5n)

TABLE A.5m
Power Unit Conversion

1 kW
= 1.3415 hp
= 738 ft-lb[a]/s
= 44,268 ft-lb/min
= 2,656,100 ft-lb/h
= 0.948 Btu/s
= 56.9 Btu/min
= 3413 Btu/h

1 hp
= 0.7455 kW
= 550 ft-lb/s
= 33,000 ft-lb/min
= 1,980,000 ft-lb/h
= 0.707 Btu/s
= 0.424 Btu/min
= 2544 Btu/h

1 ft-lb/s
= 0.001355 kW
= 0.001818 hp
= 60 ft-lb/min
= 3600 ft-lb/h
= 0.001284 Btu/s
= 0.0771 Btu/min
= 4.62 Btu/h

1 ft-lb/min
= 0.00002259 kW
= 0.0000303 hp
= 0.01667 ft-lb/s
= 60 ft-lb h
= 0.00002141 Btu/s
= 0.001284 Btu/min
= 0.0771 Btu/h

1 ft-lb/h
= 0.000000376 kW
= 0.000000505 hp
= 0.000278 ft-lb/s
= 0.01667 ft-lb/min
= 0.000000357 Btu/s
= 0.00002141 Btu/min
= 0.001284 Btu/h

1 Btu/s
= 1.055 kW
= 1.416 hp
= 778 ft-lb/s
= 46,700 ft-lb/min
= 2,802,000 ft-lb/h
= 60 Btu/min
= 3600 Btu/h

1 Btu/min
= 0.01759 kW
= 0.02359 hp
= 12.98 ft-lb/s
= 778 ft-lb/min
= 46,700 ft-lb/h
= 0.01667 Btu/s
= 60 Btu/h

1 Btu/h
= 0.0002931 kW
= 0.0003932 hp
= 0.2163 ft-lb/s
= 12.98 ft-lb/min
= 778 ft-lb/h
= 0.0002778 Btu/s
= 0.01667 Btu/min

[a] ft-lb means foot-pound, the work done in moving against one pound-force a distance of one foot.

TABLE A.5n
Power or Heat Flow Unit Definitions

Name of Unit	Symbol	Definition	Power or Heat Flow in SI Units
BTU (International Table) per hour	atm ccm	$\equiv 1$ atm $\times$ 1 cm^3/min	$=1.688\ 75 \times 10^{-3}$ W
BTU (International Table) per hour	atm ccs	$\equiv 1$ atm $\times$ 1 cm^3/s	$=0.101\ 325$ W
BTU (International Table) per hour	atm cfh	$\equiv 1$ atm $\times$ 1 cu ft/h	$=0.797\ 001\ 244\ 704$ W
BTU (International Table) per hour	atm·cfm	$\equiv 1$ atm $\times$ 1 cu ft/min	$=47.820\ 074\ 682\ 24$ W
BTU (International Table) per hour	atm cfs	$\equiv 1$ atm $\times$ 1 cu ft/s	$=2.869\ 204\ 480\ 9344 \times 10^3$ W
BTU (International Table) per hour	BTU$_{IT}$/h	$\equiv 1$ BTU$_{IT}$/h	$\approx 0.293\ 071$ W
BTU (International Table) per minute	BTU$_{IT}$/min	$\equiv 1$ BTU$_{IT}$/min	$\approx 17.584\ 264$ W
BTU (International Table) per second	BTU$_{IT}$/s	$\equiv 1$ BTU$_{IT}$/s	$=1.055\ 055\ 852\ 62 \times 10^3$ W
Calorie (International Table) per second	cal$_{IT}$/s	$\equiv 1$ cal$_{IT}$/s	$=4.1868$ W
Erg per second	erg/s	$\equiv 1$ erg/s	$=10^{-7}$ W
Foot-pound-force per hour	ft lbf/h	$\equiv 1$ ft lbf/h	$\approx 3.766\ 161 \times 10^{-4}$ W
Foot-pound-force per minute	ft lbf/min	$\equiv 1$ ft lbf/min	$=2.259\ 696\ 580\ 552\ 334 \times 10^{-2}$ W
Foot-pound-force per second	ft lbf/s	$\equiv 1$ ft lbf/s	$=1.355\ 817\ 948\ 331\ 4004$ W
Horsepower (boiler)	bhp	≈ 34.5 lb/h $\times$ 970.3 BTU$_{IT}$/lb	$\approx 9.810\ 657 \times 10^3$ W
Horsepower (European electrical)	hp	$\equiv 75$ kp·m/s	$=736$ W
Horsepower (imperial electrical)	hp	$\equiv 746$ W	$=746$ W
Horsepower (imperial mechanical)	hp	$\equiv 550$ ft lbf/s	$=745.699\ 871\ 582\ 270\ 22$ W
Horsepower (metric)	hp	$\equiv 75$ m kgf/s	$=735.498\ 75$ W
Liter-atmosphere per minute	L·atm/min	$\equiv 1$ atm $\times$ 1 L/min	$=1.688\ 75$ W
Liter-atmosphere per second	L·atm/s	$\equiv 1$ atm $\times$ 1 L/s	$=101.325$ W
Lusec	lusec	$\equiv 1$ L·µmHg/s [14]	$\approx 1.333 \times 10^{-4}$ W
Poncelet	p	$\equiv 100$ m kgf/s	$=980.665$ W
Square foot equivalent direct radiation	sq ft EDR	$\equiv 240$ BTU$_{IT}$/h	$\approx 70.337\ 057$ W
Ton of air conditioning		$\equiv 2000$ lbs of ice melted / 24 h	$\approx 3\ 504$ W
Ton of refrigeration (imperial)		$\equiv 2240$ lb $\times$ ice$_{IT}$ / 24 h: ice$_{IT}$ = 144°F $\times$ 2326 J/kg.°F	$\approx 3.938\ 875 \times 10^3$ W
Ton of refrigeration (IT)		$\equiv 200$ lbs $\times$ ice$_{IT}$ / 24 h: ice$_{IT}$ = 144° $\times$ 2326 J/kg.°F	$\approx 3.516\ 853 \times 10^3$ W
Watt (SI unit)	W	The power which in one second of time gives rise to one joule of energy[27]	$=1$ W $= 1$ J/s $= 1$ N·m/s $= 1$ kg·m^2/s^3

Pressure Unit Conversion Factors
(Tables A.5o through A.5s)

TABLE A.5o
Pressure Unit Conversion (Courtesy of DGSI)

Bar	kPa	psi	m H$_2$O	ft H$_2$O	atm	millibar	inHg
1	100	14.504	10.197	33.456	0.98692	1000	29.529
0.010	1	0.14504	0.10197	0.33455	0.0098692	10	0.2953
0.068947	6.8957	1	0.70305	2.3067	0.068046	68.948	2.036
0.098068	9.8068	1.4223	1	3.2808	0.096788	98.068	2.8959
0.029891	2.9891	0.43353	0.3048	1	0.029499	29.891	0.88267
1.0133	101.33	14.696	10.332	33.899	1	1013.3	29.921
0.001	0.1	0.014504	0.010197	0.033455	0.00098692	1	0.02953
0.033864	3.3864	0.49115	0.34566	1.1329	0.033421	33.864	1

Example of use: To convert millibars to psi, find the 1 under the millibar column. The 1 marks the row that contains multipliers to convert millibars to other units. Follow the row to the psi column. The multiplier is 0.014504. Thus you would mulitiply the millibar value by 0.014504 to convert it to psi. Note that factors related to head of water are derived from water at a temperature of 4°C (39.2°F).

TABLE A.5p
Pressure Unit Conversion

1 in. water[a]
   = 0.0833 ft water
   = 0.0735 in. Hg
   = 0.577 oz/sq in.
   = 0.831 oz/sq ft
   = 0.0361 lb/sq in.
   = 5.20 lb/sq ft

1 ft water
   = 12 in. water
   = 0.882 in. Hg
   = 6.93 oz/sq in.
   = 998 oz/sq ft
   = 0.433 lb/sq in.
   = 62.4 lb/sq ft

1 in. Hg
   = 13.61 in. water
   = 1.131 ft water
   = 7.84 oz/sq in.
   = 1129 oz/sq ft
   = 0.491 lb/sq in.
   = 70.5 lb/sq ft

1 oz/sq in.
   = 1.732 in. water
   = 0.1443 ft water
   = 0.1276 in. Hg
   = 144 oz/sq ft
   = 0.0625 lb/sq in.
   = 9 lb/sq ft

1 oz/sq ft
   = 0.01203 in. water
   = 0.001002 ft water
   = 0.000886 in. Hg
   = 0.00694 oz/sq in.
   = 0.000434 lb/sq in.
   = 0.0625 lb/sq ft

1 lb/sq in.
   = 27.71 in. water
   = 2.31 ft water
   = 2.04 in. Hg
   = 16 oz/sq in.
   = 2304 oz/sq ft
   = 144 lb/sq ft

1 lb/sq ft
   = 0.1924 in. water
   = 0.01604 ft water
   = 0.01418 in. Hg
   = 0.1111 oz/sq in.
   = 16 oz/sq ft
   = 0.00694 lb/sq in.

[a] in. water means inches of water at 60°F. in. Hg means inches head of mercury at 32°F.

TABLE A.5q

Pressure Conversion (psi to kg/cm²)

| \multicolumn{8}{c}{1 pound per square inch = 0.0703 kilogram per square centimeter} |

lb/in.²	kg/cm²	lb/in.²	kg/cm²	lb/in.	kg/cm²	lb/in.	kg/cm²
1.00	0.0703	2.25	0.1582	4.50	0.3164	8.25	0.5800
1.10	0.0773	2.30	0.1617	4.75	0.3339	8.50	0.5976
1.20	0.0844	2.40	0.1687	5.00	0.3515	8.75	0.6151
1.25	0.0879	2.50	0.1758	5.25	0.3691	9.00	0.6327
1.30	0.0914	2.60	0.1828	5.50	0.3867	9.25	0.6503
1.40	0.0984	2.70	0.1898	5.75	0.4042	9.50	0.6679
1.50	0.1055	2.75	0.1933	6.00	0.4218	9.75	0.6854
1.60	0.1125	2.80	0.1969	6.25	0.4394	10.00	0.7030
1.70	0.1195	2.90	0.2039	6.50	0.4570		
1.75	0.1230	3.00	0.2109	6.75	0.4746		
1.80	0.1265	3.25	0.2285	7.00	0.4921		
1.90	0.1336	3.50	0.2461	7.25	0.5097		
2.00	0.1406	3.75	0.2636	7.50	0.5273		
2.10	0.1476	4.00	0.2812	7.75	0.5448		
2.20	0.1547	4.25	0.2988	8.00	0.5624		

1 kg/cm² = 14.223 lb/in.²

kg/cm²	lb/in²	kg/cm²	lb/in.	kg/cm²	lb/in.	kg/cm²	lb/in.²
1.0	14.22	2.5	35.56	4.0	56.89	7.5	106.67
1.1	15.65	2.6	36.98	4.1	58.31	8.0	113.78
1.2	17.07	2.7	38.40	4.2	59.74	8.5	120.90
1.3	18.49	2.8	39.82	4.3	61.16	9.0	128.01
1.4	19.91	2.9	41.25	4.4	62.58	9.5	135.12
1.5	21.33	3.0	42.67	4.5	64.00	10.0	142.23
1.6	22.76	3.1	44.09	4.6	65.43		
1.7	24.18	3.2	45.51	4.7	66.85		
1.8	25.60	3.3	46.94	4.8	68.27		
1.9	27.02	3.4	48.36	4.9	69.69		
2.0	28.45	3.5	49.78	5.0	71.12		
2.1	29.87	3.6	51.20	5.5	78.23		
2.2	31.29	3.7	52.63	6.0	85.34		
2.3	32.71	3.8	54.05	6.5	92.45		
2.4	34.14	3.9	55.47	7.0	99.56		

TABLE A.5r
Pressure Conversion (Head Pressure)

Pressure Wanted, lb/in.²	Known Pressure or Head	Head Wanted, ft H₂O	Pressure Wanted, lb/in.²	Known Pressure or Head	Head Wanted, ft/H₂O	Pressure Wanted, lb/in.²	Known Pressure or Head	Head Wanted, ft H₂O	Pressure Wanted, lb/in.²	Known Pressure or Head	Head Wanted, ft H₂O
0.433	1	2.309	11.259	26	60.044	22.084	51	117.779	32.909	76	175.514
0.866	2	4.619	11.692	27	62.354	22.517	52	120.089	33.342	77	177.824
1.299	3	6.928	12.125	28	64.663	22.950	53	122.398	33.775	78	180.133
1.732	4	9.238	12.558	29	66.973	23.383	54	124.708	34.208	79	182.433
2.165	5	11.547	12.991	30	69.282	23.816	55	127.017	34.642	80	184.752
2.598	6	13.856	13.424	31	71.591	24.249	56	129.326	35.075	81	187.061
3.031	7	16.166	13.857	32	73.901	24.682	57	131.636	35.508	82	189.371
3.464	8	18.475	14.290	33	76.210	25.115	58	133.945	35.941	83	191.680
3.897	9	20.785	14.723	34	78.520	25.548	59	136.255	36.374	84	193.990
4.330	10	23.094	15.156	35	80.829	25.981	60	138.564	36.807	85	196.299
4.763	11	25.403	15.589	36	83.138	26.414	61	140.873	37.240	86	198.608
5.196	12	27.713	16.022	37	85.448	26.847	62	143.183	37.673	87	200.918
5.629	13	30.022	16.455	38	87.757	27.280	63	145.492	38.106	88	203.227
6.062	14	32.332	16.888	39	90.067	27.713	64	147.803	38.539	89	205.537
6.495	15	34.641	17.321	40	92.376	28.146	65	150.111	38.972	90	207.846
6.928	16	36.950	17.754	41	94.685	28.579	66	152.420	39.405	91	210.155
7.361	17	39.260	18.187	42	96.995	29.012	67	154.730	39.838	92	212.465
7.794	18	41.569	18.620	43	99.304	29.445	68	157.039	40.271	93	214.774
8.227	19	43.879	19.053	44	101.614	29.878	69	159.349	40.704	94	217.084
8.660	20	46.188	19.486	45	103.923	30.311	70	161.658	41.137	95	219.393
9.093	21	48.497	19.919	46	106.232	30.744	71	163.967	41.570	96	221.702
9.526	22	50.807	20.352	47	108.542	31.177	72	166.277	42.003	97	224.012
9.959	23	53.116	20.785	48	110.851	31.610	73	168.586	42.436	98	226.321
10.392	24	55.426	21.218	49	113.161	32.043	74	170.896	42.869	99	228.631
10.825	25	57.735	21.651	50	115.470	32.476	75	173.205	43.302	100	230.940

Note: The center column, marked "Known," may be used either for head in feet or for pressure in pounds per square inch—when used as head, the corresponding pressure is found in the column at the left designated as "Pressure Wanted"; when used as pressure, the corresponding head is found in the column at the right designated as "Head Wanted." For example, a 10 ft head has a pressure of 4.33 lb per sq in., and a 10 lb pressure has a head of 23.094 ft. By moving the decimal place, quantities larger than 100 may be used.

TABLE A.5s
Pressure Unit Definitions

Name of Unit	Symbol	Definition	Pressure in SI Units
Atmosphere (standard)	atm		$\equiv$101 325 Pa [28]
Atmosphere (technical)	at	$\equiv$1 kgf/cm^2	=9.806 65 × 10^4 Pa [28]
Bar	bar		$\equiv$10^5 Pa
Barye (cgs unit)		$\equiv$1 dyn/cm^2	=0.1 Pa
Centimeter of mercury	cmHg	$\equiv$13 595.1 kg/m^3 × 1 cm × g	$\approx$1.333 22 × 10^3 Pa [28]
Centimeter of water (4°C)	cmH$_2$O	$\approx$999.972 kg/m^3 × 1 cm × g	$\approx$98.063 8 Pa [28]
Foot of mercury (conventional)	ftHg	$\equiv$13 595.1 kg/m^3 × 1 ft × g	$\approx$40.636 66 × 10^3 Pa [28]
Foot of water (39.2°F)	ftH$_2$O	$\approx$999.972 kg/m^3 × 1 ft × g	$\approx$2.988 98 × 10^3 Pa [28]
Inch of mercury (conventional)	inHg	$\equiv$13 595.1 kg/m^3 × 1 in. × g	$\approx$3.386 389 × 10^3 Pa [28]
Inch of water (39.2°F)	inH$_2$O	$\approx$999.972 kg/m^3 × 1 in. × g	$\approx$249.082 Pa [28]
Kilogram-force per square millimeter	kgf/mm^2	$\equiv$1 kgf/mm^2	=9.806 65 × 10^6 Pa [28]
Kip per square inch	ksi	$\equiv$1 kipf/sq in.	$\approx$6.894 757 × 10^6 Pa [28]
Micron (micrometer) of mercury	µmHg	$\equiv$13 595.1 kg/m^3 × 1 µm × g $\approx$ 0.001 Torr	$\approx$0.133 322 4 Pa [28]
Millimeter of mercury	mmHg	$\equiv$13 595.1 kg/m^3 × 1 mm × g $\approx$ 1 Torr	$\approx$133.3224 Pa [28]
Millimeter of water (3.98°C)	mmH$_2$O	$\approx$999.972 kg/m^3 × 1 mm × g = 0.999 972 kgf/m^2	=9.806 38 Pa
Pascal (SI unit)	Pa	$\equiv$N/m^2 = kg/(m·s^2)	=1 Pa [29]
Pièze (mts unit)	pz	$\equiv$1000 kg/m·s^2	=1 × 10^3 Pa = 1 kPa
Pound per square foot	psf	$\equiv$1 lbf/ft^2	$\approx$47.880 26 Pa [28]
Pound per square inch	psi	$\equiv$1 lbf/in^2	$\approx$6.894 757 × 10^3 Pa [28]
Poundal per square foot	pdl/sq ft	$\equiv$1 pdl/sq ft	$\approx$1.488 164 Pa [28]
Short ton per square foot		$\equiv$1 sh tn × g / 1 sq ft	$\approx$95.760 518 × 10^3 Pa
Torr	Torr	$\equiv$101 325/760 Pa	$\approx$133.322 4 Pa [28]

Specific Enthalpy, Entropy, Volume Conversion (Tables A.5t through A.5v)

TABLE A.5t
Specific Enthalpy and Latent Heat Conversion

Joule per kilogram	J/kg	4186.8
Joule per gram	J/g	4186.8
Calorie$_{IT}$ per gram	Cal$_{IT}$/g	1
British thermal unit$_{IT}$ per pound	Btu$_{IT}$/lb	1.8
Calorie$_{th}$ per gram	Cal$_{th}$/g^1	1.000669
British thermal unit$_{th}$ per pound	Btu$_{th}$/lb	1.000669

TABLE A.5u
Specific Heat and Entropy Unit Conversion

Joule per kilogram-kelvin	J/kg/K^1	4186.8
Joule per gram-kelvin	J/g/K	4186.8
Joule per gram degree Celsius	J/g/°C	4186.8
Calorie$_{IT}$ per gram degree Celsius	cal$_{IT}$/g/°C	1
British thermal unit$_{IT}$ per pound degree Fahrenheit	Btu$_{IT}$ lb/°F	1
British thermal unit$_{IT}$ per pound degree Rankine	Btu$_{IT}$ lb/°R	1
Calorie$_{th}$ per gram degree Celsius	cal$_{th}$/g/ °C	1.000669
British thermal unit$_{th}$ per pound degree Fahrenheit	Btu$_{th}$/lb/°F	1.000669
British thermal unit$_{th}$ per pound degree Rankine	Btu$_{th}$/lb/°R^1	1.000669

TABLE A.5v
Specific Volume Conversion

Cubic centimeter per gram	cm³/g	1000
Cubic meter per kilogram	m³/kg	1
Liter per gram	l g⁻¹	1
Cubic foot per pound	ft³/lb	16.0185
Imperial gallon per pound	gal/lb	99.777
U.S. gallon per pound	gal.lb	119.83

Temperature Conversion (Table A.5w)

TABLE A.5w
Temperature Conversion

Degrees Fahrenheit to Centigrade
$-°C = 5/9\ (°F + 32)$
$+°C = 5/9\ (°F - 32)$

°F	0	25	50	75
−200	−128.9	−142.8	−156.7	−170.6
−100	−73.3	−87.2	−101.1	−115.0
−0	−17.8	−31.7	−45.6	−59.4
+0	−17.8	−3.9	+10.0	+23.9
100	+37.8	+51.7	65.6	79.4
200	93.3	107.2	121.1	135.0
300	148.9	162.8	176.7	190.6
400	204.4	218.3	232.2	246.1
500	260.0	273.9	287.8	301.7
600	315.6	329.4	343.3	357.2
700	371.1	385.0	398.9	412.8
800	426.7	440.6	454.4	468.3
900	482.2	496.1	510.0	523.9
1000	538.0	551.7	565.6	579.4
1100	593.2	607.2	621.1	635.0

Degrees Centigrade to Fahrenheit
$-°F = (9/5 \times °C) - 32$
$+°F = (9/5 \times °C) + 32$

°C	0	25	50	75
−200	−328	−373	−418	
−100	−148	−193	−238	−283
−0	+32	−13	−58	−103
+0	+32	+77	+122	+167
100	212	257	302	347
200	392	437	482	527
300	572	617	662	707
400	752	797	842	887
500	932	977	1022	1067
600	1112	1157	1202	1247
700	1292	1337	1382	1427

Time Unit Conversion (Table A.5x)

TABLE A.5x
Time Unit Conversion

1 s
= 0.01667 min
= 0.0002778 h
= 0.00001157 days
= 0.0000003805 month[a]
= 0.0000000317 year

1 day
= 86,400 s
= 1440 min
= 24 h
= 0.0329 month
= 0.002740 year

1 min
= 60 s
= 0.01667 h
= 0.000694 day
= 0.0000228 month
= 0.000001903 year

1 month
= 2,628,000 s
= 43,800 min
= 730 h
= 30.4 days
= 0.0833 year

1 h
= 3600 s
= 60 min
= 0.0417 day
= 0.001370 month
= 0.0001142 year

1 year
= 31,536,000 s
= 525,600 min
= 8760 h
= 365 days
= 12 months

[a] Month used is exactly 1/12 year.

Velocity Unit Conversion (Tables A.5y and A.5z)

TABLE A.5y
Velocity Unit Conversion

1 ft/s
= 60 ft/min
= 3600 ft/h
= 0.01136 m/min
= 0.682 mi/h

1 m/min
= 88 ft/s
= 5280 ft/min
= 316,800 ft/h
= 60 mi/h

1 ft/min
= 0.01667 ft/s
= 60 ft/h
= 0.0001894 m/min
= 0.01136 mi/h

1 mi/h
= 1.467 ft/s
= 88 ft/min
= 5280 ft/h
= 0.01667 m/min

1 ft/h
= 0.002778 ft/s
= 0.01667 ft/min
= 0.00000316 m/min
= 0.0001894 mi/h

TABLE A.5z
Velocity Unit Definitions

Name of Unit	Symbol	Definition	Velocity in SI Units
Foot per hour	fph	$\equiv$ 1 ft/h	$\approx 8.466\,667 \times 10^{-5}$ m/s
Foot per minute	fpm	$\equiv$ 1 ft/min	$= 5.08 \times 10^{-3}$ m/s
Foot per second	fps	$\equiv$ 1 ft/s	$= 3.048 \times 10^{-1}$ m/s
Furlong per fortnight		$\equiv$ 1 furlong/fortnight	$\approx 1.663\,095 \times 10^{-4}$ m/s
Inch per hour	iph	$\equiv$ 1 in./h	$\approx 7.05\,556 \times 10^{-6}$ m/s
Inch per minute	ipm	$\equiv$ 1 in./min	$\approx 4.23\,333 \times 10^{-4}$ m/s
Inch per second	ips	$\equiv$ 1 in./s	$= 2.54 \times 10^{-2}$ m/s
Kilometer per hour	km/h	$\equiv$ 1 km/h	$= 1/3.6$ m/s $\approx 2.777\,778 \times 10^{-1}$ m/s
Knot	kn	$\equiv$ 1 nmi/h = 1.852 km/h	$\approx 0.514\,444$ m/s
Knot (Admiralty)	kn	$\equiv$ 1 NM (Adm)/h = 1.853 184 km/h[citation needed]	$= 0.514\,773$ m/s
Mach number	M	Ratio of the speed to the speed of sound in the medium varies especially with temperature; about 1225 km/h (761 mph) in air at sea level to about 1062 km/h (660 mph) at jet altitudes (12,200 m [40,000 ft]);[25] unitless	≈ 340 to 295 m/s for aircraft
Meter per second (SI unit)	m/s	$\equiv$ 1 m/s	$= 1$ m/s
Mile per hour	mph	$\equiv$ 1 mile/h	$= 0.447\,04$ m/s
Mile per minute	mpm	$\equiv$ 1 mile/min	$= 26.8224$ m/s
Mile per second	mps	$\equiv$ 1 mile/s	$= 1,609.344$ m/s
Speed of light in vacuum	c	$\equiv$ 299 792 458 m/s	$= 299,792,458$ m/s
Speed of sound in air	s	Varies especially with temperature; about 1225 km/h (761 mph) in air at sea level to about 1062 km/h (660 mph) at jet altitudes	≈ 340 to 295 m/s at aircraft altitudes

Viscosity Unit Conversion (Table A.5aa and Figure A.5a)

TABLE A.5aa
Viscosity Unit Conversion

Kinematic Viscosity Centistokes (K)	Seconds Saybolt Universal	Seconds Saybolt Furol	Seconds Redwood	Seconds Redwood Admiralty	Degrees Engler	Degrees Bardey
1.00	31	—	29.1	—	1.00	6200
2.56	35	—	32.1	—	1.16	2420
4.30	40	—	36.2	5.10	1.31	1440
5.90	45	—	40.3	5.52	1.46	1050
7.40	50	—	44.3	5.83	1.58	838
8.83	55	—	48.5	6.35	1.73	702
10.20	60	—	52.3	6.77	1.88	618
11.53	65	—	56.7	7.17	2.03	538
12.83	70	12.95	60.9	7.60	2.17	483
14.10	75	13.33	65.0	8.00	2.31	440
15.35	80	13.70	69.2	8.44	2.45	404
16.58	85	14.10	73.3	8.86	2.59	374
17.80	90	14.44	77.6	9.30	2.73	348
19.00	95	14.85	81.5	9.70	2.88	326
20.20	100	15.24	85.6	10.12	3.02	307
31.80	150	19.3	128	14.48	4.48	195
43.10	200	23.5	170	18.90	5.92	144
54.30	250	28.0	212	23.45	7.35	114
65.40	300	32.5	254	28.0	8.79	95
76.50	350	35.1	296	32.5	10.25	81
87.60	400	41.9	338	37.1	11.70	70.8
98.60	450	46.8	381	41.7	13.15	62.9
110	500	51.6	423	46.2	14.60	56.4
121	550	56.6	465	50.8	16.05	51.3
132	600	61.4	508	55.4	17.50	47.0
143	650	66.2	550	60.1	19.00	43.4
154	700	71.1	592	64.6	20.45	40.3
165	750	76.0	635	69.2	21.90	37.6
176	800	81.0	677	73.8	23.35	35.2
187	850	86.0	719	78.4	24.80	33.2
198	900	91.0	762	83.0	26.30	31.3
209	950	95.8	804	87.6	27.70	29.7
220	1000	100.7	846	92.2	29.20	28.2
330	1500	150	1270	138.2	43.80	18.7
440	2000	200	1690	184.2	58.40	14.1
550	2500	250	2120	230	73.00	11.3
660	3000	300	2540	276	87.60	9.4
770	3500	350	2960	322	100.20	8.05
880	4000	400	3380	368	117.00	7.05
990	4500	450	3810	414	131.50	6.26

(Continued)

TABLE A.5aa (Continued)
Viscosity Unit Conversion

Kinematic Viscosity Centistokes (K)	Seconds Saybolt Universal	Seconds Saybolt Furol	Seconds Redwood	Seconds Redwood Admiralty	Degrees Engler	Degrees Bardey
1100	5000	500	4230	461	146.00	5.64
1210	5500	550	4650	507	160.50	5.13
1320	6000	600	5080	553	175.00	4.70
1430	6500	650	5500	559	190.00	4.34
1540	7000	700	5920	645	204.50	4.03
1650	7500	750	6350	691	219.00	3.76
1760	8000	800	6770	737	233.50	3.52
1870	8500	850	7190	783	248.00	3.32
1980	9000	900	7620	829	263.00	3.13
2090	9500	950	8040	875	277.00	2.97
2200	10000	1000	8460	921	292.00	2.82

Note: The viscosity is often expressed in terms of viscometers other than the Saybolt Universal. The formulas for the various viscometers are given opposite.

If viscosity is given at any two temperatures, the viscosity at any other temperature can be obtained by plotting the viscosity against temperature in degrees Fahrenheit on special log paper. The points for a given oil lie in a straight line.

$$\text{Kinematic viscosity} = \frac{\text{Absolute viscosity}}{\text{Specific gravity}}$$

$$\text{Redwood } K = 0.26t - \frac{180}{t} \text{ (British)}$$

$$\text{Redwood Admirality } K = 2.7t - \frac{20}{t} \text{ (British)}$$

$$\text{Saybolt Universal } K = 0.22t - \frac{195}{t} \text{ (American)}$$

$$\text{Saybolt Furol } K = 2.2t - \frac{184}{t} \text{ (American)}$$

$$\text{Engler } K = 0.147t - \frac{374}{t} \text{ (German)}$$

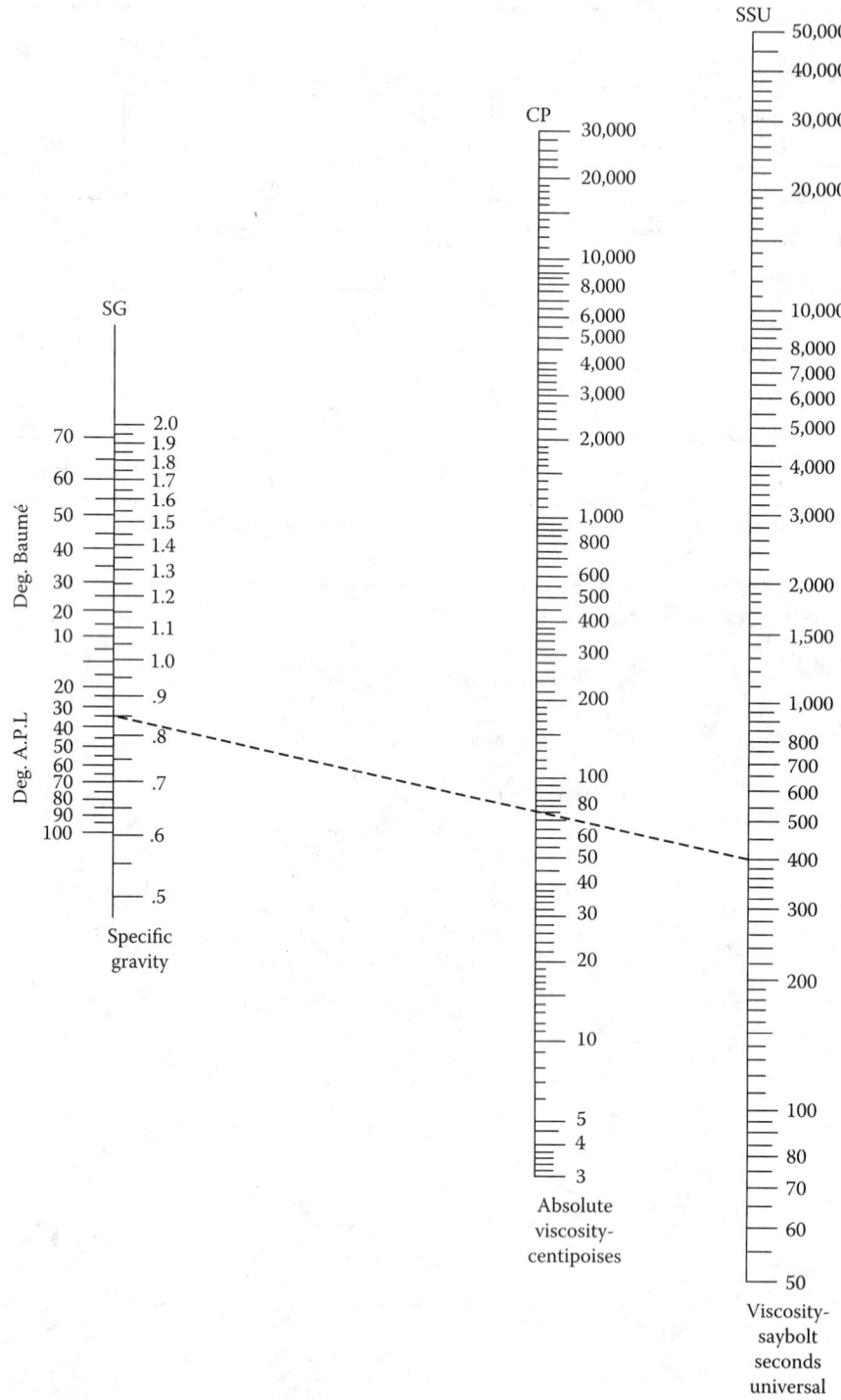

FIG. A.5a

Viscosity unit conversion chart.

This chart enables the direct conversion of a viscosity in centipoises to a SSU viscosity (Figure 2.5aa). As an example, suppose the liquid under consideration has a specific gravity of .85 and a viscosity of 75 centipoises. To determine the viscosity in SSU, lay a straight edge between 75 on the CP scale and .85 on the G scale. The viscosity in SSU can be read directly on the SSU scale. In this instance, the SSU viscosity is 400 (see dotted line).

If the viscosity value is given in centistokes (kinematic viscosity), it can be used directly on the viscosity correction nomograph. The relationship between the absolute viscosity and the kinematic viscosity is expressed by the following formula:

$$\text{Centipoises} = \text{Centistokes} \times \text{Specific Gravity}$$

Volume Unit Conversion (Tables A.5ab and A.5ac)

TABLE A.5ab
Volume Unit Conversion

1 cu in.
= 0.00433 gal
= 0.000579 cu ft
= 0.0000214 cu yd

1 gal
= 231 cub in.
= 0.1337 cu ft
= 0.00495 cu yd
= 0.00000307 acre-ft[a]

1 cu ft
= 1728 cu in.
= 7.48 gal
= 0.0370 cu yd
= 0.0000230 acre-ft

1 cu yd
= 46,656 cu in.
= 202.0 gal
= 27 cu ft

1 acre-ft
= 325,800 gal
= 43,560 cu ft
= 1613 cu yd

[a] Acre-ft of water is the volume in 1 ft of depth covering 1 acre.

TABLE A.5ac
Volume Unit Definitions

Name of Unit	Symbol	Definition	Volume in SI Units
Acre-foot	ac ft	$\equiv$1 ac × 1 ft = 43 560 ft^3	=1 233.481 837 547 52 m^3
Acre-inch		$\equiv$1 ac × 1 in.	=102.790 153 128 96 m^3
Barrel (imperial)	bl (imp)	$\equiv$36 gal (imp)	=0.163 659 24 m^3
Barrel (petroleum)	bl; bbl	$\equiv$42 gal (U.S.)	=0.158 987 294 928 m^3
Barrel (U.S. dry)	bl (U.S.)	$\equiv$105 qt (U.S.) = 105/32 bu (U.S. lvl)	=0.115 628 198 985 075 m^3
Barrel (U.S. fluid)	fl bl (U.S.)	$\equiv$31½ gal (U.S.)	=0.119 240 471 196 m^3
Board foot	fbm	$\equiv$144 cu in.	$\equiv$2.359 737 216 × 10^{-3} m^3
Bucket (imperial)	bkt	$\equiv$4 gal (imp)	=0.018 184 36 m^3
Bushel (imperial)	bu (imp)	$\equiv$8 gal (imp)	=0.036 368 72 m^3
Bushel (U.S. dry heaped)	bu (U.S.)	$\equiv$1 ¼ bu (U.S. lvl)	=0.044 048 837 7086 m^3
Bushel (U.S. dry level)	bu (U.S. lvl)	$\equiv$2 150.42 cu in.	=0.035 239 070 166 88 m^3
Butt, pipe		$\equiv$126 gal (wine)	=0.476 961 884 784 m^3
Coomb		$\equiv$4 bu (imp)	=0.145 474 88 m^3
Cord (fire wood)		$\equiv$8 ft × 4 ft × 4 ft	=3.624 556 363 776 m^3
Cord foot		$\equiv$16 cu ft	=0.453 069 545 472 m^3
Cubic fathom	cu fm	$\equiv$1 fm × 1 fm × 1 fm	=6.116 438 863 872 m^3
Cubic foot	cu ft	$\equiv$1 ft × 1 ft × 1 ft	$\equiv$0.028 316 846 592 m^3
Cubic inch	cu in.	$\equiv$1 in. × 1 in. × 1 in.	$\equiv$16.387 064 × 10^{-6} m^3
Cubic meter (SI unit)	m^3	$\equiv$1 m × 1 m × 1 m	$\equiv$1 m^3
Cubic mile	cu mi	$\equiv$1 mile × 1 mile × 1 mile	$\equiv$4 168 181 825.440 579 584 m^3
Cubic yard	cu yd	$\equiv$27 cu ft	$\equiv$0.764 554 857 984 m^3
Cup (breakfast)		$\equiv$10 fl oz (imp)	=284.130 625 × 10^{-6} m^3
Cup (Canadian)	c (CA)	$\equiv$8 fl oz (imp)	=227.3045 × 10^{-6} m^3
Cup (metric)	c	$\equiv$250.0 × 10^{-6} m^3	=250.0 × 10^{-6} m^3
Cup (U.S. customary)	c (U.S.)	$\equiv$8 U.S. fl oz $\equiv$ 1/16 gal (U.S.)	=236.588 2365 × 10^{-6} m^3
Cup (U.S. food nutrition labeling)	c (U.S.)	$\equiv$240 mL[16]	=2.4 × 10^{-4} m^3

(Continued)

TABLE A.5ac (Continued)
Volume Unit Definitions

Name of Unit	Symbol	Definition	Volume in SI Units
Dash (imperial)		$\equiv$1/384 gi (imp) = ½ pinch (imp)	=369.961 751 302 08 3 × 10^{-9} m^3
Dash (U.S.)		$\equiv$1/96 U.S. fl oz = ½ U.S. pinch	=308.057 599 609 375 × 10^{-9} m^3
Dessertspoon (imperial)		$\equiv$1/12 gi (imp)	=11.838 776 0416 × 10^{-6} m^3
Drop (imperial)	gtt	$\equiv$1/288 fl oz (imp)	=98.656 467 013 8 × 10^{-9} m^3
Drop (imperial) (alt)	gtt	$\equiv$1/1 824 gi (imp)	≈77.886 684 × 10^{-9} m^3
Drop (medical)		$\equiv$0.9964/12 mL	=83.03 × 10^{-9} m^3
Drop (medical)		$\equiv$1/12 mL	=83.3 × 10^{-9} m^3
Drop (metric)		$\equiv$1/20 mL	=50.0 × 10^{-9} m^3
Drop (U.S.)	gtt	$\equiv$1/360 U.S. fl oz	=82.148 693 22916 × 10^{-9} m^3
Drop (U.S.) (alt)	gtt	$\equiv$1/456 U.S. fl oz	≈64.854 231 496 71 × 10^{-9} m^3
Drop (U.S.) (alt)	gtt	$\equiv$1/576 U.S. fl oz	≈51.342 933 268 23 × 10^{-9} m^3
Fifth		$\equiv$1/5 U.S. gal	=757.082 3568 × 10^{-6} m^3
Firkin		$\equiv$9 gal (imp)	=0.040 914 81 m^3
Fluid drachm (imperial)	fl dr	$\equiv$⅛ fl oz (imp)	=3.551 632 8125 × 10^{-6} m^3
Fluid dram (U.S.); U.S. fluidram	fl dr	$\equiv$⅛ U.S. fl oz	=3.696 691 195 3125 × 10^{-6} m^3
Fluid scruple (imperial)	fl s	$\equiv$1/24 fl oz (imp)	=1.183 877 60416 × 10^{-6} m^3
Gallon (beer)	beer gal	$\equiv$282 cu in.	=4.621 152 048 × 10^{-3} m^3
Gallon (imperial)	gal (imp)	$\equiv$4.546 09 L	=4.546 09 × 10^{-3} m^3
Gallon (U.S. dry)	gal (U.S.)	$\equiv$⅛ bu (U.S. lvl)	=4.404 883 770 86 × 10^{-3} m^3
Gallon (U.S. fluid; wine)	gal (U.S.)	$\equiv$231 cu in.	=3.785 411 784 × 10^{-3} m^3
Gill (imperial); noggin	gi (imp); nog	$\equiv$5 fl oz (imp)	=142.065 3125 × 10^{-6} m^3
Gill (U.S.)	gi (U.S.)	$\equiv$4 U.S. fl oz	=118.294 118 25 × 10^{-6} m^3
Hogshead (imperial)	hhd (imp)	$\equiv$2 bl (imp)	=0.327 318 48 m^3
Hogshead (U.S.)	hhd (U.S.)	$\equiv$2 fl bl (U.S.)	=0.238 480 942 392 m^3
Jigger (bartending)		$\equiv$1½ U.S. fl oz	≈44.36 × 10^{-6} m^3
Kilderkin		$\equiv$18 gal (imp)	=0.081 829 62 m^3
Lambda	λ	$\equiv$1 mm^3	=1 × 10^{-9} m^3
Last		$\equiv$80 bu (imp)	=2.909 4976 m^3
Liter	L	$\equiv$1 dm^3 [17]	=0.001 m^3
Load		$\equiv$50 cu ft	=1.415 842 3296 m^3
Minim (imperial)	min	$\equiv$1/480 fl oz (imp) = 1/60 fl dr (imp)	=59.193 880 208 3 × 10^{-9} m^3
Minim (U.S.)	min	$\equiv$1/480 U.S. fl oz = 1/60 U.S. fl dr	=61.611 519 921 875 × 10^{-9} m^3
Ounce (fluid, imperial)	fl oz (imp)	$\equiv$1/160 gal (imp)	$\equiv$28.413 0625 × 10^{-6} m^3
Ounce (fluid, U.S. customary)	U.S. fl oz	$\equiv$1/128 gal (U.S.)	$\equiv$29.573 529 5625 × 10^{-6} m^3
Ounce (fluid, U.S. food nutrition labeling)	U.S. fl oz	$\equiv$30 mL [16]	$\equiv$3 × 10^{-5} m^3
Peck (imperial)	pk	$\equiv$2 gal (imp)	=9.092 18 × 10^{-3} m^3
Peck (U.S. dry)	pk	$\equiv$¼ U.S. lvl bu	=8.809 767 541 72 × 10^{-3} m^3
Perch	per	$\equiv$16½ ft × 1½ ft × 1 ft	=0.700 841 953 152 m^3
Pinch (imperial)		$\equiv$1/192 gi (imp) = ⅛ tsp (imp)	=739.923 502 60416 × 10^{-9} m^3
Pinch (U.S.)		$\equiv$1/48 U.S. fl oz = ⅛ U.S. tsp	=616.115 199 218 75 × 10^{-9} m^3
Pint (imperial)	pt (imp)	$\equiv$⅛ gal (imp)	=568.261 25 × 10^{-6} m^3
Pint (U.S. dry)	pt (U.S. dry)	$\equiv$1/64 bu (U.S. lvl) $\equiv$ ⅛ gal (U.S. dry)	=550.610 471 3575 × 10^{-6} m^3
Pint (U.S. fluid)	pt (U.S. fl)	$\equiv$⅛ gal (U.S. fl)	=473.176 473 × 10^{-6} m^3
Pony		$\equiv$3/4 U.S. fl oz	=22.180 147 171 875 × 10^{-6} m^3
Pottle; quartern		$\equiv$½ gal (imp) = 80 fl oz (imp)	=2.273 045 × 10^{-3} m^3
Quart (imperial)	qt (imp)	$\equiv$¼ gal (imp)	=1.136 5225 × 10^{-3} m^3
Quart (U.S. dry)	qt (U.S.)	$\equiv$1/32 bu (U.S. lvl) $\equiv$ ¼ gal (U.S. dry)	=1.101 220 942 715 × 10^{-3} m^3
Quart (U.S. fluid)	qt (U.S.)	$\equiv$¼ gal (U.S. fl)	=946.352 946 × 10^{-6} m^3
Quarter; pail		$\equiv$8 bu (imp)	=0.290 949 76 m^3

(Continued)

TABLE A.5ac (Continued)
Volume Unit Definitions

Name of Unit	Symbol	Definition	Volume in SI Units
Register ton		≡100 cu ft	=2.831 684 6592 m^3
Sack (imperial); bag		≡3 bu (imp)	=0.109 106 16 m^3
Sack (U.S.)		≡3 bu (U.S. lvl)	=0.105 717 210 500 64 m^3
Seam		≡8 bu (U.S. lvl)	=0.281 912 561 335 04 m^3
Shot (U.S.)		usually 1.5 U.S. fl oz[14]	≈44 × 10^{-6} m^3
Strike (imperial)		≡2 bu (imp)	=0.072 737 44 m^3
Strike (U.S.)		≡2 bu (U.S. lvl)	=0.070 478 140 333 76 m^3
Tablespoon (Australian metric)			≡20.0 × 10^{-6} m^3
Tablespoon (Canadian)	tbsp	≡½ fl oz (imp)	=14.206 531 25 × 10^{-6} m^3
Tablespoon (imperial)	tbsp	≡5/8 fl oz (imp)	=17.758 164 0625 × 10^{-6} m^3
Tablespoon (metric)			≡15.0 × 10^{-6} m^3
Tablespoon (U.S. customary)	tbsp	≡½ U.S. fl oz	=14.786 764 7825 × 10^{-6} m^3
Tablespoon (U.S. food nutrition labeling)	tbsp	≡15 mL[16]	=1.5 × 10^{-5} m^3
Teaspoon (Canadian)	tsp	≡1/6 fl oz (imp)	=4.735 510 416 × 10^{-6} m^3
Teaspoon (imperial)	tsp	≡1/24 gi (imp)	=5.919 388 02083 × 10^{-6} m^3
Teaspoon (metric)		≡5.0 × 10^{-6} m^3	=5.0 × 10^{-6} m^3
Teaspoon (U.S. customary)	tsp	≡1/6 U.S. fl oz	=4.928 921 595 × 10^{-6} m^3
Teaspoon (U.S. food nutrition labeling)	tsp	≡5 mL[16]	=5 × 10^{-6} m^3
Timber foot		≡1 cu ft	=0.028 316 846 592 m^3
Ton (displacement)		≡35 cu ft	=0.991 089 630 72 m^3
Ton (freight)		≡40 cu ft	=1.132 673 863 68 m^3
Ton (water)		≡28 bu (imp)	=1.018 324 16 m^3
Tun		≡252 gal (wine)	=0.953 923 769 568 m^3
Wey (U.S.)		≡40 bu (U.S. lvl)	=1.409 562 806 6752 m^3

Weight Unit Conversion (Tables A.5ad and A.5ae)

TABLE A.5ad
Weight Unit Conversion

lb	kg/lb	kg	lb	kg/lb	kg	lb	kg/lb	kg	lb	kg/lb	kg	lb	kg/lb	kg
2.205	1.0	0.454	6.614	3.0	1.361	11.023	5.0	2.268	15.432	7.0	3.175	19.841	9.0	4.082
2.425	1.1	0.499	6.834	3.1	1.406	11.243	5.1	2.313	15.653	7.1	3.221	20.062	9.1	4.128
2.646	1.2	0.544	7.055	3.2	1.452	11.464	5.2	2.359	15.873	7.2	3.266	20.282	9.2	4.173
2.866	1.3	0.590	7.275	3.3	1.497	11.684	5.3	2.404	16.094	7.3	3.311	20.503	9.3	4.218
3.086	1.4	0.635	7.496	3.4	1.542	11.905	5.4	2.449	16.314	7.4	3.357	20.723	9.4	4.264
3.307	1.5	0.680	7.716	3.5	1.588	12.125	5.5	2.495	16.535	7.5	3.402	20.944	9.5	4.309
3.527	1.6	0.726	7.937	3.6	1.633	12.346	5.6	2.540	16.755	7.6	3.447	21.164	9.6	4.355
3.748	1.7	0.771	8.157	3.7	1.678	12.566	5.7	2.586	16.975	7.7	3.493	21.385	9.7	4.400
3.968	1.8	0.816	8.377	3.8	1.724	12.787	5.8	2.631	17.196	7.8	3.538	21.605	9.8	4.445
4.189	1.9	0.862	8.598	3.9	1.769	13.007	5.9	2.676	17.416	7.9	3.583	21.826	9.9	4.491
4.409	2.0	0.907	8.818	4.0	1.814	13.228	6.0	2.722	17.637	8.0	3.629	22.046	10.0	4.536
4.630	2.1	0.953	9.039	4.1	1.860	13.448	6.1	2.767	17.857	8.1	3.674			
4.850	2.2	0.998	9.259	4.2	1.905	13.669	6.2	2.812	18.078	8.2	3.720			
5.071	2.3	1.043	9.480	4.3	1.950	13.889	6.3	2.858	18.298	8.3	3.765			
5.291	2.4	1.089	9.700	4.4	1.996	14.109	6.4	2.903	18.519	8.4	3.810			
5.512	2.5	1.134	9.921	4.5	2.041	14.330	6.5	2.948	18.739	8.5	3.856			
5.732	2.6	1.179	10.141	4.6	2.087	14.550	6.6	2.994	18.960	8.6	3.901			
5.952	2.7	1.225	10.362	4.7	2.132	14.771	6.7	3.039	19.180	8.7	3.946			
6.173	2.8	1.270	10.582	4.8	2.177	14.991	6.8	3.084	19.400	8.8	3.992			
6.393	2.9	1.315	10.803	4.9	2.223	15.212	6.9	3.130	19.621	8.9	4.037			

TABLE A.5ae
Weight Unit Definitions

Name of Unit	Symbol	Definition	Weight in SI Units
Atomic mass unit, unified	u; AMU		≈1.660 538 73 × 10^{-27} ± 1.3 × 10^{-36} kg
Atomic unit of mass, electron rest mass	m_e		≈9.109 382 15 × 10^{-31} ± 45 × 10^{-39} kg[18]
Bag (coffee)		≡60 kg	=60 kg
Bag (Portland cement)		≡94 lb av	=42.637 682 78 kg
Barge		≡22½ sh tn	=20 411.656 65 kg
Carat	kt	≡3 1/6 gr	≈205.196 548 333 mg
Carat (metric)	ct	≡200 mg	=200 mg
Clove		≡8 lb av	=3.628 738 96 kg
Crith			≈89.9349 mg
Dalton	Da		≈1.660 902 10 × 10^{-27} ± 1.3 × 10^{-36} kg
Dram (apothecary; troy)	dr t	≡60 g	=3.887 9346 g
Dram (avoirdupois)	dr av	≡27 11/32 g	=1.771 845 195 3125 g
Electronvolt	eV	≡1 eV (energy unit)/c^2	=1.7826 × 10^{-36} kg
Gamma	γ	≡1 µg	=1 µg
Grain	gr	≡1/7000 lb av	≡64.798 91 mg
Grave	G	Grave, the original name of kilogram	≡1 kg
Hundredweight (long)	long cwt or cwt	≡112 lb av	=50.802 345 44 kg
Hundredweight (short); central	sh cwt	≡100 lb av	=45.359 237 kg
Kilogram (kilogramme)	kg	≡mass of the prototype near Paris (≈ mass of 1 L of water)	≡1 kg (SI base unit)[8]
Kip	kip	≡1000 lb av	=453.592 37 kg
Mark		≡8 oz t	=248.827 8144 g
Mite		≡1/20 g	=3.239 9455 mg
Mite (metric)		≡1/20 g	=50 mg
Ounce (apothecary; troy)	oz t	≡1/12 lb t	=31.103 4768 g
Ounce (avoirdupois)	oz av	≡1/16 lb	=28.349 523 125 g
Ounce (U.S. food nutrition labeling)	oz	≡28 g[16]	=28 g
Pennyweight	dwt; pwt	≡1/20 oz t	=1.555 173 84 g
Point		≡1/100 ct	=2 mg
Pound	lb	≡slug·ft/s^2	=0.45359237 kg
Pound (avoirdupois)	lb av	≡0.453 592 37 kg = 7000 grains	≡0.453 592 37 kg
Pound (metric)		≡500 g	=500 g
Pound (troy)	lb t	≡5 760 grains	=0.373 241 7216 kg
Quarter (imperial)		≡1/4 long cwt = 2 st = 28 lb av	=12.700 586 36 kg
Quarter (informal)		≡¼ short tn	=226.796 185 kg
Quarter, long (informal)		≡¼ long tn	=254.011 7272 kg
Quintal (metric)	q	≡100 kg	=100 kg
Scruple (apothecary)	s ap	≡20 g	=1.295 9782 g
Sheet		≡1/700 lb av	=647.9891 mg
slug; geepound; hyl	slug	≡1 gee × 1 lb av × 1 s^2/ft	≈14.593 903 kg
Stone	st	≡14 lb av	=6.350 293 18 kg
Ton, assay (long)	AT	≡1 mg × 1 long tn ÷ 1 oz t	≈32.666 667 g
Ton, assay (short)	AT	≡1 mg × 1 sh tn ÷ 1 oz t	≈29.166 667 g
Ton, long	long tn or ton	≡2240 lb	=1 016.046 9088 kg
Ton, short	sh tn	≡2000 lb	=907.184 74 kg
Tonne (mts unit)	t	≡1000 kg	=1 000 kg
Wey		≡252 lb = 18 st	=114.305 277 24 kg (variants exist)
Zentner	Ztr.	Definitions vary (see [19] and [14])	

A.6 Chemical Resistance of Materials

Table A.6a on the pages that follow lists a variety of materials (metals, carbons, ceramics, rubbers, plastics, and woods) and indicates their suitability—from a chemical resistance point of view—to be in contact with a long, alphabetical list of chemicals. In addition to this data, chemical compatibility data can also be found on the Internet at the following web addresses:

Emerson:
http://www.documentation.emersonprocess.com/groups/public/documents/reference/d351798x012_06.pdf.

Cole-Parmer:
http://www.coleparmer.com/Chemical-Resistance.

Habonim:
http://www.habonim.com/pdf/engineering/corrosion_table.pdf.

TABLE A.6a
Chemical Compatibility Guide

Media	Chemical Formula	Aluminum	Brass	Carbon Steel	Ductile Iron / Cast Iron	316/316TI/321 S.St	17-4PH	Alloy20	Monel	Hastelloy C®	Inconel 625	Titanium	Bronze	304 Stainless Steel	Duplex	Buna N (Nitrile)	EPDM/EPR	Viton	Flexible Graphite	Delrin®	Peek®	PVDF	Teflon® and Reinforced Teflon®	PCTFE	UHMWPE®	Vespel®	PFA	KEL-F®
Acetaldehyde	C₂H₄O	B	B	C	C	A	A	A	A	A		A	C	A		D	A	D	A	A	B	D	A	A	A		A	A
Acetamine	HCl	B	B	B	B	B				A				B		A				A			A					
Acetate Solvents	—	A	A	D	B	A	A					A	B	B		D	A	D	A	D	B	A	A	A	A			B
Acetic Acid (aerated)	C₂H₄O₂	C	D	D	C	A	A	A	A	A		A	C			C	A	C	A	D	A	A	A	A	A			
Acetic Acid (air free)	C₂H₄O₂	C	D	D	D	A	A	A	A	A		A	C			D	A	D	A	D	A	A	A	A	A			
Acetic Acid (crude)	C₂H₄O₂	C	D	D	D	A	B	A	A	A		A	C			D	B	D	A	D	A	B	A		A			
Acetic Acid Glacial	C₂H₄O₂	C	D	D	D	A	A	A	A	A	A	A	C	C		D	B	C	A	D	A	C	A	A	A	A	A	A
Acetic Acid (pure)	C₂H₄O₂	C	D	D	D	A	A	A	A	A		A	C	C		D	B	D	A	B	A	A	A	A	A		A	A
Acetic Acid 10%	C₂H₄O₂	C	D	D	D	A	B	A	A	A		A	C	C		D	C	D	A	D	A	C	A	A	A		A	A
Acetic Acid 80%	C₂H₄O₂	C	D	D	D	A	A	A	A	A		A	C	D		D	A	D	A	D	A	A	A	A	A		A	A
Acetic Acid Vapors	—	B	D			C	C	B	C	A						D			A				A					
Acetic Anhydride	C₄H₆O₃	C	D	D	D	C	B	B	B	A		A	C	B		D	C	D	A	D	A	D	A	A	A		A	A
Acetone	C₃H₆O	A	A	A		A	B	B	B	A	A	A	A	B		D	A	D	A	B	A	D	A	A	A	B	A	A
Other Ketones	RC(=O)R'	A	A	A		A	A	A	A	A		A	A	A		D	B	D		B	D	D	A	A	A		A	A
Acetyl Chloride	CH₃COCl	C	B	A	C	A	A			A			D	D		D	D	D		D	B	B	A	A				
Acetylene	C₂H₂	A	C	A	A	A	A	B		A	A	B	C	A	A	B	A	A		A	B	B	A	A	A		A	
Acid Fumes	—	B	D	D	D	B				B		B				C				D			A	A				
Acrylonitrile	C₃H₃N	B	A	A	C	A	A	A	B	B	A	A	A	A		D	D	B	D	D	A	A	A			A	A	A
Air	—	A	A	A	A	A	A	A	A	A	A		A	A	A	A	A	A	A	A	A	A	A					
Alcohol, Amyl	C₅H₁₁OH	B	B	B	B	A	B	A	B	A		A	B	A		C	A	B	A	A	A	A	A	A	A	A	A	A
Alcohol, Butyl	C₄H₉OH	A	B	B	C	A	A	B	B	A		B	B	A		B	B	A	A	A	A	A	A	A	A	A	A	A
Alcohol, Diacetone	CH₃C(O)CH₂C(OH)(CH₃)₂	A	B	B	B	A	B	A	B	A		A	B	A		D	A	D	A	A	A	B	A	A	A	B	A	B
Alcohol, Ethyl	C₂H₆O	B	B	B	B	A	B	A	B	A		A	B	A		A	A	A	A	B	A	A	A	A	A	A	A	A
Alcohol, Fatty	—	B	B	B	B	A	B			A			B	B		B	A	A	A	A	A	A	A	A	A		A	
Alcohol, Isopropyl	C₃H₈O	B	B	B	B	B	B	A	B	A		B	B	B		C	A	B	A	A	A	A	A	A	A		A	A
Alcohol, Methyl	CH₄O	C	B	B	B	A	B	A	B	A		A	B	A		B	A	C	A	B	A	B	A	A	A		A	B
Alcohol, Propyl	C₃H₈O	B	B	A	B	A	B	A	B	A		A	B	A		B	A	A	A	A	A	A	A	A	A		A	A

Ratings: A-Excellent | B-Good | C-Poor | D-Do not use | Blank-No Information.
KEL-F® is a registered trademark of 3M Company, NYLATRON® is a registered trademark of Quadrant DSM Engineering Plastic Products, PEEK® is a trademark of VICTREX, TFM™ is a trademark of Dyneon, Teflon® is a registered trademark of DuPont, DELRIN® is a registered trademark of DuPont, VESPEL® is a registered trademark of DuPont.

(Continued)

TABLE A.6a (Continued)
Chemical Compatibility Guide

Media	Chemical Formula	Aluminum	Brass	Carbon Steel	Ductile Iron / Cast Iron	316/316Ti/321 S.St	17-4PH	Alloy20	Monel	Hastelloy C	Inconel 625	Titanium	Bronze	304 Stainless Steel	Duplex	Buna N (Nitrile)	EPDM/EPR	Viton	Flexible Graphite	Delrin	Peek	PVDF	Teflon and Reinforced Teflon	PCTFE	UHMWPE	Vespel	PFA	KEL-F
Alumunia	—	A								A						A	A			A			A					
Aluminum Acetate	Al(C$_2$H$_3$O$_2$)$_3$	C	C	D	D	B	B	A	B	B		C	C	D		D	A	D	A	D	A	A	A		A	A	A	A
Aluminum Chloride (dry)	AlCl$_3$	C	D	D	D	C	B	D	B	A	A	B	D	D		B	A	A	A	C	A	A	A	A	A		A	A
Aluminum Chloride Solution	—	B	D	D	D	D	C	B	B	A		C	D	D		B				D		A	A					
Aluminum Fluoride	AlF$_3$	C	D	D	D	C	C	B	A	B		C	C	D		A	A	A	A	C	A	A	A	A	A			A
Aluminum Hydroxide	AlH$_6$O$_3$	B	D	D	D	A	A	A	B	A		B	D	A		B	A	A	A	B	A	A	A		A			B
Aluminum Nitrate	AlN$_3$O$_9$	D	D	D	D	B	A	A	C	B		A	D	C		B	B	D	A	C	A	B	A					B
Aluminum Oxalate	C$_6$AlO$_{12\cdot3}$	B			B	A		A		A			D						A				A					
Aluminum Potassium Sulfate	AlKO$_8$S$_2$	B	B	C	D	B	C	B	B	C		A	C	C		B	A	B	A	C	A	A	A	A				A
Aluminum Sulfate (Alum)	Al$_2$O$_{12}$S$_3$	C	C	D	D	B	B	A	C	B		A	C	B	A	A	A	A	A	C	A	A	A	A	A			A
Amines	R$_3$-xNHx	C	B	B	C	A	A	A	B	A	A	B	C	A		D	C	D	A	D	A		A					A
Ammonia Alum	AlH$_{28}$NO$_{20}$S$_2$	C			C	A	B	A	B	A		B	C	B		B	B	A	A	C		B	A					
Ammonia, Anh. Liquid	NH$_3$	A	D	A	B	A	A	A	A	B		B	D	B		B	B	D	A	D	A	B	A	A	A			A
Ammonia Aqueous	H$_5$NO	B	C	C	C	A	A	A	B	A		B	C	A		D	B	C	A	D	A	A	A	A	A			
Ammonia Gas (hot)	H$_3$N	B	C	B	B	B	B	A	B	A		B	D	A		B	B	D	A	D	A		A					
Ammonia Liquor	—	A	C	A	B	B	A	A	A	B			D	C		C	A	D	A	D	A	B	A	A				
Ammonia Solutions	—	B	D	B	C	B	A	A	A	A		B	D			C	A	D	A	D	A		A					
Ammonium Acetate	C$_2$H$_3$O$_2$NH$_4$	A	D	D	D	A	A	A	A	A			D	B		C	A	D	A	D			A					
Ammonium Bicarbonate	NH$_4$HCO$_3$	B	B	C	C	B		B	B	B				B		B	A	A		B			A					
Ammonium Bromide 5%	NH$_4$Br	D				B		B	B			A								B			A					
Ammonium Carbonate	(NH$_4$)$_2$CO$_3$	B	D	B	B	B	B	B	B	B		A	D	B		C	A	B	A	C	A	A	A	A	A			A
Ammonium Chloride	NH$_4$Cl	C	C	D	D	C	C	B	B	B		C	D	D		B	A	A	A	C	A	A	A	A	A	A	A	A
Ammonium Hydroxide 28%	NH$_4$OH	C	D	C	B	B	A	A	B	B		A	D	B		B	B	B	A	C		A	A		A	B	A	A

(Continued)

TABLE A.6a (Continued)
Chemical Compatibility Guide

Media	Chemical Formula	Aluminum	Brass	Carbon Steel	Ductile Iron / Cast Iron	316/316Ti/321 S.St	17-4PH	Alloy20	Monel	Hastelloy C®	Inconel 625	Titanium	Bronze	304 Stainless Steel	Duplex	Buna N (Nitrile)	EPDM/EPR	Viton	Flexible Graphite	Delrin®	Peek®	PVDF	Teflon® and Reinforced Teflon	PCTFE	UHMWPE®	Vespel®	PFA	KEL-F®
Ammonium Hydroxide Conc.	NH₄OH	C	D	C	B	B	A	A	C	B		A	D	A		C	A	B	A	D	A	B	A			D		
Ammonium Monosulfate	—	D							B	B										D								
Ammonium Nitrate	(NH₄)(NO₃)	B	D	D	B	B	A	A	A	A		A	D	A		A	A	B	A	C	A	A	A					A
Ammonium Oxalate 5%	C₂H₈N₂O₄	A		A	D	A	A	A	B	A		B	D	A		D	A	B	A	B			A					
Ammonium Persulfate	H₈N₂O₈S₂	D	D	C	D	D	A	A	D	B		A	C	A		D	B	B	A	D		B	A					A
Ammonium Phosphate (mono)	(NH₄)₃PO₄	B	C	D	D	B	A	A	C	A		A	C	C		C	A	A	A	B	A	A	A					
Ammonium Phosphate Di-basic	(NH₄)₂HPO₄	B	C	D	C	B	B	B	C	A		A	D	C		A	A	A	A	B		A	A					A
Ammonium Phosphate Tri-basic	—	B	C	D	C	B	B	B	C	A		A	C	B		A	A	A	A	B			A					
Ammonium Sulfate	(NH₄)₂SO₄	C	D	C	C	C	B	A	B	B		A	D	C		B	A	C	A	C	A	A	A	A	A			A
Ammonium Sulfide	(NH₄)₂S	B	D	D	D	B	B	B	A	A			D	C		A	A	D	A	D	A	A	A		A			
Ammonium Sulfite	(NH₄)₂SO₃	D	C	C	C	A	B	B	D	A		A	B	B		B	B	C	D	D	A	A	A	A	A			A
Amyl Acetate	CH₃COO(CH₂)₄CH₃	A	A	C	C	B	A	A	A	A		A	B	C		D	B	C	D	C	A	A	A	A	A		A	B
Amyl Chloride	C₅H₁₁Cl	B	B	A	B	B	B	B	A	A		C	A	B		B	D	D	A	A	A	B	A	A	A		A	A
Aniline	C₆H₅N	C	D	C	C	A	A	B	B	B		C	A	A		D	B	C	A	A	A	A	A	A		D	A	A
Aniline Dyes	—	C	C	C	C	A	A	A	A	A			C	A		C	B	B	A	B	A		A					
Apple Juice	—	B	C	D	D	B	A	A		C			C			A	A	A	A	A	A	B	A	A	A			A
Aqua Regia (Strong Acid)	—	D	D	D	D	C	C	B	B	C		B	D	D		D	D	D	D	D	D	A	A	A	D		A	A
Aromatic Solvents	—	A	A	C	B	A	B	A	B	B			A			D	D	A	A	A	A	A	A					A
Arsenic Acid	AsH₃O₄	D	C	D	D	B	A	A	A	A		B	C	A		A	A	D	A	D	A	A	A	A				A
Asphalt Emulsion	—	C	B	B	B	A	A	A	B	B			B	A		D	B	A	A	B	A	A	A	A				A
Asphalt Liquid	—	C	B	B	B	A	A	A	B	A			A			C	D	A	A	B	A	A	A	A				A
Barium Carbonate	CBaO₃	C	B	B	C	B	A	A	A	A		A	A	B		B	A	A	A	A	A	A	A	A				A
Barium Chloride	BaCl₂	D	B	C	C	B	A	A	A	A		A	B	A		A	A	A	A	A	A	A	A	A				A

(Continued)

TABLE A.6a (Continued)
Chemical Compatibility Guide

Media	Chemical Formula	Aluminum	Brass	Carbon Steel	Ductile Iron / Cast Iron	316/316Ti/321 S.St	17-4PH	Alloy20	Monel	Hastelloy C	Inconel 625	Titanium	Bronze	304 Stainless Steel	Duplex	Buna N (Nitrile)	EPDM/EPR	Viton	Flexible Graphite	Delrin	Peek	PVDF	Teflon and Reinforced Teflon	PCTFE	UHMWPE	Vespel	PFA	KEL-F
Barium Cyanide	$Ba(CN)_2$	D	C		C	B	A	B	C	A			C	A		B	A	A		B			A					
Barium Hydrate	BaH_2O_2	D	D						B									A		A			A					
Barium Hydroxide	$Ba(OH)_2$	D	D	C	C	A	A	A	A	B		B	D	B		A	A	A	A	D	A	A	A	A	A		A	A
Barium Nitrate	$Ba(NO_3)_2$	B	D	A	A	A	B	A		A		A	D	B		A	A	A	A	B	A		A		A			A
Barium Sulfate	$BaSO_4$	C	B	B	C	B	A	A	A	A		A	C	C		A	A	A	A	B	A	B	A	A	A		A	A
Barium Sulfide	BaS	D	C	C	C	B	A	A	A	A		B	D	C		A	A	A	A	B	A		A	A	A		A	A
Beer	—	A	B	C	D	A	A	A	A	A	A	A	B	A		B	A	A	A	A	A	A	A	A	A		A	A
Beet Sugar Liquors	—	A	B	B	B	A	A	A	A	A		A		A		D	A	A	A	B	A	A	A	A	A			A
Benzaldehyde	C_7H_6O	B	B	B	C	A	A	A	B	A	A	A	A	C		D	D	D	A	A	A	A	A	A	A	A	A	B
Benzene (Benzol)	C_6H_6	B	A	B	B	B	A	A	A	A	A	A	A	B		D	D	A	A	A	A		A	A	A	A	A	A
Benzoic Acid	$C_7H_6O_2$	B	C	D	D	B	A	A	B	A		A	C	B		C	D	A	A	B	A	A	A	A	A	A	A	A
Beryllium Sulfate	BeO_4S	B	B		B	B		A	B							B	B	B		A	A		A					
Bleaching Powder Wet	—				C	C	B	B	A					B		D	B	B		B	D		A					
Blood (meat juices)	—	B	B	C	D	A	A	A	A	A	A	B	B	A		A	B	B	A	B	A		A		A	A		A
Borax (Sodium Borate)	$Na_2B_4O_7 \cdot 10H_2O$	C	B	C	A	A	A	A	A	A			B	A		A	A	A	A	B	A	A	A	A	A	A	A	A
Bordeaux Mixture	—					A		A		A		A											A					
Borax Liquors	—	C	A	C	C	B	B	A	B	A	A		B	B		A	A	A	A	B	A		A	A	A		A	A
Boric Acid	BH_3O_3	C	C	D	D	C	B	A	B	A	A		B	B		B	A	A	A	A	A	A	A	A	A	A	A	A
Brake Fluid	—	B	B	A	B	A	A	A	B	A			B	A		D	B	D	A	B	A		A	A	A	A	A	
Brines (saturated)	—	C	B	C	C	B	B	B	B	A			A	B		A	A	A	A	B	A	A	A	A	D			A
Bromine (dry)	Br	D	C	D	D	D	C	B	B	A	A	B	D	D		D	D	A	D	D	D	A	A	A		A	A	A
Bunker Oils (Fuel)	—	A	B	A	B	A		A	A	A			A	A		B	C	A	A	A	A		A					A
Butadiene	C_4H_6	A	B	B	B	B	A	A	A	B	A	A	C	A		C	C	A	A	A	A	A	A	A	A	A	A	A
Butane	C_4H_{10}	A	A	B	C	A	A	A	A	A	A	A	B	A		B	D	A	A	A	A	A	A	A	A	A	A	A
Butter	—	A		D	D	A		A	A				D	C		A	B	A		A	A		A	A				A
Buttermilk	—	A	D	D	D	D	A	B	A	A		A	D	A		B	B	D	A	A	A	A	A	A	A			A
Butyl Acetate	$C_6H_{12}O_2$	B	B	B	A	B	A	A	A	A		B	B	C		D	C	D	A	B	A	A	A	A	A		A	A
Butylene	C_4H_8	A	A	A	A	A	A	A	A	A			A	C		C	D	D	A	B	A	A	A	A	A		A	C
Butyric Acid	$C_4H_8O_2$	B	B	D	D	B	A	A	A	A		A	C	C		C	C	C	A	B	A	A	A	A	A		A	C

(Continued)

TABLE A.6a (Continued)
Chemical Compatibility Guide

Media	Chemical Formula	Aluminum	Brass	Carbon Steel	Ductile Iron / Cast Iron	316/316Ti/321 S.St	17-4PH	Alloy20	Monel	Hastelloy C®	Inconel 625	Titanium	Bronze	304 Stainless Steel	Duplex	Buna N (Nitrile)	EPDM/EPR	Viton	Flexible Graphite	Delrin®	Peek®	PVDF	Teflon® and Reinforced Teflon	PCTFE	UHMWPE®	Vespel®	PFA	KEL-F®
Calcium Bisulfite	CaH$_2$O$_6$S$_2$	C	D	D	D	B	B	A	D	B		A	C	B		A	D	A	A	D	A	A	A					A
Calcium Carbonate	CaCO$_3$	C	D	C	D	B	A	A	A	A		A	C	A		B	B	A	A	A	A	A	A					
Calcium Chlorate	Ca(ClO$_3$)$_2$	B	C	B	C	B	A	A	A	B		B	C	B		B	B	B	B	C	A	A	A		A	A	A	A
Calcium Chloride	CaCl$_2$	D	C	C	C	B	B	A	A	A	A	A	B	C		A	A	A	A	D	A	A	A	A	A	A	A	A
Calcium Hydroxide	Ca(OH)$_2$	C	D	C	C	B	A	A	A	A	A	A	C	B		B	A	A	A	D	A	A	A	A	A	D	A	B
Calcium Nitrate	Ca(NO$_3$)$_2$	C	B	B	C	B	B	A	A	B	A	B	B	C		A	B	B	A	D	A	A	A	A	A	A	A	
Calcium Phosphate	Ca$_3$(PO$_4$)$_2$	D	C	B	C	B	A	B	A							B	B	B	A	B	A	A	A					
Calcium Silicate	Ca$_2$SiO$_4$	D	C	B	C	C	A	A	A	A		A	B	B		A	A	B	A	A	A	A	A			A		A
Calcium Sulfate	CaO$_4$S	C	B	B		A	A	A	A	B		A		B	A	B		B	A	D	A	A	A	A				
Caliche Liquor	—			B		A	A	A	C											A	A	A	A	A				
Camphor	C$_{10}$H$_{16}$O	C	C	B	C	B	A	A	B	B		A	B	A		B	B	B	A	A	A	A	A	A	A			A
Cane Sugar Liquors	C$_{12}$H$_{22}$O$_{11}$	B	B	A	B	A	A	A	B	A	A	A	A	A	B	B	A	B	A	A	A	A	A	A				
Carbonated Beverages	—	B	B	D	B	B	B	B	C	B		B		B		B	B	B	A	A	A	A	A	A				
Carbonated Water	—	A	A	B	C	A	B	A	B	A		A	B	A		A	A	A	A	A	A	A	A	A	A	A		
Carbon Bisulfide	CS$_2$	B	C	B	B	B	A	A	A	A			B	A		D	D	A	B	B	A	A	A	A	B	A	A	
Carbon Dioxide (dry)	CO$_2$	B	B	B	B	B	A	A	B	A		B	A	B		C	B	B	A	A	A	A	A	A	A	A	A	A
Carbonic Acid	CH$_2$O$_3$	A	D	B	D	B	A	A	B	B		C	C	A		B	B	B	A	B	A	B	A	A	B	A	A	D
Carbon Monoxide	CO	A	A	B	B	A	A	A	A	A		A		A		B	D	B	A	B	A	A	A	A	B	A	A	
Carbon Tetrachloride (dry)	CCl$_4$	C	B	C	C	A	A	A	A	A		A	B	B		D	D	B	A	B	A	A	A	A	B	A	A	B
Carbon Tetrachloride (wet)	CCl$_4$	D	C	D	D	B	A	B	B	A		A	C	A		D	D	B	A	B	A	A	A	A	A	A	A	D
Casein	—	C	C	C	C	B	A	B	C	B				A		B	B	B	A	A	A	A	A	A	A	A		B
Castor Oil	—	A	A	A	A	A	A	A	A	A		A	B	A		A	B	A	A	A	A	A	A	A	A	A	A	
Caustic Potash	HKO	C	C	B	B	A	A	B	A	A		C	C			B		C	A	D	A	A	A	A	A			
Caustic Soda	HNaO	D	D	B	B	A	A	A	A	A		B	B	A		C	B	C	A	D	A	A	A	A	A			
Cellulose Acetate	—	B	B	B	B	B	A	A	B	B			B			D	B	D	C	C			A					
China Wood Oil (Tung)	—	A	C	C	C	A	A	A	A	A		A		A		A	D	A	A	D		A	A			A	A	D
Chlorinated Solvents	—	C	C	C	C	C	A	A	A	A			A			D	D	C	A	A	B		A					B

(Continued)

TABLE A.6a (Continued)
Chemical Compatibility Guide

Media	Chemical Formula	Aluminum	Brass	Carbon Steel	Ductile Iron / Cast Iron	316/316Ti/321 S.St	17-4PH	Alloy20	Monel	Hastelloy C	Inconel 625	Titanium	Bronze	304 Stainless Steel	Duplex	Buna N (Nitrile)	EPDM/EPR	Viton	Flexible Graphite	Delrin®	Peek®	PVDF	Teflon® and Reinforced Teflon	PCTFE	UHMWPE®	Vespel®	PFA	KEL-F®
Chlorinated Water		D	D	C	D	B	C	A	A	B		A	C	C		C	C	A	B	D	D	A	A	A				A
Chlorine Gas (dry)	Cl_2	C	D	C	B	B	B	B	C	A		A	D	C		C	D	B	A	D	D	A	A	A	B	A	A	D
Chlorobenzene (dry)	C_6H_5Cl	B	B	B	C	A	A	A	A	A			B	B		D	D	A	A	C	C	B		A		A	A	B
Chloroform (dry)	$CHCl_3$	C	B	C	C	A	A	A	A	B		B	B	A		D	D	B	A	B	C	C	A	A	B	A	A	C
Chlorophyll (dry)	—	B	B	B	B	B		A	B	A		A				B	B	B	A	B			A		D			
Chlorosulfonic Acid (dry)	HSO_3Cl	C	C		B	C	C	B	A	B			C	D		D	D	D	A	D	B	D	A	A	D	A	A	A
Chrome Alum	$CrKO_8S_2$	C	C	B	C	A	B	A	B	A						B	B	B	A	B			A	A				
Chromic Acid<50%	CrH_2O_4	D	D	D	D	C	B	B	C	C		A	D			D	C	C	C	D	A	A	A	A	A	D	A	A
Chromic Acid>50%	CrH_2O_4	D	D	D	D	C	C	B	D	C		B	D	C		D	C	C	C	D	D	A	A	A	A	D	A	B
Chromium Sulfate	$Cr_2O_{12}S_3$	B	C		D	B	B	C						B		B	B	B	A	C			A					
Cider	—	B			D	A	A	B	A	A			A	A		A	A	A	A	A	A	A	A		A			A
Citric Acid	$C_6H_8O_7$	C	D	D	D	C	B	A	A	A		A	D	B		B	B	A	A	C	A	A	A			D		
Citrus Juices	—	C	B	D	D	B	A	A	A	A		A		A		A	A	A		A		A	A	A				
Coca Cola Syrup	—					A		A								A	A	B	A	A			A					
Coconut Oil	—	B	B	C	C	A	A	A	B	A		A	B	A		B	A	B	A	A	A	A	A	A	A			
Coffee	—	A	A	C	C	A	A	A	B	A			A	A		A	A	A	A	A		A	A	A				
Coffee Extracts (hot)	—	A	A	C	C	A		A	B	A										A		A	A	A				
Coke Oven Gas	—	A	C	B	B	A	A	A	B	A		A	B			C	D	B	A	C	A	A	A	A	A			
Cooking Oil	—	B	B	B	B	B		A	B	A				A		A	D	A	A	A		A	A	A				
Copper Acetate	$C_4H_6CuO_4$	D	D	C	D	C	A	A	A	A		A	C	B		C	B	D	A	C	A	A	A	A				
Copper Carbonate	$CCuO_3$	D			D			A	B	A										A			A					
Copper Cyanide	$CCuN$	D	D	C	D	A	A	A	C	A		A	D	B		A	B	B	A	C	A	A	A	A				
Copper Nitrate	$Cu(NO_3)_2$	D	D	D	D	B		A	D	B		B	D	A		A	B	B	A	B	A	A	A	A				
Copper Sulfate	CuO_4S	D	D	B	D	B	A	A	D	A		A	D	B		A	A	A	A	D	A	A	A	A				
Corn Oil	—	B	B	B	C	B		A	B				B	A		A	A	A	A	A	A	A	A	A	A			
Cotton seed Oil	—	B	B	B	C	B	A	A	B				B			C	C	B	A	B		A	A	A				
Cresol	C_7H_8O	A	C	A	C	B	A	B	B	B		B	A	B		D	D	D	A	D	D	B	A	A	A	A		B
Creosote Oil	—	B	B	B	B	A	A	A	A	A		A	B	A		C	D	A	A	D	A	A	A	A	A			
Cresylic Acid	C_7H_8O	C	B	C	D	B	A	A	A	A		A	C	A		D	D	B	A	D	A	C	A	A			A	

(Continued)

TABLE A.6a (Continued)
Chemical Compatibility Guide

Media	Chemical Formula	Aluminum	Brass	Carbon Steel	Ductile Iron / Cast Iron	316/316Ti/321 S.St	17-4PH	Alloy20	Monel	Hastelloy C®	Inconel 625	Titanium	Bronze	304 Stainless Steel	Duplex	Buna N (Nitrile)	EPDM/EPR	Viton	Flexible Graphite	Delrin®	Peek®	PVDF	Teflon® and Reinforced Teflon	PCTFE	UHMWPE®	Vespel®	PFA	KEL-F®
Crude Oil, sour	—	A	C	C	C	B	A	A	B	A	A			B	A	A	D	A		A	A		A			A		
Crude Oil, sweet	—	A	B	A	B	A	A	A	B	A	A			B	A	A	D	A		A	A		A			A		
Cupric Nitrate	CuN_2O_6	D	A	A		A	A	A	D	A	A		A							D	A	A	A					
Cutting Oils, Water Emulsions	—	A	A	B	B	A	A	A	B	A	A					A				A	A	A	A					
Cyanide Plating Solution	—	D	D	B	D	B	A	A	D	A	A						B	B		D	A	A	A					
Cyclohexane	C_6H_{12}	A	A	A	B	A	A	A	A	A	A	A	A	A	A	B	D	A	A	A	A	C	A	A	A			A
Cyclohexanone	$C_6H_{10}O$	A	B	B	B	A	A	A	B	B	B	A	B	A	A	D	B	D	A	A	A	C	A		A			B
Detergents, Synthetic	—	B	B	B	B	B	A	A	B	A	A	A	B	A		B	B	B	A	A	A	A	A	A	A		A	
Dextrin	$C_{18}H_{32}O_{16}$	B	A	B	B	B	A	A	B	A	A	A	A	A	A	B	B	B	A	A	A	A	A	A	A			
Dichloroethane	$C_2H_4Cl_2$	B	B		C	A	B	A	B	D	A	A	B	D		D	D	B	A	B	A	A	A	A				A
Dichloroethyl Ether	$C_4H_8Cl_2O$	B	B	A	B	B	A	A	B	A	A		D	D		D	C	D	A	C	A	B	A	A	D			
Diesel Oil Fuels	—	A	A	A	A	A	A	A	A	A	A	C	A	A	A	A	D	A	A	A	A	B	A	A	A			B
Diethylamine	$C_4H_{11}N$	B	C	C	B	A	A	A	B	A	A	B	A	A		B	C	D	A	B	A	B	A	A	A			A
Diethyl Benzene	—				B	B			B					A		D	D	A	A	C	A	B	A					
Diethylene Glycol	$C_4H_{10}O_3$	B	B	A	A	B	B	A	B	B	A	A	A	B	A	A	B	B	A	B	A	A	A	A	A			A
Diethyl Sulfate	$(C_2H_5)_2SO_4$	A	B	A	B	B	A	B	B	A	A		C		A	C	C	B	A	B	A	C	A	B				
Dimethyl Formamide	C_3H_7NO	B	B	B	B	A	A	B	A	A	A			C		B	B	D	A	C	A	A	A			D		A
Dimethyl Phthalate	$C_{10}H_{10}O_4$				A	A		A								B	B	D	A	C	A	A	A	A				
Dioxane	$C_4H_8O_2$	B	A	A	B	B	A	B	A	A	A		A	A		D	C	D	A	A	A	A	A		A			
Dipentane (Pinene)	$C_{10}H_{16}$	A	A	A	A	B	A	B	A	B	A		A	A		B	D	B	A	A	C	B	A					
Disodium Phosphate	HNa_2O_4P	B	A	B	B	B	A	A	A	B	B			A		B	B	B	A	A	A	A	A	A				
Dowtherm	$C_{24}H_{20}O$	A	A	A	B	A	A	A	B	A	A			A		D	D	A	A	A	B	A	A					
Drilling Mud	—	B	B	C	B	A	A	A	B	B	B					B	B	A	D	A	A	A	B					
Dry Cleaning Fluids	—	A	C	B	B	A	A	A	A							B	A	B	D	A	A	A	A					
Drying Oil	—	C	C	C	B	B	A	A	B	A	A					B	C	D	A	A	A	A	D					
Enamel	—		A														A		B		A							
Epsom Salts	MgO_4S	B	A	A	A	B	A	B	B	A	A		B	A		B	D	A	B	A	A		A		A			A
Ethane	C_2H_6	A	B	B	A	A	A	A	B	A	A		A	A		A	A	A	A	A	A	A	A	A				

(Continued)

TABLE A.6a (Continued)
Chemical Compatibility Guide

Media	Chemical Formula	Aluminum	Brass	Carbon Steel	Ductile Iron / Cast Iron	316/316Ti/321 S.S.	17-4PH	Alloy20	Monel	Hastelloy C®	Inconel 625	Titanium	Bronze	304 Stainless Steel	Duplex	Buna N (Nitrile)	EPDM/EPR	Viton	Flexible Graphite	Delrin®	Peek®	PVDF	Teflon® and Reinforced Teflon	PCTFE	UHMWPE®	Vespel®	PFA	KEL-F®
Ethers	H_2O	B	A	B	B	A	A	A	A	B		A	B	A	A	D	C	C	A	C	A	C	A	A	A		A	C
Ethyl Acetate	$C_4H_8O_2$	B	B	B	B	B	A	A	A	B		A	B	B		D	C	D	A	A	A	C	A		A	A		B
Ethyl Acrylate	$C_5H_8O_2$	B	B	A	B	A	A	A	A	A		A	A	A		D	C	D	A	B	A	B	A					
Ethyl Benzene	C_8H_{10}	A		A	B	B	B	A	A	A			B	B		C	D	D		B	A	A	A					
Ethyl Bromide	C_2H_5Br	B	A		B	B		C	B				B			B	B	B	A	A			A					B
Ethyl Chloride (dry)	C_2H_5Cl	B	B	B	B	A	A	B	A	C		A	C	A		C	C	B	A	B	A	B	A	A				
Ethyl Chloride (wet)	C_2H_5Cl	D	C	C	D	B	B	B	B	C			C	C		C	B	B	A	B		B	A					B
Ethylene Chloride	C_2H_4Cl	B		B	C	A	B	A	A	B		B	A	B		D	D	D	A	B	A	B	A		B	A		B
Ethylene Dichloride	$C_2H_4Cl_2$	B	B	B	A	A	B	A	A	C		B	B	B		D	D	D	A	C	A	B	A		A	A		A
Ethylene Glycol	$C_2H_6O_2$	A	A	B	B	A	A	A	B	A		A	A	B	A	A	A	A	A	B	A	A	A		A	A		A
Ethylene Oxide	C_2H_4O	D	D	B	C	B	B	A	A	B			D	B		D	C	D	A	C	A	B	A					
Ethyl Ether	$C_4H_{10}O$	B	B	A	B	A	B	A	A	A		A	B	B		D	D	D	A	B	A	A	A		A			B
Ethyl Silicate	$C_8H_{20}O_4Si$	B	B		B	B		B	B	B				B		B	B	B	A	A			A					
Ethyl Sulfate	$C_4H_{10}O_4S$	A				D		B					C	D		B	C	A	A	A		A	A		A			A
Fatty Acids	R-COOH	C	C	D	D	B	A	A	A	A		A	C	B		B	D	A	A	B		A	A	A	A	A		A
Ferric Hydroxide	FeH_3O_3	C		D	D	A	A	A	A	B				B		B	A	A	A	A			A		A			
Ferric Nitrate	FeN_3O_9	D	D	D	D	B	B	A	D	A		A	D	B		A	A	A	A	D	A	A	A		A			A
Ferric Sulfate	$Fe_2O_{12}S_3$	D	C	D	D	B	B	A	D	B		A	D	B		A	A	A	A	D	A	A	A		A			A
Ferrous Ammonium Citrate	$C_6H(5+4y)Fe(x)N(y)O_7$	B			A	B		B		A						A			A	A			A	A					
Ferrous Chloride	Cl_2Fe	D	D	D	D	D	A	D	D	C		A	C	D		A	A	A	A	D	A	C	A		A			C
Ferrous Sulfate	FeH_2O_4S	C	C	D	D	B	A	D	B	B		A	C	B		A	D	A	A	D		C	A					A
Ferrous Sulfate (Saturated)	—	C	B	C	C	A	B	A	A	B		A	A			C	B	B	A	B			A		A			
Fertilizer Solutions	—	D	C	A	A	A	A	B	B	A			D	A		B	A	A	A	A	A	C	A			A	A	C
Fish Oils	—	B	B	B	B	A	A	A	A	A			A	A		A	D	A	A	D			A					
Flue Gases	CO	C	B	C	B	A	B	A	A	A			A	A		C	D	C	A	C			A					A
Fluoboric Acid	BF_4H	C		D	C	B	B	A	A	B		D	C	B		A	A	A	A	B	A	B	A			A	A	B
Fluorosilicic Acid	F_6H_2Si	D	B	D	D	B	B	A	A	A		D	B	C		C	C	C	A	B		C	A		A		A	C
Formaldehyde (cold)	CH_4	B	B	B	B	A	B	A	A	A		B	B	B		B	B	D	A	A	A	A	A	A	A	A	A	A
Formaldehyde (hot)	CH_4	B	B	C	D	C	B	B	A	B				B		B			A	A	A	A	A	A	A	A	A	A

(Continued)

TABLE A.6a (Continued)
Chemical Compatibility Guide

Media	Chemical Formula	Aluminum	Brass	Carbon Steel	Ductile Iron / Cast Iron	316/316Ti/321 S.St	17-4PH	Alloy20	Monel	Hastelloy C	Inconel 625	Titanium	Bronze	304 Stainless Steel	Duplex	Buna N (Nitrile)	EPDM/EPR	Viton	Flexible Graphite	Delrin	Peek	PVDF	Teflon and Reinforced Teflon	PCTFE	UHMWPE	Vespel	PFA	KEL-F
Formic Acid (cold)	CH_2O_2	B	D	D	D	B	A	A	B	A	B	B	C	C		D	A	B	A	C	B	A	A	A	A	A	A	B
Formic Acid (hot)	CH_2O_2	B	D	D	D	B	B	B	B	B	B	C	C	C		D	A	A	A	D	B	A	A	A	A	A	A	
Freon Gas (dry)	—	B	A	A	B	A	A	A	A	B						C	C	C	A				A					A
Freon 11, MF,112, BF	CCl_3F	D	A	B	C	A	A	A	A	A		B	B	A		C	C	C	A	D	D	D	A					C
Freon 12, 32, 114, 115	CCl_2F_2	B	A	B	B	A	A	A	B	A		B	B	B		B	B	B	A	B	A	C	A			A		
Freon 21, 31	$CHCl_2F$	B	A	B	C	A	A	A	B	A				C		D	D	D	A	A	A	A						A
Freon 22	$CHClF_2$	D	A	B	C	A	A	A	B	A		B		A		D	D	D	A	A	B	A	A					
Freon 113, TF	$C_2Cl_3F_3$	D	A	B	C	A	A	B	B	A		C	B	B		B	B	D	A	A	B	A	B					A
Freon (wet)	—	D	B	C	D	C	B	B	B	B			C			B	B	D	A	B								
Fruit Juices	—	B	D	D	D	A	A	A	B	A	A	A	B	A		A	A	A		C	A	A	D	A	A		A	A
Fuel Oil	—	B	B	A	B	A	A	A	B	A		A	A	A	A	A	D	A	A	A	A	A	B		A			A
Fumaric Acid	$C_4H_4O_4$	A	A	A		A	A	A	A	A					A	B	B	A		A	A	A	A					
Furfural	$C_5H_4O_2$	B	A	A	B	A	A	A	A	A		A	B	A		D	C	D	A	B	A	D	A					D
Gallic Acid 5%	$C_7H_6O_5$	C	C	D	D	B	A	B	B	A		B	B	B		B	C	D	B	A	B	A	A					A
Gas, Manufactured	—	B	B	B	B	B		B	B	A				A		A	D	A	A	A	A	A	A					
Gas, Natural	—	B	B	B	B	A		B	A	B				A	A	A	C	A	A	A	A		A					
Gas, Odorizers	—	A	A	A		A			B					A	A	B	D	A		B	A		A					
Gasoline (aviation)	—	A	A	A	B	A	A	A	A	A		A		A		C	D	A	A	C	A	A	A		A	A		A
Gasoline (leaded)	—	A	A	A	A	A	A	A	B	A			A	A		C	D	A	A	A	A	A	A		A	A		A
Gasoline (motor)	—	A	A	A	B	A	A	A	B	A		A		A		C	D	A	A	A	A		A		A			
Gasoline (refined)	$C_{12}H_{13}N_5O_6S_2$	A	B	B	B	A		B	B	A			B			A	D	A	A	A	A	B	A		A	A		A
Gasoline (sour)	—	A	B	B	B	A		A	C	A				A		A	A	A	A	A	A		A					
Gasoline (unleaded)	—	A	A	A	B	A	A	A	B	A		A	B	A		C	D	A	A	A	A	B	A		A			A
Gelatine	—	A	C	D	D	A		A	A	B		A	A	A		A	A	A	A	B	A	A	A	A	A		A	
Glucose	$C_6H_{12}O_6$	A	A	B	B	B	B	A	A	A		A		A		A	B	B	A	A	A	A	A		A			
Glue	—	A	A	A	B	B	A	A	A	A		A	C	A		C	A	C	A	A	A	A	A	A	A		A	A
Glycerine (Glycerol)	$C_3H_8O_3$	C	A	A	B	A	A	A	A	A		A		B		A	D	C	A	C	A	A	A		A			
Glycol Amine	—																											

(Continued)

TABLE A.6a (Continued)
Chemical Compatibility Guide

Media	Chemical Formula	Aluminum	Brass	Carbon Steel	Ductile Iron / Cast Iron	316/316Ti/321 S.St	17-4PH	Alloy20	Monel	Hastelloy C	Inconel 625	Titanium	Bronze	304 Stainless Steel	Duplex	Buna N (Nitrile)	EPDM/EPR	Viton	Flexible Graphite	Delrin®	Peek®	PVDF	Teflon® and Reinforced Teflon	PCTFE	UHMWPE®	Vespel®	PFA	KEL-F®
Glycol	$C_2H_6O_2$	B	B	B	B	B	A	A	A	B			A	A	A	B	B	A		C	A	A	A	A	A		A	
Graphite	CH_4	B	B		C	B		A	B					A		B	B	B		A			A					
Grease	—	B	B	A	A	A	A	A	A	A			B	A		A	D	A		B			A					
Helium Gas	He	B	B		B	A								A	A	B	B	B		A			A					
Heptane	C_7H_{16}	A	A	A	A	A		A	A	A			A	A		A	D	A	A	A	A	A	A					A
Hexane	C_6H_{14}	A	A	A	A	A	A	A	A	A		A	A	A		A	D	A		A	A	A	A			A		A
Hexanol, Tertiary		A	A	A	A	A	A	A	A	A		A	A	A		A	D	B		A			A					
Hydraulic Oil Petr. Base	—	A	B	A	A	A	A	A	A	A			A	A		A	D	A	B	B	A		A		A		A	
Hydrazine	H_4N_2	C	D	C	D	B	A	A	B	A		B	C	A		C	B	C	A	B	D	A	A	A	A	D	A	B
Hydrocyanic Acid	CHN	C	D	D	C	A	B	A		A			D	B		B	B	A	A	C	D	A	A		A	A	A	
Hydrofluoric Acid 20%	HF	D	D	D	D	D	C	B	A	B		D	C	D		D	D	A	A	D	D	B	B	A	C			B
Hydrofluoric Acid 50%	HF	D	D	D	D	D	D	C	B	C		D	C	D		D	D	B	A	D	D	B	B	A	C			B
Hydrofluoric Acid 75%	HF	D	D	D	D	D	D		B	C		D	C	D		D	D	B	A	D	D	B	B	A	C			B
Hydrofluoric Acid 100%	HF	D	D	D	D	D	D	B	A	B	A	D	C	D	A	D	D	B	A	D	D	A	A	A	C			A
Hydrofluosilicic Acid	H_2SiF_6	D	B	D	D	C	B	B	B	C		D	B	C		B	B	A	B	B	D	B	A					B
Hydrogen Gas (cold)	H	A	A	B	A	A	A	A	A	A	A	A	A	A		B	B	A		A	A	B	A		A	A		B
Hydrogen Gas (hot)	H	C	A	B		B		A	A	A						B	B			A			A					
Hydrogen Peroxide Conc.	H_2O_2	C	D	D	D	B	A	A	C	B		B	D	B		D	C	B	C	C	A	B	A	A	A		A	A
Hydrogen Peroxide, Dilute	H_2O_2	C	D	D	D	B	A	A	B	A	A	B	D	B		A	A	A	B	D	A	B	A	A	A	A	A	B
Hydrogen Sulfide (dry)	H_2S	C	C	C	C	C	B	B	B	B		A	A	C		C	A	D	A	C	A	B	A	A	A	A	A	A
Hydrogen Sulfide (wet)	H_2S	C	C	C	D	C	C	B	C	B		B	D	C		C	B	D	A	C	A	B	A	A	A	A	A	B

(Continued)

TABLE A.6a (Continued)
Chemical Compatibility Guide

Media	Chemical Formula	Aluminum	Brass	Carbon Steel	Ductile Iron / Cast Iron	316/316Ti/321 S.St	17-4PH	Alloy20	Monel	Hastelloy C®	Inconel 625	Titanium	Bronze	304 Stainless Steel	Duplex	Buna N (Nitrile)	EPDM/EPR	Viton	Flexible Graphite	Delrin®	Peek®	PVDF	Teflon® and Reinforced Teflon	PCTFE	UHMWPE®	Vespel®	PFA	KEL-F®
Hypo (Sodium Thiosulfate)	$Na_2S_2O_3$	B	C	D	D	B	A	B	B				B	B		C	A	A		A	A							A
Illuminating Gas	—	A	A	A	A	A	A	A	A	A						A	D	A	A	A	A		A					
Ink-Newsprint	$C_{11}H_{16}O_3P+$	C	C	D	A	A	B	A	B	A		B	B	C		A	B	A	A	B		B	A			A		B
Iodoform	CH_3	C	C	B	C	A	A	A	C	B			C	A		D	A	A	A	A			C					
Iso-Butane	C_4H_{10}	A	A	B	B	A	A	A	A	B			A	A		B	D	A	A	A	A	A	A					A
Iso-Octane	C_8H_{18}	A	A	A	A	A	A	A	A	A		A	A	A	A	D	D	D	A	A	A	B	A		A			B
Isopropyl Acetate	$C_5H_{10}O_2$	B	A	B	C	B	A	A	A	A		A	A	B		D	D	D	A	D	A	B	A		A			
Isopropyl Ether	$C_6H_{14}O$	A	A	A	B	A	A	A	A	A		A	A	A		C	D	D	A	D	A	A	A		A	A		A
JP-4 Fuel	—	A	A	A	A	A	A	A	A	A		A	A	A		A	D	A	A	A	A	A	A			A		A
JP-5 Fuel	—	A	A	A	A	A	A	A	A	A		A	A	A		B	D	A	A	A	A	A	A			A		A
JP-6 Fuel	—	A	A	A	A	A	A	A	A	A			A	A		A	D	A	A	A	A	A	A			A		A
Kerosene	—	B	A	B	B	A	A	A	A	A		A	A	A		A	D	A	A	A	A	A	A		A	A		
Ketchup	—	C	D	C	C	A	A	A					C	B				A		A	A		A		A			C
Ketones	CH_2O	B	A	A	A	B	B	A	A	A		A	B	B		D	D	D	A	D	A	D	A	A	A	A		B
Laquer (and Solvent)	—	A	C	D	C	B	A	A	A	A		B	A	B		D	B	A	A	A	A	B	A		A	A		
Lactic Acid Concentrated (cold)	$C_3H_6O_3$	C	D	D	D	A	A	A		B		A	D	B		B	C	B	A	C	C	C	A	A	A			B
Lactic Acid Concentrated (hot)	$C_3H_6O_3$	C	D	D	D	B	B	A		B		A	D	B		C	C	B	A	C	A	B	A	A	A			
Lactic Acid Dilute (cold)	$C_3H_6O_3$	B	D	D	D	A	A	A	A	A		A	C	B		B	B	A	A	B	A	B	A	A	A	A		B
Lactic Acid Dilute (hot)	$C_3H_6O_3$	B	D	D	D	A	B	A	A	B		A	C	B		C	C	B	A	B	A	C	A	A	A	A		
Lactose	$C_{12}H_{22}O_{11}$	B	B	C	C	B	A	B	B					A		B	B	B	A	A	A		A		A	A		
Lard	—	A	B	A	A	A	A	A	A	A			C	B		B	C	A	A	A	A	A	A	A	A	A		
Lard Oil	—	B	B	C	C	B	B	B	B				C	B		B	A	B		A	A		A		A	A		
Lead Acetate	$C_4H_6O_4Pb$	D	C	C	D	B	B	B	B	B		A	B	C		A	B	D	A	B	A	A	A	A	A	A	A	A
Lead Sulfate	O_4PbS	D	C	C	D	B	B	B	B	B			B	C		B	B	B	A	A	A	A	A	A	A	A		
Lecithin	$C_{46}H_{89}NO_9P+$	B	C	C	C	B	A	B	B	A			A	A		D	D	B	A	A	A	A	A		A			A

(Continued)

TABLE A.6a (Continued)
Chemical Compatibility Guide

Media	Chemical Formula	Metals														Elastomers				Polymers								
		Aluminum	Brass	Carbon Steel	Ductile Iron / Cast Iron	316/316Ti/321 S.St	17-4PH	Alloy20	Monel	Hastelloy C®	Inconel 625	Titanium	Bronze	304 Stainless Steel	Duplex	Buna N (Nitrile)	EPDM/EPR	Viton	Flexible Graphite	Delrin®	Peek®	PVDF	Teflon® and Reinforced Teflon®	PCTFE	UHMWPE®	Vespel®	PFA	KEL-F®
Linoleic Acid	$C_{18}H_{32}O_2$	B	C	C	B	B	B	A	A	A			C	B		A	D	B	A	B	A	A	A					
Linseed Oil	—	A	A	A	A	B	A	A	A	A	A		B	A		A	D	B	A	A	A	A	A		A			
Lithium Chloride	ClLi	C	B	C	B	B	A	A	A	A			B	A		B	B	B	A	A	A	A	A					
Liquid Petroleum Gas (LPG)	$C_3H_7NO_2$	A	A	A	B	B		B	B	A				A		A	D	A		A	A		A					
Lubricating Oil Petroleum Base	C_6H_6	A	A	A	A	A	A	A	A	A		A	A	A		A	D	A	A	A	A	A	A					
Ludox	O_2Si	C	D	A	B	B	A	A	A	A			C			B	B	B	A	B			A					
Magnesium Bisulfate	$HMgO_4S+$	D	B	B	B	A	A	A	B		A	A	A	A		B	B	B	B	A		A	A					
Magnesium Bisulfade	—	C	D		D	B		B	B	B						B	B	B		A			A					
Magnesium Carbonate	$CMgO_3$	A	B	B	B	A	A	A	A	B		A	B	B		B	B	B	A	A	A	A	A					
Magnesium Chloride	Cl_2Mg	D	C	C	D	C	D	C	B	A		A	B	D		A	A	A	A	A	B	A	A		A	A		
Magnesium Hydroxide	H_2MgO_2	D	B	B	B	B	A	A	B	A		A	B	B		A	A	A	A	A	A	A	A					
Magnesium Hydroxide (hot)	H_2MgO_2	D	C	B	B	A	B	A	A	B		A		B		B	A	A	A	A	A		A					
Magnesium Nitrate	MgN_2O_6	B	C	B	B	A	B	A	B	A		A	C	B		B	A	B	A	A	A	A	A					
Magnesium Sulfate	MgO_4S	B	B	B	B	A	B	A	B	B		A	C	B		A	A	A	A	A	A	A	A	A				
Maleic Acid	$C_4H_4O_4$	B	B	C	C	B	B	A	B	B		A	B	A		B	D	A	A	D	A	A	A	A	A		A	
Maleic Anhydride	$C_4H_2O_3$	B	B		D	B	B	B	B	A	A		C	A		D	D	B	B	D	A	A	A		A			
Malic Acid	$C_4H_6O_5$	B	B	D	D	B	C	B	B	B		A	C			A	D	B	B	A	A	A	A	A	A			
Malt Beverages	—					A		B	A			A		A		A	B	A	A	A	A	A	A					
Manganese Carbonate	—	B				B		A								B	B		A	A	A	A	A					
Manganese Sulfate	MnO_4S	B	D	B	D	A	B	A	A	A		A	A	A		B	B	B	A	B	A	A	A		A	A	A	B
Mayonnaise	—	D	B	D	D	A	B	A	B	B			D	B		A	B	B	A	A	A	A	A					
Meat Juices	—	B				A										B				A	A	A	A		A			
Melamine Resins	$C_3H_6N_6$		D			C	C	C	A	A			D	C		B	A	A	C	A	A	A	A	A	A			
Methanol	CH_4O	A	A	A	A	A	A	A	A	A		B	A			B	D	B	A	A	A	A	A	A	A			
Mercuric Chloride	Cl_2Hg	D	D	D	D	D	D	C	D	B		A	D	D		A	A	A	B	B	A	A	A	A	A		A	B

(Continued)

TABLE A.6a (Continued)
Chemical Compatibility Guide

Media	Chemical Formula	Aluminum	Brass	Carbon Steel	Ductile Iron / Cast Iron	316/316Ti/321 S.St	17-4PH	Alloy20	Monel	Hastelloy C®	Inconel 625	Titanium	Bronze	304 Stainless Steel	Duplex	Buna N (Nitrile)	EPDM/EPR	Viton	Flexible Graphite	Delrin®	Peek®	PVDF	Teflon® and Reinforced Teflon	PCTFE	UHMWPE®	Vespel®	PFA	KEL-F®
Mercuric Cyanide	C_2HgN_2	D	D	D	D	B	C	A	D	A		A	D	C		A	B	A		A	B	A	B					D
Mercurous Nitrate	$HgNO_3$	D	D	D	D	A	A	A	D	A		A	D	A		B		B	C	A	B	A	A					
Mercury	Hg	C	D	A	A	A	A	A	B	A		A	A	A		A	A	A	C	A	B	A	A	A	A		A	A
Methane	CH_4	A	A	A	A	A	A	A	A	A		A	A	A	A	A	A	A	B	A	A	A	A	A	A			
Methyl Acetate	$C_3H_6O_2$	A	A	B	B	A	A	A	A	A			B	A		D	B	D	B	B	A	C	A	A			A	A
Methyl Acetone	C_4H_8O	A	A	A	A	A	A	A	A	A			D	A		D	A	D	A	D		D	A					
Methyl Amine	CH_5N	B	C	B	B	B	A	B	C	B			D	A		B	B	D	A	D	A	D	A					
Methyl Bromide 100%	CH_3Br	C	C	C	D	B	A	B	B	B			D	B		B	D	D	A	D	B	A						
Methyl Cellosolve	$C_3H_8O_2$	B	A	A	B	A	A	A	A	A	A	A	A	B		C	B	D	A	D	A	A	A					
Methyl Cellulose	—					A		A	A	B						D	D	B	C	C								
Methyl Chloride	CH_3Cl	D	C	A	B	A	A	A	A	A			B	A		D	D	B	A	C	B	A	A	A				B
Methyl Ethyl Ketone	C_4H_8O	B	A	A	A	B	A	B	A	A		B	B	A		D	B	D	A	B	B		A	A	A			B
Methylene Chloride	CH_2Cl_2	C	B	B	B	A	A	B	A	A			B	B		D	D	C	A	B	A	C	A	A	B		A	A
Methyl Formate	$C_2H_4O_2$	B		C	B	A	A	A	A	A		A	A	A		D	B	D	A	A	A		A	A				
Methyl Isobutyle Ketone	$C_6H_{12}O$	B	B	A	B	B	A	A	A	A			A	A		D	C	D	A	C	B	D	A	A				A
Milk & Milk Products	—	C	C	C	D	A	A	A	A	A		A	B	A		A	A	A	A	A	A	A	A	A	A		A	A
Mineral Oils	—	A	A	A	B	A	A	A	A	A		A	A	A		A	D	A	A	A	A	A	A	A	A			
Mineral Spirits	—	A	B	B	B	B	A	B	B	B		B	B	A		A	D	A	A	A	A	A	A	A	A		A	A
Mixed Acids (cold)	—	D	D	C	C	B	A	B	C	B		B	B			D	D	B	A	D		C	A	A	B			
Molasses, crude	—	B	B	A	A	A	A	A	A	A		A	A			A	B	A	A	A	A	C	A	A	A			A
Molasses, Edible	—	A	B	A	C	A	A	A	A	A			A	A		A	B	A	A	A	A	B	A	A	A			A
Molybdic Acid	H_2MoO_4					A		A	A	A										A			A	A				
Monochloro Benzene (dry)	C_6H_5Cl	B		B	A	A	A	A	A	A		A	B	B		D	D	D	A	C	B		A	A	A			
Morphine	$C_{17}H_{19}NO_3$	B	B	B	B	A	A	A	B	A		A	A			D	B	D	A	A	A	A	A		A			
Mustard	$C_2H_8Cl_2S$	B	A	B	B	A	A	A	A	A		A	B	B		A	D	A	A	C	B	A	A	A	A		A	
Naptha	—	A	A	A	B	A	A	A	A	A		A	A	A		B	D	A	A	B	A	A	A	A	A		A	A

(Continued)

TABLE A.6a (Continued)
Chemical Compatibility Guide

Media	Chemical Formula	Aluminum	Brass	Carbon Steel	Ductile Iron / Cast Iron	316/316Ti/321 S.St	17-4PH	Alloy20	Monel	Hastelloy C	Inconel 625	Titanium	Bronze	304 Stainless Steel	Duplex	Buna N (Nitrile)	EPDM/EPR	Viton	Flexible Graphite	Delrin	Peek	PVDF	Teflon and Reinforced Teflon	PCTFE	UHMWPE	Vespel	PFA	KEL-F
Naphthalene	$C_{10}H_8$	A	B	A	B	A	A	A	A	A		A	A	A		D	D	A	A	B	A	A	A		A		A	A
Natural Gas, Sour	—	B	A	A	B	A	A	A	B	A			A	A		A	A	B	A	B	A		A					
Nickel Ammonium Sulfate	$H_8N_2NiO_8S_2$	D	D	D	D	A		A	C				D			A	B	B		C	A		A	A	B			
Nickel Chloride	Cl_2Ni	D	D	D	D	B	B	A	C	A		A	D	D		A	B	A	A	A	A	A	A	A	A			A
Nickel Nitrate	N_2NiO_6	D	D	D	D	B	B	A	C	B		B	C	B		A	A	A	A	C	A		A	A	A	D	A	A
Nickel Sulfate	NiO_4S	D	D	D	D	B	B	A	B	C		B	C	B		A	B	B	A	B	A		A	A	B		A	A
Nicotinic Acid	$C_6H_5NO_2$	A	A	C	C	A		A	B				B			D	D	B		C	A		A		A			
Nitric Acid 10%	HNO_3	C	D	D	D	B	B	A	D	B		A	D	A		C	B	A	A	D	B	B	A	A	A	D	A	B
Nitric Acid 30%	HNO_3	D	D	D	D	B	B	A	D	A		A	D	A		C	C	A	C	D	D	B	A	A	A		A	B
Nitric Acid 80%	HNO_3	D	D	D	D	B	C	A	D	B		B	D	A		D	D	B	D	D	D	B	A	A	B		A	B
Nitric Acid 100%	HNO_3	B	D	D	D	B	D	A	D	B		B	D	A		D	D	A	D	D	D	B	A	A	B		A	B
Nitric Acid Anhydrous	—	B	D	D	C	B		A	D				D	B		D	D	A	C	D	A		A		A			
Nitrobenzene	$C_6H_5NO_2$	B	D	A	B	A	B	A	B	B		A	B	B	A	D	C	C	B	B	D	B	A	A	A	A	A	B
Nitrogen	N_2	A	A	A	A	A	A	A	A	A		A	A	A		A	B	A	A	A	A	B	A	A		A	A	
Nitrous Acid 10%	HNO_2	D	D	D	D	B	B	B	D	C		A	C	B		C	A	A	A	B	B	B	A	A	A			B
Nitrous Gases	—	B	D	B	C	A	B	B	D								A			B	B	B	A		A			
Nitrous Oxide	N_2O	B	B	B	C	B	B	B	D	B		B	C	B		B	A	A	B	A	A	D	A	A				
Oils & Fats	—	B				A		A								B	B	A		A		A	A	A				
Oils & Animal	—	A	A	A	A	A	A	A	A	A				A		A	B	B		A		A	A	A				
Oils, Petroleum (refined)	C_6H_6	A	B	A	B	A	A	A	A	A			C	A	A	A	A	B	A	B	A	A	A		A			
Oils, Petroleum (sour)	C_6H_6	A	C	B	C	A	A	A	A	A			C	A	A	A	A	D	D	B	A	A	A		A			
Oils, Water Mixture	—	A	A	B	B	A	A	A	A	A			A	A		A	D	A	A	A	A	A	A		A	A	A	
Olaic Acid	—	B				B		B	A	A			A			B	C			D	A		A					
Oleic Acid	$C_{18}H_{34}O_2$	B	D	C	C	B	A	A	B	B		B	C	B		B	D	C	A	C	A	A	A	A		A		B
Oleum	H_2O_4S	C	C	B	D	B	B	B	C	B		D	D	D		D	C	A	C	D	D	D	A	A		D		A
Oleum Spirits	—	D	D		D	B	A	B	D							C	D	A		D		C	A					

(Continued)

TABLE A.6a (Continued)
Chemical Compatibility Guide

Media	Chemical Formula	Aluminum	Brass	Carbon Steel	Ductile Iron / Cast Iron	316/316Ti/321 S.St	17-4PH	Alloy20	Monel	Hastelloy C	Inconel 625	Titanium	Bronze	304 Stainless Steel	Duplex	Buna N (Nitrile)	EPDM/EPR	Viton	Flexible Graphite	Delrin	Peek	PVDF	Teflon and Reinforced Teflon	PCTFE	UHMWPE	Vespel	PFA	KEL-F
Olive Oil	—	A	B	B	A	A	A	A	A	A		A	A	A		A	B	A	A	A	A	A	A	A	A	A	A	A
Oxalic Acid	$C_2H_2O_4$	C	C	C	D	B	B	B	B	B		A	C	B		C	B	A	A	C	C	A	A	A	A	A	A	D
Oxygen	O_2	A	A	A	A	A	A	A	A	A	A	A	A		A	B	A	B	A	D	C	B	A		B	A	A	A
Ozone (dry)	O_3	A	A	A	A	A		A	A	A			A	B		D	A	B	C	C	A	B	A	A	B		A	
Ozone (wet)	O_3	B	B	B	C	A		A	A	A			B			D	B	B		C	A	B	A	A	B		A	
Paints & Solvents	—	A	A	A	A	A	A	A	A	A		A		A		D	D	A	A	A	A		A					
Palmitic Acid	$C_{16}H_{32}O_2$	B	C	C	C	B	A	B		A		A	B	B		B	D	A	A	D	A	B	A		A		A	
Palm Oil	—	A	B	C	C	B	A	A	A	A		A	C	A		B	D	B	A	D	A	A	A		A		A	
Paper Pulp	—	C	B	B	B	A		A	B							B	B	B	A	A	A	A	A					
Paraffin	CnH_{2n+2}	A	A	B	B	A	A	A	A	A		A	A	A		A	D	A	A	A	A	A	A	A	A		A	A
Paraformaldehyde	CH_2O	B	B	B	B	B		B	B	A						B	C	A		A			A					
Paraldehyde	$C_6H_{12}O_3$					B		B	B	A				B		B	D		A	A	A	A	A	A				
Pentane	C_5H_{12}	A	A	B	B	A	A	A	A	B			B	B		A	D	A	A	B	A	C	A	A	A		A	
Perchloroethylene (dry)	C_2Cl_4	B	C	B	B	A	A	A	A	A		A	B	B		D	D	A	A	B	A	A	A		A		A	A
Petrolatum (Vaseline Petr. Jelly)	—	B		C	C	B	A	A	A	A	A	B	A	A		A	A	A	A	B	A	A	A	A	A	A	A	
Phenol	C_6H_6O	A	B	C	D	A	A	A	A	A		A	A	B		D	B	B	A	C	D	B	A	A	A	A	A	B
Phosphate Ester 10%	—	C	D	A	C	A		A	A	A		A	C			D	A	A	A	A			A			A	A	
Phosphate Acid 10%	—	C	D	D	D	C		B		A		B				B	B	A	A	D			A		A	A	A	
Phosphoric Acid 50% (cold)	H_3O_4P	D	D	D	D	C	B	B	D	B		B	D	D		B	B	B	A	D	A	B	A	A	A	A	A	A
Phosphoric Acid 50% (hot)	H_3O_4P	D	D	D	D	C	C	B	D	B		C	D	D		B	B	A	A	A	A	A	A	A	A	A	A	A
Phosphoric Acid 85% (cold)	H_3O_4P	D	D	D	D	C	C	B	D	B		B	D	D		B	B	B	A	D	A	B	A	A	A	A	A	A
Phosphoric Acid 85% (hot)	H_3O_4P	D	D	D	D	D	D	B	D	B		C	D	D		C	C	B	A	D	B	B	A	A	A	A	A	A
Phosphoric Anhydride	P_2O_5	C		D		A	A	B		A		D	C	B		D	B	D	A	A	A	D	A				A	

(Continued)

TABLE A.6a (Continued)
Chemical Compatibility Guide

Media	Chemical Formula	Aluminum	Brass	Carbon Steel	Ductile Iron / Cast Iron	316/316Ti/321 S.St	17-4PH	Alloy20	Monel	Hastelloy C®	Inconel 625	Titanium	Bronze	304 Stainless Steel	Duplex	Buna N (Nitrile)	EPDM/EPR	Viton	Flexible Graphite	Delrin®	Peek®	PVDF	Teflon® and Reinforced Teflon®	PCTFE	UHMWPE®	Vespel®	PFA	KEL-F®
Phosphorus Trichloride	PCl_3	C		B	C	A	A	A	A	B		A		A		D	B	B	A	D			A		A			A
Phthalic Acid	$C_8H_6O_4$	C	B	D	C	B	A	A	B	A		A	B	B		C	C	A	B	C	A	B	A	A	A	A	A	
Phthalic Anhydride	$C_8H_4O_3$	B	B	D	C	B	A	B	A	A			B	A		C	B	A	A	C	A	A	A		A	A		
Picric Acid	$C_6H_3N_3O_7$	C	D	D	D	B	B	B	D	B		A	C	B		C	B	A	A	B	A	B	A				A	A
Pineapple Juice	—	A	C	C	C	A	A	A	A	A		A		A		A	D	A	A	A	A	A	A					
Pine Oil	—	B	B	B	B	A	A	A	A			A	C			C	D	A	B	D	A	A	A	A				
Pitch (Bitumen)	C_6H_6					A		A								B	D	A	A	A	A		A					
Polysulfide	—	C	D	B	B	B	B	B	B							B	B	B	A	C			A					A
Polyvinyl Acetate	$C_6H_{12}O_2$	B	B	C	B	B	B	A	B	A		A	B	A			B	A	A	A	A	A	A					
Polyvinyl Chloride	C_2H_3Cl	B	D	D	B	B	B	B	B	A				A			B	A	A	A			A					
Potassium Bicarbonate	$CHKO_3$	C		A	D	A	B	A	B	B		A		B		B	A	A	A	B	A	A	B	A	A		A	A
Potassium Bichromate	$Cr_2K_2O_7$	B	B	B		A	A	A	A	A	B	A	C	B		B	A	B	A	B	A	A	B	A	A	D		
Potassium Bisulfate	HKO_4S	B	C	C	C	C	A	A	C	A		A		B		B	B	A	A	A	A	A	A					A
Potassium Bisulfite	HKO_4S	C	C	D	D	B	A	B	D	A		A	C	B		A	B	A	A	A		A	A					A
Potassium Bromide	BrK	C	C	C	D	A	B	B	B	B		A	C	B		A	B	A	A	A	A		A				A	
Potassium Carbonate	CK_2O_3	C	B	B	B	A	A	A	B	A		A	C	B		A	B	A	B	B	A	A	A		A			A
Potassium Chlorate	$ClKO_3$	B	B	B	B	A	A	A	B	A		A	C	B		A	B	A	B	B	A	B	A		A	D		A
Potassium Chloride	KCl	D	C	C	B	B	B	A	A	A		A	C	B		A	A	A	A	A	A	A	B	A	A			A
Potassium Chromate	CrK_2O_4	B	B	B	B	B	A	A	B	A		A	A	B		B	A	A	A	A		A	A		A			A
Potassium Cyanide	KCN	D	D	B	B	B	A	A	A	A		A	D	B		A	B	A	A	C		A	A		A	D		A
Potassium Dichromate	$Cr_2K_2O_7$	B	D	C	C	B	A	A	C	B		A	C	B		A	B	A	B	C	A	A	A					
Potassium Ferricyanide	$C_6FeK_3N_6$	C	D	C	C	A	B	A	A	B		A	D	B		A	A	A	B	A	A	B	A			D		B
Potassium Ferrocyanide	$C_6FeK_4N_{6+4}$	C	C	C	C	B	A	A	A	B		A	C	B		A	A	A	A	A	A	A	A					A
Potassium Hydroxide Dilute (cold)	HKO	D	D	B	D	B	A	A	A	A	B	C	D	B		A	A	D	B	C	A	A	A		A	D		B

(Continued)

TABLE A.6a (Continued)
Chemical Compatibility Guide

Media	Chemical Formula	Metals														Elastomers				Polymers								
		Aluminum	Brass	Carbon Steel	Ductile Iron / Cast Iron	316/316Ti/321 S.St	17-4PH	Alloy20	Monel	Hastelloy C	Inconel 625	Titanium	Bronze	304 Stainless Steel	Duplex	Buna N (Nitrile)	EPDM/EPR	Viton	Flexible Graphite	Delrin®	Peek®	PVDF	Teflon® and Reinforced Teflon	PCTFE	UHMWPE	Vespel®	PFA	KEL-F®
Potassium Hydroxide to 70% (cold)	HKO	D	D	C	D	B	B	B	A	B		C	D	B		B	B	D		C	A		A			D		
Potassium Hydroxide Dilute (hot)	HKO	D	D	B	D	B	C	B	A	A		D	D	B		B	A			D	A		A			D		
Potassium Hydroxide to 70% (hot)	HKO	D	D	C	D	B	C	B	A	B		D	D	B		C	B			D	A	B	A			D		
Potassium Iodide	IK	B	C	B	B	B	A	A	B	B		A	D	A		A	B	A	B	A	A	B	A		A	A		B
Potassium Nitrate	KNO$_3$	B	B	B	B	B	B	A	A	A		A	B	B		A	B	A	B	B	A		A					A
Potassium Oxalate	C$_2$K$_2$O$_4$	B	B	B	A	B	B	A	A	A		A	A	B		C		A	B	C	A		A					
Potassium Permanganate	KMnO$_4$	B	B	B	B	A	A	A	A			A	B			A	B	A	C	C	A		A		A	D		
Potassium Phosphate (mono)	KH$_2$PO$_4$	D	C	B	C	B	A	B	B	B				A	A	A	A	A		A			A					
Potassium Phosphate Di-basic	K$_2$HPO$_4$	B	B	A	A	A	B	A	B	B		A				A	B	A		A			A					
Potassium Phosphate Tri-basic	K$_3$PO$_4$	D		A	A	B	A	B	B							B	B	A		A			A					
Potassium Sulfate	K$_2$O$_4$S	B	C	B	B	A	A	A	A	A		A	B	B		A	A	A	A	B	A	A	A			A		B
Potassium Sulfide	HK$_2$S+	C	D	D	C	A	B	B	D	B		A	D	B		A	B	A	A	A	A		A			A		A
Potassium Sulfite	K$_2$O$_3$S	B	B	D	C	B	B	B	C	B		A	C	C		B	A	A	A	A	A		A			A		
Producer Gas	—	B	B	B	B	B	A	B	A					A	A	A	D	A	A	A	A		A					
Propane Gas	C$_3$H$_8$	A	A	A	A	A	A	A	B	A		A	B	A	A	A	D	A	A	B	A		A			A		A
Propyl Bromide	C$_3$H$_7$Br	B	B	B	B	A	B	A	B				B			B	B	B	A	A	A		A					
Propylene Glycol	C$_3$H$_8$O$_2$	C	B	B	B	B	A	A	A	A		A	A	B		A	D	B	A	B	A		A		A	B		
Pyridine	C$_5$H$_5$N	B	B	B	B	B	A	A	C	B		B	B	A		D	B	D	A	B	B	D	A	A	A	A		B
Pyrolgalic Acid	—	B	B	B	B	B	A	A	A	A		A	B	B		A	B	D	A	D	A		A			A		
Quench Oil	—	A	A	A	A	A	B	A	B	A		A	B	A		A	A	A	A	A	A		A					A
Quinine Sulfate (dry)	C$_{40}$H$_{50}$N$_4$O$_8$S		B		A	A	A	A	A	A		A	A	B		B		B		A			A			A		
Resins & Rosins	—	A	A	A	A	A	A	A	A	A			A			B	B	A		A	A		A			A		
Resorcinal	C$_6$H$_6$O$_2$	A	B	C	C	A	A	A	A	A		A	B	A		D	B	B		B	A		A					

(Continued)

TABLE A.6a (Continued)
Chemical Compatibility Guide

Media	Chemical Formula	Aluminum	Brass	Carbon Steel	Ductile Iron / Cast Iron	316/316Ti/321 S.S.	17-4PH	Alloy20	Monel	Hastelloy C®	Inconel 625	Titanium	Bronze	304 Stainless Steel	Duplex	Buna N (Nitrile)	EPDM/EPR	Viton	Flexible Graphite	Delrin®	Peek®	PVDF	Teflon® and Reinforced Teflon	PCTFE	UHMWPE®	Vespel®	PFA	KEL-F®
Road Tar	$C_2H_4O_3$	A	A	A	A	A	A	A	A			A		A		B		A		A			A					
Roof Pitch	—	A	A	B	B	A	A	A	A			A		A				A		A			A					
Rosin Emulsion	$C_{15}H_{20}O_6$	A	A	A	A	A	A	A	B	A		A	A	A		D	B	C	A	C	A		A					
R P-1 Fuel	—	B	B	C	C	B	A	A	B				A			A		A		B	A		A	A				B
Rubber Latex Emulsions	—	C	C	D	D	A	B	A	C	A		A	B	B		A	B	B		C	A	A	A					
Rubber solvents	—	B	B	C	C	B	B	A	A	A		A	B	A		A	A	A	A	A	A	A	A		A			A
Salad Oil	—	B	B	B	D	B	B	B	B	B		A	B	B		A	A	B	A		A	A	A		A			
Salicylic Acid	$C_7H_6O_3$					B		B								A			A	C								
Salt (NaCl)	$NaCl+H_2O$	B	D	C	D	B	B	B	A	A		A		C	B	A	A	B	A	A	A	A	A	A	A		A	A
Salt Brine	NaCl	C	C	C	D	B	A	B	B	A			C	B	B	A	B	B	A	B	A		A					
Sauerkraut Brine	—	A	B	A	B	A	A	B	B	A		A	A	A		B	A	B	A	A	A		A					
Sea Water	—	B	B	B	B	B		B		B		B	A			B	A	B	A	A	A	A	A	A	A			A
Sewage	—	D				C	C	A	B	A		A	D	D				B	A	D			A		A			A
Shellac	—	D	D	D	D	B		B	B	A		A	D	D		B		B	A	D			A					
Silicone Fluids	Si	D	D	D	D	A	B	A	D	A		A	D	B		C	A	A	A	A	A	A	A	A	A			A
Silver Bromide	AgBr	B				A		A		A		A				A	A	A			A							
Silver Cyanide	CAgN	C	B	B	B	A	A	A	B	A		B	B	A		A	B	A	A	A	A		A	A	A		A	A
Silver Nitrate	CAgN	B	B	C	C	B	B	A	C	B		A	C	B		B	A	D		B	A	B	A	A			A	A
Silver Plating sol.	—	D	B	B	C	A	A	B	B	B		A	B			A	A	A	A	B	A		A	A	A				
Soap Solutions (Stearates)	$C_{21}H_{42}O_4$	B				B	B	B	A	B		A	A	A		B	A	B	A	B	A	B	A		A			
Sodium Acetate	$C_2H_4O_2$	C	B	C	C	B	A	A	A	A	A	A	B	A		A	A	A	A	A	A	A	A	A	A			A
Sodium Aluminate	$AlNaO_2$	B				A	A	A	A	B			C			D	B	A	C	A	A		A	A				
Sodium Benzoate	$C_7H_6O_2$	D	D	D	D	A	A	A	B	B		B	C	B		A	B	A	A	B	A	A	A	A				A
Sodium Bicarbonate	$CHNaO_3$	C	B	D	D	B	B	A	C	B		A	B	B		A	B	A	A	C	A	B	A		A			A
Sodium Bichromate	$Cr_2Na_2O_7$	C	B	C	C	B	B	A	B	B		B	B	B		A	B	A	A	A	A	A	A	A				A
Sodium Bisulfate 10%	H_2NaO_4S	D	B	B	D	A	B	B	B	B		A	B	C		A	B	A	A	A	A	B	A		B			B
Sodium Bisulfite 10%	$HNaO_3S$	D	B	D	D	A		B	B	B			B			A	B	A		D			A				A	

(Continued)

TABLE A.6a (Continued)
Chemical Compatibility Guide

Media	Chemical Formula	Aluminum	Brass	Carbon Steel	Ductile Iron / Cast Iron	316/316Ti/321 S.St	17-4PH	Alloy20	Monel	Hastelloy C®	Inconel 625	Titanium	Bronze	304 Stainless Steel	Duplex	Buna N (Nitrile)	EPDM/EPR	Viton	Flexible Graphite	Delrin®	Peek®	PVDF	Teflon® and Reinforced Teflon	PCTFE	UHMWPE®	Vespel®	PFA	KEL-F®
Sodium Borate	$C_{15}H_{22}O_2$	B	B	C	C	B		B	B							A	B	A	A	A	A		A		A			
Sodium Bromide 10%	BrNa	B	B	C	D	B		B	B							A	B	A	A	A				A					
Sodium Carbonate (Soda Ash)	CNa_2O_3	D	B	B	B	B	A	A	B	A		A	B	A		A	B	A	A	B	A	A	A	A	A	A	D	A	A
Sodium Chlorate	$ClNaO_3$	C	B	B	C	B	A	B	B	B		A	B	A		A	B	A	C	A	A	A	A	A		A			A
Sodium Chloride	$ClNa$	B	C	B	C	B	B	B	C	B		A	B	B		A	A	A	A	B	B	A	A	A		A	A		A
Sodium Chromate	$CrNa_2O_4$	B	C	B	B	A		A	A	B			B	B		A	B	A	A	D	D	A	A	A		A	A		A
Sodium Citrate	$C_6H_5Na_3O_7$	D		B		A		B	A	A										A	A		A	A					
Sodium Cyanide	$CNNa$	D	D	B	B	A	B	A	B	A		A	D	A		A	A	A	A	A	A	A	A	A		A	A		A
Sodium Ferricyanide	$C_6FeN_6Na_3$	B	D		C	B	B	A	B	B			C	B						A				A					A
Sodium Fluoride	FNa	B	C	D	D	B	B	A	A	A		A	B	D		A	B	A	A	B	A	A	A	A		A	A		
Sodium Hydroxide 20% (cold)	$NaOH$	D	C	A	A	A	A	B	A	A	A	A	B	B		A	B	B	A	C	A	A	A	A		A	D		A
Sodium Hydroxide 20% (hot)	$NaOH$	D	C	B	B	A	B	B	B	A	A	A	B	B		B	B	C	A	C	A	A	A	A		A	D		B
Sodium Hydroxide 50% (cold)	$NaOH$	D	D	A	B	A	A	B	A	B		A	D	B		A	B	C	B	C	A	A	A	A		A	D		A
Sodium Hydroxide 50% (hot)	$NaOH$	D	D	B	B	A	B	B	B	B		A	D	C		B	C	C	C	C	A	A	A			A	D		A
Sodium Hydroxide 70% (cold)	$NaOH$	D	D	C	C	A	A	B	A	A		A	D	C		B	C	D	A	D	A	A	A			A	D		A
Sodium Hydroxide 70% (hot)	$NaOH$	D	D	C	C	A	B	B	B	B		A	D	D		D	D	D	A	D	A	A	A			A	D		
Sodium Hypochlorite (Bleach)	$ClNaO$	D	D	D	D	D	D	D	D	A		C	D	D		B	C	B	B	D	B	A	A		A	D			A
Sodium Hyposulfite	$Na_2S_2O_4$	B				B		B	B											A	A		A						
Sodium Lactate	$C_3H_5NaO_3$	D		A		A		A	B							A	A	A	A	A				A					
Sodium Metaphosphate	$Na_6O_{18}P_6$	C	D	C	B	B		A	B	A			B	A		A	A	A	A	B	A	A	A	A					

(Continued)

TABLE A.6a (Continued)
Chemical Compatibility Guide

Media	Chemical Formula	Aluminum	Brass	Carbon Steel	Ductile Iron / Cast Iron	316/316Ti/321 S.St	17-4PH	Alloy20	Monel	Hastelloy C®	Inconel 625	Titanium	Bronze	304 Stainless Steel	Duplex	Buna N (Nitrile)	EPDM/EPR	Viton	Flexible Graphite	Delrin®	Peek®	PVDF	Teflon® and Reinforced Teflon	PCTFE	UHMWPE®	Vespel®	PFA	KEL-F®
Sodium Metasilicate (cold)	Na₂O₃Si	D	B	B	C	A	A	A	A	A			A	A		B	B	B		D	A		A					
Sodium Metasilicate (hot)	Na₂O₃Si	D	B	C	D	A	A	A	A	A				A						D	A		A					
Sodium Nitrate	NNaO₃	B	B	B	B	A	B	A	B	B		A	B	B		C	A	A	C	A	A	A	A		A	A		A
Sodium Nitrite	NNaO₂	A		B	B	B	A	A	B	B			B	A		C	A	B		B	A	A	A					
Sodium Perborate	BNaO₃	C	D	B	C	C	B	A	A	A			B	B		C	A	A	C	B	A	A	A	A	B	A	A	A
Sodium Peroxide	Na₂O₂	C	C	C	D	B	A	A	A	A			D	A		C	A	A	A	B	A	A	A					A
Sodium Phosphate (mono)	NaH₂PO₄	D	C	B	D	B	A	A	B	B		A	D	A		B	A	A		D	A	A	A					
Sodium Phosphate Di-basic	Na₂HPO₄	D	C	C	C	B	A	B	B	A		A	C	A		A	A	A	A	B	A	A	A					
Sodium Phosphate Tri-basic	Na₃PO₄	D	C	C	C	B	A	B	B	A		A	C	A		B	A	A		A	A	A	A					
Sodium Polyphosphate	Na₆O₁₈P₆	D	D	C	D	B	B	A	B	A		A	C	B		B	A	A	A	B			A					A
Sodium Salicylate	C₇H₅NaO₃					A		A		A			D	A	A		A	A	C	C	A	A	A		A			
Sodium Silicate	Na₂O₃Si	B	C	B	B	B	A	A	B	A		A	D	A		A	A	A	A	C	A	A	A	A	A		A	A
Sodium Silicate (hot)	Na₂O₃Si	C	D	C	C	B	B	B	B	B			D	B		A	B	A		C	A	A	A					
Sodium Sulfate	Na₂O₄S	A	B	B	B	A	A	A	A	A		A	B	B		A	A	A	A	B	A	A	A	A	A		A	A
Sodium Sulfide	HNaS	D	D	C	B	B	A	A	A	B		A	D	B		A	A	B	A	B	A	A	A		A		A	B
Sodium Sulfite	NaO₃S-	C	C	B	A	A	A	A	C	A		A	B	B		A	A	B	A	A	A	A	A	A	A			
Sodium Tetraborate	C₁₅H₂₂O₂	C	D	C	A	A	A	A	B	B		A	B	A		A	A	A	A	B	A	A	A	A	A			A
Sodium Thiosulfate	Na₂S₂O₃	A	D	C	C	B	A	B	B	B		A	B	A		A	A	A	A	C	A	A	A					A
Soybean Oil	—	B	B	C	C	A	A	A	A	A		A	A	A		A	B	A	A	D	A	A	A		A			
Starch	C₂₇H₄₈O₂₀	B	B	B	B	A	A	A	A	A			B	A		A	C	A	A	A	A	A	A	A	A		A	A
Steam (212°F)	—	B	B	A	A	A	A	A	A	A		A	B	A		D	B	C	A	D	A	A	A			A	A	A
Stearic Acid	C₁₈H₃₆O₂	B	C	C	C	B	A	A	A	A		A	B	B		A	B	B	A	A	A	A	A	A	A			A
Styrene	C₈H₈	A	A	A	B	A	A	A	A	A		A	B	A		D	D	B	A	D	A	A	A				A	A
Sugar Liquids	—	A	A	B	B	A	A	A	A	A		A	B	A		A	A	A	A	A	A	A	A	A	A			A

(Continued)

TABLE A.6a (Continued)
Chemical Compatibility Guide

Media	Chemical Formula	Aluminum	Brass	Carbon Steel	Ductile Iron / Cast Iron	316/316L/321 S.S.	17-4PH	Alloy20	Monel	Hastelloy C	Inconel 625	Titanium	Bronze	304 Stainless Steel	Duplex	Buna N (Nitrile)	EPDM/EPR	Viton	Flexible Graphite	Delrin	Peek	PVDF	Teflon and Reinforced Teflon	PCTFE	UHMWPE	Vespel	PFA	KEL-F
Sugar, Syrups & Jam	—	B	B	B	C	B	A	B	A	A			C	A					A	A	A	A	A	A	A		A	
Sulfate Black Liquor	H₂O₄S	D	C	C	C	B	A	B	B	B			C	B		C	A	B	A	D	A	A	A	A				
Sulfate Green Liquor	H₂O₄S	D	C	C	C	B	A	B	B	B			D	B					A	A	A	A	A	A	A			
Sulfate White Liquor	H₂O₄S	D	C	C	C	B	B	B	C	B			D	B		C	A	B	A	A	A	A	A	A	A			
Sulfur	S	C	C	C	C	B	B	B	B	B			D	B		D	C	B	C	B	A	B	A	A	A	A	A	A
Sulfur Chloride	SCl	D	D	D	D	D	D	C	D	B			D	D		D	D	A	A	D	A	A	A			A	A	A
Sulfur Dioxide (dry)	SO₂	B	B	B	B	D	B	B	D	A		A	C	C		D	B	A	A	B	A	A	A	A		A	A	A
Sulfur Dioxide (wet)	SO₂	B	D	B		B	C	B	A	A		A	B	D		D	B	B	A	B	A	B	A		D			
Sulfur Hexafluoride	SF₆	A	B			A		A		C					A	B	B	B		D	A		A		D			
Sulfur, Molten	—	A	D	C	B	B	A	A	C	A		D	C	A		B	B	B	A	D	A	D	A	A	A	A	A	A
Sulfur Trioxide	SO₃	A	D	B	B	B	B	A	B	A		A	C	D		D	B	B	C	D	D		A	A	D	A	A	A
Sulfur Trioxide (dry)	SO₃	D	D	D	D	C	C	B	C	A		C	D	D		B	C	A	D	D	D	A	A	A	A	B	A	A
Sulfuric Acid 0 to 77%	H₂SO₄	D	D	C	C	B	C	A	B	B		D	D	C		D	C	B	D	D	D	A	A	A	A	D	A	A
Sulfuric Acid 100%	H₂SO₄	C	D	D	D	B	B	A	B	B		A	D	B		C	C	A	A	D	B	B	A	A	A	C	A	A
Sulfurous Acid	H₂O₃S	C	B	B	B	A	A	A	A	A		A	B	A		B	D	A	A	A	A	A	A	A	A	A	A	A
Tall Oil	—		B	B	C	B	B	B	B	B		B	A	B		B	A	A	A	A	A	B	A	A			A	A
Tannic Acid (Tannin)	C₂₇H₂₄O₁₈	A	A	A	A	A	A	A	A	A		A	A	A		B	B	A	B	B	A	A	A	A	A	A	A	A
Tanning Liquors	C₂₇H₂₄O₁₈			A	A	A	A	A	A	A			A	A		C	D	A	B	B	A		A					
Tar & Tar Oils	C₂H₄O₃	B	A	A	A	A	A	A	A	A			A	A		C	C	A	A	A	A	A	A	A	A	A	A	A
Tartaric Acid	C₄H₆O₆	B	D	D	D	A	A	A	A	B		A	B	C		B	B	A	B	B	A	B	A	A	A	A	A	A
Tetraethyl Lead	C₈H₂₀Pb	B	B	B	C	B	A	B	A				B			B	D	A	C	C	A	A	A		B	B		B
Toluol (Toluene)	C₇H₈	A	A	A	A	A	A	A	A	A		A	A	A		D	D	A	A	A	A	B	A	A		A	A	A
Tomato Juice	—	A	C	B	C	A	A	A	A	A			A	A		A	B	A	A	A	A	A	A				A	A
Transformer Oil	—	A	B	A	B	A	A	A	A	A			B	A		A	D	A	A	A	A	A	A	A	B	A	A	A
Tributyl Phosphate	C₁₂H₂₇O₄P	A	A	A	C	B	A	B	A	B		B	B	B		D	D	D	A	A	A	B	A	A		A	A	A
Trichlorethylene	C₂HCl₃	B	B	B	C	B	B	B	B	A			B	B		D	D	B	A	A	A	B	A		B			
Trichloroacetic Acid	C₂HCl₃O₂	D	B	C	D	D	C	B	B	A		D	B	D		C	B	C	A	D	A	C	A		A	A	A	A
Triethanolamine	C₆H₁₅NO₃			C	C	B	A	C	B			C	C	A		C	A	D	A	A	A	B	A			A	A	A
Triethylamine	C₆H₁₅NO₃	B	B	A		B		B		A			A			B	A	C	A	D	A	B	A					A

(Continued)

TABLE A.6a (Continued)
Chemical Compatibility Guide

Media	Chemical Formula	Aluminum	Brass	Carbon Steel	Ductile Iron / Cast Iron	316/316Ti/321 S.S!	17-4PH	Alloy20	Monel	Hastelloy C®	Inconel 625	Titanium	Bronze	304 Stainless Steel	Duplex	Buna N (Nitrile)	EPDM/EPR	Viton	Flexible Graphite	Delrin®	Peek®	PVDF	Teflon® and Reinforced Teflon	PCTFE	UHMWPE®	Vespel®	PFA	KEL-F®
Trisodium Phosphate	Na_3O_4P	D		A	B	A	A	A	A	A			C	B		A	A	B	A	A	A	A	A			D		
Tung Oil	—	B	B	B	B	A	A	A	A	A			B	A		A	D	A		A	A	A	A					
Turpentine	$C_{10}H_{16}$	A	C	B	B	B	A	A	A	A		B	B	A		B	D	A	A	A	A	A	A	A			A	A
Urea	CH_4N_2O	B	B	C	C	B	A	A	B	B		A	C	B		C	B	C	A	A	A	A	A	A	A			
Uric Acid	$C_5H_4N_4O_3$	C				A	B	A		B		A	B	B					A	B		A	A		A		A	
Varnish	—	A	B	C		A	A	A	A	A			A	A		A	D	A	A	A	A		A	A				A
Vegetable Oils	—	A	B	A	B	A	A	A	A	A			B	A		A	C	A	A	B	A	B	A	A	A		A	A
Vinegar	$C_2H_4O_2$	D	C	D	D	A	A	A	B	A		A	C	C		C	A	C	A	C	A	B	A				A	A
Vinyl Acetate	$C_4H_6O_2$	B	B	A	B	B	A	B	B	A			C	B		D	B	C	A	B	A	B	A					
Water (distilled)	H_2O	A	B	C	D	A	A	A	A	A	A	A	A	A		A	A	B	A	B	A	A	A	A	A	B	A	A
Water (fresh)	H_2O	A	B	C	C	A	A	A	A	B		A	A	A	B	A	A	B	A	A	A	A	A	A	A	B	A	A
Water (acid mine)	—	D	D	D	D	B	A	A	B	B		A	C	B		A	A	D	A	A	A	A	A	A			A	
Waxes	—	A	A		A	A		A	A				A	A		A	C	A	A	A		A	A	A				A
Whiskey & Wines	—	C	B	D	D	A	A	A	A	A		A	B	A		A	A	A	A	A	A	A	A	A				A
Xylene, Xylol (dry)	$C_{24}H_{30}$	A	A	B	B	A	A	A	A	A			A	B		D	D	B	A	A	A	A	A	A			A	A
Zinc Bromide	Br_2Zn	D	B		D	B		B	B	B				D		B	B	B	A	A	A	A	A		B		A	A
Zinc Hydrosulfite	ZnS_2O_4	D	C	A	B	A	A	A	B	B			D			A	A	A	A	C	A	A	A					
Zinc Sulfate	$ZnSO_4$	D	C	C	D	B	A	A	B	B		A	C	A		A	A	A	A	C	A	A	A	A		A		A
Zinc	Zn	C				A				A				A			A	A					A					

A.7 Composition and Properties of Metallic and Other Materials

B. G. LIPTÁK (2017)

INTRODUCTION

This chapter concentrates on providing the reader with information concerning the composition, suppliers and properties of some of the piping, vessel, and other equipment materials. Naturally, the data is not complete, and the reader can find further information on alloys and their requirements for piping applications on the ASTM website: (http://www.wermac.org/materials/chemical_composition_part1.html.

Similarly, additional and detailed data can be found on metal and alloy compositions on the MetWeb website: http://www.matweb.com/search/CompositionSearch.aspx.

As to the Tables included in this chapter, Table A.7a provides information on the composition of metals and also provides a partial list of their manufacturers. Table A.7b provides the same information for Carbon, Ceramic, Plastic, and Rubber materials. Finally, Table A.7c lists the specific gravities, thermal expansion coefficients, and melting points of a list of alloys.

TABLE A.7a
Composition and Manufacturers of Metallic Materials

No.	Material	Manufacturer	Composition or Description
Metals			
17	Aluminum		
19	Aloyco-20	Alloy Steel Products Co.	Fe; 19–21 Cr; 28–30 Ni; 4.0–4.5 Cu; 2.5–3.0 Mo; 1.5 max. Si; 0.65–0.85 Mn; 0.07 max. C
19a	720 Alloy	General Plate	20 Mn; 20 Ni; Cu
54–60	Brass		Various commercial grades ranging 60–65 Cu; 35–40 Zn; 0.5–3.0 Pb
63	Brass, red		85 Cu; 15 Zn
66	Bronze, comm.		90 Cu; 10 Zn
73	Bronze, phosphor, 5% A		94.8–95.5 Cu; 4.3–5 Sn; P
74	Bronze, phosphor, 8% C		Cu; 7–9 Sn; 0.03–0.25 P
75	Bronze, phosphor 10% D		89.5–90 Cu; 10–10.5 Sn; P
76	Bronze, phosphor, spec. free cutting		88 Cu; 4 Zn; 4 Sn; 4 Pb
81	CA-FA20	Cooper Alloy	Fe; 19–21 Cr; 28–30 Ni; 3.5 Mo; 4–4.5 Cu; 0.07 max. C
82	CA-MM	Cooper Alloy	67 Ni; 30 Cu; 1.4 Fe; 0.1 Si; 0.15 C
86	Cast iron		Ordinary unalloyed cast iron
88	Chlorimet 2	Duriron Co.	63 Ni; 32 Mo; 3 max. Fe; 0.15 max. C; 1 Si; 1 Mn
89	Chlorimet 3	Duriron Co.	60 Ni; 18 Mo; 18 Cr; 2 Fe; 0.07 max. C; 1 Si; 1 Mn
111	Copper		99.9+ Cu
112	Copper, Be		97.5 Cu; 2.15 Be; 0.35 Ni
119	Corrosiron	Pacific Fdry.	Fe; 14.5 Si
140	Durichlor	Duriron Co.	Fe; 0.85 C; 14.5 Si; 3 Mo; 0.35 Mn
141	Durimet 20		Fe; 20 Cr; 29 Ni; 0.07 max. C; 2 Mo; 4 Cu; 1 Si
142	Durimet T	Duriron Co.	Fe; 19 Cr; 22 Ni; 0.07 max. C; 2 Mo; 1 Cu; 1 Si
143	Duriron	Duriron Co.	Fe; 0.80 C; 14.5 Si; 0.35 Mn
148	Everdur 1000	Amer. Brass	94.9 Cu; 4 Si; 1.1 Mn
149	Everdur 1010	Amer. Brass	95.8 Cu; 3.1 Si; 1.1 Mn
150	Everdur 1015	Amer. Brass	98.25 Cu; 1.5 Si; 0.25 Mn
156	Gold		99.99 Au
156a	Green gold		75% Au; 25% Ag
159	Hastelloy A	Haynes Stellite	Ni; 17–21 Mo; 17–21 Fe
160	Hastelloy B	Haynes Stellite	Ni; 24–32 Mo; 3–7 Fe; 0.02–0.12 C
161	Hastelloy C	Haynes Stellite	Ni; 14–19 Mo; 4–8 Fe; 0.04–0.15 C; 12–16 Cr; 3–5.5 W
162	Hastelloy D	Haynes Stellite	Ni; 8–11 Si; 2–5 Cu; 1 max. Al
163	Stellite 1	Haynes Stellite	Co; 28–34 Cr; 11–15 W
165	Stellite 6	Haynes Stellite	Co; 25–31 Cr; 3–6 W
184	Inconel	Int'l Nickel	79.5 Ni; 13 Cr; 6.5 Fe; 0.08 C; 0.2 Cu; 0.25 Mn
191	Lead		99.9 + Pb
192	Lead, antimonial		94 Pb; 6 Sb
193	Lead, antimonial		Pb; 4–12 Sb
196	Lead, chemical		99.93 Pb; 0.06 Cu
200	Lead, Te		99.88 Pb; 0.045 Te; 0.06 Cu
216	Monel	Int'l Nickel	67 Ni; 30 Cu; 1.4 Fe; 0.1 Si; 0.15 C
219	Muntz Metal		60 Cu; 40 Zn
224	Nickel	Int'l Nickel	99.4 Ni; 0.2 Mn; 0.1 Cu; 0.15 Fe; 0.05 Si
224a	Z-Nickel	Int'l Nickel	95 + Ni
226	Nickel–Silver 18% A		65 Cu; 18 Ni; 17 Zn
227	Nickel–Silver 18% B		55 Cu; 18 Ni; 27 Zn
227a	Ni-Span	Int'l Nickel	Ni, Ti, Cr, C, Mn, Si, Al
229	Ni-Hard	Int'l Nickel	Fe; 3.4 C; 1.5 Cr; 4.5 Ni; 0.6 Si
231	Ni-Resist	Int'l Nickel	Fe; 2.8 C; 14 or 20 Ni; 6 Cu (optional); 2 Cr; 2 Si
240	Platinum		99.99 Pt
268	Silver		99.9+ Ag
275	S.S. 301		Fe; 16–18 Cr; 6–8 Ni; 0.08–0.15 C
276	S.S. 302		Fe; 17–19 Cr; 8–10 Ni; 0.08–0.15 C
278	S.S. 303		Fe; 17–19 Cr; 8–10 Ni; 0.15 max. C; 0.07 min. P, S, Se; 0.6
279	S.S. 304		Fe; 18–20 Cr; 8–11 Ni; 0.08 max. C; 2 max. Mn
282	S.S. 310		Fe; 24–26 Cr; 19–22 Ni; 0.25 max. C

(Continued)

TABLE A.7a (Continued)
Composition and Manufacturers of Metallic Materials

No.	Material	Manufacturer	Composition or Description
Metals			
283	S.S. 316		Fe; 16–18 Cr; 10–14 Ni; 0.1 max. C; 1.75–2.75 Mo
284	S.S. 317		Fe; 17.5–20 Cr; 10–14 Ni; 0.1 max. C; 3–4 Mo
285	S.S. 321		Fe; 17–19 Cr; 8–11 Ni; Ti, 5 × C min.
286	S.S. 347		Fe; 17–19 Cr; 9–12 Ni; Cb, 10 × C min.
287	S.S. 403		Fe; 11.5–13 Cr; 0.15 max. C
290	S.S. 410		Fe; 11.5–13.5 Cr; 0.15 max. C
292	S.S. 416		Fe; 12–14 Cr; 0.15 max. C; 0.07 min. P, S, Se; 0.6 max. Zr, Mo
295	S.S. 430		Fe; 14–18 Cr; 0.12 max. C
303	S.S. 446		Fe; 23–27 Cr; 0.35 max. C; 0.25 max. N
360a	Steel		Plain carbon steel
368	Tantalum	Fansteel	99.9+ Ta
390	Worthite	Worthington Pump	Fe; 20 Cr; 24 Ni; 0.07 max. C; 3.25 Si; 3 Mo; 1.75 Cu; 0.5 Mn

TABLE A.7b
Composition and Manufacturers of Carbon, Ceramic, Plastic and Rubber Materials

No.	Material	Manufacturer	Description
Carbon and graphite			
401	Karbate (carbon)	National Carbon	Impervious carbon
402	Karbate (graphite)	National Carbon	Impervious graphite
Ceramics			
611	Lapp porcelain	Lapp Insulator Co.	Chemical porcelain
614	Pfaudler Glass lining	Pfaudler Co.	Glass-lined steel equipment
615	Plate glass		Polished plate glass, flat or bent
616	Pyrex	Corning Glass Wks.	Glass
Plastics			
700	Ace Saran	American Hard Rubber	Vinylidene chloride
710	Geon	B. F. Goodrich	Polyvinyl chloride
711	Haveg 41	Haveg Corp.	Phenolic asbestos
712	Haveg 43	Haveg Corp.	Phenolic graphite
713	Haveg 60	Haveg Corp.	Furan asbestos
714	Haveg 63	Haveg Corp.	Furan graphite
715	Heresite M 66	Heresite & Chem. Co.	Transparent molding powder
716	Heresite MF 66	Heresite & Chem. Co.	Black molding powder
717a	Kel-F	M. W. Kellogg	Polymerized trifluoroethylene
718	Koroseal	B. F. Goodrich	Plasticized polyvinyl chloride
731	Nylon FM-101	E. I. du Pont	Injection, compression, and extrusion moldings (tubing, sheeting, wire covering, gasketing)
731a	Plastisol		Polyvinyl chloride
735	Polythene	E.I. du Pont	Polyethylene
740	Saran	Dow Chemical	Vinyl chloride–vinylidene chloride copolymer
740a	Sirvene	Chicago Rawhide	Synthetic rubber
742	Teflon	E.I. du Pont	Polymerized tetrafluoroethylene
746	Tygon	U.S. Stoneware	Synthetic compounds
Rubber			
800	Ace Hard Rubber	American Hard Rubber	Vulcanized rubber
805	Butyl (GR-I)	Stanco Distributors	Solid copolymer of isobutylene and isoprene
820	Hycar (GR-A)	B. F. Goodrich	Nitrile-type synthetic rubber
829	Neoprene	E. I. du Pont	Polymer of chloroprene
836	Natural (soft)		
837	Natural (hard)		
838	GR-S (soft)		
839	GR-S (hard)		
853	Thiokol (GR-P)	Thiokol Corp.	

TABLE A.7c
Composition and Properties of Alloys*

Composition	Name	Specific Gravity	Thermal Expansion Coefficient (μ/m per °C)	Melting Point (°C)
Aluminum				
99.2Al	Aluminum 2S	2.71	23.94	660
95Al, 5Cu	Lynite, body alloy		26	
94Al, 4Cu, 0.5Mg, 0.5Mn, coated with 99.7+ Al	Alclad 17 ST	2.96	21.96	
90Al, 10Mg	Magnalium	2.50	24	600
70Al, 30Mg	Magnalium	2.00		435
98Al, 1.25Mn	Aluminum alloy 3S	2.74	23.04	640–655
95Al, 5Si	Aluminum–silicon 43	2.58	21.96	577–630
91Al, 9Zn		2.80	26	650
70Al, 30Zn			26	610
Bismuth				
53Bi, 32Pb, 15Sn	Eutectic fusible alloy			96
52Bi, 40Pb, 8Cd	Eutectic fusible alloy			91.5
50Bi, 27.1Pb, 22.9Sn	Rose metal			
50Bi, 27Pb, 13Sn, 10Cd	Eutectic fusible alloy; Lipowitz alloy			70–74
50Bi, 25Pb, 12.5Sn, 12.5Cd	Wood's metal	9.70		70
40Bi, 40Pb, 20Sn	Bismuth solder			111
54Bi, 26Sn, 20Cd	Eutectic fusible alloy			103
Cobalt				
Co, Cr, W alloy	Stellite No. 1 Alloy	8.59	14.4	1250
Co, Cr, W alloy	Stellite No. 6 Alloy	8.38	16.9	1275
Co, Cr, W alloy	Stellite No. 12 Alloy	8.40	15.8	1263
Co, Cr, W alloy	Stellite Star J-Metal	8.76	14.6	1270
Copper				
99.9 + Cu	Deoxidized copper	8.50	17.71	1082
99.90 + Cu, 0.01P	Deoxidized copper	8.91	17.71	1082
90Cu, 10Al	Aluminum bronze	7.6	16.5	1050
90Cu, 9Al, 1Fe	Resistac			1066
80–90Cu, 8–10Al, 6–7Fe	Ampco Metal	7.20		649
88–96.1Cu, 2.3–10.5Al, Fe, Sn	Aluminum bronze	7.50–8.19		1038–1071
95Cu, 5Mn	Manganese bronze	8.8		1060
82Cu, 15Mn, 3Ni	Manganin	8.5		
88.5Cu, 5Ni, 5Sn, 1.5Si	Barberite	8.80		1070
80Cu, 20Ni	Nickeline	8.5		1185
75Cu, 25Ni	Nickel coinage, U.S.A			1205
75Cu, 20Ni, 5Zn		8.58	16.40	1150
65Cu, 18Ni, 17Zn	Nickel silver 18% A	8.75	18.36	1110
60Cu, 40Ni	Constantan	8.4		1280
55Cu, 18Ni, 27Zn	Nickel silver 18% B	8.69		1055
94.8–96Cu, 3–4Si, 1–1.2Mn		8.46	16.99	1000
95.5Cu, 4.3Sn, 0.2P	Phosphor bronze 30	8.91	18.90	1050
95Cu, 4Sn, 1Zn	Coinage bronze	8.96		
91.6Cu, 8.25Sn, 0.15P	Phosphor bronze 47	8.91		
90Cu, 10Sn, trace P	Phosphor bronze 209	9.00		
90Cu, 10Sn	Bronze, gun metal	8.8	18	1000
79.7Cu, 10Sn, 9.5Sb, 0.8P	Phosphor bronze	8.8		
78Cu, 22Sn	Bell metal	8.7		870

* Table A.7c has been prepared by Oliver Seely (oseely@csudh.edu). It lists the compositions and physical properties of many common alloys. Compositions in the first column are given in mass percent units: For example, 50Bi, 25Pb, 12.5Sn, and 12.5Cd mean 50% Bi, 25% Pb, 12.5% Sn, and 12.5% Cd. As this table been prepared by an individual scientist, it can contain errors.

(Continued)

TABLE A.7c (Continued)
*Composition and Properties of Alloys**

Composition	Name	Specific Gravity	Thermal Expansion Coefficient (μ/m per °C)	Melting Point (°C)
67Cu, 33Sn	Bronze, speculum metal	8.6	18.6	745
90Cu, 10Zn	Commercial bronze; red brass	8.80	18.18	1050
89Cu, 9Zn, 2Pb	Hardware bronze	8.83	18.18	1050
85Cu, 15Zn	Red brass	8.75	18.72	1030
70Cu, 29Ni, 1Sn	Adnic.		16.29	1205
70Cu, 29Zn, 1Sn	Admiralty	8.17	20.16	935
67Cu, 33Zn	Brass, ordinary yellow	8.40	18.5	940
60Cu, 40Zn	Muntz metal			840
50Cu, 50Zn	Solder, refractory			900
Gold				
79Au, 21Al	Roberts-Austen (purple gold)			750
92Au, 8Cu	Standard gold, Great Britain			900
90Au, 10Cu	Coinage	17.17		940
84Au, 16Cu	Jewelry			895
75Au, 24Cu	Jewelry			925
50Au, 50Cu	Dark red gold			1000
75Au, 25Fe	Blue gold			1165
90Au, 10Pd	White gold, palladium gold			1265
80Au, 20Pd	Palau			1375
60Au, 40Pt	Platinum gold, white			1500
Iridium				
95Ir, 5Pt		22.38		
Iron				
99.94Fe, 0.025S, 0.017Mn, 0.012C, 0.005P	Armco Ingot Iron	7.86		1530
98.5Fe	Wrought iron	7.70		1510
80Fe, 20Al	Ferro-aluminum	6.30		1480
99Fe, 1C	Steel	7.83	12.0	1430
97Fe, 3C	Cast iron, white	7.60		1150
94Fe, 3.5C, 2.5Si	Cast iron, gray	7.0	11.2	1230
99.4Fe, 0.45Cu, 0.07Mo, 0.03C	Toncan copper; molybdenum iron	7.83	11.99	1525
90Fe, 10Cr, <0.5Mn, + 0.25C	Stainless steel	7.75		1510
90–92Fe, 8Cr, 0.4Mn, < 0.12C	Stainless steel	7.75	11.00	1450
90Fe, 8Cr, 0.4Mn, < 0.12C	Stainless iron	7.75	11.00	1450
88Fe, 16–17Cr, 0.4Mn, 0.1C max.	Stainless iron		9.99	
86–88Fe, 12–14Cr, 0.3C	Carpenter stainless steel 2	7.75		1425
86–88Fe, 12–14Cr, <0.5Mn, <0.1C, trace Ni	Defirust rustless iron	7.75		1480
86–88Fe, 12–14Cr, 0.1C	Carpenter stainless steel 1	7.78		1496
85–89 Fe, 10–14 Cr, <0.5Mn, <0.13C	Stainless iron	7.78	10.19	1490
82–86Fe, 12–16Cr, <0.5Ni, <0.05Si, <0.5 Mn, <0.12C	Ascoloy 33	7.64	10.89	1495
86.7Fe, 12.5Cr, 0.35Mn, 0.35Ni, 0.12C	Sterling stainless steel T	7.75	9.99	1430
86.4Fe, 13.5Cr, 0.1C	Stainless I	7.75	10.91	
84–86Fe, 12.5–14.5Cr, 0.5Mn max., 0.5Si max., 0.5Ni, min., 0.12–0.18C	Enduro S15	7.86	10.89	1475
84–86Fe, 12.5–14.5Cr, 0.5Mn max., 0.5Si max., 0.25Ni, 0.05–12C	Enduro S	7.86	10.89	1500
85.8Fe, 13.5Cr, 0.35Mn, 0.35C	Sterling stainless steel A	7.75	10.30	1425
85.6Fe, 14Cr, 0.35C	Stainless A	7.75	10.91	
85Fe, 13–14Cr, 2Ni max., 0.3–0.6Mn, 0.12C max.	Enduro KM1	7.75	9.99	1490
82–84Fe, 16–18Cr, <0.5Mn, C	Duraloy B	7.61		1510
82–84Fe, 16–18Cr, <0.5Mn, <0.1C, trace Ni	Special defirust rustless iron	7.75		1480
81–83Fe, 16.5–18.5Cr, 0.75Si, 0.1C max.	Enduro A	7.64	11.00	1510

(Continued)

TABLE A.7c (Continued)
Composition and Properties of Alloys*

Composition	Name	Specific Gravity	Thermal Expansion Coefficient (µ/m per °C)	Melting Point (°C)
82.5Fe, 16.5Cr, 0.65C, 0.35Mn	Sterling stainless steel B	7.72	10.91	1425
82Fe, 16–18Cr, 0.5Mn, 0.5Ni, 0.35C	Sweetaloy 16	7.83	11.00	1495
79–82Fe, 16–19Cr, <0.5Mn, <0.5Ni, <0.5Si, <0.12C	Ascoloy 66	7.64		
79–82Fe, 16.5–18.5Cr, 0.5Mn, 0.5–1.25Si, 0.25Ni, 0.1C max.	Enduro A	7.86	10.80	1490
79–81Fe, 16.5–18Cr, 1–1.1C, 0.75–1Si, 0.35–0.5Mn	Delhi hard	7.75	9.99	1500
78.7Fe, 20Cr, 1Cu, 0.3C	Carpenter stainless steel 3	7.70		1475
71–76Fe, 17–19Cr, 7–10Ni, <0.05Mn, 0.2C	Defistain rustless iron	7.83		1455
71–75Fe, 17–19Cr, 8–9Ni, <0.5Mn, 0.06–0.25C	Midvale V2A	7.89	16.99	1450
70–75Fe, 25–30Cr, <0.5Mn, 0.25C, trace Ni	Defiheat rustless iron	7.89		1595
70–75Fe, 17–20Cr, 7–10Ni, <0.5Mn, <0.5Si, 0.2C	Allegheny metal	7.86–0.95	17.30	1430–1470
69–75Fe, 16.5–19.5Cr, 7–10Ni, 0.75Si max., 0.5Mn max., 0.15C max.	Enduro KA2	7.86	15.98	1400
53.85Fe, 46Ni, 0.15C	Platinite	8.2	7.5e−6	1470
47–52Fe, 34–36Ni, 10–12Cr, 4.5–5.5Si, 0.15C	Rezistal 355C	7.81		
45Fe, 35–37Ni, 15–17Cr, 1.4–1.6Si, 0.6–0.8 ~ 1n, 0.5–0.7C	Standard Misco	7.97	13.50	1540
50Fe, 35Ni, 15Cr	Chromax castings	7.81	12.19	1480
48Fe, 35Ni, 12Cr, 5Si, 0.25C	Durimet B	7.89		1500
45Fe, 36Ni, 18Cr, 0.5Mn, 0.3C	Sweetaloy 20	7.97		1495
84.86Fe, 13.5Si, 1C, 0.4Mn, 0.18P, 0.05S	Tantiron	7.83		1315
84.3Fe, 14.5Si, 0.85C, 0.35Mn	Duriron	7.00	15.59	1265
66Fe, 17W, 10Cr, 3.5C, 2.5Mo	Cristite 1	7.61	15.59	
Lead				
94Pb, 6Sb	Battery plates			300
92–94Pb, 6–8Sb	Antimonial lead	11.0	27.00	245–290
90Pb, 10Sb	Magnolia			270
85Pb, 15Sb		10.4	19.5	250
75Pb, 19Sb, 5S-1, 1Cu	White metal	9.5		238
92Pb, 8Cd	Aluminum solder, U.S.P. 1,333,666			310
99.93Pb, 0.08Cu	Chemical lead	11.35	28.98	327
67Pb, 33Sn	Solder, plumber's	9.4	25.0	275
50Pb, 50Sn	Solder, half and half		24	225
Magnesium				
93.8Mg, 6Al, 0.2Mn	Dowmetal E	1.79	27e−6	616
90.8Mg, 6Al, 0.2Mn, 3Zn	Dowmetal H	1.83	27e−6	613
92.6Mg, 6.5Al, 0.2Mn, 0.7Zn	Dowmetal J	1.80	27e−6	618
98.5Mg, 1.5Mn	Dowmetal M	1.76	27e−6	649
90.8Mg, 8.5Al, 0.2Mn, 0.5Zn	Dowmetal O	1.80	27e−6	610
88.9Mg, 10Al, 0.1Mn, 1Zn	Dowmetal P	1.82	27e−6	596
90.2Mg, 9Al, 0.2Mn, 0.6Zn	Dowmetal R	1.81	27e−6	604
96.8Mg, 3Al, 0.2Mn, 3Zn	Dowmetal X	1.80	27e−6	635
Mercury				
80Hg, 20Bi	Bismuth amalgam			90
Nickel				
99–99.5Ni(+Co), 1–0.2C, 1–0.25Si, 1–0.3Mn 1–0.55Fe, 1–0.25Cu	Nickel	8.86		1450
Ni-Cr steel alloy of high Si content	Elcomet	8.03		
80Ni, 20Cr	Chromel A	8.4		
80Ni, 20Cr	Tophet A	8.50	13 00	1345
80Ni, 20Cr	Nichrome IV	8.50	13.21	1395

(Continued)

TABLE A.7c (Continued)
*Composition and Properties of Alloys**

Composition	Name	Specific Gravity	Thermal Expansion Coefficient (μ/m per °C)	Melting Point (°C)
58Ni, 22Cr, 6–7Cu, 4Mo, 2W, 1Mn, 6–7Fe	Illium (Illium G)	8.31	13.50	1300
60Ni, 20Cr, 10Fe, 1.75Mn, 0.5C	Firearmor	8.00	13.99	1330
73Ni, 17.5Co, 6.5Fe, 2.5Ti, 0.2Mn	Konel	8.61	10.66	1450–1500
90Ni, 3Cu, 1.5Al, 10 ± Si	Hastelloy D	7.81	11.59	1160
60–70Ni, 25–35Cu, 1–3Fe, 0.25–2Mn, 0.02–1.5Si, 0.5–0.3C	Monel metal	8.80		1330
60Ni, 33Cu, 6.5Fe	Monel metal	8.90	14	1360
61Ni, 23Fe, 16Cr	Chromel C	8.24		
60–62Ni, 23–26Fe, 10–11Cr, 2–2.5W, 1.2–1.5Mn, 0.3–0.35C	Midvale BTG	8.47		1450
G0Ni, 28Fe, 12Cr	Tophet C	8.19	13.70	1350
60Ni, 25Fe, 15Cr, 0.7C	Nichrome castings	8.08	12.10	
60Ni, 24Fe, 16Cr, 0.1C	Nichrome	8.17	13.70	1350
60Ni, 20Fe, 20Mo	Hastelloy A	8.80	10.71	1300
35Ni, 17Fe, 15Cr, 1.75Mn, 0.5C	Zorite	7.92		1300
Ni, Fe, Mo	Hastelloy C	8.91		1350
Palladium				
67Pd, 33Ag	Palladium alloy			1415
Platinum				
80–100Pt, 0–20Ir	Platinum-iridium		7.5–8.8	
90Pt, 10Ir	Platinum-iridium	21.61	8.8	
80–100Pt, 0–20Rh	Platinum-rhodium for thermocouples		8.8	
Silver				
92.5Ag, 7.5Cu	Standard (sterling) silver		18	920
92Ag, 8Cu	Silver-rupee			920
90Ag, 10Cu	Silver U.S. coins	10.3		890
80Ag, 20Cu	Jewelry		18	820
73Ag, 27Pt	Platinum solder			1160
66.7Ag, 33.3Pt	Platinum silver			1230
Tantalum				
99.5Ta	Tantalum	16.6	6.50	2850
Tin				
90Sn, 10Sb	Brittania			255
80Sn, 20Sb				320
75Sn, 12.5Sb, 12.5Cu	Antifriction	7.53		233
68Sn, 32Cd		7.70		180
97Sn, 3Cu	Rhine metal	7.35		300
67Sn, 33Pb	Solder, Tinman's			180
50Sn, 32Pb, 18Cd	Eutectic fusible alloy			145
Tungsten				
W2C	Blackor	14.0		
W, 0.5–0.75ThO$_2$	Tungsten filaments			
WC + 13% Co	Carboloy	14.10	6	
Zinc				
96Zn, 4Al, 0.05Mg	ASTM Alloy XXIII	6.7	26.9	380.9
95Zn, 4Al, 1Cu, 0.05Mg	Zamak-5	6.7	27.4	380.6
93Zn, 4Al, 3Cu, 0.05Mg	ASTM Alloy XXI	6.8	27.7	379.5
95Zn, 5Al		6.80	28	380
67Zn, 33Cu	Solder, readily fusible		20	795
60Zn, 40Cu	Solder, white		21	840

A.8 Steam and Water Tables

DEFINITIONS

Specific volume (v): It is a property of materials, defined as its volume per unit mass (Table A.8c). Specific volume is inversely proportional to density. If the density of a substance doubles, its specific volume, as expressed in the same base units, is cut in half. In the International System of Units (also referred to as SI or metric units), it is usually given as the number of cubic meters occupied by one kilogram of the particular substance (m^3/kg) or in terms of the number of cubic centimeters occupied by one gram of a substance (cm^3/g). To convert m^3/kg to cm^3/g, multiply by 1000; conversely, multiply by 0.001.

In the English System, it is most often used as the number of cubic feet occupied by one pound of the substance (ft^3/lbm). For a tabulation of the most often used specific volume units and the conversion among them, refer to Table A.8a.

TABLE A.8a
Common Specific Volume Units and the Conversions Among Them

Cubic centimeters per gram	$cm^3\ g^{-1}$	1000
Cubic meters per kilogram	$m^3\ kg^{-1}$	1
Liters per gram	$L\ g^{-1}$	1
Cubic feet per pound	$ft^3\ lb^{-1}$	16.0185
Imperial gallons per pound	$gal\ lb^{-1}$	99.777
US gallons per pound	$gal\ lb^{-1}$	119.83

Enthalpy (h): Specific enthalpy is a thermodynamic quantity equivalent to the total energy content of the unit mass of a substance relative to a reference condition. It is equal to the internal energy (u) of the system plus the product of pressure (p) and specific volume (v), in other words $h = u + pv$. In the English System, the usual unit of enthalpy is BTU/lbm, while in the SI (Metric) System, the usual unit is kilojoules per kilogram, where BTU = 1054.8 J = 1.055 kJ and lbm = 0.4536 kg and, therefore, BTU/lbm = 2.324 kJ/kg. In addition to joules and BTUs, other energy units such as ft-lbf and calories are also used. Enthalpy cannot be measured; only the difference between enthalpies can be determined if referred to a reference state at which it is given an arbitrary value, usually zero. In steam tables, the reference value ($h = 0$) is the enthalpy of saturated water at 32°F. For other substances, the reference values are different. For example for Freon-12, it is the enthalpy of saturated liquid at −4.44°C (−40°F).

For a tabulation of the most often used enthalpy units and for the conversion among them, refer to Table A.8b.

TABLE A.8b
Common Enthalpy Units and the Conversions Among Them

Joules per kilogram	$J\ kg^{-1}$	4186.8
Joules per gram	$J\ g^{-1}$	4186.8
Calories$_{IT}$ per gram	$cal_{IT}\ g^{-1}$	1
British thermal units$_{IT}$ per pound	$BTU_{IT}\ lb^{-1}$	1.8
Calories$_{th}$ per gram	$cal_{th}\ g^{-1}$	1.000669
British thermal units$_{th}$ per pound	$BTU_{th}\ lb^{-1}$	1.000669

Entropy (s): Entropy and energy are both abstract concepts, which human beings devised to aid in describing thermodynamic observations. Specific entropy (entropy per unit mass) is a thermodynamic quantity representing the unavailability of the thermal energy of a substance for conversion into mechanical work (it is often interpreted as the degree of disorder or randomness in the system). Entropy is a property of a substance, and it approaches zero at the absolute zero temperature. It is most often used quantified relative to a reference state at which it is given an arbitrary value, usually zero. In steam tables, the reference value ($s = 0$) is the entropy of saturated water at 32°F. For other substances, the reference values are different. For example for Freon-12, it is the entropy of saturated liquid at −4.44°C (−40°F). In the English System, the most often used entropy unit is BTU/(lbm°R), which in Metric Units equals 4186.8 J/(kg°K) or 4.187.8 kJ/(kg°K).

For a tabulation of the most often used entropy units and for the conversion among them, refer to Table A.8c.

TABLE A.8c
Common Entropy Units and the Conversion Among Them

Joules per kilogram kelvin	$J\ kg^{-1}\ K^{-1}$	4186.8
Joules per gram kelvin	$J\ g^{-1}\ K^{-1}$	4186.8
Joules per gram degree Celsius	$J\ g^{-1}\ °C^{-1}$	4186.8
Calories$_{IT}$ per gram degree Celsius	$cal_{IT}\ g^{-1}\ °C^{-1}$	1
British thermal units$_{IT}$ per pound degree Fahrenheit	$BTU_{IT}\ lb^{-1}\ °F^{-1}$	1
British thermal units$_{IT}$ per pound degree Rankine	$BTU_{IT}\ lb^{-1}\ °R^{-1}$	1
Calories$_{th}$ per gram degree Celsius	$cal_{th}\ g^{-1}\ °C^{-1}$	1.000669
British thermal units$_{th}$ per pound degree Fahrenheit	$BTU_{th}\ lb^{-1}\ °F^{-1}$	1.000669
British thermal units$_{th}$ per pound degree Rankine	$BTU_{th}\ lb^{-1}\ °R^{-1}$	1.000669

TABLE A.8d
Properties of Dry and Saturated Steam at Various Temperatures (Specific Volume, Enthalpy and Entropy)

Temp., °F/°C	Abs. Press., psia	Specific Volume, ft³/lbm			Enthalpy, BTU/lbm			Entropy, BTU/(lbm°R)		
		Sat. Liquid v_f	Evap. v_{fg}	Sat. Vapor v_g	Sat. Liquid h_f	Evap. h_{fg}	Sat. Vapor h_g	Sat. Liquid s_f	Evap. s_{fg}	Sat. Vapor s_g
32/0	0.08854	0.01602	3306	3306	0.00	1075.8	1075.8	0.0000	2.1877	2.1877
35/1.7	0.09995	0.01602	2947	2947	3.02	1074.1	1077.1	0.0061	2.1709	2.1770
40/4.4	0.12170	0.01602	2444	2444	8.05	1071.3	1079.3	0.0162	2.1435	2.1597
45/7.2	0.14752	0.01602	2036.4	2036.4	13.06	1068.4	1081.5	0.0262	2.1167	2.1429
50/10	0.17811	0.01603	1703.2	1703.2	18.07	1065.6	1083.7	0.0361	2.0903	2.1264
60/15.6	0.2563	0.01604	1206.6	1206.7	28.06	1059.9	1088.0	0.0555	2.0393	2.0948
70/21.1	0.3631	0.01606	867.8	867.9	38.04	1054.3	1092.3	0.0745	1.9902	2.0647
80/26.7	0.5069	0.01608	633.1	633.1	48.02	1048.6	1096.6	0.0932	1.9428	2.0360
90/32.2	0.6982	0.01610	468.0	468.0	57.99	1042.9	1100.9	0.1115	1.8972	2.0087
100/37.8	0.9492	0.01613	350.3	350.4	67.97	1037.2	1105.2	0.1295	1.8531	1.9826
110/43	1.2748	0.01617	265.3	265.4	77.94	1031.6	1109.5	0.1471	1.8106	1.9577
120/49	1.6924	0.01620	203.25	203.27	87.92	1025.8	1113.7	0.1645	1.7694	1.9339
130/54	2.2225	0.01625	157.32	157.34	97.90	1020.0	1117.9	0.1816	1.7296	1.9112
140/60	2.8886	0.01629	122.99	123.01	107.89	1014.1	1122.0	0.1984	1.6910	1.8894
150/66	3.718	0.01634	97.06	97.07	117.89	1008.2	1126.1	0.2149	1.6537	1.8685
160/71	4.741	0.01639	77.27	77.29	127.89	1002.3	1130.2	0.2311	1.6174	1.8485
170/77	5.992	0.01645	62.04	62.06	137.90	996.3	1134.2	0.2472	1.5822	1.8293
180/82	7.510	0.01651	50.21	50.23	147.92	990.2	1138.1	0.2630	1.5480	1.8109
190/88	9.339	0.01657	40.94	40.96	157.95	984.1	1142.0	0.2785	1.5147	1.7932
200/93	11.526	0.01663	33.62	33.64	167.99	977.9	1145.9	0.2938	1.4824	1.7762
210/90	14.123	0.01670	27.80	27.82	178.05	971.6	1149.7	0.3090	1.4508	1.7598
212/100	14.696	0.01672	26.78	26.80	180.07	970.3	1150.4	0.3120	1.4446	1.7566
220/104	17.186	0.01677	23.13	23.15	188.13	965.2	1153.4	0.3239	1.4201	1.7440
230/110	20.780	0.01684	19.365	19.382	198.23	958.8	1157.0	0.3387	1.3901	1.7288
240/116	24.969	0.01692	16.306	16.323	208.34	952.2	1160.5	0.3531	1.3609	1.7140
250/121	29.825	0.01700	13.804	13.821	218.48	945.5	1164.0	0.3675	1.3323	1.6998
260/127	35.429	0.01709	11.746	11.763	228.64	938.7	1167.3	0.3817	1.3043	1.6860
270/132	41.858	0.01717	10.044	10.061	238.84	931.8	1170.6	0.3958	1.2769	1.6727
280/138	49.203	0.01726	8.628	8.645	249.06	924.7	1173.8	0.4096	1.2501	1.6597
290/143	57.556	0.01735	7.444	7.461	259.31	917.5	1176.8	0.4234	1.2238	1.6472
300/149	67.013	0.01745	6.449	6.466	269.59	910.1	1179.7	0.4369	1.1980	1.6350
310/154	77.68	0.01755	5.609	5.626	279.92	902.6	1182.5	0.4504	1.1727	1.6231
320/160	89.66	0.01765	4.896	4.914	290.28	894.9	1185.2	0.4637	1.1478	1.6115
330/166	103.06	0.01776	4.289	4.307	300.68	887.0	1187.7	0.4769	1.1233	1.6002
340/171	118.01	0.01787	3.770	3.788	311.13	879.0	1190.1	0.4900	1.0992	1.5891
350/177	134.63	0.01799	3.324	3.342	321.63	870.7	1192.3	0.5029	1.0754	1.5783
360/182	153.04	0.01811	2.939	2.957	332.18	862.2	1194.4	0.5158	1.0519	1.5677
370/188	173.37	0.01823	2.606	2.625	342.79	853.5	1196.3	0.5286	1.0287	1.5573
380/193	195.77	0.01836	2.317	2.335	353.45	844.6	1198.1	0.5413	1.0059	1.5471
390/199	220.37	0.01850	2.0651	2.0836	364.17	835.4	1199.6	0.5539	0.9832	1.5371
400/204	247.31	0.01864	1.8447	1.8633	374.97	826.0	1201.0	0.5664	0.9608	1.5272
10/210	276.75	0.01878	1.6512	1.6700	385.83	816.3	1202.1	0.5788	0.9386	1.5174
420/216	308.83	0.01894	1.4811	1.5000	396.77	806.3	1203.1	0.5912	0.9166	1.5078
430/221	343.72	0.01910	1.3308	1.3499	407.79	796.0	1203.8	0.6035	0.8947	1.4982
440/227	381.59	0.01926	1.1979	1.2171	418.90	785.4	1204.3	0.6158	0.8730	1.4887

(Continued)

TABLE A.8d (Continued)
Properties of Dry and Saturated Steam at Various Temperatures (Specific Volume, Enthalpy and Entropy)

Temp., °F/°C	Abs. Press., psia	Specific Volume, ft³/lbm			Enthalpy, BTU/lbm			Entropy, BTU/(lbm°R)		
		Sat. Liquid v_f	Evap. v_{fg}	Sat. Vapor v_g	Sat. Liquid h_f	Evap. h_{fg}	Sat. Vapor h_g	Sat. Liquid s_f	Evap. s_{fg}	Sat. Vapor s_g
450/232	422.6	0.0194	1.0799	1.0993	430.1	774.5	1204.6	0.6280	0.8513	1.4793
460/238	466.9	0.0196	0.9748	0.9944	441.4	763.2	1204.6	0.6402	0.8298	1.4700
470/243	514.7	0.0198	0.8811	0.9009	452.8	751.5	1204.3	0.6523	0.8083	1.4606
480/249	566.1	0.0200	0.7972	0.8172	464.4	739.4	1203.7	0.6645	0.7868	1.4513
490/254	621.4	0.0202	0.7221	0.7423	476.0	726.8	1202.8	0.6766	0.7653	1.4419
500/260	680.8	0.0204	0.6545	0.6749	487.8	713.9	1201.7	0.6887	0.7438	1.4325
520/271	812.4	0.0209	0.5385	0.5594	511.9	686.4	1198.2	0.7130	0.7006	1.4136
540/282	962.5	0.0215	0.4434	0.4649	536.6	656.6	1193.2	0.7374	0.6568	1.3942
560/293	1133.1	0.0221	0.3647	0.3868	562.2	624.2	1186.4	0.7621	0.6121	1.3742
580/304	1325.8	0.0228	0.2989	0.3217	588.9	588.4	1177.3	0.7872	0.5659	1.3532
600/316	1542.9	0.0236	0.2432	0.2668	617.0	548.5	1165.5	0.8131	0.5176	1.3307
620/327	1786.6	0.0247	0.1955	0.2201	646.7	503.6	1150.3	0.8398	0.4664	1.3062
640/338	2059.7	0.0260	0.1538	0.1798	678.6	452.0	1130.5	0.8679	0.4110	1.2789
660/349	2365.4	0.0278	0.1165	0.1442	714.2	390.2	1104.4	0.8987	0.3485	1.2472
680/360	2708.1	0.0305	0.0810	0.1115	757.3	309.9	1067.2	0.9351	0.2719	1.2071
700/371	3093.7	0.0369	0.0392	0.0761	823.3	172.1	995.4	0.9905	0.1484	1.1389
705.4/374.1	3206.2	0.0503	0	0.0503	902.7	0	902.7	1.0580	0	1.0580

Source: Abridged from Keenan, J.H. and Keyes, F.G., *Thermodynamic Properties of Steam*, John Wiley & Sons, Inc., New York, 1936.

[a] psia = 0.069 bar (abs); ft³/lbm = 62.4 L/kg; BTU/lbm = 0.556 kcal/kg.

TABLE A.8e
Properties of Superheated Steam at Various Temperatures (Specific Volume, Enthalpy and Entropy)

Abs. Press.- psia (Sat. Temp. °F)		200/93	220/104	300/149	350/177	400/204	450/232	500/260	550/288	600/316	700/371	800/427	900/482	1000/538
1 (101.74)	v	392.6	404.5	452.3	482.2	512.0	541.8	571.6	601.4	631.2	690.8	750.4	809.9	869.5
	h	1150.4	1159.5	1195.8	1218.7	1241.7	1264.9	1288.3	1312.0	1335.7	1383.8	1432.8	1482.7	1533.5
	s	2.0512	2.0647	2.1153	2.1444	2.1720	2.1983	2.2233	2.2468	2.2702	2.3137	2.3542	2.3923	2.4283
5 (162.24)	v	78.16	80.59	90.25	96.26	102.26	108.24	114.22	120.19	126.16	138.10	150.03	161.95	173.87
	h	1148.8	1158.1	1195.0	1218.1	1241.2	1264.5	1288.0	1311.7	1335.4	1383.6	1432.7	1482.6	1533.4
	s	1.8718	1.8857	1.9370	1.9664	1.9942	2.0205	2.0456	2.0692	2.0927	2.1361	2.1767	2.2148	2.2509
10 (193.21)	v	38.85	40.09	45.00	48.03	51.04	54.05	57.05	60.04	63.03	69.01	74.98	80.95	86.92
	h	1146.6	1156.2	1193.9	1217.2	1240.6	1264.0	1287.5	1311.3	1335.1	1383.4	1432.5	1482.4	1533.1
	s	1.7927	1.8071	1.8595	1.8892	1.9172	1.9436	1.9689	1.9924	2.0160	2.0596	2.1002	2.1383	2.1744
14.696 (212.00)	v			27.15	30.53	32.62	34.68	36.73	38.78	40.82	42.86	46.94	51.00	55.07	59.13
	h			1154.4	1192.8	1216.4	1239.9	1263.5	1287.1	1310.9	1335.8	1383.2	1432.3	1482.3	1533.1
	s			1.7624	1.8160	1.8460	1.8743	1.9008	1.9261	1.9498	1.9734	2.0170	2.0576	2.0958	2.1319
20 (227.96)	v			22.36	23.91	25.43	26.95	28.46	29.97	31.47	34.47	37.46	40.45	43.44
	h			1191.6	1215.6	1239.2	1262.9	1286.6	1310.5	1334.4	1382.9	1432.1	1482.1	1533.0
	s			1.7808	1.8112	1.8396	1.8664	1.8918	1.9160	1.9392	1.9829	2.0235	2.0618	2.0978
40 (267.25)	v			11.040	11.843	12.628	13.401	14.168	14.93	15.688	17.198	18.702	20.20	21.70
	h			1186.8	1211.9	1236.5	1260.7	1284.8	1308.9	1333.1	1381.9	1431.3	1481.4	1532.4
	s			1.6994	1.7314	1.7608	1.7881	1.8140	1.8384	1.8619	1.9058	1.9467	1.9850	2.0214
60 (292.71)	v			7.259	7.818	8.357	8.884	9.403	9.916	10.427	11.441	12.449	13.452	14.454
	h			1181.6	1208.2	1233.6	1258.5	1283.0	1307.4	1331.8	1380.9	1430.5	1480.8	1531.9
	s			1.6492	1.6830	1.7135	1.7416	1.7678	1.7926	1.8162	1.8605	1.9015	1.9400	1.9762
80 (312.03)	v				5.803	6.220	6.624	7.020	7.410	7.797	8.562	9.322	10.077	10.830
	h				1204.3	1230.7	1256.1	1281.1	1305.8	1330.5	1379.9	1429.7	1480.1	1531.3
	s				1.6475	1.6791	1.7078	1.7346	1.7598	1.7836	1.8281	1.8694	1.9079	1.9442
100 (327.81)	v				4.592	4.937	5.268	5.589	5.905	6.218	6.835	7.446	8.052	8.656
	h				1200.1	1227.6	1253.7	1279.1	1304.2	1329.1	1378.9	1428.9	1479.5	1530.8
	s				1.6188	1.6518	1.6813	1.7085	1.7339	1.7581	1.8029	1.8443	1.8829	1.9193
120 (341.25)	v				3.783	4.081	4.363	4.636	4.902	5.165	5.683	6.195	6.702	7.207
	h				1195.7	1224.4	1251.3	1277.2	1302.5	1327.7	1377.8	1428.1	1478.8	1530.2
	s				1.5944	1.6287	1.6591	1.6869	1.7127	1.7370	1.7822	1.8237	1.8625	1.8990
140 (353.02)	v					3.468	3.715	3.954	4.186	4.413	4.861	5.301	5.738	6.172
	h					1221.1	1248.7	1275.2	1300.9	1326.4	1376.8	1427.3	1478.2	1529.7
	s					1.6087	1.6399	1.6683	1.6945	1.7190	1.7645	1.8063	1.8451	1.8817
160 (363.53)	v					3.008	3.230	3.443	3.648	3.849	4.244	4.631	5.015	5.396
	h					1217.6	1246.1	1273.1	1299.3	1325.0	1375.7	1426.4	1477.5	1529.1
	s					1.5908	1.6230	1.6519	1.6785	1.7033	1.7491	1.7911	1.8301	1.8667

(Continued)

TABLE A.8e (Continued)
Properties of Superheated Steam at Various Temperatures (Specific Volume, Enthalpy and Entropy)

Abs. Press.- psia (Sat. Temp. °F)		200/93	220/104	300/149	350/177	400/204	450/232	500/260	550/288	600/316	700/371	800/427	900/482	1000/538
										Temperature,°F/°C				
180	v					2.649	2.852	3.044	3.229	3.411	3.764	4.110	4.452	4.792
(373.06)	h					1214.0	1243.5	1271.0	1297.6	1323.5	1374.7	1425.6	1476.8	1528.6
	s					1.5745	1.6077	1.6373	1.6642	1.6894	1.7355	1.7776	1.8167	1.8534
200	v					2.361	2.549	2.726	2.895	3.060	3.380	3.693	4.002	4.309
(381.79)	h					1210.3	1240.7	1268.9	1295.8	1322.1	1373.6	1424.8	1476.2	1528.0
	s					1.5594	1.5937	1.6240	1.6513	1.6767	1.7232	1.7655	1.8048	1.8415
220	v					2.125	2.301	2.465	2.621	2.772	3.066	3.352	3.634	3.913
(389.86)	h					1206.5	1237.9	1266.7	1294.1	1320.7	1372.6	1424.0	1475.5	1527.5
	s					1.5453	1.5808	1.6117	1.6395	1.6652	1.7120	1.7545	1.7939	1.8308
240	v					1.9276	2.094	2.247	2.393	2.533	2.804	3.068	3.327	3.584
(397.37)	h					1202.5	1234.9	1264.5	1292.4	1319.2	1371.5	1423.2	1474.8	1526.9
	s					1.5319	1.5686	1.6003	1.6286	1.6546	1.7017	1.7444	1.7839	1.8209
260	v						1.9183	2.063	2.199	2.330	2.582	2.827	3.067	3.305
(404.42)	h						1232.0	1262.3	1290.5	1317.7	1370.4	1422.3	1474.2	1526.3
	s						1.5573	1.5897	1.6184	1.6447	1.6922	1.7352	1.7748	1.8118
280	v						1.7674	1.9047	2.033	2.156	2.392	2.621	2.845	3.066
(411.05)	h						1228.9	1260.0	1288.7	1316.2	1369.4	1421.5	1473.5	1525.8
	s						1.5464	1.5796	1.6087	1.6354	1.6834	1.7265	1.7662	1.8033
300	v						1.6364	1.7675	1.8891	2.005	2.227	2.442	2.652	2.859
(417.33)	h						1225.8	1257.6	1286.8	1314.7	1368.3	1420.6	1427.8	1525.2
	s						1.5360	1.5701	1.5998	1.6268	1.6751	1.7184	1.7582	1.7954
350	v						1.3734	1.4923	1.6010	1.7036	1.8980	2.084	2.266	2.445
(431.72)	h						1217.7	1251.5	1282.1	1310.9	1365.5	1418.5	1471.1	1523.8
	s						1.5119	1.5481	1.5792	1.6070	1.6563	1.7002	1.7403	1.7777
400	v						1.1744	1.2851	1.3843	1.4770	1.6508	1.8161	1.9767	2.134
(444.59)	h						1208.8	1245.1	1277.2	1306.9	1362.7	1416.4	1.469.4	1522.4
	s						1.4892	1.5281	1.5607	1.5894	1.6398	1.6842	1.7247	1.7623

		500/260	550/288	600/316	620/327	640/338	660/349	680/360	700/371	800/427	900/482	1000/538	1200/649	1400/760	1600/871
450	v	1.1231	1.2155	1.3005	1.3332	1.3652	1.3967	1.4278	1.4584	1.6074	1.7516	1.8928	2.170	2.443	2.714
(456.28)	h	1238.4	1272.0	1302.8	1314.6	1326.2	1337.5	1348.8	1359.9	1414.3	1467.7	1521.0	1628.6	1738.7	1851.9
	s	1.5095	1.5437	1.5735	1.5845	1.5951	1.6054	1.6153	1.6250	1.6699	1.7108	1.7486	1.8177	1.8803	1.9381
500	v	0.9927	1.0800	1.1591	1.1893	1.2188	1.2478	1.2763	1.3044	1.4405	1.5715	1.6996	1.9504	2.197	2.442
(467.01)	h	1231.3	1266.8	1298.6	1310.7	1322.6	1334.2	1345.7	1357.0	1412.1	1466.0	1519.6	1627.6	1737.9	1851.3
	s	1.4919	1.5280	1.5588	1.5701	1.5810	1.5915	1.6016	1.6115	1.6571	1.6982	1.7363	1.8056	1.8683	1.9262
550	v	0.8852	0.9686	1.0431	1.0714	1.0989	1.1259	1.1523	1.1783	1.3038	1.4241	1.5414	1.7706	1.9957	2.219
(476.94)	h	1223.7	1261.2	1294.3	1306.8	1318.9	1330.8	1342.5	1354.0	1409.9	1464.3	1518.2	1626.6	1737.1	1850.6
	s	1.4751	1.5131	1.5451	1.5568	1.5680	1.5787	1.5890	1.5991	1.6452	1.6868	1.7250	1.7946	1.8575	1.9155

(Continued)

TABLE A.8e (Continued)
Properties of Superheated Steam at Various Temperatures (Specific Volume, Enthalpy and Entropy)

Abs. Press.-psia (Sat. Temp. °F)	Temperature, °F/°C													
	500/260	550/288	600/316	620/327	640/338	660/349	680/360	700/371	800/427	900/482	1000/538	1200/649	1400/760	1600/871
v	0.7947	0.8753	0.9463	0.9729	0.9988	1.0241	1.0489	1.0732	1.1899	1.3013	1.4096	1.6208	1.8279	2.033
600 h	1215.7	1255.5	1289.9	1302.7	1315.2	1327.4	1339.3	1351.1	1407.7	1462.5	1516.7	1625.5	1736.3	1850.0
(486.21) s	1.4586	1.4990	1.5323	1.5443	1.5558	1.5667	1.5773	1.5875	1.6343	1.6762	1.7147	1.7846	1.8476	1.9056
v		0.7277	0.7934	0.8177	0.8411	0.8639	0.8860	0.9077	1.0108	1.1082	1.2024	1.3853	1.5641	1.7405
700 h		1243.2	1280.6	1294.3	1307.5	1320.3	1332.8	1345.0	1403.2	1459.0	1513.9	1623.5	1734.8	1848.8
(503.10) s		1.4722	1.5084	1.5212	1.5333	1.5449	1.5559	1.5665	1.6147	1.6573	1.6963	1.7666	1.8299	1.8881
v		0.6154	0.6779	0.7006	0.7223	0.7433	0.7635	0.7833	0.8763	0.9633	1.0470	1.2088	1.3662	1.5214
800 h		1229.8	1270.7	1285.4	1299.4	1312.9	1325.9	1338.6	1398.6	1455.4	1511.0	1621.4	1733.2	1847.5
(518.23) s		1.4467	1.4863	1.5000	1.5129	1.5250	1.5366	1.5476	1.5972	1.6407	1.6801	1.7510	1.8146	1.8729
v		0.5264	0.5873	0.6089	0.6294	0.6491	0.6680	0.6863	0.7716	0.8506	0.9262	1.0714	1.2124	1.3509
900 h		1215.0	1260.1	1275.9	1290.9	1305.1	1318.8	1332.1	1393.9	1451.8	1508.1	1619.3	1731.6	1846.3
(531.98) s		1.4216	1.4653	1.4800	1.4938	1.5066	1.5187	1.5303	1.5814	1.6257	1.6656	1.7371	1.8009	1.8595
v		0.4533	0.5140	0.5350	0.5546	0.5733	0.5912	0.6084	0.6878	0.7604	0.8294	0.9615	1.0893	1.2146
1000 h		1198.3	1248.8	1265.9	1281.9	1297.0	1311.4	1325.3	1389.2	1448.2	1505.1	1617.3	1730.0	1845.0
(544.61) s		1.3961	1.4450	1.4610	1.4757	1.4893	1.5021	1.5141	1.5670	1.6121	1.6525	1.7245	1.7886	1.8474
v			0.4532	0.4738	0.4929	0.5110	0.5281	0.5445	0.6191	0.6866	0.7503	0.8716	0.9885	1.1031
1100 h			1236.7	1255.3	1272.4	1288.5	1303.7	1318.3	1384.3	1444.5	1502.2	1615.2	1728.4	1843.8
(556.31) s			1.4251	1.4425	1.4583	1.4728	1.4862	1.4989	1.5535	1.5995	1.6405	1.7130	1.7775	1.8363
v			0.4016	0.4222	0.4410	0.4586	0.4752	0.4909	0.5617	0.6250	0.6843	0.7967	0.9046	1.0101
1200 h			1223.5	1243.9	1262.4	1279.6	1295.7	1311.0	1379.3	1440.7	1499.2	1613.1	1726.9	1842.5
(567.22) s			1.4052	1.4243	1.4413	1.4568	1.4710	1.4843	1.5409	1.5879	1.6293	1.7025	1.7672	1.8263
v			0.3174	0.3390	0.3580	0.3753	0.3912	0.4062	0.4714	0.5281	0.5805	0.6789	0.7727	0.8640
1400 h			1193.0	1218.4	1240.4	1260.3	1278.5	1295.5	1369.1	1433.1	1493.2	1608.9	1723.7	1840.0
(587.10) s			1.3639	1.3877	1.4079	1.4258	1.4419	1.4567	1.5177	1.5666	1.6093	1.6836	1.7489	1.8083
v				0.2733	0.2936	0.3112	0.3271	0.3417	0.4034	0.4553	0.5027	0.5906	0.6738	0.7545
1600 h				1187.8	1215.2	1238.7	1259.6	1278.7	1358.4	1425.3	1487.0	1604.6	1720.5	1837.5
(604.90) s				1.3489	1.3741	1.3952	1.4137	1.4303	1.4964	1.5476	1.5914	1.6669	1.7328	1.7926
v				0.2407	0.2597	0.2760	0.2907	0.3502	0.3986	0.4421	0.5218	0.5968	0.6693	
1800 h				1185.1	1214.0	1238.5	1260.3	1347.2	1417.4	1480.8	1600.4	1717.3	1835.0	
(621.03) s				1.3377	1.3638	1.3855	1.4044	1.4765	1.5301	1.5752	1.6520	1.7185	1.7786	
v					0.1936	0.2161	0.2337	0.2489	0.3074	0.3532	0.3935	0.4668	0.5352	0.6011
2000 h					1145.6	1184.9	1214.8	1240.0	1335.5	1409.2	1474.5	1596.1	1714.1	1832.5
(635.82) s					1.2945	1.3300	1.3564	1.3783	1.4576	1.5139	1.5603	1.6384	1.7055	1.7660

(Continued)

TABLE A.8e (Continued)
Properties of Superheated Steam at Various Temperatures (Specific Volume, Enthalpy and Entropy)

Abs. Press.-psia (Sat. Temp. °F)		500/260	550/288	600/316	620/327	640/338	660/349	680/360	700/371	800/427	900/482	1000/538	1200/649	1400/760	1600/871
2500	v							0.1484	0.1686	0.2294	0.2710	0.3061	0.3678	0.4244	0.4784
(668.13)	h							1132.3	1176.8	1303.6	1387.8	1458.4	1585.3	1706.1	1826.2
	s							1.2687	1.3073	1.4127	1.4772	1.5273	1.6088	1.6775	1.7389
3000	v								0.0984	0.1760	0.2159	0.2476	0.3018	0.3505	0.3966
(695.36)	h								1060.7	1267.2	1365.0	1441.8	1574.3	1698.0	1819.9
	s								1.1966	1.3690	1.4439	1.4984	1.5837	1.6540	1.7163
3206.2	v									0.1583	0.1981	0.2288	0.2806	0.3267	0.3703
(705.40)	h									1250.5	1355.2	1434.7	1569.8	1694.6	1817.2
	s									1.3508	1.4309	1.4874	1.5742	1.6452	1.7080
3500	v								0.0306	0.1364	0.1762	0.2058	0.2546	0.2977	0.3381
	h								780.5	1224.9	1340.7	1424.5	1563.3	1689.8	1813.6
	s								0.9515	1.3241	1.4127	1.4723	1.5615	1.6336	1.6968
4000	v								0.0287	0.1052	0.1462	0.1743	0.2192	0.2581	0.2943
	h								763.8	1174.8	1314.4	1406.8	1552.1	1681.7	1807.2
	s								0.9347	1.2757	1.3827	1.4482	1.5417	1.6154	1.6795
4500	v								0.0276	0.0798	0.1226	0.1500	0.1917	0.2273	0.2602
	h								753.5	1113.9	1286.5	1388.4	1540.8	1673.5	1800.9
	s								0.9235	1.2204	1.3529	1.4253	1.5235	1.5990	1.6640
5000	v								0.0268	0.0593	0.1036	0.1303	0.1696	0.2027	0.2329
	h								746.4	1047.1	1256.5	1369.5	1529.5	1665.3	1794.5
	s								0.9152	1.1622	1.3231	1.4034	1.5066	1.5839	1.6499
5500	v								0.0262	0.0463	0.0880	0.1143	0.1516	0.1825	0.2106
	h								741.3	985.0	1224.1	1349.3	1518.2	1657.0	1788.1
	s								0.9090	1.1093	1.2930	1.3821	1.4908	1.5699	1.6369

Source: Abridged from Keenan, J.H. and Keyes, F.G., *Thermodynamic Properties of Steam*, John Wiley & Sons, Inc., New York, 1936.

TABLE A.8f
Properties of Water at Various Temperatures (Specific Volume, Specific Gravity, 1bm/ft³, Vapor Pressure)

Temp., °F	Temp., °C	Specific Volume ft³/lb	Specific Gravity	Specific Weight (lb/ft³)	Vapor Pressure, psia
40	4.4	0.01602	1.0013	62.42	0.1217
50	10.0	0.01603	1.0006	62.38	0.1781
60	15.6	0.01604	1.0000	62.34	0.2563
70	21.1	0.01606	0.9987	62.27	0.3631
80	26.7	0.01608	0.9975	62.19	0.5069
90	32.2	0.01610	0.9963	62.11	0.6982
100	37.8	0.01613	0.9944	62.00	0.9492
120	48.9	0.01620	0.9901	61.73	1.692
140	60.0	0.01629	0.9846	61.39	2.889
160	71.1	0.01639	0.9786	61.01	4.741
180	82.2	0.01651	0.9715	60.57	7.510
200	93.3	0.01663	0.9645	60.13	11.526
212	100.0	0.01672	0.9593	59.81	14.696
220	104.4	0.01677	0.9565	59.63	17.186
240	115.6	0.01692	0.9480	59.10	24.97
260	126.7	0.01709	0.9386	58.51	35.43
280	137.8	0.01726	0.9293	58.00	49.20
300	148.9	0.01745	0.9192	57.31	67.01
320	160.0	0.01765	0.9088	56.66	89.66
340	171.1	0.01787	0.8976	55.96	118.01
360	182.2	0.01811	0.8857	55.22	153.04
380	193.3	0.01836	0.8736	54.47	195.77
400	204.4	0.01864	0.8605	53.65	247.31
420	215.6	0.01894	0.8469	52.80	308.83
440	226.7	0.01926	0.8328	51.92	381.59
460	237.8	0.0196	0.8183	51.02	466.9
480	248.9	0.0200	0.8020	50.00	566.1
500	260.0	0.0204	0.7863	49.02	680.8
520	271.1	0.0209	0.7674	47.85	812.4
540	282.2	0.0215	0.7460	46.51	962.5

Source: Computed from Keenan & Keyes Steam Table.

INDEX

A

AAS, *see* Atomic absorption spectrometer
Absolute viscosity, 1006
Absorption
    air quality monitoring, 101
    colorimetry, 211–212
    FOPs, 324
    mercury, 500–501
    UV analyzers, 983–984, 986–987
Accuracy
    analyzer instrumentation, 27
    ISEs, 464
    Nernst equation, 464
Acid-base titration, 1088
Acoustic sensor, 879
Acousto-optic tunable filters (AOTF), 32, 40–41, 450
Acute Exposure Guideline Levels (AEGLs), 404
Adenosine 5′-triphosphate (ATP) biometers, 118
Adiabatic calorimeters, 385
Adiabatic flame temperature, 384
Adsorption, air quality monitoring, 101–102
AEGLs, *see* Acute Exposure Guideline Levels
Agitator power viscometer, 1032–1033
Airborne-particulate counter, 737
Air filters, 103–104
Airflow calorimeter, 383
Air purge, 750–751
Air quality monitoring
    absorption, 101
    adsorption, 101–102
    air quality and meteorological parameters, 92
    audits, 98
    automatic analyzers, 98–100
    automatic monitoring, 97
    basics, 91
    carbon dioxide, 126–127
    carbon monoxide, 138
    Clean Air Act*, 93
    combustibles, 221–231
    common errors, 92
    cost of, 92
    data processing, 97–98
    data transmission, 97
    dust-fall jars, 95–96
    gas sampling, 101–103
    hydrocarbon, 393–399
    laboratory analyses, 96–97
    mercury, 498–503
    moisture, 519–535
    NAAQS, 93
    odor, 625–630
    particulate sampling
        air filters, 103–104
        electrostatic precipitators, 105
        fiberglass papers, 104
        filter papers, 103
        impactors, 104
        impingers, 104
        thermal precipitators, 105
    sampling problems, 100
    sampling site selection, 95
    sensors, 97
    single source/stacks monitoring, 94
    static monitoring, 95–96
    static sensors, 92
    urban area monitoring, 94
    vapor sampling, 101–103
Air-saturated vapor pressure analyzers, 809–810
Air temperature rise calorimeter, 383
Alloys, composition and manufacturers, 1283–1286
Amalgamation, mercury, 500
Ambient air conditions
    analyzer sampling, 100
    carbon dioxide, 126–127
    carbon monoxide, 138
    particulates, 754–759
    sulfur dioxide analyzers, 925–927
American Oil Chemists Society (AOCS), 27
American Society for Testing and Materials (ASTM)
    D 56-90, 817
    D 93-90, 817
    D 97-87, 813
    D 323-90, 808–809
    D 1160, 804
    D 1267-89, 809
    D 2386-88, 816
    D 2500-88, 814–815
    D 2533-90, 812
    D 2699, 818
    D 2700, 818
    D-2885, 818–819
    D 2887-89, 804
    D 3710-88, 804
    D 4953-90, 809
    D 86-IP-123, 803–804
Aminonaphtholsulfonic acid method, 793
Ammonia analyzers
    basics, 109
    measurement in gases
        ammonia production plants, 112
        emission monitoring systems, 112
        SNCR, 112
        TDL, 113
    measurement in water
        colorimetric method, 110–111
        ISE, 111–112
        NH_3 to NH_4^+ ratio, 110
        UV spectrophotometry method, 112
Amperometric analyzers
    chlorine
        buffers and reagents, 149
        electrode cleaners, 149
        free residual chlorine type, 147
        immersion probe-type polarographic chlorine detector, 148
        microprocessor-based chlorine analyzers, 147
        relative merits, 144
        titrators, 146
    voltammetric analyzers, 291, 295–297
Amperometric sensors, 36, 362, 629, 725–727
Analyte-specific techniques, 37
Analyzers
    accuracy and precision, 28
    analysis frequency, 28
    calibration, 28
    cost, 43–45
    data handling, 43
    location of, 42–43
    maintenance, 44
    methane monitoring, 19
    preparation phase
        at-line measurement, 26
        information gathering, 27

in-line measurement, 26
installation and check-out, 20
installation drawings, 20
invasive measurement, 26
online measurement, 26
orientation table, 20–25
preliminary quotations, 26
re-calibration requirements, 19
self-calibrating analyzers, 19
self-diagnosing analyzers, 19
specificity, 27–28
spectrometers, 29–34
Analyzer sampling, see Sampling systems
ANSI/ISA-60079-28 standard, 322
AOCS, see American Oil Chemists Society
AOTF, see Acousto-optic tunable filters
Aqueous conductivity, 958–960
Area unit definitions, 1238
Ash analysis, 204
AST, see Automatic sampling train
At-line measurement, 26
Atomic absorption spectrophotometry (AAS), 501–502, 509–510
Atomic fluorescence spectroscopy, 503
ATP biometers, see Adenosine 5′-triphosphate biometers
Attenuated total reflectance (ATR) probe, 326
Attenuation
 moisture, 571–572
 particle size, 736
 particulates, 748–751
 ultrasonic attenuation, 736
Audits, air quality monitoring, 98
Autocalibration, 532
Automatic air monitoring, 97
Automatic capillary viscometer, 1051
Automatic efflux cup viscometer, 1057
Automatic on-line designs, oxygen demand, 957–958
Automatic osmometers, 583–584
Automatic probe cleaning, 65
Automatic refractometer, 855–856
Automatic retraction, 777
Automatic sampling train (AST), 85–86, 746
Automatic stack sampling, 69–70
Autotitrator-based, closed-loop control systems, 1089
Aviation fuel, freezing-point analyzer, 816
Avogadro's law, 543

B

Backpressure viscometers, 1023
Backscatter turbidity sensors, 976
Bag break detector, 753
Bare metal electrodes, 725–726
Basis weight compensation, moisture, 571
Batch-type, automatic titrator, 1089
Beer–Lambert law, 434–435, 717, 890, 985
Bench chemistry, see Wet chemistry
Bending rheometers, 873
Bingham solid, 1028
Biological oxygen demand (BOD) analyzers
 basics, 665–666
 and COD, 677

electrolysis instrument, 671
extended BOD test, 669
features, 677
five-day BOD tests, 667–668
incubation time, 667
manometric BOD test, 669–670
measurements, 1077
nitrification, 667
pH, 666–667
robotic analyzer, 671
seed, 666
in sewage treatment plants, 670
temperature, 667
test procedures, 666
toxicity, 667
Bioluminescence ATP determination kits, 119
Biometers
 ATP measurement, 118–119
 basics, 117
 definition, 117
 soil samples activity measurement, 117
Bituminous coal analysis, 202
Blackbody
 grating spectrophotometers, 438
 radiation interference, 889
Blade consistency analyzers, 251
BOD analyzers, see Biological oxygen demand analyzers
Bomb calorimeter, 379, 381
Brazing, 319
Broken bag monitors, 753
Brookfield viscosities, 1062
Brush cleaners, 776–777
Bubble-time viscometers, 1048
Buffers, 149
Bypass filters, 63–64, 913

C

Cadmium sulfide photocell, 338
Calculation form, pitot tube, 83–84, 87
Calibration
 analyzer instrumentation, 28
 analyzer sampling, 58
 carbon monoxide, 133–134
 and cleaning system, 784–787
 cloud-point analyzers, 816
 conductivity analyzers, 241
 dissolved oxygen, 709
 flash-point analyzer, 818
 gas chromatography, 395
 hazardous and toxic gas monitoring, 371–372
 hydrocarbon analyzers, 399
 industrial viscometers, 1020
 ion-selective electrodes, 468–469
 IR analyzers, 448
 laboratory viscometers, 1049–1050
 moisture analyzers, 534–535, 555–557
 NIR analyzers, 820
 NO_x analyzers, 618–619
 octane analyzers, 819
 online distillation analyzers, 808
 oxidation–reduction potential, 660
 pH measurement, 784–787

pour-point temperature analyzers, 814
Raman process analyzer, 839
streaming current detector, 912
sulfur dioxide, 925
thermal conductivity detectors, 946–947
TOD analyzers, 675
vapor–liquid ratio analyzers, 813
vapor pressure analyzers, 811
California Air Resources Board (CARB), 98
Candles, 96
Capacitance hygrometer
 capacitance-type measuring cell, 545
 limitations, 548
 sampling systems, 546
 thin-film capacitance probes, 546–547
Capacitance moisture gauge, 572–573
Capillary extrusion viscometer, 1052–1053
Capillary plastometers, 1029–1030
Capillary rheometers, 874
Capillary viscometers
 industrial viscometers
  backpressure type, 1023
  calibration, 1020
  differential pressure type, 1021–1022
  disadvantages, 1020
  Hagen–Poiseuille law, 1020
  measurement span, 1020
  Newtonian fluids, 1020
 laboratory viscometers
  accuracy and limitations, 1049
  automatic capillary viscometer, 1051
  calibration, 1049–1050
  capillary extrusion viscometer, 1052–1053
  Hagen–Poiseuille law, 1048
  intrinsic viscosity and molecular weight, 1051
  Mark–Houwink relation, 1052
  measurement principles, 1048
  Newtonian liquid measurement, 1048
  Ostwald viscometer, 1049
CARB, see California Air Resources Board
Carbon, ceramic, plastic and rubber materials, composition and manufacturers, 1282
Carbon dioxide
 ambient air measurement, 126–127
 basics, 123–124
 emissions, 125–126
 in global warming, 124–125
 Orsat equipment, 124, 128
Carbon monoxide (CO)
 ambient air sampling, 138
 analyzers calibration techniques, 133–134
 basics, 132–133
 characteristics, 133
 in chemical processing industries, 133
 electrochemical reaction, 137
 gas chromatographs, 136
 GFC, 135
 measuring devices, 138
 mercury vapor analyzer, 135–136
 NDIR analysis, 134
 personal toxic gas monitors, 133
 portable monitors, 137–138
 threshold limit value, 133
 water vapor interference control, 135

Carrier gas flow control, 175
Carrier supply, 192–193
Cascade impactor, 104, 758
Catalytic analysis, 138
Catalytic bead (CB) sensors, 414–415
Catalytic combustion detectors
　combustibles, 222–227
　oxygen, gases, 686
Catalyzed ozone hydroxyl radical oxidation analyzer, 955–956
Cavity ring-down spectroscopy (CRDS), 526, 554, 887, 902
CB sensors, see Catalytic bead sensors
CCTV detectors, 334–335
CDTs, see Color detector tubes
Cell constant, 236–237
Cells
　electrochemical cells, 346
　electrodeless cells, 239
　electrolytic cells, 526–527
　flow-through cells, 705
　galvanic cells, 704–975
　gas cells, 446
　liquid cells, 447
　low-temperature electrochemical cells, 695
　polarographic (voltametric) cells, 703–704
　TCD analyzer, 944
　thallium cells, 707
　two-electrode cells, 238
Cellulose filters, 59
Cellulose hygrometers, 530
CFR, see Code of Federal Regulations
Charge transfer, 758–759
Chemical analyzers, 695
Chemical cleaners, 777
Chemical compatibility guide, 1258–1279
Chemical oxygen demand (COD) analyzers
　automatic on-line measurement, 672–673
　basics, 665–666
　and BOD, 677
　definition, 666
　dichromate COD test, 671–672
　oxygen quantity measurement, 1077
Chemical reactor samplers, 71
Chemiluminescence detector (CLD), 609–611
Chemiluminescence ozone monitor, 720
Chemiluminescence total nitrogen analyzer, 610–611
Chemometrics, 29, 820
Chilled-mirror hygrometer, 528
Chlorine (Cl)
　amperometric analyzers
　　buffers and reagents, 149
　　electrode cleaners, 149
　　free residual chlorine type, 147
　　immersion probe-type polarographic chlorine detector, 148
　　microprocessor-based chlorine analyzers, 147
　　relative merits, 144
　　titrators, 146
　basics, 142–143
　for bleaching and disinfection, 143
　colorimetric analyzer, 144–146

elemental chlorine, 143
　in gas and water, 144
　iodometry, 145
　quality control procedures, 150
Chromatographic calorimeter, 384
Chromatography
　gas (see Gas chromatography)
　liquid (see High-performance liquid chromatography (HPLC))
　water quality monitoring, 1080–1081
Chromium reduction, 658–659
Circular variable filter (CVF), 440
CLD, see Chemiluminescence detector
Clean Air Act*, 93
Cleaning and calibration systems
　pH sensor
　　digital pH sensors, 785
　　new pH sensor developments, 784
　　smart pH sensors, 784–785
　pH transmitters
　　asset management, 786–787
　　new pH transmitters, 785–786
　　smart pH transmitters, 786
　　wireless HART™ transmitters, 787
Cloud-point analyzers
　applications, 816
　calibration, 816
　cloud-point tests (ASTM Method D 2500-88), 814–815
　optical, 815–816
CN_F determination, see Free cyanide determination
CN_T determination, see Total cyanide determination
Coal
　ash analysis, 204
　basics, 200
　bituminous coal analysis, 202
　characteristics, 201
　composition analysis, 201
　gross calorific value, 201–203
　on-line monitors, 204–206
　property measurement, 201–202
　proximate analysis, 201
　sulfur analysis, 201
　sulfur determination, 203–204
　TG technique, 202
　types, 200
Coalescing gas filter, 59
Coaxial cylinder rheometers, 873
Coaxial cylinder viscometer, 1058–1060
COD analyzers, see Chemical oxygen demand analyzers
Code of Federal Regulations (CFR), 98
Cold box, 84
Cold vapor atomic absorption spectroscopy (CVAAS), 511
Cold vapor atomic fluorescence spectrometer (CVAFS), 512
Color-change badges, 366–368
Color detector tubes (CDTs), 368–370
Colorimeters
　absorption, 211–212
　ambient air analyzers, 925–926
　basics, 210–211
　color measurements, 211–212
　continuous color monitors

in-line liquid color measurement, 214
　multispectral analysis, 215
　online shade monitors, 214
　optical fluorescence and luminescence sensors, 214–215
　spectrophotometric analyzers, 212–213
　TOC, 961
　transmittance, 211–212
　tristimulus method, 213–214
　visual colorimetry, 211
　wet chemistry
　　applications, 1095
　　high-precision laboratory analyzers, 1092–1094
　　on/off batch colorimeter, 1094–1095
　　requirements, 1092
　　simple, continuous colorimetric analyzers, 1094
Colorimetric detection and analysis
　mercury, 502–503, 509
　ozone, 727
　phosphate, 793–795
Color measurements, 211
Columns
　HPLC, 193–195
　process gas chromatograph, 162–163
Combustible analyzers
　basics, 219–220
　catalytic combustion type
　　accessories, 225
　　advantages and disadvantages, 227
　　diffusion head design, 224
　　limitations, 222
　　measuring circuits, 222–224
　　multiple head system, 226
　　remote head system, 225–226
　　sampling system, 225
　　tube sampling system, 226–227
　detector types, 221
　FIDs, 227
　flammable materials properties, 221
　infrared types
　　combustibles monitoring instruments, 228
　　open-path instruments, 228–231
　　photometers and spectrometers, 228
　　point measurement, 231
　　point monitoring system, 228–229
　PIDs, 227–228
Compensation
　basis weight, moisture, 571
　oxygen, liquid, 709–710
　refractometry, 856
Compliance recording, 386
Composition and manufacturers
　alloys, 1283–1286
　of carbon, ceramic, plastic and rubber materials, 1282
　of metallic materials, 1281–1282
Concentration measurements, 239
Concentric viscometer, 1027–1028
Condensation, 566
Conductimetric analyzers, 926
Conductivity analyzers
　asset management, 243
　basics, 235
　calibration and maintenance, 241

cell constant, 236–237
cell dimensions, 237–238
concentration measurements, 239
corrosive and fouling applications, 240
digital conductivity sensors, 241
electrodeless cells, 239
four-electrode measurement, 238–239
high-purity water
   measurements, 239–240
new conductivity transmitter
   developments, 241–242
smart conductivity transmitters, 243
theory of operation, 236
two-electrode cells, 238
wireless HART™ transmitters, 244
Conductivity sensors, 35–36, 241–242, 291, 1075
Cone-and-plate plastometer, 1028–1029
Cone-and-plate rheometer
   conversion among rheological
      properties, 871
   dynamic mechanical analysis, 870
   flow field disturbances, 869
   flow geometry, 868
   molecular weight distribution, 870–871
   selection, 869
   shear flow properties, 868
   strain-controlled rheometer, 868
   stress-controlled rheometer, 868
   time–temperature superposition, 871
   torque ranges, 869
   viscoelastic shear moduli, 870
Cone-and-plate viscometer, 1060–1061
Consistency measurement
   basics, 248
   gravimetric method, 249
   paper industry applications, 249
   sensor design
      blade sensors, 250
      blade types, 251
      gamma attenuation devices, 249
      mechanical sensors, 250
      microwave measurement
         techniques, 249
      optical designs, 252–254
      optical sensors, 249–250
      plastic scintillation detector
         technology, 249
      probe types, 251
      rotating sensors, 250–252
      stationary sensors, 250
      ultra-high-frequency electromagnetic
         wave type detector, 250
Continuous capillary viscometers,
   see Capillary viscometers
Continuous color monitors
   in-line liquid color measurement, 214
   multispectral analysis, 215
   online shade monitors, 214
   optical fluorescence and luminescence
      sensors, 214–215
Continuous phosphorus analyzer
   experimental setup, 793–794
   orthophosphate analyzer, 794
   pumpless design, 794–795
Continuous vapor–liquid ratio analyzers,
   812–813

Controlled potential coulometry, 298–299
Controller
   data system, 893
   gas chromatography, 175–176
Cooling methods, air moisture, 528–529
Copper interference, 509
Coriolis mass flowmeters, 599–600,
   1038–1039
Correlation spectrometry, 925
Corrosion monitoring
   advantages, 258
   basics, 258
   corrosion resistance, 259
   coupon monitoring, 260–261
   DCMS, 264–265
   ER monitor, 261–262
   instrumentation for, 259
   LPR and ER combination designs, 264
   LPR monitors, 263
   mobile monitoring laboratories, 259
   nondestructive testing, 259
   wall thickness monitoring, 265
Coulometric analyzers, 292, 298–299
   ambient air analyzers, 927
   TOC, 961
Coulometric sensor
   dissolved oxygen
      advantage, 707
      differential conductivity
         detectors, 708
      multiple-anode detector, 707
      thallium cells, 707
   electrochemical oxygen
      sensors, 692
Coulter reference method, 734–735
CRDS, see Cavity ring-down spectroscopy
Critical angle of refraction, 852
Critical-angle refractometer, 33,
   854–856, 860
CVAAS, see Cold vapor atomic absorption
   spectroscopy
CVAFS, see Cold vapor atomic fluorescence
   spectrometer
CVF, see Circular variable filter
Cyanide analyzers
   basics, 269
   CN_F determination, 270
   CN_T determination, 271–272
   CN_{WAD} determination, 271
   free cyanide, 270
   measurement methods, 270
   online cyanide analysis methods, 273
   primary and alternate analytical
      methods, 272
   strong metal complexes, 270
   weak acid dissociable cyanide, 270
Cyclic voltammetry, 291

D

Data
   air quality monitoring, 97–98
   analyzers sampling, 43
   mass spectrometers, 492
   OP-FTIR spectrometry, 893
   Raman process analyzer, 838–840

Data handling, 43
Data system/controller, 893
DCMST, see Down-hole corrosion
   monitoring
Deflection-type analyzer, 684–685
Denitrification, 608
Density-based sensors, 976
Density unit definitions, 1239
Detectors
   ammonium detectors, 608–609
   backscatter-type traversing detector, 639
   bag break detector, 753
   capacitance-type water-in-oil-detectors,
      635–636
   catalytic combustion detectors,
      222–227, 686
   CCD detector, 835
   CCTV detectors, 334–335
   chemiluminescence detector, 609–611
   combustible analyzers, 221
   dielectric constant detector, 196
   differential conductivity detectors, 708
   discharge ionization detectors, 174
   electrochemical detectors, 702–708
   electron capture detector, 173–174
   flame ionization detectors, 169, 227,
      393–394, 961–962
   flame photometric detector, 169–171, 425,
      627, 795, 932–933
   gas chromatography, 166–175
   hydrogen cyanide detectors, 403–407
   hydrogen sulfide detectors, 421–425
   ICP-type detector, 305–307
   immersion probe-type polarographic
      chlorine detector, 148
   infrared detectors, 445–446
   intrusive and erosion based sand detector,
      880–881
   leak detectors, 474–482
   liquid chromatographs, 195–196
   Luft detector, 441–443
   microphone type, 436
   microwave water-in-oil detector, 636
   multiple-anode detector, 707
   NDIR detectors, 445
   open-path spectrometry, 892–893,
      903–904
   optical absorbance detector, 195
   optical flame detectors, 332–334
   optical phototransistor detectors, 528
   orifice-capillary detector, 172
   ORP detectors, 1075
   oxidation–reduction potential, 1075
   oxygen demand detectors, 1075–1077
   passive acoustic sand detector, 879
   photoconductive detectors, 446
   photo-ionization detectors, 172–173,
      227–228
   portable hydrogen leak detector, 416
   portable multigas detector, 405
   pulsed discharge detector, 174–175
   pulsed flame photometric detector,
      170–171
   pyroelectric detector, 445
   Raman instrumentation, 835
   refractive index detector, 195
   RVP detectors, 277

RVR detectors, 759
scattered light detectors, 974–975
self-cleaning suspended solids detector, 39
smoke detectors, 330–332
solid-state type, 436
standpipe detector, 481
streaming current detector, 909–915
thermal conductivity detectors, 941–947
thermistor-type detector, 946
thermocouple detector, 223
thermopile detectors, 445–446
toxic gas monitoring, 360–366
ultra-high-frequency electromagnetic wave type detector, 250
ultraviolet analyzers, 990
visual flame detectors, 335
water quality monitoring systems, 1075–1077
Device and function symbols
alarm indicators, 1217
functional diagrams, 1217–1218
instrument signal connections, 1216–1217
multipoint and multifunction instruments, 1218
PFD diagram, 1214
process control description, 1218–1219
process variable measurement devices
analyzer primary element, 1215
double process connection, 1215
final control elements, 1215–1216
instrument-to instrument connection symbols, 1185–1186, 1214
orifice meter tubes/runs, 1215
orifice plate primary elements, 1215
primary element and transmitter symbols, 1178, 1214
primary element symbols, 1180–1182
process root block valves, 1215
with sample conditioners, 1214
secondary instrument symbols, 1182
single process connection, 1215
without sample conditioners, 1215
Dew cup, 528–529
Dew-point meter, 398
Diagnostics
gas chromatography, 180
Raman instrumentation, 840
Dichotomous sampler, 756
Dichromate COD test, 671–672
Dielectric constant detector, 196
Differential pressure
physical property analyzers, 813–814
viscometers, 1021–1022
Differential refractometer sensor, 589
Differential vapor pressure
basics, 277
microprocessor-based smart instruments, 279
refrigerant characteristics, 277–278
RVP detectors, 277
temperature compensation, 279
temperature–vapor pressure relationships, 279
transmitter design and installation, 280
Diffusion head analyzers, 224
Digital conductivity sensors, 241

Diode lasers, 899
Dioxin analyzers
basics, 284
industrial processes, 284
PCDD and PCDF emissions measurement
cleanup procedure, 286
condenser and absorbent trap, 285
gas chromatograph, 286
high-resolution gas chromatography, 286
high-resolution mass spectrometry, 286
sample extraction, 286
sampling train, 285
polychlorinated biphenyls, 284
Dipole polarization effect moisture sensor, 553–554
Direct low-level photometric analyzer, 424–425
Discharge ionization detectors, 174
Dispersive Raman instrumentation, 832–833
Dissolved oxygen (DO)
applications, 711
basics, 700–701
calibration methods, 709
coulometric sensor
advantage, 707
differential conductivity detectors, 708
multiple-anode detector, 707
thallium cells, 707
electrochemical detectors
components, 702–703
flow-through cells, 705
galvanic cells, 704–975
polarographic (voltametric) cells, 703–704
probe design, 705–706
sample temperature and flow, 704
trace oxygen sensor, 707
fluorescence sensors, 708–709
installations, 711
measurement
polarographic DO sensor, 668
Winkler titration method, 668
microbes, 701
pressure effects, 710–711
salinity effects, 711
sensors, 1077
temperature compensation, 709–710
vibratory paddle-type probe cleaner, 702
water aeration, 701
wet chemistry analyzers, 709
Distillation analyzers
ASTM Method D 1160, 804
ASTM Method D 2887-89 and D 3710-88, 804
ASTM Method D 86-IP-123, 803–804
batch technique, 803
distillation curves, 803
online, 803–808
true boiling-point distillation, 803
Distribution monitors, see Particle size distribution (PSD) monitors
Dithizone, 502

DO, see Dissolved oxygen
Dosimeters
CDTs, 368–370
color-change badges, 366–368
sorption-type dosimeters, 370–371
Double junctions, 773
Down-hole corrosion monitoring (DCMST), 264–265
Drinking water disinfection, 725
Dry Reid method, 809
Dual-beam analyzer
dual-detector UV analyzer, 993
IR analyzer, 746
single-detector UV analyzer, 992–993
Dual-beam ultraviolet analyzers, 992–993
Dual-capillary viscometer, 1022
Dual-chamber Luft detector, 442–443
Duckbill samplers
analyzer sampling, 71
water quality monitoring systems, 1072
Dust-fall jars, 95–96
Dynamic liquid-pressure-type viscometer, 1033–1034
Dynamic mechanical analysis, 865–866, 870
Dynamic vapor pressure analyzers, 810–811

E

ECD, see Electron capture detector
Edge fracture, 869
EDXRF instruments, see Energy-dispersive instruments
Efflux cup viscometers, 1055–1057
Elastomeric gaskets, 319
Electric resistance (ER) monitor, 261–262
Electrochemical analyzers
ambient air analyzers, 927
basic principle, 34
basics, 290–291
conductivity sensors, 35–36
gas analyzer, 927
potentiometric sensors, 34–35
sulfur dioxide, 927
voltammetric analyzers (see Voltammetric analyzers)
Electrochemical calibrating gas generators, 372
Electrochemical cells
fluoride analyzers, 346
hydrogen sulfide, 422
Electrochemical DO detectors
components, 702–703
flow-through cells, 705
galvanic cells, 704–975
polarographic (voltametric) cells, 703–704
probe design, 705–706
sample temperature and flow, 704
trace oxygen sensor, 707
Electrochemical oxygen sensors
categories, 686
coulometric sensor, 692
galvanic sensors, 689–692
high-temperature current-mode oxygen sensors, 689

high-temperature fuel cell oxygen
  sensors, 687–689
 polarographic sensor, 692
Electrochemical ozone sensor, 719
Electrode cleaners, pH
 automatic cleaners
  brush cleaners, 776–777
  chemical cleaners, 777
  design variations, 775
  types, 775
  ultrasonic cleaners, 776
  water-jet cleaners, 776–777
 automatic retraction, 777
 manual cleaning, 777
 on-line cleaning, 777–778
 shrouds and filters, 775
Electrodeless cells, 239
Electrodes
 bare metal electrodes, 725–726
 cleaners, 149, 775–778
 ion-selective electrodes, 111–112, 344, 458–470, 1073–1079
 metallic ORP electrodes, 657
 mounting, 657–658
 pH measurement, 770–778
Electrolysis instrument, 671
Electrolytic hygrometer, 543–545
Electrolytic probes, 294–295
Electron capture detector (ECD), 173–174
Electronic nose (e-nose), 628–630
Electron microscopy
 molecular weight analyzers, 591
 particle size distribution monitors, 734
Electrophoresis, 195
Electrostatic precipitators, 105
Elemental analyzers
 arsenic limit, 304
 atomic absorption spectrometer, 304–305, 307
 basics, 303
 ICP-type detector, 305–307
 mass spectrometer, 304
 sea water elements, 304
 XRF spectrometers, 307–308
Elliptical-mirror-type airborne-particulate counter, 737
End group determination, 591
Energy and heat unit conversion, 1239
Energy-dispersive (EDXRF) instruments, 936–937
Energy unit definitions, 1240
Enthalpy (h)
 definition, 1287
 dry and saturated steam properties, 1288–1289
 superheated steam properties, 1290–1293
 units and conversions, 1287
 water properties, 1294
Entrainment removal, 68
Entropy (s)
 definition, 1287
 dry and saturated steam properties, 1288–1289
 superheated steam properties, 1290–1293
 units and conversions, 1287
 water properties, 1294

Environmental pollution sensors, 635
Environmental Protection Agency (EPA) particulate sampling system
 control unit, 84–85
 heated compartment, 84
 ice-bath compartment, 84
 microprocessor-controlled stack sampling, 80
 pitot tube assembly
  isokinetic sampling, 84
  particle collection, sampling velocity, 84
  pitot tube calculation form, 83–84, 87
  sampling point selection, 82
  traversing point locations, 82–83
  type S pitot tube manometer assembly, 81–84
 schematic diagram, 80
EPA particulate sampling system, see Environmental Protection Agency particulate sampling system
ER monitor, see Electric resistance monitor
Eutrophication, 1070
Expansion tube calorimeter, 384
Extensional flow rheometer, 873–874

F

Fabre Perot interferometer, 326
Falling ball viscometer, 1024–1025
Falling element viscometers
 drag force, 1023
 falling-piston viscometer, 1023–1024
 falling-slug/falling-ball viscometers, 1024–1025
 Stokes (creeping) flow, 1023
Falling needle viscometer, 1055
Falling-piston viscometer, 1023–1024
Falling/rolling element viscometers
 automatic falling ball viscometer
  auxiliary control unit, 1054
  electrical timer, 1054
  with variable inclination angle, 1055
  viscosity test assembly, 1054
 drag force, 1053
 gravitational force, 1053
 manual falling ball unit, 1053–1054
Falling-slug viscometers, 1024–1025
FIA, see Flow injection analysis
Fiberglass papers, 104
Fiber mist eliminator, 60–61
Fiber-optic (FO) cables
 basics, 312–313
 environmental effects, 323
 internet supporting cables, 313–315
 refractive index, 313
 safety considerations, 322–323
 total internal reflection, 313
Fiber-optic light delivery and collection, 835
Fiber-optic probes (FOPs), 38–41
 absorption measurement, 324
 ATR probe, 326
 basics, 312–313
 chemical processing applications, 317
 cross-pipe dual probe, 321

 designs, 317–318
 environmental effects, 323
 extrinsic mode, 315
 Fabre Perot interferometer, 326
 vs. fiber-optic cables, 315
 fiber-optic multiplexer, 317
 fluorescence optrode, 325
 intrinsic mode, 315
 legacy fiber-optic spectrophotometer, 315–316
 level sensors, 325
 Nephelometry*, 324
 optical coupler, 317
 process interfaces, 320
 Raman analyzers, 325
 refractive index, 325–326
 return waveguide, 316
 safety considerations, 322–323
 sealing techniques
  brazing, 319
  compression rings, 319–320
  elastomeric gaskets, 319
  epoxy adhesive, 318
 sensor tip, 316
 single-pass probe, 320
 source optical waveguide, 316
Fiber-optic Raman analyzer, 833
Fiber optic RI probes, 859
Fiber-optic spectrophotometer, 212, 316
FIDs, see Flame ionization detectors
Field-programmable autotitrator, 1089
Filters
 pH measurement, 775
 Raman instrumentation, 836
 sampling systems, 40
Flame and fire detectors
 basics, 330–331
 burner flame safeguards, 335
 CCTV detectors, 334–335
 conduction sensor rods, 336
 flame rods, 336
 flame safeguard installation, 339
 heat detection cables, 332
 heat sensor rods, 336
 multisensor smoke alarm, 332
 optical flame detectors
  cone of vision, 333
  dual and multi-spectrum IR detectors, 334
  false alarm source impact, 333
  flame detection distances, 333
  flickering frequency, 332
  IR detectors, 334
  UV detectors, 333–334
  UV/IR detectors, 334
 rectification sensor rods, 336–337
 thermal sensors, 331–332
 visible radiation, 337–338
 visual flame detectors, 335
Flame ionization detectors (FIDs), 169, 227, 393–394, 961–962
Flameless atomic absorption spectroscopy, 502
Flame photometric analyzers, 795–796, 927
Flame photometric detector (FPD), 169–171, 425, 627, 795, 932–933

Flash-point analyzer
    calibration and applications, 818
    flammable liquid, 816
    flash-point tests (ASTM Methods D 56-90 and D 93-90), 817
    high-temperature, 818
    low-temperature, 817–818
Flat glass electrodes, 770–771
Flicker photometer, 993–994
Float viscometers
    concentric viscometer, 1027–1028
    precautions, 1025
    rotameter-type flowmeter, 1025
    single-float viscometer, 1026
    two-float viscometer, 1026
Flowing junction reference electrode, 768
Flow injection analysis (FIA), 1087
    application, 1096–1097
    cyanide analysis, 271
    detectors, 1095
    industrial design, 1096
    laboratory design, 1096
    sequential injection and bead injection, 1095
Flow injection analyzers (FIAs), 1077–1078
Flow sheets
    definition, 1174
    of distillation column, 1174
    electrical schematics symbols, 1209–1211
    final control element symbols
        functional diagram symbols, 1193
        function block symbols, 1194–1200
    function and calculation symbols
        control valve failure positions, 1192
        final control element actuator symbols, 1188–1189
        final control element symbols, 1187
        regulator relief device symbols, 1190–1191
    identification code letters, 1176
    instrumentation device and function symbols, 1174, 1177
    instrument index, 1174–1176
    logic function symbols, 1200–1208
    measurement symbols
        auxiliary and accessory device symbols, 1183
        measurement related abbreviations, 1179
        primary element and transmitter symbols, 1178
        primary element symbols, 1180–1182
        secondary instrument symbols, 1182
    miscellaneous device and function symbols, 1177
    signal and connection symbols
        instrument-to instrument connection symbols, 1185–1186
        instrument to process connection symbols, 1184
Fluorescence
    fiber-optic probes, 325
    optrode, 325
    sensors, dissolved oxygen, 708–709
Fluoride analyzers
    automatic and manual methods, 344
    basics, 343–344
    commercial MS analyzers, 348
    detector tubes, 345
    electrochemical cells, 346
    fluoride compounds types, 345
    Freons, 348
    IMS, 346–347
    ion-selective electrodes, 344
    IR spectroscopy, 347
    ISE, 347–348
    laboratory methods, 348
    laser-based open-path systems, 348
    liquid fluoride analyzers, 345
    paper tape, 346
    qualitative sensors, 345
    silicon dioxide sensors, 348
FO cables, see Fiber-optic cables
FOPs, see Fiber-optic probes
Ford cup viscometers, 1056
Formazin turbidity units (FTUs), 967–968
Forward scattering/transmission type sensor
    colorimeter, 969
    dual-beam design, 969
    instrument designs summary, 971–973
    laser-type meter, 971
    ratioing, 969–971
    suspended solids and sludge density sensors, 974
Fouling, conductivity, 240
Four-electrode measurement, 238–239
Fourier-transform infrared spectroscopy (FTIR) gas analyzer, 405–406
Fourier transform Raman instrumentation, 831
Fourier transform spectrometers, 440
FPD, see Flame photometric detector
Free cyanide (CN_F) determination, 270
Free residual chlorine analyzer, 147
Freeze-out sampling, 103
Freezing-point analyzer, 816
Fritted absorber, 102–103
FTIR gas analyzer, see Fourier-transform infrared spectroscopy gas analyzer
FTUs, see Formazin turbidity units

G

Galvanic cells, 704–975
Galvanic DO sensors, 702, 704–975
Galvanic probes, 291, 294–295
Gas and vapor sampling
    bag sampling, 101–103
    samples grabbing device, 101
Gas cells, 446
Gas chromatography (GC)
    basic elements of, 156
    basics, 154
    carrier gas flow control, 175
    definition, 155
    discharge ionization detectors, 174
    ECD, 173–174
    FID, 169
    with FPD, 169–170, 932–933
    vs. HPLC, 192
    hydrocarbon analyzers
        basic apparatus, 395
        calibration, 395
        nonmethane hydrocarbons, 395
        reactive hydrocarbons, 395–396
    hydrogen, 415
    industrial emission applications, 618
    mobile phase, 155
    multidimensional capillary GC, 184
    NO_x analyzers, 618
    OCD, 172
    odor detection, 627
    online distillation analyzers, 808
    organic mercury detection, 513–514
    oxygen in gases, 695
    PCDDs and PCDFs emissions measurement, 286
    PDD, 174–175
    PFPD, 170–171
    PGC
        components of, 158–166
        features, 157
        schematic diagram, 157
    PIDs, 172–173
    programmer controller, 175–176
    retention time, 155–156
    stationary phase, 155
    TCD, 167–168
    temperature-programmed chromatography, 156–157
    workstation portals, 184–185
Gas filter correlation (GFC), 127, 135, 443
Gas flushing, see Sparging
Gas phase ozone monitoring, 715–716
Gas removal, 62–63
Gauges
    capacitance moisture gauge, 572–573
    impedance moisture gauge, 574
    resistance moisture gauge, 573
GC, see Gas chromatography
Gel-permeation chromatography, 29, 195, 588
Gel permeation columns, 195
GFC, see Gas filter correlation
Glass electrodes, 770–771
Glass impedance measurements, 783
Glass ion-selective electrodes, 464–465
Glasteel®, 771
Global carbon cycle, 124
Gold cyanidation, 404
Gold film sensor, 423
Gravimetric methods
    absorption and condensation, 566
    desiccants, 565
    disadvantages, 563
    lyophilization, 565–566
    oven drying, 563–565
    TGA, 565
Griess–Saltzman method, 620
Gross calorific value, 201–203
Gyrating-element viscometer, 1032

H

Hagen–Poiseuille law, 1003, 1020, 1048
Hair hygrometer, 529
Half-cell reactions, 655–656

Handheld corrosion monitor, 264
Handheld digital refractometer, 858
Handheld indoor air quality monitors, 100
HART®, see Highway Addressable Remote Transducer
Hazardous and toxic gas monitoring
   area detection, 372–373
   basics, 352
   calibration, 371–372
   choice of diffusion vs. extractive detection, 372
   communication and maintenance, 373
   detectors
      advantages and limitations, 365
      amperometric sensors, 362
      applications, 365
      catalytic bead detectors, 366
      electrochemical sensors, 361–363
      features, 360
      fixed detectors, 363
      open-path detectors, 366
      oxygen detectors, 361
      photo and flame ionization detectors, 366
      portable detectors, 364
      semiconductor detectors, 366
      spectrometers, 365
      typical design, 360
   dosage sensors
      CDTs, 368–370
      color-change badges, 366–368
      sorption-type dosimeters, 370–371
   duct detection, 373
   electronic monitors, 353
   environmental conditions, 372
   hazardous areas
      classification of, 356
      exposure limits, 357–359
      safety instrument performance standards, 357
      toxicity levels, 357
      zones, 356
   interference gas, 372
   location, 372
   maintenance requirements, 372
   stack detection, 373
   suppliers list, 354–355
   target gas and detection range, 372
HCN detectors, see Hydrogen cyanide detectors
Heart-cutting, process gas chromatograph, 166
Heat balance calorimetry, 385
Heated compartment, 84
Heat flow calorimetry, 385
Heating value calorimeters
   applications, 386
   basics, 378–379
   bomb calorimeter, 379, 381
   features and specifications, 379–380
   inferential calorimeters, 379
   sample conditioning system, 386
   temperature-based designs
      adiabatic calorimeters, 385
      adiabatic flame temperature, 384
      heat balance calorimetry, 385
      heat flow calorimetry, 385
      MTDSC, 385
      optical calorimeters, 385–386
      thermopile calorimeter, 385
   true calorimeters, 379
   water temperature rise calorimeter
      airflow calorimeter, 383
      air temperature rise calorimeter, 383
      calibration heater, 382
      chromatographic calorimeter, 384
      expansion tube calorimeter, 384
      residual oxygen calorimeter, 384
   Wobbe index, 381
Heating, ventilating, and air conditioning (HVAC), 523–524
Heat-of-adsorption hygrometer, 550–551
High absorption particulate arrestance (HEPA) filters, 732
High-performance liquid chromatography (HPLC)
   analysis section, 192
   applications, 196
   basics, 190–191
   carrier supply, 192–193
   columns
      applications, 193
      electrophoresis, 195
      gel permeation columns, 195
      ion exchange columns, 195
      liquid–adsorption columns, 194–195
      liquid–partition columns, 194
      selectivity and resolution, 194
   control section, 192
   definition, 191
   dielectric constant detector, 196
   vs. gas chromatography, 192
   mobile phase, 191
   optical absorbance detector, 195
   pressure and flow controls, 193
   refractive index detector, 195
   schematic diagram, 191
   stationary phase, 191
   supply pumps, 193
   valves, 193
High-pressure stream sampling, 71
High-purity water measurements, 239–240
High-temperature current-mode oxygen sensors, 689
High-temperature flash-point analyzer, 818
Highway Addressable Remote Transducer (HART®), 243, 786–787
Homogenizers, 65–66
Horizontal still distillation analyzer, 807
Hot box, 84
HPLC, see High-performance liquid chromatography
H_2S detectors, see Hydrogen sulfide detectors
Human olfactory system, 626
Humidity sensors
   electrolytic cells, 526–527
   operating ranges, 525
   spectroscopic absorption, 526
HVAC, see Heating, ventilating, and air conditioning
Hydrocarbon analyzers
   atmospheric hydrocarbon analyzers, 392
   basics, 390–391
   calibration, 399
   FIDs, 393–394
   gas chromatography
      basic apparatus, 395
      calibration, 395
      nonmethane hydrocarbons, 395
      reactive hydrocarbons, 395–396
   hydrocarbon dew-point meter, 398
   IMS–MS, 398
   spectrometric methods
      LIDAR, 396
      perimeter monitoring, 397
      VCSEL, 396–397
Hydrocarbon dew-point meter, 398
Hydrogen cyanide (HCN) detectors
   basics, 403
   cyanide chemistry, 404
   exposure limits, 404
   FTIR, 405–406
   performance standards, 406–407
   portable multigas detector, 405
Hydrogen detector transmitters, 947
Hydrogen, stream/air
   CB sensors, 414–415
   electrochemical sensors, 414
   explosions, 412
   gas chromatography, 415
   lower explosive limit, 412
   mass spectrometry, 415
   palladium-based thick film sensors, 414
   properties and safety, 412
   role, 411–412
   sampling and installations, 415–417
   solid-state sensors, 414
   TCDs, 413
   upper explosive limit, 412
Hydrogen sulfide (H_2S) detectors
   applications, 425
   basics, 421
   direct low-level photometric analyzer, 424–425
   electrochemical sensors, 422
   FPD, 425
   gold films, 423
   lead acetate tape, 423–424
   solid-state sensors, 423
Hygrometers
   capacitance hygrometer, 545–548
   cellulose hygrometers, 530
   chilled-mirror hygrometer, 528
   electrolytic hygrometer, 543–545
   hair hygrometer, 529
   heat-of-adsorption hygrometer, 550–551
   impedance hygrometer, 548–549
   microwave absorption hygrometer, 552
   piezoelectric hygrometer, 549–550

I

Ice-bath compartment, 84
ICMI, see International Cyanide Management Institute
ICP detector, see Inductively coupled plasma detector
Image analyzers, 734

Immersion probe-type polarographic chlorine detector, 148
Impaction devices
　ambient air opacity monitoring, 757–758
　particulate sampling, 104
Impactors, 104
Impedance hygrometer, 548–549
Impedance moisture gauge, 574
Impinger methods
　air quality monitoring, 104
　mercury, 498–500
IMS, see Ion mobility spectrometry
IMS–MS, see Ion-mobility spectrometry–mass spectrometry
Incubation time, oxygen demand, 667
Indigo method, ozone monitoring, 727
Inductively coupled plasma (ICP) detector, 305–307
Industrial emission applications
　chemiluminescent analyzers, 618
　electrochemical sensors, 618
　gas chromatographs, 618
　NDIR analyzers, 617
　paramagnetic analyzers, 616
　thermal conductivity analyzers, 616
　UV analyzers, 617–618
Industrial viscometers
　basics, 1016–1017
　capillary viscometers, 1020–1023
　Coriolis mass flowmeter, 1038–1039
　developments
　　communication, 1039
　　display and human machine interface, 1039
　　self-diagnosis, 1039
　　squeeze flow, 1039–1040
　　surface acoustic wave sensors, 1039
　falling element viscometers, 1023–1025
　float viscometers, 1025–1028
　vs. laboratory viscosity analyzers, 1046
　orientation table, 1018–1019
　oscillating viscometers, 1034–1036
　plastometers, 1028–1030
　rotating-spindle-type industrial viscometer, 1030–1032
　rotational element viscometers, 1030–1034
　vibrational viscometers, 1036–1038
Information gathering, 26–27
Infrared (IR) analyzers, 32
　applications, 449
　auto-calibration, 450
　basics, 429–430
　Beer–Lambert law, 434–435
　calibration, 448
　cell selection
　　gas cells, 446
　　liquid cells, 447
　　path length selection, 446
　　solid samples, 448
　continuous chemical analysis, 432
　dual-beam IR analyzer, 438
　functional group frequency chart, 434
　in situ stack gas analysis, 437–438
　intelligent units, 450

intrusive and erosion based sand detector, 880–881
　laboratory analyzers
　　filter spectrometers, 440
　　Fourier transform spectrometers, 440
　　grating spectrophotometers, 438–439
　　process control applications, 441
　　tunable lasers, 440–441
　laboratory spectrophotometers, 435–436
　MIR/ATR crystal probes, 450
　NDIR detectors, 445
　online applications
　　dual-chamber Luft detector, 442–443
　　GFC spectrometers, 443
　　multiple-component fixed filter analyzer, 444
　　narrow-band-pass filters, 443
　　portable octane analyzer, 444
　　portable, single-beam IR photometer, 444
　　programmed CVF analyzer, 444
　　single-component analyzers, 441–442
　packaging, 449
　photoconductive detectors, 446
　pyroelectric detector, 445
　single-beam configuration, 436–437
　sources, 445
　spectra, 432–433
　TAS AOTF, 450
　thermopile detectors, 445–446
Inorganic carbon (IC), 957
Installation
　impedance hygrometer, 548–549
　pH measurement, 781–783
　process gas chromatograph, 183–184
　Raman process analyzer, 841
Instrumentation
　oxygen demand, 667–668
　Raman instrumentation, 830–833, 836
Intelligent IR analyzer, 449
Interfaces
　conductivity-based interface, 638
　fiber-optic probes, 320
　process gas chromatograph, 178–180
　Raman process analyzer, 836
　smart user interfaces, 1064
　ultrasonic pipeline interface, 638
　viscometers, 1039
Interferences
　carbon monoxide, 135
　ion-selective electrodes, 467–468
　open-path spectrophotometry, 904
　oxygen demand, 676
International Cyanide Management Institute (ICMI), 270
International Society of Automation (ISA), 25
International unit symbols, 1222
Internet supporting cables, 313–315
Intrinsic viscosity
　molecular weight, 1051
　viscometers, 1004
Intrusive and erosion based sand detector, 880–881
Invasive measurement, 26
Iodometry, 145, 1109

Ionic strength adjustment buffer (ISAB), 462–463
Ionization chamber sensors, 331
Ion mobility spectrometry (IMS), 346–347
Ion-mobility spectrometry–mass spectrometry (IMS–MS), 398
Ions
　concentrations, pH, 767
　liquid chromatographs, 195
　mass spectrometry detection, 491
　separation, 489–491
Ion-selective electrodes (ISEs)
　advantages, 469
　ammonia water analyzer, 111–112
　basics, 458–459
　calibration, 468–469
　commercially available electrodes, 459
　electrode types
　　glass, 464–465
　　liquid ion exchange, 466–467
　　solid-state electrodes, 465–466
　fluoride analyzers, 344
　interference, 467–468
　measurement range, 467
　potential disadvantages, 469–470
　theory of operation
　　activity coefficient, 461–462
　　analytical concentration, 461–462
　　ISAB, 462–463
　　Nernst slope changes, 460
　　Nikolsky equation, 460
　　reference electrode, 460–461
　　system accuracy, 464
　　temperature effects, 463–464
　water quality monitoring systems
　　conductivity sensors, 1075
　　dissolved oxygen sensors, 1077
　　ion electrodes, 1073–1074
　　nephelometers, 1077
　　ORP detectors, 1075
　　oxygen demand detectors, 1075–1077
　　pH probes, 1075
　　TC and TOC analyzers, 1077
　　turbidity meters, 1077
　　VOC detection, 1078–1079
　　wet chemistry analyzers, 1077–1078
Ion-selective field effect transistor (ISFET), 772
Ion-specific electrodes (ISE), 347–348
Ion-trapping sections, mass spectrometry, 490–491
IR analyzers, see Infrared analyzers
ISA, see International Society of Automation
ISAB, see Ionic strength adjustment buffer
ISEs, see Ion-selective electrodes
ISFET, see Ion-selective field effect transistor
Isokinetic sampler, 79
Isokinetic sampling technique, 84, 752
Isopotential point, 463, 769

J

Jackson turbidity unit (JTU), 967
Jacobs–Hochheiser method, 620

K

Karl Fischer titration, 542
Kinematic viscosity, 1003, 1008
Kjeldahl nitrogen standard method, 610
Kneader plastometer, 1029

L

Lab algorithm, textile industry, 213
Laboratory equipment and methods
  air quality monitoring, 96–97
  fluoride analyzers, 348
  ion-selective electrodes, 464
  moisture analyzers, 542
Laboratory streaming current analyzer, 911
Laboratory viscometers
  basics, 1045–1046
  bubble-time viscometers, 1048
  capillary viscometers, 1048–1053
  falling/rolling element viscometers, 1053–1055
  vs. industrial viscosity analyzers, 1046
  orientation table, 1046–1047
  orifice/efflux cup viscometers, 1055–1057
  oscillating piston viscometer, 1057
  rotational viscometers, 1057–1063
  trends
    adaptability, 1064
    automation, 1063
    lab management systems, 1063–1064
    smart accessories, 1064
    smart user interfaces, 1064
  units, 1048
  vibrational viscometers, 1063
Lambert–Beer law, 434–435, 717, 985
Laser-based open-path systems, 348
Laser excitation source, 834
Laser-induced breakdown spectroscopy (LIBS), 202, 206
Laser-induced Doppler absorption radar (LIDAR), 396
Laser-type meter, 971
Lead acetate tape method, 423–424, 933–934
Lead peroxide candles, 96
Leak detectors
  aboveground detection
    combustible/toxic leaks, 476
    personal alarms, 477
    pressure-and vacuum-based testing, 476
  application, 475
  basics, 474–475
  externally based systems, 481
  halogen detectors, 478–479
  internally based systems, 481
  laser-based fiber-optic Raman sensing, 482
  sub-seas leakage detection, 482
  ultrasonic detectors, 477–478
  underground detection
    level monitoring, 479–480
    soil concentration detection, 480–481

Legacy fiber-optic spectrophotometer, 315–316
Length unit conversion, 1242
LIBS, see Laser-induced breakdown spectroscopy
Light absorption, 967
Light attenuation, 748–751
Light direction and ranging (LIDAR), 759
Light measurement, refractometers, 859
Light scattering, particle size, 735
Light-scattering photometers, 585–587
Linear polarization resistance (LPR) monitors, 263
Line-powered transmitters, 242, 244, 785, 787
Liquefied natural gas processing, 386
Liquefied petroleum gases, 809
Liquid–adsorption columns, 194–195
Liquid cells, infrared instruments, 447
Liquid chromatography,
  see High-performance liquid chromatography
Liquid–gas separation, 62
Liquid ion exchange, 466–467
Liquid–liquid separation, 62
Liquid–partition columns, 194
Loop diagram
  control valve failure positions, 1192
  definition, 1173
  final control element actuator symbols, 1188–1189
  final control element symbols, 1187
  graphic symbols, 1213
  instrument-to instrument connection symbols, 1185–1186
  instrument to process connection symbols, 1184
  ISA-5.4-1991 standard guidelines, 1211
  regulator relief device symbols, 1190–1191
  required informations, 1211–1212
  terminal, power supply and action symbols, 1212–1213
Low-temperature flash-point analyzer, 817–818
LPR monitors, see Linear polarization resistance monitors
Luft-type detector, 441–442
Lyophilization, 565–566

M

Magnetically coupled viscometers, 1032
Magnetic sectors, mass spectrometry, 489
Maintenance
  analyzer instrumentation, 44
  conductivity analyzers, 241
  oxidation–reduction potential, 660
  Raman process analyzer, 840–841
Manometric BOD test, 669–670
Manual stack samplers, 752
Mark–Houwink relation, 1052
Mass flow unit conversion, 1241
Mass spectrometers
  basics, 486
  data reduction and presentation, 492

ion-trapping sections, 490–491
magnetic sectors, 489
operation principles
  automatic metering valves, 487
  computer-based data reduction system, 492
  ion detection, 491
  ion separation, 489–491
  online spectrometer applications, 487
  preanalysis sample conditioning, 487
  process stream applications, 487
  sample ionization, 487–488
  sintered metal leak device, 487
  thin diaphragms, 487
  vacuum environment, 491–492
quadrupole filter, 489–490
RGA, 492
time-of-flight filter, 491
Mass spectroscopy, oxygen in gases, 693
Measurement ranges
  ion-selective electrodes, 467
  pH measurement, 766
  thermal conductivity detectors, 942–943
Measuring cells, ultraviolet analyzers, 990
Measuring circuits, combustible analyzers, 222–224
Median selector, pH, 783
Membrane-covered probes, 295
Membrane osmometer, 583–584
Membrane probes, chlorine, 147–148
Membrane-type amperometric ozone monitors, 726–727
Memosens® sensors, 785
Mercury measurement
  in ambient air
    activated absorption, 500–501
    amalgamation, 500
    atomic fluorescence flame spectrophotometric methods, 503
    basics, 497–498
    colorimetric methods, 502–503
    impinger collection methods, 498–500
    mercury vapors conversion, 501
    organic mercurials, 503
    radiochemical procedure, 503
    regulations, 503
    sample collections, 498
    ultraviolet absorption, 501–502
    wet chemical procedure, 498
  in water
    atomic absorption spectrophotometry, 509–510
    basics, 507
    colorimetric detection, 509
    direct/thermal decomposition, 515
    online measurement, 511–512
    organic mercury detection, 513–514
    sample treatment, 508–509
Mercury vapor analyzer, carbon monoxide, 135–136
Metallic materials, composition and manufacturers, 1281–1282
Metal oxide pH electrodes, 771–772
Metal oxide sensors, odor detection, 629
Methane and carbon monoxide analyzer, 136, 395

Micronose, *see* Electronic nose
Microorganisms quantity and activity measurement, *see* Biometers
Microprocessor-controlled portable infrared spectrometer, 98–100
Microwaves
    absorption hygrometer, 552
    attenuation, 571–572
    oil measurements, 636–638
    and radio frequency analyzers, 33
Microwave sensors, 254, 552
Microwave-type solids moisture analyzers, 571–572
Modern and International Units, 1220
Modulated-temperature differential scanning calorimeters (MTDSC), 385
Moisture analyzers
    in air
        airline environment, 524
        autocalibration, 532
        basics, 519–521
        breathing air, 524
        calibration methods, 534–535
        cellulose hygrometers, 530
        chilled-mirror hygrometer, 528
        clean rooms, 524
        combustion, 524
        condensing atmospheres, 531–532
        dew cup, 528–529
        duct installation, 531
        enthalpy, 522–523
        hair/synthetic fiber element, 529
        handheld calibration, 532–533
        humidity sensors, 525–527
        HVAC, 523–524
        ISO standard, 533
        NIST, 533
        optical phototransistor detectors, 528
        psychrometric chart, 521–523
        relative humidity, 521–525
        sling psychrometer, 530
        thin-film capacitance probes, 527–530
        two-pressure humidity generator, 533–534
    in gases and liquids
        basics, 540–541
        calibration, 555–557
        capacitance hygrometer, 545–548
        CRDS, 554
        dipole polarization effect moisture sensor, 553–554
        electrolytic hygrometer, 543–545
        heat-of-adsorption hygrometer, 550–551
        impedance hygrometer, 548–549
        laboratory analyzers, 542
        microwave absorption hygrometer, 552
        neutron backscatter moisture online analyzer, 554–555
        piezoelectric hygrometer, 549–550
        refraction-based moisture analyzer, 555
        sampling system, 543
        spectroscopy, 551–552
    in solids
        basics, 561–562
        capacitance measurement, 573
        capacitance moisture gauge, 572–573
        chemical methods, 563
        combined neutron/gamma probes, 567–568
        fast neutron moderation, 566–567
        gravimetric methods, 563–566
        impedance moisture gauge, 574
        infrared absorption and reflection, 569
        Karl Fischer type units, 563
        microwave attenuation, 571–572
        NMR apparatus, 574–575
        radio frequency absorption, 575
        resistance moisture gauge, 573
        TDR, 570
Molecular weight analyzers, liquid
    advantages and limitations, 591–592
    basics, 580–581
    from chromatogram, 589–591
    differential refractometer sensor, 589
    electron microscope, 591
    end group determination, 591
    free-standing assembly, 589
    gel-permeation chromatography, 588
    light-scattering photometers, 585–587
    membrane osmometer, 583–584
    osmotic pressure, 582
    photometer, 586–587
    sedimentation equilibrium method, 592
    sedimentation velocity method, 592
    subassemblies, 589
    TEMs, 591
    ultracentrifuge, 592
    vapor pressure osmometers, 584–585
    viscometers, 587–588
    Zimm plot, 586
Monochromatic wavelength-dispersive (MWDXRF) instruments, 936–937
Mooney unit, 1028–1029
Motorized self-cleaning filters, 59
Mounting
    oxygen in liquids, 711
    refractometers, 856–857
MTDSC, *see* Modulated-temperature differential scanning calorimeters
Multiple-anode oxygen detector, 707
Multiple-component fixed filter analyzer, 444
Multiple head system, combustible analyzers, 226
Multiple junction references, 774
Multipliers to convert engineering units, 1123–1237
Multisensor smoke alarm, 332
Multistream analysis, gas chromatograph, 183
Multistream sampling, 74–75
Multivariable predictive analysis, 838–839
MWDXRF instruments, *see* Monochromatic wavelength-dispersive instruments

N

Narrow-IR-band-pass filters, 440, 443
National Ambient Air Quality Standards (NAAQS), 93
National Institute for Standards and Technology (NIST), 533
Natural gas measurements
    flow measurement, 597–598
    orifice plates, 598
    turbine flowmeters, 599
    ultrasonic flow meters
        accuracy, 599
        advantages, 599
        analyzers, 600
        bidirectional, multipath, LNG flowmeter, 599
        chromatographs, 601–602
        Coriolis mass flowmeters, 599–600
        mass spectrometers, 602
        sampling systems, 601
        venturi tube, 598–599
NDIR analyzers, *see* Nondispersive infrared analyzers
Near-infrared (NIR) analyzers
    absorption bands, 32–33, 451
    basics, 430–431
    calibration and applications, 820
    calibration transfer, 453
    chemometrics, 820
    fiber-optics, 451–452
    gases, 452–453
    liquids, 453
    moisture measurement, 32
    new developments, 820
    online design, 820
    optical sources, 452
    property and radiation absorption, 819
    sample temperature control, 451
    solids, 453
    types, 452
Near-infrared (NIR) photometers, 983, 997
Nephelometers
    ambient air opacity monitoring, 757
    oil-on-water measurements, 641–642
    scattered light detectors, 974–975
    water quality monitoring systems, 1077
Nephelometric turbidity units (NTUs), 967–969, 1077
Nephelometry*, 324
Nernst temperature effect, 769
NeSSIT, *see* New Sampling/Sensor Initiative
Neutron backscatter moisture online analyzer, 554–555
New Sampling/Sensor Initiative (NeSSIT), 75
Newtonian fluids, 1004–1005, 1020
NIR analyzers, *see* Near-infrared analyzers
NIST, *see* National Institute for Standards and Technology
Nitrification, 608
Nitrogen, ammonia, nitrite and nitrate
    ammonification, 608
    ammonium detectors, 608–609
    basics, 606–607
    CLD, 609–611

denitrification, 608
fertilizer, 608
Kjeldahl nitrogen standard method, 610
methemoglobinemia, 608
nitrate measurement, 609–610
nitrification, 608
nitrite measurement, 609
nitrogen cycle, 607
oxidation states, 608
total nitrogen measurement, 610–611
Nitrogen oxide (NO_x) analyzers
  ambient air monitoring
    calibration methods, 618–619
    colorimetric analyzers, 620
    NO–NO_2 combination analysis, 619
  basics, 615–616
  in combustion products, 616
  industrial emission applications
    chemiluminescent analyzers, 618
    electrochemical sensors, 618
    gas chromatographs, 618
    NDIR analyzers, 617
    paramagnetic analyzers, 616
    thermal conductivity analyzers, 616
    UV analyzers, 617–618
  portable monitors, 620
NMR apparatus, *see* Nuclear magnetic resonance apparatus
Nonaqueous solutions, pH measurement, 780
Nondispersive infrared (NDIR) analyzers
  carbon monoxide analyzer, 134
  NO concentration determination, 617
  SO_2, 923–924
  TOC, 954–955
Nondispersive infrared sensors, 126–127
Nonmethane hydrocarbons, 395
Non-Newtonian behavior, 867
Non-Newtonian fluids, 1004–1005
NO_x analyzers, *see* Nitrogen oxide analyzers
NTUs, *see* Nephelometric turbidity units
Nuclear magnetic resonance (NMR) apparatus, 36–37, 574–575
Nuclear techniques, spectrometers, 34

O

Occupational Safety and Health Administration (OSHA), 353
OCD, *see* Orifice-capillary detector
Octane analyzers
  applications, 819
  calibration, 819
  octane tests (ASTM Methods D 2699 and D 2700), 818
  online comparator analyzer (ASTM Method D-2885), 818–819
  reactor-tube continuous, 819
Odor detection
  amperometric sensors, 629
  basics, 625
  calorimetric sensors, 629
  chemical/chromatographic methods, 628
  chemocapacitors, 629
  e-nose/micronose, 628–630
  flame photometric detectors, 627
  gas chromatographs, 627
  gravimetric sensors, 628
  human nose, 628
  human olfactory system, 626
  hydrocarbon odor, 629
  metal oxide sensors, 629
  odorant substances, 625
  odor panels, 627
  organoleptic methods, 626
  polymeric film sensors, 629
  portable chromatograph, 628
  potentiometric sensors, 629
  relative odor intensities, 625
  sniff odor test system, 626–627
Oil-in-water measurements, 640–641
Oil measurements
  basics, 634–635
  environmental pollution sensors, 635
  oil-in-water measurements, 640–641
  oil-on-water measurements, 641–642
  water-in-oil measurements
    backscatter-type traversing detector, 639
    capacitance, 635–636
    conductivity-based interface, 638
    near-infrared meters, 639–640
    radiation-based flexible sensor, 639
    radio frequency/microwave sensor, 636–638
    strip-type nuclear source, 639
    ultrasonic pipeline interface, 638
Oil-on-water measurements, 641–642
Oil slick thickness detection, 642
Olfactory system, human, 626
OLR, *see* Online rheometer
Online comparator analyzer, 818–819
Online cyanide analysis methods, 273
Online distillation analyzers
  applications, 808
  calibration, 803
  calibration and sampling, 808
  vs. continuous plant distillation analyzers, 804
  end-point distillation analyzers, 805
  horizontal still distillation analyzer, 807
  simulated distillation, gas chromatography, 808
  vacuum distillation analyzer, 806
Online measurements
  analyzer instrumentation, 26
  mercury measurement, water, 511–512
  particle size, 736–737
On-line monitors
  coal analyzers, 204–206
  colorimeters, 214
Online particle size monitor, 738
Online rheometer (OLR), 866
Online streaming current monitor, 915
On/off batch colorimeter, 1094–1095
Opacity measurement
  dust loading, 745
  light attenuation and transmissometers, 748–751
  percent opacity, 745
  photopic and incandescent light sources, 744–745
  Ringelmann card numbers, 745–746
  transmittance, 745
  turbidity attenuation coefficient, 745
OP-CRDC spectroscopy, 902
Open-path absorption, 889–890
Open-path combustible hydrocarbon (OP-HC) detectors, 903–904
Open-path FTIR spectrometry
  components, 891
  data system/controller, 893
  instrumental configurations, 893–896
  interferometer, 891–892
  open-path air monitor, 891
  transfer optics and detector types, 892–893
Open-path (OP) monitoring
  applications
    combustible vapor measurement, 903–904
    OP-HC detectors, 903–904
    toxic GAS measurement, 903
  basics, 886–887
  design types
    blackbody radiation interference, 889
    instrument designs, 887
    interferometers, 889
    IR, UV, and TDLAS, 888
  long-path monitoring, 887
  OP-CRDC spectroscopy, 902
  OP-FTIR spectrometry
    components, 891
    data system/controller, 893
    instrumental configurations, 893–896
    interferometer, 891–892
    open-path air monitor, 891
    transfer optics and detector types, 892–893
  OP-TDLAS spectrometry
    applications, 900
    diode lasers, 899
    lead salt lasers, 898
    portable gas emission analyzer, 898
    principle of operation, 900
    wavelength modulation spectrometry, 901–902
  OP-UV spectromctry, 897–898
Open-path TDLAS spectrometry
  applications, 900
  diode lasers, 899
  lead salt lasers, 898
  portable gas emission analyzer, 898
  principle of operation, 900
  wavelength modulation spectrometry, 901–902
Open-path UV spectrometry, 897–898
Operator interface, 178–180
OP-HC detectors, *see* Open-path combustible hydrocarbon detectors
OP monitoring, *see* Open-path monitoring
Optical absorbance detector, 195
Optical and digital refractometers, 858
Optical calorimeters, 385–386
Optical cloud-point analyzers, 815–816
Optical flame detectors
  cone of vision, 333
  dual and multi-spectrum IR detectors, 334
  false alarm source impact, 333
  flame detection distances, 333

flickering frequency, 332
IR detectors, 334
UV detectors, 333–334
UV/IR detectors, 334
Optical fluorescence and luminescence sensors, 214–215
Optical particle analyzer, 736–737
Optical scattering (single particle), 735
Optical sensors
consistency measurement, 249–250
odor detection, 629
Optical sources, near-infrared analyzers, 452
Optrode, 325
Organic fluoride analyzers, 348
Organic mercury detection, 513–514
Organoleptic methods, 626
Orifice-capillary detector (OCD), 172
Orifice/efflux cup viscometers
automatic efflux cup viscometer, 1057
efflux time, 1055
Ford cup viscometers, 1056
Saybolt viscometer, 1055–1056
Zahn cup viscometer, 1056
ORP, see Oxidation–reduction potential
Orsat equipment, 124, 128
Oscillating blade viscometer, 1034
Oscillating piston viscometer
industrial viscometer, 1035–1036
laboratory viscometer, 1057
Oscillating viscometers
oscillating blade viscometer, 1034
oscillating piston viscometer, 1035–1036
OSHA, see Occupational Safety and Health Administration
Osmometers
membrane osmometer, 583–584
vapor pressure osmometers, 584–585
Ostwald viscometer, 1049
Oven, gas chromatograph, 159–160
Oxidation–reduction potential (ORP)
applications, 653–654
basics, 652–653
biological batch reactor control, 661
calibration, 660
chromium reaction, 658–659
control, 660–661
corrosion protection, 659–660
cylindrical ORP electrode cell, 657
detectors, 1075
electrode mounting, 657–658
maintenance, 660
measurement principles
cell potential, 656
half-cell reactions, 655–656
microprocessor-based units, 656
oxidation and reduction, 655
pH, 766
redox couples and reversibility, 655
scales, 654
metallic ORP electrodes, 657
Nernst equation, 657
reaction constituents monitoring, pharmaceuticals, 661
smart/digital sensor, 661
Oxygen demand
basics, 665–666
BOD analyzers

electrolysis instrument, 671
extended BOD test, 669
features, 677
five-day BOD tests, 667–668
incubation time, 667
manometric BOD test, 669–670
measurements, 1077
nitrification, 667
pH, 666–667
robotic analyzer, 671
seed, 666
in sewage treatment plants, 670
temperature, 667
test procedures, 666
toxicity, 667
COD analyzers
automatic on-line measurement, 672–673
definition, 666
dichromate COD test, 671–672
oxygen quantity measurement, 1077
TOD analyzers
applications, 676–677
basic components, 674
calibration, 675
definition, 666
interferences, 676
manual injection, 674
on-line monitoring, 676
organic and inorganic impurities detection, 666
oxygen demand measurement, 1077
platinum–lead fuel cell, 675
rotary sample valve, 675
sliding plate valves, 674
yttrium-doped zirconium oxide ceramic tube, 675
Oxygen in gases
basics, 682–683
catalytic combustion oxygen sensors, 686
chemical analyzers, 695
diatomic molecule, 683
in dry atmospheric air, 683
electrochemical oxygen sensors
categories, 686
coulometric sensor, 692
galvanic sensors, 689–692
high-temperature current-mode oxygen sensors, 689
high-temperature fuel cell oxygen sensors, 687–689
polarographic sensor, 692
gas chromatography, 695
low-temperature electrochemical cells, 695
paramagnetic oxygen sensors
deflection-type analyzer, 684–685
dual-gas analyzer, 686
magnetic susceptibility, 684
thermal analyzer, 685
portable and pocket oxygen analyzers, 695
spectroscopic oxygen detection
mass spectroscopy, 693
NIR spectroscopy, 693–694
zirconia-type oxygen analyzers, 695
Oxygen in liquids, see Dissolved oxygen

Ozone
in gas
amperometric electrochemical cell, 719
applications, 715, 725
basics, 714–715
chemiluminescent design, 720
concentration, 715
generators, 715
monitoring and analysis, 715–716
thin-film semiconductor design, 719–720
UV analyzer, 716–718
in water
amperometric sensors, 725–727
basics, 724
colorimetric method, 727
drinking water disinfection, 725
indigo method, 727
ozone contactors, 725
stripping and gas phase monitors, 727
UV absorption, 727

P

Packaging
infrared analyzers, 449
Raman process analyzer, 840–841
thermal conductivity detectors, 946–947
Packed columns, 162
Packed towers, 60–61
PAD, see Pulsed amperometric detection
Palladium-based thick film sensors, 414
Paper tape, fluoride analyzers, 346
Parallel disk rheometers, 871–872
Parallel plate viscometer, 1061
Paramagnetic analyzers, 616
Paramagnetic oxygen sensors
deflection-type analyzer, 684–685
dual-gas analyzer, 686
magnetic susceptibility, 684
thermal analyzer, 685
Particle charge analyzer, see Streaming current detector
Particle size distribution (PSD) monitors
applications
objectives, 733–734
range of, 733
basics, 731–732
HEPA filters, 732
laboratory measurements
Coulter reference method, 734–735
electron microscopy, 734
image analyzers, 734
light scattering (multiple particle), 735
optical methods, large samples, 735
optical microscopy, 734
optical scattering (single particle), 735
photo sedimentation, 735
sieving method, 735
ultrasonic attenuation, 736
laser diffraction, 737–738
online measurement
airborne-particulate counter, 737
optical analyzer, 736–737

Particulate sampling
  air filters, 103–104
  electrostatic precipitators, 105
  fiberglass papers, 104
  filter papers, 103
  impactors, 104
  impingers, 104
  thermal precipitators, 105
Passive acoustic sand detector, 879
Path integrated concentrations (PIC), 890
Path length, infrared analyzers, 446
PCDDs and PCDFs emissions measurement,
    see Polychlorinated dibenzo-
    p-dioxins and polychlorinated
    dibenzofurans emissions
    measurement
PDA spectrophotometers, see Photodiode
    array spectrophotometers
PDD, see Pulsed discharge detector
Peak processor, gas chromatography, 176
Permeable tube sample dryer, 68
PFPD, see Pulsed flame photometric detector
PGC, see Process gas chromatograph
PGNAA, see Prompt-gamma neutron
    activation analysis
pH measurement
  application problems
    acid errors, 779
    nonaqueous solutions, 780
    probe coating and low conductivity,
      780–781
    salt errors, 778–779
    solution temperature effects, 779
    water concentration errors, 779–780
  applications, 766
  basics, 764–765
  calibration
    and automatic cleaning systems,
      784–787
    buffer errors, 784
  electrodes
    cleaners, 775–778
    flat glass electrodes, 770–771
    Glasteel®, 771
    metal oxide pH electrodes, 771–772
    reference electrodes, 773–775
    shroud designs, 770
  flowing junction reference
    electrode, 768
  functional layers, glass membrane, 768
  glass measurement electrode, 768
  hydrogen ion activity, 766
  installation methods
    median selector, 783
    principles, 781
    retractable units, 782
    self-diagnostics, 783
    submersion assemblies, 781
  ion concentrations, 767
  measurement error, 765
  measurement range, 766
  measurement, reference, and combination
    electrodes, 769
  neutralization, 767
  sensors
    digital pH sensors, 785
    fault signaling, 783
    impedance diagnostic
      measurements, 783
    ISFET, 772
    new pH sensor developments, 784
    on-line cleaning systems, 778
    relative performance, 772–773
    smart pH sensor data, 788
    smart pH sensors, 784–785
  temperature effects, 769
  transmitters, 788
    asset management, 786–787
    new pH transmitters, 785–786
    smart pH transmitters, 786
    wireless HART™ transmitters, 787
Phosphate
  basics, 792
  chromatography, 796
  continuous colorimeter
    experimental setup, 793–794
    orthophosphate analyzer, 794
    pumpless design, 794–795
  flame photometric analysis, 795–796
  laboratory analysis, 793
  orthophosphate, organic form, 793
  phosphorus in wastewater, 793
  sample handling systems, 796
  total phosphate analysis, 795
Photoconductive detectors, 446
Photodiode array (PDA) spectrophotometers,
    994–995
Photoelectric sensors, 332
Photo-ionization detectors (PIDs)
  combustible analyzers, 227–228
  gas chromatography, 172–173
Photopic and incandescent light sources,
    744–745
Photo sedimentation, 735
Physical property analyzers, petroleum
    products
  basics, 800–801
  cloud-point analyzers
    applications, 816
    calibration, 816
    cloud-point tests (ASTM Method D
      2500-88), 814–815
    optical, 815–816
  continuous analyzers, advantages, 803
  distillation analyzers
    ASTM Method D 1160, 804
    ASTM Method D 2887-89 and D
      3710-88, 804
    ASTM Method D 86-IP-123, 803–804
    batch technique, 803
    distillation curves, 803
    online, 803–808
    true boiling-point distillation, 803
  flash-point analyzer
    calibration and applications, 818
    flammable liquid, 816
    flash-point tests (ASTM Methods D
      56-90 and D 93-90), 817
    high-temperature, 818
    low-temperature, 817–818
  freezing-point analyzer, 816
  NIR analyzers
    calibration and applications, 820
    chemometrics, 820
    new developments, 820
    online design, 820
    property and radiation absorption, 819
  octane analyzers
    applications, 819
    calibration, 819
    octane tests (ASTM Methods D 2699
      and D 2700), 818
    online comparator analyzer (ASTM
      Method D-2885), 818–819
    reactor-tube continuous, 819
  orientation, 801–802
  pour-point temperature analyzers
    applications, 814
    calibration, 814
    by differential pressure, 813–814
    pour-point test (ASTM Method D
      97-87), 813
    viscous-drag pour pointer, 814
  vapor–liquid ratio analyzers
    applications, 813
    calibration, 813
    continuous, 812–813
    volatility, 811–812
  vapor pressure analyzers
    air-saturated, 809–810
    applications, 811
    calibration, 811
    dry Reid method (ASTM Method D
      4953-90), 809
    dynamic, 810–811
    liquefied petroleum gases (ASTM
      Method D 1267-89), 809
    Reid method (ASTM Method D
      323-90), 808–809
Picric acid method, 271
PIDs, see Photo-ionization detectors
Piezoelectric crystal mass balance, 757
Piezoelectric effect, 36
Piezoelectric hygrometer, 549–550
Piping and instrument diagram (P&ID),
    see Flow sheets
Pitot tube assembly, 81–84
Plastometers
  advantages and disadvantages, 1030
  capillary plastometer, 1029–1030
  cone-and-plate plastometer, 1028–1029
  kneader plastometer, 1029
  plastic behavior, 1028
Point measurement, combustible
    analyzers, 231
Point monitoring system, 228–229
Polarographic analyzers, 291, 298
Polarographic (voltametric) cells, 702–704
Polarographic sensor, 686–687, 692,
    702, 726
Polychlorinated dibenzo-p-dioxins (PCDDs)
    and polychlorinated dibenzofurans
    (PCDFs) emissions measurement
  cleanup procedure, 286
  condenser and absorbent trap, 285
  gas chromatograph, 286
  high-resolution gas
    chromatography, 286
  high-resolution mass spectrometry, 286
  sample extraction, 286
  sampling train, 285

Polymer
    chain-like structure, 581
    molecular weight analyzers
        advantages and limitations, 591–592
        from chromatogram, 589–591
        differential refractometer sensor, 589
        electron microscope, 591
        end group determination, 591
        free-standing assembly, 589
        gel-permeation chromatography, 588
        light-scattering photometers, 585–587
        membrane osmometer, 583–584
        osmotic pressure, 582
        sedimentation equilibrium method, 592
        sedimentation velocity method, 592
        subassemblies, 589
        TEMs, 591
        ultracentrifuge, 592
        vapor pressure osmometers, 584–585
        viscometers, 587–588
        Zimm plot, 586
    size and shape, 582
Polymeric film sensors, 629
Polymerization, 838
Pope cell, 527
Portable gas emission analyzer, 898
Portable hydrogen leak detector, 416
Portable monitors
    carbon monoxide, 137–138
    NO_x analyzers, 620
Portable multigas detector, 405
Portable spectrophotometer, 1081
Potentiometric analyzers, 291, 293–294
Potentiometric sensors, 34–35
Pour-point temperature analyzers
    applications, 814
    calibration, 814
    by differential pressure, 813–814
    pour-point test (ASTM Method D 97-87), 813
    viscous-drag pour pointer, 814
Power/heat flow unit definitions, 1243
Power unit conversion, 1242
Pressure controls, 193
Pressure unit conversion, 1243–1256
Pressure unit definitions, 1247
Probe cleaners
    analyzer instrumentation, 41–42
    analyzer sampling systems, 65
    vibratory paddle-type probe cleaner, 702
Probe coating, pH measurement, 780–781
Probe consistency analyzers, 251
Probes
    analyzer sampling, 68–69
    chlorine, 147–148
    cleaners, 41–42, 65, 702
    consistency measurement, 251
    design, 705–706
    electrochemical detectors, 705–706
    electrolytic probes, 294–295
    galvanic probes, 291, 294–295
    gas chromatography, 180
    gas sampling probe, 68–69
    membrane probes, 147–148, 295
    moisture analyzers, 527–530
    nitrogen, ammonia, nitrite and nitrate, 608–609
    oil measurements, 637–638
    oxygen, 705–706
    pH probes, 1075
    Raman instrumentation, 836–838, 842
    refractometers, 859
    thin-film capacitance probes, 546–547
Probe-type critical angle refractometer, 856
Probe-type sensors, 39
Probe-type stack gas analyzers, 40
Probe-type turbidity transmitters, 975
Process gas chromatograph (PGC)
    analyzer, 158–159
    back-flush, 165
    chromatograph transmitter, 158
    columns
        functions, 163
        packed columns, 162
        WCOT columns, 162–163
    features, 157
    heart-cutting, 166
    installation, 183–184
    microprocessor operated PGC, 176–178
    operator interface, 178–180
    oven, 159–160
    pneumatic transmitter, 158
    sample injection, 164–165
    schematic diagram, 157
    SHS/SCS
        multistream analysis, 183
        sample conditioning, 181–182
        sample disposal, 183
        sample probe, 180
        sample transport, 181
    valveless/Deans switching, 164
    valves
        diaphragm valve, 161–162
        function, 160
        rotary valve, 161
        six-and ten-port valves, 164
        sliding plate valve, 161
Programmed circular variable filter (CVF) analyzer, 444
Prompt-gamma neutron activation analysis (PGNAA), 202, 204–205
PSD monitors, see Particle size distribution monitors
Pseudoplastics, 1005
Pulsed amperometric detection (PAD), 291, 296
Pulsed discharge detector (PDD), 174–175
Pulsed flame photometric detector (PFPD), 170–171, 933
Pyridine/barbituric acid method, 271
Pyroelectric detector, 445
Pyrolysis gas sample fractionation and conditioning unit, 56

Q

Quadrupole filter, mass spectrometry, 489–490
Quartz crystal microbalances, 36, 529

R

Radiation
    sources, 988
    spectrum, 984
    wavelength, 983
Radiochemical method, 503
Radio frequency
    absorption, 575
    analyzer instrumentation, 33
Radio frequency sensor, 636–638
Radiometric devices, 758
Rag layer profiler, 637
Raman process analyzer
    advantages, 842–843
    applications, 843
    CCD detector, 835
    data analysis
        calibration model, 839
        calibration transfer, 840
        multivariable predictive analysis, 838–839
        polymerization application, 838
        spectral processing, 838
    diagnostics and maintenance, 840
    disadvantages, 842–843
    fiber-optic light delivery and collection, 835
    fiber-optic Raman analyzer, 833
    installation and maintenance, 841
    laser and Raman filters, 836
    laser excitation source, 834
    laser safety, 840
    legacy Raman gas sampling design, 325
    NIR and IR absorption spectroscopy methods, 842
    outputs and communication, 840
    packaging, 840–841
    probe designs, 842
    Raman probes, 836–838
    requirements, 833
    sample interface, 836
    spectrometer, 834–835
Raman scattering
    basics, 825–826
    definition, 826
    elastic scattering, 827
    energy-level diagram, 828–829
    inelastic scattering, 827–828
    instrumentation
        dispersive Raman instrumentation, 832–833
        fluorescence signal, 830
        FT spectrometer, 831
        silver-doped solgel film, 830
        visible laser excitation, 830
    intensity, 829
    LASERs, 826
    mid-IR absorption, 829
    online Raman spectrum, 827
    pictorial representation, 828
    Raman shift, 828–829
    Stokes and anti-Stokes, 827
Raman spectroscopy, 33
    advantages and disadvantages, 826
    applications, 826, 843

1310 Index

vs. IR spectroscopy, 831–832
principles
 elastic scattering, 827
 inelastic scattering, 827
 optical spectroscopy, 827
 Raman scattering, 827–829
 spectral information, 829–830
Rayleigh scattering
 definition, 827
 energy-level diagram, 828–829
Reactive hydrocarbons, 395–396
Reactor-tube continuous octane
 analyzers, 819
Reagents, chlorine, 149
Recombination effect, 544
Rectangular torsion rheometers, 872
Rectilinear simple shear flow, 867
Redox titrations, 1088
Reference electrodes
 impedance, 783
 ion-selective electrodes, 460–461
 pH measurement, 773–775
Reference filaments, 946
Reflectance, colorimeters, 213–214
Reflected light detecting refractometer, 859
Reflection liquid cells
Refraction-based moisture analyzer, 555
Refractive index
 fiber-optic cables, 313
 fiber-optic probes, 325–326
 liquid chromatographs, 195
Refractometers
 angle of refraction, 850
 automatic refractometer, 855–856
 basics, 849–850
 critical angle of refraction, 852
 critical-angle refractometer, 854–855
 design variations
  handheld and benchtop designs, 858
  mounting and cleaning options, 857
  probe-type refractometer, 856
 fiber optic RI probes, 859
 mounting and compensation, 856
 on-line refractometer, 857, 860
 phase velocity, 850
 reflected light measurement, 859
 refraction
  definition, 850
  terms, 852
 refractive index, 850–851, 860
 single-pass differential refractometer, 853
 two-pass differential refractometer, 853–854
Refractometry (RI), 33
Reid method, 808–809
Reid vapor pressure (RVP) detectors, 277
Remote head system, combustibles, 225–226
Remote sensing, particulates, 759
Residence time, 661
Residual gas analyzers (RGA), 490, 492, 693
Residual oxygen calorimeter, 384
Resistance moisture gauge, 573
Retroreflector probes, 996–997
RGA, see Residual gas analyzers
Rheometers
 basics, 864
 design selection, 867

detector designs
 capillary, 874
 coaxial cylinder, 873
 cone-and-plate flow geometry, 868–871
 extensional flow rheometer, 873–874
 parallel disk, 871–872
 rectangular torsion, 872
 tension/compression and bending, 873
extensional rheometers, 865
nonlinear shear flow, 867
OLR, 866
polymer characterization, 865–866
rotational/shear rheometers, 865
shear creep test, 867
shear-strain tests, 866
viscoelastic properties, 865
Rheopectic substances, 1005
RI, see Refractometry
Ringelmann card numbers, 745–746
Rotameter-type flowmeter, 1025
Rotary disc filter, 64
Rotary valve, gas chromatographs, 161
Rotating sensors, consistency analyzers, 250–252
Rotating-spindle-type industrial viscometer, 1030–1032
Rotational element viscometers
 agitator power viscometer, 1032–1033
 dynamic liquid-pressure-type viscometer, 1033–1034
 gyrating-element viscometer, 1032
 rotating-spindle-type industrial viscometer, 1030–1032
 rotation rate, solid shape, 1030
Rotational/shear rheometers, 865
Rotational viscometers
 advantage, 1057
 advantages and precautions, 1062
 Brookfield viscosities, 1062
 coaxial cylinder viscometer, 1058–1059
 cone-and-plate viscometer, 1060–1061
 features, 1058
 geometries, 1057
 non-Newtonian fluids, 1059–1060
 operating principle, 1057
 parallel plate viscometer, 1061
 performance, 1060
 spindle viscometer, 1061–1062
 Stabinger viscometer, 1062–1063
Runway visual range (RVR) detectors, 759
RVP detectors, see Reid vapor pressure (RVP) detectors

S

Safety
 fiber-optic cables, 322–323
 fiber-optic probes, 322–323
 hazardous and toxic gas monitoring, 357
 Raman instrumentation, 840
Salinity, 711
Salt errors, 778–779
Sampling systems
 air quality monitoring, 92–107
 ambient effects, 58

AOTF, 40
applications
 automatic stack sampling, 69–70
 chemical reactor samplers, 71
 Duckbill samplers, 72
 gas sampling, 68–69
 high-pressure stream sampling, 71
 stack gas sampling, 69
automatic float drain, 60
automatic, online analyzer sampling system, 54
automatic probe cleaning, 65
basics, 54
bypass-return fast loop, 57
bypass-stream transport, 57
cellulose filters, 59
coalescing gas filter, 59
component materials selection, 68
component selection, 58–59
conditioning system, 66–67
disposal, 57
entrained gas separator, 62
entrainment removal, 57
feasibility evaluation, 54–55
fiber mist eliminator, 60–61
FOPs, 38–41
heated sample probe assembly, 56
homogenizers, 65–66
liquid–gas separation, 62
liquid–liquid separation, 62
mass spectrometers, 75
multistream sampling, 74–75
NeSSIT, 75
packed towers, 60
particles and water removal, 37
point location, 54
probe-type sensors, 39
probe-type stack gas analyzers, 40
self-cleaning filters, 59, 63–65
self-cleaning suspended solids detector, 39
single-line transport, 56–57
sintered metallic filters, 59
slipstream/bypass filters, 63
slipstreaming plus coalescing filter, 63
solid sampling, 72–73
sparger, 60
spray-stripping chamber, 61
stack monitoring, 79, 87–89
temperature-compensated RI-detecting fiber-optic probe, 38
test and calibration, 58
time lag calculation, 57
trace analysis sampling, 73–74
transportation delay, 54
vaporizing samples, 67–68
warm air purge, 61
Sand detector
 basics, 878
 intrusive and erosion based sand detector, 880–881
 passive acoustic sand detector, 879
 sand monitoring system/data flow, 882
Sand production, 878
Saybolt viscometer, 1055–1056
Scanning spectrophotometers, 995–996

Scattered light detectors, 974–975
SCD, *see* Streaming current detector
SCM, *see* Streaming current monitor
Screen-printed electrochemical sensor (SPEC), 291
SCV, *see* Streaming current value
Sedimentation
    molecular weight, 592
    particle size, 735
Seed, oxygen demand, 666
Selective noncatalytic reductions (SNCR), 112
Selenium sulfide, 502
Self-cleaning filters, 63–65
Self-cleaning suspended solids detector, 39
Self-diagnostics, 688–689
Semiconductor sensors, 224
Sensor fault signaling, 783
Sensors
    acoustic sensor, 879
    air quality monitoring, 97
    amperometric sensors, 36, 362, 629
    backscatter turbidity sensors, 976
    calorimetric sensors, 629
    catalytic bead sensors, 414–415
    catalytic combustion oxygen sensors, 686
    conductivity sensors, 35–36, 241–242, 291, 1075
    coulometric sensors, 693, 707–708
    density-based sensors, 976
    differential refractometer sensor, 589
    digital conductivity sensors, 241
    dipole polarization effect moisture sensor, 553–554
    dosage sensors, 366–371
    electrochemical oxygen sensors, 686–692
    electrochemical ozone sensor, 719
    environmental pollution sensors, 635
    galvanic DO sensors, 702, 704–975
    gold film sensor, 423
    gravimetric sensors, 628
    high-temperature current-mode oxygen sensors, 689
    humidity sensors, 525–527
    ionization chamber sensors, 331
    Memosens^r sensors, 785
    metal oxide sensors, 629
    microwave sensors, 254, 552, 636–638
    nondispersive infrared sensors, 126–127
    optical fluorescence and luminescence sensors, 214–215
    optical sensors, 249–250, 629
    oxidation–reduction potential, 661
    palladium-based thick film sensors, 414
    paramagnetic oxygen sensors, 684–686
    pH measurements, 772–773, 784–785
    photoelectric sensors, 332
    polarographic DO sensor, 668
    polymeric film sensors, 629
    potentiometric sensors, 34–35
    probe-type sensors, 39
    radiation-based flexible sensor, 639
    radio frequency sensor, 636–638
    screen-printed electrochemical sensor, 291
    semiconductor sensors, 224
    silicon dioxide sensors, 348
    solid-state sensors, 414, 423
    surface acoustic wave sensors, 1039
    suspended solids and sludge density sensors, 974
    thermal sensors, 331–332
    toxic gas monitoring, 361–363, 366–371
    trace oxygen sensor, 707
    water quality monitoring systems, 1075, 1077
Shear flow properties, rheometers, 868
Shear-strain tests, rheometers, 866
Shear stress relaxation, 866
Shirtpocket-size ozone sensor, 720
Shrouds, pH, 770, 775
Silicon dioxide sensors, 348
Simulated distillation, 808
Single-beam UV analyzer, 991
Single-component analyzers, 441–442
Single-float viscometer, 1026
Single-pass differential refractometer, 853
Single source/stacks monitoring, air quality monitoring, 94
Sintered metallic filters, 59
SLAMS, *see* State and Local Air Monitoring Stations
Sliding plate valves
    gas chromatograph, 8
    oxygen demand, 674
Sling psychrometer, 530
Slipstream filters, 63
Slipstreaming plus coalescing filter, 63
SMA* connectors, *see* Subminiature version A connectors
Smart conductivity transmitters, 243
Smoke and dust density measurement
    ambient air opacity monitoring
        airborne material, 754
        ambient air sampling, 754–756
        charge transfer (triboelectricity), 758–759
        dichotomous sampler, 756
        high-volume sampler, 755–756
        impaction devices, 757–758
        particulate concentration, 754
        piezoelectric crystal mass balance, 757
        radiometric devices, 758
        remote sensing, 759
        surface ionization, 759
        tape sampler, 757
        visual observation, 759
    attenuation, 744
    light absorption, reflection, and scattering radiation, 744
    light scattering designs, 751–752
    opacity measurement
        dust loading, 745
        light attenuation and transmissometers, 748–751
        percent opacity, 745
        photopic and incandescent light sources, 744–745
        Ringelmann card numbers, 745–746
        transmittance, 745
        turbidity attenuation coefficient, 745
    stack emission monitoring
        particulate sampling, 746–747
        stack gas analysis and sampling, 747–748
        stack samplers, 752–753
Smoke detectors
    basics, 330–331
    central fire alarm system, 331
    ionization chamber sensors, 331
    photoelectric sensors, 332
    in U.S., 331
SNCR, *see* Selective noncatalytic reductions
Sniff odor test system, 626–627
Sodium ion error, 778
Soiling index, 757
Solid electrodes, 296–297
Solid polymer measurements, 865–866
Solids
    analyzer sampling, 72–73
    infrared instruments, 448
Solid-state instruments
    hydrogen, stream/air, 414
    hydrogen sulfide detectors, 423
    ion-selective electrodes, 465–466
Solid-state sensors, 414, 423
Sorption-type dosimeters, 370–371
Spargers
    analyzer sampling, 60–61
    water quality monitoring systems, 1079–1080
SPEC, *see* Screen-printed electrochemical sensor
Specific enthalpy and latent heat conversion, 1247
Specific heat and entropy unit conversion, 1247
Specificity, analyzer instrumentation, 27
Specific volume (v)
    definition, 1293
    dry and saturated steam properties, 1288–1289
    superheated steam properties, 1290–1293
    units and conversions, 1248, 1287
    water properties, 1294
Spectral information, 829–830
Spectrometers
    electromagnetic energy, absorption/emission, 30
    fluorescent, 31–32
    frequency and wavelength ranges, 30
    hazardous and toxic gas monitoring, 365
    hydrocarbon analyzers, 396–397
    infrared and near-infrared, 32
    microwave and radio frequency, 33
    nuclear techniques, 34
    open-path monitoring, 887–904
    Raman spectrometers, 33
    refractometry, 33
    turbidity and particle size, 33–34
    ultraviolet and visible spectrums, 30–31
Spectrometric perimeter monitoring, 397
Spectrophotometer, 211
Spectroscopic oxygen detection
    mass spectroscopy, 693
    NIR spectroscopy, 693–694
Spindle viscometer, 1031, 1061–1062
Split-beam UV analyzer, 992
Spot sampling, carbon monoxide, 138
Squarewave voltammetry, 291, 294
Stabinger viscometer, 1062–1063
Stack gas sampling, 69, 747–748

1312 Index

Stack monitoring
　AST, 85–86
　basics, 79
　EPA particulate sampling system
　　control unit, 84–85
　　heated compartment, 84
　　ice-bath compartment, 84
　　microprocessor-controlled stack sampling, 80
　　pitot tube assembly, 81–84
　　schematic diagram, 80
　isokinetic sampler, 79
Standpipe detector, 481
Stannous chloride method, 793
State and Local Air Monitoring Stations (SLAMS), 98
Static methods, 96
Stiff gel references, 774
Stokes' law, 1003
Strain-rate-controlled cone-and-plate rheometer, 868
Streaming current detector (SCD)
　applications
　　control system, 914–915
　　sampling, 913
　　titration, 913
　basics, 909
　calibration, 912
　operating principle
　　streaming current, 910–911
　　treatment chemical selection, 912
Streaming current monitor (SCM), 910, 915
Streaming current particle charge analyzer, see Streaming current detector
Streaming current value (SCV), 910
Stress-controlled cone-and-plate rheometer, 868
Strippers, 60–61
Submarine internet cables, 313–314
Submersion assemblies, pH, 781
Subminiature version A (SMA) connectors, 323
Sub-seas leakage detection, 482
Sugar industry applications, 572
Sulfur analyzers
　automotive catalytic converters, 932
　basics, 931–932
　chemiluminescence analyzer, 934
　gas chromatography with FPD, 932–933
　hydrogenolysis and rateometric colorimetry, 933–934
　pyrolysis and ultraviolet fluorescence, 935
　x-ray fluorescence, 936–937
Sulfur dioxide (SO_2)
　ambient air analyzers
　　automated wet chemical devices, 925
　　calibration, 925
　　colorimetric analyzers, 925–926
　　conductimetric analyzers, 926
　　coulometric analyzers, 927
　　electrochemical analyzers, 927
　　flame photometric analyzers, 927
　applications, 922–923
　basics, 920–921
　industrial analyzer types
　　correlation spectrometry, 925
　　NDIR analyzers, 923–924
　　thermal conductivity analyzers, 925
　　UV analyzers, 924
　NAAQS history, 922
　production, 921
　sources, 922
Sulfur trioxide (SO_3), 920–921
Supply pumps, liquid chromatography, 193
Surface ionization technique, 759
Surface Water Treatment Rules, 725
Suspended solids and sludge density sensors, 974
Synergy 2 SL luminescence microplate reader, 119

T

Tape sampler, 757
TAPPI, see Technical Association of the Pulp and Paper Industry
TAS AOTF, see Thallium–arsenic–selenide acousto-optic tunable filters
TC analyzers, see Total carbon analyzers
TCDs, see Thermal conductivity detectors
TDL, see Tunable diode laser
TDLS, see Tunable diode laser spectrometer
TDR, see Time domain reflectometry
Technical Association of the Pulp and Paper Industry (TAPPI), 27
Temperature
　infrared instruments, 451
　ion-selective electrodes, 463–464
　oxygen demand, 667
　oxygen, liquid, 709–710
　pH measurements, 769
　refractometry, 854
　thermal conductivity detectors, 945–946
Temperature-compensated RI-detecting fiber-optic probe, 39
Temperature conversion, 1248
Temperature-programmed chromatography, 156–157
TEMs, see Transmission electron microscopes
Tension/compression rheometers, 873
Testing
　analyzer sampling, 58
　rheometers, 866–867
TGA, see Thermogravimetry
Thallium–arsenic–selenide acousto-optic tunable filters (TAS AOTF), 450
Thallium cells, 707
Thermal analyzer, oxygen, 685
Thermal conductivity
　composition measurement, 942
　definition, 942
　factors, 942
　katharometer/catharometer, 942
　measurement ranges, 942–943
　TCD analyzer
　　basics, 941
　　case temperature control, 945–946
　　cells, 944
　　components, 943
　　detectors, 943–944
　　operation, 946
　　packaging and calibration, 946–947
　　Wheatstone bridge circuits, 944–945
　unit systems, 942
Thermal conductivity detectors (TCD) analyzer
　basics, 941
　case temperature control, 945–946
　cells, 944
　components, 943
　detectors, 943–944
　operation, 946
　packaging and calibration, 946–947
　Wheatstone bridge circuits, 944–945
Thermal conductivity sulfur dioxide analyzers, 925
Thermal precipitators, 105
Thermal sensors, 331–332
Thermistor-type detector, 946
Thermocouple detector, 223
Thermogravimetry (TGA), 202, 565
Thermopile calorimeter, 385
Thermopile detectors, 445–446
Thin-film capacitance
　air moisture, 527–530
　probes, 527–530, 546–547
Thin-film semiconductor ozone monitor, 719–720
Thin-layer chromatography, 514
Thixotropic materials, 1005
TIC, see Total inorganic carbon
Time domain reflectometry (TDR), 570
Time-of-flight filter, mass spectrometry, 491
Time unit conversion, 1248
TOC analyzers, see Total organic carbon analyzers
TOD analyzers, see Total oxygen demand analyzers
Torsional oscillation viscometers, 1036–1037
Total carbon (TC) analyzers, 1077
　aqueous conductivity, 958
　automatic on-line design, 957
　carbon measurement techniques, 952–953
　catalyzed ozone hydroxyl radical oxidation analyzers, 955–956
Total cyanide (CN_T) determination, 271–272
Total inorganic carbon (TIC)
　carbon measurement techniques, 952–953
　catalyzed ozone hydroxyl radical oxidation analyzer, 953, 955–957
　colorimetric analyzer, 961
Total organic carbon (TOC) analyzers, 1077
　aqueous conductivity
　　chemometric equation, 960
　　conductivity cell temperature, 960
　　laboratory and in-line applications, 958
　　online conductivity TOC analyzer, 959
　　semiconductor and power applications, 959–960
　　semipermeable hydrophobic membrane, 960
　　ultraviolet-persulfate oxidation, 958
　basics, 951–952
　BOD and COD, 952
　carbon measurement techniques, 952–953

colorimetric analyzer, 961
coulometric analyzer, 961
determination methods, 954
development, 953
FID analyzers, 961–962
nondispersive infrared analyzers, 954–955
oxidation methods
    automatic on-line design, 957–958
    catalyzed ozone hydroxyl radical oxidation analyzer, 955–956
    high-temperature combustion analyzer, 956–957
    inorganic carbon, 957
    wet chemical oxidation, 957
Total oxygen demand (TOD) analyzers, 1075, 1077
    applications, 676–677
    basic components, 674
    basics, 665–666
    calibration, 675
    definition, 666
    interferences, 676
    manual injection, 674
    on-line monitoring, 676
    organic and inorganic impurities detection, 666
    platinum–lead fuel cell, 675
    rotary sample valve, 675
    sliding plate valves, 674
    yttrium-doped zirconium oxide ceramic tube, 675
Total phosphate analysis, 795
Toxic gas monitoring, *see* Hazardous and toxic gas monitoring
Toxicity, oxygen demand, 667
Trace analysis sampling, 73–74
Transfer optics, open-path spectrometry, 892–893
Transmission electron microscopes (TEMs), 591
Transmission liquid cells, 447
Transmissometers
    air purge, 750–751
    atmospheric extinction, 749
    double-beam, double-pass transmissometer, 749–750
    optical characteristics, 750
    optical density, 748–749
    single-pass transmissometer, 748–750
    spectral characteristics, 750
Transmittance colorimeters, 211–212
Transport
    analyzer sampling, 56–57
    gas chromatography, 181
    water quality monitoring systems, 1071–1073
Transport lag, 57
Traversing point locations, 82–83
Triboelectric effect, 758
Triboelectricity, 758–759
Tristimulus method, 213–214
Tube sampling, combustibles, 226–227
Tube torsional movement, 1038–1039
Tunable diode laser (TDL), 113
Tunable diode laser spectrometer (TDLS), 396
Tunable lasers, 440–441

Turbidity
    attenuation coefficient, 745
    backscatter turbidity sensors, 976
    basics, 966
    density-based sensors, 976
    forward scattering/transmission type sensor
        colorimeter, 969
        dual-beam design, 969
        instrument designs summary, 971–973
        laser-type meter, 971
        ratioing, 969–971
        suspended solids and sludge density sensors, 974
    meters
        cloudiness detection, 967
        light absorption and scattering, 967
        water quality monitoring systems, 1077
    scattered light detectors (nephelometers), 974–975
    units
        conversion, 968
        FTU, 967–968
        interferences, 968–969
        JTU, 967
        NTU, 967–969
Turbidity meters, 969, 971, 974–976, 1077
Turbine flowmeters, 599
Two-electrode cells, conductivity analyzers, 238
Two-float viscometer, 1026
Two-liquid phase separation, analyzer sampling, 62
Two-pass differential refractometer, 853–854
Tyndall effect, 974
Tyndall light, 974

U

Ultracentrifuge, 592
Ultrasonic attenuation, particle size, 736
Ultrasonic flow meters
    accuracy, 599
    advantages, 599
    analyzers, 600
    bidirectional, multipath, LNG flowmeter, 599
    chromatographs, 601–602
    Coriolis mass flowmeters, 599–600
    mass spectrometers, 602
    sampling systems, 601
Ultrasonic pipe thickness monitoring, 265
Ultraviolet absorption, mercury, 501–502
Ultraviolet (UV) analyzers, 924
    absorption
        absorbance characteristics, 983–984
        compounds, 986–987
        spectrum, 987
    basics, 981–982
    Beer–Lambert law, 985
    broad emission sources, 988
    components, 982, 987
    detectors, 990
    discrete line emission sources, 988–989

    dual-beam, dual-detector, 993
    dual-beam, single-detector, 992–993
    measuring cell, 990
    monochromator, 989
    optical dispersion, 990
    photometers
        absorbance/transmittance measurement, 982
        applications, 987
        flicker photometer, 993–994
    radiation source, 988
    radiation spectrum, 984
    radiation wavelength, 983
    readout devices, 991
    retroreflector probes, 996–997
    single-beam, 991
    spectrophotometers, 983
        PDA, 994–995
        scanning, 995–996
    split-beam, 992
    wavelength selection, 989
Ultraviolet and visible spectrums, 30–31
Ultraviolet (UV) ozone analyzer
    double-beam design, 718
    ozone absorption coefficient value, 716
    single-beam design, 717
Unit of Mooney, 1028–1029
Units of area, 1238
Units of density, 1239
Units of mass, 1220
Urban area, air quality monitoring, 94
U.S. Environmental Protection Agency (EPA), 925
UV analyzers, *see* Ultraviolet analyzers

V

Vacuum distillation analyzer, 806
Vacuum environment, mass spectrometry, 491–492
Valves
    gas chromatography, 160–162, 164
    liquid chromatography, 193
Vaporizing, analyzer sampling, 67–68
Vaporizing regulator assembly, 67
Vapor–liquid ratio analyzers
    applications, 813
    calibration, 813
    continuous, 812–813
    volatility, 811–812
Vapor pressure analyzers
    air-saturated, 809–810
    applications, 811
    calibration, 811
    dry Reid method (ASTM Method D 4953-90), 809
    dynamic, 810–811
    liquefied petroleum gases (ASTM Method D 1267-89), 809
    Reid method (ASTM Method D 323-90), 808–809
Vapor pressure osmometers, 584–585
Vapor sampling
    air quality monitoring, 101–103
    analyzer sampling, stack monitoring, 86
VCSEL, *see* Vertical cavity surface emitting laser

Velocity unit conversion, 1248
Velocity unit definitions, 1249
Venturi tube, 598–599
Vertical cavity surface emitting laser (VCSEL), 396–397
Vibrating-reed/rod viscometer, 1037–1038
Vibrational viscometers
   industrial viscometer
      resonator's damping measurement methods, 1036
      torsional oscillation viscometers, 1036–1037
      vibrating-reed/rod viscometer, 1037–1038
   laboratory viscometer, 1063
VIS analyzers, *see* Visible analyzers
Viscometers
   applications, 1010–1011
   basics, 1002
   orientation table, 1008–1009
   selection, 1010
   vibrational viscometers
      industrial viscometer, 1036–1038
      laboratory viscometer, 1063
Viscometers, industrial, *see* Industrial viscometers
Viscometers, laboratory, *see* Laboratory viscometers
Viscosity
   absolute, 1006
   definition, 1002
   Hagen–Poiseuille law, 1003
   intrinsic, 1004
   kinematic, 1003, 1008
   Newtonian and non-Newtonian fluids, 1004–1005
   shear rate, 1003
   shear stress, 1003
   SI and non-SI units, 1018
   Stokes' law, 1003
   two-plate model, 1002
   unit conversion, 1250–1252
   units, 1006–1008
   viscometer
      applications, 1010–1011
      orientation table, 1008–1009
      selection, 1010
Viscous-drag pour pointer, 814
Visible (VIS) analyzers
   basics, 981–982
   components, 982
   photometers, 982
      NIR photometers, 997
      visible light photometers, 997
      spectrophotometer, 983
Visible light photometers, 997
Visual flame detectors, 335
Volatile organic compounds (VOC) detection, 1078–1079
Volatility test, 812
Voltammetric analyzers
   amperometric analyzers, 291, 295–297
   characteristic response curves, 293
   conductivity sensors, 291
   coulometric analyzers, 292, 298–299
   cyclic voltammetry, 291
   description, 291
   developments, 299
   electrolytic probes, 294–295
   galvanic analyzers, 291, 294–295
   limitations, 299
   PAD, 291
   polarographic analyzers, 291, 298
   potentiometric analyzers, 291, 293–294
   SPEC, 291
   squarewave voltammetry, 291, 294
Volumetric flow unit conversion, 1241
Volumetric flow unit definitions, 1242
Volumetric titrators
   applications, 1091
   high-precision automatic titrator, 1089–1090
   pH and colorimetric end point types, 1089
   simple, continuous automatic titrator, 1091
Volume unit conversion, 1253
Volume unit definitions, 1253–1255

W

Waste water treatment
   coagulation, 910
   SCD, 910
   SCM, 910
   SCV, 910
Water concentration errors, 779–780
Water-in-oil measurements
   backscatter-type traversing detector, 639
   capacitance, 635–636
   conductivity-based interface, 638
   near-infrared meters, 639–640
   radiation-based flexible sensor, 639
   radio frequency/microwave
      microwave water-in-oil detector, 636
      rag layer profiler, 637
      RF probes, 637–638
      water dump control, 636–637
   strip-type nuclear source, 639
   ultrasonic pipeline interface, 638
Water-jet cleaners, pH, 776–777
Water, ozone monitoring
   amperometric sensors
      bare metal electrodes, 725–726
      membrane-type designs, 726–727
   colorimetric method, 727
   drinking water disinfection, 725
   indigo method, 727
   ozone contactors, 725
   stripping and gas phase monitors, 727
   UV absorption, 727
Water quality monitoring systems
   basics, 1069
   chromatographs
      multivariable, unattended systems, 1080–1081
      sparging, 1079–1080
   components in water, 1073
   eutrophication, 1070
   industrial systems, 1071
   information sources, 1071
   ion-selective electrodes
      conductivity sensors, 1075
      dissolved oxygen sensors, 1077
      ion electrodes, 1073–1074
      nephelometers, 1077
      ORP detectors, 1075
      oxygen demand detectors, 1075–1077
      pH probes, 1075
      TC and TOC analyzers, 1077
      turbidity meters, 1077
      VOC detection, 1078–1079
      wet chemistry analyzers, 1077–1078
   nitrate concentration, 1070
   portable spectrophotometer, 1081
   regulatory systems, 1071
   sampling systems
      Duckbill samplers, 1072
      permanent installation, 1072
      sample transport system, 1071–1073
      submersible sensor housing, 1072
   water quality standards, 1082
Water temperature rise calorimeter
   airflow calorimeter, 383
   air temperature rise calorimeter, 383
   calibration heater, 382
   chromatographic calorimeter, 384
   expansion tube calorimeter, 384
   residual oxygen calorimeter, 384
Wavelength modulation spectrometry, 901–902
Wavelength selection, 989
WCOT columns, 162–163
Weight unit conversion, 1255
Weight unit definitions, 1256
Wet chemistry
   analyzers and autotitrators
      colorimetric analyzers, 1092–1095
      FIA, 1087, 1095–1097
      in-process measurement, 36
      titration end points, 1088–1089
      volumetric titrators, 1089–1092
      water quality monitoring systems, 1077–1078
   basics, 1087
   liquid phase observation, 1087
Wettable metals, 500
Wheatstone bridge
   combustibles, 223–224
   thermal conductivity detectors, 944–945
Winkler titration method, 668
Wireless corrosion monitor, 264
Wireless HART™ transmitters, 244
Wobbe index, 381
Wood-free pulp measurement, 252

X

XMTC thermal conductivity binary gas transmitter, 947
X-ray fluorescence (XRF), 34
X-ray fluorescence (XRF) spectrometer, 307–308

Z

Zahn cup viscometer, 1056
Zeta potential titrations, 1088
Zimm plot, 586
Zirconia-type oxygen analyzers, 695
Zirconium dioxide fuel cells, high-temperature, 687–689